Transmission Performance of Evolving Telecommunications Networks

For a complete listing of the Artech House Telecommunciations Library,
turn to the back of this book

Transmission Performance of Evolving Telecommunications Networks

John Gruber
Godfrey Williams

Artech House
Boston • London

Library of Congress Cataloging-in-Publication Data

Gruber, John.
    Transmission performance of evolving telecommunications / John
Gruber, Godfrey Williams.
      p.   cm.
    Includes bibliographical references and index.
    ISBN 0-89006-591-8
    1. Telecommunication systems--Evaluation. I. Williams, Godfrey.
II. Title.
TK5102.5.G753 1992   92-3283
621.382--dc20   CIP

British Library Cataloguing in Publication Data

Gruber, John J.
    Transmission Performance of Evolving
    Telecommunications Networks
    I. Title II. Williams, Godfrey
    621.382

    ISBN 0-89006-591-8

© 1992 ARTECH HOUSE, INC.
685 Canton Street
Norwood, MA 02062

All rights reserved. Printed and bound in the United States of America. No part of this book may be reproduced or utilized in any form or by any means, electronic or mechanical, including photocopying, recording, or by any information storage and retrieval system, without permission in writing from the publisher.

International Standard Book Number: 0-89006-591-8
Library of Congress Catalog Card Number: 92-3283

10 9 8 7 6 5 4 3 2 1

CONTENTS

Preface	xv
Acknowledgments	xix
Chapter 1 Introduction	1
1.1 Definition of Transmission Performance	1
1.1.1 Relationship of Transmission Performance to Categories of Performance and to Phases of the Communications Process	1
1.1.2 Relationship of Transmission Performance to Performance Criteria	3
1.2 Scope of the Book	5
1.2.1 Relevant Telecommunications Services	5
1.2.2 Relevant Telecommunications Networks	7
1.3 Historical, Contemporary, and Future Perspectives	9
1.3.1 Regulated, Predivestiture Environment	9
1.3.2 The Competitive Environment	10
1.3.2.1 The United States	10
1.3.2.2 Canada and Other Countries	12
1.4 Introduction to the Overall Performance Process	12
1.5 References	15
Chapter 2 User-Oriented Performance	19
2.1 Definition and Background	19
2.2 Voice and Data Performance Considerations	20
2.3 User-Oriented Performance Parameters	23
2.4 Performance Assessment Techniques and Implications	23
2.4.1 Customer Reports and Surveys	23
2.4.2 Laboratory and Modeling Techniques	24
2.4.3 Factors Influencing User Expectations	24
2.5 Grade-of-Service Methodology	25
2.5.1 Subjective Testing of Speech Quality	26
2.5.2 Transmission Rating Models	28

	2.5.3	Opinion Models and Opinion Data Bases	29
	2.5.4	Performance Classes	33
	2.5.5	Behavioral Models and Behavioral Data Bases	34
2.6	Potential for Modeling Nonvoice Parameters		36
2.7	References		36

Chapter 3 Translating User-Oriented Performance into End-to-End Network-Oriented Performance 39

- 3.1 Introduction 39
- 3.2 User- and Network-Oriented Performance for Voice Service 44
  - 3.2.1 Speed of Service Performance Criterion 44
    - 3.2.1.1 Fixed Delay 44
    - 3.2.1.2 Variable Delay 46
  - 3.2.2 Accuracy of Service Performance Criterion 50
    - 3.2.2.1 Loudness Loss 50
    - 3.2.2.2 Idle Circuit Noise 54
    - 3.2.2.3 Talker Echo 54
    - 3.2.2.4 Listener Echo 58
    - 3.2.2.5 Quantization Noise 60
    - 3.2.2.6 Acoustic (Room) Noise 62
    - 3.2.2.7 Sidetone 64
    - 3.2.2.8 Bandwidth and Slope 67
    - 3.2.2.9 Crosstalk and Interference 71
    - 3.2.2.10 Nonlinear Distortion 72
    - 3.2.2.11 Analog Transients 75
    - 3.2.2.12 Transmission Errors 76
    - 3.2.2.13 Short Interruptions 82
    - 3.2.2.14 Speech Clipping 84
  - 3.2.3 Dependability of Service Performance Criterion 89
    - 3.2.3.1 Long Interruptions 89
    - 3.2.3.2 Call Cutoff 89
  - 3.2.4 Nontransmission Parameters 90
- 3.3 User- and Network-Oriented Performance for Data Service 90
  - 3.3.1 General 90
  - 3.3.2 Speed of Service Performance Criterion 90
    - 3.3.2.1 Transfer Delay (Response Time) 90
    - 3.3.2.2 Transfer Rate (Throughput Efficiency) 91
  - 3.3.3 Accuracy of Service Performance Criterion 92
    - 3.3.3.1 Error Probability 92
    - 3.3.3.2 Residual Error Probability 93
  - 3.3.4 Dependability of Service Performance Criterion 94
    - 3.3.4.1 Transfer Denial 94
    - 3.3.4.2 Premature Disengagement 95

3.4	Summary			95
3.5	References			96

Chapter 4 End-to-End Network-Oriented Performance — 105
- 4.1 Introduction — 105
- 4.2 Network-Oriented Parameters — 105
- 4.3 Voiceband Network Performance — 107
  - 4.3.1 Connection Models — 107
  - 4.3.2 End-to-End Performance Information — 112
    - 4.3.2.1 Survey Data — 112
    - 4.3.2.2 Performance Objectives/Requirements — 113
  - 4.3.3 Characterized Connections — 114
  - 4.3.4 End-to-End Network Performance Characterization — 117
    - 4.3.4.1 Loss and Related Parameters — 117
    - 4.3.4.2 Noise and Related Parameters — 120
    - 4.3.4.3 Echo Path Loss Parameters — 122
    - 4.3.4.4 Echo Path Delay Parameters — 126
    - 4.3.4.5 Distortion and Incidental Modulation Parameters — 129
    - 4.3.4.6 Analog Transient Parameters — 133
    - 4.3.4.7 Error Performance Parameters — 135
- 4.4 Digital Network Performance — 138
  - 4.4.1 CCITT — 138
  - 4.4.2 North America — 141
    - 4.4.2.1 Performance Objectives — 141
    - 4.4.2.2 Network Performance Estimates — 147
    - 4.4.2.3 Note on Jitter and Wander — 151
- 4.5 Packet Network Performance — 151
  - 4.5.1 CCITT — 151
  - 4.5.2 North America (U.S.) — 156
    - 4.5.2.1 Performance Objectives — 156
    - 4.5.2.2 Network Performance Measurements — 156
- 4.6 References — 160

Chapter 5 Allocation of End-to-End Performance to Networks and Network Components — 165
- 5.1 Introduction — 165
  - 5.1.1 Purpose of Hypothetical Reference Connections — 168
- 5.2 Networks and Network Components — 168
  - 5.2.1 CCITT—International Networks — 168
  - 5.2.2 North American Networks—A Comparison — 171
    - 5.2.2.1 Canadian Networks — 172
    - 5.2.2.2 U.S. Networks — 172
  - 5.2.3 Network Components — 176

5.3	Performance Allocation Criteria		178
	5.3.1	Distance	178
	5.3.2	Technology	178
	5.3.3	U.S. Postdivestiture and Evolutionary/Competitive Factors	179
5.4	Performance Allocation Methodology		179
	5.4.1	General	179
	5.4.2	Addition Laws for Serial Connection Configurations and for Fixed Parameter Values	180
		5.4.2.1 Algebraic Addition Laws	180
		5.4.2.2 Voltage, Power, and Related Addition Laws	197
	5.4.3	Addition Laws for Serial Connection Configurations and for Distributed Parameter Values	211
		5.4.3.1 Introduction	211
		5.4.3.2 Monte-Carlo Simulation	211
		5.4.3.3 The Normal Distribution and Related Moments	213
		5.4.3.4 Algebraic Addition Laws	217
		5.4.3.5 Voltage, Power, and Related Addition Laws	226
	5.4.4	Addition Laws for Parallel Configurations	245
		5.4.4.1 General	245
		5.4.4.2 Methodology	245
		5.4.4.3 Examples of Parallel Paths with Normally Distributed Parameters	249
		5.4.4.4 Examples of Parallel Paths with Log-Normally Distributed Parameters	253
5.5	References		260
Appendix A.5.1	Conversion of Q to Subjectively Equivalent Idle Circuit Noise		265
Appendix A.5.2	Relationship Between Normal and Log-Normal Distribution		267
Chapter 6	Performance of Network Components		271
6.1	Introduction		271
6.2	Performance of Voiceband Network Components		272
	6.2.1	Description of Main Connection Components	272
	6.2.2	Performance Objectives, Requirements, and Limits	275
	6.2.3	Network Component Performance Characterization	275
		6.2.3.1 Loss and Related Parameters	275
		6.2.3.2 Signal Levels	279
		6.2.3.3 Noise and Related Parameters	283
		6.2.3.4 Return Loss Parameters	288
		6.2.3.5 Delay Parameter	294
		6.2.3.6 Distortion-Related Parameters	296

		6.2.3.7	Incidental Modulation-Related Parameters	301
		6.2.3.8	Analog Transient Parameters	302
		6.2.3.9	Error Performance	303

6.3 Error Performance of Digital Network Components ... 314
    6.3.1 Allocation of BER and ES Objectives ... 314
    6.3.2 Measured Performance of Network Components ... 318
6.4 Jitter and Wander in Digital Networks ... 322
    6.4.1 Causes and Effects of Jitter and Wander ... 322
        6.4.1.1 Jitter ... 322
        6.4.1.2 Wander ... 326
    6.4.2 Accumulation of Jitter and Wander ... 330
        6.4.2.1 Regenerative Repeater Jitter ... 330
        6.4.2.2 Multiplex/Demultiplex Waiting Time Jitter ... 335
    6.4.3 Jitter and Wander Performance ... 337
        6.4.3.1 Output Jitter and Wander at Network Interfaces ... 337
        6.4.3.2 Input Jitter and Wander Tolerances for Digital Equipment ... 338
        6.4.3.3 Jitter and Wander Transfer Characteristics of Digital Equipment ... 341
        6.4.3.4 Jitter and Wander Generated by Digital Equipment ... 341
        6.4.3.5 Clock and Synchronous Node Timing Requirements ... 343
6.5 References ... 349
Appendix A.6.1 Objective Interpretation of Subjective "Compandor Penalty" Associated with Analog Compandors ... 357
Appendix A.6.2 Breakdown of Digital Switching Delays ... 360

Chapter 7 Network Transmission Planning ... 363
7.1 Introduction ... 363
    7.1.1 Purpose ... 363
    7.1.2 Application ... 364
    7.1.3 Historical Perspective ... 364
        7.1.3.1 The Via Net Loss Era ... 365
        7.1.3.2 The Fixed Loss Plan Era ... 365
        7.1.3.3 The All-Digital Network Era ... 369
7.2 Transmission Planning Focus ... 370
7.3 Transmission Planning: Basic Principles and Process ... 371
    7.3.1 Summary of Main Transmission Planning Parameters/ Impairments ... 372
    7.3.2 Summary of User-Oriented Performance Goals ... 374
    7.3.3 Methods of Controlling Impairments ... 374

7.4	The Fixed Loss Plan		378
	7.4.1	Implementation and Control of Loss	378
	7.4.2	Intra-Office Considerations	379
7.5	The All-Digital Network Era—The Future		380
	7.5.1	End-to-End All-Digital Network (Including ISDN)	381
	7.5.2	Interworking of All-Digital and FLP-Based Networks	384
	7.5.3	Simplification of the FLP	391
7.6	Loop Carrier Systems		394
	7.6.1	Digital Loop Carrier Systems	394
	7.6.2	Integrated Digital Loop Carrier (IDLC) Systems	396
	7.6.3	Extension of IDLC to the Customer's Premise	397
7.7	Remote Switching Systems		399
	7.7.1	Transmission Planning Objectives and Parameters	399
	7.7.2	RSS Planning Methodology	401
	7.7.3	The Future	401
7.8	Private Networks		402
	7.8.1	PBX Loss Plan Approach	406
	7.8.2	Existing PBX Loss Plan	408
	7.8.3	All-Digital Network and ISDN Loss Plan	410
7.9	Transmission Delay and Echo Control Considerations		410
	7.9.1	Echo Control	410
	7.9.2	The Management of Increasing Delay	412
	7.9.3	Industry Guidelines	418
7.10	32 kb/s ADPCM Application Rules		418
7.11	References		422
Appendix A.7.1	The Via Net Loss Plan		426
Appendix A.7.2	The Fixed Loss Plan		428
	A.7.2.1	Basic Attributes of the FLP	428
	A.7.2.2	Loss Objectives	429
Appendix A.7.3	Transmission Level Points and Digital Reference Signals		432
	A.7.3.1	Transmission Level Point Methodology	432
	A.7.3.2	Representations of Digital Signal Levels	434
	A.7.3.3	Application of ELP and DLP in Digital Switches	436
Chapter 8	**Transmission Performance Standards**		**439**
8.1	Standards Need		439
8.2	U.S. Standards Bodies		443
	8.2.1	Standards Committee T1—Telecommunications	443
	8.2.2	TIA—TR.30—TR.41	447
	8.2.3	IEEE	447
	8.2.4	Other U.S. Performance-Related Standards Organizations	448

8.3	Canadian Standards Bodies		449
8.4	International Forums		451
	8.4.1	CCITT	451
	8.4.2	CCIR	452
	8.4.3	Private Network Forums	453
8.5	Regional Bodies		453
8.6	Performance Standards Development		454
	8.6.1	End User/Top-Down Driven Standards	455
	8.6.2	Operations-Related/Bottom-Up Driven Standards	458
8.7	Relationships Among Standards Forums		459
	8.7.1	United States	459
	8.7.2	Canada	462
	8.7.3	International (CCITT)	463
8.8	References		465
Appendix A.8.1		Activities of the Standards Forums in CCITT, T1, TIA, and IEEE	468
	A.8.1.1	CCITT (International Telegraph and Telephone Consultative Committee) Study Groups (SGs)	468
	A.8.1.2	Committee T1—Telecommunications:Technical Subcommittees	468
	A.8.1.3	Institute of Electrical and Electronics Engineers: Telecommunications-Related Societies (IEEE)	469
	A.8.1.4	Telecommunications Industry Association	469
Appendix A.8.2		CCITT Study Group Questions on Performance, 1989 to 1992 Study Period	471
	A.8.2.1	CCITT SG XII—Transmission Performance of Telephone Networks and Terminals	471
	A.8.2.2	CCITT SG XVIII—ISDN	472
Appendix A.8.3		List of CCITT Recommendations and Supplements Which Address Transmission Performance	473
	A.8.3.1	Recommendations P.10-P.79 and Supplements— Telephone Transmission Quality	473
	A.8.3.2	Recommendations G.101-G.181 and Supplements— General Characteristics of International Telephone Connections and Circuits	475
	A.8.3.3	Recommendations in G.700 and G.800 Series on Digital Networks Performance	477
	A.8.3.4	Recommendations I.350 to I.352 on Integrated Services Digital Network Performance Objectives (Addressed by CCITT SG XVIII)	478
	A.8.3.5	Recommendations X.130 to X.141 on Data Networks Performance	478

Appendix A.8.4		IEEE, TIA, and T1 Transmission Performance Standards	479
	A.8.4.1	IEEE Standards Addressing Transmission Performance and Its Measurement	479
	A.8.4.2	TIA (EIA) TR-41 Performance Standards	479
	A.8.4.3	ANSI T1 Performance Standards	480
Chapter 9	Maintenance Considerations		481
9.1	Introduction		481
9.2	Maintenance Processes		482
	9.2.1	Preventative Maintenance Process and Functions	482
	9.2.2	Corrective Maintenance Process and Functions	484
9.3	Embedded Maintenance Support Systems		486
	9.3.1	Circuit Maintenance	486
		9.3.1.1 Customer Premises Equipment Maintenance	488
		9.3.1.2 Loop Maintenance	489
		9.3.1.3 Switching System Maintenance	489
		9.3.1.4 Trunk Maintenance	490
		9.3.1.5 Special Service Circuit Maintenance	490
	9.3.2	Carrier Transmission Facility Maintenance	491
9.4	Telecommunications Management Network		493
	9.4.1	Motivation for TMN	493
	9.4.2	TMN Maintenance Functions	495
9.5	Examples of TMN-Like Maintenance Support Systems		496
	9.5.1	Test Access for Analog and Digital Circuits	496
	9.5.2	Quality Assurance Measurements of Circuits	498
	9.5.3	Performance Monitoring and Alarm Surveillance for Digital Networks and Services	501
9.6	Performance Monitoring—Network Considerations		504
	9.6.1	Operations and Transport Network Architecture	504
	9.6.2	Network Elements and Standardized Hierarchical Rates	504
	9.6.3	Layered Approach to Performance Monitoring	509
	9.6.4	Layer 1 Performance Monitoring	512
		9.6.4.1 Section, Line, and Path Monitoring	512
		9.6.4.2 Monitoring at Near-End, Far-End, and Intermediate Locations	514
	9.6.5	In-Service Monitoring for Specific Types of Digital Signals	517
		9.6.5.1 Line Code Oriented Signals	517
		9.6.5.2 Frame-Oriented Signals	518
		9.6.5.3 Other Signals	520
9.7	Generic Performance Monitoring Functions		522
	9.7.1	Nature of Digital Impairments	523
	9.7.2	Performance Primitives	526

		9.7.2.1	Transmission Error Primitives	528
		9.7.2.2	Frame Synchronization Primitives	530
		9.7.2.3	Time Synchronization Primitives	530
		9.7.2.4	Failure-Related Primitives and Alarm and Status Indications	531
		9.7.2.5	Protection Switching Primitives	534
	9.7.3	Performance Parameters		534
		9.7.3.1	Transmission Error Parameters	535
		9.7.3.2	Frame Synchronization Parameter	536
		9.7.3.3	Time Synchronization Parameters	537
		9.7.3.4	Failure-Related Parameter	537
		9.7.3.5	Protection Switching Parameters	538
		9.7.3.6	Far-End Performance Parameters	538
	9.7.4	Accuracy of Performance Assessment		539
		9.7.4.1	Limitations of Block Error Detection Codes	539
		9.7.4.2	Other Accuracy Considerations	543
	9.7.5	Interaction Between Available and Unavailable States		546
	9.7.6	Storing, Thresholding, and Reporting of Performance Information		549
		9.7.6.1	Memory	549
		9.7.6.2	Thresholding	549
		9.7.6.3	Reporting	551
9.8	Review of Maintenance Capabilities Based on Continuous, In-Service Performance Monitoring			552
	9.8.1	Basic Maintenance Capabilities		552
	9.8.2	Enhanced Maintenance Capabilities		553
9.9	References			554

Epilogue	571
List of Abbreviations and Acronyms	573
Index	587

PREFACE

The goal of this book is to address the task of achieving high-quality service in telecommunications networks in the face of many changes currently underway in the telecommunications industry. These changes include the evolution from current mixed analog-digital networks to eventual end-to-end all-digital networks. The focus is on transmission performance for voiceband (voice and voiceband data) services and for digital data and packet data services. Higher rate services such as high-speed data, video applications, and multimedia services are also considered, but mainly in the context of wideband and broadband digital network performance; that is the underlying performance needed at the open systems interconnection (OSI) layer 1 level to support the most critical of such services. Broadband digital networks are capable of supporting broadband access (B-ISDN) and broadband transport using asynchronous transfer mode (ATM) techniques. In this regard, consideration is given to the effect of ATM on the performance of voiceband services.

A systematic approach to quality of service is particularly required when addressing the performance of public switched networks. In the U.S., this is motivated largely by the divestiture of AT&T and the resulting competitive multivendor environment. In this environment, a single telecommunication administration can no longer completely determine performance on an end-to-end basis. To some considerable degree, this responsibility now lies with standards forums; in particular, the American National Standards Institute (ANSI) Accredited Standards Committee T1—Telecommunications in the public networks domain, and the Telecommunications Industry Association (TIA) in the private and mobile networks domain. This trend is also visible in Europe where, for example, the European Telecommunications Standards Institute (ETSI) is addressing basic network performance requirements.

To achieve the goal of this book, the concept of an overall performance process is used; this concept provides a logical and structured framework for discussing transmission performance. In particular, this book provides a formalized approach to achieving high transmission quality of service in terms of two principle aspects of the performance process:

- A top-down aspect in which quality of service is designed into networks by individual network vendors in cooperation with coordinating standards bodies.
- A bottom-up aspect in which quality of service is verified and controlled under actual network operating conditions.

This approach remains largely consistent with methods used previously by telecommunications administrations and standards bodies, and, in addition, it reflects changing present-day economic, technological, and competitive constraints.

This book is intended for telecommunications professionals (engineers, managers, and consultants) who would have occasion to use it in situations involving network design, performance assessment, planning, maintenance, and equipment specification. This book would also serve as a useful reference for overseas organizations seeking alignment with North American standards, operating companies interworking with North American counterparts to support international traffic, and manufacturers entering the North American telecommunications market. Finally, the book can also serve as a reference for graduate and undergraduate electrical engineering students studying communications or telecommunications. As prerequisites, some background in electrical engineering or in communications would be helpful.

This book attempts to bridge the gap between industry-oriented books on transmission, which tend to take a particular operating company perspective and require specific background and insight to understand, and more academic books, which tend to cover a broader range of more theoretical topics with generally less depth and/or less emphasis on practical matters. In bridging this gap, this book focuses on transmission performance from a contemporary industry perspective. Background and practical insights are included to make the book more readily useful to nonexperts on the subject. Each chapter includes an extensive list of references, and some chapters include appendices.

This book is organized around the overall performance process noted above, which links all the chapters together as follows: the performance process (introduced in Ch. 1) provides an appreciation of performance ranging from user-oriented to network component-oriented (Chs. 2 to 6). These chapters are followed by associated network planning (Ch. 7), industry standards (Ch. 8), and maintenance considerations (Ch. 9). However, the reader is not limited to following the chapters in sequence. That is, most chapters may be referenced individually. Also, an alternative sequence of usage is possible. The sequence from Chapters 2 to 6 noted above represents a top-down view of how end-to-end user-oriented performance (Ch. 2) is translated (Ch. 3) into end-to-end network-oriented performance (Ch. 4), and is then allocated (Ch. 5) to network components (Ch. 6). An equally instructive alternative is the reverse sequence from Chapters 6 to 2. This represents a bottom-up view of how network component-oriented performance (Ch. 6) can be accumulated (Ch. 5) to obtain end-to-end network-oriented performance (Ch. 4), and finally translated (Ch. 3) into end-to-end user-oriented performance (Ch. 2).

In addition to the performance content of each chapter, some chapters also provide the necessary background on relevant architectural aspects of networks, digital transmission equipment, and operations systems. Specifically:

- Chapters 3, 4, and 5 include analog and digital connection models and hypothetical reference connections (HRX).
- Chapter 6 includes functional aspects of digital regenerators and digital multiplexers, and of synchronization networks.
- Chapter 7 includes architectural aspects of access network evolution from a mixed analog-digital, to an eventual all-digital state.
- Chapter 9 includes fundamentals of digital signal hierarchies (including the synchronous optical network (SONET) hierarchy) and related equipment, as well as the functions of embedded and future operations systems for circuit and facility maintenance.

ACKNOWLEDGMENTS

The content of this book has been influenced by the long and fruitful association the authors have had with their many colleagues at Bell Northern Research (BNR), as well as at Bell Canada and at Northern Telecom. Those colleagues who coauthored the papers referenced herein with the authors are especially acknowledged.

The authors appreciate the managerial encouragement and administrative support in clearing the way for this book to proceed, provided in particular by Roy Mills and Roly Champagne of BNR, by Stu Erskine (formerly of BNR), by Doug Carruthers of Bell Canada, and by Bob Kenedi of Northern Telecom Canada.

The authors are grateful for the efforts of those who reviewed and commented on the drafts of this book: in particular, Bernie Murphy and Joe Zebarth of Bell Canada, Dick Skillen of Northern Telecom Canada, and Roy Mills, Bob Brett, and Jim Barry of BNR.

Finally, the authors are indebted to Amy Ciancibello, Eileen Devlin, and Eluned Davies for their excellent work in preparing most of the text, figures, and tables for the book.

Chapter 1
INTRODUCTION

1.1 DEFINITION OF TRANSMISSION PERFORMANCE

1.1.1 Relationship of Transmission Performance to Categories of Performance and to Phases of the Communications Process

In basic terms, transmission refers to the transport of information, such as voice signals or data messages, by electronic means from one location (source or origin) to another location (sink or destination). Transmission performance refers to the fidelity with which the information is carried, or the extent to which the integrity of the information is preserved.

In more formal terms, transmission performance is associated with one of the many categories of performance that determine the overall quality of telecommunication services as perceived by end users. The various user-oriented performance categories as defined by the International Telegraph and Telephone Consultative Committee (CCITT) are shown in Figure 1.1 [1–3, 4]. The main categories for which performance parameters have been determined are service accessibility, service integrity, and service retainability [5, 6]. As shown in Table 1.1, these user-oriented categories correspond to network-oriented traffic, transmission, and availability performance, respectively. These categories are also associated with the various phases of the communications process as discussed below [7–10]. In particular:

- *Service Accessibility* is the ease with which users gain access to telecommunication services or resources. This aspect of performance relates to the access phase of the communications process, and is governed mainly by traffic considerations. Service accessibility relates to the functions of connection establishment (such as for voice calls or data sessions) and associated data base access, protocol conversion, etc., as required to support a particular service.
- *Service Integrity (or Suitability)* is users' general opinion of, or level of satisfaction with, a service after gaining access to it. This aspect of performance relates to the transfer phase of the communications process, and is governed mainly

Figure 1.1 Performance categories contributing to overall quality of service. (The focus in this book is on the Service Integrity category, which relates to network transmission performance.)

Table 1.1
Main User- and Network-Oriented Performance Categories

User/Service Perspective	Network Performance Perspective
Service Accessibility	⟶ Traffic Performance
Service Integrity	⟶ Transmission Performance
Service Retainability	⟶ Availability Performance

by transmission considerations. **Service integrity is the performance category emphasized in this book.**

- *Service Retainability (or Robustness)* is the ability of the communication service to remain functional for the duration required by users. This aspect of performance is related to all phases of the communications process, and is governed mainly by availability considerations. Note that availability is affected by equipment failure, as well as by degradations in traffic and/or transmission performance that are severe and persistent enough to render the service unusable. Such degradations in transmission performance are emphasized here.

Figure 1.2 illustrates the phases of the communications process referred to above, namely:

- The access (or connection setup) phase.
- **The information transfer (or information exchange) phase.**
- The disengagement (or connection clearing) phase.

Figure 1.2 Phases of the communication process. (The focus of this book is on the Information Transfer phase for which transmission performance is manifested.)

Thus, for example, if the communications process is a telephone voice call, then transmission performance (associated with the information transfer phase) relates to speech quality during the conversation. Alternatively, if the communications process is a data session, transmission performance relates to the accuracy and timeliness of the data messages exchanged.

It should be noted that there are dependencies among the service accessibility (traffic), integrity (transmission), and retainability (availability) categories. For example, the effect of severe and persistent traffic and transmission performance degradation on availability was noted above. Further, for certain "packet"-oriented services, congestion (a traffic aspect of performance), can be manifested not only during the access and disengagement phases, but also during the information transfer phase, and thereby affect the integrity of information transfer.

1.1.2 Relationship of Transmission Performance to Performance Criteria

For each phase of the communications process, three performance criteria have been defined, namely, the speed, accuracy, and dependability of service criteria. As indicated in Table 1.2, performance parameters are associated with each pair-wise (communications phase/performance criterion) combination, and these parameters are provided later in Chapter 2. For transmission performance, the various performance criteria have the following meaning:

- The **speed** criterion (how "fast" information is transferred) relates mainly to delay for voice service and throughput and response time for data services.

- The **accuracy** criteria (how "well" information is transferred) relates mainly to numerous analog impairments for voice service and to error performance for all services.
- Finally, the **dependability** criterion (how "reliably" information is transferred) relates to loss of information (or possibly loss of service) for reasons related to transmission performance degradation.

Table 1.2
Communication Phases and Performance Criteria Framework (The focus in this book is on transmission performance parameters associated with the information transfer phase.)

Communication Phase \ Performance Criteria	Speed	Accuracy	Dependability
Access			
Information Transfer		Primary Performance Parameters*	
Disengagement			

* Primary parameters are more-or-less directly assessable or measureable.
 Secondary parameters (i.e., relating to availability) can be derived from the primary parameters.

In summary, transmission performance is associated with the information transfer phase of the communications process and relates to the integrity of the information exchange as embodied in the speed, accuracy, and dependability performance criteria. These aspects of performance are combined in Table 1.3 to provide a framework and focus for dealing with transmission performance throughout this book. The performance parameters associated with Table 1.3 are provided in Chapter 2. There are primary parameters (i.e., more or less directly assessable or measurable parameters) for each of the above criteria. In addition, secondary parameters can be derived from the primary parameters. In particular, performance degradations below a threshold for some combination of the primary parameters serve to define unavailability. Some aspects of unavailability from a performance standpoint are included in Chapters 3 and 4, and aspects from a maintenance standpoint are in Chapter 9.

Table 1.3
Overall Framework for Performance Parameters
(The focus in this book is on transmission performance parameters associated with the service integrity performance category and the information transfer phase.)

Performance Category		Communications Phase	Performance Criteria		
User-oriented/ Quality-of-service Perspective	Network Performance Perspective		Speed	Accuracy	Dependability
Accessibility	Traffic	Access and Disengagement			
Integrity/ Suitability	Transmission	Information Transfer		Performance Parameters *	
Retainability/ Robustness	Availability	All			

* These parameters will include some related to unavailability, as contributed to mainly by excessively degraded transmission performance.

1.2 SCOPE OF THE BOOK

The scope of this book, as already indicated, centers on transmission performance. At this point, the scope is also outlined in terms of other relevant factors, such as the types of services and networks for which the performance work herein is applicable.

1.2.1 Relevant Telecommunications Services

The existing and new services that a fully integrated telecommunications network may support are

- Voice:
  - Plain ordinary telephone service (POTS)
  - Teleconferencing/hands-free telephony
  - High-quality telephony
  - Voice response applications
  - Sound programming
- Data:
  - Private line (mainly for bulk/file transfer, batch mode, and remote job entry applications)
  - Packet-switched (mainly for real-time, interactive, time-sharing, transaction, and inquiry/response applications)

- Circuit switched; analog voiceband data or digital data (for many of the above applications)
- Image:
  - Facsimile
  - Broadcast television
  - Video conferencing
  - Audio-graphics conferencing
  - Videotex
  - Video telephony
  - Advanced/enhanced/improved television
  - High-definition television (HDTV)
  - High-resolution imaging
- Signaling:
  - End-to-end user signaling
  - Data base services
- Multimedia:
  - Combinations of voice, data, and image services made possible by "fast" packet techniques (e.g., asynchronous transfer mode (ATM))

The focus of this book is on the transmission performance of evolving telecommunication networks supporting mainly conversational voice service, voiceband data service, and digital data and packet data services. To some degree, voiceband "data" also encompasses certain other voiceband services as well, such as facsimile and signaling.

With regard to the higher bandwidth data, audio, video, and multimedia services, etc., this book is concerned with the underlying transmission performance of wideband and broadband all-digital transport networks. The goal of such networks is to eventually support the most stringent (and hence all) of the above services. In terms of the open systems interconnection (OSI) model, this corresponds mainly to layer 1 (the service-independent physical layer). Chapter 9 includes architectural aspects of digital transport networks in a maintenance context.

The above paragraph presents a possible solution to the problem of how an integrated services network can satisfy the different performance requirements of different services; that is, satisfy the most stringent requirements. The following paragraphs discuss this and other alternatives, including some of their advantages and disadvantages.

A traditional approach is to require end users (e.g., data users) to enhance performance of their particular service by means of suitable line speeds, block lengths, end-to-end protocols, and associated terminal equipment Examples include user-provided error control by retransmission, and protocol translators (e.g., from stop and wait BISYNC to continuous synchronous data link control (SDLC) for satellite connections).

A related alternative (from the network rather than the user's perspective) is based on the class of service concept, whereby a data user, for example, is provided with a range of performance levels (at a corresponding range of prices). This approach is applicable to current and future networks above OSI layer 1; for example, to the integrated services digital network (ISDN) and the service-dependent ATM adaptation layer (AAL) for broadband ISDN (B-ISDN) access [11–16].

Another traditional network-based approach is to segregate services, with more critical services being supported by specially designed and maintained overlay networks, typically for digital data and packet data services. This approach tends to run contrary to the integrated services concept, and also tends toward more administrative cost and complexity. On the other hand, this approach will continue to be viable in situations where total integration of services into one network is not yet practical for technological or economic reasons. A practical manifestation of this approach is integrated access (ISDN), but segregated (e.g., circuit, packet) transport (see Fig. 9.14 in Chapter 9).

A variation on the above segregated services approach arises from evolution of existing public and private networks, wherein subcomponents are replaced or upgraded with the introduction of new technology, to the point where more stringent services can be supported. This new technology includes

- Inherently higher performance fiber optic transmission.
- New transport modes such as ATM, which provide the basis for services integration.
- Distribution of network intelligence between network elements and operations systems to better enable evolving network capabilities to support more stringent services.

1.2.2 Relevant Telecommunications Networks

Networks can be classified in a number of ways. The following classification is largely application-oriented, but also has some relation to technology and geography:

- General purpose (public) and special purpose (private and military).
- Local (local-exchange or intra-LATA) and toll (inter-exchange or inter-LATA).
- Segregated and integrated (with regard to transmission, switching, and services).
- Switched, slow-switched, and nonswitched.

The following network classification is largely geographically-oriented, but also has some relation to application and technology:

- Local, metropolitan, and wide area networks (LAN, MAN, WAN).
- Land-based and mobile (including personal communication networks).

- Terrestrial, undersea, and satellite.
- National and international.

The following network classification is largely technology-oriented, but also has some relation to application:

- Two-wire and four-wire.
- All-analog, mixed analog-digital, and all-digital.
- Circuit, packet, hybrid, fast circuit, and fast packet.

The intent of this book is to provide a sufficiently generic treatment of transmission performance that can be applied to most of the above networks. However, the focus will be on evolving nationwide public switched networks supporting voice and data services, taking into account the constraints of international connections. To a large extent, the information also applies to switched and nonswitched private networks. The following discussion provides additional background on technological and evolutionary aspects of such networks [17]. Chapters 4, 5, and 9 include network connection models and aspects of network architecture.

Voice networks are largely based on a slowly evolving hybrid combination of analog and digital technology. Data networks, on the other hand, have evolved rapidly, but less consistently, over the last few decades, resulting in the following data network capabilities:

- Use of voiceband modems to transmit data over the telephone network at speeds of up to 9.6 kb/s, and more recently at higher speeds of up to 19.2 kb/s.
- Use of digital data sets and dedicated facilities to support end-to-end data transmission at speeds of up to 56 kb/s, and more recently up to 64 kb/s.
- Use of circuit- and packet-switched data networks to support a wide range of data speeds and terminal protocols.
- Use of emerging "fast" packet techniques such as frame relay [18], switched multimegabit data service (SMDS) [19–21], and ATM [11–16] to support high-speed data transfer.

The trend is to use the emerging digital telephone network to support voice, data, video, and eventually multimedia services. The digital telephone network is the result of the deployment and integration of digital transmission and digital switching systems, and it is commonly referred to as the integrated digital network (IDN). The benefits of the IDN for carrying voice are lower cost, improved end-to-end performance, high availability, and sophisticated services based on stored program control.

Given the inherent digital capability of the IDN and the general ubiquity of the telephone network, it is appropriate to also consider the IDN for support of existing and new data services. Transparent data services at rates of up to 64 kb/s in the digital circuit-switched form are intrinsic. In addition, the IDN offers advanced routing, signaling, maintenance, and administration capabilities to support both voice and data services. This concept of voice and data services supported in the IDN forms the basis of the emerging ISDN.

The early stages of this integration process, from the user-service perspective, will enable users to gain access to a variety of telecommunications services by way of a small number of standardized user-network interfaces, including Basic Rate, Primary Rate, and Broadband Access. In the core, or backbone, portion of the network, the switching scheme used will not necessarily be pure circuit or packet switching, but for a considerable transition period will consist of both schemes. Eventually, an appropriate hybrid combination of schemes [22] or a single scheme such as ATM is envisaged [11-16].

In summary, the discussion of performance for voice service in this book addresses the transmission performance of existing mixed analog-digital networks, as well as the effects on performance of evolution toward the all-digital ISDN. Network performance for data service necessarily reflects an evolutionary process that is less consistent than for voice. The discussions of voice and data performance apply largely to a wide variety of appropriate network designs, transmission schemes, and switching architectures. Both traditional transmission performance parameters, as well as more contemporary parameters associated with new and evolving systems are considered [23, 24]. These systems include digital circuit multiplication equipment (DCME) such as low bit rate codecs (LBRC) [25-28], digital speech interpolation (DSI) systems [29, 30], as well as fast packet [31] and ATM networks. Additional references on these systems are provided in Chapter 3.

1.3 HISTORICAL, CONTEMPORARY, AND FUTURE PERSPECTIVES

1.3.1 Regulated, Predivestiture Environment

In the U.S., the regulated telephone companies of the predivestiture era provided "good" network performance, and hence "good" quality of service for end users. This was driven through performance standards as well as equipment and system specifications set by the Bell System (e.g., see [32]). In this period, the Bell System was also effective in influencing world standards through their participation in CCITT affairs (e.g., see [33]).

The decision to proceed with deregulation in the U.S was not based on quality, but rather on the possible disadvantage of the monopolistic Bell System in terms of pricing and competitiveness [34-37]. A forerunner of deregulation was the emergence of a number of specialized, or "other common carriers" (OCC), such as Microwave Communications Inc. (MCI), which began by providing private network services [38].

The situation in Canada during this period was different [39]. Telecom Canada (formerly the Trans Canada Telephone System—TCTS) and its member operating companies such as Bell Canada, proceeded at a relatively fast pace to modernize the nationwide network [40, 41]. In many respects, conversion of the Canadian network to digital working has proceeded at a faster rate than in the U.S. There are many reasons for this, including the part that the Canadian telecommunications industry has played in developing and utilizing digital communication technology [42]. Telecom Canada continues to influence world standards through its participation in CCITT.

1.3.2 The Competitive Environment

1.3.2.1 The United States

An important factor influencing network performance in the U.S. is the increasingly deregulated, competitive, and market-oriented nature of the telecommunications industry. Central to this was the divestiture of AT&T and the resulting multivendor environment. The impact of divestiture in the U.S. has been considerable, especially in the performance area in terms of setting performance objectives and establishing necessary agreements between network providers [43, 44]. In this respect, three aspects of divestiture in the U.S. are addressed as follows:

  (1) Local-exchange carrier and inter-exchange carrier partitioning.
  (2) Technological advances.
  (3) End-user impact.

Local-Exchange Carrier and Inter-Exchange Carrier Partitioning

Divestiture divided the Bell System into a number of parts consisting of AT&T, which retained the "long lines" division, and the seven regional bell operating companies (RBOC) supported by Bell Communications Research (Bellcore). AT&T is now a major inter-exchange carrier (IEC), and the RBOCS and Bellcore constitute the local-exchange carrier (LEC) part of the industry. Given this situation, responsibility for setting standards, specifying equipment, and assuring end user quality no longer resides with one entity [34, 35, 45]. In addition, a large number of small IECs arose to take advantage of the rapidly emerging competitive environment. It was recognized early in this process that a U.S. forum for setting standards and addressing standards issues among the various carriers was required, and in 1986 the Committee T1-Telecommunications was established for this purpose [46]. Within this forum, a number of technical committees were organized with, for example, T1Q1 addressing performance requirements, and T1M1 addressing maintenance. (Ch. 8 addresses various standards forums in detail.)

T1 is now fully established and has already generated a number of ANSI accredited standards. These standards tend to involve minimal levels of performance, as dictated by existing network capabilities and technologies, and by economic factors.

Technological Advances

Technological advances in processing power and memory capacity and in digital switching and transmission have influenced network performance. These advances have altered technological limits such that telecommunications equipment manufacturers can produce competitive products with higher levels of performance and increased functionality. This in turn provides an opportunity for network vendors to offer a range of performance at appropriate prices consistent with the range of services to be supported. For example, triggered by end user needs, network providers could offer higher performance on a competitive basis and use this as a means of differentiating their service offerings from those of other network providers.

In the early days of divestiture, because of the technological limitations imposed by the access arrangement between the LECs and some of the IECs, overall performance in the U.S. was generally lower than in the predivestiture era [47]. However, because some of the IECs offered service at a substantially lower price than AT&T, it is conceivable that the end user was able to trade this price against the lower performance and so accept the overall situation. Pricing of services will remain a significant factor in determining the viability of many of the IECs [48, 49].

In the years following divestiture, "equal access" has minimized the performance differentials between the IECs, and there has been a consolidation of IECs, many of the smaller ones having gone out of business. Many of the remaining IECs are proceeding with the implementation of new "high-performance" technology such as fiber-optic transmission systems [50]. The ongoing competition with AT&T is an example of the use of performance as an advertising and promotional tool.

With the emergence of "equal access" arrangements for connection to the IECs, the price differential between AT&T and the other IECs for services is being reduced. This leaves performance as a potentially significant factor in influencing the end user's choice of IEC. In summary, performance is becoming, and will continue to be, a factor in determining the viability of many of the IECs [51–57].

End User Impact

During the current evolutionary phase of telecommunication networks, network performance is influenced by the increasing awareness, expectations, and sophistication of the end users. End users include individual business and residential users, as well as large corporate users of private networks. The needs of these various users are not necessarily the same.

Large business users provide the greatest percentage of revenue and, as such, are sought after by the network providers. For the large business, both price and performance are significant factors. Price is important because it has an impact on corporate economics (overhead) and competitive advantage. Performance is important in terms of employee and customer satisfaction; that is, for voice service, many individual business users expect improved performance. AT&T currently retains a large number of business customers, a contributory factor being the high overall performance (of transmission and traffic, etc.) offered by the AT&T network.

For the residential user, there is still some confusion over the proliferation of carriers and billing procedures [58], although equal access has made it easier for the end user to choose a carrier. It remains to be seen how this and performance-related media advertising will affect residential users.

1.3.2.2 Canada and Other Countries

In Canada, competition is currently limited to terminals and private networks [39, 59, 60, 61]; thus, the need for a forum analogous to T1 is not required. Telecom Canada currently participates in the U.S. standards forums and in CCITT to address cross-border interworking issues, and also participates in CCITT in developing performance and maintenance-related recommendations. Should competition in the public network domain evolve in Canada, it is likely that standards analogous to those developed in T1 or CCITT will be required.

In European countries, although telecommunications is still largely controlled by government agencies with little or no competition, there are indications of a trend toward deregulation and competition. For example, in the United Kingdom Mercury Communications competes with British Telecom [62]. There are similar situations in various other countries, as reflected in [63].

1.4 INTRODUCTION TO THE OVERALL PERFORMANCE PROCESS

In the long run, the combined effect of such factors as competition, new service offerings, user awareness, and technological advances will be an evolution toward networks with generally higher average performance. However, there will be a transition phase during which networks will exhibit a broad range of performance capabilities, only the highest of which can support the most stringent service requirements.

The purpose of this book is to provide a framework for achieving overall quality of service for evolving telecommunications networks in the face of the changing conditions noted above. In order for this framework to reflect the evolutionary aspects of the telecommunications industry, it should retain a large measure of consistency with previous approaches used by telecommunication administrations (e.g., The Bell System) and standards bodies (e.g., CCITT), and be flexible enough to incorporate evolutionary changes.

The framework is illustrated in Figure 1.3, and it has the following two principle components [8–10]:

- A *top-down* process in which the required quality of service is designed into networks. This process consists of three traditional steps: (1) translation of subjective user-oriented performance measures into quantitative network-ori-

ented performance parameters, (2) allocation of the end-to-end network-oriented performance to network components, and (3) network planning (including a performance assessment plan) to ensure that network implementations achieve intended performance [64]. Associated with each of these top-down steps are practical considerations (dotted lines in Fig. 1.3) which reflect the effects of economic, technological, and competitive realities. The top-down aspects of the performance process, as shown in Figure 1.3, are addressed in Chapters 2 to 8 of this book.
- An overall *bottom-up* validation process consisting of quality assurance and performance control. Quality assurance and performance control are specific aspects of network operations administration and maintenance (OAM). Quality assurance, in this context, is the process of verifying that intended network performance is achieved with actual implementations and under day-to-day network operating conditions. The emphasis is on continuous in-service performance monitoring and surveillance-based maintenance (rather than trouble-driven maintenance) as a means of identifying performance problems. When performance deficiencies are identified by the quality assurance process, performance control then provides the ability to maintain the network on an essentially continuous and real-time basis so that satisfactory end user performance is achieved [65]. The emphasis is on the use of network intelligence (rather than fixed rules or human intervention) to control performance on a dynamic basis. Network intelligence can be used to select network resources to tune or customize the network in accordance with specific service requirements (recall the various services listed in Sec. 1.2.1). However, from the standpoint of operational simplicity, it is preferable that the network be able to support the most stringent requirements. The bottom-up process, as shown in Figure 1.3, is addressed in terms of operations and maintenance considerations in Chapter 9 of this book.

The following is a brief synopsis of each chapter (with reference to Fig. 1.3):

- Chapter 1 defines transmission performance in the context of a broad quality-of-service framework, outlines the scope of the book in terms of technical and contemporary considerations, and provides an overall performance process within which transmission performance is examined throughout the book.
- Chapter 2 addresses end-to-end, user-oriented performance and parameters primarily for voice and data services, with a focus on performance assessment techniques and grade-of-service methodology.
- Chapter 3 addresses the specific relationships between subjective user-oriented performance and objective network-oriented performance in terms of an expanded list of transmission performance parameters for voice and data services, and in terms of detailed grade-of-service results for voice service.
- Chapter 4 gives more detailed consideration of end-to-end network-oriented performance on a per parameter basis for voiceband, digital, and packet-type

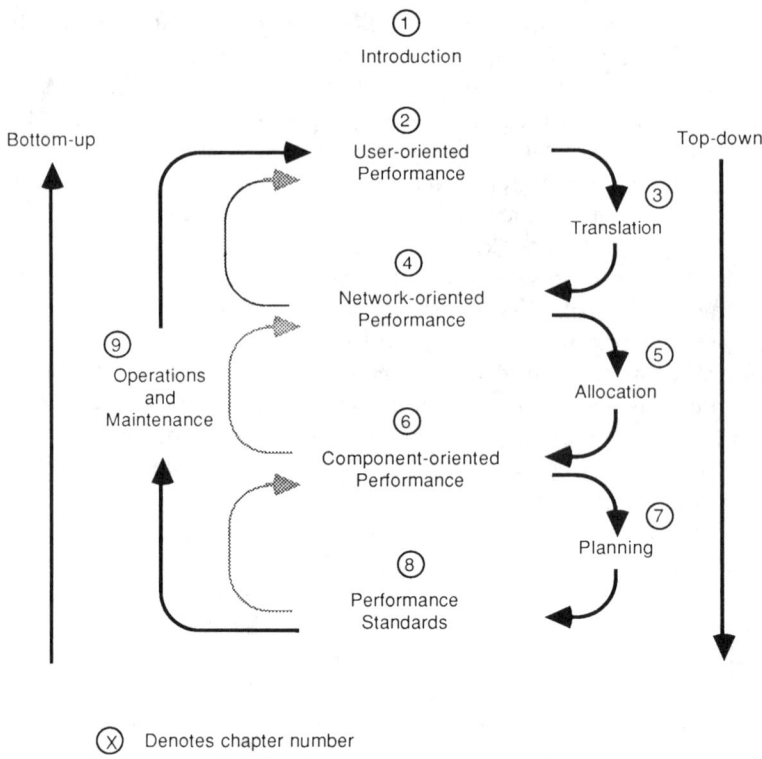

Figure 1.3 The overall performance process. (Introduced in Chapter 1)

services, and in terms of connection models, survey data, and typical performance objectives.
- Chapter 5 addresses the allocation of end-to-end network-oriented performance among various networks and network components that may make up an end-to-end connection. The allocation processes include algebraic and power-like addition laws, deterministic and statistical (normal and log-normal) techniques, and serial and parallel configurations, for various analog and digital parameters.
- Chapter 6 addresses the performance of network components (or, in some cases, the performance of portions of network connections) on a per parameter basis for voiceband services (including 32-kb/s adaptive differential pulse code modulation (ADPCM) performance), digital transport networks (error performance, and jitter, wander, and synchronization-related performance) in terms of survey data, and typical performance limits.
- Chapter 7 addresses transmission planning, that is, the rules and guidelines for network implementation (e.g., of loss and echo control) and equipment

deployment (e.g., of ADPCM codecs) to ensure that end-to-end performance (as allocated to network components) is achieved, while capital expense and operational costs are minimized. This chapter includes transmission planning considerations for access network evolution by way of subscriber carrier, remote switching and PBX systems, to provide eventual end-to-end all-digital connectivity. Also included are transmission planning considerations for mobile systems and for ATM based transport networks.
- Chapter 8 addresses standards for telecommunications networks, which ensure that networks involving many different administrations and systems will interwork, can be maintained, and will provide satisfactory end-to-end performance. This chapter focuses on transmission-related standards activities in North America (i.e., Committee—T1, EIA/TIA, and IEEE) and CCITT recommendations internationally. This chapter also touches on regional standards activities in Europe as reflected in ETSI.
- Finally, Chapter 9 addresses the overall bottom-up process of verifying and controlling quality of service. This process may be referred to as quality assurance, which consists of verifying that intended network performance is achieved under day-to-day operating conditions, and of initiating appropriate action when performance is not as intended. This chapter includes fundamentals of digital signal heirarchies (including the synchronous optical network (SONET) hierarchy) and related equipment, and outlines the maintenance functions of embedded and future operations systems for circuit and facility maintenance. The focus is on performance monitoring and alarm surveillance of digital transport systems to ensure the satisfactory support of emerging broadband services.

1.5 REFERENCES

1. CCITT Rec. E.800, "Terms and Definitions Related to the Quality of Service of Telecommunications Services," *Blue Book,* Vol. II, Fas. II.3, Geneva, 1989, pp. 257–268. Also, Supp. No. 1, "Terms and Definitions for Quality of Service, Network Performance, Dependability and Trafficability Studies," pp. 317–365.
2. CCITT Rec. I.350, "General Aspects of Quality of Service and Network Performance in Digital Networks, Including ISDN," *Blue Book,* Vol. III, Fas. III.I, Geneva, 1989, pp. 129–139.
3. F. Riciniello, "The CCITT and the Quality of Service," *ICC Rec.,* Vol. 1, Boston, June 1989, pp. 12.5.1–12.5.5.
4. *Handbook on Quality of Service and Network Performance,* Preliminary Draft, Issue 1, CCITT WP II/4 Q.21, June 1991.
5. N.B. Seitz, "Performance Standards for Packet-Switched Services," *ICC Rec.,* Vol. 1, Boston, June 1989, pp. 3.1.1–3.1.8.
6. *American National Standard for Telecommunications—Performance Parameters for Packet Switched Data Communication Service,* ANSI T1.504-1989.
7. J.S. Richters and C.A. Dvorak, "A Framework for Defining the Quality of Communications Services," *IEEE Comm. Mag.,* Vol. 26, No. 10, Oct. 1988, pp. 17–23.
8. J. Gruber, E. Abdou, P. Richards and G. Williams, "Quality-of-Service in Evolving Telecommu-

nications Networks," *IEEE Journal on Sel. Areas in Comm.*, Issue on Quality Assurance for the Communications Community, Vol. SAC-4, No. 7, Oct. 1986, pp. 1084–1089.
9. J. Gruber, "From Performance to Planning: An Overview," *NCF Rec.*, Vol. 39, Chicago, Oct. 1985, pp. 551–555, 565, 566.
10. D.R. Jones and H.A. Malec, "Communications Systems Performability: New Horizons," *ICC Rec.*, Vol. 1, Boston, June 1989, pp. 1.4.1–1.4.9.
11. *IEEE Network: The Magazine of Computer Communications*, Special Issue on "Broadband Networks," Vol. 3, No. 1, Jan. 1989.
12. "Preliminary Report on Broadband ISDN Transfer Protocols," Bellcore SR-NWT-001763, Issue 1, Dec. 1990.
13. CCITT Recommendations Drafted by Working Party 8 (General B-ISDN Aspects), Report COM XVIII-R 34-E, June 1990.
14. "B-ISDN: High Performance Transport," *IEEE Mag.*, Vol. 29, No. 9, Sept. 1991.
15. "Broadband ISDN Transport Network Elements Framework Generic Criteria," Bellcore FA-1109, Issue 1, Dec. 1990.
16. "Broadband ISDN Switching System Framework Generic Criteria," Bellcore FA-1110, Issue 1, Dec. 1990.
17. J. Gruber and N.H. Le, "Performance Requirement for Integrated Voice and Data Networks," *IEEE Journal on Sel. Areas in Comm.*, Vol. SAC-1, No. 6, pp. 981–1005, Dec. 1983. Also, GLOBECOM Rec. Vol. 1, San Diego, Nov./Dec. 1983, pp. 23.5.1–23.5.7. Also, "Perspectives on Packetized Voice and Data Communications," IEEE Press, 1991, pp. 37–61.
18. American National Standard for Telecommunications—ISDN—Architectural Framework and Service Description for Frame-Relaying Bearer Service, ANSI T1.606-1990.
19. "Generic Systems Requirements in Support of Switched Multi-megabit Data Service," Bellcore TR-772, Issue 1, May 1991.
20. "Local Access System Generic Requirements, Objectives, and Interface in Support of SMDS Service," Bellcore TR-773, Issue 1, June 1991.
21. "Switched Multi-megabit Data Service (SMDS) Operations Technology Network Element Generic Requirements," Bellcore TA-774, Issue 3, Feb. 1991.
22. J. Gruber, "Performance Considerations for Integrated Voice and Data Networks: A State-of-the-Art Review of Traffic Management Strategies," *Comp. Comm.*, Vol. 4, June 1981, pp. 106–126. Also, *IEEE Trans. Comm.*, Vol. COM-29, June 1981, pp. 786–800. Also, *Proc. 6th Data Comm. Symp.*, Pacific Grove, California, Nov. 27–29, 1979, pp. 166–180.
23. J. Gruber and L. Strawczynski, "Subjective Effects of Variable Delay and Speech Clipping in Dynamically Managed Voice Systems," *IEEE Trans. Comm.*, Vol. COM-33, No. 8, Aug. 1985, pp. 801–808. Also, *GLOBECOM Rec.*, Vol. 3, Miami, Nov.–Dec. 1982, pp. F7.3.1–F7.3.5. Also, *Telesis*, Vol. 10, No. 2, BNR, 1983, pp. 30–34.
24. J. Gruber, "A Comparison of Measured and Calculated Speech Temporal Parameters," *IEEE Trans. Comm.*, Vol. COM-30, Apr. 1982, pp. 728–738. Also, Conf. Rec., 15th HICSS, Hawaii, Jan. 1982, pp. 671–680.
25. N.S. Jayant, "Coding Speech at Low Bit Rates," *IEEE Spectrum*, Vol. 23, No. 8, Aug. 1986, pp. 58–63.
26. S. Lipoff, "New Techniques in Voice Compression," *Bus. Comm. Rev.*, Vol. 19, No. 5, May 1989, pp. 81–83.
27. P. Strauss, "Squeeze Play: Carriers Quietly Trying Compressed Voice; Users Could Soon Squeeze Big Savings," *Data Comm.*, Vol. 16, No. 8, July 1987, pp. 66, 70, 72, 74, 78, 80.
28. N. Muller, "ADPCM Offers Practical Methods for Doubling T1 Capacity," *Data Comm.*, Vol. 16, No. 2, Feb. 1987, pp. 213–216.
29. W.G. Schmidt, "Digital Circuit Multiplication Gets 120 Speech Channels from T1," *Data Comm.*, Vol. 16, No. 2, Feb. 1987, pp. 197, 198, 201–206, 209, 210.

30. E. Looney, "Using DSI Technology for Optimizing T1 Performance," *Telecomm.*, Vol. 21, No. 12, Dec. 1987, pp. 62, 65, 67.
31. D. Sparrell, "Wideband Packet Technology," *GLOBECOM Rec.*, Vol. 3, Hollywood, Florida., Nov./Dec. 1988, pp. 49.2.1–49.2.5.
32. *Catalog of Communications Technical Publications*, PUB 10000, AT&T, April 1989.
33. CCITT Supp. No. 3 (to Rec. P.11) on "Transmission Rating Models," pp. 248–261, and Annex A (to Supp. No. 3) on "Opinion Ratings of Transmission Impairments," pp. 299–306, Bellcore. *Blue Book*, Vol. V, Geneva, 1989.
34. *IEEE Comm. Mag.*, Vol. 23, No. 12, Dec. 1985. Issue on "Divestiture: Two Years Later."
35. *IEEE Spectrum*, Vol. 22, No. 11, Nov. 1985. Issue on "Telecommunications."
36. P. Temin and L. Galambos, "The Fall of the Bell System," Cambridge Univ. Press, 1988.
37. S.A. Simon and M. Whelan, *After Divestiture: What the AT&T Settlement Means for Business and Residential Telephone Service*, Knowledge Industry Publications, 1985.
38. E.C. Gaines, "Specialized Common Carriers—Competition and Alternative," *Telecomm.*, Vol. 7, No. 9, Sept. 1973, pp. 15–18, 20–23, 26, 27.
39. R.M. Fenson, "Diverging Regulations—In North America, Europe and Japan: Canada," *GLOBECOM Rec.*, Vol. 2, New Orleans, Dec. 1985, pp. 26.4.1–26.4.3.
40. B. Stoffels, "How They Ring the Bell in Canada," *TE&M*, Vol. 92, No. 10, May 15, 1988, pp. 137–138.
41. J.D. Fahey, C.T. Seager, and E.O. Capes, "Bell Canada Studies Its Digital Experience—and Is Pleased," *Telephony*, Vol. 207, No. 2, July 9, 1984, pp. 85, 86, 91, 92, 94, 120.
42. "20 Years of Innovation," *Telesis*, Issue 92, BNR, July 1991.
43. R.E. Collins, "Service Quality—Measuring the Network and the Customer," *Telephony*, Vol. 209, No. 24, Dec. 9, 1985, pp. 32–34.
44. M.N. Woinsky, "National Performance Standards for Telecommunications Services," *IEEE Comm. Mag.*, Vol. 26, No. 10, Oct. 1988, pp. 70–81.
45. T.E. Bell, "Technical Challanges to the Decentralized Phone System," *IEEE Spectrum*, Vol. 27, No. 9, Sept. 1990, pp. 32–37.
46. E.J. Cohen, "Exchange Carrier Standards Primer," *TE&M*, Vol. 88, No. 19, Oct. 1 1984, pp. 116, 117, 120.
47. F.D. Richter, "Usage Measurements on Feature Group Lines," *TE&M*, Vol. 90, No. 4, Feb. 15, 1986, pp. 107–110.
48. "Long-Distance Phoning: Time to Get Fickle," *Consumer Reports*, Vol. 51, No. 5, May 1986, pp. 302–304.
49. G.D. Solomon, "This Coupon Good for One Free Phone Call," *TE&M*, Vol. 90, No. 6, Mar. 1986, p. 81.
50. E. Booker and C. Inan, "Interexchange Carriers Campaign for Market Share in 1988," *Telephony*, Vol. 214, No. 11, Mar. 14, 1988, pp. 22, 23, 26, 27.
51. "Ratings Reveal Carriers are Worlds Apart," *Data Comm.*, Vol. 17, No. 8, Aug. 1988, pp. 103–106, 108.
52. "Report Cards Out: Carriers Rated on Service, Quality," *Data Comm.*, Vol. 16, No. 9, Aug. 1987, pp. 127–130, 133, 134.
53. A.A. Burgess and H.S. Bott Jr., "Telephone Customers Sound Off About Customer Service," *Telephony*, Nov. 12 1990, pp. 46, 48, 50, 52, 54.
54. D.H. Sculman and C.G. Savolaine, "AT&T's International Competitive Advantage—Quality-of-Service," *AT&T Tech.*, Vol. 5, No. 2, 1990, pp. 4–9.
55. M. Jander, "Users Rate Long-Distance Carriers," *Data Comm.*, Vol. 19, No. 10, Aug. 1990, pp. 91–96.
56. J.M. Cummins and E.C. Stonebaker, "Total Quality Management of Telecommunications," *Bus. Comm. Rev.*, Vol. 19, No. 12, Dec. 1989, pp. 36–41.

57. "Users Say Carriers Quality Varies with Services," *Data Comm.,* Vol. 18, No. 10, Aug. 1989, pp. 117–119.
58. J.G. Reed, "Effective Customer Relations," *TE&M,* Vol. 88, No. 10, May 15, 1984, pp. 102, 103, 105.
59. P.O. Tegelberg, "Voice Concentration on Leased Private-Line Circuits Reduces Communications Costs," *Comm. Systems,* Vol. 3, No. 3, July 1983, pp. 10, 11, 34.
60. D. McKendry, "Regulation: The Call for Competition," *Canadian Consumer,* Vol. 15, No. 2, Feb. 1985, pp. 40, 41.
61. E.V. Hird, "Equal Rights for Equal Access," *TE&M,* Vol. 95, No. 14, July 15, 1991, pp. 42, 44, 45, 47.
62. J. Williamson and P. Porton, "David vs. Goliath in the U.K. The U.K.'s Competitive War: Who Is Winning?," *Telephony,* Vol. 214, No. 4, Jan. 25, 1988, pp. 28.30, 34, 36, 38, 40.
63. D. Hammer, "Privatization: It's Blowing in the Wind," *TE&M,* Vol. 94, No. 17, Sept. 1, 1990, pp. 64, 66, 68, 71.
64. Sessions in Seminar NET-08 on "Telecommunication Network Performance and Planning," *NCF Proc.,* Vol. 39, Chicago, Oct. 1985, pp. 550–568.
65. Sessions in Seminar NET-10 on "Network Management and Control: Challenge and Response," *NCF Proc.,* Vol. 39, Chicago, Oct. 1985, pp. 587–600.

Chapter 2
USER-ORIENTED PERFORMANCE

2.1 DEFINITION AND BACKGROUND

This chapter addresses end-to-end, user-oriented performance primarily for voice and data services. User-oriented performance is the starting point for the top-down portion of the overall performance process introduced in Figure 1.3 of Chapter 1. The term "user-oriented" refers to performance that is directly experienced by, perceived by, or relevant to the user, irrespective of the network involved. User-oriented performance also represents the highest level of consideration in a top-down approach to determining network performance requirements. Lower levels of consideration, such as mapping to network-oriented performance and the allocation of impairments to subcomponents of the network, are equally important and are addressed in Chapters 3 to 6.

User-oriented performance considerations for voice and data networks have been based on two different but related approaches: (1) the grade-of-service (GOS) concept for voice, and (2) end-to-end service requirements for data. The GOS concept accounts for the subjective effects of network impairments on voice users. The service requirement approach attempts to arrive at a suitable set of performance criteria and parameters applicable to the many different types of data networks and data circuits.

Extensive work has been carried out in recent years to develop performance criteria and parameters, largely for data networks, and several approaches have been examined [1]. In choosing appropriate performance criteria and parameters, consideration has been given to their intended use and applicability. The desirable attributes of performance criteria that cover both subjective and objective effects on users are that they be

- User-oriented; that is, they refer to the user's experience with the network and not to the technical performance of the network.
- Inclusive of the user's significant performance factors.
- Assessable by an experienced user.

- Suitable for evaluating and comparing the performance of different networks.
- Sufficiently sensitive to deterioration in performance to aid the user in applying corrective measures, such as lowering data speeds or initiating diagnostic tests.

As noted in Chapter 1 (where transmission performance was defined), the performance criteria now being used in the International Telegraph and Telephone Consultative Committee (CCITT) and North American standards forums (Committee T1—Telecommunications) are the speed, accuracy, and dependability of service criteria [2, 3, 4]. The meaning of these criteria for transmission performance was also briefly discussed in Chapter 1.

These criteria were also combined with the performance categories (accessibility, integrity, and retainability) and call phases (access, information transfer, and disengagement) to provide an overall framework for identifying voice and data performance parameters. This framework was shown in Table 1.3 of Chapter 1, and constitutes a common approach to establishing performance criteria for voice and data. In this chapter, Table 1.3 is filled in with specific user-oriented performance parameters for voice and data services (see Table 2.1). First, however, some of the philosophical differences between voice and data, which have a bearing on performance, are noted [1]. These differences become apparent when the identification of the performance parameters for these services is attempted.

2.2 VOICE AND DATA PERFORMANCE CONSIDERATIONS

The first difference relates to the way in which voice and data users experience a certain level of performance. For voice users, performance is subjectively perceived, while for data users performance is experienced in mainly objective terms, although subjective effects are also apparent in some applications, such as response time for transaction services.

A possible user-oriented "parameter" for voice service is simply a single GOS indication representing a composite subjective assessment of all of the individually user-perceived network impairments. This approach would be consistent with the fact that users are generally not consciously aware (in a detailed technical sense) of the many individually perceived impairments that influence their overall subjective assessment. However, this single GOS indicator approach is not amenable to the subsequent need to map between user-oriented and network-oriented performance and to allocate impairments to subcomponents of the network (Chs. 3 to 6). For this reason user-oriented voice parameters are defined here as those end-to-end network-oriented parameters that are directly perceived by the voice user. This approach provides objectivity for user-oriented voice performance, which puts it on a more common footing with user-oriented data performance.

A second difference between voice and data performance considerations relates to network evolution. For voice service, regardless of the state of network evolution,

Table 2.1
User-Oriented Performance Parameters

Performance Category & Communication Phases	Performance Criteria	Parameter	
		Voice	Data
Accessibility (Traffic)	Speed	Dial tone delay Post dialing delay Clear down delay	Pre-selection delay Access delay Disengagement delay
Access/ Disengagement Phases	Accuracy	Ineffective/errored attempts Cut-offs	Incorrect access Incorrect disengagement
	Dependability	Blocking Failure to clear down	Access denial Disengagement denial
Integrity/ Suitability (Transmission) Information Transfer Phase	Speed	Fixed delay Variable delay	Transfer delay (response time) Transfer rate (throughput efficiency)
	Accuracy	Analog * · Continuous · Transient Digital · Transmission errors · Short interruptions	Digital · Error probability · Residual error probability Errored Information ! Extra Information #
Retainability/ Robustness (Availability) All phases	Dependability	Speech clipping Long Interruptions Call cut-off User isolation	Lost information @ Reset, Transfer denial Premature disengagement & Service outage Mean Time To Service Outage Mean Time To Repair

* A number of analog parameters, such as loss and noise, are itemized later in Chapter 3.
! May include out of sequence informaton.
\# May include duplicated and misdelivered information.
@ May include misdelivered informaton. Some standards regard this as a dependability parameter, but here it is grouped with related accuracy parameters.
& Some standards regard this parameter as belonging strictly to the disengagement phase, but here it is considered to also affect dependability of the information transfer phase by unexpectedly terminating this phase.

an element of analog technology will always prevail, since even in an end-to-end "all-digital" network, the voice terminal transducers will be partly analog in nature. For this reason, many performance parameters, such as loudness loss, idle circuit noise, bandwidth, etc., will also prevail, even though associated impairment levels should be much lower and have much less variability than in current, more analog-based networks.

On the other hand, for digital data users, there are no analog voiceband impairments involved. Even for voiceband data users, analog impairments are not directly perceived, but rather are indirectly perceived in the form of incorrect data resulting from transmission errors. However, when error correcting protocols are used, even the effects of transmission errors are not directly perceived, but rather the indirect effects of errors are experienced in terms of residual (uncorrected) errors and anomalies, and related effects such as increased response time or reduced throughput.

Thus, user-oriented parameters for voice service correspond directly to many end-to-end network-oriented parameters. Conversely, user-oriented parameters for data service involve fewer and somewhat different end-to-end network-oriented parameters than for voice service.

Even in cases where there are common user-oriented parameters for voice and data (e.g., delay), the requirements may still be different for each service. This is because of other differences between voice and data performance related to traffic characteristics and to information integrity. Figure 2.1 is the classical illustration of these differences between voice and data [5, 6]. Conversational voice signals give rise to fairly continuous (not very bursty) traffic with high throughput and low delay requirements. Information integrity, on the other hand, is not critical as long as intelligibility and subjectively perceived quality are not unduly affected. This may be contrasted with (1) interactive data services, which produce bursty traffic and require relatively low throughput and low delay, and with (2) batch mode data services, which also produce continuous traffic and require relatively high throughput, but allow higher delays. In addition, information integrity requirements for data are generally much more critical than for voice service.

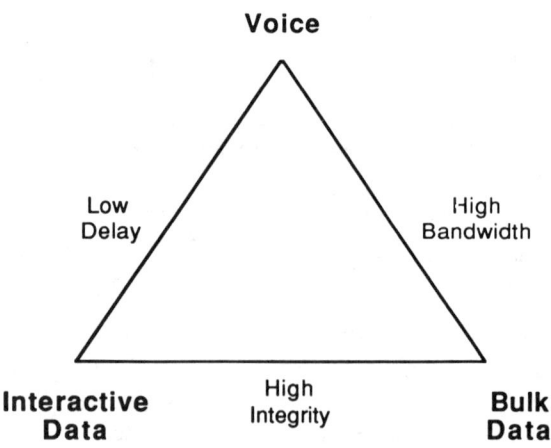

Figure 2.1 Fundamental differences between voice and data performance requirements.

In summary, when considering end-to-end user-oriented performance requirements for voice and data services, it can be expected that (1) a different set of parameters may be involved for each service, and (2) even with specific common parameters, different requirements may be necessary for each service. It is emphasized that user-oriented parameters are here defined as those network-oriented parameters that are directly perceived by users. Performance parameters of this nature are provided in the next section for voice and data services, and are organized according to the overall framework provided in Chapter 1.

2.3 USER-ORIENTED PERFORMANCE PARAMETERS

In accordance with the user-oriented perspective, the parameters in Table 2.1 are viewed as an essentially complete set without regard to specific types of networks. It is to be understood that certain networks may not require all parameters; for example, private line networks do not require all traffic parameters, and circuit-switched data networks do not require certain transmission parameters.

Again, the focus in this book is on the service integrity performance category, which relates to network transmission performance as manifested during the information transfer phase of the communications process. To some degree there is an overlap between the service integrity (transmission) and service retainability (availability) categories shown in Table 2.1. Here the retainability (availability) category is only considered in terms of seriously degraded transmission (during the transfer phase), and not in terms of hard equipment or facility failures. In addition, certain parameters such as call cut-off, while associated with the disengagement phase, also have a bearing on the transfer phase by unexpectedly terminating it.

Chapter 3 will expand upon the transmission-oriented parameters in Table 2.1. The remainder of Chapter 2 will deal with the evaluation of user-oriented performance, as well as related GOS models for voice service.

2.4 PERFORMANCE ASSESSMENT TECHNIQUES AND IMPLICATIONS

2.4.1 Customer Reports and Surveys

The most direct and accurate approach to assessing quality of service is by using subjective means. Thus, user-oriented performance has traditionally been evaluated using the following techniques [7–9]:
- Analysis of customer complaints, trouble reports and requests for credit.
- Localized customer attitude surveys based on callback interviews after users have completed the communication process.

- Formal, networkwide customer opinion surveys, again based on callback interviews.

Examples of opinion surveys include those involving transmission performance of terrestrial and satellite-based telephone connections [10, 11], and a survey of the traffic and transmission performance of the predivestiture AT&T network [12]. With the emergence of a more competitive (e.g., postdivestiture) telecommunication business environment, a number of consumer-oriented surveys have also been conducted to assess end user satisfaction with regard to circuit quality and other communication service attributes, such as planning assistance, reliability of operation, and response to trouble reports [13–22].

2.4.2 Laboratory and Modeling Techniques

Laboratory tests and network surveys use questionnaires that assess users' opinions of service quality [23–28]. For particular combinations of objective parameters (such as loss, noise, and delay), GOS indicators can be determined, such as the percentage of users who rate the service as Good or Better (% GOB). Such techniques have been developed mainly to evaluate performance for voice service [29–31].

GOS models are derived based on laboratory subjective tests, as well as on networkwide surveys in which GOS indications are determined under a variety of known objective conditions. These surveys also serve in bottom-up validation of the GOS models. For example, objective surveys, which assess objective performance parameters, track the performance of networks as they evolve in response to new service demands and as they take advantage of cost/performance trade-offs made possible with the introduction of new technology [32–35]. Similarly, subjective (or opinion) surveys, referenced earlier in this section, serve to track user expectations. For example, in a competitive environment, users' opinions are probably influenced by the price of alternative service offerings, and their changing expectations in this regard are reflected in survey results.

2.4.3 Factors Influencing User Expectations

In practice, performance, as perceived by end users, is influenced by many factors [36–40], including:

- The needs and sophistication of the user.
- The type and price of the service (price/quality trade-offs).
- The availability of competitive alternatives.

Users of performance-sensitive (e.g., high-speed data) services are generally the most performance-conscious; but corporate users, in particular, are also influenced by economic considerations. That is, users' opinions of lower performance may be mitigated by lower price, as was evidenced by the emergence of the other common carriers (OCC) prior to and after divestiture. Users' acceptance of a given level of performance may also depend on the nature of the telecommunications business environment (regulated or competitive), and on the response of service providers and regulators to users' complaints. Generally, the existence of competitive alternatives enables users to exercise price/quality trade-offs that better suit their needs. However, in the long run, migration toward higher performance can be expected because of the nature of new technology and the need to support both critical and noncritical services in an operationally simple and economic manner.

Other factors that influence user expectations are

- Distance, as manifested by local and toll (or intra- and inter-LATA) connections and by national and international connections.
- Application, as manifested by public and private networks.

Generally, there is a tendency for users to be more tolerant in their opinions when using long-distance connections. This is illustrated later in this chapter when opinion data bases are discussed. Corporate users may also be more inclined to accept lower performance in a private network environment, since a lower price for the service may have been a strong motivating factor for deciding to use a private network in the first place.

These are some of the considerations to be noted when using GOS models to map between subjective user-oriented performance and objective network-oriented performance (recall Fig. 1.3 of Ch. 1). A given GOS model is only valid under certain assumptions about users' expectations in a particular environment. For this reason, models do not necessarily give accurate indications of absolute GOS. Rather, under given conditions, the models establish a link between performance as perceived by end users and the network impairments, and are useful for assessing the relative performance of alternative systems, networks, and transmission plans. Transmission planning is presented in Chapter 7.

2.5 GRADE-OF-SERVICE METHODOLOGY

Chapter 3 will address the specific relationships between user- and network-oriented performance. This chapter outlines the principle methodology used in North America to establish these relationships for voice service. The mapping between user- and network-oriented performance for voice service in terms of two types of GOS models [41]:

- Opinion models, which address users' perception or expectation of a particular service.
- Behavioral models, which address how users react if they are not satisfied with a particular service.

These GOS models are based on subjective testing and on the derivation of transmission rating models, which are discussed next (see [42] for other approaches).

2.5.1 Subjective Testing of Speech Quality

The transmission GOS models discussed below are empirical models based on the results of subjective tests. Most of these are listening tests, but conversational tests are also used to evaluate impairments likely to affect the dynamics of conversations (e.g., delay). In either case, test conditions consisting of simulated impairments are presented to a suitably large number of "naive" test subjects in a controlled acoustical and electrical environment. (Naive subjects are those unfamiliar with the technical aspects of the communications process, and, in this sense, are considered unbiased in their opinions; the detailed methodology of subjective testing can be found in [23–26, 43–45].) For listening tests, Figures 2.2 and 2.3 illustrate typical ways to simulate impairments and present them to subjects. Reference conditions of previ-

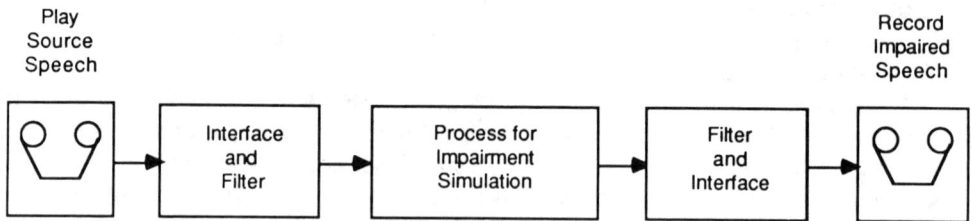

Figure 2.2 Typical functional arrangement for impairment simulation.

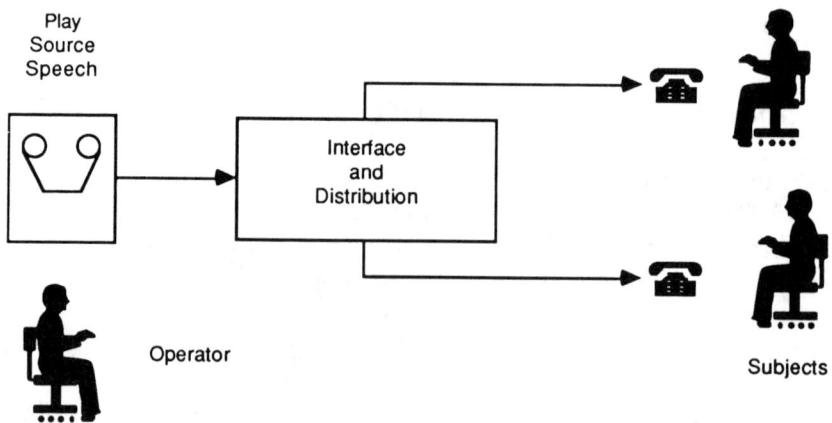

Figure 2.3 Typical functional arrangement for listening tests.

ously studied impairment levels are also included in the test to enable comparison with results of previous tests and previously established models, and to enable the combination of results from different tests for further modeling purposes.

Speech quality testing as described above is a well-known and established subjective evaluation procedure for telephony applications. For such applications, speech intelligibility is generally high, and speech quality tests provide a finer measure of performance than a speech intelligibility test. Quality is a measure of a listener's acceptance of the speech, whereas intelligibility is only a measure of the listener's ability to understand what has been said. Although intelligibility is affected by some impairments (e.g., speech clipping), existing information can be used as a guide to ensure high intelligibility [46]. Speech clipping and related parameters are addressed in Chapter 3.

The test subjects rate the speech quality test conditions, often using a 5-point opinion rating scale of quality, very similar to the CCITT scale [24]. This scale consists of opinion ratings and numerical weights as follows:

Five-Point Opinion Rating Scale

Opinion Rating	Numerical Weight (i)	Proportion of Ratings P_i
Excellent (E)	5	P_5
Good (G)	4	P_4
Fair (F)	3	P_3
Poor (P)	2	P_2
Bad (B)	1	P_1

where $\sum_{i=1}^{5} P_i = 1$

Often in North America, the Unsatisfactory (U) rating has been used instead of the Bad (B) rating, but it has the same numerical weight of 1 [43]. In some cases, such as for field opinion surveys, a 4-point scale is used consisting only of Excellent, Good, Fair, and Poor ratings with weights of 4, 3, 2, and 1, respectively. The reason is that extreme conditions at the low end of the quality scale in field situations are rarely encountered [11]. In addition, a 4-point scale is easier for a subject to remember, since a printed rating scale is not used with callback interviews.

A well-known subjective measure of quality is the mean opinion score (MOS). For each test condition, the MOS is the weighted average rating and is given by the sum of the weights (i), each multiplied by the corresponding proportion of the subjects' ratings ($P_i = n_i/N$, where n_i is the number of subject ratings of weight i, and N is the total number of subjects); thus,

$$\text{MOS} = \frac{1}{N} \sum_{i=1}^{5} i n_i = \sum_{i=1}^{5} i P_i \qquad (2.1)$$

Also of interest, mainly for modeling purposes, is the standard deviation (SD) of the MOS, which is given by

$$\text{MOSSD} = \sqrt{\frac{1}{N}\sum_{i=1}^{5}(i - \text{MOS})^2 n_i} = \sqrt{\sum_{i=1}^{5} i^2 P_i - \text{MOS}^2} \qquad (2.2)$$

With the MOS and MOSSD, it is possible to fit a normal (Gaussian) probability density function (pdf) to the actual frequency of occurrence of subject ratings for each parameter value. This is part of the process leading to transmission rating models, which are described in the next section.

Another subjective measure used for transmission GOS purposes is the percentage of subjects who rate a test condition as % GOB (good or excellent); thus,

$$\% \text{ GOB} = 100\left(\frac{n_4 + n_5}{N}\right) = 100(P_4 + P_5) \qquad (2.3)$$

Similarly, the percentage of subjects who rate a test condition as Poor or Worse (poor or bad) is

$$\% \text{ POW} = 100\left(\frac{n_1 + n_2}{N}\right) = 100(P_1 + P_2) \qquad (2.4)$$

The measures in Equations (2.1), (2.3), and (2.4) give GOS indications directly from the results of subjective tests. For example, it is possible to plot MOS versus a range of values for a given parameter (such as noise), and thereby obtain an empirical MOS "model" for that parameter in terms of a smooth analytical function. Such a model then enables user opinion to be estimated for a continuous range of parameter values.

However, this "modeling" process can be formalized to account for the effects of more than one parameter and for differences among subjective tests. This topic is addressed in the next section.

2.5.2 Transmission Rating Models

One difficulty with the modeling approach noted above is that the results of subjective tests (i.e., the test subjects' opinion ratings) can depend significantly on a number of factors other than the transmission performance parameter values being investigated. These factors include

- The application environment (laboratory or more natural setting).
- Type of test (listening or conversational).
- Range of test conditions (range of parameter values).
- Types of subjects (age, sex, occupation, previous test experience, etc.).

For this reason, determining GOS modeling consists of two stages:
(1) Determining a normalized model that is only a function of transmission parameter values and not the more secondary factors noted above. This model is called the *transmission rating model* and is generically denoted by R.
(2) Determining an "absolute" model that is a function of the transmission rating model but also reflects the other factors noted above, principally, the application environment (most of the other factors should be controllable by the proper experimental design of subjective tests). The "absolute" model is called the opinion model; it is often expressed in terms of, for example, the percentage of users that rate a connection as % GOB.

Background, rationale, and details associated with determining R models can be found in [31, 41, 43]. It is simply noted here that an R model can be viewed as a normalization (or linear transformation) of a so-called "fit" MOS model, where the "fit" MOSs are first moments (averages) of normal pdfs, fitted to the frequencies of occurrence of opinion ratings as observed from subjective tests or opinion surveys. The normalization is with respect to objective test conditions that span a representative range of performance, and to which arbitrary values of R are assigned. The equations for the R models are well known and are summarized in Table 3.4 of Chapter 3; they are also referred to later in that chapter when the GOS curves for specific transmission parameters are presented.

However, to illustrate for introductory purposes, Figure 2.4 shows R for the commonly known loss/noise transmission rating model. Note that the transmission rating scale is rather arbitrary and is not constrained to a range from 0 to 100. In practice, R is almost always considerably greater than 1, but can also be greater than 100 for certain impairments, such as echo.

2.5.3 Opinion Models and Opinion Data Bases

Details associated with determining opinion models can also be found in [31, 41, 43]. It is simply noted here that opinion models can be viewed as areas under a normal pdf fitted to the frequency of occurrence of opinion ratings, with the "fit" MOS and "fit" MOSSD as the pdf parameters; the areas are therefore a function of the "fit" MOS and "fit" MOSSD. Since R is also a function of the "fit" MOS through the linear transformation discussed earlier, the areas and hence opinion ratings can be expressed in terms of R. For example, the relationship for % GOB and % POW are

$$\% \text{ GOB} = 100\Phi(A), \text{ and } \% \text{ POW} = 100[1 - \Phi(B)] \tag{2.5}$$

where $\Phi(x)$ is the standard normal error function given by

$$\Phi(x) = \frac{1}{\sqrt{2\pi}} \int_{-\infty}^{x} e^{-t^2/2} dt \tag{2.6}$$

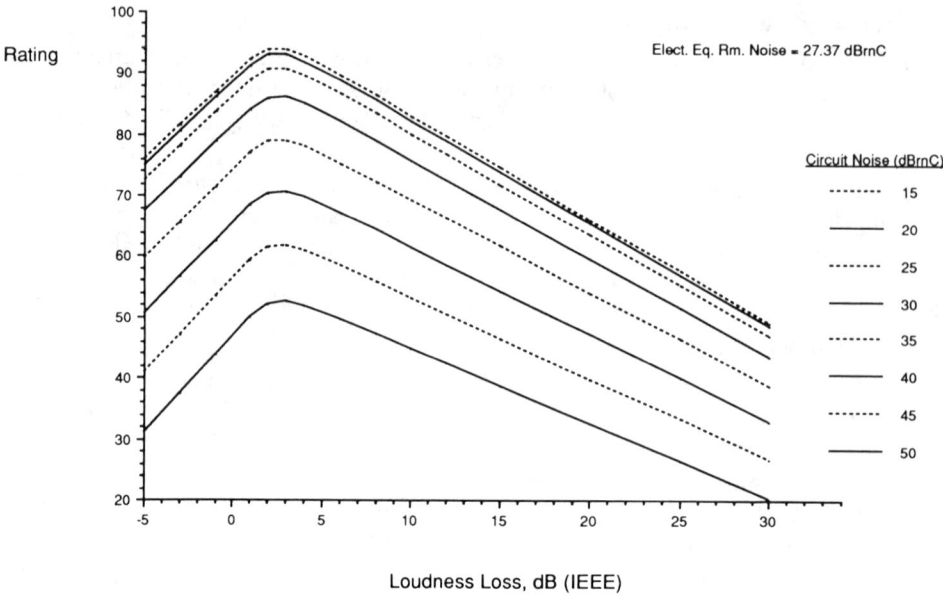

Figure 2.4 Loss/noise transmission rating.

and where

$$A = (R - \alpha)/\gamma, \text{ and}$$
$$B = (R - \beta)/\gamma \quad (2.7)$$

The terms α, β, and γ are opinion data base parameters determined for the particular environments in which experiments are conducted. (In the nonstandard form of a normal pdf, A and B are relative deviations, R is the random variable, α and β are means, and γ is the SD; see Section 5.4.3.3 of Chapter 5.) Three representative experiments and the corresponding values of α, β, and γ, based on the 5-point opinion scale (unless otherwise noted), are

Opinion Data Base Parameters

Experiment	α	β	γ
1965 Murray-Hill SYBLE Test	64.07	51.87	17.57
CCITT Conversational Tests	62	43	15
Long Toll Interviews	51.5	40.98	15.7
(4-point scale)	(51.5)	(23.69)	(15.85)

Figure 2.5 gives the relationship between GOS and transmission rating for each of the above data bases. Figure 2.6 is an example of loss/noise GOS based on the CCITT opinion data base, and it corresponds directly to the loss/noise transmission ratings given in Figure 2.4. Loss/noise GOS curves are also presented in Figure 2.7 to illustrate the effect of the different data bases on GOS; this figure is for a specific example value of noise (30 dBrnC) selected from Figures 2.4 and 2.6.

Figure 2.5 Grade of service versus transmission rating for various opinion data bases.

In Figure 2.7, the Murray-Hill data base predicts the most stringent opinions, apparently because the data is from the results of intra-building business calls where there was a general expectation of good quality. The CCITT data base is a composite of the results of various conversational tests reported to CCITT and predicts somewhat less stringent opinions, apparently owing to the different environment and lower expectations of good quality associated with these tests. Finally, the Long Toll data base is from the results of interviews with North American customers following calls on long toll connections. This data base predicts considerably less stringent opinions than the other two, evidently because expectations of good quality are not as high with such an application (i.e., long-distance calls).

Thus, depending on the application being investigated, the most appropriate data base can be selected. For illustrative purposes, the CCITT opinion data base is used nearly exclusively in this book because it is perhaps the most generally repre-

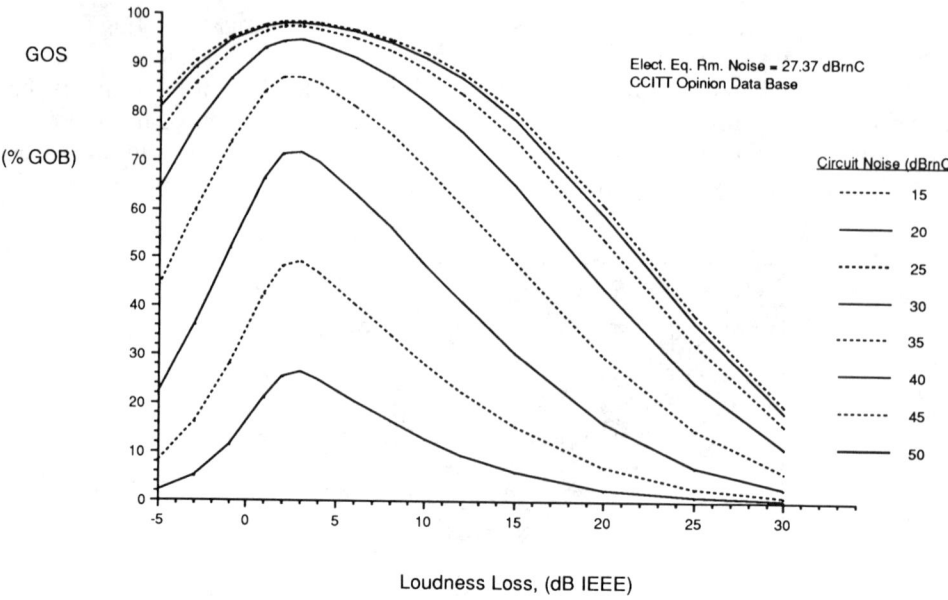

Figure 2.6 Loss/noise grade of service.

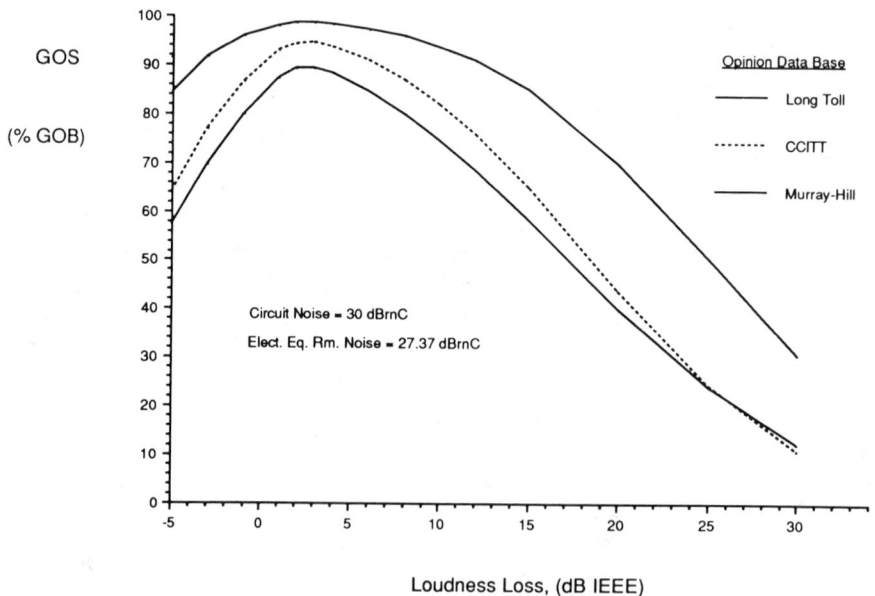

Figure 2.7 Loss/noise grade of service. (Effect of opinion data base.)

sentative. However, the opinions predicted by any of the data bases should not be viewed as absolutely accurate indications of the particular problem being assessed. Rather, the usefulness of opinion models for a selected data base is in the relative changes of opinion predicted for different parameter values associated with the different connection characteristics being investigated.

2.5.4 Performance Classes

Given that a GOS model is developed, it can be used to assess user-oriented performance and to provide end users with the methodology for evaluating performance. This applies particularly to large corporate users of private networks. Because of the competitive era in which end users are dependent on several vendors for the provision of end-to-end service, the IEEE voiceband standard was developed to provide methodologies for assessing basic end-to-end transmission performance [29, 30]. As part of this process, the voiceband standard characterizes the range of transmission performance by defining the following five classes of performance:

Performance Classes

Performance Class	Description	Performance (% GOB) (CCITT Opinion Data Base)
A1:	Typical local call (metropolitan network)	≥ 80
A2:	Typical public network toll call (average distance)	≥ 60
A3:	Limiting public network toll call (nation wide)	≥ 30
A4:	Limiting private network call (tandem tie trunk network)	≥ 15
A5:	Unsuitable	< 15

Given that the range of performance has been characterized, a possible further step would be to associate a price with each performance classification. Such a price/quality relationship would represent a new criterion for performance assessment and provide guidance on how to use the broad range of telecommunications network performance capabilities, in accordance with the base of embedded technology, the service requirements of end users, and the competitive pressures from other network vendors.

This end user perspective contrasts with past practice, in which network performance modeling was used by telecommunications administrations (e.g., the Bell System) and standards bodies (CCITT) to optimize overall end-to-end network performance from a network cost point of view. This end-to-end function now falls more into the domain of standards bodies and reflects more directly the needs of end users. Individual network providers now tend to be more concerned about the per-

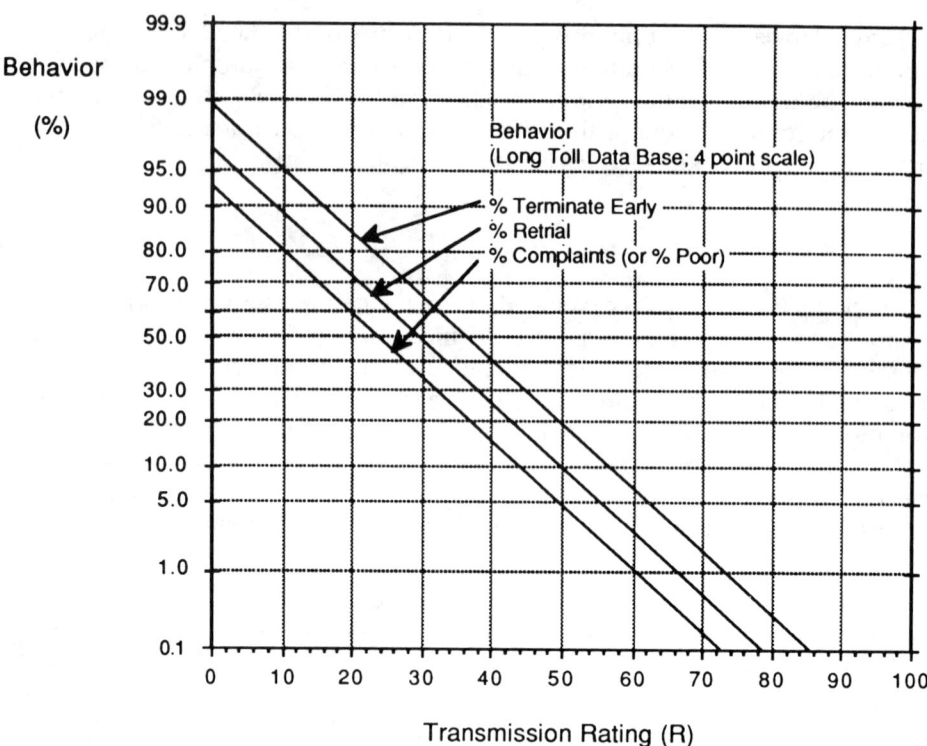

Figure 2.8 Behavior versus transmission rating for behavioral data base.

formance and economical well-being of their own particular networks, and they use modeling to address these specific concerns.

2.5.5 Behavioral Models and Behavioral Data Bases

As noted earlier, customer behavior models reflect how users react to the level of performance they perceive [11, 41, 47, 48]. For transmission performance, overt user reaction has been assessed in terms of

- Call abandonment (terminating a call early).
- Retrial (dialing again).
- Complaining to the operator.

The probabilities of users taking these actions have been modeled as a function of transmission rating in a manner similar to that for opinion rating models described previously. The relationships are all of the form

$$\% \text{ of Users Taking a Certain Action} = 100[1 - \Phi(B)] \tag{2.8}$$

where $B = (R - \beta)/\gamma$, as defined in Equation (2.7), and where $\Phi(x)$ is as defined in Equation (2.6).

The values of β and γ are behavioral data base parameters determined for the particular environment in which the behavioral investigation was done. In this case, the results are all based on the AT&T Long Toll interviews using the 4-point opinion rating scale, for which the values of β and γ are as follows:

Behavioral Data Base Parameters

Behavioral Reaction	β	γ
Terminate Early	36.30	15.85
Retrial	29.50	15.85
Complaints (or POW)	23.69	15.85

Figure 2.8 gives the relationship between behavior and transmission rating for each behavioral reaction. In addition, using Figures 2.5 and 2.8, it is possible to relate the well-known % GOB measure to behavior as illustrated in Figure 2.9. Note that

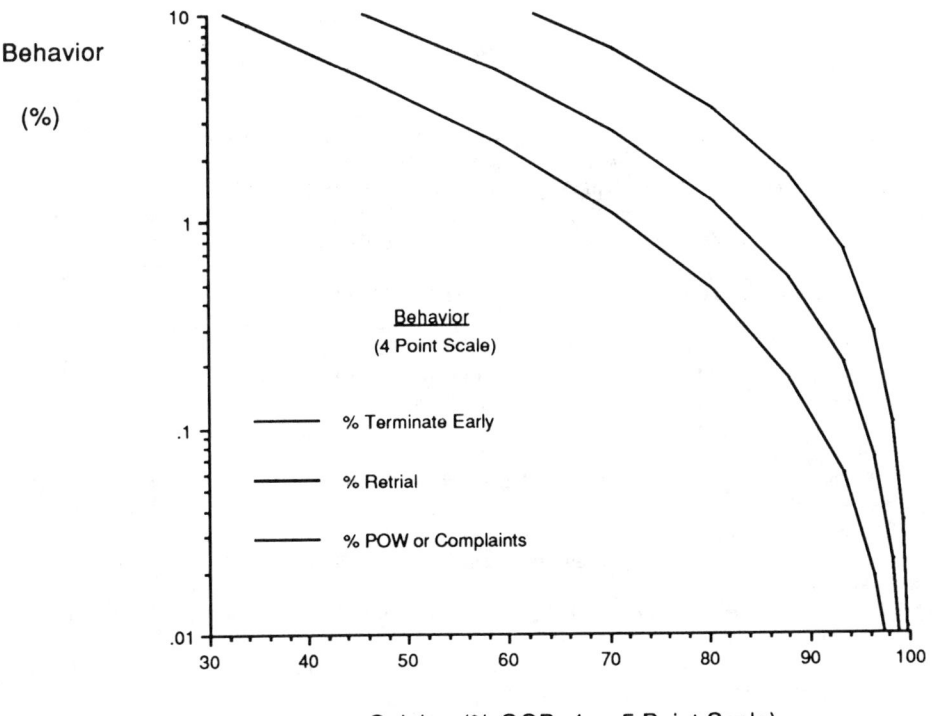

Figure 2.9 Behavior versus opinion.

in Figure 2.5 the relationship for % GOB using the Long Toll data base for the 5-point opinion scale is nearly the same as that for the 4-point scale. Thus, Figure 2.8 applies for either the 4- or 5-point opinion rating scale.

2.6 POTENTIAL FOR MODELING NONVOICE PARAMETERS

The GOS techniques discussed in this chapter have been developed mainly to evaluate performance for voice service. However, similar concepts and models could also be applied to other services; for example, to assess user opinion of response time for interactive data services [49], and to assess picture quality for image applications [50–55]. The reason for the lack of formal models for data and image services is likely due to the dominance to date of voice over other services. Models of a more network-oriented nature exist for voiceband data; these models relate modem error performance to lower level network impairments (e.g., noise) [56, 57]. The error performance data can then be used to assess such user-oriented performance measures as data throughput and response time.

2.7 REFERENCES

1. J. Gruber and N.H. Le, "Performance Requirements for Integrated Voice and Data Networks," *IEEE Journal on Sel. Areas in Comm.*, Vol. SAC-1, No. 6, Dec. 1983, pp. 981–1005. Also, *GLOBECOM Rec.*, Vol. 1, San Diego, Nov./Dec. 1983, pp. 2.2.1–2.2.9.
2. CCITT Rec. I.350, "General Aspects of Quality of Service and Network Performance in Digital Networks, Including ISDN," *Blue Book*, Vol. III, Fas. III.1, Geneva 1989, pp. 129–139.
3. American National Standard for Telecommunications: Performance Parameters for Packet Switched Data Communication Service, ANSI T1.504-1989.
4. CCITT Rec. X.140, "General Quality-of-Service Parameters for Communication Via Public Data Networks," *Blue Book*, Vol. VIII, Fas. VIII.3, Geneva 1989, pp. 422–446.
5. S.L. Casner, E.R. Mader, and E.R. Cole, "Some Initial Measurements of ARPANET Packet Voice Transmission," *NTC Rec.*, Vol. 1, Birmingham, Alabama, Dec. 1978, pp. 12.2.1–12.2.5.
6. D. Cohen, "Packet Communication of Online Speech," *AFIPS NCC Proc.*, Vol. 50, Chicago, May 1981, pp. 169–176.
7. "Telecommunications Transmission Engineering," Chapter 19 on "Measurements and Indices," Vol. 3, *Networks and Services*, 2nd ed., Bell Telephone Labs., Inc., 1977. Also, Chapter 20 on "Measurements," in 3rd ed., Bellcore, 1990.
8. D. Riggenbach, " 'Listening' Programs Keep Customers Happy," *TE&M*, Vol. 85, No. 1, Jan. 1, 1981, pp. 68–70.
9. J.L. Sullivan, "Is Transmission Satisfactory? Telephone Customers Help Us Decide," *Bell Labs. Rec.*, Vol. 52, No. 3, Mar. 1974, pp. 91–98.
10. J.A. Post and R.D. Silverthorn, "Results of a Subjective Comparison of Echo Control Devices in Terrestrial and Satellite Trunks," *NTC Rec.*, Vol. 3, Nov. 1979, pp. 48.4.1–48.4.5.
11. L.S. DiBiaso, "Satellite User Reaction Tests: A Subjective Evaluation of Echo Control Methods," *NTC Rec.*, Vol. 3, Nov. 1979, pp. 48.6.1–48.6.6.

12. J.P. Healey, J.M. Kelly, D.G. Leeper, M.G. Linnell, and J.A. Maher, "How's Service? Ask the Customer," *Bell Labs. Rec.*, Vol. 60, Feb. 1982, pp. 31–34.
13. "Ratings Reveal Carriers Are Worlds Apart," *Data Comm.*, Vol. 17, No. 8, Aug. 1988, pp. 103–106, 108.
14. "Report Cards Out: Carriers Rated on Service, Quality," *Data Comm.*, Vol. 16, No. 9, Aug. 1987, pp. 127–130, 133, 134.
15. "Communications Lines Score Low," *Data Comm.*, Vol. 13, No. 8, July 1984, pp. 76, 77.
16. G. Harris, "How Do Users Rate Vendor Service? On What Do They Base Their Opinions?" *Data Comm.*, Vol. 15, No. 5, May 1986, pp. 159, 160, 163, 164, 167, 168.
17. "19,000 Ear Survey Proves Digital Better than Analog in Voice Networks," *Data Comm.* Vol. 15, No. 10, Sept. 1986, pp. 84, 86, 88.
18. "Users Say Carrier Quality Varies with Services," *Data Comm.*, Vol. 18, No. 10, Aug. 1989, pp. 117–119.
19. M. Jander, "Users Rate Long-Distance Carriers," *Data Comm.*, Vol. 19, No. 10, Aug. 1990, pp. 91–96.
20. A.A. Burgess and H.S. Bott Jr., "Telephone Customers Sound Off About Customer Service," *Telephoney*, Nov. 12, 1990, pp. 46, 48, 50, 52, 54.
21. J.M. Cummins and E.C. Stonebaker, "Total Quality Management of Telecommunications," *Bus. Comm. Rev.*, Vol. 19, No. 12, Dec. 1989, pp. 36–41.
22. J.T Johnson, "Users Rate Long-Distance Carriers," *Data Comm.*, Vol. 20, No. 10, Aug. 1991, pp. 91–94, 96, 98.
23. CCITT Rec. P.82 "Method for Evaluation of Service from the Standpoint of Speech Transmission Quality," *Blue Book*, Vol. 5, Pt. I, Section 7, Geneva 1989, pp. 203–207.
24. CCITT Rec. P.84, "Subjective Listening Test Method for Evaluating Digital Circuit Multiplication and Packetized Voice Systems," *Blue Book*, Vol. 5, Pt. I, Section 7, Geneva 1989, pp. 208–236.
25. CCITT Supp. No. 2 (to Rec. P.80), "Methods Used for Assessing Telephony Transmission Performance," *Blue Book*, Vol. 5, Pt. II, Geneva 1989, pp. 237–247.
26. CCITT Supp. No. 14 (to Rec. P.81), "Subjective Performance Assessment of Digital Processes Using the Modulated Noise Reference Unit (MNRU)," *Blue Book*, Vol. 5, Pt. II, Geneva 1989, pp. 341–360.
27. A. Stewart, "Grade-of-Service in a Digital Network," *TE&M*, Vol. 92, No. 3, Feb. 1988, pp. 54, 55, 58, 59–60.
28. J.G. Reed, "Effective Customer Relations," *TE&M*, Vol. 88, No. 10, May 15, 1984, pp. 102, 103, 105.
29. Papers in Session on "Standards on Telecommunication Network Transmission Performance," *GLOBECOM Rec.*, Vol. 2, San Diego, Nov./Dec. 1983, pp. 676–714.
30. "Methodologies for Specifying Voice Grade Channel Transmission Parameters and Evaluating Transmission Performance for Speech Telephony," ANSI/IEEE Standard 823-1989.
31. CCITT Supp. No. 3 (to Rec. P.11) on "Transmission Rating Models," pp. 248–261, and Annex A (to Supp. No. 3) on "Opinion Ratings of Transmission Impairments," Bellcore, *Blue Book*, Vol. V, Geneva 1989, pp. 299–306.
32. M.B. Carey, H.T. Cheu, A. Descloux, J.F. Ingle, and K.I. Park, "1982/83 End Office Connection Study; Analog Voice and Voiceband Data Transmission Performance Characterization of the Public Switched Network," *BLTJ*, Vol. 63, No. 9, Nov. 1984, pp. 2059–2119.
33. J. Bush, "A Quality Comparison: Which Carrier Tests Best," *Data Comm.* Vol. 18, No. 3, Mar. 1989, pp. 96–98, 101, 102, 105, 106, 109.
34. E.E. Mier, "Comparing the Long-Distance Carriers," *Data Comm.*, Vol. 15, No. 9, Aug. 1986, pp. 90–113.
35. E.E. Mier and J. Bush, "Rating the Long-Distance Carriers," *Data Comm.*, Vol. 13, No. 9, Aug. 1984, pp. 103–114.

36. L. Anderson, "Users Sound off on Service," *Telephony,* Vol. 214, No. 7, Feb. 15, 1988, pp. 42, 46.
37. L.C. Lasota and D.M. Maxwell, "Custom Calling Helps Sell Network Service," *TE&M,* Vol. 90, No. 6, Mar. 15, 1986, pp. 53, 54, 57.
38. R.E. Collins, "Service Quality—Measuring the Network and the Customer," *Telephony,* Vol. 209, No. 24, Dec. 9, 1985, pp. 32–34.
39. P.K. Pitsh, "The Impact of Divestiture on Quality—A Consumer Based View," *GLOBECOM Rec.,* Vol. 2, Atlanta, Nov. 1984, p. 17.1.1.
40. G.T. Ray, "Quality, A Commitment," *GLOBECOM Rec.,* Vol. 2, Atlanta, Nov. 1984, pp. 17.2.1.–17.2.4
41. B.W. Kort, "Models and Methods for Evaluating Customer Acceptance of Telephone Connections," *GLOBECOM Rec.,* Vol. 2, San Diego, Nov./Dec., 1983, pp. 20.6.1.–20.6.9.
42. "Technology Independent, User-Oriented, Objective Classification of Voice Transmission Quality," Committee T1—Telecommunications Draft Technical Report T1Y1/LB91-05, Oct. 1991, T1Y1/91-083, T1Y1.2/91-015R2, July 1991.
43. J.R. Cavanaugh, R.W. Hatch, and J.L. Sullivan, "Models for the Subjective Effects of Loss, Noise and Talker Echo on Telephone Connections," *BSTJ,* Vol. 55, Nov. 1976, pp. 1319–1371.
44. H.S. Cohen, "Measuring and Modeling User Satisfaction with Telephone Switching and Transmission Performance," *Proc. of 9th Int. Symp. on Human Factors in Telecomm.,* Red Bank, N.J., Sept./Oct. 1980, pp. 237–242.
45. H.S. Cohen and B.W. Kort, "Psychological Evaluation of Telephone Switching and Transmission Performance," *Proc. of 21st Human Factors Society Meeting,* 1977, pp. 505–507.
46. J. Gruber and L. Strawczynski, "Subjective Effects of Variable Delay and Speech Clipping Loss in Dynamically Managed Voice Systems," *IEEE Trans. Comm.,* Vol. COM-33, No. 8, Aug. 1985, pp. 801–808; also, *GLOBECOM Rec.,* Vol. 3, Miami, Nov./Dec. 1982, pp. F7.3.1–F7.3.5.
47. L.S. DiBiaso, "User Reaction to Transmission Quality: A Cost Model," *ICC Rec.,* Vol. 2, Seattle, June 1987, pp. 34.6.1–34.5.5.
48. L.S. DiBiaso, "Telecommunications System Performance Needs for Standards," *GLOBECOM Rec.* Vol. 2, San Diego, Nov./Dec. 1983, pp. 20.1.1–20.1.7.
49. S.S. Katz, "Introduction of New Network Services," *Proc. 1 of 10th Int. Teletraffic Cong.,* Montreal Quebec, 1983.
50. M. Miyahara, "Quality Assessment of Visual Services," *IEEE Comm. Mag.,* Vol. 26, No. 10, Oct. 1988, pp. 51–60.
51. H. Murakami, H. Hashimoto, and Y. Hatari, "Quality of Band-Compressed TV Services," *IEEE Comm. Mag.,* Vol. 26, No. 10, Oct. 1988, pp. 61–69.
52. N. Ohta, M. Nomura, and T. Fujii, "Variable Rate Video Coding Using Motion-Compensated DCT for Asynchronous Transfer Mode Networks," *ICC Rec.,* Vol. 3, Philadelphia, June 1988, pp. 38.6.1–38.6.5.
53. T.A. White et al., "Double-Stimulus Quality Rating Method for Television Digital Codecs," *Electronic Letters,* Vol. 16, No. 18, Aug. 1980, pp. 714–716.
54. M. Inove, "Noise Specification Methods for Mixed Analog-Digital Video Transmission," *Rev. of the Elect. Comm. Labs.,* Vol. 37, No. 2, 1989, pp. 139–148.
55. M. Nomura et al., "Basic Characteristics of Variable Rate Video Coding in ATM Environment," *IEEE Journal on Sel. Areas in Comm.,* Vol. 7, No. 5, June 1989, pp. 752–760.
56. B. Kort and M. Kalb, "Models and Methods for Evaluating and Assuring the Performance of Voice and Voiceband Data Services," *NCF Rec.,* Vol. 39, Chicago, Oct. 1985, pp. 555–557.
57. M. Kalb, C.H. Mortin, and J.J. Shynk, "DATACAL—A Voiceband Data Communications Connection Performance Model," *Proc. of 2nd Int. Network Planning Symp.,* Conf. Pub. No. 215, Univ. of Sussex, Brighton, U.K., Mar. 1983, pp. 171–177.

Chapter 3
TRANSLATING USER-ORIENTED PERFORMANCE INTO END-TO-END NETWORK-ORIENTED PERFORMANCE

3.1 INTRODUCTION

This chapter addresses the next step in the overall performance process introduced in Figure 1.3 of Chapter 1: the specific relationships between subjective user-oriented performance and objective network-oriented performance. In this regard, Table 2.1 of Chapter 2 provided end-to-end parameters for each of the three user-oriented (and network-oriented) performance categories, accessibility (traffic), integrity (transmission), and retainability (availability). Here, Table 3.1 provides an expanded list of transmission performance parameters corresponding mainly to the service integrity category and covering the speed, accuracy, and dependability of service performance criteria. These transmission parameters are in accordance with the focus of this book and apply to various types of public and private voice and data networks; a subset of the data parameters apply to circuit-switched and nonswitched data networks.

This chapter will first address end-to-end user-oriented transmission performance parameters for voice service. In Table 3.1, many of these parameters may be characterized according to the nature of the impairment they produce and according to the state of network evolution. This characterization is shown conceptually in Table 3.2, and in more detail in Table 3.3 [1]. It is apparent that as networks evolve from an all-analog to an all-digital state, the corresponding general trend is for impairments to also evolve from being mostly continuous, to mostly transient in nature. Another general trend will be from the highly variable performance of all-analog networks toward the more consistently high performance of all-digital networks. The reason for this trend is that in digital networks, those parameters which correspond to continuous impairments (such as loudness loss and idle circuit noise) can be readily controlled to levels which yield high performance [2]. In addition, those parameters that correspond to transient impairments (such as errors and inter-

Table 3.1
Detailed Classification of User-Oriented Voice and Data Transmission Performance Parameters
(Mainly for the Information Transfer Phase)

Performance Criteria	Parameters	
	Voice	Data
Speed	Fixed Delay Variable Delay	Transfer Delay (Response Time) Transfer Rate (Throughput efficiency)
Accuracy	Analog – Continuous: - Loudness Loss - Idle Circuit Noise - Echo (Talker and Listener) - Quantization Noise - Room (Acoustic) Noise - Sidetone - Bandwidth and Slope - Cross-talk/Interference - Non-Linear Distortion Analog – Transient @: Digital - Transmission Errors - Short Interruptions	Digital - Error Probability Residual Error Probability * - Errored Information - Extra Information
Dependability	Speech Clipping Long Interruptions Call Cut-off	- Lost Information # Reset, Transfer Denial Premature Disengagement #

* See Fig. 3.24 for notes on further breakdown of these parameters.
\# See Table 2.1 of Chapter 2 for explanatory notes on these parameters.
@ Analog transient parameters are gain and phase hits, signal dropouts,
and impulse noise (see Table 3.14). While perceptable for voice service, these
parameters are controlled by voiceband data requirements (see Ch. 4 and 6).

Table 3.2
Conceptual Characterization of Transmission Performance Parameters for Voice Service

Table 3.3
More Detailed Characterization of Voice Parameters

Nature of Impairments	Parameters	Comments on Parameters in Terms of Network Evolution		
		Analog	Mixed	Digital
Continuous	Delay	Fixed per call	Potentially larger and mainly fixed per call	Still larger and fixed or variable per call
	Loss	Variable due to access and backbone network	Variable mainly due to access network	Fixed
	Noise	Stationary in nature	Stationary and speech correlated in nature	Mainly speech correlated in nature
	Echo	Governed by variable loss and echo control	Governed by fixed loss and echo control	Governed by incidental acoustic and crosstalk paths
	Sidetone	Variable due to access network	Variable due to access network and its evolution	Fixed
	Bandwidth/ Slope	Governed by access and backbone network	Governed mainly by access network	Governed mainly by voice terminals
	Cross-talk/ Interference	Governed by access and backbone network	Governed mainly by access network	Governed by incidental crosstalk paths
	Non-linear Distortion	Governed by voice terminals and backbone network	Governed mainly by voice terminals and signal processing	Governed mainly by signal processing
Transient	Transmission Errors	–	Random and bursty	Mainly bursty
	Interruptions	Hits, dropouts, impulses, and speech clipping	Hits, dropouts, impulses, frame slips, frame loss, and speech clipping	Frame slips, frame loss and speech clipping

ruptions) can have their frequencies of occurrence limited to levels where these impairments are rarely perceived.

The approach here uses nationwide telephone network evolution as a basis for discussing performance. This evolution is taking place in North America, as well as in other countries (e.g., see [3]), and is the result of the ongoing deployment of digital transmission and switching technology. Network evolution from a mainly all-analog state to a mixed analog/digital state generally gives rise to improved performance, although in some cases impairments of a transitional nature are introduced. Short-term evolution relates to the situation in which the access network is mainly analog, and the backbone network is mixed. Longer term evolution relates to the situation in which the access network is mixed, and the backbone network is mainly digital. Ultimately, long-term evolution relates to the end-to-end all-digital network. In this case, performance tends to correspond either to very little impairment, or to the average or preferred performance level associated with the mixed analog/digital case.

The basis for the discussion of many of the voice parameters in Tables 3.1 and 3.3 is grade of service (GOS), as determined from the well-known transmission rating models introduced in Chapter 2 and summarized here in Tables 3.4(a), (b), and (c). Also provided in this chapter are definitions of the parameters and relevant discussions relating to the performance of evolving networks. For this purpose, Figure 3.1 illustrates a typical voice connection and a number of the main voice parameters to be referred to in this and later chapters.

Table 3.4(a)
Main Transmission Rating Models for Loudness Loss, Circuit Noise, and Talker Echo

1. <u>Loss/Noise</u>

$$R_{LN} = 147.76 - 2.257 \sqrt{(L_e - 2.2)^2 + 1} - 1.907 N_F + 0.02037 (L_e) N_F \qquad (1)$$

where:

- L_e = Loudness loss (in dB IEEE) of the overall telephone connection,
- N_F = Total effective noise (in dBrnC) referred to the input of a reference receiving system with ROLR* = 46 dB,
    - = N_C "+" N_R "+" N_Q (where "+" signifies power addition)
- N_C = Circuit noise (in dBrnC) referred to ROLR = 46 dB,
- N_R = Circuit noise equivalent (in dBrnC) of room noise referred to ROLR = 46 dB (e.g., N_R = 27.37 dBrnC), and
- N_Q = Circuit noise equivalent (in dBrnC) of quantization noise referred to ROLR = 46 dB.

2. <u>Talker Echo</u>

$$R_E = 106.4 - 53.45 \log_{10}\left[\frac{1 + D}{\sqrt{1 + (D/480)^2}}\right] + 2.277E \qquad (2a)$$

where:

- E = Loudness loss (in dB IEEE) of the talker echo path, and
- D = Round-trip talker echo path delay (in ms)

<u>Loss/Noise/Talker Echo</u>

$$R_{LNE} = \frac{R_{LN} + R_E}{2} - \sqrt{\left[\frac{R_{LN} - R_E}{2}\right]^2 + 100} \qquad (2b)$$

---

* ROLR: Receiving Objective Loudness Rating, the sensitivity of the receiving system in converting an electrical signal to an acoustic signal.

Table 3.4(b)
Extension of Transmission Rating Models for Listener Echo, Quantization Noise, and Room Noise

3. <u>Listener Echo</u>

R_{LE} = 9.3 (WEPL + 7) $(D_L - 0.4)^{-0.229}$ AT&T model   (3a)

   = 10.5 (WEPL + 7) $(D_L + 1)^{-0.25}$ Compromise model   (3b)

where:

WEPL = Weighted (Listener) Echo Path Loss (in dB)

D_L = Round-trip listener echo path delay (ms).

<u>Loss/Noise/Listener Echo</u>

$$R_{LNLE} = \frac{R_{LN} + R_{LE}}{2} - \sqrt{\left[\frac{R_{LN} - R_{LE}}{2}\right]^2 + 169}$$ (3c)

4. <u>Quantization Noise</u>

N_Q = 77.0 - L_e - 2.36 Q   AT&T model   (4a)
    = 76.1 - L_e - 2.02 Q   Compromise model   (4b)
    = 76.1 - L_e - (2.403 - 0.0503L_e) Q   Compromise model that accounts for received speech level variations   (4c)

where:

L_e and N_Q are defined in Table 3.4(a)

Q = Speech to speech-correlated-noise ratio (dB)

5. <u>Room Noise</u>

$$N_R = N_A - 35 + 0.0078 (N_A - 35)^2 + 10\log\left[1 + 10^{\frac{1-SOLR}{10}}\right]$$ (5)

where:

N_R is defined in Table 3.4(a)

N_A = Room noise in dB(A), and

SOLR = Sidetone Objective Loudness Rating (in dB IEEE) of the telephone set sidetone path

Table 3.4(c)
Extension of Transmission Rating Models for Bandwidth/Slope and Sidetone

6. <u>Bandwidth/Slope</u>

R_{LNBW} = k_{BW} (R_{LN} - 22.8) + 22.8   (6a)

where:

R_{LN} is given in Table 3.4(a)

k_{BW} = $k_1 k_2 k_3 k_4$,   (6b)

k_1 = 1 - 0.00148 (F_1 - 310),   (6c)

Table 3.4(c) continued

k_2	=	$1 + 0.000429\,(F_u - 3200)$,	(6d)
k_3	=	$1 + 0.0372\,(S_1 - 2) + 0.00215\,(S_1 - 2)^2$,	(6e)
k_4	=	$1 + 0.0119\,(S_u - 3) - 0.000532\,(S_u - 3)^2 - 0.00366\,(S_u - 3)(S_1 - 2)$,	(6f)
F_l, F_u	=	Lower and upper band limits (in Hz) at which the acoustic-to-acoustic response is 10 dB lower than the response at 1000 Hz (for F_u > 3200 Hz, use 3200 Hz), and	
S_l, S_u	=	Lower and upper inband response slopes (in dB/octave) below and above 1000 Hz, respectively, which would have the same loudness loss as the actual response shapes.	

7. <u>Sidetone</u>

R_{LNST}	=	$K_{ST}\,R_{LN}$	(7a)

where:

R_{LN} is given in Table 3.4(a)

K_{ST}	=	$1.021 - 0.002\,(SOLR - 10)^2 + 0.001\,(SR - 2)^2\,(SOLR - 10)$	(7b)
SOLR	=	Sidetone Objective Loudness Rating (in dB IEEE),	
SR	=	Sidetone response slope in dB/octave below 1000 Hz. (The slope above 1000 Hz is 1.5 times that below 1000 Hz.)	
R_{EST}	=	$R_E + 2.6\,(7 - SOLR) - 1.5\,(4.5 - SR)^2 + 3.38$	(7c)

where:

R_E is given in Table 3.4(a)

3.2 USER- AND NETWORK-ORIENTED PERFORMANCE FOR VOICE SERVICE

3.2.1 Speed of Service Performance Criterion

3.2.1.1 Fixed Delay

Fixed delay (also called *absolute delay*) is the deterministic time for a signal to travel from the mouth of the talker to the ear of the listener, and is contributed to by processing, switching, and transmission delays, particularly in the backbone network. Survey data for North American terrestrial network toll connections indicates one-way delays of at most 30 ms for the backbone network [4, 5] (see also Table 4.15 in Ch. 4). However, as noted later (in Sec. 3.2.2.3 on talker echo), network evolution and the introduction of certain technology can cause delays to increase. In the absence of echo control, connection delays must be constrained in order to minimize the talker echo impairment. On the other hand, with adequate echo control, absolute fixed delay for terrestrial connections is not critical and could readily be increased to 40 ms. This delay is consistent with maximum-length (4000 mi) national connec-

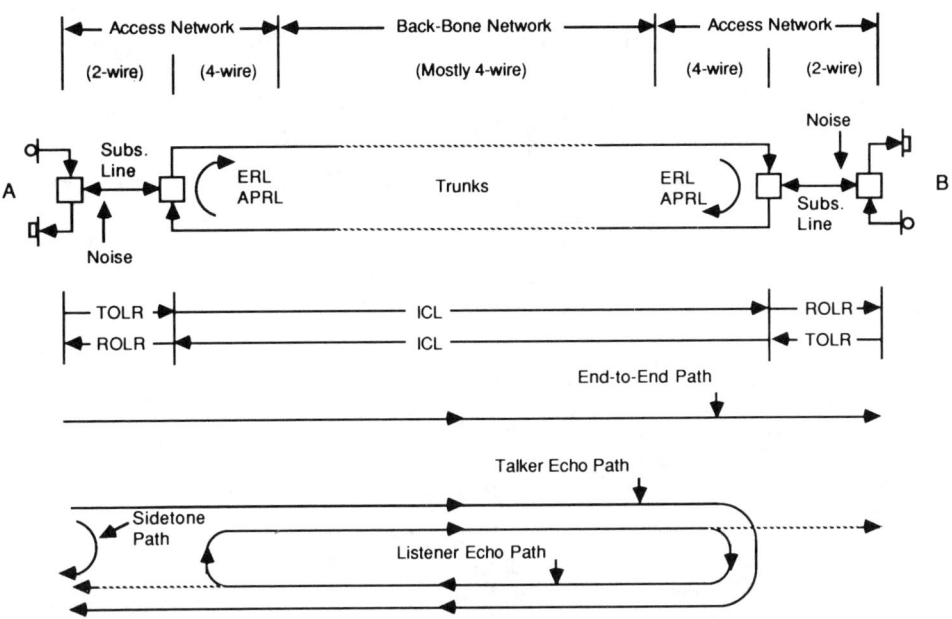

Figure 3.1 Typical mixed analog digital connection and related transmission parameters.

tions using mixed analog/digital technology, and can be estimated from [6]. In addition, with the introduction of newer, more delay-prone technology, such as asynchronous transfer mode (ATM) systems, larger delays of 100 to 150 ms may be encountered [7].

Over the years, a number of investigations have assessed the subjective effect of fixed delay on conversational speech [8–13]. Perhaps the most comprehensive and representative results are given in [8], and these are summarized in Figure 3.2. These results cover terrestrial as well as half-hop, one-hop, and two-hop satellite connections. The corresponding one-way delays are about 25 ms (estimated), 150 ms (average), 300 ms, and 600 ms, respectively. Also included are the effects of echo cancellers, and analog and digital echo suppressors.

These results suggest that with adequate echo control (preferably echo cancellers), considerable delay, such as that encountered on satellite links, can be tolerated, although with some degradation. Also, although double-hop satellite connections should generally be avoided [6], the results indicate that if equipped with echo cancellers, they can be used (i.e., in exceptional situations where no economic alternatives exist, such as alternate routing on international connections [8], emergency restoration for disaster recovery/network survivability, and access to remote regions).

However, in recent years the use of satellite transmission for voice service has been reduced owing to user sophistication and the availability of competitive alter-

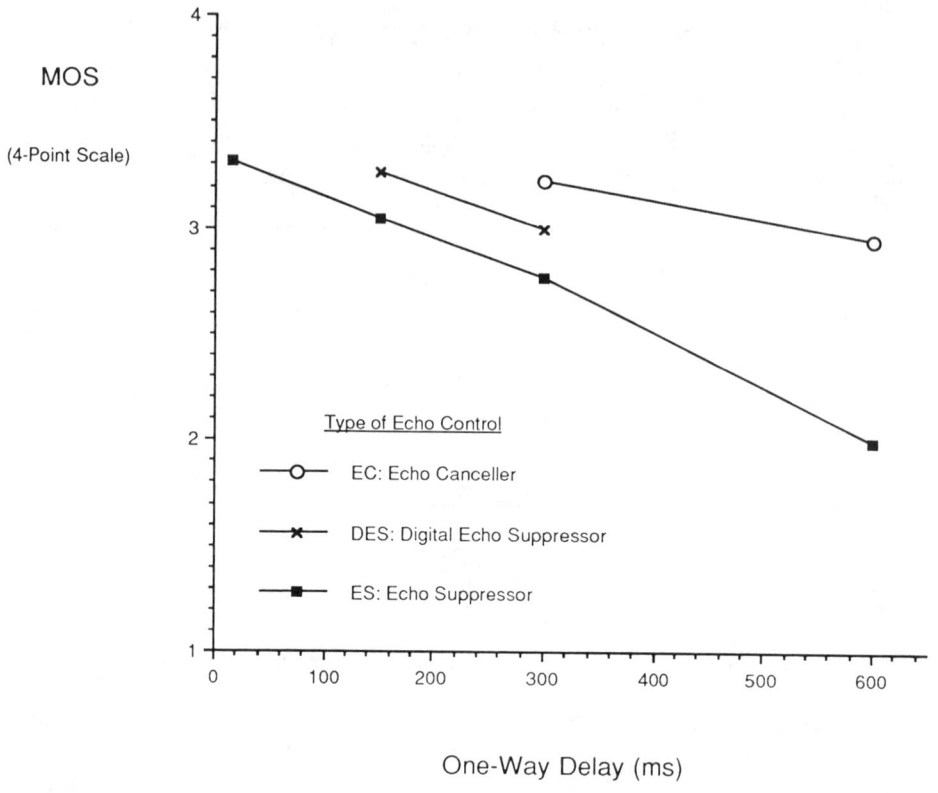

Figure 3.2 Subjective effect of fixed delay.

natives. Today, the delay of a single satellite link with cancellers is generally believed to significantly affect the dynamics of conversational speech [7]. Knowledgeable users may be able to adapt to this situation. However, if users are unknowingly exposed to both terrestrial and satellite routes, the contrast from call to call is likely to be very noticeable and possibly objectionable. Also, in the competitive U.S. environment, some users (business users in particular), realizing that they have terrestrial alternatives, may decline the use of satellite transmission. These considerations give rise to the fixed-delay performance levels summarized in Table 3.5.

3.2.1.2 Variable Delay

Variable delay is the random buffering or queueing delay imposed on speech burst traffic. This delay arises in certain advanced communication systems that use speech

Table 3.5
Performance for Fixed Delay

Transmission Medium	Fixed One-Way Delay (ms)	Subjective Effect of Delay		Echo Control Device
		Pre-Divestiture Non-Competitive Environment	Post-Divestiture Competitive Environment	
Terrestrial	≤ 40*	None	None	Cancellers or Suppressors #
Single Hop Satellite	≤ 300	Little or none	Significant, especially with suppressors	Cancellers or Suppressors #
Double Hop Satellite	≤ 600	Some, but generally not objectionable	Objectionable	Cancellers

* Higher delays are anticipated which depend on the type of newer, more delay prone technology (see text).
\# Tandeming of echo suppressors should be avoided.

activity detection to exploit speech idle time, and that manage the resulting speech burst traffic on a delayed (or buffered) basis. These systems are a subset of those called *digital circuit multiplication equipment* (DCME) [14–20]. Examples of systems that introduce speech burst delay include buffered speech interpolator-type digital speech interpolation (DSI) systems (e.g., see [15] and references in [21]), packet voice networks (e.g., see [22, 23] and references in [1, 6, 24]), and various integrated services systems, including ATM [25–34, 75].

With such systems, variable delay can be smoothed by buffering the speech at the destination node. In this case, variable delay is absorbed into a larger fixed delay (as addressed in Sec. 3.2.1.1). Depending on the type of system, the smoothing may also involve time stamping of transmitted packets to enable proper reconstitution of the speech into a smooth stream at the destination. Various anomalies must also be contended with, such as missing (late) packets and buffer overflow and underflow.

An alternative to the fixed-delay approach is to permit some variable delay to occur. This can be done by appropriately controlling the effective speech burst duration and speech burst rate at the originating node. The resulting variable delay mechanism is illustrated in Figure 3.3. The subjective effect of variable delay is highly dependent on the speech detector characteristics, that is, the speech detector threshold and hangover time [21]. These two characteristics determine the effective speech burst rate and hence the rate of delay occurrence and corresponding sensitivity to the variable delay impairment. Figure 3.4 illustrates how the speech detector hangover serves to increase effective speech burst duration and reduce effective speech burst rate (and thereby reduce the rate of delay occurrences).

Figure 3.3 Illustration of variable speech burst delay.

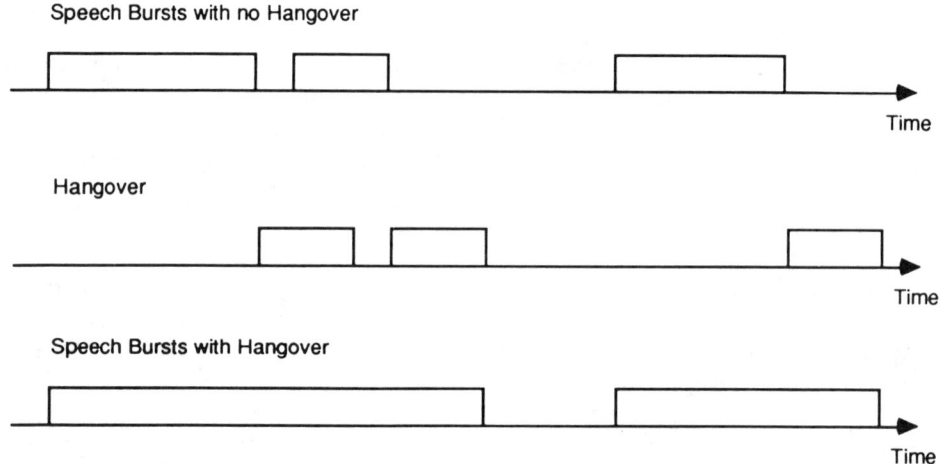

Figure 3.4 Illustration of how speech detector hangover increases effective speech burst duration and reduces effective speech burst rate.

The results of subjective testing are shown in Figure 3.5 in terms of mean opinion score (MOS) versus the "average" variable delay imposed on speech bursts [35]. Note, that the "average" is the first moment of the speech burst delay distribution. Since in these tests the distribution was a truncated exponential, the "average" is also approximately equal to the standard deviation and, as such, is a useful measure of delay variability. The parameter in Figure 3.5 is the speech detector hang-

Figure 3.5 Subjective effect of variable speech burst delay.

over, which is selected pair-wise with speech detector threshold to maintain high speech quality with no speech burst delay [36]. Figure 3.5 suggests the following guidelines concerning variable delay:

- Speech detector hangovers of 100 ms or less cause considerable GOS degradation to all but the shortest "average" variable delays (<50 ms). The reason for this behavior is that small hangovers result in many short speech bursts, and, hence, a high exposure rate to variable delays, including the highly perceptible delays between syllables and words.
- Hangovers of 150 ms or more are preferred because they provide nearly constant GOS for "average" variable delays on the order of the hangover value or less. These longer hangovers result in fewer and longer speech bursts, and, hence, a lower exposure rate to variable delay. Also, the delays, being between phrases and sentences, are less perceptible. To minimize effective speech activity, hangovers should not exceed 250 or 300 ms. Overall, 200 ms is a good design value.

Variable delay performance is summarized in Table 3.6. These results should be interpreted from the point of view of a special-purpose (e.g., private) network

Table 3.6
Performance for Variable Speech Burst Delay (Speech Detector Hangover = 200 ms)

"Average" Variable Speech Burst Delay (ms)	Subjective Effect	Grade-of-Service *	
		MOS	Remark
≤ 200	Perceptible threshold	4	Good
≤ 300	Perceptible but not objectionable	3.5	Fair to Good

* In the absence of echo (see text).

application taking advantage of the economic saving provided by DCME-like equipment [15, 21]. For a general-purpose public network, the results in Table 3.6 may be optimistic. Also, unless connections encountering these delays are 4-wire end-to-end, echo control would be required (see Sec. 3.2.2.3).

3.2.2 Accuracy of Service Performance Criterion

3.2.2.1 Loudness Loss

General: Three perceptible aspects of loudness loss are end-to-end loss, loss asymmetry, and loss contrast. The first is the most important and is addressed subsequently in some detail. The remaining two are then briefly considered.

End-to-end loudness loss: This loss is sustained by an acoustic signal from the mouth of the talker to the ear of the listener (see Fig. 3.1). For an end-to-end connection, it is given approximately by [37]

$$\text{OOLR} = -20 \, \text{Log}_{10}(S_E/S_M) = \text{TOLR} + \text{ICL} + \text{ROLR} \quad (3.1)$$

where OOLR is the overall objective loudness rating in decibels, and S_E and S_M are the sound pressure levels at the ear and mouth reference points, respectively (in Pascals).

Also, TOLR and ROLR are the transmit and receive objective loudness ratings (or sensitivities) in decibels, respectively, of the voice terminals given by

$$\text{TOLR} = -20 \log_{10}(V_T/S_M), \text{ and } \text{ROLR} = -20 \log_{10}(2S_E/V_W) \quad (3.2)$$

where V_t and V_w are the transmit and (open circuit) receive voltages in millivolts, respectively. Finally, OOLR also includes the 1-kHz inserted connection loss (ICL) between the switching offices or between remote switching systems (see Fig. 3.1).

Typical values of TOLR and ROLR are in the region of −47 to −43 dB and 46 to 50 dB, respectively [37] (see also Table 6.2 of Ch. 6). In the case of telephone networks, TOLR plus ROLR includes the loudness loss of the telephone sets and of the two-wire subscriber lines, which connect the sets to the switching offices or to the remote switching systems (i.e., the two-wire portions of the access network shown in Fig. 3.1).

For North American public switched telephone networks, ICL is nominally about 0, 3, or 6 dB, depending on the type of connection; 0 dB is for intra-office connections (i.e., through one switching office), 3 dB is for inter-office connections (i.e., metropolitan network connections through two offices), and 6 dB is for toll connections (i.e., long-distance connections typically through more than two offices). These connections are illustrated later in Figures 4.1 through 4.4 of Chapter 4, and loss planning is detailed in Chapter 7.

Note that the definitions of loudness loss in terms of OOLR, TOLR, and ROLR are based on the IEEE loudness measurement method now commonly used in North America [38]. However, there are other loudness loss definitions, such as those based on the ElectroAcoustic Rating System (EARS) and the International Telegraph and Telephone Consultative Committee (CCITT) methods [39–41]. Therefore, the approximate conversions among loudness loss measures are included for reference in Table 3.7.

In past years, the perception of loudness loss has been assessed by various subjective tests and GOS models. Figure 3.6 is based on the equations for the loss/noise transmission rating model in Table 3.4(a) and is a repeat of Figure 2.6 presented in

Table 3.7
Approximate Conversions Among Loudness Loss Measures

Sensitivities	IEEE	EARS	CCITT	
	Objective Loudness Rating (OLR)	Rating (R)	Reference Equivalent (RE)	Corrected Reference Equivalent (CRE)
Transmitting Loop (TL), Transmitting (T), or Sending (S)	TOLR	TLR = TOLR + 25	SRE = TOLR + 52	CSRE = TOLR + 55
Receiving Loop (RL) or Receiving (R)	ROLR	RLR = ROLR − 20	RRE = ROLR − 46	CRRE = ROLR − 45
Overall (O) *	OOLR	OR = OOLR + 5	ORE = OOLR + 6	CORE = OOLR + 6 **

* For End-to-End, Sidetone, and Echo Path Loudness Losses
** CORE = OOLR + 10 for Echo Path Loudness Loss

Figure 3.6 Loss/noise grade of service.

Chapter 2. Figure 3.6 represents the model used in North America for the transmission performance assessment of existing and evolving telephone networks [42, 43]. This model predicts an optimum loudness loss of about 3 dB. However, for low noise, this optimum is not critical, and higher losses up to about 9 dB cause little GOS degradation. These losses correspond to receive speech levels, which are regarded as neither too loud nor too soft. In North America, 3 dB corresponds approximately to the average loudness loss encountered on local telephone calls between subscribers connected to the same switching office [44, 45] (i.e., intra-office calls within the access network). Thus, from an optimum performance standpoint, an ultimate long-term nominal value of loudness loss for end-to-end all-digital connections is 3 dB. However, 6 dB is recommended; this is for the practical reasons of interworking with existing networks and minimizing contrast with existing nominal losses (of about 3, 6, and 9 dB for intra-office, inter-office, and toll calls, respectively) [46, 47] (see Ch. 7).

For analog and mixed analog/digital telephone connections, there is considerable end-to-end loudness loss variability, which is contributed to by telephone sets, subscriber lines in the access network, and by transmission facilities between offices in the backbone network. For mainly analog connections, it can be estimated from survey information [5, 37, 48, 49] that at the 5% level loudness loss may be as low as about −3 dB for short intra-office connections, and at the 95% level as high as about 19 dB for long toll connections (see Table 4.8 of Ch. 4). As can be seen in Figure 3.6, this range produces considerable GOS degradation from the optimum loss, but still results in acceptable performance for a large percentage of connections.

With mixed analog/digital connections, there is considerable scope for reducing loudness loss variability. For example, the use of digital transmission facilities in the backbone network [50, 51] can reduce the high loudness loss of long toll connections from 19 to about 17 dB. Also, the introduction of digital technology in the form of subscriber carrier and remote switching systems into the access network, as well as the use of gain equalization for excessively long subscriber lines, can further reduce the range of loudness losses [52–54]. In addition, certain access network applications, such as those involving digital private branch exchanges (DPBX), use loss equalization to increase the low loudness loss of associated short subscriber lines [55]; this increases the lower loudness loss from −3 to about 0 dB. Thus, in the longer term, the above considerations indicate a reduced loudness loss range of about 0 to 14 dB for mixed analog/digital networks. As noted earlier, in an all-digital connection, the loudness loss range will be considerably reduced, and the loss will center on a fixed value of about 6 dB [47] (see Ch. 7). Loudness loss performance in terms of network evolution is summarized in Table 3.8, after the idle circuit noise parameter is discussed.

Loss asymmetry: Loss asymmetry is the difference between the end-to-end loudness losses in the two directions of transmission as experienced during a given telephone conversation. As might be expected, subjective tests indicate that users' opinions are most affected when loss variations are introduced in the direction toward, rather than away from, the user (i.e., listening is affected more than talking) [56]. However, if transmission performance is otherwise not excessively degraded, loss asymmetry is not considered critical and can be as much as a few decibels. This is reasonable, given that the speech levels of the two parties in a conversation may readily differ by more than a few decibels [57]. Thus, there is generally no strong need to specifically control this impairment. Accordingly, as a guideline, national networks are allocated 4 dB of maximum loss asymmetry, so that 8 dB is not exceeded on international connections [58].

Loss asymmetry is caused by differences in the way telephone-set transmit and receive sensitivities vary with the length of subscriber loops [37, 59], and also by loss differences in each direction of transmission on 4-wire trunks [60]. Loss asymmetry may also occur as a result of the interworking of different types of networks (e.g., public and private networks). Nominally, loss asymmetry is usually 0 dB by design. However, owing to the loss variations noted above, an analysis of survey data indicates that at the 95% worst-case level, there could be as much as about 3, 4, and 5 dB of loss asymmetry on intra-office, inter-office, and toll connections, respectively (see Table 4.8 of Ch. 4). As networks become more digital, loss asymmetry will reduce, since there will be progressively fewer points of loss insertion, and therefore less opportunity for loss variability. Ultimately, 4-wire digital telephones will be used, resulting in elimination of analog subscriber loops, and the virtual elimination of loss asymmetry.

Loss contrast: Loss contrast is the difference in loss perceived from one telephone call to the next. It is expected that loss contrast would be most noticeable on successive calls between the same parties. As with loss asymmetry, if transmission performance is otherwise not excessively degraded, loss contrast is not critical, and as much as 5 dB would be just perceptible [58].

Loss contrast is caused, for example, by alternate or dynamic routing procedures in either metropolitan or toll networks, where these networks are either of an analog or a mixed analog/digital nature. In either case, the alternate route will typically involve at least one additional trunk, and hence additional loss in the connection. For metropolitan network connections involving from one to four trunks, the loss contrast has been estimated to be less than 5 dB for 90% of calls [61]. However, loss contrast will reduce as networks become more digital (i.e., use digital trunks with 0 dB loss).

3.2.2.2 Idle Circuit Noise

Idle circuit noise is the continuous background noise level perceived mainly in the absence of speech. This noise is generally stationary and independent of the speech level, and is also assumed to have a uniform frequency response and a Gaussian (normal) amplitude density. This noise is assessed objectively in terms of its short-term average power as measured across the telephone voice frequency band in conjunction with an appropriate frequency response weighting; in North America, C-message weighting is used. For GOS purposes, the noise is expressed in decibels reference noise C-weighted (dBrnC) at the input to a reference receiving system (i.e., telephone set plus subscriber line having a sensitivity of ROLR = 46 dB [37]).

The perception of noise may be assessed using the GOS model considered in Figure 3.6. Note that, in this figure, about 27 dBrnC of electrically equivalent room noise (as coupled in through the telephone set sidetone path) is power summed with the idle circuit noise [37]. (Under nominal sidetone conditions, this 27-dBrnC figure corresponds to about 57.5 dB(A) of acoustic room noise, which is typical of that found in a business office environment [62]; the sidetone parameter is discussed later in this chapter.) Under these room noise conditions, it is evident from Figure 3.6 that idle circuit noise levels less than 20 dBrnC have a negligible effect on performance. This noise level of 20 dBrnC also corresponds to the near maximum idle channel noise level likely to be produced by 64 kb/s analog-digital (A/D) plus digital-analog (D/A) converters [63, 64]. Thus, from a purely performance standpoint, 20 dBrnC may be regarded as an ultimate long-term lowest practical value for idle circuit noise on end-to-end all-digital connections. It should also be noted that the entire elimination of noise is unnecessary and possibly undesirable because some low level of noise serves to provide feedback to the user that a connection has been established or is functioning properly. However, a requirement for such low-level noise is generally always automatically satisfied because connections are never entirely noise free.

For analog and mixed analog/digital connections, an analysis of survey information [5, 37] shows that received noise levels vary considerably and can exceed 35 dBrnC (see Table 4.9 of Ch. 4). From Figure 3.6 this noise level, together with the 19-dB maximum loudness loss noted earlier, can result in considerable GOS degradation. However, this is acceptable as a lower performance limit, and in fact is somewhat conservative. This is because statistical independence between the surveyed values of loudness loss and idle circuit noise has been assumed in the above discussion. There is some degree of dependency such that high (low) loss connections will tend to reduce (increase) end-to-end noise by way of more (less) attenuation; the reverse is true as indicated in parentheses. A more detailed consideration of these dependencies is given in Chapter 5.

Mixed analog/digital connections can produce considerably lower noise levels. For example, replacement of the long-haul noise-controlling analog carrier transmission facilities by corresponding digital facilities makes it reasonable to expect an overall idle circuit noise level of about 30 dBrnC or less. Eventual conversion of the entire backbone network to digital operation only leaves the noise contribution of one A/D plus D/A conversion (20 dBrnC), plus the access network noise, which is generally less than 20 dBrnC [51, 65] for each end of the connection. This corresponds to an overall noise level of about 25 dBrnC. Loudness loss and idle circuit noise performance is summarized in Table 3.8.

3.2.2.3 Talker Echo

As shown in Figure 3.1, talker echo is the result of an acoustic signal originating from the mouth of the talker being reflected by (usually) one major discontinuity in the connection, and returned to the ear of the talker. Multiple discontinuities are considered later in Chapters 5 [66]. Talker echo is characterized primarily by its amplitude and delay relative to the originating signal. Talker echo path loss (TEPL) is the loss in decibels sustained by the acoustic signal from the mouth to the ear of the talker, and for the A end of the connection in Figure 3.1, is given approximately by

$$\text{TEPL} = -20 \log_{10}(S_E/S_M) = \text{OOLR for the talker echo path}$$
$$= \text{TOLR}_A + \text{ROLR}_A + 2\,\text{ICL} + \text{ERLB}_B \quad (3.3)$$

All of the above terms have been previously defined (see Eqs. (3.1) and (3.2)), except for echo return loss (ERL). From Figure 3.1, ERL is the loudness loss of the

Table 3.8
Loss/Noise Performance

Network		Loudness Loss Range (dB, IEEE)	Idle Circuit Noise (dBrnC)	Worst Case Grade-of-Service (Percent GOB)
Mainly Analog		-3 to 19 *	≤ 35	~ 35
Mixed	Short Term	-3 to 17 *	≤ 30	~ 55
	Long Term	0 to 14 *	≤ 25	~ 80
Digital		~ 6 #	≤ 20	~ 95

* Nominal values are about 3, 6, and 9 dB for intra-office, inter-office, and toll connections, respectively.
\# Nominal value.

reflected signal as determined at the reflection point, and usually over the 500- to 2500-Hz region of the voice band [67].

In North America, the perception of talker echo is assessed using the GOS model illustrated in Figure 3.7 (this is based on the talker echo transmission rating model in Table 3.4(a) [42]). This figure is determined for loudness loss and noise

Figure 3.7 Loss/noise/talker echo grade of service.

levels that exemplify a mixed analog/digital toll connection. From this figure, it is evident that, for the largest talker echo path delays likely to be encountered (e.g., 600 and 1200 ms on single- and double-hop satellite connections, respectively), about 50 to 55 dB of TEPL is required in order to give negligible talker echo impairment. Therefore, in terms of performance, this range indicates ultimate long-term minimum values of TEPL necessary for end-to-end all-digital connections; for such connections, only an acoustic echo path and perhaps an incidental electrical path due to crosstalk will be associated with the acoustic and analog electrical portions of the far-end voice terminal. Depending on the performance of voice terminals [68], echo control may still be required on satellite connections, but should not be required on terrestrial connections.

However, for analog and mixed connections, electrical echo paths are predominantly due to 4- to 2-wire interfaces between the backbone and access network, as shown in Figure 3.1 (see Figs. 6.1 and 6.2 of Ch. 6 for more detail). Analysis of survey information [4, 49] shows that such echo paths in mainly analog toll connections can have TEPLs lower than 15 dB (see Table 4.13 of Ch. 4). Thus, to provide satisfactory GOS on high-delay connections, echo control devices (i.e., echo suppressors or echo cancellers) are required to effectively increase the TEPL to high values [69].

However, in practice, a small percentage of terrestrial connections with TEPLs as low as about 15 dB, and with echo path delays exceeding about 35 ms, may not encounter echo control [4]. As can be seen in Figure 3.7, this results in considerably degraded performance, but only for a small percentage of connections. This percentage tends to be small because, in practice, echo control also appears on some connections with delays as low as 25 ms [4]. In addition, in mixed analog/digital connections, there is some degree of dependency such that higher delay connections tend to also have higher TEPL.

Echo control is generally intended, by network design, to be associated with connections having echo path delays ≥ 45 ms. However, in practice, the deployment of echo control devices is such that they are usually encountered on connections with delays higher than about 35 ms. Echo control results in TEPLs of at least 55 dB, which is achieved by echo suppressors [70], or echo cancellers with center clippers [71, 72].

With network evolution, general technological improvements in the access portion, as well as more sophisticated balance networks in backbone/access interfaces, provide higher ERLs and, therefore, higher performance for connections not encountering echo control [52–54, 73] (see Tables 6.10 and 6.12 of Ch. 6). Thus, TEPLs ≥ 20 dB for connections with echo path delays ≤ 35 ms can be expected in the short term. Also, recent developments in echo canceller technology tend to make their use more economically feasible on lower delay (shorter) connections.

Echo control on shorter connections may also be necessitated by recent developments in the industry, which are resulting in increased network delay [74]. These developments include the introduction of higher delay technology such as fiber optic transmission systems, digital cross-connect systems, and possibly low bit rate coding schemes. Other developments include architectural changes to networks, such as access network evolution with fiber-based remote switching systems, additional switching and trunking in exchange access networks, dynamic/time of day routing in backbone networks, and, eventually, some form of wideband transport system such as ATM [25–28, 75–80]. Some of these developments are discussed in more detail in Chapter 7. Accounting for these developments, plus the fact that some connections with delays as low as 25 ms already have echo control, TEPLs ≥ 20 dB for connections with echo path delays ≤ 30 ms may be expected in the longer term. Based on the above discussion, talker echo performance is summarized in Table 3.9.

Table 3.9
Loss/Noise Talker Echo Performance

Network		Talker Echo Path Loss (dB IEEE)	Talker Echo Path Delay (ms)	Echo Control Devices	Worst Case Grade-of-Service (percent GOB)
Analog		≥ 15	≤ 35	No	30
		≥ 45	> 35	Yes	55
Mixed	Short Term	≥ 20	≤ 35	No	55
		≥ 45	> 35	Yes	55
	Long Term	≥ 20	≤ 30	No	70
		≥ 55	> 30	Yes	> 90
Digital		≥ 55	≤ 1200 *	No **	> 90 *

   * Note 1200 ms delay corresponds to two tandemed satellite links, which, while feasible, is not recommended. Note also that the GOS refers to loss/noise talker echo GOS only, and does not account for degradations due to excessive delays that affect the dynamics of conversations (see Sec. 3.2.1.1).
** Echo control devices are not required for terrestrial connections but may still be required for satellite connections.

3.2.2.4 Listener Echo

Listener echo results from an acoustic signal, originating from the talker's mouth, being received more than once at the listener's (and talker's) ear because of a principal reflection point at each end of the connection, as shown in Figure 3.1. Note that a listener echo signal is generally received more than once because it circulates in the echo path (until reduced to zero by the listener echo path loss (LEPL)).

Listener echo is characterized primarily by the delay and loss in the listener echo path. Since this echo path is prone to positive feedback and potential instability, LEPL has been specified in terms of the singing margin (SM), which is the minimum loss of the listener echo path at a single frequency in the voiceband. SM must be greater than 0 dB to ensure stability, and considerably greater than 0 dB to provide satisfactory performance.

However, more recent listener echo studies also account for the frequency response of the listener echo path which, from a subjective standpoint, provides a more accurate measure of LEPL [66, 81–83]. This measure is called *weighted echo path loss* (WEPL) and is not as stringent as the SM measure. A conservative estimate of WEPL in decibels is given by

$$\text{WEPL} = \text{APRL}_A + 2 \text{ ICL} + \text{APRL}_B \tag{3.4}$$

where the average power return loss (APRL_X) is the return loss of the reflected signal as determined at a reflection point (near the X end of the connection), and averaged over the 200- to 3200-Hz region of the voice frequency band [51, 65].

Owing to the averaging aspect of the WEPL measure, it is possible for the LEPL at specific frequencies to be significantly lower than the WEPL value, and in extreme cases to be ≤0 dB. To protect against such unstable situations, in addition to meeting WEPL objectives, the SM of each connection should exceed 4 dB.

The perception of listener echo may be assessed using the GOS model illustrated in Figure 3.8 (based on the compromise listener echo transmission rating

Figure 3.8 Loss/noise/listener echo grade of service.

model in Table 3.4(b) [84]). This figure applies, for example, to contemporary applications where digital technology, in the form of remote switching, is used in the access network; digital switching introduces some delay relative to the older analog switching [10, 52, 53, 85, 189]. However, in these applications, intra-office connections (i.e., connections using the access network) are such that ICL is usually at or near 0 dB to be consistent with the previous analog environment [54, 65]. This low loss corresponds to the lowest values of WEPL and, with the higher delay, to the most degradation in GOS.

In this regard, it is noted that intra-office connections that use all-analog technology usually do not have an obvious listener echo path, although there may be 2-wire impedance discontinuities, which effectively produce such a path. However, the delay of such a path would be minimal, and hence there is no significant listener echo impairment. Also, inter-office and toll connections (i.e., connections involving both the access and backbone networks) generally have nonzero ICLs to control talker echo, and, hence, WEPLs large enough that GOS degradations due to listener echo are minimal. Therefore, these connections are not of interest in terms of listener echo.

Echo path delays of contemporary mixed analog/digital intra-office connections are usually less than 5 ms, and survey data indicates that most WEPLs exceed 10 dB [52, 82, 86] (see Table 4.11 of Ch. 4). Higher WEPLs, usually exceeding 16 dB, are possible with more sophisticated balance networks at the interfaces between analog subscriber lines and local digital switching or transmission systems in the access network. On the other hand, some of the newer technologies noted in Section 3.2.2.3, could increase echo path delays on intra-office connections to about 10 ms. However, as can be seen in Figure 3.8, it is evident that the above WEPLs will still result in little or no listener echo impairment for voice service. In fact, nonvoice services such as voiceband data (VBD) are controlling as far as listener echo is concerned [87]; VBD service is addressed in the context of listener echo in Chapter 4.

It is also evident from Figure 3.8 that sufficiently high WEPLs will produce negligible impairment regardless of the echo path delay. This scenario corresponds to eventual end-to-end all-digital connections for which WEPLs will be very large. Another way to view this is to note that the ultimate long-term value for TEPL of 55 dB, considered earlier, will serve to eliminate any possibility of a listener echo impairment. Listener echo performance is summarized in Table 3.10.

3.2.2.5 Quantization Noise

Quantization noise is correlated with the speech signal amplitude and is the result of speech transmission through one or more A/D plus D/A conversion processes (also called *encoder plus decoder pairs,* or *codecs*). Examples of such codecs are 64-kb/s μ-law pulse code modulation (PCM) [88] and 32-kb/s adaptive differential PCM (ADPCM) [16, 20, 89–97].

Table 3.10
Loss/Noise Listener Echo Performance

Access Network		WEPL	Listener Echo Path Delay (ms)	Grade-of-Service (Percent GOB)
Mixed	Short Term	≥ 10	≤ 5	> 90
	Long Term	≥ 16	≤ 10	> 90
Digital		–	–	> 90

There are a number of objective measures of quantization noise or of signal-to-quantization noise ratio. However, most of these measures do not provide an accurate indication of the subjective effect of this impairment, particularly for more than one codec in tandem. For this reason, a measure of signal-to-quantization noise ratio called Q (in decibels) is determined based on subjective tests [98–108] and related GOS modeling [109]. Such a model is shown in Figure 3.9 [43, 99]. For mod-

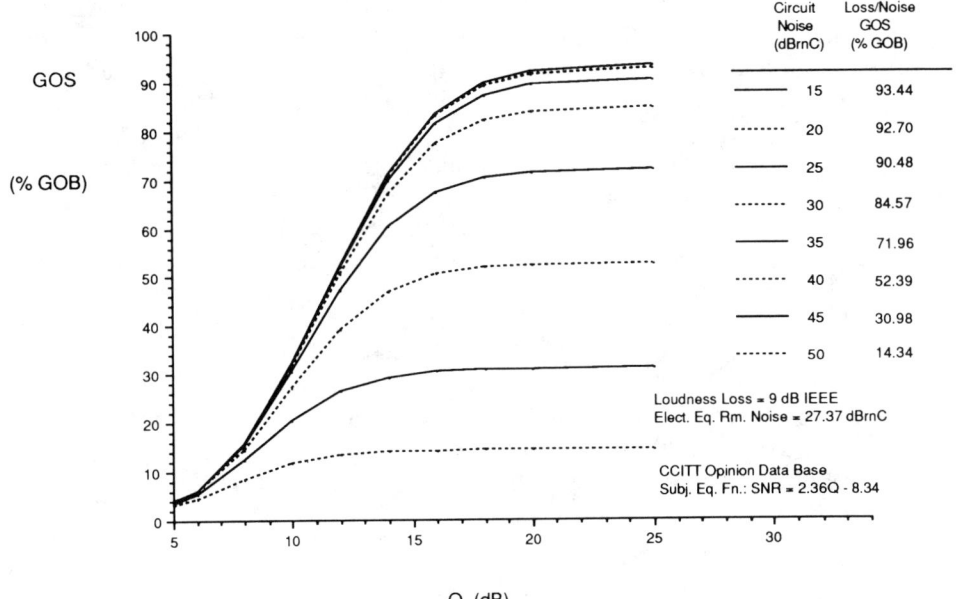

Figure 3.9 Loss/noise/Q grade of service.

eling purposes, Q is converted to subjectively equivalent uncorrelated (idle circuit) noise, which adds on a power basis to the idle circuit noise already present in the connection. See Table 3.4(b) for the way in which quantization noise is incorporated into the loss/noise transmission rating model (further detail is provided in Appendix A.5.1 of Ch. 5). The transmission rating model accounts for the fact that quantization noise, being correlated with the speech signal amplitude, is less objectionable than idle circuit noise. Various estimates of the degree of subjective improvement are given in Table 3.4(b); the results in Figure 3.9 are based on the AT&T equation, which to date (while not the most recent or necessarily the most accurate) is the most commonly used (e.g., see [37]).

Q is approximately related to the number of tandemed codecs by

$$Q_N = -15 \log_{10} \left\{ \sum_{i=1}^{N} 10^{Q_i/15} \right\} \text{ (dB)} \qquad (3.5)$$

where Q_N is the resultant Q for N tandemed codecs, and Q_i is the Q for each individual codec. For identical codecs,

$$Q_i = Q_1 \text{ for all } i, \text{ and } Q_N = Q_1 - 15 \log_{10}(N) \qquad (3.6)$$

For example, $Q_1 \approx 37$ dB for 64-kb/s PCM and $Q_1 \approx 29$ dB for 32-kb/s ADPCM [110] (see Table 6.9 of Ch. 6 for other values of Q). The slope of Equation (3.6) represents the degradation in Q of $15 \log_{10}(2) = 4.5$ dB for each doubling of the number of tandemed codecs, N. On a worst-case basis, considerable tandeming of PCM or ADPCM codecs could be permitted until Q_N reduces to about 15 dB. Most users would perceive this level of impairment because it is comparable to the average performance of long-toll analog network connections [111–114].

As can be seen in Figure 3.9, there is little GOS degradation for Q values larger than 20 to 25 dB. This is the range corresponding to the approximate perceptible threshold of Q for the average user. $Q \geq 20$ dB and $Q \geq 25$ dB correspond to progressively higher performance levels in the short and longer term, respectively, and to correspondingly fewer tandemed codecs as the networks evolve toward an end-to-end all-digital state. Ultimately, only one codec is required; when this stage is reached, it is expected that sub-64-kb/s codecs will be able to achieve the ultimate long-term value of $Q \geq 30$ dB. Most users would not perceive any impairment for this range of Q. Performance for quantization noise is summarized in Table 3.11.

3.2.2.6 Acoustic (Room) Noise

Acoustic noise is the background ambient level of room noise perceived by telephone users. This noise is assessed objectively in terms of its short-term average intensity as

Table 3.11
Quantization Noise Performance

Network		Total Signal to Quantization Noise Ratio Q_N (dB)	Number of Tandemed A/D plus D/A Processes (N)		Grade-of-Service for 25 dBrnC of noise (percent GOB)
			PCM	ADPCM	
Mixed	Worst Case	≥ 15	≤ 29[a]	≤ 8	75
	Short Term	≥ 20	≤ 13	≤ 4	90
	Long Term	≥ 25	≤ 6	≤ 2	> 90
Digital		≥ 30	≤ 3[b]	1	> 90

[a] Note this number of PCM codecs is only allowable from a quantization noise standpoint. Other impairments, such as attenuation distortion, limit the total number to ≤ 14 [192].
[b] Hypothetical, since only one codec is required for all-digital connections.

measured by a sound pressure meter with an appropriate frequency response weighting. In North America "A" weighting is used, and the noise is expressed in decibels(A), or dB(A), which is the sound pressure relative to the threshold of hearing intensity of 10^{-12} W/m². In general, sound intensities may span a range from 0 dB(A) (threshold of hearing) to about 120 dB(A) (threshold of pain). However, for telephony purposes, relevant levels are 40 and 55 dB(A), which correspond approximately to ambient noise levels in an average residence and an average office environment, respectively [62]. For comparison, average conversational speech at a distance of three feet has an intensity of about 70 dB(A).

Room noise affects telephone users directly (i.e., acoustically) and indirectly by the coupling of the acoustic noise into the telephone set electrical sidetone path. As indicated in Figure 3.1, and as discussed in more detail later in this chapter, electrical sidetone is the result of the acoustic signal in the vicinity of the talker's mouth being returned through the voice terminal to the talker's ear. In the case of an acoustic noise signal, the overall result is considered as electrically equivalent room noise which adds (on a power basis) to idle circuit noise. See Table 3.4(b) for the way in which room noise is incorporated into the loss/noise transmission rating model.

Thus, the perception of acoustic noise may be assessed using the GOS model in Figure 3.10 [43, 84]. This figure shows that acoustic noise levels up to about 55

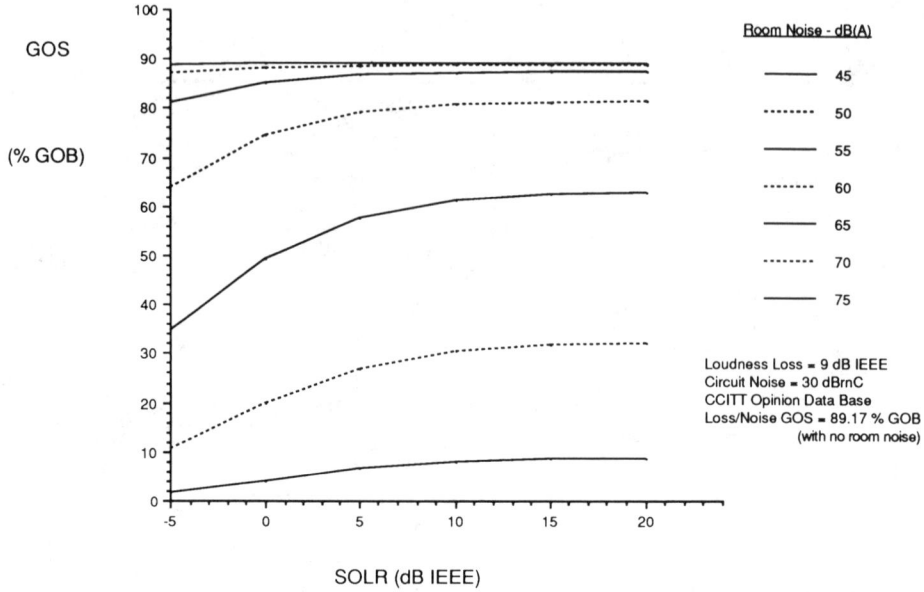

Figure 3.10 Loss/noise/room noise grade of service.

dB(A) have little effect on GOS. This is especially true since the sidetone path loss (defined in the next section as sidetone objective loudness rating (SOLR)) is usually higher than 0 dB.

For room noise much above 55 dB(A), GOS is significantly degraded. However, there is little dependence of this degradation on sidetone path loss for values of SOLR ≥ 10 dB. This implies that considerable room noise impairment is perceived directly, and not through the sidetone path.

3.2.2.7 Sidetone

As indicated in Figure 3.1, electrical sidetone is the result of the acoustic signal originating from the talker's mouth being returned through the voice terminal to the talker's ear (there are also various acoustic sidetone paths, i.e., through the air, through the telephone handset, and through the user's head; however, these acoustic paths are not considered here). Sidetone is characterized by its amplitude, and to a lesser degree by its frequency response relative to the originating signal. Side tone path loss (STPL) is measured in terms of SOLR, and is the loss in decibels sustained by the acoustic signal from the mouth to the ear of the talker. It is given by

$$\text{SOLR} = -20 \log_{10}(S_E/S_M) = \text{OOLR for the electrical sidetone path} \quad (3.7)$$

where the above terms have been previously defined in relation to Equation (3.1). Thus, low values of SOLR result in high sidetone signal levels, and vice versa.

The perception of sidetone is manifested in a number of ways. One example is the coupling of ambient acoustic noise through the sidetone path to the ear, as previously demonstrated by the model in Figure 3.10. This figure indicates that, for values of SOLR of about 10 dB or higher, there is little adverse enhancement of the effects of ambient acoustic noise.

In fact, some level of sidetone is desirable to serve as feedback to the talker and provide assurance that the telephone set is functioning. Considering this factor, together with ambient noise, an SOLR of about 10 dB is near optimum. This is supported by the model illustrated in Figure 3.11, which indicates maximum or near maximum GOS at SOLR = 10 dB (see Table 3.4(c)).

The dependency of sidetone perception on the frequency response of the sidetone path is also illustrated in Figure 3.11. The dependency is minimal for SOLR = 10 dB (the range of responses considered in the model is shown in Figure 3.12; note that slopes above 1 kHz are constrained to be 1.5 times the slopes below 1 kHz). Slopes from 1 to 3 dB/octave below 1 kHz (with corresponding slopes of 1.5- to 4.5-dB/octave above 1 kHz) give the highest performance over most of the range of SOLR.

Figure 3.11 Loss/noise/sidetone grade of service.

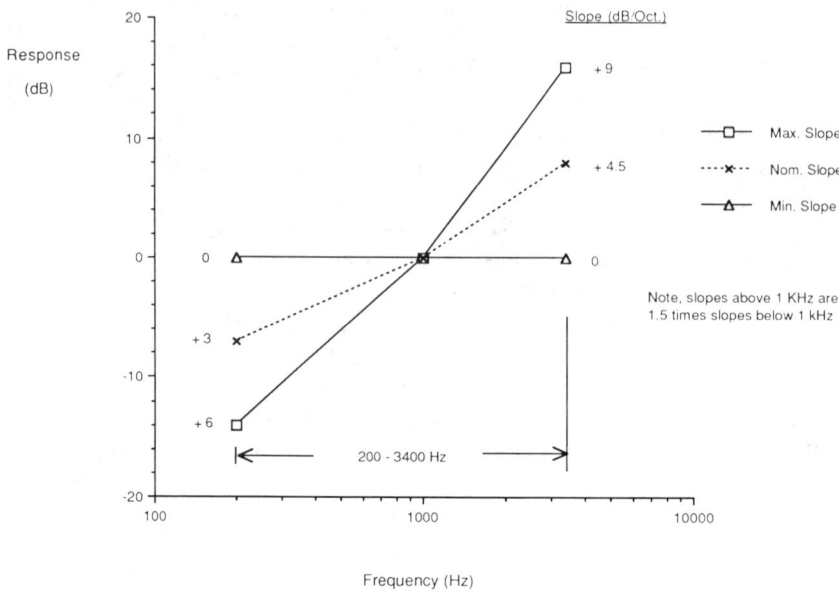

Figure 3.12 Range of slopes for sidetone response.

Sidetone perception is also affected by the presence of echo as shown by the model in Figure 3.13, which compares sidetone GOS in the absence and presence of an echo impairment (again, see Table 3.4(c)). The model shows that in the presence of significant echo there is a considerable penalty for high sidetone path loss. This implies that some level of sidetone is desirable in order to mask the effect of echo. In the presence of significant echo, the optimum SOLR is somewhat lower than 10 dB (i.e., about 7 dB).

Another aspect of sidetone is its potential effect on talker volume [115]. For example, sufficiently low SOLR may cause talkers to speak softer, or to move the telephone handset away from the head to obtain relief from the loud sidetone signal. This in turn can make it difficult for listeners to hear, particularly on connections with high loss and noise. This particular phenomenon is not accounted for in the above models, but has been addressed in other, more sophisticated models [116].

Most of the above-noted effects related to sidetone should not be manifested in eventual all-digital connections having optimum loss and low noise. A reasonable ultimate long-term nominal value for such connections is SOLR = 10 dB. This is consistent with the center of the range of SOLR values specified for digital telephones in [47].

For analog and mixed connections, there is considerable variability in SOLR because of its critical dependency on numerous factors related to telephone set

Figure 3.13 Loss/noise/sidetone grade of service.

design, loop plant characteristics, and the input impedance of central office switches. SOLRs can readily range from about 0 to 20 dB [117]. This range is also representative of the effects of introducing digital technology into the access network, which also causes SOLR variability [118]. Reduction of SOLR and response variability is possible with appropriate telephone set design and loop plant evolution. A corresponding SOLR range is from about 5 to 15 dB under these circumstances. This is consistent with the 3- to 15-dB range specified in [47]. Performance for sidetone is summarized in Table 3.12.

3.2.2.8 Bandwidth and Slope

Bandwidth in hertz is the range of frequencies bounded by a frequency f_u, above 1 kHz, and a frequency f_l, below 1 kHz (i.e., $f_u - f_l$). Note that the amplitude response of the voice channel at frequencies f_u and f_l, is 10 dB below the response at 1 kHz.

Slope in decibels per octave is an idealized measure of the amplitude response of the voiceband channel above (S_u) and below (S_l) 1 kHz. Note that an amplitude response characterized by these idealized slopes has the same loudness loss as the amplitude response of the actual voice channel under consideration. Figure 3.14

Table 3.12
Sidetone Performance

Network		SOLR Range (dB)	Worst Case Grade-of-Service (Figure 3.13)	
			No Echo	Echo
Analog and Mixed	Short Term	0 - 20	≥ 45	–
	Long Term	5 - 15	≥ 80	> 45
Digital		10	> 85	

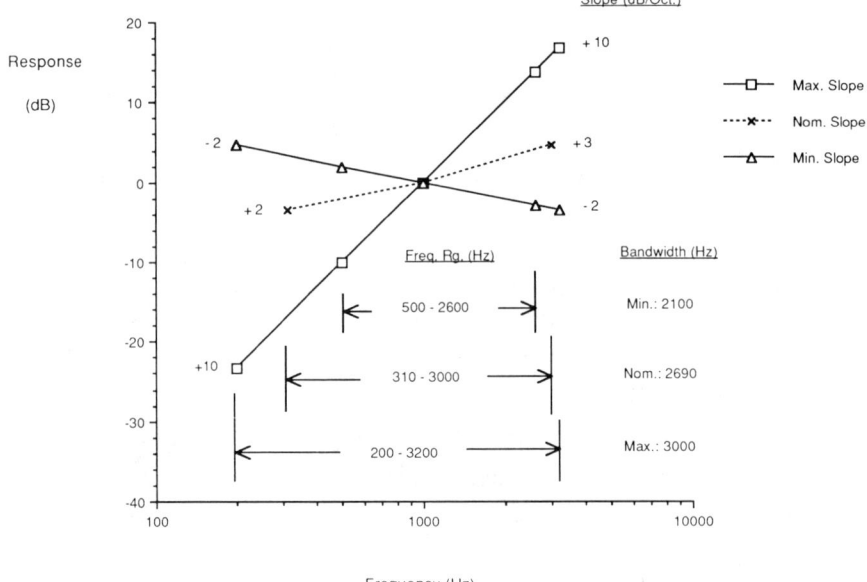

Figure 3.14 Range of bandwidths and slopes.

illustrates the applicable range of bandwidths and slopes for the bandwidth/slope model discussed below.

In general, bandwidth is determined by the frequency response of the access network (i.e., the loop plant and the subscribers' terminals), as well as by the response of the backbone network (i.e., carrier systems and associated tandeming of voiceband filters). An assessment of backbone network survey information [5] and the frequency response limits of North American telephone sets with long subscriber lines [119, 120] indicates that most connection bandwidths exceed 3000 − 400 = 2600 Hz. This minimum bandwidth results in considerably degraded performance, as estimated from Figure 3.15, which illustrates a GOS model that may be used to assess the perception of bandwidth [43, 84] (see Table 3.4(c)).

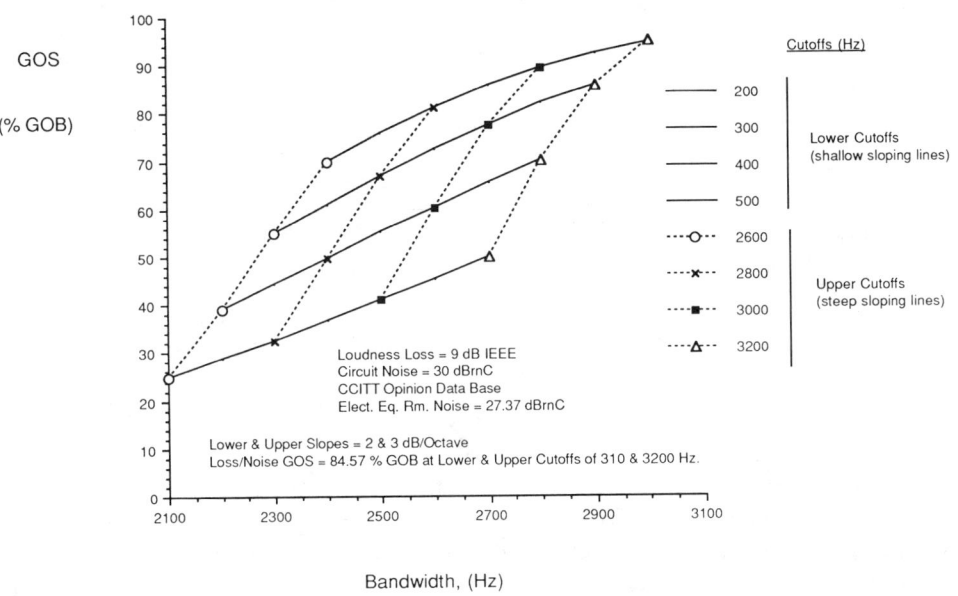

Figure 3.15 Loss/noise/bandwidth grade of service.

Degradations also depend on the slope of the amplitude response, as shown in Figure 3.16. Generally, the steeper positive slopes are preferred. Because of the transmit frequency response of telephone sets, positive slopes in excess of 3 dB/octave tend to be encountered in practice, which, as can be seen in Figure 3.16, tends to mitigate adverse effects of slope [121, 122].

With the further deployment of both digital transmission and switching, considerable reduction in bandwidth variability can be expected [98, 121]; this gives rise to values for bandwidth of 3300 − 300 = 3000 Hz, or 3200 − 200 = 3000 Hz [123].

It is evident from Figure 3.15 that for telephony purposes a bandwidth of 3000 Hz is sufficient to provide very high performance. Thus, there would appear to be little motivation for increasing the bandwidth. However, since wider bandwidths are

Figure 3.16 Loss/noise/slope grade of service.

achievable on end-to-end all-digital connections, an ultimate long-term value of 3500 − 200 = 3300 Hz is reasonable on the basis that it would provide somewhat enhanced naturalness, as well as somewhat enhanced intelligibility of unvoiced sounds [124, 125]. Performance as a function of bandwidth is summarized in Table 3.13.

Table 3.13
Bandwidth/Slope Performance

Network		Bandwidth (Hz)	Worst Case Grade-of-Service for Nominal Slope (percent GOB)
Mixed	Short Term	400 → 3000 = 2600	60
	Long Term	300 → 3300 = 3000	85
Digital		200 → 3500 = 3300	> 90

It should be noted that the above discussion applies to normal telephony speech signals, nominally referred to as 4-kHz speech. Recently, standardization efforts are underway for high-quality speech signals, nominally referred to as 7-kHz speech [126, 127]. This is a new telephony application, for which the above bandwidth/slope model does not specifically apply, and for which no GOS model currently exists.

3.2.2.9 Crosstalk and Interference

Crosstalk is undesired energy from one or more signals, unintentionally coupled from one or more disturbing circuits into a disturbed circuit where it is perceived. In general, crosstalk may consist of intelligible or nonintelligible speech signals, or of nonspeech signals such as interfering signaling tones or noise. Crosstalk is caused by a number of factors, including electromagnetic coupling between circuits, nonlinearities in multiplexing and modulation systems, and erroneous time-slot interchange in digital systems.

Crosstalk is controlled by the shielding of conductors and equipment, physical separation of conductors, and by designing and maintaining systems to control crosstalk. For example, the latter is done by ensuring balanced-to-ground operation of voice frequency facilities. Also, judicious use of gain and control of the number and level of signals avoids overload and associated nonlinearities. In the case of speech signals, loss of privacy is the overriding concern, and for this reason intelligible crosstalk is controlled by limiting it to a small probability of occurrence (also called the *crosstalk index*).

With reference to Figure 3.17, the probability of crosstalk accounts for a number of random factors, including

- The probability of a conversation on the potentially disturbing and disturbed circuits.
- The talker's volume on the disturbing circuit.
- The loss on the disturbing circuit from the talker to the point of crosstalk.
- The crosstalk coupling loss between the disturbing and disturbed circuits.
- The loss on the disturbed circuit from the point of crosstalk to the listener.
- The received noise on the disturbed circuit.
- The hearing acuity of the listener on the disturbed circuit.

The crosstalk probability is 1% for intertoll trunks and 0.5% for toll-connecting (and all other metropolitan inter-office) trunks in the core network, and 0.1% for subscriber loops in the access parts of the network [49, 128]. One reason for lower probabilities (more stringent requirements) for transmission facilities nearer the customer is that there is an increased chance of users being acquainted. Another reason is that higher loss and noise in the core network tend to mitigate crosstalk effects.

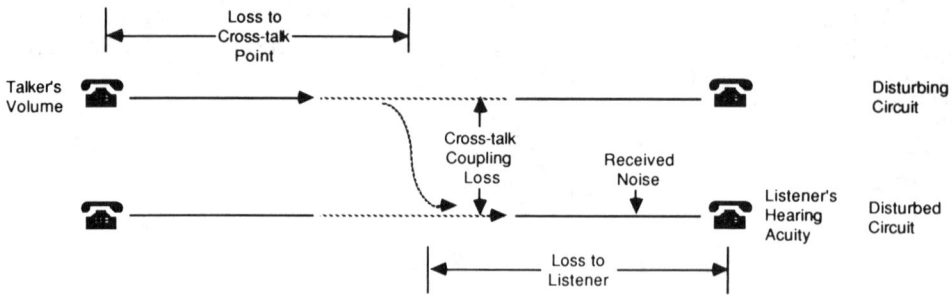

Figure 3.17 Illustration of crosstalk path.

Although loss and noise will be less with digital technology, and thus tend to allow enhanced perception of any crosstalk signals, digital technology is less prone to generation of crosstalk, and crosstalk can be controlled by equipment design. Thus, network evolution from analog to digital can be expected to reduce the incidence of crosstalk.

Unintelligible crosstalk results from one or more speech signals that are not understandable, but that are nevertheless syllabic in nature. Thus, unintelligible crosstalk also has connotations regarding privacy, and is therefore more disturbing than noise. For this reason such signals should be controlled to levels lower than that of noise so as to be masked by the noise. Interference from nonspeech signals (such as signaling tones) is also more perceptible than noise, and is controlled by keeping the level of such signals below the level of noise by at least 10 to 12 dB [129]. When interference is bursty, such as from signaling tones, the level of such interference should be at least 3 dB below the level of noise [130].

3.2.2.10 *Nonlinear Distortion*

General: Nonlinear distortion occurs in systems for which the input/output (transfer) characteristic is nonlinear. The nonlinearity may take on many forms, including those with and without discontinuities. An example of a discontinuous nonlinearity is a "hard" overload characteristic of an amplifier or codec. In this case the output level will be linearly related to the input up to some input level, beyond which the output will be constant (saturated); that is, the input signal will appear amplitude-clipped at the output. This is called *overload distortion.*

An example of a continuous nonlinearity is that of the carbon microphone transmitter commonly used in telephone handsets (although in recent years these devices are being replaced by linear microphones). The acoustic-to-electric efficiency (gain) of carbon microphones tends to be higher for higher acoustic input levels. This type of nonlinearity can be represented by a power series expansion consisting of linear, quadratic, cubic, and possibly higher order terms. For sinusoidal input signals, the resulting distortion is called *harmonic distortion*. For more complex input signals, like speech, it is called *intermodulation distortion*.

Over-load distortion: Figure 3.18 shows the subjective effect of overload distortion for three types of PCM codecs [125]. In this figure the peak level of the speech signal corresponds to 0-dB amplitude clipping on the abscissa (i.e., no amplitude clipping). The 1 to 10 rating scale on the ordinate is arbitrary, covering the range from least to most acceptable performance. (With reference to the more common 5-point rating scale discussed in Ch. 2, the ratings 9, 7, 5, 3, and 1 on the 1 to 10 scale, may correspond approximately to the excellent, good, fair, poor, and bad ratings on the 5-point scale [131].)

The results in Figure 3.18 do not provide details for the low levels of amplitude clipping, but suggest that considerable clipping is possible without serious impair-

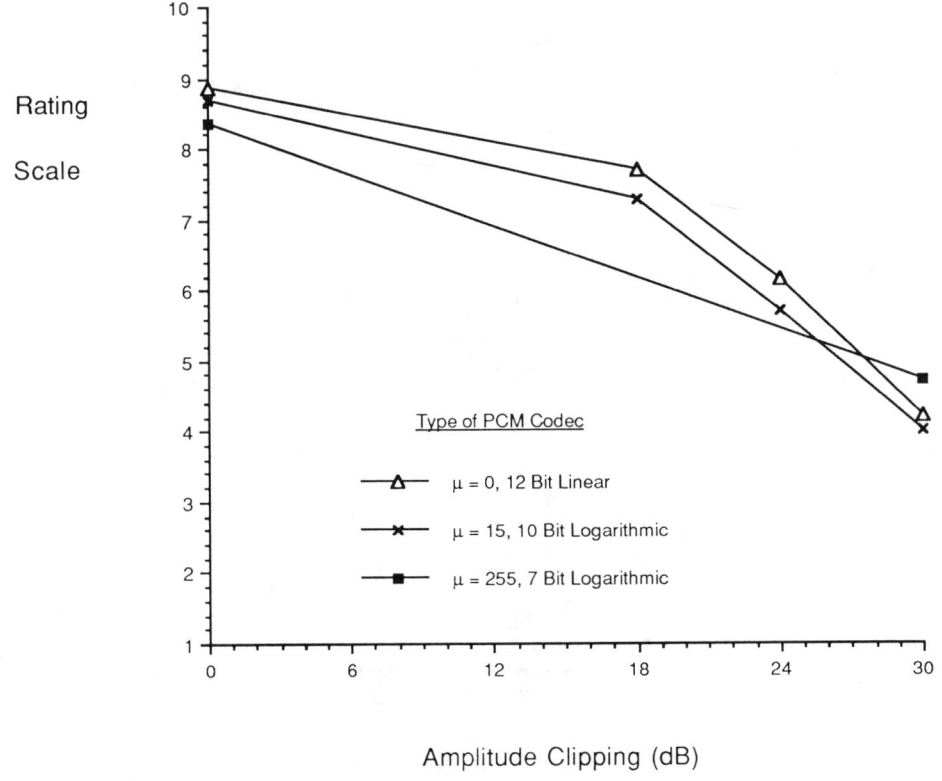

Figure 3.18 Subjective effect of overload distortion.

ment. This is supported by the fact that, for the linear codec and 6, 12, 18, 24, and 30 dB of amplitude clipping, about 0.7%, 1.7%, 13%, 25%, and 40%, respectively, of speech samples are clipped [125]. This suggests that up to about 12 dB of amplitude clipping (corresponding to only 1.7% of samples being clipped) could be tolerated with little degradation. In fact, overload distortion is a much more serious impairment for nonspeech signals such as voiceband data and signaling tones, as discussed later in Chapter 6.

Intermodulation distortion: Figure 3.19 shows the subjective effect of quadratic and cubic distortion [132]. In this figure the distortion level on the abscissa is expressed in terms of the percentage of harmonic distortion of a sinusoidal signal having the same root mean square (rms) level as the speech signal.

As may be expected intuitively, the results show that the subjective effect of a cubic nonlinearity is more severe than for a quadratic nonlinearity. What is perhaps less expected is the degree to which cubic distortion is more serious than quadratic.

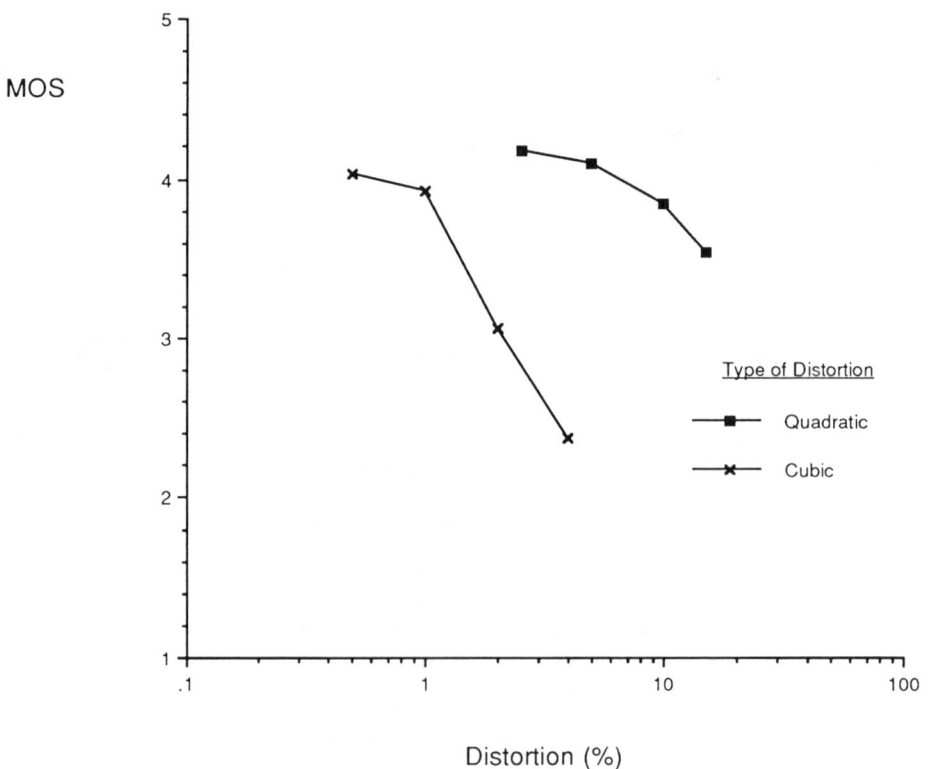

Figure 3.19 Subjective effect of intermodulation distortion.

For example, about 1% and 4% of cubic and quadratic distortion, respectively, are approximately equivalent in terms of subjective impairment.

However, as with overload distortion, discussed above, intermodulation distortion of the type described here is generally a more significant impairment for voiceband data transmission, as discussed in Chapters 4 to 6. This applies to voiceband data transmission using acoustic couplers for transmission through telephone handsets. Other systems such as syllabic compandors also introduce intermodulation distortion.

3.2.2.11 Analog Transients

Various analog transient disturbances such as impulse noise, signal dropouts, gain hits, and phase hits can arise on voiceband channels and affect speech and other voiceband signals. However, these are primarily network-oriented parameters which most seriously affect nonspeech signals, such as voiceband data (discussed in Chs. 4 and 6). Impairment levels for each of these parameters are summarized in Table 3.14 [51, 65].

Table 3.14
Summary of Voiceband Data-Oriented Performance Levels for Various Analog Transient Disturbances

Parameter	Impulse Noise	Drop-outs	Gain Hits	Phase Hits
Impairment Level	> -18 dBm	< -12 dB	> +3 dB	> ±20°
Impairment Duration (ms)	≥ 10	≥ 10	> 4	> 4
Impairment Frequency (occurrences/min)	1 for 50% of connections	1/30	8/15	8/15

As far as their effect on voice service is concerned, impairments arising from these parameters may all be regarded as perceptibly similar to transmission errors or to short durations of speech clipping, to be described later in Section 3.2.2.14. On this basis, the durations and frequencies of occurrence of transient impairments, as shown in Table 3.14, can be compared to those of interest for speech clipping shown later in Figure 3.23. Thus, speech clipping performance also accounts for these parameters.

3.2.2.12 Transmission Errors

Transmission errors are state discrepancies between the transmitted and received bits of digital signals. These errors are produced mainly by digital regenerative repeaters located at various points in a digital transmission system, and may result from numerous factors such as fading, interference, noise, and transient disturbances. Errors can occur randomly or can be bursty in nature. For voice service, errors are characterized in terms of bit error ratio (BER), which is the ratio of the number of bits received in error to the total number of bits received during a test period of suitable duration [133]. For data service, error performance in terms of the percentage of error-free intervals is addressed later in this chapter.

Random errors: The subjective effect of random bit errors on 56 kb/s PCM codecs is qualitatively indicated in Table 3.15 [134]. More quantitative subjective results for 64-kbit/s PCM codecs and 32-kbit/s ADPCM codecs (more robust to transmission errors than PCM) are shown in Figure 3.20 [112]; similar results can also be found in [20]. In the longer term, performance should be readily sustainable at these or possibly higher levels (higher MOSs), consistent with the capabilities of fiber transmission systems and the prospects of more advanced, high-quality (e.g., wideband) speech coding schemes. Generally, higher error performance for speech service also tends to narrow the gap between voice requirements and the generally more stringent requirements for data.

Table 3.15
Subjective Effect of Random Errors on 56 kb/s A-Law PCM Speech

BER	Subjective Effect
$\leq 10^{-6}$	Not detectable
$\leq 10^{-5}$	Just discernible on low level speech
$\leq 10^{-4}$	Annoying on low level speech
$\leq 10^{-3}$	Serious over whole speech range

Based in part on the results in Table 3.15, the current CCITT BER Recommendation G.821 is as summarized in Table 3.16 [133]. In this table, BERs $> 10^{-6}$ over 1 min are called *degraded minutes* (DM), and BERs $> 10^{-3}$ over 1 sec are called *severely errored seconds* (SES); note that SESs are excluded from DMs. The CCIR BER recommendations for current and future satellite links are also summarized in Table 3.17 [135, 136] for comparison with CCITT Recommendation G.821.

Low BERs of $\leq 10^{-6}$ for $\geq 90\%$ of the time, as given in CCITT Recommendation G.821, translates to about $\geq 93\%$ of the time for national networks based on

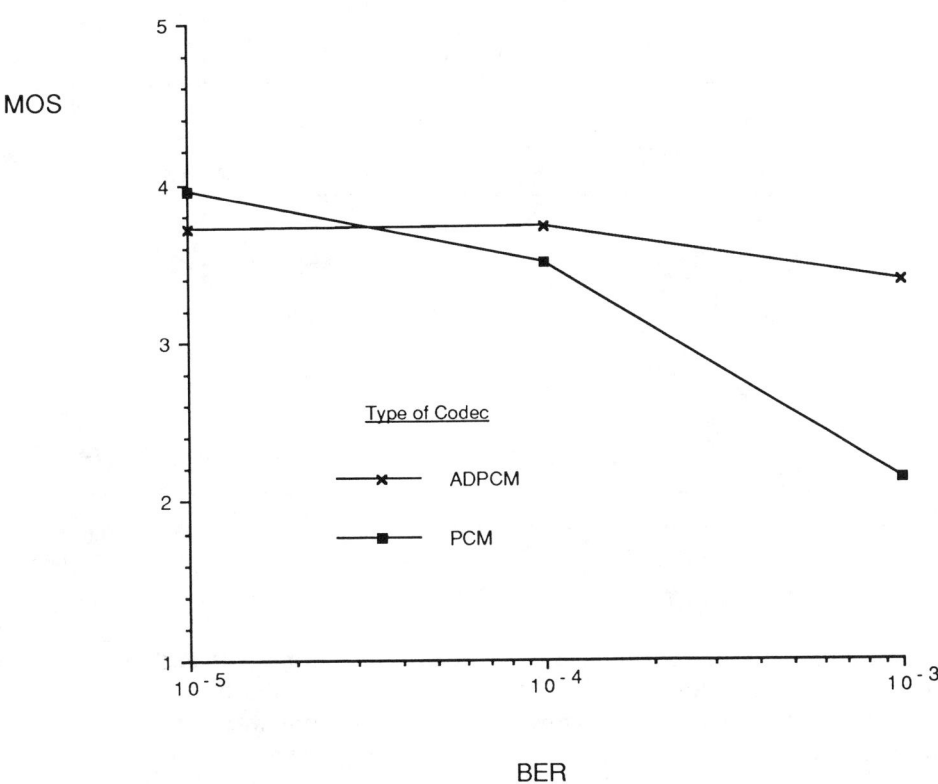

Figure 3.20 Subjective effect of transmission errors.

Table 3.16
CCITT BER Recommendation G.821 for Voice Service on 64 kbit/s International ISDN Connections

BER	Performance
$\leq 10^{-6}$ for $\geq$ 90% of 1 min. periods	Satisfactory
$> 10^{-6}$ for < 10% of 1 min. periods *	Degraded
$> 10^{-3}$ for < 0.2% of 1 sec. periods #	Unacceptable

*   < 10% Degraded Minutes (DM). A DM is a 1 minute period with BER > 10^{-6} and no SESs. Note that the DM parameter may be removed from a future version of G.821.
\#   < 0.2% Severely Errored Seconds (SES). A SES is a 1 second period with BER > 10^{-3}.

Table 3.17
CCIR BER Recommendations for Satellite Links

BER	Proportion of Time (%) for:	
	Non-ISDN Connections Providing PCM Telephony	ISDN Connections
$\leq 10^{-7}$	($\geq$ 45)	$\geq$ 90 #
$\leq 10^{-6}$	$\geq$ 80 *	$\geq$ 98 #
$\leq 10^{-5}$	($\geq$ 96.5)	($\geq$ 99.4)
$\leq 10^{-4}$	$\geq$ 99.7 *	($\geq$ 99.9)
$\leq 10^{-3}$	$\geq$ 99.95 *	$\geq$ 99.97 #
$> 10^{-3}$	< 0.05 *	< 0.03 #

( ) Estimated based on log normally distributed BERs [190, 191].
* From CCIR Rec. 522-2 [135].
\# From CCIR Rec. 614 [136].

national/international percentage of time allocation for BER; depending on the length of national networks, the national/international allocation can vary from about 60/40 to 70/30 [133]. Figure 3.21 illustrates near-worst international and national hypothetical reference connections (HRX), for which the allocation is 68/32; thus, $100 - (100 - 90)*68/100 = 93.2\%$. (HRXs and allocation procedures are addressed in more detail in Chapter 5.)

BERs $\leq 10^{-6}$ for $\geq$93% of the time are somewhat less stringent than a national (AT&T) objective of BER $\leq 10^{-6}$ for 95% of the time [51]; both levels of performance are consistent with the capabilities of most connections with current terrestrial digital facilities [137, 138]; see Chapter 6 for additional results. However, these BERs may not be consistent with the capabilities of older generation satellite links, as reflected by CCIR Recommendation 522-2 [135]. Thus, performance somewhat poorer than in CCITT Recommendation G.821 can be expected when these satellites are involved until higher performance satellites with forward error control prevail. Higher performance satellite links are considered in CCIR Recommendation 614 [136], and, as shown here in Table 3.17, these satellite systems satisfy Integrated Services Digital Network (ISDN) performance requirements. Owing to the above variations in performance capabilities, short-, medium-, and long-term error performance levels are considered as summarized at the end of this subsection on random errors.

At high BERs, $> 10^{-3}$, the CCITT performance level of <0.2% of 1-sec periods so affected corresponds to an average of roughly one such 1-sec period every 8 min.

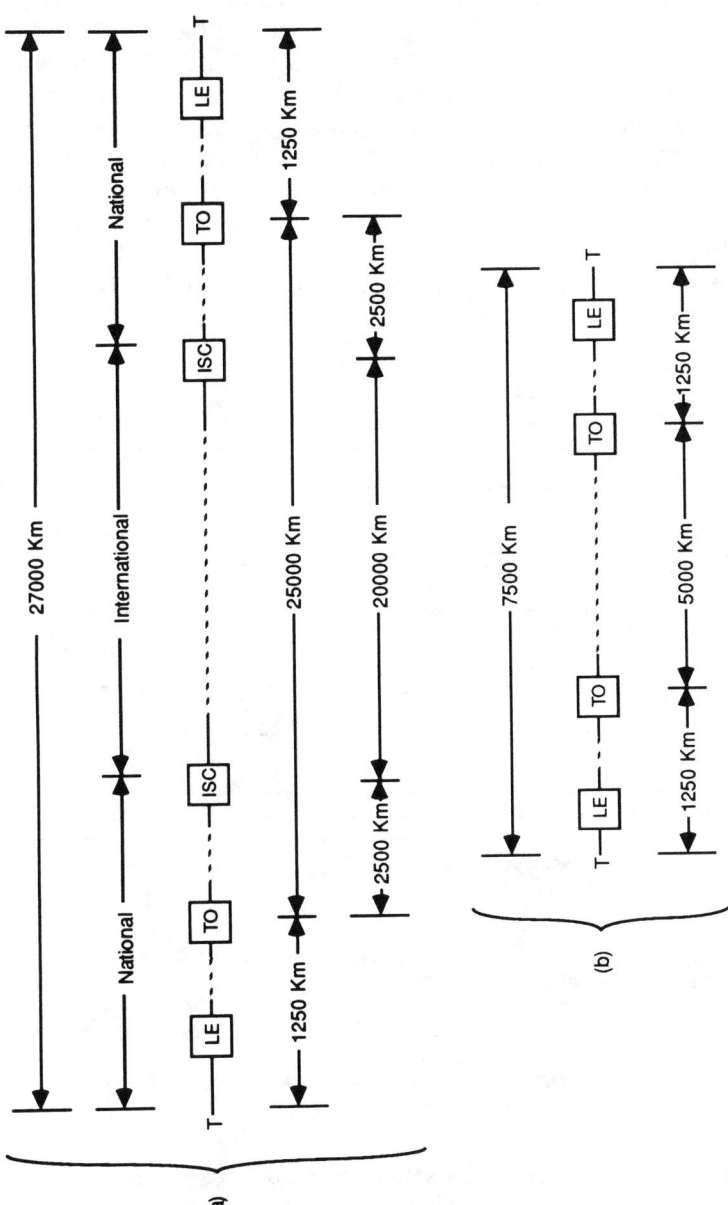

Figure 3.21 Near-worst hypothetical reference connections (HRXs) for error performance: (a) international HRX; (b) representative national HRX.

This performance level is consistent with the subjective effect of short, randomly occurring error burst and speech clipping events, as given later in this section (Table 3.19) [35, 139]. The CCITT value of <0.2% translates to about <0.14% being allocated to national networks (although note that, strictly, according to Recommendation G.821, only one-half of the 0.2% is allocated, while the other one-half can be assigned for adverse conditions—e.g., for radio fading). The above discussion assumes that the 1-sec periods with BERs $>10^{-3}$ occur independently in time. If clustered in time, 10 or more such consecutive periods (i.e., consecutive SESs) are regarded as unavailable time and are called *available seconds* (UAS) [133]; unavailable time ends after 10 consecutive non-SESs. Unavailable time is discussed again later in this chapter under the dependability of service criterion. Random BER performance for voice is summarized in Table 3.18.

Table 3.18
Summary of BER Performance for Near-Worst National and International Connections

BER	Proportion of Time (%) for:					Performance
	Short Term Non-ISDN Connections @		Medium Term ISDN Connections		Long Term Conservative Estimate	
	International $	National &	International	National #	National	
$\leq 10^{-7}$	(≥ 29.6)	(≥ 31.4)	(≥ 77.3)	(≥ 84.6)	≥ 87.5	Satisfactory
$\leq 10^{-6}$	≥ 73.2	≥ 74.0	≥ 90.0 *	≥ 93.2	≥ 95	
$\leq 10^{-5}$	(≥ 94.1)	(≥ 94.4)	(≥ 96.4)	(≥ 97.6)	≥ 98.3	Degraded
$\leq 10^{-4}$	≥ 99.2	≥ 99.3	(≥ 99.2)	(≥ 99.5)	≥ 99.5	
$\leq 10^{-3}$	≥ 99.814	≥ 99.856	≥ 99.8 *	≥ 99.864	≥ 99.9	
$> 10^{-3}$!	< 0.186	< 0.144	< 0.2 *	< 0.136	< 0.1	Unacceptable

( ) Estimated based on log normally distributed BERs [190, 191].
* From CCITT Rec. G.821 (Table 3.16) [133].
\# Based on 68/32 National/International proportion of time allocation.
@ Based on CCIR Rec. 522-2 (Table 3.17) [135].
$ Based on sum of the proportions of time, BER is higher than indicated for: National ISDN Connection, and Satellite link as per CCIR Rec. 522-2.
& Based on sum of the proportions of time, BER is higher than indicated for: local plus medium grade of National ISDN Connection, and Satellite link as per CCIR Rec. 522-2 (see Ch. 4 and 5).
! Also applies to burst error performance and to the out-of-frame type of interruption performance (Sec. 3.2.2.13).

Burst errors: Burst errors are transmission errors that occur clustered rather than randomly distributed in time, and produce relatively high BERs (e.g., $>10^{-3}$) for relatively short durations (e.g., ≤1 sec), that is, SESs. From subjective tests with random errors, BERs $>10^{-3}$ are considered to provide unacceptable performance [134].

However, since error burst durations are usually quite short, the distribution of a given number of error occurrences within a 1-sec interval may, to a first approximation, be ignored. In other words, performance may still be regarded as unacceptable if an entire 1-sec interval has a BER of 10^{-3}, or if, for example, 10% of a 1-sec interval has a BER of 10^{-2}. In this respect, the $>10^{-3}$ random BER threshold for unacceptable performance also adequately reflects the effect of error bursts. Therefore, on this basis, error bursts correspond to unacceptable performance and should only be allowed to affect suitably small percentages of speech.

A guideline for this percentage can be determined based on informal qualitative comparisons of short, randomly occurring error burst and speech clipping events, as summarized in Table 3.19 [139]. (In this table, the BER for the error bursts was fixed at 10^{-2}.) These results indicate that, generally, error bursts are less severe than speech clipping. Therefore, since the acceptable level of speech clipping performance given later in this chapter (Sec. 3.2.2.14), range from $<0.2\%$ to $<0.5\%$, the more stringent level of 0.2% also serves as a conservative performance limit for error bursts. Again, the $<0.2\%$ CCITT value for BERs $> 10^{-3}$ translates to about $<0.14\%$ when allocated to the national portion of connections.

Table 3.19
Qualitative Comparison of Error Bursts and Midspeech Burst Clipping Occurrences

Performance Measure	Error Bursts (BER = 10^{-2})	Speech Clipping
Intelligibility	Unaffected	Affected
Quality at low rates *	Low to medium energy impulse noise-like clicks	Higher energy clicks
Quality at high rates *	Grainy, nearly continuous noise	Raspy quantization-like noise
Overall assessment	Preferred	Not preferred

* Rates refer to the relative frequency of occurrence of the impairment.

Error burst performance limits for speech are summarized in the last row of Table 3.18. The percentage of speech affected by short, randomly occurring error burst events (with BERs $>10^{-3}$, in ≤ 1 sec) is regarded as equivalent to the percentage of time the BER exceeds 10^{-3}; that is, short error bursts occurring independently in time is a mechanism whereby BERs may exceed 10^{-3}. Again, if clustered in time, 10 consecutive 1-sec periods containing error bursts may be regarded as unavailable time. In practice, BERs $>10^{-3}$, whether random or bursty, would cause protection switching of carrier facilities (for facilities equipped with protection). Otherwise, such BERs, if of sufficiently long duration (a few seconds) will cause the termination of service (e.g., cutoff of calls) by terminal equipment. This is an aspect of transmission availability that will be discussed later in this chapter.

Note that there is currently a trend away from BER as a performance parameter, with more emphasis now being placed on time interval-based parameters such as errored seconds (ES) and SESs. This is evidenced by the intent to remove the BER-based DM parameter from CCITT Recommendation G.821 [140], and also by the directions being taken in developing the new CCITT draft Recommendation G.82x on performance of systems at rates above 64 kb/s [141]. One reason for the above trend is that, generally, BERs cannot be readily measured on an in-service basis (see Ch. 9).

3.2.2.13 Short Interruptions

General: Short interruptions are momentary events during which there is an absence of speech or a burst of noise (the noise may either be during active speech or during silent periods). These events can arise in conventional carrier communication systems, and may be caused by a variety of factors. These factors are categorized as follows:

- Frequency instability of incoming and local clocks, and subsequent misalignment of incoming and local frames in synchronous digital carrier systems; this causes controlled frame slips.
- Various factors in digital carrier systems that cause loss of frame synchronization, followed momentarily by reframe; this causes uncontrolled frame slips.

Controlled frame slips: A controlled slip occurs when the timing sources for the digital (usually PCM) equipment at the two ends of a transmission system are not exactly frequency synchronized. When this happens, an entire frame's worth of information can be lost (or duplicated) even though the individual bits may be correct. Controlled slips usually only occur when the digital equipment at the receiving end of the transmission system uses a different clock from that at the transmitting end (i.e., different than the clock that can be derived at the receiving end from the incoming signal sent by the transmitting end).

The case of a frame deletion corresponds to a very minor example of speech clipping to be discussed later (in Sec. 3.2.2.14). Frame deletions involve very small speech clipping durations (≈ 125 μsec), and, in a properly designed network, small frequencies of occurrence (<1 in 2 min, at most) [142]. Therefore, there are correspondingly small amounts of clipped speech ($<10^{-4}$ %; see the next section on speech clipping). In fact, speech is very robust to frame slips as indicated by informal listening tests summarized in Table 3.20 [139]. As a result, controlled frame slip rates are determined entirely by other service requirements, such as those for digital data, as indicated by CCITT Recommendation G.811, and shown here in Table 3.21

Table 3.20
Perceptible Effects of Controlled Frame Slips on PCM Speech

- Single slips are not perceptible.
- High slip rates are perceived as a quantization distortion effect.
- Frame repetitions are essentially indistinguishable from frame deletions.

Slip rate	Subjective Effect
20/sec.	Just perceptible
80/sec.	Objectionable
200/sec.	Unacceptable

Table 3.21
CCITT Rec. G. 811 Controlled Frame Slip Performance for
64-kb/s National and International Connections

Slip rate	Proportion of Time (%)		Performance
	International	National	
≤ 5 in 1 day	> 98.9	> 98.988	Satisfactory
> 5 in 1 day and ≤ 30 in 1 hour	< 1.0	< 0.92	Acceptable but Degraded
> 30 in 1 hour	< 0.1	< 0.092	Unacceptable

[142]. In North America the slip objectives are similar and are discussed with synchronization performance in Chapter 6.

Out-of-frame events: Out-of-frame (OOF) events (also called *uncontrolled frame slips,* or, more recently, *severely errored frame* (SEF) events) are periods during which many bits, bytes, or frames of information may be in error due to loss of framing information and subsequent reframe activity. In most cases an impairment occurs when a certain pattern of framing bit errors are detected and the framer seeks for, finds, and applies the correct frame position (i.e., reframes). Different types of framers are discussed from a maintenance perspective in Chapter 9.

The errors that cause OOF events in digital carrier systems can result from a number of factors, including

- Interference from unwanted signals (power surges/AC induction, lightning, and other static discharges).
- Maintenance activity (e.g., automatic or manual protection switching).

Therefore, OOF events are transient disturbances, and may have durations up to a few tens of ms (e.g., 50 ms or more at the DS1 rate of 1.544 Mb/s). As such, they may be more readily perceived than the analog transients, random transmission errors, and controlled slips discussed in Sections 3.2.2.11–3.2.2.13. OOF events fall into the category of error bursts, since at the 64-kb/s rate they will generally result in BERs $>10^{-3}$ over a 1-sec period. Thus, these events contribute to SESs and should be included in the performance limits given in the last row of Table 3.18. However, in practice, these events tend to be rare enough that they do not materially affect speech quality. As with controlled frame slips, the performance limits for OOF events are determined by data requirements; these requirements are further addressed in Chapters 4 and 6.

3.2.2.14 Speech Clipping

General: Speech clipping is the complete loss of speech energy over a given time duration. Manifestations of this parameter arise in certain advanced communication systems (DCMEs) that use speech activity detection to exploit speech idle time, and that resolve contention among the resulting speech bursts on a speech clipping basis [14–19, 143, 144].

Contention among speech bursts from different sources can be resolved by discarding the first part of the most recent speech burst to arrive; this is called *front-end clipping* (FEC) and is often associated with speech interpolators [36]. Alternatively, short speech durations in the middle of one or more contending speech bursts can be discarded until more resources become available; this is called *midspeech burst clipping* (MSC) and can arise in packetized voice systems [22].

A variety of techniques have also been investigated to mitigate the effects of MSC, including replacement of the missing speech with noise or the previous speech segment, displacement of the missing speech with subsequent speech, and estimation of the missing speech. In general, these techniques all improve performance relative to the case of simple MSC [145, 146]. Thus, the results given here for simple MSC serve as a worst case bound on performance for the above techniques.

Figure 3.22 illustrates the main speech clipping mechanisms (FEC and MSC), and these are discussed presently. Regardless of the clipping mechanism, subjective performance depends on the following four factors [35]:

- The duration of each speech clip (L, in sec).
- The proportion of active speech that is clipped (P).
- The frequency of speech clipping occurrences (F, in clips/sec).
- The speech activity (α, $0 < \alpha < 1$).

These factors are related by the expression

$$P = F^*L/\alpha \tag{3.7}$$

Figure 3.22 Illustration of speech clipping mechanisms.

Note that the clipping only applies to active speech, not to the silent intervals. (The statistics of speech active and silent intervals are discussed in [21, 147, 148].)

Because the results presented below are in part based on tests with monolog speech, inferences about conversational speech are made from monolog speech statistics as follows [21]. By ignoring the subtleties of conversational dynamics (e.g., double talk and mutual silence), the main difference between monolog and conversational speech is that the distribution of silence durations for a conversation accounts for the additional time that one talker spends listening to the other talker; this additional time consists of the other talker's talking plus silence time. Thus, for a given talker, speech level, and speech detector, the following approximate relationships should apply:

Parameter	Monolog (m)	Conversation (c)	
Mean Speech Burst Duration	T_m	$T_c = T_m$	
Mean Silence Duration	S_m	$S_c = 2S_m + T_m$	(3.8)
Speech Burst Rate	R_m	$R_c = R_m/2$	
Speech Activity	α_m	$\alpha_c = \alpha_m/2$	

Note that, in general, the speech burst rate (R) and speech activity (α) are related to the mean speech burst (T) and silence (S) durations as follows:

$$R = 1/(T+S) \tag{3.9}$$
$$\alpha = T/(T+S) = T*R \tag{3.10}$$

Equations (3.7) and (3.8) imply that for a given proportion (P) of speech clipping and duration (L) of each speech clip, the frequency (F) of speech clipping is reduced by about a factor of two in a conversation, owing to the reduced speech activity (α). This means that (although the percentage is the same) the absolute amount of clipped speech in a conversation is about half that for a monolog. It is therefore hypothesized that the performance for conversations may be somewhat higher than for a monolog. Thus, the results for monolog speech may serve as a conservative estimate for conversational speech.

Front-end clipping: FEC is a well-known impairment for which various performance guidelines have been established. For example, an FEC duration of 15 ms is the approximate threshold of perceptibility of this impairment [149], and the CCIR percentage of speech loss limit is ≤0.5% [150]. Based on Equation (3.7) and subjective test results with high-activity monolog source speech [35], these limits, when taken together, correspond to a frequency of 18 FECs/min, and to good to excellent performance as indicated by an MOS exceeding 4; see Table 3.22. These limits are probably somewhat stringent; it is estimated that the percentage of speech loss could be increased to about 1%, and the speech loss duration could be increased to about 30 ms (with corresponding frequency remaining at 18 clips per min), and still provide good performance (MOS of about 4 [35]).

Loss durations >50 ms should generally be avoided, since they have a potentially adverse effect on intelligibility [149]. Accordingly, the Campanella criterion states that no more than 2% of FECs should exceed 50 ms [151]. Based on subjective tests, it is conservatively estimated that this criterion should result in at least a fair to good performance as indicated by an MOS of about 3.5 or more, depending on the distribution and range of the random loss durations.

Since FEC affects the start of each speech burst, the subjective effect of this impairment depends on the speech detector design. The above qualitative observations pertain to a speech detector with a 200-ms hangover, which generates nearly 20 speech bursts/min for high-activity monolog speech. FEC performance is summarized in Table 3.22.

Finally, from [35] it is noted that in terms of speech quality, FEC is preferred to MSC (discussed below). This is intuitively reasonable since, unlike MSC, FEC is preceded by silence, so the listener is anticipating the subsequent speech to a lesser degree than with MSC. Also, there is only one discontinuous amplitude transition with each FEC event, and two with each MSC event. It is reasonable to expect that

Table 3.22
Speech Clipping Performance—FEC

Front-End Clipping (FEC)			Grade-of-Service	
Duration	Proportion	Frequency [e]		
L (ms)	P (%)	F(Clips/min)	MOS [f]	Remarks
≤ 15 [a]	0.5 [b]	18	> 4	Good to Excellent
≤ 30	1.0 [b]	18	≈ 4	Good
≤ 50	≤ 1.6	≤ 17.6	≥ 3.5	Fair to Good
> 50 [c]	> 0.04	> 0.4 [d]	–	–

[a] Threshold of perceptibility [149].
[b] CCIR guideline [150].
[c,d] Campanella criterion [151].
[d] ≤ 2 percent of 18 clips/min. ≈ ≤ 0.4 clips/min.
[e] Based on eqn. (3.7) for about 18 speech bursts/min. and high activity ($\alpha \approx 0.9$) monolog speech [35].
[f] For about 18 clips/min. and high activity ($\alpha \approx 0.9$) monolog speech [35]

the perception of FEC would approach that of MSC when there are many more than 20 clips/min. This would apply to a speech detector with a hangover of much less than 200 ms.

Midspeech burst clipping: The perception of MSC can be assessed from Figure 3.23, which is based on formal subjective listening tests using high-activity monolog source speech [35, 144]. The general behavior of the results in Figure 3.23 is similar to other results for speech clipping [152], but the ratings in Figure 3.23 are somewhat lower (more conservative). Generally, the results in Figure 3.23 are also more conservative than those observed in [153, 154], because in those cases the application was specialized, and in some instances involved very low bit rate coding, for which the distortion impairment can mask that arising from speech clipping. This is in contrast to the less specialized private (and perhaps public) network application emphasized here, where there is an expectation of relatively higher speech quality. Still other results can be found in [143, 145, 146, 155]. Figure 3.23 suggests the following guidelines on MSC:

- Little impairment results from ≤1% clipping with durations ≤5 ms.
- Speech clipping durations from 10 to 50 ms produce noticeable GOS degradations unless the percentage of speech clipping is quite low (≤0.2%).
- GOS degrades for high percentages of speech clipping and ≤50-ms durations. However, it is expected that intelligibility is not adversely affected; therefore, this range for MSC is useful for transient overload situations.

Figure 3.23 Subjective effect of midspeech burst clipping.

- Speech clipping durations >50 ms cause GOS as well as potential intelligibility degradations. These conditions should generally be avoided for telephony use.

Based on these observations, performance for MSC is summarized in Table 3.23.

Table 3.23
Speech Clipping Performance—MSC

Midspeech Burst Clipping (MSC)			Grade-of-Service	
Duration	Proportion	Frequency		
L (ms)	P (%)	(Clips/min)[a]	MOS[b]	Remarks
50	0.2	2.2	≥ 4	Good to Excellent
50	0.5	5.4	3.5	Fair to Good

[a] Based on eqn. (3.7) for high activity ($\alpha \approx 0.9$) monolog speech.
[b] See Fig. 3.23.

3.2.3 Dependability of Service Performance Criterion

3.2.3.1 Long Interruptions

Long interruptions are associated with periods of excessively degraded transmission performance. Such degraded performance could be associated with one or more of the various parameters considered above. For example, according to CCITT Recommendation G.821 [133], BERs $> 10^{-3}$ in each of 10 or more consecutive seconds render such periods as unavailable time (in other words, periods of 10 or more SESs are UASs). Little information is available pertaining to users' reactions to such "unavailable times." However, some results exist for long interruptions [156], which are essentially speech clipping occurrences with durations much longer than those considered earlier in this chapter (recall Fig. 3.23). Such long interruptions may be regarded as an extreme worst-case manifestation of excessively degraded transmission performance. When such degradations arise after a voice call is in progress, it may be deemed necessary by either or both users to terminate the call because of the excessively degraded performance.

The results in [156] for long speech clipping interruptions indicate that users may terminate an established call with an interruption as short as about 5 sec. Thus, a reasonable approach would be to permit durations of a few seconds of excessively degraded performance for a suitably small percentage of voice calls. This is consistent with current practice, in which calls are effectively cut off by network switching equipment after long interruptions exceeding about 2 to 3 sec (see the next section). Although the percentage of calls so affected has yet to be determined, suggested values are in the range of $<0.15\%$ to $<0.5\%$, corresponding to call cutoff rates discussed in the next section [157]. For less severe interruptions (e.g., BERs $> 10^{-3}$), users may tolerate longer durations of the interruption before terminating the call, but such information is currently unknown.

3.2.3.2 Call Cutoff

After a call is in progress, it may be inadvertently terminated (cut off) by the network as a result of a network outage; such an outage may in turn be caused by a failure, congestion, or a degradation of transmission performance below some threshold level. For example, BERs exceeding 10^{-3}, or an inability to acquire frame synchronization for a period of 2 to 3 sec, can cause the cutoff of calls by switching equipment.

Based on survey results and studies on availability, it has been estimated that the performance of North American networks is such that about $<0.5\%$ of all calls experience cutoff [158]. A longer term cutoff rate of $<0.15\%$ has also been proposed on the basis that the percentage of cutoff should be about an order of magnitude lower than the percentage of blocked calls, which in turn should nominally be less than 1% [157].

3.2.4 Nontransmission Parameters

In general, the GOS concept applies not only to transmission performance, but also to traffic performance. The definitive work in this area is in [159, 160], which describe a call setup rating model. This model is a function of such traffic-oriented performance parameters as dial-tone delay and postdialing delay (recall Table 2.1 in Ch. 2). Furthermore, the call setup model is combined with the transmission rating model to give an overall call completion rating model. Finally, the call completion rating model is expressed in terms of opinion rating and behavior in a manner similar to that described earlier for transmission performance in Chapter 2. Since the focus in this book is on transmission performance, the modeling details for traffic performance are not covered here, but may be pursued further in [159].

3.3 USER- AND NETWORK-ORIENTED PERFORMANCE FOR DATA SERVICE

3.3.1 General

As noted in Chapter 2, end-to-end user-oriented performance considerations for data service are somewhat different from voice service, even in the context of ISDN. Among other things, the differences are in terms of performance parameters. Many voice performance parameters have been well developed and researched as described above. On the other hand, user-oriented data performance parameters have been more recently developed, and are only now beginning to be more widely accepted even though the user-oriented versus network-oriented concept has been well understood [1]. The outcome of extensive work in this area results in the data performance parameters summarized earlier in Table 3.1 [161–163]. These parameters are intended to be network-independent and to cover all types of data services. Also shown in brackets in Table 3.1 are terminologies for the parameters that are more familiar to both network users and network providers.

It should be noted that similar but somewhat more technology-specific performance parameters are also being developed for new frame- and cell-based systems (e.g., for frame relay [164], SMDS [165], and ATM [166]).

3.3.2 Speed of Service Performance Criterion

3.3.2.1 Transfer Delay (Response Time)

Transfer delay is end-to-end delay, and is the elapsed time between the start of transfer of a unit of information (e.g., a block or packet of data) from the source to the successful reception of the information unit at the destination. The transfer is declared successful if it is completed within a specified time-out period, and the received information unit has exactly the same form and content as transmitted by the source. Transfer delay is a function of transmission error performance, user and network protocols, and of processing, switching, and propagation delays. These dependencies may be further characterized according to the following:

- The nature of the data service; that is, the user information unit size, access transmission rates, and user protocols will influence transfer delay.
- The nature of the data network; that is, the switching scheme, if packet-based, will result in queueing delays that depend on network loading; the topology and architecture will affect delay through the distance and the number of nodes traversed. The transmission rates, transmission performance, and related network protocols will contribute to store and forward, and to retransmission delays.

Response time is a round-trip delay related to transfer delay. It is the elapsed time between the start of transfer of an information unit from the source, to the successful reception of a response (from the destination) by the source. Response time is a function of the same factors as transfer delay, plus the processing time of the application from which a response is being requested.

Response time is a relevant user-oriented performance parameter for any type of real-time, interactive, transaction, or inquiry/response data service. Requirements may be expressed in terms of a mean, and a worst-case (e.g., 90th or 95th percentile) delay [167].

Reaction to variation in response time has been assessed for inexperienced terminal users using the 5-point rating scale introduced earlier [168]. Based on these studies, a proposed objective is to achieve user acceptance at the 80% Good-or-Better (GOB) level. To accomplish this, the average response time must be less than 3 sec. A reasonable worst-case response time may be of the order of 10 sec. In practical networks, depending on the numerous factors noted earlier, response times can vary from as low as a few tens of ms, to many seconds [169–181] (see Sec. 4.5 of Ch. 4 for further detail).

3.3.2.2 Transfer Rate (Throughput Efficiency)

Transfer rate is the number of information units successfully transferred from source to destination between start of transmission of the first unit and reception of the last unit. The term "successful" means that none of the information units are errored, missing, or out of sequence, and that there are no extra (e.g., duplicated) units. As with transfer delay, transfer rate is a function of transmission error performance, user and network data protocols, and of processing, switching, and propagation delays.

Throughput efficiency is related to transfer delay and is the ratio of the number of information units successfully transferred from source to destination, to the maximum number of units that could ideally be transferred. Throughput efficiency is a relevant user-oriented performance parameter for file/bulk data transfer and remote job entry data services, and manifests itself in terms of the time it takes to transfer a file.

As with response time, throughput may be expressed in terms of a mean and a worst case [167]. There appears to be no known subjective user-oriented performance guidelines for throughput, except that users are probably most sensitive to the time it takes to transfer a file or to process a job; the lower the throughput, the longer the transfer/processing time. In practice, depending on the throughput-dependent factors noted earlier, values can range from as low as a few percent to as high as nearly 100% [172–174, 176–178, 181] (see Sec. 4.5 of Ch. 4 for details).

3.3.3 Accuracy of Service Performance Criterion

3.3.3.1 Error Probability

Error probability is the ratio of total incorrect user information units to total successfully transferred plus total incorrect user information units during the measurement time [182]. An information unit can be a single bit or a block of bits. The former case results in BER, defined earlier for voice. The latter case results in block error ratio (BLER). Different users can use different block lengths, and BLER is specified as a function of block length. The alternative is to choose a time interval that approximates a typical block length at a certain speed; 1 sec is a well-known example of such a time interval. An ES is declared when one or more bits in that second is in error. BLER then becomes "ES ratio" and the complement of errored second ratio is more commonly known as the percentage of error-free seconds (PEFS).

BER was used earlier for assessing error performance for voice. For data applications, BLER or PEFS is more meaningful, since users usually transmit data in some form of blocks. For random errors, PEFS can be derived from BER as follows:

$$\text{PEFS} \approx 100(1 - R \times \text{BER}) \qquad (3.11)$$

where R is the bit rate in bits/s.

Equation (3.11) applies, for example, to satellite links without forward error correction. For the majority of transmission systems, including terrestrial radio and cable facilities used in both long-haul or short-haul systems, and especially loop access systems, errors have been found to occur in bursts [137, 138, 183, 184]. Since burstiness is not readily modeled, however, it has been found useful to retain both BER and PEFS for characterizing error performance.

For a 64-kb/s international ISDN connection, the PEFS performance limit is given in CCITT Recommendation G.821 as 92% of the available time [133]. Of the allowable 8% ESs, about 5.5% is allocated to national networks (based on the 68/32 national/international allocation used earlier for BER performance in Sec. 3.2.2.12).

This means that the corresponding PEFS performance level for a national network should be about 94.5% of the available time. It is evident that this performance level can be achieved in practice, for example, from performance objectives for switched and nonswitched digital data services in North America, which call for 95 and 99.5 PEFS, respectively [179, 180, 185–187] (see Sec. 4.4 of Ch. 4 for details).

3.3.3.2 Residual Error Probability

This parameter applies to applications where forward error correction (FEC) or error recovery protocols based on retransmission techniques are used. Owing to imperfections in these systems, a residual (uncorrected) error probability (REP) will generally prevail, and is the ratio of errored (N_E), lost (N_L), and extra (N_X) user information units, to total (N) transmitted or received information units:

$$\text{REP} = (N_E + N_L + N_X)/N \tag{3.12}$$

where $N = N_C + N_E + N_L + N_X$ and where N_C is the number of correct information units. On a worst-case basis, REP should be ≤ 2 to 5×10^{-10}, assuming user information units are in bits [182]. From Equation (3.11), this corresponds to about 99.999 to 99.997 PEFSs at the 64-kb/s rate.

The various components of residual error probability are illustrated in Figure 3.24. The errored information probability (P_E) is similar to error probability as defined in the previous section, except that P_E applies to the residual error case and is the ratio of total errored to total correct plus errored user information units, given by

$$P_E = N_E/(N_C + N_E) \tag{3.13}$$

User information loss probability (P_L) can result from buffer overflow, misdelivery due to misrouting, protocol anomalies, and equipment failure. P_L is the ratio of total lost (N_L) to total transmitted user information units, and is given by

$$P_L = N_L/(N_C + N_E + N_L) \tag{3.14}$$

The extra user information probability (P_X) is the ratio of total unrequested extra (N_X) to total received information units, and may be due to misdelivered or duplicated information units; it is given by

$$P_X = N_X/(N_C + N_E + N_X) \tag{3.15}$$

More detailed information on these parameters is provided in [173, 176, 178] and in Section 4.5 of Chapter 4.

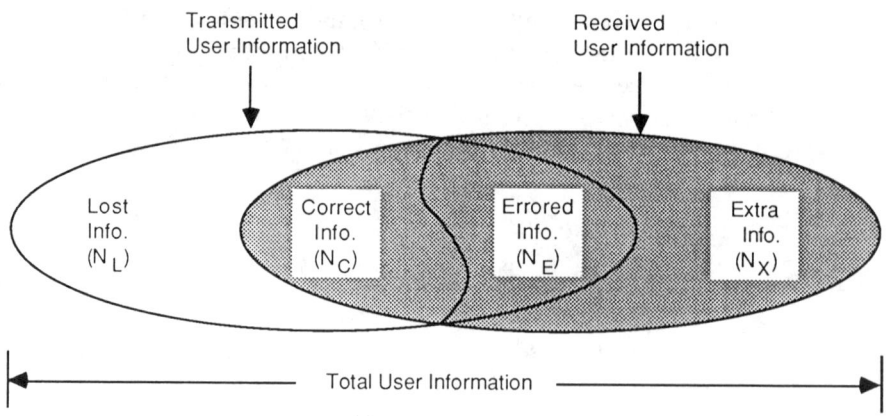

Figure 3.24 Components of residual error probability.

3.3.4 Dependability of Service Performance Criterion

3.3.4.1 Transfer Denial

Transfer denial probability is the ratio of total transfer denials to total transfer samples (blocks of user information) observed during a measurement period [161, 188]. A transfer sample is a discrete observation of network performance in transferring user information between a specified source and destination. A transfer denial is a transfer sample in which the observed performance is worse than a minimum acceptable level during the information transfer phase. This degradation may be due to a number of factors, such as unacceptable transmission performance (i.e., high BER or low PEFS).

As noted earlier with respect to voice service (in Secs. 3.2.2.12 and 3.2.3.1), CCITT Recommendation G.821 considers that a period of unavailable time begins when the BER in each second is worse than 10^{-3} for a period of 10 consecutive seconds [133]. The period of unavailable time ends when the BER in each second is better than 10^{-3} for a period of 10 consecutive seconds.

Error performance is only one of the transmission factors that may contribute to the total unavailable time. Transfer denial is also identified by comparing the measured values for certain other quality-of-service parameters to specified transfer denial thresholds. Thus, transfer denial may also occur because of unacceptably high response time or low throughput, or because of excessive residual error probability [161]. As a result, transfer denial may manifest itself in practice by protocol reset, which means that a time out is exceeded or an out-of-sequence condition cannot be recovered from.

Transfer denial affects transmission availability, which in turn is contributed to by equipment failure and excessive performance degradation. Availability objectives and achieved levels of availability vary considerably depending on the type of network, that is, from 97% to values exceeding 99.9% [173, 174, 176, 179, 180, 185, 187, 188]. Details are provided in Chapter 4 (Tables 4.24, 4.27–4.29).

3.3.4.2 Premature Disengagement

Premature disengagement (or premature disconnection) for data services is analogous to call cutoff, which was discussed earlier for voice service in Section 3.2.3.2. The call cutoff rate is typically contributed to by network failures, congestion, and excessively degraded transmission performance. Call cutoff rate can therefore be used as an unavailability indicator for switched data applications. The lowest call cutoff rate noted previously for voice service is 0.15%. This could be used as a worst-case performance limit for switched data services.

3.4 SUMMARY

This chapter has addressed top-down, end-to-end, user-oriented performance requirements for integrated voice/data networks. First, a common framework for both voice and data parameters was provided. Secondly, various philosophical differences between voice and data performance were recognized, relating to the ways in which performance is experienced, to evolutionary considerations, and to traffic and information integrity. As a result, different sets of parameters are involved for voice and data and, even for those parameters which are common, different requirements are generally necessary for each service.

Next, performance levels for the various parameters were presented. The orientation was toward nationwide public and private network applications, and the approach was to discuss basic underlying performance that applies to a variety of network designs, transmission schemes, switching architectures, and technological implementations. For voice service the approach is also strongly tied to an essentially continuous and consistent evolutionary process, as well as to the prevailing nature of an element of analog technology. However, for data service, the evolutionary process is less continuous.

Finally, the intent in this chapter was to discuss performance as reflected by up-to-date information and positions adopted by various organizations involved with performance standards, particularly the North American T1 forum and CCITT. In some instances, these positions are provisional. More detailed information on performance levels is provided in Chapters 4 and 6.

3.5 REFERENCES

1. J. Gruber and N.H. Le, "Performance Requirements for Integrated Voice and Data Networks," *IEEE Journal on Sel. Areas in Comm.,* Vol. SAC-1, No. 6, Dec. 1983, pp. 981–1005. Also, *GLOBECOM Rec.,* Vol. 1, San Diago, Nov./Dec. 1983, pp. 2.2.1–2.2.9.
2. J. Gruber, J. Bhullar, and G. Williams, "Acoustic Factors Determining Telephone Speech Quality," *Canadian Acoust.,* Vol. 10, No. 2, Apr. 1982, pp. 29–39. Also, abstract, J. Acoust. Soc. Amer., Supp. 1, Vol. 69, Spring 1981, p. S76.
3. K. Takahashi, "Transmission Quality of Evolving Telephone Services," *IEEE Comm. Mag.,* Vol. 26, No. 10, Oct. 1988, pp. 24–35.
4. F.P. Duffy, G.K. McNees, J. Nasell, and T.W. Thatcher, J., "Echo Performance of Toll Telephone Connections in the United States," *BSTJ,* Vol. 54, No. 2, Feb. 1975, pp. 209–243.
5. M.B. Carey, H.T. Cheu, A. Descloux, J.F. Ingle, and K.I. Park, "1982/83 End Office Connection Study; Analog Voice and Voiceband Data Transmission Performance Characterization of the Public Switched Network," *BLTJ,* Vol. 63, No. 9, Nov. 1984, pp. 2059–2119.
6. CCITT Rec. G.114 on "Mean One-Way Propagation Time," *Blue Book,* Vol. III, Fas. III.1, Geneva 1989, pp. 84–94.
7. Contrib. T1Q1.1/89-017 on "Effects of Echo-Free Delay on Conversations," AT&T, Feb. 7, 1989.
8. L.S. DiBiaso, "Satellite User Reaction Tests: A Subjective Evaluation of Echo Control Methods," *NTC Rec.,* Nov. 1979, Vol. 3, pp. 48.6.1–48.6.6. Also, CCITT Contrib. COM XII-No. 165-E to Quest. 6, AT&T, June 1979.
9. J.A. Post and R.D. Silverthorn, "Results of a Subjective Comparison of Echo Control Devices in Terrestrial and Satellite Trunks," *NTC Rec.,* Nov. 1979, Vol. 3, pp. 48.4.1–48.4.5.
10. J. Gruber and N. Psimenatos, "Subjective Testing for Echo Evaluation in the Evolving Network," *TELESIS,* Vol. 9, No. 2, 1982, pp. 12–19.
11. G. Williams and L.S. Moye, "Subjective Evaluation of Unsuppressed Echo in Simulated Long-Delay Telephone Communications," *Proc. IEEE,* Vol. 118, Mar.–Apr. 1971, pp. 401–408; also, *Proc. of 5th Int. Symp. on Human Factors in Telecomm.,* London, Paper No. 22, Sept. 1970.
12. Contrib. T1Q1.1/89-002, "Long Propagation Delay and Echo Related Considerations for Telephone Circuits (CCITT Rec. G.114, Annex A), COMSAT, Dec. 22, 1988.
13. N. Kitawaki and K. Itoh, "Pure Delay Effects on Speech Quality in Telecommunications," *IEEE Journal on Sel. Areas in Telecomm.,* Vol. 9, No. 4, May 1991, pp. 586–593.
14. American National Standard for Telecommunications: DCME—Interface Functions and Performance Specification, ANSI T1.309-1991.
15. D.H.A. Black, "PLC-1: A TASI System for Small Trunk Groups," *IEEE Trans. on Comm.,* Vol. COM-30, No. 4, Apr. 1982, pp. 786–791. Also, *ICC Rec.,* Vol. 3, Denver, June 1981, pp. 49.1.1–49.1.5.
16. CCITT Rev. Draft Rec. G.763 on "DCME Using 32 kbit/s ADPCM and Digital Speech Interpolation," COM XV-R 39-E, Aug. 1990.
17. CCITT G.764 on "Voice Packetization—Packetized Voice Protocols," Geneva, 1990.
18. American National Standard for Telecommunications—Voice Packetization—Packetized Voice Protocols, ANSI T1.312-1990.
19. M. Onufry and S.J. Engelman, "Subjective Evaluation of the Intelsat TDMA/DSI System," *COMSAT Tech. Rev.,* Vol. 16, No. 1, Spring 1986, pp. 239–251.
20. Y. Yatsuzuka et al., "High-Performance ADPCM Codec for Voice and Voiceband Data and its Application to DCME," *ICC Rec.,* Vol. 1, Philadelphia, June 1988, pp. 13.4.1–13.4.8.

21. J. Gruber, "A Comparison of Measured and Calculated Speech Temporal Parameters," *IEEE Trans. on Comm.*, Vol. COM-30, No. 4, Apr. 1982, pp. 728–738. Also, *15th HICSS Rec.*, Honolulu, Jan. 1982, pp. 671–680.
22. B. Gold, "Digital Speech Networks," *Proc. IEEE*, Vol. 65, No. 12, Dec. 1977, pp. 1636–1658.
23. W.S. Mosley, "Low Bit-Rate Packet Voice Systems," *Telecomm.*, Vol. 20, No. 9, Sept. 1986, pp. 93, 94, 96, 98, 108.
24. J.G. Gruber, "Delay Related Issues in Integrated Voice and Data Networks," *IEEE Trans. on Comm.*, Vol. COM-29, No. 6, June 1981, pp. 786–800. Also, *Comp. Comm.*, Vol. 4, No. 3, June 1981, pp. 106–126. Also, *6th Data Comm. Symp.*, Pac. Grove, California, Nov. 1979, pp. 166–180.
25. R.P. Singh and S. Singhal, "Voice Transmission over Packet- and Circuit-Switched Networks," *INFOCOM Rec.*, New Orleans, Mar. 1988, pp. 2A.3.1–2A.3.9.
26. L.T. Wu and M. Kerner, "Emulating Circuits in a Broadband Packet Network," *GLOBECOM Rec.*, Vol. 2, Hollywood, Florida, Nov./Dec. 1988, pp. 21.6.1–21.6.10.
27. D. Sparrell, "Wideband Packet Technology," *GLOBECOM Rec.*, Vol. 3, Hollywood, Florida, Nov./Dec. 1988, pp. 49.2.1–49.2.5.
28. W.A. Williamson, "Voice Delay and Echo Considerations for B-ISDN" Contrib. T1S1.1/89-216, Bell South Services, May 15, 1989.
29. N. Kitawaki et al., "Speech Coding Technology for ATM Networks," *IEEE Comm. Mag.*, Vol. 28, No. 1, Jan. 1990, pp. 21–27.
30. M.M. Buchner, Jr. et al., "Evolution of a Universal Information System Environment," *ISSLS Rec.*, Tokyo, Sept./Oct. 1986, pp. 65–70.
31. R.W. Muise et al., "Experiments in Wideband Packet Technology," *Int. Zurich Sem. on Digital Comm.*, 1986, pp. D4.1–D4.5.
32. K.M. Hafner and D.W. Campt, "Packetized Voice Could Move out of the Laboratory Soon," *Data Comm.*, Vol. 14, No. 2, Feb. 1985, pp. 45, 46.
33. E.E. Mier, "How AT&T Plans to Conquer Voice-Data Integration," *Data Comm.*, Vol. 15, No. 8, July 1986, pp. 51–56.
34. K. Sriram, "Voice Packetization and Compression in Broadband ATM Networks," *IEEE Journal on Sel. Areas in Telecomm.*, Vol. 9, No. 3, April 1991, pp. 294–304.
35. J. Gruber and L. Strawczynski, "Subjective Effects of Variable Delay and Speech Clipping in Dynamically Managed Voice Systems," *IEEE Trans. on Comm.*, Vol. COM-33, No. 8, Aug. 1985, pp. 801–808. Also, *GLOBECOM Rec.*, Vol. 3, Miami, Nov./Dec. 1982, pp. F7.3.1–F7.3.5.
36. H. Miedema and M.G. Schachtman, "TASI Quality—Effect of Speech Detectors and Interpolation," *BSTJ*, Vol. 51, No. 4, July 1962, pp. 1455–1473.
37. "Methodologies for Specifying Voice Grade Channel Transmission Parameters and Evaluating Transmission Performance for Speech Telephony," ANSI/IEEE Standard 823-1989.
38. "IEEE Standard Method for Determining Objective Loudness Ratings of Telephone Connections," IEEE Standard 661-1979.
39. J.L. Sullivan, "A Laboratory System for Measuring Loudness Loss of Telephone Connections," *BSTJ*, Vol. 50, No. 8, Oct. 1971, pp. 2663–2739.
40. Annex of CCITT Contrib. COM XII, No. R 4-E, "Conversion from EARS Loudness Ratings to Corrected Reference Equivalents," AT&T, June 1982, pp.14–15.
41. "Quality Measures and the Design of Telecommunications Systems," J. Fennick, Artech House, 1988.
42. J.R. Cavanaugh, R.W. Hatch, and J.L. Sullivan, "Models for the Subjective Effects of Loss, Noise and Talker Echo on Telephone Connections," *BSTJ*, Vol. 55, No. 9, Nov. 1976, pp. 1319–1371.
43. J.R. Cavanagh, R.W. Hatch, and J.L. Sullivan, "Transmission Rating Model for Use in Planning of Telephone Networks," *GLOBECOM Rec.*, Vol. 2, San Diego, Nov./Dec. 1983, pp. 20.2.1–20.2.6.

44. L.S. DiBiaso, "Transmission Considerations for Local Switched Digital Network," *Telephony*, Vol. 133, No. 17, Oct. 24, 1977, pp. 40–45, 52.
45. A.M. Lessman, and J.R. Rosenberger, "ISDN Voice CPE Performance Compatibility Information," *GLOBECOM Rec.*, Vol. 1, Houston TX., Dec. 1986, pp. 14.5.1–14.5.10.
46. Draft of American National Standard for Telecommunications—Network Performance—Loss Plan for Evolving Digital Networks, T1Q1/91-026, July 1991.
47. "Acoustic-to-Digital and Digital-to-Acoustic Transmission Requirements for ISDN Terminals," EIA/TIA—579, 1990.
48. L.M. Manhire, "Physical and Transmission Characteristics of Customer Loop Plant," *BSTJ*, Vol. 57, No. 1, Jan. 1978, pp. 35–59.
49. "Engineering and Operations in the Bell System," Bell Telephone Labs., Inc., 1977, Chapter 6, Section 6.3.
50. J.E. Abate, L.A. Brandenburg, J.C. Lawson, and W.L. Ross, "The Switched Digital Network Plan," *BSTJ*, Vol. 56, No. 7, Sept. 1977, pp. 1297–1321. Also, *Telephony*, Vol. 113, No. 17, Oct. 24 1977, pp. 25, 28, 29, 32, 33, 36–39.
51. Notes on the Network, AT&T, 1980, Section 7 on "Transmission Considerations."
52. J.L. Neigh, "Transmission Planning for an Evolving Local Switched Digital Network," *IEEE Trans. on Comm.*, Vol. COM-27, No. 7, July 1979, pp. 1019–1024.
53. P.N. Burgess, J.L. Neigh, and R.G. Sparker, "No. 10A Remote Switching System: Transmission Plan," *BSTJ*, Vol. 61, No. 4, April 1982, pp. 627–646. Also, *ICC Rec.*, Vol. 1, Seattle, June 1980, pp. 1.2.1–1.2.5.
54. "LATA Switching Systems Generic Requirements (LSSGR)," Bellcore TR-507 Issue 3, Mar. 1989, Section 7 on "Transmission."
55. Annex (to CCITT Rec. G.171) on "Digital Private Network Loss Plan/Performance," *Blue Book*, Vol. III, Fas. III.1, Geneva 1989, pp. 251–269.
56. CCITT Contrib. COM XII-No. 107-E to Quest. 3 on "Effect of Asymmetry Between the Two Directions of Transmission," AT&T, Sept. 1971.
57. W.C. Ahern, F.P. Duffy, and J.A. Maher, "Speech Signal Power in the Switched Message Network," *BSTJ*, Vol. 57, No. 7, Sept. 1978, pp. 2695–2726.
58. CCITT Rec. G.121 on "Corrected Reference Equivalents (CREs) and Loudness Ratings (LRs) of National Systems," *Blue Book*, Vol. III, Fas. III.1, Geneva, 1989, pp. 109–125.
59. L.S DiBiaso, "Subscriber Line Impact on System Transmission Performance: A Principle of Asymmetry," *NTC Rec.*, Vol. II, Dallas TX., Dec. 1976, pp. 23.6-1 to 23.6-3.
60. F.P. Duffy and T.W. Thatcher, Jr., "Analog Transmission Performance of the Switched Telecommunication Network," *BSTJ*, Vol. 50, No. 4, Apr. 1971, pp. 1311–1347.
61. *Telecommunications Transmission Engineering*, Vol. 3, Networks and Services, 2nd Ed. AT&T 1977, Chapter 2 on "Metropolitan Network Plans," p. 36.
62. *Reference Data for Radio Engineers*, 5th Ed., ITT, 1969, Chapter 35 on "Electroacoustics."
63. G.K. McNees, "Transmission Performance of m = 225 Quantization in a Local Digital Office," *BSTJ*, Vol. 59, No. 10, Dec. 1980, pp. 1943–1964.
64. R.W. Hatch and J.L. Sullivan, "Recent Trends in Transmission Planning for Local Networks," *ISSLS Rec.*, GA Inst. of Tech., Mar. 1978, pp. 195–199.
65. "Notes on the BOC INTRA-LATA Networks," Bellcore TR-275, Issue 1, Apr. 1986, Section 7 on "Transmission."
66. J. Gruber, "Estimating the Transmission Grade-of-Service of Telephone Connections with Multiple Echo Paths," *GLOBECOM Rec.*, Vol. 1, Hollywood, Florida, Nov./Dec. 1988, pp. 14.6.1–14.6.5.
67. *Telecommunications Transmission Engineering*, Vol. 1: Principles, 2nd ed., AT&T, 1977, Chapter 20 on "Echo in Telephone Channels." Also, 3rd. ed., Bellcore 1990, pp. 578.

68. Contrib. T1Q1.1/89-102 on "Talker Echo on All Digital Connections Including ISDN," Telecom Canada, Sept. 1989.
69. T.C. Spang, "Loss-Noise-Echo Study of the Direct Distance Dialing Network," *BSTJ*, Vol. 55, No. 1, Jan. 1976, pp. 1-36.
70. CCITT Rec. G.164, "Echo Suppressors," *Blue Book*, Vol. III, Fas. III.1, Geneva, 1989, pp. 186-221.
71. CCITT Rec. G.165, "Echo Cancellers," *Blue Book*, Vol. III, Fas. III.1, Geneva, 1989, pp. 221-243.
72. K. Doherty, "Digital Echo Cancellation: A Tutorial," *Telecomm.*, Vol. 25, No. 9, Sept. 1991, pp. 37, 38, 40, 42, 44.
73. CCITT Contrib. COM XII-No. 23-E to Quest. 16/XII, "Return Loss Variations in Subscriber Lines and Telephone Sets—Balancing Circuit Considerations," AT&T, Mar. 1977.
74. G. Williams and J. Gruber, "Transmission Planning for Digital Connectivity in the Post Divestiture Environment," *NCF Rec.*, Vol. 40, Bk. 2, Chic. ILL., Oct. 1986, pp. 675, 677-682.
75. D.O. Bowker and C.A. Dvorak, "Speech Transmission Quality of Wideband Packet Technology," *GLOBECOM Rec.*, Vol. 3, Tokyo, Nov. 1987, pp. 47.7.1-47.7.3.
76. G.R. Ash, A.H. Katker, and K.R. Krishnan, "Intercity Dynamic Routing Architecture and Feasibility," *Proc. 1 of 10th Int. Teletraffic Cong.*, Montreal, Quebec, 1983, Session #3.2, ITC-10, Paper #2.
77. American National Standard for Telecommunications—Network Performance—Supplement to Transmission Specifications for Switched Exchange Access Network, Draft ANSI T1.506a, Supp. to ANSI T1.506-1990.
78. *Telecommunications Transmission Engineering*, Vol. 2: Facilities, 3rd. ed., Bellcore 1990, Chapter 13 on "PCM Transmission Systems."
79. *IEEE Comm. Mag.*, Issue on "Advanced Traffic Control Methods for Circuit Switched Telecommunications Networks," Vol. 28, No. 10, Oct. 1990.
80. "Generic Requirements for the Switched DS1/Switched Fractional DS1 Service Capability From an ISDN Interface (SWF—DS1/ISDN)," Bellcore TA-1203, Issue 1, April 1991.
81. J.R. Cavanaugh, R.W. Hatch, and J.L. Neigh, "A Model for the Subjective Effects of Listener Echo on Telephone Connections," *BSTJ*, Vol. 59, No. 6, July/Aug. 1980, pp. 1009-1060.
82. O.P. Mahajan and E.J. Isganitis, "Average Power Return Loss (APRL)—A Single-Ended Parameter for Specifying Listener Echo Objectives," *ICC Rec.*, Vol. 2, Seattle, June 1980, pp. 2041-2046.
83. CCITT Contrib. COM XII, No. 14-E, "Listener Echo Subjective Test Results for Non-Uniform Echo Path Losses," Canada: BNR, Apr. 1981.
84. CCITT Supp. No. 3 (to Rec. P.11) on "Transmission Rating Models," pp. 248-261, and Annex A (to Supp. No. 3) on "Opinion Ratings of Transmission Impairments," Bellcore, pp. 299-306. *Blue Book*, Vol. V, Geneva, 1989.
85. G. Williams, J. Gruber, and N. Psimenatos, "Transmission Planning for Local Digital Networks," in *Proc. 2nd IEE Int. Conf. on Telecom. Transmission*, London, Mar. 17-20, 1981, pp. 205-208. Also, *ICC Rec.*, Vol. 2, Seattle, June 1980, pp. 20.1.1-20.1.6.
86. J.B. Terry, D.R. Younge, and R.T. Matsunaga, "A Subscriber Line Interface for the DMS-100 Digital Switch," *NTC Rec.*, Wash. D.C., Nov. 1979, Vol. 2, pp. 28.3.1-28.3.6.
87. J. Gruber, "Performance Analysis and Design Guidelines for Multiple 4-wire Path Telephone Connections," *IEEE Trans. on Comm.*, Vol. COM-34, No. 7, July 1986, pp. 647-653. Also, ICC Rec. Vol. 1, Toronto Ontario, June 1986, pp. 7.4.1-7.4.6. Also, CCITT Contrib. COM XVI, No. 42-E to Quest. 5, "Some analytical results for multiple 4-wire paths in connections supporting voiceband data," Canada: BNR, Mar. 1983.
88. CCITT Rec. G.711 on "64-kbit/s PCM Algorithm," *Blue Book*, Vol. III, Fas. III.3, Geneva 1989, pp. 175-184.
89. CCITT Rec. G.726 on "40, 32, 24, 16 kbit/s ADPCM," Geneva, 1990.

90. "Digital Processing of Voiceband Signals—32-kbit/s ADPCM Line Format Standard," ANSI T1.302-1989.
91. J.C. Bellamy, *Digital Telephony*, Wiley, 1982. Also, 2nd ed., 1991.
92. American National Standard for Telecommunications—Digital Processing of Voiceband Signals—Algorithm for 24, 32 and 40 kbit/s ADPCM, ANSI T1.303-1989.
93. CCITT Rec. G.727 on "5-, 4-, 3- and 2-Bits Sampled Embedded ADPCM," Geneva, 1990.
94. American National Standard for Telecommunications—Digital Processing of Voiceband Signals—Algorithm for 5-, 4-, 3- and 2-Bits per Sample Embedded ADPCM, ANSI T1.310-1991.
95. N.S. Jayant, "High Quality Coding of Telephone Speech and Wideband Audio," *ICC Rec.*, Vol. 3, Atlanta, April 1990, pp. 322.1.1-322.1.5. Also, *IEEE Comm. Mag.*, Vol. 28, No. 1, Jan. 1990, pp. 10-20.
96. M.H. Sherif et al., "Overview of CCITT Embedded ADPCM Algorithms," *ICC Rec.*, Vol. 3, Atlanta, April 1990, pp. 324.5.1-324.5.5.
97. S. Lipoff, "New Techniques in Voice Compression," *Bus. Comm. Rev.*, Vol. 19, No. 15, May 1989, pp. 81-83.
98. W.R. Daumer and J.R. Cavanaugh, "A Subjective Comparison of Selected Digital Codecs for Speech," *BSTJ*, Vol. 57, No. 9, Nov. 1978, pp. 3119-3165.
99. W.R. Daumer and J.L. Sullivan, "Subjective Quality of Several 9.6-32 kbit/s Speech Coders," *ICASSP Rec.*, Paris, France, May 1982, Vol. 3, pp. 1709-1712. Also, *IEEE Trans. on Comm.*, Vol. COM-30, No. 4, Apr. 1982, pp. 655-662.
100. D.J. Goodman and R.D. Nash, "Subjective Quality of the Same Speech Transmission Conditions in Seven Different Countries," *IEEE Trans. on Comm.*, Vol. COM-30, No. 4, Apr. 1982, pp. 642-654. Also, *ICASSP Rec.*, Paris, France, May 1982, Vol. 2, pp. 984-987.
101. CCITT Contrib. COM XII, No. 148-E, "Determination of Subjectively Equivalent Noise, and Threshold of Speech Correlated Noise," Canada: BNR, Apr. 1979.
102. CCITT Contrib. COM XII, No. 61-E, "Subjectively Equivalent Noise for Linear and Carbon Microphone Originated Speech Signals," Canada: BNR, Apr. 1982.
103. V. Gupta, H. Suyderhoud, K. Viropaksha, and M. Onufry, "Subjective Equivalence of Speech-Correlated and Stationary Noise," *COMSAT Tech. Rev.*, Vol. 13, No. 1, Spring 1983, pp. 165-173.
104. T. Watanabe, H. Nagabuchi, and N. Kitawaki, "Law of Addition for Subjective Quality of Voiceband Codecs," *Rev. of the Elect. Comm. Labs.*, Vol. 35, No. 4, 1987, pp. 435-441.
105. CCITT Contrib. COM XII, No. 130-E, "Subjective Equivalence Functions for Consideration in a Quantization Distortion Opinion Model," Canada: BNR, Mar. 1983.
106. N. Kitawaki, "Speech-Quality Assessment Methods for Speech-Coding Systems," *IEEE Comm. Mag.*, Vol. 22, No. 10, Oct. 1984, pp. 26-33.
107. CCITT SG XII Contrib. SQ-15.88 on "Definition of QDU and Use of MNRU," July 1988.
108. N. Kitawaki and H. Nagabuchi, "Quality Assessment of Speech Coding and Speech Synthesis Systems," *IEEE Comm. Mag.*, Vol. 26, No. 10, Oct. 1988, pp. 36-44.
109. J.R. Cavanaugh, "A Model for the Subjective Effects of Quantization Distortion in Digital Transmission of Speech," *ICC Rec.*, Seattle, Vol. 2, June 1980, pp. 20.5.1-10.5.5.
110. D.W. Petr, "32 kbit/s ADPCM-DLQ Coding for Network Applications," *GLOBECOM Rec.*, Miami, Florida, Nov./Dec. 1982, Vol. 1, pp. A8.3.1-A8.3.5. Also, Annex 2 of CCITT Contrib. COM XVI-No. R2-E, "32 kbit/s ADPCM-DLQ coding," AT&T, June 1982, pp. 17-28.
111. P. Mermelstein and G. Williams, "Network Performance Issues in 32 kbit/s Coding of Speech and Voiceband Data," *GLOBECOM Rec.*, Vol. 1, Miami, Florida, Nov./Dec. 1982, pp. A8.2.1-A8.2.5.
112. G. Williams and H. Suyderhoud, "Subjective Performance of the 32 kbit/s ADPCM Algorithm," *GLOBECOM Rec.*, Vol. 2, Nov. 1984, pp. 23.2.1-23.2.8.
113. N. Psimenatos, J. Gruber, G. Goddard, and D. Mondor, "The Application of 32 kb/s ADPCM

Systems in Telecommunications Networks," *GLOBECOM Rec.*, Vol. 2, Atlanta, Nov. 1984, pp. 23.5.1–23.5.7. Also, *3rd IEE Int. Conf. on Telecomm. Transmission*, London, Mar. 1985, pp. 27–30.
114. CCITT Contrib. COM XVI, No. 41-E, "An Assessment of Recommendation G.113 in Relation to 32 kbit/s ADPCM Performance," Canada: BNR, Mar. 1983.
115. D.L. Richards, "Telecommunications by Speech," London, Butterworth, 1973.
116. CCITT Supp. No. 3 (to Rec. P.11) on "Prediction of Transmission Qualities from Objective Measurements," pp. 259–271, and Annex B (to Supp. No. 3) on "Calculated Transmission Performance of Telephone Networks," pp. 307–317, British Telecom., *Blue Book*, Vol. V, Geneva, 1989.
117. P.V. Coverdale, "Taking Some of the Black Magic Out of Sidetone," *ICC Rec.*, Vol. III, Philadelphia, June 1982, pp. 61.3.1–61.3.5.
118. G.I. Riley and P.K. Webb, "Transmission Performance of 4-Wire PABX's," *Proc. Int. Conf. Private Electron. Switching syst.*, London, Apr. 1978, pp. 165–168.
119. F.T. Andrews, Jr. and R.W. Hatch, "National Telephone Network Transmission Planning in the American Telephone and Telegraph Company," *IEEE Trans. on Comm.*, Vol. COM-19, No. 3, June 1971, pp. 302–314.
120. A.H. Inglis and W.L. Tuffnell, "An Improved Telephone Set," *BSTJ*, Vol. 29, No. 4, Apr. 1951, pp. 239–270.
121. CCITT Contrib. COM XII-No. 84-E to Quest. 14, on "Effect of Attenuation Distortion," AT&T, June 1982.
122. Annex B (to CCITT Rec. P.11) on "Effects of Attenuation Distortion on Transmission Performance," *Blue Book*, Vol. 5, Geneva 1989, pp. 30–37.
123. *Telecommunications Transmission Engineering*, Vol. 1: Principles, 3rd. ed., Bellcore 1990, Chapter 26 on "Transmission Objectives and Requirements."
124. P.J. Patrick, R. Steele, and C.S. Xydeas, "Voiced/Unvoiced Band Switching System for Transmission of 6 kHz Speech Over 3.4 kHz Telephone Channels," *Radio Eng.*, Vol. 51, No. 9, May 1981, pp. 233–235.
125. D.J. Goodman, B.J. McDermott, and L. Nakatani, "Subjective Evaluation of PCM Coded Speech," *BSTJ*, Vol. 55, No.8, Oct. 1976, pp. 1087–1109.
126. CCITT Rec. G.722 on "7 kHz Audio-Coding Within 64 kb/s," *Blue Book*, Vol. III, Fas. III.3, Geneva, 1989, pp. 269–341.
127. P. Mermelstein, "G.722, A New CCITT Coding Standard for Digital Transmission of Wideband Audio Signals," *IEEE Comm. Mag.*, Vol. 26, No. 1, Jan. 1988, pp. 8–15.
128. *Telecommunications Transmission Engineering*, Vol. 1: Principles, 2nd ed., AT&T, 1977, Chapter 17 on "Noise and Crosstalk." Also, 3rd. ed., Bellcore, 1990.
129. *Telecommunications Transmission Engineering*, Vol. 1: Principles, 2nd ed., AT&T, 1977, Chapter 26 on "Transmission Objectives," pp. 610, 611. Also, 3rd. ed., Bellcore, 1990.
130. *Telecommunications Transmission Engineering*, Vol. 3, Networks and Services, 2nd ed. AT&T 1977, Chapter 16 on "Short-Haul Systems," pp. 424, 425.
131. Annex E (to CCITT Rec. P.84) on "Examples of Other Subjective Scales," *Blue Book*, Vol. 5, Pt. I, Section 7, Geneva 1989, p. 32.
132. CCITT Annex F (to Rec. P.11) on "Effects of Nonlinear Distortion on Transmission Performance," *Blue Book*, Vol. V, Geneva 1989, p. 43.
133. CCITT Rec. G.821, "Error Performance on an International Digital Connection Forming Part of an Integrated Services Digital Network," *Blue Book*, Vol. III, Fas. III.3, Geneva, 1989, pp. 26–35.
134. CCIR Doc. 4/75-E, "Effects of Bit Errors on the Transmission of Speech Through PCM Systems," U.K., Feb. 17, 1976.
135. CCIR Rec. 522-2, "Allowable Bit Error Rate at the Output of the Hypothetical Reference Circuit for System in the Fixed Satellite Service Using Pulse-Code Modulation for Telephony," Vol. IV, Pt. 1, 1986, pp. 64–65.

136. CCIR Rec. 614, "Allowable Error Performance for a Hypothetical Reference Digital Path in the Fixed-Satellite Service Operating Below 15 GHz when forming part of an International Connection in an Integrated Services Network," Vol. IV, Pt. 1, 1986, pp. 65,66.
137. G.R. Ritchie and P.E. Scheffer, "Projecting the Error Performance of the Bell System Digital Network," *ICC Rec.*, Philadelphia, June 1982, Vol. 1, pp. 2D.2.1–2D.2.6.
138. M.A. Rezny, "SDC Error Performance Considerations," *ICC Rec.*, Vol. 2, Philadelphia, June 1982, pp. 3D.5.1–3D.5.6.
139. J. Gruber, R. Vickers, and D. Cuddy, "Impact of Errors and Slips on Voice Service," *ICC Rec.*, Vol. 1, Philadelphia, June 1982, pp. 2D.5.1–2D.5.7.
140. CCITT SG XVIII TD30 to Quest.6, "Report on Error Performance Issues," Geneva, June 1991.
141. CCITT SG XVIII Draft Rec. G.82x on "Error Performance Parameters and Objectives for Continuous Bit Rate International Digital Connections at or above the Primary Rate," Annex 1 to CCITT SG XVIII TD30 to Quest.6, "Report on Error Performance Issues," Geneva, June 1991.
142. CCITT Rec. G.822, "Controlled Slip Rate Objectives on an International Digital Connection," *Blue Book*, Vol. III, Fas. III.3, Geneva, 1989, pp. 36–38.
143. D.J. Goodman et al., "Packet Reservation Multiple Access for Local Wireless Communications," *IEEE Trans. on Comm.*, Vol. 37, No. 8, Aug. 1989, pp. 885–890.
144. C.M. Carbalis, "The Challenge of Voice by the Packet," *Telecomm.*, Vol. 20, No. 10, Oct. 1986, pp. 47, 48, 50, 52, 54.
145. O.J. Wasem, D.J. Goodman, C.A. Dvorak, and H.G. Page, "The Effect of Waveform Substitution on the Quality of PCM Packet Communications," *IEEE Trans. on Ac., Sp. and Sig. Proc.*, Vol. 36, No. 2, Mar. 1988, pp. 342–348.
146. J. Suzuki and M. Taka, "Missing Packet Recovery Techniques for Low-Bit-Rate Coded Speech," *IEEE Journal on Sel. Areas in Comm.*, Vol. 7, No. 5, June 1989, pp. 707–717.
147. W. Yeh et al., "Sensitivity of Bandwidth Allocation Algorithms for Unidirectional Networks to Voice Parameters," *ICC Rec.* Vol. 4, Atlanta, April 1990, pp. 333.5.1–333.5.5.
148. H.H. Lee and C.K. Un, "A Study of On-Off Characteristics of Conversational Speech," *IEEE Trans. on Comm.*, Vol. COM-34, No. 6, June 1986, pp. 630–637.
149. J.A. Jankowski, Jr., "A New Digital Voice-Activated Switch," *COMSAT Tech. Rev.*, Vol. 6, No. 1, Spring 1976, pp. 159–178.
150. CCIR Doc. 4/29-E, "Proposed Amendment to Annex to Report 211-3: Speech Interpolation Technique in TDMA Systems," Japan, Jan. 19, 1976.
151. S.J. Campanella, "Digital Speech Interpolation," *COMSAT Tech. Rev.*, Vol. 6, No. 1, Spring 1976, pp. 127–158.
152. CCITT Contrib. COM CMBD, No. 36-E, "Quality of Speech Impaired by Short Interruptions," KDD, June 1982.
153. H. Tanaku et al., "Design of a Packet Voice Transmission System," in *NTC Rec.*, Vol. 1, Wash. D.C., Dec. 1979, pp. 13.1.1–13.1.5.
154. N.S. Jayant and S.W. Christensen, "Effects of Packet Losses in Waveform Coded Speech and Improvements Due to an Odd-Event Sample-Interpolation Procedure," *IEEE Trans. on Comm.*, Vol. COM-29, No. 2, Feb. 1981, pp. 101–109.
155. "Voice Services Internetworking Between ATM Networks and PSTN," CCITT SGXV Liaison to SGXVIII, Annex 16 to Report COM XV-R 64 of WP XV/8, Geneva, Feb. 1991.
156. CCITT Contrib. COM CMBD, No. R2-E, Annex 5, "Tolerability of Telephone Users to Transmission Interruptions Lasting Several Seconds or More," Oct. 1982, pp. 40–41.
157. CCITT Contrib. COM CMBD, No. 20-E, "The Use of Call Cutoff Rate as Measure of Telephone Service Reliability," AT&T, May 1982.
158. J.P. Healey, J.M. Kelly, D.G. Leeper, M.G. Linnell, and J.A. Maher, "How's Service? Ask the Customer," *Bell Lab. Rec.*, Vol. 60, No. 2, Feb. 1982., pp. 31–34

159. B.W. Kort, "Models and Methods for Evaluating Customer Acceptance of Telephone Connections," *GLOBECOM Rec.*, Vol. 2, San Diego, Nov./Dec. 1983, pp. 20.6.1.–20.6.9.
160. *Telecommunications Transmission Engineering,* Vol. 3: Networks and Services, Chapter 20 on "Measurements," 3rd ed., Bellcore 1990.
161. CCITT Rec. X.140 "General Quality of Service Parameters for Communication via Public Data Networks," *Blue Book,* Vol. VIII, Fas. VIII.3, Geneva, 1989, pp. 422–446.
162. CCITT Rec. I.350, "General Aspects of Quality of Service and Network Performance in Digital Networks, Including ISDN," *Blue Book,* Vol. III, Fas. III.1, Geneva, 1989, pp. 129–139.
163. American National Standard for Telecommunications: Performance Parameters for Packet Switched Data Communication Service, ANSI T1.504-1989.
164. American National Standard for Telecommunications—IDSN—Architectural Framework and Service Description for Frame Relaying Bearer Service, ANSI T1.606-1990.
165. "Generic System Requirements in Support of SMDS," Bellcore TR-772, Issue 1, May 1991.
166. CCITT Draft Rec. I-35B on "Broadband ISDN Performance," TD 27 (XVIII/6), Geneva, June 1991.
167. CCITT Rec. X.135 on "Speed of Service (Delay and Throughput) Performance Values for Public Data Networks When Providing International Packet-Switched Service," *Blue Book,* Vol. VIII, Fas. VIII.3, Geneva, 1989, pp. 368–391.
168. S.S. Katz, "Introduction of New Network Services," *Proc. 1 of 10th Int. Teletraffic Cong.*, Montreal, Quebec, 1983.
169. R.S. Bloomfield and K.P. Spies, "User-Oriented Performance Evaluation of Data Communications Services: Measurement Results," *ICC Rec.*, Vol. 1, Boston, June 1989, pp. 3.5.1–3.5.7.
170. M.T. Neibert, "International Packet Switched Correction Performance Measurements," *ICC Rec.*, Vol. 1, Boston, June 1989, pp. 3.6.1–3.6.3.
171. M.B. Hidalgo, W.S. Keely, and T.N. Washburn, "Verification/Validation of Performance Simulation Using X.3102 Measurements," *IEEE Network,* Vol. 2, No. 6, Nov. 1988, pp. 15, 16, 23–27.
172. S.H. Richman, "A Comparison of Field Performance Measurements on Multi-National Packet Switching Networks," *GLOBECOM Rec.*, Vol. 3, Tokyo, Nov. 1987, pp. 41.6.1–41.6.8.
173. L.S. Newman, "Early Alternatives to the ISDN," *Telephony,* Vol. 211, No. 19, Nov. 10, 1986, pp. 44, 46, 47, 74.
174. K.E. Carmichael, W.K. Pehlert, and M. Ulema, "Maintenance of the ACCUNET Packet Service Network," *ICC Rec.,* Vol. 1, Toronto Ontario, June 1986, pp. 18.2.1–18.2.6.
175. N.B. Seitz, D.R. Wortendyke, and K.P. Spies, "User-Oriented Performance Measurement on Public Data Networks," *ICC Rec.*, Vol. 1, Chicago, June 1985, pp. 12.6.1–12.6.7.
176. N.B. Seitz, D.R. Wortendyke, and K.P. Spies, "User-Oriented Performance Measurement on the ARPANET," *IEEE Comm. Mag.,* Vol. 21, No. 5, Aug. 1983, pp. 28–44.
177. K.P. Spies, W.R. Wortendyke, E.L. Crow, M.J. Miles, E.A. Quincy, and N.B. Seitz, "User-Oriented Performance Evaluation of Data Communication Service: Measurement Design, Conduct and Results," NTIA Report 88–238, U.S. Dept. of Commerce, Aug. 1988.
178. D.R. Wortendyke, N.B. Seitz, K.P. Spies, E.L. Crow, and D.S. Grubb, "User-Oriented Performance Measurement on the ARPANET: The Testing of a Proposed Federal Standard," NTIA Rep. 82–112, U.S. Dept. of Commerce, Nov. 1982.
179. J.J. Mahoney Jr., J.J. Mansell and R.C. Metlack, "Digital Data System: User's View of the Network," *BSTJ,* Vol. 54, No. 5, May/June 1975, pp. 833–884.
180. "International ACCUNET Digital Service Description and Network Interface Specifications," AT&T PUB 54019, Issue 1, Sept. 1987. Also, Adden 1, April 1988.
181. D.S. Gokhale et al., "Quality-of-Service Measurement of Satellite-Based International Packet-Switched Networks," *COMSAT Tech. Rev.*, Vol. 19, No. 2, Fall 1989, pp. 235–258.
182. CCITT Rec. X.136 on "Accuracy and Dependability Performance Values for Public Data Net-

works When Providing International Packet-Switched Service," *Blue Book,* Vol. VIII, Fas. VIII.3, Geneva, 1989, pp. 392–410.
183. E.C. Bender, J.G. Kneuer, and W.J. Lawless, "Digital Data System: Local Distribution System," *BSTJ,* Vol. 54, No. 5, May/June 1975, pp. 915–942.
184. M.B. Brilliant, "Observations of Errors and Error Rates on T1 Digital Repeatered Lines," *BSTJ,* Vol. 57, No. 3, Mar. 1978, pp. 711–746.
185. N.E. Snow and K. Knapp Jr., "Digital Data System: System Overview," *BSTJ,* Vol. 54, No. 5, May/June 1975, pp. 811–832.
186. "ACCUNET Switched 56 Service to Public Switched Digital Service Interface Advisory," AT&T PUB 61330, Sept. 1986.
187. "Digital Data System Channel Interface Specification," AT&T PUB62310, Issue 2, Nov. 1987. Also, Adden. 3 Dec. 1989.
188. CCITT Rec. X.137 on "Availability Performance Values for Public Data Networks when Providing International Packet Switched Service," *Blue Book,* Vol. VIII, Fas. VIII.3, Geneva, 1989, pp. 410–421.
189. J.P. Taylor, "Making the Switch to an All Digital Network," *Telephony,* Vol. 196, No. 10, Mar. 5, 1979, pp. 24–20, 32, 33, 52.
190. CCITT Contrib. COM XVIII to Quests. 1 and 2, "Measures of Error Performance for Digital Systems," AT&T, June 1977.
191. CCITT Contrib. COM XVIII to Quest. 1, "Error Ratio Parameters for IDN /ISDN," Canada: CTCA, Mar. 1977.
192. CCITT Rec. G.113, "Transmission Impairments," *Blue Book,* Vol. III, Fas. III.1, Geneva, 1989, pp. 63–84.

Chapter 4
END-TO-END NETWORK-ORIENTED PERFORMANCE

4.1 INTRODUCTION

This chapter addresses the next step in the overall performance process introduced in Chapter 1 (Fig. 1.3). More detailed consideration is given to end-to-end network-oriented performance than provided in Chapter 3, where the focus was on the relationships between user- and network-oriented performance. In particular, this chapter includes information on those voiceband network-oriented performance parameters not directly perceived by end users of data services, but which ultimately affect voiceband data error performance. In general, this chapter estimates the performance capabilities of networks based on connection models and on survey data, and compares this capability with typical performance objectives. This is done on a parameter-by-parameter basis for voiceband-, digital-, and packet-type services.

4.2 NETWORK-ORIENTED PARAMETERS

Table 4.1 provides a detailed classification of network-oriented voice and data transmission performance parameters analogous to the user-oriented parameters given earlier in Table 3.1 of Chapter 3. A comparison of Tables 3.1 and 4.1 reveals the following similarities and differences between user- and network-oriented parameters:

- For voice service, the user- and network-oriented parameters in the respective tables are similar. The reason, as noted in Chapter 3, is that most network-oriented voice parameters are directly perceived by voice users, and as such are also user-oriented parameters.
- For digital data and packet data services, the user- and network-oriented parameters in the respective tables are also similar. The reason is that most network-oriented digital data and packet data parameters are perceived in some sense by these data users, and as such are also user-oriented parameters.

Table 4.1
Detailed Classification of Network-Oriented Voice and Data Transmission Performance Parameters
(Mainly for the Information Transfer Phase)

Performance Criteria	Parameters			
	Voice	Data		
		Circuit Switched/Dedicated		Packet Switched
		Voiceband	Digital	
Speed	Fixed Delay Variable Delay	Propagation Delay	Propagation Delay	Throughput Cap. Transfer Delay
Accuracy	*Analog: Continuous*			
	Loudness Loss	Electrical Loss	—	—
	Idle Circuit Noise	C-message Noise	—	—
	—	3 kHz Flat Noise	—	—
	TEPL	TEPL	—	—
	LEPL (WEPL)	LEPL (SM)	—	—
	Sig./Quant. Dist. (Q)	Sig./C-Notched Noise	—	—
	Quantization Noise	C-Notched Noise	—	—
	—	3 kHz Flat-Notched Noise	—	—
	Acoustic Noise	—	—	—
	Sidetone	—	—	—
	Bandwidth/Slope	Attenuation Distortion	—	—
	—	Envelope Delay Dist.	—	—
	—	Peak/Average Ratio	—	—
	Crosstalk/Interference	—	—	—
	Non-Linear Distortion	Intermodulation Dist.	—	—
	—	Amp. & Phase Jitter	—	—
	—	Frequency Shift	—	—
	Analog: Transients			
	Impulse Noise	Impulse Noise	—	—
	Gain & Phase Hits	Gain & Phase Hits	—	—
	Dropouts	Dropouts	—	—
	Speech Clipping	—	—	—
	Error Performance #			
	BER DM, SES Short Interruptions SEFS, CSS	BER, BLER	BER, BLER, DM, ES, SES	Residual Error Rate
Dependability	Long Ints. (UASs) Cutoff * (Carrier Alarm)	Long Ints. (UASs) Cutoff * (Carrier Alarm)	UASs Premature Disc. *	Reset Premature Disc. *

\# See Chapter 9 for detailed error performance terminology and other parameters needed for maintenance.
* For switched services.

- However, for voiceband data service, the user- and network-oriented parameters for the accuracy of service category in the respective tables are not similar. The reason, as also noted in Chapter 3, is that most network-oriented analog parameters for voiceband data are not directly perceived by voiceband data users.

A comparison of the voice and data network-oriented parameters in Table 4.1 reveals the following similarities and differences:

- The voice and voiceband data parameters are similar in many instances, since both are associated with the voiceband.
- However, voiceband parameters for the accuracy of service category differ from those for digital data and packet data. The reason is that analog voiceband parameters are not relevant to digital data services, nor to packet data services supported by a digital transport system. It is possible for packet data service to be supported by an analog, or mixed analog/digital transport system. In this case, the packet service is viewed here as a higher layer application carried by voiceband data-like circuits, and these circuits must satisfy voiceband data requirements. As such, the packet layer parameters are disassociated from the analog voiceband parameters.

This chapter will address network performance in terms of end-to-end performance based on survey data, as well as end-to-end performance objectives. For this purpose, various network connection models are considered in the next section.

4.3 VOICEBAND NETWORK PERFORMANCE

4.3.1 Connection Models

The types of intra-LATA and inter-LATA network connections considered here are listed in Table 4.2 and illustrated in Figures 4.1 and 4.2, respectively (a more detailed discussion of the nature of these networks is provided in Ch. 5). Notice that these figures represent more specific instances of the generalized connection model introduced earlier in Chapter 3 (Fig. 3.1). The backbone portion of the intra- and inter-LATA networks are further characterized in Figures 4.3 and 4.4, respectively. Generally, routing alternatives are available in both cases, and estimates of the probabilities of occurrence of these routes are given in Table 4.3 [1–3]. These estimates, expressed in terms of the number of switches likely to be encountered in a route, are regarded as upper bounds, since in recent years there has been a trend toward higher network connectivity, fewer hierarchical levels, and a corresponding move toward dynamic routing (e.g., of the two-link type [4, 5]). Two-link dynamic routing means that most traffic within an intra-LATA network, or within the inter-exchange carrier (IEC) portion of an inter-LATA network, is carried on a direct route, and the remainder on a route with, at most, two links.

Regarding access from the customer premises equipment (CPE) across the network interface (NI) to the end office (EO), a combination of analog 2-wire loop for the distribution portion, plus 4-wire digital carrier for the feeder portion, is assumed in Figures 4.1 to 4.4. However, eventually, the distribution portion will also be dig-

Table 4.2
Types of Voiceband Connections

Intra-LATA (Metropolitan)	Intra-Office:	Without Distributed Switching With Distributed Switching *
	Inter-Office:	Direct Route Tandem Route Inter-tandem Route
Inter-LATA (Toll)	LEC Portion:	Direct Route (FGC, FGD) Tandem Route (FGD)
	IEC Portion:	Short Distance Medium Distance Long Distance

* Distributed switching means distribution of local (end-office) switching functions by means of remote switching or subscriber carrier units, and may also be associated with all other types of connections.

(a)

(b)

CPE : Customer Premise Equipment
NI   : Network Interface
EO : End-Office Switch

Figure 4.1 Intra-LATA (metropolitan) end-to-end connection models: (a) intra-office; (b) inter-office.

Figure 4.2 Inter-LATA (toll) end-to-end connection model.

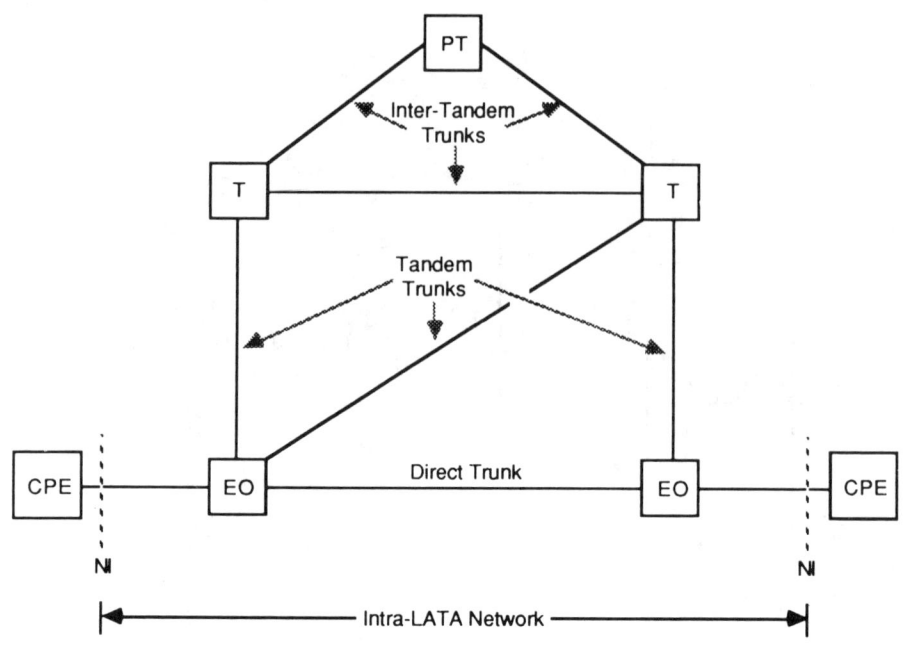

Figure 4.3 Intra-LATA (inter-office) end-to-end network connection model.

ital, resulting in end-to-end all-digital connections. Network performance and planning considerations for this situation are addressed in [6–8], and also in Chapter 7.

Regarding the local-exchange carrier (LEC) portion of inter-LATA connections, a variety of arrangements provide access to the IEC portion, as discussed in Chapter 5. However, the trend is toward the Feature Group D (FGD) arrangement shown here in Figure 4.4 [9–11]. In this figure, it is expected that the exchange access arrangement via the tandem switch will be most often used, since this route provides more economical access to many end offices than direct routing to each end office. In fact, for a given LEC end office and IEC point-of-presence (POP) switch, there may often be only one route (e.g., via an access tandem switch), so in this sense there

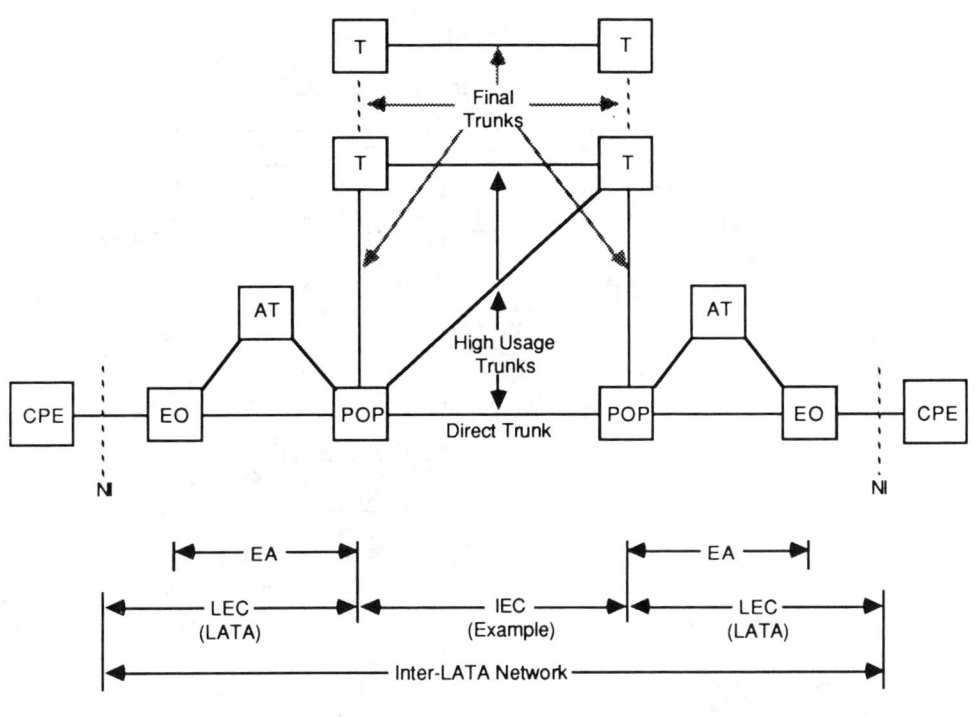

Figure 4.4 Inter-LATA (toll) end-to-end network connection model.

is no alternate routing capability. However, across many connections, it can be viewed that there are two access alternatives. In addition, for reasons of survivability, more instances of both direct and tandem routes will likely occur in the future. Under the principle of "equal access," the performance of these two routing alternatives will be very similar [10, 12]. Regarding the IEC portion of inter-LATA connections, it should be noted that many hierarchical levels are possible as implied in Figure 4.4 (i.e., there were four levels in the predivestiture AT&T inter-toll network [13, 14]). However, today the AT&T network essentially has only a two-level hierarchy, owing to nonhierarchical dynamic routing [4]. Also, non-AT&T IEC networks typically have a two-level hierarchy [15].

Table 4.3
Upper Bound Estimates of the Number of Trunks on Various Types of Connections

Routing	Number	Trunks in Connection				
		Probability (%)				
		Intra-LATA Inter-Office	Inter-LATA, IEC Portion *			
			Short	Medium	Long	Overall
Hierarchical	1	70	64	28	23	56
	2	20	30	45	38	33
	3	7	5	23	29	9
	4	3	1	4	10	2
	Ave. # Trunks	1.4	1.4	2.0	2.3	1.6
2-Link Dynamic	1	70	64	28	23	56
	2	30	36	72	77	44
	Ave. # Trunks	1.3	1.4	1.7	1.8	1.4

	*Distance Category		
	Short	Medium	Long
Probability of Occurrence (%)	77.8	13.4	8.8
Airline Distance Ranges (mi)	0 – 180	180 – 725	725 – 2600
Route Distance Ranges (mi)	0 – 340	340 – 1065	1065 – 3380
Approx. Mean Airline Distance (mi)	45	385	1525
Approx. Mean Route Distance (mi)	85	655	2135
Approx. Route/Airline Ratio	1.9	1.7	1.4

4.3.2 End-to-End Performance Information

4.3.2.1 Survey Data

For the purposes of characterizing the end-to-end performance of the networks just described, principal sources of survey data are used as indicated in Table 4.4. In general, the applicable survey data consists of two components: loop survey data [16], which is common to all connections, and trunk (or backbone network) survey data [1, 17, 18]. In this chapter, the loop survey data for certain parameters is concatenated with the trunk-related survey data using the analytical impairment accumulation techniques (addition laws) presented in Chapter 5 (Sec. 5.4). This process provides an estimate of the end-to-end performance. The process applies largely to end-to-end loss and related parameters (such as attenuation distortion), and to talker echo path loss (TEPL). It also applies to noise, although in most cases loop noise is negligible. For most other parameters, the impairment contribution of the loop plant is negligible, and in these cases the end-to-end performance is based on the relevant trunk-related survey data (adjusted by loop loss where necessary).

Regarding trunk-related survey data, the exchange access study (EAS) [17] is regarded here as applicable to most intra-LATA trunks (Figs. 4.1 and 4.3). This is because of the similarity of the transmission plan, as well as the short-haul distances and technologies associated with both exchange access and inter-office trunking. The EAS is directly applicable to the exchange access portion of inter-LATA connections (Figs. 4.2 and 4.4); but for the purpose of end-to-end connection characterization, it is recognized that the exchange access is included in the end-office survey (EOS) data [1]. The EOS includes the entire backbone connection from end office to end office, excluding loops. The EOS only includes exchange access by a direct route (e.g., to the AT&T IEC network). However, owing to the concept of "equal access," the performance via an access tandem is in most instances intended to be very similar to that of the direct route. There is an exception for the delay parameter when non-AT&T IEC networks are involved in inter-LATA connections; in this case, the trunking from the access tandem to the IEC POP may be substantial in distance, whereas with the AT&T IEC network, the access tandem and IEC POP may often be coincident. The additional delay involved in access tandem to IEC POP trunking is treated separately.

The format of the survey data presented here for continuous analog parameters is most often in terms of a mean value and a standard deviation (SD), with the underlying assumption that the distributions are approximately normally distributed. This enables worst-case values to be estimated using percentiles, as described in Chapter 5. For transient parameters (analog and digital), the distribution for the count of transient events is most often log-normal, and performance is expressed in terms of a median rate (event count per unit time) and a maximum rate (percentile value), with corresponding probability.

4.3.2.2 *Performance Objectives/Requirements*

To provide further appreciation for end-to-end performance, performance objectives for various parameters are included for comparison with the survey data. Performance objectives are levels of performance suitable for support of existing and likely future services, accounting for technical, economic, and operational trends and planned evolution. Performance requirements (or limits), on the other hand, are minimum levels of performance that must be met to satisfy current voiceband services. However, such a clear distinction between these terms is generally not maintained in the literature; i.e., a given performance level may be referred to as an objective in one reference, and as a limit in another. Thus, no attempt is made in this book to maintain a rigorous distinction, and the term "objective" is most often used throughout, since the book addresses evolving networks.

The principal sources of information for performance objectives are included in Table 4.4. Note that, in addition to these sources, similar and in some instances more detailed information can be found in earlier, pre- and post-divestiture sources such as the *Blue Book* [20], the *Red Books* (including the recent updates) [21–23], the *Green Book* [24] and the *Brown Book* [25], as well as various AT&T Publications (PUB) and Practices [26], Bellcore Technical Advisories (TA) and References (TR) [27], *Bell System Technical Journal* (BSTJ) and *Bell Laboratories Technical Journal* (BLTJ) articles, and other published papers. These sources are referenced as required in the following sections of this chapter. In some instances, end-to-end performance objectives are not available directly from the above or other sources; in these cases, derived objectives or performance guidelines can be inferred from objectives for network components and accumulated using the techniques outlined in Chapter 5.

Table 4.4
Principle Sources of Basic Survey Data and Performance Objectives Used to Characterize End-to-End Network Connections

Connection		Sources of Survey Data	Sources of Performance Objectives *
Type	Description		
Intra-LATA	Intra-office: Without Distributed Switching With Distributed Switching	Loop Survey [16] Loop + EAS [17, 18]	LSSGR [19]
	Inter-office: Direct and Tandem Routes Intertandem Routes	Loop + EAS Loop + EAS or Loop + EOS (short distance) [1]	BOC Intra-LATA Notes on the Network [9]
Inter-LATA	Toll: Short, Medium and Long Distances	Loop + EOS [1]	AT&T Notes on the Network [13]

* Guidance on end-to-end loudness loss for evolving mixed analog/digital networks, and for all digital networks, is also provided in the ANSI draft loss planning standard [7], and in the EIA/TIA ISDN Terminal standard [8].

The format for performance objectives may be in terms of a mean and SD, or in terms of a maximum limit; in some cases the limit is to be met with a certain probability, where the probability may be interpreted as either the percentage of connections meeting the limit, or the percentage of time that the limit is met. Tables 4.5 to 4.7 provide a collection of representative performance limits for parameters most relevant to voiceband data performance [28–36]. These and other objectives are noted in relation to the survey data throughout this chapter.

4.3.3 Characterized Connections

Owing to the large number of different possible connections arising from different routing alternatives, distances, and technologies, etc., attention is focused here on

Table 4.5
Summary of End-to-End Network Performance Objectives for Continuous Analog Voiceband Transmission Performance Parameters

Parameter	Units	Condition	Public Network		Basic and C-Conditioned Private Lines	
			[29]	[35]	[10, 36]	[28-30]
1 kHz Loss	dB	Range Nominal	≤ 30		≤ 28	12 - 20 16
1 kHz Loss Deviation	dB	Long Term Short Term Installation				±4 ±3 ±1
Attenuation Distortion relative to 1 kHz	dB	0.4 to 2.8 kHz	≤14	≤ 14	≤ 14	*
C-Message Noise	dBrnC					28 - 44 !
C-Notched Noise	dBrnC					≤ 48
Sig./C-Notched Noise (SNR)	dB		≥ 24	≥ 24	≥ 28	≥ 24 a
Single Freq. Interference	dB					≥ 3 d
Talker Echo Path Loss	dB				≥ 15	
Listener Echo Path Loss	dB					≥12
Env. Delay Dist. (EDD) relative to 1.8 kHz	ms	0.6 to 2.8 kHz 1.0 to 2.4 kHz 1.0 to 2.2 kHz	≤ 2.6 ≤ 0.8	≤ 2.6 ≤ 0.6	≤ 2.6 ≤ 0.8	*
Peak/Average Ratio (P/AR)	—		≥ 48	≥ 50#		
InterMod. Dist. Ratio (IMD)	dB	2nd harmonic 3rd harmonic	≥ 27 ≥ 32	≥ 27 ≥ 32	≥ 37 ≥ 36	≥ 27 b ≥ 32 c
Phase Jitter	° p-p	20 - 300 Hz 4 - 300 Hz 4 - 20 Hz	≤ 10 ≤ 15 ≤ 5	≤ 10 ≤ 15	≤ 10 ≤ 15	≤ 2 - 10 @ ≤ 7 - 15 @
Frequency Shift	Hz		± 5	± 3	± 2	± 3, ± 5

* See Table 4.6 for detailed objectives.
\# 70 for 4.8 kb/s voiceband data.
@ Increases in steps of 2° p-p for facility lengths of 0-250, 250-500, 500-1000, 1000-2000 and 2000-4000 miles.
! 28, 31, 34, 38, 40, 42, 44 dBrnC for facility lengths of 0-50, 50-100, 100-400, 400-1000, 1000-1500, 1500-2500 and 2500-4000 miles.
a ≥ 28 dB for D-conditioned private lines.
b ≥ 35 dB for D-conditioned private lines.
c ≥ 40 dB for D-conditioned private lines.
d ≥ 3 dB below C-Message Noise.

Table 4.6
Detailed End-to-End Network Performance Limits for Attenuation Distortion, and Envelope Delay Distortion Parameters

Parameter	Frequency Range (kHz)	Limit					
		Public Network	Private Leased/Dedicated Lines: Conditioning				
			Basic	C1	C2	C4	C5
Attenuation Distortion (dB) relative to 1 kHz	1 - 2.4 .5 - 2.5 .5 - 2.8 .5 - 3.0 .4 - 2.8 .3 - 2.7 .3 - 3.0 .3 - 3.2	≤ 14	- 2 to + 8 - 2 to + 10 - 3 to + 12	- 1 to + 3 - 2 to + 6 - 3 to + 12	-1 to +3 -2 to +6	-2 to +3 -2 to +6	-0.5 to +1.5 -1 to +3
Envelope Delay Distortion (ms) relative to 1.8 kHz	1 - 2.4 1 - 2.6 .8 - 2.6 .8 - 2.8 .6 - 2.6 .6 - 2.8 .6 - 3.0 .5 - 2.8 .5 - 3.0	≤ 0.8 ≤ 2.6	≤ 1.75	≤ 1 ≤ 1.75	≤ .5 ≤ 1.5 ≤3.0	≤ 0.3 ≤0.5 ≤ 1.5 ≤ 3.0	≤ 0.1 ≤ 0.3 ≤ 0.6

the three main types of representative connections itemized below. Note that although described in postdivestiture terms, these connections are also largely representative of predivestiture connections and North American connections in general.

(1) Intra-LATA (or metropolitan), intra-office connections, including those with distributed digital switching (e.g., remote switching systems) and with nominally 0 dB inserted loss; see Figure 4.1(a). The relevant survey data is the EAS [17, 18], with appropriate adjustments for loss (0 rather than 3 dB), and technology (T-carrier rather than a technology mix).

(2) Intra-LATA (or metropolitan), inter-office connections with direct routing and nominal 3-dB inserted loss; see Figure 4.1(b), and also the direct trunking case in Figure 4.3. The relevant survey data is the EAS [17, 18].

(3) Inter-LATA (or toll), short, medium, and long connections with nominal 6-dB inserted loss; see Figures 4.2 and 4.4. The relevant survey data is the EOS [1]. As noted earlier, this data applies directly to the case of direct trunking in the exchange access portions of Figure 4.4. However, owing to the "equal access" concept (i.e., with Feature Group D access), it is also largely applicable

Table 4.7
Summary of End-to-End Network Performance Limits for Transient Analog Voiceband Data Transmission Performance Parameters

Parameter	Condition	Limit @
Gain Hits	≥ 3 dB	≤ 8 *
Phase Hits	≥ 20°	≤ 8 *
Dropouts	≥ 12 dB	≤ 2 *
Impulse Noise	#	≤ 15 *
BER		$\leq 10^{-5}$
BLER	1000 bit block	$\leq 10^{-2}$

@ For public networks and private lines.
* Event counts are in 15 minute intervals, except for dropouts for switched data service, for which the interval is 1 hour.
\# For the limit count, the noise threshold is 6 dB below the received signal. Counts are 9 and 5 for thresholds 2 dB below, and 2 dB above, the received signal, respectively.

to the access tandem route for all parameters, except delay, which is addressed separately. The short distance category (0 to 180 miles) for toll connections is also indicative of performance for intra-LATA connections with tandem or inter-tandem routing (and distances ≤200 miles): compare Figure 4.3 with tandem or inter-tandem routing to Figure 4.4 with direct routing in the exchange access portions and one or two trunks in the IEC portion.

4.3.4 End-to-End Network Performance Characterization

4.3.4.1 Loss and Related Parameters

The end-to-end performance of these parameters (which were introduced in Ch. 3) is provided in Table 4.8 in terms of surveyed performance and inferred performance objectives. The distinction between acoustic and electrical loss in Table 4.8 is as follows.

Acoustic Loss

Acoustic loss is an end-to-end voice-oriented loudness measure which includes the sensitivity of the telephone sets, as well as the variation in this sensitivity, as discussed

Table 4.8
End-to-End Network Performance for Loss and Related Parameters

Loss Parameter			Type of Connection					
Type	Description/ Condition		Value	Intra-Office	Inter-Office	Toll @		
						Short	Medium	Long	Overall
Acoustic	End-to-End Loss		Mean (dB)	3.7	6.7	10.1	10.9	11.4	10.3
			SD (dB)	4.3	4.3	4.5	4.4	4.6	4.5
			Objective (dB) *	≤ 12.0	≤ 15.0	≤ 18			
			% ≤ Objective	97.5	97.5	96.0	94.5	92.5	95.5
	Loss Asymmetry		Mean (dB)	0.0	0.0	0.0	0.0	0.0	0.0
			SD (dB)	3.7	3.7	4.1	3.9	4.2	4.2
			Guideline (dB)	≤ 4					
			% ≤ Guideline	72.0	72.0	67.0	69.5	66.0	66.0
Electric	End-to-End Loss		Mean (dB)	7.5	10.5	13.9	14.7	15.2	14.1
			SD (dB) #	4.1	4.1	4.3	4.2	4.4	4.2
			Objective (dB) *	≤17.0	≤ 20.0	≤ 23			
			% ≤ Objective	99.0	99.0	98.5	97.5	96.0	98.0
	Attenuation Distortion	0.4 kHz	Mean (dB)	-1.0	-0.9	-0.3	-0.1	-0.2	-0.2
			SD (dB)	2.5	2.5	2.5	2.5	2.7	2.5
		2.8 kHz	Mean (dB)	7.0	7.2	8.2	7.6	7.6	8.0
			SD (dB)	5.0	5.1	5.4	5.2	5.1	5.3
		0.4, 2.8 kHz	Objective (dB)	≤14					
			% ≤ Objective	92.0	91.0	86.0	89.0	89.5	87.0

@ From Table 4.3, short, medium and long connection probabilities in % are: 77.8, 13.4 and 8.8, respectively
* Inferred objective: see text
# Also indicative of end-to-end loss deviation (mean deviation = 0 dB)

in Chapter 6 on the performance of individual network components. In Table 4.8, the common 500 type telephone set, or equivalent, is assumed [37–39], and it is evident that a large percentage of connections meet the inferred performance objectives. The performance objectives for end-to-end loss are inferred from maximum loop loss objectives, together with the nominal inserted connection losses of 0, 3, and 6 dB on intra-office, inter-office, and toll connections, respectively. For electrical loss, the loop limits are 8.5 dB each, while for acoustic loss the limits are TOLR = -38 dB and ROLR = 50 dB [9]. Somewhat more relaxed limits are reflected in [7].

Note that the contribution to loss by the loop plant can vary considerably depending on the type of plant. Table 4.8 is derived from general loop plant loss statistics [16], and is considered to be representative in this regard [1]. However, it is possible for somewhat longer (more lossy) [40], as well as shorter (less lossy), loop populations to be encountered. Longer loops are encountered in rural applications, but treatment of the loops with gain is then used to prevent excessive loss [41]. Shorter loops can apply particularly with distributed switching arrangements (e.g., remote switching systems) where the feeder portion of the loop plant is a digital carrier, and the distribution portion may consist of loops shorter than those of the general loop plant (see Fig. 4.1(a)). Eventually, with end-to-end all-digital telephony (e.g., ISDN), analog loops will disappear, and an essentially fixed "loop" loss will be manifested in the sensitivity of the voice terminal [6–8, 42, 43]. Evolution to this situation is discussed in Chapter 7.

Loss Asymmetry

Loss asymmetry is a relevant aspect of end-to-end acoustic loss for speech service, and is the difference in loss in the two directions of transmission as experienced during a conversation. It is seen in Table 4.8 that only about 65% to 70% of connections meet the 4-dB guideline; but, as discussed in Chapter 3, loss asymmetry is not a critical parameter.

Electrical Loss

Electrical loss is a single-frequency (1 kHz) measure from the output of the transmitting terminal (e.g., voiceband data modem) to the input of the receiving terminal; this includes the backbone network and loop plant losses up to the NIs, as shown in Figures 4.1 to 4.4. This loss by itself is not a sensitive parameter for voiceband data, because most modems use automatic gain control (AGC) to compensate for loss variations on connections. This explains the relatively nonstringent 30-dB loss limit given in Table 4.5. However, excessive losses can contribute to signal-to-noise ratio (SNR) degradation, which is important and is discussed later. In any case, most connections meet the inferred performance objectives for electrical loss in Table 4.8. Note that the inferred objectives are more stringent than the ≤ 30-dB limit given in Table 4.5. The ≤ 30-dB limit is readily met by virtually all connections; the reason is that less loss is required for satisfactory support of voice service.

Loss Deviation

Loss deviation is relative to a fixed electrical loss—such as mean loss—at 1 kHz, and is similar to the variation in electrical loss. There are no end-to-end loss deviation objectives for public networks, although it can be seen in Table 4.5 that ±4 dB is used for private lines; in Table 4.8 only about 65% of public network connections meet this objective, owing to less control of loss in the loop plant of public networks.

Attenuation Distortion

Attenuation distortion is the difference between the electrical losses at 0.4 kHz and 1 kHz, and between the electrical losses at 2.8 kHz and 1 kHz, and is a parameter most relevant to voiceband data. (The general mechanisms whereby voiceband data performance is degraded are discussed later in this chapter under distortion and incidental modulation parameters in Sec. 4.3.4.5.) Fairly large percentages of connections meet the attenuation distortion objective (see Table 4.8). The related parameter for voice is bandwidth and slope, as discussed in Chapter 3.

4.3.4.2 Noise and Related Parameters

Certain of these parameters were also introduced in Chapter 3.

Idle Circuit Noise

Idle circuit noise (or C-message noise) performance is shown in Table 4.9. This noise accounts for loop loss, but does not include the metallic component of loop noise. As discussed in Chapters 5 and 6, loop noise tends to have a very low mean and high SD, and therefore does not readily lend itself to accumulation with backbone network noise using analytical techniques; simulation techniques are required as discussed in Chapter 5. In any case, except for high noise values in the tails of loop metallic noise distributions, loop noise is negligible compared to network noise [44]. Idle circuit noise is shown referred to the input of a reference receiving system, which is appropriate for voice grade-of-service (GOS) assessment, as discussed in Chapter 3; Chapter 5 shows how the noise referral is done. Idle circuit noise is also shown at the NI as appropriate for voiceband data-like services.

In Table 4.9, the NI-to-NI performance objectives for idle circuit noise are shown for short-, medium-, and long-toll connections [13]. Using the analytical techniques described in Chapter 5, similar objectives are derived for intra- and interoffice connections using switch [19] and short-haul trunk [9] noise objectives, respectively, together with the trunk and loop losses provided in Chapter 6. The objectives for a reference receiving system were determined in a similar manner. Generally, it is seen that the surveyed performance aligns favorably with the objectives.

Three-kiloHertz Flat Noise

Three-kiloHertz flat noise also applies at the NI, and, owing to the flat weighting, includes low-frequency noise, for example, 60-Hz induction and related harmonics. The 3-kHz flat noise measured balanced to ground is generally lower than 3-kHz flat

Table 4.9
Network Performance for End-to-End Idle Circuit Noise
(Loop Noise not Included)

Noise Parameter			Type of Connection					
Name	Condition	Value	Intra-Office	Inter-Office	Toll @			
					Short	Medium	Long	Overall
C-Message Noise (dBrnC)	At Reference ROLR	Mean	18.6	16.1	23.9	27.8	30.7	25.0
		SD	5.5	6.1	5.3	4.6	4.3	5.6
		Obj.:* Mean	19.3	24.3	22.0	26.0	30.0	23.2
		SD	2.2	4.4	5.6	4.5	3.4	5.9
	At Network Interface	Mean	15.6	13.1	20.9	24.8	27.7	22.0
		SD	5.9	6.4	5.7	5.1	4.8	6.0
		Obj.: Mean	16.3*	21.3*	19.0	23.0	27.0	20.2
		SD	3.1*	4.9*	6.0	5.0	4.0	6.2
3 kHz Flat Noise (dBrn)	Balanced measurement	Mean	32.6	30.0	40.2	40.3	41.7	40.4
		SD	15.7	15.3	8.4	7.7	7.4	8.2
	To ground measurement	Mean	54.7#	52.0#	48.3	45.9	47.1	47.9
		SD	8.6#	8.9#	10.3	7.7	7.1	9.8

@ From Table 4.3, short, medium and long connection probabilities in % are: 77.8, 13.4 and 8.8, respectively
* Derived objective (see text)
# Estimated from distributions in [17] using techniques described in Chapter 5.
Obj. Objective

noise measured directly to ground owing to the partial cancellation of noise from the longitudinal balancing process (see Table 4.9).

Signal-Dependent Noise and SNR

Signal-dependent noise and SNR performance is shown in Table 4.10 in a manner similar to Table 4.9. The objectives for signal-dependent noise (also called *C-notched noise*) are determined from the voiceband data 24-dB SNR objective in Table 4.6, the transmitted voiceband data signal level, and from connection losses (see Chs. 5 and 6). Essentially all connections meet these objectives. In fact, large percentages would also meet the more stringent 28-dB SNR objective for D-conditioned private lines, noted in Table 4.6.

Signal-to-Quantization Distortion Ratio

Signal-to-quantization distortion ratio (Q) performance is estimated in Table 4.10 by inferring from signal-to-C-notched noise ratio results that the equivalent of one

Table 4.10
Network Performance for End-to-End Signal-Dependent Noise and Signal-to-Noise Ratios
(Loop Noise not Included)

Parameter		Type of Connection				
Name	Value	Intra-Office	Inter-Office	Toll @		
				Short	Medium	Long	Overall
C-Notched Noise !	Mean (dBrnC)	35.5	29.0	31.8	32.7	33.2	32.0
	SD (dB)	4.0	8.8	4.6	3.9	3.9	4.5
	Objective (dBrnC) *	≤49.3	≤46.3	≤43.3			
	% ≤ Objective	99.975	97.5	98.7	99.3	99.0	98.75
3 kHz Flat Notched Noise !	Mean (dBrn)	42.7	38.0	41.5	41.3	42.3	41.5
	SD (dB)	8.1	10.1	6.3	6.0	6.1	6.3
Signal-to-C-Notched Noise Ratio !	Mean (dB)	38.1	41.8	34.9	33.3	32.3	34.4
	SD (dB)	1.7	8.4	2.6	2.6	2.3	2.7
	Objective (dB)	≥24					
	% ≤ Objective	100	98.75	100	99.98	99.985	99.995
Q #	Mean (dB)	35.0	35.0	30.4	27.7	25.8	29.6
	SD (dB)	1.7	1.7	1.2	1.0	0.9	1.9
	Obj. *: Mean (dB)			24.3 } ≤22.3 with 99% Probability			
	SD (dB)			0.8			

@ From Table 4.3, short, medium and long connection probabilities in % are: 77.8, 13.4 and 8.8, respectively
* Derived objectives; see text.
# Estimated; see text.
! Holding tone level = −13 dBm0

pulse code modulation (PCM) coder-decoder (codec) is encountered for intra- and inter-office connections, and that the equivalent of two, three, and four PCM codecs are encountered on short, medium, and long toll connections, respectively. This number of codecs is less than the maximum number of five allocated by CCITT Rec. G.113 for national connections [45]. As such, it is expected that an inferred performance objective for Q based on five tandemed PCM codecs is met by essentially all connections. The values of Q per codec are discussed in Chapters 3 and 6, and the degradation of Q with tandeming is discussed in Chapters 3 and 5.

4.3.4.3 Echo Path Loss Parameters

Listener Echo Path Loss

Listener echo path loss (LEPL) parameters are summarized in Table 4.11 in terms of weighted echo path loss (WEPL) and singing margin (SM) as estimated from dis-

Table 4.11
Network Listener Echo Path Loss Performance for Intra-Office Connections
(Effect of Loop Plant and Loop Balance Network)

Listener Echo Path Loss Parameter				Type of Loop Balance Network		
Name	Description/ Condition		Value (dB)	Series	Parallel	Double
Weighted Echo Path Loss (WEPL)	Loop Plant	General	Mean SD	19.8 3.4	24.5 3.7	27.0 3.7
		Rural	Mean SD	17.5 4.4	18.5 4.3	24.0 4.6
		Rural (no design violations)	Mean SD	— —	— —	27.6 4.1
	Objective		Mean SD	$14 + D_L$, for D_L = LEPD = 2 to 8 ms * 3 dB		
Singing Margin (SM)	General Loop Plant		Mean SD	11.2 1.7	16.6 3.4	21.4 3.7
	Objective	For Voice	Mean SD	$\left.\begin{array}{c}15\\3\end{array}\right\}$ ≥ 10 dB for 95% of Connections		
		For Voice-band Data	Limit	≥ 12 dB		
	Flat SM Guideline for voice-band data	≤ 2.4 kb/s	Limit	≥ 18 dB		
		≤ 4.8 kb/s	Limit	≥ 25 dB		

— No data available
* E.g., for a 4 ms LEPD, the mean WEPL objective is 18 dB.

tributions of these parameter values [46–49] (these parameters were defined earlier in Ch. 3). The results are seen to depend significantly on the nature of the loop plant and on the loop balance network, which forms part of the interface between the 4-wire trunking and the 2-wire loop. Table 4.11 shows that, in comparison to the general loop plant, the rural loop plant (which consists of longer loops, as well as loops with design violations) provides significantly lower LEPLs. As discussed in Chapter 3, Table 4.11 also shows that LEPL performance can be improved by use of a loop balance network, which is more sophisticated than the simple series network (900 Ω plus 2.16 μf). That is, the parallel network (1100 Ω in parallel with 0.03 μf) provides a somewhat higher LEPL for the general and rural loop populations. The double balance network consists of two parallel networks, one for nonloaded loops (800 Ω in parallel with 100 Ω plus 0.05 μf) and one for loaded loops (1650 Ω in parallel with 100 Ω plus 0.005 μf). Thus, owing to the application of a different balance network to loops segregated in this manner, still better LEPL performance can be achieved, as shown. The mean WEPL objective is a function of listener echo path delay (LEPD) [9, 13, 50, 51]; for a representative delay such as 4 ms, the mean WEPL objective is $\geq$ 18 dB with SD = 3 dB, which aligns favorably with most of the WEPL results in Table 4.11.

Also shown in Table 4.11, are results and objectives for the SM parameter. This parameter is no longer used to characterize voice performance (it has been replaced by WEPL). However, SM is still used for voiceband data performance characterization [28, 30]. It is seen that the 12-dB SM objective is met by a larger percentage of connections if a parallel or double balance network is used. (The objective is strictly a signal-to-echo ratio, but for high values this is similar to SM [52].) The 12-dB objective generally applies to (and depends on) nonuniform (or nonflat) echo path loss frequency responses. This means that the return loss in the center of the voiceband where the voiceband signal is located, is usually higher than at the band edges. Table 4.11 provides an indication of the SM required if the echo path frequency response is uniform (flat), or, alternatively, what the equivalent uniform echo path loss objective might be [53]. In this regard, since WEPL is an equivalent uniform measure of echo path loss, it may also be useful for voiceband data performance characterization.

Results similar to those in Table 4.11 can also be determined for inter-office and toll connections by accounting for the nominal 3 and 6 dB of inserted loss, respectively, in these connections. However, the resulting LEPL on such connections is high enough to render listener echo a negligible impairment. On the other hand, talker echo can be a significant impairment on all types of connections.

Talker Echo Path Loss

TEPL results are summarized in Table 4.12 for intra- and inter-office connections. These results are derived from far-end loop balance return loss [48], plus switch or trunk loss (for the intra-office or inter-office cases, respectively) [17], plus near-end loop or loop plus set loss (for electrical [16] or acoustic [54] loss, respectively); see Chapter 5. Again, the results generally improve with more sophisticated balance networks. In a manner similar to that for the end-to-end loss parameter described earlier, both acoustic and electric TEPLs are shown and are relevant for voice and non-voice services, respectively. Numerically, there is an average of nearly 4 dB less loss for the acoustic connections owing to the net effect of the 500 type telephone set sensitivity in conjunction with the loop plant loss. Notice that TEPL for inter-office connections is about 6 dB higher than for intra-office connections owing to the nominal 3 dB of inserted loss on inter-office connections, which contributes an additional 6 dB to the TEPL. Finally, the variations in TEPL in Table 4.12 are contributed to mostly by variation in loop loss and telephone set sensitivities, and the fact that there is considerable directional dependence for loop plus telephone set loss (the effect of

Table 4.12
Network Talker Echo Path Loss Performance for Intra- and Inter-Office Connections
(General Loop Plant, Effect of Loop Balance Network)

Type of Connection	Parameter		Type of Loop Balance Network		
	Type	Value (dB)	Series	Parallel	Double
Intra-Office	Acoustic	Mean SD	15.1 6.4	18.3 6.5	18.3 6.1
	Electric	Mean SD	18.9 6.5	22.1 6.7	22.1 6.2
Inter-Office	Acoustic	Mean SD	21.1 6.4	24.3 6.6	24.3 6.1
	Electric	Mean SD	24.9 6.5	28.1 6.7	28.1 6.2

this dependence and how it increases TEPL variations is illustrated in detail in Ch. 5).

Table 4.13 shows TEPLs for short-, medium-, and long-toll connections [16, 55]. It is expected that, for the most part, the echo path loss survey results in [55] are indicative of performance for the general loop plant and for a single balance network.

Table 4.13
Network Talker Echo Path Loss Performance for Toll Connections
(General Loop Plant, Series Balance Network, Effect of Echo Control)

Echo Path Loss Parameter			Type of Toll Connection @			
Type	Condition of Echo Control	Value (dB)	Short *	Medium	Long	Overall
Acoustic	Disabled	Mean SD	25.4 8.2	27.2 8.3	27.6 8.4	25.8 8.3
	Enabled	Mean SD	25.4 8.2	27.3 8.3	43.3 15.5	27.2 10.4
Electric	Disabled	Mean SD	29.2 8.3	31.0 8.4	31.4 8.5	29.6 8.4
	Enabled	Mean SD	29.2 8.3	31.1 8.4	47.1 15.5	31.0 10.5
Estimated Percentage of Connections Encountering Echo Control			0.0	0.2	55.6	4.9

@ From Table 4.3, short, medium and long connection probabilities in % are: 77.8, 13.4 and 8.8, respectively
* Estimated Talker Echo Path Losses; see text.

(Note also that echo path loss survey results are not available for the short-distance category, but are estimated in Table 4.13 from a consideration of estimated return losses and short-distance trunk losses [55, 56]; details are provided in Sec. 6.2.3.4 of Ch. 6). The toll TEPLs are about 12 and 6 dB higher than the intra- and inter-office TEPLs, respectively, owing to the nominal 6 dB of inserted loss on toll connections. The variation in TEPL in Table 4.13 is large due both to the backbone network and to the loop plant and telephone sets. The core network portion of the variation is larger than would be expected from a straightforward accumulation of return loss and trunk loss variations, even when assuming directional dependencies of the trunk losses (see Ch. 5). The larger than expected TEPL variations are also evident from [57]. The reason for such large TEPL variation may be due to a variety of factors, including:

- The trunk losses contributing to the surveyed echo path losses are not 1-kHz losses, but are, rather, losses across the 500- to 2500-Hz portion of the voice frequency band corresponding to the echo return loss (ERL) filter bandwidth [55].
- The TEPLs include multiple echo paths in the connection, principally due to the main reflection at the class 5 end office, and possibly a secondary reflection at the class 4 toll office (both offices being at the far end of the connection).
- There may be some directional dependence in trunk loss variations (inadvertently introduced by trunk maintenance procedures).

Finally, the results in Table 4.13 show the effect of echo control on TEPL. (In general, echo control may have consisted of a mixture of echo suppressors and echo cancellers, although it is expected that mostly echo suppressors were encountered in the survey data in [55].) The survey indicates that echo control is essentially only encountered on somewhat more than half of the long-toll connections (it is expected that echo control would be encountered on the longest of the long connections), and that, overall, only about 5% of all connections encounter echo control. It should be noted that the overall distribution of TEPLs for toll connections with enabled echo control will tend to be distinctly nonnormal (bimodal); this is because those connections without echo control will tend to have TEPLs on the order indicated in Table 4.13 with echo control disabled, and those connections with echo control will have very high TEPLs (on the order of at least 55 dB).

4.3.4.4 Echo Path Delay Parameters

Table 4.14 provides echo path delay information for intra- and inter-office connections, while Table 4.15 provides the same for toll connections. The results within the tables are for the trunking in the backbone network, plus the 4-wire portions of the access networks. These values are applicable to listener echo paths, and also approximately to talker echo paths. The relatively low delay of 2-wire loops only applies to

Table 4.14
Network Echo Path Delay Performance for Intra- and Inter-Office Connections
(Effect of Type of Local Switching)

Type of Connection	Echo Path Delay Parameter Value (ms)	Type of Local Switching		
		Analog	Digital	
			Non-Distributed	Distributed
Intra-Office [19]	Mean *	~ 0	1.9	3.3
	SD	~ 0	0.21	0.45
	Objective Mean	—	1.9	5.5
	SD	—	0.21	0.43
	99%	—	≤ 2.4	≤ 6.5
Inter-Office [9, 19] #	Mean *	1.6	3.2	4.6
	SD	0.66	0.29	0.46
	Objective Limit	≤ 10	≤ 10	≤ 14

* For talker echo path delay, add 0.4 ms/mile of 2-wire subscriber loop at the talker end of the connection (see text).
\# See also Section 6.2.3.5 of Chapter 6.

Table 4.15
Network Talker Echo Path Delay Performance for Toll Connections

Talker Echo Path Delay Parameter Value (ms)	Type of Toll Connection @			
	Short	Medium	Long	Overall
Mean *	9.6	16.5	32.6	12.5
SD	4.2	4.5	9.0	8.2
Objective Limit	≤ 45 ms with no echo control			
Estimated % of Connections with Delay ≥ 45 ms	0	0.05	6.0	0.53
Estimated % of Connections with Echo Control	0	0.2	55.6	4.9

@ From Table 4.3, short, medium and long connection probabilities in % are: 77.8, 13.4 and 8.8, respectively
* As noted in the text:
for talker end of the connection add - 0.4 ms/mile of 2-wire subscriber loop
for each end of the connection add - 0.7 ms for remote switching and 16 µs/mile of access tandem to IEC POP trunking

the talker (not listener) echo impairment, and is considered separately as follows. The round-trip delay of 2-wire loops is included by adding this delay at the talker end of the connection to the TEPDs in Tables 4.14 and 4.15 to obtain better estimates of TEPD. The round-trip delay of 2-wire loops is about $2 \times 0.2 = 0.4$ ms/mile [58], and can vary from near zero to about 2 ms for loops with working lengths from near zero to 5 miles, respectively; an average loop of about 2 miles of working length would have a round-trip delay of about 0.8 ms.

The results of mean echo path delay Tables 4.14 and 4.15 may also be used to estimate mean end-to-end delay by dividing by a factor of 2. The SD of end-to-end delay may be obtained by dividing the echo path delay SD by 2 if there is total directional dependency of delay, or by $\sqrt{2}$ if there is directional independence. The actual end-to-end delay variation should be bounded by the above extremes. Situations where partial directional dependencies occur are addressed in Chapter 5.

For the intra- and inter-office connections in Table 4.14, no survey data exists for the delay parameters; thus, the delays in this table are based on specified delays for analog switching systems, digital switching systems including distributed switching [19], and on knowledge of delays for transmission systems (for details of the following, see Sec. 6.2.3.5 of Ch. 6). For transmission systems, both terminal and distance-dependent delays are estimated [59–63]. With digital switching, T-carrier technology is assumed for distributed switching applications and for inter-office trunking. With analog switching, a technology mix in accordance with the EAS is assumed for inter-office trunking. For this technology mix, the average distances considered (based on survey information [64, 65]) are about 21 miles, with a SD of about 14 miles. For T-carrier technology, the average distances considered is also about 21 miles, but with a SD of about 7 miles. The same T-carrier distance is used for distributed switching application (i.e., distance from central office to each remote switching unit). This may somewhat overestimate the distributed switching distances. However, these distances are appropriate as a conservative estimate, in that the maximum planning distance for inter-office connections is on the order of 150 to 200 miles with 3 or 4 trunks (50 miles per trunk) [9, 66], and the maximum design distance for remote switching is at least 50 miles [59].

In Table 4.14, the delay objectives for intra-office connections are based on listener echo considerations [50, 51]. The delay objectives for inter-office connections are based on an intra-LATA round-trip delay limit of 10 ms for the maximum length short-haul connection of 150 to 200 miles noted above [9]. The delay objectives for distributed switching are adjusted (based on the intra-office delays) to reflect an additional maximum expected delay of 4 ms. Note that the inter-office delays in Table 4.14 are generally consistent with those reflected in the recent addendum in [10].

Table 4.15 provides the echo path delay information for short-, medium-, and long-toll connections based on survey data [1]. These results include both analog and digital switching, but not remote switching. As can be seen in Table 4.14, the incremental delay due to remote switching at both ends of the connection can be accounted for by adding about 1.4 ms to the mean delays in Table 4.15; the SD would not change significantly. These results also do not include the effect of delay for exchange access via a tandem switch, where trunking from the tandem to the IEC POP is significant. Additional echo path delay due to this factor can be accounted for by adding about $2 \times 8 = 16$ μs per mile of the additional trunking in question. Thus, for example, if the trunk in question was 100 miles long for each exchange access network, the total incremental round-trip delay would be about 3.2 ms.

It is of interest to note from Table 4.15 that only about 6% of long connections (about 0.5% of all connections) are estimated to have echo path delays exceeding 45 ms, the value of echo path delay beyond which echo control is required by design [58]. However, it can be estimated from survey data that over 50% of long connections (about 5% of all connections) encounter echo control [55]. This suggests that echo control is effectively deployed on connections having less than 45 ms of echo path delay, and that the echo delay performance of the surveyed network can be expected to exceed that corresponding to the 45-ms design limit. The reason for the above situation is perhaps due to the practical aspects of providing echo control. Until recently, echo control has been provided on inter-toll trunks exceeding 1,850 miles in length [9, 13]. As noted in Chapter 3, this provisioning rule evidently results in echo control on connections with echo path delays exceeding 35 ms. This aspect of performance, including more recent echo control deployment guidelines, is discussed in more detail in Chapter 7 on transmission planning.

4.3.4.5 Distortion and Incidental Modulation Parameters

Tables 4.16 and 4.17 summarize results for a number of parameters that have a direct bearing on voiceband data performance. These parameters, as well as attenuation distortion discussed earlier in Section 4.6.1, impair voiceband data performance by reducing the noise margin at which errors are produced. This occurs as a result of the impairments partially closing the "eye" or "window" within which the voiceband data signal regenerator must sample the received signal to determine if the signal is currently above or below a threshold (i.e., to determine if a "1" or "0" has been transmitted; see Fig. 6.7 of Ch. 6). The sampling must be done at or near the correct time, and the signal amplitude must be sufficiently high and free of noise, distortion, and interference to enable a correct decision to be made. Generally, the distortion and jitter impairments all tend to obscure the sampling time and/or the signal amplitude in such a way that there is an increased probability of error. Thus, these impairments must be controlled to levels that enable voiceband data bit error ratio (BER) and block error ratio (BLER) objectives to be met, as discussed later in this chapter. These impairments are contributed to mainly by the backbone (trunk part) of connections where various carrier modulation schemes and associated voiceband channel filtering give rise to the degradations noted above. The following elaborates somewhat on

Table 4.16
Network Performance for Certain End-to-End Distortion-Related Parameters
(Contribution from Loops Assumed Negligible)

Parameter			Type of Connection				
Name	Condition	Value	Intra-Office	Inter-Office	Toll @			
					Short	Medium	Long	Overall
Envelope Delay Distortion	0.6 kHz	Mean (μsec)	222	279	853	1150	1337	935
		SD (μsec)	98	211	352	471	613	430
	2.8 kHz	Mean (μsec)	156	173	599	701	727	624
		SD (μsec)	53	121	230	275	272	245
	0.6, 2.8 kHz	Objective (μsec)	*	*	≤ 2600			
		% ≤ obj.	100	100	100	99.9	98.0	~100
Peak to Average Ratio (P/AR)		Mean	97.0	95.9	87.5	85.4	84.2	86.8
		SD	1.2	3.1	6.2	7.6	9.0	6.8
		Objective	*	*	≥ 48			
		% ≥ obj.	100	100	100	100	100	100
Signal to Intermodulation Distortion Ratio	2nd Harmonic	Mean (dB)	58.1	60.5	53.3	52.2	51.6	53.0
		SD (dB)	5.9	8.8	6.6	6.2	6.1	6.5
		Objective (dB)	*	*	≥ 27			
		% ≥ obj.	100	100	100	100	100	100
	3rd Harmonic	Mean (dB)	60.0	61.4	53.4	51.0	50.2	52.8
		SD (dB)	5.1	7.7	6.9	4.7	6.9	6.7
		Objective (dB)	*	*	≥ 32			
		% ≥ obj.	100	100	99.9	100	99.6	~99.9

@ From Table 4.3, short, medium and long connection probabilities in % are: 77.8, 13.4 and 8.8, respectively
* More stringent objectives than those indicated for toll connections could conceivably be applied for all connections, especially for intra- and inter-office connections (see text).

each of the parameters noted in Tables 4.16 and 4.17. The survey results for these parameters are based on the EOS and EAS [1, 17]. A general discussion of the parameters can be found in [34, 45].

Envelope Delay Distortion

Envelope delay distortion (EDD) is the difference between the envelope delay at 0.6 kHz and 1.7 kHz, and the difference between the envelope delay at 2.8 kHz and 1.7 kHz, as shown in Table 4.16. Envelope delay is the slope of the channel phase characteristic versus frequency, and 1.7 kHz is a reference frequency at or near the frequency for minimum envelope delay. This reference is usually between 1.6 and 1.8 kHz; 1.7 kHz applies to the EDD results in Table 4.16, except for the objective,

Table 4.17
Network Performance for End-to-End Incidental Modulation Type Parameters
(Contribution from Loops Assumed Negligible)

Parameter			Connection					
Name	Condition	Value	Intra-Office	Inter-Office	Toll @			
					Short	Medium	Long	Overall
Amplitude Jitter	In 2-300 Hz Band	Mean (%) SD (%)	2.0 0.7	1.7 1.1	2.8 1.4	3.2 1.4	3.6 1.6	2.9 1.4
	In 20-300 Hz Band	Mean (%) SD (%)	2.0 0.7	1.7 1.0	2.6 1.1	2.8 1.0	3.2 1.2	2.7 1.1
		Guideline (%) % ≤ Guideline	10 ~100					
Phase Jitter	In 2-300 Hz Band	Mean (° p-p) # SD (° p-p)	1.6 0.6	1.8 1.5	5.5 3.7	6.4 3.5	7.6 3.8	5.8 3.7
		Objective (° p-p) % ≤ obj.	≤ 15					
			100	100	99.5	99.3	97.5	99.35
	In 20-300 Hz Band	Mean (° p-p) SD (° p-p)	1.4 0.6	1.3 1.0	3.1 1.4	3.9 1.7	4.9 2.1	3.3 1.6
		Objective (° p-p) * % ≤ obj.	≤ 10					
			100	100	100	99.98	99.25	~100
Frequency Shift	N/A	% ≥ 2 Hz % ≥ 1 Hz	N/A	N/A	0.19 0.24			
		Objective (Hz) % ≤ obj.			± 5 100			

@ From Table 4.3, short, medium and long connection probabilities in % are: 77.8, 13.4 and 8.8, respectively
N/A Not Applicable
* More stringent objectives could conceivably be applied, especially for intra- and inter-office connections (see text).
\# Strictly, the objective applies to the 4-300 Hz band.

which relates to 1.8 kHz. Essentially all connections meet the 2600-μs EDD objective for public networks given in Table 4.5. Thus, a considerable percentage probably also meets the more stringent objectives for private networks in Table 4.6.

Peak-to-Average Ratio

Peak-to-average ratio (P/AR) is a relatively simple measurement technique used to indicate the degree of EDD as well as attenuation distortion [67]. A transmitter generates an approximate representation of a voiceband data signal with known P/AR. The channel distortion will modify the P/AR in a way that can be related to intersymbol interference, which tends to "close the eye" of voiceband data signals. A P/AR value of <100 implies signal degradation. Table 4.16 indicates that essentially all connections meet the P/AR objective of ≥ 48. This is also true of more stringent objectives such as P/AR ≥ 70 for private network voiceband data service at 4.8 kb/s (see Table 4.5). Also, based on expected switch performance [19], more stringent objectives could conceivably be applied to intra- and inter-office connections.

Intemodulation Distortion

Intermodulation distortion (IMD) is a measure of channel nonlinearities as determined by the four-tone measurement method. Table 4.16 shows the results expressed as a signal-to-intermodulation distortion ratio in decibels. Generally, there is little dependence on distance, and nearly all connections meet the objectives indicated. In fact, large percentages of connections would also meet tighter objectives for D-conditioned private lines (Table 4.5). Most intra- and inter-office connections would also meet more stringent objectives based on expected switch performance [19].

Phase Jitter

Phase jitter is the deviation of zero crossings of a 1-kHz tone from their nominal or steady state position in time, and is caused by incidental modulation of the signal by noise and other unwanted or interfering signals [68]. Phase jitter is measured by using a phase-locked loop to compare the instantaneous and average phase of the received signal. The jitter is usually measured in both a "narrow" (20 to 300 Hz) and "wide" band (2 or 4 to 300 Hz). Table 4.17 shows that for toll connections there is some dependence of jitter on distance, and that large percentages of connections meet the phase jitter objectives.

Amplitude Jitter

Amplitude jitter is the deviation of the peak of a 1-kHz tone from its nominal value, and increases somewhat with distance. Amplitude jitter is caused by factors similar to those that produce phase jitter. There appear to be no objectives for amplitude jitter, although a guideline corresponding to $<10\%$ amplitude modulation may be inferred from [34]. Essentially all connections readily satisfy this value, as shown in Table 4.17. For this and other reasons [35], this parameter is not critical for voiceband services.

Frequency Shift

Frequency shift (also called *frequency offset* or *frequency translation*) is due to oscillator drift which, in certain analog carrier systems, results in the shifting of all frequency components of the voiceband signal. This parameter can affect the perception of speech signals and data performance where narrowband filters are involved. Performance objectives have accordingly been in the range of ± 2 Hz to ± 5 Hz. However, in more recent years this parameter has also required control so as not to impair the performance of echo cancellers, which can only track frequency shifts of about ± 1 Hz. On the other hand, as analog carrier technology is replaced by digital carrier systems in a frequency-synchronized network, the frequency shift impairment will be less prevalent. Table 4.17 indicates relatively few incidents of significant frequency shifts.

4.3.4.6 Analog Transient Parameters

These parameters were introduced in Chapter 3 and include gain and phase hits, dropouts, and impulse noise. The resulting impairments mainly affect voiceband data services in that any transient event of this type is likely to cause one or more bit errors or a block error. The process by which bit errors are produced is similar to that described for distortion and jitter in the previous section, except that disturbances to the sampling time and signal amplitude are of a momentary rather than continuous nature. Nevertheless, the degradation is likely to be sufficient to cause errors during these events.

Network performance for gain and phase hits and dropouts is summarized in Table 4.18 and for impulse noise in Table 4.19. In most cases, observed (surveyed) performance for the count of these events was over a 5-min period, whereas objectives are specified over a 15-min period. Thus, the observed 5-min results are assumed to be stationary over 15 min and the counts are multiplied by three to enable comparison with objectives. Also, in all cases the count is approximately lognormally distributed. As discussed in Chapter 5, this means that the distribution of event counts is normal in the logarithmic domain, and log-normal in the inverse logarithmic domain. Therefore, the results in Tables 4.18 and 4.19 are based on more or less worst-case normal approximations of surveyed performance as presented in the log domain [1, 17].

Note that in accordance with the treatment of log-normally distributed parameters in Chapter 5, all percentile values (e.g., median = 50% value) on the normal distribution in the log domain, can be determined in the inverse log domain by simply taking the inverse logarithm. Thus, the mean (or median) and SD of the normal distribution can be used in the usual way to estimate percentile values in the log domain, which are then readily convertible to the inverse log domain. For this reason, the results in Tables 4.15 and 4.16 are expressed in terms of the median number of events counted in 15 min, the SD (in decades in the log domain), and the percentage of connections for which the counts exceed a value corresponding to the end-to-end performance objective.

Table 4.18
Network Performance for Analog Transients
(Contributions from Loops Assumed Negligible)

Parameter			Type of Connection	
Name	Threshold Condition	Value * (Count in 15 minutes)	Inter-Office	Toll
Gain Hits **	3 dB	Median SD (decades) % ≤ 8 [a]	— — 99.96 [#]	7.1×10^{-3} 1.2 99.4
Phase Hits **	20°	Median SD (decades) % ≤ 8 [a]	— — 99.95 [@]	7.1×10^{-3} 1.6 97.5
Dropouts	12 dB	Median SD (decades) % ≤ 2 [b]	5.4×10^{-3} 1.4 96.7	1.1×10^{-1} 0.91 91.7

* Note, the medians and SDs are not necessarily accurate in themselves, but can be used to estimate higher percentile values.
— Cannot be estimated from available data.
** No gain or phase hits observed on inter-office metallic cable facilities.
\# Dominated by radio performance.
@ Dominated by N-carrier performance.
a The objective for gain and phase hits is a count of ≤ 8 hits in 15 min. for each type of connection.
b The objective for dropouts is a count of ≤ 2 dropouts in 15 min. for each type of connection.

Table 4.19
Network Performance for Impulse Noise (Noise Threshold 6 dB Below Received Signal, Effect of Type of Local Switch, Contribution from Loops Not Included)

Impulse Noise Count in 15 Minutes	Type of Connection				
	Inter-Office	Toll			
		Type of Local Switch and Percentage Encountered			
		Analog		Digital	Overall
		Step-by-Step (12%)	Cross-Bar (30%)	(58%)	(100%)
Median	0.62	8.7	6.0	0.75	1.9
SD (decades)	0.95	0.69	0.81	0.89	0.97
Percentage of Connections with ≤ 15 Counts *	92.8	63.0	68.0	92.5	82.3

* The requirement is that 50% of connections have no more than 15 impulses in 15 minutes.

Note that median values of less than 1 count in 15 min, as well as the associated SDs, are not necessarily accurate. That is, the median and SD are based on extrapolations of the normal approximation, and the median rate is only meaningful to the extent that the event process is stationary for longer periods of time. For example, a median value of 0.1 counts in 15 min implies a long-term stationary process, with a median of 1 count in 15/0.1 = 150 min or 2.5 hours. The event processes may not be stationary over such long periods. Nevertheless, the median and SD can be used to estimate higher percentile values of more than 1 count in 15 min; using the techniques provided in Chapter 5, the general expression for the count (C) at some percentile (p) is C_p given by

$$C_p = \text{inverse log } [\log(C_{50}) + z \times \text{SD in decades}] \qquad (4.1)$$

For example, if the median rate for a parameter is $C_{50} = 0.1$ counts in 15 min with SD of 1 decade, then the 95% rate is

$$C_{95} = \text{inverse log } [\log(0.1) + 1.645 \times 1] = 4.4 \text{ counts in 15 min}$$

where $z = 1.645$ is the relative deviation on the normal distribution at the 95% level; see Table 5.10 of Chapter 5.

The results in Tables 4.18 and 4.19 do not include contributions from the loop plant, whether metallic, carrier, or both; hence, no results are presented for intra-office connections. In some instances (i.e., for gain and phase hits) the events are specific to carrier systems, and there should be few if any such events on metallic loops. Regarding the carrier in the loop plant, it could be inferred that impulse noise performance is similar to that of inter-office connections using the same type of carrier (i.e., T-carrier). However, cable engineering is not as stringent in the loop plant as in the trunk network; thus, there is more susceptibility to interfering signals in the loop plant, which may lead to different (probably lower) performance [69].

In general, fairly high percentages of connections meet the phase and gain hit and dropout objectives in Table 4.18. Regarding impulse noise in Table 4.19, the results are a function of switching technology rather than distance. Owing to the trend toward digital switching, the results indicate that impulse noise performance can be expected to improve. In any case, current impulse noise requirements are met by a comfortable margin; that is, more than 50% of connections have $\leq$ 15 counts in 15 min.

4.3.4.7 Error Performance Parameters

Error performance for voiceband data connections is given in Table 4.20. Again, observed (surveyed) results tend to be approximately log-normally distributed. Thus, the representation of the error performance information in Table 4.20 is similar to

Table 4.20
Network Performance for Voiceband Data BER and BLER
(1000-Bit Blocks, Artificial Loops)

Voiceband Data Rate (kb/s)	Parameter		Type of Toll Connection @			
	Name	Value *	Short	Medium	Long	Overall
1.2	BER	Median	1.2×10^{-8}	4.8×10^{-8}	3.7×10^{-7}	2.0×10^{-8}
		SD (in decades)	2.7	2.0	1.8	2.6
		% ≤ 10^{-5} #	86.0	87.5	78.0	85.0
	BLER	Median	1.6×10^{-7}	5.1×10^{-5}	4.4×10^{-4}	7.0×10^{-7}
		SD (in decades)	3.6	1.5	1.2	3.5
		% ≤ 10^{-2} !	91.0	94.0	87.5	88.2
4.8	BER	Median	1.5×10^{-7}	1.2×10^{-7}	2.6×10^{-7}	1.5×10^{-7}
		SD (in decades)	1.5	1.6	1.6	1.5
		% ≤ 10^{-5} #	88.8	88.5	84.0	88.8
	BLER	Median	1.1×10^{-4}	3.0×10^{-4}	3.5×10^{-4}	2.7×10^{-4}
		SD (in decades)	0.92	0.73	0.87	0.89
		% ≤ 10^{-2} !	98.3	98.1	95.3	96.1

@ From Table 4.3, short, medium and long connection probabilities in % are: 77.8, 13.4 and 8.8, respectively
* Note, the medians and SDs are not necessarily accurate in themselves, but can be used to estimate higher percentile values (i.e., BERs > 10^{-6} and BLERs > 10^{-3}; see text).
# BER objective is ≤ 10^{-5} for 85% of connections
! BLER objective is ≤ 10^{-2} for 85% of connections

that described in the previous section for analog transients. Note again, however, that low median values of BER and BLER (i.e., much less than 10^{-6} and 10^{-3}, respectively, in Table 4.20) are based on extrapolations of the normal approximations; as such, these medians are not necessarily accurate, but can be used with the SD in the log domain to estimate higher percentile values (i.e., greater than 10^{-6} and 10^{-3} for BER and BLER, respectively). The general expression for a BER or BLER at some percentile (p) is similar to Equation (4.1); that is,

$$\text{BER}_p \text{ or } \text{BLER}_p = \text{inverse log } [\log(\text{BER}_{50} \text{ or } \text{BLER}_{50}) + z \times \text{SD in decades}] \quad (4.2)$$

For example, if a median BER is 10^{-7} with SD of 2 decades, then from Equation (4.2), the 85% BER is

$$\text{BER}_{85} = \text{inverse log } [\log(10^{-7}) + 1.036 \times 2] = 1.2 \times 10^{-5}$$

The voiceband data results in Table 4.20 are based on [1] for which two different modems were evaluated; the results are indicative of the lowest performance of the two modems. The results tend to indicate a general degradation in BER and BLER with distance, but generally BER and BLER performance objectives of 10^{-5} and 10^{-2}, respectively, are met by a higher than required percentage of connections (i.e., by more than 85% of connections [70]). The BLER objective assumes random errors and a data block size of 1000 bits.

An interesting by-product of BER and BLER measurements is that an indication of the randomness or burstiness of the error process can be determined (i.e., the number of bit errors per block error). This is given by

Burst Factor = Block Size × BER / BLER   (4.3)

Theoretically, the burst factor can vary from as low as one error for random errors, to as high as half the block size being in error for very bursty error conditions. The BLER results in Table 4.21 are for 1000-bit blocks. The burst factor is determined from the mean BER and mean BLER to give the mean number of errors per block using Equation (4.3). Mean BERs and BLERs are in turn obtained from the mean of log-normal distributions using the medians and SDs in Table 4.20 (see Eq. (5.73)

Table 4.21
Estimated Error Burst Factor Based on Voiceband Data BER and BLER Performance
(For Mean BERs and BLERs, and for 1000-Bit Blocks)

Voiceband Data Rate (kb/s)	Parameter	Type of Toll Connection @			
		Short	Medium	Long	Overall !
1.2	Mean BER	*	1.9×10^{-3}	2.0×10^{-3}	2.0×10^{-3}
	Mean BLER	*	2.0×10^{-2}	2.0×10^{-2}	2.0×10^{-2}
	Burst Factor #	*	95	100	100
4.8	Mean BER	5.8×10^{-5}	1.1×10^{-4}	2.3×10^{-4}	8.0×10^{-5}
	Mean BLER	1.0×10^{-3}	1.2×10^{-3}	2.6×10^{-3}	1.2×10^{-3}
	Burst Factor #	58	92	88	67

@ From Table 4.3, short, medium and long connection probabilities in % are: 77.8, 13.4 and 8.8, respectively
* Cannot be estimated (see text)
# Burst Factor = Block Size x BER/BLER, in bits
! Does not include short connections for 1.2 kb/s data rate

or App. A.5.2 of Ch. 5). That is,

$$\text{BER}_{mean} = \text{inv. log }[\log(\text{BER}_{med}) + (\text{BER}_{SD}\text{ in decades})^2(\ln 10)/2]$$
$$\text{BLER}_{mean} = \text{inv. log }[\log(\text{BLER}_{med}) + (\text{BLER}_{SD}\text{ in decades})^2(\ln 10)/2] \quad (4.4)$$

The results in Table 4.21 indicate that relative to the maximum of 500 errors per block, there is considerable burstiness, with about 60 to 100 errors per block. Note that in Table 4.21, the means cannot be estimated from the survey data for short connections and for the 1.2-kb/s data rate. This is because of the relatively large SDs, which, while providing meaningful percentile values over the range of interest, result in unmeaningful mean BER and BLER values larger than one. A truncated log-normal distribution would have to be used to overcome this problem. In fact, the means for the other conditions in Table 4.21 may for similar reasons also be too large, but since Equation (4.3) involves a ratio BER/BLER, the error should be mitigated.

4.4 DIGITAL NETWORK PERFORMANCE

4.4.1 CCITT

End-to-end digital network performance was discussed in some detail in Chapter 3, largely from the standpoint of CCITT and CCIR recommendations. Here, additional information is provided in terms of the end-to-end performance of international and national connections, as well as the allocation to major components of national networks. The relevant longest length of international hypothetical reference connections (HRX) is shown in Figure 4.5, and a representative HRX suitable for North American national networks is shown in Figure 4.6 (these HRXs are adapted from Fig. 3.21 of Ch. 3; further information on HRXs is also provided in Ch. 5). Note the local-, medium-, and high-quality grades to which various percentages of end-to-end performance are allocated. The end-to-end performance is specified in terms of the percentage of time performance is below a certain limit; the limits are determined by the definition of the following CCITT standardized performance parameters at the DS0 (64-kb/s) rate [71]:

- Errored seconds (ES): Seconds with ≥ 1 error.
- Severely errored seconds (SES): Seconds with BER $> 10^{-3}$.
- Degraded minutes (DM): Minutes with BER $> 10^{-6}$, excluding SESs.

The relative and absolute performance allocations for these parameters are shown in Table 4.22; also shown are the overall ratios of national to international performance allocations, some of which were referred to in Section 3.2.2.12 of Chapter 3. Similarly, performance allocations for the controlled slip parameter is provided in Table 4.23 (referred to in Sec. 3.2.2.13 of Ch. 3 [72]).

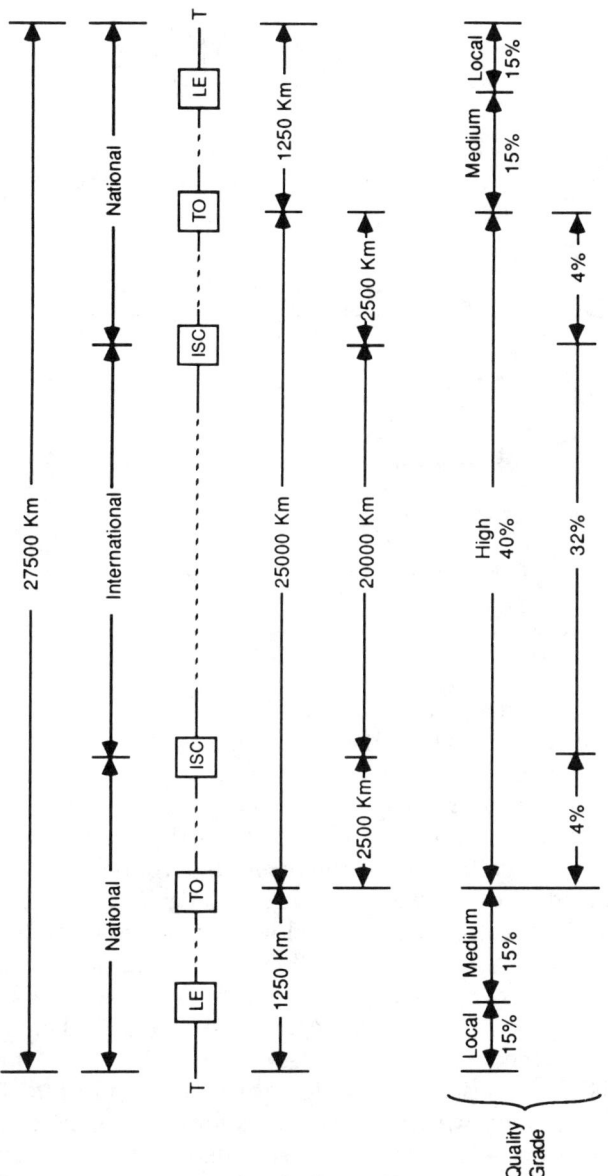

Figure 4.5 Near-worst international HRX for error performance.

Figure 4.6 Near-worst representative national HRX for error performance.

From Figures 4.5 and 4.6, and Tables 4.22 and 4.23, it is apparent that the national HRX is a subset of the international HRX. For error and slip performance, about 68% and 92%, respectively, is allocated to national networks. The reason for the relatively high allocation to national networks is owing to the generally lower performance capabilities of certain portions of national networks; i.e., the local and medium grades portions for error performance, and the local portion for slip performance. With continued introduction of more recent technology (e.g., digital cable and fiber carrier in the access network), the relatively lower performance of the local grade will become less applicable, as is evident in the performance objectives currently being developed for rates above 64 kb/s [73].

Notice that for the SES parameter in Table 4.22, there is a CCITT-allocated component that is transmission media-independent, as well as a transmission media-dependent component [71]; the latter is not allocated by CCITT and is intended to cover adverse conditions associated with radio transmission systems used in the medium- and high-grade portions of connections. To enable equitable use of this component of SES performance, a possible allocation for this component is included in Table 4.22.

Table 4.22
CCITT Error Performance Allocation for 64-kb/s Digital Connections

Type of Allocation	Parameter		Percentage of Time Allocation					
			Parts of Connections and Quality Grades				End-End Total	Total Nat./ International Ratio
	Name	Qualifier	Each National			International	@	!
			Local	Medium	High	High		
Relative	DM, ES, SES	Media Indep. #	15	15	4	32	100	68/32
	SES	Media Dep. *	0	17	17	32	100	68/32
		Total*	7.5	16	10.5	32	100	68/32
Absolute	DM &	-	1.5	1.5	0.4	3.2	10	6.8/3.2
	ES	-	1.2	1.2	.32	2.56	8	5.44/2.56
		Media Indep.	.015	.015	.004	.032	.1	.068/.032
	SES	Media Dep. *	0	.017	.017	.032	.1	.068/.032
		Total*	.015	.032	.021	.064	.2	.136/.064

\# In this Table, the SES parameter has a media independent, and a media dependent, allocation.
   Note that in CCITT Rec. G.821, the media dependent part is not allocated.
   Allocations for the other parameters are media independent.
* Possible allocation for media dependent portion of SES impairment,
   and resulting media independent plus media dependent total.
& The DM parameter may be removed from future issues of G.821 [113].
@ End-to-end performance of an international connection is contributed to by two national portions,
   and the international portion.
! End-to-end performance of a national connection is contributed to by two national portions,
   and is given by the numerators in the ratios shown.

4.4.2 North America

4.4.2.1 *Performance Objectives*

Table 4.24(a) summarizes specified performance levels for dedicated digital circuits operating at DS0 (56 or 64 kb/s), DS1 (1.544 Mb/s), and DS3 (44.736 Mb/s) rates, based on the earlier versions of [74–78]. Table 4.24(b) provides similar, but more stringent, performance levels based on the more recent versions of these references, which in turn are based in part on recent measurements of access link performance objectives [79]. Generally, it is expected that performance levels may become even more stringent with time, consistent with the introduction of higher-performance technology and more performance-sensitive services. Performance levels similar to

Table 4.23
CCITT Slip Performance Allocation for 64-kb/s Digital Connections

Type of Allocation	Slip Performance Limits	Percentage of Time Allocation				
		Parts of Connections			End-End Total @	Total Nat./ International Ratio !
		Each National		International		
		Local	Transit	Transit		
Relative	All	40	6	8	100	92/8
Absolute	> 5/day	.44	.066	.088	1.1	1.012/.088
	> 5/day and ≤ 30/hour	.4	.06	.08	1.0	.92/.08
	> 30/hour	.04	.006	.008	0.1	.092/.008

@ End-to-end performance of an international connection is contributed to by two national portions, and the international portion.
! End-to-end performance of a national connection is contributed to by two national portions, and is given by the numerators in the ratios shown.

those for DS1 in Table 4.24(a) can also be found for fractional DS1 ($N \times$ DS0) in [63]. Also, performance levels for international connections can be found in [80].

Note that the access links referred to in Table 4.24 connect each customer directly to the IEC network. The performance of these links is also indicative of the performance required of LEC networks providing comparable access service; in this case, the access arrangement would typically consist of a loop portion and one or two short-haul trunks (see Fig. 4.2). Generally, dedicated digital circuits are intended to support data, video, and other services, some data services being at lower rates than those indicated in Table 4.24, but time division multiplexed with similar and other services so as to occupy the higher rates. Thus, the delay specification in Table 4.24 will include that of such multiplexing equipment, and is relevant in terms of the impact on throughput and response time for data services, as discussed in Chapter 3.

The following discussion relates mainly to the earlier performance levels in Table 4.24(a), since these can be compared with CCITT levels and estimates of actual network performance evident during the same time frame (see Sec. 4.4.2.2).

Also relevant to data services in Table 4.24(a) are the error performance specifications in terms of percentage error free seconds (PEFS). Notice that the 99.5 PEFS value in Table 4.24(a) for 64-kb/s circuits is about an order of magnitude higher

Table 4.24(a)
Summary of Early Performance Objectives for North American Dedicated Digital Services

Service Rate	Parameter	Each Access	IEC # Short	IEC # Medium	IEC # Long	End-to-End # @ Short	End-to-End # @ Medium	End-to-End # @ Long
DS0 (56, 64 kb/s)	1-way Delay (ms)	- -	- -	- -		≤ 50 2-point ≤ 70 * multipoint		
	PEFS/day ES/day BER (x 10^{-8}) &	99.875 108 2.2, 2.0		99.75 216 4.5, 3.9			99.5 432 8.9, 7.8	
	% Availability/year Outage time/yr (hrs)	- -		- -			99.9 8.76	
DS1 (1.544 Mb/s)	1-way Delay (ms)	-		-			≤ 100	
	PEFS/day ES/day BER (x 10^{-9}) &	98.75 1080 8.1	99.1 778 5.8	98.5 1296 9.7	97.5 2160 16	96.6 2938 22	96.0 3456 26	95.0 4320 32
	% Availability/year Outage time/yr (hrs)	99.925 6.57		99.85 13.1			99.7 26.3	
DS3 (44.736 Mb/s)	1-way Delay (ms)	10		60			80	
	PEFS/day ES/day BER (x 10^{-10}) &	99.0 864 2.2	97.1 2506 6.4	95.2 4147 11	92.0 6912 18	95.1 4234 11	93.2 5875 15	90.0 8640 22
	% Availability/year Outage time/yr (hrs)	99.925 6.57		99.85 13.1			99.7 26.3	

* Guideline
\# Distances (miles): Short = 0 – 250, Medium = 250 – 1000, Long $\geq$ 1000.
  Single column entries apply to all distances.
@ End-to-end performance is contributed to by two access portions, and the IEC portion.
& Calculated BERs corresponding to PEFSs and ESs, assuming random errors; see eqn. (4.5).
  For DS0, the two entries per column correspond to 56 and 64 kb/s, respectively.

Table 4.24(b)
Summary of Recent Performance Objectives for North American Dedicated Digital Services

Service Rate	Parameter	Each Access	IEC # Short	IEC # Medium	IEC # Long	End-to-End # @ Short	End-to-End # @ Medium	End-to-End # @ Long
DS0 (56, 64 kb/s)	1-way Delay (ms)	- -	-	-		≤ 50 2-point ≤ 70 * multipoint		
	PEFS/day ES/day BER (x 10^{-8}) &	99.96 35 .72, .63	99.97 25 .52, .45	99.97 30 .62, .54	99.96 35 .72, .63	99.89 95 2.0, 1.7	99.88 100 2.1, 1.8	99.88 105 2.2, 1.9
	SES/day	4	10	25	30	18	33	38
	% Availability/year Outage time/yr (hrs)	99.975 2.19		99.95 4.38			99.90 8.76	
DS1 (1.544 Mb/s)	1-way Delay (ms)	5		50			60	
	PEFS/day ES/day BER (x 10^{-9}) &	99.88 100 .75	99.96 35 .26	99.93 60 .45	99.90 90 .67	99.80 175 1.3	99.77 195 1.5	99.75 215 1.6
	SES/day	4	10	25	30	18	33	38
	% Availability/year Outage time/yr (hrs)	99.975 2.19	99.97 2.63	99.95 4.38	99.93 6.14	99.87 11.4	99.85 13.1	99.83 14.9
	% Avail./Quarter Outage time/Qtr (hrs)	99.975 .55	99.97 .66	99.94 1.31	99.92 1.75	99.82 3.94	99.80 4.38	99.77 5.04
DS3 (44.736 Mb/s)	1-way Delay (ms)	5		50			60	
	PEFS/day ES/day BER (x 10^{-10}) &	99.54 400 1.0	99.92 70 .18	99.86 120 .31	99.79 180 .47	99.00 865 2.2	98.94 920 2.4	98.87 980 2.5
	SES/day	4	10	25	30	18	33	38
	% Availability/year Outage time/yr (hrs)	99.975 2.19	99.97 2.63	99.95 4.38	99.93 6.14	99.87 11.4	99.85 13.1	99.83 14.9
	% Avail./Quarter Outage time/Qtr (hrs)	99.975 .55	99.97 .66	99.94 1.31	99.92 1.75	99.82 3.94	99.80 4.38	99.77 5.04

* Guideline
\# Distances (miles): Short = 0 – 250, Medium = 250 – 1000, Long ≥ 1000.
   Single column entries apply to all distances.
@ End-to-end performance is contributed to by two access portions, and the IEC portion.
   For PEFS and ES, DS0 and DS3 performance is based on worst case addition, while DS1
   performance is based on a statistical approach [75].
   For availability, DS0 performance is based on worst case addition, while DS1 and DS3
   performance is based on service provider circuit management considerations [75].
& Calculated BERs corresponding to PEFSs and ESs, assuming random errors; see eqn. (4.5).
   For DS0, the two entries per column correspond to 56 and 64 kb/s, respectively.

(more stringent) than the corresponding allocated CCITT value in Table 4.22 of 100 − 5.44 ∼ 94.5% for national networks. This can be explained by two factors. First, the North American objectives for switched (not dedicated) data service at 56 kb/s is 95 PEFS [81, 82], consistent with the CCITT values. Second, since 95 PEFS also applies to the DS1 rate (1.544 Mb/s in North America), there is potential for higher PEFS performance at DS0, depending on the error producing mechanism at DS1. If DS1 errors are very bursty, then about the same number of ESs will occur at DS1 and at DS0 rates. (The ultimate manifestation of this is for SESs, which tend to be the same in number at all rates; see Table 4.24(b).) However, if DS1 errors are random or moderately bursty, then fewer ESs will occur at DS0 rates. Based on early experience in North America, an ES at DS1 will evidently affect only about 2 or 3 of the 24 DS0s, thus resulting on average in about 10 times fewer ESs at DS0 than at DS1 [83]. In principle, similar behavior occurs when errors at DS3 affect embedded DS1s [84]. The relationship between ESs at DS1 and DS0 rates is discussed further in the next section.

Calculated BERs corresponding to the PEFS values are also shown in Table 4.24(a). These BERs assume random errors (one error per ES), for which the following approximate relationship applies:

$$\text{BER} = \text{PES}/(100 \times R) \tag{4.5}$$

where PES = (100 − PEFS) is the percentage of errored seconds, and R is the circuit transmission rate in bits per second. Note that although the PEFS requirements are relaxed with higher transmission rates, the corresponding random BER values are more stringent. However, if errors are bursty (more than one error per ES), higher BERs can correspond to the same PEFS values. For example, if, on the average, there were 10 errors per ES, then the BERs can increase by a factor of 10 and result in the same PEFS values.

Finally, although not emphasized in this book, but included for completeness, Table 4.24 also includes availability figures appropriate for dedicated circuits. Also shown are the total outage times in hours per year corresponding to the percentage availability values. (Availability parameters are discussed in more detail in the next section of this chapter.)

Table 4.25 provides performance objectives and limits associated with end-to-end DS1 connections supporting switched services. In addition to PEFS values, BER, framing and controlled slip values are also provided [9, 13, 60, 66, 85]. The BER objective is such that BERs may only exceed 10^{-6} for 5% of calls. Thus, the nominal BER must be considerably lower than 10^{-6} (consistent with the lower BERs noted in Table 4.24(a)).

Regarding framing, the out-of-frame (OOF) events are known to correspond to BERs $> 10^{-3}$ in one second [86]. These events, being relatively rare, are assumed here to occur only once in a given second, and as such contribute directly to the number of SESs. The SES objective ($\sim 0.01\%$) is more stringent than the corresponding

Table 4.25
DS1 End-to-End Digital Performance Objectives and Limits for Support of Switched Services

Parameter	End-To-End DS1 Performance		
	Objective	Limit	
		Maintenance (Minor alarm)	Out-of-Service (Major alarm)
BER	10^{-6}/95% of calls	10^{-6}/10 minutes	10^{-3}/10 seconds
PEFS	95%/day	66%/15 minutes	-
SES #	8/day (~ 0.01%)	-	-
OOF* #	4/day	17/day	511/day
Slip	1/5 hours	4/day	255/day

* Out-of-Frame (OOF) events are assumed to correspond to a subset of SESs
( ) % SES/day
\# For short-haul (intra-LATA) connections

nationally allocated CCITT values (~0.07%) shown in Table 4.22. However, the SES objective in Table 4.25 is only for short-haul intra-LATA connections. For longer (inter-LATA) connections, see Table 4.24(b) for which the SES objectives range from ~0.02 to 0.04%, more consistent with the CCITT allocation of ~0.07%.

Finally, the end-to-end controlled slip objective of one slip in five hours is comparable to the CCITT value of five slips per day. The CCITT value is allowed to be exceeded only about 1% of the time [72]. No such percentage of time value is associated with the North American objective, but slip rates this high would only occur under abnormal (e.g., timing failure, or excessive jitter or wander) conditions, which should be relatively rare and therefore occupy a small percentage of time.

Notice that in Table 4.25 the maintenance limit for slips is similar to that for the end-to-end objective [85]. The reasoning is as follows. The performance limits in Table 4.25 apply to the monitoring of performance at a particular nodal network element (e.g., digital circuit switch, or DS1/DS0-type digital cross-connect system), and reflect the performance of the DS1 signals at that network element. In particular, the limits for slips reflect the stability of the DS1 rate clocks local and incoming to the network element. The limits for slips at DS1 do not necessarily reflect the total end-to-end performance impact of slips on DS0 signals, since there may be other nodes in the DS0 connection where slips were encountered. However, this would imply simultaneously abnormal conditions at two or more nodes, which is not likely.

Thus, in effect, the limits for slips at DS1 should be indicative of end-to-end performance at DS0.

As discussed in Chapter 3, the effect of controlled slips on telephony voice service is minimal [87], although the effect on higher quality speech for such services as sound programming may be more significant. However, the main concern is for data services. The effect of slips on voiceband data is to cause interruptions in the demodulated data signal, during which the error rate on this signal will be very high [88–90]. The duration of these interruptions can vary from a few milliseconds to a few seconds and depends largely on the type of modem and modem speed (1.2 to 9.6 kb/s). The durations are not strongly related to the level of analog impairments associated with the voiceband channel (impairments such as SNR, attenuation, and phase distortion, and amplitude and phase jitter).

Slips also affect voiceband facsimile transmission by causing up to eight scan lines (0.01 inch per line) to be missing [91]. Depending on the nature of the facsimile original, this may or may not warrant the retransmission of the affected page. The "errors" associated with the slips as detected by the facsimile receive terminal are evidently not sufficient to cause the receive terminal to signal the transmit terminal to reduce the transmit speed (from 9.6 to 4.8, or from 4.8 to 2.4 kb/s). This is probably appropriate since slips are rare events, even though the speed reduction would result in improved immunity to slips.

4.4.2.2 Network Performance Estimates

In this section additional information is provided to illustrate end-to-end digital network performance as projected from the performance of subcomponents of a digital connection. From [69], which examines short- and long-toll connections, it is also possible to estimate performance for nontoll connections as included in Figure 4.7. The performance estimates for all connections are provided in Table 4.26 and are based on the performance of network components as itemized in Table 6.32 of Chapter 6. These estimates are also based on the data in [69], but re-interpreted to be more consistent with CCITT performance parameter definitions. That is, long-term BER as interpreted here does not include large numbers (thousands) of errors occurring over short periods of time due to such events as protection switching activity on transmission facilities. Rather, these highly errored events are included in ES and in SES counts. Similarly, here, the estimated number of minutes having BERs $>10^{-6}$ do not include the above highly errored events, in order to be consistent with the CCITT definition of a DM; that is, a minute with BER $>10^{-6}$, but excluding SESs.

The results in Table 4.26 apply to the DS1 hierarchical transmission rate of 1.544 Mb/s, but for most parameters are also indicative of performance at the DS0 embedded tributary rates of 64 kb/s. The exception is for ESs at DS0 which, depend-

148

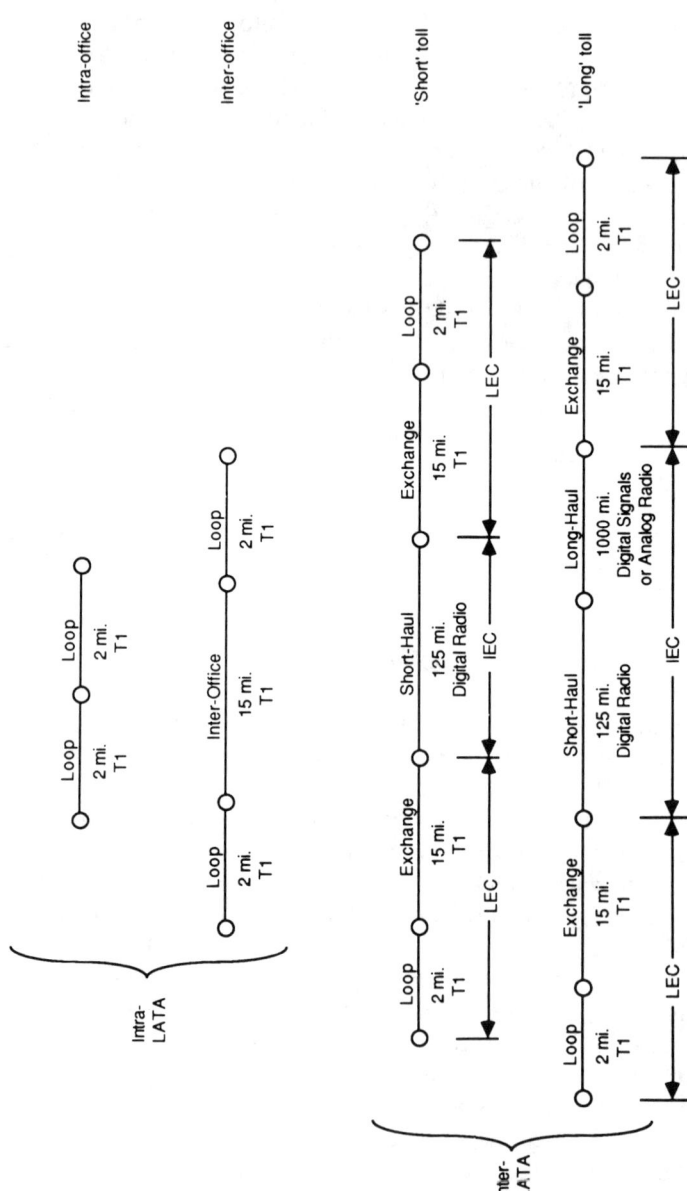

Figure 4.7 Examples of digital network connections (all facilities, including "loops," are digital [69]).

Table 4.26
Example of Digital Network Performance

Parameter		Type of Connection #					
		Intra-Office	Inter-Office	Toll			
				Short		Long	
Name	Value			Non-Fading	Fading	Non-Fading	Fading
Long Term BER	Median %$\leq 10^{-6}$ *	8.0×10^{-9} 94	8.1×10^{-9} 91	8.2×10^{-9} 88			
	NA Obj. (%$\leq 10^{-6}$)	$\geq 98 - 99$	$\geq 96.5 - 97.5$	$\geq 94.5 - 95.5$		$\geq 93 - 94$	
% DMs	Median 95% Level *	.01 0.6	.015 0.9	.02 1.2			
	CCITT Obj.	≤ 3	≤ 4.5	≤ 6.8			
% ESs	DS1 Median 95% Level *	.03 3	.0315 4	.0332 5.0016	.035 5.016	.0348 5.0146	.051 5.146
	DS1 NA Obj.	≤ 1	≤ 1.375	≤ 2.65		≤ 5.15	
	DS0 Median 95% Level *	.003 .3	.00315 .4	.0035 .5016	.0053 .516	.0051 .5146	.021 .646
	DS0 NA Obj. CCITT Obj.	$\leq .1$ ≤ 2.4	$\leq .1375$ ≤ 3.6	$\leq .265$		$\leq .515$	
				≤ 5.44			
% SESs	Median 95% Level	– –	– –	.0002 .0016	.002 .016	.0018 .0146	.018 .146
	CCITT Obj.	$\leq .03$	$\leq .062$	≤ 0.136			

\# See Fig. 4.7
* Conservative Estimates [69]. See text.
NA Obj.: North American (U.S.) Connection Specific Objective. See Table 4.25 for overall connection objectives.

ing on the cause of ESs at DS1, are, in Table 4.26, an adjusted or translated version of the ESs at DS1. As noted in the previous section, ESs due to transmission errors at DS1 will generally result in fewer ESs for any one DS0; that is, if there are less than 24 bit errors in an ES at DS1, then fewer than 24 DS0s will be affected (evidently about 10 times fewer [83]). On the other hand, ESs due to protection switching activity or OOF events at DS1 or higher rates will cause the same number of ESs on all DS0s (similarly, OOFs at DS3 will cause ESs on all 28 DS1s [84], thereby affecting all DS0s). These factors have been accounted for in the estimation of the percentage of ESs at DS0 in Table 4.26. That is, ESs at DS0 are estimated to consist of one tenth of the ESs at DS1 due to transmission errors, plus all of the ESs at DS1 due to OOF-like events. Table 4.26 also includes connection-specific performance objectives as derived in Chapter 6 for the specific North American connections in Figure 4.7 and based on CCITT Recommendations as discussed earlier in this chapter and in Chapter 3. These specific objectives are in addition to the overall objectives presented earlier in Tables 4.24 and 4.25, and enable comparison with projected performance for specific connections. With the above background in mind, the following is a discussion of the performance levels in Table 4.26.

The long-term BER and DM results are due solely to T-carrier transmission for all connections. This transmission system technology is directly applicable to the intra- and inter-office connections, while, for the toll connections, the contribution to BER and DMs from the radio systems is assumed to be negligible once the highly errored events due to protection switching activity are removed (but included instead in ESs and SESs). The estimated percentage of connections achieving a long-term BER $\leq 10^{-6}$ is based on the simple rule that $n(100 - 97) = 3n\%$ of connections will have BERs $> 10^{-6}$, where 3% applies to one T-carrier link [69], and n is the number of T-carrier links in the connection. This rule may be conservative, but Section 5.4.3.4 of Chapter 5 shows that the rule agrees reasonably well with the results of convolution; see Equations (5.91) and (5.92), and Table 5.15. Thus, as seen in Table 4.26, significant percentages of connections do not achieve a BER $\leq 10^{-6}$ [69]. This implies selection of T-carrier facilities, as discussed further in Section 6.3.1 of Chapter 6 [92]. However, generally, the overall 95% $\leq 10^{-6}$ BER objective is more difficult to achieve when measured continuously over longer time intervals, as opposed to over many short contiguous intervals, such as 1 min. The same is true for the connection-specific 10^{-6} BER objectives shown in Table 4.26 (as derived in Ch. 6). This is supported by the DM results, which readily satisfy CCITT recommended performance levels. For simplicity, the median results in Table 4.26 are all based on the addition of the median for each network component. This may be optimistic, and more detailed relationships are discussed in Chapter 5; see Equations (5.93) and (5.96).

Consistent with [69], this simple approach is also applied to the higher percentile values for the DM, ES, and SES parameters, which amounts to adding the percentile values for each network component. Again, this may not be the most accurate approach, and alternatives are presented in Chapter 5 (see Tables 5.15 and 5.16 and the related discussion). ESs for intra- and inter-office connections are based on T-carrier performance. For toll connections, a small percentage of ESs due to protection switching activity is also included, and, according to [69], this percentage is 10 times higher under fading conditions, as shown in Table 4.26. As discussed earlier, these different causes of ESs at higher rates are accounted for in estimating the ESs at DS0. ES performance is marginal relative to the overall North American objectives of 95 and 99.5 PEFS for DS1 and DS0 services, respectively (Table 4.24(a)); the same is true for the connection-specific PEFS objectives in Table 4.26 as derived in Chapter 6. At DS0, the less stringent CCITT ES levels are readily met.

No specific data on SESs for T-carrier facilities is provided in [69]. However, some SESs are expected in practice due to OOF events, and error bursts are expected due to lightning hits, power surges, and maintenance activity on such systems. For example, it is evident from more recent results for access circuits [79] that a relatively small number of SESs of less than two per day occur at the 97% level (<0.002% SESs/day). In relation to the percentiles used in Table 4.26, it can be estimated that less than one SES in two days occurs at the 95% level ($\approx$0.0005% SESs/day). These results are not included in Table 4.26, because they are from different systems and from a different time frame than in [69]. However, these recent results are presented in more detail in Section 6.3.2 of Chapter 6. For toll connections, SESs correspond to protection switching activity, which causes about 10 times as many SESs under fading conditions as opposed to nonfading conditions. SES performance relative to CCITT levels is readily met under nonfading conditions [69]. Performance is marginal under fading conditions, but is acceptable when averaged over fading and nonfading periods, because most fades occur during a few days per month (e.g., half in 1 day, 70% in 3 days [69]). It should also be noted that more recent technology can be expected to provide improved performance; examples are provided in Chapter 6.

4.4.2.3 Note on Jitter and Wander

Jitter and wander are important lower-level digital network impairments which have a bearing on synchronization performance and, in turn, on error performance. Owing to the specialized nature of this subject, it is treated in its entirety in Chapter 6 (rather than distributed through Chs. 4, 5, and 6 according to end-to-end values, accumulation techniques, and component values, respectively).

4.5 PACKET NETWORK PERFORMANCE

4.5.1 CCITT

In a manner analogous to Figures 4.5 and 4.6 presented previously for digital network connections, Figures 4.8 and 4.9 provide international and national packet network reference connections, respectively. Each connection consists of packet networks interconnected by circuit (nonpacket) sections [93]. Tables 4.27(a) and (b) provide minimum (worst case) performance requirements for the national and international portions, respectively, as well as for representative end-to-end connections composed of national and international portions. The parameters include those given earlier for packet networks in Table 4.1, as well as availability-related parameters.

The transfer delay and throughput requirements apply to 9.6-kb/s signaling rates and 128-byte packets. (In Table 4.27, transfer delay is also shown as a general function of transmission rates [94].) The delay requirements for connection portions

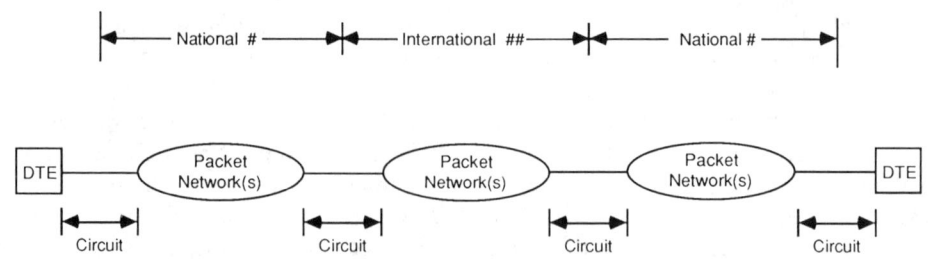

Figure 4.8 Simplified international virtual connection showing national and international portions, as well as packet network sections and circuit sections.

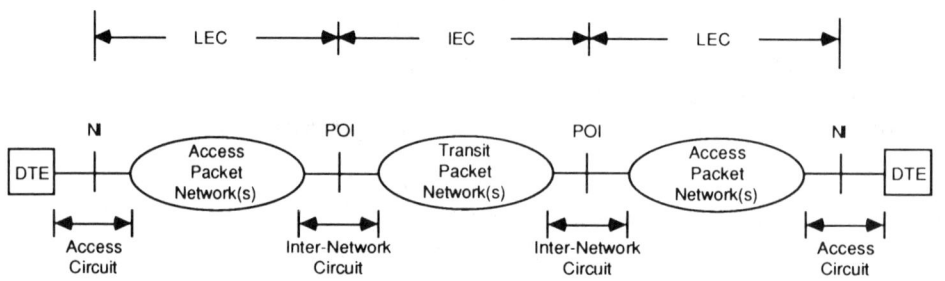

Figure 4.9 Simplified national virtual connection showing local and inter-exchange carrier portions, as well as packet network sections and circuit sections.

are accumulated algebraically, assuming independence from portion to portion, to obtain the requirements for end-to-end connections. The resulting end-to-end requirements represent rather large delays, generally on the order of 1 sec or more. Throughput efficiency requirements for connections are limited by the efficiency of the poorest performing portion of the connection. The resulting end-to-end requirements correspond to rather low efficiencies on the order of 20% or less, owing to the considerable overhead associated with conventional packet switching.

The connection requirements for residual error, reset, and premature disconnect probabilities are determined by algebraic addition of the values for the connection components [95]. As discussed in Chapter 3, residual error probability is the resulting error probability after application of packet network error control by retransmission. In general, residual errors include not only errored packets, but also lost, misdelivered, out-of-sequence, and multiple (e.g., duplicate) packets. In Table 4.27, the reset and premature disconnect probabilities do not include the reset stimulus and premature disconnect stimulus probabilities, respectively, which are typically an order of magnitude smaller.

Criteria for defining the availability requirements in Table 4.27 [96, 97] include limits on

- Throughput capacity of ≤ 80 b/s.
- Residual error rate of $\geq 10^{-3}$.
- Reset and premature disconnect probabilities of $\geq 1.5 \times 10^{-2}$ and $\geq 10^{-2}$, respectively.

Table 4.27(a)
Worst-Case CCITT Performance for National Virtual Packet Data Connections

Parameter		Type of National Portion		Type of End-to-End National Connection	
Name	Value	Nat. A	Nat. B	Nat. Type 1 (A + A)	Nat. Type 2 (A + B)
Transfer Delay (ms)	Mean* SD 95 %* Mean # 95 % #	350 + y 106 525 + y 463 638	650 + y 106 825 + y 763 938	700 + 2y 150 947 + 2y 926 1173	1000 + 2y 150 1397 + 2y 1225 1622
Throughput Efficiency # (%)	Mean 95 %	31 25	25 21		
Residual Error	Prob.	10^{-10}	2×10^{-10}	2×10^{-10}	3×10^{-10}
Reset	Prob.	10^{-5}	2×10^{-5}	$\sim 2 \times 10^{-5}$ @	$\sim 3 \times 10^{-5}$ @
Premature Disc.	Prob.	5×10^{-6}	10^{-5}	$\sim 10^{-5}$ @	$\sim 1.5 \times 10^{-5}$ @
Availability	%	99.5	99	99	98.5
MTBSO	Hours	1200	800	600	480
MTTSR	Hours	6	8	6.1	7.3

* y = 1088/R (ms), R = Signaling rate (kb/s), 128 byte packets.
\# R = 9.6 kb/s, 128 byte packets.
@ Ignores ~ negligible Reset Stimlulus and Premature Disconnect Stimulus probabilites, respectively.

Table 4.27(b)
Worst-Case CCITT Performance for International Virtual Packet Data Connections

Parameter		Type of International Portion		Type of End-to-End International Connection &	
Name	Value	Int. A	Int. B	Nat. Type 1 + Int. A	Nat. Type 2 + Int. B
Transfer Delay (ms)	Mean *	215	950	915 + 2y	1950 + 2y
	SD	0	106	150	184
	95 % *	215	1125	1162 + 2y	2253 + 2y
	Mean #	215	950	1141	2175
	95 % #	215	1125	1388	2479
Throughput Efficiency # (%)	Mean	21	19	~ 21	~ 19
	95 %	19	16	19	16
Residual Error	Prob.	~ 0	2×10^{-10}	2×10^{-10}	5×10^{-10}
Reset	Prob.	~ 0	2×10^{-5}	$\sim 2 \times 10^{-5}$ @	$\sim 5 \times 10^{-5}$ @
Premature Disc.	Prob.	~ 0	10^{-5}	$\sim 10^{-5}$ @	$\sim 2.5 \times 10^{-5}$ @
Availability	%	99.5	99	98.5	97.5
MTBSO	Hours	1600	800	436	300
MTTSR	Hours	8	8	6.5	7.5

* y = 1088/R (ms), R = Signaling rate (kb/s), 128 byte packets.
\# R = 9.6 kb/s, 128 byte packets.
@ Ignores ~ negligible Reset Stimlulus and Premature Disconnect Stimulus probabilites, respectively.
& See Table 4.27(a) for National Type 1 and Type 2 Portions.

These criteria are also referred to for North American packet networks [98] and are included later with other criteria in Table 4.28 [99]. The well-known relationships for availability performance are summarized as follows:

$$\text{MTBSO} = \text{Mean Time Before Service Outage}$$
$$= 1/\lambda = \text{inverse of service outage rate}$$
$$\text{MTTSR} = \text{Mean Time To Service Restoral}$$
$$= 1/\mu = \text{inverse of service restoral rate}$$

$$\text{Availability} = \frac{\text{MTBSO}}{\text{MTTSR} + \text{MTBSO}} = \frac{\mu}{\lambda + \mu} \qquad (4.6)$$

$$\text{Unavailability} = 1 - \text{Availability} = \text{transfer denial probability}$$
$$= \frac{\text{MTTSR}}{\text{MTTSR} + \text{MTBSO}} = \frac{\lambda}{\lambda + \mu} \qquad (4.7)$$

Table 4.28
Performance Goals and Requirements for North American (U.S.) Packet Data Networks

Parameter		Performance			
		Typical Goal @ [100 - 102]	SMDS [103]	Worst Case [99]	Transfer Denial Criteria [99]
Block Transfer Time	Total	150 ms ave. * 200 ms ave. #	20 to 220 ms &	2.5 sec.	7.5 sec.
	User Fraction	0.1 * 0.2 #			0.13
Bit Transfer Efficiency		95 %		1 kb/s	333 b/s {80 b/s}
Bit Error Probability					2.1×10^{-2} {10^{-3}}
Block Probability	Error	10^{-8}	5×10^{-13}	10^{-5}	
	Misdelivery	0.5 to 1×10^{-7}	5×10^{-8}		
	Extra	10^{-9} !	5×10^{-8}	10^{-5}	
	Out of Sequence	10^{-9}			
	Lost	5×10^{-8}	1×10^{-4}	10^{-5}	
Reset Probability **					{> 0.015}
Premature Disconnect Prob. **		2×10^{-6}			{>0.01}
Transfer Denial Probability				10^{-3}	-
Availability		97 to 99.7 $	99.7		-
MTBSO (hours)		162 to 1662 $	1092		-
MTTSR (hours)		5	3.5		-

@   ~ 56 and 64 kb/s rates
*   Average delay for 128 byte packets
#   Average delay for 256 byte packets
!   Duplicated
$ Specifically, 98.5%, 436 hours [97]
** Includes respective stimulus probabilities
-   Not Applicable
{ } From [96-98]
&   Depends on access arrangement and addressing scheme

Connection unavailability (i.e., the complement of availability), is determined by algebraically adding the unavailability for each connection portion. The connection MTBSO is the harmonic mean (inverse of the sum of inverses) of the MTBSOs for the various portions of the connection. From these results, the connection MTTSR can then be determined from either Equation (4.6) or (4.7).

4.5.2 North America (U.S.)

4.5.2.1 Performance Objectives

Table 4.28 provides typical performance goals and requirements for packet data networks used in North America [96–103]. Also included are criteria for defining unavailability (transfer denial). The typically expected block transfer delays are much less than the worst-case requirements shown in Table 4.28, and less than the worst-case CCITT requirements in Table 4.27. The user fraction of block transfer delay enables the network portion and total user portion of the delay to be determined (i.e., 135 and 15 ms, respectively, for 128-byte packets, and 160 and 40 ms, respectively, for 256-byte packets). Expected throughput efficiency in Table 4.28 is also much superior to that of the CCITT requirements in Table 4.27. On the other hand, residual error probability (and possibly availability) requirements are somewhat lower for North American networks, relative to the CCITT requirements. The reason for the above noted differences in North American versus CCITT performance specifications is expected to be due to the higher speed and more streamlined, less processing- and protocol-intensive nature of the North American networks envisaged in Table 4.28. This results in improved throughput and delay performance, but lower error (and possibly availability) performance for these North American networks. However, lower error and availability performance can be compensated for by end-to-end network, or user, protocols. This is especially true of such recent transport schemes as frame relay [104] and cell-based asynchronous transfer mode (ATM) [105], which do not provide error correction by retransmission. A particular example of such systems is included in Table 4.28 in the form of cell-based switched multimegabit data service (SMDS), which supports high-speed data applications at DS1 and DS3 access rates [103]. SMDS performance in Table 4.28 is considered indicative of that achievable by ATM for the support of future broadband services.

4.5.2.2 Network Performance Measurements

Table 4.29 presents measured data network performance for a variety of terrestrially based packet networks and for the public circuit-switched network; for performance

Table 4.29
Measured Performance of North American (U.S.) Data Networks

Parameter		Network					
		Packet				Circuit	
		ARPANET		TELENET/TYMNET/UNINET **			
Name	Value	Mean	Min. to Max. Range	Mean	95 % Conf. Int.	Mean	95 % Conf. Int.
Block Transfer Time (ms):	Total User-Fraction	.262 *, .709 # .13	.216-2.86 *, .569-3.95 # -	3.79-5.86 .089-.105	3.61-7.63 .075-.134	1.41 .093	1.39-1.42 .093-.094
Bit Transfer Efficiency (%)	-	9.7 #&	-	67.8-75.5	35.1-100	79.3	79.3-79.3
Bit Probability	Error Extra Lost	0 0 .006 !	0-3 x 10^{-6} @ 0-3 x 10^{-6} @ -	0-7 x 10^{-6} 0 0	0-3 x 10^{-5} 0-3 x 10^{-5} 0-3 x 10^{-5}	0 0 0	0-5 x 10^{-5} 0-5 x 10^{-5} 0-5 x 10^{-5}
Block Probability	Error Extra Lost	0 0 .006	- - -	0-.002 0 0	0-.03 0-.03 0-.03	0 0 0	0-.04 0-.04 0-.04
Tranfer Denial Probability (Unavailability)	-	.047	.028-.067	0	0-.05	0	0-.08

* 64 byte blocks.
# 512 byte blocks.
& User-Fraction = 0.013, relevant condition: 50 kb/s rate.
@ Upper limits were not measured, but are statistical estimates.
! 0.002 excluding losses due to transfer timeouts.
** Relevant conditions: 128 byte packets, 1.2 kb/s rates.

of a satellite based packet network, see [106]. The Advanced Research Project Agency Network (ARPANET) exhibits large variation in block transfer delay and rather low throughput efficiency, presumably because of the considerable overhead and processing intensiveness of this network [107, 108]. The user fraction of mean block transfer delay is relatively small at 0.13, which, for example, with 512-byte packets, corresponds to a network portion of 617 ms and a total user portion of 92 ms for an end-to-end total of 709 ms. The user fraction of throughput efficiency of 0.13 means that, owing to user portion restrictions, the network portion of 11.2% is degraded by about 1.5%, for a net efficiency of 9.7%. No bit or block errors or extra bits or blocks were observed in the ARPANET measurements, which is indicative of effective error control by retransmission (the 95% confidence limits indicated for these parameters in Table 4.29 are not measured, but are worst-case estimates). Bit and packet losses were observed, most being due to transfer denial caused by excessive transfer delay, and the rest probably due to host computer delays. Bit and packet misdeliveries were not measured. Transfer denial is largely due to host computer and host interface message processor (IMP) behavior, and corresponds to mean unavailabilities during transfer of 3.1% and 1.6%, respectively, for an overall mean availability during transfer of 95.3%.

In contrast to the ARPANET, the collective results representative of the TELENET, TYMNET, and UNINET networks in Table 4.29 have high delays on the order of a few seconds, but with much lower delay variability; throughput efficiency is also considerably higher (on the order of 70%), although there is considerable variability [109–111]. These characteristics are expected to be the result of both the lower (1.2 kb/s) link rates and less intensive processing of these networks relative to the ARPANET. The error performance shown for TELENET, TYMNET, and UNINET in Table 4.29 is poorer than expected from packet networks, owing to the inclusion of a nonpacket connection to measurement equipment, which probably produced most of the errors. No extra or lost packets were observed. Transfer denial is superior to that of ARPANET, probably because of fewer host computer problems and, perhaps, less tendency to congest.

Finally, in contrast to all the packet networks, the circuit-switched network with 1.2-kb/s rates provides moderate and nearly constant delay of about 1.4 sec and throughput efficiency of about 80%. Again, no extra or lost packets were observed, and error performance is reasonable for a circuit switched network. Transfer denial is comparable to that of the TELENET, TYMNET, and UNINET packet networks.

Table 4.30 provides additional measurement results for the throughput and delay performance of a variety of packet switches and packet networks, as well as pairs of packet networks operating in series [112]. (In Table 4.30 there is not necessarily a one-to-one correspondence between any of the switches, networks, or pairs of networks.) The packet switches and packet network services originate mostly in North America, but also in Europe and elsewhere. The objectives and performance results shown for the packet switches and networks are with reference to 128-byte

Table 4.30
Measured Transfer Delay and Throughput Efficiency of Packet Switches and Networks

Parameter		Type of Packet Switch #					
Name	Objective	1	2	3	4	5	6
Transfer * Delay (ms)	40-70	28-45	20-50	55-75	95	220	370
Throughput * Efficiency (%)	> 95	>97.9	> 99	≤ 95.8	86.5	> 42.7	87.5

Parameter		Type of Packet Network #				
Name	Objective	1	2	3	4	5
Transfer * Delay (ms)	160	90-105	115-180	220	400-800	175-200
Throughput * Efficiency (%)	95	> 98.5	98.2	42.7	84.4-90.6	68.8-78.1

	Type of Packet Network Pair #					
Connection	1	2	3	4	5	6
Distance Range	Short		Medium		Long	
Distance (mi.)	750	850	3500		10500	
No. of Switches	2	5	3	2	3	
9.6 kb/s Gateway						
Window Size (packets)	3		3	2	2	
Packet Size (bytes)	256		256	128	128	
Parameter						
Transfer * Delay (ms)	382	289	252	951	426	361
Throughput * Efficiency (%)	88.3	98.2	98.2	12.7	42.9	74.3

* With respect to 128 byte packets and 9.6 kb/s rates.
\# Note, there is not necessarily a one to one correspondence between any of the packet switches, networks, or network pairs.

packets and 9.6-kb/s service rates. The same is true for the network connections involving pairs of packet networks between X.25 data terminals, except that the X.75 gateway between the packet networks also uses 256-byte packets and operates with various flow control window sizes, as shown in Table 4.30.

The results indicate considerable variability in performance of the packet switches and networks, some of which meet the objectives and some of which do not. Although not specifically identified as such, it is expected that performance of individual networks would in large measure be influenced by switch performance. Similarly, the performance of the connections involving pairs of packet networks is probably dominated by the more poorly performing individual networks.

4.6 REFERENCES

1. M.B. Carey, H.T. Cheu, A. Descloux, J.F. Ingle, and K.I. Park, "1982/83 End Office Connection Study; Analog Voice and Voiceband Data Transmission Performance Characterization of the Public Switched Network," *BLTJ,* Vol. 63, No. 9, Nov. 1984, pp. 2059–2119.
2. F.T. Andrews, Jr., and R.W. Hatch, "National Telephone Network Transmission Planning in the American Telephone and Telegraph Company," *IEEE Trans. Comm.,* Vol. COM-19, No. 3, June 1971, pp. 302–314.
3. "Telecommunications Transmission Engineering," Vol. 3: Networks and Services, 2nd ed., Chapter 2 on "Metropolitan Network Plans," AT&T, 1977. Also, 3rd ed., Bellcore 1990.
4. G.R. Ash, A.H. Kafker, and K.R. Krishnan, "Intercity Dynamic Routing Architecture and Feasibility," *Proc. 1 of 10th Int. Teletraffic Cong.,* Montreal, Quebec, 1983, Session #3.2, paper #2.
5. W.H. Cameron, J. Regnier, P. Galloy, and A.M. Savoie, "Dynamic Routing for Intercity Telephone Networks," *Proc. 1 of 10th Int. Teletraffic Cong.,* Montreal, Quebec, 1983, Session #3.2, paper #3.
6. A. Niedzwiecki, R.T. Matsunaga, and P.V. Coverdale, "All-Digital Network Transmission Planning and Terminal Requirements," *IEEE Journal on Sel. Areas in Comm.,* Vol. 7, No. 8, Oct. 1989, pp. 1166–1171.
7. Draft of American National Standard for Telecommunications—Network Performance—Loss Plan for Evolving Digital Networks, T1Q1/91-026, July 1991.
8. "Acoustic-to-Digital and Digital-to-Acoustic Transmission Requirements for ISDN Terminals," EIA/TIA-579, 1990.
9. "Notes on the BOC INTRA-LATA Networks," Bellcore TR-275, Issue 1, Section 7 on "Transmission," April 1986.
10. American National Standard for Telecommunications—Network Performance Standards—Switched Exchange Access Network Transmission Performance Standard, ANSI T1.506-1989. Also, Supplement to Transmission Specifications for Switched Exchange Access Network, Draft ANSI T1.506a.
11. F.D. Richter, "Usage Measurements on Feature Group Lines," *TE&M,* Vol. 90, No. 4, Feb. 15, 1986, pp. 107–110.
12. R.F. Downing, "Engineering Equal Access," *Telecom.,* Vol. 19, No. 9, Sept. 1985, pp. 88–90, 92.
13. *Notes on the Network,* AT&T, Section 7 on "Transmission Considerations," 1980.
14. "Telecommunications Transmission Engineering," Vol. 3: Networks and Services, 2nd ed., AT&T, 1977, Chapter 1 on "The Network Plan for Distance Dialing." Also, 3rd ed., Bellcore, 1990.
15. R. Havens, "How a 'Different' Carrier Works," *TE&M,* Vol. 88, No. 18, Sept. 15, 1984, pp. 141, 142, 147, 149.
16. L.M. Manhire, "Physical and Transmission Characteristics of the Customer Loop Plan," *BSTJ,* Vol. 57, No. 1, Jan. 1978, pp. 35–59.
17. J.F. Ingle, K.I. Park, and T.C. Redman, "1983 Exchange Access Study: Analog Voice-Frequency Transmission Performance Characterization of the Exchange Access Plant," *IEEE Trans. on Comm.,* Vol. COM-35, No. 1, Jan. 1987, pp. 104–112. Also, Bellcore TR-037, Issue 1, Nov. 1984.
18. "Estimated Transmission Performance of Switched Access Service Feature Group-D," Bellcore TR-002, Issue 1, July 1984.
19. LATA Switching Systems Generic Requirements (LSSGR): Section 7 on "Transmission," Bellcore TR-507, Issue 3, Mar. 1989.

20. "Transmission Systems For Communications," Bell Telephone Labs. Inc., 5th ed., 1982.
21. "Telecommunications Transmission Engineering," Vol. 1: Principles, 2nd ed., AT&T, 1977. Also, 3rd ed., Bellcore, 1990.
22. "Telecommunications Transmission Engineering," Vol. 2: Facilities, 1st ed., AT&T, 1977. Also, 3rd ed., Bellcore, 1990.
23. "Telecommunications Transmission Engineering," Vol. 3: Networks and Services, 2nd ed., AT&T, 1977. Also, 3rd ed., Bellcore, 1990.
24. "Engineering and Operations in the Bell System," Bell Telephone Labs. Inc., 1977.
25. "Engineering and Operations in the Bell System," AT&T Bell Labs. Inc., 1983.
26. "Catalog of Communications Technical Publications," PUB 10000, AT&T, April 1989.
27. *1990–1991 Catalog of Technical Information,* Bellcore SR-264, Issue 6, July 1990–July 1991.
28. R.L. Fredrick, "Measurement of Voice Frequency Transmission Impairments," *Telecomm.,* Vol. 19, No. 3, Mar. 1985, pp. 85, 86, 88, 88b, 88e, 88f.
29. W. Hand, "How to Find Phone-Line Faults and What to Do About Them," *Data Comm.,* Vol. 17, No. 10, Sept. 1988, pp. 179–182, 185, 186, 189–192, 195–197.
30. "Data Communications Using Voiceband Private Line Channels," AT&T PUB. 41004, Oct. 1973.
31. "AT&T Analog Voice Total and Coordinated Services," Preliminary PUB 43202, May 1985.
32. "Data Communications Using the Switched Telecommunication Network," AT&T PUB 41005, May 1971.
33. "'DATAPHONE' Data Communications Service Minimum Acceptable Performance (MAP) Criteria: Data Systems on the Public Switched Network," AT&T Co. Standard, Section 314-205-503, Issue 4, Nov. 1981.
34. "Transmission Parameters Affecting Voiceband Data Transmission—Description of Parameters," AT&T PUB. 41008, July 1974.
35. "'DATAPHONE' Service and Data Access Arrangements: Analysis of Analog Parameters and Their Effects on Voiceband Data Transmission," Bell System Practice Section 880-440-101, Issue 3, Nov. 1974.
36. Standard EIA-496-A: "Interface Between Circuit-Terminating Equipment (DCE) and the Public Switched Telephone Network (PSTN)," 1991.
37. Standard EIA-470-A: "Telephone Instruments with Loop Signaling," July 1987.
38. A.H. Inglis and W.L. Toffnell, "An Improved Telephone Set," *BSTJ,* Vol. 29, No. 4, April 1951, pp. 239–270.
39. Bell System Practice on "500 Series Combined Telephone Sets: Transmission Performance; C and D Series," Section AB22.066-3, Issue 2, Oct. 1956.
40. J.O. Norback, "Transmission Analysis of GT&E Customer Loop Plant," *IEEE Trans. on Comm.,* Vol. COM-19, No. 3, June 1971, pp. 326–332.
41. G.S. Schrage, "Rural Subscriber Loop Performance: Some Favourable Results from the 1986 FEA Subscriber Loop Study," *Telephony,* Vol. 92, No. 2. *TE&M,* Jan. 15, 1988, pp. 73, 76–79.
42. G. Williams and J. Gruber, "Transmission Planning for Digital Connectivity in the Post Divestiture Environment," *NCF Rec.,* Vol. 40, Bk. 2, Chicago, Oct. 1986, pp. 675, 677–682.
43. A.M. Lessman and J.R. Rosenberger, "ISDN Voice CPE Performance Compatibility Information," *GLOBECOM Rec.,* Vol. 1, Houston, Dec. 1986, pp. 14.5.1–14.5.10.
44. D.V. Batorsky and M.E. Burke, "1980 Bell System Noise Survey of the Loop Plant," *BLTJ,* Vol. 63, No. 5, May–June 1984, pp. 775–818.
45. CCITT Rec. G.113, "Transmission Impairments," *Blue Book,* Vol. III, Fas. III.1, Geneva, 1989, pp. 63–84.
46. J.L. Neigh, "Transmission Planning for an Evolving Local Switched Digital Network," *IEEE Trans. on Comm.,* Vol. COM-27, No. 7, July 1979, pp. 1019–1024.
47. E.J. Isganitis, "Listener Echo Considerations for a Digital Central Office," AT&T/BTL—USITA "Digital Transmission Planning Workshop," AT&T, Basking Ridge, N.J., April 15, 1980.

48. CCITT Contrib. COM XII-No. 23-E to Quest. 16, on "Return Loss Variations in Subscriber Lines and Telephone Sets—Balancing Circuit Considerations," AT&T, Mar. 1977.
49. L.S. DiBiaso, "Transmission Considerations for Local Switched Digital Network," *Telephony*, Vol. 133, No. 17, Oct. 24, 1977, pp. 40–45, 52.
50. O.P. Mahajan and E.J. Isganitis, "Average Power Return Loss (APRL)—A Single-Ended Parameter for Specifying Listener Echo Objectives," *ICC Rec.*, Seattle, June 1980, Vol. 2, pp. 2041–2046.
51. J.R. Cavanaugh, R.W. Hatch, and J.L. Neigh, "A Model for the Subjective Effects of Listener Echo on Telephone Connections," *BSTJ*, Vol. 59, No. 6, July/Aug. 1980, pp. 1009–1060.
52. CCITT Contrib. COM XVI-No. 42-E to Quest. 5 on "Some Analytical Results for Multiple 4-wire Paths, In Connections Supporting Voiceband Data," Canada: BNR, Mar. 1983.
53. J. Gruber, "Performance Analysis and Design Guidelines for Multiple 4-wire Path Telephone Connections," *IEEE Trans. on Comm.*, Vol. COM-34, No. 7, July 1986, pp. 647–653. Also, *ICC Rec.* Vol. 1, Toronto, June 1986, pp. 7.4.1–7.4.6. Also, [46].
54. "Engineering and Operations in the Bell System," Bell Telephone Labs. Inc., Chapter 6 on "Transmission," 1977.
55. F.P. Duffy, G.K. McNees, J. Nasell, and T.W. Thatcher, "Echo Performance of Toll Telephone Connections in the United States," *BSTJ*, Vol. 54, No. 2, Feb. 1975, pp. 209–243.
56. F.P. Duffy and T.W. Thatcher, Jr., "Analog Transmission Performance of the Switched Telecommunication Network," *BSTJ*, Vol. 50, No. 4, Apr. 1971, pp. 1311–1347.
57. T.C. Spang, "Loss-Noise-Echo Study of the Direct Distance Dialing Network," *BSTJ*, Vol. 55, No. 1, Jan. 1976, pp. 1–36.
58. "Telecommunications Transmission Engineering," Vol. 1: Principles, 2nd ed., Chapter 26 on "Transmission Objectives," AT&T, 1977. Also, 3rd ed., Bellcore, 1990.
59. P.N. Burgess, J.L. Neigh, and R.G. Sparker, "The No. 10A Remote Switching System: Transmission Plan," *BSTJ*, Vol. 61, No. 4, April 1982, pp. 627–646.
60. LATA Switching Systems Generic Requirements (LSSGR): Section 32 on "Data Services," Bellcore TR-534, Issue 2, July 1987.
61. "Telecommunications Transmission Engineering," Vol. 3: Networks and Services, 2nd ed., Chapter 15 on "Private Switched Networks," AT&T, 1977.
62. "Telecommunications Transmission Engineering," Vol. 2: Facilities, 3rd ed., Chapter 13 on "PCM Transmission Systems," Bellcore, 1990.
63. "Generic Requirements for the Switched DS1/Switched Fractional DS1 Service Capability From an ISDN Interface (SWF—DS1/ISDN)," Bellcore TA-1203, Issue 1, April 1991.
64. J.E. Kessler, "The Transmission Performance of Bell System Toll Connecting Trunks," *BSTJ*, Vol. 50, No. 8, Oct. 1971, pp. 2741–2777.
65. J.R. Davis, L.E. Forman, and L.T. Nguyen, "The Metropolitan Digital Trunk Plant," *BSTJ*, Vol. 60, No. 6, July/Aug. 1981, pp. 933–964.
66. L.T. Wu and M. Kerner, "Emulating Circuits in a Broadband Packet Network," *GLOBECOM Rec.*, Vol. II, Hollywood, Florida, Nov./Dec. 1988, pp. 21.6.1–21.6.10.
67. H.D. Wright, "Is Your Network up to P/AR?," *Telecomm.*, Vol. 16, No. 7, June 1982, pp. 71, 72.
68. J. Kelly, "The Relationship of Phase Jitter to the Measurement of Other Voice Channel Parameters," *Telecomm.*, Vol. 6, No. 10, Oct. 1972, pp. 46–48, 50.
69. G.R. Ritchie and P.E. Scheffer, "Projecting the Error Performance of the Bell System Digital Network," *ICC Rec.*, Vol. 1, Philadelphia, June 1982, pp. 2D.2.1–2D.2.6.
70. H.S. Merrill, "Digital Performance Parameters for Switched Voice Grade Services," *NCF Rec.*, Vol. 1, Chicago, Oct. 1986, pp. 179–197.
71. CCITT Rec. G.821, "Error Performance on an International Digital Connection Forming Part of an Integrated Services Digital Network," *Blue Book*, Vol. III, Fas. III.3, Geneva, 1989, pp. 25–65.
72. CCITT Rec. G.822, "Controlled Slip Objectives on an International Digital Connection," *Blue Book*, Vol. III, Fas. III.3, Geneva, 1989, pp. 36–38.

73. CCITT SG XVIII Draft Rec. G.82x on "Error Performance Parameters and Objectives for Continuous Bit Rate International Digital Connections at or above the Primary Rate," Annex 1 to [113].
74. "Digital Data System Channel Interface Specification," AT&T PUB. 62310, Issue 2, Nov. 1987. Also, Adden. 3, Dec. 1989.
75. "ACCUNET T1.5 Service Description and Interface Specifications," AT&T PUB 62411, Issue 2, Dec. 1988. Also, Adden. 1, Mar. 1990. Also, Issue 3, Dec. 1990.
76. "ACCUNET T45 Service Description and Interface Specifications," AT&T PUB 54014, Issue 1, June 1987. Also, Adden. 3, Mar. 1990.
77. "High-Capacity Digital Special Access Service: Transmission Parameter Limits and Interface Combinations," Bellcore TR-342, Issue 1, Feb. 1991.
78. "Access Specification for High Capacity (DS1/DS3) Dedicated Digital Services," AT&T PUB 62415, June 1989.
79. "Access Performance Objectives for DS1 Services," Contributions T1Q1.4/88-024, T1Q1.4/89-010, T1Q1.4/90-004, and T1Q1.4/90-014, AT&T, Aug. 17, 1988, Feb. 8, 1989, Jan. 31, 1990, and April 25, 1990, respectively.
80. "International ACCUNET Digital Service Description and Network Interface Specifications," AT&T PUB 54019, Issue 1, Sept. 1987. Also, Adden. 1, April 1988.
81. "ACCUNET Switched 56 Service to Public Switched Digital Service Interface Advisory," AT&T PUB 61330, Sept. 1986.
82. "Network Transmission Interface and Performance Specifications Supporting ISDN," Bellcore TA-938, Issue 1, Feb. 1990.
83. J.J. Mahoney Jr., J.J. Mansell, and R.C. Metlack, "Digital Data System: User's View of the Network," *BSTJ*, Vol. 54, No. 5, May/June 1975, pp. 833–884.
84. S.D. Akers, S.C. Donn, S.R. Gregg, and S. Ravikumar, "Uses of In-Service Monitoring Information to Estimate Customer Service Quality," *GLOBECOM Rec.*, Vol. 3, Dallas, Nov./Dec. 1989, pp. 51.3.1–51.3.5.
85. LATA Switching Systems Generic Requirements (LSSGR): Section 18 on "Synchronization," Bellcore TR-518, Issue 2, July 1987.
86. R.E. Mallon, "Detection of Bursty Error Conditions through Analysis of Performance Information," *GLOBECOM Rec.*, Vol. 3, Tokyo, Nov. 1987, pp. 51.1.1–51.1.5.
87. J. Gruber, R. Vickers, and D. Cuddy, "Impact of Errors and Slips on Voice Service," *ICC Rec.*, Vol. 1, Philadelphia, June 1982, pp. 2D.5.1–2D.5.7.
88. H. Drucker and A.C. Morton, "The Effect of Slips on Data Modems," *ICC Rec.*, Vol. 1, Seattle, June 1987, pp. 12.4.1–12.4.5.
89. J.F. Ingle, "Identification of Digital Impairments in a Voiceband Channel," *ICC Rec.*, Vol. 1, Boston, June 1989, pp. 12.3.1–12.3.5.
90. EIA/TIA "Digital Private Network Synchronization Standard," PN-2198, 1991.
91. J.E. Abate and H. Drucker, "The Effect of Slips on Facsimile Transmission," *ICC Rec.*, Vol. 2, Philadelphia, June 1988, pp. 32.3.1–32.3.4.
92. S.E. Miller and I.P. Kaminow, *Optical Fiber Telecommunications II*, Chapter 22 on "Interoffice Transmission Systems," Academic Press, 1988, p. 844.
93. CCITT Rec. X.140 "General Quality of Service Parameters for Communication via Public Data Networks," *Blue Book*, Vol. VIII, Fas. VIII.3, Geneva, 1989, pp. 422–446.
94. CCITT Rec. X.135 on "Speed of Service (Delay and Throughput) Performance Values for Public Data Networks When Providing International Packet-Switched Service," *Blue Book*, Vol. VIII, Fas. VIII.3, Geneva, 1989, pp. 368–391.
95. CCITT Rec. X.136 on "Accuracy and Dependability Performance Values for Public Data Networks When Providing International Packet-Switched Service," *Blue Book*, Vol. VIII, Fas. VIII.3, Geneva, 1989, pp. 392–410.

96. CCITT Rec. X.137 on "Availability Performance Values for Public Data Networks when Providing International Packet Switched Service," *Blue Book*, Vol. VIII, Fas. VIII.3, Geneva, 1989, pp. 410–421.
97. CCITT Draft Rec. I.35A on "Availability Performance for 64 kb/s ISDN Connection Types," Annex 1 of TD 22 (XVIII/6), June 1991.
98. American National Standard for Telecommunications—Packet Switched Data Communications Service—Performance Parameters, ANSI T1.504-1989.
99. American National Standard for Information Systems—Data Communications Systems and Services—Measurement Methods for User-Oriented Performance Evaluation, ANSI X3.141-1987.
100. L.S. Newman, "Early Alternatives to the ISDN," *Telephony*, Vol. 211, No. 19, Nov. 10, 1986, pp. 44, 46, 47, 74.
101. K.E. Carmichael, W.K. Pehlert, and M. Ulema, "Maintenance of the ACCUNET Packet Service Network," *ICC Rec.*, Vol. 1, Toronto, June 1986, pp. 18.2.1–18.2.6.
102. L.S. Newman and E.G. Turman, "Public Packet Switching Networks for the BOCs: Technical Requirements," *GLOBECOM Rec.*, Vol. 2, Atlanta, Nov. 1984, pp. 21.4.1–21.4.4.
103. "Generic Systems Requirements in Support of Switched Multi-megabit Data Service," Bellcore TR-772, Issue 1, May 1991.
104. American National Standard for Telecommunications—IDSN—Architectural Framework and Service Description for Frame Relaying Bearer Service, ANSI T1.606-1990.
105. CCITT Draft Rec. I-35B on "Broadband ISDN Performance," TD 27 (XVIII/6), Geneva, June 1991.
106. D.S. Gokhale et al., "Quality-of-Service Measurement of Satellite-Based International Packet-Switched Networks," *COMSAT Tech. Rev.*, Vol. 19, No. 2, Fall 1989, pp. 235–258.
107. N.B. Seitz, D.R. Wortendyke, and K.P. Spies, "User-Oriented Performance Measurement on the ARPANET," *IEEE Comm. Mag.*, Vol. 21, No. 5, Aug. 1983, pp. 28–44.
108. D.R. Wortendyke, N.B. Seitz, K.P. Spies, E.L. Crow, and D.S. Grubb, "User-Oriented Performance Measurement on the ARPANET: The Testing of a Proposed Federal Standard," NTIA Rep. 82-112, U.S. Dept. of Commerce, Nov. 1982.
109. R.S. Bloomfield and K.P. Spies, "User-Oriented Performance Evaluation of Data Communications Services: Measurement Results," *ICC Rec.*, Vol. 1, Boston, June 1989, pp. 3.5.1–3.5.7.
110. N.B. Seitz, D.R. Wortendyke, and K.P. Spies, "User-Oriented Performance Measurement on Public Data Networks," *ICC Rec.*, Vol. 1, Chicago, June 1985, pp. 12.6.1–12.6.7.
111. K.P. Spies, W.R. Wortendyke, E.L. Crow, M.J. Miles, E.A. Quincy, and N.B. Seitz, "User-Oriented Performance Evaluation of Data Communication Service: Measurement Design, Conduct and Results," NTIA Report 88-238, U.S. Dept. of Commerce, Aug. 1988.
112. S.H. Richman, "A Comparison of Field Performance Measurements on Multi-National Packet Switching Networks," *GLOBECOM Rec.*, Vol. 3, Tokyo, Nov. 1987, pp. 41.6.1–41.6.8.
113. CCITT SG XVIII TD 30 to Quest. 6, "Report on Error Performance Issues," Geneva, June 1991.

Chapter 5
ALLOCATION OF END-TO-END PERFORMANCE TO NETWORKS AND NETWORK COMPONENTS

5.1 INTRODUCTION

Following the consideration of end-to-end network-oriented performance, as addressed in Chapter 4, the next step in the performance process is the allocation of end-to-end performance among the various networks and network components that make up an end-to-end connection (recall Fig. 1.3 of Ch. 1). This allocation process is required for network design and network maintenance purposes within networks, and on an interworking basis between networks to ensure that end-to-end performance levels are achieved. Allocation is based on a top-down process, taking into account bottom-up design, maintenance, and operational considerations. Designers of network components must know what level of performance each component should support to ensure that end-to-end performance can be achieved. Similarly, network maintenance personnel must know the performance level that network components should meet in order to determine, upon monitoring or testing of the components, whether maintenance action is required. Also, after repair, performance of the components must be restored to a known level, again, to ensure that end-to-end performance is achieved. (Aspects of maintenance are addressed in Ch. 9.)

Figure 5.1 is a simple illustration of a network (or interconnected networks) terminated at each end by customer premise equipment (CPE). As an example, consider the loudness loss parameter. Both the network and CPE contribute to the end-to-end loudness loss. Therefore, to specify each network component, a process is required to allocate end-to-end performance between, in this case, the CPE and network. In addition, connections can involve more than one network, each consisting of a number of network components; thus, allocation to these networks, and to components within each network is required. For example, in the U.S., divestiture has resulted in a number of different network jurisdictions. In addition, for international

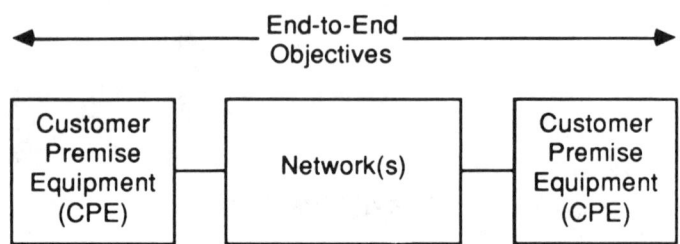

Figure 5.1 End-to-end objectives: network and CPE.

connections, there is a need to consider allocation among the international and national parts of the connection.

Thus, from a technical standpoint, allocating end-to-end network-oriented performance to lower level entities involves three main steps [1, 2]:

- Allocation to national and international networks.
- Allocation to different networks within a nation (e.g., to multivendor networks).
- Allocation to components within each national network.

Various types of connections and associated administrative boundaries are illustrated in Figure 5.2. Figure 5.2(a) illustrates an international connection made up of two national extensions that interface to the international part at the gateways. In this figure, the CPE is considered to be part of the national network. Allocation to national and international networks is dealt with largely in the International Telegraph and Telephone Consultative Committee (CCITT). For example, Study Group XVIII addresses digital network performance [3], particularly in terms of how much impairment is allowed for each national part. This step is a useful starting point, because any national network (public or private) that supports international traffic should comply with international standards in order to provide satisfactory end-to-end performance.

Within North America, the Canadian and U.S. situations regarding allocation of performance are different. In Canada, Telecom Canada is responsible for ensuring that national extensions meet CCITT requirements, while, in the U.S., divestiture has resulted in a division of responsibility within the national extension. Figures 5.2(b) and (c) illustrate the U.S. predivestiture and postdivestiture situations, respectively. Prior to divestiture, both the core and access parts of the network were managed by the Bell System. However, with the advent of terminal interconnect, CPE performance was considered separately. This required collaboration between the Bell System and the Electrical Industries Association (EIA, now the Telecommunications Industries Association (TIA)) standards body, which now provides CPE requirements.

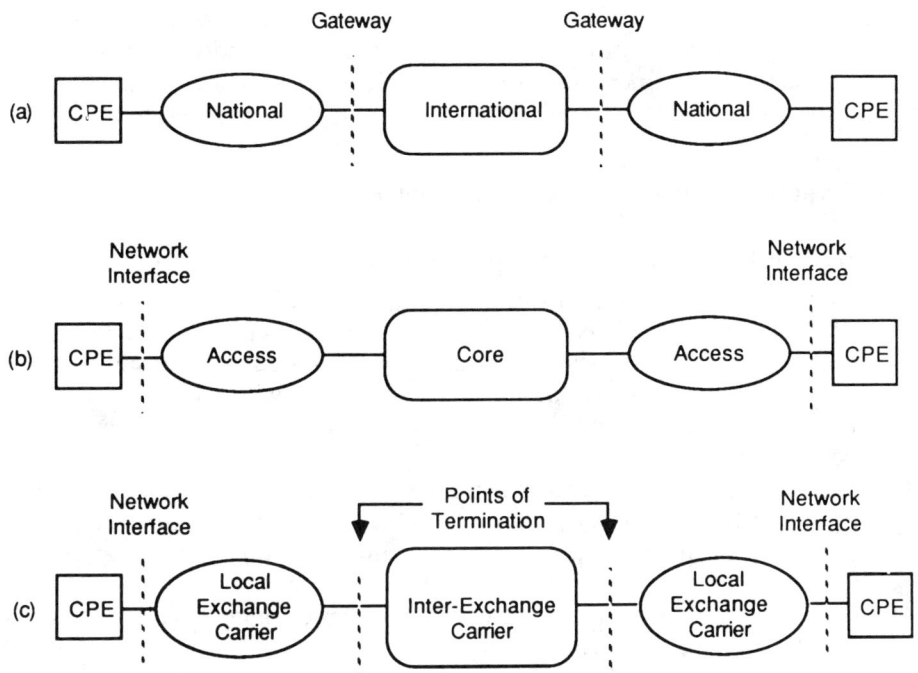

Figure 5.2 End-to-end connections and network boundaries: (a) national and international; (b) U.S. national: predivestiture; (c) U.S. national: postdivestiture.

Figure 5.2(c) illustrates the status following divestiture with three distinct types of network entities: CPE, local-exchange carriers (LEC) and inter-exchange carriers (IEC). The TIA TR 41forum addresses CPE requirements (including those of private networks), T1Q1 addresses performance allocation between IECs and LECs, and CCITT addresses allocation for international connections. These standards forums are discussed in detail in Chapter 8.

The final step in the performance allocation process addresses components within specific national networks, including subnetworks such as allocation to the access and exchange access portions within LECs (recall Figs 4.2 and 4.4 of Ch. 4). Such allocations are based largely on the performance capabilities of the technology (which are mainly transmission facilities and associated terminal equipment) and subject to the constraints of achieving the required level of end-to-end performance. However, to some degree, such allocation may also be at the discretion of the network owner, depending on experience, as well as economic and competitive factors.

Individual networks consist of network components such as switching, multiplexing, and transmission equipment (facilities and associated terminals). Therefore, network models or hypothetical reference connections (HRX) are required as a basis for establishing requirements for these components. Such models account for various factors, such as network architecture, technology, and distance; in the U.S. environment these models also account for LEC and IEC networks, and the CPE.

5.1.1 Purpose of Hypothetical Reference Connections

HRXs for both analog and digital connections serve the following purposes:

(1) To ensure end-to-end performance objectives are met in order to support a range of services, such as 4-kHz voiceband services and digital data services.
(2) To aid national administrations in examining the effect on transmission quality of possible changes to performance allocation within national networks; that is, to test compliance with national/international performance requirements.
(3) Nationally, in the U.S., to aid network providers in establishing performance objectives for their networks, and to aid the Standards Committee T1-Telecommunications and TIA standards committees in ensuring that end user needs are met.
(4) To provide a basis for allocating end-to-end performance objectives to design objectives for network components.
(5) To assist in establishing guidelines for the deployment of new technology (such as ADPCM coders-decoders (codecs)).

HRXs should also account for connection length and administrative boundaries between networks and between jurisdictions having different responsibilities. In addition, recognizing that it may be economically undesirable to design for the worst possible connection, the HRX is usually chosen to represent a near worst configuration (i.e., near longest connection).

5.2 NETWORKS AND NETWORK COMPONENTS

5.2.1 CCITT—International Networks

The main parts of an international connection are illustrated in Figure 5.3 [4]. These parts include national switches and international switching centers (ISC) that carry international traffic. For large countries such as Canada, the U.S., and Australia, it is important that the number of switching nodes and connection length be dependent on country size because of the general accumulation of impairments with the amount of equipment in the connection, and, hence, with distance. For this reason, HRXs are generally classified to cover connections of near longest length and of moderate length. Figures 5.4 and 5.5 illustrate long- and moderate-length digital HRXs [3].

Figure 5.3 Principle parts of an international connection.

Figure 5.4 Standard digital hypothetical reference connection (longest length).

Note - See Fig. 5.4 for legend.

Figure 5.5 Standard digital hypothetical reference connection (moderate length).

Similar HRXs have been agreed upon for analog impairments, and Figure 5.6 illustrates a moderate-length HRX for the loss and noise parameters [5]. In this figure, only one national part is illustrated (the connection is assumed to be symmetrical with respect to national parts). Figure 5.7 also illustrates how CPE characteristics are accounted for in specifying allocation of loudness loss to a large country

ISC International Switching Center
PC Primary Center
LE Local Exchange

Figure 5.6 Example of an international connection of moderate length showing one national portion and including loss and noise allocations.

Figure 5.7 Example of the allocation of maximum objective loudness rating values for an international connection in a large country.

[6]. Note that the loudness terminology in Figure 5.7 is in terms of objective loudness ratings obtained by applying the appropriate correction factor (from Table 3.7 of Ch. 3) to the corrected reference equivalent values given in [6].

5.2.2 North American Networks—A Comparison

In Canada competitive influence is currently limited to private networks and to the provision of CPE, and the concept of a national HRX is well established. Also, although Canadian and U.S. networks have technological similarities, in some cases there are significant architectural and topological differences; that is, the Canadian nationwide network is essentially "long" and "thin," as are certain networks in the U.S., such as those of MCI and US Sprint, whereas the AT&T long-haul network is more "mesh" like, with higher connectivity. For these reasons HRXs in Canada and the U.S. are treated separately.

In addition, divestiture in the U.S. has influenced network architecture and spheres of responsibility, and these factors need to be accounted for in a network model applicable to this environment. For example, an administrative boundary is required between the main network providers (LEC and IEC), and some agreement on reference connections is emerging in the T1Q1 standards forum to provide the basis from which to proceed in the future.

5.2.2.1 Canadian Networks

Canada has typically long and relatively thin telecommunications network topologies, which makes it fall under the CCITT large-country category. These networks run east to west and encompass the ten provinces. For example, Telecom Canada has developed HRXs to represent nationwide connections. Figure 5.8 illustrates an HRX for voice and voiceband data services covering interprovincial and provincial intertoll trunks, toll connecting trunks, and the access network [7]. The access network includes remote units and CPE (e.g., private branch exchanges (PBX)). In terms of the future and the higher penetration of fiber transport systems, more emphasis in the HRX on alternate routing or network reconfiguration may be appropriate to account for network survivability [8, 9].

5.2.2.2 U.S. Networks

The predivestiture Bell System network was hierarchical in nature. This network is well defined and documented and the allocation of performance followed traditional

Figure 5.8 Example of a Canadian hypothetical reference connection.

methodology (e.g., see [13] and [20] to [25] of Ch. 4). Following divestiture, the resolution of performance issues, including end-to-end performance and its allocation, became the responsibility of the T1Q1 standards committee (Ch. 8 describes this committee in detail). However, there are as yet no detailed HRXs agreed on in the U.S.

Divestiture resulted in two main types of network—LECs and IECs—which serve two different functions. The LECs support intra-local access and transport area (LATA) traffic, and provide access to IECs, and to international networks via IECs. LECs therefore support intra-LATA, inter-LATA, and international traffic. This situation is illustrated in Figure 5.9, which corresponds to the reference connection agreed to by the T1Q1 committee in its studies [10]. This figure shows the main network components of a U.S national HRX. Note that the LEC part only extends to the network interface (NI) between the LEC and CPE, and excludes the CPE (unlike in CCITT where the CPE is included in national HRXs). There is little published

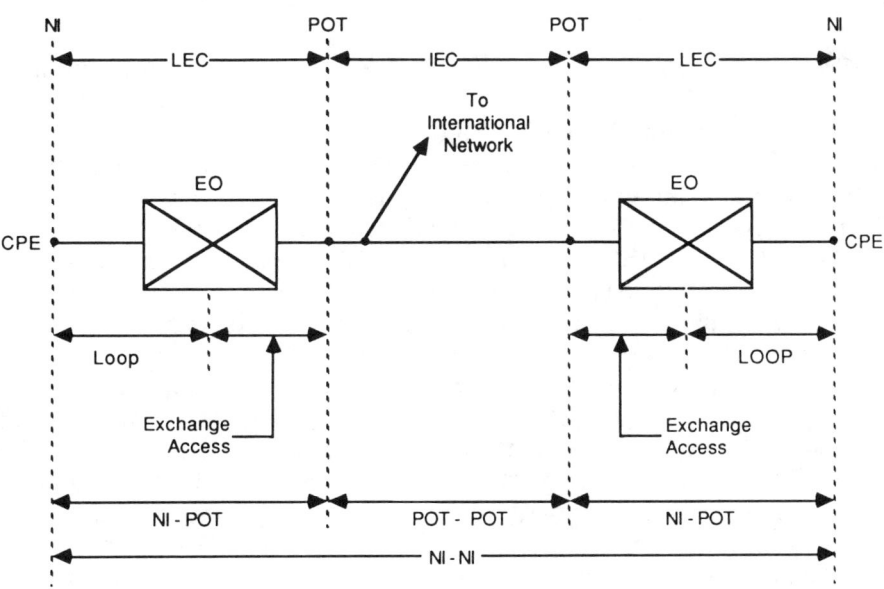

CPE - Customer Premises Equipment (e.g.,Telephone Set)
NI - Network Interface
EO - End Office
POT - Point of Termination

Figure 5.9 Principle parts of a U.S. network connection.

information on the recent nature of IEC networks, owing to competition among the IECs. Competition has limited public discussion of IEC network architecture in standard forums, because doing so would be construed as revealing proprietary information. Therefore, the industry is left with only the broad functional view of U.S. networks as given in Figure 5.9.

IEC Networks

The little information that does exist on the status of IEC networks reveals that most tend to be a two-level hierarchy [11], and many are currently engaged in extensive deployment of fiber-optic transmission [12–15]. Figure 5.10 illustrates the general layout of the US Sprint network, including the deployment of echo control devices [16]. Some information about the AT&T IEC network can be inferred from previous survey data [17]. The AT&T network is now nonhierarchical in nature, with time of day routing being used for traffic equalization purposes [18, 19].

LEC Networks

"Notes on the BOC Intra-LATA Network" provides an architectural view of intra-LATA networks and a number of reference connections [20]. However, these reference connections do not provide distance information. Figure 5.11 illustrates a hypothetical local (intra-LATA) network, including an inter-LATA connection provided by an IEC. Each IEC providing inter-LATA connectivity must have its own point of presence (POP) in each LATA. From Figure 5.11, the intra-LATA architecture provides for alternate routing between many of the end offices (EO).

An important part of the LEC network is the exchange access, which provides the connection between the POP and EO, as shown in expanded form in Figure 5.12. The possible access arrangements are tariffed offerings, and are described in [20]. As tariffed offerings, they come under the headings of feature group (FG) A, B, C, and D, with FGA providing line side access at the EO, and the others providing trunk side access. Following divestiture, to meet short-term needs, the line side arrangement (FGA) was implemented to provide access to IEC networks. The longer term (trunk side) arrangements, which are intended to provide more equitable performance for access to every IEC, are as illustrated in Figure 5.12. The industry objective is to universally provide the trunk side FGD equal access arrangement.

Performance of these arrangements has been the subject of standardization in the T1Q1 forum, resulting in the "Switched Exchange Access Network Transmission Performance Standard," which covers both voice and voiceband data services [21]. A key aspect of the FGD arrangement in Figure 5.12 is that the performance specification is the same for either a direct connection or a connection via an access tandem between a POP and a given EO (i.e., equal access principle). FGD ensures

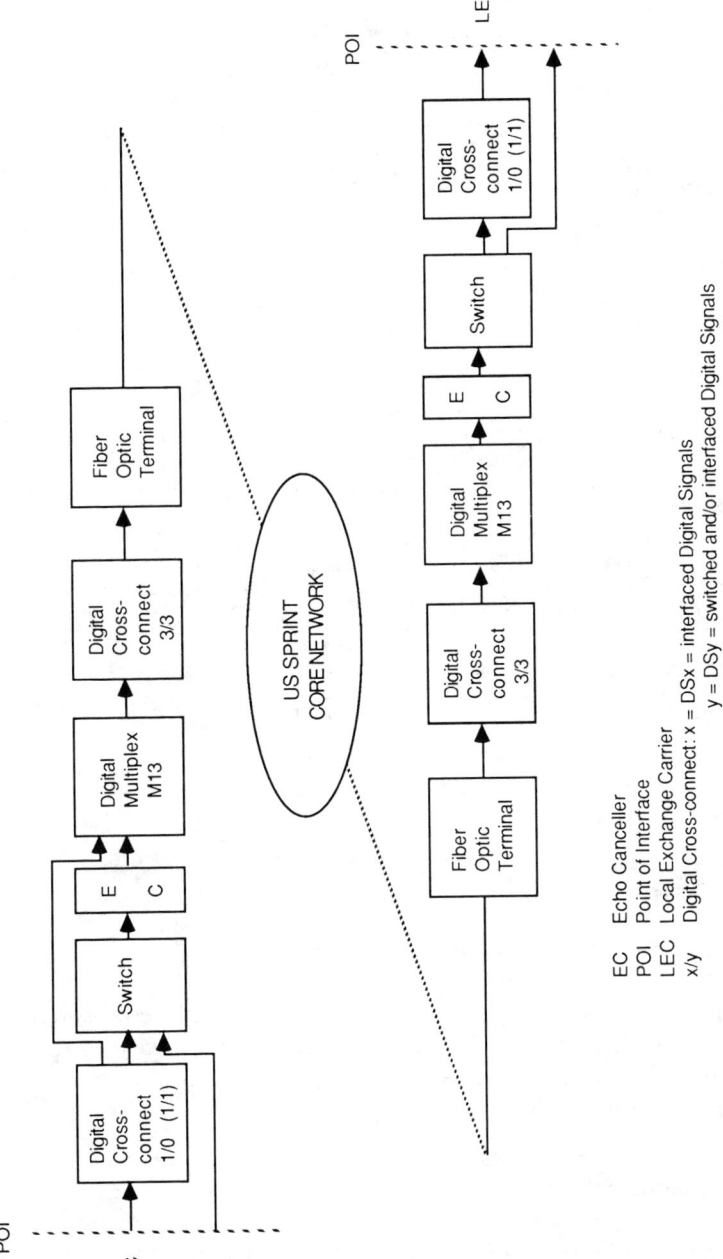

Figure 5.10 General layout of the US Sprint network (one direction of transmission shown[16]).

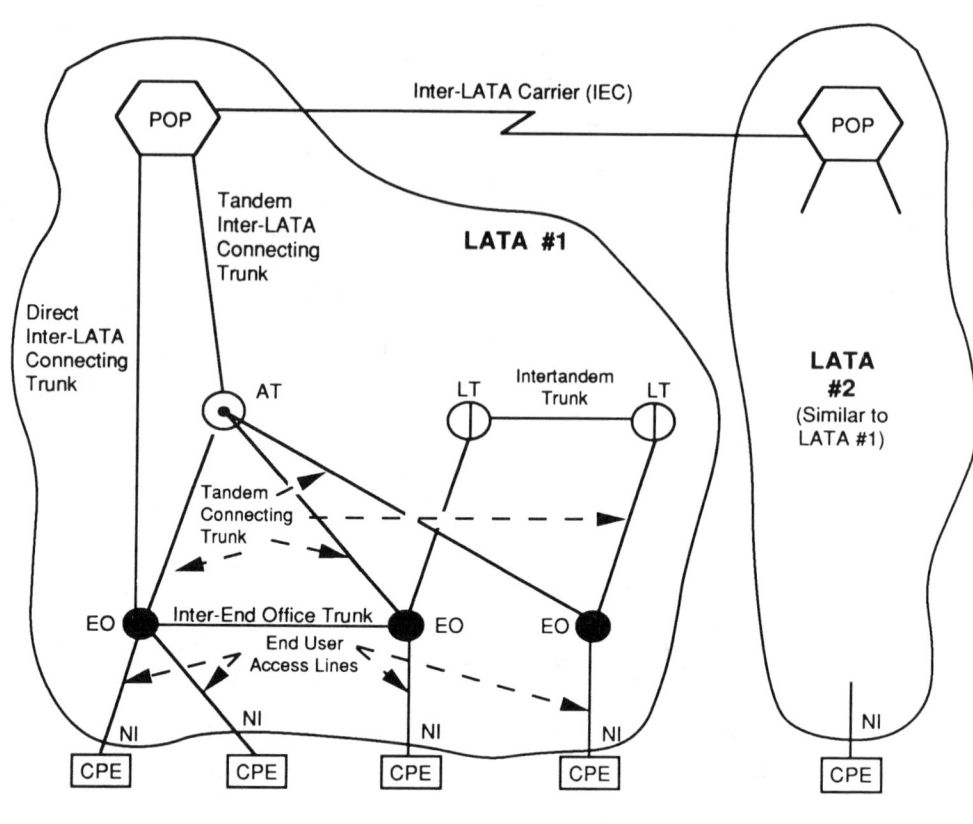

Figure 5.11 Hypothetical intra-LATA network (enclosed).

that the IECs using an access tandem are provided performance equivalent to that for IECs with direct access. In the context of allocation, the process in developing this industry standard was to use measured survey data as the basis, taking into account end user requirements [22].

5.2.3 Network Components

To facilitate the study of digital transmission performance at the equipment level, it is necessary to define network models comprising a combination of different types

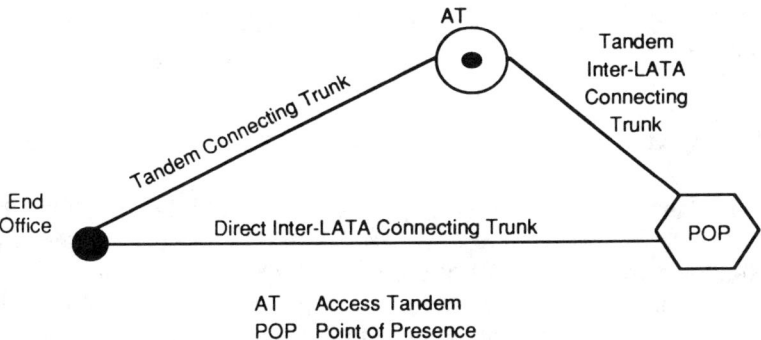

AT   Access Tandem
POP Point of Presence

Figure 5.12 Exchange access reference connection.

of network components. In some models (i.e., for circuit switching) the components consist of subscriber lines, switches, and trunks. For transport networks, the components consist of multiplexers and transmission systems. In the latter case, the model constitutes a hypothetical reference digital path (HRDP) with a specified number of network components, as shown in Figure 5.13. A digital path is the means

▶◀ Digital regenerator
ME Maintenance entity
LTE Line terminating equipment
X Digital distribution frame or equivalent point
Mux Digital line multiplexer

Figure 5.13 Digital path and maintenance entities.

of transmission of a digital signal of specified rate between two digital distribution frames [23, 24].

5.3 PERFORMANCE ALLOCATION CRITERIA

From the previous sections in this chapter, a number of different allocation criteria for allocating performance were noted in passing. Before proceeding with performance allocation methodology, this section will review the various allocation criteria. The more classical/traditional criteria are distance and technology, while in the U.S. the more contemporary criteria are a function of postdivestiture and evolutionary/competitive factors.

5.3.1 Distance

Continuous analog impairments such as loss and noise, when associated with analog facilities, accumulate as a function of distance. However, networks are continually evolving to become all digital, and many of the traditional analog impairments are now a function of transmission planning requirements (e.g., the fixed loss transmission plan discussed in Ch. 7) and terminal equipment characteristics. One analog impairment that is distance-dependent and that will remain so, even with end-to-end digital connectivity, is absolute delay. This delay is contributed to by transmission facilities and terminals, as well as by switching equipment. The effect of the increase in network delay on transmission planning requirements is addressed in Chapter 7. Digital facilities, while minimizing impairments such as loss and noise, are characterized by transient impairments such as digital errors, which generally also accumulate as a function of distance, owing to the larger numbers of equipment components (such as regenerators and transmission terminals) in longer connections.

5.3.2 Technology

An example of a technology-related performance allocation criterion applies to digital circuit multiplication equipment (DCME). In Chapter 3, and later in this chapter, it is shown that the quantizing distortion impairment is directly related to the number of DCME systems (e.g., 32-kb/s ADPCM codecs) in a connection. Therefore, on an international connection, the allowable number must be allocated between the international and the two national portions. Another example relates to the nature of transmission systems; fiber transmission systems are generally capable of providing higher transmission performance than older type radio systems, and thus could conceivably be assigned a more stringent (smaller) allocation of the end-to-end performance budget. However, for administrative simplicity, it is preferable to allocate performance in a technology-independent manner.

5.3.3 U.S. Postdivestiture and Evolutionary/Competitive Factors

Given that network components in a connection contribute to the accumulation of end-to-end impairments, and given the competitive nature of the U.S. telecommunications industry, the allocation or specification of performance for the LEC or IEC networks is a sensitive issue. Examples of recent industry standards agreements show that there are basically two approaches to addressing this issue.

The first is an approach based on the current performance of the network, as used in the "Switched Exchange Access Network Transmission Performance Standard" [21]. Here, basic network survey data was used, tempered with a view of the end-to-end performance requirements. The second is an approach based on an allocation of the number of allowable systems to each network portion, as used in the ADPCM performance standard [25]. In this case, with no more than three to four ADPCM systems in a connection, end user requirements can still be supported, and, on this basis, one ADPCM system was allocated to each LEC network and one to the IEC network (as discussed in Ch. 7).

Also an important factor in the U.S. is the position of the LECs between the IECs and end users (CPE), and as providers of service to both parties. As such, the LECs are a "necessary" part of every connection and would prefer an equitable performance allocation based on top-down principles. However, some IECs, while not wanting to reveal their network characteristics, prefer more stringent exchange access network performance according to a technology-based allocation approach. The IECs have the alternative of bypassing the LEC networks. Similarly, the CPE vendors want more stringent NI-to-NI requirements to ease performance requirements for CPE. Thus, the LECs, IECs, and CPE vendors all support different allocation approaches, and one of the main tasks of the U.S. standards forums is to resolve these differences in an equitable fashion.

5.4 PERFORMANCE ALLOCATION METHODOLOGY

5.4.1 General

The term "allocation" is a top-down term implying assignment of end-to-end network performance to smaller network entities. Conversely, the term "accumulation" may be regarded as a bottom-up term implying the combination of smaller component-oriented performance quantities into the end-to-end value. Thus, "allocation" and "accumulation" procedures are inverses of one another, and either or both may be used. However, as a practical matter, "accumulation" is often a simpler procedure than "allocation," and for this reason will be most often used here. The actual portion of end-to-end performance that is allocated to particular components is often a complicated function of the performance allocation criteria discussed earlier (Sec. 5.3), as well as many other factors and engineering judgments, including those related to service requirements, economics, backward compatibility. Some of these factors fall under transmission planning as addressed in Chapter 7.

In the remainder of this chapter, the mathematical methods associated with the "allocation" or "accumulation" procedures are provided. These are also sometimes referred to as the "addition laws," where the "addition" may be on a voltage, power, or some other basis. Some of the simpler addition laws involving fixed values and serial (i.e., single route) connection configurations have been introduced previously (e.g., see Eqs. (3.1) to (3.7) for loudness loss and echo-path delay, etc., in Sec. 3.2.2 of Ch. 3). Here these and other laws are formalized, and the more general situations involving asymmetries, distributions of values, as well as parallel (e.g., alternate route) connection configurations, are also considered. Examples and practical insights into the use of the laws are also provided.

5.4.2 Addition Laws for Serial Connection Configurations and for Fixed Parameter Values

For convenience and for reference purposes, the relevant aspects of these laws are summarized in Tables 5.1 to 5.6. Tables 5.1 and 5.2 indicate the parameters for which algebraic and power-like addition laws apply, respectively [21, 26]. Table 5.3 defines the symbols and other relationships associated with the parameters. Tables 5.4 and 5.5 summarize the specific algebraic and power-like laws, respectively. Finally, Table 5.6 indicates another less common addition law for voiceband jitter. (Note: the specialized addition laws for digital network jitter are treated separately in Ch. 6.) The following elaborates on the addition laws in Tables 5.4 and 5.5 by considering asymmetrical situations and various examples.

5.4.2.1 Algebraic Addition Laws

The two basic techniques that may be used for performance allocation/accumulation purposes are analytical and simulation. The analytical technique is particularly useful in situations where only fixed parameter values are involved. In this case, relatively simple addition laws apply. (In fact, these same simple laws are also used for each set of parameter values sampled as part of the Monte-Carlo simulation technique; this technique is described later when distributed parameter values are addressed.) For example, in serial connection configurations, accumulated loss in decibels is the algebraic sum of all component losses in decibels; this applies to end-to-end loss and echo path loss situations. The same is true of other parameters that

Table 5.1
Parameters for Which Algebraic and Related Addition Laws Apply (Fixed Parameter Values)

Generalized Algebraic Addition Law

$$X = \sum_{i=1}^{n} x_i$$

Continuous Analog Parameters

    Basic Parameters:
        Loss and Loss Deviation
        Attenuation Distortion
        Delay
        Envelope Delay Distortion

    Parameters Derived from Basic Parameters:
        Output Signal Level
        Talker and Listener Echo Path Loss and Delay
        Loss, Noise, and Talker Echo Path Loss and Delay Asymmetry

Independent, Transient Event-Like Parameters

    Analog Events:
        Noise Impulses
        Gain and Phase Hits
        Drop outs
        Speech Clips
        Variable Delays

    Digital Events: *
        Anomalies: Errors (measured by BER, BLER, CV, ES, SES parameters)
                          Controlled Slips (measured by CSS, ES parameters)
        Defects: LOS, SEF, LOP, AIS, FERF (measured by LOSS, SEFS, AISS,
                          SAS, ALS, ES, SES parameters)
        Threshold Crossings (indicated by alert messages)

    Digital Failures (Alarm and Status Conditions) and Related Events: *
        Periods of $\geq$ 10 consecutive SES (measured by UAS parameter)
        Loss-of-Signal (LOS) #
        Loss-of-Frame (LOF) #
        Loss-of-Pointer LOP) #
        Alarm Indication Signal (AIS) #
        Remote Alarm Indication (RAI) @
        Far-End Receive Failure (FERF) @
        Protection Switch Occurrences (measured by PSC parameter)
        Protection Switch Durations (measured by PSD parameter)

---

*   See Chapter 9 for terminology, and definitions of digital events, parameters, failures, etc.
\#   Indicated by alarm or status message and included in UAS parameter
@ Indicated by alarm or status message and included in far-end UAS parameter

Table 5.2
Parameters for Which Power and Related Addition Laws Apply (Fixed Parameter Values)

Generalized "Power-Like" Addition Law

$$X = \sum_{i=1}^{n}{}_{Kp} x_i = K\log_{10}\left\{\sum_{i=1}^{n} 10^{x_i/K}\right\}$$

Continuous Analog Parameters	K
Idle Circuit Noise: *	
C-Message, and 3 and 15 kHz Flat	+ 10
Signal Dependent Noise: *	
C, and 3 and 15 kHz Flat Notched	+ 10
Intermodulation Distortion:	
2nd Harmonic	+ 10
3rd Harmonic	+ 16
Multiple Echo Path Losses:	
Talker and Listener Echo Path Losses	− 10
Signal-to-Noise Ratio:	
Idle Circuit and Signal Dependent Noise	− 10
Signal-to-Quantization Noise Ratio (Q):	0 to −15 #
Signal-to-Intermodulation Distortion Ratio:	
2nd Harmonic	− 10
3rd Harmonic	− 16

* Noise components are appropriately adjusted for loss (see text in Sec. 5.4.2.2).
\# Depending on the type of codec tandeming law (see Table 5.8)

Table 5.3
Relevant Parameter Definitions for Fixed Parameter Values and for the ith Network Component

Parameter	Unit	Symbol	Definition/Description
Specified Loss	dB	L_{si}	1 kHz Design Loss
Loss	dB	L_i	1 kHz inserted loss
Loss Deviation	dB	L_{di}	$L_{si} - L_i$
Frequency Response			
at 400 Hz	dB	$L_{0.4i}$	Loss at 400 Hz
at 2800 Hz	dB	$L_{2.8i}$	Loss at 2800 Hz
Attenuation Distortion			
at 400 Hz	dB	$L_{0.4ai}$	$L_{0.4i} - L_i$
at 2800 Hz	dB	$L_{2.8ai}$	$L_{2.8i} - L_i$

Table 5.3
continued

Parameter	Unit	Symbol	Definition/Description		
Input Signal Level	dBm	S_{i-1}	e.g., at 1 kHz		
Output Signal Level	dBm	S_i	$S_{i-1} - L_i$		
Output Noise	dBrnC	N_i	$\alpha_i S_{i-1} + \beta_i - L_i = \alpha_i S_i + \beta_i - (1-\alpha_i)L_i$		
Companding Factor		α_i	Compandor None PCM Analog	α_i 0 1 1/2 (example)	
Signal-to-Noise Ratio	dB	SNR_i	$S_i - N_i + F = (1-\alpha_i)(S_i + L_i) - \beta_i + F$		
Unit Conversion Factor	dB	F	Noise Idle Circuit Idle Circuit Idle Circuit Signal Dependent	Weighting C 3 kHz Flat 15 kHz Flat All	F 88.0 88.8 90.0 90.0

Table 5.4
Summary of Specific Algebraic Addition Laws for Fixed Parameter Values and Directional Symmetry of the Connection

Parameter	Unit	Network Component Symbol	Network Connection Symbol	Law
Loss	dB	L_i	L	$= \sum_{i=1}^{n} L_i$
Loss Deviation	dB	L_{di}	L_d	$= \sum_{i=1}^{n} L_{di}$
Attenuation Distortion	dB	L_{ai}	L_a	$= \sum_{i=1}^{n} L_{ai}$
Delay	ms	d_i	d	$= \sum_{i=1}^{n} d_i$
Envelope Delay Distortion	μs	d_{ei}	d_e	$= \sum_{i=1}^{n} d_{ei}$
Connection Input Signal	dBm	-	S_I	-
Component Output Signal	dBm	$S_i = S_{i-1} - L_i$	S_i	$= S_I = \sum_{j=1}^{i} L_i$
Connection Output Signal	dBm	-	S_O	$= S_I - L$
Talker Echo Path Loss	dB	L_i, ERL_p	TEPL	$= 2\sum_{i=1}^{p} L_i + ERL_p$
Listener Echo Path Loss	dB	L_i, RL_q, RL_p	LEPL	$= RL_q + 2\sum_{i=q}^{p} L_i + RL_p$

Table 5.4
continued

Parameter	Unit	Network Component Symbol	Network Connection Symbol	Law
Talker Echo Path Delay	ms	d_i	TEPD	$= 2 \sum_{i=1}^{p} d_i$
Listener Echo Path Delay	ms	d_i	LEPL	$= 2 \sum_{i=q}^{p} d_i$
General Independent events	-	e_i	e	$= \sum_{i=1}^{n} e_i$

Table 5.5
Summary of Specific Power-Like Addition Laws for Fixed Parameter Values and for Directional Symmetry of the Connection

Parameter	Unit	Network Component Symbol	Network Connection Symbol	Law
Noise	e.g., dBrnC	N_i	N	$= \sum_{i=1}^{n} 10p \left(N_i - \sum_{j=i+1}^{n} L_j \right)$
2nd Harm. Intermod. Dist.	dBm	N_{d2i}	N_{d2}	$= \sum_{i=1}^{n} 10p \, N_{d2i}$
3rd Harm. Intermod. Dist.	dBm	N_{d3i}	N_{d3}	$= \sum_{i=1}^{n} 16p \, N_{d3i}$
Multiple Talker Echo Path Losses	dB	$TEPL_i$	TEPL	$= \sum_{i=1}^{n} {}_{-10p} TEPL_i$
Multiple Listener Echo Path Losses	dB	$LEPL_i$	LEPL	$= \sum_{i=1}^{n} {}_{-10p} LEPL_i$
Signal-to-Noise Ratio	dB	SNR_i	SNR	$= \sum_{i=1}^{n} {}_{-10p} SNR_i$
Signal-to-Quantization Noise Ratio (Q)	dB	Q_i	e.g., Q	$= \sum_{i=1}^{n} {}_{-15p} Q_i$
Signal-to-Intermod. Dist. Ratio				
2nd Harm.	dB	r_{2i}	r_2	$= \sum_{i=1}^{n} {}_{-10p} r_{2i}$
3rd Harm.	dB	r_{3i}	r_3	$= \sum_{i=1}^{n} {}_{-16p} r_{3i}$

Harm.:   Harmonic

Intermod. Dist. Ratio: Intermodulation Distortion (IMD) Ratio

$\sum_{i=1}^{n} {}_{\pm xp}$:   Notation denoting global power sum.

Table 5.6
Addition Laws for Voiceband Jitter

Sum of Variables raised to a Fixed Power, Law:

$$X = \left\{ \sum_{i=1}^{n} x_i^q \right\}^{1/q}$$

E.g.: For peak amplitude or phase jitter (x_i) on n = 2 tandemed trunks:

n = 2,
q = 1.33 for wideband (2 - 300 Hz) jitter, and
q = 1.45 for narrowband (20 - 300 Hz) jitter.

Therefore: $X_a = (x_{a_1}^q + x_{a_2}^q)^{1/q}$ for total amplitude jitter, and

$X_p = (x_{p_1}^q + x_{p_2}^q)^{1/q}$ for total phase jitter.

Note: Specialized Addition Laws for Digital Network Jitter are included in Chapter 6.

accumulate algebraically, such as fixed delay in milliseconds. These and other cases are addressed in detail below.

End-to-End Loss

In the case where *electrical connection loss* is of interest (e.g., for voiceband data service), the losses of the trunks and of the subscriber loops are all 1-kHz losses in dB. This situation is illustrated in Figure 5.14, where for n components in a connection, each with loss L'_i dB in the transmit direction and L''_i dB in the receive direction, the accumulated losses for the connection are

$$L_{AB} = \sum_{i=1}^{n} L'_i \quad \text{and} \quad L_{BA} = \sum_{i=1}^{n} L''_i \text{ (dB)} \tag{5.1}$$

where A and B denote the A and B ends of the connection in Figure 5.14, AB denotes the A to B direction, and BA denotes the B to A direction.

In cases where *acoustic connection loss* is of interest (e.g., for speech service), the losses of the subscriber loops plus station sets are in terms of transmit and receive loudness ratings (TOLR and ROLR, respectively), as introduced in Section 3.2.2.1 of Chapter 3; in this case,

$$\text{ICL}_{AB} = \sum_{i=2}^{n-1} L'_i \quad \text{and} \quad \text{ICL}_{BA} \sum_{i=2}^{n-1} L''_i \tag{5.2a}$$

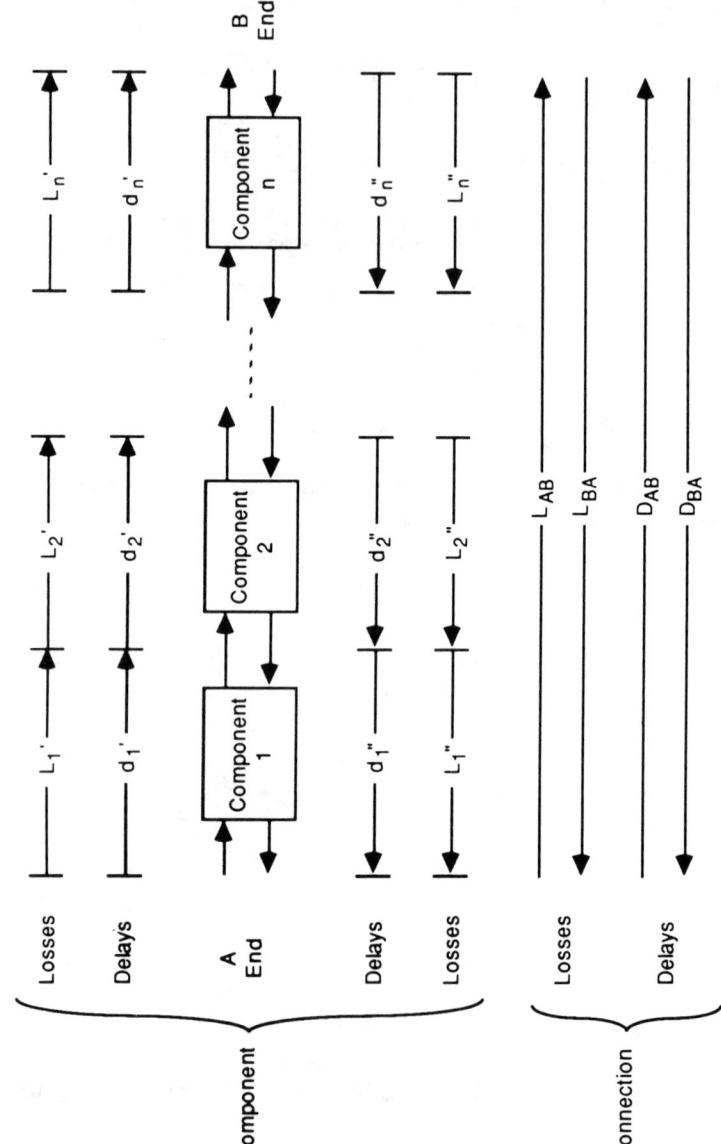

Figure 5.14 Illustration of losses and delays for connection components and for end-to-end connection.

$$L_{AB} = \text{OOLR}_{AB} = \text{TOLR}_A + \text{ICL}_{AB} + \text{ROLR}_B \text{ and}$$
$$L_{BA} = \text{OOLR}_{BA} = \text{TOLR}_B + \text{ICL}_{BA} + \text{ROLR}_A \qquad (5.2b)$$

where ICL is the inserted connection loss of the trunk part of the connection, and OOLR is the overall objective loudness rating of the connection, as also given earlier in Chapter 3. The loss values in Equation (5.2b) would be used for grade-of-service (GOS) purposes in the loss/noise transmission rating equations in Table 3.4(a) of Chapter 3.

As will be apparent from Chapter 7, it is possible for the loss of each trunk component to be the same in each direction; this is the case in all-analog networks. Also, owing to the 2-wire nature of subscriber loops, their electrical loss will be the same in each direction. Thus, in this specific type of symmetrical situation,

$$L'_i = L''_i = L_i \qquad (5.3)$$

For mixed analog/digital or all-digital trunking in the core portion of networks, the loss of certain trunks will be different in the two directions, so that Equation (5.3) does not apply to all components. However, the *total* connection loss can still be the same in both directions. In this less specific type of symmetrical situation,

$$\sum_{i=1}^{n} L'_i = \sum_{i=1}^{n} L''_i = \sum_{i=1}^{n} L_i \qquad (5.4)$$

Applying either of the symmetrical situations in Equations (5.3) or (5.4) to Equation (5.1), the end-to-end *electrical* connection loss simplifies to

$$L = L_{AB} = L_{BA} = \sum_{i=1}^{n} L_i \qquad (5.5)$$

For *acoustic* connections, the situation is different with regard to subscriber loop plus station set components. First, the transmit and receive sensitivities (TOLR and ROLR) are different. Secondly, the variation in these sensitivities differs with different loop lengths, so that symmetry is only achieved when loop lengths are the same at both ends of a connection. This is discussed further below.

Loss Asymmetry

Loss asymmetry is a relevant aspect of end-to-end loss for speech service, and is the difference in loss in the two directions of transmission as experienced during a conversation. That is, using Equation (5.2b), the general expression for loss asymmetry is

$$\Delta L = L_{AB} - L_{BA}$$
$$= \text{TOLR}_A + \text{ROLR}_B + \text{ICL}_{AB} - \text{ICL}_{BA} - (\text{TOLR}_B + \text{ROLR}_A) \quad (5.6)$$

If either the individual or total trunk losses are the same for each direction (see Eqs. (5.3) and (5.4), respectively), then the loss is symmetrical for the trunk part of the connection, and (with reference to Eq. (5.2a):

$$\text{ICL}_{AB} = \text{ICL}_{BA} = \text{ICL} = \sum_{i=2}^{n-1} L_i \quad (5.7)$$

and from Equations (5.6) and (5.7), the connection asymmetry reduces to

$$\Delta L = \text{TOLR}_A + \text{ROLR}_B - (\text{TOLR}_B + \text{ROLR}_A) \quad (5.8)$$

which is loss asymmetry due only to the subscriber loop plus station set sensitivities. Generally, this component of loss asymmetry is only eliminated if the loop plant is the same at both ends of the connection. This is because TOLR is more sensitive to changes in loop length than is ROLR, so that if the loop plant is not the same at both ends, then $\text{TOLR}_A + \text{ROLR}_B \neq \text{TOLR}_B + \text{ROLR}_A$, and therefore $\Delta L \neq 0$ dB. This is illustrated by an example from [27], as shown in Table 5.7.

This example shows that the end user with the shortest loop (at the A end of the connection in this example) will experience the most end-to-end loss (L_{BA} in this example). Again, the reason is that TOLR is more sensitive to change in loop length than is ROLR. That is, TOLR_B decreases considerably due to the long loop at end B, which results in a reduced signal (more loss experienced) at end A. In addition, as discussed later in this chapter, the user at the A end of this example connection also experiences higher echo and noise due to asymmetry for these impairments. Such asymmetries are significant in interworking situations (e.g., between private and public networks). End users served by the lowest loss access arrangement will generally experience the most performance degradation [28].

End-to-End Delay

In a manner similar to end-to-end loss in Figure 5.14 for the fixed delay parameter, given n components of delay in milliseconds of d'_i in one direction and d''_i in the other direction, then the connection delays are

$$d_{AB} = \sum_{i=1}^{n} d'_i, \quad \text{and} \quad d_{BA} = \sum_{i=1}^{n} d''_i \text{ (ms)} \quad (5.9)$$

Table 5.7
Example of Asymmetries for End-to-End Loss, Talker Echo Path Loss, and Noise,
Due to Differences in Subscriber Loop Plus Station Set Sensitivities

Example Loops	At End #	Loop				Acoustic Loss (dB)					Asymmetries (dB)		
		Length (kft)	Electrical Loss (dB)	TOLR	ROLR	End-to-End OOLR		Talker-Echo TEPL		Loss	TEPL *	Noise *
						L_{AB}	L_{BA}	E_A	E_B	$\Delta L = L_{AB} - L_{BA}$	$\Delta E = E_A - E_B$	$\Delta N = ROLR_A - ROLR_B$
Identical	A	9.5	4	-42.6	47.9	5.3	5.3	5.3	5.3	0	0	0
	B	9.5	4	-42.6	47.9							
Non-Identical	A	4.9	2	-46.2	46.4	3.4	7.3	0.2	10.5	-3.9	-10.3	-3.2
	B	13.5	6	-39.1	49.6							

* Asymmetries for Talker Echo Path Loss and Noise are discussed in subsequent sections of this chapter.
# - "End" refers to the loop at the A or B end of the connection.

Very often in practice, either the individual component delays or the total delays in each direction will be the same so that

$$d = d_{AB} = d_{BA} = \sum_{i=1}^{n} d_i \text{ (ms)} \qquad (5.10)$$

One may also define delay asymmetry, but ordinarily little or no delay asymmetry is encountered in practice. Even in rare instances where there is significant delay asymmetry (e.g., satellite route in one direction, and terrestrial route in the other), this asymmetry per se is not a significant factor in affecting the dynamics of conversations. In fact, it is well known that, in terms of conversational dynamics, the effect is the same if the delay is distributed between the directions or all lumped in one direction [29, 30].

Talker Echo Path Loss

Figure 5.15 illustrates a connection with a single talker echo path as observed from the A end of the connection. Usually, at least one talker echo path would also be observed from the B end of the connection. Given n component losses in the transmit direction L'_i and in the receive direction L''_i, and a reflection after the pth component, the *electrical* talker echo path loss (TEPL) is

$$\text{TEPL}_A = \sum_{i=1}^{p} (L'_i + L''_i) + \text{ERL}_p \text{ (dB)} \qquad (5.11)$$

where ERL_p is an echo return loss in decibels, as defined in Section 3.2.2.3 of Chapter 3. Although not shown in Figure 5.15, a talker echo path as observed from the B end of the connection and having a reflection after the qth component would similarly have loss:

$$\text{TEPL}_B = \sum_{i=q}^{n} (L'_i + L''_i) + \text{ERL}_q \text{ (dB)} \qquad (5.12)$$

Again, electrically, the L'_i and L''_i terms are 1-kHz losses. Acoustically, the losses of the subscriber loop plus station set components are in terms of TOLR and ROLR. In the *acoustic* case,

$$\text{TEPL}_A = E_A = \text{TOLR}_A + \text{ROLR}_A + \sum_{i=2}^{p} (L'_i + L''_i) + \text{ERL}_p \text{ and}$$

$$\text{TEPL}_B = E_B = \text{TOLR}_B + \text{ROLR}_B + \sum_{i=q}^{n-1} (L'_i + L''_i) + \text{ERL}_q \qquad (5.13)$$

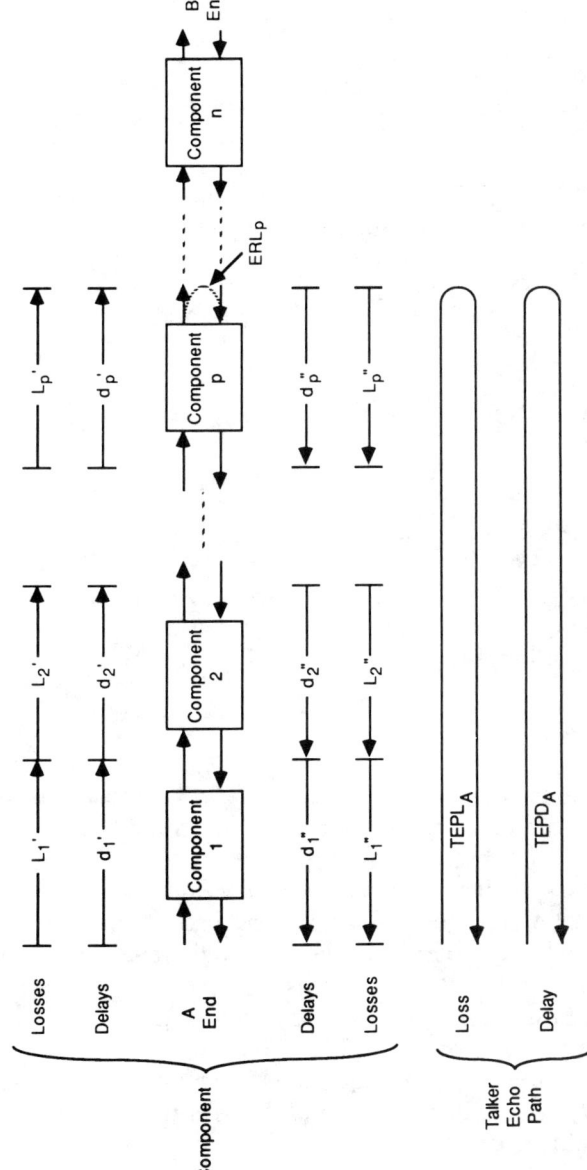

Figure 5.15 Illustration of losses and delays for connection components and for a single talker echo path.

These are the TEPLs that would be used for GOS purposes in the talker echo transmission rating equation given in Table 3.4(a) in Chapter 3. If the total *electrical* loss is the same for each direction, which often happens in practice, then, using Equation (5.3) in Equations (5.11) and (5.12),

$$\text{TEPL}_A = 2 \sum_{i=1}^{p} L_i + \text{ERL}_p \quad \text{and} \quad \text{TEPL}_B = 2 \sum_{i=q}^{n} L_i + \text{ERL}_q \quad (5.14)$$

However, *acoustically,* this symmetry would only apply to trunk components; therefore, using Equation (5.3) in Equation (5.13),

$$E_A = \text{TOLR}_A + \text{ROLR}_A + 2 \sum_{i=2}^{p} L_i + \text{ERL}_p \text{ and}$$

$$E_B = \text{TOLR}_B + \text{ROLR}_B + 2 \sum_{i=q}^{n-1} L_i + \text{ERL}_q \quad (5.15)$$

Talker Echo Path Loss Asymmetry

TEPL asymmetry is relevant for speech service and is the difference in TEPL as observed by users at each end of the connection during a conversation; therefore, with reference to Equations (5.13) or (5.15), TEPL asymmetry is

$$\Delta E = E_A - E_B \quad (5.16)$$

From the previous discussions of Equations (5.11) to (5.15), it is apparent that TEPLs as observed from each end of the connection could be completely different, owing to different ERLs, subscriber loops plus station set sensitivities, and even different trunks that are spanned by the echo paths. For speech service, this generally results in different echo GOS for the two subscribers. A common situation is that illustrated in Figure 3.1 of Chapter 3, in which the main reflections occur at the interface between the trunk and loop plant. In this case, the reflections are located at $p = n - 1$, and $q = 2$ in Equations (5.11) to (5.15), and the total trunk losses (ICL_{AB} and ICL_{BA}) are then as shown in Equation (5.2a). Also, the total trunk loss will often be the same in each direction, and then Equation (5.7) applies. Therefore, using Equations (5.2a) and (5.7) in Equations (5.15) and (5.16) eliminates the trunk losses, and the asymmetry simplifies to

$$\Delta E = \text{TOLR}_A + \text{ROLR}_A + \text{ERL}_{n-1} - \text{ERL}_2 - (\text{TOLR}_B + \text{ROLR}_B) \quad (5.17)$$

Apart from the obvious effect on TEPL asymmetry of different ERLs, it is also important to note the contribution to asymmetry from the different subscriber loop plus station set sensitivities. The contribution to TEPL asymmetry by the loop plus set sensitivities is illustrated by the example given in Table 5.7 (where equal ERLs are assumed). Again, asymmetry is only eliminated if the loop plus station set sensitivities are the same at both ends of the connection. Furthermore, the table shows that the subscriber on the short loop is exposed to the least TEPL (lower by 10.3 dB in this example), and, therefore, the most talker echo GOS degradation (other factors, such as ERLs, being equal). Thus far it has been shown that the user at the A end of the example connection in Table 5.7 experiences more loss and echo degradation. Later in this chapter it will be seen that this user also experiences more noise.

Talker Echo Path Delay and Delay Asymmetry

In a manner similar to that for TEPL, talker echo path delays (TEPD) for the connection in Figure 5.15 are

$$\text{TEPD}_A = D_A = \sum_{i=1}^{p} (d'_i + d''_i) \quad \text{and} \quad \text{TEPD}_B = D_B = \sum_{i=n}^{q} (d'_i + d''_i) \qquad (5.18)$$

These are the values of TEPD that would be used for GOS purposes in the talker echo transmission rating equation in Table 3.4(a) in Chapter 3. Often in practice the total delays will be the same in each direction, so that $d'_i = d''_i$ for all i, and, therefore, from Equation (5.18),

$$D_A = 2 \sum_{i=1}^{p} d_i \quad \text{and} \quad D_B = 2 \sum_{i=n}^{q} d_i \qquad (5.19)$$

In general, TEPD asymmetry will exist and is given by

$$\Delta D = D_A - D_B \qquad (5.20)$$

For the common case where the main reflections are at the 4- to 2-wire interfaces to the subscriber loops (i.e., $p = n - 1$ and $q = 2$ in Eqs. (5.18) and (5.19)), the delays will tend to be similar in each direction for the trunk part of the network. Also, the delays for the loop plant are usually negligible compared to other delays, so that, for this common case, there is little or no delay asymmetry (i.e., $\Delta D \approx 0$ in Eq. (5.20)).

Listener Echo Path Loss

Figure 5.16 illustrates a connection with a single listener echo path. Thus, following a discussion similar to that for talker echo, the listener echo path loss (LEPL) is

$$\text{LEPL} = \text{RL}_q + \sum_{i=q}^{p}(L'_i + L''_i) + \text{RL}_p \qquad (5.21)$$

where RL_x is a measure of stability return loss near the X end of the connection. RL_x may be expressed in a number of ways, including single-frequency values and weighted average values across one or more portions of the voiceband. Since the listener echo path is prone to positive feedback and potential instability, the LEPL is often specified in terms of singing margin (SM), which is the minimum loss of the listener echo path at a single frequency in the voiceband. SM must be greater than 0 dB to ensure stability, and considerably greater than 0 dB to minimize attenuation and phase (also called *near-singing*) distortion, and thereby provide satisfactory performance.

For single-frequency values (f) of RL_x, SM is given by

$$S = \underset{f}{\text{Min}}(\text{LEPL}) = \underset{f}{\text{Min}}\{\text{RL}(f)_q + \text{RL}(f)_p\} + \sum_{i=q}^{p}(L'_i + L''_i) \qquad (5.22)$$

On the other hand, when RL_x is in terms of weighted average values across some portion of the voiceband, an estimate of SM is given by

$$\text{SM} = \underset{f}{\text{Min}}(\text{LEPL}) \approx \text{SRL}_q + \sum_{i=q}^{p}(L'_i + L''_i) + \text{SRL}_p \qquad (5.23)$$

where the singing return loss (SRL_x) is the weighted average loss of the reflected signal as determined at a reflection point near the X end of the connection, and usually over the 200- to 500-Hz and 2500- to 3200-Hz regions of the voice frequency band [31, 32]. The regions correspond to SRL-LOw and SRL-HIgh, respectively. Thus, in Equation (5.23),

$$\text{SRL}_q + \text{SRL}_p = \text{Min}\{\text{SRL-LO}_q + \text{SRL-LO}_p, \text{SRL-HI}_q + \text{SRL-HI}_p\} \qquad (5.24)$$

The above measures of LEPL have been used for speech service, but are now used mostly to estimate the objective performance of listener echo paths for the purpose of supporting nonspeech services such as voiceband data.

For speech service, another, more recent, measure of return loss is now used, called average power return loss (APRL), introduced in Chapter 3 [20]. As the name implies, it is the return loss power averaged across the voiceband (from 200 to 3400 Hz). As a practical matter, APRL can also be estimated from a frequency weighed power sum of SRL-LO and -HI, and ERL as follows [33] (see the example provided later in relation to Eq. (5.89)):

$$\text{APRL} \approx -10 \log_{10}\{0.1 \times 10^{-\text{SRL-LO}/10} + 0.5 \times 10^{-\text{ERL}/10} + 0.4 \times 10^{-\text{SRL-HI}/10}\} \quad (5.25)$$

When the APRL measure of return loss is used, the LEPL is called *weighted echo path loss* (WEPL) [34], and is estimated by

$$\text{LEPL} = \text{WEPL} = \text{APRL}_q + \sum_{i=q}^{p} (L'_i + L''_i) + \text{APRL}_p \quad (5.26)$$

WEPL is the parameter used for GOS purposes in the listener echo transmission rating equation in Table 3.4(b) in Chapter 3.

A common situation (similar to that for talker echo discussed earlier) is that the return losses are at the 4- to 2-wire conversion points between the trunks and the loop plant; in this case, $q = 2$ and $p = n - 1$, and Equation (5.2a) applies. Also, for all of the above measures of LEPL, the total trunk loss will usually be the same in each direction, so that Equation (5.7) also applies. Thus, using Equations (5.2a) and (5.7) in Equations (5.21), (5.23), and (5.26) gives

$$\begin{aligned} \text{LEPL} &= \text{RL}_q + 2\,\text{ICL} + \text{RL}_p \\ \text{SM} &= \text{SRL}_q + 2\,\text{ICL} + \text{SRL}_p \\ \text{WEPL} &= \text{APRL}_q + 2\,\text{ICL} + \text{APRL}_p \end{aligned} \quad (5.27)$$

Further, as far as speech service is concerned, listener echo is most prevalent on zero loss intra-office connections (i.e., for ICL = 0 in Eq. (5.27)). Thus, loss in the listener echo path is often governed only by return losses. Finally, it is noted that listener echo is considered to be equally perceived at both ends of the connection, so that the notion of asymmetry does not apply in this case.

Listener Echo Path Delay

In a manner similar to that for LEPL, the listener echo path delay (LEPD) parameter illustrated in Figure 5.16 is given by

$$\text{LEPD} = D_L = \sum_{i=q}^{p} (d'_i + d''_i) \quad (5.28)$$

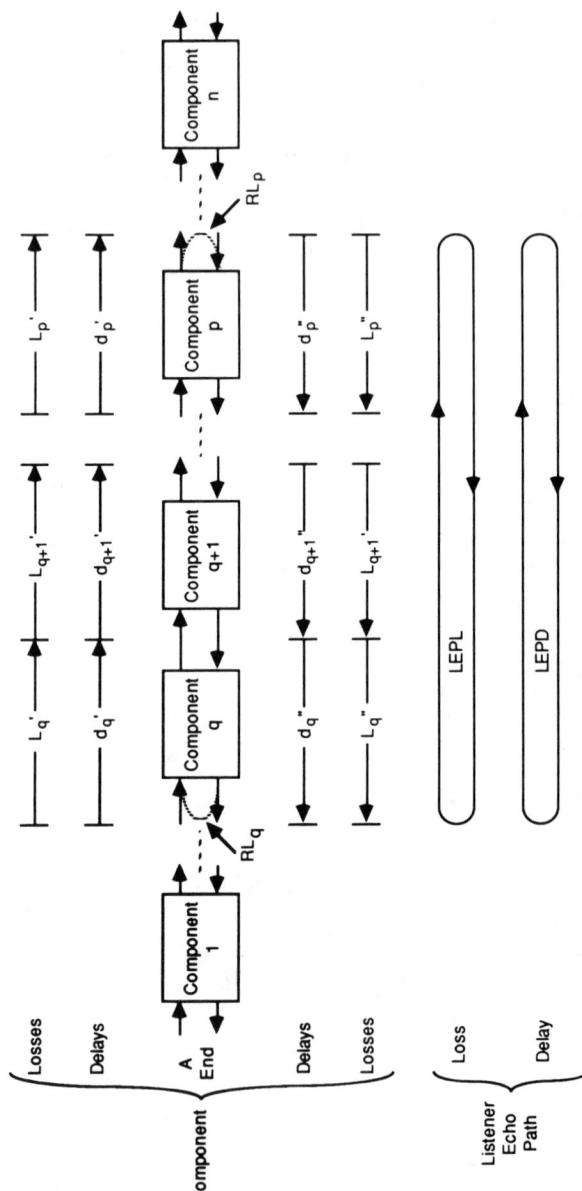

Figure 5.16 Illustration of losses and delays for connection components and for a single listener echo path.

This is the value of LEPD that would be used for GOS purposes in the listener echo transmission equation in Table 3.4(b) in Chapter 3. Again, as with LEPL, return losses are usually at the end of the connection ($q = 2$, and $p = n - 1$), and the total delays will often be the same in each direction ($d'_i = d''_i$). Thus, from Equation (5.28),

$$\text{LEPD} = D_L = 2 \sum_{i=2}^{n-1} d_i \tag{5.29}$$

Finally, it is noted that the notion of delay asymmetry does not apply to listener echo.

5.4.2.2 *Voltage, Power, and Related Addition Laws*

The various loss and delay parameters considered in the previous sections all add algebraically in the decibel domain. However, this is not the case for other parameters such as noise, signal-to-noise ratio, and multiple echo path loss. In these cases, the impairment accumulates in the power domain. Also, in the case of noise, before a noise component (say, i) is accumulated with another downstream value of noise (say, j), noise component i (in decibels reference noice C-weighted (dBrnC)) is first reduced by the downstream loss (in decibels) between i and j; this accounts for the attenuation of noise by downstream losses. Tables 5.2, 5.3, and 5.5 summarize information relevant to power-like addition laws. These laws are addressed in more detail below for the most important parameters.

End-to-End Noise

This parameter is illustrated in Figure 5.17 where for n network components in a connection, each component has a loss and noise value in the transmit and receive directions as shown. Note, that for each network component, the noise is considered to be at the output of that component (i.e., after the loss of that component). The types of noise covered by this methodology include idle circuit and signal dependent noise (where each may be C or flat weighted, in dBrnC or dBrn, respectively).

For fixed parameter values there are two equivalent ways to formulate the accumulation of noise. The most general way is to take the noise from the first component, reduce it by the loss of the second component, and power sum it with the noise of the second component; this noise sum is then reduced by the loss of the third component and power summed with the noise of the third component, etc. The result is as follows:

$$N = [\{(N_1 - L_2) \text{``+''} N_2\} - L_3] \text{``+''} N_3 \text{ ---- etc.} \tag{5.30}$$

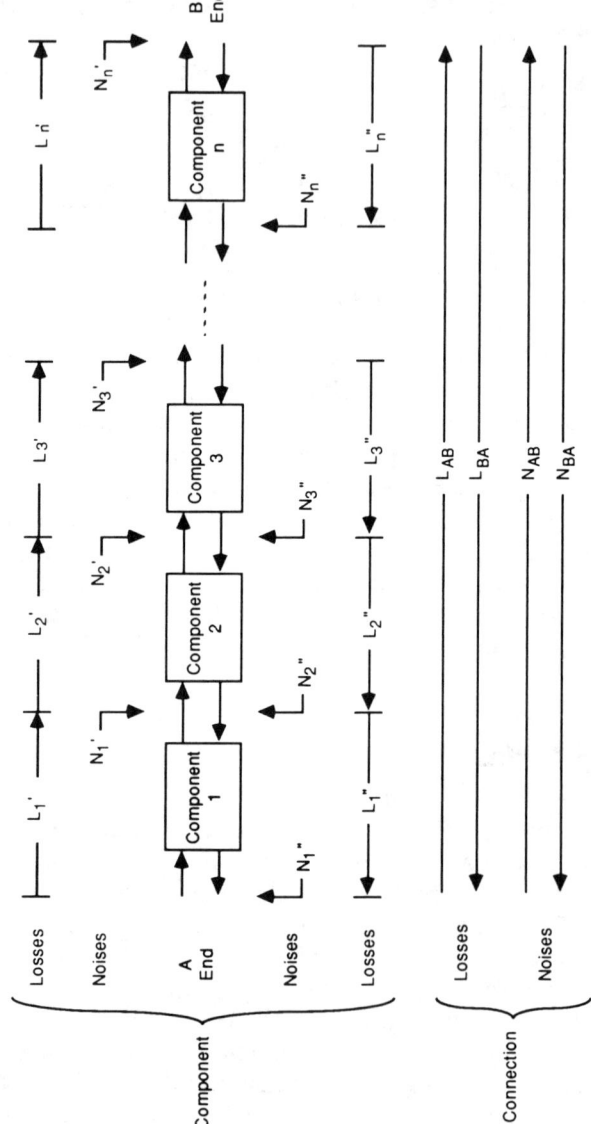

Figure 5.17 Illustration of losses and noises for connection components and for end-to-end connection.

The method in Equation (5.30) is required for analytical accumulation techniques involving distributions of values of loss as discussed later in this chapter. The reason is that Equation (5.30) accounts for dependencies; for example, the loss L_3 above is common to both noise components N_1 and N_2. Thus, in a Monte-Carlo simulation (described later), the same loss sample for L_3 is applied to both N_1 and N_2.

The symbol "+" above denotes a power sum; for example, if N_x and N_y are two noise values in the decibel domain (e.g., in dBrnC) and are referred to the same point in a connection, the total noise at that point is

$$N = N_x \text{``+''} N_y = 10 \log_{10}\{10^{N_x/10} + 10^{N_y/10}\} \quad (5.31)$$

When strictly fixed values of loss are involved, an equivalent but more compact formulation for accumulated noise is obtained by referring all noise values to the end of the connection before they are power summed; that is, reduce each noise by the total downstream loss it will be attenuated by, and then power sum the resulting components. The result is as follows:

$$N = (N_1 - \sum_{j=2}^{n} L_j) \text{``+''} (N_2 - \sum_{j=3}^{n} L_j) \text{``+''} (N_3 - \sum_{j=4}^{n} L_j) \text{``+''} \cdots \text{``+''} N_n \quad (5.32)$$

This may be simplified to

$$N = \sum_{i=1}^{n}{}_{10p} (N_i - \sum_{j=i+1}^{n} L_j), \, j \leq n \quad (5.33)$$

where Σ_{10p} denotes a global power sum of noise components.

Equations (5.30) to (5.33) are for a given direction of transmission. In general, the noise can be different in each direction; thus, referring to Figure 5.17 (and for fixed parameter values),

$$N_{AB} = \sum_{i=1}^{n}{}_{10p} (N'_i - \sum_{j=i+1}^{n} L'_j), \, j \leq n \quad \text{and}$$

$$N_{BA} = \sum_{i=1}^{n}{}_{10p} (N''_i - \sum_{j=1}^{i-1} L''_j), \, j \geq 1 \quad (5.34)$$

It is important to note that noise from network components near the end of the connection will tend to be more significant in the sense that this noise will generally not be attenuated as much as will noise originating nearer the beginning of the connection. An example of noise accumulation is provided in Figure 5.19, after the discussion of signal-to-noise ratios in this section.

Referral of Noise to the Input of a Reference Receiving System

For GOS purposes, the total accumulated noise on a connection must be referred to the input of a reference receiving system for which the GOS models were developed. A reference (r) receiving system is a subscriber loop plus station set with receive sensitivity ROLR = ROLR_r = 46 dB. The referral process is illustrated in Figure 5.18 for the B end of the connection; of course, a similar situation also applies for the A end.

ROLR_B	=	Receive Objective Loudness Rating of Actual Loop + Station Set
ROLR_r	=	Receive Objective Loudness Rating of Reference Loop + Station Set
ROLR_s	=	Receive Objective Loudness Rating of Station Set

Figure 5.18 Illustration of how total noise is referred to the input of a reference receiving system.

The noise referral consists of two parts: one for the noise accumulated up to the input of loop B, and one for the noise of loop B itself at the input to station set B (e.g., N'_n in Fig. 5.17). For the first part, the accumulated noise is first referred to the output of the station set by subtracting the sensitivity of the actual receiving system, ROLR_B, and then referred back to the input of the reference receiving system by adding the sensitivity of the reference system, ROLR_r. Similarly, for the second part, the noise of loop B is first referred to the output of the station set by subtracting the set sensitivity, ROLR_s, and then referred back to the input of the reference system by adding ROLR$_r$. Thus, the total referred noise is

$$N_{ABr} = \left\{ \sum_{i=1}^{n-1} {}_{10p}(N'_i - \sum_{j=i+1}^{n-1} L'_j) - \text{ROLR}_B + \text{ROLR}_r \right\}$$
$$\text{``+''} (N'_n - \text{ROLR}_s + \text{ROLR}_r)$$
(5.35)

where $j \leq n - 1$. Similarly, for the other direction,

$$N_{BAr} = \left\{ \sum_{i=2}^{n} {}_{10p}(N''_i - \sum_{j=2}^{i-1} L''_j) - \text{ROLR}_A + \text{ROLR}_r \right\}$$
$$\text{``+''} (N''_1 - \text{ROLR}_s + \text{ROLR}_r)$$
(5.36)

where $j \geq 2$. Note that for the common 500 type telephone set, ROLR$_s$ is approximately 42.2 dB [35]; thus, in this case, in Equations (5.35) and (5.36), $N_x - \text{ROLR}_s + \text{ROLR}_r = N_x - 46 + 42.2 = N_x + 3.8$ dB. N_{ABr} and N_{BAr} in these equations are the idle circuit noises that would be used for GOS purposes in the loss/noise transmission rating equation in Table 3.4(a) of Chapter 3.

Noise Asymmetry

It is evident from the previous discussions that the total accumulated noise may be different for the two directions of transmission, owing to different component losses, noises, or both, in each direction. This is called *noise asymmetry,* which is relevant from a GOS point of view, and, with reference to Equations (5.35) and (5.36), is given by

$$\Delta N_r = N_{ABr} - N_{BAr} \quad (5.37)$$

Often, the accumulated losses and noises for the trunk part of the network may be similar in each direction. If so, there would be little noise asymmetry from these factors. Also, loop noise is usually low enough to be negligible compared to other noise sources. Applying these condition to Equations (5.35) and (5.36), the noise asymmetry from Equation (5.37) simplifies to

$$\Delta N_r = \text{ROLR}_A - \text{ROLR}_B \quad (5.38)$$

which is entirely dependent on the loop plus station set receive sensitivities. As with loss and echo asymmetry discussed earlier in relation to Table 5.7, the loop plant and station sets would have to be the same at both ends of the connection to eliminate this source of asymmetry. The example for various asymmetries considered in Table 5.7 shows that the subscriber on the short loop will be most affected by noise asymmetry. The reason is that the short loop will attenuate the accumulated trunk noise less than would a longer loop (by 3.2 dB in this example).

Thus, overall, subscribers on short loops are affected the most from all three asymmetries considered (due to end-to-end loss, recall Sec. 5.4.2.1; talker echo path loss, recall Sec. 5.4.2.1; and idle circuit noise differences). Of these, TEPL is the most significant (all other echo path loss factors being equal). As noted earlier, such asymmetries are significant in interworking situations (e.g., between private and public networks). End users served by the lowest loss access arrangement will generally experience the most performance degradation [28].

Signal-to-Noise Ratio

For various types of noise (e.g., idle circuit noise or signal-dependent noise), signal-to-noise ratios (SNR) can be determined for each component in the connection and for the entire connection. The signal-dependent noise of most interest here is also called *C-notched noise* (used for voiceband data performance characterization purposes), although, in addition to C-weighting, 3-kHz and 15-kHz flat weighting can also be used. Similarly, idle circuit noise is called *C-message noise.* To determine SNR, the signal and noise levels at the output of each component must be known (where noise levels have been previously considered in this section).

With reference to Figure 5.17, and the example presented in Figure 5.19, consider connection input (S_{IA}, S_{IB}) and output (S_{OA}, S_{OB}) signals in decibels milliwatt (dBm), where I and O denote connection input and output, and where A and B denote the A and B ends of the connection. Further, consider individual network component output signals S'_i and S''_i in dBm for each direction of transmission (these signals are at the same location as the noises N'_i and N''_i in Figs. 5.17 and 5.19). Thus, the SNRs in decibels for each direction for each component are:

$$\text{SNR}'_i = S'_i - N'_i + F \quad \text{and} \quad \text{SNR}''_i = S''_i - N''_i + F \text{ (dB)} \tag{5.39}$$

where F is a constant to convert the noise units (e.g., in dBrnC) to that of the signal (i.e., in dBm) (see Table 5.3).

The component output noises will generally be known (e.g., from survey data), and the output signals for each component can be determined from the connection input signal and the component losses as follows:

$$S'_1 = S_{IA} - L'_1, \; S'_2 = S'_1 - L'_2 = S_{IA} - L'_1 - L'_2, \; \text{---- etc.} \tag{5.40}$$

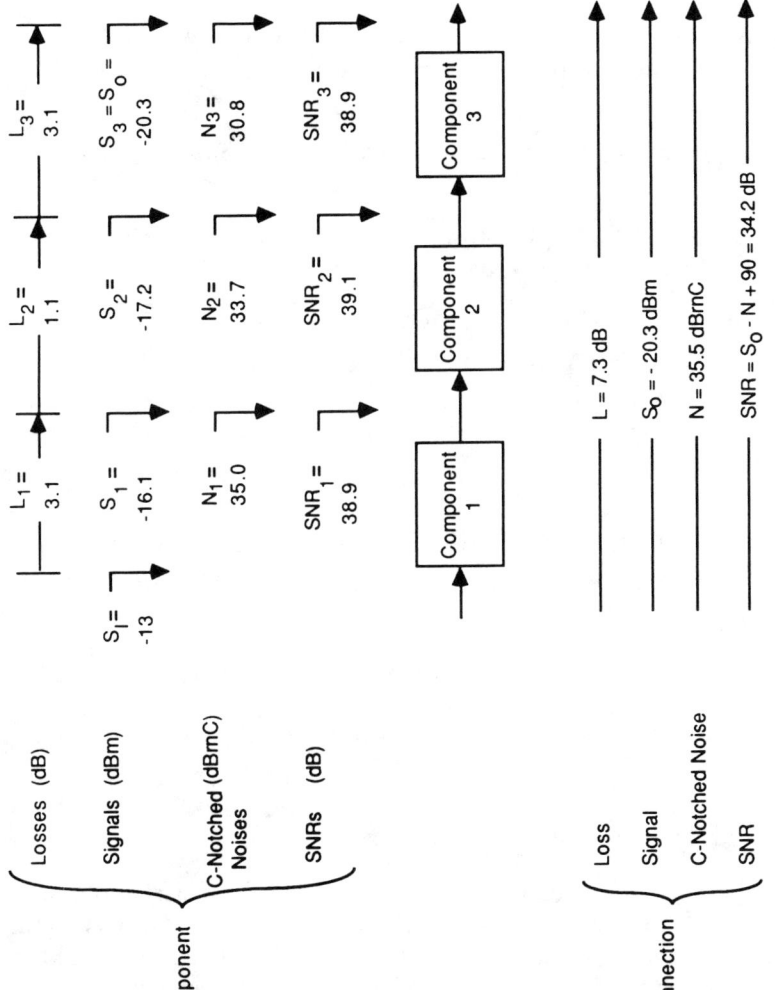

Figure 5.19 Example illustration of loss, noise, and SNR accumulation (fixed values).

In general, for the jth component the output signal is

$$S'_j = S_{IA} - \sum_{i=1}^{j} L'_i \quad \text{and} \quad S''_j = S_{IB} - \sum_{i=j}^{n} L''_i \tag{5.41}$$

For the last component ($j = n$) in the A to B direction, and ($j = 1$) in the B to A direction, the connection output signals are determined as

$$S'_n = S_{OB} = S_{IA} - \sum_{i=1}^{n} L'_i = S_{IA} - L_{AB} \quad \text{and similarly,} \quad S_{OA} = S_{IB} - L_{BA} \tag{5.42}$$

The SNR for the connection is the power sum of the component SNRs:

$$\text{SNR}_{AB} = \text{SNR}'_1 \text{``+''} \text{SNR}'_2 \text{``+''} \cdots \text{``+''} \text{SNR}'_n$$

$$= -10 \log_{10} \left\{ \sum_{i=1}^{n} 10^{-\text{SNR}'_i/10} \right\} = \sum_{i=1}^{n} {}_{-10p} \text{SNR}'_i \tag{5.43}$$

Similarly, $\text{SNR}_{BA} = \sum_{i=1}^{n} {}_{-10p} \text{SNR}''_i$

Alternatively (if fixed values of loss are involved), the connection SNR may also be determined from the connection output signal and the total connection output noise as follows:

$$\text{SNR}_{AB} = S_{OB} - N_{AB} + F \quad \text{and} \quad \text{SNR}_{BA} = S_{OA} - N_{BA} + F \tag{5.44}$$

Often, signal levels and accumulated noises will be the same for each direction, in which case

$$\text{SNR} = \text{SNR}_{AB} = \text{SNR}_{BA} \tag{5.45}$$

SNRs of the type just described are most relevant to voiceband data performance. See the simple example in Figure 5.19 for one direction of transmission. This example is representative of the trunk part of a medium-length end office to end office toll connection with T-carrier facilities used for trunk components 1 and 3 [17, 22, 26].

Signal-to-Quantization Distortion Ratio (Q)

As introduced in Section 3.2.2.6 of Chapter 3, the Q parameter is associated with digital codecs and degrades with the tandeming of codecs. From an objective standpoint, this parameter is analogous to the signal-to- (C- or flat-) notched noise ratio parameter which degrades according to a power law as discussed previously in this chapter. However, subjectively, Q may degrade differently depending on the type of tandeming. In general, the degradation is of the following form:

$$Q = -K \log_{10} \left\{ \sum_{i=1}^{n} 10^{-Q_i/K} \right\} \text{ (dB)} \tag{5.46}$$

where n is the total number of tandemed codecs, Q_i is the Q of the ith codec, and K depends on the types of tandeming. If the codecs are identical ($Q_i = Q_1$ for all i), then

$$Q = Q_1 - K \log_{10}(n) \text{ (dB)} \tag{5.47}$$

and the degradation per doubling of the number of tandem codecs is

$$\Delta Q = K \log_{10}(2) \text{ (dB)} \tag{5.48}$$

The worst-case type of tandeming is called *analog* or *asynchronous tandeming* where the signals at the output of each codec are in analog form (e.g., to pass through an analog switch) before being tandemed with the next codec. In this case, the analog speech signal is resampled for each tandem operation and generally at different time instants than previously sampled (i.e., asynchronously with respect to previous sampling). This causes the most increase in quantization noise. In this case, $K = 15$ and, from Equation (5.48), $\Delta Q = 4.5$ dB, as given in Chapter 3 [35].

The best-case type of tandeming is *digital tandeming* in the *synchronous mode* where signals in digital form at the output of one device are used identically by the next device without any further manipulation of the bit stream. This only applies to identical devices having the synchronous encoding property (e.g., all PCM codecs of this type, or all ADPCM codecs of this type). In this case, there is no degradation and $K = \Delta Q = 0$ dB.

Between the worst and best cases above is *digital tandeming* involving either resampling of the digital stream or other manipulation of the stream (*semisynchronous mode*). Degradations in such cases are not well characterized but are estimated to be such that $K = 5$ to 10 dB, and correspondingly, from Equation (5.48), $\Delta Q = 1.5$ to 3 dB. For example, it is estimated that for digitally tandemed PCM codecs that use inband robbed bit signaling (RBS) and zero code suppression (ZCS) for bit sequence independence $K = 5$ and $\Delta Q = 1.5$ dB. Similarly, for digitally tandemed ADPCM codecs of this type, a somewhat worse degradation corresponding to $K = 10$ and $\Delta Q = 3$ dB is estimated. These estimates are summarized in Table 5.8 and illustrated in Figure 5.20.

As an example, consider a three-trunk connection (similar to that of Fig. 5.19) consisting of asynchronously tandemed PCM, ADPCM, and PCM codecs with Qs

Table 5.8
Examples of Approximate Addition Laws for Q

Tandeming Mode	Codec with ZCS + BRS?	Tandeming Law *						Comment
		64 kb/s PCM			32 kb/s ADPCM			
		Q_1 (dB)	K (dB)	ΔQ (dB)	Q_1 (dB)	K (dB)	ΔQ (dB)	
Analog !	No	37	15	4.5	28.7	15	4.5	Worst Case
	Yes	35	15	4.5	25.8	15	4.5	
Digital	No #	37	0	0	28.7	0	0	Best Case
	Yes @	35	5	1.5	25.8	10	3	Estimated

* See eqns. (5.47) and (5.48)
! Asynchronous tandeming
\# Synchronous tandeming
@ Semi-Synchronous tandeming

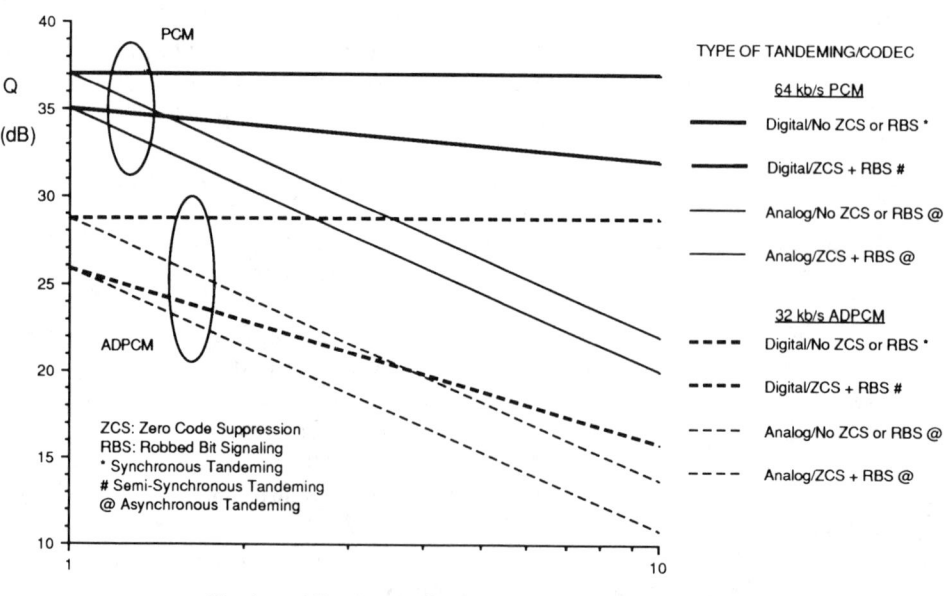

Figure 5.20 Degradation of Q with tandeming.

of 35, 28.7, and 35 dB, respectively. From Equation (5.46), the total Q is

$$Q = -15 \log_{10}\{10^{-35/15} + 10^{-28.7/15} + 10^{-35/15}\} = 25 \text{ dB}$$

It is also of interest to compare the total value of Q to the equivalent number of PCM codecs with the synchronous encoding property (i.e., with $Q_1 = 37$ dB). This is done by solving Equation (5.47) for n to give

$$n = 10^{(Q_1-Q)/K} \tag{5.49}$$

For the above example, $n = 10^{(37-25)/15} = 6.3$, which corresponds to at least 6 PCM codecs. The term n is also referred to as the number of quantization distortion units (QDUs) [36].

In order to incorporate the subjective effect of Q in the GOS models presented in Chapter 3 (Table 3.4(b)), Q is first converted to subjectively equivalent idle circuit noise to be power summed with other idle circuit noise components. Details of the conversion are provided in Appendix A.5.1 [35].

Multiple Echo Paths

In general, connections may contain more than one talker-echo path or more than one listener echo path, owing to a mix of 2-wire and 4-wire technologies, and because of the interworking of different networks [33, 37–39]. Consider the simple examples in Figure 5.21. From an *objective* standpoint, the net effect of multiple echo paths is reduced overall echo path loss. The overall combined echo path loss is the power sum of the individual echo path losses. Thus, for a connection with k talker-echo paths, as observed from one end of the connection, the overall TEPL is

$$\text{TEPL} = -10 \log_{10}\left\{\sum_{i=1}^{k} 10^{-\text{TEPL}_i/10}\right\} = \sum_{i=1}^{k}{}_{-10p} \text{TEPL}_i \tag{5.50}$$

For listener echo, the situation is somewhat different in that for k physical listener echo paths, there are $m = k(k+1)/2$ electrical listener echo paths [37]. Thus, the overall LEPL is

$$\text{LEPL} = -10 \log_{10}\left\{\sum_{i=1}^{m} 10^{-\text{LEPL}_i/10}\right\} = \sum_{i=1}^{m}{}_{-10p} \text{LEPL}_i \tag{5.51}$$

From a *subjective* standpoint, it is not meaningful to simply power sum echo path losses in the above manner, unless the echo path delay is the same (or nearly

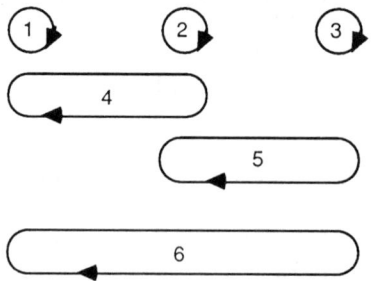

Figure 5.21 Example enumeration of multiple talker and listener echo paths.

the same) for each path. This will not always be the case in a practical connection. However, it is possible to adjust all the echo path delays to be the same (i.e., equal to a reference (r) delay such as, for example, the delay of the echo path with the lowest echo performance—the dominant echo path [33]). To compensate for each echo path delay adjustment, each echo path loss is also adjusted so that with the reference delay the same echo transmission rating is maintained as with the original delay. For example, if a low echo path delay is increased to equal the reference delay, the echo path loss would also be increased so as to maintain the same echo transmission rating.

The talker and listener echo transmission rating equations given in Chapter 3 (Table 3.4(a) and (b), respectively) are used for this purpose. For talker echo, the adjusted echo path losses are $TEPL_{ia} = E_{ia}$, given by

$$E_{ai} = E_r + (R_{Ei} - R_{Er})/2.277 \tag{5.52}$$

where E_r is the TEPL of the echo path with the reference delay, and R_{Ei} and R_{Er} are the echo transmission ratings for the ith path and for the path having the reference delay, respectively. The resulting adjusted echo path losses can then be power summed to obtain an overall combined (c) "subjectively equivalent" echo path loss $TEPL_c = E_c$ for the connection, given by

$$E_c = -10 \log_{10} \left\{ \sum_{i=1}^{k} 10^{-E_{ia}/10} \right\} = \sum_{i=1}^{k}{}_{-10p} E_{ia} \tag{5.53}$$

Finally, the combined echo path loss can then be used in determining the overall echo transmission rating for the connection as follows:

$$R_{Ec} = R_{Er} - 2.277(E_r - E_c) \tag{5.54}$$

For listener echo, a procedure can be followed entirely similar to talker echo above [33]. In particular, the adjusted echo path losses are $LEPL_{ia} = WEPL_{ia}$ and are given by

$$WEPL_{ai} = (R_{LEi}/R_{LEr})(WEPL_r + 7) - 7 \tag{5.55a}$$

where $WEPL_r$ is the LEPL of the echo path with the reference delay, and R_{LEi} and R_{LEr} are the listener echo transmission ratings for the ith path and for the path having the reference delay, respectively. The overall combined LEPL is

$$WEPL_c = -10 \log_{10} \left\{ \sum_{i=1}^{m} 10^{-WEPL_{ia}/10} \right\} = \sum_{i=1}^{m}{}_{-10p} WEPL_{ia} \tag{5.55b}$$

Finally, the overall listener echo transmission rating is

$$R_{LEc} = R_{LEr}(\text{WEPL}_c + 7)/(\text{WEPL}_r + 7) \qquad (5.55c)$$

The above procedure is discussed in detail in [33] and is illustrated by examples in Table 5.9. Notice that accumulations of the echo path losses subjectively adjusted for delay are considerably larger than the accumulation of the objective (unadjusted) echo path losses. This means that not adjusting the echo path losses would result in an overly conservative estimate of echo GOS (i.e., the GOS estimate would be too low).

Table 5.9
Example Accumulations of Multiple Talker and Listener Echo Path Losses, Including the Subjective Effect of Echo Path Delay

(a) Talker Echo (n = 5 echo paths)

Echo Path No., i	TEPD, D_i (ms)	TEPL, E_i (dB)	Transmission Rating, R_{Ei}	Rank Ordering of most Dominant Paths	TEPLs Adjusted For D_r and R_{Ei}, E_{ia} (dB) #
1	4.3	21.0	115.5	4	38.7
2	14.3	38.0	129.6	5	44.9
3	27.4	36.2	111.2	3	36.8
4	29.0	35.2	107.6	1 *	35.2
5	31.4	36.2	108.1	2	35.4
		TEPL$_c$ = 19.6 dB			E_c = 30.1 dB

* Dominant (reference) path i = r = 4
D_r = 29.0 ms, E_r = 35.2 dB, R_{Er} = 107.6
R_{Ec} = f(D_r, E_c) = 96.0
Adjustment is according to subjective criteria (see text).

(b) Listener Echo (n = 2 physical echo paths, m = 3 electrical echo paths)

Echo Path No., i	LEPD, D_{Li} (ms)	LEPL, WEPL$_i$ (dB)	Transmission Rating, R_{LEi}	Rank Ordering of most Dominant Paths	LEPLs Adjusted For D_{Lr} and R_{LEi}, WEPL$_{ia}$ (dB) #
1	4.0	14	147.5	4	16.2
2	2.0	12	151.6	3	16.8
3	6.4	15	140.1	1 *	15.0
		LEPL$_c$ = 8.7dB			WEPL$_c$ = 11.2 dB

* Dominant (reference) path i = r = 3
D_{Lr} = 6.4 ms, WEPL$_r$ = 15 dB, R_{LEr} = 140.1
R_{LEc} = f(D_rLr, WEPL$_c$) = 115.9
Adjustment is according to subjective criteria (see text).

5.4.3 Addition Laws for Serial Connection Configurations and for Distributed Parameter Values

5.4.3.1 Introduction

This section considers the more general situation where parameter values are distributed. Distributions of parameter values can be handled by simulation and by analytical techniques, presented below. Analytical techniques may be used to determine the moments, that is, the mean (μ) and standard deviation (SD, σ) of certain accumulated (connection) parameter values. In this case, the parameter mean and SD for each network component in the connection must be known (as they often are from survey data, as illustrated in Ch. 6). In general, whether there is dependency among the components or between the directions of transmission must also be known. Usually, independence can be assumed among components, but not always between directions. For example, because delay is distance-dependent, the delay in one direction of an echo path will be correlated with the delay in the opposite direction. Similarly, for 2-wire facilities without gain (e.g., most subscriber loops), the loss will be distance-dependent and therefore related in the two directions of transmission for a given loop; this is relevant when determining talker echo path loss.

5.4.3.2 Monte-Carlo Simulation

The role of Monte-Carlo simulation in determining end-to-end performance is illustrated in Figure 5.22. Monte-Carlo simulation is based on the use of a distribution for each parameter value for each network component in a connection. The distribution can be analytical (e.g., normal), empirical (e.g., based on actual samples of field data), or modeled (e.g., based on histograms of field data or on distributions fitted to field data). The simulation essentially models numerous connections, each of which has a particular set of fixed parameter values as randomly sampled for each network component from the above noted distributions. For each connection and each parameter, the randomly sampled fixed values for each component are combined using the algebraic and power-like addition laws described earlier in this chapter. As alluded to previously, it is important that various dependencies, where they exist, be accounted for in the simulation. For example, if a component delay in one direction of transmission is the same as the delay in the other direction, then the same single random sample must be used for each direction.

After simulating numerous connections (typically, at least 1,000), distributions of end-to-end connection parameter values can be provided, as well as related statistics (e.g., mean, median, SD, minimum, and maximum percentile values, etc.). In addition, each end-to-end connection parameter value can be used in the transmission rating and customer opinion models introduced in Chapter 3 to determine the GOS of each connection. Thus, distributions and related statistics of end-to-end

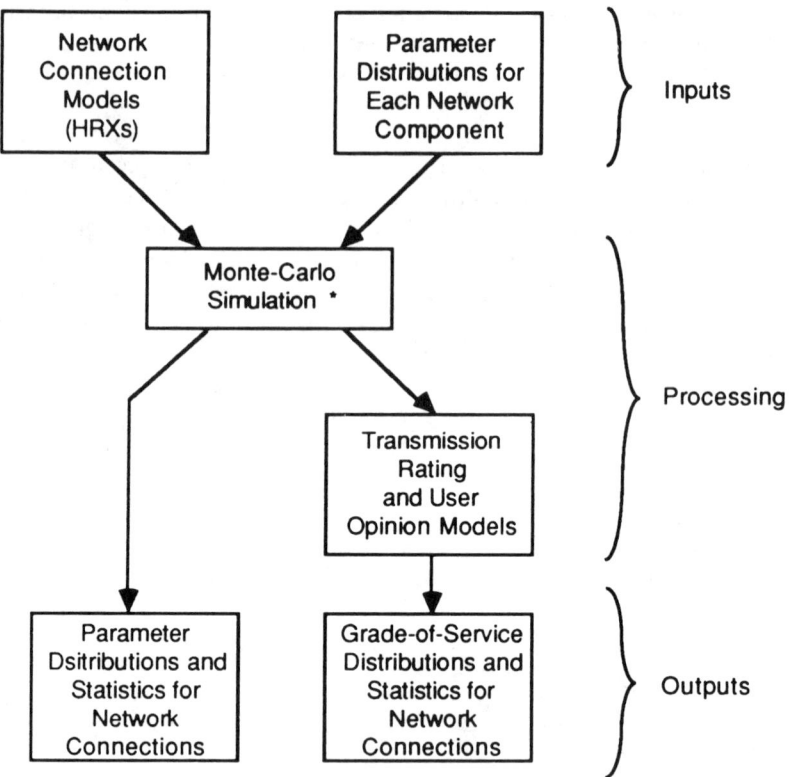

* Uses algebraic and power-like addition laws for fixed (sampled) parameter values.

HRX: Hypothetical Reference Connection

Figure 5.22 Role of Monte-Carlo simulation in determining end-to-end performance.

GOS are determined. Monte-Carlo simulation in the context of GOS tools for network transmission planning purposes is discussed in [40, 41].

The next section in this chapter outlines the analytical treatment of distributions of parameter values. The purpose of this section is to

- Enable estimation of connection performance from distributions and/or simple statistics of survey data (in the absence of the detailed data base of survey information).
- Enable estimation of connection performance in the absence of a Monte-Carlo simulation.

- Provide insight into the equivalent analytical operation of a Monte-Carlo simulation, and thereby into the meaning and interpretation of simulation results.

5.4.3.3 The Normal Distribution and Related Moments

Determination of Percentiles from Moments, and an Example

When a parameter value is distributed, it is of interest to know not only the mean and SD, but also the near worst maximum (and possibly the near minimum) value(s). This can be done by using percentile values (e.g., 5% minimum and 95% maximum values). However, this requires that the distributions of the parameter values be known. Often the distribution is approximately normal, or is assumed to be normal, in which case the near worst-case value can be readily estimated [41, 42]. This is most applicable to algebraically accumulated parameters which, owing to the central limit theorem, will tend to be normally distributed, even if individual component distributions are not normally distributed. For a normal distribution, the relationship among the mean, SD, and minimum or maximum value is as follows (see Fig. 5.23):

$$z = [x(\%) - \mu]/\sigma \tag{5.56}$$

$$f(x) = \frac{1}{\sqrt{2\pi}\sigma} e^{-(x-\mu)^2/2\sigma^2}$$

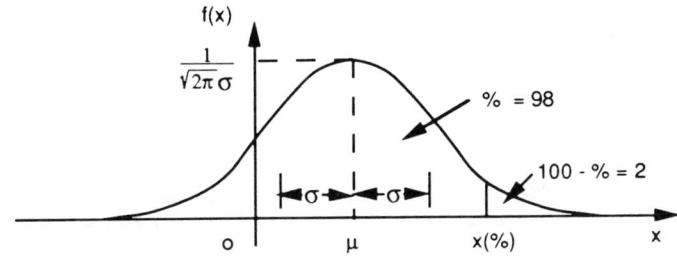

Figure 5.23 Illustration of percentiles for normal probability density function.

Min. Percentile Value
e.g., x(2)

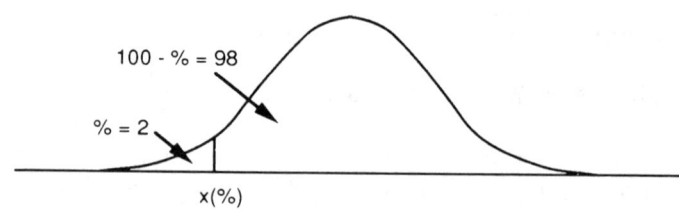

Min. and Max. Percentile
Values with Equal Deviation
from Mean
e.g., x(2) and x(98)

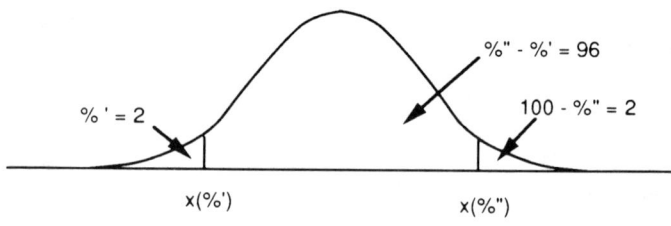

Min. and Max. Percentile
Values with Unequal
Deviation from Mean
e.g., x(2) and x(95)

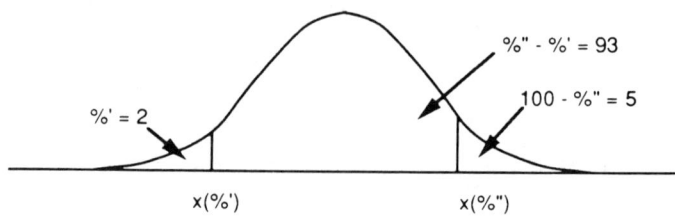

Figure 5.23 Continued

where $x(\%)$ denotes the minimum or maximum percentile value of the random variable x under consideration at the percentile of interest (e.g., 5% minimum or 95% maximum), and z is a factor corresponding to the number of SDs that the maximum or minimum value deviates from the mean value, μ. Values of z are available from tables of normal distributions, and common instances are summarized here in Table 5.10 [43]. For example, this table shows that positive deviations of $z = 0, 1, 2,$ and 3 SDs correspond approximately to 50%, 84%, 98%, and 99.9% maximum percentile values, respectively. Thus, from a knowledge of the mean and SD for a normal distribution, near worst minimum and maximum percentile values can be determined from Equation (5.56) and information of the type in Table 5.10.

For example, consider an inter-office or toll connecting trunk with mean loss $\mu = 3.1$ dB and loss SD $\sigma = 0.9$ dB [33]. As can be seen from Equation (5.56) and

Table 5.10
Examples of Minimum and Maximum Deviation from the Mean for the Normal Distribution

Percent (%)		Deviation (z) *
Minimum	Maximum	(no. of SDs)
50.00	50.00	0
15.87	84.13	± 1
15.00	85.00	± 1.036
10.00	90.00	± 1.282
9.00	91.00	± 1.341
8.00	92.00	± 1.405
7.00	93.00	± 1.476
6.00	94.00	± 1.555
5.00	95.00	± 1.645
4.00	96.00	± 1.751
3.00	97.00	± 1.881
2.28	97.72	± 2
2.00	98.00	± 2.054
1.00	99.00	± 2.326
0.50	99.50	± 2.576
0.13	99.87	± 3
0.10	99.90	± 3.090
0.01	99.99	± 3.719

* The +ve sign corresponds to maximum deviations.
The -ve sign corresponds to minimum deviations.

Table 5.10, the 5% minimum and 95% maximum values of loss are

$$x(\%) = z\sigma + \mu \tag{5.56a}$$

and, therefore, $x(5) = -1.645 \times 0.9 + 3.1 = 1.6$ dB and

$$x(95) = 1.645 \times 0.9 + 3.1 = 4.6 \text{ dB}$$

and the loss lies between the minimum and maximum values with relative probability of $95 - 5 = 90\%$.

Determination of Moments from Percentiles, and an Example

Given any two percentile values from a normal distribution, say $x(\%')$ and $x(\%'')$, the corresponding mean and SD can be determined by first obtaining the corresponding deviations z' and z'' (e.g., from Table 5.10), and then, based on Equation (5.56), solving the following equations:

$$z' = [x(\%') - \mu]/\sigma, \text{ and } z'' = [x(\%'') - \mu]/\sigma \text{ to give:} \tag{5.57}$$

$$\sigma = [x(\%'') - x(\%')]/(z'' - z'), \text{ and } \mu = x(\%') - z'\sigma = x(\%'') - z''\sigma \tag{5.58}$$

For example, consider the 90th and 99th percentile values for subscriber loop noise as measured for Bell Systems loops, and as illustrated here in Table 5.11 [44]. As can be seen in Table 5.10, the corresponding values of z are $+1.282$ and $+2.326$, respectively. Thus, using Equations (5.58), the μ and σ values are as given in Table 5.11.

Table 5.11
"Equivalent" Mean and SD for Subscriber Loop Noise in the Bell System

Location of Noise	Surveyed Percentile Values #		Equivalent	
	x (90) (dBrnC)	x (99) (dBrnC)	Mean (dBrnC)	SD (dB)
At Subscriber's Station Set *	13	31.2	-9.3	17.4
At Central Office (Station Set off-hook)	6	28	-21.0	21.1

* With quiet termination at central office
\# From [44]

It is important to note that in this example the noise distribution may not be normal throughout the entire range of the noise variable, particularly for the low values. However, from a practical standpoint, this is not important because the low values of noise will contribute negligibly to the overall noise of the connection. In other words, it is not important if the mean, the deviations below the mean, and the deviations up to a point above the mean (i.e., up to about 15 to 20 dBrnC) are accurate, since these values of noise will not materially affect the overall connection noise. What is important is that the right-hand tail of the distribution represent the high values of loop noise. Normal distributions with statistics as in Table 5.11 accomplish this, and can be used in a Monte-Carlo simulation to represent loop noise.

5.4.3.4 Algebraic Addition Laws

Addition and Subtraction of Independent Random Variables

For *addition* of independent random variables, it is well known that mean values add algebraically just as fixed values do, as was considered in Section 5.4.2.1 (see Eqs. (5.1) to (5.5)). Similarly, it is well known that SDs accumulate algebraically on a sum-of-squares basis. In summary,

$$\mu = \sum_{i=1}^{n} \mu_i \quad \text{and} \quad \sigma = \sqrt{\sum_{i=1}^{n} \sigma_i^2} \qquad (5.59)$$

where μ_i and σ_i are the parameter mean and SD, respectively, for the ith network component.

Equation (5.59) applies to parameters such as end-to-end loss, for which component values add algebraically and for which there is no reason to expect values to be related from one component to the next. This also applies largely to end-to-end delay, although if delay is contributed to by traffic loading, then this aspect of delay may be at least partially related from component to component. Also, for parameters involving both directions of transmission, such as talker echo path loss and talker echo path delay, dependencies between the directions must be considered. These situations are addressed in more detail later in this chapter.

Subtraction of independent random variables arises with loss asymmetry, if all component losses are independent, for each direction of transmission. The mean losses subtract just as fixed values do, as was considered in Equations (5.6) and (5.16). However, owing to independence, SDs still add on a sum-of-squares basis. In summary, for the difference of two independent random variables x and y,

$$\Delta\mu = \mu_x - \mu_y \quad \text{and} \quad \Delta\sigma = \sqrt{\sigma_x^2 + \sigma_y^2} \qquad (5.60)$$

Addition of Dependent Random Variables

In this case there is no difference regarding the mean values; the means still accumulate algebraically, as indicated in Equation (5.59). However, if there is complete dependence among *all* the components, the SDs accumulate on a direct algebraic basis (not on a sum-of-squares basis); that is,

$$\sigma = \sum_{i=1}^{n} \sigma_i \tag{5.61}$$

If there is complete dependence among *some* network components and not others, a combination of algebraic and sum-of-squares accumulation can be used as appropriate:

$$\sigma = \sqrt{\sigma_D^2 + \sigma_I^2} \tag{5.62}$$

where the total dependent (D) and independent (I) variations are given, respectively, by

$$\sigma_D = \sum_i \sigma_i \quad \text{and} \quad \sigma_I = \sqrt{\sum_j \sigma_j^2} \tag{5.63}$$

(in which the σ_is and σ_js are the individual SDs for the dependent and independent components, respectively). This situation may be relevant to certain end-to-end calculations (e.g., for delay).

Another such situation occurs if there is *partial* dependence from one component to another. In this case, separate SDs are estimated for the dependent (σ_{iD}) and independent (σ_{iI}) parts of each network component such that the total SD for each component is

$$\sigma_i = \sqrt{\sigma_{iD}^2 + \sigma_{iI}^2} \tag{5.64}$$

Then the total SD is given by Equation (5.62), where, in this case,

$$\sigma_D = \sum_{i=1}^{n} \sigma_{iD} \quad \text{and} \quad \sigma_I = \sqrt{\sum_{i=1}^{n} \sigma_{iI}^2} \tag{5.65}$$

Finally, there may be complete or partial dependence between the direction of transmission for some components and not for others. This is relevant to round-trip calculations (e.g., for talker echo path loss (TEPL)). For the most general case of partial dependence, the total dependent and independent portions can be combined using Equation (5.62), where, in this case,

$$\sigma_D = \sqrt{\sum_i (\sigma'_{iD} + \sigma''_{iD})^2} \quad \text{and} \quad \sigma_I = \sqrt{\sum_j [(\sigma'_{jI})^2 + (\sigma''_{jI})^2]} \quad (5.66)$$

and where, again, i and j index the components with dependent and independent variations, respectively.

To illustrate some of the above relationships, consider the example connection in Figure 5.24 for a medium-distance toll connection involving various loss parameters [26, 45, 46]. In this example, complete independence of loss is assumed from component to component in a given direction. However, partial dependence is assumed for direction as indicated in Figure 5.24 for the variations in transmit and receive sensitivities (TOLR and ROLR) of the loop plus station set at each end of the connection. The assumed independent component of this variation is relatively small and reflects incidental design tolerances in the send and receive sensitivities [46]. The assumed dependent component of this variation reflects the common aspects of both the physical loop and the regulation of sensitivity for the two directions of transmission [45] (see Secs. 6.2.3.1 of Ch. 6 for more detail on this point).

End-to-End Loss Example. Since the connection in Figure 5.24 is symmetrical, from Equation (5.59)—or more specifically, Equation (5.2)—the mean end-to-end loss is

$$L = L_{AB} = L_{BA} = \text{TOLR} + \sum_{i=2}^{4} L_i + \text{ROLR} = 10.9 \text{ dB}$$

For end-to-end loss, directional dependencies are irrelevant, and, from Equation (5.64), the total variation for each of the send and receive sensitivities is

$$\sigma'_1 = \sigma''_5 = \sqrt{\sigma^2_{iD} + \sigma^2_{iI}} = \sqrt{3.6^2 + 1.3^2} = 3.8 \text{ dB for each TOLR}$$

and similarly, $\sigma''_1 = \sigma'_5 = \sqrt{1.7^2 + 0.6^2} = 1.8$ dB for each ROLR

Thus, from Equation (5.59), the total end-to-end loss variation is

$$\sigma_L = \sqrt{\sum_{i=1}^{5} \sigma_i^2} = \sqrt{3.8^2 + 0.97^2 + 0.28^2 + 0.97^2 + 1.8^2} = 4.4 \text{ dB}$$

As an aside, it is noted that the loss variation σ_L above is similar to the loss deviation parameter, which is the variation in loss relative to a fixed loss (e.g., mean

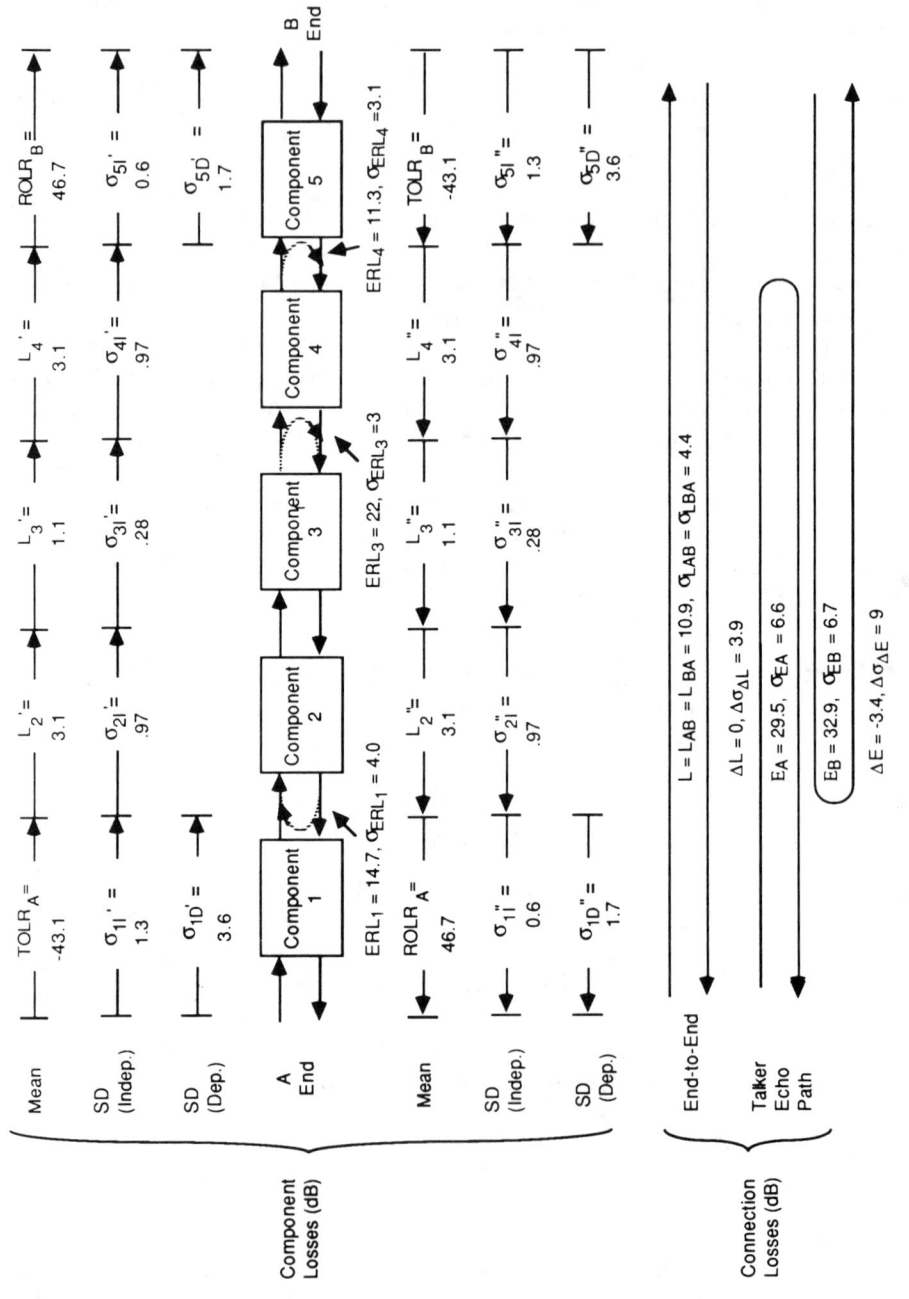

Figure 5.24 Example illustration of end-to-end loss, talker echo path loss, and loss asymmetries involving distributions of losses.

loss) at 1 kHz. However, loss deviation is ordinarily an electrical, rather than acoustical, performance measure.

Talker Echo Path Loss Example. For the A end of the connection in Figure 5.24, the mean TEPL—using Equation (5.15)—is

$$E_A = \text{TOLR}_A + \text{ROLR}_A + 2\sum_{i=2}^{4} L_i + \text{ERL}_4 = 29.5 \text{ dB}$$

For TEPL, the directional dependencies for the transmit and receive sensitivities are relevant, and, for $i = 1$, are accounted for in the loss variations from Equation (5.66) as follows:

$$\sigma_D = \sigma'_{1D} + \sigma''_{1D} = 3.6 + 1.7 = 5.3 \text{ dB}$$

$$\sigma_I = \sqrt{(\sigma'_{1I})^2 + (\sigma''_{1I})^2 + 2\sum_{j=2}^{4}(\sigma_{jI})^2}$$
$$= \sqrt{1.3^2 + 0.6^2 + 2(0.97^2 + 0.28^2 + 0.97^2)} = 2.4 \text{ dB}$$

Thus, the total variation in TEPL, including the ERL variation, is, from an equation similar to (5.62),

$$\sigma_{EA} = \sqrt{\sigma_D^2 + \sigma_I^2 + \sigma_{\text{ERL4}}^2} = \sqrt{5.3^2 + 2.4^2 + 3.1^2} = 6.6 \text{ dB}$$

Notice that a great deal of the total variation is due to the directionally dependent variations of the subscriber loop plus station set sensitivity of 5.3 dB.

Subtraction of Dependent Random Variables

This situation arises, for example, with end-to-end loss asymmetry and with talker echo path loss asymmetry, where, again, there are generally some dependencies to be accounted for. Again, mean values of loss subtract algebraically, just as fixed values do, as was considered in Equation (5.60). Also, for variations that are independent, the SDs will add on a sum-of-squares basis, as in Equation (5.60). However, for variations that are dependent, the SDs will subtract algebraically. In summary, for the difference of two partially dependent random variables x and y,

$$\Delta u = u_x - u_y, \quad \Delta\sigma_I = \sqrt{\sigma_x^2 + \sigma_y^2}, \quad \Delta\sigma_D = |\sigma_x - \sigma_y| \quad (5.67)$$

and based on Equation (5.62), the total variation is

$$\Delta\sigma = \sqrt{\Delta\sigma_I^2 + \Delta\sigma_D^2} \qquad (5.68)$$

End-to-End Loss Asymmetry Example. Consider again the example in Figure 5.24. From Equation (5.6), the mean loss asymmetry in this case is

$$\Delta L = L_{AB} - L_{BA} = 10.9 - 10.9 = 0 \text{ dB}$$

This is due to the symmetric nature of the total connection loss (even though some individual component losses are not symmetric—i.e., TOLR and ROLR differ). Regarding the variation in loss asymmetry, the component variations involving directional dependencies will subtract algebraically. Thus, based on the last part of Equation (5.67), for the TOLR plus ROLR pair at each end of the connection,

$$\Delta\sigma_{1D} = |\sigma'_{1D} - \sigma''_{1D}| = 3.6 - 1.7 = 1.9 \text{ dB, and similarly, } \Delta\sigma_{5D} = 1.9 \text{ dB}$$

Note, however, that $\Delta\sigma_{1D}$ and $\Delta\sigma_{5D}$, being at different ends of the connection, are independent of one another. Therefore, from the first part of Equation (5.66), the total contribution to asymmetry from dependent variations is

$$\Delta\sigma_D = \sqrt{\Delta\sigma_{1D}^2 + \Delta\sigma_{5D}^2} = \sqrt{2(1.9)^2} = 2.7 \text{ dB}$$

From the last part of Equation (5.66), the total variation in loss asymmetry due to independent variations is the sum of squares of all independent variations as follows:

$$\Delta\sigma_I = \sqrt{\sum_{j=1}^{5} [(\sigma'_{jI})^2 + (\sigma''_{jI})^2]}$$
$$= \sqrt{2(1.3^2 + 0.97^2 + 0.28^2 + 0.97^2 + 0.6^2)} = 2.8 \text{ dB}$$

Thus, from Equation (5.68), the total variation in loss asymmetry is

$$\Delta\sigma_{\Delta L} = \sqrt{\Delta\sigma_D^2 + \Delta\sigma_I^2} = \sqrt{2.7^2 + 2.8^2} = 3.9 \text{ dB}$$

Talker Echo Path Loss Asymmetry Example. The talker echo path from the A end of the connection has already been discussed in Figure 5.24. Consider also the talker echo path from the B end with a reflection at component 2 having mean $ERL_2 =$

14.7 dB and SD $\sigma_{ERL2} = 4.0$ dB. Thus, from Equation (5.15), the TEPL for this path is

$$E_B = \text{TOLR}_B + \text{ROLR}_B + 2\sum_{i=2}^{4} L_i + \text{ERL}_2 = 32.9 \text{ dB}$$

which, except for the difference in ERL, is the same as E_A for the A end of the connection given earlier for the talker echo path loss example.

Similarly, the variation in TEPL_B is the same as for TEPL_A, except for the difference in the variation of ERL. Thus, using Equation (5.66) in a manner similar to the previous example, $\sigma_D = 5.3$ dB, $\sigma_I = 2.4$ dB, and, from Equation (5.26),

$$\sigma_{EB} = \sqrt{\sigma_D^2 + \sigma_I^2 + \sigma_{ERL2}^2} = \sqrt{5.3^2 + 2.4^2 + 4.0^2} = 6.7 \text{ dB}$$

For this example, E_A and E_B are symmetric in the mean except for the mean ERLs; thus, from Equations (5.16) and (5.17), the mean TEPL asymmetry is

$$\Delta E = E_A - E_B = \text{ERL}_4 - \text{ERL}_2 = 11.3 - 14.7 = -3.4 \text{ dB}$$

This means that the user at the A end of the connection will experience an average of 3.4 dB less TEPL, and hence poorer echo performance, than the user at the B end.

Regarding asymmetry due to TEPL variation, various dependencies must be considered. First, notice that for this example in Figure 5.24, all trunk losses (L_2, L_3, and L_4) are symmetric *and* common to both of the echo paths considered above, and therefore will not contribute to asymmetry variations. Secondly, notice that the TOLR plus ROLR terms at each end of the connection are the same, but are *not* common to each talker echo path, and are therefore independent. Finally, notice that the ERLs are not common to each echo path, and so are also independent. Thus, the total variations for the TOLR plus ROLR and the ERL terms will add on a sum-of-squares basis as determined below.

Specifically, from Equation (5.66) with $i = 1$ or 5 and $j = 1$ or 5, consider variations for the TOLR plus ROLR pair at each end of the connection:

The dependent part is

$$\sigma_{1D} = \sigma'_{1D} + \sigma''_{1D} = 3.6 + 1.7 = 5.3 \text{ dB}$$

Similarly, $\sigma_{5D} = 5.3$ dB

the independent part is:

$$\sigma_{1I} = \sqrt{(\sigma'_{1I})^2 + (\sigma''_{1I})^2} = \sqrt{1.3^2 + 0.6^2} = 1.4 \text{ dB}$$

Similarly, $\sigma_{5l} = 1.4$ dB

Thus, from Equation (5.64), the total variation for the TOLR plus ROLR term at each end of the connection is

$$\sigma_1 = \sqrt{\sigma_{1D}^2 + \sigma_{1l}^2} = \sqrt{5.3^2 + 1.4^2} = 5.5 \text{ dB}$$

Similarly, $\sigma_5 = 5.5$ dB

Overall, the total variation in TEPL asymmetry due to telephone plus loop plant and ERL variations is

$$\Delta\sigma_{\Delta E} = \sqrt{\sigma_1^2 + \sigma_{ERL4}^2 + \sigma_5^2 + \sigma_{ERL2}^2} = \sqrt{5.5^2 + 3.1^2 + 5.5^2 + 3.4^2} = 9.0 \text{ dB}$$

This example illustrates that TEPL asymmetry is quite large, and is contributed to most by directionally dependent loss variations associated with the subscriber loop plus station set sensitivities.

Attenuation Distortion and Example. Attenuation distortion is the difference between the losses at 0.4 kHz and 1 kHz, and the difference between the losses at 2.8 kHz and 1 kHz. If the loss statistics (mean and SD) are known for each of the above frequencies, the mean and SD can be estimated for attenuation distortion. The mean attenuation distortion (*ad*) is the difference in the mean losses as follows:

$$L_{0.4ad} = L_{0.4} - L_{1.0} \quad \text{and} \quad L_{2.8ad} = L_{2.8} - L_{1.0} \tag{5.69}$$

The variation in attenuation distortion depends on the part of the network being assessed. For the *trunk plant* having voiceband filters, there is a partial dependency between the losses at 1 kHz and 0.4 kHz, and between the losses at 1 kHz and 2.8 kHz, that needs to be considered. Generally, the losses and their variations at 0.4 kHz and 2.8 kHz, will be larger than at 1 kHz. The larger variations at 0.4 and 2.8 kHz may be viewed as consisting of variations on the order of, and dependent on, those at 1 kHz, plus additional independent variations owing to larger tolerances in the frequency response near the edges of the voiceband. Thus,

$$\sigma_{0.4} = \sqrt{\sigma_{1.0}^2 + \sigma_{0.4ad}^2} \quad \text{and} \quad \sigma_{2.8} = \sqrt{\sigma_{1.0}^2 + \sigma_{2.8ad}^2} \tag{5.70}$$

and, accordingly, the variation in attenuation distortion is estimated by rearranging Equation (5.70) to give

$$\sigma_{0.4ad} = \sqrt{\sigma_{0.4}^2 - \sigma_{1.0}^2} \quad \text{and} \quad \sigma_{2.8ad} = \sqrt{\sigma_{2.8}^2 - \sigma_{1.0}^2} \tag{5.71}$$

As an example, consider the results from survey data for medium-length toll connections in Table 5.12(a). In this example the statistics for attenuation distortion are estimated from surveyed 0.4-, 1.0-, and 2.8-kHz loss statistics using Equations (5.69) and (5.71), and compared with the exact statistics as computed sample by sample from the survey data [17]. These results demonstrate that the above estimates are useful for engineering purposes.

For the *loop plant,* the situation is somewhat different at the low-frequency end of the voiceband. In this case, owing to the nature of increased loop attenuation with frequency, the loss and its variations at 1 kHz will generally be larger than at 0.4 kHz

Table 5.12(a)
Comparison of Estimated and Surveyed Attenuation Distortion Statistics for the Trunk Plant
(Example from Survey Data for Medium Length End-Office to End-Office Toll Connections [17])

Frequency	Loss (dB)		Attenuation Distortion (dB)			
			Estimated		Surveyed	
kHz	Mean	SD	Mean	SD	Mean	SD
0.4	8.75	1.57	1.46	0.67	1.4	0.7
1.0	7.29	1.43				
2.8	8.09	1.96	0.80	1.34	0.8	1.2

Table 5.12(b)
Estimated Attenuation Distortion Statistics for the Loop Plant
(Example from Survey Data for General Loop Distribution [47])

Frequency	Loss (dB)		Estimated Attenuation Distortion (dB)	
kHz	Mean	SD	Mean	SD
0.4	2.9 *	2.2 *	-0.8	1.7
1.0	3.7	2.8		
2.8	7.1 #	4.5 #	3.4	3.5

* Linearily interpolated from values at 0.3 and 0.5 kHz
# Linearily interpolated from values at 2.5 and 3.0 kHz

(opposite to that in the trunk plant). The larger variation at 1 kHz consists of variations on the order of, and dependent on, those at 0.4 kHz, plus an additional independent variation owing to the larger tolerance in the frequency response at higher frequencies. Thus, loop plant attenuation distortion at 0.4 kHz is determined as follows:

$$\sigma_{1.0} = \sqrt{\sigma_{0.4}^2 + \sigma_{0.4ad}^2} \quad \text{and therefore} \quad \sigma_{0.4ad} = \sqrt{\sigma_{1.0}^2 - \sigma_{0.4}^2} \tag{5.72}$$

which differs from the 0.4-kHz portions of Equations (5.70) and (5.71). However, the 2.8-kHz portion of Equations (5.70) and (5.71) also apply to the loop plant.

As an example, consider the results from survey data for the general loop plant in Table 5.12(b) [47]. The estimated attenuation distortion results are obtained from surveyed 0.4-, 1.0-, and 2.8-kHz loss statistics using Equations (5.69) and (5.72); the 0.4- and 2.8-kHz losses are interpolated as shown in the table. In this case there are no exact attenuation distortion statistics provided for comparison with the estimates.

5.4.3.5 *Voltage, Power, and Related Addition Laws*

There is no exact closed form analytical solution for the accumulation of values for parameters such as noise, SNRs, and multiple echo path losses, which are distributed and which accumulate other than algebraically (e.g., accumulate on a power basis). It is necessary to rely on simulation techniques (i.e., Monte-Carlo, discussed earlier) or on approximate analytical solutions [48]. When values are normally distributed in the logarithmic (log) or decibel domain, a relatively simple approximate accumulation technique is called *Wilkinson's method*, which provides the mean and SD for the accumulated parameter values and assumes that the resulting distribution is also normal [49–51]. Comparison with simulation results has shown that this method is reasonably accurate, provided that individual component SDs are not too large (i.e., less than 4 to 6 dB), and provided that not too many random variables are accumulated. For transmission performance purposes, most SDs fall within the range indicated, and usually it is not required to combine a large number of values; therefore, Wilkinson's method is useful. One exception is for subscriber loop noise, which typically has a very large SD (see Table 5.11 [44]). In general, the error is such that the accumulated mean and SD are under and overestimated, respectively, by Wilkinson's method. As a result, when using Equation (5.56a), this error will tend to be reduced somewhat when estimating maximum percentile values (e.g., at 95%), and increased somewhat when estimating minimum percentile values (e.g., at 5%). Owing to its reasonable accuracy and simplicity, Wilkinson's method is used here. Readers are referred to [49] for a more accurate (but more complex) analytical technique.

The procedure for using Wilkinson's method is as follows. The component means and SDs in the log (or decibel) domain are used to determine the corresponding means and SDs in the inverse log (or inverse decibel) domain (i.e., in the log-normal domain). The log-normal means and SDs are then accumulated according to voltage, power, or other similar addition laws as required, taking into account dependencies as discussed earlier in this chapter. The resulting accumulated mean and SD in the log-normal domain are then used to determine the corresponding mean and SD back in the log (or decibel) domain. This procedure is now summarized in general mathematical terms, and then illustrated specifically for various relevant parameters. (Note that for reference purposes Appendix A.5.2 provides a detailed comparison of normal and log-normal distributions and the relevant statistics associated with each.)

Step 1: Consider n network components in series, each with a parameter value characterized by a normal random variable with mean μ_i and SD σ_i in the log (or decibel) domain. The corresponding means and SDs in the log-normal domain are

$$U_i = 10^{\pm \mu_i/K + (\sigma_i/K)^2 (\ln 10)/2} \quad \text{and} \quad V_i = U_i \sqrt{10^{(\sigma_i/K)^2 (\ln 10)} - 1} \quad (5.73)$$

Where:

- K is a constant that depends on the type of accumulation (e.g., for power and voltage accumulations, $K = 10$ and 20, respectively; see Tables 5.2 and 5.13 for other examples).
- The $\pm$ sign depends on whether the accumulated mean increases or decreases (e.g., for noise the mean increases, so the $+$ve sign is used, whereas for SNR, Q, and multiple echo path losses, the mean decreases, so the $-$ve sign is used).

Step 2: Next, the means and SDs in the log-normal domain are accumulated as follows:

$$U = \sum_{i=1}^{n} (\pm \omega_i U_i) \quad \text{and} \quad V = \left(\sum_{i=1}^{n} (\pm \omega_i V_i^y) \right)^{1/y} \quad (5.74)$$

Where:

- ω_i is a weighting term such that usually $\omega_i = 1$ for all i. However, for some applications, $0 < \omega_i < 1$, and $\sum_{i=1}^{n} \omega_i = 1$.

  For example, frequency weighting can be used to power sum distributions of singing and echo return losses (SRL-LO, ERL, and SRL-HI) to estimate APRL as discussed elsewhere in this chapter (see the frequency weights in Eq. (5.25) of Sec. 5.4.2.1 or in Eq. (5.89) of Sec. 5.4.3.4).
- A $+$ve or $-$ve sign applies separately to each term depending on whether addition or subtraction of the random variable is being performed.
- The y accounts for the dependencies among the random variables. For independent random variables, $y = 2$, which corresponds to a sum of squares, while for dependent random variables, $y = 1$, which corresponds to an algebraic sum.

Table 5.13
Example Applications of Wilkinson's Method

Parameter to be Accumulated	Addition Law K (dB)	Weighting (ω) for:		Exponent for SD y	
		Added RVs	Subtracted RVs #	Indep. RVs	Dep. RVs
Noise	10	+1	-1	2	N/A
SNR	-10	+1	-1	2	N/A
Q	0 to -15*	+1	-1	2	N/A
Multiple EPL	-10	+1	-1	##	##
APRL**	10	$+\omega_i$	$-\omega_i$	2	N/A
IMD					
2nd Harmonic	10	+1	-1	2	N/A
3rd Harmonic	16	+1	-1	2	N/A
Signal to IMD Ratio					
2nd Harmonic	-10	+1	-1	2	N/A
3rd Harmonic	-16	+1	-1	2	N/A
Events	@	+1	-1	2	N/A

*   Depends on type of tandeming (see Table 5.8)
** APRL estimate from SRL-LO, ERL and SRL-HI statistics
\#   Subtraction can be used, for example, to determine the difference between an accumulated value and a component value; in this case the component value would have a -ve weight.
\#\# Echo path losses are usually partially dependent on one another owing to commonality of certain network component losses to more than one echo path. Thus, an appropriate combination of y = 1 and 2 is required to include both dependent and independent losses (see text, Sec. 5.4.3.4).
@   Sign is +ve. Magnitude depends on type of event. Empirically, for BERs and Percentage of BERs > 10^{-6} K ≈ 2, and for Percentage Errored Seconds (PESs) K ≈ 1.5 (see Tables 5.15 and 5.16). For other events, assume K = 1.
N/A Not Applicable, in that the accumulated random variables are usually independent; if dependent, y = 1.

Step 3: Finally, the accumulated mean μ, and SD σ, back in the log (or dB) domain are

$$\mu = \pm K\{\log U - (\tfrac{1}{2})\log(1 + V^2/U^2)\} \quad \text{and} \quad (5.75)$$
$$\sigma = K\sqrt{(1/\ln 10)\log(1 + V^2/U^2)}$$

Notice from Equations (5.73) to (5.75) that if the original component SDs σ_i, are all zero, then $V_i = 0$ for all i; therefore, $V = 0$ and $\sigma = 0$, and Wilkinson's method reduces to an ordinary power-like sum for fixed parameter values as in Table 5.2.

Examples for the application of Wilkinson's method are shown in Table 5.13 and are addressed in more detail below.

End-to-End Noise and Example

To illustrate, consider Figure 5.17, which has n network components each of which generates noise mean N_i and SD σ_{Ni} (SDs are not shown in Fig. 5.17 but would appear at the same locations as the noise values; also, for convenience, consider one direction of transmission only). As noted earlier, before the noise components can be accumulated, they must be appropriately attenuated by subsequent losses in the connection. Note that in this case dependencies of the noise on loss variations must be accounted for. That is, some trunk losses will be common to more than one noise component. For this reason, the general formulation for the accumulation of noise in Equation (5.30) must be used as presented earlier in Section 5.4.2.2; note, however, that in this case "+" denotes a power sum using Wilkinson's method in Equations (5.73) to (5.75).

Let the losses have means L_i, and SDs σ_{Li}. Then, as described earlier in Section 5.4.2.2, the noise of the first component is reduced by the loss of the second component and power summed (using Wilkinson's method) with the noise of the second component; this noise sum is then reduced by the loss of the third component and power summed with the noise of the third component, and so on. Thus, the noise, say at the B end of the connection in Figure 5.17, is

For the mean: $N_{AB} = [\{(N_1 - L_2) \text{ "+" } N_2\} - L_3] \text{ "+" } N_3$ ---- etc. (5.76)

for the SD: $\sigma_{NAB} = \sqrt{[\{(\sigma_{N1}^2 + \sigma_{L2}^2) \text{ "+" } \sigma_{N2}^2\} + \sigma_{L3}^2] \text{ "+" } \sigma_{N3}^2}$ ---- etc. (5.77)

Again, the "+" associated with the noise means and SDs in the decibel domain in Equations (5.76) and (5.77), respectively, denotes a power law accumulation using Wilkinson's method.

To illustrate more specifically, refer to Table 5.13, and consider the example in Figure 5.25 for the accumulation of distributions of noise (and loss) from a medium-length end office to end office toll connection [26]. Notice that, as expected, Wilkinson's method somewhat under and overestimates the mean and SD, respectively, of the connection noise, relative to the result determined by simulation.

Noise Asymmetry and Example

Noise asymmetry can be determined by extending the approach outlined for fixed values in Section 5.4.2.2. For distributed values, the *mean* asymmetry is the difference in the mean noises in the two directions as given by Equation (5.37); that is,

$\Delta N_r = N_{ABr} - N_{BAr}$ (5.78)

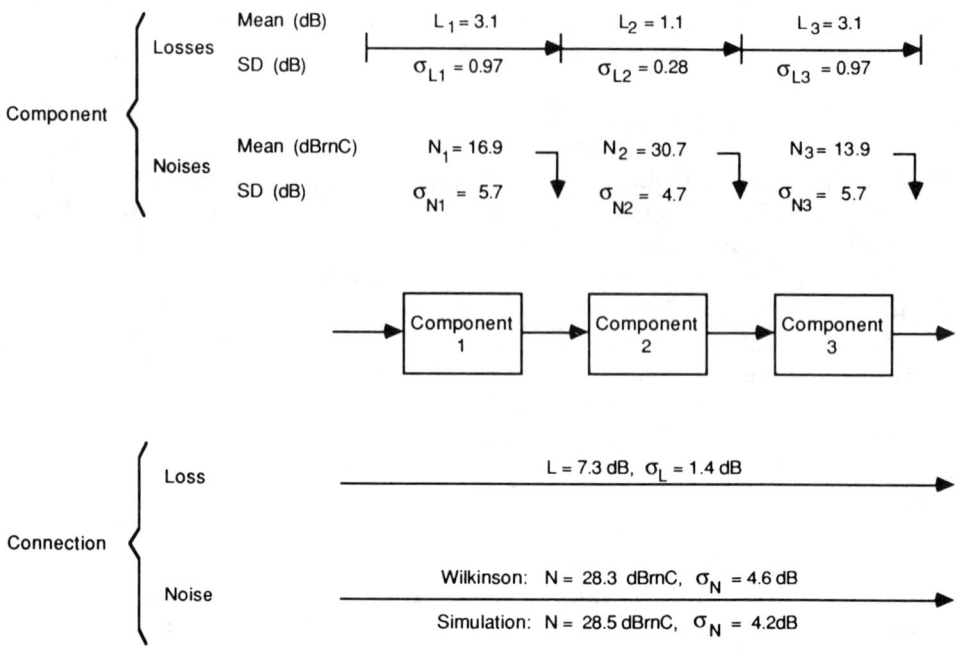

Figure 5.25 Example for accumulation of noise distributions: comparison of Wilkinson with simulation method.

where the subscript r denotes referral of the noise to a reference receiving system (which is readily done by an extension of process for fixed values described in Sec. 5.4.2.2). Assuming directional independence of the noises, the *variation* in asymmetry is

$$\Delta \sigma_{\Delta Nr} = \sqrt{\sigma_{NABr}^2 + \sigma_{NBAr}^2} \tag{5.79}$$

This assumption of directional independence may not always be true. For example, the noise of long-haul analog transmission systems tends to be distance-dependent both in the mean and the SD [52]. Thus, there may also be some directional dependence. This effect is not considered here, but to the extent that such dependencies can be quantified, they can be accounted for using the technique outlined in this chapter. There should be no distance or directional dependencies for noise on digital transmission systems.

Again, as in Section 5.4.2.2, ignoring loop noise and assuming the noise statistics for trunks are the same for each direction, the noise asymmetry simplifies as follows:

For the mean (recall Eq. (5.38)): $\Delta N_r = \text{ROLR}_A - \text{ROLR}_B$ (5.80)

for the SD: $\Delta \sigma_{\Delta Nr} = \sqrt{2\sigma_{Nr}^2 + \sigma_{\text{ROLR}A}^2 + \sigma_{\text{ROLR}B}^2}$ (5.81)

where σ_{Nr} is the SD of the total trunk noise in one direction.

If, in addition, the subscriber loop plus station set statistics are the same at both ends of the connection, then, from Equations (5.80) and (5.81),

$\Delta Nr = 0$ and $\Delta \sigma_{\Delta Nr} = \sqrt{2(\sigma_{Nr}^2 + \sigma_{\text{ROLR}}^2)}$ (5.82)

Consider an example where the subscriber loop plus station set receive sensitivity (ROLR) has mean of 46.7 dB and SD of 1.8 dB (e.g., as in Fig. 5.24). Consider also the trunk noise statistics from the example in Figure 5.25 for which the mean is 28.5 dBrnC and the SD is 4.2 dB. Thus, from Equation (5.82), the mean loss asymmetry is $\Delta N_R = 0$ dB, with SD $\Delta \sigma_{\Delta Nr} = 6.5$ dB. The variation in noise asymmetry is dominated by noise variations.

Signal-to-Noise Ratios and Example

Consider the example in Figure 5.19 and the related SNR analysis given in Section 5.4.2.2 (also, consider one direction of transmission only). Component output signal, noise, and SNR formulae are summarized in Table 5.14, and discussed below. When distributions are considered, the connection input signal has mean S_I, and SD σ_{SI}, and the signal level of the output of each component is

For the mean (recall Eq. (5.40)): $S_i = S_{i-1} - L_i = S_I - \sum_{j=1}^{i} L_j$ (5.83)

For the SD: $\sigma_{Si} = \sqrt{\sigma_{Si-1}^2 + \sigma_{Li}^2} = \sqrt{\sigma_{SI}^2 + \sum_{j=1}^{i} \sigma_{Lj}^2}$ (5.84)

Similarly, each network component produces noise with mean N_i, and SD σ_{Ni}. Thus, the mean SNR at the output of each component is similar to Equation (5.39):

$\text{SNR}_i = S_i - N_i + F$ (5.85)

For the SD of the SNR, the situation is more involved, since various dependencies must be accounted for. Consider first *signal-independent* (idle circuit) *noise* (e.g., C-message noise). In this case, it is necessary to consider whether both the signal and the noise of the *i*th component in the connection are reduced by the loss of *i*th component. For facilities with digital loss or for 2-wire facilities (without gain or without fixed inserted loss, e.g., subscriber loops with distributed structural loss), the output noise is not necessarily reduced by the loss of such a facility, and independence between the facility loss and noise can be assumed. In this case, when loss *does not* reduce noise, the SD for the SNR is

$$\sigma_{\text{SNR}i} = \sqrt{\sigma_{Si}^2 + \sigma_{Ni}^2} \qquad (5.86)$$

For the other types of facilities in Table 5.14 with fixed inserted loss, both the signal and noise will be identically reduced by this loss. Thus, the SNR mean and SD as determined before the loss are identical to the mean and SD after the loss. Therefore, at the output of the *i*th component (after the loss), the SNR variation due to the loss of the *i*th component, being identical for both the output signal and noise, is eliminated from the SD for the output SNR. In this case, where loss *does* reduce noise,

$$\sigma_{\text{SNR}i} = \sqrt{\sigma_{Si-1}^2 + \sigma_{\beta i}^2} = \sqrt{\sigma_{Si}^2 + \sigma_{Ni}^2 - 2\sigma_{Li}^2} \qquad (5.87)$$

Of more interest is the case of *signal dependent* (e.g., C-notched) *noise*, which occurs, for example, on digital (PCM-based T-carrier) and analog (N-carrier) compandored transmission facilities [26]. Formulae for component output noises and SNRs for this signal-dependent case were given in Table 5.3 for fixed values. The same formulae apply for mean values as shown here in Table 5.14. However, both loss and signal dependencies need to be accounted for in the SD for noise and SNR, as included in Table 5.14. Again, the SNR mean and SD are the same before and after the loss; thus, signal and loss variations, being identical for both the output signal and noise, are eliminated from the SD for the output SNR as follows:

$$\sigma_{\text{SNR}i} = \sqrt{(1 - \alpha_i^2)\sigma_{Si-1}^2 + \sigma_{\beta i}^2} = \sqrt{\sigma_{Si}^2 + \sigma_{Ni}^2 - 2\sigma_{Li}^2 - 2\alpha_i^2(\sigma_{Si}^2 - \sigma_{Li}^2)} \qquad (5.88)$$

Note from Table 5.14 that for no companding ($\alpha_i = 0$), the noise and SNR mean and SD reduce to the idle circuit noise case (e.g., compare Eq. (5.88) to Eq. (5.87)). Note also that for PCM companding ($\alpha_i = 1$) the SNR mean and SD, when expressed in terms of the component input signal level, are independent of the input signal; this is a property of PCM companding. However, for analog companding ($\alpha_i = \frac{1}{2}$) the SNR mean and SD are a function of the input signal level.

For a given facility type, the appropriate SNR means and SDs in Table 5.14 can be accumulated using Wilkinson's method. Figure 5.26 is an example in which the components are assumed to be compandored T-carrier transmission facilities ($\alpha_i = 1$) [26]. Notice in the figure that, in all cases, the SD of any output SNR is always less than the SD of the corresponding output noise. This is also generally true from an examination of survey data for both trunks [22] and connections [17]. The

Table 5.14
Characterization of Component Noise and SNR

Noise Type →	Idle-Circuit			Signal Dependent	
Medium Type →	Metallic	Carrier		Carrier	
Loss Type →	Structural *	Digital	Analog #	Digital or Analog #	
Output Signal Mean: S_i	$S_{i-1} \cdot L_i$		$S_{i-1} \cdot L_i$	$S_{i-1} \cdot L_i$	
SD: σ_{Si}	$\sqrt{\sigma_{Si-1}^2 + \sigma_{Li}^2}$		$\sqrt{\sigma_{Si-1}^2 + \sigma_{Li}^2}$	$\sqrt{\sigma_{Si-1}^2 + \sigma_{Li}^2}$	
Output Noise Mean: N_i	β_i		$\beta_i \cdot L_i$	$\alpha_i S_{i-1} + \beta_i \cdot L_i$	
SD: σ_{Ni}	$\sigma_{\beta i}$		$\sqrt{\sigma_{\beta i}^2 + \sigma_{Li}^2}$	$\sqrt{\alpha^2 \sigma_{s-i}^2 + \sigma_{\beta i}^2 + \sigma_{Li}^2}$	
Output SNR Mean: $SNR_i = S_i - N_i + F$	$S_{i-1} - L_i - \beta_i + F$		$S_{i-1} - \beta_i + F$	$S_{i-1} - \alpha_i S_{i-1} - \beta_i + F$	
SD: σ_{SNR_i}	$\sqrt{\sigma_{Si-1}^2 + \sigma_{Li}^2 + \sigma_{\beta i}^2}$ $= \sqrt{\sigma_{Si}^2 + \sigma_{Ni}^2}$		$\sqrt{\sigma_{S-i}^2 + \sigma_{\beta i}^2}$ $= \sqrt{\sigma_{Si}^2 + \sigma_{Ni}^2 - 2\sigma_{Li}^2}$	$\sqrt{\sigma_{Si-1}^2 - \alpha_i \sigma_{Si-1}^2 + \sigma_{\beta i}^2}$ $= \sqrt{\sigma_{Si}^2 + \sigma_{Ni}^2 - 2\sigma_{Li}^2 - 2\alpha_i^2(\sigma_{Si}^2 - \sigma_{Li}^2)}$	

* No fixed loss or gain

\# These equations also apply to metallic medium with fixed loss or with fixed gain

234

Figure 5.26 Examples for derivation and accumulation of SNR distributions.

reason is that loss (and in some cases signal) dependencies are common to the output signal and noise, and therefore reduce the SDs for output SNRs relative to the SD for the output noise; this observation is consistent with Equations (5.87) and (5.88) and with Table 5.14.

Loss/Noise Dependencies on GOS and Example

The accumulated loss and C-message noise at the end of a connection are dependent, since component noises are reduced by downstream losses. When Monte-Carlo simulation is used to determine GOS, the end-to-end loss, noise, and GOS are determined on a sample connection by sample connection basis, so that the dependencies are accounted for. However, in the absence of a Monte-Carlo simulator, it is possible to estimate the average GOS using the mean connection loss and noise in the loss/noise transmission rating equation. It is also possible to estimate the near worst-case GOS by using percentiles from the connection loss and noise distributions. However, so as not to be too conservative in the worst-case GOS estimate, the dependencies between the connection loss and noise distributions can be removed. This is warranted if loss variations contribute significantly to noise variations [53]; this is perhaps more likely to happen in a digital environment where noise variations are small and more readily influenced by loss variations.

Dependencies are removed by not including any loss variations in the SDs of the network component noises before the component noises are accumulated. Thus, each component noise SD σ_{Ni} is reduced by the SD of the component loss σ_{Li} (using an approach similar to that in Equation (5.71)), and is not increased by the SDs of subsequent component losses. The component noise means, loss means, and SDs are otherwise accumulated in the usual way, as described earlier. Consider the example in Figure 5.25. By applying the above procedure, the SD of the connection noise is reduced slightly (by about 0.1 dB). Thus, in this example, the effect of the connection loss/noise dependence is negligible owing to the inherently large noise SDs as compared to loss SDs. However, the effect is greater if loss variations contribute significantly to noise variations [53]. As noted above, this is more likely in a digital environment, although the overall GOS would also be higher than in an analog environment.

Multiple Echo Path Losses and Example

For multiple echo paths, at least some network components will be common to more than one echo path (recall Fig. 5.21). Thus, generally, there are dependencies among the echo paths that must be considered when accumulating the echo path losses.

More specifically, recall Figure 5.24, which shows a talker echo path for the A end of the connection resulting from the reflection at component 4 with mean of $ERL_4 = 11.3$ dB and SD $\sigma_{ERL} = 3.1$ dB (this is a typical value of loop balance return loss as discussed in Ch. 6 [47]). As an example, consider another talker echo path for the A end of the connection resulting from a reflection at component 3 with mean of $ERL_3 = 22$ dB and SD $\sigma_{ERL3} = 3$ dB (this is a typical value of terminal balance return loss as discussed in Ch. 6 [50]). Thus, in Figure 5.24, the losses of components 1, 2, and 3 are common to both of these particular echo paths. In a Monte-Carlo simulation, the same loss samples for components 1, 2, and 3 would be used in the calculation of each of the echo path losses, and in this manner account for the dependency of each echo path on these losses. The resulting two TEPLs are then power summed (for subjective applications, the TEPLs can also be adjusted for delay differences before power summing, as described in Sec. 5.4.2.2 [33]).

Analytically, the approach to accounting for dependencies is to first combine those aspects of the talker echo paths that are independent. Referring to Figure 5.24, the reflections (ERLs) at components 3 and 4, as well as the loss of component 4, are all independent. Thus, the reflection at component 4 can be referred to, and combined with, the reflection at component 3. The referred reflection is

For the mean: $ERL_{4r} = ERL_4 + 2L_4 = 11.3 + 2 \times 3.1 = 17.5$ dB

For the SD: $\sigma_{ERL4r} = \sqrt{\sigma_{ERL4}^2 + 2\sigma_{L4}^2} = \sqrt{3.1^2 + 2(0.97)^2} = 3.4$ dB

where the subscript r denotes referral of the parameter value. Note that, for subjective applications, the referred reflection can also be adjusted for the delay of component 4, which would reduce ERL_{4r} somewhat. This is not addressed here, since, in the context of this example, component 4 would be a toll connecting trunk for which the delay is relatively small (in general the effect of delay can be included using a procedure similar to that in Sec. 5.4.2.2 [33]). Thus, ignoring the delay of component 4, the referred reflection is now power summed with the reflection already at component 3 (i.e., $ERL_3 = 22$ dB, $\sigma_{ERL3} = 3$ dB). This is done using Wilkinson's method (see Eqs. (5.73) to (5.75) and Table 5.13), and the combined reflection at component 3 is

For the mean: $ERL_{3c} = ERL_3$ "+" $ERL_{4r} = 22$ "+" $17.5 = 15.8$ dB

For the SD: $\sigma_{ERL3c} = \sigma_{ERL3}$ "+" $\sigma_{ERL4r} = 3$ "+" $3.4 = 2.8$ dB

where the subscript c denotes combined values.

This reflection is now added to the losses of components 1, 2, and 3 in Figure 5.23, which are common to both echo paths, to give an overall combined TEPL. Thus, following the Figure 5.24 example computed in the talker echo path loss example, the mean TEPL is

$$E_{Ac} = \text{TOLR}_A + \text{ROLR}_A + 2\sum_{i=2}^{3} L_i + \text{ERL}_{3c}$$

$$= -43.1 + 46.7 + 2(3.1 + 1.1) + 15.8 = 27.8 \text{ dB}$$

which, as expected, is lower than the 29.5 dB previously computed for the single A end echo path in Figure 5.24.

Accounting for direction dependencies in a manner similar to that in the talker echo path loss example, for a single echo path, the TEPL SD for two echo paths is determined from dependent and independent contributors as follows:

$$\sigma_D = \sigma_{1D} + \sigma_{1D} = 3.6 + 1.7 = 5.3 \text{ dB (as before), and}$$

$$\sigma_I = \sqrt{(\sigma'_{1I})^2 + (\sigma''_{1I})^2 + 2\sum_{i=2}^{3}(\sigma_{iI})^2}$$

$$= \sqrt{1.3^2 + 0.6^2 + 2(0.97^2 + 0.28^2)} = 2.0 \text{ dB}$$

which is lower than the 2.4-dB value obtained previously because the trunk loss SD for component 4 was included elsewhere in the combined SD for ERL ($\sigma_{\text{ERL}3c}$). Thus, the total variation in TEPL including the combined ERL variation is

$$\sigma_{EAc} = \sqrt{\sigma_D^2 + \sigma_I^2 + \sigma_{\text{ERL}3c}^2} = \sqrt{5.3^2 + 2.0^2 + 2.8^2} = 6.3 \text{ dB}$$

which is lower than the 6.6 dB obtained with a single echo path. Thus, as expected, the overall combined TEPL mean and SD for two talker echo paths are somewhat lower than for the single talker echo path in Figure 5.24. As a practical matter, the difference is expected to be small for this example, since terminal balance return loss should not significantly affect connection performance; connection performance is dominated by loop balance return loss (see Ch. 6).

Estimation of APRL and an Example

As discussed in Section 5.4.2.1, APRL can be estimated from SRL and ERL measures [33]. For fixed values, APRL is given by Equation (5.25), repeated here as

$$\text{APRL} \approx -10 \log_{10}\{0.1 \times 10^{-\text{SRL-LO}/10} + 0.5 \times 10^{-\text{ERL}/10} + 0.4 \times 10^{-\text{SRL-HI}/10}\} \quad (5.89)$$

For normally distributed values, APRL can be estimated using Wilkinson's method (see Eqs. (5.73) to (5.75) and Table 5.13). In this case, frequency weights are involved and are $\omega_1 = 0.1$, $\omega_2 = 0.5$, and $\omega_3 = 0.4$, as shown in Equation (5.89). The following is an example in which APRL is estimated from objectives for SRL-LO, ERL, and SRL-HI [34].

Moments	SRL-LO	ERL	SRL-HI	APRL
Mean (dB)	11	11	7	9.6
SD (dB)	2	3	3	3.1

The mean value of APRL is a frequency weighted average of the ERL and SRL values, while the SD for APRL is somewhat larger than the ERL and SRL SDs. If all SDs were zero in the above example, then, from Equation (5.89), APRL = 8.9 dB. Thus, the nonzero SDs have the effect of somewhat increasing the mean value of APRL.

Bit and Block Error Ratios and Example

Bit and block error ratios (BER and BLER) are usually expressed in the form $a \times 10^b$, where the multiplier a is a positive real number, and the exponent b is a negative integer (e.g., BER = 5×10^{-6}). Fixed ratios of this form add algebraically (e.g., $5 \times 10^{-6} + 4 \times 10^{-5} = 4.5 \times 10^{-5}$ (see Table 5.1)). However, survey data indicates that, as a rough approximation, the logarithm of BER is normally distributed [54, 55]. Therefore, BERs are approximately log-normally distributed, and Wilkinson's method can be used to estimate the accumulated value of BERs for systems operated in series in a connection (see Table 5.13).

For example, consider the following percentile values from the surveyed BER of T-carrier transmission systems (as in Table 5.15 and Figure 5.27): 50% of systems surveyed had BERs $\leq 10^{-10}$ (i.e., median BER = 10^{-10}), and 97% had near worst BERs $\leq 10^{-6}$ [55]. To estimate the results for two such systems in series in a connection using Wilkinson's method first, BERs are expressed in the equivalent form: BER = $a \times 10^b = 10^d$, where $d = \log(\text{BER}) = \log(a \times 10^b) = \log(a) + b$. The exponent d is the approximately normally distributed random variable typically used in the presentation of survey data. More formally, exponents d, or percentile values in the log domain, are

$$x_i(\%) = d_i = \log(\text{BER}_{\%-i}) \qquad (5.90)$$

Thus, for the above example,

$$x_i(50) = \log(10^{-10}) = -10, \quad \text{and} \quad x_i(97) = \log(10^{-6}) = -6$$

Table 5.15
Example of Accumulation of Log-Normal BER Statistics: Comparison of
Wilkinson and Empirical Methods with Convolution

For n = 1 Link: Median BER = 10^{-10}, 97th percentile BER = 10^{-6}
∴ x(50) = -10, x(97) = -6, and μ = -10, σ = 2.13

For n = 2 Links: See the table below, and Fig. 5.27

BER Statistic	Accumulation Method		
	Convolution	Estimated	
		Wilkinson	Empirical
Median BER	10^{-9}	7.9×10^{-10} ! $	9.8×10^{-10} @
SD σ (decades)	2.03 *	2.00 !	1.94 *
Percentile @ 10^{-6} BER	93	94 #	94 @
BER @ 93rd Percentile	10^{-6}	7.1×10^{-7} $	7.1×10^{-7} $

See text for further explanation of the following notes:

! From eqns. (5.73) to (5.75) with K = 2
\# From eqn. (5.56) and Table 5.10, using log median BER and SD in decades
$ From eqn. (5.90a) and Table 5.10, using log median BER and SD in decades.
@ Median from eqn. (5.94) with x = 3.3, and higher percentile from eqn. (5.92)
* From eqn. (5.58) and Table 5.10, using log median BER and log BER of 10^{-6}

Next, to apply Wilkinson's method, the means and SDs for d are needed and can be estimated from the survey data using the previously presented technique for relating moments and percentiles (Eq. (5.58) and Table 5.10 in Sec. 5.4.3.2); the SDs and means (or medians) for d are

$$\sigma_{di} = [x_i(97) - x_i(50)]/(z''_i - z'_i) = [-6 - (-10)]/(1.881 - 0) = 2.13$$

and

$$\mu_{di} = x_i(50) - z''_i \sigma_{di} = -10 - 0 \times 2.13 = -10$$

Combining these statistics for two such systems in series using Wilkinson method (Eqs. (5.73) to (5.75) and with $K = 2$ from Table 5.13) gives

$$\text{mean (or median) } \mu_d = -9.1 \quad \text{and} \quad \text{SD } \sigma_d = 2.00$$

Figure 5.27 Example of accumulation of log-normal BER CDFs for two identical links in series. (See Table 5.15 for Link Statistics.)

Thus, using Equation (5.56a) and Table 5.10, various percentile values for two systems in series can now be determined from $x(\%) = z\sigma_d + \mu_d$, and then by taking the inverse logarithm of Equation (5.90), these percentile values are converted to the BERs shown in Table 5.15. That is (similar in form to Eqs. (4.1) and (4.2) used in Chapter 4),

$$\text{BER}_\% = \text{inverse log}[x(\%)] = \text{inverse log}[z\sigma_d + \mu_d] \quad (5.90a)$$
$$\text{BER}_{50} = \text{inverse log}[0 \times 2.00 + (-9.1)]$$
$$= \text{inverse log}(-9.1) = 7.9 \times 10^{-10}$$
$$\text{BER}_{93} = \text{inverse log}[1.476 \times 2.00 + (-9.1)]$$
$$= \text{inverse log}(-6.15) = 7.1 \times 10^{-7}$$

Also of interest is the percentile at a BER of 10^{-6}; this is obtained from the relative deviation in Equation (5.56):

$$z = [x(\%) - \mu_d]/\sigma_d = [-6 - (-9.1)]/2.00 = 1.55$$

for which the percentile using Table 5.10 is 94% (see Table 5.15 and Fig. 5.27).

As shown in Table 5.15 and Figure 5.27, one convolution of the BER field data results in a median BER for two systems in series of about 10^{-9}, and about 93% of systems in series have a BER $<10^{-6}$. Thus, comparing the results just estimated from Wilkinson's method with those obtained from convolution shows that for engineering purposes the agreement is reasonable. From another example in [56], it can be shown that the estimated percentile at a BER of 10^{-6} for three links in series (using Wilkinson's method with $K = 2$) also agrees reasonably well with the results of convolution. Similarly, from another example in [55], it can be shown that the estimated accumulation of the percentage of 5-min periods with BERs $> 10^{-6}$ for two links in series (using $K = 2$) agrees reasonably well with the results of convolution.

However, it should be noted that (from Table 5.13) the value of $K = 2$ used in Wilkinson's Equations (5.73) and (5.75) was specifically selected to give the reasonable agreement with the results from convolution. In the absence of other information (i.e., the convolution results), a value of $K = 1$ would be used, but this significantly underestimates the accumulated BERs for systems in series. The reason may be that the various BER distributions used in these examples are only roughly lognormal.

Estimated results using a simple, alternative, *empirical rule* for the accumulation of BERs are also shown in Table 5.15 and Figure 5.27. The rule for *high percentiles* (i.e., $>95\%$) is

$$\%_n = 100 - \sum_{i=1}^{n} (100 - \%_i) \tag{5.91}$$

If the systems are identical, then $i = 1$ for all i, and

$$\%_n = 100 - n(100 - \%_1) \tag{5.92}$$

where, for example, in Table 5.15 at BER $= 10^{-6}$, the percentile for a single link is $\%_1 = 97$, and, therefore, from Equation (5.92) at BER $= 10^{-6}$, the percentile for $n = 2$ two links in series is $\%_2 = 100 - 2(100 - 97) = 94$. It can be shown that Equation (5.92) also applies approximately to accumulation of percentiles for BERs on three links [56]. Rules such as those in Equations (5.91) and (5.92) have been used in [55] to estimate accumulation of percentiles for BER. These rules were also used in Section 4.4.2.2 of Chapter 4 to estimate end-to-end digital performance (see Table 4.26); the rules were applied to the network component BERs in Table 6.32 of Chapter 6.

The empirical rule for certain BER *percentile values* is assumed to be

$$\text{BER}_{\%-n} = n^{x-1} \sum_{i=1}^{n} \text{BER}_{\%-i} \tag{5.93}$$

where the exponent x is discussed below, and Equation (5.93) is such that if the systems are identical, then $i = 1$ for all i, and

$$BER_{\%-n} = n^x \, BER_{\%-1} \qquad (5.94)$$

Empirical rules of the nature of Equation (5.92) and Equation (5.94) with exponent for n of $x = 1$ were used earlier in Section 4.4.2.2 of Chapter 4 to estimate end-to-end digital network performance. Here, an exponent of $x = 3.3$ in Equation (5.94) was selected to give good agreement with the median BER results from convolution (see Table 5.15 and Fig. 5.27). For example, in Table 5.15 the median BER for a single link is $BER_{50-1} = 10^{-10}$, and therefore, from Equation (5.94), the median BER for $n = 2$ two links in series is

$$BER_{50-2} = 2^{3.3} \times 10^{-10} = 9.8 \times 10^{-10}$$

It can be shown that Equations (5.92) and (5.94) also apply approximately to the accumulation of the percentages of 5-min periods with BER $> 10^{-6}$ on two links [55].

Note that the agreement with $x = 3.3$ is only known to apply for identical systems in Equation (5.94); it is not known if there is good agreement for nonidentical systems in Equation (5.93), since there are no results from convolution to compare with. Also, this agreement is only known to apply to median values, and not necessarily to other percentile values. However, in general, other percentile values can be estimated by first using Equation (5.58) and Table 5.10 to obtain the SD σ, in decades. This SD is determined from any two known percentiles, such as the following two which have already been computed:

- The log median value, $\mu = x(50) = \log(BER_{50-2}) = \log(9.8 \times 10^{-10}) = -9.01$, corresponding to Equation (5.94).
- The log percentile value, $x(94) = \log(BER_{94-2}) = \log(10^{-6}) = -6$, corresponding to Equation (5.92). Using these values in Equation (5.58) gives

$$\sigma = [x(94) - x(50)]/(z'' - z') = [-6 - (-9.01)]/(1.55 - 0) = 1.94$$

Then Equation (5.90a) is used with the log median BER, $\mu = x(50)$, to obtain any other percentile values. For example, consider the BER at the 93rd percentile in Table 5.15 and Figure 5.27. From Equation (5.90a), $BER_{93} =$ inverse $\log[z\sigma + \mu]$, and, therefore,

$$BER_{93} = \text{inverse } \log[1.476 \times 1.94 + (-9.01)]$$
$$= \text{inverse } \log[-6.15] = 7.1 \times 10^{-7}$$

Percent Errored Seconds and Example

Percent errored seconds (PES) performance can also be estimated for more than one link in series in a connection in a manner similar to that outlined above for BERs. An example is shown in Table 5.16 and Figure 5.28 for $n = 3$ and $n = 6$ links [57]. It can be seen that both the Wilkinson and empirical methods provide somewhat optimistic estimates for median PES performance and conservative estimates for near worst PES performance at the higher percentile. Note that (from Table 5.13) a value of $K = 1.5$ was used for Wilkinson's method, again to obtain reasonable agreement with the results from convolution.

Table 5.16
Example of Accumulation of Log-Normal PES Statistics: Comparison of Wilkinson and Empirical Methods with Convolution

For n = 1 Link: Median PES = .0006%, 98.5th percentile PES = 5%
∴ x(50) = -5.2, x(98.5) = -1.3, and μ = -5.2, σ = 1.80

For n = 3 and 6 Links: See the table below, and Fig. 5.28

PES Statistic	Accumulation Method					
	n = 3			n = 6		
	Convolution	Estimated		Convolution	Estimated	
		Wilkinson	Empirical		Wilkinson	Empirical
Median PES	.008	.0074 !$	.0075 @	.04	.035 !$	.037 @
SD σ (decades)	1.54 *	1.67 !	1.67 *	1.46 *	1.59 !	1.58 *
Percentile @ 5 PES	96.5	95.5 #	95.5 @	92.5	91.4 #	91 @
PES @ 96.5th percentile	5	7.9 $	8.0 $	-	-	-
PES @ 92.5th percentile	-	-	-	5	6.8 $	7.0 $

   ! From eqns. (5.73) to (5.75) with K = 1.5
   # From eqn. (5.56) and Table 5.10, using log median PES and SD in decades
   $ From eqn. (5.90a) and Table 5.10, using log median PES and SD in decades
   @ Median from eqn. (5.96) with y = 2.3, and higher percentile from eqn. (5.92)
   * From eqn. (5.58) and Table 5.10, using log median PES and log PES of 5 %

Figure 5.28 Example of accumulation of log-normal PES CDFs for identical links in series. (See Table 5.16 for Link Statistics.)

Also, for the empirical results, the estimates of higher percentiles were based on Equation (5.92), while the rule for percentile values of PES is assumed to be

$$\text{PES}_{\%-n} = n^{y-1} \sum_{i=1}^{n} \text{PES}_{\%-i} \qquad (5.95)$$

such that if the systems are identical, then $i = 1$ for all i, and

$$\text{PES}_{\%-n} = n^{y} \text{PES}_{\%-1} \qquad (5.96)$$

where in this case an exponent for n of $y = 2.3$ was selected to give good agreement with the median PES results from convolution (see Table 5.16). It can also be shown that Wilkinson's method (with $K = 1.5$) and the empirical method (Eqs. (5.92) and (5.96)) will provide reasonable PES performance estimates for $n = 2$ links [55].

Again, the agreement with $y = 2.3$ is known to apply to the identical systems in Equation (5.96), but not necessarily to the nonidentical systems in Equation (5.95). Also, this agreement is only known to apply to median values. Often, for simplicity, higher percentile values are simply added ($y = 1$ in Eqs. (5.95) and (5.96)).

This is the approach taken in [55] and, for consistency with this reference, the same approach was followed in Section 4.4.2.2 of Chapter 4. However, in a manner similar to that for BER in Table 5.15 and Figure 5.27, other percentile values can be estimated by first using Equation (5.58) and Table 5.10 to obtain the SD in decades. Then Equation (5.90a) is used with the log median PES to obtain other percentile values; examples are shown in Table 5.16 and Figure 5.28.

5.4.4 Addition Laws for Parallel Configurations

5.4.4.1 General

It is useful to combine performance across parallel configurations. These configurations can arise because of a variety of situations, including multiple routes, multiple distance categories, and multiple technologies. Multiple (i.e., alternate or dynamic) routing systems usually involve one direct route between two locations, which is selected with relatively high probability, and one or more other indirect routes, which are selected with lower probabilities [58]. Multiple routing arises in international networks owing to the generally hierarchical nature of these networks [59]. Nationally, in North America, multiple routing occurs in both metropolitan (intra-LATA) networks and in long distance (inter-LATA) networks as illustrated earlier in Figures 4.3 and 4.4 of Chapter 4 [20]. These networks have traditionally been hierarchical, although nonhierarchical dynamic routing is now used in some instances (e.g., for the AT&T IEC network [18]).

Multiple distance categories were also noted in Chapter 4, where, for example, the surveyed transmission performance of short, medium and long end office to end office connections carrying various percentages of traffic was used in the estimate of overall end-to-end network performance [17]. Finally, an example of multiple technologies is that of surveyed exchange access network transmission performance for trunks using T-carrier, N-carrier, radio, and metallic cable types of transmission facilities, which are encountered with various probabilities [22].

In all of the above cases, by knowing the performance and probabilities for each route, technology, or distance category (as the case may be), it is possible to estimate the overall combined performance of the parallel configuration using either Monte-Carlo simulation (see Sec. 5.4.3.2) or analytical techniques. Again, both fixed and distributed parameter values and various dependencies may be involved, and must be accounted for in simulation and in analytical techniques. Below, the analytical methodology is provided and followed by various examples.

5.4.4.2 Methodology

Consider the network configuration in Figure 5.29, which consists of two serial portions and a parallel portion (in general, a network configuration may consist of one

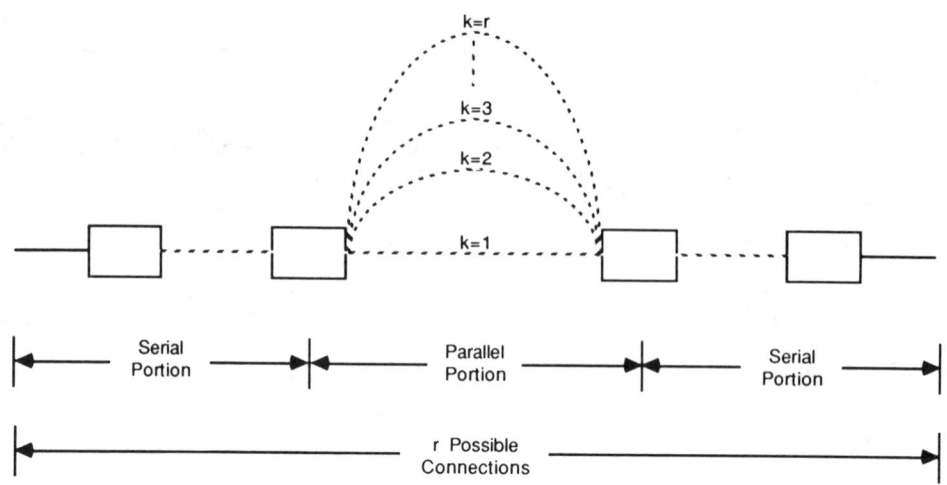

Figure 5.29 General illustration of a network configuration with serial and parallel portions.

or more serial and/or parallel portions). On a connection-by-connection basis, the serial portions are common to specific individual paths in the parallel portion. The methodology for combining parallel portions with serial portions is to first determine all possible end-to-end parallel connections, each consisting of the serial portions and one of the possible paths in the parallel portion. The end-to-end performance of each of the resulting parallel connections is then determined by combining all the resulting serial portions in each connection using the methods outlined previously in this chapter (this will not be repeated here). Next, the parallel connections are "collapsed" into a single equivalent serial connection using the methodology outlined below.

Note that parallel connections are assumed to be independent, even though each serial portion is common to all the parallel connections. The reason for this assumption is that parallel connections as manifested in practice by routing, technology, and distance alternatives imply that the analysis is across a large number of connections all set up at different times, and therefore involving an ensemble for each of the series portions within these connections. Therefore, serial portions are assumed not to introduce significant dependencies among the parallel connections. The following specifically addresses the process of collapsing a general set of independent parallel paths into an equivalent serial path.

As in Figure 5.29, consider r independent parallel paths, each with probability of occurrence P_k, where

$$\sum_{k=1}^{r} P_k = 1 \tag{5.97}$$

For a fixed parameter value (x_k) for each path (k), the overall combined parameter value across the parallel paths is simply the weighted sum of the values for each path:

$$x = \sum_{k=1}^{r} P_k x_k \tag{5.98}$$

Similarly, for distributed parameter values, if each path has mean μ_k, the combined mean value is

$$\mu = \sum_{k=1}^{r} P_k \mu_k \tag{5.99}$$

and the combined second moment for distributed values is

$$\overline{x^2} = \sum_{k=1}^{r} P_k \overline{x_k^2} \quad \text{where} \quad \overline{x_k^2} = \mu_k^2 + \sigma_k^2 \tag{5.100}$$

Also, since $\overline{x^2} = \mu^2 + \sigma^2$ for the combined path, then the combined SD is

$$\sigma = \sqrt{\overline{x^2} - \mu^2} \tag{5.101}$$

Notice that the overall combined SD in Equation (5.101) is not expressed directly in terms of the SDs σ_k, for each path. The reason is that it is rather complex to do so for more than two parallel paths. However, it is possible to do so iteratively by considering two parallel paths at a time. Thus, for $r = 2$ in Equations (5.99) to (5.101):

$$\mu = P_1 \mu_1 + P_2 \mu_2 \tag{5.102}$$

$$\overline{x^2} = P_1 \overline{x_1^2} + P_2 \overline{x_2^2} = P_1(\mu_1^2 + \sigma_{x1}^2) + P_2(\mu_2^2 + \sigma_{x2}^2) \tag{5.103}$$

$$\sigma = \sqrt{\overline{x^2} - \mu^2} = \sqrt{P_1 \sigma_1^2 + P_2 \sigma_2^2 + P_1 P_2 (\mu_1 - \mu_2)^2} \tag{5.104}$$

The above combined mean and SD for the two parallel paths in Equations (5.102) and (5.104), respectively, can now be viewed as a single equivalent path and can be combined with a third parallel path using the same procedure as above. In this iterative manner, any number of parallel paths characterized in terms of their parameter means μ_k and SD σ_k can be combined.

Notice from Equation (5.104) that for any two parallel paths the overall SD σ consists of two parts, the weighted sum of squares of the path SDs, plus a weighted term related to the difference in the path means. Thus, even if the individual path

SDs are zero (fixed parameter values), there is still an overall SD, owing to the difference of the path means.

To determine percentiles for the combined paths requires that the overall distribution for the combined paths be known. If the probability density for each path is $f_k(x)$, then the combined density is a weighted sum of the path densities:

$$f(x) = \sum_{k=1}^{r} P_k f_k(x) \tag{5.105}$$

In general, $f(x)$ will tend to be multimodal (with potentially one mode or peak corresponding to each path), unless the modes for each path are quite broad and/or close together, in which case the modes will tend to merge into fewer than r, and possibly only one mode. More specifically, if the individual path densities are normally distributed, the combined distribution will not be normal, although it may be approximately normal if multiple modes are not too evident. See the example in Figure 5.30, which is discussed in more detail later.

Figure 5.30 Example of combined normal pdfs for losses on two parallel routes.

In the remainder of this chapter, examples of parallel combinations will be considered for each of two cases:

- Parameters that are approximately normally distributed. These include most continuous parameters, such as loss, noise, and distortion, many of which are expressed in the log or decibel domain.
- Parameters that are approximately log-normally distributed. These include most transient parameters, such as analog transients (hits, dropouts, impulse noise) and digital events (BERs, BLERs, PESs, etc.), many of which are expressed in the inverse log domain. It is significant and of practical use later in this chapter that log-normally distributed parameters are normally distributed in the logarithmic domain. For this reason, and for reference purposes, the relationship between normal and log-normal distributions is included in Appendix A.5.2.

5.4.4.3 Examples of Parallel Paths with Normally Distributed Parameters

Multiple Routes

To illustrate, consider the example in Figure 5.30 for $r = 2$ parallel routes, for which the individual and combined route information in Table 5.17 applies. The direct route ($k = 1$) consists of one link carrying 70% of traffic, and the alternate route ($k = 2$) consists of two tandemed links carrying 30% of traffic. The percentages of traffic are not necessarily representative of all applications, but this example is otherwise indicative of a metropolitan or exchange access network with a direct route and an alternate route through a tandem location (e.g., see Fig. 4.3 of Ch. 4, or Fig. 5.12).

In Figure 5.30, the individual route loss densities are assumed to be normal and the exact combined density obtained from Equation (5.105), while not bimodal, shows evidence of two modes. The combined density is also approximated by a combined normal density having the same mean and SD as the exact density. See Figure 5.31 for the corresponding distributions. An indication of the accuracy of the normal approximation can be obtained by comparing a particular percentile from the normal approximation with that from the exact combined density. The results are illustrated in Table 5.17 and Figure 5.31, and the procedure is described below.

Given a percentile value, $x(\%)$, for the normal approximation, then the corresponding deviations (z_k) for each of the individual routes are computed from Equation (5.56), i.e., $z_k = [x(\%) - \mu k]/\sigma_k$. Next, from information on deviations and percentiles in Table 5.10, the corresponding percentile for each route, $\%_k$, is obtained. Finally, the exact combined percentile is

$$\% = \sum_{k=1}^{r} P_k \%_k \qquad (5.106)$$

Table 5.17
Example to Illustrate Combination of Normal Statistics for Losses on Two Parallel Routes

Parameter	Route			Equations #	
	1	2	Combined		
Probability	0.7	0.3	1.0	5.97	
Mean Loss (dB)	3	6	3.9	5.102	
Loss SD (dB)	1	1.5	1.8	5.104	
			Normal Approx. / Exact	Normal Approx. / Exact	
Median Loss (dB)*	3	6	3.9 $ / 3.5	5.102 / 5.106@	
95% Max Loss (dB)*	4.65	8.47	6.86 / 7.45	5.56a ! / 5.106@	
% ≤ 6.86 dB *	99.994	71.7	95 / 91.5	5.56 ! / 5.106	
% ≤ 7.45 dB *	~100	83.3	97.5 / 95	5.56 ! / 5.106	

\#   Apply to results in "Combined" column
*   Assumes normal pdfs for Route 1 and Route 2 losses
@   Iteratively with eqn. (5.56) and Table 5.10
!   With Table 5.10
$ Median loss ~ mean loss

For example, from Equation (5.56a), consider the percentile value $x(95) = 3.9 + 1.645 \times 1.8 = 6.86$ dB on the normal approximation in Figure 5.30. Applying the above procedure, the *percentile* on the exact combined distribution is about 91.5%. This is illustrated as follows. From Equation (5.56),

For route $k = 1$, $z_1 = [x(95) - \mu_1]/\sigma_1 = [6.86 - 3]/1 = 3.86$
For route $k = r = 2$, $z_2 = [x(95) - \mu_2]/\sigma_2 = [6.86 - 6]/1.5 = 0.5733$

From information as in Table 5.10 [43], the percentiles for each route are $\%_1 = 99.994$, and $\%_2 = 71.7$. Therefore, from Equation (5.106), the combined percentile is

$$\% = 99.994 \times .7 + 71.7 \times .3 = 91.5$$

Thus, the normal approximation is somewhat optimistic in this example, since it indicates that 95% of losses are less than 6.86 dB, whereas this is true for only 91.5% of losses.

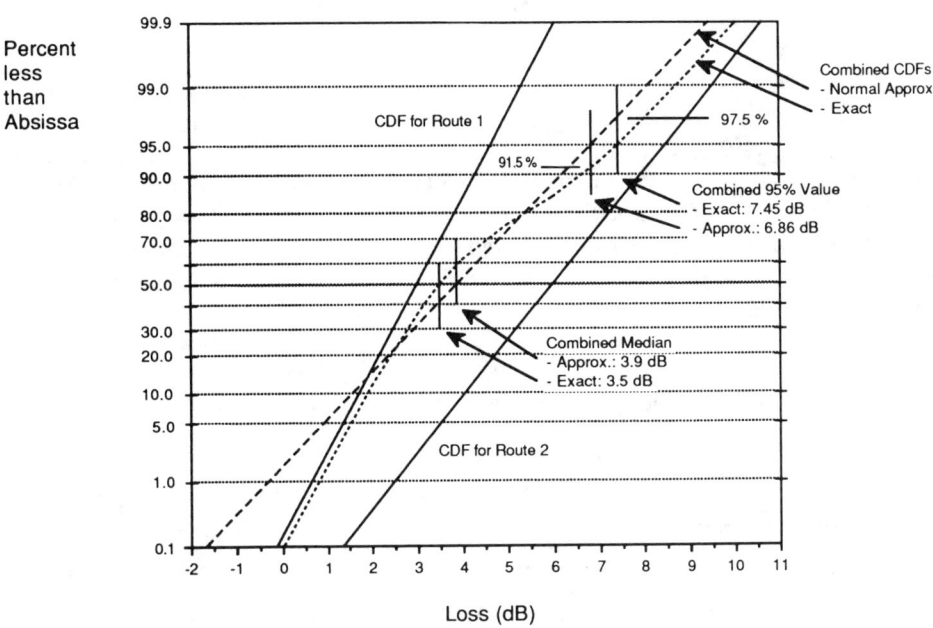

Figure 5.31 Example of combining normal CDFs for losses on two parallel routes. (See Figure 5.30 for Corresponding pdfs, and Table 5.17 for Route Statistics.)

Carrying the example further, a *percentile value* on the exact combined distribution can be determined iteratively by assuming a combined percentile value, $x(\%)$, computing the corresponding individual percentiles, $\%_1$ and $\%_2$ from Equation (5.56) and Table 5.10, and, finally, checking if these percentiles satisfy Equation (5.106). Depending on the percentile value of interest, iteration may not be necessary; for instance, for a percentile value, $x(95)$, it can be seen by inspection of Figure 5.30 that iteration is not necessary, since percentile values greater than $x(91.5) = 6.86$ dB are large enough that the corresponding percentile for the first route is known at essentially 100%. Thus, in such cases, the percentile for the second route can be solved directly from Equation (5.106); that is, for the above example with $r = 2$,

$$\% = \sum_{k=1}^{2} P_k \%_k = 0.7 \times 100 + 0.3 \times \%_2 = 95$$

Thus, $\%_2 = 83.3\%$; therefore, from information in Table 5.10 [43], $z_2 = 0.966$, and from Equation (5.56a), $x_2(83.3) = z_2\sigma_2 + \mu_2 = 0.966 \times 1.5 + 6 = 7.45$ dB. These and various other percentiles and deviations relevant to this example are summarized in Table 5.17 and Figure 5.31.

Multiple Distance Categories;

To combine performance across distance categories requires knowledge of the probability of occurrence of these categories. These estimated probabilities are provided in Table 5.18 [17, 60]. With this information, the combined performance of short, medium and long end office to end office toll connections is shown for connection loss in Table 5.19 [17, 60], and for TEPL and TEPD in Table 5.20 [61]. In each case, the combined performance as determined analytically using Equations (5.102) and (5.104), is compared with that from actual measurements and is generally found to be in good agreement.

Table 5.18
Estimated Pre-Divestiture Distance Distribution of Toll Calls

Airline Distance Between End-Offices (miles)		Percentage of Calls in Distance Category	
Course	Fine	Course [17]	Fine *
0 - 180 (short)	0 - 23 23 - 45 45 - 90 90 - 180	77.8	31.3 18.6 16.9 11.0
180 - 725 (medium)	180 - 360 360 - 725	13.4	7.7 5.7
725 - 2600 (long)	725 - 1450 1450 - 2600	8.8	4.7 4.1

* Estimated from earlier survey data [60].
Note: For additional distance information, see Table 4.3 of Chapter 4.

In some instances, the analytical estimates are somewhat larger than those from measurements (i.e., for TEPL with echo control and for TEPD in Table 5.28). This may be due to various dependencies in the data (e.g., common trunks in the various distance categories) not accounted for in the analytical approach, which assumes complete independence of performance among the distance categories. The error may also be due to the estimated probabilities associated with the various distance categories (Table 5.19).

Multiple Technologies;

As a final example, Table 5.21 illustrates how trunk loss and noise [26] are combined across various trunk technologies, and shows that the analytical results using Equations (5.102) and (5.104) compare favorably with those from simulation [22].

Table 5.19
Combination of Moments for Loss of Pre-Divestiture End-Office to End-Office Connections

Survey	Parameter	Distance Category (miles)			Combined	
		0-180	180-725	725-2600 #	Estimated	Surveyed
1969/70 [61]	Probability (%) * Loss: mean (dB) SD (dB)	83.7 6.45 2.05	12.4 7.20 2.20	3.9 7.55 2.25	100 6.59 2.10	100 6.65 2.10
1983 [17]	Probability (%) Loss: mean (dB) SD (dB)	77.8 6.53 1.63	13.4 7.29 1.43	8.8 7.84 1.79	100 6.75 1.67	100 6.7 1.3 **

* From [60].
** Believed to be a typographical error (too low) in [17].
\# Range is 725 - 2900 miles for 1969/70 survey.

5.4.4.4 Examples of Parallel Paths with Log-Normally Distributed Parameters

Bit Error Ratio

In practice, field survey data that is approximately log-normally distributed in the inverse log domain is usually presented in terms of approximately normal distributions in the log domain. For example, a data sample value of BER $= 10^{-6}$ in the inverse log domain would be presented as log(BER) $= -6$ in the log domain. It can be shown that if $f(x)$ in Equation (5.105) is the exact combined density for parallel normal densities in the log domain, then the counterpart in the inverse log domain is the combined log-normal density:

$$f(y) = \sum_{k=1}^{r} P_k(f_k(y)) \qquad (5.107)$$

where $f_k(y)$ is the log-normal density of y for the kth parallel path, and

$$y = 10^{x/K} \qquad (5.108)$$

is the log-normal variable (parameter value) in the inverse log domain, x is the corresponding normal variable in the log domain, and K is constant depending on the type of parameter (see Table 5.13). As described for the example involving losses on parallel routes in Section 5.4.4.3, for any percentile value in the log domain, $x(\%)$, the corresponding percentile for the exact combined distribution in the log domain—Equation (5.105)—is obtained from Equation (5.56), Table 5.10, and Equation (5.106). The significant point is that the same percentile applies to the exact

Table 5.20
Combination of Moments for Talker Echo Path Loss and Delay of Pre-Divestiture End-Office to End-Office Connections

Talker Echo Path		Distance Category (miles)					Combined	
Parameter	Value	180-360	360-725	725-1450	1450-2900		Estimated	Surveyed #
Probability *	(%)	34.7	25.6	21.2	18.5		100	100
Loss (No Echo Control)	Mean (dB) SD (dB)	23.1 5.7	24.3 6.8	24.6 6.3	23.3 6.4		23.76 6.29	23.8 6.3
Loss (Echo Control)	Mean (dB) SD (dB)	23.2 5.6	24.3 6.8	31.4 11.9	49.2 10.9		30.03 12.95	28.9 12.1
Delay	Mean (ms) SD (ms)	11.7 3.4	16.4 3.8	24.8 5.0	37.3 6.1		20.42 10.35	19.5 9.5

* Derived from Table 5.16 [60]
# [38]

Table 5.21
Combination of Moments for Pre-Divestiture Toll Connecting Trunk Loss and
Idle Circuit Noise for Multiple Technologies

Parameters			Technology @				Combined	
			T - Carrier	N - Carrier	Cable	Radio	Estimated	Simulated #
Probability		(%)	71.1	4.1	15.8	9.0	100	100
Loss	Mean	(dB)	3.14	2.63	3.04	3.03	3.09	3.1
	SD	(dB)	1.05	0.75	0.54	0.86	0.96	0.97
Noise*	Mean	(dBrnC)	13.26	14.67	11.36	22.47	13.85	13.9
	SD	(dB)	5.19	6.17	5.05	3.39	5.81	5.7

* at End-Office
@ [22]
# [26]

combined distribution in the inverse log domain (Equation (5.107)), and the corresponding percentile value is given by Equation (5.108). Thus, apart from the application of Equation (5.108), the approach for combining parallel paths with log-normally distributed parameters is the same as that for normally distributed parameters.

Table 5.22 and Figure 5.32 illustrate an example involving approximately log-normal BER statistics on two parallel routes, where the direct route ($k = 1$) consists of one T-carrier facility (with 70% probability), and the alternate route ($k = 2$) consists of two tandemed T-carrier facilities (with 30% probability). Again, this example is indicative of a metropolitan or exchange access network with a direct route, and an alternate route through a tandem location (e.g., see Fig. 4.3 of Ch. 4, or Fig. 5.12).

From Table 5.15, the principle BER statistics for each route are as follows: the median BERs are 10^{-10} and 10^{-9} for routes 1 and 2, respectively, and 97% and 93% of connections have BERs $\leq 10^{-6}$ on routes 1 and 2, respectively [55]. In Table 5.22 and Figure 5.32, the parallel combination of the two parallel routes results in a median BER of 2.04×10^{-10} (or about 2×10^{-10}) and 95.8% (or about 96%) of connections with BERs $\leq 10^{-6}$.

These results are illustrated as follows. The combined *percentile* for a BER of 10^{-6} is obtained directly from Equation (5.106); that is, % = $0.7 \times \%_1 + 0.3 \times \%_2$ = $0.7 \times 97 + 0.3 \times 93 = 95.8$. However, *percentile values* such as the median BER are obtained by iteration, for example, by assuming a combined log median percentile value, $\mu = x(50)$, computing the corresponding individual percentiles $\%_1$ and $\%_2$ from Equation (5.56) and Table 5.10, and, finally, checking if these percentiles satisfy Equation (5.106). Note that, from Table 5.22, a good initial estimate of the log median BER, is the log mean BER = -9.7. The last (or near last) step in the iterative

Table 5.22
Example to Illustrate Combination of Log-Normal BER Statistics on Two Parallel Routes

Parameter	Route 1	Route 2	Combined		Equations #	
Probability	0.7	0.3	1.0		5.97	
Log median BER	-10	-9	-9.69 *		5.106 !	
Log mean BER	-10	-9	-9.7		5.99 $	
BER SD (decades)	2.13	2.03	2.15		5.101 $$	
			Log-Normal Approx.	Exact	Log-Normal Approx.	Exact
Median BER	10^{-10}	10^{-9}	2.00×10^{-7}	2.04×10^{-10}	5.108 &	5.108 @
95% Max BER	3.19×10^{-7}	2.18×10^{-6}	6.87×10^{-7}	6.68×10^{-7}	5.90a !!	5.106 **
% $\leq 6.68 \times 10^{-7}$	96.37	91.80	94.94	95.00	5.56 !!	5.106
% $\leq 6.87 \times 10^{-7}$	96.41	91.88	95.00	95.05	5.56 !!	5.106
% $\leq 10^{-6}$	97	93	95.7	95.8	5.56 !!	5.106

\#   Apply to results in "Combined" column
*   Log median BER ~ log mean BER
!   Iteratively with eqn. (5.56) and Table 5.10 (see text)
@   With K = 1, and applied to log median BER
$ Or pairwise using eqn. (5.102)
$$ Or pairwise using eqn. (5.104)
&   With K = 1, and applied to log mean of BER
!!   With Table 5.10
**   Iteratively with eqn. (5.56), Table 5.10 and eqn. (5.108) with K = 1

Figure 5.32 Example of combining log-normal CDFs for BERs on two parallel routes. (See Table 5.22 for Route Statistics.)

process is illustrated as follows:

Assume $\mu = x(50) = -9.69$, then, using Equation (5.56),
For route $k = 1$, $z_1 = [x(50) - \mu_1]/\sigma_1 = [-9.69 - (-10)]/2.13 = 0.1455$
For route $k = r = 2$, $z_2 = [x(50) - \mu_2]/\sigma_2 = [-9.69 - (-9)]/2.03 = -0.3399$

Then, from information in Table 5.10 [43], the percentiles for each route are $\%_1 = 55.79$, and $\%_2 = 36.70$. Finally, from Equation (5.106), the combined percentile is $\% = 55.79 \times 0.7 + 36.70 \times 0.3 = 50.06 \sim 50\%$. A similar procedure is used for other percentile values, such as the 95% maximum BER in Table 5.22.

Also included in Table 5.22 and Figure 5.32 is an approximate log-normal solution, which is easier to obtain than the exact one just outlined that centers on Equation (5.106). That is, the combined density in the log domain is approximated by a combined normal density having the same log mean BER and BER SD in decades as the exact density in the log domain. This enables all percentiles to be estimated by way of a normal approximation in the log domain using Equation (5.56) and Table 5.10. Owing to the properties of normal and log-normal distributions, the

Table 5.23
Example to Illustrate Combination of Log-Normal Statistics for Pre-Divestiture Toll Connecting Trunk Impulse Noise for Multiple Technologies

Parameter	Technology				Combined		Equations #	
	T- Carrier	N- Carrier	Cable	Radio				
Probability	.711	.041	.158	.09	1		5.97	
Log Median Impulse Noise Count	-.296	-.0468	-.0522	.136	-.170 *		5.106 !	
Log Mean Impulse Noise Count	-.296	-.0468	-.0522	.136	-.2084		5.99 $	
SD (decades)	1.037	.980	.570	.521	.9491		5.101 $$	
					Log-Normal Approx.	Exact	Log-Normal Approx.	Exact
Median Impulse Noise Count	.506	.898	.887	1.37	.6189	.6761	5.108 &	5.108 @
% ≤ 15	92.2	89.4	98.4	97.7	92.76	93.57	5.56 !!	5.106

\#   Apply to results in "Combined" column
*   Log median Impulse Noise Count ~ log mean Impulse Noise Count
!   Iteratively with eqn. (5.56) and Table 5.10
@   With K = 1, and applied to log median Impulse Noise Count
$ Or pairwise using eqn. (5.102)
$$ Or pairwise using eqn. (5.104)
&   With K = 1, and applied to log mean Impulse Noise Count
!!   With Table 5.10

normal *percentiles* are directly applicable in the inverse log domain (see App. A.5.2). For example, consider BER = 6.68×10^{-7}; from Equation (5.90),

$x(\%) = \log(\text{BER}) = \log(6.68 \times 10^{-7}) = -6.175$ and, from Equation (5.56),
$z = [x(\%) - \mu_1]/\sigma_1 = [-6.175 - (-9.7)]/2.15 = 1.640$.

Thus, from information in Table 5.10 [43], % = 94.94.

Also, all *percentile values* can be estimated in the log domain using Equation (5.90a) and Table 5.10. For example, consider the 95% BER. From Equation (5.90a), BER_{95} = inverse log[$z\sigma + \mu$], and, therefore,

$$\text{BER}_{95} = \text{inverse log}[1.645 \times 2.15 + (-9.7)]$$
$$= \text{inverse log}[-6.163] = 6.87 \times 10^{-7}$$

For the example in Table 5.22 and Figure 5.32, the approximate solution agrees very well with the exact solution. The reason is that the means of the log BERs are not very different, and the BER SDs (in decades) are large enough to cause considerable overlap in the normal distribution for each route and, therefore, little deviation of the exact combined distribution from normal.

Figure 5.33 Example of combining log-normal CDFs for impulse noise across different transmission technologies. (See Table 5.23 for Statistics for each Technology.)

Impulse Noise

A final example is shown in Table 5.23 and in Figure 5.33 involving approximately log-normally distributed impulse noise counts over 15 min for four different trunk technologies [22]. Further discussion of this specific example is provided in Section 6.2.3.8 of Chapter 6 (see Table 6.22). The methodology involved is the same as that for BERs in Table 5.22 and Figure 5.32 above. Table 5.23 shows that, across the technologies, percentages of connections ranging from about 90% to 98% have less than 15 counts in 15 min, and, overall, about 93% of connections have less than 15 counts in 15 min. In this case, owing to the significant differences in performance for each technology, the approximate combined results deviate somewhat from the exact combined results.

5.5 REFERENCES

1. J. Gruber, E. Abdou, P. Richards, and G. Williams, "Quality-of-Service in Evolving Telecommunications Networks," *IEEE Journal on Sel. Areas in Comm.*, Issue on Quality Assurance for the Communications Community, Vol. SAC-4, No. 7, Oct. 1986, pp. 1084–1089.
2. J. Gruber, "From Performance to Planning: An Overview," *NCF Rec.*, Vol. 39, Chicago, Oct. 1985, pp. 551–555, 565, 566.
3. CCITT Rec. G.801, "Digital Transmission Models," *Blue Book*, Vol. III, Fas. III.5, Geneva, 1989, pp. 3–8.
4. CCITT Rec. G.101, "General Characteristics for International Telephone Connections and International Telephone Circuits," *Blue Book*, Vol. III, Fas. III.1, Geneva, 1989, pp. 9–25.
5. CCITT Rec. G.103, "Hypothetical Reference Connections," *Blue Book*, Vol. III, Fas. III.1, Geneva, 1989, pp. 28–37.
6. CCITT Rec. G.121, "Corrected Reference Equivalents (CREs) and Loudness Ratings (LRs) of National Systems," *Blue Book*, Vol. III, Fas. III.1, Geneva, 1989, pp. 109–125.
7. "Digital Network Notes," Telecom Canada, 1983. Issued by Technical Planning and Standards, Network Development and Operations, 410 Laurier Avenue, West, Ottawa, Ontario, Canada, K1P 6H5.
8. *IEEE Comm. Mag.* Special Issue on "Surviving Disasters," Vol. 28, No. 6, June 1990.
9. T. Wu and M.E. Burrowes, "Feasibility of a High-Speed SONET Self-Healing Ring Architecture in Future Interoffice Networks," *IEEE Comm. Mag.*, Vol. 28, No. 11, Nov. 1990, pp. 33–51.
10. M.N. Woinsky, "National Performance Standards for Telecommunications Services," *IEEE Comm. Mag.*, Vol. 26, No. 10, Oct. 1988, pp. 70–81.
11. R. Havens, "How a 'Different' Carrier Works," *TE&M*, Vol. 88, No. 18, Sept. 15, 1984, pp. 141, 142, 147, 149.
12. A. Almlie et al., "Design Issues in the Development of a National Fiber-Optic Transmission Network," *ICC Rec.*, Vol. 1, Philadelphia, June 1988, pp. 2.4.1–2.4.6.
13. G.P. Abbott, "Crossing the Country with Fiber Optics," *TE&M*, Vol. 90, No. 4, Feb. 15, 1986, pp. 90, 91.
14. L.G. Cohen, "Trends in U.S. Broad-Band Fiber Optic Transmission Systems," *IEEE Journal on Sel. Areas in Comm.*, Vol. SAC-4, No. 4, July 1986, pp. 488–497.

15. T. Volovic, "Fiber-Optic Deployment Among the Interexchange Carriers," *Telecomm.*, Vol. 21, No. 5, May 1987, pp. 40, 43, 44, 46, 48, 53.
16. "Propagation Delay Distributions in the IC Network," Contrib. T1Q1.1/89-048, US Sprint, April 1989.
17. M.B. Carey, H.T. Cheu, A. Descloux, J.F. Ingle, and K.I. Park, "1982/83 End Office Connection Study; Analog Voice and Voiceband Data Transmission Performance Characterization of the Public Switched Network," *BLTJ*, Vol. 63, No. 9, Nov. 1984, pp. 2059–2119.
18. G.R. Ash, A.H. Kafker, and K.R. Krishnan, "Intercity Dynamic Routing Architecture and Feasibility," *Proc. 1 of 10th Int. Teletraffic Cong.*, Montreal, Quebec, 1983, Session #3.2, paper #2.
19. G.R. Ash, R.G. Crafton, and P.E. Timm, "Building a Dynamic Routing Architecture," *Telephony*, Vol. 204, No. 26, June 27, 1983, pp. 42, 46, 50, 51, 53.
20. "Notes on the BOC INTRA-LATA Networks," Bellcore TR-NPL-000275, Issue 1, April 1986, Section 7 on "Transmission."
21. American National Standard for Telecommunications—Network Performance Standards—"Switched Exchange Access Network Transmission Performance Standard," ANSI T1.506-1989.
22. J.F. Ingle, K.I. Park, and T.C. Redman, "1983 Exchange Access Study: Analog Voice-Frequency Transmission Performance Characterization of the Exchange Access Plant," *IEEE Trans. on Comm.*, Vol. COM-35, No. 1, Jan. 1987, pp. 104–112. Also, Bellcore TR-037, Issue 1, Nov. 1984.
23. CCITT Rec. G.701, "Vocabulary of Digital Transmission and Multiplexing and Pulse Code Modulation (PCM) Terms," *Blue Book*, Vol. III, Fas. III.4, Geneva, 1989, pp. 7–40.
24. CCITT Rec. G.803, "Maintenance of Digital Networks," *Red Book*, Vol. III, Fas. III.3, Geneva, 1985, pp. 294–296.
25. American National Standard for Telecommunications—Network Performance—Tandem Encoding Limits for 32 kb/s ADPCM, ANSI-T1.501-1988.
26. Estimated Transmission Performance of Switched Access Service Feature Group-D, Bellcore TR-002, Issue 1, July 1984.
27. L.S. DiBiaso, "Subscriber Line Impact on System Transmission Performance: A Principle of Asymmetry," *NTC Rec.*, Vol. II, Dallas, Dec. 1976, pp. 23.6-1 to 23.6-3.
28. F.T. El-Mokedem, "Effect of Imperfect Echo Cancellation and Loop Variations on Voice Quality of Satellite Circuits," *ICC Rec.*, Vol. 3, Seattle, June 1987, pp. 42.6.1–42.6.6.
29. G.K. Helder, "Customer Evaluation of Telephone Circuits with Delay," *BSTJ*, Vol. 45, No. 7, Sept. 1966, pp. 1157–1191.
30. P.T. Brady, "Effects of Transmission Delay on Conversational Behaviour on Echo-Free Telephone Circuits," *BSTJ*, Vol. 50, No. 1, Jan. 1971, pp. 115–134.
31. *Telecommunications Transmission Engineering*, Vol. 1: Principles, 2nd ed., AT&T, 1977. Also, 3rd ed., Bellcore, 1990.
32. IEEE Standard 743, "Methods and Equipment for Measuring the Transmission Characteristics of Analog Voice Frequency Circuits," Nov. 1984.
33. J. Gruber, "Estimating the Transmission Grade-of-Service of Telephone Connections with Multiple Echo Paths," *GLOBECOM Rec.*, Vol. 1, Hollywood, Florida, Nov./Dec. 1988, pp. 14.6.1–14.6.5.
34. LATA Switching Systems Generic Requirements (LSSGR): Section 7 on "Transmission," Bellcore TR-507, Issue 3, Mar. 1989.
35. J.R. Cavanaugh, "A Model for the Subjective Effects of Quantization Distortion in Digital Transmission of Speech," *ICC Rec.*, Seattle, Vol. 2, June 1980, pp. 20.5.1–10.5.5.
36. CCITT Rec. G.113, "Transmission Impairments," *Blue Book*, Vol. III Fas. III.1, Geneva, 1989, pp. 68–84.
37. J. Gruber, "Performance Analysis and Design Guidelines for Multiple 4-wire Path Telephone Connections," *IEEE Trans. on Comm.*, Vol. COM-34, No. 7, July 1986, pp. 647–653. Also, *ICC Rec.*,

Vol. 1, Toronto, June 1986, pp. 7.4.1–7.4.6. Also, CCITT Contrib. COM XVI, No. 42-E to Quest. 5, Canada: BNR, Mar. 1983.
38. F.P. Duffy, G.K. McNees, I. Nasell, and T.W. Thatcher, Jr., "Echo Performance of Toll Telephone Connections in the United States," *BSTJ*, Vol. 54, No. 2, Feb. 1975, pp. 209–243.
39. P.H. Wittke, S.R. Penstone, and R.J. Keightley, "Measurement of Echo Parameters Pertinent to High-Speed Full-Duplex Data Transmission on Telephone Circuits," *IEEE Journal on Sel. Areas in Comm.*, Vol. SAC-2, No. 5, Sept. 1984, pp. 703–710. Also, *ICC Rec.*, Vol. 1, Amsterdam, May 1984, pp. 365–368.
40. Papers in Session 23 on "Decision Making Models in Transmission Management," *NTC Rec.*, Vol. II, Nov./Dec. 1976, Dallas, pp. 23.1-1 to 23.6-3.
41. L.S. DiBiaso, "User Reaction to Transmission Quality: A Cost Model," *ICC Rec.*, Vol. 2, Seattle, June 1987, pp. 34.6.1–34.6.5.
42. L.S. DiBiaso, "Telecommunications System Performance Needs for Standards," *GLOBECOM Rec.*, Vol. 2, San Diego, Nov./Dec. 1983, pp. 20.1.1–20.1.7.
43. E. Kreyszig, "Advanced Engineering Mathematics," 3rd ed., Wiley, 1972, p. 839.
44. D.V. Batorsky and M.E. Burke, "1980 Bell System Noise Survey of the Loop Plant," *BLTJ*, Vol. 63, No. 5, May/June 1984, pp. 775–818.
45. "Engineering in the Bell System," Bell Telephone Labs. Inc., Chapter 6, 1977, p. 167.
46. F.T. Andrews, Jr., and R.W. Hatch, "National Telephone Network Transmission Planning in the American Telephone and Telegraph Company," *IEEE Trans. Comm.*, Vol. COM-19, June 1971, pp. 302–314.
47. L.M. Manhire, "Physical and Transmission Characteristics of the Customer Loop Plan," *BSTJ*, Vol. 57, No. 1, Jan. 1978, pp. 35–59.
48. "Methodologies for Specifying Voice Grade Channel Transmission Parameters and Evaluating Transmission Performance for Speech Telephony," ANSI/IEEE Standard 823-1989.
49. S.C. Schwartz and Y.S. Yeh, "On the Distribution Function and Moments of Power Sums with Log-Normal Components," *BSTJ*, Vol. 61, No. 7, Sept. 1982, pp. 1441–1462.
50. N.A. Marlow, "A Normal Limit Theorem for Power Sums of Independent Random Variables," *BSTJ*, Vol. 46, No. 9, Nov. 1967, pp. 2081–2089.
51. I. Nasell, "Some Properties of Power Sums of Truncated Normal Random Variables," *BSTJ*, Vol. 46, No. 9, Nov. 1967, pp. 2091–2110.
52. I. Nasell, C.R. Ellison, Jr., and R. Holmstrom, "The Transmission Performance of Bell System Intertoll Trunks," *BSTJ*, Vol. 47, No. 8, Oct. 1968, pp. 1561–1613.
53. B.W. Kort, "Models and Methods for Evaluating Customer Acceptance of Telephone Connections," *GLOBECOM Rec.*, Vol. 2, San Diego, Nov./Dec. 1983, pp. 20.6.1–20.6.9.
54. M.B. Brilliant, "Observations of Errors and Error Rates in T1 Digital Repeatered Lines," *BSTJ*, Vol. 57, Mar. 1978, p. 711–746.
55. G.R. Ritchie and P.E. Scheffer, "Projecting the Error Performance of the Bell System Digital Network," *ICC Rec.*, Vol. 1, Philadelphia, June 1982, pp. 2D.2.1–2D.2.6.
56. H.S. Merrill, "Digital Performance Parameters for Switched Voice Grade Services," *NCF Rec.*, Vol. 1, Chicago, Oct. 1986, pp. 179, 184–197.
57. R.J. Schweizer, "DS1 Performance," Contribution T1Q1.4/88-012, Bellcore, May 12, 1988.
58. "Telecommunications Transmission Engineering," Vol. 3: Networks and Services, 2nd ed., AT&T, Chapter 2 on "Metropolitan Network Plans," 1977. Also, 3rd ed., Bellcore 1990.
59. *Telecommunications Transmission Engineering,* Vol. 1: Principles, 2nd ed., Chapter 28 on "International Communications," AT&T, 1977.
60. I. Nasell, "Some Transmission Characteristics of Bell System Toll Connections," *BSTJ*, Vol. 47, No. 6, Jul./Aug. 1968, pp. 1001–1018.
61. F.P. Duffy and T.W. Thatcher, Jr., "1969–70 Connection Survey: Analog Transmission Perfor-

mance on the Switched Telecommunications Network," *BSTJ,* Vol. 50, No. 4, April 1971, pp. 1311–1347.
62. T.C. Spang, "Loss-Noise-Echo Study of the Direct Distance Dialing Network," *BSTJ,* Vol. 55, Jan. 1976, pp. 1–36.
63. J.R. Cavanaugh, R.W. Hatch, and J.L. Neigh, "A Model for the Subjective Effects of Listener Echo on Telephone Connections," *BSTJ,* Vol. 59, July/Aug. 1980, pp. 1009–1060.

Appendix A.5.1
CONVERSION OF Q TO SUBJECTIVELY EQUIVALENT IDLE CIRCUIT NOISE

In order to incorporate the subjective effect of Q in the GOS models presented in Chapter 3 (Table 3.4(b)), Q is first converted to subjectively equivalent idle circuit noise to be power summed with other idle circuit noise components. The conversion is as follows [35]. Based on the results of subjective tests, the relationship between SNR (speech to idle circuit noise ratio) and Q (speech to speech correlated noise ratio) is determined and given in simplified terms by

$$\text{SNR} = aQ + b \text{ (dB)} \tag{A.5.1}$$

where the coefficients 'a' and 'b' are here considered to be 2.36 and -8.34, respectively. Other examples are as discussed in Chapter 3 (see Table 3.4 (b)).

From a relationship similar to that of Equation (5.39),

$$\text{SNR} = V_r - N_q + F^* \text{ (dB)} \tag{A.5.2}$$

where V_r is the speech level in volume units (Vu) at the input to a reference receiving system (with $\text{ROLR}_r = 46$ dB), N_q is the subjectively equivalent idle circuit noise in dBrnC, and F^* is a constant to account for the different units for speech and noise levels; in this case, $F^* = F + 1 = 88 + 1 = 89$ dB, where, evidently, the 1 dB converts the units for speech level from dBm to Vu.

For GOS purposes, it is also possible to express the speech level in terms of an equivalent end-to-end loudness loss ($\text{OOLR} = L_e$ in decibels). This is done by first considering a representative (average) transmit loop plus 500 type station set with $\text{TOLR}_a = -44.1$ dB [63]. Thus, the end-to-end loudness loss due to the loop plant is $\text{TOLR}_a + \text{ROLR}_r = -44.1 + 46 = 1.9$ dB.

Secondly, the loss of the trunk part of the network (ICL) is estimated by the difference in speech level V_r into the reference receiving system (i.e., at the equivalent of a receiving class 5 office), and the average speech level at a sending class 5 office of $V_s = -22.2$ Vu [63]. Thus,

$$\text{ICL} = V_s - V_r \qquad (A.5.3)$$

and, from Equation (5.2b),

$$L_e = \text{TOLR}_a + \text{ICL} + \text{ROLR}_r \qquad (A.5.4)$$

Therefore, from Equations (A.5.3) and (A.5.4),

$$V_r = \text{TOLR}_a + \text{ROLR}_r + V_s - L_e \qquad (A.5.5)$$

Finally, combining Equations (A.5.1), (A.5.2), (A.5.5), and the various numerical values determined above gives

$$\begin{aligned} N_q &= \text{TOLR}_a + \text{ROLR}_r + V_s - L_e + F^* - aQ - b \\ &= 77.0 - 2.36Q - L_e \end{aligned} \qquad (A.5.6)$$

which is the idle circuit noise that is subjectively equivalent to Q, and is power summed with other components of idle circuit noise. This and other equations of this nature were noted in Table 3.4(b) of Chapter 3.

Appendix A.5.2
RELATIONSHIP BETWEEN NORMAL AND LOG-NORMAL DISTRIBUTIONS

An example of a normal density and corresponding distribution in the log domain is shown in Figures A.5.2.1 and A.5.2.3, respectively. The log-normal counterpart in the inverse log or power domain, is shown in Fig. A.5.2.2. Table A.5.2.1 summarizes the various statistics for normal and log-normal distributions.

The most useful of these statistics are the median and other percentile values in the log domain, because these are readily determined by inspection from distri-

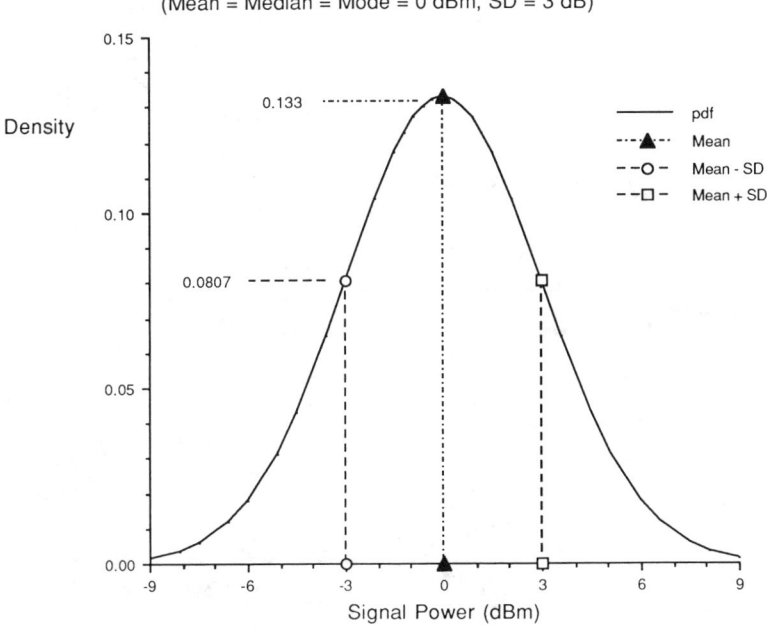

Figure A.5.2.1 Example of a normal pdf.

Figure A.5.2.2 Example of a log-normal pdf.

Figure A.5.2.3 Example of normal CDF.

Table A.5.2.1
Attributes of Normal and Log-Normal Distributions

Random Variable Description	Normal		Log-Normal	
	Random Variable	Density	Random Variable	Density
General	x	$f(x) = Ae^{-(x-\mu)^2 B}$	$y = 10^{x/K}$	$f(y) = (C/y)e^{-(K \log y - \mu)^2 B}$
Minimum	$-\infty$	0	0	0
Maximum	$+\infty$	0	$+\infty$	0
Mean	μ	A	$U = 10^{\mu/K} + D/2$	$(C/U)e^{-D(\ln 10)/8}$
Median	μ	A	$M = 10^{\mu/K} = U^2/\sqrt{U^2 + V^2}$	C/M
Mode	μ	A	$M\,10^{-D} = MU^2/(U^2 + V^2)$ $= M^2/\sqrt{U^2 + V^2}$	$CUe^{-2\mu/K} \ln 10$ $= (C/M)e^{(D/2)\ln 10}$
SD	$\pm \sigma$	$f(\pm\sigma)$	$\pm V = \pm U\sqrt{10^D - 1}$	$f(\pm V)$
N SDs about mean	$\mu \pm n\sigma$	$Ae^{-n^2/2}$	$10^{(\mu \pm n\sigma)/K}$ Also, $y = U \pm nV$	$Ce^{-n^2/2}10^{-(\mu + n\sigma)/K}$ $f(U \pm nV)$
Percentile Value	X	$\%* = 100\int_0^X f(x)\,dx$	$Y = 10^{X/K}$	$\%* = 100\int_0^Y f(y)\,dx$

$A = 1/(\sqrt{2\pi}\sigma)$, $B = 1/(2\sigma^2)$, $C = 1/(\sqrt{2\pi}(\sigma/K)\ln 10) = KA/\ln 10$, $D = (\sigma/K)^2 \ln 10$, K is a constant (see Table 5.13)

* Percentiles, % = 100 times the area under the density; the normal and log-normal areas are equal.

butions of survey data. From the median and one other percentile value, the SD can be determined from Equation (5.58), and all other percentile values in the log domain can then be estimated using Equation (5.56a) and Table 5.10. A similar process is followed for parallel combinations of distributions, as discussed in Section 5.4.4.3. For most continuous analog parameters, which are manifested in the log domain, it is only necessary to deal with parallel distributions in the log domain.

For transient analog parameters that are manifested in the inverse log domain, the process is similar to the above except that percentile values in the inverse log domain are determined by simply taking the inverse log of those in the log domain using Equation (5.108). The same also applies to parallel combinations of such parameters, as discussed in Section 5.4.4.4.

Chapter 6
PERFORMANCE OF NETWORK COMPONENTS

6.1 INTRODUCTION

This chapter addresses the next step in the overall performance process introduced in Chapter 1 (Fig. 1.3); that is, the performance of network components (or, in some cases, the performance of portions of network connections). In general, the performance of network components as addressed in this chapter is related to the end-to-end network performance discussed in Chapter 4 through the performance allocation methodologies presented in Chapter 5.

In this chapter, performance of network components is also addressed in terms of the network-oriented performance parameters introduced in Table 4.1 of Chapter 4, and in terms of the connection models introduced in Figures 4.1 to 4.4 of Chapter 4. For voiceband services, these connections typically consist of the following network components: customer premises equipment (CPE) (e.g., station set or modem), 2-wire analog subscriber loops, switching equipment, and the trunking that interconnects the switches. The focus for voiceband services is on loops and various types of trunks (as described in the next section). The sources of survey data and performance objectives identified in Table 4.4 of Chapter 4 are also applicable in this chapter. See [1, 7–9, 13, 16–19] in Chapter 4. Additional sources of performance requirements and maintenance limits are cited in this chapter as required (the main ones are outlined later in Table 6.1).

In addition to voiceband services, digital data and other existing and future services, such as video, will continue to be supported by transport systems. These systems (carrier transmission facilities and associated terminal equipment) are rapidly evolving toward all-digital technology. Accordingly, the error performance of digital fiber, as well as radio and twisted pair carrier systems is also included in this chapter, to complement the end-to-end performance provided by such systems as presented in Chapter 4. In addition, this chapter addresses jitter and wander in digital networks in terms of their effect on error performance, and in terms of jitter accumulation and jitter and wander specifications.

Table 6.1
Main Sources of Basic Survey Data and Objectives, Requirements and Limits for Characterizing Connection Components

Connection Component	Sources of Survey Data	Sources of Performance Objectives, Requirements and Limits
Station Sets		IEEE Terminal Standard [1] IEEE Voiceband Standard [14] ANSI Draft Loss Plan Standard [7] EIA/TIA ISDN Terminal Standard [8]
Analog Loop	Loop Surveys [12, 13]	IEEE Voiceband Standard [14] ANSI Draft Loss Plan Standard [7] BOC Notes on the Network [11]
Digital Switch	Included in EAS [16]	LSSGR [15]
Short-Haul Trunking	EAS [16]	ANSI Exchange Access Standard [21] BOC Notes on the Network [11] Estimated FGD Performance [19]
Long-Haul Trunking	EAS and EOS [16, 23] Inter-toll Trunk Survey [22] *	AT&T Notes on the Network [18]
End-Office to End-Office	EOS [23]	BOC Notes on the Network [11] AT&T Notes on the Network [18] ANSI Draft Loss Plan Standard [7]

* Now out of date

For data services (packet data in particular), performance requirements and measured performance results were presented in Chapter 4 for end-to-end performance of various packet networks, as well as for packet switches. It appears that there is little additional performance data available specifically for packet network components to warrant further discussion in this chapter.

6.2 PERFORMANCE OF VOICEBAND NETWORK COMPONENTS

6.2.1 Description of Main Connection Components

For voiceband services, the following briefly describes the main components encountered in a connection. Associated with these components are the sources of survey data and performance objectives, etc., as summarized in Table 6.1. For reference, recall also the connection models in Chapter 4 (Figs. 4.1 to 4.4). The connection components are as follows:

Station Set: Typical, conventional 500 type telephone set or equivalent satisfying industry standards [1-3]. For acoustic parameters, the set is typically associated with the analog subscriber loop, as described below [4–6]. For digital subscriber "loops," 4-wire digital telephones (e.g., ISDN terminals) are applicable [7–9].

Modem: Devices that support voiceband data applications by converting transmitted binary data signals into voiceband signals (MODulation), and converting voiceband signals into received binary data signals (DEModulation). A variety of modems are considered in relation to the voiceband data error performance results presented in Section 6.2.3.9 of this chapter.

Analog Loop: 2-wire connection from the end office line side distribution frame (or from the line side of a remote switching terminal when distributed switching is used) to the customer/network interface (NI) [10]. For acoustic parameters, the customer premise wiring and station set performance are included in the parameter values [11]. Loops typically involve a -48 volt/400Ω DC power feeding arrangement and 900Ω source and terminating impedances. Generally these factors all have some bearing on performance for acoustic parameters. The focus here is mainly on the general Bell System loop plant for which the above values apply [12, 13]. Performance for various other types of loop plant, feeding arrangements and impedances, etc., is discussed in the IEEE voiceband standard [14]. Performance and planning considerations for digital subscriber "loops," in conjunction with 4-wire digital telephones are discussed in [7–9]; see also Chapter 7.

Digital Switch: Refers principally to port-to-port performance through an end office switch. Ports may be analog, remote-analog, or digital, depending on the type of connection and the types of lines or trunks interfacing to the switch [15]. Accounting for the difference in nominal switch loss (0 dB) and trunk loss (3 dB), the digital switch performance (analog port to analog port) is estimated to be the same as for the digital (T-carrier) trunking measurements included in the exchange access study (EAS) [16]. Since some of this T-carrier technology may be rather old, these results (especially for standard deviations (SD)) may be somewhat poorer (larger in value) than achieved by more modern day digital switches, and are probably conservative.

Short-Haul Trunking

Predivestiture: Toll connecting trunk from (and including) the end office switch up to (but not including) the Class 4 toll switch; also indicative of inter-office and tandem trunking in metropolitan networks [17, 18].

Postdivestiture: Feature group (FG) C or FGD exchange access direct trunking from (and including) the end office switch up to (but not including) the inter-exchange carrier (IEC) point-of-presence (POP) switch (see exchange access portion of Fig. 4.4) [11, 19]; also indicative of FGD nondirect trunking (owing to the equal access principle), as well as inter-office and tandem trunking in intra-LATA networks (see Fig. 4.3).

Short-haul distances are less than 200 miles (typically less than 50 miles), and short-haul trunks usually operate with a nominal 3-dB loss. The performance results presented here for short-haul trunking are based on the technology mix measured in the (EAS) [16], and should be representative of a variety of nominal 3-dB loss trunks, such as the intra-LATA exchange access trunks, as well as the metropolitan direct and tandem trunks noted above.

Long-Haul Trunking

Predivestiture: Inter-toll trunking from Class 4 toll switch to Class 4 toll switch (including the Class 4 switches) [18, 20]. "Short"-distance inter-toll trunking is also indicative of inter-tandem trunking in metropolitan networks.

Postdivestiture: IEC trunking from POP switch to POP switch (including the POP switches; see IEC portion of Fig. 4.4 of Ch. 4) [11, 21]. "Short"-distance IEC trunking is also indicative of inter-tandem trunking in intra-LATA networks (see Fig. 4.3).

"Long-haul" strictly signifies distances longer than 200 miles, although many "short" inter-toll or IEC trunks are less than 200 miles long. However, all such trunks operate at or near 0 dB loss (that is, 0 dB for digital trunks, 1 dB for combination—mixed analog/digital—trunks, and via net loss (VNL) for analog trunks; the transmission loss plan for trunks is discussed in detail in Ch. 7).

Performance for long-haul trunking is cited in an inter-toll trunk survey [22]. However, this survey is now quite old and is probably no longer representative in that more advanced technology has since been introduced, which means performance has likely improved considerably compared to that reflected in this survey. Thus, where possible, long-haul trunk performance is estimated from the more recent EAS [16] and end office survey (EOS) [23] by subtracting short-haul (e.g., toll-connecting or exchange access) trunking performance from the end office to end office performance. In some cases, this procedure is not possible owing to limitations of analytical techniques and other considerations as discussed later with each specific parameter.

End Office to End Office

Connections from Class 5 end office switch to Class 5 end office switch (including the end office switches) as assessed in the EOS [23], and consisting of the following:

Predivestiture: Two toll connecting (short-haul) trunks, plus intertoll (long-haul) trunking [18, 20].

Postdivestiture: Two exchange access (short-haul) arrangements, plus IEC (long-haul) trunking [11, 21]; see Figure 4.4.

6.2.2 Performance Objectives, Requirements, and Limits

Table 6.1 provides the sources of these performance levels. There are a variety of ways of specifying the performance of network components, varying from highest to lowest performance as follows: design objectives and requirements, acceptance (or pre-service, or restoral) limits, maintenance limits, and immediate action limits. These performance levels are briefly described as follows. The design objectives and requirements are similar in principle and intent to the end-to-end objectives and requirements outlined in Chapter 4.

Objectives are design goals achievable with new, if not current, technology.

Requirements are minimum achievable levels necessary for viable support of current services.

Acceptance limits (AL) are performance levels above which a network component can initially support a service, or can be returned to service after repair [21, 24, 25]. Thus, ALs are sometimes referred to as preservice limits (PSL), and are also similar in value to restoral limits (RL).

Maintenance limits represent degradations in performance to a level warranting maintenance action of an investigative or perhaps corrective nature.

Immediate action limits (IAL) are performance levels below which the network component must be removed from service and subjected to corrective maintenance [21, 24, 25].

6.2.3 Network Component Performance Characterization

6.2.3.1 *Loss and Related Parameters*

Acoustic Loss

Acoustic loss (sensitivity) of station sets and of station sets plus subscriber loops is provided in Table 6.2. The mean set plus loop sensitivities are based on models of loop survey distributions [26] and are within a decibel of other similar estimates not shown in Table 6.2 (i.e., other transmit (TOLR) and receive objective loudness rating (ROLR) means are -43.9 and 47.7 dB, respectively [14]). The SDs in Table 6.2 [26] are, however, somewhat larger than estimated elsewhere (i.e., other TOLR and ROLR SDs are 3.04 and 1.28 dB, respectively [14]), and, as such, are perhaps conservative. From a practical standpoint, a conservative approach is probably warranted given the general uncertainty about the loop plant relative to the comparatively well-characterized and maintained trunk part of the network, plus the wide variety of telephones in use since the advent of terminal interconnect. With increasing penetration of digital technology into the loop plant, less variation in sensitivity

Table 6.2
Estimated Acoustic Loss of Station Sets and Subscriber Loops

Acoustic Loss Parameter		Connection Component		
Description	Value	Station Set	Station Set Plus Loop	
			With No Set Variation	With Set Variation
TOLR	Mean (dB)	-47.1	-43.1	-43.1
	SD (dB)	1.3	3.6	3.8
	Objective (dB)	-46 &	≤ -38 #	≤ -38 #
	% ≤ Objective	-	92.2	91
ROLR	Mean (dB)	42.7	46.7	46.7
	SD (dB)	0.6	1.7	1.8
	Objective (dB)	48 &	≤ 50 #	≤ 50 #
	% ≤ Objective	-	97.4	96.7
Contribution to Asymmetry	Mean (dB)	*	*	*
	SD (dB)	1.4	1.9	2.4
Contribution to TEPL	Mean (dB)	-4.4	3.6	3.6
	SD (dB)	1.4	5.3	5.5

is expected, although the range specified for digital sets is still considerable [8]. Thus, overall, the loop plus set variations in Table 6.2 are considered representative.

Note that the sensitivity variations for analog sets (excluding the independent variations within the set alone, as discussed below) are correlated with respect to direction of transmission for a given loop, owing to the obvious commonality of the physical loop and its loss in each direction, plus the dependence of the set transmit and receive regulation on loop length. For the station set plus loop, this dependence accounts for the relatively low contribution to asymmetry variation and high contribution to echo path loss variation in Table 6.2 (methodologies for computing loss asymmetry and echo path losses are provided in Sec. 5.4.3.4 of Ch. 5).

The mean sensitivity of the set alone is estimated by subtracting the mean electrical loop loss of about 4 dB (see Table 6.3) from the mean set plus loop sensitivities. The variation in sensitivities of the set alone is based on independent SDs of about 1 dB each for transmit and receive sensitivities [27]; however, because of the generally poorer stability of transmit sensitivities (particularly with carbon transmitters), a larger SD (by a factor of approximately two) is assigned to the transmit, relative to the receive, sensitivity variation, such that the net transmit plus receive variation (of $\sqrt{2}$ dB) is the same as if 1 dB were assigned to each of transmit and receive.

The variations of only the set are assumed to be directionally independent, and therefore contribute to overall variations, as well to asymmetry and echo path vari-

Table 6.3
Electrical Loss Parameters for Connection Components

Loss Parameter			Connection Component										
			Analog Loop	Digital Switch	Short-Haul Trunking	Long-Haul Trunking &				End-Office to End-Office &			
Name	Condition	Value (dB)				Short	Medium	Long[b]	Overall	Short	Medium	Long[b]	Overall
End-to-End Loss	1.0 kHz	Mean SD	3.7 2.8	0.1 1.1	3.1 0.97	0.35 0.82	1.11 0.28	1.6 1.17	0.5 1.0	6.5 1.6	7.3 1.4	7.8 1.8	6.7 1.7
		Des. Obj. Des. Limit	~4 ≤8.5	0 ≤1	3.0 4, 5 @	0.5 ≤0.8	1.4 ≤1.7	2.3, 0, 1.1 ≤2.9, 0, 1.4	0.8 1.1	6.5 ≤6.8	7.4 ≤7.8	8.3, 6, 7.1 ≤8.9, 6, 7.4	6.8 7.1
Loss Deviation	1.0 kHz	Mean SD	0 2.8	0 1.1	0 1.0	0 0.82	0 0.28	0 1.17	0 0.81	0 1.6	0 1.4	0 1.8	0 1.7
		AL IAL	-	±0.3 * -.5 to +2 #	±0.7 ±2.5	-	-	-	-	-	-	-	-
Attenuation	0.4 kHz	Mean SD	-0.8 1.7	0.61 0.30	0.71 0.39	-0.09 0.24	0.04 0.38	-0.07 1.18	-0.07 0.45	1.33 0.60	1.46 0.67	1.35 1.30	1.35 0.71
	2.8 kHz	Mean SD	3.4 3.5	0.22 0.11	0.45 0.77	0.45 1.76	-0.10 0.78	-0.14 0.48	-0.14 1.54	1.35 2.07	0.80 1.34	0.76 1.19	1.22 1.89
Distortion	0.4/2.8 kHz	AL IAL	-1/3 a -	-.5/1.0 ! -	-1.5/+3.5 x, -1.5/+5.0 y -2.0/+4.0 x, -2.0/+5.5 y		~1a				~8a		
Contribution to Loss Asymmetry		Mean SD	0 0	0 1.6	0 1.4	0 1.16	0 0.4	0 1.65	0 1.14	0 2.3	0 2.0	0 2.5	0 2.4
Contribution to TEPL		Mean SD	7.4 5.6	0.2 1.6	6.2 1.4	0.6 1.16	2.2 0.4	3.2 1.65	1.04 1.43	13.0 2.3	14.6 2.0	15.6 2.5	13.4 2.4

* 99% Objective (Obj.).
\# 99% Requirement (Req.).
a Allocated from end-to-end limit of 14 dB (see Table 4.5 of Ch. 4).
b Triple entries account approximately for long trunks without echo control, with echo control, and the combined case (see text).
! Min./Max. Req. at 0.3 and 3.0 kHz, respectively.
@ 4 dB for exchange access, 5 dB for inter-office.
x, y Values for 4-wire and 2-wire IEC switches, respectively.
& From Table 4.3 of Chapter 4, short, medium and long connection probabilities in % are: 77.8, 13.4 and 8.8, respectively

ations, in an independent manner as shown in Table 6.2 (again, recall the methodologies in Ch. 5).

Electrical Loss

Electrical loss for various network components is shown in Table 6.3. As discussed previously, the loop losses are directionally dependent. For all other network components, the losses are assumed to be directionally independent.

Based on measured T-carrier trunk loss, loss through a digital switch (analog port to analog port) is estimated to be nominally about 0 dB with SD of about 1 dB. Again, this SD may be somewhat high owing to dated T-carrier technology (T-carrier measurements from the EAS with mean loss reduced by 3 dB were used as the estimated switch loss).

Loss for short-haul trunks is nominally about 3 dB with SD of about 1 dB. For short-haul trunks, the ALs and IALs are determined from the T1 Exchange Access Standard [21] (when provided; otherwise limits are from other relevant documents such as those noted in Table 6.1 [21, 24, 25]). For other network components, the AL and IAL designations may have other interpretations, as shown in the footnotes in Table 6.3.

The loss for the long-haul (e.g., IEC) trunking in Table 6.3 is derived by subtracting twice the short-haul (e.g., exchange access) trunking loss from the end office to end office loss [19]. The end office to end office loss is determined directly from the EOS [23]. The long-haul loss design objectives and limits correspond approximately to the VNL rule [11, 18], which stipulates that loss from 0.5 to 2.9 dB is assigned to analog trunks in steps of 0.3 dB according to the relationship:

$$\text{VNL} = 0.0015 \times \text{Avg. trunk length} + 0.4 \text{ dB} \qquad (6.1)$$

For VNL planning purposes, the overall trunk length ranges from 0 to 1,850 miles, while the average in Equation (6.1) ranges from about 67 to 1,670 miles in steps of approximated 200 miles. For the design *objectives* in Table 6.3, the estimated *mean* route distances of about 85 and 655 miles were used in Equation (6.1) for the short and medium distance categories, respectively (see Table 4.3 in Ch. 4). However, in the long-distance category, for trunk lengths longer than 1,850 miles, the loss is 0 dB and echo control is applied to the trunk. Thus, the mean route distance of trunks with no echo control in the long-distance category is estimated from Table 5.18 of Chapter 5 to be about 1,215 miles, and is used in Equation (6.1). The resulting VNL losses are shown in Table 6.3 to the nearest 0.3 dB, according to the intended usage of Equation (6.1). Similarly, for the design *limits* in Table 6.3, the *maximum* route distances of 340 and 1,065 miles were used in Equation (6.1) for the short- and medium-distance categories, respectively (see Table 4.3 in Ch. 4), while for the long-distance category, the maximum route distance of trunks with no echo control is

1,850 miles and is used in Equation (6.1). Again, the resulting losses in Table 6.3 are to the nearest 0.3 dB. Also, trunk losses with and without echo control and for the mixed case are shown for the long-distance category in Table 6.3. The mixed case is estimated using the probability of echo control on long trunks (see Table 4.15 of Ch. 4, or Table 6.13 later in this chapter).

When long-haul trunking is all digital, the loss will be 0 dB. During the transition from analog to digital, however, some trunks will use digital transmission technology but may still be terminated by an analog switch at one end, in which case they are called *combination* (analog/digital) trunks, and the loss is 1 dB (see Ch, 7 for further discussion of transmission loss planning). Thus, it is evident from Table 6.3 that a mixture of analog, combination, and some digital technology was probably in use for long-haul trunking at the time of the EOS [23].

Attenuation Distortion

Attenuation distortion of the long-haul trunking is also estimated in Table 6.3 by subtracting twice the attenuation distortion for the short haul from the end office to end office values. Judging from some of the negative mean values, this approach may result in an underestimate of the long-haul attenuation distortion. The reason is probably due to the wide range of technologies covered in the EAS, which may not have prevailed for the samples used in the EOS. Thus, the short-haul attenuation distortion values used here may be somewhat larger than those encountered as part of the EOS. The attenuation distortion limit of 1 dB shown in Table 6.3 for long-haul trunking is estimated by subtracting from the end-to-end limit of (14 dB—see Table 4.5 of Ch. 4) the allocation to loops (3 dB each [11, 18]), and the AL for short-haul trunks (3.5 dB each [21]), which leaves 1 dB. This value of 1 dB is consistent with Table 6.4, which provides addition attenuation distortion limits applicable for private line trunks [28–30].

Finally, Table 6.3 shows the contribution of each network component to electrical loss asymmetry and to echo path loss. These contributions are affected by directional dependence of loop loss and directional independence of trunk losses, and are determined using the methodologies in Chapter 5.

6.2.3.2 Signal Levels

The voiceband signals of interest include those for speech, voiceband data, and facsimile services, as well as signaling tones, test tones, and call progress tones. The levels of these signals are important for a number of reasons, and they must not be too low or too high. Levels that are too low will be adversely corrupted by noise and distortion, and therefore will not be properly detected by receiving equipment, or heard by listeners in the case of speech signals. Levels that are too high will result in

Table 6.4
Private Access Line and Trunk Performance Limits for Attenuation,
and Envelope Delay Distortion Parameters

Parameter	Frequency Range (kHz)	Limit for C3 Conditioned:	
		Access Line	Trunk
Attenuation Distortion (dB) w.r.t. 1 kHz	.5 - 2.8 .3 - 3.0	-0.5 to +1.5 -0.8 to +3.0	-0.5 to +1 -0.8 to +2
Envelope Delay Distortion (ms) w.r.t. 1.8 kHz	1 - 2.6 .6 - 2.6 .5 - 2.8	.11 .30 .65	.08 .26 .50

w.r.t.: with respect to

excessive crosstalk coupling between metallic (noncarrier) facilities; that is, between voice frequency loops and between voice frequency trunks. Levels that are too high will also cause crosstalk interference (intermodulation distortion) on analog carrier systems, and overload (nonlinear) distortion in both analog and digital carrier systems.

Thus, to avoid the above problems, signal levels must have an appropriate operating range. Generally, the approach is to keep signals as high as possible without causing the adverse crosstalk or distortion effects noted above.

Table 6.5 summarizes the specified levels of a variety of voiceband signals, and these levels are discussed as follows. For most carrier equipment, the overload level (at which amplitude clipping of the signal occurs) is +6 dBm0. Thus, the maximum sinewave test tone level is about +3 dBm0 (sinewave peak to rms ratio is about 3 dB). Levels as high as 0 dBm0 for test tones are permissible on the basis that they are generally of short duration. However, owing to crosstalk considerations, for tests of long duration, −10 dBm0 levels are currently used, and eventually −16 dBm0 levels are intended to be used. Similarly, levels for voiceband data signals are required to be no more than −13 dBm0. A number of call progress tones also have levels of −13 dBm per frequency (see [31, 32] for other levels for these tones).

The above noted levels are related to the send level from the trunk side of an originating end office switch. Levels at other downstream locations would typically be lower owing to network losses. For example, receiving levels at the line side of a terminating end office switch would be nominally 0, 3, or 6 dB lower than those above for intra-office, inter-office, and toll calls, respectively. Conversely, levels upstream from the originating end office would typically be higher than those sent

Table 6.5
Summary of Voiceband Signal Levels

Signal		Level (dBm0)
Type	Description	
Loop Signaling	Dual Tone Multi-Frequency Dial Tone	-8 to +4 * -13 to +3 *
Trunk Signaling	Single Frequency Muti-Frequency	-20 to -8 -7 to -6 *
Test Tone	Sinewave	-16, -10, 0
Call Progress Tone	Sinewave	@
Voiceband Data	-	-13
Speech #	-	-16

* Levels are per frequency and increase by 10 log (no. of frequencies).
  There are two frequencies in each of these signals which corresponds to a 3 dB increase.
\# See Table 6.6 for surveyed speech levels.
@ Many call progress tone levels are -13 dBm per frequency; see [31, 32] for other levels.

from the originating end office. For example, a typical send level from a voiceband data modem is about -9 dBm, such that with an average loop with 4 dB loss, the level received at the end office is -13 dBm. Levels for dual tone multifrequency (DTMF) signaling tones sent from telephone sets can vary considerably depending on the set powering and the loop current. However, send levels are such as to be received at less than 0 dBm at the end office [1].

For speech signals, crosstalk and distortion are not as perceptible as with tones, but owing to the large peak-to-rms ratio of about 15 dB for speech, the maximum level should not exceed -16 dBm0 to minimize amplitude clipping (-13 dBm0 is permitted for short speech durations up to 3 sec [11, 18].

A survey of speech levels (see Table 6.6) indicates that mean near-end speech levels received at the line side of originating end office switches are about -25 dBm0 with SD of about 5 dB, so that most signals are lower than -16 dBm0 [33]. (Near-end signals are those from the "near" subscriber and are subjected to loop loss only.) These near-end levels are also largely independent of the class of service (business or residential, Table 6.6(a)), and the type of call (local or toll, Table 6.6(b)).

Table 6.6(a)
Summary of Speech Level Statistics: Class of Service

Speech Level (dBm)	Class of Service and Percentage									
	Business (59 %)						Residential (41 %)		Overall (100 %)	
	Type of Originating Station Set and Percentage									
	Bell System (53 %)		Non-Bell System (47 %)		Overall (100 %)					
	Mean	SD	Mean	SD	Mean	SD	Mean	SD	Mean	SD
Near-End	-24.9	5.2	-27.5	5.4	-25.0	5.2	-25.7	5.4	-25.3	5.3
Far-End	-28.4	5.5	-29.3	5.1	-28.4	5.5	-28.2	5.6	-28.3	5.6
Difference in dB	3.4 (3.5)	6.7 (7.6)	1.6 (1.8)	6.6 (7.4)	3.4 (3.4)	6.7 (7.6)	2.5 (2.5)	6.6 (7.8)	2.9 (3.0)	6.7 (7.7)

( ) Estimated difference between Near-End and Far-End speech levels (see text).

Table 6.6(b)
Summary of Speech Level Statistics: Type of Call

Speech Level (dBm)	Type of Call and Percentage Business/Residential									
	Local*				Toll*				Overall (59/41 %)	
	Intrabuilding (34/66 %)		Interbuilding (50/50 %)		Within NPA (59/41 %)		Between NPA (69/31 %)			
	Mean	SD	Mean	SD	Mean	SD	Mean	SD	Mean	SD
Near-End	-25.3	5.3	-27.5	5.5	-24.7	5.2	-24.2	4.8	-25.3	5.3
Far-End	-26.5	5.4	-29.7	5.2	-29.3	5.7	-31.1	4.9	-28.3	5.6
Difference in dB	1.0 (1.2)	6.2 (7.6)	3.9 (4.0)	6.6 (7.6)	4.9 (4.6)	6.5 (7.7)	7.0 (6.9)	6.7 (6.9)	2.9 (3.0)	6.7 (7.7)

( ) Estimated difference between Near-End and Far-End speech levels (see text).
NPA Number Plan Area.
* Intrabuilding and Interbuilding correspond to Intra-office and Inter-office, respectively.
Within NPA and Between NPA correspond to medium toll and long toll, respectively. (See text.)

An exception to the above is when non-Bell System station equipment is used for business services (Table 6.6(a)); in this case, near-end levels are about 2 to 3 dB lower than in other cases. This is most likely due to lower transmit sensitivities of non-Bell System station sets.

Regarding the type of call (Table 6.6(b)), the "intra-building" and "inter-building" calls can be viewed as corresponding to intra-office and inter-office connections discussed throughout this book (particularly in Chapters 4 to 6). Similarly, for the toll calls, the "within number plan area (NPA)" calls correspond to short-toll connections, and the "between NPA" calls correspond to medium- and long-toll connections.

Finally, also shown in Table 6.6 are the far-end speech levels received at the line side of the same end office switch and on the same lines where near-end speech was measured. (Far-end signals are those from the "far" subscriber and are subjected to loop plus network loss.) The differences between the near- and far-end mean speech levels are an indication of the apparent network loss on connections between end offices. Over all connections, this apparent loss is about 3 dB; this is expected, since the nominal design loss is 0, 3, and 6 dB for intra-office, inter-office, and toll calls, respectively, which, for more or less equally sampled connections across all types of calls, averages to about 3 dB. The effect of call type on the apparent loss is shown in Table 6.6(b) where this loss is about 1 and 4 dB for intra-office and inter-office connections, respectively, about 5 dB for short toll connections, and about 7 dB for medium- and long-toll connections. This corresponds reasonably well with expected nominal values of 0, 3, and 6 dB, for short, medium, and long connections, respectively.

The SD of the apparent losses is large mainly due to speech level variation, and is not indicative of network loss variations. Also shown in Table 6.6 are estimates of apparent losses based on subtracting the aggregate near-end and far-end statistics, assuming they are independent. The estimated means agree closely with those computed from the difference between the pair-wise near-end and far-end survey measurements [33]. The estimated SDs, however, are somewhat larger than those computed from the difference between the pair-wise measurements. This implies some dependency between the measured pair-wise near-end and far-end speech levels, which is not reflected in the estimated SDs.

6.2.3.3 Noise and Related Parameters

Idle Circuit Noise

Idle circuit noise (or signal-independent, C-message noise) performance is provided in Table 6.7. The C-message loop noise is at the station set for the general Bell System loop plant, and is estimated (modeled) from the tails of measured noise distributions [13]. (Also, not shown in Table 6.7 is the estimated loop noise at the central office end of the loop, with an off-hook station set at the customer end of the loop; mean = -21 dBrnC and SD = 21.1 dB; see the methodology and discussion for these estimates in Sec. 5.4.3.3 and Table 5.11 of Ch. 5.)

The C-message noise for the digital switch is based on T-carrier trunk measurements [16] and, as discussed in Section 6.2 of this chapter, may be somewhat high in this regard. Note that the switch noise is before any loss that may be inserted

Table 6.7
Idle Circuit Noise for Connection Components

Noise Parameter			Connection Component									
		Analog Loop	Digital Switch	Short-Haul Trunking !	Long-Haul Trunking &				End-Office to End-Office &			
Name	Value				Short	Medium	Long	Overall	Short	Medium	Long	Overall
C-Message Noise (dBrnC)	Mean	-9.3 *	19.3	13.8	24.9	30.7	34.1	26.5	24.6	28.5	31.4	25.7
	SD	17.4 *	5.2	5.8	6.6	4.7	4.1	6.9	4.8	4.2	3.9	5.3
	Obj. Mean	-	-	22	25-31	26.3-31.3	31.3-36.3	-	22.7	26.7	30.7	23.9
	Obj. SD	-	-	4	4	4	4	-	5.3	4.1	2.5	5.6
	AL	≤ 20 #	≤ 20 a (≤ 21) c	≤ 22 e, ≤ 24 e, ≤ 28 f	≤ 25 e, ≤ 28-34g ≤ 31 e, ≤ 37-39g	≤ 25 e, ≤ 34-37g ≤ 31 e, ≤ 39-41g	≤ 25 e, ≤ 39-42g ≤ 31 e, ≤ 43-47g	-				
	IAL	≤ 30 @	≤ 23 b (≤ 25) d					-				
3 kHz Flat Noise (dBm)	Balanced Mean	-	36.3	30.7					43.9	44.0	45.4	44.1
	SD	-	15.5	15.1					7.9	7.2	6.9	7.7
	Balanced to Ground Mean	91	58.4	52.7					52.0	49.6	50.8	51.6
	SD	12	8.1	8.5					9.9	7.1	6.6	9.4

a Objective average
b 99% Requirement
c Satifactory level
d Unsatisfactory level (interpretation depends on the number of measurements)
e Limits for digital trunks
f Limits for 0 to 50 mile analog trunks (see text)
g Limits for analog trunks (see text)
* Noise at station set (see text)
! Mean noise is at the end-office for toll connecting or exchange access trunking (see text)
Objective
@ Unacceptable limit
( ) End-office noise (including office wiring, etc.)
& From Table 4.3 of Chapter 4, short, medium and long connection probabilities in % are: 77.8, 13.4 and 8.8, respectively

into a connection on the line side of the switch (nominally, 0, 3, or 6 dB for intra-office, inter-office, and toll connections, respectively). Thus, depending on the type of connection, the switch noise is reduced by these values of loss.

The mean C-message noise for short-haul trunking is at the end office for a toll connecting or exchange access trunk (after 6 dB of connection loss). Not shown in Table 6.7 is the noise at the POP and at end offices for inter-office trunks (after 3 dB of loss); therefore, add 3 dB to the end office noise to reflect the transmission loss plan for these cases. The short-haul mean noise objectives are such that the mean is 22 dBrnC for distances from 0 to 60 miles [11, 18], assumed to be applicable here. For distances between 60 and 250 miles, the general short-haul mean objective is

$$\text{Noise} \leq 22 + 3 \log_2(\text{distance}/60) \quad \text{in dBrnC} \tag{6.2}$$

The noise objective SD is 4 dB, independent of length. The C-message noise limits for digital and analog short-haul trunking are based on the T1 Exchange Access Standard [21]. In general, the analog limits vary with distances ranging from 0 to 400 miles. However, here it is assumed that, for the most part, short-haul distances are from 0 to 50 miles.

C-message noise for the end office to end office portion of a connection in Table 6.7 is obtained directly from the EOS [23], and the long-haul noise as estimated elsewhere by simulation [19] is the end office to end office noise, less twice the short-haul exchange access noise (accounting for trunk losses). An example and discussion of this process using analytical techniques was presented in Section 5.4.3.5 and Figure 5.25 of Chapter 5. The long-haul noise objectives in Table 6.7 account for distance dependence owing to the apparently significant presence of analog technology in the long-haul network. For distances between 250 and 4,000 miles, the general long-haul mean noise objective is

$$\text{Noise} \leq 25 + 3 \log_2(\text{distance}/250) \quad \text{in dBrnC} \tag{6.3}$$

The noise objective SD is 4 dB independent of length [11, 18].

The 3-kHz flat noise results in Table 6.7 are directly from survey data, where available [16, 23]. For long-haul trunking, no up-to-date survey data exists. Also, because of the large variability of the end office to end office and short-haul measurements, estimates based on the analytical techniques of Chapter 5 cannot be made; simulation results for same are also not available.

C-Notched Noise

C-notched noise (or signal-dependent noise) performance is provided in Table 6.8, and much of the previous discussion for C-message noise applies here as well. The

Table 6.8
Signal-Dependent Noise and Signal-to-Noise Ratios for Connection Components

Noise Parameter		Connection Component									
		Analog Loop	Digital Switch	Short-Haul Trunking !	Long-Haul Trunking &				End-Office to End-Office &		
Name	Value				Short	Medium	Long	Overall	Short	Medium	Long	Overall
C-Notched Noise @ (dBrnC)	Mean	-	39.2	29.7					36.5	37.4	37.9	36.7
	SD	-	2.9	8.3					3.4	2.7	2.7	3.5
	AL	≤ 28 dBrnC0	-	≤ 38.5, ≤ 40 **								
	IAL	-	-	≤ 42.5, ≤ 43 **								
3 kHz Flat Notched Noise @ (dBm)	Mean	-	46.4	38.7					46.2	46.0	47.0	46.2
	SD	-	7.6	9.7					5.7	5.3	5.4	5.6
Signal-to--C-Notched Noise Ratio @ (dB)	Mean	-	38.1	41.8					34.9	33.3	32.3	34.4
	SD	-	1.7	8.4					2.6	2.6	2.3	2.7
	AL	≥ 49	≥ 37 #*	≥ 32.5, ≥ 31 **			≥ 26.2			≥ 24		
	IAL	-	≥ 33 ##*	≥ 28.5, ≥ 28 **								
Q (dB)	Mean	-	35.0	35.0	35.0	30.4	27.7	33.7	30.4	27.7	25.8	29.6
	SD	-	1.7	1.7	1.7	1.2	1.0	2.9	1.2	1.0	0.9	1.9

\# Objective average
\#\# 99% Requirement
@ -13 dBm0 holding tone
** For digital end-offices with digital loss; Service Affecting Limit (SAL) = 31 dB (see text)
* Subtract 1.1 dB for systems using Robbed Bit Signaling
! Mean noise is at the end-office for toll connecting or exchange access trunking (see text)
& From Table 4.3 of Chapter 4, short, medium and long connection probabilities in % are: 77.8, 13.4 and 8.8, respectively

distinction is that C-Notched noise is generally higher than C-message noise, owing to the increase with signal level of distortion in digital systems and noise enhancement with signal level in analog compandored systems.

The estimated subjective signal-to-quantization noise ratios (Q) in Table 6.8 assume the equivalent of 1 pulse code modulation (PCM) codec for all short-haul trunking in a connection, and 1, 2, and 3 codecs for the short, medium, and long categories of long-haul trunking, respectively. Table 6.9 provides values of Q for various other digital processes [34]. Also shown in Table 6.9 are the associated quantization distortion units (QDU) associated with each process. As discussed in Section

Table 6.9
Subjective Signal-to-Quantization Distortion Ratio (Q) for Digital Processes

Digital Process	Q (dB)	QDU
Processes involving A/D Conversion		

8-bit (64 kb/s) PCM A- or µ-law
- synchronous encoding
- semi-synchronous encoding
  * with Zero Code Suppression (ZCS)
  * Trans-multiplexer
  * Echo Canceller
	37	1

8-bit (64 kb/s) PCM µ-law - semi-synchronous encoding * with Robbed Bit Signaling (RBS)	~35	~1.4
7-bit (56 kb/s) PCM A- or µ-law	29.8	3

4-bit (32 kb/s) ADPCM
- synchronous encoding
- semi-synchronous encoding
  * with Zero Code Suppression (ZCS)
  * with Robbed Bit Signaling (RBS)
  * with ZCS plus RBS
	28.8
	
	~28.3
	~26.3
	~25.8
		3.5
		
		~3.8
		~5.1
		~5.5

Digital Processes		
µ/A/µ-law Tandem Conversion	46	.25
A/µ-law or µ/A-law Tandem Conversion A/µ/A-law Tandem Conversion	41.5 41.5	.5 .5
8-bit PCM Pad, A- or µ-law	39.3	.7
PCM/ADPCM/PCM Tandem	31	2.5
8-7-8 bit PCM, A- or µ-law	29.8	3

QDU: Quantization Distortion Unit; see eqn.(6.4)

5.4.2.2, and in terms of Equation (5.49), of Chapter 5:

$$QDU = 10^{(Q_1-Q)/K} \tag{6.4}$$

where $Q_1 = 37$ dB is the Q for a PCM codec with the synchronous encoding property, and where K is a factor corresponding to the type of codec tandeming.

For those processes in Table 6.9 involving analog to digital (A/D) conversion, tandeming is in accordance with the analog $K = 15$ tandeming law outlined in Chapter 5 (see Fig. 5.19). For the digital processes in Table 6.9, there will generally be less degradation with tandeming, and, in fact, no degradation ($K = 0$) for the PCM or ADPCM codecs having the "synchronous encoding" property (this property is discussed later in this chapter under error performance for voiceband data services). However, in most other digital cases, there is likely to be degradation corresponding to $K = 5$ to 10. Generally, these processes have not been precisely characterized for speech service; performance is estimated in Figure 5.19 of Chapter 5.

With regard to analog compandored systems, there is also a subjective advantage due to enhancement of noise during speech signal presence and reduction of noise in the absence of speech. However, this subjective effect is expressed somewhat differently than for digital systems, as either "compandor advantage" [35, 36], or as "compandor penalty" [22]. "Compandor penalty" is perhaps the simplest of the two approaches, and is the amount (in decibels) by which the compandor output idle circuit noise (output noise with no input signal) is subjectively enhanced by a speech signal during a conversation; this enhancement is evidently about 5 dB [22]. Thus, for example, if compandor output idle circuit noise is 20 dBrnC, the subjectively perceived noise during conversation is about 25 dBrnC. Appendix A.6.1 provides an objective interpretation of compandor penalty. Field results involving analog compandored systems are presented in [37].

6.2.3.4 Return Loss Parameters

Return loss as measured at a single frequency is illustrated in Figure 6.1. Based on single-frequency measurements, weighted average measures across various portions of the voice frequency band give echo return loss (ERL), singing return loss (SRL) (low and high) and average power return loss (APRL), etc., as previously defined and discussed in Chapters 3 and 5. Later in this section, measured results and performance limits are provided in terms of these parameters.

It is evident from Figure 6.1, that return loss is a sensitive function of the 2-wire path impedance and the balance network impedance. The 2-wire path impedance (Z_2) will be determined by the structural nature of the 2-wire line or trunk (gauge of wire, gauge changes, loading, bridged taps, irregularities, and design violations), as well as the terminating impedance at the far end of the 2-wire line or trunk. Ideally, the balance network impedance (Z_1) should equal the 2-wire path impedance (Z_2),

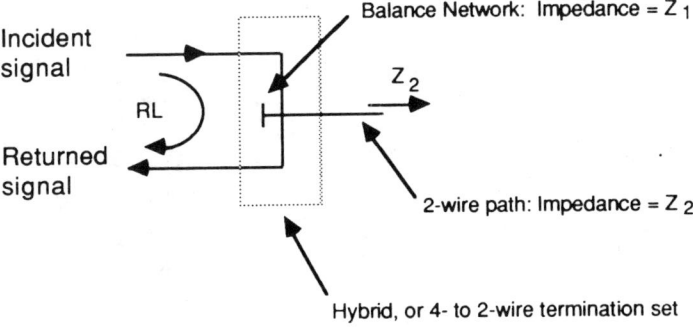

Return Loss (RL) in dB = Incident signal (dBV) - Returned signal (dBV)

$$= 20 \log_{10} \left| \frac{Z_1 + Z_2}{Z_1 - Z_2} \right|$$

dBV = $20 \log_{10}$ (Signal Voltage)

Figure 6.1 General illustration of return loss measurement.

which would result in infinite return loss (no echo). However, since this is impossible to achieve in practice, return losses will always be finite and must be controlled to ensure satisfactory network performance. Network return losses are assessed more specifically in terms of the configurations illustrated in Figure 6.2, and described as follows:

Loop Balance. The loop balance is the return loss of a loop as measured from the line (i.e., loop) side of an end office switch toward the customer end of the loop; the customer end of the loop is terminated by an off-hook station set. In the case of voiceband data transmission the customer end of the loop may be terminated by a modem or equivalent fixed impedance, such as 600Ω or 900Ω. However, in most instances, measurements are reported with an off-hook station as the termination. Loop balance return loss is generally responsible for the principal echo occurring in voiceband connections, and will exist as long as the 2-wire loop plant is in use.

As shown in Table 6.10, loop balance return loss is a rather sensitive function of the type of loop plant and the balance network. The discussion provided in Chapter 4 (Table 4.11) with respect to talker and listener echo for intra-office connections

Figure 6.2 Illustration of loop, terminal, and through balance return loss measurement.

Table 6.10
Loop Balance Return Loss Performance
(Effect of Loop Plant and Balance Network)

Return Loss Parameter				Type of Loop Balance Network		
Name	Description/Condition		Value (dB)	Series	Parallel	Double
Average Power Return Loss (APRL)	Loop plant	General	Mean SD	9.9 * 2.4 *	11.2 2.1	13.1 2.1
		Rural	Mean SD	8.8 * 3.1 *	8.7 2.0	11.5 2.0
		Rural (no design violations)	Mean SD	- -	- -	13.8 * 2.9 *
	Objective		Mean SD	$(14 + D_L)/2$ for D_L = LEPL = 2 to 8 ms # 2.1		
Echo Return Loss (ERL)	General loop plant		Mean SD	11.3 3.0	14.4 3.3	14.4 2.1
	Objective		Mean SD	11 3	14 3	
Singing Return Loss (SRL)	General loop plant			Mean SD	5.6 1.2	8.3 2.4	10.7 2.6
	Objective	For Voice	SRL-LO	Mean SD	11 2	
			SRL-HI	Mean SD	7 3	
			Overall **	Mean SD	7.5 $\geq$ 4 dB for 95 % of Connections 2.1		
		For Voice-band data		Limit	≥ 6		
	Flat SRL guideline for voiceband data	$\leq$ 2.4 kb/s		Limit	≥ 9		
		$\leq$ 4.8 kb/s		Limit	≥ 12.5		

- No data available
\# E.g., for a 4 ms LEPD, the mean APRL objective is 9 dB.
* No data available; determined from WEPL values in Table 4.11 of Chapter 4 (see text).
** Determined from WEPL objectives in Table 4.11 of Chapter 4 (see text).

is largely applicable to Table 6.10 as well (see [9, 13, 28, 30, 46–53] of Chapter 4). In fact, the single-ended return loss-oriented performance measurements and limits in Table 6.10 are approximately related to the echo path-oriented measures in Table 4.11 by dividing the means and SDs in Table 4.11 by 2 and $\sqrt{2}$, respectively. Deviations from this behavior are likely due to the return loss statistics (means and SDs) in Table 6.10 being estimated from return loss distributions, while the echo path loss statistics in Table 4.11 were estimated from echo path loss distributions.

Terminal Balance. Predivestiture: The return loss of a toll connecting trunk as measured from the inter-toll side of a Class 4 toll switch toward the end office; the trunk is terminated in 900Ω + 2.16 μf on the line side of the end office switch [18, 38–41].

Postdivestiture: The return loss of exchange access trunking as measured from the IEC side of an access tandem or IEC POP switch toward the end office; again the trunk terminates in 900Ω + 2.16 μf on the line side of the end office switch [11, 21].

Terminal balance also applies in a similar manner to intra-LATA or metropolitan tandem trunks between tandem switches and end office switches (see Fig. 4.3 of Ch. 3). Relative to loop balance return loss, terminal balance return loss is generally responsible for the next most important echo in a voiceband connection, and, again, will exist as long as 2-wire loops are in use; this is true even if all core network trunking and switching is digital (4-wire). Examples of terminal balance objectives for various types of trunks are included in Table 6.11 [11, 18, 21].

Through Balance. Through balance is the return loss of a 2-wire Class 3 or higher toll switch, or IEC equivalent switch, or principal tandem metropolitan switch. The outgoing side of the switch is usually terminated in 600Ω. Through balance return loss is generally responsible for the least important of all echoes in a voiceband connection, and will be irrelevant when all toll switches are digital (4-wire). See Table 6.11 for limits on through balance return loss [11, 18, 21].

Talker Echo Path Loss

Table 6.12 provides the electrical core network contribution to talker echo path loss (TEPL) for intra- and inter-office connections. These contributions to TEPL consist of the loop balance return loss (ERL) plus switch loss in each direction for intra-office connections, and the ERL plus inter-office trunk loss in each direction for inter-office connections [12, 15, 16]. Similarly, Tables 6.13 provides the same information for toll connections, including the effect of enabled and disabled echo control [42]. By adding loop losses to the values in Tables 6.12 and 6.13, TEPLs for complete connections are determined as discussed in Tables 4.12 and 4.13 of Chapter 4. Much of the discussion of these tables in Chapter 4 also applies to Tables 6.12 and 6.13 in this chapter.

Table 6.11
Terminal and Through Balance Return Losses

Balance Return Loss		Type of Balance	
Parameter	Value (dB)	Terminal	Through
ERL	Mean SD AL IAL	26.5 5.9 $18_a^a, 18_b^b, 22_c^c$ $13^a, 16^b, 16^c$	- - 27 21
SRL-LO	Mean SD	26.0 6.7	- -
SRL-HI	Mean SD	22.6 6.0	- -
SRL	Mean SD	$10_a^a, 13_b^b, 15_c^c$ $6^a, 11^b, 11^c$	20 14

- No data available
a 2-wire toll or POP switch, and 2-wire trunk (same values as in Exchange Access Standard [21])
b 4-wire toll or POP switch, and 2-wire trunk
c 2- or 4-wire toll or POP switch, and 4-wire trunk
   (Exchange Access mean ERL and SRL ALs are 21 and 14 dB, respectively [21])

Table 6.12
Core Network Contribution to Talker Echo Path Loss Performance for Intra- and Inter-Office Connections (Effect of Loop Balance Network)

Type of Connection	Echo Path Loss (dB)	Type of Loop Balance Network		
		Series	Parallel	Double
Intra-Office	Mean SD	11.5 3.3	14.7 3.6	14.7 2.6
Inter-Office	Mean SD	17.5 3.3	20.7 3.6	20.7 2.6

Table 6.13
Core Network (End-Office to End-Office) Contribution to Talker Echo Path Loss Performance for Toll Connections (Series Balance Network, Effect of Echo Control)

Condition of Echo Control	Echo Path Loss (dB)	Type of Toll Connection @			
		Short *	Medium	Long	Overall
Disabled	Mean SD	21.8 6.1	23.6 6.2	24.0 6.4	22.2 6.2
Enabled	Mean SD	21.8 6.1	23.7 6.2	39.7 14.5	23.6 8.8
Estimated Percentage of Connections With Echo Control		0.0	0.2	55.6	4.9

@ From Table 4.3 of Chapter 4, short, medium and long connection probabilities in % are: 77.8, 13.4 and 8.8, respectively

* Estimated Talker Echo Path Losses; see text.

Note that, as mentioned in Section 4.3.4.3 of Chapter 4, echo path loss survey results are not available for the short-distance category, but are estimated in Table 6.13 as follows. For the medium and long connections, trunk loss statistics [43] are subtracted from echo path loss statistics [42] to obtain an overall estimate of the loop return loss statistics encountered in the echo path loss survey (of mean $\sim$ 8.8 dB and SD $\sim$ 5.3dB). These estimated echo path loss statistics were then added to short trunk loss statistics [43] to estimate echo path loss statistics for short-toll connections. The techniques used are in accordance with Section 5.4.3.4 of Chapter 5.

6.2.3.5 *Delay Parameter*

One-way delay is contributed to by propagation delay of transmission media and processing delay of network nodes. Table 6.14 summarizes propagation delays for a variety of radio, fiber, (nonfiber) cable, and satellite transmission systems [44, 45]. These delays are not all purely due to propagation, but, in the case of carrier systems include repeater delays and delays of high rate multiplexers at various intermediate and end terminals. It is significant that, with evolution from digital radio to fiber transmission, network delay will increase somewhat.

Tables 6.15 and 6.16 summarize delays of various network nodal equipment [15, 46–51]; delays for certain other equipment and network arrangements are included in [21, 52, 53]. Table 6.15 focuses on delays of digital switching and digital cross-connect systems (DCS). Appendix A.6.2 (Figs. A.6.2.1 and A.6.2.2) estimates a detailed breakdown of digital switching delays to account for various distributed

Table 6.14
One-Way Transmission Facility Delays for Carrier Systems, Including Repeater, Regenerator, and High-Rate Terminal Delays

Type	Facility Description			Propagation Delay		Propagation Velocity	
				μs/km	μs/mi #	km/ms	mi/ms
Radio	Analog			4.7	7.5	218	136
	Digital			3.3	5.4 (5.3)	300	186
Fiber				5	8 (8.4)	200	125
Non-Fiber Cable	Coaxial	Terrestrial		4	6.4	250	156
		Submarine		6	9.6	167	104
	Twisted Pair *	Carrier		4.7	7.5 (7.8)	218	136
		Metallic	4-wire	53	8.5	18.8	11.8
			2-wire	125	200	8	5
Satellite	Non-Synchronous 14000 km = 8750 mi alt.			Delay = 110 ms			
	Synchronous 36000 km = 22500 mi alt.			Delay = 260 ms (250 ms)			

\# A conservative estimate for carrier systems is 10 μs/mi [53].
( ) Other more recent estimates [21, 52, 53].
* Delays derived from Via Net Loss (VNL) Factors

and nondistributed switching situations, and thereby enables appropriate switching delays to be accounted for in various applications [15].

Note that DCS delay decreases with higher rates, particularly when slip control buffers for synchronization purposes are not required—that is, for synchronous optical network (SONET)-based DCSs. (DCS definitions, typical standardized hierarchical digital signals, and related applications are discussed in Ch. 9.)

"Pure" propagation delays of transmission facilities and related terminal delays can be used to estimate the total delay of trunks, portions of connections, and complete connections. This is illustrated in Table 6.17 for short-haul trunking, for which there is generally no published survey data that is specific to delay. However, as discussed in Section 4.3.4.4 of Chapter 4, survey data on trunk lengths, together with Table 6.17, enable estimates of total round-trip delays to be determined for short-haul trunking consisting of various transmission technologies, as summarized

Table 6.15
One-Way Network Equipment Delays for Digital Switches and Cross-Connect Systems

Equipment		Delay (ms)	
Type	Ports	Average	99% Maximum
Digital Switch	Digital - Digital *	.5 (.45 - 1.2)	.7 (.75 -1.7) {1}
	Digital - Analog *	.7 (.975 - 1.35)	.95 (1.35 -1.95) {1.2}
	Analog - Analog *	.95 (1.5)	1.2 (1.95) {1.5}
	Digital or Analog - Remote Analog **	1.75	2.0
	Remote Analog - Remote Analog **	2.75	3.25
Digital Cross-connect System (DCS)	1/0, 3/1/0		.5 - .7 [46]
	3/1 with slip control		.125 - .25 [47]
	3/1 with no slip control		.025 - .125 [45, 47]
	SONET STS/VT or STS/STS		.125 [47]
	DS1/VT/DS1 mapping		.02 - .16 @ [51]

   * Includes Intra-Remote Switching configuration [15]
  ** Includes Inter-Remote Switching configuration [15], with 100 mi. of T-carrier facility from switch to each remote unit
  @ Depending on the type of mapping
  ( ) CCITT values; Maximums are 95% values [50]
  { } PBX values; Maximums are 95% values [49]
  [ ] Reference

in Table 6.18 (see [59–61, 64, 65] of Chapter 4). Overall, the distances have a mean and SD of about 21 and 14 miles, respectively, and corresponding round-trip delay mean and SD of about 1.6 and 0.7 ms, respectively. With short-haul trunking, it is important to have a good estimate of terminal delays, since these delays tend to dominate the overall delay of such trunks.

Round-trip delays for long-haul trunking can be estimated by subtracting the estimated short-haul trunking round-trip delays in Table 6.18 from the round-trip talker echo path delays (TEPD) provided in Table 4.15 of Chapter 4. The resulting delays for long-haul trunking are shown in Table 6.19.

6.2.3.6 *Distortion-Related Parameters*

Table 6.20 provides performance measurements and limits for envelope delay distortion (EDD), peak-to-average ratio (P/AR) and signal-to-intermodulation distortion (IMD) ratio. Generally, analog loops contribute little to these impairments, and the bulk of the impairment is caused by voiceband filters in trunk terminating equipment. Therefore, the digital switch, short-haul trunking, and end office to end office

Table 6.16
One-Way Network Delays for Other Equipment

Equipment		Delay
Type	Description	
Channel Bank (pair)	Digital - PCM Analog - Companded - Non-Companded	0.6 ms # 1.0 ms 1.5 ms
Transcoder	PCM-ADPCM-PCM	0.5 ms @
Transmultiplexer	Analog-Digital	1.5 ms *
Echo Canceller	Digital Analog	1.0 ms 1.6 ms
Device Pairs for Bit Sequence Transparency	ZBTSI B8ZS	0.5 ms 10 μs
First to Third Level Multiplexers	M13 (DS1/DS3) SONET (VT/STS) SONET VT ADM [48]	10-20 μs 15-30 μs 50 μs
Third to Higher Level Carrier Facility Multiplexers	DS3/NxDS3 SONET STS/STS-N	5-10 μs 10-20 μs
DCME and DSI Devices	Non-Packet	30-40 ms

\# .37 to .97 ms, depending on type of channel bank (D4 or D5), and type of analog interface (2-wire or 4-wire) [52].
@ .25 to 1 ms depending on type of ADPCM system [52].
* 3.3 ms if at digital satellite earth station

results in Table 6.20 are essentially the same as those for intra-office, inter-office, and toll connections, respectively, as presented in Table 4.16 of Chapter 4. Table 6.4 also provides EDD limits for private access lines and trunks.

Where possible, the performance for long-haul trunking in Table 6.20 is estimated by subtracting twice the performance of short-haul trunking from that of end office to end office performance (again, using the methodologies in Ch. 5). This procedure is straight forward for EDD. However, for P/AR, there is evidently no known techniques for accumulating this impairment. For high mean values of P/AR, it is assumed that the complements of the mean P/AR values ($100 - P/AR$) accumulate

Table 6.17
Breakdown of One-Way Carrier Facility Transmission Delays

Transmission Medium		Carrier Terminals	Propagation				Delay of Terminals *
			Delay		Velocity		
			μs/km	μs/mi	km/ms	mi/ms	ms
Radio		T N2 N3, N4, A5, A6	3.3	5.4	300	186	0.6 0.5 1.2
Cable	Twisted Pair	T N1 N2 ON, N3, N4	4.4 6.7	7.1 10.7	225 150	141 93.5	0.6 0.68 0.38 1.2
	Coaxial	A5 A6	4.3	6.8	235	147	0.85 1.3

T-carrier is digital, N-, O- and A-carriers are analog
* Total delay of terminals at both ends of the facility

Table 6.18
Estimate of Toll Connecting or Direct Exchange Access Trunking Distances and Round Trip Delays

	Trunk				Round Trip Delay (ms)				
Technology	Probability	Distance (mi)		Propagation Velocity	Propagation		Terminals	Total	
		Mean	SD		Mean	SD		Mean	SD
	(%)			(μs/mi)					
T-carrier	71.1	21	6.8	17.1	0.31	0.10	1.2	1.51	0.10
N-carrier	4.1	24	21.4	10.7	0.51	0.46	2.4	2.91	0.46
Cable	15.8	6.9	6.1	85	1.17	1.0	0	1.17	0.096
Radio	9.0	41.5	28.5	5.4	0.45	0.31	2.4	2.85	0.31
Combined	100	20.7	14.1	-	0.46	0.53	1.17	1.63	0.66

Table 6.19
Estimate of Intertoll or IEC Trunking Round Trip Delays

Probability (%)	Trunking			Round Trip Delay (ms)	
	Category	Distance		Mean	SD
		Airline (mi)	Route (mi)		
77.8	Short	0 - 180	0 - 340	6.3	4.1
13.4	Medium	180 - 725	340 - 1065	13.2	4.4
8.8	Long	725 - 2600	1065 - 3380	29.3	9.0
100	Overall			9.2	8.2

algebraically, and that the SDs accumulate in an independent manner, as follows:

$$P/AR_n = 100 - \sum_{i=1}^{n} (100 - P/AR_i) \tag{6.5}$$

$$\sigma_{P/AR_n} \approx \sqrt{\sum_{i=1}^{n} \sigma_{P/AR_i}^2}$$

Equation (6.5) results in what appears to be reasonable estimates of P/AR for long-haul trunking in Table 6.20. For second and third harmonic signal-to-IMD ratios, accumulation of these impairments involves 10 log and 16 log power additions, respectively, of sample values from the distributions of these parameters; see Table 6.20 [19]. The corresponding approximate analytical approach (Wilkinson's Method) outlined in Chapter 5 cannot be used for second harmonic IMD, owing to the large SDs involved for the short-haul values. Wilkinson's method is used for the third harmonic IMD in Table 6.20, but the values obtained are probably only a rough estimate, owing to the fairly large SDs involved.

Table 6.20
Distortion-Related Parameters for Connection Components

Parameter			Connection Component								
			Analog Loop	Digital Switch	Short-Haul Trunking	Long-Haul Trunking &			End-Office to End-Office &		
Name	Condition	Value				Short	Medium	Long	Overall	Short	Medium	Long	Overall
EDD (μsec.) w.r.t. 1.8 kHz	0.6 kHz	Mean SD AL IAL	- - - -	222 98 - -	279 211 ≤1300 * ≤1450	295 187	592 364	779 535	377 310	853 352	1150 471	1337 613 ~2600 -	935 430
	2.8 kHz	Mean SD AL IAL	- - ≤100 -	156 53 - -	173 121 ≤800 # ≤1100	253 154	355 215 ~800 -	381 211	278 172	599 230	701 275	727 272 ~2600 -	624 245
P/AR		Mean SD AL IAL	- - ≥90 -	97.0 1.2 ≥96 ≥94	95.9 3.1 - -	95.7 4.4	93.6 6.2	92.4 7.9	95.1 5.2	87.5 6.2	85.4 7.6	84.2 9.0 ~48 -	86.8 6.8
Signal-to-IMD Ratio (dB)	2nd Harmonic	Mean SD Obj. Ave. AL IAL	- - - - -	58.1 5.9 ≥48 ≥44 ≥43	60.5 8.8 - ≥37 @ ≥33					53.3 6.6	52.2 6.2	51.6 6.1 ~27 -	53.0 6.5
	3rd Harmonic	Mean SD Obj. Ave. AL IAL	- - - - -	60.0 5.1 ≥49 ≥45 ≥44	61.4 7.7 - ≥34 @ ≥31	67.8 11.3	58.2 4.4	57.0 9.2	65.6 11.3	53.4 6.9	51.0 4.7	50.2 6.9 ~32 -	52.8 6.7

w.r.t. with respect to
* SAL = 1500 μsec.
# SAL = 1000 μsec.
@ SALs = 40 dB
& From Table 4.3 of Chapter 4, short, medium and long connection probabilities in % are: 77.8, 13.4 and 8.8, respectively

6.2.3.7 Incidental Modulation-Related Parameters

Table 6.21 summarizes survey results and limits for amplitude and phase jitter, and for frequency shift. Again, these parameters are not manifested on analog loops, and therefore the digital switch, short-haul trunking, and end office to end office results in Table 6.21 are essentially the same as those for intra-office, inter-office, and toll connections, respectively, presented in Table 4.17 of Chapter 4. Various limits on these parameters are provided for short-haul trunking, as shown in Table 6.21.

Although formulae exist for the accumulation of fixed values of jitter (see Table 5.6 of Ch. 5), evidently no estimates for these parameter values have been made for long-haul trunking (by "subtracting" twice the short-haul values from the end office to end office values). Also, evidently no analytical methodology is available for accumulating distributions of these parameters. Thus, no estimates of long-haul trunking performance for these parameters are provided in Table 6.21.

Table 6.21
Incidental Modulation-Related Parameters for Connection Components

Parameter			Connection Component					
			Digital Switch	Short-Haul Trunking	End-Office to End-Office &			
Name	Condition	Value			Short	Medium	Long	Overall
Amplitude Jitter (%)	In 2-300 Hz Band	Mean SD AL IAL	2.0 0.7 - -	1.7 1.1 ≤ 4.7 # ≤ 8.9	2.8 1.4	3.2 1.4	3.6 1.6 - -	2.9 1.4
	In 20-300 Hz Band	Mean SD AL IAL	2.0 0.7 - -	1.7 1.0 ≤ 3.6 @ ≤ 6.2	2.6 1.1	2.8 1.0	3.2 1.2 - -	2.7 1.1
Phase Jitter (° p-p)	In 2-300 Hz Band	Mean SD AL IAL	1.6 0.6 - -	1.8 1.5 ≤ 11.0 # ≤ 14.4	5.5 3.7	6.4 3.5	7.6 3.8 ≤ 15 -	5.8 3.7
	In 20-300 Hz Band	Mean SD AL IAL	1.4 0.6 - -	1.3 1.0 ≥ 6.1 @ ≥ 7.7	3.1 1.4	3.9 1.7	4.9 2.1 ≤ 10 -	3.3 1.6
Frequency Shift (Hz)	N/A	% ≥ 2 % ≥ 1 AL	N/A	- - ± 2			0.19 0.24 ± 5	

\# SALs = 6 % and 10 ° p-p in 4-300 Hz band.
@ SALs = 5 % and 6.5 ° p-p.
& From Table 4.3 of Chapter 4, short, medium and long connection probabilities in % are: 77.8, 13.4 and 8.8, respectively

6.2.3.8 Analog Transient Parameters

Table 6.22 provides surveyed performance and limits for gain and phase hits, dropouts, and impulse noise. As discussed in Table 4.18 of Chapter 4, hits and dropouts are not expected in the loop plant (and are also not measured or specified for switching equipment). Thus, short-haul and end office to end office results for the hit and dropout parameters in Table 6.22, are essentially the same as those for inter-office and toll connections, respectively, presented in Table 4.18 of Chapter 4.

Table 6.22
Estimated Analog Transient Performance of Connection Components

Parameter			Connection Component		
Name	Condition	Value (Count in 15 min.)	Short-Haul Trunking	Long-Haul Trunking	End-Office to End-Office
Gain Hits	3 dB Threshold	Median SD in decades Limit	@ @ 0-2 *	- - 2	7.1×10^{-3} 1.2 ≤8
Phase Hits	20° Threshold	Median SD in decades Limit	@ @ 0-2 *	- - 2	7.1×10^{-3} 1.6 ≤8
Dropouts	12 dB Threshold	Median SD in decades Limit	5.4×10^{-3} 1.4 1/30 min. *	- - 1/30 min.	1.1×10^{-1} 0.91 ≤2
Impulse Noise	Noise Threshold 6 dB below received signal	Median SD in decades Limit !	.62 .95 5/5 min. *	- - 5/5 min.	1.9 0.97 ~15

- No data available
@ Cannot be estimated from available data [16]
! See Table 6.23 for detailed impulse noise limits
* Proposed Service Affecting Limits (SALs) are 1, 3, 1, and 15 counts in 15 min. for gain and phase hits, dropouts, and impulse noise, respectively [173].

This last statement also applies for impulse noise (compare Table 6.22 and Table 4.19 of Ch. 4), but for other reasons; that is, measured impulse noise performance for loops is not readily available. However, impulse noise for loops and switches is not necessarily negligible, as is evident from detailed limits for impulse noise as provided in Table 6.23 [11].

Note that in Table 6.23 the impulse noise threshold is usually expressed in absolute terms (i.e., N, in decibels reference noise C-weighted at zero level point

Table 6.23
Impulse Noise Limits for Network Components

Network Component		Impulse Noise Limit		
Type	Description	Threshold (dBrnC0)	Count AL	IAL
Loop	- Analog	59	≤ 15/15 min.	-
Switching Office	- Cross Bar - Step-by-Step - Electronic	54 59 47	≤ 5/5 min. for 50% of calls	≥ 20/5 min.
Short-Haul Trunking	- Voice Frequency - Companored or Mixed - T-carrier - Noncompandored (0 - 125 mi)	54 66 62 58	≤ 5/5 min. for 50% of calls	≥ 20/5 min.
Long-Haul Trunking (Noncompandored)	- 125 - 1000 mi - 1000 - 2000 mi - > 2000 mi	59 61 64	≤ 5/5 min. for 50% of calls	≥ 20/5 min.
Connection	- Intra-LATA - Inter-LATA	6 dB below signal level 5 dB below signal level	≤ 15/15 min. for 85% of calls ≤ 15/15 min. for 50% of calls	- -

(dBrnC0)) for network components, and in relative terms (i.e., analogous to a signal-to-noise ratio (SNR) T, in decibels) for network connections. With respect to a signal level S (e.g., of -13 decibels milliwatt at zero level point (dBm0)) and a factor $F = 90$ to convert from dBrnC0 to dBm0, the relationship between the relative and absolute thresholds is given by [54]

$$T = S - N + F \tag{6.6}$$

6.2.3.9 Error Performance

Introduction to Voiceband Codec Performance

In Chapter 4, end-to-end error performance based on survey data was presented for 1.2- and 4.8-kb/s voiceband data services [23]. Transmission errors are produced by the receiver (demodulator) portion of voiceband data modems. The errors are caused by the variety of continuous and transient analog impairments as discussed for end-to-end connections in Chapter 4, and as discussed for network components in this chapter (see also Fig. 6.7, presented later with the discussion of jitter in digital networks).

Network components that are significant contributors to certain continuous analog impairments, principally quantization noise, are PCM and ADPCM voiceband codecs. These codecs are associated with the terminal equipment for first-level carrier transmission systems, for example, T-carrier systems at 1.544 Mb/s. The terminal equipment is typically a channel bank or, in the case of PCM codecs, a digital switch.

In this chapter, voiceband data error performance is examined when a number of 64-kb/s PCM or 32-kb/s ADPCM codecs are tandemed in the voiceband connection, as shown in Figure 6.3 (one tandem consists of one codec pair, as shown in the figure). The performance results are based on laboratory tests in terms of modem error performance versus the modem input SNR [55, 56]. Also present in the simulated laboratory connections are representative levels of continuous analog impairments, as shown in Table 6.24, to realistically evaluate practical modem performance [57, 58]. Here, the most relevant information to be gleaned from the test results is whether desired levels of error performance can be achieved with modem input SNRs of ≤ 24 dB for switched network or basic C-conditioned private line applications, and ≤ 28 dB for D-conditioned private line applications (recall Table 4.5 of Ch. 4). Thus, the results are summarized in Tables 6.25 and 6.26 in terms of the minimum test SNRs required to achieve the desired levels of error performance (nominally, these performance levels are bit error ratios (BER) $\leq 10^{-5}$ and block error ratios (BLER) $\leq 10^{-2}$ for 1,000-bit blocks [59]).

A variety of representative 2.4-, 4.8-, and 9.6-kb/s modems have been evaluated, as shown in Tables 6.25 to 6.27. Table 6.25 provides results for PCM and ADPCM codec algorithms as standardized in CCITT Recommendations G.711 and G.721, respectively [56]. These codecs possess the so-called "synchronous encoding" property. This means that, in the absence of transmission errors on the carrier facility (i.e., at 1.544 Mb/s), quantization distortion will not accumulate when these codecs are digitally tandemed (digital tandeming means the channel signals remain in a digital form between the codec pairs—see Fig. 6.3). With this form of digital tandeming, there is essentially no performance degradation, and the results are similar to those for one tandem. Thus, the results in Table 6.26 are for the case of analog (or asynchronous) tandeming; that is, when the channel signals are converted to analog between codec pairs. In this case there is a performance degradation with tandeming.

Tables 6.26 and 6.27 provide results for a North American (ANSI) 32-kb/s ADPCM algorithm which is based on Recommendation G.721, but also embodies the zero code suppression (ZCS) feature [55]. As a practical matter, ZCS is one of the techniques used in North America to satisfy bit sequence restrictions on digital T-

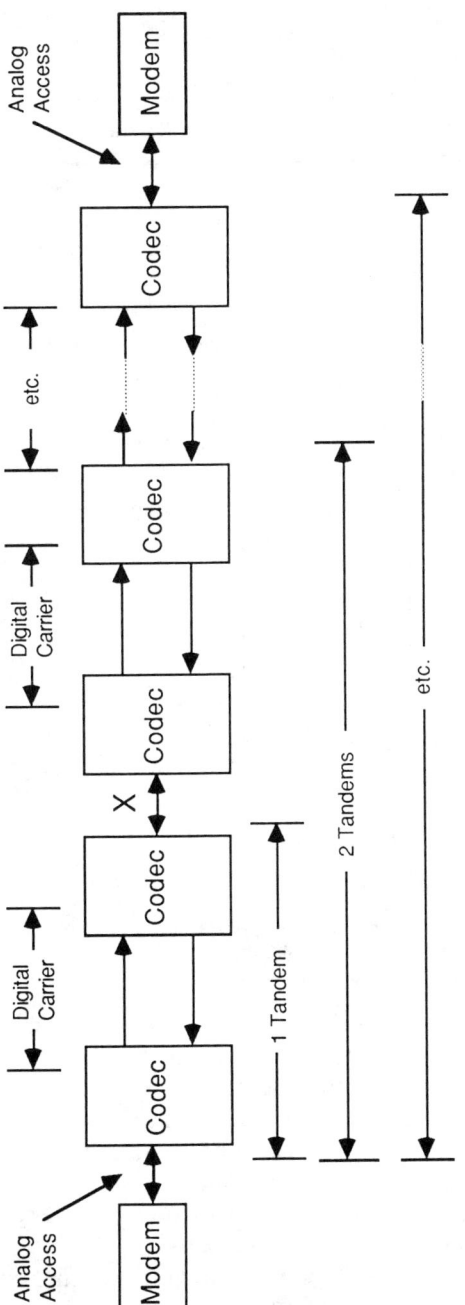

Figure 6.3 Illustration of types of tandeming.

Table 6.24
Analog Impairments Associated with PCM and ADPCM Error Performance
Evaluations for Voiceband Data Service

Parameter *	Units	Condition	Value
Signal Level	dBm0	-	-15
C-Message Noise	dBrnC0	-	Varied to achieve SNRs #
Attenuation Distortion w.r.t. 1 kHz	dB	0.4 kHz 2.8 kHz	0.9 3.9
Envelope Delay Distortion w.r.t. 1.8 kHz	ms	0.6 kHz 2.8 kHz	1.2 0.934
Signal-to Non-Linear Distortion Ratio	dB	2nd order 3rd order	36 38
Phase Jitter	°p-p	120 Hz	±3

w.r.t. with respect to
* Impairments are applied at the input to the Codec.
# The Signal (S) and Noise (N) levels determine SNRs, i.e., SNR = S - N + 88 (dB).
Network SNR Requirements are:
   SNR ≥ 24 dB for switched network, and for C-conditioned private line applications, and
   SNR ≥ 28 dB for D-conditioned private line applications.
Thus, the modem and codec test SNRs of significance are ≤ 24 and ≤ 28 dB such
that error performance requirements for the above applications are satisfied.

carrier transmission facilities; that is, avoid long strings of "zeros," which disrupt timing recovery. This algorithm is sometimes referred to as the 15-level algorithm, since, of the $2^4 = 16$ possible quantizer levels, the "zero" level is not used, leaving 15 levels. As a result of the ZCS feature, this algorithm provides somewhat degraded performance relative to the G.721 algorithm, and, in addition, does not retain the synchronous encoding property. This gives rise to so-called "semi-synchronous" tandeming in which the 32-kb/s channels undergo an impairment with tandeming, even though these channels remain digital between codec pairs.

Tables 6.26 and 6.27 also provide results for a 32-kb/s algorithm that, in addition to ZCS, supports inband robbed bit signaling (RBS). Again, as a practice matter,

Table 6.25
Minimum Signal-to-Noise Ratios to Support Voiceband Data Services with CCITT Asynchronously Tandemed PCM and ADPCM Codecs

			Minimum SNRs (dB) to Achieve BLERs # or MSURs for:					
	Modem		No. of PCM Tandems			No. of ADPCM Tandems		
Speed (kb/s)	Mode of Operation *	Type	1	2	4	1	2	4
2.4	Continuous Carrier (BLER # $\leq 5 \times 10^{-3}$)	V.26 Ter.!	13.4	13.8	14.1	14.4	14.2	15.0
		V.22 Bis.!	15.7	15.8	16.1	16.4	17.3	21.0
4.8	Continuous Carrier (BLER # $\leq 5 \times 10^{-3}$)	IBM 3864-1	16.4	16.2	16.4	17.5	19.6	23.7
		V.32 @	17.4	17.3	17.5	18.0	18.4	21.3
		V.27 Bis.	18.1	18.1	17.9	18.7	19.0	20.5
4.8	Switched Carrier (MSUR $\leq 10^{-2}$)	IBM 3864-1	~ 17.3	~ 17.5	18.0	18.4	19.2	23.7
		Codex V.29	18.1	17.9	18.3	19.0	19.2	20.5
		AT&T 2048	19.7	20.0	19.7	20.0	20.7	22.7
9.6	Continuous Carrier (BLER # $\leq 5 \times 10^{-3}$)	V.32 @	20.4	20.2	20.7	25.0	> 29	-
		V.29	25.4	26.3	27.6	» 30	-	-

\# 511 bit blocks; corresponding BLER $\leq 10^{-2}$ for 1000 bit blocks, & BER ~ 10^{-5}.
* All modems were subjected to the analog impairments in Table 6.24
! Signal to Echo (S/E) Ratio = 6 dB.   @ Signal to Echo (S/E) Ratio = 12 dB.
- Required BLER cannot be achieved with any SNR.

RBS is required in North America for compatibility with current inband signaling schemes. Eventually, with widespread common channel signaling (CCS), RBS will not be necessary. The RBS algorithm is sometimes referred to as the 15/7-level algorithm, since in every sixth frame (of the 1.544-Mb/s signal carrying 48 32-kb/s channels) the least significant bit is "robbed" for signaling purposes; this leaves $2^3 = 8$, less one for ZCS = 7 levels for the robbed frames, and 15 levels for the other frames. Like the ZCS algorithm, RBS also destroys the synchronous encoding property, and is therefore characterized as another variation of semi-synchronous tandeming. Bit robbing causes a significant degradation relative to the G.721 and ZCS algorithms.

The following is a detailed interpretation of the results in Tables 6.25 to 6.27. Results pertaining to other, more specialized codecs (e.g., at 24 and 40 kb/s) and to other modems are addressed elsewhere [60–63]. The latest codec-related ANSI standards and CCITT Recommendations were referenced in Chapter 3 (see [88–90, 92–94, 126]).

Table 6.26
Minimum Signal-to-Noise Ratios to Support Voiceband Data BERs with ANSI Tandemed ADPCM Codecs

Modem		Tandeming	Minimum SNRs (dB) to Achieve BERs $\leq 10^{-5}$ for No. of ADPCM Tandems								
Speed (kb/s)	Type		1		2			3		4	
			A	B	A*2	B*2		B*A*B	A*4	B*(A*2)*B	B*4
2.4	V.22	Async.	18.4	19.2	18.4	19.3		20.2	18.4	20.0	20.2
	V.32	Async.	18.1	18.7	18.1	18.7		19.6	20.0	22.7	25.1
		Semi-Sync.			18.3	20.1		19.8	18.2	22.6	23.3
	2048	Async.	21.6	22.2	21.2	22.3		23.2	23.1	24.5	26.6
		Semi-Sync.			22.2	22.6		23.3	21.5	23.8	27.0
4.8	V.29	Async.	18.0	18.3	17.8	19.2		25.1	23.1	29.3	>30
	V.27	Async.	20.3	21.4	20.4	22.2		22.6	24.1	29.0	>30
		Semi-Sync.			20.4	27.3		>30	20.8	26.1	*30
	QAM	Async.	19.6	20.9	19.7	21.5		>28	24.4	*28	*28
	208	Async.	24.7	25.4	25.0	27.2		*28	*28	*28	*28

A: ANSI Zero Code Suppression ADPCM algorithm.
B: ANSI (T1Y1) Robbed Bit Signaling ADPCM algorithm.
* Signifies tandeming operation with like, or different codecs (see text)

Table 6.27
Minimum Signal-to-Noise Ratios to Support Voiceband Data BLERs with ANSI Tandemed ADPCM Codecs

Modem			Minimum SNRs (dB) to Achieve BLERs** $\leq 10^{-2}$ for No. of ADPCM Tandems								
Speed (kb/s)	Type	Tandeming	1		2		3		4			
			A	B	A*2	B*2	B*A*B	A*4	B*(A*2)*B	B*4		
2.4	V.22	Async.	18.0	18.6	18.0	18.6	18.7	18.1	18.8	19.5		
	V.32	Async.	16.8	17.3	17.1	17.8	19.4	18.2	19.0	20.0		
		Semi-Sync.			17.8	19.0	19.0	17.5	20.2	22.1		
	2048	Async.	20.6	21.6	20.5	21.6	22.3	22.1	23.0	25.0		
		Semi-Sync.			21.1	22.0	22.5	21.1	20.0	26.0		
4.8	V.29	Async.	17.1	17.4	16.8	17.0	19.4	19.4	21.7	25.1		
	V.27	Async.	20.0	20.5	20.0	20.5	21.0	21.7	22.8	25.4		
		Semi-Sync.			20.0	21.0	22.6	19.9	21.8	>30		
	QAM	Async.	18.5	19.6	18.5	20.0	25.9	23.2	>28	>28		
	208	Async.	23.5	25.0	23.5	25.0	≈28	≈28	≈28	≈28		

A: ANSI Zero Code Suppression ADPCM algorithm
B: ANSI (T1Y1) Robbed Bit Signaling ADPCM algorithm
* Signifies tandeming operation with like, or different codecs (see text)
** 1000 bit blocks.

Performance With CCITT Codecs

The results in Table 6.25 apply to asynchronously tandemed PCM and ADPCM codecs [56]. The results are rank ordered as to the modem speed and type which gives the highest performance (at the top of the table), to the lowest performance (at the bottom of the table). "Highest performance" means the minimum modem input SNR necessary to achieve the required error performance. Accordingly, the results in Table 6.25 are in terms of the minimum SNRs necessary to achieve BLERs $\leq 5 \times 10^{-3}$ with block size (BS) of 511 bits. Based on the approximate relationship

$$\text{BLER} = \text{BER} \times \text{BS}/K \qquad (6.7)$$

which applies to random errors if $K = 1$ bit error per block error, the same SNRs can also be applied to BLERs $\leq 10^{-2}$ for 1,000-bit blocks. However, to the extent that the codec tests may have involved some error burstiness (i.e., $K > 1$ bit error per block error), the SNRs are conservative; that is, they are somewhat higher than required to achieve BLERs $\leq 10^{-2}$ with 1,000-bit blocks. The reason is that error burstiness provides an advantage for larger sized blocks.

For random errors as per Equation (6.7) with $K = 1$, the SNR results in Table 6.25 also apply to BERs $\leq 10^{-5}$. However, if the errors were bursty, the SNR results in Table 6.25 are probably optimistic; that is, somewhat lower than required to achieve BER $\leq 10^{-5}$. (A later comparison of Tables 6.26 and 6.27 indicates that the SNR required to achieve BER $\leq 10^{-5}$, is at least about 1 dB higher than the SNR to achieve BLER $\leq 10^{-2}$ with 1,000-bit blocks.)

The BER and BLER measures apply to continuous carrier voiceband data operation, where, regardless of the data traffic to be carried, the modem transmitted carrier signal is continuously present on the voiceband channel. Conversely, with switched carrier operation, the modem transmitted carrier is switched on when there is a data message to send, and switched off when the message ends. In this case an important performance factor is the ability of the modem receiver to start up and properly demodulate the carrier when it is switched on. This is measured by the missed startup rate (MSUR), which, to be consistent with BLER requirements, should not exceed 10^{-2} (i.e., ≤ 1 missed startup in 100 messages).

With the above background in mind, the following are specific observations about the results in Table 6.25. Again, the results are interpreted in terms of the following two applications: (1) switched network and basic C-conditioned private line network applications for which the required minimum SNR is ≥ 24 dB, and (2) higher performing D-conditioned private line applications for which the required minimum SNR is ≥ 28 dB (recall Table 4.5 of Ch. 4).

- As expected, performance degrades (i.e., higher SNRs are necessary) with more codecs in tandem. Also, PCM codecs perform better than the ADPCM codecs.

For the 2.4- and 4.8-kb/s modems, the improvement with PCM depends largely on tandeming and varies from ≤ 1 dB with one tandem to a few dB with four tandems.
- All of the 2.4- and 4.8-kb/s modems can provide the required error performance for both applications with at least four asynchronous PCM or ADPCM tandems. Performance at 2.4 kb/s is very robust.
- For the 9.6-kb/s modems, there are some limitations:
  - The V.32 modem will support both applications with at least four asynchronous PCM tandems, but will only operate with one ADPCM tandem, and then only on D-conditioned private lines.
  - The V.29 modem will operate on D-conditioned private lines with at least four PCM tandems, but will not operate with ADPCM.
- A final observation is that, comparing the 4.8-kb/s IBM modem performance in the continuous and switched carrier modes, typically somewhat lower performance is indicated (i.e., up to about 1.5 dB higher SNR is necessary) in the switched mode.

Performance With ANSI Codecs

The results for BER and BLER in Tables 6.26 and 6.27, respectively, all apply to the continuous carrier mode of modem operation [55]. The results also apply to both asynchronous and semi-synchronous ZCS and RBS ADPCM tandems (designated A and B, respectively, in the tables).

Included in the results are mixes of the ZCS and RBS codecs for the connections with three and four tandems, as shown in the tables. Again, the results are rank ordered as to the modem speed and type, which gives the highest to lowest performance. In this case however, the performance criteria for ranking, while based on SNR, is further qualified by the specific ability to achieve required error performance with ≤ 24- and ≤ 28-dB SNRs across all tandeming conditions. Thus, modems near the top of the table satisfy these conditions for all or nearly all tandeming conditions, while modems near the bottom of the page do so for fewer tandeming conditions. This behavior is summarized in Table 6.28 in terms of the minimum ADPCM tandeming capabilities that will support the indicated voiceband data services.

As shown by the footnote (*) in this table, depending on the types of codecs used in the connection, more tandems may be supported than indicated by the strict unqualified minimum numbers in Table 6.28. For example, Table 6.28 shows that with up to three asynchronous ADPCM tandems, the V.32 modem can achieve a BER ≤ 10^{-5} with ≤ 24-dB SNR. However, from Table 6.26, the same is also true with certain (but not all) arrangements of four ADPCM tandems. Generally, the more RBS codecs, as opposed to ZCS codecs, in the connection, the more difficult it is to achieve required error performance. The tandeming limits in the range of three

Table 6.28
Minimum Tandeming Capability of ANSI ADPCM Codecs

Modem			Tandeming	BERs ≤ 10⁻⁵		BLERs ≤ 10⁻²	
Speed (kb/s)	Type	#		Switched Network (Test SNR ≤ 24 dB)	Private Line (Test SNR ≤ 28 dB)	Switched Network (Test SNR ≤ 24 dB)	Private Line (Test SNR ≤ 28 dB)
2.4	V.22		Async.	4	4	4	4
	V.32		Async.	3*	4	4	4
			Semi-Sync.	4	4	4	4
	2048		Async.	3*	4	3*	4
			Semi-Sync.	3*	4	3*	4
4.8	V.29		Async.	2*	3*	3*	4
	V.27		Async.	3	3*	3*	4
			Semi-Sync.	1*	2*	3*	3*
	QAM		Async.	2	2*	2*	3*
	208		Async.	0	2	1*	2

* Tandeming capability is higher for certain selected Codec tandeming arrangements (see Tables 6.26 and 6.27).
# All modems operating in the continuous carrier mode.

or four are consistent with those recommended in [69]. The specific planning rules for the deployment of ADPCM are discussed in Section 7.10 of Chapter 7.

The following are specific observations about the results in Tables 6.26 and 6.27.

- Again, as expected, performance degrades with more codecs in tandem. Also, ZCS codecs (designated "A" in the tables) perform better than RBS codecs (designated "B"), and the difference increases both with increased tandeming and with the poorer performing modems. The difference ranges from less than 1 dB in SNR for one tandem and high performing modems, to a few dB for four tandems and poorer performing modems.
- It is perhaps expected that asynchronous tandeming would not perform as well as semi-synchronous tandeming. However, a comparison of the asynchronous and semi-synchronous results shows that the difference in performance is generally small (often < 1 or 2 dB) and is not consistent over all tandeming arrangements. In fact, for one to three tandems, semi-synchronous performance is most often somewhat lower than the asynchronous performance, while for four tandems, the reverse is true. This suggests that, overall, for voiceband data services, impairment due to semi-synchronous tandeming is similar to that of asynchronous tandeming.
- Finally, in comparing the BER and BLER results in Tables 6.26 and 6.27, the BLER results are better in terms of lower required SNRs, as expected. The difference in performance is due to the advantage provided by block transmission when errors are bursty. The difference tends to increase somewhat with tandeming and with the poorer performing modems. The difference ranges from about 1 dB for one tandem and higher performing modems, to a few dB with four tandems and poorer performing modems.

The following are more applications-oriented observations about the results in Tables 6.25 to 6.28.

- The 2.4-kb/s modem and, with a minor exception, the V.32 4.8-kb/s modem can provide the required error performance with at least four ADPCM tandems; again, 2.4-kb/s performance is very robust. (An "exception" is considered "minor" when the test SNR exceeds the required network value by only about 1 dB or so.)
- For the 2048 4.8-kb/s modem, performance is somewhat poorer than the V.32 4.8-kb/s modem for the switched or basic private line application when four tandems are used. However, this modem will operate satisfactorily on D-conditioned private lines.
- The remaining four modems have considerable difficulty with certain tandeming arrangements. This is signified by SNRs >28 or >30 dB in Tables 6.26 and 6.27, which means that, even with very high SNRs, achieving the required error performance would likely be marginal at best. The V.29 modem indicates this

behavior for one of the arrangements of four tandems, while the remaining three modems also indicate this behavior for three tandems.
- The last three modems definitely cannot meet required error performance with certain tandeming arrangements, regardless of SNR. This is indicated by SNRs $\gg 28$ or $\gg 30$ dB in Tables 6.26 and 6.27. The V.27 modem indicates this behavior for one of the arrangements of four tandems, and the remaining two modems also have this behavior for three tandems.
- It is evident from the above observations that the impairments due to RBS, in particular, are significant in limiting the application of ADPCM codecs for 4.8-kb/s voiceband data. Recall from Table 6.25 that no such limitations were evident with the non-RBS CCITT ADPCM algorithms.

Signaling and Facsimile Performance

The previous sections have dealt exclusively with voiceband data performance. Other voiceband services and signals that may encounter codecs are dual tone multifrequency (DTMF) and multifrequency (MF) signaling tones, and facsimile signals.

MF signaling is usually node to node, and would therefore generally only encounter one tandem. Similarly, DTMF signals usually only apply to the access portion of networks, and, in this case, would usually only encounter one tandem. However, there are also applications where DTMF signals may traverse end-to-end connections, in which case more tandems may be encountered. Similarly, facsimile signals apply to end-to-end connections. Generally, under reasonable carrier error rate conditions (BER $\ll 10^{-4}$), DTMF signaling and Group 1 and Group 2 facsimile applications can be supported with up to four ADPCM tandems [55, 56].

6.3 ERROR PERFORMANCE OF DIGITAL NETWORK COMPONENTS

6.3.1 Allocation of BER and ES Objectives

In Chapter 4, end-to-end performance was discussed for digital services. Also, performance objectives were allocated to national and international portions of connections, as well as to various portions of national connections (see Tables 4.22 to 4.24 of Ch. 4). In this chapter, a more detailed performance allocation to network components is provided for the common BER and errored second (ES) parameters, and mainly for the DS0 and DS1 hierarchical rates. These allocated objectives are used for the following purposes:

(1) To compare with the measured or estimated performance of various network components (see Table 6.32, discussed later in this chapter).

(2) To derive specific end-to-end objectives for each of the example digital connections introduced in Chapter 4 (see Fig. 4.7). Each of these connections was evaluated in Chapter 4 by comparing the specific end-to-end objectives to the projected end-to-end performance based on the measurements or estimates of performance for the components noted in item (1) above (see Table 4.26).

Based on [70], the allocation for the BER parameter is illustrated in Table 6.29 in terms of the percentage of time that the BER is $> 10^{-6}$, where the overall end-to-end objective is such that the BER can be $> 10^{-6}$ for $\leq 5\%$ of time. The table shows that, with the given allocations to components, the derived end-to-end objectives for specific intra-office, inter-office, and some "short"-toll connections fall within the overall end-to-end objective. The same is perhaps not so for the longer toll connections. Part of the uncertainty comes from the allocation to analog loops, which are difficult to characterize and maintain, and therefore do not have a definitive objec-

Table 6.29
Example Allocations of North American (U.S.) BER Objectives for Switched Digital Data Service

Network Component	BER Allocation (% > 10^{-6})		Example Connections #			
	Per Component	Per Connection (multiplier)	Intra-Office	Intra-Office	Toll	
					Short	Long
Digital Loop	0.5 to 1	x 2	1.0 - 2.0	1.0 - 2.0	1.0 - 2.0	1.0 - 2.0
Circuit Switch *	.005	x n	x 1 = .005	x 2 = .01	x 4 = .02	x 5 = .025
Short-Haul Trunking	0.1 per mi.	x 15 x m	-	1.5 x 1 = 1.5	1.5 x 2 = 3	1.5 x 2 = 3
Long-Haul Trunking 'Short'	1.0 per 250 mi.	x 125 mi.	-	-	0.5	0.5
Long	1.0 per 4000 mi.	x 1000 mi.	-	-	-	0.25
Entrance Link Associated with Long Trunks	0.6	x 2	-	-	-	1.2
Connection Totals			~ 1 - 2	~ 2.5 - 3.5	~ 4.5 - 5.5	~ 6 - 7

* Typical Digital Switch BER Requirement is $< 10^{-9}$ [15].
\# See Fig. 4.7 of Chapter 4
n, m   The number of circuit switches, and short-haul trunks in the connection, respectively.

tive [70]. Regarding the other components, the switch allocation is negligible, and the trunk allocations are distance related with the per-mile allocation being much more stringent for the longer trunks; this is consistent in principle with the CCITT medium- and high-grade categories, which correspond to shorter and longer network portions, respectively [71]. Finally, the entrance link leading to long-haul radio systems uses a considerable portion of the overall BER budget.

Table 6.30 is a somewhat more detailed allocation of the DS0, DS1, and DS3 ES objectives considered earlier in Chapter 4 [64, 72–75], including the overall end-to-end connection objectives. Note that the allocations are based on objectives for dedicated digital services as specified in the earlier versions of [64, 72, 73], which relate allocated and end-to-end objectives using worst-case addition; see Table 4.24(a) of Chapter 4. More stringent objectives have been specified in the later ver-

Table 6.30
Allocations of Early North American (U.S.) ES Objectives for Dedicated Digital Services

Network Component		% ES per Day Allocation		
		DS0	DS1	DS3
Digital Loop		.05	0.5	(0.4)
Short-Haul Trunk		.0375	0.375	(0.3)
Access *		.125	1.25	1.0
Long-Haul Trunking	'Short' Medium Long	(.09) (.15) .25	0.9 1.5 2.5	2.9 4.8 8.0
Connection **	'Short' Medium Long	(.34) (.40) .50	3.4 4.0 5.0	4.9 6.8 10.0

  * Assumed to consist of digital loop plus two short-haul trunks.
 ** Consists of two access arrangements (*) plus long-haul trunking.
 ( ) Estimated (in proportion to DS1 objectives).

sions of [64, 72, 73], and in some cases relate allocated and end-to-end objectives using statistical and other criteria; see Table 4.24(b) of Chapter 4. These criteria are not considered here.

Essentially, the additional detail provided in Table 6.30, over that in Chapter 4, is for the loop and short-haul trunking making up each access link [65, 66]. Also, certain distance-dependent long-haul trunking allocations are estimated for DS0 services based on the relationship of 1 ES at DS1 affecting 2 or 3 (about one tenth) of the 24 DS0s in the DS1 stream (i.e., 1 ES at DS1 corresponds to an average of about 0.1 ES at DS0; this relationship is discussed at some length in Chs. 4 and 9 [65]). For the access, this same relationship does not necessarily hold between DS3 and DS1 rates, because different technologies may be used for DS3 and DS1 rates. However, for the long-haul trunking where DS1s and DS3s would both be transported by the same carrier technology, it can be inferred from Table 6.30 that 1 ES at DS3 evidently affects about 9 (about one third) of the 28 DS1s in the DS3 (i.e., 1 ES at DS3 corresponds to an average of about 0.3 ES at DS1). This is conservative when compared with measurements and models of error performance at DS1 and DS3 rates where 1 ES at DS3 corresponds to an average of about 0.12 ES at DS1 [76].

The allocated ES objectives in Table 6.30 are used in Table 6.31 to derive the specific end-to-end ES objectives for the various example connections considered in

Table 6.31
Example Allocations of Early North American (U.S.) ES Objectives for Dedicated Digital Services

Network Component		DS0 % ES Allocation *		Example Connections #			
		Per Component	Per Connection (multiplier)	Intra-Office	Intra-Office	Toll	
						Short	Long
Digital Loop		0.05	x 2	0.1	0.1	0.1	0.1
Short-Haul Trunking		.0375	x m	-	x 1 = .0375	x 2 = .075	x 2 = .075
Long-Haul Trunking	'Short'	0.09	x 1	-	-	0.09	0.09
	Long	0.25	x 1	-	-	-	0.25
Connection Totals				0.1	0.1375	0.265	0.515

* Multiply by 10 for DS1 % ES allocations and totals.
\# See Fig. 4.7 of Chapter 4
m The number of short-haul trunks in the connection.

Chapter 4 (Fig. 4.7), and they were also used for comparison with estimated end-to-end ES performance in Chapter 4 (Table 4.26). The Table 6.31 objectives are for DS0 rates, but can be readily multiplied by a factor of 10 to determine DS1 objectives. From this table, the connection-specific objectives all essentially fall within the overall end-to-end objectives of $\leq 0.5\%$ ES per day at DS0 (5% ES per day at DS1).

6.3.2 Measured Performance of Network Components

Here information is provided that is indicative of component performance achieved in practice [77]. This information applies to the components in the various digital connections shown in Figure 4.7 of Chapter 4. Also, the end-to-end performance results for these connections was provided in Table 4.26 of Chapter 4. Finally, the earlier discussion in Section 4.4.2.2 of Chapter 4 relating to interpretation of the data in [77] and to the nature of performance parameters, etc., also applies here. With this background in mind, the following is a discussion of the performance levels in Table 6.32.

The long-term BER and DM results are caused by T-carrier transmission. This transmission system technology is directly applicable to the digital loops and to the short-haul trunking used in the exchange access network. For the toll connections, the contribution to BER and DMs from the radio systems is assumed to be negligible once highly errored events due to protection switching activity are removed. However, these errors are included in ESs and severely errored (SES), as discussed later. The estimated percentages of trunks achieving a long-term BER $\leq 10^{-6}$ generally fall somewhat short of the connection-specific objectives (as derived in Table 6.29). This implies selection of facilities (i.e., a small percentage of T-carrier spans would have to be either upgraded or rejected to support digital services [79]). Generally, when measured over relatively long time intervals, BER objectives are more difficult to achieve than when measured over shorter time intervals. This is supported by the DM results, which readily satisfy both North American (U.S.) and CCITT recommended performance levels.

ESs for digital loops and short-haul trunks are based on T-carrier performance. For long-haul trunks, a small percentage of ESs due to protection switching activity is also included, and this percentage is 10 times higher under fading conditions, as shown. These different causes of ESs are accounted for in estimating the ESs at DS0, as discussed in Chapter 4. ES performance for T-carrier technology is lower than the North American (U.S.) objectives (as derived in Table 6.30). Again this implies that a degree of selection of T-carrier spans would be necessary to achieve required performance for digital services [79]. ES performance is readily met by the long-haul technology. At DS0, the less stringent CCITT ES levels are readily met by all components.

Table 6.32
Example of Digital Performance of Network Components

Parameter		Network Component					
Name	Value	Digital Loop	Short-Haul Trunking (Exchange)	Long-Haul Trunking			
				'Short'		Long	
				Non-Fading	Fading	Non-Fading	Fading
Long Term BER	Median $\% \leq 10^{-6}$	4.0×10^{-9} 97	10^{-10} 97	–		–	
	NA Obj. $(\% \leq 10^{-6})$	$\geq 99 - 99.5$	≥ 98.5	≥ 99.5		≥ 98.55	
% DMs	Median 95% Level	.005 0.3	.005 0.3	–		–	
	CCITT Obj.	≤ 1.5	≤ 1.5	≤ 0.8			
% ESs	DS1 Median 95% Level	.015 1.5	.0015 1	.0002 .001625	.002 .01625	.0016 .013	.016 .13
	DS1 NA Obj.	$\leq .5$	$\leq .375$	≤ 0.9		≤ 2.5	
	DS0 Median 95% Level	.0015 .15	.00015 .1	.0002 .001625	.002 .01625	.0016 .013	.016 .13
	DS0 NA Obj.	$\leq .05$	$\leq .0375$	≤ 0.09		$\leq .25$	
	CCITT Obj.	≤ 1.2	≤ 1.2	≤ 0.64			
% SESs	Median 95% Level	– –	– –	.0002 .001625	.002 .01625	.0016 .013	.016 .13
	CCITT Obj.	$\leq .015$	$\leq .032$	≤ 0.042			

NA Obj.: North American (U.S.) Objective

Regarding SESs, no specific data is provided in [77] for T-carrier facilities, although, as noted in Chapter 4, at least a small percentage of SESs is bound to occur in practice, and more recent results are available as summarized later in this chapter [60] and as specified in the later versions of [64, 72, 73]. For toll connections, SESs correspond to protection switching activity, which causes about 10 times as many SESs under fading conditions as opposed to nonfading conditions. Performance is marginal under fading conditions, but is acceptable when averaged over fading and nonfading periods, since most fades take place during a few days per month (e.g., half in one day, 70% in three days [77]).

It should be noted that newer technology, particularly fiber-optic and advanced radio transmission systems, will eventually replace most older long-haul systems, as well as T-carrier systems currently used for digital loops, short-haul trunks, and

Table 6.33
Transport System Digital Performance Requirements/Expectations

System			BER		Parameter Req. PEFS		SES/day #		Availability	
Type	Condition		Req.	Exp.	DS1	DS3	Req.	Exp.	Req.	Exp.
Fiber Short-Haul (250 mi.) [80, 84]	-		$\leq 10^{-9}$ *	10^{-11} @	≥ 99.86	≥ 99.0	≤ 4 &	-	≥ 99.98 !	-
Radio 90 Mb/s (~ 480 mi.) [82]	Non-fading		-	10^{-9}	(99.85)	(99.55)			$	99.99
	Fading		-	10^{-8}	(98.46)	(95.53)				
Satellite 60 Mb/s with FEC [82]	Non-fading		-	10^{-9}	(99.85)	(99.55)			$	99.99
	Fading		-	10^{-7}	(84.56)	(55.26)				
Generic [83]	Repeater Span (25 mi)		$\leq 10^{-9}$	10^{-11}	(≥ 99.85)	(≥ 99.55)	-	< 1	-	-
	Short-Haul (250 mi)		$\leq 10^{-8}$	-	(≥ 98.46)	(≥ 95.53)	≤ 4			
	Long-Haul (4000 mi)		$\leq 10^{-8}$	-	(≥ 98.46)	(≥ 95.53)	≤ 8			

- Radio uses 6 and 11 GHz, and is subject to rainy weather and multipath fading degradations.
- Satellite uses 4/6 GHz, QPSK modulation, 7/8 rate convolutional Forward Error Correction (FEC), and is subject to rainy weather and sun transit degradations.

@ For ~ 400 Mb/s, 480 mi. system [82].
\# SESs have > 100 errors per second. • Excludes error bursts (i.e., SESs). & Excludes protection switching events.
! 99.98 % for DS1 with ~ 44/56 allocation to equipment/fiber media failures.
  99.985% for DS3, and 99.98% for SONET, both with 25/75 allocation [84].
$ Generally designed to meet ≥ 99.98 %, and allocated 50/50 to equipment failure/propagation effects.
( ) Conservative estimates computed from BERs assuming random errors; see eqn. (6.10).
- No information in references.

entrance links, etc. This will result in much improved performance, as already in evidence from the later versions of [64, 72, 73]. More specifically, specified and expected performance at DS1 and DS3 rates when supported by various transport system technologies is shown in Table 6.33 [80–84]. Expected performance of fiber systems is on the order of BER $\leq 10^{-11}$ for 400-mile systems [82]. Also, nationwide (4,000-mile) performance goals are on the order of BER $\leq 10^{-10}$ [85]. Indications are that these performance levels are achievable when considering the capabilities of recent technologies, as exemplified in Table 6.34 [86–90]. These measured performance levels in terms of percent error-free seconds (PEFS), if linearly prorated to the short-haul distance, readily meet the levels specified in Table 6.33 for short-haul systems. Also, assuming random errors, computed BERs corresponding to the PEFS values (if prorated to the short-haul distance) readily meet the BER $\leq 10^{-9}$ requirement in Table 6.33.

In fact, prorating these BERs to long-haul distances (of 4,000 miles) also shows that a BER $\leq 10^{-10}$ objective is achieved in most instances. The BERs are computed

Table 6.34
Typical Measured Performance of High-Speed Digital Carrier Transport Systems

System		Measured Performance					Computed BER &
Type	Length (mi)	Duration (days)	Rate	PEFS per day	Prot. Sw. per day *	Outages #	
Fiber (417 Mb/s) [86, 87]	950	59	DS1	99.998	.24	0	1.3×10^{-11}
	100	91	DS1	99.999	.028	0	6.5×10^{-12}
	100	11	417 Mb/s	99.9998	.18	-	5.0×10^{-15}
Radio (90 Mb/s) [88]	650	152 (Aug.-Dec.)	DS1	99.9989	-	-	7.1×10^{-12}
	79	69 (July-Sep.) @	DS3	99.825	-	-	3.9×10^{-11}
	79	21 (Aug.-Sep.) @	DS3	99.998	-	-	4.5×10^{-13}
Radio (135 Mb/s) [88]	76	78 (June-Aug.)	DS1	> 99.9999 !	-	-	$< 6.5 \times 10^{-13}$
		78 (June-Aug.)	DS1	99.9999 !	-	-	6.5×10^{-13}
	365	74 (Sep.-Dec.)	DS1	99.9992 !	-	-	5.2×10^{-12}
		116 (Sep.-Dec.)	DS1	99.9998 !	-	-	1.3×10^{-12}

* Protection switching of the carrier will cause Out-of-Frame (OOF) events, and hence ESs at the DS1 rate.
\# Defined as $\geq$ 10 consecutive seconds with BER > 10^{-3}
@ Last 21 days tested were with adjustments to radio system to optimize performance.
! Measurements shown for each direction of transmission.
& Assuming random errors (see text).

from PEFS as follows:

$$\text{BER} = (\text{No. errors in time } T)/(R \times T) \tag{6.8}$$

$$\text{PEFS} = 100[1 - (\text{No. ES in time } T)/T] \tag{6.9}$$

where R is the transmission rate in bits per second, and T is the measurement time in seconds. For random errors, the "No. errors in time T" = "No. ES in time T"; thus, combining Equations (6.8) and (6.9) gives

$$\text{BER} = (1 - \text{PEFS}/100)/R \tag{6.10}$$

In practice, for a given PEFS value, the corresponding BER could be larger than in Equation (6.10), owing to burstiness. That is, for bursty errors, assuming the "No. errors in time T" = $K \times$ "No. ES in time T", then

$$\text{BER} = K(1 - \text{PEFS}/100)/R \tag{6.11}$$

Finally, Table 6.35 shows recent measured performance of DS1 access circuits, probably of short distances and using T-carrier and other (fiber, radio) technologies [91]. It is seen that the performance is considerably improved from that of older technology (Table 6.32) and also readily satisfies the objectives in Table 6.30; as already mentioned, more stringent objectives have been proposed as footnoted in Table 6.35 [91], and in the later versions of [64, 72, 73].

6.4 JITTER AND WANDER IN DIGITAL NETWORKS

6.4.1 Causes and Effects of Jitter and Wander

Jitter and wander are digital network impairments that, if excessive, can lead to higher level impairments, such as the transmission errors and short interruptions (frame loss events and slips) discussed previously in this chapter and in Chapters 3 and 4. Jitter may also directly affect end user perceived performance for certain applications, such as video. Measurement of these parameters is dealt with in [92–94].

6.4.1.1 Jitter

In the context of digital transmission systems, jitter (also called *timing jitter*) is the unwanted phase modulation of digital signals at modulation rates (jitter frequencies) above 10 Hz.

Table 6.35
Recent Measured Performance of DS1 Access Circuits

Number of 2-way Circuits *	Measurement Duration (Days)	Parameter		Performance #	
		Name	Statistic	Number of Events	Percentage
25	75	ES per day	Mean	87	0.10
			95 %	< 170	< 0.20
		SES per day	Mean	2	0.002
			95 %	< 2	< 0.002
86	117	ES per day	Mean	67	0.078
			95 %	< 125	< 0.14
474	132	ES per day	Mean	21.4	0.025
			95 %	< 30	< 0.035
			98.5 %	< 170	< 0.20
756	71	ES per day	Mean	18.5	0.021
			95 %	< 30	< 0.035
			99 %	< 170	< 0.20

* Circuits are expected to be short-haul (~ 10-20 miles), and use various carrier technologies (fiber, radio, twisted pair cable). Each direction of transmission was separately measured.

\# Proposed objectives in [91] based on this data are: ≤ 170 ESs per day and ≤ 2 SESs per day, where the ESs per day may exceed 170, one day per month. Later objectives are 100 ESs per day [73, 75].

Regenerative Repeater Jitter

A principal source of timing jitter is from imperfections in the timing recovery process in each repeater in repeated transmission lines (digital repeaters are also called *regenerative repeaters,* or *regenerators*). The timing recovery process introduces pattern-dependent jitter, which accumulates from repeater to repeater [78, 89, 95–103]. Figures 6.4 and 6.5 show a chain of digital repeaters and the functional makeup of a digital repeater, respectively. Figure 6.6 illustrates the nature of the most significant type of repeater jitter, which is discussed in more detail later. In general, excessive accumulation of repeater jitter can result in errors in downstream timing recovery circuits, as well as errors and interruptions due to buffer overflows and underflows in downstream digital terminal equipment [104–109]. Errors in timing

Figure 6.4 Digital transmission facility with chain of regenerators.

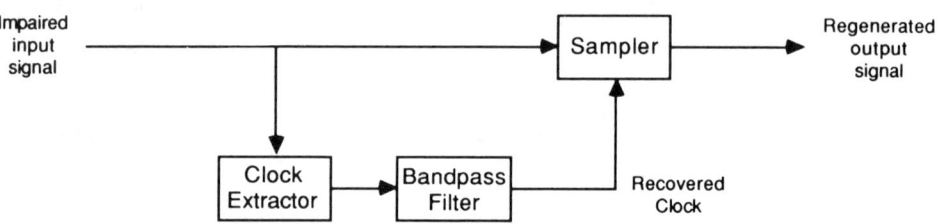

Figure 6.5 Simplified block diagram of a digital regenerator.

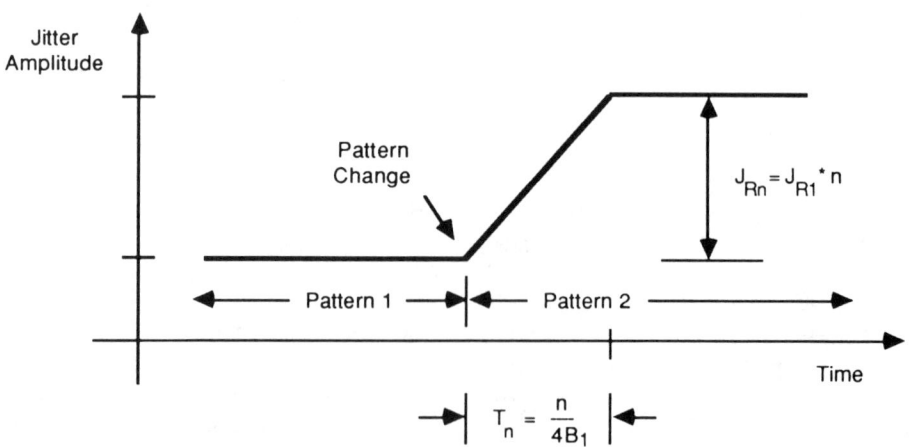

Figure 6.6 Illustration of jitter due to change in repetitive pattern, after n digital regenerators.

recovery circuits are due to misalignment of the data and clock signals, as shown in Figure 6.7. This misalignment is caused by jitter as well as other impairments, and has the effect of closing the digital data signal "eye" [110], not unlike the case of impaired voiceband data signals as discussed in Chapter 4 and earlier in this chapter.

Figure 6.7 Idealized illustration of received data signals and recovered clocks: (a) unimpaired; (b) impaired.

Multiplex/Demultiplex Jitter

Another principal source of timing jitter is from the stuffing synchronization process used in digital time division multiplexer/demultiplexer (mux/demux) terminal equipment [111, 112]. This process results in a low-frequency waiting time component of stuffing jitter which accumulates from terminal to terminal [67, 113–118]. Figures 6.8 to 6.11 show the terminal equipment, and Figure 6.12 illustrates the nature of stuffing jitter. This jitter applies largely to positive pulse stuffing in the asynchronous mux/demux equipment shown in Figure 6.8(a). With synchronous mux/demux equipment such as that shown for SONET in Figure 6.8(b), positive/negative byte stuffing is used to phase align signals that are already synchronous or nearly synchronous (plesiochronous) in frequency [119–122]. Byte stuffing with SONET also produces a form of low-frequency waiting time jitter. For both asynchronous and synchronous systems, excessive accumulation of waiting time jitter can result in the overflow or underflow of synchronizer and/or desynchronizer buffers in mux/demux equipment (Figs. 6.10 and 6.11). This in turn causes transmission errors, frame loss events, and slips in downstream digital signals.

6.4.1.2 Wander

Wander is unwanted phase modulation of digital signals at frequencies less than 10 Hz [123–128]. It is caused by slowly varying changes in digital signal phase due to clock frequency offset and drift, changes in propagation delay induced by temperature changes in terrestrial media, and by deviations of satellite distance from the earth. Wander is also contributed to by low-frequency jitter, such as waiting time jitter components at frequencies <10 Hz and quantization effects in the output of slaved clocks used in master-slave synchronization systems. Figure 6.13 shows the architecture of such synchronization systems.

Wander is not particularly relevant to the previously discussed conventional asynchronous transmission systems, which use a positive pulse stuffing synchronization scheme, since such systems can readily accommodate wander and low-frequency jitter by treating these components as small frequency offsets (Fig. 6.8). However, wander is relevant to synchronous transmission systems such as SONET and to DS1-based master-slave synchronized networks. In the case of SONET, phase changes due to wander (and jitter) must be accommodated by the positive/negative byte stuffing process and by storage in the equivalent of synchronizer and desynchronizer buffers.

For DS1 synchronization, the phase changes must be stored in DS1 slip buffers. Excessive wander (plus jitter) causes the slip buffers to either fill up or empty out, thereby causing controlled slips (deletion or repetition of DS1 frames, respectively). Wander also accumulates to some degree in the chain of timing links and clocks in the synchronization hierarchy (Fig. 6.13).

327

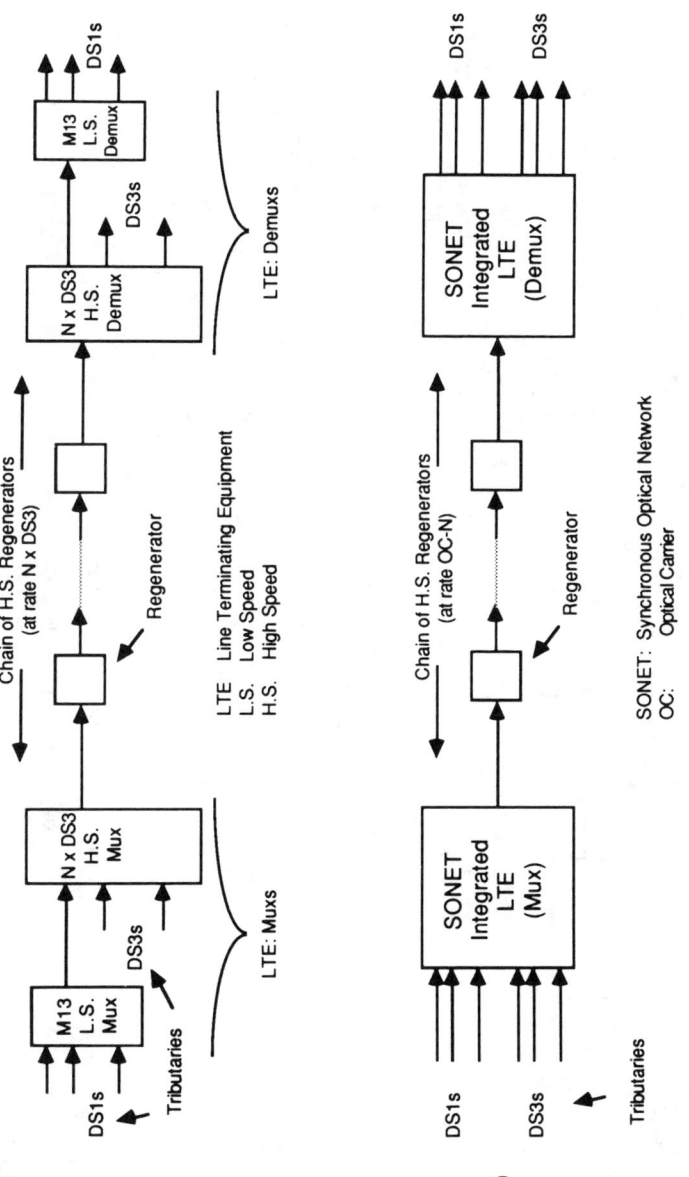

Figure 6.8 Examples of digital transmission facilities with typical line terminating equipment (one direction of transmission shown): (a) asynchronous; (b) synchronous.

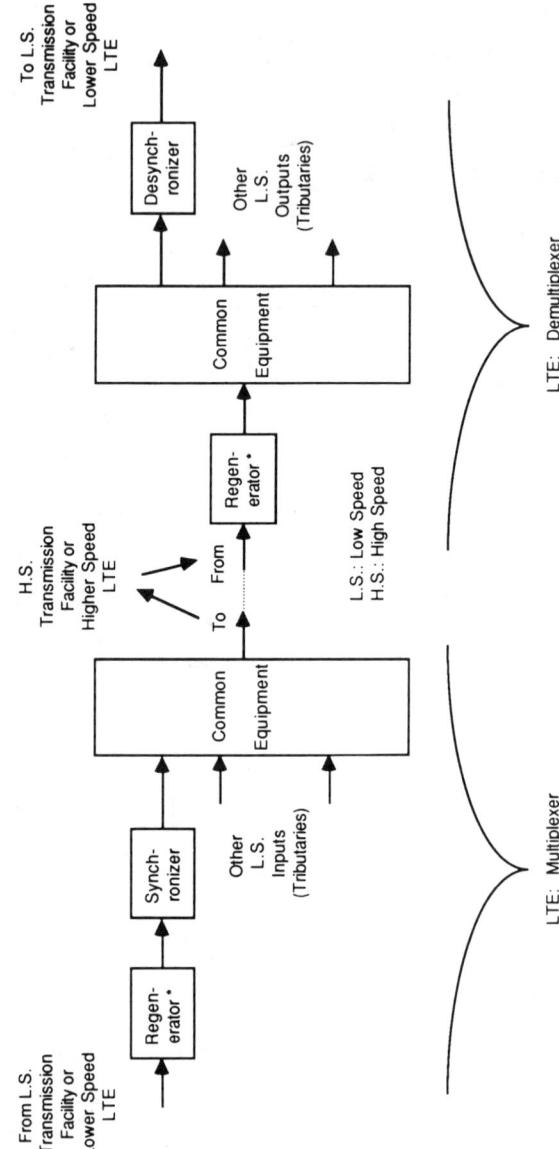

Figure 6.9 Simplified block diagram of mux/demux LTEs.

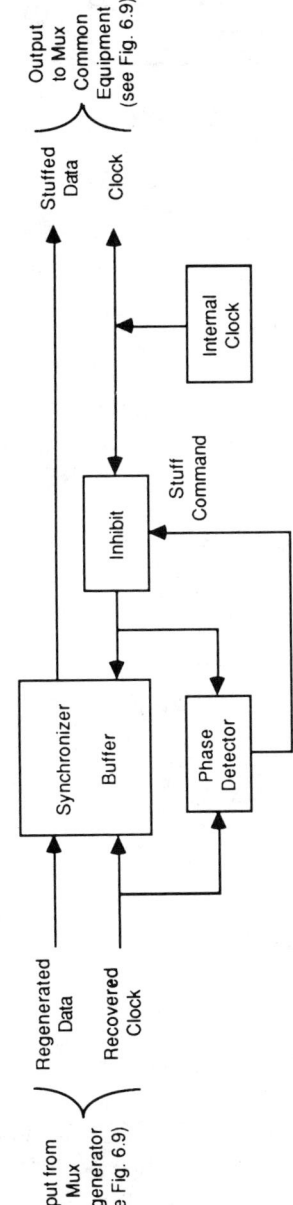

Figure 6.10 Simplified block diagram of mux synchronizer.

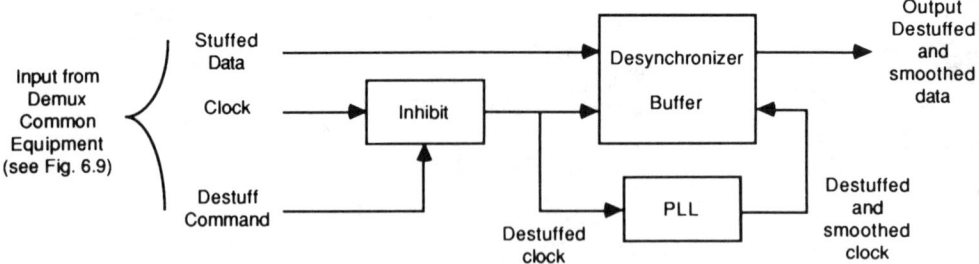

Figure 6.11 Simplified block diagram of demux desynchronizer.

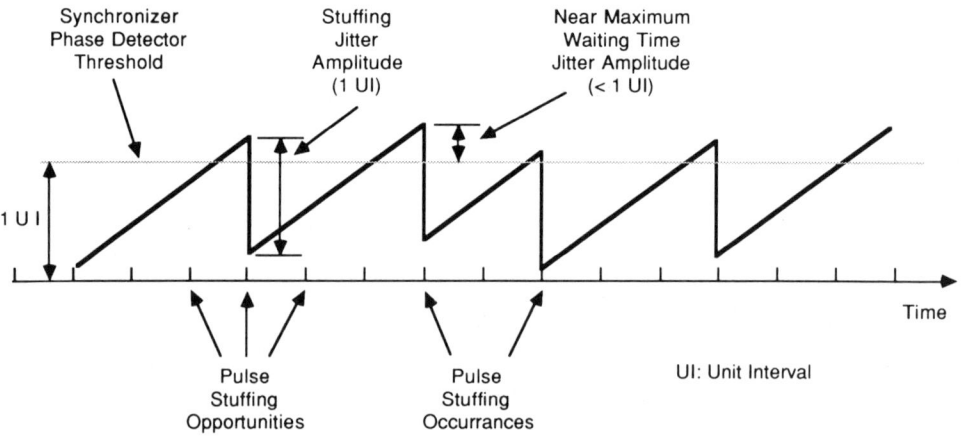

Figure 6.12 Illustration of positive pulse stuffing jitter.

6.4.2 Accumulation of Jitter and Wander

6.4.2.1 Regenerative Repeater Jitter

Thermal Sources

Repeater jitter can arise from thermal sources and from pattern-dependent sources. Thermal sources of jitter are independent from repeater to repeater, and after a chain of n identical repeaters will produce total rms jitter, J_{Tn}, of about

$$J_{Tn} = J_{T1} \times n^{1/4} \qquad (6.12)$$

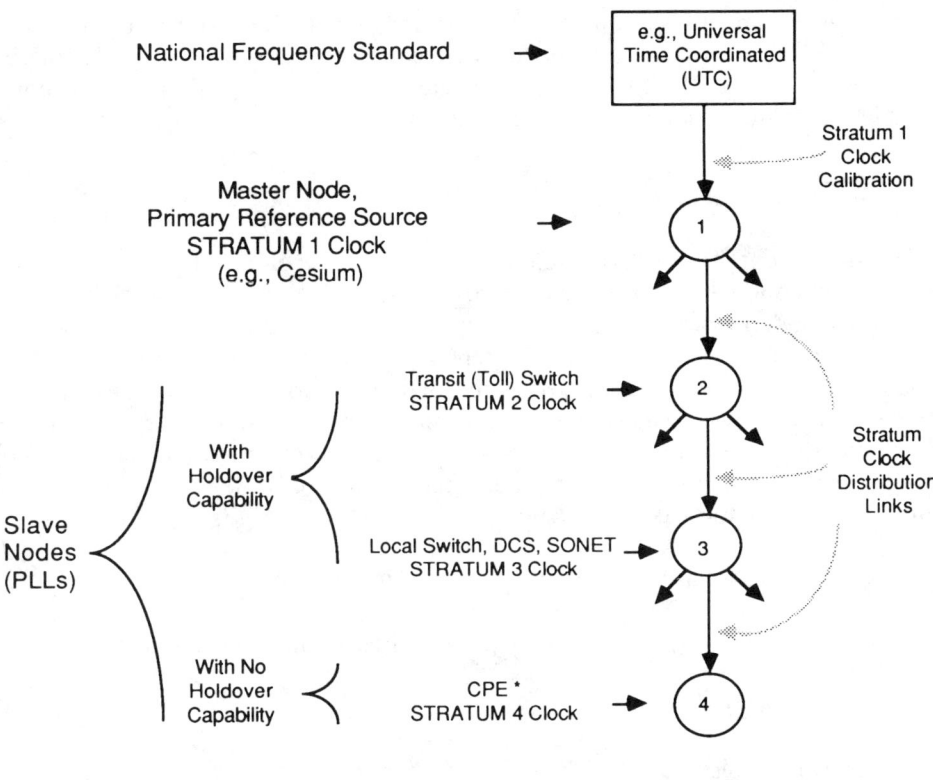

CPE: Customer Premise Equipment
  * May be Stratum 4 Enhanced Clock, or
    Non-duplicated Stratum 3 Clock

DCS: Digital Cross-connect System

SONET: Synchronous Optical Network (terminals)

PLL: Phase Locked Loop

Figure 6.13 National hierarchical master-slave synchronization architecture and timing distribution.

where J_{T1} is the rms jitter of one repeater (recall Fig. 6.4.). The fourth root relationship is the result of two factors: independence of the jitter sources and filtering by downstream repeaters [96, 97]. Thermally caused jitter is typically negligible compared to pattern-dependent jitter, discussed below.

Pattern-Dependent Sources

Pattern dependent jitter is correlated from repeater to repeater. The term "pattern" refers to the sequence of "ones" and "zeros" making up the digital signal. The jitter differs greatly, depending on how the pattern changes with time. With *repetitive patterns,* meaning a change from one fixed repetitive pattern to another fixed repetitive pattern, the resulting jitter increases approximately linearly starting at the instant of the pattern change, and persisting for a time related to the number of repeaters in the chain, and the time constant of the repeater clock recovery circuits (see Fig. 6.6) [102]. Thus, the repetitive pattern jitter after a chain of n identical repeaters is a "ramp" with peak-to-peak value J_{Rn} of

$$J_{Rn} = J_{R1} \times n \quad \text{and duration} \quad T_n = n/(4B_1) \tag{6.13}$$

where J_{R1} is the peak-to-peak repetitive pattern jitter of one repeater, and

$$B_1 = \pi f_o/(4Q) \tag{6.14}$$

is the one-sided noise bandwidth in Hertz of each repeater clock recovery filter (assumed to be a simple single tuned filter; see Fig. 6.5). In turn, f_o is the center frequency of the clock (and digital) signal in Hertz, and

$$Q = f_o/B_{3dB} \tag{6.15}$$

is a measure of the selectivity of the clock recovery circuit (B_{3dB} is the two-sided 3-dB bandwidth in Hertz of the clock recovery circuit). Thus, from Equation (6.13), both the amplitude and duration of repetitive pattern jitter are directly proportional to the number of repeaters. Correspondingly, from Equation (6.13), the slope of the ramp of repetitive pattern jitter is

$$\text{Slope} = J_{Rn}/T_n = 4J_{R1}B_1 \tag{6.16}$$

that is, the narrower the repeater clock recovery bandwidth, the lower the slope, and the longer it takes to reach the maximum jitter value. Knowledge of this slope is important in the design of downstream terminal equipment, since the slope effectively represents a frequency offset that must be accommodated by this equipment.

Example 1: Consider a T-carrier system with f_o = 1.544 Mb/s, Q = 100, and n = 50 repeaters. Then, from Equations (6.14) and (6.15), B_1 = 12.1 kHz, and B_{3dB} = 15.44 kHz, respectively. Consider also a change from a repetitive ⅛ pattern (one "one" followed by seven "zeros" = 10000000) to a repetitive all "ones" pattern (eight "ones" = 11111111), for which the jitter per repeater is evidently about J_{R1} = 1.8° peak to peak [102]. Then, for n = 50 repeaters, the total jitter from Equation (6.13) is J_{Rn} = 90° peak to peak, with duration T_n of about 1 ms. Therefore, from Equation (6.16), the slope is about 90°/ms (or 90 × 1000/360 = 250 Hz). Some pattern changes can cause much higher jitter. For example, a change from an alternating "one"-"zero" pattern (10101010) to an all "ones" pattern results in 11.2° of jitter per repeater [102]; on a 100 repeater system this corresponds to 1120° or over three unit intervals (UI) of jitter. (Note: one UI is one time slot which, for example, at DS1, corresponds to 1/1.544 MHz ≈ 648 ns; in relative terms, one UI is also equivalent to one clock cycle, 360°, or 2π radians.)

With *random patterns*, rms jitter accumulates as

$$J_{rn} = J_{r1} \sqrt{2p(n)} \qquad (6.17)$$

where J_{r1} is the rms jitter of one repeater, and $P(n)$ is a rather complicated function of the number of repeaters [95, 102]; however, $P(n)$ = ½ for n = 1, and asymptotically $P(n)$ approaches n for $n >$ 100. Thus, conservatively, it is assumed here that $p(n) = n$, and therefore

$$J_{rn} \sim J_{r1} \sqrt{2n} \qquad (6.18)$$

Equation (6.18) assumes zero or negligible positive jitter gain in the timing recovery filters; this is usually the case in practice for tuned circuit-type filters, and is also achievable with phase-locked loop (PLL)-type filters [129]. Note that Equation (6.18) represents a significantly reduced rate of random pattern jitter accumulation relative to the repetitive pattern jitter in Equation (6.12).

Owing to the cascading of timing recovery filters in a chain of n repeaters, the one-sided jitter bandwidth (B_n in Hertz) reduces with n as follows:

$$B_n = B_1 \times 2P(n)/n^2 \sim 2B_1/n \qquad (6.19)$$

Knowledge of the jitter bandwidth is important in determining jitter reduction by either jitter reducers or downstream demux terminal equipment (discussed later).

Example 2: Consider the T-carrier system discussed in Example 1, with n = 50 repeaters and B_1 = 12.1 kHz. Assume also jitter of J_{r1} = 1° rms per repeater [102]; then, from Equation (6.18), the total jitter is about $J_{rn} \approx$ 10° rms, and, from Equation (6.19), the one-sided noise bandwidth is $B_n \approx$ 485 Hz.

Rms jitter is only one aspect of the jitter process. It is also necessary to consider the peak value of the the jitter, which may affect the design of buffers in downstream terminal equipment. For this purpose it is generally assumed that random pattern jitter is normally distributed with zero mean and SD = J_{rn}; see Equation (6.18). The peak jitter can therefore be represented by some number (K) of SDs. The peak jitter of $K \times J_{rn}$ must be such that it does not cause buffers in downstream terminal equipment to over or underflow more often than a certain rate (typically very rarely; e.g., less than once per year). This topic is not addressed in detail here, but can be intuitively appreciated by considering a buffer overflow rate of

$$A \alpha p \times B_n \text{ (overflows/sec)} \tag{6.20}$$

where α signifies "is proportional to," p is the probability of overflow, and the jitter bandwidth B_n is an indication of the maximum rate that jitter could exceed a given threshold. Clearly, if the overflow rate A is required to be small (e.g., corresponds to, say, one overflow per year), then the overflow probability p must also be small; that is,

$$p \alpha A/B_n \tag{6.21}$$

Since the jitter is normally distributed with mean 0 and SD J_{rn}, the number of SDs (K) corresponding to the probability p can be determined from the normal distribution (as discussed in Sec. 5.4.3.3 of Ch. 5). Typically, K will be quite large (e.g., >10), and the corresponding buffer size required to achieve the desired overflow rate is $2K \times J_{rn}$ to accommodate both positive and negative peaks. Buffer size allocation for line jitter usually corresponds to a few UIs. However, in practice, the buffer size may be governed by repetitive pattern jitter—see Equation (6.13)—which is typically larger than random pattern jitter.

Effect of Scramblers

If scramblers are used in repeaters, pattern dependence from repeater to repeater would be removed and the resulting random jitter would accumulate at a lower rate according to the fourth root of the number of repeaters (similar to accumulation of jitter due to the independent thermal jitter sources as discussed earlier; see Equation (6.12)).

Effect of Dejitterizing

Phase jitter due to random patterns on repeatered lines can be filtered by downstream standalone dejitterizers [130] and/or by demux equipment. If the dejitterizer one-sided noise bandwidth is B_d Hz, then the dejitterizer output rms jitter is approximately

$$J_d = J_{rn} \sqrt{B_d/B_n} = n \times J_{r1} \sqrt{B_d/B_1} \quad \text{for } B_d < B_n \quad (6.22)$$
$$= J_{rn} \quad \text{for } B_d \geq B_n$$

Thus, jitter is reduced if the dejitterizer noise bandwidth is less than that of the jitter.

Example 3: Consider Example 2 above with $n = 50$, $J_{rn} = 10°$ rms, and $B_n = 485$ Hz. If the dejitterizer bandwidth $B_d = 100$ Hz, then from Equation (6.22), the dejitterizer output jitter is $J_d = 4.5°$ rms.

The situation is similar when filtering is done by desynchronizer PLLs in demux equipment (Fig. 6.12), except that in this case particular attention must be paid to the units used to describe the jitter parameters. If the units are in terms of absolute time (e.g., in nanoseconds), then Equation (6.22) is valid. However, if the units are in terms of relative time (i.e., UIs, degrees, or radians) then the difference in demux input and output rates must be accounted for. The reason is that a UI, being one clock cycle, is larger at the lower rates than at higher rates, and this in itself reduces the relative magnitude of the demux input jitter. Thus, for demux filtering in terms of relative units,

$$J_d = J_{rn}(R_o/R_i) \sqrt{B_d/B_n} \quad \text{for } B_d < B_n \quad (6.23)$$
$$= J_{rn} R_o/R_i \quad \text{for } B_d \geq B_n$$

where R_i and R_o are the demux input and output bit rates, respectively. For example, with an M13-type demux [111], the ratio $R_o/R_i = (1.544 \text{ Mb/s})/(44.736 \text{ Mb/s}) \sim 1/28$.

6.4.2.2 Multiplex/Demultiplex Waiting Time Jitter

The nature of jitter resulting from the positive pulse stuffing synchronization process is shown in Figure 6.12 and consists of two main components: a periodic high-frequency stuffing component and a random low-frequency waiting time component [113]. The amplitude of the stuffing jitter component is one UI peak to peak with single pulse stuffing (the usual case), and proportionally higher in systems that stuff more than one pulse at a time. The stuffing component occurs at a frequency equal to the difference between the low-speed mux input frequency and a higher speed internal mux clock frequency to which the input frequencies are all increased by the stuffing synchronization process. The stuffing frequency is typically high enough that stuffing jitter, although high in amplitude, can be readily attenuated to negligible levels by PLL circuits in the desynchronizer portion of demuxs (see Fig. 6.11).

The weighting time component of jitter occurs at much lower frequencies (as low as 0 Hz) and is random in frequency, owing to clock frequency drifts and phase offsets. Waiting time jitter results from the fact that stuff requests are random, whereas stuffing opportunities are periodic. Thus, stuffing must usually "wait" a random period of time until the next stuffing opportunity occurs, during which time the jitter increases beyond that of the stuffing component and gives rise to the waiting time component, as illustrated in Figure 6.12.

Waiting time jitter amplitude is related to the ratio of the nominal stuffing rate to the maximum stuffing rate made possible by all stuffing opportunities. Typical ratios for some muxs are ⅓ and ⅕, and these ratios are the maximum peak-to-peak waiting time jitter amplitude in UIs for these muxs [111, 131, 132]. (Note: since the waiting time jitter waveform is a "saw tooth," the rms jitter is determined by dividing the peak-to-peak value by $2\sqrt{3}$.)

Owing to the low frequency of waiting time jitter, it is typically not reduced substantially by PLL filtering in demuxs, and therefore accumulates from demux to demux. The accumulation is that of independent sources, since the waiting time jitter tends to be different in phase at each mux due to clock frequency and phase offsets. Thus, ignoring PLL filtering, the total rms waiting time jitter J_{wm} after m demuxs is:

$$J_{wm} = J_{w1} \sqrt{m} \qquad (6.24)$$

where J_{w1} is the rms waiting time jitter after one demux. If PLL filtering is substantial, then the total jitter would accumulate at a lower rate, corresponding to the fourth root of m. Thus, in general, the total rms waiting time jitter is bounded [96, 97] by

$$J_{w1} \times m^{1/4} \leq J_{wm} \leq J_{w1} \sqrt{m} \qquad (6.25)$$

but, to be conservative, the right-hand bound is assumed here. Again, it is assumed that there is negligible jitter gain in demux PLL closed loop response [129].

Finally, an estimate of the total peak jitter is required. Conservatively, this is less than m times the peak-peak jitter after one demux, that is,

$$J_{pm} < J_{p1} \times m \qquad (6.26)$$

where J_{p1} is the peak-to-peak waiting time jitter after one demux. A more accurate analysis of peak jitter would account for overflow rate, and would generally result in a somewhat lower estimate of peak jitter.

Example 4: Assume the waiting time jitter after one demux is $J_{p1} = ⅓$ UI peak to peak (equivalent to $1/(3 \times 2\sqrt{3}) \approx 0.1$ UI rms), then, after $m = 10$ demuxs, the total rms jitter from Equation (6.24) is about 0.3 UIs. The total peak-to-peak jitter from Equation (6.26) is less than 3⅛ UIs.

It should be noted that with SONET, which uses positive/negative byte stuffing, the stuffing frequency is ideally zero, but in practice is random between zero and a frequency slightly higher than zero to track jitter, wander, and small frequency offsets. Thus, SONET jitter may be viewed as a form of waiting time jitter which is relatively large in amplitude for two reasons: (1) byte (not bit) stuffing is used, and (2) the ratio of nominal to maximum stuffing frequency is zero, so that when a stuff does occur, it does so with maximum (1 byte) amplitude. The effects of this relatively large jitter amplitude are mitigated by desynchronizer PLLs which, in addition to linearly filtering the jitter somewhat, also further smooth the 1-byte phase transition by adjusting the PLL voltage-controlled oscillator in steps of 1 bit at a time, until the entire 8-bit (1 byte) phase difference is compensated for. Because of these factors (discussed in more detail in [133–135]), it is generally not straightforward to analytically determine how SONET jitter accumulates from system to system, and simulation techniques are typically relied on for this purpose [135, 136].

6.4.3 Jitter and Wander Performance

Since jitter and wander from various sources can accumulate (as discussed above), limits are required for the maximum jitter and wander that can appear at network interfaces and at the input and output of network equipment. In addition, limits are required for the jitter and wander generated by equipment and the jitter gain imposed by equipment. A number of these requirements are discussed in the subsequent sections.

6.4.3.1 Output Jitter and Wander at Network Interfaces

Jitter at network interfaces can be illustrated in terms of the frequency bands in Figure 6.14, and is specified in Table 6.36 for the common DS1 (1.544 Mb/s) and DS3 (44.736 Mb/s) hierarchical rates [137–139]. Also included are output jitter specifications for SONET carrier systems operating at various optical carrier (OC) rates of OC-N($N \times$ 51.84 Mb/s) [140–142]. The jitter is measured over an entire band of jitter frequencies and over a high-frequency band. This specification of jitter is useful in terms of the jitter requirements that should be met at interfaces between different digital networks. Under these conditions, the output jitter from one network should produce essentially no errors in other networks it interfaces with and vice-versa.

Output wander at network interfaces is specified in a different manner than in Figure 6.14 and Table 6.36, as discussed later under clock and synchronous node timing requirements in Section 6.4.3.5.

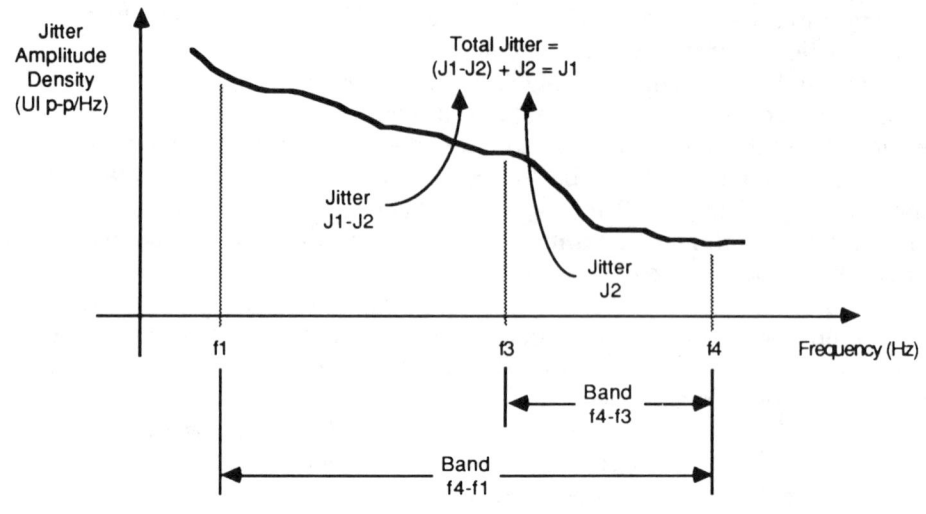

Figure 6.14 Illustration of output jitter limits at network interfaces.

6.4.3.2 Input Jitter and Wander Tolerances for Digital Equipment

Input jitter and wander is illustrated in Figure 6.15 in terms of a generalized mask, and is specified in Table 6.37 [111, 137, 138, 143, 144]. Input jitter tolerances (for frequencies above 10 Hz) apply to asynchronous equipment such as pulse stuffing mux/demuxs, and to synchronous equipment such as digital switches and certain DCSs (see Ch. 9 for a discussion of DCSs). Input jitter tolerances also apply to SONET mux/demuxs, as shown in Table 6.37; these specifications are consistent with [141, 144]. Input wander tolerances (for frequencies below 10 Hz) apply to synchronous equipment only. In general, input jitter and wander tolerances are also an indication of the worst-case output jitter and wander expected from networks. In fact, the input tolerances, being in terms of sinusoids rather than across a band of frequencies, can accommodate the network output jitter specified in Table 6.36. Equipment that can tolerate the input jitter and wander specified in Table 6.37 can be expected to operate essentially error free as far as jitter and wander are concerned.

Table 6.36
Maximum Output Jitter at Hierarchical Network Interfaces

Interface and Bit Rate (Mb/s)	Jitter (UI * p-p)		Measurement Filter Cutoff Frequencies		
	J1 in Band f4-f1	J2 in Band f4-f3	Lower		Upper
			f1 (Hz)	f3 @ (kHz)	f4 (kHz)
DS1: 1.544	5	0.1 #	10	8	40
DS3: 44.736	5	0.1	10	30	400
SONET OC-3: 155.52 !	1.5	0.15	0.5	65	1300
SONET OC-12: 622.08 !	1.5	0.15	1	250	5000
SONET OC-48: 2488.32 !	1.5	0.15	5	1000	20000

Note: see Fig. 6.14, which relates to this table.
* 1 IU = 648 ns at DS1
      22.3 ns at DS3
      19.3/N ns at SONET OC-N
\# Provisional. Equipment with autonomous (e.g., stratum 4) clocks may have more stringent requirements (e.g., 0.01 UI rms as per CCITT Rec. G.743 [131]).
@ f3 may be the one-sided bandwidth of the timing recovery filter; i.e., f3 = f0/(2Q), where f0 is the filter center frequency (Hz) corresponding to the bit rate, and Q is the selectivity of the filter (typically ranging from 50 to 100, for DS1 and DS3).
! Proposed for SONET carrier systems [140, 141].

Figure 6.15 General illustration of maximum sinusoidal input jitter and wander tolerances for digital equipment.

Table 6.37
Minimum Input Jitter and Wander Tolerances for Digital Equipment

Interface and Bit Rate (Mb/s)	Equipment Type	Jitter Amplitude (UI p-p)			Frequency				
		A0	A1	A2	Hz			kHz	
					f0	f1	f2	f3	f4
DS1: 1.544 !	Timing Recovery*	-	10	0.3	-	10	#	6.43	40
	Mux./Demux.*	-	5	0.1	-	10	0.5	8	40
	Sync. and Async.	28	5	0.1	.000012	10	0.12	6	40
DS3: 44.736	Timing Recovery*	-	20	0.3	-	10	#	22.3	300
	Mux./Demux.*	-	5	0.1	-	10	2.3	60	300-400
	Sync. and Async.	805	5	0.1	.000012	10	0.6	30	400
N x DS3: ~ N x 44.736	Timing Recovery*	-	20	0.3	-	10	#	N x 22.3	TBD
SONET OC-3: 155.52	Timing Recovery	15	1.5	0.15	30 &	300	6.5	65	#
	Mux./Demux.	-	1.5	0.15	-	#	6.5	65	#
SONET OC-12: 622.08	Timing Recovery	15	1.5	0.15	30 &	300	25	250	#
	Mux./Demux.	-	1.5	0.15	-	#	35	250	#
SONET OC-48: 2488.32	Timing Recovery	15	1.5	0.15	600 &	6000	100	1000	#
	Mux./Demux.	-	1.5	0.15	-	#	100	1000	#

Note: see Fig. 6.15, which relates to this table.
! Some of these tolerance have recently been made more stringent (see [73]).
* Asynchronous Equipment
- Not applicable
& Slope is 0 dB/octave from 10 Hz to these frequencies.
\# Not specified
TBD: To Be Determined

6.4.3.3 Jitter and Wander Transfer Characteristics of Digital Equipment

A generalized jitter transfer characteristic is illustrated in Figure 6.16, and the requirements are specified in Table 6.38 [137, 141, 143, 144]. The transfer characteristic represents the attenuation that timing recovery filters and desynchronizer PLLs apply to jitter signals. In some cases, the characteristic is simpler than that shown in Figure 6.16, and consists of a single -20-dB/decade slope starting at either frequency f_2 (for DS3 demuxs) of f_3 (for timing recovery circuits), as indicated in Table 6.38. Of particular importance is the gain between frequencies f_1 and f_2, which must be kept small (preferably 0 dB) to avoid jitter enhancement when equipment is tandemed.

Transfer characteristics for wander are not included in Table 6.38, but rather are implicit in the synchronous node timing requirements discussed later in this chapter. Typical implemented wander transfer characteristics provide small gain at low frequencies (on the order of <0.2 dB), a very low cutoff frequency (on the order of 10^{-4} to 10^{-3} Hz), and a slope of at least -20 dB/decade [145, 146]. The above noted cutoffs readily satisfy those specified at <0.1 Hz in [50, 147].

6.4.3.4 Jitter and Wander Generated by Digital Equipment

Equipment output jitter with no input jitter is provided in Table 6.39, and differs depending on the type of equipment, as shown [74, 131, 137, 141, 144, 147]. The largest values of jitter are caused by waiting time-like components appearing at the output of demux equipment. In some cases the output jitter of synchronous nodal equipment is as yet unspecified, owing to quantization effects, oscillator noise, and other factors requiring further investigation (see Table 6.40).

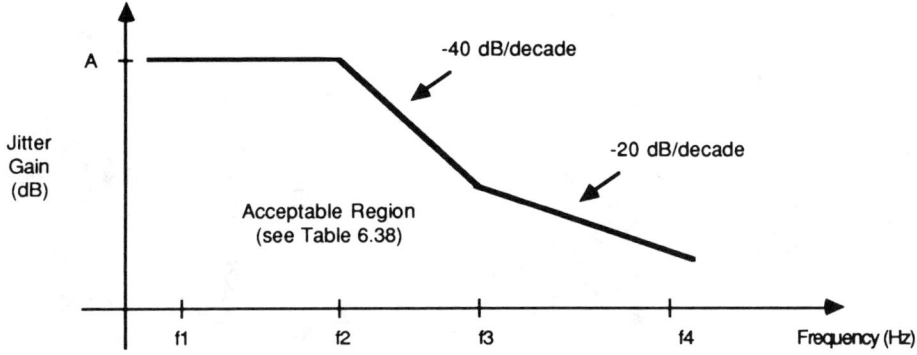

Figure 6.16 General illustration of jitter transfer characteristic.

Table 6.38
Maximum Jitter Gain for Jitter Transfer Characteristic

Bit Rate (Mb/s)	Equipment Type	Jitter Gain A (dB)	Frequency			
			Hz		kHz	
			f1	f2	f3	f4
DS1: 1.544 DS3 to DS1	Timing Recovery Demux.	0.1 0.1	10 10	# .175 - .35	9.65 2.5	- 15
DS3: 44.736 N x DS3 to DS3	Timing Recovery Demux.	0.1 0.1	10 10	# 1	5.96 •	- 15
N x DS3: ~ N x 44.736	Timing Recovery	0.1	10	#	N x 28	-
SONET OC-3: 155.52	Timing Recovery	0.1	@	#	130	-
SONET OC-12: 622.08	Timing Recovery	0.1	@	#	500	-
SONET OC-48: 2488.32	Timing Recovery	0.1	@	#	2000	-

Note: see Fig. 6.16, which relates to this table.
  • Slope between f2 and f4 is -20 dB/decade
  # Slope between f1 and f3 is 0 dB/decade
  - Not specified
  @ Not specified, but probably 10 Hz

Table 6.39
Equipment Output Jitter with No Input Jitter

Equipment		Jitter	
Type	Example	Cause	Amplitude
Asynchronous	Mux. with autonomous clock	Oscillator noise	≤ 0.05 ns rms @ [131, 132, 143]
	Demux.	Waiting time	≤ 0.3 UI rms * ≤ 1.0 UI p-p [74, 137]
Synchronous	Switch or DCS with slave clock	Oscillator noise, digital PLL quantizing effects	TBD [147]
	SONET demux.	Pointer adjustments	≤ 1.5 UI p-p [144] #

*   Also applies to SONET demux. equipment for the condition of no pointer adjustments [144].
\#   Includes the effect of pointer adjustments.
@   ≤ 0.01 UI rms for SONET regenerators, based on [141].

Wander generation, which relates largely to synchronous equipment, is discussed in the next section.

6.4.3.5 Clock and Synchronous Node Timing Requirements

The hierarchical master-slave network synchronization architecture was introduced in Figure 6.13. The following is a brief description of the architecture in relation to the slip rates in Table 6.40 and the timing requirements outlined in Figure 6.17 and Table 6.41. The main components of the architecture are the stratum 1, 2, 3, and 4 clocks, which have decreasing accuracy with increasing stratum numbers, as shown in the tables [147–152]. The stratum 1 clock is calibrated with respect to a national reference, such as Universal Time Coordinated (UTC), Long-Range Navigation (LORAN), or, eventually, the Global Positioning System (GPS) [153–155]. Therefore, stratum 1 clocks have a long-term relative frequency accuracy or offset of $\pm 1 \times 10^{-11}$ [156, 157]. (Absolute frequency offset in Hertz is determined by multiplying the relative offset by the nominal frequency in question, usually the DS1 rate of 1.544 MHz.) National references such as UTC are typically about two orders of magnitude more accurate than a stratum 1, or about $\pm 1 \times 10^{-13}$. The stratum 2, 3, and 4 clocks are in turn slaved (phase locked) to a stratum 1 through timing distribution links, as shown in Figure 6.13 [148, 149, 158–163].

Table 6.40
Range of Clock-Relative Frequency Offsets and Corresponding Slip Rates at the DS1 Rate

Clock/Offset Description	Offset	Time to Slip #	Slip Rate #
AT&T Primary Node: per week	1×10^{-13}	40 years !	2.5 per 100 years !
AT&T Secondary Node: per week	3×10^{-13}	13 years !	7.6 per 100 years !
AT&T Primary Node: per day	7×10^{-13}	5.7 years !	1.8 per 10 years !
AT&T Secondary Node: per day	8×10^{-13}	5.0 years !	2.0 per 10 years !
Stratum 1: long term (free run)	1×10^{-11} *	4.8 months	2.5 per year
Two Pleisiochronous Stratum 1s	2×10^{-11}	2.4 months (~ 70 days) *	5.0 per year
Stratum 2: holdover per first day	1×10^{-10} *	2.1 weeks	2.1 per month
Stratum 2: initial **	5×10^{-10}	2.9 days	1.5 per week
Stratum 2: drift per day **	1×10^{-9}	1.5 days	4.8 per week
NA allocated slip objective	3.5×10^{-9} *	10 hours *	2.4 per day
NA slip maintenance limit	5.8×10^{-9}	6 hours	4 per day *
NA end-to-end slip objective	6.9×10^{-9}	5 hours *	4.8 per day
CCITT degraded slip limit	7.2×10^{-9}	4.8 hours	5 per day *
Stratum 3: initial **	1×10^{-8}	3.5 hours	9.8 per day
Stratum 2: long term (free run)	1.6×10^{-8} *	2.2 hours	11 per day (~ 10 per day)
Stratum 3: drift per day **	2×10^{-8}	1.7 hours	14 per day
Stratum 3: holdover per first day @	3.7×10^{-7} *	5.7 minutes	11 per hour (~ 255 per day) *
CCITT unacceptable slip limit	1×10^{-6}	2 minutes	30 per hour *
Stratum 3: long term (free run)	4.6×10^{-6} *	27 seconds	2.2 per minute
20 Parts Per Million (PPM)	2×10^{-5}	6.3 seconds	9.6 per minute
Stratum 4: long term (free run)	3.2×10^{-5} *	3.9 seconds	15 per minute

! These values are shown for completeness only; the weekly and daily offsets would have to be sustained for very long periods for these values to have practical significance.

\# Note, time to slip and slip rates are means based on the frequency offsets in the second column from the left, using eqn. (6.27). Under normal conditions (no clock or timing link failures), slips are very rare.

* Commonly specified values.

@ North American (NA) slip out-of-service limit.

** CCITT Rec. G.812 values [158].

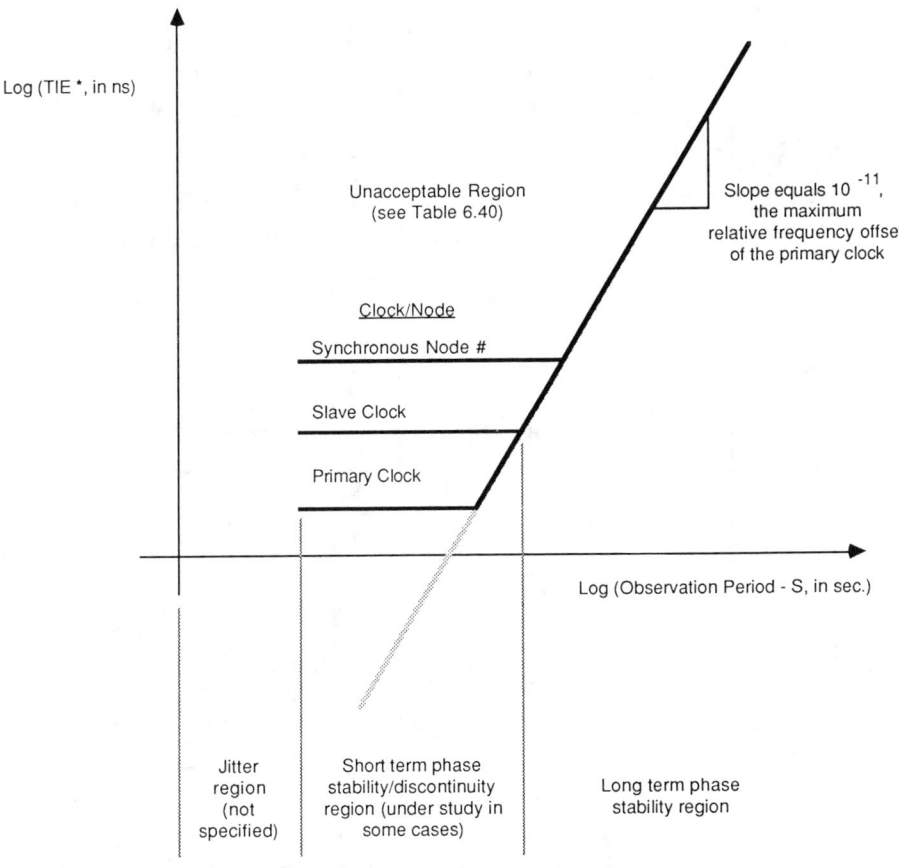

Figure 6.17 Simplified illustration of clock and synchronous node output timing requirements.

Recently, there have been attempts to improve the accuracy of stratum 1 (primary) and stratum 2 (secondary) clocks to bring them more in line with the accuracy of national references, and thereby better ensure that required levels of synchronization performance are achieved throughout the clock hierarchy [164–166]. These more stringent accuracies are included in Table 6.40. There is also a perceived need

Table 6.41
Clock and Synchronous Node Output Timing Requirements (DS1 Rate)

Parameter	Clock				Synchronous Node With input jitter and wander?	
	Primary (G.811, T1.101) Stratum 1	Slave (G.812, T1.101)		Slave $$ Stratum 4 e.g., CPE	No (Q.541)	Yes (G.824, T1.101) (Network Interface)
		Stratum 2 e.g., Transit Switch	Stratum 3 e.g., Local Switch			
Frequency Offset						
- Long term	1×10^{-11}	1.6×10^{-8}	4.6×10^{-6}	3.2×10^{-5}	N/A	
- Initial $	-	5×10^{-10}	1×10^{-8}	-	N/A	
- Drift (per day) $	-	1×10^{-9}	2×10^{-8}	-	N/A	
- Holdover (per first day)	N/A	1×10^{-10}	3.7×10^{-7}	N/A	-	
Phase Discontinuity (ns)						
- Normal @	81	81 per ≤ 1.326 ms, to max. of 1000		N/A	81	
- Under test/rearrangement	-	-			81 per ≤ 1.326 ms, to max. of 1000	
Phase Variation (ns)						
- Short term (jitter)	TBD	TBD	TBD ** : .05 < S < 100	TBD	See Table 6.39	See Table 6.36
- Long term (wander)			$1000 ** : S \geq 100$			
* Ideal !	100S: .05 < S ≤ 5 5S + 500: 5 < S ≤ 500 .01S + 3000 #: S > 500	Transit: .5S + 5.8 x 10^{-6}S^2 + 1000: S ≥ 100 Local: 10S + 1.15 x 10^{-4}S^2 + 1000: S ≥ 100		TBD	100S + 81: S < 10 1000: S ≥ 10	TBD: .05 < S < 10000 & .01S + 10000: S > 10000 &
* Holdover	N/A			N/A		

Note: see Fig. 6.17, which relates to this table.   S Measurement Interval in seconds.   TBD To Be Determined.   N/A Not Applicable.   - Not Specified.
3000 is provisional; revisions to T1.101 are: 29S + 10 ns for .05 < S ≤ 1000, and .01S + 300 ns for S > 1000 [157].
** MRTIE (Maximum Relative Time Interval Error); 10 ns for 0.1 ≤ S ≤ 1000, and TBD for S > 1000 for SDH [172].
! 3S ns rms for 1 < S ≤ 100, 300 ns rms for S > 100 for SONET timing links; 20 ns for .1 < S < 1, 20S .477 ns for 1 < S < 10, and 60 ns for S > 10 for SONET slave clocks [144];
  other proposals are under consideration [169]. $ CCITT values [158]. $$ PN-2198 [160].
@ 300 ns rms for 100 < S ≤ 1000 for SONET slave clocks [144].
& Revisions to T1.101 are: 2.5S + 300 ns for .05 < S ≤ 280, and .01S + 1000 ns for S > 280,
  and for SONET: 3.32S + 70 ns for .05 < S ≤ 280, and .01S + 999 ns for S > 280 [157].

to improve stratum 4 clock stability, and this can be done either by way of an enhanced stratum 4 clock, or by the use of a nonredundant stratum 3 clock in customer premise equipment (CPE) [160]. Notice also in Figure 6.13 that for the most part SONET equipment (when externally synchronized) is expected to have stratum 3 (or better) clocks [120]. This is to ensure that, if the external synchronization signal is lost (in which case the stratum 3 clock provides timing), SONET output jitter specifications can still be met. An exception is for SONET regenerators, which may use $\pm 20 \times 10^{-6}$ (i.e., ± 20 parts per million (PPM)) clocks, since, in the event of loss of the regenerator input signal or regenerator failure, it is only necessary to produce a regenerator output signal stable enough to carry maintenance signals. Also, for near-term point-to-point applications (analogous to asynchronous system operation), external synchronization is not required, and 20 PPM SONET clocks are sufficient.

Table 6.40 lists, for reference purposes, the range of clock stabilities, ranging from most to least stringent (including stratums 1, 2, 3, and 4), and also indicates the slip rates associated with these stabilities. However, it should be noted that, under normal conditions of no synchronization related failures, networks should operate essentially slip free [167].

Included in Table 6.40 are the North American *end-to-end slip rate objectives* of 1 slip in 5 hours, with 1 slip in 10 hours allocated to jitter, wander, and delay variation in transmission facilities (including the timing distribution links, as shown in Fig. 6.13), and to certain synchronization network failure conditions (principally, failure of incoming timing links to stratum 3 clocks in local switching nodes) [147]. Also included in Table 6.40 are the North American *slip rate maintenance and out-of-service limits* of 4 and 255 slips per day, respectively [147].

The following is a discussion of the various modes of operation of slave clocks, including those related to synchronization-related failures. Also discussed are the clock and synchronous node timing requirements, as summarized in Table 6.40. In this regard, Figure 6.17 is a generalized illustration of how timing requirements are specified in terms of time interval error (TIE) versus an observation period. TIE is the difference in time between a given timing signal and an ideal reference signal, as observed over the observation period.

Ideal Operation

Ideal operation means no interruptions of the input timing reference to a slave clock (e.g., due to switching from a failed timing link to a redundant timing link), no switching of clock modes (e.g., to and from the "holdover" mode; see below), and no switching from a failed clock to a redundant clock. Under these conditions, all clocks run with a long-term average frequency accuracy essentially equivalent to a stratum 1 ($\pm 1 \times 10^{-11}$). This corresponds to slip-free operation within a given network where usually all nodes are synchronized (slaved) to a single stratum 1. For two networks, each with its own stratum 1, the maximum frequency difference between the two networks is $\pm 2 \times 10^{-11}$. This is the case for national and international networks, and also within nations where, for example, some local-exchange and interexchange carriers may each have their own stratum 1 clock. Such situations with more than one stratum 1 is known as plesiochronous operation, and corresponds to at most 1 slip in about 70 days [156].

In general, slip rate (SR) is determined by

$$SR = R \times \Delta/B \qquad (6.27)$$

where: B is the size of the slip buffer in clock cycles (or bits) allocated to frequency offset,
R is the clock rate in Hertz, and
Δ is the relative frequency offset between the slip buffer input write clock and output read clock [168].

Thus, for the above example of plesiochronous operation, let $B = 193$ bits (the DS1 frame size), $R = 1.544$ MHz, and $\Delta = 2 \times 10^{-11}$; then, from Equation (6.27), SR ≈ 1 slip/70 days.

Under ideal operation, the long-term phase variation at the output of clocks is of the form shown in Figure 6.17, and specified in Table 6.41 in terms of maximum time interval error (MTIE). MTIE is the maximum difference in time between a given timing signal and an ideal reference signal as observed over a particular time period. The long-term phase variation includes a component due to frequency offset for which the limit increases (ramps up) with observation time, and a random component for which the limit is more or less fixed with time (as illustrated in Fig. 6.17).

Stressed Operation

Stressed operation means operation of slave clocks under realistic network conditions; that is, interruptions or failure of incoming timing links and failures of upstream clocks or the local clock. To counter failures, redundant links and clocks are provided to supply backup timing. For example, failure of an incoming timing link to a stratum 3 clock should cause an automatic switch to the redundant timing link. Switching from one timing link to another or from one clock to another causes a phase discontinuity, which under normal (nontest) conditions is required to be less than ⅛ UI, or less than about 81 ns. Under test conditions, discontinuities are allowed to accumulate in time at the rate of 81 ns per 1.3 ms, to a maximum of 1 ms, as shown in Table 6.41 [156–158].

Holdover Operation

Under a failure condition where, for example, the incoming timing link to a stratum 2 clock fails, the stratum 2 clock will enter the holdover mode. Also, under more severe failure conditions where both timing links to a stratum 3 clock fail, the stratum 3 clock will enter the holdover mode. A clock in the holdover mode coasts from the last known phase position before the loss of timing. In this mode, the long-term phase variation is as given in Table 6.41 and depends on the initial frequency offset and the drift in frequency per day (also included in the table).

Of most importance is the holdover offset, which occurs in the first day following the failure, since this will determine the slip rate encountered during this period [147]. The slip rate for a stratum 2 clock in holdover is about 2 slips/month, which allows considerable operational flexibility in the repair of a failed timing link. For this reason, a redundant incoming timing link to a stratum 2 clock is not strictly required for performance reasons, although it is for other (operational) reasons. However, the slip rate for a stratum 3 clock in holdover is about 255 slips/day, which is an out-of-service condition. For this reason a redundant incoming timing link to a stratum 3 clock is required, which serves to minimize such out-of-service situations.

Eventually, if not repaired, a slave clock would settle to a frequency offset corresponding to its long-term (free-run) offset specification (Table 6.41). The slip rates corresponding to these offsets are shown in Table 6.40. In particular, the long-term free-run slip rates for stratum 2 and 3 clocks are about 10 slips/day and 2 slips/min, respectively.

Stratum 4 clocks generally do not have a holdover mode, but simply free run under complete failure of incoming timing. In this case, slip rates are much higher at 15/min owing to the lower long-term stability of stratum 4 clocks. Generally, output phase variation for stratum 4 clocks remains to be determined [159, 160].

Also, in general, the output short term phase variation (jitter) of slave clocks is currently under study (e.g., see [157]). This jitter will likely exceed the output jitter of autonomous clocks (see Table 6.39) because of quantizing effects, etc. Because of the direct effect of short-term phase variations on the operation of SONET equipment, requirements for these variations have been specified for SONET, as noted in Table 6.41 [134, 140, 144], but are under revision (e.g., see [169]).

Finally, Table 6.41 includes the long-term phase variation at the output of synchronous nodes (e.g., switches), both with and without input jitter and wander [138, 157, 170]. Some of these values have been recently revised (e.g., based on [171]).

6.5 REFERENCES

1. Standard on "Telephone Instruments with Loop Signaling," EIA-470-A, July 1987.
2. A.H. Inglis, and W.L. Toffnell, "An Improved Telephone Set," *BSTJ*, Vol. 29, No. 4, April 1951, pp. 239–270.
3. Bell System Practice on "500 Series Combined Telephone Sets: Transmission Performance; C and D Series," Section AB22.066-3, Issue 2, Oct. 1956.
4. ANSI/IEEE Standard 820-1984, "IEEE Standard Telephone Loop Performance Characteristics," Dec. 15, 1984.

5. J.R. Palladino, and W.B. Wilkens, "Standard on Voice Frequency Telephone Loop Performance Characteristics," *GLOBECOM Rec.*, Vol. 2, San Diego, Nov./Dec. 1983.
6. "Method for Determining Objective Loudness Ratings of Telephone Connections," IEEE Standard 661-1979.
7. Draft of American National Standard for Telecommunications—Network Performance—Loss Plan for Evolving Digital Networks, T1Q1/91-026, July 1991.
8. "Acoustic-to-Digital and Digital-to-Acoustic Transmission Requirements for ISDN Terminals," EIA/TIA-579, 1990.
9. A. Niedzwiecki, R.T. Matsunaga, and P.V. Coverdale, "All-Digital Network Transmission Planning and Terminal Requirements," *IEEE Journal on Sel. Areas in Comm.*, Vol. 7, No. 8, Oct. 1989, pp. 1166-1171.
10. *Telecommunications Transmission Engineering*, Vol. 3: Networks and Services, 2nd ed., Section 2, AT&T, 1977. Also, 3rd ed., Bellcore, 1990.
11. "Notes on the BOC INTRA-LATA Networks," Bellcore TR-NPL-000275, Issue 1, Section 7 on "Transmission," April 1986.
12. L.M. Manhire, "Physical and Transmission Characteristics of the Customer Loop Plan," *BSTJ*, Vol. 57, No. 1, Jan. 1978, pp. 35-59.
13. D.V. Batorsky, and M.E. Burke, "1980 Bell System Noise Survey of the Loop Plant," *BLTJ*, Vol. 63, No. 5, May/June 1984, pp. 775-818.
14. "Methodologies for Specifying Voice Grade Channel Transmission Parameters and Evaluating Transmission Performance for Speech Telephony," ANSI/IEEE Standard 823, 1989.
15. LATA Switching Systems Generic Requirements (LSSGR): Section 7 on "Transmission," Bellcore TR-507, Issue 3, Mar. 1989.
16. J.F. Ingle, K.I. Park, and T.C. Redman, "1983 Exchange Access Study: Analog Voice-Frequency Transmission Performance Characterization of the Exchange Access Plant," *IEEE Trans. on Comm.*, Vol. COM-35, No. 1, Jan. 1987, pp. 104-112. Also, Bellcore TR-037, Issue 1, Nov. 1984.
17. *Telecommunications Transmission Engineering*, Vol. 3: Networks and Services, 2nd ed., Chapters 2, 6, and 8, AT&T, 1977. Also, 3rd ed., Bellcore, 1990.
18. Notes on the Network, AT&T, Section 7 on "Transmission Considerations," 1980.
19. Estimated Transmission Performance of Switched Access Service Feature Group-D, Bellcore TR-002, Issue 1, July 1984.
20. *Telecommunications Transmission Engineering*, Vol. 3: Networks and Services, 2nd ed., Chapters 2 & 9, AT&T, 1977. Also, 3rd ed., Bellcore, 1990.
21. American National Standard for Telecommunications—Network Performance Standards—Switched Exchange Access Network Transmission Performance Standard, ANSI T1.506-1989. Also, Supplement to Transmission Specifications for Switched Exchange Access Network, Draft ANSI T1.506a.
22. I. Nasell, C.R. Ellison, Jr., and R. Holmstrom, "The Transmission Performance of Bell System Intertoll Trunks," *BSTJ*, Vol. 47, No. 8, Oct. 1968, pp. 1561-1613.
23. M.B. Carey, H.T. Cheu, A. Descloux, J.F. Ingle, and K.I. Park, "1982/83 End Office Connection Study; Analog Voice and Voiceband Data Transmission Performance Characterization of the Public Switched Network," *BLTJ*, Vol. 63, No. 9, pp. 2059-2119, Nov. 1984.
24. "Voice Grade Switched Access Service: Transmission Parameter Limits and Interface Combinations," Bellcore TR-334, Issue 1, June 1986, Rev. 1, Nov. 1987, Rev. 2, Feb. 1990.
25. "Voice Grade Special Access Service: Transmission Parameter Limits and Interface Combination," Bellcore TR-335, Issue 1, June 1986, Rev. 1, Feb. 1987, Rev. 2, Nov. 1987. TA-335, Issue 1, Dec. 1989.
26. "Engineering and Operations in the Bell System," Bell Telephone Labs. Inc., Chapter 6 on "Transmission," 1977.
27. F.T. Andrews, Jr., and R.W. Hatch, "National Telephone Network Transmission Planning in the

American Telephone and Telegraph Company," *IEEE Trans. Comm.,* Vol. COM-19, No. 3, June 1971, pp. 302–314.
28. W. Hand, "How to Find Phone-Line Faults and What to Do About Them," *Data Comm.,* Vol. 17, No. 10, Sept. 1988, pp. 179–182, 185, 186, 189–192, 195–197.
29. "Data Communications Using Voiceband Private Line Channels," AT&T PUB. 41004, Oct. 1973.
30. "AT&T Analog Voice Total and Coordinated Services," Preliminary PUB 43202, May 1985.
31. "Notes on the BOC INTRA-LATA Networks," Bellcore TR-NPL-000275, Issue 1, Section 5 on "Signalling," April 1986.
32. Notes on the Network, AT&T, Section 5 on "Signalling," 1980.
33. W.C. Ahern, F.P. Duffy, and J.A. Maher, "Speech Signal Power in the Switched Message Network," *BSTJ,* Vol. 57, No. 7, Sept. 1978, pp. 2695–2726.
34. CCITT Rec. G.113, "Transmission Impairments," *Blue Book,* Vol. III, Fas. III.1, Geneva, 1989, pp. 63–84.
35. "Transmission Systems For Communications," Bell Telephone Labs. Inc., 5th ed., 1982, Chapter 16, Sections 16.1 on "Speech Signal Characteristics," and 16.6 on "Syllabic Companding."
36. G.G. Szarvas, and H.G. Suyderhoud, "Enhancement of FDM-FM Satellite Capacity by use of Compandors," *COMSAT Tech. Rev.,* Vol. 11, No. 1, Spring 1981, pp. 1–58.
37. E. Wallace, et al., "A Field Test for Companded SSB Modulation: Implications for Capacity Enhancement and Transmission Planning," *COMSAT Tech. Rev.,* Vol. 15, No. 1, Spring 1985, pp. 89–112.
38. T.C. Spang, "Loss-Noise-Echo Study of the Direct Distance Dialing Network," *BSTJ,* Vol. 55, No. 1, pp. 1–36, Jan. 1976.
39. *Telecommunications Transmission Engineering,* Vol. 3: Networks and Services, 2nd ed., Chapter 9 on "Toll Trunk Design," AT&T.
40. "Through and Terminal Balance General Engineering Considerations," Bell System Practice Section 853-500-100, Issue 1, AT&T, Nov. 1973.
41. "Transmission Design Considerations and Objectives: Switched Special Services and PBX Services," Bell System Practice Section 851-300-100, Issue 2, AT&T, Feb. 1966.
42. F.P. Duffy, G.K. McNees, J. Nasell, and T.W. Thatcher, J., "Echo Performance of Toll Telephone Connections in the United States," *BSTJ,* Vol. 54, No. 2, Feb. 1975, pp. 209–243.
43. F.P. Duffy, and T.W. Thatcher, Jr., "Analog Transmission Performance of the Switched Telecommunication Network," *BSTJ,* Vol. 50, No. 4, Apr. 1971, pp. 1311–1347.
44. CCITT Rec. G.114 on "Mean One-way Propagation Time," *Blue Book,* Vol. III, Fas. III.1, Geneva, 1989, pp. 84–94.
45. B.J. Rabinovich, "Propagation Delay Distribution in the IC Network," Contribution T1Q1.1/89-048, US Sprint, April 17, 1989.
46. "Digital Cross-Connect Systems Requirements and Objectives," Bellcore, TR-170, Issue 1, Nov. 1985.
47. "Wideband and Broadband Digital Cross-Connect Systems Generic Requirements and Objectives," Bellcore TR-233, Issue 2, Sept. 1989.
48. "SONET Add-Drop Multiplex Equipment (SONET ADM) Generic Criteria," Bellcore TR-496, Issue 2, Sept. 1989.
49. EIA/TIA Standard—464-A, "Private Branch Exchange (PBX) Switching Equipment for Voiceband Applications," Section 4.8 on "Transmission Parameters," Feb. 1989. Also, Draft Adden. EIA/TIA-464-A-1, Aug. 1991.
50. CCITT Rec. Q.551 on "Transmission Characteristics of Digital Exchanges," *Blue Book,* Vol. VI, Fas. VI.5, Geneva, 1989, pp. 121–134.
51. R.W. Cubbage, and P.A. Littlewood, "To Float or Not to Float? That is the Question for SONET Mapping," *Telephony,* Vol. 217, No. 5, Oct. 2, 1989, pp. 52, 54, 56, 58.
52. Telecommunications Transmission Engineering, Vol. 2: Facilities, 3rd ed., Chapter 13 on " PCM Transmission Systems," Bellcore, 1990.

53. "Generic Requirements for the Switched DS1/Switched Fractional DS1 Service Capability From an ISDN Interface (SWF—DS1/ISDN)," Bellcore TA-1203, Issue 1, April 1991.
54. *Telecommunications Transmission Engineering,* Vol. 3: Networks and Services, 2nd ed., Chapter 16 on "Private Line Channels," AT&T, 1977, p. 424.
55. "Liaison Statement from T1Y1 to T1Q1 Concerning ADPCM," T1Y1.2/88-035R1, T1Y1/88-017, T1Q1.1/88-0035, May 4, 1988
56. J.M., Ravlin, W.R. Belfield, and T. Nishitani, "Objective Test Results for the 32 kb/s ADPCM Codec," *GLOBECOM Rec.,* Vol. 2, Atlanta, Nov. 1984, pp. 23.3.1–23.3.5.
57. Annex 1 (to report R28 of the meeting of Working Party XVIII/2 concerning Question 7) "Report on the Work of the Ad-Hoc Group on 32 kbit/s ADPCM," Section 6 on Non-Voice Results, Tables 10 and 11. Geneva, Nov. 1983.
58. D.W. Petr, "32 kbps ADPCM—DLQ Coding for Network Applications," *GLOBECOM Rec.,* Vol. 1, Miami, Nov./Dec. 1982, pp. A.8.3.1–A.8.3.5. Also CCITT Contribution COM XVI—No 94-E to Question 18, AT&T, June 1982.
59. H. S. Merrill, "Digital Performance Parameters for Switched Voice Grade Services," *NCF Rec.,* Vol. 1, Chicago, Oct. 1986, pp. 179-197.
60. CCITT Contrib. COM XVIII—No. DM83/XVIII to Quests. 25, 31, "Summary of Objective Measurements on ADPCM Codec Performance reported to Quest. 25/XVIII DCME Ad-Hoc Group," COMSAT Corp., June 1987.
61. N. Benvenuto, and G. Bertocci, "Enhancement in the 32 kbit/s ADPCM Algorithm for Voice-Band Data Signals," *IEEE Trans. on Comm.,* Vol. 36, No. 3, May 1988, pp. 626–628. Also, "FSK Enhancements for Proposed T1Y1 ADPCM Algorithm," T1Y1.2/86-006, AT&T Bell Labs, Jan. 1986.
62. "Test Bed Pushes Modems to the Limit, Revealing Performance Differences," *Data Comm.,* Vol. 16, No. 1, Jan. 1987, pp. 75–79.
63. N. Benvenuto, and W.R. Daumer, "Wave Form Coding of 9.6 kb/s Voiceband Data Signals at 32 kb/s," *ICC Rec.,* Vol. 3, Seattle, June 1987, pp. 49.1.1–49.1.4.
64. "ACCUNET T45 Service Description and Interface Specifications," AT&T PUB 54014, Issue 1, June 1987. Also, Adden. 3, Mar. 1990.
65. J.J. Mahoney, Jr., J.J. Mansell, and R.C. Metlack, "Digital Data System: User's View of the Network," *BSTJ,* Vol. 54, No. 5, May/June 1975, pp. 833–884.
66. A. Tabalabai, and M. Molek-Zavarei, "Generic Design of Digital Service Circuits," *GLOBECOM Rec.,* Vol. 1, Atlanta, Nov. 1984, pp. 13.6.1–13.6.3.
67. S. Kozuka, "Phase Controlled Oscillator for Pulse Stuffing Synchronization System," *Rev. of the Elect. Comm. Lab.,* Vol. 17, Nos. 5,6, May/June 1969, pp. 376–387.
68. J.E. Kessler, "The Transmission Performance of Bell System Toll Connecting Trunks," *BSTJ,* Vol. 50, No. 8, Oct. 1971, pp. 2741–2777.
69. American National Standard for Telecommunications—Network Performance Standards—32 kb/s ADPCM Tandem Encoding Limits, ANSI T1.501-1988.
70. M.A. Rezny, "SDC Error Performance Considerations," *ICC Rec.,* Vol. 2, Philadelphia, June 1982, pp. 3D.5.1–3D.5.6.
71. CCITT Rec. G.821, "Error Performance on an International Digital Connection Forming Part of an Integrated Services Network," *Blue Book,* Vol. III, Fas. III.3, pp. 26–35, Geneva 1989.
72. "Digital Data System Channel Interface Specification," AT&T PUB. 62310, Issue 2, Nov. 1987. Also, Adden. 3, Dec. 1989.
73. "ACCUNET T1.5 Service Description and Interface Specifications," AT&T PUB 62411, Issue 2, Dec. 1988. Also, Adden. 1, Mar. 1990. Also, Issue 3, Dec. 1990.
74. "High Capacity Digital Special Access Service: Transmission Parameter Limits and Interface Combinations," Bellcore TR-342, Issue 2, Jan. 1991.
75. "Access Specification for High Capacity (DS1/DS3) Dedicated Digital Services," AT&T PUB 62415, June 1989.

76. S.D. Akers, S.C. Donn, S.R. Gregg, and S. Ravikumar, "Uses of In-Service Monitoring Information to Estimate Customer Service Quality," *GLOBECOM Rec.*, Vol. 3, Dallas, Nov./Dec. 1989, pp. 51.3.1–51.3.5.
77. G.R. Ritchie, and P.E. Scheffer, "Projecting the Error Performance of the Bell System Digital Network," *ICC Rec.*, Philadelphia, June 1982, Vol. 1, pp. 2D.2.1–2D.2.6.
78. R.J.S. Bates, "A Model for Jitter Accumulation in Digital Networks," *GLOBECOM Rec.*, Vol. 1, San Diego, Nov./Dec. 1983. pp. 4.7.1–4.7.5.
79. S.E. Miller, and I.P. Kaminow, "Optical Fiber Telecommunications II," Chapter 22 on "Interoffice Transmission Systems," Academic Press, 1988, p. 844.
80. "Transport Systems Generic Requirements (TSGR): Common Requirements," Bellcore TR-499, Issue 3, Section 4 on "Error Performance," Dec. 1989.
81. L.T. Wu, and M. Kerner, "Emulating Circuits in a Broadband Packet Network," *GLOBECOM Rec.*, Vol. II, Hollywood, Florida, Nov./Dec. 1988, pp. 21.6.1–21.6.10.
82. C.M. Siperko, "Lasernet—A Fiber Optic Interstate Network (Planning and Engineering Considerations)," *IEEE Comm. Mag.*, Vol. 23, No. 5, May 1985, pp. 31–45.
83. S.E. Miller, and I.P. Kaminow, *Optical Fiber Telecommunications II*, Chapter 23 on "Terrestrial Intercity Transmission Systems," Academic Press, 1988, pp. 857–860.
84. "Generic Reliability Assurance Requirements for Fiber Optic Transport Systems," Bellcore TA-418, Issue 2, Sept. 1989. Also, Supp. 1, Mar. 1991.
85. K. Nagaraj, J. Gruber, J. Leeson, and B. Fleury, "Improvements in Availability and Error Performance of SONET Compared to Asynchronous Transport System," *ICC Rec.* Vol. 1, Atlanta, April 1990, pp. 224.1.1 to 224.1.7. Also, *Telephony*, Vol. 218, No. 26A, June 25 1990, pp. 34, 36, 38, 40, 42, 44.
86. K.A. Tse, R.M. O'Conner, and N. Fatseas, "DS1 Facility Performance: Optical Fiber as a Long-Haul Technology," *ICC Rec.*, Vol. 2, Seattle, June 1987, pp. 19B.2.1–19B.2.4.
87. K.A. Tse, and D.E. Smith, "Transmission Performance Monitoring of the AT&T FT Series G Lightguide System," *ICC Rec.* Vol. 3, Toronto, June 1986, pp. 48.2.1–48.2.4.
88. M.J. Zinnikas, et al., "DS1 Facility Performance: Digital Radio as Long-Haul Technology," *ICC Rec.*, Vol. 2, Seattle, June 1987, pp. 19.B.4.1–19.B.4.4.
89. S. Lumish, et al., "1.7 Gb/s Lightwave Transmission Field Experiment," *ICC Rec.*, Vol. 3, Toronto, June 1986, pp. 48.1.1–48.1.4.
90. M. LeGall, and R. Auffret, "Transmission Systems on Optical Fibers Used in Digital Broadband Networks," *ICC Rec.*, Vol. 3, Toronto, June 1986, pp. 53.6.1–56.3.4.
91. "Access Performance Objectives for DS1 Services," Contributions T1Q1.4/88-024, T1Q1.4/89-010, T1Q1.4/90-004 and T1Q1.4/90-014, AT&T, Aug. 17 1988, Feb. 8 1989, Jan. 31 1990 and April 25 1990, respectively.
92. "Slave Stratum Clock Performance Measurement Guidelines," Committee T1—Telecommunications TR No. 6, June 1990.
93. "Jitter Measurement Methodology," Committee T1—Telecommunications TR No. 8, Feb. 1991.
94. "A Test Set for Field Measurement of Synchronization Parameters," NTI Contrib. T1X1.3/91-050, May 1991.
95. "Transmission Systems For Communications," Bell Telephone Labs. Inc., 5th ed., Chapter 30 on "Baseband Digital Transmission," 1982.
96. CCITT Annex B (to Rec. G.823) on "Jitter Accumulation in Digital Networks," *Blue Book* Vol. III, Fas. III.3, Geneva, 1989, pp. 46–50.
97. B.N., Kearnsey, and R.W. McLintock "Jitter in Digital Telecommunications Networks," *The Radio and Electronic Eng.*, Vol. 54, No. 2, Feb. 1984, pp. 70–78.
98. P.R. Trischitta, et al., "A Circulating Loop Experimental Technique to Simulate the Jitter Accumulation of a Chain of Fiber-Optic Regenerators," *IEEE Trans. on Comm.*, Vol. 36, No. 2, Feb. 1988, pp. 205–213.
99. B.I. McNally, S.F. Carter, and S.D. Walker, "Jitter Accumulation in a Simulated 591.2 Mbit/s,

6000 km Optical Transmission System," *Elect. Letters,* Vol. 24, No. 11, May 26, 1988, pp. 676–678.
100. M. Amenikya, K. Sato, and M. Aiki, "Design of 1.55 mm Gigabit/s Submarine Repeatered Transmission System," *ICC Rec.,* Vol. 3, Boston, June 1989, pp. 50.3.1–50.3.5.
101. C.K. Korgel, "Timing Jitter Accumulation Through a Chain of Phase Locked Loop Repeaters," *ICC Rec.,* June 1975, San Francisco, pp 20-28 to 20-31.
102. C.J. Byrne, B.J. Karafin, and D.B. Robinson, Jr., "Systematic Jitter in a Chain of Digital Regenerators," *BSTJ,* Vol 42, No. 6, Nov. 1963, pp. 2679–2714.
103. L.E. Zegers, "The Reduction of Systematic Jitter in a Transmission Chain with Digital Regenerators," *IEEE Trans, Comm.,* Vol. COM-15, Aug. 1967, pp. 542–551.
104. A. Lesea, "T1 Link Problem: Jitter," *TE&M,* Vol. 92, No. 1, Jan. 1, 1988, pp. 66, 67.
105. S. Ishimatsu, "How to Detect Mistuned PCM Repeaters," *TE&M,* Vol. 93, No. 12, June 15, 1989, pp. 92, 97–100.
106. R. Bridge, and R. Taylor, "ISDN Is More Analog Than You Thought," *Telephony,* Vol. 216, No. 15, April 10, 1989, pp. 37–39.
107. R.B. Flynn, "What You Don't Know About Jitter Can Give You the Jitters," *Data Comm.,* Vol. 16, No. 5, May 1987, pp. 151, 152, 155–158, 161, 162, 165, 166.
108. B. Gewin, "Jitter Testing of T1 Equipment and Networks," *Telecomm.,* Vol. 20, No. 7, July 1986, pp. 77, 78, 80, 82, 84.
109. W.W. Rollins, "Jitter in Digital Transmission Systems," *Telecomm.,* Vol. 18, No. 1, Jan. 1984, pp. 56, 58, 61, 65.
110. V. Brands, "Error-Free Digital Signals Through Precise Jitter Detection," *Telecomm.,* Vol. 24, No. 3, Mar. 1990, pp. 69–72.
111. "Asynchronous Digital Multiplexes: Requirements and Objectives," Bellcore TR-09, Issue 1, May 1986.
112. B. Fleury, "Asynchronous High Speed Digital Multiplexing," *IEEE Comm. Mag.,* Vol. 24, No. 8, Aug. 1986, pp. 17–25.
113. "Transmission Systems For Communications," Bell Telephone Labs. Inc., 5th ed., Chapter 29 on "Digital Multiplex Impairments," 1982.
114. D. Choi, "Waiting Time Jitter Reduction," *IEEE Trans. on Comm.,* Vol. 37, No. 11, Nov. 1989, pp. 1231–1236.
115. D.L. Duttweiler, "Waiting Time Jitter," *BSTJ,* Vol. 51, No. 1, Jan. 1972, pp. 165–207.
116. P.E.K. Chow, "Jitter Due to Pulse Stuffing Synchronization," *IEEE Trans. on Comm.* Vol. COM-21, No. 7, July 1973, pp. 854–859.
117. P.E.K. Chow, "Generalized Formula for Pulse Stuffing Jitter," *Elect. Letters,* Vol. 22, No. 24, Nov. 20, 1986, pp. 1313–1314.
118. T.E. Moore, R.G. Kusyk, and W.D. Grover, "A Computer Simulation Model for Analysis of Waiting Time Jitter," *Canadian Conf. on Elect. and Comp. Eng. Rec.,* Montreal, Sept. 1989, pp. 553–556.
119. "Synchronous Optical Network (SONET) Transport Systems: Common Generic Criteria," Bellcore TR-253, Issue 1, Sept. 1989. Also, TA-253, Issue 6, Sept. 1990. Also, update, Bellcore Digest, Aug. 1991.
120. American National Standard for Telecommunications—Digital Hierarchy—Optical Interface Rates and Formats Specifications, ANSI T1.105-1988. Also, Draft ANSI T1.105-199x, T1LB-213, April/May 1991.
121. R. Ballart, and Y. Ching, "SONET: Now It's the Standard Optical Network," *IEEE Comm. Mag.,* Vol. 27, No. 3, Mar. 1989, pp. 8–15.
122. C.E. Holborow, and R. Mediavilla, "Synchronization of Transmission Systems," *ICC Rec.,* Vol. 2, Philadelphia, June 1988, pp. 28.6.1–28.6.4.
123. J. Philips, "Analyzing Timing Problems in T1 Networks," *Telecomm.,* Vol. 22, No. 7, July 1988, pp. 26–28, 30, 32, 36.

124. A. Lesea, "T1 Link Problem: Wander," *TE&M,* Vol. 91, No. 21, Nov. 1, 1987, pp. 53, 54.
125. C. Griffith, "Synchronizing Digital Networks—Here's How," *Telephony,* April 21, 1980, pp. 31–33, 36, 37.
126. R. Smith, and L.J. Millott, "Synchronization and Slip Performance in a Digital Network," *The Radio and Electronic Eng.,* Vol. 54, No. 2, Feb. 1984, pp. 87–96.
127. A. Lesea, "T1 Problems—Drift and Synchronization," *TE&M,* Vol. 92, No. 3, Feb. 1, 1988, pp. 72, 73, 76.
128. J.C. Hart, and P.H. Fredette, "Timing Considerations in Hybrid Networks," *Telecomm.,* Vol. 20, No. 11, Nov. 1986, pp. 55, 58, 67.
129. C.J. Byrne, "Properties and Design of the Phase-Controlled Oscillator with a Sawtooth Comparator," *BSTJ,* Vol. 41, No. 2, Mar. 1962, pp. 559–602.
130. R.F. Bridge, S. Biley, J. Klass, and R. Taylor, "Jitter Attenuation in T1 Networks," *ICC Rec.,* Vol. 2, Atlanta, April 1990, pp. 314.1–314.1.5.
131. CCITT Rec. G.743 on "Second Order Digital Multiplex Equipment Operating at 6312 kbit/s and Using Positive Justification," *Blue Book,* Vol. III, Fas. III.3, Geneva, 1989, pp. 427–431.
132. CCITT Rec. G.752 on "Characteristics of Digital Multiplex Equipments Based on a Second Order Bit Rate of 6312 kbit/s and Using Positive Justification," *Blue Book,* Vol. III, Fas. III.3, Geneva, 1989, pp. 460–469.
133. CCITT Rec. G.782, on "Types and General Characteristics of SDH Multiplexing Equipment," Section 4.1 on "Timing and Synchronization Overview," Geneva, 1990, pp. 22–24.
134. CCITT Rec. G.783, on "Characteristics of SDH Multiplexing Equipment Functional Blocks," Section 6 on "Timing Functions," and Section 7 on "Specification of Jitter and Wander," Geneva, 1990, pp. 40–47.
135. Draft Committee T1—Telecommunications "Technical Report on the Effects of SONET on Jitter, Wander and Synchronization," July 1991.
136. CCITT SGXVIII Draft Recs. G.sna1 on "Architectures of Transport Networks based on SDH," Section 6 on "Architecture of Synchronization Networks," Annex 3 of TD 35, Geneva, June 1991; and G.sna2 on "Performance and Management Capabilities of Transport Networks Based on the SDH," Section 6 on "Jitter and Wander in SDH Networks," Annex 4 of TD 35, Geneva, June 1991.
137. "Transport Systems Generic Requirements (TSGR): Common Requirements," Bellcore TR-499, Issue 3, Section 7 on "Jitter," Dec. 1989.
138. CCITT Rec. G.824, "The Control of Jitter and Wander Within Digital Networks Which Are Based on the 1544 kbit/s Hierarchy," *Blue Book,* Vol. III, Fas. III.3, Geneva, 1989, pp. 50–55.
139. American National Standard for Telecommunications: Digital Heirarchy—Electrical Interfaces," ANSI T1.102-1987. Also, Draft ANSI T1.102-199x, T1X1.4/91-004R1, July 1991.
140. CCITT Draft Rec. G.82j on "The Control of Jitter and Wander within Digital Networks Which Are Based on the SDH," Annex B of TD-26 (XVIII/6), Geneva, June 1991.
141. CCITT Rec. G.958 on "Digital Line Systems Based on the SDH for Use on Optical Cables," Section 6 on "General Characteristics of Synchronous Optical Line Systems," Geneva, 1990.
142. American National Standard for Telecommunications—Optical Interface Specifications: Single Mode, ANSI T1.106-1988.
143. CCITT Rec. G.755 on "Digital Multiplex Equipment Operating at 139264 kbit/s and Multiplexing Three Tributaries at 44736 kbit/s," *Blue Book,* Vol. III, Fas. III.3, Geneva, 1989, pp. 476–481.
144. "Synchronous Optical Network (SONET) Transport Systems: Common Generic Criteria," Bellcore TR-253, Issue 1, Sections 5.4 and 5.6 on "Network Synchronization" and "Jitter," Sept. 1989.
145. E.A. Munter, "Synchronized Clock for DMS-100 Family," *IEEE Trans. on Comm.,* Vol. COM-28, No. 8, Aug. 1980, pp. 1276–1284.
146. M. Kihara, "Performance Aspects of Reference Clock Distribution for Evolving Digital Networks," *IEEE Comm. Mag.,* Vol. 27, No. 4, April 1989, pp. 24–34.

147. Synchronization—LATA Switching Systems Generic Requirements (LSSGR): Section 18, Bellcore TR-518, Issue 2, July 1987.
148. "Digital Synchronization Network Plan," Bellcore TA-436, Issue 1, Nov. 1986.
149. J. Porter, "New ANSI Standards for Network Synchronization," *Telecomm.*, Vol. 23, No. 12, Dec. 1989, pp. 56–58.
150. R. Avery, "Loose Slips Sync'ed by BITS," *TE&M*, Vol. 93, No. 3, Feb. 1, 1989, pp. 39–41.
151. T.C. Warren, "Do You Have The Time?" *TE&M*, Vol. 93, No. 1, Jan. 1, 1989, pp. 46, 47.
152. P.R. Strauss, "Bad Timing on T1 Circuits Gives More and More Users the Jitters," *Data Comm.*, Vol. 16, No. 12, Nov. 1987, pp. 54, 56, 58.
153. D.R. Smith, *Digital Transmission Systems,* Van Nostrand, Chapter 10 on "Network Timing and Synchronization," 1985.
154. H. Meyr, and G. Ascheid, *Synchronization in Digital Communications,* Vol. 1, Phase-, Frequency Locked Loops, and Amplitude Control, Chapter 3, Section 3.5 on "Time and Frequency Stability of Signal Generators," Wiley, 1990, pp. 133–147.
155. IEEE Proc. Special Issue on "Time and Frequency," Vol. 79, No. 7, July 1991.
156. CCITT, Rec. G.811 "Timing Requirements at the Outputs of Primary Reference Clocks Suitable for Plesiochronous Operation of International Digital Links," *Blue Book,* Vol. III, Fas. III.3, Geneva, 1989, pp. 16–21.
157. American National Standard for Telecommunications, "Synchronization Interface Standards for Digital Networks," ANSI T1.101-1987. Also, Draft ANSI T1.101-199x, T1X1.3/90-026R7, Sept. 1991.
158. CCITT Rec. G.812, "Timing Requirements at the Outputs of Slave Clocks Suitable for Plesiochronous Operation of International Digital Links," *Blue Book,* Vol. III, Fas.III.3, Geneva, 1989, pp. 21–26.
159. EIA/TIA Standard −464-A, "Private Branch Exchange (PBX) Switching Equipment for Voiceband Applications," Section 4.11 on "Synchronization," Feb. 1989.
160. Draft EIA/TIA "Private Digital Network Synchronization Standard," SP 2198-A, 1991.
161. AT&T PUB 60110, "Digital Synchronization Network Plan," Dec. 1983.
162. "Timing Signal Generator (TSG) Requirements and Objectives" Bellcore TA-378, Issue 1, April 1986.
163. "Synchronized Clocks: Common Generic Criteria," Bellcore TA-1244, Issue 1, Sept. 1991.
164. J.E. Abate, et al., "AT&T's New Approach to the Synchronization of Telecommunications Networks," *IEEE Comm. Mag.,* Vol. 27, No. 4, Apr. 1989, pp. 35–45.
165. A. Zampatti, "Provisional Stratum 2 Clock Stability Proposal," Contribution T1X1.3/88-040, AT&T, Nov. 1988.
166. J.T. Mulgreen, "Carriers Upgrade Synchronization," *Data Comm.,* Vol. 19, No. 6, May 1990, pp. 54, 56, 58.
167. CCITT Rec. G.822, "Controlled Slip Rate Objectives on an International Digital Connection," *Blue Book,* Vol. III, Fas III.3, Geneva, 1989, pp. 36–38.
168. J. Bellamy, "Digital Telephony," Chapter 7 on "Network Synchronization, Control and Management," Wiley, 1982. Also, 2nd ed., 1991.
169. "Proposals for SONET NE Clock Short-Term Stability Requirement," NTI Contrib. T1X1.3/91-050, July 1991.
170. CCITT Rec. Q.541 on Digital Exchange Design Objectives—General," *Blue Book,* Vol. VI, Fas. VI.5, Geneva, 1989, p. 41–46.
171. "Proposal for MTIE at Network Timing Reference Interfaces," NTI Contrib. T1X1.3/91-087, July 1991.
172. CCITT Draft Rec. G.81s on "Timing Characteristics of Slave Clocks Suitable for Operation in SDH Equipments," Annex A of TD-26 (XVIII/6), Geneva, June 1991.
173. M. Kalb, "Service Affecting Limits (SALs) for Switched Exchange Access Service," Contribution T1Q1.1/90-042, AT&T, April 23 1990.

Appendix A.6.1
OBJECTIVE INTERPRETATION OF SUBJECTIVE "COMPANDOR PENALTY" ASSOCIATED WITH ANALOG COMPANDORS

For trunks that use analog compandors, the subjectively perceived noise and the corresponding subjective penalty can be estimated in objective terms as follows. From the compandor compression characteristic (typically 2:1 [35]), measurements of compandor output idle circuit noise and of compandor output noise with an input tone signal [16, 68], the compandored trunk output noise versus input signal level characteristic can be modeled as shown in Figure A.6.1.1. Next, the compandor input speech level is obtained from speech level survey data [33], and knowledge of the network transmission loss plan [11, 18]. This enables the compandor output noise with an input speech signal to be estimated using the model in Figure A.6.1.1.

Thus, the perceived compandor output noise is then estimated as the sum of the idle circuit noise and the noise with active input speech, each appropriately weighted by the speech activity factor. Finally, the compandor penalty is the difference between the perceived noise and the idle circuit noise. As shown by the example below, this approach, when applied to compandored trunk data, results in an objective estimate of compandor penalty that agrees well with the subjective value of 5 dB [22].

Mathematically, the above procedure is as follows: based on the model in Figure A.6.1.1 (see also Tables 5.3 or 5.14 of Ch. 5), the compandor output noise in dBrnC is

$$N = N_I \qquad \text{for } S < (N_I - b)/a \qquad (A.6.1.1)$$

$$= aS + b \qquad \text{for } S \geq (N_I - b)/a \qquad (A.6.1.2)$$

where N_I is the compandor output idle circuit noise (dBrnC) with no input signal; S is the compandor input signal (dBm); a is the compandor compression characteristic

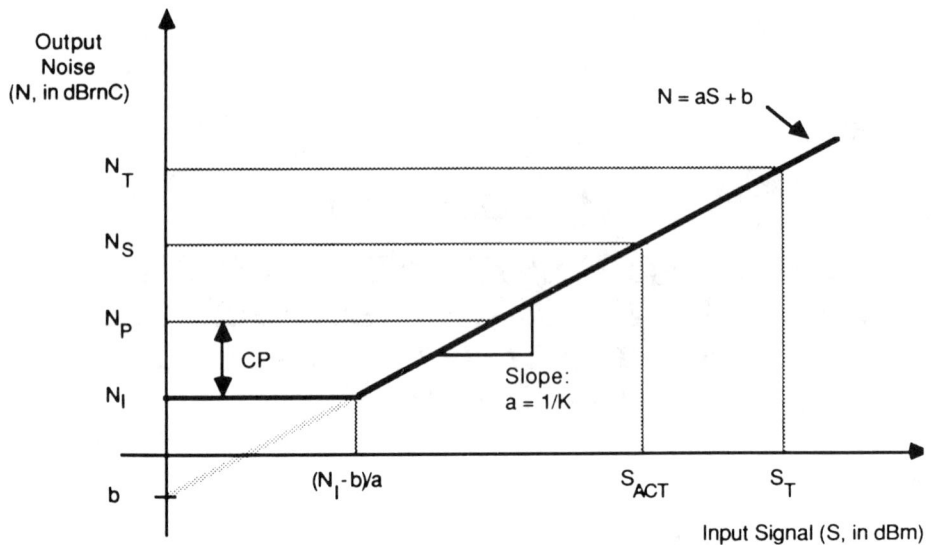

CP: Compandor Penalty = $N_P - N_I$ (dB)

See text - eqns. (A.6.1.1) to (A.6.1.7) - for definitions of terminology.

Figure A.6.1.1 Simplified model of analog compandor output noise versus input signal characteristic.

(i.e., for a K:1 compandor, $a = 1/K$; typically $K = 2$ [22]); and b is the signal-independent part of the compandor output noise (which includes a factor $F = 90$ dB to convert from dBm to dBrnC).

From measurements of compandor output idle circuit noise ($N = N_I$ with $S < (N_I - a)/b$), and compandor output noise with input signal tone ($N = N_T$ with $S = S_T$), the factor b is determined from Equation (A.6.1.2) as

$$b = N_T - aS_T \tag{A.6.1.3}$$

From measurements of average speech power (S_{AVE}), the active speech power (S_{ACT}) is

$$S_{ACT} = S_{AVE} - 10\log_{10}(t) \tag{A.6.1.4}$$

where t is the speech activity factor (typically $t = 0.35$ for a conversation [22]).

Thus, from Equations (A.6.1.2) to (A.6.1.4), the corresponding output noise ($N = N_S$) during active speech ($S = S_{ACT}$), is

$$N_S = aS_{ACT} + b = N_T - a(S_T - S_{ACT}) \quad (A.6.1.5)$$

The perceived noise (N_P) during a conversation is the sum of the idle circuit noise—N_I from Equation (A.6.1.1)—and the noise with active speech—N_S from Equation (A.6.1.5)—each weighted by the speech activity factor (t) as follows:

$$N_P = (1 - t)N_I + tN_S \quad (A.6.1.6)$$

Finally, the compandor penalty (CP) is the difference between the perceived noise—N_P from Equation (A.6.1.6)—and the idle circuit noise—N_I from Equation (A.6.1.1):

$$CP = N_P - N_I = t(N_S - N_I) \quad (A.6.1.7)$$

Example: Consider a 2:1 compandored N-carrier analog trunk ($a = \frac{1}{2}$) having the following measured output noise performance [16] (note: levels are at a 0-level point so that trunk loss does not need to be explicitly accounted for):

for $S < (N_I - b)/a$, $\quad N = N_I = 20.3$ dBrnC0
for $S = S_T = -13$ dBm0, $\quad N = N_T = 37.6$ dBrnC0

For compandor input speech, such that

$S_{AVE} = -25.3$ dBm0 [33] and $t = 0.35$ [22], then from Eq. (A.6.1.4),
$S_{ACT} = -25.3 - 10 \log_{10}(0.35) = -20.7$ dBm0

From Equation (A.6.1.5), the compandor output noise with active input speech is

$$N_S = 37.6 - (-13 - \{-20.7\})/2 = 33.8 \text{ dBrnC0}$$

Finally, from Equation (A.6.1.6), the perceived compandor output noise is

$$N_P = (1 - 0.35) \times 20.3 + 0.35 \times 33.8 = 25.0 \text{ dBrnC0}$$

and, from Equation (A.6.1.7), the compandor penalty is

$$CP = 25.0 - 20.3 = 4.7 \text{ dB}$$

which agrees closely with the subjective value of 5 dB [22].

Appendix A.6.2
BREAKDOWN OF DIGITAL SWITCHING DELAYS

This appendix provides a breakdown of approximate average and 99% maximum digital switching system delays, based largely on [15]. The total average delay of port-to-port connections through the switch is the sum of the average delays of each switching component in the connection. As a worst case, and for simplicity, the same approach is used for the maximum delays.

These delays are illustrated in Figures A.6.2.1 and A.6.2.2 for intra- and inter-local switching arrangements, respectively. The terminology used in these figures has the following meanings:

- Nondistributed switching: a central office (or host) switch only.
- Distributed switching: a central office (or host) switch, and a remote switching or subscriber carrier system.
- Intra-local switching: a remote switching or subscriber carrier system whose traffic is switched by the remote switching or subscriber carrier system (included in Fig. A.6.2.1).
- Inter-local switching: a remote switching or subscriber carrier system whose traffic is switched by the host switch (included in Fig. A.6.2.2).

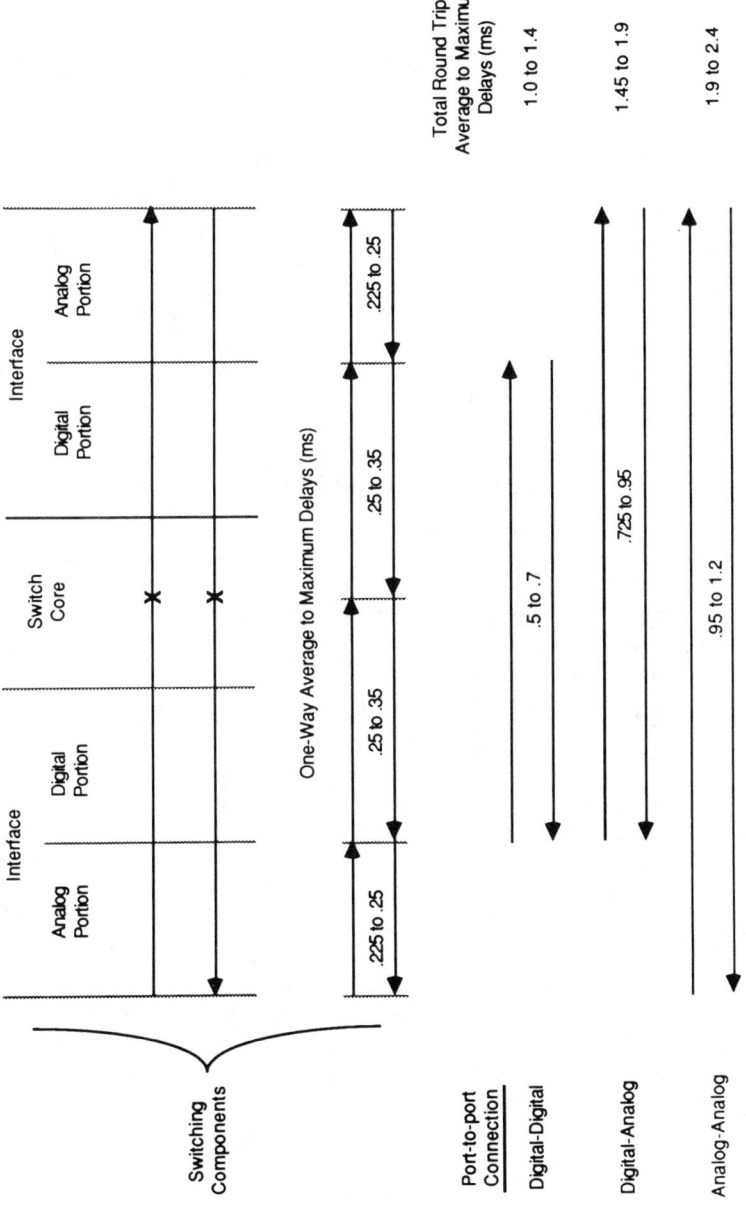

Figure A.6.2.1 Illustration of digital switching system components and delays (non-distributed and distributed intra-local switching).

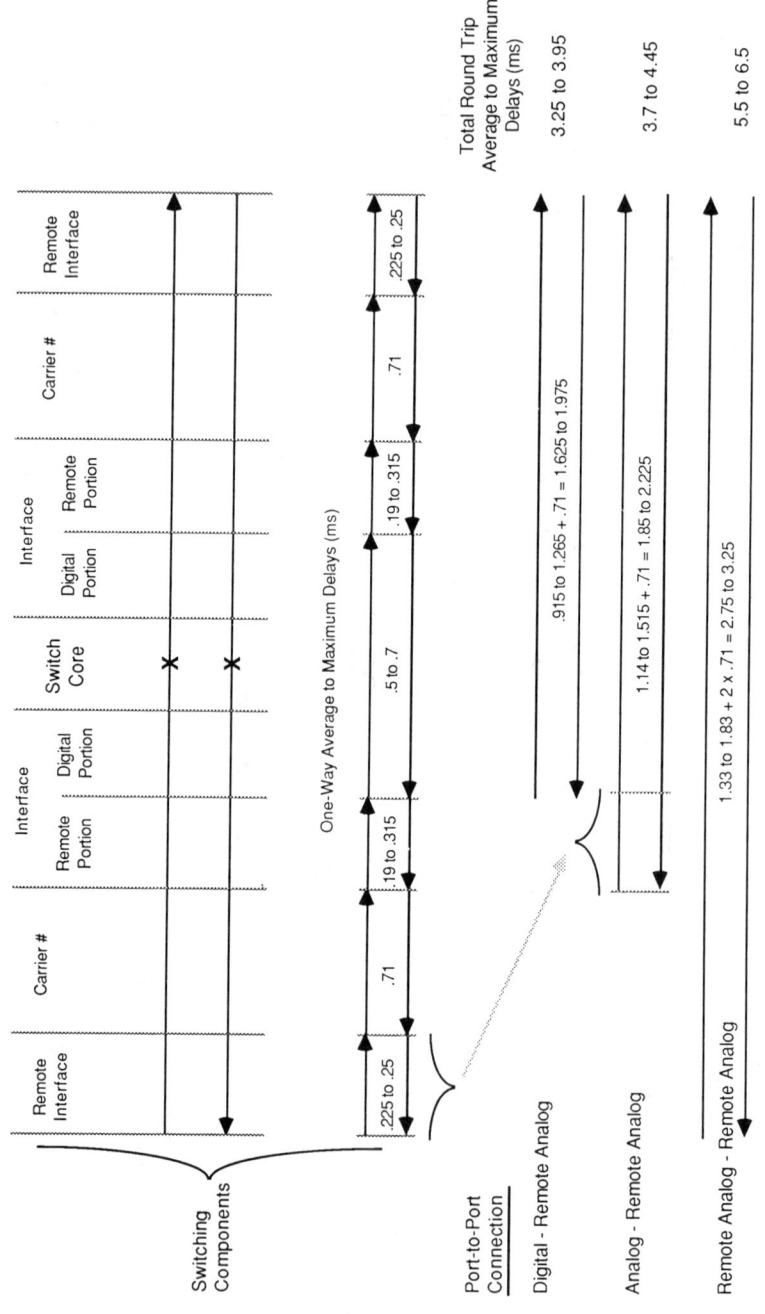

Figure A.6.2.2 Illustration of digital switching system components and delays (non-distributed and distributed inter-local switching).

Chapter 7
NETWORK TRANSMISSION PLANNING

7.1 INTRODUCTION

The final step in the top-down process in Figure 1.3 of Chapter 1 is network planning, leading to network standards. In general, the purpose of network planning is to provide the rules and guidelines for implementation to ensure that end-to-end performance (as allocated to network components) is achieved. Guidelines are also required for the deployment of the various technologies that affect network performance. Examples of such technologies include echo control devices, low bit rate coding systems, such as 32-kb/s adaptive differential pulse code modulation (ADPCM), asynchronous transfer mode (ATM), and remote switching systems. The network planning challenge is to utilize the performance capabilities of evolving networks to satisfy end user requirements while minimizing capital expense and operational cost. Thus, network planners assess the performance capabilities of systems and technologies in terms of end users' needs and service requirements (Chs. 2 and 3) as translated and allocated to the equipment-level components of the network (Chs. 4 to 6). From the results of this assessment, network standards are developed and updated (Ch. 8). The focus in this chapter is on the transmission aspects of network planning.

7.1.1 Purpose

In general, the purpose of a transmission plan is to provide acceptable grade of service (GOS) to end users according to agreed-on criteria or objectives (see Chs. 2 and 3), but subject to economic and technical constraints. This is achieved through a set of specifications applicable to each network element making up the end-to-end connection. These specifications control impairments such as loss, noise, and echo, and account for signal levels as well as maintenance and operational considerations. Transmission planning is primarily associated with 4-kHz voiceband services and, to a lesser extent, with digital services. An example of a non-4-kHz service that requires transmission planning is 7-kHz high-quality voice. This is currently under study in CCITT SG XII (see Ch. 8, App. A.8.2).

7.1.2 Application

Telephone networks are generally in a constant state of evolution driven by changing end user needs and technological advances. A distinction, however, may be drawn between telephone networks that are in an early stage of evolution, such as those in third-world countries, and networks that are in an advanced stage, such as those in North America.

In the former case, it may be possible to redesign the entire network using modern switching and transmission technology, in that backward compatibility with the existing network is not as important a criterion as it is in the latter case. However, for the latter, the key element of transmission planning is the assurance of high quality as networks evolve, that is, maintaining backward compatibility with an existing network already in a relatively high technological state. End user expectations are already at a high level, and the introduction of new technology must not negatively impact these high expectations. New network responsibilities and architectures brought about by divestiture in the U.S. are an additional consideration.

Transmission planning of evolving North American networks to meet changing end user requirements and new network architectures is the main focus of this chapter, although the principles outlined are generally applicable on a global basis.

7.1.3 Historical Perspective

Since the early 1950s, the North American telephone network has undergone a number of fundamental transmission planning changes mainly associated with the introduction of new technology and the continuous need to streamline operational procedures. In addition, since 1984, divestiture has had a significant impact in the U.S., not so much on the basic transmission planning principles, but more on the interrelationships between the different network providers and the need to assure end-to-end quality of service on connections involving different networks. Consequently, standards (as discussed in detail in Ch. 8) play an increasingly important role in providing industry agreements for interworking between different networks.

In this chapter three stages or eras in network evolution are considered, and a perspective on the corresponding transmission planning philosophy in each case is given.

The three stages, or eras, are

1. The via net loss (VNL) era.
2. The fixed loss plan (FLP) era.
3. The all-digital network (ADN) era.

7.1.3.1 The Via Net Loss Era

The VNL plan was developed in the early 1950s in North America to control echo by introducing loss as a function of distance. Section 13 of "AT&T Notes on the Network" [1] includes a number of references on the VNL plan. Echo, together with loss and noise, are the major performance parameters that need to be considered in large countries, such as those in North America. The VNL plan was the first approach used to control the effect of echo. This was the time of analog facilities and analog switches, and under this plan (based on controlling echo) the loss associated with facilities was allowed to increase as a function of connection distance, up to a maximum loss. The maximum loss was determined from customer opinion models (as in Chs. 2 and 3), and beyond this distance, echo control in the form of echo suppressors was used. Appendix A.7.1 provides a brief summary of the basic principles of the VNL plan. In smaller countries, such as those in Europe, VNL concepts were not required. "Transmission Planning of Switched Telephone Networks" [2] provides information on the planning principles that were generally applicable in Europe in this era.

7.1.3.2 The Fixed Loss Plan Era

In the early 1970s, with the introduction of digital switches into networks already containing some digital facilities, it was clear that in the long term, the VNL concept would not be appropriate for this new technology. Specifically, digital loss would be required at intermediate points in connections, that is, using digital attenuators (also called digital pads). Alternatively, signals would have to be converted to analog at these points, and the loss implemented with analog loss pads. Both techniques introduce added impairment in the form of increased quantization distortion (see Sec. 6.2.3.3 of Ch. 6) and are not compatible with the concept of digital bit transparency, which was already recognized as a necessary future requirement to support digital services. Thus was born the FLP [3-6] which introduced fixed values of analog loss in the end office (Class 5 switch) as a function of connection type. This plan accounts for the loss, noise, and echo impairments, and provides appropriate echo control rules.

The FLP was developed with the objective of ensuring smooth evolution from the VNL concept, both in terms of maintaining acceptable transmission GOS and of meeting associated operational and maintenance requirements. Indeed, the operational requirements and the need to ensure compatibility with existing VNL, or analog practices, played a key role in establishing the FLP planning concept. Appendix A.7.2 outlines the basic principles of the FLP.

Figures 7.1(a) and (b) illustrate the VNL and FLP concepts. In Figure 7.1(a), the increase in loss with distance is illustrated. In Figure 7.1(b) the FLP with loss

Figure 7.1(a) VNL transmission plan illustration (mainly analog switching/transmission).

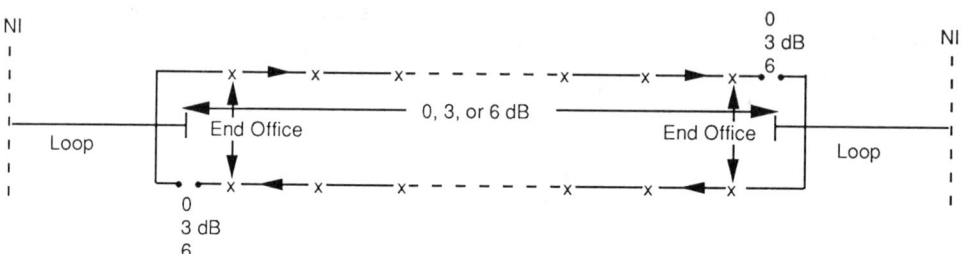

Figure 7.1(b) Fixed loss plan illustration (mainly digital switching/transmission).

implemented in the end office (EO) is illustrated; in this case the EO-to-EO loss is 0, 3, or 6 dB for intra-office, inter-office, and toll digital connections, respectively.

The FLP was initially conceived out of work carried out by both the AT&T Bell System and Telecom Canada and later agreed to by AT&T and Telecom Canada [4], as the North American Transmission Plan. The FLP supports all-digital connectivity from EO to EO as shown in Figure 7.1(c), and interworking with the analog loop plant as shown in Figure 7.1(d).

The introduction of digital technology during the FLP era also brought about the introduction of new access network architectures. These architectures take

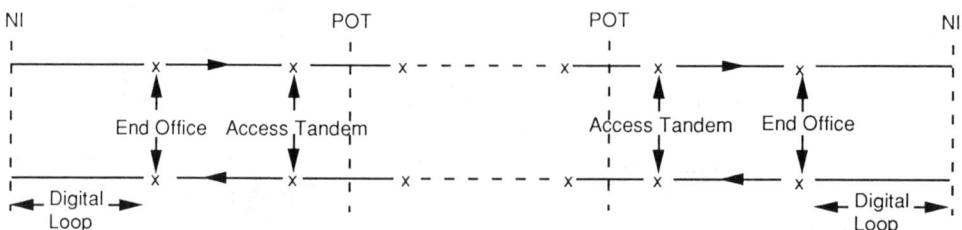

Figure 7.1(c) All-digital connection illustration (postdivestiture era).

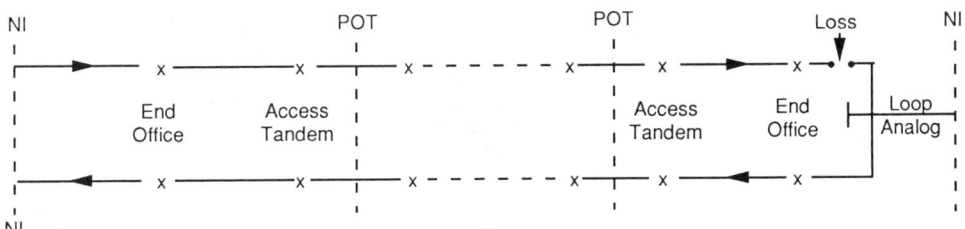

Figure 7.1(d) All-digital/FLP interworking illustration (postdivestiture era).

advantage of the signaling and control capabilities of digital technology to offer economic access to the EO switch through digital loop carrier (DLC) and remote switching systems (RSS) [7]. Figures 7.2(a) and (b) illustrate both of these access configurations.

DLC systems perform a concentrating function by replacing the feeder section(s) of the loop plant with a multiplexed carrier. RSSs, on the other hand, are an extension of the EO, or "host" switch. The "host" provides the operations and administrative support, and the remote unit is located usually less than 50 miles, but possibly as far as 100 miles from the host. This provides economic access as well as, in some cases, a local switching function, that is, intra-local switching. Note that when the DLC is integrated with the digital Class 5 switch, it is called integrated DLC (IDLC) [8]. Both configurations are subject to specific transmission planning guidelines and Section 7.6 of this chapter provides further details.

During the FLP era, the increasing need to interconnect public switched networks with private networks and digital private branch exchanges (PBX) was also recognized by the PBX manufacturers. The focus for developing transmission standards in this area was Electrical Industries Association (EIA) TR41. This function is still carried out by the same forum under its new title, Telecommunications Industry Association (TIA) TR41. In 1974, a standards project was agreed upon to develop requirements for digital PBXs, including, in particular, interfaces with digital end offices that were designed to meet FLP requirements. More details of the function of TIA TR41.1, its relationship with the EIA, and its projects are given in Chapter 8.

Impact of Divestiture

With the breakup of the Bell System after divestiture, the focus for transmission planning in the U.S. was lost, and this, coupled with the emergence of a multitude of competitive inter-exchange carriers (IEC), resulted in a period of slowdown of transmission planning in the industry.

Under its operating mandate, T1Q1.1 (in Standards Committee T1—Telecommunications; see Sec. 8.2.1 of Ch. 8) took on the task of resolving industry needs,

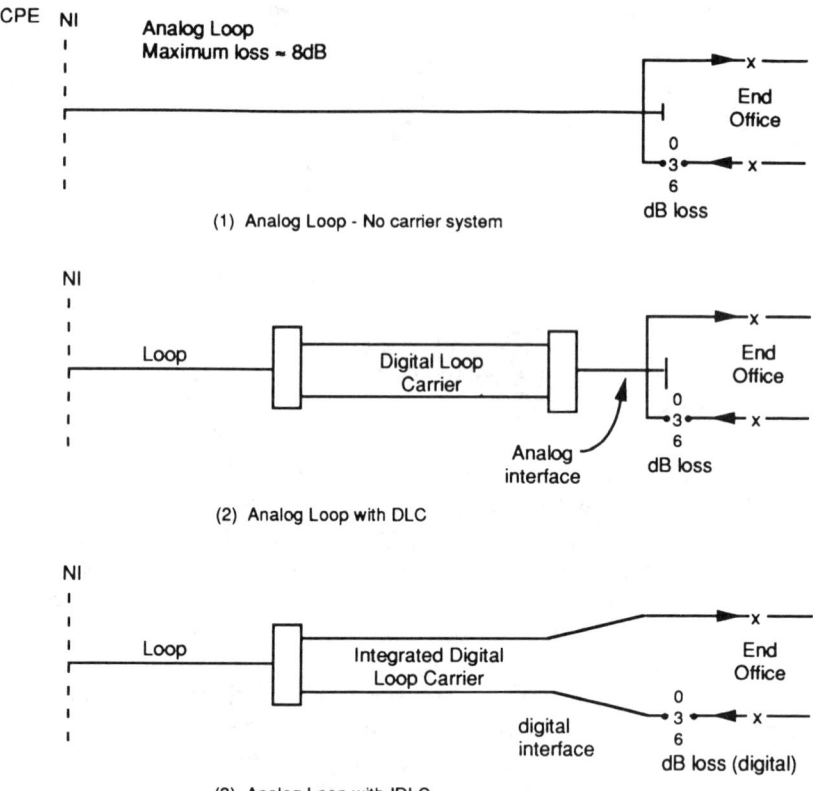

Figure 7.2(a) Illustration of access network evolution using loop carrier systems.

Figure 7.2(b) Illustration of remote switching system.

and it is the current focus for network transmission planning issues spanning network interfaces (NI to NI). TIA TR41 is the focus for private networks and PBX requirements on the customer side of the NI. Divestiture also slowed progress on the EIA/TIA standard on digital PBXs, a key reason being the lack of industry guidance over implementation of the FLP in end offices that serve PBXs, and specifically the lack of planning guidelines for this application.

Recognizing this, T1Q1.1 and TIA TR41.1 set up a joint liaison activity addressing interworking issues having a bearing on the EIA/TIA standard. The EIA/TIA standard was completed, received ANSI accreditation in December 1988, and was released in late 1989 [9]. Chapter 8 gives more information on the activities of these standards forums. Transmission planning activities in Canada on a national and international interworking basis have progressed smoothly mainly under the influence of the Telecom Canada group of operating companies and associated CCITT activities.

International Activities

CCITT Study Group (SG) XII is the focus for developing recommendations for the interworking of national/international networks. Generally, all developed countries have progressed to a stage similar to the FLP era in North America. In smaller countries, the loss plan can be a simple variation of the FLP, since echo is not as significant a problem, and very often advantage can be taken of different loop plant characteristics. For example, in the U.K. [10], a single loss of 7 dB between end offices covers local, inter-office, and toll digital connections.

7.1.3.3 The All-Digital Network Era

With the introduction of an increasing number of digital PBXs, it became clear in the industry that the FLP concept required revision for the digital interconnection of digital end offices and digital PBXs. The reason is the same as that behind the development of the FLP in the first place: performance and digital connectivity. With digital connectivity to digital PBXs, the end office is an intermediate point in the connection and not an appropriate point to implement loss. Also, following divestiture, increased impetus was being given in the access network to field trials of Integrated Services Digital Network (ISDN) technological concepts. It was clear that the FLP was being overtaken by industry needs for digital connectivity up to and beyond the NI.

Recognizing this, both T1Q1.1 and TIA TR41.1 agreed in 1986 to develop transmission planning standards covering all-digital networks in their respective public and private network domains. This is discussed in Section 8.6.1.1 of Chapter 8.

A transmission plan for an all-digital connection between the NI interfaces, as illustrated in Figure 7.1(c), is simple in concept, since the network is effectively an all-digital pipe. Note that the network interfaces in Figure 7.1(c) reflect the current postdivestiture situation.

However, the challenge in developing a transmission plan covering the evolution of the network to the ideal in Figure 7.1(c) is the assurance of backward compatibility. This was previously the case for the FLP, and is also the case for the plan currently being developed. Indeed, the title of the T1Q1.1 standard [11] reflects this situation. This compatibility need is shown in Figure 7.1(d), which illustrates a "mixed" connection of the type the network will have to support into the foreseeable future. On the left of the figure, the access arrangement is digital from the NI to the end office. On the right of the figure, the access from the NI is analog to an end office that implements the FLP.

Clearly, in Figure 7.1(d), the solution must provide acceptable quality of service to end users at each end of such connections, the end user on the left using a digital terminal, and the user on the right, an analog terminal. Given other considerations such as preservation of bit transparency, a solution encompassing the customer premises equipment (CPE) characteristics is required. This topic is covered in detail in Section 7.5.2 of this chapter.

In addition, other solutions to cover network evolution based on a simplification of the FLP rules are also being considered, as covered in Section 7.5.3 of this chapter.

7.2 TRANSMISSION PLANNING FOCUS

Given the above historical perspective and current industry emphasis on the evolution to all-digital networks and ISDN, the focus of the detailed part of this chapter will be on transmission planning for all-digital networks using the FLP as it currently exists as the basis.

The basic concepts of the FLP will be described in terms of the main transmission performance parameters: loss, noise, and echo/delay. Although noise is of secondary importance in a mainly digital environment, loss requirements must still be considered in the context of a network in which delay is increasing; that is, loss is still required to compensate for the echo impairment. The increase in delay results from the introduction of delay-prone digital technology stimulated by the processing/switching potential of a digital network and the bandwidth and error performance benefits of fiber transmission. This technology trend will be discussed and analyzed with respect to its impact on the evolving network. A simplification of the FLP is also addressed, which is a key step in the evolutionary process. In addition, the operational aspects of the FLP will be presented, and simplification of operational needs as the network evolves to all-digital working will be discussed.

Finally, private network considerations will be included, taking into account particular interworking needs and the potential application of digital circuit multiplication equipment (DCME) such as 32-kb/s ADPCM in public, private, and international networks.

7.3 TRANSMISSION PLANNING: BASIC PRINCIPLES AND PROCESS

As mentioned at the beginning of this chapter, the purpose of a network transmission plan is to provide acceptable GOS to end users according to certain criteria or objectives, but subject to economic and technical constraints. These criteria or objectives include the need to minimize end user complaints when using any network connection. This is achieved through control of relevant transmission impairments as provided by the transmission plan and as dictated by economic and technical constraints. Thus, with reference to Figure 7.3, transmission planning is essentially an optimization process governed by the tradeoff between end user need, as reflected in user-oriented performance objectives, and network simplicity in controlling impairments (to minimize end user complaints), as reflected in corresponding network-oriented performance objectives.

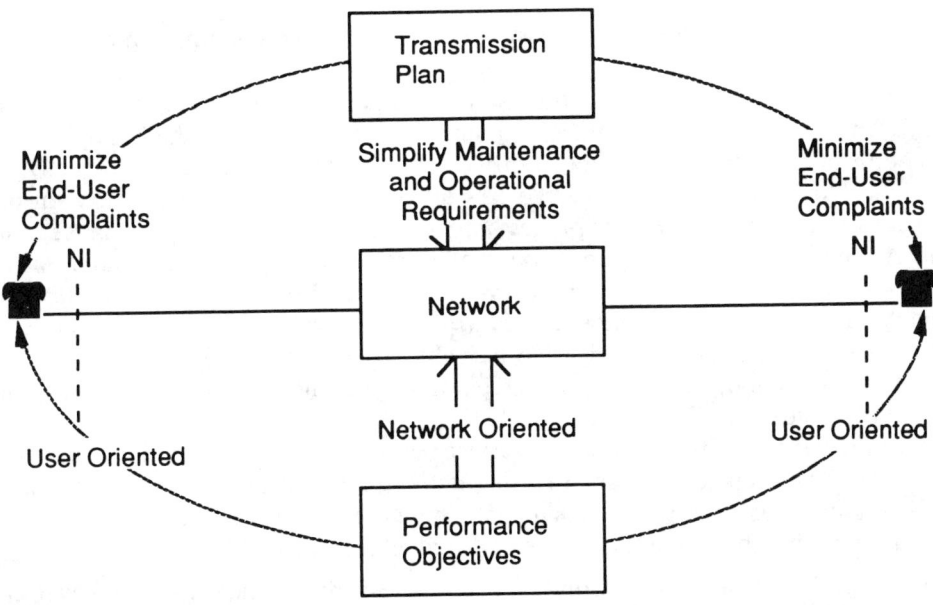

Figure 7.3 Illustration of the transmission planning process in relation to performance objectives.

In developing a transmission plan, the following aspects must be accounted for:
1. The user-oriented performance goals.
2. The main transmission parameters/impairments.
3. The method of controlling the impairments to meet user-oriented performance.
4. The technological and economic impact of tightening or relaxing the control of impairments.

Coupled with the above is the allocation of the end-to-end network-oriented objectives among the network components. This allocation is also a function of technological and economic constraints, and in the current U.S. internetworking environment, is dependent upon changing network architecture (such as the shift from hierarchical to dynamic routing) and network demarcation responsibilities.

In addition, a transmission plan must include

- Design rules covering, for example, the implementation of loss and the application of echo control.
- Maintenance limits on parameters such as loss, noise, and echo return loss.
- Consideration of the deployment of certain technologies, such as low bit rate codecs.
- Consideration of routing and certain interworking arrangements.

7.3.1 Summary of Main Transmission Planning Parameters/Impairments

A distinction is made between a transmission impairment such as noise, which generally impairs performance, and a parameter such as added loss, which is inserted in a connection for optimization or improvement of overall performance.

Table 2.1 of Chapter 2 categorizes all parameters according to the various phases of the communications process and according to various performance criteria. A transmission plan only addresses the information transfer phase, and the accuracy-of-service (transmission) performance criteria. For voice service, there are many user-oriented parameters that can impact accuracy of service, and these are discussed in Section 3.2.2 and listed in Table 3.1 of Chapter 3. Correspondingly, there are also a many network-oriented parameters as discussed in Section 4.3.4 and listed in Table 4.1 of Chapter 4.

Although all of the accuracy-of-service parameters in Table 4.1 affect perceived end user performance, for transmission planning purposes, only the main parameters/impairments need to be considered. These are as illustrated in Figure 7.4 and described as follows. Note that Figure 7.4 is an elaboration of Figure 3.1.

End-to-End Loudness Loss: Consists of the one-way electrical loss plus terminal (telephone) acoustic losses in the transmit and receive directions.

Figure 7.4 Network connections and key transmission planning impairments associated with the network and terminal (CPE).

Noise: Consists of idle circuit noise (C-message weighted) and, where applicable, quantizing noise. Quantizing noise will be present when digital or DCME devices are used in a connection. Usually, a small number of digital pulse code modulation (PCM) codecs are encountered, and quantization noise can be ignored. Chapter 6 discusses the effect of tandemed PCM and ADPCM codecs on the most sensitive services (e.g., voiceband data).

Delay: Consists of delay through network elements (transmission facilities, switching equipment, terminal equipment, and other digital or analog processing devices). In the absence of echo, relatively high values of delay are perceived as a degradation for voice service because of the impact on conversational speech dynamics. Delay can also impact other services, such as interactive voice/video.

Talker Echo Path Loss: Results from a combination of connection loss and hybrid mismatch loss, and is generally dominated by the 4-wire to 2-wire termination at the 2-wire loop interface, as illustrated in Figure 7.4.

The remaining accuracy impairments listed in Table 4.1 of Chapter 4 are not considered in the transmission planning process because in general their effect is minimized through appropriate equipment and network design. For example, crosstalk interference in the loop plant is reduced to acceptable levels through feeder cable design and control of signal levels (as discussed in Sec. 3.2.2.9 of Ch. 3). Amplifier and codec design, together with control of signal levels, also minimizes the effects of many distortion-related parameters.

7.3.2 Summary of User-Oriented Performance Goals

Chapter 3 discusses the translation of user-oriented performance to network-oriented performance and presents performance values (and ranges) for the network-oriented parameters/impairments. For the key transmission planning parameters/impairments, these are as follows:

End-to-End Loudness Loss: From Table 3.8 of Chapter 3, an acceptable range is from about 0 to 14 dB across all connections (intra-office, inter-office, and toll) with optimum values of about 3, 6, and 9 dB, respectively. An ultimate long-term optimum for an all-digital network is about 3 to 6 dB for all connections.

Noise: From Table 3.8 of Chapter 3, a value of noise less than 25 dBrnC is acceptable. A value less than 20 dBrnC has minimum impact on performance and is a reasonable objective for an all-digital network.

Delay: As discussed in Section 3.2.1.1 of Chapter 3, in the absence of echo (either on a complete 4-wire connection or on connections with echo cancellers), end-to-end one-way delay should be limited to $\leq$ 150 ms for terrestrial connections [12], and to $\leq$ 400 ms for satellite connections [13].

Talker Echo Path Loss (TEPL): As shown in Figure 7.4, TEPL is dependent upon the hybrid mismatch at the 4- to 2-wire conversion points. As given in Table 3.9 of Chapter 3, TEPL should at least exceed 20 dB on connections with echo path delay less than 30 ms and with no echo control devices. For higher delay connections with echo control devices present, at least 55 dB of TEPL should be achieved.

7.3.3 Methods of Controlling Impairments

The methods of controlling impairments are closely related to the technology used and the needs of evolving all-digital networks, such as providing bit transparency for all services. These methods are illustrated using the connection diagram in Figure 7.5.

Control of End-to-End Loudness Loss

As networks evolved from the VNL era to the FLP era, the range of loop plus terminal loudness loss characteristics have remained essentially unchanged, with values as shown in Figure 7.5(a) (see Sec. 6.2.3.1 of Ch. 6). However, control of loss within the trunk part of the network has changed from a distributed (VNL) methodology to a discrete methodology, as in the FLP, with loss implemented and controlled in the end office, as illustrated in Figure 7.5(a). Ideally, this loss should be implemented

Figure 7.5(a) VNL and FLP loss control.

on the analog side of the digital-to-analog (D/A) converter, and this is generally the case. Digital loss implementation (on the digital side of the D/A converter) is also possible, but is not compatible with the evolution to bit-transparent networks. However, in some systems the only solution is to provide digital loss until such time as the network evolves to all-digital working. Section 7.6 discusses some remote switching systems where digital loss is employed.

As the network evolves to all-digital working, loss will migrate to the digital terminal (telephone set) as shown in Figure 7.5(b). Section 7.5 discusses the development of the transmission plan to cover the ADN era.

Figure 7.5(b) All-digital network loss control.

Figure 7.5(c) Talker echo path loss (TEPL) control.

Control of Noise

Noise is largely a function of the technology used for transmission facilities. In the VNL era with analog facilities, idle circuit noise was directly related to connection/facility length. Noise on analog facilities can exceed 35 dBrnC (see Sec. 3.2.2.2 of Ch. 3) and compandors were sometimes used to improve the signal-to-noise ratio (SNR) (see Sec. 6.2.3.3 of Ch. 6). Digital facilities have the advantage of not generating excessive amounts of idle circuit noise. On a connection with a single codec, the end-to-end noise is generally on the order of 20 dBrnC. The introduction of other digital processing, such as echo cancellers, can contribute to the overall noise level, but not to an appreciable extent.

However, digital PCM systems as well as other processing systems, such as DCMEs, can contribute to significant levels of quantization distortion (see Sec. 6.2.3.4 of Ch. 6), and, for this reason, rules are developed to control the number of such systems in a connection. Section 7.10 discusses the case of 32-kb/s ADPCM and the deployment rules developed and agreed to by the industry to minimize the impact these devices can have on voiceband services, particularly voiceband data at 4.8 kb/s and higher rates.

Generally, DCME systems are not used in public networks because of the administrative problem associated with the control of the number of such devices that can occur in a connection, particularly in the North American multinetwork environment. This would require interworking agreements among the various network providers. However, such devices offer economic advantages in private/special service networks and on satellite-based systems, and are used for this purpose. They are also used on some international links forming part of an international connection.

Control of Delay

Delay is a function of the transmission facility (and physical distance) and switching node technology (see Sec. 6.2.3.5 of Ch. 6). From a planning perspective, control of delay can be achieved either through a choice of low-delay technology, or through routing restrictions to minimize connection distances. However, low-delay technology may have economic or capacity disadvantages. Also, routing restrictions are not always possible because of practical network constraints. Therefore, unlike loss and noise, there is little that a transmission planner can do to control delay, except to impose overall delay objectives and take note of the delay contributions from the various network elements. The subjective effect of delay on voice is closely related to talker echo path loss, which is controlled as discussed subsequently. Delay can also impact nonvoice services such as digital data, in which throughput is delay sensitive, depending on facility error performance. Transmission delay considerations are covered in Section 7.9.

Control of Talker Echo Path Loudness Loss

Talker echo path loudness loss is a function of network loss, echo return loss, and the terminal characteristics, as shown in Figure 7.5(c). In this example, with a digital telephone set at the A end of the connection, and an analog set at the B end, the TEPL values as observed at the two ends can be significantly different.

For end A, TEPL = 2 × network loss + echo return loss + TOLR + ROLR. The network loss is fixed (according to the FLP), and the echo return loss (ERL) is a function of the loop characteristics. There is a tendency for echo return loss to improve with the deployment of remote digital switching and subscriber carrier systems, as discussed in Sections 7.6 and 7.7. Control of TEPL by the digital set transmit (TOLR) and receive objective loudness rating (ROLR) characteristics is discussed in detail in Section 7.5.

For end B, there are no 4- to 2-wire conversion points to complete an electrical echo path. However, there does remain an acoustic echo path in the digital terminal due mainly to acoustic coupling between the transmitter and receiver. There may also be some crosstalk leakage in the digital set electronics and handset cord. However, these echo signals can generally be controlled by appropriate digital set design to ensure they do not significantly degrade end user performance (see Sec. 7.5.1 of this chapter).

7.4 THE FIXED LOSS PLAN

Appendix A.7.2 provides an overview of the basic attributes of the FLP and the planning process used to develop the loss values. However, the key attributes, namely, the implementation and control of loss aspects, are summarized here.

7.4.1 Implementation and Control of Loss

The main attribute of the FLP is that loss is controlled in the receive direction by the terminating end office. Generally, this loss is analog, is inserted on a per line card basis, and is a function of the connection type, as illustrated in Figure 7.6. Loss is

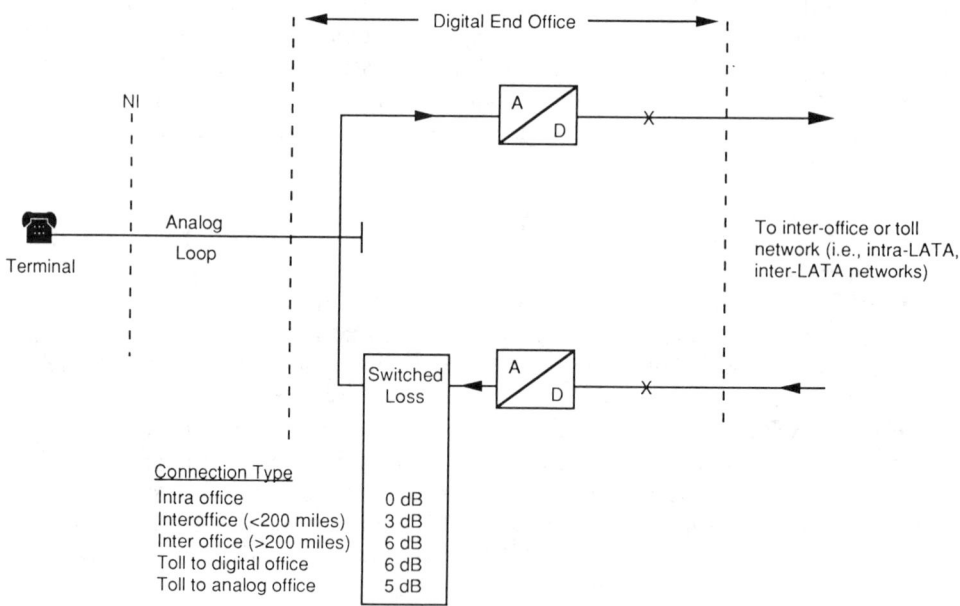

Figure 7.6 The FLP: control and implementation of loss.

therefore fixed as a function of connection type, but is switched depending upon the connection type. The loss, as illustrated in the figure, varies between 0 dB for an intra-office connection, to 6 dB for a toll connection. Appendix A.7.2 discusses the process leading to this choice of values in North America.

Generally, smaller countries use a fixed value of loss for all connection types. CCITT Rec. G.121 [10] contains a table of loss values in terms of the receive and transmit pads that are inserted on a connection basis.

7.4.2 Intra-Office Considerations

With the introduction of digital switches, it was necessary to pay particular attention to the listener echo impairment arising from echos circulating in the 4-wire transmission paths in the connection. This impairment is controlled by listener echo path loss (LEPL), as discussed in Chapters 3 and 4 [14].

An intra-office connection, as illustrated in Figure 7.7, is the most sensitive in this regard, because the inserted connection loss is 0 dB. Therefore, such connections rely on good loop balance return loss (RL) at each 4- to 2-wire conversion point (see Sec. 6.2.3.4 of Ch. 6). This is required not only to maintain stability, but also to support voiceband services, in particular voiceband data.

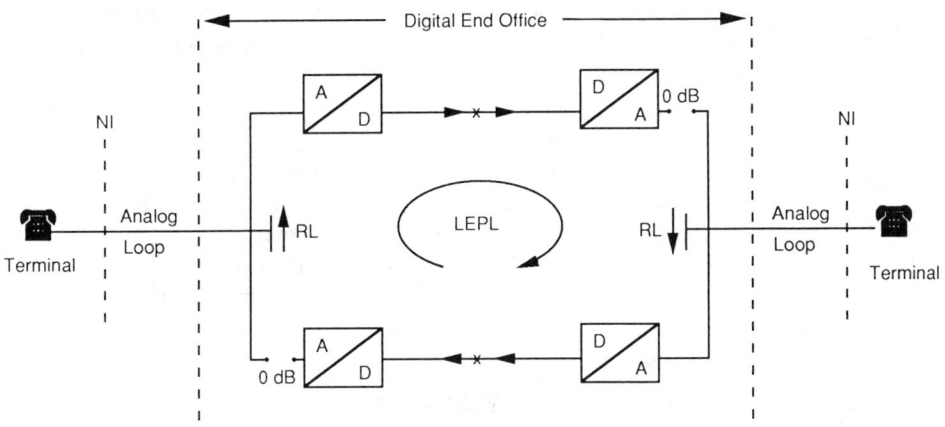

Figure 7.7 Intra-office connection through digital switch.

To ensure connection stability, a LEPL criterion was developed. This criterion imposes a minimum singing margin (SM) of 4 dB, and ensures satisfactory voice performance by imposing a minimum weighted echo path loss (WEPL) of, for example, 18 dB (with standard deviation (SD) of 3 dB) for a listener echo path delay of 4 ms.

LEPL is a function of the loop balance return loss obtained at each 4- to 2-wire conversion point. In turn, the return losses are a function of the loop characteristics. To ensure meeting the LEPL criterion, a double balance network design has been generally adopted in which loaded and nonloaded loops are assigned a different balance network. This maintains the balance return loss statistic of 11 dB, with 3 dB SD achieved with analog end offices, and can result in improvements in return loss under some loop conditions. In addition, some switching (EO) systems were developed that can automatically, on a routine per call basis, determine the return loss stability and insert 2 dB of loss if this is required to meet the 4-dB overall minimum singing margin requirement. The switched balance network capability is specified in Section 7.0 of the LATA Switching Systems Generic Requirements (LSSGR) [15] and is a standard feature in most digital end office switch designs [16].

Note that connections in smaller countries that have a single fixed (nonzero) loss independent of connection type should not be subject to listener echo impairment.

7.5 THE ALL-DIGITAL NETWORK ERA—THE FUTURE

Ensuring end-to-end performance across multivendor networks in the future will be achieved through end-to-end digital connectivity. The ultimate all-digital network will be relatively simple in transmission planning terms. However, as eluded to earlier in this chapter, evolution towards this ultimate objective presents a number of transmission planning challenges.

This section discusses the likely stages in this evolution process, but in reverse chronological order. That is, the ultimate objective of an all-digital network is addressed first, followed by the interim period involving evolution from the current FLP situation discussed in Section 7.4 to the all-digital objective.

The stages to be covered (in reverse chronological order) are as follows. The evolutionary period is covered by items (2) and (3).

1. End-to-end all-digital network (including ISDN).
2. Interworking of "all-digital" and FLP-based networks.
3. Simplification of the FLP.

These stages are a reasonable representation of the future. Industry agreement on a transmission plan(s) for digital public and private networks and ISDN are under development within the U.S. in the following standards forums:

T1Q1.1 — Public Network Domain [11]

TIA TR41.1 — Private Network Domain [17]

In addition, liaison activity is in place between these forums to ensure compatible interworking between public and private networks, specifically, in terms of CPE and network equipment requirements as discussed in Section 8.6.1.1. The development of a transmission plan covering international digital connections is the subject of study in CCITT SG XII (see Sec. 8.4.1 of Ch. 8).

7.5.1 End-to-End All-Digital Network (Including ISDN)

Figure 7.8 illustrates an all-digital connection and the main transmission planning parameters. This connection consists of an all-digital network terminating at each end in a digital telephone (e.g., ISDN) set that provides the A/D and D/A conversion processes (i.e., the codec functions). The connection is bit transparent.

The key transmission planning parameters are

- Overall objective loudness rating (OOLR).
- Transmit and receive objective loudness ratings.
- Talker echo path, and acoustic echo path losses.
- End-to-end delay.

Note that idle channel noise associated with the codec process is assumed to be negligible, and is therefore not included in the above list. Each of the listed parameters is now examined in turn.

Figure 7.8 All-digital connection and key transmission planning parameters.

Overall Objective Loudness Rating

In Chapter 3, curves are given to illustrate the relationship between overall objective loudness rating and GOS. These GOS curves show that there is a broad optimum range of acceptable loudness rating in the absence of other impairments, such as talker echo.

Currently in North America, a value of 5 dB OOLR is being considered both in T1Q1.1 [11] and TIA TR41.3 [19] as a desirable overall objective for all types of connections. This is also compatible with CCITT SGXII Recommendation P.31 covering transmission characteristics for digital telephones [18] on international connections.

Transmit and Receive Objective Loudness Ratings

In an all-digital transparent connection, loss is entirely a function of the digital telephone set transmit and receive sensitivities. The current view of nominal TOLR and ROLR values in North America [19] is as follows:

$$\text{TOLR} = -46 \text{ dB (spread } -51 \text{ to } -38)$$
$$\text{ROLR} = 51 \text{ dB (spread } 46 \text{ to } 56)$$

These values result in an average of 5 dB OOLR. The rationale for the respective TOLR and ROLR values is based on compatibility with existing network return losses as discussed later in Section 7.5.2.

Talker Echo Path and Acoustic Echo Path Losses

As noted earlier, in current network connections, talker echo results mainly from reflections at 4- to 2-wire conversion points. In an all-digital connection, there are no such conversion points, and essentially no echo, except for such factors as the acoustic echo path between the telephone set receiver and transmitter.

Acoustic echo manifests itself by the transmitter sensing the acoustic signal as generated by loud speaker action of the receiver. Therefore, in Figure 7.8, at end A of the connection, a low-level echo signal will be present because of the acoustic echo path at end B, and vice-versa.

This echo is negligible compared to that produced by 4- to 2-wire conversion in current networks. Measurements of telephone sets indicate that the acoustic echo path loss is on the order of 40 to 50 dB [20–22]. Currently, echo cancellers are used on long delay connections to control echo. However, for all-digital and ISDN connections, echo cancellers are not compatible with network transparency for the transmission of digital data; that is, the cancellers would have to be disabled using appropriate signaling. An alternative approach currently being considered by the industry is to specify a minimum acoustic echo path requirement for digital telephones that is sufficient to control echo on long delay connections.

This approach is being addressed in the TIA TR41.1 forum as part of the development of an ISDN digital telephone set standard, and acoustic echo path losses of ≥ 45 dB are being proposed as a requirement, with 50 dB as a desirable objective [19]. Such a specification will also include other potential echo paths, such as those caused by crosstalk within the handset and electronics. The specification will also include a measurement methodology that accounts for the acoustic/electrical nature of the signal and its low level. TIA SP-1920A [19] also specifies the acoustic echo measurement methodology based on a weighted acoustic echo path response.

End-to-End Delay

In Tables 6.14 to 6.17 of Chapter 6, values of delay are given for various transmission facilities and nodal network elements. Delay is a function not only of these network components, but also of the network architecture. Network delay will increase in the future because of such factors as changes in access network architecture to include fiber-based remote switching systems and survivable ring structures [23-25]. In an all-digital connection the impact of nominal delay increases will not be significant if acoustic echo can be controlled. However, there is a need to consider the effect of high delay on conversation dynamics.

This happens on connections with one or more satellite hops. Although the use of one satellite hop is acceptable, the use of more than one is not encouraged, because with two hops, the round-trip delay of 1200 ms seriously affects conversational dynamics [26, 27]. Owing to this experience and to the trend towards increased network delays, T1Q1 has developed guidelines for the maximum delay for terrestrial connections, which are incorporated in the loss plan for evolving digital networks [11]. An additional consideration is the proliferation of relatively high-delay technology on the CPE side of the NI (e.g., mobile digital cellular radio [28]). Voiceband data services can also be impacted by end-to-end delay [29].

The current view is that the maximum terrestrial one-way delay should be between about 100 and 150 ms, sufficient to accommodate foreseeable delay increases, but low enough to minimize the impact on end user performance [11, 30].

Determining this delay objective, required a subjective test procedure that is more sensitive than the conventional mean opinion score (MOS) test described in Chapter 2. For example, the subjective criterion can be based on meeting the needs of end users who are most sensitive to delay, and by taking into account the effect on interruptability during a conversation [30].

For private networks (including mobile cellular systems) and their interworking with public networks, delay may extend beyond the limit proposed above for terrestrial connections. This situation will have to be accounted for by either relaxing the objectives to accommodate interworking, or by developing guidelines that indicate to the end user the impact of such interworking. Fortunately, as delay increases, the impairment is gradual so that conversation will still be possible but under a more structured format, with end user awareness that performance may be degraded.

Other aspects of transmission delay and echo control requirements are discussed in Section 7.9 of this chapter.

7.5.2 Interworking of All-Digital and FLP-Based Networks

Figure 7.9 illustrates a connection involving networking, and the related transmission planning parameters. For this connection, a digital telephone set at one end interworks with an analog set and loop at the other, a configuration that will be very common as networks evolve to all-digital working.

The digital telephone set provides the A/D and D/A conversion processes. The key transmission planning parameters are

- Overall objective loudness rating as determined by digital set transmit and receive loudness ratings.
- End-to-end loudness loss asymmetry.
- Talker echo path loss.
- End-to-end delay.

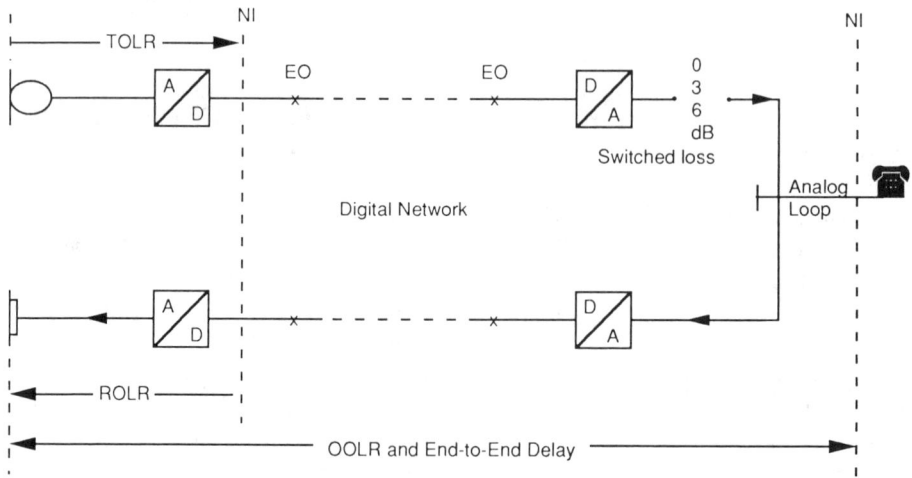

Figure 7.9 All-digital FLP interworking connection and key transmission planning parameters.

The rapid technological changes occurring in telephone networks require careful transmission planning to ensure compatibility with existing public switched telephone networks (PSTN). At the same time, consideration must be given to the future all-digital environment from both the network and terminal perspectives [31, 32].

Thus, the key requirements of an all-digital (including ISDN) network transmission plan are

1. Provide optimal performance on all-digital and ISDN connections.
2. Ensure end-to-end transparency for digital services.
3. Be compatible with current network(s).

The following subsections examine a transmission loss plan that encompasses both network and terminal considerations. This plan satisfies the above requirements in near optimum fashion, and has been agreed to by the U.S. industry [11, 19].

Loss Plan for Digital Access Network

In all-digital connections, talker echo will only be a potential source of degradation on high-delay connections because of acoustic echo and crosstalk in the terminal. This is less of a problem than in current mixed analog/digital connections, and echo on satellite and very long terrestrial connections will probably be controlled through appropriate acoustic echo specification, as discussed in Section 7.5.1. However, mixed analog/digital connections with significant echo generated primarily at 4- to 2-wire interfaces will predominate for the foreseeable future. Therefore, a transmission plan for future all-digital and ISDN connections must be compatible with the need to control echo in the existing mixed analog/digital network environment.

In general, three different options may be considered for the future all-digital network transmission plan. These options involve the use of

1. Switched digital loss pads in the network (i.e., in the end office).
2. Switched analog loss pads in the digital telephone.
3. Compromise, fixed analog loss pad in the digital telephone.

The first option is essentially an extension of the existing FLP. In this case, the sensitivities for the digital telephone would be designed to be the same as those for an average analog telephone set plus loop. According to EIA Standard RS-470 [33], these sensitivities are

$$\text{Transmit Objective Loudness Rating (TOLR)} = -46 \text{ dB}$$
$$\text{Receive Objective Loudness Rating (ROLR)} = 48 \text{ dB}$$

Thus, interworking of the all-digital ISDN network with the existing network would be straightforward. However, a loss plan based on providing digital switched pads in the end office is incompatible with the transparency requirement for ISDN and digital services. This approach is also complex in terms of interworking between private and public networks and in supporting integrated voice/data services, since signaling is required to disable the loss when other than voiceband services are carried.

The second option is not desirable since it would require signaling between the end office and digital telephone to control the insertion of the required loss on a per-call basis, and therefore increases the complexity and cost of the digital telephone.

This leaves the third and most preferred option of using fixed loss in digital sets. To determine the appropriate value of fixed loss to use in the digital set, a loss/noise and echo/delay transmission GOS analysis was conducted [31, 34]. In addition, the effect of loss asymmetry was investigated.

The results are in terms of a performance comparison between (1) mixed network connections under investigation with fixed loss in the digital set at one end, and a 500-type analog loop at the other end, as in Figure 7.10(a), and (2) reference connections indicative of the existing network with a 500-type set and analog loop at both ends, as in Figure 7.10(b). The core portions are digital in both cases. The fixed network parameters used in the study were as follows:

$$\text{TOLR} = -45.5 \text{ dB} \quad (\text{SD} = 3.2 \text{ dB})$$
$$\text{ROLR} = 47.5 \text{ dB} \quad (\text{SD} = 1.8 \text{ dB})$$
$$\text{ERL} = 11 \text{ dB} \quad (\text{SD} = 3 \text{ dB})$$
$$\text{Noise} = 20 \text{ dBrnC}$$

For each of the above noted transmission planning options, three types of connections were analyzed: intra-office, inter-office, and toll, as shown in Figures 7.10(a) and (b). Each of the mixed connections under investigation in Figure 7.10(a) was examined with three values of compromise fixed receive loss (i.e., 0, 3, and 6 dB). The results of the GOS analysis are shown in Table 7.1 [31, 34]. The results are given for the digital end of the mixed connections, which is the worst-case end from an echo/delay standpoint.

The results in Table 7.1 indicate that an additional 3 dB of receive loss in the digital telephone of the mixed connection provides almost comparable performance with the reference connections using the fixed loss plan. A more detailed discussion of the comparison in terms of connection type follows.

Figure 7.10(a) Mixed network connections under investigation having digital and analog access arrangements.

388

Figure 7.10(b) Existing reference network connections having analog access arrangements at both ends.

Table 7.1
Summary of Grade of Service Analysis Between Reference and Mixed Network Connections
(Results given for the digital end)

Type of Connection	Receive loss in Digital Terminal (dB)	Grade of service of mixed connections relative to references		
		Loss/Noise	Echo/Delay	Loss Assymetry
Intra-office	0	Equal	Equal	Equal
	3	Equal	Equal	Equal
	6	**Worse**	Equal	**Worse**
Inter-office	0	Equal	**Worse**	Equal
	3	Equal	Equal	Equal
	6	**Worse**	Equal	**Worse**
Toll	0	Equal	**Worse**	Equal
	3	Equal	Note	Equal
	6	Equal	Equal	Equal

Equal: Grade of service of mixed connections equal or better than reference connections.
Worse: Grade of service of mixed connections worse than reference connections.
Note: Average grade of service of mixed connections worse for delays >12 ms worstcase grade of service of mixed connections better than reference connections for all delays.

Mixed Intra-Office Connections

In this case, the use of 3 dB of receive loss in the digital terminal results in negligible degradation in loss/noise performance and an improvement in echo/delay performance relative to the reference connection with 0 dB loss. The negative effect of the 3-dB inserted loss is offset by the much smaller variation in sensitivity associated with a digital telephone, as compared to an analog telephone and analog loop. In particular, the effect of long, high loss analog loops is eliminated at the digital end of the connection.

Regarding loss asymmetry, 6 dB loss in the digital terminals for the mixed intra-office connections results in significantly greater loss difference in the two directions as compared to the loss difference on the corresponding reference connections. However, 3 dB loss in the digital terminal, together with the smaller variability in sensitivity of the digital set, results in smaller and acceptable loss differences in the two directions.

Mixed Inter-Office Connections

In this case, the 3-dB receive loss in the digital terminal results in no degradation in performance, since this loss is the same as in the reference connection. In fact, there is some improvement in performance because of the smaller variation in sensitivity at the digital end of the connections, as discussed above. For similar reasons, loss asymmetry is also improved.

Mixed Toll Connections

In this case, the 3-dB receive loss in the digital terminal results in lower average echo/delay performance than for the reference connections. This is illustrated in the two top curves in Figure 7.11. For one-way delays less than about 19 ms, there is a degradation in GOS of <2% Good or Better. This is considered acceptable, since a 2% degradation is not expected to be perceived by end users [35]. For the worst-case connections (the two bottom curves in Fig. 7.11), performance of the mixed connection is better than for the reference connection, again because of less variation in sensitivity in the digital terminal.

Figure 7.11 Comparison of % GOB grade of service for mixed and reference toll connections.

When the one-way delay is longer than 19 ms, the deterioration for the average connection becomes significant. However, this will be controlled under existing North American echo control rules where echo control devices are inserted for one-way delays longer than about 20 ms (as discussed in Sec. 3.2.2.3 of Ch. 3). In addition, the trend towards better loop plant return loss characteristics (i.e., use of double balance networks) significantly improves performance. This results from the approximately 3-dB improvement in average return loss from the use of a double balance network (see Sec. 6.2.3.4 of Ch. 6). For example, although not shown in the figure with a double balance network, the performance cross-over point in the upper curves in Figure 7.11 increases from about 12 ms one-way delay to about 18 ms, with the 2% degradation point occurring beyond 25 ms.

Use of 6 dB of loss in the digital terminal would obviously be preferable from the standpoint of toll connection echo/delay and asymmetry performance. However, 6 dB would adversely degrade intra-office loss/noise and asymmetry performance. Thus, overall, 3 dB is judged to be the best compromise.

Performance at the Analog End of Mixed Connections

Table 7.1 applies to the user at the digital end of the mixed connection. For the user at the analog end, echo/delay performance will not be a significant factor because the return loss at the digital end will be considerably improved (a function of the acoustic echo characteristic of the digital telephone). Performance as perceived at the analog end will, therefore, depend mainly on the loss/noise factors. Loss/noise performance of mixed intra-office connections will be essentially as good as the reference intra-office connections, because the digital telephone considerably minimizes overall loss variability. For the same reasons, loss/noise performance of the inter-office and toll connections will be better than the corresponding reference connections.

All-Digital Connection Performance

The use of digital terminals with 3 dB loss at both ends of connections results in an overall connection loss of 5 dB OOLR. This is a desirable objective, as discussed in Section 7.5.1.

7.5.3 Simplification of the FLP

In Sections 7.5.1 and 7.5.2 evolutionary transmission planning considerations have been presented addressing compatibility with the current FLP as the network evolves to the ideal all-digital state. This section will discuss an additional nearer term step that has been agreed upon by the industry.

The considerations of importance to operating companies for migrating loss implementation from the network to the terminal are to achieve transparent digital connectivity, and to eliminate the switched loss currently required by the FLP. The latter consideration has operational benefits.

The FLP was developed taking into account analog loop loss distributions and the need to ensure acceptable GOS to the end user, given the mix of analog and digital network facilities. The analog loop plant was the main consideration, in particular, the large spread of loop lengths and losses. With the increasing penetration of digital remotes and digital loop carrier systems (discussed in detail in Secs. 7.6 and 7.7 of this chapter), there is a clear trend to migrate the line card (which includes the 4- to 2-wire conversion, connection loss, and A/D and D/A converters) nearer to the NI, with a resulting decrease in the analog loop plant length and loss (see Fig. 7.12). This raises the possibility of simplifying the FLP to account for these changing network statistics.

In Figure 7.12(a) the conventional FLP is illustrated with the line card located at the EO (host switch). In Figure 7.12(b), the line card is located at a position immediately adjacent to the NI, resulting in a near zero loss loop. This arrangement is achieved through the use of subscriber carrier or remote switching systems in the access network. These systems are discussed in detail in Sections 7.6 and 7.7 of this chapter. These discussions include the rationale for using a single nonswitched network loss to approximate the sensitivity of an average loop plus telephone set.

Given this evolutionary trend (of increased penetration of digital technology in the access network), there will not only be an increasing percentage of short loops, but also an increasing number of digital telephones that use a fixed compromise receive loss of 3 dB, as described in Section 7.5.1. This raises the question of whether the three main switched loss values of 0, 3, and 6 dB will still be required at the analog access end of the connection.

Although there is some justification from an operational viewpoint [36] for collapsing the 3- and 6-dB switched loss values into a single value for inter-office and toll connections, there will, however, remain a substantial percentage of end offices serving the general analog loop plant (i.e., without digital access) for which 0 dB may remain a requirement for intra-office connections. For this reason, proposals have been made for a simplification of the FLP [36–39]. These proposals contain two main options as follows:

Option A	Intra-local connection	—0 dB (or another value)
	All other connections (including remote-remote)	—6 dB (or another value)
or		
Option B	All connections	—6 dB (or another value)

Based on these proposals and the need to maintain flexibility as the network evolves, Technical Subcommittee T1Q1 of Committee T1 has developed a draft loss planning standard [11], which at the time of publication was going through the approval process.

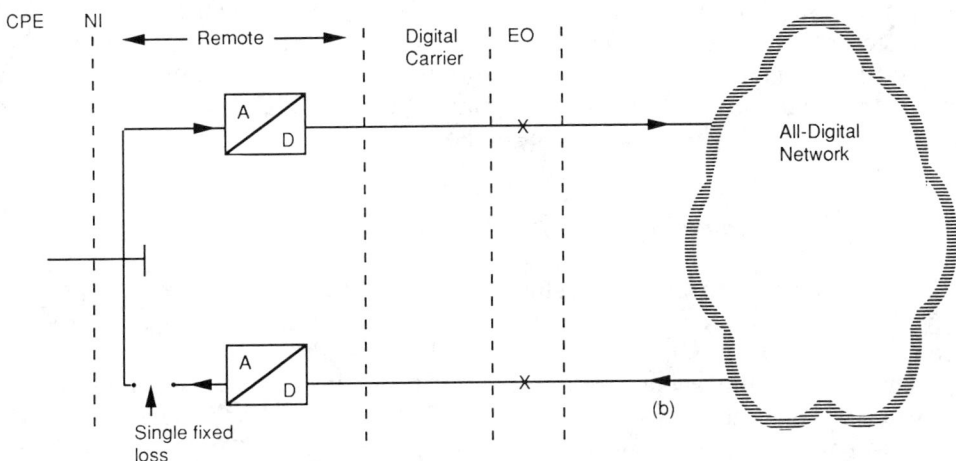

Figure 7.12 FLP implementation: (a) conventional (b) simplification.

The network loss plan may be summarized as follows:

Standard Values: These cover interconnections between different network providers.

- *Inter-LATA and International:* 6 dB to meet talker echo considerations and international requirements.
- *Intra-LATA (different ECs):* 3 or 6 dB in the interim with 6 dB a long-term preferred objective.

Guidelines: These cover connections provided by a single network provider

Intra-LATA (same EC) covering connections to remote switching systems	0 dB up to 3.2 msec round trip delay 3 dB up to 8 msec round trip delay 6 dB up to 12 msec round trip delay

7.6 LOOP CARRIER SYSTEMS

7.6.1 Digital Loop Carrier (DLC) System

In the 1970s, DLC systems [7] were introduced as an economic alternative to metallic pair facilities for long feeder routes (i.e., as pair gain systems). The basic system architecture is illustrated in Figure 7.13. This consists of a remote terminal (RT) at the subscriber end connected to the distribution part of the loop plant. The RT is also connected via a digital (usually T-carrier) facility to a central office terminal (CT) in the end office, and then to the end office switch.

Transmission Planning Criteria, Parameters, and Methodology

The transmission planning criteria for the DLC application, is to provide subscribers terminated on the RT with the same or better transmission GOS as provided with conventional analog loops; see Figure 7.13(a) and (b). The main transmission parameter, therefore, is network loss. However, this application also serves to demonstrate the presence of listener echo—see Figure 7.13(c)—as specified by WEPL and SM parameters discussed in Sections 3.2.2.4 of Chapter 3 and 4.3.4.3 of Chapter 4.

Methodology

As illustrated in Figure 7.13(c), the DLC system introduces a listener echo path between the major sources of echo located at the RT and CT 4- to 2-wire conversion

Figure 7.13 Access (loop) network arrangements: (a) analog loop access; (b) digital loop carrier (DLC); (c) listener echo path (LEPL) for DLC system.

points. To protect against instability or near singing conditions, a SM of at least 4 dB is desirable under all operating conditions. This can be achieved by inserting 2 dB fixed loss, as in the listener echo path indicated in Figure 7.13(c). With this approach, the maximum loss of the physical loop must be limited to 6 dB, given that the maximum designed loop loss is about 8 dB for analog access, as in Figure 7.13(a). Other loss tradeoffs may be possible [15]. An alternative approach is to provide zero loss within the carrier system, and allow 8 dB maximum loss for the physical loop. However, steps then must be taken through selection of appropriate balance networks to ensure that unstable conditions do not occur. A tradeoff therefore exists between loop range limitation and ensuring stability. Systems with 0 dB inserted loss and systems with 2 dB inserted loss are currently in use in the network. For these systems, the end office switch provides the additional switched network loss according to the FLP, and in the same way as for conventional loop access; see Figure 7.13(c).

Balance networks used in DLC systems generally follow those used in the EO, although other networks may be used provided they meet the WEPL objectives [15].

In summary, the planning criteria are to provide access network loss consistent with the maximum loss of conventional analog loops, and to ensure stability.

7.6.2 Integrated Digital Loop Carrier (IDLC) Systems

With the integration of digital switching and digital transmission, the CT at the end office can be interfaced directly with the digital switch, thereby avoiding the D/A and A/D conversion between the CT and the switch. Technical requirements for IDLC systems are given in Bellcore Technical Reference (TR)-303 [8]. This results in an IDLC system as illustrated in Figure 7.14(a). From a transmission planning perspective, this eliminates the listener echo path problem in the access network, although listener echo can still occur, and must be controlled for intra-office connections. In addition, control of network loss must still be provided. Two alternatives are given in Figure 7.14(b). Firstly, switched digital loss can be provided in the end office switch. Secondly, switched analog loss can be provided in the RT. Therefore, there is a tradeoff between nontransparency caused by a digital loss implementation in the first instance, and a more complex RT configuration with the additional signaling required to switch the analog loss in the second instance. However, with the trend toward digital network transparency and, possibly, a simplified two loss (0 or 6 dB) network, IDLC systems could evolve to a configuration in which the RT implements a single fixed analog loss (i.e., 6 dB), independent of the type of connection [40]; that

Figure 7.14 Integrated digital loop carrier (IDLC) system: (a) general configuration; (b) network loss alternatives.

is, IDLC-IDLC connections would be like a remote-remote connection with 6 dB loss implemented.

7.6.3 Extension of IDLC to the Customer's Premise

As noted earlier, the trend toward all-digital networks and ISDN is causing migration of the line card functionality closer to the NI interface, and current systems are being developed in which the RT will be located very close to this interface. Although there are regulatory issues to be resolved with respect to whether the RT will be designated as CPE equipment or network equipment, the transmission planning issues are the same.

In Figure 7.15 the RT is shown located adjacent to the network interface. In most instances, the served loop plant will be very short, with a range of from zero to about 300 feet. This raises a number of transmission planning issues related to—

1. Compensation of loop loss (for short loops).
2. Network connection loss requirements (x dB).
3. RT line card balance impedance Z_B.
4. RT line card input impedance Z_{in}.

These issues have been discussed by the industry [41], and the following factors and values have been presented for future consideration:

Compensation of Loop Loss

This loss is analogous to the fixed loss specified in the PBX standard developed by EIA [9]. The loss value is 3 dB and serves to compensate for short loops so that each short loop plus telephone set is approximately equivalent in sensitivity to an average set plus an average loop. As shown in Figure 7.15, the 3-dB loss is implemented in both the transmit and receive paths, and provides an additional 6 dB of return loss. This provides very significant improvements of talker echo path loss and ensures connection stability.

Network Connection Loss Requirement (x dB)

In addition to loop loss compensation, a fixed network loss is required for compatible interconnection with the FLP network and for ensuring satisfactory echo performance. Current values being considered for x are 3 dB and 6 dB.

RT Line Card Balance Impedance Z_B

Given the short loop plant and no loaded loops, Z_B could be made equal to the balance network impedance for conventional nonloaded loops. Alternatively, a more

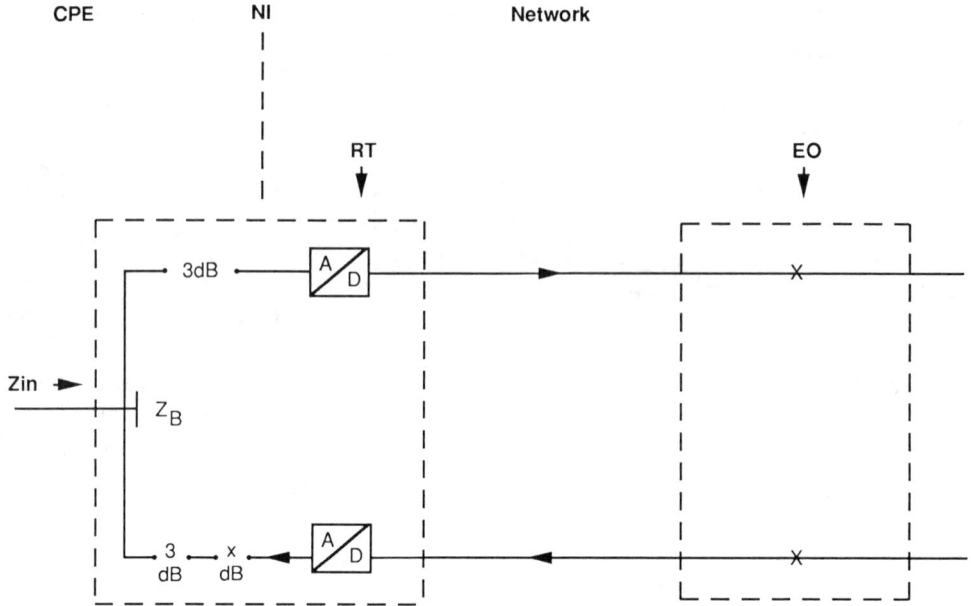

Figure 7.15 RT located adjacent to the network interface (NI), and transmission planning factors.
Zin = input impedance
Z_B = balance impedance
3dB pads provided for short loop compensation
xdB is required to meet network end-to-end connection loss requirements.

optimized network could be developed that is specific to short loops. In addition, if the IDLC system is designated to be CPE, that is, on the customer premises, the PBX balance network specification [9] could be used.

RT Line Card Input Impedance (Z_{in})

The input impedance requirements presents an interesting challenge in that the conventional value (e.g., 900 + 2 μF) for an end office is not optimum, particularly if the RT provides reduced constant current feed to the telephone set. Given the need to minimize power consumption in the RT as much as possible, reduced current feed is highly likely in this environment.

With reduced current feed and a short loop, sidetone balance in the telephone can be adversely affected [42]. This imbalance can be corrected by an appropriate complex input impedance design for the RT line card, and this approach is currently being considered by the industry. The use of a complex input impedance Z_{in} will require nonconventional test measurement equipment. However, this problem can be overcome based on developments in CCITT, where the European trend toward a complex impedance Z_{in} has been accommodated by suitable test procedures [43].

7.7 REMOTE SWITCHING SYSTEMS

A central office switch may serve as a host switch for two types of access network applications. In Section 7.6, loop carrier applications were discussed that provide an economic replacement for the existing feeder portion of the loop plant (i.e., as pair gain systems) and serve a somewhat shorter portion of the loop plant.

In this section, RSSs are addressed, which serve approximately the same general loop population as the host. Therefore, although the transmission planning considerations are similar in principle to the DLC case, the RSS solution is more difficult because a tradeoff does not exist between RSS loss and loss of the distribution portion of the loop plant.

7.7.1 Transmission Planning Objectives and Parameters

When RSSs were initially conceived, the planning objective [44] was to provide end users within the EO and carrier serving area (CSA) environments with essentially the same GOS, because the RSSs are considered to be an extension of the EO (or host switch). Figure 7.16(a) illustrates a RSS connected via the host switch to the network

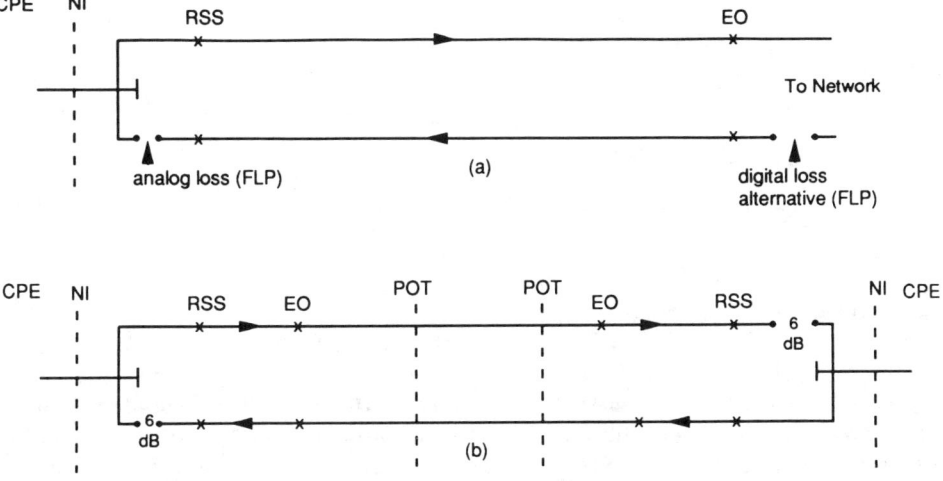

Figure 7.16 Examples of connections between RSS and the public network: (a) RSS to PSTN; (b) RSS to PSTN to RSS.

and providing intra-office, inter-office, and toll (inter-LATA) connections. On inter-office and toll connections, conventional 3 and 6 dB of loss, respectively, is inserted, and generally these losses are sufficient to accommodate the additional transmission delay of the RSS application, especially when double balance networks are used in the line card, as discussed below. This is illustrated in Figure 7.16(b) for a toll (inter-LATA) connection with 6 dB of receive loss.

The challenge with the RSS application is to provide 0-dB inter-office connections, particularly when the RSS only supports inter-local switching. That is, RSSs can be of two types, either with intra-local or inter-local switching capability. Intra-local switching means that calls between users connected to the same RSS are switched by the RSS (less delay), and not by the EO switch, as with inter-local switching (more delay). Although the RSSs with inter-local switching are less complex, they have the disadvantage that, if the digital carrier to the EO fails, all users served by the attached RSS cannot call either to the network at large, or, more importantly, within their own community (i.e., to other users on the same RSS). For this reason, RSSs are generally provisioned with intra-local switching. However, it is still necessary to consider transmission planning for RSSs without intra-local switching, because they represent a worst case, and the same rules are required in any event for more general intra-office connections (i.e., not involving the same RSS).

The inter-local intra-office connection of interest is illustrated in Figure 7.17, with two end users A and B connected to the same RSS and to one another through the host switch. Note that the connection is stretched out linearly in Figure 7.17, for explanation purposes. As such, Figure 7.17 also reflect the case where users A and B are connected to different RSSs.

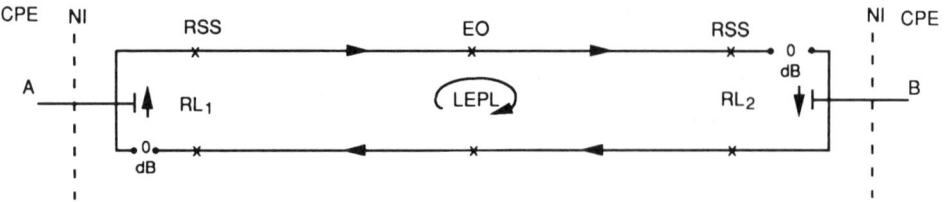

Figure 7.17 Intra-office connection using remotes.

As in the case of loop carrier (pair gain) systems, listener echo is a key planning parameter, but more so in this case, because, as noted earlier, the loop plant (and its loss) is not restricted, and RSSs are intended for operation up to 50 to 100 miles from the EO switch. The carrier facility transmission delay and RSS processing delay coupled with the zero inserted loss creates the potential for listener echo. Overall loss and circuit noise are generally not the controlling parameter for this type of connection.

7.7.2 RSS Planning Methodology

As discussed in Section 3.2.2.4 of Chapter 3, listener echo results from finite return losses at RL1 and RL2 (see Fig. 7.17), creating voiceband attenuation and phase distortion (also called near singing distortion), which manifests itself subjectively as a "rain barrel" or "hollow" sounding effect. This distortion also affects voiceband data performance (see Sec. 4.3.4.3 of Ch. 4).

Results from a GOS model on the subjective effect of listener echo were shown in Figure 3.8 of Chapter 3, which illustrates performance in percentage Good or Better (% GOB) versus WEPL, with round-trip delay as a parameter. From Table 4.11 of Section 4.3.4.3, the WEPL objective for a 4-ms delay may be expressed in terms of a mean of 18 dB (SD = 3 dB). Table 4.11 also shows that this objective is met for all types of loop plant when a double balance network is used. Therefore, from Figure 3.8, with an average round-trip delay of 4 ms (resulting from a RSS to host to connection of 100 miles), and the WEPL objective of 18 dB, listener echo will not be a problem since performance under these conditions is limited by the loss and noise; that is, performance is asymptotic to over 95% GOB.

This brings out the relationship between two important transmission planning criteria—remoting distance and GOS. For RSS to host switch distances beyond 50 miles, say to 100 miles, additional loss of 1 or 2 dB can, if necessary, be used to offset the listener echo effect. This would correspond to a round-trip delay of about 5 ms (see Sec. 6.2.3.5 of Ch. 6 for a detailed breakdown of access network delays). Transmission planning rules applicable for intra-office connections are

RSS to Host Distance (miles)	Loss (dB)	RSS to Host to RSS Round Trip Delay (ms)	
		Objective/Average	Requirement/Max.
0–50	0	4	5
50–100	1–2 (if necessary)	5.5	6.5

The additional 1 to 2 dB of loss violates the planning objective of providing the same GOS to users on RSSs as to users on the host switch. However, this additional loss is not sufficient to be perceived by the majority of end users, and is preferable to a listener echo impairment. An additional factor is that since the initial introduction of RSSs in the 1970s, most RSSs now incorporate the intra-local switching feature that eliminates the long RSS to host to RSS connection for users in the same community of interest (connected to the same RSS). Finally, for users connected to different RSSs, it may be argued that for a RSS to host distance of 100 miles, the connection is essentially equivalent to a 3-dB inter-office connection.

7.7.3 The Future

Since the development of the above planning rules for RSSs, the industry has considered and agreed upon a simplification of the FLP, as discused in Section 7.5.3

[11]. This simplification also accounts for the use of RSSs, given the important role they plan in evolving network architectures [40].

The guidelines covering RSSs may be summarized as follows:

- 0 dB can be used with up to 3.2-ms round-trip delay (approx. 80 miles).
- 3 dB can be used with up to 8-ms round-trip delay (approx. 360 miles).
- 6 dB can be used with up to 12-ms round-trip delay (approx. 600 miles).

These guidelines can also include RSSs in tandem as well as other configurations involving, for example, PBXs. A fixed loss approach, say of 6 dB, offers distinct operational/implementation advantages for RSSs, in that distance will no longer be a major planning consideration. Note that in countries in which loop plant characteristics and geographical distance allow for the use of a fixed loss independent of distance (e.g., the U.K.) [10], the application of remotes is considerably simplified.

7.8 PRIVATE NETWORKS

In North America, the TIA TR41.1 forum is the focal point for developing industry agreements on private network transmission planning, and has been since the conception of the work in 1974 (in EIA TR41.1) to develop requirements for digital PBXs, including values of loss.

This committee, under Project PN-1993, is addressing the all-digital network and ISDN requirements, and has ongoing liaison with T1Q1.1 on interworking considerations between private and public networks.

Generally, in developing a private network transmission plan, the same planning criteria and methodology are followed as for public networks. However, there are some key differences, such as the nature of the loop plant and interworking with public networks, which must be accounted for. ANSI Standard EIA/TIA-464A, released in final form in January 1989, addresses transmission requirements for digital PBX [9] in a mixed analog/digital environment.

Transmission Planning Objectives and Parameters

Figure 7.18 illustrates a private network connection and its key transmission planning parameters. It also serves to illustrate the main differences from a public network connection, such as the fixed transmit and receive loss to compensate for short loops.

As private networks evolve to all-digital working, the core part will consist of digital tie trunks and tandem PBXs. Tandem PBXs can either terminate a connection or, as in Figure 7.18, provide a tandem function. Currently, analog and digital tie trunks can coexist, although the trend is toward higher penetration of digital trunk technology. Analog trunks follow the VNL plan as in the public network domain.

Figure 7.18 Private network connections.

Apart from possible hierarchical differences in the core portion, the main difference between private and public networks is the role of the terminating switching machine, which in the case of private networks is a terminating PBX. Generally, PBXs serve a significantly different loop population than do public network end office switches. In the majority of cases, loops are short (e.g., within a building) and these loops are called on-premises station (ONS) extensions. However, in large installations such as airports, long loops can also occur, and these are called off-premises station (OPS) extensions.

The ONS extensions are similar to restricted loops served by DLC systems in the public network, and similar transmission planning considerations apply, that is, provision of loop compensation loss in addition to the network loss, as shown in Figure 7.18. A single series balance impedance Z_B also applies in the PBX, although the impedance value [9] is somewhat different than for the DLC case.

The private network loss requirements are similar to those provided by the EO in the public network, but a more extensive range of loss values is required because of the need to interface with both analog and digital facilities and to interwork with the public network. Examples of private to public network inter-connections are shown in Figure 7.19 for

1. Terminating PBX—EO, using digital interconnecting facilities.
2. Terminating PBX—EO, using analog interconnecting facilities.
3. Tandem PBX—EO or toll office, using digital interconnecting facilities.

The complete range of private and private/public network connections is illustrated in Figure 7.20. The private/public connections are those that connect to

- ATO—analog toll office
- DTO—digital toll office
- DEO—digital end office
- AEO—analog end office

Figure 7.19 Examples of digital PBX interworking arrangements between private and public networks.

These abbreviations and others used in Figure 7.20 are listed in Table 7.2. Note that, as shown in Figure 7.20, a satellite PBX (not to be confused with satellite transmission) receives all calls from the public network through the PBX on which it is homed. However, outgoing calls to the public network can be made directly from the satellite PBX. Apart from this special "satellite PBX" designation, a digital PBX can generally terminate a private network connection or act in a tandem switching capacity at intermediate points within the private network connection.

Figure 7.20 Representative PBX network connections both within the private domain and with the public domain.

Table 7.2
PBX Interface Types and Abbreviations

Interface Type	Abbreviations
Line interface to on-premise station	ONS
Line interface to off-premise station	OPS
Analog trunk interface to tie trunk (TT)	A/TT
Digital trunk interface to digital tie trunk, combination tie trunk or any other tie trunk with a digital termination at the PBX	D/TT
Analog trunk interface to analog satellite PBX tie trunk	S/ATT
Digital trunk interface to digital satellite PBX tie trunk	S/DTT
Analog trunk interface to analog end office (EO) trunk	A/EO
Digital trunk interface to digital EO trunk, combination EO trunk or any other EO trunk with a digital termination at the PBX	D/EO
Analog trunk interface to analog toll office (To) trunk	A/TO
Digital trunk interface to digital To trunk, combination To trunk, or any other To trunk with a digital termination at the PBX	D/TO

7.8.1 PBX Loss Plan Approach

The private network loss plan was developed along the same lines as the FLP. Figure 7.21 illustrates the main aspects for private network connections. Firstly, for ONS extensions a 3-dB fixed loss is implemented on the line card in both send and receive directions to compensate for the low loss in the short extensions. The 3-dB fixed loss is always associated with ONS extensions irrespective of the connections leaving the PBX. For OPS extensions, this loss is zero because there is significant loss in the long extensions. Therefore, as shown in Figure 7.21(a), the loss for intra-PBX connections is 6 dB with ONS extensions and 0 dB with OPS extensions. This is similar to the plan discussed in Section 7.6.1 for DLC systems serving restricted loops.

For inter-PBX private network connections using digital tie trunks (DTT), an additional fixed loss of 6 dB is implemented on receive paths, as shown in Figure 7.21(b). This loss is the same as that used in the FLP for public network toll connec-

Figure 7.21 Examples of private network connections.

tions, and may be implemented either on the line or trunk side of the PBX. Other types of connections using analog facilities are also possible, as discussed in the PBX standard [9]. In addition TIA TR41.1 has developed a technical bulletin containing a set of application guidelines covering the loss plan specification contained in the PBX standard. This technical bulletin [45] also provides a rationale on which the loss planning aspects of the standard are based.

For connections to the public network, the transmission plan assumes that the serving end office would implement the necessary FLP loss value, as shown in Figure 7.22(a). Therefore, in a digital connection, digital loss pads are required in the EO, which destroys digital network transparency. Although this is an assumption in RS-464 [9], current studies in TIA TR41.1 and T1Q1.1 are revisiting the transmission plan for these types of connections to achieve digital transparency to support digital data and ISDN services.

Figure 7.22(b) illustrates a tandem tie trunk connection to the public network. Because the terminating PBX implements a 6-dB fixed receive loss and the EO also implements 6 dB loss as part of the FLP, the tandem PBX is required to implement 6 dB in the send direction to retain symmetry. Again, although this connection is specified in RS-464, these digital loss requirements will likely be eliminated in the standard addressing all-digital connections and ISDN. An option is also shown in Figure 7.22(b) in the tandem PBX to implement a 3-dB loss and 3-dB gain in the send and receive paths, respectively. This option is provided to offset the relatively large end-to-end losses that otherwise occur in such connections (e.g., in the right to left direction in Figure 7.22(b), $6 + 0 + 6 = 12$ dB of network loss is reduced to $6 - 3 + 6 = 9$ dB).

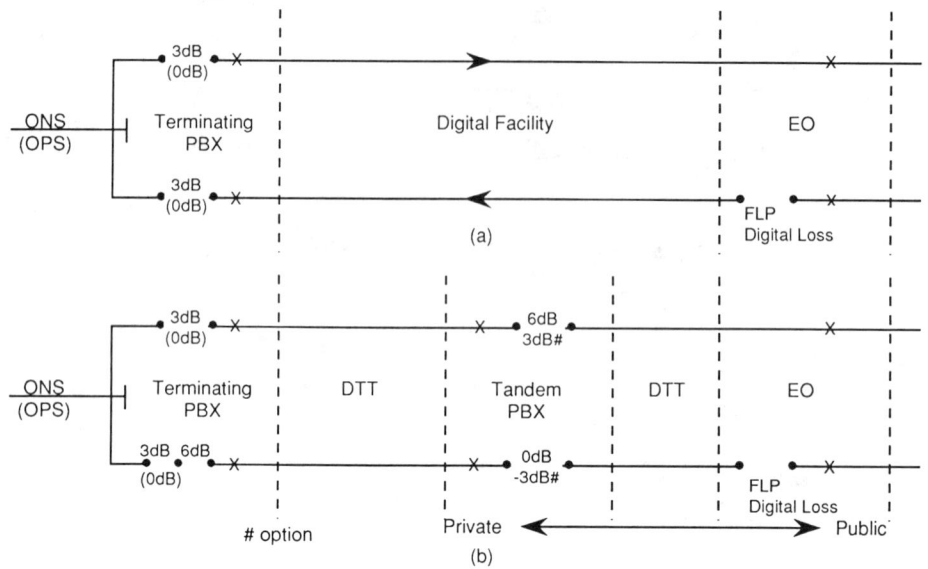

Figure 7.22 Example of connections between private and public networks: (a) PBX to public network; (b) PBX to tandem PBX to public network.

7.8.2 Existing PBX Loss Plan

Table 7.3 lists the loss values (port to port) that a digital PBX is required to implement depending on the type of line or trunk both entering and leaving the PBX. Table 7.2 lists the abbreviations used in Table 7.3.

These port-to-port losses cover the following PBX interfaces:

- On and off promise extensions (ONS, OPS).
- Analog and digital end offices (AEO, DEO).
- Analog and digital tie trunks (A/TT, D/TT).
- Analog and digital toll offices (A/TO, D/TO).
- Satellite analog and satellite digital tie trunks (S/ATT, S/DTT).

With reference to the connections in Figure 7.22(b), the required port-to-port losses can be determined from Table 7.3 by proceeding from left to right in the connection as follows. For the terminating PBX, the table column and row entries are ONS (OPS) to D/TT, respectively. For the tandem PBX the table column and row entries are D/TT to D/EO, respectively.

Table 7.3
PBX Port-to-Port Loss (see Table 7.2 for Terminology)

Port ↓	A ONS ↑↓	B OPS ↑↓	C A/TT ↑↓	D D/TT ↑↓	E S/ATT S/DTT ↑↓
1 → ONS ←	6 6	3 6	3 3	3 9	3 9
2 → OPS ←	3 3	0 0	2 2	0 6	2 2
3 → A/TT ←	3 3	2 2	0 0	-3 3	0 0
4 → D/TT ←	9 3	6 0	3 -3	0 0	6 0
5 → S/ATT S/OTT ←	3 3	2 2	0 0	0 6	0 0
6 → A/EO ←	0 0	0 0	0/2 (Note 2) 0/2	-3/0 (Notes 1,2) 3/6	0 0
7 → D/EO ←	3 3	0 0	2 2	0/-3 (Note 3) 6/3	0 0
8 → A/TO ←	6 6	3 3	0 0	-3 3	3 3
9 → D/TO ←	9 3	6 0	3 -3	0 0	6 0

Notes for Table 7.3 [9]
(1) The -3/3-dB value pair should be provided for connections between an A/EO port and a D/TT port serving as the interface to a combination tie trunk to a satellite PBX.
(2) It is desirable that the low-loss option (0/0 or -3/3) be used when the PBX-E0 trunk loss is greater than or equal to 2 dB and the ERL ≥ {18, 13} and SRL ≥ {10, 6} measured into a 900 ohm + 2.16 uF termination at the CO. {The notation {M, L} signifies that the median value is M and the lower limit is L.}
(3) The 0/6 dB loss pair shall always be provided. The -3/3 dB loss pair is a desirable option to be used for internetwork applications in which no significant configuration will encounter echo, stability, or overload problems because of the reduced loss. With the -3/3 dB loss pair, subscriber station DTMF signals transmitted through the DEO into the private network might experience nonrecoverable digit mutilation in secondary signaling applications (DTMF signaling after the connection has been established; e.g., order entry) because of the 3dB gain.
(4) Details on PBX loss plan and port connections may be found in the EIA "Digital PBX Loss Plan and Implementation Guide" [45].

7.8.3 All-Digital Network and ISDN Loss Plan

As for the public network, there is a great deal of interest in the industry to agreement on a PBX transmission plan that will support digital services. To this end, TIA TR41.1 has developed requirements that will enable PBXs to provide digital connectivity not only within private networks, but also when interworking with the public networks.

As pointed out earlier in this chapter, events in the industry have overtaken EIA/TIA Standard 464A [9], in that network providers see the advantage in providing transparent digital connectivity as opposed to providing the nontransparent type of connection illustrated in Figure 7.22(b). Indeed, the provision of digital loss in the EO in Figure 7.22(b), an assumption in RS-464A, is contrary to current industry thinking on the future evolution of all-digital network(s).

Discussion in TR41.1 has resulted in agreement on port-to-port losses for digital PBXs to support connectivity to ISDN compatible stations (ICS), and to ISDN compatible trunks (ICT). These values will be provided as an addendum to EIA/T1A-464A [17].

7.9 TRANSMISSION DELAY AND ECHO CONTROL CONSIDERATIONS

Network evolution toward digital connectivity results in a reduction of loss and noise to levels where they no longer significantly affect end users. On the other hand, transmission delay is increasing, and talker echo will remain a dominant transmission impairment until 4- to 2-wire conversion points are eliminated from the network.

The effect of talker echo and related transmission planning considerations have been addressed in Chapter 3 and in previous sections of this chapter, respectively. The aim here is to summarize and provide a future view of the impact of increasing network delay and how it may be controlled.

7.9.1 Echo Control

Echo cancellers are widely deployed to control the effects of talker echo on long-toll connections. The application of cancellers and their location in the network depends on two main criteria, as illustrated in Figure 7.23.

These criterion are end delay and application distance. End delay is the maximum round-trip delay within which the canceller can cancel reflected echos. End delay capability is a function of the echo canceller design, and can generally be provided for values typically between 16 ms and 48 ms [46], although higher delay capability may be required for some applications. Longer end delay capability requires a longer shift register length (memory) and is more costly. The echo canceller application distance and location is determined from loss/noise echo/delay GOS (see Sec.

Figure 7.23 Application of echo cancellers.

3.2.2.3 of Ch. 3) as well as network architecture/trunk length distributions and practical echo control deployment and cost considerations. Generally, cancellers are deployed on trunks that would result in connection talker echo path delay exceeding about 35 ms; that is, approximately the round-trip delay beyond which the 6-dB FLP inserted loss (inter-LATA connection) is not sufficient to attenuate the echo signal to acceptable levels (see Sec. 4.3.4.4 of Ch. 4).

Therefore, substantial increase in network delay can impact overall transmission quality through increase in end delay and application distance, unless the echo canceller deployment rules are changed. Given the current deployment rules adopted by the IECs, increased delay can introduce significant economic penalties and means other than increased deployment of echo control need to be considered (as discussed later in this chapter) to manage increases in network delay.

Figure 7.24 illustrates the functional effect of introducing a process or system which increases delay in the U.S. network, either in the LEC part or the IEC part of

Figure 7.24 Impact of increase in network delay.

a connection. An increase in delay in the LEC part will increase end delay and application distance. If the canceller's end delay capability is exceeded by the network end-delay, the cancellers will not cancel those high-delay echoes originating at the far end 4- to 2-wire conversion in the canceller's tail circuit. This situation results in failure of the canceller to control echo, and a corresponding serious echo/delay impairment. Increased LEC delay also impacts the application distance even if end delay is not exceeded. However, this results in a graceful degradation in echo/delay performance. The same degradation occurs for a delay increase in the IEC part of the connection. This must be compensated for by reducing the echo canceller application distance to compensate for increased delay.

Either reducing the application distance or increasing the canceller end delay capability are decisions that are not taken lightly because of the economic consequences. Likely solutions in the short term to cope with increased network delay will involve application of additional system-specific cancellers. These solutions are discussed below.

End-User Tolerance to Increased Delay

With the industry awareness of increasing delay, attention has been given to incremental network delay above that which already exists, and which end users could tolerate without network providers having to change existing echo control application rules.

A related study was carried out by the T1Q1.1 forum as part of their assessment of the impact of introducing certain delay-prone technology into the network [47, 48]. In this instance, the particular technology was zero byte time slot interchange (ZBTSI), a technique for providing 64 kb/s clear channel capability on DS1 circuits [49]. However, the analysis applies to any other technology that introduces delay. This study was based on the criterion that average loss/noise, echo/delay GOS could be allowed to degrade by up to 2% GOB before end users became aware of an effect. This study determined that network delay could increase by a 5-ms round trip (2.5 ms one way) without end users perceiving a significant difference in performance [50]. This value of incremental delay has become an industry guideline. Although this delay allows some flexibility in deploying delay-prone technology, it is still fairly restrictive in terms of, for example, inter-LATA connections that involve two LECs and an IEC, as well as two CPE environments, and possibly interworking with private networks. Thus, there are numerous opportunities for increased delay, with a very limited budget.

The following section provides examples of how to manage increasing delay beyond the 5-ms round-trip limit.

7.9.2 The Management of Increasing Delay

This section gives three examples of systems that require echo control management technique because of their inherent delay characteristics, that is, mobile radio, ATM, and personal communication systems (PCS). In addition, this section covers delay increase in IEC networks.

Mobile Radio Systems

An example of a network application that requires specialized echo control treatment is mobile radio. Low bit rate coding systems, which are being introduced to more efficiently utilize the radio spectrum, introduce delays on the order of 10 to 50 ms in each direction [28]. But mobile systems are 4-wire, and for calls between mobile users the connection is, therefore, 4-wire end-to-end, so that echo/delay performance is generally not an issue, and echo control is still not required (given that acoustic echo is sufficiently controlled (see Sec. 7.5.1)).

However, for connections between the mobile and public network users, attention must be given to the impact of the mobile low bit rate coding delay and associated enhanced echo on the mobile subscriber. In Figure 7.25, the mobile subscriber at end A of the mobile/public network connection will be subject to the varying conditions in the public network, and will, for example, encounter connections with and without echo cancellers; that is, the echo control characteristics of each connection cannot be predetermined. Therefore, given the significant increase in delay of the mobile system noted earlier (exceeding the incremental 5-ms round-trip criterion), steps must be taken to control echo generated at the 4- to 2-wire conversion point at end B of the connection in Figure 7.25. However, the user at end B in the public network will not require special attention, since the mobile system is 4-wire and generates no significant echo. This assumes acoustic echo in the mobile terminal is appropriately controlled and specified [19] as discussed in Section 7.5.1.

Returning to the mobile end A, because of uncertainty as to echo canceller deployment in the public network, an echo canceller is required for all mobile connections to the public network. The role of this canceller, as shown in Figure 7.25, is to cancel any and all echos up to end B of the connection, where these connections could be intra-LATA (intra-office or inter-office) or inter-LATA (toll). Therefore, with no knowledge of whether cancellers are already present in a given connection,

Figure 7.25 Application of echo cancellers in mobile to public network connections.

the mobile canceller must accommodate the largest expected end delay; this corresponds to maximum length inter-LATA connections not having echo cancellers.

Based on knowledge of public network delays and echo control deployment rules (see Sec. 4.3.4.4 of Ch. 4 and 6.2.3.5 of Ch. 6), cancellers are expected to require an end delay capability of at least 48 ms [51]. This is based on a reasonable allocation of round-trip delay of 8 to 15 ms to the LEC portion (not including mobile delay), and 36- to 40-ms round trip for the IEC portions not equipped with cancellers. In summary, the mobile radio application requires an integrated echo canceller with at least 48 ms end delay capability. Although in this discussion we considered the U.S. network environment, the same rules will apply throughout North America and, to a lesser extent, in smaller countries that incorporate delay-prone access technology.

Asynchronous Transfer Mode Systems

For some time, the industry has been addressing the problem of how to integrate the provision of services to the end user in a future broadband ISDN (B-ISDN) environment. Such services will range from voice and low-speed data at one end of the spectrum, to high definition television (HDTV), image transfer, and high-speed data at the other. Of the many techniques considered, including fast circuit and fast packet techniques, the industry focus is now on ATM as the transport vehicle of choice [52–55]. ATM attempts to combine the advantage of circuit switching (simplicity, high throughput, and low delay) with the advantages of packet switching (efficiency and flexibility of bandwidth utilization). In so doing, ATM assembles the user information (payload) into labeled cells (similar to packets). However, unlike a packet-switched system, the cells occur in fixed time slots and are relatively small, with a 5-byte header (label), and with the user information contained in a 48-byte information field [54].

For 64-kb/s voice service, the assembly (filling) of these cells with user information at an originating ATM node creates a delay in a connection of 125 μs/byte; that is, for a 48-byte information field, the delay in each direction would be 6 or 12 ms round trip. Additional but lower delay will also occur due to congestion control at intermediate ATM nodes and to buffering (smoothing) at the destination ATM mode.

Figures 7.26(a) and (b) illustrate an inter-LATA connection with a conventional, and an ATM-based LEC network, respectively. The ATM system provides access and includes the EO and access tandem (AT) switching functions. Figure 7.26(b) represents a likely introduction scenario for ATM. This figure also depicts both digital (i.e., 4-wire) terminal access and conventional analog access to the ATM-based LEC. Digital terminal access is more likely in this future environment; however, ATM interworking with existing analog access must also be considered during the transition. Note that, with digital terminal access, Figure 7.26(b) is very similar with respect to delay to the mobile radio connection in Figure 7.25. With a 48-byte cell size, the round-trip delay resulting from the ATM process will be at least 12 ms.

(a) Inter-LATA connection with conventional LEC networks

(b) Inter-LATA connection with ATM LEC network

Figure 7.26 Inter-LATA connection illustrating conventional and ATM-based LEC network and echo control requirements.

By accounting for the delay values of the network elements (in Fig. 7.26a) which the ATM system replaces, the incremental increase in delay over the conventional LEC network will be less than 12 ms, by about 3 ms [56]. Therefore, the incremental increase in delay is about 9 ms for a 48-byte cell size, which exceeds the 5 ms of allowable increment. Thus, to control talker echo on connections not having echo control, the ATM applications require an echo canceller with 48-ms end delay capability, as in the mobile radio application.

In addition to the need for a canceller facing the IEC network as discussed above, a canceller is also required on the terminal side for the case of analog access to the ATM network; see Figure 7.26(b). This echo control is required because of the possibility of the ATM network either exceeding the end delay capability of cancellers already in the existing IEC network, or of exceeding the 40-ms round-trip delay (application distance) criterion discussed in Sections 7.5 and 7.9. The end delay capability of the canceller on the terminal side of the ATM network can be less than for the canceller on the network side; a 16-ms end delay should suffice and account for interworking with private networks, given that private networks should provide some degree of echo control within its domain.

However, there are alternative ways of managing voice traffic in which only part of the cell is used by a given voice source, and therefore the cell assembly delay is less. One way is called *partial cell fill.* For example, if only 16 bytes are used, then the ATM-associated round-trip delay is about 4 ms, with a delay increment over conventional networks of about 1 ms. The U.S. contribution to SGXVIII [56] discusses this approach in detail and develops the following round-trip delay increments as a function of cell fill:

Cell Fill		ATM Associated Round Trip Delay	Round Trip Delay Increment*
16 bytes	—	4 ms	1.2 ms
32 bytes	—	8 ms	5.2 ms
48 bytes	—	12 ms	9.2 ms

*Compared with conventional LEC network

Using the 5-ms tolerable incremental round-trip network delay criterion discussed earlier, the 16-byte cell fill for voice suggests additional echo control is not required. This would also apply if, in addition, the LEC on the right in Figure 7.26(b) was also ATM-based. For 32 bytes cell fill, echo control is not required for one ATM network, as shown in Figure 7.26(b), but would be for two ATM networks supporting analog access at the end of the connection. For 48-byte cell fill, echo control is required in all cases where analog access is supported.

The disadvantage of the partial cell technique is that part of the cell bandwidth is wasted (traded for lower assembly delay). Another technique which overcomes this disadvantage and still reduces cell assembly delay is the shared, or composite, cell approach. In this case more than one user's information is put in each cell (e.g., 16 bytes from each of three users). The delay performance considerations are similar to those for the partial fill technique, but raise other issues regarding cell processing complexity and restrictions in cell routing. Covington and Vitella [57] discuss these and other techniques for managing voice traffic.

The application of cancellers on a system basis, as in the mobile radio and ATM examples discussed above, relies on the fact that cancellers can operate in tandem [58]. This is essential, since in these examples there may already be echo control deployed in the existing network whose operation must not be interfered with.

As ATM penetrates the network, the stage will eventually be reached where complete end-to-end all digital connections with digital terminal access will be achieved, and where echo control is no longer required (e.g., end-to-end all-digital ISDN connections).

Personal Communication Systems

In the late 1990s, the telecommunications industry started taking interest in personal communication systems [59, 60], and in 1991 Committee T1 set up Technical Subcommittee T1P1 to manage the standardization process (as discussed in Ch. 8).

Through a combination of wireless technology and low bit rate coding, PCS provides the end user with the capability of accessing public and private networks with a wireless portable unit.

The coding rate in PCS will likely vary from 32 kb/s down to 8 kb/s depending on application and quality considerations. Therefore, as for mobile systems, discussed earlier in this section, PCS will introduce delays which, for lower coding rates, will likely reach the same value. This will require echo control arrangements similar to that illustrated in Figure 7.25 for mobile systems.

Delay Increase in IEC Networks

Increased delay within the IEC part of inter-LATA connections can be addressed through application of cancellers on connections of shorter distance. Conventionally, with networks using hierarchical routing, echo control has been provided on connections in which the end-to-end delay exceeds about 20 ms (i.e., 40 ms round trip). This corresponds to distances of IEC or toll trunks above 1,850 miles in length. With nonhierarchical routing and the introduction of higher delay fiber transmission systems, these rules have changed. In the AT&T network, the echo control application distance is about 1,200 miles to accommodate nonhierarchical routing [61].

Echo control application distances are affected by delay increases, regardless of whether these increases are in LEC, IEC, or both networks. Technically, most delay increase can be accommodated by echo canceller deployment on shorter trunks within IEC networks along the lines discussed above. However, this generally means increased echo canceller deployment, and therefore increased cost. For this reason, IECs are generally not anxious to alter their application rules to accommodate increased delay in LEC networks. The increase in cost depends on IEC network architecture.

For example, the AT&T network, being mesh-like, is such that most inter-LATA connections (on a traffic basis) are not expected to be subject to echo control. However, for the US SPRINT network, being a long, thin architecture, most inter-LATA connections are already subject to echo control [62]. Therefore, the US SPRINT network would not be affected as much as the AT&T network in this respect.

An alternative is for delay increase to be accommodated in LEC networks. This is feasible by deploying echo cancellers on the IEC side of access tandem (AT) switches. It is discussed briefly in Section 9.5.2 of Chapter 9.

7.9.3 Industry Guidelines

Apart from the delay increases associated with specific systems, such as mobile, ATM, and PCS as discussed previously in this section, changing technology and network architectures are also bringing about general delay increases in the network. Recognizing this, Technical Subcommittee T1Q1 of Committee T1—Telecommunications has been working on the development of delay guidelines for the exchange access network. Several of the references [24, 25, 62–67] reflect industry awareness and the need for such guidelines. This work in T1Q1 culminated in 1991 with a draft supplement [12] to ANSI T1.506-1990 [68].

Based on an access reference model, these guidelines provided estimated delay maximum values, typical values, and minimum values as a function of NI to point-of-termination (POT) route distance. For example, at 600 miles (one way), the estimated maximum is 20 ms round-trip delay (with 10 ms associated with the network elements).

These guidelines also indicate that network providers should cooperate to seek a solution when delay is anticipated to be a problem (e.g., delay going beyond the guideline maximum). These guidelines, in conjunction with the absolute round-trip delay limit of 300 ms on terrestrial corrections [11], provide the North American industry with planning information necessary to accommodate technological and architectural network evolution with minimum impact on end user performance.

Although subject to similar technological and architectural changes, networks in smaller geographical areas, such as the European countries, will not be subject to the same delay constraints. However, on an international basis consideration must be given to these evolutionary changes.

CCITT Rec. G.114 [13] covers the limitation of delay on connections that include satellite-based facilities. (See also Section 6.2.3.5.)

7.10 32-kb/s ADPCM APPLICATION RULES

Digital circuit multiplication equipment (DCME) such as low bit rate codecs (i.e., less than 64 kb/s) and digital speech interpolation (DSI) systems provide the means for at least doubling the traffic carrying capacity of, for example, T-carrier systems operating at the DS1 rate (1.544 Mb/s). Examples of such systems are referenced in Chapter 1 ([23–30] of Ch. 1). In addition, there is current interest in the application of low bit rate coding systems in PCS for increasing the capacity of the available radio spectrum [59]. However, the increase in capacity is traded for a reduction in end-to-end performance associated with the performance characteristics. In the case of low bit rate codecs, quantization distortion increases with the number of codecs in a connection (see Sec. 3.2.2.5 of Ch. 3 and 5.4.2.2 of Ch. 5). For DSI systems, the performance degradation manifests itself mainly in the form of speech clipping (see Sec. 3.2.2.4 of Ch. 3). Transmission planning application rules are therefore required to ensure that end-to-end service needs are not significantly affected by the application of DCME. These rules include the maximum number of DCME systems allowed in a connection, where this number is a function of service type.

As an example of the need for DCME application rules, this section focuses on 32-kb/s ADPCM. ADPCM offers a practical method of doubling capacity at DS1 rates, and standards covering the ADPCM algorithms and line formats exist both under ANSI standards [69, 79] and in CCITT Recommendations [70, 71]. In particular, specific reference is made here to ANSI Standard T1-501-1988 [72], which provides tandem encoding limits for ADPCM systems incorporating algorithms as specified in ANSI Standard T1-301-1987 [69]. DSI systems will not be discussed in this section, insofar as they are not well suited for voiceband data service and are generally limited to specialized applications, such as satellite, undersea cable, and point-to-point private networks.

The development of application rules for DCME 32-kb/s ADPCM is discussed for three service categories of

1. Voice.
2. Voiceband data.
3. Signaling and facsimile.

A single ADPCM link (or ADPCM tandem; see Fig. 6.3 of Ch. 6) is a transmission system, or part of a transmission system, which performs one PCM-ADPCM transcoding followed by a corresponding ADPCM-PCM transcoding. Asynchronous tandeming as discussed in Section 5.4.2.2 of Chapter 5 is considered here as the appropriate worst-case performance scenario for planning purposes.

Voice

The criteria for determining the 32-kb/s tandeming limit is based on a measure called Q (signal-to-quantizing distortion ratio), as discussed in Section 3.2.2.5 of Chapter 3, 4.3.4.2 of Chapter 4 and 5.4.2.2 of Chapter 5. From subjective tests, the approximate perceptible threshold of Q is in the range of 20 to 25 dB. This is consistent with Figure 3.9 of Chapter 3, which illustrates Q in terms of a combined loss/noise/Q GOS model.

The laws of addition for Q are discussed in Section 5.4.2.2 of Chapter 5. Based on the value of Q for a single ADPCM link, as given in Table 6.9 of Chapter 6, the combined Q for any number of tandemed links can be determined.

Using a minimum value (threshold) of Q of 20 dB as the minimum end-to-end performance criterion, Table 3.11 of Chapter 3 shows that up to four ADPCM links without zero code suppression (ZCS) or robbed bit signaling (RBS) can be tandemed. Q is significantly degraded if the ADPCM algorithm includes ZCS and RBS as illustrated in Figure 5.20 of Chapter 5 and Table 6.9 of Chapter 6. In this case, only two ADPCM links meet the Q criteria of 20 dB.

For an ANSI T1-301-based ADPCM algorithm (which includes only ZCS), the guidelines in ANSI T1.501 specify a maximum of three ADPCM tandems, although four tandems would meet the Q of 20 dB criteria. This takes into account the presence of other coding systems such as 64-kb/s codecs and echo cancellers.

In the U.S. environment this provides an equitable allocation of one ADPCM tandem (asynchronously connected) in each of the LECs and one in the IEC making up an inter-LATA connection (see Figs. 4.2 and 4.4 of Chapter 4, or Figs. 5.2 and 5.9 of Chapter 5). Table 6.9 of Chapter 6 also lists the quantizing distortion units (QDU) associated with digital processes including ADPCM. One QDU corresponds to the distortion of one 8-bit 64-kb/s codec, as discussed in Section 5.4.2.2, and the net effect of tandemed digital processes in a connection can be approximated by adding the QDUs for each process. The CCITT in Rec. G.113 [73] has established a maximum provisional planning value of 14 QDUs for end-to-end international connections. This corresponds to a Q of about 20 dB and, in terms of the QDU values for ADPCM in Table 6.9 of Chapter 6, allows two to four ADPCM tandems, depending on the algorithm. As in the case of inter-LATA connection within the U.S., fourteen QDUs allows for one ADPCM in each national part and one in the international part of an international connection. Note that the worst-case ADPCM tandem having ZCS plus RBS would only apply to the North American part of the international connection.

With increasing digital connectivity there will be a trend for ADPCM links, either in public or private networks, to be synchronously (or semi-synchronously) tandemed (see Sec. 5.4.2.2 of Ch. 5). In this case, the performance improves and the planning limit of three tandems for voice service may be relaxed.

Voiceband Data

Section 6.2.3.9 of Chapter 6 provides a comprehensive view of the effect of both tandemed PCM and 32-kb/s ADPCM on a range of voiceband data modems. The modem data rates covered are between 2.4 and 9.6 kb/s.

The performance criteria for voiceband data are twofold, namely, a block error ratio (BLER) equal to or less than 10^{-2} for a block size of 1,000 bits, and a bit error ratio (BER) equal to or less than 10^{-5}. In addition, modem error performance is in the presence of representative values of continuous analog impairment, as discussed in Section 6.2.3.9 of Chapter 6. In Chapter 6, Table 6.25 summarizes results for CCITT asynchronously tandemed PCM and ADPCM codecs, and Table 6.28 summarizes the results for ANSI ADPCM codecs and includes the effect of ZCS and RBS.

In terms of an inter-LATA connection, from Table 6.28, one ADPCM tandem in each LEC and one in the IEC part will readily support voiceband service with a 2.4-kb/s (V22) modem type. However, at 4.8 kb/s, although generally three ADPCM tandems will support this speed, for many of the modems there are instances where this is not the case (e.g., for 4.8-kb/s QAM modem types). Although 9.6-kb/s is not included in Table 6.28, from Table 6.25, which provides results for the CCITT ADPCM codec, it is clear that 9.6-kb/s voiceband data cannot be supported over three asynchronous ADPCM tandems.

These voiceband data results illustrate the difficulty of developing easily administered application rules for 32-kb/s ADPCM in a multivendor network environment like the one in the U.S. However, because performance is service-dependent, T1Q1, in developing ANSI-T1.501-1988, formulated a solution in terms of two signal or service categories in which the number of ADPCM tandems are limited to ensure required end-to-end performance.

These service categories are as follows:

Category 1: Limit of three asynchronous T1.301-type ADPCM tandems.

Category 2: Limit is zero T1.301 ADPCM tandems.

The implication of category 1 is that, to support these services, each LEC and IEC can have one ADPCM tandem in their part of an inter-LATA connection. For category 2, no ADPCM tandems are allowed.

Therefore, from earlier discussion, voice service is in category 1 together with 2.4-kb/s voiceband data service (Table 6.28 of Ch. 6). On the other hand, 4.8-kb/s fall into category 1 in most cases, but with some exceptions. Finally, 9.6-kb/s voiceband data is in category 2.

Signaling and Facsimile

From Section 6.2.3.9 it is concluded that DTMF signaling and Group 1 and Group 2 facsimile can be supported with up to four ADPCM tandems. These therefore fall into category 1. However, Group 2 facsimile is generally considered to be in category 2 [72].

Other Deployment Considerations

In concluding this section it is important to consider that 32-kb/s ADPCM, like other digital processing systems, such as echo cancellers and A/μ conversion, is not strictly compatible with the provisions of end-to-end digital connection transparency. With the introduction of more fiber transmission facilities and ISDN, the network is clearly evolving to provide this end-to-end capability.

Therefore, the provision of 32-kb/s ADPCM must be considered as part of a network evolution plan in which, for example, either ADPCM is restricted to parts of the network which will not, for some time to come, provide digital transparency, or the ADPCM systems must be capable of being disabled when transparency is needed. Several references [74–76] discuss typical application scenarios in which these and other ADPCM planning considerations are taken into account. Note that the 32 kb/s ADPCM planning guidelines based on QDUs are generally not applicable at lower coding rates. Objective measurement methodologies are currently under study for such systems.

7.11 REFERENCES

1. AT&T Notes on the Network, 1980.
2. "Transmission Planning of Switched Telephone Networks," ITU, Geneva, 1976.
3. T.C. Spang, "Loss-Noise-Echo Study of the DDD Network," *BSTJ,* Vol 55, No 1, Jan. 1976, pp. 1–36.
4. CCITT Com XVI, "Transmission Plan for the North American Digital Network," No 68/69E, June 1978.
5. L.S. Dibiaso, "Transmission Consideration for Local Switched Digital Networks," *Telephony,* Oct. 1977, pp. 40–46.
6. J.E. Abate, and J.C. Lawson, "The Switched Digital Network Plan," *Telephony,* Oct. 1977, pp. 25–39.
7. Bellcore, "Telecommunications Transmission Engineering: Vol 2—Facilities," 3rd ed., 1990.
8. Bellcore TR-TSY-000303, "Integrated Digital Loop Carrier Systems Generic Requirements, Objectives and Interfaces," Issue 1, Sep. 1986.
9. ANSI/EIA/TIA-464A-1989, "Private Branch Exchange (PBX) Switching Equipment for Voiceband Applications."
10. CCITT Rec. G.121 (Annex c), "Loudness Ratings (RLs) of National Systems," *Blue Book,* Vol III, Fascicle III.I, Geneva, 1989, pp. 109–125.
11. T1 Advisory Group Doc. No. T1AG/91-189, "Draft Proposed American National Standard—Network Performance—Loss Plan for Evolving Digital Networks," T1 Letter Ballot—LB 272, Nov. 1991.
12. American National Standard—Supplement to ANSI T1.506, Network Performance—Transmission Specifications for Switched Exchange Access Network, ANSI T1.506a-1992.
13. CCITT Rec G.114, "Mean One-Way Propagation Time," *Blue Book,* Vol III, Fascicle III.I, Geneva, 1989, pp. 84–93.
14. J.R. Cavanaugh, R.W. Hatch, and J.L. Neigh, "A Model for the Subjective Effect of Listener Echo on Telephone Connections," *BSTJ,* Vol 59, July/Aug. 1980, pp. 1009–1060.
15. Bellcore R-TSY-000507 LSSGR (LATA Switching System Generic Requirements), Issue 3, March, 1989.
16. J.B, Terry, D.R. Younge, and R.T. Matsunaga, "A Subscriber Line Interface for the DMS-100 Digital Switch," NTC Rec., Vol 2, Washington D.C., Nov. 1979.
17. EIA Engineering Standards Proposal No. 1993-A, Proposed Addendum to EIA/T1A-464A, October 1991.
18. CCITT Rec. P. 31 "Transmission Characteristics for Digital Telephones," *Blue Book,* Vol V, Geneva, 1989, pp. 55–56.
19. TIA SP-1920A "Acoustic-to-Digital and Digital-to-Acoustic Transmission Requirements for ISDN Terminal," July 1990.
20. Bellcore, "Weighted Acoustic Echo Path Loss Requirements," Doc. No. TIA TR41.3/90-05-011, May 1990.

21. Northern Telecom, "Measurement of Handset Acoustic Echo," Doc. No. TIA TR41.3/90-06-011, May 1990.
22. AT&T Technologies, "Acoustic Echo Path Loss," TIA Doc. No. TR41.3/90-06-012, May 1990.
23. Bell South, "Round-Trip Delay in a Digital Switched Access Network," Doc. No. T1Q1.1/89-011, Feb. 1989.
24. Seiscor Technologies, "Echo Path Delay in Seiscor (Fiber Traq)," Doc. No. T1Q1.1/90-046, April 1990.
25. GTE Telephone Operations, "A Reference Model for POT-NI Delay," Doc. No. T1Q1.1/90-046, April 1990.
26. AT&T, "Effects of Echo-Free Delay on Conversations," Doc. No. T1Q1.1/89-017, Feb. 1989.
27. Telecom Canada, "Subjective Impact of Long Delays in the Absence of Echo," Doc. No. T1Q1.1/90-012, Feb. 1990.
28. J.E. Natvig, S. Hansen, and J. de Brato, "Speech Processing in the Pan-American Digital Mobile Radio System (GSM)—System Overview," *Globecom Rec.*, pp. 1060–1064.
29. General Datacom, "Effects of Network Propagation Delay on Multipoint Data Networks," Doc. No. T1Q1.1/90-060, April 1990.
30. Telecom Canada, "Subjective Impact of Long Delays in the Absence of Echo," Doc. No. T1Q1.1/90-012, Jan. 1990.
31. Northern Telecom, "Transmission Loss Plan for ISDN—Impact on the Network," Doc. No. T1Q1.1/86-086, Nov. 1986
32. Northern Telecom, "Transmission Loss Plan for ISDN—Impact on the Digital Terminal," Doc. No. TIA TR41.3/86-011 Sep. 1987.
33. ANSI/EIA RS-470-A-1987, "Telephone Instruments with Loop Signaling," 1987.
34. A. Niedzwiecki, R.T. Matsunaga, and P.V. Coverdale, "All-Digital Network Transmission Planning and Terminal Requirements," *IEEE JSAC*, Vol 7, No 8, Oct. 1989, pp. 1166–1171.
35. Bellcore, "The Effects on Voice Service of Delay Introduced by ATM," Doc. No. T1Q1.1/88-070, Oct. 1988.
36. Bellcore, "RDSN (Regional Digital Switched Network) Transmission Plan," Doc. No. T1Q1.1/88-007, March 1988.
37. Telecom Canada, "Telecom Canada Fixed Loss Plan," Doc. No. T1Q1.1/88-032, June 1989.
38. Pacific Bell, "Modern Fixed Loss Transmission Plan," Doc. No. T1Q1.1/88-080, Nov. 1988.
39. AT&T, "Loss on Inter LATA Connections," Doc. No. T1Q1.1/88-081, Nov. 1988.
40. US West Communication, "US West Recommendation on Loss/Level Planning," Doc. No. T1Q1.1/89-061, July 1989.
41. Northern Telecom Inc., "Transmission Planning Consideration for Customer Premise Collocated Loop Carrier Systems," Doc. No. T1Q1.1/89-052, May 1989.
42. G.I. Riley, and P.K. Webb, "Transmission Performance of 4-wire PABXs," *Proc Int. Conf. Private Electronic Switching Systems,* London, April 1978, pp. 165–168.
43. CCITT Rec. Q.552, "Transmission Characteristics at 2-Wire Analog Interfaces of Digital Exchanges," *Blue Book,* Vol VI, Fascicle VI.5., Geneva, 1989, pp. 134–158.
44. G. Williams, and D.J. Kavanagh, "Application of the Fixed Loss Plan in Local Distributed Switching Networks," ICC Rec., Seattle, 1980.
45. EIA/TIA-TSB-32, "DPBX Loss Plan Application Guide," August 1990.
46. Bellcore TA-TSY-000434, "Digital Echo Canceller Requirements," Issue 1, May 1987.
47. S.S. Gorshe, and G.J. Beveridge, "A Robust Technique for Providing Clear Channel Capability over Existing and Future Digital Transmission Facilities," *ICC Rec.,* Vol 3, Seattle, June 1987, pp. 42.5.1–42.5.10.
48. Northern Telecom Inc., "Evaluation of ZBS/TSI Method for 64 kbit/s Clear Channel Operation," Doc. No. T1Q1.1/85-016, 1985.
49. GTE Telecom Systems, "New Algorithm Which Offers Reduced Delay in the Zero-Byte Time Slot

Interchange (2BTSI) Clear Channel Schemes, and a Method for Improving Robustness," Doc. No. T1X1.4/85-DD, 1985.
50. T1Q1, "Liaison Letter from T1Q1.1 to T1X1.4 on 64 kbit/s Clear Channel Performance," Doc. No. T1Q1.1/85-046R1, 1985.
51. Northern Telecom Inc., "Transmission Specification for the Digital Mobile Network," Doc. No. T1Q1.1/90-069, July 1990.
52. T1S1, "T1S1 Technical Sub Committee, Broadband Aspects of ISDN, Baseline Document," Doc. No. T1S1.1/89-200, May 1989.
53. CCITT Rec. I.121, "Broadband Aspects of ISDN," *Blue Book,* Vol III, Fascicle III.7, Geneva, 1989, pp. 34–61.
54. CCITT Rec. I.150, "B-ISDN Asynchronous Transfer Mode Functional Characteristics," SG XVIII TD-39, Melbourne, Dec. 1991.
55. A. Kaser, and S. Rao, "International Standard of Broadband Aspects of ISDN," Globecom Rec. 1989.
56. United States Contribution to SGXVIII, "Echo Control for ATM Voice Services Interworking with PSTN," Doc. No. T1Q1.1/89-056R1.
57. W.O. Covington, and M.E. Vitella, "Voice Transport on an ATM Broadband Network," *ICC Rec.,* 1989, pp. 1921–1925.
58. AT&T, "Summary of Recent Canceller Studies," Doc. No. T1Q1.1/89-018, Feb. 1989.
59. Bellcore FA-NWT-001013, "Generic Framework Criteria for Universal Digital Personal Communications Systems (PCS)," Issue 2, Dec. 1990.
60. TIA MCD Microcell/PCS Ad-Hoc Committee, Final Report, April 1991.
61. G.R. Ash, A.H. Kafkerand, and K.R. Krushman, "Intercity Dynamic Routing Architecture and Feasibility," *Proc. 1 of 10th Int. Teletraffic Cong. Montreal,* Quebec, 1983, Session 3.2, paper 2.
62. US Sprint, "Propagation Delay Distribution in the IC Network," Doc. No. T1Q1.1/89-048, April 1989.
63. Bellcore, "Delay in Public Switched Telephone Networks," Doc. No. T1Q1.1/88-028, May 1988.
64. AT&T, "Propagation Delay in the Switched Access Network," Doc. No. T1Q1.1/88-036, May 1988.
65. US WEST Communications, "Round Trip Delay in Distributed Local Switching Networks with Integrated Transport," Doc. No. T1Q1.1/89-003, Feb. 1989.
66. US WEST Communications, "Switching Architectures of Network Providers and the Effects on Round-Trip Delay," Doc. No. T1Q1.1/89-013, Feb. 1989.
67. Bellcore, "A Reference Model for POT-NI Delay," Doc. No. T1Q1.1/90-010, Feb. 1990.
68. American National Standard for Telecommunications—Network Performance—Transmission Specification for Switched Exchange Access Network, ANSI T1-506-1990.
69. American National Standard for Telecommunications—Digital Processing of Voice-Band Signals—Algorithms and Line Format for 32 kbit/s ADPCM," ANSI T1.301-1987.
70. CCITT Rec. G.721, "32 kbit/s Adaptive Differential Pulse Code Modulations (ADPCM)," *Blue Book,* Vol III, Fascicle III.4, Geneva, 1989, pp. 231–268.
71. CCITT Rec. G.722, "Extensions of Recommendation G.721 ADPCM to 24 and 40 kbit/s for DCME Applications," *Blue Book,* Vol III, Fascicle III.4, Geneva 1989, pp. 269–340.
72. American National Standard for Telecommunication—Network Performance Standards 32 kbit/s ADPCM Tandem Encoding Limits. ANSI T1.501-1988.
73. CCITT Rec. G113, "Transmission Impairments," *Blue Book,* Vol III, Fascicle III.I, Geneva, 1989, pp. 63–84.
74. N. Psimenatos, J. Gruber, G. Goddard, and D. Mondor, "The Application of 32 kb/s ADPCM Systems in Telecommunication Network," *Globecom Rec.,* Nov. 1984, pp. 796–801.
75. T. Kilm, and D. Sparrell, "Introduction of 32 kb/s ADPCM Equipment into the North American Network," *Globecom Rec.,* Nov. 1984, pp. 803–807.

76. J. Gruber, and N. Psimenatos "Network Performance Assessment and Transmission Planning for 32 kbit/s ADPCM," IEE Conference Publication No. 246—Telecommunication Transmission, March 1985, pp. 27–30.
77. CCITT Rec. G.101, "The Transmission Plan," *Blue Book,* Vol III, Fascicle III.I, Geneva, 1989.
78. CCITT Rec. G.711, "Pulse Code Modulations (PCM) of Voice Frequencies," *Blue Book,* Vol III, Fascicle III.I, Geneva, 1989.
79. American National Standard for Telecommunications—Digital Processing of Voice-Band Signals—32 kbit/s ADPCM Line Format Standard, ANSI T1.302-1989.
80. American National Standard for Telecommunications—Digital Processing of Voice-Band Signals—Algorithms for 24, 32 and 40 kbit/s ADPCM, ANSI T1.303-1989.

Appendix A.7.1
THE VIA NET LOSS PLAN

The objective of the via net loss (VNL) plan, developed in the early 1950s, was to control talker echo by insertion of loss in connections, commensurate with near optimum loss/noise echo/delay performance [1]. The inserted loss was implemented on trunks, with the loss value being proportionate to the trunks' length.

This plan, which was based on subjective echo tolerance curves, accounted for the number of trunks in a connection, the random deviation of trunk losses from design values, and the expected echo return loss at the end office. The plan assumed that dominant echos arose at end offices. Lesser echos arose at intermediate switching offices because of more stringent balance requirements at these locations (see Ch. 6).

At this time, and until the early 1980s, the North American network employed a five-level switching hierarchy consisting of a Class 5 (end office) and four levels of toll offices (see Ch. 4). The trunks between the local exchange and the Class 4 (lowest) level of toll office were called *toll connecting trunks* (analogous to the exchange access EO-POT trunks in today's U.S. environment). Trunks between toll offices were called *intertoll trunks* (analogous to the IEC trunks in today's U.S. environment).

The basic loss objectives for these trunks under the VNL plan are given in Table A.7.1.1.

The velocity of propagation depends on the type of transmission facilities (see Ch. 6). A maximum inserted connection loss under the VNL plan of approximately 9 dB was judged to be optimum for about 45 ms of round-trip delay. Longer connections were equipped with echo suppressors, in which case the intertoll trunk loss was 0 dB.

Also, although the VNL formula is a continuous function of distance, it was administratively desirable to use a nine-step approximation to the formula. Thus, intermediate nodes, facilities, and equipment were designed to provide the required VNL loss dependent on the technology and distance covered. This approximate characteristic of the VNL plan was a factor in the development of the FLP, appropriate for the introduction of digital facilities and switches.

Table A.7.1.1
VNL Plan Trunk Loss Objectives

Trunk Type	Loss (dB)	Maximum Loss (dB)	Echo Suppressor Application Distances
Toll Connecting	$2.5 + \text{VNLF} \times D + 0.4$ dB	4	Not applicable
Intertoll	$\text{VNLF} \times D + 0.4$ dB	2.9 (0 with echo suppressors)	1850 miles (~3000 km)

D = one way distance in miles
VNLF = Via Net Loss Factor
$= \dfrac{2 \times .102}{\text{Velocity of propagation in miles/ms}}$

Appendix A.7.2
THE FIXED LOSS PLAN

In the early 1970s, the envisaged deployment of digital switches with associated digital transmission facilities brought into question the appropriateness of the VNL plan, which, in the long term, would not be compatible with the potential benefits of a digital network, that is, simplification with respect to loss, and bit transparency. Consequently, a new plan, the fixed loss plan (FLP), was developed to exploit the benefits of digital transmission. This appendix discusses these aspects of the FLP, some of which are still relevant in today's environment.

A.7.2.1 BASIC ATTRIBUTES OF THE FLP

Loss Control/Implementation

The FLP requires that loss be controlled by the digital end office (switch) and implemented in the receive direction only. Generally, this loss is analog, inserted on the analog side of the D/A converter. Analog (as opposed to digital) loss provides for bit transparency in the digital part of the network for the carriage of digital services. Absence of digital loss is a key attribute and is also applicable in evolving all-digital networks. With in-band (associated) signaling, this enabled the carriage of services up to 56 kb/s. With out-band (common channel) signaling, 64 kb/s can be carried.

Overall Transmission Quality

The VNL plan was optimized for the analog network. The FLP, although optimized for the digital network (up to the end office), was designed to be compatible with the VNL plan, insofar as mixed analog/digital network connections would perform as well as their all-analog or all-digital counterparts.

In addition, the FLP has the following quality attributes:
- Provides symmetrical loss on individual connections.
- Maintains talker echo performance on international connections
- Provides stability margin.

Operational Considerations

The FLP was designed to be compatible with existing North American maintenance and administrative procedures. In addition, it retained existing lineup procedures for PCM multiplex equipment and their conventional transmission level points (TLP). TLPs are discussed in Appendix A.7.3.

A.7.2.2 LOSS OBJECTIVES

The intent of the FLP is to approximate the variable loss of the VNL plan with a fixed loss approximation. End-to-end loss (OOLR = L_e) and talker echo path loss (TEPL = E) parameters are discussed in Chapter 3 (see Eqn. (3.1) and Fig. 3.6, and Eqn. (3.3) and Fig. 3.7, respectively). Equations (3.1) and (3.3) show that both OOLR and TEPL are a function of the loss inserted in the connection (i.e., ICL according to the VNL plan or the FLP). Figure 3.6 and Equation (1) in Table 3.4 (a) show that, for a given value of noise, loss/noise performance degrades for increasing values of loss (L_e) greater than about 3 dB. However, Figure 3.7 and Equation (2a) in Table 3.4(a) show that, for a given value of echo path delay, echo/delay performance improves with increased loss (E). Thus, there is an optimum value of ICL such that maximum loss/noise, echo/delay performance is achieved. This is illustrated conceptually in Figure A.7.2.1 (detailed versions of such figures can be found in [3]).

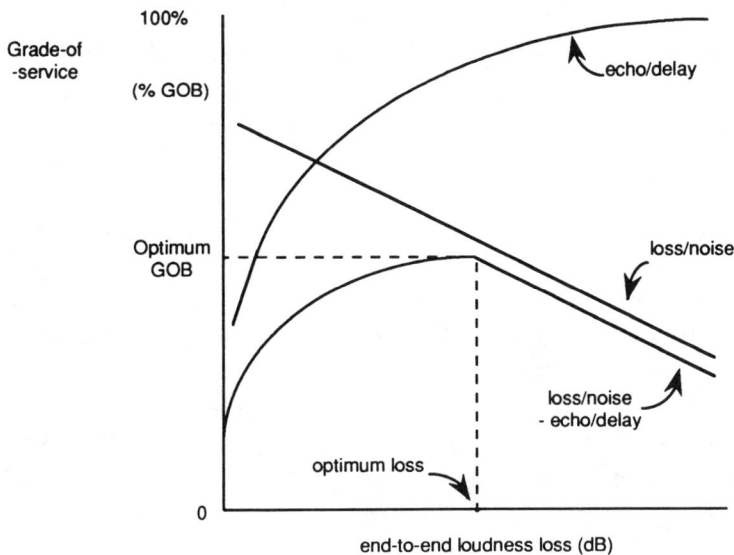

Figure A.7.2.1 Optimum loss/noise-echo/delay performance (GOB).

Given representative connection models, an optimum loss versus airline distance can be determined, as shown conceptually in Figure A.7.2.2 [3]. It can be shown that following this approach the VNL plan is close to the optimum [3]. Furthermore, it has been recommended that, for the emerging digital network, fixed loss approximation to the VNL plan be implemented. Thus was born the basis of the FLP.

Figure A.7.2.2 Grade-of-service (GOS) modeling process to determine fixed loss approximation to optimum loss.

Loss Objectives for Different Types of Trunk Technology

When the FLP was developed, it was necessary to identify three types of trunks to cover the co-existence of analog and digital equipment. To a certain extent, this situation still applies in the current network.

Analog Trunk: A trunk using analog (or derived analog) transmission and analog interfaces at both ends. Very few of this type of trunk exists currently (see App. A.7.1 for loss values).

Combination Trunk: A trunk using digital transmission, an analog switch interface at one end, and a digital switch interface at the other.

Digital Trunk: A trunk using digital transmission and digital switch interfaces at both ends.

The loss objectives for combination and digital trunks, which are still implemented using the FLP, are given in Table A.7.2.1.

Table A.7.2.1
FLP Trunk Loss Objectives (dB)

Trunk Type	Inter-Office (dB)	Toll Connecting	Toll End-to-End	Inter-Toll
Digital	3	3*	6	0
Combination	3	3*	6	1

*The 3 dB loss objective for toll connecting trunks is for test measurement purposes only. In practice, loss on an all-digital end-to-end toll connection would be implemented on the receive side at each end-office switch and be equal to the sum of these objectives, i.e., 6 dB.

Operational and Administrative Aspects of the FLP

Generally, for analog trunks, loss in both directions is symmetrical and maintenance and operations practices were developed based on this fundamental principle. With the introduction of digital facilities, and loss insertion only on the receive side of trunks (e.g., 6 dB on toll connecting trunks for toll connections), it was necessary to develop a maintenance and operations plan that was consistent with the "analog" practices, while at the same time meeting the needs of the FLP. It is worth noting that this aspect of the FLP is still not well understood and has resulted in some confusion in the industry.

The key considerations in this regard are as follows:

1. *Measured* digital trunk loss should be symmetrical (the same in each direction), as opposed to *actual* loss, which for toll connecting trunks will not be symmetrical.
2. Test tone levels for analog carrier systems should be retained.
3. Transmission levels for analog offices should remain at 0 dBm and -2 dBm (or at 0 TLP and -2 TLP) for end offices and toll offices, respectively.

The loss of circuits in the network is administered through the use of the transmission level point (TLP) concept. TLPs are discussed in Appendix A.7.3, with examples given for the FLP.

Appendix A.7.3
TRANSMISSION LEVEL POINTS AND DIGITAL REFERENCE SIGNALS

In an analog network and in the current mixed analog/digital network, it is necessary to define transmission level points (TLP) to which losses or gains can be referred. TLPs were used to facilitate operation and maintenance procedures associated with installation, maintenance, and testing of analog voiceband trunks. TLPs are therefore essentially analog measurements of test tone levels.

In an all-digital connection between, for example, end offices, the need to relate analog levels to reference points is no longer required since the signal is digital, and operations- and maintenance-related functions can also be digital in nature. This appendix reviews the basics of TLPs for analog systems and digital reference signals (DRS) for digital systems.

A.7.3.1 TRANSMISSION LEVEL POINT METHODOLOGY

TLP Definition

The TLP at any point on a transmission system is defined as the decibel value of the ratio of the power of a signal at that point, to the power of the same signal at a reference point; the reference point is called the 0 TLP point [15, Sec. 7].

Use of TLPs on Analog Connections

Figure A.7.3.1 illustrates the use of TLPs on part of an analog connection, using as an example the old VNL plan. The outgoing trunk of each switch is designated as a reference point. However, as in Figure A.7.3.1, the absolute signal level at these points are not the same. For the EO (Class 5) switch, the outgoing trunk is at 0 TLP point (or 0 dBm signal level). For a toll (Class 4) switch, the outgoing trunk is at -2

Figure A.7.3.1 Use of TLPs on part of an analog connection (only one direction of transmission illustrated).

TLP point (or -2 dBm signal level). These TLPs are related to the average signal levels at the reference points, the average signal level at a toll reference point being about 2 dB lower than that at the EO reference point. In addition, when testing trunks between switches, it is necessary to exercise the equipment over a range of signal levels without incurring overload. The above test levels (in brackets) satisfy the overload criteria.

In addition to TLPs for analog switches, voice frequency ports on carrier transmission system also have standard reference levels based, as in analog switches, on a dynamic range criterion. For 4-wire trunk carrier systems, the input is a -16 TLP and the output is a $+7$ TLP, as illustrated in Figure A.7.3.2 (i.e., a -16-dBm input level results in a $+7$-dBm output level). An example of the use of TLPs on carrier systems in a connection between two analog end offices is given in Figure A.7.3.3. In this instance, the trunk loss design objective (referred to as the *inserted connection loss* (ICL)) is 3 dB. This loss is achieved through the use of appropriate loss pads, as shown.

Figure A.7.3.2 Trunk carrier system TLPs (only one direction of transmission illustrated).

Figure A.7.3.3 Example of an interoffice 3-dB trunk implemented with a 4-wire carrier system.

Use of TLPs in Mixed Analog/Digital Connections and Switches

With the introduction of digital facilities and switches, the TLP concept had to be extended in digital switches to include the encoding (A/D conversion) and decoding (D/A conversion) characteristics of North American 64-kb/s μ-law PCM codecs.

A.7.3.2 REPRESENTATIONS OF DIGITAL SIGNAL LEVELS

With reference to Figure A.7.3.4, the encoder output and decoder input level representation is in terms of the DRS [7, 77]. A 1,000-Hz DRS is the digital representation of an analog 1,000-Hz sinewave, and is precisely defined in terms of a PCM bit sequence [78]. However, in practice, a 1,004-Hz sinewave is used for test purposes to minimize the effect on level measurements of interference of the test signal with submultiples of the codec 8-kHz sampling frequency. The 1,004-Hz test signal results in a PCM bit sequence that is much longer than the 1,000-Hz test signal. The dynamic range limitation of encoders and decoders (i.e., codecs) is accommodated by ensuring that signal levels operate below the clipping level of +3.17 dBm0.

The encoder and decoder in Figure A.7.3.4 have 0-dB encode levels and decode levels, respectively, and are referred to as a 0-level encoder and a 0-level decoder. Figure A.7.3.5 illustrates examples of a +6-level encoder and a −6-level decoder. The encoder includes the equivalent of a 6-dB input loss pad, which means that, to develop the DRS level, a +6-dBm, 1,004-Hz input analog sinewave is required. In this case, clipping will occur at an input level of +9.17 dBm (see Fig. A.7.3.5).

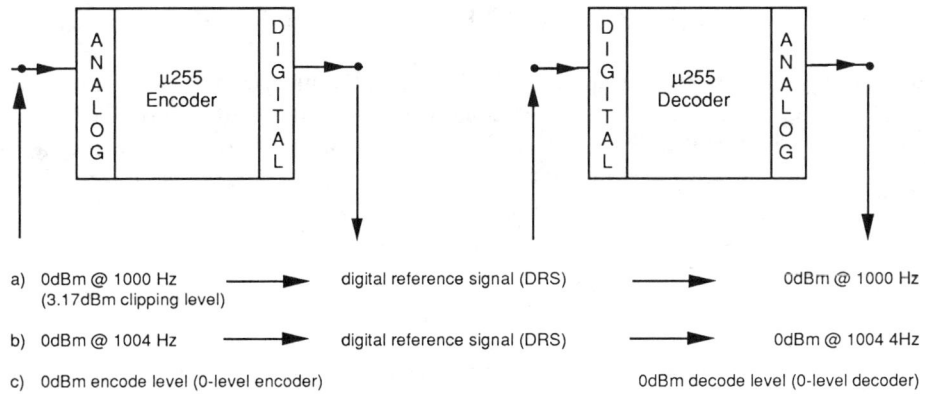

a) 0dBm @ 1000 Hz → digital reference signal (DRS) → 0dBm @ 1000 Hz
   (3.17dBm clipping level)

b) 0dBm @ 1004 Hz → digital reference signal (DRS) → 0dBm @ 1004 4Hz

c) 0dBm encode level (0-level encoder)   0dBm decode level (0-level decoder)

Figure A.7.3.4 Encoding/decoding levels representations.

- +6dBm @ 1004 Hz → DRS   DRS → -6dBM @ 1004 Hz
  (+9.17 dBm clipping level)

- 6dBm encode level (6-level encoder)   -6dBm decode level (-6-level decoder)

Figure A.7.3.5 Example of a +6 level encoder and −6 level decoder.

For the decoder, the 6-dB loss pad results in the DRS level at the input generating a −6-dBm, 1,004-Hz analog sinewave at the output. In practice, the equivalent of these loss pads can be integrated into the codec processing algorithms and not necessarily implemented as separate functions. Conversely, gain devices can be associated with the codec processes, resulting in different encode/decode levels. This illustrates how various loss values can be implemented digitally.

The encode and decode levels can also be referred to as *encode level points* (ELP) and *decode level points* (DLP) and are defined as follows:

1. ELP—The encode level at any point is defined as the decibel ratio of the power of a signal at that point that results in a DRS, to a power of 1 mW. Thus, in Figure A.7.3.5, the ELP is +6 dB.
2. DLP—The decode level at any point is defined as the decibel ratio of the power of a signal at that point, which results from decoding a DRS, to a power of 1 mW. Thus, in Figure A.7.3.5 the DLP is −6 dB.

A.7.3.3 APPLICATION OF ELP AND DLP IN DIGITAL SWITCHES

To illustrate the application of TLP, DLP, and ELP concepts, Figure A.7.3.6 shows a combination trunk between an analog end office and digital end office implemented according to the FLP rules, as discussed in Appendix A.7.2. The analog end office is connected via a digital carrier system to the digital end office to provide a 3-dB loss.

Figure A.7.3.6 Inter-office combination 3 dB loss trunk.

Analog EO to Digital EO Direction: The PCM channel bank develops a DRS for a -8.5-dBm input; that is, the channel bank is a -8.5 encoder, or the input is a -8.5 ELP. The DRS at the digital EO develops a -3-dBm output through the 3-dB FLP loss pad; that is, the digital EO is a -3 decoder, or the output is a -3 DLP. The resulting trunk loss is 3 dB.

Digital EO to Analog EO Direction: At the digital EO, 0 dBm at the line card input generates a DRS; that is, the EO is a 0 encoder, or the input is a 0 ELP. At the channel bank decoder, the DRS generates $+4$ dBm ($+4$ TLP); that is, the channel bank is a $+4$ decoder, or the output is a $+4$ DLP. Finally, the 7-dB pad results in a -3-dBm output level. Again, the resulting trunk loss is 3 dB.

Test Requirements: Test access points are provided in the digital EO for 0 ELP and -3 DLP levels, as shown in Figure A.7.3.6. The injection of a 0-dBm signal level at one end of the connection results in a -3-dBm signal level at the other end, and vice versa. Note that, by replacing the analog EO with a digital EO, a 3-dB trunk loss is still implemented and can be tested.

Figure A.7.3.7 illustrates a combination tandem (or toll) connecting trunk providing 3 dB loss in each direction and illustrating the following:

1. Application of the 5-dB FLP loss pad value.
2. -2 TLP analog tandem (or toll) switch and corresponding test levels.
3. A channel bank with a 4-wire analog interface.

Figure A.7.3.7 Tandem (or toll) combination trunk with 3 dB loss.

Note, in particular, that for test purposes, a -2 TLP signal is injected at the analog tandem (or toll) switch. In practice, the injected signal level always corresponds in level to the TLP designation of the switch, as discussed previously (e.g., see Figs. A.7.3.1 and A.7.3.5).

ELPs and DLPs in All-Digital Connections

In an all-digital network, TLPs are not required, and ELPs and DLPs are sufficient to define connection losses. With a conventional 0-dB ELP at the transmit end of a connection, the DLP at the receive end will determine the loss of the connection. This is illustrated in Figure A.7.3.8 for a 6-dB inter-LATA or toll connection. In each

Figure A.7.3.8 All-digital inter-LATA connection.

direction, a 0-dBm input signal develop a DRS, and DRS is decoded to -6 dBm at the output (i.e., a -6 DLP).

Other connection losses can be implemented by changing the DLP pads. For the FLP, the DLP values are as follows:

Table A.7.3.1
Fixed Loss Plan (FLP) Decode Level Points

Network	Connection Type	DLP
Intra-LATA	Intra-office	0
	Inter-office	-3
	Toll	-6
Inter-LATA	Toll	-6

Chapter 8
TRANSMISSION PERFORMANCE STANDARDS

8.1 STANDARDS NEED

Telecommunications networks consist of many systems that must work in harmony to ensure end user satisfaction. This requires that individual equipment and systems have mutually compatible interface specifications, and the necessary performance capability to satisfy end-to-end performance objectives. Within a single network administered by a single telecommunications operating company, both the network end-to-end performance objectives, as well as the performance of individual equipment and systems can, in principle, constitute an internal company-developed set of standards. Generally, however, in the current North American and international environments, such objectives and system specifications are the subject of consensus among the various network providers and equipment manufacturers. This standardization process is the subject of this chapter.

In addition to network end-to-end performance, equipment and system performance, and interface standards, a transmission plan is also required to ensure that, from a network implementation standpoint, the required end-to-end performance is achieved, and that appropriate transmission levels are met. Also, operational standards are required to ensure that appropriate transmission performance parameter measurement techniques are used, and that effective operations and maintenance procedures such as monitoring, testing, and repair/replacement are in place.

The transmission performance of a network therefore depends on standards in the following four areas:

1. End-to-end performance (to meet end user needs).
2. Equipment and system performance requirements.
3. Transmission plans.
4. Operations and maintenance procedures.

A prime example of a single company-controlled network was the predivestiture Bell System in the U.S. Through standards generated in the above areas, the Bell System planned and maintained its network to a performance level considered to be among the best in the world in terms of overall quality. A similar situation exists currently in Canada, within the Telecom Canada network.

In the predivestiture era, in terms of external standards in the U.S., the Electronic Industries Association (EIA), established in 1924, served the electronic equipment and component manufacturing industry (e.g., private branch exchanges (PBX) and terminal equipment). From 1988, this role was taken over by the Telecommunications Industry Association (TIA), formed by a merger between the telecommunications sector of the EIA and the United States Telecommunications Suppliers Association. In addition, the Institute of Electrical and Electronics Engineers (IEEE), established in the 1960s, is one of the most active and prominent professional organizations in the area of measurement methodologies and related standards.

On an international basis, the International Telegraph and Telephone Consultative Committee (CCITT) (formed in 1956 out of a merger of the CCIF-Telephone and CCIT-Telegraph) is the key forum for developing standards/recommendations. These address technical, operating, and tariff questions related to telegraph and telephone services required to ensure compatibility among the public international networks across countries. The CCITT is a main organ of the International Telecommunication Union (ITU), established in 1865, the main purpose of which is to promote cooperation and development in the provision of worldwide telecommunications services. The ITU also cooperates with ISO (International Organization for Standardization), another active global standards organization, which, among other things, addresses standards for information systems. Specifically, the Joint Technical Committee on Information Technology (JTCI) of ISO and the International Electrotechnical Commission (ISO/IEC) deals with information technology. Rutkowski [1], Cohen [2], and the IEEE special magazine issue on "Telecommunications Standards" [3] provide an overview and details on these major standards organizations.

In the predivestiture era, the EIA/TIA and IEEE included "performance" as a general study area, and have remained active in the U.S. through divestiture. However, divestiture brought with it a variety of alternative inter-exchange carriers (IEC) in addition to AT&T Communications (formerly AT&T Long Lines). Also, the resulting competition in the equipment field encouraged the growth of a multitude of equipment manufacturers (some off-shore based), all competing for the attention of the Bell Operating Company (BOC), local-exchange carrier (LEC), and IEC markets. In this environment, it was realized that, with divestiture, a standards forum was required to replace the performance and equipment compatibility standards functions previously provided "in-house" by the Bell System. Such a standards forum would have to operate in an analogous fashion to the CCITT, but within the

U.S. environment, and address interworking between the LECs and IECs, as well as associated equipment standards. Thus was born the Exchange Carriers Standards Association (ECSA), established in 1983 by members of the wireline exchange carrier industry to address, in particular, exchange access interconnection standards. Such standards are developed under the ECSA-sponsored Standards Committee T1—Telecommunications (established in 1984), an American National Standards Institute (ANSI)-accredited standards body [4]. In this chapter this standards body is referred to as "T1."

At the time of publication of this book, telecommunications competition is not a major factor in Canada. However, standards do play a role in the evolution of Canadian telecommunications networks. For example, part of the role of the Canadian Standards Association (CSA) [5] is to ensure terminal compatibility with the network(s) that fall under the responsibility of the telecommunications carriers. This is achieved through the issue of CSA-developed voluntary standards. The CSA is accredited by the Standards Council of Canada (SCC) [6], which performs an analogous function to that of ANSI in the U.S. Section 8.3 of this chapter discusses the current Canadian standards organizations.

Subsequent sections of this chapter will address performance standards activities under three bodies as follows:

1. National in U.S. and Canada (in Secs. 8.2 and 8.3 respectively).
2. International (or Global) (in Sec. 8.4).
3. Regional (in Sec. 8.5).

Table 8.1 illustrates this breakdown, in which standards forums in each category are further divided into those associated with *Telecommunication* standards and those associated with *Information Technology* standards. Information Technology standards address transport of computer-based information, including the necessary data networks. In addition, the corresponding umbrella organization with which these standards forums are associated or accredited are tabulated in Table 8.1. Some of these standards forums have already been touched upon in the introduction to this chapter. Table 8.1 includes others that will be discussed in subsequent sections of this chapter.

However, within this overall structure, the focus in this chapter will be on Telecommunication transmission performance standards, particularly those that fall within the heavy lined boxed areas of Table 8.1. Information Technology standards activities and other activities outside of these focused areas will be covered to the extent that they impact the overall performance standards process.

The standards developed by the organizations included in Table 8.1 are generally voluntary in nature. That is, there is no obligation on behalf of network providers or equipment manufacturers to adhere to these standards. However, a class of standards that is mandatory in nature also exist which has an indirect bearing on

Table 8.1
Standards Organizational Structure

	Telecommunication Standards	Information Technology Standards	Umbrella Organization
National - U.S.	T1 TIA (EIA) IEEE (T1Q1) (TR41) (TR45)	X3 (X3S3)	ANSI
National - Canada	CSA SCOT	CSA SCIT	SCC
International or Global	CCITT CCIR (SG XII) (CMTT) (SG XVIII)	——— JTC1	ITU ISO/IEC
Regional - Europe	ETSI	CEN/CENELEC + ECMA	———

T1	= Standards Committee T1 - Telecommunication
TIA	= Telecommunication Industry Association
X3	= Committee X3 - Information Processing Systems
ANSI	= American National Standards Institute
IEEE	= Institute of Electrical and Electronics Engineers
CSA SCOT	= Canadian Standards Association - Steering Committee on Telecommunications
CSA SCIT	= Canadian Standards Association - Steering Committee on Information Technology
CCITT	= International Telegraph and Telephone Consultative Committee
CCIR	= International Radio Consultative Committee
JTC1	= Joint Technical Committee on Information Technology
ITU	= International Telecommunication Union
ISO	= International Organization for Standardization
IEC	= International Electrotechnical Commission
ETSI	= European Telecommuncations Standards Institute
ECMA	= European Computer Manufacturers Associations
EIA	= Electrical Industries Association
CEN/CENELEC	= European Committee of Standardization/European Committee for Electro Technical Standardization
SCC	= Standards Council of Canada

performance. Mandatory standards are generally developed by government or regulatory agencies, and examples of such standards will be given in subsequent sections of this chapter.

In the case of regional standards bodies, developments in Europe are touched upon as an example of changes that are occurring in the global standards scene, and which could have a significant impact both on national standards in North America and international standards in CCITT.

Appendix A.8.1 summarizes the activities of the CCITT, T1, TIA and IEEE in terms of their study groups, working groups and technical sub-committees which generally address performance standards and related areas.

8.2 U.S. STANDARDS BODIES

Within the U.S. environment, performance-related standards are addressed primarily by three standards bodies. These are as follows:

1. T1 (through its Technical Subcommittee T1Q1).
2. TIA (through its Technical Requirements (TR) groups).
3. IEEE (through its various technical committees).

These bodies, apart from the TIA[1], are accredited organizations under ANSI. ANSI was founded in 1918 as a nonprofit organization that coordinates voluntary standards activities in the U.S. Performance-related standards developed in T1 and IEEE are designated ANSI standards and are subject to the ANSI rules and procedures for establishing industry consensus.

The various responsibilities of T1, TIA, and IEEE are shown in Figure 8.1. This figure illustrates a connection traversing LEC and IEC networks and providing a communication link to end users between the customer/network interfaces (designated NI in Fig. 8.1). The TIA mandate is to address performance and equipment specifications on the end user side of each NI. The T1 mandate is to address corresponding performance issues on the network side of each NI, including network-to-network (i.e., LEC to IEC) interworking.

The IEEE continues to provide a professional venue for the dissemination of technical information through publications and meetings and, in addition, engages in significant standards development [7]. In the performance area, the current role of the IEEE is mainly to research and agree upon new measurement methodologies in support of TIA and T1 performance-related standards.

Participation of the U.S. in CCITT comes under the auspices of the U.S. Department of State (see Sec. 8.4 of this chapter).

8.2.1 Standards Committee T1—Telecommunications

T1 [4] is the focal point for the generation of U.S. network-related standards in the general areas of switching, signaling, transmission, and performance, as well as operations, administration, maintenance and provisioning. The scope and responsibilities are laid out in Article II of the Bylaws of T1 [8]. This committee consists of a

---

[1]At the time of this publication, the TIA was applying for ANSI accreditation.

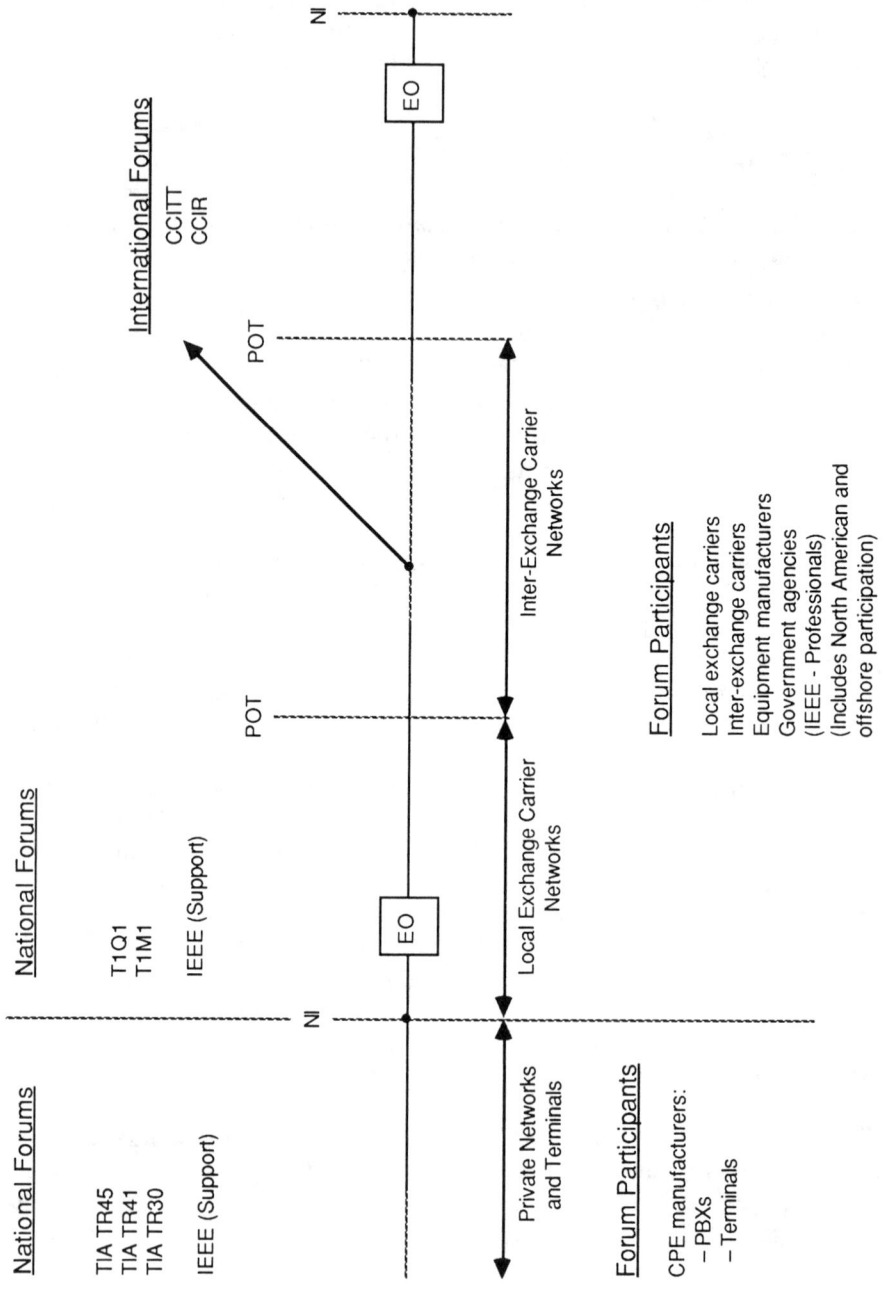

Figure 8.1 Responsibilities of the T1, TIA and IEEE forums.

number of technical subcommittees (listed in App. A.8.1.2), each of which has a number of working groups, which in turn address a particular area requiring standardization.

Figure 8.2 details the activities carried out by all of the T1 subcommittees and associated working groups at the time of this publication.[2] The focal point for performance issues in T1 is technical subcommittee "T1Q1—Performance" [9]. This subcommittee covers the following areas under five working groups:

T1Q1.1—Voice and Voiceband Data.

- Traffic and availability.
- Transmission loss planning.

T1Q1.2—Survivability.

T1Q1.3—Digital Packet and ISDN.

T1Q1.4—Digital Dedicated and Circuit Switched.

T1Q1.5—Wideband Program/Audio.

- Program audio.
- Commercial TV.
- Digital TV.
- Video teleconferencing.

Appendix A.8.4.3 lists the standards already developed in T1Q1.

Being the focal point for addressing performance issues, there is a need for T1Q1 to interact with other T1 technical subcommittees to ensure that the performance aspects of the activities in these subcommittees are addressed. For example, there is strong liaison with T1M1 on operations and maintenance, and with T1Y1 on new transmission techniques such as sub-64 kb/s voiceband encoding and 45 mb/s digital video encoding. In the area of new transport mode performance, such as asynchronous transfer mode (ATM), there is liaison with T1S1 on such aspects as transmission delay and bit error mitigation techniques. With T1X1 there is, for example, liaison on the performance requirements for synchronous optical network (SONET) survivable ring architectures. More recently, liaison has been established with T1P1 on the performance aspects of personal communications systems and services.

---

[2]Note: In October 1991, Committee T1 agreed upon a reorganization of T1Q1, T1Y1, and T1E1, to take effect in February 1992. Under this reorganization, T1Q1 and T1Y1 are terminated, and a new technical subcommittee, T1A1, will address "Performance and Signal Processing"; i.e., T1Q1 and part of T1Y1. The remaining part of T1Y1 is merged with T1E1, to address interface and environmental standards.

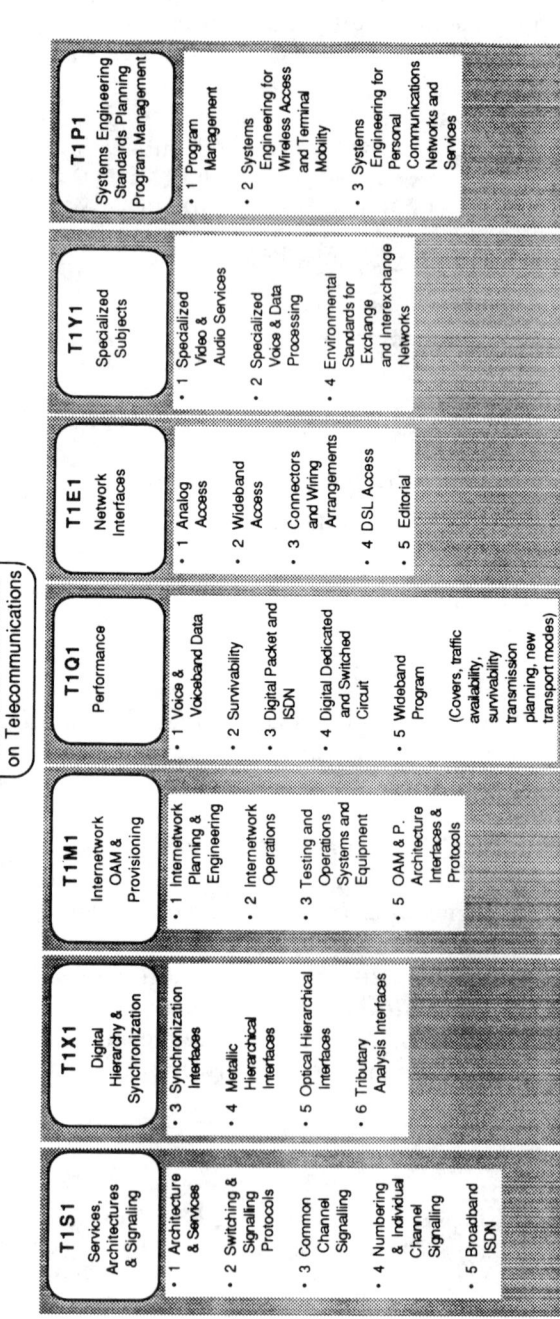

Figure 8.2 T1 technical subcommittees and their working groups.

8.2.2 TIA—TR.30—TR.41

Appendix A.8.1.4 briefly describes the structure of the TIA and the structure of the EIA before 1988. The main groups in TIA addressing performance issues are as follows:
- TR-30 Data Transmission Systems and Equipment.
    TR30.2 Data Transmission Interfaces.
    TR30.3 Data Interfaces to Leased Lines and Dial-Up Networks.
- TR-41 Telephone Terminals.
    TR41.1 PBX and Primary Rate Access.
    TR41.2 Key Telephones and Basic Rate Access.
    TR41.3 Telephone and Acoustic Transfer Characteristics.
- TIATR-45 Cellular Systems.
    TR45.1 Cellular Radio.
    TR45.2 Cellular System Operations.
    TR45.3 Digital Cellular.

These groups in TIA address end user needs for both data and voice services in terms of terminal requirements and private network loss/level plans. Appendix A.8.4.2 lists EIA/TIA performance-related standards.

8.2.3 IEEE

The focal point in the IEEE [7] for generating transmission performance-related standards is the Transmission Systems Committee (TRANSYSCOM) of the IEEE Communications Society. Other committees in IEEE are listed in Appendix A.8.1.3. TRANSYSCOM, in turn, has a number of subcommittees addressing performance, with the emphasis on performance measurement techniques. These are as follows:
1. Subcommittee on Telephone Instrument Testing.
2. Subcommittee on Telecommunication System Performance Characterization.

Subcommittee 2 has been involved with two major activities in the performance area, carried out by the following working groups:
- IEEE Working Group on Voiceband Channel Performance Criteria.
- IEEE Working Group on Loop Performance Criteria.

Appendix A.8.4.1 lists the standards that have been developed by TRANSYSCOM in the general area of performance, some of which are referred to in this section. With the issuance of IEEE Standards 820 on Loop Performance Characteristics and 823 on Voiceband Channel Performance Criteria in 1990, work on performance characteristics and criteria in the Subcommittee on Telecommunications System Performance Characterization has more or less come to an end. The main reason for this is the emergence of the T1 committees, in particular T1Q1, as the focus in the U.S. for network performance standards activities. For example, consider the Loop Performance Characteristics Standard 820. This standard addresses the performance characteristics of analog loops, in addition to the measurement procedures. When this standard was completed, the group turned its attention to digital loop characteristics. However, performance requirements, etc., of digital loops fall under the mandate of T1Q1.3, and in 1987, following completion of a recommended practice covering digital parameters (Standard 1051), this work in IEEE was terminated.

On the other hand, the IEEE subcommittee on telephone instruments testing continues to be active, addressing, in particular, updates to Standard 269 on measurement of telephone set characteristics to address digital sets and headsets. This IEEE subcommittee has a strong liaison with TIA TR-41 on telephone terminal equipment.

In addition to the specific IEEE activities on performance, IEEE Project 802 (under the auspices of the IEEE Computer Society) is addressing a family of standards for local and metropolitan area networks, which include specific performance-related protocol needs. For example, 802.6 is addressing the distributed queue dual bus (DQDB) subnetwork of a metropolitan network, and includes a means of identifying in the protocol quality-of-service attributes of delay and cell loss (arising from contention or congestion) [10].

8.2.4 Other U.S. Performance-Related Standards Organizations

A number of other committees under the auspices of the ECSA also exist to identify and resolve transmission issues between the LECs and IECs. Although these committees do not develop standards, their function includes identifying the need for standards, ensuring compatibility between carriers, and putting in place programs to carry out field performance measurements. These committees are as follows:

CLC—Carrier Liaison Committee [11].

ICCF—Interexchange Carriers Compatibility Forum [12]. In the past, this forum has addressed compatibility issues such as echo control and transmission quality measurements.

NOF—Networks Operations Forum (including a testing committee) [13].

The Federal Communications Commission (FCC), in its capacity as the government agency with responsibilities for the regulation of telecommunications in the U.S., has issued directives, some of which have a bearing on performance. For example, FCC Standards Part 68 (Rules and Regulations) [14] covers the interconnection of terminating equipment with the network, and, as such, prescribes maximum signal levels to ensure no harm comes to the network. Signal levels have a bearing on performance in that, if too high, they cause excessive crosstalk and interference, as well as nonlinear distortion due to overload (see Sec. 6.2.3.2 of Ch. 6).

Under Information Technology standards activities (see Table 8.1 in Sec. 8.1.1), Committee X3—"Information Processing Systems," also accredited under ANSI, is a large standards forum [15] dealing with a wide range of information processing systems standards. Committee X3 is sponsored by the Computer and Business Equipment Manufacturers' Association (CBEMA) [15]. X3 is involved with performance-related standards development, particularly through the group on data communications—Technical Subcommittee X3S3.

In the 1980s, this committee generated three standards on user-oriented performance as follows:

X3.102-1983 "Data Communications Systems and Services—User-Oriented Performance Parameters."

X3.105-1985 "Data Communications Systems and Services—User-Oriented Performance Measurement Methods."

X3.141 "Data Communications and Services—Measurement Methods for User-Oriented Performance Evaluations."

These standards address data communications in terms of the three main performance criteria: speed, accuracy, and reliability (see Chs. 1, 2, and 3).

X3S3 also has a technical advisory group (TAG) function in the U.S. for developing contributions to the ISO/IEC Joint Technical Committee on Information Technology—ISO/IEC JTC1/SC6 (see Sec. 8.4.3).

8.3 CANADIAN STANDARDS BODIES

Within the North American continent, the Canadian telecommunications network is, from an administrative and regulatory viewpoint, wholly Canadian and separate from the U.S. telecommunications network. However, there is interconnection with the U.S. Within the Canadian public switched telephone network, competition (at the time of publication) is not a factor, although in networks exclusively used for data transport, competition is allowed. In the customer premises equipment (CPE) area, interconnection has been permitted since the 1970s and has progressed to the point where there are currently a large number of competing CPE manufacturers. Regulation of the federally regulated part of the telecommunication industry in Canada is the responsibility of the Canadian Radio-Television and Telecommunications Commission (CRTC) [16].

The Canadian telecommunication network consists of a number of separately administered networks (e.g., Bell Canada in Ontario and Quebec) and coordinated (with the exception of a few small telephone administrations) on a national basis by Telecom Canada [16]. In terms of technical direction, Telecom Canada is to a degree analogous to the predivestiture Bell System, developing technical procedures [17] and guidelines with the intent of ensuring end user quality of service on an interprovincial, nationwide basis. Such guidelines are applied by the individual provincially based operating companies, which are members of Telecom Canada.

Telecom Canada is a voting member of the T1 Committee and generally participates in all of the technical subcommittees thereby ensuring strong coupling with U.S. standards development. Canada is a member of CCITT and to establish Canadian positions, national study groups (NSG) under the auspices of the Canadian Department of Communications (DOC) meet on a regular basis (see Sec. 8.4.1).

In the CPE area, following the CRTC's decision in the 1970s to allow increased CPE interconnections, there was a need for a voluntary terminal/CPE standards process. The CSA [5], Canada's foremost standards development, verification, and testing organization, took on this role, with the objective of ensuring terminal compatibility with the network, in conjunction with the telecommunications carriers, which are responsible and accountable to the CRTC. Following a DOC study, the Steering Committee on Telecommunications (SCOT) [18] was established in 1983, and a number of technical subcommittees under SCOT have been established. To date the role of the CSA-SCOT in Canada has been similar to the role of the TIA TR-41 in the U.S. In this respect, joint liaison activities have been established between these two organizations.

The terms of reference of SCOT are as follows:
- To ensure that all interests of end users and equipment providers are served in a timely manner through the development of telecommunications standards.
- To maintain appropriate liaison with national and international organizations concerned with telecommunications standards.
- To establish work priorities and identify the availability of human and financial resources required to carry out this program.

An example of a CSA activity that includes a performance aspect is CSA-SCOT Project T512.2-M, entitled "Functional and Compatibility Requirements for Digital PBXs, and Analog PBXs with Digital Interfaces." This activity started in August 1987 and is analogous to a similar activity in TIA TR 41.1 in the U.S. on the development of standard ANSI/EIA/TIA-464A-1989 covering PBXs (see App. 8.4.2).

In addition, in 1989 CSA-SCOT accepted the responsibility of developing standards in the areas of Integrated Services Digital Network (ISDN) terminal-to-network interfaces, and terminal equipment performance and functionality (excluding network protection standards). CSA-SCOT also agreed, whenever possible, to adopt T1 and TIA standards, only introducing changes, when required, to address particular Canadian policy, regulatory, and legal requirements. This is a key step towards rationalizing standards development from a North American perspective.

The CSA steering committee on information technology (CSA SCIT) plays a role in the development of information technology standards (see Table 8.1 in Sec. 8.1.1).

In Canada, mandatory network protection standards are developed under the auspices of the DOC-sponsored Terminal Attachment Program Advisory Committee (TAPAC). Document CS-03 contains "Standards for Terminal Equipment and Connection Arrangement Systems, Network Protective Device."

8.4 INTERNATIONAL FORUMS

8.4.1 CCITT

Appendix A.8.1.1 lists the CCITT Study Groups (SG). Of these, the following SGs and respective Working Parties (WP) focus on performance-related issues, SG XII and SG XVIII being the prime Study Groups on transmission performance:

SG II Network Operations
- WP II/5 Quality of Service, Network Management and Mobile Service.
- WP II/6 Availability and Reliability.

SG XII Transmission Performance of Telephone Networks and Terminals
- WP XII/1 Telephonometry.
- WP XII/2 Terminals.
- WP XII/3 Transmission Performance and Models.
- WP XII/4 Transmission Planning.

SG XVIII ISDN
- WP XVIII/6 Performance Aspects.

In addition, these Study Groups interact with other Study Groups to ensure that performance issues are addressed. In particular, relevant groups include SG I on Quality of Service Aspects, SG IV on Maintenance, SG XVII on Data Transmission, and SGXV on Transmission Equipment and System Specification.

Each Study Group has a number of questions which are addressed during the study period, and the responses to these questions and resulting documentation usually result in CCITT Recommendations. In the case of SG XII and SG XVIII, Appendix A.8.2 lists the questions (issues being addressed) for the 1989–1992 study period, and Appendix A.8.3 provides a list of existing Recommendations and Supplements pertaining to transmission performance covering G, P, I, and X Series Recommendations. SG II is also developing a "Handbook on Quality of Service and Network Performance" [19], which contains a comprehensive reference section on

CCITT Recommendations which not only cover transmission performance, but also trafficability, dependability, and serviceability performance. With advances in technology and changes in end user needs, etc., these Recommendations are also subject to change. Such changes are generally made as required at the end of each four-year study period. However, with the rapid evolution now occurring in communication technology, the CCITT frequently adopts a mode of accelerated approval of Recommendations within the four-year study period, under CCITT Resolution No. 2 [20].

As the CCITT is an international body in which national positions are debated, organizations within each country meet to agree on national positions and/or contributions before presenting these to CCITT. In the U.S., four SGs have been set up under the auspices of the U.S. Department of State to agree on U.S. contributions and positions. These, with their respective Study Group responsibilities are as follows:

Study Group	Subject	CCITT Study Group Responsibility
SG-A	Services, Operations, Tariffs	I, II, III, IX
SG-B	ISDN	XI, XVIII, all ISDN
SG-C	Telephone network	IV, V, VI, X, XII, XV
SG-D	Data Communications	VII, VIII, XVII

As an example of this process in the U.S., the T1 Committee includes in its scope the development of U.S. contributions/positions. However, before becoming approved U.S. positions, they must be subject to review in one of the above Study Groups [21].

In Canada, a similar process of approval exists in which NSGs meet on a regular basis under the auspices of the Canadian Department of Communications to approve contributions and agree on positions. There is one NSG for each CCITT Study Group, administered under a Canadian National Organization (CNO/CCITT) [22].

8.4.2 CCIR

The International Radio Consultative Committee (CCIR) is an organ of the ITU like the CCITT. The work of this group includes radio propagation, transmission and techniques, in general. On the performance side, there is necessarily strong liaison with CCITT, particularly CCITT SG XVIII. For example, the CMTT is a joint CCIR/CCITT Study Group on Transmission of Sound Broadcasting and Television Signals over Long Distances. With increasing interest in new transport modes of communications for broadband services such as video, CMTT is addressing and providing input to CCITT SG XVIII on end user requirements for these services.

In addition, with increasing interest in personal communication systems the CCIR is turning its attention towards future public land mobile telecommunications systems (FPLMTS), with particular regard to overall objectives (including performance), suitable frequency bands, and the degrees of compatibility or commonality among these systems [23].

The establishment of hypothetical reference connections and performance allocation covering radio technology is also a subject in which the CCIR has a great interest and ongoing dialog with the CCITT.

8.4.3 Private Network Forums

With the increasing use of private networks on an international basis, an international committee has been established under ISO/IEC JTC1 (Joint Technical Committee on Information Technology) to address network interworking and performance related issues. This committee, ISO/IEC JTC1/SC6/WG6 [24], referred to as Private Integrated Services Networking (PISN), addresses those aspects of private networks not covered by CCITT, including the transmission planning aspects. U.S. input into PISN is coordinated through X3S3 (U.S. TAG; see Sec. 8.2.4), and will, in particular, cover work areas currently being addressed in TIA TR41 on private network/PBX performance-related standards and synchronization. T1S1 also provides input to X3S3 on PISN.

8.5 REGIONAL BODIES

CEPT (European Conference of Postal and Telecommunications Administrations) was established in 1959 [25]. Part of its work was to provide the means for establishing European consensus on telecommunication standards and specifications. More recently, CEPT has played a key part in ensuring that when the single European market emerges in 1992, a standardization organization and process will exist to serve European needs both internally and externally on a global basis, in general, and in CCITT, in particular.

For example, in 1985 CEPT agreed to introduce a special family of telecommunication specifications for connecting with European Post Telegraph and Telephone (PTT) organizations. These mandatory specifications are referred to as NETS (Normes Europeene de Telecommunications). There are a total of 33 NETS addressing different technological and service aspects of telecommunications [26].

Examples of NETS that have performance content are as follows:

NET-4	Analog subscriber interface
NET-7	ISDN terminal adapter
NET-11	Mobile terminal

NETS 20–25   Modems

NEC-33   Telephonic characteristics of terminal equipment for ISDN

More recently, in 1988, CEPT established the European Telecommunications Standards Institute (ETSI) [27, 28], which has the responsibility for the development of voluntary European Telecommunications Standards (ETS). In 1990, the role of ETSI in the development of mandatory standards or NETS was agreed upon [29]. Firstly NETS are replaced by Common Technical Regulations (CTR). Secondly, ETSI is mandated to develop draft CTRs that will be passed on to a Committee of Member States for consideration and ratification as mandatory CTRs. Thus, ETSI has become the focus in Europe for the development of voluntary standards (ETS) and the development of draft mandatory standards (draft CTRs).

Within ETSI the focus for the development of network (public and private) and equipment performance standards is ETSI—Sub Technical Committee—Business Telecommunications 2 (ETSI STC BT2). This has five working partys (A-E), with working party A addressing transmission [30].

Clearly, equipment manufacturers in North America will need to meet NETS or CTR requirements in order to market their equipment in Europe, and European manufacturers will need to satisfy CSA, TIA, and T1 requirements for the North American market. Recognizing this and the desirable objective of unifying standards on a global basis, high-level conferences between delegations from T1, ETSI, and the Telecommunications Technology Committee (TTC) of Japan have already taken place on the subject of standards harmonization [31–33]. These Interregional Telecom Standards Conferences (ITSC) addressed particularly the process of unifying standards in general, and, in this respect, agreed that the CCITT is the preeminent forum for achieving final global agreement on standards. Dialogue is continuing on how to ensure timely development of global standards and necessary liaison between the three standards bodies, T1, ETSI, and TTC. This covers not only performance but all standards currently being addressed by these bodies.

Information Technology standards in Europe come under the umbrella of the European Committee of Standardization/European Committee for Electro Technical Standardization (CEN/CENELEC). CEN/CENELEC activities include standards addressing private networks, which it addresses in collaboration with manufacturing associations, such as the European Computer Manufacturers Association (ECMA) [26]. Currently there is little being undertaken in the performance area.

8.6 PERFORMANCE STANDARDS DEVELOPMENT

The main steps involved in the development of a performance-related standard parallel those given in the top-down approach used as a framework for the chapters of this book (see Fig. 1.1 of Ch. 1). This framework is used here so that the standards process can be related to the relevant chapters of this book.

Also, in order to explain the complexities and interactions resulting from divestiture, the standards development process description begins in terms of an end-to-end network, analogous to the process carried out by the predivestiture Bell System. Then, the added dimensions and interworking considerations of the postdivestiture environment are introduced. Both national and international standards are considered, with an indication of which technical committee or Study Group is involved at each stage of the process. Examples of standards activities are provided.

8.6.1 End User/Top-Down Driven Standards

Transmission quality of a telecommunications network is influenced by three main considerations as follows:

1. The end users' needs (service requirements).
2. Technological capabilities.
3. Economic factors.

For example, after a telephone call is established, the end user may become aware of voice impairments such as loudness loss, noise, echo, delay, and distortion. Similarly, for data transmission, the end user may become aware of degradations in throughput and response time caused by transmission delays and errors. However, provided the impairments do not exceed certain threshold values, the user will consider the connection satisfactory. By relating the user's tolerance of network impairments to the technology available and the cost of providing this technology, cost-effective networks can be designed.

To design a network based on end users' needs, subjective, user-oriented performance parameters must be translated into objective, measurable, network-oriented parameters that the designer can use (recall Chs. 2 to 4). Transmission performance and planning standards, which define the quality of transmission and provide planning rules, respectively, bridge the gap between user- and network-oriented parameters.

To develop transmission standards, network planners go through three main stages as illustrated in Figure 8.3:

- Translation.
- Allocation.
- Transmission planning.

A consensus stage is also required to establish interworking among multivendor national networks, as well as international networks, as discussed later. Although economic factors are considered at each stage (in a broad sense to ensure that the network is not over designed, and that services can be offered at competitive prices), these factors are not explicitly agreed upon as part of the standard.

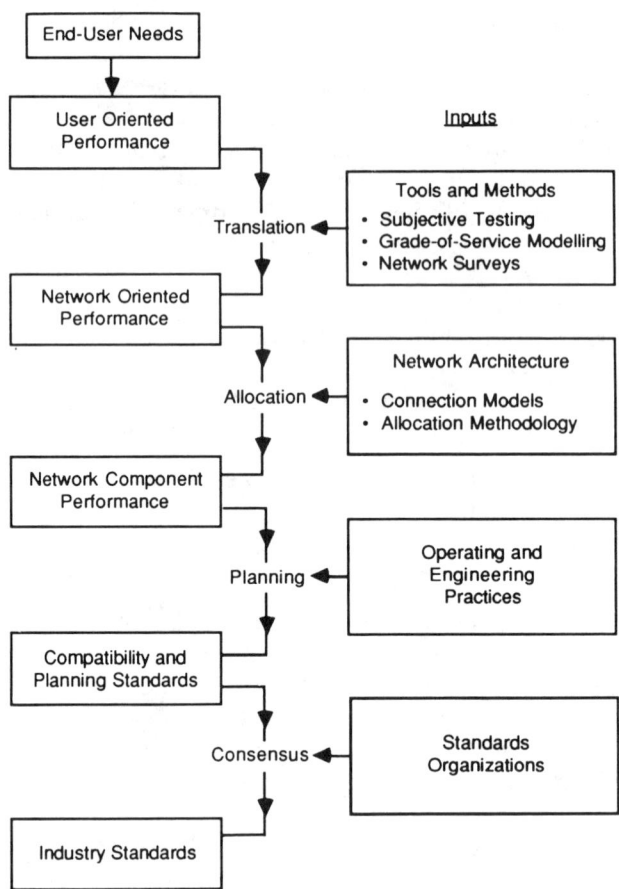

Figure 8.3 Top-down standards development stages.

In the first stage (translation), the user-oriented performance (or quality expectation) is converted into the end-to-end network-oriented transmission performance requirements. For example, consider the dial tone of a telephone. The pitch and volume of that tone meet a standard that was developed by relating available technology to people's answers to subjective questions. These questions included: Can you hear the tone? Do you find it annoying? Should it be higher or lower? Similar examples can be given for digital data transport in terms of error rate and throughput. The translation between end user needs and network-oriented performance is an area in which the TIA (representing manufacturers of end user equipment and the development of related equipment performance standards) and T1 (developing network-oriented performance standards) are finding it necessary to cooperate closely. This is discussed in more detail later in this section.

Tools and methods used in the translation process include subjective testing, grade-of-service (GOS) models and network surveys, as discussed in Chapters 2 and 3. Users' opinions of a particular service are gathered through subjective experiments in the laboratory and through networkwide end user opinion surveys, such as those referenced in Chapter 2 (e.g., [34] in this chapter). Through laboratory experiments and user opinion surveys, subjective transmission GOS models are derived, as also discussed in Chapter 2. GOS models relate the transmission impairments of a connection to the subjective rating of that connection. To establish the subjective rating, listeners judge the quality of speech transmission as excellent, good, fair, poor, or bad, or other scales may be used. GOS models as applied to the performance of networks using mature technology are discussed in Chapter 3.

In the second stage (allocation in Figure 8.3), overall network performance requirements are partitioned into requirements for individual network components such as transmission systems (see Chs. 5 and 6). Performance allocation is also determined within a network by considering economic factors as well as the particular network's architecture. On an international basis, hypothetical reference connections (HRX) are used to determine equitable allocations to national portions (accounting for the size of the countries) and to international portions of connections. A similar situation and approach is required to address the U.S. network environment in terms of performance allocation to the LEC and IEC networks, and in terms of performance allocation to CPE and the network. However, in this case, the competitive nature of the U.S. environment makes it more difficult to agree on the allocation. This is discussed in Section 5.3 of Chapter 5.

In the third stage (planning in Figure 8.3), network planners develop deployment rules for the various transmission systems and other components and technologies in a network (see Ch. 7), including, in particular, transmission plans. These rules ensure that end users' needs are met, and that different networks can be interconnected. For example, the rules ensure that speech volume levels stay within an acceptable range, noise levels are not excessive, and echo control is sufficient for all connections that involve one or more telecommunications networks. Operating and engineering practice must also be accounted for during the planning stage.

Finally, through consensus on an industrywide basis, both national and international standards are set. In particular, the TIA and T1 are the main forums within the U.S., CSA in Canada, and within both countries the CCITT is the main forum for international network performance standards.

Although the above standards development process has been described largely in terms of voiceband analog performance, the process is equally applicable to digital performance at standardized hierarchical rates (e.g., DS1, DS3, SONET, and synchronous digital hierarchy (SDH)). That is, the same steps of translation, allocation and planning are required to relate, in this case, end user digital performance requirements to digital network performance standards.

8.6.2 Operations-Related/Bottom-Up Driven Standards

Thus far in the performance standards development process, consideration has been given to the top-down translation, allocation, and planning stages required to link user-oriented performance to the necessary standards. This process contributes greatly to ensuring that end users' needs are met. However, there is also a need to provide more direct (bottom-up) assurance that service quality is being achieved, as shown in Figure 8.4. This is accomplished by maintenance-related functions such as performance monitoring, testing, and surveillance (as discussed in Ch. 9), and by associated standards.

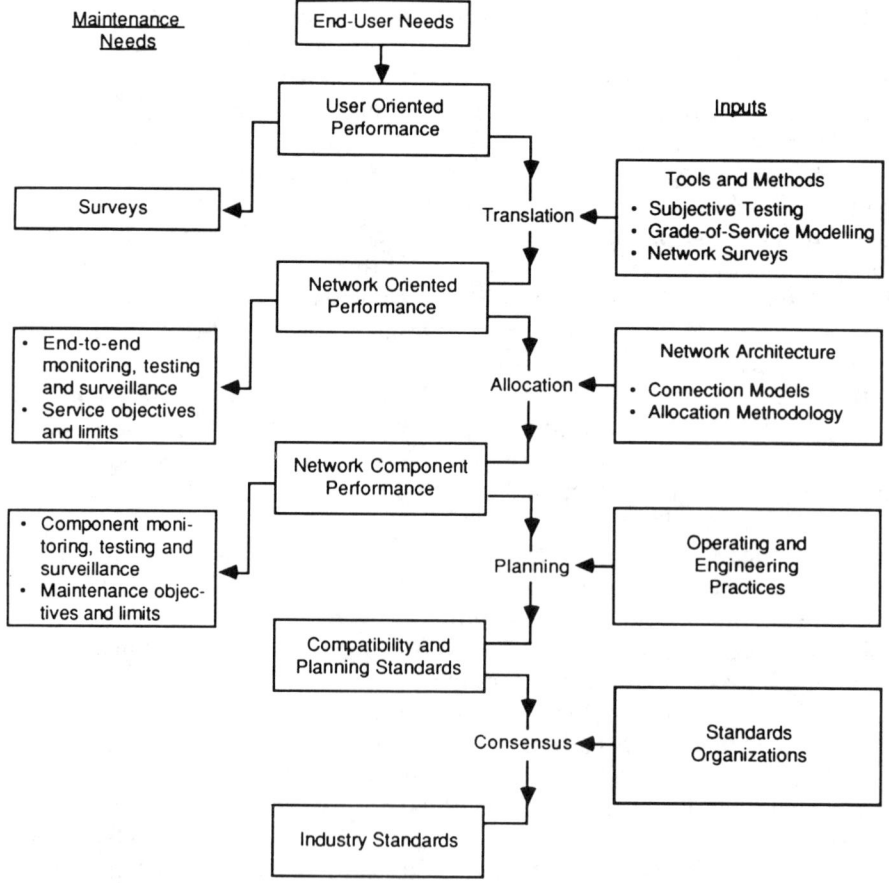

Figure 8.4 Standards development stages including bottom-up maintenance needs.

At the network-oriented (end-to-end) performance level, maintenance-related equipment can monitor large portions of the network, including complete end-to-end connections. In the context of digital performance, this is referred to as *path monitoring*. An example is end-to-end DS1 path monitoring using the cyclical redundancy check (CRC) block error code built into the DS1 extended super-frame format (ESF) to support continuous in-service performance monitoring. At the analog voiceband level, monitoring is usually carried out on an "out of service" basis, although in-service monitoring is becoming available at this level as well, as discussed in Chapter 9.

At the network component performance level, appropriate monitoring, testing, and surveillance procedures are required to ensure that the performance levels allocated to component parts of the connection are achieved. In the context of digital performance, an important aspect assuring performance at the network component level is referred to as transmission *line monitoring* (see Ch. 9). Allocated performance is expressed in terms of appropriate maintenance objectives and maintenance limits. An example is line monitoring of T-carrier transmission facilities using the bipolar violation (BPV) line error code, which also enables continuous in-service monitoring of these facilities.

The role of maintenance-related capabilities and agreement on standards to specify their functionality (i.e., in the T1M1 forum) is becoming important in the U.S. competitive environment in terms of

- Replacing reactive maintenance as driven by end user complaints with a more pro-active surveillance based approach.
- Resolving administrative responsibility in the event of overall performance degradation on multivendor connections.
- Comparing the network performance provided by different network vendors.
- Relating service quality to the price paid by end users.

8.7 RELATIONSHIPS AMONG STANDARDS FORUMS

Having discussed the national and international standards bodies that address performance issues, and having described the process of developing performance standards within these bodies, the relationships among these bodies can now be illustrated. This is done in terms of current activities, and the complexities brought about by divestiture and interworking situations in the U.S. postdivestiture environment can be seen.

8.7.1 United States

Figure 8.5 illustrates the relationships among the main standards bodies, T1, TIA, and IEEE in developing public network(s) standards. End user performance needs are addressed by the TIA in terms of terminal equipment specifications, which ensure performance compatibility with the network (e.g., sensitivities for voice terminals and BER capabilities for data modems). Corresponding network standards involving performance translation, allocation, and planning are then addressed by T1Q1.1 and by T1Q1.3/.4 for analog and digital services, respectively. Finally, the consensus process is achieved in accordance with ANSI rules, which promotes participation by all members of the industry.

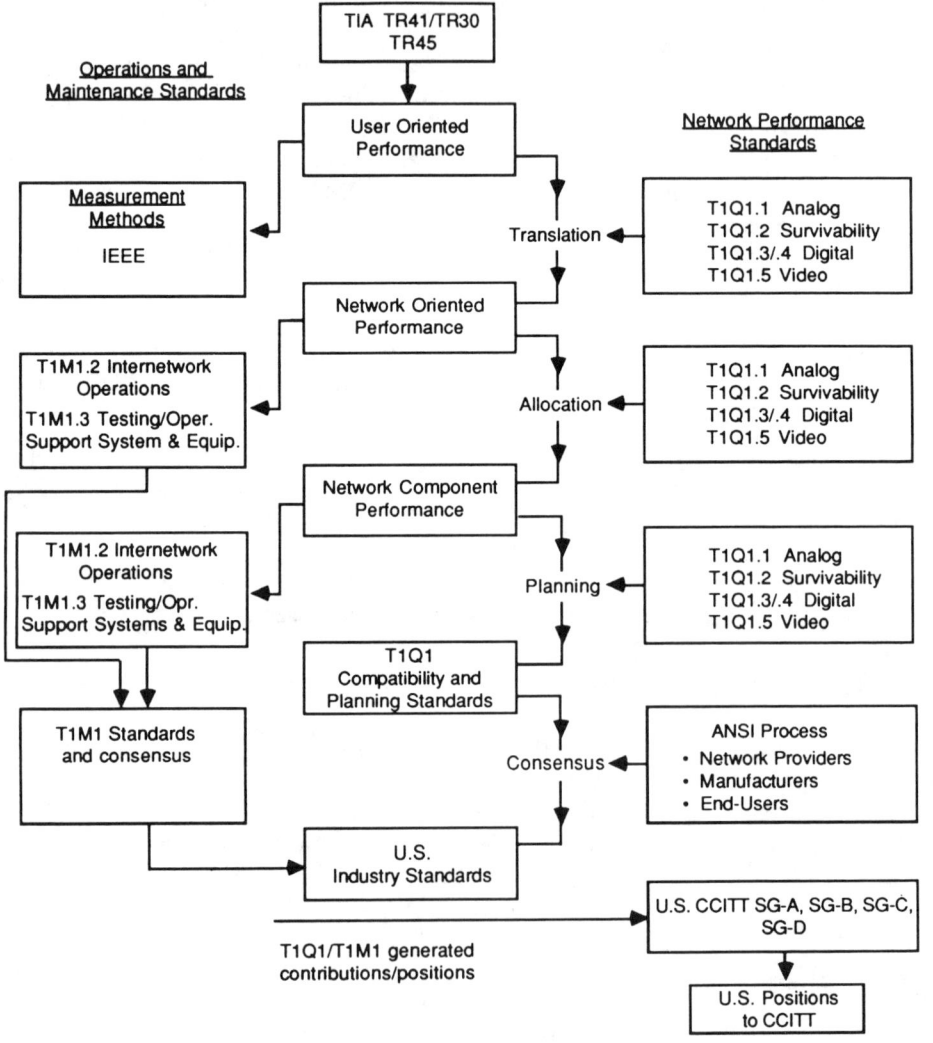

Figure 8.5 Relationships among U.S. forums (T1, TIA, IEEE) in the development of public network transmission performance standards.

The relevant TIA and T1Q1 committees cooperate by way of liaison at all stages in the development of a standard, to ensure that the resulting standard satisfies industry needs. Cost is also a significant underlying factor in this process; for example, relaxed requirements for terminals should not impose severe economic penalties in terms of more stringent specifications for the network, and vice-versa. That is, although it may be technically possible to achieve more stringent network performance, the cost may be prohibitive. On the other hand, network performance must be sufficiently high to support the desired services.

Considering operations and maintenance, T1M1 addresses the development of relevant standards in this area, as shown in Figure 8.5. Both at the network end-to-end level and at the network component level, the need for performance measurement and maintenance procedures and limits provides the impetus for T1M1 to develop appropriate maintenance-related standards. The consensus process within T1M1 mainly involves equipment manufacturers and network providers, in which the manufacturers tend to support the provision of a minimum set of equipment features, as opposed to network providers that tend to support a more comprehensive range of features.

Following the consensus process in Figure 8.5, an ANSI-accredited standard is established. These are subject to review and update on a five-year basis. As mentioned previously, positions agreed upon within T1 are often used as the basis for developing U.S. positions for CCITT. This requires prior agreement in one of the U.S. Study Groups (A, B, C, or D). as illustrated in Figure 8.5, and as described in Section 8.4.1.

The work of T1Q1 in the digital performance area can be used to provide two examples of the development of U.S. positions into CCITT SG XVIII, as illustrated in Figure 8.5. In the circuit switched digital performance area, T1Q1.4 is actively developing U.S. positions on draft CCITT Rec. G.82x [35], and, through liaison with T1M1.3, ensuring that the new performance parameters in Rec. G.82x are compatible with future network element performance monitoring requirements. In the broadband ISDN performance area, T1Q1.3 is actively developing U.S. positions on the I series recommendations in parallel with an activity to develop a national performance standard covering ISDN [36].

With reference to Figure 8.5, liaison not only takes place between T1Q1 and T1M1 on performance monitoring and maintenance aspects, but also between T1Q1 and T1X1 on performance aspects of new systems and network architectures (including synchronization and survivability), with T1Y1 on performance aspects of new coding systems, T1E1 on interface performance requirements, T1S1 on new transport modes, and with T1P1 on personal communications performance.

The role of the IEEE in the postdivestiture environment is generally to investigate and develop new measurement techniques in support of both TIA terminal requirements and standards, and T1Q1 network requirements and standards. Generally, some IEEE participants also participate in TIA and T1Q1 activities, and this contributes toward necessary liaison.

Although Figure 8.5 has focused on the development of standards for public network(s), a similar process would be followed within TIA in developing a private/corporate network or CPE standard. In this case, T1Q1 would provide input related to interfacing with, and performance requirements of, public networks. Therefore, the process of developing a private network standard would also be one of interworking and close liaison, and would also involve the IEEE in the area of measurement methods as required.

Two specific US transmission performance standards activities which serve to illustrate the process and interaction between T1Q1 and TIA are:

1. The activity leading to ANSI Standard T1.501-1988, "32 kbit/s ADPCM Tandem Encoding Limits" [37], and
2. The activity to develop a transmission planning standard covering evolving all-digital networks and ISDN [38, 39].

The first standard provides guidelines on how many ADPCM systems may be tandemed dependent upon the services to be carried [37]; (this topic is discussed in detail in Sec. 7.10 of Ch. 7). An important service in this respect is voiceband data, in terms of modulation rates and methods, and end user service requirements. To ensure these factors were accounted for, liaison was established between T1Q1.1 (involving data performance) and TR 30.3 (involving data interfaces) in which TR 30.3 provided information reflecting user service requirements over a range of modem speeds and modulation methods. In addition, TR 30.3 delegates actively participated in T1Q1.1 meetings and this contributed strongly to the timely concensus and development of the ANSI Standard T1.501.

Regarding development of a transmission plan for evolving all-digital networks and ISDN, both TR 41 and T1Q1.1 have developed draft standards covering private [38] and public networks [39], respectively. The TIA activity is under TIA TR41.1 Standard Project PN-1993. The T1Q1 activity is under the general T1Q1 Standard Project T1Q1-11. These activities included liaison between the two groups as a means of ensuring future interworking compatibility between private and public networks in the evolving all-digital environment.

Liaison has also been established between T1Q1 and TR 45 in the area of cellular mobile radio performance and interworking to cover aspects of public to mobile network interworking covered in the draft loss planning standard [39].

8.7.2 Canada

Compared with the U.S., the standards bodies' relationships are not as complex. CSA SCOT develops terminal standards much like the TIA in the U.S. However, network guidelines are developed internally for example by Telecom Canada for use by the individual provincial telephone operating companies which belong to the Telecom Canada consortium.

However, the need for Canadian positions does arise when developing positions for CCITT. In this case, Telecom Canada is one player among others, e.g., UNITEL and Teleglobe Canada who meet on a regular basis in National Study Groups (NSGs) to agree on Canadian positions (see Sec. 8.4). Participation also includes representatives of the Department of Communications, terminal manufacturers, and scientific or industrial organizations such as Bell Northern Research.

Through the interaction between Canadian and US standards forums, there is a positive trend towards developing a unified set of North American standards, particularly for ISDN. This is occurring through:

1. CSA-SCOT agreement to adopt ANSI-T1 and ANSI-EIA/TIA standards as the basis for Canadian standards.
2. Participation of Canadian organizations such as Telecom Canada and Mitel in the US standard forum arenas of T1 and TIA.

Although there will in some cases be differences in terms of regulatory and legal factors between US and Canadian standards, it is anticipated that differences at the technical level will be minimal.

8.7.3 International (CCITT)

Figure 8.6 illustrates the relationship among the CCITT Study Groups which address terminals, network, and maintenance related issues, in terms of the overall process of developing performance related CCITT Recommendations.

Starting with user-oriented performance, a number of Study Groups provide necessary input apart from that generated internally in SG XII by terminal manufacturers. Within the CCITT, and in SG XII in particular, the end user/terminal performance needs, and the public network performance needs are addressed jointly. This has the advantage of ensuring compatibility of requirements, and illustrates an important distinction between the operation of the CCITT, and operation of the TIA and T1 in the U.S. That is, in the U.S., end user and terminal performance needs are addressed in TIA, separate from the network performance needs addressed in T1.

SGs I and II (in conjunction with SG XVIII for ISDN) address "quality-of-service" issues (see Ch. 1 and 2) which are becoming more important with the evolution toward narrowband and broadband ISDN. These issues relate to performance measures which end users can relate to, such as data throughput and response time. From the data terminal perspective, both SG VII and SG XVIII contribute to, and liaise with, SG XII on the needs of data users and on requirements for new data terminals. For packet data liaison is with SG VII while for digital data, liaison is with SG XVIII. In addition, the CCIR through the CMTT, has a role to play in providing user-oriented performance requirements for broadband services, such as, video which is becoming an increasingly important aspect of the broadband ISDN studies in SG XVIII. SG I is also involved in studying "quality of service" issues for video services.

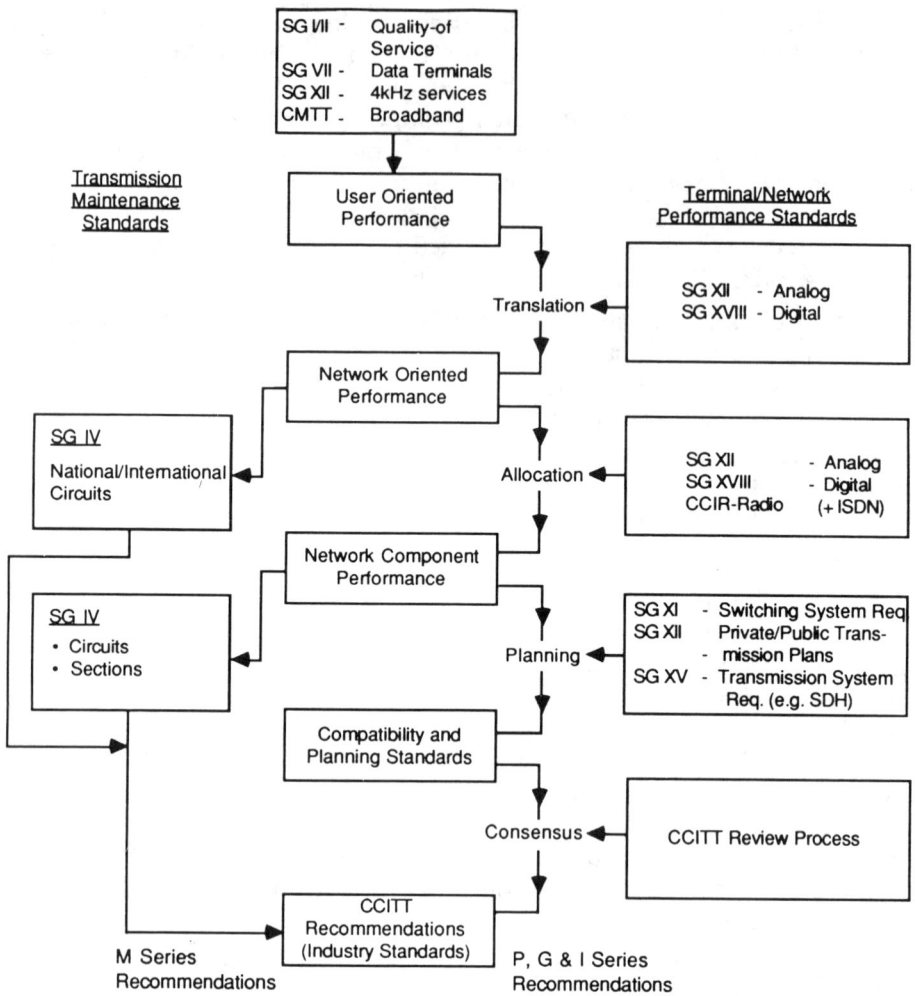

Figure 8.6 CCITT Study Group relationships arising in the development of recommendations covering performance and maintenance.

The performance translation and allocation phases are addressed by SG XII and SG XVIII, covering 4 kHz voiceband channels and digital channels, respectively. This includes agreement on HRXs and the levels of analog and digital performance to be provided by the national and international portions of international connections. The CCIR also plays a key role through liaison with CCITT SG XVIII in addressing digital radio relay system performance, HRXs and allocation.

At the planning stage, equipment requirements are taken into account through interaction with SG XI on switching equipment requirements, and SG XV on general transmission systems requirements. SG XV not only addresses specific transmission equipment specifications such as echo cancellers, but also the much broader requirements associated with the synchronous digital hierarchy (SDH).

Regarding maintenance aspects in Figure 8.6, SGIV provides the focus, addressing the maintenance of national and international circuits, and in addition, addressing the requirements of smaller connection elements such as circuits and sections. These activities are related to the end-to-end network standards and the network component standards as illustrated in Figure 8.6.

Finally, subject to the consensus process, recommendations covering performance (P,G and I Series Recommendations) and maintenance (M Series Recommendations) are developed in SG XII, SG XVIII and SG IV respectively.

Appendix A.8.3 lists existing Recommendations pertaining to performance and appendix A.8.2 lists current questions under study in SG XII (WP XII/1-4) and SG XVIII (WP XVIII/6).

Examples of Recommendations which pertain to digital transmission performance and under development or revision in SG XVIII [40] in the 1989-1992 Study Period are as follows:

I.350R   General aspects of quality of service and network performance in digital networks, including ISDN
I.35B   Performance aspects of B-ISDN
G.82x   Error Performance in high-speed digital networks.

G.82x is fundamental for future high-speed digital networks and services as it defines transmission parameters and objectives for rates at DS1 and above. This will in particular impact service definition and network performance monitoring and is also the subject of debate on a national level in T1Q1 (see Sec. 8.7.1).

8.8 REFERENCES

1. A.M. Rutkowski, "An overview of the Forums for Standards and Regulations for Digital Networks," Telecomm. Vol. 20, No. 10, Oct. 1986, pp. 84–96.
2. E.J. Cohen, "A Standards Primer". TE&M" Vol. 89. No. 22, Nov. 1985, pp. 133–136.
3. IEEE Comm. Mag. Special Issue on "Telecommunications Standards," Vol. 23, No. 1, Jan. 1985.
4. I.M. Lifchus, "Standards Committee T1-Telecommunications," IEEE Comm. Mag., Vol. 23, No. 1, Jan. 1985, pp. 34–37.
5. R.E. Olley, "The Role of Canadian Standards Association in the Development of Telecommunications Standards in Canada," Canadian Business Telecomm. Alliance Annual Conf., Sep. 1985, Edmonton, Alberta
6. S. Brasier, "The National Standards System: A Historical Perspective Standards Council of Canada CONSENSUS, Vol. 15, No. 4, Autumn 1988.
7. J.L. Koepfinger, "IEEE: The pivot of standards work in the United States," Telephony, Vol. 27, No. 2, July 1984, pp. 100–105.

8. T1. Secretariat "Standards Committee T1-Telecommunications Procedures Manual," Sixth Issue, June 1991.
9. M.N. Woinsky, "National Performance Standards for Telecommunications Services," IEEE Comm. Mag., Vol. 26, No. 10, Oct. 1988, pp. 70–81.
10. IEEE P802.6 "Proposed Standard—Distributed Queue Dual Bus (DQDB) Subnetwork of a Metropolitan Area Network Doc. P802-6/D14 July, 1990.
11. "Carrier Liaison Committee Established," ECSA Tielines, Spring 1985.
12. H. McElveen, "The ICCF: Past Present, Future," ECSA Tielines, Third Quarter 1988.
13. J. Johnson, and B. Elliott, "The NOF: Success through Consensus," ECSA Tielines, Fourth Quarter 1985.
14. W.H. Von Alven, "Part 68: A 'Living' Document." Telephony Vol. No. September 1980, pp 50–62.
15. T.E. Bell, "Technical Challenges to a Decentralized Phone System," IEEE Spectrum, pp 32–37, Sep. 1990.
16. M. Winfield, and P. von Lipinski "Regulations and Competition—Canadian Style" Telephony Vol. 209, No. 7., July 1985, pp. 56–63.
17. Telecom Canada "Digital Network Notes" 1983.
18. Management Briefing on the Canadian Standards Association. Telecomm. Report No. 1, July 1983.
19. CCITT SG II "Handbook on Quality of Service and Network Performance" Preliminary Draft, CCITT WP 11/4 Q.21, Issue 1, January 1991.
20. CCITT Resolution No. 2 "Approval of new and revised Recommendations between Plenary Assemblies." Blue Book, Vol. 1, Fascicle 1.2, Geneva 1989.
21. D.M. Cerni, "The United States Organization for the CCITT," IEEE Comm. Mag. Vol. 23, No. 1, Jan. 1985, pp. 38–42.
22. Canadian National Organization [CNO/CCITT] "Manual" August 1990.
23. CCIR Question 39-2/8 "Future Public Land Mobile Telecommunicative Systems." Questions of the CCIR, 1990, Vol. XV-2. SG-8 Mobile Radio Determinations, Amateur and Related Satellite Services. CCIR, Geneva, 1990.
24. ISO/IEC JTC1/SC6 "Terms of Reference of WG6 in Private Integrated Services Networking (PISN)," Standards Council of Canada, Doc. N 5562, June 1989.
25. M. Toutan, "CEPT Recommendations," IEEE Comm. Mag., Vol. 23, No. 1, Jan. 1985, pp. 28–30.
26. D. Gagliardi, F. Lucas, J. Remy "European Cooperation in Implementing ISDN," Computation and Transmission, No 1., 1989, pp. 39–66.
27. Lonberg Ib, "ETSI—The European Telecommunication Standard Institute," Teleteknik, No. 2., 1988, pp. 68–73.
28. J. Williamson, "Raising the European Standard" Telephony, Vol. 220, No. 22, June 1991, pp 46–50.
29. European Telecommunication Standards Institute Technical Assembly ACTE/TRAC/ETSI Developments. 7th Technical Assembly Document No. ETSI/TA 7(90) 24, London, June 1990.
30. ETSI Sub Technical Committee BT2. Report of Meeting No. 5 Oslo 13–16 May 1991.
31. "Interregional Telecom Standards Conference." Telecomm. Report, Feb. 26, 1990.
32. I. Dorros, "The Standards Slowdown" Telephony, Vol. 218, No. 9, Feb. 1990. pp. 46–52.
33. ETSI "Summary Report of ITSC2 meeting," 11th Technical Assembley, Nice, 18–20 September 1991 Temp. Doc. No. 48.
34. J.A. Post, and R.D. Silverthorn "Results of a Subjective Comparison of Echo Control Devices in Terrestrial and Satellite Trunks," NTC Rec. Vol. 3, Wash. D.C., Nov. 1979, pp. 48.4.1–48.4.5.
35. Technical Subcomittee T1Q1. Proposed contribution to CCITT Study Group XVIII "Proposals for CCITT Recommendation G.82x," August 1991.

36. Technical Subcommittee T1Q1 "Baseline Text for Proposed American National Standard T1.1 (Telecommunications—Performance Parameters, Objectives, and Measurement Methods for Integrated Services Digital Networks). Doc. No. T1Q1.3/91-020R2, August 1991.
37. American National Standard for Telecommunications—Network Performance Tandem Encoding Limits for 32 kbit/s Adaptive Differential Pulse-code Modulation (ADPCM), ANSI T1.501-1988.
38. EIA Engineering Standards Proposal No. 1993-A, Proposed Addendum to EIA/TIA-464-A, October 1991.
39. Draft American National Standard for Telecommunications—Network Performance—Loss Plan for Evolving Digital Networks, T1Q1/91-026, August 1991.
40. CCITT Study Group XVIII — Report R54 "Report of Working Party XVIII/6 — Performance, January 1991.
41. CCITT COM XII — 1, "List of Questions proposed for study during the 1989-1992 Study Period," Melbourne, Jan. 1989.
42. CCITT P Series Recs "Telephone Transmission Quality" Blue Book Vol. V, Geneva, 1989.
43. CCITT Recs G.101 — G.181, "General Characteristics of International Telephone Connections and Circuits" Blue Book, Vol. III, Fascicle III.1, Geneva, 1989.
44. CCITT Recs G.700 — G.772. "General Aspects of Digital Transmission Systems: Terminal Equipment" Blue Book, Vol. III, Fascicle III. 4, Geneva, 1989.
45. CCITT Recs G.801 — G.906 "Digital Networks, Digital Sections and Digital Line Systems" Blue Book, Vol. III, Fascicle III.5, Geneva, 1989.
46. CCITT Recs I.350-I.352. "Integrated Services Digital Network (ISDN). Overall Network Aspects and Functions, ISDN User-Network Interfaces," Blue Book, Vol. III, Fascicle III.8, Geneva, 1989.
47. CCITT Recs. X.40 — X.181 "Data Communications Networks Transmission, Signalling and Switching, Network Aspects, Maintenance and Administrative Arrangements" Blue Book, Vol. VIII, Fascicle VIII.3, Geneva, 1989.
48. CCITT COM XVIII — 1, "Questions allocated to Study Group XVIII (ISDN) Melbourne, Jan. 1989.

Appendix A.8.1
ACTIVITIES OF THE STANDARDS FORUMS IN CCITT, T1, TIA, AND IEEE

A.8.1.1 CCITT (INTERNATIONAL TELEGRAPH AND TELEPHONE CONSULTATIVE COMMITTEE) STUDY GROUPS (SGs)

SG No.	Title
I	Services.
II	Network Operations.
III	Tariff and accounting principles.
IV	Maintenance.
V	Protection against electromagnetic effects.
VI	Outside Plant.
VII	Data communications networks.
VIII	Terminals for telematic services.
IX	Telegraph networks and telegraph terminal equipment.
X	Languages for telecommunications applications.
XI	Switching and signaling.
XII	Transmission performance of telephone networks and terminals.
XV	Transmission systems and equipment.
XVII	Data transmission over the telephone network.
XVIII	ISDN.

A.8.1.2 COMMITTEE T1-TELECOMMUNICATIONS: TECHNICAL SUBCOMMITTEES

T1E1	"Network Interfaces"
T1M1	"Internetwork Operations, Administration, Maintenance and Provisioning"
T1P1	"Systems Engineering, Standards Planning and Program Management"
T1Q1	"Performance"

T1S1	"Services, Architectures and Signalling"
T1X1	"Digital Hierarchy & Synchronization"
T1Y1	"Specialized Subjects"

A.8.1.3 INSTITUTE OF ELECTRICAL AND ELECTRONICS ENGINEERS: TELECOMMUNICATIONS RELATED SOCIETIES (IEEE)

Acoustics, Speech & Signal Processing Society

Aerospace & Electronic Systems Society

Antennas & Propagation Society

Circuits & Systems Society

Communications Society

Transmission Systems Committee (TRANSYSCOM)

Computer Society

Electromagnetic Compatibility Society

Instrumentation & Measurement Society

Microwave Theory & Techniques Society

Professional Communication Society

Reliability Society

Standards Coordination Committees

A.8.1.4 TELECOMMUNICATIONS INDUSTRY ASSOCIATION

The EIA before 1988 was a conglomeration of a number of more or less independent groups as follows:

1. Consumer Electronics Group (CEG)
2. Information and Telecommunications Group (ITG)
3. Government Division (GD)
4. Parts Division (PD)
5. Electronic Display and Tube Division (TD)
6. Solid State Products Division (SSPD)
7. Industrial and Electronics Group (IEG)

The TIA was the result of a merger in 1988 between the Information and Telecommunications Group (ITG) and the United States Telecommunications Suppliers Association.

The ITG was made up of 4 divisions which now constitute the TIA work areas as follows:

1. Mobile Communications Division
2. Fiber Optics Division
3. Network Equipment Division
4. Users Premises Equipment Division

Specifically a number of committees within these divisions include the performance areas of TR-30 and TR-41 (and TR-45 covering cellular standards). The complete set of committees is as follows:

1. TR-8—Land Mobile Services
2. TR-14—Point-to-Point Microwave
3. TR-29—Facsimile Systems
4. TR-30—Data Transmission Systems and Equipment
5. TR-32—Personal Radio Equipment
6. TR-34—Satellite Equipment and Systems
7. TR-41—Telephone Terminal Equipment
8. TR-45—Cellular Standards
9. FO-2—Optical Communications
10. FO-6—Fiber Optics

Appendix A.8.2
CCITT STUDY GROUP QUESTIONS ON PERFORMANCE, 1989 TO 1992 STUDY PERIOD

A.8.2.1 CCITT SG XII—TRANSMISSION PERFORMANCE OF TELEPHONE NETWORKS AND TERMINALS [41]

General

Question 1	Future Program of Work
Question 3	Definitions in the Field of Telephonometry and of Characteristics of International Connections and Circuits
Question 4	Updating of the CCITT Telephonometric and Transmission Planning Handbook
Question 6	Harmonization of G.100 Recommendations

Working Party 1: Telephonometry

Question 8	Improvements of the Methods for the Determination of Loudness Rating
Question 9	Sidetone
Question 12	Artificial Voices, Mouths and Ears
Question 14	Applications for the Artificial Voice

Working Party 2: Terminals

Question 2	Handsfree Telephony
Question 10	Speech Transmission Characteristics for Digital Handset Terminals
Question 20	Wideband Telephony
Question 23	Coupling of Hearing Aids to Telephone Receivers
Question 30	Methods for Evaluating the Transmission Performance of Digital Telephone Sets
Question 31	Speech Quality in Multimedia Terminals

Working Party 3: Transmission Performance and Modelling

Question 5	Speech Synthesis/Recognition Systems
Question 7	Models for Predicting Transmission Quality from Objective Measurements
Question 11	Degradation Introduced by Interaction Between Voice-Operated Devices
Question 13	Methods for the Evaluation of Non-Linear Distortion
Question 15	Loudness Rating, Algorithm and Application Rules
Question 18	Transmission Performance of Digital Systems
Question 27	Talker Echo, Propagation Time and Stability in Telephone Networks, ISDN and Interconnection with ISDN
Question 28	Listener Echo (Receive end Echo) in the Public Switched Telephone Networks
Question 32	In-service Non-intrusive Assessment of Telephone Channel Transmission Performance

Working Party 4: Transmission Planning

Question 16	Impedance Strategy in the Local Network
Question 17	Actual and Preferred Speech Levels in Telephone Connections
Question 19	Recommended Values for Loudness Ratings
Question 21	Relative Levels at the Boundary Between National Systems and the International Chain
Question 22	International Telephone Conferencing
Question 24	Integration of Mobile Systems into the Public Switched Network
Question 25	Transmission Impairments in the Evolving Mixed Analog/Digital and ISDN Networks
Question 26	Setting Objectives for Mixed Analog/Digital Circuits
Question 29	Transmission Plan Aspects of the Interworking Between PSTN and ISDN in the Evolving Network

A.8.2.2 CCITT SG XVIII – ISDN [48]

Working Party XVIII/6

Question 5	General Aspects of Quality of Service and Network Performance in Digital Networks including ISDNs
Question 6	Network Performance Objectives for ISDN Circuit Mode Information Transfer
Question 7	Performance Objectives for Timing and Controlled Slips (Synchronization), Filter, Wander and Propagation Delay
Question 8	Network Performance Objectives for ISDN Connection, Processing and Packet Mode Information Transfer
Question 9	Performance Objectives for ISDN Availability

Appendix A.8.3
LIST OF CCITT RECOMMENDATIONS AND SUPPLEMENTS WHICH ADDRESS TRANSMISSION PERFORMANCE

A.8.3.1 RECOMMENDATIONS P.10-P.79 AND SUPPLEMENTS— TELEPHONE TRANSMISSION QUALITY [42]

Vocabulary and Effects of Transmission Parameters on Customer Opinion of Transmission Quality and Their Assessment

Rec. P.10   Vocabulary of terms on telephone transmission and telephone sets
Rec. P.11   Effect of transmission impairments
Rec. P.16   Subjective effects of direct crosstalk; thresholds of audibility and intelligibility

Subscribers' Lines and Sets

Rec. P.33   Subscriber telephone sets containing either loudspeaking receivers or microphones associated with amplifiers
Rec. P.34   Transmission characteristics of loudspeaker telephones
Rec. P.35   Handset telephones
Rec. P.36   Efficiency of devices for preventing the occurrence of excessive acoustic pressure by telephone receivers
Rec. P.37   Magnetic field strength around the earcap of telephone handsets which provide for coupling to hearing aids

Transmission Standards

Rec. P.42   Systems for the determination of reference equivalents
Rec. P.43A Instructions for forwarding to the CCITT Laboratory primary and working standard systems for calibration and commercial telephone systems for determining equivalent values (and other parameters, if requested)
Rec. P.48   Specification for an intermediate reference system

Objective Measuring Apparatus

Rec. P.51	Artificial voices, artificial mouths, artificial ears
Rec. P.52	Volume meters
Rec. P.53	Psophometers (apparatus for the objective measurement of circuit noise)
Rec. P.54	Sound level meters (apparatus for the objective measurement of room noise)
Rec. P.55	Apparatus for the measurement of impulsive noise

Objective Electro-Acoustical Measurements

Rec. P.61	Methods for the calibration of condenser microphones
Rec. P.62	Measurements on subscribers' telephone equipment
Rec. P.63	Methods for the evaluation of transmission quality on the basis of objective measurements
Rec. P.64	Determination of sensitivity/frequency characteristics of local telephone systems to permit calculation of their loudness ratings
Rec. P.65	Objective instrumentation for the determination of loudness ratings

Subjective Voice-Ear Measurements

Rec. P.70	Modulated noise reference unit (MNRU)
Rec. P.71	Measurement of speech volume
Rec. P.72	Measurement of reference equivalents and relative equivalents
Rec. P.73	Measurement of the sidetone reference equivalent
Rec. P.74	Methods of subjective determination of transmission quality
Rec. P.75	Standard conditioning method for handsets with carbon microphones
Rec. P.76	Determination of loudness ratings; fundamental principles
Rec. P.77	Method for evaluation of service from the standpoint of speech transmission quality
Rec. P.78	Subjective testing method for determination of loudness ratings in accordance with Recommendation P.76
Rec. P.79	Calculation of loudness ratings

Supplements to P. Series Recommendations

Supp. 1	Precautions to be taken for correct installation and maintenance of IRS
Supp. 2	Methods used for assessing telephony transmission performance
Supp. 3	Transmission rating models
Supp. 4	Prediction of transmission qualities from objective measurements
Supp. 5	The SIBYL method of subjective testing
Supp. 6	Attenuation of the electro-acoustic efficiency of telephone sets in view of protection against acoustic shocks

Supp. 7   Generation of the artificial voice
Supp. 8   Objective measurement of active speech level
Supp. 9   Provisional methods of testing headsets so that P-Series Recommendations may be used
Supp. 10   Considerations relating to characteristics for subscriber sets (handset telephones)
Supp. 11   Some effects of sidetone on the subscriber when talking
Supp. 12   Composition and alignment of the NOSFER system
Supp. 13   Noise spectra
Supp. 14   Subjective performance assessment of digital processes using the modulated noise reference unit (MNRU)

A.8.3.2 RECOMMENDATIONS G.101-G.181 AND SUPPLEMENTS—GENERAL CHARACTERISTICS OF INTERNATIONAL TELEPHONE CONNECTIONS AND CIRCUITS [43]

General Characteristics for International Telephone Connections and International Telephone Circuits

G.101   The transmission plan
G.102   Transmission performance objectives and recommendations
G.103   Hypothetical reference connections
G.104   Hypothetical reference connections (digital network)
G.105   Hypothetical reference connection for crosstalk studies
G.106   Terms and definitions related to quality of service, availability and reliability
G.107   General considerations and model of a basic telephone call
G.108   Models for the allocation of international telephone connection retainability

General Recommendations on the Transmission Quality for an Entire International Telephone Connection

G.111   Corrected reference equivalents (CREs) and loudness ratings (LRs) in an international connection
G.113   Transmission impairments
G.114   Mean one-way propagation time
G.117   Transmission aspects of unbalance about earth (definitions and methods)
G.120   Transmission characteristics of national networks
G.121   Corrected reference equivalents (CREs) and loudness ratings (LRs) of national systems
G.122   Influence of national systems on stability, talker echo, and listener echo in international connections

G.123   Circuit noise in national networks
G.125   Characteristics of national circuits on carrier systems

General Characteristics of the 4-Wire Chain Formed by the International Circuits and National Extension Circuits

G.131   Stability and echo
G.132   Attenuation distortion
G.133   Group-delay distortion
G.134   Linear crosstalk
G.135   Error on the reconstituted frequency

General Characteristics of the 4-Wire Chain of International Circuits; International Transit

G.141   Transmission losses, relative levels and attenuation distortion
G.142   Transmission characteristics of exchanges
G.143   Circuit noise and the use of compandors

General Characteristics of International Telephone Circuits and National Extension Circuits

G.151   General performance objectives applicable to all modern international circuits and national extension circuits
G.152   Characteristics appropriate to long-distance circuits of a length not exceeding 2500km
G.153   Characteristics appropriate to international circuits more than 2500 km in length

Apparatus Associated with Long-Distance Telephone Circuits

G.161   Echo-suppressors suitable for circuits having either short or long propagation times
G.162   Characteristics of compandors for telephony
G.163   Call concentrating systems
G.164   Echo suppressors
G.165   Echo cancellers
G.166   Characteristics of syllabic compandors for telephony on high capacity long distance systems

Transmission Plan Aspects of Special Circuits and Connections Using the International Telephone Connection Network

G.171   Transmission plan aspects of privately operated networks
G.172   Transmission plan aspects of international conference calls

General Recommendations on Availability and Reliability of Entire International Telephone Connections

G.180   Connection accessibility objective for the international telephone service
G.181   Connection retainability objective for the international telephone service

Supplements to G. Series Recommendations 101 to 181

Supp. 1   Talker echo on international connections
Supp. 20   Possible combinations of basic transmission impairments in hypothetical reference connections
Supp. 21   The use of quantizing distortion units in the planning of international connections (contribution of Bell-Northern Research)
Supp. 24   Considerations concerning quantizing distortion units of some digital devices that process encoded signals
Supp. 25   Guidelines for placement of microphones and loudspeakers in telephone conference rooms

A.8.3.3 RECOMMENDATIONS IN G.700 AND G.800 SERIES ON DIGITAL NETWORKS PERFORMANCE

Recommendations in G.700 Series [44]

G.700   Framework of the series G.700, G.800 and G.900 Recommendations
G.701   Vocabulary of digital transmission and multiplexing, and pulse code modulation (PCM) terms
G.711   Pulse code modulation (PCM) of voice frequencies
G.721   32 kbit/s adaptive differential pulse code modulation (ADPCM)
G.731   Primary PCM multiplex equipment for voice frequencies
G.737   Characteristics of an external access equipment operating at 2048 kbit/s offering digital access at 384 kbit/s and/or synchronous digital access at 64 kbit/s

Recommendations in G.800 Series [45]

G.801   Digital transmission models
G.802   Interconnection of digital paths using different techniques
G.803   Maintenance of digital networks
G.811   Timing requirements at the outputs of reference clocks and network nodes suitable for plesiochronous operation of international digital links
G.821   Error performance of an international digital connection forming part of an integrated services digital network
G.822   Controlled slip rate objectives on an international digital connection
G.823   The control of jitter and wander within digital networks which are based on the 2048 kbit/s hierarchy
G.824   The control of jitter and wander within digital networks which are based on the 1544 kbit/s hierarchy

A.8.3.4 RECOMMENDATIONS I.350 TO I.352 ON INTEGRATED SERVICES DIGITAL NETWORK PERFORMANCE OBJECTIVES (ADDRESSED BY CCITT SG XVIII) [46]

I.350   General aspects of quality of service performance in digital networks including ISDN

I.351   Recommendation in other series concerning network performance objectives that apply at reference points T of an ISDN

I.352   Network performance objectives for connection processing delays in an ISDN

A.8.3.5 RECOMMENDATIONS X.130 TO X.141 ON DATA NETWORKS PERFORMANCE [47]

X.130   Call processing delays in public data networks when providing international synchronous circuit-switched data services

X.131   Call blocking in public data networks when providing international synchronous circuit-switched data services

X.134   Portion boundaries and packet layer reference events: basis for defining packet-switched performance parameters

X.135   Speed of service (delay and throughput) performance values for public data networks when providing international packet-switched services

X.136   Accuracy and dependability performance values for public data networks when providing international packet-switched services

X.137   Availability performance values for public data networks when providing international packet-switched services

X.140   General quality of service parameters for communication via public data networks

X.141   General principles for the detection and correction of errors in public data networks.

Appendix A.8.4
IEEE, TIA, AND T1 TRANSMISSION PERFORMANCE STANDARDS

A.8.4.1 IEEE STANDARDS ADDRESSING TRANSMISSION PERFORMANCE AND ITS MEASUREMENT

297-1969 IEEE Recommended Practice for Speech Quality Measurements
661-1979 IEEE Standard for Determining Objective Loudness Ratings of Telephone Connection
269-1983 IEEE Standard Method for Measuring Transmission Performance of Telephone Sets
743-1984 Standard Methods and Equipment for Measuring the Transmission Characteristics of Analog Voice Frequency Circuits
752-1987 Functional Requirement for Methods and Equipment for Measuring the Performance of addressing Signaling Systems
820-1987 IEEE Standard on Loop Performance Characteristics
823-1988 IEEE Standard on Voiceband Channel Performance Criteria

Definitions for Switching System Performance

1051-1988 IEEE Recommended Practice for Parameters to characterize Digital Loop Performance
P1007 Methods and Equipment Standard for Measuring the Transmission Characteristics of PCM Telecommunications Circuits and Systems

A.8.4.2 TIA (EIA) TR-41 PERFORMANCE STANDARDS

ANSI/EIA/TIA-464A-1989
    Telephone instruments with loop signalling of voiceband applications.
RS 464 Private Branch Exchange (PBX) Switching Equipment for voiceband applications 1989
RS 334 Signal Quality at interface between data terminal equipment and synchronous data communication equipment for serial data transmission

A.8.4.3 ANSI T1 PERFORMANCE STANDARDS

501-1988	32 Kbit/s ADPCM Tandem Encoding Limits
502-1989	System M-NTSC Television Signals Transmission Specification
503-1989	Dedicated Service Parameter Definitions and Measurement Method
504-1989	Packet Switching Parameter Definitions
505-1989	Advanced Digital Program Audio Services Specification
506-1990	4 kHz Switched Exchange Access Network
506a-1992	Supplement to ANSI T1.506-1990 Addressing Network Delay

Chapter 9
MAINTENANCE CONSIDERATIONS

9.1 INTRODUCTION

So far in this book, consideration has been given to how quality of service can be designed and planned into telecommunications networks by individual network vendors, in cooperation with coordinating standards bodies. Both the top-down steps, and various bottom-up economic, technological, and competitive considerations associated with each step, have been discussed. The remainder of the book deals with the overall bottom-up process of verifying and controlling quality of service (see Fig. 1.3 of Chapter 1). This process may be referred to as quality assurance, which consists of verifying that intended network performance is achieved under day-to-day operating conditions, and of initiating appropriate action when performance is not as intended [1, 2].

Broadly speaking, quality assurance includes procedures ranging from

- Manual, out-of-service (intrusive) testing of specific network components, to
- Automatic and continuous in-service (nonintrusive) monitoring and centralized surveillance of networks and their components, to
- Direct end-to-end subjective assessment based on user-opinion surveys.

The last procedure noted above is the high-level subjective aspect of quality assurance which has been discussed in Chapter 2 in terms of opinion surveys associated with user-oriented performance. In addition to the subjectively oriented methods of assessing quality of service, the objective methods listed above are also required. These methods are broadly referred to here as maintenance procedures, which range from out-of-service testing, to in-service monitoring and centralized surveillance.

This chapter provides an overview of testing procedures, but emphasizes the continuous in-service monitoring and surveillance procedure [3, 4]. This emphasis is consistent with the trend away from dependence on trouble reports and out-of-service testing (i.e., trouble-driven maintenance), and the increasingly important role of in-service monitoring as a surveillance-based maintenance tool; this tool may also serve as a means of supporting higher level functions, such as provisioning, and restoration of degraded or failed circuits and facilities. The emphasis on in-service monitoring and surveillance is also consistent with the overall maintenance objective of safe-guarding network performance at a reasonable cost. In this regard, maintenance plans for modern telecommunications networks are taking advantage of new technologies in both network components and in operations support systems (OS), and are evolving toward further automation and centralization of maintenance procedures [5–22].

9.2 MAINTENANCE PROCESSES

9.2.1 Preventative Maintenance Process and Functions

The preventative maintenance process and related maintenance functions are illustrated in Figure 9.1 [23, 24]. Preventative maintenance as defined here is the detection of potential performance degradations before they affect service and are noticed by end users. The process therefore involves the continuous surveillance and analysis of performance data to determine the extent of any degradations, and what subsequent preventative action should be taken. Preventative maintenance as defined here also embodies "controlled" maintenance as defined in [23].

Until recently, data gathering has been largely on the basis of periodic routine testing, usually on an out-of-service basis during off-peak traffic hours. As discussed later in this chapter (Sec. 9.3 on embedded maintenance support systems), this type of testing is still carried out on trunks, for example, and in a mechanized manner. When measured performance deviates significantly from the expected range, exception reports are issued so that follow-up preventative (or corrective) action can be initiated.

With the introduction in recent years of new digital telecommunication equipment, as well as the advent of lower cost processing and memory technologies, the trend is toward continuous in-service performance monitoring (performance monitoring is discussed in detail in Secs. 9.6 and 9.7 of this chapter). The data reported from monitoring processes can be in its entirety, or in terms of degradations below some threshold limit, that is, in the form of threshold alert messages. Owing to the potential volume of data, generally some form of automated data analysis is required, for example, to assess the extent and source of degradations, to predict impending, more serious degradations through trend analysis, and perhaps to generate alarms in situations deemed to be serious enough to warrant corrective action (see Fig. 9.2, presented with the next topic). There are a variety of maintenance philosophies relating to the analysis of performance data and alerts, but the overall intent is to proactively address performance degradations before they affect end users of telecommunications services [25].

One approach along these lines is to compare the reported performance data with performance objectives and limits, such as those discussed in Chapters 4 and 6

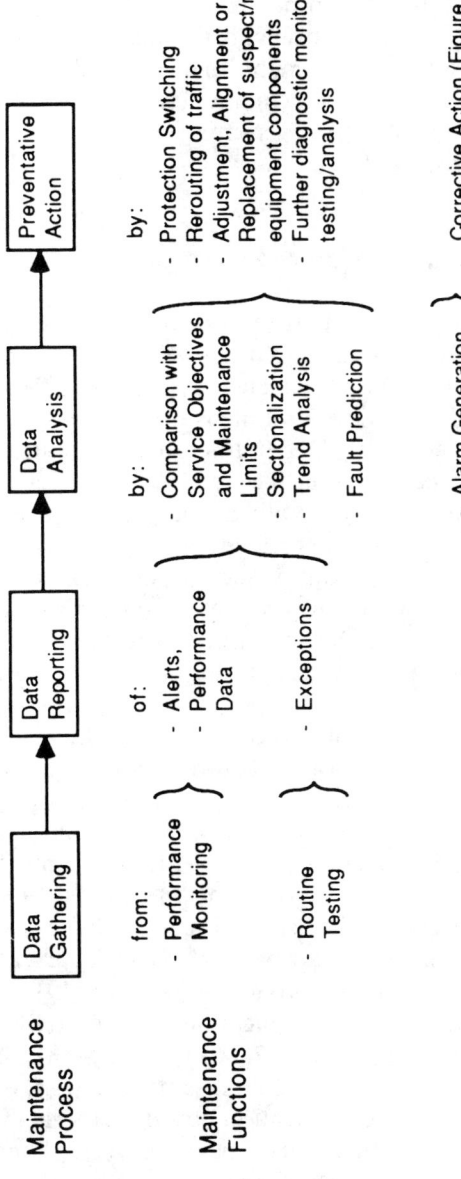

Figure 9.1 Preventative (proactive) maintenance process and functions.

[4, 26]. If the measured performance is on a connection basis, then the end-to-end performance objectives in Chapter 4 are applicable. On the other hand, if the measured performance is from network components (e.g., trunks or transmission facilities), then the performance limits in Chapter 6 are applicable.

The preventative actions that can be taken in the event of significant performance degradations range from further diagnostic analysis to actual substitution of marginal, intermittent, or suspect equipment, so as to avoid a subsequent failure or further degradation. Other actions include manual (forced) or automatic protection switching to standby equipment, or the rerouting of performance-sensitive traffic onto better performing transmission facilities.

9.2.2 Corrective Maintenance Process and Functions

The corrective maintenance process and related maintenance functions are illustrated in Figure 9.2 [23, 24]. Corrective maintenance involves all of the steps necessary from the moment that trouble is detected, to when the affected equipment has been repaired and returned to service. Trouble is usually detected through the surveillance of service-affecting alarms. Depending on the maintenance philosophy, trouble detection may also be based on the number and severity of performance monitoring alerts. (Note that alarms generally denote failures or serious performance troubles, while alerts denote less serious performance degradations.) Finally, trouble detection can be from the results of routine testing and from customer complaints, although the trend is away from these techniques, toward in-service monitoring.

As soon as trouble is detected, the first priority is to avoid, or at least minimize, service outages. When sufficient amounts of traffic are involved, this is often accomplished by automatic protection switching to redundant equipment, as soon as the trouble is detected. However, if redundant equipment is already being used, or has previously failed, then route restoration is required. In the past this has been done largely by manual patching at cross-connect panels, and is known as *emergency restoration*. However, more recent technology enables restoration to be carried out automatically with advanced protection switching arrangements (e.g., bidirectional rings), or by remotely controlled digital cross-connect systems [19, 21, 27–47]. Automatic restoration of large traffic cross sections is central to achieving network survivability. Such large cross sections occur particularly with current-day high-capacity fiber systems with transmission rates up to 2.4 Gb/s [48, 49].

After service is protected, trouble verification is achieved by the built-in alarm and performance monitoring capabilities [12, 16–18,22, 50–58], which should automatically indicate whether the trouble is still present; otherwise, manual or mechanized testing is required for trouble verification purposes. (Note that there may also be a separate and rapid trouble verification or persistence check made prior to automatic protection/restoration activity.) In the case of manual or mechanized testing,

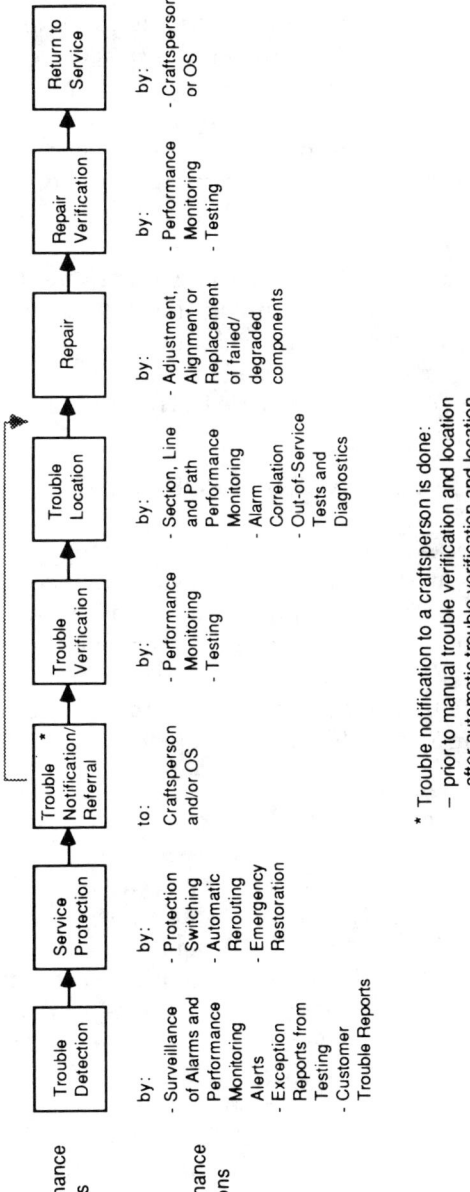

Figure 9.2 Corrective (reactive) maintenance process and functions.

craftspersons are notified of the trouble so they can perform the trouble verification function.

Similarly, depending on operational capabilities, sectionalization and correction of the trouble may be automatic or manual, or possibly a combination of both. For digital carrier transmission systems, it is expected that through correlation of the alarms and of the alerts gathered from the alarm and alert surveillance processes, trouble can in most instances be automatically sectionalized to the responsible maintenance jurisdiction, and to the regenerator section, transmission line, or associated carrier terminal equipment within the jurisdiction. Sectionalization is discussed in more detail in Section 9.6.4 of this chapter.

Finally, if not already informed, craftspersons are notified to carry out the repair process and possibly repair verification using manual or mechanized testing. However, with newer technology, repair verification can also be done with built-in performance monitoring capabilities, which should automatically indicate to a craftsperson or OS whether or not the trouble has been corrected.

9.3 EMBEDDED MAINTENANCE SUPPORT SYSTEMS

This section provides a brief overview of examples of existing maintenance support systems which assist in carrying out the preventative and corrective maintenance functions discussed above [24, 59–65]. In many instances, these support systems, although mechanized, still involve a degree of manual or out-of-service testing. As a result, these systems are viewed as precursors of, and an introduction to, the future, more automated systems built around the telecommunications management network (TMN) described in Section 9.4 of this chapter.

9.3.1 Circuit Maintenance

Circuit maintenance applies to voiceband circuits carrying services such as voice, voiceband data, and facsimile. These circuits are implicit in public switched telephone networks as well as in private networks, and in switched or nonswitched Special Services arrangements for business users. Circuits may be analog or digital. The discussions below will address maintenance of analog circuits (and analog circuits derived from digital facilities). Maintenance of digital circuits involving more modern digital test access techniques is described in Section 9.5.1 of this chapter.

Circuit maintenance involves a number of jurisdictions such as the loop plant and trunks as shown in Figure 9.3 [24]. See Figures 4.1 to 4.4 of Chapter 4 for correspondence with end-to-end connection models. Also shown in Figure 9.3 are examples of existing maintenance support systems used for each of these jurisdictions. Certain of these systems are expanded on from an architectural point of view in Figure 9.4.

Jurisdiction:	Customer Premise	Loop Plant	Switching System	Trunks	Special Service Circuits	End-to-End Circuits
Examples:	- Automatic Data Test System (ADTS) - Trouble Reports - Repair Visits - CPE and Centralized diagnostic tests	Loop Maintenance Operation System (LMOS)	Switching Center Control System (SCCS)	Centralized Automatic Reporting on Trunks (CAROT)	Switched Access Remote Test System (SARTS)	In-Service Non-Intrusive Measurement Device (INMD) *

* New capability [146]

Figure 9.3 Circuit maintenance jurisdictions and examples of maintenance support systems.

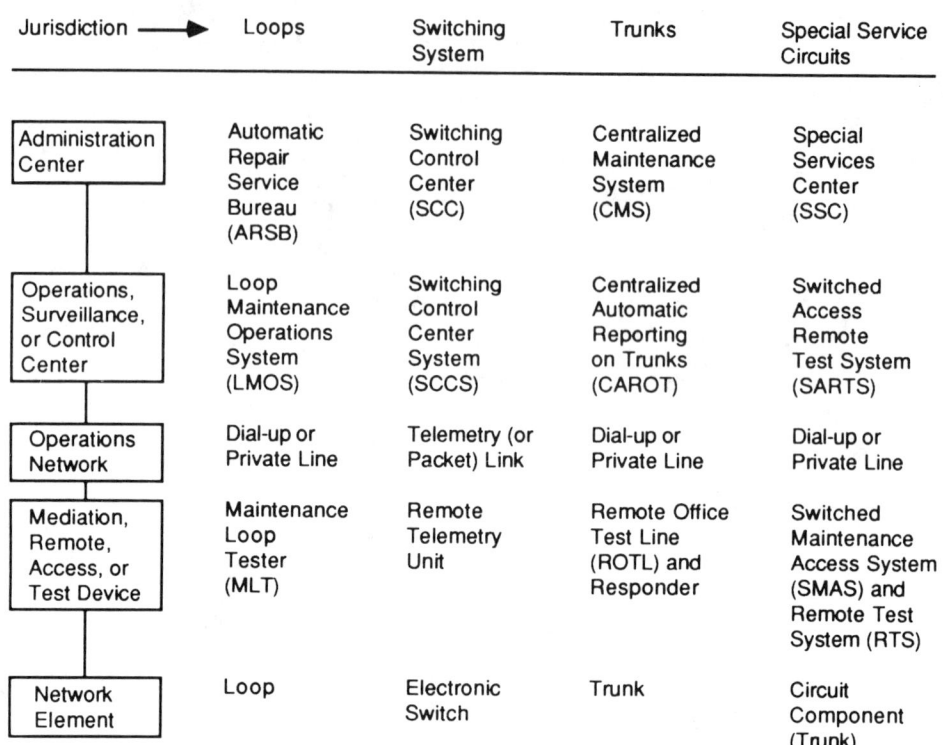

Figure 9.4 Operations architecture for examples of existing circuit maintenance support systems.

9.3.1.1 Customer Premises Equipment Maintenance

Customer premises equipment (CPE) refers to equipment on the customer's premises, including connections to the telecommunications service provider at the network interface (NI). The equipment may be customer-owned or leased from the service provider. Generally, the service provider is only responsible for CPE maintenance in the latter case. The CPE may consist of station sets (i.e., residential, coin and business telephones) as well as key telephone systems, private branch exchange (PBX) systems, data sets (modems) and terminals, and local area networks (LAN).

At present, CPE maintenance is often initiated by customer trouble reports. To some extent, trouble can be diagnosed by the service provider from a central location. For example, the automatic data test system (ADTS) is capable of testing remote data sets used in both public and private network data connections [62, 65]. Currently it is often necessary to dispatch a craftsperson for both diagnostic and repair purposes. In some cases, the CPE has diagnostic capabilities or built-in maintenance support (e.g., in PBXs). In other cases, diagnostic tests can be run by dialing special numbers (e.g., for verifying the correct functioning of telephones).

In the future, with ISDN access, more enhanced CPE maintenance capabilities will be possible in conjunction with TMN. This includes the possibility of mutual access to certain performance information by both the end user and the service provider. These capabilities are discussed in Section 9.6 of this chapter.

9.3.1.2 Loop Maintenance

Trouble conditions occurring on the analog loop facility connecting the CPE to the central office are detected by loop tests or by customer trouble reports to a repair service bureau. The automated repair service bureau (ARSB) supports testing and trouble analysis (see Fig. 9.4). It consists of, among other things, the loop maintenance operations system (LMOS) and the maintenance loop testing (MLT) system [66]. LMOS functions include data analysis relevant to trouble reports and control of automated testing. The MLT is an automated testing system controlled by LMOS. The MLT accesses and tests customers' loops using a test algorithm tailored to the expected electrical characteristics of the loop. MLT outputs include analog-measurement results, pass/fail indications, and recommended maintenance actions.

In more recent years, advances in loop testing and maintenance capabilities have evolved to reflect the needs of business users and new ISDN services, as well as the need to control maintenance costs and account for demarcations between service providers and end users [58, 83, 377–383].

In addition, with the advent of digital loop carrier technology, carrier facility maintenance is also required for the loop plant [67–70]. This is addressed in Section 9.5.1 of this chapter in the context of analog and digital test access to circuits, and in Sections 9.6 and 9.7 on performance monitoring of digital facilities.

9.3.1.3 Switching System Maintenance

Switching system maintenance involves a variety of functions, including surveillance of alarms for the detection of switching system troubles, control of scheduled routine testing of switching equipment, control of testing for trouble verification and sectionalization, analysis of alarms and test results, and repair of switching equipment [71–73].

The switching control center (SCC) is the hub of centralized switching system maintenance for a geographical area (see Fig. 9.4). The SCC can remotely administer, monitor and control switching equipment. The SCC also surveys and analyzes fault detection indications and the periodic results of automated checks for hardware and software errors. When an alarm is received, action is taken to protect service and locate and correct the problem, either from the SCC or by dispatch of a craftsperson to the switching office in question. The switching control center system (SCCS) supports SCCs by providing remote alarms and logging and analysis of trouble indicators, as well as enabling remote control of the switch from the SCC by means of a central telemetry unit at the SCCS, and a remote telemetry unit at the switch location.

Generally, switching system maintenance is only considered in this chapter to the extent that it involves transmission performance; that is, trunk testing and monitoring of certain carrier facilities (i.e., T-carrier) can be done in conjunction with a switch, particularly a modern digital switch. These maintenance functions are described later in this chapter, in relation to test access and performance monitoring. Recent requirements for other aspects of switching system maintenance can be found in [74, 75].

9.3.1.4 Trunk Maintenance

Trunk maintenance depends primarily on routine or demand transmission and operational testing of trunks, usually by mechanized means [76, 77]. The centralized automatic reporting on trunks (CAROT) system carries out such tests on trunks within a given geographical area (see Fig. 9.4). The CAROT system can be connected to a circuit maintenance system (CMS) to facilitate maintenance of trunks terminating on electronic switching machines. The CAROT system consists of a controller that enables trunks to be seized by a remote office test line (ROTL), and test equipment (e.g., responders) to be connected to each end of the trunk. In this manner, a variety of analog transmission performance (and other) tests can be conducted on the trunks, including trunk loss, C-message and C-notched noise, terminal balance return loss, and attenuation distortion (these parameters are discussed in detail in Ch. 6).

It should be noted that with the advent of digital transmission and switching technology, CAROT-based testing of trunks is not as necessary, and is also less useful, since the remaining analog technology is no longer associated with the trunk being tested, but rather is at the end points of a digital circuit. Trunk testing may be included as part of in-service nonintrusive circuit monitoring, as discussed later in Section 9.5.2 of this chapter.

9.3.1.5 Special Service Circuit Maintenance

Special Services differ from ordinary services, such as public telephone service to residential and business users. Special Services include foreign exchange (FX), wide area telephone service (WATS), Centrex (CTX), private line, and switched private network services, and are used mostly by business customers. Special service circuits require different treatment in a number of respects, including maintenance [78–86]. For the local plant involving loops and short haul trunks, maintenance is based largely on customer trouble reports to a repair service bureau or Special Services center (SSC). This is followed by manual and mechanized testing such as with the automatic data test system (ADTS) for modems. For the toll plant, the SCC provides centralized supervision and control of maintenance, remote circuit access, and transmission testing (see Fig. 9.4). The access and testing is carried out by the switched access remote test system (SARTS), which enables access to circuits through the switched maintenance access system (SMAS), and testing by means of the remote test system (RTS). The RTS can carry out, among other things, transmission tests, including circuit loss, noise, and distortion.

In recent years Special Service circuits have been provided by digital cross-connect systems (DCS), or "slow-switches" [87]. DCSs enable remote provisioning of, and test access to, Special Service circuits. DCSs also enable performance monitoring of T-carrier facilities carrying digital Special Service circuits, and terminating on DCSs. Test access and performance monitoring are discussed later in this chapter.

9.3.2 Carrier Transmission Facility Maintenance

Maintenance of carrier transmission systems differs considerably from circuit maintenance. This is because of the generally large numbers of circuits that are multiplexed for transport over a single carrier facility. As a result, equipment redundancy and surveillance of alarms are required to ensure that service to a large number of users is protected in the event of failure or serious degradation to a single facility. In addition, routine periodic tests can be performed for preventative maintenance purposes. However, recently the trend has been to replace routine testing by in-service performance monitoring, particularly for digital carrier systems. Surveillance of performance monitoring alerts also serves as a basis for proactive preventative maintenance (see Fig. 9.1).

The architectural arrangement of various existing carrier facility maintenance support systems is shown in Figure 9.5 [88–93]. Other examples can be found in [94, 95]. The facility maintenance and administration center (FMAC) centralizes and coordinates alarm surveillance, preventative maintenance, and repair and control of transmission facility networks consisting of nonfiber cable, radio, and fiber carrier systems. The FMAC depends on surveillance and control of transmission systems (SCOTS) or telecommunications alarm surveillance and control (TASC) systems, both of which are essentially alarm surveillance systems. SCOTS is used to monitor equipment associated with metallic and coaxial cables as well as microwave radio and fiber-optic transmission systems. TASC provides centralized alarm and status surveillance and reporting of data as gathered from certain transmission and switching equipment. TASC also allows remote control of certain parts of the monitored plant and automates many operations, including responding to alarms. A carrier transmission maintenance system (CTMS) is used to provide automated performance monitoring of long-haul analog carrier systems and associated multiplex terminal equipment. These measurements are specific to the analog carrier system and are carried out on a daily and in-service basis.

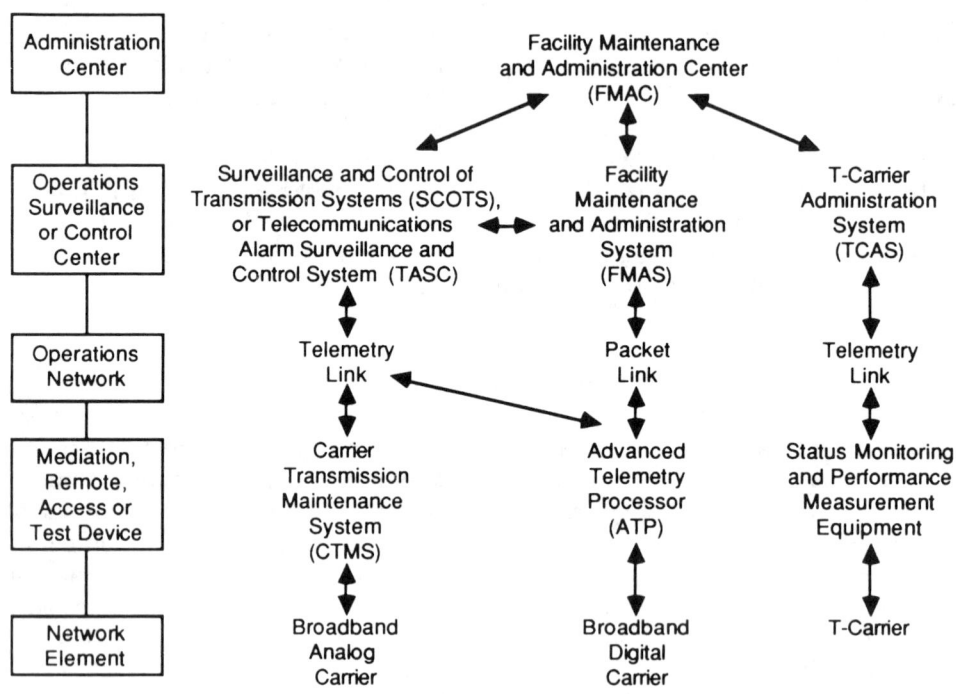

Figure 9.5 Operations architecture for examples of existing facility maintenance support systems.

For other carrier systems, notably, digital radio and fiber systems, the advanced telemetry processor (ATP) reports alarm and status information to SCOTS. It also provides in-service performance monitoring and remote fault locating down to the level of a regenerator; the information is reported to a facility maintenance and administration system (FMAS). SCOTS and FMAS also communicate with each other to coordinate the alarm and performance monitoring information. The performance monitoring information consists of counts of block error detection events (e.g., parity violations), out-of-frame (OOF) events, and a classification of different severities of errored seconds (i.e., Errored Second—ES type A with 1 error, type B with 2 to X errors, and type C with greater than X errors [92]. X depends on the rate being monitored; for example, X = 44 for an estimated random bit error ratio (BER) of about 10^{-6} at the DS3 rate of about 45 Mb/s). The monitored events are collected for 15-min intervals and stored for up to 24 hours. The counts are also thresholded and reported upon threshold crossing, to alert the FMAS of performance degradations.

In a manner similar to FMAS, the T-carrier administration system (TCAS) uses real-time surveillance and telemetry to monitor the status of T-carrier systems (rather than broadband systems, as with SCOTS and TASC). TCAS makes routine periodic digital transmission performance measurements on T-carrier systems. TCAS can automatically detect T-carrier failures and categorize the nature of the alarm as an outage or as an intermittent occurrence. Finally, it can also identify major route failures and automatically sectionalize failures to a defective span or terminal.

In recent years there have been improvements in maintenance of digital carrier transmission facilities through enhanced digital signal formats (e.g., DS1-ESF, DS3 C-Bit, and SONET), as well as in terms of enhanced monitoring capabilities (e.g., customer serving units (CSU), bridge-on, and integrated types of performance monitors), and through more widespread use of surveillance-based maintenance. These aspects are discussed in detail later in this chapter.

9.4 TELECOMMUNICATIONS MANAGEMENT NETWORK

9.4.1 Motivation for TMN

Currently in the North American telecommunications industry, there are a variety of approaches to operations, administration, maintenance, and provisioning (OAM&P), and a variety of interfaces between network elements (NE) and OSs. This is apparent, for example, from the range of maintenance support systems just reviewed. These systems tend to be vertically integrated; that is, they tend to be standalone systems from a single supplier, and they generally do not cooperate well with other systems (i.e., are not horizontally integrated). It is desirable that certain functional maintenance capabilities (i.e., for testing and performance monitoring) and associated operations interfaces be standardized.

Standardization in these areas will provide a uniform understanding of performance measures in the industry; this is important given the multivendor environment and the need to resolve responsibilities for end-to-end performance degradations at interfaces between different networks, and at interfaces between users and networks. Maintenance for ISDN basic and primary rate access is one area in which industry standardization is completed [58, 96–98]. Standardization is also at an advanced stage for general transport networks [50, 99], and is underway for broadband ISDN (B-ISDN) [100–103]. In addition, standardized operations interfaces will enable NEs procured from different manufacturers to be compatible with network providers' OSs [104-108]. Finally, standardization of interfaces and maintenance functions will enable improved cooperation among (horizontal integration of) different maintenance-related applications.

Accordingly, the purpose of the TMN is to provide an overall operations architecture to enable the interconnection of various different types of OSs with telecommunications equipment (NEs) using a set of standardized protocols and interfaces (see Fig. 9.6 [109-123]). In addition, the purpose of TMN is to provide a framework for communications among OSs, and to thereby enable the integration of OAM&P capabilities. To fulfill these purposes, it is necessary to have a standardized generic network model and OAM&P functions, as well as standardized OS/OS and OS/NE interfaces, communications protocols for the data communications (operations) network, and maintenance-related messages (e.g., see [20]).

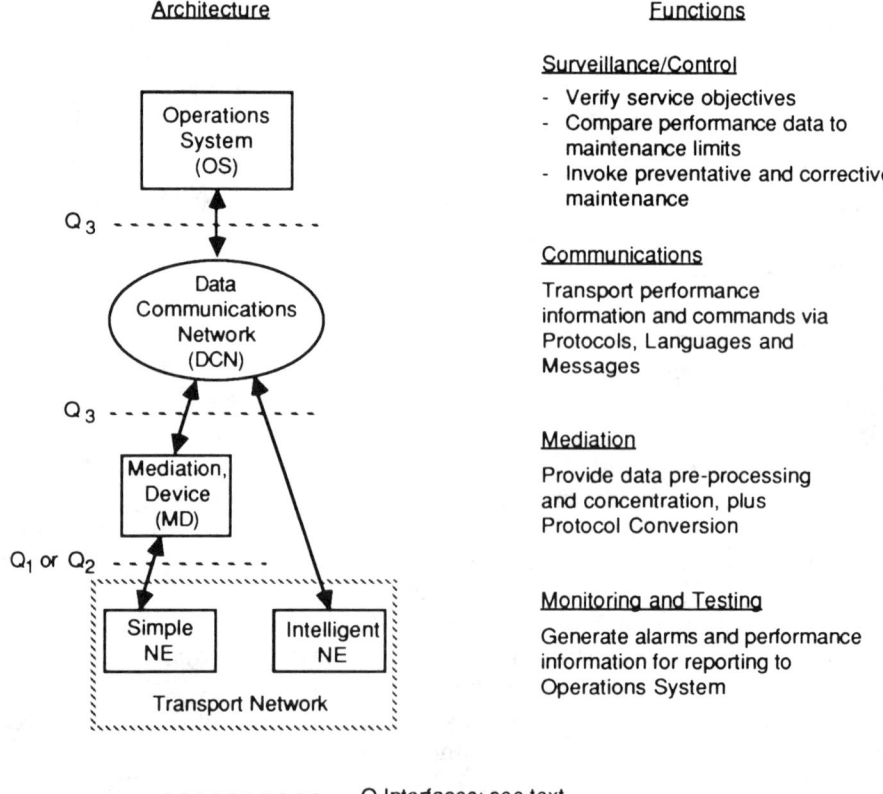

Figure 9.6 Generalized TMN surveillance-based maintenance architecture and maintenance-related functions.

9.4.2 TMN Maintenance Functions

From the perspective of the remainder of this chapter which focuses on maintaining digital transmission performance, the relevant functions of each of the main components of the TMN are as outlined in Figure 9.6 [4, 118]. The NEs, possibly in cooperation with mediation devices (MD), are responsible for detecting and reporting alarms, and providing test access and/or performance monitoring for collecting, storing, thresholding, and reporting performance-related information to a surveillance center (OS), across standard interfaces. These interfaces can either be between the NE and the surveillance center, or between an intermediate location (MD, remote interface unit) and the surveillance center. The various interfaces between NEs, MDs, and OSs, etc., depend on the type of NE and the OAM&P functions to be supported.

The main interface types associated with TMN are as follows:

- The Q1 interface supports a small set of relatively simple functions, such as the transport of binary indications representing alarm states, loop back requests, etc. Q1 is used between an NE and an MD, and is appropriate when large numbers of simple, small-scale NEs are supported by a MD. Such NEs may include certain digital multiplexing and transmission equipment, particularly embedded systems. Protocols for Q1 involve parts of layers 1 and 2 of the open systems interconnection (OSI) model.
- The Q2 interface supports a larger set of more complex OAM&P functions, such as provisioning, alarm surveillance, and performance monitoring. Q2 is also distinguished by the ability to transport data rather than only binary indications; the data may also be concentrated. Q2 may be used between a more intelligent, larger scale NE and a MD. Such NEs may include certain new generation multiplexing and transmission equipment. Protocols for Q2 involve parts of layers 1, 2, and 3 of the OSI model.
- The Q3 interface is a full-capability interface which can support all OAM&P functions, and is appropriate for large-scale intelligent NEs, such as digital switches and DCSs. Q3 is intended for use between an NE and OS or between a MD and OS via an operations network. Protocols for Q3 require layers 1 to 7 of the OSI model.

For the purposes of the maintenance considerations being addressed in this chapter, Q1 and Q2 are the sufficient and applicable interfaces.

After performance data is communicated from NEs to an OS via the interfaces just noted, the data can be analyzed in a number of ways. For example, it can be compared to allocated performance objectives (maintenance limits), and appropriate maintenance actions can then be initiated according to a proactive maintenance strategy. The data can also be used to verify that end-to-end service objectives are satisfied (see Fig. 9.12, discussed later). Of the various OAM&P functions supported by TMN, the following is an elaboration of the maintenance functions relevant to this chapter. Broadly, the functions include those outlined earlier in relation to preventative and corrective maintenance (Figs. 9.1 and 9.2).

- Alarm surveillance to rapidly capture out-of-service conditions due to equipment failures or severe performance degradations, so that service continuity is preserved, and immediate corrective action can be taken.
- Performance monitoring to detect and verify on a continuous in-service basis less serious degradations before end users are affected.
- Fault location for both the causes of out-of-service conditions, as well as the less serious degradations. This involves sectionalizing the trouble to the responsible maintenance administration or jurisdiction, and locating the trouble within these regions of responsibility. Sectionalization is supported by various levels of performance monitoring data as gathered by different types of NEs; that is, those that terminate regenerator sections, transmission lines, and digital paths as discussed in Section 9.6.4.1 of this chapter [3, 50, 51, 124].
- Testing, for detailed diagnostics, and for sectionalization and fault location (when this cannot be accomplished automatically by performance monitoring). Testing can also be used for repair verification of performance following corrective maintenance and service restoration (although, preferably, this can also be done by performance monitoring). In general, testing can be in-service (nonintrusive), or out-of-service (intrusive) if necessary, the former being preferred.
- Status and Control, for checking and changing the state of an NE on demand (e.g., in/out-of-service, protection switch status, service restore), initiating maintenance related activities (e.g., loop back, testing, data gathering and reporting), changing the attributes associated with performance monitoring (e.g., threshold change, enabling and disabling of alerts, etc.).

The above maintenance capabilities, when used with performance objectives and maintenance limits, provide the tools to enable meaningful and cost-effective maintenance actions (see Fig. 9.12, presented later).

9.5 EXAMPLES OF TMN-LIKE MAINTENANCE SUPPORT SYSTEMS

9.5.1 Test Access for Analog and Digital Circuits

Recent years have brought significant developments in the area of analog and digital testing [20, 125–137]. The TMN-like operations architecture for circuit testing is representative, and is shown in Figure 9.7. The integrated test system (ITS) is a new, forward-looking maintenance support system for testing of both analog and digital circuits. These circuits include those associated with Plain Old/Ordinary Telephone Service (POTS) and Special Services. Also, what is meant by digital circuits is not only DS0s but also higher hierarchical rates such as DS1 and DS3 circuits. It is also the intent of ITS to integrate the diverse individual test systems shown in Figure 9.4.

Figure 9.7 Example of TMN-like architecture for testing of digital and analog circuits.

The test system controller (TSC) receives test requests from ITS to test analog or digital Special Service circuits, message network trunks, loops, or any combination of these. The TSC decides which remote test units (RTU) are to be used to test the circuits in question, and, in cooperation with ITS, which tests are to be performed. The TSC will also interpret the test results and send them to ITS for further evaluation.

The RTU is the remotely controlled test equipment that measures the transmission (and signaling) parameters of a circuit. The transmission parameters include analog and digital circuit loss, idle circuit noise, impulse noise, and digital error rate measurements. These circuit measurements may be made point-to-point, or in a looped-around mode by setting up a loop-back at the far end of the circuit.

The digital test access unit (DTAU) is the means by which an RTU accesses a digital circuit such as a DS0 or DS1. The DTAU may be integrated into NEs such as DCSs. The DTAU can be used where DS1, DS0, voiceband, or subrate (e.g., 2.4, 4.8,

9.6, 56 kb/s) test access is required. There are two basic test access modes, the first being the monitor or bridge-on mode, which provides a hitless (nonintrusive) connection onto the circuit under test. That is, the circuit under test continues normal transmission during monitoring, with the test access connection transparent to the circuit. The second mode is the split (disruptive or intrusive) mode in which the circuit under test is opened and may be tested in either or both directions of transmission from the access point. Thus, the split mode enables the RTS to both transmit signals and to receive (or terminate) incoming signals. The incoming signals may be the result of a test signal being applied at the far end of the circuit, or the result of a loop-back being set up at the far end. The ability to access and loop circuits at various locations enables faults to be sectionalized. This is also aided by the application of various test signals (stress patterns) which, for example, can identify such troubles as those relating to timing misalignment, faulty equalization, faulty termination, excessive jitter, low power, etc., on T-carrier (twisted pair) facilities [138–142]. (For fiber facilities, optical time domain reflectometers (OTDR) are used to isolate cable defects such as breaks, poor splices, micro-bending, etc. [69, 143–145].)

The Analog (or metallic) test access unit (ATAU) is the means for accessing and testing an analog (or metallic) circuit. Again, the basic access modes are the monitor and split modes, similar to those noted above for the DTAU. However, for the ATAU, analog test signals (e.g., 1-kHz tones, bands of noise) and quiet terminations are used to measure loss, return loss, and noise, etc., as noted in Section 9.3.1.

9.5.2 Quality Assurance Measurements of Circuits

The circuit testing capability discussed in the previous section, when applied to analog circuits or to digital DS0s, must often be conducted in an out-of-service or intrusive mode. This is because of the need to insert test signals to measure loss, and to terminate circuits to measure noise, etc. More recently however, an in-service technique for estimating transmission parameters such as loss and noise, as well as echo path losses and delays, has been made possible through the use of echo canceller-like technology. This technique can be applied to circuits carrying voice traffic [146–151]. An example of a maintenance support system employing such a capability is illustrated in Figure 9.8 [152]. Future versions of this technology may also evaluate other voiceband signals.

The network service evaluation system (NSES) collects loss, noise, and echo path loss and delay information from the in-service nonintrusive measurement device (INMD), also called a "smart" canceller. The NSES can evaluate this information in terms of the GOS models discussed in Chapter 3 of this book. These models include both opinion and behavioral models. In this manner, it is possible for the NSES to provide useful information from both an overall connection performance point of view, as well as from a maintenance or trouble sectionalization point of view. To appreciate this, it is useful to consider how INMD devices may be deployed in a typical inter-LATA network connection. Figure 9.9 shows such a connection

Figure 9.8 Example of TMN-like architecture for performance monitoring and surveillance of digital or analog circuits carrying voice traffic.

(similar to that discussed in Chs. 4 and 5). In this figure, the INMD devices can be usefully deployed between the local- and inter-exchange carriers, to enable sectionalization of troubles to these networks. The INMD deployment is shown in more detail in Figure 9.10.

In general, the INMD is capable of measuring the analog performance of DS0s carrying voice traffic, as well as the digital performance in terms of error rates, etc., of DS1 signals which transport the DS0s (details of digital performance monitoring is addressed in Secs. 9.6 and 9.7 of this chapter). In addition, by accessing inband signaling information (or by way of a signaling interface), the INMD can collect traffic information relevant to the voice signals being carried in the DS0s. This traffic information includes call origination and destination numbers as well as interexchange carrier codes, which assist in the sectionalization and location of transmission troubles that may be identified by the INMD. If INMDs are integrated into digital circuit switches, the traffic information may also include call disposition information such as call completions, percentage of calls terminated early, percentage of retrials, etc.; this enables higher priority to be attached to transmission troubles associated with unfavorable (revenue-affecting) call dispositions.

More specifically, in the area of transmission performance of voice carrying DS0s, the INMD assessment capabilities are as follows. The INMD, owing to its echo

Figure 9.9 Inter-LATA connection.

Figure 9.10 Local exchange carrier and INMD deployment.

canceller technology, employs a speech detector; when this detector is on (speech is present), it can be used to measure speech levels received at the INMD, from which inferences can be made about the loss on the connection from the end user's station set to the INMD. Similarly, when the speech detector is off (speech is absent), the noise can be measured. In addition, implicit in echo-canceller technology is the ability to measure echo path loss and echo path delay information, including that for more than one echo path [153]. Figure 9.10 shows that the INMD would measure a principal echo occurring at the end office. There may be additional echoes, for example, at the access tandem, if it is a 2-wire switch, and also in the access network if there is a subscriber carrier system or PBX (not shown in Fig. 9.10). Similarly, by operating in the reverse direction, the INMD can assess loss, noise, and echo path performance from the IEC part of the connection (see Fig. 9.10). Finally, by combining the information from the two directions, overall connection-based performance can be estimated and further assessed, either locally or by the NSES, in terms of GOS. One of the possible outcomes of this analysis is the ability to decide, using GOS criteria, on the assignment of echo control to connections on a dynamic, per-call basis. Another possible outcome is the assessment of the effectiveness of maintenance in terms of reduced complaints, and thereby reduced maintenance costs [154].

9.5.3 Performance Monitoring and Alarm Surveillance for Digital Networks and Services

Recent developments in the telecommunications industry require increased emphasis on performance monitoring, surveillance, and maintenance of networks [3]. One of these developments is the trend toward integrated services networks, including B-ISDN. These networks must support new services, such as high-speed digital data, high-resolution imaging, high-quality video and video conferencing, and sound programming, all of which can be much more sensitive to digital network impairments than ordinary telephone voice service. Very often these impairments are not simple "hard" failures, but "soft" degradations, or are of a transient and intermittent nature. (The nature of digital signal impairments is discussed in more detail in Sec. 9.7.1 of this chapter.) An enhanced monitoring scheme is required to capture these effects and provide the basis for effective network maintenance. In turn, this will enable a network provider to address the increasing performance sensitivity of end users, particularly large business users, in the increasingly more competitive telecommunications industry.

The objective of performance monitoring is to enable early detection and correction of troubles, preferably before the end user is affected. This requires monitoring to be done continuously or nearly continuously, and on an in-service basis (as opposed to out-of-service testing in response to customer trouble reports). It also requires a degree of centralized surveillance by an operations center to collect the monitoring information and enable automatic trouble location and initiation of appropriate corrective action.

The TMN-like operations architecture for performance monitoring of digital transmission systems is shown in Figure 9.11. Network monitoring and analysis (NMA) is an example of a new, forward-looking maintenance support system for alarm surveillance and performance monitoring of digital networks and services; this includes surveillance and monitoring of all digital network components (or NEs) making up the network [6, 11, 13, 16, 18, 93]. The intent of NMA is to automate many of the maintenance functions associated with the preventative and corrective maintenance processes outlined earlier in this chapter with respect to Figures 9.1 and 9.2. In particular, NMA will gather performance data, detect, verify, and locate degraded and trouble conditions, and initiate appropriate maintenance actions, ideally before end users are affected. The intent of NMA is to also integrate maintenance functions across access, inter-office, and switching boundaries, and in so doing to accommodate both new and embedded technologies. This will serve to simplify (horizontally integrate) the variety of existing facility maintenance systems (for example, those discussed in Fig. 9.5).

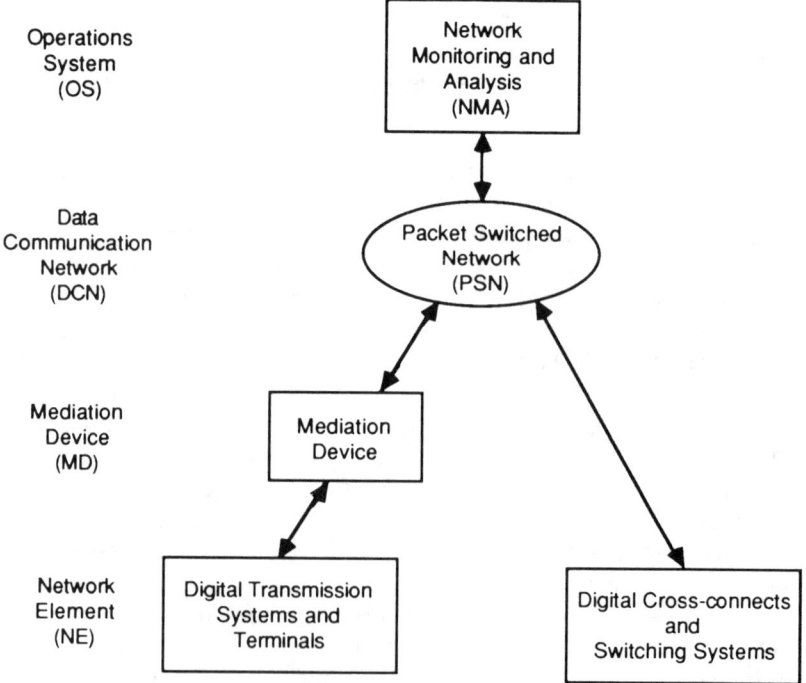

Figure 9.11 Example of a TMN-like architecture for performance monitoring and surveillance of digital networks.

The general advantages of the approach in Figure 9.11, as well as the role of mediation devices therein, were noted earlier in the TMN discussion (recall Sec. 9.4 and Fig. 9.6 of this chapter [115–119,155]). The remainder of this chapter will expand on performance monitoring and surveillance functions, the intent of these functions being to support network maintenance activities which are related to, and thereby address, end user performance needs [5–14, 50–57]. This approach is consistent with an integrated view of top-down network performance objectives and bottom-up performance monitoring capabilities, which aligns with the performance process followed throughout this book (Fig. 1.3 of Ch. 1). This performance process can be given a maintenance context as shown in Figure 9.12. To summarize, there are two main aspects of the resulting maintenance process, as follows. The first is top-down/planning oriented, and principally involves the following steps:

- Determination of end user needs in terms of long-term, end-to-end digital transmission performance objectives.
- Allocation of these objectives to maintenance entities; this may include the expression of objectives as shorter term limits consistent with maintenance needs.

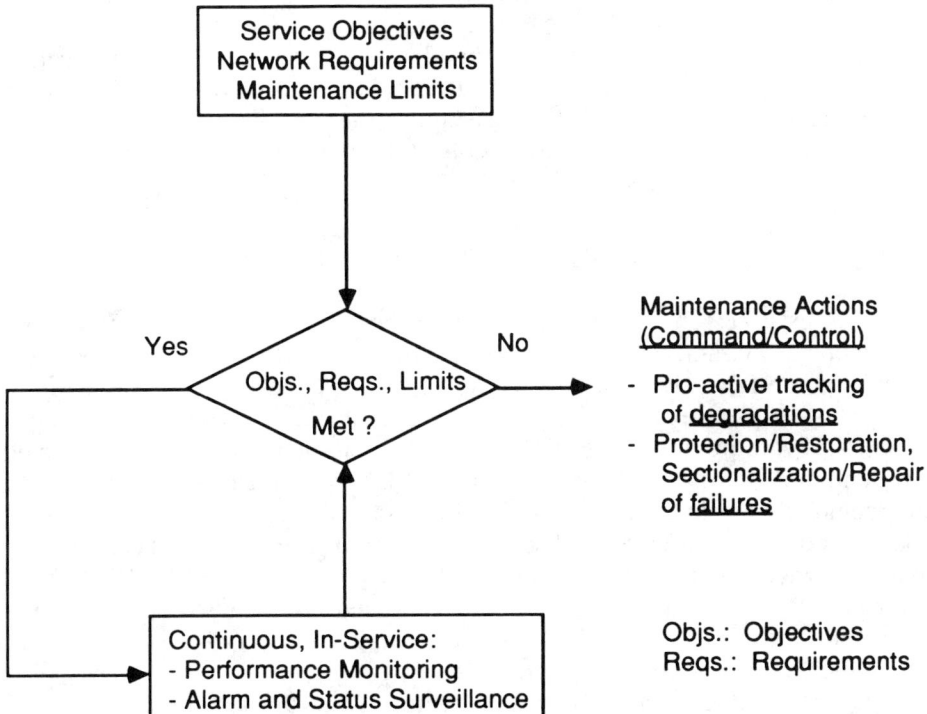

Figure 9.12 Network maintenance process.

The second part of the process is bottom-up/operations oriented, and involves the following steps:

- Continuous in-service performance monitoring of NEs, and reporting of the resulting performance data to a centralized surveillance center.
- Comparison of this data to the performance objectives and maintenance limits determined from the top-down process, and initiation of maintenance action when required.

9.6 PERFORMANCE MONITORING—NETWORK CONSIDERATIONS

9.6.1 Operations and Transport Network Architecture

Performance monitoring is discussed here in the broad context of the operations and transport network architecture shown in Figure 9.13. In this figure, each transport network (local- and inter-exchange carrier) is monitored and controlled by a TMN-based maintenance support system (note that for large networks there may be more than one such support system). The TMN for each LEC is responsible for maintenance from the NI to the point of interface (POI), and the TMN for the IEC is responsible for maintenance from POI to POI. In addition, each CPE may also have a maintenance support system. To the extent that these support systems communicate with one another, either directly or via embedded operations channels (EOC) associated with transport signals, this will assist in enabling the troubles detected on end-to-end connections, to be coordinated and resolved among the multivendors involved in providing the end-to-end service.

9.6.2 Network Elements and Standardized Hierarchical Rates

Table 9.1 lists the general types of NEs and their functions as encountered in the overall operations and network architecture in Figure 9.13. A further breakdown of the transport and switching functions can be provided in the context of Figure 9.14, which shows different types of access arrangements and core networks anticipated during the evolution toward ISDN. Eventually, in a totally integrated services environment, one access arrangement such as B-ISDN is envisaged which can functionally include the Basic and Primary rates [55, 100, 101, 156–166]. Also, eventually, one integrated switching technology is envisaged for the access and core networks, such as asynchronous transfer mode (ATM) [102, 156, 160, 163, 165, 167, 170–176]. However, there will be a long transition period during which the different access and core capabilities will prevail and be supported by a variety of standardized hierarchical rates and associated NEs.

Details related to the ISDN access arrangements are shown in Table 9.2, and the main standardized hierarchical levels used in both access and core portions of

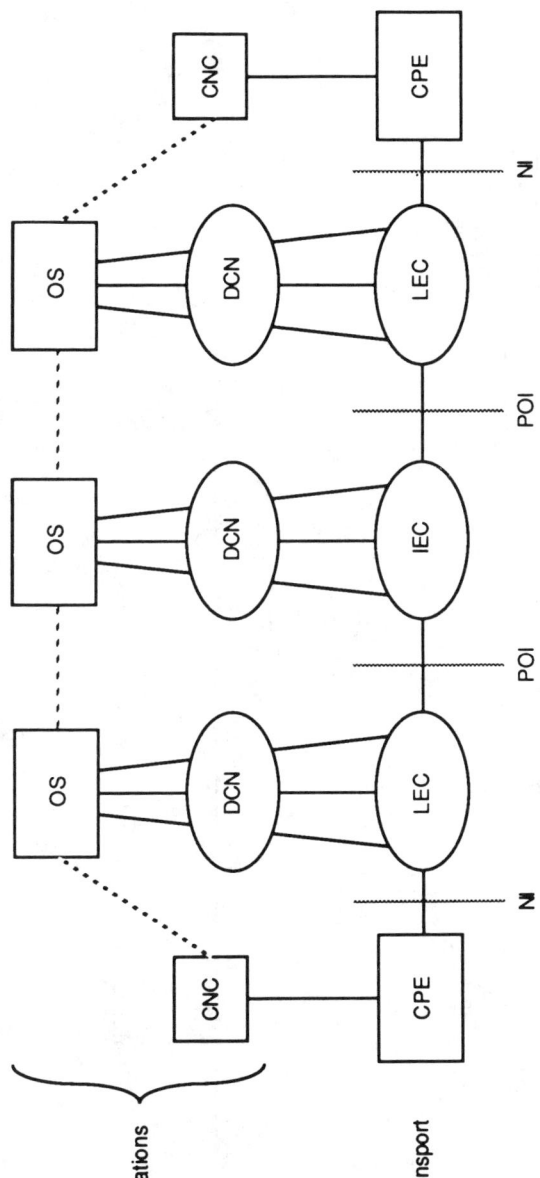

Figure 9.13 Overall operations and transport network architecture.

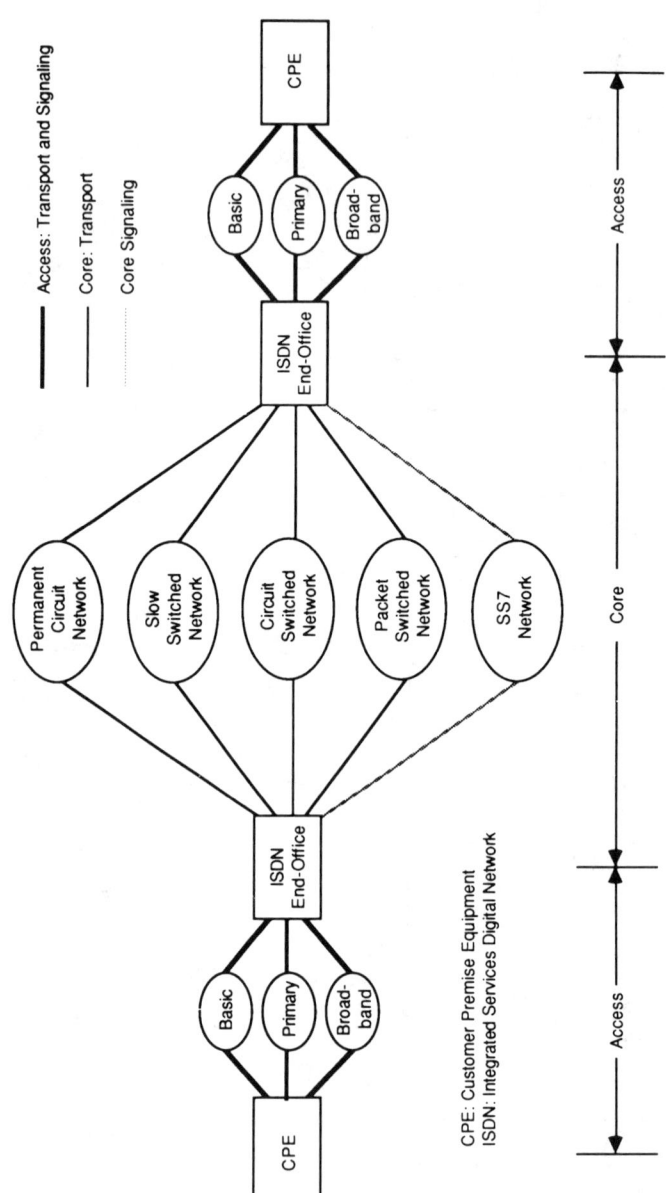

Figure 9.14 ISDN access arrangements and core network alternatives.

Table 9.1
Types of Network Elements and Their Functions

Type	Element	Function
Operations	Network operations center (Operations System - OS)	• Maintenance and provisioning • Surveillance and control
	Operations network (Data Communications Network - DCN)	Carriage of operations information • Performance data and alarms • Control messages
Transport	Transmission facilities and terminals (Line Terminating Elements - LTEs)	Multiplexing and carriage of services at multiples of: 64 kb/s (DS0),1.5 Mb/s (DS1), 1.7 Mb/s (VT-1.5), 45 Mb/s (DS3), 55 Mb/s (STS)
	Add-Drop Muxes (ADMs), and Digital Cross-connects Systems (DCSs)	Drop and insert, and slow switching for network rearrangement at multiples of DS0, DS1, VT, DS3, STS
Switching *	Circuit switches	Circuit switched services typically at DS0, also DS1
	Packet switches	Packet switched services typically at ≤ DS0, also DS1
	Broadband switches	Packet-like switching at rates > DS1

* Note, switching type network elements form part of the transport network.

Table 9.2
ISDN Access Arrangements

Type	Structure	Level	Payload (Mb/s)	Overhead (kb/s)	Total (Mb/s)
Basic	2B + D + OH	Subrate	.144	16	.160
Primary	23B + D + OH	DS1	1.536	8	1.544
Broadband	PL + OH	DS3	~ 44.210 *	~ 526	44.736
		STS-1	49.536	2304	51.84
		STS-3c	149.76 #	5760	155.52
		STS-12c	600.192	21888	622.08

DSn   nth level Digital Signal
PL   Payload
OH   Overhead
B   64 kb/s Bearer channel for circuit and packet traffic
D   16 kb/s (Basic) and 64kb/s (Primary) D channel for signaling and packet data
2B + D   Basic rate PL of 144 kb/s
23B + D   Primary rate PL which supports H11 channel of 1.536 Mb/s
STS-Nc   Nth level SONET Synchronous Transport (path) Signal - concatenated (Nc x 51.84 Mb/s)
*   Supports H22 channel of 44.16 Mb/s
#   Supports H4 channel of 132.48 to 138.24 Mb/s

networks are shown in Table 9.3. These levels are shown in order of increasing transmission rate and include conventional hierarchical rates of DS0, DS1, and DS3, as well as the more recent SONET virtual tributary (VT) and synchronous transport signal (STS) rates [17, 43, 54, 124, 177–205]. The higher rates are derived by conventional digital multiplexing techniques [54, 206–209], as well as more recently by other technologies, such as add drop multiplexer (ADMs) and DCSs (see Table 9.4) [210–213]. ADMs and DCSs are a class of nodal NEs which, together with switching and other transmission NEs, form the basis of modern digital transport networks [34, 37, 39, 40, 129, 158, 190, 195, 197, 199, 214–232]. As discussed later, such NEs also play a central role in the performance monitoring of digital networks. DCSs also enable test access for maintenance purposes, as well as grooming of traffic according to different services (i.e., switched public network service versus private line or Special Services), consolidation of traffic for increased utilization of higher transport rates, reconfiguration for customer network control [233–238], and network survivability [19, 27, 28, 30, 31, 35, 40, 239, 240–243].

Table 9.3
Main Standardized Hierarchical Levels

Level	Rate (Mb/s)	Number of				
		DS0s	DS1s	VTs	DS3s	STSs
DS0	.064	1	–	–	–	–
DS1	1.544	24	1	–	–	–
VT	1.728	28	1	1	–	–
DS3	44.736	672	28	–	1	–
STS-N	N x 51.84	N x 672	N x 28	N x 28	N	N

N: e.g., 3, 12, 48

Figure 9.15 shows a possible functional configuration of DCS and switching NEs which supports all hierarchical rates. It should be noted that, in the future, each DCS or switch will not necessarily be a separate physical entity, but rather, two or more functions may be integrated into a single physical entity. For example, a circuit switch and a 1/0 type DCS may be implemented in an integrated manner; an integrated circuit switch, 1/0 DCS, and 3/1 DCS is another possibility. SONET, particularly with direct optical interfaces to a switch, also enables a high degree of functional integration [214, 244, 245].

Table 9.4
Types of Digital Cross-Connect Systems

Type	Interfaced Levels	Slow Switched Levels
1/0	DS1	DS0 (64 kb/s)
3/1	DS1, DS3	DS1 (1.544 Mb/s)
STS/VT	DS1, DS3, STS-N	VT-1.5 (1.728 Mb/s)
3/3	DS3	DS3 (44.736 Mb/s)
STS/STS	DS3, STS-N	STS (51.84 Mb/s)

N: e.g., 3, 12, 48

9.6.3 Layered Approach to Performance Monitoring

In general terms, performance monitoring is required at three layers as follows:

Layer 3   Network, or Packet Layer
Layer 2   Link, or Frame Layer
Layer 1   Physical, or Line Code and Frame Format Layer

These layers are illustrated for the example of ISDN access as shown in Figure 9.16, and correspond to the lower (first three) protocol layers of the OSI model [55–57, 96–98, 109, 118, 185, 246, 247]. A similar arrangement exists for the core portion of networks operating in the frame and/or "packet" mode. These modes include not only classical packet switching in the context of, for example, ARPANET, TELENET, or DATAPAC, but also various new transfer modes [156, 168, 170, 175, 248–259] such as fast-packet, frame relay, and, in particular, ATM (where in this case the transmitted information units are called "cells," rather than frames or packets). The main reason for layer 2 and layer 3 monitoring is to detect protocol abnormalities occurring at the link and network layers, respectively. These anomalies include residual error rate due to lost, extra, out-of-sequence, or misrouted packets, etc. Also, performance in terms of throughput and delay can be monitored to assess quality of service (see Sec. 3.3 of Ch. 3, and Sec. 4.5 of Ch. 4 for discussions of packet network performance).

Monitoring to assess information integrity in terms of residual error performance is particularly important with new simplified transport schemes such as frame

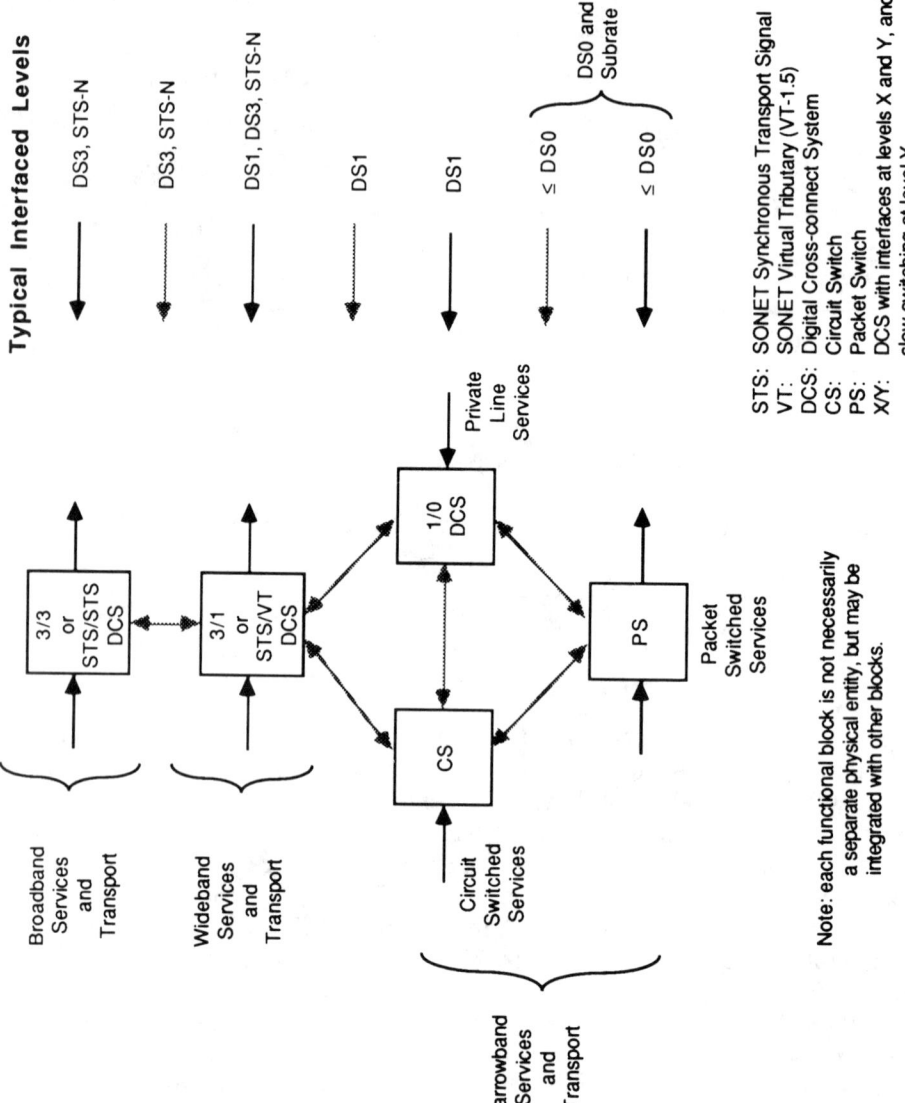

Figure 9.15 Nodal network switching and cross-connect functions, and related services.

511

ET : Exchange Termination (e.g., ISDN switch)
LT : Line Termination (may be integrated with ET)
NT1 : Network Termination 1 (e.g., Customer Serving Unit - CSU)
NT2 : Network Termination 2 (e.g., PBX, see Fig. 9.17)
TE1 : Terminal Element 1 (ISDN Terminal, see Fig. 9.17)

TA : Terminal Adapter (for Non-ISDN Terminals, see Fig. 9.17)
L_i : Layer i Protocol and Performance Monitoring
N : Network Interface (between service provider and ISDN user)
T, U, V : Other interface points

* Depending on their capabilities, the NT1 and LT may not provide full layer 1 monitoring.

Figure 9.16 Illustration of layered performance monitoring for ISDN access.

relay and ATM, since these schemes do not provide node-to-node error correction by retransmission, but rather, they depend on the high performance at the layer 1 level of fiber-optic or new generation radio transmission systems [214].

Layer 2 and layer 3 performance is largely dependent on the more fundamental aspects of transmission performance at the physical, or layer 1 level. Layer 1 maintenance is most fundamental since layer 1 transport is common to, and supports, all transport modes illustrated in Figure 9.14, whether circuit, packet, or dedicated, and ranging from narrowband to broadband. The scope of the remainder of this chapter is limited to maintenance (performance monitoring and alarm surveillance) at the layer 1 level.

9.6.4 Layer 1 Performance Monitoring

9.6.4.1 Section, Line, and Path Monitoring

Within Layer 1 there are a number of possible "sublayers" of monitoring; for convenience, these sublayers are considered here as layers within the overall OSI layer 1, but are not to be confused with OSI layers. The main layers within layer 1 are those for regenerator section monitoring, transmission line monitoring, and digital path monitoring. These layers are illustrated for ISDN access in Figure 9.17, and in generic terms for the core network in Figure 9.18 [3, 12, 14, 51, 52, 54, 55, 124, 196, 260].

Figure 9.17 Layer 1 performance monitoring for the ISDN subscriber access reference configuration.

STE: Section Terminating Element
LTE: Line Terminating Element (will usually also include STE)
PTE: Path Terminating Element (will usually also include LTE and STE)

 Regenerator

Figure 9.18 Generic layer 1 performance monitoring.

Figure 9.18 illustrates the main NEs that perform the monitoring function. These are section terminating elements (STE), that is, regenerators, line terminating elements (LTE), such as line multiplexers and protection switching equipment, and path terminating elements (PTE), such as digital switches and DCSs. These NEs monitor regenerator sections, transmission lines, and digital paths, respectively, as also shown in Figure 9.18. Note that some NEs may perform more than one layer of monitoring; for example, digital switches and DCSs may serve as combined LTE/PTEs for some digital signals, and SONET equipment will serve as combined STE/LTE/PTEs.

Regenerator Section Monitoring is used for the portion of a digital transmission facility between two regenerators, or between an LTE and a regenerator. The purpose is to sectionalize line troubles to the responsible regenerator section. Regenerator section monitoring and related processing is often specific to (proprietary for) particular transmission systems [261–263]. An exception is for new systems using the SONET format which enables regenerator section monitoring in a standardized manner [3, 50, 54, 124, 264]. Thus, error primitives in the SONET section signal can be monitored, and the resulting detected errors can be reported by each regenerator, for example, to the downstream LTE in a SONET overhead or embedded operations channel (EOC). Also, the section error primitives are corrected at each regenerator prior to downstream signal transmission to the next regenerator. This enables the LTE to identify regenerator sections that are producing errors by examining the errors from each regenerator as reported in the EOC.

For nonstandard regenerator section monitoring, there will usually not be a dedicated section error primitive available for this purpose. In these cases, the error primitive at the transmission line layer may be monitored (but not corrected) by regenerators, and reported to the LTE in a manner similar to the above. Although not as straightforward as with SONET, the LTE can still infer from these reports which regenerators are producing errors (by taking differences between the error counts reported by the different regenerators), and also determine the performance of the entire transmission line.

Transmission Line Monitoring is used for a digital transmission facility between two consecutive LTEs [3, 52]. For example, a common form of line monitoring is bipolar violations (BPV) used for DS1 and DS3 cable carrier transmission [265–268]. Line monitoring may also be applied to protection switching sections within a transmission facility. Line monitoring is performed for maintenance purposes; that is, for sectionalization of digital path problems to the responsible transmission facility or protection switching section. With SONET, the line error primitive enables measurement of line performance by the downstream LTE, independent of the regenerators [50, 54]. As mentioned above, one mode of operation for nonstandard systems is for the error primitive in the line format not to be corrected by regenerators within the line prior to transmission to the downstream LTE, so as to enable direct measurement of transmission line performance by the LTE.

Digital Path Monitoring is used for a connection between two PTEs [3, 52]. The purpose of path monitoring is to provide an overall measure of the quality of service of a path, and (together with line monitoring) to aid in sectionalization of path troubles [10, 92]. The latter function is illustrated by the example shown in Figure 9.19. For instance, if a performance degradation is detected on the end-to-end DS1 path, the source of the degradation may be rapidly sectionalized by examining performance monitoring data from a higher rate path. For example, if monitoring of the SONET STS path shows no degradation, the problem is likely in one of the two DS1 lines, and the separate line monitoring for these DS1 lines will indicate which line is responsible.

The format for a path signal is a well-defined standard; examples are DS1 with superframe format (SF) or extended superframe format (ESF) [183], DS3, DS3 C-Bit [178, 269], or DS3S (SYNTRAN) [179], and SONET synchronous transport signals (STS) and virtual tributaries (VT) [124]. A digital path signal format is continuous (unchanged/transparent) throughout the path.

9.6.4.2 *Monitoring at Near-End, Far-End, and Intermediate Locations*

Sections, lines, and paths are normally monitored for the incoming (receive) direction only, which is called *near-end monitoring*. However, when line and path formats provide for the far-end block error (FEBE) function, provision can be made to mon-

Figure 9.19 Example connection with digital cross-connects showing transmission lines (facilities) and digital paths.

itor far-end performance as well. The FEBE function enables performance monitored at the far end of a path to be reported to the near end, in the path overhead. This feature is provided by the DS1-ESF, DS3 C-Bit, DS3S, SONET line, and SONET STS and VT path signals [124, 178, 179, 183]. The feature is useful when, for example, the path crosses interfaces between two different networks (e.g., LEC/IEC, public/private, national/international networks) or between a network and end user. In all of these cases each network provider (or end user) only has physical access to one (their) end of the system, but nevertheless can assess performance in both directions of transmission from that one end using the FEBE function. This is referred to as single-ended monitoring (or single-ended maintenance [67]).

As an example, consider the case of performance assessment between a network and end user as illustrated in Figure 9.20 for ISDN primary rate access (PRA) using the DS1-ESF signal [97, 56]. Taking the perspective of the ISDN switch (or exchange termination (ET)) in Figure 9.20, near-end performance is monitored in the usual way using the cyclical redundancy check (CRC) block error code, which is generated at the end user location (or network termination 2 (NT2)) and checked at the ET. Near-end performance from the perspective of the NT2 is monitored in a similar way using the CRC generated at the ET and checked at the NT2. However, the result of the CRC check at the NT2 can also be returned to the ET in the DS1-

ESF overhead using the FEBE capability (the FEBE is called a *performance report message* (PRM) in this case [180]). Thus, by checking the received CRC and FEBE, the ET is able to determine performance in both directions of the PRA path from a single end (the ET end). The NT2 has a similar (optional) capability. Thus, in this manner, both parties can each have access to the same performance information and are better able to verify the existence of troubles and their direction.

M :	Monitor CRC and FEBE	NE :	Near-End (Performance)
G :	Generate CRC and FEBE	FE :	Far-End (Performance)
		CRC :	Cyclical Redundancy Check
* :	Pass NE Performance to FE (FEBE functionality)	FEBE :	Far-End Block Error

See Fig. 9.16 for other terminology.

Figure 9.20 Near-end and far-end path monitoring for ISDN PRA.

However, this capability, while useful in detecting troubles, does not ordinarily determine on which side of the NI the trouble originated. To provide this additional capability requires the ability to not only monitor paths at the point where they terminate, but also at intermediate locations. This can be done with a bridge-on monitoring function [270–274]. Again, such an intermediate location may be the interface between two different networks, or between a network and an end user.

This situation is illustrated in Figure 9.21, which shows an intermediate monitoring unit (IMU), which, ideally, is situated at the NI in Figure 9.20. In practice, the IMU may be close to (if not right at) the NI; it may be a standalone unit (e.g., see

[275]), or integrated into telecommunications equipment such as a remote switching system or transmission terminal [67]. IMU operation is as follows: The upper monitor in Figure 9.21 checks both the incoming CRC and FEBE from the ET. This provides a measure of near-end performance from the ET to the NI, and a measure of far-end performance from the NT2 to the ET. Similarly, the bottom monitor in Figure 9.21 provides near-end performance from the NT2 to the NI, and far-end performance from the ET to the NT2. Thus, by comparing the results of both monitors, it is readily possible to sectionalize trouble with respect to direction of transmission, and the side of the NI that is the source of trouble.

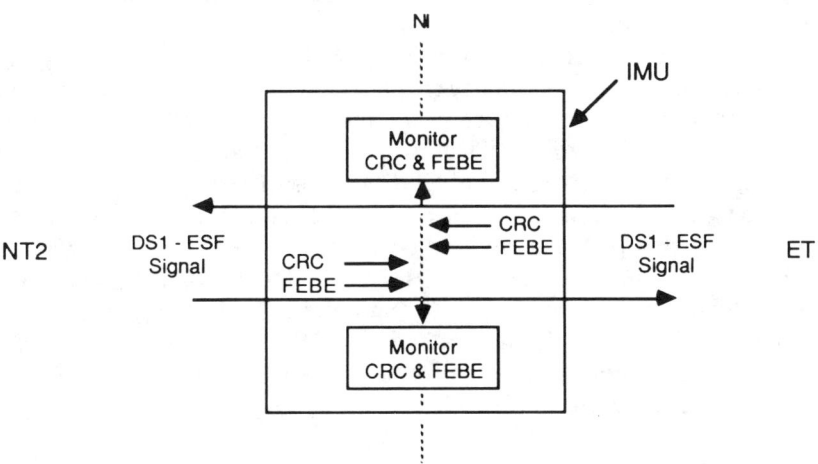

Intermediate Monotoring Unit (IMU) Provides Sectionalization of Troubles:
- Direction of trouble
- Which side of NI trouble is on

ESF: Extended Superframe Format

See Figs. 9.16 and 9.20 for other terminology

Figure 9.21 Intermediate monitoring unit (IMU) functionality.

9.6.5 In-Service Monitoring for Specific Types of Digital Signals

9.6.5.1 Line Code Oriented Signals

Layer 1 monitoring depends on the type of digital signal under consideration. One type is characterized by a line code which, apart from making the signal suitable for transmission over cable carrier facilities, also enables error detection. The most common such code in North America is the alternate mark inversion (AMI) code [177, 276]. This is a three-level code ($+1, 0, -1$), where a positive pulse ($+1$) and a negative pulse (-1) each represent a logical "1", and no pulse (0) represents a logical "0". The coding is such that the $+1$ and -1 pulses must alternate. Thus, if two consecutive pulses are of the same polarity, an error has occurred, and is detected as such and called a *bipolar violation* (BPV). While this code has some imperfections for error detection, it is known to be sufficiently accurate over an applicably wide range of BERs, and is used on DS1 and DS3 line signals. (Performance monitoring accuracy is considered in Sec. 9.7.4.) A basic characteristic of this code is that it only applies for the length of a continuous cable carrier transmission system. Thus, if any other type of equipment is used in the connection, this code cannot be used for end-to-end performance monitoring.

A variation on the AMI code is the bipolar N zeros substitution (BNZS) code, which is used to provide bit sequence independence [276–282]. That is, long strings of zeros (which are detrimental to clock recovery on cable carrier systems) are avoided by substituting strings of N zeros at the line transmitter, by a specific code word with a known BPV pattern. This code word is then detected at the line receiver and replaced by zeros. For DS1 and DS3, B8ZS and B3ZS code words are used, respectively, for this purpose, and, when detected, are not included in the error count. Apart from this, error detection is similar to that for AMI.

Note that an alternative to the B8ZS method of providing bit sequence independence is zero byte time slot interchange (ZBTSI), which is a DS1 path (not line) oriented technique. The various performance and operational implications of B8ZS versus ZBTSI are discussed in [283–286].

9.6.5.2 *Frame-Oriented Signals*

For most signals, layer 1 performance monitoring depends on the frame structure of the digital signals to be monitored (see Fig. 9.22(a) and (b)). That is, most digital signals are frame-oriented, which means that one or more payloads plus overhead are contained in the signal. The overhead determines where the payloads are located, provides for block error checking of the payloads, and may also provide an EOC to carry operations control messages and performance information. As discussed above in Section 9.6.4.2, such signal structures often enable the monitoring of both near-end and far-end performance of digital signals, as well as monitoring at intermediate locations, and some ability to report performance information from one location to another in an EOC. Examples of signals with all or some of these capabilities include the DS1-ESF path signals, DS3 path signals, and SONET section line and path signals. In some instances, path signals will traverse many different types of equipment transparently, and as such can be used for end-to-end performance monitoring, or for monitoring of portions of a connection as spanned by the various paths (e.g., see Fig. 9.19).

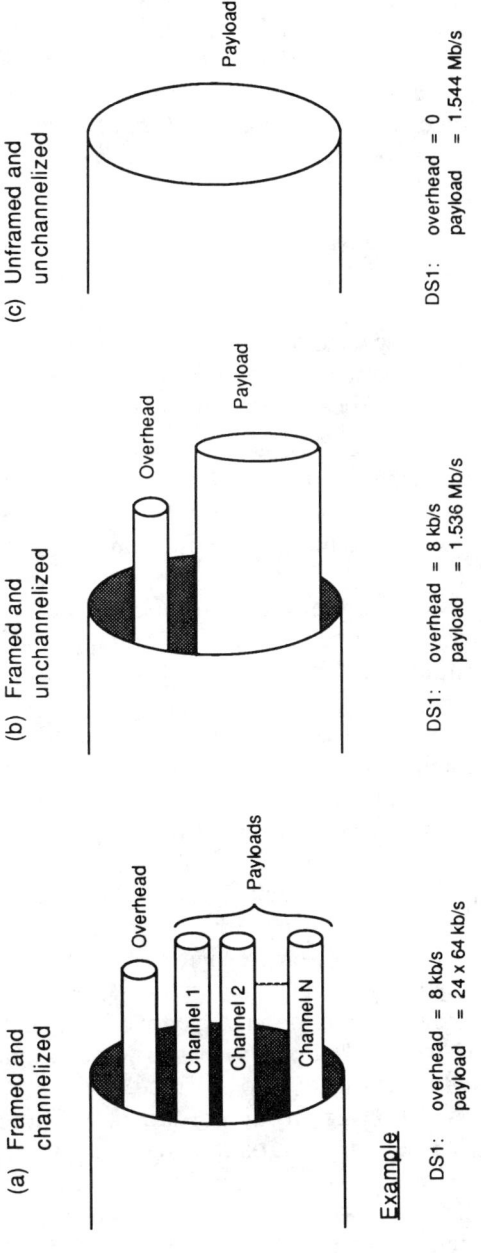

Figure 9.22 Digital signal structures for a given hierarchical level.

In recent years, considerable industry attention has been given to the use of the DS1-ESF for monitoring of end-to-end DS1 paths carrying performance-sensitive services [287, 288], as well as for intermediate monitoring for sectionalization purposes. The DS1-ESF and its capability for performance monitoring are addressed in [180, 183, 289–293]. The applications of the DS1-ESF in terms of end-to-end and bridge-on monitoring are discussed in [56, 97, 218, 275, 294–310].

9.6.5.3 Other Signals

"Other signals" refers to signals which do not satisfy the conditions upon which in-service performance monitoring of digital paths depends, that is,

(1) The presence and knowledge of overhead or redundancy in the digital path signal which can be used for error detection.
(2) Transparency of this overhead from path source to path destination.

The most common example of a digital path signal that does not satisfy the first of the above requirements is the standard DS0 signal; that is, this signal has no redundancy or overhead for error detection. Certain specialized DS0 signals, such as those used for data services have redundancy due to the replication of low-speed data rates of 2.4, 4.8, and 9.6 kb/s [311–315]. Thus, a majority vote can be used to detect errors for these data rates, and hence estimate errors for the DS0 carrying these rates. Other variations on this approach include dedicating part of the DS0 for error detection [316] and using an additional DS0 for error detection and correction purposes [317, 318].

Another example of a digital signal not satisfying the first of the above requirements is an unframed/unchannelized DS1 signal (e.g., see Fig. 9.22(c)). Again, with no overhead, monitoring is not possible except with specific knowledge of the payload. In some cases there may be overhead within the payload that can be used for monitoring purposes. However, in general, a public network provider would not be able to use this capability because of a lack of knowledge of (or jurisdiction over) the payload content. Proprietary monitoring techniques are therefore most appropriate for end user terminals, or for special-purpose networks such as private line or packet networks with proprietary frame formats or protocols, respectively. An alternative for the public network provider is to map the payload into a network-defined signal envelope having its own overhead, to enable performance monitoring and possibly other operations-oriented functions to be performed. This technique, and an example involving a SONET VT-1.5 carrying an unframed, unchannelized DS1, is illustrated in Figure 9.23 (a VT-1.5 can be used to carry any type of DS1 signal). In this case the standardized SONET VT-1.5 overhead can serve to effectively monitor the DS1 payload [187, 319].

An example of a digital signal not satisfying the second of the above requirements is a DS1 signal whose 24 DS0s are all switched intact (transparently) through a circuit switch or a 1/0 DCS, but the overhead portion of the DS1 signal is not. In

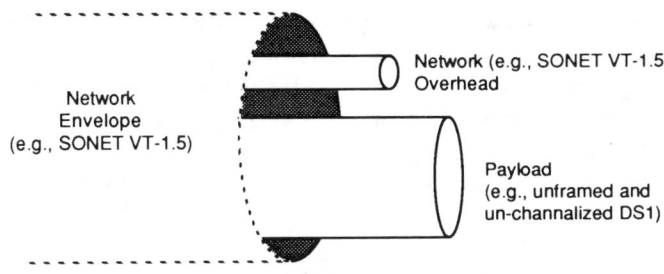

VT: SONET Virtual Tributary
SONET: Synchronous Optical Network

Figure 9.23 Network handling of unframed and unchannelized payload signal.

such cases, end-to-end DS1 path monitoring is not possible on connections composed of intermediate switches or DCSs. In these cases the DS1 digital path, from the standpoint of the overhead, is effectively terminated and originated again at every node.

A technique that overcomes the above noted limitation is illustrated in Figure 9.24. Depending on the type of signal, this technique may be used for all or a portion

Figure 9.24 Generalized method of monitoring digital signals having no error detection overhead, or non-transparent error detection overhead (one direction shown).

of a signal that transparently traverses points A and B in the figure, a special case being when A and B represent an end-to-end connection. In the case of a framed DS1 signal, the sending test unit can bridge onto a DS0, N DS0s [320–325], or the entire DS1, and generates periodic block error check codes; this can also apply to an entire unframed DS1 [326, 327]. The codes are sent to the receive end in a separate channel; for example, in the EOC of another DS1 signal if it is transparent from A to B, another associated DS0, or any other disassociated channel between A and B (e.g., voiceband data link). At the receiving test unit, the received check codes are synchronized and compared with codes generated locally from the received portion of the signal under test, and the differences are counted as block errors. This technique can also be used to monitor far-end performance.

An approach conceptually similar to that in Figure 9.24 will be used to monitor the performance of cell-based broadband transport networks at the ATM cell layer [100, 328–330]. Since at this layer ATM cells are processed and monitored on a node-to-node basis, this monitoring does not give an end-to-end view of ATM network performance. This can be overcome by computing an error check on each of transmitted strings of contiguous cells, and transmitting each resulting error check code in a special OAM cell following each string of contiguous cells. At the receiver, the same error check is computed on each of the received strings of cells, and each result is compared with its corresponding error check in received OAM cells. Any discrepancies are counted as cell block errors. Lost cells can also be estimated by this method [331]. This technique can also be used for far-end monitoring, and for monitoring at intermediate locations.

9.7 GENERIC PERFORMANCE MONITORING FUNCTIONS

A survey of a variety of in-service and out-of-service performance monitoring techniques is provided in [332]. The focus and purpose of the remainder of this chapter is to present in-service digital transmission performance monitoring functions for NEs. These functions are being developed in industry standards [50, 99], and support centralized surveillance and maintenance of mainly the core portion of digital transport networks. However, subject to economic and other considerations, the core network monitoring philosophy is also applicable to the access network [96, 97, 55, 67, 195] (see Ch. 8 for a discussion of T1Q1 and CCITT activities).

The focus is on line monitoring of physical transmission systems for maintenance purposes, and on monitoring of digital paths for quality-of-service purposes. Included in the functional description are performance primitives, performance parameters, memory requirements, thresholding and alerting, and the reporting of performance information to a surveillance center. Performance failures (alarm and status indications) are also included, as described in Section 9.7.2.

Within the constraints of in-service monitoring capabilities, this approach is consistent with T1Q1 and CCITT digital transmission performance parameters

[333–335]. This approach also supports a variety of operating company maintenance strategies ranging from basic (reactive) to enhanced (proactive), and enables verification of end user service objectives.

9.7.1 Nature of Digital Impairments

For a better understanding of the functions related to performance monitoring, it is useful to appreciate the nature of digital impairments. In this regard, Figure 9.25 illustrates three principal impairment modes, some or all of which may occur during the operation of a given system. The simplest mode is the hard/abrupt failure, which can be represented by a simple binary (on-off) indication. In general, such indications are referred to as failures (alarm or status indications), of which there are a number of different kinds, such as loss-of-signal (LOS) energy, and loss-of-frame (LOF) synchronization, etc. Alarms have been provided by NEs, and gathered by maintenance support systems for many years (recall Sec. 9.3.2 and Fig. 9.5 [208, 210, 336]). Usually, the maintenance response to alarms is immediate corrective action, either autonomously by NEs by switching to redundant capabilities, or by automatic or manual route restoration techniques, to avoid or minimize loss of service. Alarms continue to be an important aspect of maintaining network performance from the standpoint of service availability.

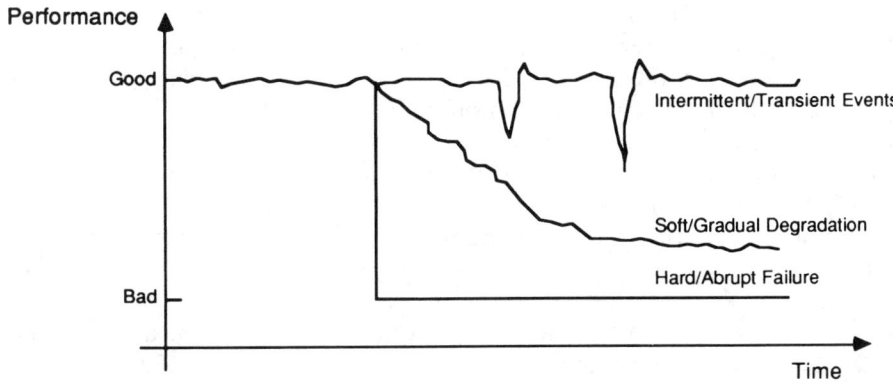

Figure 9.25 Modes of digital performance impairment.

The next impairment mode considered in Figure 9.25 is the soft/gradual degradation. In practice, this degradation may manifest itself as a steady increase with time in the number of transmission errors or other events. Such events may occur, for example, due to a deteriorating or drifting component in a NE, such as a timing recovery circuit in a regenerative repeater or line interface module. Such degradations can be detected by counting the errors or other events over time, and comparing them to one or more threshold limits. When a threshold is crossed, the maintenance support system is alerted to the situation and can take preventative or corrective action as required, depending on which threshold is crossed, or on how frequently thresholds are crossed. An important aspect of the soft/gradual degradation mode is that unlike the hard/abrupt case, it lends itself to trend analysis techniques and to the application of preventative maintenance actions, to avoid further degradation or failure.

The final impairment considered in Figure 9.25 is that of intermittent/transient events. This mode tends to be very common with digital transmission systems, and, depending on the system, can arise due to a number of factors such as power surges, lightning and other static discharges, maintenance/protection switching activity, fading, interference from other signals and systems, synchronization problems, etc. Depending on the cause, the resulting events may manifest themselves as bursts of transmission errors, out-of-frame events, or controlled slips. Such impairment events can be characterized by their

- Depth (or intensity).
- Duration.
- Frequency of occurrence.

From both a performance and maintenance perspective, the *depth or intensity* of the event may be quantified by a threshold. For example, a severely errored second (SES) at the DS0 rate is as a second for which the BER exceeds a threshold of 10^{-3} [335]. The *duration* of the event is often simplified by considering one second to be the smallest time interval necessary for maintenance and performance assessment purposes, and during which, in fact, more than one event may have occurred. For example, a second with one or more errors is called an *errored second* (ES), and a second with one or more out-of-frame events is called a *severely errored frame second* (SEFS) [50]. Finally, the *frequency of occurrence* is determined by counting the events (or simplified one-second representations of them, as discussed above) over longer time intervals, such as 15 minutes or a day. These longer term counts can also be thresholded to indicate degradations on a longer term basis. For example, the number of ESs in a day is a commonly used performance level (e.g., at DS0 and DS1 rates, 0.5% and 5% ESs are allowed per day, which corresponds to 432 and 4,320 ESs per day, respectively [287, 288, 320, 337, 338]). Thus, daily thresholds can be set at the tariffed performance level to indicate when this level is reached or exceeded. Further, for maintenance purposes, 15-min thresholds can be set to forewarn that daily tariffed levels may be exceeded, and thereby enable appropriate preventative action to be taken.

The transient/intermittent case in Figure 9.25 warrants further elaboration as illustrated in Figure 9.26. In this figure, the CCITT percentage of degraded time performance assessment methodology is illustrated using as an example the BER parameter. (Note that, from the standpoint of in-service monitoring, BER is usually estimated from block error checks—e.g., DS3 parity, and by assuming a random error distribution; see Sec. 9.7.4.) This methodology accounts for the fact that, with digital systems, background performance is generally at a relatively high level most of the time, but there are occasional short-term durations in which performance is significantly degraded [4, 214]. The approach is to capture the periods of degradation by counting, for example, the number of 15-min periods during which the BER exceeds a threshold of 10^{-8}. This count may also be expressed as a percentage of the total number of 15-min intervals in the total evaluation period, and compared to a threshold of acceptable performance, such as 5% of (or about 5) 15-min intervals per day.

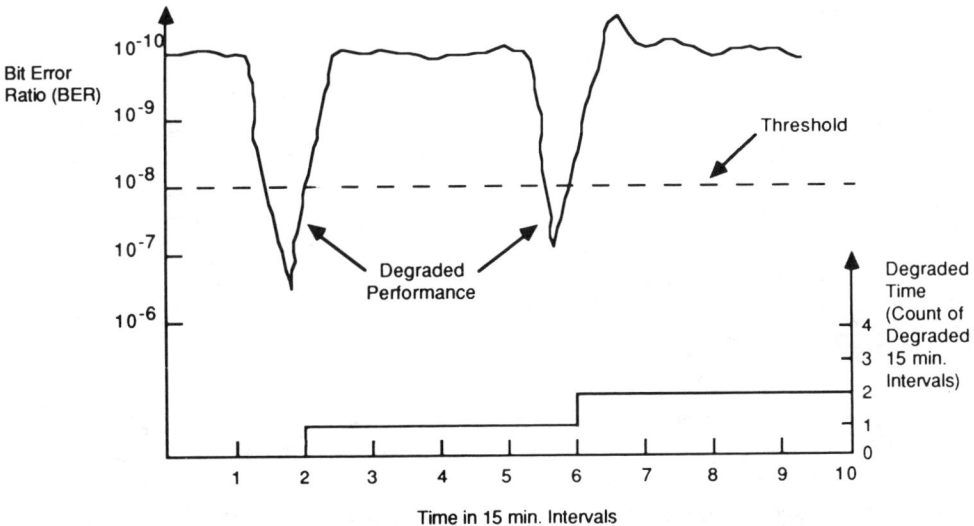

Figure 9.26 Example of degraded time accumulation.

A similar approach can be used for other parameters. For example, a certain number of ESs per 15 minutes may be thresholded, and those 15-min periods, having an equal or greater number of ESs, can be counted to accumulate the total number of 15-min time intervals that are impaired in this manner.

Table 9.5 provides a framework in which the various types of impairments discussed above can be addressed from a maintenance perspective. Broadly, performance can be addressed in terms of two basic maintenance states: available (in-service) and unavailable (out-of-service). As discussed below, the in-service state is

Table 9.5
Performance Categories and Impairments in Relation to Maintenance Indicators

Performance Category (or Maintenance State)	Type/Area of Impairment	Maintenance Assessment Indicator
Available (In-Service)	– Transmission Errors – Frame Synchronization – Time Synchronization	Performance Primitives and Parameters
Unavailable (Out-of-Service)	– Failures (severe degradation) – Protection Switching *	Failure Related Primitives, and Alarms and Status

* Protection Switching does not contribute to, but is used to avoid, unavailable time.

characterized in terms of performance primitives and parameters associated with various digital signal formats, and captures all types of impairments having a bearing on quality of service and on maintenance (i.e., errors, framing, and timing). The out-of-service state is characterized by alarm and status indications in accordance with traditional practice, as well as by a parameter (unavailable seconds, as discussed later) that logs total unavailable time.

Monitoring of in-service performance can be accomplished in terms of the functions illustrated in the generic NE performance monitoring process shown in Figure 9.27. The functions represent a logical flow of performance-related information as follows [3, 50]:

- The detection of performance primitives associated with digital signals.
- Derivation of performance parameters and alarm and status indications from the primitives.
- Provision of memory for performance data (to store counts of events for each parameter).
- Generation of alerts messages by comparing parameter counts to thresholds.
- Reporting of performance data and alerts to a surveillance center across a standard NE/OS interface.

The following sections address these functions in more detail.

9.7.2 Performance Primitives

Performance primitives are associated with specific regenerator section, transmission line, and digital path signals. The primitives discussed here are for all previously noted signal formats (e.g., DS1, DS3, and SONET), and capture all modes of impair-

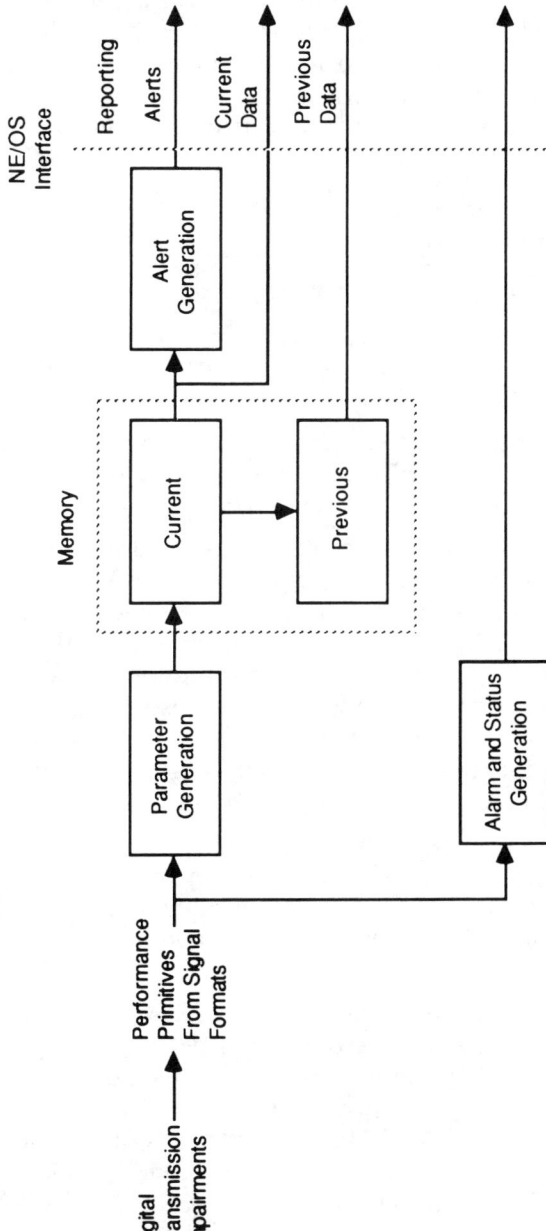

Figure 9.27 Generic NE monitoring process.

ment, as discussed in Section 9.7.1. Performance primitives and typical examples of their use are discussed in general terms as follows, and in accordance with the framework in Table 9.5 (see [50] for detailed definitions for each digital signal). Note that, typically, a given signal format only provides a subset of all primitives as illustrated in the discussion below and as summarized in Table 9.6. Also, accuracy considerations associated with the transmission error oriented primitives are discussed later in Section 9.7.4.

9.7.2.1 Transmission Error Primitives

- Bipolar violations (BPV): Added or deleted pulses, or changes in pulse polarity, that violate the pulse alternating property of a bipolar transmission line code, and thereby enable error monitoring. BPVs are typically used on DS1 and DS3 line signals (recall Sec. 9.6.5.1) [177].

Related to this primitive are excessive zeros (EXZ), which are occurrences of longer than expected strings of zero pulses (i.e., >15 zeros for DS1 AMI, >7 zeros for DS1 B8ZS, and >2 zeros for DS3 B3ZS).

- Frame bit errors (FBE): The bits of a frame-oriented signal that have a known pattern for framing purposes, and that enable frame loss detection. FBEs can also be used for limited error monitoring of any frame-oriented signal, but are usually only specifically used for the DS1 superframe format (SF) path signal, since there is no other primitive for error monitoring of this signal [339]. Limitations of the FBE monitoring method are discussed later in Section 9.7.4.
- A variation on FBEs is frame word errors (FWE), which are known words of a frame-oriented signal used for framing purposes, such as for ISDN basic rate access (BRA) [340], and for SONET section signals [124].

The following primitives are block-oriented error detection codes (parity, CRC, and BIP). In each case, the code is computed over a typically large number of information bits (called the *block size*) at the signal origination point, and recomputed at a downstream point for block error monitoring purposes (recall Sec. 9.6.5.2).

- Parity errors (PE): Usually a single bit (or effectively a single bit) of certain frame-oriented signals that is used for error detection. Specifically, parity is used by DS3 paths [178] and by various nonstandard line signal formats for radio and fiber-based digital transmission systems.
- Cyclical redundancy check (CRC-X): The X bits of certain frame-oriented path signals, which are used for error detection. Specifically, ISDN BRA [340], DS1-ESF [180], and DS3S [179], use CRC-12, CRC-6, and CRC-9, respectively.

Table 9.6
Summary of Performance and Failure Primitives, and Related Digital Signals

Impairment	Primitive	Digital Signal						
		DS1		DS3		SONET		
		Line	Path & ──── SF ESF	Line	Path # ──── A S C	Sect.	Line	Path ──── STS VT
Transmission Errors	BPV/EXZ FBE/FWE CRC-X PE BIP-X FEBE	•	• • • • •	•	• • • • • • • • • •	• •	• •	• • • •
Frame Sync.	SEF		• •		• • •	•		
Time Sync.	CS		• •		•			
Failure Related @	LOS LOF LOP AIS FERF	•	• • • • • •	•	• • • • • •	• •	•	• • • •
Prot. Sw.	PS	•	•		• • •		•	

  & SF: Superframe Format
    ESF: Extended Superframe Format
  # A: Asynchronous DS3 format using P-bit parity.
    S: DS3S (SYNTRAN) format using P-bit parity and CRC-9.
    C: DS3 C-Bit format using P-bit parity and C-bit parity.
  @ When failure related primitives persist, they result in alarm or status indications
    with the same names as the primitives; see Table 9.7 and text.
    Failure related primitives also contribute to various performance parameters; see Table 9.8 and text.

FBE/FWE: Frame Bit Error (DS1 and DS3)/Frame Word Error (SONET)
CRC-X: Cyclical Redundancy Check (see Table 9.10 and Fig. 9.29 for values of X)
PE: Parity Error (P-bit for all DS3 paths, and C-bit for DS3 C-bit path)
BIP-X: Bit Interleaved Parity (see Table 9.10 and Fig. 9.29 for values of X)

- Bit interleaved parity (BIP-X): The X bits of SONET signals, which are used for error detection. Specifically, there are separate BIP-8 indicators for section, line, and path monitoring of synchronous transport signals (STS), and also a BIP-2 indicator for virtual tributary (VT) signals [124]
- Far-end block error (FEBE): The bit or bits in the overhead of certain digital signals, which enable performance as measured at one end to be reported to the opposite end. FEBE-like functions are supported by ISDN BRA [340], DS1-ESF [180], DS3 C-Bit [178], DS3S [179], and SONET line and SONET STS and VT path signals [124] (recall Sec. 9.6.4.2).

9.7.2.2 Frame Synchronization Primitives

- Out-of-frame (OOF) event: Detection of a certain pattern of framing bit errors (e.g., two or more errors in four consecutive framing bits at DS1), followed by an "on-line" search of the bit stream for the correct frame position [52]. This search destroys the old frame position and causes an impairment. The OOF condition and associated impairment will persist until the proper frame position is located. OOF usually signifies a momentary condition lasting for a few tens of microseconds (e.g., about 250 μs for SONET sections), to a few milliseconds (e.g., 2 to 3 ms for DS3 paths), to a few tens of milliseconds (e.g., 50 ms for DS1 paths), depending on the type of signal as shown [49, 336, 341]. If the condition persists long enough, it results in a loss-of-frame alarm as discussed later.
- Change-of-frame alignment (COFA) event: Detection of a pattern of framing bit errors as with OOF above, but followed by an "off-line" search of the bit stream for the correct frame position [52]. This search does *not* affect the old frame position. Thus, at the end of the search, if the old frame position turns out to be correct (framing bits were just in error), no change occurs and no impairment is introduced by the search. However, if a new frame position is found, a COFA occurs, and an impairment is encountered. COFA also signifies a momentary condition, as with OOF. If a new frame position cannot be found, a loss-of-frame alarm results, as discussed later.

Off-line framing giving rise to COFAs (an alternative to on-line framing giving rise to OOFs) has the advantage that it only causes a reframe (impairment) when there is a true loss of frame alignment, and not when there are only errors in the frame bits. However, for performance monitoring consistency in interworking situations where different equipment may use on- or off-line framers, it is the intent that off-line framers include OOFs with the COFA primitive [50] (effectively, an off-line framer need only provide OOFs, since a subset of OOFs will result in COFAs).

- Severely errored framing (SEF) event: A unifying primitive representative of OOF and COFA but in a standardized manner for performance monitoring consistency [50]; this term is used subsequently in this chapter.

9.7.2.3 Time Synchronization Primitives

- Controlled Slips: Loss or duplication of an entire frame's worth of data, for example, 192 information (payload) bits at DS1, even though all of these bits may be error-free [342]. Controlled slips will usually only occur when a PTE uses a different clock than that derived from incoming signals. That is, a controlled slip can occur when the timing sources for the PTEs at the two ends of

a transmission path are not exactly frequency synchronized. Controlled slips apply to DS1 and DS3S path signals (synchronization is discussed in Sec. 6.4.3 of Ch. 6).
- Pointer justification (PJ): Change of the address of (or pointer to) the start of STS and VT path payloads in SONET signals [124]. PJs occur due to differences in incoming and local path signal rates. Ordinarily, PJs in themselves do not cause any error impairment. However, if PJs occur at an excessive rate, they may be indicative of a clock synchronization problem that could result in controlled slips and associated impairments at downstream locations where DS1 paths (carried by SONET) terminate.

The limitations of PJs for monitoring purposes are that PJs related to common synchronization troubles are difficult to distinguish from those due to normal operation; and further, if synchronization trouble is detected, it is difficult to locate the source of trouble without considerable processing and global network information about SONET paths [343]. See the pointer justification count (PJC) parameter discussed later in Section 9.7.3.

9.7.2.4 Failure-Related Primitives and Alarm and Status Indications

Failure-related primitives such as LOS and LOF are included in Table 9.6 and are discussed below (see also Fig. 9.27). Typically, these primitives are detected in a relatively short time period (e.g., in a few frames), and then checked for persistence (e.g., for 2 to 3 sec). If the primitive persists, a corresponding failure (alarm or status indication) is declared (activated). If the primitive is intermittent, it is either cleared, or, in some cases, specific integration algorithms may be used to decide whether to declare the failure or not [54]. Note also that failure-related primitives contribute to various performance parameters as discussed in Section 9.7.3. Once in the failure state, the normal (nonfailure) condition is also detected rapidly and checked for persistence (e.g., for 15 sec), after which the failure is cleared (deactivated). The above failure activate and deactivate persistence times of 2 to 3 sec and 15 sec, respectively, are typical values, whereas the specified ranges of values are 2 to 10 sec and 20 or fewer sec, respectively [50].

Failures are reported from an NE to an OS as follows: an "alarm" is reported automatically, while a "status" is reported on demand by the OS (see Fig. 9.27). Any failure can be provisioned as an "alarm" or as a "status." (Certain failures that cause trunk conditioning on carrier terminals or on switching machines are called *carrier failure alarms* (CFA) [210].) For trouble diagnosis purposes, failures indications as observed at a particular location are categorized as follows:

- Local "red" indication (e.g., LOS, LOF): Signifies a local failure requiring local corrective action; usually provisioned as "alarm."

- Remote (formerly "blue") indication (alarm indication signal (AIS): Signifies an upstream failure (red alarm) requiring upstream corrective action; usually provisioned as "status."
- Remote "yellow" indication (e.g., remote alarm indication (RAI)): Signifies a downstream failure (red alarm) requiring downstream corrective action; usually provisioned as "status."

The following is a general discussion of common failure-related primitives and their associated alarm or status indications. See Table 9.7 for a complete list of failures, and see [50] for detailed definitions for each digital signal.

Table 9.7
Summary of Failures (Alarm and Status Indications), and Related Digital Signals

Failure		Digital Signal						
Type	Name	DS1		DS3		SONET		
		Line	Path & SF ESF	Line	Path # A S C	Sect.	Line	Path STS VT
Local 'Red'	LOS	•		•		•		
	LOF		• •		• • •	•		
	LOP							• •
Remote Upstream 'Blue'	AIS		• •		• • •		•	• •
Remote Downstream 'Yellow'	FERF						•	
	RAI		• •		• •			• •

&, # See corresponding footnotes in Table 9.6.

- Loss-of-signal (LOS): This primitive occurs when a certain number of consecutive zeros (e.g., 175 ± 75 zeros at DS1 and DS3) are received on an incoming line signal. The LOS primitive ends when valid signal framing is detected, or (if frame processing is not done) when the average "ones" density of the digital line signal meets a given threshold, depending on the type of interface (12.5% "ones" density for DS1, 33% for DS3). The LOS primitive contributes to ES, SES, and LOS second (LOSS) parameters for the signal being monitored.

When frame processing is not done (e.g., by an LTE terminating a DS1 line signal only), a persistent LOS primitive will cause an LOS alarm. If frame processing is done (e.g., by a combined LTE/PTE terminating DS1 line and path signals), the LOS primitive will lead to a LOF primitive, and, if persistent, will in turn lead to an LOF alarm as discussed below.

- Loss-of-frame (LOF): This primitive occurs when the NE loses framing on an incoming signal and cannot find a valid framing pattern. That is, the LOF primitive is an SEF (OOF/COFA) event which persists for a certain period of time, depending on the type of signal. For DS1 and DS3 paths this time should not exceed 1.5 times the maximum average reframe time, but could be as short as an SEF event [336]. For SONET sections, the time is 3 ms [124]. The SEF primitive contributes to ES, SES, SEF second (SEFS), and SEF/AIS second (SAS) parameters, as applicable for the signal being monitored (the same is implicitly true for the LOF primitive, since the SEF primitive leads to an LOF primitive).

If the LOF primitive persists, it causes an LOF alarm. Note that, as discussed above, in some cases a persistent LOS primitive will also result in an LOF alarm. An LOF alarm contributes to the unavailable second (UAS) parameter for the signal being monitored (see Sect. 9.7.3, and Fig. 9.28).

- Loss-of-pointer (LOP): This primitive occurs when a valid pointer to a SONET STS or VT path payload cannot be found within a certain period of time (e.g., within eight consecutive frames for SONET STS path signals [124]). This primitive contributes to ES, SES, and AIS/LOP second (ALS) parameters for the signal being monitored.

If this primitive persists, it causes an LOP alarm. An LOP alarm contributes to the UAS parameter for the signal being monitored.

- Alarm indication signal (AIS), formerly called "blue" signal: A signal sent in the downstream (forward) direction in response to a failure-related primitive being detected on the incoming or received signal. When an upstream failure-related primitive is first detected, all the affected downstream path signals are immediately replaced by AIS. A simple "all ones" pattern is used as AIS for the DS1 signal. For DS3 and SONET signals, more sophisticated AIS patterns are used so as to preserve framing and certain overhead information. This primitive contributes to ES, SES, AIS second (AISS), SAS and ALS parameters, as applicable for the signal being monitored.

Reception of a persistent AIS signal causes local red alarms at the reception point to be inhibited and replaced by an AIS status indication to signify that a remote (upstream) failure has occurred [124, 336]. An AIS status contributes to the UAS parameter for the signal being monitored.

- Far-end receive failure (FERF): A signal sent in the upstream (reverse) direction in response to a failure-related *primitive* on the incoming or received signal. FERF signals are carried in SONET line and in SONET STS and VT path signal overheads [124]. This primitive contributes to far-end ES, SES, AISS, and ALS parameters, as applicable for the signal being monitored.

If this primitive persists, it causes a FERF alarm or status. For SONET lines, a FERF alarm or status contributes to the far-end UAS parameter. SONET STS and VT paths use RAI; see below.

- Remote alarm indication (RAI), also called "yellow" signal: A signal sent in the upstream (reverse) direction in response to a *failure* on the incoming or received signal (note that in this regard there is no RAI primitive, but rather a direct indication of a remote downstream failure). RAI signals are usually carried in the digital signal overhead; that is, this is the case for the DS1-ESF signal [180], DS3 C-bit and DS3S signals [178, 179], and for SONET STS and VT path signals [124].

Reception of an RAI signal causes an immediate RAI status indication to signify that a remote (downstream) failure has occurred. An RAI status contributes to the far-end UAS parameter for the signal being monitored.

9.7.2.5 Protection Switching Primitives

- Protection switch (PS) event: Automatic or manual transfer of a signal from a working (but possibly failed or degraded) line (or path), to a protection or spare line (or path), and vice versa. Usually switching is proprietary and is associated with transmission lines (facilities); SONET is an exception where protection switching is standardized [114, 124]. Protection switching can also be associated with DS1 and DS3 paths. For DS1, code words in the DS1-ESF overhead can be used [180, 344, 345], while for DS3 the X-bits [178] can be used on a proprietary basis, and for SONET specific line overhead ("K" bytes) are used [124]. The PS event contributes to the PS count (PSC) and duration (PSD) parameters discussed later.

All of the above performance primitives and failures are summarized and related to the applicable digital signals in Tables 9.6 and 9.7, respectively.

9.7.3 Performance Parameters

With reference to the generic NE monitoring process in Figure 9.27, performance parameters are standard measures of performance as derived from the system-specific performance primitives discussed above. The measure for each parameter is, in most cases, a count. The parameters are discussed in general terms as follows, and are summarized along with associated digital signals in Table 9.8. See [50] for detailed parameter definitions specific to each digital signal.

9.7.3.1 Transmission Error Parameters

- Code violation (CV): A generic term to indicate the count of one of the error primitives: BPVs, FBEs, PEs, CRC, or BIP violations, depending on the signal being monitored. CVs exclude the possibly large numbers of errors associated with SEF and failure primitives (i.e., the monitoring processor is provisioned to inhibit CV counts during seconds containing these events; inhibiting these errors enables a more meaningful/uncorrupted estimate of the background CV count, while the SEF and failure primitives are captured in other parameters as discussed below).

CV counts enable an estimate of BER, but unprocessed CV counts can be reported to the OS, and BERs can then be estimated by the OS. Similarly, CV-related alarm or alert thresholds need only be based on CV counts corresponding to certain BERs. The relationship between BER and CV counts and the accuracy of the BER estimates is discussed in Section 9.7.4 [3].

Table 9.8
Summary of Performance Parameters and Related Digital Signals

Impairment	Parameter @	DS1			DS3			SONET			
		Line	Path &		Line	Path #		Sect.	Line	Path	
			SF	ESF		A S C				STS	VT
Transmission Errors	CV	•	•	•	•	• • •		•	•	•	•
	ES	•	•	•	•	• • •		•	•	•	•
	ES-A			•	•	• • •		•	•	•	•
	ES-B			•	•	• • •		•	•	•	•
	SES	•	•	•	•	• • •		•	•	•	•
Frame Sync.	SEFS							•			
	SAS		•	•		• • •					
Time Sync.	CSS		•	•		•					
Pre-Failure Related	LOSS	•			•			•			
	ALS									•	•
	AISS		•	•		• • •			•	•	•
Failure	UAS !		•	•		• • •		•	•	•	
Prot. Sw.	PSC	•		•		• • •		•			
	PSD	•		•		• • •		•			

&, # See corresponding footnotes in Table 9.6.
 @ See [50] for specific guidance as to which parameters are required, optional, etc.
 ! Failures are also separately reported as alarm and status indications (see Table 9.7).

- Errored second (ES): Second containing one or more CVs, SEFs, CSs, or failure primitives (i.e., LOS, LOP, or AIS, depending on the digital signal; see Table 9.7). This definition, when used to monitor at DS1 and higher rates, captures all the impairments that cause errors at the rate being monitored and at the lower, embedded tributary (or payload) rates; recall Figure 9.22(a).

An ES can start with detection of an error event (called an *event-triggered*, or *synchronous* ES), or an ES can be independent of the occurrence of an error event and governed by a system clock (called an *asynchronous* ES) [346]. The accuracy associated with these approaches is discussed in the next section. However, for practical, maintenance purposes, either synchronous or asynchronous ESs are acceptable.

Depending on the distribution of errors among the tributaries of a higher rate signal, the ES count monitored at the higher rate may be larger than the ES count on any given tributary within the high-rate signal. To approximately correct for this difference, ESs may be translated from the rate being monitored to lower tributary rates using appropriate algorithms [335]. This is also discussed further in the next section on accuracy.

Finally, it is noted that additional parameters can be defined that provide a breakdown of the contents of ESs. These parameters are ES-A (second containing only a single CV), ES-B (second containing only two or more, but fewer than X CVs, where X corresponds to the number of CVs in a severely errored second; see below) [50, 92, 274].

In addition, it is possible to provide separate counts of seconds containing specific failure primitives (or groups of failure primitives), that is, LOSS, AISS, and ALS [50].

- Severely errored second (SES): Second containing CVs exceeding a certain BER threshold (e.g., 10^{-3} at DS1, 10^{-6} at DS3), one or more SEFs [347], or applicable failure primitives [50]. For higher rates, lower BER thresholds are used for SESs to reflect the generally higher performance of higher rate transmission systems, as well as the possibly more critical nature of some services at these rates (e.g., video). Note that an SES is a subset of an ES.
- Degraded minute (DM): Minute containing CVs exceeding a certain BER threshold (e.g., 10^{-3} at DS0 and 10^{-6} at DS1) and no SESs. Again, for higher rates, the BER threshold may be lowered (e.g., to 10^{-8}). It is currently envisioned that the DM parameter will not be used for monitoring at rates above DS0, and may also be unnecessary for out-of-service evaluation of performance at DS0 [335, 348, 349].

9.7.3.2 Frame Synchronization Parameter

- Severely errored frame second (SEFS): Second containing one or more SEF events. Note that at DS1 and DS3, this parameter is modified to include AIS primitives, and is called the *SEF/AIS Second* (SAS) parameter [50].

9.7.3.3 Time Synchronization Parameters

- Controlled slip second (CSS): Second containing one or more CS events.
- Pointer justification count (PJC): A count of PJ events. Note that PJC may or may not become an industry standardized parameter because of difficulty in distinguishing counts due to normal operation from those due to common synchronization trouble, and difficulty in locating the cause of trouble [343]. However, PJCs may be useful for internal diagnosis of certain synchronization failures within NEs [54], or in detecting trouble due to excessive jitter accumulation.

An alternative to PJCs for SONET synchronization network maintenance is the use of synchronization status messages. Based on fundamental synchronization failure or other trouble indications at an upstream node, the synchronization status (e.g., stratum clock currently in use) is communicated to downstream subtending nodes in the overhead of the (DS1 path or SONET line) timing signal. Upon reception of the status message at the downstream node, corrective self-healing action can be taken if necessary (e.g., switch to an alternative higher stratum clock). Synchronization status messages are currently being standardized in the industry (e.g., see [350]).

9.7.3.4 Failure-Related Parameter

- Unavailable second (UAS): Period starting with the onset of ten or more consecutive SESs. Once unavailable, service becomes available only at the onset of ten consecutive seconds containing no SESs [335].

Depending on the digital signal, UAS counts are also incremented during applicable failure conditions (i.e., LOF, LOP alarms and AIS status for near-end UAS, and FERF and RAI for far-end UAS; see Fig. 9.28 [351]). The time to detect ten consec-

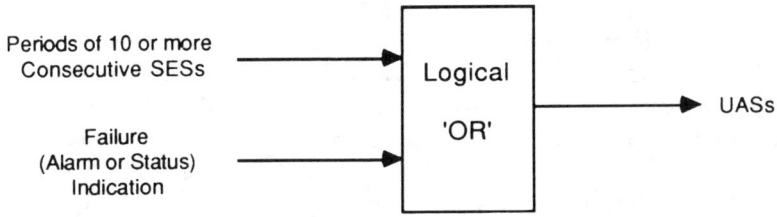

See Fig. 9.30 for timing, and text for definition details [9.20].

SES: Severely Errored Second
UAS: Unavailable Second

Figure 9.28 Contributions to unavailable second parameter.

utive SESs, or a failure (e.g., 2 to 3 sec) is included in UAS. The time to detect ten consecutive non-SESs or clearance of a failure (e.g., 15 sec) is excluded from UAS. This timing is illustrated later in Figure 9.30 and detailed definitions are provided in [50].

In addition to being included in UAS parameters, failures are also separately reported to an OS as alarm or status indications.

9.7.3.5 Protection Switching Parameters

- Protection switching count (PSC): A count of PS events.
- Protection switching duration (PSD): A parameter that counts the total duration, in seconds, for which service was automatically removed from the working line (or path) being monitored. This parameter only applies to revertive protection switching; that is, to a working line (or path) which the system automatically switches back (reverts) to when the trouble clears.

9.7.3.6 Far-End Performance Parameters

These parameters are based on the FEBE, FERF, and RAI functions provided in whole or in part by certain digital signals (see Table 9.6). Subject to the limitation of these functions, far-end parameters are intended to be the same as their near-end counterparts. Generally, far-end CV, ES, SES, and UAS parameters can be derived from these functions. That is, far-end block errors and failure primitives are used to derive far-end CV, ES, and SES parameters. Also, far-end SESs and far-end failures are used to derive far-end UAS. See [50] for specific details for each digital signal.

Table 9.8 summarized the section, line and path performance parameters just discussed. A specific signal format and NE will usually involve a reduced set of Table 9.8. For example, slips only apply to DS1 and DS3S signals. Also, some signals are not frame-oriented (e.g., DS1 and DS3 line signals) and so do not require SEF monitoring. In addition, it may not be necessary for some smaller-scale NEs to implement the full set of parameters in Table 9.8. Finally, embedded NEs may not provide all of these parameters, but some NEs may do so through planned upgrades.

For these reasons it is useful to partition performance monitoring parameters into those that are required, those that are optional (or objectives), and those that are application-specific. Requirements correspond to minimum monitoring capabilities necessary primarily for basic maintenance and quality-of-service assessment, while objectives correspond to enhanced monitoring suitable for more detailed network performance characterization. Application-specific parameters are those that apply under specific circumstances. For example, SONET section CV, ES, and SES parameters are only relevant if there are regenerators in the SONET transmission line, and, as such, are application-specific parameters. Specific guidance as to which parameters are required, optional, etc., is provided in [50]. Generally, most NEs must report failures and a subset of performance parameters to ensure that errors, framing, and time synchronization is functioning properly, and that unavailable time is accounted for.

9.7.4 Accuracy of Performance Assessment

9.7.4.1 Limitations of Block Error Detection Codes

Block error detection codes such as parity, CRC, and BIP have limitations, particularly, for BER estimation. These limitations affect the accuracy of BERs in general, as well as in specific instances such as for severely errored second thresholds (e.g., BER threshold of 10^{-3} at DS1). The asymptotic relationships between BER and CVs (the count of parity, CRC or BIP violations) are shown in terms of formulae in Table 9.9, and illustrated by examples in Table 9.10 and in Figure 9.29 [3]. Although the

Table 9.9
Asymptotic Formulae for Relating BLER, BER, and CV Counts

Performance Primitive	Formula Type	Formula < Saturation BER #	Formula ≥ Saturation BER
BPV	BLER	BER	1/2
	BER	CV / (BR * T)	1/2
	CV	BER * BR * T	BR * T / 2
FBE	BLER	BER	1/2
	BER	CV * FS / (BR * T)	1/2
	CV	BER * BR * T / FS	BR * T / (2 * FS)
PE, BIP *	BLER	BS * BER	1/2
	BER	CV / (BR * T)	1 / (2 * BS)
	CV	BER * BR * T	BR * T / (2 * BS)
CRC	BLER	BS * BER	1
	BER	CV / (BR * T)	1 / BS
	CV	BER * BR * T	BR * T / BS

   * Individual BIP parity bits (SONET)
   # Assumes Random Errors
   T: Measurement time in sec.
   BS: Block Size in bits for Saturation BER estimation. For SONET BIP, this is the effective BS corresponding to each individual parity bit (see Table 9.10)
   FS: Frame Size in bits
   BR: Bit Rate per channel in bits/sec.
See Table 9.10 and Fig. 9.29 for examples.

Table 9.10
Estimated BER Range for Various Performance Primitives

Performance Primitive				Maximum (Saturation) BER @	Bit Rate, BR (Mb/s)	Minimum BER in Time T ***		
Type	Usage	Block Size, BS (Bits) &	Frame Dur., FD (µs)			15 min.		Day
BPV	DS1	1*	.648	1/2	1.544	7.2×10^{-10}		7.5×10^{-12}
	DS3	1*	.02235	1/2	44.736	2.5×10^{-11}		2.6×10^{-13}
FBE	e.g., DS1-SF	193 **	125	1/2	44.736	1.4×10^{-7}		1.4×10^{-9}
PE	DS3	4760	106.4	1.1×10^{-4}	44.736	2.5×10^{-11}		2.6×10^{-13}
CRC-X	CRC-6: DS1-ESF	4632	3000	2.2×10^{-4}	1.544	7.2×10^{-10}		7.5×10^{-12}
	CRC-9: DS3S	14280	319.21	7.0×10^{-5}	44.736	2.5×10^{-11}		2.6×10^{-13}
	CRC-12: BRA	480	3000	2.1×10^{-3}	0.160	6.9×10^{-9}		7.2×10^{-11}
BIP-X (SONET)	BIP-8: STS-N Sect.	6480N/8 #	125/8	6.2×10^{-4}/N	N x 51.84	2.1×10^{-11}/N		2.2×10^{-13}/N
	BIP-8: STS-N Line	6480/8 #	125/(8N)	6.2×10^{-4}	N x 51.84	2.1×10^{-11}/N		2.2×10^{-13}/N
	BIP-8: STS-Nc Path	6480Nc/8 #	125/8	6.4×10^{-4}/Nc	Nc x 51.84	2.1×10^{-11}/Nc		2.2×10^{-13}/Nc
	BIP-2: VT-Nc Path	864Nc/2 #	500/2	1.2×10^{-3}/Nc	Nc x 1.728	6.4×10^{-10}/Nc		6.7×10^{-12}/Nc

& In general, Block Size (BS) = Bit Rate (BR) X Frame Duration (FD)
@ See right most column of Table 9.9 for formulae.
* In these cases there is no block error detection scheme, but rather BPVs, which result in an effective BS of 1 bit.
** In this case there is no block error detection scheme, but rather periodic frame bits which define the Frame Size (FS): In this example, every frame bit is monitored (i.e., 1 bit in 193).
*** CV Count = 1, random error assumption; see second column from right in Table 9.9 for formulae.
# For SONET, the BSs and FDs shown, effectively correspond to each of the X individual BIP-X parity bits. The entire BS and FD over which an entire BIP-X is computed, is X times the BS and FD shown.

exact relationships are somewhat more complex (e.g., see [192, 290, 352, 353]), for maintenance purposes, the asymptotic behavior as presented in Tables 9.9 and 9.10 and in Figure 9.29 is sufficient.

Figure 9.29 Asymptotic BLER vs. BER behavior for block error detection codes.

The formulae are based on an assumption of random errors, which means that for sufficiently low BERs there will be only one bit error per block error (i.e., one bit error per CV). Thus, the estimated BER is simply the number of CVs divided by the transmitted bits in the measurement time, T. For higher BERs, however, a point is reached where there is more than one bit error per block error, and where eventually, all blocks are in error. At this point, higher BERs cannot be estimated from CV counts. The BER in this region is called the *saturation BER* (BER_{SAT}); this BER is inversely proportional to the block size (BS) of the error detection code. That is, the larger the block size, the lower the BER at which all blocks become errored. The block sizes and saturation BERs for various codes are also shown in Table 9.10.

To illustrate, consider an *ideal block error code;* assuming random errors, the CV ratio or Block Error Ratio (BLER) is related to BER by

$$BLER = 1 - (1 - BER)^{BS} \qquad (9.1)$$

Below saturation BER $\ll 1$, and the relationship simplifies to

$$\text{BLER} \approx 1 - (1 - \text{BS} \times \text{BER})$$
$$= \text{BS} \times \text{BER}, \quad \text{for BER} < \text{BER}_{SAT} \quad (9.2)$$

which is the asymptotic behavior below saturation. At and above saturation, all blocks are in error (BLER = 1) and, therefore, from Equation (9.2), $\text{BER}_{SAT} = 1/\text{BS}$; thus,

$$\text{BLER} = 1, \quad \text{for BER} \geq \text{BER}_{SAT} \quad (9.3)$$

Smaller block sizes are preferred since they enable estimation of higher BERs (i.e., ideally, estimation of BERs as high as 10^{-3} is desirable). However, block sizes are usually governed by other factors related to the digital signal format efficiency, and are therefore often not as small as desired.

Note that for a *nonideal block error code,* although all blocks may be in error at high BER, the block error detection scheme may not indicate this, owing to its limitations. For example, with parity check codes, BLER = ½ under high error conditions, owing to errors in the parity bit itself, so that in this case, from Equation (9.2), $\text{BER}_{SAT} = 1/(2\text{BS})$. Thus, for parity check codes (including the individual parity bits for the bit interleaved parity (BIP) code used with SONET),

$$\text{BLER} = \tfrac{1}{2}, \quad \text{for BER} \geq \text{BER}_{SAT} \quad (9.4)$$

As another example, consider CRC-X and BIP-X codes for which $\text{BLER} = 1 - 2^{-x}$ due to imperfect detection of errors under high error conditions [290, 354]; here X is the number of bits used in the CRC and BIP codes. In these cases,

$$\text{Error Detection Capability (\%)} = 100(1 - 2^{-x}) \quad (9.5)$$

For example:

CRC-X/BIP-X	Error Detection Capability (%)
CRC-6 (DS1-ESF)	99.44
CRC-9 (DS3S)	99.80
CRC-12 (ISDN BRA)	99.98
BIP-8 (SONET STS)	99.61
BIP-2 (SONET VT)	75.00

For maintenance purposes, CRC-X and BIP-8 are considered to approximate an ideal code, so that Equation (9.3) approximately applies as follows:

$$\text{BLER} = 1 - 2^{-x} \approx 1, \quad \text{for BER} \geq \text{BER}_{SAT} \tag{9.6}$$

The above considerations are accounted for in Tables 9.9 and 9.10 and further illustrated in Figure 9.29. Notice from Table 9.10 and Figure 9.29 that one of the practical consequences of BER saturation is the inability of some codes to accurately estimate SESs having a BER threshold of 10^{-3}. In these cases, $\text{BER}_{SAT} < 10^{-3}$, and conservative approximations are required; that is, assume that $\text{BER}_{SAT} = 10^{-3}$. On the other hand, this is mitigated at DS3 and SONET rates where lower SES thresholds in the region of 10^{-6} and 10^{-7} are used [50].

Finally, notice also the special case where (effectively) the block size equals unity (BS = 1), which applies for line codes using BPVs, and for frame bit errors. In these cases, the limitations related to block size do not occur. However, these detection schemes have other limitations, as discussed later.

In the case of nonrandom (i.e., bursty) errors, the limitation of block codes is the uncertainty as to the number of bit errors in an errored block. This uncertainty increases with increased block size, and ranges from 1 to BS/2 bit errors per errored block. In this regard, smaller block sizes are preferred. As an example, for the BIP-X code used in SONET, X individual parity bits are observed, each effectively corresponding to an X times smaller block size than for CRC-X. Since Equation (9.4) applies to each of the X individual BIP-X parity bits, effectively, the uncertainty in bit errors per block error is only reduced by X/2 relative to CRC-X. This consideration also has a bearing on the random error case, and has been accounted for in Tables 9.9 and 9.10, and in Figure 9.29.

Another consequence of burst errors in relation to parity codes is the likelihood of the parity bit being in error and failing to indicate a block error [353]; see Equation (9.4). The practical significance of this limitation is an underestimation of the count of ESs. Less likely is the possibility of a parity bit error falsely indicating a block error. These limitations regarding ES counts are minimal for CRC-X and BIP-X codes, since, effectively, X parity bits are jointly considered, and errors in all of them are unlikely (i.e., probability of 2^{-x}, the complement of the detection capabilities in the table following Eqn. (9.5)).

9.7.4.2 Other Accuracy Considerations

Frame Bit Errors: Framing bits typically represent a very small percentage (i.e., <1%) of a digital signal, and, as such, reduce the confidence of the error measurement. For example, FBEs may be monitored for DS1-SF signals where there is no other path monitoring capability [210]. In this case, the frame bits only span $100 \times \frac{1}{193} \approx 0.5\%$ of the DS1 signal. The result of this limitation is that BER and ES counts are likely to be underestimated [339]. The confidence can be improved by measuring over sufficiently long periods of time [281, 355, 356]. However, in some cases these times are too long to be practical for maintenance purposes. This limitation is removed by new digital signal formats which have more powerful block detection codes, such as CRC-6 for DS1-ESF, and which obviate the need to depend on FBEs.

Bipolar Violations: These codes have some imperfections for error detection in that it is possible for certain error patterns to result in too few or too many errors being detected [284, 281]. However, for practical maintenance purposes, the codes are known to provide a nearly one-to-one correspondence with BER over an applicably wide range of BERs [357] (i.e., to as high as 10^{-2}). Also, these codes are peculiar to cable carrier systems, and are generally not applicable to radio and fiber signal formats, where block codes tend to be used.

ES Estimation; As noted in the previous discussion of the ES parameter in Section 9.7.2, ESs may be detected using either the asynchronous or synchronous method. The asynchronous method uses a 1-sec clock not synchronized to bit errors, while with the synchronous method, the 1-sec clock is synchronized to the first bit error in the second. Generally, the asynchronous method may indicate a higher (conservative) number of ESs. Also, measurement of the same error pattern by two different "asynchronous" monitors may not give the same ES count because of possible differences in the phases of the monitors' 1-sec clocks in relation to the error occurrence [346]. Conversely, the synchronous method provides an exact (minimum) and consistent ES count.

More specifically, under the condition of *random errors,* when there are very few errors (much less than one per second), the synchronous and asynchronous methods will give essentially the same answer. This is because each second will tend to have either one error or no errors, and, therefore, the total ESs will be similar regardless of the clock phase. Also, for a very high error rate (many errors per second), both methods will tend to give the same answer, since most seconds will be in error, regardless of the clock phase. The worst case is when there is about one or two errors per second. Then differences in clock phase with respect to errors will most affect the number of ESs. In this worst case, the asynchronous method will indicate nearly 20% more ESs than the synchronous method [346]. However, across all random BER conditions, the difference would be much less than 20% and on the conservative (high) side.

For *error bursts* with durations from near zero to near 1 sec, the number of ESs using the asynchronous method can vary from 1 to twice that given by the synchronous method [346]. The best (most accurate) case is for very short bursts which are very likely to all fall within a second regardless of clock phase (this results in no increase in the number of ESs using the asynchronous method). The worst (least accurate) case occurs for bursts somewhat shorter than 1 sec, since, conceivably, each burst could either be within a second, or within 2 sec (i.e., split by the boundary between 2 sec); this corresponds to a 100% increase. For longer bursts, the worst-case error is less, since, for example, with a burst somewhat shorter than 3 sec, there can either be 3 or 4 ESs depending on the clock phase, which represents 33% increase.

However, across all types of error bursts, the error (which is on the conservative side) would be less than the worst-case values indicated above, and is considered adequate for maintenance purposes.

Hierarchical Translation of ESs: The count of ESs as monitored at a given hierarchical rate (e.g., at DS3) may not be the same as the count of ESs on a given tributary rate carried within the higher rate (e.g., at DS1 or DS0). The counts depend on the cause of ESs. If an ES is due to SEF events at the higher rate, then each tributary will also encounter an ES. Also, if the ES is due to a controlled slip event at the higher rate (e.g., at DS1), then each tributary (DS0), will again encounter an ES. Thus, in these cases the ES counts are the same at all rates. However, if an ES at a higher rate is due to transmission errors, then, depending on the number and distribution of errors in the ES, as few as one, to as many as all, tributary rates will encounter an ES (Note: no tributary may encounter an ES if only the higher rate overhead is in error, which is unlikely to happen). Thus, with transmission errors, the ES counts at higher, and at tributary rates, may differ; in general, the count at the lower rates will tend to be less than at the higher rates [352, 358, 359].

The significance of this difference in ES counts at higher and at tributary rates depends on services carried at these rates and how performance-sensitive these services are. For example, if certain higher rate services (such as video) are most sensitive, then a uniform (service-independent) maintenance philosophy which satisfies the most sensitive services at higher rates will also satisfy services at tributary rates (i.e., ES counts at higher rates are an upper bound on these at tributary rates). In this case, ES translation to tributary rates is unnecessary. On the other hand, if tributary rate services are most performance sensitive and difficult to achieve with overbound estimates at higher rates, then ES translation may be useful.

An estimate of ES counts at lower rates can be made using the following CCITT-based algorithm [335]. A count (n) of the number of CVs is made for each ES at the higher rate, and the CVs are assumed to be randomly distributed in time within each second. If the number of CVs equals or exceeds the number of tributaries (N), it is assumed that at least one CV will affect each tributary, and, therefore, each tributary is considered to have encountered an ES. However, if the number of CVs in the ES at the higher rate is less than the number of tributaries, then the number of tributaries encountering an ES is assumed to equal the number of CVs. Overall, in each second, the average number of ESs across all tributaries is between 0 and 1 (i.e., the ratio—n/N—of the number of CVs per second to the number of tributaries, truncated to a maximum value of 1). By carrying out this processing for each second, the average number of ESs at tributary rates is estimated. The estimate is conservative, in the sense that CVs may be distributed such as to affect fewer than the maximum number of tributaries. On the other hand, there may also be more than one bit error per CV which could affect more than one tributary.

If estimates of the average number (m) of bit errors per CV are known (e.g., from network characterization studies), this number can be multiplied by the number of CVs per second to estimate the average number of bit errors per second, and used in the algorithm discussed above (i.e., the ratio—$m \times n/N$—of the estimated number of bit errors per second to the number of tributaries, truncated to a maximum value of 1).

Miscellaneous; All parameter counts will generally be exact, except when there is register overflow, in which case registers should remain (peg) at the maximum value until they are read and reset at the end of the reporting interval (after the data is stored).

For CV counts the register sizes could ideally be sufficient to accommodate a BER of 10^{-3} for DS1 and SONET VT signals, and a BER of 10^{-4} for DS3 and SONET STS signals, in 15-min or day periods; in practice, smaller register sizes may be used. For other parameters the register sizes should ideally be sufficient to accommodate essentially all possible intervals in 15-min or day periods. Register sizes are discussed in more detail in Section 9.7.6.

For certain signals, approximations due to BER saturation and exceptions due to too few errors in the reporting interval need to be accounted for as discussed earlier in this section (see Tables 9.9 and 9.10, and Fig. 9.29). The resolution in terms of BER is an integer ($1 \leq c \leq 9$) times the decade value (d), that is, $c \times 10^{-d}$, where values of d satisfy the BER ranges discussed in Section 9.7.6.

Time interval durations and their alignment with hours and days, as well as absolute time of day and time stamping of threshold alert messages and performance data reports, should all be based on the most accurate network clock available in NEs with performance monitoring functions. Accuracy of time of day synchronization of NE clocks is also important. However, due to the limitations of various existing time-of-day architectures and distribution schemes, this accuracy is not yet determined in the industry.

9.7.5 Interaction Between Available and Unavailable States

Returning to the summary of performance parameters in Table 9.8, it is important to distinguish between the two performance categories: available (or in-service), and unavailable (or out-of-service). The parameters associated with the available state are intended to be monitored during this state only, as shown in Figure 9.30. This applies especially to quality-of-service parameters (ESs and SESs) and also to most other parameters. It is not considered meaningful to monitor these parameters under conditions of unavailability; that is, periods of 10 or more consecutive SESs and/or failures [50, 351].

The specific inhibiting rules are as follows (further details can be found in [50]):

(i) CVs are inhibited during all seconds containing SEF (if applicable) and failure primitives, and during UAS.

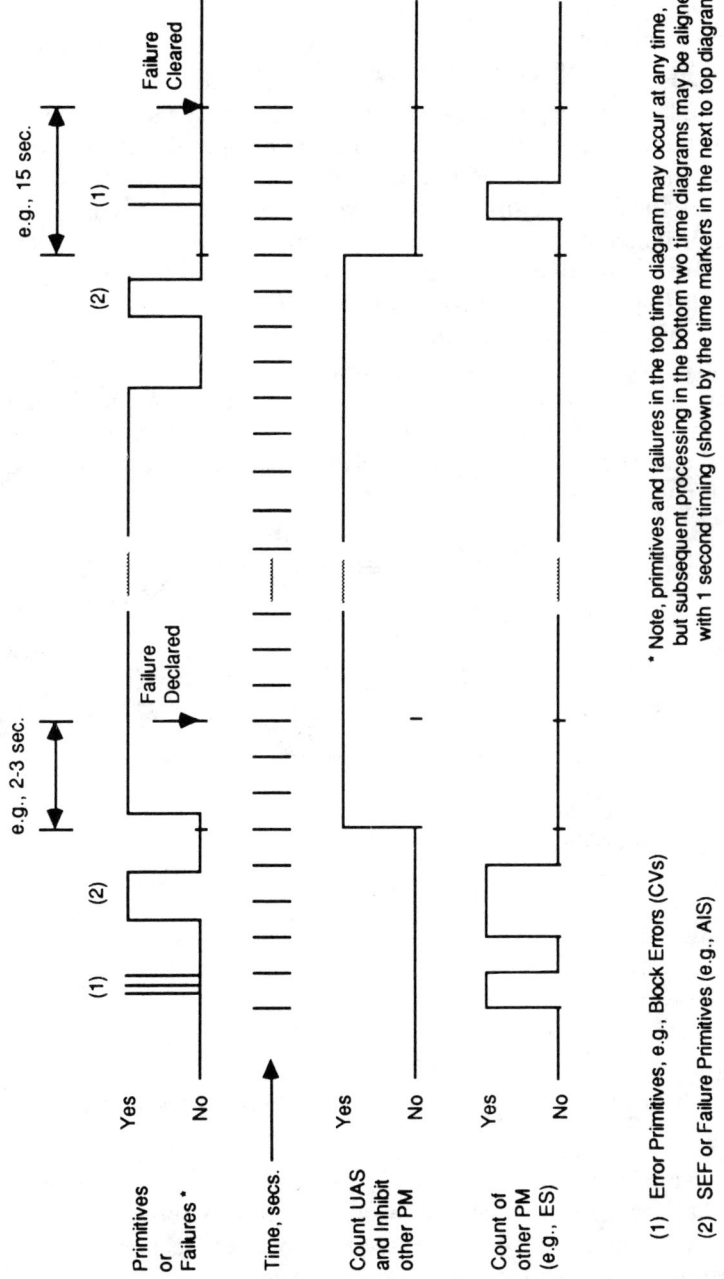

Figure 9.30 Inhibiting of performance monitoring (PM) during failures.

(ii) When the UAS parameter is provided, all parameters are inhibited during UAS, except UAS, PSC, and PSD.
(iii) When UAS is not provided, only CVs are inhibited during all seconds containing SEF (if applicable) and failure primitives.

In the context of layered performance monitoring, it is important to note the dependencies among the various sublayers of layer 1 monitoring as illustrated by the example for SONET in Figure 9.31. In this figure, impairments propagate from the functionally lower, to the functionally higher sublayers, and the reverse is not true. This applies to performance primitives, to failure primitives, and to failures (alarm and status indications). Primitives are propagated implicitly, for example, regenerator section block (BIP-8) errors cause transmission line block (BIP-8) errors, which in turn cause STS path block (BIP-8) errors, etc. (recall Figs. 9.17 and 9.18). Failure primitives (and hence failures) are also propagated explicitly through AIS. For example, section failure primitives (LOS or LOF for SONET) will result in the application of line AIS at the point where the failure primitive is first detected (i.e., at the STE). This AIS will then be detected at the line level (at the LTE), and a corresponding line status indication will be raised.

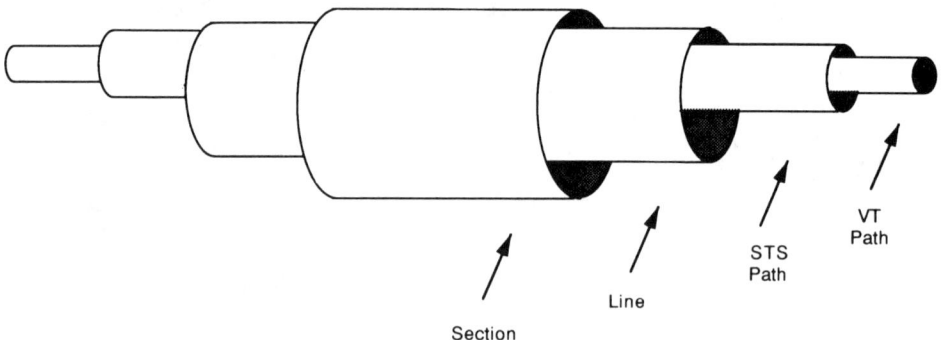

- Section troubles affect Line & Paths
- Line troubles affect Paths
- STS Path troubles affect VT Paths
- Reverse of any of the above is <u>NOT</u> true

Figure 9.31 Dependencies among heirarchical signals as illustrated by example for SONET.

The reverse of the above processes is not true. Therefore, since path performance does not affect lines, line parameters can continue to be counted during path troubles. Similarly, since line performance does not affect sections, section parameters can continue to be counted during line troubles.

9.7.6 Storing, Thresholding, and Reporting of Performance Information

With reference to Figure 9.27, the remaining generic performance monitoring functions are the provision of memory for current and previous counts of each parameter, comparison of each current parameter count to alert thresholds, and modes of reporting performance data and alerts across the NE/OS interface [3, 50, 52]. These functions are illustrated in more detail in Figure 9.32.

9.7.6.1 Memory

Memory is required in the NE (or associated mediation device) to store the counts for each section, line, and path parameter that is monitored. Where applicable, separate memory is provided for each direction of transmission (i.e., for near-end and far-end performance). This memory is useful for diagnostic purposes to verify customer-reported troubles, and to enable the OS or local maintenance personnel to examine past history leading up to a trouble condition, and thereby aid in identifying the cause of trouble, and in sectionalization. This memory also enables an OS to compile history of past performance to verify long-term performance objectives, and to perform trend analysis and other performance characterizations. As shown in Figure 9.32, 32 individual 15-min periods and at least one day period of performance data is a suitable amount of previous history required for the above purposes.

In addition to history, current 15-min and current day counts are also required. These current counts can be queried at frequent intervals (e.g., every minute) to aid in diagnostics of a particular section, line, or path that is exhibiting a degraded condition. The current counts are also used for alert generation, as discussed below.

For most parameters, the sufficient size for the memory counters is 10 bits for the 15-min counters and 16 bits for the daily counters. An exception is for CV counts where larger counters of 14 and 20 bits are specified for 15-min and day counts of CVs, respectively [50]. If the maximum value of a counter is reached, the counter should remain at the value until it is read, the data is stored, and the counter is reset.

9.7.6.2 Thresholding

As shown in Figure 9.32, the current 15-min and day counts are each compared to a programmable threshold level. Separate thresholding is provided for each direction of transmission, although the threshold level is usually the same for each direction. The comparison is done frequently relative to the 15-min and day time intervals (ideally, once per second, but at least once per minute for the 15-min threshold, and at least once per 15 min for the day threshold). When a count reaches or exceeds the threshold level, an alert message is generated and reported to the OS (and local maintenance interface). When a threshold is reached or crossed, the NE continues count-

Figure 9.32 Detailed storing, thresholding, and reporting functions for each parameter.

Tf, Td: 15 min. and daily threshold values, respectively.
@: Locally and remotely settable thresholds with default values.
Y: Yes
N: No

#: Reset every 15 min. and aligned with hours
##: Reset every day (e.g., at midnight).
&: 1 day is required; additional days (e.g., up to 7, are optional)

ing to the end of the 15-min or day period. At most, one alert message (per parameter, per monitored signal, per direction of transmission) is sent during any 15-min or day period. Alert threshold values are programmable either from the OS or from the local maintenance interface.

Ideally, for most parameters, the range of threshold values should be from 1 event to the maximum number of events in a 15-min or day period. This corresponds to ranges from 1 up to 2^{10} and 2^{16} for 15-min and day threshold values, respectively. The exception is for CVs, for which the upper values for the ranges of 2^{14} and 2^{20} are specified in [50]. For DS1 and VT signals, this enables BERs to range from as high as about 10^{-5}, to as low as about 10^{-9} and 10^{-11} for the 15-min and day threshold values, respectively (see Table 9.10). Similarly, for DS3 and STS signals, BERs range from as high as about 3×10^{-7}, to as low as about 10^{-10} and 10^{-12} for the 15-min and day threshold values, respectively.

Alert messages notify maintenance forces that performance has degraded below some level. An example of how the 15-min and day thresholds may be used is as follows. For the ES parameter, the day threshold can be set to the tariffed performance level (e.g., for a DS1 path, 95% EFS/day corresponds to a threshold level of 4320 ESs/day [287, 288]). In addition, the 15-min threshold can be set to correspond to a lower ES count (e.g., $4320/(24 \times 4) = 45$ ES/15 min). Thus, by observing the number of 15-min alerts, maintenance forces can be forewarned that the daily ES performance level may be exceeded. This enables the maintenance forces to take proactive steps to correct the condition before the end user is adversely affected. Examples of maintenance limits are provided in [26], and default alert thresholds in [52, 67, 254].

9.7.6.3 *Reporting*

Having discussed how performance data and alert messages are generated, the final performance monitoring function in Figure 9.32 addresses the reporting of this information to an OS. Flexibility in the reporting of performance data is useful in order to accommodate existing and future maintenance requirements and OS capabilities. All the data can be periodically reportable autonomously by the NE (e.g., every 8 hours), or as requested (on demand) by the OS. The OS may also request reporting of current data continuously with a start time to stop time command, or for a specified (scheduled) period of time. The current data can also be reportable with alerts, or at the end of 15-min or day intervals in response to events (where an event is a nonzero count of one or more parameters in the reporting interval). Continuous reporting of current data is the simplest mode, but requires the most operations network bandwidth. Reporting in response to alerts reduces the bandwidth requirements, but only results in reporting of performance that is worse than threshold level. On the other hand, event-based reporting provides a complete performance record (e.g., for verifying service objectives and enabling proactive maintenance), while also reducing operations network bandwidth requirements. The reporting intervals are nominally 15 min and a day, but can be shorter (e.g., 1 min) for diagnostic purposes on selected ports.

When a threshold for a given parameter is reached or crossed, the NE immediately initiates sending of the corresponding alert message and continues the parameter count to the end of the reporting interval, when the count is stored and reset. Alert messages are time stamped. No more than one threshold alert message (per parameter, per monitored signal, per direction of transmission) is sent during any reporting interval.

There should be a per-port and per-NE disabling feature for selected performance parameters and/or alerts. This serves to eliminate data from ports or NEs where, for example, there is an already known problem, or to ignore certain parameters and/or alerts, for example, on ports carrying noncritical traffic. The OS should be notified of such disabling.

9.8 REVIEW OF MAINTENANCE CAPABILITIES BASED ON CONTINUOUS, IN-SERVICE PERFORMANCE MONITORING

9.8.1 Basic Maintenance Capabilities

The performance monitoring functions presented in Section 9.7 are sufficient to enable network providers to maintain their digital transport networks to required levels of transmission performance, and thereby demonstrate to end users and other network providers that allocated portions of end-to-end path quality-of-service objectives are satisfied [3, 360]. These functions, which support centralized surveillance and maintenance, are summarized as follows:

Automatic Trouble Identification: The performance data and alert messages provided by in-service monitoring of sections, lines, and paths enable an OS to identify troubles (ideally before end users are affected), verify if the trouble is persisting, sectionalize the trouble to the responsible administration or jurisdiction, and locate the trouble within an administration.

Trouble Correction: The first priority after trouble detection is to preserve service continuity (avoid an outage); based on the performance information, either automatic or manual protection switching, or route restoration can be invoked. Once service continuity is ensured, corrective maintenance is undertaken. After the trouble is corrected, performance monitoring then enables automatic repair verification before a repaired system is restored to service.

Monitoring During Trouble-Free Periods: The role of performance monitoring in trouble-free periods is to replace the need for routine preventative maintenance; this is possible since all facilities can be continuously monitored. The resulting information provides performance history that can be used for the prediction of impending troubles (proactive maintenance) or reactive (after the fact) maintenance, and for the verification of end-to-end and allocated service objectives [361].

The use of performance information for verification of allocated service objectives is particularly useful in terms of assessing performance at interfaces between different network vendors, and between end users and network vendors, in order to resolve responsibility for end-to-end degradations. This information can also enable end users to directly assess the performance of their virtual networks, and to relate service quality to the price of service.

Finally, performance information can be used to support higher level network functions, such as provisioning, network reconfiguration, and rerouting of performance-sensitive traffic [1]; these enhanced maintenance functions are discussed in the following section.

9.8.2 Enhanced Maintenance Capabilities

Performance monitoring provides information whereby network vendors can determine the extent to which intended network performance is being achieved. This information is also used directly to maintain the network as described above. The same information can also be used together with expert systems and database images of the network, to dynamically manage and control network resources on a broader scale [4, 39, 307, 362–368]. This is achieved with building blocks similar to those outlined earlier in the overall operations and network architecture (Fig. 9.13); that is,

- Digital transmission links operating at multiples of DS0, DS1, VT, DS3. or STS rates.
- Digital switches which control traffic flow (typically at DS0 rates).
- Transport ADMs and DCSs which enable reconfiguration of logical circuits in near real time at DS0 and higher rates.
- Networkwide operations support systems which use the switches and transport equipment to monitor and control the network.

The priority is to ensure the survivability of networks, especially when nodes and links carry the large amounts of traffic made possible by fiber-optic transmission. Network survivability is a network's ability to cope with random failures, traffic overloads, and serious transmission performance degradation. By using route diversity (e.g., ring technology, or by establishing a certain proportion of redundant channels at DS1 and higher rates that can be used universally), the network can enable dynamic restoration of service and therefore achieve the desired overall service availability [19, 21, 27–47, 237]. Another alternative is to use media diversity (e.g., radio in place of a failed fiber link [369]). Similarly, dynamic routing at DS0 rates adapts in near real time to unexpected traffic demands, which also improves network survivability [370–373].

Network performance can also be improved in terms of responsiveness to changes in demand, and in terms of increasing network efficiency [40, 233–235, 238, 360, 374–376]. For example, dynamic bandwidth control permits an OS to allocate facilities or bandwidth on a near real-time basis, thus improving network responsiveness (i.e., decreased response time and increased throughput). The cell structure of ATM is also suitable for this purpose [102, 156, 160, 163, 168–170, 173, 175, 176].

Dynamic control can also improve network efficiency through near real-time provisioning. An example is when a network determines that an excessive number of calls between the same nodes are using alternate routes through other nodes. The controller can improve the efficiency of the network by indicating the need for increased capacity on the direct (high usage) route between the two nodes involved.

Finally, performance control facilitates adapting network performance to satisfy end users' changing needs. This can be achieved in a user's virtual network in which performance is monitored and, to a degree, can be varied as required. The network vendor can, for example, respond by assigning network resources that satisfy the end users' particular service or performance requirements. In this manner, the network vendor can also exploit the performance range resulting from a technology base consisting of both new and embedded equipment.

9.9 REFERENCES

1. J. Gruber, E. Abdou, P. Richards, and G. Williams, "Quality-of-Service in Evolving Telecommunications Networks," *IEEE Journal on Sel. Areas in Comm.*, Vol. SAC-4, No. 7, Oct. 1986, pp. 1084–1089.
2. J. Gruber, "From Performance to Planning: An Overview," *NCF Rec.* Vol. 39, Chicago, Oct. 1985, pp. 551–555, 565, 566.
3. J. Gruber, "Performance Monitoring and Surveillance in Integrated Services Networks," *IEEE Journal of Sel. Areas in Comm.*, Vol. 6, No. 8, Oct. 1988, pp. 1378–1383.
4. J. Leeson, J. Gruber, A. Sand, and D. Lachance, "Safeguarding End User Performance Through Digital Network Maintenance," *ICC Rec.*, Vol. 1, Atlanta, April 1990, pp. 224.1.1–224.1.7. Also, *NCF Rec.*, Vol. 42, Bk. 2, Chicago, Oct. 1988, pp. 1114–1120.
5. H.M. Dib *et al.*, "AT&T International Digital Service: Operations and Maintenance," *ICC Rec.*, Vol. 1, Atlanta, June 1990, pp. 218.3.1–218.3.8.
6. E.V. Hird, "The Network: How They Manage It," *TE&M*, Vol. 94, No. 9, May 1, 1990, pp. 43, 46–48.
7. "NetView Shows Highs and Lows of Net Management," *Data Comm.*, Vol. 18, No. 9, July 1989, pp. 85–87.
8. E.V. Hird, "Network Management with Supermasters," *TE&M*, Vol. 92, No. 21, Nov. 1, 1988, pp. 57–61.
9. K.R. Blake, "An Insider's View of Network Maintenance," *Telephony*, Vol. 212, No. 24, June 15, 1987, pp. 82, 83.
10. P.A. Wolfmeyer and R.P. Guenther, "Maintaining Quality Performance," *ICC Rec.*, Vol. 2, Seattle, June 1987, pp. 198.7.1–198.7.4.
11. H. Sticker, "Network Monitoring and Analysis (NMA)—An Advanced Network Monitoring Tool," *GLOBECOM Rec.*, Vol. 3, Houston, Dec. 1986, pp. 36.4.1–36.4.5.
12. S.D. Chikte *et al.*, "An Overview of Surveillance-Based Maintenance of the Transmission Network," *GLOBECOM Rec.*, Vol. 3, Houston, Dec. 1986, pp. 36.5.1–36.5.5.
13. H.T. Stump, "Surveillance-Based Maintenance: A Strategy for Future Network Operation," *ISSLS Rec.*, Tokyo, Sept./Oct. 1986, pp. 271–276.

14. E. Turner, "Fiber Optic Maintenance," *ICC Rec.*, Vol. 1, Seattle, June 1987, pp. 13.7.1–13.7.8
15. *TE&M*, Vol. 94, No. 14, July 15 1990. Issue on "Operations Support Systems."
16. T. Kiernan, "Net Management—From an Alarming Point of View," *TE&M*, Vol. 95, No. 5, Mar. 1, 1991, pp. 64–67.
17. R. Holter, "SONET—A Network Management Viewpoint," *NCF Rec.*, Vol. 44, Chicago, Oct. 1990, pp. 181–191.
18. G.M. Durochik, "Network Monitoring and Analysis (NMA)," *NCF Rec.*, Vol. 44, Chicago, Oct. 1990, pp. 977–979.
19. Papers in Session 301 on "Network Configuration Management," *GLOBECOM Rec.*, Vol. 1, San Diego, Dec. 1990.
20. P.E. Prozeller, "A Model for OSI-Based Message Set for the Unified Management of Testing in Communications Networks," *GLOBECOM Rec.*, Vol. 2, San Diego, Dec. 1990, pp. 701.6.1–701.6.6.
21. Papers in Session 802 on "Integrated OAM for Network Management and Control" *GLOBECOM Rec.*, Vol. 3, San Diego, Dec. 1990.
22. D. Desikan, "Bringing Strategic Value to Network Management Systems," *Telecomm.*, Vol. 24, No. 12, Dec. 1990, pp. 60–62.
23. CCITT Rec. M.20, "Maintenance Philosophy for Telecommunications Networks," *Blue Book*, Vol. IV, Fas. IV.1, Geneva, 1989, pp. 4–20.
24. "Engineering and Operation of the Bell System," AT&T Bell Labs. Inc.. Chapter 13 on "Overview of Telephone Company Operations," 1983.
25. CCITT Rec. M.34, "Performance Monitoring on International Transmission System and Equipment," *Blue Book*, Vol. IV, Fac. IV.1, Geneva, 1989, pp. 65–68.
26. CCITT Rec. M.550, "Performance Limits for Bringing into Service and Maintenance of Digital Paths, Sections and Line Sections," *Blue Book*, Vol. IV, Fas. IV.1, Geneva, 1989, pp. 314–328. Also, Rev., TD No. 178, Tokyo, April 1991.
27. G.S. Disher, "Planning for Survivability in Bell Canada's Interexchange Network," *Conform Conf. on Survivability*, Phoenix, Mar. 1990, pp. D4-1 to D4-8.
28. W.D. Grover *et al.*, "Development and Performance Assessment of a Distributed Asynchronous Protocol for Real-Time Network Restoration," *IEEE Journal on Sel. Areas in Comm.*, Vol. 9, No. 1, Jan. 1991, pp. 112–125. Also, *GLOBECOM Rec.*, Vol. 1, San Diego, Dec. 1990, pp. 403.3.1–403.3.7.
29. T. Flanagan, "Fiber Network Survivability," *IEEE Comm. Mag.*, Vol. 28, No. 6, June 1990, pp. 46–53.
30. IEEE Comm. Mag. Special Issue on "Surviving Disaster," Vol. 28, No. 6, June 1990.
31. D.J. Holste, "Creating a Network That Heals Itself," AT&T Tech., Vol. 5, No. 2 1990, pp. 42–47.
32. T. Helmes, "Crash-Proof Fiber Rings," *Telephony*, Vol. 217, No. 9, Aug. 21 1989, pp. 22, 23.
33. W. Ergle, "A Network Survival Game Plan," *Telephony*, Vol. 217, No. 9, Aug 21, 1989, pp. 24, 25, 28–31.
34. T. Flanagan, "Planning a SONET Network," *ICC Rec.*, Vol. 3, Atlanta, April 1990, pp. 321.2.1–321.2.6.
35. B. Pekarske, "Restoration in a Flash—Using DS3 Cross-Connects," *Telephony*, Vol. 219, No. 12, Sept. 10 1990, pp. 35, 36, 38, 40.
36. J.W. Schladweiler, "Evolution of Business Continuity Services," *Telecomm.*, Vol. 24, No. 8, Aug. 1990, pp. 63–66.
37. S. Fleming, "Better to be Safe than SONET-ized," *TE&M*, Vol. 95, No. 9, May 1, 1991, pp. 54–57, 59.
38. J. Sosnosky and T. Wu, "SONET Ring Applications for Survivable Fiber Loop Networks," *IEEE Comm. Mag.*, Vol. 29, No. 6, June 1991, pp. 51–58.
39. Papers in Session 304 on "Connecting the Future—Operations Considerations," *GLOBECOM Rec.*, Vol. 1, San Diego, Dec. 1990.

40. Papers in Session 705B on "Digital Cross-Connects Systems—Changing Applications and Technology," *GLOBECOM Rec.*, Vol. 2, San Diego, Dec. 1990.
41. T. Wu and M.E. Burrowes, "Feasibility Study of a High-Speed SONET Self-Healing Ring Architecture in Future Interoffice Networks," *IEEE Comm. Mag.*, Vol. 28, No. 11, Nov. 1990, pp. 33–51.
42. D. Kidd and D. Hiscoe, "Getting to the Source," *Telephony*, Vol. 219, No. 19, Oct. 29, 1990, pp. 26, 30, 32, 34.
43. Telephony Special Issue on "SONET," Vol. 220, No. 4, Jan. 28, 1991.
44. C.W. Bowen, "Lightning Bolt Jolts Contel of Kentucky," *Telephony*, Vol. 219, No. 17, Oct. 15, 1990, pp. 26, 27, 30, 31.
45. A. Stewart, "How Fiber Survives," *TE&M*, Vol. 95, No. 6, Mar. 15, 1991, pp. 120, 123–125.
46. D. Borg, G. May and T. Flanagan, "SONET Rings for Network Management and Survivability," *NCF Rec.*, Vol. 44, Chicago, Oct. 1990, pp. 160–169.
47. M. Mikolas, "RBOC Strategies for Local Loop Reliability," *Bus. Comm. Rev.*, Vol. 20, No. 10, Oct. 1990, pp. 48–53.
48. P. Heywood, "Extending a 'Global Fiber Highway' Across the Pacific," *Data Comm.* Vol. 20, No. 3, pp. 62, 64.
49. D.T. Kong, "2.488 Gb/s SONET Multiplexer/Demultiplexer with Frame Detection Capability," *IEEE Journal on Sel. Areas in Comm.*, Vol. 9, No. 5, June 1991, pp. 726–731.
50. Draft American National Standard for Telecomm.—Layer 1 In-Service Digital Transmission Performance Monitoring, T1M1.3/91-003R3, T1M1/91-056, T1M1/LB91-06, June 1991.
51. H. Walker, "Ins and Outs of DS1 and DS3 Measurement," *Telephony*, Vol. 217, No. 12, Sept. 11, 1989, pp. 36, 40, 44, 46.
52. "Operations Technology Generic Requirements (OTGR): Network Maintenance: Transport Surveillance," Bellcore, TR-475, Issue 3, June 1990.
53. C.E. Pittman, "Transmission Performance Monitoring for Digital Services," *GLOBECOM Rec.*, Vol. 3, Houston, Dec. 1986, pp. 36.2.1 to 36.2.4.
54. "SONET Transport Systems: Common Generic Criteria," Bellcore TR-253, Issue 1, Sept. 1989. Also, TA-253, Issue 6, Sept. 1990. Also, update, Bellcore Digest, Aug. 1991.
55. H.R. Salloum and S. Yoneda, "B-ISDN User Network Interface: Implementation of Performance Monitoring Functions Using SONET Overhead," *ICC Rec.*, Vol. 3, Atlanta, April 1990, pp. 318.2.1–318.2.4.
56. S. Banerjee and R.C. Roman, "ISDN Primary Rate Access Maintenance," *ICC Rec.*, Vol. 1, Boston, June 1989, pp. 2.6.1–2.6.5.
57. L.W. Campbell, S.G. Pelosi, and R.C. Roman, "ISDN Basic Access Line Transmission Performance Maintenance," *GLOBECOM Rec.*, Vol. 3, Houston, Dec. 1986, pp. 45.1.1–45.1.5.
58. T.J. Starr, "OAM&P for ISDN Basic Access DSL Based on ANSI Physical Layer Standards," *IEEE Comm. Mag.*, Vol. 29, No. 6, June 1991, pp. 59–63.
59. "Notes on the BOC INTRA-LATA Networks," Bellcore TR-275, Issue 1, Section 8 on "Maintenance Considerations," April 1986.
60. "Notes on the Network," AT&T, Section 8 on "Maintenance Considerations," 1980.
61. W.O. Fleckenstein, "Operation Support Systems: Computer Aids for the Local Exchange," *Bell Labs. Rec.*, Sept. 1982, p. 185–193.
62. "Engineering and Operations in the Bell System," Bell Telephone Labs. Inc., Chapter 16 on "Testing and Maintenance," 1977.
63. "Telecommunication Transmission Engineering," Vol. 3: Networks and Services, 2nd ed., AT&T, Chapter 20 on "Maintenance and Reliability," 1977. Also, 3rd ed., Bellcore, 1990.
64. "Transmissions System for Communications," Bell Telephone Labs. Inc., 5th ed., Chapter 36 on "Maintenance," 1982.
65. "Telecommunications Transmission Engineering," Vol. 2: Facilities, 1st ed., AT&T Chapter 24 on "Maintenance Systems," 1977. Also, 3rd ed., Bellcore, 1990.

66. P.S. Boggs and J.R. Wright, "Knocking Problems for a Loop," *AT&T Bell Labs. Rec.*, Jan. 1985, pp. 22–26.
67. "Integrated Digital Loop Carrier System Generic Requirements Objectives, and Interface," Bellcore TR-303, Issue 1 Sept. 1986. Also, Revs. 1 (Dec. 1987), 2 (July 1989) and 3 (Mar. 1990). Also, Supps. 2 (Oct. 1989), and 3 (Mar. 1990). Also, TA-303 Issue 2, July 1987.
68. Functional Criteria for Digital Loop Carrier Systems," Bellcore TR-057, Issue 1, April 1987. Also, Rev. 1 (Nov. 1988). Also, Bull. No. 1, Mar. 1990. Also, TA-057, Issue 4, June 1988.
69. D. Pfeiffer, "Finding Fault with The Local Loop," Special Supp. on "Transmission," *Telephony*, Nov. 1990, pp. 36, 37.
70. R. Maciel, "Pacific Bell Expects Big Things from Small Packages," *Telephony*, Vol. 219, No. 20, Nov. 5, 1990, pp. 30, 34, 36.
71. R. Ahmari and R.J. Healey, "Helping Switches to Account for Themselves," *AT&T Bell Labs. Rec.*, Mar. 1985, pp. 16–22.
72. R.E. Machol, "Evolution of No. 2 Switching Control Center System," *ICC Rec.*, Vol. 3, Chicago, June 1985, pp. 42.1.1–42.1.5.
73. K.D. Meola and E.C. Wenning, "Tracking Trouble in the Switching Network," *AT&T Bell Labs. Rec.*, Jan 1984, pp. 20–26.
74. "LATA Switching System Generic Requirements (LSSGR)," Section 9 on "Maintenance," Bellcore TR-509, Issue 2, July 1987. Also, Rev. 1, Dec. 1988.
75. B. Stoffels, "Telcos Search for Self-Sufficient Switches," *TE&M*, Vol. 95, No. 6, Mar. 15 1991, pp. 116, 119.
76. J.J. Plato and B.S. Robb, "CAROT's New Features Expand Maintenance and Administrative Function," *AT&T Bell Labs. Rec.*, July/Aug. 1983, pp. 18–24.
77. J.R. Hemmert, "Every Communications Manager Should Know Network Testing," *Telecomm.*, Vol. 17, No. 7, July 1983, pp. 95, 96, 108.
78. W.G. Albert *et al.*, "Special-Service Circuits: D5 Delivers Them Faster and Easier," *AT&T Bell Labs. Rec.*, Mar. 1985, pp. 24–28.
79. B.T. Barish, "SARTS COER Provides Automatic Analysis of Traffic Data," *AT&T Bell Labs. Rec.*, Vol. 63, No. 4, July 1985, p. 11.
80. E. Howard, "Remote Access Testing Reaches Small Remote Offices," *Telephony*, Vol. 206, No. 19, May 7, 1984, pp. 30–34.
81. I.R. Feuster and D.C. Radeschi, "Mechanizing the Administration of Special Service Circuits," *ICC Rec.*, Vol. 3, Boston, June 1985, pp. E3.4.1–E3.4.5.
82. P.F. Lynn, "Switched Access Remote Test System—The Test System for the Evolving ISSN," *ICC Rec.*, Vol. 3, Boston, June 1985, pp. E.3.5.1–E.3.5.5.
83. G.P. Outzen, "A Smart Way to Test Loops," *TE&M*, Vol. 92, No. 21, Nov. 1 1988, pp. 62, 63, 66.
84. J. Bourgois, "Seeing Clearly with Windows: "Fine Tuning for Special Services; What's Unique about Special Services?," *TE&M*, Vol. 92, No. 10, May 15 1988, pp. 98–101, 104.
85. N.J. Hess, "Application of Advanced Digital Terminal Systems in a Telephone Company Environment," *GLOBECOM Rec.*, Vol. 3, Houston, Dec. 1986, pp. 45.7.1–45.7.5.
86. M.E. Koblentz, "Generic Approach to Testing Using Standard Interfaces—Workstation Prototype," *GLOBECOM Rec.*, Vol. 1, New Orleans, Dec. 1985, pp. 10.3.1–10.3.4.
87. D.K. Doherty *et al.*, "High Capacity Digital Network Management and Control," paper in [19], pp. 301.3.1–301.3.5.
88. M.L. Romane and B.A. McLeod, "SCOTS Generic 2: A Sharper Eye on Transmission Facilities," *AT&T Bell Labs. Rec.*, Jan. 1985, pp. 16–21.
89. R.J. Boyd and P.J. Schachter, "AT&T Communication Maintenance Interface for Digital Transmission Systems," *GLOBECOM Rec.*, Vol. 2, Atlanta, Nov. 1984, pp. 31.5.1–31.5.4.
90. W.G. Robinson, "Maintenance of Digital Radio Using Remoted Performance Monitoring," *ICC Rec.*, Vol. 1, Amsterdam, May 1984, pp. 456–461.

91. M. Dinkes and D.W. Zoccoli, "Measuring the Performance of the Digital Network," *AT&T Bell Labs. Rec.*, Jul./Aug. 1983, pp. 25–29.
92. Compatibility Bulletin (CB) No. 149, "Maintenance Standards for Digital Transmission Systems," Issue 4, 1989, AT&T Com, Nov. 1.
93. E.A. Jones Jr. and D.S. Walsh, "Monitoring and Analysis Strategies for Digital Networks," paper in [122], pp. 715–721.
94. J.R. Chaley and J.T. Reid, "Surveillance and Maintenance of the DSR-8 Long Haul Digital Microwave Radio System," *IEE 3rd Int. Conf. on Telecomm. Transmission*, London, Mar. 1985, pp. 36–39.
95. "Digital Network Notes," Telecom Canada, 1983. Issued by: Technical Planning and Standards, Network Development and Operations, 410 Laurier Avenue, West, Ottawa, Ontario, Canada, K1P 6H5.
96. American National Standard For Telecomm.—ISDN Management—Basic Rate Physical Layer, ANSI T1.216-1991.
97. American National Standard for Telecomm.—ISDN Management—Primary Rate Physical Layer, ANSI T1.217-1991.
98. CCITT I.600 Series Recs. on Maintenance Principles, *Blue Book*, Vol. III, Fas. III.9, Geneva, 1989.
99. CCITT Rec. G.784, on "Synchronous Digital Hierarchy (SDH) Management," Geneva, 1990.
100. "Broadband ISDN Operations and Maintenance Principles," Draft ANSI Tech. Rep. T1S1.5/91-006R1, May 1991.
101. Draft American National Standard for Telecomm.—Broadband ISDN—Customer Installation Interfaces, Physical Media Dependent Specification, T1M1.2/91-021, Feb. 1991.
102. Draft American National Standard for Telecomm.—Broadband ISDN ATM Aspects—ATM Layer Functionality and Specification, T1M1.2/91-023, Aug. 1990.
103. CCITT Draft Rec. I.610 on "OAM Principles of B-ISDN Access," Mar. 1991.
104. M. Blake-Knox and K.G. Knightson, "Network Operations Protocol—OSI Protocols for Operations, Administration and Maintenance," *GLOBECOM Rec.*, Vol. 1, Houston, Texas, Dec. 1986, pp. 28.6.1–28.6.6.
105. W.H. Cameron, C. LaCerte, and J.F. Noyer, "Integrated Network Operations Architecture and its Application to Network Maintenance," *IEEE Comm. Mag.*, Vol. 25, No. 8, Aug. 1987, pp. 48–53. Also, ISSLS Rec. Tokyo, Sept./Oct. 1986, pp. 261–266.
106. J.J. Appel, "Future Operations Strategy for Network Elements and Their Operation Systems," *ICC Rec.*, Vol. 1, Chicago, June 1985, pp. 6.4.1–6.4.6.
107. C.J. Byrne *et al.*, "Standard Provisioning, System Maintenance and Access/Testing Functions and Commands at the Operation System/Network Terminal Equipment Interface," *ICC Rec.*, Vol. 1, Chicago, June 1985, pp. 6.5.1–6.5.5.
108. W.P. Arvidson, *et al*, "A New Interface for Communication Between Network Terminal Equipment and its Operations System," *GLOBECOM Rec.*, Vol. 2, Atlanta, Nov. 1984, pp. 31.4.1–31.4.8.
109. American National Standard for Telecomm.—OAM&P—Lower Layer Protocols for Interfaces Between OSs and NEs, ANSI T1.204-1991.
110. American National Standard for Telecomm.—OAM&P—Upper Layer Protocols for Interfaces Between OSs and NEs, ANSI T1.208-1991.
111. American National Standard for Telecomm.—OAM&P—Generic Network Model for Interfaces Between OSs and NEs, ANSI T1.214-1990. Also, Draft T1.214a-199x, T1M1/LB-275, Nov. 1991.
112. CCITT Rec. G.771, "Q Interfaces and Associated Protocols for Transmission Equipment in the TMN," *Blue Book*, Vol. III, Fas. III.4, Geneva, 1989, pp. 544–570.
113. CCITT Rec. G.773, "Protocol Suites for Q-Interfaces for Management of Transmission Systems," Geneva, 1990.
114. Draft American National Standard for Telecomm.—SONET OAM&P Communications. T1X1.5/91-001R2, June 1991.

115. CCITT Rec. M.30, "Principles for a Telecommunications Management Network," *Blue Book,* Vol. IV, Fas. IV.1, Geneva, 1989, pp. 22–61.
116. M.S. Real, "An Overview of the CCITT and ANSI T1 With Emphasis on OSI Network Management Activities," *ICC Rec.,* Vol. 1, Boston, June 1989, pp. 11.1.1–11.1.7.
117. J.R. Scherer and T.A. Murray, "Management of the Intelligent Network," *IEEE Comm. Mag.,* Vol. 26, No. 12, Dec. 1988, pp. 21–24.
118. B. Higgins, B. Hurley and R. Luxford, "Operation, Administration and Maintenance of ISDN and its Relationship to TMN," *GLOBECOM Rec.,* Vol. 1, Hollywood, Florida, Nov./Dec. 1988, pp 6.6.1–6.6.8.
119. H. Walker, "Testing Future Telecom Networks," *TE&M,* Vol. 92, No. 21, Nov. 1, 1988, pp. 72, 73, 76, 76-A, 76-C.
120. American National Standard for Telecomm.—OAM&P—Fault Management Messages for Interface Between Operations System and Network Elements," ANSI T1.215-1990.
121. American National Standard for Telecomm.—OAM&P—Principles of Functions, Architecture and Protocols for Interfaces between OSs and NEs, ANSI T1.210-1989. Also, Draft T1.210-199x, T1M1/91-053R1, T1M1/LB91-07, July 1991.
122. *IEEE Journal on Sel. Area in Comm.,* Issue on Telecommunications Network Operations and Management, Vol. 6, No. 4, May 1988.
123. American National Standard for Telecomm.—OAM&P—Performance Management Functional Area Services for Interfaces Between OSs and NEs, T1LB 256, Nov. 1991.
124. American National Standard for Telecomm.—Digital Hierarchy—Optical Interface Rates and Formats Specifications, ANSI T1.105-1988. Also, Draft T1.105-199x, T1LB 213; April/May 1991.
125. G. Denman, "LiTel Takes 'Remote' Control," *Telephony,* Vol. 215, No. 3, July 18 1988, pp. 36, 38, 40.
126. T.J. Zebo, "Circuit Maintenance Operations on Digital Loop Carrier Channels in an Integrated Network Environment," *GLOBECOM Rec.,* Vol. 1, New Orleans, Dec. 1985, pp. 10.2.1–10.2.3.
127. R. Gross, "The Changes in Network Tests," *TE&M,* Vol. 91, No. 1, Jan. 1 1987, pp. 63–65.
128. R. Cox, "PCM Channel Access and Testing," *Telecomm.,* Vol. 18, No. 7, July 1984, pp. 98a, 98b, 102b.
129. L. Bohn, "Digital Cross-Connect Keeps Network Honest," *TE&M,* Vol. 92, No. 21, Nov. 15, 1986, pp. 102–104.
130. P. Marshall, "Maintaining the CCS Network," *TE&M,* Vol. 92, No. 21, Nov. 1 1988, pp. 78–80.
131. T.H. Murray, "A CAT that can Scratch Trouble," *TE&M,* Vol. 92, No. 21, Nov. 1 1988, pp. 91–94.
132. "Specification for Generic Test System Controller/Remote Test Unit," Bellcore TA-099, Issue 2, Feb. 1987.
133. "Operations Technology Generic Requirement (OTGR): Network Maintenance: Access and Testing," Bellcore TR-476, Issue 2, Feb. 88. Also, Section 6 of Revs. 1 (Mar. 1989), 2 (May 1989), and 3 (Nov. 1989).
134. "Generic Requirements for Circuit Maintenance of POTS and Special Services on Digital Loop Carrier Systems," Bellcore TA-313, Issue 2, July 1988.
135. "Basic Testing Functions for Digital Networks and Services," Bellcore TA-055, Issue 3, April 1987.
136. "Digital Test Access Unit Functional Specification," Bellcore TA-044, Issue 3, April 1987.
137. A Skaggs *et al.,* "Centralized Testing Gives IXC a Competitive Advantage," *Telephony,* Vol. 218, No. 6A, Feb. 5, 1990, pp. 34, 35, 38, 39.
138. H. Walker, "What T1 Test Patterns Tell You," *TE&M,* Vol. 95, No. 1, Jan. 1, 1991, pp. 35, 36, 38, 39.
139. D.P. Haas, "Detecting Hidden Bridge Taps with T-1 Test Patterns," *Telephony,* Vol. 218, No. 19, May 7, 1990, pp. 28–30, 32.
140. A.H. Miller, "Fast and Efficient DSX Template Testing," *Telephony,* Vol. 218, No. 15, April 9, 1990, pp. 70, 72, 74.

141. Committee T1-Telecommunications "A Technical Report on Test Patterns for DS1 Circuits," Draft T1M1.3/91-002R2, Nov. 1991.
142. D.P. Haas, "Intelligent TDRs Score High on T-1 Tests," *Telephony*, Vol. 221, No. 12, Sept. 16 1991, pp. 18–20, 23, 24.
143. W.R. Campbell, "No Small Change," *TE&M*, Vol. 94, No. 23, Dec. 1, 1990, pp. 54–56.
144. V. Vittore, "OTDRs Offer Three Tools in One," *TE&M*, Vol. 94, No. 23, Dec. 1, 1990, pp. 64, 67.
145. T. Yount, "OTDRs for Common Folk," *TE&M*, Vol. 94, No. 21, Nov. 1, 1990, pp. 56, 58, 60.
146. American National Standard for Telecomm.—In-service Nonintrusive Measurement Device (INMD) Voice Service Measurement, ANSI T1.122-1991.
147. J. O'Neill, "A New Monitoring Technology For All Networks," *Telephony*, Vol. 216, No. 10, Mar. 6 1989, pp. 52–54.
148. R.A. Sutton and J.P. Boyles, "Get More Control Over The Loop," *Telephony*, Vol. 216, No. 8, Feb. 1989, pp. 46–48.
149. R.A. Sutton et al., "Tracing Quality with Live Calls," *TE&M*, Vol. 90, No. 22, Nov. 15 1986, pp. 88–90, 94.
150. R. Avery, "The Line-Call Test System Does Variety of Jobs," *TE&M*, Vol. 90, No. 22, Nov. 15 1986, pp. 94–96.
151. E.C. Mosley, "DOBS Keeps the Echo out of LATA Switched Access Network," *Telephony*, Vol. 210, No. 20, May 19 1986, pp. 66–68, 72, 74, 76, 78, 82, 84.
152. L.S. DiBiaso, Technical and Product Information Package on DB Telco Service Evaluation System, DB Telco Inc., Telecommunications Network Services, at Village West, Box 385, Gilford, NH 03247, (606) 524-6133.
153. J. Gruber, "Estimating the Transmission Grade-of-Service of Telephone Connections with Multiple Echo Paths" *GLOBECOM Rec.*, Vol. 1, Hollywood, Florida, Nov./Dec. 1988, pp. 14.6.1–14.6.5.
154. L.S. DiBiaso, "User Reaction to Transmission Quality: A Cost Model," *ICC Rec.*, Vol. 2, Seattle, June 1987, pp. 34.6.1–34.6.5.
155. V. Sahin et al., "Telecommunications Management Network (TMN) Architecture and Interworking Designs," Paper in ref. [122], pp. 685–696.
156. W. Stallings, "CCITT Standards Foreshadow Broadband ISDN," *Telecomm.*, Vol. 24, No. 3, Mar. 1990, pp. 29, 30, 32, 34, 41.
157. IEEE Journal on Sel. Areas in Comm., Issue on Broad-band Communications Systems, Vol. SAC-4, No. 4, July 1986.
158. T.C. Miller, "SONET and BISDN: A Marriage of Technologies," *Telephony*, Vol. 216, No. 20, May 15, 1989, pp. 32–35, 38.
159. *TE&M*, Vol. 94, No. 15, Aug. 1 1990. Issue on "Transmission Trends: Deploying Broadband in the Local Loop."
160. S.E Minzer, "Broadband ISDN and Asynchronous Transfer Mode (ATM)," *IEEE Comm. Mag.*, Vol. 26, No. 6, June 1988, pp. 16–26.
161. D. Fisher, P. Maynard and S. Alexander, "Broadband ISDN," *Comp. Comm.*, Vol. 11, No. 4, Aug. 1988, pp. 185–190.
162. *IEEE Network*, Vol. 3, No. 1, Jan. 1989. Special Issue on Broadband Networks.
163. M. De Prycker, "Evolution from ISDN to BISDN: A Logical Step Towards ATM," *Comp. Comm.*, Vol. 12, No. 3, June 1989, pp. 141–146.
164. M. Lee, "What Will Drive Residential Broadband?," *Telephony*, Vol. 219, No. 8, Aug. 13, 1990, pp. 20, 21.
165. *TE&M*, Vol. 94, No. 16, Aug. 15, 1990. Issue on Fiber in the Local Loop.
166. T.J. Robe and K.A. Walsh, "A SONET STS-3c User Network Interface Integrated Circuit," *IEEE Journal on Sel. Areas in Comm.*, Vol. 9, No. 5, June 1991, pp. 732–740.
167. G.M. Woodruff and R. Kositpaiboon, "Multimedia Traffic Management Principle for Guaranteed

ATM Network Performance," *IEEE Journal on Sel. Area in Comm.*, Vol. 8, No. 3, April 1990, pp. 437-446.
168. G. Fioretto *et al.*, "ATM Based Network Transport Service," *ICC Rec.*, Vol. 3, Atlanta, April 1990, pp. 318.3.1-318.3.5.
169. A. Forcina *et al.*, "A Strategy for ATM Introduction into Public Networks," *ICC Rec.*, Vol. 4, Atlanta, April 1990, pp. 342.5.1-342.5.6.
170. K. Asatani and M. Taka, "Quality Design Approaches for Present and Future Telecommunication Services," *ICC Rec.*, Vol. 4, Atlanta, April 1990, pp. 334.6.1-334.6.5.
171. N. Kitawaki *et al.*, "Speech Coding Technology for ATM Networks," *IEEE Comm. Mag.*, Vol. 28, No. 1, Jan. 1990, pp. 21-27.
172. M. Nomura *et al.*, "Basic Characteristics of Variable Rate Video Coding in ATM Environment," *IEEE Journal on Sel. Areas in Comm.*, Vol. 7, No. 5, June 1989, pp. 752-760.
173. J.F. Mallenauer, "Metropolitan Area Network Update: The Global LAN is Getting Closer," *Data Comm.*, Vol. 18, No. 16, Dec. 1989, pp. 111, 112, 115, 119, 120.
174. G.P. Smith, "Foundations for Tomorrow's Private Network," *Bus. Comm. Rev.*, Vol. 20, No. 7, July 1990, pp. 35-40.
175. T. Nguyen *et al.*, "Evolution Path to ATM," *Telecomm.*, Vol. 24, No. 7, July 1990, pp. 26, 28, 30, 31.
176. C.A. Johnson and H.J. Chao, "The ATM Layer Chip: An ASIC for B-ISDN Applications," *IEEE Journal on Sel. Areas in Comm.*, Vol. 9, No. 5, June 1991, pp. 741-750.
177. American National Standard for Telecomm.: Digital Hierarchy—Electrical Interfaces," ANSI T1.102 −1987. Also, Draft T1.102-199x.
178. American National Standard for Telecomm.: Digital Hierarchy—Formats Specifications," ANSI T1.107-1988. Also, Supp. ANSI T1.107a-1990.
179. American National Standard for Telecomm.: Digital Hierarchy—Synchronous DS3—Formats Specifications," T1.103-1987. Also, Supp. T1.103a-1990.
180. American National Standard for Telecomm.—Carrier-to-Customer Installation—DS1 Metallic Interface, ANSI T1.403-1989.
181. American National Standard for Telecomm.—Carrier-to-Customer Installation—DS3 Metallic Interface, ANSI T1.404 −1989.
182. CCITT Recs. G.702 to G.709 on interfaces and hierarchical bit rates, Blue Book, Vol. III, Fas. III.3, Geneva 1989. Also, update, COM XVIII-R 33-E, 1991.
183. "Extended Superframe Format (ESF) Interface Specification," Bellcore TR-194, Issue 1, Dec. 1987.
184. "Synchronous DS3 Format Interface Specification," Bellcore TR-021, Issue 1, June 1984.
185. CCITT I.400 Series Recs. on ISDN User-Network Interfaces, *Blue Book*, Vol. III, Fas. III.8, Geneva, 1989.
186. G.R. Werth, "The Fundamentals of DS3," *Telecomm.*, Vol. 23, Part I, No. 8, Aug. 1989, pp. 67, 68, 70, 72-75; Part II, No. 9, Sept. 1989, pp. 73, 75, 76, 82.
187. S. Fleming, "To Know SONET, Know Your VTs," *TE&M*, Vol. 93, No. 12, June 15, 1989, pp. 62, 66-69.
188. K. H. Molloy, "ABCs of Digital Formats," *TE&M*, Vol. 93, No. 12, June 15 1989, pp. 70, 74, 75, 141.
189. R. Ballart and Y. Ching, "SONET: Now It's the Standard Optical Network," *IEEE Comm. Mag.*, Vol. 27, No. 3, Mar. 1989, pp. 8-15.
190. R.J. Boehm, "The SONET Interface and Network Applications," *GLOBECOM Rec.*, Vol. 2, Hollywood, Florida, Nov./Dec. 1988, pp. 30.4.1-30.4.5.
191. T.P.J. Flanagan, B.E. Allen and G.C. Copley, "Application Benefits of SONET Networking," *GLOBECOM Rec.*, Vol. 2, Hollywood, Florida, Nov./Dec. 1988, pp. 30.3.1-30.3.5.
192. N.B. Sardesara *et al.*, "SONET Intra-Office Interconnect Signal," *GLOBECOM Rec.*, Vol. 2, Hollywood, Florida, Nov./Dec. 1988, pp. 30.5.1-30.5.7

193. IEEE Comm. Mag., Vol. 18, No. 8, Aug. 1990. Issue on Global Deployment of SDH-Compliant Networks.
194. Telephony, Vol. 218, No. 26A, June 1990. Special Supp. on SONET: Metamorphosis of the Network.
195. R.B. Misra and H.M. Jablecki, "IDLC and SONET: A Winning Combination," *Telephony*, Vol. 216, No. 11, Mar, 13 1989, pp. 44, 46, 48–50.
196. H. Nakajima, "SONET: Progress Report from Japan," *Telecomm.*, Vol. 24, No. 8, Aug. 1990, pp. 58–61.
197. B. Stoffels and A. Stewart, "SONET: Building Connectivity," pp. 4–7, Supp. to *TE&M*, Vol. 95, No. 9, May 1, 1991.
198. R.W. Cubbage and F.R. Martin, "A Progress Report: SONET Standards," pp. 10, 11. Supp. to *TE&M*, Vol. 95, No. 9, May 1, 1991.
199. G.T. Hawley and F.S. Nabavi, SONET: This Isn't OZ Anymore," *TE&M*, Vol. 95, No. 9, May 1, 1991, pp. 45, 46, 51, 52.
200. K. Head, "A SONET Milestone for the History Books," *Telephony*, Vol. 219, No. 25, Dec. 10, 1990, pp. 24, 25, 27.
201. R.J. Boehm, "Basic Concepts of the SONET Interface," *NCF Rec.*, Vol. 44, Chicago, Oct. 1990, pp. 128–131.
202. E.V. Hird, "Sacrifices in the Name of Simplification," *TE&M*, Vol. 95, No. 1, Jan. 1, 1991, pp. 43–47.
203. P. Purton, "SDH Breaks the Network Barrier," Special Supp. on "Transmission," *Telephony*, Nov. 1990, pp. 24, 25.
204. R.W. Cubbage, "SONET's Many Phases: A Progress Report," *TE&M*, Vol. 94, No. 19, Oct. 1 1990, pp. 79, 80, 85, 87.
205. F.J. Martin, "SONET Makes Cents," *Telecomm.*, Vol. 25, No. 7, July 1991, pp. 33, 34, 38.
206. CCITT Recs. G.711 to G.755 on codecs and multiplexers, *Blue Book*, Vol. III, Fas. III.3, Geneva, 1989.
207. CCITT Recs. G.781 to G.784 on "General Aspects of Digital Transmission Systems Terminal Equipments," Geneva, 1990.
208. "Asynchronous Digital Multiplexes: Requirements and Objectives," Bellcore TR-09, Issue 1, May 1986.
209. B. Fleury, "Asychronous High Speed Digital Multiplexing," *IEEE Comm. Mag.*, Vol. 24, No. 8, Aug. 1986, pp. 17–25.
210. "Digital Cross-Connect Systems Requirements and Objectives," Bellcore TR-170, Issue 1, Nov. 1985.
211. "Wideband and Broadband Digital Cross-Connect Systems Generic Requirements and Objectives," Bellcore TR-233, Issue 2, Sept. 1989.
212. "SONET Add-Drop Multiplex Equipment (SONET ADM) Generic Criteria for a Self-Healing Ring Implementation," Bellcore TR-496, Issue 3, Aug. 1990.
213. "Synchronous DS3 Add-Drop Multiplex (ADM 3/X) Requirements and Objectives," Bellcore TR-010, Issue 1, Feb. 1988.
214. K. Nagaraj, J. Gruber, J. Leeson and B. Fleury, "Improvements in Availability and Error Performance of SONET Compared to Asynchronous Transport System," *ICC Rec.*, Vol. 1, Atlanta, April 1990, pp. 224.1.1 to 224.1.7. Also, *Telephony*, Vol. 218, No. 26A, June 25 1990, pp. 34, 36, 38, 40, 42, 44.
215. S. Fleming, "Getting Ready for T3 Networking," *Data Comm.*, Vol. 18, No. 11, Sept. 1989, pp. 82–86, 88, 90, 92, 94.
216. G. Power, "Seize Control with a Digital Cross-Connect," *Telephony*, Vol. 215, No. 8, Aug. 22, 1988, pp. 38, 42, 46.
217. E.V. Hird, "Emerging High Capacity Digital Cross-Connect T1 Systems," *TE&M*, Vol. 92, No. 1, Jan. 1 1988, pp. 51, 54–58.

218. P.F. Thompson, "Tomorrow is Today in DS3 Networking," *Telephony,* Vol. 212, No. 24, June 15 1987, pp. 54, 55, 59, 60, 64.
219. K.D. Cooley et al., "Wideband Virtual Networks," *Telecomm.,* Vol. 21, No. 2, Feb. 1987, pp. 65, 66, 70, 72, 84.
220. E.V. Hird, "Saving $$$ with T1 DCS," *TE&M,* Vol. 91, No. 1, Jan. 1, 1987, pp. 66, 71, 74–76.
221. M.W. Harvey and J.E. Rathmann, "Network Planning for the SYNTRAN Add/Drop Multiplex (ADM)" *GLOBECOM Rec.,* Vol. 3, Houston, Dec. 1986, pp. 45.3.1–45.3.5.
222. R.J. Conway and T.R. Crawford, "Corporate Network Design Options," *Telecomm.,* Vol. 20, No. 11, Nov. 1986, pp. 38–40, 44, 46, 48, 53.
223. Papers in Session 18 on "Services Provided by Digital Cross Connect Systems," *GLOBECOM Rec.,* Vol. 2, New Orleans, Dec. 1985.
224. L.M. Martz and A.J. Osofsky, "Operations Impact of New Technology: A Case Study of the Digital Access and Cross-Connect System," *ICC Rec.,* Vol. 1, Philadelphia, June 1982, pp. 10.2.1–10.2.5.
225. D.R. Gibson and J.M. Curry, "New Techniques for Digital Transmission," *Telecomm.,* Vol. 22, No. 1, Jan. 1988, pp. 68, 69, 71.
226. R. Ballart and J.M. Rydzak, "Synchronous Transmission at DS3," *Telephony,* Vol. 211, No. 26, Dec. 29 1986, pp. 36, 40–42. Also, *Telephony,* Vol. 210, No. 20, May 19 1986, pp. 98, 100, 104, 106.
227. J.K. Conlisk, "What Will Tomorrow Bring?," *Telephony,* Vol. 216, No. 25, June 19, 1989, pp. 22, 23, 26, 30, 31.
228. P. Runyon, "Fiber Moves into the Feeder Plant," *Telephony,* Vol. 216, No. 1, Jan. 2, 1989, pp. 40, 41, 44, 45.
229. J.S. McConnell III, "The Power of a SONET ADM," *Telephony,* Vol. 218, No. 16, April 16, 1990, pp. 130, 132, 134, 136.
230. H.C. Blackmon, "Digital Cross-Connects: A Gateway to the Synchronous Optical Network," *Telecomm.,* Vol. 24, No. 7, July 1990, pp. 50–54.
231. S. Makgill, "Utilizing the Power of SONET Add/Drop Multiplexers," *GLOBECOM Rec.,* Vol. 1, San Diego, Dec. 1990, pp. 303.7.1–303.7.6.
232. "The DCS Debate—Parts 1, 2 and 3," *Telephony,* Vol. 221, No. 8, Aug. 19 1991, pp. 22, 23, 26, 30, 32–34, 36.
233. M. Gawdon, "Customer-Controlled Network Management," *Telecomm.,* Vol. 21, No. 7, July 1987, pp. 36–40.
234. P.A. Barth, "Customer Controlled Reconfiguration," *Telecomm.,* Vol. 20, No. 11, Nov. 1986, pp. 71, 72, 74.
235. R.K. Berman and R.W. Lawrence, "Customer Network Management and Control Services," *ISSLS Rec.,* Sept./Oct. 1986, Tokyo, pp. 94–98.
236. J. Kane, "Taking Full Advantage of the New T1 Network Services," *Telecomm.,* Vol. 19, No. 9, Sept. 1985, pp. 50k, 50p, 56t.
237. E.V. Hird, "The New DCS Buzz Words: Restoration, Rerouting," *TE&M,* Vol. 93, No. 12, June 15 1989, pp. 50–52.
238. C. Wilson and M. Warr, "The Search for Customer Control," *Telephony,* Vol. 218, No. 6A, Feb. 5 1990, pp. 28–30.
239. Papers in Session 403 on "Transport Reliability Enhancement," *GLOBECOM Rec.,* Vol. 1, San Diego, Dec. 1990.
240. W.D. Grover and T.E. Moore, "Precision Time-Transfer in Transport Networks Using Digital Cross-Connect Systems," *IEEE Trans. on Comm.,* Vol. 38, No. 9. Sept. 1990, pp. 1325–1332. Also, *GLOBECOM Rec.,* Vol. 3 Hollywood, Florida, Nov./Dec. 1988, pp. 47.2.1–47.2.5.
241. W. Beller, "Telcos in the Trenches: Defensive Strategies for Offensive Actions," *TE&M,* Vol. 95, No. 14, July 15 1991, pp. 37–40.
242. A. Prest, "Built-In Reliability," *TE&M,* Vol. 95, No. 17, Sept. 1 1991, pp. 74, 76, 79, 80.

243. D.S Eckel, "Network Survivability: Must Reliability Cost More?" *Telephony*, Vol. 221, No. 13, Sept. 23, 1991, pp. 30, 32, 34, 36.
244. S. Bootman, "The Integration of SONET into Switching Equipment" The Merger of Transmission and Switching," *ICC Rec.*, Vol. 1, Boston, June 1989, pp. 14.3.1–14.3.6.
245. "SONET Digital Switch Trunk Interface Criteria," Bellcore TR-782, Issue 2, Sept. 1989.
246. American National Standard for Telecomm.—ISDN Management—Data Link and Network Layer, ANSI T1.218-1991.
247. J.I. Falek and M.A. Johnston, "Standards Makers Cementing ISDN Subnetwork Layers," *Data Comm.*, Vol. 16, No. 11, Oct. 1987, pp. 237, 238, 241, 242, 245, 246, 249, 250, 253–255.
248. T.T. Bradley and T. Suda, "Performance of Error Recovery Schemes in a Fast Packet Switching Network," *ICC Rec.*, Vol. 2, Atlanta, April 1990, pp. 308.2.1–308.2.6.
249. S. Mazumdar and A.A. Lazar, "Monitoring Integrated Networks for Performance Management," *ICC. Rec.*, Vol. 2, Atlanta, April 1990, pp. 302.1.1–302.1.6.
250. K.A. Roa, "ATM Adaptation Layer and Its Functions," Contribution T1S1.1/88-482, AT&T Tech., Oct. 1988.
251. P.J. Severino, "T1 for LAN/WAN Links," *Telecomm*, Vol. 22, No. 10, Oct. 1988, pp. 39–42. Also, *Telephony*, Vol. 215, No. 12, Sept. 19 1988, pp. 46, 48–50.
252. K. Takami and T. Takanaka, "Architecture and Functional Aspects of a Multi-Media Packet Switched Network," *ISS Rec.*, Vol. 2, Phoenix, Mar. 1987, pp. B6.4.1–B.6.4.5.
253. "Generic Systems Requirements in Support of Switched Multi-megabit Data Service," Bellcore TA-772, Issue 3, Oct. 1989. Also, Supp. 1, Dec. 1990.
254. "Local Access System Generic Requirements, Objectives, and Interface in Support of SMDS Service," Bellcore TA-773, Issue 2, Mar. 1990. Also, Supp. 1, Dec. 1990.
255. "Switched Multi-megabit Data Service (SMDS) Operations Technology Network Element Generic Requirements," Bellcore TA-774, Issue 3, Feb. 1991.
256. T.M. Chen and D.G. Messerschmitt, "Integrated Voice/Data Switching," *IEEE Comm. Mag.*, Vol. 26, No. 6, June 1988, pp. 16–26.
257. ISS Procs., Phoenix AZ., Mar. 1987. Vol. 2 Sessions B5 and B6 on Fast Packet Switching, and C5 on Switching/Network Projections. Vol. 3 Sessions A7 and A8 on Network Architectures and Broadband Networks, and A9 and A10 on Broadband Switching.
258. N. Lippis, "Frame Relay Redraws the Map for Wide-Area Networks," *Data Comm.*, Vol. 19, No. 9, July 1990, pp. 80–82, 84, 88, 90, 92, 94.
259. B. Bhushan, "User's Guide to Frame Relay," *Telecomm.*, Vol. 24, No. 7, July 1990, pp. 39, 40, 42, 43.
260. L. Bishop, "Use SONET for In-Service Testing," *TE&M*, Vol. 93, No. 21, Nov. 1, 1989, pp. 53–55, 58.
261. "Centralized Fault Location for Digital Transmission Systems," Bellcore TA-053, April 1985. Also, *GLOBECOM Rec.*, Vol. 1, Atlanta, Nov. 1984, pp. 13.2.1–13.2.8.
262. E. Elkins *et al.*, "Support Systems Play Catch-up," *Telephony*, Vol. 218, No. 8, Feb. 19, 1990, pp. 24, 25, 28, 30.
263. W. Voisin and J. Ferri, "System Predicts Repeater Faults," *TE&M*, Vol. 91, No. 1, Jan. 1, 1987, pp. 58, 59.
264. "SONET Regenerator (SONET RGTR) Equipment," Bellcore TA-917, Issue 2, Mar. 1990.
265. C.T. Prok, "When Bipolar Violations Tests Won't Do," *TE&M*, Vol. 90, No. 22, Nov. 15, 1986, pp. 105, 106.
266. S.C. Taylor, "Error Detection in T1 Transmission," *Telecomm.*, Vol. 20, No. 10, Oct. 1986, pp. 78, 80–82.
267. R.C. Troutman, "T1 Circuits—In-Service Monitoring and Out-of-Service Testing," *Telecomm.*, Vol. 20, No. 7, July 1986, pp. 42, 43, 44, 46, 48.
268. A. Bortz, "Monitoring System Keeps T-Carrier Network Functioning and in Good Shape," *Telephony*, Vol. 204, No. 20, May 16 1983, pp. 34, 36, 38.

269. B. Nelson and B. Sharp, "C-Bit Parity Frames T3 for Maximum Management," *Data Comm.*, Vol. 20, No. 10, Aug. 1991, pp. 79, 80, 82, 84, 86.
270. "DS1 Rate Digital Service Monitoring Unit Functional Specifications," Bellcore TA-147, Issue 1, Oct. 1987.
271. K. Cadieux et al., "A New Network Element for Performance Monitoring and Test Access Spanning the Digital Hierarchy," *ICC Rec.*, Vol. 4, Atlanta, April 1990, pp. 334.4.1–334.4.5.
272. M.A. Wright, "Don't Leave Your Control Logic in the Dark," *Telephony*, Vol. 217, No. 12, Sept. 11 1989, pp. 57, 61, 62, 64.
273. F.S. Knight, "T1 Surveillance System From Clear Communications," *Bus. Comm. Rev.*, Vol. 19, No. 3, Mar. 1989, pp. 70, 71.
274. "Requirements for Interfacing Digital Terminal Equipment to Services Employing the Extended Superframe Format," AT&T PUB 54016, Sept. 1989.
275. C.M. Brown, "Sizing Up ESF Service Units," *Data Comm.*, Vol. 19, No. 6, May 1990, pp. 96–99.
276. "Transmission Systems for Communications," Bell Telephone Labs. Inc., 5th ed., Chapter 30 on "Baseband Digital Transmission," 1982, pp. 742–746.
277. P.A. Johnson and D.R. Walker, "A Standard for Clear Channel Capability," *TE&M*, Vol. 90, No. 16, Aug. 15, 1986, pp. 112, 115.
278. E.H. Lee, "A BOC Explains Where the Bits Went," *Data Comm.*, Vol. 15, No. 7, June 1986, pp. 141–143.
279. J. Parsons, "Are You Using the Right Stress Patterns?," *Telephony*, Vol. 218, No. 16, April 16, 1990, pp. 150, 152, 155, 156, 158.
280. "High-Capacity Digital Service (1.544 Mb/s) Interface Generic Requirements for End Users," Bellcore TR-054, Issue 1, April 1989.
281. J.A. Philips, "Ten Things You Should Know about T1 Testing," *Telecomm.*, Vol. 24, No. 8, Aug. 1990, pp. 31–33, 35, 36.
282. Committee T1—Telecommunications, "The Performance of AMI Signals Through B8ZS Optional Equipment Across Network Boundaries," TR No. 2, Mar.1989.
283. P.R. Strauss, "Big T1 Giveaway: New Option Adds 192 kbit/s of Bandwidth," *Data Comm.*, Vol. 18, No. 2, Feb. 1989, pp. 49–52.
284. S.S. Gorshe and G.J. Beveridge, "A Robust Technique for Providing Clear Channel Capability over Existing and Future Digital Transmission Facilities," *ICC Rec.*, Vol. 3, Seattle, June 1987, pp. 42.5.1–42.5.10. Also, *TE&M*, Vol. 90, No. 5, Mar. 1, 1986, pp. 63–67, 68; also, No. 4, Feb. 15, 1986, pp. 111–117.
285. E.E. Mier, "Binary Dilemma: How Many 'Ones' Does T1 Need?" *Data Comm.*, Vol. 15, No. 7, June 1986, pp. 53, 54, 56.
286. A.A. Knapp and J.W. Miller, "Implementing ISDN: Issues, Trials, and Answers," *Telecomm.*, Vol. 20, No. 5, May 1986, pp. 48–50, 52, 80.
287. "ACCUNET T1.5 Service Description and Interface Specifications," AT&T PUB 62411, Issue 2, Dec. 1988. Also, Adden. 1, Mar. 1990. Also, Issue 3, Dec. 1990.
288. D.D. Ferguson and W.D. Hutcheson, "New Digital Network Service Capabilities," *ICC Rec.*, Vol. 2, Seattle, June 1987, pp. 19B.8.1–19B.8.5.
289. S.C. Taylor, "Why Fe?," *Telecomm.*, Vol. 19, No. 9, Sept. 1985, pp. 56y, 56z, 56bb.
290. A.B. Sripad and J.W. Smith, "Performance of the Cyclic Redundancy Check Code in the New 1.544 Mb/s Extended Framing Format," *GLOBECOM Rec.*, Vol. 2, Atlanta, Nov. 1984, pp. 31.3.1–31.3.3.
291. J.W. Smith, "Signaling of the T1-Carrier System," *NCF Rec*, Vol. 37, Oct. 1983, pp. 174–177.
292. W. Radwill, "Fe Format: A Common Link for the Change from 56 to 64 kb/s Digital Network Capability," *Telephony*, Vol. 202, No. 25, June 21 1982, pp. 34–36.
293. D.M. McWhorter and E.L. Varma, "A New 'Health' Check for the Digital Network," *Bell Labs. Rec.*, Vol. 60, No. 2, Feb. 1982, pp. 37–39.

294. P.T. Tux, "T1 Testing and the Standard Answer," *TE&M,* Vol. 94, No. 1, Jan 1 1990, pp. 50, 53, 57.
295. W.J. Buckley, "Which Standard for the ESF Data Link?" *TE&M,* Vol. 93, No. 20, Oct. 1985.
296. M. Lefkowitz, "A Tale of Two Standards—the T1 ESF (R)Evolution," *Data Comm.,* Vol. 18, No. 13, Oct. 1989, pp. 143, 144, 147,148, 151.
297. K. Stauffer and A. Brajkovic, "DS1 Extended Superframe Format and Related Performance Issues," *IEEE Comm. Mag.,* Vol. 27, No. 4, April 1989, pp. 19–23.
298. E.V. Hird, "A Decision is Needed on T1 Performance," *TE&M,* Vol. 93, No. 1, Jan. 1, 1989, pp. 39, 40, 44, 45.
299. S. Bootman, "Some Things Never go out of Style," *Telephony,* Vol. 215, No. 23, Dec 5, 1988, pp. 26, 29, 30.
300. P. Marshall, "T1-Liftoff," *Telephony,* Vol. 215, No. 23, Dec. 5, 1988, pp. 38–40.
301. "ESF Making a Hit in T1 Diagnostics; New Standard Could Make it Bigger," pp. 58, 60. "T1 Muxes in State of Flux Over Bringing ESF to Users," pp. 62, 64. *Data Comm.,* Vol. 17, No. 6, June 1988.
302. S.C. Taylor, "ESF: Its Time Has Come," *TE&M,* Vol. 91, No. 1, Jan. 1, 1987, pp. 98, 99.
303. E.E. Mier, "Controversy Brews over AT&T T1 Issue," *Data Comm.,* Vol. 15, No. 12, Nov. 1988, pp. 53, 54, 56, 58.
304. B.E. Worne; "An HMO for T1 Circuits," *TE&M,* Vol. 90, No. 22, Nov. 15, 1986, pp. 111–114.
305. F. Bradley; "The Hidden Treasures of ESF," *Data Comm.,* Vol. 15, No. 10, Sept. 1986, pp. 204–206, 209, 210, 213.
306. "Application Guidelines for Use of the DS1 Extended Superframe Format Data Link," ANSI Draft Tech. Rep. T1E1/90-057R2, T1E1.2/90-030R4.
307. J.M. Gregory, "AIN Poses Maintenance Challenges," *Telephony,* Vol. 218, No. 16, April 16, 1990, pp. 146–148.
308. G. Mauk, "DS1 Testing Sees the Light," *Telephony,* Vol 219, No. 5, July 30, 1990, pp. 41, 42.
309. P.T. Rux and C.V. Liles, "ESF: Rx for Healthy T1 Nets," *Data Comm.,* Vol. 19, No. 6, May 1990, pp. 81, 82, 84, 86, 88, 90, 92, 94.
310. P. Corr, "ESF: Functionality and Benefits," *Telecomm.,* Vol. 24, No. 12, Dec. 1990, pp. 63–65.
311. P. Benowitz *et al.,* "Digital Data System: Digital Multiplexers," *BSTJ,* Vol. 54, No. 5, May/June 1975, pp. 893–918.
312. "Digital Data System—Private Line Service—Overall Description," Bell System Practice, Section 314-500-100, Issue 3, July 1979. Section 3 on DDS Signal Structure, pp. 3–7.
313. T.H. Murray, "The Evolution of DDS Networks: Part 1," *Telecomm.,* Vol. 23, No. 2, Feb. 1989, pp. 39, 40, 42, 44, 46, 47.
314. S.R. Blackwell *et al.,* "Switched Digital Services: How the RBOCs Make Them Go," *Bus. Comm. Rev.,* Vol. 19, No. 1, Jan. 1989, pp. 74–77.
315. E.E. Mier, "AT&T Planning to Put DDS into T1," *Data Comm.,* Vol. 15, No. 4, April 1986, pp. 55, 56, 58, 60, 62.
316. F.M. Gilvinson, S.S. Corkovic and J.A. Kennedy, "Monitoring and Management Systems for DATAROUTE," *NTC Record,* Vol. 3, New Orleans, Nov./Dec. 1981, pp. F6.4.1–F6.4.6.
317. J. Paremba, "ADN: The Next Step on the DDS Ladder," *TE&M,* Vol. 93, No. 18, Sept. 15, 1989, pp. 48, 49, 51, 52.
318. P.R. Strauss, "From the Lamented Victoria to a Highly Cunning DSU," *Data Comm.,* Vol. 18, No. 3, Mar. 1989, pp. 54, 56, 58.
319. R.W. Cubbage and P.A. Littlewood, "To Float or not to Float? That is the Question for SONET Mapping," *Telephony,* Vol. 217, No. 5, Oct. 2, 1989, pp. 52, 54, 56, 58.
320. "Generic Requirements for the Public Switched DS1/Switched Fractional DS1 Service Capability," Bellcore TA-1068, Issue 2, April 1991.
321. M.L. McLoughlin and N.J. Muller, "Fusing ISDN and Fractional T-1: Will it be Another By-Pass Threat?," *Telephony,* Vol. 218, No. 6A, Feb. 5, 1990, pp. 42, 44, 46, 48.

322. N. Lippis, "At AT&T, Much Ado About Switched Services," *Data Comm.*, Vol. 19, No. 10, Aug. 1990, pp. 29, 30.
323. J.T. Reynolds, "Optimizing Performance for Fractional T1," *Telecomm.*, Vol. 24, No. 8, Aug. 1990, pp. 27, 28.
324. T.H. Murray, "Fractional T1: New Twists in Old Operations," *Telecomm.*, Vol. 24, No. 6, June 1990, pp. 63, 65, 66, 68, 70.
325. "Generic Requirements for the Switched DS1/Switched Fractional DS1 Service Capability From an ISDN Interface (SWF-DS1/ISDN)," Bellcore TA-1203, Issue 1, April 1991.
326. T.H. Murray "Line Data Analysis for T1 Testing," *Telecomm.*, Vol. 23, No. 1, Jan. 1989, pp. 78–83.
327. B.E. Worne, A. Darling and T. Murray, "Installation and Maintenance Measurement Parameters and Techniques for Digital Circuits, Facilities and Networks" Contribution T1M1.3/88-055, T1E1.2/89 −043, Sept. 5 1988.
328. US CCITT SGXVIII Contrib., "A VPC/VCC Performance Monitoring Scheme," T1S1.5/91-141, June 1991.
329. H. Breuer, "ATM-Layer OAM: Principles and Open Issues," *IEEE Comm. Mag.*, Vol. 29, No. 9, Sept. 1991, pp. 75–78.
330. J. Anderson, and M.D. Nguyen, "ATM—Layer OAM Implementation Issues," *IEEE Comm. Mag.*, Vol. 29, No. 9, Sept. 1991, pp. 79–81.
331. "Relation of Performance Monitoring Based on OAM Cells to the I.35B Performance Parameters," Bellcore Contrib. T1Q1.3/91-042, April 1991.
332. E.A. Newcombe and S. Paspathy, "Error Rate Monitoring For Digital Communications," *Proc. IEEE*, Vol. 70, No. 8, Aug. 1982, pp. 805–868.
333. American National Standard for Telecomm.—Network Performance Parameters for Dedicated Digital Service—Definitions and Measurement Methods. ANSI T1.503-1989.
334. American National Standard for Telecomm.—Network Performance Parameter for Circuit-Switched Digital Services—Definition and Measurement Methods. ANSI T1.507-1990.
335. CCITT Rec. G.821, "Error Performance on an International Digital Connection Forming Part of an ISDN," *Blue Book*, Vol. III, Fas. III.3, Geneva, 1989, pp. 26–35.
336. "Alarm Indication Signal Requirements and Objectives," Bellcore TR-191, Issue 1, May 1986.
337. "Digital Data System Channel Interface Specification," AT&T PUB. 62310, Issue 2, Nov. 1987. Also, update Dec. 1989.
338. A. Tabatabai and M. Malek-Zavarei, "Generic Design of Digital Service Circuits," *GLOBECOM Rec.*, Vol. 1, Atlanta, Nov. 1984, pp. 13.6.1–13.6.5.
339. J.H Eu, "An Evaluation of Error Performance Estimation Schemes for DS1 Transmission Systems Carrying Live Traffic," *IEEE Trans. on Comm.*, Vol. 38, No. 3, Mar. 1990, pp 384–391.
340. American National Standard for Telecomm.—ISDN—Basic Access Interface for Use on Metallic Loops for Application on the Network Side of the NT (Layer/Specification), ANSI T1.601-1988.
341. J.H. Eu and W.W. Rollins, "A Performance Review of Out-of-Frame Detection Schemes for DS1 Signals," *IEEE Trans. on Comm.*, Vol. 39, No. 6, June 1991, pp. 1004–1009.
342. Synchronization—LATA Switching System Generic Requirements (LSSGR): Section 18, Bellcore TR-518, Issue 2, July 1987.
343. J. Gruber, "Alternative to Monitoring of SONET Pointer Justifications," Contribution T1X1.5/89-220R1, T1M1.3/89-096, NTI, Jan. 1990.
344. "DS1 Automatic Facility Protection Switching (AFPS) Features for Digital Terminal Systems: Requirements and Objectives," Bellcore TA-435, Issue 1, Feb. 1987.
345. A.H. Miller, "Take it Easy—T1 Protection's Here," *TE&M*, Vol. 95, No. 3, Feb. 1 1991, pp. 72, 74, 76–78.
346. W.W. Rollins; "Error Second Measurements as Performance Indicators," *Telecomm.*, Vol. 14, No. 10, Oct. 1980, pp. 80, 82, 132.
347. R.E. Mallon, "Detection of Bursty Error Conditions through Analysis of Performance Information," *GLOBECOM Rec.*, Vol. 3, Tokyo, Nov. 1987, pp. 51.1.1–51.1.5.

348. CCITT SG XVIII TD 30 to Quest. 6, "Report on Error Performance Issues," Geneva, June 1991.
349. CCITT SG XVIII Draft Rec. G.82x on "Error Performance Parameters and Objectives for Continuous Bit Rate International Digital Connections at or above the Primary Rate," Annex 1 to ref. [369].
350. "Proposal for Node-to-Node Synchronization Status Messages to Compliment OS Surveillance of a Synchronization Network," NTI Contrib. T1X1.3/91-015, Feb. 1991.
351. J. Gruber, "Proposal for Monitoring Unavailable Time and Inhibiting Performance Monitoring Parameter Counts at DS1, DS3 and SONET," Contribution T1M1.3/90-007, NTI, Mar. 1990.
352. F. Inumaro et al., "Error Performance Estimation for Digital Circuits Using In-Service Monitoring Information," *GLOBECOM Rec.*, Vol. 3, Dallas, Nov. 1989, pp. 51.2.1–51.2.5.
353. G.S. Fang; "Alarm Statistics of the Violation Monitor and Remover," *BSTJ*, Vol. 55, No. 8, Oct. 1976, pp. 1197–1217.
354. K.A. Witzke and C. Leung, "A Comparison of Some Error Detecting CRC Code Standards," *IEEE Trans. on Comm.*, Vol. COM-33, No. 9, Sept. 1985, pp. 956–998.
355. W.W. Rollins; "Confidence Level in Bit Error Rate Measurement," *Telecomm.*, Vol. 11, No. 12, Dec. 1977, pp. 67, 68.
356. A. Haime; "The Chernoff Bound Applied to Error Rate Measurements," *Australian Telecomm. Rev.*, Vol. 15, No. 2, 1981, pp. 13.16.
357. J.S. Mayo; "A Bipolar Repeater for Pulse Code Modulation Signals," *BSTJ*, Vol. 41, No. 1, Jan. 1962, pp. 25–97.
358. S.D. Akers, S.C. Donn, S.R. Gregg and S. Ravikumar, "Uses of In-Service Monitoring Information to Estimate Customer Service Quality," *GLOBECOM Rec.*, Vol. 3, Dallas, Nov./Dec 1989, pp. 51.3.1–51.3.5.
359. R.A. Court et al., "The Multiplexing Factor in Digital Transmission Systems," *Australian Telecomm. Rev.*, Vol. 15, No. 2, 1981, pp. 58–63.
360. V.E. Charlton, "Maintenance Management System: A Strategic Option," *TE&M*, Vol. 92, No. 21, Nov. 1, 1988, pp. 82, 84, 86.
361. M. Swaringen, "Predictive Maintenance Keeps USAir in the Air," *Telephony*, Vol 219, No. 5, July 30 1990, pp. 30, 31, 34.
362. A.C Stark, Jr. and P.M. Villiere, "Harnessing Intelligent Networks," *AT&T Tech.*, Vol. 2, No. 3, 1987, pp. 14–21.
363. S. Rabie et al., "DRESS: Diagnostic Repair Expert System Shell," *ICC Rec.*, Vol. 3, Philadelphia, June 1988, pp. 45.4.1–45.4.6.
364. L. Bernstein, "Expert System in Field Operations," *TE&M*, Vol. 92, No. 8, April 15 1988, pp. 117–119.
365. L. Cynar and D. Mueller, "Expert Systems Solve Network Problems and Share the Information," *Data Comm.*, Vol. 15, No. 5, May 1986, pp. 187, 188, 191, 192.
366. T. Miyazaki et al., "Dynamic Operation and Maintenance Systems for Switching Networks," *IEEE Comm. Mag.*, Vol. 28, No. 9, Sept. 1990, pp. 34–39.
367. R. Karpinski, "Rhyme and Reason: Artificial Intelligence in the Public Network," Special Supp. on "Horizons of Technology," *Telephony*, Feb. 1991, pp. 32, 33.
368. Y. Lirov and O. Yue, "Expert Maintenance System in Telecommunications," *GLOBECOM Rec.*, Vol. 2, San Diego, Dec. 1990, pp. 707B.2.1–707B.2.7.
369. W. Cooper, "Private Microwave for Route Diversity," *Bus. Comm. Rev.*, Vol. 20, No. 7, July 1990, pp. 51–54.
370. G.R. Ash, A.H. Kafker and K.R. Krishnan, "Intercity Dynamic Routing Architecture and Feasibility," *Proc. 1 of 10th Int. Teletraffic Cong.*, Montreal, 1983, Session 3.2, paper 2.
371. G.R. Ash, R.G. Crafton and P.E. Timm, "Building a Dynamic Routing Architecture," *Telephony*, Vol. 204, No. 26, June 27 1983, pp. 42, 46, 50, 51, 53.

372. W.H. Cameron *et al.,* "Dynamic Routing for Intercity Telephone Networks," *10th Int. Teletraffic Cong., Proc. 1,* Montreal, 1983, Session 3.2, paper 3.
373. *IEEE Comm. Mag.* Issue on "Advanced Traffic Control Methods for Circuit Switched Telecommunications Networks," Vol. 28, No. 10, Oct. 1990.
374. I. Ebert, E. Abdou and P. Richards, "Network Flexibility Through Dynamic Transport Control," *Telecomm.,* Vol. 22, No. 1, Jan. 1988, pp. 60, 63, 65, 67.
375. S.E. Aidarous *et al.,* "A Distributed Architecture for Customer Network Control," *ICC Rec.,* Vol. 2, Philadelphia, June 1988, pp. 35.4.1–35.4.6.
376. S.S. Cheng and L.T. Wu, "Reconfiguration in a Flash!," *Telephony,* Vol. 212, No. 20, May 18, 1987, pp. 76–80.
377. R.M. Lefkowits, "ISDN Basic Rate Line Testing," *Telephony,* Vol. 214, No. 7, Feb. 15 1988, pp. 28–30.
378. P. Black, "ISDN—A New Challenge for Test Instrumentation," *Telecomm.,* Vol. 18, No. 7, July 1984, pp. 49, 50, 53.
379. T.E. Hagersick, "Line Test System Must 'Answer the Call'," *TE&M,* Vol. 89, No. 22, Nov. 15 1985, pp. 108–111.
380. R.L. Fredrick, "Measurement of Voice Frequency Transmission Impairments," *Telecomm.,* Vol. 19, No. 3, Mar. 1985, pp. 85, 86, 88, 88b, 88e, 88f; also, Part II, No 7, July 1985, pp. 40, 44, 48a.
381. C.E. White, "Trends for Test and Measurement," *Telecomm.,* Vol. 22, No. 7, July 1988, pp. 39, 40, 43, 47.
382. J.L. Crowley, "ISDN Operations Objectives and Their Implementation for Loop Maintenance," *ICC Rec.,* Vol. 1, Boston, June 1989, pp. 2.4.1–2.4.6.
383. B. Dentinger *et al.,* "The Problem is the Problem," *TE&M,* Vol. 95, No. 5, Mar. 1, 1991, pp. 57–61.

EPILOGUE

In this book, an approach to achieving overall transmission quality of service of evolving telecommunications networks has been presented. From a technical standpoint, the principles involved are largely consistent with those used previously by telecommunication administrations and in standards forums. However, the approach also reflects various economic, technological, and competitive realities, as well as increasing expectations associated with today's telecommunications industry.

The top-down technical steps (performance translation and allocation, network planning and standards) of designing quality of service into networks are each weighted by a number of bottom-up considerations. For example, different types of end users have different needs and expectations which influence their perception of service quality. Different user expectations and business environments must be recognized as variables and limitations in the process of modeling the relationships between user-oriented and network-oriented performance.

From a network standpoint, no single provider is responsible for end-to-end performance of multivendor network connections. This presents a challenge to Standards Committee T1-Telecommunications and other standards forums to equitably allocate end-to-end performance among multivendor networks, and to arrive at the necessary planning guidelines and related standards to enable interworking and compatibility both on a national/regional and international basis.

The bottom-up technical steps of verifying and controlling quality under network operating conditions reflect recent technological and competitive trends in the industry. In addition to its role in trouble detection and maintenance support, quality assurance based on continuous in-service performance monitoring can also be used to characterize each vendor's internal network performance, resolve responsibility for performance degradations in multivendor connections, provide end users with a performance assessment capability, and support various higher level network performance control functions.

The requirements for new service and technology-driven standards plus the presence of multivendor networks has dictated the need for a dynamic network that can address the present and future needs and performance sensitivity of end users. Performance control coordinates and uses the information provided by monitoring and surveillance systems to manage and automatically control network resources so as to address performance deficiencies in a productive manner. This is made possible on a dynamic basis owing to the use of networkwide intelligence rather than fixed rules and human intervention.

LIST OF ABBREVIATIONS AND ACRONYMS

A-A	analog (port-to)-analog (port)
AAL	ATM adaptation layer
AC	alternating current
A/CO	analog central office
ACT	active
A/D	analog to digital
Adden.	addendum
ADM	add-drop multiplexer
ADN	all digital network
ADPCM	adaptive differential pulse code modulation
ADTS	automatic data test system (AT&T)
AGC	automatic gain control
AIS	alarm indication signal (formerly, blue signal)
AISS	alarm indication signal second
AL	acceptance limit
ALM	alarm
ALS	AIS/LOP second
AMI	alternate mark inversion (DS1, DS3)
ANSI	American National Standards Institute
APC	adaptive predictive coding
APRL	average power return loss
APS	automatic protection switching
ARPANET	advanced research project agency network
ARSB	automatic repair service bureau (AT&T)
AT	access tandem
AT&T	American Telephone and Telegraph
ATAU	analog (or metallic) test access unit (Bellcore)
ATP	advanced telemetry processor (AT&T)
A/TO	analog toll office
A/TT	analog tie trunk

ATDM	asynchronous time division multiplexing
ATM	asynchronous transfer mode
AVE or Ave.	average
B-DCS	broadband digital cross-connect system
Bellcore	Bell Communications Research
BER	bit error ratio
BIP-N	bit interleaved parity -N (SONET, e.g., BIP-2, BIP-8)
B-ISDN	broadband ISDN
BITS	building integrated timing supply
BLER	block error ratio
BLTJ	Bell Labs. Tech. Journal (formerly BSTJ, now AT&T Tech. Journal)
BNR	Bell Northern Research
BNZS	bipolar with N zero substitution (e.g., B3ZS, B6ZS, B8ZS)
BOC	Bell operating company
BPV	bipolar violation (DS1, DS3)
BRA	basic rate access (ISDN)
BS	block size
BSC	BISYNC (IBM binary synchronous stop and wait link protocol)
BSTJ	Bell System Technical Journal
Bus.	business
CAROT	centralized automatic reporting on trunks (AT&T)
CB	compatibility bulletin (AT&T)
CBEMA	Computer and Business Equipment Manufacturers Association
CCIR	International Radio Consultative Committee
CCITT	The International Telegraph and Telephone Consultative Committee
CCS	common channel signaling
CDF	cumulative distribution function
CEPT	European Conference of Postal and Telecommun. Administrations
CFA	carrier failure alarm
CI	customer installation
CGA	carrier group alarm (DS1)
CLC	carrier liason committee
CMS	central maintenance system (AT&T)
CMTT	Joint CCITT/CCIR study group on transmission of sound broadcasting and television over long distances

CNC	customer network control
CNCP	Canadian National Canadian Pacific (Telecom., now UNITEL)
CNO	Canadian National Organization
CO	central office
codec	coder-decoder
COFA	change of frame alignment (event)
Comm.	communications
Comp.	computer
CORE	corrected overall reference equivalent (CCITT)
CP	compandor penalty
CPE	customer premise equipment
CRC	cyclical redundancy check
CRTC	Canadian Radio-Television Telecommunications Commission
CS	circuit switch(-ed, -ing)
CSA	Canadian Standards Association
CT	central terminal
CTMS	carrier terminal maintenance system (AT&T)
CTX	centrex
CSRE	corrected send reference equivalent (CCITT)
CTR	common technical regulation
CV	code violation
D/A	digital to analog
dB	decibel
dB(A)	decibels acoustic
dBm0	decibels milliwatt at zero level point
dBrnC0	decibels reference noise C-weighted at zero level point
dBV	decibels Voltage
DC	direct current
DCME	digital circuit multiplication equipment
DCN	digital communications network
D/CO	digital central office
DCS	digital cross-connect system
D-D	digital (port-to)-digital (port)
DDS	digital data system, dataphone data service (AT&T)
demux	demultiplexer
des.	design
DES	digital echo suppressor
DIC	direct inter-LATA connecting (trunk)
DLC	digital loop carrier

DLP	decode level point
DM	degraded minute (CCITT)
DMW	digital milliwatt
DOBS	dynamic office balance system
DOC	Department of Communications (Canadian Federal)
DPBX	digital private branch exchange
DQDB	distributed queue dual bus
DR	dynamic routing
DRS	digital reference signal
DS	digital signal
DSn	digital signal, level n (e.g., DS1, DS3)
DSI	digital speech interpolation
DSP	digital signal processing
DTAU	digital test access unit (Bellcore)
DTMF	dual tone multi-frequency (signaling)
D/TO	digital toll office
D/TT	digital tie trunk
DT	digital terminal
DTS	digital test signal
EA	exchange access
EARS	electro-acoustic rating system (Bell System)
EAS	exchange access study
EC	echo control, echo canceller, exchange carrier
ECMA	European Computer Manufacturers Association
ECSA	Exchange Carriers Standards Association
Ed.	edition
EDD	envelope delay distortion
EFS	error free second
EIA	Electrical Industries Association
elect.	electrical
ELP	encode level point
EO	end-office
EOC	embedded operations channel
EOS	end-office survey
eq.	equivalent
eqn.	equation
ERL	echo return loss
ERLE	echo return loss enhancement
ES	errored second, echo suppressor
ESF	extended superframe format (DS1)
ET	exchange termination (ISDN)
ETS	European telecommunications standards

ETSI	European Telecommunications Standards Institute
exp.	expected
EXZ	excessive zeros
FA	framework advisory (Bellcore)
FBE	frame bit error
FCC	Federal Communications Commission (U.S.)
FD	frame duration
FDL	facility data link (e.g., for DS1)
FEBE	far-end block-error
FEC	forward error correction, front end clipping
FEPR	far-end performance report
FERF	far-end receive failure (SONET)
FGD	feature group D (Access)
Fig.	figure
FLP	fixed loss plan
FLS	frame loss second (now SEFS)
FMAC	facility maintenance and administration center (AT&T)
FMAS	facility maintenance and administration system (AT&T)
fn.	function
FO	fiber optics
FPLMTS	future public land mobile telecommunications system
freq.	frequency
FS	frame size
FWE	frame word error
FX	foreign exchange
Gb/s	gigabits per second
GLOBECOM	Global Communications (Conference)
GHz	gigahertz
GOB	good-or-better
GOS	grade-of-service
GPS	global positioning system
HDLC	high level data link control (ISO continuous link protocol)
HDTV	high definition television
HRX	hypothetical reference connection
H.S.	high speed
Hz	Hertz (cycles per second)
IAL	immediate action limit
IBM	International Business Machines
ICB	intelligent channel bank

ICC	International Conference on Communications
ICCF	inter-exchange carrier compatibility forum
ICL	inserted connection loss
ICS	ISDN compatible station
ICT	ISDN compatible trunk
IDLC	integrated digital loop carrier (Bellcore)
IDN	integrated digital network
IEC	inter-exchange carrier, International Electrotechnical Commission (standards)
IEEE	Institute of Electrical and Electronic Engineers
IMD	intermodulation distortion
IMP	interface message processor (ARPANET)
IMT	inter-machine trunk
IMU	intermediate monitoring unit
INE	intelligent network element
INMD	in-service nonintrusive measurement device
info.	information
Int.	international
IRS	intermediate reference system (CCITT)
ISDN	integrated services digital network
ISO	International Organization for Standardization
ISSLS	International Symposium on Subscriber Loop Systems
ITS	integrated test system (Bellcore)
ITT	inter-toll trunk
ITU	International Telecommunications Union
JTCI	Joint Technical Committee on Information Technology
KAS	keep alive signal (now AIS)
kb/s	kilobits per second
km	kilometer
kHz	kiloHertz
KTS	key telephone system
Lab.	laboratory
LAN	local area network
LAPB	link access procedure balanced
LAPD	link access procedure delta (D) channel
LATA	local access and transport area
LBRC	low bit rate coding (or codec)
LEC	local-exchange carrier
LEPD	listener echo path delay

LEPL	listener echo path loss
LH	long-haul
LMOS	loop maintenance operation system (AT&T)
LOC	loss of clock
LOF	loss-of-frame (persistent OOF condition)
log	logarithm
LOP	loss-of-pointer (SONET)
LORAN	long range navigation
LOS	loss-of-signal
LOSS	loss-of-signal second
LR	loudness rating
L.S.	low speed
LSSGR	LATA switching system generic requirements (Bellcore)
LT	line termination (ISDN)
LTE	line terminating equipment
M13	DS1 to DS3 multiplex
Mag.	magazine
MAN	metropolitan area network
max.	maximum
Mb/s	megabit per second
MCI	Microwave Communications Incorporated
MD	mediation device
metro	metropolitan (network)
MF	multi-frequency (signaling)
MHz	megaHertz
mi.	mile
min.	minute, minimum
ML	maintenance limit
MLT	maintenance loop tester (AT&T)
MML	machine-machine language
MNRU	modulated noise reference unit
modem	modulator demodulator
MOS	mean opinion score
MRTIE	maximum relative time interval error
ms	millisecond
MSC	mid-speech burst clipping
MSUR	missed start up rate
MTAU	metallic (or analog) test access unit (Bellcore)
MTBSO	mean time before service outage
MTCE	maintenance
MTIE	maximum time interval error

MTTSR	mean time to service restoral
mux	multiplexer
NA	North American
Nat.	national
NBS	National Bureau of Standards (U.S.)
NDF	new data flag (SONET)
NDL	network data link
NE	network element
NETS	Normes Europeane Telecommunications
NHDR	non-hierarchical dynamic routing (AT&T)
NI	network interface
NIC	near instantaneous companding
NMA	network monitoring and analysis (Bellcore)
NNI	network-node interface
No.	number
NOF	networks operations forum
nom.	nominal
NOS	network operations center
NOSFER	the new fundamental system for the determination of ref. equivalents
NPA	number plan area
ns	nanosecond
NSA	non-service affecting
NSES	network service evaluation system
NSG	National Study Group (for CCITT in Canada)
N-STD	non-standard
NT1, NT2	network termination types 1 and 2 (ISDN)
NTE	network terminating equipment
NTI	Northern Telecom Incorporated
NTIA	National Telecommunications and Information Administration (U.S.)
OAM or OA&M	operations, administration and maintenance
OAM&P	operations, administration, maintenance and provisioning
Obj.	objective
OC	optical carrier
OCC	other common carrier
OC-N	optical carrier level N (SONET, e.g., OC-1, OC-3, OC-12, OC-48)
ONS	on premise station
OOF	out-of-frame (momentary event or defect)

OOLR	overall objective loudness rating (IEEE)
OPS	off premise station
OR	overall rating (EARS)
ORE	overall reference equivalent (CCITT)
OS	operation system
OSI	Open Systems Interconnection
OSS	operations support system
OTDR	optical time domain reflectometer
OTGR	operations technology generic requirements (Bellcore)
p.	page
P/AR	peak to average ratio
PBX	private branch exchange
PCM	pulse code modulation
PCS	personal communications systems
pdf	probability density function
PE	parity error
PEFS	percent error free seconds
PES	percent errored seconds
PISN	private integrated service networks
PJ	pointer justification (SONET)
PJC	pointer justification count (SONET)
PLL	phase locked loop
POI	point of interface
POP	point of presence
POT	point of termination
POTS	plain old (ordinary) telephone service
POW	poor or worse
pp.	pages
PPM	parts per million
PRA	primary rate access (ISDN)
Prot.	protection
PS	protection switch(-ed, -ing)
PSC	protection switch count
PSD	protection switch duration
PSDN	packet-switched data network
PSL	pre-service limit
PSN	packet-switched network
PSTN	public switched telephone network
PTE	path terminating element
PTT	European Post Telegraph and Telephone
PUB	publication (AT&T)

Q	signal-to-quantization distortion ratio
QDU	quantization distortion unit
QOS	quality-of-service
Qtr	quarter
Quest.	question
RAI	remote alarm indication (yellow signal)
RBOC	Regional Bell Operating Company
RBS	robbed bit signaling
REA	Rural Electrification Association (U.S.)
Rec.	record, recommendation (CCITT)
ref.	reference
REMA-REMA	remote analog (port-to)-remote analog (port)
REP	residual error probability
Req.	required
Rev.	review, revised, revision
Rg	range
RL	return loss, restoral limit
RLR	receive loop rating (EARS)
rm.	room
rms	root mean square
ROLR	receive objective loudness rating (IEEE)
ROTL	remote office test line (AT&T)
RRE	receive reference equivalent (CCITT)
RT	remote terminal, round trip
RTS	remote test system (AT&T)
RTU	remote test unit (Bellcore)
SA	service-affecting
SAL	service-affecting limit
SAS	SEF/AIS second
SAT.	satellite, saturated
SARTS	switched access remote test system (AT&T)
S/ATT	satellite analog tie trunk
S/C	surveillance and control
SCC	switching control center (AT&T)
	Standards Council of Canada (standards)
SCCS	switching center control system (AT&T)
SCOT	Steering Committee on Telecommunications (Canada)
SCOTS	surveillance and control of transmission systems (AT&T)
SD	standard deviation
SDH	synchronous digital hierarchy (CCITT)

SDLC	synchronous data link control (IBM continuous link protocol)
S/DTT	satellite digital tie trunk
S/E	signal to echo (ratio)
sec. or s	second
SEF	severely errored frame
SEFS	severely errored frame second
Sel.	selected
SES	severely errored second (CCITT)
SF	superframe format (DS1)
SG	study group (CCITT)
SH	short-haul
SM	singing margin
SMAS	switched maintenance access system (AT&T)
SMDS	switched multi-megabit data service (Bellcore)
SNR	signal-to-noise ratio
SOLR	sidetone objective loudness rating (IEEE)
SONET	synchronous optical network
SPE	synchronous payload envelope
SR	slip rate
SRE	send reference equivalent (CCITT)
SRL-HI, -LO	singing return loss-high, -low
SS	special services
STE	section terminating element (SONET)
STPL	sidetone path loss
STS	synchronous transport signal (SONET)
STS-N	synchronous transport signal level N (e.g., STS-1, STS-3, STS-12)
STS-Nc	synchronous transport signal level N concatenated (e.g., STS-3c)
subj.	subjective
SW	switch
Symp.	symposium
Sync.	synchronization
SYNTRAN	synchronous transmission (DS3)
T	tandem
T1	Committee T1-Telecommunications
T1AG	T1 advisory group
TA	technical advisory (Bellcore), terminal adapter (ISDN)
TABS	telemetry asynchronous block serial (AT&T Protocol)
TAPAC	terminal attachment program advisory committee

TASC	telecommunications alarm surveillance and control (AT&T)
TBD	to be determined
TBOS	telemetry byte oriented serial (AT&T Protocol)
TC	tandem connecting (Trunk)
TCAS	T-carrier administration system (AT&T)
TCT	toll connecting trunk
TCTS	Trans Canada Telephone System (now Telecom Canada)
TD	temporary document (CCITT)
TE1	ISDN terminal element
TE2	non-ISDN terminal element
Telecomm.	Telecommunications (journal)
TE&M	Telecommunications Engineering and Management (journal)
TEPD	talker echo path delay
TEPL	talker echo path loss
TIA	Telecommunications Industry Association
TIC	tandem inter-LATA connecting (Trunk)
TIE	time interval error
TLP	transmission level point
TLR	transmit loop rating (EARS)
TMN	telecommunications management network (CCITT)
TOLR	transmit objective loudness rating (IEEE)
TP	telemetry processor (AT&T)
TQTF	transmission quality task force
TR	technical reference, technical requirement (Bellcore)
TRANSYSCOM	Transmission Systems Committee (IEEE)
TRIF	Telecommunications Requirements Industry Forum (Bellcore)
Trk.	trunk
TSC	test system controller (Bellcore)
TSGR	transmission system generic requirements (Bellcore)
TTC	Telecommunications Technology Committee (Japan)
TX	transmission
UAS	unavailable seconds (CCITT)
UI	unit interval
UNI	user-network interface
μs or μsec.	microsecond
US(A)	United States (of America)
USITA	United States Independent Telecommun. Association (now USTA)

USTA	United States Telecommunications Association
UTC	universal time coordinated
VBD	voiceband data
Vol.	volume
VNL	via net loss
VNLF	via net loss factor
VT	virtual tributary (SONET)
Vu	volume units
WAN	wide area network
WARL	weighted average return loss
WATS	wide area telephone service
WATTC	World Administration Telephone and Telegraph Conference
W-DCS	wideband digital cross-connect system
WEPL	weighted echo path loss
WP	working party (CCITT)
w.r.t.	with respect to
yr	year
ZBTSI	zero byte time slot interchange
ZCS	zero code suppression

INDEX

A/D converter, 55, 61
Adaptive differential pulse code modulation. See ADPCM.
Add drop multiplexer. See ADM.
Addition laws
  algebraic, 180–197, 217–226
  distributed values, 211–245, 246–259
  fixed values, 180–211
  parallel configurations, 245–259
  power, 197–211, 226–245
  serial configurations, 180–245
ADM, 380, 508, 553
ADPCM, 14, 60, 62, 76, 178, 205, 288, 304–314, 462
  ANSI MT1-502-1988, 419
  application rules, 419
  effect on DTMF, 421
  effect on voiceband data, 420
  Q, 419
  transcoding, 419
Advanced telemetry processor (ATP), 492
AEPL, 382
AIS second (AISS), 534, 536
AIS, 532, 533, 537, 538, 548
AIS/LOP second (ALS), 533, 534, 536
Alarm and status indications, 492, 522, 526, 531, 538, 548
Alarm indication signal. See AIS.
Alarms. See AIS, LOF, LOP, LOS.
Alternate mark inversion (AMI), 518
Amplitude clipping, 73
Analog loop. See Loop plant.
Analog test access unit (ATAU), 496
Analog transients
  dropouts, 133, 247, 302
  gain hits, 75, 133, 249, 302
  impulse noise, 75, 133, 249, 302
  phase hits, 75, 133, 249, 302
  signal dropouts, 75
ANSI, 441, 462
APRL, 59, 194, 227, 238, 288
ARPANET, 158
Asynchronous transfer mode. See ATM.
ATM, 6–9, 45, 48, 90, 158, 414, 504, 522, 554
  cell size, 415
  cells, 414
  delay, 414
  echo control, 415
ATM adaption layer (AAL), 7
Automated data test system (ADTS), 488, 491
Automated repair service bureau (ARSB), 489
Availability, 1, 95, 151, 153
Average power return loss. See APRL.

B-ISDN, 7, 493, 501, 504
B-ISDN standards
  T1Q1.3, 462
Balance impedance, 288, 397
Balance network, 123, 125, 288, 391
  double, 124, 379
  parallel, 124
  PBX, 402
  single, 124, 125
Bandwidth, 67
Behavioral data base
  parameters, 35
Bell System, 10, 12, 440
Bellcore, 10, 114
  Technical Ref (TR)-303, 396
BER, 76, 80, 84, 92, 94, 129, 136–151, 238–245, 249, 253–260, 304, 311, 314, 318, 493, 518, 524, 536–537, 539–542, 543, 546, 551
Bipolar N zeros substitution (BNZS), 518, 528
Bipolar violation, See B-ISDN.
Bit error ratio, See BER.
Bit interleaved parity (BIP), 528, 536, 539, 548
Bits
  errored, 158
  misdelivery, 158
BLER, 92, 129, 136, 137, 238, 249, 304, 310, 311, 541
Block error ratio. See BLER.
Block size, 137, 310
BPV, 459, 514, 518, 528, 536, 543, 544
Burst factor, 137

Call cutoff, 89
Canadian standards
  CRTC, 449
  CSA, 441, 450
  CSA-SCOT, 450, 463
  NSG, 452, 463
  SCC, 441
  Telecom Canada, 450
Carrier transmission maintenance system (CTMS), 492
Carriers
  AT&T, 10–12, 24, 111, 114, 174, 245
  IEC, 11, 117, 129, 142, 167, 173, 174, 273, 278, 292, 501, 504, 515
  LEC, 11, 142, 167, 173, 174, 504, 515
  MCI, 9
  Mercury, 12
  OCCs, 9, 25
  Telecom Canada, 9, 166
CBEMA, 449
CCIF, 440

CCIR, 452
   FPLMTS, 453
CCITT, 440
   Rec. G.121, 379
   Rec. P.31, 382
   SG I, 451
   SG II, 451
   SG IV, 451
   SG VII, 463
   SG XI, 465
   SG XII, 381, 451
   SG XV, 451, 465
   SG XVIII, 451
CCS, 307
CEN/CENELEC, 454
Center clipper, 57
Centralized automatic reporting on trunks (CAROT), 490
CEPT, 454
Change-of-frame alignment (COFA), 530–533
Channel bank, 436
Circuit maintenance system (CMS), 490
CLC, 448
Clock operation
   holdover, 348
   ideal, 347
   stressed, 348
CMTT, 452
Code violation (CV), 536, 538, 539, 546, 549
Codecs, 303. *See also* ADPCM, DCME, PCM.
   ANSI, 311
   CCITT, 310
Common channel signaling. *See* CCS.
Communication phases
   access, 2, 20
   disengagement, 2, 20, 23
   transfer, 2, 20, 23
Compandors,
   advantage, 288, 357
   analog, 288, 357
   compression characteristic, 357
   penalty, 359
Competition, 10, 12
Connection characterization, 114
Connection models, 107
Connections
   analog, 53, 57
   digital, 53, 55, 57
   interoffice, 51, 117, 124, 126, 135, 150
   intraoffice, 51, 60, 67, 114, 124, 126
   mixed, 53, 55, 57
   toll, 51, 116, 117, 126, 147
Controlled slips, 82, 322, 326, 347, 349, 530, 537
   effect on facsimile, 147
   effect on voice, 146
   effect on voiceband data, 146
Conversational dynamics, 383
Crosstalk, 280, 378
   index, 71
   intelligible, 71
   unintelligible, 71
CRTC, 449
CSA, 441, 450
CTR, 454

Customer premise equipment. *See* Equipment, CPE.
Customer serving unit (CSU), 493
Cylincal redundancy check (CRC), 515, 522, 528, 537, 539, 543

D/A converter, 55, 61
DCME, 9, 47, 84, 178, 376
   32 kbit/s ADPCM, 371, 419
   ADPCM, 419
   ADPCM Standard, 462
   application rules, 419
   DSI, 418
   low bit rate codecs, 418
   PCS, 418
   speech clipping, 418
DCS, 294, 338, 491, 495, 497, 508, 513, 520, 553
Decode level point (DLP), 434
Degraded minute. *See* DM.
Dejitterizer, 334
Delay, 188, 294, 360
   ATM, 414
   asymmetry, 193
   CCITT Rec. G.114, 418
   control of, 377
   echo path/round trip, 128
   end, 410
   end-to-end, 383
   fixed, 45
   guidelines, 417
   heirarchical routing, 416
   IEC, 411, 417
   LEC, 411
   listener echo path, 124, 195
   mobile radio systems, 413
   PCS, 417
   RSS round trip, 401
   satellite, 45, 57
   supplement to ANSI T1-506-1990, 418
   talker echo path, 55, 193
   terrestrial, 45
   transfer, 90
   variability, 158
   variable, 46
Densities
   log-normal, 133, 137, 253–260, 267
   normal, 133, 213–217, 249–253, 267
Digital circuit multiplication equipment. *See* DCME.
Digital cross-connect system. *See* DCS.
Digital impairments, 523
Digital loop carrier (DLC), 365
Digital network performance
   estimates, 147
   objectives, 141
Digital reference signal (DRS), 434–435
Digital signals
   DS0, 141, 149, 317
   DS1-ESF, 493, 515, 520, 529, 534, 542, 543
   DS1-SF, 543
   DS1, 141, 149, 317, 322, 326, 333, 337, 343, 348, 457
   DS3, 141, 149, 317, 321, 337, 457
   DS3 C-Bit, 493, 515, 529
   DS3 SYNTRAN, 514

fractional DS1, 142
SONET, 457
SONET line, 515, 529, 534, 537
SONET section, 514, 520, 529, 530, 533, 538
SONET STS, 515, 529, 533, 534, 546
SONET VT, 515, 529, 533, 534, 546
Digital speech interpolation. *See* DSI.
Digital switch, 273, 278, 283, 294, 304, 338, 360
Digital test access unit (DTAU), 497
Distortion, 249
    attenuation, 279
        loop plant, 225
        trunk plant, 224
    envelope delay, 296
    intermodulation, 280, 296
        cubic, 74
        quadratic, 74
    near-singing, 194
    nonlinear, 72
    overload, 72
    quantizing, 60, 121, 178, 365, 376, 419
    quantizing distortion units, 420
Divestiture, 12, 116, 166, 273, 292, 440
DM, 76, 82, 138, 147, 150, 318, 537
Domain
    dB, 197, 199, 217, 229
    inverse log, 249, 253, 267
    log, 238, 255, 259, 267
    log/dB, 227, 228, 249
DPBX, 53
DQDB, 448
DS1, 9, 47, 459
DTMF, 281, 314, 421
Dual tone multifrequency. *See* DTMF.

Echo, 123–129
    acoustic, 57, 378
    cancellers, 410
    control, 391, 410
    cross talk, 57
    listener, 58, 123, 400
    talker, 55, 124
Echo canceller, 383
    tandem operation, 416
Echo control, 44, 126, 129, 252, 278, 294
    application distance, 410
    cancellers, 45, 57
    suppressors, 45, 57
Echo return loss. *See* ERL.
ECMA, 454
ECSA, 441, 448
    CLC, 448
    ICCF, 448
    NOF, 448
EDD, 130, 296
EIA, 440
    Standard RS-470, 385
Embedded operations channel (EOC), 504, 517, 518, 522
Empirical rule, 241, 243
Encode level point (ELP), 435
End office survey. *See* Survey data, EOS.
Envelope delay distortion. *See* EDD.
Equal access, 11, 113, 117, 174

Equipment
    2-wire, 107, 124, 128, 187, 195, 207, 211, 232, 271, 292
    4-wire, 108, 123, 128, 207, 292
    CPE, 107, 165, 172, 347, 370, 488, 504
    ERL, 55, 58, 126, 190, 195, 221, 222, 227, 236, 237, 288, 292
Error performance, 135, 238, 271
    digital, 314, 322
    voiceband data, 303–314
Error probability, 92
    residual, 93, 153, 156
Errored second. *See* ES.
ES, 82, 92, 138, 150, 314, 318, 524, 532, 534, 536–538, 543, 545, 551
ES type A, 492
ES type B, 492
ETSI
    STC BT2, 454
European standards
    CEPT, 454
    CTR, 454
    ETS, 454
    ETSI, 454
    NETS, 454
Excessive zeroes (EXLs), 528
Exchange access survey. *See* Survey data, EAS.
Extended superframe format. *See* Digital signals; DS1-ESF.

Facilities
    fiber, 271, 294
    N-carrier, 232, 359
    nonfiber cable, 294
    radio, 271, 294
    T-carrier, 117, 128, 135, 150, 204, 232, 238, 273, 283, 304, 318, 490, 491, 493, 498
    twisted pair, 291
Facility maintenance and administration center (FMAC), 491
Facility maintenance and administration system (FMAS), 492
Facsimile
    effect on 32 kbit/s ADPCM, 421
Far-end block error (FEBE), 514, 529, 538
Far-end receive failure (FERF), 534, 538
FCC standards
    Part 68, 449
Feature Group D. *See* FGD.
FGA, 174
FGD, 174, 273
FLP
    analog trunk, 430
    combination trunk, 430
    digital trunk, 430
    loss control, 428
    loss objectives, 429
    operational administrative aspects, 431
    operational considerations, 429
    overall transmission quality, 428
    simplification, 329
    transmission plan, 365
    trunk loss objectives, 431
FPLMTS, 453

Frame bit error (FBE), 528, 536, 544
Frame relay, 91, 156
Frame word error (FWE), 528
Frequency shift, 132

Global positioning system (GPS), 343
Global standards, 441, 443
Good or better. See Subjective measure, GOB.
Grade-of-service, 13, 19, 20, 23, 41, 91, 193, 195, 197, 210, 211, 235, 457, 501
   analysis, 386
   good or better, 389
   loss/noise, 52, 54
   loss/noise, bandwidth/slope, 69
   loss/noise, echo/delay, 55
   loss/noise, listener echo/delay, 60
   loss/noise/Q, 60
   loss/noise/sidetone, 63, 65

HDTV, 414
Hierarchical rates. See Digital signals.
High-speed data, 414
Hypothetical reference connection, (HRX), 78, 138, 168
Hypothetical reference digital path (HRDP), 177

ICCF, 448
ICL, 51, 55, 60, 187, 192, 195, 266
IEEE, 440, 443, 447
   Project 802, 448
   Project 802.6, 448
IMD, 132, 296
IMP, 158
Impedance
   balance, 397
   input, 397
   input, complex, 398
In-service nonintrusive measurement device (INMD), 498
Information technology standards, 441
Input impedance, 397
   complex, 398
Integrated digital loop carrier, 369
Integrated digital network (IDN), 8
Integrated services network. See ISDN.
Integrated test system (ITS), 496
Intelligibility, 27
Inter-exchange carrier. See Carriers, IEC.
Interfaces
   network
      NI, 107, 120, 488, 504, 516
      POI, 504
   operations
      Q1, 495
      Q2, 495
      Q3, 495
Interference, 71
Intermediate monitoring unit (IMU), 516
Intermodulation distortion. See IMD.
Interruptions
   long, 89
   short, 82
ISDN, 90, 489, 493, 504, 509, 515–516
   digital telephone set, 383

ISDN-compatible stations (ICS), 410
ISDN-compatible trunks (ICT), 410
ISO, 440
ISO/IEC JTC1/SC6, 449
ITSC, 454
ITU, 440

Jitter, 151
   amplitude, 132, 147, 301
   multiplex, 326
   phase, 132, 147, 301
   repeater, 323, 330, 333
   waiting time, 335
Jitter and wander, 271, 322–349
   accumulation, 330
   causes and effects, 322
   performance, 337
      input, 338
      intrinsic, 341
      output, 341
      transfer, 341

LATA, 394
LEPD, 124, 195
LEPL, 58, 59, 122, 123, 194, 195, 207
   criterion, 379
Limits
   acceptance, 275
   immediate action, 275
   maintenance, 275, 347
   out of service, 347
   preservice. See Limits, acceptance.
   restoral. See Limits, accpetance.
Line terminating element. See LTE.
Listener echo, 401
Listener echo path loss. See LEPL.
Local-exchange carrier. See Carriers, LEC.
LOF, 523, 531, 533, 538, 548
Long range navigation (LORAN), 343
Loop carrier systems
   digital, 394
   integrated, 396
Loop maintenance operations system (LMOS), 489
Loop plant, 117, 137, 489
   analog, 271, 273, 275, 283, 289, 315
   digital, 319
LOP, 533, 536, 538
LOS, 523, 531–532, 536, 548
LOS second (LOSS), 531, 536
Loss, 117–119, 275–279
   acoustic, 119, 185, 187, 219, 275
   acoustic echo path, 382
   analog, 375, 378
   asymmetry, 53, 119, 188, 222, 276
   contrast, 54
   control of, 378
   deviation, 119
   digital, 375
   electrical, 119, 185, 187, 278
   end-to-end, 50
   fixed, 53
   ICL, 51
   implementation, 372
   listener echo path, 58, 123, 197

loop, 281
loudness, 50
multiple echo path, 235
    objective, 207
    subjective, 207
network, 283
optimum, 53
pads, 385
sidetown path, 63
switch, 273, 285
switched, 51
talker echo path, 55, 124, 292, 382
    acoustic, 190, 221
    asymmetry, 192, 223
    electrical, 190, 192
    trunk, 273
Loss pads
    analog, 385
    digital, 385
Loss-of-framed. See LOF.
Loss-of-pointer. See LOP.
Loss-of-signal. See LOS.
Loudness
    CCITT, 51
    EARS, 51
    IEEE, 51
Loudness rating
    OOLR, 50
    receive reference, 54
    ROLR, 50
    SOLR, 64
    TOLR, 50
Low bit rate coding (LBRC), 9
LSSGR, 380
LTE, 532, 548

Maintenance capabilities, 552–554
    basic, 552
    enhanced, 553
Maintenance loop testing (MLT), 489
Maintenance processes
    corrective, 484
    preventative, 482
Maintenance standards
    CCITT SG IV, 465
Maintenance systems
    circuit, 486–491
        CPE, 488
        loop, 489
        special service, 490
        switching, 490
        trunk, 490
    embedded, 486–493
    facility, 491–493
Mandatory standards, 443
Maximum time interval error (MTIE), 348
Mean opinion score. See MOS and Subjective
    measures, MOS.
Mean time before service outage (MTBSO), 156
Mean time to service restoral (MTTSR), 156
Mediation device (MD), 495
Metallic test access unit. See Analog test access
    unit.
Mobile digital cellular radio, 383

Mobile radio systems, 413
    echo canceller, 413
    echo control, 413
Model
    behavioral, 25, 34
    data performance, 35
    grade of service, 25
    loss/noise, 29
    opinion, 25, 29
    transmission rating, 25, 28
Modem, 281, 289, 304, 310, 313
Monte Carlo simulation, 180, 199, 211, 217, 235,
    245
MOS, 48, 76, 86
Multifrequency (MF), 281, 314

National standards, 441
NETS, 453
Network
    all-digital, 380, 384
    architecture, 504
    circuit-switched, 23
    components, 176
    data, 9
    delay, 417
    digital access, 385
    evolution, 20, 39, 113
    exchange access, 174
    IEC, 417
    intelligence, 13
    inter-LATA, 107, 117, 146, 174, 245, 394
    international, 168
    intra-LATA, 107, 117, 146, 174, 245, 274,
        288, 292, 394
    metropolitan, 54
    North American, 172
    packet voice, 46
    private line, 23, 370, 402
Network components, 272–274
Network elements (NES), 494, 504, 508, 513,
    538
Network interface. See Interfaces, network.
Network monitoring and analysis (NMA), 502
Network performance
    digital, 138–151
        CCITT, 138
        North America, 141
    packet, 156–160
        CCITT, 156
        North America, 156
    voiceband, 107–138
Network service evaluation system (NSES), 492
New transport modes
    T1S1, 461
NOF, 448
Noise, 197, 229, 283–288
    3-kHz flat, 120
    acoustic/room, 63
    analog compandor, 357
    asymmetry, 201, 229
    C-notched, 121, 202, 285
    control of, 376
    idle circuit, 54, 120, 283, 373, 376
    impulse, 75, 133, 135, 249, 260, 302

Q, 121, 204
quantizing, 60, 204, 288, 304
referred, 200
SNR, 202, 231
subjectively equivalent, 60, 204
weighting
   15-kHz flat, 202
   3-kHz flat, 120, 202, 285
   acoustic-A, 63
   C-message, 54, 120, 202, 283
Normal pdf, 28, 29
Number plan area, 283

OAM, 17
Off-premise station (OPS), 403
On-premise station (ONS), 403
OOF, 145, 149, 492, 530, 533
OOLR, 50, 51, 55, 64, 265, 381
Open systems interconnection. See OSI.
Operations administration, maintenance and provisioning (OAM&P), 493
Operations system (OS), 494
Opinion data base
   CCITT, 33
   long toll, 31, 35
   Murray-Hill, 31
   parameters, 31
Opinion scale
   4-point, 27, 35
   5-point, 27, 73
   10-point, 73
Opinion surveys, 23, 457
Optical time domain reflectometer (OTDR), 498
OSI, 6, 495, 509, 512
Other common carriers. See Carriers, OCC.
Out-of-frame. See OOF.
Overall objective loudness rating. See OOLR.

P/AR, 131, 296
Packet data performance standards
   CCITT SG VII, 463
Packet network performance, 272
   measurements, 156
   objectives, 156
Packets, 151, 156, 158
   errored, 158
   extra, 158
   lost, 158
   misdelivery, 158
Parameter
   delay, 372
   noise, 376
   talker echo path loss, 374
Parity error (PE), 528, 536
Path monitoring, 459
Path terminating element. See PTE.
PBX, 398
   digital, 402
   ONS, 403
   OPS, 403
   tandem, 402
   transmission requirements, 402
PCM, 60, 62, 76, 122, 205, 232, 288, 304–306
PCS, 417

DCME, 418
Peak to average ratio. See P/AR.
PEFS, 92, 145, 150, 321
Percent error free seconds. See PEFS.
Performance
   assessment, 24
   categories
      accessibility, 1, 20
      integrity, 2, 39
      retainability, 2, 20
   classes, 33
   component-oriented, 14, 271
   criteria
      accuracy, 3, 20, 50, 92
      dependability, 3, 20, 89, 94
      speed, 3, 20, 44, 90
   data, 20, 90
   digital, 314–349
   end-to-end, 13
   network-oriented, 13, 39, 105, 165
   objectives, 113, 275
   parameters
      data, 20
      voice, 20
   requirements, 275
   subjective, 13, 20, 24
   traffic, 1
   transmission, 1, 39
   user-oriented goals, 1, 14, 19, 39, 105, 374
   voice, 20, 44
   voiceband, 271–314
Performance allocation, 14, 165, 178, 457
   criteria
      competitive, 179
      distance, 178
      network portion, 156–158
      technology, 179
   methodology, 179
   objectives, 371
   user fraction, 156–158
Performance monitoring
   accuracy, 539–548
      bipolar violations, 544
      block errors, 539
      errored seconds, 544
      frame bit errors, 543
      hierarchical translation, 545
      miscellaneous, 546
   far-end, 514
   functions, 522–552
   intermediate, 516
   layer, 512
   layered, 509
   line, 512
   memory, 549
   near-end, 514
   network considerations, 504–522
   parameters, 535–538
      error, 536
      failure, 538
      far-end, 538
      framing, 537
      protection switching, 538
      synchronization, 537

path, 514
primitives, 526-535
  error, 528
  failure, 532
  framing, 530
  protection switching, 535
  synchronization, 531
  reporting, 551
  section, 513
  signals
    framed, 518
    line code, 517
    other, 520
  thresholding, 549
Performance parameters, 105
  accuracy of services, 372
  AEPL, 382
  analog, 50
  delay, 374, 383
  digital, 75-83
  end-to-end loudness loss, 372, 374
  listener echo, 401
  network-oriented, 372
  noise, 373, 374
  nontransmission, 90
  OOLR, 381
  ROLR, 382
  SM, 394
  TEPL, 382
  TOLR, 382
  user-oriented, 23, 372
  WEPL, 394, 401
Performance process, 39, 95, 105, 165, 271, 481
  bottom-up, 13
  top-down, 12
Performance report message (PRM), 516
Personal communication standards
  T1P1, 462
PES, 145, 244, 249
Phase locked loop. See PLL.
PISN, 453
PLL, 341
Point of interface. See Interface network, POI.
Point of presence (POP), 110, 174
Pointer justification (PJ), 531, 537
Pointer justification count (PJC), 531, 537
Poor-or-worse. See Subjective measure, POW.
Power feed, 273
Premature disconnect probability, 153
Premature disengagement, 95
Price/quality, 33
Private line
  C-conditioned, 304
  D-conditioned, 304, 311
Protection standards
  TAPAC, 451
Protection switch (PS), 535, 538
Protection switching count (PSC), 535, 538, 548
Protection switching duration (PSD), 535, 538, 548
Protocols
  X.75, 159
PTE, 513, 514, 530, 532
Pulse code modulation. See PCM.

Q, 204, 227
  laws of addition, 61, 419
CS, 419
QDU, 207, 287
  CCITT Rec. G.113, 420
  end-to-end, 420
  IEC, 420
  LEC, 420
Quality assurance, 481, 498
Quality of service, 9, 25
Quality of service standards
  CCITT SG I, 463
  CCITT SG II, 463
Quantization distortion unit. See QDU.
Quantizing distortion, 62, 12, 178, 287, 304, 365, 376

Random variables
  dependent
    addition, 217
    subtraction, 221
  independent
    addition, 217
    subtraction, 217
RBS, 306, 311, 314, 419
Receive objective loudness rating. See ROLR.
Recommmednation
  CCIR, 522-2, 78
  CCITT, G.113, 122
  CCITT, G.114, 418
  CCITT, G.121, 379
  CCITT, G.711, 304
  CCITT, G.721, 304
  CCITT, G.811, 82
  CCITT, G.821, 78, 82, 89, 93, 94
  CCITT, G.82x, 82
  CCITT, P.31, 381
Reference conditions, 27
Regenerator. See Section terminating element.
Regional Bell Operating Company (RBOC), 12
Regional standards, 441, 442
Regulation
  CRTC, 449
  FCC, 448
Remote alarm indication (RAI), 532, 534, 538
Remote office test line (ROTL), 490
Remote switching systems (RSS), 367, 399
Remote test system (RTS), 491, 497
Remote test unit (RTU), 497
Repeater. See STE.
Reset probability, 153
Response time, 91
Return loss, 288-294
  loop balance, 289
  terminal balance, 292
  through balance, 292
Robbed bit signaling. See RBS.
ROLR, 50, 54, 55, 118, 185-192, 200, 219, 222, 223, 231, 265, 275, 382
Routing
  direct, 110, 117
  dynamic, 111
  hierarchical, 111
  intertandem, 117

tandem, 117
RSS
    round trip delay, 401
SCC, 441
Scrambler, 334
SDH standards
    CCITT SG XV, 463
SDH, 459
Section terminating element. See STE.
SEF/AIS second (SAS), 533, 537
SEFS, 524, 533, 537
Service
    digital data, 5, 20, 36, 271
    facsimile, 314
    image, 5, 36
    multimedia, 5
    POTS, 496
    segregated, 7
    signaling, 5, 314
    special, 490, 496, 508
    voice, 5, 20
    voiceband, 271
SES, 76, 80, 82, 84, 89, 138–150, 318–319, 524,
    532, 534, 536, 538
Severely-errored frame second. See SEFS.
Severely-errored framing (SEF), 530, 533,
    536–537, 545–546
Severely-errored second. See SES.
Sidetone, 64
Signal levels, 279–283
Signal-to-noise ratio. See SNR.
Singing return loss. See SRL.
SM, 59, 121, 122, 194, 394
SMDS, 8, 90, 156
SNR, 202, 204, 226, 227, 231, 232, 265, 303–304,
    310, 313
SOLR, 64–67
SONET, 15, 295, 326, 337–338, 347, 508, 513,
    535, 537, 548
Special services center (SSC), 490
Speech
    activity, 85, 358
    burst, 85
    conversational, 85
    detection, 85
    level, 281, 358
    monolog, 85
Speech clipping, 27, 84
    FEC, 84, 86
    MSC, 84, 87
Speech detector
    hangover, 49
    threshold, 49
Speech quality, 27
SRL, 195, 227, 238, 288
Standards
    CCITT, 12, 20, 33, 95, 138, 151, 166, 522,
        545
    CCITT SG XII, 369
    EIA/TIA, 15, 167
    ETSI, 15
    IEEE, 15, 33
    T1, 15, 20, 95, 168

T1M1, 10
T1Q1, 10, 167, 172, 369, 522
TIA, 168
Standards forums
    T1P1, 417
    T1Q1, 383
    T1Q1.1, 380, 402
    TIA, 402
    TIA TR41.1, 380
Station set. See Telephones.
Statistics
    moments
        mean, 211
        SD, 211
    percentiles, 213, 235, 239, 241, 244, 248, 250,
        255
STE, 513, 548
STPL, 64
Statum clocks, 343
Subjective measure
    GOB, 24, 28, 29, 33
    MOS, 27, 29
    POW, 28, 29
Subjective testing
    conversational, 27
    listening, 27
Subscriber loop. See Loop plant.
Superframe format. See Digital signals, DS1-
    SF.
Surveillance and control of transmission systems
    (SCOTS), 491
Survey data, 112, 271
    EAS, 113, 116, 273, 274, 278
    EOS, 113, 117, 274, 278, 285
    intertoll trunk, 274
    loop, 113
    noise, 283
    speech level, 357
    trunk, 113
    trunk lengths, 295
    voiceband data error performance, 303
Survivability performance standards
    T1Q1.2, 462
Survivable ring structures, 383
Switched access remote test system (SARTS),
    491
Switched maintenance access system (SMAS),
    491
Switched multi-megabit data service. See
    SMDS.
Switching control center (SCC), 490, 491
Switching control center system (SCCS), 490
Switching equipment standards
    CCITT SG XI, 464
Synchronization standards
    T1X1, 461
Synchronization, 343–349
Synchronous optical network. See SONET.

T-Carrier administration system (TCAS), 493
T1, 441, 443
T1M1, 448
    T1M1.2, 461
T1P1, 417, 445

T1Q1, 383
T1Q1.1, 369, 380, 402, 445
T1Q1.2, 445
T1Q1.3, 445
T1Q1.4, 445
T1Q1.5, 445
T1S1, 445
T1X1, 445
T1Y1, 445
Talker echo path loss. *See* TEPL.
Tandeming
   analog/asynchronous, 207, 288
   digital/semisynchronous, 207, 288
   digital/synchronous, 207, 288
TAPAC, 451
Technology, 10–13, 252
   analog, 39
   delay prone, 44, 57
   digital, 39
Telecom Canada, 450
Telecommunications
   networks, 7
   services, 5
   standards, 15
Telecommunications alarm surveillance and control (TASC), 491
Telecommunications management network. *See* TMN.
TELENET, 158
Telephones
   500 type, 118, 124, 201, 272, 385
   analog, 52, 67–68, 275
   digital, 54, 67, 271, 275, 384
   microphones, 73
TEPD, 193, 252
TEPL, 55, 57–58, 60, 124, 125, 190–193, 202, 207, 221, 224, 226, 252, 292, 382
   control of, 377
Terminating impedance, 288
Test signal
   1000 Hz, 434
   1004 Hz, 434
Test subjects, 28
Test system controller (TSC), 497
Throughput efficiency, 91, 152, 158
TIA, 440, 443
   TR-30, 447
   TR-30.3, 463
   TR-41, 447
   TR41.1, 402
   TR-45, 447
TIA TR41, 369
TIA TR41.1, 380
Tie trunk
   digital, 406
   tandem, 407
Time interval error (TIE), 347
TLP
   definition, 432
   on analog connections, 432
   on mixed analog/digital connection, 433
   switches, 433
TMN, 489, 493–503

examples
   circuit quality assurance, 498
   performance and alarm monitoring, 501
   testing, 496
maintenance functions, 495
motivation for, 493
TOLR, 50–51, 55, 118, 185–192, 219, 222, 223, 265, 275, 382
Transfer
   delay, 151, 156, 158
   denial, 94, 156, 158
   rate, 91
Transmission errors
   bursty, 80
   random, 76
Transmission performance
   definition, 1, 5, 15
Transmission plan
   Canada, 369
   FLP, 365, 428
   North American, 366
   UK, 369
   VNL, 364, 426
Transmission planning standards
   T1Q1.1, 462
   TIA TR-30.3, 462
Transmission planning, 15
Transmit objective loudess rating. *See* TOLR.
TRANSYSCOM, 447
Trunking
   intertoll, 274
   long-haul, 274, 278, 285, 296, 317, 318
   short-haul, 273, 278, 285, 296, 302, 317
TTC, 454
TYMNET, 158

UAS, 80, 89, 533, 538, 546
Unavailability, 546. *See also* Availability *and* UAS.
Unavailable second. *See* UAS.
UNINET, 158
Universal time coordinated, (UTC), 343
US study groups
   SG-A, 452
   SG-B, 452
   SG-C, 452
   SG-D, 452
User expectations, 24, 25

Via net loss, *See* VNL.
Video performance standards
   CCITT SG XVIII, 463
   CMTT, 463
VNL, 274, 278, 426
   transmission plan, 364
   trunk loss objectives, 426
Voice
   32 kbit/s encoding limits, 419
   7-kHz high quality, 364
Voiceband data, 60, 75, 107, 303, 377, 383
   effect of 32 kbit/s ADPCM, 420
   performance criteria, 420
Voluntary standards, 443

Wander, 151, 326

WEPL, 59, 60, 122, 195, 209, 383, 394, 401
Wilkinson's method, 226, 229, 232, 236, 238, 239, 241, 243, 299

X3, 449
X3S3, 449
   US TAG, 453
X3.102-1983, 449
X3.105-1985, 449
X3.141, 449

ZBTSI, 412, 518
ZCS, 304, 311
Zero byte time slot interchange. *See* ZBTSI.
Zero code suppression. *See* ZCS.

The Artech House Telecommunications Library

Vinton G. Cerf, Series Editor

Advances in Computer Communications and Networking, Wesley W. Chu, ed.

Advances in Computer Systems Security, Rein Turn, ed.

Analysis and Synthesis of Logic Systems, Daniel Mange

A Bibliography of Telecommunications and Socio-Economic Development, Heather E. Hudson

Codes for Error Control and Synchronization, Djimitri Wiggert

Communication Satellites in the Geostationary Orbit, Donald M. Jansky and Michel C. Jeruchim

Communications Directory, Manus Egan, ed.

The Complete Guide to Buying a Telephone System, Paul Daubitz

Current Advances in LANs, MANs, and ISDN, B.G. Kim, ed.

Design and Prospects for the ISDN, G. Dicenet

Digital Cellular Radio, George Calhoun

Digital Hardware Testing: Transistor-Level Fault Modeling and Testing, Rochit Rajsuman, ed.

Digital Signal Processing, Murat Kunt

Digital Switching Control Architectures, Giuseppe Fantauzzi

Digital Transmission Design and Jitter Analysis, Yoshitaka Takasaki

Distributed Processing Systems, Volume I, Wesley W. Chu, ed.

Disaster Recovery Planning for Telecommunications, Leo A. Wrobel

E-Mail, Stephen A. Caswell

Expert Systems Applications in Integrated Network Management, E.C. Ericson, L.T. Ericson, and D. Minoli, eds.

FAX: Digital Facsimile Technology and Applications, Second Edition, Dennis Bodson, Kenneth McConnell, and Richard Schaphorst

Fiber Network Service Survivability, Tsong-Ho Wu

Fiber Optics and CATV Business Strategy, Robert K. Yates et al.

A Guide to Fractional T1, J.E. Trulove

Handbook of Satellite Telecommunications and Broadcasting, L.Ya. Kantor, ed.

Implementing X.400 and X.500: The PP and QUIPU Systems, Steve Kille

Information Superhighways: The Economics of Advanced Public Communication Networks, Bruce Egan

Integrated Broadband Networks, Amit Bhargava

Integrated Services Digital Networks, Anthony M. Rutkowski

International Telecommunications Management, Bruce R. Elbert

International Telecommunication Standards Organizations, Andrew Macpherson

Internetworking LANs: Operation, Design, and Management, Robert Davidson and Nathan Muller

Introduction to Satellite Communication, Burce R. Elbert

Introduction to Telecommunication Electronics, A. Michael Noll

Introduction to Telephones and Telephone Systems, Second Edition, A. Michael Noll

The ITU in a Changing World, George A. Codding, Jr. and Anthony M. Rutkowski

Jitter in Digital Transmission Systems, Patrick R. Trischitta and Eve L. Varma

LAN/WAN Optimization Techniques, Harrell Van Norman

LANs to WANs: Network Management in the 1990s, Nathan J. Muller and Robert P. Davidson

The Law and Regulation of International Space Communication, Harold M. White, Jr. and Rita Lauria White

Long Distance Services: A Buyer's Guide, Daniel D. Briere

Mathematical Methods of Information Transmission, K. Arbenz and J.C. Martin

Measurement of Optical Fibers and Devices, G. Cancellieri and U. Ravaioli

Meteor Burst Communication, Jacob Z. Schanker

Minimum Risk Strategy for Acquiring Communications Equipment and Services, Nathan J. Muller

Mobile Information Systems, John Walker

Networking Strategies for Information Technology, Bruce Elbert

Numerical Analysis of Linear Networks and Systems, Hermann Kremer et al.

Optimization of Digital Transmission Systems, K. Trondle and Gunter Soder

The PP and QUIPU Implementation of X.400 and X.500, Stephen Kille

Packet Switching Evolution from Narrowband to Broadband ISDN, M. Smouts

Principles of Secure Communication Systems, Second Edition, Don J. Torrieri

Principles of Signals and Systems: Deterministic Signals, B. Picinbono

Private Telecommunication Networks, Bruce Elbert

Radiodetermination Satellite Services and Standards, Martin Rothblatt

Residential Fiber Optic Networks: An Engineering and Economic Analysis, David Reed

Setting Global Telecommunication Standards: The Stakes, The Players, and The Process, Gerd Wallenstein

Signal Processing with Lapped Transforms, Henrique S. Malvar

The Telecommunications Deregulation Sourcebook, Stuart N. Brotman, ed.

Television Technology: Fundamentals and Future Prospects, A. Michael Noll

Telecommunications Technology Handbook, Daniel Minoli

Telephone Company and Cable Television Competition, Stuart N. Brotman

Terrestrial Digital Microwave Communciations, Ferdo Ivanek, ed.

Transmission Networking: SONET and the SDH, Mike Sexton and Andy Reid

Transmission Performance of Evolving Telecommunications Networks, John Gruber and Godfrey Williams

Troposcatter Radio Linds, G. Roda

Virtual Networks: A Buyer's Guide, Daniel D. Briere

Voice Processing, Walt Tetschner

Voice Teletraffic System Engineering, James R. Boucher

Wireless Access and the Local Telephone Network, George Calhoun

Applied Biochemistry and Biotechnology

Volumes 70–72, Complete, Spring 1998

Copyright ©1998 Humana Press Inc.

All Rights Reserved.

No part of this publication may be reproduced or transmitted in any form or by any means, electronic or mechanical, including photocopy, recording, or any information storage and retrieval system, without permission in writing from the copyright owner.

Applied Biochemistry and Biotechnology is abstracted or indexed regularly in *Chemical Abstracts, Biological Abstracts, Current Contents, Science Citation Index, Excerpta Medica, Index Medicus,* and appropriate related compendia.

TP
248.3
.J68
1998
v. 70-72

Biotechnology for Fuels and Chemicals

Proceedings of the Nineteenth Symposium on Biotechnology for Fuels and Chemicals Held May 4–8, 1997, at Colorado Springs, Colorado

Editors

Mark Finkelstein

National Renewable Energy Laboratory

Brian H. Davison

Oak Ridge National Laboratory

Sponsored by

US Department of Energy's Office of Fuels Development
and the Office of Industrial Technologies
National Renewable Energy Laboratory
Oak Ridge National Laboratory
Idaho National Environmental and Engineering Laboratory
Lockheed Martin Energy Systems, Inc.
A. E. Staley Manufacturing Company
Arkenol
Bio-Technical Resources, L. P.
Cargill Inc.
Chronopol, Inc.
Colorado Institute of Research for Biotechnology
Coors Brewing Company
E. I. Du Pont de Nemours and Company
Grain Processing Corporation
Henkel, Corp.
Raphael Katzen Associates International, Inc.
Nedalco B. V.
American Chemical Society's Division of Biochemical Technology

BUCKMAN LABORATORIES, INTL.
TECHNICAL INFORMATION CENTER

Humana Press • Totowa, New Jersey

Introduction

MARK FINKELSTEIN

National Renewable Energy Laboratory

BRIAN H. DAVISON

Oak Ridge National Laboratory

The proceedings of the 19th symposium on *Biotechnology for Fuels and Chemicals,* held in Colorado Springs, Colorado, May 4–8, 1997, had over 200 attendees. This meeting continues to provide a unique forum for the presentation of new applications and recent research advances in the production of fuels and chemicals through biotechnology. The utilization of renewable resources, and in particular cellulosic biomass, has broad implications in today's world of greenhouse gases, global warming, ozone layers, climate change, energy sustainability, and carbon emissions. It also has relevance to the chemical industry's continuing need to both lower current chemical production costs and produce novel chemicals. Biotechnology and bioprocessing are now making it possible to convert this biomass to fuels and chemicals in a commercially attractive fashion.

The 19th Symposium captures a wide range of technical topics from an academic, industrial, or government perspective. A variety of biomass feedstocks are discussed in Session 1, along with several updated and innovative pretreatment processing approaches. The ability to turn lignocellulosic materials into simple sugars offers great opportunities to generate cost-effective feedstocks to be used in biotechnological processes for the production of fuels and chemicals.

Through the advent of genetic engineering, the development of a series of exciting new biocatalysts and microbes were presented in Session 2. The ability to enhance the metabolic capabilities of microorganisms and further refine them into minifactories is becoming increasingly routine. Enzymes are being dissected, re-assembled, and altered for heightened kinetic properties.

In Session 3, new reactor designs and innovative conversion processes were the focus. These were further refined in Session 4 as past, current, and future integrated operations were discussed in the context of state and industrial needs, as well as estimated and actual costs. This session was a particularly exciting forum for discussion of the near-term feasibility of biomass-based ethanol plants.

The application of biotechnology to the production of bulk and specialty chemicals was the spotlight of Session 5. A number of microbes or plants were genetically modified to produce enzymes, key chemical intermediates, or perform novel bioconversions. The paper and pulp industry was the beneficiary of the power of biotechnology as a variety of enzymes were exploited in Session 6.

The technical program consisted of 37 oral and 138 poster presentations covering topics in feedstock supply and processing, applied biological research, bioprocessing research, commercialization and process economics, specialty chemicals with emphasis on environmentally benign products and processes, and biotechnology in the paper and pulp industry. Two special topics sessions discussed "Extremophiles and Their Enzymes" and "Climate Change and Global Warming." The productive interactions of the 200 attendees between technical sessions in addition to the banquet and tour of the National Renewable Energy Laboratory's ethanol Process Development Unit helped continue the fond tradition of this Symposium.

The Nineteenth Symposium was sponsored by the Office of Fuels Development (DOE), Office of Industrial Technologies (DOE), the National Renewable Energy Laboratory, Oak Ridge National Laboratory, Idaho National Engineering Laboratory, Lockheed Martin Energy Research, Arkenol, Nedalco B.V., American Chemical Society-Division of BioChem Technology, Chronopol Inc., Raphael Katzen Associates, Colorado Institute of Research for Biotechnology, A. E. Staley, DuPont, Henkel Corporation, Cargill, Inc., Bio-Technical Resources, L.P., Grain Processing Corporation, and the Coors Brewing Company.

Organization of the Symposium was as follows:

Organizing Committee

Mark Finkelstein, Conference Chairman, National Renewable Energy Laboratory

Brian H. Davison, Conference Co-chairman, Oak Ridge National Laboratory

David Boron, US Department of Energy

Ting Carlson, Cargill

Kathleen Clarkson, Genencor

Bruce Dale, Michigan State University

James Doncheck, Bio-Technical Resources, L.P.

Introduction

Mark Donnelly, Argonne National Laboratory
Tom Jeffries, USDA-Forest Products Laboratory
Don Johnson, Grain Processing Corporation
Raphael Katzen, Raphael Katzen Associates International
Lee Lynd, Dartmouth College
Mark Reeves, Oak Ridge National Laboratory
Jack Saddler, University of British Columbia
Valerie Sarisky-Reed, U.S. Department of Energy
Sharon Shoemaker, University of California
Liz Willson, National Renewable Energy Laboratory
Charles Wyman, BC International Corporation

Session Chairpersons and Cochairpersons

Session 1: Feedstock Supply and Processing
Michael R. Ladisch, *Purdue University*
Lynn Wright, *Oak Ridge National Laboratory*

Session 2: Applied Biological Research
Tom Jeffries, *University of Wisconsin, Forest Products Laboratory*
Mike Himmel, *National Renewable Energy Laboratory*

Session 3: Bioprocessing Research
Rakesh Bajpai, *University of Missouri*
David Glassner, *National Renewable Energy Laboratory*

Session 4: Industrial Needs, Commercialization and Process Economics
Patrick Foody, *Iogen Corporation*
J. Russell Miller, *Arkenol Energy, Inc.*

Session 5: Specialty Chemicals with Emphasis on Environmentally Benign Products and Processes
Robert Dorsch, *DuPont*
Nhuan Nghiem, *Oak Ridge National Laboratory*

Session 6: Biotechnology in the Paper and Pulp Industry
Jack Saddler, *University of British Columbia*
Esteban Chornet, *University of Sherbrooke*

Acknowledgments

The able assistance of Joan Ross, as Conference Coordinator, Liz Willson, Symposium Secretary and Proceedings Editor, assisted by Heather Bulmer, Renae Humphrey, Netta Ingle, Gail McNabb, Sally NeuDeld, and Cricket Pierce.

The National Renewable Energy Laboratory is operated by Midwest Research Institute, for the US Department of Energy under contract DE-AC36-83CH190093.

Oak Ridge National Laboratory is managed by Lockheed Martin Energy Systems, Inc., for the US Department of Energy under contract DE-AC05960R22464.

The submitted manuscript has been authored by a contractor of the US Government under contract DE-AC05-960R22464. Accordingly, the US Government retains a nonexclusive, royalty-free license to publish or reproduce the published form of this contribution, or allow others to do so, for US Government purposes.

Other Proceedings in This Series

1. "Proceedings of the First Symposium on Biotechnology in Energy Production and Conservation" (1978), *Biotechnol. Bioeng. Symp.* **8.**
2. "Proceedings of the Second Symposium on Biotechnology in Energy Production and Conservation" (1980), *Biotechnol. Bioeng. Symp.* **10.**
3. "Proceedings of the Third Symposium on Biotechnology in Energy Production and Conservation" (1981), *Biotechnol. Bioeng. Symp.* **11.**
4. "Proceedings of the Fourth Symposium on Biotechnology in Energy Production and Conservation" (1982), *Biotechnol. Bioeng. Symp.* **12.**
5. "Proceedings of the Fifth Symposium on Biotechnology for Fuels and Chemicals" (1983), *Biotechnol. Bioeng. Symp.* **13.**
6. "Proceedings of the Sixth Symposium on Biotechnology for Fuels and Chemicals" (1984), *Biotechnol. Bioeng. Symp.* **14.**
7. "Proceedings of the Seventh Symposium on Biotechnology for Fuels and Chemicals" (1985), *Biotechnol. Bioeng. Symp.* **15.**
8. "Proceedings of the Eigth Symposium on Biotechnology for Fuels and Chemicals" (1986), *Biotechnol. Bioeng. Symp.* **17.**
9. "Proceedings of the Ninth Symposium on Biotechnology for Fuels and Chemicals" (1988), *Appl. Biochem. Biotechnol.* **17,18.**
10. "Proceedings of the Tenth Symposium on Biotechnology for Fuels and Chemicals" (1989), *Appl. Biochem. Biotechnol.* **20,21.**

Introduction

11. "Proceedings of the Eleventh Symposium on Biotechnology for Fuels and Chemicals" (1990), *Appl. Biochem. Biotechnol.* **24,25.**
12. "Proceedings of the Twelfth Symposium on Biotechnology for Fuels and Chemicals" (1991), *Appl. Biochem. Biotechnol.* **28,29.**
13. "Proceedings of the Thirteenth Symposium on Biotechnology for Fuels and Chemicals" (1992), *Appl. Biochem. Biotechnol.* **34,35.**
14. "Proceedings of the Fourteenth Symposium on Biotechnology for Fuels and Chemicals" (1993), *Appl. Biochem. Biotechnol.* **39,40.**
15. "Proceedings of the Fifteenth Symposium on Biotechnology for Fuels and Chemicals" (1994), *Appl. Biochem. Biotechnol.* **45,46.**
16. "Proceedings of the Sixteenth Symposium on Biotechnology for Fuels and Chemicals" (1995), *Appl. Biochem. Biotechnol.* **51/52.**
17. "Proceedings of the Seventeenth Symposium on Biotechnology for Fuels and Chemicals" (1996), *Appl. Biochem. Biotechnol.* **57/58.**
18. "Proceedings of the Eighteenth Symposium on Biotechnology for Fuels and Chemicals" (1997), *Appl. Biochem. Biotechnol.* **63–65.**

This symposium has been held annually since 1978. We are pleased to have the proceedings of the Nineteenth Symposium currently published in this special issue to continue the tradition of providing a record of the contributions made.

The Twentieth Symposium is planned for May 3–7, 1998 in Gatlinburg, Tennessee. We encourage comments or discussions relevant to the format or content of the meetings.

Applied Biochemistry and Biotechnology
Vols. 70–72, Spring 1998

CONTENTS

Introduction
Mark Finkelstein and Brian Davison iii

SESSION 1—FEEDSTOCK SUPPLY AND PROCESSING

Introduction to Session 1
Michael R. Ladisch and Lynn Wright 1

Comparison of SO_2 and H_2SO_4 Impregnation of Softwood Prior to Steam Pretreatment on Ethanol Production
Charlotte Tengborg, Kerstin Stenberg, Mats Galbe, Guido Zacchi,* Simona Larsson, Eva Palmqvist, and Bärbel Hahn-Hägerdal 3

Pretreatment of Softwood by Acid-Catalyzed Steam Explosion Followed by Alkali Extraction
Daniel Schell,* Quang Nguyen, Melvin Tucker, and Brian Boynton 17

Comparison of Yellow Poplar Pretreatment Between NREL Digester and Sunds Hydrolyzer
M. P. Tucker,* J. D. Farmer, F. A. Keller, D. J. Schell, and Q. A. Nguyen 25

Shrinking-Bed Model for Percolation Process Applied to Dilute-Acid Pretreatment/Hydrolysis of Cellulosic Biomass
Rongfu Chen, Zhangwen Wu, and Y. Y. Lee* 37

Cost Estimates and Sensitivity Analyses for the Ammonia Fiber Explosion Process
Lin Wang, Bruce E. Dale,* Lale Yurttas, and I. Goldwasser 51

Selective Polarity- and Adsorption-Guided Extraction/Purification of *Annona* sp. Polar Acetogenins and Biological Assay Against Agricultural Pests
J. D. Fontana,* F. M. Lanças, M. Passos, E. Cappelaro, J. Vilegas, M. Baron, M. Noseda, A. B. Pomílio, A. Vitale, A. C. Webber, A. A. Maul, W. A. Peres, and L. A. Foerster 67

Dilute Acid Pretreatment of Softwoods: *Scientific Note*
Q. A. Nguyen,* M. P. Tucker, B. L. Boynton, F. A. Keller, and D. J. Schell 77

Pretreatment of Sugarcane Bagasse Hemicellulose Hydrolysate for Xylitol Production by *Candida guilliermondii*
Lourdes A. Alves,* Maria G. A. Felipe, João B. Almeida E. Silva, Silvio S. Silva, and Arnaldo M. R. Prata 89

Continuous pH Monitoring During Pretreatment of Yellow Poplar Wood Sawdust by Pressure Cooking in Water
Joseph Weil, Mark Brewer, Richard Hendrickson, Ayda Sarikaya, and Michael R. Ladisch* 99

---

*For papers with multiple authorship, the asterisk identifies the author to whom correspondence and reprint requests should be addressed.

SESSION 2—APPLIED BIOLOGICAL RESEARCH

Introduction to Session 2
  Tom Jeffries and Mike Himmel 113

Fuel Ethanol Production from Corn Fiber: *Current Status and Technical Prospects*
  Badal C. Saha,* Bruce S. Dien, and Rodney J. Bothast 115

Xylose Reductase Production by *Candida guilliermondii*
  S. M. A. Rosa, M. G. A. Felipe, S. S. Silva, and M. Vitolo* 127

Yeast Adaptation of Softwood Prehydrolysate
  Fred A. Keller,* Delicia Bates, Ray Ruiz, and Quang Nguyen 137

Production of Xylitol by *Candida mogii* from Rice Straw Hydrolysate: *Study of Environmental Effects Using Statistical Design*
  Z. D. V. L. Mayerhoff, I. C. Roberto,* and S. S. Silva 149

Improving Fermentation Performance of Recombinant *Zymomonas* in Acetic Acid-Containing Media
  Hugh G. Lawford* and Joyce D. Rousseau 161

Conditions that Promote Production of Lactic Acid by *Zymomonas mobilis* in Batch and Continuous Culture
  Hugh G. Lawford* and Joyce D. Rousseau 173

A Novel Fermentation Pathway in an *Escherichia coli* Mutant Producing Succinic Acid, Acetic Acid, and Ethanol
  Mark I. Donnelly,* Cynthia Sanville Millard, David P. Clark, Michael J. Chen, and Jerome W. Rathke 187

Cloned *Bacillus subtilis* Alkaline Protease (*aprA*) Gene Showing High Level of Keratinolytic Activity
  Taha I. Zaghloul 199

Isolation, Identification, and Keratinolytic Activity of Several Feather-Degrading Bacterial Isolates
  Taha I. Zaghloul,* M. Al-Bahra, and H. Al-Azmeh 207

Acetamide Degradation by a Continuous-Fed Batch Culture of *Bacillus sphaericus*
  F. Ramirez, O. Monroy,* E. Favela, J. P. Guyot, and F. Cruz 215

Use of Hemicellulose Hydrolysate for β-Glucosidase Fermentation
  K. Réczey, A. Brumbauer, M. Bollók, Zs. Szengyel,* and G. Zacchi 225

Production of a Novel Pyranose 2-Oxidase by the Basidiomycete *Trametes multicolor*
  Christian Leitner, Dietmar Haltrich,* Bernd Nidetzky, Hansjörg Prillinger, and Klaus D. Kulbe 237

Broad Spectrum and Mode of Action of an Antibiotic Produced
by *Scytonema* sp. TISTR 8208 in a Seaweed-Type Bioreactor
Aparat Chetsumon, Fusako Umeda, Isamu Maeda, Kiyohito Yagi,*
Tadashi Mizoguchi, and Yoshiharu Miura 249

Comparative Study of Xylanase Kinetics Using Dinitrosalicylic,
Arsenomolybdate, and Ion Chromatographic Assays
Thomas W. Jeffries,* Vina W. Yang, and Mark W. Davis 257

Production and Purification of CGTase of Alkalophylic *Bacillus*
Isolated from Brazilian Soil
*Graciette Matioli, Gisella M. Zanin, Manoel F. Guimarães,
and Flávio F. de Moraes** 267

Submerged Culture Screening of Two Strains of *Streptomyces* sp.
with High Keratinolytic Activity
O. Garcia-Kirchner, Ma. E. Bautista-Ramirez,
and M. Segura-Granados* 277

Cofermentation of Glucose, Xylose, and Arabinose by Mixed Cultures
of Two Genetically Engineered *Zymomonas mobilis* Strains
Ali Mohagheghi,* Kent Evans, Mark Finkelstein, and Min Zhang 285

Improvement of Substrate Conversion to Molecular Hydrogen
by Three-Stage Cultivation of a Photosynthetic Bacterium,
Rhodovulum sulfidophilum
Isamu Maeda, Wasimul Q. Chowdhury, Kenji Idehara, Kiyohito Yagi,*
*Tadashi Mizoguchi, Toru Akano, Hitoshi Miyasaka, Toshio Furutani,
Yoshiaki Ikuta, Norio Shioji, and Yoshiharu Miura* 301

Effect of Drying on Bioremediation Bacteria Properties
*F. Weekers, Ph. Jacques, D. Springael, M. Mergeay, L. Diels,
and Ph. Thonart** 311

Production of L-Lactic Acid by *Rhizopus oryzae* in a Bubble Column
Fermenter
Jianxin Du,* Ningjun Cao, Cheng S. Gong, and George T. Tsao 323

Factors that Affect the Biosynthesis of Xylitol by Xylose-Fermenting
Yeasts: A Review
Silvio S. Silva,* Maria G. A. Felipe, and Ismael M. Mancilha 331

Cloning and Sequence Analysis of the Poly(3-Hydroxyalkanoic Acid)-
Synthesis Genes of *Pseudomonas acidophila*
Fusako Umeda, Yoshiharu Kitano, Yuki Murakami, Kiyohito Yagi,*
Yoshiharu Miura, and Tadashi Mizoguchi 341

Continuous Culture Studies of Xylose-Fermenting *Zymomonas mobilis*
*Hugh G. Lawford, Joyce D. Rousseau, Ali Mohagheghi,
and James D. McMillan** 353

Effect of Surfactants and Zeolites on Simultaneous Saccharification
and Fermentation of Steam-Exploded Poplar Biomass to Ethanol
*I. Ballesteros, J. M. Oliva, J. Carrasco, A. Cabañas, A. A. Navarro,
and M. Ballesteros** 369

Thermal Stability and Energy of Deactivation of Free and Immobilized
Amyloglucosidase in the Saccharification of Liquefied Cassava Starch
Gisella M. Zanin* and Flavio F. De Moraes 383

Hydrolysis of Cellulose Using Ternary Mixtures of Purified Cellulases
John O. Baker, Christine I. Ehrman, William S. Adney,
Steven R. Thomas, and Michael E. Himmel* 395

The Production and Properties of a New Xylose Reductase
from Fungus: *Neurospora crassa*
Xin Zhao,* Peiji Gao, and Zunong Wang 405

SESSION 3—BIOPROCESSING RESEARCH

Introduction to Session 3
Rakesh Bajpai and David A. Glassner 415

An Integrated Bioconversion Process for Production of L-Lactic
Acid from Starchy Potato Feedstocks
S. P. Tsai* and S.-H. Moon 417

Production of Ethanol from Starch by Co-Immobilized *Zymomonas
mobilis*–Glucoamylase in a Fluidized-Bed Reactor
May Y. Sun, Nhuan P. Nghiem, Brian H. Davison, Oren F. Webb,
and Paul R. Bienkowski* 429

Ethanol Production from AFEX-Treated Forages and Agricultural Residues
Khaled Belkacemi, Ginette Turcotte, Damien de Halleux,
and Philippe Savoie* 441

Adaptive Optimal Control of Fed-Batch Alcoholic Fermentation
T. L. M. Alves,* A. C. Costa, A. W. S. Henriques, and E. L. Lima 463

Nonisothermal Simultaneous Saccharification and Fermentation
for Direct Conversion of Lignocellulosic Biomass to Ethanol
Zhangwen Wu and Y. Y. Lee* 479

Comparison Between Experimental and Theoretical Values
of Effectiveness Factor in Cephalosporin C Production
Process with Immobilized Cells
M. Lucia G. C. Araujo, Roberto C. Giordano, and Carlos O. Hokka* 493

Downstream Processing of Inulinase: *Comparison of Different Techniques*
Adalberto Pessoa, Jr.* and Michele Vitolo 505

Improvement of Lactic Cell Production
S. Desmons,* H. Krhouz, P. Evrard, and P. Thonart 513

In Situ Global Method for Measurement of Oxygen Demand
and Mass Transfer
K. Thomas Klasson, Karin M. O. Lundbäck, Edgar C. Clausen,
and James L. Gaddy* 527

Improvement of Oxygen Transfer Coefficient During *Penicillium canescens* Culture: *Influence of Turbine Design, Agitation Speed, and Air Flow Rate on Xylanase Production*
A. Gaspar,* L. Strodiot, and Ph. Thonart 535

Batch Foam Recovery of Sporamin from Sweet Potato
Samuel Ko, Veara Loha, Aleš Prokop, and Robert D. Tanner* 547

Batch Foam Fractionation of Kudzu (*Pueraria lobata*) Vine Retting Solution
Jirawat Eiamwat, Veara Loha, Aleš Prokop, and Robert D. Tanner* 559

Development of a Novel, Two-Step Process for Treating Municipal
Biosolids for Beneficial Reuse
Christopher J. Rivard,* Brian W. Duff, and Nicholas J. Nagle 569

Phenomenological and Neural-Network Modeling of Cephalosporin
C Production Bioprocess
A. J. G. Cruz, M. L. G. C. Araujo, R. C. Giordano, and C. O. Hokka* 579

Biosorption of Nickel Using Filamentous Fungi
L. Mogollón,* R. Rodríguez, W. Larrota, N. Ramirez, and R. Torres 593

Conversion of Food Industrial Wastes into Bioplastics
P. H. Yu,* H. Chua, A. L. Huang, W. Lo, and G. Q. Chen 603

Improved Oxygen Delivery in a Continuous-Roller-Bottle Reactor
*R. Eric Berson, Trupti V. Mane, C. Kurt Svihla,
and Thomas R. Hanley** 615

Xylanase Recovery: *Effect of Extraction Conditions on the Aqueous
Two-Phase System Using Experimental Design*
Silgia A. Costa, Adalberto Pessoa, Jr.,* and Inês C. Roberto 629

Extracellular Proteolytic Processing of *Aspergillus awamori* GAI
into GAII is Supported by Physico-Chemical Evidence
Hilton J. Nascimento, Valeria F. Soares, Elba P. S. Bon,
and José G. Silva, Jr.* 641

Technical and Economic Evaluation of Different Reactors
for Methanotrophic Cultures for Propylene Oxide Production
Bhupendra K. Soni,* Robert L. Kelley, and Vipul J. Srivastava 651

Xylanase Recovery by Ethanol and Na$_2$SO$_4$ Precipitation
  Ely V. Cortez, Adalberto Pessoa, Jr.,* and Adilson N. Assis 661

Production of Citronellyl Acetate in a Fed-Batch System Using
  Immobilized Lipase: *Scientific Note*
  Heizir F. de Castro, Diovana A. S. Napoleão,
  and Pedro C. Oliveira* 667

Production and Purification of Tartrate Dehydrogenase:
  Role of Aqueous Two-Phase Extraction
  R. Harve and R. K. Bajpai* 677

Demonstration-Scale Evaluation of a Novel High-Solids Anaerobic
  Digestion Process for Converting Organic Wastes to Fuel Gas
  and Compost
  Christopher J. Rivard, Brian W. Duff, James H. Dickow,
  Carlton C. Wiles, Nicholas J. Nagle, James L. Gaddy,
  and Edgar C. Clausen* 687

Recycling of Process Streams in Ethanol Production from Softwoods
  Based on Enzymatic Hydrolysis
  Kerstin Stenberg, Charlotte Tengborg, Mats Galbe, Guido Zacchi,
  Eva Palmqvist, and Bärbel Hahn-Hägerdal* 697

Biological-Chemical Treatment of Soils Contaminated
  with Exploration and Production Wastes
  Bhupendra K. Soni, J. Robert Paterek, Salil Pradhan,
  and Vipul J. Srivastava* 709

Recovery and Refining of Au by Gold-Cyanide Ion Biosorption
  Using Animal Fibrous Proteins
  Shin-Ichi Ishikawa* and Kyozo Suyama 719

Effect of Temperature and Pressure on Growth and Methane
  Utilization by Several Methanotrophic Cultures
  B. K. Soni,* John Conrad, Robert L. Kelley, and Vipul J. Srivastava 729

The Production of Hydrocarbons from Photoautotrophic Growth
  of *Dunaliella salina* 1650
  Don-Hee Park, Hwa-Won Ruy, Ki-Young Lee, Choon-Hyoung Kang,
  Tae-Ho Kim, and Hyeon-Yong Lee* 739

Fouling and Protein Adsorption: *Effect of Low-Temperature Plasma
  Treatment of Membrane Surfaces*
  J. Johansson, H. K. Yasuda, and R. K. Bajpai* 747

Biocatalytic Removal of Nickel and Vanadium from Petroporphyrins
  and Asphaltenes
  L. Mogollón, R. Rodríguez, W. Larrota, C. Ortiz, and R. Torres* 765

Expanded-Bed Adsorption Utilizing Ion-Exchange Resin to Purify
  Extracellular β-Galactosidase
  *José Antonio Marques Pereira, Paulo De Tarso Vieira, E. Rosa,
  Glaucia Maria Pastore, and Cesar Costapinto Santana**...................... 779

β-Cyclodextrin Production by Simultaneous Fermentation
  and Cyclization
  Heron O. S. Lima, Flavio F. De Moraes, and Gisella M. Zanin*.............. 789

SESSION 4—INDUSTRIAL NEEDS, COMMERCIALIZATION, AND PROCESS ECONOMICS

Introduction to Session 4
  Patrick Foody and J. Russ Miller 805

Use of Net Present Value Analysis to Evaluate a Publicly Funded
  Biomass-to-Ethanol Research, Development, and Demonstration
  Program and Valuate Expected Private Sector Participation
  Norman D. Hinman* and Mark A. Yancey 807

Coordinating California's Efforts to Promote Waste to Alcohol Production
  William J. Blackburn* and Jonathan M. Teague 821

SESSION 5—SPECIALTY CHEMICALS WITH EMPHASIS ON ENVIRONMENTALLY BENIGN PRODUCTS AND PROCESSES

Introduction to Session 5
  Robert Dorsch and Nhuan Nghiem 843

Production of L-Malic Acid via Biocatalysis Employing Wild-Type
  and Respiratory-Deficient Yeasts
  Xiaohai Wang, C. S. Gong, and George T. Tsao* 845

Fate of Branched-Chain Fatty Acids in Anaerobic Environment
  of River Sediment
  H. Chua,* W. Lo, and P. H. F. Yu 853

Simultaneous Enzymatic Synthesis of Gluconic Acid and Sorbitol:
  *Continuous Process Development Using Glucose-Fructose
  Oxidoreductase from* Zymomonas mobilis
  *Marisol Silva-Martinez, Dietmar Haltrich, Senad Novalic,
  Klaus D. Kulbe, and Bernd Nidetzky** 863

Biotechnological Production of Xylitol from Agroindustrial Residues:
  Evaluation of Bioprocesses
  Denise C. G. A. Rodrigues, Silvio S. Silva, Arnaldo Márcio R. Prata,
  and Maria das Gracas A. Felipe* 869

Ethanol from Babassu Coconut Starch: *Technical and Economical Aspects*
  Edmond A. Baruque Filho,* Maria da Graça A. Baruque,
  Denise M. G. Freire, and Geraldo L. Sant'Anna, Jr. 877

Biotechnological Production of Acrylic Acid from Biomass
  H. Danner,* M. Ürmös, M. Gartner, and R. Braun 887

Bacillus stearothermophilus for the Thermophilic Production of L-Lactic Acid
  H. Danner,* M. Neureiter, L. Madzingaidzo, M. Gartner,
  and R. Braun 895

In Situ Mutagenesis and Chemotactic Selection of Microorganisms
  in a Diffusion Gradient Chamber
  Mark R. Mikola, Mark T. Widman, and R. Mark Worden* 905

Bioconversion of Fumaric Acid to Succinic Acid by Recombinant *E. coli*
  Xiaohai Wang, C. S. Gong, and George T. Tsao* 919

Accumulation of Biodegradable Copolyesters of 3-Hydroxy-Butyrate
  and 3-Hydroxyvalerate in *Alcaligenes eutrophus*
  H. Chua,* P. H. F. Yu, and W. Lo 929

SESSION 6—BIOTECHNOLOGY IN THE PULP AND PAPER INDUSTRY

Introduction to Session 6
  Jack Saddler and E. Chornet 937

Efficient Production of Mannan-Degrading Enzymes
  by the Basidiomycete *Sclerotium rolfsii*
  Alois Sachslehner, Dietmar Haltrich,* Georg Gübitz,
  Bernd Nidetzky, and Klaus D. Kulbe 939

Production and Characterization of *Phanerochaete chrysosporium*
  Lignin Peroxidases for Pulp Bleaching
  M. E. A. de Carvalho,* M. C. Monteiro, E. P. S. Bon,
  and G. L. Sant'Anna, Jr. 955

In Vitro Degradation of Insoluble Lignin in Aqueous Media
  by Lignin Peroxidase and Manganese Peroxidase
  David N. Thompson,* Bonnie R. Hames, C. A. Reddy,
  and Hans E. Grethlein 967

Pulp Bleaching Using Laccase from *Trametes versicolor* Under High
  Temperature and Alkaline Conditions
  M. C. Monteiro and M. E. A. de Carvalho* 983

Microbial Oxidation of Mixtures of Methylmercaptan
  and Hydrogen Sulfide
  Anbu Subramaniyan, Ravindra Kolhatkar, K. L. Sublette,*
  and Robert Beitle 995

Author Index 1007

Subject Index 1011

Session 1

Feedstock Supply and Processing

MICHAEL R. LADISCH[1] AND LYNN WRIGHT[2]

[1] *Laboratory of Renewable Resources Engineering and Department of Agricultural and Biological Engineering West Lafayette, IN* and
[2] *Oak Ridge National Laboratory, Oak Ridge, TN*

The availability of biomass feedstocks' as well as efficient and cost-effective processing technology, are key determinants of the economic viability of obtaining fuel additives and oxygenated chemicals from biomass. This series of papers examines biomass feedstock availability, and the impact of advances in hydrolysis and pretreatment methods on improving extents of conversion of cellulose to glucose. Significant progress has been made, but the economic impact of these advances will not be fully realized or documented until the first biomass conversion plants are built. In the meantime, these papers provide insights into the developments which enhance the prospects for implementation.

Biomass feedstocks will be available for properly located conversion plants - if the price is right. Based on the assumption of a constant biomass feedstock supply, economic analysis shows that the price of biomass will increase as this resource becomes scarce. Low-cost wood residues, for instance, are a limited supply. Availability of the feedstock will be a key factor, along with infrastructure considerations, in determining size and location of a conversion facility. One type of cellulosic material which could become widely available is switchgrass, a native grass which can be grown in most cropland regions, and has the potential to produce annual yields on the order of 6 to 8 dry tons per acre. Current analysis suggests that biomass feedstocks are likely to cost on average, $25 to 40/dry ton at the farmgate, depending on location and size of the industry.

Thorough examination of the shrinking bed hydrolysis concept shows that a stream of pentoses and hexoses can be obtained by carrying out hydrolysis in a reactor which facilitates compression of the biomass as it mass decreases during the conversion of the insoluble cellulose and soluble hemicelluloses to monosaccharides via acid hydrolysis. An alternate approach is to pretreat the biomass, and then follow pretreatment with enzyme hydrolysis. Aqueous based, hydrothermal pretreatments are attractive since these entail little or no addition of extraneous reagents, while making the cellulose more susceptible to enzyme based hydrolysis. The approaches discussed in this session are steam explosion with added SO_2 or H_2SO_4 to pretreat residue from Douglas fir, use of hot liquid water and steam to obtain reactive, readily hydrolyzed fibers from sugarcane bagasse and Aspen; and pretreatment of wood sawdust by pressure cooking it in water in an agitated batch reactor with a specially designed pH monitoring system. The water in this system was controlled to near neutral pH, in order to minimize autohydrolysis during pretreatment.

These papers give an overview of the state-of-the-art, and experimental results based on a number of different lignocellulosic substrates. This session of the 19th Biotechnology Symposium shows that significant progress is being made in the generation, front end treatment, and hydrolysis of lignocellulosic substrates.

Comparison of SO₂ and H₂SO₄ Impregnation of Softwood Prior to Steam Pretreatment on Ethanol Production

CHARLOTTE TENGBORG,[1] KERSTIN STENBERG,[1] MATS GALBE,[1] GUIDO ZACCHI,*,[1] SIMONA LARSSON,[2] EVA PALMQVIST,[2] AND BÄRBEL HAHN-HÄGERDAL[2]

[1]*Department of Chemical Engineering I and* [2]*Department of Applied Microbiology, Lund University, P.O. Box 124, SE-221 00 Lund, Sweden*

ABSTRACT

The pretreatment of softwood with sulfuric acid impregnation in the production of ethanol, based on enzymatic hydrolysis, has been investigated. The parameters investigated were: H_2SO_4 concentration (0.5–4.4% w/w liquid), temperature (180 – 240°C), and residence time (1–20 minutes). The combined severity (log Ro-pH) was used to combine the parameters into a single reaction ordinate. The highest yields of fermentable sugars, i.e., glucose and mannose, were obtained at a combined severity of 3. At this severity, however, the fermentability declined and the ethanol yield decreased. In a comparison with previous results, SO_2 impregnation was found to be preferable, since it resulted in approximately the same sugar yields, but better fermentability.

Index Entries: Softwood, H_2SO_4; SO_2; steam pretreatment, ethanol production

INTRODUCTION

Several countries, including Sweden, have decided to reduce or to allow no increase in the present level of carbon dioxide discharge to the atmosphere. Interest in fuel ethanol produced from renewable resources has therefore increased during recent years, since ethanol produced from biomass results in no net contribution of carbon dioxide to the atmosphere.

In Sweden, the most abundant raw materials for ethanol production are softwoods in the form of logging waste and waste from the forest industry. In contrast to hardwoods and agricultural residues, which have

---

*Author to whom all correspondence and reprint requests should be addressed.

been thoroughly investigated *(1–3)*, softwoods are not easily converted to sugars *(4–6)*.

The enzymatic digestibility of wood is improved when it is impregnated with sulfuric acid or sulfur dioxide prior to steam pretreatment *(5,7–11)*. Acid-catalyzed steam pretreatment of softwood has been utilized to increase sugar yields, but pretreatment conditions resulting in high sugar yields also result in the formation of degradation products *(12)*. When the pretreatment conditions for SO_2-impregnated steam-pretreated softwood were optimized regarding ethanol yield, it was observed that the ethanol production rate was also affected by the pretreatment conditions *(13)*.

One of the most important factors for the economic outcome of ethanol production from lignocellulose is the overall ethanol yield, which requires a high degree of utilization of the raw material *(14)*. In the present study on steam pretreatment of H_2SO_4-impregnated spruce, the effects of the following parameters were investigated: H_2SO_4 concentration (0.5–4.4% w/w liquid), pretreatment temperature (180–240°C), and residence time (1–20 min). To compare the different pretreatment conditions in a normalized way, a single reaction ordinate, the combined severity factor, was used. The combined severity factor is an extension of the severity factor, which has been used in various studies of the pretreatment conditions of lignocellulose *(3,15–18)*. The effects of pretreatment were assessed by the sugar yields after pretreatment and after enzymatic hydrolysis of the fibrous material, and the ethanol yield following fermentation of the sugar solution after pretreatment. The effect on the ethanol productivity was also determined. The results were compared with the results obtained in a previous study using SO_2-impregnated mixed softwoods *(13)*.

MATERIALS AND METHODS

Raw Material

Fresh-chipped spruce, free from bark, was provided by a sawmill, Höörsågen AB (Höör, Sweden). The chip size was less than 30 mm, the material had a dry matter (DM) content of 43%, and was stored in plastic bags at 4°C. The composition, analyzed by STORA AB (Säffle, Sweden), is given in Table 1.

Pretreatment

The following parameters were investigated: sulfuric acid concentrations of 0.5, 2.4, and 4.4% w/w liquid; temperatures of 180, 200, 210, 225, and 240°C, and times of 1, 5, 10, 15, and 20 min (Table 2). Pretreatment was performed in equipment previously described *(13)*. Wood chips corresponding to 200 g DM were mixed, in a plastic bag, with 100 mL dilute

Table 1
Composition of Raw Material

Component	% of DM
Extractives	1.0
Galactan	1.8
Glucan	43.4
Mannan	12.0
Arabinan	1.1
Xylan	4.9
Lignin	28.1

sulfuric acid, and stored at room temperature overnight. The pretreatment vessel (volume 2 L) was preheated with steam prior to loading of the impregnated wood chips. The wood chips were heated by steam to the desired temperature and, when the preset pretreatment time (the heating-up time, less than 10 s, excluded) had elapsed, the material was discharged into a flash drum. The material was then separated, by filtration, into a solid fraction and a filtrate. The solid fraction was washed and the yield of fibrous material determined. The filtrate was analyzed for solubilized glucose, mannose, arabinose, galactose, and xylose, and the byproducts acetic acid, 5-hydroxy-methyl-2-furaldehyde (HMF), and furfural.

Enzymatic Hydrolysis

The washed fibrous material was enzymatically hydrolyzed to determine the maximum-obtainable sugar yield. Hydrolysis was performed at 2% wDM/w in a 0.1 M sodium acetate buffer (pH 4.8), supplemented with 0.2 g Celluclast 2L and 0.05 g Novozym per g dry substrate. The activity of Celluclast was 75 FPU/g *(19)*. The β-glucosidase activity in Celluclast was 12 IU/g *(20)*, and in Novozym it was 392 IU/g. The enzyme preparations were gifts from Novo Industri A/S (Bagsvaerd, Denmark). The hydrolysis vessel, containing a total amount of 500 g material, was maintained at 40°C, and the hydrolysis time was 96 h. Samples withdrawn after 0, 6, 24, 48, 72, and 96 h were analyzed for glucose and mannose content.

Fermentation

The filtrates, i.e., the liquid fractions after pretreatment, were fermented using compressed baker's yeast, *Saccharomyces cerevisiae*, Jästbolaget AB (Rotebro, Sweden). The pH was adjusted to 5.5 with 20% w/w $Ca(OH)_2$, the filtrates were then centrifuged for 10 min at 5000g, and the supernatants were fermented. Fermentation was carried out in 25-mL glass flasks containing a total volume of 20 mL (18.5 mL filtrate, 0.5 mL nutrients, 1 mL inoculum), and sealed with rubber stoppers with cannulas for the

Table 2
Experimental Design of Pretreatment Stage and Dry Matter Content After Pretreatment

Run Id	Combined severity (log Ro-pH)	Temperature (°C)	Time (min)	H_2SO_4 (% w/w Liquid)	DM content after pretreatment (%)
1	1.4	180	1	0.5	17.3
2	2.0	200	1	0.5	11.9
3	2.1	180	5	0.5	14.5
4	2.3	210	1	0.5	10.3
5	2.3	180	1	4.5	14.4
6	2.6	200	1	2.2	10.6
7	2.7	200	5	0.5	10.8
8	2.7	180	20	0.5	10.8
9	2.7	225	1	0.5	9.5
10	2.9	200	1	4.4	11.0
11	2.9	210	1	2.4	8.5
12	3.0	210	5	0.5	8.9
13	3.0	200	10	0.5	9.0
14	3.0	180	5	4.5	10.2
15	3.1	180	10	2.4	10.3
16	3.1	240	1	0.5	6.9
17	3.3	210	10	0.5	7.5
18	3.3	200	20	0.5	7.9
19	3.3	200	5	2.3	6.5
20	3.4	225	1	2.4	6.0
21	3.4	225	5	0.5	7.8
22	3.6	200	10	2.3	5.7
23	3.6	180	20	4.4	8.7
24	3.6	210	5	2.4	6.4
25	3.9	210	10	2.4	7.5
26	3.9	210	10	2.4	7.3
27	3.9	210	10	2.4	6.5
28	4.0	210	10	2.5	7.5
29	4.0	225	20	0.5	7.0
30	4.1	240	1	4.3	5.8
31	4.1	225	5	2.5	4.2
32	4.2	210	10	4.3	6.6
33	4.2	210	20	2.4	7.0
34	4.3	240	15	0.5	8.6
35	4.5	240	20	0.5	6.5
36	4.8	240	10	2.4	6.0
37	4.9	225	20	4.3	5.4
38	5.3	240	15	4.3	4.1
39	5.4	240	20	4.6	3.1

removal of carbon dioxide. The concentration of fermentable sugars was adjusted to 60 g/L with glucose. This addition was performed to obtain fermentation results that were as comparable as possible. The final concentrations of nutrients were yeast extract (1 g/L), $(NH_4)_2HPO_4$ (0.5 g/L), $MgSO_4 \cdot 7H_2O$ (0.025 g/L), and NaH_2PO_4 (0.1 M). The filtrates were inoculated with compressed baker's yeast to a cell mass of 10 g DM/L, incubated at 30°C, and stirred with a magnetic stirrer. Two reference fermentations containing 60 g/L glucose and nutrients were included at every fermentation occasion. Samples were collected at the start, before yeast addition, and after 2, 4, 6, 8, 12, and 24 h of fermentation. The sampling frequency was slightly adjusted, depending on the fermentability of the hydrolysates. The samples were analyzed for glucose, ethanol, lactic acid, acetic acid, and glycerol. Fermentation was continued until a glucose stick (Boehringer Mannheim, Mannheim, Germany) was negative, or for a maximum of 24 h. To estimate the influence of pH on the fermentability of the filtrates, fermentation was performed at pH 5.5 and 5.7.

Analyses

Filtrates and samples from hydrolysis and fermentation were analyzed on an HPLC (Shimadzu, Kyoto, Japan) with refractive index detection (Shimadzu). The samples were filtered (0.20 μm) prior to HPLC analysis. Glucose, mannose, arabinose, galactose, and xylose were separated on an Aminex HPX-87P column (Bio-Rad, Hercules, CA) at 80°C, using ultrapure water as eluent, at a flow rate of 0.5 mL/min. Glucose, ethanol, lactic acid, acetic acid, glycerol, HMF, and furfural were separated on the same equipment, at 65°C, using 5 mM H_2SO_4 as eluent, at a flow rate of 0.5 mL/min. The amount of glucose obtained in the pretreatment step was determined from the HPX-87H chromatograms because of background interference in the chromatograms on the HPX-87P column, whereas glucose from the enzymatic hydrolysis was determined from the HPX-87P chromatograms.

RESULTS

The severity of different pretreatment conditions was compared by calculating a severity parameter, in which the temperature and residence time variables were combined into a single reaction ordinate. The severity factor, log Ro, is defined by

$$Ro = t \cdot \exp[(T_r - T_b)/14.75] \qquad (1)$$

where t is the reaction time (min), T_r the reaction temperature (°C) and T_b a reference temperature, which was set to 100°C (15). Because the pretreatment was performed under acidic conditions, the effect of pH was taken into consideration by the combined severity (16) defined as

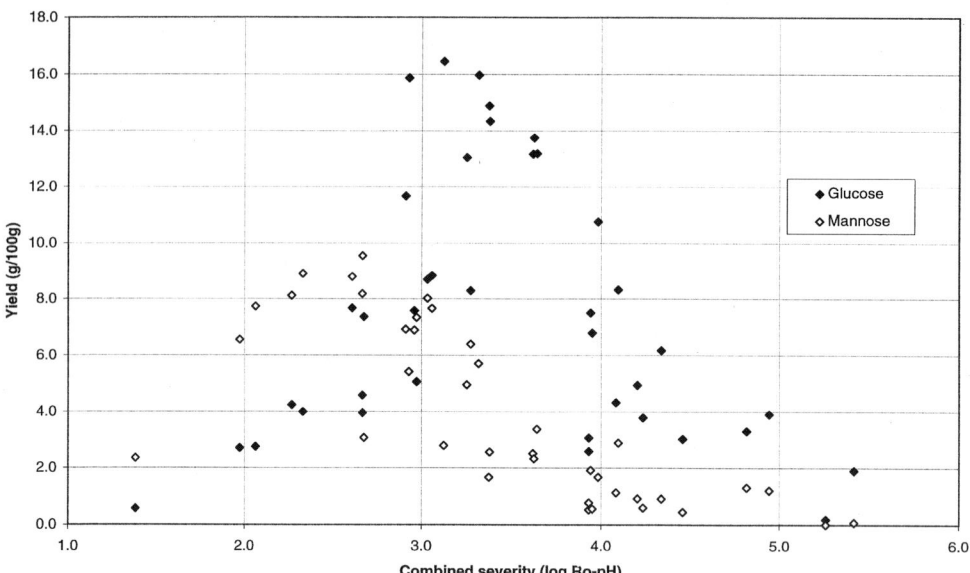

Fig. 1. The yield of glucose and mannose after pretreatment as a function of the combined severity.

$$\text{Combined severity} = \log Ro - pH \quad (2)$$

The pH was calculated from the amount of sulfuric acid added, and the combined severity parameter was calculated for each set of pretreatment conditions (Table 2).

Unless otherwise stated, the yields following pretreatment and hydrolysis are expressed as g/100 g of original dry wood. The yields following fermentation, on the other hand, are expressed as the amount of ethanol divided by the amount of total fermentable sugars, i.e., glucose and mannose, in the fermentation broth.

Arabinose and xylose are not fermentable with *S. cerevisiae*. Galactose was present at very low concentrations in the filtrates, and was not fermented by the strain of *S. cerevisiae* used in the present investigation. Therefore, only glucose and mannose, the predominant sugars in softwoods, were used for the evaluation of the yields. The maximum yields of mannose and glucose following pretreatment were achieved in different ranges of combined severity: 2.3–2.7 and 2.9–3.4, respectively (Fig. 1). The highest glucose yield, about 16 g/100 g, was obtained for runs 11, 16, and 19 (Table 2); the highest mannose yield, about 9 g/100 g, was achieved for runs 5, 6, and 8.

The maximum glucose yield in the enzymatic hydrolysis, about 19 g/100 g, was obtained in the same range of combined severity (2.9–3.4) as for the maximum glucose yield in the pretreatment (Fig. 2). One single run deviates from the general trend, resulting in a glucose yield of

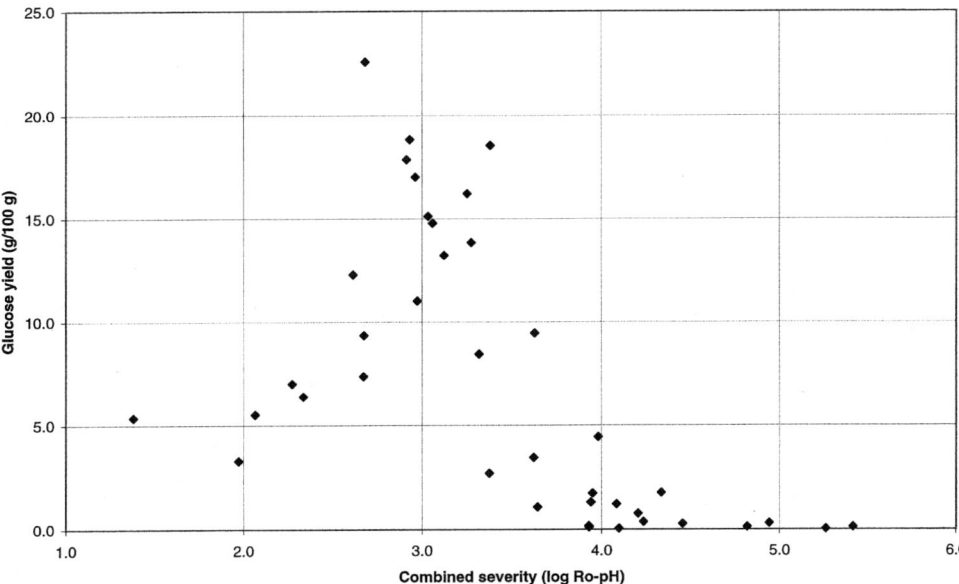

Fig. 2. The yield of glucose following enzymatic hydrolysis as a function of the combined severity.

23 g/100 g at a combined severity of 2.7 (corresponding to 225°C, 1 min, and 0.5% H_2SO_4). For all acid concentrations, the highest yield following hydrolysis was obtained at a residence time of 1 min, but at different temperatures. For 2.4% H_2SO_4, the maximum yield, 19 g/100 g, was obtained at 210°C, and for 4.4% H_2SO_4 the maximum yield of 18 g/100 g was obtained at 200°C.

The total yield of fermentable sugars after both pretreatment and hydrolysis is shown in Fig. 3, together with information on whether or not the filtrates were fermentable. The fermentability was defined as an ethanol yield of more than 50% of the theoretical yield after 24 h. The highest total yield of fermentable sugars, 40 g/100 g, was obtained following pretreatment at 210°C, for 1 min with 2.4% H_2SO_4 (Fig. 3). These conditions also resulted in the highest total glucose yield after pretreatment and hydrolysis, i.e., 35 g/100 g, corresponding to 72% of the theoretical yield. Among the fermentable samples, the highest fermentable sugar yield was 35 g/100 g, obtained for the pretreatment conditions 225°C, 5 min, and 0.5% H_2SO_4.

Only the filtrates after pretreatment were fermented, since no additional byproduct formation occurs in the enzymatic hydrolysis. The ethanol yield showed a sharp decline at a combined severity of about 3, from 0.3–0.35 g/g to 0–0.05 g/g (Fig. 4). The ethanol yield in the reference fermentation was 0.40 ± 0.02 g/g. For the samples obtained with a combined severity of about 3, a small variation in pH, such as 0.2 units, influ-

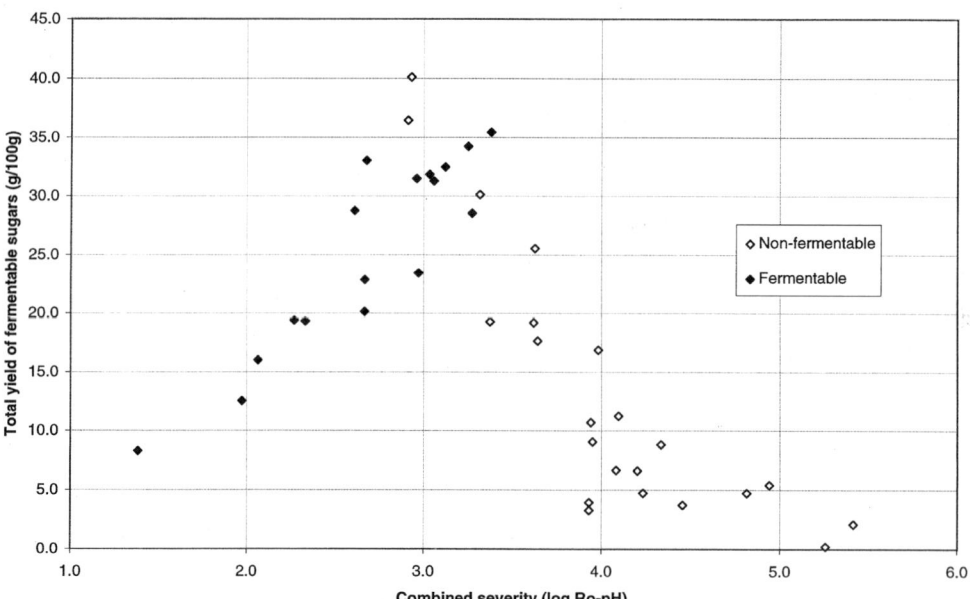

Fig. 3. The total yield of fermentable sugars as a function of the combined severity.

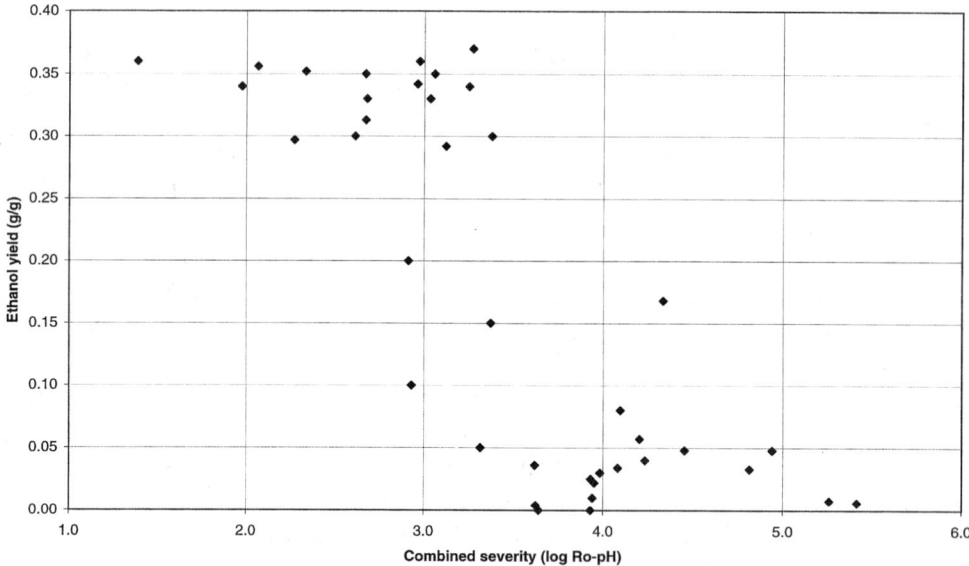

Fig. 4. The ethanol yield at pH 5.5 as a function of the combined severity.

enced the yield. For example, for run 10 the yield in the fermentation increased from 0.21 g/g at pH 5.5 to 0.35 g/g at pH 5.7. However, the sample that gave the highest yield of fermentable sugars did not ferment. The ethanol yield was 0.1 g/g at pH 5.5, and, when the pH was adjusted

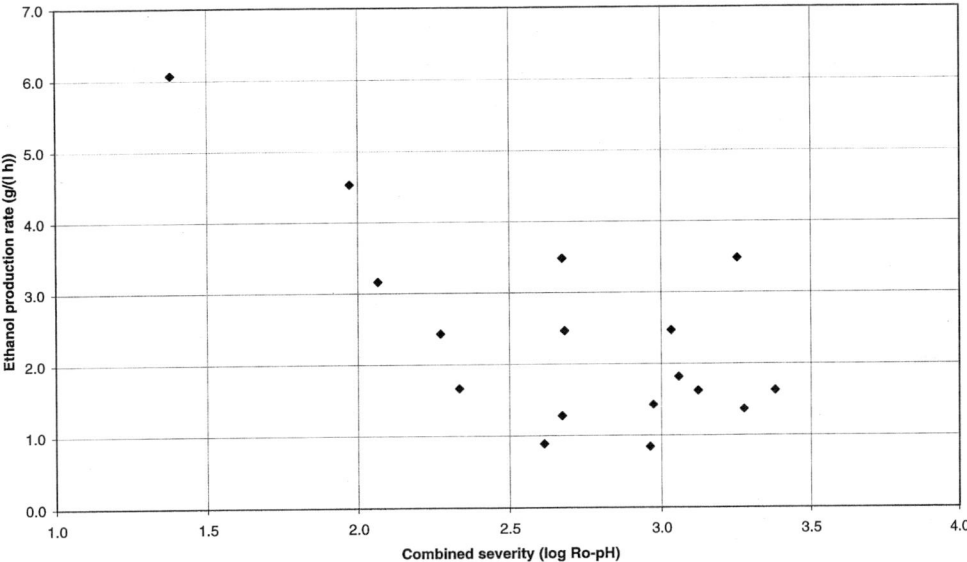

Fig. 5. The average ethanol production rate after 2 h (excluding the lag phase), r_{2h}, as a function of the combined severity.

to 5.7, the yield increased only marginally, to 0.14 g/g. The average ethanol production rate during the first 2 h, r_{2h}, was determined excluding the lag phase, which was less than 2 h. If the lag phase was included in the ethanol production rate, no distinct relationship between the production rate and the combined severity could be seen. For the fermentable filtrates, r_{2h} varied between 0.8 and 6.1 g/(L h), and decreased with increasing combined severity (Fig. 5). The mean value of the ethanol production rate in the reference fermentations was 5.9 g/(L h). The dry wt of the pretreated material varied between 3 and 17% (Table 2). The samples treated under the most severe conditions had the lowest DM contents. A long residence time and high pretreatment temperature increased the heat losses in the pretreatment reactor, which resulted in the formation of more condensate and a reduction in the DM content, since direct steam is used in the pretreatment unit. Therefore, the amount of compounds solubilized in the pretreatment stage were unequally diluted in the liquid after pretreatment. Increased degradation of the material also resulted in a reduced DM content because of an increase in the formation of soluble byproducts. The fermentability was low for the samples with the highest combined severity, despite the higher dilution.

The maximum concentrations of the byproducts furfural and HMF were found to be approx 2 g/L and 5 g/L, respectively, at a combined severity between 3 and 4 (Fig. 6). These concentrations correspond to a yield of about 1.5 g/100 g and 3 g/100 g, respectively. The concentration

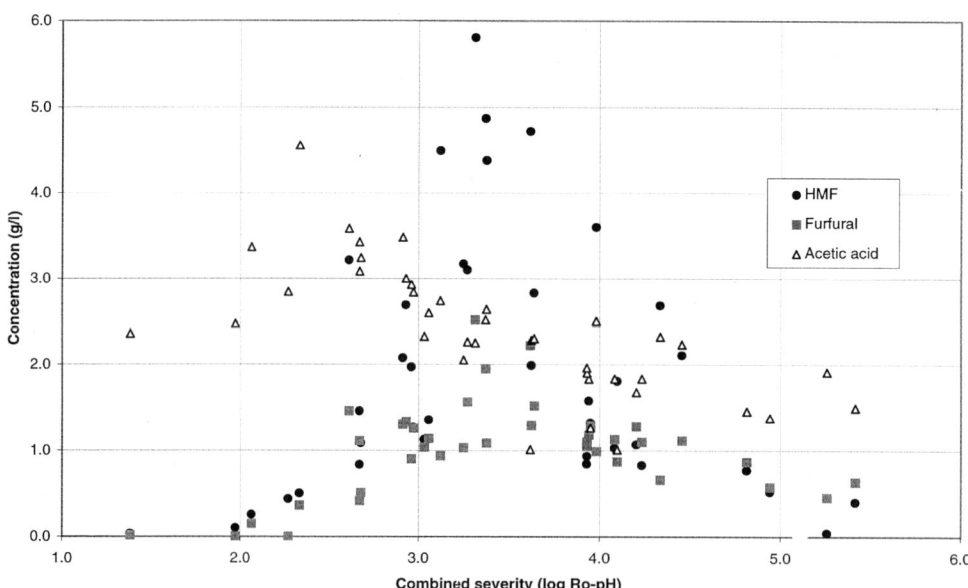

Fig. 6. The concentration of the byproducts HMF, furfural, and acetic acid, as a function of the combined severity.

of acetic acid reached a maximum at a combined severity of 2.3, then the concentration decreased slightly from 3.0 to 1.5 g/L (Fig. 6), corresponding to yields of 2 g/100 g and 1 g/100 g, respectively.

DISCUSSION

Impregnation with SO_2 or H_2SO_4 prior to steam pretreatment and enzymatic hydrolysis has been investigated in several studies (5,7–9). It is difficult to compare the efficiencies of the two acid catalysts, since different equipment and raw materials have been used. However, comparisons between SO_2 and H_2SO_4 have been performed on willow (21) and aspen (7). In the study on willow, the sugar yields after pretreatment and after enzymatic hydrolysis were compared; in the study on aspen, the enzymatic hydrolysis was replaced by combined hydrolysis and fermentation. The latter investigation was based on a limited amount of data. In the present study, a comparison for softwood was made possible by comparing the results obtained for H_2SO_4 impregnation with the results from a previous study on SO_2 impregnation, performed with the same equipment, and using the same procedure (13). In the SO_2 study, the following pretreatment conditions were investigated: temperature (190–230°C) and residence time (2–15 min). The SO_2 concentration absorbed by the wood was less than 1% w/w DM. The sugar yields after pretreatment and after enzymatic hydrolysis, as well as the fermentability, were investigated. To the best of the authors' knowledge, this is the first time both hydrolysis yield and fermentation yield have been considered in a comparison between SO_2 and H_2SO_4 impregnation of softwood. The most significant differences between H_2SO_4 and SO_2 impregnation were the step in which the sugars were formed (pretreatment or enzymatic hydrolysis) and the difference in fermentability.

H_2SO_4 impregnation resulted in high yields of both glucose and mannose after the pretreatment step; the glucose yield following enzymatic hydrolysis was low. With SO_2 impregnation, the situation was the opposite: Most of the fermentable sugars were released during hydrolysis. However, SO_2 impregnation resulted in approximately the same total yield of fermentable sugars, based on theoretical yield (66%), as H_2SO_4 impregnation (65%). This was also observed in a similar study on willow, in which H_2SO_4 impregnation resulted in high yields of hemicellulose sugars, but the glucose yield after pretreatment and enzymatic hydrolysis was low in comparison with the yield obtained with SO_2 impregnation (21).

As shown in the present study, there is a relation between the yields of sugars and ethanol, and the combined severity in the region investigated. In consequence, the maximum yield was the same for all acid concentrations, but at different temperatures and residence times. In contrast, an increase in SO_2 concentration in the impregnation step, up to about 3%, increased the yield of glucose (9,12). The optimal sugar yield was achieved at the same temperature and the same residence time for different levels of SO_2. In the latter studies, no fermentation was performed, and it is therefore not possible to conclude whether the overall ethanol yield increased or not. However, another study showed that softwood pretreated with SO_2 at a concentration of 4% w/w DM could be fermented (22).

An obvious difference in fermentability was observed between H_2SO_4 and SO_2 impregnation. The sharp decline in ethanol yield at a combined severity of about 3 for H_2SO_4-impregnated samples was not observed for SO_2 impregnation, in which all samples were fermentable (13). Despite the fact that only the fermentable filtrates were considered, the average ethanol production rate for H_2SO_4-impregnated samples was low compared with SO_2-impregnated samples. The ethanol production rate, r_{2h}, for H_2SO_4-impregnated samples was about 2 g/(L h), while it was about 5 g/(L h) for the SO_2-impregnated samples. In H_2SO_4 impregnation, both the pretreatment temperature and the residence time influenced the ethanol production rate; for SO_2 impregnation the pretreatment temperature had a major influence, but the effect of the residence time was negligible. The difference in fermentability between H_2SO_4- and SO_2-impregnated material indicates that H_2SO_4 impregnation either leads to higher concentrations of byproducts or to the formation of additional inhibitory byproducts.

The sharp decline in ethanol yield for H_2SO_4 impregnation did not correlate with the concentration of the analyzed byproducts HMF, furfural, and acetic acid (Fig. 6), which decreased at combined severities higher than 3. However, the decline in ethanol yield was obtained at the same combined severity as the maximum concentration of the analyzed byproducts. At higher values of combined severity, the ethanol yield was still low, despite the decrease in the concentration of the analyzed byproducts caused by dilution and further degradation (HMF and furfural). This indicates that the production of byproducts not analyzed increased, e.g., levulinic acid and formic acid, which are known to be secondary degradation products from HMF *(23,24)*, and lignin degradation products *(25)*.

The maximum yield of fermentable sugars and the decline in ethanol yield were observed at the same combined severity, emphasizing the importance of considering the ethanol yield when optimizing the pretreatment conditions. In this region, the fermentability could be improved by a small increase in pH, which indicates that the fermentability was affected by weak acids *(26,27)*. However, other byproducts, such as lignin degradation products, were probably also responsible for the inhibition *(25)*, since an increase in pH did not improve the fermentability for all filtrates.

To recover the maximum amount of sugars from both hemicellulose and cellulose, different pretreatment conditions are required. Two-step, countercurrent pretreatment of hardwood, using dilute sulfuric acid at different temperatures, resulted in high yields of both xylose and glucose *(28)*. The different conditions for maximum mannose yield and maximum glucose yield obtained in the present study indicate that a two-step pretreatment procedure would also be advantageous for softwoods. Another alternative would be to recover the hemicellulose sugars in the first step, using H_2SO_4 under mild conditions, and, in the following step, to increase the reactivity of the cellulose, using SO_2 at a higher temperature. In such a process configuration, the increased consumption of chemicals would have to be evaluated in relation to a potentially higher yield of ethanol.

ACKNOWLEDGMENTS

The Swedish National Board for Industrial and Technical Development (NUTEK) is gratefully acknowledged for its financial support. The authors are grateful to Nina Fällstrand Larsson for her invaluable help with the pretreatment of the samples.

REFERENCES

1. Torget, R., Werdene, P., Himmel, M., and Grohmann, K. (1990), *Appl. Biochem. Biotechnol.* **24/25**, 115–126.
2. Martinez, J., Negro, M. J., Saez, F., Manero, J., Saez, R., and Martin, C. (1990), *Appl. Biochem. Biotechnol.* **24/25**, 127–134.
3. Ropars, M., Marchal, R., Pourquié, J., and Vandecasteele, J. P. (1992) *Bioresource. Technol.* **42**, 197–204.
4. Grethlein, H. E. and Converse, A. O. (1991), *Bioresource Technol.* **36**, 77–82.
5. Ramos, L. P., Breuil, C., and Saddler, J. N. (1992), *Appl. Biochem. Biotechnol.* **34/35**, 37–48.
6. Gregg, D. J. and Saddler, J. N. (1996), *Biotechnol. Bioeng.* **51**, 375–383.

7. Mackie, K. L., Brownell, H. H., West, K. L., and Saddler, J. N. (1985), *J. Wood Chem. Technol.* **5,** 405–425.
8. Schwald, W., Breuil, C., Brownell, H. H., Chan M., and Saddler, J. N. (1989), *Appl. Biochem. Biotechnol.* **20/21,** 29–44.
9. Schwald, W., Smaridge, T., Chan, M., Breuil, C., and Saddler, J. N. (1989), in *Enzyme Systems for Lignocellulose Degradation,* Coughlan, M. P., ed., Elsevier, New York, pp. 231–242.
10. Mamers, H. and Menz, D. N. J. (1984), *Appita* **37,** 644–649.
11. Clark, T. A., Mackie, K. L., Dare, P. H., and McDonald, A. G. (1989), *J. Wood Chem. Technol.* **9** 135–166.
12. Clark, T. A. and Mackie, K. L. (1987), *J. Wood Chem. Technol.* **7,** 373–403.
13. Stenberg, K., Tengborg, C., Galbe, M., and Zacchi, G. (1997) *J. Chem. Tech. Biotechnol.,* accepted.
14. von Sivers, M. and Zacchi, G. (1996), *Bioresource. Technol.* **56,** 131–140.
15. Overend, R. P. and Chornet, E. (1987), *Phil. Trans. R. Soc. London,* Series A, **321,** 523–536.
16. Chum, H. L., Johnson, D. K., Black, S. K., and Overend, R. P. (1990), *Appl. Biochem. Biotechnol.* **24/25,** 1–14.
17. Chum, H. L., Johnson, D. K., and Black, S. K. (1990), *Ind. Eng. Chem. Res.* **29,** 156–162.
18. van Walsum, G. P., Allen, S. G., Spencer, M. J., Laser, M. S., Antal, M. J., and Lynd, L. R. (1996), *Appl. Biochem. Biotechnol.* **57/58,** 157–170.
19. Mandels, M., Andreotti, R., and Roche, C. (1976), *Biotechnol. Bioeng. Symp.* **6,** 21–33.
20. Berghem, L. E. R. and Petterson, L. G. (1974), *Eur. J. Biochem.* **46,** 295–305.
21. Eklund, R., Galbe, M., and Zacchi G. (1995), *Bioresource. Eng.* **52,** 225–229.
22. Stenberg, K., Tengborg, C., Galbe, M., Zacchi, G., Palmqvist, E., and Hahn-Hägerdal, B., (1997), *Appl. Biochem. Biotechnol.,* accepted.
23. Nilvebrant, N.-O., Reimann, A., de Sousa, F., Kleen, M., and Palmqvist, E. (1997) Proc. 9th International Symposium on Wood and Pulping Chemistry, Poster no. 79, Montréal, Canada, June 9–12.
24. Baugh, K. D. and McCarty, P. L. (1988), *Biotechnol. Bioeng.* **31,** 50–61.
25. Clark, T. A. and Mackie, K. L. (1984), *J. Chem. Tech. Biotechnol.* **34B,** 101–110.
26. Russel, N. J. and Gould, G. W. (1991), *Food Preservatives,* Blackie and Son, Glasgow and London.
27. Taherzadeh, M. J., Niklasson, C. and Lidén, G. (1997), *Chem. Eng. Sci.* **52,** 2653–2659.
28. Torget, R., Hatzis, C., Hayward, T. K., Hsu, T.-A., and Philippidis, G. P. (1996), *Appl. Biochem. Biotechnol.* **57/58,** 85–101.

Pretreatment of Softwood by Acid-Catalyzed Steam Explosion Followed by Alkali Extraction

DANIEL SCHELL,* QUANG NGUYEN, MELVIN TUCKER, AND BRIAN BOYNTON

National Renewable Energy Laboratory, Golden, CO 80401

ABSTRACT

A process for converting lignocellulosic biomass to ethanol hydrolyzes the hemicellulosic fraction to soluble sugars (i.e., pretreatment), followed by acid- or enzyme-catalyzed hydrolysis of the cellulosic fraction. Enzymatic hydrolysis may be improved by using an alkali to extract a fraction of the lignin from the pretreated material. The removal of the lignin may increase the accessibility of the cellulose to enzymatic attack, and thus improve overall economics of the process, if the alkali-treated material can still be effectively converted to ethanol.

Pretreated Douglas fir produced by a sulfuric-acid-catalyzed steam explosion was treated with NaOH, NH_4OH, and lime to extract some of the lignin. The treated material, along with an untreated control sample, was tested by an enzymatic-digestion procedure, and converted to ethanol by simultaneous saccharification and fermentation using a glucose-fermenting yeast. NaOH was most effective at removing lignin (removed 29%), followed by NH_4OH and lime. However, the susceptibility of the treated material to enzymatic digestion was lower than the control and decreased with increasing lignin removal. Ethanol production was similar for the control and NaOH-treated material, and lower for NH_4OH- and lime-treated material.

Index Entries: Ethanol; lignocellulosic biomass; lignin extraction; cellulose hydrolysis; alkali.

INTRODUCTION

The production of ethanol from lignocellulosic biomass has received considerable attention, because of the potential of producing large quantities of ethanol for use as a transportation fuel *(1)*. The process involves

---

*Author to whom all correspondence and reprint requests should be addressed.

hydrolyzing the hemicellulosic and cellulosic fractions of biomass to their component sugars for subsequent conversion to ethanol by a fermentative process. Hemicellulose is typically hydrolyzed using a chemical process (e.g., by acid or caustic treatment, commonly referred to as "pretreatment"); cellulose is hydrolyzed by chemical (acid) or biological (enzyme) attack. The economic success of these processes will depend on their ability to obtain good sugar conversion, and to successfully convert the sugars to ethanol at high yields.

The enzymatic approach to hydrolyzing cellulose to glucose is receiving attention because enzymes can achieve high yields, since they do not catalyze glucose degradation reactions common for the acid process. However, the cellulose must be accessible to the enzyme. The accessibility depends on the severity of the pretreatment process. A greater degree of hemicellulose and lignin removal during pretreatment increases the accessibility of the cellulose, and thus the efficacy of cellulose hydrolysis.

One potential method of converting wood to ethanol involves removing the hemicellulosic sugars by an acid process, then, after washing the sugars from the solids, the solids are subjected to a lignin-extraction step. After separating the lignin-containing liquor from the solids, the soluble lignin is precipitated, and this liquor is combined with the hemicellulosic liquor (rich in six- and five-carbon sugars) and the solids in a simultaneous saccharification and fermentation (SSF) process. Combining the liquor streams reduces the amount of dilution water needed in the process, and thus increases ethanol concentrations. This should improve glucose conversion from enzymatic hydrolysis of cellulose, when compared to single treatment.

Heitz et al. *(2)* steam treated aspen wood *(Populus tremloides)* in a Stake II reactor (Stake Technology, Norval, Ontario, Canada). The treated material was delignified with NaOH at 5% by weight consistency, pH 13.0, and at 100°C for 30 min. Lignin in the caustic solution was recovered (precipitated) from solution by adjusting to pH 1.5–2.0, and holding at 80°C for 10 min. The amount of lignin recovered in the solution was strongly dependent on steam-treatment conditions ranging from 5 to 90% as treatment severity increased (residence time and steam temperature). Although, enzymatic glucose yields increased with steam treatment severity, delignification may or may not have increased yields.

Parajo et al. *(3)* delignified eucalpytus wood with a 95% acetic acid–0.2% HCl solution (boiling temperature for 1 h), and then treated the residue with varying concentrations (4, 12, and 20 wt%) of NH_4OH at 60°C for 3 h. NH_4OH caused little change in treated material composition, but more than doubled enzymatic hydrolysis yields. Varying NH_4OH concentration had no effect on yields.

Schwald et al. *(4)* investigated steam explosion (240°C, 60–140 s) of aspen wood chips and steam explosion of SO_2-impregnated chips (1.6% SO_2 dry basis, 220°C, 100 s). The treated chips were washed with water, and then twice washed with 0.4% NaOH. There were no significant differences in enzymatic hydrolysis yields and rates between untreated and alkali-washed material.

Steam-exploded aspen wood with no catalyst, and with impregnation by SO_2 and sulfuric acid, was investigated by Mackie et al. (5). The treated chips were washed with water, and then washed with 0.4% NaOH at room temperature. Extracted material (as percent of original wood weight) was approx 15% for no acid and SO_2-treated wood, and approx 10% for sulfuric-acid-treated wood. The effectiveness of the various treatments on the enzymatic susceptibility of the treated wood was tested by performing an SSF procedure. However, no comparison was made between alkali-washed and unwashed material.

The goal of this work was to test the enzymatic digestibility and SSF performance of a softwood feedstock (Douglas fir) subjected to sulfuric-acid-catalyzed steam explosion, followed by alkali extraction. Various alkalis (NaOH, NH_4OH, and lime) were tested for their ability to remove lignin, increase enzymatic cellulose hydrolysis yields, and not significantly affect fermentation performance. Although NaOH is a stronger caustic and should be a more effective delignification agent, NH_4OH may have some nutritive effect, and lime is known to detoxify hydrolysates (6).

METHODS

Pretreatment

Debarked Douglas fir logs were chipped and milled in a knife mill equipped with a 9.5-mm screen. Particles smaller than 2 mm were discarded. Screened particles were soaked in 0.35% sulfuric acid at 60°C for 6 h. They were then placed in a steam explosion apparatus previously described (7), and heated to 215°C by direct steam injections. After heating for 140 s, the particles were explosively decompressed into a collection tank. The solids were collected and frozen for later use.

Solids concentrations were determined by oven-drying a sample of solids or liquid at 105°C. The composition of the pretreated biomass was determined by methods previously described (8).

Alkali Extraction

Figure 1 shows a flow diagram of the process used to produce the alkali-treated material used in this study. Pretreated wood (at 31% solids) was diluted to 20% solids with 60°C water (~100 g of water) and filtered, then washed with another 100 g of water to extract most of the soluble sugars. This low-pH sugar solution was saved, and the solids were extensively washed with fresh 60°C water. The washed solids were diluted to a 15% solids concentration with an alkali solution and held at 60°C in either 0.4% NaOH (0.1 M), 10% NH_4OH (2.8 M), or 0.06% lime (0.008 M).

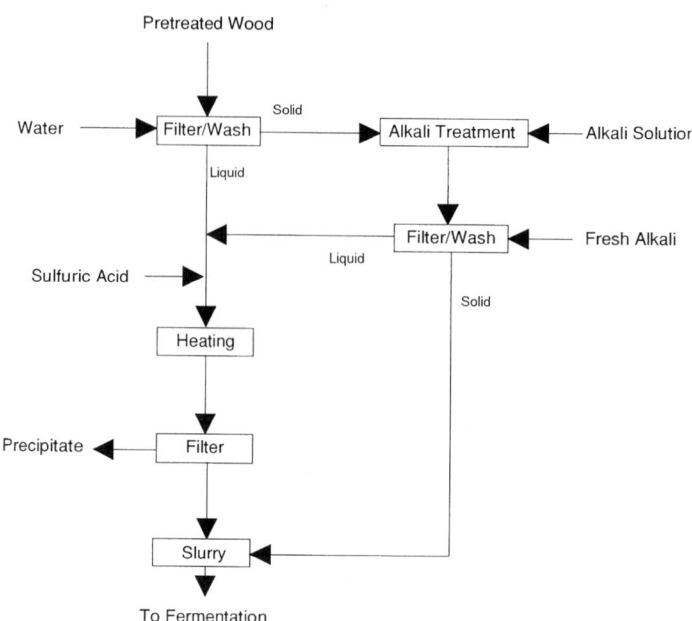

Fig. 1. Alkali treatment experimental procedure.

The lime solution was at its solubility limit at 60°C. After 1 h, the solids were filtered and washed with the fresh 60°C alkali solution. This alkali solution was saved, and the solids were again extensively washed with fresh alkali solution. The treated washed solids were analyzed for enzymatic digestibility. The total solids concentrations were measured on all saved and discarded solutions for tracking solids.

The low-pH sugar solution was combined with the alkali solution, and was adjusted to pH 2.0–2.5 with sulfuric acid. The solution was heated to 80–90°C and held for 15 min, to precipitate the lignin solubilized during the alkali treatment. The solution was filtered to remove precipitated lignin, and later combined with the treated solids to test performance during SSF. When NH_4OH was used, the alkali solution was heated to 80–90°C and held until the pH stabilized as the ammonia vapor was driven off. This additional step simulates an ammonia recovery step. A control sample (nonalkali-treated) was generated by following only the first filter-and-wash step, and saving the produced solids and liquid.

Enzyme Digestibility

Enzymatic digestibilities (defined as glucose produced divided by potential glucose) were performed on the extensively washed alkali-treated and control samples, using the following procedure. The amount of washed solids required to give 0.1 g cellulose in 10 mL was added to a vial. The buffer for the digestion was 50 mM citrate, pH 4.8, containing

40 µg/mL tetracycline and 30 µg/mL cycloheximide. The Iogen (Ottawa, Ontario, Canada) cellulase enzyme (Iogen Super Clean cellulase, lot #BRC 191095) loading was adjusted to 60 FPU/g of cellulose in the vials. The contents of the vials were prewarmed to 50°C before enzyme was added. Digestibility assays were carried out in duplicate at 50°C on a Roto-Torque rotaton (Cole Parmer, Niles, IL), with rotation at a 45-degree angle from the horizontal and 120 rpm, and compared to an identical no-enzyme blank. Duplicate Solka Floc (grade NF-FCC, lot #1016, Fiber Sales and Development, Urbana, OH 43078) digestions were used as controls. One-half mL samples were removed and analyzed for glucose, using a YSI Model 2700 Select Biochemistry Analyzer equipped with immobilized glucose oxidase membranes (Yellow Springs Instruments, Yellow Springs, OH). Samples were centrifuged at 12,000g for 5 min and diluted to keep the glucose readings below the 2.5 g/L level used for calibrating the instrument.

Simultaneous Saccharification and Fermentation

Alkali-treated solids and liquor were thawed and recombined. The pH levels of the NaOH and lime-treated slurries were near the required 5.0, after the acidified liquor was combined with the high-pH solids. The ammonia-treated sample was near pH 10.0 and was adjusted to 5.0 with sulfuric acid. The control sample was raised to pH 5.0 with lime. Because of dilution caused by adding liquid during the treatment steps described above, the total solids concentration of the combined slurries was 6–7%. The solids concentration of the control sample was adjusted to be in this range. The concentration of Difco (Detroit, MI) yeast extract and peptone in each SSF flask was 2.5 g/L and 5.0 g/L, respectively. The enzyme loadings were 70–75 FPU/g cellulose, so that adequate glucose would be available from cellulose hydrolysis. Each flask was inoculated with a 10% (w/w) *Saccharomyces cerevisiae* D_5A (9), a glucose-fermenting yeast, grown on 50 g/L glucose at 30°C. Inoculated flasks were placed on a rotary shaker that operated at 30°C and 150 rpm.

Flasks were analyzed for sugars with a Hewlett-Packard (Palo Alto, CA) 1090L high-pressure liquid chromatograph (HPLC) equipped with a Bio-Rad (Hercules, CA) HPX-87P carbohydrate-analysis column that operated at 85°C. The mobile phase was deionized water at a flow rate of 0.6 mL/min. Fermentation products were quantified using the same samples on another HPLC equipped with a Bio-Rad HPX-87H organic-acid analysis column operating at 65°C. The mobile phase was 0.01 N sulfuric acid at a flow rate of 0.6 mL/min.

RESULTS

Alkali Extraction

After pretreatment, the wood slurry is 31.1% total solids and 21.6% insoluble solids, and approx one-third of the wood was solubilized by the treatment. The composition is 46% cellulose and 45% lignin. One goal

Table 1
Amount of Lignin Dissolved from Pretreated Douglas Fir

Alkali	Lignin dissolved during alkali treatment step[a]	
	% original pretreated wood	% lignin in pretreated wood
NaOH	13.3	29.5
NH_4OH	4.7	10.5
Lime	0.2	0.5

[a] Assumes only lignin is dissolved during the alkali extraction step, and includes lignin in the discarded alkali-washed liquor.

of this work was to determine the ability of each alkali to solubilize lignin that remained in the pretreated wood. The results are presented in Table 1.

As expected, NaOH was the most effective delignification agent. It solubilized 13.3% of the pretreated wood (after water washing), or 29.5% of the lignin. This is comparable to the results of Mackie et al. (5), who achieve 15% solubilization of the pretreated wood. Ten percent NH_4OH solubilized only 10.5% of the lignin, about one-third of the solubilization achieved by NaOH, and was inferior to NaOH as a delignification agent. Lime was clearly not effective as a delignification agent, particularly with its limited solubility.

Enzymatic Digestibility

Enzymatic digestibilities as functions of time for each alkali are shown in Figure 2. For calculation purposes, the authors assumed that the cellulose content of all the samples was the same as the control; this is not true, because some lignin was dissolved during the alkali treatment (cellulose content was not measured for the treated material). This would increase the potential glucose, and bias the reported digestibilities higher than they should be, which makes it even more clear that the control was superior to any of the alkali-treated material. The results indicate that, as the extent of delignification increased, digestibilities decreased, which was counter to the expected trend. This phenomena has been reported before (10), and researchers have suggested that either redistribution of the lignin or alteration in the crystalline cellulose structure is responsible for loss of hydrolysis performance.

Simultaneous Saccharification and Fermentation

Figure 3 shows ethanol production during SSF for the control and each of the three alkali-treated samples. The initial glucose concentration for each sample was 7, 5, 3, and 1 g/L for the control, NaOH-, lime-, and

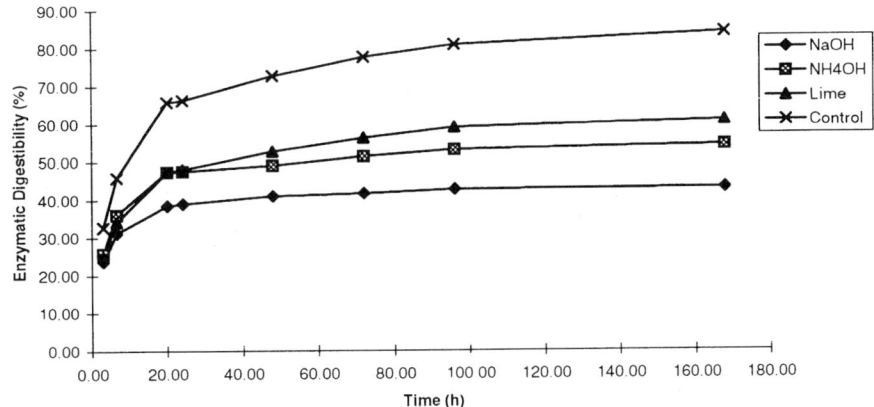

Fig. 2. Enzymatic digestibility as a function of time for each of the alkali-treated samples and the untreated control sample.

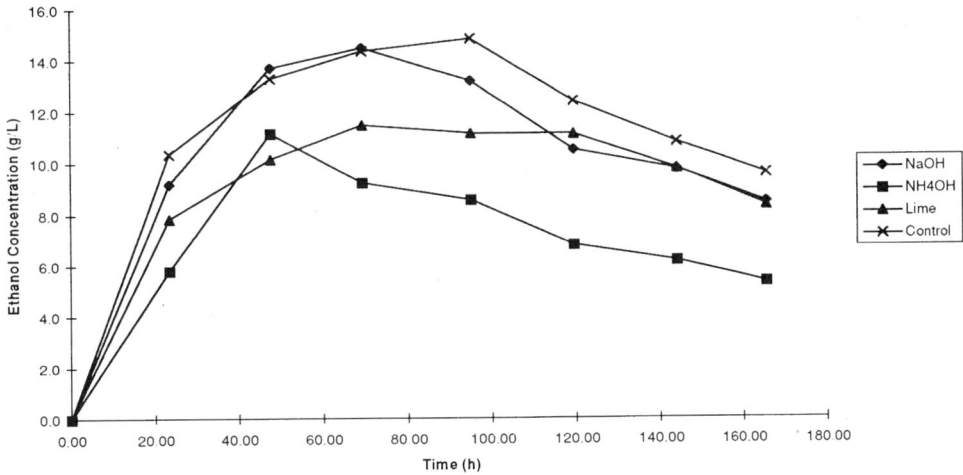

Fig. 3. Ethanol concentration as a function of time for each of the alkali-treated samples and the untreated control sample.

NH_4OH-treated samples, respectively. The alkali treatment is apparently responsible for some sugar loss, particularly with NH_4OH. However, the data in Fig. 3 includes the contributions from the initial glucose and glucose produced during enzymatic hydrolysis of the cellulose, since this reflects the overall performance of the system.

The SSF results show that the NaOH treatment had performance similar to the control (both peaked at about 80% ethanol yield); both NH_4OH and lime treatments produced significantly less ethanol. A small part of this can be attributed to the lower amount of initial glucose in the samples. But reduced enzymatic digestibility is probably the primary factor responsible for the reduced yields. The enzymatic digestibility measured above for the NaOH sample was significantly lower than the control, but both samples had similar performance during SSF. The reason for this behavior is not known. But it does illustrate the importance of evaluating processes based on fermentation performance of treated material and not extrapolating fermentation performance based on digestibility measurements.

SUMMARY

This work has shown that NaOH is the most effective delignification agent, compared to NH_4OH and lime. Relatively low concentrations (0.1 M) of NaOH can dissolve 30% of the lignin in acid-pretreated Douglas fir. But this alkali treatment does not improve either enzymatic digestibility or SSF performance compared to an untreated control. The fractionation process would have to show a significant performance improvement to justify the increased cost and complexity that would be added to a biomass-to-ethanol process.

ACKNOWLEDGMENTS

This work was funded by the Biochemical Conversion Element of the Office of Fuels Development of the US Department of Energy. The assistance of Fannie Posey-Eddy and Jim Hora in performing the HPLC analysis on fermentation samples is appreciated.

REFERENCES

1. Schell, D. J., McMillan, J. D., Philippidis, G. P., Hinman, N. D., and Riley, C. (1992), in *Advances in Solar Energy*, vol. 7, Boer, K. W., ed., American Solar Energy Society, Boulder, CO, pp. 373–448.
2. Heitz, M., Capek-Menard, E., Koeberle, P. G., Gagne, E., Chornet, E., Overend, R. P., Taylor, J. D., and Yu, E. (1991), *Bioresource Technol.* **35,** 23–32.
3. Parajo, J. C., Alonso, J. L., and Santos, V. (1995), *Process Biochem.* **30,** 537–545.
4. Schwald, W., Breuil, C., Brownell, H. H., Chan, M., and Saddler, J. N. (1989), *Appl. Biochem. Biotechnol.* **20/21,** 29–44.
5. Mackie, K. L., Brownell, H. H., West, K. L., and Saddler, J. N. (1985), *J. Wood Chem. Technol.* **5,** 405–425.
6. McMillan, J. D. (1994), in *Enzymatic Conversion of Biomass for Fuels Production*, Himmel, M. E., Baker, J. O., and Overend, R. P., eds, American Chemical Society, Washington, DC, pp. 411–437.
7. Nguyen, Q. A., Tucker, M. P., Boynton, B. L., Keller, F. A., and Schell, D. J. (1998), to be published in *Appl. Biochem. Biotechnol.*, accepted.
8. Vinzant, T. B., Ponfick, L., Nagle, N. J., Ehrman, C. I., Reynolds, J. B., and Himmel, M. E. (1994), *Appl. Biochem. Biotechnol.* **45/46,** 611–626.
9. Spindler, D. D., Wyman, C. E., Grohman, K., and Philippidis, C. P. (1992), *Biotechnol. Lett.* **14,** 403–407.
10. Ramos, L. P., Breuil, C., and Saddler, J. N. (1992), *Appl. Biochem. Biotechnol.* **34/35,** 37–47.

Copyright © 1998 by Humana Press Inc.
All rights of any nature whatsoever reserved.
0273-2289/98/70-72—0025$10.75

Comparison of Yellow Poplar Pretreatment Between NREL Digester and Sunds Hydrolyzer

M. P. Tucker,* J. D. Farmer, F. A. Keller, D. J. Schell, and Q. A. Nguyen

National Renewable Energy Laboratory, Golden, CO 80401

ABSTRACT

Single-stage cocurrent dilute acid pretreatments were carried out on yellow poplar *(Liriodendron tulipifera)* sawdust using an as-installed and short residence time modified pilot-scale Sunds hydrolyzer and a 4-L bench-scale NREL digester (steam explosion reactor). Pretreatment conditions for the Sunds hydrolyzer, installed in the NREL process development unit (PDU), which operates at 1 t/d (bone-dry t) feed rate, spanned the temperature range of 160–210°C, 0.1–1.0% (w/w) sulfuric acid, and 4–10-min residence times. The batch pretreatments of yellow poplar sawdust in the bench-scale digester were carried out at 210 and 230°C, 0.26% (w/w) sulfuric acid, and 1-, 3-, and 4-min residence times. The dilute acid prehydrolysis solubilized more than 90% of the hemicellulose, and increased the enzymatic digestibility of the cellulose that remained in the solids. Compositional analysis of the pretreated solids and liquors and mass balance data show that the two pretreatment devices had similar pretreatment performance.

Index Entries: Biomass; ethanol; dilute-acid pretreatment; bioconversion.

INTRODUCTION

The enzymatic utilization of lignocellulosic biomass for ethanol production requires steps to make the recalcitrant cellulose more accessible to cellulase enzymes. This may be accomplished by chemically hydrolyzing the bonds in the lignin–hemicellulose matrix by dilute acid hydrolysis, which allows the subsequent enzymatic conversion of cellulose to glucose to proceed at reasonable rates. The glucose formed can then be fermented

*Author to whom all correspondence and reprint requests should be addressed.

into useful products. Prehydrolysis of lignocellulosic biomass with dilute sulfuric acid, at temperatures higher than 160°C, effectively hydrolyzes the glycosidic bonds in the hemicellulosic component, and increases the enzymatic digestibilities of the cellulosic component in the remaining pretreated residues (1–6).

Dilute acid prehydrolysis of yellow poplar (*Liriodendron tulipifera*) sawdust was carried out in a series of experiments using the Sunds CD-300 Laboratory Hydrolyzer (Sunds Defibrator, Norcross, GA) installed in the NREL PDU. To conduct experiments at residence times shorter than the unit was designed for, the Sunds hydrolyzer was modified with a custom-built displacement cylinder. High-temperature, short-residence-time pretreatments were investigated at low acid concentrations that would result in savings in catalyst and lime costs, and lower costs for materials of construction for the reactor. It would also reduce the problems and costs associated with handling the gypsum formed during neutralization. Because of the considerable investment in human resources, operating expenses, and quantities of feedstock involved in operating the Sunds hydrolyzer (7), a 4-L-batch steam-explosion reactor (NREL Digester) was designed and installed for conducting optimization experiments at a smaller scale. A series of experiments to test the pretreatment performance of the NREL digester were conducted to compare the possibility of scale-up of this reactor to the pilot-scale Sunds hydrolyzer. Applicability of the bench-scale pretreatment results to the larger-scale reactor would result in considerable savings in resources and feedstock.

This paper reports the compositional analysis of hydrolysate liquors and solids, xylose and carbon mass balance closures, and enzymatic digestibility results from several experimental runs that used dilute acid prehydrolysis of yellow poplar sawdust pretreated in the pilot-scale Sunds hydrolyzer and the NREL digester.

MATERIALS AND METHODS

Feedstock Preparation

Yellow poplar sawdust was obtained from SawMiller (Haydenville, OH) and stored at −20°C in plastic-lined cardboard totes of approx 1000 lbs (~50–55% solids). The totes were thawed for several days before being pretreated in the Sunds hydrolyzer. Thawed totes of sawdust were dumped into a feedhopper bin and metered onto a weighbelt for feedback control of the feed, acid, water, and steam addition rates to the Sunds hydrolyzer. The sawdust used in the NREL digester pretreatment experiments was collected by combining and mixing several buckets-full each of material removed from totes during the experimental runs for the Sunds hydrolyzer.

Large chips, twigs, and branches in the poplar sawdust collected for pretreatment experiments with the NREL digester were removed by screening through a 7.4-mm screen. The screened sawdust was soaked in 0.3% (w/w) sulfuric acid solution at 60°C for 3 h, then drained overnight to approx 38% solids before pretreatment. The concentration of acid that remained in the chips to be pretreated was determined by titration. A small sample of the impregnated chips was placed in deionized water and blended with a Waring blender for 30 min. Titrations with standardized NaOH solutions (J. T. Baker, Phillipsburg, NJ) gave 0.26% (w/w) sulfuric acid concentration in the chips, as reported in the tables.

Pretreatment Reactors

The Sunds CD-300 Laboratory Hydrolyzer used in the pretreatment section of the NREL PDU can process as much as 1 t/d (bone-dry t) of various lignocellulosic feedstocks, using dilute acid prehydrolysis at temperatures as high as 230°C and residence times as long as 60 min (7). Acid and water are added in a stainless steel pug-mill mixer to achieve the desired acid and solids concentration in the hydrolyzer. A stainless steel plug-flow screw feeder compresses the acid-wetted biomass feed into a plug solid enough to resist the steam pressure (maximum 400 psig). A blow-back preventer valve is also present, in case the biomass plug fails. Liquid stream produced during the compression process (squeezate) is pumped back into the vertical impregnator section of the hydrolyzer to maintain the desired solids and acid concentrations. The acid impregnated feedstock stream in the reactor is heated by direct steam injection with ports near the bottom and top of the hydrolyzer. Temperature is measured near the bottom, middle, and top, with feedback control via temperature or pressure sensors. Noncondensable gases entrained in the feedstock are bled off the top of the reactor with volatile components (furfural, HMF, and acetic acid), condensed, and analysis performed for use in mass-balance calculations. Residence time is a function of the various feed rates and level within the hydrolyzer, and is controlled by a level sensor that controls the operation of a double-reciprocating valve isolating the high pressure of the reactor from the low-pressure (5 psig) flash tank. The pretreated biomass is flash-cooled in the flash tank, and flash vapor is condensed and components analyzed. A high-volume open-faced centrifugal pump (Discflo, San Diego, CA) is used to recirculate the pretreated slurry (approx 25% total solids) between the bottom and top of the flash tank, to keep the solids in the slurry suspended. Lime may be added to the flash tank for either neutralization or overliming of the slurry. Metering off the recirculation loop allows the pretreated slurry to enter the fermentation train.

The NREL digester consists of a 4-L-batch steam-jacketed reactor equipped with a 4-in (10-cm) top and 2-in (5-cm) bottom ball valve, two direct steam injection ports near the top and bottom, with K-type thermocouples inserted near the top and bottom of the reactor used for temperature measurements. At the end of pretreatment, the contents of the reactor are discharged into a cooled flash tank and the flash vapor is condensed in a separate condenser, as is the case with the Sunds pretreatments. Noncondensable gases are scrubbed by a dilute alkali solution prior to exhausting to atmosphere. Noncondensable gases can be vented from the top of the digester during pretreatment, if temperature and resulting steam pressure differ significantly.

Pretreatment

Pretreatment experiments with the Sunds hydrolyzer were carried out at temperatures that ranged from 160 to 210°C, acid concentrations of 0.1–1%, and residence times of 4–10 min. The feed rate of sawdust into the Sunds hydrolyzer was set at 37 or 40 bone-dry kg/h. The higher feed rate was needed to consistently achieve 4-min residence times. The reactor was brought to steady state, residence times measured, and then operated for a minimum of 1 h before being sampled. Before the sample was taken, the flash tank was emptied, and material was allowed to collect for at least 15 min with recirculation before sampling. The solids loadings in the Sunds hydrolyzer and flash tank were controlled to approx 20–25% total solids. Steady-state conditions could be changed 2–3× in a long day.

Pretreatment with the batch NREL digester was accomplished by pre-warming the reactor to the desired pretreatment temperature, then loading it with a batch of pre-weighed, acid-soaked chips. Steam was introduced at the desired pressure and pretreatment allowed to proceed for the desired residence time. At the end of pretreatment, the contents of the reactor were discharged into the flash tank. Typically, two 4-L batches are pretreated at each temperature, after which the contents of the flash tank were emptied and cone-blended, and the slurry subjected to analysis. The flash tank surfaces were rinsed down and the rinsate collected separately for analysis. Portions of the pretreated solids were stored at −20°C for later digestibility and fermentation assays. The pretreated material (approx 20–25% total solids) was processed into a liquor fraction by vacuum filtration (Whatman No. 1 paper), and a water-insoluble fraction (after washing with 10 vol of H_2O) followed by chemical analysis of each fraction.

Analysis of Wood and Solid Residues

Dry wt (by oven drying at 105°C to constant weight) *(8)* and Klason acid-soluble and acid-insoluble lignin were determined by standard methods *(9)*. Anhydrosugars in the whole-wood and pretreated solids were determined by a procedure slightly modified from that developed at the US Forest Products Laboratory *(4,10)*. Ash in the wood and pretreated solid residues were analyzed by standard gravimetric methods *(11)*.

Analysis of Liquid Residues

Organic acids, glycerol, hydroxymethyl furfural (HMF), and furfural in the filtrate and rinsate fractions were determined by high-performance liquid chromatography (HPLC), using Bio-Rad Aminex HPX-87H columns (Bio-Rad, Hercules, CA) *(4,10,12)*. Monomeric sugars were also determined by HPLC using Bio-Rad Aminex HPX-87P columns *(4,10,12)*. Subsequent to the HPLC analysis above, the oligomeric sugars in the liquors and rinsate fractions were converted to monomers using 4% H_2SO_4 hydrolysis at 121°C for 1 h, and the monomeric sugars analyzed using the Bio-Rad HPX-87P column, and corrected for sugar losses, using sugar recovery standards *(10,12)*. Acetic acid released by hydrolysis of the oligomeric xylan in the liquor fraction, following the 4% posthydrolysis analysis, was determined by HPLC with the Bio-Rad HPX-87H column.

Enzymatic Analysis

Extensively washed (10 vol of H_2O) biomass residues were tested for enzymatic digestibilities with Iogen cellulase (Iogen, Ottawa, ON, Can) enzyme (lot no. BRC 191095, assayed at 91 FPU/mL and 198 IU/mL β-D-glucosidase activity) *(13)* at a loading equivalent to 60 FPU/g of cellulose in prewarmed (50°C) 10-mL reaction cocktails that contained solids equivalent to 1% (w/v) cellulose, 50 mM citrate buffer, pH 4.8, and 40 µg/mL tetracycline and 30 µg/mL cycloheximide, to minimize contamination *(6)*. Duplicate reaction vials were incubated at 50°C with 120 rpm rotation at a 45-degree angle, and compared to controls that contained 1% (w/v) Solka Floc, grade NF-FCC (Fiber Sales and Development, Urbana, OH), and enzyme blanks. One-half mL samples were removed and glucose concentrations determined with a YSI Model 2700 Select Biochemistry Analyzer equipped with an immobilized glucose oxidase membrane (Yellow Springs Instruments, Yellow Springs, OH) calibrated with YSI-supplied 2.50 g/L glucose calibration standards. Samples were centrifuged at 12,000g for 5 min, with the supernatant diluted to keep the glucose readings below the 2.50 g/L level used for calibrating the YSI instrument.

RESULTS

Composition of hydrolysate liquors from the various pretreatments are shown in Table 1. The data show that, as expected, at harsher pretreatment conditions (i.e., higher temperatures and acid concentrations), more glucose is produced from the cellulosic portion of the biomass. The data indicate that at low acid concentrations (below 0.21%) and short residence times, the Sunds hydrolyzer incompletely hydrolyzes the hemicellulose to monomeric sugars, even at temperatures of 200 and 210°C (rows 4 through 8). Xylose concentrations of 30–34 g/L result from pretreatments at temperatures of 200°C and higher in both reactors; concentrations between 39 and 40 g/L are found at 160–170°C (rows 9 and 10). Because the furfural concentrations are comparable for the various pretreatment conditions listed, xylose is apparently converted to other products that are not measured (*see* losses of xylose listed in Table 3). The lower acid concentrations

Table 1
Compositional Analysis of Hydrolysate Liquors of Pretreated Yellow Poplar

Experiment number	Reactor	Temperature °C	Acid (%-w/w)	Residence time (min)	Cellobiose (g/L)	Glucose (g/L)	Xylose (g/L)	Mannose (g/L)	Acetic acid (g/L)	Furfural (g/L)	HMF (g/L)
1	Digester	210	0.26	3	0.9	13.9 (15.8)[a]	34 (33.9)[a]	7.5 (8.9)[a]	10.5 (14)[a]	1.8	0.4
2	Digester	210	0.26	4	0.8	14.6 (16.4)	31.1 (30)	7.2 (8.2)	10.5 (14.2)	2.8	0.7
3	Digester	230	0.26	1	1.1	15.2 (17.9)	31.7 (32)	6.7 (8.7)	8.3 (13.5)	1.3	0.4
4	Sunds[b]	161	1.01	9.2		11.7 (11.7)	39.4 (39.7)	6.4 (9.8)	12.5	2	nd[d]
5	Sunds[b]	170	0.72	10.5		9.2 (9.7)	38.6 (39.6)	6.7 (8.2)	12.9	1.7	nd
6	Sunds[c]	190	0.2	4	4.5	3.2 (6.7)	14.5 (35.3)	2.7 (7.6)	3.8 (12.1)	0.4	0.1
7		200	0.1	4	4.2	1.6 (5.8)	6.9 (32.2)	1.4 (6.5)	2.9 (11.1)	0.4	0.3
8		200	0.15	4	4.7	2.8 (6.5)	13 (34.5)	2.6 (7.5)	4 (12.9)	0.6	0.3
9		200	0.21	4	3.9	3.2 (6.1)	14.6 (31.1)	2.9 (6.9)	3.7 (11.6)	0.7	0.2
10		210	0.1	4	4	1.7 (5.7)	7.4 (30.9)	1.6 (6.3)	3.2 (11.5)	0.4	0.3
11	Sunds[c]	200	0.32	4.6	0.9	8.9 (10.7)	32.2 (33.6)	6.7 (9.9)	10.5	2.1	0.5

[a] Parenthesis indicates component analysis following 4% hydrolysis at 121°C for 1 h.
[b] As-installed Sunds hydrolyzer.
[c] Sunds hydrolyzer modified with displacement cylinder.
[d] nd = not determined.

(rows 4 through 8) result in lower furfural concentrations, presumably because the solubilized oligomeric xylans are partially protecting the xylose incorporated in the oligomers. HMF concentrations (not shown) were less than 0.8 g/L. Galactose concentrations (not shown) are near 3.5–4 g/L, and arabinose concentrations were near 2.2 g/L for all the liquors. The galactose and arabinose were usually hydrolyzed to monomeric sugars, except for the low-acid concentration experiments with the Sunds hydrolyzer, in which they were mixtures with monomer/total sugar ratio of approx 50%. Mannose concentrations, except in the low-acid concentration pretreatment experiments in the Sunds hydrolyzer (in which they are associated with the solubilized oligomeric xylan in a monomer/total sugar ratio of approx 50%), were in the range of 6.4–9.9 g/L. Acetic acid concentrations varied between 8 and 14 g/L, except at the low-acid concentration pretreatments in the Sunds hydrolyzer (rows 4 through 8). The total solids of the pretreated material and the compositional analysis of the liquors are similar, which suggests that the two reactors have similar performance under similar pretreatment conditions.

The solids compositional analysis (based on a 105°C dry wt) of the pretreated solids and the starting yellow poplar sawdust is shown in Table 2. The values for yellow poplar sawdust are averages of two separate determinations of representative sawdust samples. Data for glucan and xylan components (presented as glucose and xylose) in the pretreated solids show that, at the lower temperatures (161 and 170°C), the percent glucan is near 60%, and the xylan remaining is near 2–3%; thus about 90% of the xylan is solubilized. The pretreatment at 0.21% acid and 200°C in the Sunds hydrolyzer leaves about 4% of the residue composed of xylan, which represents less solubilization (~87%). Increasing the acid concentration to 0.32% at 200°C increased the solubilization and decreased the amount of xylan in the solid residue to 0.7%, and decreased the percentage of mannose in the residue from 0.68 to 0.33%. Galactose and arabinose (not shown) in the sawdust were 0.25 and 0.7%, respectively, but very little of the sugars were found in the pretreated residues (less than 0.06%). Klason lignin in the sawdust was approx 25% of the sawdust, which increased to approx 33 to 35% of the water-insoluble solids following pretreatment. The acid-soluble lignin decreased to approx 1.5% following pretreatment, except in the lower temperature, higher-acid concentration experiments, in which approx 4.4–4.8% of the pretreated residue was acid-soluble lignin. Total ash in all cases decreased from 0.9% in the sawdust to 0.5% in the pretreated solids.

The xylose and carbon closures (14) for the various pretreatments are shown in Table 3. Conditions of low acid (rows 6 and 7) and lower temperatures (rows 4 and 5) result in 9 and 15%, respectively of the pretreated solids composed of xylan. In addition, low temperatures (rows 4 and 5) result in higher percentages of xylose recovered in the hydrolysate liquor, in which recoveries near 85% are possible. Xylose closure, based

Table 2
Solids Compositional Analysis of Yellow Poplar[a]
Components (% dry wt)

Experiment number	Reactor/ feedstock	Temperature °C	Acid conc. (%-w/w)	Residence time (min)	Glucose	Xylose	Mannose	Klason lignin	Acid soluble lignin	Total ash
	Yellow poplar sawdust				46.3	20.2	3.6	23.5	2.6	0.9
1	Digester	210	0.26	3	69.2	0.4	0.08	34.2	1.5	0.5
2	Digester	210	0.26	4	69.3	0.5	0.11	34.8	1.5	0.5
3	Digester	230	0.26	1	67.5	0.4	0.05	35.3	1.6	0.5
4	Sunds	161	1.01	9.2	61	3	0	33.1	4.4	nd[b]
5	Sunds	170	0.72	10.5	62.6	2.4	0	34.2	4.8	nd
9	Sunds	200	0.21	4	65.7	3.9	0.68	31.4	1.7	0.5
11	Sunds	200	0.32	4.6	65.8	0.7	0.33	35.2	1.3	0.5

[a] Sawdust and water-washed pretreated sawdust solids.
[b] nd = Not determined.

Table 3
Xylose Recovery and Cellulose Digestibility

Experiment number	Reactor	Temperature °C	Acid conc. (% w/w)	Residence time (min)	% Xylose in solids	% Xylose in liquor	% Xylose to furfural	% Xylose mass closure	% Carbon mass closure	Digestibility[a] %
1	Digester	210	0.26	3	0.38	21.9	3.2	25.5	72	79.8
2	Digester	210	0.26	4	0.5	66.3	11.1	77.9	101	80.5
3	Digester	230	0.26	1	0.36	38.7	1.5	40	89	83.9
4	Sunds	161	1.01	9.2	9.9	84.8	9.8	101.5	94.6	nd[b]
5	Sunds	170	0.72	10.5	9	74.8	11.1	88.9	98.6	nd
7	Sunds	200	0.1	6.6	13.8	66.5	12.2	86.4	93.2	nd
9	Sunds	200	0.21	4	15.4	50.2	7	76.7	105.4	67.6
11	Sunds	200	0.32	4.6	2.3	60.5	25.8	81	107	82.9

[a] Digestibility determined with 60 FPU/g cellulose at 50°C, pH 4.8.
[b] nd = Not determined.

on xylose recoveries in the hydrolysate liquor, the solid residue, and xylose converted to furfural, are calculated to be 25–102% for the various pretreatments. Higher temperatures, lower acid concentrations, and shorter residence times (rows 6 and 7) decreased the recoveries. Presumably, the lost xylose is converted to furfural and other degradation products. Increasing the acid concentration in the Sunds hydrolyzer from 0.21 to 0.32% at 200°C, with a 4.6-min residence time, increases the amount of xylose converted to furfural (row 8). The results for the NREL digester pretreatment at 210°C, 4-min residence time (row 2) compares favorably with the Sunds hydrolyzer experiment carried out at 200°C and 4.6-min residence time (row 8). Increasing the temperature in the NREL digester to 230°C, with 1-min residence time (row 3), decreases the amount of xylose recovered in the liquor to approx 40%, and the %-xylose mass-balance closure to 40%, which suggests that these pretreatment conditions are severe. The total mass balance closures for xylose (25.5%) and carbon (72%) are low for the NREL digester experiment summarized in row 1, because of problems in recovering all of the hydrolysate liquor. With the above exception, the carbon mass-balance closure *(14)* for the other pretreatments ranged from 89 to 107%.

The enzymatic digestibility of washed pretreated solids in Table 3 was approx 80–84% after 24 h, at a loading of 60 FPU/g of cellulose and 50°C. The digestibility decreased to 55–60% at a loading of 25 FPU/g of cellulose and 37°C (data not shown). The digestibility of the NREL digester pretreated washed solids were near that obtained for the Sunds hydrolyzer. The high-temperature, short-residence-time, low-acid (0.21%) pretreatment in the Sunds hydrolyzer produced solids with a lower digestibility of approx 68% (row 7), which suggests the pretreatment conditions were not as severe.

CONCLUSIONS

The Sunds hydrolyzer very effectively solubilizes the hemicellulosic portion of biomass, and increases the enzymatic digestibility of the cellulose that remains in the solid residues for temperatures higher than 160°C, at which recoveries of xylose in the hydrolysate liquor approaches 85% (Table 3, rows 4 and 5). Increasing the temperature to 200°C with 0.32% (w/w) sulfuric acid concentrations, and the short residence time of 4.6 min, increases the amount of hemicellulose solubilized; however, a considerable amount of xylose is lost to unaccounted-for degradation products, and the amount of xylose recovered in the hydrolysate liquor decreases to approx 61%. Acid concentrations below 0.21% in the Sunds hydrolyzer pretreatments do not effectively solubilize the hemicellulose or increase digestibility of the cellulose in the residues, even at 200–210°C (Table 3, rows 6 and 7), which suggests that the ash within the sawdust is neutralizing a considerable amount of the acid catalyst. The pH measured between 2.3

and 2.9 for the low-acid experiments for hydrolysate liquors from the Sunds hydrolyzer (data not shown), which suggests that dilute acid hydrolysis with pH higher than 2.2, and temperatures of 200°C and 4.5-min residence times, is less effective for solubilizing hemicellulose in yellow poplar sawdusts. The results demonstrate that high-temperature (~200°C), short-residence-time, dilute-acid prehydrolysis with the Sunds hydrolyzer is an effective pretreatment that lowers the catalyst requirements (sulfuric acid) and gypsum formed when neutralizing the pretreated residues with lime, before they enter the fermentation train. The results presented for the NREL digester give similar levels of solubilization of hemicellulose and xylose recoveries in the liquors, compared to the Sunds hydrolyzer at short residence times. The results show only minor differences between the bench- and pilot-scale reactors in the pretreatment of yellow poplar sawdusts. The use of the bench-scale digester would result in shorter turnaround times for optimization experiments, and reduce requirements for feedstock and resources. The pretreatment parameters developed in the digester can then be applied to the Sunds hydrolyzer when pilot plant demonstration is required.

ACKNOWLEDGMENTS

This work was funded by the Biochemical Conversion Element of the Office of Fuels Development of the US Department of Energy. Chemical analysis of some of the Sunds hydrolysate liquors were performed by John Brigham, whose help was greatly appreciated. Solids compositional analysis were performed by Hauser Laboratories (Boulder, CO).

REFERENCES

1. Grethlein, H. E. (1980), US Patent 4, 237,226.
2. Knappert, D., Grethlein, H. E. and Converse, A. (1980), *Biotech. Bioeng.* **22**, 1449.
3. Knappert, D., Grethlein, H. E. and Converse, A. (1981), *Biotech. Bioeng. Symp.* **11**, 67.
4. Grohmann, K., Himmel, M., Rivard, C., Tucker, M., Baker, J., Torget, R., and Graboski, M. (1984), *Biotech. Bioeng. Symp.* **14**, 137.
5. Grohmann, K., Torget, R., and Himmel, M. (1985), *Biotech. Bioeng. Symp.* **15**, 59.
6. Grohmann, K., Torget, R., and Himmel, M. (1986), *Biotech. Bioeng. Symp.* **17**, 135–151.
7. Nguyen, Q. A., Dickow, J. H., Duff, B. W., Farmer, J. D., Glassner, D. A., Ibsen, K. N., et al. (1996), *Bioresource Technol.* **59**, 189–196.
8. TAPPI Test Methods (1991), T210 cm-86, *Weighing, Sampling and Testing Pulp for Moisture,* TAPPI, Atlanta, GA.
9. TAPPI Test Methods (1994–1995), T222 om-88, *Acid-Insoluble Lignin in Wood and Pulp,* TAPPI, Atlanta, GA.
10. Moore, W. E., and Johnson D. B. (1967), *Procedures for the Chemical Analysis of Wood and Wood Products,* Forest Products Laboratory, U.S. Department of Agriculture, Madison, WI.
11. TAPPI Test Methods (1991), T211 om-85, *Ash in Wood and Pulp,* TAPPI, Atlanta, GA.
12. Ehrman, C. I. and Himmel, M. E. (1994), *Biotechnol. Techniques* **87**, 99–104.
13. Ghose, T. K. (1987), *Pure Appl. Chem.* **59**, 257–268.
14. Hatzis, C., Riley, C., and Philippidis, G. (1996), *Appl. Biochem. and Biotech.* **57** and **58**, 443–459.

Shrinking-Bed Model for Percolation Process Applied to Dilute-Acid Pretreatment/Hydrolysis of Cellulosic Biomass

RONGFU CHEN, ZHANGWEN WU, AND Y. Y. LEE*

Department of Chemical Engineering, Auburn University, AL 36849

ABSTRACT

For many lignocellulosic substrates, hemicellulose is biphasic upon dilute-acid hydrolysis, which led to a modified percolation process employing simulated two-stage reverse-flow. This process has been proven to attain substantially higher sugar yields and concentrations over the conventional single-stage percolation process. The dilute-acid pretreatment of biomass solubilizes the hemicellulose fraction in the solid biomass, leaving less solid biomass in the reactor and reducing the bed. Therefore, a bed-shrinking mathematic kinetic model was developed to describe the two-stage reverse-flow reactor operated for hydrolyzing biphasic substrates, including hemicellulose, in corn cob/stover mixture (CCSM). The simulation indicates that the shrinking-bed operation increases the sugar yield by about 5%, compared to the nonshrinking bed operation in which 1 reactor volume of liquid passes through the reactor (i.e., $\tau = 1.0$). A simulated optimal run further reveals that the fast portion of hemicellulose is almost completely hydrolyzed in the first stage, and the slow portion of hemicellulose is hydrolyzed in the second stage. Under optimal conditions, the bed shrank 27% (a near-maximum value), and a sugar yield over 95% was attained.

Index Entries: Dilute-acid pretreatment; modeling; shrinking bed; percolation; corn cobs/stover mixture.

INTRODUCTION

Pretreatment is necessary to bioconvert lignocellulosic biomass into fuels and chemicals. Pretreatment with dilute sulfuric acid is a viable process option; however, a concern is that the sugars decompose under high temperature and low pH to form undesirable components that are toxic in the subsequent fermentation. To avoid this, it is important to select

*Author to whom all correspondence and reprint requests should be addressed.

proper reaction conditions, reactor configurations, and operation conditions. Previous studies (1–4) have established that a percolation reactor (packed-bed, flow-through type) is most suitable for biomass pretreatment, because the sugar products are discharged from the reactor as they form, thus reducing sugar decomposition. High sugar concentrations can also be attained because of the high solid-to-liquid ratio that exists in a packed-bed reactor. In the previous bench-scale modeling and experimental work (5,6), it was demonstrated that a two-stage, two-temperature, reverse-flow scheme significantly enhances the overall performance of the percolation reactor. It simulates countercurrent flow of the biomass solid and hydrolysis liquor, and it exploits the fact that hemicellulose exhibits a biphasic behavior upon dilute acid hydrolysis. A low temperature is applied in the first stage to hydrolyze the labile xylan, then a high temperature is applied in the second stage to hydrolyze the resilient xylan.

Acid pretreatment solubilizes the hemicellulose in biomass. In previous modeling work (5), to retain the linearity of the governing equations, the authors assumed that the bulk-packing volume of the solid biomass in the reactor remained constant during the hydrolysis. This was done to retain the linearity of the governing equations. However, in actual operation, the bulk-packing density of the solid biomass in the reactor changes because of the solubilization of the hemicellulose, which leaves less solid biomass in the reactor. To further optimize the pretreatment process, a shrinking-bed reactor was proposed by NREL (7). It was designed to keep a constant bulk-packing density of the solid biomass in the reactor, so that the high solid-to-liquid ratio would allow a high product concentration to be obtained. Figure 1 shows a simplified diagram of a bench-scale, shrinking-bed, percolation reactor. The reactor has a fixed and a movable end; the movable end is supported by a compressed spring. As the reaction progresses, the gradual depletion of the packed-solid biomass causes the particle structure to be less dense. To overcome this limitation, the spring-attached movable end presses the loose biomass particles closer, allowing the bulk-packing density of the lignocellulosic biomass to be maintained constant. Although pilot-scale reactor designs, which allow for the continual shrinkage of the biomass-bed as a function of hydrolysis, will most likely not resemble the bench-scale design (Fig. 1), it is instructive to model the shrinking-bed concept for subsequent design. In this work, process modeling and simulation were performed for this shrinking-bed reactor as it applies to the acid hydrolysis of lignocellulosic.

This investigation establishes a process model for the shrinking-bed reactor, which prehydrolyzes the hemicellulose in a corn cobs/stover mixture (CCSM). The modeling work is directed toward optimal operation of the shrinking-bed reactor and analysis on the bed-shrinking phenomena. The issues addressed in the modeling and simulation are the extent and the effect of bed-shrinking, substrate variation in the reactor, and product yield and concentration.

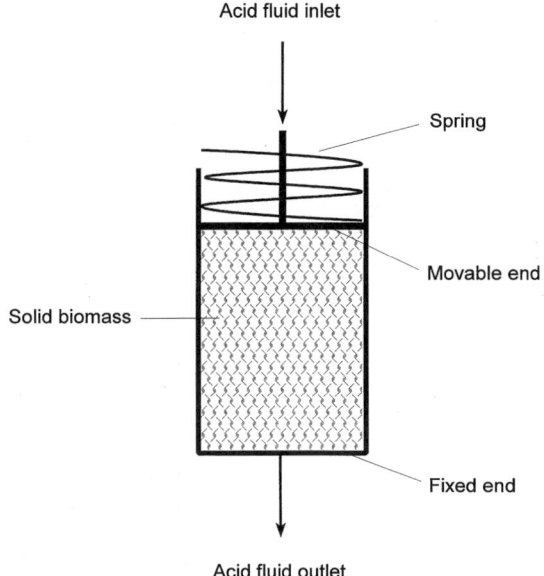

Fig. 1. A simplified diagram of a shrinking-bed reactor applied in pretreatment.

MODEL DEVELOPMENT

The shrinking-bed percolation reactor has a solid bed that shrinks during hydrolysis because of hemicellulose solubilization (Fig. 1). The effective reactor volume (or the length of solid bed of biomass in the percolation reactor) is related to hemicellulose conversion. For a differential value of τ^i, or a differential amount of acid fluid, the percolation process is assumed to be a nonshrinking-bed reactor, because a differential amount of hemicellulose is removed during that time span. The nonshrinking process is then followed by a compression stage, in which the biomass is compressed to its original packing density. It is assumed that no reaction occurs in this stage. The model for the shrinking-bed reactor can be decoupled, as shown in Figure 2, by repeating the operation of nonshrinking-bed reaction with a differential amount of liquid τ^i followed by a compression process. Adoption of this method allows the governing partial differential equations to be linear; therefore, the analytical solution previously obtained for nonshrinking-bed operation (5) directly applies to the present case.

A shrinking factor (ξ^i) is defined as the ratio of the reactor volume after the ith compression operation to that before the ith compression operation.

$$\xi^i = V^i/V^{i-1} \qquad (1)$$

For the cylindrical reactor, Eq. 1 can also be expressed in terms of reactor length:

Fig. 2. Decomposition of a shrinking-bed reactor operation.

$$\xi^i = L^i/L^{i-1} \quad (2)$$

The shrinking factor ξ^i (≤ 1) is a function of hemicellulose conversion only.

The conversion (θ^i) of hemicellulose during one differential value of τ^i is

$$\theta^i = H^{i-1} - H^i/H^{i-1} \quad (3)$$

which is valid only for the nonshrinking-bed operation. The hemicellulose concentration in the reactor H^{i-} and H^i are determined by Eq. 8 in ref. 5.

The composition (η^i) of hemicellulose in the solid biomass changes during the hydrolysis, as hemicellulose and some lignin are solubilized into the hydrolyzate. It can, based on material balance, be expressed as

$$\eta^i = \frac{\eta^{i-1}(1 - \theta^i)}{1 - \eta^{i-1}\theta^i(1 + r)} \quad (4)$$

where γ is the ratio of solubilized lignin to solubilized hemicellulose and cellulose during the pretreatment. It is assumed to be constant throughout the reaction. For the sample substrate of CCSM, with 20.0% xylan, 39.2% glucan, and 23.3% lignin (8), it was assumed that 80% of total lignin was dissolved with the solubilization of hemicellulose and cellulose. Therefore, the γ value is calculated to be 0.315 from the composition of the feedstock.

The change of reactor volume resulting from the solubilization of hemicellulose and lignin is expressed as

$$V^i = V^{i-1}[1 - \eta^{i-1} \theta^{i-1} (1 + \gamma)] \quad (5)$$

Therefore, the shrinking factor is determined by

$$\xi^i = 1 - \eta^{i-1} \theta^i (1 + \gamma) \quad (6)$$

for one stage of compression operation.

The overall reaction conversion (θ_{overall}) after the nth nonshrinking and compression operations, based on the initial reaction conditions, is expressed as:

$$\theta_{overall} = 1 - \frac{H_j}{H_0} \prod_{i=1}^{j} \xi^i \tag{7}$$

Similarly, the overall length of the shrinking-bed reactor during the reactions is:

$$L = L_0 \prod_{i=1}^{j} \xi^i \tag{8}$$

The overall yield is determined by summing the yields from the nonshrinking-bed model with consideration of the compression operation. It is then expressed as:

$$Y_\Theta = \frac{1}{H_0} \sum_{j=1}^{j} (Y^j_{ns} H^j \prod_{i=1}^{j} \xi^j \tag{9}$$

where Y_s = sugar yield from shrinking-bed operation,
Y_{ns} = sugar yield from nonshrinking-bed operation (5),
H_0 = initial concentration of hemicellulose as xylose in the reactor,
j = 1, 2, ... n, and
i = 1, 2, ... j.

RESULTS AND DISCUSSION

The shrinking-bed operation is similar to the nonshrinking-bed operation, except there is a compression step between each differential reaction period τ^i. The compression process increases the solid-packing density to its previous level, but does not change the biomass composition; therefore, the optimum temperature profile previously determined for the nonshrinking operation is also applied to the shrinking-bed operation. In the shrinking-bed reactor, the bed depth changes during the reaction; therefore, the optimum flow rate must be adjusted to respond to the reduced liquid residence time. The computational modeling work was done for CCSM. The two-stage, reverse-flow operation (Fig. 3A) is identical in theory to the process shown in Figure 3B (5), so the shrinking-bed process was modeled based on the scheme of Figure 3B.

Effect of Acid Flow Rate and τ

In the previous nonshrinking-bed model, the optimal flow rate is obtained for a given reactor length at particular temperature (5). The dimensionless reaction rate β_i ($= l_i L/u$) is an optimized operational parameter. If the length of the reactor is reduced, the flow rate of the liquid u should also be reduced to maintain β at its optimum. However, this opera-

Fig. 3. Schematics of temperature step-change, two-stage, reverse flow, and shrinking-bed operation.

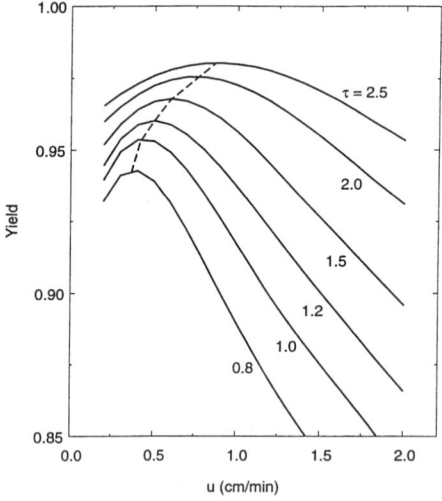

Fig. 4. Effect of acid flow rate on sugar yield at various τ values. (Two-stage, reverse-flow, shrinking-bed mode, T_1 = 140°C, T_2 = 170°C, L_o = 15.24 cm, C_o = 3.33 w/v%, and acid concentration = 0.8 wt%.)

tion is not convenient in practice. The authors have studied the effect of acid flow rate on sugar yield at a constant flow-rate operation. Figure 4 shows the effect of flow rate on sugar yield at the optimum temperature step change (140–170°C) (5). For a given τ, there is an optimum u to obtain

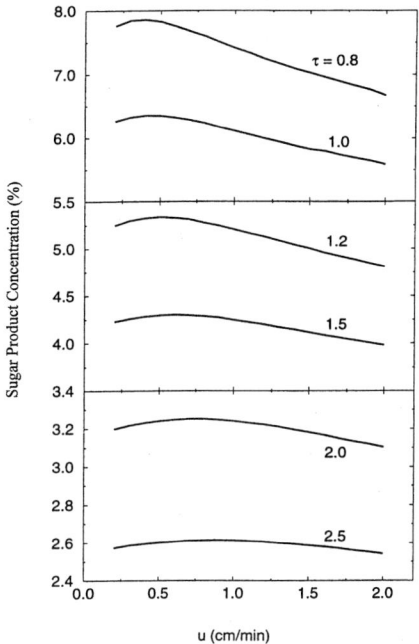

Fig. 5. Effect of acid flow rate on sugar product concentration (Cp) at various τ values. (Two-stage, reverse-flow, shrinking-bed mode, $T_1 = 140°C$, $T_2 = 170°C$, $L_o = 15.24$ cm, $C_o = 3.333$ w/v%, and acid concentration = 0.8%).

the maximum yield. For $\tau = 0.8$, the maximum yield of 0.94 is obtained at u = 0.4 cm/min; at $\tau = 2.5$, the maximum yield of 0.98 is obtained at u = 0.8. High τ-values give high sugar yields; however, high τ lowers the sugar concentration in the hydrolysate, as clearly shown in Figure 5. With application of the respective optimum flow rates, the maximum sugar concentrations are 7.8% at $\tau = 0.8$ and 2.6% at $\tau = 2.5$. Obviously, there is a trade-off between yield and sugar concentration. A proper choice of τ can only be made considering the overall process economics.

Comparison Between Shrinking-Bed and Nonshrinking-Bed Operations

One advantage of the shrinking-bed operation, over the nonshrinking one, is that τ becomes higher, even for the same amount of liquid throughput. Consider a shrinking-bed reactor with an initial volume the same as that of a nonshrinking-bed reactor. For $\tau = 1$, this means one reactor volume of liquid has passed through the nonshrinking bed reactor during the reaction. For the shrinking-bed operation, if, for example, 50% of the biomass dissolves, 1 reactor volume of liquid at the initial phase of the reaction will become 2 reactor volumes at the latter phase. Because yield increases with τ (fluid input into the reactor) at a given temperature and

Fig. 6. Yield and sugar product concentration under various operation modes. (Acid concentration = 0.8%, C_o = 3.333 w/v%, u = optimum values for shrinking-bed operation).

flow rate, the shrinking-bed operation is expected to give a higher yield. Figure 6 compares the shrinking-bed and nonshrinking-bed operations, both with temperature-step-change, reverse-flow operations, and uniform temperature operations. It shows that the shrinking-bed operation gives the highest sugar yield for a given τ. The sugar yield increases about 5% at τ = 1.0, 2% at τ = 1.5, and 0.2% at τ = 3.0 over those from nonshrinking-bed, step-change, reverse-flow operation modes. The sugar concentration from the shrinking-bed operation is slightly higher than that of the nonshrinking-bed operation.

Overall Hemicellulose Conversion

Figure 7 shows the profile of hemicellulose conversion during a two-stage shrinking-bed operation with τ = 1.5 for various flow rates. As shown in Figure 4, the flow rates of 0.2 and 2.0 cm/min are the lower and upper limits of this work, and a flow rate of 0.6 cm/min is the optimal value for τ = 1.5. Figure 7 indicates that a low flow-rate of 0.2 cm/min induces an excessive residence time, causing overreaction and significant decomposition. The reaction achieves near-complete conversion at about τ = 1; however, the sugar yield is about 0.95, and sugar loss, about 5%. On the other hand, with the flow rate of 2.0 cm/min, the hemicellulose conversion is only 91%, and the sugar yield is 0.90. Loss of sugar because of decomposition is therefore about 1%. With the flow rate of 0.6, the hemicellulose conversion is in excess of 99%, and the yield is 0.97, thus

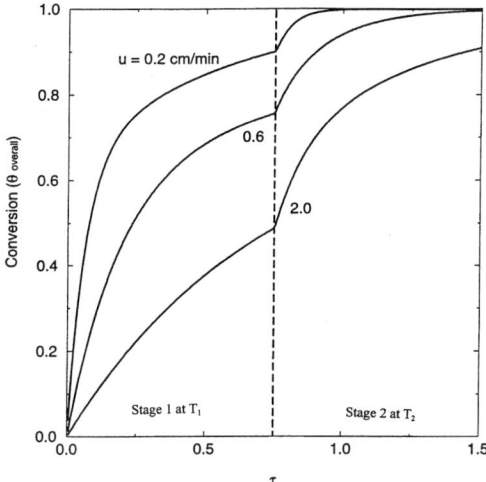

Fig. 7. The extent of hemicellulose conversion ($\theta_{overall}$) at various flow rate (u). ($T_1 = 140°C$, $T_2 = 170°C$, $L_o = 15.24$ cm, $C_o = 3.333$ w/v%, and acid concentration = 0.8%).

causing about 2% sugar loss. It is also seen that 75% conversion of hemicellulose was achieved in the first stage alone (low-temperature stage).

Distribution of Fast and Slow Portions of Hemicellulose During Hydrolysis

A two-stage temperature-step-change operation is particularly beneficial for hydrolysis of biphasic hemicellulose. In the overall process scheme, the fast portion of the hemicellulose is hydrolyzed at low temperature and the slow portion of hemicellulose at a high temperature. In a previous kinetic study (5), it was determined that the fast portion of hemicellulose in CCSM is 65%. Figure 8 shows the distribution of fast and slow portions of hemicellulose in the two-stage reactor at optimum temperature-step-change (140–170°C) condition (5). Three flow-rate conditions of 0.2, 0.6, and 2.0 cm/min, at $\tau = 1.5$, were applied in the study. For the case of u = 0.2 cm/min, the fast portion of hemicellulose was quickly dissolved at the early stage, at τ less than 0.5, and about 60% of the slow portion of hemicellulose was hydrolyzed after the first stage. The remaining slow portion of hemicellulose was completely hydrolyzed in the second stage, with a final sugar yield of 0.95. For u = 2.0, only about 60% of the fast portion and no slow portion of hemicellulose were hydrolyzed after the first stage. The remaining fast portion and about 61% of the slow portion of hemicellulose were dissolved after the second stage, giving a sugar yield of only 0.90, because of incomplete hydrolysis. With u = 0.6 and $\tau = 1.5$ (the optimum point), about 97% of the fast portion and 19% of the slow portion of were hydrolyzed after the first stage. The remainder of the fast portion and 99% of the slow portion were hydrolyzed after the second stage, to give the total of 97% yield.

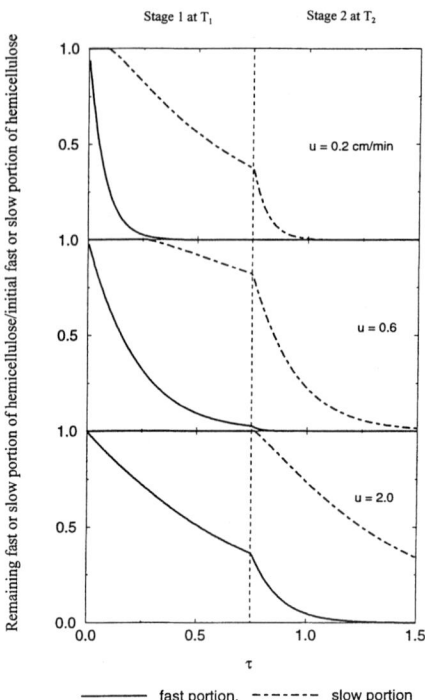

Fig. 8. Effect of acid flow rate (u) on the remaining fast or slow portion of hemicellulose. (T_1 = 140°C, T_2 = 170°C, L_o = 15.24 cm, C_o = 3.333 w/v%, and acid concentration = 0.8%).

Bed-Shrinking During Hydrolysis

In the bed-shrinking model, the shrinking process is terminated when the hydrolysis of hemicellulose is completed. Figure 9 shows the extent of shrinkage for the period of τ = 1.5 at various liquid flow rates (u). For the flow rate of 0.2 cm/min, the shrinking process was terminated at about τ = 1.0, because of complete hydrolysis. At the termination, the total bed shrinkage was 27% at the completion of the hydrolysis, and the shrinkage was about 24% after the first stage. For the flow rate of 2.0 cm/min, the final shrinkage was only 13% after the first stage, and 24% after the second stage. The reaction was not completed after the second stage. At the flow rate of 0.6 cm/min, the solid bed shrank 20% after the first stage and 27% after second stage, which was the maximum observed. The reaction was indeed complete at the end of the second stage.

Analysis of Sugar Loss During Hydrolysis of Hemicellulose

Sugar produced from hemicellulose hydrolysis is subjected to decomposition. Figures 10 and 11 show how the sugar product is decomposed in each stage. The overall τ is 1.5 for the two-stage operation (τ = 0.75 for each stage). With u = 0.2 cm/min, about 2.3% sugar is decomposed in

Shrinking-Bed Model for Percolation

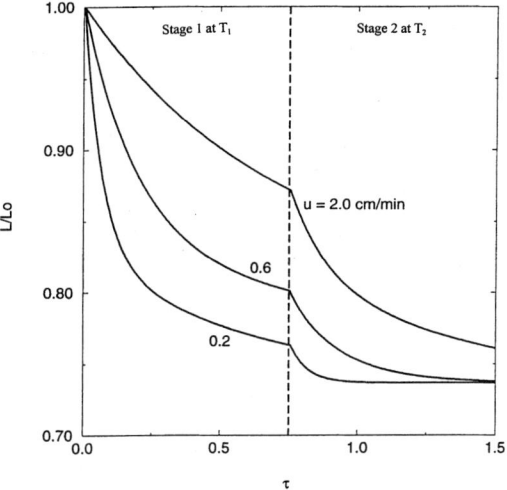

Fig. 9. Effect of acid flow rate (u) on bed shrinking (L/L_o). (T_1 = 140°C, T_2 = 170°C, L_o = 15.24 cm, C_o = 3.333 w/v%, and acid concentration = 0.8%).

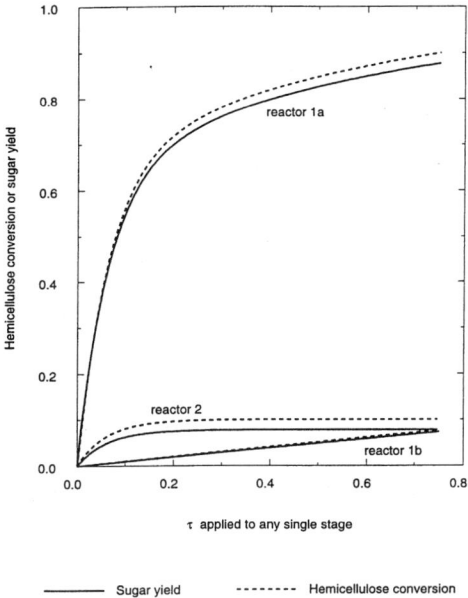

Fig. 10. Sugar decomposition at each stage during hydrolysis of hemicellulose. (T_1 = 140°C, T_2 = 170°C, L_o = 15.24 cm, C_o = 3.333 w/v%, u = 0.2 cm/min, and acid concentration = 0.8%).

Fig. 11. Sugar decomposition at each stage during hydrolysis of hemicellulose. (T_1 = 140°C, T_2 = 170°C, L_o = 15.24 cm, C_o = 3.333 w/v%, u = 0.6 cm/min, and acid concentration = 0.8%).

Reactor A1, a reactor packed with fresh biomass at low temperature (refer to Fig. 3B for Reactors A1, A2, and B). About 2.7% sugar is decomposed in Reactor B, a reactor packed with treated biomass at high temperature. There is only a trace amount of sugar loss from Reactor A2, an artificial reactor without solid biomass. For the optimum run of u = 0.6 cm/min, there is about 2% sugar loss from Reactor B, and much less in Reactors A1 and A2. The improved performance is attributed to the low-temperature condition in Reactor A1.

CONCLUSIONS

A modeling and simulation was performed on the shrinking-bed, two-stage, reverse-flow reactor operating for dilute-acid pretreatment of CCSM. The simulation results showed that the shrinking-bed operation increases the sugar yield by about 5%, compared to the nonshrinking-bed operation at a representative τ value of 1.0. The flow rate has emerged as an important parameter acutely affecting the performance of the two-stage reactor. A simulated optimal run at τ = 1.5 reveals that fast portion of hemicellulose is almost completely hydrolyzed after the first-stage reaction. Most of the slow portion of hemicellulose is hydrolyzed in the second stage. The simulation results further prove that the two-stage operation is well suited for hydrolysis of biphasic substrates, including hemicellulose in CCSM. With application of optimal flow rates, the bed-shrinkage reached a maximum of 27%, giving almost complete conversion of hemicellulose in

CCSM. The corresponding yield was over 95%. About three-quarters of the total shrinkage occurred after the first stage. Almost all the sugar decomposition occurred at the second stage (the high-temperature reactor).

ACKNOWLEDGMENTS

This research was conducted as a part a subcontract from NREL (No. XAW-3-13441-01). Additional support was provided by the Engineering Experiment Station of Auburn University.

NOMENCLATURE

C_0	initial xylan concentration in percolation, %
H	concentration of hemicellulose as xylose in the reactor
k_{0i}	frequency factor for k_i
k_i	rate constant = $A^{n_i} k_{0i} \exp(E_i/RT)$, min^{-1}. A is acid concentration.
L	reactor length, cm
t	time, min
u	velocity inside percolation reactor, cm/min
V	reactor volume
Y	yield
β_i	dimensionless reaction rate, $k_i L/u$
γ	ratio of solubilized lignin to solubilized hemicellulose
η	composition of hemicellulose in solid biomass
θ	conversion of hemicellulose during hydrolysis
ξ	shrinking factor, the ratio of the reactor volume after a compression operation to that of before the compression operation
τ	dimensionless residence time, tu/L
ρ	biomass density, g/mL
0	value at t = 0
ns	nonshrinking-bed operation
overall	overall value based on initial condition
s	shrinking-bed operation
i	*i*th operation
j	*j*th operation

REFERENCES

1. Lee, Y. Y., Lin, C. M., Johnson, T., and Chambers, R. P. (1978), *Biotechnol. Bioeng. Symp.* **8**, 75–88.
2. Limbaugh, M. L. (1980), MS Thesis, Auburn University, AL.
3. Cahela, D. R., Lee, Y. Y., and Chambers, R. P. (1983), *Biotechnol. Bioeng.* **25**, 3–17.
4. Kim, B. J., Lee, Y. Y., and Torget, R. W. (1993), *Appl. Biotechnol. Bioeng.* **39**, 119–129.
5. Chen, R., Lee, Y. Y., and Torget, R. W. (1996), *Appl. Biotechnol. Bioeng.* **57/58**, 133–146.
6. Torget, R. W., Hayward, T. K., Hatzis, C., and Philippidis, G. P. (1996), *Appl. Biotechnol. Bioeng.* **57/58**, 119–129.
7. Torget, R. W., Hayward, T. K., and Elander, R. (1997), *Nineteenth Symposium on Biotechnology for Fuels and Chemicals*, paper No. 4 Colorado Springs, CO.
8. Chen, R. (1997), PhD Dissertation, Auburn University, AL.

Copyright © 1998 by Humana Press Inc.
All rights of any nature whatsoever reserved.
0273-2289/98/70-72—0051$12.00

Cost Estimates and Sensitivity Analyses for the Ammonia Fiber Explosion Process

LIN WANG,[1] BRUCE E. DALE,[2,*] LALE YURTTAS,[1] and I. GOLDWASSER[1]

[1] Texas A&M University, College Station, Texas 77843-3122; and
[2] Michigan State University, East Lansing, Michigan 48824-1226

ABSTRACT

Process designs were conducted for each unit of the conceptual ammonia fiber explosion (AFEX) process, and fixed capital investment and operating costs were estimated. AFEX costs about $20–40/t of dry biomass treated. Several promising areas for reducing process costs exist. Return on investment (ROI) calculations were also done for AFEX-treated materials (as digestibility-enhanced animal feeds), in conjunction with sensitivity analyses on the overall processing costs. Estimated ROIs range from over 100%/y to negative, depending on the system variables. The most important variables are the cost of corn and corn fiber, ammonia loading, and whether or not drying is required.

Index Entries: Biomass pretreatment; cost estimation and economic analysis; process design; computer simulation; ammonia fiber explosion.

INTRODUCTION

The production of fuels and chemicals by fermentation of renewable resources has received increasing attention over the past several decades. Biomass is the only renewable resource that can be directly converted to liquid fuels. The dominant forms of biomass are grain starches and lignocellulosic crop and forest materials. The most viable carbohydrate substrates for fermentation to produce fuels and chemicals at very large scale will probably be derived from lignocellulosic crop and forest materials, rather than from grain or other food materials, because of economic considerations and the volume of raw material available. These lignocellulosic crop and forest materials have also been investigated for many years as potential animal feedstuffs.

*Author to whom all correspondence and reprint requests should be addressed.

A key problem in utilization of these lignocellulosics for either feedstuffs or fuel/chemical production is the relatively unreactive nature of the cellulose, in particular the poor yields of glucose from cellulose by acids or enzymes. The AFEX pretreatment increases biomass digestibility significantly (1); yields of fermentable sugars by enzymatic hydrolysis, following AFEX are typically 4–5X greater than those obtained on untreated materials. Ammonia, a volatile chemical, is easy to recover, compared with other chemical pretreatments. Few, if any, fermentation inhibitors or sugar degradation products are formed during the AFEX process (2). In addition, the energy requirements and capital costs for AFEX are expected to be relatively modest (3).

The purpose of this study was to develop a computer simulation program that could be used to carry out engineering and economic analyses of the current conceptual AFEX process, under different scenarios, using existing laboratory data.

METHODS

Process Design

Process Description

The current conceptual design for the AFEX process is simple and straightforward, but has two major variations: the so-called wet product option (Fig. 1) and the dry product option (Fig. 2). Under both options, a V-Ram pump feeder takes a slug of biomass from the hopper to the high pressure AFEX pretreatment reactor, in which the liquid ammonia is added. When the reaction is complete, a large valve at the bottom of the reactor opens, allowing the biomass to explode into a flash tank. The high pressure is suddenly released, and most of the ammonia flashes at that moment. The amount of ammonia flashed depends on reaction temperature, pressure, and, especially, the water and ammonia loading. The ammonia vapor from the flash tank is taken to a precondenser, which concentrates ammonia vapor to 99.8% (w/w), and then the ammonia vapor from the precondenser enters a total condenser, which condenses ammonia vapor to 99.8% (wt) liquid ammonia. The liquid ammonia from the total condenser is pumped and stored in an ammonia drum.

The liquid phase from the precondenser is taken to a distillation column. Ammonia vapor with a concentration of 99.8% (wt) is obtained from the top product of the distillation column, which enters a total condenser, and condenses to liquid ammonia. Through the liquid ammonia pump, pressurized liquid ammonia is transferred and stored in the ammonia drum. The recovered 99.8% (wt) liquid ammonia can be recycled to the AFEX pretreatment. The bottom product of the distillation column is water with 0.01% (wt) ammonia, which is stored in a water tank.

For the wet option, as shown in Fig. 1, the pretreated biomass from the flash tank is transferred to a unit that uses water to wash out enough

Computer Analysis of AFEX Process

Fig. 1. The AFEX pretreatment process flow sheet diagram for the wet option.

Fig. 2. The AFEX pretreatment process flow sheet diagram for the dry option.

of the remaining ammonia in the pretreated biomass to match the design requirement, which specifies how much ammonia can remain in the pretreated biomass as a nitrogen source for fermentation or for animal feeding. For the dry option, as shown in Fig. 2, the pretreated biomass from the flash tank is transferred to a dryer, which uses heated ammonia vapor to evaporate the extra ammonia remaining in the pretreated biomass, to meet the design requirement. This evaporated ammonia is absorbed in water and distilled as above. The pretreated biomass is now ready for hydrolysis, fermentation, or animal feeding. Dry AFEX-treated biomass can be easily stored and transported, but the wet, AFEX-treated biomass is presumed to be used within a few hours, and is not assumed to be transported significant distances.

Model Description

This computer simulation work was based on published AFEX experimental results *(1-4)*, as-yet unpublished results, personal communications *(5,6)* and the known vapor-liquid equilibrium of the ammonia-water system *(7,8)*. Since the strong polar character of aqueous ammonia solution causes severe nonideality problems, equilibrium models, which correlate other systems very well, often fail to correlate this strong polar and water-containing system. A set of equations developed by Schulz *(8)* was used to correlate vapor-liquid equilibria and other thermodynamic data for the ammonia-water system over relatively wide temperature and pressure ranges.

The most difficult work is in the modeling of the distillation columns and flash tank, because the strong polarity of the ammonia-water system causes difficulties with convergence of the model equations. Different numerical methods are used to solve the model equations *(9)*. Other areas of considerable uncertainty, and potential research subjects, are the drying curves for ammonia-laden biomass, and the equilibrium relationships between ammonia, biomass, and water in the biomass. Standard design equations are also used for equipment sizing *(10)*.

The calculation procedure was programmed in MS (i.e., Microsoft) Fortran, a Fortran programming language based on Fortran 77, which has some advanced features of Fortran 90. To enhance the portability of the program, an IBM-compatible PC was used in the programming. Microsoft Fortran 5.1 was used as the development tool, which does the compiling and linking of the program.

Summary Conditions: Base Case:

- Costs — 1995 dollars
- Plant location — Unspecified
- On-stream time — 8000 h/yr
- Feed — Corn fiber or rice straw
- Ammonia loading (lb ammonia/lb dry biomass) — 1.0 (corn fiber), 1.5 (rice straw)
- Water loading (lb water/lb dry biomass) — 1.5 (corn fiber), 1.5 (rice straw)
- Treatment pressure (kpa) — 1135 (corn fiber), 2413 (rice straw)
- Treatment temperature (°C) — 90 (corn fiber), 95 (rice straw)
- Treatment time (min) — 30 (corn fiber), 20 (rice straw)
- Nominal capacity — 10 t dry biomass/h
- Product — Pretreated biomass
- Utilities — Provided by existing facilities

Economics

The investment costs for different scenarios were developed by determining bare equipment costs for each piece of equipment. Costs of major pieces of equipment were obtained from recent vendor quotes *(11)*, as well as from other sources *(12)*. From the bare equipment costs, the fixed capital investment was estimated, using Lang multiplication factors *(12)* to calculate the capital necessary for the installed process equipment with all auxiliaries needed for complete process operation. Assuming the AFEX pretreatment plant is similar to an average chemical plant that processes both solids and liquids, the Lang multiplication factor used here is 4.1 *(12)*. The working capital for the AFEX process is estimated as 15% of the total capital investment *(12)*. The total capital investment is the sum of fixed capital investment and the working capital. All expenses directly connected with the manufacturing operation or the physical equipment of a process plant itself are included in the total manufacturing costs.

The assumed selling price for the AFEX-treated biomass was determined by measuring the digestibility of the treated material in rumen fluid, and assuming that the selling price was proportional to the digestibility, given the known digestibility of the corn fiber after treatment (90%), the digestibility and moisture content of grain (80 and 10%, respectively), and the average selling price of corn grain in the first three-quarters of 1995 *(13)*, or approx $3.80/bu. This approach tends to underestimate the value of the AFEX-treated corn fiber, since corn fiber also contains over 10% protein and some nonprotein nitrogen from the AFEX process, which will increase the feed value of the material. Since recent corn prices have been somewhat higher than historic levels ($2.50–$3.00/bu), a sensitivity analysis was done to determine the effect of corn prices at historical levels of about $2.50/bu on the calculated ROIs.

The return on investment is then calculated as the following:

$$ROI = (\text{Selling price} - \text{Feedstock price} - \text{Total manufacturing cost}) \cdot \text{ton of production}/\text{Total capital investment}$$

All cost estimates were performed in the spreadsheet program.

RESULTS AND ANALYSES

Capital Investment

Purchased-capital investment breakdowns for corn fiber are shown in Table 1 for the wet option, and in Table 2 for the dry-product option. The fixed capital investment for corn fiber in Table 1 is estimated at about $2.2 million for the wet option, and at about $14 million for the dry option (Table 2), reflecting the much higher costs for the dryers needed in the second option. Since there is little or no information on the rate of ammonia evaporation from biomass, a very conservative approach to cost estimation was taken, and this probably contributed to the high estimated drying costs.

Table 1
Purchased Equipment Cost for Corn Fiber Base Case: Wet Option

Purpose	Equipment	Unit price ($)	Total cost ($)
Feeding and flashing	V-Ram pump feeder	17,300	69,200
	Reactor agitator	11,900	47,600
	Reactor	46,610	186,440
	Flash tanks	50,430	50,430
	Flash tank cyclone	3020	3020
	Flash tank screw Conveyor	4420	4420
	Sub total		361,100
Ammonia recovery from treated corn fiber and flash vapor	Washer	100,000	100,000
	Distillation column	36,730	36,730
	Distillation condenser	8320	8320
	Reboiler	4620	4620
	Reflux pump	2080	2080
	Reboiler pump	4890	4890
	First condenser	1670	1670
	Total condenser	2600	2600
	Liquid compressor	3200	3200
	Accumulator	4620	4620
	Sub total		169,400
	Total		530,500
	Total updated to 1995		910,000
	Fixed capital investment		3,730,000

Note: Costs of major pieces of equipment were obtained from recent vendor quotes (11), as well as from other sources (12). From the bare equipment costs, the fixed capital investment was estimated using Lang factor (12), which represents the capital necessary for the installed process equipment with all auxiliaries needed for complete process operation. Assuming the AFEX pretreatment plant is similar to an average chemical plant that processes both solids and liquids, the Lang multiplication factor used here is 4.1 (12). Marshall and Swift equipment cost indexes were used to update cost data to 1995. The Marsh and Swift all-industry equipment index for 1995 is 1042.9.

Production Costs

Summaries of the costs of production of AFEX-treated corn fiber base case for both options are shown in Tables 3 and 4. The feedstock cost at $50/t (13) is the largest component of the cost for both options, representing 53% of the total costs for the wet option and 32% of the costs for the dry option. Utility costs are high for both options, and fixed charges are nearly as significant as utility costs for the dry option.

SENSITIVITY ANALYSES

Using available experimental data (1,2,6), a variety of simulations were obtained from the computer simulation program and spreadsheet

Table 2
Purchased-Equipment Cost for Corn Fiber Base Case: Dry Option

Purpose	Equipment	Unit price ($)	Total cost ($)
Feeding and flashing	Feeder	17,300	69,200
	Reactor agitator	11,900	47,600
	Reactor	46,610	186,440
	Flash tanks	50,430	50,430
	Flash tank cyclone	3020	3020
	Flash tank screw Conveyor	4420	4420
	Sub total		361,100
Flash vapor recovery	First condenser	1670	1670
	Distillation column	33,830	33,830
	Distillation condenser	11,890	11,890
	Reflux pump	2070	2070
	Distillation reboiler	2920	2920
	Reboiler pump	3520	3520
	Total condenser	2600	2600
	Liquid compressor	3200	3200
	Accumulator	4620	4620
	Sub total		66,320
Ammonia recovery from treated corn fiber	Dryer screw conveyor	4420	75,140
	Dryer	49,550	842,350
	Cyclone	11,400	193,800
	Blower	88,170	1,499,000
	Heat exchanger	6030	102,510
	Sub total		2,712,790
	Total		3,140,000
	Total updated to 1995		5,340,000
	Fixed capital investment		21,900,000

Note: See corresponding note for Table 1.

program. Selected results from these simulations are presented in Fig. 3A–D to 5A–C.

Effect of Ammonia Loading and Reaction Pressure

Figure 3A–C shows that when the ammonia loading is increased (which also increases the reaction pressure), the total capital investment (TCI) increases, because larger equipment is required to recover the extra ammonia. The value of TCI increases very quickly for the dry option, because the number of dryers that is required for recovering extra ammonia increases dramatically, and these dryers are a dominant part of the TCI for this option. As the ammonia loading increases, the reaction pressure also increases. This also raises the TCI, because the higher reaction pressure requires more robust equipment.

Table 3
Preliminary Cost Estimate for Corn Fiber Base Case: Wet Option

Selling price ($/t)	$202	
Feedstock price ($/t)	$50	
H/yr	8000	
Tonnes/hr feedstock	10	
Number of operators per hr	2	
Direct costs:		
Equipment costs:	$910,000	
Lang factor (solid/fluids processing plant)	4.1	
Fixed capital investment (FCI)	$3,731,000	
Working capital (15% of FCI)	$560,000	
Total capital investment (TCI)	$4,291,000	
Manufacturing costs:	$/t	$/yr
Raw materials ($0.10/lb of ammonia)	0.66	53,000
Operating labor ($16/h)	3.20	256,000
Direct supervisory and clerical labor (18% of operating labor)	0.58	46,000
Maintenance and repairs (6% of FCI)	2.80	224,000
Operating supplies (15% of maintenance)	0.42	33,600
Lab charges (15% of operating labor)	0.48	38,400
Utilities:	$/t	$/yr
Coal ($3/million BTU)	10.09	810,000
Cooling cost ($2.00/t–d (288,000 BTU removed)	20.46	1,640,000
Fixed charges:	$/t	$/yr
Depreciation (10-yr life, straight line, no salvage value)	4.62	370,000
Local taxes (2.5% of FCI)	1.16	93,000
Insurance (0.7% of FCI)	0.32	26,000
Total manufacturing cost:	44.80	3,590,000

The ammonia loading also affects operating costs, especially for dryer operation. When the ammonia loading is increased from 0.10 to 2.0 (kg ammonia/kg biomass), the TCI increased by almost $3 million for the wet option, and by $13 million for the dry option. The utility cost increased almost $16/t biomass in the wet option, and $30/t biomass in the dry option.

However, the ammonia loading effect on ROI does not parallel the effect on TCI and utilities, because improvements in expected rumen digestibility increase the value of the treated material, and digestibility does not change in the same way with the changing ammonia loading (6). Better data on the effect of ammonia loading on digestibility are needed, especially at low ammonia loadings, but it is obvious that the less ammonia, the better the ROI, at least down to the point at which the effectiveness of the treatment is significantly reduced. Even at 0.1 kg ammonia/kg dry

Table 4
Preliminary Cost Estimate for Corn Fiber Base Case: Dry Option

Selling price ($/t)	$202	
Feedstock price ($/t)	$50	
H/yr	8000	
Tonnes/hr feedstock	10	
Number of operators per h	3	
Direct costs		
Equipment costs	$5,340,000	
Lang factor (new technology and solid/fluids)	4.1	
Fixed capital investment (FCI)	$21,900,000	
Working capital (15% of FCI)	$3,870,000	
Total capital investment (TCI)	$25,770,000	
Manufacturing costs	$/t	$/yr
Raw materials ($0.10/lb of ammonia)	1.10	88,000
Operating labor ($16/h)	4.80	384,000
Direct supervisory and clerical labor (18% of operating labor)	0.86	69,120
Maintenance and repairs (6% of FCI)	16.43	1,314,000
Operating supplies (15% of maintenance)	2.46	197,100
Lab charges (15% of operating labor)	0.72	57,600
Utilities	$/t	$/yr
Coal ($3/million BTU)	15.15	1,212,000
Cooling cost ($2.00/t/d (288,000 BTU removed)	29.06	2,324,800
Fixed charges	$/t	$/yr
Depreciation (10-yr life, straight line, no salvage value)	27.38	2,190,000
Local taxes (2.5% of FCI)	6.84	547,500
Insurance (0.7% of FCI)	1.92	153,300
Total manufacturing cost	107.00	8,540,000

corn fiber, very substantial improvements in digestibility were obtained. Significant, although not nearly as large, improvements in rice straw digestibility were also obtained at low ammonia loadings.

The results for the different simulation cases are summarized in Fig. 3C. Although the calculated ROI values for the dry option are still economically attractive, they are much less attractive than those for the wet option. Although it is perhaps somewhat obvious, it is still worth knowing explicitly that one should not attempt to evaporate ammonia twice, once in the dryer and a second time in the distillation column, otherwise, process economics suffer.

Effect of Treatment Temperature

These simulations were done using experimental results obtained on rice straw, since there is not a complete corresponding data set for corn

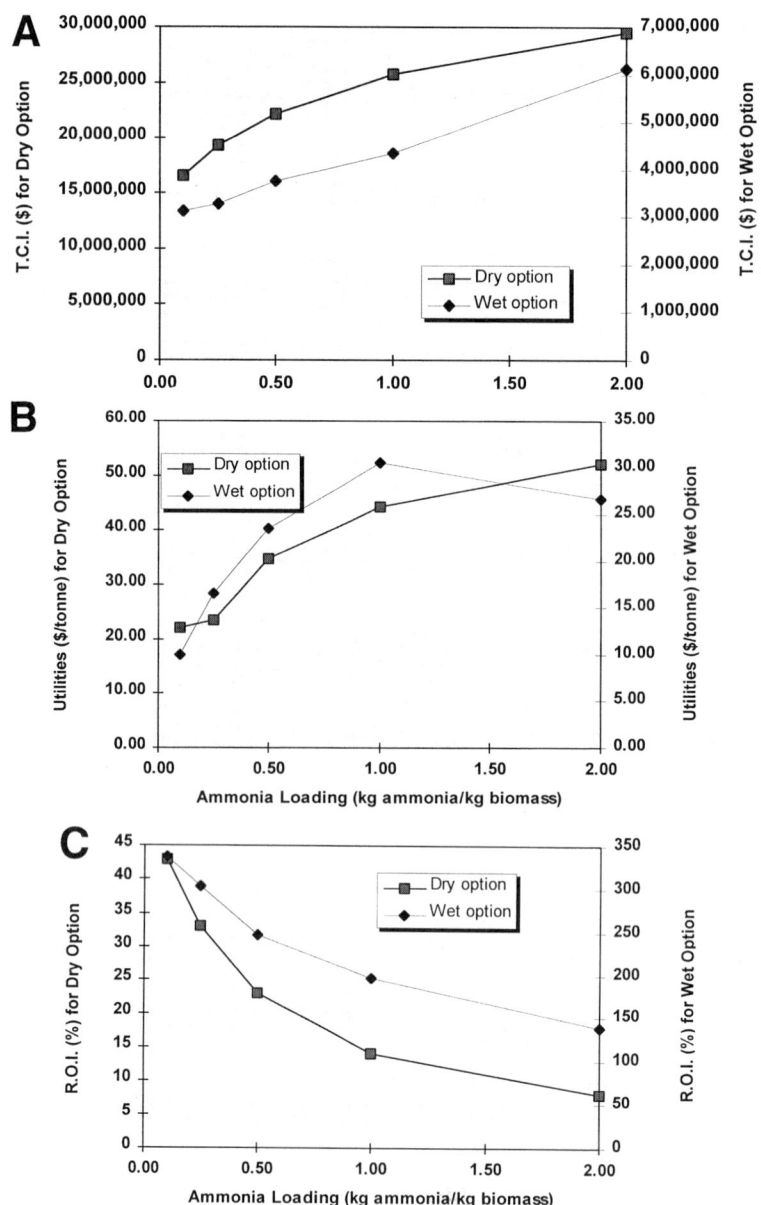

Fig. 3. **(A)** Ammonia loading effect on TCI for corn fiber. **(B)** Ammonia loading effect on utilities for corn fiber. **(C)** Ammonia loading effect on ROI for corn fiber.

fiber. From Fig. 4A–C, the treatment temperature strongly affects the values of TCI and utilities in the dry option. When the treatment temperature increased from 80 to 95°C, the value of TCI for the dry option increased about $20 million, and utilities increased by about $24/t of biomass. The dryers made a significant contribution to the TCI and utilities increases

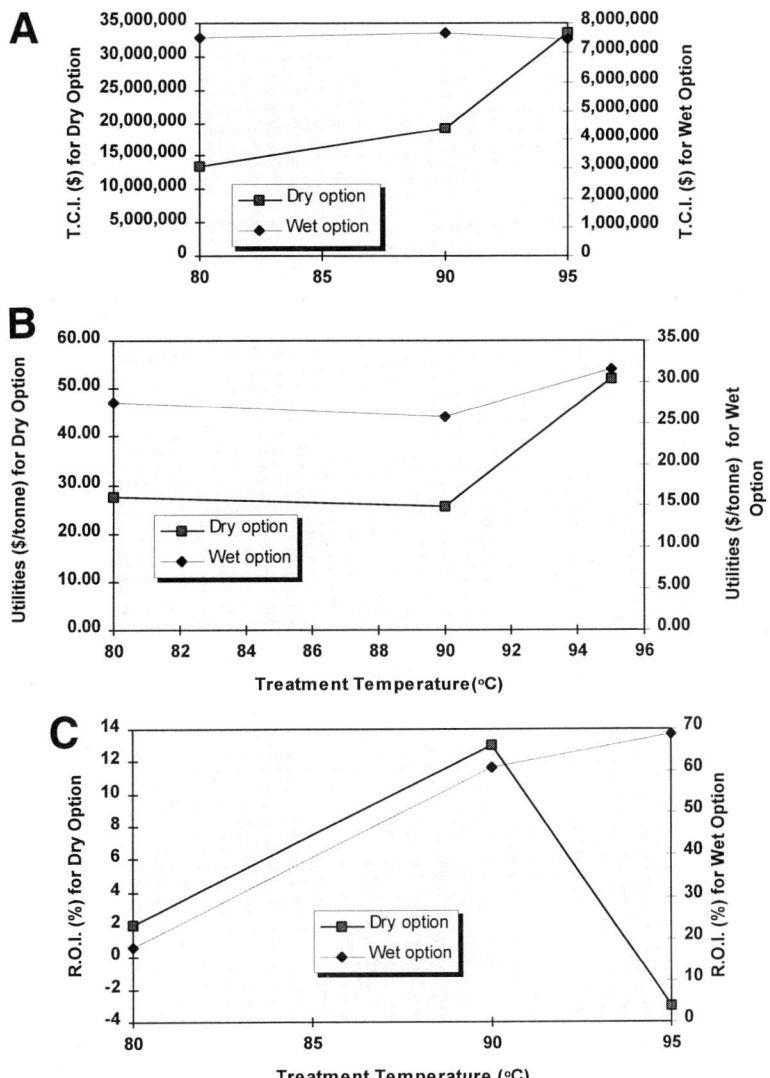

Fig. 4. **(A)** Treatment temperature effect on TCI for rice straw. **(B)** Treatment temperature effect on utilities for rice straw. **(C)** Treatment temperature effect on ROI for rice straw.

because the number of dryers increased very fast when the treatment temperature increased.

In the wet option, the treatment temperature did not affect TCI significantly, because the reaction pressure was not as strong a function of temperature, because of the presence of water. In the wet option, utilities were not significantly affected by the temperature because the wet option requires much less energy than the dry option. Because the digestibility was significantly improved when the treatment temperature increased, the value of ROI also improved significantly for the wet option. However, the dry option costs much more in TCI and utilities, and as a result the digestibility improvement did not much affect the ROI. However, as mentioned, drying costs are probably substantially overestimated because of lack of information on the rate of ammonia evaporation from biomass, and the temperatures required.

In general, the wet option seems to be preferred to the dry option for the AFEX pretreatment process, because it does not cost nearly as much in both TCI and utilities, and both TCI and utility costs are not strongly affected when the ammonia loading and treatment temperature are increased in order to increase the effectiveness of the treatment. However, there may be cases in which a wet AFEX-treated material is not practical for subsequent use, and, at least under some conditions, the dry option still provides attractive economics (e.g., Fig. 3C). Note also that calculations of ROI depend on product values determined from estimates of ruminant digestibility, and that these estimates are still quite preliminary and incomplete. Notwithstanding the uncertainties surrounding ROI calculations, the relative merits of the wet and dry options, and the relative impacts of various system parameters on the system economics, are quite clear. Clarifying the relative importance of these factors was the major purpose of this study.

Effect of Reaction Time

From Fig. 5A–C, one sees that, when the reaction time increases from 4 to 20 min, the TCI increases by about 50% in the wet option. Reaction time does not affect the TCI significantly for the dry product option because the TCI for this option is dominated by the dryer costs. The TCI for the dry option depends on the ammonia loading, water loading, and reaction pressure. Reaction time does not affect the utility cost in either option, because the utility costs are most strongly determined by the ammonia loading and the water loading. Reaction time does affect the ROI, because the digestibilities of treated biomass are significantly different at different reaction times.

Effect of Feedstock Price, Cost of Cooling, Cost of Energy, and Cost of Corn

As the results in Fig. 6, indicate the feedstock price has a large impact on ROI When the feedstock prices for corn fiber are 30, 40, 50, and $60/dry t, the values of ROI are 234, 216, 197, and 179%, respectively—a decline of 56% for a doubling in the feedstock price. In Fig. 7, one sees that, when the cost of cooling varies from $1.00–2.50/t/d for the corn-fiber base case, the ROI declines from 216 to 188%, a decline of 19%, for a doubling in cooling cost. In Fig. 8, when the cost of energy doubles from $2 to $4/million BTU for corn fiber base case, the ROI decreases from 203 to 191%,

Computer Analysis of AFEX Process

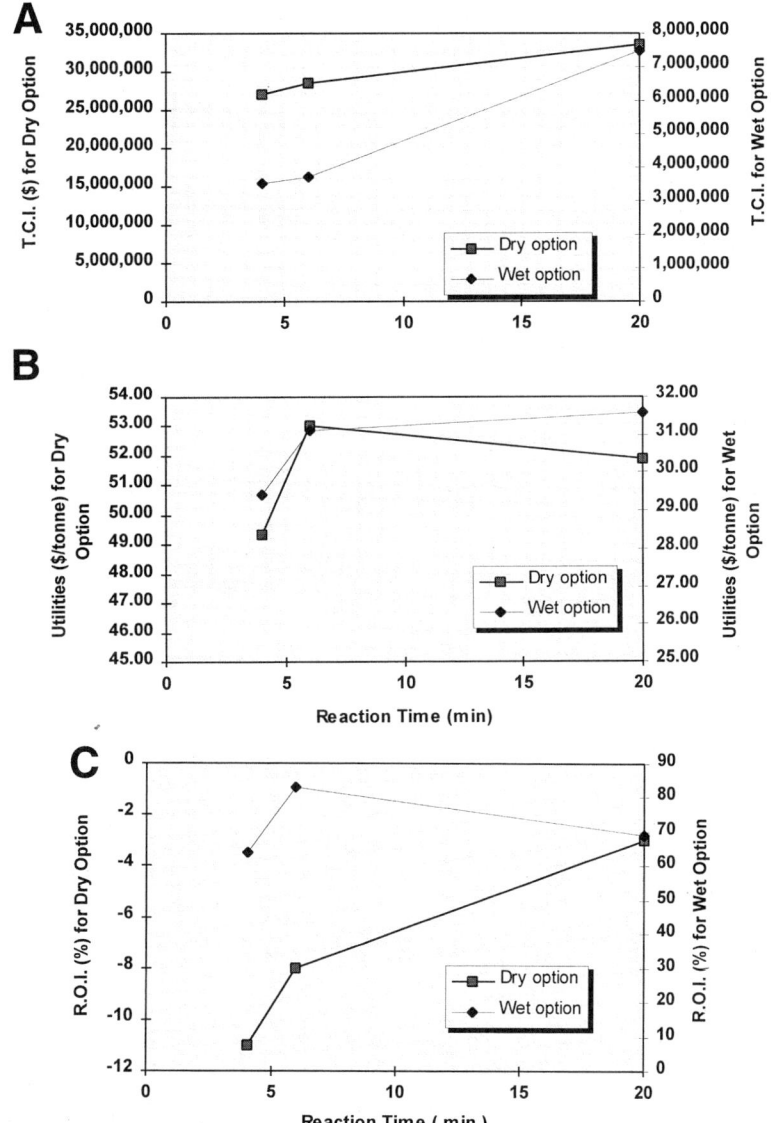

Fig. 5. **(A)** Reaction time effect on TCI for rice straw. **(B)** Reaction time effect on utilities for rice straw. **(C)** Reaction time effect on ROI for rice straw.

a decline of 12%. Therefore, the costs of cooling and energy do significantly affect the ROI value, but they are not nearly as important as the feedstock cost in their impact on ROI, as expected.

However, by far the most significant effect of all is the effect of corn price on the ROI. Figure 9 shows that, under the corn fiber base case conditions for the wet product option, a 50% increase in corn selling price (from about \$2.50/bu to about \$3.80/bu) increases the ROI from 74 to 197%, or

Fig. 6. Feedstock price effect on ROI for corn-fiber base case.

Fig. 7. Cooling cost effect on ROI for corn-fiber base case.

Fig. 8. Energy cost effect on ROI for corn-fiber base case.

an increase in ROI of 266%. Similar percentage changes are observed for the dry-product option. Obviously, profitability depends very strongly on the spread between the feedstock price and the price of the competing digestible material, which is corn in this case. In countries with ample grass and other cellulosic feedstocks, such as tropical areas, but where grain is costly, the AFEX process might have an exceptional impact.

Fig. 9. Effect of corn selling price on ROI for corn-fiber base case.

CONCLUSIONS

A computer simulation program has been developed for the AFEX pretreatment process. This program can be used to assist design of the process, and to predict the economics for the AFEX pretreatment under different scenarios, thereby helping guide further research to reduce costs and increase profitability. Although some of the data on which the program relies are incomplete (e.g., the evaporation rate of ammonia from biomass), a number of important conclusions have resulted from this first generation model and simulation study, namely:

1. The cost of AFEX treatment is about $20–$40/t of dry biomass treated for the wet option, and significant opportunities exist for further reducing process costs.
2. Very attractive ROIs result for AFEX-treated biomass as an animal feed in these studies, and treated corn fiber has significantly better ROIs than rice straw, because of its very high digestibilities.
3. The wet option is to be preferred to the dry option under most conceivable circumstances, because it has much lower TCI and utility costs.
4. The cost of corn is the dominant factor in ROI, when AFEX-treated biomass is considered as a replacement for corn grain in ruminant animal diets. Cost of the biomass itself is the next most important effect on ROI.
5. Higher reaction pressures (and temperatures) will increase both TCI and utility costs, but much less for the wet option, compared to the dry option.
6. Ammonia loading does not affect TCI greatly, but low ammonia loadings do improve the ROI through lower utility costs.
7. Increased water-loading significantly increases TCI and utility costs, and decreases the ROI, especially for the dry option.

8. Increased reaction time increases TCI significantly for the wet product option. It also increases the ROI somewhat because of the increased digestibility.
9. Finally, the AFEX process conditions must be tuned, both technically and economically, for a given feedstock and a given intended final use of the sugars made available by the pretreatment. Optimal conditions are not obvious beforehand.

Parallel cost–benefit studies on preparing AFEX-treated material for enzymatic hydrolysis to fermentable sugars are strongly suggested by these results, in order to optimize the treatment for such applications, and to suggest future research priorities.

ACKNOWLEDGMENTS

Partial support for this work by the United States Department of Agriculture under Specific Cooperative Agreement No. 58-3620-4-131 is gratefully acknowledged.

REFERENCES

1. Dale, B. E. and Moreira, M. J. (1982), *Biotechnology and Bioengineering Symposium*, **12**, 31–43, Wiley, New York.
2. Dale, B. E., Henk, L. L., and Shiang, M. (1985), *Dev. Ind. Microbiol.* **26**, 223–233.
3. Holtzapple, M. T., Jun, J. H., Ashok, G., Patibandla, S., and Dale, B. E. (1991), *Appl. Biochem. Biotechnol.* **28/29**, 59–74.
4. Dale, B. E. and Tsao, G. T. (1982), *J. Appl. Polymer Sci.* **27**, 1233–1251.
5. Eubank, P. T. (1995), Personal Communication, Department of Chemical Engineering, Texas A&M University.
6. Byers, F. M. (1996), Personal Communication, Department of Animal Science, Texas A&M University.
7. Wang, P. Y., Bolker, H. I., and Purves, C. B. (1964), *Can. J. Chem.* **42**, 2434–2456.
8. Schulz, S. C. G. (1973), *Progr. Refrig. Sci. Technol., Proc. Int. Congr. Refrig., 13th*, **2**, 431–446.
9. Raghu, R. (1985), *Chemical Process Computation*, Elsevier, New York.
10. Perry, R. H. and Chilton, H. C. (1984), *Chemical Engineer's Handbook*, 6th ed., McGraw-Hill, New York.
11. Holtzapple, M. (1987), *Ethanol Production Module No. 4*, Texas A&M University, from Beaumont Birch Company.
12. Peters, M. S. and Timmerhaus, K. D. (1979), *Plant Design and Economics for Chemical Engineers*, McGraw-Hill, New York.
13. *Feedstuffs*, **67**, 42.

Selective Polarity- and Adsorption-Guided Extraction/Purification of *Annona* sp. Polar Acetogenins and Biological Assay Against Agricultural Pests

J. D. FONTANA,*,[1] F. M. LANÇAS,[2] M. PASSOS,[1]
E. CAPPELARO,[2] J. VILEGAS,[2] M. BARON,[1] M. NOSEDA,[1]
A. B. POMILIO,[3] A. VITALE,[3] A. C. WEBBER,[4] A. A. MAUL,[5]
W. A. PERES,[6] and L. A. FOERSTER[6]

[1]LQBB~Biomass Chemio/Biotechnology Lab–UFPR (Federal University of Parana) PO Box-19046 (81531–990) Curitiba-PR-Brazil; [2]CROMA-USP-São Carlos; [3]PROPLAME–UBA/Buenos Aires; [4]Pharmaceutical Sciences/USP; [5]Biology/FUAM/Manaus; [6]Zoology Department UFPR

ABSTRACT

Annonaceae acetogenins (AG) comprise a family of natural chemical modifications of long-chain fatty acids (C_{35-37}) bearing one to several hydroxyls (less often oxo), middle-chain tetrahydrofuran rings, and a γ-lactonized, α/β-unsaturated carboxyl group. Acetogenins' strong biological activity as larvicides, pesticides, and antitumorals is dependent on these structural variations. The hydroxylation degree is particularly important for these effects. Seeds, albeit rich in fats (mostly triacylglycerols, [TAG]), are a nonpredatory source of these drugs as compared to other botanical parts such as roots and stems. Conventional lipid extractions lead to quantitative lipid recovery and then the unfavorable natural ratio of TAG:AG in the range >90:<0.1 These extracts thus require, for instance, partitions and extensive sílica gel column chromatographic steps, in order to enrich or purify the AG fraction(s). Great operational difficulties result from the similar polarity and mol. wt. range of TAG and AG when carrying out these purification steps. An alternative fast two-step procedure to obtain polar acetogenin (pAG)-enriched preparations was developed. The extraction procedure for *Annona* spp. seeds pAG was carried out with acetonitrile ($E° = 0.65$; $\log P = -0.33$) as a polar organosolvent, followed by the adsorption of the solvent-free extract on activated charcoal, then washed with hexane and/or chloroform ($E° = 0.0$

*Author to whom all correspondence and reprint requests should be addressed.

and 0.40: log P = 3.5 and 2.0) for most of the contaminating TAG removal, and then with acetone (E° = 0.56; log P = −0.23) to the desorption of an enriched **pAG** fraction. An alternative procedure for pAG extraction was supercritical fluid extraction (SFE) at moderate thermopressurization conditions (65–82°C; 120–130 atm) using CO_2, with 10% acetonitrile as the polarity modifier. The pAG fractions' bioactivity was evaluated with the brine-shrimp test (BST), and for feed deterrance, growth inhibition, and lethality against the high-impact agricultural pests *Anticarsia gemmatalis* and *Pseudaletia sequax* caterpillars feeding on soya or grass leaves sprayed with a 10% alcohol-stabilized emulsion of pAG.

Index Entries: *Annonaceae;* polar acetogenins; acetonitrile; downstream; caterpillars.

INTRODUCTION

Acetogenins from *Annonaceae*, and particularly from the richest genera, *Annona, Rollinia, Asimina, Uvaria,* and *Goniothalamus*, display strong biological activity against several organisms because of their particular mode of action, that is, inhibition of the mitochondrial NADH: ubiquinone reductase. Effective doses as low as $ED_{50} = 10^{-2}$ µg/mL were found using VERO as a tumoral cell line model. Sixty-one new structural acetogenin variations were recently described, comprised of three main groups according to the number of tetrahydrofuran (THF) rings or their precursors (epoxides) in the biosynthetic route *(1)*. Important potential applications of acetogenins in the oncology field are being reported (one example is in human prostate tumor *[2]*).

Annona muricata (soursop; "guanabana," "graviola") is one of the most-studied models as a mono-THF acetogenin source, and as many as 16 bioactive fatty acid lactones were found in its leaves *(3–6)*. Among them, gigantetrocin A, muricatetrocins A and B, annonacin A, goniothalamicin, and annonacin-10-one were also found in the seeds, along with the lesser hydroxylated forms like murisolin, solamin, corossolone, and diepomuricanin. The seed acetogenins bearing 4 hydroxyls (gigantetrocins and muricatetroxins) are particularly bioactive, since increased polarity correlates with bioactivity *(1)*. The chemical structure for one of these compounds is shown in Fig. 1.

Downstream processing of two di-THF acetogenins (bullatacin and bullatacinone) from the bark of *Annona bullata* was based on sequential partitions of a crude ethanol extract using biphasic systems, namely chloroform/methanol and hexane/90% methanol. Final purification steps were chromatography and rechromatography on silica gel columns eluted with hexane/chloroform/methanol and chloroform/ethyl acetate/methanol gradients. About 195 mg of pure bullatacin, a tetrahydroxylated C_{37}-(THF)

Fig. 1. Gigantetrocin A from *A. muricata* seeds (example of a pAS structure used with permision from ref. *1*).

$_2$ acetogenin, was thus obtained from approx. 4 kg of *A. bullata* bark, as well as 13 mg of bullatacinone, the related oxo-trihydroxylated form *(7)*.

Interested in the highly hydroxylated and more bioactive acetogenins (polar acetogenins [pAG]), the authors adopted *A. muricata* seeds as a nondestructive working source for an alternative and faster downstream process. The proposed methodology took in account the high neutral lipid (TAG) content (~ 30%) previously reported for these particular seeds *(8)*.

MATERIALS AND METHODS

Seeds and Lipid Material Extraction by Organosolvents or Supercritical Fluid Extraction (SFE)

Ripened fruits from *A. muricata* ("graviola"), *Annona* sp. (probably *squamosa*; "fruta-do-conde"), and *Annona coriacea* ("pinha-do-cerrado" [PR]) were respectively collected at Feira de Santana (Brasfrut, Bahia State, Brazil), Porecatu, and Jaguariaiva (Parana State), Brazil. Washed and sundried seeds were powdered in a Waring blender (Dynamics, New Hartford, CT; 30-mesh sieve), and stored in plastic bags with an average 10% moisture content. Lipid material extraction was carried out with 3 vol of each organosolvent overnight at 28°C in a rotatory shaker. Acetonitrile (polarity, E° = 0.65, Riedel scale; hydrophobicity, log P = −0.33) was the preferable solvent. Following filtration by a frit glass screen, extract solvent was removed in a rotatory vacuum evaporator. Acetonitrile extracts were designated as AE. Guidelines on solvent polarity were taken from a published log-P hydrophobicity-ordered list *(9)*, and from a Riedel-de-Haen table (Seelze, Germany).

Supercritical fluid extraction (SFE) was carried out in a steel apparatus (reactor working vol = 10 mL) specially designed by one of the authors (F. M. Lanças) from the Chromatography Laboratory at the University of São Paulo (USP) at São Carlos-SP *(10)*. The SFE apparatus consists of a crushed-ice bath involving a 500-mL high-pressure gas-mix chamber (carbonic anhydride and nitrogen), with or without addition of 0.1 vol of modifier solvent (acetonitrile or hexane), an extraction vessel (sample-containing reactor) in a thermostat-equipped water bath, a restrictor outlet, and a collection vessel containing methanol in a cryogenic bath. Moderated ranges of temperature (maximum 82°C) and pressure (120–130 atm) were employed. Gas flow was set at 45 mL/min for dynamic SFE of solutes, after an equilibration time of 10 min at the lowest temperature (65°C; static SFE; no gas flow). During the average time of 90 min for each run, temperature increased from 65 to 82°C, and fractions were collected at 10 min intervals. Sample solvent of the SFE fractions from powdered seeds was evaporated in a vacuum oven, and each residue was dissolved in chloroform:methanol (2:1) or acetonitrile:water (85:15) for analyses by thin layer chromatography (TLC) or high-presssure liquid chromatography (HPLC).

Acetogenins Enrichment and Purification

Acetogenin extract (light brown oil) was suspended in petroleum ether or hexane, thoroughly mixed with 4–5 parts activated charcoal (Sigma C-3585; St. Louis, MO), and left in contact for 1 h. Filtration was carried out in a frited glass filter/Buchner flask assembly using a double layer of filter paper and applying 2 column vol of each solvent from an eluotropic series: hexane ($E° = 0.0$; glycerides and less polar acetogenins); chloroform ($E° = 0.40$; remanent glycerides and midpolar acetogenins); acetone ($E° = 0.56$; polar acetogenins, pAG), and finally, methanol ($E° = 0.95$). A triglyceride-free pAG standard was further obtained on neutral alumina or silica gel 60 columns (gradient elution with hexane:chloroform up to 70:30, and then acetone).

Analytical Procedures

Thin-layer chromatography was carried out in Merck (Darmstadt, Germany) silica gel chromatoplates G 60 (art. 105553) with hexane:chloroform:nitroethane:ethyl acetate:acetone:methanol:acetonitrile:water, 120:20:40:40:10:20:16:1 as mobile phase. Lipids were detected with four different reagents (11): iodine vapor (unsaturated lipids); phosphomolibdic acid, saturated solution in ethanol (general lipids) at 120°C, 10 min; 1% anisaldehyde in sulfuric acid:methanol (5:95) for the differentiation between TAG and phytosterols (deep lilac to wine color) and acetogenins (yellow to olive color) at 105°C for 5 min; and the Kedde reagent, 3-5-dinitrobenzoic acid, followed by KOH, both as 5% methanolic solutions, as a selective reagent sequence for acetogenins (lactones; pink to wine color, fast-fading).

HPLC was run on a Spectra Physics C-18 reverse-phase column assembled in a Waters multimodule (SC 600E/WISP 712) with double monitoring with a 410 DRI or 484 UV detector at 215 nm, using acetonitrile:water (85:15) as mobile phase at a 0.75 mL/min flow rate. Sample solvent mix (chloroform:methanol:water or aqueous acetonitrile, detected as the initial chromatographic peaks) was adjusted to ensure complete lipid solubilization.

Biological Assays

The brine shrimp test (BST) *(12)*, affording the LD_{50} for *Artemia salina* nauplii (24 h after the cyst hatching in artificial sea water), was performed as a routine cytotoxic activity assay. Feed deterrance, growth inhibition, and lethality were assayed against *Pseudaletia sequax* larvae (V instar) feeding on fresh leaves of *Penisetum clandestinum* ("kikuyo grass," 7-cm length and 0.8-cm width leaf pieces), or against *Anticarsia gemmatalis* caterpillars feeding on fresh leaves of *Glycine max* (soya; 2.5- to 3.5-cm length young leaves). Plant leaves were sprayed (10 s) with a fine mist of a milky, homogeneous, and stable suspension of pAG (fraction eluted from active charcoal with acetone) in 10% ethanol, in order to ensure an 18 (low dose), 36 (medium dose), and 54 µg (high dose)/leaf as determined in grass leaf pieces. Control leaves were sprayed with the same solvent or water. Following solvent evaporation, leaves were offered *ad libitum* for the growing caterpillars for 2 d. From d 3 untreated leaves were the exclusive food in all experiments.

RESULTS AND DISCUSSION

The current fractionation flow sheet is depicted in Fig. 2. Progressive enrichment of pAG was obtained in the sequence from acetonitrile extraction to the activated charcoal fractionation step. Compared to an ethanol extract Fig. 3A, B; (lanes EE-1–3) from *A. muricata* seeds, whose AG content is barely visible with both the chromogenic anisaldehyde and Kedde reagent sprays, because of the dominance of TAG (triglycerides), the alternative acetonitrile extract (same Fig., lane EA) displayed an approx 1:2–1:3 AG:TAG ratio (color profiles in Fig. 3A, B). Higher TAG content in the ethanolic extracts, compared to the acetonitrile extract, was qualitatively confirmed using the lipase/glycerokinase/glycerophosphate oxidase/peroxidase/benzoquinone-monoimido-phenazone kit (Diagnostic Systems, Holzheim, Germany). Gravimetric measurement of total fat content of *A. muricata* seeds, after exhaustive ethanol or acetonitrile extractions, corresponded to 33 and 26%, respectively. The corresponding figures for *A. coriacea* were 20 and 13%. The acetone eluate, when processing the acetonitrile extract on an activated charcoal bed (lane AC-A), was progressively enriched in pAG because of the efficient and previous elution of TAG (and also most of the less polar acetogenins) with hexane, followed by chloroform. In fact, considering the TLC fast-moving spot as TAG, since a triolein standard cochromatographed with the same Rf, the AC-A fraction is purer than the equivalent step of the previously patented procedure *(7)*, namely, the aqueous methanolic phase from the hexane/90% methanol partition (lane M[H]; Fig. 3A, B). For the sake of illustration, the complete purification of the dominant and most polar acetogenin(s), attained on an alumina pad washed with a chloroform:acetone gradient (up to 70:30,

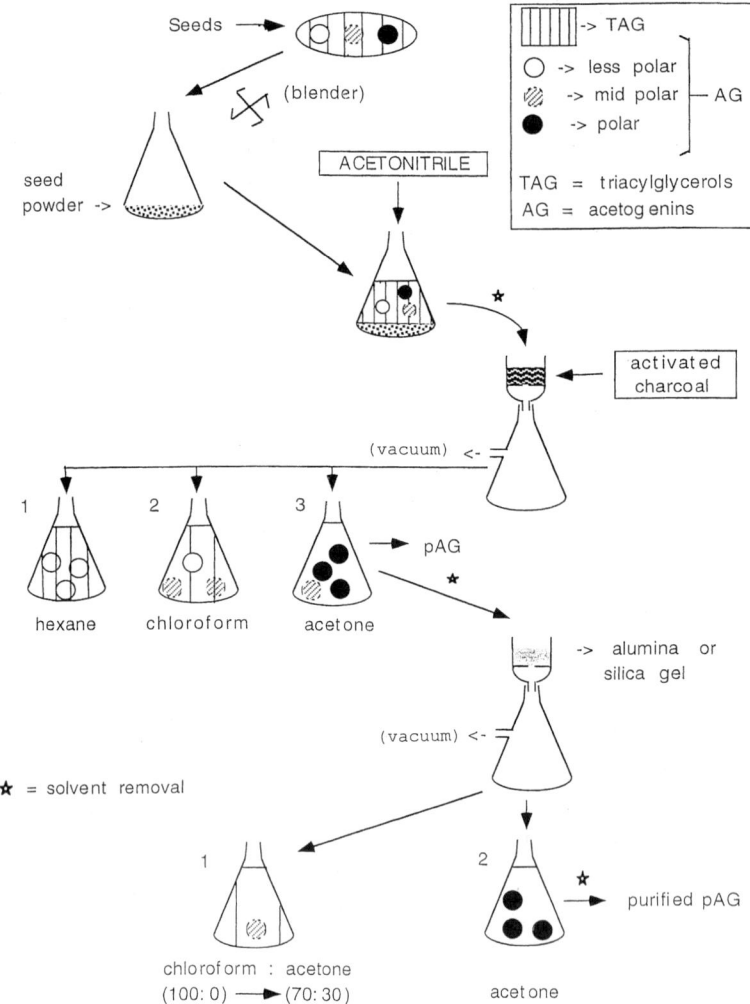

Fig. 2. Downstreaming for pAGs using acetonitrile as extractant and activated charcoal for fractionation and purification.

v/v, and then pure acetone), is also shown (pure pAG fraction in Fig. 2, and lane AL-A in Fig. 3A, B). The relative profiles for polar acetogenins (pAG; peaks with Rt = 9.7 and 11.0 min as monitored at 215 nm) were also verified in the same samples by HPLC (Fig. 4A–C, referring to initial alternative acetonitrile extract, acetone eluate from charcoal, and acetone eluate from alumina, respectively Fig. 4D refers to the initial classic ethanol extract). Quantitation indicated that the sum of the two main peaks (a resolution not attained in TLC) corresponded to 43% of all 215 nm-detectable material. Scanning the TLC plates (or their derived color pictures) on a Shimadzu CS-9301-PC Dual Wavelength Flying Spot Densitometer (results not shown), using the most suitable wavelength (545 nm) for both anisalde-

Fig. 3. TLC monitoring for *A. muricata* seed acetogenins extraction and purification. **(A)** p-anisaldehyde spray. **(B)** Kedde's reagent spray. EE, ethanol extraction and derived fractions; EE-1, EE-2, EE-3, sequential initial ethanol extracts; M(H), aqueous methanolic phase from the second partition of the EE mix; AE, acetonitrile extract; AC-A, acetone eluate from the activated charcoal bed; AL-A, acetone fraction from the alumina bed.*, **, and ***, less polar, midpolar, and pAG, respectively.

hyde and Kedde reagent chromogenic sprays, indicated that pAG contributions were 11 and 42% of the colored components, respectively. It may be recalled, as shown in Fig. 3A, B, that anisaldehyde is an efficient chromogen for both TAG (wine to violet) and AG (yellow to olive), compared to the Kedde reagent, specific for AG (light to deep pink; no color for TAG and other related or unrelated lipids). Efficiency of acetonitrile as a selective extractant for all acetogenins, irrespective of their polarity, was also successfully assessed using other species of *Annona*. The TLC results for the sp. *coriacea*, for instance, indicated an even better ratio of AG:TAG, but this may be attributed to the lesser relative content of TAG in this seed type (J. D. Fontana, unpublished data).

Pursuing the same goal, that is, the preferential extraction of polar acetogenins, SFE was carried on *A. muricata* powdered seeds using either 10% hexane or 10% acetonitrile as polarity modifiers for the carbonic anhydride thermopressurized flowing stream. TLC analysis indicated best results for the CO_2 acetonitrile mix, whose earlier fractions, 1–4 among 10, were also particularly enriched on pAG, most of TAG being collected in fractions 5–9 in the higher temperature range. The CO_2–hexane mix revealed a preferential extractant for TAG, thus concentrated in the initial five fractions, no significant lipid amounts being detectable in fractions 6–10. The encouraging results obtained with the inclusion of acetonitrile in SFE is being better explored and the complete results will be published elsewhere.

An emulsifier agent is needed for the application of pAG in agricultural plague control. The choice for this purpose was 10% ethanol. Fig. 5A–C show the comparative effectiveness of this low-cost emulsifying mixture on whole oil (WO), acetonitrile extract (AE), and semipurified polar

Fig. 4. Comparative HPLC analyses for the extracts or purified fractions of *A. muricata* seed polar acetogenins (pAG). **(A)** Initial acetonitrile extract (AE); **(B)** acetone eluate from AE purification on an activated charcoal bed (AC-A); **(C)** acetone eluate from AC-A final purification on an alumina bed (AL-A); **(D)** initial ethanol extract. (UV monitoring at 215 nm with an AUFS = 2.0. Earlier peaks correspond to different solvent sample mix. Samples at 50 mg/mL; injection volume of 15 µL for samples A and D, and 5 µL for samples B and C. A sample of a triolein standard presented no detection in the 25 min run).

acetogenins (AC-A) samples at a 0.25% concentration (w/v). Pictures were taken just after mixing (A), 3 h (B), and 24 h (C). As is clearly visible, the progressive enrichment of *A. muricata* seed lipomaterial in polar acetogenins leads to partial (Fig. 5B; sample AE), or even complete, emulsion stabilization (Fig. 5C, sample AC-A), when 10% aqueous ethanol replaced pure water or other emulsifying agents. This latter preparation of pAG was then used as a fine mist for spraying of legume (soya) or grass (wheat and "kykuio") leaves (18, 36, and 54 µg of pAG/leaf) in order to check acetogenin bioactivity against the caterpillar pests *Anticarsia gemmatalis* and *Pseudaletia sequax*. The use of low and medium doses led to 10 and 30% mortality, respectively, reduced food consumption (feed deterrance), enlargement of the larval period, and generation of a surplus instar (designed as the 7th, not existing in the controls). Results of this triplicate experience are detailed in Table 1. Wheat caterpillar (*P. sequax*) proved to be more resistant to pAG inhibitory action, since the highest dose (54 µg/leaf cut)

Annona sp. Polar Acetogenin Extraction

Table 1
Food Consumption by *Anticarsia gemmatalis* Caterpillars for the Three Final Instar Periods, Larval Period Delay, and Mortality Rate Following a 2 d-Period Feeding on *Glycine max* Leaves Treated or Untreated with Polar Acetogenins (pAG).

Treatment	Average food consumption (%)			Larval Period (d)	Mortality (%)
	5th instar	6th instar	7th instar[a]		
Control; water	51.3 ± 8.7	77.5 ± 7.0	—	19.9 ± 1.3	0%
Control; 10% ethanol	55.8 ± 13.2	75.2 ± 9.4	—	19.3 ± 1.8	0%
pAG[b]; low dose	18.3 ± 3.4	35.3 ± 10.4	74.7 ± 13.7	25.0 ± 2.5	10%
pAG[b]; medium dose	20.9 ± 7.6	49.2 ± 22.9	72.8 ± 15.5	29.3 ± 3.0	30%

[a] Surplus instar resulting from the cytotoxic action of acetogenins actually seen.
[b] pAG, polar acetogenins (acetone eluate from activated charcoal; doses detailed in Materials and Methods).

resulted only in delayed larval growth because of the feed deterrance, which persisted until the fifth d, fresh untreated food allowing further recovery for the initial effect (results not shown).

Fig. 5. *A. muricata* crude and semipurified acetogenins forrnulation in 10% ethanol as emulsifier. WO, seed whole oil; AE, seed acetonitrile extract; AC-A, acetone eluate from the activated charcoal bed. (All samples at 0.25 g% were vigorously mixed in a Vortex for 1 min and pictures from the laid test tubes were taken just after mix **(A)**, after 3 h **(B)**, and after 24 h **(C)**.

CONCLUSIONS

As a nonpredatory source for polar and more bioactive pesticidal and antitumoral acetogenins, *Annona muricata* seeds were conveniently and quickly processed using acetonitrile as a more selective extractant, compared to ethanol. The time-consuming steps of partitions were replaced by adsorption and desorption of the polar acetogenins on activated charcoal. A second promising extraction method was supercritical fluid extraction, provided 10% acetonitrile was used as polarity modifier for the moderately thermopressurized carbonic anhydride supercritical stream. Polar acetogenins semipurified fraction produced stable emulsions in 10% ethanol, even at a 0.25% concentration, thus facilitating their uniform application for agricultural pests control. The soya caterpillar was significantly affected.

ACKNOWLEDGMENTS

The authors thank the Brazilian funding agencies (CABBIO–Argentine/Brazilian Center for Biotechnology, CNPq, and CAPES bench fees), as well as the kind provision of *A. muricata* seeds by Brasfrut Co. The help of Rafael F. Martin with drawings is acknowledged. Data on acetonitrile extraction and activated charcoal purification will be addressed to lNPl–Brazil (patent request).

REFERENCES

1. Fang, X., Rieser, M.-J., Gu, Z., Zhao, G. and McLaughlin, J. L. (1993), *Phytochem. Anal* **4**, 49–67.
2. Hopp, D. C., Zeng, L., Gu, Z. M., and Laughlin, J. L. (1996), *J. Nat. Prod.* **59**, 97–99.
3. Wu, F. E., Gu, Z. M., Zeng, L., Zhao, G. X, Zhang, Y., McLaughlin, J. L., and Sastrodihardjo, S. (1995), *J. Nat. Prod.* **58**, 830–836.
4. Wu, F. E., Gu, Z. M., Zhao, G. X., Zhang, Y., Schwedler, J. T., McLaughlin, J. L., and Sastrodihardjo, S. (1995), *J. Nat. Prod.* **58**, 902–908.
5. Wu, F. E., Gu, Z. M., Zhao, G. X., Zhang, Y., Schwedler, J. T., McLaughlin, J. L., and Sastrodihardjo, S. (1995), *J. Nat. Prod.* **58**, 909–915.
6. Wu, F. E., Gu, Z. M., Zhao, G. X., Zhang, Y., Schwedler, J. T., McLaughlin, J. L., and Sastrodihardjo, S. (1995), *J. Nat. Prod.* **58**, 1430–1437.
7. McLauglin, J. L. and Hui, Y. (1993), US Patent 5229419.
8. Fontana, J. D., Almeida, E. R. A., Baron, M., Guimaraes, M. F., Deschamps, F. C., Schwartsmann, G., et al. (1994), *Appl. Biochem. Biotechnol.* **45/46**, 295–313.
9. Laane, C., Boeren, S., Vos, K., and Veeger, C. (1987), *Biotechnol. Bioeng.* **30**, 84.
10. Sargenti, S. R. and Lanças, F. M. (1994), *J. Chromatogr.* **667**, 213–218.
11. Dawson, R. M. C., Elliot, D. C., Elliot, W. H., and Jones, K. M., eds. (1986), *Data for Biochemical Research*, 3rd. ed., Clarendon, Oxford.
12. Meyer, B. N., Ferrigini, N. R., Putnam, J. E., Jacobsen, L. B., Nichols, D.E., and McLauglin, J. L. (1982), *Planta Medica* **45**, 31–34.

Dilute Acid Pretreatment of Softwoods

Scientific Note

Q. A. NGUYEN,* M. P. TUCKER, B. L. BOYNTON, F. A. KELLER, AND D. J. SCHELL

Biotechnology Center for Fuels and Chemicals, National Renewable Energy Laboratory, Golden, CO 80401

ABSTRACT

Selective thinning of forests in the western United States will generate a large, sustainable quantity of softwood residues that can be an attractive feedstock for fuel ethanol production. The major species available from thinning of forests in northern California and the eastern Rocky Mountains include white fir *(Abies concolor)*, Douglas fir *(Pseudotsuga menziesii)*, and Ponderosa pine *(Pinus ponderosa)*. Douglas fir chips were soaked in 0.4% sulfuric acid solution, then pretreated with steam at 200–230°C for 1–5 min. After pretreatment, 90–95% of the hemicellulose and as much as 20% of the cellulose was solubilized in water, and 90% of the remaining cellulose can be hydrolyzed to glucose by cellulase enzyme. The prehydrolysates, at as high as 10% total solid concentration, can be readily fermented by the unadapted yeast *Saccharomyces cerevisiae* D_5A.

Index Entries: Biomass; ethanol; pretreatment; softwood; bioconversion.

INTRODUCTION

After decades of human intervention in suppressing forest fires, large quantities of small-diameter trees and underbrush have overcrowded forests in the western United States, and created a severe fuel-loading problem. The resulting forest fires are often catastrophic. They not only destroy important natural resources, but endanger property, pollute the air, and can lead to soil erosion and flooding. Recognizing the need to reduce this severe fuel loading, the United States and various state forest service agencies have started selective thinning operations in national and state forests. Thinning millions of acres of overgrown forest areas would be expensive,

*Author to whom all correspondence and reprint requests should be addressed.

and the small-diameter trees to be removed are of low economic value to the current forest-products industry. Furthermore, the volume of wood generated from forest-thinning operations would be too large to be absorbed by traditional uses such as producing wood composite products, firewood, compost, and mulch.

A potential use of forest thinnings is to convert it into fuel ethanol and cogenerated electricity. This option has merit in that it can convert the enormous quantity of available biomass to an excellent fuel oxygenate such as ethanol. Additionally, there are synergistic benefits to locating biomass ethanol plants next to biomass power generation facilities. Such combinations would strengthen the process economics of both technologies and provide an ecologically sound solution to alleviate the forest fuel-loading problem.

In contrast to hardwoods and herbaceous materials, published data on softwood conversion to ethanol are limited. Little attention has been paid to softwoods, perhaps because of softwoods have high lignin contents that cannot be fermented to ethanol, and pretreatment processes successfully developed for herbaceous materials do not necessarily work well for softwoods. Sulfur dioxide-catalyzed steam pretreatment of spruce *(1)* and *Radiata* pine *(2)* was reported to greatly enhance the separation of major wood components. Most of the hemicellulose is rendered water soluble, and the cellulose can be enzymatically hydrolyzed to glucose.

The predominant softwood species available from forest thinning operations in the western United States are white fir *(Abies concolor)*, Douglas fir *(Pseudotsuga menziesii)*, and Ponderosa pine *(Pinus ponderosa)*. The objective of this study was to determine the potential ethanol yield from softwood via dilute sulfuric acid pretreatment and enzymatic hydrolysis. We determined sugar recoveries after pretreatment, glucose yields from enzymatic hydrolysis of water-washed substrates, and fermentability of the prehydrolysate. Douglas fir was selected as the model feedstock in our preliminary work.

MATERIALS AND METHODS

Feedstock Preparation

Softwood forest thinnings of Douglas fir and Ponderosa pine of 7–20-cm diameter were selected and harvested by Colorado State Forest Service agents (Golden district) from a north-facing slope in Golden Gate Canyon near Golden, Colorado. After being harvested and transported, the segregated logs were debranched, manually debarked, and chipped using a 65-hp Brush Bandit mobile knife chipper (Foremost, Remus, MI). The wood chips were then milled in a rotary knife mill (model 10 × 12, Mitts and Merrill, Saginaw, MI) equipped with a 3/8-in (9.5-mm) rejection screen, and cone/quartered blended. The fines were removed by screening through a 2-mm screen. The final chips were packed in polyethylene-lined drums and stored at $-20°C$.

Milled and screened Douglas fir chips were soaked in dilute sulfuric acid (0.35–0.4%) solution at 60°C for 4 h, and drained overnight to approx 40% solids before being pretreated.

Pretreatment

All pretreatment experiments were performed using a 4-L steam-explosion reactor equipped with a steam jacket, a 4-in (10-cm) ball valve at the top for loading biomass, a 2-in (5-cm) ball valve at the bottom for discharging the contents of the reactor, two steam-injection ports near the top and bottom, and K-type thermocouples inserted near the top and bottom for reactor-temperature measurements. The reactor was made of Hastelloy C-22™ to resist corrosion. The reactor temperature was controlled at or near the desired value by using a pressure-control valve to control the steam-supply pressure. At the beginning of each experiment, the reactor was preheated to near the desired operating temperature by admitting steam into the jacket and cycling steam repeatedly through the reactor. A batch of preweighed acid-soaked chips (approx 1 kg wet weight) was then loaded into the reactor and saturated steam admitted (defined as time zero). After a predetermined cooking time, the steam was shut off, then the contents of the reactor (cooked wood chips, condensate, and steam) were discharged into a cooled flash tank with the flash vapor condensed in a separate condenser. The contents of the flash tank were emptied and blended, then the flash-tank surfaces were rinsed with water and collected for analysis. Portions of the pretreated solids were stored at $-20°C$ for digestibility and fermentation assays. The pretreated materials were processed into liquor samples (obtained by pressing the liquid from the wet samples) and water-insoluble solid samples (obtained by extensively washing the samples) for chemical analyses and enzyme digestibility assays.

Analysis of Wood and Water-Insoluble Solids

Dry weights were determined by oven drying at 105°C to constant weight *(3)* and Klason acid-insoluble lignin and acid-soluble lignin were determined by standard methods *(4,5)*. Anhydrosugars in the whole wood and pretreated solids were determined by a procedure slightly modified from that developed at the U.S. Forest Products Laboratory *(4,6)*. Ash in the wood and pretreated solid residues was analyzed by standard gravimetric methods *(7)*.

Analysis of Liquor

Organic acids, glycerol, hydroxymethyl furfural (HMF), and furfural in the liquor and rinsate fractions were determined by ion-moderated partition chromatography using Bio-Rad Aminex HPX-87H columns (Bio-Rad, Hercules, CA) *(4,8)*. Monomeric sugars were determined by HPLC (high-pressure liquor chromatography) using Bio-Rad Aminex HPX-87P columns. Oligomeric sugars in the liquor and rinsate fractions were converted to monomers using 4% H_2SO_4 hydrolysis at 121°C for 1 h, followed by determining monomeric sugars with the Bio-Rad HPX-87P column and correcting for sugar losses *(4,8)*.

Enzymatic Hydrolysis

Extensively washed pretreated wood samples were tested for enzymatic digestibility with Iogen cellulase (Iogen Super Clean cellulase, lot no. BRC 191095, Iogen, Ottawa, Ontario, Canada). The filter paper activity of the enzyme was assayed at 91 FPU/mL and the β-1 glucosidase activity at 198 IU/mL. The digestibility assays were perform at pH 4.8 with an enzyme loading of 60 FPU/g of cellulose in prewarmed (50°C) 10-mL reaction cocktails that contained solids equivalent to 1% cellulose, 50 mM citrate buffer, 40 mg/mL tetracycline, and 30 mg/mL cycloheximide. The antibiotics were to minimize contamination *(9,10)*. Duplicate reaction vials were incubated at 50°C with mixing at 120-rpm rotation at a 45° angle and compared to controls that contained 1% Solka-Floc, grade NF-FCC (Fiber Sales and Development, Urbana, OH) and enzyme blanks. One-half milliliter samples were removed at time zero, 2, 4, 6, 18, 24, 48, 72, 96, and 168 h. Glucose levels were determined with a YSI Model 2700 Select Biochemistry Analyzer equipped with immobilized glucose oxidase membranes (Yellow Springs Instruments, Yellow Springs, OH). Samples were centrifuged at 12,000g for 5 min and diluted to keep the glucose readings below the 2.50 g/L level used to calibrate the instrument.

Fermentability and Simultaneous Saccharification and Fermentation Performance

A fermentability test was carried out on whole slurry of pretreated wood samples. Thawed samples of pretreated wood were adjusted to pH 5.5 with $Ca(OH)_2$ then autoclaved at 121°C for 30 min in 250-mL DeLong (Fisher Scientific, Pittsburgh, PA) culture flasks. Sterile 10X solutions of yeast extract, peptone, and glucose (50 g/L, 100 g/L, and 500 g/L, respectively) were added to each flask to bring the total solids to 10 or 15% in a total of 150 g. The flasks were inoculated with a 15-h culture of *Saccharomyces cerevisiae* D_5A at a 10% level (w/w). Inoculated flasks were placed on a rotary shaker that operated at 30°C and 150 rpm. Sugar levels in samples were analyzed by HPLC using an HPX-87P column with deionized water as the eluant and refractive-index monitoring. Ethanol and fermentation byproducts were analyzed using an HPX-87H column with 0.01 N H_2SO_4 as the mobile phase and refractive-index monitoring *(8)*.

Standard simultaneous saccharification and fermentation (SSF) assays were carried out on exhaustively washed pretreated solids to determine the potential ethanol yield from the cellulose fraction. Cellulase enzyme (Iogen) at 25 IU/g of cellulose was added to each flask, which also contained 10 g/L yeast extract and 5 g/L peptone (Difco Laboratories, Detroit, MI) and enough solids to give 30 g/L cellulose in the 100-mL cocktail. The D_5A pre-inoculum was started in yeast extract peptone dextrose media (YPD), pH 5.0, from a frozen culture and incubated at 37°C for 6 h. A 10% volume of the preinoculum was used to start an overnight inoculum (12–16 h) in YPD, pH 5.0, mixed at 150 rpm and 37°C. The SSF was initiated by adding a 10% volume of the overnight inoculum and enzyme to flasks sealed with bubble traps. Samples were taken at time 0 and at 24-h intervals. SSF assays of whole slurry samples at 10% solid concentrations were also carried out to determine the effect of inhibitors present in the hydrolysate on ethanol yield.

RESULTS

Pretreatment Results

Five pretreatment conditions tested during the first run ranged from temperatures of 200 to 230°C and residence times of 125 to 305 s. The chips were impregnated with 0.4% sulfuric acid at 60°C for 4 h and then drained overnight to approx 40% solid before pretreatment. Depending on the pretreatment temperature and time, the solid contents of pretreated materials ranged from 25 to 30% based on wet weight. Compositional data were collected only on the liquid fraction of the pretreated material. Water used to rinse the flash tank (rinsate) was not collected during these experiments. A summary of the results is shown in Table 1. Solids recovery is the fraction of the original wood weight recovered in the pretreated material and should be less than 100% because of loss of volatile components. Recovery will also be low because rinsate was not collected. Solids solubilized is the fraction of the original wood lost as volatile or as water-soluble components, such as sugars, soluble lignin, and nonvolatile decomposition products. The sugar and byproduct yields and solids solubilized are also low because rinsate was not collected, however, the error will be small (6–9%, based on mass balance experiments, data not shown) and will be assumed to not affect a relative comparison of the results.

The results show that as pretreatment severity increases, the solids recovery decreases, and the solid solubilized increases. Both results are expected. Using xylose yield as an indication of hemicellulose hydrolysis, the maximum value occurred at 201°C and 305 s. The higher temperature conditions produced lower xylose yields and higher furfural yields, which indicated that the material was overcooked. However, it is not possible from this data to identify the maximum yield conditions. The glucose yields are difficult to interpret because glucose is produced from both

Table 1
Sugar Yields from Liquid Fractions of Pretreated Douglas Fir[a]

Exp. #	Temp. (°C)	Time (s)	Solids Recovery (%)	Solids solubilized (%)	Glucose[2] Yield (%)	Xylose[2] Yield (%)	HMF Yield (%)	Furfural Yield (%)
1	201	125	93.4	30.9	6.8	35.0	0.5	2.7
2	201	305	88.0	32.4	20.9	61.7	1.6	7.1
3	230	305	60.2	60.3	29.0	12.8	4.9	10.2
4	231	125	70.2	59.0	40.8	30.2	4.7	11.2
5	216	215	83.0	44.2	33.2	43.0	3.1	9.4

[a] Wood chips impregnated with 0.4% H_2SO_4.
[b] % of theoretical amount in wood, based on total sugars (monomeric + oligomers).

the cellulose and hemicellulose. Assuming a mannan to glucan ratio of 3:1 in the hemicellulose, the maximum glucose yield from the hemicellulose glucan is approx 8% based on wood weight. Glucose yields higher than this are from cellulose.

The experiment was set up as a two-factor (temperature and time), two-level factorial experiment with one center point, so it can be analyzed using the techniques of experimental design. The effects of temperature and time on solids solubilized (similar for hydrolysis of hemicellulose) and furfural yield (similar for hydroxymethyl furfural [HMF] yield) are shown in Figs. 1 and 2, respectively. The results indicate that temperature has a significant effect on all responses and that time has little or no effect on solids solubilized and hemicellulose yields. Time does become a more significant factor for HMF and furfural yields, but temperature still dominates. These results are true only within the tested range of variables.

An additional experiment was performed to obtain complete mass balance information on one set of pretreatment conditions. Ten batches of Douglas fir chips that had been impregnated with 0.35% sulfuric acid were pretreated at 212°C for 105 s. Rinsate was collected and analyzed in this experiment to complete the material balance. Table 2 presents the feedstock composition and component yield calculations based on the measured component concentrations in the pretreated solids, pretreated liquor, and flash vapor. The only decomposition products measured were HMF and furfural. The unaccounted-for fraction is material not accounted for by the indicated components. Additionally, it is assumed that HMF was produced only from glucan-derived glucose and furfural only from xylose. The large unaccounted-for fraction for most of the sugars is partially from: not accounting for byproduct production that is HMF or furfural; from production of byproducts not measured; and from experimental error.

Fig. 1. Effects of pretreatment temperature and time on solids solubilized (%).

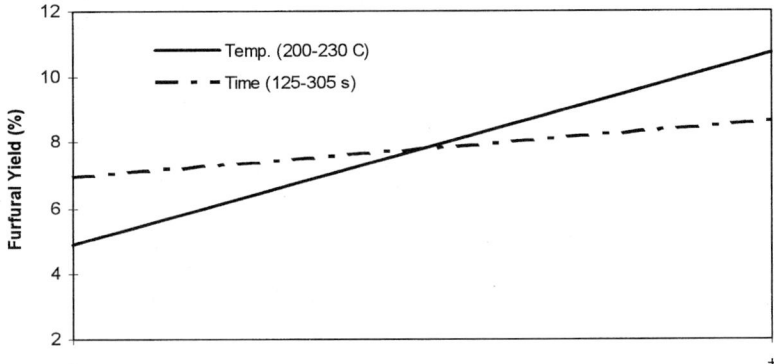

Fig. 2. Effects of pretreatment temperature and time on HMF and furfural yields (%).

The results show that more than 90% of the hemicellulose was solubilized during the pretreatment. The hemicellulosic sugar yields ranged from 56 to 63%, except for arabinose at 36%. A large fraction of the remaining hemicellulose is unaccounted for reasons discussed above.

Seventy percent of the glucan, which consisted of hemicellulosic glucan and cellulose, was not converted during the pretreatment. Twenty-three percent was converted to soluble sugars, and the rest to HMF or was unaccounted for. The mass recovery (mass accounted for by known products divided by mass input) based on individual component measurements for glucan was good at 96.8% and was 86.5% for all measured components (carbohydrates, lignin, and ash). The value is low because of the low recoveries for the hemicellulosic sugars (approx 60%). However, because they are only a small fraction of the feedstock, errors in these mass balances contribute less to the overall mass balance.

Enzymatic Hydrolysis Results

The effect of pretreatment time on the enzymatic hydrolysis glucose yield from Douglas fir is presented in Fig. 3. For a pretreatment tempera-

Table 2
Feedstock Composition and Component Yields of Pretreated
Douglas Fir (Pretreatment at 212°C, 105 s, 0.35% H_2SO_4)

	Feedstock Composition (%)	Yields after Pretreatment (%)
Glucan	44.1	
Unconverted		70.6
To Monomeric Glucose		22.0
To Oligomeric Glucose		1.1
To HMF		3.1
Unaccounted For		3.2
Mannan	13.2	
Unconverted		3.4
To Monomeric Mannose		55.8
To Oligomeric Mannose		7.2
Unaccounted For		33.6
Galactan	3.7	
Unconverted		1.6
To Monomeric Galactose		63.4
To Oligomeric Galactose		4.3
Unaccounted For		30.6
Xylose	6.0	
Unconverted		0.4
To Monomeric Xylose		59.4
To Oligomeric Xylose		0
To Furfural		13.5
Unaccounted-For		30.0
Arabinan	3.0	
Unconverted		0.8
To Monomeric Arabinose		36.3
To Oligomeric Arabinose		0
Unaccounted For		64.4

ture of 212°C and sulfuric acid concentration of 0.4%, the 65-s pretreatment time resulted in 32% solubilization of wood and 91% solubilization of hemicellulose in water. Longer pretreatment times (105–185 s) resulted in 37 to 40% solubilization of wood and more than 95% of hemicellulose. When 91% solubilization of hemicellulose was achieved, the enzymatic digestibility of cellulose in pretreated Douglas fir was 65% of theoretical. With 95% solubilization of hemicellulose, the cellulose digestibility increased to 85%.

Fermentability and SSF Results

The results of fermentability tests of pretreated Douglas fir samples from the first pretreatment run are shown in Table 3. Ethanol yields were calculated by dividing the net ethanol produced by the theoretical amount based on total hexose sugars (glucose, mannose, and galactose). Fermentation performance was best for the less severe pretreatments (samples 1, 2,

Table 3
Effects of Solid Concentration on Ethanol Yield from Pretreated Douglas Fir[a]

Sample	Pretreatment Temperature (C)	Pretreatment Duration (s)	Solids Conc. (wt. %)	Hexose[2] Conc. (g/L) 0h	9h	46h	Ethanol Produced (g/L) 9h	46h	Ethanol Yield % of theo. 9h	46h
1	201	125	10	74.2	10.8	0.9	25.6	23.1	68	61
			15	82.5	45.4	1.7	25.2	33.3	60	79
			20	97.5	84.8	6.4	2.5	35.3	5	71
2	201	305	10	66.9	15.2	1.0	25.8	27.3	76	80
			15	87.8	40.2	4.7	20.2	32.4	45	72
			20	103.2	69.5	7.0	13.7	38.5	26	73
3	230	305	10	62.0	68.9	69.4	0.2	0.0	1	0
			15	80.0	84.3	84.7	0.0	0.0	0	0
			20	87.7	87.0	93.1	0.3	0.0	1	0
4	231	125	10	76.2	75.4	72.6	0.8	0.4	2	1
			15	95.0	94.4	94.8	0.0	0.0	0	0
			20	109.8	102.0	111.9	0.0	0.0	0	0
5	216	215	10	74.3	38.6	2.5	15.4	28.2	40	74
			15	87.2	79.2	3.9	3.4	36.4	8	82
			20	106.3	105.0	140.3	1.0	5.1	2	9
Glucose Control	N/A	N/A	N/A	44.4	0.0	0.0	14.3	12.4	63	55

[a] Wood chips impregnated with 0.4% H_2SO_4.
[b] Hexose includes glucose added and hexose (glucose, mannose, and galactose) present in the wood hydrolysate.

Fig. 3. Enzymatic hydrolysis glucose yield of pretreated Douglas fir (pretreatment at 212°C, 65–185 s, 0.4% H_2SO_4; enzymatic hydrolysis at 1% cellulose, 60 FPU/g cellulose, and 50°C).

Fig. 4 Ethanol yield during SSF of pretreated Douglas fir pretreatment at 212°C, 105 s, 0.35% H_2SO_4; extensively washed with water; SSF at 3% cellulose, 25 FPU/g cellulose, *S. cerevisiae* D_5A yeast, and 37°C.

and 5) and at solids concentrations lower than 15%. For the more severe pretreatments (i.e., high temperatures and longer residence times), samples 3 and 4, the yeast was severely inhibited even at 10 wt% total solids loading.

Figure 4 shows the ethanol yield from SSF of the Douglas fir pretreated at 212°C for 105 s. With more than 95% of the hemicellulose removed after pretreatment, the ethanol yield from the cellulose portion of the pretreated wood sample was comparable to that of Solka-Floc (Fiber Sales and Development, Urbana, OH) (i.e., 80–85% of theoretical). Enzymatic hydrolysis and SSF of whole-slurry-pretreated Douglas fir at 10% solid concentrations (data not shown) indicated inhibition of both enzyme activity and fermentation, which resulted in lower ethanol yields (i.e., 60–70% of theoretical). We have since successfully adapted the D_5A yeast to a 21% solids hydrolysate from Douglas fir pretreated at 212°C, 0.35% sulfuric acid, and 105 s *(11)*.

DISCUSSION

The results of our preliminary work show that dilute sulfuric acid pretreatment effectively renders most of the hemicellulose fraction of Douglas fir soluble in water and the cellulose fraction highly digestible by cellulase enzyme. However, the pretreatment conditions that favor high enzyme digestibility substantially degrade the hemicellulosic sugar. To maximize hemicellulose recovery and minimize formation of inhibitors such as furfural and HMF, the pretreatment temperatures need to be lower than those used in these experiments. This suggests a two-stage pretreatment method, with the first stage operating at a lower temperature (e.g., 170–190°C), and a washing step between stages to remove the hemicellulosic sugars before carrying out the second-stage pretreatment at a higher temperature (e.g., 210–230°C). However, this method, would increase the complexity and cost of the pretreatment process.

Another key issue is the presence of bark and needles. In forest-thinning operations, the normal practice is to chip the whole tree, which results in high bark and needle content in the chips. Because bark and needles contain high levels of extractives that may inhibit enzymes and the fermenting organism, whole tree chips may not be readily converted. We are currently investigating the effect of bark and needle content on the performance of pretreatment and fermentation. The trade-off between the additional cost of debarking and reduced ethanol yield caused by bark and needles will be evaluated.

ACKNOWLEDGMENT

Chemical analyses of hydrolysis and fermentation samples by F. Posey-Eddy, J. Hora, L. W. Brown, R. Ruiz, and C. I. Ehrman are greatly appreciated. Solid compositional analyses were performed by Hauser Laboratories (Boulder, CO). This work was funded by the Biochemical Conversion Element of the Office of Fuels Development of the US Department of Energy.

REFERENCES

1. Schwald, W., Smaridge, T., Chan, M., Bruil, C., and Saddler, J. N. (1989), in *Enzyme Systems for Lignocellulose Degradation*, Coughlan, M. P., ed., Elsevier, New York, pp. 231–242.
2. Clark, T. A. and Mackie, K. L. (1987), *J. Wood Chem. Technol.* **7(3),** 373–403.
3. TAPPI Test Methods (1991), T 210 cm-86, *Weighing, Sampling, and Testing Pulp for Moisture*, TAPPI, Atlanta, GA.
4. Moore, W. E. and Johnson, D. B. (1967), *Procedures for the Chemical Analysis of Wood and Wood Products*, Forest Products Laboratory, US Department of Agriculture, Madison, WI.
5. TAPPI Test Methods (1994–1995), T 222 om-88, *Acid-Insuluble Lignin in Wood and Pulp*, TAPPI, Atlanta, GA.
6. Grohmann, K., Himmel, M., Rivard, C., Tucker, M., Baker, J., Torget, R., and Graboski, M. (1984), *Biotechnol. Bioeng. Symp.* **14,** 137.
7. TAPPI Test Methods (1991), T 211 om-85, *Ash in Wood and Pulp*, TAPPI, Atlanta, GA.
8. Ehrman, C. I. and Himmel, M. E. (1994), *Biotechnol. Techniques* **87(2),** 99–104.
9. Ghose, T. K. (1987), *Pure Appl. Chem.* **59,** 257–268.
10. Grohmann, K., Torget, R., and Himmel, M. (1986), *Biotechnolo. Bioeng. Symp.* **17,** 135–151.
11. Keller, F. A., Bates, D., Ruiz R., and Nguyen, Q., *Nineteenth Symposium on Biotechnology for Fuels and Chemicals*, Colorado Springs, Colorado, May 4–8, 1997.

Pretreatment of Sugarcane Bagasse Hemicellulose Hydrolysate for Xylitol Production by *Candida guilliermondii*

LOURDES A. ALVES,* MARIA G. A. FELIPE, JOÃO B. ALMEIDA E. SILVA, SILVIO S. SILVA, ARNALDO M. R. PRATA

Departamento de Biotecnologia, FAENQUIL, Rod. Itajubá-Lorena, km 74,5 12600–000, Lorena, SP, Brazil

ABSTRACT

In order to remove or reduce the concentrations of toxic substances present in the sugarcane bagasse hemicellulose hydrolysate for xylose-to-xylitol bioconversion, the hydrolysate was pretreated by changing the initial pH level through the combination of different bases and acids with or without the subsequent addition of activated charcoal. Attention was given to the influence of the fermentation time as well.

The experiments were based on multivariate statistical concepts, with the application of fractional factorial design techniques to identify the important variables in the process. Subsequently, the levels of these variables were quantified by the response surface methodology, which permitted the establishment of the best pretreatment procedure with a xylose-to-xylitol bioconversion efficiency of 86.2%. This procedure consisted in increasing the pH of the hydrolysate from 0.5 to 7.0 with CaO and reducing it to 5.5 with H_3PO_4. Next, the hydrolysate was treated with activated charcoal (2.4%). The highest xylitol yield (0.79 g/g) corresponded to a productivity of 0.47 g/L/h.

Index Entries: Sugarcane bagasse; hydrolysate; pretreatment; xylitol; *Candida guilliermondii*.

INTRODUCTION

Sugarcane bagasse consists of 25 to 35% hemicellulose, of which D-xylose is the major component. This pentose can be directly fermented to xylitol by *Candida guilliermondii* as an alternative to the production of xylitol

---

*Author to whom all correspondence and reprint requests should be addressed. E-mail: feqlps@eu.ansp.br

(1,2), which today is obtained by chemical process (3). However, various toxic compounds such as furfural, hydroxymethylfurfural, and acetic acid are formed during hydrolysis of hemicellulose. Once present in the hydrolysate, these compounds are potential inhibitors of the microbial metabolism (1,4).

Xylose in high concentrations favors the xylitol production by the yeast (5,6). This could be achieved by concentrating the hydrolysate before using it as a culture medium. However, the more concentrated the hydrolysate, the lower the xylose consumption and xylitol production, because of the simultaneous increase in the concentrations of toxic compounds (7). According to these authors, the inhibitory effect could be overcome by using an adequate hydrolysate pretreatment (8), a hydrolysate-adapted yeast strain (1), and/or bioreactors inoculated with a high cell concentration (2).

Different procedures for pretreating the bagasse hemicellulose hydrolysate were tested with a view to improving the xylose-to-xylitol fermentation efficiency.

MATERIALS AND METHODS

Microorganism and Inoculum Preparation

Cells of *Candida guilliermondii* FTI 20037 (9) maintained at 4°C on malt-extract agar slant were inoculated in the culture medium containing 30 g of xylose supplemented with the following nutrients (g/L): $(NH_4)_2SO_4$ (2.0), $CaCl_2 \cdot 2H_4O$ (0.1), and rice bran (20.0). Fifty milliliters of this medium was placed into 125-mL Erlenmeyer flasks and agitated at 200 rpm, at 30°C for 24 h. The cells were then centrifuged at $2200g$ for 15 min and washed in sterile distilled water. A suspension was prepared with the cell mass in sterile distilled water and utilized as the inoculum. For the experiments, the initial cell concentration was 0.45 g/L.

Preparation and Treatment of Hemicellulose Hydrolysate

The sugarcane bagasse was introduced into a 250-L reactor and mixed with concentrated H_2SO_4 (100 mg of acid per gram of dry matter) with a solid:liquid ratio of 1:10. After hydrolysis (121°C, 10 min), the hydrolysate containing (g/L) glucose (2.1), xylose (15.7), arabinose (2.3), acetic acid (3.9), furfural (0.06), and hydroximethylfurfural (0.05) was filtered and concentrated under vacuum at 70°C to increase the xylose concentration threefold. The hydrolysate thus obtained had the following composition (g/L): glucose (5.9), xylose (50.2), L-arabinose (6.5), acetic acid (6.9), furfural (0.03), hydroxymethylfurfural (0.15). The hydrolysate was treated as follows: the initial pH (0.5) was raised to 7.0 or 10.0 with CaO or $Ca(OH)_2$ (commercial powder); the hydrolysate was then acidified with concentrated H_2SO_4 or H_3PO_4 to pH 5.5, with or without the subsequent addition of 3% w/v activated charcoal (refined powder). The charcoal was mixed with the hydrolysate under agitation (200 rpm) at 30°C for 1 h. In all the treatments the precipitate resulting from pH adjustment and from addition of activated charcoal was removed by vacuum filtration. These treatments resulted in several hydrolysates that were autoclaved at 111°C, 0.5 atm, to be used as culture media.

Medium and Fermentation Conditions

The hydrolysates containing 43–48 g/L xylose were supplemented with the same nutrients described in the inoculum preparation. Fermentations were carried out in 125-mL Erlenmeyer flasks containing 50 mL of the culture medium (pH 5.5), on a rotary shaker at 200 rpm at 30°C. The influence of the fermentation time was also evaluated (45 or 63 h of incubation).

Analytical Methods

The concentrations of glucose, L-arabinose, xylose, xylitol, acetic acid, furfural, and hydroxymethylfurfural were determined by high-pressure liquid chromatography *(2)*. Cell concentration was determined by optical density at 600 nm.

Experimental Design

The effects of the treatments and fermentation time were appraised with the application of 2^{5-1} fractional factorial design with two replicates *(11,12)*, to identify the variables important to the process (Table 1). The factors and their respective levels represented by the signs (−) and (+), were as follows: [A] acid (H_2SO_4, H_3PO_4), [B] base (CaO, Ca(OH)$_2$), [pH] pH (7.0, 10.0), [CA] activated charcoal (0.3%), and [FT] fermentation time (45 or 63 h). After statistical analysis, the factors with significant effects were used for additional experiments (2^2 full factorial design with a centered face and three replicates at the center point). The intermediary level for the selected factors was 1.5% of activated charcoal and 54 h of fermentation time.

The statistical analysis was performed using the STATGRAPHICS statistical software version 6.0 and STATISTICA program version 5.0.

RESULTS AND DISCUSSION

Table 1 presents the experimental matrix as well as the results achieved for the fermentation parameters of the xylose-to-xylitol bioconversion by *C. guilliermondii* as a function of different pretreatments of the sugarcane bagasse hemicellulose hydrolysate. The analysis of the estimated effect (Table 2) shows that for xylitol yield (Y p/s), the significant main effects ($p < 0.05$) are: fermentation time, interaction between acid

Table 1
Experimental Design and Fermentative Parameters in the
Bioconversion Xylose-to-Xylitol by *C. guilliermondii* as a Function of
Different Treatments of the Sugarcane Bagasse Hemicellulose
Hydrolysate for the 2^{5-1} Fractional Fatorial Design[a]

Treatment	[A]	[B]	[pH]	[CA]	[FT]	$Y_{p/s}$[a]	Q_p[b]
1	-	-	-	-	+	0.58	0.35
2	+	-	-	-	-	0.55	0.42
3	-	+	-	-	-	0.56	0.44
4	+	+	-	-	+	0.51	0.34
5	-	-	+	-	-	0.55	0.46
6	+	-	+	-	+	0.53	0.36
7	-	+	+	-	+	0.56	0.35
8	+	+	+	-	-	0.57	0.37
9	-	-	-	+	-	0.55	0.50
10	+	-	-	+	+	0.51	0.37
11	-	+	-	+	+	0.47	0.34
12	+	+	-	+	-	0.70	0.54
13	-	-	+	+	+	0.58	0.39
14	+	-	+	+	-	0.62	0.51
15	-	+	+	+	-	0.61	0.49
16	+	+	+	+	+	0.60	0.35

[a] Average of two replicate experiments.
[b] $Y_{p/s}$, xylitol produced (g)/xylose consumed (g).
[c] Q_p, volumetric productivity (g/L/h).

and activated charcoal, and the interaction between fermentation time and activated charcoal. With respect to productivity, charcoal and fermentation time have significant main effects ($p < 0.05$). Also significant is the effect of the interaction between these two factors (Table 2).

Despite the results of other researchers (1,2,8,13,14,15), our experiments did not show a significant improvement in fermentation with overlining to pH 10.0. The base and pH did not present significant effects (at the 5% level) for any of the parameters evaluated. The increase in the pH of the hydrolysate to 7.0 or 10.0 during the treatment with base did not affect the fermentative parameters. Raising the pH to 7.0 implied a reduction in the amounts of base and acid to be added as well as in the time required for the treatment, which consequently affects the cost of the process. As for the base factor, according to Roberto et al. (8), treating the sugarcane bagasse hydrolysate with CaO and $Ca(OH)_2$ increased the xylose consumption to 95 and 98%, respectively, whereas with KOH the consumption was only 56%. Increasing the pH of the hydrolysate with Ca^{2+} ions may cause precipitation of the toxic compounds present in the

Table 2
Effect Estimates, Standard Errors, *T*-Test for the Yield and Productivity in the Xylose-to-Xylitol Bioconversion by *C. guilliermondii*, in the 2^{5-1} Fractional Factorial Design

Factors and Interactions	Yield			Productivity		
	estimate	Standard errors	t	Estimate	Standard errors	t
Average	0.562	± 0.008	-	0.408	± 0.006	-
[A]	0.017	± 0.016	1.064	-0.006	± 0.012	0.468
[B]	0.016	± 0.016	0.985	-0.017	± 0.012	1.403
[pH]	0.027	± 0.016	1.695	-0.003	± 0.012	0.260
[CA]	0.028	± 0.016	1.774	0.049	± 0.012	4.104*
[FT]	-0.046	± 0.016	2.877*	-0.109	± 0.012	9.090*
[A] x [B]	0.028	± 0.016	1.774	0.003	± 0.012	0.260
[A] x [pH]	-0.011	± 0.016	0.670	-0.018	± 0.012	1.506
[A] x [CA]	0.038	± 0.016	2.404*	0.019	± 0.012	1.610
[A] x [FT]	-0.026	± 0.016	1.616	0.006	± 0.012	0.468
[B] x [pH]	0.003	± 0.016	0.197	-0.022	± 0.012	1.818
[B] x [CA]	0.017	± 0.016	1.064	0.006	± 0.012	0.468
[B] x [FT]	-0.029	± 0.016	1.852	-0.008	± 0.012	0.675
[pH] x [CA]	0.021	± 0.016	1.301	-0.001	± 0.012	0.052
[pH] x [FT]	0.027	± 0.016	1.695	0.013	± 0.012	1.091
[CA] x [FT]	-0.034	± 0.016	2.168*	-0.039	± 0.012	3.273*

[a] Significant at 5% probability level.

hydrolysate, which can then be removed by filtration *(16)*. Since the base did not have a significant effect, the treatment with CaO was preferred, for its low cost.

As is evident from Table 2, the activated charcoal has a positive main effect and the effect of the interaction between the charcoal and the acid is also positive. Thus, H_3PO_4 was selected for the next experiments. A 2^2 factorial design with a centered face and three replicates at the center point was used to obtain the mathematical model representing this fermentative process by the response-surface methodology. The matrix referring to this design and to the fermentative parameters Y p/s and Qp) is shown in Table 3. As can be seen, treatment 4 provided the highest xylitol yield (0.79 g/g), whereas treatment 1 provided the lowest yield (0.62 g/g). This difference is equivalent to 22%. With respect to productivity, the variation between maximum and minimum values (0.52 g/L/h for treatment 2 and 0.40 g/L/h for treatment 8), corresponded to an increase of 23%, which confirms the

Table 3
Experimental Design and Fermentative Parameters of the Xylose-to-Xylitol Bioconversion by *C. guilliermondii* as a Function of Different Treatments of Sugarcane Bagasse Hemicellulose Hydrolysate for the 2^2 Factorial Design with a Centered Face and Three Replicates at the Center Point

Treatment	[CA]	[FT]	$Y_{p/s}$	Q_p
1	-1	-1	0.62	0.44
2	+1	-1	0.75	0.52
3	-1	1	0.64	0.41
4	+1	1	0.79	0.47
5	-1	0	0.63	0.44
6	+1	0	0.74	0.50
7	0	-1	0.73	0.49
8	0	1	0.76	0.40
9	0	0	0.76	0.47
10	0	0	0.75	0.44
11	0	0	0.76	0.51

Table 4
Regression Coefficients, Standard Errors, *T*-Test, and Significance Level for the Model Representing the Yield in the Xylose-to-Xylitol Bioconversion by *C. guilliermondii* in the 2^2 Factorial Design with a Centered Face and Three Replicates at the Center Point

Factors	coefficients	Standard errors	t	P
Constant	0.7511	±0.00655	114.6194	-
X_1	0.065	±0.00523	12.4648	0.0001*
X_2	0.015	±0.00523	2.8765	0.0347*
X_1X_2	0.005	0.00639	0.7829	0.4691
X_1^2	-0.0576	±0.00803	-7.1813	0.0008*
X_2^2	0.0024	0.00803	0.2951	0.7798

[a] Significant at 5% probability level.

influence of the treatments on the xylose-to-xylitol bioconversion. The maximum value of xylitol yield (0.79 g/g) was obtained when 3% w/v activated charcoal was utilized. Parajo et al. *(17)* found that the treatment of eucalyptus hydrolysate with activated charcoal (0.5%) increased the xylitol yield from 0.4 to 0.6 g/g, as compared with the untreated hydrolysate.

Table 4 presents the regression coefficients, standard errors, *t* values, and significance levels for the model representing xylitol yield. At 5% probability level, the activated charcoal, the fermentation time, and the charcoal quadratic term presented significant effects. The conversion efficiency increased by 17% when charcoal was used and by 4% when the fermentation time was changed from 45 to 63 h. The linear and quadratic

Fig. 1. Response surface and contour lines described by the Y_1 model representing the yield in the xylose-to-xylitol bioconversion by *C. guilliermondii* in sugarcane bagasse hemicellulosic hydrolysate.

terms presented an optimum significance level. The regression variance analysis shows that the mathematical model is significant. This is confirmed by the determination coefficient ($R^2 = 0.98$), which indicates that the selected model is suitable for the process and allows an estimation of 98% variance as a function of the activated charcoal concentration and fermentation time. The mathematical model for the xylitol yield is represented by the equation (1):

$$Y_1 = 0.754 + 0.070\, X_1 + 0.015\, X_2 - 0.057\, X_1^2 \qquad (1)$$

where Y_1 represents the yield, X_1 the activated charcoal ratio, and X_2 the fermentation time.

Solving this mathematical model for the optimal conditions predicts a yield of 0.79 g/g using 2.4% charcoal and a fermentation time of 63 h. This optimal region can be observed in Fig. 1, which depicts the response surface and contour lines described by the Y_1 model.

Table 5 presents the regression coefficient, standard errors, t values, and significance level for the model representing the xylose-to-xylitol bioconversion productivity at 5% significance level. In this case, charcoal and fermentation time have significant effects. The results lead to the conclusion that using charcoal in the hydrolysate treatment increases the xylitol productivity by 14%, whereas changing the fermentation time from 45 to 63 h reduces it by 12%. The regression variance analysis and the coefficient determination of the model ($R^2 = 0.78$) reveal that the mathematical model is significant and can be expressed by the equation (2):

Table 5
Regression Coefficients, Standard Errors, T-Test, and Significance Level for the Model Representing the Productivity of the Xylose-to-Xylitol Bioconversion by C. guilliermondii in the 2^2 Fatorial Design with a Centered Face and Three Replicates at the Center Point

Factors	Coefficients	standard errors	t	P
Constant	0.4695	± 0.0137	34.3671	-
X_1	0.0333	± 0.0109	3.0662	0.0279*
X_2	-0.0283	± 0.0109	-2.6062	0.0479*
$X_1 X_2$	-0.005	± 0.0133	-0.3755	0.7227
X_1^2	0.0063	± 0.0167	0.3775	0.7213
X_2^2	-0.0187	± 0.0167	-1.1168	0.3149

[a] Significant at 5% probability level.

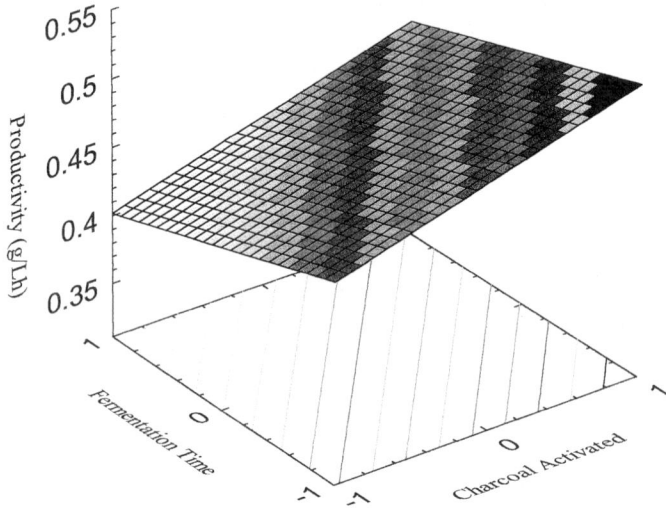

Fig. 2. Response surface and contour lines described by the Y_2 model representing the productivity in the xylose-to-xylitol bioconversion by C. guilliermondii in sugarcane bagasse hemicelullosic hydrolysate.

$$Y_2 = 0.463 + 0.033\ X_1 - 0.031\ X_2 \qquad (2)$$

where Y_2 represents the productivity, X_1 the activated charcoal ratio, and X_2 the fermentation time.

Figure 2 shows the response surface and the contour lines described by the Y_2 model. The best results were predicted from coded values 1 and −1, corresponding to 3% charcoal and 45 h of fermentation time, respectively. The mathematical model makes it possible to obtain the maximum point for these factors, resulting in a productivity of 0.52 g/L/h.

CONCLUSION

Developing a new methodology for the pretreatment of sugarcane bagasse hemicellulose hydrolysate is indispensable for increasing xylose-to-xylitol conversion efficiency. An extensive factorial design showed that the best conditions for this bioconversion were: increasing the pH level to 7.0 with CaO, reducing it to 5.5 with H_3PO_4, and adding 2.4% of activated charcoal. Under these conditions the highest xylitol yield (0.79 g/g) and productivity (0.52 g/L/h) were obtained after 63 and 45 h of fermentation time, respectively. The charcoal reduced the concentration of the toxic compounds furfural and hydroxymethylfurfural in the hydrolysate, whereas the concentration of acetic acid was not significantly affected. From previous studies conducted in our laboratory *(1,2,10)*, which are in agreement with other reports *(4,16,18)*, it is evident that the acetic acid inhibits the xylose metabolism of the yeast. The effect of this acid mainly depends on its concentration level *(10)* and the pH of the fermentation. Phenolic compounds, derived from the breakdown of lignin, have also been considered as inhibitors of various bioconversion processes employing hydrolysates obtained from lignocellulosic materials *(8,17,19)*. Although the concentration levels of these compounds were not measured in our experiments, activated charcoal has been reported by several authors as able to remove phenolics from hydrolysate *(8,17,20)*.

ACKNOWLEDGMENTS

The authors acknowledge financial assistance from FAPESP/Brazil and Maria Eunice M. Coelho for the helpful revision on this paper.

REFERENCES

1. Felipe, M. G. A., Vitolo, M., and Mancilha, I. M. (1996A), *Acta Biotechnologica* **16,** 73–79.
2. Felipe, M. G. A., Alves, L. A., Silva, S. S., Roberto, I. C., Mancilha, I. M., and Almeida e Silva, J. B., (1996B) *Bioresource Technol.* **56,** 281–283.
3. Melaja, A. J. and Hämäläinen, L. (1997), *Process for Making Xylitol.* US Patent no. 4.008.285. 1975, 15 fev.
4. Domingues, J. M., Cheng, S. G., and Tsao, G. T. (1996), *Applied Biochem. Biotechnol.* **57/58,** 49–56.
5. Meyrial, V., Delgenes, J. P. Molletta, R., and Navarro, J. M. (1991), *Biotechnol. Lett.* **13,** 281–286.
6. Nolleau, V., Preziosi-Belloy, L., Delgenes, J. P., and Delgenes, J. M. (1993), *Curr. Microbiol.* **27,** 191–197.
7. Felipe, M. G. A., Vitolo, Mancilha, I. M., and Silva, S. S. (1997), *J. Indust. Microbiol.* **18,** 251–254.
8. Roberto, I. C., Felipe, M. G. A., Lacis, L. S., Silva, S. S., and Mancilha, I. M. (1991), *Bioresource Technol.* **36,** 271–275.
9. Barbosa, M. F. S., Medeiros, M. B., Mancilha, I. M., Schneider, H., and Lee, H. (1988), *J. Indust. Microbiol.* **3,** 241–251.
10. Felipe, M. G. A., Vieira, D. C., Vitolo. M., Silva, S. S., Roberto, I. C., Mancilha, I. M., and Rosa, S. A. M. (1995), *J. Basic Microbiol.* **35,** 171–177.

11. Box, G. E. P., Hunter, W. G., and Hunter, J. S. (1978), *Statistics for Experimenters: An Introduction to Design, Data Analysis and Model Building*. Wiley, New York.
12. Barros Neto, B., Scarminio, I. S., and Bruns, R. E. (1995), *Planejamento e Otimização de Experimentos*, 1 ed., Editora da UNICAMP, Campinas.
13. Chen, L. F. and Gong, C. S. (1985), *J. Food Sci.* **50**, 226–228.
14. Felipe, M. G. A., Mancilha, I. M., Vitolo, M., Roberto, I. C., Silva, S. S., and Rosa, S. A. M. (1993), *Arquivos de Biologia e Technologia*, **36**, 103–114.
15. Roberto I. C., Mancilha, I. M., Souza, C. M. A., Felipe, M. G. A., Sato, S., and Castro, H. F. (1994), *Biotechnol. Lett.* **16**, 1211–1216.
16. van Zyl, C., Prior, B. A., and Du Preez, J. C. (1988), *Appl. Biochem. Biotechnol.* **17**, 357–369.
17. Parajó, J. C., Domingues, H., and Domingues, J. M. (1996B) *Bioresource Technol.* **57**, 179–185.
18. Parajó, J. C., Domingues, H., and Domingues, J. M. (1996A), *Biotechnol. Lett.* **18**, 593–598.
19. Chung, I. S., and Lee, Y. Y. (1985), *Biotechnol. Bioengineer.* **27**, 308–315.
20. Wang, R. C., Kuo, C. C., and Shyu, C. C. (1997), *J. Chem. Tech. Biotechnol.* **68**, 187–194.

Continuous pH Monitoring During Pretreatment of Yellow Poplar Wood Sawdust by Pressure Cooking in Water

JOSEPH WEIL, MARK BREWER, RICHARD HENDRICKSON, AYDA SARIKAYA, AND MICHAEL R. LADISCH*

Laboratory of Renewable Resources Engineering and Department of Agricultural and Biological Engineering, Purdue University, West Lafayette, IN 47907

ABSTRACT

Yellow poplar wood sawdust consists of 41% cellulose and 19% hemicellulose. The goal of pressure cooking this material in water is to hydrate the more chemically resistive regions of cellulose in order to enhance enzymatic conversion to glucose. Pretreatment can generate organic acids through acid-catalyzed degradation of monosaccharides formed because of acids released from the biomass material or the inherent acidity of the water at temperatures above 160°C. The resulting acids will further promote the acid-catalyzed degradation of monomers that cause both a reduction in the yield and the formation of fermentation inhibitors such as hydroxymethyl furfural and furfural. A continuous pH-monitoring system was developed to help characterize the trends in pH during pretreatment and to assist in the development of a base (2.0 M KOH) addition profile to help keep the pH within a specified range in order to reduce any catalytic degradation and the formation of any monosaccharide degradation products during pretreatment. The results of this work are discussed.

Index Entries: pH monitoring; aqueous pretreatment; pH control; hydrolysis; cellulose; hydrothermal.

INTRODUCTION

Cellulose is a linear polymer of glucose that is found in herbaceous and woody plants. It is associated with another polysaccharide, hemicellulose, and both are sealed with lignin, a complex three-dimensional polyaro-

*Author to whom all correspondence and reprint requests should be addressed.

matic compound that is resistant to enzyme and acid hydrolysis *(1)*. Another challenge to hydrolysis is the crystallinity of the cellulose itself. The β-1,4 orientation of the glucosidic bonds result in the formation of six hydrogen bonds per anhydro-glucose molecule, thus, allowing the cellulose to form a tight crystalline structure *(2)*. One way to overcome these challenges is to pretreat the lignocellulosic material to increase the accessible surface area of the cellulose and enhance conversion of the cellulose to glucose. There are many methods of pretreating lignocellulosic material to this end *(3)*. Recent work with aqueous pretreatment of yellow poplar wood sawdust have shown improved hydrolysis of the cellulose to glucose *(4)*, and diminished formation of degradation products during pretreatment when the pH was controlled. Hydrolysis refers to the formation of monosaccharides from polysaccharides and oligosaccharides from hemicellulose and cellulose. Degradation refers to the formation of aldehydes such as hydroxymethyl furfural (HMF) and furfural, and organic acids such as levulinic and formic acid from monosaccharides. These aldehydes and organic acids are also referred to as degradation products.

The goal of aqueous pretreatment of lignocellulosic material is to hydrate the crystalline structure of the cellulose and modify it to a form that is more accessible and susceptible to enzyme hydrolysis, thereby enhancing conversion of cellulose to glucose *(5)*. During the pretreatment, organic acids can form and dissociate hydrogen ions, which will promote acid-catalyzed hydrolysis of the cellulose to glucose and, also, the acid-catalyzed degradation of the glucose to HMF, and levulinic and formic acids. This process competes with the physical hydration of the cellulose by the water. Hydrogen-ion formation from water and from organic acids is an important factor during aqueous pretreatment of lignocellulosic material since the wood sawdust and water mixture will reach temperatures as high as 260°C and pressures as high as 700 psi during the pretreatment. These high temperatures and pressures will accelerate the acid-catalyzed hydrolysis of the cellulose and hemicellulose as well as the acid-catalyzed degradation of the glucose and xylose. The loss of substrate to degradation products will decrease the extent of enzymatic conversion of the polysaccharides and oligosaccharides to monosaccharides. Monitoring and control of the pH of this system is necessary to minimize the hydrogen-ion concentration during pretreatment and help prevent hydrolysis of the cellulose and hemicellulose to oligosaccharides and monosaccharides and, more importantly, the acid-catalyzed degradation of the monosaccharides to degradation products *(5)*.

Because of the accumulation of organic acids and the liberation of hydrogen ions during pretreatment, pH is an important parameter to monitor. Conventional laboratory-scale pH probes are too delicate to read the pH of solutions at temperatures of 100°C or greater. These probes will lose their accuracy and will be permanently damaged. Probes for use at temperatures above 100°C are still experimental and internal referencing does not give completely accurate results (6). One way to overcome this challenge is to continuously pull a stream of the pretreatment supernatant from the vessel, cool it down to room temperature where the pH probe is most accurate, read the pH, and then pump the supernatant back to the pretreatment vessel.

This paper explains the method in which the pH was monitored and the procedure for which the pH was modified during the aqueous pretreatment of wood sawdust. This work will also discuss the effects of controlling pH on the polysaccharides in the wood sawdust and compare those results to the pretreatments in which the pH was not controlled.

MATERIALS AND METHODS

Materials

The yellow poplar wood sawdust was provided by National Renewable Energy Laboratory (NREL) in 1 × 5-mm average particles. The untreated wood sawdust consisted of 41% cellulose and 19% hemicellulose that consisted of predominantly xylose (4). The wood sawdust was immediately frozen upon arrival and removed from the freezer as needed.

Pretreatment Reactor

A Parr, 304 SS reactor (Model 4843; Moline, IL) was used to pretreat yellow poplar wood sawdust in water at selected temperatures and pH values (Fig. 1). The reactor has a total volume of 2 L, with three turbine propeller agitators, and a proportional integral derivative (PID) temperature controller (± 1.0°C). Cooling water was circulated through a serpentine coil to cool the reactor contents at the end of each run. A bottom port and two inlet ports allow sampling of the pretreated material and addition of reagents to the reactor.

Pretreatment Procedures

The wood sawdust was pretreated under six conditions. The three final set-point temperatures were 220, 240 and 260°C. Two runs were made at each set-point temperature. One run pretreated the sawdust with the addition of KOH to help control the pH, and another was carried out without any pH control.

The wood sawdust was first soaked in water at room temperature for approximately 14–18 h. The vessel loading for the wood sawdust was between 53 and 63 g/L (dry weight basis) depending on the dry weight of the wood sawdust. The working volume of the pretreatment vessel was 1.5 L for the 220 and 240°C pretreatments, and 1.0 L for the 260°C pretreatments. The volumes were selected to maintain adequate head space since pretreatment at these temperatures caused expansion of the liquid phase $\rho = 0.845$ g/mL at 220°C vs $\rho = 0.785$ g/mL at 260°C). The ratio of solids to liquids was kept at a constant level.

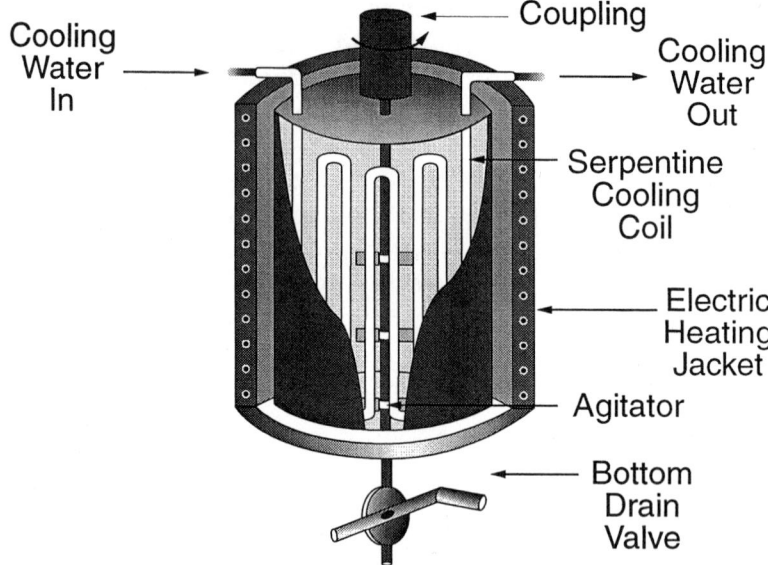

Fig. 1. Schematic diagram of the pretreatment vessel.

Fig. 2. Schematic diagram of the pH-monitoring system.

Prior to pretreatment, the pumps and tubing for the pH-monitoring system were primed with deionized water (Fig. 2). This was done to calibrate the two pumps to the same flow rate and to evaluate the system for leaks and for plugging.

The pretreatment slurry was agitated using 3 turbine impellars (1 cm wide by 6 cm diameter) set at 115 rpm. The reactor contents were initially at ambient temperature. The heater set-point temperature was set at 220,

240 or 260°C, and heat-up was initiated by turning the heat switch to high. The heat-up rate for the wood sawdust was between 3.5 and 4.0°C/min. The control of temperature and heat-up was achieved using the PID controller supplied by Parr (Moline, Illinois).

A continuous pH-monitoring system was developed to sample the supernatant without loss during the pretreatment. The system consisted of two Milton Roy pumps (Mini-pump; 46/460 mL/hr) (Fig. 2). pH monitoring was initiated when the pretreatment temperature reached 150°C, with a vessel pressure of 50 psi (± 3 psi). Pump 1 continually metered liquid supernatant from the reactor through a 0.2-micron filter stone (Alltech, Deerfield, IL.), then through 1/8-in od (0.085-in id) stainless-steel tubing. The vessel pressure ranged from 0 to 700 psi over the course of the pretreatment and hence, provided the back pressure required for the supernatant to pass through the filter stone. The supernatant passed through an ice bath, then through pump 1. From pump 1, the supernatant went through a water bath, at ambient temperature, and was pumped through a check valve (set at 750 psi) into a pH probe/flow cell at a rate of 8.5 to 9.0 mL/min. The total time for supernatant to travel from the filter stone in the pretreatment vessel to the pH flow cell was approx 1.5 min.

The flow cell, which housed the pH probe, was fabricated from a 50-mL graduated cylinder with two approx 1 mm id (3/8-in od) glass ports fitted at the top and bottom of the cylinder by the Purdue University Chemistry Glass Shop. The temperature of the supernatant entering the flow cell was 24°C. The fluid entered a port at the bottom of the cylinder and flowed out through a port at the top into a reservoir. The reservoir was a graduated beaker that held approx 40 mL of fluid. The volume of the reservoir was checked regularly to calibrate the flow rates of the two pumps in the monitoring line. A second Milton Roy pump (Mini-pump; 46/460 mL/h) then returned the supernatant from the reservoir to the pretreatment vessel at approximately the same flow rate as pump 1. The total time for supernatant to flow through this circuit was approx 10 min.

A Markson pH meter (Markson Science, Phoenix, AZ) and pH electrode (Markson V-830) were used to determine the pH of the supernatant in the flow cell. The pH electrode was calibrated using standardized buffer solutions (Fischer, Pittsburgh, PA) of pH 4.0 and 7.0. The electrode was stored in pH 4.0 buffer when it was not being used.

For the pretreatments in which the pH was controlled by the addition of KOH, the base was metered into the reactor by manual control using a third pump, Dynamax SD-200 (Woburn, MA), pumping at 10 mL/min, shown in Fig. 2. The base was added directly to the pretreatment vessel to reduce the time between reading the pH meter and the physical contact of the base with the contents of the pretreatment vessel. This minimized the possibility of alkaline degradation reactions that might otherwise occur if the base were added to the reservoir, where a small portion of the slurry can sit at a high pH (7.0–12.0) at room temperature for a period of 5 min.

Table 1
pH Results for Wood Sawdust Pretreatment Without pH Control

Temperature (°C)	Initial pH	Final pH
220	5.14	3.03
240	5.27	2.81
260	5.30	2.91

Once the predetermined set-point temperature was obtained, the heater was turned off and cooling water was charged through the serpentine cooling coil. The contents of the reactor cooled down to 180°C in approx 2 min, and to 150°C in approx 5 min. The reactor was kept sealed, and the slurry agitated until the reactor headplate had cooled to approx 50°C. The agitator drive was then disconnected and the reactor was then physically removed from the heating jacket and placed on the bench. The reactor was then opened, and the contents removed for further analysis and testing.

RESULTS AND DISCUSSION

Pretreatments of Wood Sawdust Without pH Control

The pH ranges for the wood sawdust pretreatments in which pH was not controlled are summarized in Table 1. The initial pH of the wood sawdust and water slurry after sitting 16 h was between 5.14 and 5.30. The pH reaches a final reading between 2.8 and 3.0 upon the completion of pretreatment and cool-down of the slurry to room temperature. There does not appear to be a strong correlation between final pretreatment temperature and the final pH. For the 240 and 260°C pretreatments, the final pH approached 2.8, where autohydrolysis of the cellulose may occur (7).

A typical pH profile that occurs during pretreatment to 240°C without pH control is illustrated in Fig. 3. Between 180 and 200°C, the pH decreased significantly from 5.0 to 3.5. From the 200°C point to the end of the pretreatment, where the slurry was cooled to 180°C, the pH continued dropping to approx 2.8.

Wood Sawdust Pretreatments with pH Control

Table 2 shows the experimental pH ranges observed for the wood sawdust pretreatments with KOH added to help control the pH. Before pretreatment, the pH of the slurry was measured directly from the vessel before pretreatment and was approx 5.2 for all three pretreatments. Approximately 0.3 mL of 2.0 M KOH were added to the vessel to bring the

Table 2
pH Results for Wood Sawdust Pretreatments with pH Control

Temperature (°C)	Initial pH	Final pH	pH range	KOH added (mls)
220	7.10	5.91	4.50 - 7.10	88.6
240	8.20	5.71	5.02 - 8.20	80.4
260	7.02	6.13	4.95 - 7.02	108.0

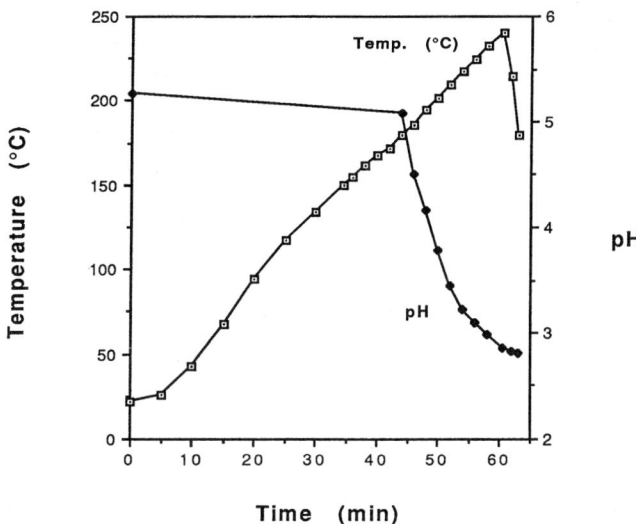

Fig. 3. pH profile for wood sawdust pretreated to 240°C without pH control.

pH to more neutral values, between 7.0 and 8.2, before the start of the pretreatment, Table 2. The target range for the pH was calculated to be between 5.0 and 7.0 and the volume of KOH added to control the pH was not to exceed 10% of the working volume. For the pretreatments to 220 and 240°C, the volume of 2.0 M KOH added was between 80 and 90 mL. A more significant volume of 2.0 M KOH, 108 mL, was needed to keep the pH between 5.0 and 7.0 for the pretreatment to 260°C.

The KOH addition profile was developed from the trend of the pH during pretreatment in the absence of pH control as well as the observation of the instantaneous change in pH caused by the addition of KOH during that particular pretreatment. The KOH addition profile for the pretreatment to 240°C with pH control is shown in Fig. 4. Prior to the pretreatment, 0.3 mL of 2.0 M KOH were added to the pretreatment vessel and changed the pH from approx 5.2 to 8.2. Typically, an addition of 0.3 mL 2.0 M KOH added prior to the pretreatment would raise the pH from 5.0 to approx 7.0 (Table 2, initial pH). As the pretreatment temperature rose to 200°C,

Fig. 4. pH profile for wood sawdust pretreated to 240°C with KOH added to help control the pH.

corresponding to 42 min into the pretreatment, 40 mL 2.0 M KOH were added to bring the pH from 6.5 to 7.5. As the pretreatment temperature rose from 200 to 240°C, another 40 mL 2.0 M KOH were added to keep the pH from dropping below 5.0. The pH profile and volume of KOH added over this temperature are consistent with the release of an acidic component from the sawdust, and may include significant quantities of acetic acid. Overall, the pH remained between 5.0 and 7.5 with the final pH being 5.7 during the hottest parts of the pretreatments, where temperatures ranged from between 170 and 240°C, Fig. 4.

Effects of Aqueous Pretreatment on Polysaccharides in Wood Sawdust Total Solids

The material balances for each pretreatment are shown in Table 3. From the overall material balance, the amount of solid retention was determined by comparing the amount of dry material remaining after the pretreatment to the amount of dry material initially being pretreated. The quantity of solid material remaining after pretreatment ranged from 56 to 84% with the assumption that the solid material not accounted for was solubilized during pretreatment, Table 3 and Fig. 5. The solubilization process may include the partial hydrolysis of the cellulosic and hemicellulosic polysaccharides into soluble oligosaccharides and their corresponding monosaccharides, glucose and xylose.

The addition of KOH helps to control the pH during the pretreatment of wood sawdust as well as reduce the quantity of total solids being solubilized. The range of solid solubilization for the three pretreatments with

Table 3
Material Balances for Each Pretreatment Condition

	220°C	220°C KOH	240°C	240°C KOH	260°C	260°C KOH
Material In:						
Dry	86.3	80.6	88.2	63.8	60.8	63.2
Liquid	1513.7	1519.4	1511.8	1003.2	1006.2	1003.8
KOH	0.00	88.6	0.00	80.5	0.00	107.9
Total In	1600.0	1688.6	1600.0	1147.5	1067.0	1174.9
Material Out:						
Dry	62.2	67.9	58.2	53.7	33.8	44.3
Liquid	1425.9	1559.5	1453.0	1012.8	998.7	899.5
Samples	20.1	20.4	25.1	25.3	30.2	30.5
Losses	30.1	10.2	20.1	20.2	10.1	173.1
Total Out	1538.3	1658.0	1556.4	1112.0	1072.8	1147.4
% Recovery	96.1	98.2	97.3	96.9	101.0	97.7

KOH addition was from 16 to 30%. This is less than the extent of solubilization for the pretreatments carried out in the absence of pH control where the solid solubilization was between 28 and 44%, depending on the temperature (Fig. 5). The greatest solubilization of solid material occurred at 240 and 260°C, both in the absence of pH control, where 34 and 44% of the original material was solubilized, respectively, Table 3 and Fig. 5. For the pretreatments that use KOH to control pH, both the 220 and 240°C pretreatments solubilized only 16% of the original material, whereas the 260°C pretreatment solubilized 30% of the original material, Fig. 5. For the pretreatments with and without the addition of KOH, the quantity of solid material solubilized increases as the final pretreatment temperature increases. This trend is more apparent for the pretreatments in which the

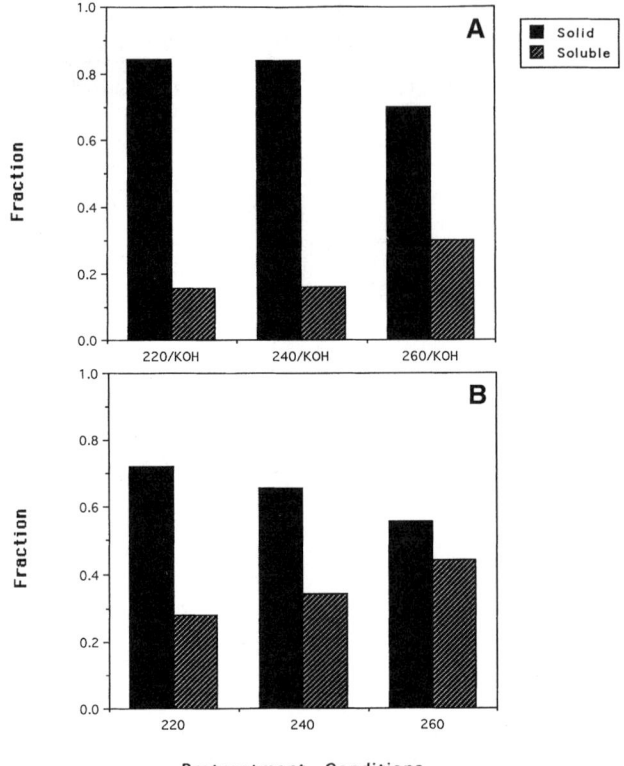

Fig. 5. Fraction of the wood sawdust remaining in the solid phase and that which has been dissolved based on pretreatment conditions; **(A)** with KOH added to control pH; **(B)** without pH control.

pH is not controlled, where the extent of solubilization increases with final pretreatment temperature, Fig. 5.

Cellulose

Significant hydrolysis of cellulose and degradation of the glucose occurs if the pH is below 3.5 and the temperature above 200°C (7–9). Released organic acids may facilitate polymeric hydrolysis and degradation of hemicellulose and cellulose fractions, whereas hydroxyl-mediated degradation of the released saccharides leads to low yields. Therefore a goal of this study was to maintain pH in a range in which the production of organic acids from degradation of monosaccharides are minimized and organic acids are in a salt form that minimizes their activity as acid catalysts and thereby keeps the carbohydrates in an oligomer form at conditions whose goal is to hydrate the cellulose fraction.

Treatment of wood sawdust using water at high temperatures and pressures and the addition of KOH for pH control may help the conversion

Table 4
Composition of Treated and Untreated Solid Wood Sawdust (% Basis)

	220°C	220°C KOH	240°C	240°C KOH	260°C	260°C KOH	Untrtd
Cell.	65.6 (± 1.2)	47.0 (± 1.3)	60.9 (± 1.2)	66.3 (± 1.7)	54.8 (± 1.8)	67.6 (± 2.0)	41.0 (± 0.3)
Xylans & Arabans	0.00	0.00	0.00	0.00	0.00	0.00	19.4 (± 0.1)
Klason Lignin*	29.4 (± 1.6)	25.3 (± 0.1)	32.8 (± 0.5)	24.0 (± 0.3)	38.1 (± 0.6)	19.7 (± 0.7)	22.7 (± 0.3)
Acid Lignin	9.0 (± 0.9)	13.9 (± 0.8)	12.7 (± 0.4)	1.9 (± 0.0)	3.2 (± 0.5)	1.7 (± 0.3)	4.3 (± 0.0)
Ash	0.2 (± 0.0)	2.9 (± 0.1)	0.2 (± 0.0)	3.8 (± 0.0)	0.3 (± 0.2)	5.3 (± 0.1)	0.5 (± 0.0)
Total	104.2	89.1	106.4	96.0	96.4	94.3	87.9

[a] Contains components other than Klason Lignin (possibly some protein).

of cellulose to glucose by concentrating the cellulose found in the treated solid material by removing other components. The pretreatments with KOH addition have the highest percentage of cellulose in the remaining solids. The cellulose content of the remaining solids for the pretreatment to 260°C with KOH addition was approx 68% and for the 240°C pretreatment with KOH the percentage of cellulose in the remaining solids was 66%, Table 4. Pretreatments to 240 and 260°C without KOH addition showed a lower cellulose content in the remaining solid material at 61 and 55%, respectively, Table 4. The pretreatments to 220°C were the exception in this case. The cellulose content for the remaining solids after pretreatment to 220°C with no KOH was approx 66%, whereas the cellulose content for the remaining solids pretreated to 220°C with KOH was 47%. These values compare to a cellulose content of 41% for untreated wood sawdust, Table 4.

The prevention of cellulose solubilization by controlling the pH with KOH may be one way to increase the cellulose content after pretreatment. The most notable indication that the addition of KOH helps prevent cellulose solubilization can be shown by the pretreatments to 260°C. Pretreatment to 260°C without KOH solubilized the most cellulose, approx 26%, whereas the pretreatment to 260°C with KOH addition showed all of the cellulose remained with the solids after pretreatment conditions were removed, Fig. 6.

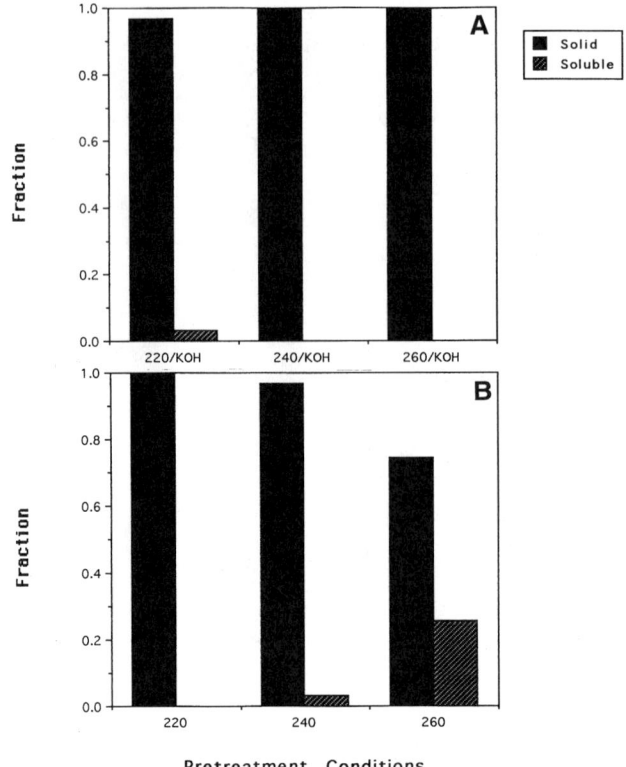

Fig. 6. Fraction of cellulose remaining in the solid and that fraction solubilized based on pretreatment conditions; **(A)** with KOH added to control pH; **(B)** without pH control.

Hemicellulose

Another polysaccharide present in the wood sawdust is hemicellulose, and is composed of xylose, a small amount of arabinose, and acetate. The xylan and araban polysaccharide composes approx 20% of the dry-wood sawdust. The hemicellulosic polymer was completely dissolved from the solid at all the pretreatment conditions. Xylans and arabans were not detected in the remaining solids from any pretreatment, Table 4. Liquid chromatography analysis indicates that the majority of the solubilized hemicellulosic material appeared as oligosaccharides. Xylose was found in quantities that corresponded to less than 10% of the total xylan balance for the 240 and 260°C pretreatments with no pH control. Furfural was only found in trace quantities, less than 0.3 g/L, in the supernatant of the 260°C pretreatment carried out in the absence of pH control. There were no detectable quantities of xylose or furfural in the supernatant for all of the pretreatments where KOH was used to control pH.

CONCLUSIONS

pH control is necessary for aqueous pretreatment of yellow poplar wood sawdust, especially at the high temperatures and pressures used for treating the sawdust. The addition of base to the pretreatment vessel prevents the accumulation of acids that will promote the hydrolysis of the cellulose to soluble oligosaccharides and glucose, and the degradation of the glucose to HMF and levulinic and formic acid. The pH monitoring system described here, when combined with a previously determined, well-characterized, pH profile enables the development of a base addition profile that will keep the pH within a specified range.

ACKNOWLEDGMENTS

The authors wish to acknowledge The National Renewable Energy Laboratories (NREL) for its financial support (contract XAC-4-13511-01) of this work, and thank Bob Torget, Mark Finkelstein, and Tammy Hayward for their helpful review and comments on this work. We thank Kyle Beery and Hong Li for their review of this manuscript and Doris Rau and Joan Goetz for technical assistance in carrying out NREL standard procedures for the analysis of the compositions of untreated and pretreated wood sawdust.

REFERENCES

1. Kirk, T. K. and Farrell, R. L. (1987), *Ann. Rev. Microbiol.* **41,** 465–505.
2. Zhbankov, R. G. (1992), *J. Mol. Structure* **270,** 523–539.
3. Weil, J., Westgate, P. J., Kohlmann, K., and Ladisch, M. R. (1994), *Enzyme Microb. Technol.* **16,** 1002–1004.
4. Weil, J., Sarikaya, A., Rau, D., Goetz, J., Ladisch, C., Brewer, M., Hendrickson, R., and Ladisch, M. R., (1997), *Appl. Biochem. Biotechnol.* **68,** 41–60.
5. Kohlmann, K. L., Sarikaya, A., Westgate, P. J., Weil, J., Velayudhan, A., Hendrickson, R., and Ladisch, M. R. (1995), in *Enzymatic Degradation of Insoluble Polysaccharides.* M. Penner and J. Saddler, eds., American Chemical Society, Washington, DC, ACS Symp. Ser. No. 618, 237–255.
6. Macdonald, D. D., Hettiarachchi, S., Song, H., Makela, K., Emerson, R., and Ben-Haim, M. (1992), *J. Solution Chem.* **21(8),** 849–881.
7. Baugh, K. D. and McCarty, P. L. (1988), *Biotechnol. Bioeng.* **31,** 50–61.
8. Ladisch, M. R. (1989), in *Biomass Handbook,* C. W. Hall and O. Kitani, eds., Gordon and Breach, pp. 434–451.
9. Bienkowski, P. R., Ladisch, M. R. Narayan, R., Tsao, G. T., and Eckert, R. (1987), *Chem. Eng. Comm.* **51,** 179–192.

Session 2
Applied Biological Research

TOM JEFFRIES[1] AND MIKE HIMMEL[2]
[1]FLP Madison, WI and [2]NREL, Golden, CO

The applied biological research session has traditionally emphasized novel approaches to long-standing problems. This year most of the novelty emerged from metabolic and enzyme engineering. Increasingly, genetically engineered enzymes or organisms with altered traits are being integrated into processes. Fermentation of xylose by genetically engineered bacteria and yeast continues to attract attention. The high ethanol yield *Zymomonas mobilis* characteristically attains on glucose was shown to extend to the fermentation of glucose/xylose sugar mixtures in continuous cultures with strains that had been genetically engineered to express a xylose metabolic pathway. This consisted of xylose isomerase, xylulokinase, transketolase, and transaldolase. These studies indicate that there is no inherent technical barrier to further scale-up of continuous fermentations with this organism. Other research in this field has resulted in the development of an engineered *Z. mobilis* strain that will coferment arabinose. Increasing attention is being given to the use of hydrolysates by *Z. mobilis* and other ethanologens. Also, a number of yeasts have been screened and adapted to the fermentation of softwood prehydrolysates. The nature of the adaptation is unknown, but fermentation performance increases with successive batch transfers.

Genetic engineering of xylose fermentations by yeasts is also progressing. In the case of mixed sugar fermentations by recombinant *Saccharomyces,* two of the principal barriers have been the excess production of xylitol, and the diminished consumption of xylose under anaerobic conditions. Both of these factors could be related to redox imbalances arising from cofactor preferences. One of the key enzymes enabling xylose fermentation by *Saccharomyces* is the xylose reductase of *Pichia stipitis*. This enzyme is capable of using either NADH or NADPH for the reduction of xylose to xylitol, but NADPH is strongly favored. The next enzyme in the pathway, xylitol dehydrogenase, uses NADH exclusively, so the excess consumption of NADPH can lead to cofactor can block assimilation. One paper described genetic engineering of xylose reductase to use NADH preferentially. In other research, xylose reductases from *Neurospora crassa* and *Candida guillierondii* were characterized.

To date, the bulk of genetic approaches to improved xylose fermentations have been carried out in *Escherichia coli, Z mobilis* and *Saccharomyces sp.,* but the development of an effective genetic system for P. *stipitis* is beginning to result in improvements there as well. Even though wild type strains of P. *stipitis, Candida shehatne,* and *Pachysolen tannophilus* will ferment xylose to ethanol, their product yields have been limited by an oxygen requirement for growth. The biochemical basis for that requirement was poorly understood, because it cannot be satisfied simply by supplying essential lipids (as in the case of *Saccharomyces*). However, *Saccharomyces* has adapted to anaerobic growth in other ways. One of the most critical, is its ability to produce uracil under anaerobic conditions. This capacity is conferred by a novel form of the enzyme, dihydroorotate dehydrogenase. In most eukaryotic organisms, activity of this enzyme is necessarily tied to respiratory metabolism. In *Saccharomyces*, it can be coupled to the reduction of fumarate. By transforming P. *stipitis* with the *Saccharomyces* gene for this enzyme, anaerobic growth was conferred on this pentose-fermenting yeast.

Pretreatment and saccharification of lignocellulosic substrates remain significant problems in bioconversion. Steam pretreatment is an inexpensive means to prepare low- lignin content materials, such as willow, for enzymatic dissolution. Steam pretreatment releases hemicellulose as oligomers and opens up the wood pore structure. Enzymatic saccharification is rapid on the "amorphous" cellulose, but the bulk of the substrate is found in crystalline structures. Cellulose binding domains (CBD) are an integral part of most cellulases, but since they do not actively participate in the hydrolytic reaction, one can ask what role they play in saccharification. Removal of the CBD by proteolytic cleavage decreased the hydrolytic capacity of the enzyme on polymeric substrates. Presumably, the CBD keeps the catalytic domain in the vicinity of the substrate as it progresses along the chain. Cellulolytic microbes almost always form cellulases in mixtures of beta-glucosidases, endoglucanases, and exoglucanases. The relative portions of these enzymes vary from one organism to another, and it has been unclear as to whether the native mixtures maximize cellulose hydrolysis. Research reported in this section shows that synergism among various enzyme preparations can be increased, but that aside from deficiencies in beta-glucosidase, the native mixtures are nearly optimal for natural (untreated) substrates.

Lignocellulosic substrates for initial commercial use will most likely come from low-value agricultural byproducts. Corn fiber is one of the most interesting because it is present in large quantities in most ethanol and starch processing plants, and it has few competing uses. However, bioconversion of corn fiber faces several challenges. Most particularly, its hemicellulose is highly substituted and resistant to enzymatic degradation. Likewise, pretreatment must be carried out in an exacting manner, because the bulk of the hemicellulosic sugars, xylose and arabinose, are subject to degradation. Current research reported here has identified optimal conditions for pretreatment and several novel enzyme systems for saccharification.

While the bulk of bioconversion research reported in this symposium has focused on ethanol production, many other products are attracting increasing interest. These include bioplastics, enzymes, extractives, oils, xylitol, and oligosaccharides. Transglycosylation products are often problematic during starch saccharification, because they reduce ethanol yields. Cyclodextrins; however, are among the more valuable and versatile transglycosylation products from starch. They form hydrophobic molecular cages into which various small molecules can be entrapped for later diffusion and slow release. They are formed by cyclodextrin glycosyltransferases (CGTase). Cyclodextrins have a market of several thousand metric tons per year and a value of around $5.00/kg. Cyclodextrin glycosyltransferase is produced by various alkalophilic *Bacillus sp.* and research reported in this session identified a CGTase that can convert starch to cyclodextrins in a 21% yield.

Overall, this symposium session remains one of the most important in the field.

Fuel Ethanol Production from Corn Fiber

Current Status and Technical Prospects

BADAL C. SAHA,* BRUCE S. DIEN, AND RODNEY J. BOTHAST

Fermentation Biochemistry Research Unit, National Center for Agricultural Utilization Research, Agricultural Research Service, US Department of Agriculture, Peoria, IL 61604**

ABSTRACT

Corn fiber, which consists of about 20% starch, 14% cellulose, and 35% hemicellulose, has the potential to serve as a low cost feedstock for production of fuel ethanol. Currently, the use of corn fiber to produce fuel ethanol faces significant technical and economic challenges. Its success depends largely on the development of environmentally friendly pretreatment procedures, highly effective enzyme systems for conversion of pretreated corn fiber to fermentable sugars, and efficient microorganisms to convert multiple sugars to ethanol. Several promising pretreatment and enzymatic processes for conversion of corn fiber cellulose, hemicellulose, and remaining starch to fermentable sugars were evaluated. These hydrolyzates were then examined for ethanol production in bioreactors, using genetically modified bacteria and yeast. Several novel enzymes were also developed for use in pretreated corn fiber saccharification.

Index Entries: Fuel ethanol; corn fiber; pretreatment; enzymatic saccharification; fermentation.

INTRODUCTION

In the United States, over 1.3 billion gallons of ethanol are produced annually, primarily from corn starch. Various agricultural residues, such as corn stover, straw, and bagasse, can also serve as low-value and abundant feedstocks for production of fuel ethanol. In general, these contain about 35–50% cellulose, 20–35% hemicellulose, and 5–25% lignin. Corn fiber

---

* Author to whom all correspondence and reprint requests should be addressed.
** Names are necessary to report factually on available data; however, the USDA neither guarantees nor warrants the standard of the product, and the use of the name by USDA implies no approval of the product to the exclusion of others that may also be suitable.

represents a renewable resource that is available in sufficient quantities from the corn wet-milling industries to serve as a low-cost feedstock for production of fuel ethanol. Currently, the utilization of corn fiber to produce fuel ethanol presents significant technical and economic challenges, and its success depends largely on the development of environmentally friendly pretreatment procedures, highly effective enzyme systems for conversion of pretreated corn-fiber substrate to fermentable sugars, and efficient microorganisms to ferment mixed sugars to ethanol. In this paper, a comprehensive progress report on this endeavor is presented.

Corn Fiber as Feedstock

In the United States, both wet- and dry-milling processes are currently used to produce ethanol from corn. Wet milling accounts for about 60% of total ethanol production. In a typical wet-milling process, cleaned corn is soaked in circulating water, slightly acidified with 0.1–0.2% SO_2, at 51–54°C for 24–48 h to soften the kernel, loosen the germ, and hull and swell the endosperm. Corn fiber is a mixture of corn hulls and residual starch not extracted during the milling process, making up 11% of the dry wt of the corn kernel (1). Presently, corn fiber is marketed in corn-gluten feed. A typical composition of corn fiber is given in Table 1. It contains about 70% fermentable sugars, of which approx 20% comes from starch. Typically, 32 lb of starch is obtained from a bu of corn (56 lb), which yields about 2.5 gal of ethanol in industrial practice. From the same bu of corn, about 4.5 lb of corn fiber is obtained, which produces about 3.15 lb of fermentable sugars. These fermentable sugars will theoretically yield about 0.3 gal of ethanol. In practice, it may yield about 0.24 gal of ethanol per bu of corn. The low cost and high carbohydrate content of corn fiber makes it an attractive feedstock for conversion to fuel ethanol. Gulati et al. (2) estimated that a wet-milling facility that currently produces 100 million gal of ethanol per yr from starch could generate an additional $4–8 million of annual income, if the fiber components were processed into ethanol. The conversion of corn fiber to ethanol involves four basic steps: pretreatment, enzymatic saccharification, fermentation of hydrolyzate, and ethanol recovery. The enzymatic saccharification and fermentation can also be performed together in a process known as simultaneous saccharification and fermentation (SSF).

Pretreatment and Enzymatic Saccharification of Corn Fiber

Native corn-fiber cellulose and hemicellulose are resistant to enzymatic hydrolysis. The pretreatment of corn fiber is thus crucial before enzymatic hydrolysis. Various pretreatment options can be used to solubilize, hydrolyze, and separate starch, cellulose, hemicellulose, and lignin components of corn fiber. A number of pretreatment procedures, such as dilute acid, alkali, ammonia, alkaline peroxide, and liquid hot-water treatments,

Table 1
Composition of Corn Fiber

Component[a,b]	%[c]
Carbohydrate	
Crude fiber	14.1 (0.97)
Starch	19.68 (0.91)
Glucose	37.19 (1.86)
Xylose	17.58 (1.76)
Arabinose	11.25 (1.46)
Galactose	3.59 (0.336)
Total sugars	69.6 (5.03)
Other components	
Protein	10.98 (0.52)
Klason lignin	7.78 (0.74)
Acetyl groups	1.71 (0.13)
Ash	0.6 (0.05)
Crude fat	2.53 (0.31)
Unknown	6.79

[a] Average concentration of each component is expressed in wt% of total dry solids.
[b] The sugar content is expressed on anhydrous basis.
[c] Numbers in parentheses refer to standard deviations.
From Ref. 1.

have been tried for pretreatment of corn fiber *(1,3–5)*. Each of them has distinct advantages and disadvantages. Osborn and Chen *(6)* reported that the starch fraction of the corn hull was completely hydrolyzed by glucoamylase after the corn hull was heated with steam for 5 min, and the destarched hemicellulose fraction of the corn hull was readily hydrolyzed with dilute sulfuric acid at 135°C. Grohmann and Bothast *(1)* investigated saccharification of corn fiber by treating first with dilute sulfuric acid (100–160°C), and then with enzymes (cellulase and glucoamylase, 45°C), after partial neutralization. The sequential treatment achieved a high (~85%) conversion of corn-fiber polysaccharides to monomeric sugars. The formation of compounds inhibitory to fermentative microorganisms was evident for all pretreatments tested at 140 and 160°C. Moniruzzaman et al. *(4)* optimized the conditions necessary for pretreatment of corn fiber with high moisture content (150% moisture, dry wt basis) by ammonia fiber explosion (AFEX). The best results were obtained at 90°C with an ammonia and dry corn fiber ratio of 1:1, and a residence time of 30 min (reactor pressure ~200 psig). More than 80% of the theoretical sugar yield was obtained during enzymatic hydrolysis of the AFEX-pretreated corn fiber with a combined mixture of commercial enzyme preparations (α-amylase, glucoamylase, cellulase, hemicellulase, and β-glucosidase).

However, xylooligosaccharides represented about 30–40% of xylan-degradation products, and very little xylose was produced (7). Recently, Leathers and Gupta (5) reported that pretreatment of corn fiber with alkaline peroxide nearly doubled the susceptibility of corn-fiber hemicellulose to enzymatic digestion.

The hydrolysis of cellulose to glucose, which requires the synergistic actions of endoglucanase, cellobiohydrolase, and β-glucosidase, is very slow because of product and substrate inhibitions (8). In addition, cellulases are expensive because of high production costs, and are difficult to reuse. Both endoglucanase and cellobiohydrolase are inhibited by cellobiose. Although β-glucosidase hydrolyzes cellobiose to glucose, most β-glucosidases are inhibited by glucose, as well as by cellobiose (9). Freer (10,11) purified and characterized a unique β-glucosidase from *Candida wickerhamii*, and studied the kinetics of the enzyme in detail. The enzyme demonstrated optimal activity between pH 4.0 and 5.0, and was stable below 40°C. Cellodextrins were good substrates for the enzyme. No inhibition of *p*-nitrophenyl β-D-glucoside hydrolysis by the enzyme was detected with 50 mM glucose. Skory and Freer (12) have isolated and cloned the gene encoding this glucose-insensitive β-glucosidase in *Escherichia coli*. The introduction of this gene in *Saccharomyces cerevisiae* has the potential to yield a cellodextrins-fermenting yeast (13). In a continuing search for a thermophilic β-glucosidase insensitive to glucose inhibition, Saha and Bothast (14) screened a variety of yeasts from the Agricultural Research Service (ARS) culture collection. Enzymes from 15 yeast strains showed very high glucose tolerance (<50% inhibition at 30% glucose). Optimal temperature and pH for these 15 β-glucosidase preparations varied from 30 to 65°C and pH 4.5 to 6.5. The β-glucosidase from *Debaryomyces yamadae* NRRL Y-11714 showed the highest optimal temperature at 65°C, followed by enzymes produced by *Candida chilensis* NRRL Y-17141 and *Kluyveromyces marxianus* NRRL Y-1195 at 60°C. The optimal pHs of these three enzyme preparations were 6.5, 6.0, and 6.5, respectively. The β-glucosidase from all these strains hydrolyzed cellobiose. The β-glucosidase from *Candida peltata* NRRL Y-6888 was purified and characterized (15). Enzyme production was not repressed by glucose, making glucose a good carbon source for production of the enzyme. The optimal pH and temperature for the action of the purified enzyme were 5.0 and 50°C, respectively. The enzyme hydrolyzed cellobiose and cellooligosaccharides very well, and was highly tolerant to glucose, with a K_i of 1.4 M (252 mg/mL). The β-glucosidase was not inhibited by cellobiose (15%). Cellobiose (10%) was almost completely hydrolyzed to glucose by the purified β-glucosidase, in both the absence and presence of 6% glucose (Fig. 1). The β-glucosidase from a color-variant strain of *Aureobasidium pullulans* was purified, characterized, and found to be highly thermostable and optimally active at 75°C (16). The half-life of the crude enzyme was 72 h at 75°C and 24 h at 80°C. The enzyme hydrolyzed both cellobiose and cellooligosaccharides very well, but was strongly inhibited by glucose with a K_i of 5.65 mM.

Progress in Fuel Ethanol from Corn Fiber

Fig. 1. Time-course of cellobiose (10%, w/v) hydrolysis by purified β-glucosidase (1.5 U/mL) from *C. peltata* Y-6888 in the absence and presence of glucose (6%, w/v) at pH 5.0 and 50°C *(15)*. Symbols: (○), cellobiose only; (□), glucose formed from cellobiose, (△), cellobiose with glucose; (▽), glucose formed from cellobiose in the presence of glucose (from Ref. 15).

The total hydrolysis of xylan requires endoxylanase, β-xylosidase, and several accessory enzymes, such as α-L-arabinofuranosidase, α-glucuronidase, acetyl xylan esterase, feruloyl esterase, and *p*-coumaroyl esterase, which are necessary for hydrolyzing various substituted xylans. Many xylanases do not cleave glycosidic bonds between xylose units that are substituted. Therefore, the side chains must be cleaved before the xylan backbone can be completely hydrolyzed *(17)*. On the other hand, several accessory enzymes only remove side chains from xylooligosaccharides. These enzymes require xylanases to partially hydrolyze hemicellulose before side chains can be cleaved *(18)*. The corn-fiber hemicellulose was resistant to hydrolysis to fermentable sugars by commercial hemicellulase preparations *(7)*. Structural analysis of the corn-fiber xylan suggests that over 70% of the xylose backbone residues have one or more arabinose, 4-O-methylglucuronic acid, or other side chains *(19)*. As a result, there are few regions in corn-fiber xylan where several contiguous xylose residues are unsubstituted, which makes enzymatic hydrolysis of xylan difficult.

Fermentation of Corn Fiber Hydrolyzates

Most of the xylose- and glucose-fermenting yeasts, such as *Pichia stipitis*, *Candida shehatae*, and *Pachysolen tannophilus*, do not have the ability to produce ethanol from L-arabinose *(20–22)*. Since corn fiber contains 12% L-arabinose, Dien et al. *(23)* screened 116 different L-arabinose-utilizing yeasts for production of ethanol from L-arabinose, and found the following species able to ferment the sugar: *Ambrosiozyma monospora* (NRRL Y-1484), *Candida* sp. (NRRL YB-2248), *Candida auringiensis* (NRRL Y-11848), and *Candida succiphila* (NRRL Y-11998). These yeasts produced low levels (up to 4.1 g/L) of ethanol, and are potential candidates for mutational or other genetic enhancements for increased ethanol production. Most yeasts are inefficient in the regeneration of the co-factor required for the conversion of L-arabinose to ethanol. Saha and Bothast *(24)* studied the production of L-arabitol from L-arabinose by two superior L-arabitol producers *(Candida entomaea* NRRL Y-7785 and *Pichia guilliermondii* NRRL Y-2075). These two strains produced L-arabitol (0.70 g/g) from L-arabinose (50 g/L) at 34°C and pH 5.0 and 4.0, respectively. Both yeasts produced ethanol (0.32–0.33 g/g) from glucose (50 g/L), and only xylitol (0.43–0.51 g/g) from xylose (50 g/L). The yeasts co-utilized xylose (6.2–6.5 g/L) and L-arabinose (4.9–5.0 g/L) from a corn-fiber acid hydrolyzate simultaneously, and produced xylitol (0.10 g/g xylose) and L-arabinose (0.53–0.54 g/g L-arabinose). *Klebsiella oxytoca* strain P2 is a recombinant organism in which the pyruvate decarboxylase *(pdc)* and alcohol dehydrogenase *(adh B)* genes from *Zymomonas mobilis* have been integrated into the *pfl* gene within the chromosome of *K. oxytoca* M5A1, and expressed at high levels *(25)*. This strain diverts the metabolism of pyruvate to ethanol, and produces a lesser quantity of acetic acid. Bothast et al. *(26)* evaluated this organism for its ability to ferment L-arabinose, xylose, and glucose, alone and in mixtures, in pH-controlled batch fermentations. The recombinant organism produced 0.34–0.43 g ethanol/g sugar at pH 6.0 and 30°C on 8% sugar substrate, and utilized L-arabinose very well. With the mixture of sugars, *K. oxytoca* strain P2 showed a preference for glucose. Xylose utilization was slow, and approx 47 and 29% of supplied xylose were left unutilized in mixtures A (glucose:xylose:arabinose, 1:1:1) and B (glucose:xylose:arabinose, 1:2:1), respectively, even after 114 h fermentation. Sugar utilization was glucose > arabinose > xylose, and ethanol production was xylose > glucose > arabinose.

Grohmann and Bothast *(1)* investigated the fermentation of dilute acid hydrolyzates of 15% corn fiber slurries with recombinant *E. coli* K011. *E. coli* K011 is a recombinant derivative of *E. coli* B, in which the *Z. mobilis* genes encoding *pdc* and *adh B* genes have been integrated *(27)*. The fermentations were conducted in 2 L magnetically stirred bioreactors (Multigen™ 2000 fermenter, New Brunswick, Edison, NJ) or covered beakers (Fleaker,™ Corning, NY) equipped with pH, temperature, and agitation controls *(28)*. Final ethanol concentrations exceeded 3% w/v in 3 d, and yields ranged from 19–62% of theoretical. As mentioned earlier, the formation of inhibitory compounds became readily apparent for all pretreatments tested at 140 and 160°C. Asghari et al. *(29)* reported that hemicellulose hydrolyzates of the agricultural residues bagasse, corn stover, and corn hulls plus fibers were readily fermented to ethanol by recombinant *E. coli*

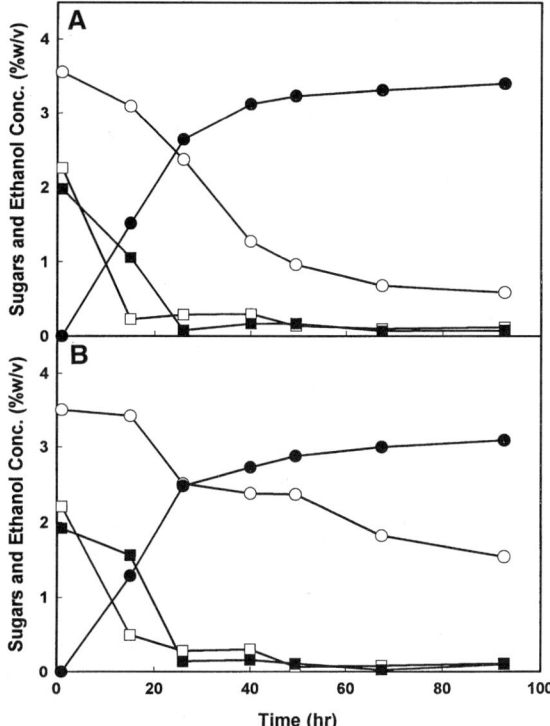

Fig. 2. Fermentation of corn fiber acid hydrolysate with recombinant *e. coli* **(A)** strain K011 and **(B)** strain SL40. Symbols: □, glucose; ■, L-arabinose; ○, xylose and galactose; ●, ethanol. Xylose and galactose comigrated during HPLC analysis (from Ref. 3).

strain K011. Corn-steep liquor and crude yeast autolyzate served as excellent nutrients. Fermentations were essentially complete within 48 h, often achieving over 40 g ethanol/L.

Dien et al. *(3)* treated corn fiber with dilute sulfuric acid (1%, v/v H_2SO_4) at 121°C for 1 h. Following hydrolysis, insoluble materials were removed by filtration, and the hydrolyzates were neutralized to pH 6.5 with $Ca(OH)_2$. Insoluble $CaSO_4$ was then removed by filtration. The total reducing sugar present in a typical hydrolyzate preparation was 89.8 g/L with glucose (26.1 g/L), xylose and galactose (40.9 g/L), and L-arabinose (22.8 g/L). These hydrolyzate preparations were then fermented to ethanol using recombinant *E. coli* strains K011 and SL40. *E. coli* SL 40 is a phosphomycin-resistant mutant of strain K011 that has been reported to produce 60 g/L ethanol from 120 g/L xylose in 60 h, 20% more than K011 under identical conditions *(30)*. Ethanol yields were 0.38–0.41 g/g of sugar consumed, and fermentations were complete within 60 h. Both strains fermented glucose and L-arabinose well, but the fermentation of xylose was slow and incomplete (Fig. 2). Both strains produced minor amounts of organic acids such as lactic acid (2.1–2.2 g/L), acetic acid (0.5–0.8 g/L),

Table 2
Fermentation of AFEX-Pretreated Corn Fiber Enzymatic Hydrolyzate (EH) and Simulated Sugar Mixtures (SM) by Recombinant Bacteria[a]

Strain	Substrate	Cell mass (g/L)	Base consumed[b] (mmol/L)	Maximum ethanol[c] (g/L)	Ethanol yield[d] (g/g)
E. coli strain K011	EH	3.2	65	27.1	0.47
	SM	3.2	69	27.1	0.47
E. coli strain SL40	EH	2.9	61	26.6	0.46
	SM	3.2	69	27.2	0.47
K. oxytoca strain P2	EH	2.1	57	20.0	0.35
	SM	2.9	62	20.3	0.35

[a] Values reported are from duplicate experiments.
[b] Base (KOH) refers to that added automatically to maintain pH at 6.0 during fermentation.
[c] Ethanol yields are corrected for dilution by base addition.
[d] Ethanol yields in g/g of substrate available for fermentation.
From Ref. 31.

and succinic acid (0.4–0.5 g/L). The amount of base (2 M KOH) added to the cultures to maintain at pH 6.5 was 163 mmol/L in the case of strain K011 and 120 mmol/L in the case of strain SL40. Overall, strain K011 tended to use sugars better than strain SL40. The presence of glucose in the fermentation medium inhibited xylose utilization by E. coli strain K011 (28).

Moniruzzaman et al. (31) studied ethanol production from AFEX-pretreated corn fiber by recombinant E. coli strains K011 and SL40, and by K. oxytoca strain P2, under pH-controlled conditions. Enzymatic digestion of AFEX-pretreated corn fiber produced a hydrolyzate containing (g/L): glucose, 30.2; xylose, 3.1; L-arabinose, 4.0; and galactose, 2.0. The hydrolyzate was supplemented with additional xylose and arabinose, so that theoretical concentrations of xylose (15.2 g/L) and L-arabinose (10.5 g/L) were realized. Both E. coli strains (K011 and SL40) efficiently utilized most of the sugars contained in the hydrolyzate and produced a maximum of 27.1 and 26.6 g/L ethanol, respectively, equivalent to 92 and 90% of the theoretical yield. Very little difference was observed in cell growth and ethanol production between the fermentation of the enzymatic hydrolyzate and mixtures of pure sugars used to simulate the hydrolyzate (Table 2). These results indicate good compatibility of AFEX-pretreatment, and subsequent fermentation with recombinant bacteria.

Moniruzzaman et al. (32) also studied ethanol production from AFEX-pretreated corn fiber by recombinant Saccharomyces strain 1400 (pLNH32). Saccharomyces strain 1400 (pLNH32) was genetically engineered to ferment xylose by expressing genes encoding a xylose reductase, a xylitol dehydrogenase, and a xylulose kinase (33). The fermentation experiment with the recombinant yeast was performed in a pH 5.0-controlled flask under anaerobic conditions with an inoculum level of 2.0 g/L (dry wt). The recombinant yeast fermented the AFEX-pretreated corn-fiber enzymatic hydrolyzate containing glucose (33.5 g/L), xylose (7.5 g/L), arabinose (5.0 g/L), and galactose (1.0 g/L), and produced ethanol.

Although genetically engineered bacteria and yeast hold tremendous potential for the fermentative conversion of multiple substrates to ethanol, questions still remain concerning the stability or hardiness of these organisms and their ability to perform in a large-scale industrial process. Ingram et al. (34) were the first to report a metabolically engineered *E. coli* for high alcohol production. The resultant recombinant strain produced more than 4% w/v ethanol from glucose in media containing ampicillin, with positive selection pressure for the plasmid. A considerably more stable strain was developed by Ohta et al. (27), by integrating the PET operon (gene cluster producing ethanol) and chloramphenicol (cm) resistance gene into the *E. coli* chromosome. The resultant *E. coli* KO11 strain did not require cm in the growth media for retention of the PET operon, but, in the absence of cm, ethanol production was lower, presumably because of reduced PET gene copy number. When mutants were selected for resistance to high levels (600 µg/mL) of cm, high ethanol production was restored. Hespell et al. (35) have undertaken an alternative approach to eliminate the requirement for antibiotics. A lactate dehydrogenase (*ldh*)–pyruvate formate lyase (*pfl*) double mutant of *E. coli* was used as the cloning host (36). Although capable of aerobic growth, this mutant strain FMJ39 is incapable of anaerobic growth, because of its inability to regenerate oxidized pyridine nucleotides by reduction of pyruvate to lactate. Clones having recombinant plasmids containing a *ldh* gene can be isolated by complementing for anaerobic growth by this strain. It was reasoned that an alternative complementation for anaerobic growth would be the PET-operon-containing plasmids, because expression of the *pdc* and *adh* genes would convert pyruvate to ethanol and regenerate oxidized pyridine nucleotides. If so, the resultant strains should be quite anaerobically stable for ethanol production. The strains should also not require antibiotics (ampicillin or tetracycline) in the growth media to maintain positive selective pressure for cells containing the PET operon plasmid, because loss of the plasmid would be a conditionally lethal event under anaerobic growth conditions. *E. coli* strains FBR1 and FBR2 were created by transforming *E. coli* FMJ39 with the PET operon plasmids pLO1295 and pLO1297, respectively. Both strains were capable of anaerobic growth and displayed no apparent PET plasmid losses after 60 generations in serially transferred (9×) anaerobic batch cultures. In contrast, similar aerobic cultures rapidly lost the PET operon plasmids. In high-cell-density batch fermentations, up to 4.4% w/v ethanol was produced from 10% glucose. In anaerobic, glucose-limited continuous culture, one strain grown for 20 d (51 generations, 23 with tetracycline, and then 28 after tetracycline was removed) showed no loss of antibiotic resistance. Anaerobic, serially transferred batch cultures and high-density fermentations were inoculated with cells taken at 57 generations from the previous continuous culture. Both cultures continued high ethanol production in the absence of tetracycline. The genetic stability conferred by selective pressure for PET-containing cells, without requirement for antibiotics, suggests potential commercial suitability for these *E. coli* strains.

CONCLUSION

It is now possible to convert corn fiber hydrolyzates containing mixed sugars (glucose, xylose, L-arabinose, and galactose) to ethanol by using recombinant microorganisms. However, many fundamental problems need to be overcome before this becomes a commercial reality. They include:

1. High cost of cellulase enzymes.
2. Ineffectiveness of commercial hemicellulases to degrade corn fiber xylans.
3. Production of inhibitory substances to microbial fermentation during dilute acid pretreatment.
4. Stability and ethanol tolerance of the recombinant organisms.
5. Recovery of dilute ethanol from fermentation broth.
6. Disposal of waste containing recombinant organisms.

REFERENCES

1. Grohmann, K. and Bothast, R. J. (1997), *Process Biochem.* **32**, 405–415.
2. Gulati, M., Kohlmann, K., Ladisch, M. R., Hespell, R. B., and Bothast, R. J. (1997), *Bioresource Technol.* **56**, 253–264.
3. Dien, B. S., Hespell, R. B., Ingram, L. O., and Bothast, R. J. (1998), *World J. Microbiol. Biotechnol.* in press.
4. Moniruzzaman, M., Dale, B. E., Hespell, R. B., and Bothast, R. J. (1997), *Appl. Biochem. Biotechnol.* **67**, 113–126.
5. Leather, T. D. and Gupta, S. C. (1996), *Appl. Biochem. Biotechnol.* **59**, 337–347.
6. Osborn, D. and Chen, L. F. (1984), *Starch/Starke* **36**, 393–395.
7. Hespell, R. B., O'Bryan, P. J., Moniruzzaman, M., and Bothast, R. J. (1997), *Appl. Biochem. Biotechnol.* **62**, 87–97.
8. Saha, B. C. and Bothast, R. J. (1997), in *Fuels and Chemicals from Biomass*, Saha, B. C. and Woodward, J., eds., American Chemical Society, Washington, DC, pp. 46–56.
9. Saha, B. C., Freer, S. N., and Bothast, R. J. (1995), in *Enzymatic Degradation of Insoluble Polysaccharides*, Saddler, J. N. and Penner, M. H., eds., American Chemical Society, Washington, DC, pp. 197–206.
10. Freer, S. N. (1985), *Arch. Biochem. Biophys.* **243**, 515–522.
11. Freer, S. N. (1993), *J. Biol. Chem.* **268**, 9337–9342.
12. Skory, C. D. and Freer, S. N. (1995), *Appl. Environ. Microbiol.* **61**, 518–525.
13. Skory, C. D., Freer, S. N., and Bothast, R. J. (1996), *Curr. Genet.* **30**, 417–422.

14. Saha, B. C. and Bothast, R. J. (1996), *Biotechnol. Lett.* **18**, 155–158.
15. Saha, B. C. and Bothast, R. J. (1996), *Appl. Environ. Microbiol.* **62**, 3165–3170.
16. Saha, B. C, Freer, S. N., and Bothast, R. J. (1994), *Appl. Environ. Microbiol.* **60**, 3774–3780.
17. Lee, S. F. and Forsberg, C. W. (1987), *Can J. Microbiol.* **33**, 1011–1016.
18. Poutanen, K., Tenkanen, M., Korte, H., and Puls, J. (1991), in *Enzymes in Biomass Conversion*, Leathers, G. F. and Himmel, M. E., eds., American Chemical Society, Washington, DC, pp. 426–436.
19. Montgomery, R. and Smith, F. (1957), *J. Amer. Chem. Soc.* **79**, 695–697.
20. Gong, C.-S., Claypool, T. A., McCraken, L. D., Maun, C. M., Ueng, P. P., and Tsao, G. T. (1983), *Biotechnol. Bioeng.* **25**, 85–102.
21. Slininger, P. J., Bolen, P. L., and Kurtzman, C. P. (1987), *Enzyme Microbiol. Technol.* **9**, 5–15.
22. McMillan, J. D. and Boynton, B. L. (1994), *Appl. Biochem. Biotechnol.* **45/46**, 569–584.
23. Dien, B. S., Kurtzman, C. P., Saha, B. C., and Bothast, R. J. (1996), *Appl. Biochem. Biotechnol.* **57/58**, 233–242.
24. Saha, B. C. and Bothast, R. J. (1996), *Appl. Microbiol. Biotechnol.* **45**, 299–306.
25. Ohta, K., Beall, D. S., Meija, J. P., Shanmugam, K. T., and Ingram, L. O. (1991), *Appl. Environ. Microbiol.* **57**, 2810–2815.
26. Bothast, R. J., Saha, B. C., Flosenzier, A. V., and Ingram, L. O. (1994), *Biotechnol. Lett.* **16**, 401–406.
27. Ohta, K., Beall, D. S., Meija, J. P., Shanmugam, K. T., and Ingram, L. O. (1991), *Appl. Environ. Microbiol.* **57**, 893–900.
28. Beall, D. S., Ohta, K., and Ingram, L. O. (1991), *Biotechnol. Bioeng.* **38**, 296–303.
29. Asghari, A., Bothast, R. J., Duran, J. B., and Ingram, L. O. (1996), *J. Ind. Microbiol.* **16**, 42–47.
30. Lindsay, S. E., Bothast, R. J., and Ingram, L. O. (1995), *Appl. Microbiol. Biotechnol.* **43**, 70–75.
31. Moniruzzaman, M., Dien, B. S., Ferrer, B., Hespell, R. B., Dale, B. E., Ingram, L. O., and Bothast, R. J. (1996), *Biotechnol. Lett.* **18**, 985–990.
32. Moniruzzaman, M., Dien, B. S., Skory, C. D., Chen, Z. D., Hespell, R. B., Ho, N. W. Y., Dale, B. E., and Bothast, R. J. (1997), *World J. Microbiol. Biotechnol.* **13**, 341–346.
33. Ho, N. W. Y. and Tsao, G. T. (1995), *PCT Patent* No. WO95/13362.
34. Ingram, L. O., Conway, T., Clark, D. P., Sewell, G. W., and Preston, J. F. (1987), *Appl. Environ. Microbiol.* **54**, 2420–2425.
35. Hespell, R. B., Wyckoff, H., Dien, B. S., and Bothast, R. J. (1996), *Appl. Environ. Microbiol.* **62**, 4594–4597.
36. Mat-Jan, F., Alam, K. Y., and Clark, D. P. (1989), *J. Bacteriol.* **171**, 342–348.

Xylose Reductase Production by *Candida guilliermondii*

S. M. A. Rosa,[1] M. G. A. Felipe,[2] S. S. Silva,[2] and M. Vitolo*[1]

[1]Depto. Tecnologia Bioquímico-Farmacêutica, FCF, University of São Paulo.
P.B. 66083, 05389-970, São Paulo, SP, Brasil. FAX: 0055-11-8156386;
[2]Faculdade de Engenharia Química de Lorena, Depto. Biotecnologia,
12600-000, Lorena, SP, Brasil

ABSTRACT

The effect of pH, time of fermentation, and xylose and glucose concentration on xylitol production, cell growth, xylose reductase (XR), and xylitol dehydrogenase (XD) activities of *Candida guilliermondii* FTI 20037 were determined. For attaining XR and XD activities of 129–2190 U/mg of protein and 24–917 U/mg of protein, respectively, the cited parameters could vary as follows: initial pH: 3.0–5.0; xylose: 15–60 g/L; glucose: 0–5 g/L; and fermentation time: 12–24 h. Moreover, the high XR and XD activities occurred when the xylitol production by the yeast was less than 19.0 g/L.

Index Entries: Xylose reductase; *Candida guilliermondii*; xylitol dehydrogenase.

INTRODUCTION

The xylose reductase (EC.1.1.1.21) (XR), an enzyme found mainly in yeasts, catalyzes the conversion of xylose into xylitol, a product with sweetening and anticarious properties *(1)*.

The xylose–xylitol bioconversion is an alternative of economic interest to the catalytic hydrogenation of pure xylose, which is currently obtained from wood hydrolysates. Two reasons can explain this. First, some yeasts convert xylose into xylitol at high yield, either in sugarcane bagasse or in wood hydrolysates, without previous purification of xylose *(2)*. Second, the bagasse is a plentiful residue available from ethanol distilleries, so it is by far cheaper than wood. Nevertheless, even the fermentative process could lead to an unsuitable xylitol production, because of perturbations on the microbial metabolism caused by factors such as oxygen limitation

---

*Author to whom all correspondence and reprint requests should be addressed.

(3), pH, and temperature *(4)*; presence of inhibitors (furfural and acetic acid, among others); and/or sugars other than xylose *(5)*. Thus, carrying out the xylose–xylitol conversion directly with the XR, those problems would be overcome. In this case, however, the enzyme must be readily available and inexpensive.

Of course, the main sources of XR would probably be the xylose-fermenting yeasts belonging to the genera *Pachysolen*, *Pichia*, *Candida*, *Hansenula*, and *Debaryomyces (6)*.

Although XR production and characterization are well-documented for the species *Pichia stipitis (7)*, *Pachysolen tannophilus (8)*, and *Candida shehatae (9)*, little information is available for the XR of *Candida guilliermondii*, despite the fact that this yeast is a good xylitol producer *(10)*. Indeed, the strain *C. guilliermondii* FTI 20037 was already fully described and adapted to the sugar cane and wood hemicellulosic hydrolysates, which are useful raw materials for bioconversions *(2,10,11,12)*.

This paper deals with the effect of initial pH, fermentation time, and xylose and glucose concentrations on the xylose reductase production by *C. guilliermondii* FTI 20037. Conditions under which low xylitol dehydrogenase activity occurred were also considered.

METHODS

Microorganism and Inoculum Preparation

C. guilliermondii FTI 20037, described by Barbosa et al. *(10)*, was used in all experiments. The stock culture was maintained at 4°C on malt-extract agar slants. A loopful of the stock culture was transferred into a 125-mL Erlenmeyer flask containing 50 mL of the following medium: xylose, 30.0 g/L; rice bran, 20.0 g/L; $(NH_4)_2SO_4$, 2.0 g/L; $CaCl_2 \cdot 2H_2O$, 0.1 g/L. The initial pH was adjusted to 5.0. The flask was incubated in a rotary shaker at 200 rpm and 30°C for 24 h. After that, the cells were separated by centrifugation (2000 g; 10 min), rinsed twice with distilled water, and the cell cake was resuspended in an adequate volume of distilled water to attain a final concentration of 5 g dry cell/L.

Growth Conditions and Cell Disruption

Five mL of the cell suspension was introduced into a 125-mL Erlenmeyer flask containing 45 mL of the medium: rice bran, 20 g/L; $(NH_4)_2SO_4$, 2.0 g/L; $CaCl_2 \cdot 2H_2O$, 0.1 g/L; xylose (15, 30, 60, 90, 120, or 240 g/L) and glucose (0, 5, 10, 20 or 30 g/L). The initial pH was adjusted to 3.0, 5.0, or 6.5. The flasks were left at 30°C for 12 h, 24 h and 36 h in a rotary shaker (200 rpm). At each time three flasks were drawn, and the suspensions centrifuged (5000g; 10 min). Using high-pressure liquid chromatography (HPLC) the xylose, xylitol, and glucose concentrations were measured in the supernatant. The cell cake was washed twice with distilled water, resuspended in 0.1 M phosphate buffer (pH 7.2), and submitted to disruption in a vortex, in the presence of 0.5-mm glass beads for 6 min. Then the suspension was left resting in an ice bath for 3 min. The vortexing/resting cycle was repeated six times, leading to a cell disruption of about 70%. This determination was accomplished through microscope observation, and by counting into a Neubauer chamber (area = 1/400 mm^2; height = 0.100 mm) the number of intact cells before and after vortexing/resting. The cell-free extract was employed for xylose reductase and xylitol dehydrogenase activities measurement.

Analytical Methods

Xylose, xylitol, and glucose were measured by HPLC (Shimadzu [Kyoto, Japan] LC-10AD), using a refractive index (RI) detector and a Bio-Rad (Hercules, CA) HPX87H (300 × 7.8 mm) column under the following conditions: 0.01 N H$_2$SO$_4$ as the eluent; 0.6 mL/min flow rate; column temperature 45°C; detector attenuation: 16X; sample volume 20 μL. Cell concentration, expressed as g dry matter/L, was measured by filtering 5 mL of cell suspension through a Millipore (Bedford, MA) membrane (0.22-μm pore diameter), followed by drying in an oven at 105°C for 2 h.

The measurement of xylose reductase activity (XR) was carried out in the following reaction medium: 5mM xylose, 0.1 M phosphate buffer (pH 7.2), 3.0 μM NADPH, and 0.1 M β-mercaptoethanol. For xylitol dehydrogenase (XD), the reaction medium employed was: 5mM xylitol, 0.5 M Tris buffer (pH 8.6), 2.5 μM NADP, and 0.1 M β-mercaptoethanol. The reactions were initiated by the addition of xylose or xylitol, and the variation of absorbance was spectrometrically monitored (340 nm) against the reaction time at 25°C.

Protein in the cell-free extract was determined by the dye-binding technique, using Coomassie blue reagent *(13)* and bovine serum albumin (Sigma, St. Louis, MO >99% purity) as the standard protein. One XR or XD unit (U) was defined as the amount of enzyme catalyzing, respectively, the formation of 1 μmol NADP/min or 1 μmol NADPH/min. The specific activity was expressed as U/mg of protein.

RESULTS AND DISCUSSION

Table 1 shows that 66.4 g/L of xylose was completely consumed after 48 h of fermentation, the xylitol concentration was equal to 43 g/L, which led to a volumetric productivity (VP) of 0.90 g/Lh. When this VP is compared to that attained by Meyrial et al. *(14)* for *C. guilliermondii* NRC 5578, which was equal to 0.30 g/Lh, it is concluded that *C. guilliermondii* FTI 20037 has a superior xylitol producing capability. It is also observed that the pH of the medium decreased 3 pH units (from 5.0 to 2.0) after 48 h of fermentation. This behavior is quite common among yeasts, because of the high permeability to cations and hydrogen ions presented by the cell

Table 1
Xylose Consumption, Xylitol and Biomass Production, and pH
Variation of Medium for *C. guilliermondii* FTI 20037, Cultivated
Batchwise in a Synthetic Medium

Time (h)	Xylose (g/L)	Biomass (g/L)	Xylitol (g/L)	pH
0	66.4	0.50	0	5.0
16	48.5	2.80	17.6	2.8
24	31.0	3.76	27.4	2.5
48	0	5.29	43.4	2.2
51	0	5.50	46.8	2.1
62	0	6.20	47.6	2.0

60 g/L of xylose; 20 g/L of rice bran; 2 g/L of ammonium sulfate; and 0.1 g/L of calcium chloride.

membrane (15). Moreover, the cell and xylitol concentrations increased, respectively, 14.6 and 8.8% between 48 h and 62 h (Table 1). Probably, some reserve substance synthesized during the high cell-growth rate phase was lately converted into biomass. Meanwhile the extra xylitol into the medium would result from a late excretion of this substance still accumulated inside the cells. However, the fermentative parameters for xylitol production are affected by factors such as pH, temperature, dissolved oxygen, and xylose and glucose concentration (3–5).

Tables 2 and 3 show that the xylitol yield factor ($Y_{p/s}$) and VP diminished markedly for initial xylose and glucose concentrations higher than 90.0 g/L and 5.0 g/L, respectively. Regarding xylose, the result is in accordance with Silva et al. (16), who set 100 g/L as the upper limit for the initial xylose concentration without affecting xylitol production by *C. guilliermondii*. Furthermore, according to Du Preez et al. (17) and Silva et al. (16), under high initial xylose concentration (higher than 100 g/L), the yeast would be submitted either to the dissolved oxygen limitation or to the increased osmotic pressure of the culture medium. In addition, the cell yield factor ($Y_{x/s}$) diminished 71% as the initial xylose concentration varied from 15.0 g/L to 240 g/L (Table 2), but increased 75% in a medium containing 60.0 g/L of xylose, and the initial glucose concentration was varied from 0 to 30.0 g/L (Table 3). In this case, it is clear that glucose is a more suitable carbon source for cell growth than xylose.

As the reaction sequence xylose/xylitol/xylulose (the initial steps of the pentose pathway in the yeast) are catalyzed by XR and XD, the effect of pH, xylose, and glucose on their production by *C. guilliermondii* FTI 20037 were evaluated.

From Table 4, it is clear that the initial pH of the medium affects the XR and XD activities. At pH 3.0 and 6.5, the XR activity decreased during the fermentation, but the opposite occurred at pH 5.0. The same occurred

Table 2
Xylose Conversion into Xylitol ($Y_{p/s}$), Xylitol Volumetric
Productivity (VP), and Xylose Conversion into Biomass ($Y_{x/s}$)
for *C. guilliermondii* FTI 20037 Cultivated in a Synthetic
Medium for 36 h at Different Xylose Concentrations

Xylose (g/L)	$Y_{p/s}$ (g/g)	$Y_{x/s}$ (g/g)	VP (g/Lh)
15	0.23	0.63	0.096
30	0.42	0.22	0.34
60	0.57	0.15	0.71
90	0.58	0.19	0.80
120	0.33	0.13	0.44
240	0.00	0.096	0.00

Table 3
Effect of Initial Glucose Concentration on Xylitol Production by
C. guilliermondii FTI 20037 Cultivated in a Synthetic Medium
for 36 h and in Presence of Xylose at Initial Concentration
of 60 g/L

Glucose (g/L)	$Y_{p/s}$ (g/g)	$Y_{x/s}$ (g/g)	VP (g/Lh)
0	0.50	0.14	0.69
5	0.56	0.16	0.64
10	0.49	0.22	0.41
20	0.20	0.32	0.10
30	0.30	0.55	0.11

Table 4
Effect of Initial pH of Culture Medium on Xylose Reductase (XR) and Xylitol
Dehydrogenase (XD) Activities of *C. guilliermondii* FTI 20037.

pH (Initial)	Time (h)	Biomass (g/L)	Xylose (residual) (g/L)	Xylitol (g/L)	U/mg protein	
					XR	XD
	12	3.4	53.5	0.00	1097	184.5
3.0	24	4.7	38.5	9.52	665.3	167.0
	36	5.1	23.4	20.2	572.2	142.3
	12	5.0	52.4	0.00	327.9	47.30
5.0	24	6.7	34.2	11.6	673.8	191
	36	7.0	12.6	25.2	801.7	171.2
	12	2.1	58.1	0.00	889.2	255.1
6.5	24	8.7	32.3	10.6	690.5	211.7
	36	9.6	10.5	22.3	505.0	183.5

In all tests the initial biomass and xylose concentrations were 0.5 g dry matter/L and 60.0 g/L, respectively.

Table 5
Effect of Initial Xylose Concentration on Xylose Reductase (XR) and Xylitol Dehydrogenase (XD) Activities of *C. guilliermondii* FTI 20037

Xylose (initial) (g/L)	Time (h)	Biomass (g/L)	Xylose (residual) (g/L)	Xylitol (g/L)	U/mg protein XR	U/mg protein XD
	12	7.2	6.8	0.0	319.6	135.2
		(0.34 g/Lh)[a]				
15	24	8.6	0.15	0.0	2189.5	917.2
	36	9.9	0.0	3.4	213.0	145.2
	12	6.4	24.8	2.3	452.9	152.8
		(0.25 g/Lh)[a]				
30	24	6.5	11.8	19.0	2033.5	777.0
	36	7.1	0.45	12.4	288.3	136.0
	12	5.9	50.7	1.7	824.9	117.4
		(0.26 g/Lh)[a]				
60	24	6.7	36.9	20.4	694.2	242.0
	36	7.1	15.3	25.5	365.3	275.6
	12	5.6	75.4	0.0	1201.7	393.0
		(0.23 g/Lh)[a]				
90	24	6.1	54.6	21.3	528.1	224.7
	36	6.9	40.9	28.7	320.9	303.2
	12	5.1	100	0.0	1172.3	493.5
		(0.22 g/Lh)[a]				
120	24	5.8	75.3	11.7	554.1	217.3
	36	5.9	71.5	15.8	292.3	327.0
	12	3.9	226	0.0	213.9	140.4
		(0.20 g/Lh)[a]				
240	24	5.2	214	0.0	258.8	185.1
	36	5.8	210	0.0	264.7	296.4

In all tests $X_o = 0.5$ g dry matter/L and initial pH = 5.0.
[a] Growth rate calculated as follows: $(X_{24} - X_o)/24$, where X_{24} and X_o were the biomass concentration (g dry matter/L) at $t = 24$ h and $t = 0$ h, respectively.

for XD, except that at pH 5.0 its activity increased until $t = 24$ h, and decreased after that. Furthermore, at pH 3.0, 5.0, and 6.5, the xylose consumption between $t = 12$ h and $t = 36$ h were 56, 76, and 82%, respectively. It must be pointed out that until $t = 24$ h the xylitol production was low, but XR (665.3 U/mg of protein) and XD (167.0 U/mg of protein) activities were high. This is evidence that xylose was directed to biomass formation. When $t > 24$ h, the growing rate slows down and the xylitol accumulates in the medium, reaching a final concentration of 25.2 g/L at pH 5.0 (Table 4). The XR:XD ratio always diminishes as the fermentation time or the pH increases (Table 7). Hence, to attain a good yield for XR, the initial pH should be between 3.0 and 5.0, and the fermentation carried out up to 12 h.

Table 6
Effect of Initial Glucose Concentration on Xylose Reductase (XR) and Xylitol Dehydrogenase (XD) Activities of *C. guilliermondii* FTI 20037

Glucose (initial) (g/L)	Time (h)	Xylose (residual) (g/L)	Biomass (g/L)	Xylitol (g/L)	U/mg protein XR	U/mg protein XD
0	12	50.7	5.9	1.7	824.9	117.4
	24	36.9	6.7	20.4	694.2	242.0
	36	15.3	7.1	25.5	365.3	275.6
	(1.4 g/Lh)^a					
5	12	53.3	3.9	0.0	158.4	27.6
	24	32.9	7.2	10.1	782.5	70.5
	36	18.4	7.9	23.2	300.6	62.5
	(1.2 g/Lh)^a					
10	12	49.3	4.2	0.0	122.4	46.8
	24	40.9	7.0	5.9	285.2	64.0
	36	29.8	7.1	14.9	396.9	68.7
	(0.84 g/Lh)^a					
20	12	56.5	5.0	0.0	411.8	45.4
	24	52.7	6.5	0.0	135.7	46.3
	36	41.3	7.3	3.7	246.2	45.4
	(0.52 g/Lh)^a					
30	12	55.2	6.5	0.0	178.9	29.6
	24	46.6	7.8	0.0	138.0	35.5
	36	46.6	8.4	4.0	61.8	38.6
	(0.37 g/Lh)^a					

In all tests the initial xylose concentration (XY_o) was equal to 60 g/L. The initial pH and cell concentration were 5.0 and 0.5 g dry matter/L, respectively.

^{a} Xylose consumption rate calculated as follows: ($XY_{36} - XY_o$)/36, where XY_{36} and XY_o were the xylose concentrations at $t = 36$ h and $t = 0$ h, respectively.

The highest activities of XR (2189.5 U/mg of protein) and XD (917.2 U/mg of protein) occurred after 24 h of fermentation, at an initial xylose concentration of 15.0 g/L (Table 5). Taking into account that, under these conditions, no xylitol was formed, and that the highest growth rate (0.34 g/Lh) occurred, it can be assumed that the biosynthesis of XR and XD is linked to the cell growth. This is confirmed by the fact that, at 90.0 g/L of xylose, when xylitol is formed (21.32 g/L) and the growth rate diminished to 0.24 g/Lh, the XR and XD activities decreased 76%. Table 5 shows also that, in the 30.0 g/L of the initial xylose-test, the xylitol concentration decreases by 35%, probably because of its consumption by the cell as an alternative carbon source, since after $t = 24$ h the xylose concentration in the medium was low.

It is quite important to establish the effect of glucose on the XR and XD activities of *C. guilliermondii*, because this hexose is always present in lignocellulosic hydrolysates, which presently is the main constituent of the

Table 7
Effect of pH, Xylose, and Glucose Concentration on XR:XD
Ratio at Different Fermentation Times

Parameter		Fermentation time (h)		
		12	24	36
pH	3.0	5.95	3.98	4.02
	5.0	6.93	3.52	4.68
	6.5	3.49	3.26	2.75
Xylose (g/L)	15	2.36	2.38	1.47
	30	2.96	2.62	2.12
	60	7.03	2.87	1.33
	90	3.06	2.35	1.06
	120	2.38	2.55	0.89
	240	1.52	1.40	0.89
Glucose (g/L)	0	5.40	8.20	3.76
	5	5.74	11.1	4.81
	10	2.62	4.46	5.78
	20	9.07	2.93	5.42
	30	6.04	3.89	1.60

fermentative medium for xylitol production by yeasts (16). Therefore, it is clear from Table 6 that XR and XD activities of *C. guilliermondii* were sensitive to the glucose present in the culture medium. The highest XR (782.5 U/mg of protein) was attained at 5.0 g/L of glucose, but the high XD activity (81.2 U/mg of protein) occurred in the absence of glucose. Regarding the xylitol formation, a negative effect occurs when the initial glucose concentration is higher than 20.0 g/L. The authors can see that, between $t = 12$ h and $t = 36$ h, the xylose consumption rate, which was 1.4 g/Lh (absence of glucose) or 1.2 g/Lh (in the presence of 5.0 g/L of glucose), decreased at least 45% for glucose concentration above 10.0 g/L. Taking into account that XR and XD are inducible enzymes (18,19), and that glucose can interfere in the xylose-transport mechanism across the cell membrane (16), the authors hypothesize that the diminution of xylose uptake by *C. guilliermondii* FTI 20037 led to the observed diminution of XR and XD activities.

According to the data presented, the production of XR should take into account the conditions under which XR:XD ratio is between 2.36 and 11.1 (Table 7). Thereby, a suitable XR production could be attained by setting the parameters studied as follows: initial pH: 3.0–5.0; xylose: 15–60 g/L; glucose: 0–5 g/L; and fermentation time: 12–24 h.

ACKNOWLEDGMENTS

To Conselho Nacional De Desenvolvimento Científico E Tecnológico for sponsoring this work.

REFERENCES

1. Emodi, A. (1978), *Food Technol.* **32,** 28–32.
2. Felipe, M. G. A., Vitolo, M., and Mancilha, I. M. (1996), *Acta Biotechnol.* **16,** 73–79.
3. Delgenes, J. P., Moletta, R., and Navarro, J. M. (1989), *Biotechnol. Bioeng.* **34,** 398–402.
4. Forage, R. G., Harrison, D. E. F., Pitt, D. E. (1985), in *Comprehensive Biotechnology,* vol. 1, Moo-Young, M., ed., Pergamon, New York, pp. 251–280.
5. Nolleau, V., Preziosi, B. L., Delgenes, J. P., and Delgenes, J. M. (1993), *Curr. Microbiol.* **27,** 191–197.
6. Onishi, H. and Suzuki, T. (1966), *Agricultural Biol. Chem.* **30,** 1139–1144.
7. Steven, R. and Lee, H. (1992), *J. Gen. Microbiol.* **138,** 1857–1863.
8. Smiley, K. L. and Bolen, P. L. (1982), *Biotechnol. Lett.* **4,** 607–610.
9. Du Preez, J. C., Driessel, B., and Prior, B. A. (1989), *Arch. Microbiol.* **152,** 143–147.
10. Barbosa, M. F. S., Medeiros, M. B., Mancilha, I. M., Schneider, H., and Lee, H. (1988), *J. Ind. Microbiol.* **3,** 241–251.
11. Roberto, I. C., Felipe, M. G. A., Mancilha, I. M., Vitolo, M., and Silva, S. S. (1995), *Bioresource Technol.* **51,** 255–257.
12. Silva, S. S., Dimas, J., Felipe, M. G. A., and Vitolo, M. (1997), *J. Ind. Microbiol.* **63/65,** 557–563.
13. Bradford, M. M. (1976), *Anal. Biochem.* **72,** 248–254.
14. Meyrial, V., Delgenes, J. P., Moletta, R., and Navarro, J. M. (1991), *Biotechnol. Lett.* **13,** 281–286.
15. Weitzel, G., Pilatus, U., and Rensing, I. (1987), *Exp. Cell Res.* **170,** 64–79.
16. Silva, S. S., Vitolo, M., Pessoa, A., Jr, and Felipe, M. G. A. (1996), *J. Basic Microbiol.* **36,** 187–191.
17. Du Preez, J. C., Bosch, M., and Prior, B. A. (1986), *Enzyme Microb. Technol.* **8,** 360-364.
18. Bicho, P. A., Runnals, P. L., Cunninghan, J. D., and Lee, H. (1988), *Appl. Environ. Microbiol.* **54,** 50–54.
19. Lee, H. (1992), *FEMS Microbiol. Lett.* **92,** 1–4.

Yeast Adaptation on Softwood Prehydrolysate

FRED A. KELLER,* DELICIA BATES, RAY RUIZ, AND QUANG NGUYEN

Biotechnology Center for Fuels and Chemicals, National Renewable Energy Laboratory, Golden, CO 80401

ABSTRACT

Several strains and genera of yeast, including *Saccharomyces cerevisiae* D_5A, *Pachysolen tannophilus*, *S. cerevisiae* K-1, *Brettanomyces custersii*, *Candida shehatae*, and *Candida acidothermophilum*, are screened for growth on dilute acid-pretreated softwood prehydrolysate. Selected softwood species found in forest underbrush of the western United States, which contain predominantly hexosan hemicellulose, were studied. This phase of the work emphasized debarked Douglas fir. The two best initial isolates were gradually selected for improved growth by adaptation to increasing prehydrolysate concentrations in batch culture, with due consideration of nutrient requirements. Microaerophilic conditions were evaluated to encourage tolerance of pretreatment hydrolysate, as well as ethanol product. Adaptation and simultaneous saccharification and fermentation (SSF) results are used to illustrate improved performance with an adapted strain, compared to the wild type.

Index Entries: Yeast; adaptation; softwood; ethanol; fermentation.

INTRODUCTION

Effort to produce renewable alternative sources of transportation fuels from biomass have resulted in considerable progress in the conversion of hardwood and agricultural waste into ethanol. Added yields are expected if the pentosan hemicellulose, in addition to the cellulose, is effectively fermented to ethanol. Except for sulfite waste liquor, reports of the conversion of softwood (SW) materials have been limited. Furthermore, the hemicellulose for most SWs studied is primarily hexosan, composed mainly of mannose with smaller amounts of glucose and galactose, as well as some pentoses. Traditional *Saccharomyces cerevisiae* yeast cultures ferment these

*Author to whom all correspondence and reprint requests should be addressed.

hexoses to ethanol very well, and are expected to produce high ethanol yields if they can tolerate low concentrations of countless toxins present in prehydrolysates generated from SW. Prehydrolysates from SW may be more toxic than those from hardwood biomass sources, because Sws usually contain more extractives and often more bark than do hardwoods. Potential toxic substances include biomass components themselves, particularly extractives such as terpenes, aldehydes, and polyhydroxy aromatics. Other sources of toxins are prehydrolysis products and degradation products including acetic acid from acetylated sugars, furfural, and hydroxymethyl furfural, the initial degradation products from pentose and hexose sugars, respectively, and oligomers formed by reaction of the furfurals with sugars. Degradation of coniferous lignin yields complex guaiacyl propyl units. Corrosion products from equipment also can be toxic, or the metallic ions can behave as catalysts to produce additional products. Fortunately most of the toxins in well-prepared prehydrolysate are present at less than one g/L, and only a very few, such as furfural, are present at a few g/L. However, over time, yeast can adapt themselves to tolerate many of these substances in the presence of glucose sugar. These toxins prevent or greatly reduce growth and ethanol production. The authors intend to test the capability of yeast to gradually adapt to their environment and encourage them to eventually grow and produce very near stoichiometric ethanol, while the unadapted (wild-type) yeast can do neither.

Traditional wastes, such as SW trimmings and sawdust, are available sources of SW biomass. Recently, however, SW forest underbrush of selected regions of western United States has been harvested in an effort to prevent extremely hot forest fires which can destroy all vegetation, along with the trees. Serious land erosion or floods can follow. The authors intend to test the capability of yeasts to gradually adapt to their environment and encourage them to eventually grow and produce very near stoichiometric ethanol, when the unadapted (wild-type) yeast could do neither.

OBJECTIVES

Fermentation processes are usually both capitals and operating-cost-intensive. The authors' ultimate objectives are to determine whether these costs can be significantly reduced by demonstrating improved yeast fermentation performance through adaptation of selected yeast cultures. Specifically, high fermentation yield, process efficiency, and volumetric productivity in the presence of real prehydrolysates, while supplying minimal nutritional requirements is wanted. The high yield and efficiency ensures good conversion of saccharides to ethanol with a minimum of by products. The high productivity and low nutrient charge minimizes the fermentor size (and cost) and keeps cost of supplies down, respectively. Another objective factor is the concentration of the broth or slurry used. Higher concentrations minimize the fermenter size for a given product capacity. But perhaps what is more important, they also increase the potential final ethanol concentration in the spent beer, significantly improving the economics of ethanol recovery. Consequently, one of the first direct benefits of culture adaptation to higher prehydrolysate slurry concentrations, and the objective of this phase of the work, is to achieve significantly improved ethanol production at desired elevated slurry prehydrolysate concentrations, in which an unadapted strain does poorly, or cannot survive at all.

EXPERIMENTAL METHODS, MATERIALS, AND MICROORGANISMS

Freshly cut Douglas fir tree trunks that measured 3–8 in. in diameter were debarked, chipped (65 HP Brush Bandit, Foremost, Remiss, MI), and shear-milled (Mitts Merrily, Sagging, MI) through a 9-mM screen. The milled wood was then soaked in 60°C acidified water, using 0.35% (w/w) sulfuric acid, for 4 h. The wood particles were allowed to drain and air-dry to about 40–50% (w/w) solids, and then were prehydrolyzed in a batch digester (1). The resulting semisolid prehydrolysate contained ~20–30% (w/w) dry solids.

Our culturing goals were to minimize the requirement for prehydrolysate and retain or recycle all yeast cells from the prehydrolysate, while maximizing the number of generation times accumulated. Achieving the latter two goals improves the probability of obtaining and keeping desirable spontaneous mutants. Therefore, small batch cultures were used. In late log phase, or early stationary phase, cultures to be transferred were split into succeeding cultures. Initially, to facilitate culture-handling, sampling, and analysis, prehydrolysate filtrate was used. It was sterile-filtered before use. To maximize the number of generations the adapting cultures underwent in a limited time, full-strength (~25% w/w) prehydrolysate filtrate was diluted only slightly, using nutrients and inocula, to growth permissive strengths. These final concentrations of diluted liquors are referred to as the equivalent solids (ES) concentration, which is the solids concentration (dry basis) from which they were prepared. This helped in measuring progress of adaptation. This reference also permits direct estimation of the maximum ethanol concentration possible (at maximum theoretical yield and efficiency) from a given total equivalent saccharide concentration in whole prehydrolysate slurry. At the time of dilution, supplements were added and figured into the dilution.

Prehydrolysate Media

A simulated Douglas fir acid prehydrolysate was also used by mixing the five main sugars of Douglas fir hemicellulose, all of research-grade purity. Typically, they were mixed to simulate hydrolyzed hemicellulose, supplemented with significant cellulose hydrolysis. The synthetic, simulated Douglas fir acid prehydrolysate, under these conditions, consisted of 5% (w/v) glucose, 2.5% (w/v) xylose, 0.5% (w/v) arabinose, 4.5% (w/v) mannose, 0.5% (w/v) galactose, 1% (w/v) yeast extract, and 2% peptone (latter replaced by 1.5% [w/v] ammonium sulfate), allowing room for a 10% (w/v) inoculum. These sugars were filter-sterilized, except xylose, which was autoclaved. Yeast extract and peptone were autoclaved; the ammonium sulfate was filter-sterilized. Chemicals were purchased from Baker (Windsor, UK) or Sigma (St. Louis, MO); yeast extract and peptone from Difco, (Detroit, MI). Oxygen absorption rates, to provide microaerophilic conditions of at least 0.075 mM oxygen/L/min, as determined by sulfite oxidation, were provided by controlling liquid volume to flask volume, baffling, and shaker (Model 4000 NBS, Edison, NJ) speed. They were grown at either 30°C, or, for culture 6, at 38°C, and at 150 rpm shaker speed.

Adaptation media were prepared by adding 1.5% (w/v) corn-steep liquor (CSL), 0.5% (w/v) yeast extract, and 1.5% (w/v) ammonium sulfate to debarked Douglas fir hydrolysate filtrate, then adjusting to pH 5.5 ±0.1 with calcium hydroxide (~6 g). CSL (Grain Processing, Muscatine, IA), at a 50% solids (w/v), was sterile-filtered after centrifuging and prefiltering, or was cross-flow-filtered.

SSF slurry was prepared with pretreated debarked Douglas fir. The SW was digested at 215°C for 100 s, after being soaked with 0.35% (w/v) sulfuric acid. The final cake was pH-adjusted with calcium hydroxide to pH 5.4 ±0.1, blended, and added to 250-mL shake flasks. After autoclaving, each flask was supplemented with filter-sterilized nutrient solution to provide 1.5% (w/v) filter-sterilized CSL, 0.5% (w/v) yeast extract, and 1.5% (w/v) ammonium sulfate. Filter-sterilized Iogen cellulase enzyme (Ottawa, Ontario, Canada) was added at a concentration of 36 IFPU/g cellulose just before inoculation. The usual ratio of enzyme IFPU/g cellulose was increased from 25 to 36, because of prehydrolysate inhibition of cellulase enzyme. Cultures were inoculated with either an adapted or wild-type strain of *S. cerevisiae* D_5A at proportional concentrations. The adapted strain, growing on prehydrolysate, required 3× the broth volume to yield the same amount of cells (dry wt basis) as did the wild-type growing on synthetic prehydrolysate. To achieve 3× the concentration, one-third was added with spent media; the other two-thirds was centrifuged and resuspended in the fresh liquor of each flask. Cultures were incubated at 30°C, and sampled daily for 7 d.

Chemical Analyses

Ethanol was analyzed with a YSI instrument (Yellow Springs, OH), or HP Series II gas chromatograph (GC) using a Porapak Q column (HP, Palo Alto, CA). Dilutions were made to stay below the YSI standard value of 1.03 g ethanol/L, or 2.5 g glucose/L. An HPLC (HP # 1090 with refractive index detector) was used for hexose concentrations, sampled at selected intervals. During adaptation, absorbances were read with a Genesys 5

Spectronic (Milton Roy, Rochester, NY) spectrophotometer at 600 nM. Dry wt were obtained by carefully washing 10 mL of culture once with 10 mL of sterile nanopure water. A final volume of 5 mL was weighed and dried at 105°C overnight. Viability of the cultures, obtained at sampling intervals, was determined by adding Wolford viable stain (2) to the culture, at concentrations of 3:1 for liquid cultures and 9:1 for slurry cultures, since the solids consume a lot of stain. Viable cells were able to keep the stain from entering, and appeared colorless. Expired cells took up the stain and appeared purple or blue. Cells were counted using a Pettroff Hauser hemocytometer.

Microorganisms

The yeast strains included *Brettanomyces custersii*, an NREL culture available as #34447 from the American Type Culture Collection (ATCC Rockville, MD). *S. cerevisiae* (D_5A) and (K-1) are NREL strains genetically derived from Commercial Red Star brewers yeast, and from Lallemand brewers yeast, respectively. *Candida shehatae* was obtained from ATCC, # 22984, and *Candida acidothermophilum* from ATCC, # 20381. *Pachysolen tannophilus* was obtained as NRRL Y-2460, from the US Dept. of Agriculture (USDA), National Center for Agricultural Utilization Research, formerly the Northern Regional Research Laboratory.

Wild-type cultures were plated on YPD: 1% (w/v) yeast extract (Y), 2% (w/v) peptone (P), 2% (w/v) dextrose (D), and 2% (w/v) agar; or RMG: 1% (w/v) yeast extract, 0.2% (w/v) monobasic potassium hydrogen phosphate, 2% (w/v) glucose, and 2% (w/v) agar. Adapted strains were plated on prehydrolysate plates 15% (w/v) filter-sterilized prehydrolysate, 1.5% (w/v) agar. All sugars were sterilized separately.

RESULTS AND DISCUSSION

Six species of yeast, which were preserved frozen in glycerol or sucrose, were grown on YPD broth and plated on YPD agar for short-term supply. Baseline performances of the yeast were initially determined by growing them in a simulated Douglas fir acid prehydrolysate that contained YP at concentrations as in YPD. During this time, they were acclimated to grow in medium formulated with 1.5% ammonium sulfate, in place of peptone, and then yeast extract reduced to one-half strength and supplemented with 1.5% (v/v) CSL, in preparation for leaving out the yeast extract entirely.

Late log-phase cultures on synthetic prehydrolysate were initially divided in half or quarters and yeast incubated after being spun down and reconstituted with actual hydrolysate filtrate back to the original volumes. Growth was monitored by measuring absorbance at 600 nM, glucose by (Yellow Springs Instrument, Yellow Springs, OH) YSI, and ethanol by GC or YSI. They were subcultured similarly when the glucose became low

Table 1
Growth of Six Cultures on Prehydrolysate After 48 h

Culture	Equivalent solids concentration	A_{600}
S. cerevisiae, D$_5$A	**17 b**	**12.10**
S. cerevisiae, D$_5$A	**21 a**	**11.50**
P. tannophilus	15 b	2.85
P. tannophilus	17 a	3.15
S. cerevisiae, K-1	**17 b**	**12.85**
S. cerevisiae, K-1	**21 a**	**12.00**
B. custersii	15 b	3.40
B. custersii	17 a	3.70
C. shehatea	15 b	2.95
C. shehatea	17 a	2.80
C. acidothermophilum	15 b	3.45
C. acidothermophilum	17 a	3.90

Best three cultures are in bold.

or the absorbance approached doubling or quadrupling, depending on whether they were divided into halves or into quarters. However, after splitting, each half or quarter could be reconstituted with a higher concentration prehydrolysate, and then the other half with the same initial hydrolysate concentration. This provided culture backup in the event of loss of the culture at the higher concentration. Carryover of spent, conditioned media was minimized by aseptically centrifuging the cultures, discarding the supernatant, and then resuspending all of the cells in fresh prehydrolysate. This procedure was more toxic (less-conditioned medium was used), and required additional performance from the culture.

Screening for the Most Resistant Cultures

The number of cultures that were continuously maintained, by dropping (and storing) those that adapted more slowly than others. This should not imply that they would not adapt over time, or that, with patience, they would not even be better than those that adapt more quickly; it is simply a matter of time. Cell density (Table 1) for all six yeast strains is measured by the culture absorbance at 600 nM, after ~ 2 wk of acclimating the six strains in prehydrolysate filtrate at 15, 17, and 21% ES. The initial absorbance was usually between 2.5 and 3.0; the double-digit numbers are the ES levels. As they were able to develop, they were split into the next higher ES, and a repeat ES. S. cerevisiae D$_5$A and K-1 grew sufficiently well to be transferred into higher and higher ES levels. They did well at 21%, and slightly better at 17% ES; the other strains struggled at 15 and 17% ES.

In this way, we dropped all strains, except for the latter two, shortly after testing them on 15 and 17% ES hydrolysate. The adaptation sequence was continued with the *S. cerevisiae* strains for 3 wk more, until they were able to tolerate 23% ES, or better. Prehydrolysate concentrations of 23% reduced generation times (doublings) to less than one every 3 d. At that rate, sufficient doublings to maintain reasonable probabilities of getting spontaneous mutants were not obtained. Some progress was made however, because these cultures did better than those kept at lower concentrations (17%), but concentrations that initially would not support growth. Selection at the lower concentrations for less than 2 wk was maintained at which time we compared an adapted culture of *S. cerevisiae* D_5A with an unadapted (wild) strain in the simultaneous saccharification and fermentation (SSF) process.

Strain Comparison in Simultaneous Saccharification and Fermentation

This SSF test was run in duplicate with three levels of whole prehydrolysate slurry at 6.5, 10, and 15% total solids (dry basis), consisting of 2.1, 3.2, and 4.8% cellulose per flask. Table 2 shows the results of the first test. All data are averages of duplicates. The standard deviations for glucose and ethanol values are all less than +/−5% of the concentrations. An overview of Table 2 shows immediately that the adapted strain in culture sets 3 (10% solids) and 5 (15% solids) took off rapidly in the presence of initial sugar, and quadrupled the ethanol concentration within 1 d, from 0 to 24 h. The wild-type performed similarly, but only at the low 6.5% solids. The wild-type caught up to the adapted strain in 10% solids, at 72 h, but only after a long lag, from 0 to 72 h. At 15% solids, which is near the minimum solids concentration for economic ethanol production, the wild-type produced no ethanol, even in the presence of at least 2% (20 g/L) sugar concentration, as noted by the glucose level, from the very beginning through 168 h. The ethanol volumetric productivity of the adapted strain at 15% solids was at least about 0.41 g/Lh over the first 24 h, including a lag. There was some prehydrolysate inhibition of enzyme, but even at 15% solids, and feedback inhibition from greater than 2% sugar, the enzyme slowly produced about another 1% glucose (column 6). The enzyme levels decreased with solids level, because they were dosed on the basis of units of enzyme/g cellulose. Enzyme inhibition is probably the reason cultures 5 and 3 became sugar (i.e., enzyme rate)-limiting at about 48 h. Cultures 4 almost catch up to 5, albeit about 3 d later and 5% less solids. All adapted strains carried over some hydrolysate with the 10% (v/v) inoculum. Therefore, prehydrolysate inhibition of the enzyme is probably the reason the adapted culture pair became sugar-limiting sooner than the wild-strain pairs. Table 2 shows that the adapted culture viability is very good at 96 h. This shows quite an improvement over that observed for the unadapted yeast under similar low-sugar conditions, as indicated in Table 3. The actual conditions of SSF may not be as severe as sugar

Table 2
Adapted vs Wild S. cerevisiae D_5A SSF Performance on Prehydrolysate

Time (h)	Analyte (g/L)	6.50% Solids		10.00% Solids		15.00% Solids	
		Adapted culture 1	Wild culture 2	Adapted culture 3	Wild culture 4	Adapted culture 5	Wild culture 6
0	EtOH	3.60	5.20	3.65	5.65	3.80	6.00
	Glucose	10.85	9.15	14.45	15.15	23.90	23.50
t-24	EtOH	9.60	13.20	12.90	7.90	13.20	6.25
	Glucose	0.47	0.12	3.50	11.30	16.30	25.25
t-48	EtOH	11.00	14.65	15.65	11.40	20.50	6.45
	Glucose	0.26	0.14	0.77	7.22	4.00	26.85
t-72	EtOH	10.15	14.35	16.15	16.55	19.90	6.15
	Glucose	1.56	0.37	2.97	1.36	7.09	31.03
t-96	EtOH	10.05	12.30	16.35	17.80	22.15	5.05
	Glucose	1.51	0.32	2.80	0.67	5.99	32.42
	Viability, %	88	83.00	74.00		98.00	72.00
t-120	EtOH	9.30	12.25	15.90	20.20	22.60	5.70
	Glucose	1.49	0.31	2.53	0.58	5.27	34.98
t-144	EtOH	7.75	10.85	14.40	20.40	22.55	5.30
	Glucose	0.14	0.06	0.23	0.17	0.81	33.80
t-168	EtOH	0.15	0.06	0.25	0.14	0.77	36.60
	Glucose						

% Solids = % insoluble solids of whole prehydrolysate, dry basis.

Table 3
Yeast Viability at Low Sugar Concentration

Expt.	Yeast	Viability	Notes
1 a,b	S. cerevisiae D_5A	2.3, 4.0	Wolford Stain
2 a,b	P. tannophilus	< 10/mL	On plates
3 a,b	S. cerevisiae K-1	50.7, 54.9	Wolford Stain
4 a,b	B. custersii	~ 0.005%	On plates
5 a,b	C. shehatae	55.0, 64.7	Wolford Stain
6 a,b	C. acidothermophilum	< 10/mL	On plates

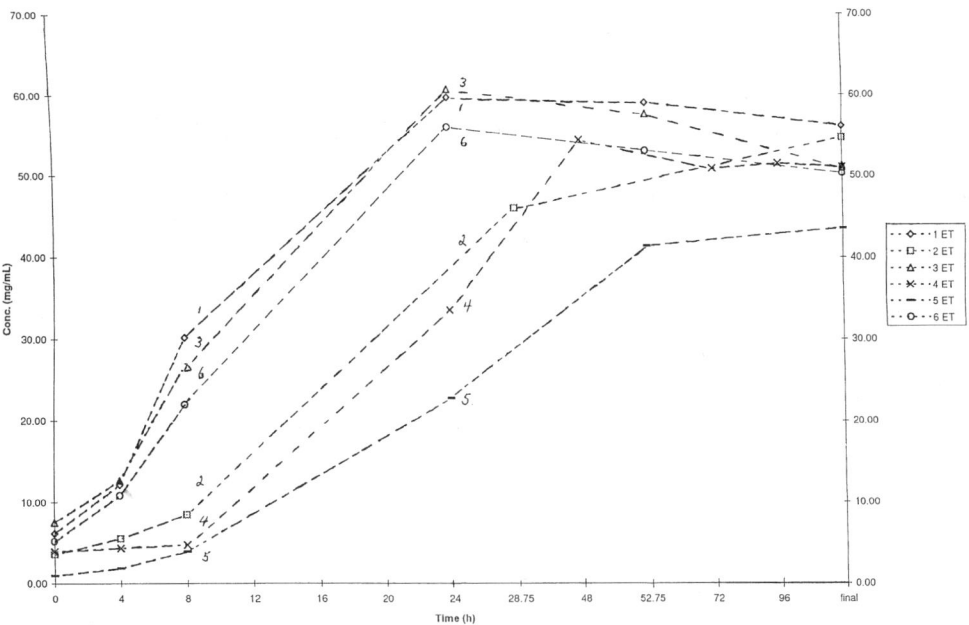

Fig. 1. Ethanol production by six yeast species grown on simulated hydrolysate.

starvation, because even though the sugar concentration is low in SSF, it usually is still being produced, but is consumed by yeast as rapidly as it is generated by the enzyme.

Characterization of Strains on Simulated Prehydrolysate

Characterization of culture performance on simulated prehydrolysate in the absence of inhibitors is expected to indicate the ultimate performance that may be expected from the strains after adaptation to real prehydrolysate containing inhibitors. Figure 1 illustrates ethanol concentrations and productivities achieved by the six cultures grown on rich nitrogen and nutrient sources, using late log-phase inocula. The cultures that per-

Table 4
Ethanol Performance Parameters

Culture	Ethanol concn @ 24 h (%, w/v)	Ethanol concn max (%, w/v)	Ethanol dP/dt max (g/L-h)	Ethanol eff. max (% theor., w/w)	Ethanol yield max (% theor., w/w)
1 S. cerevisiae D_5A	6.0	6.0	6.7	89.0	85.1
2 P. tannophilus	4.6	5.4	2.2	84.1	78.2
3 S. cerevisiae K-1	6.1	6.1	4.5	88.2	86.3
4 B. custersii	2.1	5.4	1.1	83.8	77.6
5 C. shehatae	2.3	4.3	1.4	66.9	62.2
6 C. acidothermophilum	5.6	5.6	4.4	83.1	75.3

Table 5
Performance Data for Evaluated Yeasts Grown on Simulated Hydrolysate

	Xylitol (g/L)	HOLac (g/L)	Gly (g/L)	HOAc (g/L)	Cell mass (%, g/100 mL)	Cell N (%, w/w)	Beer N (%, g/100 mL)	μ (1/h)	t_g (h)
1 S. cerevisiae, D₅A	6.03	–	3.31	2.37	0.68	8.19	0.0611	0.26	2.7
2 P. tannophilus	2.23	–	–	1.01	0.69			0.104	6.6
3 S. cerevisiae, K-1	6.12	0.87	2.88	2.31	0.95	8.62	0.0819	0.23	3.0
4 B. custersii	1.17	–	0.45	2.77	1.16			0.066	10.6
5 C. shehatea	4.84	–	5.21	0.74	0.70			0.49	1.4
6 C. acidothermophilum	1.86	0.20	3.55	0.71	0.47	11.3	0.0530	0.24	2.9

formed the best in actual prehydrolysate 1 and 3 (Table 1), also achieved the highest ethanol concentrations and the highest ethanol productivities (Fig. 1) when cultured on simulated prehydrolysate. The highest values observed are noted (Table 4) when sampled at 4, 8, 24, 48, and 72 h of fermentation. Lower efficiency is usually a result of formation of by products such as xylitol and glycerol (Table 5). Table 5 also reports cell mass produced, maximum specific growth rates observed (or minimum generation times), and cell mass nitrogen as percent (w/w) in dry cell mass, as well as cell nitrogen per culture volume, expressed as nitrogen in beer because of nitrogen contained by the yeast. For example, although culture 6 contained high relative nitrogen concentration in its cell mass, because of relatively low cell mass per culture volume, the nitrogen consumed by the cell mass per volume was actually lower than cultures 1 and 3.

It was also determined that the apparent cell viabilities after 48–72 h incubation in spent beer with low hexose sugar, as shown in Table 3. This may be thought of as simulating conditions during enzyme rate-limiting SSF. In Table 3, culture 1 had relatively low viability, 2.3–4%. However, in Table 2, after prehydrolysate adaptation, the cell viability is very good (96–100%) at (96 h) with low sugar.

ACKNOWLEDGMENTS

The authors wish to thank Fannie Posey Eddy and Jim Hora for their superb and timely analytical support, and Brian Boynton for useful discussion. This work was funded by the Biochemical Conversion Element of the Office of Fuels Development of the US Department of Energy.

REFERENCES

1. Nguyen, Q., Tucker, M. P., Boynton, B. L., Keller, F. A., and Schell, D. J. (1997), 19th Symposium on Biotechnology for Fuels and Chemicals, Colorado Springs, CO, May 4–8.
2. McDonald, V. R. (1963), *J. Food Sci.* **28,** 135.

Production of Xylitol by *Candida mogii* from Rice Straw Hydrolysate

Study of Environmental Effects Using Statistical Design

Z. D. V. L. Mayerhoff, I. C. Roberto,* S. S. Silva

Department of Biotechnology, Faculty of Chemical Engineering of Lorena, 12600-000, Lorena/SP, Brazil

ABSTRACT

The influence of aeration level, initial pH, initial cell concentration, and fermentation time on the xylitol production from rice straw hemicellulose hydrolysate by *Candida mogii* was studied. A multifactorial experimental design was adopted to evaluate this influence. A statistical analysis of the results showed that the aeration level and the initial pH had significant effects on yield factor, volumetric productivity, and xylose consumption. For the latter, fermentation time was also a significant variable. Based on the response surface methodology, models for the range investigated were proposed. The maximum values for the yield factor ($Y_{p/s}$) and volumetric productivity (Q_p) were, respectively, 0.71 g/g and 0.46 g(Lh).

Index Entries: Rice straw; *Candida mogii*; hemicellulose hydrolysate; factorial design; xylitol.

INTRODUCTION

The fermentation conditions for the biotechnological production of xylitol by several yeast strains have been the target of many investigations. The literature reports that this bioconversion is influenced by a number of variables, such as aeration, initial pH, and initial cell concentration. The aeration level interferes with the level of oxidized co-factor NAD^+, blocking or promoting the activity of the enzyme xylitol dehydrogenase, responsible for the oxidation of xylitol into xylulose. As frequently observed, decreasing the oxygen supply increases the xylitol formation. Nevertheless, the optimal oxygenation levels appear to be specific for each yeast strain (1–4). Likewise, the pH level promoting the highest xylitol

---

*Author to whom all correspondence and reprint requests should be addressed.

accumulation is different for each yeast strain *(1,3)* and each fermentation medium *(5)*. The influence of the initial cell concentration also showed to be dependent on the yeast strain employed *(1,6,7)*.

In a previous study, *Candida mogii* NRRL Y-17032 was selected from among 31 yeast strains as a promising xylitol producer from rice straw hemicellulose hydrolysate *(8)*. However, no investigation on the effect of the fermentation conditions has been conducted so far. In this work, a multifactorial experimental design was adopted to evaluate the influence of aeration, initial pH, initial cell concentration, and fermentation time on the xylose consumption, yield factor, and volumetric productivity during the xylitol production by this yeast strain. A response surface methodology (RSM) was used to determine a statistical model for these parameters. Statistical designs have already been employed in the evaluation of the influence of different factors on the xylitol production by yeasts *(9,10)*.

MATERIALS AND METHODS

Preparation of Hemicellulose Hydrolysate

Rice straw hemicelullose hydrolysate was obtained by acid hydrolysis of the rice straw in an AISI 316 stainless steel 25-L stirred tank reactor (10 g of H_2SO_4 0.7% w/v per g of dry matter). The hydrolysis (145°C, 21 min) was followed by vacuum filtration. The hydrolysate was then concentrated under vacuum, at 70°C, to increase the initial xylose content 5×. The pH of the concentrate was raised with NaOH pellets to 9.5, and then lowered to 5.4 with H_2SO_4 72% (w/w). Each time the pH level was changed, the precipitate was removed by centrifugation (1000g, 15 min). Next, the hydrolysate was autoclaved under free-flowing steam for 15 min.

Microorganism and Inoculum Preparation

C. mogii NRRL Y-17032 obtained from Northern Regional Research Laboratory (Peoria, IL) was maintained at 4°C on malt-extract agar slants. Inoculum was prepared by cultivating cells in 125-mL Erlenmeyer flasks containing 25 mL of medium composed of 20 g/L rice bran extract, 3 g/L $(NH_4)_2SO_4$, 0.1 g/L $CaCl_2 \cdot 2H_2O$, and 50% v/v rice straw hemicellulose hydrolysate. The flasks were incubated for 48 h at 30°C, under 200 rpm.

Fermentation Conditions

A fermentation medium with a composition similar to that used for the inoculum preparation was placed into 125-mL Erlenmeyer flasks, which were incubated on a rotary shaker at 30°C, under 200 rpm. Different aeration levels were obtained by varying the volume of the medium in the flasks. Three different volumes of medium (60, 40, and 20 mL), as well as three initial pH values (4.50, 5.75, and 7.00), three initial cell concentration levels (1.0, 2.5, and 4.0 g/L), and three fermentation times (44, 58, and 72 h), were employed. The minimum, intermediate, and maximum values of each variable correspond to the coded levels $-1, 0,$ and $+1$, respectively.

Statistical Analysis

A 2^4 experimental design (11) was used to evaluate the effect of the variables studied. Statistical models for xylose consumption, yield factor, and volumetric productivity, employing the significant variables, were determined by the response surface regression procedure. A 2^3 factorial with face-centered design was performed. The models determined are expressed by the equation:

$$\hat{Y} = b_0 + \sum b_i X_i + \sum b_{ii} X_i^2 + \sum b_{ij} X_i X_j$$

where $\hat{Y}$ is the response variable; b, the regression coefficients, and X, the experimental factors coded levels.

Analytical Methods

Concentrations of D-xylose and xylitol were determined by high performance liquid cromatography (HPLC) using a Shimadzu C-R7A cromatograph (Kyoto, Japan) equipped with a refractive index (RI) detector and a Bio-Rad (Hercules, CA) Aminex HPX-87H column under the following conditions: temperature: 45°C; eluent: 0.01 N sulfuric acid; flow: 0.6 mL/min; sample volume: 20 µL.

RESULTS AND DISCUSSION

The effect estimates, standard errors, and Student's t-test for xylose consumption (X_C), yield factor ($Y_{p/s}$), and volumetric productivity (Q_P) in xylitol obtained from the 2^4 factorial design are shown in Table 1. The initial cell concentration did not show a significant effect for any of the response variables in the range studied. Only the first-order effects of aeration, initial pH, and fermentation time were significant (at 95% confidence level) for xylose consumption. For the yield factor, the most significant effects were the initial pH and the interaction between aeration and initial pH (90% confidence level). On the other hand, for volumetric productivity, the aeration and pH factors were significant at 95% confidence level. However, no interaction effect was found for this response.

The results of the 2^3 factorial with face centered design, performed using the variables aeration, initial pH, and fermentation time to determine the statistical models from RSM, are shown in Table 2. Based on these results, an analysis of variance for each response, considering all effects, was carried out (data not shown). In order to increase the degrees of free-

Table 1
Effects Estimates, Standard Errors, and Student's (t_3) test for Xylose Consumption (X_C), Yield Factor ($Y_{P/S}$), and Volumetric Productivity (Q_P) in Xylitol Obtained from 2^4 Factorial Design

Effects	Estimates				Standard errors				t		
	X_C (g/L)	$Y_{P/S}$ (g/g)	Q_P g/(L h)		X_C (g/L)	$Y_{P/S}$ (g/g)	Q_P g/(L h)		X_C (g/L)	$Y_{P/S}$ (g/g)	Q_P g/(L h)
Average	26.668	0.620	0.301		±1.739	±0.024	±0.025		15.34	26.08	12.09
X_1: Aeration	14.200	0.070	0.169		±3.789	±0.052	±0.054		3.75[a]	1.35	3.12[a]
X_2: pH	20.025	0.103	0.246		±3.789	±0.052	±0.054		5.29[a]	1.98[b]	4.55[a]
X_3: Cell conc.	7.325	−0.035	0.076		±3.789	±0.052	±0.054		1.93	0.68	1.41
X_4: Time	12.075	−0.023	−0.021		±3.789	±0.052	±0.054		3.19[a]	0.43	0.39
$X_1 X_2$	−1.675	−0.110	−0.021		±3.789	±0.052	±0.054		0.44	2.12[b]	0.39
$X_1 X_3$	3.075	0.028	0.044		±3.789	±0.052	±0.054		0.81	0.53	0.81
$X_1 X_4$	2.275	0.025	−0.014		±3.789	±0.052	±0.054		0.60	0.48	0.25
$X_2 X_3$	−2.250	−0.005	−0.019		±3.789	±0.052	±0.054		0.59	0.10	0.35
$X_2 X_4$	0.200	−0.063	−0.086		±3.789	±0.052	±0.054		0.05	1.21	1.59
$X_3 X_4$	−0.100	−0.055	−0.031		±3.789	±0.052	±0.054		0.03	1.06	0.58

Standard error from total error with 8 d.f. 95% confidence level ($t = 2.30665$).
Standard error from total error with 8 d.f. at 90% confidence level ($t = 1.860$).
[a] Significant at 95% confidence level.
[b] Significant at 90% confidence level.

Table 2
Matrix of 2^3 with Face-Centered Factorial Design and Experimental Responses

Assay	Coded variables			Responses		
	Aeration	pH	Time	X_C (g/L)	$Y_{P/S}$ (g/g)	Q_P (g/(L h)
1	−1	−1	−1	6.0	0.46	0.06
2	+1	−1	−1	12.9	0.59	0.18
3	−1	+1	−1	20.8	0.71	0.33
4	+1	+1	−1	37.8	0.68	0.58
5	−1	−1	+1	9.0	0.45	0.06
6	+1	−1	+1	33.8	0.68	0.32
7	−1	+1	+1	37.5	0.63	0.32
8	+1	+1	+1	45.6	0.58	0.37
9	−1	0	0	19.3	0.70	0.24
10	+1	0	0	42.0	0.67	0.48
11	0	−1	0	17.0	0.53	0.16
12	0	+1	0	36.8	0.62	0.39
13	0	0	−1	17.4	0.62	0.24
14	0	0	+1	34.2	0.66	0.32
15	0	0	0	34.3	0.74	0.44
16	0	0	0	32.7	0.73	0.41
17	0	0	0	32.9	0.72	0.41
18	0	0	0	26.9	0.69	0.32
19	0	0	0	27.8	0.65	0.31

Xylose consumption (X_C) = g/L.
$Y_{P/S}$ = g xylitol produced per g xylose consumed.
Q_P = g xylitol produced per L/h.

dom for the estimate of the effects, a second analysis was made using only the terms that were significant in the first. The models determined are described by the following equations:

$$\hat{Y}_1 = 27.62 + 7.95\ X_1 + 9.98\ X_2 + 6.52\ X_3 \quad (1)$$

$$\hat{Y}_2 = 0.688 + 0.049\ X_2 - 0.052\ X_1X_2 - 0.097\ X_2^2 \quad (2)$$

$$\hat{Y}_3 = 0.352 + 0.090\ X_1 + 0.122\ X_2 - 0.042\ X_2X_3 - 0.074\ X_2^2 \quad (3)$$

where $\hat{Y}_1$ = predicted xylose consumption (g/L), $\hat{Y}_2$ = predicted yield factor (g/g), $\hat{Y}_3$ = predicted volumetric g(Lh), X_1 = coded aeration level, X_2 = coded initial pH, and X_3 = coded fermentation time.

Deriving the Eqs. 2 and 3 and making it equal to zero, the maximum values for yield factor and volumetric productivity obtained in the range studied were 0.71 g/g and 0.46 g/(Lh), respectively. The value for yield factor was calculated considering the coded levels −1 for aeration and 0.5

Table 3
Anova for Xylose Consumption as Function of Parameters Aeration, pH, and Fermentation Time

Effect	Sum of squares	Degrees of freedom	P-value
A: Aeration	632.02500	1	0.0017
B: pH	996.00400	1	0.0007
C: Time	425.10400	1	0.0035
Lack-of-fit	262.16426	11	0.2402
Pure error	44.40800	4	
Total (corr.)	2359.70526	18	

R-squared = 0.87008.

Table 4
F-values from Estimate of Effects in 2^3 Factorial with Face-Centered Design

Statistical parameter	F-value		
	X_C	$Y_{P/S}$	Q_P
Regression	33.49	10.99	13.59
Lack-of-fit	2.15	0.68	1.21

for pH. For volumetric productivity, the coded levels were +1 and 0.54 for aeration and pH, respectively. These aeration levels were employed because the points of maximum related to this variable are out of the range studied for both responses.

The analysis of variance for xylose consumption, with nonsignificant effects eliminated, is presented in Table 3. The percentage of variance explained in this analysis (R^2 = 87%) indicates that the selected model is likely to be adequate for describing the xylose consumption behavior as a function of the factors within the range studied. The significance level (P-values) found for aeration, pH, and fermentation time confirmed the significant effect of these factors on xylose consumption. Lack-of-fit was not found to be significant for this model. The validity of the model can be also confirmed by the F-test for regression model (Table 4).

Table 5 shows the analysis of variance for the yield factor. As can be concluded from the P-values, the first-order effect of the initial pH, the interaction between aeration and pH, and the second-order effect of pH were once more significant terms for the yield factor. The percentage of variance explained (R^2) was 76% out of 81% of variance explainable. This demonstrates that the model is adequate for this response, which can be confirmed by the F test using the values of Table 4, and by the P-value found for lack-of-fit in Table 5.

Table 5
Anova for Yield Factor as Function of Parameters Aeration and pH

Effect	Sum of squares	Degrees of freedom	P-value
A: Aeration	0.0057600	1	0.1527
B: pH	0.0240100	1	0.0102
AB	0.0220500	1	0.0127
B^2	0.0443650	1	0.0016
Lack-of-fit	0.0065827	4	0.6187
Pure error	0.0240429	10	
Total (corr.)	0.12681053		

R-squared = 0.758494.

Table 6
Anova for Volumetric Productivity as Function of Parameters Aeration, pH, and Fermentation Time

Effect	Sum of squares	Degrees of freedom	P-value
A: Aeration	0.0810000	1	0.0085
B: pH	0.1488400	1	0.0028
C: Time	0.0000000	1	1.0000
BC	0.0144500	1	0.1109
B^2	0.0260950	1	0.0518
Lack-of-fit	0.0377456	9	0.4600
Pure error	0.0138800	4	
Total (corr.)	0.32201053		

R-squared = 0.839677.

The analysis of variance for volumetric productivity is presented in Table 6. P-values show the significance of the first-order effects of aeration and pH, and of the second-order effect of pH at 95% confidence level. The interaction between pH and fermentation time, significant at 90% confidence, was also considered for the model of volumetric productivity. Likewise, for this response, the percentage of variance explained ($R^2 = 84\%$), the F-test in Table 5 and the P-value for the lack-of-fit indicate the adequacy of the selected model.

The analysis of variance showing the value of R^2 for full regression in the three models determined is presented in Table 7.

The response surfaces for the models determined are shown in Figs. 1–3. For plotting the response surfaces for xylose consumption and volumetric productivity, the fermentation time was set at the coded value +1 because of its positive effect in the utilization of the xylose present in the

Table 7
Analysis of Variance for Full Regression of Models Determined

Statistical parameter	Xylose consumption		Yield factor		Volumetric productivity	
	Model	Error	Model	Error	Model	Error
Sum of squares	2053.13	306.572	0.0904	0.0364	0.2704	0.0516
Degrees of freedom	3	15	3	15	4	14
Mean square	684.3780	20.4382	0.0301	0.0024	0.0676	0.0037
F-ratio	33.4853		12.4259		18.3310	
P-value	0.0000		0.0002		0.0000	
R^2	0.8701		0.7131		0.8397	

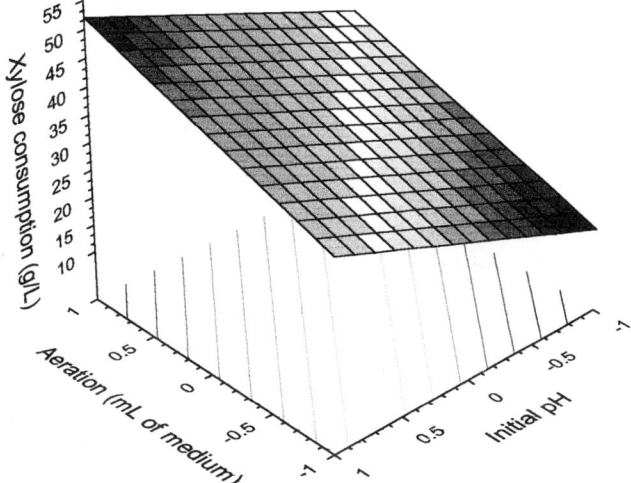

Fig. 1. Response surface and contour plot for the predicted model for xylose consumption.

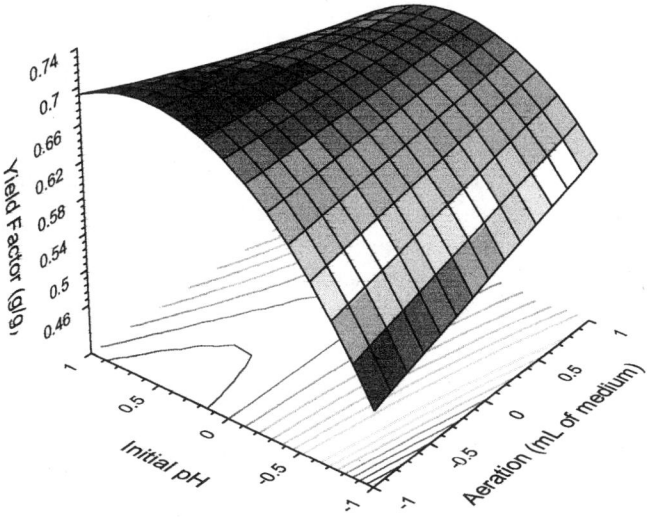

Fig. 2. Response surface and contour plot for the predicted model for yield factor.

hydrolysate. As can be seen from Fig. 1, xylose consumption augments linearly with the increase of both aeration and initial pH. Within the ranges studied, the maximum value for the yield factor was predicted with the lowest value of aeration (60 mL of medium in the flask); maximum volumetric productivity was predicted with the highest aeration (20 mL of medium in the flask, corresponding to the coded level +1). For maximum values for both responses, the initial pH was approx 6.5, corresponding to the coded level 0.5. This opposite effect of the aeration for the yield

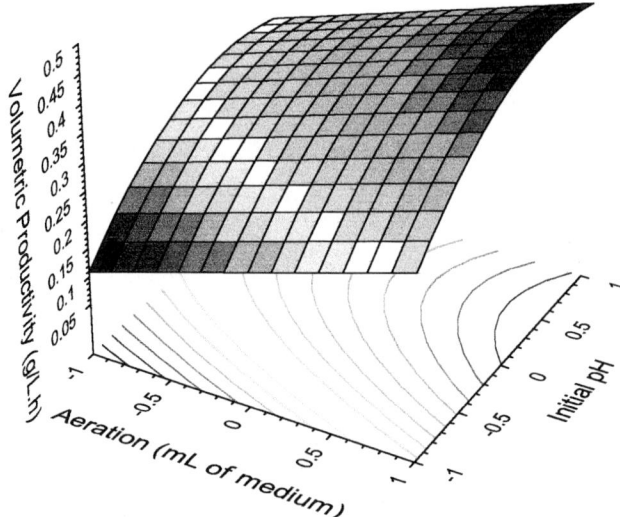

Fig. 3. Response surface and contour plot for the predicted model for volumetric productivity.

factor and productivity in xylitol was also observed for *C. mogii* ATCC 18364 by Sirisansaneyakul et al. (*12*). According to these authors, this was probably caused by the effect of accumulation of intracellular xylose, which affects its own transport across the cell membrane. The most suitable pH value for xylitol yield and productivity found in this study was also found in a study using a strain of *Candida boidinii* (*1*). Nevertheless, the influence of the initial pH on this bioconversion seems to be dependent on the yeast strain employed, and on the composition of the fermentation medium (*3,5*). The latter has been associated with the presence of acetic acid, which occurs in hydrolysate-based media. Indeed, the xylitol production using *Candida* sp B-22 in synthetic medium without acetic acid was not influenced by the initial pH (*6*). Although the initial cell concentration has been found to influence this bioconversion (*1,6*), no influence was observed in the present work.

CONCLUSION

According to the statistical analysis of the results, it can be concluded that aeration and initial pH significantly affect xylose consumption, yield factor, and volumetric productivity. The fermentation time also affects the xylose consumption, and its interaction with initial pH influences the volumetric productivity. Highest yield factor is achieved using 60 mL of medium and initial pH of 6.2, and highest volumetric productivity, using 20 mL of medium and initial pH of 6.4. Further experiments will be conducted in a bench-scale fermentor, employing the significant variables selected in the present study.

ACKNOWLEDGMENTS

ZDVLM acknowledges the financial support from the Fundação de Amparo à Pesquisa do Estado de São Paulo (FAPESP), Brazil. The authors wish to thank Maria Eunice M. Coelho for reviewing this manuscript.

REFERENCES

1. Vandeska, E., Amartey, S., Kuzmanova, S., and Jeffries, T. W, (1995), *World J. Microbiol. Biotechnol.* **11,** 213–218.
2. Roseiro, J. C., Peito, M. A., Girio, F. M., and Amaral-Collaço, M. T. (1991), *Arch. Microbiol.*, **156,** 484–490.
3. Noulleau, V., Preziosi-Belloy, L., and Navarro, J. M. (1995), *Biotechnol. Lett.* **17,** 417–422.
4. Furlan, S. A., Bouilloud, P., and de Castro, H. F. (1994), *Process Biochem.* **29,** 657–662.
5. Roberto, I. C., Silva, S. S., Felipe, M. G. A., Mancilha, I. M., and Sato, S. (1996), *Appl. Biochem. Biotechnol.* **57/58,** 339–347.
6. Cao, N.-J., Tang, R., Gong, C.-S., and Chen, L. F. (1994), *Appl. Biochem. Biotechnol.* **45/46,** 515–519.
7. Roberto, I. C., Sato, S., and Mancilha, I. M. (1996), *J. Ind. Microbiol.* **16,** 348–350.
8. Mayerhoff, Z. D. V. L., Roberto, I. C., and Silva, S. S. (1997), *Biotechnol. Lett.* **19,** 407–409.
9. Parajó, J. C., Domínguez, H., and Domínguez, J. M. (1996), *Bioresource Technol.* **57,** 179–185.
10. Roberto, I. C., Sato, S., Mancilha, I. M., and Taqueda, M. E. S. (1995), *Biotechnol. Lett.* **17,** 1223–1228.
11. Box, G. E. P., Hunter, W. G., and Hunter, J. S. (1978), *Statistic for Experimenters: an Introduction to Design, Data Analysis and Model Building*, John Wiley, New York.
12. Sirisansaneeyakul, S., Rizzi, M., and Reuβ, M. In *DECHEMA Biotechnology Conferences 5*. VCH Verlogsgemeinschaft, Weinheim (1992), pp. 541–544.

Improving Fermentation Performance of Recombinant *Zymomonas* in Acetic Acid-Containing Media

HUGH G. LAWFORD* AND JOYCE D. ROUSSEAU

Bio-engineering Laboratory Department of Biochemistry, University of Toronto, Toronto, Ontario, Canada M5S 1A8

ABSTRACT

In the production of ethanol from lignocellulosic biomass, the hydrolysis of the acetylated pentosans in hemicellulose during pretreatment produces acetic acid in the prehydrolysate. The National Renewable Energy Laboratory (NREL) is currently investigating a simultaneous saccharification and cofermentation (SSCF) process that uses a proprietary metabolically engineered strain of *Zymomonas mobilis* that can coferment glucose and xylose. Acetic acid toxicity represents a major limitation to bioconversion, and cost-effective means of reducing the inhibitory effects of acetic acid represent an opportunity for significant increased productivity and reduced cost of producing fermentation fuel ethanol from biomass. In this study, the fermentation performance of recombinant *Z. mobilis* 39676:pZB4L, using a synthetic hardwood prehydrolysate containing 1% (w/v) yeast extract, 0.2% KH_2PO_4, 4% (w/v) xylose, and 0.8% (w/v) glucose, with varying amounts of acetic acid was examine. To minimize the concentration of the inhibitory undissociated form of acetic acid, the pH was controlled at 6.0. The final cell mass concentration decreased linearly with increasing level of acetic acid over the range 0–0.75% (w/v), with a 50% reduction at about 0.5% (w/v) acetic acid. The conversion efficiency was relatively unaffected, decreasing from 98 to 92%. In the absence of acetic acid, batch fermentations were complete at 24 h. In a batch fermentation with 0.75% (w/v) acetic acid, about two-thirds of the xylose was not metabolized after 48 h. In batch fermentations with 0.75% (w/v) acetic acid, increasing the initial glucose concentration did not have an enhancing effect on the rate of xylose fermentation. However, nearly complete xylose fermentation was achieved in 48 h when the bioreactor was fed glucose. In the fed-batch system, the rate of glucose feeding (0.5 g/h) was designed to simulate the rate of cellulolytic diges-

*Author to whom all correspondence and reprint requests should be addressed.

tion that had been observed in a modeled SSCF process with recombinant Zymomonas. In the absence of acetic acid, this rate of glucose feeding did not inhibit xylose utilization. It is concluded that the inhibitory effect of acetic acid on xylose utilization in the SSCF biomass-to-ethanol process will be partially ameliorated because of the simultaneous saccharification of the cellulose.

Index Entries: Recombinant *Zymomonas*; acetic acid; xylose; ethanol; SSCF; cofermentation; synthetic biomass prehydrolyzate; glucose fed-batch.

INTRODUCTION

In a survey of industrial biocatalysts, designed to identify promising host strains for genetic transformation directed to rapid and efficient ethanologenic pentose metabolism, the Gram-negative bacterium *Zymomonas mobilis* met the selection criteria, which were based on several fermentation performance characteristics considered to be essential, as well as a number of secondary traits considered to be desirable, for a commercial biomass-to-ethanol process *(1,2)*.

Scientists at the National Renewable Energy Laboratory (NREL) (Golden, CO) have constructed a series of transformation vectors, consisting of a marker gene for tetracycline resistance and four xylose metabolism genes (xylose isomerase, xylulokinase, transaldolase, and transketolase) cloned from *Escherichia coli (3)*, and the proprietary xylose-fermenting *Z. mobilis* recombinants *(4)* are among several biocatalysts for the production of ethanol from biomass that are currently under investigation. The cellulose component of lignocellulosic biomass is recalcitrant to enzymic digestion, unless the impediments caused by the acetylated pentosans (hemicellulose) and lignin are removed by pretreatment *(5)*. Thermochemical pretreatment of lignocellulosic biomass is efficient and cost-effective *(6–8)*, and acetic is a well-known byproduct of dilute-acid prehydrolysis *(9,10)*. The acetic acid concentration of lignocellulosic prehydrolysate can be predicted from the structure and composition of the biomass feedstock *(11)*. NREL is currently working with yellow poplar wood as a feedstock for cellulosic ethanol. The prehydrolysate produced from this hardwood by dilute-acid pretreatment contains about 4% (w/v) xylose, 0.8% (w/v) glucose, and acetic acid in the range of 1.2–1.5% (w/v) *(11–14)*.

Acetic acid is known to be an effective antimicrobial agent, and is used as such in the food and beverage industries *(15)*. The inhibitory effect of acetic acid on ethanologenic biocatalysts is well-documented (for review, *see* ref. *14*). The mechanism of acetic acid toxicity is also well-understood (for review, *see* ref. *16*), and relates to the ability of the undissociated (protonated) form of the weak acid (pK_a = 4.75) to traverse the cell membrane and to act as a membrane protonophore (i.e., proton transporter), thereby causing an acidification of the cytoplasm. Hence, the inhibitory effect of acetic acid is pH-dependent *(17–19)*, and derives from its ability to interfere with the homeostatic mechanisms related to the maintenance of a constant intracellular pH *(18,19)*.

The authors have studied the effect of acetic acid on ethanologenic recombinant *E. coli* *(13,18,20–22)*, and both wild-type *(19,23)* and recombinant *Z. mobilis* *(24)*. Using recombinant *E. coli*, observed that the energetic uncoupling effect of acetic acid, as reflected in the decreased growth yield, is more pronounced with xylose as energy source, compared to glucose *(18)*. For *Z. mobilis* ATCC 29191 growing in a glucose-based semisynthetic medium, with the pH controlled at 6.0, the final cell mass concentration was reduced about 25% at an acetic acid concentration of 0.5% (w/v) *(19)*.

At this symposium last year, the authors reported on the growth and fermentation characteristics of recombinant *Zymomonas* CP4:pZB5 using synthetic hardwood prehydrolysate media, and in a systematic factorial analysis showed that acetic acid was the most statistically significant limiting factor for xylose utilization and seed production from a corn-steep liquor-based medium in which xylose was the principal sugar *(24)*. With the pH controlled at 6.0, the final cell mass concentration was reduced about 50% at an acetic acid concentration of 0.5% (w/v) *(24)*.

NREL is currently assessing a variety of bioconversion processes for converting lignocellulosic biomass to ethanol at an industrial scale. In the general process design, feedstock comminution is followed by a dilute-acid pretreatment process. Economic sensitivity analysis of several process designs has demonstrated the substantial cost reduction that accompanies modifying the design from one of sequential hydrolysis and fermentation (SHF) to an (simulataneous saccharification and fermentation (SSF) process *(1,25–27)*. Furthermore, there is potential for additional cost reduction (capital and operating costs) by combining the pentose fermentation and cellulose conversion (SSF) unit operations of the process *(2)*, using a biocatalyst with broad substrate specificity. At the seventeenth symposium in 1995, S. Picataggio presented a paper on NREL's proposed simultaneous saccharification co-fermentation (SSCF) process, which employed a proprietary *(28)* recombinant *Z. mobilis* strain CP4:pZB5. Those preliminary observations on the co-conversion efficiency of the recombinant *Zymomonas* were based on a model fermentation medium containing 4% (w/v) xylose, 0.8% (w/v) glucose, 6% (w/v) cellulose and cellulase, but there was no acetic acid in the medium *(28)*.

Using a hardwood prehydrolysate medium, NREL has screened several *Zymomonas* isolates as potential hosts for transformation *(1)*. Ongoing *Zymomonas*-related research and development at NREL is now focused on another similarly engineered recombinant in which *Z. mobilis* ATCC 39676 was transformed with a plasmid designated as "pZB4L". For the past year, our research at the University of Toronto, in collaboration with NREL, has involved studies on the physiological characteristics of this recombinant *Zymomonas* variant, and in a separate paper presented at this symposium, the authors are reporting on the continuous fermentation performance of this same recombinant culture (29).

The purpose of this study was to examine the effect of simultaneous cellulose saccharification on xylose fermentation, using an acetic acid-containing synthetic biomass prehydrolysate medium. The author used a fed-batch system in which glucose feeding mimicked a constant rate of cellulose hydrolysis. The synthetic prehydrolysate medium and the glucose feed medium were balanced with respect to acetic acid concentration. It was observed that, at relatively low rates of glucose loading, the inhibitory effect of acetic acid was apparently reduced, as reflected in the reduced amount of time required for nearly complete xylose utilization.

MATERIALS AND METHODS

Organism

The xylose-utilizing recombinant Z. mobilis strain ATCC 39676, carrying the plasmid pZB4L (designated as Zm 39676:pZB4L), was received from M. Zhang (NREL, Golden, CO). Stock cultures were stored in glycerol at −70°C, and precultures were prepared as previously described (24).

Fermentation Medium and Equipment

The fermentation medium was a nutrient-rich, complex medium (designated as RM) consisting of distilled water with 10g/L Difco Yeast Extract (YE) (Difco, Detroit, MI) and 2g/L KH_2PO_4 (30). In all experiments, the medium also contained 4% (w/v) xylose and tetracycline (10 mg/L), but the amount of glucose and acetic acid was variable. Batch and fed-batch fermentations were conducted in 2-L bioreactors (model F2000 MultiGen, New Brunswick Scientific, Edison, NJ) fitted with agitation (100 RPM), pH, and temperature control (30°C). The pH was monitored using a sterilizable combination pH electrode (Ingold), and was controlled at a set-point of 6.0 by automatic titration with 4 N KOH (NBS model pH-40 controller). Fermentations were initiated by directly transferring about 100 mL of a flask preculture into the bioreactor containing 1400 mL of sterilized medium. The initial cell density was monitored spectrophotometrically to give an OD_{600} (1-cm light path) in the range 0.1–0.2, corresponding to 25–50 mg dry wt cells/L. In fed-batch fermentations, the concentration of acetic acid in the feed medium was identical to that in the fermentation medium. A peristaltic pump was used to deliver sterile medium at a constant rate through the central agitator shaft of the bioreactor, and the flow rate (range 8–10 mL/h) was determined with the aid of an in-line pipet.

Analytical Procedures, Growth, and Fermentation Parameters

Growth was measured turbidometrically at 600 nm (1-cm light path) (Unicam spectrophotometer, model SP1800). In all cases, the blank cuvet contained dH$_2$O. Dry cell mass (DCM) was determined by microfiltration of an aliquot of culture, followed by washing and drying of the filter to constant weight under an infrared heat lamp. Neither the OD or the final cell mass were corrected for dilution in fed-batch experiments. Fermentation media and cell-free spent media were compositionally analyzed by HPLC as previously described (24). The ethanol yield ($Y_{p/s}$) was calculated as the mass of ethanol produced per mass of sugar added to the medium. In the plots of fermentation time-courses, the values given for the concentrations of xylose, glucose, and ethanol were those of the sample medium, and they were not corrected for dilution in the case of fed-batch fermentations.

RESULTS AND DISCUSSION

Effect of Acetic Acid in pH-Stat Batch Fermentations

Since much of the previous work with another similarly engineered NREL recombinant, namely *Z. mobilis* CP4:pZB5 (3,24), had been performed using RM medium (30), for purposes of direct comparison, this same RM medium was selected for use in this study. The RM medium was supplemented with 4% (w/v) xylose and 0.8% (w/v) glucose, to mimic the composition of a dilute-acid hardwood hemicellulose hydrolysate which is known to also contain about 1.5% (w/v) acetic acid (14). The pH-dependent effect of acetic acid on ethanologenic recombinant *E. coli* (18), was previously examined as well as wild-type *Z. mobilis* (19) and recombinant Zm CP4:pZB5 (24). Because the causative agent in acetic acid toxicity is the undissociated form of the acid, the inhibitory effect can be reduced by operating the fermentor at a pH control set-point that is elevated above the pK_a value of 4.75 (19,24). The pH optimum for both growth rate and cell mass yield for wild-type *Z. mobilis* is close to 6.0 (31,32), and therefore this pH value was selected as the control set-point in this study.

Figure 1 (Expt. B1) shows a typical growth and fermentation time-course, using the recombinant culture Zm 39676:pZB4L in the nutrient-rich RM medium. The final cell mass concentration was 1.44 gDCM/L, and the ethanol yield was 0.50 g/g, equivalent to a conversion efficiency of 98% of theoretical maximum (Table 1). Under these conditions, the growth and fermentation performance of NREL recombinants Zm CP4:pZB5 and 39676:pZB4L are identical (24). The addition of 0.4% (w/v) acetic acid to the RM medium (Expt. B2) resulted in a decrease in both growth rate and yield, but the ethanol yield remained unaffected (Fig. 1 and Table 1). With 0.4% acetic acid in the medium (at pH 6), the final cell mass decreased from 1.44 to 0.91 gDCM/L (Table 1); the time required for complete xylose utilization increased from 24 to 34 h (Fig. 1B). These observations on the effect of acetic acid on recombinant Zm 39676:pZB4L are very similar to those previously reported for recombinant Zm CP4:pZB5 using a medium with corn-steep liquor as the sole nutritional supplement (24).

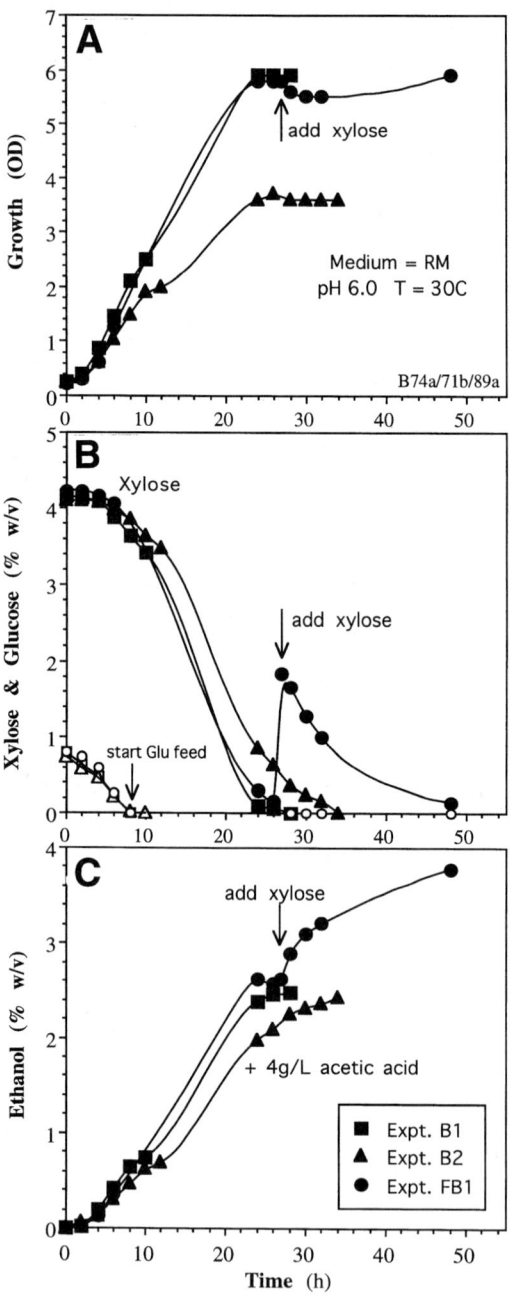

Fig. 1. Time-course of pH-stat batch and fed-batch fermentations with recombinant *Z. mobilis* 39676:pZB4L. **(A)** Growth, **(B)** glucose and xylose utilization, and **(C)** ethanol production. The medium was RM and was supplemented with 4% (w/v) xylose and 0.8% (w/v) glucose. The temperature was kept constant at 30°C. The pH-control setpoint was 6.0. The observed final cell mass concentrations, ethanol yield values, and the conditions for the fed-batch Expt. FB1, are given in Table 1.

Table 1
Effect of Acetic Acid on Cofermentation of Xylose and Glucose by Recombinant *Zymomonas* 39676:pZB4L in Batch and Fed-Batch Systems

Expt.	Medium composition					Fermentation parameters					
	Xylose % (w/v)	Glucose % (w/v)	Ac acid % (w/v)	Glucose in feed % (w/v)	Ac in feed % (w/v)	Feed rate (ml/h)	Feed interval (h)	Total Glu (g)	Final cell mass (gDCM/L)	Ethanol yield (g/g)	Conversion efficiency % (theoret. max)
Batch fermentations											
B1	4.0	0.8	0	—	—	—	—	12	1.44	0.50	98
B2	4.0	0.8	0.40	—	—	—	—	12	0.91	0.50	98
B3	4.0	0.8	0.75	—	—	—	—	12	0.43	0.47	92
B4	4.0	2.2	0.75	—	—	—	—	33	0.49	0.49	96
Fed-batch fermentations											
FB1	4.0[a]	0.8	0	5.2	0	9.6	8–48	32	1.40	0.50	98
FB2	4.0	0.8	0.75	0	0.75	8.9	0–48	12	0.31	0.49	96
FB3	4.0	0.8	0.75	5.0	0.75	8.5	0–48	33	0.53	0.48	94
FB4	4.0	0.8	0.75	13.6	0.75	8.7	0–48	69	0.70	0.48	94

Note: The base medium (RM) contained 10g/L Difco yeast extract and 2 g/L KH_2PO_4. The temperature of the pH-stat stirred-tank fermentors was maintained constant at 30°C, and the pH was controlled at 6.0. Maximum theoretical ethanol yield = 0.51 g/g. DCM was determined by ultrafiltration (see Methods), and was not corrected for dilution in fed-batch fermentations.

Abbreviations: Ac, acetic acid; B, batch fermentation; FB, fed-batch fermentation

[a] In Expt. FB1, xylose was added at 27 h, and the concentration was 1.8% (w/v) (see Fig. 1).

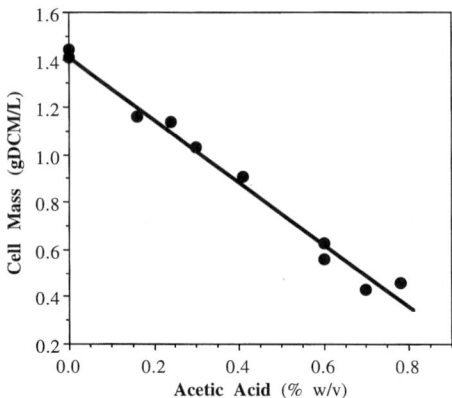

Fig. 2. Final cell mass concentration for recombinant Zm 39676:pZB4L as a function of the acetic acid concentration. These data represent a series of batch fermentations similar to Expt. B2 shown in Fig. 1. DCM was determined by ultrafiltration, as described in Materials and Methods. The RM medium contained 4% (w/v) xylose and 0.8% (w/v) glucose, and varying amounts of acetic acid. The temperature was 30°C and the pH was controlled at 6.0.

The acetic acid concentration of the dilute-acid poplar wood prehydrolysate produced by NREL is approx 1.5% (w/v) (14). However, based on observations with recombinant Zm CP4:pZB5 in acetic acid-containing synthetic biomass prehydrolysate media, the recommended upper limit for acetic acid for the purpose of seed production (i.e., preparation of inoculum) was 0.75% (w/v), equivalent to that of a 50% prehydrolysate medium (24). Figure 2 shows a plot of the final cell mass concentration as a function of the acetic acid concentration (range 0–0.8%) in RM medium containing 4% (w/v) xylose and 0.8% (w/v) glucose. The results shown in Figure 2 for recombinant Zm 39676:pZB4L are very similar to those previously observed with recombinant Zm CP4:pZB5 under the same operating conditions (unpublished results). In Fig. 2, the line created by linear regression analysis is represented by the following relationship: DCM concentration (g/L) = 1.42 − 1.33 (% w/v acetic acid) (with a regression coefficient of 0.985).

Effect of Acetic Acid in pH-Stat Fed-Batch Fermentations

One of several process designs currently under assessment at NREL for the conversion of biomass to ethanol on an industrial scale is simultaneous saccharification and cofermentation (SSCF) (2). In such a process, the saccharifying enzymes release glucose from cellulose and cellobiose. In this study, the effect of simultaneous saccharification of cellulose on the efficiency of xylose fermentation by recombinant Zm 39676:pZB4L was examined using a fed-batch system in which glucose is fed to the pH-stat bioreactor at rates commensurate with anticipated rates of cellulose hydrolysis by the exogenous enzymes in the SSCF, as proposed by NREL (28). In the fed-batch fermentation experiment, designated as FB1, 5.2% (w/v) glucose was fed at a rate of 9.6 mL/h, commencing after 8 h of batch fermentation. To prevent complete exhaustion of xylose from the medium, additional xylose (1.8%) was added after 27 h (Fig. 1). In this experiment, there was no acetic acid in the medium. This level of glucose feeding did not appear to affect the rate of xylose conversion (Fig. 1B). After 48 h, the final ethanol concentration was 3.7% (w/v), representing an ethanol yield from both xylose and glucose of 0.50 g/g (98% conversion efficiency) (Fig. 1C). The tailing off that is observed in the xylose concentration trajectory in the latter stages of the fermentation (Fig. 1B) is probably caused by the competition by glucose and xylose for uptake by the common transporter (33). Based on a cell mass concentration of about 1.4 gDCM/L, the specific productivity associated with cofermentation by recombinant 39676:pZB4L over the fed-batch fermentation interval of 28–34 h is estimated at approx 0.7 g ethanol/gDCM/h (Fig. 1C).

In this study, the highest level of acetic acid tested was 0.75% (w/v), which represents about a 50% dilution of the level anticipated in hardwood prehydrolysate. The effect of this level of acetic acid in batch fermentation, with the pH controlled at 6.0, is represented by Exp. B3 (Fig. 3). The final cell mass concentration is only about one-third of that observed in the absence of acetic acid (Fig. 3), and the ethanol yield (based on sugar consumed) is also reduced from 0.50 to 0.47 g/g (Table 1). With 0.75% acetic acid in the medium, only about one-third of the xylose is consumed after 48 h (Fig. 3B). However, when a solution of 5% glucose and 0.75% acetic acid is fed to the bioreactor (Expt FB3), at a similar rate to that used in experiment FB1, the xylose was almost completely fermented after 48 h (Fig. 3B). Since the xylose concentration trajectories in Fig. 3B were not corrected for the volumetric dilution effect produced by glucose feeding, a control experiment (Expt. FB2) was performed in which the bioreactor was fed with a solution of 0.75% acetic acid (Fig. 3B). Fed-batch experiment FB2 shows that the dilution of xylose, caused by feeding, produced an artificial or apparent improvement in the rate of xylose utilization; however, this was not significant compared to the enhancing effect produced by the presence of glucose in the feed (Fig. 3B).

Supplementing the medium with glucose can be expected to result in an increase in the cell mass concentration, and the rate of xylose utilization can be expected to be proportional to the cell concentration. Increasing the concentration of glucose in the feed about 2.5-fold (Expt. FB4) did not decrease the time required for nearly complete xylose fermentation (Fig. 3), despite the increase in the observed final cell density from 0.53 to 0.70 g DCM/L (Fig. 2A and Table 1). In *Zymomonas*, the uptake of glucose and xylose is by a common transporter that exhibits much higher affinity for glucose than xylose (33). The failure of extra glucose to further improve the rate of xylose utilization is probably caused by the preference of

Fig. 3. Time-course of pH-stat batch and fed-batch fermentations with recombinant *Z. mobilis* 39676:pZB4L. **(A)** Growth, **(B)** glucose and xylose utilization, and **(C)** ethanol production. The experimental conditions, as well as the values for final cell mass concentrations and ethanol yields, are presented in Table 1.

the common transporter for glucose (33). Hence, the accumulation of glucose over the initial 10 h interval could be expected to have a retarding effect on xylose uptake during that same interval. To examine this phenomenon further, a separate batch fermentation was performed (Expt. B4), in which the initial concentration of glucose in the medium was increased from 0.8 to 2.2% (w/v). Although there was an increase in growth (Fig. 3A), with the final cell mass increasing from 0.43 to 0.49 gDCM/L (Table 1), xylose utilization was significantly decreased with only about 10% of the xylose consumed after 48 h (Fig. 3B). This observation emphasizes the importance of the proper balance between the concentration of the two sugars during cofermentation.

It was concluded that, under certain well-defined conditions, the detrimental effect of acetic acid on xylose utilization by recombinant *Zymomonas* can be reduced, if glucose is supplied for cofermentation at a rate that does not result in glucose accumulation, since this appears to interfere with xylose uptake. Consequently, it is expected that the inhibitory effect of acetic acid in the SSCF process will be partially ameliorated because of the continuous supply of glucose provided by the action of the exogenous cellulolytic enzymes.

ACKNOWLEDGMENTS

This work was funded by the Biochemical Conversion Element of the Office of Fuels Development of the US Department of Energy. Research conducted at the University of Toronto was part of Phase III of a Subcontract AAP-4-11195-03 from NREL. The authors are grateful to M. Zhang for providing the recombinant *Z. mobilis* 39676:pZB4L, and to J. McMillan for helpful discussions.

REFERENCES

1. Zhang, M., Franden, M. A., Newman, M., McMillan, J., Finkelstein, M., and Picataggio, S. (1995), *Appl. Biochem. Biotechnol.* **51/52,** 527–536.
2. Picataggio, S. K., Zhang, M., and Finkelstein, M. (1994), in *Enzymatic Conversion of Biomass for Fuels Production*, Himmel, M. E., Baker, J. O., and Overend, R. A., eds., American Chemical Society, Washington, DC, *ACS Symposium Series* 566, pp. 342–362.
3. Zhang, M., Eddy, C., Deanda, K., Finkelstein, M., and Picataggio, S. K. (1995), *Science*, **267,** 240–243.
4. Picataggio, S. K., Zhang, M., Eddy, C. K., Deanda, K. A., and Finkelstein, M. (1996), US Patent 5,514,583.
5. Wright, J. D. (1988), *Chem. Eng. Prog.* **84,** 62–74.
6. Grethlein, H. E. (1985), *BioTechnology* **3,** 155–160.
7. Grethlein, H. E., Allen, D. C., and Converse, A. O. (1984), *Biotech. Bioeng.* **26,** 1498–1505.
8. Torget, R., Werdene, P., Himmel, M., and Grohmann, K. (1990), *Appl. Biochem. Biotechnol.* **24/25,** 115–126.
9. Grohmann, K., Himmel, M., Rivard, C., Tucker, M., Baker, T. Torget, R., and Graboski, M. (1984), *Biotechnol. Bioeng. Symp.* **14,** 139–157.
10. Kong, F., Engler, C. R., and Soltes, E. (1992), *Appl Biochem. Biotechnol.* **34/35,** 23–35.

11. Timell, T. E. (1964), *Adv. Carbohydr. Chem.* **19,** 247–302.
12. Lawford, H. G. and Rousseau, J. D. (1993), in *Energy from Biomass and Wastes XVI.* (1992) Klass, D. L., ed., Institute of Gas Technology, Chicago, pp. 559–597.
13. Lawford, H. G. and Rousseau, J. D. (1993), *Appl. Biochem. Biotechnol.* **39/40,** 667–685.
14. McMillan, J. D. (1994), in *Enzymatic Conversion of Biomass for Fuels Production,* Himmel, M. E., Baker, J. O., and Overend, R. A., eds., American Chemical Society, Washington, DC, *ACS Symposium Series* 566, pp. 411–437.
15. Freese, E., Sheu, C. W., and Galliers, E. (1973), *Nature* **241,** 321–326.
16. Booth, I. R. (1985), *Microbiol. Rev.* **49,** 63–91.
17. Prior, B. A., Killan, S. G., and du Preez, J. C. (1989), *Process Biochem.* **24,** 21–32.
18. Lawford, H. G. and Rousseau, J. D. (1993), *Appl. Biochem. Biotechnol.* **39/40,** 301–322.
19. Lawford, H. G. and Rousseau, J. D. (1994), *Appl. Biochem. Biotechnol.* **45/46,** 437–448.
20. Lawford, H. G. and Rousseau, J. D. (1991), *Biotechnol. Letts.* **13,** 191–196.
21. Lawford, H. G. and Rousseau, J. D. (1993), in: *Energy from Biomass and Wastes XVI,* Klass, D. L., ed., Institute of Gas Technology, Chicago, IL, pp. 559–597.
22. Lawford, H. G. and Rousseau, J. D. (1992), *Appl. Biochem. Biotechnol.* **34/35,** 185–204.
23. Lawford, H. G. and Rousseau, J. D. (1993), *Appl. Biochem. Biotechnol.* **39/40,** 687–699.
24. Lawford, H. G., Rousseau, J. D., and McMillan, J. D. (1997), *Appl. Biochem. Biotechnol.* **63–65,** 269–286.
25. Hinman, N. D., Wright, J. D., Hoagland, W., and Wyman, C. E. (1989), *Appl. Biochem. Biotechnol.* **20/21,** 391–401.
26. Hinman, N. D., Schell, D. J., Riley, C. J., Bergeron, P. W., and Walter, P. J. (1992), *Appl. Biochem. Biotechnol.* **34/35,** 639–649.
27. Lynd, L. R. (1990), *Appl. Biochem. Biotechnol.* **24/25,** 695–719.
28. Picataggio, S. K., Eddy, C., Deanda, K., Franden, M. A., Finkelstein, M., and Zhang, M. (1996), *Seventeenth Symposium on Biotechnology for Fuels & Chemicals* (Paper #9).
29. Lawford, H. G., Rousseau, J. D., Mohagheghi, A., and McMillan, J. D. (1998), *Appl. Biochem. Biotechnol.,* (19th Symp) **70–72**.
30. Goodman, A. E., Rogers, P. L., and Skotnicki, M. L. (1982), *Appl. Environ. Microbiol.* **44,** 496–498.
31. Lawford, H. G. and Ruggiero, A. (1990), *Biotechnol. Appl. Biochem.* **12,** 206–211.
32. Lawford, H. G., Holloway, P., and Ruggiero, A. (1988), *Biotechnol. Letts.* **10,** 809–814.
33. DiMarco, A. and Romano, A. H. (1985), *Appl. Environ. Microbiol.* **49,** 151–157.

Conditions that Promote Production of Lactic Acid by *Zymomonas mobilis* in Batch and Continuous Culture

Hugh G. Lawford* and Joyce D. Rousseau

Bio-engineering Laboratory, Department of Biochemistry, University of Toronto, Toronto, Ontario, Canada M5S 1A8

ABSTRACT

This study documents the similar pH-dependent shift in pyruvate metabolism exhibited by *Zymomonas mobilis* ATCC 29191 and ATCC 39676 in response to controlled changes in their steady-state growth environment. The usual high degree of ethanol selectivity associated with glucose fermentation by *Z. mobilis* is associated with conditions that promote rapid and robust growth, with about 95% of the substrate (5% w/v glucose) being converted to ethanol and CO_2, and the remaining 5% being used for the synthesis of cell mass. Conditions that promote energetic uncoupling cause the conversion efficiency to increase to 98% as a result of the reduction in growth yield (cell mass production). Under conditions of glucose-limited growth in a chemostat, with the pH controlled at 6.0, the conversion efficiency was observed to decrease from 95% at a specific growth rate of 0.2/h to only 80% at 0.042/h. The decrease in ethanol yield was solely attributable to the pH-dependent shift in pyruvate metabolism, resulting in the production of lactic acid as a fermentation byproduct. At a dilution rate (D) of 0.042/h, decreasing from pH 6.0 to 5.5 resulted in a decrease in lactic acid from 10.8 to 7.5 g/L. Lactic acid synthesis depended on the presence of yeast extract (YE) or tryptone in the 5% (w/v) glucose–mineral salts medium. At D = 0.15/h, reduction in the level of YE from 3 to 1 g/L caused a threefold decrease in the steady-state concentration of lactic acid at pH 6. No lactic acid was produced with the same mineral salts medium, with ammonium chloride as the sole source of assimilable nitrogen. With the defined salts medium, the conversion efficiency was 98% of theoretical maximum. When chemostat cultures were used as seed for pH-stat batch fermentations, the amount of lactic acid produced correlated well with the activity of the chemostat culture; however, the mechanism of this prolonged induction

*Author to whom all correspondence and reprint requests should be addressed.

effect is unknown. The levels of lactic acid produced by Z. *mobilis* in this study have not been previously reported. *Zymomonas* is Gram-negative, and at no time did microscopic inspection of lactic-acid-producing cultures indicate the presence of Gram-positive organisms. Although these observations are very preliminary in nature, they have implications for the regulation of glycolytic flux in *Zymomonas*, and demonstrate the possibility of an alternative fate for pyruvate previously presumed not to exist.

Index Entries: *Zymomonas*; lactic acid; lactic acid dehydrogenase; ethanol yield; continuous fermentation.

INTRODUCTION

Cost sensitivity analyses associated with the industrial-scale production of fermentation fuel ethanol have identified product yield as the most important factor affecting production costs *(1–3)*. The bacterium *Zymomonas mobilis* (for reviews, *see* refs. 4–6) exhibits fermentation performance characteristics *(7)* that have placed it high on the candidacy list of ethanologenic biocatalysts currently being investigated for their bioconversion efficiency in proposed fermentation processes utilizing diverse feedstocks *(8)*. Given the economic importance of ethanol yield *(3)*, a particularly attractive feature of *Zymomonas* is its high degree of ethanol selectivity, with a sugar-to-ethanol conversion efficiency typically in the range of 94–98% of theoretical maximum *(7,9–10)*. The genes for both pyruvate decarboxylase and alcohol dehydrogenase II have been cloned from *Zymomonas* and used to transform bacteria such as *Escherichia coli (11–14)* and *Klebsiella oxytoca (15)* which are capable of utilizing the different hexose and pentose sugars found in lignocellulosic hydrolysates, into highly ethanologenic recombinants. In a survey conducted by the National Renewable Energy Laboratory (NREL) (Golden, CO), *Zymomonas* was selected as the most promising host for metabolic engineering directed to the utilization of pentose sugars *(8)*. NREL used a hardwood prehydrolysate medium to test the hardiness of several *Zymomonas* strains as part of its screening for superior hosts *(16)* for genetic transformation with their proprietary *(17)* xylose assimilation and utilization plasmid *(18)*.

The authors previously, reported on the fermentation performance characteristics of one of NREL's recombinant *Zymomonas (18)* in a study that focused on the optimization of seed production for a proposed simultaneous saccharification cofermentation biomass-to-ethanol process *(19)*. In extending the investigation from batch to continuous cofermentations, it was observed that certain operating conditions appeared to promote a decrease in the ethanol yield because of the production of lactic acid. In their review of the biology of *Zymomonas*, Swings and DeLey *(4)* point out the conflicting reports in the early literature pertaining to lactic acid production by *Zymomonas* (previously known as *Termobacterium mobile* and *Pseudomonas lindneri*). Swings and DeLey *(4)* stated that "One of the important conclusions on the carbohydrate metabolism of *Zymomonas* is that, when growing in a complex medium, 98% of the glucose is converted to ethanol, CO_2, ATP, and heat, and only 2% is used as building material for the cell."

The concentrations of lactic acid that were observed in the author's work with recombinant *Zymomonas* were considerably greater than those reported by others for various fermentations using a variety of different *Z. mobilis* isolates (20–24) in which lactic acid production did not appreciably alter the ethanol yield. With respect to other fermentation byproducts, *Zymomonas* is known to produce mannitol, glycerol, and dihydroxyacetone from fructose *(21)*, and fructose, levan, and sorbitol from sucrose *(26–28)*. The production of these byproducts can cause the ethanol yield to decrease from 0.50 to 0.35 g/g *(29)*. Therefore, most of the reports in the literature relating to the effect of environmental conditions on ethanol yield in *Zymomonas* are based on sucrose fermentations *(23,24)*. The reduction of fructose to sorbitol is dependent on the dilution rate in continuous fermentations, with very little byproduct formation observed at low dilution rates *(23)*. In a study of nutritional effects on glucose conversion efficiency by chemostat cultures of *Z. mobilis*, Cromie and Doelle *(30)* concluded that "the conversion efficiency of the glucose taken up to ethanol is not affected at all by environmental conditions." Glucose is metabolized by the Entner-Doudoroff pathway in *Zymomonas*, and there is no synthesis of fructose *bis*-phosphate *(4–6)*, but the fate of glyceraldehyde-3-phosphate is the same as in the yeasts commonly used in ethanol fermentations, with the exception that in *Zymomonas* there are two isozymes of alcohol dehydrogenase *(31)*. The regulation of glycolytic flux in *Z. mobilis* has been investigated *(32,33)*, but there was no mention of lactic acid dehydrogenase. In the author's literature search concerning lactic acid and *Zymomonas* it was discovered that a segment of DNA that lies between the genes for phosphoglycerate mutase *(pgm)* and alcohol dehydrogenase I *(adhA)* has recently been assigned as the gene for a D-isomer-specific 2-hydroxyacid dehydrogenase *(ddh)* *(34)*. The designation of this gene as *ddh* was made on the basis of the correlation in both identity (31.5%) and similarity (55.4%) between its transcription product (331 amino acid polypetide with a NAD-binding domain) and the D-isomer lactic acid dehydrogenase from *Lactobacillus plantarum* *(34)*. However, the precise function of this gene in *Zymomonas* remains unknown *(34)*. Fructose *bis*-phosphate is known to play a regulatory role in lactic acid dehydrogenase activity in certain bacteria *(35)*, and, although it is not an intermediate of the glucose dissimilation pathway in *Zymomonas*, fructose-6-phosphate is produced during xylose metabolism as a product of the transketolase activity in recombinant *Z. mobilis* *(18)*.

The present study was undertaken with wild-type cultures of *Z. mobilis* to ascertain if the shift in metabolism, which was observed in continuous fermentations with recombinant *Z. mobilis* at relatively low dilution rates, was a response common to natural isolates, or a response that was a consequence of the genetic transformation—specifically, the operation of the metabolically engineered pentose metabolism pathway, by virtue of the expression of the four plasmid-encoded xylose utilization genes.

MATERIALS AND METHODS

Organisms

Z. mobilis strains ATCC 29191 and ATCC 39676 were obtained from the American Type Culture Collection (Rockland, MD).

Fermentation Media, Equipment, and Operating Conditions

The composition of the different media used in this study are described in Table 1. Bacto Yeast Extract (YE) and Bacto Tryptone were obtained from Difco (Detroit, MI). Other chemicals were laboratory-grade purity. Glass-distilled water was used to prepare all media.

Batch fermentations were conducted in 2-L MultiGen stirred-tank bioreactors (Model F2000, New Brunswick Scientific, Edison, NJ) fitted with agitation (100 RPM), pH, and temperature control (30°C). The working volume was 1500 mL, and the pH was controlled by the addition of 4 N KOH (NBS model pH-40 controller). Continuous fermentations were conducted with 750 mL MultiGen bioreactors (Model F1000, New Brunswick Scientific), except that the glass vessel had an overflow outlet. The working volume of the chemostats was about 350 mL. The flow rate of the feed medium was determined by collecting the effluent into a graduated cylinder for a specified period of time. Sampling was effected in a similar fashion. Steady state was assumed only after a minimum of 3 vol had exchanged. In all fermentations, the amount of preculture (inoculum) added was sufficient to produce an OD_{600nm} reading (1-cm light path) of 0.1–0.2 (equivalent to an initial cell density of approx 30–60 mg dry cell mass/L).

Analytical Procedures

Growth was measured turbidometrically at 600nm (1-cm light path) (Unicam spectrophotometer, model SP1800). In all cases, the blank cuvet contained dH_2O. Dry cell mass (DCM) was determined by microfiltration of an aliquot of culture, followed by washing and drying of the filter to constant weight under an infrared heat lamp. Fermentation media and cell-free spent media were compositionally analyzed by HPLC with a refractive index monitor and computer-interfaced controller/integrator (Bio-Rad, Hercules, CA). Separations were performed at 65°C, using an HPX-87H column (300 × 7.8 mm) (Bio-Rad), as previously described (19). The ethanol yield ($Y_{p/s}$) was calculated as the final concentration of ethanol divided by the concentration of glucose determined to be in the medium prior to inoculation. The $Y_{p/s}$ was not corrected for the dilution caused by

Table 1
Zymomonas Media Formulations

Ingredient (g/L)	Complex RM[a]	Semisynthetic ZM			Defined salts DS[b]	
Glucose	100	50	50	100	50	100
Yeast Extract (Difco)	10.0	3.0	–	3.0	–	–
Tryptone (Difco)	–	–	2.1	–	–	–
NH_4Cl	–	1.6	1.6	2.4	1.6	2.4
KH_2PO_4	2.0	3.48	3.48	3.48	3.48	3.48
$MgSO_4$	–	0.49	0.49	0.49	0.49	0.49
$FeSO_4 \cdot 7H_2O$	–	0.01	0.01	0.01	0.01	0.01
Citric acid	–	0.21	0.21	0.21	0.21	0.21
Ca Pantothenate (mg)	–	–	–	–	1.0	1.0
Biotin (mg)	–	–	–	–	1.0	1.0

[a] Goodman et al. (1982) *Appl. Environ. Microbiol.* **44:** 496–498.
[b] Ref. 43.

adding alkali during the fermentation. For the purpose of carbon balancing (% C recovery), the carbon content of the cell mass (48.7%) was considered to be constant. Carbon dioxide was not measured, but was assumed to be produced at a molar equivalent to ethanol.

RESULTS AND DISCUSSION

Z. mobilis ATCC 39676 is one of the strains selected by NREL for metabolic engineering directed to xylose fermentation, using their proprietary transformation vector (17) carrying E. coli genes for xylose isomerase, xylulose kinase, transaldolase, and transketolase (18). Recombinant Z. mobilis 39676:pZB4L is the subject of ongoing research conducted in collaboration with NREL. The fermentation performance of this recombinant in continuous cofermentations is reported elsewhere (36).

Z. mobilis ATCC 39676 is a patent strain (37) with claimed superior sucrose fermentation characteristics, which has been developed from the designated neotype strain Z. mobilis ATCC 29191 (NCIB 11199) by Doelle and his research colleagues at the University of Queensland (25,29,38). The performance of strain 39676 has been tested in large-scale fermentations (17,000 gal) using both milo (sorghum) (39) and corn starch (40). Jones and Doelle (41) used continuous cultures of strain 39676 in a laboratory investigation on the effects of pH and nutrient limitation on fermentation performance. Since Z. mobilis 39676 is a patent culture for which the physiological characteristics are largely unknown, were interested in comparing its fermentation characteristics to that of strain ATCC 29191, which has been the subject of extensive research in their laboratory over the past 18 years (9,10,42–45).

Under conditions of glucose-limited growth in continuous culture, using a semisynthetic medium containing 50 g/L glucose and 3 g/L yeast extract (YE) (Table 1), and with the pH controlled at 6.0, both *Z. mobilis* 39676 and 29191 produce lactic acid as a fermentation byproduct at dilution rates (i.e., growth rates) <0.2/h (Fig. 1A). Over the range of dilution rates tested (0.048 to 0.2/h), the steady-state concentration of lactic acid increased as the dilution rate decreased, and the pattern exhibited by both cultures, as a function of dilution rate, appeared similar (Fig. 1A). With strain 39676 at D = 0.042/h, the steady-state level of lactic acid decreased from 10.8 to 7.5 g/L when the set-point for the pH controller was changed from 6.0 to 5.5 (Fig. 1A). There was good closure of the carbon balance (results not shown), indicating that, apart from cell mass and carbon dioxide, lactic acid, and ethanol were the only products of glucose fermentation.

Accordingly, the decrease in ethanol yield from 0.484 g ethanol/g glucose (i.e., 95% conversion efficiency) at D = 0.2/h to 0.41 g/g (80% conversion efficiency) at D = 0.042/h, was assumed to be a direct consequence of the redirection of pyruvate metabolism from ethanol to lactic acid (Fig. 1B). *Z. mobilis* is Gram-negative (4), and at no time in this study did microscopic inspection of lactic acid-producing cultures reveal the presence of Gram-positive organisms such as lactobacilli. To the authors' knowledge, the synthesis of such high levels of lactic acid by *Z. mobilis* has not been previously reported in the literature. Viikari (23,27) reported that *Z. mobilis* strain VTT-2-78042 produced 0.2 g/L lactic acid (D- and L-isomers) from 15% (w/v) fructose, at pH 5.5. Schmidt and Schügerl (21) reported that *Z. mobilis* ATCC 29191 produced a maximum of 0.2g/L lactic acid from 10% (w/v) glucose in chemostat culture, at a dilution rate of about 0.04/h. What was particularly interesting about the observations of Schmidt and Schügerl (21) was the decrease in lactic acid concentration from a maximum at low dilution rates to a minimum at D = 0.2/h. Although the absolute concentrations were vastly different, the pattern of acidogenesis as a function of dilution rate was similar to the authors' results (Fig. 1A).

Figure 1 shows that replacing the YE-based semisynthetic medium in the chemostat reservoir with a defined salts (DS) medium, in which ammonium chloride is the sole source of assimilable nitrogen results in a virtual eradication of the lactic acid synthesis (with the possible exception of relatively low dilution rates [< 0.08/h]). Although Figure 1 shows only the results for *Z. mobilis* 29191, identical results were obtained with strain 39676. Even at the theoretical maximum ethanol yield (0.51 g/g), the ethanol in these fermentations is not expected to be at a growth-inhibitory concentration, and the value of the maximum dilution rate, with either the semisynthetic or defined salts medium, is known to be greater than 0.2/h (45). The addition of YE to strain 29191 grown in DS feed medium (1g/L), when the chemostat was being operated at a dilution rate of 0.15/h, resulted in the steady-state lactic acid concentration of 1.6g/L (Fig. 1A) which was about one-third the level of lactic acid produced by the same medium

Fig. 1. Glucose-limited continuous culture of Z. *mobilis* 29191 and 39676. **(A)** Steady-state concentration of lactic acid, **(B)** ethanol yield (g ethanol/g glucose), and **(C)** steady-state DCM concentration, as a function of dilution (growth) rate. All media contained 5% (w/v) glucose, the temperature was maintained at 30°C, and the pH was controlled at 6.0 by the addition of 4 N KOH. For Zm 39676 in ZM medium, the pH was controlled at 5.75. Details of media composition are given in Table 1. Also shown are separate experiments with Zm 39676 in which the pH was 5.5, and with Zm 29191 in which the level of YE in the ZM medium was 1 g/L. The arrow in B indicates the theoretical maximum ethanol yield of 0.51 g/g.

containing 3× as much YE. Apart from the obvious stoichiometric correlation between the YE concentration of the medium and the level of lactic acid, this observation suggested that a component of the YE was the causative agent that was somehow responsible for lactic acid synthesis. Yeast extract is a complex nutritional adjunct, and is generally regarded as a source of inorganic elements, vitamins, and organic nitrogen. Bacto tryptone has a composition similar to Bacto yeast extract (YE), but it lacks the vitamins and growth factors that are present in YE. Replacing the YE (3 g/L) in the semisynthetic medium with an equivalent amount of Bacto Tryptone *(46)* (2.1 g/L, based on specifications provided by Difco with respect to the total nitrogen content of their products) (Table 1) had no appreciable effect on the pattern of lactic acid production and ethanol yield as a function of dilution rate (Fig. 1). Collectively, these observations with different media suggest that the causative agent might be an amino acid(s). In this context, it is interesting to note that in continuous fermentations with Z. mobilis ATCC 31821, using defined mineral salts media containing mixtures of different amino acids *(47)*, Beyeler et al. *(48)* did not observe any appreciable difference in ethanol yield relative to a complex YE-based medium; however, the relatively high dilution rate at which the glucose-limited chemostats were operated in the work of Beyeler et al. *(48)* compromises the significance of their observations in terms of the proposed effect of amino acids on *Zymomonas* metabolism at low growth rates. Work is ongoing in our laboratory to ascertain the chemical nature of the causative agent and the mechanism for the regulation of pyruvate metabolism.

Batch Fermentations

There are two probable reasons why lactic acid has not been commonly observed as a product of *Zymomonas* metabolism. First, lactic acid is not produced in batch fermentations, which the inoculum has been produced by the usual procedure of flask culture. Under these conditions, growth is near maximal. In order to maximize productivity, most work with *Zymomonas* with continuous fermentation systems has been performed at relative high dilution (growth) rates. In systems in which cell recycle or cell retention (immobilized) has been employed, and in which growth is likely to be less than maximal, the ethanol yield is often lower than in comparable cell-free systems *(49)*. Second, the majority of *Zymomonas* fermentations have been done at pH 5.0–5.5 *(7)*, and it is known that the ethanol yield is increased at pH <6.0 *(44,50)*

Figure 2 shows a typical time-course for two batch fermentations with Z. mobilis 29191 in a stirred-tank fermentor, with the pH controlled at 6.0. Two different media were used: DS medium, with 2.4 g/L NH_4Cl as sole source of assimilable nitrogen (open squares); and a semisynthetic medium (ZM) of identical composition, except that it was supplemented with 3 g/L YE (filled squares) (Table 1 and Fig. 2). For these batch fermentations, the inoculum was generated in a glucose-limited chemostat operating at D = 0.12/h, with a DS medium (Table 1). The observed specific growth rates

Lactic Acid from Zymomonas mobilis

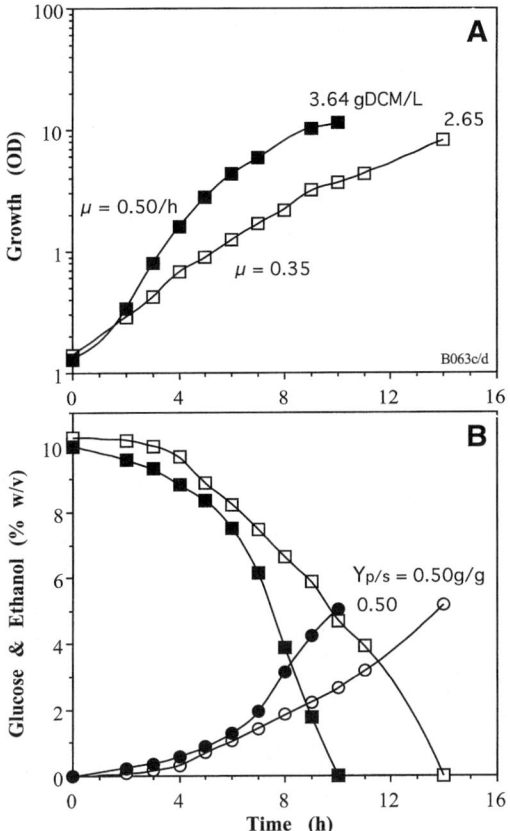

Fig. 2. Time-course of pH-stat batch fermentations with Z. mobilis 29191. **(A)** Growth, and **(B)** glucose utilization and ethanol production. (□) defined mineral salts medium (DS) with 10% (w/v) glucose; (■) DS medium with 3 g/L YE and 10% glucose. Details of the composition of the media are given in Table 1. The temperature was 30°C and the pH was 6.0. Values for both the specific growth rate (μ) and final cell mass (dry w-DCM) are given in A. The value of the ethanol yield ($Y_{p/s}$) is indicated in B.

with the DS and ZM media were 0.35/h and 0.50/h, respectively (Fig. 2A); however, with both media, the ethanol yield ($Y_{p/s}$) was 0.50 g ethanol/g glucose (i.e., 98% theoretical maximum conversion efficiency), and no lactic acid was detected (Fig. 2B). More typically, for batch fermentations, the inoculum (preculture or seed) is prepared as a flask culture using a nutrient-rich, complex medium, and a typical pH-stat batch fermentation would exhibit a growth and fermentation time-course very similar to the one shown in Figure 2. (results not shown).

Effect of pH and Inoculum History

Table 2 shows the results of a series of batch fermentations using four pH-stat batch fermentors containing RM medium with 10% (w/v) glucose (Table 1) were inoculated with Z. mobilis 29191 cultures produced in two

Table 2
Effect of Physiological History of Inoculum on Production of Lactic Acid by Z. *mobilis* 29191 in Batch Fermentation with pH Controlled at Either 5.5 or 6.0

Growth of inoculum			Batch fermentation parameters				
						Ethanol yield	
Chemostat feed medium	Dilution rate (1/h)	pH	Final cell mass (gDCM/L)	Lactic acid (g/L)		(gEtOH/g Glu)	Conversion efficiency (% theoret. max)
ZM [DS + 3g/L YE]	0.08	5.5	3.34	6.79		0.472	92.4
		6.0	3.55	9.09		0.465	91.0
DS	0.08	5.5	3.21	1.77		0.50	97.8
		6.0	3.25	2.56		0.495	96.9

Note: The defined mineral salts medium (DS) contained 2.4 g/L NH_4Cl and the semisynthetic medium (ZM) was the same composition, except that it contained 3 g/L Difco YE. Both media contained 10% (w/v) glucose. The temperature of the pH-stat stirred-tank fermentors was maintained constant at 30°C.

glucose-limited chemostats operating at D = 0.08/h (Fig. 1) with either YE-based ZM or a DS (Table 1). The pH was controlled at either 5.5 or 6.0. The YE-based complex medium (RM) was selected for these experiments, because it has been used extensively by others as a reference medium in research with *Zymomonas* directed to comparative fermentation performance and nutritional requirements in formulating cost-effective media for use in industrial-scale operations (46). At D = 0.08/h, the chemostat culture growing in a DS medium produced very little lactic acid (Fig. 1). However, when transferred to the batch fermentor, with the pH controlled at 6.0, it produced 2.56 g/L lactic acid, with an ethanol yield of 0.495 g/g (96.9% conversion efficiency) (Table 2). When the pH was controlled at 5.5, the final concentration of lactic acid produced from 10% (w/v) glucose decreased to 1.77 g/L, and the ethanol yield was correspondingly increased to 0.50 g/g (97.8%) (Table 2). At D = 0.08/h, the chemostat culture growing in YE-based ZM medium produced about 8 g/L lactic acid (Fig. 1), and when transferred to the batch fermentor, with the pH controlled at 6.0, it produced 9.09 g/L lactic acid, with an ethanol yield of 0.465 g/g (91%) (Table 2). When the pH was controlled at 5.5, the lactic acid level decreased to 6.79 g/L and the ethanol yield increased to 0.472 g/g (92.4%) (Table 2). These observations clearly demonstrate the beneficial effect of pH on ethanol yield whereby the synthesis of the coproduct is reduced by controlling the pH <6.0.

If lactic acid synthesis is dependent on a reduced glycolytic flux, as is achieved under a condition of glucose-limited growth in a chemostat operating at relatively low dilution rates, it is expected that lactic acid production would be minimized in batch fermentations, growth and glucose catabolism are maximal. In these experiments, growth in batch culture represents only a few generations (about five or six); however, the persistent nature of lactic acid production exhibited in pH-stat batch fermentations, in which the inoculum (seed) came from a lactic acid-producing chemostat, suggests a mechanism of enzyme induction rather than allosteric regulation, the latter being more instantaneous with respect to stopping acidogenesis.

These experiments are only exploratory in nature. Further research will be required to properly describe both the causative agent(s) and conditions that are responsible for promoting lactic acid synthesis by *Zymomonas*. Collectively these observations suggest that the characteristic high ethanol selectivity of *Zymomonas* can be compromised by the production of lactic acid. The mechanism of the metabolic shift from solventogenesis to acidogenesis appears to be exacerbated by elevated pH and a consequence of the low glycolytic flux achieved through glucose-limited growth in complex media containing organic nitrogen (amino acids). The results of this study are consistent with results of continuous fermentations using a xylose-utilizing recombinant *Z. mobilis*, in which the pH was controlled at 6.0 to minimize the acetic acid toxicity of the lignocellulosic hemicellulose hydrolysate medium (36).

This study did not include an investigation into either strain specificity or the stereoisomeric specificity of lactic acid synthesis. This work has established the operating parameters for a systematic screening of our *Zymomonas* culture collection for strain specificity with respect to lactic acid production.

ACKNOWLEDGMENTS

This work was funded by the University of Toronto.

REFERENCES

1. Wright, J. D. (1988), *Chem. Eng. Prog.* **84,** 62–74.
2. Hinman, N. D., Wright, J. D., Hoagland, W., and Wyman, C. E. (1989), *Appl. Biochem. Biotechnol.* **20/21,** 391–401.
3. Wyman, C. E. and Hinman, N. D. (1990), *Appl. Biochem. Biotechnol.* **24/25,** 735–753.
4. Swings, J. and DeLey, J. (1977), *Bacteriol. Rev.* **41,** 1–46.
5. Montenecourt, B. S. (1985), in *Biology of Industrial Microorganisms*, Demain, A. L. and Simon N. A., eds., Benjamin/Cummings, Meno Park, C, pp. 216–287.
6. Doelle, H. W., Kirk, L., Crittenden, R., Toh, H., and Doelle, M. (1993) *Crit. Rev. Biotechnol.* **13,** 57–98.
7. Rogers, P. L., Lee, K. J., Skotnicki, M. L., and Tribe, D. E. (1982), *Adv. Biochem. Eng.* **23,** 37–84.
8. Picataggio, S. K., Zhang, M., and Finkelstein, M. (1994), in *Enzymatic Conversion of Biomass for Fuels Production*, Himmel, M. E., Baker, J. O., and Overend, R. A., eds., American Chemical Society, Washington, DC, *ACS Symposium Series* 566, pp. 342–362.
9. Lawford, H. G. (1988), *Appl. Biochem. Biotechnol.* **17,** 203–219.
10. Beavan, M., Zawadzki, B., Droiniuk, R., Fein, J. E. and Lawford, H. G. (1989), *Appl. Biochem. Biotechnol.* **20/21,** 319–326.
11. Alterthum, F. and Ingram, L. O. (1989), *Appl. Environ. Microbiol.* **54,** 397–404.
12. Ingram, L. O., Conway, T., and Alterthum, F. (1991), US Patent 5,000,000.
13. Ohta, K., Alterthum, F., and Ingram, L. O. (1990), *Appl. Environ. Microbiol.* **56,** 463–465.
14. Beall, D. S., Ohta, K., and Ingram, L. O. (1991), *Biotechnol. Bioeng.* **38,** 296–303.
15. Doran, J. B., Aldrich, H. C., and Ingram, L. O. (1994), *Biotechnol. Bioeng.* **44,** 240–247.
16. Zhang, M., Franden, M. A., Newman, M., McMillan, J., Finkelstein, M., and Picataggio, S. K. (1995), *Appl. Biochem. Biotechnol.* **51/52,** 527–536.
17. Picataggio, S. K., Zhang, M., Eddy, C. K., Deanda, K. A., and Finkelstein, M. (1996), US Patent 5,514,583.
18. Zhang, M., Eddy, C., Deanda, K., Finkelstein, M., and Picataggio, S. K. (1995), *Science*, **267,** 240–243.
19. Lawford, H. G. and Rousseau, J. D. (1997), *Appl. Biochem. Biotechnol.* **63–65,** 269–286.
20. Jöbses, I. M. L., Egberts, G. T. C., van Baalen, A., and Roels, J. A. (1985), *Biotechnol. Bioeng.* **27,** 984–995.
21. Schmidt, W. and Schügerl, K. (1987), *Chem. Eng. J.* **36,** B39–B48.
22. Wecker, M. S. A. and Zall, R. R. (1987), *Appl. Environ. Microbiol.* **53,** 2815–2822.
23. Viikari, L. (1988), *Crit. Rev. Biotechnol.* **7,** 237–261.
24. Johns, M. R., Greenfield, P. F., and Doelle, H. W. (1991), *Adv. Biochem. Eng. Biotechnol.* **44,** 97–121.
25. Doelle, H. W. and Greenfield, P. F. (1985), *Appl. Microbiol. Biotechnol.* **22,** 405–410.
26. Viikari, L. (1984), *Appl. Microbiol. Biotechnol.* **19,** 252–255.
27. Viikari, L. (1984), *Appl. Microbiol. Biotechnol.* **20,** 118–123.

28. Viikari, L. and Korhola, M. (1986), *Appl. Microbiol. Biotechnol.* **24,** 471–476.
29. Suntinanalert, P., Pemberton, J. P., and Doelle, H. W. (1986), *Biotechnol. Lett.* **8,** 351–356.
30. Cromie, S. and Doelle, H. W. (1982), *Eur. J. Appl. Microbiol.* **14,** 69–73.
31. Keshew, K. F., Yamano, L. P., An, H., and Ingram, L. O. (1990), *J. Bacteriol.* **172,** 2491–2497.
32. Osman, Y. A., Conway, T., Bonetti, S. J., and Ingram, L. O. (1987), *J. Bacteriol.* **169,** 3726–3736.
33. Algar, E. M. and Scopes, R. K. (1985), *J. Biotechnol.* **2,** 275–281.
34. Yomano, L. P., Scopes, R. K., and Ingram, L. O. (1993), *J. Bacteriol.* **175,** 3926–3933.
35. Garvie, E. I. (1980), *Microbiol. Rev.* **44,** 106–139.
36. Lawford, H. G., Rousseau, J. D., Mohagheghi, A., and McMillan, J. D. (1998), 19th Symp. *Appl. Biochem. Biotechnol.* **70–72.**
37. Doelle, H. W. (1989), US Patent 4,797,360.
38. Doelle, M. B., Greenfield, P. F., and Doelle, H. W. (1990), *Process Biochem.* **25,** 151–156.
39. Millichip, R. J. and Doelle, H. W. (1989), *Process Biochem.* **24,** 141–145.
40. Doelle, M. B., Millichip, R. J., and Doelle, H. W. (1989), *Process Biochem.* **24,** 137–140.
41. Jones, C. W. and Doelle, H. W. (1991), *Appl. Microbiol. Biotechnol.* **35,** 4–9.
42. Fein, J. E., Lawford, H. G., Lawford, G. R., Zawadski, B. C., and Charley, R. C. (1983), *Biotechnol. Lett.* **5,** 19–24.
43. Lawford, H. G. and Stevnsborg, N. (1986), *Bioeng. Biotechnol. Symp.* **17,** 209–219.
44. Lawford, H. G. (1989), US Patent 4,840,902.
45. Fein, J. E., Charley, R. C., Hopkins, K., Lavers, B., and Lawford, H. G. (1983), *Biotechnol. Lett.* **5,** 1–6.
46. Lawford, H. G. and Rousseau, J. D. (1996), *Appl. Biochem. Biotechnol.* **57/58,** 307–326.
47. Nipkow, A., Beyeler, W., and Fiechter, A. (1984), *Appl. Microbiol. Biotechnol.* **19,** 237–240.
48. Beyeler, W., Rogers, P. L., and Fiechter, A. (1984), *Appl. Microbiol. Biotechnol.* **19,** 277–280.
49. Prince, I. G. and Barford, J. P. (1982), *Biotechnol. Lett.* **8,** 525–530.
50. Lawford, H. G., Holloway, P., and Ruggiero, A. (1988), *Biotechnol. Lett.* **10,** 809–814.

A Novel Fermentation Pathway in an *Escherichia coli* Mutant Producing Succinic Acid, Acetic Acid, and Ethanol

MARK I. DONNELLY,[*,1] CYNTHIA SANVILLE MILLARD,[1] DAVID P. CLARK,[3] MICHAEL J. CHEN,[2] AND JEROME W. RATHKE[2]

[1]*Environmental Research Division and* [2]*Chemical Technology Division, Argonne National Laboratory, Argonne, IL 60439; and* [3]*Department of Microbiology, Southern Illinois University, Carbondale, IL 62901*

ABSTRACT

Escherichia coli strain NZN111, which is unable to grow fermentatively because of insertional inactivation of the genes encoding pyruvate: formate lyase and the fermentative lactate dehydrogenase, gave rise spontaneously to a chromosomal mutation that restored its ability to ferment glucose. The mutant strain, named AFP111, fermented glucose more slowly than did its wild-type ancestor, strain W1485, and generated a very different spectrum of products. AFP111 produced succinic acid, acetic acid, and ethanol in proportions of approx 2:1:1. Calculations of carbon and electron balances accounted fully for the observed products; 1 mol of glucose was converted to 1 mol of succinic acid and 0.5 mol each of acetic acid and ethanol. The data support the emergence in *E. coli* of a novel succinic acid:acetic acid:ethanol fermentation pathway.

Index Entries: *Escherichia coli*; fermentation; succinic acid.

INTRODUCTION

In microbial fermentations of organic growth substrates, energy-yielding oxidative reactions generate a pool of partially oxidized intermediates and a surplus of reduced cofactors *(1)*. The excess reductant generated, primarily NADH, must be dissipated for metabolism to continue, and in most cases this is accomplished by reducing the available organic intermediates to end products that are excreted into the medium. The nature and proportions of these end products vary widely *(1)*. *Escherichia coli* produces several end products from the fermentation of sugars,

---

*Author to whom all correspondence and reprint requests should be addressed.

principally ethanol and acetic, formic, and lactic acids *(2,3)*. Under some conditions, formic acid is further metabolized to hydrogen and CO_2. Smaller amounts of succinic acid are also formed by carboxylation of phosphoenolpyruvate (PEP) and reductive reactions of the tricarboxylic acid (TCA) cycle. The ratio of these products varies, depending on the strain and the growth conditions, but is always adjusted to balance the overall metabolism by consuming exactly the reducing equivalents generated in glycolysis *(2)*.

Some fermentative organisms produce succinic acid in higher amounts. Certain propionic acid bacteria make large amounts of succinic acid by fermentation of lactate *(1)*. In other cases, succinic acid may be formed from glucose. At low pH, the obligate anaerobe *Anaerobiospirillum succiniciproducens* ferments glucose plus CO_2 to a mixture of succinic and acetic acids in a 2:1 molar ratio *(4)*. Recently, the strain *Bacterium* 130Z has been shown to produce succinic acid as its major fermentation product, along with smaller amounts of acetic, formic, propionic, and pyruvic acids *(5)*. Wild-type *E. coli*, on the other hand, consistently forms only small amounts of succinic acid under a wide variety of conditions.

The authors have previously shown that overexpression of PEP carboxylase, or, in the appropriate genetic background, of malic enzyme, can significantly increase *E. coli*'s production of succinic acid *(6,7)*. A spontaneous mutation in a nonfermenting strain of *E. coli* that restored its ability to ferment glucose was recently discovered by the authors. Succinic acid was the major product. In addition, substantial but lower amounts of acetic acid and ethanol were formed. Here the discovery of the mutant, the quantitative characterization of its fermentative metabolism, and aspects of electron balance that affect product distribution is described. The pathway constitutes the first known instance of a succinic acid:acetic acid:ethanol fermentation.

MATERIALS AND METHODS

Bacterial Strains and Culture Methods

Strains of *E. coli* (Table 1) were routinely cultured in Luria broth (LB) or M9 medium at 37°C *(8)*. Anaerobic minimal medium was supplemented with Fe, Se, Mo, and Mn *(9)*. For plates, anaerobic conditions were established by use of a GasPak (Becton Dickinson, Cockeysville, MD) and grown under a atmosphere of $H_2:CO_2$. Liquid anaerobic cultures were grown in stoppered serum tubes under an atmosphere of sterile, anaerobic CO_2 at 14 psi, which was established by use of a gassing manifold *(10)*. During the fermentation of higher concentrations of glucose in rich medium, the pH was controlled by the addition of solid $MgCO_3$ to the serum tubes. To monitor the growth of anaerobic cultures, cells were grown in serum tubes containing minimal medium without $MgCO_3$, and absorbances were moni-

Table 1
Bacterial Strains and Plasmids

Strain	Relevant markers	Source or ref.
W1485	F$^+$ wild type	CGSC 5024
FMJ123	W1485 pfl::Cam	Bunch et al. (11)
NZN111	FMJ123 ldh::Kan	Bunch et al. (11)
AFP111	Spontaneous mutant of NZN111	This work
pMDH14	E. coli mdh (A80P/R81Q) in pTRC99a (Pharmacia, Piscataway, NJ)	Boernke et al. (12)

tored with a Spectronic 20 (Bausch & Lomb, Rochester, NY). When appropriate, antibiotics were added at the following concentrations: ampicillin, 100 μg/mL; carbenicillin, 100 μg/mL; tetracycline, 10 μg/mL; kanamycin, 30 μg/mL; chloramphenicol, 30 μg/mL. Strain NZN111 *(11)*, containing a plasmid encoding a mutant malate dehydrogenase that possessed lactate dehydrogenase activity *(12)*, was cultured in sealed serum tubes in 10 mL of LB medium, supplemented with 100 μg/mL carbenicillin, 12 g/L glucose, 0.1 mM isopropyl-β-D-thiogalactopyranoside (IPTG), and 0.5 g MgCO$_3$.

Analytical Methods and Enzyme Assays

Glucose consumption and product formation were determined by high-performance liquid chromatography (HPLC) using a Bio-Rad (Hercules, CA) Aminex HPX-87H column (7.8 × 300 mm) and a Shimadzu LC-10A chromatography system equipped with UV absorbance and refractive index detectors. The column was eluted isocratically with 5 mM H$_2$SO$_4$, and data were analyzed with an EZChrom™ data system (Scientific Software, San Ramon, CA). Glucose levels were also monitored enzymatically using the commercial kit from Stanbio, (San Antonio, TX).

Analysis of Products Formed from [1-^{13}C]-D-Glucose

Quantitative measurement of the products formed from [1-^{13}C]-D-glucose (99.5 atom %, Cambridge Stable Isotopes, Andover, MA) was achieved by nuclear magnetic resonance (NMR) performed in the presence of the chemical relaxation reagent, erbium chloride. Because erbium salts of the products and of bicarbonate precipitated at neutral pH, the samples were acidified and degassed prior to its addition. Following the anaerobic conversion of 18.3 g/L [1-^{13}C]-D-glucose in LB medium under the conditions described above, cultures were clarified by filtration and acidified by treatment with Dowex AG50W-X4 (H$^+$ form) (Bio-rad, Hercules, CA), to convert the bicarbonate present to CO$_2$. The Dowex was removed by filtration and the sample was degassed under vacuum to completely remove the CO$_2$. Erbium chloride (ErCl$_3$·6H$_2$O) was added to a final concentration of 7.9 (±0.3) mM. Spectra were obtained with a Varian (Palo Alto, CA) Unity 400 spectrometer at a probe frequency of 100.577 MHz. Transients (300 to 1000) of spectral width 25,000 Hz were collected at ambient temperature, with an acquisition time of 0.6 s, a relaxation time of 25 s, and a pulse width of 22 µs (PW90 = 25 µs). Protons were decoupled during acquisition, and the field was locked on D_2O. Chemical shifts are reported in parts per million, relative to tetramethylsilane. Reference spectra of succinic acid, acetic acid, and ethanol were also obtained, and the chemical shifts observed were within 0.1 ppm of published values.

RESULTS

Selection of Mutant E. coli Producing Succinic Acid

Previous site-directed mutagenesis of *E. coli* malate dehydrogenase performed in our laboratory generated a series of mutants with activity toward pyruvate, instead of the natural substrate, oxaloacetate (12). The lactate dehydrogenase activity of these mutants was low, but sufficient to complement a chromosomal inactivation in *E. coli* of the fermentative lactate dehydrogenase gene, as illustrated in Fig. 1. When the mutant malate dehydrogenase encoded by plasmid pMDH14 (Table 1) was expressed in the nonfermenting *E. coli* host NZN111, slow fermentative production of lactic acid occurred (Table 2).

Since fermentative growth in NZN111(pMDH14) appeared to be limited by the low lactate dehydrogenase activity of the mutant malate dehydrogenase, experiments in directed evolution in an attempt to select a variant enzyme with improved lactate dehydrogenase activity were initiated. After several cycles of subculturing without mutagenesis, accelerated glucose catabolism was detected but found that other products, in addition to lactic acid, were formed (Table 2). Succinic acid was the major product observed, as well as lactic acid, acetic acid, and ethanol. To establish the source of the additional products, single colonies were isolated anaerobically on LB plates containing glucose, as well as kanamycin and chloramphenicol, the antibiotics encoded by the cassettes used to inactivate *pfl* and *ldhA* in the construction of NZN111 (11). Isolated colonies were restreaked twice in the absence of antibiotic, then evaluated for antibiotic sensitivity. All colonies were resistant to kanamycin and chloramphenicol, but sensitive to ampicillin, indicating that they arose from NZN111 and no longer possessed the plasmid pMDH14. The pattern of fermentation products generated by several colonies was evaluated and found to be identical, indicating that the colonies were probably siblings. Succinic acid was the major fermentation product formed, with lower amounts of acetic acid and ethanol (Table 2).

Characterization of Mutant Strain

The purified mutant, named AFP111, was compared to its ancestral strains. The pattern of fermentation products it generated was very different from those formed by NZN111, its immediate precursor, FMJ123, the

Fermentation Pathway in E. coli Mutant

Fig. 1. Complementation of mutations in *E. coli* NZN111 by plasmid pMDH14. Inactivation (−) of the pyruvate:formate lyase and lactate dehydrogenase genes, *pfl* and *ldhA*, respectively, in strain NZN111 blocks the conversion of pyruvate and eliminates the ability to ferment glucose. Plasmid pMDH14 directs the expression of a mutant *E. coli* malate dehydrogenase with lactate dehydrogenase activity. Expression of this protein allowed NZN111 to metabolize glucose slowly by a homolactic acid fermentation.

Table 2
Emergence of Mutant Strain During Enzyme Selection Experiments[a]

Strain	Glucose remaining and products formed (g/L)				
	Glucose	Succinic acid	Lactic acid	Acetic acid	Ethanol
NZN111[b]	10.5	0.5	0	0	0
NZN111(pMDH14)[c]	3.8	0.8	5.7	0	0
Selection experiment[d]	0	5.8	3.0	2.6	1.7
Plasmid-free mutant[e]	0	8.4	0	2.2	1.4

[a] Products formed overnight from 12 g/L glucose in 10 mL of LB with $MgCO_3$ under an atmosphere of CO_2.
[b] Unable to ferment glucose. Pyruvate produced at 0.2–0.8 g/L (2–9 mM).
[c] pMDH14, which encodes *E. coli* MDH mutant with LDH activity, was induced with IPTG.
[d] Products formed by induced NZN111(pMDH14) after six cycles of subculturing.
[e] Named AFP111. Isolated as Amp^s, Kan^r, Cam^r single colonies.

parent of NZN111, which contains a functional *ldh* gene *(11)*, or W1485, the wild-type ancestor (Table 3). AFP111 fermented glucose to a mixture of succinic acid, acetic acid, and ethanol; FMJ123 metabolized glucose by a homolactic acid fermentation; and W1485 generated the typical spectrum of mixed acid fermentation products. NZN111 failed to ferment glucose and excreted pyruvate into the medium.

The anaerobic growth of AFP111 on glucose in minimal medium supplemented with casamino acids was compared to that of NZN111 and W1485 (Fig. 2). Strain NZN111 failed to grow, as reported earlier *(11,12)*;

Table 3
Products Formed by Strains in Lineage of Mutant AFP111[a]

Strain	Product (g/L)				
	Succinic acid	Lactic acid	Formic acid	Acetic acid	Ethanol
W1485	2.1	1.8	4.9	4.5	4.1
FMJ123	1.1	13.2	0.3	1.0	0.7
NZN111[b]	0.1	0.2	0.2	0.2	0.4
AFP111	11.3	0.6	0	3.9	2.3

[a] Products formed from 18.4 g/L glucose in 10 mL of LB broth incubated in sealed tubes under an atmosphere of CO_2 with 1 g $MgCO_3$ to maintain pH.
[b] Unable to ferment glucose. Pyruvate also produced at 1 g/L.

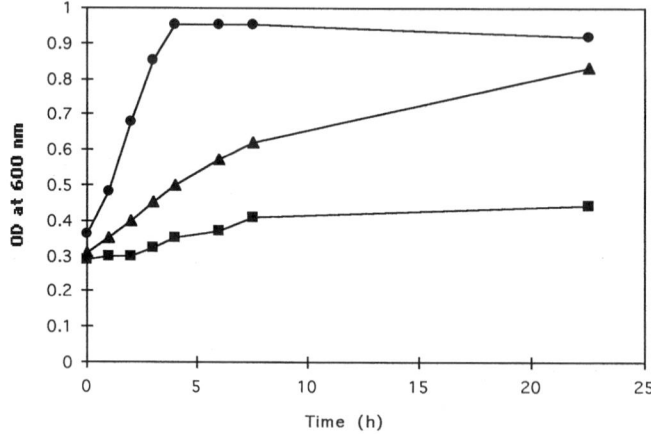

Fig. 2. Anaerobic growth of strains on glucose in minimal medium. Strains AFP111 (▲), its immediate parent NZN111 (■), and the wild-type ancestor W1485 (●) were grown anaerobically at 37°C, in 10 mL of M9 medium containing 2.9 g/L glucose, in sealed vials under an atmosphere of CO_2. Optical densities were measured with a Spectronic 20.

the small amount of growth observed was most likely caused by carryover of metabolites and oxygen in the inoculum. Strain AFP111 grew much more slowly than W1485, but eventually attained approximately the same final cell density.

Fermentation in minimal medium allowed quantitative evaluation of the fermentative metabolism of AFP111 (Table 4). Approximately 1 mol of succinic acid was generated per mol of glucose consumed. Acetic acid and ethanol were formed in lower, approximately equal amounts; in total, their amount approached 2 mol/mol of glucose. A trace of lactic acid was also formed. Analysis of the products in terms of carbon and electron

Table 4
Carbon and Electron Balances Glucose Fermentation by AFP111[a]

	Growth substrate and products		Carbon balance mol product/ 100 mol glucose	Electron balance mol NADH/ 100 mol glucose
	g/L	mM		
Glucose (substrate)	2.92	16		200
Succinic acid	1.90	16	99	−199
Lactic acid	0.13	1	9	−9
Acetic acid	0.47	8	48	48
Ethanol	0.29	6	39	−39
Cell mass[b]	0.16			
Total products[c]	2.95		195	1

[a] Products formed from 2.9 g/L glucose in M9 medium containing 1% casamino acids in sealed tubes under an atmosphere of CO_2.

[b] Cell mass estimated from the established relationship between cell mass and optical density.

balance indicated tight closure of the metabolic pathway. For the mixed acid fermentation pathway, a maximum of 200 mol of products (excluding formic acid) are anticipated per 100 mol of glucose (1). The authors accounted for 195 mol as excreted products. Electron balance calculations, based here on reductions and oxidations required by established metabolic pathways for glucose fermentation in *E. coli*, supported tight coupling of reductant generation and consumption.

Metabolism of [1-^{13}C]-D-Glucose

To establish that all the products observed in rich medium arose from glucose, as opposed to other components of the medium, the products formed by AFP111 in the fermentation of [1-^{13}C]-D-glucose were analyzed. The chemical relaxation reagent erbium chloride was added to the treated broths, to permit quantitative integration of the resonances and determination of the relative abundances of the products. For clarity, a single set of assignments is used for both spectra (Fig. 3).

The spectrum of products generated by W1485 (Fig. 3A) reveals ethanol, lactic acid, acetic acid, and succinic acid methyl and methylene carbons (resonances a–d, respectively), the ethanol hydroxymethyl carbon (e), formic acid (f), and the acetic acid and succinic acid carboxyl carbons (g and h, respectively). No resonance for CO_2 is observed, because the samples were acidified and degassed prior to analysis. In the AFP111 spectrum (Fig. 3B), the increased abundance of the succinic acid resonances (d and h), relative to those in the W1485 spectrum, is apparent.

Fig. 3. ^{13}C-NMR spectra of fermentation products. Products of fermentation of 18.3 g/L of [1-^{13}C]-D-glucose by **(A)** W1485 and **(B)** AFP111. Peaks are assigned as follows: a, ethanol methyl carbon; b, lactic acid methyl carbon; c, acetic acid methyl carbon; d, succinic acid methylene carbon; e, ethanol hydroxymethyl carbon; f, formic acid carboxyl carbon; g, acetic acid carboxyl carbon; and h, succinic acid carboxyl carbon. The methyl and methylene carbon signals are much larger because of enrichment from [1-^{13}C]-D-glucose.

The relative abundance of the products formed from [1-^{13}C]-D-glucose can be determined from the integrated intensities of the alkyl resonances of each product—assuming that each species was formed from a single PEP. Calculation, in this manner, of the mole fraction of each product formed by AFP111 agreed reasonably well with the product distribution determined by HPLC analysis of the same experiment (Table 5). The estimates of succinic acid and ethanol are not fully in agreement, however; NMR estimates indicate slightly greater than stoichiometric formation of succinic acid from glucose, at the expense of reduced ethanol formation.

Table 5
Products Formed During Growth on [1-^{13}C]-Glucose in Rich Medium^a

Product	Analysis by HPLC		Analysis by NMR	
	g/L	mol fraction^b	Integrated intensity^c	mol fraction^d
Succinic acid	12.8	0.51	55.00	0.59
Lactic acid	0.2	0.01	0.53	0.01
Acetic acid	3.3	0.26	25.09	0.27
Ethanol	2.2	0.23	13.14	0.14
Total products	18.5	1.0	93.76	1.0

^{a} Products formed from 18.3 g/L [1-^{13}C]-glucose (99 atom %) in 10 mL of LB broth incubated in sealed tubes under an atmosphere of CO_2 with 1 g $MgCO_3$ to maintain pH.
^{b} Based on molarity of products formed, as calculated from g/L and mol wt.
^{c} Integrated intensity of methyl carbon of lactic acid, acetic acid, and ethanol, and of methylene carbon of succinic acid.
^{d} Based on summation of integrated intensities.

Possibly, a small amount of ethanol was lost by degassing during filtration of the sample following acidification and treatment with Dowex.

DISCUSSION

The results described above establish that the E. coli mutant AFP111 carries out a balanced fermentation of 1 mol glucose to 1 mol succinic acid, one-half mol of acetic acid, and approx one-half mol of ethanol (Fig. 4). These stoichiometries were obtained consistently, both in minimal supplemented with casamino acids and in rich media. They were substantiated by the distribution of ^{13}C derived from metabolism of [1-^{13}C]-D-glucose. The pathway represents the first example of an E. coli strain that produces succinic acid as its major fermentation product. Figure 4 satisfies the requirements of fermentations for both the balanced production and consumption of reducing equivalents and the generation of ATP. The ability of AFP111 is attributed to ferment glucose to a spontaneous chromosomal mutation that became enriched under the selective conditions of the directed evolution experiment.

The pathway can be envisioned as the conversion of glucose to two PEP, followed by reduction of one PEP to succinic acid, using the reducing power generated in glycolysis (two reductive steps are required for succinic acid formation), and conversion of the second PEP to an equal mixture of acetic acid and ethanol. The reducing equivalents generated in the oxidation of pyruvate are consumed in the reduction of acetylCoA to ethanol. The obligatory coupling of acetic acid and ethanol production documented earlier for E. coli (13) persists in AFP111.

The pathway proposed in Fig. 4 differs from a similar fermentation carried out under acidic conditions by A. succiniciproducens. This later path-

Fig. 4. Proposed fermentative pathway of AFP111. Because of the use of the PTS for the uptake of glucose, glycolysis in *E. coli* can be considered to yield one PEP, one pyruvate, and two reducing equivalents as NADH. Metabolism of PEP via the reductive arm of the TCA cycle would consume two reducing equivalents. Oxidation of pyruvate to acetylCoA would generate additional reducing equivalents, which can be consumed in the reduction of half of the acetylCoA to ethanol. The remaining acetylCoA would be hydrolyzed to acetic acid to complete the conversion of 1 mol glucose to 1 mol succinic acid, 0.5 mol acetic acid, and 0.5 mol ethanol.

way converts glucose to a 2:1 mixture of succinic and acetic acids *(4)*. It also satisfies the need for redox balance; two-thirds of the PEP formed in glycolysis is converted to succinic acid; oxidation of the remaining third to acetic acid provides the reducing equivalents needed for succinic acid formation. Such a pathway may not be possible in *E. coli* because of the use of the phosphotransferase system (PTS) for glucose uptake. Use of the PTS effectively mandates that half of the PEP generated glycolytically be converted to pyruvate *(14)*, which is then dissimilated to acetic acid and ethanol. Such a restriction would make the production of greater than stoichiometric amounts of succinic acid impossible.

Although the proposed pathway fits the data very well, and would be totally satisfactory for a less well-characterized organism, many aspects of it are not satisfactory in the face of the voluminous knowledge available for *E. coli*. Most obviously, the enzyme that catalyzes the conversion of pyruvate to acetylCoA needs to be determined. Pyruvate:formate lyase is not responsible, since this enzyme has been inactivated by disruption of the *pfl* gene. The locations of ^{13}C observed in the products generated from [1-^{13}C]-glucose suggest that conventional reactions of known enzymes are used, but the well-established regulation of these enzymes argues that they should not be operative under these conditions. Preliminary genetic mapping and transductions of genetic knock-outs of specific genes indicates that the mutation that created AFP111 was not in any of the structural genes of the pyruvate dehydrogenase complex. Neither was the mutation in the gene *fdx*, which encodes the ferrodoxin that participates in pyruvate:ferrodoxin oxidoreductase, nor in the global regulatory genes *arcB* or *fnr*.

Regardless of which enzyme catalyzes pyruvate conversion, the abundant production of succinic acid by AFP111 must be explained. Fermentation by AFP111 is slower than that of the wild type or of mutants lacking either lactate dehydrogenase or pyruvate:formate lyase alone, which metabolize glucose to a reduced spectrum of products. Possibly the slower metabolism in AFP111 results in the accumulation of both pyruvate and PEP to higher concentrations than normally occur in other strains, and thereby opens a metabolic channel normally of minor consequence in the presence of efficient competing pathways.

The authors are currently investigating the nature of the mutation, through a combination of genetic mapping, enzymatic analyses, and two-dimensional gel electrophoresis.

ACKNOWLEDGMENTS

This work was supported by the Alternative Feedstocks Program, Office of Industrial Technology, US Department of Energy, Assistant Secretary for Energy Efficiency and Renewable Energy, under contract W-31-109-Eng-38.

REFERENCES

1. Gottschalk, G. (1985), *Bacterial Metabolism*, 2nd ed. Springer series in microbiology, Springer-Verlag, New York.
2. Clark, D. P. (1989), *FEMS Microb. Rev.* **63**, 223–234.
3. Bock, A. and Sawers, G. (1996), in *Escherichia coli and Salmonella*, Neidhardt F. C., et al., eds. American Society for Microbiology, Washington, DC, pp. 262–282.
4. Samuelov, N. S., Lamed, R., Lowe, S., and Zeikus, J. G. (1991), *Appl. Environ. Microbiol.* **57**, 3013–3019.
5. Guettler, M. V., Jain, M. K., and Rumler, D. (1996), in *US Patent and Trademark Office, patent no. 5573931*, Michigan Biotechnology Institute.

6. Millard, C. S., Chao, Y. -P., Liao, J. C., and Donnelly, M. I. (1996), *Appl. Environ. Microbiol.* **62,** 1808–1810.
7. Stols, L. and Donnelly, M. I. (1997), *Appl. Environ. Microbiol.* **63,** 2695–2701.
8. Sambrook, J., Fritsch, E. F., and Maniatis, T. (1989), *Molecular Cloning: A Laboratory Manual*, 2nd ed., Cold Spring Harbor Press, Cold Spring Harbor, NY.
9. Mat-Jan, F., Kiswar, A. Y., and Clark, D. P. (1989), *J. Bacteriol.* **171,** 342–348.
10. Balch, W. and Wolfe, R. S. (1976), *Appl. Environ. Microbiol.* **32,** 781–791.
11. Bunch, P. K., Mat-Jan, F., Lee, N., and Clark, D. P. (1997), *Microbiology* **143,** 187–195.
12. Boernke, W. E., Millard, C. S., Stevens, P. W., Kakar, S. N., Stevens, F. J., and Donnelly, M. I. (1995), *Arch. Biochem. Biophys.* **322,** 43–52.
13. Gupta, S. and Clark, D. P. (1989), *J. Bacteriol.* **171,** 3650–3655.
14. Postma, P. W., Lengeler, J. W., and Jacobsen, G. R. (1996), in *Escherichia coli and Salmonella*, Neidhardt F. C., et al., eds. American Society for Microbiology, Washington, DC, pp. 1149–1174.

Cloned *Bacillus subtilis* Alkaline Protease (*aprA*) Gene Showing High Level of Keratinolytic Activity

Taha I. Zaghloul

Department of Bioscience and Technology, Institute of Graduate Studies and Research, University of Alexandria, Alexandria, Egypt

ABSTRACT

The *Bacillus subtilis* alkaline protease (*aprA*) gene was previously cloned on a pUB110-derivative plasmid. High levels of expression and gene stability were demonstrated when *B. subtilis* cells were grown on the laboratory medium 2XSG. *B. subtilis* cells harboring the multicopy *aprA* gene were grown on basal medium, supplemented with 1% chicken feather as a source of energy, carbon, and nitrogen. Proteolytic and keratinolytic activities were monitored throughout the cultivation time. A high level of keratinolytic activity was obtained, and this indicates that alkaline protease is acting as a keratinase. Furthermore, considerable amounts of soluble proteins and free amino acids were obtained as a result of the enzymatic hydrolysis of feather. Biodegradation of feather waste using these cells represents an alternative way to improve the nutritional value of feather, since feather waste is currently utilized on a limited basis as a dietary protein supplement for animal feedstuffs. Moreover, the release of free amino acids from feather and the secreted keratinase enzyme would promote industries based on feather waste.

Index Entries: *Bacillus subtilis*; alkaline protease gene; keratinolytic activity.

INTRODUCTION

Recently, our research group is focusing on the aspect of the utilization of some environmental wastes, such as chicken feather and tomato processing waste, using genetically engineered *Bacillus subtilis* strains. Although environmental wastes are found in great quantities, and many are rich in proteins and various carbon compounds, little attention is given to utilizing or recycling these wastes. Additionally, the accumulation of some of these wastes in nature is considered to be a serious source of pollution and health hazards. Upon the directed utilization by genetically modified microorganisms, several valuable products are produced, such as extracellular enzymes, soluble proteins, peptides, amino acids, and other compounds, which can satisfy part of the animal feed and/or human need (1,2). These products can also promote and enhance several new industries in many countries.

The *B. subtilis* alkaline protease *(aprA)* gene was cloned on a pUB110-derivative plasmid. The expression of the *aprA* gene occurred late in the stationary phase of growth, and the multicopy plasmid that carries the *aprA* gene was segregationally and structurally stable (3). *B. subtilis* cells carrying the multicopy *aprA* gene were grown on basal medium supplemented with 1% whole chicken feather (nearly pure keratin protein) as a source of carbon and nitrogen. High level of keratinolytic activity was developed in the culture, and, consequently, considerable amounts of soluble proteins and amino acids were obtained as a result of the biodegradation of feather. On the other hand, the keratinase *(kerA)* gene of *B. licheniformis* PWD-1 was cloned and found to have 98 and 97% sequence homology with genes encoding subtilisin Carlsberg of *B. licheniformis* and subtilisin of *B. licheniformis* NCIB, respectively (4).

MATERIALS AND METHODS

Bacterial Strains and Plasmids

B. subtilis DB100 his$^-$ met$^-$ (pS1) was used in this study. The pS1 plasmid (6.7 kbp) is a pUB110-derivative plasmid carrying the complete *B. subtilis (aprA)* gene, as well as a kanamycin resistance gene (selectable marker) (3).

Media

Bacterial strain was activated and grown on PY medium (5) (Bacto peptone, 10 g; Difco [East Moseley, Surrey, UK] yeast extract, 5 g; and NaCl 5 g/L). PA medium is PY supplemented with 1.5% agar agar. Basal medium II (6) (NH$_4$Cl, 0.5 g; NaCl, 0.5 g; K$_2$HPO$_4$, 0.3 g; KH$_2$PO$_4$, 0.4 g; MgCl$_2$·6H$_2$O, 0.1 g; and yeast extract 0.1 g/L) supplemented with 1% (w/v) chicken feather, was used in the fermentation process. Kanamycin was added to a final concentration of 5 µg/mL medium.

Monitoring Proteolytic and Keratinolytic Activity

The proteolytic and keratinolytic activity was monitored throughout the growth of the bacterial strain. Cells were activated by growing them overnight on PA plates at 37°C. Fresh colonies were then transferred to 10 mL PY kanamycin medium, and the culture was allowed to grow at 37°C, with shaking, for two h. One hundred mL preautoclaved basal medium II, supplemented with 1% whole chicken feather and kanamycin

(5 μg/mL), was inoculated with 5 mL of the above culture. Cells were allowed to grow at 37°C, with shaking at 200 rpm, to the indicated time (1–5 d). At the indicated time, 1 mL culture was taken to determine bacterial growth in the form of colony-forming units, as described before (7). Another 2 mL culture were taken and centrifuged at 4500g in a microcentrifuge for 2 min. The supernatant was then used as crude solution to determine the extracellular alkaline protease activity, as well as the keratinolytic activity.

Proteolytic Activity

Proteolytic activity was measured according to the method of Cliffe and Law (8), using hide powder azure (HPA). One unit of enzyme is the amount of enzyme that develops a change of absorbance against control reaction, at 595 nm/30 min at 37°C. Additionally, the proteolytic activity was determined according to the method of Lin et al. (9), using skim milk agar plate.

Keratinolytic Activity

Keratinolytic activity was determined based on the free amino (NH_2 groups) that were released as a result of the biodegradation of feather by bacterial cells. Free amino groups were determined as described earlier (9), and a standard curve for leucine (0.01–0.1 μmol) was established. Additionally, the physical appearance of chicken feather in the culture was observed.

Analysis of Soluble Proteins by SDS-Polyacrylamide Gel Electrophoresis (SDS-PAGE)

Cell-free supernatant of the feather culture was precipitated with solid ammonium sulfate to reach 70% saturation, and was kept on an ice bath for 2 h. The mixture was centrifuged at 7000g for 30 min at 4°C. The pellet was suspended in a small volume of Tris-HCl buffer, pH 8.0, and dialyzed overnight against the same buffer. A second precipitation by acetone (3 vol) was carried out, and the precipitate was collected by centrifugation at 3500g for 5 min, using microcentrifuge. The pellet was directly suspended in a small volume of sample application buffer (SAB) and applied to 10% SDS-polyacrylamide gel. The SDS-PAGE was carried out according to the method of Laemmli (10).

Amino Acid Analysis

Amino acids were analyzed using Beckman 119 CL amino acid analyzer, (Beckman Instruments, Palo Alto, CA), using standard amino acids. The analysis was performed at the Central Laboratory, Faculty of Agriculture, University of Alexandria.

Fig. 1. Monitoring proteolytic and keratinolytic activities of the bacterial strain throughout the cultivation time. Activated *B. subtilis* DB100 (pS1) cells were grown on basal medium II, supplemented with 1% whole chicken feather and 5 µg/mL kanamycin, to the indicated time. Proteolytic activity (+) is expressed as the change in absorbance at 595 nm, of the released blue color from HPA, per mL supernatant per 30 min at 37°C. Free amino groups, expressed as µmol of leucine per mL cell-free supernatant, represent the keratinolytic activity of the alkaline protease.

RESULTS AND DISCUSSION

Monitoring Proteolytic and Keratinolytic Activity of *aprA* Gene

Earlier the author reported the expression of the cloned *aprA* gene throughout the growth of *B. subtilis* DB100 (pS1) strain in 2×SG (sporulation medium) (3). This medium was used to give a short exponential phase and an extended stationary phase. Expression of the *aprA* gene started late, after 2 h from the end of the exponential phase (3). Proteolytic and keratinolytic activity of the alkaline protease was monitored throughout the growth of the same strain that harbors the *aprA* gene, on basal medium II, supplemented with 1% whole feather. Proteolytic activity, as measured rapidly and accurately by the synthetic substrate HPA, increased as the cultivation time was increased, up to d 2, then remained constant (Fig. 1). Data of skim milk agar plate gave the same pattern (not shown). On the other hand, keratinolytic activity, as determined by the free amino group method, using standard curve for leucine, increased gradually as the cultivation time was increased up to d 5, when it reached about 95 µmol free ($-NH_2$ group) per mL culture (Fig. 1).

The physical appearance of chicken feather in the culture was observed each day and compared to the control (d 0), in order to evaluate the level of keratinolytic activity of the culture (Fig. 2). Most of the feather

Fig. 2. Keratinolytic activity of the alkaline protease enzyme secreted by *B. subtilis* DB100 (pS1) cells in a culture containing 1% whole feather. A–C represent feather samples of the above culture after 0, 1, and 2 d after inoculation, respectively.

Fig. 3. SDS-polyacrylamide gel to analyze possible soluble proteins resulting from the biodegradation of whole feather at d 4. Preparation of samples and gel condition was described in Materials and Methods. Lanes 1, 2, and 3 represent BSA, mol wt marker, and 100 µg of soluble proteins sample, respectively. Arrow points at the location of alkaline protease enzyme (about 27 kDa protein). Note that considerable amounts of small mol wt proteins appear at the tail of the gel.

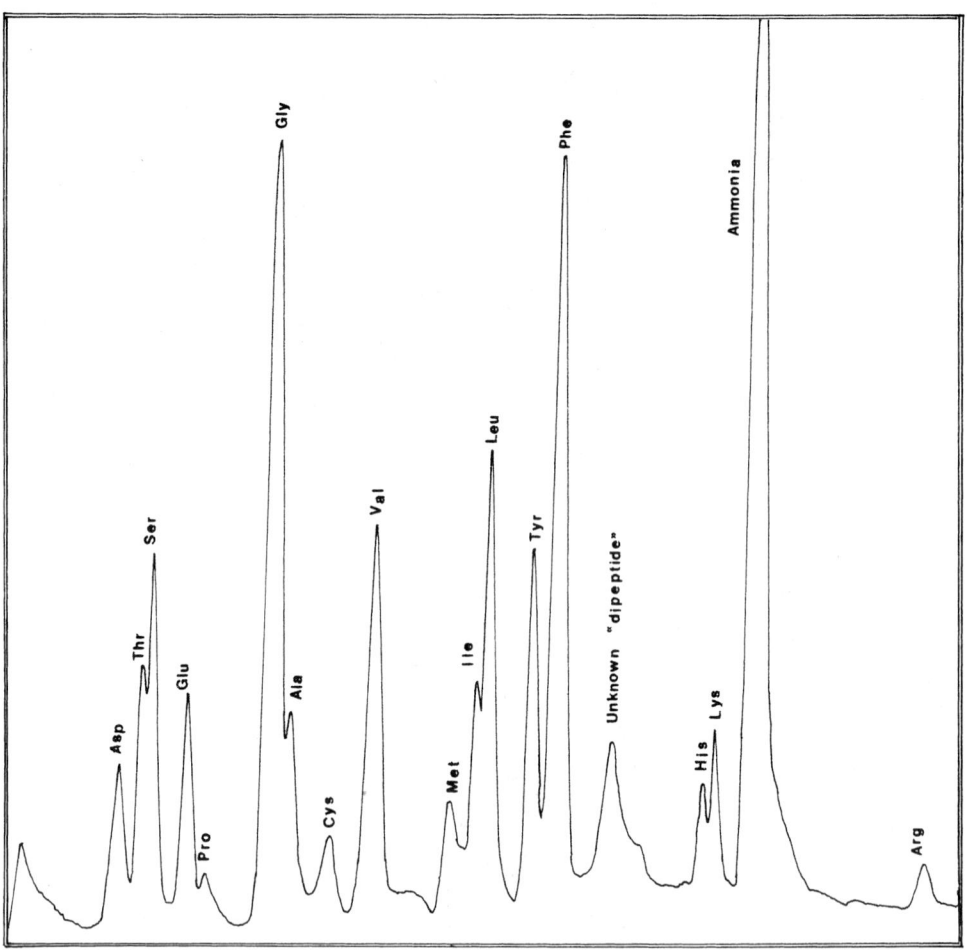

Fig. 4. Amino acid analysis of cell-free and protein-free sample derived from d 4. Analysis was carried out as described in the text, using amino acid analyzer. Several amino acids, such as gly, phe, leu, val, ser, and tyr, resulted from the utilization process.

was degraded after 48 h except the feather shaft, which was completely degraded after 72 h of cultivation.

Alkaline Protease is Acting as a Keratinase

Data shown above would indicate that the cloned *aprA* gene encoded a protease that has a high level of keratinolytic activity. Several *Bacillus* serine proteases have been cloned, sequenced, and characterized *(4,11–13)*. Members of the subtilisin family of serine proteases share extensive sequence homology. The keratinase (*kerA*) gene of *B. licheniformis* shares 98 and 97% sequence homology with genes encoding subtilisin Carlsberg of *B. licheniformis* and subtilisin of *B. licheniformis* NCIB *(4)*. Moreover, the alkaline protease of *Bacillus* sp AH101, which shows high hydrolyzing activity against keratin, has extensive sequence homology with subtilisin BPN$^-$, subtilisin Carlsberg 221 and Ya-B alkaline protease *(14)*. Additionally, the activity of the keratinase enzyme secreted by *B. licheniformis* PWD1 was inhibited by PMSF, which inhibits serine proteases *(4)*. The alkaline protease enzyme (in this study) is also inhibited by PMSF (not shown).

On the other hand, considerable amounts of soluble proteins and free amino acids were obtained as a result of the enzymatic hydrolysis of feather (Figs. 3 and 4). Biodegradation of feather waste using *B. subtilis* DB100 (pS1) strain represents an alternative way to improve the nutritional value of feather, since feather waste is currently utilized on a limited basis as a dietary protein supplement for animal feedstuffs. Moreover, the release of several amino acids from feather, as well as secreted keratinase, would promote industries based on feather waste.

REFERENCES

1. Zaghloul, T., Ibrahim, N., El-Sewedy, S., and Helmi, S. (1992), *First Egyptian–Italian Symposium on Biotechnology*, Assiut, Egypt, Abstract 38.
2. Zaghloul, T., Abdel Aziz, A., and Mostafa, M. (1992), *First Egyptian–Italian Symposium on Biotechnology*, Assiut, Egypt, Abstract 64.
3. Zaghloul, T., Abdel Aziz, A., and Mostafa, M. (1994), *Enzyme Micro. Technol.* **16**, 534–537.
4. Lin, X., Kelemen, D. W., Miller, E. S., and Shih, J. C. (1995), *Appl. Environ. Microbiol.* **61**, 1469–1474.
5. Bernhard, K., Schrempf, H., and Goebel, W. (1978), *J. Bacteriol.* **133**, 897–903.
6. Williams, C. M., Richter, C. S., Mackenzie, J. M., and Shih, J. C. (1990), *Appl. Environ. Microbiol.* **56**, 1509–1515.
7. Pelczar, M. J. and Chan, E. C. (1977), *Laboratory Exercises in Microbiology*, 4th ed., McGraw-Hill, New York.
8. Cliffe, A. J. and Law, B. A. (1982), *J. Dairy Res.* **49**, 209–219.
9. Lin, X., Lee, C. G., Casale, E. S., and Shih, J. C. (1992), *Appl. Environ. Microbiol.* **58**, 3271–3275.
10. Laemmli, U. K. (1970), *Nature (London)* **227**, 680–685.
11. Jacobs, M., Eliasson, M., Uhlen, M., and Flock, J. (1985), *Nucleic Acids Res.* **13**, 8913–8926.
12. Stahl, M. L. and Ferrari, E. (1984), *J. Bacteriol.* **158**, 411–418.
13. Wells, J. A., Ferrari, E., Henner, D. J., Estell, D. A., and Chen, E. Y. (1983), *Nucleic Acids Res.* **11**, 7911–7925.
14. Takami, H., Akiba, T., and Horikoshi, K. (1990), *Appl. Microbiol. Biotechnol.* **33**, 519–523.

Isolation, Identification, and Keratinolytic Activity of Several Feather-Degrading Bacterial Isolates

Taha I. Zaghloul,*,[1] M. Al-Bahra,[2] and H. Al-Azmeh[2]

[1]*Department of Bioscience and Technology, Institute of Graduate Studies and Research, University of Alexandria, Alexandria, Egypt; and* [2]*Department of Chemistry, Faculty of Science, Damascus University, Syria*

ABSTRACT

Several feather-degrading bacterial isolates were isolated from Egyptian soil. These isolates were able to degrade chicken feather, when grown on basal medium containing 1% native feather as a source of energy, carbon, and nitrogen. Feather waste, generated in large quantities as a byproduct of commercial poultry processing, is nearly pure keratin, which is not easily degradable by common proteolytic enzymes. The isolates were identified according to the morphological characteristics, biochemical tests, and API 50 CHB *Bacillus* system. Proteolytic and keratinolytic activities of these isolates were monitored throughout the cultivation of the bacterial isolates on feather. Resulting soluble proteins, which were released as a result of the biodegradation of feather, were demonstrated by SDS-PAGE.

Index Entries: *Bacillus*; isolation; identification; keratinolytic activity.

INTRODUCTION

Environmental wastes are found in large quantities in many countries. Although some of them contain a considerable amount of protein and various carbon compounds, little attention is given to using them in a technological way. Recently, the authors have focused on the utilization of some environmental wastes, mainly feather waste. Feather waste, generated in large quantities as a byproduct of commercial poultry processing, is nearly pure keratin *(1)*. Because of the high degree of disulfide bonds, hydrophobic interactions, and hydrogen bonds, keratin in its native state is not degradable by common proteolytic enzymes, such as pepsin, trypsin,

---

* Author to whom all correspondence and reprint requests should be addressed.

and papain (2). Feather waste is utilized on a limited basis as a dietary protein supplement for animal feedstuffs. Generally, the feather is steam-pressure-cooked or chemically treated before use (3). Biodegradation of feather by microorganisms represents an alternative method to improve the nutritional value of feather waste.

The present work reports the isolation of several feather-degrading bacterial isolates from soil. These isolates are able to degrade whole chicken feather when grown on basal medium containing 1% native feather. The isolates were classified as members of the genus *Bacillus*. Proteolytic and keratinolytic activities of these bacterial isolates were monitored throughout the biodegradation process.

MATERIALS AND METHODS

Media

Bacterial isolates were activated and grown on PY medium (4) (Bacto peptone, 10 g; Difco (East Molesley, Surrey, UK) yeast extract, 5 g; and NaCl, 5 g/L). PA medium is PY supplemented with 1.5% agar agar. Basal medium II (5) (NH_4Cl, 0.5 g; NaCl, 0.5 g; K_2HPO_4, 0.3 g; KH_2PO_4, 0.4 g; $MgCl_2 \cdot 6H_2O$, 0.1 g; and yeast extract, 0.1 g/L), supplemented with 1% (w/v) whole chicken feather, was used to check the proteolytic and keratinolytic activity of the bacterial isolates. Feather plates were made as follows: 2 g of whole feather were treated with about 5 mL of NaOH (5 N), to the point that feather was converted to a soft paste. The volume was adjusted to 100 mL with basal medium II, after which the pH was adjusted to pH 7.0, and 1.5 g agar was added. The medium was autoclaved at 110°C for 10 min.

Isolation of Microorganisms

Several pieces of whole chicken feather were placed into wet soil in plastic containers. The containers were watered every 2 d for 2 wk, after which the partially degraded feather pieces were used to inoculate PY medium. The new culture was allowed to grow at 37°C for 24 h, with shaking. Suitable dilutions of this culture were plated on PA plates. Single colonies were screened for their ability to hydrolyze milk and keratin, by patching them into milk agar plates and feather plates, respectively. Alternatively, 100 mL of PY medium was inoculated with 1% soil suspension, and the culture was allowed to grow overnight, with shaking, at 37°C. Basal medium II, containing 1% whole feather, was inoculated with 1 mL of the above culture, and the new culture was allowed to grow at 37°C for 2, 4, and 6 d. A sample was taken every 2 d, diluted, and plated on PA plates. Single colonies were checked on milk plates and feather plates, as described above.

Determination of Bacterial Viable Count

Colony-forming units (CFU) were determined as described earlier *(6)*.

Identification of Feather-Degrading Isolates

Feather-degrading isolates were identified according to morphological examinations and several biochemical tests *(7,8)*. Identification was mainly based on the use of API 50 Carbohydrates *Bacillus* (CHB) strips (API Laboratory, Basingstoke, Hants, UK). The strips contain 49 *Bacillus*-specific carbohydrates tests. Bacterial isolates were prepared and treated as described earlier *(9)*.

Monitoring Proteolytic and Keratinolytic Activity

Bacterial isolates were activated by growing them overnight on PA plates at 37°C. Fresh colonies of each isolate were then transferred to 10 mL PY medium, and cultures were allowed to grow at 37°C for 2 h. One hundred mL preautoclaved basal medium II, containing 1% untreated whole feather, was inoculated with 5 mL of the above culture. Cells were allowed to grow at 37°C, with shaking, to the indicated time. At the indicated time, 1 mL culture was taken to determine the CFU, as described earlier. Two other mL were taken, centrifuged at 4500g for 2 min, using a microcentrifuge, and the supernatants were used as crude solution to determine proteolytic and keratinolytic activity.

Proteolytic Activity

Proteolytic activity was measured as described by Cliffe and Law *(10)*, using hide powder azure (HPA). One unit of enzyme is the amount of enzyme that develops a change of absorbance, against control reaction, at 595 nm/30 min at 37°C. Additionally, the proteolytic activity was determined according to the method of Lin et al. *(11)*, using skim milk agar plate.

Keratinolytic Activity

Keratinolytic activity was determined based on the free amino ($-NH_2$ groups) that were released as a result of the biodegradation of feather by bacterial cells. Free amino groups were determined as described earlier *(11)*, and a standard curve for leucine (0.01–0.1 μmol) was established. Additionally, the physical appearance of chicken feather in the culture was observed.

Analysis of Soluble Proteins by SDS-Polyacrylamide Gel Electrophoresis (SDS-PAGE)

Cell-free supernatant of each feather culture was precipitated with solid ammonium sulfate to reach 70% saturation, and was kept on an ice bath for 2 h. The mixture was centrifuged at 7000q for 30 min at 4°C.

The pellet was suspended in a small volume of Tris-HCl buffer, pH 8.0, and dialyzed overnight against this buffer.

A second precipitation by acetone (3 vol) was carried out, and the precipitate was collected by centrifugation at 3500g for 5 min using microcentrifuge. The pellet was directly suspended in a small volume of sample application buffer (SAB) and applied to a 10% SDS-polyacrylamide gel. The SDS-PAGE was carried out according to the method of Laemmli *(12)*.

RESULTS AND DISCUSSION

Identification of Feather-Degrading Bacterial Isolates

Feather, in its native state, is not degradable by common proteolytic enzymes such as trypsin, pepsin, and papain. However, feather or keratin-containing compounds do not accumulate in nature (especially in soil) *(5)*. Isolation of the feather-degrading bacterial isolates was carried out using soil sample and/or embedded feather in soil. The identification of these bacterial isolates was based on cell morphology, colony morphology, growth characteristics, several biochemical and supplementary tests, and, finally, the use of API 50 CHB strips that were designed to identify members of the genus *Bacillus*. API 50 CHB system has been shown to be rapid, accurate, and more reproducible than the classical tests. A taxonomy based on API 50 CHB tests is in good agreement with those obtained by other methods *(9)*.

Throughout the identification steps, *B. subtilis* DB100 $his^-\,met^-$ strain *(13)* was used as a reference strain. All feather-degrading bacterial isolates were Gram-positive, endospore-forming, rod-shaped bacilli. Bacterial isolate Bio 8/2 was identified as *B. pumilus*, isolates Bio 27/1, Bio 1/2, Bio 23, and Bio 6/2 were identified as *B. cereus*, and the bacterial isolate Bio 23' was identified as *B. subtilis*.

Monitoring Proteolytic and Keratinolytic Activity

Bacterial isolates were activated by growing them on PY medium for 2 h, to get vegetative active cells. Proteolytic activity was rapidly and accurately determined, using the synthetic substrate HPA, as well as the skim milk agar plate method. Data from the latter method (not shown) were consistent with those of HPA. Figure 1 shows two patterns for the production of extracellular proteases regarding the time of secretion, taking into consideration viable count of each isolate (log CFU). The isolates Bio 8/2, Bio 23, and Bio 23' secreted their proteases early (early genes), after which the proteolytic activity was decreased with increasing the cultivation time. In contrast, bacterial isolates Bio 27/1 and Bio 1/2 secreted their proteases late (late genes). Since members of the genus *Bacillus* are characterized by the formation of endospores, some genes encoding pro-

Fig. 1. Monitoring proteolytic activity (U/log CFU) of the bacterial isolates throughout the cultivation time. Isolates were activated on PY medium for 2 h, after which they were grown on basal medium II, supplemented with 1% whole chicken feather.

teases are expressed early during the log phase (i.e., neutral protease of *B. subtilis*), while other protease genes are known to be expressed late, during the stationary phase, or just before sporulation (i.e., the alkaline protease gene of *B. subtilis*) *(14)*. Unlike the above two patterns, the bacterial isolate Bio 6/2 showed constant production of their extracellular protease.

The keratinolytic activity of the bacterial isolates was determined based on measuring the free amino groups that were released in the cultures as a result of the biodegradation of feather. Generally, keratinolytic activity (μmol/mL culture/log CFU) increased with cultivation time (Fig. 2). Bacterial isolates Bio 23 and Bio 23' showed high level of both proteolytic and keratinolytic activities. This would suggest that their proteases may be acting as keratinases, and this is in agreement with some previous reports *(15–17)*. In contrast, the bacterial isolate Bio 27 showed high level of proteolytic activity, but a moderate level of keratinolytic activity.

The authors reported earlier the cloning of the *B. subtilis* alkaline protease (*aprA*) gene on a pUB110-derivative plasmid. *(14)* High levels of expression and gene stability were demonstrated when *B. subtilis* cells were grown on the laboratory medium 2XSG *(14)*. When these cells were grown on basal medium, supplemented with 1% whole chicken feather, high levels of proteolytic and keratinolytic activity were obtained. Data would indicate that the *aprA* gene is acting as a keratinase gene *(17)*. In the present study, *B. subtilis* DB100 (pS1) cells carrying the multicopy *aprA* gene were used to compare the proteolytic and the keratinolytic activity of these cells

Fig. 2. Monitoring keratinolytic activity of the bacterial isolates throughout the cultivation time. Isolates were activated on PY medium, then grown on basal medium II, supplemented with 1% whole chicken feather. The keratinolytic activity is expressed as μmol of -NH_2 groups per mL of cell-free supernatant per log CFU.

Fig. 3. SDS-polyacrylamide gel to analyze soluble proteins resulted from the biodegradation of whole feather at d 4. Preparation of samples and gel conditions were described in Materials and Methods. Lanes A–F represent 100 μg soluble proteins of bacterial isolates Bio 8/2, 23, 23′, 6/2, 1/2, and Bio 27/1, respectively.

with those of the bacterial isolates. The proteolytic activity of *B. subtilis* DB100 (pS1) strain was several-fold higher than that of any of the bacterial isolates; the keratinolytic activity (specific activity) was at the level of isolates Bio 23, Bio 23' and Bio 1/2 (6.7 μmol free NH_2 group/mL/log CFU) *(17)*.

Analysis of Soluble Proteins

On the biodegradation of feather using the bacterial isolates, considerable amounts of soluble proteins were obtained. These proteins were analyzed by SDS-polyacrylamide gel as described in Materials and Methods (Fig. 3). A zymogram technique, using gelatin in the polyacrylamide gel, is needed to know which protein band(s) might have a proteolytic activity.

The biodegradation of feather waste using these bacterial isolates represents an alternative way to improve the nutritional value of feather, since feather waste is currently utilized on a limited basis as a dietary protein supplement for animal feedstuffs. Research is being carried out to improve the ability of some of these isolates to degrade feather by transforming these bacterial isolates with the multicopy (pS1) plasmid that carries the *aprA* gene *(14)*.

REFERENCES

1. Goddard, D. R. and Michaelis, L. (1934), *J. Biol. Chem.* **106,** 604–614.
2. Harrap, B. S. and Woods, E. F. (1964), *Biochem. J.* **92,** 19–26.
3. Steiner, R. J., Kellems, R. O., and Church, D. C. (1983), *J. Anim. Sci.* **57,** 495–502.
4. Bernhard, K., Schrempf, H., and Goebel, W. (1978), *J. Bacteriol.* **133,** 897–903.
5. Williams, C. M., Richter, C. S., Mackenzie, J. M., and Shih, J. C. (1990), *Appl. Environ. Microbiol.* **56,** 1509–1515.
6. Pelczar, M. J. and Chan, E. C. (1977), *Laboratory Exercises in Microbiology*, 4th ed., McGraw-Hill, New York.
7. Claus, D. and Berkeley, R. C. W. (1986), in *Bergey's Manual of Systematic Bacteriology*, vol. 2 Sneath, P. H. A., Mair, N. S., Sharpe, M. E., and Holt, J. G. (eds). Williams and Wilkins, Baltimore, MD pp. 1105–1138.
8. Cheesbrough, M. (1985), *Medical Laboratory Manual for Tropical Countries*, vol. 2, ELBS.
9. Logan, N. A. and Berkeley, R. C. W. (1984), *J. Gen. Microbiol.* **130,** 1871–1882.
10. Cliffe, A. J. and Law, B. A. (1982), *J. Dairy Res.* **49,** 209–219.
11. Lin, X., Lee, C. G., Casale, E. S., and Shih, J. C. (1992), *Appl. Environ. Microbiol.* **58,** 3271–3275.
12. Laemmli, U. K. (1970), *Nature (London)* **227,** 680–685.
13. Sadaie, Y. and Kada, T. (1983), *J. Bacteriol.* **153,** 813–821.
14. Zaghloul, T., Abdel Aziz, A., and Mostafa, M. (1994), *Enzyme Micro. Technol.* **16,** 534–537.
15. Lin, X., Kelemen, D. W., Miller, E. S., and Shih, J. C. (1995), *Appl. Environ. Microbiol.* **61,** 1469–1474.
16. Takami, H., Akiba, T., and Horikoshi, K. (1990), *Appl. Microbiol. Biotechnol.* **33,** 519–523.
17. Zaghloul, T. (1998), *19th Symposium on Biotechnology for Fuels and Chemicals*, ABAB paper no. 2-08.

Acetamide Degradation by a Continuous-Fed Batch Culture of *Bacillus sphaericus*

F. Ramirez, O. Monroy,* E. Favela, J. P. Guyot, and F. Cruz

Departamento de Biotecnología, Universidad Autónoma Metropolitana-Iztapalapa. A P 55-535, 09340, Iztapalapa, D.F., MEXICO ¹Invited professor from ORSTOM, present address PMC, 911 Avenue Agropolis, BP 5045, 34032 Montpellier Cedex 1, France.

ABSTRACT

The methanogenesis of acetamide occurs through a two-step reaction in methanogenic sludges. First, acetamide is hydrolyzed to acetate and ammonia by a strict aerobic bacterium *(Bacillus sphaericus)*, then acetate is used by *Bacillus* as carbon source or converted to methane by methanogens. In this work, the kinetics of acetamide degradation by *B. sphaericus* was studied in a continuous reactor with biomass accumulation, fed with acetamide. The oxygen supplied was dissolved in the feed (6.4 mg/L) to resemble conditions in an anaerobic wastewater treatment reactor. A reaction in series model (acetamide → acetate → biomass) was used to find the kinetic parameters. Results show that *B. sphaericus* can hydrolyze acetamide in a second-order reaction with $K_1 = 1.1$ L/g/d, implying that the amount of biomass determines the rate and that no reaction will take place at specific loading rates greater than 35 gAm/gX/d. Growth parameters on acetate, as carbon source, under limiting O_2 conditions, are $\mu_{max} = 0.102/d$, $K_s = 37$ mg/L, $Y = 0.081$ gX/gAm.

Index Entries: Acetamide; hydrolysis; *Bacillus sphaericus*; continuous culture with biomass accumulation; UASB reactor.

INTRODUCTION

Acetamide is a highly toxic xenobiotic compound widely used in the lacquer, cosmetic, explosive, textile, and pharmaceutical industries *(1)*, and it is also produced by acetonitrile biodegradation *(2)*.

Acetamide degradation has been studied in an upflow anaerobic sludge blanket (UASB) reactor *(7)*. It was found that at low acetamide loading rates ($B_v = 1$ g. L/d), 86% removal efficiencies were obtained;

* Author to whom all correspondence and reprint requests should be addressed.

Table 1
Experimental Design for Growth of *B. sphaericus* in Continuous Culture

Am_o (mg/L)	1000	1500	2000	3000	3000	3000	3000
HRT (d)	1	1	1	1	2	3	0.5

Am_o inlet acetamide concentration; HRT, hydraulic retention time.

Table 2
Kinetic Equations for Acetamide Uptake, Acetate Production, and Biomass Growth

$$Am \xrightarrow{v_1}_{MN} Ac \xrightarrow{v_2}_{MN} X \qquad \text{Sequence reaction (2)}$$

$$\frac{dAm}{dt} = rAm = -K_1 \cdot Am\, X \qquad \text{Second-order reaction (3)}$$
Acetamide Hydrolysis

$$\frac{dAc}{dt} = +K_1 \cdot Am\, X \frac{\mu_{max} \cdot Ac}{K_{Ao} + Ac} \frac{X}{Y} \qquad \text{Acetate production (4) and uptake}$$

$$\frac{d\chi}{dt} = \frac{\mu_{max} \cdot Ac}{K_{Ac} + Ac} X \qquad \text{Monod (5)}$$

increasing B_v caused lower removal efficiencies. It was also noticed that acetamide caused inhibition of acetate methanization.

It was later found *(6)* that this degradation was possible through a synergistic association between a sporulating, Gram-positive, strictly aerobic rod, which transformed acetamide to acetate, and ammonia and methanogens, which transformed acetate to methane *(6)*.

$$CH_3 CONH_2 + H_2O \quad \textit{Bacillus sphaericus} \quad CH_3COO^- + NH_4^+ \quad (1)$$
$$CH_3COO^- + H_2O \quad \textit{Methanogens} \quad CH_4 + HCO_3^-$$

The coexistence of facultative aerobes with strict anaerobes is possible when aerobic bacteria take up the available oxygen in the media *(3–8)*.

Acetamide degradation can be modeled in a two-step consecutive reaction, in order to find out if the rate limitations are caused by dissolved oxygen limitations or by the acetamide substrate inhibition. Therefore, the acetamide hydrolysis kinetics by *B. sphaericus* were studied in a continuous tubular reactor, with biomass accumulation, under limited amounts of oxygen.

Acetamide Degradation by B. sphaericus

Fig. 1. Tubular reactor with syntered glass to retain the *B. sphaericus* biomass, and a gas separator.

MATERIAL AND METHODS

Continuous Cultivation

A 0.8-L continuous-operation volume tubular reactor (Fig. 1) was operated at 34–36°C to hydrolyze acetamide. It was inoculated with 80 mL of a pure strain of *B. sphaericus* in the exponential growth phase isolated from a UASB reactor (7).

The feed was a solution consisting of Balch medium (9), oligoelements, 0.5 g/L yeast extract, 0.1 g/L of casein peptone and acetamide, and saturated with air to get an oxygen concentration of 6.4 mg/L. Acetamide concentration and hydraulic retention times were varied, as shown in Table 1, to assess the kinetics constants. The media was adjusted to pH 7.0, and sterilized in an autoclave.

Analyses

Acetamide and acetate were determined in a Varian gas chromatograph with a flame ionization detector, using a capillary column (0.22 mm × 30 m) (At-1000, Altech), with helium as carrier gas. One-mL samples were pretreated with 50 µL of formic acid by centrifugation at 3000 rpm

Table 3
Transient Mass Balances in *B. sphaericus* Growth in Continuous Culture with Biomass Accumulation

$$\frac{dAm}{dt} = D(Amo - Am) - K1 Am X \quad (6)$$

$$\frac{dAc}{dt} = -Dac + AmX - \frac{\mu^* Ac}{K_{Ac} + Ac} \frac{X}{Y_{XS}} \quad (7)$$

$$\frac{dX}{dt} = \mu X \quad (8)$$

for 15 min. Biomass was measured by $OD_{600\,nm}$ (Bausch & Lomb, Spectronic 20), and calibrated against known concentrations of *B. sphaericus*. Oxygen concentration was measured with a HACH meter (HACH, Loveland, CO).

Kinetic Characterization of the Strain

The affinity (K_s), the maximum specific growth rate (μ_m), and the yield (Y) constants were determined with the sequence reaction kinetic equations and mass balances shown in Tables 2 and 3, respectively. The acetamide hydrolysis is modeled as a second-order reaction (Eq. 3), in which acetate is an intermediate product in a sequence reaction (Eq. 2 and 4). Biomass growth follows Monod kinetics (Eq. 5). The liquid phase is in continuous flow while biomass is retained within the vessel.

RESULTS AND DISCUSSION

Figure 2 shows the acetamide loading rates (B_v) applied to the reactor, together with the outgoing rates of acetamide and acetate. The largest acetamide uptake efficiency was observed at a $B_v = 1$ gAm L/d. At Bv = 1.5, acetate accumulates and the biomass remains constant (Fig. 3). It is only in the $B_v = 2$ when biomass starts to accumulate, while acetate and acetamide start to be consumed. By $B_v = 3$, acetate is being totally consumed, while biomass growth rate is the highest. When the B_v is again reduced to 1.5 by increasing the HRT, biomass builds up to 900 mg/L. With this high biomass concentration, acetamide and acetate are both at almost zero. During this period, there is no acetamide hydrolysis and no growth, probably because of the sevenfold specific acetamide load to the bacteria (Fig. 4).

Although a limited amount of oxygen is being fed to the reactor, Table 4 shows that the reaction itself is not limited by oxygen, because the amount available for the bacteria is inversely proportional to the biomass concentration.

Fig. 2. Continuous culture of *B. sphaericus*; acetamide loading rate (X), acetamide (□), and acetate (♦) accumulation rates. The acetamide predicted behavior (−) as described in Eq. 3 with K_S = 37 mg/L and μm = 0.10 per d.

Fig. 3. *B. sphaericus* biomass accumulation in the tubular reactor exposed to several loading rates. The brokes line is a reinoculation of the reactor.

The hydrolysis rate of acetamide follows a second-order reaction, as shown in Eq. 3. Reaction constants were evaluated by both algebraic (selecting steady states) and differential methods (taking discrete increments in differential Eq. 6), in order to take all the experimental points into ac-

Table 4
Acetamide Hydrolysis by *B. sphaericus*

B_v (g/l/d)	Am_i (mg/L)	η_m (%)	Ac (mg/L)	η_c (%)	X (mg/L)	B_{OD} (mgOD/gxd)	B_x (gAm/gxd)
1	1000	88	411	41	157	32.6	6.4
1.5	1500	44	536	36	163.25	31.4	9.2
2	2000	37	354	18	288.25	18	7
3	3000	28	31.2	1	413.25	12.4	7.26
1.5	3000	43	0	0	788.25	3.2	1.9
1	3000	78	0	0	913	2	1.1
6	3000	4.4	0	0	207	49.46	29

B, acetamide loading rate; Am, acetamide concentration; η, conversion efficiencies; Ac, acetate concentration; X, biomass; subindex: v = volumetric; x = specific; m = acetamide; c = acetate; and OD = oxygen dissolved.

Fig. 4. Specific acetamide loading rate (gAm/gX/d).

count. The results obtained by these two methods were similar: $K_i = 1.1$ and 0.8 L/gX/d for the algebraic are differential methods, respectively. The continuous line in Fig. 4 shows the predicted values obtained by Eq. 6.

Solving Eqs. 7 and 8 simultaneously, the growth yield coefficient was found to be $Y_{X/Ac} = 0.018$ gX/gAc.

Figure 3 shows a period of exponential growth between d 30 and 48. It was used to estimate the kinetic parameters of Monod equation. By equating Eqs. 5 and 8, and solving for a linear regression, the values of μ_m and K_s were estimated to be 0.102 per d and 0.0367 gAc/L, respectively.

Fig. 5. Hydrolysis efficiency as a function of B_v. Large efficiencies are obtained at low rates.

Fig. 6. The hydrolysis efficiency, as a function of the biomass loading rate, shows the same tendency as the B_v effect.

Table 4 shows the steady-state values of each run. From these data, Fig. 5 shows that the hydrolysis efficiency drops as B_v increases. This is also associated to the biomass loading rate (B_x = gAm /gX/d) (Fig. 6).

The specific acetamide hydrolysis rate (gAm/gX/d) is a negative function of acetamide concentration, thus suggesting acetamide substrate inhibition (Fig. 7). Two runs are out of this tendency, because the reaction rate

Fig. 7. The specific acetamide hydrolysis rate, as a function of the concentration, suggests a decreasing rate with increasing concentration. The two points out of the tendency are limited by loading rate.

Fig. 8. Specific acetate consumption rate shows the negative effect of the B_{ox} at three different levels of B_x.

(gAc/L/d) is limited by the loading rate (gAmo/L/d), and the specific oxygen loading rate (B_{ox}).

The acetate uptake rate is negatively influenced by the level of specific oxygen loading rate (B_{ox} = mg O_2/g X/d). Figure 8 shows the specific acetate consumption rate as a function of both the B_{ox} and the B_x. It suggests that increasing B_{ox} adversely affects the acetate consumption rate. Three B_x levels can be distinguished here, one at the level 1–2 gAm/gX/d, another at around 7, and a third one of $B_x \geq 9$, suggesting that increasing specific acetamide loads need more oxygen.

CONCLUSIONS

These results show that *B. sphaericus* can degrade acetamide and does not need the participation of methanogens for it to use acetate as substrate for growth. At 6.4 mg oxygen/L, *B. sphaericus* exhibited a K_s for acetate of 37 mg/L, which makes it a better scavenger than methanogens that have K_s = 160–300 mg/L. Under these conditions, the coculture is possible, because *B. sphaericus* grows slowly with a μ_m = 0.10 per d; methanogens exhibit twice this value *(10)*.

Given that the hydrolysis of acetamide is intended as a first step for its methanization, the present study orients toward the coculture conditions. Results suggest that, in order to promote high acetamide hydrolysis and keep a low acetate consumption rate by *B. sphaericus*, the biomass loading rates (B_x) should be about 1 or 2 gAm/gX/d at low Am concentrations and low oxygen-loading rates (B_{ox} = 3 mgO_2/gAm/d). Under these conditions there would be enough acetate for the methanogens.

REFERENCES

1. Moretti, T. A. (1978), in *Kirk-Othmer Encyclopedia of Chemical Technology* vol. 1. Mark H. F., McKettea J. J., and Othmer D. F., eds. John Wiley, New york, pp. 148–151.
2. DiGeronimo, M. J. and Antoine, A. D. (1967), *Appl. Environ. Microbiol.* **31,** 900–906.
3. Gerritse, J. and Gottschal J. C., (1993), *J. Gen. Microbiol.* **139,** 1853–1860.
4. Guyot, J. P. and Fajardo, C. (1993), *Biotechnol. Lett.* **15,** 743–748.
5. Guyot, J. P., Gutierrez, G., and Rojas, M. G. (1993), *Appl. Microbiol. Biotechnol.* **40,** 139–142.
6. Guyot, J. P., Ramírez, F., and Ollivier, B. (1994), *Appl. Microbiol. Biotechnol.* **42,** 452–456.
7. Guyot, J. P., Ferrer, H., and Ramírez F. (1995), *Appl. Microbiol. Biotechnol.* **43,** 1107–1111.
8. Wu, W., Hu, J., Gu, X., Zhao, Y., Zhang, H., and Gu, G. (1987), *Water Res.* **21,** 789–799.
9. Balch, W. E., Fox, G. E., Magrum, L. J., Woese, C. R., and Wolfe, R. S. (1979), *Microbiol. Rev.* **43,** 260–296.
10. Pavlostathis, S. G. and Giraldo Gomez, E. (1991), *Water Sci. Tech.* **24,** 35–39.

Use of Hemicellulose Hydrolysate for β-Glucosidase Fermentation

K. Réczey,[1] A. Brumbauer,[1] M. Bollók,[1] Zs. Szengyel,*,[2] and G. Zacchi[2]

[1]*Department of Agricultural Chemical Technology, Technical University of Budapest, Szent Gellért tér 4, Budapest, H-1521, Hungary; and* [2]*Department of Chemical Engineering I, University of Lund, P.O. Box 124, S-221 00 Lund, Sweden*

ABSTRACT

Hydrolysis of cellulose by *Trichoderma* cellulases often results in a mixture of glucose, cellobiose, and low-mol-wt cellodextrins. Cellobiose is nonfermentable for most yeasts, and therefore it has to be hydrolyzed to glucose by β-glucosidase prior to ethanol fermentation. In the present study, the β-glucosidase production of one *Penicillium* and three *Aspergillus* strains, which were previously selected out of 24 strains, was investigated on steam pretreated willow. Both steam-pretreated willow and hemicellulose hydrolysate, released during steam explosion of willow, were used as carbon sources. Reference cultivation runs were performed using prehydrolyzed Solka Floc and glucose. The four strains were compared with *Trichoderma reesei* regarding sugar consumption and β-glucosidase production. *Aspergillus niger* and *Aspergillus phoenicis* proved to be the best enzyme producers on hemicellulose hydrolysate. The maximum β-glucosidase activity, 4.60 IU/mL, was obtained when *A. phoenicis* was cultivated on the mixture of hemicellulose hydrolysate and steam-pretreated willow. The maximum yield of enzyme activity, 502 IU/g total carbohydrate, was obtained when *Aspergillus foetidus* was cultivated on the hemicellulose hydrolysate.

Index Entries: β-glucosidase production; hemicellulose hydrolysate of willow; *Trichoderma; Penicillium; Aspergillus.*

INTRODUCTION

Different techniques are available for the conversion of lignocellulosic materials to fuel ethanol. During the past decade, process alternatives based on enzymatic hydrolysis have been the focus of interest, showing

* Author to whom all correspondence and reprint requests should be addressed.

good yields of fermentable sugars from both soft- and hardwoods (1–3). Most of these processes consist of the following five major process steps: pretreatment of raw material, including chopping, screening, and prehydrolysis of hemicellulose; cellulase enzyme production; enzymatic hydrolysis of the cellulose fraction; fermentation of the hydrolysate, using a suitable microorganism; and ethanol refining (4). One method for prehydrolysis of the hemicellulose fraction, which has been used for a large variety of woody materials, is high-pressure steam pretreatment (5). During the steam-pretreatment of wood, the hemicellulose part of the material is degraded mostly to monosaccharides (xylose, mannose, galactose, glucose, and arabinose). The fibrous material remaining after pretreatment consists of cellulose and structurally modified lignin. As a result of the steam explosion, the pretreated material is more accessible to enzymatic attack.

The cellulose fraction of the fibrous material is further hydrolyzed to cellobiose and glucose by means of cellulolytic enzymes. Mainly, three groups of hydrolytic enzymes are involved in the hydrolysis of cellulose to glucose: endoglucanases (EG), cellobiohydrolases (CBH), and β-glucosidase/cellobiase (6,7). The cellobiase enzyme has an important role in the hydrolysis of cellulose. It cleaves the β-glucosidic bond in the cellobiose, which is released by EG and CBH from the cellulose. The conversion of cellobiose, which is inhibitory to CBH and EG (8), increases the yield considerably (9).

The β-glucosidase activity in most *Trichoderma* cellulase preparations has been shown to be much lower than what is required for an efficient conversion of cellulose to glucose (10). Although the cellobiase production of *Trichoderma reesei* Rut C 30 can be enhanced by changing the pH and temperature profile of the fermentation (11), addition of external cellobiase from other microorganisms is required. The optimal ratio of β-glucosidase: cellulase activity required for sufficient hydrolysis of cellulose was reported to be in the range of 0.8–1.5 IU of β-glucosidase per filter paper unit of cellulase activity (12).

When hardwood, in which the xylan content varies between 15–23% based on dry matter (13), is used for ethanol production, the xylose fraction leaves the process unchanged. This reduces the overall ethanol yield based on carbohydrates, since it cannot be fermented by ordinary baker's yeast. There are other pentose fermenting microorganisms, which are capable of converting C5 carbohydrates to ethanol, but these are sensitive to inhibitory compounds formed during the steam pretreatment (14,15). Another option for utilizing the xylose-rich liquid is to use it for cellulase enzyme production. In a previous study, in which *T. reesei* Rut C 30 was cultivated on the hemicellulose hydrolysate, supplemented with fibrous pretreated willow at a total carbohydrate concentration of 20 g/L, a cellulolytic enzyme activity of 1.79 filter paper activity (FPU)/mL, and only 0.43 IU/mL β-glucosidase activity, was obtained (16). An alternative is to utilize the hydrolyzed hemicellulose fraction for cellobiase production, which can be used for supplementing the *Trichoderma* fermentation broth with β-glucosidase. When *Aspergillus niger* VKMF-2092 was cultivated on a 50–50% mixture of glucose and wheat bran at 10 g/L total carbohydrate concentration, an activity of 3.25 IU/mL was obtained after 148 h residence time *(17)*. *Aspergillus wentii* cultivated on various carbon sources at 30 g/L carbohydrate concentration yielded β-glucosidase activities in the range of 0.35–1.10 IU/mL *(18)*.

In the present study, four different β-glucosidase producers, *A. niger*, *Aspergillus phoenicis*, *Aspergillus foetidus*, and *Penicillium ochro-chloron* were investigated. The yield of β-glucosidase, using these fungal strains grown on glucose, prehydrolyzed Solka Floc 200 (FS&D, Urbana, OH), and hemicellulose hydrolysate from willow, were determined and compared with the yield obtained with *T. reesei*.

MATERIALS AND METHODS

Pretreatment of Willow

Salix caprea, a fast-growing willow species, was used for enzyme production. The raw material contained 36.8% cellulose, 23.0% hemicellulose, 20.7% lignin, and 19.5% other compounds. The willow was chipped and sieved, and the fraction between 1 and 3.5 mm was used. First, the willow chips were presteamed for 40 min with 1 bar saturated steam. The hot presteamed material was immediately transferred into plastic bags, impregnated with 1% SO_2, based on oven-dried material (ODM), and stored overnight at room temperature. The steam pretreatment of the impregnated material was performed at 207°C for 5 min with saturated steam *(13,19)*. The pretreated material was diluted to approx 5% ODM with hot tap water, stirred for 30 min, and filtered using a PF 0.1H2 (Larox OY, Finland) filter press unit *(4)*. The filtrate comprised of hydrolyzed hemicellulose, was divided into two fractions. One fraction was used without further treatment. The other fraction was concentrated by vacuum evaporation, removing 75% of the water at 80°C and pH 2.8–3.0, using a Büchi RE 121 rotavapor (Büchi Labortechnik AG, Switzerland), thus removing the volatile compounds that have been shown to be inhibitory to *T. reesei* *(16)*. The composition of the steam-pretreated willow (SPW), regarding cellulose and lignin content, was determined by Hägglund's method *(20)*, with the modification that the sugar content of the acid hydrolysate was analyzed as well for total sugars, which was used to calculate the cellulose content. The steam-pretreated willow contained 50% cellulose and 37% lignin, based on ODM. Both the original filtrate (F) and the concentrated filtrate (CF) were analyzed for sugar content, using phenol sulfuric acid method *(21)*. The total sugar concentration was 64 g/L and 267 g/L for F and CF, respectively. The fibrous material (SPW), the F, and the CF were all used as carbon (C) sources for β-glucosidase production. The amounts of SPW, F, and CF added to the medium were based on the cellulose and sugar contents given above.

Prehydrolysis of Cellulose

The prehydrolysis of cellulose powder (Solka Floc) was carried out in an NBS G24 (New Brunswick, NJ) rotary shaker using 500 mL E-flasks. A total volume of liquid of 300 mL, containing 20 g/L Solka Floc, 8 mL/L Celluclast 1.5 L enzyme preparation (Novo A/S, Denmark), and tap water, was incubated at 50°C, pH 4.8, and 150 rpm for 20 h. The obtained slurry, i.e., the prehydrolyzed Solka Floc (PSF), was diluted twice with tap water, and used as a C source for β-glucosidase production.

Fungal Strains

The following five fungal strains were used for cellobiase production: *A. niger* BKM F-1305, *A. phoenicis* QM329, *A. foetidus* Biogal 39 (strain collection Dr. G. Szakacs, Technical University of Budapest, Dept. of Agricultural Chemical Technology, Budapest, Hungary), *P. ochro-chloron* WFPL 175A, and *T. reesei* Rut C 30, respectively. The strains were obtained from the Department of Agricultural Chemical Technology, Technical University of Budapest.

Inoculum Preparation

Prior to use, all fungi were stored on agar slants containing (in g/L) 20 malt extract, 5 glucose, 1 proteose peptone, and 20 bacto agar. After 2–3 wk at 30°C, the conidia were used to initiate growth in 750-mL E-flasks containing 150 mL medium. The original Mandels medium *(22)*, in which the concentration of nutrients were (in g/L) 0.3 urea, 1.4 $(NH_4)_2SO_4$, 2.0 KH_2PO_4, 0.3 $CaCl_2$, 0.3 $MgSO_4$, 0.25 yeast extract, and 0.75 proteose peptone, together with 10 g/L Solka Floc cellulose powder, was used for *Trichoderma* inoculum preparation. Trace elements were also added (mg/mL): 5 $FeSO_4 \cdot 7H_2O$, 20 $CoCl_2$, 1.6 $MnSO_4$, and 1.4 $ZnSO_4$. The pH before sterilization was adjusted to 5.5–6.0. The inoculum was ready after 4 d at 30°C and 300 rpm. For Aspergilli; and *Penicillium*, the medium contained only 5% malt extract in tap water. Prior to sterilization, the pH was set to 5.5–6.0. After 2 d at 30°C and 300 rpm, the inoculum was ready.

Shake-Flask Cultivation

The mycelia obtained from the inoculum were used to initiate growth in 750-mL E-flasks containing 150 mL Mandels medium, including the various C sources, i.e., glucose, PSF, F, CF, and F, supplemented with fibrous SPW. Each strain was cultivated on each C source. The total concentration of the carbohydrates available in the medium was set to 10 g/L, except for F, in which it was 5 g/L. At this sugar concentration, the filtrate proved to be not inhibiting *(16)*. The inoculum constituted 10% of the medium. The enzyme production was performed in a rotary shaker at 300 rpm and 30°C. Samples were withdrawn once a day, and, at the same time, the cultivation was adjusted to pH 6.0 with addition of either 10% NaOH or 10% H_2SO_4 solutions, when required. The samples were centrifuged using a Janetzki T24 (Leipzig, Germany) centrifuge at C 6100g for 10 min. The supernatant was analyzed for enzyme activity and sugar content.

ANALYSIS

The enzyme activity of the samples from *Aspergillus* and *Penicillium* cultivation was determined as β-glucosidase activity, using Berghem's method *(23)*; the samples from the cultivation with *Trichoderma* were analyzed for both β-glucosidase activity, using Berghem's method, and FPU, using Mandels procedure *(24)*. The reducing-sugar content of the samples was determined with the DNS method *(25)*.

RESULTS AND DISCUSSION

The four strains of *Aspergillus* and *Penicillium* were selected after a screening of 24 strains of *Aspergillus, Penicillium, Trichoderma, Chaetomium, Geotrichum,* and *Paecilomyces,* using PSF as a C source. The four selected strains yielded the highest β-glucosidase activity after inoculation and 10 d of cultivation (results not shown). The aim of the present study was to utilize the pentose rich hydrolysate containing sugars only in form of monosaccharides for enzyme production.

Does β-Glucosidase Production Need to be Induced?

The biosynthesis, not only of the cellulases but of β-glucosidase in microorganisms, is, in most cases, subjected to catabolite repression *(26,27)*. The production of β-glucosidase is initiated only after the glucose or other easily metabolizable monosaccharides in the medium have been utilized. To investigate whether the β-glucosidase synthesis needed to be induced, all strains were cultivated in shake flasks, both on PSF and on glucose-containing media, at a total C source concentration of 10 g/L. For each experimental condition three fermentation runs were performed in parallel, and the mean values of the enzyme activity and the sugar concentration were calculated. The standard deviation (SD) was 0.05 IU/mL for the enzyme activity.

When *A. niger* was cultivated on PSF, a β-glucosidase activity of 2.46 IU/mL was obtained, compared with 2.80 IU/mL reached, when cultivated on glucose medium (Fig. 1A). After 2 d of cultivation, most of the sugars present in both media were consumed. The higher yield on glucose indicates that the β-glucosidase secretion of *A. niger* is not repressed by glucose.

For *A. phoenicis,* the maximum β-glucosidase activities obtained after 6 d were 2.80 IU/mL and 3.60 IU/mL, when cultivated on glucose and on PSF, respectively (Fig. 1B). In both cases, the soluble sugar content of the medium was reduced to 0 g/L after 2 d, and, at the same time, the enzyme production was started. The somewhat lower enzyme activity

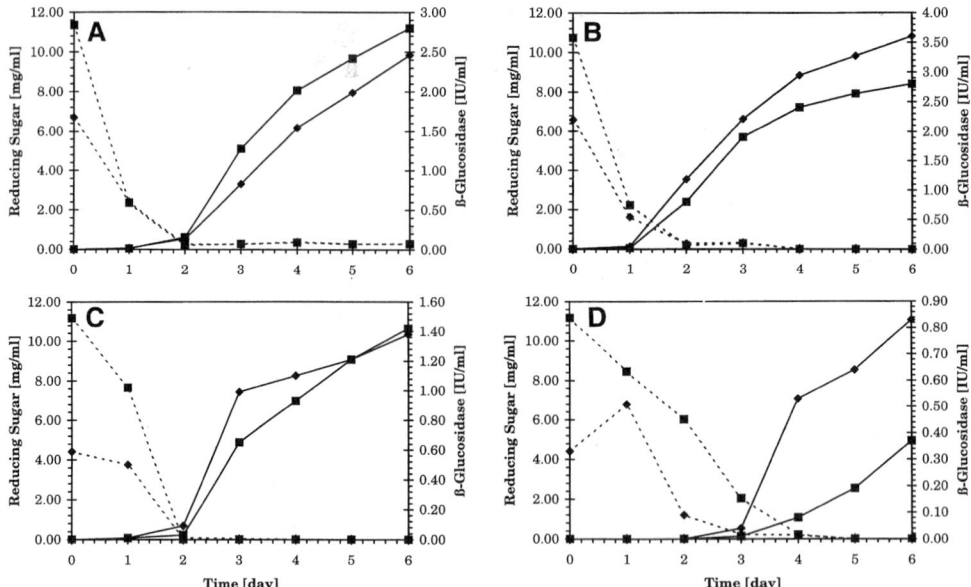

Fig. 1. β-Glucosidase activity and reducing sugar concentration vs time for cultivation of **(A)** *A. niger*, **(B)** *A. phoenicis*, **(C)** *A. foetidus*, and **(D)** *P. ochro-chloron*, on glucose medium (■), and on PSF medium (♦). Solid lines = enzyme activity, broken lines = sugar concentration.

reached after 6 d on glucose cannot be explained with catabolite repression. A significant amount of enzyme was produced on glucose medium, with an initial production rate equal to that obtained with PSF. After 4 d, the production rate started to decline (Fig. 1B), presumably because of depletion of nutrients. The cell mass produced on glucose was higher than that produced on PSF, resulting in a higher consumption of nutrients.

There was no significant difference between the two C sources when *A. foetidus* was cultivated, but only 1.40 IU/mL activity was reached on both media (Fig. 1C).

For the *P. ochro-chloron*, the possibility of catabolite repression cannot be excluded (Fig. 1D). There was a significant difference in enzyme activity after 5 d of cultivation for the two C sources. The enzyme activity reached was 237% higher on PSF than on glucose. The final activity obtained on glucose was 0.37 IU/mL, which is 55% less than that reached on PSF. Furthermore, the sugar consumption rate was lower on glucose than on PSF, which indicates catabolite repression.

Enzyme Production on Concentrated Filtrate

The enzyme production on concentrated filtrate was investigated at a carbohydrate concentration of 10 g/L. The highest β-glucosidase activity after 6 d cultivation, 3.75 IU/mL, was obtained with *A. phoenicis*, and the

β-Glucosidase Fermentation

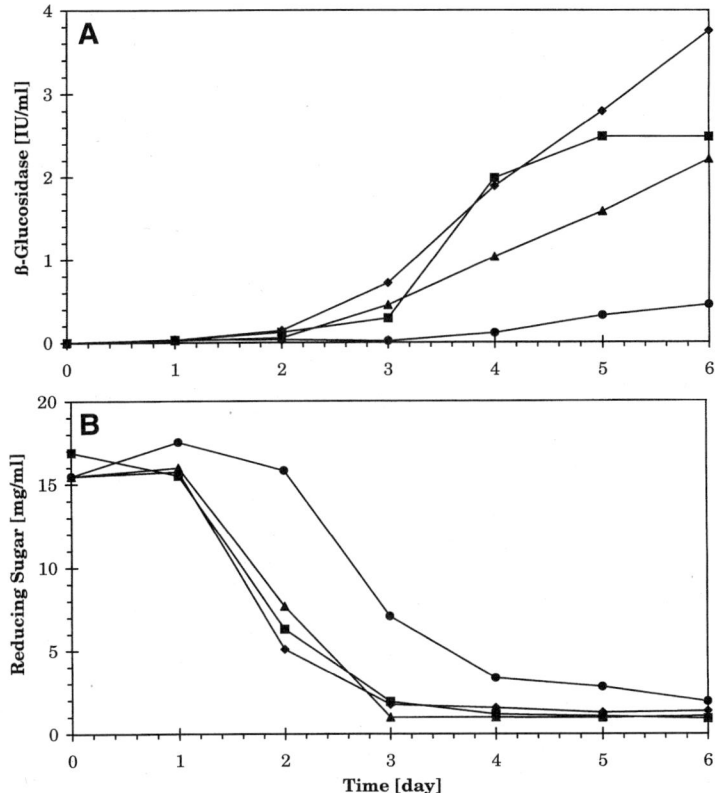

Fig. 2. β-Glucosidase activity and reducing sugar concentration vs time for various fungal strains cultivated on CF at a sugar content of 10 g/L. **(A)** β-Glucosidase activity. **(B)** Sugar concentration. *A. niger* (■), *A. phoenicis* (♦), *A. foetidus* (▲), *P. ochro-chloron* (●).

lowest, 0.45 IU/mL, for *P. ochro-chloron* (Fig. 2). The enzyme yield of *A. niger* obtained on CF was somewhat lower than that reached on pure glucose medium. All the other strains produced significantly more enzyme on CF than on glucose (Table 1), even though the sugar consumption rates were lower on the CF medium. Although the yield of enzyme activity for *P. ochro-chloron* reached on CF was increased with 22%, compared to that obtained on glucose, it was much lower than for all the other strains (Fig. 2).

Enzyme Production Using Original Filtrate

Two main series of experiments were performed to study the β-glucosidase production, using F as a C source. In the first series, the total concentration of sugars in the medium was set to 5 g/L by dilution with tap water. In the second series, the same diluted F was used, but it was

Table 1
β-Glucosidase Activities and Yields Based of Total Carbohydrates After 6 d of Cultivation Using Glucose, PSF, and F- and CF-Containing Media

Strain	Glucose		PSF		F		CF	
	Activity (IU/mL)	Yield (IU/g)	Activity (IU/mL)	Yield (IU/g)	Activity (IU/mL)	Yield (IU/g)	Activity (IU/mL)	Yield (IU/g)
A. niger	2.80	280	2.46	246	1.90	380	2.48	248
A. phoenicis	2.80	280	3.60	360	2.11	422	3.75	375
A. foetidus	1.42	142	1.38	138	2.51	502	2.21	221
P. ochro-chloron	0.37	37	0.83	83	0.24	48	0.45	45

Table 2
β-Glucosidase Activities Obtained After 6 d of Cultivation Using the Mixture of Diluted Filtrate and Steam Pretreated Willow

Strain	Activity (IU/mL)	Yield (IU/g soluble CH)	Yield (IU/g total CH)
A. niger	2.90	580	290
A. phoenicis	4.60	920	460
A. foetidus	2.02	404	202
P. ochro-chloron	0.69	138	69

supplemented with fibrous SPW at a cellulose concentration corresponding to 5 g/L. For each strain, three parallel runs were performed. The average SD in enzyme activity was 0.05 IU/mL. The results are presented as average values in Table 1 and Table 2.

The β-glucosidase activities and yields after 6 d of cultivation with the F-containing medium are shown in Table 1. For comparison, data obtained on glucose and PSF are also shown. The highest β-glucosidase activity, 2.51 IU/mL, was obtained with A. foetidus. Also, for this C source, P. ochro-chloron showed the lowest production of β-glucosidase, resulting in a final activity of 0.24 IU/mL. Although the sugar consumption rate on d 1 was lower for all strains, compared to those obtained on glucose and PSF, 95% of the available carbohydrates were consumed within 2 d. The enzyme yields based on added carbohydrates were used for comparison of the β-glucosidase production on different C sources (Table 1). All strains gave the highest yields when cultivated with the F-containing medium. The highest yield of 502 IU/g carbohydrate was obtained with A. foetidus, which was 3.5× higher than that reached on pure glucose, or on PSF. The second best yield, 422 IU/mL, was obtained with A. phoenicis on F, which is only slightly higher than that obtained on PSF.

In the second series of experiments using the mixture of F and SPW, A. niger, A phoenicis, and P. ochro-chloron gave higher β-glucosidase activities than when cultivated on the filtrate (Table 2). For A. phoenicis, an activity of 4.60 IU/mL was obtained, which is more than twice that obtained on F. The yield based on total available carbohydrate (Table 2) is higher than that obtained on F, which indicates that most of the cellulose was utilized for β-glucosidase production. When A. niger was cultivated on the mixed carbon source, a β-glucosidase activity of 2.90 IU/mL was obtained, which is 1.5-fold higher than when cultivated on F, indicating that only a part of the cellulose was utilized for β-glucosidase production. The β-glucosidase production of A. foetidus was lower, 2.02 IU/mL, on the mixed C source than on F.

Table 3
Cellulase and β-Glucosidase Activities Obtained with *T. reesei*
Cultivated on Different C Sources

Carbon source[a]	FPU (FPU/mL)	β-Glucosidase (IU/mL)
Glucose	0.30	0.11
PSF	1.06	0.40
CF	0.61	0.11
F	0.27	0.05
F + SPW	0.83	0.15

[a] For abbreviations, see text.

β-Glucosidase Production of *T. reesei*

For comparison, *T. reesei* Rut C 30 was cultivated on all five C sources. The obtained β-glucosidase and FPU activities after 6 d of cultivation on the different media are summarized in Table 3. The highest β-glucosidase and cellulase activities, 0.40 IU/mL and 1.06 FPU/mL, respectively, were obtained on PSF. This gives an enzyme activity ratio of 0.38 β-glucosidase activity to FPU activity. Although relatively high cellulase activity, 0.83 FPU/mL, was reached on the mixture of F and SPW, the β-glucosidase concentration was low, 0.15 IU/mL, resulting in a β-glucosidase:FPU ratio of 0.18. This ratio is considerably lower than that required for efficient hydrolysis of cellulose. On the media containing only soluble sugars (glucose, F, and CF), very low β-glucosidase activities were obtained.

CONCLUSIONS

The hemicellulose hydrolysate of willow obtained after steam pretreatment of the raw material proved to be a very good substrate for β-glucosidase production with three different Aspergilli strains. The highest β-glucosidase activity of 4.60 IU/mL was obtained when *A. phoenicis* was cultivated on the mixture of filtrate and steam-pretreated willow at a total carbohydrate concentration of 10 g/L. This activity is higher than shown by other reported data on *Aspergillus* strains cultivated on different C sources *(17,18)*. Cultivation of *A. foetidus* on filtrate alone, at a carbohydrate concentration of 5 g/L, resulted in a β-glucosidase activity of 2.51 IU/mL, with the highest yield of enzyme activity, 502 IU/g carbohydrate. All Aspergilli strains performed well on the concentrated filtrate, with a maximum β-glucosidase activity of 3.75 IU/mL reached in the cultivation of *A. phoenicis*.

ACKNOWLEDGMENT

The Swedish National Board for Industrial and Technical Development (NUTEK), and the National Research Fund of Hungary (OTKA-T0117201) are gratefully acknowledged for their financial support.

REFERENCES

1. Jurasek, L. (1979), *Develop. Ind. Microbiol.* **20**, 177–183.
2. Saddler, J. N., Brownell, H. H., Clermont, L. P., and Levitin, N. (1982), *Biotech. Bioeng.* **24**, 1389–1402.
3. Grethlein, H. E., Allen, D. C., and Converse, A. O. (1984), *Biotech. Bioeng.* **26**, 1498–1505.
4. Palmqvist, E., Hahn-Hägerdal, B., Galbe, M., Larsson, M., Stenberg, K., Szengyel, Zs., Tenborg, C., and Zacchi, G. (1996), *Bioresource Technol.* **58**, 171–179.
5. Saddler, J. N. (1993), in *Bioconversion of Forest and Agricultural Plant Residues*, CAB International, Wallingford, UK.
6. Wood, T. M. (1989), *Enzyme Syst. Lignocellul. Degrad., Proc. Workshop Prod., Charact. Appl. Cellul.-, Hemicellul.-, Lignin-Degrading Enzyme Syst.*, Coughlan, M. P., ed., Elsevier; London, UK, pp. 5–16.
7. Wood, T. M. and Garcia-Campayo, V. (1990), *Biodegradation* **1**, 147–161.
8. Mandels, M., and Reese, E. T. (1963), in *Advances in Enzymatic Hydrolysis of Cellulose and Related Materials*, Reese, E. T., ed., Pergamon, London. pp. 115–157.
9. Eklund, R., Galbe, M., and Zacchi, G. (1990), *Enzyme Microb. Technol.* **12**, 225–228.
10. Sternberg, D., Vijayakumar, P., and Reese, E. T. (1977), *Can. J. Microbiol.* **23**, 139–147.
11. Tangnu, S. K., Blanche, H. W., and Wilke, C. R. (1981), *Biotech. Bioeng.* **23**, 1837–1849.
12. Duff, S. J. B. (1985), *Biotechol. Lett.* **7**, 185–190.
13. Eklund, R., Galbe, M., and Zacchi, G. (1995), *Bioresource Technol.* **51**, 225–229.
14. Palmqvist, E., Hahn-Hägerdal, B., Galbe, and Zacchi, G. (1996), *Enzyme Microb. Technol.* **19**, 470–476.
15. Olsson, L. and Hahn-Hägerdal, B. (1993), *Process Biochem.* **28**, 249–257.
16. Szengyel, Zs., Zacchi, G., and Réczey, K. (1997), *Appl. Biochem. Biotechnol.*, **63–65**, 351–362.
17. Kerns, G., Dalchow, E., Klappach, G., and Meyer, D. (1986), *Acta Biotechnol.* **6**, 355–359.
18. Srivastava, S. K., Gopalkrishnan, K. S., and Ramachandran, K. B. (1987), *J. Fermentation Technol.*, **65**, 95–99.
19. Eklund, R., Galbe, M., and Zacchi, G. (1988), *International Symposium on Alcohol Fuels*, VIII, 101–105.
20. Hägglund, E. (1951), in *Chemistry of Wood*, Academic, New York, NY., pp 324–332.
21. Dubois M., Gilles K. A., Hamilton J. K., Rebers P. A., and Smith F. (1956), *Anal. Chem.* **28**, 350–356.
22. Mandels, M. and Weber, J. (1969), *Adv. Chem. Ser.* **95**, 391–414.
23. Berghem, L. E. R. and Petterson, L. G. (1974), *Eur. J. Biochem.* **46**, 295–305.
24. Mandels, M., Andreotti, R., and Roche, C. (1976), *Biotechnol. Bioeng. Symp.* **6**, 21–33.
25. Miller, G. (1959), *Anal. Chem.* **31**, 426–428.
26. Kerns, G., Okunev, O. N., Ananin, V. M., and Golovlev, E. L. (1987), *Acta Biotechnol.* **6**, 535–545.
27. Kubicek, C. P., Messner, R., Gruber, F., Mach, R. L., and Kubicek-Pranz, E. M. (1993), *Enzyme Microb. Technol.* **15**, 90-99.

Production of a Novel Pyranose 2-Oxidase by Basidiomycete *Trametes multicolor*

CHRISTIAN LEITNER,[1] DIETMAR HALTRICH,*,[1] BERND NIDETZKY,[1] HANSJÖRG PRILLINGER,[2] AND KLAUS D. KULBE[1]

[1]*Abteilung Biochemische Technologie, Institut für Lebensmitteltechnologie and* [2]*Institut für Angewandte Mikrobiologie, Universität für Bodenkultur BOKU Wien (University of Agricultural Sciences Vienna), Muthgasse 18, A-1190 Wien, Austria*

ABSTRACT

During a screening for the enzyme pyranose 2-oxidase (P2O) which has a great potential as a biocatalyst for carbohydrate transformations, *Trametes multicolor* was identified as a promising, not-yet-described producer of this particular enzyme activity. Furthermore, it was found in this screening that the enzyme frequently occurs in basidiomycetes. Intracellular P2O was produced in a growth-associated manner by *T. multicolor* during growth on various substrates, including mono-, oligo-, and polysaccharides. Highest levels of this enzyme activity were formed when lactose or whey were used as substrates. Peptones from casein and other casein hydrolysates were found to be the most favorable nitrogen sources for the formation of P2O. By applying an appropriate feeding strategy for the substrate lactose, which ensured an elevated concentration of the carbon source during the entire cultivation, levels of P2O activity obtained in laboratory fermentations, as well as the productivity of these bioprocess experiments, could be enhanced more than 2.5-fold.

Index Entries: *Trametes multicolor*; pyranose 2-oxidase; screening; culture medium development

INTRODUCTION

Pyranose 2-oxidase (P2O, glucose 2-oxidase, pyranose:oxygen 2-oxidoreductase, EC 1.1.3.10) catalyzes the C-2 oxidation of several aldopyranoses to form the corresponding 2-keto derivatives, with the preferred

*Author to whom all correspondence and reprint requests should be addressed.

substrate being D-glucose, which is oxidized to D-*arabino*-2-hexosulose (2-keto-D-glucose, D-glucosone). During this oxidation, electrons are transferred to molecular oxygen to yield hydrogen peroxide. P2O has been purified and characterized from several sources *(1–6)*. Typically, it is a rather large glycoprotein that contains covalently bound flavin adenine dinucleotide.

There is very good indication of a likely involvement of P2O in lignocellulose degradation. A possible role of this enzyme could be as a major source of hydrogen peroxide *(7)*. This view is supported by the fact that P2Os typically are rather unspecific, accepting all major sugars found in lignocellulose as substrates, and that their main localization is in the periplasmic space, or even extracellular, which has been demonstrated *in situ* during wood decay for several organisms *(7,8)*.

A completely different physiological function of P2O has been suggested by Baute et al. *(9,10)*, who showed that glucose can be converted to the antibiotic cortalcerone, via glucosone, by the fungus *Corticium caeruleum*. The formation of this antibiotic substance could also be proven in a number of other fungi (13.65% of the species tested in a screening). Moreover, the existence of the enzyme pyranosone dehydratase, which transforms glucosone into cortalcerone, has been shown in *Phanerochaete chrysosporium* *(11,12)*. However, a number of organisms that exert P2O activity do not possess this ability to convert glucosone into cortalcerone, apparently lacking the second enzyme of this pathway. These findings indicate that glucosone may play a more significant role as an intermediate in glucose metabolism in fungi, cortalcerone biosynthesis being probably one of the possible alternative pathways *(11)*.

P2O offers an attractive potential as a biocatalyst for the specific oxidation of unprotected sugars. The dicarbonyl compounds formed by these enzymatic transformations can be used as fine chemicals, or as building blocks in synthetic carbohydrate chemistry *(13,14)*. Because chemical syntheses of these compounds are laborious and result in relatively low yields, together with a number of byproducts, the enzymatic conversion represents an interesting alternative to the chemical route. Because of the high regioselectivity of the enzyme employed, a close-to-complete formation of three different dicarbonyl sugars, with yields of 85–99% by using P2O from *Peniophora gigantea*, has been recently reported *(15)*. Furthermore, some of the dicarbonyl sugars have proposed applications in food technology. 2-Ketoglucose can be used as the key intermediate in the production of fructose, mannitol, or sorbitol, and has attracted considerable attention in this respect *(16–18)*.

It was the objective of our work to identify a suitable producing strain of the enzyme P2O, which has a great potential for carbohydrate transformations. Additionally, growth conditions favoring the enhanced formation of this enzyme in an appropriate organism should be investigated in detail.

MATERIAL AND METHODS

Chemicals

All chemicals were of the highest purity available, and were obtained from Merck, Darmstadt, Germany, unless otherwise stated. 2,2'-Azino-bis(3-ethylbenzthiazoline-6-sulfonic acid) (ABTS), syringaldazine, malt extract, corn-steep liquor, avicel (microcrystalline cellulose), cellobiose, and xylitol were obtained from Sigma (St. Louis, MO); horseradish peroxidase, grade I, was from Boehringer Mannheim (Mannheim, Germany); tryptone was from Oxoid (Basingstoke, UK); soybean meal (Provasoy™) was from Vamo Mills (Izegem, Belgium); beechwood xylan was from Lenzing AG (Lenzing, Austria); casamino acids were from Marcor (Hackensack, NJ); casein, casein hydrolysate, and NH_4Cl were from Fluka (Buchs, Switzerland). Sunflower oil and skim milk were purchased at a local supermarket. Whey powder was a gift from Bundeslehranstalt für Alpenländische Milchwirtschaft (Rotholz, Austria).

Microbial Strains and Culture Conditions

All fungal organisms were from the culture collection of the Institute of Applied Microbiology, Universität für Bodenkultur Wien, where they are deposited under the indicated strain numbers. The wild-type strain of *T. multicolor* (= *T. zonata*) MB 49, which was used for most parts of this study, was isolated from hardwood in southern Germany. Stock cultures were maintained on glucose-maltose Sabouraud agar, and were transferred every 8 wk. Inoculated plates were incubated at 25°C for 4–6 d, and then stored at 4°C.

For the initial screening tests, two different culture media were used. Medium A contained (in g/L): maltose, 20; D-glucose, 10; $MgSO_4 \cdot 7H_2O$, 0.5; inositol, 0.05; peptone from casein, 0.2; yeast extract 2.0; KH_2PO_4, 0.4; $ZnSO_4 \cdot 7H_2O$, 0.001; $FeCl_3 \cdot 6H_2O$, 0.01; $MnSO_4 \cdot H_2O$, 0.005. Medium B contained (in g/L): D-glucose, 5.0; D-galactose, 5.0; D-xylose, 5.0, L-sorbose, 5.0, yeast extract, 5.0, malt extract, 10.0 *(19)*. For the initial screening performed on agar plates, medium B was supplemented with agar (20 g/L) and ABTS (0.25 g/L), before autoclaving. Peroxidase (1,000 IU/L) was added by sterile filtration, when the autoclaved medium had cooled to 50°C *(19)*. All optimization experiments were done in 300-mL baffled conical flasks containing 100 mL of medium. These were inoculated with a piece (1 cm²) from an actively growing, 4–6-day-old colony of *T. multicolor* on Sabouraud agar. The inoculated flasks were continuously shaken on an orbital shaker at 110 rpm (stroke 25 mm) and 25°C for 12 d. Mycelia were harvested by centrifugation, washed twice using saline, and then disrupted for the determination of P2O activity (*see* Enzyme Assay).

Bioprocess Experiments

Fermentation studies were carried out in a 20-L laboratory fermenter (MBR Bio Reactor, Wetzikon, Switzerland) with a working volume of 15 L, and equipped with four disk turbine impellers, each with six flat blades. The basal culture medium for these cultivations contained whey powder, which was added so that the final lactose concentration was 25 g/L, peptone from casein (10 g/L), and KH_2PO_4 (1.0 g/L). The temperature was controlled at 25°C, and the pH was allowed to float. Aeration was automatically varied from 0.1 to 1.0 vol of air/fluid vol/min to maintain a pO_2 of 40% of air saturation. Foaming was controlled by using 10% v/v aqueous polypropylene glycol P2000 (Fluka).

Enzyme Assay

Mycelia (2 g wet wt) were resuspended in 10 mL of potassium phosphate buffer (50 mM, pH 6.5, containing 10 mM EDTA), homogenized, and then disrupted using a French press operating at 1,310 bar (19,000 lb/in^2) cell pressure. Following three passages through the French press cell (4°C), debris were removed by centrifugation (30,000g, 4°C, 20 min). The clear supernatant thus obtained was used for enzyme activity determination. P2O activity was spectrophotometrically determined at 30°C, using ABTS as described by Danneel et al. *(6)*. One IU of P2O activity is defined as the amount of enzyme necessary for the oxidation of 2 μmol of ABTS/min under the given conditions. Laccase activity was determined by oxidation of syringaldazine *(20)*. β-Galactosidase activity was assayed using *p*-nitrophenyl-β-D-galactopyranoside (8 mM final concentration) as the substrate.

Other Analyses

Protein determinations were done according to the dye-binding method of Bradford *(21)*, using bovine serum albumin (fraction V; United States Biochemical Corp., [USB] Cleveland, OH) as standard. Lactose was assayed by using a commercial kit (Boehringer Mannheim).

RESULTS

Screening

In an initial screening procedure, in which a number of fungi were tested for the ability to produce carbohydrate-oxidizing enzyme activities, agar plates containing the chromogen ABTS (2,2' azino-*bis*[3-ethylbenzthiazoline-6-sulfonic acid]) and peroxidase were employed. By using this method, positive strains could be easily identified by a characteristic purple-blue dye, which is formed by the peroxidase-catalyzed reaction of the chromogen and H_2O_2. It has been previously shown that this screening method is suitable for both extra- and intracellular carbohydrate oxidases *(19)*. This preliminary screening procedure included a total of 38 species of basidiomycetes from 24 genera. Approximately two-thirds of those basidiomycetous fungi tested were found to form carbohydrate oxidases under the conditions selected for the screening. However, the incubation time, which was necessary for the first color response to be detectable, varied greatly from 1 to 7 d for the different strains. Carbohydrate oxidizing enzyme activities were detected among several *Armillaria* sp, *Phanerochaete* sp, *Phellinus* sp, *Pholiota* sp, *Trametes* sp, as well as in *Asterophora lycoperdoides, Daedaleopsis tricolor, Flammulina velutipes, Fomes fomentarius, Ganoderma applanatum, Lenzites betulina, Marasmius alliaceus, Oudemansiella mucida, Phlebiopsis gigantea,* and *Schizophyllum commune*. Based on these results, selected organisms were grown in shaken-flask cultures, using different growth media, and P2O activity was assayed from the mycelial extracts. In this experiment, 19 species were investigated, 11 of which formed intracellular P2O activity (Table 1). Only organisms with unequivocally positive P2O activity are listed. Gluconic acid formed by the activity of glucose 1-oxidase, which also yields hydrogen peroxide and thus interferes with the P2O assay employed, could only be detected in *P. chrysosporium* MB 59; hence, this organism is not shown in Table 1.

Culture Medium Development

T. multicolor, which formed both the highest specific, as well as total, activity of P2O in the screening experiment, was selected for further studies. To investigate the effect of the carbon (C) source on P2O formation in this organism, various carbohydrate substrates, including mono-, oligo-, and polysaccharides, were added to a basal medium containing yeast extract (2.0 g/L), malt extract (1.0 g/L), and peptone from casein (3.0 g/L). The concentration of the latter complex nitrogen (N) source was increased, compared to the initial screening experiment, since it was found to favorably affect P2O formation. The results of these experiments are given in Table 2. Growth of *T. multicolor* was good on most of the substrates tested. Intracellular P2O activity was formed by the organism during growth on all of the various C sources employed, albeit to a greatly varying extent. Best results pertaining to the total activity of P2O were obtained for whey powder, which contains approx 80% lactose, as well as for D-xylose, lactose, and cellobiose. Since whey powder not only showed excellent results, but also is a cheap, technical substrate, it was selected for further experiments.

Since one of the objectives of this work was to identify medium components that positively influence P2O formation in *T. multicolor*, the effect of various, mainly complex N sources, which were shown to significantly affect the formation of this enzyme in another fungus *(22)*, was studied as well. The indicated N sources were added to a medium containing whey powder (30 g/L) and KH_2PO_4 (1.0 g/L). Their concentrations were based on their total N concentration, as estimated by the Kjeldahl method, thus

Table 1
Formation of Pyranose 2-Oxidase Activity in Different Fungi

Organism	Medium	Growth[a] (g/L)	Time (h)	Enzyme activity		
				(IU/mL)[b]	(IU/mg)[c]	(IU/L)[d]
F. fomentarius MB 79	B	26.8	474	0.008	0.083	1.3
L. betulina MB 78	B	31.2	526	0.209	0.230	39.1
O. mucida MB 121	B	23.2	386	0.615	0.346	84.9
O. mucida MB 122	B	20.4	553	0.416	1.30	50.9
P. chrysosporium MB 56	A	20.8	233	0.037	0.023	4.6
P. chrysosporium MB 56	B	49.3	334	ND	ND	ND
P. gigantea MB 70	A	25.6	526	0.005	0.052	0.7
P. gigantea MB 70	B	35.2	261	0.187	0.245	39.5
P. mixta MB 116	B	10.4	549	0.006	0.006	0.4
S. commune MB 145	B	113.6	214	0.023	0.015	15.7
T. hirsuta MB 50	B	82.4	236	0.018	0.165	8.9
T. multicolor MB 49	A	104.0	286	0.267	2.67	167.0
T. multicolor MB 49	B	48.7	168	0.574	0.892	167.8
T. pubescens MB 88k	B	188.2	387	0.025	0.096	28.4
T. pubescens MB 90	B	217.6	214	0.032	0.178	40.8
Trametes versicolor MB 53	A	43.6	187	0.208	0.186	54.7
T. versicolor MB 53	B	119.2	236	0.053	0.363	37.9
T. versicolor MB 54	A	41.9	186	0.197	0.277	49.6
T. versicolor MB 54	B	249.2	261	0.010	0.048	15.0

All organisms were incubated at 25°C for the time indicated.
[a] Mycelial wet wt.
[b] Volumetric activity determined in the crude enzyme preparation (mycelial extract).
[c] Specific activity (IU/mg protein) in the crude enzyme.
[d] Total activity (IU/L fermentation medium).
ND, not detectable.

ensuring a constant concentration of 1.26 g/L total N in each medium. Results for growth and P2O formation are summarized in Table 3. All of the complex, organic nutrients tested sustained good growth of the organism. Addition of these organic N sources seems to be necessary, since they could not be substituted by NH_4Cl, which gave only poor results. Clear differences pertaining to P2O formation in *T. multicolor* exist for the different N sources examined in this experiment. Especially, various enzymatic digests of casein, e.g., peptone from casein, tryptone, or casein hydrolysate, significantly stimulate formation of total P2O activity. Furthermore, increasing the concentration of peptone from casein to 9.1 g/L, compared to the initial experiments, enhanced both growth and P2O formation considerably.

Table 2
Effect of Various Carbon Sources on Growth and Formation of Pyranose 2-Oxidase in T. multicolor

Growth substrate	Biomass[a] (g/L)	Enzyme activity		
		(IU/mL)[b]	(IU/mg)[c]	(IU/L)[d]
Medium B	48.5	0.57	0.76	166
L-Arabinose	21.4	0.01	0.02	1.3
D-Xylose	294.7	0.21	0.88	362
D-Galactose	51.6	0.49	0.96	152
D-Glucose	195.9	0.04	0.22	41
L-Sorbose	189.2	0.01	0.03	11
Mannitol	112.3	0.30	0.68	199
Xylitol	29.1	1.06	1.07	184
Cellobiose	147.2	0.29	0.36	261
Lactose	234.9	0.22	0.76	315
Maltose	145.1	0.10	0.32	89
Sucrose	199.1	0.04	0.57	48
Whey powder	101.2	0.81	0.59	494
Sunflower oil	341.9	0.01	0.12	29
Xylan from beechwood	NA	0.96	2.31	158
Cellulose microcrystalline	NA	0.11	0.33	23

Substrates were added in equal concentrations (20 g/L) to a medium that further contained peptone from casein (3.0 g/L), yeast extract (2.0 g/L), malt extract (1.0 g/L), and KH_2PO_4 (1.0 g/L). Medium B contained D-glucose, D-galactose, D-xylose, L-sorbose (each at a concentration of 5.0 g/L).

[a] Mycelial wet wt.
[b] Volumetric activity determined in the crude enzyme preparation.
[c] Specific activity (IU/mg protein).
[d] Total activity (IU/L fermentation medium).
NA, not available.

Bioprocess Experiments

Production of P2O was studied in a 20-L laboratory fermenter; the working volume was 15 L of a medium based on whey powder (final lactose concentration, 25 g/L) and peptone from casein, which were both found to be optimal for enhanced P2O formation in previous experiments. The time-course of this bioprocess experiment is shown in Fig. 1. The stationary phase of growth was reached after approx 150 h. At this time, 6.7 g/L of dry wt biomass, corresponding to a mycelial wet wt of 42 g/L, was formed. It is interesting to note that lactose was not depleted at the beginning of the stationary phase, but still could be found in concentrations of 8.8 g/L. Extracellular β-galactosidase activity could not be detected at any time in this cultivation. The pH value, which was initially 5.8, and was allowed to float, continuously dropped during growth of the organism, to

Table 3
Effect of Various Nitrogen Sources on the Formation of Pyranose 2-Oxidase in
T. multicolor When Grown in Shaken Flask Cultures

N source	Concentration (g/L)	Biomass[a] (g/L)	Enzyme activity		
			(IU/mL)[b]	(IU/mg)[c]	(IU/L)[d]
Peptone from casein	9.1	289.4	0.89	1.24	1550
Tryptone	9.9	354.7	0.54	1.01	1150
Casein hydrolysate	10.0	370.5	0.47	0.78	1040
Casamino acid	10.9	62.0	1.15	0.73	429
Yeast extract	12.0	112.2	0.59	0.57	402
Peptone from meat	10.7	97.5	0.57	0.71	332
Soybean meal	15.7	118.7	0.37	1.07	262
Peptone from soybean	13.1	139.0	0.31	0.91	257
Cornsteep liquor	38.2	92.8	0.44	0.28	243
Casein	10.0	334.3	0.11	0.21	228
Meat extract	10.5	108.4	0.18	0.28	118
Skim milk	40% (v/v)	233.3	0.05	0.13	73.3
NH_4Cl	4.8	29.5	0.34	0.91	60.7
None	–	37.1	0.03	0.27	5.5

The N sources were added on the basis of an equivalent N concentration of 1.26 g/L total N. Whey powder (20 g/L) was used as growth substrate.
[a] Mycelial wet wt.
[b] Volumetric activity determined in the crude enzyme preparation.
[c] Specific activity (IU/mg protein).
[d] Total activity (IU/L fermentation medium).

reach a minimum of 4.1. At the transition of the exponential to the stationary phase of growth, it started to increase slightly. As can be clearly seen in Fig. 1, formation of P2O is growth-associated in *T. multicolor*, but laccase

Fig. 1. Time-course of a batch fermentation of *T. multicolor* on a medium containing whey powder as the substrate. Symbols: (■), mycelial dry wt; (▲), lactose; (X), total pyranose 2-oxidase activity (IU/L medium); (♦), laccase activity.

Fig. 2. Time-course of a fed-batch fermentation of *T. multicolor* on a whey-based medium. The arrows mark the addition of a 1 M solution of lactose, so that the lactose concentration in the medium was increased by approx 10 g/L. Symbols: (■), mycelial dry wt; (▲), lactose; (X), total pyranose 2-oxidase activity (IU/L medium).

activity peaked at the early growth phase and thereafter quickly decreased. The maximum of total P2O activity (588 IU/L of culture medium, corresponding to 87.8 IU/g dry biomass) was found after 200 h of growth; thereafter, P2O activity slightly decreased.

Since both growth of *T. multicolor* and formation of total P2O activity were only observed at elevated lactose concentrations in the batch cultivation, a fed-batch fermentation was performed. The initial growth medium was identical to the batch cultivation. At 107, 192, and 264 h cultivation time, a 1 M lactose solution was added by sterile filtration to increase the lactose concentration in the medium by approx 10 g/L, thus ensuring an elevated concentration of this carbohydrate during the entire course of the cultivation. Results are shown in Fig. 2. Contrary to the batch cultivation, an increase in biomass was observed for more than 300 h, reaching a maximum of 14.9 g/L dry wt (corresponding to 85.4 g/L wet wt). Simultaneously, total P2O activity increased to a maximum value of 1,650 IU/L, which corresponds to 111 IU/g dry biomass. Again, the initial culture pH of 5.8 slowly dropped to approx 4.2, and then stayed constant for the entire cultivation.

DISCUSSION

In an initial screening procedure, which was done on agar plates containing the chromophore ABTS and peroxidase, it was found that carbohydrate-oxidizing enzyme activities are widespread among basidiomycetes. This confirms the results of previous screenings for these enzyme activities (10,19,23,24). Out of 38 different species belonging to 24 genera, 25 organisms that formed a carbohydrate oxidase under the growth conditions selected for this experiment were detected by a characteristic dye formation. The employed screening method proved to be very efficient and convenient. However, since the agar medium used in this screening contained the carbohydrates D-galactose, D-glucose, L-sorbose and D-xylose, several other sugar oxidases would be detected by this method, in addition to P2O (EC 1.1.3.10). These could include glucose oxidase (EC 1.1.3.4), hexose oxidase (EC 1.1.3.5), galactose oxidase (EC 1.1.3.9), or L-sorbose oxidase (EC 1.1.3.11).

To unequivocally identify suitable P2O-producing strains, several of the carbohydrate–oxidase-positive organisms were cultivated in shaken flasks, and P2O activity was then assayed in the mycelial extracts. Formation of P2O activity was commonly found in these selected basidiomycetes. Eleven out of 19 organisms formed P2O activity under the chosen growth conditions. Because of this frequent occurrence, it seems likely that this enzyme activity has a more general importance in the metabolism of basidiomycetous fungi than has been suggested before (24). The effects of the two different media used for the cultivation of the basidiomycetes on the formation of P2O activity are not unambiguous. Especially with *P. chrysosporium* MB 56, this enzyme activity could not be detected when the organism was cultivated on medium B, but growth on medium A clearly resulted in the formation of P2O. It can be concluded that the growth conditions used for the screening were certainly not optimal for P2O synthesis, and that, because of this fact, detection of several P2O-positive strains could have been missed. As a result of the screening, *T. multicolor* MB 49 was identified as a promising, not-yet-described producing strain of intracellular P2O activity, and was selected for further investigations.

The most suitable substrate for the efficient production of P2O by *T. multicolor* was found to be whey powder, which contains approx 80% lactose. Although several more readily metabolized carbohydrates, when employed as C and energy source, resulted in higher biomass formation, levels of P2O activities were found to be significantly lower with these substrates than those obtained with whey powder or lactose. It seems likely that the relatively slow utilization of lactose by this basidiomycetous fungus, which does not secrete β-galactosidase activity into the extracellular environment under these growth conditions, is favorable for P2O synthesis, since higher concentrations of monosaccharides, i.e., glucose or galactose, which typically exert catabolite repression in microbial cells, are avoided. This is further corroborated by the fact that an appropriate feeding, which ensured a certain lactose concentration during the entire cultivation, greatly enhanced not only growth of the organism, but also the formation of total P2O activity, as well as of the units of P2O formed per g biomass.

Synthesis of P2O under apparently derepressed conditions is in agreement with several other reports on different fungal organisms. Its synthesis is growth-associated in an unidentified basidiomycete (no. 52), when lactose is used as a substrate that was also poorly utilized (22). P2O is only formed in the idiophase after the depletion of the C source, when using glucose- or polyol-based media by several organisms, including *Coriolopsis occidentalis*, *P. chrysosporium*, or *P. gigantea* (6,24,25). Contrarily, its formation by *Polyporus obtusus* and *Oudemansiella mucida* was reported to be associated with growth, even when glucose was employed as the C source. With these two organisms, the maximum P2O activity coincided with the depletion of the carbohydrate substrate (2,26). However, these physiological aspects of P2O synthesis in fungi have not so far been studied in detail.

Both growth of *T. multicolor* and P2O formation considerably decreased when the organism was grown in a stirred-tank laboratory fermenter, compared to the cultivations in shaken flasks. A possible explanation for this could be damage to the mycelium caused by high shear stress, which probably will occur near the tips of the impellers. These negative effects of shear stress and mechanical forces on filamentous fungi, which can cause breakage of the hyphae and leakage of intracellular material, have recently been reviewed (27). Moreover, a characteristic change in the morphology was observed when comparing growth in shaken flasks and stirred-tank reactors. *T. multicolor* formed small pellets (2–3 mm diameter) in shaken flask cultures, but it grew in filamentous form in the fermenter cultivations. Presumably, conditions of low shear, e.g., as characteristic for air lift fermenters, will be favorable for the enhanced production of P2O in *T. multicolor*. Further investigations in this respect will be carried out in the authors' laboratory.

ACKNOWLEDGMENT

The authors would like to thank Elisabeth Mayr for the gift of whey powder. This work was supported by a grant from the Austrian Science Foundation (FWF) P11459-MOB.

REFERENCES

1. Janssen, F. W. and Ruelius, H. W. (1975), *Methods Enzymol.* **41**, 170–173.
2. Volc, J., Sedmera, P., and Musílek, V. (1978), *Folia Microbiol.* **23**, 292–298.
3. Machida, Y. and Nakanishi, T. (1984), *Agric. Biol. Chem.* **48**, 2463–2470.
4. Volc, J. and Eriksson, K.-E. (1988), *Methods Enzymol.* **161**, 316–322.
5. Izumi, Y., Furuya, Y, and Yamada, H. (1990), *Agric. Biol. Chem.* **54**, 1393–1399.
6. Danneel, H.-J., Rössner, E., Zeeck, A., and Giffhorn, F. (1993), *Eur. J. Biochem.* **214**, 795–802.
7. Daniel, G., Volc, J., and Kubátová, E. (1994), *Appl. Environ. Microbiol.* **60**, 2524–2532.
8. Daniel, G. (1994), *FEMS Microbiol. Rev.* **13**, 199–233.
9. Baute, M.-A., Baute, R., Deffieux, G., and Filleau, M.-J. (1977), *Phytochemistry* **16**, 1895–1897.
10. Baute, M.-A. and Baute, R. (1984), *Phytochemistry* **23**, 271–274.
11. Volc, J., Kubátová, E., Sedmera, P., Daniel, G., and Gabriel, J. (1991), *Arch. Microbiol.* **156**, 297–301.

12. Gabriel, J., Volc, J., Sedmera, P., Daniel, G., and Kubátová, E. (1993), *Arch. Microbiol.* **160,** 27–34.
13. Röper, H. (1991), in *Carbohydrates as Organic Raw Materials,* Lichtenthaler, F. W., ed., VCH, Weinheim, Germany, pp. 267–288.
14. Crueger, A. and Crueger, W. (1990), in *Microbial Enzymes and Biotechnology,* Fogarty, W. M. and Kelly, C. T., eds., Elsevier Applied Science, London, pp. 177–226.
15. Huwig, A., Danneel, H.-J., and Giffhorn, F. (1994), *J. Biotechnol.* **32,** 309–315.
16. Neidleman, S. L., Amon, W. F., and Geigert, J. (1981), U.S. Patent 4246347.
17. Liu, T.-N. E., Wolf, B., Geigert, J., Neidleman, S. L., Chin, J. D., and Hirano, D. S. (1983), *Carbohydr. Res.* **113,** 151–157.
18. Geigert, J., Neidleman, S. L., and Hirano, D. S. (1983), *Carbohydr. Res.* **113,** 159–162.
19. Danneel, H.-J., Ullrich, M., and Giffhorn, F. (1992), *Enzyme Microb. Technol.* **14,** 898–903.
20. Petroski, R. J., Peczynska-Czoch, W., and Rosazza, J. P. (1980), *Appl. Environ. Microbiol.* **40,** 1003–1006.
21. Bradford, M. M. (1976), *Anal. Biochem.* **72,** 248–254.
22. Furuya, Y., Yamada, H., and Izumi, Y. (1993), *J. Ferment. Bioeng.* **76,** 532–534.
23. Gancedo, J. M., Gancedo, C., and Asensio, C. (1967), *Arch. Biochem. Biophys.* **119,** 588–590.
24. Volc, J., Denisova, N. P., Nerud, F., and Musílek, V. (1985), *Folia Microbiol.* **30,** 141–147.
25. Eriksson, K.-E., Pettersson, B., Volc, J., and Musilek, V. (1986), *Appl. Microbiol. Biotechnol.* **23,** 257–262.
26. Ruelius, H. W., Kerwin, R. M., and Janssen, F. W. (1968), *Biochim. Biophys. Acta* **167,** 493–500.
27. Thomas, C. R. (1990), in *Chemical Engineering Problems in Biotechnology,* Winkler, M. A., ed., Elsevier Applied Science, London, pp. 23–93.

Broad Spectrum and Mode of Action of an Antibiotic Produced by *Scytonema* sp. TISTR 8208 in a Seaweed-Type Bioreactor

APARAT CHETSUMON, FUSAKO UMEDA, ISAMU MAEDA,
KIYOHITO YAGI,* TADASHI MIZOGUCHI AND YOSHIHARU MIURA

Faculty of Pharmaceutical Sciences, Osaka University, 1–6 Yamada-oka, Suita, Osaka 565, Japan

ABSTRACT

A photobioreactor was constructed using anchored polyurethane foam strips (1 × 1 × 40 cm) fixed onto a stainless-steel ring to prevent flotation, as a biomass support material (BSM). This type of reactor was named a seaweed-type bioreactor. A filamentous cyanobacterium, *Scytonema* sp. TISTR 8208, which produces a novel cyclic dodecapeptide antibiotic, was immobilized in seaweed-type photobioreactor and cultivated with air containing 5% CO_2 sparged at a gas flow rate of 250 mL/min under illumination at a light intensity of 200 μmol photon $m^{-2}s^{-1}$. The antibiotic produced in the seaweed-type photobioreactor was purified by HPLC and examined regarding its spectrum and mode of action. The antibiotic effectively inhibited the growth of Gram-positive bacteria, pathogenic yeasts, and filamentous fungi, but it had only a weak effect on Gram-negative bacteria. Scanning electron micrograph analysis showed that the most characteristic change was swelling of the cells after exposure to the antibiotic. The antibiotic seems to alter the conformation of the microbial cell membrane, thereby changing its permeability, leading to osmotic shock.

Index Entries: Cyanobacteria; *Scytonema*; photobioreactor, antibiotic; polyurethane foam.

INTRODUCTION

Prokaryotic cyanobacteria can be grown photoautotrophically using light energy and CO_2, which is a major greenhouse gas partially responsible for global warming. Recently developed technology enables CO_2 to be

---

*Author to whom all correspondence and reprint requests should be addressed.

recovered from the emission gases of steam power plants. It is environmentally and economically important to produce valuable substances photoautotrophically from CO_2 by cyanobacteria. Various strains of cyanobacteria are known to produce intracellular and extracellular metabolites, with diverse biological activities, which are classified into two groups (1). The first group contains lactones, phenols, and acids such as cyanobacterin, a γ-lactone antialgal from *Scytonema hofmanni* (2), an antibacterial, brominated phenol from *Calothrix brevissima* (3), and an antimicrobial, O-methyl acid from various shallow-water varieties (4). The second group, which is the major one, consists of nitrogen-containing substances, such as a cytotoxin, malyngamide D from *Lyngbya majuscula* (5), and an antialgal and antimycotic, hapalindole A from *Hapalosiphon fontinalis* (6). The authors previously screened nine strains and five genera of cyanobacteria for antibiotic production, and a filamentous cyanobacterium, *Scytonema* sp. TISTR 8208, had the strongest activity against the bacteria tested (7). Stable immobilization of the cyanobacterium, which was shown to produce a cyclic peptide antibiotic, could be established by utilizing a fibrous biomass support material. The optimal medium composition for the production of the antibiotic was determined (8), and a seaweed-type photobioreactor was then constructed for the continuous cultivation of the cyanobacterium (9). Stable production of the antibiotic was achieved in the bioreactor for 16 d.

The aim of the present study was to characterize the antibiotic produced by immobilized *Scytonema* sp. TISTR 8208, to determine its spectrum and mode of action toward susceptible microorganisms.

MATERIALS AND METHODS

Cultivation of cyanobacterium

Scytonema sp. TISTR 8208, which was obtained from the culture collection of the Thailand Institute of Scientific and Technological Research Center (TISTR), was cultivated in modified BGA medium (MBGA) (7). The seaweed-type photobioreactor used is illustrated in Fig. 1. Details of the bioreactor dimensions were given previously (9). Cyanobacterial cells of about 1.6 g dry wt were inoculated into the 2.3-L bioreactor containing 2.0 L of MBGA medium. The bioreactor was incubated in a $30 \pm 1°C$ incubation room. The basal conditions of light illumination and gas (air containing 5% CO_2) flow rate were 200 μmol photon $m^{-2}s^{-1}$ and 250 mL/min, respectively, unless otherwise stated. A linear bank of fluorescent lamps was used on one side of the bioreactor.

Purification of Antibiotic

The culture supernatant was concentrated under reduced pressure at 30°C, and the antibiotic was extracted with methanol. After vacuum drying, the sample was then partitioned into a solvent system of $CHCl_3$:

Fig. 1. Structure of seaweed-type photobioreactor. (1), air inlet; (2), air filter; (3) and (4), sampling ports; (5), air outlet; (6), polyurethane foam strips; (7), stainless-steel ring.

methanol:H_2O at a ratio of 2:1:1. The lower phase was applied onto a silica gel 60 column and eluted with $CHCl_3$:methanol:H_2O at a ratio of 10:3:1. Fractions of 2 mL were collected, and this aliquot volume was used for antibiotic assay against *Bacillus subtilis* ATCC6633. The active fractions were collected and further purified by reverse-phase HPLC, using a YMC ODS AM-302 column (10 × 250 mm, Yamamura Chemical Laboratories, Japan). The elution was carried out with H_2O:CH_3CN:TFA in a ratio of 650:350:1, at a flow rate of 4 mL/min. The concentration of the purified sample was 0.001% (w/v).

Determination of Minimum Inhibitory Concentrations (MICs)

The MIC test against bacterial strains was done by the microtube broth-dilution technique *(10)*, and the MICs against yeasts and filamentous fungi were determined by the tube broth-dilution method *(11)*.

Analysis of Mode of Antibiotic Action

The mode of action of the antibiotic was observed by a scanning electron micrograph (SEM) (JEOL, JEM-2000 EX). Microbial cells were treated and observed with the SEM according to the methods described by Mallie et al. *(12)*.

RESULTS AND DISCUSSION

The filamentous cyanobacterium *Scytonema* sp. TISTR 8208 was cultivated photoautotrophically in the seaweed-type bioreactor for 2 wk. Secreted antibiotic was purified as described in Materials and Methods. As

Fig. 2. Purification of the antibiotic secreted from *Scytonema* sp. TISTR 8208 by reverse-phase HPLC. A partially purified sample was injected into a YMC ODS AM-302 column, and the antibiotic was eluted as described in Materials and Methods.

shown in Fig. 2, four peaks were obtained in reverse-phase HPLC, and the antibiotic activity was found in peaks A, B, and C. The major fraction, C, eluted at a retention time of 16 min, was used for the subsequent characterization. NMR and FAB mass analysis revealed that the antibiotic was a novel cyclic dodecapeptide with a mol wt of 1490 (data not shown).

The MICs of the antibiotic against various microorganisms were determined and compared with those of representative antibiotics. The antibiotic was shown to have strong activity toward Gram-positive bacteria, such as *B. subtilis*, at the same level as streptomycin and tetracycline (Table 1). However, the action was weak toward Gram-negative bacteria, especially when compared with gentamicin. Although the antibiotic activity against pathogenic yeasts varied, depending on the strain tested, it was stronger than that of imidazole, which is the nucleus of antifungal imidazole derivatives (Table 2). The antibiotic was also shown to have stronger activity against filamentous fungi than cycloheximide (Table 3).

The mode of action of the antibiotic against a bacterium (*B. subtilis* ATCC 6633), a yeast (*Pichia membranaefaciens* TISTR 5107), and a dermatophyte (*Microsporum audouini* IFO 8147), which were the strains most sus-

Table 1
MICs of Antibiotics Against Bacteria

Test organism	Antibiotic (µg/mL)					
	Sample	Ampicillin	Gentamicin	Kanamycin	Streptomycin	Tetracycline
Bacteria						
B. subtilis ATCC 6633	4.0	0.13	0.25	1.0	4.0	4.0
E. coli ATCC 8739	64.0	2.0	8.0	32.0	32.0	0.5
P. aeruginosa ATCC 9027	128.0	128.0	1.0	64.0	32.0	8.0
P. alkanolyticum IFO 12319	16.0	16.0	0.5	16.0	2.0	0.5
P. putida ATCC 15175	64.0	64.0	0.13	32.0	1.0	32.0
P. reptilivora IFO 3461	128.0	64.0	0.06	32.0	1.0	16.0

Table 2
MICs of Antibiotics Against Yeasts

Test organism	Antibiotic (µg/mL)	
	Sample	Imidazole
Yeasts		
C. albicans IFO 0579	32.0	64.0
C. albicans TISTR 5239	32.0	64.0
C. krusei TISTR 5099	8.0	32.0
C. tropicalis TISTR 5045	16.0	16.0
P. kluyveri TISTR 5150	16.0	64.0
P. membranaefaciens TISTR 5107	4.0	32.0

Table 3
MICs of Antibiotics Against Fungi

Test organism	Antibiotic (µg/mL)	
	Sample	Cycloheximide
Filamentous fungi		
A. fumigatus IFO 31952	16.0	32.0
A. fumigatus TISTR 3108	64.0	32.0
A. niger TISTR 3390	16.0	16.0
F. solani IFO 31093	16.0	16.0
M. audouini IFO 8147	4.0	4.0
T. mentagrophytes IFO 32412	16.0	64.0

ceptible to the antibiotic, was investigated. Cells were treated with 40 µg/mL antibiotic for 2 h in the case of *B. subtilis*, and for 18 h in the case of *P. membranaefaciens* and *M. audouini*. Figure 3 shows photographs of treated and nontreated cells taken under SEM observation. Swelling occurred in *B. subtilis* within 2 h of the commencement of antibiotic treatment, as a result of cell membrane damage. In *P. membranaefaciens* TISTR 5107, swollen cells with wrinkled, knob-like portions were observed within 24 h prior to cell lysis. Cell membrane damage was also evident in *M. audouini* as a result of antibiotic treatment. These changes in cell conformation were dose- and time-dependent (data not shown). The findings clearly showed that the antibiotic altered the microbial cell membrane, thereby changing its permeability. This led to osmotic shock, thus allowing leakage of the intracellular cell contents. Well-known antifungal antibiotics, such as polyene antibiotics and imidazoles, have the same kind of actions to the susceptible microorganisms *(11)*. The polyene antibiotics interact with sterols in the cell membrane, to give either membrane fragmentation or lipid rearrangement *(13)*. Imidazole and its derivatives interact with unsaturated

Fig. 3. Scanning electron micrographs of microorganisms before and after treatment by the antibiotic. The bacterium *B. subtilis* ATCC 6633 (**A** and **B**), yeast *P. membranaefaciens* TISTR 5107 (**C** and **D**), and filamentous fungus *M. audouini* TISTR 8147 (**E** and **F**) were used to analyze the mode of action of the antibiotic. A, C, and E are nontreated cells. B, D, and E are cells treated as described in the text. Bars represent 1 μm.

phospholipids in the cell membrane of susceptible organisms, thereby altering its permeability *(14)*. Further investigation is now under way to elucidate the exact mechanism of the action of the cyclic peptide antibiotic produced by *Scytonema* sp. TISTR 8208. *Scytonema* sp. strain U-3-3, which appears to be similar to TISTR 8208, has been reported to produce an cyclic peptide named scytonemin A *(15)*, the structure of which is different from the antibiotic purified in this work. Since scytonemin A has calcium-antagonistic properties, the substance purified from *Scytonema* sp. TISTR 8208 might have some pharmacological activity. The continuous production of such a valuable product in a photobioreactor, using solar energy, appears to be a very promising means of recycling CO_2.

REFERENCES

1. Glombitza, K.-W. and Koch, M. (1989), in *Algal and Cyanobacterial Biotechnology*, Cresswell, R. C. and Shah, N. eds., Longman, Essex, p. 161–238.
2. Mason, C. P., Edwards, K. R., Carlson, R. E., Pignatello, J., Gleason, F. K., and Wood, J. M. (1982), *Science* **215,** 400–402.
3. Pedersen, M. and DaSilva, E. J. (1973), *Planta* **115,** 83–86.
4. Cardellina, J. H., II, Moore, R. E., Arnold, E. V., and Clardy, J. (1979), *J. Org. Chem.* **44,** 4039–4042.
5. Kashiwagi, M., Mynderse, J. S., Moore, R. E., and Norton, T. R. (1980), *J. Pharm. Sci*. **69,** 735–738.
6. Moore, R. E., Cheuk, C., and Patterson, G. M. L. (1984), *J. Am. Chem. Soc.* **106,** 6456–6457.
7. Chetsumon, A., Fujieda, K., Hirata, K., Yagi, K., and Miura, Y. (1993), *J. Appl. Phycol.* **5,** 615–622.
8. Chetsumon, A., Maeda, I, Umeda, U., Yagi, K., Miura, Y., and Mizoguchi, T. (1994), *J. Appl. Phycol.* **6,** 539–543.
9. Chetsumon, A., Maeda, I., Umeda, F., Yagi, K., Miura, Y., and Mizoguchi, T. (1995), *J. Appl. Phycol.* **7,** 135–139.
10. Keneman, E. W., Allen, S. D., Dowell, V. R., and Sommess, H. M. (1979), in *Colar Atlas and Textbook of Diagnostic Microbiology* J. B. Lippincott, Philadelphia, p. 319–346.
11. Herman, C. and Donald, B. L. (1980), in *Antibiotic in Laboratory Medicine*, Lorian, V., eds, J. B. Lippincott, Philadelphia, p. 170–192.
12. Mallie, M., Butty, P., Montes, B., Jouvert, S., and Bastide, J.-M. (1991), *Can. J. Microbiol.* **37,** 964–970.
13. Hamilton-Miller, J. M. T. (1973), *Bacteriol. Rev.* **37,** 166–196.
14. Yamaguchi, H. (1978), *Antimicrob. Agents Chemother.* **13,** 423–426.
15. Helms, G. L., Moore, R. E., Niemczura, W. P., Patterson, G. M. L., Tomer, K. B., and Gross, M. L. (1988), *J. Org. Chem.* **53,** 1298–1307.

Comparative Study of Xylanase Kinetics Using Dinitrosalicylic, Arsenomolybdate, and Ion Chromatographic Assays

Thomas W. Jeffries,* Vina W. Yang, and Mark W. Davis

USDA, Forest Service, Forest Products Laboratory, One Gifford Pinchot Drive, Madison, WI 53705

ABSTRACT

Xylanases are commonly assayed by the dinitrosalicylic acid (DNS) or the arsenomolybdate (ARS) method. However, specific activities are many times higher with DNS than with ARS. This is because the DNS assay is more reactive and the ARS assay is less reactive with xylooligosaccharides than with xylose. Xylose is often used as a standard, even though oligosaccharides are prevalent, so the DNS method overestimates and the ARS method underestimates specific activity. Ion chromatography, with pulsed amperometric detection, separates and measures all products and intermediates, but quantitation on a molar basis is difficult, because few xylooligosaccharide response factors are known. This report directly compares these three assay methods for the assay of xylanase activities.

Index Entries: Oligosaccharides; degradation; pulsed amperometric detection; xylanase; chromatography.

INTRODUCTION

Many xylanases have been isolated and characterized in recent years to determine their potential usefulness in bleaching kraft pulps *(1–5)*. High productivity is essential for commercial manufacture, so accurate measures of activities are important. However, xylanase activity, as measured by reducing-sugar release, does not always correlate well with increased pulp brightness or reduced chemical demand following enzyme treatment. The question remains whether some xylanases have inherent characteristics that make them better than others. The more critical fundamental question

*Author to whom all correspondence and reprint requests should be addressed.

is, how does one measure the molar turnover number of an enzyme if reducing-sugar assays give such radically different values?

The first step in assessing relative efficacy is to determine how much enzyme is required to obtain a desired result. From a practical perspective, this is done on a volumetric or cost basis, but fundamental studies are needed that can relate the effect to specific enzymatic activities or molar turnover numbers. Despite interlaboratory studies (6), there is no well-defined, reliable means of comparing the activity of one xylanase to another. The dinitrosalicylic acid (DNS) method has been recommended (7), but several researchers have shown that this assay does not give a consistent response when used with various substrates (8,9). Moreover, activities obtained by DNS do not correlate with those obtained with other reducing-sugar assays, such as the arsenomolybdate method (ARS) of Nelson and Somogyi (10). This problem was first noted in trying to use the DNS assay for starch hydrolysis (11). It is apparently a result of the partial hydrolysis of oligosaccharides by the DNS reagent.

The objective of the present research was to reassess the relative merits of the DNS and ARS methods in light of more contemporary ion chromatograph (IC) methods that separate product sugars. Reducing sugar assays can be used as broad indicators of enzymatic activity, but DNS tends to overestimate and ARS tends to underestimate oligosaccharide production, and HPLC-based IC, combined with pulsed amperometric detection (PAD), is a better means to determine enzyme kinetics.

MATERIALS AND METHODS

Xylanase Assay

Xylanase activity was determined by measuring the release of reducing sugars and oligosaccharides from an approx 1% (w/v) solution of water-soluble birch wood xylan, using ARS (10), DNS (12) and IC/PAD (13) methods. Aliquots for ARS, DNS, and IC/PAD assays were removed from a single reaction mixture, and either transferred directly into the ARS or DNS reagents, or heat-inactivated for subsequent IC/PAD analysis. The ARS assays used 1.0 mL aliquots; the DNS assay used 0.5 mL; IC/PAD used 10 µL. Substrate preparation and assay conditions have been previously described (14). SP342* xylanase was obtained from Novo Nordisk (Franklinton, NC, Ecozyme was obtained from Zeneca Bio Products (Mississauga, ON, Can). Each stock enzyme preparation was diluted as necessary to obtain appropriate levels of product formation for the assay employed (see Fig. 2 legend below).

---

*The use of trade or firm names in this publication is for reader information and does not imply endorsement by the US Department of Agriculture of any product or service.

Sugar Standards

Analytical grade xylose (Aldrich, Milwaukee, WI) and xylooligosaccharides from Megazyme (North Rocks, Australia) were used as standards for ARS, DNS, and IC/PAD.

Chromatographic Conditions

Because of artifacts created by the presence of phosphate buffer in the reaction mixtures, samples for IC were diluted to less than 10 mM PO_4 prior to analysis. The chromatography system consisted of a 738 autosampler (Alcott Chromatography, Norcross, GA), a GP40 quaternary gradient high-pressure pump (Dionex, Sunnyvale, CA) and a pulsed amperometric detector (PAD, Dionex). Product separation was performed with two Dionex Carbo Pac PA1 guard columns and a single analytical column connected in series. These were eluted at a flow rate of 1.0 mL/min, with the following sodium acetate gradient, in 100 mM NaOH: 50 mM for 7 min; linear ramp to 200 mM at 25 min. Following each 10-μL injection, column cleanup was performed by elution (2 min) with 800 mM acetate in 100 mM NaOH, and column equilibration was performed by elution (11 min) with 50 mM acetate in 100 mM NaOH. To remove hydrophobic components that have the potential to foul these columns, on-line solid-phase extraction with a Dionex IonPac NG1 guard column was employed. The NG1 guard column and the autosampler were removed from the flow path 1.2 min after each sample injection. Detector settings were as follows: E1 = 0.1 V (300 ms), E2 = 0.6 (120 ms), E3 = −9 V (300 ms).

RESULTS AND DISCUSSION

Although both the ARS and DNS assays are based on reducing group formation, they do not give similar responses with different xylose oligosaccharide standards (Table 1). The ARS method is almost 12× more sensitive than the DNS method when assaying the presence of xylose (X). With xylobiose (X2) as a standard, reactivity of ARS is less, but the reactivity of DNS is greater, so the ratio of sensitivities drops to 4.9. With xylotriose (X3), the reactivity of the DNS assay is still greater, and ARS method is only 4.4× as sensitive. Thus, for the ARS assay, the molar response factor decreases, and, for the DNS assay, the molar response factor increases, with increasing degree of polymerization (DP), for the X, X2, and X3 series.

This means that the apparent enzyme activity depends on the sugar assay method and the prevailing oligosaccharide present in the hydrolysis mixture, and that the DPs of the product sugars affect the assays in inverse ways. Moreover, because xylose is commonly used as a standard (even though most xylanases produce little xylose), the ARS assay tends to underestimate, and the DNS assay to overestimate, enzyme activity when it is expressed as international units (i.e., μ/min) of reducing sugars released (as xylose equivalents).

Table 1
Molar Response Factors for Xylose, Xylobiose, and Xylotriose as
Determined by ARS and DNS Methods

Analyte	Molar response factor		
	ARS	DNS	Ratio
Xylose	1.038	0.087	12
Xylobiose	0.675	0.130	4.9
Xylotriose	0.642	0.152	4.4

The difficulty in determining relative efficacy of various enzyme preparations is compounded further when capacity for reducing-sugar production is compared at slightly different activities. The rate of reducing-sugar release is linear only when enzyme activities are very low, or when reaction times are short. If the reaction is allowed to proceed, the reaction rate falls off rapidly (data not shown).

Because most researchers like to report the highest possible activities for enzyme production by their isolates, the usual practice is to compare enzymes based on apparent activities following only limited hydrolysis, i.e., while the reaction is in the linear region. However, this approach does not reveal differences in action patterns or substrate specificities, which can only be discerned by separating and quantifying the oligosaccharides.

Figure 1A,B illustrates the product profiles formed from birch glucuronoxylan by SP342 and Ecozyme. The oligosaccharide products, ranging from X1 to X11, were separated into well-resolved peaks between 4 and 19 min. The linear xylan series is resolved to even higher DP, but the products are masked by a second series of oligosaccharides that begin to elute after 19 min. Their longer retention on the IC column suggests that these consist of 4-O-methylglucuronosides of xylan oligomers. A 4-O-methylglucuronic acid standard elutes at or immediately prior to the first member of this series of peaks, depending on the chromatographic conditions employed. Compositional analysis of the soluble xylan used as substrate indicated a minimal concentration of 11% (w/w) 4-O-methylglucuronic anhydride. Following prolonged incubations, late-eluting products disappeared and xylooligosaccharides accumulated, along with other earlier eluting members of the presumed acidic series (data not shown). But no standards are available, so the identities of the late-eluting products as acidic xylan oligosaccharides cannot be confirmed.

In examining these neutral and acidic products, it is first notable that almost no xylose is present in either reaction mixture at the 30-min timepoint (when the reaction is still linear), so an estimate of enzyme activity based on a xylose standard is meaningless. Second, there is no apparent transition from an accumulation of higher DP oligosaccharides to an accumulation of lower DP oligosaccharides during the time frame examined.

Fig. 1. Xylan oligosaccharide and acidic oligosaccharide products of **(A)** SP342 and **(B)** Ecozyme reaction mixtures at 30, 60, and 120 min. Reaction mixtures received 1.0 and 2.0 µL/mL of stock enzyme solution for SP342 and Ecozyme, respectively.

This is to say, the xylanases do not form intermediate DP (X5 to X11) products in a transient manner. Rather, what one observes is the progressive accumulation of X2, X3, and X4, plus acidic oligosaccharides, in the case of Ecozyme, and the accumulation of a wide range of oligosaccharides, in the case of SP342. These results suggest that the greatest affinity remains with the high-mol-wt xylan, because the ratios of the DP × to X11 and acidic xylooligosaccharides do not change much in the course of this reaction.

It is informative to estimate sugar release based on the total carbohydrate present in the solubilized oligosaccharides. This can be done only if response factors are available for each of the products. Quantitative estimates of PAD responses, with standard mixtures of xylan oligosaccharides having a degree of polymerization of up to 5, show that the detector response is more closely related to the oligosaccharide mass than to its molarity (Table 2), but since oxidation at the electrode is incomplete and varies with the product, no definitive quantitation of higher DP products can be obtained without proper standards. This is particularly true with respect to the analysis of acidic oligosaccharides. The molar reactivities of uronic acids are known to be less than those of xylose, and the reactivities of xylan oligosaccharides with uronic acid moieties attached are probably lower than the reactivities of xylooligosaccharides, so the PAD response probably underestimates the accumulation of acidic oligo saccharides.

On a molar basis, the PAD response is similar for xylose and X2, but, on a weight basis, the PAD response drops dramatically. For X3 and X5, the molar PAD response factor increases significantly, but the change in the weight-response factor is less dramatic. A similar PAD response was previously reported for amylodextrins up to DP 7 *(15)*, so our observation is consistent with previous results.

After the sharp drop from xylose to xylobiose, the PAD mass-response factor seems to change only slightly (Table 2). If one assumes that the areas under the peaks are representative of the masses of these products, it is possible to obtain a rough estimate of the amounts of sugars present in the higher oligosaccharide series. To make the results comparable to ARS and DNS assays, these products have been expressed as xylose equivalents. For example, the amounts of xylotriose present in the reactions are reported as three xylose equivalents.

Using actual molar response factors for DP 2, 3, and 5, linear regression was used to estimate molar response factors for DP 4 and DP 6 to 11 xylooligomers (Response factor = [966563 * DP] + 948743; r^2 = 0.993 (3 pts).

The total products detected by IC/PAD on a mass basis were compared to reducing sugars measured by the DNS and ARS methods, and the DP 1 to X11 products, as calculated from their molar response factors (Fig. 2). By this comparison on a mass basis, the IC/PAD analysis detects almost twice as much product as the DNS method. The IC/PAD data in the mass analysis include materials eluted during the column wash that are not shown in Fig. 1. The total molar concentration of DP 1 to 11 xylooligosaccharides approximates, or is slightly less than, the molar xylose equivalents assayed by the ARS method. Because the molar response factors are unknown for the higher acidic oligosaccharide series, they are not

Table 2
Molar- and Mass-Based Response Factors for
Xylooligosaccharides Relative to that Observed for
Xylose when Analyzed by the IC/PAD Method

Analyte	Relative response factor	
	Molar	Mass
Xylose	1	1
Xylobiose	0.92	0.49
Xylotriose	1.31	0.47
Xylopentaose	1.88	0.42

Fig. 2. Sugar production from birch xylan by (**A**) SP342 and (**B**) Ecozyme as assayed by ARS, DNS, and IC/PAD. ARS (●) and DNS (▲) reducing-sugar concentrations were calculated using xylose as a standard. For the IC/PAD assays, total oligosaccharide concentrations were estimated either by using a mass response factor and converting into xylose equivalents (■), or by using molar response factors and expressing the X1 to X11 series as total mM of reducing equivalents (□). SP342 and Ecozyme reaction mixtures received 0.1 µL/mL and 0.2 µL/mL of stock enzyme solution, respectively.

included in this latter calculation. This could account for the smaller amount of product observed by IC/PAD.

Under the conditions employed, kinetic assays based on the ARS method can be represented as international units (i.e., μmoles of xylooligosaccharide product/min), but kinetic assays based on the DNS method would be better represented as mg (or μg) of xylan solubilized. However, the extent to which the ARS or DNS methods measure the amounts of acidic oligosaccharides cannot be determined from these data.

CONCLUSIONS

In comparing these data, it is reassuring to note that the relative activities of the two enzyme preparations are similar by each of the three assay methods. However, DNS greatly overestimates reducing-group formation. The actual value for μmol of oligosaccharide products formed is better approximated by the ARS assay, but until more accurate response factors can be determined for xylooligosaccharides, and particularly for those in the acidic series, meaningful analysis of the kinetics, such as determination of specific molar turnover numbers, cannot be carried out, even by the IC/PAD method. Given the large difference in reactivities between xylose and X2, the declining difference in reactivities of higher oligosaccharides, and the prevalence of X2 or X3 as ultimate reaction products for most xylanases, it is better to employ X2 or X3, rather than xylose as a standard, when estimating activities and molar turnover numbers. Product profiles from pulp samples have many more complexities than reported here, so a sound basis for understanding xylanase bleaching in terms of enzyme action requires further study.

REFERENCES

1. Viikari, L., Ranua, M., Kantelinen, A., Sundquist, J., and Linko, M. (1986), Proc. 3rd Int. Conf. Biotechnol. Pulp Paper Indus., Stockholm, pp. 67–69.
2. Viikari, L., Kantelinen, A., Poutanen, K., and Ranua, M. (1990), in *Biotechnology in Pulp and Paper Manufacture,* Kirk, T. K. and Chang, H.-M., eds., Butterworth-Heinemann, Boston, pp. 145–151.
3. Clark, T. A., McDonald, A. G., Senior, D. J., and Mayers, P. R. (1990), in *Biotechnology in Pulp and Paper Manufacture,* Kirk, T. K. and Chang, H.-M., eds., Butterworth-Heinemann, Boston, pp. 153–167.
4. Paice, M. G., Bernier, R. Jr., and Jurasek, L. (1988), *Biotechnol. Bioeng.* 32, 235–239.
5. Kirk, T. K. and Jeffries, T. W. (1996), in *Enzymes for Pulp and Paper Processing,* Jeffries, T. W. and Viikari, L., eds., ACS Symp. Ser. 655, pp. 2–14.
6. Ghose, T. K. and Bisaria, V. S. (1987), *Pure Appl. Chem.* **59**, 1739–1752.
7. Bailey, M. J., Biely, P., and Poutanen, K. (1992), *J. Biotechnol.* **23**, 257–270.
8. Bruell, C. and Saddler, J. N. (1985), *Enzyme Microb. Technol.* **7**, 327–332
9. Royer, J. C. and Nakas, J. P. (1989), *Enzyme Microb. Technol.* **11**, 405–410.

10. Somogyi, M. (1952), *J. Biol. Chem.* **195,** 19–23.
11. Roybt, J. F. and Whelan, W. H. (1972), *Anal. Biochem.* **45,** 510–516.
12. Miller, G. L. (1959), *Anal. Chem.* **31,** 426–429.
13. Rocklin, R. D., and Pohl, C. A. (1983), *J. Liquid Chromatogf.* **6,** 1577–1590.
14. Grabski, A. C. and Jeffries, T.W. (1990), *Appl. Environ. Microbiol.* **57,** 987–992.
15. Wong, K. S., and Jane, J. (1997), *J. Liq. Chrom. Rel. Technol.* **20,** 297–310.

Production and Purification of CGTase of Alkalophylic *Bacillus* Isolated from Brazilian Soil

GRACIETTE MATIOLI,[1] GISELLA M. ZANIN,[1] MANOEL F. GUIMARÃES,[2] AND FLÁVIO F. DE MORAES*,[1]

[1]*State University of Maringá, Chemical Engineering Department, Av. Colombo, 5790 - BL. E46 - 09 87020-900, Maringá, PR, Brazil;*
[2]*Federal University of Paraná, P. O. Box 19046, Biochemistry Department, 81531-990-Curitiba - PR, Brazil*

ABSTRACT

Alkalophylic bacilli that produce cyclodextringlycosyltransferase (CGTase) were isolated from Brazilian soil, with a scheme of two plating steps. In the first step, the bacterial isolate forms a halo in the cultivation medium that contains γ-cyclodextrin (CD) complexing dyes. The CGTase of an isolate was purified 157-fold by biospecific affinity chromatography, with β-CD showing a mol wt of 77,580 Daltons. It produces a γ- to β-CD ratio of 0.156 and a small amount of α-CD, using maltodextrin 10% as substrate, at 50°C, pH 8.0 and 22 h reaction time, reaching 21.4% conversion of the substrate to cyclodextrins. In the second screening step, the isolates chosen give larger halos with β-CD complexing dyes, and smaller halos with β-CD complexing dyes, leading to a 30% improvement in γ-CD selectivity, although at lower total yield for cyclodextrins (11.5%).

Index Entries: Cyclodextringlycosyltransferase; CGTase; cyclodextrin; alkalophylic bacillus, screening.

INTRODUCTION

Cyclodextrins (CDs) are cyclic oligosaccharides formed by D-glucosyl residues linked by α,1–4 bonds. The most common are the α-, β-, and γ-CD, containing 6, 7, and 8 glucosyl residues, respectively. They are produced by reacting liquefied starch with the enzyme cyclodextringlycosyltransferase

---

*Author to whom all correspondence and reprint requests should be addressed.
e-mail: flavio@cybertelecom.com.br

(CGTase). Normally, a mixture of CDs is formed and the concentration ratio depends on the enzyme source. Frequently, β-CD is produced in a greater amount, in some cases α-CD, but γ-CD is very seldom produced in high yields. Depending on which cyclodextrin, α-, β-, or γ-CD, is the main product, the enzyme is called an α-, β-, or γ-CGTase, respectively. Only two nonrecombinant strains are known to produce relatively high yields of γ-CD, those of Englbrecht et al. *(1)*, and Mori et al. *(2)*. However, these research groups have put an extreme effort to select those strains that were isolated from so many CGTase-positive microorganisms.

The CD rings are highly hydrophilic externally, and relatively hydrophobic internally *(3)*. In aqueous solution, this structure thermodynamically favors the inclusion of nonpolar molecules that can be fitted entirely or partially inside the ring *(4)*. This encapsulation at the molecular level has been amply used for protecting labile molecules against chemical or physical damage in diverse industrial products, such as food, drugs, agrochemicals, cosmetics, and perfumes *(5)*.

More recently, the pharmaceutical industry has shown renewed interest in encapsulation with γ-CD, because this CD can accommodate large drug molecules. This has stimulated research to aim at better ways of producing larger quantities of γ-CD. Three different routes have been taken, as follows:

1. The development of superior processes for separating γ-CD from other CDs in the reaction mixture *(6)*.
2. Prevention of γ-CD destruction by reversible reactions. This is achieved through the formation of a stable γ-CD complex with an appropriate substance *(7)*.
3. Alternatively, the search for new strains whose CGTase enzyme kinetically favors the production of γ-CD at the beginning of the reaction *(1,2)*. However, it is worthy of note that, at equilibrium, reached after a long time, β-CD will always be in a greater concentration *(8)*.

This article presents a scheme of plate screening, with two steps that help in selecting new bacterial isolates that produce CGTase with higher γ-CD selectivity. Results of the application of this scheme to regional soil samples, as well as data on production of cyclodextrins by the CGTase-producing isolates, are shown.

MATERIALS AND METHODS

First Screening Step

The first screening technique to isolate microorganisms that produce CGTase was based on the Nakamura and Horikoshi screening medium II *(9)*, for alkalophylic microorganisms. The medium was supplemented with dyes to detect CGTase-positive isolates. The isolates produce a clear halo around the colonies, since the complexation of these dyes with CDs changes their light absorption characteristics. Dyes that were appropriate for complexing with γ-CD were Congo red and xylene cyanole FF. These dyes have been used by Hamaker and Tao with Luria-Bertoni neutral media (pH 7.0) containing 1% starch to detect γ-CD production by recombinant *Escherichia coli* cells *(10)*. The final composition of the modified plate medium was (w/v): soluble starch (1%), peptone (0.5%), yeast extract (0.5%), K_2HPO_4 (0.1%), $MgSO_4 \cdot 7H_2O$ (0.02%), Na_2CO_3 (1%), Congo red (0.01%), xylene cyanole FF (0.001%), and agar (1.5%). The pH of this medium was 10.3. A drop of solution containing either α-, β-, or γ-CD, deposited over a plate containing this medium, produced a clear halo only with γ-CD.

Microorganisms were isolated from soil samples collected from farm fields used for cultivation of wheat, corn, potato, and cassava. Each sample was suspended in dH_2O and screening plates containing the above modified medium were inoculated with appropriate volumes of these suspensions. Plates were incubated at 37°C for 24 h, after which high-enzyme-producing isolates, associated with larger halos, were chosen.

Second Screening Step

In this case, a second plating medium was used with the isolates of the first screening step. For the second plate medium, the dyes were changed, to be specific for detecting the presence of β-CD. These dyes are phenolphthalein mixed with methyl orange, and have been used by Park et al. *(11)*.

CGTase Activity

The enzymatic activity of the CGTase produced by the bacterial isolates was determined through the cultivation of these microorganisms in liquid cultures containing a medium with the same composition as the plate medium, except the dyes, and agar. Cultures were grown at 37°C up to 5 d, after which cell-free supernatant was used to assay for CGTase activity. Assay tubes with 0.5 mL of the CGTase-containing sample were mixed with 0.5 mL of substrate solution: soluble starch 1% (w/v), Tris-HCl buffer, pH 8.0 (0.01 M), and $CaCl_2$ (5 mM). Enzyme samples and substrate solution were warmed to 50°C before mixing, and left to react at this temperature for 20 min. Reactions were stopped by boiling the tubes for 5 min, and the CDs produced were measured by colorimetric assays as described below. One unit of CGTase is the amount of enzyme that produces one μmol of β-CD/min/mL of enzyme in these conditions. CGTase enzyme activity was also assayed by the method of successive dilutions of the supernatant. The amount of CD produced in this case was assayed through the precipitation with (trichloroethylene) (TCE) *(12)*.

β-CD concentration was measured, based on the discoloration of phenolphthalein solutions at 550 nm, which occurs after complexation with β-CD, following the method of Vikmon *(13)* as modified by Hamon and

Moraes *(14)*. γ-CD concentration was determined based on increased absorption at 620 nm of bromocresol green solutions upon complexation with γ-CD, according to the method of Kato and Horikoshi *(15)*, as modified by Hamon and Moraes *(14)*. Soluble proteins were determined by the method of Bradford *(16)*, using BSA as standard protein.

Enzyme Purification

Selected microorganisms were cultivated in 5-L cultures (medium described in CGTase Activity) for 5 d at 37°C, with agitation. Each culture was centrifuged at 8800/g for 15 min to pellet cells. The cell-free supernatant was mixed with ammonium sulfate (80% saturation) and left in a refrigerator overnight. A second centrifugation at 8800/g for 30 min was conducted, and the precipitate was dissolved in Tris-HCl buffer, pH 8.0 (0.01 M). The enzyme was further purified by biospecific affinity chromatography column using Sepharose 6B gel, and both β-CD *(17)* and γ-CD *(1)* were independently tried as affinants. Purified CGTase was concentrated by ultrafiltration, dialyzed 5 × with Tris-HCl buffer, pH 8.0 (0.01 M), to remove molecules smaller than 30 kD, and stored in a refrigerator for later use.

Mol Wt Determination

The mol wt of the CGTase was determined according to Weber and Osborn *(18)*, by SDS-PAGE, using a mol wt reference kit (Pharmacia, Uppsala, Sweden) of six standard proteins with mol wt in the range of 14,400 to 94,000 Daltons. The relation between log mol wt and relative mobility was established, and the mol wt of the CGTase was determined through this relation.

Cyclodextrin Production Tests

A batch reactor test was used for the production of cyclodextrins with the purified CGTase. Conditions were: 50°C, about 1 mg/L of pure enzyme (80.1 U), and the substrate solution contained maltodextrin 10% (w/v) (Dextrin 10, Fluka [Buchs, Switzerland] article 31412), Tris-HCl buffer, pH 8.0 (0.01 M), and 5 mM CaCl$_2$. The test was run for a period of 24 h, and samples were taken at regular intervals and boiled for 5 min, after which they were assayed for β- and γ-CD with the colorimetric methods above described. α-CD was measured by HPLC, using C-18 hydrophobic interaction column and water–methanol (92–8%) eluent *(19)*.

RESULTS AND DISCUSSION

First Screening Step

Application of the first screening step has screened for 57 strains that produce CGTase. The cell-free supernatant of isolate number 37, up to a dilution of 2^8, was able to produce CDs in sufficient amount to be precipitated by TCE. This isolate has shown the highest enzyme activity in this group. Media presented in Table 1 were tested with isolate 37. The highest activity observed in the cell-free supernatant occurred with medium A, which contained peptone and yeast extract as source of nitrogen, reaching 0.1155 µmol of β-CD/(min·mL). Therefore, isolate 37 has shown a divergent behavior in comparison with the alkalophylic bacillus isolated by Nakamura and Horikoshi (9) which produces much higher CGTase activity in medium B, which contains corn-steep liquor as source of nitrogen.

Enzyme Purification

CGTase enzyme from isolate 37 was purified as described in Materials and Methods, giving a electrophoretically homogeneous enzyme solution with 0.171 mg of protein/mL, activity of 13.7 µmol of β-CD/(min·mL) and 157-fold purification. The specific activity of this enzyme was 80.1 µmol of β-CD/(min·mg of enzyme). This value is very close to the specific activity determined by Hamon and Moraes (14) for the CGTase enzyme from alkalophylic bacillus 1-1 described by Schmid et al. (23). The purification obtained in this work is superior to that obtained by László et al. (17), who obtained 120-fold, and also, greater than that obtained by Saha et al. (24), who purified the enzyme by agarose and Sephacryl 200 column, and obtained 129-fold purification.

Enzyme Mol Wt

The mol wt of the CGTase enzyme of isolate 37 was determined by SDS-PAGE, and found to be 77,580 Daltons. This value is within the usual range of mol wt obtained for CGTases from different microorganisms (66,000 to 80,000 Daltons).

Production of Cyclodextrins

The production of CDs from maltodextrins upon using purified CGTase of isolate 37, in the 24 h batch test, is shown in Fig. 1. As can be seen, β-CD concentration, at any time, is higher than γ-CD concentration, indicating that this enzyme is a β-CGTase enzyme. β-CD concentration reached about 16 mM, and the ratio of γ- to β-CD concentration was 0.156; that is, a maximum of 2.5 mM of γ-CD was obtained. Production of α-CD was very low, reaching only 0.085 mM. This lower value would reduce separation requirements in an industrial production process of CDs. Moreover, total conversion of the substrate, maltodextrin at 10% (w/v), reached 21.4% after a period of 22 h, with 1 mg/mL of CGTase enzyme of isolate 37, at 50°C and pH 8.0.

Second Screening Step

The second screening step was applied to the isolates already screened through the first step. The idea of this new screening step is based on the fact that, although the first screening technique spotted the colonies that

Table 1
Production of CGTase by Isolate 37 with Alternative Liquid Cultivation Media

Medium component composition % w/v	Ref. 9 Medium A	9 B	20 C	21 D	22 E	2 F
Soluble starch	2	1	2	2	6	2[a]
Peptone	0.5	–	–	–	0.5	–
Yeast extract	0.5	–	–	0.5	0.15	1.0
Corn-steep liquor	–	5[b]	2	–	–	–
MgSO$_4$·7H$_2$O	0.02	0.02	0.02	0.02	0.02	0.02
K$_2$HPO$_4$	0.1	0.1	0.1	0.1	0.6	0.1[a]
Na$_2$CO$_3$	1	1	1	1	1	1
Total protein in the cell-free supernatant (µg/mL)	1488	1616	1343	1372	1442	1376
CGTase protein in the cell-free supernatant (µg/mL)	1.44	0.42	0.58	1.06	0.44	1.16
CGTase activity (µmol β-CD/min mL)	0.1155	0.0333	0.0463	0.0849	0.0355	0.0929

[a] Mori et al. (2) used half these values.
[b] In v/v.

Fig. 1. Production of cyclodextrins by the CGTase from isolates 37 and 41, using maltodextrins 10% (w/v) as substrate, pH 8.0, at 50°C, and about 1 mg/mL of pure enzyme.

produce γ-CD, they could still be large producers of β-CD. Through the detection of small halos in the second plate medium, this technique aims at selecting isolates that would produce γ-CD at higher selectivity, that is, at higher ratios of γ- to β-CD.

Two new isolates, 41 and 47, were chosen to work with. Production level of CDs by the purified enzyme of isolate 41 is also shown in Fig. 1, for comparison with the results of isolate 37. Both β-CD and γ-CD production by the CGTase of isolate 41 are proportionally smaller than that of isolate 37, but the ratio of γ- to β-CD concentration produced by this isolate is higher, and reached 0.204. Therefore, isolate 41 produces γ-CD at about 30% higher selectively than isolate 37, although with lower total CD yield (11.5%). The γ-CD selectivity improvements, so far, are modest. The production of CDs by the purified enzyme of isolate 47 was very similar to that of isolate 41, and the ratio of γ- to β-CD concentration was 0.187.

Although the purification by affinity chromatography of the CGTase enzyme of isolates 37, 41, and 47 was attempted in two different columns, one containing β-CD as affinant and the other γ-CD, purification was achieved only with the β-CD column, showing that these CGTase are in fact β-CGTase enzymes.

Sabioni and Park *(21,25)* have exclusively used the phenolphthalein screening technique of Park et al. *(11)*, with Brazilian soil from Campinas-SP, and isolated an alkalophylic β-CGTase strain of *Bacillus lentus*, which formed α-, β-, and γ-CD in the proportion of 1:67:1.6, respectively. The ratio of γ- to β-CD in this case is much lower: 0.015. This confirms that the application of the screening scheme with two steps proposed in this work helps to isolate CGTase with higher selectivities for γ-CD. Since the γ-CGTase isolated by Mori et al. *(2)* can reach a γ- to β-CD ratio of 1.4 with 10% starch *(26)*, there is still much scope for further improvements

CONCLUSIONS

The isolation of alkalophylic microorganisms that secrete CGTase enzyme has become relatively easy with the application of the screening scheme presented in this paper.

The second screening step introduced was able to screen for isolates whose CGTase leads to slightly better selectively for γ-CD. A 30% higher selectivity over step one was achieved, although at lower overall yield for cyclodextrins.

New, improved schemes should still be developed for facilitating the isolation of strains that are good producers of γ-CGTase, since only two are known presently, and these were obtained by highly laborious and time-consuming procedures.

The interest of the pharmaceutical industry in using γ-CD for complexing large drugs warrants further research in this area.

ACKNOWLEDGMENTS

The authors thank Conselho Nacional de Desenvolvimento Científico e Tecnológico (CNPq), Fundação Coordenacão de Aperfeiçoamento de Pessóal de Nível Superior (CAPES), Programa de Apoio ao Desenvolvimento Científico e Tecnológico (PADCT), and the State University of Maringá for financial support.

REFERENCES

1. Englbrecht, A., Harrer, G., Lebert, M., and Schmid, G. (1990), in *Minutes of the 5th International Symposium on Cyclodextrins*, Duchêne, D. ed., Editions. de Santé, Paris, pp. 25–31.
2. Mori, S., Hirose, S., Oya, T., and Kitahata, S. (1994), *Biosci. Biotechnol. Biochem.* **58,** 1968–1972.
3. Duchêne, D., Debruères, B., and Brétillon, A. (1984), *Labo-Pharma-Probl. Tech.* **32,** 843–850.
4. Szejtli, J. (1988), in *Cyclodextrin Technology*, Szejtli, J. ed., Kluwer, Dordrecht, The Netherlands, pp. 79–185.
5. Lee, J. H., Choi, K. H., Choi, J. Y., Lee, Y. S., Kwon, I. B., and Yu, J. H. (1992), *Enzyme Microb. Technol.* **14,** 1017–1020.
6. Mattsson, P., Mäkelä, M., and Korpela, T. (1988), in *Proceedings of the Fourth International Symposium on Cyclodextrin*, Huber, O. and Szejtli, J., eds., Kluwer, Dordrecht, The Netherlands, pp. 65–70.
7. Schmid, G., Huber, O. S., and Eberle, H. J. (1988). in *Proceedings of the Fourth International Symposium on Cyclodextrin*, Huber, O. and Szejtli, J, eds., Kluwer, Dordrecht, The Netherlands pp. 87–92.

8. French, D. (1957), *Adv. Carbohydr. Chem.* **12,** 189–260.
9. Nakamura, N. and Horikoshi, K. (1976), *Agric. Biol. Chem.* **40,** 753–757.
10. Hamaker, K. and Tao, B. Y. (1993), *Starch/Stärke* **45,** 181, 182.
11. Park, C. S., Park, K. H., and Kim, S. H. (1989), *Agric. Biol. Chem.* **53,** 1167–1169.
12. Nomoto, M., Shew, D. C., Chen, S. J., Yen, C. W. L., and Yang, C. P. (1984), *Agric. Biol. Chem.* **48,** 1337, 1338.
13. Vikmon, M. (1981), in *The First International Symposium on Cyclodextrin.* Szejtli, J. ed., D. Riedel, Dordrecht, The Netherlands, pp. 69–74.
14. Hamon, V. and de Moraes, F. F. (1990), in *Etude Preliminaire a L'immobilsation de L'enzyme CGTase WACKER.* Laboratoire de Tecnologie Enzymatique. Université de Tecnologie de Compiègne.
15. Kato, T. and Horikoshi, K. (1984), *Anal. Chem.* **56,** 1738–1740.
16. Bradford, M. (1976), *Anal. Biochem.* **72,** 248.
17. László, E., Bánky, B., Seres, G., and Szejtli, J. (1981), *Starch/Stärke* **33,** 281–283.
18. Weber, K. and Osborn, M. (1969), *J. Biol. Chem.* **244,** 4406–4412.
19. Chatjigakis, A. K., Cardot, P. J. P., Coleman, A. W., and Parrot-Lopez, H. (1993), *Chromatographia* **36,** 174–178.
20. Sabioni, J. G. (1991), PhD Thesis, Universidade Estadual de Campinas, Campinas-SP, Brasil.
21. Sabioni, J. G. and Park, Y. K. (1992), *Revista de Microbiologia* **23,** 128–132.
22. Tulfahi, A. (1991), PhD Thesis, Université de Technologie de Compiègne, Compiègne, France.
23. Schmid, G., Englbrecht, A., and Schmid, D. (1988), in *Proceedings of the Fourth International Symposium on Cyclodextrins,* Huber, O and Szetjtli, J., eds., Kluwer, Dordrecht, Netherlands, pp. 71–76.
24. Saha, B. C., Freer, S. N., and Bothast, R. J. (1994), *Appl. Environ. Microbiol.* **60,** 3774–3780.
25. Sabioni, J. G. and Park, Y. K. (1992), *Starch/Stärke* **44,** 225–229.
26. Mori, S., Goto, M., Mase, T., Matsuura, A., Oya, T., and Kitahata, S. (1995), *Biosci. Biotechnol. Biochem.* **59,** 1012–1015.

Submerged Culture Screening of Two Strains of *Streptomyces* sp. with High Keratinolytic Activity

O. GARCIA-KIRCHNER*, MA. E. BAUTISTA-RAMIREZ, AND M. SEGURA-GRANADOS

Departamento de Bioprocesos, Unidad Profesional Interdisciplinaria de Biotechnología (UPIBI/IPN) del Instituto Politécnico Nacional, Av. Acueducto s/n Barrio La Laguna; Ticomán, Zacatenco. 07340-México, D.F.

ABSTRACT

Keratinases can be used for the production of potentially important hydrolyzed proteins and chemicals. This study investigated the keratinolytic activity of *Streptomyces* sp on keratinaceous materials like wool. High levels of proteolytic and keratinolytic activity were obtained after 96 h of culture when two *Streptomyces* sp strains were grown on basal medium containing mineral salts and 3% (w/v) of defatted wool as a source of energy, carbon, and nitrogen. The cell-free culture filtrates exhibited rapid proteolytic digestion of keratin powder. Currently, the authors are testing whether the enzymatic activity obtained is in fact keratinolytic, and not only an alkaline protease activity.

Index Entries: Keratinases; enzymatic hydrolysis; *Streptomyces* sp.; M-Zyme; alkaline proteases.

INTRODUCTION

Keratin forms mammalian outer tissues such as hair, nails, feathers, and horn. Keratinous proteins are insoluble and resistant to degradation by common proteolytic enzymes, because of their extensive crosslinking by disulfide bonds, hydrogen bonds, and hydrophobic interactions. This insolubility and resistance to proteolytic enzymes makes them inaccessible to most living organisms.

However, keratin does not accumulate in nature. Despite the unusual stability of keratinous proteins, several microbial keratinolytic proteases have been reported, such as the keratinase of *Trichophyton mentagrophytes*

---

*Author to whom correspondence and reprint requests should be addressed.

(1). Keratinolytic activity has also been reported for some species of *Aspergillus* (2), proteinase K of *Tritirachium album* (3), the alkaline protease of *Streptomyces* sp (4), the thermostabile alkaline protease of *Bacillus* sp (5), and a purified keratinase from *Streptomyces fradie*, which solubilizes wool keratin (6).

Laboratory research efforts are focusing on the processing of waste animal protein into either value-added byproducts or derivatives that are more quickly recycled in the environment. Biodegradation by microorganisms possessing keratinolytic activity is a possible method for improving the nutritional value of keratinaceous waste. Microbial keratinases can be used as food additives to improve the digestibility and nutritious value of proteins, since keratin contains all common amino acids, and mainly differs from other structural fibrous proteins in its high cysteine content. Enhanced production and processing of keratinase enzymes would benefit applications such as converting keratineous materials and waste proteins into usable amino acids.

In this article the authors report the production of alkaline protease activity and keratinolytic activity by submerged culture of two different strains of *Streptomyces* sp using defatted wool as an inducer in a simple mineral salts medium.

MATERIALS AND METHODS

Microorganisms

Streptomyces sp KER p-08 and *Streptomyces* sp KER p-17 were isolated in our laboratory from samples of soil from slaughterhouses that were enriched with strands of wool and pieces of cattle hoof. These strains have been shown to possess the ability to digest native keratin rapidly, as previously reported (7,8).

The actinomycete strains were grown at $29 \pm 1°C$ for 5 d on potato dextrose slants, and maintained on the same medium at 4°C. Subcultures were prepared at monthly intervals.

Alkaline Protease and Keratinase Production

Six grams of defatted wool were placed in 500-mL Erlenmeyer flasks containing 180 mL of fermentation medium. All chemicals used in this culture medium were industrial grade, and tap water was used according to the basal salt formula, indicated in Table 1. In our experiments, keratin was neither autoclaved nor sterilized. It was washed with water, extracted with chloroform–methanol (9), and soaked several times in 0.05 M phosphate buffer, 1 mM Mg^{2+}, pH 7.8, then rinsed with water and methanol, and air-dried at room temperature. The wool was either cut with scissors to a length of 3–5 mm, or with a knife to pieces of 1–3 mm length.

Table 1
Culture Medium Composition Utilized for the
Production of Keratinolytic Activity by
Submerged Fermentation

	Industrial-grade medium (G/L)
Defatted wool[a] (size: 3 mm)	30.0
K_2HPO_4	1.2
KH_2PO_4	1.0
KNO_3	3.0
$MgSO_4 \cdot 7H_2O$	0.6
$MnCl_2$	0.3
$ZnSO_4 \cdot 7H_2O$	0.6
$FeSO_4 \cdot 2H_2O$	0.6
$CaCl_2$	0.5
pH	8.0
Tap water	1.0 L

All the constituents are industrial grade chemicals.

[a] The wool was neither autoclaved nor sterilized. It was either cut with scissors or knife to pieces of 1–3 and 3–5-mm long, washed with water, extracted with chloroform–methanol and soaked several times in 0.05 M phosphate buffer, 1 mM $MgCl_2$, pH 7.0, then rinsed with water and methanol, and air-dried at room temperature. The wool was treated prior to being added to the medium previously sterilized by autoclaving.

Inocula were prepared by harvesting spores from 1-wk-old PDA slants of the two strains in the proportion 3:1 in sterile distilled water containing 0.1–0.3 mL of Tween-80.

Fermentation flasks were inoculated to give 10^6–10^7 spores/mL. Incubation was at $29 \pm 1°C$. The pH value was initially adjusted to 8.0 before sterilization.

During the culturing, samples were removed aseptically and centrifuged at 3200 g, to obtain cell-free filtrates.

Enzyme Activity Assays and Soluble Protein Determination

These assays were performed with the supernatants after removing the cell growth and residual wool. Removal of the filamentous growth from cultures of *Streptomyces* grown in shake flasks on wool–salts medium was readily accomplished by vacuum filtration through Whatman (Maidstone, UK) No. 1 paper.

Protein content was determined according to Lowry *(10)* in aliquots of cell-free culture filtrates after $\pm 4°C$ overnight dialysis, with bovine serum albumin as a standard. Culture filtrates of both strains of *Streptomyces* sp.

were concentrated 5× by ultrafiltration (mol wt cutoff >10,000), with respect to the initial soluble protein content, or until both proteins reached approximately the same concentration.

Alkaline protease activity was measured by the Anson's method modified by Keay et al. (11,12). The concentrated culture filtrates were incubated with 5% casein in 1 mL buffer 50 mM Tris-HCl (pH 7.8), containing 1 mM CaCl$_2$ for 30 min at 37°C. The reaction was stopped by adding 2 mL of TCA solution (0.11 M trichloroacetic acid, 0.22 M CH$_3$COONa, 0.33 M CH$_3$COOH). After 30 min of incubation at 30°C, the precipitate was removed by centrifugation and the absorption of the supernatant was measured at 280 nm. One unit of protease activity is defined as the amount of enzyme required to release 1 μmol of tyrosine in 30 min at 37°C, using casein as substrate.

The determination of keratinolytic activity was estimated by incubating concentrated culture filtrates with 10 mg of keratin powder in 1 mL of buffer A (50 mM Tris-HCl buffer, pH 7.8, containing 1 mM CaCl$_2$) with vigorous shaking for 1 h at 37°C. The reaction was stopped by adding 2 mL of TCA solution (0.11 M trichloroacetic acid, 0.22 M CH$_3$COONa, 0.33 M CH$_3$COOH). After 30 min of incubation at 30°C, the substrate was removed by centrifugation and the absorption of the supernatant was measured at 280 nm, as described by Shu-Wen et al. (13). One unit of keratinase activity is defined as the amount of enzyme required to release 1 μmol of tyrosine in 60 min at 37°C, using keratin powder as substrate.

Enzymatic Hydrolysis

A keratin-digestion assay was devised, in which the amount of protein solubilized during the incubation period was estimated spectrophotometrically at 280 nm. Keratin powder was hydrolyzed by a commercial purified keratinase, the M-Zyme-(Merck) (14), by the cell-free culture filtrates obtained from *Streptomyces* sp strains, and by other microbial alkaline proteases. The enzymes were incubated with 20 mg substrate (keratin powder, Merck, Rahway, NJ) in 1 mL buffer (0.05 M Tris-HCl containing 1mM MgSO$_4$, pH 7.8) with shaking during 3 h at 37°C, and filtered through a Whatman No. 2 paper. Absorption of the supernatant at 280 nm was measured after (TCA) precipitation.

RESULTS AND DISCUSSION

As previously reported (8), some preliminary tests were carried out at the authors' laboratory using dyed keratin powder (keratin-azure, Merck). After a final screening in Petri dishes using casein medium, many strains of filamentous fungi and some actinomycetes were obtained. In the present work, the production of alkaline protease and keratinolytic activities by two different strains of *Streptomyces*, chosen for the previously indicated reasons was determined by submerged culture in a cheap and simplified

Fig. 1. Production of caseinolytic activity and soluble protein in cell-free culture filtrates of KER p-08 (-■-), (-▲-) and KER p-17 (-□-), (-○-) *Streptomyces* sp. strains, respectively, as well as pH variation (-●-) on liquid fermentation using defatted wool.

medium. Figure 1 shows caseinolytic activity, protein content in cell-free culture filtrates, and pH variation over 120 h of growth of *Streptomyces* sp under liquid fermentation on defatted wool.

Considerable keratinolytic activity was produced by the *Streptomyces* sp strains (KER p-08 and KER p-17) after 72 h.

It is important to notice in Figure 2 that maximal enzyme production was achieved in a medium containing 3.0% (w/v) defatted wool, as a sole carbon and organic nitrogen source, and mineral salts at pH 7.7 and 30°C for 96 h.

Table 1 shows the gross composition of the culture medium used for all the fermentations. The use of tap water did not show negative effects on the production of alkaline protease and keratinolytic activities. Table 2 compares enzymatic activities evaluated at maximal production time with reference to the soluble protein content present in culture filtrates obtained during fermentation.

Keratin that has been denatured and degraded by treatments such as sterilization and ball milling can be decomposed by organisms that are incapable of attacking native keratin *(15)*. The pretreatments mentioned drastically decrease the cysteine content of wool *(16)*. Upon destruction of the disulfide bonds, keratin becomes easily denatured, and then loses its natural insolubility and resistance to proteases.

As indicated by Noval and Nickerson *(17)*, a basic question involved in studies on microbial decomposition of keratin is this: Do the microorganisms decompose native keratin during their growth on keratinaceous substrates, or do they grow at the expense of nonkeratinaceous nutrients (fats,

Table 2
Caseinolytic and Keratinolytic Activities Obtained in
Submerged Fermentation for 96 h by the Strains of
Streptomyces sp

Enzymatic activities (U/mL)	KER p-08	KER p-17
CASEINOLYTIC	26.2	15.2
KERATINOLYTIC	19.8	9.4

The proteolytic activity was measured by Anson's method modified by Keay et al. *(11,12)*. One unit of protease activity is defined as the amount of enzyme required to release 1 µmol of tyrosine in 30 min at 37°C, using casein as substrate.

The keratinolytic activity was estimated according Shu-Wen *(13)*. One unit of keratinase activity is defined as the amount of enzyme required to release 1 µmol of tyrosine in 60 min. at 37°C, using keratin powder as substrate.

Fig. 2. Production of keratinolytic activity and soluble protein in cell-free culture filtrates of KER p-08 (-■-), (-▲-) and KER p-17 (-□-), (-○-) *Streptomyces* sp. strains, respectively, as well as pH variation (-●-) on liquid fermentation using defatted wool.

carbohydrates, and proteins) that are usually present as minor constituents of such substrates? In seeking an answer to this question, the following experimental requirements must be met: Nondenatured keratin must be used as the substrate, and the evidence that keratin has been decomposed must be unequivocal.

To evaluate if the enzymatic activity obtained in our experiments really represents keratinolytic activity, several microbial enzymes, includ-

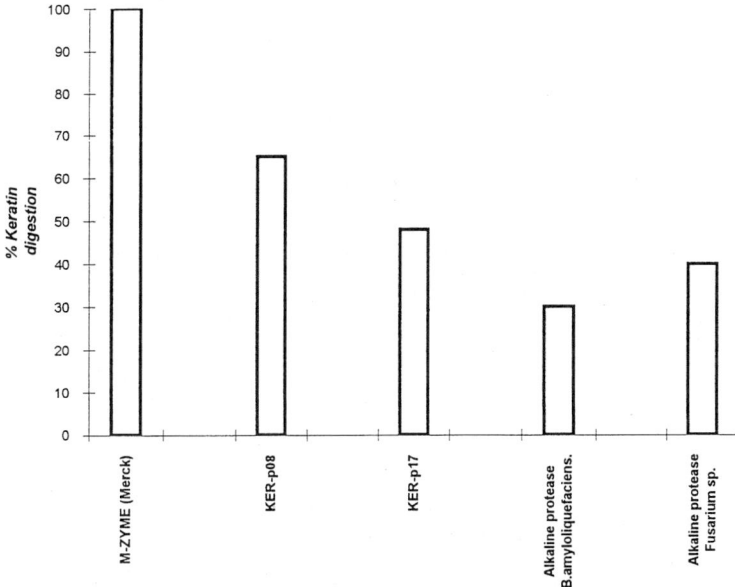

Fig. 3. Hydrolysis of keratin powder with a purified keratinase, two alkaline proteases and the cell-free culture filtrates obtained from *Streptomyces* sp. The enzymes were incubated with 20 mg substrate (keratin powder, Merck) in 1 mL of buffer, with shaking for 3 h at 37°C. Absorption at 280 nm of supernatant was measured after TCA precipitation.

ing a commercial preparation, were tested. Figure 3 shows the effects of the cell-free culture filtrates on keratin powder.

The percentage of keratin digestion of the filtrates with respect to purified keratinase was about 65% for the KER p-08 enzyme and 48% for the KER p-17 enzyme. This result partially supports hypothesis that the obtained enzymatic activity is keratinolytic, because the alkaline protease activity obtained was much lower.

In conclusion, a good level of keratinolytic activity was obtained with the KER p-08 strain of *Streptomyces* sp after 96 h by submerged culture at 37°C in the evaluated culture medium. The keratinolytic activity observed with this strain was compared with that of some microbial proteases. Keratinolytic activity in cell-free culture filtrates from *Streptomyces* sp hydrolyzes several types of substrates, and exhibits high proteolytic activity, compared to other alkaline proteases. The highest hydrolysis level was obtained with keratin powder, but it is necessary to test other keratinaceous materials like feathers, wool, hide, and so on, in order to demonstrate real keratinolytic activity, and not the presence of an alkaline protease in the culture filtrates.

Not much research is presently done in Mexico on the use of native keratinaceous materials that are considered agroindustrial or cattle wastes. Enhanced production of *Streptomyces* sp KER p-08 and processing of keratinase will benefit applications such as recycling hides, feathers, wool, and waste proteins as usable peptides. These molecules have nutritional value and can be used as animal food supplements, or in other industrial applications. Microbial proteases have extensive commercial applications, and their use in agricultural applications could be increased.

There is considerable interest in producing chemicals from renewable resources. This increasing interest is justified by recent technological advances that can create economically competitive processes. These technologies must be environmentally acceptable and produce acceptable marketable materials. For these reasons, the authors are working to turn keratinaceous materials (an abundant renewable feedstock) into a wide variety of industrial chemicals. However, numerous technical improvements must be achieved before the potential global benefits of these biological processes can be fully realized.

ACKNOWLEDGMENTS

Support from Polytecnic National Institute through their Research Division for the project no. 923001 is gratefully acknowledged. The authors thank Francisco Martinez-Arce for the English translation; Julieta Cortez, who typed the manuscript; and Waldo Toledo, who corrected the final technical version. Isabel Pérez Montfort corrected the English version of the manuscript.

REFERENCES

1. Yu, R. J., Harmon, S. R., and Blank, F. (1968), *J. Bacteriol.* **96,** 1435–1436.
2. Koh, W., Santro, A., and Messing, R. (1958), *Bacteriol. Proc.* 18.
3. Ebeling, W., Hennrich, N. Klocknow, M., Metz, H., Orth, H. D., and Lang, H. (1974), *Eur. J. Biochem.* **47,** 91–97.
4. Nakanishi, T. and Yamamoto, T. (1974), *Agric. Biol. Chem.* **38,** 2391–2397.
5. Takami, H., Nakamura, S. Aono, R., and Horikoshi, K. (1992), *Biosci. Biotech. Biochem.* **56,** 1667–1669.
6. Yu, R. J., Ragot, J., and Blank, F. (1972), *Specialia*: **15,** 1512–1513.
7. García-Kirchner, O. (1992), Memorias del IX Congreso Nacional de Ing. Bioquímica, 10–13 Nov. 1992, Oaxtepec, Mor. México.
8. García-Kirchner, O. and Gómez-Montes, M. E. (1993), *Biotechnología*. vol. 3, Revista de La Sociedad Mexicana de Biotechnología y Bioingeniería, A. C. México.
9. Yu, R. J., Harmon, S. R., and Blank, F. (1968), *J. Bacteriol.* **96,** 1435–1436.
10. Lowry, O. H., Rosebrough, J., Farr, A. L., and Randall, R. J. (1951), *J. Biol. Chem.* **96,** 193, 265–275.
11. Keay, L., Moses, P. W., and Wildi, B. S. (1970), *Biotechnol. Bioeng.* **12,** 213–215.
12. Keay, L. and Wildi, B. S. (1970), *Biotechnol. Bioeng.* **12,** 179–185.
13. Shu-Wen, Ch., Hsien-Ming, H., Shu-Whe, S., Hiroshi, T., Minao, A., and Ying-Chieh, T. (1995), *Biosci. Biotech. Biochem.* **59,** 2239–2243.
14. The Merck Index. (1983), An Encyclopedia of Chemicals, Drugs, and Biologicals, 10th ed. Merck, Rahway, NY, p. 760.
15. White, W. L., Mandels, G. R., and Siu, R. G. H. (1950), *Mycologia* **42,** 199–223.
16. Stall, W. H., McQue, B., and Siu, R. G. H. (1949b), *J. Biol. Chem.* **177,** 69–73.
17. Noval, J. J. and Nickerson, W. J. (1958), *J. Bacteriol.* **77,** 251–263.

Cofermentation of Glucose, Xylose, and Arabinose by Mixed Cultures of Two Genetically Engineered *Zymomonas mobilis* Strains

Ali Mohagheghi,* Kent Evans, Mark Finkelstein, and Min Zhang

Biotechnology Center for Fuels and Chemicals, National Renewable Energy Laboratory, Golden CO 80401

ABSTRACT

Cofermentation of xylose and arabinose, in addition to glucose, is critical for complete bioconversion of lignocellulosic biomass, such as agricultural residues and herbaceous energy crops, to ethanol. A factorial design experiment was used to evaluate the cofermentation of glucose, xylose, and arabinose with mixed cultures of two genetically engineered *Zymomonas mobilis* strains (one ferments xylose and the other arabinose). The pH range studied was 5.0–6.0, and the temperature range was 30–37°C. The individual sugar concentrations used were 30 g/L glucose, 30 g/L xylose, and 20 g/L arabinose. The optimal cofermentation conditions obtained by data analysis, using Design Expert software, were pH 5.85 and temperature 31.5°C. The cofermentation process yield at optimal conditions was 72.5% of theoritical maximum. The results showed that neither the arabinose strain nor arabinose affected the performance of the xylose strain; however, both xylose strain and xylose had a significant effect on the performance of the arabinose strain. Although cofermentation of all three sugars is achieved by the mixed cultures, there is a preferential order of sugar utilization. Glucose is used rapidly, then xylose, followed by arabinose.

Index Entries: Recombinant *Zymomonas*; ethanol; cofermentation; xylose; arabinose; mixed culture fermentation

INTRODUCTION

Lignocellulosic feedstocks are composed predominantly of cellulose, hemicellulose, and lignin, and are naturally resistant to chemical and bio-

*Author to whom all correspondence and reprint requests should be addressed.

logical conversion. Because the feedstock can represent more than 40% of all process costs *(1)*, an economical biomass-to-ethanol process critically depends on the rapid and efficient conversion of all sugars (e.g., glucose, xylose, and arabinose) present in their cellulosic and hemicellulosic fractions *(2)*. Many organisms can ferment the glucose component of cellulose to ethanol, but efficient conversion of the pentose sugars, particularly xylose and arabinose, in the hemicellulose fraction has been hindered by the lack of a suitable biocatalyst *(3)*. Xylose is the predominant pentose sugar derived from the hemicellulose of most hardwood feedstocks, but arabinose can constitute a significant amount of the pentose sugars derived from various agricultural residues and other herbaceous crops, e.g. switchgrass, that are being considered for use as dedicated energy crops. Although arabinose makes up only 2–4% of the total pentoses in hardwoods, it represents 10–20% of the total pentoses in many herbaceous crops *(4)*. Arabinose contents can be as high as 30–40% of the total pentoses in corn fiber, a byproduct of corn processing *(5)*.

Zymomonas mobilis is a unique organism with the ability to anaerobically ferment sucrose, glucose, and fructose via the Entner-Doudoroff pathway, which is used primarily by aerobic organisms *(6)*. Highly expressed pyruvate decarboxylase and ethanol dehydrogenase genes rapidly convert pyruvate to ethanol *(7)*. Comparisons of high glucose (25% w/v) fermentations between *Z. mobilis* and *Saccharomyces carlsbergenesis* show that *Z. mobilis* is 2.5× faster in specific glucose uptake rate and 3× faster in specific ethanol productivity (8). Higher theoretical ethanol yields are also achieved because of lower cell biomass formation. Although *Z. mobilis* demonstrates many characteristics needed for an ideal ethanol-producing strain, its substrate utilization range is restricted to the fermentation of glucose, fructose, and sucrose *(9)*. Wild-type strains are not naturally suited for fermenting the xylose found in lignocellulosic feedstocks, because they lack the essential xylose assimilation and pentose metabolism pathways *(10)*.

Enzymatic analyses have confirmed that *Z. mobilis* completely lacks xylose isomerase, xylulokinase, and transaldolase activities, and contains insufficient levels of transketolase activity *(10)*. Consequently, introducing and expressing xylose isomerase, xylulokinase, transaldolase, and transketolase into *Z. mobilis* allowed the authors to complete a functional metabolic pathway that converts xylose to central intermediates of the Entner-Doudoroff pathway, and enables *Z. mobilis* to ferment xylose to ethanol *(10)*. Introducing and expressing L-arabinose isomerase, L-ribulokinase, L-ribulose-5-phosphate-4-epimerase, transaldolase, and transketolase from *Esherichia coli* led to the creation of an arabinose-fermenting strain of *Z. mobilis* (11).

In this study, two full factorial experiments are designed to study cofermentation of glucose, xylose, and arabinose by mixed culture. Experiment 1 examines the effects of each strain (individual and mixed) and each sugar (glucose, xylose, and/or arabinose) on mixed sugar cofermentation performance. The parameters evaluated were process yield (Yp), % xylose, and arabinose utilization. The strains used are xylose-fermenting *Z. mobilis* ATCC 39676 (pZB4L) *(10)* and arabinose-fermenting *Z. mobilis* ATCC 39676 (pZB206) *(11)*. The ratios of sugars are similar to those found in agricultural residues such as corn fiber. The full factorial experiment examines the effect of each strain, individually and combined, by analyzing each fermentation for sugar consumption, ethanol yield, and byproduct formation. Glucose, xylose, and arabinose combinations are varied to determine any effects the sugar combination would have on the performance of the organisms. Experiment 2 examines the effects of pH and temperature on the cofermentation process. The authors hope to demonstrate that a mixed-culture fermentation, capable of fermenting glucose, xylose, and arabinose to ethanol, has potential applications to industries with feedstocks that contain hexose and mixed pentose sugars.

MATERIALS AND METHODS

Microorganisms and Media

Z. mobilis ATCC 39676 (pZB4L) (xylose-fermenting strain) *(10)* and ATCC 39676 (pZB206) (arabinose-fermenting strain) *(11)* were used for mixed-culture cofermentation. Both strains were stored in concentrated working stocks in rich media (RM) + 10% glycerol, and stored at −70°C until use. RM medium (1% yeast extract + 2 g/L KH_2PO_4) was prepared as a 10X concentrated stock solution. Glucose and xylose were prepared as 50% w/v stock solutions, and arabinose as a 25% w/v stock solution. All the solutions were filter-sterilized using a 0.2-μm filter (Nalgene, Rochester, NY).

Cultivation of Inoculum

A 500-mL Erlenmeyer flask that contained 400 mL of RM medium + 30 g/L glucose + 5 g/L xylose (or arabinose) was inoculated with 1.0 mL of thawed culture stock (one flask for each strain). Both flasks were placed in a rotary shaker (New Brunswick, Edison, NJ) at 150 rpm, at 30°C for 15 h, or until the optical density of the culture measured 2.0–3.0 at 600 nm. Cultures were transferred to sterile 150-mL centrifuge bottles and centrifuged at 2800 *g* for 10 min. The cell pellets were resuspended in 10 mL of RM medium, and used to inoculate the fermenters.

Fermentations

Fermentations were conducted in Biostat-Q fermenters from B. Braun (Allentown, PA) that contained 500 mL working volume of RM medium + 10 mg/L tetracycline + sugars. In the first factorial experiment, which was a 2-level, 5-factor full factorial, the sugar concentrations and type

Table 1
List of Fermentations and Their Results for Design 1, Performed to Study Interaction of Two Strains 39676(pZB4L) and 39676(pZB206) and Three Sugars (Glucose, Xylose, and Arabinose) at Constant pH 5.5 and T 35°C

Fermentation no.	Strains 39676 (pZB4L)	Strains 39676 (pZB206)	Glucose (g/L)	C-Source xylose (g/L)	Arabinose (g/L)	Yp (%)	Xylose consumed (%)	Arabinose consumed (%)
1	Y	Y	0.0	29.0	0.0	88.8	97.9	0.0
2	Y	N	30.7	30.5	0.0	85.0	96.4	0.0
3	N	Y	31.1	30.1	23.1	73.8	22.9	78.4
4	N	Y	33.8	0.0	0.0	82.7	0.0	0.0
5	Y	N	32.4	0.0	22.3	89.0	0.0	2.6
6	Y	Y	0.0	0.0	22.0	57.4	0.0	72.5
7	N	Y	0.0	28.8	23.4	14.7	0.9	16.0
8	Y	Y	30.9	31.4	0.0	83.8	94.2	0.0
9	Y	N	0.0	30.1	0.0	84.6	97.8	0.0
10	Y	N	0.0	0.0	22.5	0.0	0.0	0.0
11	Y	Y	32.5	0.0	22.5	74.7	0.0	74.7
12	Y	N	30.9	29.8	23.1	84.9	94.6	0.0
13	Y	N	33.1	0.0	0.0	86.4	0.0	0.0
14	Y	Y	0.0	27.2	23.2	53.5	96.6	13.8
15	N	Y	31.0	30.8	0.0	94.3	9.4	0.0
16	N	Y	33.6	0.0	22.1	79.4	0.0	98.6
17	N	Y	0.0	0.0	22.0	82.7	0.0	97.1
18	Y	Y	31.3	30.2	23.2	70.4	91.0	70.7
19	N	Y	0.0	28.6	0.0	1.7	0.3	0.0
20	Y	Y	33.8	0.0	0.0	83.2	0.0	0.0
21	Y	N	0.0	27.5	23.1	49.9	97.5	0.4

Table 2
List of Fermentations and Their Results for Design 2, Performed to Study Effect of pH and Temperature on Mixed Culture Cofermentation of Glucose, Xylose, and Arabinose by Two Strains 39676(pZB4L) and 39676(pZB206)

Fermentation no.	T (C°)	pH	Yp (%)	Xylose consumed (%)	Arabinose consumed (%)
22	30.0	5.0	68.50	84.88	51.50
23	33.5	5.0	65.90	82.42	52.83
24	37.0	5.0	57.20	60.57	47.32
25	30.0	5.5	67.80	89.41	56.23
26	33.5	5.5	69.80	87.67	66.12
27	33.5	5.5	71.60	89.45	68.13
28	37.0	5.5	62.80	71.48	57.72
29	30.0	6.0	73.20	93.57	73.70
30	33.5	6.0	70.20	90.80	70.92
31	37.0	6.0	61.50	66.12	70.07

varied with the conditions. Fermentations were controlled at 300 rpm agitation, 35°C, and pH 5.5, controlled using 2 M KOH. For the second factorial experiment, which was a 3-level, 2-factor full-factorial design with duplicate centerpoints, pH and temperature varied, depending on the condition being tested. The medium used was RM + 10 mg/L tetracycline + 30 g/L glucose + 30 g/L xylose + 20 g/L arabinose at a agitation rate of 300 rpm. All fermenters were inoculated with the concentrated cells to achieve an initial OD_{600} of 0.2. Data from 72 h fermentation were used for data analysis.

Analyses

Samples were taken periodically throughout the course of the fermentations and analyzed for sugars, ethanol, and byproducts by HPLC. HPLC analysis was conducted using a Hewlett-Packard (Wilmington, DE) series 1090 HPLC with a Bio-Rad (Hercules, CA) HPX-87H column. Process yield was calculated as ethanol concentration × 100 ÷ initial sugar concentration × 0.51.

Experimental Design and Data Analysis

The statistical design programs Design Ease and Design Expert (Stat-Ease, Minneapolis, MN) were used to design and analyze the results of the experiments. The designed experiments are shown in Tables 1 and 2.

RESULTS AND DISCUSSION

Two factorial experiments were carried out to evaluate the performance of the mixed-culture, mixed-sugar cofermentation system. First, a

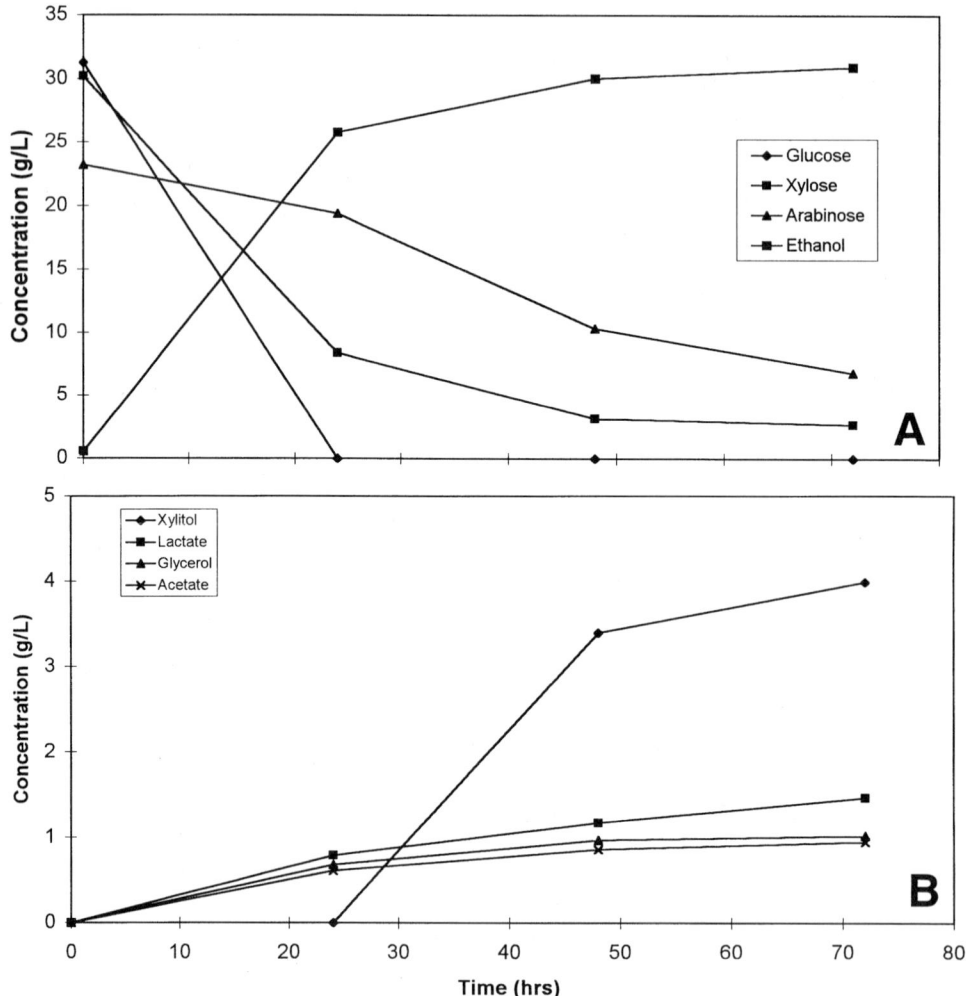

Fig. 1. Coculture fermentation profile of strains 39676(pZBAL) and 39676(pZB206) on mixed sugar of 30 g/L glucose, 30 g/L xylose, and 20 g/L arabinose at pH 5.5 and T 35°C. **(A)** Growth profile. **(B)** Byproduct profile.

two-level, five-factor full factorial (2^5) design was used to evaluate the main and two-way interaction effects of each strain (39676 (pZB4L) [xylose strain] and/or 39676 (pZB206) [arabinose strain]) and each sugar (glucose, xylose, and/or arabinose) on mixed-sugar cofermentation performance (fermentations 1–21). This first factorial experiment was run at 35°C and pH 5.5. Following this experiment, a 3-level 2-factor full-factorial (3^2) design with duplicate centerpoints was run at a fixed sugar concentration of 80 g/L (30 g/L glucose; 30 g/L xylose; and 20 g/L arabinose), to characterize the effect of pH and temperature on the cofermentation process (fermentations 22–31). In this second factorial experiment, the pH was varied

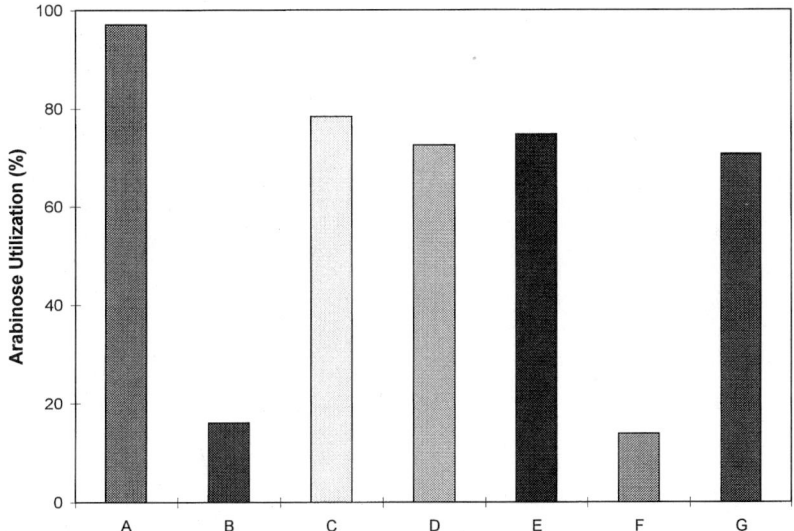

Fig. 2. Effect of glucose, xylose, and xylose strain (39676[pZB4L]) on arabinose utilization by arabinose strain (39676[pZB206]) at pH 5.5 and T 35°C. **(A)** Control (arabinose strain grown on arabinose only). **(B)** Arabinose strain grown on arabinose and xylose (effect of xylose). **(C)** Arabinose strain grown on arabinose plus xylose and glucose (effect of xylose and glucose). **(D)** Arabinose and xylose strain grown on arabinose (effect of xylose strain). **(E)** Arabinose and xylose strain grown on arabinose and glucose (effect of xylose strain and glucose). **(F)** Arabinose and xylose strain grown on arabinose and xylose (effect of xylose strain and xylose). **(G)** Arabinose and xylose strain grown on arabinose, xylose, and glucose (effect of xylose strain, xylose, and glucose).

from 5.0 to 6.0, and the temperature from 30 to 37°C (centerpoint at pH 5.5 and 33.5°C).

Evaluation of Cofermentation Performance as a Function of Sugar Mixture and Strain-Combination Fixed pH and Temperature

Figure 1 shows an example of the growth and byproduct formation profiles observed during mixed-culture co-fermentation carried out at pH 5.5 and temperature 35°C (fermentation 18). As this figure illustrates, in the presence of all three sugars, first, glucose is used most rapidly, then xylose and arabinose appear to coferment simultaneously. Another characteristic of this system is that the rate of xylose utilization is somewhat faster than that of arabinose. Finally, except for xylitol and/or arabitol, very little byproduct formation is observed, although low levels of acetate, lactate, and glycerol are produced.

The results of this 2^5 factorial experiment are summarized in bar graph form in Figs. 2,3 (fermentations 3,6,7,11,14,17,18),4, and 5 (fermentations 1,8,9,12,14,18,21), and in three-dimensional (3-D) form in Figs. 6–8 (fermen-

Fig. 3. Effect of glucose, xylose, and xylose strain (39676[pZB4L]) on ethanol process yield (Yp) for arabinose utilization by arabinose strain (39676[pZB206]) at pH 5.5 and T 35°C. **(A)** Control (arabinose strain grown on arabinose only). **(B)** Arabinose strain grown on arabinose and xylose (effect of xylose). **(C)** Arabinose strain grown on arabinose plus xylose and glucose (effect of xylose and glucose). **(D)** Arabinose and xylose strain grown on arabinose (effect of xylose strain). **(E)** Arabinose and xylose strain grown on arabinose and glucose (effect of xylose strain and glucose). **(F)** Arabinose and xylose strain grown on arabinose and xylose (effect of xylose strain and xylose). **(G)** Arabinose and xylose strain grown on arbinose, xylose, and glucose (effect of xylose strain, xylose, and glucose).

tation 1–21). A complete summary of the key performance results (process yields, xylose utilization [%], and arabinose utilization [%] is also provided in Table 1.

The key findings from this experiment are that xylose, and, to a lesser extent, the xylose strain inhibit the cofermentation performance of the arabinose strain. Figures 2 and 3 show the effects of glucose, xylose, and the xylose strain on arabinose utilization and ethanol Yp, respectively, for the arabinose strain. These figures show that the presence of both xylose and the xylose strain negatively affect the arabinose strain. Although virtually all the arabinose (97%) was utilized by the arabinose strain in medium that contained only arabinose, arabinose utilization decreased substantially when xylose or the xylose strain, or both, were present (to only 16, 72, and 14%, respectively). Similarly, the Yp was reduced in the presence of xylose and the xylose strain, decreasing from 83% in medium that contained arabinose alone to 15 and 57% in medium that contained arabinose and xylose or the xylose strain, respectively. Based on these results, xylose greatly inhibits arabinose utilization by the arabinose strain. However, experimental results show that the addition of glucose diminishes the in-

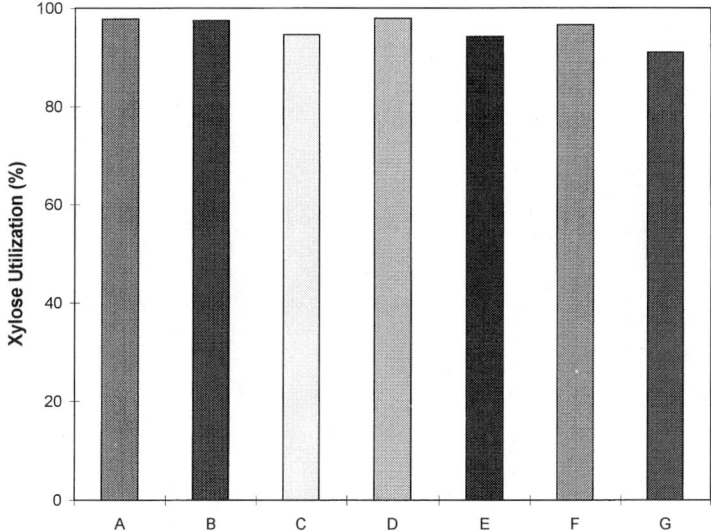

Fig. 4. Effect of glucose, arabinose, and arabinose strain (39676[pZB206]) on xylose utilization by xylose strain (39676[pZB4L]) at pH 5.5 and T 35°C. **(A)** Control (xylose strain grown on xylose only). **(B)** Xylose strain grown on xylose and arabinose (effect of arabinose). **(C)** Xylose strain grown on xylose plus arabinose and glucose (effect of arabinose and glucose). **(D)** Xylose strain and arabinose strain grown on xylose (effect of arabinose strain). **(E)** Xylose and arabinose strain grown on xylose and glucose (effect of arabinose strain and glucose). **(F)** Xylose and arabinose strain grown on arabinose and xylose (effect of arabinose strain and arabinose). **(G)** Xylose and arabinose strain grown on arabinose, xylose, and glucose (effect of Arabinose strain, arabinose, and glucose).

hibitory effect of xylose and the xylose strain on arabinose utilization. Figures 2 and 3 show that arabinose utilization and Y_p increased to 78 and 74%, respectively, when the arabinose strain was grown in the presence of all three sugars.

Figures 4 and 5 show that the neither arabinose nor the arabinose strain had a major effect on xylose utilization by the xylose strain. As Fig. 4 demonstrates, xylose utilization was above 95% on medium that contained only xylose, regardless of whether glucose or arabinose were present. Similarly, Fig. 5 shows that the Y_p achieved by the xylose strain was around 85% in all cases, except when arabinose was present (in which case, the Y_p yield decreased to about 50%, because arabinose was not utilized.

The main factor effects and the two-way factor–factor interaction effects identified in this experiment can be visualized more easily when the results are represented in 3-D form as cubic plots. Figures 6–8 show cubic representations of the results generated using the Design-Ease software. In the cubic plots shown in these figures, the three orthogonal axes (A, B, and C) represent the three factors being varied in the experiment. The

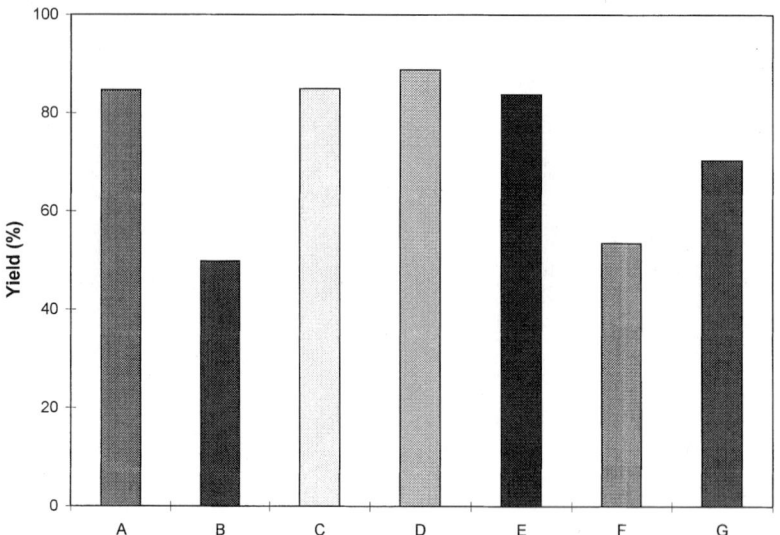

Fig. 5. Effect of glucose, arabinose, and arabinose strain (39676[pZB206]) on ethanol process yield (Yp) for xylose utilization by xylose strain (39676[pZB4L]) at pH 5.5 and T 35°C. **(A)** Control (xylose strain grown on xylose only). **(B)** Xylose strain grown on xylose and arabinose (effect of arabinose). **(C)** Xylose strain grown on xylose plus arabinose and glucose (effect of arabinose and glucose). **(D)** Xylose strain and arabinose strain grown on xylose (effect of arabinose strain). **(E)** Xylose and arabinose strain grown on xylose and glucose (effect of arabinose strain and glucose). **(F)** Xylose and arabinose strain grown on arabinose and xylose (effect of arabinose strain and arabinose). **(G)** Xylose and arabinose strain grown on arabinose, xylose, and glucose (effect of arabinose strain, arabinose, and glucose).

eight corners of the cubes correspond to the eight independent cases investigated for each possible subset of the cofermentation system (negative sign is indication of absence and positive sign indicates the presence of the factor), and the numerical values are the performance results achieved at each condition. For example, Fig. 6 summarizes the effect of the xylose strain (axis A), glucose (axis B), and xylose (axis C) on the performance of the arabinose strain in arabinose-containing media. Figure 6A graphically illustrates how the Yp varied under the conditions studied, i.e., at each corner of the cube. As Fig. 6A shows, a Yp of 83% was achieved when the fermentation was carried out using only arabinose, i.e., in the absence of glucose, xylose, and the "4L" xylose strain (A−, B−, C− lower left-hand front corner of the cube). Figure 6A readily shows that the Yp falls to 57% when the xylose strain is added (A+, B−, C− lower right-hand front corner of the cube), and to only 15% when xylose is added (A−, B−, C+ lower left hand back corner of the cube); a Yp of 54% is achieved when both xylose and the xylose strain are present (A+, B−, C+ lower right hand back corner of the cube). The cubical representation of the results makes the beneficial effect of glucose on alleviating the inhibitory effect

Fig. 6. Cubical presentation of the effect of xylose strain 3967PZB4L), xylose, and glucose on the performance of arabinose strain 39676(pZB206). The effects are predicted by the software program Design Expert. The effects of three factors (glucose, xylose, and xylose strain) on the Yp, xylose utilization, and arabinose utilization are shown in Figs. **6A–C,** respectively. Note: sign (−) means absence of the factor, and sign (+) means presence of the factor.

of xylose and the xylose strain on Yp readily apparent. All conditions under which glucose is present (i.e., the B+ corner conditions depicted in the upper plane of the cube) achieve higher Yps than when glucose is absent (i.e., the B− corners depicted in the lower plan of the cube), except for the arabinose-only control. Figure 6B,C shows similar cubic representations of the xylose and arabinose utilization performance results achieved for the arabinose–arabinose strain co-fermentation system.

Figure 7 shows similar plots that summarize the performance of the xylose–xylose strain system. Figure 8 shows a summary of cofermentation system performance as a function of the tertiary sugar mixture, when both strains are present. As Fig. 8 shows, when all three sugars were present, the mixed-culture cofermentation system achieved an ethanol Yp of 70%,

Fig. 7. Cubical presentation of the effect of arabinose strain 39676(pZB206), arabinose, and gluose on the performance of xylose strain 39676(pZB4L). The effects are predicted by the software program Design Expert. The effects of three factors (glucose, arabinose, and arabinose strain) on the Yp, xylose utilization, and arabinose utilization are shown in Fig. 7A–C, respectively. Note: sign (−) means absence of the factor, and sign (+) means presence of the factor.

with 91% xylose utilization and 71% arabinose utilization. The marginal Yp of 70% is mostly a result of incomplete arabinose utilization.

Evaluation of Mixed-Culture Cofermentation Performance as a Function of pH and Temperature

The effect of pH and temperature on the mixed culture cofermentation process was studied using a 3-level 2-factor full-factorial (3^2) experiment carried out at a fixed sugar concentration (30 g/L glucose; 30 g/L xylose; and 20 g/L arabinose). The pH ranged from 5.0 to 6.0 and temperature from 30 to 37°C. Duplicate centerpoints were run at pH 5.5 and 33.5°C. Results of this experiment are summarized in Table 2 and Fig. 9 (fermentations 22–31).

Fig. 8. Cubical presentation of the effect of glucose, xylose, and arabinose on the cofermentation by mixed cultures of xylose strain 39676(pZB4L) and arabinose strain 39676ZP206). The effects are predicted by the software program Design Expert. The effects of three factors (glucose, xylose, and arabinose) on the Yp, xylose utilization, and arabinose utilization are shown in Figs. **8A–C,** respectively. Note: sign (−) means absence of the factor, and sign (+) means presence of the factor.

As shown in both Table 2 and Fig. 9, the highest Yp and xylose and arabinose utilization were 73, 94, and 74%, respectively, and were achieved at pH 6.0 and 30°C (fermentation 29). The lowest values for these same performance metrics were 57, 61, and 47% at pH 5.0 and 37°C (fermentation 24). In general, increasing pH and decreasing temperature had positive effects on the cofermentation process.

Statistical analysis of the results showed that both Yp and xylose utilization were more sensitive to temperature than to pH. By increasing the pH from 5.0 to 6.0, with temperature ranging from 30 to 37°C, the ethanol Yp yield and xylose utilization increased only by 5 and 9%, respectively; increasing temperature from 30 to 37°C caused the ethanol Yp and xylose utilization to drop 10 and 20%, respectively. On the other hand, the relative sensitivity of arabinose utilization to pH and temperature was the

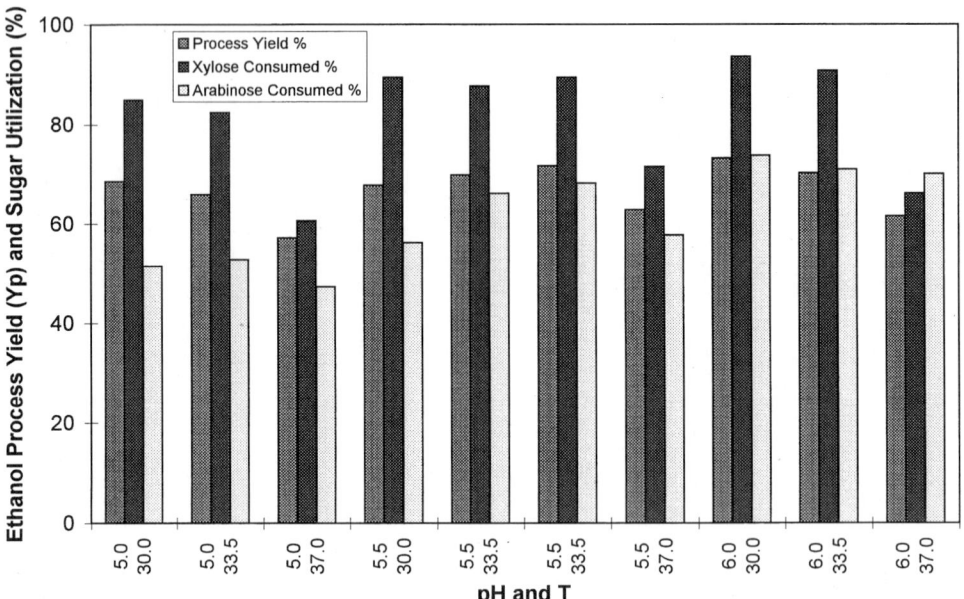

Fig. 9. Effect of pH and temperature on the ethanol process yield (Yp), xylose, and arabinose utilization by cofermentation of glucose, xylose, and arabinose, using mixed cultures of xylose strain 39676(pZB4L) and arabinose strain 39676(pZB206).

opposite, with pH exerting a more significant effect than temperature. Arabinose utilization increased by about 20% when pH was increased from 5.0 to 6.0, at temperature range of 30 to 37°C, but decreased only by 5% when temperature was raised from 30 to 37°C. The optimum conditions for mixed-culture cofermentation with sugar mixture (glucose:xylose:arabinose at 30:30:20 g/L) was estimated to be approx pH 5.85 and 31.5°C, using the Design-Expert program. Under these optimal conditions, the Yp and xylose and arabinose utilization were predicted to be 72.5, 94, and 71%, respectively.

As in previous experiments, low levels of byproducts were produced by the mixed culture system over the range of conditions tested. Major byproducts were xylitol and lactate, which were produced at concentrations as high as 4 g/L and 2 g/L, respectively. The concentrations varied, depending on the experimental conditions. Other minor byproducts formed were glycerol and acetate; both accumulated to less than 1 g/L under all conditions. *Zymomonas* generally produces few by-products *(12)*. Byproducts found in this work are similar to those reported earlier *(12)*.

ACKNOWLEDGMENTS

This work was funded by the Biochemical Conversion Element of the Office of Fuels Development of the US Department of Energy.

REFERENCES

1. Hinman, N. D., Wright, J. D., Hoagland, and W., Wyman, C. E. (1989), *Appl. Biochem. Biotechnol.* **20/21,** 391–401.
2. Wright, J. D. (1987), AIChE National Meeting, Minneapolis, MN.
3. Zhang, M., Franden, M. A., Newman, M., McMillan, J., Finkelstein, M., and Picataggio, S. (1995), *Appl. Biochem. Biotechnol.* **51/52,** 527–536.
4. McMillan, J. D. and Boynton B. L. (1994), *Appl. Biochem. Biotechnol.* 45/46, 569–584.
5. Carlson, T. (1994), *Fifth Corn Utilization Conference,* June 8–10, St. Louis, MO, National Corn Growers Association.
6. Doelle, H. W., Kirk, L., Crittenden, R., Toh, H., and Doelle M. B. (1993), *Crit Rev in Biotechnol.* **13,** 57–98.
7. Ingram, L. O., Eddy, C. K., Mackenzie, K. F., Conway, T., and Alterthum F. (1989), *Dev. Indust. Microbiol.* **30,** 53–69.
8. Rogers, P. L., Lee, K. J., and Tribe, D. E. (1979), *Biotechnol. Lett.* **1,** 165–170.
9. Eveleigh, D. E., Stokes, H. W., and Dally, E. L., (1983), *Organic Chemicals from Biomass,* Wise, D., ed., Benjamin/Cummings, Menlo Park, CA.
10. Zhang, M., Eddy, C., Deanda, K. A., Finkelstein, M., and Picataggio, S. (1995), *Science* **267,** 240–243.
11. Deanda, K. A., Eddy, C., Zhang, M., and Picataggio, S. (1996), *Appl. Environ. Microbiol.* **62,** 4465–4470.
12. Swings, J. and DeLay, J. (1977), Bacteriol. Rev. **41,** 1–46.

Improvement of Substrate Conversion to Molecular Hydrogen by Three-Stage Cultivation of a Photosynthetic Bacterium, *Rhodovulum sulfidophilum*

Isamu Maeda,[1] Wasimul Q. Chowdhury,[1] Kenji Idehara,[1] Kiyohito Yagi,*,[1] Tadashi Mizoguchi,[1] Toru Akano,[2] Hitoshi Miyasaka,[2] Toshio Furutani,[2] Yoshiaki Ikuta,[3] Norio Shioji,[3] and Yoshiharu Miura[2]

[1]*Faculty of Pharmaceutical Sciences, Osaka University, Suita, Osaka 565;* [2]*Kansai Electric Power Company, Amagasaki, Hyogo 661; and* [3]*Mitsubishi Heavy Industries, Takasago, Hyogo 676, Japan*

ABSTRACT

In photosynthetic bacteria, after transition to light-anaerobic and nitrogen-deficient conditions, hydrogen evolution starts with expression of nitrogenase activity. Until the expression of enough activity, *Rhodovulum sulfidophilum* consumed substrates and converted them to poly(3-hydroxybutyrate) (PHB), resulting in a decrease in the proportion of substrate converted into hydrogen gas. To prevent conversion to PHB during the period when nitrogenase activity is derepressed, the authors employed a cultivation method consisting of three stages: cell growth, nitrogenase derepression, and hydrogen production. Cells cultivated by this method exhibited no lag time before the commencement of hydrogen evolution and gave an improved yield of hydrogen from the algal fermentative products.

Index Entries: Hydrogen production; poly(3-hydroxybutyrate) (PHB) accumulation; nitrogenase derepression; ethanol; algal fermentative products.

INTRODUCTION

Phototrophic purple bacteria have been studied extensively because of their metabolic variety and their possible biotechnological applications.

---

*Author to whom all correspondence and reprint requests should be addressed. E-mail yagi@phs.osaka-u.ac.jp.

They cannot utilize water as an electron donor, but, instead, utilize low-mol-wt organic compounds, and reduced sulfur under light illumination (1). They also have the potential to evolve hydrogen hydrogen gas under the catalysis of nitrogenase in the light. Many scientists have proposed applying this property of purple bacteria to the production of hydrogen gas for use as a clean energy source (2).

The authors have proposed a method in which photosynthetic bacteria produce hydrogen gas from organic substrates photosynthetically produced from carbon dioxide and water by algae (3). Acetic acid, ethanol, glycerol, and trace amounts of organic acids were excreted into a minimal medium after anaerobic treatment of the green alga *Chlamydomonas* sp. MGA161, which accumulated starch (4,5). A photosynthetic bacterium, *Rhodovulum sulfidophilum* strain W-1S, continuously produced molecular hydrogen for at least 7 d from an algal fermentative broth, used as a culture supernatant after anaerobic incubation of the green alga in the dark (3). However, in a daily cycle of cell resuspension into newly prepared algal fermentative broth, the hydrogen evolution rate in the first few days was less than that in the subsequent days, resulting in a drop in the efficiency of substrate conversion to hydrogen gas.

The present study examines hydrogen evolution by the phototroph *R. sulfidophilum* strain W-1S from an algal fermentative broth and the individual organic compounds that constituted the fermentative product. The results indicate that the decrease in the efficiency of hydrogen evolution from the substrates was caused by an accumulation of poly(3-hydroxybutyrate) (PHB) that promptly occurred until the beginning of hydrogen evolution. Improvement of the method of cultivating strain W-1S resolved this problem, and enhanced the production of hydrogen gas.

METHODS

Strain Origin

Strain W-1S was isolated by enrichment culture of marine samples collected near the coast in the Kinki region of Japan (3). Although W-1S has previously been characterized as a strain of the genus *Rhodopseudomonas* (6–8), nucleotide sequence data of its 16S rRNA amplified by the PCR (LA PCR kit; Takara Shuzo, Shiga, Japan) have established that the isolate is actually a strain of *R. sulfidophilum*. The strain has been deposited at FERM (Fermentation Research Institute, Ibaraki, Japan) under accession number FERM P-15320.

Cultivation Conditions

Strain W-1S was cultured in modified Okamoto medium (MOM) (4), with the omission of vitamin B_{12} and the addition of 3% NaCl, 400 μg vitamin B_1/L, 500 μg nicotinic acid/L, 300 μg *p*-aminobenzoic acid/L, and

50 µg biotin/L. Pyruvate, succinate, malate, and acetate were simultaneously added as substrates for photoheterotrophic growth, each at a concentration of 0.1%. Cells were grown at 30°C under illumination from incandescent lamps at 140 W m^{-2}. Cultivation was done in a 1.6-L Roux bottle containing 1.0 L of the medium, with gentle agitation by a magnetic stirrer.

Experimental Procedures

Cells in the logarithmic growth phase were harvested and washed once or twice with the carbon- and nitrogen-free (CN-free) medium used for cultivation. The cells were then resuspended in the CN-free medium or an algal fermentation broth. After resuspension into the CN-free medium, the organic substrate used for each experiment was added. To prepare the algal fermentation broth, cells of *Chlamydomonas* sp. MGA161 were incubated under dark-anaerobic conditions in the CN-free medium (9). Incubation for hydrogen evolution was carried out in vessels sparged with argon at 30°C under 55-W m^{-2} illumination.

Assay of Hydrogen Evolution, and Measurement of Nitrogenase Activity and Polysaccharide Content

The hydrogen produced was measured by a gas chromatograph (model-164; Hitachi, Tokyo). The column was filled with a molecular sieve 13X, 30/60 mesh (Gasukuro Kogyo, Tokyo). Nitrogenase activity was measured by the acetylene reduction method, as described previously (6).

PHB Extraction and Analysis

PHB extracted by chloroform, as described previously (8). Extracts were identified by the 200-MHz ^{1}H NMR spectra (VXR-200; Varian, Palo Alto, CA). The polymer extracted from strain W-1S was shown to be the homopolymer PHB by the NMR analysis (8).

Conversion of the substrate into PHB was quantitated by the radioactivity of the chloroform extract from cells incubated in the presence of [1-^{14}C]acetate (Du Pont/NEN, Wilmington, DE). The radioactivities of the chloroform extract and [^{14}C]acetate remaining in the medium were determined by scintillation counting (Tri-Carb 2100TR; Packard, Meriden, CT). After chloroform was completely evaporated, the polymer in the extract was hydrolyzed by boiling in 10 N NaOH for 1 h for HPLC analysis. After adequate dilution, the hydrolysate was subjected to HPLC (model 302; Gilson, France) on a TSK-GEL column (Tosoh, Tokyo), and the radioactivities of the elutes were determined by scintillation counting (LS 3801, Beckman, Palo Alto, CA). A peak of 3-hydroxybutyric acid was identified by using the alkaline-hydrolysate of natural-origin PHB (Aldrich, Milwaukee, WI) as a standard.

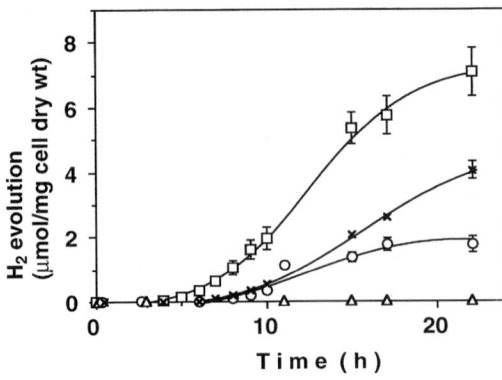

Fig. 1. Hydrogen evolution from the main organic compounds excreted during the algal fermentation. Succinate was used as a control. Each substrate was added at a concentration of 1 mM to carbon- and nitrogen-free minimal medium. △, ethanol; ○, acetate; ×, glycerol; □, succinate.

RESULTS

Hydrogen Evolution from Algal Fermentative Products

Because acetic acid, ethanol, and glycerol are the principal products excreted into the medium during anaerobic cultivation in the dark of the green alga *Chlamydomonas* sp. MGA161, the characteristics of each of these organic compounds as a substrate for hydrogen evolution in strain W-1S were assessed (Fig. 1). Hydrogen evolution was investigated during incubation of W-1S suspended in the CN-free minimal medium. Hydrogen evolution from succinate was employed as a positive control. Slower rates of hydrogen evolution and longer lag times were observed during the incubations with acetate and glycerol, compared to that with succinate; in the presence of ethanol, no hydrogen gas was detected during the incubation period. The conversion efficiencies of succinate, acetate, and glycerol to hydrogen gas were 44.1, 19.3, and 25.1%, respectively at 22 h. The uptake rates of these organic compounds in hydrogen evolving incubation were next investigated (Fig. 2). The culture of strain W-1S could consume acetate, ethanol, and glycerol immediately after transition to the hydrogen evolving condition. The uptake of acetate added to the medium was almost completed within 1 h. There was no proportional relationship between the conversion efficiency and the uptake rate. The rates of uptake of ethanol and glycerol were much lower than that of acetate.

PHB Accumulation Under the Condition for Hydrogen Evolution

The effect of incubation under the hydrogen evolving condition on the PHB content was investigated in the presence of ethanol as an external substrate (Table 1). After sufficient growth, cells accumulated PHB to the

Table 1
Conversion of Ethanol to PHB During Incubation for Hydrogen Evolution

Condition	Initial PHB mg	Initial PHB % of cell dry wt	Final PHB mg	Final PHB % of cell dry wt	PHB accumulation[a] mg	H₂ evolution[a] μmol/mg cell dry wt
Succinate[b]	52.7	44.7	29.0	29.1	−23.7	34.3
Ethanol[b]	52.7	44.7	61.5	51.5	8.8	0.46
No substrate[c]	52.7	44.7	13.0	15.4	−39.7	23.8

[a] Incubation was continued for 43 h.
[b] Each substrate was added at a concentration of 1 mM.
[c] No external substrate was added.

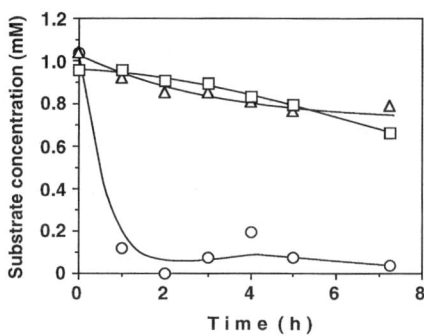

Fig. 2. Consumption of substrates in hydrogen evolving incubation in the presence of each product of algal fermentation. ○, acetate; △, ethanol; □, glycerol.

extent of 44.7% of the cell dry wt. The PHB content increased to 51.5% after the incubation for hydrogen evolution ended, as consequences of ethanol uptake and the absence of hydrogen evolution. On the other hand, extensive PHB degradation and hydrogen evolution occurred during the incubation with succinate, which is one of the substrates suitable for hydrogen evolution with strain W-1S. Similar results were obtained in the absence of an external substrate. A reciprocal relationship was observed between PHB accumulation and hydrogen evolution. To analyze this relationship further, the uptake of [^{14}C]acetate into PHB was examined under the hydrogen evolving condition. Cells were incubated anaerobically in the presence of [^{14}C]acetate, and PHB was extracted by chloroform, as described in Methods (Table 2). Cells having no nitrogenase activity incorporated about 70% of the count into the PHB fraction within 3 h. The labeled PHB fraction was hydrolyzed by alkaline-heat treatment, and the hydrolysate was analyzed by HPLC with scintillation counting, to confirm that the [^{14}C]acetate was actually incorporated into PHB. About 84% radioactivity was detected in a 3-hydroxybutyrate fraction, indicating that al-

Table 2
Effect of Cellular Condition Regarding Nitrogenase on Hydrogen Evolution
and PHB Accumulation from Acetate

	0 h	1 h	3 h	6 h
Before derepression of nitrogenase[a]				
H_2 evolution (μmol/200 mL batch)	0	N.T.[d]	0	Trace
PHB[c] (× 10^6 CPM/200 mL batch)	1.0	N.T.[d]	18.7 ± 8.8	18.2 ± 3.9
Supernatant[c] (× 10^6 CPM/200 mL batch)	26.8 ± 0.6	N.T.[d]	4.3 ± 0.5	4.1 ± 0.3
After derepression of nitrogenase[a,b]				
H_2 evolution (μmol/200 mL batch)	0	54.4	87.5	N.T.[d]
PHB[c] (× 10^6 CPM/200 mL batch)	1.0	1.3 ± 0.1	1.2 ± 0.2	N.T.[d]
Supernatant[c] (× 10^6 CPM/200 mL batch)	26.8 ± 0.6	1.8 ± 0.1	7.1 ± 0.1	N.T.[d]

[a] Nitrogenase activity was measured in a sample anaerobically taken at 0 h before addition of [^{14}C]acetate. The enzyme activities of cells before and after derepression were 0 and 1087 ± 126 nmol/mg cell dry wt/h, respectively.

[b] Cells after growth were incubated in the carbon- and nitrogen-free minimal medium supplemented with 1 mM succinate, and, 1 d later, [^{14}C]acetate was injected anaerobically.

[c] Radioactivity values are means ± SD.

[d] N.T., not tested.

most all the count incorporated was derived from PHB. Cells incubated anaerobically with succinate for 1 d had high nitrogenase activity. After the addition of [^{14}C]acetate, hydrogen evolution was observed, but no significant incorporation of the count into PHB occurred, in spite of an immediate decrease of [^{14}C]acetate in the supernatant. Therefore, the fate of the incorporated acetate could be dependent on the nitrogenase activity.

Improvement of Conversion of Substrates to Hydrogen Gas

When a three-stage cultivation method, composed of cell growth, nitrogenase derepression, and hydrogen production from the substrate, was applied to a culture for hydrogen evolution from the algal fermentative broth (Fig. 3), no lag time was observed before hydrogen evolution began, regardless of contact with oxygen during the resuspension of cells into the broth (Fig. 4). The conversion efficiency of the substrate to hydrogen gas over a period of 1 d was improved to 60.8% in this three-stage cultivation method, compared with 29.4% in a culture that lacked a stage for nitrogenase derepres-

Fig. 3. Schematic diagram of the three-stage cultivation method for hydrogen evolution.

Fig. 4. Effect of preincubation for nitrogenase derepression on hydrogen evolution from the algal fermentative broth. Cells were suspended in the broth after growth (△) or after the derepression of nitrogenase by preincubation with 1 mM succinate (○).

sion and hydrogen evolution stages was achieved by filling the incubation vessel with argon gas. The uptake of substrates during incubation for hydrogen evolution was accelerated in a culture employing three-stage cultivation, compared to that in a culture in which the substrates were supplied after cell growth (Table 3). When the strain W-1S was used for a hydrogen evolution test after growth, it could not convert ethanol to hydrogen gas because of PHB accumulation (Table 1). To establish whether or not the

Table 3
Effect of Preincubation for Nitrogenase Derepression on Uptake of Substrates

		Substrate concentration (mM)	
		Before depression of nitrogenase[a]	After depression of nitrogenase[a]
Acetate	0 h	5.64	5.64
	21 h	1.85 ± 0.03	0.45 ± 0.18
	Consumption	3.79	5.19
Ethanol	0 h	4.58	4.58
	21 h	4.23 ± 0.18	3.86 ± 0.17
	Consumption	0.35	0.72
Glycerol	0 h	5.05	5.05
	21 h	4.63 ± 0.11	3.87 ± 0.09
	Consumption	0.42	1.18

[a] Cells were suspended in the medium containing acetate, ethanol, and glycerol after growth or after the derepression of nitrogenase by preincubation with 1 mM succinate.

Fig. 5. Conversion of ethanol at various concentrations to hydrogen gas when supplied to the culture after the nitrogenase derepression stage.

three-stage cultivation method for hydrogen evolution was effective with ethanol as a substrate, the conversion of ethanol at various concentrations to hydrogen gas was investigated in cultures to which adequate quantities of ethanol were anaerobically added after the nitrogenase derepression stage (Fig. 5). Although the largest quantity of hydrogen gas was evolved at an ethanol concentration of 50 mM, the conversion efficiency deteriorated markedly at concentrations of 50 mM or more. Using the three-stage cultivation method, more than 50% of ethanol at a concentration of 10 mM could be converted into hydrogen gas.

DISCUSSION

The cultivation of photosynthetic bacteria for H production can be categorized into two basic methods; in one, the cell growth and H production are carried out concomitantly; in the other, the H production period is independent of cell growth. Although the former method is advantageous because of its simplicity, glutamate must be added to the cultivation medium as a nitrogen source *(10,11)*. Hence, the latter method is more commonly employed in H production by photosynthetic bacteria *(12,13)*, including the culture of immobilized cells *(14,15)*. The authors have investigated hydrogen production, using the fermentative broth of a marine green alga as a H-donor, of the marine photosynthetic bacterium R. *sulfidophilum* strain W-1S *(3)*. Using the above cultivation methods, however, strain W-1S could not evolve H gas with high efficiency from the fermentative products represented by acetate, ethanol, and glycerol. The rate of acetate uptake was much greater than that of ethanol or glycerol, and uptake of acetate, along with the conversion of PHB, started promptly before occurrence of H evolution. On the other hand, H evolution from fermentative products was improved in cells exposed to a nitrogenase derepression stage. This suggests that the rate-limiting stage in H evolution from fermentative products is the derepression of enzyme activity. In a period of cell growth, nitrogenase activity is repressed by ammonium ion and dissolved oxygen in the medium *(1,2)*. The purpose of the three-stage cultivation is to derepress nitrogenase sufficiently to allow H to be produced promptly after cell growth.

These results showed that the introduction of a period of nitrogenase derepression after cell growth is essential for the efficient evolution of H gas from fermentative products. Without such a period, a lag time and a low conversion efficiency to H was observed in the case of acetate, and only a small amount of H was evolved from ethanol. Competition between H evolution and PHB accumulation for the consumption of reducing equivalents has been suggested in *Rhodopseudomonas palustris (16)*, *Rhodobacter sphaeroides*, and *Rhodospirillum rubrum (17)*. As suggested previously for photosynthetic bacteria, with strain W-1S, the absence of nitrogenase activity and the occurrence of substrate uptake immediately after cell growth facilitated PHB accumulation from the substrates, resulting in a decrease of the ratio of conversion into H gas. Improvement of the conversion efficiency by the three-stage cultivation method is not only the result of a reduction of the substrates left in the medium, but also because of prevention of the conversion of substrates into PHB. This cultivation method for photosynthetic bacteria described here should enable the use of a broader range of H substrates that are not suitable at present for conversion into H gas by the bacteria.

REFERENCES

1. Vignais, P. M., Colbeau, A., Willison, J. C., and Jouanneau, Y. (1985), *Adv. Microb. Physiol.* **26,** 155–234.

2. Sasikara, K., Ramana, C. V., Raghuveer Rao, P., and Kovacs, K. L. (1993), *Advan. Appl. Microbiol.* **38,** 211–295.
3. Miura, Y., Saitoh, C., Matsuoka, S., and Miyamoto, K. (1992), *Biosci. Biotech. Biochem.* **56,** 751–754.
4. Miura, Y., Ohta, S., Mano, M., and Miyamoto, K. (1986). *Agric. Biol. Chem.* **50,** 2837–2844.
5. Maeda, I., Miyashiro, M., Hikawa, H., Yagi, K., Miura, Y., and Mizoguchi, T. (1996), *Biosci. Biotech. Biochem.* **60,** 975–978.
6. Yagi, K., Maeda, I., Idehara, K., Miura, Y., Akano, T., Fukatu, K., Ikuta, Y., and Nakamura, H. K. (1994), *Appl. Biochem. Biotechnol.* **45/46,** 429–436.
7. Miura, Y. (1995), *Process Biochem.* **30,** 1–7.
8. Chowdhury, W. Q., Idehara, K., Maeda, I., Umeda, F., Yagi, K., Miura, Y., and Mizoguchi, T. (1996), *Appl. Biochem. Biotechnol.* **57/58,** 361–366.
9. Ohta, S., Miyamoto, K., and Miura, Y. (1987), *Plant Physiol.* **83,** 1022–1026.
10. Kim, J. S., Ito, K., and Takahashi, H. (1982), *Agric. Biol. Chem.* **46,** 937–941.
11. Stevens, P., Plovie, N., De Vos, P., and De Ley, J. (1986), *Syst. Appl. Microbiol.* **8,** 19–23.
12. Nakada, E., Asada, Y., Arai, T., and Miyake, J. (1995), *J. Ferment. Bioeng.* **80,** 53–57.
13. Zürer, H. and Bachofen, R. (1979), *Appl. Environ. Microbiol.* **37,** 789–793.
14. Planchard, A., Mignot, L., Jouenne, T., and Junter, G. A. (1989), *Appl. Microbiol. Biotechnol.* **31,** 49–54.
15. Fißler, J., Kohring, G. W., and Giffhorn, F. (1995), *Appl. Microbiol. Biotechnol.* **44,** 43–46.
16. De Philippis, R., Ena, A., Guastini, M., Sili, C., and Vincenzini, M. (1992), *FEMS Microbiol. Rev* **103,** 187–194.
17. Hustede, E., Steinbühel, A., and Schlegel, H. G. (1993), *Appl. Microbiol. Biotechnol.* **39,** 87–93.

Effect of Drying on Bioremediation Bacteria Properties

F. WEEKERS,[1] PH. JACQUES,[2] D. SPRINGAEL,[3] M. MERGEAY,[3] L. DIELS,[3] AND PH. THONART,*[,1]

[1]*University of Liege, Walloon Center for Industrial Biology, Bat. B 40 - 4000 Sart-Tilman, Liege, Belgium;* [2]*Faculty of Agricultural Sciences of Gembloux, Bio-industries Unit, Passage des Déportés, 2-5030 Gembloux; and* [3]*Flemish Institute for Technological Research (VITO) Boerentag, 200 - B2400 Mol. Belgium*

ABSTRACT

Bioremediation bacteria with drought-resistance characteristics were selected and compared to a collection of 10 strains selected only for their bioremediation properties. Twenty-six strains were selected from dried diesel-polluted soil, and they exhibit a better level of survival during drying, compared to collection bioremediation strains (two orders of magnitude difference). The lyophilization process does not affect the strains' ability to grow on xenobiotic compound when measured immediately after drying. However, collection bioremediation strains selected only for their bioremediation properties lose up to 80% of their properties when stored at 25°C for 15 d, but the strains selected for their drought resistance lose their properties to a lesser extent during the same period. The maximal growth rate and the rate of xenobiotic degradation of the still-active cells are not affected by the drying process.

Index Entries: Biodegradation; drought resistance; selection; maintenance of properties.

INTRODUCTION

The estimated number of contaminated industrial sites in the European Union is significant (150,000), as well as in the rest of the world (over 1,500,000 leaking underground storage tanks estimated in the United States alone) *(1)*. Different techniques have been developed for the remediation of these sites *(2)*. *In situ* bioremediation is one that has already been applied, but that deserves further development. The use of microbial products in

* Author to whom all correspondence and reprint requests should be addressed.

the bioremediation processes, however, is controversial and, in most cases, is being abandoned. These products usually have high efficiency in vitro, but competition, predation, lag phase, heavy metals copollution, and so on, make them less competitive than autochthonous strains when used *in situ* (3,4). However, in some cases of specific recalcitrant compound pollution, the use of appropriate starter cultures can readily boost the clean-up process (5). These starter cultures are mostly available in ready-to-use dry form, commercially distributed.

In these cases, it is important to have good knowledge of suitable techniques for the production and the conditioning of the starters. The drying process and its direct influence on the properties of the product constitute a bottleneck between the production chain and the *in situ* use of the bacteria. The final product must have a high survival ratio and maintain a high level of biodegradation activity.

To ensure a high level of survival after drying, the technique must be adapted to respect the cells' integrity. The kinetics of water activity (a_w) variation is a very important factor (6–8) for the viability of bacteria subjected to a drying process. Since slower decrease of the a_w, down to a limit threshold, affords higher survival ratio, the drying methods should be designed to allow slow water depletion.

Survival after drying and stability over time of the surviving fraction are necessary, but not sufficient, conditions for a starter culture to be competitive (9). In addition, the degradation properties must be maintained in the surviving cells. Actually, ensuring genetic stability after drying and during preservation is a problem, since the viability of the cells after preservation may not correlate with the full maintenance of all properties. Plasmid-encoded degradation activities may be lost at high frequencies during drying of a culture, although little loss of viability occurs (10). Changes in various properties have been reported, especially during inadequate lyophilization (11–13). Although Lang and Malik (10) found a loss of properties in their strains, they could not detect any plasmid loss.

To quantify the biodegradation ability of a bacterium, the rate of substrate uptake ($-dS/dt$), i.e., degradation rate, is an important parameter to monitor (14). When comparing the influence of drying on biodegradation properties, one should also look at the maximal growth rate in different instances, because significant bacterial growth is of prime importance in a bioremediation process.

This laboratory specializes in large-scale drying of sensitive microorganisms of industrial interest. In this context, the fundamental phenomena accompanying the drying of the cells are studied. This paper reports the selection of bioremediation strains according to their resistance to the drying process, the characterization of their degradation properties, and the influence of the drying process on their survival, as well as on the maintenance of their degradation properties.

MATERIALS AND METHODS

Culture Medium Composition

Minimum (mineral) medium: 10 mM buffer, $Na_2HPO_4 \cdot 2H_2O/KH_2PO_4$ pH 7.0; 1 mM $FeSO_4 \cdot 7H_2O$; (100 mM) $CaCl_2 \cdot 2H_2O$; (8.5 mM) NaCl; (1 mM) $MgSO_4 \cdot 7H_2O$; (18 mM) $(NH_4)_2SO_4$; 500 µL/L of an aqueous vitamin solution (20 mg/L folic acid, 50 mg/L pentothenic acid, 50 mg/L riboflavin, and 30 mg/L pyridoxal), sterilized by filtration, are added after autoclave sterilization of the medium. One % of diesel (or other tested hydrocarbon), sterilized by filtration, is added as carbon C source before inoculation.

869 rich medium: 10 g/L Peptone from casein; 5 g/L yeast extract; 5 g/L NaCl; 1 g/L glucose and 0.345 g/L $CaCl_2$; pH 7.0.

Collection Strain Origin

The bioremediation reference strains come from Dr L. Diels' VITO (Mol, Belgium) strain collection.

Selection of Strains with Drought Resistance Characteristics

Soil samples, collected on a diesel-polluted site, were dried in a (Niro-Aeromatic AG Budendorf, CH) fluidized bed for 20 min with an inlet air temperature of 45°C and outlet temperature ranging from 21 up to 28°C. The dry matter of the final dried product reached 98%. Duplicate samples were lyophilized. The freezing occurred in liquid nitrogen ($-198°C$), and the sublimation was carried out for 24 hr in a Leybold-Heraeus Lyovac GT2 (Köln, Germany) at a pressure of 10^{-5} bar, and with a trap temperature of $-40°C$. The samples were rehydrated and plated-out on mineral medium, with diesel as sole source of C. The survivors were isolated, and were used to inoculate liquid mineral cultures with diesel as C source.

Standardized Drying Procedure

The strains were grown in rich medium (869), harvested, and washed with 2 vol of magnesium sulfate (10^{-2} M). The sample were resuspended in 1 vol of a 0.5% (w/v) trehalose solution as protective agent, frozen in liquid nitrogen ($-198°C$), and dried under conditions described above.

Other Drying Method

Slow dehydration was achieved by the method described by Mattimore *(15)*. The water activity of the atmosphere around the samples is set at 0.49 by the use of silica gel (Nerk, Darmstadt, Germany). The samples were analyzed after 48 hr and 14 d of dehydration at a temperature of 25°C.

Some samples were dried in the fluidized bed described above, with silica as a solid substratum, on which the $MgSO_4$ cell suspension is sprayed with 15% (w/v) maltodextrine as protective agent.

In the course of the determination of the maintenance of the degradation properties, the lyophilizations were made after freezing at $-60°C$. The samples were washed in the same manner as in the standard lyophilization and resuspended in 1 vol of $MgSO_4$ (10^{-2} M), either with 0.5% trehalose or without protector.

Rehydration Procedure

All dried samples were rehydrated in $MgSO_4$ (10^{-2} M) at 30°C under shaking for 30 min.

Determination of Maintenance of Biodegradation Properties

Before and after drying, cells were plated-out on plate-count agar and simultaneously on mineral medium, with decane as sole source of C supplied via the gas phase. The survival ratio was determined by the ratio of the plate-count agar (PCA) count before and after treatment. The still-active fraction was measured by the ratio between the mineral medium count and PCA count, both done at the same time after drying.

Kinetic Factor Determination

Five-mL mineral medium cultures were inoculated with 100 µL of a $MgSO_4$ suspension of cells centrifugated after 24 h of culture in 869 medium. To ensure reproducibility, hydrocarbon extractions and colony counts were done in duplicate every 24 h on the total volume of the tightly closed culture vials. To measure the kinetic factors after the lyophilization process, 5 mL of cell suspensions in $MgSO_4$ were lyophilized, rehydrated in an equal volume, and used to inoculate the mineral cultures as before drying.

The generation time (g [h]), which measures the time required to double the cell concentration of the culture, is directly correlated to the growth rate by the equation below. It was calculated from the maximal growth rate (μ_{max}) values. The μ_{max} was evaluated from the experimental data from the steepest slope of the growth curve. The rate of hydrocarbon consumption ($-dS/dt$) was measured from the steepest slope of the hydrocarbon consumption curve.

$$g = \log(2x) - \log(x)/\mu = \log(2)/\mu$$

Hydrocarbon Dosage

The hydrocarbons were extracted from the liquid culture medium at regular intervals. The extraction was made with 1 vol of cyclohexane. The organic phase was analyzed by gas chromatography on a Chrompack capillary column WCOT fused silica CP-Sil8CB. The carrier gas was helium and the head column pressure was 60 kPa. The injector temperature was maintained at 200°C and the flame ionization detector (FID) temperature at 300°C. The column temperature was raised from 50°C to 280°C, with a gradient of 7°C/min. 50 µL of a 5000 ppm heptanol standard solution were added to 500 µL of the sample before injection. One µL was injected for each measurement. The extraction efficiency was measured by the addition of a known quantity of hexadecane to the liquid culture just before extractions; its average value was 87%. The removal of hydrocarbons caused by evaporation was evaluated by the determination of the quantity left in a reference sterile, closed vial at the time of the last sampling. It was measured to be less than 5% with decane and less than 1% with diesel.

RESULTS AND DISCUSSION

Drying Resistance of Selected Collection Strains

Collection strains were selected for their ability to decompose rather recalcitrant compounds (e.g., polyaromatic hydrocarbons [PAH]), and were submitted to a standardized lyophilization procedure.

Table 1 shows the characteristics of strains representative of different genera, and the result of the lyophilization test. The bacteria have good remediation potentialities, but withstand poorly the drying procedure.

Screening for Drought-Resistant Microorganisms

In order to find drought-resistant bioremediation bacteria, soil samples from a hydrocarbon-polluted site were dried according to the procedure described above, and 26 different strains were collected. The strains named T901–T907, T981–T987, and TF1–TF7 were selected from fluidized soil samples. The strains named TFL1–TFL5 were selected from fluidized soil samples that were subsequently rehydrated and lyophilized.

Those strains selected for their ability to resist drying techniques were then tested for their growth in liquid mineral medium, with diesel as sole source of C (see Fig. 1).

Four different strains (TF5 and TFL1 were identified to be the same one) exhibited good growth levels in these conditions. They were submitted to the same standardized drying procedure as the reference strains, and identified (see Table 2).

The strains selected from a dried soil survive with a higher proportion to the same drying technique than the collection bacteria. Thus, selecting the bacteria according to their drought resistance yields new strains more suitable to the technological conditioning treatment. However, the survival values stay low. The lyophilization conditions used did not yield good survival. Although some strains (T902 and TF1) belong to the same genus (*Rhodococcus* and *Acinetobacter*, respectively) as collection strains, they do not have the same behavior during the drying process.

To improve the yield, and to verify whether the difference between the two categories stayed the same, other drying techniques were tested.

Table 1
Collection Strains Description and Their Survival Ratio After Standardized Lyophilization

Strains	Genera and species	Hydrocarbon used	Other characteristics	Survival ratio (%)
LB208	*Rhodococcus* sp	Pyrene, fluoranthene, phenanthrene		<0.01
SK15	*Arthrobacter* sp	Biphenyl		0.36
LB126	*Sphingomonas* sp	Fluorene		0.05
AEX5	*Alcaligenes eutrophus*	3-CBA, 4-CBA, 3CBP, 4-CBP, 2-CBP, BP[a]	Cd, Ni, Zn resistant	0.11
LH240	*Pseudomonas* sp	Oil	Lux marked[b]	<0.01
PaW1	*P. putida*	Biphenyl		0.09
GpO1	*P. oleovorans*	Oil		0.01
LH168	*Acinetobacter calcoaceticus*	Oil		0.16
LB400	*Pseudomonas* sp	Biphenyl		0.11
A5.1.	*Alcaligenes eutrophus*	Biphenyl, 4-CBP	pSS50[c]	—[d]

[a] 3-chloro-benzoate, 4-chloro-benzoate, 3-chloro-biphenyl, 4-chloro-biphenyl, 2-chloro-biphenyl, biphenyl, respectively.
[b] Ref. 16.
[c] Catabolic plasmid, Springael, D. personal communication.
[d] Not determined.

Fig. 1. Growth level (OD, 600 nm) of the collected soil strains in mineral liquid medium with diesel as sole source of carbon, measured after 4 d.

Improvement of Survival Ratio

The technique adapted from Mattimore *(15)* consists of a slow extraction of moisture from the cells. It affords higher survival yields than the standardized lyophilization used so far. The results are given in Table 3.

Table 2
Identification and Survival Ratio After Standardized
Lyophilization of Some Strains Collected from Dried Soil

Strains	Genera and species	Survival ratio (%)
T902	Rhodococcus erythropolis	1.54
TF1	Acinetobacter johnsonii	1.89
TF7	Micrococcus luteus	2.55
TFL1	Methylobacterium extorquens	1.03

Table 3
Survival Ratio of Different Selected Strains After Slow Drying

Strain	48 hr (%)	15 d (%)	Fluidized bed (%)
T902	68	14.84	15.5
TF1	6.61	1.36	–
TF7	2.98	2.73	12.5
TFL1	23.38	17.50	–
LB208	0.56	0.15	–
SK15	–[a]	–	–
LB126	0.21	<0.01	–
AEX5	0.90	<0.01	–
LH240	0.79	0.77	–
PaW1	0.17	0.10	–
GpO1	1.16	0.01	–
LH168	0.27	0.03	2
LB400	0.35	0.03	–

[a] Not determined.

After 48 h of drying, the first survival value is measured when the a_w of the bacteria is in equilibrium with the environment water activity, equaled to 0.49. The second value, measured after 15 d, shows the evolution of the survival ratio over time, i.e., the stability of the dry product.

Some samples were also dried in a fluidized bed. This technique allows a slow water activity depletion as well.

The strains selected for their drought-resistance characteristics exhibit a better survival level than the bioremediation collection strains, when dried with these techniques as well. Furthermore, most of the collection strains undergo a rapid decay of their surviving population with time. The dry products are not stable at 25°C. The bacteria collected from the dried soil yield much better drying resistance, and are more stable in time.

Characterization of Carbon Source Pattern of the Strains

Some soil bacteria were streaked on minimal medium Petri dishes with different C sources, in order to determine their pattern of degradation

Table 4
Description of Hydrocarbons Metabolized by Strains Collected from Dried Soil

Strains	General and species	Hydrocarbons degraded
T902	*Rhodococcus erythropolis*	Alkanes (C_{10}–C_{15}), branched alkanes, diesel
TF1	*Acinetobacter johnsonii*	Branched alkanes, diesel
TF7	*Micrococcus luteus*	Benzene, phenol, m-xylene, diesel
TFL1	*Methylobacterium extorquens*	Diesel
T901	–[a]	Diesel
T981	–	Xylenes, p-cymene, biphenyl, diesel
T982	–	Diesel
T986	–	Diesel, benzene
T987	–	Alkanes (C_{10}–C_{11}), diesel
TFL3	–	Alkanes (C_8–C_{20}), diesel

[a] Not determined.

potentialities. Their growth was evaluated after 4 d on 40 different compounds, ranging from simple light alkanes to the heaviest polyaromatic hydrocarbons. A summary of the results is given in Table 4.

Since the first screenings were done with diesel as C source, not surprisingly, the strains selected are only able to grow on the most easily degradable molecules: n-alkane, and the simplest aromatic molecules. The more recalcitrant compounds are not decomposed. The selection should, therefore, be made on the recalcitrant compounds, if strains are to be isolated with other degradation properties.

The screening test with the dried soil samples was carried out again with a mixture of PAH (naphtalene, phenanthrene, anthracene) as sole source of C. No strain was isolated from this soil in these conditions.

Maintenance of Biodegradation Properties After Drying

The influence of the drying process on the maintenance of the degradation properties of four microorganisms selected from dried soil (*Rhodococcus erythropolis, Acinetobacter johnsonii, Micrococcus luteus, Methylobacterium extorquens*) and of two of the collection strains (*Acinetobacter calcoaceticus*, LH168, and *Alcaligenes eutrophus*, A5.1.) was investigated. The effect of the different drying procedures (lyophilization and Mattimore's slow drying) on the bioremediation properties of the cells was measured by comparing their ability to grow on rich medium and on mineral medium, with a xenobiotic compound as sole source of C and energy.

The direct measurement of the degradation abilities of the bacteria after lyophilization is equal to the measurement made before drying, even for *A. eutrophus* A5.1, known to be a plasmid bearer. One hundred % of the survivors are still able to grow on decane, chosen as the xenobiotic C

Table 5
Maximal Growth Rate, Generation Time, Hydrocarbon Uptake Rate, and Lag Phase Duration of *Rhodococcus erythropolis* (T902) and *Micrococcus luteus* (TF7)

Strain	Xenobiotic compound		μ_{max} (h^{-1})	g (h)	$-dS/dt$ (ppm/h)	Lag time
T902	Decane	Before drying	0.0129	54	88	0
		After drying	0.0114	60	85	60 hr
T902	Diesel	Before drying	0.0210	33	44	0
		After drying	0.0180	38	34	48 hr
TF7	Decane	Before drying	0.0115	60	92	0
		After drying	0.0130	54	108	48–72 hr
TF7	Diesel	Before drying	0.0195	35	59	0
		After drying	0.0168	41	52	24–48 hr

All parameters were measured before and after lyophilization.

source. On the other hand, when dried with the slow technique, activity is lost during conservation. A part of the survivors have lost their ability to grow on the xenobiotic compound after a conservation of 15 d at 25°C of the dried powders (*see* Fig. 2). Loss of activity does not come from the drying process itself, but rather from the storage period. The phenomenon is common to both strain categories, but a difference appears between the drought-resistant bacteria and the collection ones. Eighty % of the survivors of the A5.1. strain already lost their degradation properties after 48 h and 75% of LH168 after 15 d, but only 10% of T902, 18% of TF7, 30% of TF1, and 43% of TFL1 lost their properties after 15 d.

In a second phase, two strains, *R. erythropolis* (T902) and *M. luteus* (TF7), showing the best rate of growth with hydrocarbons as sole source of C, were further studied. The maximum rate of growth (μ_{max}), and the rate of hydrocarbon uptake ($-dS/dt$), with decane and diesel, were measured from the growth curve and the hydrocarbon degradation curve (*see* Fig. 3). Their values were compared before and after drying. The lag phase before growth start was estimated. The values are given in Table 5.

R. erythropolis (T902) and *M. luteus* (TF7) both grow faster on diesel than on decane alone, but they retain comparable generation time, before and after lyophilization. However, the lag time before the exponential growth starts gets longer after drying, especially on decane alone as C source. The lyophilization process does not have any effect on the growth rate of the strains, as measured by this experiment.

The degradation rate of decane is higher than that of diesel, although the growth rates are in the reversed order, for both strains. Since diesel is a complex mixture of compounds, some of them might not be decomposed at all, which makes the overall degradation rate slower, but, in diesel, the bacteria find other compounds that allow them to grow faster than on decane alone. The degradation rate values, measured before and after ly-

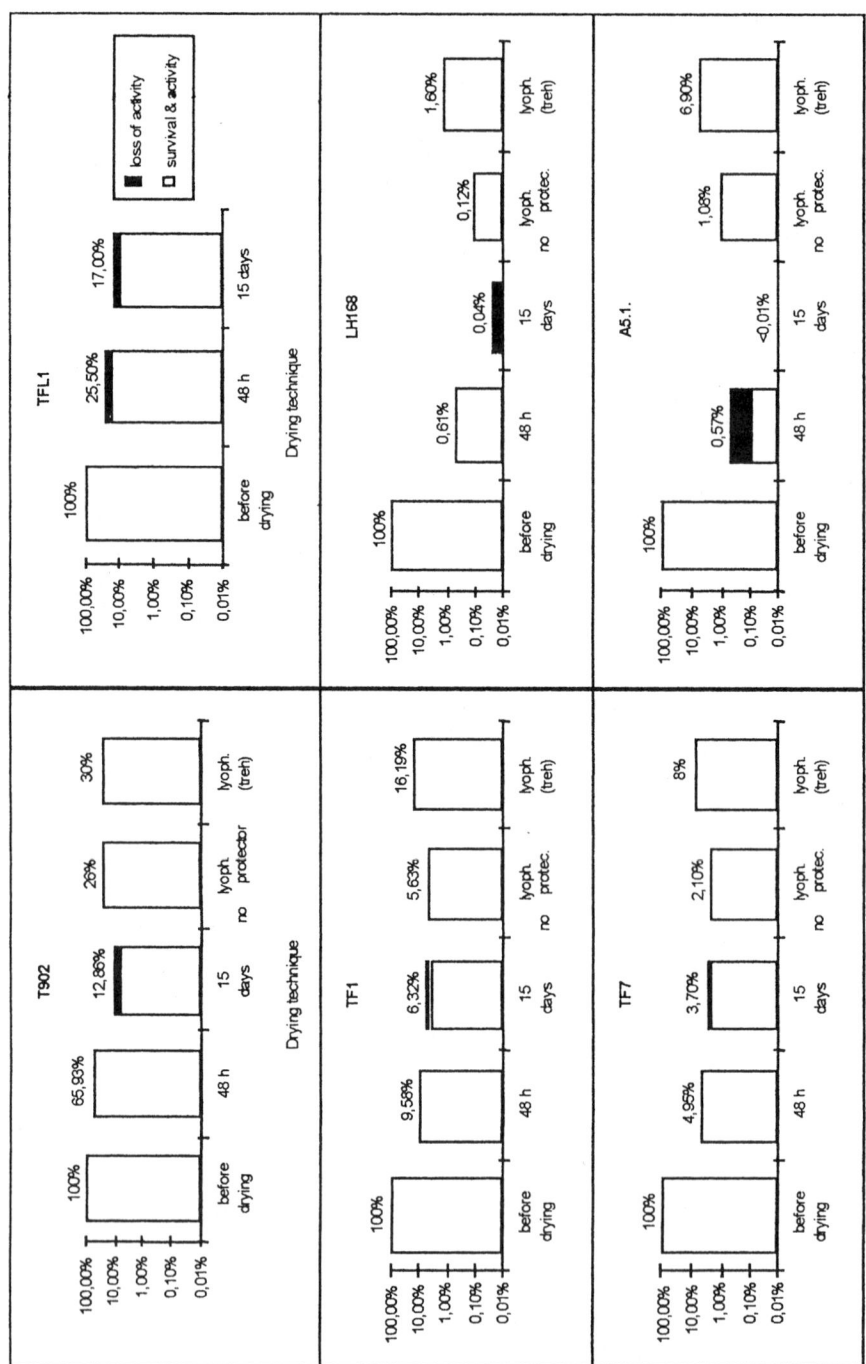

Fig. 2. Survival and maintenance of the biodegradation properties of soil strains and of two collection strains after 48 h of slow drying, 15 d of storage, lyophilization without protector, and lyophilization with 0.5% (w/v) trehalose as protector.

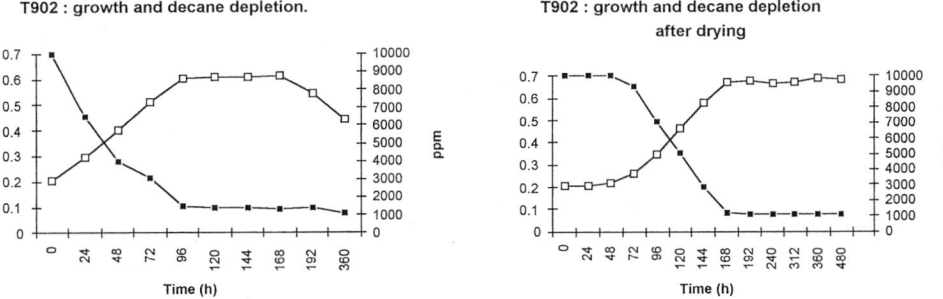

Fig. 3. Growth (O.D.) and decane consumption before and after drying of the strain T902.

ophilization, are essentially the same. The lyophilization process does not seem to have any effect on the rate of biodegradation of decane or diesel, once the lag phase is finished. However, because of the longer lag phase of the lyophilized starter, the overall speed of hydrocarbon uptake is slowed down compared to the use of fresh inoculum. This is true for both strains.

CONCLUSIONS

This new approach to the selection of potential bioremediation strains, which are suitable for the conditioning process, yields good results. The strains selected from dried soil have a survival ratio two orders of magnitude higher than the collection strains. However, since they grow only on the compound on which they were selected, their biodegradation potentialities could be improved.

ACKNOWLEDGMENTS

F. Weekers is a recipient of a F.R.I.A. (Fonds pour la Formation à la Recherche dans l'Industrie et l'Agriculture) fellowship.

REFERENCES

1. Glass, D., Risto, T., and Van Eijk, J. (1995), *Gen. Eng. News* **15,** 6–9.
2. Jacques, Ph., Bossrez, S., Mergeay, M., and Thonart, Ph. (1995), *Arch. Int. Physio.* **103,** B46.
3. Venosa, A. D., Haines, J. R., Nisamaneepong, W., Govind, R., Prahan, R., and Siddique, B. (1992), *J. Ind. Microbiol.* **10,** 13–23.
4. Dott, W., Feidieker, D., Kampfer, P., Schleibinger, H., and Strechel, S. (1989), *J. Ind. Microbiol.* **4,** 365–374.
5. Venosa, A. D., Haines, J. R., Nisamaneepong, W., Govind, R., Prahan, R., and Siddique, B. (1992), *J. Ind. Microbiol.* **10,** 1–22.
6. Poirier, I., Marechal, P.-A., and Gervais, P. (1996), *Med. Fac. Lanbouww. Univ. Gent* **61,** 1559–1564.

7. Hodzic, D., Zgoulli, S., Gilsoul, J.-J., Llabres, G., and Thonart, Ph. (1994), *Lebensm. Wiss. u. Technol.* **28,** 21–24.
8. Gervais, P., Marechal, P., and Molin, P. (1992) *Biotechnol. Bioeng.* **43,** 165–170.
9. Weekers, F., Jacques, Ph., Springael, D., Mergeay, M., Diels, L., and Thonart, Ph. (1996), *Med. Fac. Landbouww. Univ. Gent* **61,** 2161–2164.
10. Lang, E., and Malik, K. (1996), *Biodegradation* **7,** 65–71.
11. Ashwood-Smith, M., and Grant, E. (1976), *Cryobiology* **13,** 206.
12. Sakane, T., Banno, I., and Iijima, T. (1983), *IFO Res. Comm.* **11,** 14–24.
13. Simione, F. (1992), *J. Parent. Sci. Technol.* **46,** 226–232.
14. Zhou, E., and Crawford, R. (1995), *Biodegradation* **6,** 127–140.
15. Mattimore, V. and Battista J. R., (1996) *J. Bacteriol.* **178,** 633–637.
16. Springael, D., Carpels, M., Hooyberghs, D., Bastiaens, L., Kinnaer, L., and Diels L. (1995), *Med. Fac. Landbouww. Univ. Gent* **60,** 2647–2650.

Production of L-Lactic Acid by *Rhizopus oryzae* in a Bubble Column Fermenter

JIANXIN DU,* NINGJUN CAO, CHENG S. GONG, and GEORGE T. TSAO

Laboratory of Renewable Resources Engineering, Purdue University, West Lafayette, IN 47907

ABSTRACT

Two distinctive forms of growth (mycelial filamentous and mycelial pellets) of *Rhizopus oryzae* were obtained by manipulating the initial pH of the medium with the controlled addition of $CaCO_3$ in a bubble fermenter. In the presence of $CaCO_3$, diffused filamentous growth was obtained when the initial pH of the substrate was 5.5. In the absence of $CaCO_3$, mycelial pellet growth was obtained when the initial pH was 2.0. The fermentation study indicated that the mycelial growth has a shorter lag period before the onset of acid formation. Both physical forms of growth of *Rhizopus* exhibited a high yield of L-lactic acid in the bubble fermenter when the initial glucose concentration exceeded 70 g/L. A final lactic acid concentration of 62 g/L was produced by the filamentous form of *Rhizopus* from 78 g/L glucose after 27 h. This showed a weight yield of 80% of glucose consumed, with an average specific productivity of 1.46 g/h/g. Similarly, the pellet form of *Rhizopus* produced a final lactic acid concentration of 66 g/L from 76 g/L glucose after 43 h, with a weight yield of 86% and an average specific productivity of 1.53 g/h/g.

Index Entries: Bubble fermentor; L-lactic acid; mycelial pellet; *Rhizopus oryzae*.

INTRODUCTION

Lactic acid is an intermediate-volume specialty chemical that is used in a wide range of food-processing and industrial applications. Because of its chemical properties, lactic acid has the potential of becoming a large-volume, commodity-chemical intermediate that can serve as feedstock for

*Author to whom all correspondence and reprint requests should be addressed.

biodegradable polymers, oxygenated chemicals, environmentally friendly solvents, and other intermediates. Polylactate is the feedstock for the biodegradable polymer that is synthesized from lactic acid. For polylactate synthesis, L-lactic acid is preferred.

Rhizopus oryzae is known to produce L-lactic acid (1–3). Similar to many other mycelial fungal species, *Rhizopus* cultures are morphologically complex. They can grow as mycelial mats, mycelial clumps, or mycelial pellets, depending on the growth conditions and the strain of fungus. Different morphological growth forms can have a significant effect on the rheology of the fermentation broth that will affect the performance of the bioreactors. Large-scale production of lactic acid in a traditional stirred tank may be difficult, if the fungal cells were grown in the filamentous form. This is because filamentous growth results in highly viscous broth. The high viscosity will have a negative impact on the mass-transfer properties of the broth (4). The problem is more profound when the fermentations require high oxygen transfer, such as in lactic acid production by *R. oryzae*. As a rule, there are two types of growth in the submerged cultures of mycelial fungi; the mycelial pellet form, and the extended filamentous form. The size of mycelial pellet varied from 1 mm to as large as 10 mm in diameter; the filamentous growth forms a homogeneous suspension that can disperse throughout the liquid medium. The growth of *R. oryzae* in the shake flasks varied from extensive mycelial growth to small pellets to conglomerates. The physical forms of fungal growth are influenced by the strains of fungi, nutrients, pH of the medium, agitation, aeration, inoculum size, and substrate concentrations. In order to obtain a uniform size of mycelial pellets (1–2 mm in diameter) from *R. oryzae,* the authors have formulated a growth medium with xylose as the sole carbon source. The characteristics of *Rhizopus* pellets were examined and the results were documented (3).

In this report, the relationship among growth conditions, the morphology of the mycelial growth, and lactic acid production in a bubble fermenter were studied.

MATERIALS AND METHODS

Microorganism and Inoculum

Culture of *R. oryzae* ATCC 52311, purchased from American Type Culture Collection (Rockville, MD), was chosen for this study because it is a good sporangiospore producer. The culture was maintained on YMP (Difco, Detroit, MI) agar slants and propagated by growing in Erlenmeyer flasks, in YMP agar, which consists of 0.3% yeast extract, 0.3% malt extract, 0.3% peptone, 1% glycerol, and 1.5% agar to obtain sporangiospores. For fermentation study, spores were collected by washing spore-bearing cultures with sterile water and collecting this as a spore suspension.

Preculture

For each experiment, the spore suspension (1×10^6 spores/mL) was inoculated into three 2.5-L Erlenmeyer flasks containing 1 L cultivation medium. The medium consisted of 20 g glucose, 3 g urea, 0.6 g KH_2PO_4, 0.25 g $MgSO_4 \cdot 7H_2O$, and 0.088 g $ZnSO_4 \cdot 7H_2O$ in 1 L dH_2O. Incubation was carried out at 35°C and 200 rpm in an incubator–shaker for 20 h.

Bubble Column

The batch fermentations were performed in a 5-L bubble column with an operating volume of 3.5 L. The fermenter was constructed with transparent polycarbonate (Meyer Plastics, Lafayette, IN) plastic pipe. It consisted of a column having a diameter of 6 cm and a height of 90 cm, with a perforated plate (10 holes of 1 mm in diameter), located at the bottom of the column, to serve as the air sparger. Sterile air was supplied at 0.2–0.3 vvm.

Fermentation

After the germ tubes were formed (up to 4 mm in length) in the preculture step, the germinated spores were transferred aseptically into the sterilized bubble column containing only glucose in water. Depending upon the experiment, sterilized $CaCO_3$ was added whenever needed to maintain the pH at 5.5. The fermentation was operated at an aeration rate of 0.2–0.3 vvm at 31°C. Samples were taken periodically for HPLC analysis.

Analytical Methods

Lactic acid, glycerol, ethanol, and glucose were determined and quantified by high-performance liquid chromatography (Hitachi, L-6200A (Hercules, CA)), using a Bio-Rad Aminex HPX-87H ion exclusion column (300×7.8 mm) with a refractive index detector (Hitachi, L-3350 RI). The column was eluted with dilute sulfuric acid ($0.005\,M$) at a column temperature of 80°C and a flow rate of 0.8 mL/min over an 18-min period.

RESULTS

Fermentation of Lactic Acid by Diffused Filamentous Form of *Rhizopus*

When the germinated spores of *R. oryzae* obtained from preculture were transferred into the bubble column in the presence of $CaCO_3$ (pH = 5.5) and glucose, cells grew into the distinctive diffused filamentous form with extended hyphae. The length of mycelial growth extended to 10 mm

Fig. 1. Kinetics of lactic acid fermentation by filamentous form of *Rhizopus oryzae* in a bubble fermenter. (■, Glucose; □, L-Lactic acid; ✻, pH; ◇, Ethanol; ◆, Glycerol).

long. Lactic acid was detected after a 10 h lag period (Fig. 1). After 27 h of incubation, glucose was entirely consumed. An average specific lactic acid productivity of 1.46 g/h/g, and a final lactic acid concentration of 62.5 g/L, were obtained. Throughout the fermentation period, lactic acid yield was constant and reached a value of 0.8 g/g. Ethanol (3 g/L) and glycerol (2 g/L) were also detected during the fermentation.

Fermentation of Lactic Acid by Mycelial Pellet Form of *Rhizopus*

The growth of pellet form of *R. oryzae* in the bubble column was achieved after transferring the germinated spores into liquid glucose medium without $CaCO_3$. During the first 16 h of incubation with bubbling air, broth dropped from about pH 5.0 to 2.0, because of the production of a small amount of lactic acid (Fig. 2). During this period, the germinated spores grew into more or less uniform sized elliptical pellets (1–2 mm). The growth of the pellets stopped after reaching this size, presumably because of the exhaustion of the available nitrogen source. A small amount of $CaCO_3$ was then pumped into the column after the completion of pellet development, to maintain the broth in the range of 5.0 to 5.5 pH in order to enhance the lactic acid production. During the fermentation stage, aeration was increased from 0.2 to 0.3 vvm, to maintain fermentation rate.

After 43 h of incubation, glucose was consumed to produce a final lactic acid concentration of 66 g/L, with an average specific lactic acid productivity of 1.53 g/h/g. The longer incubation time was required for *Rhizopus* pellets to complete the fermentation. This is a result of the long lag period (21 h). Throughout the acid production period, the product yield was constant and reached a value of 0.8 g/g. Ethanol and glycerol were detected during later stages of fermentation.

Fig. 2. Kinetics of lactic acid fermentation by pellet form of *Rhizopus oryzae* in a bubble fermenter. (■, Glucose; □, L-Lactic acid; ✻, pH; ◇, Ethanol; ◆, Glycerol).

Cell Recycle

One of the advantages of fermentation in the bubble fermenter is that the mycelia and the pellets can be easily separated from the fermentation broth by simply allowing the cells to settle. In the cell-recycle study, fresh glucose solution was supplied after about 75% of the fermented broth was drained. The repeat fermentation was initiated by supplying oxygen. Cells of *Rhizopus* can continue to produce lactic acid from glucose under this nongrowth condition, with much shorter lag period and higher volumetric productivity than the first cycle fermentation. However, the fermentation activity declined after the second cycle. The results of cell recycle experiments are summarized in Table 1.

DISCUSSION

L-Lactic acid fermentation by *R. oryzae* can be separated into two distinctive stages, cell growth and product formation. During the growth stage, in the presence of a nitrogen source, *R. oryzae* grows with extended hyphae and forms large-sized pellets or mycelial aggregates. Once the cells are fully grown, the fermentation stage can be augmented by changing the medium to one without a nitrogen source. When fermentation was carried out in the conventional stirred-tank fermenter, the growing mycelia were adsorbed onto the heat exchanger and impellers, and formed mycelial clumps. The problem was accentuated when $CaCO_3$ was used as the neutralizing agent. Added $CaCO_3$ intermingled with mycelial aggregates and further complicated the clumping problem. Under the conditions described, the fermentation time was long, and ethanol was accumulated in place of lactic acid. One way to avoid this problem is by conducting the fermentation in a bubble fermenter. The bubble fermenter can provide better gas and mass transfer than the stirred-tank fermenter *(5)*.

Table 1
Summary of Results for Lactic Acid Production by *Rhizopus oryzae* in Bubble Column

Forms of *Rhizopus oryzae*	Pelleted forms			Filamentous forms		
Cycle no.	1	2	3	1	2	3
Initial glucose concentration (g/L)	76.5	70.8	63.0	77.9	73.2	57.1
Final lactic acid concentration (g/L)	66.2	65.8	55.6	62.5	63.2	47.2
Fermentation time (hr)	43	15	19	27	12	18
Lag phase (hr)	21	0.5	1.0	10	0.5	2.0
Biomass dry weight (g/L)	1.97	–	2.80	2.52	–	3.93
Yield (%)	86.5	92.9	88.2	80.2	86.4	82.6
Volumetric productivity (g-lactic acid/h/L)	1.54	4.39	2.93	1.69	5.06	2.36
Specific productivity (g-lactic acid/h/g biomass)	1.53	–	1.10	1.46	–	0.75

The authors' previous study on lactic acid fermentation by *R. oryzae* indicates that pellets with a loose structure and a diameter of 1–2 mm provide the best physical characteristics for producing high yields of lactic acid (3). In this study, two different physical forms (filamentous or pellet) of cell growth can be realized by controlling the timing of addition of the neutralizing agent, $CaCO_3$, to the bubble fermenter. The controlled $CaCO_3$ addition can bypass the difficulties encountered in the stirred tank fermentation. It can also avoid the complication of using a specially formulated growth medium to achieve pellet formation before subjecting the cells to the nongrowth fermentation condition.

In the authors' experience, a higher concentration (>10%) of calcium lactate can be obtained by supplying a higher initial glucose concentration. However, this is undesirable, because the limited solubility of the calcium lactate resulted in the *in situ* crystallization of the product during operation. In this study, the substrate concentration was kept at under 80 g/L to avoid the crystallization of calcium lactate.

CONCLUSION

The results of this study indicate that two distinctive forms of mycelial growth can be obtained by controlling the timing of $CaCO_3$ addition. The bubble column fermentation results indicate that both the filamentous and the pellet forms of *Rhizopus* exhibited good fermentation rates and high product yield.

ACKNOWLEDGMENTS

The authors thank Linda Liu for analysis. This research was supported in part by US Department of Agriculture Grant 96-35500-3192.

REFERENCES

1. Soccol, C. R., Stonoga, V. J., and Raimbault, M. (1994), *World J. Microbiol. Biotechnol.* **10**, 433–435.
2. Rosenberg, M. and Kristofikova, L. (1995), *Acta Biotechnol.* **15**, 367–374.
3. Yang, C. W., Lu, Z., and Tsao, G. T. (1995), *Appl. Biochem. Biotechnol.* **51/52**, 57–71.
4. Vanags, J. J., Priede, M. A., and Viesturs, U. E. (1995), *Acta Biotechnol.* **4**, 355–366.
5. Lubbert, A., Paaschen, T., and Lapin, A. (1996), *Biotechnol. Bioeng.* **52**, 248–258.

Factors that Affect the Biosynthesis of Xylitol by Xylose-Fermenting Yeasts

A Review

SILVIO S. SILVA,* MARIA G. A. FELIPE, AND ISMAEL M. MANCILHA

Department of Biotechnology, Faculty of Chemical Engineering of Lorena, P.O. Box 116, Lorena, São Paulo, Brazil

ABSTRACT

Xylitol is a sweetener with important technological properties like anticariogenicity, low caloric value, and negative dissolution heat. Because it can be used successfully in food formulations and pharmaceutical industries, its production is in great demand.

Xylitol can be obtained by microbiological process, since many yeasts and filamentous fungi synthesize the xylose reductase enzyme, which catalyses the xylose reduction into xylitol as the first step in the xylose metabolism. The xylitol production by biotechnological means has several economic advantages in comparison with the conventional process based on the chemical reduction of xylose. The efficiency and the productivity of this fermentation chiefly depends upon the microorganism and the process conditions employed. In this mini-review, the most significant upstream parameters on xylitol production by biotechnological process are described.

Index Entries: Xylose fermentation; xylitol; xylose-fermenting yeasts; hemicellulosic hydrolysates.

XYLOSE-FERMENTING YEASTS: BIOCHEMICAL CONSIDERATIONS

Traditionally, yeasts were considered nonfermentors of xylose (1,2). Nevertheless, in later years several yeasts able to ferment xylose have been identified, namely *Pachysolen tannophilus*, *Candida shehatae*, *Candida tropicalis*, *Pichia stipitis*, *Debaryomyces hansenii*, and *Candida mogii* (3–7). Xylitol is the main byproduct of these yeasts during the xylose metabolism. How-

---

*Author to whom all correspondence and reprint requests should be adressed. E-mail: silvioss@fastnet.com.br ever, the factors that regulate the production and excretion of xylitol have not been clearly established *(3,8–10)*.

There are two possible metabolic routes for the utilization of xylose by microorganisms. In prokaryotic organisms, the first step of the metabolism involves enzyme induction followed by xylose isomerization into xylulose. Thereafter, the xylulose is phosphorylated to xylulose-5-phosphate *(11)*. In eukaryotic organisms such as yeasts, the oxido-reductive xylitol pathway to xylose prevails. In these reactions, the xylose is reduced to xylitol by the xylose reductase (E.C. 1.1.1.21) enzyme linked to nicotinamide-dinucleotide phosphate in the reduced form (NADPH). Next, the xylitol is oxidized to xylulose by xylitol dehydrogenase (E.C.1.1.1.9) linked to nicotinamide-dinucleotide in the oxidized form (NAD^+). The xylulose formed is phosphorylated to xylulose 5-phosphate that can be converted into pyruvate through the connection of the phosphopentose pathway with the Embden Meyerhof Parnas pathway *(12)*. The major enzymes for the metabolic pathway utilized by xylose-fermenting-yeasts are: xylose reductase, xylitol dehydrogenase, xylulokinase, phosphopentose epimerase, phosphopentose isomerase, transaldolase, transketolase, phosphohexoisomerase, glucose 6-P dehydrogenase, lactonase, and 6-P glucose dehydrogenase.

According to Gong *(11)*, the xylose reduction to xylitol and the activation of the phosphopentose pathway in yeasts is controlled by the $NADP^+$ and NADPH disposability. The $NADP^+$ generation during the xylose reduction can stimulate the activity of the 6-P-glucose dehydrogenase which, in turn, stimulates the activation of the phosphopentose pathway. Jeffries, 1982, cited by Gong *(11)*, suggests that the xylose metabolism in yeasts operates in a coordinated and closed cycle, assuring the NADPH regeneration necessary to reduce xylose to xylitol. The xylitol produced will be catabolized only to eventually produce glucose 6-phosphate for regeneration of NADPH and to maintain the cycle. This NADPH regeneration is essential for a economic xylitol production starting from xylose. Through this mechanism, the xylose metabolism in yeasts by the phosphopentose pathway and subsequent glucose 6-phosphate oxidation generates 2 moles of NADPH per each mole of CO_2 liberated.

Taylor et al. *(13)* observed that the xylitol production is favored by an excess production of NADPH generated during the xylose catabolism. However, few papers report the quantitative aspects of the production and consumption of NADPH by yeasts. So, the xylitol production by yeasts seems to be essentially related with the NADPH pool and the xylose bioconversion into xylitol occurs primarily for coenzyme regeneration and probably as a mechanism of cellular detoxification.

The xylose reductase and xylitol dehydrogenase enzymes play a fundamental role in the xylose metabolism by yeasts *(14–16)*. The continued understanding of mechanisms regulating the activities of these enzymes may allow the establishment of conditions that favor the production of the desired product. According to Skoog and Hahn-Hagerdal *(17)*, the xylose reductase and xylitol dehydrogenase enzymes differ in specificity by the coenzymes NADPH, NADH, NAD$^+$, and NADP$^+$. In *Candida utilis*, xylose reductase requires only NADPH as a cofactor, whereas xylitol dehydrogenase requires only NAD *(17,18)*. In other yeasts able to ferment xylose, like *Candida shehatae*, *Pichia stipitis*, and *Candida tenius*, xylose reductase requires both NADPH and NADH as cofactors *(18,19)*. The initial reactions of the xylose metabolism by yeasts are limited in different degrees because of a double specificity of xylose reductase by NADPH and NADH *(20)*. The accumulation of xylitol by yeasts in a medium containing xylose is associated with the complete absence of xylose reductase activity dependent on NADH *(19)*. The xylose reductase and xylitol dehydrogenase enzymes are inducible by cell growth in a medium containing xylose *(17)*. Bicho et al. *(21)* noticed that both in *Phachysolen tannophilus* and in *Pichia stipitis* the induction of these enzymes by the xylose was inhibited by glucose. These authors concluded that the repression of these enzymes was the principal regulating system in yeasts that metabolize xylose and that this repression can inhibit the yeasts' potential to ferment pentoses in lignocellulosic substrates.

XYLITOL-PRODUCING YEASTS

The development of an economic fermentative process for xylitol production involves the selection of microbial yeast strains with high productivity, the establishment of conditions that maximize the conversion of xylose into xylitol, and the optimization of these parameters for process scale-up. From a study of 58 species of yeasts that utilize xylose aerobically, Onishi and Suzuki *(22)* found that the best xylitol producers were *Candida polymorpha*, *Candida tropicalis*, *Candida guilliermondii*, *Pichia miso*, and *Hansenula anomala*. However, fermentation efficiency was only 40% between 7 and 10 d of cultivation. Gong et al. *(23)* also found that xylitol was the main metabolite formed during the xylose fermentation by yeasts. These authors concluded that a mutant strain of *Candida tropicalis* HXP2 is a promising xylitol producer, since its fermentation efficiency was approx 90% after 4 d cultivation. *Candida tropicalis* and *Candida guilliermondii* yeasts also proved to be suitable for xylitol production, presenting a high fermentation yield (81%) after 48 h cultivation, and an insignificant formation of byproducts *(9)*. The yeast *Debaryomyces hansenii* was able to produce xylitol with a conversion efficiency of approx 70% in only 28 h cultivation *(19)*. Although this microorganism is very promising, *Candida guilliermondii* has been considered as an outstanding xylitol producer. However, there are few studies on the xylose metabolism of this yeast, as well as on the factors regulating the xylitol production by this microorganism.

BIOTECHNOLOGICAL PARAMETERS IN XYLITOL FORMATION BY YEASTS

The xylose bioconversion into xylitol is regulated by different factors such as initial xylose concentration *(20,23–26)*, pH *(27,28)*, presence of glucose *(21,28,29)*, and aeration *(30–32)*.

The initial concentration of xylose in the fermentation medium has great influence on the xylitol production by yeasts. High concentrations of xylose in the medium further the xylose consumption by the yeasts and consequently, enhances the xylitol production *(26)*. Gong et al. *(23)* noticed that *Candida tropicalis* HXP2 produced more xylitol when the xylose concentration was increased from 5 to 20%. This behavior was also exhibited by *Candida shehatae* Y-12856, *Pachysolen tannophilus* NRRL Y-2460, and *Pichia stipitis* NRRL Y-7124 *(33)*. The correlation between xylitol accumulation and xylose concentration may be a consequence of an oxygen reduction resulting from high cell densities of highly concentrated substrates. The osmotic pressure exerted by xylose concentrations over 30% can interfere with the xylitol production *(26,34)*. This interference can be reduced or even eliminated by altering the process to fed batch *(34)*.

The extracellular pH has great influence on the metabolic process and on the product formation as well. This is evident from the behavior of the xylose-fermenting yeasts in relation to the pH of the medium. In general, the yeasts grow better in acid media at pH between 3.5 and 3.8. The tolerance limits are between 2.5 and 8.0, for several species. The optimum pH range for xylitol production by *Candida shehatae* is 3.5 to 4.0 *(35)*. For *Pachysolen tannophilus* the xylitol production is maximum and constant at pH ranging from 3.0 to 5.8 (Watson, 1983, cited by Du Preez et al., *(cf. 36)*. Silva and Afschar *(26)* detected significant xylitol production at pH 2.5 in cultures of *Candida tropicalis*.

Temperature has a significant effect on growth, metabolism, viability, and fermentative capacity of the yeasts *(11)*. The optimum temperatures for yeast growth are between 20 to 30°C, although some species grow within the range of 0 to 47°C. The xylose-to-xylitol conversion into xylitol seems to be stimulated by the temperature increase *(9)*. Indeed, Du Preez et al. *(36)* found that the xylitol production by *Candida shehatae* enhanced when the temperature was raised from 22 to 36°C. This was also observed for *Candida guilliermondii*, the maximum accumulated xylitol concentration (23 g/L) and the highest specific growth velocity (0.78/h) occurring at temperatures of 30 or 35°C *(9)*.

The presence of hexoses, such as glucose, in the fermentation medium is also a critical factor that regulates the xylitol production by yeasts. The presence of glucose may repress the activity of the key xylose reductase enzyme involved in the xylose conversion into xylitol resulting in low yields of the product *(21)*. In a fermentation medium without glucose an accentuated increase in the xylitol production by *Candida guilliermondii* was detected *(28,29)*. In fermentations performed with *Candida shehatae* and *Pichia stipitis* containing a mixture of sugars (xylose and glucose), glucose was preferably consumed *(37)*. These authors observed that a lag period was necessary to synthesize the enzymes of the xylose metabolism before this sugar was metabolized. However, this period can be significantly reduced with the inoculum growth in a medium containing xylose as a carbon source. In these conditions, the enzymes necessary for the xylose metabolism are induced and a simultaneous utilization of sugars is observed *(37)*. The simultaneous utilization of sugars, especially by yeasts, is a phenomenon still little studied and literature reports with conflicting opinions about the effect of hexoses on the utilization of xylose.

The constituents of the fermentation medium determine whether the fermentative processes are feasible. They should meet the elementary requirements for producing metabolites and forming biomass. It is known that the yeasts demand several inorganic ions such as: Ca^{+2}, Co^{+2}, Cu^{+2}, K^+, Mg^{+2}, and Mn^{+2} in micro and millimolar quantities for optimum culture growth in synthetic media *(38)*. Many of these ions activate the enzymatic reactions or participate in various biosynthetic reactions. Normally, it is necessary to supplement the fermentation medium with nutritional factors such as yeast extract, peptone, and meat extract. The yeast extract stimulates the yeast growth, mainly because it is rich in vitamins and amino acids. Increasing the concentration of this nutrient from 1 to 30 g/L resulted in the decrease of the xylitol production *(9)*, since the accumulation of xylitol by yeasts seems to be associated with growth limitation and acts as a secondary metabolite. A simple and inexpensive fermentation medium has been pursued by our research group *(31,39,40)*.

Oxygen is important for the xylose metabolism by yeasts *(17,29–31)*. The biochemical and physiological aspects of xylose metabolism requiring oxygen are not entirely known yet. This metabolism appears to be related to sugar transport, coenzyme regeneration, and the ATP production during the oxidative phosphorylation. Some yeasts need oxygen for an optimum xylose fermentation. Under aerobic conditions, a high production of cell mass occurs, whereas under anerobic conditions a great part of the xylose is converted into xylitol and the ethanol production is small *(17,31,40,41)*. A likely explanation for the oxygen demand is based upon the necessity for recuperation of the coenzymes required in the initial steps of the xylose utilization *(18)*. According to these authors, the xylose metabolism through xylose reductase linked to NADPH and xylitol dehydrogenase linked to NAD^+, under anerobic conditions, brings about an overproduction of NADH, which paralyzes the subsequent metabolism reactions, and consequently stimulates the accumulation of xylitol in the culture medium. In fact, the xylitol production by *Pachysolen tannophilus* increases by decreasing the aeration rate *(41)*. The same happens to the xylitol production by *Candida guilliermondii (9,32), Candida parapsilosis (30), Debaryomyces hansenii (6)*, and *Candida mogii (7,42)*. The agitation of the medium is also an important parameter for the xylose fermentation into xylitol, since the oxygen transference rate is favored by higher agitation speeds. According to Ojamo et al. *(34)* for xylitol production by *Candida guilliermondii*, it is essential that the oxygen supply to the yeast be restricted. High xylitol yields are obtained under appropriate agitation and aeration conditions. Silva et al. *(29)* found that the xylitol production by *Candida guilliermondii* was favored by increasing the agitation speed from 200 to 300/min, whereas increasing to 400/min promoted an increment in the xylose consumption to the disadvantage of the xylitol formation. The agitation role in this fermentative process is still little known. However, for xylitol production, it seems that a moderate agitation is necessary and that the maximum production is likely to be reached by means of an adequate agitation/aeration relationship.

The initial cell concentration also influences the xylose fermentation to xylitol by yeasts. Increasing the initial concentration of the inoculum from 0.25 to 3.0 g/L results in a greater xylose consumption *(43)*. Besides, the amount of xylitol produced by *Candida guilliermondii* increases by 20%. Similar results were obtained by Barbosa et al. *(9)* using the same yeast strain, but under distinct nutritional conditions. These authors obtained yields of approx 81% with media inoculated with high cell concentrations. This xylitol accumulation in the medium, with increased cell concentrations, is a consequence of the oxygen decrease, which favors the xylitol production by yeasts *(9,29,32)*.

SOME ASPECTS OF THE XYLITOL PRODUCTION FROM HEMICELLULOSIC HYDROLYSATES

In most studies on xylitol production by fermentative processes, xylose of analytical grade is commonly the main substrate. Despite the fact that agroforest residues are abundant, inexpensive, and contain a large proportion of xylose, few studies have reported the utilization of hemicellulosic hydrolysates coming from these materials for xylitol production. The main problem in the fermentation of these hydrolysates is the presence of toxic compounds released from the lignocellulosic structure during the hydrolytic process, as well as those originated from the sugar degradation *(44)*, which inhibit the microbial growth and the fermentative activity of the yeasts. In this way, several methods have been proposed with the purpose for minimizing this effect. According to Felipe et al. *(31)* for an effective xylitol production from sugarcane bagasse hydrolysate is very important a previous adaptation of the cells to the hydrolysate. Alves *(45)* verified that *Candida guilliermondii* produced more xylitol when the sugarcane bagasse hydrolysate is first treated with CaO to adjust the pH to 7.0 and subsequently treated with H_3PO_4 to lower the pH to 5.5, adding 2.4% of activated charcoal. Under these conditions, 90% of the initial xylose (48 g/L) contained in the hydrolysate was consumed after 40 h of batch fermentation, and the xylitol concentration was 24.2 g/L, which corresponds to a 67% conversion efficiency. As for eucalyptus hemmicellulosic hydrolysate, Silva et al. *(46)* observed that the maximum xylitol production (54 g/L) occurs when the hydrolysate is first treated with CaO until reaching pH 8.4 and then treated with H_3PO_4 until the pH decreases to 6.0. The pH of fermentation is another important factor in the fermentation of hemicellulosic hydrolysates. Its effect is related to the acetic acid concentration in the hydrolysate. Acetic acid concentration higher than 3.0 g/L inhibits the *C. guillermondii* capability to convert xylose into xylitol *(47)*. The nonionized acetic acid, which is found in the medium at pH < 7.0, probably acts as an inhibitor of the yeast metabolism. Therefore, the inhibition of xylose/xylitol bioconversion can be related to the coupled effect of low pH and undissociated acetic acid concentration over 5.0 g/L *(48)*. These results show that the hemmicellulosic hydrolysates from agroforest residues can be efficiently utilized in fermentative processes for xylitol production after an initial treatment designed to remove or reduce the compounds known to be toxic to cell metabolism.

LARGE-SCALE XYLITOL PRODUCTION: SCALE-UP CONSIDERATIONS

Knowing the factors that regulate the xylose metabolism to xylitol and determining the fermentative parameters that maximize this bioconversion is fundamental for establishing an economic process for xylitol production. The biotechnological process, whose efficiently has been demonstrated on a laboratory scale, demands optimization and scale-up studies before transfer to the productive sector. The scale-up of fermentative processes, particularly of aerobic fermentations involving non-newtonian fluids, is difficult to achieve. The methods employed are mostly empirical and based upon similitude concepts and nondimensional groups. The existing correlations are limited and, in general, the variables considered to be important to the process are the fermenter geometry, the agitation frequency, the potency transmitted to the system, the oxygen transfer, and the rheological properties of the fluid. For these empirical methods to be successfully utilized, it is necessary to identify the factor having the strongest influence on the process efficiency and then use it as a scale-up criterion. As mentioned before, the dissolved oxygen concentration in the medium is one of the factors that regulates the bioconversion of xylose into xylitol. The volumetric coefficient of oxygen transfer ($K_L a$) should be used as a criterion to scale-up processes requiring oxygen. Thus, when the $K_L a$ is maintained constant, in both scales (laboratory and pilot plant), other variables (potency, agitation frequency, and diameter of the tank) can be determined for a larger scale, based upon empirical correlations already developed.

No literature was found on the scale-up of xylose fermentation to xylitol. Considering that the existing correlations were developed for particular cases under different operational conditions, it is necessary to determine suitable correlations for the development of new technologies for xylitol production on a large scale.

CONCLUDING REMARKS

Several studies have been published on xylitol production from xylose by yeasts. However, to determine the best fermentation conditions, most authors have used Erlenmeyer flasks in rotary shakers. Also, in most cases the discontinuous fermentation system has been employed. Few studies specify the conditions for xylitol production in large- or laboratory-scale bioreactors. Likewise, few studies exist on the utilization of hemicellulosic hydrolysates from agroforest residues as the raw material for this bioconversion.

The number of research groups interested in a new technology for xylitol production using the biotechnological processes has grown considerably.

Among the environmental factors that exert influence on the xylitol production by xylose-fermenting yeasts, the dissolved oxygen concentration is noteworthy and must be carefully controlled.

ACKNOWLEDGMENTS

The authors acknowledge the financial support of "Fundação de Amparo à Pesquisa do Estado de São Paulo" FAPESP and "Conselho Nacional de Desenvolvimento Científico e Tecnológico" CNPq/Brazil. The authors are also grateful to Maria Eunice Machado Coelho for the revision of this paper.

REFERENCES

1. Barnett, J. A. (1976), *Adv. Carbohyd. Chem. Biochem.* **32,** 125–234.
2. Jeffries, T. W. (1983), *Adv. Biochem. Eng./Biotechnol.* **27,** 1–32.
3. Schneider, H., Mahmourides, G., Labelle, J. L., Lee, H., Maki, N., and Mc Neill, H. J. (1985), *Biotechnol. Lett.* **7(53),** 61–64.
4. Delgenes, J. P., Moletta, R., and Navarro, J. M. (1988), *J. Ferment. Technol.* **66(4),** 417–422.
5. Ligthelm, M. E., Prior, B. A., and du Preez, J. C. (1988), *Appl. Microbiol. Biotechnol.* **28(1),** 63–68.
6. Roseiro, J. C., Peito, M. A., Gírio, F. M., and AmaraL-Collaço (1991), *Arch. Microbiol.* **156,** 484–490.
7. Mayerhoff, Z. D. L., Roberto, I. C., and Silva, S. S. (1997), *Biotechnol. Lett.* **19(5),** 407–409.
8. Hahn-Hägerdal, B., Jínsson, B., and Vogel, E. L. (1985), *Appl. Microbiol. Biotechnol.* **21,** 173–175.
9. Barbosa, M. F. S., Medeiros, M. B., Mancilha, I. M., Scheneider, H., and Lee, H. (1988), *J. Ind. Microbiol.* **3,** 241–251.
10. Pfeifer, M. J., Silva, S. S., Felipe, M. G. A., Roberto, I. C., and Mancilha, I. M. (1996), *Appl. Biochem. Biotechnol.* **57/58,** 423–430.
11. Gong, C. H., Glaypool, T. A., Mccracken, L. D., Maun, C. M., Ueng, P. P., and Tsao, G. T. (1983), *Biotechnol. Bioeng.* **25,** 85–102.
12. Hahn-Hagerdal, B., Jeppson, H., Skoog, K, and Prior, B. A. (1994), *Enzyme Microb. Technol.* **16,** 933–943.
13. Taylor, K. B., Beck, M. J., Huang, D. H., and Sakai, T. T. (1990), *J. Ind. Microbiol.* **6,** 29–41.
14. Bolen, P. L., and Detroy, R. W. (1985), *Biotechnol. Bioeng.* **27,** 302–307.

15. Alexander, N. J. (1985), *Biotechnol. Bioeng.* **27,** 1739–1744.
16. Silva, S. S., Vitolo, M., Pessoa-Junior, A., and Felipe, M. G. A. (1996), *J. Basic Microbiol.* **36(3),** 187–191.
17. Skoog, K. and Hahn-Hägerdal, B. (1988), *Enzyme Microb. Technol.* **10(2),** 66–80.
18. Du Preez, J. C., Van Driessel, B., and Prior, B. A. (1989), *Biotechnol. Lett.* **11(2),** 131–136.
19. Gírio, F. M., Peito, A. M., and Amaral-Collaço, M. T. (1989), *Appl. Microbiol. Biotechnol.* **32,** 199–204.
20. Prior, B. A., Kilian, S. G., and Du Preez, J. C. (1989), *Process Biochem.* 21–32.
21. Bicho, P. A., Runnals, P. L., Cunningham, J. D., and Lee, H. (1988), *Appl. Environ. Microbiol.* **54(1),** 50–54.
22. Onisch, O. and Suzuki, T. (1966), *Agricultural Biol. Chem.* **30(11),** 1139–1144.
23. Gong, C., Chen, L. F., and Tsao, G. T. (1981), *Biotechnol. Lett.* **3(3),** 125–130.
24. Meyral, V., Delgenes, J. P., Molleta, R., and Navarro, J. M. (1991), *Biotechnol. Lett.* **13(4),** 281–286.
25. Nolleau, V., Preziosi-Belloy, L., Delgenes, J. P., and Delgenes, J. M. (1993), *Curr. Microbiol.* **27,** 191–197.
26. Silva, S. S. and Afschar, A. S. (1994), *Bioprocess Eng.* **11,** 129–134.
27. Slininger, P. J., Bothast, R. J., Ladisch, M. R., and Okos, M. R. (1990), *Biotechnol. Bioeng.* **35,** 727–31.
28. Silva, S. S., Quesada-Chanto, A., and Vitolo, M. (1997), *Zeitschrift fur Naturforschung,* **52 C,** 359–363.
29. Silva, S. S., Roberto, I. C., Felipe, M. G. A., and Mancilha, I. M. (1996), *Process Biochem.* **31(6),** 549–553.
30. Furlan, S. A., Bouilloud, P., Strehaino, P., and Riba, J. P. (1991), *Biotechnol. Lett.* **40,** 203–206.
31. Felipe, M. G., Vitolo, M., and Mancilha, I. M. (1996), *Acta Biotechnol.* **16(1),** 73–79.
32. Silva, S. S., Ribeiro, J. D., Vitolo, M., and Felipe, M. G. A. (1997), *Appl. Biochem. Biotechnol.* **63–65,** 557–563.
33. Slininger, P. J., Bothast, R. J., and Okos, M. R. (1985), *Biotechnol. Lett.* **7(6),** 431–436.
34. Ojamo, H., Ylinen, L., and Linko, M. (1988), Process for the preparation of xylitol from xylose by cultivating *Candida guilliermondii* US Patent WO 88/05467.
35. Jeffries, T. W., Fady, J. H., and Lightfoot, E. (1985), *Biotechnol. Bioeng.* **27,** 171–176.
36. du Preez, J. C., Bosch, M., and Prior, B. (1986), *Enzyme Microb. Technol.* **8,** 360–364.
37. Kastner, J. R. and Roberts, R. S. (1990), *Biotechnol. Lett.* **12(1),** 57–60.
38. Jones, R. and Greenfield, P. F. (1984), *Process Biochem.* **4,** 48–60.
39. Roberto, I. C., Sato, S., and Mancilha, I. M. (1996), *J. Ind. Microbiol.* **16,** 348–350.
40. Molwitz, M., Silva, S. S., Ribeiro, J. D., Felipe, M. G. A., Prata, A. M. R., and Mancilha, I. M. (1996), *J. Bioscience* **51C,** 404–408.
41. Woods, M. A. and Millis, N. F. (1985), *Biotechnol. Lett.* **7(9),** 679–682.
42. Sirisansaneeyakul, S., Staniszewski, M., and Rizzi, M. (1995), *J. Fermentation Bioeng.* **80(6),** 565–570.
43. Felipe, M. G. A., Vitolo, M., Mancilha, I. M., and Silva, S. S. (1997), *J. Ind. Microbiol. Biotechnol.* **18,** 251–254.
44. Mes-Hartree, M. and Saddler, J. N. (1983), *Biotechnol. Lett.* **5(8),** 531–536.
45. Alves, L. A. (1997), MSc. Thesis. Faculdade de Engenharia Química de Lorena, Lorena, São Paulo, Brasil.
46. Silva, S. S., Queiroz, M. A., Felipe, M. G. A., Roberto, I. C., and Mancilha, I. M. (1991), *Symposium on Biotechnology for Fuels and Chemicals, Book of Abstracts.* NREL, Golden, CO p. 44.
47. Felipe, M. G. A., Vieira, D. C., Vitolo, M., Silva, S. S., Roberto, I. C., and Mancilha, I. M. (1995), *J. Basic. Microbiol.* **35(3),** 171–177.
48. Felipe, M. G. A., Vitolo, M., Mancilha, I. M., and Silva, S. S. (1997), *Biomass Bioenergy,* **13($\frac{1}{2}$)** 11–14.

Cloning and Sequence Analysis of the Poly(3-Hydroxyalkanoic Acid)-Synthesis Genes of *Pseudomonas acidophila*

Fusako Umeda, Yoshiharu Kitano, Yuki Murakami, Kiyohito Yagi,* Yoshiharu Miura, and Tadashi Mizoguchi

Faculty of Pharmaceutical Sciences, Osaka University, 1-6 Yamada-oka, Suita, Osaka 565 Japan

ABSTRACT

Pseudomonas acidophila can grow with CO_2 as a sole carbon source by the possession of a recombinant plasmid that clones genes that confer chemolithoautotrophic growth ability derived from the H_2-oxidizing bacterium *Alcaligenes hydrogenophilus*. H_2-oxidizing bacteria produce poly(3-hydroxybutyric acid) (PHB) from CO_2, but recombinant *P. acidophila* can produce the more useful biopolymer poly(3-hydroxyalkanoic acid) (PHA). In this study, the *pha* genes of *P. acidophila* were cloned and a sequence analysis was carried out. A gene library was constructed using the cosmid vector pVK102. A recombinant cosmid carrying the *pha* genes was selected by the complementation of a PHB-negative mutant of *Alcaligenes eutrophus* H16. The resulting recombinant cosmid pIK7 contained a 14.8-kb DNA insert. Subcloning was done, and the recombinant plasmid pEH74 was selected by hybridization with the *A. eutrophus* H16 *pha* genes. *Escherichia coli* possessing pEH74 produced PHB, indicating that pEH74 contained the *pha* genes of *P. acidophila*. The nucleotide sequences of the PHA-synthesis genes *phaA* (β-ketothiolase), *phaB* (acetoacetyl-CoA reductase), and *phaC* (PHA synthase) in pEH74 were determined. The homologies of *phaA*, *phaB*, and *phaC* between *P. acidophila* and *A. eutrophus* H16 were 64.7, 76.1, and 56.6%, respectively.

Index Entries: polyhydroxyalkanoate; polyhydroxybutyrate; nucleotide sequence; *Pseudomonas acidophila*.

INTRODUCTION

A large variety of bacteria accumulate polyhydroxyalkanoates (PHAs) in their cells under nutrient-limited conditions. PHAs are synthe-

---

*Author to whom all correspondence and reprint requests should be addressed.

sized as a carbon and energy storage compound or as a sink for reducing equivalents *(1,2)*. Since these polyesters are thermoplastic and biodegradable in natural environments, they are of interest to the chemical industry for the biotechnological production of PHAs for various applications *(3)*. Today, PHA production from renewable feedstock is becoming prevalent. In particular, CO_2, which is increasing in the atmosphere and causing the greenhouse effect, is seen as a promising carbon source for PHA production. It will be very useful if a usable product that harmonizes with the environment can be produced from an environment-damaging material.

H_2-oxidizing bacteria, which are autotrophs, grow with CO_2 as a sole carbon source using H_2-oxidizing energy. H_2-oxidizing bacteria are rapid growers and reach a high cell concentration under chemolithoautotrophic conditions *(4)*, characteristics that make these bacteria excellent candidates for use as CO_2 utilizers/fixers/consumers. A cluster of genes from the H_2-oxidizing bacterium *Alcaligenes hydrogenophilus* encoding its chemolithoautotrophic growth ability was cloned in vivo using a transferable R-plasmid, R68.45, as a cloning vector *(5)*. The ability to grow chemolithoautotrophically was transferred to a Gram-negative bacterium, *Pseudomonas acidophila*, using the recombinant plasmid pFUS *(6)*. *P. acidophila*, which accumulates PHA copolymers from low-carbon-number organic compounds such as formate and acetate, could grow under chemolithoautotrophic conditions as a consequence of the possession of pFUS, and synthesized PHA copolymers from CO_2 *(7)*. This result is considered very significant because PHA production from CO_2 by H_2-oxidizing bacteria had previously been restricted to the homopolymer poly(3-hydroxybutyric acid) (PHB), which is one of the PHAs *(8)*. PHA copolymers are worth producing because they can confer distinct properties on polyesters *(9)*.

In our previous study, a gene library of *P. acidophila* genomic DNA was constructed and the *pha* genes were cloned to obtain more information on PHA production from CO_2 *(7)*. In the present work, the nucleotide sequences of the *pha* genes were determined and the three structural genes of the PHA synthetic pathway *(phaA, phaB,* and *phaC)* were analyzed.

MATERIALS AND METHODS

Bacterial Strain and PHA Accumulation Conditions

PHA synthesis was carried out by a two-step cultivation of recombinant *Escherichia coli* JM109 carrying the *pha* genes from *P. acidophila* IFO13774. Recombinant cells were first grown in Luria-Bertani (LB) broth under air at 37°C overnight. The cells were then harvested and washed twice with sterilized water. To promote PHA synthesis, cells were inoculated into 300 mL of a nitrogen-free mineral salts medium *(10)* supplemented with a carbon source at 1% (w/v), at an initial cell concentration of 1 optical density at 660 nm. Cultivation was carried out aerobically at

37°C for 48 h. Ampicillin was added to the medium at a final concentration of 100 µg/mL for the maintenance of the plasmid.

Analysis of PHA

The polymer was isolated from lyophilized cells, and the composition of bacterial PHA was determined by NMR analysis as described previously *(7)*.

Transformation

For transformation, *E. coli* was cultivated in LB broth containing 20 mM each $MgCl_2$ and $MgSO_4$ at 37°C. Competent cells were prepared and transformed by the calcium chloride procedure *(11)*.

Nucleotide Sequence Analysis

DNA sequencing was performed by the dideoxy-chain-termination method of Sanger et al. *(12)* with alkaline-denatured double-stranded plasmid DNA *(13)* and with [α-^{32}P]dCTP using a Δ Tth polymerase DNA sequencing PRO kit (Toyobo, Japan) according to the manufacturer's protocol. Subclonings were performed by standard procedures *(11)*. Deletion mutants were prepared using a kilosequence deletion kit (Takara Shuzo, Japan).

Analysis of Nucleotide and Amino Acid Sequences

Nucleic acid sequence data and deduced amino acid sequences were analyzed with the Genetyx-Mac program (Software Development, Japan). Homology searches were performed using the Genbank (release 3/96) database.

RESULTS

Subcloning of *pha* Genes

In our previous study, a gene library of *P. acidophila* IFO-13774 genomic DNA was constructed using the cosmid vector pVK102. A recombinant cosmid, pIK7, containing a 14.8-kb *Hind*III fragment, was selected by heterologous complementation of a PHB-negative mutant, *A. eutrophus* PHB-4, which lacked active PHB synthase *(7)*. The 14.8-kb *Hind*III fragment was hybridized with a probe containing the *phbA, phbB,* and *phbC* genes from *A. eutrophus* H16. The 14.8-kb *Hind*III fragment was digested with *Eco*RI, and a 7.4-kb *Eco*R1-*Hind*III fragment that was hybridized with the probe was subcloned using plasmid pUC19 as a vector. The resulting recombinant plasmid pEH74 contained a sequence of three *Sal*I fragments, of 0.8,

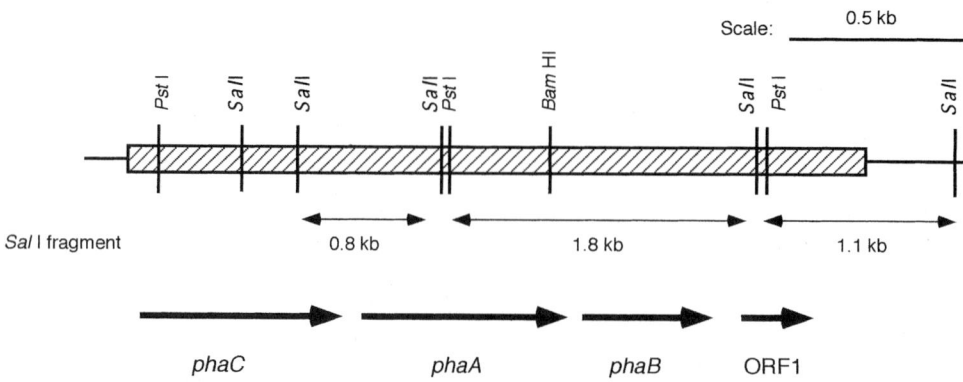

Fig. 1. Physical map of the *P. acidophila pha* gene locus and adjacent region in pEH74. The nucleotide sequence of the shaded region is given in Fig. 2. The locations and directions of the β-ketothiolase *(phaA)*, acetoacetyl-coenzymeA reductase *(phaB)*, and PHA polymerase genes *(phaC)* are indicated by arrows.

1.8, and 1.1-kb, which were hybridized with the probe (Fig. 1). For the sequencing of this region, subclonings of two of the *Sal*I fragments (those of 0.8 and 1.1-kb) and two *Bam*HI-*Sal*I fragments derived from the 1.8-kb *Sal*I fragment digested with *Bam*HI were performed using pUC19 as a vector. If necessary, deletion mutants were constructed.

Nucleotide Sequence Analysis of *pha* Genes and Their Flanking Regions

A 4169-bp region of the locus of the *P. acidophila pha* genes was sequenced using the subfragments mentioned above (Fig. 2). Four open reading frames (ORFs) were found and were identified by homology searching. First ORF: the 1152-bp structural gene of *P. acidophila* PHA synthase *(phaC)* mapped from position 64 to 1216. It encoded for a protein of 384 amino acids, and the M_r of the putative translational product was 42,779. Second ORF: the 1152-bp structural gene of *P. acidophila* β-ketothiolase *(phaA)* running from position 1350 to 2502. It encoded for a protein of 384 amino acids, and the M_r of the putative translational product was 40,200. Third ORF: the 726-bp structural gene of *P. acidophila* acetoacetyl-CoA reductase *(phaB)* starting at position 2601 and ending at 3327. It encoded for a protein of 242 amino acids, and the M_r of the putative translational product was 25,860. Fourth ORF (ORF1): this began at position 3443 and ended at 3825. It encoded for a protein of 144 amino acids, and the M_r of the putative translational product was 16,634. ORF1 identified downstream of *phaB* was compared with known DNA sequences to establish its function. Significant DNA sequence homology to known DNA sequences was not found. The overall GC content of the 4169-bp region was 63.1% (mol/mol). The four

Fig. 2. Nucleotide sequence of the *P. acidophila pha* gene locus and the adjacent region. Amino acids deduced from the nucleotide sequence of the tentative genes are specified by the standard one-letter abbreviations. Putative ribosome binding sites (Shine-Dalgarno sequences, S/D) are underlined.

ORFs were preceded by tentative ribosome-binding sites upstream of the respective ATG start codons. These data show that the three enzymes of the *P. acidophila* PHA synthetic pathway are encoded by the three genes organized as *phaC-phaA-phaB*, as illustrated in Fig. 1.

Fig. 2. *(continued)*

Comparison of PHA Polymerase, β-Ketothiolase, Acetoacetyl-CoA Reductase, and ORF1 Product

The deduced amino acid sequences of the *phaC*, *phaA*, *phaB*, and ORF1 genes from *P. acidophila* were compared with those from other microorganisms.

In an alignment of the sequences, the *phaC* product from *P. acidophila* showed 20.5 to 56.6% homology with the PHA polymerases from *A. eutro*-

phus *(14)*, *Methylobacterium extorquens (15)*, *Rhodococcus ruber (16)*, *Pseudomonas oleovorans* (1 and 2) *(17)*, *Pseudomonas aeruginosa* (1 and 2) *(18)*, *Chromatium vinosum (19)*, and *Thiocystis violacea (20)* (Fig. 3).

The deduced amino acid sequence of the *phaA* gene from *P. acidophila* exhibited 64.7, 47.3, and 51.9% homology with the β-ketothiolases from *A. eutrophus (21)*, *Zoogloea ramigera (22)*, and *C. vinosum (19)*, respectively (Fig. 4).

The deduced amino acid sequence of the *phaB* gene from *P. acidophila* was 76.1, 46.3, and 48.8% homologous with the acetoacetyl-CoA reductases from *A. eutrophus (21)*, *Z. ramigera (22)*, and *C. vinosum (19)*, respectively (Fig. 5).

The deduced amino acid sequence of ORF1 of *P. acidophila* was compared with those of ORF4 of *C. vinosum (19)* and *T. violacea (20)* (Fig. 6). The homology was 49.3 and 53.4% with the ORF4 sequences from *C. vinosum* and *T. violacea*, respectively.

Heterologous Expression of *pha* Genes in *E. coli*

E. coli JM109 was transformed with pEH74, which contained *phaA*, *phaB*, and *phaC* from *P. acidophila*. Polymer accumulation from various carbon sources was tested in the recombinant *E. coli* JM109 carrying pEH74. The *pha* genes from *P. acidophila* were expressed in *E. coli* and conferred on it the ability to synthesize polymer. The polymer content of the cells in *E. coli* varied between 2.9 and 62.1% of the cellular dry mass (Table 1). The polymer type produced by *E. coli* was PHB homopolymer with all carbon sources.

DISCUSSION

The recombinant cosmid pIK7 selected by the complementation experiment in our previous study was confirmed to contain the three structural genes of PHA synthesis. The deduced amino acid sequences of the *P. acidophila pha* genes were highly homologous with those from *A. eutrophus*: 56.6% for PHA polymerase, 64.7% for β-ketothiolase, and 76.1% for acetoacetyl-CoA reductase (Figs. 3–5).

It has been proposed that the mechanism for PHB polymerase involves two partial reactions: the formation of an acyl-S-enzyme intermediate as a first step followed by the transesterification of a primer acceptor *(23)*. Two cysteine residues conserved in PHA polymerase appear to be important in acyl-S-enzyme intermediate formation and transesterification *(14)*. In this study, *P. acidophila* PHA polymerase contained three cysteine residues, of which Cys165 and Cys297 were appropriate cysteine residues. Two highly conserved segments that seemed to be important for the polymerization reaction *(24)* and a sequence, NXXGXCXGG, which incorporates the lipase consensus sequence (lipase-box), were identified in *P. acidophila* PHA polymerase.

Fig. 3. Alignment of various PHA polymerases. The PHA polymerases are as follows: A. e., *A. eutrophus*; M. e., *M. extorquens*; R.r., *R. ruber*; P.o.1 and P.o.2, *P. oleovorans*; P.a.1 and P.a.2, *P. aeruginosa*; C.v., *C. vinosum*; T.v., *T. violacea*; P.a., *P. acidophila*. Dots indicate amino acids identical to the *A. eutrophus* sequence; dashes signify gaps introduced to maximize the alignment of the sequences; shading shows identical residues present in all the sequences. The cysteine residues at positions 319 and 459 in the *A. eutrophus* sequence, which have been proposed as candidates for the residues involved in the formation of an acyl-S-enzyme intermediate and in transesterification, were marked by asterisks. Lipase-box like sequences are indicated by a double line above the sequences. Highly conserved segments are boxed. Cysteine residues that have been proposed to be involved in the polymerization reaction are marked by asterisks.

Poly(3-Hydroxyalkanoic Acid)-Synthesis Genes 349

```
P.a.   1   MT-DVVIVSAG--PVGKFGGT-CEDRGAGAGPMVIRAVLERAGVKPEQVSEVIPGGRL----GQNPAPQSLIKAGLPSAVPGMTINKVCG
A.e.   1   ..-......ARTA......SLAKIPAPEL.AV..K.A...............M.QV.TAGS.....R.AA......AM..A......
Z.r.   1   .STPSIVIASARTA...S.N.AFANTPAHEL.AT...S........AAGE.N...L.QV.PAGE.....R.AAM...V.QEATAWGM.QL.
C.V.   1   .SENI...D..RSAI.T...SLSSSLSATEI.TA.LKGL.A.T.LA....ID...L.QV.TAGV....R.TTLH....HS..A......

P.a.  83   SGLKAVMLARTAIIAGEADIVIAGGQENMSAARRA-AGSRNGFRMGDSKLVDTMIVDGLWDVYNQYHMGITAENVAKEYGITREEQDAFA
A.e.  90   .........AN..M..D.E..V.........PHVLP...D.......A..............................A..E..
Z.r.  91   ...R..A.GMQQ.AT.D.S.IV...M.S..M.PHC-.H-LA.VK...F.MI....K...T.AFYG...T.......QWQLS.D....
C.V.  91   ......H..MQ..AC.D........S..QSSHVLPR..D.Q....WSMK..........AF.N....T.....I.QK..F...Q....

P.a. 172   RCRRTRRKGAE-GGRFNDEIVPVGDPAEEGRAAAVR-DRRVRTRRDGGRAGGLKPAFAKDGTVTAANESGLNDGLNTSDDVSYSPEATTD
A.e. 180   VGSQNKAEA.QKA.K.DE.....LI.QRK.DPV.FKT.EF..QGATLDSMS......D.A......A.....AAAVVM.AAKAKELG
Z.r. 179   VASQNKAEA.QKD...K......FIVKGRK.DITV-DA.EYI.HGATLDSMAK.R...D.E.....G.A.....AAAALLM.EAEASRRG
C.V. 181   AASQQKTEA.QKA....Q...I.IEI.QRK.DPKVFDA.EFP.HGTTAESL.K.R....S.D.S...G.A..I....AAMVVVMKE.KAKELG

P.a. 260   -THTAG-SSY-RT-LDPSVMGMGPVPASRSGAWRSGWTPGDWTPSDLDLMEINES-SRQALTVHKQMGWDTSKVNVNDGAIAIGHPIGAS
A.e. 270   L.PL.TIK..ANAGV..K.........K-----.ALSRAE...Q.........AFAA..A..Q..............G..........
Z.r. 268   IQPLGRIV.WATVGV..K....T...IPA--..--RK-ALERAG.KIG....V.A..AFAA..CA.N.DL...P.I....G..........
C.V. 271   LKPM.RLVAFASAGV...AI..T...IPA--..--TK-CLEKAG...A....I.A..AFAA..MS.NQD....L.....G..........

P.a. 345   GCRILVTLLHEMVKRDGTRGMASLCIGGGMGVALAVERPS      384
A.e. 355   ...........KR..AKK.L....Ⅰ...........K-         393
Z.r. 353   .A...N...F..KR.GARK.L.T.Ⅰ.......MCI.SL-        391
C.V. 356   .A.V.....Y..Q...AKK.L.T.Ⅰ...Q........M-       394
```

Fig. 4. Alignment of various β-ketothiolases. The β-ketothiolases are as follows: P.a., *P. acidophila*; A. e., *A. eutrophus*, Z. r., *Z. ramigera* C. v., *C. vinosum*. Dots indicate amino acids identical to the *P. acidophila* sequence; dashes signify gaps introduced to maximize the alignment of the sequences. The cysteine residue at position 81, which functions as the first step in the thiolase reaction, and that at position 369, which functions as the active-site base in deprotonation in the condensation reaction, are shaded.

```
P.a.   1   MSQRIAYVTGGMGGIGTSICQRLSKDGFRVVAG-DAH-VRRLASVNARGEKGRFDFIASEGNVADWDSTKEAFDKVKAEVGEIDVLVNNA
A.e.   1   .T......A.........A..........CGPNSP..EKWLEQQKAL.-................T.......S.....V...I...
Z.r.   1   .-S.V.L.....SR....AA.SIA.KAA..YK.A.S-Y.GNDDAAKPFK-A-..T-G--IAVYKWD.SSYEACV.GIA...E.DL.P.......
C.v.   1   .-A...L......I......T..A...CT....NCHPSEAAAAEEWKQARAAEG..IAVFTAD.SSF.DSARMVREITEQ..P..I...C.

P.a.  89   GITRDVVFRKMTREDWTAVIDTNLTSLFNVTKQVIDGMVERGWGDHQHLVGERPE-GQFGQTNYSTAKAGIHGFTMSVAQEVATKGVTVN
A.e.  90   ..............A..D..................AD....RIVNISSVNGQK..............L....AL...........
Z.r.  84   ...K.AM.H...PDQ.N...N....G...M.HP.WS..RD.SF.RIVNISSINGQK...M..A....DL...KAL..G.A..I...
C.v.  90   .....KT.K..EQAH.E...NV..N.V....R..W....L...F.RIINISSVNGQR.....A...A.....M.....AL...G.S......

P.a. 178   TVSPGYIGTDMVKAIRPDVL-LKIVATIPVRRLGAPEEIGSIVA-CSSNDSA-SDGADFSLNGGLHMG       242
A.e. 180   ........A........Q...-D........K....L....A..C.WL..EE.GF.T............   246
Z.r. 174   AIC.......E..R..PEK..NER.IPQ...G....E.D..AR..VFLA.DEAGFIT.STI.A...QFFV     241
C.v. 180   .I....VE.A.TL.MND..-RNS.ISG..M..MAQ.N...AAAI.FLAGDE.GYMT...NLPV....F.H     246
```

Fig. 5. Alignment of various acetoacetyl-coenzymeA reductases. The acetoacetyl-coenzymeA reductases are as follows: P. a., *P. acidophila*; A. e., *A. eutrophus*; Z. r., *Z. ramigera*; C. v., *C. vinosum*. Dots indicate amino acids identical to the *P. acidophila* sequence; dashes signify gaps introduced to maximize the alignment of the sequences. The NAD(P) binding region is shaded.

The catalytically essential cysteine residues discussed for the *Z. ramigera* enzyme (19) were conserved at Cys81 and Cys369 in *P. acidophila* β-ketothiolase. The former would participate in the formation of the acyl-enzyme intermediate as the first step in the β-ketothiolase reaction, whereas the latter would function as an active-site base in deprotonation in the condensation reaction.

The amino acid sequence TGGXXG has been found in acetoacetyl-CoA reductases, and is thought to participate in binding the ADP moiety

Table 1
Heterologous Expression of *Pha* Genes from *P. acidophila* in *E. coli*

Carbon source	Polymer weight (mg/liter)	Polymer content (% of dry wt)
Formate	76	23.6
Acetate	67	20.5
Propionate	18	5.4
Malate	18	4.9
Arabinose	46	14.2
Succinate	132	29.7
Gluconate	83	20.7
Glucose	184	41.9
Heptanoate	10	2.9
Octanoate	59	19.3
Dodecanoate	205	62.1

E. coli JM109 carrying recombinant plasmid pEH74 was used. For the polymer accumulation conditions, see MATERIALS AND METHODS.

```
P.a.   1   MTTTKKTAERLIKKYPNRRLYDTETSTYITLTDVKQLVLEQEDFKVVDAKSSEDLTRSIL
C.v.   1   .----N-S..I.............V.R....A..RD..MSGQP.R.L.SANDS.I.....
T.v.   1   .----N-SD.I.............V.R....A..RN..MDCTS.....TANES.I.....

P.a.  61   LQIILEEESGGVPMFSSSMLSQIIRFYGHAMLGMMGTYLEKNIQAFI-DIQ--NK-LADQ
C.v.  56   ...M....T..Q.L..AN..A.......GTL..TFAR...SSLDL.AKQQ.EVT.A.T.N
T.v.  56   ...M....T..E.L..A...A.......GTL...FAR...SSLDL.AKQQ.DMT.T.G.N

P.a. 117   S-KNLYENNAMNPEIWS--Q--FM---NMQ-AP-MMQA    144
C.v. 116   PFGTVTRLTQK.V...ADL.DEL.RAAGFPV..RKKKE    153
T.v. 116   PFEAMTRMTQK.V...ADM.EE------F---------    138
```

Fig. 6. Alignment of the deduced amino acid sequences of ORF1 of *P. acidophila* and ORF4 from *C. vinosum* and *T. violacea*. Dots indicate amino acids identical to *P. acidophila*; dashes signify gaps introduced to maximize the alignment of the sequences. Highly conserved segments are boxed.

of NAD *(25)*. This sequence was found near the N-terminus of *P. acidophila* acetoacetyl-CoA reductase.

ORF1-PHB-binding proteins are divided into 4 groups *(26)*: PHA polymerase, intracellular PHA depolymerase, a protein called phasin that stabilizes the structure of PHA, and other proteins. The surface of the PHA granule is considered to be covered with a phospholipid, but the amount is not sufficient to cover the surface perfectly. Phasin seems to partially complement the phospholipid deficiency. The deduced amino acid sequence of ORF1 from *P. acidophila* was compared with the ORF4 sequences from *C. vinosum* and *T. violacea*, the functions of which have not been clarified. They showed high homology, and a highly conserved segment was observed in all the sequences. Further studies are necessary to determine whether these ORFs are phasin-encoding ORFs.

In *A. eutrophus*, three *pha* genes are organized in one operon *(phaC-phaA-phaB) (14)*. The transcription start point is mapped 307-bp upstream from the translational initiation point of the *phbC* gene *(27)*. The −35 region (TTGACA) and the −10 region (AACAAT) identified directly upstream of the transcription start site of *phbC* were identical (TTGACA) or very similar (TATAAT) to the corresponding sequences of the *E. coli* σ^{70} consensus promoter sequences *(28)*. The order of the *pha* genes in *P. acidophila* was *phaC-phaA-phaB*, as it is in *A. eutrophus*. No promoter-like sequence was detected in the 4169-bp of the *P. acidophila pha* locus. The length of the upstream region of *phbC* analyzed was approx 50-bp. It might be necessary to conduct further sequencing to detect the promoter of the *pha* genes. In *A. eutrophus*, the expression of all three genes of the pathway *(phaC-phaA-phaB)* in *E. coli* results in the accumulation of significant levels of PHB in this bacterium. The expression of *phbC* alone in *E. coli* produces neither PHB nor significant levels of PHB polymerase activity *(14)*. In this study, *E. coli* carrying pEH74 produced considerable amounts of PHB (2.6–62.1%) (Table 1). It is probable that the 7.4-kb *Eco*RI-*Hin*dIII fragment inserted in pEH74 contains the promoter that works for the effective expression of the three *pha* genes in *E. coli*.

P. acidophila has novel characteristics that enable it to produce PHA copolymers from CO_2. However, the deduced amino acid sequences of PHA synthetic enzymes were not specific to the bacterium, but similar to those reported for other PHB-producing bacteria. In heterologous expression in *E. coli* and *A. eutrophus (7)*, the products were PHB homopolymer. Future studies will show the factors affecting the polymer type.

ACKNOWLEDGMENT

This work was sponsored by New Energy and Industrial Technology Development Organization (NEDO)/Research Institute of Innovative Technology for the Earth(RITE).

REFERENCES

1. Anderson, A. J. and Dawes, E. A. (1990), *Microbiol. Rev.* **54**, 450–472.
2. Steinbüchel, A. (1991), *Biomaterials*, Byrom, D., ed., MacMillan, Basingstoke, UK, pp. 123–213.
3. Holmes, P. A. (1985), *Phys. Technol.* **16**, 32–36.
4. Repaske, R. and Mayer, R. (1976), *Appl. Environ. Microbiol.* **32**, 592–597.
5. Umeda, F., Tanaka, N., Kimura, N., Nishie, H., Yagi, K., and Miura, Y. (1991), *J. Ferment. Bioeng.* **71**, 379–383.
6. Miura, Y. and Umeda, F. (1994), *Yakugaku Zasshi* **114**, 63–72.
7. Yagi, K., Miyawaki, I., Kayashita, A., Kondo, M., Kitano, Y., Murakami, Y., Maeda, I., Umeda, F., Miura, Y., Kawase, M., and Mizoguchi, T. (1996), *Appl. Environ. Microbiol.* **62**, 1004–1007.
8. Ishizaki, A. and Tanaka, K. (1991), *J. Ferment. Bioeng.* **71**, 254–257.
9. Doi, Y., Tamaki, A., Kunioka, M., and Doga, K. (1988), *Appl. Microbiol. Biotechnol.* **28**, 330–334.
10. Ohi, K., Takada, N., Komemushi, S., Okazaki, M., and Miura, Y. (1979), *J. Gen. Appl. Microbiol.* **25**, 53–58.
11. Maniatis, T., Fritsch, E. F., and Sambrook, J. (1982), *Molecular Cloning: A Laboratory Manual*. Cold Spring Harbor Laboratory Press, Cold Spring Harbor, NY.
12. Sanger, F., Nicklen, S., and Coulson, A. R. (1977), *Proc. Natl. Acad. Sci. USA* **74**, 5463–5467.
13. Chen, E. J. and Seeburg, P. H. (1985), *DNA* **4**, 165–170.
14. Peoples, O. P. and Sinskey, A. J. (1989), *J. Biol. Chem.* **264**, 15,298–15,303.
15. Valentin, H. and Steinbüchel, A. (1993), *Appl. Microbiol. Biotechnol.* **39**, 309–317.
16. Pieper U. and Steinbüchel, A. (1992), *FEMS Microbiol. Lett.* **96**, 73–80.
17. Huisman, G. W., Wonink, E., Meima, R., Kazemier, B., Terpstra, P., and Witholt, B. (1991), *J. Biol. Chem.* **266**, 2191–2198.
18. Timm, A. and Steinbüchel, A. (1992), *Eur. J. Biochem.* **209**, 15–30.
19. Liebergesell M. and Steinbüchel, A. (1992), *J. Biochem.* **209**, 135–150.
20. Liebergesell M. and Steinbüchel, A. (1993), *Appl. Microbiol. Biotechnol.* **38**, 493–501.
21. Peoples, O. P. and Sinskey, A. J. (1989), *J. Biol. Chem.* **264**, 15,293–15,297.
22. Peoples O. P. and Sinskey, A. J. (1989), *Mol. Microbiol.* **3**, 349–357.
23. Griebel, R. and Merrick, J. M. (1971), *J. Bacteriol.* **108**, 782–789.
24. Persson, B., B.-Olivecrona, G., Enerback, S., Olivecrona, T., and Jornvall. H. (1989), *Eur. J. Biochem.* **179**, 39–45.
25. Ploux, O., Masamune, S., and Walsh, C. T. (1988), *Eur. J. Biochem.* **174**, 177–182.
26. Steinbüchel, A., Aerts, K., Babel, W., Föllner, C., Liebergesell, M., Madkour, M. H., Mayer, F., Pieper-Furst, U., Pries, A., Valentin, H. E., and Wieczorek, R. (1995), *Can. J. Microbiol.* **41**, 94–105.
27. Schubert, P., Krüger, N., and Steinbüchel, A. (1991), *J. Bacteriol.* **173**, 168–175.
28. Hawley, D. K. and McClure, W. R. (1988), *Nucleic Acids Res.* **11**, 2237–2255.

Continuous Culture Studies of Xylose-Fermenting *Zymomonas mobilis*

Hugh G. Lawford,[1] Joyce D. Rousseau,[1] Ali Mohagheghi,[2] and James D. McMillan*,[2]

[1]*Bioengineering Laboratory, Department of Biochemistry, University of Toronto, Toronto, Ontario, Canada M5S 1A8,* [2]*Biotechnology Center for Fuels and Chemicals, National Renewable Energy Laboratory 1617 Cole Boulevard, Golden, CO 80401*

ABSTRACT

The continuous cofermentation performance of xylose-fermenting *Zymomonas mobilis* at 30°C and pH 5.5 was characterized using a pure-sugar feed solution that contained 8 g/L glucose and 40 g/L xylose. Successful chemostat start up resulted in complete utilization of glucose and greater than 85% utilization of xylose, but was only reproducibly achieved using initial dilution rates at or less than 0.04/h; once initiated, cofermentation could be maintained at dilution rates of 0.04 to 0.10/h. Whereas xylose and cell-mass concentrations increased gradually with increasing dilution rate, ethanol concentrations and ethanol yields on available sugars remained approximately constant at 20–22 g/L and 80–90% of theoretical, respectively. Volumetric and specific ethanol productivities increased linearly with increasing dilution rate, rising from approx 1.0 each (g/L/h or g/g/h) at a dilution rate of 0.04/h to approx 2.0 each (g/L/h or g/g/h) at a dilution rate of 0.10/h. Similarly, specific sugar-utilization rates increased from approx 2.0 g/g/h at dilution rate 0.04/h to approx 3.5 g/g/h at dilution rate 0.10/h. The estimated values of 0.042 g/g for the maximum *Z. mobilis* cell-mass yield on substrate and 1.13 g/g/h for the minimum specific substrate utilization rate required for cellular maintenance energy are within the range of values reported in the literature. Results are also presented which suggest that long-term adaptation in continuous culture is a powerful technique for developing strains with higher tolerance to inhibitory hemicellulose hydrolyzates.

Index Entries: Adaptation; continuous cofermentation; ethanol; xylose; *Zymomonas mobilis*.

---

* Author to whom all correspondence and reprint requests should be addressed.

INTRODUCTION

Fermentation fuel ethanol is currently produced almost exclusively from either sucrose (sugarcane or beets) or starchy feedstocks (principally corn) using *Saccharomyces* yeast *(1,2)* and the largest plants operate in a continuous mode using a cascade of several fermentors in series *(3)*. Lignocellulosic biomass (energy crops) and wastes (forest, agricultural, and municipal) represent a vast potential alternative resource for ethanol production *(4)*. However, there are a number of features that make lignocellulosic feedstocks incompatible with current bioconversion process technologies. According to Lynd et al. *(5)*, it is the lack of appreciation of the enormous resource potential as well as of the differences in processing technologies that has fostered misconceptions about the potential for cellulosic ethanol and hindered the development of emerging biomass-to-ethanol process technologies.

A biomass-to-ethanol process must achieve efficient conversion of both the cellulose and hemicellulose components to be economical *(6,7)*. Unlike starch, the cellulose in woody and herbaceous biomass feedstocks is highly resistant to enzymatic hydrolysis unless the biomass is first pretreated *(8,9)*. Dilute acid-catalyzed thermochemical hemicellulose hydrolysis is an efficient and cost-effective pretreatment method *(10–12)*. The prehydrolyzate liquor (hydrolyzate) fraction is composed of a mixture of monomeric sugars, acetic acid, furfural, and a variety of lignin-derived phenolic compounds known to be inhibitory to fermentative microorganisms *(13–15)*. The sugar component is principally the pentose sugar D-xylose in the case of hydrolysis of hardwood or herbaceous hemicellulose. There have been a variety of procedures proposed for detoxifying biomass hydrolyzates (for review, *see* ref. *15*), but their efficacy is variable and their economic impact remains largely unexplored.

Yeasts presently used in sucrose and starch-based ethanol fermentations ferment only hexose sugars to ethanol. The inability of these yeasts to utilize pentose sugars is driving considerable research to develop alternative biocatalysts that exhibit performance characteristics better suited to conversion of lignocellulosic biomass feedstocks. The search for efficient xylose-utilizing ethanologens includes selection of natural isolates and various genetic manipulations of yeast and bacteria (for reviews, *see* refs. *14–16*).

Continuous fermentation systems offer several potential advantages such as increased productivity through cell-retention or cell-recycling configurations *(17)*. The engineering strategy for increasing overall productivity and reducing manufacturing costs by switching from batch to continuous operation is similar for both starch- and cellulose-based ethanol production processes (for review, *see* ref. *17*). However, the biocatalysts (both hydrolytic enzymes and fermentative microorganisms) needed to achieve high levels of conversion are quite feedstock specific *(6,18–20)*.

Thus, whereas continuous large-scale production of ethanol from starch is a relatively mature technology, many challenges remain to achieving continuous high-yield production of ethanol from lignocellulosic feedstocks. Many researchers believe that the greatest potential for improvement of biomass-to-ethanol processes is to overcome the biological constraints associated with the conversion biocatalysts *(5,21,22).*

Techno-economic analyses of projected commercial-scale biomass-to-ethanol processes indicate that major cost reductions can be achieved through technological advances in both process configuration and biocatalyst performance. Reflecting this, over the past decade, technical breakthroughs have led the projected cost of fuel ethanol to decrease progressively from $2.66/gal to approx $1.06–1.22/gal with current technology *(6,7,14,23).* Provided significant efforts to advance biomass-to-ethanol process technology continue, projected production costs are expected to continue to fall to the point where fuel ethanol becomes cost competitive with petroleum-derived gasoline *(5,22).* It is proposed that the significant additional processing cost reductions required to achieve this goal can be achieved through a combination of process consolidation *(22),* economies of scale *(23),* and improved energy utilization *(21,22).* Advances in process consolidation would involve reducing the number of bioreactors in a design, for example, by moving from a process based on sequential hydrolysis and fermentation to one using simultaneous hydrolysis and fermentation *(18,19).* Similarly, advanced designs would employ continuous rather than batch operation. The ultimate biocatalyst for an advanced process would be an ethanologenic microorganism capable of synthesizing cellulolytic enzymes in a process called direct microbial conversion or DMC *(22,24).*

One of several biomass-to-ethanol processes currently under investigation at the National Renewable Energy Laboratory (NREL) is the simultaneous saccharification and cofermentation (SSCF) process that is based on the use of genetically engineered *Zymomonas mobilis* transformed with NREL's proprietary xylose assimilation and utilization plasmid *(25–29).* Xylose-utilizing recombinant *Z. mobilis* has been shown to exhibit excellent cofermentation performance with respect to both yield and productivity in laboratory scale pH-controlled bioreactors using a synthetic hardwood dilute-acid hydrolyzate medium with corn-steep liquor as the sole nutritional supplement *(30).* The results of successful batch SSCF trials with recombinant *Z. mobilis* also have been reported *(27,28).* More recently, we reported the effect of acetic acid on the performance of recombinant *Z. mobilis* in both batch *(30)* and glucose fed-batch systems *(31).*

This paper reports initial results from an ongoing study of continuous conversion of hardwood hemicellulose hydrolyzate to ethanol. This study focuses on the fermentation performance of the xylose-utilizing recombinant *Z. mobilis* in a pH-controlled continuous-flow bioreactor (chemostat) using both combinations of pure sugars and real hardwood hydrolyzates.

Our objective was to exploit the selective pressure provided by the continuous growth environment of a chemostat to achieve strain improvement through adaptation resulting from the long-term exposure of the recombinant microorganism to incremental increases in the level of inhibitory hydrolyzate in the feed medium. Before initiating the long-term strain adaptation effort, it was necessary to characterize the fermentation performance of the recombinant strain in chemostat culture in order to identify an appropriate dilution rate for continuous feeding of dilute-acid hydrolyzate.

MATERIALS AND METHODS

Microorganism

Zymomonas mobilis 39676:pZB4L, i.e., *Z. mobilis* host strain ATCC 39676 transformed with a derivative of the pZB5 plasmid conferring xylose assimilation and fermentation capability, as reported by Zhang et al. *(26)*. Cryovials of frozen concentrated stock culture were maintained in RM medium (10 g/L yeast extract and 2 g/L KH_2PO_4) supplemented 10 mg/L tetracycline and 10% w/w glycerol at $-70°C$ *(30)*.

Inoculum Cultivation

For the pure-sugar study, 0.25 mL of thawed frozen stock culture was inoculated into a 125-mL culture flask containing 100 mL of sterile RM supplemented with 25 g/L xylose, 25 g/L glucose, and 10 mg/L tetracycline antibiotic. For the hydrolyzate adaptation study, the medium for seed culture production was 10% (v/v) conditioned yellow poplar hemicellulose hydrolyzate (hydrolyzate; *see* Preparation of Conditioned Hydrolyzate) supplemented with 10 g/L glucose, 15 mL/L clarified corn-steep liquor (CSL) (GPC International, Muscatine, IA), and 10 mg/L tetracycline; initial medium pH was 5.8. Inoculum cultures were incubated overnight at a temperature of 30°C and an agitation rate of 150 rpm.

Chemostat Studies

Experiments were carried out in 500-mL MultiGen fermentors (New Brunswick Scientific, Edison, NJ) using a working volume of 300 mL. Pure-sugar experiments were performed at a temperature of 30°C, an agitation rate of 150 rpm and at pH 5.5 unless stated otherwise; pH was controlled through the automatic addition of 2 *N* KOH.

Pure-Sugar Chemostat Study

Fermentation was initiated batchwise by directly inoculating at a level of 10% (v/v) with overnight grown seed culture; this corresponds to an initial cell concentration of approx 0.10 g dry cell mass per liter (g DCM/L). The batch-fermentation medium and the subsequent chemostat-feed medium contained 8 g/L glucose, 40 g/L xylose, 10 mL/L clarified CSL, and 10 mg/L tetracycline. Continuous feeding was begun at a dilution rate of 0.03/h after the xylose concentration had decreased to between 15 and 20 g/L. The fermentor was sampled daily for offline analysis of cell mass and soluble metabolite concentrations, particularly for glucose, xylose, and ethanol.

Our objective was to characterize an effective operating region for performing chemostat adaptation of the culture to inhibitory hydrolyzate. We were interested in determining the approximate range of dilution rates over which effective utilization of both glucose and xylose could be achieved, i.e., conditions in which complete utilization of glucose and greater than 75% utilization of xylose occurred. Our approach was to carry out chemostat operation at a particular dilution rate until "pseudo steady-state" performance that met our criteria for effective cofermentation was achieved, i.e., sequential daily samples had approximately constant ethanol concentrations as well as constant residual xylose concentrations below 10 g/L (and constant residual glucose concentrations near zero). Once these criteria were achieved, the dilution rate was increased. The dilution rate was then maintained at this new higher value until the next steady state that met our criteria was achieved, whereupon the dilution rate was again increased. This procedure was repeated until a dilution rate of 0.10/h was reached. Efforts to operate the chemostat at a dilution rate at and above 0.11/h were ultimately abandoned because at a dilution rate of 0.11/h, the strain began to flocculate and wall growth increased dramatically; both of these factors compromised the stability of the chemostat as well as our ability to interpret the experimental results.

When chemostat operation was attempted at a dilution rate of 0.11/h, we observed increased variability in the apparent steady state with respect to culture turbidity and xylose and ethanol concentrations. We feared that wash out of the culture might occur and therefore reduced the dilution rate to 0.04/h. We also replaced the fermentor vessel with a clean vessel to reduce the possibility of wall growth confounding the results. The experiment was terminated after re-establishing a steady state at a dilution rate of 0.04/h.

Hydrolyzate-Adaptation Chemostat Study

The fermentation was started up similarly to the pure-sugar fermentation described above. In this experiment, however, the pH was controlled at 5.8 and the composition of the initial batch and feed media consisted of 10% (v/v) conditioned hydrolyzate containing 15 mL/L clarified CSL, 10 mg/L tetracycline, and supplemented with sugars to achieve final concentrations of 8 g/L glucose and 40 g/L xylose. As in the pure-sugar experiment, the chemostat was sampled daily for analysis of glucose, xylose, and ethanol concentrations; determination of cell-mass concentration was not attempted in this experiment because of the high turbidity of the hydrolyzate-based medium. After 7 d, the dilution rate was increased from 0.03 to 0.04/h. After 15 d, the concentration of CSL in the feed medium was raised to 20 mL/L to reduce the possibility of nutrient limitations under the stressed adaptation conditions.

Our objective in this experiment was to determine if a chemostat fed with increasing concentrations of hydrolyzate could be used to develop a culture capable of tolerating higher levels of inhibitory hydrolyzate. The dilution rate was fixed at 0.04/h after 7 d, since the results of the pure-sugar study showed that this dilution rate enabled high-yield cofermentation of glucose and xylose. Our approach was otherwise similar to that used in the pure-sugar study. The chemostat was operated at a particular feed hydrolyzate concentration (initially 10% v/v) until sequential samples had approximately constant ethanol concentrations as well as constant residual xylose concentrations below approx 12.5 g/L (and residual glucose concentrations near zero). Once stable performance meeting these criteria was achieved, the concentration of hydrolyzate in the feed was increased by 5% v/v. Regardless of the hydrolyzate concentration, the feed hydrolyzate solution was supplemented to achieve final concentrations of glucose and xylose of 8 g/L and 40 g/L, respectively. Chemostat operation was maintained at this new higher hydrolyzate concentration until stable performance meeting our performance criteria was once again achieved, whereupon the concentration of hydrolyzate in the feed was again increased by 5% v/v. This procedure was repeated until a hydrolyzate concentration of 35% v/v was reached.

Preparation of Conditioned Hydrolyzate

Acidic-hydrolyzate liquor was obtained by dilute sulfuric acid treatment of yellow poplar sawdust in NREL's pilot scale Sunds hydrolyzer pretreatment reactor (32). The pH of the as-received xylose-rich hemicellulose hydrolyzate liquor was in the range of 1.0–2.0. Conditioning was required to reduce the toxicity of the hydrolyzate to levels that enabled fermentation to occur (33–35). Prior to and following conditioning, the hydrolyzate liquor was stored at 4°C. The so-called "overliming" process used to condition the hydrolyzate was as follows: while continuously agitating the solution, the pH of the as-received hydrolyzate liquor was raised to 10.0–10.5 using $Ca(OH)_2$ (solid powder). The solution was then heated to 50°C, held at this temperature for 30 min, and then cooled. Concentrated H_2SO_4 (96% w/w) was then slowly added until the solution reached pH 7.0. This overlimed hydrolyzate was sterilized by filtering through a 0.2 μm sterile filter.

Analytical Methods

Cell Concentration

Cell-mass concentration was determined by measuring turbidity (OD 600 nm) using a Milton Roy Spectronic 601 spectrophotometer (Spectronic,

Rochester, NY) zeroed with distilled water. Turbidity measurements were made using test tubes (100 mm × 13 mm) with an approximate light path length of 1.2 cm; whole broth samples were diluted into the linear range of the instrument, i.e., to achieve readings below 0.5 turbidity units. Turbidity values were converted to cell concentrations using a previously established correlation factor for this strain of 0.38 gDCM/L per turbidity unit.

Glucose, Xylose, and Ethanol Concentrations

Sample supernatants were analyzed to determine the concentrations of glucose, xylose, and ethanol (and other metabolites) using high-performance liquid chromatography (HPLC), as described previously *(30)*. Mixed-component concentration verification (CV) standards were periodically run (in duplicate) to verify calibration accuracy, and analysis was repeated if the reported concentrations of CV standards deviated from their actual values by greater than ± 2.5%.

Calculations

Ethanol yield on total available sugars, also referred to as process yield, or Y_P is the most important indicator of overall process performance. Process yields were calculated as the net grams of ethanol produced per grams of glucose and xylose available in the feed. Consumed sugar yields were calculated as the net grams of ethanol produced per grams of glucose and xylose consumed. The maximum cell-mass growth yield (i.e., corrected for maintenance), $Y_{X/S}^{max}$ (g DCM per g sugar), was determined as the inverse of the slope of the best fit linear regression for specific sugar utilization rate, q_S (g sugar per g DCM per h), as a function of dilution rate, D *(36)*. The maintenance energy coefficient, m_E (g sugar per g DCM per h), was determined as the y-axis intercept of the best fit linear regression to the q_S vs D data.

RESULTS AND DISCUSSION

Continuous-culture studies were initiated to understand what range of dilution rates would support good cofermentation performance. We were interested in maintaining selective pressure for xylose fermentation during prolonged chemostat operation, and thus wanted to maintain xylose-utilization rates that were threefold greater than those for glucose. Chemostat cultures were initiated using a feed solution containing 8 g/L glucose and 40 g/L xylose, i.e., concentrations similar to those in full-strength hydrolyzate liquors obtained by dilute acid pretreatment of hardwood feedstocks *(32)*. As a minimum level of cofermentation to indicate successful chemostat start up, we chose complete utilization of glucose and over 75% utilization of xylose. This level of conversion ensures that xylose:glucose utilization ratios remain greater than 3.5:1. The ability to

Fig. 1. Time courses during start up of recombinant Z. *mobilis* continuous cofermentations at three different initial conditions: (□), D = 0.04 h and pH = 5.5; (▲), D = 0.03 h and pH = 6.0; (○), D = 0.05 h and pH = 5.5. The medium for these experiments was RM (10 g Difco yeast extract and 2 g KH_2PO_4 per L distilled water) supplemented with 40 g/L xylose, 8 g/L glucose, and 10 mg/L tetracycline. The arrow indicates when continuous feeding was initiated.

maintain selective pressure for xylose fermentation during prolonged continuous culture should ensure that this trait is maintained and may also provide a method for long-term development of strains with improved xylose utilization capability.

Fermentations were started batchwise. Continuous feeding was initiated when the xylose concentration decreased to between 15 and 20 g/L. Successful chemostat start up resulted in complete utilization of glucose and greater than 85% utilization of xylose. Compositional analysis of successive daily samples showed stable glucose concentrations of zero and xylose concentrations below 5 g/L.

Start Up

Successful start up requires using initial feed (dilution) rates of less than or equal to 0.04/h. Figure 1 shows effluent xylose concentrations during start up of continuous cofermentations of glucose and xylose using dilution rates of 0.03, 0.04, and 0.05/h. During the week following start up, the effluent xylose concentrations generally remain low for the cultures initiated at dilution rates of 0.03 and 0.04/h. These cultures achieve stable effective cofermentation within 8 d. The steady-state xylose concentrations in these successful start ups, i.e., the xylose concentrations from 8 d onward, are below 4 g/L. In contrast, poor performance results when continuous feeding is started at a dilution rate of 0.05/h. In this case, the effluent xylose concentration remains low for the first 3 d following the initiation

Xylose-Fermenting Z. mobilis

Fig. 2. Pseudo steady-state concentrations as a function of dilution rate for pure-sugar continuous cofermentation using recombinant Z. *mobilis*. The arrows and reverse symbol shading show the steady-state measurements obtained just before the experiment was terminated.

of continuous feeding, but then begins to rise. After 6 d, less than half of the feed xylose is being consumed and the xylose concentration remains above 20 g/L.

Characterization of Continuous Cofermentation

Figure 2 shows steady-state xylose, cell mass (dry cell mass), and ethanol concentrations as a function of dilution rate; no glucose was detected at any of the steady states. There is some scatter in the data but good cofermentation performance was achieved up to a dilution rate of 0.10/h. As discussed in the subheading MATERIALS AND METHODS, the system became unstable at a dilution rate of 0.11/h; the data obtained at a dilution rate of 0.11/h is shown in Fig. 2 for comparative purposes but is not considered a reliable indicator of true steady-state performance. As Fig. 2 illustrates, with the exception of the data obtained at a dilution rate of 0.11/h, the xylose concentration increased in an approximately linear fashion with increasing dilution rate, increasing from approx 3 g/L at a dilution rate of 0.04/h to approx 7 g/L at a dilution rate of 0.10/h. Similarly, the cell-mass concentration increased with increasing dilution rate, but only very slightly, rising from approx 0.9 g DCM/L at a dilution rate of 0.04/h to approx 1.1 g DCM/L at a dilution rate of 0.10/h. In contrast, the ethanol concentration remained at 21–22 g/L independent of dilution rate between dilution rates of 0.04 and 0.10/h.

Volumetric productivity increased linearly with dilution rate, rising from approx 0.8 g/L/h at a dilution rate of 0.04/h to approx 2.1 g/L/h at a dilu-

Fig. 3. Ethanol yield as a function of dilution rate. The arrows and reverse symbol shading indicate the steady-state measurements obtained just before the experiment was terminated.

Fig. 4. Specific sugar-utilization rate and specific productivity as a function of dilution rate.

tion rate of 0.10/h (not shown). Figure 3 shows that process ethanol yields (ethanol yield based on available glucose and xylose) were approx 85% of theoretical over the entire range of dilution rates tested. Calculated metabolic yields (ethanol yield based on consumed sugars) were above 90% of theoretical and trended higher with increasing dilution rate (Fig. 3).

Figure 4 shows specific ethanol productivity and specific sugar utilization rate (i.e., rates on a per-gram DCM basis) as a function of dilution

Fig. 5. Adaptation to increasing concentrations of overlimed biomass hydrolyzate.

rate. The specific rate of ethanol production increased linearly with dilution rate, increasing from a low of approx 1.0 g/g/h at dilution rate 0.04/h to a high of approx 1.8 at dilution rate 0.10/h. The specific rate of sugar utilization increased in a similar fashion, rising from a low of 2.0 g/g/h at dilution rate 0.04/h to a high of approx 3.5 g/g/h at dilution rate 0.10/h.

Adaptation to Inhibitory Hydrolyzate

The pure-sugar study results indicate that excellent xylose utilization is achieved if cofermentation is carried out at a dilution rate of 0.04/h. For this reason, this dilution rate was chosen for running the long-term adaptation experiment. Figure 5 shows the time course of this experiment, which was begun using a hydrolyzate concentration of 10% v/v. This initial hydrolyzate level was chosen because batch studies showed that 10% v/v hydrolyzate is not inhibitory (not shown); the hydrolyzate significantly inhibits batch fermentation performance by xylose-fermenting *Zymomonas mobilis* at levels of 30% v/v or higher (35).

As in the pure-sugar study, cofermentation was initiated batchwise, with continuous feeding of 10% v/v hydrolyzate (supplemented with pure sugars) starting after the effluent xylose concentration fell to 15–20 g/L. When good cofermentation performance was achieved at a given hydrolyzate level, i.e., when no residual glucose was detected and the residual xylose concentration was below 10 g/L (or near 10 g/L and trending downward), the concentration of hydrolyzate in the feed was increased by 5% v/v. Thus, the feed concentration was increased from 10 to 15% v/v at day 15, and then increased from 15 to 20% v/v at day 24; and so on. As Fig. 5

shows, after 48 d of progressive adaptation, the culture achieved good cofermentation in the presence of 35% v/v hydrolyzate. This hydrolyzate level corresponds to an acetic acid level of approx 5 g/L, a level of acetic acid inhibitory to batch cofermentation performance *(30)*.

These results suggest that there is tremendous potential for strain improvement through continuous adaptation to higher levels of hydrolyzate. Further adaptation work clearly needs to be performed, however, to develop a strain that can tolerate full-strength, or near full-strength, hydrolyzates. The results nonetheless support the widely held view that the selective pressure of a controlled-chemostat growth environment provides a powerful technique for strain adaptation. For this reason, this chemostat adaptation work is continuing with the hope that strains exhibiting greater tolerance to acetic acid and other inhibitory components found in pretreatment hydrolyzates will be developed.

DISCUSSION

In the pure sugar study, the lowest metabolic ethanol yields occurred at the beginning of the experiment, i.e., the yield based on consumed sugars was only approx 85% when the first steady state was obtained at a dilution rate of 0.04/h (Fig. 3). In contrast, just before terminating the experiment, the measured metabolic ethanol yield at dilution rate 0.04/h was approx 95% (data points indicated by arrows shown in Fig. 3). The yields on consumed sugars actually remained above 95% after the dilution rate was raised to 0.08/h (at day 30) to the point the experiment was terminated (at day 50). The roughly 10% apparent increase in measured metabolic efficiency of the culture at dilution rate 0.04/h from the beginning to the end of the experiment possibly demonstrates an improvement in the strain's performance in continuous culture. From our perspective, the potential opportunities for developing a superior culture through long-term chemostat adaptation are real.

Regression analysis of the specific sugar-utilization-rate data gives values of 0.042 g/g and 1.13 g/g/h for $Y_{X/S}^{max}$ and m_E, respectively. However, there is scatter in the data and the value for the correlation coefficient is relatively low ($R^2 = 0.86$). It is therefore unwarranted to assign too much confidence to the accuracy of these parameter estimates. This said, Table 1 shows that these values are in the range of those reported by other investigators for nonrecombinant (i.e., glucose-fermenting) strains of *Z. mobilis*. In addition to being strain specific, these physiological and bioenergetic parameters are influenced by many other factors such as medium composition, pH, and temperature. It is thus difficult to draw significant conclusions from a comparison of reported literature values, i.e., the absence of uniformity with respect to strains and environmental growth conditions confounds clear interpretation of such comparisons *(40)*. Moreover,

Table 1
Maximum Cell-Growth Yields and Maintenance Coefficients Reported for *Zymomonas*

Z. mobilis strain	Sugar Substrate	$Y_{X/S}^{max}$ (gDCM/g substrate)	m_E (g substrate/g DCM-h)	Ref.
29191	glucose	0.037-0.060	1.0-2.5	37-40
31821	glucose	0.035	2.2	41
Z-1-81	glucose	0.024	2.5	42
10988	glucose		0.5	43
39676(pZB4L)	glucose-xylose	0.042	1.1	this work

plots of qs vs D are often observed to be biphasic (37,43,44), which further complicates the interpretation of such data as originally proposed by Pirt (36).

It is generally postulated that the requirement for maintenance metabolism is because of a combination of growth-dependent energy-requiring processes and energy-wasting processes (e.g., ATPase activity or its energetic equivalent) (36). Z. mobilis is relatively unique among ethanologenic microorganisms in using the Entner-Doudoroff (ED) pathway to convert glucose to pyruvate, thus achieving a net yield of only 1 mole ATP per mole glucose fermented to ethanol (45). In this context, it is interesting to note that the reported ATPase activity of crude homogenates of Z. mobilis is 7.5 mmol ATP/gDCM/h, which is equivalent to 1.35 g glucose/gDCM/h assuming a net ATP yield per mole glucose of 1.

Our results suggest that the maintenance requirement of the xylose-fermenting strain may be a bit lower than that of the typical glucose-fermenting strain, but additional measurements need to be made to confirm this. Additional chemostat studies are planned to assess more thoroughly how $Y_{X/S}^{max}$ and m_E vary as a function of key operating variables, such as temperature, pH, and feed-sugar concentrations.

ACKNOWLEDGMENTS

This work was funded by the Biochemical Conversion Element of the Office of Fuels Development of the US Department of Energy. Research conducted at the University of Toronto was part of Phase III of a Subcontract AAP-4-11195-03 from NREL. For the studies performed at NREL, AM and JDM gratefully acknowledge technical assistance provided by M. Newman and S. Toon.

REFERENCES

1. Wright, J. D., Wyman, C. E., and Grohmann, K. (1988), *Appl. Biochem. Biotechnol.* **18**, 75–90.
2. Wyman, C. E. and Hinman, N. D. (1990), *Appl. Biochem. Biotechnol.* **24/25**, 735–753.
3. Elander, R. T. and Putsche, V. L. (1996), in: *Handbook on Bioethanol: Production and Utilization,* Wyman, C. E., ed., Taylor and Francis, pp. 329–349.
4. Lynd, L. R. (1989). *Adv. Biochem. Eng. Biotechnol.* **38**, 1–52.
5. Lynd, L. R., Cushman, J. H., Nichols, R. J., and Wyman, C. E. (1991), *Science* **251**, 1318–1323.
6. Wright, J. D. (1988), *Chem. Eng. Prog.* **84**, 62–74.
7. Hinman, N. D., Wright, J. D., Hoagland, W., and Wyman, C. E. (1989), *Appl. Biochem. Biotechnol.* **20/21**, 391–401.
8. Grohmann, K., Himmel, M., Rivard, C., Tucker, M., Baker, T, Torget, R., and Graboski, M. (1984), *Biotechnol. Bioeng. Symp.* **14**, 139–157.
9. Kong, F., Engler, C. R., and Soltes, E. (1992), *Appl. Biochem. Biotechnol.* **34/35**, 23–35.
10. Grethlein, H. E. (1985), *Bio/Technology* **3**, 155–160.
11. Grethlein, H. E., Allen, D. C., and Converse, A. O. (1984), *Biotech. Bioeng.* **26**, 1498–1505.
12. Torget, R., Werdene, P., Himmel, M., and Grohmann, K. (1990), *Appl. Biochem. Biotechnol.* **24/25**, 115–126.
13. Timell, T. E. (1964), *Adv. Carbohydrate Chem.* **19**, 247–302.
14. Lawford, H. G. and Rousseau, J. D. (1993), in *Energy from Biomass and Wastes XVI, (March 1992),* Klass, D. L., ed., Institute of Gas Technology, Chicago, IL, pp. 559–597.
15. McMillan, J. D. (1994), in *Enzymatic Conversion of Biomass for Fuels Production,* Himmel, M. E., Baker, J. O., and Overend, R. A., eds., American Chemical Society, Washington, DC, *ACS Symposium Series 566,* pp. 411–437.
16. Hahn-Hägerdal, B., Hallborn, J., Jeppsson, H., Olsson, L., Skoog, K., and Walfridsson, M. (1993), in *Bioconversion of Forest and Agricultural Plant Residues.* Saddler, J. N., ed., C.A.B. International, Wallingford, UK, pp. 231–290.
17. Godia, F., Sasas, C., and Sola, C. (1987), *Process Biochem.* **22**, 43–50.
18. Wright, J. D., Wyman, C. E., and Grohmann, K. (1988), *Appl. Biochem. Biotechnol.* **18**, 75–90.
19. Vallander, L. and Erikson, K-E. L. (1990), *Adv. Biochem. Eng.* **42**, 63–95.
20. Qureshi, N. and Manderson, G. J. (1995), *Energy Sources* **17**, 241–265.
21. Lynd, L. R. (1996), *Ann. Rev. Energy Environ.* **21**, 403–465.
22. Lynd, L. R., Elander, R. T., and Wyman, C. E. (1996), *Appl. Biochem. Biotechnol.* **57/58**, 741–761.
23. Hinman, N. D., Schell, D. J., Riley, C. J., Bergeron P. W., and Walter, P. J. (1992), *Appl. Biochem. Biotechnol.* **34/35**, 639–650.
24. South, C. R., Hogsett, D. A., and Lynd, L. R. (1993), *Appl. Biochem. Biotechnol.* **39/40**, 587–600.
25. Picataggio, S. K., Zhang, M., Eddy, C. K., Deanda, K. A., and Finkelstein, M. (1996), U.S. Patent 5,514,583.
26. Zhang, M., Eddy, C., Deanda, K., Finkelstein, M., and Picataggio, S. K. (1995), *Science* **267**, 240–243.
27. Picataggio, S. K., Eddy, C., Deanda, K., Franden, M. A., Finkelstein, M., and Zhang, M. (1995), Paper 9 presented at the Seventeenth Symposium on Biotechnology for Fuels and Chemicals, Vail, CO, May 7–11.
28. McMillan, J. D., Mohagheghi, A., Newman, M. M., and Picataggio, S. (1995), Paper 216c presented at the Annual Meeting of the American Institute of Chemical Engineers, Miami, FL, November 12–17.
29. McMillan, J. D. (1997), *Renewable Energy* **10**, 295–302.

30. Lawford, H. G., Rousseau, J. D., and McMillan, J. D. (1997), *Appl. Biochem. Biotechnol.* **63/65,** 269–286.
31. Lawford, H. G. and Rousseau, J. D. (1997), *Appl. Biochem. Biotechnol.* in press.
32. Nguyen, Q. A., Dickow, J. H., Duff, B. W., Farmer, J. D., Glassner, D. A., Ibsen, K. N., Ruth, M. F., Schell, D. J., Thompson, I. B., and Tucker, M. P. (1996), *Bioresource Technol.* **58,** 189–196.
33. Leonard, R. H. and Hajny, G. J. (1945), *Ind. Eng. Chem.* **37,** 390–395.
34. Strickland, R. C. and Beck, M. J. (1984), Proceeding 6th International Symposium on Alcohol Fuels Technology, Ottawa, Canada, Vol 2, pp. 220–226.
35. Ranatunga, T. D., Jervis, J., Helm, R. F., McMillan, J. D., and Hatzis, C. (1997), *Appl. Biochem. Biotechnol.* **67,** 185–198.
36. Pirt, S. J. (1975), in *Principles of Microbe and Cell Cultivation*, Blackwell Scientific, London, UK, pp. 66–68.
37. Lavers, B. H., Pang, P., MacKenzie, C. R., Lawford, G. R., Pik, J. R., and Lawford, H. G. (1982), in *Advances in Biotechnology*, Moo-Young, M., ed., Pergamon, Toronto, pp. 195–200.
38. Lawford, H. G. and Stevnsborg, N. (1986), *Biotechnol. Lett.* **8,** 345–350.
39. Lawford, H. G. (1988), *Appl. Biochem. Biotechnol.* **17,** 203–209.
40. Lawford, H. G. and Ruggiero, A. (1990), *Biotechnol. Appl. Biochem.* **12,** 206–211.
41. Rogers, P. L., Lee, K. J., Skotnicki, M. L., and Tribe, D. E. (1982), *Adv. Biochem. Eng.* **23,** 37–84.
42. Olivera, E. G., Morais, J. O., and Pereira, N. (1992), *Biotechnol. Lett.* **14,** 1081–1084.
43. Feischko, J. and Humphrey, A. E. (1983), *Biotechnol. Bioeng.* **25,** 1655.
44. Jobses, I. M. L. and Roels, J. A. (1985), *Biotechnol. Bioeng.* **28,** 554–563.
45. Swings, J. and De Ley, J. (1977), *Bacteriol. Rev.* **41,** 1–46.
46. Lazdunski, A. and Belaich, J. P. (1972), *J. Gen. Microbiol.* **70,** 187–197.

Effect of Surfactants and Zeolites on Simultaneous Saccharification and Fermentation of Steam-Exploded Poplar Biomass to Ethanol

I. BALLESTEROS, J. M. OLIVA, J. CARRASCO, A. CABAÑAS, A. A. NAVARRO, AND M. BALLESTEROS*

Instituto de Energís Renovables-CIEMAT, Avda. Complutense, 22 28040-Madrid-Spain

ABSTRACT

In this work, the effect of the addition of different concentrations of Tween-80 and three different zeolite-like products on enzymatic hydrolysis, ethanol fermentation, and simultaneous saccharification and fermentation (SSF) process has been investigated. The ability of these products to enhance the effectiveness of the SSF process to ethanol of steam-exploded poplar biomass using the thermotolerant strain *Kluyveromyces marxianus* EMS-26 has been tested.

Tween-80 (0.4 g/L) increased enzymatic hydrolysis yield by 20% when compared to results obtained in hydrolysis in absence of the additive. Zeolite-like products (ZESEP-56 and ZECER-56) (2.5 g/L) improved rates of conversion and ethanol yields in the fermentation of liquid fraction recovered from steam-exploded poplar. The periods required for the completion of fermentation were approx 10 h in the presence of zeolite-like products and 24 h in the absence of additives. The probable mode of action is through lowered levels of inhibitory substances because of adsorption by the additive.

Index Entries: Simultaneous saccharification and fermentation; lignocellulose biomass; zeolite; surfactant.

INTRODUCTION

The increase of the cellulose-conversion level and reduction of SSF fermentation time have significant impact on estimated production cost of fuel-ethanol *(1)*. One means of increasing cellulose-conversion yield is to

---

*Author to whom all correspondence and reprint requests should be addressed.

carry out the SSF process at the cellulase optimum temperature. Increasing hydrolysis temperature should decrease reaction time, assuming that yeasts are capable of performing well at higher temperatures.

The inhibitory action of ethanol produced in the course of fermentation has effects on the cell growth, cell viability, and fermentation of yeasts. With increasing temperature, some of the effects of ethanol on yeast cells may become more severe (2–4). Ethanol enhances thermal death by acting in a nonspecific way on membrane lipids. It has been shown that the target of thermal death and of ethanol-enhanced thermal death in *Saccharomyces cerevisiae* is a macromolecular site in the inner mitochondrial membrane (5,6). There is considerable evidence in the pertinent literature (7–10) indicating that unsaturated fatty acids and sterols in the membrane are important for enhancing ethanol tolerance in yeasts, and several groups have reported improvements in alcoholic fermentation and final ethanol concentration by broth supplementation with lipids, proteins, and vitamins (11–15).

During pretreatment of the poplar biomass by steam explosion, aromatic monomers that have inhibiting effects on the bioconversion process to ethanol are produced (16). The elimination of these toxic compounds would be desirable in order to increase the yield of the process of enzymatic hydrolysis and simultaneous fermentation.

Some natural (17) and synthetic (18) zeolites have been reported to enhance the fermentation rate of sugarcane molasses to ethanol through protection against the inhibitory effects of substrates and products. Zeolites are crystalline, hydrated aluminosilicates, and all applications of zeolites are related to their physical and chemical properties: ion-exchange, adsorption, and related molecular sieve properties. The three-dimensional crystal structure of zeolites contains pores and channels of uniform size on a molecular scale. Their shape selectivity seems to play a role in removing inhibitory substances from the fermentation medium and in favoring yeast sorption onto zeolite crystals. Their characteristics in cation-exchange capacity and cation selectivity have led to their use in waste-water treatments (19,20).

In this work, the effect of media supplementation with unsaturated fatty acids and zeolite-like products on the SSF process of steam-exploded poplar biomass to ethanol has been tested.

MATERIALS AND METHODS

Substrate and Pretreatment

Poplar biomass was ground at 5-mm mesh size. The biomass was pretreated in a steam-explosion pilot unit operated by batches and equipped with a reaction vessel of a 2-L working volume which was filled with 200 g of dry biomass. The plant description and working methodology are described in a previous paper (21). The temperature (210°C) and residence time (4 min) conditions of the biomass pretreatments were selected with regard to the maximum glucose recovery after 72 h of enzymatic hydrolysis.

Cellulase Source

The cellulolytic complex employed (Celluclast 1.5 L) and β-glucosidase (Novozyme 188) were a donation from NOVO Nordisk (Bagsvaerd, Denmark).

Surfactants and Zeolitic Products

Tween-80 (polyoxyethylenesorbitan monooleate) was purchased from Sigma (St. Louis, MO). The zeolite clinoptilolite is a natural mineral with ion-exchange and gas-adsorption properties, and it was obtained from Mineral Research (Clarkson, NY). The zeolite-like products ZESEP-56 (SiO_2/Al_2O_3) (R = 8.3) and ZECER-56 (SiO_2/Al_2O_3) (R = 12.1) were prepared by the University of Cadiz (Spain) from Sepiolite 120NF and ceramic residues, respectively, after alkaline treatment. In ceramic residues, there is a great quantity of B_2O_3 along with TiO_2. These compounds, when subjected to the alkaline treatment proper to zeolitization, can give rise to zeolitic structures in which silica and aluminum are partially substituted by boron and titanium in the crystalline framework (22). Sepiolite is a magnesium silicate with a low aluminum content. Its structure includes internal channels that provide zeolitic properties.

Microorganisms and Growth Conditions

Kluyveromyces marxianus EMS-26, a thermotolerant mutant yeast strain obtained in our laboratory (23) was used in fermentations and SSF experiments. Active cultures for inoculation were prepared by growing the organism on a rotary shaker at 180 rpm for 16 h at 42°C in a growth medium containing (g/L): yeast extract (Difco, East Molesley, Surrey, UK), 5; peptone (Oxoid), 5; NH_4Cl, 2; KH_2PO_4, 1; $MgSO_4\ 7H_2O$, 0.3; and glucose, 30.

Enzymatic Hydrolysis, Simultaneous Saccharification, and Fermentation Assays

Experiments were carried out in 100-mL Erlenmeyer flasks, each containing 50 mL of the fermentation medium (initial pH 4.1) described above, which were agitated at 150 rpm. Glucose was substituted with the lignocellulose biomasses at 10% (w/v) substrate concentration, and the cellulolytic complex (Celluclast 1.5 L at 15 FPU/g substrate and Novozyme 188 at 12.6 IU/g substrate enzyme loading, respectively,) was also added.

Tween-80 and the zeolite-like products were added at the stated levels at the beginning of fermentation.

Enzymatic hydrolysis, SSF, and fermentation assays were conducted at 42°C for 72, 144, and 24 h respectively pH was not controlled throughout assays.

In the SSF experiments, flasks were inoculated with 10% (v/v) of yeast cultures and periodically checked during the assays for ethanol and glucose.

Initial enzyme concentration (Eo) was measured as the amount of protein in the reaction medium before substrate addition. By taking the difference between the initial concentration and the concentration at a particular time during the hydrolysis, the amount of adsorbed enzyme was calculated.

Analytical Procedures

Composition of substrate and residues in potential glucose and lignin have been determined by total hydrolysis with H_2SO_4 *(24)*.

Enzymatic activities (filter paper and β-glucosidase) were measured according to the methods described by Mandels et al. *(25)* and Bailey and Nevalainen *(26)*, respectively.

Glucose was quantified by HPLC in a 1081B Hewlett Packard (HP) (Palo Alto, CA) apparatus with differential refractometer detector at the following conditions: column, AMINEX HPX-87P (Bio-Rad, Hercules, CA); temperature, 85°C; eluent, water at 0.1 mL/min.

Ethanol was measured by gas chromatography, using a HP 5890 Series II apparatus, with flame ionization detector and a column of Carbowax 20 *M* (2 m × 1/8 in) at 95°C. Injector and detector temperature: 150°C.

Protein was measured using the Lowry reaction *(27)*.

RESULTS

To analyze the effects that the supplementation of culture mediums with Tween-80 and zeolite-like products have on the process of enzymatic hydrolysis and simultaneous fermentation, some preliminary experiments about the effects of these additives on each one of these processes were carried out separately.

Effect of Tween-80 and Zeolite-Like Products on Hydrolysis Efficiency

Figure 1 shows the effect of different concentrations of Tween-80 and zeolite-like products on the enzymatic hydrolysis yield of steam-exploded poplar biomass. The results are expressed as percentages referring to the control set-up. The 100% hydrolysis yield corresponds to the glucose produced in the control experiment in which no supplementation was effected.

As may be observed, on the addition of Tween-80 to the medium in which enzymatic hydrolysis is effected, the saccharification yield increases with the corresponding incrementation of the concentration of surfactant,

Fig. 1. Effect of varying levels of Tween-80 (0.4, 2, and 4 g/L), clinoptilolite (1, 2, and 5 g/L), ZESEP-56 (1, 2, and 5 g/L), and ZECER-56 (1, 2.5, and 5 g/L) on the final hydrolysis yield of 10% steam-exploded poplar biomass. Hydrolysis yields refer to control without supplementation

achieving gains of 6, 18, and 19% in hydrolysis yield when supplements of Tween-80 at 0.4, 2, and 4 g/L respectively, are applied.

The addition of zeolite-like products has uneven effects on the enzymatic hydrolysis yield of pretreated poplar. The addition of a natural zeolite such as clinoptilolite does not improve hydrolysis yields, lower glucose productions being obtained in these tests by comparison with the control experiments. Zeolite-like products prepared from sepiolite (ZESEP-56) and ceramic residues (ZECER-56) increase the yield of hydrolysis by 5 and 15% when concentrations of 2.5 and 5 g/L, respectively, are used.

For the purpose of determining if the supplementation of culture media with Tween-80 (4 g/L), clinoptilolite (5 g/L), and zeolite-like products (5 g/L) modify the adsorption of cellulolytic enzymes, the concentration of soluble protein throughout the process of enzymatic hydrolysis was analyzed. The results of these tests are shown in Fig. 2.

As can be seen, the extent of cellulase adsorption was significantly affected by the addition of the different products tested.

The course of enzyme adsorption during the first 8 h of fermentation followed a similar profile for all cases tested. A rapid initial adsorption of cellulose was observed followed by an equally rapid desorption.

In the media that were used as controls (no supplements added) it is observed that, 8 h after the beginning of hydrolysis, a slow but constant increase of soluble protein is generated. The addition of Tween-80 and clinoptilolite produces an increase in free-protein concentration. However,

Fig. 2. Protein concentration in solution during the hydrolysis of 10% (w/v) steam-exploded poplar biomass supplementation with Tween-80 (4 g/L), clinoptilolite (5 g/L), ZESEP-56 (5 g/L), and ZECER-56 (5 g/L). Eo = 3.81 g/L.

with the presence of the zeolite-like products ZESEP-56 and ZECER-56, the adsorption profiles during enzymatic hydrolysis changed significantly and the amount of cellulose adsorbed increased with time.

Effect of Tween-80 and Zeolitic Products on Ethanol Fermentation

The effects of the supplementation of Tween-80, clinoptilolite, ZESEP-56, and ZECER-56 at different loadings on the ethanol yield of *Kluyveromyces marxianus* EMS-26 fermentations (growth conditions described in the subheading MATERIALS AND METHODS) are shown in Fig. 3.

No positive effects are observed with regard to the supplementation of Tween-80 on the production of ethanol at the assayed concentrations. The addition of zeolite-like products, ZESEP-56 and ZECER-56, likewise had no positive effect on the yield of ethanol in the fermentations. On the contrary, by supplementing the culture media with a natural zeolite such as clinoptilolite slightly increased the production of ethanol (between 5 and 7%) in the fermentation process.

For the purpose of evaluating the effect on the production of ethanol of the addition of zeolites to a natural medium containing the toxic products that are generated during the pretreatment of lignocellulose biomass, some supplementation tests (5 g/L) were carried out using as fermentation media those liquids obtained during the pretreatment of the poplar biomass by steam explosion, to which glucose was added in order to complete the initial 55 g/L. The results of these experiments are shown in Fig. 4.

Fig. 3. Effect of varying levels of Tween-80 (0.4, 2, and 4 g/L), clinoptilolite (1, 2, and 5 g/L), ZESEP-56 (1, 2, and 5 g/L), and ZECER-56 (1, 2, and 5 g/L), on the ethanol yield of growth medium containing 55 g/L initial glucose. Ethanol yields refer to control without supplementation.

Fig. 4. Effect of 5 g/L supplementation of clinoptilolite, ZESEP-56, and ZECER-56 on the progress of batch fermentation of liquid medium from steam-explosion pretreatment of poplar biomass. Initial glucose: 55 g/L.

Fig. 5. Effect of varying levels of Tween-80 (0.4, 2, and 4 g/L), clinoptilolite (1, 2, and 5 g/L), ZESEP-56 (1, 2, and 5 g/L), and ZECER-56 (1, 2, and 5 g/L) on ethanol yield in the SSF process of 10% (w/v) steam-exploded biomass. Yields refer to control experiments without supplementation.

Supplementation with zeolite-like products resulted in markedly enhanced ethanol production and in a shortening of fermentation times in all the cases tested, as compared to those obtained in the control. The supplementing of the fermentation with a natural zeolite such as clinoptilolite did not produce improvements in ethanol production as spectacular as those obtained in the experiments in which ZESEP-56 and ZECER-56 were added.

In the control tests and in the tests supplemented with clinoptilolite, maximum concentrations of ethanol of approx 18 g/L are obtained after 24 h of fermentation. The presence of zeolite-like products increases the production rate of ethanol in a drastic manner. After 10 h, 23 g/L of ethanol had been produced and almost all of the substrate had been consumed.

Effect of Tween-80 and Zeolitic Products on the Simultaneous Saccharification and Fermentation Process of Steam-Exploded Poplar Biomass to Ethanol

The addition of Tween-80 increases the yield of the SSF process by approx 6% in all the concentrations assayed (Fig. 5). The yield results of ethanol obtained in the SSF process when zeolite-like products are added do not follow a defined tendency. The best results are obtained in the following conditions: clinoptilolite, 1 g/L (7% increase); ZESEP-56, 2.5 g/L (11% increase); and ZECER-56, 5 g/L (2% increase).

The different zeolite-like products, at these concentrations, along with 0.4 g/L of Tween-80 were assayed for the purpose of establishing the effect that the simultaneous addition of surfactants and zeolite-like products

Fig. 6. Effect of different supplementations on ethanol yield in the SSF process of 10% (w/v) steam-exploded biomass. Yields refer to control experiments without supplementation.

would have on the SSF process. The results of these experiments are shown in Fig. 6. In the experiment using clinoptilolite and Tween-80, better results were not obtained than those yielded by the experiments in which each one of these products was added separately. The synthetic zeolitic products ZESEP-56 and ZECER-56, along with Tween-80, produce increases of 20 and 14% in the yields of ethanol in comparison with control. These increases are superior to those obtained by supplementing separately with each one of these products.

DISCUSSION

The simultaneous saccharification and fermentation of cellulose to ethanol is a heterogeneous reaction with a soluble catalyst and an insoluble substrate. The first step in this reaction is the adsorption of cellulolytic enzymes from the reaction medium on the surface of the cellulose substrate. Endoglucanases and exoglucanases adsorb tightly to the cellulose substrate and do not readily desorb until the substrate is degraded. Thus, large amounts of cellulases become bound to inactive sites and, consequently, this severely limits the extent of saccharification (28). Because hydrolysis involves transport of enzymes and soluble sugars between the solid substrate and the reaction medium, modifications of interfacial energy may have some impact on this transfer. Surfactants alter the surface and interfacial properties of the reaction system, and therefore the addition of surfactants would result in more intimate contact between the substrate and enzyme.

The increase in hydrolysis yield obtained in tests carried out on pretreated poplar biomass with the addition of Tween-80 suggests that the adsorption-desorption of enzymes on the substrate is influenced by the presence of this product. Surfactant also affects the disruption of cellulose structure, making the cellulose more accessible to the enzymes (29).

Results obtained are in concordance with those obtained by Helle et al. (29), in which the cellulose hydrolysis yield of steam-exploded poplar was increased by 67% in the presence of surfactants.

The course of enzyme adsorption during the first 24 h of hydrolysis was in agreement with Lee et al. (30). A rapid initial adsorption on steam-pretreated wheat straw was observed, peaking at 3 h. This was followed by an equally rapid desorption of enzymes up to 7–8 h.

The fact that in the experiments with Tween-80 greater hydrolysis was observed with less protein adsorbed (Figs. 1 and 2) indicates that the surfactants somehow protect the enzyme from nonproductive binding. Since free protein cannot be responsible for cellulose hydrolysis, an explanation postulated by Helle et al. (29) is that, without surfactant present, much of the enzyme is adsorbed on inactive sites and does not participate in the hydrolysis of cellulose.

In a previous work (31) we observed that lipid addition has a negative effect on the enzymatic hydrolysis using Solka-flock as substrate. These results are different from those obtained on lignocellulosic biomass in this work. Lee et al (30) studied the adsorption of cellulases on several different substrates. When physicochemically pretreated substrates are used, enzymes become rapidly adsorbed onto the substrate. This is because of good accessibility and is followed by a continuous decrease in adsorption. If the substrate is Solka Floc, the enzyme accessibility is low, and the initial enzyme adsorption is slow but increases with time as the continuing hydrolysis makes the substrate more accessible. The lipid addition on enzymatic hydrolysis with Solka Floc could result in a more limited adsorption of enzymes during the first hours of hydrolysis, decreasing hydrolysis yields. However, when preteated poplar biomass is used, the lipid addition improves enzymatic hydrolysis by decreasing nonproductive binding of enzymes onto substrate.

The addition of the zeolite-like products ZESEP-56 and ZECER-56 provokes changes in the pH of the media (Fig. 7) because of the properties of ion exchange that are characteristic of these materials. The increase in pH that is produced upon increasing the concentration of zeolites could be responsible for the increases in hydrolysis yield. The addition of clinoptilolite does not produce changes either in the pH of the media or increases in hydrolysis yield as compared with the control.

It is known that part of the polysaccharides and lignin are degraded during steam-explosion pretreatment. These compounds inhibit the action of cellulases and the yields of the fermentative process. The removal of inhibitors formed during pretreatment of poplar biomass by zeolites could also improve the enzymatic hydrolysis and fermentation processes.

The presence of ZESEP-56 and ZECER-56 also increased enzymatic hydrolysis yields but, contrary to surfactant addition, a decrease of free

Fig. 7. Effect of varying levels of clinoptilolite (1, 2, and 5 g/L), ZESEP-56 (1, 2, and 5 g/L), and ZECER-56 (1, 2.5, and 5 g/L) on the pH of 10% steam-exploded poplar biomass. pH refers to control without supplementation.

protein in liquid phase was observed. The immobilization of glucosidase on zeolite could be the reason for the lower free-protein content.

The addition of zeolites to the fermentations (5 g/L) which were effected using the liquid fraction obtained in the pretreatment of the lignocellulose biomass favors the production of ethanol. The shortening of fermentation times (10 h vs the 24 h of the control set-up) and the increase in the production of ethanol (superior by 30%) indicate the great capacity of ZESEP-56 and ZECER-56 for adsorbing inhibitors produced during the biomass pretreatment.

The addition of Tween-80, clinoptilolite, and zeolite-like products to the SSF process does not have effects as beneficial as those observed separately in the process of enzymatic hydrolysis and fermentation. This could be because in the SSF process, the presence of yeasts along with the enzyme already favors an increase in the rate and in the yield of saccharification.

When the joint effects of Tween-80 and the zeolites on the SSF process are analyzed, increases of 20 and 14% are obtained with ZESEP-56 and ZECER-56 plus Tween-80, respectively.

CONCLUSIONS

The adsorption of cellulases on pretreated lignocellulose biomass is observed to be modified by the addition of Tween-80, increasing the rate of enzymatic hydrolysis.

The presence of the zeolite-like products ZESEP-56 and ZECER-56 in the fermentation media of the liquid fraction obtained after the pretreatment of poplar biomass by steam explosion significantly increases the rate and yield of the alcoholic fermentation process. These effects could be caused by the elimination of inhibitors, which in turn is caused by the characteristics of ionic interchange of these products.

Surfactants and zeolite-like products would be useful additives to increase ethanol yields in the SSF process of pretreated wood biomass to ethanol, mainly if the liquid stream recovered from the steam explosion pretreatment is used.

ACKNOWLEDGMENTS

The authors wish to thank J. López Ruiz and his coworkers of the Grupo de Investigación Zeolitas-Acuicultura, Departamento de Construcciones Navales of the University of Cadiz (Spain) for supplying the zeolite samples.

REFERENCES

1. Hinman, N. D., Schell, D. J., Riley, C. J., Bergeron, P. W., and Walter P. J. (1992), *Appl. Biochem. Biotechnol.* **34/35,** 639–649.
2. Nagodawithana, T. W. and Steinkraus, K. H. (1976), *Appl. Environ. Microbiol.* **31,** 158–162.
3. Navarro, J. M. and Durand G. (1978), *Ann. Microbiol.* **129B,** 215–224.
4. Brown. S. W. and Oliver, S. G. (1982), *Biotech. Lett.* **4,** 269–274.
5. Leao, C., and Van Uden, N. (1982), *Biotechnol. Bioeng.* **24,** 1581–1590.
6. Sa-Correia, I. and Van Uden, N. (1986), *Biotechnol. Bioeng.* **28,** 301–303.
7. Ingram, L. O. and Buttke, T. M. (1984), *Adv. Microb. Physiol.* **25,** 253–300.
8. Beaven, M. J., Charpentier, C., and Rose A. H. (1982), *J. Gen. Microb.* **128,** 1447–1455.
9. Thomas, A. S., Hossack, J. A., and Rose A. H. (1978), *Arch. Microbiol.* **117,** 239–245.
10. Ingram, L. O. (1986), *Trends Biotechnol.* **4,** 40–44.
11. Casey, G. P., Magnus, C. A., and Ingledew, W. M. (1984), *Appl. Environ. Microbiol.* **48,** 639–646.
12. Casey, G. P., Magnus, C. A., and Ingledew, W. M. (1983), *Biotechnol. Lett.* **5,** 429–434.
13. Panchal, C. J. and Stewart, G. G. (1981), *Dev. Ind. Microbiol.* **22,** 711–717.
14. Viegas, C. A., Sa-Correia, I., and Novais J. M. (1985), *Biotechnol. Lett.* **7,** 515–520.
15. Viegas, C. A., Sa-Correia, I., and Novais J. M. (1985), *Appl. Environ. Microbiol.* **50,** 1333–1335.
16. Ando, S., Arai, I., Kiyoto, K., and Hanai, S. (1986), *J. Fermentation Technol.* **64(6),** 567–578.
17. Bernal, M. P. and Lopez-Real, J. M. (1993), *Bioresource Technol.* **43,** 27–33.
18. SivaRaman, H., Chandwadkar, A., Baliga, S. A., and Prabhune, A. A. (1994), *Enzyme Microb. Technol.* **16,** 719722.
19. Tarasevich, I. I. (1988), *Sov. J. Water Chem. Tech.* **10,** 22–32.
20. Liverti, L., Boari, G., Petruzzelli, D., and Passino, R. (1981), *Water Res.* **15,** 337–342.
21. Carrasco, J. E., Martínez, J. M., Negro, M. J., Manero, J. Mazon, P., Saez, F. Cabañas A. and Martín, C. (1989), in *Biomass for Energy and Industry. Fifth E.C. Conference*, Elsevier, New York pp. 38–44.
22. Ferreiro M. S., Ramos Lopez, S. R., and Lopez J. (1994), *Afinidad*, 454–461.
23. Ballesteros I., Oliva, J. M., Ballesteros M., and Carrasco J. (1993), *Appl. Biochem. Biotech.* **39/40,** 201–211.

24. Puls, J., Poutanen, K., Körner, H. U., and Viikari, L. (1985), *Appl. Microbiol. Biotechnol.* **22,** 416–423.
25. Mandels, M., Andreotti, R. and Roche, C. (1976), *Biotech. Bioeng.* **6,** 21–23.
26. Bayle, M. J. and Nevalainem, K. H. H. (1981), *Enzyme Microbiol. Technol.* **3,** 153–158.
27. Lowry, O. H., Rosebrough, N. J., Farr, A. L., and Randall, R. J. (1951), *J. Biol. Chem.* **23,** 139.
28. Howell, J. A. and Mangat, M. (1978), *Biotech. Bioeng.* **26,** 936–941.
29. Helle, S. S., Duff, S. J. B., and Cooper, D. G. (1993), *Biotech. Bioeng.* **42,** 611–617.
30. Lee, S. B., Shin, H. S., Ryu, D. D. Y., and Mandels, M. (1982), *Biotech. Bioeng.* **24,** 21–37.
31. Ballesteros, I., Oliva, J. M., Carrasco, J. E., and Ballesteros, M. (1994), *Appl. Biochem. Biotech.* **45/46,** 283–294.

Thermal Stability and Energy of Deactivation of Free and Immobilized Amyloglucosidase in the Saccharification of Liquefied Cassava Starch

GISELLA M. ZANIN* AND FLAVIO F. DE MORAES

State University of Maringá, Chemical Engineering Department, Av. Colombo, 5790 - BL. E46 - 09, 87020-900, Maringá, PR, Brazil

ABSTRACT

Amyloglucosidase from Novo (Copenhagen, Denmark) was immobilized in controlled pore silica particles with the silane-glutaraldehyde covalent method. Thermal stability of the free and immobilized enzyme (IE) was determined with 30% (w/v) α-amylase liquefied cassava starch, pH 4.5, temperatures from 35 to 75°C. Free amyloglucosidase maintained its activity practically constant for 240 min and temperatures up to 50°C. The IE has shown higher stability retaining its activity for the same period up to 60°C. Half-life for free enzyme was 20.6, 6.44, 2.07, 0.69, and 0.24 h for 55, 60, 65, 70, and 75°C, respectively, whereas the IE at the same temperatures had half-lives of 116.4, 30.88, 8.52, 2.44, and 0.73 h. The energy of thermal deactivation was thus 50.6 and 57.6 kcal/mol, respectively for the free and IE, confirming stabilization by immobilization.

Index Entries: Cassava starch; amyloglucosidase; immobilized enzyme; stability; half-life.

INTRODUCTION

Cassava starch, known as "fécula" in Brazil, is a plentiful, renewable, cheap resource that may be used for the production of modified starches of large application into textile and paper industries. It has also been considered as a source of glucose syrups produced by saccharification with amyloglucosidase. These syrups could be used directly in the food industry, or converted to ethanol of high quality to be used in perfumes or alcoholic beverages such as liqueurs and spirits. Fuel ethanol that is used

---

*Author to whom all correspondence and reprint requests should be addressed. E-mail: gisellazanin@cybertelecom.com.br

in Brazil in anhydrous form for blending with gasoline, or straight as in the azeotropic ethanol-water mixture used in ethanol cars, could be produced from cassava starch.

The potential of these applications have stimulated the development of a research program in our department aimed at studying the saccharification of liquefied cassava starch with free and immobilized amyloglucosidase *(1)*. The research program included: modeling of batch reactors with free enzyme *(2)* and of fixed- and fluidized-bed reactors with immobilized enzyme *(3,4)*, hydrodynamics *(5)*, axial dispersion *(6)*, internal *(4)* and external *(3)* mass transfer, and operational stability of the immobilized enzyme *(7)*; characterization of Novo Nordisk amyloglucosidase with respect to its activity in the saccharification of preliquefied cassava starch at different pHs and temperatures *(8)*; and determinations of the density, viscosity, and initial glucose contents of the Novo α-amylase liquefied starch, measured as a function of total dried matter in solution *(5)*.

This article covers experimental determination, of free and immobilized Novo Nordisk amyloglucosidase thermal deactivation at temperatures from 35 to 75°C during the saccharification of α-amylase preliquefied cassava starch.

DEACTIVATION HYPOTHESIS AND MODELING

As temperature is raised in a reaction catalyzed by free or immobilized enzyme, two opposing effects are observed: The enzyme activity increases with temperature, and this increases the reaction rate, but with higher temperatures the enzyme stability decreases by thermal denaturation and this reduces the concentration of active enzyme that reduces the reaction rate *(9–11)*. Usually in the range of 30 to 50°C, activation of the enzyme prevails and the enzymatic activity increases with temperature but above that, enzyme denaturation overtakes activation, and enzymatic activity begins to decline. Industrial processes that use enzyme are usually set at 60°C to avoid microbial contamination, and because of that the knowledge of the rate of deactivation of the enzyme at different temperatures is an important consideration.

The energy of activation of the thermal denaturation reaction, or as it is usually called: energy of deactivation, can be obtained from thermal stability data obtained by carrying out an experiment for various temperatures, in which the enzyme is incubated for a certain time within specified conditions, and then its residual activity is assayed. It is normally assumed that enzyme thermal denaturation is a reaction with the rate of enzyme deactivation (r_d) being first order in relation to the concentration of the active enzyme (E):

$$r_d = - K_d E \qquad (1)$$

and the deactivation constant (K_d) being a function of temperature as given by the Arrhenius equation:

$$K_d = K_d° \exp(-E_d/RT) \qquad (2)$$

where E_d is the energy of deactivation, R the universal gas constant (1.987 cal/mol K), and T the absolute reaction temperature. It should be clearly understood that the energy of deactivation is the preferred name for the energy of activation of the deactivation reaction.

For a batch reactor of constant liquid density the rate of reaction equals the time derivative of the concentration, and therefore it follows from equation (1) that:

$$(dE/dt) = -K_d E \qquad (3)$$

which integrated with the initial condition: $E = E_0$ for $t = 0$, gives:

$$E = E_0 \exp(-K_d t) \qquad (4)$$

where E_0 is the initial active enzyme concentration, and t is time elapsed during reaction.

When the enzyme is present in catalytical quantities, that is, in low concentration, the residual enzyme activity (A_r) is directly proportional to the concentration of the active enzyme (E):

$$(A_r/A_0) = (E/E_0) \qquad (5)$$

where A_0 is the initial enzyme activity observed with the initial enzyme concentration (E_0).

Combining equations (4) and (5), the residual enzyme activity results as:

$$A_r = A_0 \exp(-K_d t) \qquad (6)$$

This result is known as the exponential decay model. Therefore, by plotting residual activity data in the form of log of A_1/A_0 against time, the deactivation constant (K_r) is obtained as the angular coefficient of the adjusted straight line.

From equation (2), and, as observed experimentally, it can be seen that the deactivation constant increases with temperature. Values obtained for K_d for various test temperatures are plotted in the form of Arrhenius plot, that is log of K_d against the inverse of absolute temperature, yielding the energy of deactivation (E_d), as the angular coefficient of the adjusted straight line, times R, the universal gas constant.

In a given set of experimental conditions, another important parameter related to enzyme stability is the enzyme half-life $(t_{1/2})$, that corresponds to the period of time taken by the residual enzyme activity to decrease to

50% of its initial value. From equation (6), it results that the half-life ($t_{1/2}$) can be calculated by:

$$t_{1/2} = -\ln(0.5)/K_d = 0.693/K_d \quad (7)$$

showing that $t_{1/2}$ is inversely proportional to the deactivation constant (K_d).

The energy of deactivation (E_d) of most enzymes is normally within 47 to 96 kcal/mol, clearly higher than the energy of activation (E_a) that is normally smaller than 25 kcal/mol (12). E_a is the activation energy associated with the normal reaction catalyzed by the enzyme in which the substrate is transformed to products. These different ranges for the energy of activation and energy of deactivation, have as result the observed fact that enzymes are activated at lower temperatures and deactivated at high temperatures. This is a direct consequence of the rule that applies to competing reactions with different energies of activation, namely: high temperatures favor reactions with high energies of activation (13).

MATERIALS AND METHODS

Enzyme

Aspergillus niger amyloglucosidase kindly donated by Novo (AMG 150L, with 130 mg of protein/mL) was used as free enzyme in solution and immobilized.

Support

Controlled pore silica (CPS) was supplied by Corning Glass Works (Corning, NY) with a particle mean diameter of 0.436 mm, average pore size of 37.5 nm, and internal porosity of 56.6%.

Substrate

The substrate was cassava starch (Copagra-PR) at a final concentration of 30% (w/v), liquefied with α-amylase (Thermamyl 120L, Novo) in the presence of 70 ppm $CaCl_2$, pH 6.0, for 1 h at 95 to 100°C. After cooling to room temperature, the pH of this solution was adjusted to 4.5 with sodium acetate buffer to a final concentration of 0.02 M. The DE (dextrose equivalent) of this solution was measured with the DNS method (14), giving 38.6.

Enzyme Immobilization

Amyloglucosidase was immobilized in CPS with the silane-glutaraldehyde covalent method of Weetall (15) that consists in the following steps: silanization of the support with 0.5 (v/v) γ-aminopropyltrietoxisilane for 3 h at 75°C; wash the silanized support with distilled water and dry it for 15 h at 105°C; activate the dried silanized support with a solution of 2.5% (v/v) glutaraldehyde, pH 7.0 for 45 min at 20°C; wash with water; contact the activated support with the enzyme solution for 15 h at 20°C; wash the immobilized enzyme with water and stock it in buffer solution in a refrigerator for later use (15,16). The immobilized amyloglucosidase was kept in 0.02 M acetate buffer, pH 4.5, at 4°C.

Thermal Deactivation Test

Free or immobilized amyloglucosidase was incubated at the following temperatures: 35, 40, 45, 50, 55, 60, 65, 70, and 75°C, in the substrate solution, for a period of 4 h. In the case of the thermal deactivation test with the enzyme free in solution 10 mL of the substrate solution contained 6.2 mg of protein/mL, and each 40 min a 1 mL sample of the incubated enzyme was taken and the residual enzymatic activity was determined. For the immobilized enzyme (IE), seven stainless steel baskets containing 1,000 g wet weight of IE were used, six being immersed in 200 mL of substrate at the temperature of the thermal deactivation test, and one used for determining the initial enzymatic activity (A_0). Each 40 min, one of the baskets with the incubated IE was taken for determining the residual activity. The value of the humidity of the immobilized enzyme, which was necessary to calculate the exact IE dry weight in the basket, was determined in a parallel experiment at 105°C.

Determination of Amyloglucosidase Activity

The residual enzymatic activity was determined by the method of initial velocities (10). Samples of free or immobilized enzyme taken from the thermal deactivation test were put in contact with 50 mL of the substrate solution in a batch reactor maintained at the selected temperature, and aliquots were taken at regular intervals for determining the glucose produced by the saccharification reaction. Total reaction time was 30 min in this test to give maximum substrate conversion below 30%, usually much lower than this, and 0.5-mL samples were collected every 5 min. In these conditions the rate of glucose production as a function of time (μmol/min) is linear, and the residual enzyme-specific activity is obtained by dividing this rate by the quantity of protein present (mg). Therefore the residual specific activity was expressed in μmol/(min/mg of protein), that is, U/mg of protein.

Enzyme Activity

One unit (U) of enzyme activity corresponds to the quantity of enzyme that produces 1 μmol of glucose/min with the aforementioned substrate at the specified temperature.

Assay Methods

The glucose produced in the test for determination of the enzymatic activity was measured by the orthotoluidine method (17), and the total protein contents of the enzymatic solution was assayed by the method of Lowry et al. (18).

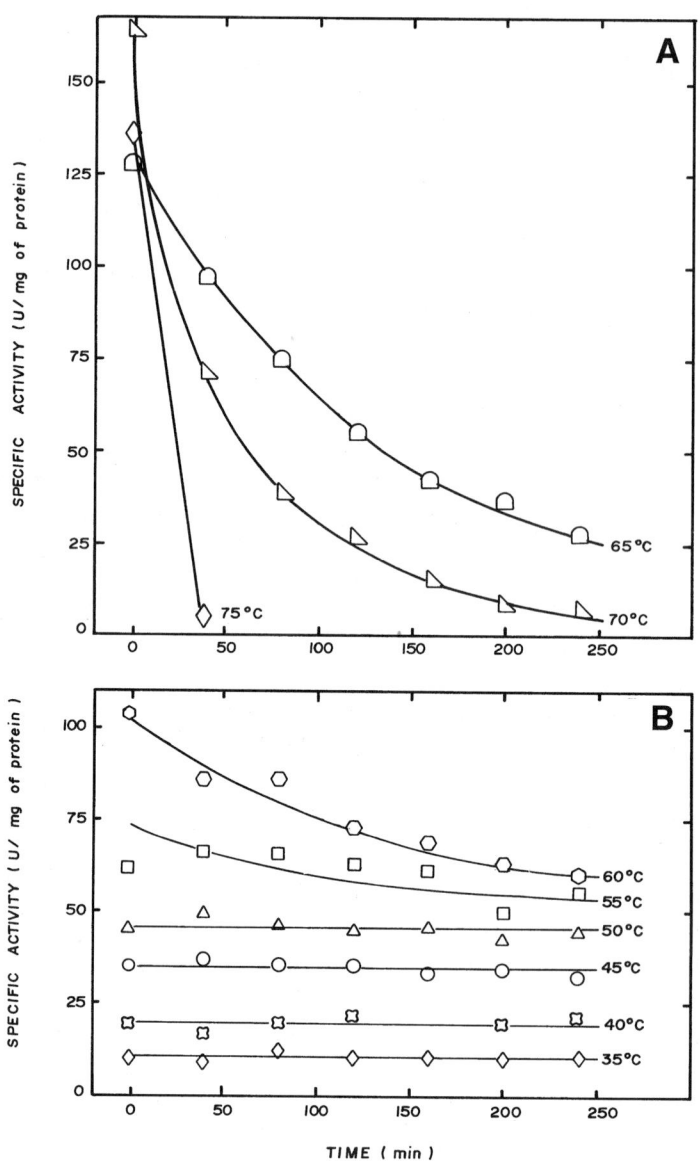

Fig. 1. Thermal deactivation data for free Novo Nordisk amyloglucosidase incubated in liquefied cassava starch, 30% (w/v), pH 4.5. Each point in the figure corresponds to the amyloglucosidase residual activity measured with 1 mL of enzyme taken from the thermal deactivation test and added to 50 mL of substrate solution giving 0.124 mg of protein/mL of solution.

RESULTS AND DISCUSSION

Immobilized amyloglucosidase was produced offering 4.48 mg/g of protein of support, and 4.12 were retained, giving an enzyme yield of 92.0%, with 22.4 U/mg of protein at 45°C.

Free and Immobilized Amyloglucosidase

Fig. 2. Thermal deactivation data for Novo Nordisk amyloglucosidase immobilized in controlled pore silica, and incubated in liquefied cassava starch, 30% (w/v), pH 4.5. Each curve corresponds to the residual activity of the IE, measured with six baskets incubated in 200 mL of substrate solution at the specific temperature. The IE had 4.12 mg/g of protein of support and each basket contained about 0.564 g of IE dry weight.

Experimental data obtained from the thermal deactivation test for free amyloglucosidase are shown in Fig. 1, and for the immobilized enzyme in Fig. 2. It can be observed that the immobilized enzyme loses very little of its activity in the period of 4 h at the temperatures of 35 to 60°C, whereas for the free enzyme this applies only for 35 to 50°C. The free enzyme thermal denaturation at 60°C is already quite noticeable, retaining only 59% of its activity after 4 h in contact with the substrate. At the higher temperatures of 65 to 75°C, the denaturation of the free amyloglucosidase is much more pronounced than that of the immobilized enzyme.

Equation *(6)* adjusted to data obtained at the higher temperatures gave the results shown in Table 1, in which the half-lives were calculated with equation *(7)*. Missing values in Table 1 reflect the difficulties of obtaining good fit at these points. For 55°C, the IE deactivation is too slow, and experimental errors are greater than the observed reduction in activity.

Table 1
Exponential Decay Model, Equation (6), Adjusted to Thermal
Deactivation of Novo Nordisk Amyloglucosidase Thermally
Denatured in Liquefied Cassava Starch 30% (w/v), pH 4.5, at Different
Temperatures

Temperature (°C)	Free Enzyme			Immobilized Enzyme		
	k_d (h^{-1})	r	$t_{1/2}$ (h)	k_d (h^{-1})	r	$t_{1/2}$ (h)
75	-	-	-	1.1484	0.9943	0.60
70	0.8712	0.9943	0.80	0.2017	0.9928	3.44
65	0.3764	0.9960	1.84	0.0933	0.9988	7.43
60	0.1297	0.9765	5.34	0.0229	0.9933	30.24
55	0.0287	0.8742	24.1	-	-	-

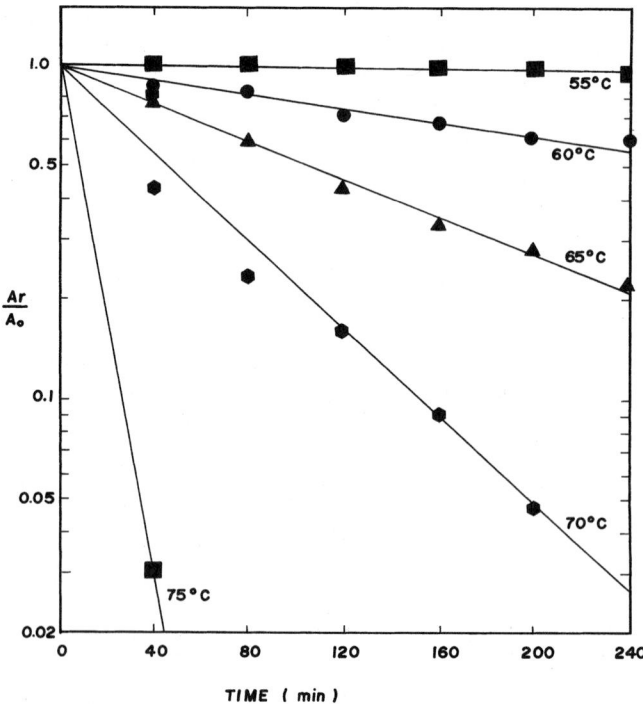

Fig. 3. Free Novo Nordisk amyloglucosidase thermal deactivation data compared with the exponential decay model, equation (6).

Fig. 4. Immobilized Novo Nordisk amyloglucosidase thermal deactivation data compared with the exponential decay model, equation *(6)*.

For 75°C, the free enzyme deactivation is too fast, and only one point was not enough to give a good measurement of K_d.

Figures 3 and 4 compare the exponential decay model with the deactivation data obtained at the higher temperatures for free and immobilized enzyme, respectively. It can be observed that there is a general good fit. Exception was seen for 75°C which is a difficult experiment owing to the fast deactivation observed at this temperature. The good fit lends support to the application of the exponential decay model for the thermal deactivation of Novo Nordisk amyloglucosidase thermally denatured in liquefied cassava starch solutions.

If industrial cassava saccharification would be run at 60°C as practiced by the corn industry in the batch process with free amyloglucosidase *(19)*, then equation *(6)* with K_d for 60°C taken from Table 1 predicts that at the end of reaction, within 48 h, practically all of the enzyme would be thermally deactivated, namely: 99.8%. After 24 h, 95.6% of the amyloglucosidase is already deactivated. Therefore, in the 48-h process, during the last 24 h, very little active enzyme is present. This suggests the staged addition of the enzyme to improve the process.

Figure 5 compares the Arrhenius plot of the deactivation constant (K_d) for free and immobilized amyloglucosidase. Equation *(2)* adjusted for this data gives:

free amyloglucosidase:

$$K_d = 1.771 \times 10^{32} \exp(-50{,}604/R\,T); \quad r = 0.993 \qquad (8)$$

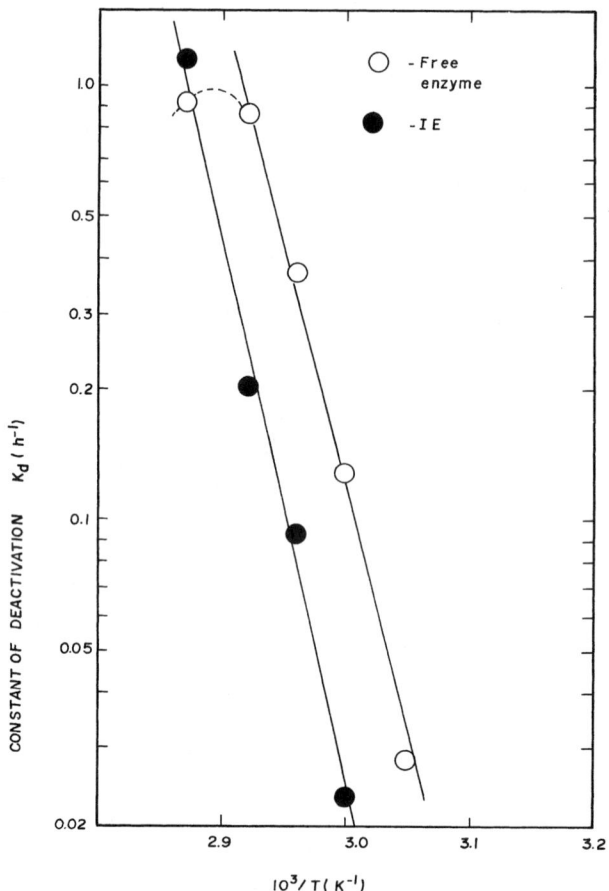

Fig. 5. Deactivation constant as a function of the inverse of absolute temperature (Arrhenius plot), for free and immobilized Novo Nordisk amyloglucosidase.

immobilized enzyme:

$$K_d = 1.409 \times 10^{36} \exp(-57{,}587/RT); \; r = 0.989 \qquad (9)$$

Therefore, the experimentally observed energy of deactivation (E_d) is approx 50.6 and 57.6 kcal/mol, respectively, for the free and immobilized Novo Nordisk amyloglucosidase thermally denatured in the liquefied cassava-starch substrate. There is a 13.8% increase in the energy of deactivation of the immobilized enzyme, confirming that immobilization confers more stability to the enzyme as observed by previous work (20–26). Also the value for the energy of deactivation of the free enzyme in contact with the liquefied cassava starch is comparable to results obtained with maltose: E_d = 50 kcal/mol (27), and soluble starch: E_d = 54.2 kcal/mol (28), and E_d = 57 kcal/mol (20).

Table 2
Comparison of Experimental and Predicted Values for Novo Nordisk
Amyloglucosidase Half-Life ($t_{1/2}$) in the Saccharification of Liquefied
Cassava Starch 30% (w/v), pH 4.5

Temperature (°C)	Free Enzyme $t_{1/2}$ (h)		Immobilized Enzyme $t_{1/2}$ (h)	
	Experimental	Predicted	Experimental	Predicted
75	-	0.24	0.60	0.73
70	0.80	0.69	3.44	2.44
65	1.84	2.07	7.43	8.52
60	5.34	6.41	30.24	30.88
55	24.1	20.6	-	116.4

Equations *(8)* and *(9)* allow one to calculate predicted values for the deactivation constant and then, with equation *(7)*, the predicted half-life is obtained. Experimental and predicted values for amyloglucosidase half-life are compared in Table 2. Agreement is satisfactory, and predicted half-lives deviate from the experimental values, on average, 16%.

CONCLUSIONS

1. Novo Nordisk amyloglucosidase when thermally denatured in a solution of liquefied cassava starch, 30% (w/v), pH 4.5, is stable up to 50°C for a period of 4 h.
2. The same enzyme, immobilized in controlled pore silica by the silane-glutaraldehyde covalent method, is stable up to 60°C, in the same conditions.
3. For higher temperatures, the enzyme deactivation reasonably follows the exponential decay model and half-lives are satisfactorily predicted with equation *(7)* and *(8)* or *(9)*, for free or immobilized enzyme, respectively.
4. The immobilized amyloglucosidase is more stable than the free enzyme, and their energy of deactivation is 57.6 and 50.6 kcal/mol, respectively.

ACKNOWLEDGMENTS

The authors thank the financial support received from CNPq, FINEP, and the State University of Maringá. The companies that supplied materials (Copagra, Novo, Corning Glass Works) are also acknowledged.

REFERENCES

1. Zanin, G. M. (1989), Ph.D. thesis, Universidade Estadual de Campinas, Campinas-SP, Brasil.
2. Zanin, G. M. and de Moraes, F. F. (1996), *Appl. Biochem. Biotechnol.* **57/58,** 617–625.
3. Zanin, G. M. and de Moraes, F. F. (1994), *Appl. Biochem. Biotechnol.* **45/46,** 627–640.
4. Zanin, G. M. and de Moraes, F. F. (1997), *Appl. Biochem. Biotechnol.* **63/65,** 527–540.
5. Zanin, G. M. and de Moraes, F. F. (1984), in *Anais do XII Encontro sobre Escoamento em Meios Porosos,* vol I, Maringá - PR, Brasil, pp. 267–285.
6. Zanin, G. M., Neitzel, I. and de Moraes, F. F. (1993), *Appl. Biochem. Biotechnol.* **39/40,** 477–489.
7. Zanin, G. M. and de Moraes, F. F. (1995), *Appl. Biochem. Biotechnol.* **51/52,** 253–262.
8. Zanin, G. M. and de Moraes, F. F. (1989), *Revista de Microbiologia* **20,** 367–371.
9. Chibata, I. (1978) *Immobilized Enzymes—Research and Development,* Wiley, New York.
10. Dixon, M. and Webb, E. C. (1979), *Enzymes,* 3rd ed., Longman Group, London, pp. 7–22.
11. Messing, R. A. (1975), *Immobilized Enzymes for Industrial Reactors,* Academic, New York.
12. Hartmeier, W. (1986), *Immobilized Biocatalysts—an Introduction,* Springer-Verlag, Berlin, pp. 54–56.
13. Levenspiel, O. (1972), *Chemical Reaction Engineering,* 2nd ed., Wiley, New York, p. 239.
14. Miller, G. L. (1959), *Anal. Chem.* **31,** 426.
15. Weetall, H. H. (1993), *Appl. Biochem. Biotechnol.,* **41,** 157–187.
16. Weetall, H. H., Vann, W. P., Pitcher, W. H. Jr., Lee, D. D., Lee, Y. Y., and Tsao, G. T. (1976), in *Methods in Enzymology,* vol. XLIV, *Immobilized Enzymes,* Mosbach, K. ed., Academic, New York, pp. 776–792.
17. Cooper, G. R. and McDaniel, V. (1970), *Clin. Chem.,* **6,** 159–170.
18. Lowry, O. H., Rosebrough, N. J., Faar, A. L., and Randall, R. J. (1951), *J. Biol. Chem.,* **193,** 265–275.
19. Pedersen, S. (1993), in *Industrial Application of Immobilized Biocatalysts,* Tanaka, A., Tosa, T., and Kabayashi, T. eds., Marcel Dekker, New York, p. 188.
20. Araujo, E. H. and Schimdell Netto, W. (1987), *Revista de Microbiologia,* **18,** 46–51.
21. Brillouet, J. M., Coulet, P. R. and Gautheron, D. C. (1977), *Biotechnol. Bioeng.,* **19,** 125–142.
22. Cabral, J. M. S. (1982), PhD thesis, Universidade Técnica de Lisboa, Lisboa, Portugal.
23. Cabral, J. M. S., Kennedy, J. F., and Novais, J. M. (1982), *Enz. Microbiol. Technol.,* **4,** 343–348.
24. Johnson, J. C. (1979), *Immobilized Enzymes—Preparation and Engineering—Recent Advances,* Noyes Data Corp., Park Ridge.
25. Johnson, D. B. and Costelloe, M. (1976), *Biotechnol. Bioeng.,* **18,** 421–424.
26. Maeda, H. and Suzuki, H. (1972), *Arch. Biol. Chem.* **36,** 1581–1593.
27. Marc, A. (1985), PhD thesis, Inst. Nat. Polytechnique de Lorraine, Lorraine, France.
28. Cardoso, J. P. (1977), PhD thesis, University of Birmingham, UK.

Hydrolysis of Cellulose Using Ternary Mixtures of Purified Cellulases

JOHN O. BAKER,* CHRISTINE I. EHRMAN, WILLIAM S. ADNEY, STEVEN R. THOMAS, AND MICHAEL E. HIMMEL

Biotechnology Center for Fuels and Chemicals, National Renewable Energy Laboratory, Golden, CO, 80401

ABSTRACT

The saccharification of microcrystalline cellulose by reconstituted ternary mixtures of purified cellulases (one endoglucanase and two cellobiohydrolases) has been studied over the entire range of mixture compositions. Ternary plots are used to compare the performance of five synthetic mixtures drawn from the cellulase systems of *Acidothermus cellulolyticus*, *Trichoderma reesei*, *Thermomonospora fusca*, and *Thermotoga neapolitana*. Results reveal that at least one synthetic mixture utilizing enzymes from three different organisms delivers performance competitive with that of a "native" (i.e., co-evolved) ternary system drawn exclusively from *T. reesei*. This heterologous system, consisting of the endoglucanase E1 from *A. cellulolyticus* and the exoglucanases CBHI from *T. ressei* and E_3 from *T. fusca*, is forgiving from the system-design point of view, in that it delivers high saccharification rates over a wide range of mixture compositions.

Index Entries: Cellulases; *Trichoderma reesei*; *Aspergillus niger*; *Acidothermus cellulolyticus*; *Thermomonospora fusca*; *Microbispora bispora*.

INTRODUCTION

The potential of lignocellulose to provide fermentable sugars as a carbon source in the production of fuels and chemical feedstocks is now well appreciated (1,2). It has long been recognized that cellulase activity is not the activity of a single enzyme, but the result of multiple activities working cooperatively, or synergistically, to efficiently solubilize crystalline cellulose (3,4). There are three categories of cellulolytic enzymes

*Author to whom all correspondence and reprint requests should be addressed.

required for this process: endo-1,4-β-D-glucanases or endoglucanases (1,4-β-D-glucan-4-glucano-hydrolases; EC 3.2.1.74), which cleave glycosidic bonds randomly in the interior of the cellulose polymer chain; the exo-1,4-β-glucosidases, which include the 1,4-β-D-glucan glucohydrolases (EC 3.2.1.74), which liberate D-glucose from 1,4β-D-glucans and hydrolyze cellobiose slowly, and the 1,4-β-D-glucan cellobiohydrolases (EC 3.2.1.91), which liberate D-cellobiose from 1,4-β-glucans; and β-D-glucosidases, or β-D-glucoside glucohydrolases (EC 3.2.1.91), which relieve end-product inhibition of the endoglucanases and cellobiohydrolases by hydrolyzing the penultimate product, cellobiose, to the less inhibitory final product, glucose.

The development during the past two decades of powerful molecular-biological methods for identifying, transferring and/or modifying, and overexpressing the genetic material that encodes specific proteins, combined with the understanding that cellulase action is a multi-activity process, suggests that cellulase mixtures might be engineered by combining enzymes that originate from various organisms to form effective, and possibly improved, hybrid cellulase systems. Among the potential benefits offered by such heterologous systems are reductions in the production cost that result from fewer enzymes being expressed, higher specific activity of the protein expressed, and the ability to tailor enzyme systems in accordance with the demands of specific processes (5).

In the present study, we have assembled an array of three endoglucanases (one fungal and two bacterial) from three organisms, plus three exoglucanases (two fungal and one bacterial), and have measured soluble-sugar production from ternary mixtures of these enzymes acting on microcrystalline cellulose. Each ternary system studied consisted of one endoglucanase, one exoglucanase (R) specific for the reducing cellulose terminus, and one exoglucanase (NR) specific for the nonreducing cellulose terminus. This paper presents the performance of selected cellulase mixtures using ternary contour diagrams to describe glucose production.

MATERIALS AND METHODS

Enzyme Purifications

Only highly purified enzymes were used to construct the cellulase cocktails evaluated in this study. The *T. fusca* enzymes (6), rE_3 and rE_3 (the prefix "r" denotes a recombinant enzyme), were purified in the laboratory of D. Wilson at Cornell University from cell lysates of *Streptomyces liyidans* TK24 that contained a plasmid carrying the appropriate *T. fusca* gene (7). rEndoglucanase A (rEndoA) from *Microbispora bispora*, expressed from a genomic fragment cloned in *Escherichia coli*, was purified in the laboratory of D. Eveleigh at Rutgers University (8).

Enzymes purified at NREL for this study included *A. cellulolyticus* endoglucanase I, EI, *(9)* expressed from *S. lividans* TK24, plus a truncated form of this enzyme (rEIcd) produced by subjecting the *S. lividans*-expressed enzyme to proteolytic treatment to remove the cellulose-binding domain *(10)*, *T. reesei* endoglucanase EG I, *T. reesei* exoglucanases CBH I and CBH II, and *Aspergillus niger* β-glucosidase. *A. niger* β-glucosidase was purified chromatographically from samples of Novozym 188 Cellobiase (Novo Nordisk, Franklinton, NC), and the *T. reesei* enzymes were prepared from samples of Laminex Cellulase, Lot 13-90091-01, code 6-5960 (Genencor International, Palo Alto, CA) according to the procedures described by Baker et al. *(10,11)*.

The purity of each enzyme preparation used in this study was verified by silver-stained, sodium-dodecylsulfate polyacrylamide gel electrophoresis. In addition, the purity of the *T. reesei* cellulases was verified by Western blot analysis using monoclonal antibodies *(12)*. These preparations were found to have compositions of at least 98% of the protein intended for evaluation.

Cellulose Digestions

Enzyme digestions of crystalline cellulose were carried out at 50°C in 50 mM acetate buffer, pH 5.0, to which 0.004% (w/v) sodium azide had been added to prevent microbial growth. Total cellulase loadings were held constant at 0.36 μM in the 1.0-mL reaction mixtures, which contained as substrate 5% (w/v) microcrystalline cellulose (Sigmacell, Type 20, Product number S-3504, lot 79F-0454, Sigma, St. Louis, MO). The total weight of protein added varied slightly with the molecular weights of the components involved, but in all cases the total cellulase loading (endoglucanase plus exoglucanase) was close to 20 μg protein per mL of digestion mixture, or 0.4 mg cellulase per g cellulose. Sufficient purified *A. niger* β-glucosidase (4.17 μg/mL of digestion mixture, equal to 0.61 units/mL or 12.2 units/mg of cellulose) was added to eliminate the problem of cellobiose inhibition. As described previously *(11)*, the sufficiency of the β-glucosidase loading was established (for the individual enzymes and for representative mixtures) by means of experiments in which reducing sugar output was measured in the presence of various loadings of β-glucosidase, to determine loadings above which further increases in β-glucosidase loading produced no further increases in yield of reducing sugar. The standard loading described above, equal to 83.33 μg β-glucosidase per g cellulose, was approx 10 times the minimum loading required to render the reducing-sugar output independent of the β-glucosidase loading *(11)*.

To reduce the consumption of purified enzyme, a miniaturized digestion apparatus was devised, using 1.5-mL Wheaton autoinjector vials as reaction vessels. The enzyme cocktails (0.3 mL total for each digestion of endoglucanase, exoglucanase, and β-glucosidase) for the various digestion mixtures were first placed in the vials, then substrate was added (as

0.7 mL of a 7.15% stirred slurry) to initiate the reaction. The vials were sealed with aluminum crimp-caps (PTFE-faced silicone septa, Kimble Glass, Vineland, NJ), placed in a custom-built rotator head immersed in a 50°C water bath, and continuously mixed by inversion at 10 rpm. After a standard 120-h digestion period, representative 0.04-mL aliquots of each (well-dispersed) digestion mixture were withdrawn, diluted to 2.0 mL with deionized water, and centrifuged to remove all solid substrate.

Analysis of Soluble Sugars Released

Sugar concentrations in aliquots harvested from the reaction vials were determined by ion-moderated partition chromatography on a Bio-Rad HPX-87P (lead-form; Hercules, CA) carbohydrate analysis column also equipped with a Bio-Rad deashing precolumn. This column system was installed in a Hewlett-Packard 1090 chromatograph and operated at 85°C with deionized water as mobile phase at a flow rate of 0.6 mL/min. The amount of glucose and cellobiose present in each sample was quantified by comparing the area of the peak against a linear calibration curve.

RESULTS AND DISCUSSION

Selection of Enzymes

The seven enzymes selected for this study included four endoglucanases along with examples of exoglucanases specific for either the reducing terminus or the nonreducing termini of the cellulose chain. Five ternary systems were developed using these purified cellulases. Table 1 shows some fundamental characteristics of the cellulases chosen for this study. Although many other cellulase component enzymes have been reported in the literature, the enzymes used in this study are well-characterized and either readily available in high purity as single-gene, recombinant products, or, in the case of the *T. reesei* enzymes, readily purified from commercial preparations. The choice of 50°C, and pH 5.0 as digestion conditions was made to accomodate the less thermostable *T. reesei* enzymes *(13)*, which constitute the reference system for our comparisons. Several other enzymes in this study are significantly more thermostable than the three *T. reesei* enzymes used. The *T. fusca* endoglucanase E_5 and the NR cellulose-terminus-specific exoglucanase, E_3 from the same organism have, respectively, temperature optima of 55 to 60° and 65°C in the hydrolysis of filter paper (D. Wilson, personal communication). *A. cellulolyticus* EI is even more thermostable, being optimally active against carboxymethylcellulose between 80 and 85°C *(9)*. Should a more thermostable reducing-terminus-specific (i.e., CBH I-like) exoglucanase be made available, either by characterizing a new enzyme or modifying CBH I itself, experiments run at higher temperatures will probably show EI, E_3, and E_3 to even greater advantage than is the case in the present study.

Table 1
Characteristics of the Exo- and Endoglucanases Used in This Study

Enzymes Studied	Glycosyl Hydrolase Family[a]	Cellulose Terminal Specificity	Product Stereo-chemistry	Form of Enzyme in Study	Thermal Tolerant
T. reesei CBH I	7	reducing	retaining	native	no
T. reesei CBH II	6	non-reducing	inverting	native	no
T. fusca E_3	6[b]	non-reducing[b]	inverting[b]	recombinant	moderately
T. reesei EG I	7	none	retaining	native	no
A. cellulolyticus EI	5	none	retaining	recombinant	yes
T. fusca E_5	5	none	retaining	recombinant	moderately
M. bispora Endo A	6	none	inverting	recombinant	yes

[a] From Bairoch (14).
[b] From D. Wilson (personal communication, 1997).

Summary of Saccharification Performance

The saccharification of microcrystalline cellulose by reconstituted ternary mixtures of purified cellulases (one endoglucanase and two exoglucanases) has been studied over the entire range of mixture compositions, and the results are shown in Figures 1–5 as ternary contour plots of the percentage of cellulose converted to soluble sugars. The actual compositions used in the experiments are indicated by the nodes (small circles). The contour lines in these figures represent a surface fitted to the experimental results using the program Statistica (Statsoft, Tulsa, OK) with the cubic spline option chosen. Assays measuring the activities of mixtures of the three *T. reesei* enzymes (Fig. 1) and those involving rEIcd and the two *T. reesei* exoglucanases (Fig. 3) were carried out in duplicate; the other experiments relied on single assays. For the assays run in duplicate, the average deviation of the duplicates from the mean was 4.2%.

The results shown in Figs. 1–5 can be compared in at least two ways. The first and simplest comparison is on the basis of maximum sugar yields, as shown in Table 2. The second way of comparing the systems considers the extent to which the system is forgiving in terms of deviations of the composition from the optimum. A useful measure of the forgiving nature of a system is the area on the ternary plot over which the sugar release is at least 85% of the maximum. For all plots shown, this area is fairly closely approximated by the two highest contour zones. In viewing these plots, the reader should bear in mind that the detail of the plots is limited by

Table 2
Maximum Saccharification from Selected Ternary Cellulase Digestions of Sigmacell-20

Ternary Cellulase System	Ratio Giving Highest Sugar Release (%)	Highest Sugar Release (%)
CBH I:CBH II:EG I	60:20:20	16.5
CBH I:CBH II:EI	40:20:40	16.2
CBH I:E_3:EI	40:20:40	14.1
CBH I:E_3:E_5	60:20:20	11.3
CBH I:E_3:Endo A	40:40:20	11.6

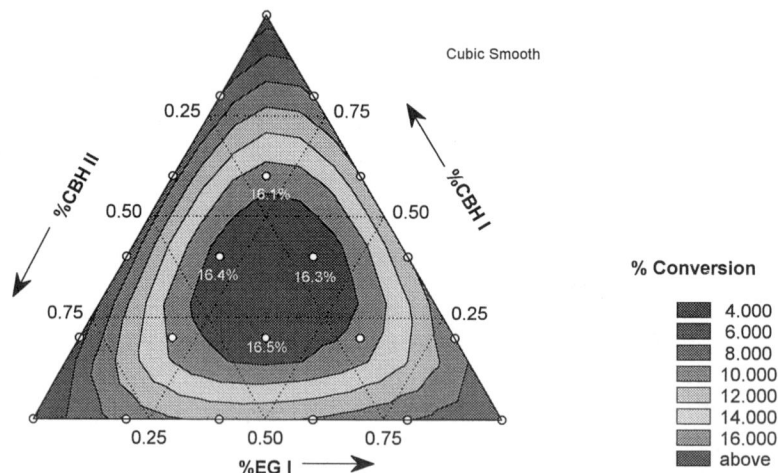

Fig. 1. Ternary plot of the saccharification of Sigmacell-20 using purified *T. reesei* CBH I, *T. reesei* CBH II, and *T. reesei* EG I. The apex indicated by each arrow represents a composition of 100% of the indicated cellulase; data points on the opposite side of the triangle represent assays that have the concentration of that cellulase set at zero.

both the relatively low resolution of the distributions of compositions (the "step" in composition being 20% in all plots), and by the relatively "stiff" surface-fitting routine employed. For instance, the authors have partial ternary diagrams run at higher resolution (data not shown) revealing that the high-yield plateau for the system of EG I, CBH II, and CBH I (Fig. 1) actually extends farther into the 100%-CBH I apex of the plot than is shown by the present plots.

The low-resolution but full-range plots shown here do, however, support some potentially useful observations. Comparison of the performance of five synthetic mixtures drawn from the cellulase systems of *A. cellulolyti-*

Hydrolysis of Cellulose

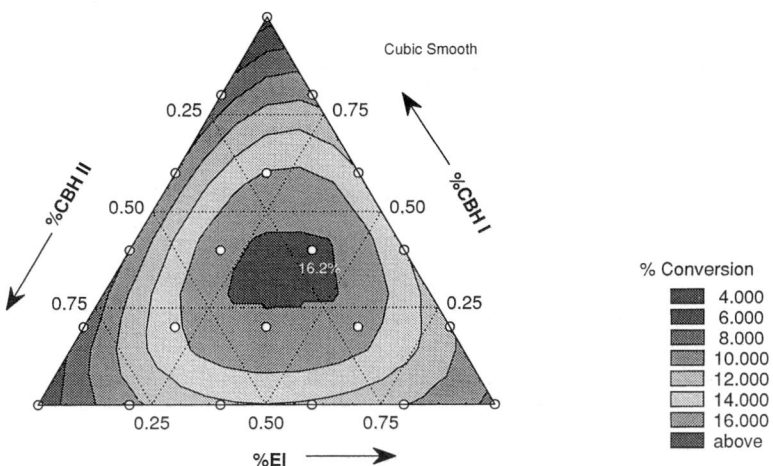

Fig. 2. Ternary plot of the saccharification of Sigmacell-20 using purified *T. reesei* CBH I, *T. reesei* CBH II, and *A. cellulolyticus* EI.

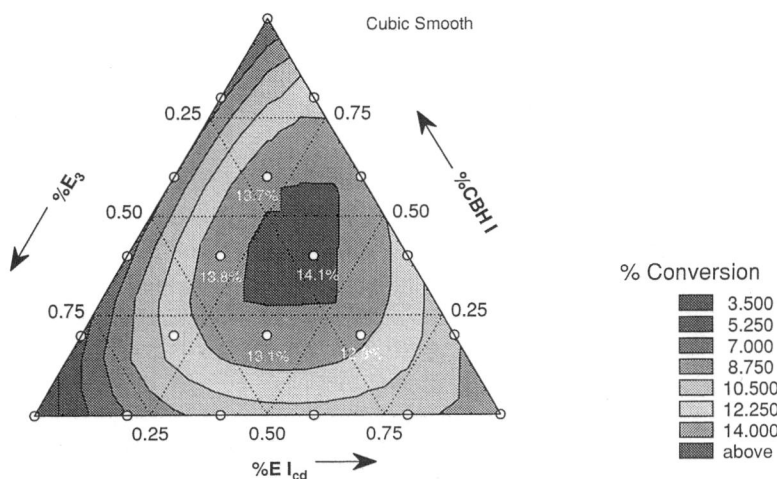

Fig. 3. Ternary plot of the saccharification of Sigmacell-20 using purified *T. reesei* CBH I, *A. cellulolyticus* rEIcd, and *T. fusca* F_3.

cus, *T. reesei*, *T. fusca*, and *M. bispora*, reveals that the performance of at least one completely heterologous synthetic mixture utilizing enzymes from three different microorganisms is competitive in both senses with the performance of a native (i.e., evolved) ternary system drawn exclusively from *T. reesei*. This heterologous system, which consists of the catalytic domain of endoglucanase EI from *A. cellulolyticus* (EIcd) and the exoglucanases CBH I from *T. reesei* and E_3 from *T. fusca*, is a forgiving system from the system-design point of view in that it delivers high saccharification rates over a wide range of mixture compositions. A

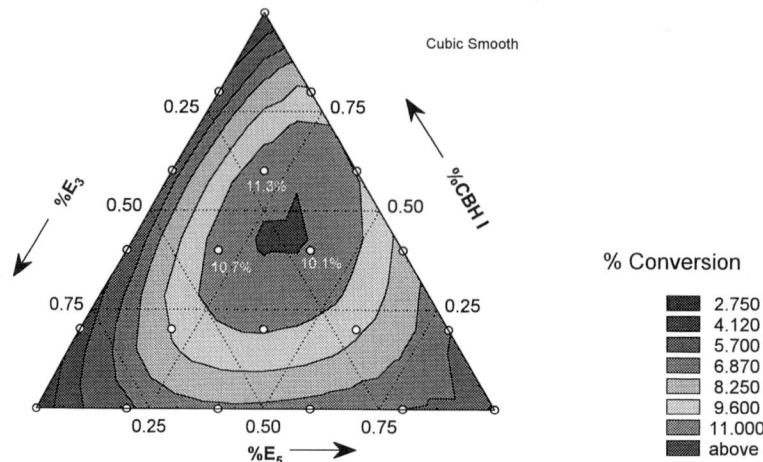

Fig. 4. Ternary plot of the saccharification of Sigmacell-20 using purified *T. reesei* CBH I, *T. fusca* E_3, and *T. fusca* E_3.

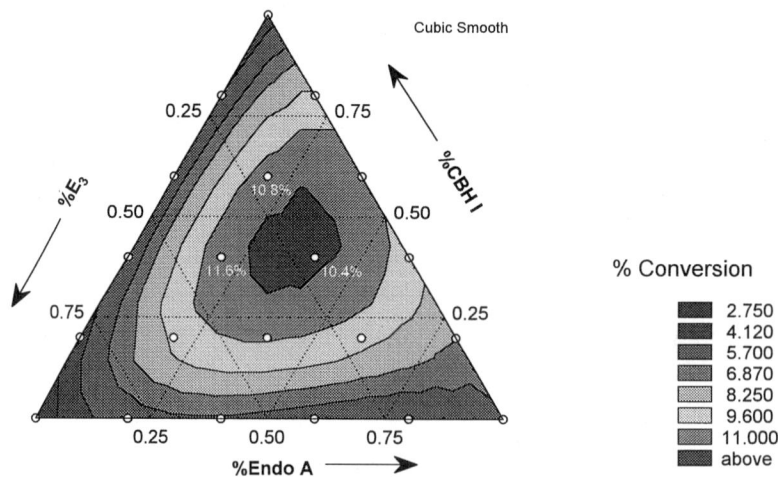

Fig. 5. Ternary plot of the saccharification of Sigmacell-20 using purified *T. reesei* CBH I, *T. fusca* F_3, and *M. bispora* EndoA.

two-organism system drawn from *A. cellulolyticus* (EI) and *T. reesei* (CBH I and CBH II) is also competitive with the native coevolved *T. reesei* system. The other completely heterologous system (*M. bispora* endoglucanase-A with *T. fusca* E_3 and CBH I, Fig. 5) and the other two-organism system (endoglucanase E_5 and exoglucanase E_3 from *T. fusca* with *T. reesei* CBH I, Fig. 4) are both less efficient and less forgiving in terms of composition than the systems of Figs. 1–3, but still illustrate that enzymes from different organisms, and indeed from different phylogenetic kingdoms, can work effectively together.

ACKNOWLEDGMENTS

This work was funded by the Ethanol from Biomass Program of the Biofuels System Division of the US Department of Energy. The authors wish to thank Douglas Eveleigh and David Wilson for the samples of rEndo A and rE_5 and rE_3 respectively, used in this study.

REFERENCES

1. Coughlan, M. P. (1990), in *Microbial Enzymes and Biotechnology*, Fogarty, W. M. and Kelley, C. T., eds., Elsevier, London, UK, pp. 1–36.
2. Grohmann, K., Wyman, C. E., and Himmel, M. E. (1992), in *Emerging Technologies for Materials and Chemicals from Biomass*, Rowell, R. M., Schultz, T. P., and Narayan, R., eds., ACS Series 476; American Chemical Society, Washington, DC, pp. 354–392.
3. Reese, E. T., Siu, R. G. H., and Levinson, H. S. (1950), *J. Bacteriol.* **59,** 485–497.
4. Gilligan, W. and Reese, E. T. (1954), *Can. J. Microbiol.* **1,** 90–107.
5. Eveleigh, D. E. (1987), *Phil. Trans. Roy. Soc. Lond.* **A321,** 435–447.
6. Wilson, D. B. (1988), *Methods Enzymol.* **160A,** 314–323.
7. Walker, L. P., Wilson, D. B., Irwin, D. C., McQuire, C., and Price, M. (1992), *Biotechnol. Bioeng.* **40,** 1019–1026.
8. Yablonsky, M. D., Bartley, T., Elliston, K. O., Kahrs, S. K., Shalita, Z. P., and Eveleigh, D. E. (1988), in *Biochemistry and Genetics of Cellulose Degradation*, Aubert, J.-P., Beguin, P., and Millet, J. eds., Academic, London, UK, pp. 249–266.
9. Himmel, M. E., Adney, W. S., Grohmann, K., and Tucker, M. P. (1994), *Thermostable Purified Endoglucanase from* Acidothermus cellulolyticus, US patent no. 5,275,944.
10. Baker, J. O., Adney, W. S., Nieves, R. A., Thomas, S. R., and Himmel, M. E. (1995), in *Enzymatic Degradation of Insoluble Polysaccharides*, Saddler, J. N. and Penner, M. H., eds., ACS Series 618, American Chemical Society, Washington, DC, pp. 113–141.
11. Baker, J. O., Adney, W. S., Nieves, R. N., Thomas, S. T., Wilson, D. B., and Himmel, M. E. (1994), *Applied Biochem. Biotechnol.* **45/46,** 245–256.
12. Nieves, R. A., Himmel, M. E., Todd, R. J., Ellis, R. P. (1990), *Appl. Environ. Microbiol.* **56,** 1103–1108.
13. Schulein, M. (1988), *Methods Enzymol.* **160,** 221–234.
14. Bairoch, A. (1996), SWISS-PROT Protein Sequence Data Bank (http://expasy.hcuge.ch/cgi-bin/lists?glycosid.text).

The Production and Properties of a New Xylose Reductase from Fungus *Neurospora crassa*

XIN ZHAO,* PEIJI GAO, AND ZUNONG WANG

Institute of Microbiology, Shandong University, 250100, Jinan, Shandong, P.R. China

ABSTRACT

Neurospora crassa X1 was found to ferment xylose and glucose simultaneously. Xylose was the appropriate inducer for the production of xylose reductase that had two isoenzymes designated as EI and EII. Both EI and EII, which were purified by affinity chromatography, had NADPH-dependent xylose reductase activities. EII also had NADH-dependent activity, and EI is the only xylose reductase found so far without any NADH-dependent activity. EI and EII had MWs of 30 kDa and 27 kDa, and pIs of 5.6 and 5.2, respectively. The specifities of EI and EII against triose, pentoses, and hexoses were studied. The K_ms against xylose for EI and EII were 2.3 mM and 1.1 mM respectively, which were much lower than those of the xylose reductase from yeast.

Index Entries: Xylose; xylose reductase; purification; characterization; *Neurospora crassa*.

INTRODUCTION

Xylose is the main sugar component of hemicellulose which comprises 30–40% in some agriculture residues such as corn stover and sugarcane [1]. Bacteria [2] and yeasts [3] have quite different ways to use xylose. Since xylose-isomerase gene from bacteria did not express sufficiently in yeasts, the studies of xylose utilization have been focused on the xylose metabolic pathway in yeasts such as *Candida*, *Pichia*, and *Pachysolen*. The characteristics of xylose reductase (EC 1.1.1.21) in yeasts vary between species and strains. *Pichia stipitis* [5] and *Candida* [6] species have only

---

*Author to whom all correspondence and reprint requests should be addressed. Current address is Agricultural and Biological Engineering, Purdue University, W. Lafayette, IN 47907

a single xylose reductase that has both NADPH- and NADH-dependent activities. The xylose reductase from *Pachysolen tannophilus* has two isoenzymes. Verduyn (7) indicated that one of these isoenzymes had both NADPH- and NADH-dependent xylose reductase activities, and the other one had NADPH-dependent activity with very low NADH-dependent activity. However, Ditzelmuler (8) reported that both of the isoenzymes from *Pachysolen tannophilus* had NADPH- and NADH-dependent activities. But all these yeasts can not utilize xylose in cellulosic hydrolysate efficiently.

Efforts have been made to look for microorganisms that can directly convert cellulosic materials to ethanol. *Neurospora crassa*, a well-studied fungus in genetic and molecular biology, is found to have the ability to not only produce cellulase and hemicellulase, but also to use xylose and glucose when it grows on cellulosic materials. The studies of the cellulase and hemicellulase from *N. crassa* has been reported (9,10), but its ability to use xylose and the enzymes involved in its xylose metabolism have not been investigated. Our studies focused on the xylose fermentation as well as the production and characterization of xylose reductase, the first enzyme, in the xylose metabolic pathway in *N. crassa* X1.

MATERIALS AND METHODS

Microorganism

Neurospora crassa X1 was isolated from decayed corn stover.

Medium

Nutrient salts for liquid medium: KH_2PO_4 2.0 g, $(NH_4)_2SO_4$ 1.4 g, urea 0.3 g, $MgSO_4 \cdot 7H_2O$ 0.3 g, $CaCl_2$ 0.3 g, $FeSO_4 \cdot 7H_2O$ 5.0 mg, $MnSO \cdot H_2O$ 1.5 mg, $ZnSO_4$ 1.4 mg, $CoCl_2$ 2.0 mg, dissolved in 1000 mL dH_2O. pH was adjusted to 5.5 with HCl or NaOH.

Liquid medium for glucose and xylose utilization contained 2.4% glucose, 1.6% xylose, and nutrient salts. Medium for the production of xylose reductase contained 1% xylose and nutrient salts.

To study the influence of sugars on the production of xylose reductase, 2% glucose, xylose, galactose, mannose or arabinose was used respectively as carbon source, and 0.1% one of the other four sugars was supplemented for the study of the induction of this sugar to xylose reductase. Sugars were dissolved in the nutrient salts solution described above.

Culture Condition

About 10^6 fresh spores of *N. crassa* X1 from 5 days' potato-agar slant were inoculated in 50 mL liquid medium, cultivated at 30°C with shaking under 100 rpm.

Preparation of Cell-Free Extract

Mycelia of *N. crassa* X1 were collected from 50 ml, 3 days' liquid culture, washed twice with water, resuspended in 5 mL of pH 7.2, 0.05 M phosphate buffer containing 1 mM 2-mercaptoethanol, then ground with quartz sand in a T-shaped motor grinder tube for 5 min at 10g. Whole cells and debris were removed by centrifugation at 10,000g, for 30 min, at 4°C. The supernatant was collected as crude xylose reductase.

Determination

Enzyme Assay (6)

Xylose reductase activity was assayed by mixing 0.1 mL diluted enzyme solution with 0.1 mL of 1 M xylose and 0.7 mL of 0.1 M, pH 7.2 phosphate buffer, then adding 0.1 mL of 3.4-mM NADPH (or NADH). This solution was immediately mixed and the absorbance (OD) at 340 nm vs time of reaction was recorded at 20°C. One unit (U) of the enzyme activity was defined as one μmol NADPH (or NADH) oxidized in one minute. Specific activity was expressed as U/mg protein.

For the studies of substrate specificities of EI and EII, xylose in the assay was replaced by D/L-glyceride, arabinose, ribose, mannose, galactose, glucose, or 2-D-deoxy-glucose, respectively.

Glycoprotein (12)

Diluted enzyme solution (1 mL) was mixed with 1 mL 2.5% phenol, 5 mL 98% H_2SO_4, and kept at room temperature for 30 min. Then the absorbance of this solution at 490 nm was determined. Glucose was used as a standard, and the concentration of sugar in protein was calculated as glucose.

Protein (13)

Soluble protein was determined by Folin-phenol method with bovine serum albumin as a standard.

Amino Acid (14)

Tyrosine and tryptophan contents (M) were determined by keeping a mixture containing 1.0 mL enzyme solution and 1.0 mL, 0.2 N NaOH at 30°C for 3 h. Then the absorbances (A) at both 294.4 nm and 280 nm were recorded for the calculations of tyrosine and tryptophan as follows:

$$M_{Tyrosine} = (0.592 \times A_{294.4} - 0.263 \times A_{280}) \times 10^{-3}$$

$$M_{Tryptophan} = (0.263 \times A_{280} - 0.170 \times A_{294.4}) \times 10^{-3}$$

Other amino acids were determined on the amino acid analyzer. The protein was hydrolyzed by mixing 1 mL protein solution with 1 mL, 6 N HCL, and hydrolyzing at 110°C for 20 h in N_2. Then the hydrolysate was evaporated, redissolved in 0.02 N HCL, and put on the amino acid analyzer (Beckman, HPLC equipped with Waters, amino acid analysis column).

Enzyme Purification

Affinity Chromatography

A column packed with reactive red agarose-120 gel (Type 3000-CL, Sigma) was equilibrated with pH 7.2, 0.05 M potassium phosphate buffer before crude xylose reductase was loaded. The loaded column was eluted with equilibration buffer, 0.5 M and 2.0 M KCl salt solutions, subsequently. The two salt-eluted fractions were collected, dialyzed against 0.05-M, pH 7.2 potassium phosphate buffer and assayed for xylose reductase activity. The purification was carried out at room temperature.

HPLC

Purified xylose reductase fractions from affinity chromatography were applied to a pre-equilibrated ion-exchange column on Waters-991 HPLC. Elution was carried out with pH 7.2, 0.05 M potassium phosphate buffer and then with a 0.1–1.0 M NaCl gradient at a rate of 1.0 mL/min. Peaks were detected at 280 nm.

Glucose and xylose were also determined by HPLC.

SDS-Gradient Ultra-Thin PAGE

Polyacrylamide gradient gel (3.9–22.5%) was used for SDS-electrophoresis on Pharmacia (Uppsala, Sweden) LKB MultiDive XL Type-Multiphor II Type equipment, under 100 V, 15′ and 600 V, 20A. Standard proteins were bovin (67,000), egg albumin (45,000), glycerol aldehyde 3-p-dehydragenase (36,000), carbonic anhydrase (30,000), trypsinogen (23,000), trypsin inhibitor (22,500). and α-lactalbumin (14,000) (Sigma).

IEF

Gradient gel (pH 3.5–9.5) was used in isoelectrofocusing at 50 mA, 20 W. The same equipment was used as in SDS-PAGE. Standard proteins were soybean trypsin inhibitor (4.55), β-lactoglobulin A (5.2), bovin carbonic anhydrase B (5.85), human carbonicanhydrase B (6.55), horse myoglobin (6.85, 7.35), lenti (8.15, 8.65), and trypsinogen (9.3) (Sigma).

RESULTS AND DISCUSSIONS

Utilization of Xylose in the Presence of Glucose

Assimilation of glucose and xylose by *N. crassa* X1 was studied in liquid medium containing glucose and xylose. Results shown in Fig. 1

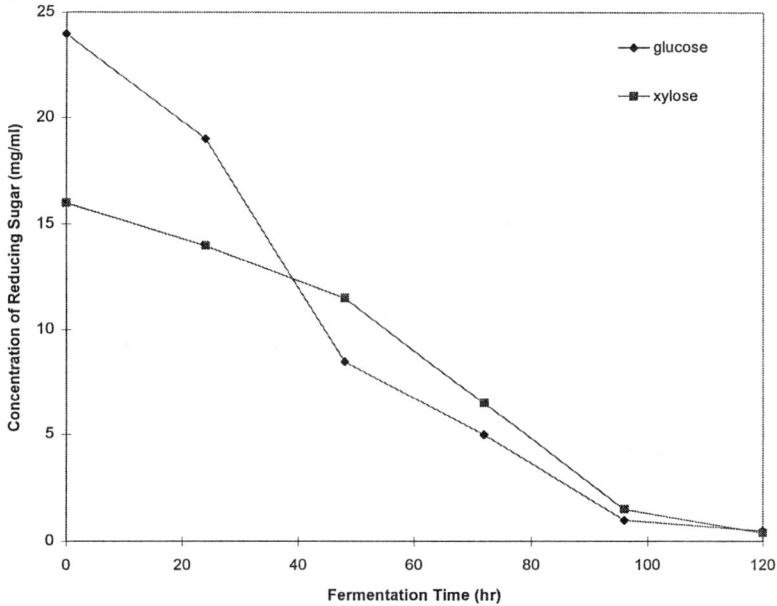

Fig. 1. Utilization of xylose in the presence of glucose by *N. crassa* X1.

indicated that *N. crassa* X1 had quite different characteristics from those of yeasts. *N. crassa* X1 can use glucose and xylose simultaneously according to Fig. 1, although the utilization of xylose was slower than that of glucose. However, in yeasts (6) such as *Candida Pichia*, and *Pachysolen*, when glucose and xylose presented in the media at the same time, xylose will not be consumed until glucose was used up, and a retention time of approx 10 h was needed before the assimilation of xylose. Similar to yeasts, *N. crassa* X1 converted glucose and xylose to ethanol and xylitol.

Production of Xylose Reductase

N. crassa X1 was cultivated in 50 mL, 1% xylose liquid medium. Samples were taken to have the xylose reductase activities and remained xylose examined. Results in Fig. 2 showed that the xylose in medium was almost used up after 96 h, and xylose reductase reached to its highest levels of approx 220 U/mL for NADPH-dependent activity, and 100 U/mL for NADH-dependent activity.

Influence of Carbon Sources on the Production of Xylose Reductase

Results of the influence of carbon sources on the production of xylose reductase from *N. crassa* X1 are presented in Table 1. Comparing the enzyme activities from cells growing in xylose with those growing in glucose or arabinose, the highest levels of xylose reductase specific activities were obtained only from the cells growing in xylose, which were 111 U/mg

Table 1
Effects of Different Carbon Sources on the
Production of Xylose Reductase from N. crassa X1

Carbon Source	Specific Xylose Reductase Activity			
	NADPH dependent		NADH dependent	
	U/ml	U/mg.protein	U/ml	U/mg.protein
2% glucose	72	34	24	11
+0.1% galactose	80	33	44	18
+0.1% arabinose	128	58	40	18
+0.1% mannose	36	37	80	33
+0.1% xylose	88	44	64	32
2% xylose	344	111	68	21
+0.1% galactose	310	111	61	22
+0.1% arabinose	352	117	48	16
+0.1% mannose	368	119	80	25
+0.1% glucose	302	104	60	21
2% arabinose	136	62	44	20

Fig. 2. The production of xylose reductase from N. crassa X1 in 1% xylose liquid culture.

protein for NADPH-dependent and 21 U/mg protein for NADH-dependent activities. The NADPH-dependent activity from cells growing in arabinose was about half of that in xylose. However, in Pachysolen (15) it was reported that the xylose reductase activity from cells growing in arabinose was two times higher than that from cells growing in xylose.

From *N. crassa* X1 growing in 2% glucose, 34 U/mg protein of NADPH-dependent as well as 11 U/mg protein of NADH-dependent reductase activities against xylose were obtained. However, it is reported for *Pachysolen* (15) that no xylose reductase activities were detected from cells growing in glucose.

From the results above, the production of xylose reductase from *N. crassa* X1 seemed not to be strictly inducible by pentose, and this may result in its simultaneous utilization of glucose and xylose (Fig. 1). This characteristic is quite different from yeasts, and might be useful in the fermentation of hydrolysate from cellulosic materials that contain glucose and xylose.

Purification of Xylose Reductase from *N. crassa* X1

Two fractions of xylose reductase, designated as EI and EII, were obtained from *N. crassa* X1 cell extract after the purification with affinity chromatography (reactive red agarose). EI, which was the fraction eluted with 0.5 M KCL, only had NADPH-dependent xylose reductase activity. EII, which was eluted with 2.0 M KCL, had both NADPH-dependent and NADH-dependent activities. HPLC was also used for further purification of EI and EII. The molecular weights of EI and EII, which were obtained from SDS-PAGE, were 30 kDa and 27 kDa, respectively. The pIs of EI and EII, which were obtained from electrofocusing, were 5.3 and 5.6, respectively.

Some of these characteristics of xylose reductase from *N. crassa* X1 were different from those reported from yeasts. First, all yeasts except *Pachysolen* had a single xylose reductase component; *Pachysolen* had two isoenzymes *(7,8)*, both of which had NADPH-dependent and NADH-dependent activities. However, we found two isoenzymes (EI and EII) in *N. crassa* X1. EII had both NADPH- and NADH-dependent activities, but EI had no NADH-dependent activity. Since EI was the only xylose reductase found so far without NADH-dependent activity, it may be used in studying the mechanism of NADH regeneration in pentose pathway to improve the enthanol production. Second, as reported, the molecular weights of xylose reductase from yeasts ranged from 31 kDa *(18)* to 70 kDa *(6)*, whereas EI and EII had MWs below 30 kDa. Finally, the pIs for yeasts were reported as 4.6 *(8)* and 5.2 *(18)*, which were lower than those (5.3 and 5.6) of EI and EII.

Determination of Glycoproteins and Amino Acid Composition of EI and EII

Both EI and EII were glycoproteins. The sugar contents (calculated as glucose) of EI and EII was 7.57% (w/w) and 11.3% (w/w) respectively.

The amino acid composition of EI and EII is shown in Table 2. The results indicate that the acidic amino acid contents of EI and EII were 20.5

Table 2
Amino Acid Composition of the Xylose
Reductase EI and EII from *N. crassa* X1

Amino acid[a]	EI	EII	Amino acid	EI	EII
	(mol%)			(mol%)	
Asx[b]	9.55	9.31	Leu	10.00	10.99
Thr	5.66	4.75	Tyr[d]	3.00	3.37
Ser	5.69	6.06	Phe	3.38	4.49
Glx[c]	10.94	11.27	Trp[d]	1.22	1.36
Gly	9.00	8.17	Lys	7.23	6.81
Ala	8.93	8.37	His	2.20	2.14
Val	7.34	6.37	Arg	4.49	5.24
Met	0.45	0.125	Pro	4.89	5.10
Ile	5.57	5.81			

[a] Trace Cys was detected.
[b] Asx = Asn + Asp.
[c] Glx = Gln + Glu.
[d] Tyr and Trp were analyzed by spectrophotometer.

and 20.6%, respectively, which are much lower than that of xylose reductase from *Candida shehatae* (6), which is 35%.

Substrate Specificity of EI and EII

Both EI and EII exhibited Michaelis-Menten kinetics with respect to their substrates. K_m value was calculated from the double reciprocal plots of v^{-1} vs $[\text{xylose}]^{-1}$, v^{-1} vs $[\text{NADPH}]^{-1}$, and v^{-1} vs $[\text{NADH}]^{-1}$ which were based on Michaelis-Menten equation (19). [xylose] (M), [NADPH] (mM) and [NADH] (mM) represent the concentrations of xylose, NADPH and NADH, respectively. Velocity (v) is expressed as the decrease of [NADPH] or [NADH] (mM) per minute. The concentration of substrate tested was up to 0.5 M xylose and 0.5 mM NADPH or NADH.

Eight sugars, listed in Table 3, were employed to determine the substrate affinity (K_m) of NADPH-dependent xylose reductase from *N. crassa* X1. The results were also compared with those of yeasts. In Table 3, the lowest K_m of 1.1 mM against xylose was obtained from EII. However, EI and the xylose reductases from yeasts had the lowest K_ms against D/L-glyceride instead of xylose. This indicated that only EII had the highest binding affinity on xylose. Therefore, EII might be considered a strict xylose reductase, and the other xylose reductases in Table 3 might be only regarded as aldose reductases.

Also from Table 3, when xylose was used as the substrate, EI and EII had K_ms of 2.3 mM and 1.1 mM against xylose, respectively, which were much lower than those of yeasts. This implies that the xylose reductases from *N. crassa* X1 (EI, EII) have a much higher binding affinity on xylose than those from yeasts.

Table 3
Comparison of Substrate Specificity of NADPH-Dependent Xylose Reductase
from N. crassa X1, P. stipitis, P. quercunm, C. utilis, and P. tannophilus

	Km (mM) (NADPH-dependent activity)					
	N. crassa EI	N. crassa EII	Pi. stipitis	Pi. quercunm	C. utilis	Pa. tannophilus
Substrate						
D.L-glyceride	0.7	36	18			1.7
xylose	2.3	1.1	42	78	28	12
arabinose	53	184	40			18
ribose	95	122	310		67	
mannose	>1000	NU				200
galactose	>1000	>1000	140			25
glucose	>1000	>1000	420			
2-D-deoxy-glucose	>1000	NU				33
Reference	This work	This work	7	5	5	16

Note: NU=no utilization

Note: NU = no utilization.

CONCLUSION

In conclusion, N. crassa X1 assimilated glucose and xylose simultaneously, and xylose reductase can be detected from cells growing in both pentose and hexose. Its combined characteristics of the utilization of cellulose/hemicellulose and glucose/xylose in a single strain will be very useful in investigating the mechanism of direct conversion of cellulosic materials to ethanol.

The xylose reductase from N. crassa X1 has two isoenzymes (EI and EII). EI is the only xylose reductase found so far without any NADH-dependent activity. EII has the highest binding affinity on xylose, and may be considered a strict xylose reductase. Further kinetics studies on EI and EII will be carried out to reveal the binding sites of xylose, NADPH, and NADH on the enzyme.

REFERENCES

1. Ladish, M. R., Lin, K. W., Voloch, M., and Tsao, G. T. (1985), *Enzyme Microb. Technol.* **5,** 82–100.
2. Asther, M. and Khan, A. W. (1985), *Appl. Microbiol. Biotechnol.* **22,** 318–324.
3. Schneider, H., Wang, P. Y., Chen, Y. K., and Malaszka, R. (1981), *Biotechnol. Lett.* **2,** 89–92.
4. Ho, W. Y. N., Stevis, P., Rosenfeld, S., Huang, J., and Tsao, G. T. (1983), *Biotechnol. Bioeng. Symp.* **13,** 245–250.
5. Jeffris, T. W. (1983), *Adv. Biochem. Eng. Biotechnol.* **27,** 1–32.
6. Toicola, A. Yarrow, D., Van den Bosch, E., Van Dejken, J. P., and Scheffers, W. A. (1984), *Appl. Environ. Microbiol* **47,** 1221–1223.
7. Verduyn, C., Van kleef, R., Frank, J., Scheruder, H., Van Dijken J. P., and Scheffers W. A. (1985), *Biochem. J.* **226,** 669–677.

8. Ditzelmuller, E. M., Kubicek-Pranz, E. M., Rihr, M., and Kubick, C. P. (1985), *Appl. Microbiol. Technol.* **22,** 297–299.
9. Yazdi, M. T., Woodward, J. R., and Radford, A. (1990), *J. General. Microbiol.* **136,** 1313–1319.
10. Yazdi, M. T., Woodward, J. R., and Radford, A. (1990), *Enzym. Microb. Technol.* **12,** 116–119.
11. Despande, V., Sulbha, K., Mishra, C., and Rao, M. (1986), *Enzym. Microb. Technol.* **8,** 149–152.
12. Dobois, M. (1959), *Anal. Chem.* **28,** 350–356.
13. Biochemistry group. (1986), *Direction on the Biotechnology*, Chinese Educational Press, Beijing, China.
14. Guo R. (1987), *The Application of Spectrometer Analysis in the Molecular Biology*, Science Press, Beijing, China, pp. 225–226.
15. Bolen, P. L. and Detroy, R. W. (1985), *Biotechnol. Bioeng.* **27,** 302–307.
16. Morimoto, S., Tawaratani, T., Azuma, K., Oshima, T., and Sinsky, A. J. (1987), *J. Ferment. Technol.* **65,** 17–21.
17. Egle, P. C. (1987), (Translated by K. W. Zhang), *Enzyme Dynamics*, Science Press, Beijing, pp. 225–226.
18. Ho, N., Lin, F. P., Huang, S., Andrews, P. C., and Tsao, F. T. (1980), *Enzyme. Microb. Technol.* **12,** 33–39.
19. Purich, D. L. (1980), *Methods Enzymol.* **63,** 8–14.

Session 3

Bioprocessing Research

RAKESH BAJPAI[1] AND DAVID A. GLASSNER[2]

[1]University of Missouri, Columbia, MO and
[2]National Renewable Energy Laboratory, Golden, CO

Bioprocessing research integrates microbiology, biochemistry, chemistry and engineering with a focus on conversion of renewable resources into value added products. Biocatalysis, and the processing of biological derived streams are typical research areas. Another important research area is the integration of biocatalysis and processing unit operations with the objective of producing a pure product.

The papers presented in the Bioprocessing Research session at the Nineteenth Symposium on Biotechnology for Fuels and Chemicals address important bioprocessing topics. Products specifically addressed in this session included bioplastics, lactic acid and ethanol. Processing technologies including mass transfer and separations were also addressed in the session.

The conversion of food processing streams to monomeric intermediates such as lactic acid and subsequent conversion to bioplastics was covered in this session. Conversion of starch to lactic acid via an integrated bioprocessing system was investigated. Process configurations for the enzymatic production of ethanol were reviewed and compared. The continuous roller bottle reactor for oxygen mass transfer in bioreactions was explored in one presentation. A patented process for the recycle and separation of acid and biomass derived sugars was reviewed. Finally, the characteristics of fouling and protein adsorption to plasma treated membranes was discussed.

An Integrated Bioconversion Process for Production of L-Lactic Acid from Starchy Potato Feedstocks

S.P. TSAI* AND S.-H. MOON

Argonne National Laboratory, 9700 South Cass Avenue, Argonne, Illinois 60439–4815

ABSTRACT

The potential market for lactic acid as the feedstock for biodegradable polymers, oxygenated chemicals, and specialty chemicals is significant. L-lactic acid is often the desired enantiomer for such applications. However, stereospecific lactobacilli do not metabolize starch efficiently. In this work, Argonne researchers have developed a process to convert starchy feedstocks into L-lactic acid. The processing steps include starch recovery, continuous liquefaction, and simultaneous saccharification and fermentation. Over 100 g/L of lactic acid was produced in less than 48 h. The optical purity of the product was greater than 95%. This process has potential economical advantages over the conventional process.

Index Entries: Lactic acid; starch; simultaneous saccharification and fermentation.

INTRODUCTION

Lactic acid (2-hydroxypropionic acid), being an acid and alcohol, is a versatile organic chemical. In addition to its current uses (mostly in food and food-related applications), lactic acid has a huge potential market as a feedstock for the synthesis of specialty and commodity biodegradable plastics, oxychemicals, and "green" solvents *(1)*. Lactic acid exists as two enantiomers: L(+)-lactic acid and D(−)-lactic acid. L-lactic acid is the natural form in human metabolism. D-lactic acid is metabolized differently by humans and has been reported to cause illness in infants *(2)*. Lactic acid can be made chemically from hydrogen cyanide and acetaldehyde, or via fermentation of carbohydrates. The chemical synthesis routes make only the racemic lactic acid, but a whole range of product optical purity (from

*Author to whom all correspondence and reprint requests should be addressed.

nearly 100% D to racemic to nearly 100% L) can be made via fermentation *(3)*, although the L-form is the desired product in most commercial lactic acid fermentation processes.

Carbohydrate metabolism of microorganisms for lactic acid production has been reviewed recently *(4)*. Although lactobacilli use simple sugars (such as glucose and lactose) efficiently, the preferred industrial *Lactobacillus* strains do not metabolize starch effectively. Although amylolytic lactobacilli and other lactic acid organisms, such as *Rhizopus oryzae* and *Bacillus laevolacticus*, have been reported, they typically suffer from such problems as lower fermentation rates, lower product yields, lower product concentration, and undesirable product stereospecificity. When starch is used as the carbon source for lactic acid fermentation using lactobacilli, the starch can be hydrolyzed into glucose before fermentation by means of well-established processes, such as the two-step enzymatic hydrolysis process widely practiced by the corn wet-milling industry at a very large scale to produce glucose syrup from corn starch. Such a process may not always be suitable for other starchy feedstocks, especially if a dedicated hydrolysis plant is needed at a smaller scale. Simultaneous saccharification and fermentation (SSF) has been used widely for ethanol production from starchy or cellulosic feedstocks *(5)*. Recently, a simultaneous liquefaction, saccharification, and fermentation process was reported for L-lactic acid production from barley starch by using *Lactobacillus casei* and a medium containing yeast extract and peptone as nutrients *(6)*. In this work, an economical process to produce L-lactic acid from starchy feedstocks has been developed by using potato as a model starchy feedstock. This process consists of simplified starch-recovery steps, continuous liquefaction of the starch, and SSF with high-performance *Lactobacillus* species.

MATERIALS AND METHODS

Enzymes and Microorganisms

α-Amylase (G-ZYME G995) and glucoamylase (G-ZYME G-990) were obtained from Enzyme Bio-Systems (Englewood Cliffs, NJ). The *Lactobacillus* culture LBM5 has been described previously *(3,7)*.

Starch Recovery

Idaho potatoes were purchased from a local grocery store, and were peeled and cut into small pieces (about $4 \times 4 \times 4$ cm). Potato wastes were obtained from a potato processing plant. The potato wastes consisted of variable sizes of cut pieces, and were partially gelatinized because of steam peeling in the plant. The raw potato materials were disintegrated by using a Waring blender (Model CB-3 Waring Products, New Hartford, CT), a hammer-mill type disintegrater (Dynacrush Soil Crusher, Custom Laboratory Equipment, Orange City, FL), or pestle and mortar. One to two L of water were added to 1 kg of potatoes for disintegration in the Waring blender. The blended potato and water were screened through a 20-mesh sieve and transferred to a 4-L beaker for starch settling. Potato starch powder was prepared by drying the wet starch under vacuum ($p = 140$ mmHg) at 50°C overnight, and crushing the material by using a mortar and pestle. The potato starch recovered in our laboratory was compared with a commercial potato starch (Sigma, St. Louis, MO Catalog No. S 4251) by using enzymatic hydrolysis, as described in Assays.

Starch Liquefaction

Continuous liquefaction experiments, using either the commercial potato starch or the potato starch recovered in our laboratory from Idaho potatoes, were performed in a continuously stirred, 500-mL tank reactor (working volume). One L of starch solution containing 248.4 g/L dry substance (DS) supplemented with 70 mg/L $CaCl_2$, was prepared. The starch solution was adjusted to pH 6.5, and 675 U of α-amylase was added. The starch solution was pumped at a flow rate of 25 mL/min into a 4.5-mm id × 91-cm stainless-steel tube immersed in a heating medium at a temperature of 102°C, and subsequently transferred into the reactor. The residence time in the gelatinization tubing was 35 s. The reaction temperature and residence time for liquefaction were controlled at 95°C and 20 min, respectively.

In the plug flow liquefaction experiment, a jacketed glass column (2.5 cm id × 30 cm) was used as the reactor, instead of the stirred tank reactor. This reactor was operated with or without packing materials. A dilute starch slurry recovered from potato waste without centrifugation was used in this experiment. One L of starch solution containing 120 g/L DS, supplemented with 70 mg/L $CaCl_2$, was prepared. The starch solution was adjusted to pH 6.0, and 338 U of α-amylase was added. The starch solution was pumped at a flow rate of 7 mL/min into a 4.5-mm id × 91-cm stainless-steel tube immersed in a heating medium at a temperature of 103.5°C, and subsequently transferred into the reactor. The reaction temperature was controlled at 103.5°C.

Samples were collected from the reactor effluent. The samples were quenched in ice water and diluted (1:10) with 0.1 N HCl solution to deactivate enzymatic activities. Concentrations of total reducing sugar in the samples were measured. Dextrose equivalent (DE), defined as the concentration of total reducing sugar (as glucose) as a percentage of DS, was calculated.

Simultaneous Saccharification and Fermentation

Precipitated starch granules recovered from potatoes as described above were used. A combination of 218 mL of water and 435 U of α-amylase were added to 363 g of a 40.2% DS potato starch suspension, in a 2-L BioFlo IIc fermenter (New Brunswick, Edison, NJ). The fermenter was autoclaved at 110°C for 40 min for liquefaction. Nutrient solution, which was autoclaved separately, was prepared by mixing 67.5 mL of corn-steep liquor, 254.5 mL of water, and 11.5 mL of 10 N sodium hydroxide to adjust to pH 7.0. A 400 mL sterile water and the nutrient solution were added to the fermenter containing the liquefied starch. The *Lactobacillus* inoculum, LBM5, was grown overnight at 42°C. At the time of the inoculation, and at 24 h, 50 U each of glucoamylase was injected to the fermenter. Initially, the pH of the medium was 6.4, and was later controlled at 5.5 by adding 5 N sodium hydroxide. The temperature and agitation speed were set at 42°C and 200 rpm, respectively.

Assays

The DS of the starch products was measured as % (w/w) solid remaining after drying overnight at 110°C. Total reducing sugars (TRS) were measured by the dinitrosalicyclic acid method *(8)*, and expressed as glucose. Methods for measuring glucose, lactic acid, and L-lactic acid were described previously *(3)*. To compare the recovered starch products with the commercial potato starch, starch solutions were prepared as 1.656% DS in phosphate buffer. Three hundred forty U of α-amylase were added to 3 mL of starch solution. Reaction conditions for liquefaction were pH 6.5 and 65°C for 3 min. For saccharification of the starch, the liquefied starch was adjusted to pH 5.0 by adding 0.5 N HCl. Four U of glucoamylase were added to the liquefied starch, and saccharification was carried out at 65°C for 1 h. Notice that these hydrolysis conditions were for assay and comparison purposes only, and were different from the conditions used in experiments for processing the materials.

RESULTS AND DISCUSSION

Starch Recovery

Results of starch recovery by using the Waring blender for disintegration are summarized in Table 1. Gelatinization was found to affect starch recovery. Without gelatinization, most of the starch precipitated as starch granules in 1–2 h, forming a bottom layer of about 40% (w/w) DS. When gelatinization was significant, either before or during potato disintegration, starch did not settle quickly, and, instead of forming starch granules, a 10–15% (w/w) DS starch slurry was formed in the settling tank. This starch slurry could be concentrated to 25–45% (w/w) DS by using a centrifuge. Also, the mechanical energy input during disintegration was found to be an important process parameter. An optimum range of the energy input, at 0.02–0.03 kWh/kg potato for the blender, led to complete disintegration and a high settling rate. When energy input was too low, disinte-

Table 1
Results of Starch Recovery Using a Blender for Disintegration

Raw materials	Fresh potatoes		Potato wastes		Commercial potato starch
Energy input during blending (kWh/kg potato)	0.053	0.024	0.015	0.024	N/A
Form of recovered starch				
w/o centrifugation	Starch slurry at 11% (w/w)	Precipitated starch granules at 40% (w/w)	Starch slurry at 12% (w/w)	Starch slurry at 13% (w/w)	N/A
w/ centrifugation	Starch mud at 30% (w/w)	N/A	Starch mud at 25% (w/w)	Starch mud at 26% (w/w)	N/A
Sugar concentrations after hydrolysis				
Glucose (g/L)	N/A	21.3	17.6	22.0	21.1
TRS (g/L)	N/A	23.6	20.3	24.2	22.9
Glucose/TRS	N/A	0.91	0.87	0.91	0.92

gration was incomplete; when energy input was too high, the temperature increase during disintegration was more noticeable, partial gelatinization was observed, and the starch did not settle quickly.

Blending 1 kg of fresh cut and peeled potato and 2 L of water at high speed for 5 min (i.e., energy input = 0.053 kWh/kg) resulted in partial gelatinization, and no starch granules were precipitated after 2 h of settling. Instead, 1.6 L of a dilute starch slurry was obtained, which was concentrated to 0.58 L of starch mud at 30% (w/w) DS after centrifugation at 1,400g for 15 min. When the same potato and water mixture was blended for 1 min at low speed and for 2 min at medium speed (i.e., total energy input = 0.024 kWh/kg), the blended mixture settled and formed three layers. The top layer (2 L) was a dark brown liquid (including some floats). The middle layer (1 L) was a dark brown slurry (10–12% [w/w] DS). The middle layer seemed to contain proteins and a gelatinized portion of starch. The bottom layer contained white starch granules (40% [w/w] DS). The volume of the starch layer was 175 mL, and its density was 1.2 g/mL. When the starch product from the bottom layer was diluted with water to 25% (w/w) DS, the slurry could be pumped by a peristaltic pump. The starch recovery yield from the bottom layer was 62%, assuming 12% starch in potatoes. More starch could be recovered from the middle layer as slurry.

When 1 kg of potato waste was blended with 1 L of water for 1 min at low speed, followed by 1 min at medium speed (i.e., total energy input = 0.015 kWh/kg), in 2 h of settling, 1500 mL of slurry was obtained. In the subsequent centrifugation of the slurry at 1400g for 30 min, 800 g of the concentrated starch mud was recovered. The DS of the starch mud was measured as 25% (w/w). When the same potato waste and water mixture was blended for 1 min at low speed, and then for 2 min at medium speed (i.e., total energy input = 0.024 kWh/kg), in 2 h of settling, 1200 mL of potato slurry was obtained. In the subsequent centrifugation at 1400g for 30 min, 658 g of the concentrated starch mud was recovered. The DS of the starch mud was measured as 26.0% (w/w). After enzymatic hydrolysis, the starch samples obtained from potato waste at 0.015 kWh/kg of blending energy input yielded less glucose, indicating insufficient disintegration. The hydrolysis results of all other samples of recovered starch were comparable with those of the commercial potato starch.

Gelatinization should be avoided for efficient recovery of starch as precipitated granules in high yields. Normal starch granules are composed of linear (amylose) and branched (amylopectin) starch molecules associated by hydrogen bonding. Undamaged starch granules are insoluble in cold water, but have a limited capacity to absorb water, and swell reversibly. If an aqueous suspension of starch is heated to a certain critical temperature (the lower limit of the gelatinization temperature), the hydrogen bonds within the granules are disrupted and gelatinization occurs. The amorphous regions of the granule are solvated first and the granule swells rapidly, accompanied by leaching out of the linear amylose molecules into the solution. When the solution containing amylose molecules, swollen granules, and granule fragments is cooled, it thickens. Therefore, gelatinization not only causes amylose to leach out, but also prevents the remaining granules from settling.

Too high a mechanical input can cause gelatinization by localized temperature increase and excessive mechanical shear. During the disintegration experiments, a temperature increase of the potato slurry was noticed. Although the maximum temperature increase in the bulk is estimated to be about 15°C (i.e., to a bulk temperature of about 35°C), poor heat transfer within the mixture can result in a localized higher temperature exceeding the gelatinization temperature of potato starch (62–68°C). In addition, excessive mechanical shear can disrupt the granular integrity and cause the starch to gelatinize more easily.

Adding water, from 1 to 2 L/kg potato, to the blender improved the starch recovery yield, compared with blending without additional water. Water probably worked as a lubricating agent and a heat sink to dissipate the heat generated by mechanical energy. Also, diluting the starch concentration might have favored retrogradation of amylose molecules and precipitation of amylose clusters.

Using the hammer-mill-type crusher and grinding the potato with mortar and pestle both produced comparable starch yields and concentrations, but with two apparent advantages: Foaming was reduced and no antifoam was needed in the settling tank; and the crushed or ground potato passed through the sieve faster than the blended potato, with effective removal of fibrous materials.

Starch Liquefaction

For continuous liquefaction in the stirred-tank reactor, at a 20-min residence time, a steady state was achieved using the commercial potato starch in 140 min (obtaining 25 g/L TRS or 10 DE); by using the potato starch prepared in our laboratory, the steady-state sugar concentration was 20 g/L TRS or 8 DE. For continuous liquefaction of starch recovered from potato waste in the plug flow reactor (PFR), the use of packing material (ceramic Berl saddle or Rashig ring) partially clogged the reactor with gelatinized starch. Even without packing material, bubbles of the boiling starch solution helped achieve efficient mixing. Continuous liquefaction was, thus, performed without packing material, and a liquefied starch of 10 DE was produced at the steady state at a 21-min residence time. In this experiment, the feed (starch slurry recovered from potato wastes) contained greater than 10^5 colony-forming-units/mL microbial contaminants that were able to grow at 37°C and 42°C on *Lactobacilli* MRS (Difco, Detroit, MI) plates. The liquefied starch (a residence time of 20 min in the reactor at 103.5°C) contained some microbes that grew at 37°C, but no colony growth was observed at 42°C on MRS plates. These findings suggest that the liquefied starch prepared by using the above procedure can be used for lactic fermentation at 42°C, without further sterilization.

Simultaneous Saccharification and Fermentation

The time-course of the SSF is shown in Fig. 1. During the run, the liquefied starch was saccharified to glucose by glucoamylase, and the glucose was metabolized by microorganisms and converted into lactic acid. Initially, the saccharification rate was greater than the fermentation rate, resulting in an accumulation of sugars. The fermentation ended at around 30 h, at which time the concentration of glucose was lower than 0.5 g/L, and the lactic acid concentration was 100 g/L. The fermentation rate was comparable with those of lactic acid by the same organisms under similar conditions using higher purity glucose. Using the crude starch prepared in this study seemed to yield a slightly higher fermentation rate. The starch recovered by blending and settling in this work contained about 0.025 g protein/g starch. This finding suggests that the simplified starch-recovery process, which generates a starch product that retains some of the potato proteins, may be more suitable for providing the carbon substrate for lactic acid fermentation than the typical intensive starch-recovery processes that make highly purified starch.

The lactic acid yield from starch was essentially 100% of theoretical yield, as predicted from the DS of the raw material, assuming that the corn-steep liquor initially contained 25% (dry basis) lactic acid. The optical purity of lactic acid at the end of fermentation was 95% L-lactic acid. This LBM5 culture has routinely produced greater than 98% L-lactic acid in our laboratory in fermentations using a well-defined medium not containing corn steep-liquor (3). The D-lactic acid introduced by the corn-steep liquor used in the fermentation medium caused the slight decrease in product stereospecificity in this run.

The glucoamylase used in this process had an optimal pH of 4.3, and is typically used at 60°C in industrial dextrose production processes. These conditions are considerably different from the optimal pH, 5.5, and temperature, 42°C (9), for lactic acid fermentation by the LBM5 culture. The enzyme supplier's literature indicates a 85% activity at pH 5.5 relative to pH 4.3 and a less than 30% activity at 42°C relative to 60°C, if all other conditions are the same. Nevertheless, the glucoamylase exhibited satisfactory activity at the process conditions. Furthermore, the glucoamylase activity was not appreciably inhibited by the constituents of corn-steep liquor or impurities in the recovered crude starch. Although the total glucoamylase dosage in this run was 0.56 U/g·lactic acid produced, the second dosage of glucoamylase added at 24 h could be reduced or eliminated.

Commercial-Scale-Process Scheme

On the basis of the laboratory results, an integrated process suitable for commercial-scale production of L-lactic acid from starchy feedstocks was designed; the process flow diagram is shown in Fig. 2. The process consists of starch recovery, continuous liquefaction, and SSF. Although the

Fig. 1. Time course of a batch SSF run.

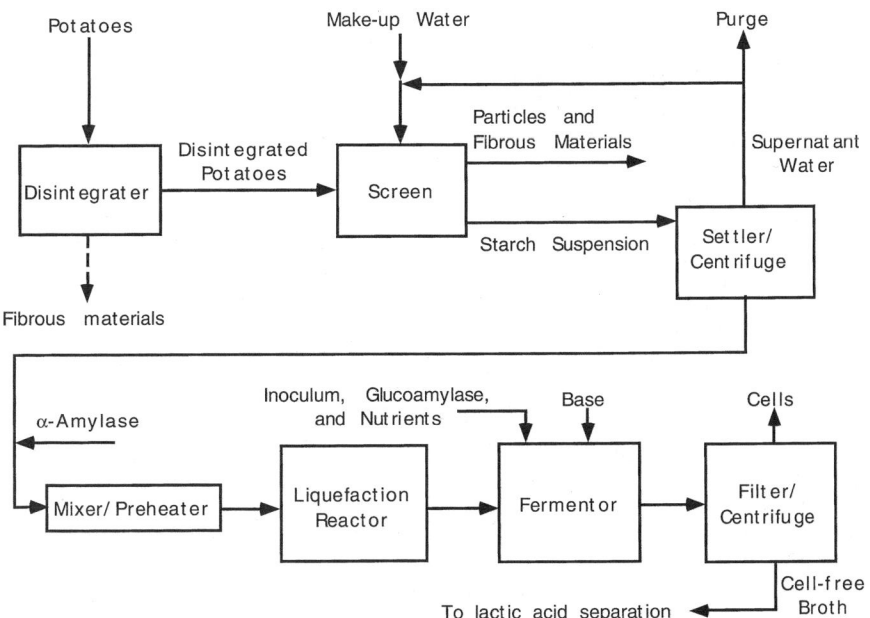

Fig. 2. Process flow diagram of the integrated process for L-lactic acid production from starchy feedstocks.

process described below involves using potatoes, other root and tuberous, starchy crops, such as cassava, could also be used.

The potato raw material is fed into a disintegrator (such as a ball mill or a hammer mill). When a high-fiber potato feed is used, a Rietz mill (a type of hammer mill) is preferred for separation of a large portion of fibrous materials as a secondary discharge. The disintegrated potato is screened to remove large particles on a mechanically vibrated woven wire or plastic texture, with the Tyler equivalent from 10- to 100-mesh. Water is sprayed over the screen to entrain starch. The starch is concentrated from the dilute starch suspension in a settling tank or a centrifuge. Although a settling tank is used for precipitating starch granules from normal (unprocessed) potato starch in a 1–2-h holding time, a centrifuge is used for concentrating the starch slurry from gelatinized starch wastes. A centrifuge can even be used for the normal potato starch to enhance the separation rate. Most of the supernatant water is recycled to the screen, and part of it is purged. If wet milling is desired, the make-up water is fed into the mill. The starch concentrator (i.e., settling tank or centrifuge) generates a crude starch product at a DS concentration from 10 to 45% (w/w) as needed. The starch products are diluted to 15–25% DS for gelatinization and liquefaction.

Complete gelatinization is achieved at a minimal ratio of water to starch of 4. Gelatinization causes the change of the retrograded starch structure, which is resistant to enzyme attack, into an amorphous pattern that is accessible to the enzymes required for degradation. Gelatinization is not completed for a starch concentration exceeding 25% (w/w DS). If the pH of the starch is far from the optimum pH of the α-amylase to be used for liquefaction, the pH should be adjusted before gelatinization. The recovered starch is mixed with α-amylase and then preheated at 90–120°C for no longer than 1 min in a line-mixer. Gelatinization occurs during the preheating. Mixing and preheating can be practiced in a single unit, but the fluid must not be heated before mixing is completed, to ensure that α-amylase is distributed uniformly. The gelatinized starch is fed into a liquefaction reactor, which can be a PFR or preferably, a continuous stirred-tank reactor (CSTR). The residence time for starch liquefaction is from 15 min to 3 h, preferably, 20 min when 3 U α-amylase/g DS are used. Calcium concentrations of 30–75 ppm on a DS basis are required for α-amylase use. The optimum residence time varies with the dosage of the enzyme. With a suitable residence time, the continuous liquefaction process should generate a solution of short-chain starch molecules, with a typical DE value of 10. The liquefied starch is cooled or heat-exchanged with the feed and fed into the fermenter without further sterilization.

The fermenter is operated in a batch mode. Initially, the fermenter is filled with the liquefied starch, sterilized corn-steep liquor, and additional water, if the total carbohydrate concentration needs to be adjusted. Inoculum culture (at 5–10% of the working volume of the production fermenter) and corn-steep liquor (ranging from 10 to 50 g/L [dry basis], preferably,

10g/L) are added. The higher concentration of corn-steep liquor increases the fermentation rate, but it also increases the concentration of impurities in the fermentation broth, which increases the purification cost in the downstream process. Suitable microorganisms include *Lactobacillus delbrueckii* subsp. *lactis*, the LBM5 culture used in this work, and many other L-specific lactobacilli reported in the literature. Fermentation can be carried out at 37–45°C (preferably, at 42°C). The medium is initially adjusted to pH 6.3. The pH is allowed to drop, and is controlled at 5.5 by adding 5–10 N sodium hydroxide. Because of the high temperature and acidic pH, this fermentation is highly resistant to microbial contamination. At the time of inoculation, glucoamylase is added to the fermenter aseptically to effect SSF. Glucoamylase dosage is typically 0.3 U/g starch. If necessary, more glucoamylase is added at timed intervals. The fermentation is completed in 25–40 h, if 10% carbohydrates are initially present in the medium. Cell mass is separated from the fermentation broth in a filter or a centrifuge. The cell-free broth containing lactate and impurities is transferred to a downstream process, in which lactic acid is recovered and purified.

CONCLUSIONS

An integrated process, consisting of starch recovery, continuous liquefaction, and SSF, has been developed for L-lactic acid production from starchy feedstocks, using potato as a model feedstock. The feasibility of using an inexpensive nutrient source, corn-steep liquor, for such a process is demonstrated, and a high optical purity (95% L) of the lactic acid product was obtained by using a stereospecific *Lactobacillus* culture. This process can use various types of potato feedstocks. Although starch recovery from the unprocessed potatoes is easier, the starch recovery steps in this process are capable of generating a suitable starch stream from partially gelatinized potato wastes. The simplified starch-recovery step used in this process, which generates a crude starch, is more suitable for supplying a low-cost carbohydrate for fermentation than the conventional potato starch manufacturing processes that involve many processing steps and generate highly purified starch *(10,11)*. The crude starch prepared in this process is expected be less expensive and can provide part of the fermentation nutrient requirements. This process uses proprietary *Lactobacillus* cultures that give a high fermentation rate (3 g/L·h), a high product concentration (over 100 g/L), a high product stereospecificity (95% L), and a high product yield (nearly 100%) in SSF.

The costs of enzyme required for this process are low: They are estimated at less than 0.5¢/lb of lactic acid produced. Although the α-amylase and glucoamylase are not used at their optimal conditions, and the enzyme dosages used in this process are slightly higher than those used in typical corn wet-milling operations, the costs of enzymes in this process are controlled at a reasonable level. This process is expected to be more economical by combining hydrolysis with fermentation. Instead of the conventional two-step hydrolysis of starch followed by fermentation, the liquefied starch is fed into the fermenter for SSF, eliminating the sterilizer and the saccharification tank, and reducing the overall process time, the capital investment, and the operating costs.

ACKNOWLEDGMENT

The work reported here was supported by the US Department of Energy, Assistant Secretary for Energy Efficiency and Renewable Energy, under Contract W-31-109-Eng-38.

REFERENCES

1. Datta, R., Tsai, S. P., Bonsignore, P., Moon, S.-H., and Frank, J. R. (1995), *FEMS Microbiol. Rev.* **16,** 221–231.
2. Benninga, H. (1990), *A History of Lactic Acid Making,* Kluwer, Dordrecht, the Netherlands.
3. Tsai, S. P., Coleman, R. D., Moon, S.-H., Schneider, K. A., and Sanville Millard, C. (1993), *Appl. Biochem. Biotechnol.* **39/40,** 323–335.
4. Bigelis, R., and Tsai, S. P. (1995), in *Food Biotechnology—Microorganisms,* Hui, Y. H. and Khachatourians, G. G., eds., VCH, New York, pp. 239–280.
5. Keim, C. R. (1983), *Enzyme Microb. Technol.* **5,** 103–114.
6. Linko, Y.-Y. and Javanainen, P. (1996), *Enzyme Microb. Technol.* **19,** 118–123.
7. Tsai, S. P., Moon, S.-H., and Coleman, R. (1995), US Patent 5,464,760.
8. Miller, G. L. (1959), *Anal. Chem.* **31,** 426–428.
9. Anonymous, *G-Zyme G990 Glucoamylase,* Enzyme Bio-Systems, Englewood Cliffs, NJ.
10. Howerton, W. W. and Treadway, R. H. (1948), *Ind. Eng. Chem.* **40,** 1402–1407.
11. Treadway, R. H. (1967), in *Starch: Chemistry and Technology,* vol. 2, Whistler, R. L. and Paschall, E. F., eds., Academic, New York, pp. 87–101.

Production of Ethanol from Starch by Co-Immobilized *Zymomonas mobilis*–Glucoamylase in a Fluidized-Bed Reactor**

May Y. Sun,[1,2] Nhuan P. Nghiem,*,[1] Brian H. Davison,[1,2] Oren F. Webb,[1] Paul R. Bienkowski[1,2]

[1]*Bioprocessing Research and Development Center, Oak Ridge National Laboratory, P.O. Box 2008, Oak Ridge, TN 37831; and* [2]*Department of Chemical Engineering, University of Tennessee, Knoxville, Tennessee 37916*

ABSTRACT

The production of ethanol from starch was studied in a fluidized-bed reactor (FBR) using co-immobilized *Zymomonas mobilis* and glucoamylase. The FBR was a glass column of 2.54 cm in diameter and 120 cm in length. The *Z. mobilis* and glucoamylase were co-immobilized within small uniform beads (1.2–2.5 mm diameter) of κ-carrageenan. The substrate for ethanol production was a soluble starch. Light steep water was used as the complex nutrient source. The experiments were performed at 35°C and pH range of 4.0–5.5. The substrate concentrations ranged from 40 to 185 g/L, and the feed rates from 10 to 37 mL/min. Under relaxed sterility conditions, the FBR was successfully operated for a period of 22 d, during which no contamination or structural failure of the biocatalyst beads was observed. Volumetric productivity as high as 38 g ethanol/(Lh), which was 74% of the maximum expected value, was obtained. Typical ethanol volumetric productivity was in the range of 15–20 g/(Lh). The average yield was 0.49 g ethanol/g substrate consumed, which was 90% of the theoretical yield. Very low levels of glucose were observed in the reactor, indicating that starch hydrolysis was the rate-limiting step.

** The submitted manuscript has been authored by a contractor of the US government under contract DE-AC05-96OR22464. Accordingly, the US government retains a nonexclusive, royalty-free license to publish or reproduce the published form of the contribution, or allow others to do so, for US government purposes.

* Author to whom all correspondence and reprint requests should be addressed.

Index Entries: Ethanol; starch; *Zymomonas mobilis*; glucoamylase; simultaneous saccharification and fermentation; fluidized-bed reactor.

INTRODUCTION

Despite the fact that ethanol fermentation is a very old process, attempts have constantly been made to improve it. Many of these efforts have aimed at the development of a continuous process. A highly efficient continuous fermentation process requires some form of cell retention to prevent wash-out and maintain sufficient cell concentrations for high substrate conversion rates. One method of cell retention is immobilization. *Zymomonas mobilis* immobilized in κ-carrageenan beads has been used in a fluidized-bed reactor (FBR) to produce ethanol from glucose at productivity as high as 120 g/(Lh) *(1)*. This was a significant improvement over other continuous systems for ethanol production, for example, 2–5 g/(Lh) in typical batch and fed-batch processes, 6–8 g/(Lh) for a free-cell continuous stirred tank reactor (CSTR), 10–16 g/(Lh) for an immobilized-cell CSTR, 10–30 g/(Lh) for a hollow-fiber reactor, and 16–40 g/(Lh) for a vertical packed bed with immobilized cells *(2)*. An FBR has distinct advantages over a mixed reactor. It allows a faster approach to reaction completion, thanks to its plug-flow characteristic, which maintains high substrate concentrations throughout most of the reactor and localization of product inhibition to the section near the exit. An FBR also offers minimization to mass transfer restrictions caused by channeling and CO_2 build-up, which are the two common problems of packed-bed reactor (PBR).

Starch is the most abundant renewable carbon source and has been used extensively in ethanol production. Unfortunately, the two best ethanol-producing organisms, *Saccharomyces cerevisiae* and *Z. mobilis*, cannot use starch. Therefore, the conversion of starch to ethanol normally requires two stages: hydrolysis of starch to glucose by acid or enzyme, and its subsequent conversion to ethanol. Siginificant savings on capital costs can be realized if both steps are carried out simultaneously in a single reactor. This has been the focus of research on starch-to-ethanol fermentation. One approach involved the integration of a gene encoding the enzyme glucoamylase (GA) into the chromosome of an ethanol-producing strain. The recombinant organism had the capability of converting soluble starch directly to ethanol *(3)*. Another approach involved the co-immobilization of GA, or an organism possessing that enzymatic function, and an ethanol-producing organism in small beads. The biocatalyst then was used to convert soluble starch to ethanol in a single reactor *(4–7)*. In addition to savings on capital costs, the co-immobilization system offers savings on enzyme costs, since, in this system, GA is retained in the reactor for repeated uses.

In a previous study, the performance of a GA-*S. cerevisiae* system was evaluated in an FBR *(6)*. In the present study, an FBR also was used to evaluate the performance of a GA–Z. *mobilis* system. The intial results are reported in this paper.

MATERIALS AND METHODS

Microorganism

Z. *mobilis* NRRL-B-14023 was used. The stock culture was maintained in 25% glycerol and kept at $-70°C$. To provide the cells for the preparation of the biocatalyst beads, Z. *mobilis* was grown in a 75-L fermenter (New Brunswick, Edison, NJ). The inoculum was prepared in a 3-L fernbach containing 2 L medium. The fernbach medium contained 50 g/L glucose, 5 g/L Tastetone 900AG yeast extract (Red Star, Juneau, WI), and 6 g/L KH_2PO_4. The medium was adjusted to pH 5.0 with concentrated phosphoric acid, sterilized by autoclaving at 121°C for 20 min, and allowed to cool to ambient temperature prior to inoculation. One stock vial containing 1.5 mL culture was thawed and used to inoculate the fernbach. The fernbach was incubated at 30°C, with gentle mixing, for 36 h before its entire contents was used to inoculate the fermenter. The fermenter medium had the same composition as the fernbach medium. The fermenter was maintained at 30°C. The pH was maintained at 5.0 using 2 N NaOH. After 20 h, when 90% of the glucose in the medium had been consumed, the cells were recovered by centrifugation (Sharples Super-Centrifuge AS26 NF). The concentrated biomass was stored at 4°C until its immobilization was carried out.

Enzyme

GA immobilized on a solid matrix having average particle diameter of 1.0–1.5 mm was provided by Genencor. Since the particle size was too large for the preparation of the co-immobilization of the enzyme with the Z. *mobilis* cells (*see* Preparation of Biocatalyst Beads), the immobilized GA was ground in a ceramic mortar placed in an ice bath, until the particle diameter was less than 0.1 mm. After the grinding process was complete, a small sample was taken and centrifuged. The supernatant was tested for activity at 35°C, using maltodextrin at 80 g/L as substrate. In this test, no glucose was released, which indicated that no enzyme was lost from the support during the grinding process. The activity of the ground, immobilized GA was measured at 35°C and pH 5.0. One hundred mL buffer (6 g/L KH_2PO_4 adjusted to pH 5.0) was used to wash 2 mL ground GA into a 250-mL flask. The flask then was placed in a 35°C water bath and its contents was mixed by a small stir bar. When the temperature in the flask was the same as the temperature of the water bath, 100 mL of a 80 g/L maltodextrin in the same buffer, which had been kept at 35°C, was quickly added to the flask. Samples were taken at intervals, quickly centrifuged to separate the immobilized enzyme, and the glucose concentration in the supernatant was measured with a YSI glucose analyzer (Yellow Springs Instruments, Yellow Springs, OH). The specific activity of the ground, immobilized GA was calculated as 1.16 g glucose/mL enzyme-h.

Preparation of Biocatalyst Beads

To prepare the biocatalyst beads, 40 g κ-carrageenan (Type NSAL 798 from FMC) was slowly dissolved in about 600 mL de-ionized water kept in a water bath at 35°C. Mixing was provided by a stir bar. Then 40 g wet wt Z. mobilis cell paste was added, followed by 150 mL ground, immobilized GA. Finally, de-ionized water was added to the final volume of 1 L. In the original procedure for the preparation of immobilized Z. mobilis in κ-carrageenan, Fe_2O_3 was added to the mixture to increase the specific density of the biocatalyst beads, to prevent them from floating (1). In this work, the addition of Fe_2O_3 was not necessary, and therefore omitted, because the solid matrix onto which the GA was immobilized by the manufacturer already made the specific density of the beads sufficiently high to allow them to sink. The slurry was pumped through a 100-mL pipet tip to form small droplets, which were allowed to drop into a solution of 0.3 M KCl. This fixing solution was stirred gently to prevent contact between droplets before they solidified. The beads were recovered, screened to remove those bigger than 2.5 mm in diameter, and stored in 0.3 M KCl at 4°C until ready for use. The total volume of the beads recovered was 1 L.

Fermentation Procedure

The FBR was a jacketed glass column of 2.54 cm id and 120 cm in length. The total working volume was 0.6 L. It consisted of four sections joined together. A schematic diagram of the FBR is shown in Fig. 1. Sample ports were installed at 0, 3, 9, and 12 cm. Temperature of the reactor contents was maintained at 35°C by circulation of water from a water bath through the jacket. The pH of the medium in the reactor was not controlled.

The feed solutions contained StarDri 100 starch (A.E. Staley, Decatur, IL) at various concentrations, from 40 to 185 g/L, 0.05 M KCl, 0.1% (w/v) Antifoam B (Dow Corning, Midland, MI), and 25% (v/v) light steep water (LSW), which was provided by the A. E. Staley corn processing plant in Loudon, TN). Feeds were sterilized by autoclaving at 121°C for 1 h.

Prior to the loading of the biocatalyst beads, the reactor was decontaminated by rinsing with hot water and 75% ethanol. The beads then were placed into the reactor. The volume occupied by the beads was 350 mL. The rest of the reactor was reserved for expansion during its operation. Feeds were pumped into the reactor at flow rates varied from 10 to 37 mL/min. Minimal efforts were made to mitigate contaminant growth. In addition to sterilization of the feed solutions and the cleaning of the reactor prior to biocatalyst loading, the feed lines and containers were changed

Fig. 1. FBR schematic. The reactor consisted of an expanded 30-cm inlet section (1.27–2.5 cm id), three 30-cm sections of 2.5 cm id jacketed glass pipe, and a-10 cm disengagement section of 9 cm id with a screened sidearm for disengagement of beads from the reactor effluent. One pH probe was inserted at 90 cm position for pH observing.

with each new charge of feedstocks. Previous experience demonstrated that sterile operation was not necessary (6). For each set of experimental conditions (substrate concentration and feed flow rate), at least six residence times were allowed before samples were taken for analyses of starch, glucose, and ethanol. These analyses were performed with a high-performance liquid chromatography (HPLC) system consisting of a Waters 410 differential refractometer detector, a Waters 717Plus Autosampler autoinjector and an Alltech 425 HPLC pump. The column was an Aminex HPX-87H (BioRad, Hercules, CA) column using a 5 mM H_2SO_4 solution as the mobile phase. Data acquisition and analysis were performed with the Waters Millenium software.

RESULTS AND DISCUSSION

The operability of the reactor was good throughout the course of the experiments. The biocatalyst beads were used continuously for 22 d in the FBR without the need of recharging. There was no noticeable loss of biocatalyst from the reactor. There also were no obvious signs of structural failure of the beads.

Inside the biocatalyst matrix, starch was hydrolyzed by GA to glucose; the glucose then produced was converted to ethanol and CO_2 by *Z. mobilis*. Fig. 2 shows an example of concentration profiles of starch, glucose, and ethanol along the reactor. Starch concentration decreased with vertical position in the reactor as GA conversion proceeded. In this particular experiment, 65% conversion of starch was obtained. Glucose was an

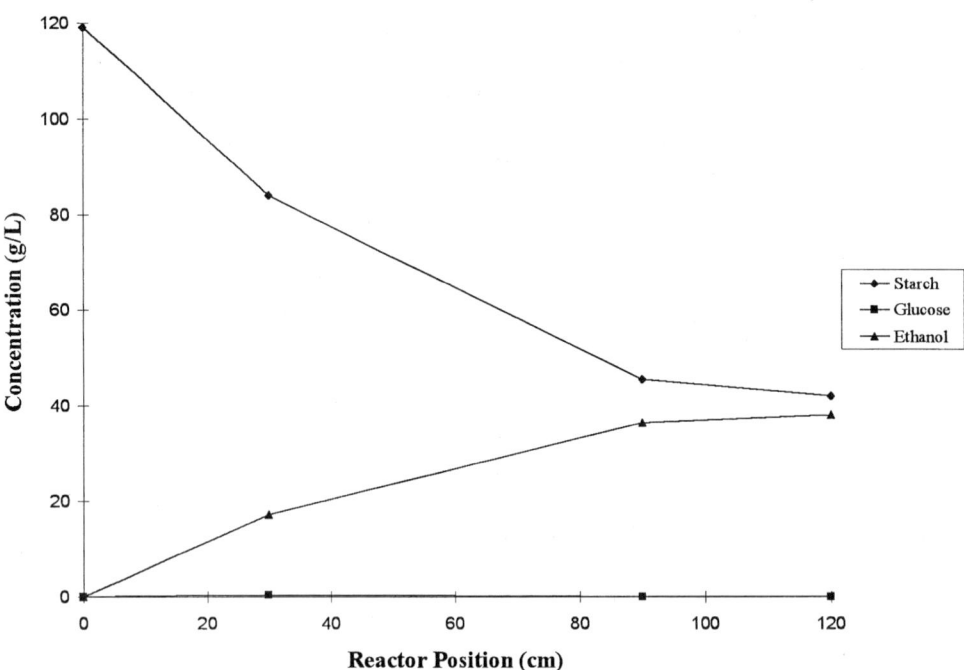

Fig. 2. Example of concentration profiles of starch, glucose, and ethanol. Starch concentration decreased with reactor position as GA conversion proceeded. Glucose was an intermediate between the GA and Z. mobilis reactions, and its concentrations were very low in the reactor. Ethanol concentration increased as the glucose released was converted.

intermediate of the series of two consecutive reactions, and therefore, its concentration was a function of both GA and Z. mobilis conversion rates. Glucose concentrations were extremely low (near zero) at all positions in the reactor. This indicated that glucose production rate was much slower than its consumption rate. In other words, the hydrolysis of starch by GA was the rate-limiting step. Glucose was converted to ethanol and CO_2 immediately after it was produced, and therefore did not accumulate in the reactor. These results were different from those obtained with the GA–S. cerevisiae system in a previous study (6). In the GA–S. cerevisiae system, starch conversion rate was much faster than glucose consumption rate, and glucose accumulated in the middle section of the reactor. Starch diffusion accounted for this difference. In the GA–S. cerevisiae system, GA was in its free form before it was co-immobilized with the yeast. In the current system, the enzyme was immobilized on a solid matrix before it was co-immobilized with Z. mobilis. This double immobilization increased the restriction of starch diffusion into the active sites of the enzyme. Consequently, the concentrations of starch at these locations were low, and, therefore, its conversion rate to glucose was slow.

Ethanol from Starch in FBR 435

Fig. 3. Ethanol volumetric productivity at different sections of the reactor. The first section (1–30 cm) of the reactor had highest productivity. The third section had lowest productivity.

Ethanol volumetric productivity was calculated for the three sections of the reactor. The results are shown in Fig. 3. In the bottom section of the reactor, the ethanol volumetric productivity was 50 g/(Lh). The productivity decreased to 17 and 3 g/(Lh) for the middle and the top sections, respectively. This decrease was related to fluid dynamics, phase hold-up, and reaction kinetics in the reactor. The fluidization of the biocatalyst bed changed rapidly with the axial position because of significant changes in fluid flow rates and physical properties (8,9). At the entrance to the reactor, the biocatalyst beads were fluidized by the liquid, and in the middle section by both liquid and the CO_2 gas. The uppermost section was characterized by high gas hold-up. The reaction rates in the reactor were a strong function of the biocatalyst concentration. This concentration was reduced in the upper part of the reactor by larger gas hold-up and liquid dispersion. The bottom section had the highest biocatalyst concentration, and consequently had the highest productivity. An example of the pH profile along the reactor is shown in Fig. 4. The pH gradually decreased from 5.5 at the reactor entrance to 4.0 at the exit. The production of ethanol also generated CO_2, some of which dissolved in the liquid and decreased the pH. Since the feed solutions were not buffered, the decrease of the liquid pH along the reactor was directly proportional to the quantities of CO_2 dissolved in it. In the bottom section of the reactor, where the rates of ethanol and CO_2 production were highest, the decreasing rate of pH also was fastest. For the same reason, the decreasing rate of pH in the middle section was faster than that in the top section. Although the pH dropped more than 1 U in the reactor, at the exit, the pH still was within the range suitable for ethanol production by *Z. mobilis* (10).

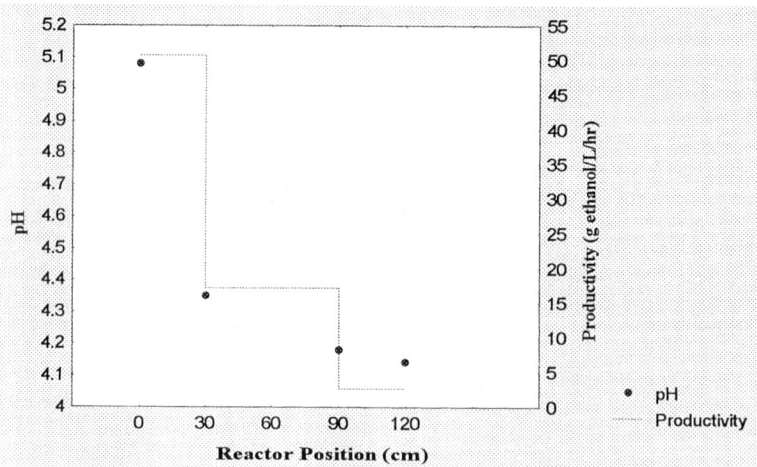

Fig. 4. pH and productivity changes along the reactor position. In the first section of the reactor, the pH decreased faster than in other sections. The rates of pH decrease was directly related to the rates of production of ethanol and CO_2.

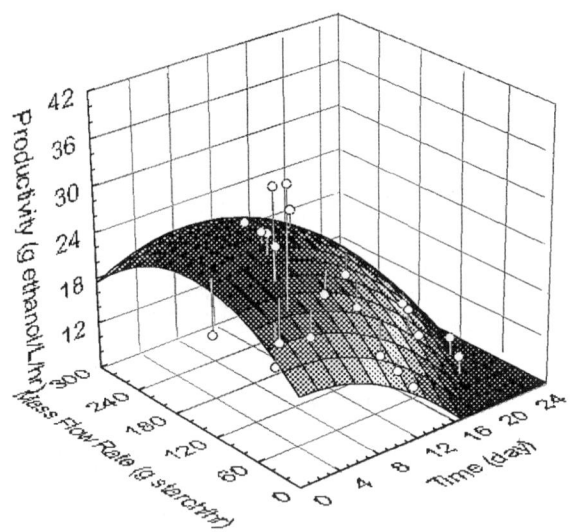

Fig. 5. The influence of starch mass feed rate and time of operation on productivity. Productivity decreased with time.

Figure 5 illustrates the effects of starch mass loading and operation time on the reactor performance. The surface was generated with the Statistica software (StatSoft, Tulsa, OK) using a method similar to that of McLain (11). The results showed that the ethanol volumetric productivity decreased with time. For example, the ethanol volumetric productivity obtained on d 4, at a mass flow rate of 71.4 g/h, was 38.5 g/(Lh); it dropped to 17.3 g/(Lh) on d 16, at a similar mass flow rate of 66 g/h. During the course of the experiments, no significant loss of the biocatalyst beads was observed. Therefore, the decrease of productivity could only be caused by lower activities of the beads. In all of the experiments, glucose concentrations stayed very close to zero in the reactor. Therefore, under the conditions of decreasing volumetric productivity, starch hydrolysis still was the rate-limiting step. The decrease in the rate of starch hydrolysis with time was probably the result of cell growth. At the beginning of the experiment, the cell concentrations inside the beads were low, and starch molecules could easily diffuse to the active sites of the immobilized GA. When the cells grew, they covered some of the active sites of the enzyme inside the beads. They could also form an outer layer, which would then severely restrict the diffusion of starch molecules into the beads. It has been shown by mathematical modeling (12) that even without a surface layer of microbial cells, starch concentration decreased rapidly along the radial direction, toward the center of the beads. Near the surface of the beads, more starch was available, which resulted in higher glucose formation and more cell growth. In the opposite, near the center of the beads, starch concentration dropped to zero and no glucose was produced. The highest volumetric producitivity of ethanol obtained was 38.5 g/(Lh). Since 150 mL ground, immobilized GA, having a specific activity of 1.16 g glucose/mL-h, was used to prepare 1 L of biocatalyst beads, and 350 mL of beads was placed in the reactor, having a working volume of 0.6 L, the ethanol volumetric productivity that could be expected in the reactor would be 51.8 g/(Lh). This was calculated assuming a yield of 0.51 g ethanol/g glucose consumed, and that the co-immobilization of the ground, immobilized GA with *Z. mobilis* did not increase restriction to starch diffusion into the biocatalyst beads. The highest ethanol volumetric productivity obtained, therefore, was 74% of the calculated value. This highest volumetric productivity obtained was slightly lower than the productivity obtained for co-immobilized GA-*Z. mobilis* in Na-alginate in a PBR, but significantly higher than the productivity obtained with the same system co-immobilized in κ-carrageenan (5). Figure 6 is a contour plot of the starch mass feed rate, operation time, and ethanol volumetric productivity. It can be seen that the optimum mass feed rate was between 120 and 180 g/h.

Figure 7 shows a plot of ethanol production vs starch consumption. The slope of the best-fit line gives an average yield of 0.49 g ethanol/g starch consumed. HPLC analysis of the substrate indicated that it contained 95 + % maltotetraose. The theoretical yield would be 0.55 g ethanol/g starch consumed. The average yield obtained, therefore, was 89% of the theoretical value. This was slightly lower than the yields obtained in other studies using immobilized *Z. mobilis* with glucose as substrate in an FBR (96% of the theoretical value) (1).

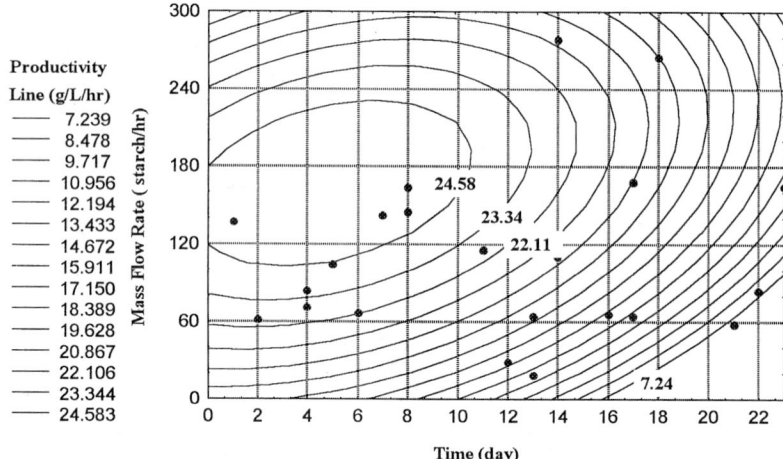

Fig. 6. The contour plot of productivity vs starch mass feed rate and time. The optimum mass feed rate was 120–180 g starch/h.

Fig. 7. Determination of average ethanol yield.

CONCLUSIONS

It has been demonstrated that co-immobilized GA and *Z. mobilis* could be used for ethanol production from starch in a single reactor. The FBR used was easy to operate and required no pH control. Under relaxed sterile conditions, no contamination was observed. The biocatalyst beads were quite structurally stable over the 22-d period of operation. Between the two consecutive conversion steps, starch hydrolysis was shown to be the rate-limiting one. The application of the FBR technology with co-immobilized GA–*Z. mobilis* will be extended to the use of real industrial feedstocks such as hydrolyzed dry-milled corn flour.

ACKNOWLEDGMENT

This research was supported by the Office of Transportation Technologies of the US Department of Energy under contract DE-AC05–96OR22464 with Lockheed Martin Energy Research. Immobilized GA and light steep water were generously donated by Genencor International Indiana, and A. E. Staley corn processing plant in Loudon, TN, respectively. The inputs from Bruce J. Jordan of Morris Ag-Energy Company, and O. J. Lantero of Genencor International Indiana, are greatly appreciated. The authors also would like to thank Bruce E. Suttle, Tommy L. Metheney, and Maria M. Blanco-Rivera for technical assistance.

REFERENCES

1. Davison, B. H. and Scott, C. D. (1988), *Appl. Biochem. Biotech.* **18,** 19–34.
2. Godia, F., Casas, C., and Sola, C. (1987), *Process Biochem.,* **22,** 43–48.
3. Nakamura, Y., Kobayashi, F., Ohnaga, M., and Sawada, T. (1997), *Biotechnol. Bioeng.* **53,** 21–25.
4. Lee, G. M., Kim, C. H., Mohammed, Z. A., Han, M. H., and Rhee, S. K. (1987), *J. Chem. Tech. Biotechnol.,* **38,** 235–242.
5. Kim, C. H., Lee, G. M., Abidin, Z., Han, M. H., and Rhee, S.K. (1988), *Enzyme Microb. Technol.,* **10,** 426–430.
6. Sun, M. Y., Bienkowski, P. R., Davison, B. H., Spurrier, M. A., and Webb, O. F. (1997), *Appl. Biochem. Biotech.* **63–67,** 483–493.
7. Tanaka, H., Kurosawa, H., and Murakami, H. (1986), *Biotechnol. Bioeng.* **28,** 1761–1768.
8. Petersen, J. N. and Davison, B. H. (1995), *Biotechnol. Bioeng.* **46,** 139–146.
9. Webb, O. F., Davison, B. H., Scott, T. C., and Scott, C. D. (1995), *Appl. Biochem. Biotech.* **51/52,** 559–568.
10. Doelle, H. W., Kirk, L., Crittenden, R., and Toh, H. (1993), *Crit. Rev. Biotechnol.* **13,** 57–98.
11. McLain, D. H., (1974), *Comput. J.* **17,** 318–324.
12. Sun, M. Y., Davison, B. H., Bienkowski, P. R., Nghiem, N. P., and Webb, O. F., (1997), Poster presented at the 213th National Meeting of the American Chemical Society, San Francisco, CA.

Copyright © 1998 by Humana Press Inc.
All rights of any nature whatsoever reserved.
0273-2289/98/70-72—0441$13.50

Ethanol Production from AFEX-Treated Forages and Agricultural Residues

Khaled Belkacemi,[1] Ginette Turcotte,*,[1] Damien de Halleux,[2] and Philippe Savoie[2,3]

[1]*Department of Food Science and Nutrition;* [2]*Department of Soil and Agri-Food Engineering University Laval, Québec, G1K 7P4, Canada;* [3]*Agriculture and Agri-Food Canada, Ste-Foy, Québec, G1V 2J3, Canada*

ABSTRACT

Lignocellulosic materials derived from forages, namely timothy grass, alfalfa, reed canary grass, and agricultural residues, such as corn stalks and barley straw, were pretreated using ammonia fiber explosion (AFEX) process. The pretreated materials were directly saccharified by cellulolytic enzymes. Sixty to 80% of theoretical yield of sugars were obtained from the pretreated biomasses. Subsequent ethanolic fermentation of the hydrolysates by *Pachysolen tannophilus* ATCC 32691 resulted in 40–60% of theoretical yield after 24 h, based on the sugars present in the hydrolysates. The uptake of sugars was not complete, indicating a possible inhibitory effect on *P. tannophilus* during the fermentation of these substrates.

Index Entries: Forages; agricultural residues; AFEX; enzymatic hydrolysis; ethanolic fermentation; biofuel.

INTRODUCTION

Perennial grasses are widely available in eastern Canada. Their potential was examined with respect to the context of Quebec's transport sector *(1).* The production of fuel ethanol from lignocellulosic materials, such as forages and agricultural residues, offers many potential economic and environmental benefits. Several methods were developed to make the production of ethanol based on lignocellulosic biomass technologically feasible, but the challenge of developing and commercializing a cost-effective process remains. One of the major problems in utilizing lignocellulosics as fermentation substrates is their resistance to hydrolysis. A wide variety

*Author to whom correspondence and reprint requests should be addressed. E-mail: ginette.turcotte@aln.ulaval.ca

of pretreatments have been used to reduce the particle size, lignin content *(2)*, and cellulose cristallinity to improve the surface area and the porosity of such materials, as well as their accessibility for the enzymes *(3–5)*. No single method has yet found widespread commercial application. Mechanical comminution by a combination of chipping, grinding, and milling techniques is a costly process because of its power requirements *(6)*. Various chemical treatments employing acids and bases are performed at high temperatures and pressure, and are costly. The chemicals are difficult to recover and recycle, and are often toxic or inhibitory to the subsequent fermentation step *(7)*. Steam explosion pretreatment can induce extensive hemicellulose degradation to furfural and its derivatives and lignin modification at high severity (temperature >220°C, and long reaction times) *(8–10)*. Impregnation prior to steam explosion with H_2SO_4 or SO_2 has been shown to satisfactorily enhance the selectivity of the process and favor hydrolysis over pyrolysis and degradation reactions *(8,11)*. Aqueous/steam fractionation of lignocellulosics was recently studied, and ethanol production from cellulosic fines derived from fractionated forages was evaluated and could be integrated within the "biorefinery" concept *(12)*. Novel pretreatments for biomass using ammonia were recently proposed. The ammonia recycled percolation (ARP) process *(13,14)* uses aqueous ammonia ($NH_3 \cdot H_2O$); ARP-H, a variant of the ARP process, combines the action of hydrogen peroxide to that of aqueous ammonia to modify structural features of herbaceous biomass *(15)*. Good results were obtained with ARP-H-treated corn cobs/stover mixture and switchgrass, but the efficiency of this process with other types of biomass remains to be demonstrated. The ammonia fiber explosion (AFEX) process was reported as a pretreatment method for improving the reactivity of lignocellulosics *(16)*. This technique, which combines ammoniacal hydrolysis and freezing to shatter and split plant material in relatively mild conditions, has been demonstrated to markedly improve the saccharification rates of several herbaceous crops and grasses *(17–23)*.

Within the mandate of the Canadian Green Plan Ethanol Program, the performance of the AFEX pretreatment on local biomasses was evaluated. The subsequent hydrolysis and fermentation steps of three hays (alfalfa, timothy, and reed canary grass) and two grain residues (barley straw and corn stalks) are reported here.

METHODS

Feedstock

Mature timothy grass (*Phleum pratense*, Basho cultivar), alfalfa (*Medicago sativa*, Apica cv.), and reed canary grass (*Phalaris arundinacea*, Vantage cv.) were mowed in July 1995 and stored as baled hay. Barley straw

(*Hordeum vulgare*) and corn stalks were baled in August and October 1995, respectively. Bales were stored at a low and stable moisture content (8% dry wt) until samples were needed for the AFEX treatment.

Enzymes

Two enzyme complexes derived from the fermentation of selected strains of *Trichoderma longibrachiatum* were purchased from Genencor International (Rochester, NY). Multifect Cellulase 300 was a soluble powder containing cellulases, β-D-glucanase, and pentosanase with mainly arabinoxylanase activity. The total declared activity was 180–190 Genencor Cellulase Units (GCU)/g of powder. Spezyme CP was a liquid solution containing cellulase activity and other combined activities (hemicellulases and pectinases), with a declared global activity of 90 GCU/mL. One GCU of Cellulase 300 was found to be equivalent to 0.8 FPU (filter paper unit).

Microorganism and Inocula

Pachysolen tannophilus ATCC 32691 was obtained from the American Type Culture Collection (Rockville, MD). Working stock cultures were grown at 30°C on agar slants containing 3g/L yeast extract, 3 g/L malt extract, 5 g/L peptone, 20 g/L D-glucose, 20 g/L D-xylose, and 20 g/L agar. They were kept at 4°C or stored in 4% (w/v) glycerol at $-80°C$.

A loopful of cells from the agar slants was inoculated in 200 mL of the culture medium described previously, but without agar, autoclaved, and adjusted to pH 5.0. The 250-mL Erlenmeyer flasks were agitated at 150 rpm for 48 h at 25–27°C in a rotary incubator (Queue Orbital Shaker, Queue Systems, Parkersburg, WV). The concentration of yeasts in the culture reached 6–7 g/L. Cells were harvested by centrifugation at 16,320g for 10 min with an RC5C model centrifuge (Sorvall® Instruments, Dupont Canada, Mississauga, ON). The resulting pellet was resuspended in sterile dH_2O to obtain the same concentration as the final concentration of the culture broth.

AFEX Treatment

Five hundred g of previously chopped biomass (9% moisture; chopped to 4–5 mm) were humidified to 15% dry wt with dH_2O and placed in a 7.6-L stainless steel Packless autoclave (Autoclave Engineers, Erie, PA). Five hundred g of liquid ammonia (99.5% purity, obtained from Prodair, Québec) was then added. The temperature was increased to 90°C, while agitating, resulting in a pressure inside the vessel of about 3.45 MPa. After maintaining these conditions for 30 min, the pressure was suddenly released. This treatment induces structural changes within the lignocellulosic matrix and increases the specific surface area of the biomass *(16)*. The treated biomass was removed from the vessel and left overnight under a fume hood to evaporate ammonia. A maximum of 2% w/w of ammonia used was left in the biomass. The evaporated biomass was placed in plastic bags until its hydrolysis.

Enzymatic Hydrolysis

Hydrolysis was carried out with 5 and 10% (dry basis) of nontreated and AFEX-treated material in 250-mL Erlenmeyer flasks with a working volume of 100 mL, or in 2-L Erlenmeyer flasks with a working volume of 1 L, in 0.05 M acetate buffer (pH 4.85). After addition of enzymes, flasks were capped and placed in a rotary incubator (Queue Systems) at 50°C and 290 rpm. The media were kept sterile with NaN_3 (0.005% w/w), except when they were followed by larger-scale fermentations. A separate flask was prepared for each sampling with the 250-mL Erlenmeyer flasks. Two-mL samples were withdrawn at specific time intervals, placed in a boiling water bath for 15 min to deactivate the enzymes, then passed through a 0.2-μm filter. A portion of 0.2 mL was used for measurement of the total reducing sugars by the DNS method (24) and compared to a glucose standard. Remaining filtrates were stored at −30°C for subsequent sugar analyses by gas chromatography (GC). Hydrolysis of AFEX-treated biomass was carried out by adding Multifect Cellulase 300 at 3, 5, 6.8, 10, and 15 GCU/g dry fiber (0.017, 0.028, 0.038, 0.056, and 0.083 g of Cellulase 300/g dry fiber). Concentration of Spezyme CP was varied at 0, 1, 2, and 3 μL/g dry matter (0.00, 0.09, 0.18, and 0.27 GCU/g dry fiber).

The potential maximum yield of total sugars (TS_{max}) was evaluated by strong acid hydrolysis with 24.1 N H_2SO_4 at 30°C for 30 min, followed by weak acid hydrolysis with 0.82 N H_2SO_4 at 120°C for 55 min. The sugar content was analyzed by GC, after neutralization with saturated $Ba(OH)_2$ solution and centrifugation at 12,062g for 10 min (25). The efficiency of enzymatic hydrolysis was expressed as a percentage of saccharification, i.e., the ratio of actual sugars released (TS_t) over potential maximum yield of sugars, corrected for soluble sugars initially present (TS_i) in the nonhydrolyzed substrates. In all cases, TS_i was found negligible.

$$\text{Saccharification (\%)} = TS_t - TS_i/TS_{max} - TS_i * 100\% \qquad (1)$$

When larger quantities of hydrolysates were required for subsequent fermentation, the initial suspensions of AFEX-treated biomass were not supplemented with NaN_3. The AFEX-treated biomasses were hydrolyzed at a working volume of 8–10 L in 20-L 316 stainless steel mechanically agitated bioreactors (New Brunswick Scientific, New Brunswick, NJ), placed in a water bath at 50°C with agitation at 300 rpm. The enzyme loading were 5 GCU of Cellulase 300 and 2 mL of Spezyme CP/g of dry fiber. Solid loading was fixed at 5% w/v. After 35 h of reaction, the obtained slurry was autoclaved at 100°C for 15 min to deactivate the enzymes, and allowed to settle overnight at 4°C. The supernatant (hydrolysate) was carefully withdrawn. The remaining solid residues were freeze-dried and stored at room temperature for a subsequent evaluation in sacco and in vitro as an animal feed (26).

Fermentation

Prior to the fermentation experiments, hydrolysates obtained from 20-L 316 stainless steel bioreactors were centrifuged by Alfa-Laval centrifuge (LAPX 202 Model), at 10,000 rpm, 160 mL/min, in order to remove small amounts of solid particles. The hydrolysates of AFEX-treated forages and agricultural residues contained 20–24 g/L total sugars. They were supplemented with 0.25 g/L of KCl and 0.4 g/L of H_3PO_4, then autoclaved at 120°C for 20 min. After cooling to 30°C, 2 g/L of urea and 1 mL/L of a vitamin solution (1.0 g/L thiamin HCl, 1.0 g/L calcium pantothenate, 0.6 g/L biotin, HCl 0.05 N) were aseptically added to the media. These media were then inoculated with a concentrated yeast solution (6–7 g/L), so that the final concentration of *P. tannophilus* in the fermentation broth was 0.5 g/L. The amounts of added nutrients were those found optimum and satisfactory for ethanol production by Beck (27). The composition of the resulting medium was such that it acted as a buffer, maintaining at pH 5.0 ± 0.1 during the fermentation. Separate 150-mL Erlenmeyer flasks were prepared for each sample and filled with a working volume of 100 mL. Flasks were placed in a Model G-53 rotary shaker (New Brunswick Scientific, New Brunswick, NJ) at 75 rpm and ambient temperature (25°C). Fermentations were monitored for 2–5 d by removing 5-mL samples, which were centrifuged with an RC5C model centrifuge (Sorvall Instruments, Dupont Canada) at 16,320*g* for 10 min at 4°C. Pellets were washed with 8% w/v saline (NaCl) solution, resuspended in dH_2O, centrifuged again, and dried at 105°C until constant weight. Supernatants from the first centrifugation were analyzed for ethanol, acetaldehyde, and carbohydrates.

Analytical Methods

AFEX-treated and nontreated biomasses were analyzed for their lignin, cellulose, ash, and hemicellulose contents using methods reported elsewhere (25).

Sugar analysis was performed as silylated sugars using HP 5890 GC system equipped with an FID detector, at 350°C, with He at 30 mL/min. STOX and TMSI (Pierce, Rockford, IL) were used as derivatization reagents; myo-inositol was used as internal standard. The total reducing sugars of these solutions was also determined by the DNS method of Miller (24).

Ethanol and acetaldehyde were determined by GC with an FID detector, an HP 19395A headspace injector, an auto-sampler system, and an HP DB-WAX 30 m × 0.25 mm capillary column (J & W Scientific, Rancho

RESULTS

As seen in Table 1, the cellulose content of all AFEX-treated forages and agricultural residues did not change substantially from that of nontreated materials. However, the hemicelluloses lost simple sugars, or methyl and acetyl groups, during the treatment, but these could be accounted for in the water extracts. The organic degradation products from lignin could be accounted for in the ethanol–toluene extracts. The proportion of pentosans in the hemicellulose fraction of nontreated biomasses represented 62% for corn stalks, 69% for barley straw, 80% for alfalfa, 88% for timothy, and 59% for reed canary grass. The residual portion of hemicelluloses usually contains sugars such as galactose, fructose, mannose, and glucose, with glucuronic acid and its methylated derivative. Except for alfalfa, the pentosan fraction increased in the hemicelluloses of AFEX-treated biomasses (for example, to 86% for reed canary grass, and to 95% for timothy), although its proportion in the biomasses remained constant. It thus seems that the AFEX treatment mainly solubilized the sugars making up the highly branched heteropolymeric hemicelluloses. With alfalfa, the treatment conditions seem to have hydrolyzed part of the pentosans as well. The fact that the potential sugars of AFEX-treated materials were 7–11% lower than in the nontreated biomasses might reflect condensation of the soluble sugars with other molecules. The potential sugars represented more than 97.8% of the theoretical amount of glucose and xylose in the cellulose and hemicellulose fractions of all materials.

Five % (w/v) of AFEX-treated materials were subjected to hydrolysis with varying amounts of enzymes at the temperature and pH recommended by the manufacturer of enzymes. Figure 1 shows the hydrolysis profiles for AFEX-treated timothy. When the hydrolysis took place with 5 GCU Cellulase 300/g dry fiber, the addition of Spezyme CP at 2 μL/g dry fiber was found essential for an acceptable 80% of saccharification of the sugars in a reasonable reaction time of 30 h. Varying the concentration of Cellulase 300 when 2 μL of Spezyme CP was present/g of dry fiber did not improve the saccharification, except at the level of 15 GCU, at which a substantial increase in early productivity of the released sugars was found: More than 50% saccharification was obtained after only 4 h. However, this level was not judged economically viable, since it would represent a cost of $360 per metric ton of AFEX-treated biomass if the price of Cellulase 300 (commercial grade) was $6/kg. Timothy not treated by AFEX was also hydrolyzed with 5 GCU Cellulase 300 and 2 μL of Spezyme CP/g dry fiber. As in the case of all other biomasses, only 35% saccharification was obtained in about 30 h (25). This proves that the AFEX treatment did provide better accessibility for the enzymes.

Table 1
Composition of Forages and Agricultural Residues Before or After AFEX-Treatment[a]

	Nontreated biomass (% dry matter)					AFEX-treated biomass (% dry matter)				
	Corn stalks	Barley straw	Alfalfa	Timothy	Reed canary grass	Corn stalks	Barley straw	Alfalfa	Timothy	Reed canary grass
α-Cellulose	35.2	32.3	27.6	34.0	27.1	33.8	32.5	28.4	34.9	28.2
Hemicelluloses	32.5	35.0	26.4	23.2	28.1	26.7	24.2	19.0	21.2	19.4
Klason lignin	8.1	6.4	6.5	11.5	15.9	5.1	3.5	5.5	8.9	11.3
Hot Water extracts	13.1	12.2	20.0	21.1	14.8	17.3	24.1	28.8	23.2	24.7
Ethanol-toluene extracts	9.0	8.4	12.6	6.2	8.1	13.2	9.9	11.4	8.0	10.6
Ash	4.8	5.9	9.0	5.9	7.9	4.6	5.6	6.6	5.4	7.0
Total	102.7	100.2	102.1	101.9	101.9	100.7	99.8	99.7	101.6	101.2
Pentosans	20.0	24.0	21.0	20.4	16.6	20.7	21.3	15.2	20.2	16.6
Potential sugars[b]	75.2	74.4	59.9	63.5	61.3	67.2	63.0	52.6	62.9	52.8

[a] Average of three samples.
[b] From acid hydrolysis.

Fig. 1. Enzymatic hydrolysis of AFEX-treated timothy (5% dry basis in 250-mL Erlenmeyer flasks) with 5 GCU Cellulase 300 and Spezyme CP at: (○), 0 μL; (■), 0.5 μL; (△), 1 μL; (◇), 2 μL; (▲), 3 μL/g dry fiber; or with 2 μL Spezyme CP and Cellulase 300 at: (○), 3 GCU; (■) 5 GCU; (△), 6.8 GCU; (□), 15 GCU/g dry fiber.

Maximum saccharification of AFEX-treated alfalfa also occurred around 30 h (Fig. 2). When 2 GCU Cellulase 300 were present/g of dry fiber, saccharification reached a maximum of 65% at 2–3 μL of Spezyme CP. This was slightly superior to AFEX-treated timothy at 2 μL of Spezyme CP with 3 GCU Cellulase 300 (see Fig. 1). Increasing the concentration of Cellulase 300 in AFEX-treated alfalfa to 5 GCU/g dry fiber showed only a small improvement. Saccharification levels of 80% similar to those for AFEX-treated timothy were finally reached with 8 GCU Cellulase 300 and 2–3 μL of Spezyme CP.

Table 2 shows results of hydrolysis after 24 and 48 h for the other studied AFEX-treated biomasses. Hydrolysis was usually faster with 2 and 3 μL of Spezyme CP at all concentrations of Cellulase 300, but their hydrolysis profile was similar. Corn stalks seemed to be more resistant to hydrolysis, with maximum saccharification never surpassing 70%, but barley straw and reed canary grass attained more than 80%.

Simple sugars expected from the hydrolysis of AFEX-treated materials were glucose (from the cellulose portion of the fibers), and xylose, arabinose, galactose, fructose, mannose, and glucose (from the hemicellu-

Ethanol from AFEX-Treated Forages

Fig. 2. Enzymatic hydrolysis of AFEX-treated alfalfa with Cellulase 300 and: (△), 1 µL; (◇), 2 µL; (▲), 3 µL of Spezyme CP/g dry fiber. **(A)** 2 GCU Cellulase 300/g dry. **(B)** 5 GCU Cellulase 300/g dry. **(C)** 8 GCU Cellulase 300/g dry.

lose portion), with a 2:1 glucose:xylose ratio. The simple sugars released during the hydrolysis of AFEX-treated timothy are shown in Fig. 3. The concentrations shown at the start of hydrolysis in Fig. 3A represent the soluble sugars. About 12 g/L of glucose were released after 40 h of hydrolysis; 4 g/L of xylose, and an almost equal quantity of arabinose and galactose, were released. Assuming that all the glucose was released from the cellulose, one can say that 65% of the cellulose and 99% of the hemicelluloses were hydrolyzed from the AFEX-treated fibers. About 1.6 g/L of the dimer cellobiose were also produced from the cellulose in the first 30 h, and contributed to 8.4% of cellulose solubilization. Then a total of 73.4%

Table 2
Enzymatic Hydrolysis of AFEX-Treated[a] Barley Straw, Corn Stalks, and Reed Canary Grass with Cellulase 300 and Spezyme CP

Enzyme loadings	Barley straw		Corn stalks		Reed canary grass	
	Saccharification (%) after:					
	24 h	48 h	24 h	48 h	24 h	48 h
.2 GCU Cellulase 300:						
1 µL Spezyme CP	33.12	56.36	25.44	54.34	36.79	46.15
2 µL Spezyme CP	62.25	61.18	42.06	57.28	58.88	53.05
3 µL Spezyme CP	67.30	67.30	51.02	67.00	58.88	61.72
.5 GCU Cellulase 300:						
1 µL Spezyme CP	55.83	50.48	40.15	58.95	49.06	49.37
2 µL Spezyme CP	76.40	68.07	44.62	65.72	56.81	71.69
3 µL Spezyme CP	75.71	71.74	50.31	62.46	68.01	72.69
.8 GCU Cellulase 300:						
1 µL Spezyme CP	67.34	62.25	51.72	62.48	41.84	64.25
2 µL Spezyme CP	69.67	76.48	52.23	67.00	57.73	84.96
3 µL Spezyme CP	65.39	82.60	51.59	70.20	61.18	81.89

[a] 5% (dry basis) of biomass in 250-mL Erlenmeyer flasks.

of cellulose fraction were converted after 40 h. The decrease in monomer concentration after 40 h could not be blamed on the consumption by microorganisms, since NaN_3 was present in the medium. Other researchers (28–31) have found inhibition of exoglucanases or β-D-glucosidase by cellobiose or glucose just formed and condensation of monomers into polymeric compounds. Although the presence of lignin and solvent-extractable components, such as tannins and terpenes, could also negatively affect hydrolysis, no such inhibition was found in the aqueous hemicellulose-rich liquors derived from aqueous/steam fractionation of these forages (12). Possible inhibition of hemicellulases (xylanases, β-D-xylosidase) and pectinases by xylobiose or xylose were also reported (32). Figure 3B confirms the adequacy between the DNS results for total reducing sugars and the total simple sugars detected by GC.

Increasing the solids loading to 10% (w/v) during hydrolysis (Fig. 4; full symbols) from 5% (open symbols) almost reduced the saccharification by half. The studied enzyme/substrate did not seem to follow classical saturation kinetics as demonstrated by Penner and Liaw (33) working with microcrystalline cellulose at 1–2% w/v solids loading. At a 5% solids loading, a 25% reduction in hydrolysis occurred in the 20-L vessel, as opposed to reaction in the 250-mL flasks. This was probably caused by the presence of dead volumes in the larger vessel, resulting in less contact between the enzymes and their substrates. Using a 2-L vessel, as opposed to a 250-mL flask, had a negligible impact at the 10% solids level. This was expected,

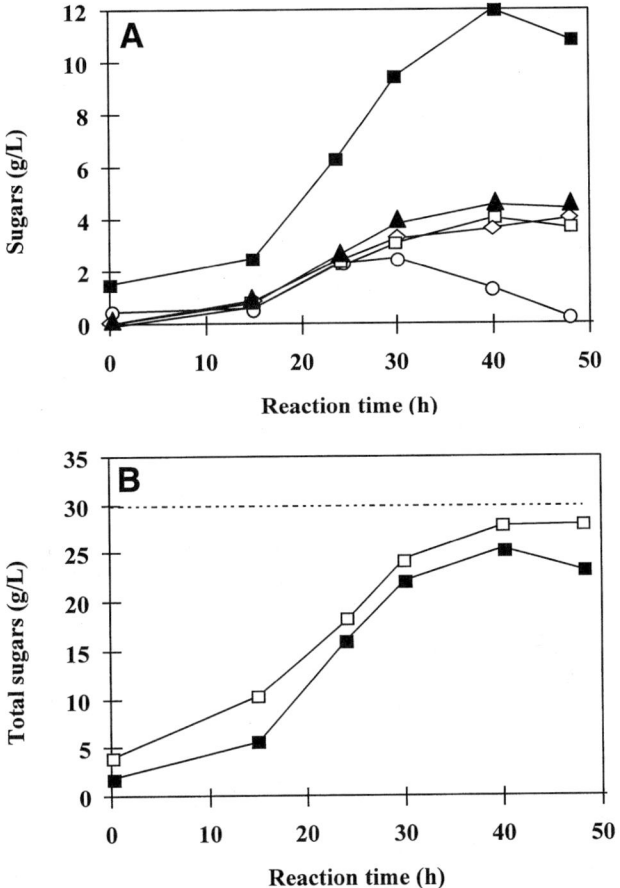

Fig. 3. Sugar profiles during hydrolysis of AFEX-treated timothy with 5 GCU of Cellulase 300 and 2 μL of Spezyme CP/g dry fiber, 5% w/v of solid loading in 250-mL flasks. Symbols: **(A)** (■), glucose; (▲), xylose; (□), arabinose; (◇), galactose; (○), cellobiose; **(B)** (□), total sugars by GC method; (■), total sugars by DNS method. Dashed line represents the potential sugars from Table 1.

since both reactors were magnetically stirred with a single bar in the bottom of the vessels.

The fermentation capabilities of *P. tannophilus* ATCC 32691 were first checked with mixtures of pure glucose and xylose (Fig. 5). When only glucose was present in the medium, the yeast converted 90% of the sugar into ethanol. Only 65% of the xylose could be converted into ethanol when xylose was the sole sugar in the medium. When xylose and glucose were mixed together, less and less ethanol could be produced out of the sugars, until a yield of 30% was obtained with 75% xylose in the mixture. This proved that catabolic repression by glucose would be present with this yeast, as investigated by Panchal et al. *(34)*. Since hydrolysates obtained from the studied AFEX-treated biomasses contained about 25 to 45% xy-

Fig. 4. Enzymatic hydrolysis of AFEX-treated timothy with 5 GCU of Cellulase 300 and 2 µL of Spezyme CP/g dry fiber. Effect of solids loading and scale of bioreactors. Symbols: (○), 5% w/v of solids in 250-mL flasks; (□), 5% w/v of solids in 2-L flasks; (△), 5% w/v of solids in 20-L bioreactors; (●), 10% w/v of solids in 250-mL flasks; (■), 10% w/v of solids in 2-L flasks.

Fig. 5. Performance of *P. tannophilus* ATCC 32691 in pure glucose/xylose mixtures at 20 g/L total sugars.

lose, expected ethanol yields of 65 to 75% were deemed acceptable for this project.

Fermentation profiles of hydrolysates from AFEX-treated barley straw and corn stalks are shown in Fig. 6. The hydrolysates contained 20 g/L of total sugars. In the case of barley straw, 13 g/L sugars were converted into a little more than 6 g/L ethanol during 48 h of fermentation. There was almost no growth of the yeast *P. tannophilus*, and almost no

Fig. 6. Fermentation profiles of hydrolysates from AFEX-treated barley straw and corn stalks using *P. tannophilus* ATCC 32691. (□), sugars; (●), ethanol; (♦), yeast; (▲), acetaldehyde.

production of acetaldehyde, a normal byproduct formed by ethanolic yeasts. Other byproducts of ethanolic fermentation were probably produced, but were not measured. Since 1.96 g sugars are theoretically needed to produce 1 g ethanol, and 2.0 g sugars for growth of the yeast, a total of 12.2 g sugars were theoretically needed to account for the ethanol production and the growth of the yeast over that 48-h period. This represented 94% of the 13 g sugars consumed/L. The performance of *P. tannophilus* with hydrolysates from alfalfa and reed canary grass was similar to that of barley straw (results not shown). In the case of corn stalks, the consumption of sugars stopped shortly after 24 h, although half of the sugars remained with less than 3 g/L of ethanol produced. Since the conversion into ethanol represented only 65% of the theoretical yield at that time, it is possible that some inhibition occurred from the 13.2% ethanol–toluene extracts (*see* Table 1; highest value of all the AFEX-treated biomasses).

Almost 3.7 g/L ethanol were formed in a hydrolysate from timothy (Fig. 7a) in the first 24 h. This represents 26 g ethanol produced/100 g sugars consumed, or 50% theoretical yield. Some growth of the yeast simul-

Fig. 7. Fermentation of hydrolysate from AFEX-treated timothy using *P. tannophilus* ATCC 32691. **(A)** Total sugars, ethanol, growth of yeast and acetaldehyde: (■), sugars; (●), ethanol; (♦), yeast; (▲), acetaldehyde. **(B)** Sugar profiles: (■), total sugars by GC method; (♦), glucose; (▲), xylose; (△), arabinose; (◇), galactose; (○), cellobiose; (●), xylitol.

taneously occurred (0.1 g/L). Fifty-one % of the 14.5 g/L sugars consumed accounted for the ethanol production and the growth of the yeast at that time. A small amount of acetaldehyde (no more than 0.1 g/L) was produced during the first 10 h of fermentation, but subsequently disappeared. Fermentation continued at a slower pace, until, at 120 h, almost all of the sugars had been consumed. The total amount of ethanol produced at that time was about 5.4 g/L, resulting in 53% of theoretical yield. The fate of the simple sugars during fermentation of the timothy hydrolysate of Fig. 7A is shown in Fig. 7B. Glucose was almost completely taken up in the first 24 h, and was largely responsible for the decrease in total sugars. More than one-third of the galactose was used in the same 24 h, but was not expected to be converted into ethanol *(35,36)*. Xylose was slowly metabolized throughout fermentation, but was not converted into ethanol. In the first 24 h, 0.4 g/L xylitol was produced from xylose and represented half of the xylose decrease in the medium. Arabinose was also slowly catabolized throughout fermentation, but was not expected to be converted into ethanol *(37)*. If it were converted into arabitol, the concentration was expected to be much less than that of xylitol *(38)*, thus, not detectable. On the other hand, cellobiose was not utilized by the yeast. This behavior was similar to that of another common ethanolic yeast, *Saccharomyces cerevisiae*.

Table 3 compares the performance of *P. tannophilus* for all studied biomasses. After 24 h of fermentation, sugar uptake was the fastest in timothy hydrolysates (66% of initial sugars had been used). However, the yeast performed at less than acceptable yield (only 53% of its capacity, based on the sugars consumed). Barley straw was found to be the most performant biomass after 24 h, with almost 50% of the initial sugars converted to 4.5 g/L ethanol, and working very near full capacity (92%). Similar results were found after 48 h of fermentation. Barley straw also gave the fastest average production of ethanol: 0.19 g ethanol/(L·h) after 24 h, and 0.13 g/(L·h) after 48 h of fermentation.

DISCUSSION

Whole forages and agricultural residues were treated by AFEX, enzymatically hydrolyzed, and fermented to ethanol with a nonengineered yeast. These steps represent the goal of the Canadian Green Plan Ethanol Program of putting forward a process as environmentally friendly as possible. This placed constraints on the process: The pretreatment must not generate compounds potentially inhibitory to either the enzymes or the ethanolic yeast; the milder conditions of enzymatic hydrolysis require longer reaction contact than for acid hydrolysis, and the ethanolic yeast must be able to convert, in a single step, both hexoses and pentoses released from the enzymatic hydrolysis.

AFEX Pretreatment

All biomasses studied in this project reacted similarly when treated with liquid ammonia under pressure: The pretreatment solubilized part of the hemicellulose and partially degraded the lignin into compounds that seemed not to be detrimental to either enzymes or yeast in the conditions used. It is not known why alfalfa's hemicelluloses were more degraded than that of other biomasses. Its lignin content was initially similar to that of barley straw and was even less degraded than its barley counterpart; its hemicellulose content was similar to those of timothy and reed canary grass. The composition and structure of xylan are more complicated than those of cellulose, and can vary qualitatively and quantitatively in various grasses and cereals *(39)*. They are highly branched and substituted

Table 3
Performance of *Pachysolen tannophilus* ATCC 32691 in Hydrolysates from AFEX-Treated Forages and Agricultural Residues after 24 h and 45 h[a]

Biomass	Sugar uptake (% of initial content)	Ethanol (g/L)	Yield[b] (% theoretical)	Yield[c] (% theoretical)	Productivity (g EtOH/L.h)
Corn stalks	46.6 (*48.9*)[d]	3.8 (*3.7*)[d]	37 (*37*)[d]	79 (*75*)[d]	0.16 (*0.08*)[d]
Barley straw	47.7 (*65.0*)	4.5 (*6.3*)	44 (*62*)	92 (*95*)	0.19 (*0.13*)
Alfalfa	37.2 (*54.0*)	3.8 (*5.7*)	38 (*56*)	100 (*100*)	0.16 (*0.12*)
Timothy	66.1 (*67.9*)[d]	4.1 (*4.4*)[d]	35 (*37*)[d]	53 (*55*)[d]	0.15 (*0.08*)[d]
Reed canary grass	35.4 (*50.4*)	3.1 (*5.2*)	31 (*51*)	87 (*100*)	0.13 (*0.11*)

[a] Values in parentheses and bold characters.
[b] Based on total initial sugars available for the fermentation.
[c] Based on consumed sugars.
[d] Value after 48 h of reaction.

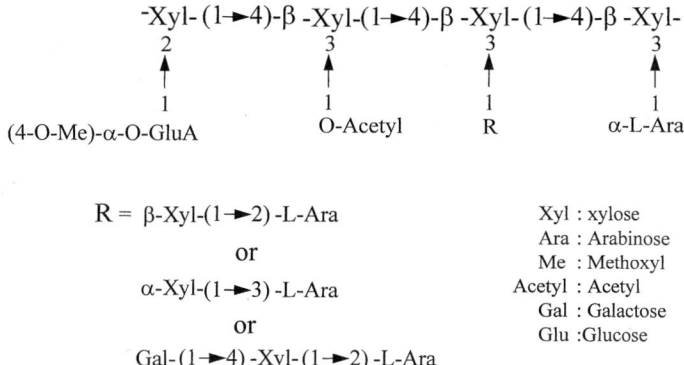

Fig. 8. Idealized structure of hemicellulose showing predominant structural features and modes of linkage.

as indicated in Fig. 8. These substitutions allow various reactions to occur in hemicelluloses.

Hemicellulases were found necessary to hydrolyze the AFEX-treated biomasses, as evidenced by the need for Spezyme CP in Fig. 1. The enzymatic activities of the enzymes used in this work was reported elsewhere *(12)*. Thus, the AFEX treatment might not have succeeded in physically separating the cellulose backbone from the hemicelluloses, or this separation occurred only at the ends of the cut fibers. Alfalfa, despite a more degraded hemicelluloses, was harder to hydrolyze than timothy or even barley straw. This seems to prove the preceeding comment.

Preliminary studies showed that the average size of the AFEX-treated biomasses' particles (in the range of 1 to 4-5 mm) had little effect on the yield of hydrolysis *(25)*. The AFEX pretreatment increased the protein degradability of the biomasses caused in part by a 1 wt% residual ammonia in them *(26)*. More than 98% of the liquid ammonia is released to the atmosphere after the AFEX treatment. It could be recompressed and recycled to make the process even more environmentally friendly. This means that the economics of recovering and reusing the 99% released after treatment should be investigated.

Hydrolysis

The AFEX pretreatment was found essential for the hydrolysis of at least 50% of the sugars in the biomasses. Alvo and Belkacemi *(40)* proved, however, that milling non-AFEX-treated biomasses down to 50 μm could yield up to 60% of saccharification. The cost of this alternative should be assessed before rejecting it. Since Cellulase 300 was found to possess a more pronounced filter paper activity than the Spezyme CP, but less β-D-xylosidase activity (result not shown), it is postulated that when Spezyme CP was added, the hemicelluloses could be opened up somewhat, giving access to the glucanases of the Cellulase 300 enzymatic complex. However, only a little xylose was released by the process in the first 15 h of reaction, as can be seen in Fig. 3, in which its concentration was half that of glucose.

During hydrolysis of timothy, a very slight inhibition by cellobiose could be demonstrated by the slight change of slope in the concentration of glucose produced at the peak of the cellobiose release, and, possibly, by the sharp decrease in glucose prior to the complete exhaustion of the cellobiose in the hydrolysate. The similar aspect of the xylose and arabinose curves during the same period might infer a possible inhibition by xylobiose, although this compound was not assayed for in the hydrolysate. These findings are in agreement with the literature (28,31,41,42) although condensation of these simple sugars could have also occurred.

Late harvesting of mature forages for ethanol production could be a complementary activity for livestock farmers who usually harvest forages early because of the higher feed value of immature hay. However, the increased lignin content of mature forages makes them more resistant to hydrolysis, as shown in Table 4, in which our results are compared with that of the literature.

The solid residue after hydrolysis was evaluated by Chiquette (26) as a potential animal feed. *In sacco* and in vivo studies seemed to indicate improved degradability and digestibility of the AFEX-treated and partially hydrolyzed forages. The increased N-ADF content seems to prove that enzymes remained mostly with the fibers. These fibers could be recycled into the hydrolysis step to reduce the amount and cost of enzymes required. But more studies are needed to evaluate the inhibition of the enzymes during hydrolysis and their reactivation with fresh AFEX-treated forages and medium.

Fermentation

No engineered microorganisms were used in this work, despite numerous potential bacteria and yeasts in the literature (43–47) since the mandate included an environmentally benign aspect. *P. tannophilus* ATCC 32691 was chosen solely on the basis of its description in the catalog as being able to convert wheat straw cellulose/hemicellulose to ethanol. Many publications reported also that *P. tannophilus* was able to ferment pure xylose (48–50). Its ethanol yield in enzymatic hydrolysates from AFEX-treated herbaceous crops varied from 40 to 60% of theoretical, based on total sugars. These values were similar to those obtained with many other ethanolic microorganisms capable of converting both glucose and xylose (51,52). All pentose-fermenting microorganisms usually produce ethanol at a slower rate than *S. cerevisiae* (53,54). No fermentation with *S. cerevisiae* was done, since this yeast was known not to be able to convert xylose into ethanol.

Growth of the yeast did not usually occur during the first 48 h of fermentation in the hydrolysates, whether a consumption of sugars oc-

Table 4
Relationship Between Maximum Saccharification of AFEX-Treated
Lignocellulosic Substrates and Their Lignin Content Prior to Treatment

Substrate	Lignin content (% dry wt)	Maximum saccharification (% of theoretical)	Ref.
Corn stalks	8.1	67	This work
Barley straw	6.4	82	This work
Alfalfa	6.5	82	This work
Timothy	11.5	80	This work
Reed canary grass	15.9	82	This work
Bermuda costal grass	6.4	94	(20)
Switchgrass[a]	5.5	96	(19)
Switchgrass[b]	12.0	68	(19)
Corn fiber	7.8	>85	(17)
Corn fiber	1.4	100	(19)
Newspaper	30.0	<40	(21)

[a] Spring harvest.
[b] Fall harvest.

curred or not. This was a bit unexpected, since the initial inoculum concentration was only 0.5 g/L. On the other hand, reduced growth meant that more energy could be channeled into ethanol production. Hydrolysates from timothy showed growth of the yeast after the first 48 h, coupled with a slow increase in ethanol production. This showed that the AFEX treatment did not produce substantial amounts of inhibitory products. However, a scale-up of the whole process might have to include a concentration step of the hydrolysates, to decrease costs associated with the separation of dilute ethanol from the fermentation broth. The influence of the AFEX conditions on the production of certain inhibitory compounds should thus be evaluated.

Xylose was shown to be slowly consumed, with a weak production of xylitol in the early stages of fermentation. No attempts were made to measure the amount of ethanol actually produced from the conversion of xylose through the xylitol metabolic pathway.

Fermentation generated 10–15 L of stillage per L of ethanol recovered. The chemical oxygen demand value was 25,000 mg O_2/L, comparable to stillage from breweries (20,000 mg O_2/L) (55).

CONCLUSIONS

Late harvesting of mature forages for ethanol production in eastern Canada could be a complementary activity for livestock farmers, who usually harvest herbaceous crops early because of the higher feed value of immature hay. However, the increased lignin content of mature forages makes them more resistant to hydrolysis.

Based on our results with barley straw (63% sugars in the AFEX-treated feedstock; 76.4% saccharification yield after 24 h of hydrolysis; 62% theoretical yield after 48 h of fermentation), an expected yield of 190 L ethanol per ton of dry biomass could be achieved with AFEX-treated and enzymatically hydrolyzed forages and agricultural residues. Between 40 and 50% of the original biomasses were hydrolyzed. The remaining portion could be used as value-added residue for animal feed.

Legislation for ethanolic microorganisms, engineered or not, is desperately needed. Without it, no recycling of the solid residue from the fermentation step would be allowed as animal feed. The nutritional value of the studied yeast, *P. tannophilus*, should be evaluated.

ACKNOWLEDGMENTS

This work was supported by a research contract from the Canadian Green Plan Ethanol Program under the supervision of Agriculture and Agri-Food Canada. The authors gratefully acknowledge Réal Michaud for advice and supervision of the contract, and Claude Gosselin for valuable assistance in GC analyses.

REFERENCES

1. Alvo, P., Savoie, P., Tremblay, D., Émond, J. P., and Turcotte, G. (1996), *Bioresource Technol.* **56,** 61–68.
2. Stone, J. E., Scallan, A. M., Danefer, E., and Ahlgren, E. (1969), *Adv. Chem. Serv.* **95,** 219–223.
3. Fan, L. T., Lee, Y. H., and Gharpuray, M. M. (1982), *Adv. Biochem. Eng. Biotechnol.* **23,** 157–187.
4. Fan, L. T., Lee, Y. H., and Beardmore, D. H. (1980), *Biotechnol. Bioeng.* **22,** 177–199.
5. Sinitsyn, A. P., Gusakov, A. V., and Vlasenko, E. Y. (1991), *Appl. Biochem. Biotechnol.* **30,** 43–59.
6. Himmel, M., Tucker, M., Baker, J., Rivard C., Oh, K., and Grohmann, K. (1985), *Biotechnol. Bioeng. Symp.* **15,** 39–58.
7. Millett, M. A., Baker, A. J., and Satter, L. D. (1976), *Biotechnol. Bioeng. Symp.* **6,** 125–153.
8. Brownell, H. H. and Saddler, J. N. (1984), *Biotechnol. Bioeng. Symp.* **14,** 55–68.
9. Chua, M. G. S. and Wayman, M. (1979), *Can. J. Chem.* **57,** 1141–1146.
10. Shultz, T. P., Biermann, C. J., and Mc Ginnis, G. D. (1983), *Ind. Eng. Chem. Prod. Res. Dev.* **22,** 344–348.
11. Mackie, K. L., Brownell, H. H., West, K. L., and Saddler, J. N. (1985), *J. Wood Chem. Tech.* **5/3,** 405–425.
12. Belkacemi, K., Turcotte, G., Savoie, P., and Chornet, E. (1997), *Ind. Eng. Chem. Res.* **36,** 4572–4580.
13. Iyer, P. V., Wu, Z. W., Kim, S. B., and Yoon, Y. L. (1996), *Appl. Biochem. Biotechnol.* **57/58,** 121–132.
14. Yoon, H. H., Wu, Z. W., and Lee, Y. Y. (1994), *Appl. Biochem. Biotechnol.* **51/52,** 5–19.
15. Kim, S. B. and Lee, Y. Y. (1996), *Appl. Biochem. Biotechnol.* **57/58,** 147–156.
16. Dale, B. E. and Moreira, M. J. (1982), *Biotechnol. Bioeng. Symp.* **12,** 31–43.
17. Bothast, R. J., Dien, B. S., Iten, L. B., Hespell, R. B., and Lawton, J. W. (1996), in *Liquid Fuels and Industrial Products from Renewable Resources*, Proceedings of the Liquid Fuel Conference, Nashville, TN, ASAE St-Joseph, pp. 241–252.

18. Dale, B. E. and de la Rosa, L. (1992), in *Liquid Fuels from Renewable Resources*, Proceedings of an Alternative Energy Conference, Nashville, TN, ASAE St-Joseph, pp. 162–170.
19. Dale, B. E., Leong, C. K., Pham, T. K., Esquivel, V. M., Rios, I., and Latimer, V. M. (1994), in *Liquid Fuels, Lubricants and Additives from Biomass*, Proceedings of an Alternative Energy Conference, Nashville, TN, ASAE St-Joseph, pp. 104–111.
20. De la Rosa, L., Reshamwala, S., Latimer, V. M., Shawky, B. T., Dale, B. E., and Stuwart, E. D. (1994), *Appl. Biochem. Biotechnol.* **45/46,** 483–497.
21. Holzapple, M. T., Jun, J. H., Ashok, G., Patibandla, S. L., and Dale, B. E. (1991), *Appl. Biochem. Biotechnol.* **28/29,** 59–74.
22. Holzapple, M. T., Ripley, E. P., and Nikolaou, M. (1994), *Biotechnol. Bioeng.* **44,** 1122–1131.
23. Moniruzzaman, M., Dien, B. S., Ferrer, B., Hespell, R. B., Dale, B. E., Ingram, L. O., and Bothast, R. J. (1996), *Biotechnol. Lett.* **18,** 985–990.
24. Miller, G. L. (1959), *Anal. Chem.* **31,** 426–428.
25. Belkacemi, K., Turcotte, G., de Halleux, D., and Savoie, P. (1996), in *Liquid Fuels and Industrial Products from Renewable Resources*, Proceedings of the Liquid Fuel Conference, Nashville, TN, ASAE St-Joseph, pp 232–240.
26. Chiquette, J. (1997), in *Proceedings of the 1997 Ethanol Research and Development Workshop*, Ottawa, ON, pp 111–114.
27. Beck, M. J. (1986), *Biotechnol Bioeng. Symp.* **17,** 615–627.
28. Coughlan, M. P. (1985), *Biotechnol Gen. Eng. Rev.* **3,** 39–109.
29. Ramos, L. P., Breuil, C., and Saddler, J. N. (1993), *Enzyme Microbiol. Technol.* **15,** 19–25.
30. Saddler, J. N. (1986), *Microbiol. Sci.* **3,** 84–87.
31. Wood, T. M. (1989), in *Enzymes Systems for Lignocellulose Degradation*, Coughlan, M. P., ed., Elsevier, London, pp. 17–35.
32. Christov, L. and Prior, B. (1993), *Enzyme Microbiol. Technol.* **15,** 460–475.
33. Penner, M. H. and Liaw, E. T. (1994), in *Enzymatic Conversion of Biomass for Fuels Production*, Himmel, M. E., Baker, J. O., and Overend, R. P., eds., ACS Symposium Series 566, pp. 363–371.
34. Panchal, C. J., Bast, L., Russell, I., and Stewart, G. G. (1988), *Can. J. Microbiol.* **34,** 1316–1320.
35. Dubus, D., Methner, H., Shulze, D., and Dellweg, H. (1983), *Eur. J Appl. Microbiol. Biotechnol.* **17,** 287–291.
36. Jeffries, T. W. (1990), in *Yeast Biotechnology and Biocatalysis*, Verachtert, H. and De Mot, R., eds. Marcel Dekker, Louvain, Belgium, pp. 349–393.
37. Dien, B. S., Kurtzman, C. P., Saha, B. C., and Bothast, R. J. (1996), *Appl. Biochem. Biotechnol.* **57/58,** 233–242.
38. Ligthelm, M. E., Prior, B. A., and du Preez, J. C. (1988), *Appl. Microbiol. Biotechnol.* **28,** 293–296.
39. Wilkie, K. C. B. (1979), in *Advances Carbohydrate Chemistry and Biochemistry*, Vol. 36 Tipson, R. S., and Horton, D., eds., Academic, New York pp. 215–264.
40. Alvo, P. and Belkacemi, K. (1997), *Bioresource Technol.* in press.
41. Fan, L. T., Gharpuray, M. M., and Lee, Y. H. (1987), in *Cellulose Hydrolysis*, Springer-Verlag, New York, pp. 5–120.
42. Poutanen, K. and Puls, J. (1989), in *Biogenesis and Biodegradation of Plant Cell Wall Polymers*, Lewis, G. and Paice, M., eds., American Chemical Society, Washington D. C., pp. 456–467.
43. Doran, J. B. and Ingram, L. O. (1993), *Biotechnol. Prog.* **9,** 533–538.
44. Ohta, K., Beall, D. S., Mejia, J. P., Shanmugam, K. T., and Ingram, L. O. (1991), *Appl. Environ. Microbiol.* **57,** 893–900.
45. Takahashi, D. F., Carvalhal, M. L., and Alterthum, F. (1994), *Biotechnol. Lett.* **16,** 747–750.
46. York, S. W. and Ingram, L. O. (1996), *Biotechnol. Lett.* **18,** 683–688.

47. Zhang, M., Eddy, C., Deanda, K., Finkelstein, M., and Picataggio, S. (1995), *Science* **267**, 240–243.
48. Schneider, H., Wang, P. Y., Chan, Y. K., and Maleska, R. (1981), *Biotechnol. Lett.* **3**, 89–92.
49. Slininger, P. J., Bothast, R. J., Van Cauwenberge, J. E., and Kurtzman, C. P. (1982), *Biotechnol. Bioeng.* **24**, 371–384.
50. Slininger, P. J., Bolen, P. L., and Kurtzman, C. P. (1987), *Enzyme Microbiol. Technol.* **9**, 5–15.
51. Delgenes, J. P., Moletta, R., and Navarro, J. M. (1986), *Biotechnol. Lett.* **8**, 897–900.
52. Woods, M. A. and Millis, N. F. (1985), *Biotechnol. Lett.* **7**, 679–682.
53. Du Preez, J. C. and Prior, B. A. (1985), *Biotechnol. Lett.* **7**, 241–246.
54. Olsson, L. and Hahn-Hägerdal, B. (1996), *Enzyme Microbiol. Technol.* **18**, 312–331.
55. Zhang, M., Eddy, C., Deanda, K., Finkelstein, M., and Picataggio, S. (1995), *Science* **267**, 240.

Adaptive Optimal Control of Fed-Batch Alcoholic Fermentation

T. L. M. Alves,* A. C. Costa, A. W. S. Henriques, and E. L. Lima

PEQ/COPPE/UFRJ, Cx. Postal 68502, CEP 21945–970, Rio de Janeiro, RJ, Brazil

ABSTRACT

An adaptive control scheme is developed for the optimization of a fed-batch ethanol production process. The fermentation process is modeled by an hybrid neural model combining mass balance equations and neural networks, used to represent the kinetic rates. The networks used, the functional link networks (FLN), allow the linear estimation of their parameters; this enables the re-estimation of the parameters at each sampling time, and thus the development of an adaptive optimal control scheme.

Index Entries: Ethanol production; optimal control; hybrid neural modeling; adaptive control.

INTRODUCTION

Brazil is the main ethanol producer in the world. This position is the result of a political strategy initiated in 1975 by the Brazilian government to cope with the sharp increase in petroleum oil prices at that time. The objective of the Brazilian National Program (PROALCOOL) was to encourage the traditional alcohol industries to increase their ethanol production in order to replace gasoline with ethanol as a main automobile fuel. Because the main objective was production, and not necessarily efficiency, a significant sector of this industry, mainly those companies with a family-based structure, have not been investing in technology. Currently, because of the stabilization of the petroleum prices at a low level and the necessity of increasing the efficiency of public administration, the government has withdrawn most of the incentives to the alcohol industry. With this new reality, only those industries that are technically efficient will continue in

*Author to whom all correspondence and reprint requests should be addressed.

the market. Therefore, there is an increased interest in optimizing all the steps of the ethanol production process.

In the optimization of fed-batch processes, the objective is to determine the substrate addition strategy that maximizes the product of interest at the end of the batch cycle. This is an optimal control problem and there are various methods proposed for its solution. Palanki et al. *(1)* used Pontryagin's maximum principle and the singular control theory. Cuthrell and Biegler *(2)* proposed a simultaneous optimization and solution strategy based on successive quadratic programming (SQP) and orthogonal collocation on finite elements. Luus *(3)* utilized iterative dynamic programming to provide piecewise linear continuous control policies. Costa *(4)* proposed a methodology to obtain optimal trajectories analytically and in feedback mode.

Once the optimization problem is solved off-line, the calculated profile of the control variable can be implemented in an open-loop fashion. This approach is appropriate in situations in which process model is accurately known and there are no external disturbances to the process.

Because fed-batch processes are transient in nature and their variables undergo significant changes during the fermentation, it is desirable that the precalculated trajectories be corrected on-line. To this end, several authors have studied different approachs. Terwiesch and Agarwal *(5)* developed a method for on-line correction of preoptimized input profiles. Lee and Ramirez *(6)* used an extended Kalman filter to estimate state variables and parameters, and implemented an on-line optimal control scheme in a process for induced foreign protein production.

However, in most real-time applications, the on-line correction of preoptimized profiles is infeasible because of prohibitive computing-power requirements *(5)*. The great difficulty is usually associated with the on-line estimation of the kinetic parameters, because they lead to the nonlinearity of the model equations.

In this work, an hybrid neural model of the process is developed. This model combines *a priori* knowledge of the process and neural networks, which are used for the estimation of the kinetic rates. Psichogios and Ungar *(7)* were the first to investigate the application of this modeling procedure to a fed-batch bioreactor. More recently, Schubert et al. *(8)* applied an improved technique of hybrid modeling to a fed-batch baker's yeast production process, and Fu and Barford *(9)* developed a hybrid neural model for a complex simulation of hybridoma cell cultivation for monoclonal antibody (Mab) production.

The goal of the hybrid neural modeling in this case is to deal with the problem of the on-line estimation. Thus, the kind of neural networks chosen to describe the kinetic rates are the functional link networks (FLN), whose great advantage is that the estimation of the network parameters is linear and it ensures convergence *(10)*. Besides, a modification is applied to these networks to increase the nonlinear approximation ability *(11)*.

A method based on orthogonal least-squares is used to eliminate nonsignificant nodes during the training of the network. This method was originally used for the identification of a multivariable non-linear systems by Billings et al. *(12)*. The use of this method significantly reduces the size and complexity of the network, and avoids overfitting of the data.

The hybrid neural model allied to the structure chosen for the neural networks make the on-line estimation of the kinetic rates very easy, and enables the development of an adaptive optimal control scheme in which the trajectory of the control variable calculated off-line is corrected at each sampling time.

The ethanol production via fementation in a fed-batch bioreactor is a typical example of a process in which the proposed strategy could be beneficial. The microorganism most used in this process is *Sacharomyces cerevisae*. However, ethanol production by *Zymomonas mobilis* can be considered a good alternative, because of its advantages when compared to *S. cerevisae*, including higher ethanol yield (to 97% of theoretical value), higher ethanol productivity, and better tolerance to acid and high concentrations of sugar and alcohol *(13)*.

This research is part of an effort to develop a new technology for ethanol production by *Z. mobilis*. At a first stage, the algorithm is tested on the best-known *S. cerevisae* process. However, a more complex process will not modify the main results, but only require more involved details.

METHODS

Hybrid Neural Model

The hybrid neural model is a combination of the mass balance equations and neural networks, used to calculate the kinetic rates.

Mass Balance Equations

The mass balance equations for a fed-batch ethanol fermentation are

$$dX/dt = (\mu - D)X \tag{1}$$

$$dS/dt = -\sigma X + D(S_F - S) \tag{2}$$

$$dP/dt = \pi X - DP \tag{3}$$

$$dV/dt = DV \tag{4}$$

where X, S, and P are the biomass, substrate, and ethanol concentrations; μ, σ, and π are the specific rates of growth, substrate consumption, and ethanol formation, respectively; S_F is the feed substrate concentration, D is the dilution rate, defined as $D = F/V$; F is the volumetric substrate feed rate, V is the volume of the reactor mixture, and t is the elapsed time.

Functional Link Networks

A neural network typically consists of many simple computational elements or nodes arranged in layers and operating in parallel. The weights, which define the strength of connection between the nodes, are estimated to yield good performance. Usually, in the training of neural networks, the inputs to a node are linearly weighted before the sum is passed through some nonlinear activation function that ultimately gives the network its nonlinear approximation ability. The same nonlinearity, however, creates problems in learning the network weights, because nonlinear learning rules must be used. The learning rate is often unacceptably slow and local minima may cause problems (10).

One way of avoiding nonlinear learning is the use of FLNs. In these networks, a nonlinear functional transform or expansion of the network inputs is initially performed, and then the resulting terms are combined linearly. The resulting structure has a good nonlinear approximation capability, and the estimation of the network weights is linear.

The general structure of a FLN is shown in Fig. 1, where x_e is the input vector and $y_i(x_e)$ is an output. The hidden layer performs a functional expansion on the inputs, in terms of nonlinear functions $h_j(x_e)$, which maps the input space, of dimension n_1, onto a new space of increased dimension, M ($M > n_1$). The output layer consists of m nodes, and each node is, in fact, a linear combiner. The input–output relationship of the FLN is

$$y_i(x_e) = \sum_{j=1}^{M} w_{ij} h_j(x_e), \quad 1 \leq i \leq m \quad (5)$$

Henrique (11) proposed a modification in the structure of the FLNs, in which the output given by Eq. 5 is passed through an invertible nonlinear activation function. The new output is

$$y_i(x_e) = f_i\left(\sum_{j=1}^{M} w_{ij} h_j(x_e)\right), \quad 1 \leq i \leq m \quad (6)$$

where f_i is an invertible nonlinear function, as for example:

$$f_i(y_i) = \tanh(y_i) = \left(1 - \exp(-2y_i)\right) \Big/ \left(1 + \exp(-2y_i)\right) \quad (7)$$

The proposed modification increases the nonlinear approximation ability of the network, and yet the estimation of the parameters remains a linear problem.

The network inputs(xe), in this case, are the biomass, substrate, and ethanol concentrations, which are supposed to be known in different sampling times. The network output (y), is the kinetic rate to be calculated. In order to obtain output values for the network training, a discrete model can be obtained from Eqs. 1–3:

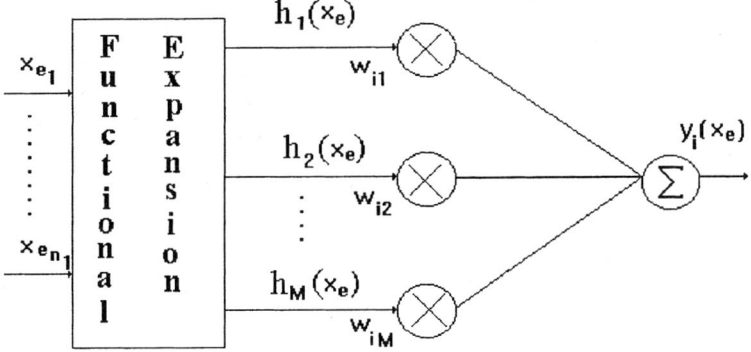

Fig. 1. General structure of a FLN.

$$\mu(t) = \left(X(t + \Delta t) - X(t)\right)\bigg/\left(\Delta t\, X(t)\right) + F(t)/V(t) \tag{8}$$

$$\sigma(t) = \left(F(t)/V(t)\right)\left((S_F - S(t))/X(t)\right) - \left(S(t + \Delta t) - S(t)\right)\bigg/\left(\Delta t\, X(t)\right) \tag{9}$$

$$\pi(t) = \left(\left((P(t + \Delta t) - P(t))\bigg/\left(\Delta t\, X(t)\right)\right)\right) - \left(F(t)/V(t)\right)\left(P(t)/X(t)\right) \tag{10}$$

where Δt is the time interval between two sampling times.

After the functional expansion of the network inputs is performed, the orthogonal least squares estimator is used to calculate the network weights (w_{ij}) and eliminate nonsignificant nodes (11).

Optimization Problem

Once the process model is determined, Pontryagin's maximum principle (14) and the singular control theory are used to solve the optimization problem.

The first step for the solution of this problem is the choice of the control variable. In the literature, the control variable usually chosen is the volumetric feed rate. Alves (15), however, showed that the dilution rate is a more suitable choice. If written as a function of this variable, the mass balance equations for biomass, substrate, and ethanol (Eqs. 1–3) are not explicit functions of the volume of the reactor. Thus, although the global balance equation, Eq. 4, is used to calculate the volume of the reactor, it is not used in the solution of the optimization problem. The dimension of the system is reduced in one order and processes described by four or less mass balance equations can be solved analytically.

Eqs. 1–3, written in a vectorial form, are

$$dx/dt = f(x) + g(x)u \qquad (11)$$

with

$$x = \begin{bmatrix} X \\ S \\ P \end{bmatrix}; \quad f(x) = \begin{bmatrix} \mu X \\ -\sigma X \\ \pi X \end{bmatrix}; \quad g(x) = \begin{bmatrix} -X \\ S_F - S \\ -P \end{bmatrix} \text{ and } u = D$$

The following constraints are imposed on the dilution rate and the final (t_f) fermenter volume

$$0 \leq D \leq F_{max}/V(t) \qquad (12)$$
$$V(t_f) = V_{max} \qquad (13)$$

The optimization problem consists of finding the optimal temporal profile of the dilution rate that leads the reactor from a given initial state to the final state that maximizes the performance index given below:

$$J = P(t_f) \qquad (14)$$

This problem can be solved by using Pontryagin's maximum principle. According to this principle, the optimal solution must maximize the Hamiltonian, which is defined as

$$H = \lambda^T(f(x) + g(x)u) \qquad (15)$$

where $\lambda(t)$ is the adjoint vector which satisfies

$$d\lambda/dt = -\partial H/\partial x \qquad (16)$$

The Hamiltonian can be rewritten as

$$H = \phi(t)u + H_0(t) \qquad (17)$$

where

$$\phi(t) = \lambda^T g(x) \qquad (18)$$
$$H_0(t) = \lambda^T f(x) \qquad (19)$$

Since the Hamiltonian is linear in the control variable, it is easy to determine $u(t)$ which maximizes this Hamiltonian by examining the sign of the function $\phi(t)$.

$$\text{if } \phi(t) > 0 \quad u = u_{max} \qquad (20)$$
$$\text{if } \phi(t) < 0 \quad u = u_{min} \qquad (21)$$

However, if $\phi(t)$ is equal to zero over a finite time interval (t_1, t_2), Pontryagin's maximum principle fails to give $u(t)$ during this interval and the singular control theory has to be used. This is called the singular interval.

The optimal temporal profile of the control variable has singular ($u = u_{sing}$) and nonsingular ($u = u_{min}$ or $u = u_{max}$) intervals. The complete solution of the optimization problem consists of the determination of the singular dilution rate expression, the switching times between the different values of this variable, and the sequence in which these values appear.

Singular Dilution Rate

During the singular interval, $\phi(t)$ is equal to zero, and hence its derivatives must also vanish. According to the singular control theory, the expression of the control variable in the singular interval can be determined by deriving $\phi(t)$ until the control variable appears explicitly. In the cases studied in this article, the singular dilution rate is determined by making the second derivative of $\phi(t)$ equal to zero.

The following equations can be written in the singular interval:

$$\phi = 0 \tag{22}$$

$$d\phi/dt = 0 \tag{23}$$

$$d^2\phi/dt^2 = 0 \tag{24}$$

In addition, if the fermentation time (t_f) is not specified a priori, the Hamiltonian on the optimal trajectory is equal to zero. Then, during the singular interval,

$$H_0(t) = 0 \tag{25}$$

In this work, Lie brackets (16) are used to obtain the singular arc and the singular dilution rate expressions. By using this differential geometric tool, the first derivative of ϕ can be written as

$$d\phi(t)/dt = \lambda^T[f,g](x) \tag{26}$$

where [f,g] is the Lie bracket, which is a vector field defined by

$$[f,g](x) = \left(\partial g(x)/\partial x\right) f(x) - \left(\partial f(x)/\partial x\right) g(x) \tag{27}$$

Its components are given by

$$([f,g](x))_i = \sum_{j=1}^{n} \left(f_j \left(\partial g_i/\partial x_j\right) - g_j \left(\partial f_i/\partial x_j\right)\right) \tag{28}$$

The second derivative of $\phi(t)$ can be written as

$$d^2\phi/dt^2 = \lambda^T[f,[f,g]] + \lambda^T[g,[f,g]]u = \lambda^T ad_f^2 g(x) + \lambda^T ad_g^2 f(x) \qquad (29)$$

where $ad_f^2 g(x)$ and $ad_g^2 f(x)$ are the notations used for the iterated Lie brackets $[f,[f,g]]$ and $[g,[f,g]]$, respectively.

The singular control is then given by

$$u_s = -\lambda^T ad_f^2 g(x) \big/ \left(\lambda^T ad_g^2 f(x)\right) \qquad (30)$$

For systems described by four mass balance equations, because the global balance equation is not used, the effective state variables vector, and thus the adjoint variables vector, are three-dimensional (the volume is not used as a state variable). In this case, it is possible to write two adjoint variables as a function of the third one and of the state variables. From Eq. 22 one can write

$$\lambda_1 = A(x)\lambda_3 \qquad (31)$$

And from Eq. 25

$$\lambda_2 = B(x)\lambda_3 \qquad (32)$$

The substitution of Eqs. 31 and 32 into Eq. 26 leads to

$$d\phi/dt = \lambda_3 G(x) \qquad (33)$$

Then, from Eq. 23, it follows that, in the singular interval

$$G(x) = 0 \qquad (34)$$

Also, substituting Eqs. 31 and 32 into Eq. 29, the following equation is obtained:

$$d^2\phi/dt^2 = \lambda_3(a(x) - b(x)u) \qquad (35)$$

From Eq. 24 it follows that, in the singular interval

$$u_s = a(x)/b(x) \qquad (36)$$

which is the equation of the control variable acting during this interval as a function only of the state variables.

In Eqs. 31–36, $A(x)$, $B(x)$, $G(x)$, $a(x)$, and $b(x)$ are functions of the state variables.

Switching Times

In the literature, the fermentation process studied in this work is considered to present two switching times: The time at which the singular arc is reached (Eq. 34 is valid), and the time at which the maximum volume of the reactor is attained *(17)*.

In this work, the final time of the fermentation is free, and, in this case, at the end of the fermentation the Hamiltonian must be equal to zero. Because at this stage, the reactor is operated in batch mode, the final time is determined from the following equation:

$$H_0(t_f) = 0 \qquad (37)$$

Also, because in the problem studied, the terminal point is not fixed a priori, the final value of the adjoint variables is given by

$$\lambda_i(tf) = \left(\partial J/\partial x_i\right)\Big|_{tf} \qquad (38)$$

The substitution of Eq. 38 in Eq. 37 enables the determination of the final time of the fermentation.

Control Variable Values Sequence

The sequence in which the values of the control variable appear is determined by the initial conditions. If at the beginning of the fermentation $\phi(0) < 0$, the sequence is ($u = 0$; $u = u_{sing}$; $u = 0$); if $\phi(0) > 0$, it is ($u = u_{max}$; $u = u_{sing}$; $u = 0$) and if $\phi(0) = 0$, the sequence is ($u = u_{sing}$; $u = 0$).

At this point, the optimal temporal profile of the control variable is completely determined. Initially, the dilution rate assumes the maximum or minimum value (depending on the initial conditions) until Eq. 34 is valid. Then, the fermenter is fed with the singular dilution rate until it is full, and after this point it is operated in batch mode until the final condition, given by Eq. 37, is attained.

Adaptive Optimal Control

Once the optimization problem has been solved, the optimal temporal profile is known in terms of the state variables. Actually, the dilution rate expression is a function of the state variables, the specific kinetic rates, and their derivatives, which are functions of the state variables. Then, the following expression can be written:

$$D = D(X, S, P, \mu, \sigma, \pi, \text{derivatives of } \mu, \sigma, \pi) \qquad (39)$$

To implement adaptive optimal control, this expression must be recalculated and the kinetic parameters re-estimated at each sampling time. If experimental measures of the concentrations of biomass, substrate, and ethanol are available, Eqs. 8–10 can be used to determine the specific rates of growth, substrate consumption, and ethanol formation, and these values used to re-estimate the network parameters at each sampling time. Because the network is linear in the parameters, the re-estimation is easy and rapid. As for the kinetic rate derivatives, in this work they are approximated by the derivatives of the corresponding neural networks, with very good results.

RESULTS

Ethanol Production by *S. cerevisae*

This process has been studied by Aiba et al. *(18)* and Modak and Lim *(17)*. The proposed kinetic model is given below:

$$\mu = \left(0.408S/(0.22 + S)\right)\exp(-0.028P) \tag{40}$$

$$\sigma = \mu/0.1 \tag{41}$$

$$\pi = \left(S/0.44 + S)\right)\exp(-0.015P) \tag{42}$$

In this work, this model is used to generate data for the modified FLN training.

Modified Functional Link Network

A neural network will be trained to estimate the specific growth rate:

$$y = [\mu] \tag{43}$$

using the following input

$$x_e = [X\ S\ P]^T \tag{44}$$

Training is based on sets of values of the specific growth rate and biomass, substrate, and ethanol concentrations at different sampling times.

The nonlinear functions $h_j(x_e)$ are obtained through a functional expansion of the inputs in the form of a polynomial expansion of degree six.

The activation function used was

$$f(y) = 1/y \tag{45}$$

After the elimination of monomials with negligible effect, the resulting neural network expression is

$$\mu_{FLN} = 1/\left(2.40 + 0.69/S + 0.082P + 0.00057\ P^2/S + 0.000030P^3\right) \tag{46}$$

The neural network performance was measured by a coefficient denoted by "cor" and defined as follows (19):

$$\text{cor} = (1 - \text{SEE}/\text{S}_{TT})100\% \qquad (47)$$

where $\text{SEE} = \sum_{k=1}^{N}(y_e(k) - y(k))^2$, $\text{S}_{TT} = \sum_{k=1}^{N}(y_e(k) - \overline{y_e})^2$, $y_e(k)$ is the k experimental output, $y(k)$ is the corresponding network output, $\overline{y_e}$ is the mean value of the experimental outputs, and N is the number of experimental data.

The FLN described the experimental data with a correlation of 99.97%.

Optimization Problem

The results of the optimization problem for the conditions: $X(0) = 0.2$ g/L, $S(0) = 100$ g/L, $P(0) = 0$ g/L and $V(0) = 5$ L, $S_F = 100$ g/L and $V_{max} = 20$ L, are shown in Figs. 2 and 3. In Fig. 2, the temporal profiles of the dilution rate calculated using the kinetic model of Aiba et al. (18) and the FLN to describe the specific growth rate are shown. Figure 3 shows the temporal profiles of ethanol concentration.

Initially, as $\phi < 0$, the fermenter is operated with the minimum dilution rate ($D = 0$) until Eq. 34 is valid. At this point, the dilution rate assumes its singular value until the reactor is full. Then, the reactor is operated in batch mode until the final condition is attained, in this case, substrate concentration equal to zero.

In Figs. 2 and 3, it can be seen that the result obtained using the modified neural network to estimate the specific growth rate is similar to the one obtained when this kinetic rate is expressed by the model equation. These results show that the derivative of the neural network can be used to describe the derivative of the model.

Adaptive Optimal Control

To show the applicability of the proposed adaptive optimal control strategy, its results are compared to the results of the open-loop implementation of the precalculated temporal profile in situations in which there is a plant–model mismatch.

Case 1

In this case, it is considered that the ethanol inhibits the specific growth rate less than the model used to calculate the optimal dilution rate profile predicts. In other words, the neural network was trained based on Eq. 40, an imperfect model, but the equation that really describes the specific growth rate is

$$\mu = \left(0.408S/(0.22 + S)\right)\exp(-0.02P) \qquad (48)$$

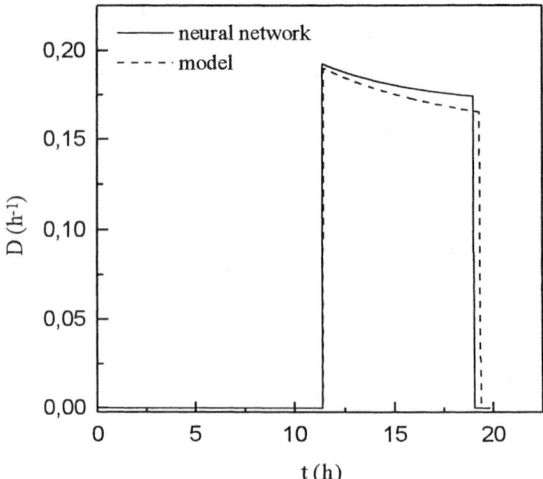

Fig. 2. Temporal profiles of the dilution rate obtained using Aiba et al. *(18)* kinetic model and the FLN to describe the specific growth rate.

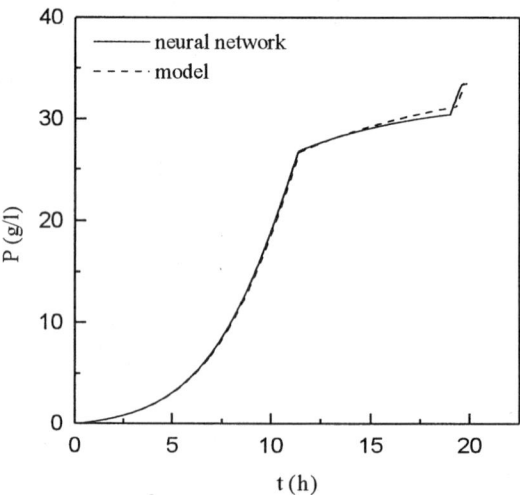

Fig. 3. Temporal profiles of ethanol concentration obtained using Aiba et al. *(18)* kinetic model and the FLN to describe the specific growth rate.

The results of the implementation of the open-loop, the adaptive control, and that which would be obtained if the perfect model (Eq. 48) were used to solve the optimization problem are shown in the Figs. 4 and 5. Figure 4 shows the results in terms of the volumetric substrate feed rate, which is easily calculated from the dilution rate (F = D/V).

Because the real inhibition of the specific growth rate by the ethanol is lower than calculated by the imperfect model, the real substrate con-

Fig. 4. Temporal profiles of the volumetric substrate feed rate for perfect model, open-loop implementation of the profile based on the imperfect model, and adaptive optimal control. Case 1.

Fig. 5. Temporal profiles of ethanol concentration for perfect model, open-loop implementation of the profile based on the imperfect model, and adaptive optimal control. Case 1.

sumption will be higher than this model predicts. In Fig. 4, it can be seen that the volumetric substrate feed rate calculated using the perfect model is the highest. When the profile calculated based on the imperfect model is implemented in an open-loop fashion, the resulting feed rate is much lower than necessary to obtain the optimal final ethanol concentration. However, when this profile is implemented in the adaptive optimal control

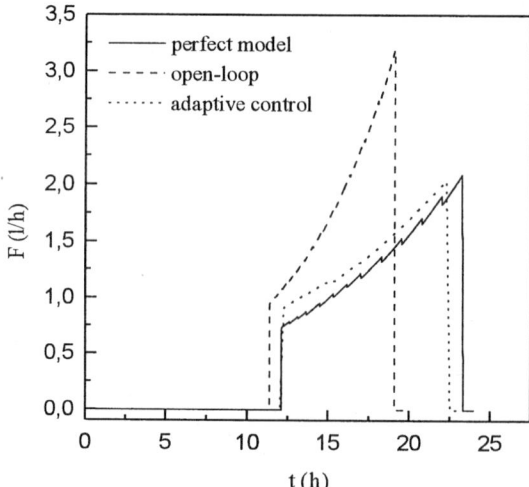

Fig. 6. Temporal profiles of the volumetric substrate feed rate for perfect model, open-loop implementation of the profile based on the imperfect model, and adaptive optimal control. Case 2.

scheme, the resulting dilution rate is similar to the optimal because of the re-estimation of the network parameters.

Figure 5 shows that the result obtained with the adaptive optimal "control scheme is the nearest to the optimal result, obtained by the use of the perfect model. The results obtained with the open-loop implementation, the adaptive optimal scheme, and the perfect model are 22 g/L, 26 g/L, and 26.6 g/L, respectively.

Case 2

In this case it was considered that the ethanol inhibits the specific growth rate more than the model used to calculate the optimal dilution rate profile predicts. The neural network was trained based on Eq. 40, an imperfect model, but the equation which really describes the specific growth rate is given below:

$$\mu = \left(0.408S/(0.22 + S)\right) \exp(-0.033P) \qquad (49)$$

In this case, the real substrate consumption is lower that the calculated by the model, and then the dilution rate, or the volumetric substrate feed rate, calculated based on the imperfect model, is higher than that required to obtain the optimal final ethanol concentration. In Figs. 6 and 7, it can be seen that the profiles obtained by the adaptive optimal control are very similar to that obtained by the perfect model. The results of the open-loop implementation, the adaptive optimal scheme, and the perfect model are 32 g/L, 41.4 g/L, and 41.9 g/L, respectively.

Fig. 7. Temporal profiles of ethanol concentration for perfect model, open-loop implementation of the profile based on the imperfect model, and adaptive optimal control. Case 2.

DISCUSSION

In most cases, the models used to determine the optimal feed policy for fed-batch biochemical processes are developed based on steady state data and work only qualitatively under transient conditions. Besides, strain modification caused by microbial adaptation or a change in the quality of nutrient medium can cause large variations in the values of parameters used to model a system (20). Thus, the development of an adaptive scheme to the optimization of these systems is of extreme importance.

In this work, an adaptive control scheme was developed for the optimization of an ethanol fermentation process. It was based on the development of a hybrid neural model for the process to be optimized. The model, composed by the mass balance equations of the process and modified FLNs, which described the kinetic rates, was shown to reproduce the process dynamics accurately.

The use of the FLNs enabled the re-estimation of the network parameters at each sampling time without much computational effort, because, in these networks, the parameter estimation is linear.

The results of the open-loop implementation when there is a plant–model mismatch showed that the final concentration of ethanol attained in the cases studied was very far from the optimal. However, the adaptive optimal control scheme led to better results.

ACKNOWLEDGMENT

The authors acknowledge CNPq for financial support.

REFERENCES

1. Palanki, S., Kravaris, C., and Wang, H. Y. (1994), *Chem. Eng. Sci.* **49,** 85–97.
2. Cuthrell, J. E. and Biegler, L. T. (1989), *Comput. Chem. Eng.* **13,** 49–62.
3. Luus, R. (1993), *Ind. Eng. Chem. Res.* **32,** 859–865.
4. Costa, A. C.(1996), M.Sc. Thesis, Federal University of Rio de Janeiro, Rio de Janeiro, Brazil.
5. Terwiesch, P. and Agarwal, M. (1994), *Comput. Chem. Eng.* **18,** S433–S437.
6. Lee, J. and Ramirez, W. F. (1996), *Chem. Eng. Sci.* **51,** 521–534.
7. Psichogios, D. C. and Ungar, L. H. (1992), *AICHE J.* **38,** 1499–1511.
8. Schubert, J., Simutis, R., Dors, M., Havlik, I., and Lübbert, A. (1994), *J. Biotechnol.* **35,** 51–68.
9. Fu, P. and Barford, J. P. (1994), *Proceedings of PSE '94,* 571–574.
10. Chen, S. and Billings, S. A. (1992), *Int. J. Control* **56,** 319–346.
11. Henrique, H. M. (1997), Personal communication, COPPE/UFRJ.
12. Billings, S. A., Chen, S., and Koremberg, M. J. (1989), *Int. J. Control* **49,** 2157–2189.
13. Veeramallu, U. and Agrawal, P. (1990), *Biotechnol. Bioeng.* **36,** 694–704.
14. Pontryagin, L. S., Boltyanskii, G. R. V., and Mischenko, E. F. (1962), *Mathematical Theory of Optimal Processes,* Wiley-Interscience, New York.
15. Alves, T. L. M. (1993), DSc. thesis, Federal University of Rio de Janeiro, Rio de Janeiro, Brazil.
16. Kravaris, C. and Kantor, J. C. (1990), *Ind. Eng. Chem. Res.* **29,** 2295–2310.
17. Modak, J. M. and Lim, H. C. (1987), *Biotechnol. Bioeng.* **30,** 528–540.
18. Aiba, S., Shoda, N., and Nagatani, N. (1968), *Biotechnol. Bioeng.* **10,** 845–853.
19. Milton, J. S. and Arnold, J. C. (1990), *Introduction to Probability and Statistics,* McGraw Hill, New York.
20. Agrawal, P., Koshy, G., and Ramseier, M. (1989), *Biotechnol. Bioeng.* **33,** 115–125.

Nonisothermal Simultaneous Saccharification and Fermentation for Direct Conversion of Lignocellulosic Biomass to Ethanol

ZHANGWEN WU AND Y. Y. LEE*

Department of Chemical Engineering, Auburn University, Auburn, AL 36849

ABSTRACT

The enzymatic reaction in the simultaneous saccharification and fermentation (SSF) is operated at a temperature much lower than its optimum level. This forces the enzyme activity to be far below its potential, consequently raising the enzyme requirement. To alleviate this problem, a nonisothermal simultaneous saccharification and fermentation process (NSSF) was investigated. The NSSF is devised so that saccharification and fermentation occur simultaneously, yet in two separate reactors that are maintained at different temperatures. Lignocellulosic biomass is retained inside a column reactor and hydrolyzed at the optimum temperature for the enzymatic reaction (50°C). The effluent from the column reactor is recirculated through a fermenter, which runs at its optimum temperature (20–30°C). The cellulase enzyme activity is increased by a factor of 2–3 when the hydrolysis temperature is raised from 30 to 50°C. The NSSF process has improved the enzymatic reaction in the SSF to the extent that it reduces the overall enzyme requirement by 30–40%. The effect of temperature on β-glucosidase activity was the most significant among the individual cellulase compounds. Both ethanol yield and productivity in the NSSF are substantially higher than those in the SSF at the enzyme loading of 5 IFPU/g glucan. With 10 IFPU/g glucan, improvement in productivity was more discernible for the NSSF. The terminal yield attainable in 4 d with the SSF was reachable in 40 h with the NSSF.

Index Entries: Simultaneous saccharification and fermentation (SSF); nonisothermal; column reactor; hydrolysis; ethanol; fermentation.

---

*Author to whom all correspondence and reprint requests should be addressed. E-mail: yylee@eng.auburn.edu

INTRODUCTION

One of the most advanced bioprocesses in fuel ethanol production is the simultaneous saccharification and fermentation (SSF) of cellulosic materials. The SSF is a single-step process in which enzymatic hydrolysis and alcoholic fermentation are carried out in a single vessel. In the SSF, the rate of hydrolysis is much lower than the rate at which the microorganism can consume glucose. The SSF therefore proceeds under glucose-limitation, and the inhibition caused by glucose and cellobiose is eliminated. Consequently, a lower enzyme loading is required *(1–4)*.

The SSF, however, has inherent problems that need to be addressed. The most significant one is the mismatch in optimum temperatures for hydrolysis and fermentation. The saccharification requires temperature of 45–50°C, and the fermentation is most efficient at 20–30°C. Since the two stages are carried out simultaneously, an SSF process is normally operated at a compromised temperature of 35–38°C. This trade-off in the temperature precludes the possibility of achieving maximum enzyme activity and the highest possible fermentation efficiency. This has been a very well-recognized problem. Substantial research effort has indeed been put forward to improve the process. Most of the research work has been focused on the identification and improvement of thermotolerant yeast or bacteria that can produce ethanol at higher temperatures *(5–10)*. This would allow hydrolysis to proceed at higher rates. From these studies, new strains that can withstand temperatures as high as 41°C have been identified. However, results from various studies indicate that thermotolerant microorganisms are less tolerant against ethanol and exhibit low productivity *(11–13)*. Furthermore the temperature range of 38–41°C is still lower than the optimum temperature for cellulases (45–50°C). Research efforts from different angles have also been made. They include changing the temperature profile *(14–15)*, varying the recipe of the enzyme, i.e., supplementing β-glucosidase *(3,16)*, further verifying of the kinetics *(17)*, increasing substrate digestibility by employing novel pretreatment methods *(18–20)*, and developing oligomeric fermenting microorganism *(21,22)*. Each proposed method has its own merits and limitations. None of them, however, has provided a feasible solution for the stated problem.

In this study, a novel bioconversion scheme that is designed to overcome the problem of temperature mismatch is introduced. A nonisothermal bioreactor is introduced to modify the SSF process. The process is named nonisothermal simultaneous saccharification and fermentation (NSSF). The key point of the NSSF is that the hydrolysis and fermentation are carried out in separate zones, yet simultaneously at their respective optimum temperatures. One zone is a column reactor in which biomass is contained and hydrolyzed at elevated temperature, and the other zone is a fermenter. Yeast cells are retained in the fermenter either by using flocculent cells or immobilized cells. A cell-free stream is recirculated through the column reactor and the fermenter. The focus of this work was on the enzymatic hydrolysis in a packed-column reactor and the overall performance of the NSSF.

MATERIALS AND METHODS

Materials

Switchgrass supplied by the National Renewable Energy Laboratory was used as the main feedstocks. The switchgrass was pretreated with 0.078 wt% sulfuric acid at 175°C for 45 min. On dry basis, the dilute-acid pretreated substrate was analyzed to contain 59% glucan, 5% xylan, and 35% lignin and ash. A commercial α-cellulose from Sigma (St. Louis, MO) (Lot 69F-0373, 93% glucan and 5% xylan) was used as a reference substrate. The cellulase enzyme, Spezyme-CP (Lot No. 41-95034-004) was obtained from Environmental Biotechnologies, Menlo Park, CA. The specific activity of the enzyme as determined by the supplier is: filter paper activity = 64.5 FPU/mL, β-glucosidase activity = 57.6 p-NPGU/mL.

Microorganism and Media

A flocculent yeast cell, *Saccharomyces cerevisiae* (ATCC 26603), was used throughout this study. The stock culture was maintained at 4°C on agar slants containing 20 g glucose, 2 g yeast extract, 3 g malt extract, 5 g peptone, and 15 g agar/L.

SSF Operation

The flocculent yeast was transferred to a growth medium twice. The yeast cells were then grown in a liquid medium containing 2.5 g/L yeast extract, 2.5 g/L malt extract, 4.0 g/L peptone, and 20 g/L glucose at 30°C, pH 5.0, and shaken at 150 rpm. The batch SSFs were run in 250-mL Erlenmeyer flasks containing 100 mL of the fermentation medium, with shaking at 150 rpm and 35°C. The flasks were attached to a water trap to maintain anaerobic condition and to vent CO_2. The fermentation medium contained 1.5 g/L yeast extract, 1.5 g/L malt extract, and 2.5 g/L peptone. Fermentation was initiated with 5% (v/v) inocula. The same media and amount of inoculum were also applied to the NSSFs. The ethanol yield is quantified as percent of the theoretical yield. The theoretical ethanol yield is defined as 0.51 (=92/180) g ethanol/g glucose.

NSSF Design and Operation

The NSSF system consists of a fermenter, a cell settler, a heat exchanger, and a column reactor for enzymatic hydrolysis (Fig. 1). The hydrolysis column reactor is made of glass, with an internal volume of 395 mL (1.24 in. id × 14 in. L). The column was jacketed and maintained at 50°C.

Fig. 1. Schematic diagram of NSSF process.

The cell is a steeply tapered glass tube settler (internal volume 90 mL), which resembles an inverted cone. Fermentation broth is pumped into the cell settler by a peristaltic pump in which flocculent yeast is naturally settled down and returns to the fermenter. A heat exchanger is inserted into the recirculation stream. The heat exchange occurs between the output stream from the cell setter (cold stream) and the output stream from the column reactor (hot stream). The cell-free supernatant containing cellulase enzymes is sucked into the bottom of the hydrolysis column. Lignocellulosic materials are hydrolyzed as the stream slowly passes through the column. The formed glucose is brought back to the fermenter, where it is consumed to produce ethanol. The glucose concentration is kept at a very low level in the entire system. The overall NSSF process scheme therefore retains the same feature as the SSF in eliminating the product inhibition.

Digestibility Test

Enzymatic hydrolysis was performed in 250 mL glass bottles at 50°C, pH 4.8 (0.05 M sodium citrate buffer). It was agitated at 150 rpm on a shaker incubator. The enzymatic digestibility is defined as (total amount of glucose released) × 0.9/total glucan. A dehydration factor of 0.9 is used to convert the glucose to glucan.

Analytical Methods

Sugars and fermentation products were analyzed by HPLC using Bio-Rad (Hercules, CA) Aminex HPX-87H column. The HPLC was operated with refractive index detector under conditions of: 0.005 M sulfuric acid as eluent, 0.6 mL/min, and 65°C.

Conversion of Lignocellulosic Biomass

Fig. 2. Enzymatic hydrolysis of cellulose at 38°C and 50°C at various enzyme loadings. The enzyme loading is based on per gram glucan. Hydrolysis condition: 3% (w/v) cellulose, pH 4.8.

RESULTS AND DISCUSSION

Effect of Temperature on Enzymatic Hydrolysis

The effect of temperature on enzymatic hydrolysis was investigated under a typical SSF cellulose loading of 3% (w/v). Two levels of temperature were selected, 38°C and 50°C, representing the optima for the cellulases and the yeast. The effect of enzyme loading on the digestibility was tested within the range of 25–100 IFPU/g glucan (Fig. 2). At 38°C, the digestibility has increased with the enzyme loading. The maximum observed digestibility at 38°C is about 58%, occurring at 120 h, with 100 IFPU/g glucan. At 50°C, the digestibility is consistently higher than those of 38°C by 20–25% for all levels of enzyme loading. For example, at

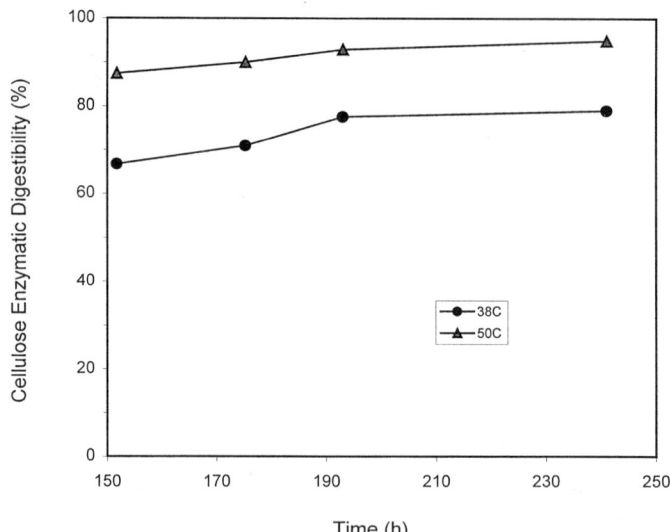

Fig. 3. Cellulose terminal enzymatic digestibility at 38°C and 50°C. Hydrolysis condition: 25 IFPU/g glucan, 3% (w/v) cellulose, pH 4.8.

120 h point, with 100 IFPU/g glucan, the yield reached 80%, a 22% increase over that of 38°C. These results indicate that reaction temperature is a prime factor determining the enzyme loading for a given conversion level. Taking a numeric example in these runs, to achieve 60% conversion at 50°C, a little less than 20 IFPU/g cellulose is needed. To achieve the same at 38°C, 100 IFPU/g cellulose is needed.

Figure 3 shows the effect of temperature on the terminal digestibility of cellulose at an enzyme loading of 25 IFPU/g glucan. At 10 d, about 90% of cellulose has been hydrolyzed at 50°C, at 38°C the digestibility is 10% lower. These results reaffirm that the temperature is an important factor controling the terminal yield of glucose as well.

From a kinetic standpoint, the initial hydrolysis rate may be a better parameter in the study of the temperature effect on the enzyme activity. The enzymatic hydrolysis was carried out under a straight batch mode at pH 4.8 using filter paper (Whatman [Clifton, NJ] No. 1) as the substrate. The initial hydrolysis rate was measured by glucose production in the first 3 h. The results in Table 1 show that the initial hydrolysis rates increase with temperature and with the enzyme loading. The initial hydrolysis rate was shown to increase by a factor of 2–3 when the temperature was raised from 30°C to 50°C. The temperature of 50°C was judged to be near the optimum for Spezyme because it was found that, at 55°C and 60°C, both hydrolysis rate and digestibility rapidly decrease.

Performance of Enzymatic Hydrolysis in a Packed-Column Reactor

The performance of the enzymatic hydrolysis in a column reactor in a recirculation mode is a key element in the NSSF process. There is no

Table 1
Effect of Temperature on Initial Hydrolysis Rate at Various Enzyme Loadings[a]

Hydrolysis temperature (°C)	Enyme loading (IFPU/g glucan)			
	5	15	25	60
30	0.25	0.51	0.85	1.36
38	0.25	0.56	0.93	1.56
45	0.70	1.39	1.56	2.03
50	0.79	1.56	1.66	2.10

[a] Hydrolysis: 1 (w/v)% glucan, pH 4.8, and Whatman No. 1 filter paper as substrate. The initial rate was defined as the average released rate of glucose and cellobiose during the first 90 min; Unit: (g/L/h).

Fig. 4. Effect of recirculating liquid flow rate on cellulose enzymatic hydrolysis. Hydrolysis condition; 25 IFPU/g glucan, 50°C, pH 4.8, 3% (w/v) cellulose.

evidence in the literature that the enzymatic hydrolysis has ever been performed in a packed-column reactor in a recirculation mode. The enzymatic hydrolysis of biomass in a packed column can be affected not only by the temperature, enzyme loading, and substrate concentration, but also by the fluid dynamic conditions in the bed.

Similar to the function of agitation in the enzymatic batch hydrolysis, a certain level of liquid flow rate in the column reactor is required to facilitate the mass transfer. The effect of liquid recirculating rate on the hydrolysis was investigated over the range of 13–60 mL/min. As shown in Fig. 4, there was little difference in digestibility among the three different flow rates applied. The mass transfer or hydrodynamic characteristics appears to be consistent over this flow rate range. Also to be noted that the

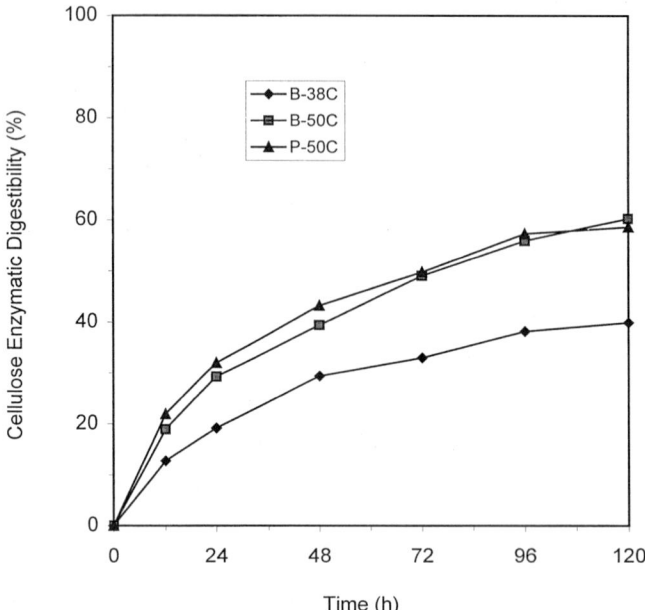

Fig. 5. Comparison enzymatic hydrolysis of cellulose in a packed-column reactor and in a straight batch process. Hydrolysis condition: 25 IFPU/g glucan, 3% (w/v) cellulose, and pH 4.8. B-38C: enzymatic hydrolysis at 38°C in a batch process. B-50C: enzymatic hydrolysis at 50°C in a batch process. P-50C: enzymatic hydrolysis at 50°C in a packed-column reactor.

recirculation needs to be applied in such a way that the liquid is sucked out of the column reactor instead of pumping the liquid through the column reactor to avoid plugging.

The hydrolysis in a packed column was first carried out using α-cellulose as the substrate. A straight batch hydrolysis at 38°C and 50°C were included as references. Average of the results from five runs was shown in Fig. 5. As seen in the figure, the enzymatic hydrolysis of cellulose at 50°C in both the packed-column reactor and the batch process showed little difference in digestibility. However, the effect of temperature was more discernible in these runs.

The performance of hydrolysis in the packed column was further tested using dilute-acid-treated hybrid poplar as the substrate. The hybrid poplar was pretreated with 0.75 wt% sulfuric acid in two stages (150°C for 10 min and 185°C for 10 min) in a percolation reactor. On dry basis, the pretreated hybrid poplar contained 64% cellulose and 33% lignin. The enzymatic hydrolysis was performed with a solid concentration of 1.55% (w/v), equivalent to 1% glucan, 50°C, pH 4.8, and an enzyme loading of 60 IFPU/g glucan. The results of hydrolysis (Fig. 6) are similar for the two modes of operation. The enzymatic hydrolysis in the column reactor has shown a higher initial rate. However, the hydrolysis in the batch mode

Fig. 6. Comparison of enzymatic hydrolysis of dilute-acid-treated poplar in a packed-column reactor and in a straight batch process. Pretreatment condition: 0.75 wt% sulfuric acid, 150°C, 10 min, and 180°C, 10 min, in a percolation reactor. Hydrolysis condition: 25 IFPU/g glucan, 50°C, pH 4.8, and 1% (w/v) glucan.

regains the activity at the latter phase, eventually yielding higher overall conversion at the end of the run.

Nonisothermal Simultaneous Saccharification and Fermentation

The NSSF runs have been made with enzyme loading of 5, 10, and 25 IFPU/glucan, using dilute-acid-pretreated switchgrass as the substrate. The SSF runs were carried out simultaneously as a reference under the same conditions, except the temperature. In the broth, cellobiose, xylose, and glucose were identifiable. A small amount of glycerol was also identified along with ethanol. At each level of enzyme loading, glucose was accumulated during the first 10–20 h in both process (Fig. 7). In the initial phase of the process, yeast could not consume glucose at the rate it was released by the enzyme. An exception is in the case of the SSF at 5 IFPU/g glucan enzyme loading, in which no glucose was accumulated during this period. An enzyme loading of 5 IFPU/g glucan is obviously too low oversupply glucose even in the initial phase of the SSF. At about 20 h, as the yeast cell mass increases, the glucose concentration in both SSF and NSSF declined to near zero level. From this point on, the process is limited by enzymatic hydrolysis, not by the microbial action. Xylan was also hydrolyzed into xylose. More xylan was converted in the NSSF than in the SSF process.

The advantage of the NSSF is evident in the cellobiose profile. At the enzyme loading of 25 IFPU/g glucan, the profiles are similar, the maximum

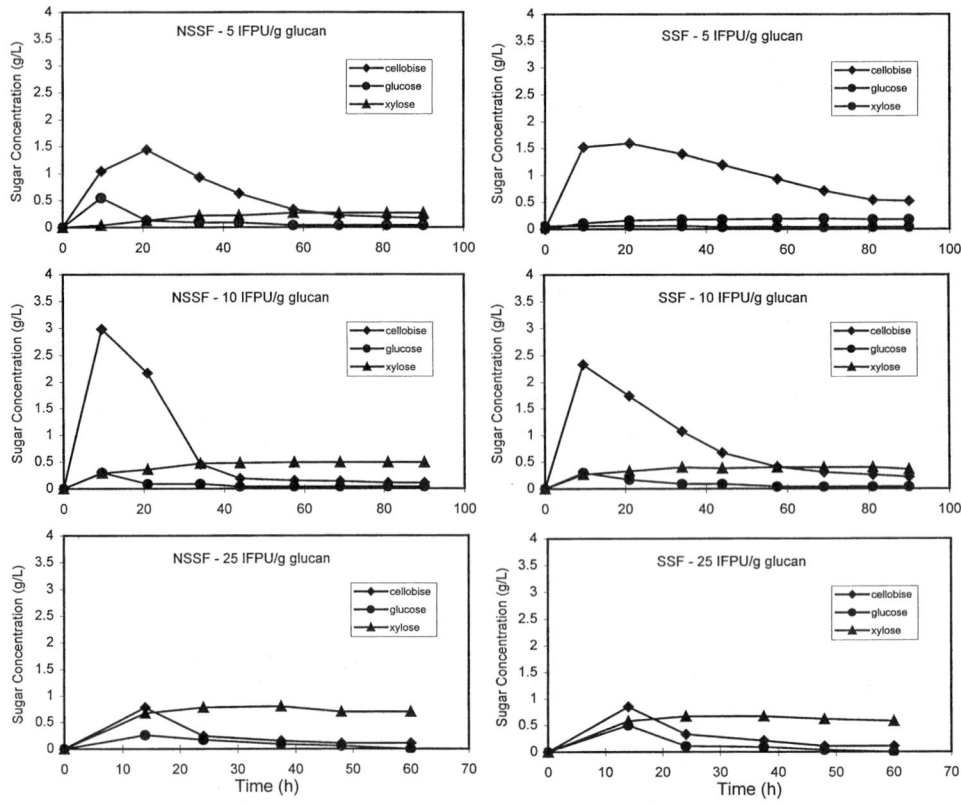

Fig. 7. Concentration profiles of sugars during NSSF and SSF of dilute-acid-pretreated switchgrass. Pretreatment condition: 0.074 wt% sulfuric acid, 170°C, and 45 min, in a batch reactor.

cellobiose concentration being less than 1 g/L for both processes. This indicates that 25 IFPU/g glucan is sufficient for both cases. With 10 and 5 IFPU/g glucan, cellobiose accumulates for both cases, the maximum cellobiose concentration reaching 3 g/L. For the NSSF process, the cellobiose concentration reached its peak value in 10–20 h, and then declined quickly to a much lower level (<0.2 g/L) at 45 h. For the SSF process, however, the cellobiose concentration profile declined gradually, then remained at a relatively high concentration for the rest of the process. Accumulation of cellobiose occurs because of insufficient β-glucosidase activity. Cellobiose is a strong inhibitor to the cellulase enzymes. It can also disrupt the synergism among the individual components of cellulase enzymes. Maintaining a low cellobiose concentration is therefore extremely important to the SSF process. Reduction of the cellobiose concentration in the SSF process can be achieved by increasing the enzyme loading or supplementing β-glucosidase. Either method, however, is uneconomical (24). The obvious benefit of the NSSF is that it enhances the overall enzyme activity, a merit of

Fig. 8. Ethanol yields of dilute-acid-pretreated switchgrass at different time in NSSF and SSF. Pretreatment condition: 0.074 wt% sulfuric acid, 170°C, and 45 min, in a batch reactor.

temperature adjustment. It is also meaningful that it raises the activity of β-glucosidase to the extent that the supplementation of it is unnecessary, even at low enzyme loading.

An advantage of the NSSF over the conventional SSF is also found in the yield of ethanol (Fig. 8). At the enzyme loading of 5 IFPU/g glucan, both ethanol yield and productivity of the NSSF process were substantially higher than those of the SSF process. The difference of ethanol yield between the NSSF and SSF process varies with time. It reached to the maximum of 17% at 67 h. Unfortunately, after that point the ethanol yield in the NSSF process started to decline. It is believed that the microorganism under extremely glucose-limiting condition utilizes ethanol as a carbon

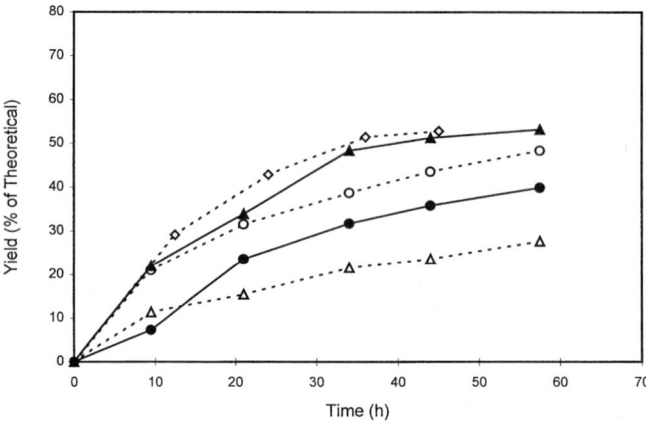

Fig. 9. Comparison of ethanol yields of dilute-acid-pretreated switchgrass in NSSF and SSF.

source. Similar phenomena were also reported by Philippidis and Smith (25). In the author's experience, this behavior is observed when the microorganism is under glucose limitation, and when air enters into the culture environment during the sampling process. Under this particular situation, the microorganism seems to switch its metabolism toward aerobic uptake of ethanol. Without the accidental metabolic pathway change, the difference in ethanol yield between the NSSF and SSF processes has been much higher. We also speculate that the microorganism may utilize CO_2 as a carbon source, along with ethanol. This comes from the observation that there is a back siphoning of liquid from the sealing trap into the fermenter that coincides with the microorganism starting to consume ethanol.

When the enzyme loading was increased to 10 IFPU/g glucan, the difference in ethanol yield between the SSF and the NSSF became insignificant. However, the productivity was much higher for the NSSF. For example, it took about 40 h for the NSSF to reach its terminal yield; it took the SSF process 4 d to reach the same conversion. The consumption of ethanol in the NSSF process also occurred after 70 h fermentation with the enzyme loading of 10 IFPU/g glucan. As the enzyme loading increased to 25 IFPU/g glucan, the advantage of the NSSF diminished.

The enzymatic hydrolysis is a limiting step in both SSF and NSSF process. The NSSF processes substantially increase the enzyme efficiency and consequently the productivity and product yield. To achieve a same-level conversion of lignocellulosic biomass, less enzyme is required in the NSSF than in the SSF. The evidence is shown in Fig. 9. The NSSF at an enzyme loading of 10 IFPU/g glucan achieves the ethanol yield achievable by the SSF with 15 IFPU/g glucan. Similarly, a 5 IFPU/g glucan enzyme loading in the NSSF exhibits enzyme activity equivalent to 8 IFPU/g glucan enzyme loading in the SSF. In general, the NSSF process reduces the enzyme requirement by 30–40%. The increase in the initial hydrolysis rate is far greater (2–3×) when the reaction temperature is increased from 30 to 50°C. The data collectively support the original claim of the NSSF that the enzyme efficiency and the productivity could significantly improved. Occasional difficulty was encountered in operation of the NSSF using powder cellulose as the substrate. The fine particles have a tendency of forming dense slurry cake, eventually causing channeling or clogging. Special care needs to be taken to cope with this problem in the design and operation of a NSSF reactor, or a recirculation packed-bed column reactor, in general.

CONCLUSION

The focus of this work was placed on the validation of two main items: the enzymatic hydrolysis in a column reactor under a recirculation mode, and the advantage of the nonisothermal SSF process. The data on enzymatic hydrolysis of lignocellulosic substrates prove that the performance of column hydrolysis is essentially the same as the stirred-batch process. The initial enzymatic hydrolysis rate is increased by a factor of 2–3 when the hydrolysis temperature is raised from 30 to 50°C. The optimum temperature for enzymatic hydrolysis of cellulose is about 50°C. Further increase in temperature beyond 50°C resulted in decrease of glucose yield.

The NSSF reactor system proposed in this work can be operated under a stable condition. Experimental results on the NSSF prove that it is a feasible engineering solution to a well-known problem associated with the SSF; the mismatch of temperatures for hydrolysis and fermentation. The NSSF process improves the enzymatic reaction in the SSF. It reduces the enzyme requirement by 30–40%. The increase in the β-glucosidase activities in the NSSF is particularly meaningful, in that it negates the cellobiose inhibition.

REFERENCES

1. Abe, S. and Takagi, M. (1991), *Biotechnol. Bioeng.* **37,** 93–96.
2. Grohmann, K. (1993), in *Bioconversion of Forest and Agricultural Plant Residues*, Saddler, J. N., ed., CAB International, Wallingford, pp. 183–210.
3. Spindler, D. D., Wyman, C. E., and Grohmann, K. (1991), *Appl. Biochem. Biotechnol.* **28/29,** 773–786.
4. Hinman, N. D., Schell, D. J., Riley, C. J., Bergeron, P. W., and Walter, P. J. (1992), *Appl. Biochem. Biotechnol.* **34/35,** 639–649.
5. Ballesteros, I., Ballesteros, M., Cabanas, A., Carrasco, J., Martic, C., Negro, M. J., Saez, F., and Saez, R. (1991), *Appl. Biochem. Biotechnol.* **28/29,** 307–315.
6. Ballesteros, I., Oliva, J. M., Ballesteros, M., and Carrasco, J. (1993), *Appl. Biochem. Biotechnol.* **39/40,** 201–211.
7. Spindler, D. D., Wyman, C. E., Mohagheghi, A., and Grohmann, K. (1988), *Appl. Biochem. Biotechnol.* **17,** 279–293.
8. Spindler, D. D., Wyman, C. E., and Grohmann, K. (1989), *Biotechnol. Bioeng.* **34,** 189–195.
9. Barron, N., Marchant, R., McHale, L., and McHale, A. P. (1995), *Appl. Microbiol. Biotechnol.* **43,** 518–520.

10. Ward, C., Nolan, A. M., O'Hanlon, K., McAree, T., Barron, N., McHale, L., and McHale, A. P. (1995), *Appl. Microbiol. Biotechnol.* **43**, 408–411.
11. Viikari, V. L., Nybergh, P., and Linko, M. (1980), In *Advances in Biotechnology*, vol. 2, Moo-Yooung M., ed., Pergamon, New York, 137–142.
12. Spindler, D. D. and Emert, G. H. (1986), *Biotechnol. Bioeng.* **28**, 115–118.
13. Saddler, J. N., Mes-Hartree, M., Yu, E. K. C., and Brownell, H. H. (1983), *Biotechnol. Bioeng. Symp.* **13**, 225–238.
14. Oh, K. K., Kim, T. Y., Jeong, Y. S., and Hong, S. I. (1996), in *Renewable Energy*, vol. 2, Sayigh, A. A. A., ed., Pergamon, New York, 962–970.
15. Huang, S. Y. Chen, C. J. (1988), *J. Fer. Technol.* **66**, 509–516.
16. Spindler, D. D., Wyman, C. E., Grohmann, K, and Mohagheghi A. (1989), *Appl. Biochem. Biotechnol.* **20/21**, 529–540.
17. Philippidis, G. P., Spindler, D. D., and Wyman, C. E. (1992), *Appl. Biochem. Biotechnol.* **34/35**, 543–556.
18. Shah, M. M., Song, S. K., Lee, Y. Y., and Torget, R. (1991), *Appl. Biochem. Biotechnol.* **28/29**, 99–109.
19. Torget, R., Hatzis, C., Hayward, T. K., Hsu, T.-A., and Philippidis, G. P. (1996), *Appl. Biochem. Biotechnol.* **58/59**, 85–101.
20. Wu, Z. and Lee, Y. Y. (1997), *Appl. Biochem. Biotechnol.* **63/65**, 21–34.
21. Mamma, D., Koullas, D., Fountoukids, G., Kekos, D., Macris, B. J., and Koukios, E. (1995), AIChE Annual Meeting, Miami Beech, FL.
22. Katzen, R. and Fowler, D. E. (1994), *Appl. Biochem. Biotechnol.* **45/46**, 697–707.
23. Wright, J. D., Power, A. J., and Douglas, L. J. (1987), *Biotechnol. Bioeng. Symp.* **17**, 285–302.
24. Hinman, N. D., Schell, D. J., Riley, C., J., Bergeron, P. W., and Walter, P. J. (1992), *Appl. Biochem. Biotechnol.* **34/35**, 639–649.
25. Philippidis, G. P. and Smith, T. K. (1995), *Appl. Biochem. Biotechnol.* **51/52**, 117–124.
26. Iyer, P. V., Wu, Z., Kim, S. B., and Lee, Y. Y. (1996), *Appl. Biochem. Biotechnol.* **57/58**, 121–132.

Comparison Between Experimental and Theoretical Values of Effectiveness Factor in Cephalosporin C Production Process with Immobilized Cells

M. LUCIA G. C. ARAUJO, ROBERTO C. GIORDANO, AND CARLOS O. HOKKA*

Universidade Federal de São Carlos, Departamento de Engenharia Química, P. O. Box 676, P.Code 13565-905, São Carlos-S.P., Brazil

ABSTRACT

Cells of *Cephalosporium acremonium* ATCC 48272 immobilized in calcium alginate beads were utilized for cephalosporin C production and the results were compared with those obtained with free cells. The experiments were performed with synthetic medium containing glucose and sucrose as carbon and energy sources. Experimental effectiveness factor values were obtained at various cell and dissolved-oxygen concentrations, considering Monod kinetics for the respiration rate, and were compared with the values calculated with zero-order kinetics in spherical bioparticle. The results showed that the assumption of oxygen limitation by diffusion in the bioparticle was correct, and that cephalosporin C production with immobilized cells is perfectly viable, although a slightly lower rate than that obtained in the free cell process was observed.

Index Entries: Cephalosporin C production; immobilized cells; respiration rate; effectiveness factor

INTRODUCTION

Cephalosporin C is a β-lactam antibiotic with some biological activity; its importance lies in the fact that it is the raw material for obtaining several semisynthetic antibiotics. Industrially, high-yield strains of the strictly aerobic filamentous fungus *Cephalosporium acremonium* are utilized through submerged cultures in aerated stirred-tank bioreactors, resulting in highly viscous non-Newtonian fermentation broths. The use of immobilized filamentous fungi in fluidized-bed tower bioreactors is a promising alternative

*Author to whom all correspondence and reprint requests should be addressed.

for minimizing gas–liquid oxygen mass transfer problems, since viscosity reduction is brought about in fermentation broth (1,2). Despite the fact that these unconventional bioprocesses present many advantages when compared to processes utilizing free cells, a rigorous examination of the intraparticle mass transfer limitations, mainly concerning oxygen transport, should be carried out in order to get a better appraisal of the advantages of these processes.

The effect of diffusion limitation of several essential nutrients in gel matrices has been investigated by many authors (3–6). However, most of these studies were conducted using gel either with entrapped dead cells or without cells, and in several studies transfer phenomena involving these biocatalysts have been evaluated only through numerical techniques and modeling, without experimental observations.

In this work, cephalosporin C production processes with both free and immobilized cells were compared through the investigation of the limitation created by oxygen diffusion into gel beads on the process rate. Kinetic parameters for the respiration rate, considering Monod kinetics model, were previously estimated by using experimental data of free cell assays. The effective oxygen diffusivity during the immobilized cell process was estimated considering a dead core model for bioparticle (7,8). These data allowed calculation of experimental effectiveness factor values (8–10), which were compared with a theoretical zero order effectiveness factor vs Thiele modulus curve. Furthermore, the influence of sugar type and concentration on the respiration rate was investigated for the process carried out with free cells.

MATERIALS AND METHODS

Microorganism

C. acremonium ATCC 48272 (C-10).

Culture Media

Inocula cultures were prepared in a synthetic medium containing (in g/L): glucose (30.0), ammonium acetate (8.8), DL-methionine (5.0), oleic acid (1.5), K_2HPO_4 (5.8), KH_2PO_4 (2.3), $CaCO_3$ (2.0), $Fe(NH_4)_2(SO_4)_2 \cdot 6H_2O$ (0.16), Na_2SO_4 (0.81), $MgSO_4 \cdot 7H_2O$ (0.384), $CaCl_2 \cdot 2H_2O$ (0.08), $MnSO_4 \cdot H_2O$ (0.032), $ZnSO_4 \cdot 7H_2O$ (0.032), $CuSO_4 \cdot 5\ H_2O$ (0.002), pH 7.0 ± 0.1.

The main fermentation medium was based on Shen et al. (11), and adapted to the nutritional requirements of the current strain. Its composition was (in g/L): glucose (27.0), sucrose (36.0), ammonium acetate (8.8), DL-methionine (5.0), oleic acid (1.5), K_2HPO_4 (2.97), KH_2PO_4 (1.08), $CaCO_3$ (2.0), with the same inorganic salts as the inoculum medium composition, pH 7.0 ± 0.1. For the immobilization cell process, the same medium was used, but, to maintain the integrity of the gel beads, K_2HPO_4 was suppressed, KH_2PO_4 concentration was changed to 1.5 g/L, and $CaCl_2 \cdot 2H_2O$ (1.0 g/L) was supplied to the medium.

Analytical Methods

Cell Concentration

The cell mass was represented as volatile suspended solids (VSS), in g/L. For the measurement of biomass in the gel beads, they were previously dissolved in a 0.5 M K_2HPO_4 solution (~20 mL/50 gel beads), according to the procedure used by Khang et al. *(12)*.

Sugar Concentration

To determine glucose concentration, GOD-PAP enzymatic method was used. For sucrose measurement, samples were previously hydrolyzed in an acid medium and GOD-PAP method was carried out.

Antibiotic Concentration

Cephalosporin C titers were determined by an agar diffusion bioassay using *Alcaligenes faecalis* ATCC 8750 *(13)*.

Experimental Procedure

Cephalosporin C Production Processes

The fermentation runs were performed in shaken flasks at 250 rpm, 26°C. The inoculum preparation procedure was the same for both processes (free and immobilized cells). Initially, resuspended lyophilized cells of *C. acremonium* were cultivated in slants containing complete medium *(11)* for 7 d at 26°C. The main fermentation inoculum was obtained after two consecutive precultures carried out in shaken flasks, for 48 h (germination) and 24 h, respectively. A percentage of 10% vol inoculum/vol culture medium was maintained between the several stages. Main fermentation periods with free and immobilized cells were of 144 and 168 h, respectively. Samples were taken every 24 h to measure pH, cell concentration, sugars, cephalosporin C, and respiration rate.

Gel bead preparation was similar to that proposed by Khang et al. *(12)*, but for the present work some modifications were made. Initially, a suspension, composed of 17% volume of the final preculture broth per total desired volume mixture of gel, alumina, and cells, was centrifuged for 2 min at 1106g, 6°C. Part of the supernatant was then discarded, so that the resulting concentrated cell suspension volume was approximately half the initial one. This suspension was added to a mixture of sodium alginate (20.0 g/L) and alumina of less than 325 mesh (10.0 g/L), to complete the desired volume. Afterwards, the mixture of gel, alumina, and cells was dripped through a needle into a stirred solution $CaCl_2 \cdot 2H_2O$

0.1 M (~600 mL solution for each 150 mL of the mixture). After preparation, the gel beads were cured in this solution for approx 1.5 h at room temperature, and then washed twice with 300 mL deionized water. The initial average diameter of the gel beads obtained was 2.2 ± 0.1 mm. For the immobilization process, the main fermentation broth was inoculated at a ratio of 15% vol gel beads/volume culture medium, which corresponded approximately to the amount of inocula from the fermentation with free cells. The diameter of the gel beads was determined with a magnifying glass throughout fermentation. An average diameter of 2.7 ± 0.15 mm was obtained during the whole process.

Measurements of Respiration Rate

For measurement of respiration rate, Yellow Springs Instrument model 5300 was used, attached to a recorder and to a dissolved oxygen analyzer. The equipment is comprised of controlled temperature (26°C) water bath, with appropriate ports in which measuring flasks (3 mL working volume), supplied with magnetic stirrers, can be inserted. To these flasks a dissolved oxygen electrode was attached (9).

Investigation of sugar influence on respiration rate was carried out with free cells. For the measurements, 10–20 mL fermentation broth was centrifuged (1106g, 6°C), the supernatant was discarded, and the compacted cells were diluted to 1:10 in fresh production medium and added to the measuring flasks. Seven glucose concentration values, between 0.2 and 19.0 g/L, and eight sucrose concentration values, between 0.5 and 31.0 g/L were utilized during free cell measurements. Blank measurements were carried out, with no sugar medium. The biomass varied from 0.9 to 1.45 g cells/L. Measurements were replicated 4–6×.

For experimental effectiveness factor determination, free and immobilized cell respiration measurements were performed in a fresh production medium containing 10.0 g/L glucose. For immobilized cell measurements, a given amount of gel beads (30–35 bioparticles) were introduced into the flasks containing fresh production medium. The free and immobilized cell concentrations, C_x and $C_{x\,biop}$, were determined following respiration rate measurements. Their values were between 0.9 and 1.8 g cells/L for the free cells, and 1.0 and 6.8 g cells/L for the immobilized cells. Replicate measurements were made, as in sugar influence on respiration study.

Estimation of Oxygen Intraparticular Diffusion Coefficient

The oxygen diffusivity, De_{O_2}, in Ca-alginate gel beads containing viable cells was determined by utilizing apparatus similar to that used by Kurosawa et al. (14). By means of this system, fresh medium saturated with oxygen passed continuously at a controled flow rate through a measuring flask of 220 mL containing a given number of gel beads. This flask was equipped with a magnetic stirrer and a water bath at 26°C. The measurements were made with gel beads taken after 92 h of the main fermentation *(8)*. By cutting the particle with a blade, a biolayer of cells inside the gel bead of about 200 μm was observed through a microscope with scaled eye-pieces, evidencing the existence of a dead core in the bioparticle. The measured values of the bioparticle radius, R_p, and the critical radius, R_{cr}, were 1.35 ± 0.05 and 1.15 ± 0.05 mm, respectively.

The reaction–diffusion model was then developed, considering the existence of a dead core in the bioparticle. This model was obtained through differential mass balances for oxygen in quasisteady state over the measuring flask and inside the gel bead along r, from R_p to R_{cr}. The oxygen diffusivity coefficient was estimated by nonlinear regression method, according to Marquardt's numerical method *(15)*, with 95% confidence interval *(8)*.

Determination of Effectiveness Factor

The effectiveness factor is defined as follows *(7,16)*:

$$\eta = \text{rate of reaction with pore diffusion resistance} / \text{rate of reaction with surface conditions} \tag{1}$$

Applied to the respiration rate at a cell concentration C_x, the effectiveness factor definition (Eq. 1) results in Eq. 2:

$$\eta = r_{obs} / R_{O_2} \cdot C_x \tag{2}$$

where r_{obs} and $R_{O_2} \cdot C_x$ refer to respiration rate with and without oxygen diffusion limitation, and C_x is the free-cell concentration expressed as biomass contained in the gel bead by culture medium volume.

In order to develop a mathematical model of the immobilized cell respiration rate, r_{obs}, the following assumptions were made:

1. The respiration rate follows Monod kinetics.
2. The gel beads are approximately spherical (mean radius R_p).
3. The biomass is homogeneously distributed in the bioparticle.
4. The medium is perfectly stirred in relation to the liquid phase.
5. Resistance to oxygen transfer through the external film around gel bead is negligible.
6. Radial oxygen flow and oxygen diffusion rate follow Fick's law.
7. The effective intraparticular diffusivity of oxygen remains constant during the process.

In quasisteady state, the global oxygen consumption rate of the encapsulated cells equals the diffusive flow of oxygen through the surface and the variation in oxygen concentration with time at the surface of the gel bead, expressed as:

$$r_{obs} = A_p/V_p \cdot De_{O_2} \cdot (dC_{O_2}/dr|_{r=R_p}) = dC_{O_2}/dt|_{r=R_p} \qquad (3)$$

where V_p and A_p are the particle volume and external surface area, respectively.

Values of r_{obs} were determined in several oxygen and cell concentrations, by means of the derivative of the curves obtained during respiration rate measurements of the immobilized cells. The experimental effectiveness factors were then calculated using Eq. 2.

By applying zero order reaction equation to the Monod kinetics, the Thiele modulus, ϕ_0, can be calculated as follows:

$$\phi_0 = \frac{R_p}{3} \sqrt{R_{max} \cdot C_x^{biop}/2 \cdot De_{O_2} \cdot C_{O_2}|_{r=R_p}} \qquad (4)$$

where R_{max} is the maximum specific respiration rate and C_x^{biop} is the gel bead cell density in g cell/L gel.

RESULTS AND DISCUSSION

Comparison Between Cephalosporin C Fermentation by Free and Immobilized Cells

The experimental results obtained in processes by either free or immobilized cells, using synthetic medium containing glucose and sucrose as the main carbon and energy sources, are shown in Figs. 1 and 2, respectively. As can be observed, both processes present two consecutive and distinct phases, as in most secondary metabolite production processes. The first phase is characterized by high cellular growth rate with the assimilation of the rapidly metabolizable sugar (glucose), resulting, however, in low production rate. In the second phase, the consumption of the more slowly assimilated carbohydrate (sucrose) results in higher production rates and insignificant biomass increase. This diauxic behavior occurs in such a way that, in the initial fermentation period, the synthesis of enzymes responsible for antibiotic production is repressed by glucose. When this sugar concentration value becomes sufficiently low, the β-lactam synthetases production is released and the microorganism is able to synthesize cephalosporin C at high rates (17).

The process with immobilized cells (Fig. 2) showed slower global rate and productivity: about 75%, compared with the process with free cells (Fig. 1). However, the productivity in relation to sucrose consumption obtained in fermentation with bioparticles was around 1.4 times higher than that obtained with free cells, showing that the new process is a promising alternative to be carried out in continuous or semicontinuous tower type bioreactors for long periods of time.

Cephalosporin C Production Comparisons 499

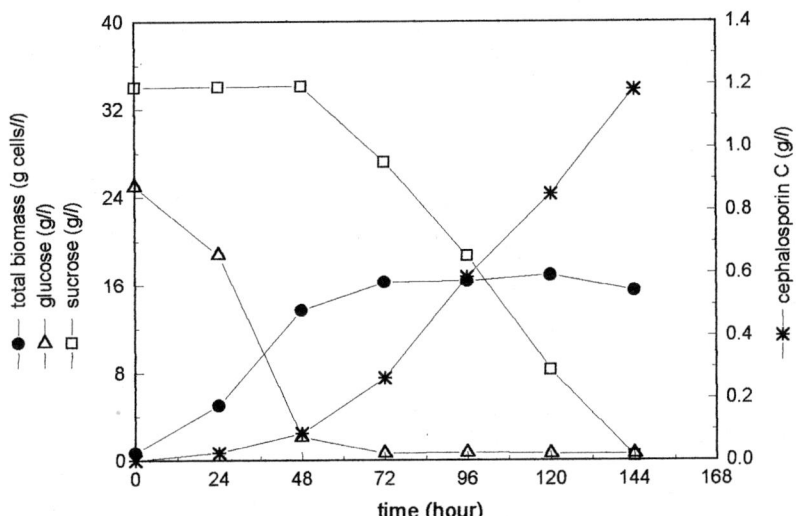

Fig. 1. Time-course of cephalosporin C batch process with synthetic medium in shaken flasks by free cells of *C. acremonium* ATCC 48272.

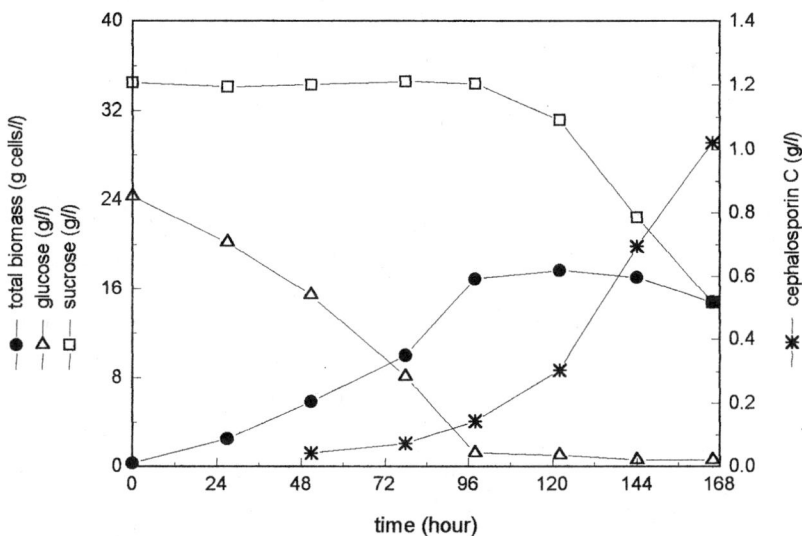

Fig. 2. Time-course of cephalosporin C batch process with synthetic medium in shaken flasks by immobilized cells of *C. acremonium* ATCC 48272 in gel beads of Ca-alginate containing alumina.

Free and Immobilized Cell Respiration Rate Measurements

Because the diauxic phenomenon strongly affects the production characteristics of this process, it was initially supposed that the specific respiration rate would be influenced by the oxygen concentration, and also by the type and concentration of sugar according to Monod equation.

The following equation represents the differential mass balance of oxygen inside the measuring flask during measurements with free cells:

$$-dC_{O_2}/dt = R_{O_2} \cdot C_x = R_{max} \cdot [C_{s_i}/(k_{s_i} + C_{s_i})] \cdot [C_{O_2}/(k_{O_2} + C_{O_2})] \cdot C_x \qquad (5)$$

where C_{O_2} is the dissolved oxygen concentration, R_{O_2} is the specific respiration rate, C_x is free cell concentration, C_{s_i} is the specific sugar concentration (glucose or sucrose), and k_{O_2} and k_{s_i} are the Monod saturation constants for oxygen and type of sugar, respectively.

Figure 3 shows the specific respiration rate data calculated from analysis of the 72 h cultivation time measurements with free cells. The oxygen consumption behavior was observed to follow a zero order kinetics in relation to glucose as well as sucrose, so that the respiration rate was not affected by the type and concentration of sugar. Furthermore, the same behavior described above was observed in blank measurements, proving that oxygen is the sole respiration rate limiting reagent.

Therefore, based on the above results, respiration rate expression can be simplified to:

$$-dC_{O_2}/dt = R_{O_2} \cdot C_x = R_{max} \cdot [C_{O_2}/(k_{O_2} + C_{O_2})] \cdot C_x \qquad (6)$$

Typical graphics of oxygen concentration vs time obtained with free and immobilized cells are shown in Fig. 4. In the free-cell systems, oxygen consumption is observed to be directly associated with the respiration rate, which seems to follow zero-order type reaction kinetics; the oxygen profile observed during measurements with immobilized cells shows that diffusion rate remarkably affects oxygen consumption rate.

Free cell measurements allowed estimation of the kinetic parameters, R_{max} and k_{O_2}, as 3.254 ± 0.608 (molO$_2$/g cell. s) $\times 10^7$ and 2.771 ± 0.730 (molO$_2$/L) $\times 10^6$, respectively, through classical nonlinear least-square regression method, according to Marquardt's procedure (15), with 95% confidence intervals (9). Immobilized cell respiration rates were determined through the derivatives calculated along the oxygen profiles.

Determination of Experimental Effectiveness Factor

The experimental effectiveness factor values were calculated using the estimated kinetic parameters, and effective oxygen diffusivity, De$_{O_2}$,

Fig. 3. Typical data of specific respiration rate of *C. acremonium* ATCC 48272 vs sugar concentration obtained during free cells process, after 72 h fermentation time.

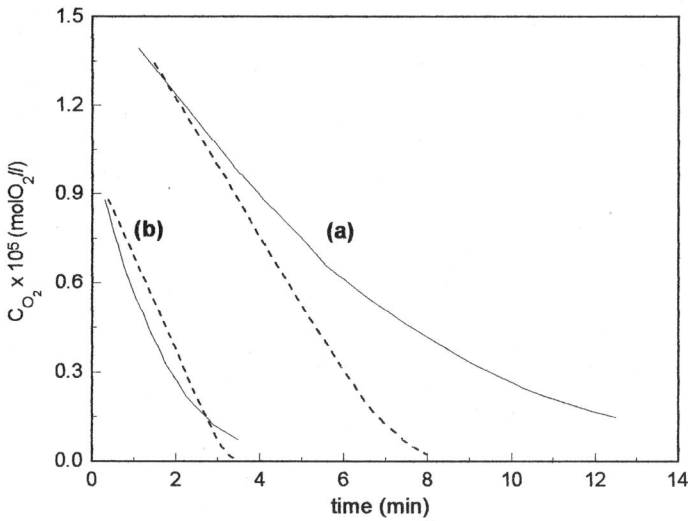

Fig. 4. Typical curves of respiration measurements with free cells (---) and immobilized cells (—) obtained after (a) 24 h cultivation time (C_x = 1.04 g cell/L; $C_{x\ biop}$ = 1.0 g cell/L) and (b) 96 h cultivation time (C_x = 1.89 g cell/L; $C_{x\ biop}$ = 6.3 g cell/L).

estimated as 1.913 ± 0.227 (m²/s) × 10⁹ (8). The effectiveness factor data are presented in Table 1.

The relationship of the dissolved oxygen concentration variation values vs cell mass concentration is shown in Fig. 5. From the slopes of the

Table 1
Experimental Effectiveness Factor Values Obtained Through Respiration Rate Measurements with Free and Immobilized Cells

C_{O_2} (molO_2/L) $\times 10^5$	8.8	7.7	6.6	5.5	4.4	3.3	2.2
C_x^{biop} (g cell/L gel)				η_{exp}			
9.4	0.685	0.648	0.594	0.514	0.441	0.363	0.325
21.3	0.509	0.471	0.432	0.377	0.317	0.258	0.228
33.0	0.565	0.535	0.474	0.405	0.317	0.283	0.253
58.4	0.533	0.509	0.445	0.370	0.313	0.258	0.223
66.4	0.501	0.471	0.410	0.347	0.296	0.247	0.218

Fig. 5. Experimental respiration rate data obtained at various dissolved oxygen concentration and cell mass concentration values.

straight lines corresponding to each oxygen concentration, the plot of Fig. 6 was obtained, showing a pseudo first order type relationship. This clearly suggests that the process with the bioparticles is limited by intraparticular oxygen mass transfer. This has already been shown that, for free cells, oxygen consumption rate behaves as a zero-order type reaction most of the time.

Plot of experimental effectiveness factor values vs calculated Thiele modulus for zero-order reaction (ϕ_0), compared with theoretical effectiveness factor η vs ϕ curve, is presented in Fig. 7.

In this figure, the experimental data obtained with low cell density bioparticles (Table 1), withdrawn at the beginning of the process, agree satisfactorily with the theoretical zero-order curve. However, the effective-

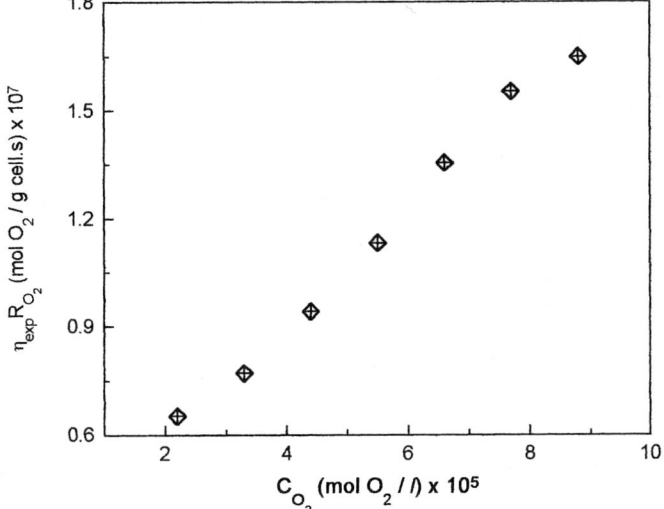

Fig. 6. Plot of immobilized cell specific respiration rate vs dissolved oxygen concentration.

Fig. 7. Plot of the experimental effectiveness factor data vs calculated ϕ_o, superimposed on theoretical η vs ϕ_o curve.

ness factor experimental values obtained with higher cell density particles, withdrawn after approx 48 h process time, begin to deviate from the theoretical values. The cause of this behavior probably lies in the fact that, although the intraparticular cell density remains at low values, the oxygen diffusion rate is sufficiently high to support maximum respiration rate, and, when the gel bead density increases in the course of the process, the diffusional limitations predominate, leading to more difficult oxygen penetration into the particle, and resulting in the reaction rate becoming oxygen-concentration dependent. Furthermore, the observed biolayer presence near the internal bioparticle surface, where total biomass was concentrated, results in a higher gel bead cell mass density, and a modified Thiele modulus based on dead core model must be utilized. At present, a study considering the effectiveness factor calculations for such a biolayer inside the particle is being carried out to compare with the experimental effectiveness factor values obtained in this work.

ACKNOWLEDGMENTS

FAPESP (São Paulo State Research Support Foundation) is acknowledged for the post doctorate scholarship granted (Process number 96/05908–3). The authors also acknowledge Paula Matvienko-Sikar for the English revision.

REFERENCES

1. Gbewonyo, K. and Wang, D. I. C. (1983), *Biotechnol. Bioeng.* **25,** 2873–2887.
2. Schügerl, K. (1990), *J. Biotechnol.* **13,** 251–256.
3. Dalili, M. and Chau, P. C. (1987), *Appl. Microbiol. Biotechnol.* **26,** 500–506.
4. Andrews, G. (1988), *Chem. Eng. J.* **37,** B31–B37.
5. Yamané, T. (1981), *J. Ferment. Technol.* **59,** 375–381.
6. Chang, H. N. and Moo-Young, M. (1988), *Appl. Microbiol. Biotechnol.* **29,** 107–112.
7. Froment, G. F. and Bischoff, K. B. (1990), *Chemical Reactor Analysis and Design,* 2nd ed., John Wiley & Sons, Singapore.
8. Araujo, M. L. G. C., Oliveira, R. P., Giordano, R. C., and Hokka, C. O. (1996), *Chem. Eng. Sci.* 14th ISCRE, **51,** 2835–2840.
9. Araujo, M. L. G. C. (1996), PhD Thesis, Chemical Engineering Program, Federal University of S. Carlos, São Carlos, SP, Brazil.
10. Araujo, M. L. G. C., Giordano, R. C., and Hokka, C. O. (1995), Proceedings of the 23th National Meeting of Porous Media (ENEMP) (1995), **2,** Maringá, PR, Brazil, pp. 1111–1123.
11. Shen., Y.-Q., Wolfe, S., and Demain, A. L. (1986), *Bio/Technology,* **4,** 61–63.
12. Khang, Y.-H., Shankar, H., and Senatore, F. (1988), *Biotechnol. Lett.* **10,** 719–724.
13. Chu, W.-B. Z., and Constantinides, A. (1988), *Biotechnol. Bioeng.* **32,** 277–288.
14. Kurosawa, H., Matsumura, M., and Tanaka, H. (1989), *Biotechnol. Bioeng.* **34,** 926–932.
15. Marquardt, D. W. (1963), *J. Soc. Indust. Appl. Math.* **11,** 431–441.
16. Aris, R. (1975), *Mathematical Theory of Diffusion and Reaction in Permeable Catalysts,* vol. 1, Oxford University Press, London.
17. Behmer, C. J., and Demain, A. L. (1983), *Curr. Microbiol.* **8,** 107–114.

Downstream Processing of Inulinase

Comparison of Different Techniques

ADALBERTO PESSOA, JR.,*,1 AND MICHELE VITOLO2

1Departamento de Biotecnologia/FAENQUIL-CP116, CEP. 12.600-000-Lorena/SP-Brazil; and 2Biochemical and Pharmaceutical Technology Departmento/FCF/USP-P.O. Box 66083- CEP 05315-970-São Paulo/SP-Brazil

ABSTRACT

Candida kefyr DSM 70106 was cultivated in a medium containing inulin as a carbon source. About 92% of the inulinase was recovered directly from the medium. Different concentration (C_f) and enrichment (E_f) factors were obtained, using the following methods: Cross-flow filtration (microfiltration and cell diafiltration were carried out using a rotary filter; enzyme ultrafiltration and diafiltration were performed using a cassette module): $C_f = 7.5$ and $E_f = 2.2$; liquid–liquid extraction of *N*-Benzyl-*N*-Dodecyl-*N-bis*[2-hydroxyethyl] ammonium chloride (BDBAC) reversed micelles: $C_f = 2.5$ and $E_f = 2.7$; and expanded-bed adsorption: $C_f = 2.8$ and $E_f = 4.3$.

Index Entries: Inulinase; downstream; cross-flow filtration; expanded-bed adsorption; reversed micelles; *Candida kefyr*.

Nomenclature: E_f; Enrichment Factor; C_f; Concentration Factor.

INTRODUCTION

Fructose has drawn much attention because of its organoleptic and sweetening properties *(1)*. This sugar is generally obtained in the form of syrup, by isomerization of glucose coming from maize hydrolysis, which normally involves the activities of several enzymes *(2)*. An alternative to this process is the enzymatic hydrolysis of inulin, a polyfructosan composed of a linear chain of D-fructofuranosides.

Inulinase (2,1 β-D-fructanfructanohydrolyse EC 3.2.1.7) is an enzyme that hydrolyzes inulin (β-D-fructofuranoside bearing a single α-D-glucopyranosyl unit) *(3)*. It is mainly produced by yeasts *(4)* and plants *(5)*.

*Author to whom all correspondence and reprint requests should be addressed.

Candida kefyr has high potential for producing this enzyme, since it reaches enzymatic activity levels comparable to those of the best strains of other species tested *(6,7)*. The industrial application of inulinase will only be viable if this enzyme is available on the market in large quantities at competitive prices *(8)*. Thus, to enhance inulinase production appropriate and feasible recovery techniques would be needed.

The costs of the extracellular enzyme recovery directly from the medium are significant. Among several separation techniques described in the current literature, with potential for scaling-up the process, cross-flow filtration, liquid–liquid extraction by reversed micelles, and expanded-bed adsorption are promising *(7,9)*.

In the case of cross-flow filtration, some fluid-dynamic characteristics have to be considered, as well as the kind of membrane, the filter geometry, and the filtration modules (rotary and cassette). The liquid–liquid extraction by reversed micelles is highly selective, specific, and, in general, compatible with proteins *(10)*. This method is able to provide desirable levels of protein concentration and purification *(11)*. Expanded-bed adsorption of proteins using Streamline™ is a recent method developed by Pharmacia (Uppsala, Sweden). It permits the recuperation of proteins without previous cell removal from the cultivated media, and efficiently replaces unitary operations such as centrifugation, extraction, and filtration *(12)*.

This work compares the yields and the purification and concentration factors of inulinase recovered directly from the medium.

MATERIALS AND METHODS

Inulinase Production

For inulinase production, *C. kefyr* DSM 70106 was grown in a 300 L fermenter containing 200 L of medium: inulin (10.0 g/L); yeast extract (2.8 g/L); peptone (6.5 g/L); $MgSO_4 \cdot 7H_2O$ (0.05 g/L); urea (2.25 g/L); KH_2PO_4 (0.30 g/L); and $CaCO_3$ (0.01 g/L). The culture was carried out batchwise, for 72 h under the following conditions: 30°C, pH 5.0, and $K_L a/43$ h. The inulinase content in the supernatant was in the range of 30–40 U/mL, and the total protein concentration was 4–5 mg/mL.

Inulinase Activity Measurement

Assays were made to measure the inulinase (2,1 β-D-fructan-fructanohydrolase, EC 3.2.1.7) in the fermentation broth, as follows: 0.20 mL of enzyme solution was mixed in a test tube with 0.80 mL of 4.0%(w/v) buffered inulin solution (0.10 *M* acetic acid/acetate buffer, pH 5.0), and the mixture was maintained at 50°C for 10 min. The reaction was stopped by immersing the test tube in boiling water. The amount of fructose was determined by the Boehringer enzymatic assay (Boehringer Mannheim,

Mannheim, FRG). One inulinase unit was defined as the amount of enzyme that catalyzes the formation of one µmol of fructose per minute under the test conditions.

Protein Determination

The amount of total protein in the filtration and adsorption tests was measured according to the Comassie blue method described by Bradford *(13)*; Lowry's method *(14)* was employed in the reversed-micelle liquid–liquid extraction. Bovine serum albumin was used as a protein concentration standard.

Cell Mass Determination

The cell concentration was obtained by means of a calibration curve to correlate optical density (OD) with dry wt (g/L).

Cross-Flow Microfiltration (Concentration and Diafiltration)

Cell harvesting by microfiltration was carried out batchwise using the rotary filter, Biodruckfilter (BDF), a prototype developed by Sulzer AG, Winterthur, Switzerland. The filter unit consists of two coaxial cylinders made of stainless steel, of which the inner is rotary and the outer cylinder is fixed. The inner cylinder is driven by an electrical motor. From a reservoir, the suspension is fed to the inlet at the lower end by a metering pump, and the concentrate is discharged by a second metering pump from the outlet at the top. The filtrate leaves the machine through a central filtrate channel at the bottom of the rotor. The outer housing and the filtrate line are cooled during processing. The suspension is pumped through the module. The retentate was continuously concentrated by repeated returning to the reservoir. The BDF technical specifications and operational conditions used in the experiments were: total membrane (PTFE 0.2 µm) area, 0.04 m^2; annular distance,: 4 mm; length of rotary cylinder, 200 mm; radius of rotary cylinder, 33 mm; radius of cylinder mantle, 37 mm; rotation speed, 3300 rpm (peripheral velocity of 11.6 m/s); pressure regime (transmembrane), ~0.5 bar; and feed rate of 30 L/h. The operation pressure was controlled by valves used for adjusting the transmembrane pressure.

Cross-Flow Ultrafiltration and Diafiltration

Plate and frame cassette devices from Filtron (Karlstein, Germany) were employed, respectively, for enzyme concentration by ultrafiltration and removal of low molecular compounds by diafiltration. Cassettes are multiple layers of ultrafiltration membranes placed between polymeric retentate separators. This type of filter consists of plates and frames arranged alternately and supported by a pair of rails. The plate has a ribbed surface. The feed channel is formed by a hole that admits feed into the frame, and at the bottom there is an outlet for the filtrate. The surface area of the PTFE membrane was 0.07 m^2, the "cutoff" 100 kDa, the transmembrane pressure about 1.0 bar, and the feed flow 225 L/h. The temperature was maintained constant at 25°C during the filtration process. Retentate and filtrate samples were withdrawn and analyzed at intervals. At the end of the process, the membranes were rinsed with 0.10 M sodium hydroxide, and with detergent solutions (including proteases).

Liquid–Liquid Extraction by Reversed Micelles

The enzyme was extracted from the whole clarified fermentation broth by N-Benzyl-N-Dodecyl-N-bis[2-hydroxyethyl] ammonium chloride (BDBAC) reversed micelles in iso-octane by a two-step procedure. The first step (forward extraction) was carried out by mixing 5.0 mL of the aqueous inulinase solution with an equal volume of micellar microemulsion (BDBAC in iso-octane/hexanol), the aqueous phase solution being adjusted to pH 6.0 by adding phosphate buffer. The phase equilibrium was obtained by intense agitation on a vortex (type Bender and Hobein AG, Zürich) for 1 min. Then the phases were separated by centrifugation at 2800g for 5 min. After that, 4.0 mL of inulinase-BDBAC-micellar phase was mixed with 4.0 mL of fresh aqueous phase (0.10 M acetate/acetic acid buffer containing 0.5 M NaCl, pH 4.0), in order to transfer the enzyme from the micelles to this aqueous phase (backward extraction), which was finally collected by centrifugation (2800g; 5 min) after intense agitation. Both aqueous phases were assayed as for enzyme activity. The extraction results are reported in terms of total recovered-activity (%) in the strip phase, using the inulinase content in the initial aqueous phase as a reference.

Expanded-Bed Adsorption Process

A 5.0-cm id chromatography column Streamline C50 (Pharmacia) was filled with 200 g Streamline DEAE. The bed was prepared by pumping the loading buffer (20 mM Tris-HCl, pH 6.5) upwards and allowing the bed height to stabilize at a twofold expansion. The cell suspension (viscosity: 2.2 mPas; density: 1.014 g/cm) was then introduced into the column from the bottom. Protein and enzyme activity in the effluent were subsequently measured. The OD (at 280 nm) of the stream leaving the bed was continuously monitored. After the adsorption step, the bed was washed with buffer flowing upwards. Then the adaptor was set down after settling of the gel bed to the initial volume of 200 mL, i.e., the expanded bed was transformed into a new fixed bed. Afterwards, the column was washed in reversed flow (downwards) with 1 bed-vol of buffer. Subsequently, elution was carried out using 20 mM Tris-HCl buffer, pH 6.5. After that, the bed was washed with 20 mM Tris-HCl containing 1.0 M NaCl, pH 6.5, before cleaning in place and reequilibration with the loading buffer. The flow rates through the columns working in the expanded-bed modes were 4 L/h (= 200 cm/h) and 1.2 L/h (= 60 cm/h), respectively.

RESULTS AND DISCUSSION

Filtration Process

Cell concentration, using microfiltration, was performed in the BDF rotary filter, and the filtrate flux obtained was higher than 100 L/h m^{-2}. Thus, this filtration process can be considered economically feasible (15). The concentrated cells were washed with acetate/acetic acid buffer (0.1 M, pH 5.0), and the inulinase recovery increased by about 6.5%. The enzymes present in the microfiltrate were ultrafiltered using a 100 kDa cutoff membrane, which provided a flux of 106.1 L/hm^{-2} and 99% of enzyme retention. Thereby, the concentration and enrichment factors for inulinase were 7.5 and 2.2, respectively (Table 1).

Liquid–Liquid Extraction by BDBAC-Reversed Micelles

Of the variables studied, those that promoted the highest extraction yields were: forward extraction: pH 6.5, 30 mM phosphate buffer, 37°C, 0.15 M BDBAC 7.5% hexanol, 92.5% isooctane, and ~5.0 mS/cm electrical conductivity; and backward extraction: pH 4.5, 100 mM acetate buffer, 25°C, and ~50 mS/cm electrical conductivity. After the forward and backward extractions, 91.0% of the inulinase was recovered from the original medium, without previous cell removal. The enrichment and concentration factors were 2.8 and 1.6, respectively (Table 1).

Expanded-Bed Adsorption

During inulinase adsorption, most of the contaminant proteins were eliminated from the column by flowing through 2 bed vol of acetate buffer. Approximately 2.8 expanded-bed volumes were used to completely remove the particulate materials from the column. Total protein concentration and optical density were used to monitor such variables. The passage of cells through the expanded bed neither blockaded nor destabilized the column and/or bed. The cells showed low affinity with the adsorbents and did not affect the enzyme adsorption. After that, the enzyme was completely eluted with 6.2 bed vol. After this chromatographic step, the inulinase enrichment and concentration factors were 4.3 and 2.8, respectively. Total recovered inulinase activity, compared to that injected into the column, was 93.1% (Table 1).

As can be seen in Table 1, the highest enrichment factor was obtained during the enzyme purification by expanded-bed adsorption, followed by the liquid–liquid extraction by reversed micelles and cross-flow filtration. The selectivity and specificity of the chromatographic technique were

Table 1
Enrichment Factor, Concentration Factor, and Enzyme Recovered by Different Processes of Enzyme Separation

Separation process	Enrichment factor (E_f)	Concentration factor (C_f)	Enzyme recovered (%)
Cross-flow filtration	2.2	7.5	92.7
BDBAD Reversed micelles liquid–liquid extraction	2.8	1.6	91.0
Expanded-bed adsorption	4.3	2.8	93.1

higher than those of the other techniques tested. However, compared to the cross-flow filtration, expanded-bed adsorption presented low capacity of enzyme concentration. Reversed-micelles liquid–liquid extraction showed neither high selectivity nor high concentration factor. To select the most appropriate technique to purify inulinase, directly from the fermented medium, both the potential industrial applications of this enzyme and their economic implications have to be considered. Through a rough analysis of the three techniques employed, it could be said that cross-flow filtration is the most adequate technique, since this technique provides a substantial increase in the concentration factor, and it exclusively depends on the initial volume employed.

CONCLUSIONS

To recover and purify inulinase, at least three techniques can be employed: cross-flow filtration, reversed-micelle liquid–liquid extraction, and expanded-bed adsorption. The recovery factor varies in the range of 91–93% for all of them. To choose the most appropriate technique for enzyme separation, two parameters were used: concentration and purification factors. Thereby, cross-flow microfiltration/ultrafiltration/diafiltration was considered the best method to produce inulinase for industrial application.

ACKNOWLEDGMENTS

A. P. acknowledges the financial support of CNPq/Brazil in the form of a Doctor of Science fellowship, and the authors thank Maria Eunice M. Coelho for revising this text.

REFERENCES

1. Kierstan, M. (1980), *Process Biochem.* **15,** 2–4.
2. Kaur, N., Kaur, M., Gupta, A. K., and Singh, R. (1992), *J. Chem. Technol. Biotechnol.* **53,** 279–284.

3. Hauly, M. C. O., Bracht, A., Beck, R., and Fontana, J. D. (1992), *Appl. Biochem. Biotechnol.* **34/35,** 297-308.
4. Manzoni, M. and Cavazzoni, V. (1992), *J. Chem. Technol. Biotechnol.* **54,** 311–315.
5. Claessens, G., Van Laere, A., and De Proft, M. (1990), *J. Plant Physiol.* **136,** 35–39.
6. Manzoni, M. and Cavazzoni, V. (1988), *Lebensm. Wiss. Technol.* **21,** 271–274.
7. Pessoa Jr., A; Vitolo, M., and Hustedt, H. (1996c) *Appl. Biotechnol. Biochem.* **57/58,** 699–709.
8. Parekh, S. and Margaritis, A. (1986), *Agric. Biol. Chem.* **50,** 1085–1087.
9. Krei, G., Meyer, U., Börner, B. and Hustedt, H. (1994), *Bioseparation* **5,** 175–183.
10. Luisi, P. L., Giomini, M., Pileni, M. P., and Robinson, B. H. (1988), *Biochim. Biophys. Acta* **947,** 209–246.
11. Krei, G. A. and Hustedt, H. (1989), *BioEngineering* **3/4,** p.32–41.
12. Frej, A.-K., B., Hjorth, R., and Hammarström, A. (1994), *Biotechnol. Bioeng.* **44,** 922–929.
13. Bradford, M. A. (1976), *Anal. Biochem.* **72,** 248–254.
14. Lowry, O. H., Rosebrough, N. J., Farr, A. L., and Randall, R. J. (1951), *J. Biol. Chem.* **193,** 265–275.
15. Kroner, K. H. and Schütte, H. (1989), in *International Technical Conference on Membrane Separation Processes* pp. 279–290.

Improvement of Lactic Cell Production

S. Desmons,* H. Krhouz, P. Evrard, and P. Thonart

Centre Wallon de Biologie Industrielle, Faculté universitaire des sciences agronomiques de Gembloux, 2, Passage des Déportés, B-5030 Belgium

ABSTRACT

The production of *Lactobacillus brevis* has been improved by changing the medium composition and the physiological conditions. The cellular concentration reaches 8.2×10^{10} cells/mL in 21 h with a fed-batch technique in MRS medium. Freeze-drying has been used as technology for drying the cells and survival rates have been improved with different additives such as glycerol, $CaCO_3$, and skimmed-milk powder up to 70%. A model has been developed to predict the stability of freeze-dried cultures during long-term conservation. This model, based on Arrhenius equation, has been confirmed by experimental data.

Index Entries: *L. brevis*; freeze-drying; fermentation; accelerated storage test.

INTRODUCTION

Lactic acid bacteria are largely employed in the food industry, especially for the fermentation of milk, meat, fruit, vegetables, and bread products *(1)*. Culture concentrates of microorganisms are then used.

The preparation of those culture concentrates requires production and maintenance techniques that maximize the storage stability, viability, and activity of the bacterial cells *(2–4)*. The most common technologies used for cell conditioning are fluidization, spray-drying, freezing, and freeze-drying. The choice depends on technological views, but also economical and practical constraints. In literature, lyophilization is frequently reported in preserving and distributing lactic starter cultures *(5,6)*.

Several factors, including the nature of the suspending solution, influence the ability of lactic organisms to survive after lyophilization *(5)*. Different cryoprotectors are commonly used to improve this stability. Glycerol, adonitol, dimethyl sulfoxide, carbohydrates, milk and its pro-

---

*Author to whom all correspondence and reprint requests should be addressed. E-mail BIOINDUS@fsagx.ac.be ducts (nonfat dry solids of milk for example), serum, peptone, dextran, sodium glutamate, and yeast extract are some of them (7,8).

The storage stability of a lactic starter is a very important factor for its industrial use. Optimal residual moisture of conservation differs in function of the culture medium composition, the protective compounds, the drying technique, and the condition of conservation used (9). De Valdez et al. (10) proposed a tool for predicting the long-term conservation of freeze-dried powder of *Lactobacillus reuteri*. It is an accelerated storage test based on the Arrhenius' law.

Among lactic acid bacteria found in industry, *Lactobacillus brevis*, a heterofermentative microorganism, appeared to be very important because of its acidification property and, especially, for its production of flavors. It is therefore used in sausage as a fermentative agent. It also improves organoleptic qualities of bread. This microorganism is then interesting to study.

The purpose of this study is to improve the production of a starter culture of *L. brevis* and to develop a predicting model for long-term conservation of this powder at low temperature. The optimization of the production involves the optimization of the growth of the lactic bacteria in fermentor and of the drying procedure by lyophilization.

MATERIALS AND METHODS

Source and Maintenance of Microorganism

Lactobacillus brevis ATCC8287 was obtained from the culture of the Laboratory of Microbiology of Gent (Belgium). The strain was maintained on MRS agar at 4°C after incubation at 30°C for 48 h.

Media

MRS broth contains 10 g of casein peptone, 5 g of yeast extract, 5 g of meat extract, 20 g of glucose, 1 mL Tween-80, 5 g of CH_3COONa, 2 g of diammonium citrate, 0.1 g of $MgSO_4 \cdot 7H_2O$, 0.1 g of $MnSO_4 \cdot H_2O$ and 2 g of K_2HPO_4 in 1 L distilled water (11). MRS agar was prepared by adding 16 g of agar, 5 g of $CaCO_3$ and 0.5 mL of an ethylic solution of bromocresol purple per liter. CSL broth contains 50 g corn-steep liquor, 21 g of yeast extract, 50 g of glucose, 0.0264 g of $MgSO_4 \cdot 7H_2O$, 0.0264 g of $MnSO_4 \cdot H_2O$ and 0.01264 g of $Fe_2(SO_4)_3$ per L of distilled water. The culture medium is adjusted to pH 6.8 with KOH 6 N before sterilization (121°C for 20 min).

Improvement of *L. brevis* Growth

Assays in Flasks

Assays were performed in flasks (300 mL) containing 150 mL broth, using a 1% inoculum and stirring at 140 rpm. Different temperatures of incubation were tested (25, 30, and 36°C). Before experimental use, microorganisms were developed once in MRS broth. The microorganism growth was followed by measurement of the optical density at 540 nm (spectrophotometer Beckman, Fullerton, USA). Viable cell count was estimated by spreading 0.1 mL of appropriate dilutions over the surface of MRS agar plates. Dilutions were performed in peptone water containing 1 g of casein peptone, 2 g of Tween-80, and 5 g of NaCl per L distilled water.

Assays in Fermentor

Bacteria were propagated twice in MRS broth at 30°C with stirring at 140 rpm (in 100 mL for 16 to 18 h, and then in 500 mL for 8 h using 2% inoculum). Bacteria were next produced in a 20-L (Biolaffite Poissy, France) fermentor containing 12 L of culture medium, using 5% inoculum (stirring at 150 rpm, aeration rate of 0.2 v/v/min).

Initial cell concentration was approx 5×10^7 CFU/mL in all cases. Evolution of the bacterial growth was evaluated by an optical density measurement at 540 nm (spectrophotometer, Beckman) and an MRS agar count. Glucose concentration was also measured by enzymatic method (Biochemistry Analyzer YSI, OH).

Freeze-Drying

After the fermentation, cells were harvested by centrifugation at 14,000g (Sharpless MV15521IHC, UK). The pellet was suspended in peptone water (ratio of the mixture 1:1). Experiments were carried out with the following protective media:

1. 10% (w/v) nonfat dry milk solids and glycerol at different concentrations (2, 3, 4, and 5% [v/v]).
2. 10% (w/v) nonfat dry milk solids, 5% (v/v) glycerol, and 0.1% (w/v) $CaCO_3$.
3. 10% (w/v) nonfat dry milk solids, 5% (v/v) glycerol, and 0.1% (w/v) $MnSO_4$.
4. 10% (w/v) nonfat dry milk solids and 5% (w/v) maltose.
5. 10% (w/v) nonfat dry milk solids and 5% (w/v) saccharose.
6. 10% (w/v) nonfat dry milk solids and 5% (w/v) sodium glutamate.

After homogenization, concentrated samples were frozen at $-20°C$ for 14 to 16 h and placed at $-50°C$ for 24 h. Samples were then freeze-dried at $+30°C$, at a vacuum of 0.5 atm for 40 h (CHRIST, alpha I-21, Osterode am Harz, Germany).

Powders obtained were grinded, distributed in plastic bags and stored at $+4°C$. Survival rate of *L. brevis* was estimated by MRS agar count after centrifugation, lyophilization, and during storage.

Accelerated Storage Test

Samples were incubated at 30, 46, and 60°C. Viability assays were performed on each sample removed at constant intervals of time by MRS agar count. Samples were removed every 24 h at 30°C, every 3 h at 46°C, and every 1 h at 60°C.

Calculation of the results was done with the equation used in the fundamental studies of Greiff and Rightsel (12). The method is resumed below. Thermal degradation of the microorganisms should follow the logarithmic form of the Arrhenius equation with respect to absolute temperature.

$$\log k = -(\Delta Ha/2303R) \cdot 1/T \tag{1}$$

where k is the specific rate of degradation (h^{-1}), ΔHa is the heat of activation (J/mole), R is the gas constant (8.32 J/mole.°K), and T is the absolute temperature (°K).

This equation indicates that any value proportional to the specific rate k would permit calculation of the slope of the log k vs $1/T$, and therefore the heat of activation. If such a relation among several values determined at high temperature is reasonably linear, the degradation rate at storage temperature can be calculated from the experimentally determined degradation rates at high temperature. The equation for a pseudo-order reaction to determine the rate of degradation (k) appeared appropriate.

$$\log N = \log N_o - k \cdot t \tag{2}$$

where N_o is the initial cell concentration (CFU/mL), N is the cell concentration at any time (t) (CFU/mL), k is the specific rate of degradation (h^{-1}) and t is the time (h).

Determination of k for each temperature tested (30, 46, and 60°C) is performed using Eq. 2. The construction of Arrhenius plot (k vs $1/T$) allows the estimation of k at lower temperatures. Cell concentration after long-term storage at low temperature can then be estimated.

Freeze-dried samples were also kept at 4°C to determine the percentage of survival after a few months and verify the model.

RESULTS

Improvement of *L. brevis* Growth

Assays in Flasks

Optimal temperature and initial pH of the culture medium have been first determined in MRS broth. Cellular growth has been followed by turbidimetric measurement (optical density at 540 nm). A correlation of

Fig. 1. **A** Effect of temperature on *L. brevis* growth in flask (MRS broth, 140 g-rpm of stirring). **B** Effect of initial pH on *L. brevis* growth in flask (MRS broth, 30°C, 140 rpm of stirring).

$N = -1.6 \times 10^9 + 9.6 \times 10^8 \cdot OD$ ($r = 0.97$) was obtained. Figures 1A and B show results obtained.

Optimal temperature for *L. brevis* growth is 30°C. *L. brevis* growth is more important when initial pH of the medium is greater than 6.0. It also seems that the growth is slower for initial medium of pH 7.0 and 7.5. A pH of 6.5 will therefore be retained as optimal initial pH medium for *L. brevis* growth.

The effect of the concentration of glucose, meat extract, and yeast extract was also tested. The final biomass is higher with increasing initial glucose concentrations (15–20 g/L). The optimal concentration of meat and yeast extracts is 3 and 5 g/L, respectively in MRS broth (data not shown).

The same final cell concentration has also been observed in MRS broth and CSL broth (6.9 × 10^9 CFU/mL in MRS broth after 23 h fermentation vs 6.3 × 10^9 CFU/mL in CSL broth).

Assays in Fermentor

During its growth, *L. brevis* generates lactic acid, which causes a diminution of the pH medium (from 6.5 at the beginning of fermentation to 4.0 at the end of the microorganism's growth). As *L. brevis* is sensitive to low pH, it is preferable to regulate this parameter. This regulation is not

Table 1
Influence of pH on the Growth of L. brevis in Fermentor

pH	Neutralizer used	Final cell concentration (cfu/ml)*10^{10}	Duration of fermentation (h)
6	KOH 6N	1,4 ±0,08	19h50'
6,2	KOH 6N	1,8 ±0,12	18h30'
6,5	KOH 6N	1,7 ±0,11	19h00
6,8	KOH 6N	1,5 ±0,16	21h00

CSL, 30°C, 150 rpm of agitation, 0.2 v/v/m in aeration rate.

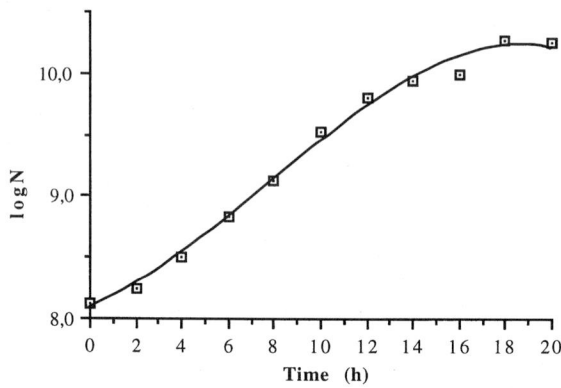

Fig. 2. Evolution of the growth of L. brevis in a 20-L fermentor. (MRS broth, 30°C, pH 6.5 regulated by KOH 6 N, 150 rpm of agitation, 0.2 v/v/min aeration rate).

possible in a flask but is in a fermentor. So, the first assays in the fermentor will concern detection of optimal pH. Fermentation is followed by glucose concentration determination and was stopped when glucose concentration was lower than 2 g/L. Cell concentration is then estimated by MRS agar count. Table 1 shows results obtained in CSL broth.

Final cell concentrations are similar in all cases. As we refer to fermentation time, it appears that optimal pH is ranging between 6.2 and 6.5.

Figure 2 shows the evolution of the growth curve of L. brevis in a 20-L fermentor (MRS broth, 30°C, pH 6.5, 150 rmp of agitation, aeration rate of 0.2 v/v/min). As assays in flasks have shown an identical growth of L. brevis in MRS and CSL broth, and as CSL broth is a turbid medium, growth curve was established in MRS broth to permit an optical density measurement.

The generation time (g = 1 h 50 min) and the maximal growth rate (μ_m = 0.39 h^{-1}) have been calculated from this growth curve.

The next step was the improvement of the final cell concentration, so two different media (CSL broth and modified CSL broth) were tested. Modified CSL broth (CSLm1 broth) consisted of CSL broth with a supple-

Table 2
Influence of the Medium on the *L. brevis* Growth

Medium	Duration of fermentation (h)	Final cell concentration (cfu/ml)*10^{10}
MRS	19	1,8 ±0,10
CSL	18	1,5 ±0,15
CSLm$_1$	18	1,7 ±0,12

30°C, 150 rpm of agitation, 0.2 v/v/m in aeration rate, pH 6.2 regulated by KOH 6 N.

ment of meat extract (2 g/L). Fermentation was followed by glucose concentration measurement and was stopped when glucose concentration became lower than 2 g/L. Final cell concentration was then measured by MRS agar count. Results are shown in Table 2. A fermentation in MRS broth is given for comparison.

In all cases, final cell concentrations reached the same value. Differences appeared in terms of medium cost. CSL broth is cheapest. Adding meat extract to this medium does not significantly improve *L. brevis* growth.

In order to improve the production of biomass, batch and fed-batch techniques were compared using MRS broth, CSL broth, and CSLm2 broth (containing 37.5 g/L of glucose instead of 50 g/L in CSL broth). Final cell concentration was determined at the end of the fermentation, when glucose concentration was lower than 2 g/L. The different assays and results are summarized in Table 3.

Fed-batch technique with CSL broth leads to a final cell concentration higher than batch technique (approx 2 times).

In the assays with MRS broth, a fed-batch technique with the whole medium gives no more growth than with the glucose alone.

Freeze-Drying

Assays of freeze-drying were performed with cells produced in CSL broth, at 30°C, 150 rpm of agitation, 0.2 v/v/min of aeration rate and pH regulated at 6.5 by KOH 6 N. Fermentation was stopped when glucose concentration was lower than 2 g/L. Under these conditions, the final cell concentration reached 1.5×10^{10} CFU/mL after 18 h of fermentation.

Loss of cells can occur at three steps, i.e., the centrifugation, the lyophilization, and the conservation. Cellular loss coming from the centrifugation process ranges from 10 to 20% (data not shown). The effect of the different protective media on the percentage of survival after lyophilization and the residual moisture content of the powders are shown in Table 4.

Table 3
Influence of Method of Carbohydrate Addition on the Growth of *L. brevis*

Medium used	Carbohydrates adding technique used	Amount of glucose added at the begining of the fermentation (g/l)	Amount and moment of the second glucose addition	Final cell concentration (cfu/ml) *10^{10}
CSL broth	batch	50	none	1,5 ±0,12
$CSLm_2$ broth	batch	37,5	none	1,9 ±0,3
CSL broth	fed-batch	25	25g/l after 11h of fermentation	2,8 ±0,18
MRS broth	fed-batch (glucose)	20	20 g/l after 12h of fermentation	8,2 ±0,8
MRS broth	fed-batch (MRS*)	20	addition of a modified MRS broth* after 12 h of fermentation (*this modified medium contains the same elements than MRS broth but at different concentration (1/2) excepted for glucose)	8,8 ±0,9

MRS and CSL broth, 30°C, 150 rpm of agitation, 0.2 v/v/m in aeration rate, pH 6.2 regulated by KOH 6 N.

Survival rates are relatively high (67–86%) and the best results concern the assay with nonfat dry milk solids, glycerol, and $CaCO_3$ (85.9%). Good percentages are also observed with glycerol and maltose (respectively, 73–82 and 79.6%).

Samples were stored at 4°C. MRS agar counts were performed over a few months. Results are shown in Fig. 3.

Percentage of survival rapidly decreases after the first weeks of storage. The decrease is less important after 1 month, and the survival concentration tends to stabilize.

Among the different cryoprotective media used, glycerol gives the best results in terms of survival rate (from 62 to 70% after 4 months of storage at 4°C).

Accelerated Storage Test

In order to elaborate a model of the stability of *L. brevis* during long-term storage, specific rate of degradation *(k)* of *L. brevis* freeze-dried in

Table 4
Effect of Cryoprotective Media on the Survival Rate and the
Percentage of Dry Matter of Lyophilized *L. brevis*.

Cryoprotective media	Survival rate (%)	Dry matter content (%)
non fat dry milk solids (NFDMS) (10%) + glycerol (5%)	79,6	87
NFDMS (10%) + glycerol (4%)	82	90
NFDMS (10%) + glycerol (3%)	79	93
NFDMS (10%) + glycerol (2%)	73	95
NFDMS (10%) +glycerol (5%) + CaCO$_3$ (0,1%)	85,9	87
NFDMS (10%) +glycerol (5%) + MnSO$_4$ (0,1%)	74,2	88
NFDMS (10%) + maltose (5%)	79,6	95
NFDMS (10%) + saccharose (5%)	67,4	93
NFDMS (10%) + sodium glutamate (5%)	67,1	92

Fig. 3. Influence of cryoprotective media on the viability of *L. brevis* freeze-dried during storage at 4°C.

presence of nonfat dry milk solids (10%) and glycerol (5%) was first determined according to Eq. 2 as described in the subheading "Materials and Methods," for three high temperatures. The viability of freeze-dried cells during short storage at 30, 46, and 60°C was evaluated by MRS agar count. The value of k was calculated for each temperature on basis of a linear regression of the different points measured (Fig. 4).

The specific rate of degradation is given by the slope of the straight lines obtained (Table 5).

Table 5
Specific Rates of Degradation at 30, 46, and 60°C of Freeze-Dried *L. brevis*.

Storage temperature (°C)	Specific rate of degradation (h⁻¹)
30	0,019 (k_{30})
46	0,32 (k_{46})
60	1,66 (k_{60})

Fig. 4. Mortality of freeze-dried *L. brevis* stored at 30, 46, and 60°C in function of time (with nonfat dry milk solids [10%] and glycerol [5%] as cryoprotectors).

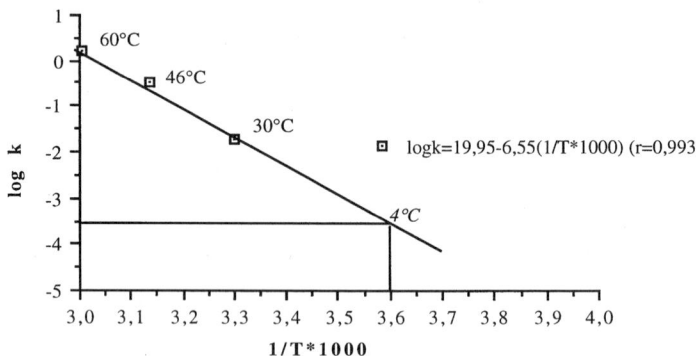

Fig. 5. Arrhenius plot of freeze-dried *L. brevis* (with nonfat dry milk solids [10%] and glycerol [5%] as cryoprotectors).

The specific rates of degradation above allow us to plot the Arrhenius graph, i.e., log k vs 1/T as shown in Fig. 5.

The specific rate of degradation at 4°C ($k_4 = 1.97 \times 10^{-4}$ h^{-1}) was estimated from this Arrhenius plot.

Table 6
Comparison of the Estimated and the Experimentally Measured
Survival Rates of Freeze-Dried *L. brevis*.

During of storage at 4°C (days)	Survival rate estimated by the accelerated storage test (%)	Survival rate experimentally measured (%)
50	58	59
137	23	16

With nonfat dry milk solids 10% and glycerol 5% as cryoprotectors.

The cell concentration of the freeze-dried *L. brevis* stored at 4°C is estimated by using Eq. 2 given in the subheading "Materials and Methods".

The survival rate of freeze-dried *L. brevis* after 50 and 137 d estimated by the accelerated storage test and experimentally measured are represented in Table 6.

After 50 days storage, *L. brevis*' survival rate estimated by accelerated storage test is quite similar to survival rate experimentally measured. After 137 d storage at 4°C, results present a difference of 30%.

DISCUSSION

Improvement of the *L. brevis* Growth

A culture medium providing an optimal growth with maximum economy would be of great interest for industrial use *(13)*.

Using the optimal operating conditions in MRS broth (30°C, pH ranging from 6.2 to 6.5 regulate with KOH 6 *N*, agitation of 150 rpm, aeration rate of 0.2 v/v/min), the generation time of *L. brevis* was fixed at 110 min. Yildiz and Westhoff *(14)* found a generation time of 130 min in filtered cabbage juice for *L. brevis*.

MRS broth is an expensive medium comparatively to CSL broth, which is based on corn steep, a byproduct of corn industry. Corn steep contains proteins, mineral salts (especially oligo-elements), and molecules with vitaminic function (phytic acid, inositol, and so on). CSL broth appeared to be a good substitute of MRS broth. It provides the same final cell concentration and costs less (it is 48% less expensive). It will be noticed than CSL broth contains more glucose (50 vs 20 g/L) and more yeast extract (21 vs 5 g/L) than MRS broth. These differences might also explain the good growth of *L. brevis* in CSL broth.

Fed-batch technique provides an appreciable gain of cell production (2.8×10^{10} CFU/mL instead of 1.5×10^{10} CFU/mL in CSL broth). A fed-batch performed in MRS broth gives a higher final cell concentration than a fed batch in CSL broth (8.2×10^{10} CFU/mL vs 2.8×10^{10} CFU/mL). As CSL broth is 48% less expensive than MRS broth, a fed-batch in MRS broth is lightly more interesting (because it allows a production of cells three times more important than CSL broth). A fed-batch realized in MRS broth with a complete medium gives no gain of cell production compared with glucose addition in two steps. This first type of fed-batch is therefore not economically interesting and won't be retained.

Some assays have been performed by replacing casein peptone in MRS broth by vegetable protein (potato protein). This modified medium gives an identical final cell concentration than MRS broth but is cheaper (data not shown). Other experiments will be needed to confirm this interesting result. Fed-batch technique with this medium had also to be performed.

Freeze-Drying

The additives for the freeze-drying procedure are very important and affect the cell survival and residual water content. The efficiency of a cryoprotective agent in the prevention of cell death during lyophilization, does not necessarily reflect its capacity to prevent loss of viability during storage (10).

Among various cryoprotective agents used in this study, glycerol exhibits the best percentage of survival after lyophilization (82 and 86%).

Addition of glycerol (5%) implies a high residual moisture of the powders (12%) compared to other cryoprotectors used (5–8%). The differences in the content of residual water observed here could be explained by a higher degree of affinity of this cryoprotective agent for water (15).

A high residual moisture content of the powders could be negative for long-term preservation. Nevertheless it had been proven that a reduction of glycerol concentration, which decreases the residual moisture, is not beneficial for the survival rate (from 85 to 73% for a glycerol concentration from 5 to 2%).

Moreover, it has been experimentally pointed out that the best conservation of the freeze-dried cells is obtained when using glycerol (from 62 to 70% after 4 months storage at 4°C).

Despite a higher residual moisture content, glycerol exhibits the best conservation results. This residual water is probably retained by glycerol and then is not available for cells. Other authors mentioned the use of 5% glycerol to decrease lethality during the drying process (18,6). Viability during storage obtained with other cryoprotectors is quite small (from 10 to 30%).

The curves of the evolution of the viability at 4°C shows a tendency for rapid death of dried cells in the early period of storage followed by a stabilization. This phenomenon has been reported by other authors (6,17–20).

It is assumed that a part of the surviving cells might be injured immediately after freeze-drying which leads to an initial postdrying lethality. Then the cell population, which is less susceptible to this postdrying inactivation mechanism might be selected to display an apparent stability in viability during the subsequent prolonged period *(20)*.

Use of $CaCO_3$ with glycerol allows a slight improve of long-term conservation.

Accelerated Storage Test

Accelerated storage test is an acceptable extrapolation tool of the stability at 4°C of freeze-dried *L. brevis* after 50 d. Indeed, comparison of predicted and experimental survival rates did not show any significant differences. Other experiments will be needed to confirm results obtained for longer storage period.

This method is advantageous because of its rapidity, with degradation rates determined at high temperatures requiring a relatively short time for their evaluation (from several hours to a couple of days). There is one restriction concerning the selection of the temperatures used for accelerated storage test, i.e., that the differences in the rates of degradation associated with each temperature must be statistically significant *(12)*.

This test can also provide a good comparison tool of efficiency of different cryprotectors for freeze-drying (in term of survival after lyophilization and stability during the storage).

It will be interesting to extend this kind of model to other cultures and other drying techniques. This model can also be applied to other fields. It has already been used to estimate thermal degradation of vitamin preparations and the measles virus *(21,12)*.

But, as the storage stability of dried cells depends on the residual moisture content, Arrhenius plot will always be associated with the residual moisture content of the sample tested.

ACKNOWLEDGMENTS

This investigation was supported by Fonds pour la Formation à la Recherche dans l'Industrie et dans l'Agriculture (FRIA). The authors also thank Roquette Society for providing corn steep liquor and potato proteins.

REFERENCES

1. Rumian, N., Angelov, M. and Tsvetkov, T.S. (1993), *Cryobiology* **30**, 438–444.
2. Wright, C. T. and Klaenhammer, T. R. (1983), *J. Food Sci.* **48**, 773–777.
3. Cox, W. A., Stanley, G. and Lewis, J. E. (1978), in *Streptococci*, Skinner, F. A. and Quesnel, L. B., eds., The Society for Applied Bacteriology, Symposium Series No. 7, Academic, New York, p. 279.
4. Gilliland, S. E. and Speck, M. L. (1974), *J. Milk Food. Technol.* **37**, 107–111.

5. Fayed E. O., Sultan N. E., Yassein N. I., and Shehata A. E. (1986), *Egypt. J. Food. Sci.* **14(2),** 313–322.
6. King, V. A.-E. and Su, J. T. (1993), *Process Biotech.* **28,** 47–52.
7. Johannsen E. (1972), *J. Appl. Bacteriol.* **35,** 423–429
8. De Antoni, G. L., Perez P., Abraham A. and Anon M. C. (1989), *Cryobiology* **26,** 149–153.
9. De Valdez, G. F. and Giori G. S. (1983), *J. Food Protection* **56(4),** 320–322.
10. De Valdez, G. F. and Diekmann, H. (1993), *Cryobiology* **30,** 185–190.
11. deMan, J. C., Rogosa, M. and Sharpe, M. E. (1960), *J. Appl. Bacteriol.* **23,** 130–135.
12. Greiff, D. and Rightsel, W. A. (1965), *J. Immunol.* **94(3),** 395–400.
13. Sinha, R. P. (1986), *J. Food Prot.* **49(4),** 260–264.
14. Yildiz, F. and Westhoff, D. (1981), *J. Food. Sci.* **46,** 962, 963.
15. Obayashi, Y., Ota, S. and Arai, S. (1961), *J. Hyg.* **59,** 77–91.
16. Bozoglu, T. F. and Gurakan, G. C. (1989), *J. Food Prot.* **52(4),** 259, 260.
17. Brennan, M., Wanismail, B., and Ray, B. (1983), *J. Food. Prot.* **46,** 887–892.
18. Cox, R. E. and Heckly, R. J. (1973), *Can. J. Microbiol.* **19,** 189–194.
19. Strange, R. E. and Cox, C. S. (1976), in *The Survival of Vegetative Microbes*, Gray T. R. G. and Postgate J. R., ed., Cambridge Univ. Press, Cambridge, UK, pp. 111–154.
20. Ishibashi N., Tatematsu, T., Shimura S., Tomita M. and Okonogi S. (1985), IIF-IIR. Commission C1, Tokyo, Japan, 1985/1.
21. Tootil, J. P. P R. (1961), *J. Pharm. Pharmacol.* **13,** 75–77.

In Situ Global Method for Measurement of Oxygen Demand and Mass Transfer

K. THOMAS KLASSON,[*,1] KARIN M. O. LUNDBÄCK,[2] EDGAR C. CLAUSEN,[2] and JAMES L. GADDY[2]

Chemical Technology Division, Oak Ridge National Laboratory,[**] Oak Ridge, Tennessee 37831; and [2]Department of Chemical Engineering, University of Arkansas, Fayetteville, Arkansas 72701

ABSTRACT

Two aerobic microorganisms, *Saccharomycopsis lipolytica* and *Brevibacterium lactofermentum*, have been used in a study of mass transfer and oxygen uptake from a global perspective, using a closed gas system. Oxygen concentrations in the gas and liquid were followed using oxygen electrodes; the results allowed for easy calculation of *in situ* oxygen transport. The cell yields on oxygen for *S. lipolytica* and *B. lactofermentum* were 1.01 and 1.53 g/g, respectively. The mass transfer coefficient was estimated as 10/h at 500 rpm for both fermentations. The advantages with this method are noticeable, since the use of model systems may be avoided, and the *in situ* measurements of oxygen demand assure reliable data for scale-up.

INTRODUCTION

In order to properly design and scale up aerobic fermentations, the oxygen requirements must be known, and appropriate methods must be developed for determining mass transfer coefficients and microbial oxygen uptake for bench-scale systems. Several methods for determining oxygen transfer rates have been reviewed *(1)*. This paper presents an alternative global method that may be applied to most bench-scale systems, either biological or chemical. The major advantage of the method is the *in situ* measurement, rather than the use of model systems, such as glucose oxidase or sulfite. Thus, problems associated with differences in medium or physical properties are effectively eliminated with this global technique. Some other methods (e.g., Warburg apparatus) rely on measuring oxygen

---

[*] Author to whom all correspondence and reprint requests should be addressed.
[**] Managed by Lockheed Martin Energy Research Corp. for the US Department of Energy under contract DE-AC05–96OR22464.

uptake parameters in systems in which the dissolved oxygen concentration may be essentially zero. Such systems are unsuitable for lysine fermentations for which oxygen limitation induces lactate production (2), thus giving a false oxygen uptake rate. Other advantages with the presented method are that only oxygen concentration measurements (partial pressure), not gas flow rates, are needed, and that changes in the mass transfer coefficient during a fermentation because of changes in surface tension, and so on, may be detected. In this paper, oxygen uptake in two aerobic fermentations is examined: citric acid production using *Saccharomycopsis lipolytica*, and lysine production using *Brevibacterium lactofermentum*.

METHODS

Microorganisms

S. (Candida) lipolytica, NRRL Y-7576, was maintained on YM-broth agar (1.5%) slants. *B. lactofermentum*, ATCC 21788, was kept on slants consisting of 1% peptone, 1% yeast extract, 0.5% NaCl, and 1.5% agar.

Medium

The medium used in the fermentation with *S. lipolytica* was a glucose, yeast extract, and salts mixture (3); a glucose, soytone, and salts medium was used in the experiment with *B. lactofermentum* (4). The fermentation jar was filled with 1 L of medium and steam-sterilized at 1 atm (15 psig) for 30 min. After cooling, each medium was inoculated using 100 mL of a 24-h-old seed culture grown in the same medium.

Equipment

The equipment used in the fermentation studies was a Biostat-M (Braun Instruments, Burlingame, CA) fermenter with a working volume of 0.7–1.5 L. The fermenter was equipped with pH, agitation, foam, and temperature control, and probes for measuring pH and dissolved oxygen tension. Cell concentration was estimated from optical density readings and a calibration curve. The measurement of oxygen uptake and *in situ* mass transfer was performed by connecting a gas bag to the setup. An external oxygen electrode and meter (Yellow Springs Instrument, Yellow Springs, OH) were used to measure the partial pressure of oxygen in the exiting gas from the fermenter. The gas was recirculated to the system using a diaphragm pump inside the gas bag. To prevent CO_2 accumulation, the gas was washed continuously with a saturated sodium hydroxide solution. A schematic of the system is displayed in Fig. 1. A single three-way valve allowed the gas to bypass the fermenter sparger; two additional valves were used to attach a flask filled with pure O_2. Both fermentations were conducted at 30°C, and the agitation rate was held constant at 500

Oxygen Demand and Mass Transfer

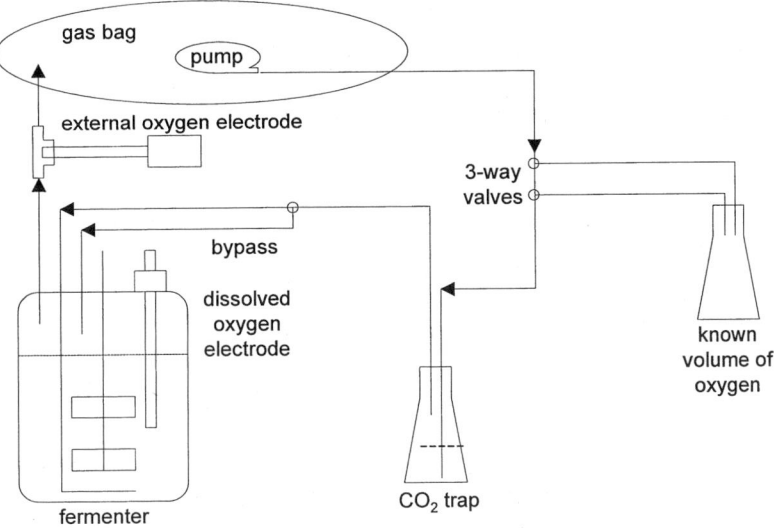

Fig. 1. Experimental setup.

rpm, which corresponds to an impeller tip speed of 0.4 m/s. The gas flow rate was maintained at approx 2 L/min.

Oxygen Balance Calculations

In an aerobic fermentation, in which nutrients and carbon substrate are plentiful, the rate at which oxygen is supplied to the active enzymes of the cell is often the limiting factor for growth and production (5,6). Since the oxygen required by the cells is supplied from a gas phase, there are a number of transport resistances, connected in series, which oxygen transport has to overcome. It is generally recognized that the main resistance to this transport, for sparingly soluble gases such as O_2, is in the liquid film (6). Assuming that Henry's law for the equilibrium between gas and liquid interfacial concentrations holds, the following equation is true:

$$\text{Oxygen transfer rate} = K_L a/H \, (p_G - p_L^*) \, V_L \quad (1)$$

At steady state (or pseudosteady state), the oxygen transfer rate equals the oxygen uptake rate by the cells.

In order to estimate the *in situ* oxygen uptake rate and mass transfer coefficient, additional equipment was attached to the fermenter to achieve a closed gas system. The additional equipment was connected and calibrated after inoculation. The method of oxygen uptake rate was then based on the decrease in the amount of oxygen present in the system.

As mentioned previously, the gas partial pressure of oxygen in the system was determined using an external oxygen electrode calibrated to give readings in partial pressure of oxygen. The total pressure in the system was assumed to be atmospheric, since a soft gasbag was used. Any carbon dioxide formed during the fermentation was removed by washing the gas with NaOH, leaving only O_2, N_2, and H_2O. This fact allowed the derivation of an expression for the oxygen present in the system at all times. It was assumed that the gas volume in the system at time zero was equal to V, and that the composition of the gas corresponded to water-saturated air at the temperature (T) and total pressure (P) of the system. After inoculation and electrode calibration, an additional known volume (V_1) of pure oxygen (at T and P) was connected to the system, and mixing was allowed. The original volume (V) can be calculated from the observed change in the partial pressure of oxygen (from p_{O_2} to p'_{O_2}) in the system. Using the ideal gas law, the following expression describing the number of moles of oxygen in the system after mixing of the two volumes may be written:

$$p_{O_2} V/RT + PV_1/RT = p'_{O_2}(V + V_1)/RT \qquad (2)$$

In addition, it is known that the amount of nitrogen present in the system is constant and described by

$$n_{N_2} = p_{N_2} V/RT \qquad (3)$$

By solving for V in Eq. 2 and substituting the result into Eq. 3, the following expression may be obtained:

$$n_{N_2} = [(P - p_{O_2} - p_{H_2O})/RT][(P - p'_{O_2})/(p'_{O_2} - p_{O_2})]V_1 \qquad (4)$$

The volume of the gas changes during the course of the fermentation, as oxygen is consumed by the culture. At any specific time, the following equation may be used to calculate this volume:

$$\text{Gas volume} = n_{O_2} RT/p_{O_2} = n_{N_2} RT/p_{N_2} = n_{N_2} RT/(P - p_{O_2} - p_{H_2O})$$

$$(5)$$

Finally Eq. 5 may be rewritten to yield

$$n_{O_2} = p_{O_2} n_{N_2}/(P - p_{O_2} - p_{H_2O}) \qquad (6)$$

in which p_{O_2} and P are measured, n_{N_2} is a constant obtained from Eq. 4, and p_{H_2O} can be estimated, assuming water saturation of the gas at all times. This derivation does not account for the oxygen present in the liquid. However, when considering the solubility of oxygen in water, it was found that the contribution from the liquid phase to the total amount of oxygen in the system was negligible.

Oxygen Demand and Mass Transfer

The procedure outlined above was used to determine the oxygen uptake rate for two different microorganisms. The mixing of the two gas volumes (pure oxygen volume and system air volume) took approx 5–10 min. To secure minimal oxygen consumption during this time, the sparger was bypassed (*see* Fig. 1), and the gas was allowed to pass over the liquid, rather than through it. Since measurements of the partial pressure of oxygen in both the gas and liquid (p_G and p_L^*), as well as the calculations of oxygen uptake from the gas, were made, the overall mass transfer coefficient could be found by rewriting Eq. 1 to yield

$$d(n_{O_2})/dt = K_L a/H \, (p_G - p_L^*) V_L \tag{7}$$

RESULTS AND DISCUSSION

The cell concentration and the amount of oxygen remaining in the system are plotted as functions of time in Figs. 2 and 3 for the two fermentations using *S. lipolytica* and *B. lactofermentum*, respectively. As is noted in the figures, the amount of oxygen (calculated from Eq. 6) decreased as the cell concentration increased in both fermentations. In the fermentation using *S. lipolytica* (Fig. 2), the data represent the initial part of the fermentation in which cell growth was more pronounced and acids were not formed. The cell yields on oxygen for the two microorganisms are represented in Fig. 4, indicating that approx 50% more cell mass was obtained in the fermentation with *B. lactofermentum* for the same amount of oxygen consumed. The combined citric and isocitric acid yield on oxygen during the latter part of the fermentation for *S. lipolytica* was 1.33 g acids/g O_2 (data not shown).

In Fig. 5, the volumetric oxygen transfer rates, calculated as the slope in data presented in Figs. 2 and 3, are plotted as a function of the driving force measured by the O_2-electrodes. As is noted, all of the data from both fermentations fell on a single straight line, indicating a constant mass transfer coefficient. This was expected, since the media and operating conditions were very similar for the fermentations. Using a value of 26.9 L atm/g O_2 for the Henry's law constant (7), the mass transfer coefficient was calculated as 10/h, based on the results obtained in Fig. 5. The fact that both model systems gave one and the same result for the mass transfer coefficient indicates that the technique employed is reliable. Validity of the results may be seen by comparing the cell yield on oxygen for *S. lipolytica* obtained in this experiment (1.01 g/g, Fig. 4) with the cell yield of 1.1 g/g obtained by Briffaud and Engasser (8) for the same organism; these results are very similar. The specific oxygen uptake rates for the two organisms were calculated as 0.32 g O_2/h, g cells for *S. lipolytica* and 0.42 g O_2/h, g cells for *B. lactofermentum* during nonoxygen-limiting conditions (in the initial part of the fermentation).

Fig. 2. Cell growth and oxygen consumption profiles for the fermentation with *S. lipolytica*.

Fig. 3. Cell growth and oxygen consumption profiles for the fermentation with *B. lactofermentum*.

Fig. 4. Cell yield on oxygen for *S. lipolytica* and *B. lactofermentum*.

Fig. 5. Estimation of the overall mass transfer coefficient in both fermentations (agitation = 500 rpm).

In conclusion, an alternative method for the measurement of the *in situ* microbial oxygen uptake and oxygen mass transfer rates was demonstrated. The advantages with this method are important, since the use of model systems may be avoided, and the accurate measurement of microbial oxygen demand assures reliable data for scale-up. The presented method may be easily adapted to any oxygen utilizing system.

REFERENCES

1. Sobotka, M., Prokop, A., Dunn, I. J., and Eisele, A. (1982), in *Annual Reports on Fermentation Processes*, vol. 5, Academic, New York, pp. 127–210.
2. Ruklisha, M., Marauska, D., Shvinka, J., Toma, M., and Galynina, N. (1981), *Biotechnol. Lett.* **3**, 465–470.
3. Klasson, K. T., Clausen, E. C., and Gaddy, J. L. (1989), *Appl. Biochem. Biotechnol.* **20/21**, 491–509.
4. Ackerson, M. D., Clausen, E. C., and Gaddy, J. L. (1989), *Appl. Biochem. Biotechnol.* **20/21**, 511–528.
5. Pirt, S. J. (1975), *Principles of Microbe and Cell Cultivation*, John Wiley, New York, pp. 81–93.
6. Bailey, J. E. and Ollis, D. F. (1986), *Biochemical Engineering Fundamentals*, 2nd ed., McGraw-Hill, New York, pp. 457–532.
7. Faust, A. S., Wenzel, L. A., Clump, C. W., Mous, L., and Andersen, L. B. (1960), in *Principles of Unit Operations*, John Wiley, New York, p. 552.
8. Briffaud, J. and Engasser, J.-M. (1979), *Biotechnol. Bioeng.*, **21**, 2083–2092.

Improvement of Oxygen Transfer Coefficient During *Penicillium canescens* Culture

Influence of Turbine Design, Agitation Speed, and Air Flow Rate on Xylanase Production

A. GASPAR,* L. STRODIOT, AND PH. THONART

Centre Wallon de Biologie Industrielle, Faculté universitaire des sciences agronomiques de Gembloux, 2, Passage Des Déportés, B-5030 Belgium

ABSTRACT

To improve xylanase productivity from *Penicillium canescens* 10–10c culture, an optimization of oxygen supply is required. Because the strain is sensitive to shear forces, leading to lower xylanase productivity as to morphological alteration, vigorous mixing is not desired. The influence of turbine design, agitation speed, and air flow rate on K_la (global mass transfer coefficient, h^{-1}) and enzyme production is discussed. K_la values increased with agitation speed and air flow rate, whatever the impeller, in our assay conditions. Agitation had more influence on K_la values than air flow, when a disk-mounted blade's impeller (DT) is used; an opposite result was obtained with a hub-mounted pitched blade's impeller (PBT). Xylanase production appeared as a function of specific power (W/m^3), and an optimum was found in 20 and 100 L STRs fitted with DT impellers. On the other hand, the use of a hub-mounted pitched blade impeller (PBT8), instead of a disk-mounted blade impeller (DT4), reduced the lag time of hemicellulase production and increased xylanase productivity 1.3-fold.

Index Entries: Xylanase; *Penicillium canescens*; K_la; specific power.

INTRODUCTION

Aerobic microorganism growth requires a continuous and efficient supply of oxygen, and hence effective gas–liquid mass transfer. This is traditionally realized by gas-sparged stirred-tank reactor *(1)*. The process of making oxygen available to a growing culture is a critical factor in

---

*Author to whom all correspondence and reprint requests should be addressed. E-Mail: BIOINDUS@fsagx.ac.be

aerobic cultivation, since agitation could create high shear environment (2–4); that may lead to mechanical disruption of cells, and hence impaired reactor performance. Examples of shear effects on biomass and product formations are abundant (2,5–11), and many authors showed a dependence of cellulase and xylanase production on agitation or aeration rate (10,12–15). Nevertheless, dependence of product formation with scaling-up parameters, such as specific power, peripheral tip speed, gas hold-up, or circulation time, are less common.

In this work, the xylanase production by *P. canescens* 10–10c was studied; earlier results showed the dependence of hyphal growth on agitation speed and oxygen supply. This suggested investigation of the influence of turbine design, agitation speed, and aeration rate on K_la in reactors fitted with one impeller. Results on K_la, specific power, and production in mono-agitated tanks are discussed.

MATERIALS AND METHODS

Strain

P. canescens 10–10c was provided by G. I. Kvesidatse, Institute of Plant Biochemistry, Academy of Sciences, Tbilisi, Georgia.

Culture Medium

The culture medium contained 30 g/L wheat straw, 30 g/L soya meal, and 5 g/L yeast extract, in mineral salt medium. The mineral salt medium contained: $Na_2HPO_4 \cdot 7H_2O$ (1.5% w/v), KCl (0.05% w/v), $MgSO_4\ 7\ H_2O$ (0.015% w/v).

Stirred-Tank Reactor (STR) (Diagram 1)

Fermentations were carried out in Biolafitte 20-L stirred-tank reactor (STR) fitted either with Rushton disk turbines with 6 or 4 blades (DT6, DT4, respectively), or pitched blade turbines with 8 or 4 blades (PBT8, PBT4, respectively). Turbine diameter (d) was 0.1 m. The working volume was 8.5 L, to have liquid height (H) equal to STR diameter (D = 0.23 m). Turbine was placed at height (h) D/3 from the bottom of the reactor. Four baffles were 0.49 m in height and 0.032 m in width. STRs were inoculated to give a spore concentration of 10^6 sp/mL, and temperature was maintained at 30°C. pH was not controlled, and varied between 5.8 and 7.5.

K_la Measurements

K_la was calculated with the static gassing-out method, in the presence of the mycelium killed by addition of sodium aside (0.5 g/L) followed by a thermal treatment at 60°C for 30 min. This procedure permitted work at

Diagram 1. Description of the STR.

real rheological culture conditions. All measurements were realized at 30°C and pH 7.0.

Xylanase Assay

Xylanase assay was carried out according to Bailey *(16)*. One unit (U) of enzyme activity is defined as the amount of sugar (in µmol) produced per min of reaction and per mL of enzyme solution, in the assay conditions.

Oxygen and Carbon Dioxide Analyzers

Oxygen and carbon dioxide were analyzed on-line with, respectively, Servomex OA 570 (Crowborough, England) and Binos (Leybold-Herous, Hanau, Germany) analyzers. This permitted the calculation of the K_la (on-line), the respiratory quotient (on-line), the oxygen uptake rate (OUR) and the carbon dioxide transfer rate (CTR).

Specific Power Calculation

The specific power (W/m^3) was calculated with the Calderbank relation *(17)*:

$$P = (1 - 12{,}6\ G/Nd^3)\ (Np\ \rho\ N^3\ d^5)\ (1/V)$$

where V is the culture volume (m^3), N is the agitation speed (h^{-1}), d is the impeller diameter (m), ρ is the volumic mass (kg/m^3) of the medium, G is the aeration rate (m^3/h) and Np is the power number.

Fig. 1. pO_2 and K_la (global transfer coefficient h^{-1}) evolution during growth of *P. canescens* 10–10c in a 20-L STR on wheat straw (3% w/v), soya meal (3% w/v), yeast extract (0.5% w/v), in mineral salt medium.

RESULTS

Evolution of pO_2 and K_la During *P. canescens* Culture

Figure 1 shows the time-course of pO_2 and K_la evolution during a cultivation of *P. canescens* on wheat straw and soya meal medium. Aeration rate was 0.75 vvm, agitation rate was 300 rpm, and the STR was fitted with one DT4.

During growth, pO_2 decreased to 0% of saturation for several hours, and K_la decreased from 60 h^{-1} to 37 h^{-1}; this could result from the increase in broth viscosity. Xylanase production started after the growth phase (after about 40 h of culture time). These results clearly indicated that there is a lack in oxygen supply, and that the culture needs an improvement of the oxygen transfer coefficient (K_la), which could lead to enhanced biomass production, and hence product formation.

Hereunder are presented results on the influence of factors acting on K_la in STR fitted with one or two impellers. Dependence of xylanase production on K_la and specific power in mono-agitated STR are also shown.

Improvement of Oxygen Transfer in *P. canescens* 10-10c Culture: Influence of Agitation and Aeration Rate on K_la in Mono-agitated Reactor

The influence of agitation speed and air flow rate on K_la in *P. canescens* culture was evaluated; the parameters studied were the turbine (DT4, DT6, and PBT8), the air flow rate (0.3, 0.5 and 0.75 vvm), and the peripheral speed (0.78, 1.05, 1.31, and 1.57 m/s). Results are shown in Figs. 2–4.

For each experiment, K_la values increased with peripheral speed; at high agitation speed (1.57 m/s), the most effective turbine for mass transfer was the DT6 one. K_la values for the PBT8 showed greater dependence

Fig. 2. K_la (h^{-1}) as a function of peripheral speed and turbine design, in a *P. canescens* culture on soya meal and wheat straw. Aeration rate is fixed at 0.3 vvm.

Fig. 3. K_la (h^{-1}) as a function of peripheral speed and turbine design, in a *P. canescens* culture on soya meal and wheat straw. Aeration rate is fixed at 0.5 vvm.

Fig. 4. K_la (h^{-1}) as a function of peripheral speed and turbine design. Aeration rate is fixed at 0.75 vvm.

Fig. 5. K_la (h^{-1}) as a function of peripheral speed and aeration rate for PBT8 impeller.

Fig. 6. K_la (h^{-1}) as a function of peripheral speed and aeration rate for DT4 impeller.

with air flow rate than with agitation speed, unlike the DT impellers (Figs. 5 and 6). The relative efficiency of PBT8 impeller grew up when air flow rate was increased and agitation speed was reduced; this means that the ratio K_la_{PBT8} on K_la_{DT} was greater at lower aeration and agitation rates than at higher gassing and mixing conditions.

Figure 7 shows K_la as a function of specific power P (W/m^3) and aeration rate G (m^3/h) for DT impellers (both DT6 and DT4). Concerning

Fig. 7. $K_l a$ as a function of specific power (W/m^3) and aeration rate (m^3/h). Data for DT6 and DT4 impellers.

PBT impeller, $K_l a$ was a function of pumping rate Qp (m^3/h) and aeration rate G (m^3/h) (Data not shown).

Influence of Turbine Design on Xylanase Production

At first, the influence of agitation speed on xylanase inactivation was evaluated: Results revealed that, in the presence of biomass, no xylanase denaturation occurred when agitation was enhanced (data not shown).

Xylanase production was carried out in STRs fitted with one impeller. Multiagitated aerated reactors were not yet evaluated: hydrodynamic currents in these reactors are complex, and correlations between results and scaling-up parameters are hazardous (19–21). Figure 8 shows the timecourse of xylanase production during batch culture of P. canescens 10-10c in a 20-L STR, fitted either with a PBT8 or a DT4 (all other culture parameters remaining unchanged). After 144 h of culture time, the batch culture with the PBT8 turbine showed relatively greater xylanase yield (844 U/mL) than the batch with DT4 (722 U/mL). Figure 8 also reveals that xylanase production started earlier with PBT8 impeller than with the DT4, but after that, production rates are similar in both cases.

Xylanase productivity was also improved when a PBT8 (844 U/mL) was used instead of PBT4 (683 U/mL) (all other culture conditions remaining unchanged).

Figure 9 shows that production after 140 h of culture time is a function of the specific power (W/m^3).

Xylanase production showed an optimum with specific power about 200–500 W/m^3. This was found in 20-L STR, fitted with DT6 or DT4 impeller, and in 100-L STR fitted with DT4.

At 200 rpm (102 W/m^3) in 20-L STR, oxygen transfer was poor, pO$_2$ was under 5% of saturation during 37 h and xylanase production reached

Fig. 8. Evolution of xylanase production (U/mL) during batch culture of *P. canescens* 10–10c in a 20-L STR fitted either with a hub-mounted pitched blade turbine (PBT8) or a disk-mounted blade turbine (DT4). Agitation speed: 300 rpm, aeration rate: 0.75 vvm.

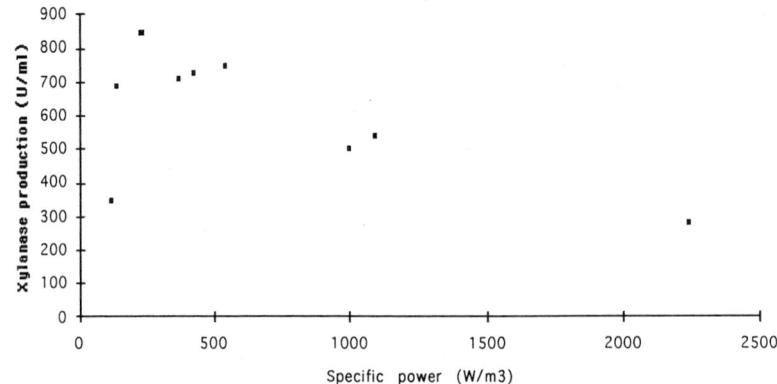

Fig. 9. Xylanase production as a function of specific power in 20-L STR. Data for DT6, DT4, PBT8, and PBT4 impellers.

a low value of 370 U/mL. On the other hand, high agitation speed produced hydrodynamic stress of hyphae; under certain conditions, hyphal breakups were observed, and this could explain the decrease in productivity above 500 W/m^3. This could explain the optimum in specific power.

In 6-L STR (working volume: 2.5 10^{-3} m^3), a continuous decrease in productivity was observed between 500 W/m^3 (450 rpm, DT6 diameter: 0.06 m) and 25,000 W/m^3 (1200 rpm, DT6 diameter: 0.07 m).

Xylanase production with PBT8 impeller reached 844 U/mL for a specific power estimated to 230 W/m^3 (Fig. 8).

DISCUSSION

Improvement of Oxygen Transfer in *P. canescens* 10-10c Culture: Influence of Agitation and Aeration Rate on K_la in Monoagitated Reactor

In preliminary experiments (22), the authors studied, in shake flasks and in STR, the optimization of culture conditions for biomass and xylanase production from *P. canescens* 10-10c. Results showed that good yields were obtained when mass transfer was optimized. Nevertheless, culture of many filamentous microorganisms in STR copes with a similar problem: poor oxygen transfer and hydrodynamic stress of hyphae (8–11).

Growth of *P. canescens* 10-10c on wheat straw, soya meal, and yeast extract in a 20-L STR developed a high-viscosity broth: This led to a drastic decrease of K_la value from 300 to $37h^{-1}$. It was concluded that an insufficient oxygen transfer limited the mycelial growth.

The influence of factors acting on K_la value in real-culture conditions was investigated. The objective is to enhance oxygen transfer and to avoid hyphal stress. The influence of mobil design on K_la and hemicellulase production, for different agitation and air flow rates, was evaluated. Performances of shearing and pumping turbines were compared in a specific medium containing straw, soya meal, and filamentous biomass.

For each experiment, K_la increased with agitation and aeration rates, whatever the impeller. K_la values were more influenced by agitation speed than by air flow rate, when DT impellers were used; an opposite result was obtained with PBT8. These observations are explained by the specific action of the two impellers, shearing (DT) or pumping (PBT8).

DT impeller, also known as radial-flow impeller (23), creates shear forces; these forces, settled by peripheral tip speed and specific power, influence bubble diameter. On the other hand, PBT8 impeller is an axial-flow impeller. It creates less shear forces and induces higher circulation time of bubbles, and hence gas hold-up. All these parameters act on the mass transfer coefficient.

Results revealed that a good compromise for *P. canescens* culture could be obtained with PBT8 impeller, because such a turbine generated lower shear forces than radial turbines, at the same agitation rate, without dramatic lost in oxygen transfer capacity.

Influence of Turbine Design on Xylanase Production

These results on K_la were correlated with xylanase production in monoagitated STR. The use of PBT8 instead of DT4 reduced the lag time of xylanase production; DT4 presented an intermediate value: at 300 rpm and 0.75 vvm, xylanase production after 50 h of culture reached, respectively, 300 U/mL with a PBT8, 150 U/mL with a DT4, and 64 U/mL with a DT6. Subsequently, after 140 h of culture time, productivity was improved by using a PBT8, instead of a DT4, by a factor 1.3.

Two things could explain these results: first, K_la values were higher with a PTB8 than with a DT4; subsequently, it could be expected to have a better biomass, and hence xylanase yields in the first case. This assumption was not verified when comparing results of DT6 and PBT8 production. Thus, another way to explain our results is the influence of stirring conditions on production: Because the specific power generated by PBT8 is lower than with DT6 impeller (all others conditions remaining unchanged), lower shear forces result from PBT8 than DT6 (19–21), and, although K_la_{DT6} was higher than K_la_{PBT8}, productivity was enhanced by using an axial impeller. The effects of high agitation speed on biomass production was already demonstrated (22), and many authors reported influence of hydrodynamic stress on filamentous microorganisms (6–9). From these results, it was concluded that a reduction in specific power is more profitable to xylanase production than an increase in K_la, beyond a critical value.

Too poor K_la is harmful to biomass growth: This was verified when production at 200 rpm with DT4 ($P = 102$ W/m^3, K_la of about 20 h^{-1}) reached only 370 U/mL. An optimum was found at 300 rpm with PBT8 ($P = $ about 232 W/m^3, $K_la = $ about 62 h^{-1}), and production reached 844 U/mL.

For similar reasons, xylanase productivity was also enhanced when STR was fitted with a PBT8, instead of a PBT4; indeed, as 8 blades furnished a higher pumping rate than 4 blades, and subsequently higher K_la values, productivity was improved.

Results also revealed that specific power could be useful for scaling-up the process: A value of 200–500 W/m^3 should be maintained to ensure best xylanase yield. Production in 500 L and 2 m^3 are planned, to verify this assumption.

CONCLUSION

P. canescens 10-10c is subject to hydrodynamic stress, and its culture in STR has to resolve contradictory aims: to increase oxygen transfer without excessive mixing. The authors propose to use a pitched blades turbine instead of a Rushton turbine; this solution permitted increased xylanase production because of lower shear forces induced. However, solutions have to be found that combine low shear and high oxygen transfer rate. Air-lift reactors could be investigated in this way.

REFERENCES

1. Mooyman, J. G. (1987), *Biotechnol. Bioeng.* **24**, 180–186.
2. Dronowat, S. N., Svihla, C. K., and Hanley, T. R. (1995), *Appl. Biochem. Biotechnol.* **51/52**, 347–354.

3. Malfait, J. L., Wilkox, D. G., Mercer, D. G., and Barker, L. D. (1981), *Biotechnol. Bioeng.* **23,** 863–877.
4. Jacques, P., Hbib, C., Vanhentenryck, F., Destain, J., Baré, G., Razafindralambo, H., Paquot, M., and Thonart, P. (1993), *Prog. Biotechnol.* **9,** 1067–1069.
5. Edwards, N., Beeton, S., Bull, A. T., and Merchuk, J. C. (1989), *Appl. Microbiol. Biotechnol.* **30,** 190–195.
6. Van Suijdam, J. C. and Metz, B. (1981), *Biotechnol. Bioeng.* **23,** 111–148.
7. Smith, J. J. and Lilly, M. D. (1990), *Biotechnol. Bioeng.* **35,** 1011–1023.
8. Gusk, T. W., Jonhson, R. D., Tyn, M. T., and Kinsella, J. E. (1990), *Biotechnol. Bioeng.* **37,** 371–374.
9. Leong-Poi, L. and Allen, D. G. (1992), *Biotechnol. Bioeng.* **40,** 403–412.
10. Hoo, M. M., Solomon, B. O., Hempei, C., Rinas, U., and Deckwer, W. P. (1995), *J. Chem. Tech. Biotechnol.* **63,** 229–236.
11. Metz, B. and Kossen, N. W. F. (1977), *Biotechnol. Bioeng.* **19,** 781–799.
12. Hoo, M. M., Hempel, C., and Deckwer, W.-D. (1994), *J. Biotechnol.* **37,** 49–58.
13. Wase, D. A., McManamey, W. J., Raymahasay, S., and Vaid, A. K. (1985), *Biotechnol. Bioeng.* **27,** 1166–1172.
14. Palma, M. B., Milagres, A. M. F., Prata, A. M. R., and de Mancilha, I. M. (1996), *Process Biochem.* **31,** 141–145.
15. Gomes, J., Purkarthofer, H., Hayn, M., Kappelmuller, J., Sinner, M., and Steiner, W. (1993), *Appl. Microbiol. Biotechnol.* **39,** 700–707.
16. Bailey, M. J., Biely, P., and Poutanen, K. (1992), *J. Biotechnol.* **23,** 257–270.
17. Calderbank, P. H. (1967), in *Mixing-Theory and Practice,* vol. 2, Academic, New York.
18. Vanderschuren, A. (1993), Thesis in Chemical Engineering, Faculte Polytechnique de Mons.
19. Bruxelmane, M. (1976), Proceedings of an International Symposium on Chemical Engineering, Louvain-La-Neuve, Belgium, pp. 321–417.
20. Bruxelmane, M. (1978), Proceedings of an International Symposium on Mixing, Facute Polytechnique de Mons, Belgium.
21. Bruxelmane, M. (1976), Proceedings of an International Symposium on Chemical Engineering, Louvain-La-Neuve, Belgium, pp. 91–177.
22. Gaspar, A., Cosson, T., Roques, C., and Thonart, Ph. (1997), *Appl. Biochem. Biotechnol.* **67,** 57–70.
23. Joshi, J. B., Pandit, A. B., and Sharma, M. M. (1982), *Chem. Eng. Sci.* **37,** 813–844.

Batch Foam Recovery of Sporamin from Sweet Potato

SAMUEL KO, VEARA LOHA, ALEŠ PROKOP
AND ROBERT D. TANNER*

Department of Chemical Engineering, Vanderbilt University, Nashville, TN 37235

ABSTRACT

The major sweet potato root protein, sporamin (which comprises about 80–90% of the total protein mass in the sweet potato) easily foams in a bubble/foam-fractionation column using air as the carrier gas. Control of that foam fractionation process is readily achieved by adjusting two variables: bulk solution pH and gas superficial velocity. Varying these parameters has an important role in the recovery of sporamin in the foam. Changes in the pH of the bulk solution can control the partitioning of sporamin in the foam phase from that in the bulk phase. A change in pH will also affect the amount of foam generated. The pH varied between 2.0 and 10.0 and the air superficial velocities (V_0) ranged between 1.5 and 4.3 cm/s. It was observed in these ranges that, as the pH increased, the total foamate volume decreased, but the foamate protein (mainly sporamin) concentration increased. On the other hand, the total foamate volume increased significantly as the air superficial velocity increased, but the foamate concentration decreased slightly. The minimum residual protein concentration occurred at pH 3.0 and $V_0 = 1.5$ cm/s. On the other hand, the maximum protein mass recovery occurred at pH 3.0 and at $V_0 = 4.3$ cm/s.

Index Entries: Batch foam fractionation; bioseparation; protein separation; sweet potato; sweet potato proteins.

INTRODUCTION

Plants usually store their food in seeds and roots for subsequent germination and budding. The storage-root plants, such as cassava, taro, and sweet potato, have been used as important food-starch sources for many centuries. Sweet potatoes are also nutritionally important for their beta-carotene, vitamin C, fiber, potassium, and protein content *(1,2)*. The sweet

* Author to whom all correspondence and reprint requests should be addressed.

potato protein level generally varies according to the way it was cultivated *(3)*, as well as the environmental growing conditions, genetic factors *(4,5)*, and storage conditions *(6)*. The protein level typically ranges between 1.3 and 10% (on a dry basis) *(2)*. The water content is generally 80% by weight. The protein distribution in a sweet potato is spread quite evenly throughout the stored root *(7)*. Sporamin (ipomoein), a globulin protein, is a major protein comprising about 80–90% of the total protein content in the sweet potato *(1,8)*. One interesting property of sporamin is that it can act as a strong surfactant, creating foam in water when air bubbles are introduced to a sweet potato extract solution. Since sporamin foams easily and naturally by itself, a batch foam fractionation process may be a promising technique for collecting, concentrating, and separating it from the other proteins. The other proteins are mostly α and β amylase, which comprise about 5% of the total soluble proteins *(4)*.

A batch foam-fractionation process generally uses air or an inert gas as a carrier gas. The gas is introduced to the process through a nozzle or a sparger at the bottom of the foam-fractionation column. The gas then creates bubbles in the liquid (mainly water) solution. If one of the proteins in the solution foams, a foam layer can be developed above the bulk-liquid phase. Because of its high surface activity, sporamin can create a high protein concentration at the air–liquid interface above the bulk liquid and below the foam layer *(9)*. Typically, sporamin from the bulk solution moves by diffusion to the air–liquid bubble interface and attaches to a rising bubble. Once bubbles reach the liquid surface, they are transformed into a foam. In the foam phase, foam cells move upward to the top of the column, and then to the foam collector as foamate (collapsed foam) *(10)*. From experience and reported work in the literature *(10–12)*, the foamate protein concentration for this foam-fractionation process is controlled by the initial solution pH, the initial bulk-protein concentration, the air superficial velocity *(11)*, and the processing time (τ) taken to generate the foam, referred to here as the batch-processing time. The initial bulk solution pH plays an important role in controlling the ionic charge balance in the sweet potato extract solution. When the net ionic charge in the protein solution is equal to zero (called the isoelectric point [pI]), the protein coagulates and becomes easier to separate from the bulk solution. The pI condition is generally a desired setting for separating proteins from a water solution and is expected to play the same role for sporamin in a sweet potato extract solution.

MATERIALS AND METHODS

Sample Preparation

Sweet potatoes (marketed by Dixie Lamb Alabama Sweet Potato Company, AL) were purchased from a local grocery store. This stored sweet potato root was cleaned with water and air-dried at room temperature. A large section was removed from the sweet potato perpendicular

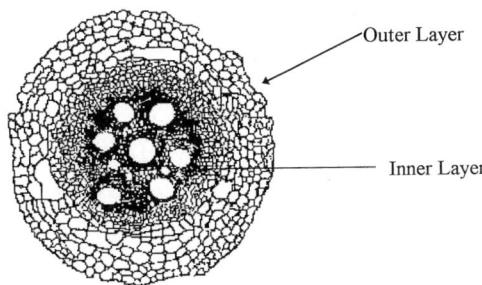

Fig. 1. Anatomy of the cross-section of sweet potato storage root.

to the longitudinal axis (approx 10 cm long) and used for the experimental samples. Both of the tapered ends of the sweet potato stored root were discarded. The typical anatomy of the cross-section of the sweet potato root is shown in Fig. 1. For the purpose of this study, the slice was dissected into two zones: the outer layer and the inner layer. Typically, the outer layer is comprised of the epidermis, cortex, and lacuna, and the inner layer is comprised of the parenchyma, phloem, protoxylem, primary cambium, secondary xylem, metaxylem, and endodermis *(13)*. The diameter, outer layer, and inner layer thickness of a sweet potato were determined to be approx 4–6 cm, 4–5 mm, and 4–6 cm, respectively. Outer layer, inner layer, and combined layer samples were analyzed for protein *(14)*, starch, fiber, reducing sugar *(15)*, moisture content, and others (tannins, beta-carotene, and so on).

The combined layer samples were used for the batch foam-fractionation experiments. A large center section of the sweet potato (about 5 cm in diameter, 1 cm thick, and ca. 20 g) was used as the experimental sample, and chopped into small pieces. These pieces were prepared for the foam fractionation run by combining them with 200 mL of deionized water in a 350-W motor food blender (Blend Master 10, Hamilton Beach/Proctor-Silex, Washington, NC). The blender chopped up these pieces for 4 min at the "liquefy" speed setting. The produced liquefied sweet potato extract solution was then filtered through Whatman (Clifton, NJ) No. 4 filter paper. The filter cake was washed several times, until only residual fiber remained on the filter paper. The filtered solution was then decanted and the starch was removed from the solution in a centrifuge (Marathon 21K, Fisher Scientific, Pittsburgh, PA) at a speed of $1073g$ for 4 min. The supernatant was collected and the starch was discarded. Water was added to the starch-free supernatant to increase the volume of the solution to 1 L. The protein-rich solution was stored in the refrigerator at 10°C, until used for the foam-fractionation experiment (typically within 2 d).

Experimental Procedure

The batch foam-fractionation apparatus was comprised of a glass cylinder 2.5 cm in diameter and 36-cm long, as shown in Fig. 2. The top port

Fig. 2. Schematic of the batch foam-fractionation process.

was used for loading the sample, and closed with a rubber stopper during foam fractionation. The opened side port allowed the foam to flow into the foam collector. A porous ceramic disk sparger was inserted into the bottom of the foam-fractionation column, and connected to the air supply copper tube by aluminum flanges. The air-flow rate was measured using an air-flow meter (rotameter type). The air supply was prehumidified by bubbling it through water to minimize both column water loss and protein contamination from the influent air. The incoming air was therefore held close to constant relative humidity, since its temperature was 23 ± 2°C. No attempt was made to determine the effect of influent air humidity on foam formation in these experiments.

Sweet potato protein-extract solution was adjusted initially to the desired pH (between 2.0 and 10.0) by adding hydrochloric acid or sodium hydroxide. This initial volume of extract solution used for batch experiment in the foam-fractionation column was 100 mL. The air (superficial velocity ranged between 1.5 and 4.3 cm/s) was introduced to the porous sparger at the bottom of the column. The bubbles entered the bulk-protein mixture, rose up to the liquid surface, and then generated foam above the liquid surface. Foams were continuously carried up to the top of the column in the open space above the liquid, eventually passing through the side port to the foam collector. Foam collapsed on its own (~10 min) and created liquid high-protein concentrate (foamate). All of the experiments were carried out at room temperature (23 ± 2°C). Typically, the bubble diameter is a function of the liquid surface tension, viscosity, and density, and these in turn are functions of temperature. Keeping the temperature fixed in these experiments helped maintain the bubble at a constant diameter. No attempt was made to vary the temperature in these experiments. The batch foam-fractionation experiments were terminated when foam flow to the foam collector stopped (batch processing time, τ, ca. 2–10 min).

Both the foamate and the residual volumes and also pHs were measured. The total protein concentrations in both the foamate and the residual liquid were determined at this processing time (τ), using the Bradford Coomassie blue method *(14)*. The optical absorbance, measured spectrophotometrically, was converted to total protein concentration in mg/L, using a total protein–optical absorbance calibration curve for sweet potato proteins, originally developed by determining the foamate protein mass gravimetrically.

Electrophoresis

Gel electrophoresis was used to identify the mol wt of the two primary proteins (sporamin and β-amylase), and their respective approximate relative concentrations within both the foamate and the bulk liquid of the foam-fractionation process. α-amylase (ca. 20% of the total amylase concentration) did not show up on the gel. Sodium dodecyl sulfate-polyacrylamide gel electrophoresis (SDS-PAGE) was employed in this study, following the Laemmli procedure *(16)*. The proteins passing through the 11% polyacrylamide gel were stained with Coomassie brilliant blue R-250 to mark the protein bands. Typically, use of this stain leads to easily seen sharp bands on the gel when the protein mass in the analyzed solution ranges between 15 and 20 µg *(17)*. In these experiments, bovine serum albumin (Lot No. 41F-9300), ovalbumin (Lot No. 55F8510), and carbonic anhydrase (Lot No. 43F-8050), purchased from Sigma (St. Louis, MO), were used as the protein mol wt markers to calibrate the gel at the approximate mol wt of 68 kDa, 45 kDa, and 29 kDa, respectively.

RESULTS AND DISCUSSION

The outer, inner, and combined layers of the sweet potato used for the foam-fractionation experiments were evaluated for starch, fiber, reducing sugar *(15)*, protein *(14)*, and moisture content, as shown in Table 1. The colors of the sweet potato extract solution of the outer, combined, and inner layers were observed to be dark brown, brownish orange and bright orange, respectively. The dark brown color of the outer layer sweet potato extract solution may be caused by its high tannin content in the root epidermis. The outer layer of the root is comprised of fiber, tannin, and lignin, in order to reinforce and protect the root from environmental damage. The fiber from the outer layer was observed to be coarse and thick, and the fiber from the inner layer was much finer. The inner layer is used by the plant for storing food in the form of starch. In particular, the starch is stored in the parenchyma cells *(13)* within the inner layer. It is seen in Table 1 that the outer layer is high in fiber, but low in starch. The protein content of the outer layer is slightly higher than that in the inner layer. This observation seems reasonable, because the outer layer is the place

Table 1
Percentage of Compositions of Outer, Inner, and Combined Layers of Sweet Potato on a Fresh Root Basis

Compositions	Outer layer % wt	Inner layer % wt	Combined layer % wt
Starch	4.32	5.98	5.60
Fiber	6.37	6.03	6.13
Reducing sugar	4.86	8.23	6.36
Proteins	0.53	0.45	0.46
Moisture	80.05	79.03	79.43
Others	3.87	0.28	2.02

Fig. 3. Foamate protein concentration for V_0 1.5, 2.9, and 4.3 cm/s, and $2 \leq \tau \leq 10$ min in a batch foam fractionation.

where the sweet potato is germinated. From these results, the total protein content (~80% of which is sporamin, the storage protein) is not much different between outer and inner layer. If the percentage of sporamin content remains a constant in both layers, then any cutting of the sweet potato root would be representative and suitable for a foam fractionation.

The air superficial velocity and the initial solution pH seem to be important independent variables in the foam-fractionation process. For example, as the air superficial velocity increases, both the foamate concentration and the foamate volume decreases. It can be seen in Fig. 3 that the foamate protein concentration increases as the initial bulk-solution pH

Fig. 4. Foamate volume for $V_0 = 1.5$, 2.9, and 4.3 cm/s, and $2 \leq \tau \leq 10$ min a batch foam fractionation.

increases. In the acidic range, it was observed that proteins seem to coagulate and precipitate (denature), as evidenced by suspended particles in the foamate. In the basic range, the soluble foamate protein concentration rises, while the concentration of suspended particles (insoluble protein) in the foamate decreases. When the air superficial velocity increases from 2.9 to 4.3 cm/s, the foamate protein concentrations increase slightly, particularly at the high pHs; the foamate volume increases considerably over the entire pH range. Air bubbles at the air superficial velocities of 2.9 and 4.3 cm/s rise very fast and are turbulent in the liquid-bulk phase. Both the bubble diameters and the foam cell diameters for these velocities are approximately constant in the range of 2–4 mm. These diameters typically increase near the end of a run when the amount of generated foam (protein surfactant) decreases. At the air superficial velocity of 1.5 cm/s, both the bubble and foam cell sizes are constant, at about 2 mm in diameter, and foam cells move slowly upward in the column. Liquid drainage from the foam can be seen clearly. The high liquid drainage from the foam phase typically enhances the foamate protein concentration, particularly at an air superficial velocity of 1.5 cm/s curve, as shown in Fig. 3. Because the bulk liquid can be entrained in the foam cellular structure at high air superficial velocities, the foamate volume increases as the air superficial velocity increases, as shown in Fig. 4. On the other hand, an increase in the initial solution pH lowers the foamate volume. This decrease of the foamate volume with an increase in pH may be caused by the changes in resulting ionic charges.

This change, in turn, reflects the underlying changes in sweet potato protein-water structure, which causes the proteins to be less surface-active, and, hence, create less foamate volume.

After the experiment is terminated (when foam ceases to flow to the foam collector), the residual bulk-liquid volume, bulk-protein concentration, and bulk-solution pH are determined. It was found that the changes in bulk solution and foamate pHs were approximately: $\Delta pH_{Foamate} = 0.5$ (acidic range) and -0.5 (basic range), where $\Delta pH_{Foamate} = pH_{Foamate} - pH_{Initial}$ and also $\Delta pH_{Residue} = -0.1$ (acidic range) and 0.1 (basic range). "Initial" pH here refers to the initial bulk pH. The residual protein concentration is lowest for the case in which the initial solution pH was 3.0, as shown in Fig. 5. The effect of a change in superficial velocity is generally less significant than the effect of a change in pH. The relative response to changes in V_0 depends on pH, and it is seen in Figs. 4–6 that the response to this effect is not obvious. Figs. 5 and 6 represent the tradeoff between protein recovery and protein concentration. Typically, before air is introduced to the foam-fractionation column, the initial solution is cloudy. After the foam-fractionation experiment has been completed, the residual bulk liquid is cleared of the colloidal matter and looks like normal deionized water. Generally, when the pH of the protein in solution is at its pI, the solution is cloudy and protein precipitates as colloids or particles. The sweet potato-water solution is most cloudy at pH 3.0, indicating that the sweet potato-protein solution pI is close to this value. Because sporamin is the major protein in the sweet potato storage root, the precipitation phenomena seems to imply that the pI of sporamin is around 3. As the initial solution pHs depart from 3, the residual protein concentrations rise, and the total protein recovery decrease, the proteins become more soluble in the bulk-liquid phase and less precipitated in the foam phase.

Typically, for pHs above 3.0, an increase in the initial solution pH enhances the foamate protein concentration, but reduces the generation of foamate volume (Fig. 4). The total protein recovery (the percentage of recovered protein mass in the foamate, relative to the initial protein mass in the sweet potato extract solution) is an important variable for determining a local maximum in the different protein-separation cases. Figure 6 shows that the total protein recovery reaches a maximum at pH 3.0 (which corresponds to the lowest residual protein concentration seen in Fig. 5), and then decreases for the initial solution pHs away from 3.0. It is also observed in Fig. 6 that an increase in air superficial velocity in the studied V_0 range tends to enhance the total protein recovery. The total protein recovery (mostly sporamin) ranged between 77 and 87% over the studied range at pH 3.0.

Another important variable to consider is the batch processing time (τ) (as shown in Fig. 7 which is to be minimized. The global optimum is resolved when the pH and V_0 variables are different for these possibly conflicting objectives through a cost analysis. Here, the optimum is clearly at pH 3.0, close to the pI.

Fig. 5. Residual protein concentration for air superficial velocity 1.5, 2.9, and 4.3 cm/s in a batch foam fractionation.

Fig. 6. Total protein recovery for air superficial velocities 1.5, 2.9, and 4.3 cm/s in a batch foam fractionation.

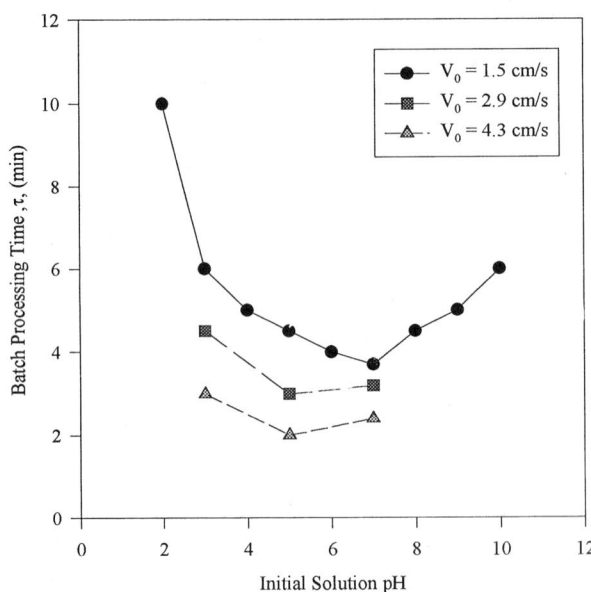

Fig. 7. Sweet potato batch processing time (τ) as a function of pH and V_0.

SDS-PAGE of foam fractionation of sweet potato proteins is illustrated in Fig. 8. Lanes 1–4 contain the three marker proteins: bovine serum albumin (68 kDa), ovalbumin (43 kDa), carbonic anhydrase (29 kDa), and lane 12 contains α-Amylase (51 kDa). Lanes 7 and 10 contain foamate proteins, and lanes 8 and 11 contain bulk-solution proteins at pH 3.0 and 5.0, respectively. Lane 9 represents the sweet potato feed stock proteins. Lanes 5 and 6 contain foamate and bulk-solution proteins, respectively. For lanes 5 and 6, the sweet potato proteins were foam-fractionated at pH 5.0, and then the residue was foam-fractionated again at pH 3.0 to obtain the foamate and the bulk solution for lane 5 and 6, respectively. From previous studies (3,8), the mol wt of sporamin was estimated at 25 kDa, which corresponds to the major strong band in lanes 5, 7, 9, and 10. As shown in lanes 5, 7, and 10, it is clear that foam fractionation can be used to concentrate sporamin from the sweet potato extract. Sporamin concentration at both pH 3.0 and 5.0 is 3–5× greater than sporamin in the original feed stock solution.

The second abundant protein, β-amylase, has a mol wt around 201 kDa, with four equal molecular subunits, each subunit being about 50 kDa (18). This protein can be observed as a small sharp band in lane 5. It is difficult to observe the band of β-amylase in lanes 7 (pH 5.0) and 10 (pH 3.0), because the sporamin is much more abundant than β-amylase, and has a stronger surface activity than β-amylase. When the sweet potato extract is foam-fractionated, the foamate containing sporamin comes out before the β-amylase. The β-amylase can be recovered at higher concentrations than that in the feedstock solution when the sporamin is depleted

Batch Foam Sporamin from Sweet Potato

Fig. 8. SDS-PAGE of marker and foam fractionated sweet potato proteins: Lane 1, bovine serum albumin; lane 2, ovalbumin; lane 3, carbonic anhydrase; lane 4, the mixture of lane 1, 2, and 3; Lane 5, foamate at pH 3.0 (run at pH 5.0 and the pH 3.0); lane 6, bulk solution at pH 3.0 (run at pH 5.0, and then pH 3.0); lane 7 foamate at pH 5.0; lane 8, bulk solution at pH 5.0; lane 9, sweet potato feed stock solution; lane 10, foamate at pH 3.0; lane 11, bulk solution at pH 3.0; and lane 12, α-Amylase.

first in the bulk solution, as shown in lane 5. One way to selectively deplete the sporamin is to first foam the bulk solution at pH 5.0, concentrating sporamin into the foamate. Then, at pH 3.0 the β-amylase can be recovered, along with residual sporamin, by foaming the bulk solution. Sporamin appears to be more surface-active than β-amylase; thus, sporamin is more easily separated by foaming. After much of the sporamin has been removed, the β-amylase attaches more readily to the adsorptive sites on the air bubbles. Generally, amylase does not foam when the air bubbles are introduced into a foam-fractionation column containing only this protein. Thus, it has been shown in this experiment that β-amylase is concentrated only in the foamate, in conjunction with a foaming protein such as sporamin. In the natural sweet potato system, both proteins are present together, and no foaming surfactant is required to recover β-amylase in the foamate.

CONCLUSIONS

Sweet potato storage protein (primarily sporamin) recovery in batch foam fractionation from a sweet potato-water extract is maximized at pH 3.0, decreasing as the initial solution pH increases. In addition, an increase in the air superficial velocity enhances the total protein recovery, which is highest (about 87%) at an air superficial velocity of 4.3 cm/s. However, the β-amylase subunit (50 kDa) can also be recovered in the foamate, at a higher concentration than the bulk solution, by first foaming the bulk solution at pH 5.0 (to remove much of the sporamin), and then foaming at pH

3.0 to recover the β-amylase, along with the residual sporamin. Foam fractionation of a sweet potato-water extract solution seems to offer a promising low-cost first step in recovering sporamin and β-amylase from sweet potato.

REFERENCES

1. Kays, S. J. (1992), in *Sweet Potato Technology for the 21st Century*, W. A. Hill and P. A. Loretan (eds.), Tuskegee University, Tuskegee, AL, pp. 201–261.
2. Collins, W. W. and Walter, Jr., W. M. (1985), in *Sweet Potato Products: A Natural Resource for the Tropics*, Bouwkamp, J. C., ed, CRC, Boca Raton FL, pp. 153–173.
3. Osujui, G. O. and Cuero, R. G. (1992), in *Sweet Potato Technology for the 21st Century*, W. A. Hill and P. A. Loretan (eds.), Tuskegee University, Tuskegee, AL, pp. 78–86.
4. Nakamura, K. (1992), in *Sweet Potato Technology for 21st Century*, W. A. Hill and P. A. Loretan (eds.), Tuskegee University, Tuskegee, AL, pp. 21–25.
5. Prakash, C. S. and Varadarajan, U. (1992), in *Sweet Potato Technology for the 21st Century*, W. A. Hill and P. A. Loretan (eds.), Tuskegee University, Tuskegee, AL, pp. 27–37.
6. Bin, W., Lu, J., Stevens, C. and Khan, V. (1995), *Tuskegee Horizons*, **6,** 19.
7. Harvey, P. J. and Boulter, D. (1993), *Phytochemistry* **22,** 1687–1693.
8. Maeshima, M., Sasaki, T., and Asahi, T. (1985), *Phytochemistry* **24,** 1899–1902.
9. Adamson, A. W. (1990), *Physical Chemistry of Surfaces*, 5th ed., John Wiley New York, pp. 525–553.
10. Lemlich, R. (1972), *Adsorptive Bubble Separation Techniques*, Academic, New York, pp. 133–143.
11. Prokop, A. and Tanner, R. D. (1993), *Starch/Stärke*, **45,** 150–154.
12. Montero, G. A. Kirschner, T. F., and Tanner, R. D. (1993), *Appl. Biochem. Biotechnol.* **39/40,** 467–475.
13. Kays, S. J. (1985), *Sweet Potato Products: A Natural Resource for the Tropics*, Bouwkamp, J. C., ed, CRC, Boca Raton FL, pp. 153–173.
14. Bradford, M. M. (1972), *Anal. Biochem.* **72,** 248–254.
15. Miller, G. L. (1959), *Anal. Chem.* **31,** 426–428.
16. Laemmli, U. K. (1970), *Anal. Biochem.* **72,** 680–685.
17. Hames, B. D. and Rickwood, D. (1990), *Gel Electrophoresis of Proteins*, Oxford University Press, Oxford, pp. 53–55.
18. Fasman, G. D. (1989), *Practical Handbook of Biochemistry and Molecular Biology*, CRC Boston, p. 173.

Batch Foam Fractionation of Kudzu (*Pueraria lobata*) Vine Retting Solution

Jirawat Eiamwat, Veara Loha, Aleš Prokop and Robert D. Tanner*

Department of Chemical Engineering, Vanderbilt University, Nashville, TN 37235

ABSTRACT

The aqueous protein solution from kudzu (*Pueraria lobata*) vine retting broth, without the addition of other surfactants, was foam-fractionated in a vertical tubular column with multiple sampling ports. Time-varying trajectories of the total protein levels were determined to describe the protein behavior at six positions along the 1-m column. The lowest two trajectories of this batch process represented a loss of proteins from the bulk liquid and tended to merge and decay together in time; the other trajectories displayed a gain in proteins in the foam phase. These upper column port protein concentration trajectories generally increased in time up to 45 min, followed by a decrease, reflecting the removal of proteins from the column ports. The foam became dryer as it passed up the column to the top port. The protein concentration was about $5-8\times$ higher in the top port foam than in the initial bulk solution, mainly as a result of liquid drainage from the foam along the column axis. This concentration increase in the collected foam was dependent on the initial pH of the bulk solution. The mol-wt profile of the proteins in the concentrated foam effluent was determined by one-dimensional gel electrophoresis. An analysis of the gel electropherograms indicated that the most abundant proteins could be cellulases and pectinases.

INTRODUCTION

Foam fractionation is an attractive technique for protein concentration and separation, because it is simple, relatively inexpensive, and easy to scale-up. Moreover, this technique has significant potential for reducing the high cost of protein recovery in commercial practice, particularly in the first recovery step in which much of the water is removed. When compared to other separation methods, such as precipitation by means of

---

*Author to whom all correspondence and reprint requests should be addressed.

salts or solvents, foam fractionation offers negligible product contamination during processing. The only additive in this processing technique is air or another gas such as carbon dioxide, with the possible exception of the addition of an acid or a base when a pH change is required. Typically, foam-fractionation technique is one in which air or an inert gas is introduced at the bottom of the fractionation column through a porous ceramic sparger or a nozzle. Above the sparger (nozzle), bubbles rise through the column and carry a surface-active substance (surfactant such as protein) on the bubbles at the air–liquid interface. When a protein, itself a surfactant, is in the solution, then no exogenous surfactant may need to be added to effect the separation. Above the liquid surface, the air bubbles may, under certain circumstances, become a foam, and the foam layer continues to rise in the column. Foam is collected and collapsed to be a liquid foamate, often using a mechanical stirrer. While the foam travels up the column, the liquid on the foam drains because of gravitational forces and the capillary forces at the complex foam interface (the plateau border). The residual surfactant material remains attached to the foam and is concentrated in the foamate (which is continuously removed).

Previous studies on foam fractionation (1–3) have demonstrated its feasibility for concentrating and separating proteins that foam. Proteins that do not foam can also be separated and concentrated by bubble fractionation alone, but to a lesser degree (4). By itself, however, this first concentration step does not generally purify proteins to their desired specifications. In the present study, the initial pH and concentration of the bulk solution are shown to play an important role in both concentrating and separating these proteins. It is found that, even though the foam and the bulk solution are comprised of the same proteins, the individual protein concentrations varied between the two different phases, indicating that fractionation occurs.

In general, foam fractionation is a very promising method for removing soluble proteins (5) from dilute solutions, e.g., potato protein wastes in the potato processing industry (6). Another candidate for foam fractionation is the extracellular protein-rich waste fermentation broth from the kudzu vine retting process (7). Although this anaerobic fermentation process usually produces a foul smell and an aesthetically unpleasant-looking liquid waste stream (similar to liquid from a swamp), the aeration of this broth can create attractive foam (similar to beer), which may have economic value as a low-cost source for cellulases and pectinases. This unappealing waste stream is markedly improved in appearance and odor by the foam-fractionation procedure.

The natural retting process (on the ground or in water) allows for development of microorganisms that have an affect on the digestion of cellulose–pectin sheath of the phloem-fiber plants, such as hemp, flex, ramie, and jute. This process is slow and may take about 2 wk. The cellulose fibers in such plants are bounded together in bundles within the phloem surrounding the stem of the plant. In order to recover fibers from such plants, the cellulose fibers must be separated from the phloem, and then from each other within the bundles. This separation of fibers requires special enzymes (cellulase and pectinase) to digest the plant. The anaerobic bacteria, by virtue of their content of cellulases and pectinases, are capable of digesting the cellulose and pectin that bind the bundles of fibers (8). The separation of fibers can subsequently be refined by using mechanical force, as for typical textile industry.

The retting process (to recover fibers from kudzu vines) promises to be of commercial interest, since the flax-like cellulosic fiber may be in demand for natural fiber clothing (9). The addition of a foam-fractionation step can serve as an attractive way of dealing with the waste, and may possibly lead to a commercial outlet for the recovered enzymes. It would be interesting to explore what enzymes are present in both the foam and the residual broth, and whether these are of sufficient value to be marketable at present.

MATERIALS AND METHODS

Retting Procedure

Kudzu vines of different diameters (0.3–0.7 cm) were harvested locally, near Gay Street and 1st Avenue S, on the west bank of the Cumberland River in downtown Nashville, Tennessee. Approximately 1.3 kg of vines were cut at a time, washed, and placed in a 10-L plastic container, which was then filled with 7 L of tap water (~ pH 5.9) (10) to give a solid concentration of 15.5% wt. These cut vines of various lengths (13–16 cm) were kept submerged in that water at ambient temperature of 23 ± 1°C. The retting was considered completed when the vines became loose and easily separated into fibers (~ 2 wk).

Sample Preparation

Sampling was performed every 1–2 wk until retting was completed. A 400-mL sample filtered through Whatman (Kent, UK) No. 4 filter paper was used for each foaming experiment. The solution to be foam-fractionated was adjusted initially to a desired pH by adding 1 N HCl or 1 N NaOH prior to aeration.

The total protein concentration was assayed using the Coomassie blue (Bradford) method (11) to measure the absorbance with a spectrophotometer, which is $A_i = K_i C_i$ (where A_i, C_i, and K_i are the absorbance, protein concentration, and the conversion value). The actual protein concentration was determined by evaporating 1 mL of sample solution in the hot-air oven for 10 hr and then weighing it. The calculated conversion value (K_i) with the concentration/absorbance is about 1.81 mL/mg.

Fig. 1. Schematic diagram of the experimental apparatus.

Experimental Procedure

The foam-fractionation experiments were carried out in a 1-m glass column (*see* Fig. 1 for the schematic) with an id of 35 mm. A porous fritted glass sparger was fitted flush with the id of the bottom of the column. The column included six sampling ports located along its length axis. Each of them was covered with a rubber septum, to allow sample collection with a hypodermic syringe. Ports were 20 cm apart, starting with the first port, which was located 2 cm above the sparger, as illustrated in Fig. 1. To eliminate evaporation, the air supply was bubbled through the water to humidify it before entering the bottom of the column. The air-flow rate was monitored by a rotameter at 0.14 vvm, or 0.056 L/min, and, throughout this study, only one air-flow rate was used. Using 10-mL hypodermic syringes, 1 mL of samples were withdrawn every 15 min throughout the experiment from each of the six side ports by inserting the needle into the septum until it entered the bulk solution or the foam. The liquid sample contents were dispensed into 10 × 65-mm test tubes for subsequent determination of the total protein concentration, for gel electrophoresis samples, and for the absorption spectral characteristic study.

Analytical Methods

Total Protein Assay

The total protein concentration was quantified spectrophotometrically at 595 nm, using the Coomassie blue (Bradford) method *(11)* with a

Spectronic 20 spectrophotometer (Bausch and Lomb, New York). The samples were assayed 2 min after the addition of the Coomassie blue dye, and were compared with a zero reference of deionized water under the same conditions.

Gel Electrophoresis

The liquid foamate was resolved to identify the primary proteins by electrophoresis on 11% (w/v) sodium dodecyl sulphate-polyacrylamide gel, according to the Laemmli procedure (12). The standard Sigma kit (Sigma, St. Louis, MO.) provided the mol wt markers. Coomassie blue R-250 (Bio-Rad, Richmond, CA) was used for staining. Protein markers (Sigma) were used for calibrating the gel and calculating R_f values for the approximation of mol wt as follows: bovine serum albumin (A-4378), mol wt 66,000; ovalbumin (A-7641), mol wt 45,000; and pectinase from *Aspergillus niger* (P-9932). The fourth marker, cellulase, was derived from *Trichoderma reesei* (MVA 1284; Gist-Brocades nv, Delft, Netherlands).

Absorption Spectral Characteristics

The absorption spectra were also used as an indicator of the characteristic proteins. The spectra of the known cellulase, known pectinase, and liquid foamate samples were measured at ambient temperature between 200 and 600 nm for the proteins in a deionized water solution in a 1-cm cuvet using a Hitachi Model 100-40 UV spectrophotometer (Hitachi, Tokyo, Japan). Each of these solutions was prepared as follows: A 0.1 mL of pectinase was added to 1.4 mL of deionized water (.067 dilution); 0.075 g of cellulase was mixed with 5 mL of deionized water, centrifuged at 1073g for 30 min, and further filtered through Whatman No. 40 filter paper; a 5-mL sample of liquid foamate was centrifuged at 1073g for 30 min, and then filtered through Whatman No. 40 filter paper.

RESULTS AND DISCUSSION

When air bubbles rise in the aqueous protein solution (from the kudzu vine retting broth) to the upper surface, foam is formed at the air–liquid interface. The enriched protein liquid leaving the bulk liquid solution is carried with the rising bubbles (bubble size ~ 0.6–1.0 mm) into the open air space as foam (foam cell ~ 0.6–2.0 mm), resulting in an enriched protein concentration in the foam phase. As the foam passes upwards, some air bubbles in the foam phase coalesce into larger bubbles, and remain attached because of capillary forces. At the same time, some unstable foam collapses. Following the operation of the foam-fractionation process over a given period of time, the amount of foam produced is lessened as the protein in the bulk solution becomes depleted. The collapsed foam has been collected; liquid volume was around 80–100 mL.

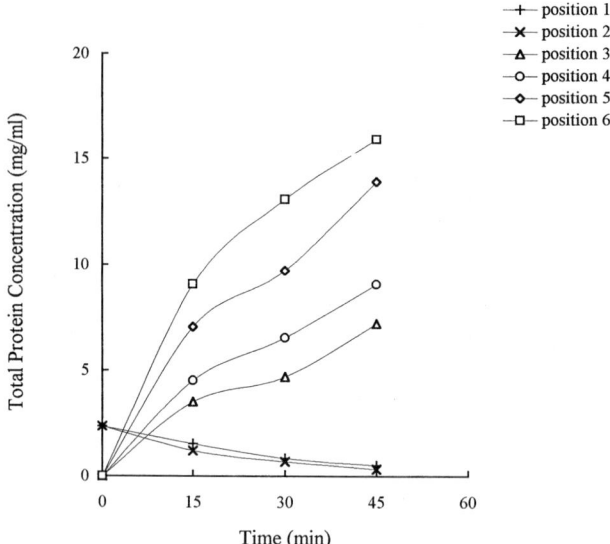

Fig. 2. Time varying trajectories of the total protein concentration at pH 3; and air flow rate 0.056 L/min.

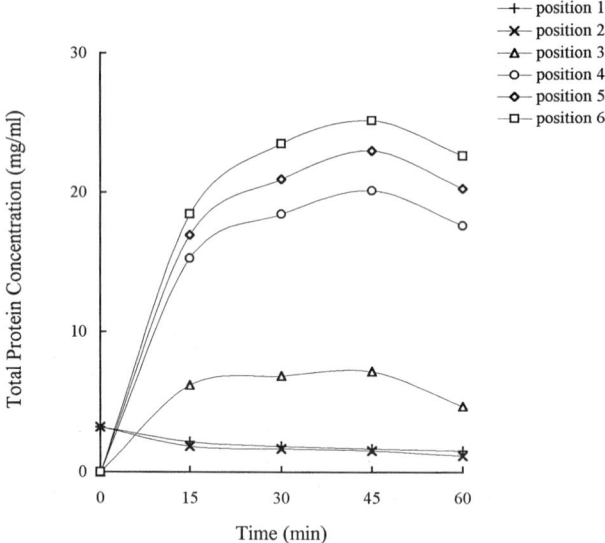

Fig. 3. Time varying trajectories of the total protein concentration at pH 5; and air flow rate 0.056 L/min.

Figures 2–4 show the protein concentration profiles for the liquid phase (positions 1 and 2) and the foam phase (positions 3 to 6) at various pHs. It is noted that since no foam is present at time zero, no protein concentrations were measured then. It could just as reasonably be argued

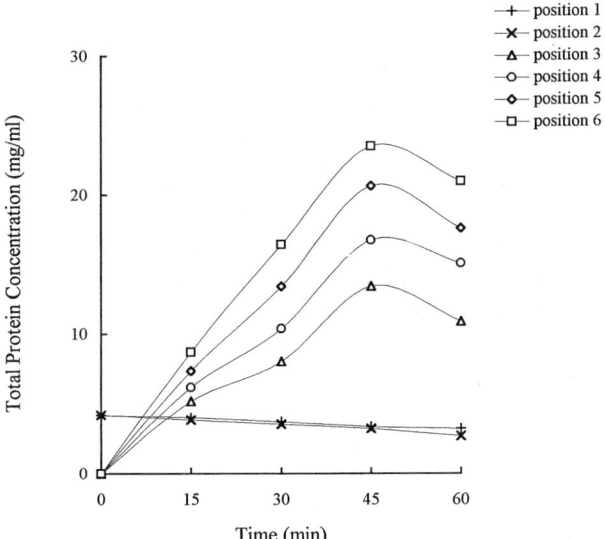

Fig. 4. Time varying trajectories of the total protein concentration at pH 7; and air flow rate 0.056 L/min.

that the concentration at time zero is the bulk concentration, rather than zero as shown on the three graphs. Although the total protein concentration of the foam phase increases appreciably as time increases, the total protein concentration of the much larger mass-liquid phase decreases gradually. This indicates that the loss of proteins from the bulk liquid is reflected in the gain of proteins in the foam phase. The total protein concentration increases up to approx 5–8× (compared to the initial concentration) at 45 min. Concomitantly, the volume of foam formed was observed to decrease beyond that time, because of the removal of foam in the samples. The pH of foamate usually differed from those of the initial pH. For the initial pH of bulk solution below 7.0, the foamate pH typically increased (by 0.2–0.3 pH units), and for the initial pH of bulk solution above 7.0, the foamate pH decreased (by about the same degree).

Analytical SDS-PAGE electrophoresis was performed in order to identify the primary proteins presented in the liquid foamate. The visualized protein bands on the gel are shown in Fig. 5. It is observed that the liquid foamate consists of at least three different proteins. The main constituent proteins may very well be cellulases and pectinases as expected, since their gel mol wt were close to the cellulase and pectinase markers. These foamate proteins, however, were very weak in color on the gel, in comparison to the protein markers. This may be the result of the relatively low concentration of proteins in the diluted vine retting solution foamate sample. The fainter bands detected on the gel apparently represent the minor protein constituents in the broth. The standard Sigma kit for mol

Fig. 5. SDS-Polyacrylamide gel electrophoresis results for markers (Lanes 1–4) and foam fractionation of kudzu vine retting solution (Lanes 5–6). Lane 1: bovine serum albumin; Lane 2: ovalbumin; Lane 3: cellulase; and Lane 4: pectinase.

Fig. 6. Comparison of UV-VIS spectra for cellulase from *T. reesei* (15 mg/ml), pectinase from *A. niger* (3.8 mg/ml), and diluted kudzu vine retting solution foamate.

wt of the unknown proteins on the SDS-PAGE gel gave mol wt estimates of 86 kDa for one apparent pectinase band and 75 kDa and 62 kDa for two apparent cellulases *(13)*.

Another direct analysis of the absorption spectra provided another means of characterization of proteins. It can be seen from Fig. 6 that the cellulase, pectinase, and liquid foamate exhibited the same absorption peaks at 330 nm wavelength. In fact, although proteins generally exhibit an absorbance at 280 nm, an absorbance at 340 nm has also been reported

(14). This, however, apparently supports the gel electrophoresis result that the liquid foamate could be composed of cellulases and pectinases.

CONCLUSION

Foam fractionation of the kudzu vine retting solution causes removal of soluble proteins present within the solution, and substantial increase ($5-8\times$) in the total protein concentration. Fractionation longer than 45 min results in a progressive decrease in the volume of the foam. The soluble proteins are apparently cellulases and pectinases. The kudzu vine retting solution foams, perhaps as a result of the foaming ability of cellulases and their hydrophobicity.

REFERENCES

1. Schutz, F. (1937), *Nature*, 629, 630.
2. Bader, R. and Schutz, F. (1954), *Nature*, 183, 184.
3. Lalchev, Z., Dimitrova, L., Tzvetkova, P., and Exerowa, D. (1982), *Biotechnol. Bioeng.* **24**, 2253–2262.
4. Loha, V., Tanner, R. D., and Prokop, A. (1997), *Appl. Biochem. Biotechnol.* **63–65**, 395–408.
5. Lemlich, R. (1972), *Adsorptive Bubble Separation Techniques*, Academic, New York, pp. 1–50.
6. Prokop, A. and Tanner, R. D. (1993), *Starch/Stärke* **45**, 154.
7. Uludag, S., Prokop, A., and Tanner, R. D. (1996), *J. Sci. Ind. Res.* **55**, 381, 382.
8. Akkawi, J. S. (1990), *US Patent*, 4,891,096.
9. Bajpai, R. K., Prokop, A., and Tanner, R. D. (1993), *Biotech. Adv.* **11**, 637.
10. Uludag, S., Loha, V., Prokop, A., and Tanner, R. D. (1996), *Appl. Biochem. Biotechnol.* **57/58**, 76.
11. Bradford, M. M. (1976), *Anal. Biochem.* **72**, 248–254.
12. Laemmli, U. K. (1970), *Nature* **227**, 680–685.
13. Reed, G. (1975), *Enzymes in Food Processing*, 2nd ed. Academic, New York, p. 102.
14. Leatham, G. F. and Himmel, M. E. (eds.) (1991), *ACS Symposium Series 460*, American Chemical Society, Washington, DC, p. 446.

Development of a Novel, Two-Step Process for Treating Municipal Biosolids for Beneficial Reuse

CHRISTOPHER J. RIVARD,*[1] BRIAN W. DUFF,[1] AND NICHOLAS J. NAGLE

[1]*Peak Treatment Systems, Inc., Golden, CO 80401; and* [2]*National Renewable Energy Laboratory, Golden, CO 80401*

ABSTRACT

Modern municipal sewage waste treatment plants use conventional mechanical and biological processes to reclaim wastewaters. This process has an overall effect of converting a water pollution problem into a solid waste disposal problem (sludges or biosolids). An estimated 10 million tons of biosolids, which require final disposal, are produced annually in the United States. Although numerous disposal options for biosolids are available, including land application, landfilling, and incineration, disposal costs have risen, partly because of increased federal and local environmental restrictions (1). A novel, thermomechanical biosolids pretreatment process, which allows for a variety of potential value-added uses, was developed. This two-step process first employs thermal explosive decompression to inactivate or kill the microbial cells and viruses. This primary step also results in the rupture of a small amount of the microbial biomass and increases the intrinsic fluidity of the biosolids. The second step uses shear to effect a near-complete rupturing of the microbial biomass, and shears the nondigested organics, which increases the overall surface area. Pretreated biosolids may be subjected to a secondary anaerobic digestion process to produce additional fuel gas, and to provide for a high-quality, easily dewatered compost product. This novel biosolids pretreatment process was recently allowed a United States patent.

Index Entries: Sewage sludge; biosolids; thermal mechanical treatment; beneficial reuse.

*Author to whom all correspondence and reprint requests should be addressed.

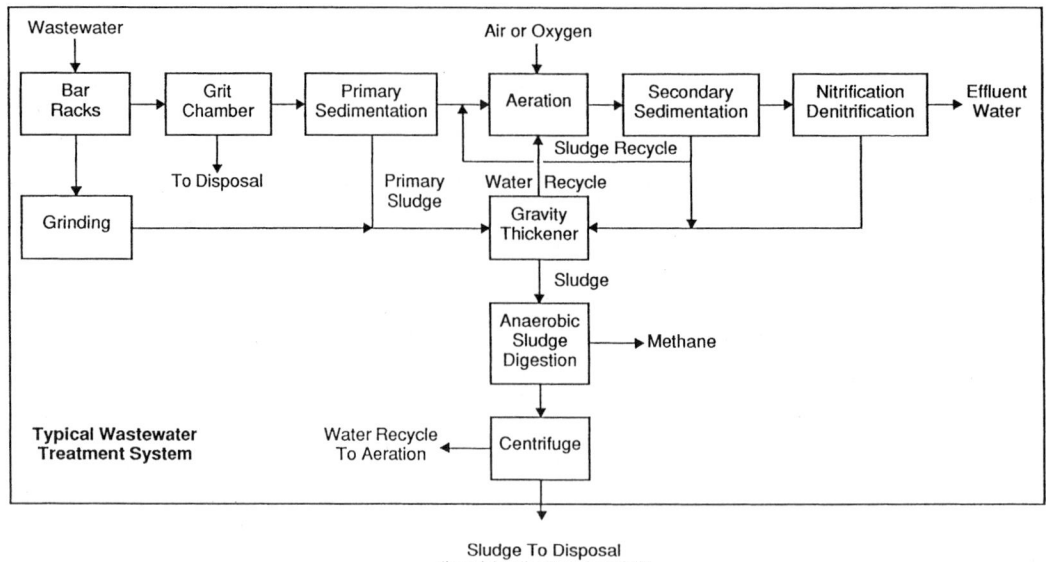

Fig. 1. Simplified process flow diagram for a conventional municipal sewage treatment plant.

INTRODUCTION

Conventional municipal sewage treatment plants use mechanical and biological processes to reclaim wastewaters (*see* Fig. 1). Disposing of microbial sludge solids that result from these treatment processes has historically been expensive, because of the extremely large volumes in which they are produced. These sludges contain high fractions of volatile solids (VS), they retain large amounts of water (70–85% before drying), and, because of the substantial bulk of the waste, the disposal costs are significant. Recently, the costs for disposing of sludges through conventional landfilling have risen dramatically, because of decreasing landfill availability. In some areas, microbial sludges are banned from landfills, because of the high pollution potential and the presence of active microbial catalysts, which further increase anaerobic bioconversion in the landfill.

An obvious solution to this disposal problem is to further reduce the organic content of the waste. This increases the potential for dewatering (reduction in bulk), and reduces both the pollution potential and pathogen load of the waste. Indeed, previous work by the authors shows that thermal mechanical pretreatment (2) effectively disrupts the macrostructure of municipal sewage sludges and renders them more amenable to secondary anaerobic bioconversion.

Although effective pretreatment was demonstrated for dilute sewage sludge, in commercial applications, sewage sludges are first dewatered to 15–30% total solids before final disposal. Therefore, pretreatment technology, which may effectively treat high-solids sludges, is required. The current study explores the use of a variety of thermal mechanical processes, such as those successfully used to pretreat biomass feedstocks before fermentation, to disrupt the macrostructure of highly dewatered sewage sludge for enhanced secondary anaerobic digestion.

MATERIALS AND METHODS

Sewage-Derived Sludge Procurement and Analysis

Municipal sewage sludge was obtained from the Denver Metropolitan Wastewater Reclamation District, Denver, CO. Dewatered sludge was collected from the belt conveyor after centrifugation, and was transferred to 40-gal plastic drums. The sludge was stored at 4°C in a cold room before use, in order to reduce unrelated breakdown of the material. The sludge was analyzed for a variety of important parameters, such as total solids, volatile solids, ash, total chemical oxygen demand (COD), soluble COD, and pH, as described previously (3,4).

Pretreatment Technology

Highly dewatered municipal sewage sludges represent a solids handling problem with respect to the application of pretreatment technologies. Therefore, various thermal mechanical processes, which have been demonstrated as effective in pretreating biomass before being biologically converted, were reviewed. Pretreatments evaluated included thermal treatment, explosive decompression, and shear, either alone or in combinations. The technologies are described here in detail, and include a Parr cell disruption bomb, a masonite gun, and a conventional hydroheater. The Parr cell disruption bomb and masonite gun represent batch-operated systems; the hydroheater was operated in a continuous mode.

Parr Cell Disruption Bomb Experiments

Experiments were performed with a Parr (Moline, IL) model #4635 cell disruption bomb constructed of 316 stainless steel. It has an internal volume of 920 mL and is rated for a maximum working pressure of 2200 psig. The disruption bomb was outfitted with an adjustable pressure-relief valve (Nupro, Denver, CO, SS-4R3A1, with adjustable spring R3A-D, rated 1500–2250 psig) in place of the normal 3000 psig rupture disk provided by the manufacturer. The pressure-relief valve was calibrated for 2100 psig activation, so as not to exceed the 2200 psig working pressure rating of the vessel.

Treatment experiments were carried out by adding a 10-g sample to a 50-mL plastic beaker and placing it in the disruption bomb. Once the headplate was secured, a compressed-gas cylinder of either ultra-high purity (UHP) nitrogen, zero air, or carbon dioxide was attached via a quick-disconnect fitting. For some of the experiments, the bomb was placed in a temperature-controlled water bath and allowed to equilibrate for 1 h before being pressurized. Treatments that used nitrogen or air were carried out at 2000 psig; those that used carbon dioxide were carried out at 900 psig (greater pressures result in liquid carbon dioxide being added to the bomb). Pressure was administered to the bomb with a high-pressure needle valve. Periodically during the treatment, additional gas was administered to maintain the target pressure, because the gas would continuously dissolve in the sample. The experiments were pressurized for 3 h, then rapidly depressurized by opening a 0.5-in. ball valve on the headplate. The sample was then removed for immediate analysis of soluble COD.

Masonite Gun Experiments

Experiments were conducted using a laboratory-scale steam-explosion process previously developed in the laboratory of Estaban Chornet at the University of Sherbrooke (Sherbrooke, PQ, Can [5]). This system consisted of a batch-loaded, steam-jacketed reaction chamber and a ball valve, followed by two flash tanks to collect the product. Experiments were conducted by adding approx 500 g of sewage sludge or sewage sludge mixed with shredded municipal solid waste (MSW) to the reactor. After the headplate was attached, the sample was subjected to pressures of 100, 211, 322, or 471 psig by the addition of steam to the reaction chamber and jacket. After 4 min of reaction at the target pressure, the contents of the reactor were expelled to atmospheric pressure into the primary flash tank through a 1-in ball valve. A representative sample of the pretreated material was removed for immediate analysis of soluble COD.

Hydroheater Experiments

The hydroheater pretreatment system used a manual stainless steel hydroheater (series M103MSX by Hydro Thermal, Waukesha, WI) to perform high-pressure steam mixing. To feed high-solids sludges at high pressures to the hydroheater, a progressing cavity pump which allowed operation at sludge-feed pressures as high as 500 psig (Moyno, model 9JKS3, Robbins Myers, Springfield, OH) was used. System capacity was 0.1–5.0 gal/min of high-solids sludge. A high-pressure boiler system provided steam to the hydroheater (18 hp, gas-fired, model HB-H44605-THK, Vapor, Chicago, IL). Following treatment with high-pressure stream in the hydroheater, the sludge was flashed to atmospheric pressure in a 40-gal stainless steel tank. The system was stopped after being operated for approx 30 min, and representative samples from the primary flash tank were obtained for immediate analysis of soluble COD.

Additional experiments that used a treatment train were performed, in which sewage sludge was first treated as described above with the hydroheater system, followed immediately by shear treatment with the Ultra Turrax (IKa Works, Staufen, Germany, model T-45-S4) for 4 min at

50% power. After this treatment train was applied, representative samples were obtained for immediate analysis of soluble COD.

Determination of Pretreatment Effectiveness

In general, pretreatment effectiveness was evaluated based on release of soluble COD from the sewage-sludge sample. Briefly, 1 g of sample was diluted with 9 mL of dH_2O, and mixed vigorously. The diluted sample was placed into a 15-mL plastic centrifuge tube and centrifuged at 1000 rpm for 5 min at room temperature using a Sorval model GLC-4 centrifuge equipped with a H1000 rotor. A 100-µL sample of the upper phase (supernatant) was added to COD test vials (Hach, Loveland, CO, high-range plus). The COD assay was incubated for 2 h at 150°C, and read at 600 nm with a spectrophotometer (Milton-Roy, Rochester, NY, model 301). Increases in soluble COD are directly related to an increase in the anaerobic biodegradation potential of the organic sample (2).

RESULTS AND DISCUSSION

Previous research identified the relative ineffectiveness of thermal, thermal–acid, thermal–alkaline, and enzymatic pretreatments in releasing soluble COD from low-solids microbial sludges (2). Additional research identified optimum pretreatment of low-solids microbial sludges (1–3%) to require thermal and shear or sonication forces for efficient release of soluble COD (2). These data were the basis for a US Patent (5380445 [1995]). However, industrial application dictates that the pretreatment technology be developed for high-solids sludges, and, if possible, in continuous mode. To this end, several approaches to high-solids sewage sludge pretreatment, including thermal treatment, explosive decompression, and shear, either alone or in combinations, were explored.

Analysis of the municipal sewage sludge obtained from the Denver Metropolitan Wastewater Reclamation Plant for this study is described in Table 1. Sludge analysis revealed a total solids content of slightly greater than 17%, and soluble COD that represented 4.9% of the total COD content.

Parr Explosive Decompression

Initial experiments that used either rapid decompression or rapid decompression with moderate thermal input are shown in Fig. 2 for microbial sewage sludge at 17.6% total solids. Compression gases included UHP nitrogen, carbon dioxide, and zero air. Data indicate that explosive decompression with nitrogen at 55°C releases the most soluble COD (17% release). Initially, the disappointing treatment effectiveness was attributed to slow gas penetration of the sludge sample, as a result of the total solids level. Therefore, experiments were also conducted with sewage sludge mixed with a processed MSW (50/50, w/w). The processed MSW consisted

Table 1
Analysis of Municipal Sewage Sludge Obtained from the Denver Metropolitan Wastewater Reclamation Plant

Parameter	Value
Total solids (TS)	17.3% ± 0.1
Volatile solids (VS, of dry wt)	66.5% ± 0.2
Ash (of dry wt)	33.5% ± 0.2
Total chemical oxygen demand (COD, mg/g wet wt)	187.5 ± 16.3
Soluble COD (mg/g wet wt)	9.2 ± 0.8
pH	8.01

Fig. 2. Efficacy of explosive decompression using the Parr reactor on the release of COD from sewage sludge. The experiments were performed with sewage sludge alone at 17.3% total solids.

of the paper and packaging fraction of residential refuse shredded with an industrial-scale knife mill to pass a 1/4-in. screen. Mixing sewage sludge with processed MSW increased the total solids content to 48.6%, and enhanced the gas permissibility of the sludge. Data shown in Fig. 3 demonstrate only marginal increases in the effectiveness of pretreatment (~2%).

Masonite Gun Pretreatment

Steam-explosion pretreatment technology imparts thermal, explosive decompression and mild shear forces to the sample. Thermal treatment is first applied in a steam-jacketed reactor, then the sample is released rapidly through a relatively small orifice (imparting shear) during rapid decompression. Results indicate that maximum treatment effectiveness was

Fig. 3. Efficacy of explosive decompression using the Parr reactor on the release of COD from sewage sludge. Experiments were performed after blending sewage sludge and shredded MSW on a 50/50 weight basis. The sewage sludge and MSW blend was 48.6% total solids.

Fig. 4. Efficacy of explosive decompression using the Masonite gun system on the release of COD from sewage sludge.

achieved with operating pressures of 211–471 psi (Fig. 4). The maximum release of soluble COD by treating sewage sludge alone was ~25% at 322 psi. Additional experiments conducted with sewage sludge mixed with processed MSW (described earlier), to increase surface area for steam penetration, had little effect on the pretreatment results. The maximum treatment effectiveness was 27% COD release at 471 psi.

Table 2
Various Operational Parameters for Hydroheater System Operation

Run	Pressure (psi) Sludge pump	Pressure (psi) Steam[a]	Temperature (°F) Steam[b]	Temperature (°F) Hydroheater	Hydroheater sludge setting	Hydroheater mix rate[c]	Moyno pump controller[d] (Hz)
1	80–100	250	410 ± 8	285 ± 6	Medium	5	14
2	75–100	250	410 ± 8	293 ± 5	Medium	10	7
3	75–100	250	410 ± 8	265 ± 5	Medium	4	20
4	80–100	250	410 ± 8	265 ± 5	High	3	20

[a] Steam pressure was attenuated from the boiler system operated at 300 psi with a steam valve.
[b] Steam temperature is listed for a thermocouple probe installed immediately preceding the hydroheater.
[c] The hydroheater mixing rate was a function of a needle valve with 11 turns (0 = no steam addition).
[d] The moyno pump rate was adjusted with a Baldor series 15 inverter control. The percentage of full pump speed was determined as a function of the maximum Hertz rating of the motor (i.e., 60 Hz).

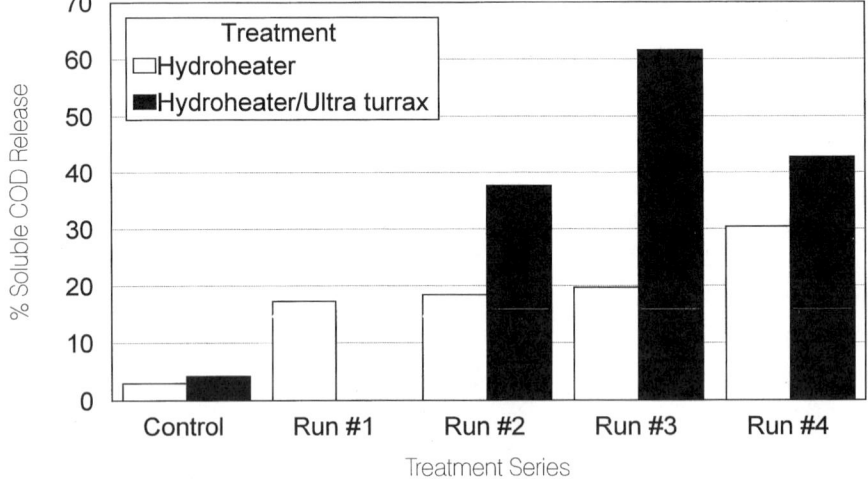

Fig. 5. Efficacy of explosive decompression using the hydroheater system on the release of COD from sewage sludge. Run conditions are as described in Table 2. For some experiments, a treatment train, including the use of shear (Ultra Turrax), was evaluated.

Hydroheater Pretreatment

Pretreatment that uses the hydroheater system is similar to steam explosion, because it imparts to the sample thermal, explosive decompression, and mild shear forces. However, the hydroheater system differs, because it represents a continuous (rather than batch) process. Several runs were conducted with the hydroheater system, with operational conditions as outlined in Table 2. Results using the hydroheater system alone demonstrate only a modest pretreatment effectiveness of 20–30% release of soluble COD (Fig. 5). However, a treatment train, which included a post-treatment with the Ultra Turrax, substantially increased soluble COD to 40–62%. Furthermore, additional experimentation may yield more improvements in treatment effectiveness. Greater than 80% release of soluble COD (similar to results obtained using low-solids pretreatment technology, [2]) may be possible.

Early analysis of energy requirements for pretreatment, compared to anticipated fuel gas yields for a secondary anaerobic digestion process, estimates the production of 237 BTU (net energy)/lb of sewage sludge treated (data not shown).

Data depicted in this paper formed the basis for a recently allowed US Patent.

ACKNOWLEDGMENTS

The authors thank Bill Morgan and Dawn Flancher (Denver Metropolitan Wastewater Reclamation Plant) for assistance in providing sewage sludge solids. This research was funded by the Director's Development Fund No. 07422031 at the National Renewable Energy Laboratory.

REFERENCES

1. Federal Register (1993), "Part 503-Standards for the Use or Disposal of Sewage Sludge," Rules and Regulations.
2. Rivard, C. J. and Nagle, N. J. (1996), *Appl. Biochem. Biotech.* **57/58,** 983–991.
3. Nagle, N. J., Rivard, C. J., Adney, W. S., and Himmel, M. E. (1992), *Appl. Biochem. Biotech.* **34/35,** 737–751.
4. APWA-AWWA-WPCF (1980), *Standard Methods for the Examination of Water and Wastewater Analysis*, 15th ed., APHA, Washington, D.C.
5. Montane, D., Farriol, X., Salvado, J., Jollez, P., and Chornet, E. (1998) *Biomass Bioenergy*, accepted.

Phenomenological and Neural-Network Modeling of Cephalosporin C Production Bioprocess

A. J. G. Cruz, M. L. G. C. Araujo, R. C. Giordano, and C. O. Hokka*

Departamento de Engenharia Química, Universidade Federal de São Carlos C. P. 676 - CEP 13565-905, São Carlos, SP, Brazil

ABSTRACT

Cephalosporin C production process with *Cephalosporium acremonium* ATCC 48272 in synthetic medium was investigated and the experimental results allowed the development of a mathematical model describing the process behavior. The model was able to explain fairly well the diauxic phenomenon, higher growth rate during the glucose-consumption phase, and the production occurring only in the sucrose-consumption phase.

Moreover, the process was simulated utilizing the neural-networks technique. Two feed-forward neural-networks with one hidden layer were employed. Both models, phenomenological and neural-networks based, satisfactorily describe the bioprocess. The difficulties in determining kinetic parameters are avoided when neural networks are utilized.

Index Entries: *Cephalosporium acremonium*; cephalosporin C, phenomenological modeling; neural network; diauxic phenomenon.

INTRODUCTION

Cephalosporin C is a β-lactam antibiotic usually produced by *Cephalosporium acremonium* in aerated and agitated tanks. They are *N*-acylated derivatives of 7-aminocephalosporanic acid (7-ACA), which is derived from the cephem nucleus. A cephem is the bicyclic ring system obtained by the fusion of a β-lactam ring with dihydrothiazine ring *(1)*. This natural product is modified by chemical or enzymatic methods to yield the semisynthetic cephalosporins, presently on a grand scale in the pharmaceutical market.

---

*Author to whom all correspondence and reprint requests should be addressed. E-mail: hokka@power.ufscar.br.

Cephalosporin C is a typical secondary metabolite and its production process presents two distinct phases. In the first phase, denominated trophophase, most growth takes place and an easily metabolized carbon source, normally glucose, is consumed. Throughout this step, low production rates are observed owing to a catabolite repression mechanism blocking the specific enzyme synthesis, as reported by Behmer and Demain (2). When the glucose is exhausted, the idiophase commences. In this phase a second carbon source, usually less easily assimilated, like sucrose, is consumed, promoting the synthesis of specific enzymes that allow higher production rates of the antibiotic. Typical diauxic phenomena are observed concerning these two carbohydrates consumption.

Because of its industrial and therapeutic importance, the cephalosporin C production process is the object of study of many research groups. Matsumura et al. (3) have proposed a kinetic model of the production process by C. acremonium ATCC 36225 (CW19). These authors have observed drastic microorganism morphological differentiation during the process time, and concluded that a certain morphology was closely associated with the production capability of the strain. They also confirmed the important role of endogenous methionine in antibiotic production and the catabolite repression by glucose. Chu and Constantinides (4) have studied this process and have developed a kinetic model aiming at the maximization of this antibiotic production by strains of C. acremonium (CW19), applying Pontryagin's maximum principle.

Nowadays the neural-networks approach is becoming popular among researchers to model many chemical and biochemical processes. The first studies in this field started in the earlier 1930s, but it was only in the 1980s that interest in the field reemerged with the publication of the back-propagation algorithm by Rumelhart et al. (5), which has established the feed-forward layered network as the major paradigm of the field.

Thibault et al. (6) have investigated the use of neural-network methodology for the modeling of a bioprocess dynamics and prediction of its variables. Di Massimo et al. (7) have reported a methodology employing neural-network techniques to infer offline assay information (primary process variables such as cellular and product concentration) with available on-line measurements. Syu and Tsao (8) have used neural networks in modeling batch microbial cell growth. In recent studies, Cruz et al. (9,10) employed this technique showing very promising results in monitoring, fault detection, and simulation of penicillin-G production bioprocess. The nonlinear behavior of the bioprocess was described satisfactorily by neural-network, and good results obtained by the authors have encouraged the study of this approach in bioprocess modeling and simulation.

In this work, experimental runs for the process were performed and the time course of cell growth, cephalosporin C production, glucose and sucrose consumption were followed during 144–166 h. For the process modeling, two approaches were utilized.

In the phenomenological approach, although the simplifying assumptions regarding the bioprocess kinetics were used, still the problem of large number of parameters to be optimized remained. To avoid this problem, the neural network approach for the bioprocess simulation was utilized by applying two feed-forward neural-networks.

MATERIALS AND METHODS

Microorganism

The strain *Cephalosporium acremonium* ATCC 48272 kindly provided by Fundação Tropical de Pesquisa e Tecnologia "André Tosello" (Campinas, SP, Brazil) was used throughout this work. This strain is able to produce more than 1000 mg cephalosporin C/L, when grown in a synthetic medium and it was kept on agar slants as proposed by Shen et al. *(11)*.

Culture media

For inoculum preparation, a synthetic medium was utilized containing (in g/L): glucose, 30.0; ammonium acetate, 8.8; *DL*-methionine, 5.0; oleic acid, 1.5; KH_2PO_4, 2.3; K_2HPO_4, 5.8; $Fe(NH_4)_2(SO_4)_2/6H_2O$, 0.16; micronutrients were provided by salt solution addition (50 mL/L). Its composition was in (g/L): Na_2SO_4, 16.2, $MgSO_4 \cdot 7H_2O$, 7.68, $CaCl_2 \cdot 2H_2O$, 1.6, $MnSO_4 H_2O$, 0.64, $ZnSO_4 \cdot 7H_2O$, 0.64, $CuSO_4 \cdot 5H_2O$, 0.04. The production medium contained (in g/L): glucose, 27.0; sucrose, 36.0; the other components were the same as the inoculum medium. The media compositions were similar to that utilized by Demain et al. *(12)*. In the case of glucose-phase parameter evaluation, sucrose was not added.

Experimental Procedure

The cell spores from agar slants were transferred to preculture medium in 250 mL shake flasks and incubated for 48 h 250 rpm at 26°C. After spore germination, a new preculture was carried out for 24 h at same conditions. The cultivated cells were then used to inoculate the main fermentation broth with free cells. The main experiments were carried out in 250-mL shake flasks, at 26°C and agitation speed of 250 rpm. The seed was 10% in volume. Runs were carried out for 144–166 h and samples were taken periodically for cell mass, the sugars, and cephalosporin C concentration.

Analysis

The cell-mass concentration was evaluated as dry weight at 105°C, in g/L. Glucose was measured utilizing enzymatic GOD-PAP method *(13)* and sucrose was measured by the same method, following acid hydrolysis. Cephalosporin C titers were determined by an agar diffusion bioassay *(14)* utilizing *Alcaligenes faecalis* ATCC 8750.

PHENOMENOLOGICAL APPROACH

Kinetic Model

The kinetic model developed was based on stoichiometric equations, representing the main steps in cell, substrates, oxygen, and product variation. In these equations, only the limiting substrate, cell, oxygen, and product are represented since the other nutrients are considered to be in excess.

The total biomass was assumed to be composed of cells X_1, submitted to catabolite repression by glucose S_1, during the growth phase (Eq. 1), and cells X_2, derepressed after glucose depletion and capable of consuming sucrose S_2 (Eq. 4). Repression and derepression refer to the regulatory mechanism of formation of the enzyme complex responsible for the cephalosporin synthesis. In glucose concentration lower than a certain critical value, Cs_{c1} (1.5 g/l), cells X_1 are transformed into cells X_2 (Eq. 3).

$$aO_2 + S_1 + X_1 \rightarrow (1 + Y_{xs1})X_1 + Y_{px1}P \tag{1}$$
$$bO_2 + X_1 \rightarrow D_1 \tag{2}$$
$$X_1 \rightarrow X_2 \tag{3}$$
$$cO_2 + S_2 + X_2 \rightarrow (1 + Y_{xs2})X_2 + Y_{px2}P \tag{4}$$
$$dO_2 + S_2 \xrightarrow{X_2} Y_{px3}P \tag{5}$$
$$eO_2 + X_2 \rightarrow D_2 \tag{6}$$
$$P \rightarrow D_3 \tag{7}$$

where total cell mass (X) includes both types of cells, defined as X_1 and X_2, D_1 and D_2 are products of cell degradation, during the glucose (S_1) and sucrose (S_2) consumption phase, respectively; and D_3 is the degradation product of P.

Rate Equations

Rate equations related to the above mentioned stoichiometric equations describing the behavior of the process were based on those proposed by Araujo et al. (15) as follows:

$$Rx_1 = \frac{dCx_1}{dt} = \left\{ \left[\left(\frac{\mu_{max1}Cs_1}{k_{x1}Cx_1 + Cs_1} \right) \cdot \left(\frac{C_{O2}}{k_{O2} + C_{O2}} \right) \right] - k_{d1} \right\} Cx_1 \\ - \left[\frac{(k_T Cx_1)}{(k_1 + Cx_1)} \right] \tag{8}$$

$$Rx_2 = \frac{dCx_2}{dt} = \left\{ \left[\left(\frac{\mu_{max2}Cs_2}{k_{s2} + Cs_2 + K_i Cs_2^2} \right) \cdot \left(\frac{C_{O2}}{k_{O2} + C_{O2}} \right) \right] - k_{d2} \right\} Cx_2 \\ + \left[\frac{(k_T Cx_1)}{(k_1 + Cx_1)} \right] \tag{9}$$

$$Rs_1 = \frac{dCs_1}{dt} = -\left[\left(\frac{1}{Y_{xs1}}\right) \cdot \left(\frac{\mu_{max1}Cs_1}{k_{x1}Cx_1 + Cs_1}\right) \cdot \left(\frac{C_{O2}}{k_{O2} + C_{O2}}\right)\right] Cx_1 \qquad (10)$$

$$Rs_2 = \frac{dCs_2}{dt} = -\left\{\left[\left(\frac{1}{Y_{xs2}}\right) \cdot \left(\frac{\mu_{max2}Cs_2}{k_{s2} + Cs_2 + K_iCs_2^2} + m\right)\right. \right.$$
$$\left.\left. \cdot \left[\left(\frac{C_{O2}}{(k_{O2} + C_{O2})}\right)\right]\right\} Cx_2 \qquad (11)$$

$$Rc_P = \frac{dCp}{dt} = \left[\left[\left\{\frac{Y_{px1}}{Y_{xs1}} \cdot \left[\frac{(\mu_{max1}Cs_1)}{(k_{x1} + Cs_1)}\right]\right\} Cx_1 \right.\right.$$
$$\left. + \left\{\left(\frac{Y_{px2}}{Y_{xs2}}\right) \cdot \left[\frac{(\mu_{max2}Cs_2)}{(k_{s2} + Cs_2 + K_iCs_2^2)}\right] + Y_{px3} \cdot m\right\} Cx_2 \right]$$
$$\left. \cdot \left[\frac{C_{O2}}{(k_{O2} + C_{O2})}\right]\right] \qquad (12)$$

$$Rc_{O2} = \frac{dC_{O2}}{dt} = k_La(C_{O2}^* - C_{O2}) - \left[\frac{(R_{max} \cdot C_{O2})}{(k_{O2} + C_{O2})}\right] \qquad (13)$$

The first term in Eqs. 8 and 9 represents growth of cells due to glucose and sucrose substrate sources. The second terms refer to the transformation of repressed cells into derepressed ones. Equation 10 is the glucose consumption rate, assuming that this substrate is used only for growth, whereas Eq. 11 represents sucrose consumption rate. The production rate is describe by Eq. 12, with three terms, related to the growth in glucose, and growth and maintenance in sucrose. Each of these terms appears associated with yield factors (or pseudo stoichiometric coefficients) Y_{Px1}, Y_{Px2} and Y_{Px3}, respectively. Equation 13 represents the time course of dissolved oxygen during the process.

Sucrose is slowly consumed during the process low activity of sucrose hydrolyzing enzyme; in this phase little cell growth is observed. Therefore in this phase it has been assumed that the process was limited by enzyme activity both in growth and maintenance reactions.

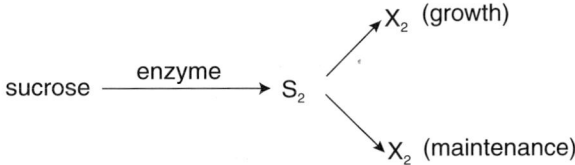

and Eq. 11 becomes:

$$Rs_2 = \frac{dCs_2}{dt} = -\left[\left(\frac{1}{Y_{x2}}\right) \cdot \left(\frac{\mu_{max2}Cs_2}{k_{s2} + Cs_2 + K_iCs_2^2}\right) \cdot \frac{C_{O2}}{k_{O2} + C_{O2}}\right] Cx_2 \qquad (14)$$

where:

$$\frac{1}{Y_{x2}} = \frac{1}{Y_{xs2}} + \frac{m}{\mu_{max2}}$$

where Y_{x2} is an overall yield coefficient which includes the amount of sucrose used by the microorganisms in growth and maintenance.

With this assumption in sucrose kinetics, the product equation becomes:

$$Rc_P = \frac{dCp}{dt} = \left[\left(\left(\frac{Y_{px1}}{Y_{xs1}} \cdot \frac{\mu_{max1}Cs_1}{k_{x1} + Cs_1}\right)\right) Cx_1 + \left(\left(\frac{Y_{px2}}{Y_{x2}}\right) \cdot \left(\frac{\mu_{max2}Cs_2}{k_{s2} + Cs_2 + K_iCs_2^2}\right)\right) Cx_2\right] \cdot \frac{C_{O2}}{k_{O2} + C_{O2}} \quad (15)$$

These equations were solved numerically using the Differential-Algebraic system solver (DASSL) algorithm developed by Petzold (16), in a 486 DX-4, 100-MHz microcomputer. The estimation of model parameters was carried out following a nonlinear least square approach, using the Marquardt method (17).

NEURAL-NETWORK APPROACH

Two feed-forward neural-networks (FNN) were employed in this work to estimate cell and antibiotic concentration. The two nets were described in Fig. 1, and consist of one input layer, one output layer, and one hidden layer. The activation function employed was sigmoidal (18). The objective function of choice for the minimization criterion in the training algorithm was the mean square error between the neural output and the desired output from all the input pattern in the data base, given by:

$$\text{Obj\_f} = \sum_p E_p \quad (16)$$

where E_p is:

$$E_p = \frac{1}{2} \|Y_p - d_p\|^2 \quad (17)$$

where y_p: value obtained from neural network, d_p: desired value from data base.

The first FNN had three hidden neurons, and an input with three variables: ($C_{s1}[k]$, $C_{s2}[k]$, $C_x[k]$), where k is the present time. The output vector was cell concentration in future time ($C_x[k + 1]$).

The second FNN was used to estimate antibiotic concentration in future time ($C_p[k + 1]$), and had four hidden neurons. Its input was composed of three variables ($C_{s1}[k]$, $C_{s2}[k]$, and $C_x[k]$) in present time.

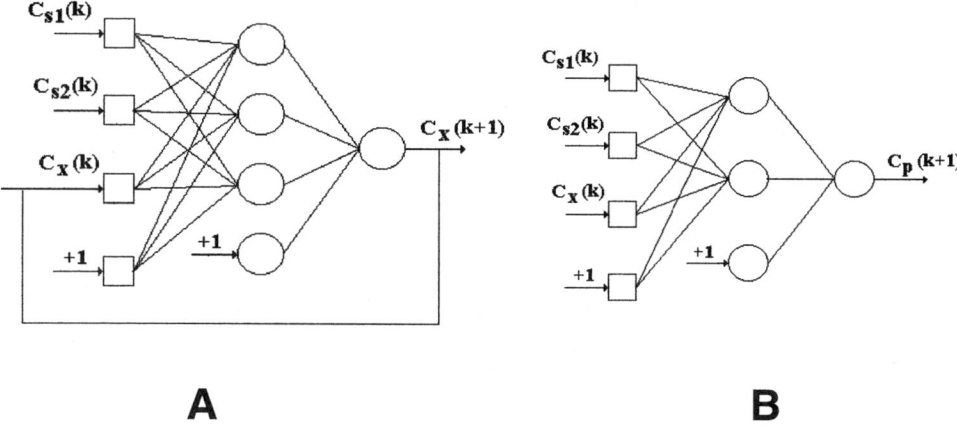

Fig. 1. Basic feed-forward neural-network architecture (FNN): **(A)** the first FNN, used to estimated cell concentration; **(B)** the second FNN, employed to infer antibiotic concentration.

Data Base

The data used to train the two FNN were obtained from 14 sets of experimental data.

RESULTS AND DISCUSSION

Parameters Estimation: Phenomenological Approach

Growth of the microorganism in a medium with glucose as the major carbon source was carried out to determine the parameters of the equation proposed by Contois *(19)*. This model was preferred as at high cell concentrations, serious diffusional limitations can be expected. This model has been utilized by Bajpai and Reuss *(20)* for describing *Penicillium chrysogenum* growth. In Fig. 2, a good fit is shown between experimental data and the model. Parameters μ_{max1}, k_x and Y_{xs1} were estimated by nonlinear regression for this experiment *(17)*; their values are shown in Table 1.

For the sucrose-utilizing phase, the study was carried out in a medium containing glucose and sucrose as major carbon sources, and the samples were taken after 48 fermentation, when it was assumed that the glucose was already depleted. Figure 3 depicts the experimental data and calculated results. Parameters μ_{max2}, k_{s2}, K_i, k_{d2}, and Y_{x2} were estimated using the Marquardt method as shown in Table 1.

Regarding the respiration rate, the influence of sucrose or glucose concentration was investigated by Araujo *(21)*, and specific respiration rate during this process was shown to be independent of sugar type or concentration. The specific respiration rate behaves as a zero-order reaction

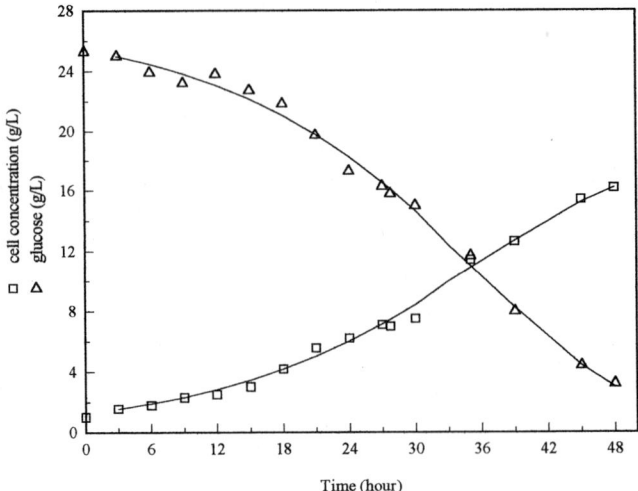

Fig. 2. Experimental data of cephalosporin C fermentation and fit of the kinetic model proposed to growth phase in glucose.

Table 1
Estimated Parameters, with 95% Confidences Intervals, Used for Simulation of Cephalosporin C Production Process by *C. acremonium* ATCC 48272 in Shake Flasks

Maximum specific growth rates	μ_{max1}	0.0671 ± 0.0029 (h^{-1})
	μ_{max2}	0.0364 ± 0.0143 (h^{-1})
Contois constant	k_{x1}	0.379 ± 0.122 (gS$_1$/gX$_1$)
saturation constant	k_{s2}	12.0 (gS$_2$/L)
inhibition constant	K_i	0.154 ± 0.105 (gS$_2$/L)$^{-1}$
death rate constant	k_{d2}	0.006 (h^{-1})
	k_{d3}	0.0015 (h^{-1})
yield coefficients (sugar consumption)	Y_{xs}	0.599 ± 0.028 (gX$_1$/gS$_1$)
	Y_{x2}	0.399 ± 0.04 (gX$_2$/gS$_2$)
yield coefficients (product formation)	Y_{px1}	0.0045 (gP/gX$_1$)
	Y_{px2}	0.035 (gP/gX$_2$)
kinetic constants	k_T	6.0 (gX$_1$/L/h)
	k_1	0.01 (gX$_1$/L)
maximum specific respiration rate	R_{max}	1.055 ± 0.219 mmolO$_2$/gX/h
kinetic constant (respiration rate)	k_{O2}	0.00277 ± 0.00073 (mmolO$_2$/L)
volumetric gas-liquid mass transfer coefficient	$k_L a$	162.0 (h^{-1})
oxygen saturation concentration	C_{O2}^*	0.22 (mmolO$_2$/L)
critical value of glucose concentration	C_{slc}	1.5 (gS$_1$/L)

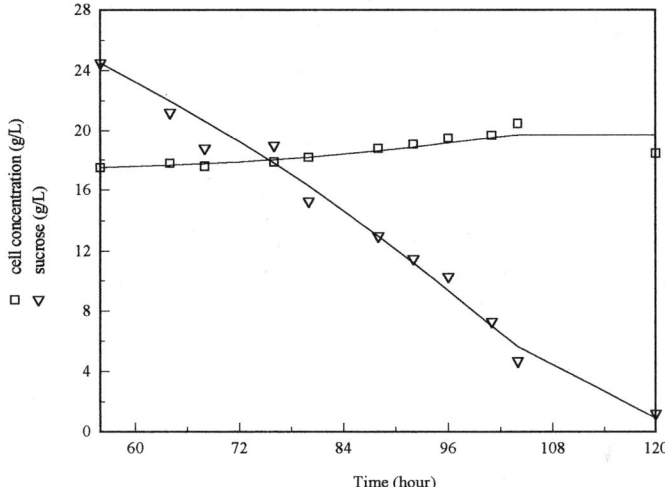

Fig. 3. Experimental and calculated data of the cephalosporin C bioprocess during sucrose consumption phase.

Fig. 4. Plot of (A) glucose, (B) sucrose, (C) cell and (D) cephalosporin C concentrations during the time course of the bioprocess. Variance and standard deviation were presented for the median values of the variables.

for both sugars. The author also measured R_{max} and k_{O2} considering that kinetics concerns oxygen concentration. The values adopted in this work are shown in Table 1.

Fourteen sets of experimental data were collected. Figure 4 presents the values of glucose, sucrose, cell, and cephalosporin C concentrations

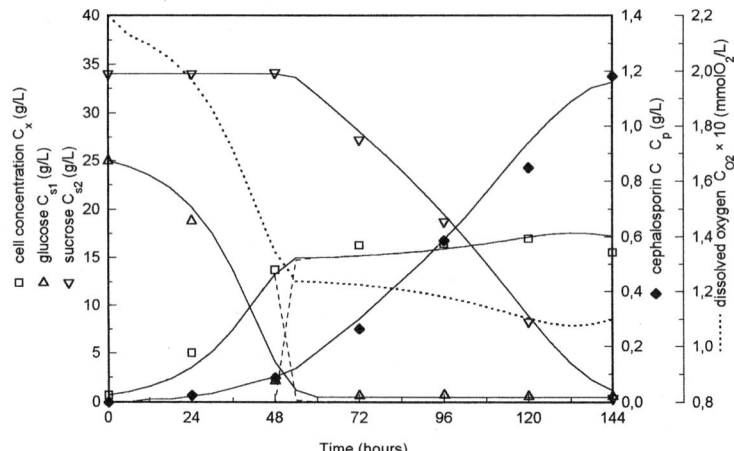

Fig. 5. Simulated results and experimental data of cephalosporin C production process by *C. acremonium* ATCC 48272 in shake flasks.

during the time course of the bioprocess. The data presented represent the average values for the 14 experimental runs.

For the remaining parameter values, the differential equation sets were solved numerically and the set of the constant values were optimized by nonlinear, least-square regression analysis following Marquadt's procedure. The values so determined are shown in Table 1.

Calculated values obtained by simulating the model are shown in Fig. 5. As can be observed, the proposed model explains very well the complex features of this bioprocess, such as diauxic growth and the higher production rate taking place mostly during the sucrose consumption phase.

A volumetric oxygen transfer coefficient was determined for this experiment. Also, dissolved oxygen behavior was simulated, and it was established that its concentration, though dropping at the beginning of cultivation in the growth phase, did not fall below 45%.

Cephalosporin C specific production rate was approx 0.53 mg/g cell/h, considering the whole process time.

Neural-Network Training

The neural-network was trained by successive presentations of input-output data pairs from the database. Three different algorithms were tested: classical back propagation (5), random search procedure (RSP) (7), and a mixed procedure combining the last two procedures (22). The second approach was most efficient in this case.

A comparison between back propagation and random search training procedures was made; the results are shown in Fig. 6. The solid line repre-

Cephalosporin C Production Bioprocess

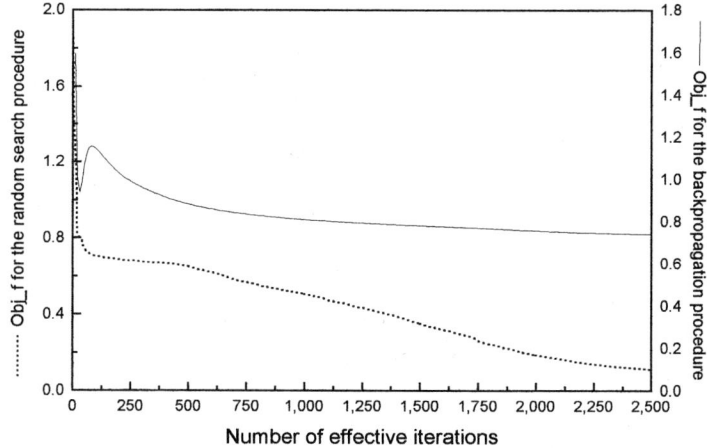

Fig. 6. Plot of error criterion for minimization of objective function based on random search algorithm and error criterion for minimization of objective function based on back propagation algorithm.

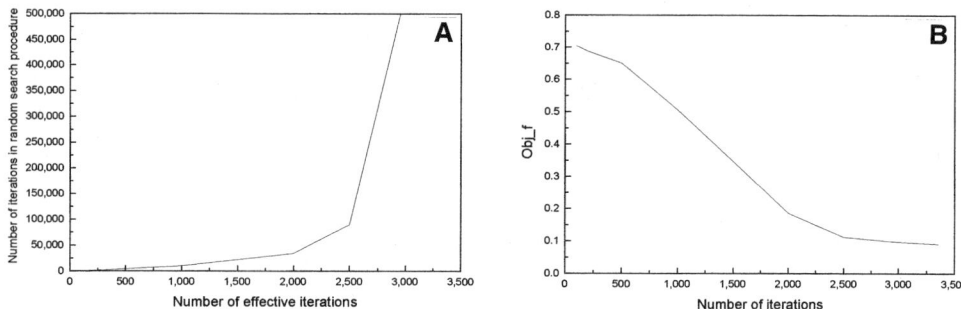

Fig. 7. **(A)** Number of effective iterations vs number of iterations in RSP algorithm; **(B)** number of iterations vs objective function values.

sents the value of objective function for the back propagation algorithm, whereas the dashed line shows its value for the random search algorithm.

In Fig. 6 only the number of effective iterations for the random search algorithm was considered. It is evident that the random search procedure is much more efficient and robust, whereas the classical backpropagation method shows a tendency to stabilize toward a local minimum.

To determine the number of iterations to be used to stop the training algorithm, the behavior of the number of iterations in RSP and the value of objective function in relation to the number of effective iterations was examined. Figure 7A, B depicts the results.

Figure 7A shows that by increasing the number of effective iterations in the random search procedure, the number of iterations become excessively high and, as a consequence, the processing time lengthens to a point

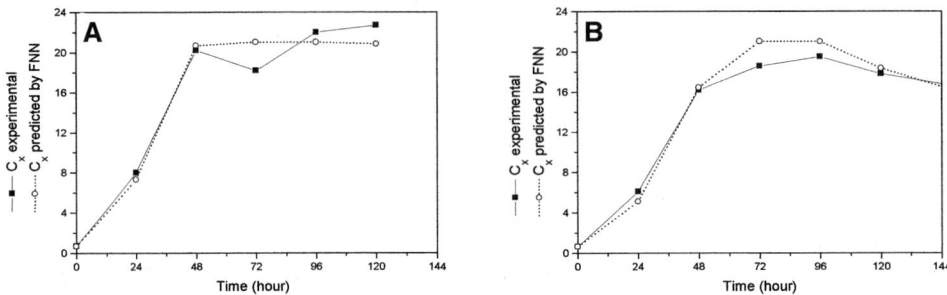

Fig. 8. Simulation results from 3 × 3 × 1 FNN: **(A)** $C_{s1} = 25$ g/L, $C_{s2} = 36$ g/L, and $C_{x1} = 0.7$ g/L initial conditions; **(B)** $C_{s1} = 25.4$ g/L, $C_{s2} = 36$ g/L, and $C_{x1} = 0.65$ g/L initial conditions.

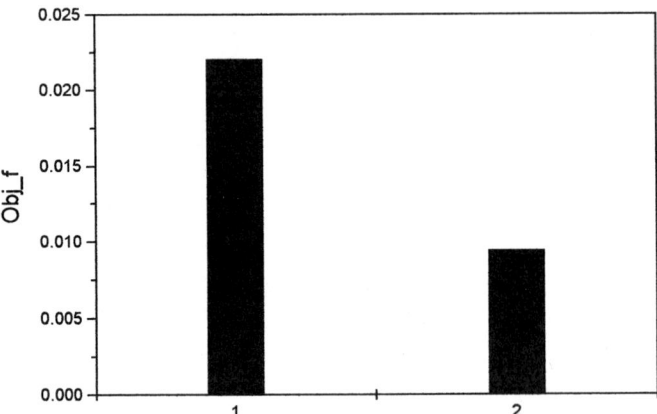

Fig. 9. Comparison between the back propagation and random-search training algorithms for 300 effective iterations.

where it becomes useless, since objective function value does not decrease significantly only above 2500 iterations, as shown by Fig. 7B. Therefore, this number of iterations was adopted as the end value for the network training process.

The neural network described in Figs. 1A, B were employed to infer the cell and cephalosporin C concentration, respectively. Figs. 8A, B show two different simulations. Figure 8A exemplifies the accuracy of the network fitting to a set of data used in the training procedure.

The good agreement is in fact already expected, because of the well-known capability of these algorithms to forecast nonlinear behaviors. Fig. 8B, on the other hand, shows an example of the FNN ability to forecast the general trend of the system, in a situation not used during the training phase. In the case depicted in Fig. 8B, the initial cell concentration is 7.14% less than in Fig. 8A (from 0.65 against 0.7 g/L). This is a typical variation

Fig. 10. Simulation results from 3 × 2 × 1 FNN: **(A)** Initial conditions: C_{s1} = 25.4 g/L; C_{s2} = 36 g/L; C_x = 0.65 g/L **(B)** Initial conditions: C_{s1} = 25 g/L; C_{s2} = 34 g/L; C_x = 0.7 g/L.

in this kind of experiment, and the FNN is able to follow the empirical observations.

For cephalosporin C estimation, sucrose, glucose and mass concentration obtained from the database were used as network inputs, whereas the cephalosporin C concentration is the network output. Best results were obtained using RSP algorithm, as depicted in Fig. 9.

Figure 10 illustrates the simulation by this technique of the cephalosporin C time course. Two data sets not used during the training are displayed in this figure; in both cases an excellent fitting to the experimental values can be noticed.

Thus, for a complex biological process, this technique is demonstrated to be very useful and promising for control and optimization when coupled with data acquisition systems.

CONCLUSIONS

The phenomenological model proposed here satisfactorily describes the main features of this bioprocess, such as diauxic effect, different sugar consumption rates and antibiotic production during the second phase.

The inherent difficulties when estimating a large number of adjustable parameters in the phenomenological approach were avoided when the neural-network approach was used.

Indeed, the neural-network method is a promising alternative to circumvent time delay in measuring cell mass and cephalosporin C concentration. Consequently, more robust control strategies are about to appear on the horizon.

Studies utilizing this technique are presently being performed in our laboratories, and indicate that this approach should be highly valuable for improvements in this process when combined with online measurements. Particularly the FNN in Fig. 1A uses only initial cell concentration to predict the cellular mass during the hole process. This feature may be useful to infer the trend of important online variables that are measured offline in the real process, such as cell and antibiotic concentrations.

ACKNOWLEDGMENTS

The authors gratefully acknowledge the financial support of CAPES (Ministry of Education Brazil), and scholarship from FAPESP (São Paulo State Foundation, Brazil) for two of the authors (A.J.G.C. and M.L.G.C.A.). The English review by Paula Matvienko-Sikar is also acknowledged.

REFERENCES

1. El-Sayed, A.-H. M. M. (1992), in *Handbook of Applied Mycology*, vol. 4, Arora, D. K., ed., Banaras Hindu University, Varanasi, India, pp. 517–564.
2. Behmer, C. J. and Demain, A. L. (1983), *Current Microbiol.* **8,** 107–114.
3. Matsumura, M., Imanaka, T., Yoshida, T., and Taguchi, H. (1981), *J. Ferment. Technol.* **59(2),** 115–123.
4. Chu, W. B. Z and Constantinides, A. (1988), *Biotechnol. Bioeng.* **32,** 277–288.
5. Rumelhart, D. E. and McClelland, J. L. (1986), in *Parallel Distributed Processing: explorations in the microstructure of cognition*, vol. 1, MIT, Cambridge, MA, pp. 318–362.
6. Thibault, J., Breusegem, Van V., and Chéruy, A. (1990), *Biotechnol. Bioeng.* **36,** 1041–1048.
7. DiMassimo, C., Montague, G. A., Willis, M. J., Tham, M. T., and Morris, A. J. (1992), *Computers Chem. Eng.* **16(4),** 283–291.
8. Syu, M.-J and Tsao, G. T. (1993), *Biotechnol. Bioeng.* **42,** 376–380.
9. Cruz, A. J. G., Hokka, C. O., and Giordano, R. C. (1996), *Proceedings of the 11st Simpósio Nacional de Fermentações (XI SINAFERM, São Carlos, SP, Brasil)*, **1,** pp. 84–89.
10. Cruz, A. J. G, Hokka, C. O., and Giordano, R. C. (1996), *Proceedings of the 3rd Congreso Interamericano de Computacion Aplicada a la Industria de Procesos (CAIP '96, Villa Maria, Cordoba, Republica da Argentina)*, pp. 115–118.
11. Shen, Y.-Q, Wolfe, S., and Demain, A. L. (1986), *Bioeng. Biotechnol.* **4,** 61–63.
12. Demain, A. L., Newkirk, J. F., and Hendlin, D. (1963), *J. Bacteriol.* **85,** 339–344.
13. Barham, D. and Trinder, P. (1972), *Analyst,* **97(2),** 142–145.
14. Claridge, C. A. and Johnson, D. L. (1962), *Antimicrobiol. Ag. Chemother.* 682–686.
15. Araujo, M. L. G. C., Oliveira, R. P., Giordano, R. C., and Hokka, C. O. (1996), *Chem. Eng. Sci.* **51(11),** 2835–2840.
16. Petzold, L. (1989), Subroutine DDASSL, Computing and Mathematics Research Division, Lawrence Livermore National Laboratory, Livermore, CA.
17. Marquardt, D. W. (1963), *J. Soc. Indust. Appl. Math.* **11,** 431–441.
18. Bhat, N. and McAvoy, T. J. (1990), *Computers Chem. Eng.* **14(4/5),** 573–583.
19. Bailey, J. E. and Ollis, D. F. (1986), *Biochemical Engineering Fundamentals*, 2nd ed., McGraw-Hill, New York.
20. Bajpai, R. K. and Reuss, M. (1981), *Biotechnol. Bioeng.* **22,** 739–763.
21. Araujo, M. L. G. C. (1996), *PhD Thesis*, Chemical Engineering Department, Federal University of São Carlos, SP, Brazil.
22. Cruz, A. J. G. and Giordano, R. C. (1996), *Proceedings of the 11st Brazilian Congress of Chemical Engineering (11 COBEQ, Rio de Janeiro, RJ)*, 571–576.

Biosorption of Nickel Using Filamentous Fungi

L. MOGOLLÓN,* R. RODRÍGUEZ, W. LARROTA, N. RAMIREZ,
AND R. TORRES

Colombian Petroleum Institute, Laboratory of Biotechnology, Km 7. Vía Piedecuesta, Bucaramanga, Colombia

ABSTRACT

Nickel (Ni) uptake capability from aqueous solutions was studied in a filamentous fungi strains group of *Rhizopus* sp., *Penicillium* sp. *Aspergillus* sp., *Trichoderma* sp., *Byschoclamyss* sp., and *Mucor* sp. The metal uptake of a *Rhizopus* sp. strain, which has the highest uptake capacity, was corroborated by electron microscopy; no Ni deposits were observed on the cell wall, but rather a homogeneous accumulation was seen on the cell surface. The influence on the capacity of metal uptake by environmental parameters such as pH, temperature, time, and the interference of other ions in the solution, was also studied. Nickel accumulation by the selected strains is fast, occurring in less than 30 min, and does not require a microorganism's active metabolism to take place. Sorption isotherms were established for the selected fungi, in order to determine the maximum metal uptake capacity. The sorption isotherms were fixed to the mathematical models of Freundlich and Langmuir, obtaining better performance on the Langmuir model.

INTRODUCTION

High quantities of toxic metals are frequently associated with effluents originating from numerous industrial operations. Because of their toxicity, they must be removed by means of different physical, chemical, or biological technologies *(1,2,3,5)*. One of the best biological options is biosorption. Biosorption refers to the passive metal uptake by different forms of biomass, which may be dead or alive. Filamentous fungi may better suit this purpose than other microbial groups, because of their high tolerance toward metals, wall binding capacity, and intracellular metal uptake capabilities. Unlike intracellular metal uptake, binding of metals to cell surface components is a very fast process. Efficiency of biosorption

*Author to whom all correspondence and reprint requests should be addressed. Email: lmogollo@infantes.ccp.com

reactions is modified by environmental conditions such as temperature, pH, ionic strength of the media, and so on *(6,8)*.

The purpose of this research was to evaluate nickel (Ni) biosorption capacity by different biomass of strains of filamentous fungi, and the role of some environmental conditions in the process efficiency. Metal uptake capabilities and other biosorbent characteristics were calculated by sorption isotherms.

MATERIALS AND METHODS

Microorganisms

For a previous work *(9)*, the biosorbents used in this research were selected from a group of Colombian native strains of filamentous fungi belonging to the genus *Penicillium* sp., *Trichoderma* sp., *Rhizopus* sp., *Mucor* sp., *Byssochlamys* sp., *Paecelomyces* sp., and *Aspergillus* sp. These strains were obtained from Andes University Culture Collection (Bogotá DC, Colombia). That work resulted in the selection of four strains with potential for Ni biosorption: *Penicillium* sp BARE1, *Mucor* sp 0620, *Rhizopus* sp 0145, and *Rhizopus* sp 0101.

Culture Media and Reagents

Maintenance and biomass growth media were Malta Extract Agar and Malta Extract Broth (Oxoid, Basingstoke, UK). Metal source was a stock obtained from a Ni tritisol standard from Merck. In washing biomass and biosorption reactions, 1 M Tris-HCl buffer, pH 7.0 was used. Changes in pH were performed by adding HCl or NaOH. For electron microscopy sample fixing, phosphate buffer (pH 7.4, plus glutaraldehyde 2.5%) was used. Ethanol solutions in increasing concentrations (30, 50, 90, 95, 100% v/v) were used to remove residual water from these samples.

Biosorption Assays

Biomass Production

Each strain was inoculated in a solid medium and the slants were cultivated for 5 d (25 +/−2°C). From these cultures, a spore solution (10^6 spores/mL) was obtained and used to inoculate flacks with 400 mL of Malta Extract Broth. These media were cultivated for 5 d in a rotary shaker (150 rpm) at 25 +/−2°C.

Biomass Preparation

The harvested biomass was washed 3× in buffer (1 M Tris-HCl, pH 7.0). Residual water was removed by vacuum filtration using 0.7-μm filters, and the biomass was stored at 4°C.

Biosorption Reaction

Metal solutions were poured in 125-mL flasks with the required wet wt of biomass. The initial concentration of metal was adjusted to give different concentrations between 10 and 300 ppm. Final volume reaction was completed with 1 M Tris-HCl buffer, pH 7.0. In all experiments except the kinetics studies, exposure time was 6 h. The reactions were carried out in a rotary shaker (150 rpm) at room temperature (20 +/− 2°C).

Separation

The biosorbent material was removed by vacuum filtration using 0.7-μm filters. The metal concentration within the biomass was calculated from the difference between metal concentration before and after metal uptake. Metal concentration was determined by adsorption spectroscopy, in a Perkin-Elmer 5100 PC, (Norwalk, CT), using an acetylene-air flame. For electron microscopy, biomass samples were taken and stored in dH_2O at 4°C [12,15].

Kinetic Biosorption Assays

These experiments were carried out using biomass of the microorganisms *Rhizopus* sp 0101, *Rhizopus* sp 0145, *Mucor* sp 0620, and *Penicillium* sp BARE1. For each microorganism, flasks were disposed with 2 g of biomass (wet wt), and the final volume was adjusted with Tris buffer, pH 7.0. Initial metal concentration was 10 ppm. Reaction was finished at different exposure times: 0.5, 1, 2, 4, 6, 8, 16, and 24 h. Each assay included an unique negative control that was sampled at 24 h.

Sorption Isotherms and Metal Uptake Capacity

Sorption isotherms were carried out using *Rhizopus* sp 0101 biomass. Harvested and washed biomass was dried at 40°C, for 12 h, to get a constant weight. Biomass particle size was adjusted to 0.85–1.16 mm in diameter. Initial metal concentrations were 10, 20, 50, 100, 200, 300, and 400 ppm. A 1 g/L biomass concentration was used. Isotherms were performed at pH 3.0, 5.0, 7.0. In pH, temperature, and sodium chloride experiments, a negative control without biomass was included.

Electron Microscopy and X-ray Elemental Analysis (SEM-EDX)

Biomass samples from isotherm assays were taken for electron microscopy and X-ray elemental analysis. Samples were dried at 40°C, until constant weight, and then covered with gold for later scanning [10,11]. Analyses were carried out in a Cambridge Instruments Stereoscan 240 electron microscope with an X-ray analytic system, EDAX.

pH Influence

pH influence in Ni uptake was assayed with pH values between 1.0 and 11.0. pH-s of the reaction media (1 M Tris-HCl) were adjusted using

HCl or NaOH. In each experiment, 2 g of biomass (wet wt) was placed in 50 mL of reaction media.

Temperature Influence

The temperature values evaluated were 4, 20, 40, and 60°C. In each assay, 2 g (wet wt) of biomass were used in 50 mL of solution.

Sodium Chloride Concentration Influence

Influence of this ion concentration in Ni uptake were performed in assays which contained 1.5 g (wet wt) and 10 ppm of Ni. Reaction media were supplemented with NaCl in the following concentrations: 100, 1000, 5000, and 10,000 ppm.

RESULTS AND DISCUSSION

Kinetics Biosorption Assays

Figure 1 shows the metal removal for each exposure time, determined as the percentage of metal removed. For both *Rhizopus* sp 0101 and 0145, Ni uptake rates were the highest between 1 and 2 h of exposure. *Rhizopus* sp 0101 reached its maximum metal uptake rate after 30 min, and *Rhizopus* sp 0145 reached equilibrium after 2 h. *Penicillium* sp Bare1 and *Mucor* sp 0620 kinetics were different: Both had lower metal uptake rates than the two strains of *Rhizopus*. *Penicillium* sp Bare1 was at equilibrium after 6 h, and *Mucor* sp 0620 reached its maximum metal uptake after 24 h.

Rhizopus sp 0101 and *Rhizopus* sp 0145 kinetics were typical of the biosorption process, and were established as passive reactions nondependent on metabolism. High metal uptake rates reached in free nutrient media are characteristic of passive biosorption (15). Metabolism-dependent intracellular uptake, in which metal ions are transported into cells across the cell wall, may be slower than passives ones. This could explain the observed *Mucor* 0620 and *Penicillium* sp Bare1 kinetics (8,12,14,15).

Sorption Isotherms and Metal Uptake Capacity

Isotherm results were adjusted to the Langmuir and Freundlich models. Figure 2 shows the typical isotherm pattern obtained with all tested materials. Table 1 shows these parameters, which were evaluated according to the least-fitting method, using the experimental Cf and q values (12,15). As seen in Table 1, correlations obtained with Langmuir model were better than Freundlich ones, except for pH 6.0. The q_{10} (uptake capacity at Cf = 10 ppm) at pH 7.0, calculated from Langmuir and Freundlich models were 21 mg/g and 6.8 mg/g, respectively; experimental q_{10} was 16 mg/g. The q_{200} (uptake capacity at Cf = 200 ppm) at pH 7.0, calculated from Langmuir and Freundlich models, were 30 mg/g and 15.6 mg/g, respectively, experimental q_{200} was 41 mg/g.

Fig. 1. Kinetics of nickel biosorption. Initial metal concentration was 10 ppm. Metal uptake was expressed as removal percentage for each exposition period. The evaluated strains were *Rhizopus 0101, Rhizopus 0145, Mucor 0620, Penicilluim BARE1*.

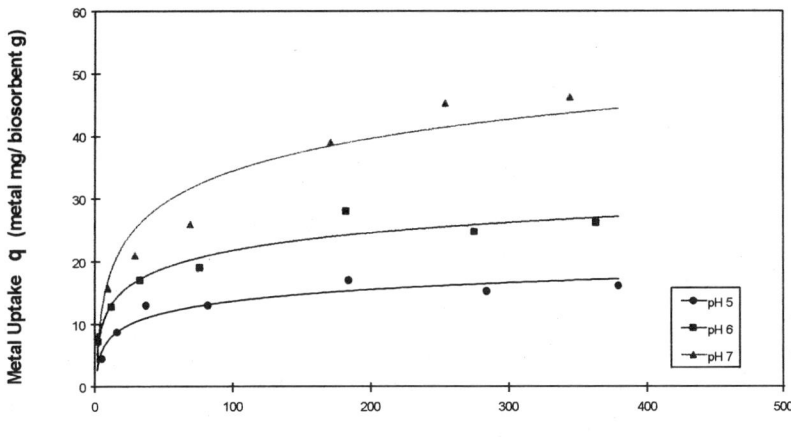

Fig. 2. Nickel Biosorpion Isotherms. Assays were performed with the strain *Rhizopus 0101*. Biomass was treated as was indicated in the text. Biomass final concentration was 1g/L dried weight.

The experimental q_{max} obtained at pH 7.0 was 45 mg/g, was greater than that reported by Fourest et al. *(17)*, who found a q_{max} of 22.2 mg/g for a strain of *Rhizopus arrhizus*. This difference may be explained by specific characteristics of the evaluated strains, although other important parameters, such as particle size, were not reported *(17)*.

Electron Microscopy and X-ray Elemental Analysis (SEM-EDX)

Electron microscopy is a powerful tool for biosorbent characterization, because it permits determination of metal accumulation by the bio-

Table 1
Nickel Biosorption Isotherm: Langmuir and Freunlich Model Parameters

pH	Langmuir model			Freundlich model		
	q_{max}	K_d	R^2	n	K	R^2
5.0	16.6	0.071	0.994	3.04	6.98	0.952
6.0	22.3	0.18	0.941	3.82	6.16	0.959
7.0	31.6	0.20	0.92	3.65	3.65	0.855

Assays were performed with the strain *Rhizopus* 0101. Biomass was treated as indicated in the text. Biomass final concentration was 1g/L dried wt.

Fig. 3. *Rhizopus sp 0101* Biosorbent Morphology. Electronic Microphotographs of *Rhizopus sp 0101* Biosorbent. **(A)** 22.7 X. **(B)** 432 X. Analysis were carried out in a Cambridge Instruments Stereoscan 240 electron microscope.

mass. EDX scan is useful for determination of metal distribution in the biosorbent *(12,13)*. Figure 3 shows the biosorbent morphology, which is an amorphous granule. The surface of the granule is irregular, with large area for metal–granule interaction. The nature of this kind of biosorbent facilitates its use in either fixed-bed or fluidized-bed reactors.

Samples from *Rhizopus* sp 0101 isotherm experiments were scanned. Nickel was found in all of them. It seems that metal deposition is homogeneous, since no precipitates were detected (Fig. 4).

Influence of pH

The pH is one of the most important factors in metal biosorption success **(4)**. pH influence was evaluated for *Rhizopus* sp 0101, *Rhizopus* sp 0145, *Mucor* sp 0620, and *Penicillium* sp Bare1. In all evaluated microorganisms, Ni biosorption is inhibited by increasing hydrogen ion concentration. With pH 5.0, reaction efficiency decreased; at lower pH, it was completely inhibited. On the other hand, neutral and alkaline pH values contributed to reaction efficiency. Decreased Ni uptake obtained at pH 3.0

Fig. 4. *Rhizopus sp 0101* Biosorbent EDX-Scan. Analysis were carried out in a Cambridge Instruments Stereoscan 240 electron microscope with an X-ray analytic system EDAX. C, O, Cl, and Ni peaks are shown.

or lower is the result of H^+ ions binding to the reactive groups in the biomass. pH values greater than 8.0 lead to ionization of amine, imidazole, and phosphate groups, increasing biomass reactivity *(4,8,17)*. pH of metal solutions were adjusted with Tris. In addition, negative controls for each pH value showed that, in the evaluated metal concentration (10 ppm), there were not interferences caused by metal precipitation. These results agree with the solubility Ni diagram, in which Ni (in concentrations greater than 8 ppm) precipitates as hydroxide at pH values higher than 9.0 *(16)*.

Influence of Temperature

For all microorganisms evaluated, the results indicate (data not shown) that the temperature does not affect Ni uptake, as reported in the literature. For instance, Brady and Duncan *(4)*, working in Cu^{+2} bioaccumulation by *S. cerevisiae*, observed that the temperature does not influence metal uptake in the range of 5 to 40°C.

Influence of Sodium Chloride on Nickel Uptake

The results obtained with these experiments are depicted in Fig. 5. High sodium chloride concentrations affected the biosoption process by decreasing metal uptake up to 20%. This effect is caused by the high ionic strength of the solution, and because sodium ions are competing with Ni for negative reactive sites in the biosorbent. This effect is very important,

Fig. 5. Sodium Chloride Effect in Biosorption Initial metal concentration was 10 ppm. Metal uptake was expressed as removal percentage for each NaCl concentration. The evaluated strains were Rhizopus 0101, Rhizopus 0145, Mucor 0620, Penicilluim BARE1.

because industrial effluents are very complex solutions with different metallic ions. These results agree with the report of Corder and Reeves (18), in which a biomass of cyanobacteria decreases its Ni uptake capacity when it was used as Ni biosorbent in media containing sodium ions (1,5,6).

ACKNOWLEDGEMENTS

This work was supported by the Colombian Petroleum Institute, ECOPETROL, Colombia. The review of the English version by Luis E. Ortiz from the Environmental Department, ICP-ECOPETROL, Colombia, is gratefully appreciated.

REFERENCES

1. Brauckmann, B. M. (1990), in *Biosorption of Heavy Metals*, Volesky, B, ed., CRC, Boca Raton, FL, pp. 52–63.
2. Hall, J. C., Raider, R. L., and Grafton, J. A. (1992), in *Water Environment and Technology*, pp. 60–63.
3. Silver, S. (1994), *Bioremediation: Scientific and Technological Issues: Tenth Forum in Microbiology*, American Society of Microbiology, Atlanta, pp. 61–66.
4. Brady, D. and Duncan, J. R. (1994), *Appl. Microbiol. Biotechnol.* **41**, 149–154.
5. Erlich, H. L. (1994), in *Biotechnology for the Treatment of Hazardous Waste*, Stoner, D. L., ed., Lewis, Boca Raton, FL, pp. 27–44.
6. Jeffers, T. H. (1991), *CA Geol.* July 1992, 155–158.
7. Gould, W. D., and McCready, R. G. L. (1994), in *Biotechnology for the Treatment of Hazardous Wastes*, Stoner, D. L. ed. Lewis, Boca Raton, FL, pp. 71–95.

8. Gadd, G. M. and White, C. (1989), in *Metal–Microbe Interaction*, Pool, R. K., ed., Oxford IRL, Oxford, UK. pp. 19–38.
9. Mogollón, L. (1996), Msc Thesis, Universidad Javeriana, Instituto Colombiano del Petróleo Bogotá, Colombia.
10. Cobaleda, G. E. and Pachón, Z. (1992), *Boletin técnico del Instituto colombiano del Petróleo*. Año 5, No. 12.
11. Cobaleda, G. E. (1995), *Boletin tecnico del Instituto colombiano del Petróleo*. Año 8, No. 3.
12. Tsezos, M. (1990), in *Microbial Mineral Recovery*, Erlich, H. L. and Brierly, C. L., eds. McGraw-Hill, New York, pp. 325–339.
13. Volesky, B. and Prastyu, I. (1994), *Biotechnol Bioeng*. **43,** 1010–1045.
14. Siegel, S. M., Galun, M., and Ziegel, B. Z. (1990). *Water Air Soil Pollu*. **53,** 335–344.
15. Volesky, B. (1990), in *Biosorption of Heavy Metals,* Volesky, B., ed., CRC, Boca Raton, FL, pp. 3–7.
16. Darimont, A. and Frenay, J. (1990), in: *Biosorption of Heavy Metals*, Volesky, B., ed., CRC, Boca Raton, FL, pp. 65–79.
17. Fourest, E., Canal, C., and Roux, J. (1994), *FEMS Microbiol. Rev*. **14,** 325–332.
18. Corder, S. L. and Reeves, M. (1994), *Appl. Biochem. Biotechnol*. **45/46** 847–859.

Conversion of Food Industrial Wastes into Bioplastics

P. H. Yu*,[1] H. Chua,[2] A. L. Huang,[1] W. Lo,[1] and G. Q. Chen[3]

[1] *Department of Applied Biology and Chemical Technology, Hong Kong Polytechnic University, Hong Kong;*
[2] *Department of Civil and Structural Engineering, Hong Kong Polytechnic University, Hong Kong; and*
[3] *Department of Biological Sciences and Biotechnology, Tsinghua University, China*

ABSTRACT

The usage of plastics in packaging and disposable products, and the generation of plastic waste, have been increasing drastically. Broader usage of biodegradable plastics in packaging and disposable products as a solution to environmental problems would heavily depend on further reduction of costs and the discovery of novel biodegradable plastics with improved properties. In the authors' laboratories, various carbohydrates in the growth media, including sucrose, lactic acid, butyric acid, valeric acid, and various combinations of butyric and valeric acids, were utilized as the carbon (c) sources for the production of bioplastics by *Alcaligenes eutrophus*. As the first step in pursuit of eventual usage of industrial food wastewater as nutrients for microorganisms to synthesize bioplastics, the authors investigated the usage of malt wastes from a beer brewery plant as the C sources for the production of bioplastics by microorganisms. Specific polymer production yield by A. Latus DSM 1124 increased to 70% polymer/cell (g/g) and 32g/L cell dry wt, using malt wastes as the C source. The results of these experiments indicated that, with the use of different types of food wastes as the C source, different polyhydroxyalkanoate copolymers could be produced with distinct polymer properties.

Index Entries: Polyhydroxyalkanoate (PHA); polyhydroxybutyrate (PHB); polyhydroxyvalerate (PHV); malt waste; Soya waste; *Alcaligenes eutrophus; Alcaligenes latus.*

---

*Author to whom all correspondence and reprint requests should be addressed. E-mail: bcpyu@hkpucc.polyu.edu.hk

INTRODUCTION

Recently, plastic usage has increased drastically *(1)*. Product-packaging plastic materials account for a large fraction of total solid-waste generation, and these are considered environmentally harmful because they are generally nonbiodegradable, or expensive for recycle usage. Of the 133 million t of solid wastes generated each year in the early 1980s, more than 30% by weight of these materials were plastics *(2)*. In Hong Kong, 9500 t of municipal solid wastes are delivered for disposal at incineration plants and landfills each day *(3)*. Approximately 11% (wet wt) of these wastes are plastics materials, including a high proportion of packaging materials and disposable products. Plastics usage and plastics-waste generation are forecast to increase at 15%/yr over the next decade *(4)*.

In the past decade, there has been much interest in the development and production of biodegradable plastics. Various biodegradable plastics have been produced by incorporating natural polymers into conventional plastics formulations, by chemical synthesis, or by microbial fermentation. In the search for a biodegradable plastic of natural or biological origin, a family of more than 40 polyhydroxyalkanoates (PHAs) and their related copolymers has been discovered, and has emerged as environmentally friendly materials. These polymers are completely biodegradable into carbon dioxide and water within a few months of burial *(5)*. Certain bacteria, such as *Escherichia coli, Clostridia* spp., and *Alcaligenes eutrophus*, produce polyhydroxybutyrate (PHB) as an intracellular metabolite and a back-up carbon (c) source when an unfavorable environment is encountered *(6)*. PHB can be synthesized intracellularly (by polymer synthase) from sugars and fatty acids by the condensation of D-3-hydroxybutyryl-CoA (catalyzed by acetoacetyl-CoA reductase), formed from acetyl-CoA via acetoacetyl-CoA (catalyzed by 3-ketothiolase) and butyryl-CoA, respectively. Microbial PHB is a biodegradable, biocompatible thermoplastic. PHB has not been commercially exploited widely because of its high price, compared with traditional thermoplastics *(7)*. Simple organic substrates, such as sucrose *(8)*, and glucose *(9)* or ethanol, propanol *(10)*, methane *(11,12)*, and so on, as C sources, and inorganic chemicals, such as ammonium or ammonia, as nitrogen (N) sources, are used in the production of PHB *(8,10)*. In a recent study, the use of microorganisms in activated sludge obtained from a wastewater treatment plant to synthesize PHAs was reported from this laboratory *(13)*. Broader usage of biodegradable plastics in packaging and disposable products as a solution to environmental problems would heavily depend on further reduction of costs and the discovery of novel biodegradable plastics with improved properties. In this study, malt refuse from a brewery was used as both C and N sources for cell growth.

MATERIALS AND METHODS

Microorganism

Alcaligenes eutrophus H16 (ATCC 17966) was purchased from American Type Culture Collection. *Alcaligenes latus* DSM 1122 and DSM 1124, gifts from George Chen of Tsinghua University, were maintained on nutrient agar slant at 4°C by monthly subculture. Mixed cultures from activated sludge were obtained from a municipal wastewater treatment plant at Shatin, Hong Kong.

Media

Preparation of Liquid Seed Medium for Fermentation

The composition of the liquid seed medium was 4 g maltose, 0.2 g K_2HPO_4, 0.4 g $(NH_4)_2SO_4$, 0.02 g $MgSO_4 \cdot 7H_2O$, 0.01 g citrate-Fe(III), 0.01 g yeast extract, 0.01 g meat peptone, and 200 mL tap water. The pH of the media was 7.0. After autoclaving and inoculating, the liquid seed media was incubated at 4g, 35°C of shaker (Forma Scientific Model 4518 Table Top Incubator Orbital Shaker) for 24 h.

Preparation of Fermentation Medium

Malt waste, mostly semisolids of spent barley and millet refuse, was obtained from a local beer brewery. Soya waste, chiefly semisolid cellular residues of soya beans (imported from the United States), was collected from a local soya milk company. The ratio of the C and N contents of the malt and soya wastes were 7:1, and 8:1, respectively, as determined by total organic carbon (TOC) and total Kjeldahl nitrogen (TKN) methods. About 300 g dry milled waste was hydrolyzed with 2500 mL 1 N HCl at 100°C for 9 h, and then the slurry was centrifuged. The malt filtrate was neutralized with NaOH to pH 7.0. In addition, 6 g K_2HPO_4, 2 g Citrate-Fe(III), and 1 g $MgSO_4 \cdot 7H_2O$ were added to the malt filtrate. Finally, the total volume of the fermentation medium was 2.4 L. The above solution was transferred to the vessel of a 5-L bioengineering fermenter. Sterilization was conducted at 121°C for 30 min.

Preparation of Nitrogen-Limited Medium

100 g sucrose, 2 g K_2HPO_4, 3g Citrate-Fe(III), and 1 g $MgSO_4 \cdot 7H_2O$ were dissolved in 300 mL water, and autoclaved for 20 min at 121°C. 100 g glucose, 2 g K_2HPO_4, 2 g $(NH_4)_2SO_4$, and 1 g $MgSO_4 \cdot 7H_2O$ were dissolved in 300 mL water, and autoclaved for 20 min at 121°C.

Fermentation

The fermentation was carried out in the computer-controlled bioengineering fermenter with the growth conditions set at DO = 20, $T = 35°C$, and pH = 7.0. The Antifoam 289 (purchased from Sigma, St. Louis, MO) was used, and 10 N NaOH was employed to adjust the pH. At each hour, 20 mL of sample were pumped out for cell dried weight, PHB, TOC, TKN analysis. Although high-density cells were obtained, the N-limited medium was fed to fermenter to increase C/N, in order to promote PHB formation in the cells.

Extraction and Precipitation of Biopolymers

After fermentation, the fermentation broth was concentrated by centrifugation at 42,110g for 25 min, washed twice, and freeze-dried. Then, 8 g of cell powder was treated with 100 mL chloroform and 100 mL of 30% sodium hypochlorite, and the mixture was agitated in a shaker at 4g at 30°C for 150 min. After the treatment, the dispersion was centrifuged at 12,282g for 10 min. The three separate phases were obtained. The upper phase was the hypochlorite solution, the middle phase contained non-PHB cell material (NPCM) and undisrupted cells, and the bottom phase was the chloroform layer, containing PHB. First, the hypochlorite solution phase was removed with a pipet, and then the chloroform phase was obtained by filtration (14,15), and further concentrated by a setup of distillation to a final volume of 20 mL. Then, the PHB material was precipitated by mixing methanol with the concentrated chloroform (methanol:chloroform, 9:1). Finally, the white precipitate was filtered by simple filtration, and then dried.

Analytical Methods

CDW Analysis

5 mL fermentation broth was centrifuged, washed with dH_2O, and then dried at 105°C for 2 h.

PHB Analysis

5 mL fermentation broth was centrifuged and washed with dH_2O. Afterward, it was mixed and vortexed with 2 mL 10% NaClO for 2 min, and then 5 mL dH_2O was added, and it was centrifuged immediately. Then, the biopolymers were washed with dH_2O and dried at 105°C for 2 h.

TOC Analysis

Adequate diluted fermentation broth supernatant was taken to measure TOC with an Astro 2000 TOC Analyzer. A 100-mg C/L sucrose solution was used as a standard. The procedure of analysis was according to APHA (4500-Norg) (16).

TKN Analysis

A 1-mL fermentation broth supernatant was analyzed with a Kjeltec Auto 1030 Analyzer. The method was according to APHA (5310C) (16).

Melting-Point (MP) Measurement

Melting point was performed by using an Electrothermal 9100 digital mp apparatus. The temperature increase rate program was set at 10°C/min.

Gas Chromatography Analysis

A 20 mg biopolymer was mixed with 1 mL chloroform and 1 mL esterification fluid (0.5 g benzoic acid, 8 mL 95–98% H_2SO_4, plus 242 mL methyl alcohol). The mixture was maintained at 100°C for 4 h. 1 mL water was added to the cooled mixture and vortexed. Standard PHB was purchased from Fluka (Buchs, Switzerland). Gas chromatographic analysis was performed on a Varian Model 3700 gas chromatograph, using a one-eight in diameter Chromosorb-WAW column with 80/100 in. mesh size, and 6 ft in length (from Supelco [Bellefonte, PA]). The recorder was a Shimadzu C-R5a Chromatopac. Nitrogen was the carrier gas, at a flow rate of 10 mL/min. The analysis started at 110°C for 3 min, whereupon the temperature was increased to 220°C at a rate of 8°C/min. After reaching 220°C, the temperature was maintained for 5 min before the analysis was terminated (17).

^1H Nuclear Magnetic Resonance (^1H NMR)

The H^1NMR analysis was carried out on a Bruker DPX-400 spectrometer using a 5-mm $^1H/^{13}C$ dual probe. ^{1}HNMR spectra were recorded at room temperature from a $CDCl_3$ solution of the extracted biopolymers with 30-degree pulse angle. Chemical shifts were referenced to the internal reference Tetramethylsilane (TMS) (18).

RESULTS

In the first part of the experiment, bioplastics were successfully biosynthesized by *A. eutrophus* H16, using organic acids as the C source. Varied concentration ratios of butyric acid and valeric acid were used as C substrates, i.e., 100% butyric acid and 0% valeric acid, mixture of 80% butyric acid and 20% valeric acid, mixture of 60% butyric acid and 40% valeric acid, mixture of 40% butyric acid and 60% valeric acid, mixture of 20% butyric acid and 80% butyric acid, and 0% butyric acid and 100% valeric acid. Analysis of the products of fermentation revealed the polymer yields and mps of the copolymers, as shown in Table 1. The polymer products may be harvested as described in Materials and Methods, and further purified by repeating the chloroform–methanol precipitation procedure. The isolated bioplastics flakes can be melted into an uniform liquid with appropriate plasticizers and additives, and molded into various types and shapes of plastic for product application, e.g., packaging films and surgical sultures and gloves.

Table 1
Polymer Yields of Bioplastics Synthesized by *Alcaligenes eutrophus* with Butyric and Valeric Acids

Sample	Carbon source $C_4:C_5$	Polymer yield (g/g)	Copolymers[a] mol ratio PHB:PHV	Melting point (°C)
1.	100:0	0.18	100:0	177.6
2.	80:20	0.41	88:12	144.0
3.	60:40	0.15	70:30	133.3
4.	40:60	0.10	65:35	127.1
5.	20:80	0.12	49:51	109.2
6.	0:100	0.06	46:54	99.2

[a] Copolymers: PHB, polyhydroxybutyrate; PHV, polyhydroxyvalerate.

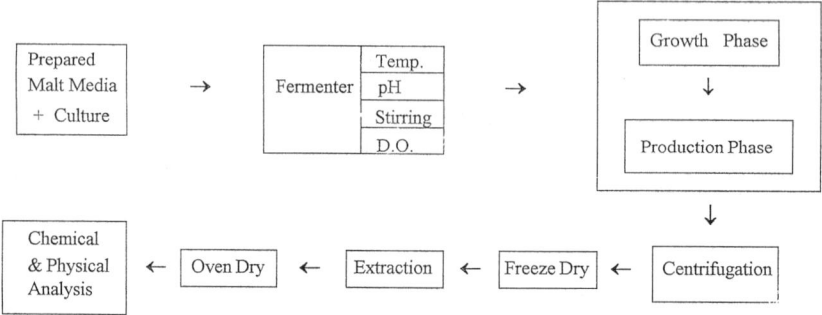

Fig. 1. Fermentation flow chart of bioplastics production.

In the second part of the experiment, a specific culture of *A. latus* DSM 1124 was selected to ferment sugar into bioplastics. Specific polymer production yield by *A. latus* DSM 1124 increased to more than 50% polymer/cell (g/g) and 30g/L cell dry wt, with increasing C/N ratio, using sucrose as the C source. The fermentation was done at 35°C for 36 h with fed-batch and dissolved oxygen controlled between 20 and 40 of calibrated units in a 5 L bioengineering fermenter.

A further experiment was conducted to investigate the conversion of malt wastes (barley), obtained from a local beer brewery, into bioplastics. A flow chart of the fermentation of the malt wastes and the isolation of the product PHAs is illustrated in Fig. 1. The results of the experiment are shown in Table 2. Specific polymer production yield by *A. latus* DSM 1124 increased to 70% polymer/cell (g/g) and 32g/L cell dry wt, using malt wastes as the C source. The data of cell growth and polymer accumulation, residual TOC and TKN in the medium during fermentation, and the changes in C/N ratio in the medium during the process of cell growth and polymer accumulation, are graphically presented in Figs. 2–4, respectively.

Food Industry Waste into Bioplastics 609

Fig. 2. Cell growth and polymer accumulation.

Fig. 3. Residual TOC and TKN in medium during fermentation.

Table 2
Production of PHB from Brewery Malt Wastes by *Alcaligenes Latus* DSM 1124

Time (h)	CDW (g/L)	Polymer (g/L)	TOC (g/L)	TKN (g/L)	C/N	Polymer/CDW (%)
0	0.26	0.00	21.45	3.86	5.56	0.00
1	0.46	0.00	21.84	3.92	5.58	0.00
2	0.34	0.00	23.81	3.96	6.02	0.00
3	0.42	0.00	20.20	3.82	5.29	0.00
4	0.38	0.00	21.85	3.77	5.80	0.00
5	0.48	0.00	21.90	3.92	5.59	0.00
7	0.46	0.00	22.52	3.87	5.81	0.00
9	0.44	0.00	21.72	3.91	5.55	0.00
10	0.54	0.00	23.14	3.91	5.92	0.00
17	0.78	0.02	21.02	3.90	5.39	2.56
21	1.20	0.06	22.66	3.81	5.95	5.00
22	1.60	0.10	23.86	3.75	6.36	6.25
24	3.38	0.86	21.20	3.55	5.98	25.44
26	6.10	2.30	18.86	3.43	5.49	37.70
29	9.60	4.70	17.02	2.90	5.86	48.96
30	10.20	5.38	16.62	2.87	5.80	52.75
32	16.68	8.50	16.46	2.75	5.99	50.96
35	21.24	10.74	21.90	2.27	9.63	50.56
41	23.10	12.42	21.38	2.19	9.76	53.77
45	28.70	17.60	25.42	2.24	11.36	61.32
46	29.28	18.98	26.46	2.23	11.89	64.82
48	30.66	21.04	25.30	2.20	11.53	68.62
50	32.34	22.80	27.00	2.23	12.10	70.50
51	32.36	22.68	26.02	2.14	12.15	70.09

Fig. 4. Changes in C/N ratio during process of cell growth and polymer accumulation.

Fig. 5. H¹ NMR spectrum of PHB produced by *Alcaligenes eutrophus* using lactic acid as the sole carbon source.

Data from the gas chromatography and ¹H NMR analysis of the biopolymers produced by fermentation of lactic acid by *A. eutrophus*, and fermentation of malt wastes by *A. latus*, are displayed in Figs. 5 and 6, respectively.

DISCUSSION

The different properties of polymers produced by fermentation of various concentration ratios of butyric and valeric acids used as C substrates are shown in Table 1. The results indicated that the higher the % of the ratio of butyrate to valerate, the higher was the polymer yield, and the mp of the extracted polymer products was also shown to be higher. The texture of the biopolymers produced from substrate of 100% of butyric acid exhibited brittleness. On the contrary, the texture of extracted polymer product fermented from substrate of 100% valeric acid appeared to be more elastic and softer than using other substrates.

The results of fermentation of biopolymers by *A. latus*, using brewery malt wastes as a C source, are shown in Table 2. During fermentation, biopolymers began to accumulate in cells after 22 h of fermentation of malt wastes, and maximized production of the biopolymers after 50 h of fermentation. The results were accorded to the scheme of permitting cell

Fig. 6. H¹ NMR spectrum of PHB produced by *Alcaligenes Latus* DSM 1124 using malt wastes as the carbon source.

growth in the initial 22 h of fermentation, using malt wastes, and maximizing biopolymer production in the second part of the fermentation, when a N-limited medium was fed to the fermenter to increase the C/N ratio in the medium. The final biomass and polymer concentration were 32.36 and 22.68 g/L dry wt, respectively, successfully yielding a production of biopolymers of 70.1% dry wt of the biomass. Since the total C consumed during fermentation (140 g C was added to a 2 L vol of medium, and 60 g residual C remained at the end of fermentation) was 80 g/2 L, and the biopolymers produced was 22.68 g/L, the mass yield of bioplastics per mass of C source consumed was calculated to be 56%. The data in Figs. 3 and 4 show that there was a gradual decrease in residual TKN, but a gradual increase of the C/N ratio (from 6 to 12) after 30 of fermentation.

In an earlier study, it was shown that higher C/N ratio (deficiency of N in the medium) would promote the production of polymers by microorganisms *(13)*. However, the fermentation of lactic acid by *A. eutrophus*, and the fermentation of malt wastes by *A. latus*, would produce only PHB, as

Table 3
Copolymer Yields of Bioplastics Biosynthesized with Various Carbon Sources

Experiment	C source	Bacteria	Copolymers[a] mol ratio PHB:PHV
1	Butyric acid	*A. eutrophus*	100:0
2	Valeric Acid	*A. eutrophus*	0:100
3	Lactic acid	*A. eutrophus*	100:0
4	Glucose	*A. latus* DSM 1124	100:0
5	Malt waste	*A. latus* DSM 1124	100:0
6	Glucose	Activated sludge[b]	55:45
7	Fructose	Activated sludge[b]	20:80
8	Malt waste	*A. latus* DSM 1122	100:0
9	Malt waste	Activated sludge[b]	92:8
10	Soya waste	Activated sludge[b]	79:21

[a] Copolymers: PHB, polyhydroxybutyric acid; PHV, polyhydroxyvaleric acid.
[b] Activated sludge, from municipal wastewater treatment plant in Shatin, Hong Kong.

shown in the data of gas chromatography and ^{1}H NMR in Figs. 5 and 6. The shapes of the mass ion peaks looked different in the NMR spectra, because of the decoupling condition of the peaks in one spectrum (Fig. 6) during the process of analysis; the results of both spectra showed only one type of polymer was present, i.e., PHB.

In another study, the authors observed that the fermentation of glucose and fructose as C source by activated sludge, and the fermentation of soya wastes as C source by *A. latus* DSM 1124, would produce PHB–polyhydroxyvolerate (PHV) copolymers. The specific effects of the types of substrates and microorganisms on the copolymer composition (e.g., PHB–PHV) of biopolymers produced, as observed in the laboratory, are summarized in Table 3. Utilizing malt wastes as a C source, fermentation results showed that biopolymers of copolymer composition of 100% PHB, and 92% PHB and 8% PHV, could be synthesized by *A. latus* DSM 1124, and mixed cultures from activated sludge, respectively. On the other hand, mixed cultures from activated sludge could utilize soya waste to produce biopolymers of copolymer composition of 79% PHB and 21% PHV (endowed with physical and thermoplastics properties different than the other bioplastics produced). Thus, specific biopolymers with copolymers of desirable physical and mechanical properties (such as flexibility, tensile strength, and melting viscosity) can be formulated in fermentation by appropriate selection of substrates (or combination of substrates), and the type of PHA-producing microorganisms. The usage of less valuable food wastes as C source in fermentation would tremendously reduce the cost of the production of bioplastics and minimize waste production, and at the same time produce environmentally friendly bioplastics.

ACKNOWLEDGMENTS

The authors wish to express their sincere gratitude to the University Research Grant Council of Hong Kong (POLYU27/96P) for the support of this research.

REFERENCES

1. Jacob, M. T. (1993), *Biocycle*, **34**, 30.
2. Johnson, R. (1988), *J. Plastic Film Sheeting* **4**, 155–170.
3. Hong Kong Environmental Protection Department (1994), in *Environment Hong Kong 1994*, Hong Kong Government Press, pp. 51–66.
4. Hong Kong Government Industry Department (1993), in *Hong Kong's Manufacturing Industries 1993*, Hong Kong Government Press, p. 85.
5. Yu, P. and Giuliany, B. (1989), *Frontiers*, **9**, 4–7.
6. Linko, S., Vaheri, H. and Seppala, J. (1993), *Enzyme Microb. Technol.* **15**, 401–406.
7. Lee, S. Y. (1996), *Trends Biotechnol.* **14**, 431–438.
8. Yamane T., Fukunagga, M., and Lee, Y. W. (1996), *Biotechnol. Bioeng.* **50**, 197–202.
9. Kim B. S. and Chang, H. S. (1995), *Biotechnol. Techniques*, **9**, 311–314.
10. Alderte, J. E., Karl, D. W., and Park, C. H. (1993), *Biotechnol. Prog.* **9**, 520–525.
11. Asenjo J. A. and Suk, J. S. (1986), *J. Ferment. Technol.* **64**, 271–278.
12. Shah N. N., Hanna, M. L., and Taylor, R. T. (1996), *Biotechnol. Bioeng.* **49**, 161–171.
13. Yu, P. H., Chua, H., and Ho, L. Y. (1996), Proceedings of the Asia-Pacific Conference on Sustainable Energy and Environmental Technology, pp. 623–630.
14. Hahn S. K., Chang, Y. K., Kim, B. S., and Chang, H. N. (1994), *Biotechnol. Bioeng* **44**, 256–261.
15. Roh, K. S., Yeom, S. H., and Yoo, Y. J. (1995), *Biotechnol. Techniques* **9**, 709–712.
16. Greenberg, A. E., Clesceri, L. S., and Eton, A. D. (1992), Standard Methods for the Examination of Water and Wastewater, 18th ed., Apha, Washington, D. C.
17. Jan, S., Roblot, C., Goethals, G., Courtois, J., Courtois, B., Saucedo, J. E. N., Seguin, J. P., and Barbotin, J. N. (1995), *Anal Biochem.* **225**, 258–263.
18. Jan S., Roblot, C., Courtois, J., Courtois, B., Barbotin, J. N., and Seguin, J. P. (1996), *Enzyme Microb. Technol.* **18**, 195–201.

Improved Oxygen Delivery in a Continuous-Roller-Bottle Reactor

R. Eric Berson, Trupti V. Mane, C. Kurt Svihla, and Thomas R. Hanley*

Department of Chemical Engineering, University of Louisville, Louisville, Kentucky 40292

ABSTRACT

Variations to the original aeration system in a continuous roller bottle reactor of novel design have been tested and compared for optimal oxygen (O) delivery. Reactor operating parameters that affect O transfer are rotation rate, liquid-volume level, fresh-feed rate, and supplementary-aeration rate. Design modifications to enhance gas–liquid O transfer include the addition of wall baffles and center baffles. The number and location of each of these baffles are compared for their effect on $k_L a$ values in the reaction chamber. The liquid feed into the system has been modified to improve the axial liquid mixing and O transfer.

Index Entries: Roller bottle reactor; continuous reactor; gas–liquid mixing.

INTRODUCTION

The capability of a cell culture to propagate can be limited if adequate oxygen (O) is not supplied efficiently throughout the system. Therefore, it is important to control the aeration and agitation system to prevent O from becoming a growth-limiting factor. Live plant or animal cells in culture are easily damaged, though, as a result of hydrodynamic stress caused by the aeration and/or agitation system used for supplying O. Hybridomas are reported to be sensitive to shear stresses beyond 1–5 N/m^2 *(1–3)*. Both O supply and hydrodynamic stress need to be considered when operating a bioreactor with cultures at high density.

Conventional roller bottles are used to grow plant or animal cells that may be too fragile to survive in the high-shear environment prevailing in conventional reactors used for submerged culture. The roller bottle may also be advantageous for growing attachment-dependent cells, since it can

* Author to whom all correspondence and reprint requests should be addressed.

provide a suitable solid surface for attachment. Conventional roller bottles are operated in batch: That is, they are filled initially with media and inoculated, and are then harvested after some set period of operation. The continuous-perfusion roller bottle reactor is designed to provide for the continuous supply of nutrients in a novel, mechanically simple manner. Continuous supply of medium, coupled with continuous or intermittent cell or product harvest, should result in a longer production phase, and in more efficient use of cells and medium. Media costs in cell culture can be extraordinarily high, and any means by which production can be extended, or media usage decreased, has the potential to offer large economic benefits.

The continuous-perfusion roller bottle reactor has already been used in the production of taxol from *Taxus* nodule culture, production of microtubers using *Solanum tuberosum*, and the growth of human carcinoma (HeLa) cells. These encouraging initial results have led to planned applications, including the production of rotavirus using epithelial cell lines, the production of differentiated colon carcinoma cells, the growth of *Francisella tularensis* (an intracellular bacterial pathogen), and other industrially important bioprocesses. One factor that has unlimited potential for development of new applications is the novelty of the design, which presents several features not found in conventional roller bottle reactors. The advantages offered by the continuous roller bottle design are primarily those that involve the supply of O and liquid nutrients to the cells. This work has therefore concentrated on establishing baseline data and quantitative models for the performance of the continuous roller bottle reactor from an engineering viewpoint, including the transfer of O from gas to liquid phase, and the supply and dispersion of liquid nutrients throughout the reaction chamber.

METHODS

Figure 1 shows the flow diagram for the various streams in the continuous perfusion roller bottle design. Salient features include the rotating drum wrapped with tubing (or spiroid), which pumps gas and liquid continuously from the fresh medium chamber to the reactor chamber; the sample loop spiroid, which provides a mean of monitoring the vessel contents, as well as enhancing the axial mixing and gas–liquid oxygen transfer in the chamber; and the outlet chamber, for harvesting product and cells. The sample loop spiroid picks up alternating plugs of gas and liquid (as it repeatedly passes through the liquid, then the headspace) from the near end of the vessel chamber, and pumps it through the external piping of the sample loop and the long pipe running through the center of the reactor, before depositing it at the far end of the chamber. The use of rotating spiroids to pump gas and liquid is mechanically simple, because the reactor needs only one moving part to turn both the reaction chamber and both

Fig. 1. Flow diagram of continuous roller bottle reactor.

spiroids. As the cells are pumped through the sample loop spiroid, low shear forces are maintained, which should pose no danger to fragile cell walls. The fresh medium chamber spiroid does deliver both gas and liquid to the cells in the reactor chamber, but supplemental gas can be introduced if O demand or removal of product gas requires higher gas throughputs than the spiroid can deliver. As shown in Fig. 1, the sample loop spiroid picks up gas and liquid from the near end of the vessel chamber and pumps it through the external piping of the sample loop and the long pipe running through the center of the reactor, before depositing it at the far end of the chamber. This loop provides a means for sampling the vessel contents, and, in addition, enhances the axial circulation of fluid in the vessel chamber.

The roller bottle chamber has a length of 0.610 m, a diameter of 0.0914 m, and a full-scale capacity of 4 L. Tests were conducted, using tap water and a liquid volume of 1.0 L, at nominal rotation rates of 2.0, 4.0, and 8.0 rpm. Each baffle has a length of 0.584 m and a width of 0.0254 m. If tests were taken with two baffles, the baffles were installed with 180-degree spacing. If three baffles were used, the baffles were installed with 120-degree spacing.

Modeling and Measuring Axial Dispersion in the Roller Bottle Reactor

These initial studies are concerned primarily with dispersion of the reactor feed, including nutrients from the feed inlet port, or the distribution of O-enriched liquid from the sample recirculation loop. Since complete dispersion is typically achieved within about 2–3 residence times (based on the recirculation loop flow rate), the inlet and outlet flows can be neglected. The recirculation flow rate is over 10× the input and output flows at the highest rotational speed. The tests are conducted by injecting a 5–10-mL pulse of fluid containing indigo disulfonate tracer (approx 5×10^{-3} M) into the sample loop at the point where the loop re-enters the reactor. Samples are taken by syringe from the same port where the sample was introduced, with sampling continued until there is no appreciable variation in the tracer concentration. Tracer concentration is detected by absorbance at 610 nm using a UV/VIS spectrophotometer. The absorbance readings are normalized by dividing by the absorbance reading after the tracer has become uniformly dispersed.

If the level in the reactor is below that of the central tube assembly, the inlet liquid sometimes drips onto the liquid surface over a distance of a few centimeters. To treat this sort of initial dispersion, the following approach was developed. The region at the far end of the reactor chamber is assumed to consist of a well-mixed zone, into which the inlet pulse of tracer is uniformly dispersed. The size of this region is estimated by visually observing the tracer introduction for each test. The rest of the reactor is assumed to be described by a dispersion model. When developing the differential equation describing the well-mixed entry region, it is important to include the term for dispersion at the edge of the boundary with the dispersion region. The boundary conditions for the dispersion region then consist of the appropriate terms for the well-mixed zone at the entry end, and a zero gradient condition at the far end of the reactor, where the sample loop pick-up is located. Flow enters the well-mixed region from the sample loop return line at a rate given by the measured recirculation loop flow rate. The initial conditions for the simulation are specified by assuming that the tracer starts by being uniformly dispersed in the well-mixed entry region, with zero initial concentration throughout the dispersion region.

Well-Mixed Entry Region:

$$dC_E/dt = F/V_E \, (C_L - C_E) + D_e A/V_E \, dC/dz|_{z = L_1} \qquad (1)$$

Initial condition:

$$C_E(t = 0) = C_{E0} \qquad (2)$$

Dispersion Region:

$$\partial C/\partial t = D_e \, \partial^2 C/\partial z^2 - u \, \partial C/\partial z \qquad (3)$$

Boundary conditions:

$$\text{At } z = L_1: \quad C(t, L_1) = C_E \tag{4}$$

$$\text{At } z = L: \quad \partial C/\partial z = 0 \tag{5}$$

Initial condition:

$$C(0, z) = 0 \tag{6}$$

Eq. 3 is solved using a finite difference approach with N equally spaced points. The resulting set of N ordinary differential equations is solved using the IMSL routine DIVPAG, which uses Gear's method, since initial value problems of this type tend to be stiff. The amount of tracer specified in the simulation is adjusted, so that the final concentration is 1. This technique provides a convenient basis for a dimensionless concentration, and allows the validity of the converged solution to be easily checked. In preliminary simulations, the number of points used in the simulation was set to 400, for which acceptable accuracy was achieved within a reasonable computation time. A typical result of the axial dispersion modeling is shown in Fig. 2, which shows both the experimental and simulated tracer response curves for a given set of conditions.

The reactor was modified to allow for multiple liquid inlet locations, as shown in Fig. 3. The modification was made to enhance axial liquid dispersion. Tests to measure the dispersion coefficient were conducted for one, two, and three wall baffles, using this modification.

Modeling Gas–Liquid Oxygen Transfer in the Roller Bottle Reactor

The sample loop increases the rate of O transfer to the liquid in the chamber, and, as a consequence, creates an axial gradient of dissolved O concentration in the vessel chamber. Therefore, in order to analyze tests conducted with the sample loop in operation, it is necessary to account for the presence of the axial concentration gradient, as well as the additional O transfer that takes place in the sample loop. The dispersion model presented to model the dispersion of a liquid-phase tracer can easily be adapted for this purpose. In the modified model, terms to account for the transfer of O from gas to liquid phase are simply added to the differential equations describing the entry region and the dispersion region. The balance on the well-mixed entry region is also modified by changing the concentration of the incoming liquid from the sample loop from C_L to C_{IN}, to reflect the additional O transfer that occurs within the sample loop. Previously, flow through the sample loop had been treated simply as a dynamic lag, which had been neglected, based on the relative time-scales of the tracer test and flow through the loop. A mass balance is then written to account for O transfer within the sample loop itself, based on the assumption that flow through the loop takes place in plug flow. The modified form of the model, which incorporates gas–liquid O transfer, is shown below.

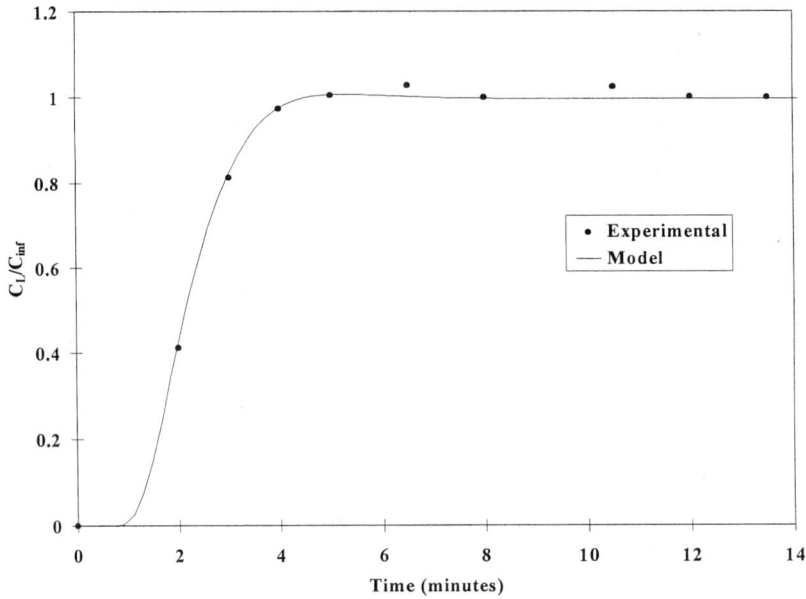

Fig. 2. Fit of dispersion model to experimental data (2 wall baffles, V = 1.0 L, 8 rpm).

Fig. 3. Modified liquid return location.

Well-Mixed Entry Region

$$dC_E/dt = F/V_E (C_{IN} - C_E) + D_e A/V_E \, dC/dz|_{z=L_1} + k_L a (\alpha C_G - C_E) \tag{7}$$

Initial condition:

$$C_E(t = 0) = C_{E0} \tag{8}$$

Dispersion Region:

$$\partial C/\partial t = D_e \, \partial^2 C/\partial z^2 - u \, \partial C/\partial z + k_L a (\alpha C_G - C) \tag{9}$$

Boundary conditions:

$$\text{At } z = L_1: C(t, L_1) = C_E \tag{10}$$

$$\text{At } z = L: \partial C/\partial z = 0 \tag{11}$$

Initial condition:

$$C(0, z) = 0 \tag{12}$$

Sample Loop Region:

$$C_{IN} = \alpha C_G - (\alpha C_G - C_L) e^{-(k_L a)_{SL} t_R} \tag{13}$$

Since the value of αC_G can be determined by noting the value of C_∞ and the residence time of fluid in the sample loop, t_R can determined by visual observation, the only two adjustable parameters in the model are the volumetric mass-transfer coefficients in the vessel and in the sample loop ($k_L a$ and $(k_L a)_{SL}$, respectively). Dissolved O concentration can be measured both in the sample loop and within the vessel, in order to facilitate the determination of the two different dynamic terms. The partial differential equations describing the simultaneous processes of axial dispersion and gas–liquid mass transfer are solved using the same approach used for the analysis of liquid mixing.

Measurement of Gas–Liquid Mass-Transfer Coefficients

Tests were initially conducted in batch, with the sample loop inlet and exit ports capped. The axial concentration profiles modeled using the dispersion equation exist only when the sample loop is functioning; therefore, analysis of the experimental data is simplified considerably when the sample loop is disabled. The balance reduces to:

$$\partial C/\partial t = k_L a (\alpha C_G - C) \tag{14}$$

Volumetric gas–liquid mass-transfer coefficients, $k_L a$, were measured

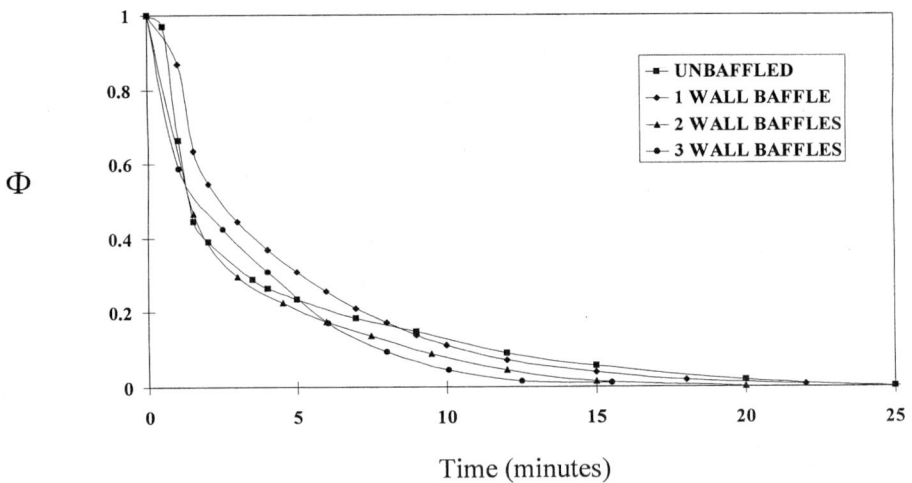

Fig. 4. Dissolved oxygen response curves (V = 1.0 L, 8 rpm).

using a dynamic technique. With the vessel held stationary, the gas headspace was sparged with O. Rotation at the desired rate was then initiated, and the resulting dissolved O response monitored until a new steady state was reached. A dissolved O electrode was inserted into the reactor chamber through the vessel inoculation port and periodically monitored until the liquid phase reached saturation. The data were fitted to a first-order response, with the time-constant from the fit taken as $1/k_L a$. R^2 values ranged from 0.906 to 0.995. The $k_L a$ values obtained in the dynamic tests are nearly linear with rotation rate, which is consistent with data reported for unbaffled horizontal rotating vessels (4,5).

Further tests were conducted with baffles installed either at the wall or in the center of the tank. Center baffles were attached to the gas and liquid inlet rods running through the center of the reaction chamber. Tests were conducted with one, two, or three baffles, and compared for $k_L a$ and effective dispersion coefficient, D_e, values.

In order to assess the significance of the O transfer in the sample loop to the overall rate of supply to the chamber, tests were also conducted with the sample loop in operation. The tests were conducted as described previously, except that the O electrode was inserted into the external piping of the sample loop, and monitored until the reading reached saturation.

RESULTS

Figure 4 shows the dissolved O response as a function of time for three wall baffle conditions. Initial concentrations are normalized at a value of 1 with no O yet transferred from the gas to the liquid. As the normalized concentration approaches zero, the concentration of O in the liquid phase

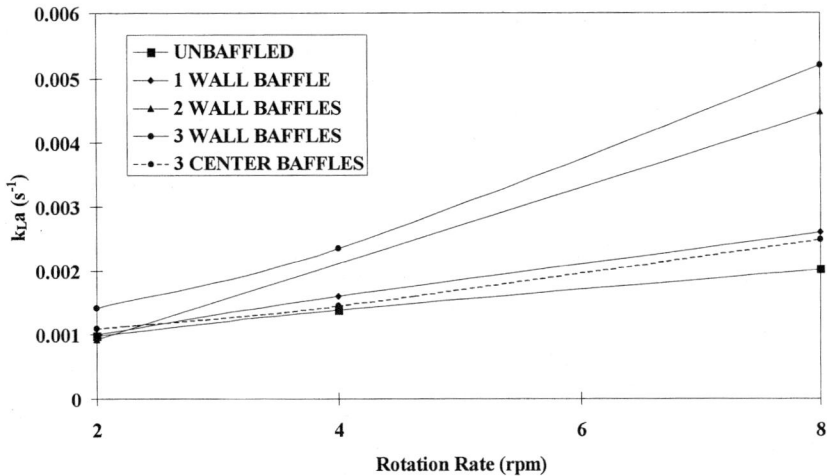

Fig. 5. Effect of baffles on $k_L a$ in the chamber (V = 1.0 L).

Fig. 6. Comparison of $k_L a$ in chamber and sampling loop (V = 1.0 L).

approaches equilibrium. As the number of baffles increase, the O concentration in the liquid reaches saturation more quickly.

Figure 5 shows how the number of baffles affects $k_L a$. As the number of wall baffles increase, $k_L a$ increases. The effect is more pronounced at higher rotation rates. The results also indicate that wall baffles are more effective than center baffles. Three center baffles resulted in a lower $k_L a$ than a single wall baffle, for the rotation rates studied.

Figure 6 shows $k_L a$ values in the sampling loop. Note the enhancement of O transfer here, compared to in the reaction chamber.

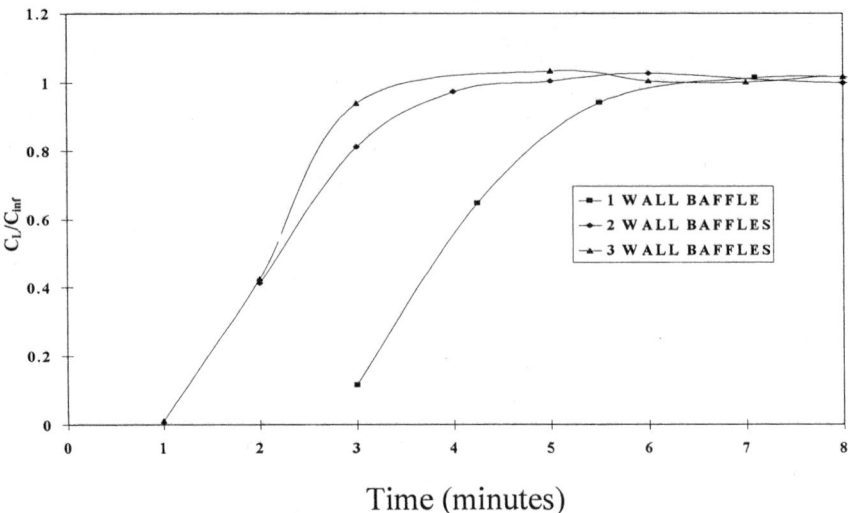

Fig. 7. Effect of baffles on tracer dispersion in the chamber (V = 1.0 L, 8 rpm).

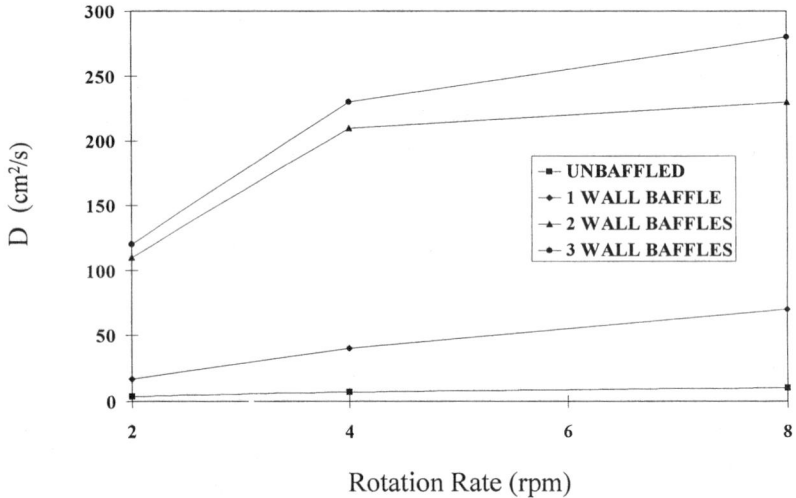

Fig. 8. Effect of wall baffles on liquid dispersion in the chamber (V = 1.0 L).

Figure 7 shows the effect of baffles on tracer dispersion in the reaction chamber. Tracer concentration is normalized by dividing by final tracer concentration. Increasing the number of wall baffles results in a quicker rise to equilibrium concentration.

Figure 8 shows the effect of the addition of wall baffles on liquid dispersion. Increasing the number of wall baffles results in higher values of D_e.

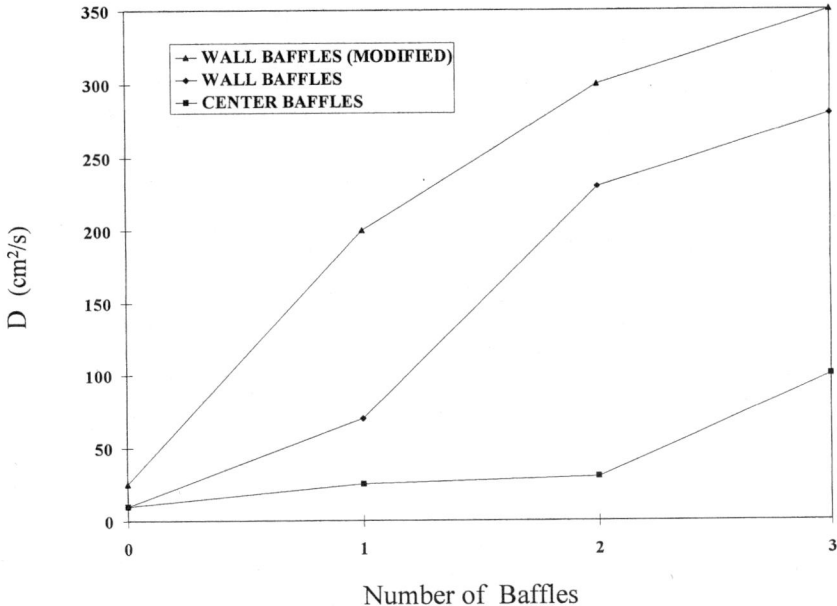

Fig. 9. Effect of baffle location and reactor modification on liquid dispersion (V = 1.0 L, 8 rpm).

Figure 9 compares D_e values for various reactor configurations. Results for center baffles, wall baffles, and wall baffles with modified liquid return inlet locations are shown. The tests with the modified liquid return inlet locations resulted in the greatest values of D_e.

DISCUSSION

The sampling loop provides a means for sampling the vessel contents, and, in addition, enhances the axial circulation of fluid in the vessel chamber. As the gas and liquid flow through the turns of the spiroid and the piping of the sample loop, the two phases are brought into close and vigorous contact. This contact results in a relatively high value of $k_L a$. The concentration of dissolved O in the liquid phase flowing through the sample loop increases substantially over the prevailing value in the reactor. Thus, in terms of monitoring the dissolved O concentration in a dynamic test, the sampling loop is not truly a sampling loop, since its dissolved concentration is greater than the average value in the reactor. The sampling loop also represents a region of enhanced gas–liquid O transfer that is not present in a conventional roller bottle.

Although $k_L a$ in the sampling loop can be $9 \times$ the $k_L a$ in the chamber, the residence time in the loop is much smaller than the residence time in the chamber. For an unbaffled tank operating at 8 rpm, the residence time in the sampling loop is 35 s but the residence time in the chamber is over 6 min. Therefore, it is important to improve O transfer in the chamber.

Baffles enhance O transfer by mixing the gas–liquid interface and increasing mass transfer, compared to transport by passive diffusion. The baffles also work to move water with dissolved gas away from the surface, increasing the concentration gradient. As the baffles move from the liquid to the air, some water droplets remain on the baffle, further expediting O transfer. Figure 5 shows that the effect of the baffles is greater at higher rotation rates. The baffles do not transfer enough energy to the liquid at 2 rpm to result in a significant effect on O transfer.

The three-tube assembly, which passes through the center of the reactor, affects the axial mixing to some extent, if the liquid level in the reactor is high enough for the tubes to be in contact with the fluid. The lowest projecting tube just touches the liquid surface at a liquid volume of approx 1.5 L. For lesser volumes, dispersion occurs relatively slowly. A pulse of tracer might be visibly dispersed throughout the reactor within 2–3 time constants, with the time constant in this case defined as the sample loop flow rate divided by the overall liquid volume. If the liquid volume in the reactor is for example 2.0 L or greater, the tracer is dispersed more rapidly, since the periodic contact of the tube assembly with the liquid surface as the vessel rotates increases the speed and extent of axial mixing in the chamber.

Baffles improve axial liquid mixing by increasing bulk flow. Although the baffles are horizontal and rotate in the radial direction, the liquid is pushed toward the surface by wall baffles (pushed toward the wall by center baffles), and the momentum is carried axially after hitting the barrier (surface or wall). Examination of Fig. 8 shows a larger increase in D_e between one and two baffles than the increase between zero and one baffle or the increase between two and three baffles. With a single baffle in operation, the baffle is only submerged throughout half of each rotation. With two or more baffles in operation, there will always be at least one baffle submerged and, hence, the difference in D_e.

Modifying the location of the liquid inlet returning from the sampling loop further increases D_e. Instead of a single entry point at one end of the chamber, the liquid enters at multiple points, thereby effectively hastening the diffusion process.

ACKNOWLEDGMENTS

The authors would like to thank the US Bioreactor company for support of this project.

NOMENCLATURE

A   Area at surface of well mixed entry region, cm^2
C_E Concentration in liquid at well mixed entry point of reactor, mol/L

C_{IN} Concentration in liquid entering reactor from sampling loop, mol/L
C_L Concentration in liquid, mol/L
C_{inf} Concentration at time = infinity
C_o Concentration at time = zero
D_e Axial dispersion coefficient, cm^2/s
F_F Volumetric flow rate through fresh feed spiroid, mL/s
F_S Volumetric flow rate through sampling spiroid, mL/s
k_La Volumetric gas–liquid mass transfer coefficient in reactor chamber, s^{-1}
k_La_{SL} Volumetric gas–liquid mass transfer coefficient in sampling loop, s^{-1}
L Reactor length, cm
t Time, s
t_R Residence time in sampling loop, s
u Velocity in sampling loop, flow rate divided by cross-sectional area, cm/s
V_L Liquid volume of reactor, mL^3
V_E Liquid volume at well-mixed entry point of reactor, mL^3
αC_G Dissolved oxygen concentration in liquid at saturation, mol/L
ϕ Dimensionless concentration, $C - C_{inf}/C_0 - C_{inf}$

REFERENCES

1. Abu-Reesh, I. and Kargi, F. (1989), *J. Biotechnol.* **9,** 167–178.
2. Petersen, J. F., McIntyre, L. V., and Papoutsakis, E. T. (1988), *J. Biotechnol.* **7,** 229–246.
3. Schurch, U., Karamer, H., Einsele, A., Widmere, F., and Eppenberger, H. M. (1988), *J. Biotechnol.* **7,** 179–184.
4. Phillips, K. L., Sallans, H. R., and J. F. T. Spencer (1961), *Ind. Eng. Chem.* **53,** 749–754.
5. Tanaka, H. (1982), *Biotechnol. Bioeng.* **24,** 425–442.

Xylanase Recovery

Effect of Extraction Conditions on the Aqueous Two-Phase System Using Experimental Design

SILGIA A. COSTA, ADALBERTO PESSOA JR,* AND INÊS C. ROBERTO

Department of Biotechnology/FAENQUIL, Rod. Itajubá-Lorena, Km 74,5. 12.600-000, Lorena/SP, Brazil

ABSTRACT

The partitioning of xylanase produced by *Penicillium janthinellum* in aqueous two-phase systems (ATPS) using poly(ethylene glycol) (PEG) and phosphate (K_2HPO_4/KH_2PO_4) was studied employing a statistical experimental design. The aim was to identify the key factors governing xylanase partitioning. The interactions of five factors (PEG concentration molecular weight, concentration of buffer K_2HPO_4/KH_2PO_4, pH, and NaCl concentration) and their main effects on the partition coefficient (K) were evaluated by means of a 2^5 full-factorial experimental design with four center points. The %PEG, %NaCl, and pH were the most important factors affecting the response variable (K). Response surface methodology (RSM) was adopted and an empirical second-order polynomial model was constructed on the basis of the results. The optimum partition conditions were pH 7.0, PEG = 8.83% and NaCl = 6.02%. Adequacy of the model for predicting optimum response value was tested under these conditions. The experimental xylanase partition coefficient (K) was 2.21, whereas its value predicted by the model was 2.33. These results indicate that the predicted model was adequate for the process. PEG molecular weight and phosphate concentration did not affect the xylanase partition coefficient.

Index Entries: Aqueous two-phase system; xylanase; *Penicillium janthinellum*; experimental design.

INTRODUCTION

Enzymes involved in lignocellulosic degradation *in situ* are of interest for their potential application to processes utilizing lignocellulosic sub-

*Author to whom all correspondence and reprint requests should be addressed.

strates. Xylanases and xylanase-producing microorganisms can potentially be applied to the production of hydrolysates from agro-industrial wastes *(1)*, to nutritional improvement of lignocellulosic feeds *(2)*, and to the processing of food *(3)*, agrofiber *(4)*, and pulp *(5)*. Efforts have been made by the pulp and paper industries to reduce the amount of chlorine needed for bleaching. Studies have been conducted on the effluent treatment and on the effectiveness of less toxic bleaching agents *(6)*. The residual lignin removal by enzymatic method is actually a very interesting research area *(7,8)*. According to Durán et al. *(9)*, xylanases produced by *Penicillium janthinellum* can be used for reducing the chlorine charge in *Eucalyptus*-pulp bleaching with a simultaneous gain in brightness. The xylanase production has been investigated *(10)*, but studies on downstream processes are still needed.

Extractive bioconversion using aqueous two-phase systems (ATPS) seems to be a very attractive method for the integration of fermentation and downstream processing of extracellular proteins *(11)*. In addition to forming a relatively mild environment suitable for cell and protein extraction, ATPS can be easily scaled up and continuously processed *(12–14)*.

The aim of this work was to identify the key factors governing xylanase partitioning using ATPS. Differences in enzyme partitioning can be ascribed to the interaction of the factors inherent in the system itself (such as choice of system components, polymer molecular weight, concentration of polymers and salts, ionic strength, and pH) with those of the target protein (such as hydrophobicity, charge, and molecular weight) *(12)*. The effects of variables like PEG molecular weight, PEG concentration, pH, and phosphate and NaCl concentration on partition coefficient (K) of xylanase were studied. Full factorial experimental designs were employed since they require a reduced number of experiments and help to identify important and interacting factors determining the enzyme partition *(15)*. Response surface methodology was used for optimizing xylanase partition coefficient (K) by ATPS and a statistical model correlating the variables was obtained.

MATERIALS AND METHODS

Microorganism and Cultivation

The microorganism *P. janthinellum*, isolated from decaying wood by Milagres *(16)*, was identified by the Biosystematic Research Center of Canada (Ottawa, Ontario, Canada) and deposited in their collection under the designation of CRC 87M-115. The strain was initially maintained in silica stocks and later on agar slants. The spore inocula were obtained after cultivation at 30°C for 5 d in medium containing 1% glucose, 0.1% yeast extract, 2% (v/v) concentrated salts solution based on Vogel's medium *(17)*, and 2% agar-agar. The final concentration of spores was 10^5 mL^{-1}. The cultivation medium for enzyme production was composed of sugarcane bagasse hemicellulosic hydrolysate (800 g of dry milled bagasse mixed with 8 L 0.25% H_2SO_4 and autoclaved for 45 min at 121°C), supplemented with 2% (v/v) concentrated salts solution based on Vogel's medium and 0.1% yeast extract. The medium was then autoclaved for 15 min at 121°C. Shake-flask cultures were grown in Erlenmeyer flasks (125 mL) containing 25 mL of medium. Standard cultivation conditions were: temperature 30°C; initial pH 5.5 (uncontrolled); cultivation time 96 h. The xylanase activity after this time was 876 nanokatals/mL^{-1}.

Determination of Enzyme Activities

Xylanase activities were determined by incubating 0.5 mL of diluted culture filtrate with 0.5 mL of a "Birchwood" xylan suspension (10 g/L^{-1}) in 0.05 M phosphate buffer (pH 5.5) for 5 min at 50°C. The released reducing equivalents were measured by a colorimetric assay *(18)* using xylose solution as a standard reference. Activity units were expressed as micromoles of reducing equivalents released per min PEG concentration level influenced this activity. The three concentration levels tested—10, 22.5, and 35%—enhanced the enzyme activity by 2.6, 10, and 18%, respectively.

Preparation of Phase Systems

Phase systems were prepared from PEG, phosphate (KH_2PO_4/K_2HPO_4), and NaCl in solid form. Three milliliters of medium containing xylanase was added to the systems and deionized water was used to adjust the desired final concentrations of the components. By varying the proportion between KH_2PO_4 and K_2HPO_4, the pH of the system was adjusted. Centrifugation (2500g for 10 min) was used after thorough vortex-mixing of the system components; the phase volumes were measured using graduated centrifuge tubes. Samples of the top and bottom phases were then assayed for enzyme activity. During all partition experiments, the temperature was ~25°C.

Experimental Designs and Statistical Analysis

To quantify the partition coefficient (K), the fraction of enzymes present in the lower and upper phases after phase separation was used. This fraction (y), which is the response factor, was measured as a function of pH (X_1), PEG MW (X_2), PEG concentration (X_3), phosphate concentration (X_4), and NaCl concentration (X_5). For each of the five factors, high (coded value: +1) and low (coded value: −1) set points were selected (Table 1). ATPSs representing all 32 (2^5), set-point combinations were made, as well as an ATPS representing the center point in which the value of all factors was in between (coded value: 0). All factors were measured twice, whereas the center point was measured times.

Table 1
Factors and Levels in the Five-Factor, Three-Level Response Surface Design
Used for Optimizing the Xylanase Partition Coefficient by ATPS

Run number	Factors	Inferior level (−1)	Superior level (+)	Center point (0)
X_1	pH	5.0	8.0	6.5
X_2	MW PEG	600	6000	4000
X_3	% PEG	10	35	22.5
X_4	%Phosphate	10	25	17.5
X_5	%NaCl	0	10	5

After statistical analysis (Tables 2 and 3) the optimization of xylanase partition coefficient (K) was achieved by three independent process variables using a 2^3-factorial experimental design with six star points (α = 1.41) and four replicates at the center point (Table 4), according to the method of Box et al. *(15)*. These independent variables (pH, %PEG, and %NaCl) acquired new values and are coded as X_1, X_3, and X_5 in the following equations:

$$X_1 = (\text{pH} - 7.4)/0.5 \quad (1)$$

$$X_3 = (\%\text{PEG} - 13.0)/5 \quad (2)$$

$$X_5 = (\%\text{NaCl} - 5.6)/0.3 \quad (3)$$

where pH, %PEG and %NaCl are true values and X_1, X_3, and X_5 are coded values.

A statistical examination of the results and a response surface study were carried out using the STATGRAPH 6.0 statistical program package. The polynomial model employed was of the form:

$$y = \beta_0 + \beta_1 X_1 + \beta_2 X_2 + \beta_3 X_3 + \beta_{1,2} X_1 X_2 + \beta_{1,3} X_1 X_3 + \beta_{2,3} X_2 X_3 \\ + \beta_{1,1} X_1^2 + \beta_{2,2} X_2^2 + \beta_{3,3} X_3^2 \quad (4)$$

where:
β_0 = constant, $\beta_{1,2}$, $\beta_{1,3}$, $\beta_{2,3}$ = cross product coefficients, β_1 β_2 β_3 = linear coefficients, X_1, X_2, X_3, = coded independent variables, and $\beta_{1,1}$, $\beta_{2,2}$, $\beta_{3,3}$ = quadratic coefficients.

Chemicals

Birchwood 4-*O*-methyl-β-D-glucoroxylan (90% xylose) was obtained from Sigma (St. Louis, MO), and poly(ethylene glycol) (PEG) from Merck (Darmstadt, Germany). All the other chemicals were of analytical grade.

Table 2
Xylanase Partition Coefficient (K) from 2^5-Full Factorial Design with Center Point under Different Treatments

Assay number	Assay sequence	Factors					K^a
		X_1	X_2	X_3	X_4	X_5	
1	25	−1	−1	−1	−1	−1	0
2	36	+1	−1	−1	−1	−1	0
3	20	−1	+1	−1	−1	−1	0
4	28	+1	+1	−1	−1	−1	0.60
5	22	−1	−1	+1	−1	−1	0
6	4	+1	−1	+1	−1	−1	1.72
7	34	−1	+1	+1	−1	−1	0
8	23	+1	+1	+1	−1	−1	0.81
9	8	0	0	0	0	0	1.69
10	14	−1	−1	−1	+1	−1	0
11	26	+1	−1	−1	+1	−1	2.03
12	6	−1	+1	−1	+1	−1	1.16
13	3	+1	+1	−1	+1	−1	1.72
14	12	−1	−1	+1	+1	−1	0
15	19	+1	−1	+1	+1	−1	0.57
16	7	−1	+1	+1	+1	−1	0
17	35	+1	+1	+1	+1	−1	0
18	16	0	0	0	0	0	1.33
19	9	−1	−1	−1	−1	+1	0
20	30	+1	−1	−1	−1	+1	2.34
21	17	−1	+1	−1	−1	+1	1.42
22	15	+1	+1	−1	−1	+1	1.94
23	5	−1	−1	+1	−1	+1	0
24	21	+1	−1	+1	−1	+1	0
25	29	−1	+1	+1	−1	+1	0
26	1	+1	+1	+1	−1	+1	0
27	24	0	0	0	0	0	1.57
28	2	−1	−1	−1	+1	+1	0
29	33	+1	−1	−1	+1	+1	1.36
30	31	−1	+1	−1	+1	+1	2.11
31	11	+1	+1	−1	+1	+1	1.33
32	18	−1	−1	+1	+1	+1	0
33	13	+1	−1	+1	+1	+1	0
34	27	−1	+1	+1	+1	+1	0
35	10	+1	+1	+1	+1	+1	0
36	32	0	0	0	0	0	1.40

$^a K = (K_1 + K_2)/2$ average partition coefficient.

Table 3
Estimated Effect, Standard Error and Student's t Test of 2^5 Factorial Design with Four Center Points

Variables	Estimated effects	Standard error	t Values
Average	0.713	±0.104	—
X_1	0.608	±0.224	2.71[a]
X_2	0.191	±0.224	0.86
X_3	−0.806	±0.224	3.36[a]
X_4	0.090	±0.224	0.41
X_5	0.118	±0.224	0.53
X_1X_2	−0.394	±0.224	1.76
X_1X_3	−0.221	±0.224	0.98
X_1X_4	−0.141	±0.224	0.63
X_1X_5	−0.187	±0.224	0.80
X_2X_3	−0.376	±0.224	1.68
X_2X_4	0.103	±0.224	0.46
X_2X_5	−0.195	±0.224	0.87
X_3X_4	−0.335	±0.224	1.50
X_3X_5	−0.505	±0.224	2.25[a]
X_4X_5	−0.203	±0.224	0.90

[a] Significant at the 5% level ($t = 2.08012$).

Table 4
Experimental Data for Xylanase Partition Coefficient (K) Under Different Treatments

Treatment	pH	%PEG	%NaCl	K
1	−1	−1	−1	2.10
2	+1	−1	−1	2.07
3	−1	+1	−1	1.75
4	+1	+1	−1	1.62
5	−1	−1	+1	2.32
6	+1	−1	+1	2.13
7	−1	+1	+1	2.14
8	+1	+1	+1	1.85
9	−1.41	0	0	2.10
10	1.41	0	0	2.27
11	0	−1.41	0	2.36
12	0	1.41	0	1.72
13	0	0	−1.41	2.27
14	0	0	1.41	2.30
15	0	0	0	2.27
16	0	0	0	2.08
17	0	0	0	2.07
18	0	0	0	2.11

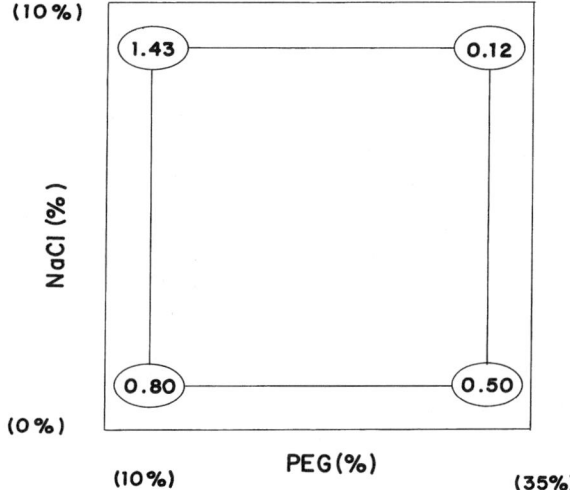

Fig. 1. Interpretation diagram of the interaction effects between %PEG and %NaCl in the 2^5 factorial design.

RESULTS AND DISCUSSION

The partition coefficients obtained after xylanase extraction by ATPS, according to the factorial design, are presented in Table 2. The factor levels were defined based on phase diagrams (MW PEG, % PEG, % phosphate and % NaCl factors) (19) and on the enzyme characteristics (pH factor) (5). The individual effects of the experimental factors and their interactions on partition coefficient (K) are shown in Table 3. As can be seen the pH (X_1), %PEG (X_3) and the interactions between %PEG (X_3) and %NaCl (X_5) presented a significative influence on partition coefficient at the 5% level. The estimated effect for the pH (X_1) was positive indicating that from level −1 to +1 the K value augments as a function of pH. The effect of this variable is independent of the other factors since there was no significant interactions among them. On the other hand, the main effect of the %PEG (X_3) cannot be interpreted separately because its interaction with %NaCl (X_5) was significant at the 5% level. As can be seen in Fig. 1, an increase in the NaCl concentration from 0 to 10% provides an increase of 0.63 U in the K value. In this variation range, at 35% NaCl, a reduction of 0.38 U in the partition coefficient is observed. On the other hand, after increasing PEG concentration from 10 to 35%, without NaCl addition the K value decreases by 0.30 U. This variation (10 to 35%) causes a decrease of 1.31 Units in the K value. As a function of these results, further optimizing experiments were performed using a 2^3 orthogonal factorial design according to the Materials and Methods section (Table 4). The regression coefficients, t values and determination coefficients (R^2) for the full quadratic

Table 5
Regression Coefficients, t Values and R^2 of Quadractic Response Surface for K Values

Term of the model	Regression coefficients	Standard error	t Values
Interception	2.175	±0.044	49.489
Factor X_1	−0.066	±0.053	1.237
Factor X_3	−0.361	±0.053	6.704[a]
Factor X_5	0.157	±0.053	2.918[b]
X_1X_3	−0.05	±0.066	0.758
X_1X_5	−0.08	±0.066	1.214
X_3X_5	0.085	±0.066	1.289
X_1X_1	−0.075	±0.066	1.138
X_3X_3	−0.220	±0.066	3.337[a]
X_5X_5	0.025	±0.066	0.379

Standard error estimated from pure error with 3 df. ($t = 3.18245$).
[a] Significant at the 5% level.
[b] Significant at the 10% level.

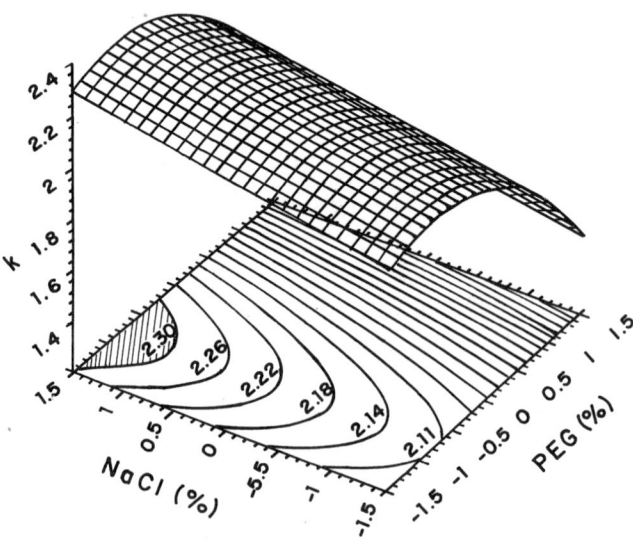

Fig. 2. Response surface of xylanase partition coefficient (K) as a function of %PEG, %NaCl, and pH.

response-surface models of K are shown in Table 5 and analysis of variance (Table 6). At the 5% probability, the pH value demonstrated no significant influence on the partition coefficient (K) as a main effect and as a second-order interaction effect, considering the new pH range (between 6.9 and 7.9). On the other hand, NaCl (%) presented a significant influence as a main effect, whereas PEG (%) also presented a second-order effect. Figure 2 illustrates the three-dimensional response surfaces and contour plots

Table 6
Analysis of Variance (ANOVA) for the Quadractic Model

Source of variations	Sum of squares	Degrees of freedom	Mean square	F Value	P value[a]
Factor X_3	0.3906	1	0.3906	31.06	0.0003
Factor X_5	0.0740	1	0.0740	5.88	0.0382
X_3^2	0.0968	1	0.0968	7.70	0.0216
Lack of fit	0.1286	5	0.0257	2.04	0.1656
Pure Error	0.1132	9	0.0126		
Total (Corr.)	0.8033	17			

$R^2 = 0.70$.
[a] $P < 0.05$.

Table 7
Regression Results: Observed Responses and Predicted Values

Observation number	Actual value (y_O)	Predicted value (y_P)	Residual ($y_O - y_P$)
1	2.10	2.15	−0.05
2	2.07	2.15	−0.08
3	1.75	1.79	−0.04
4	1.62	1.79	−0.17
5	2.32	2.31	0.01
6	2.13	2.31	−0.18
7	2.14	1.95	0.19
8	1.85	1.95	−0.10
9	2.10	2.16	−0.06
10	2.27	2.16	0.11
11	2.36	2.19	0.17
12	1.72	1.68	0.04
13	2.27	2.05	0.22
14	2.30	2.27	0.03
15	2.27	2.16	0.11
16	2.08	2.16	−0.08
17	2.07	2.16	−0.09
18	2.11	2.16	−0.05

showing the expected K values as a function of %NaCl and %PEG attained with equation 5.

$$y = 2.18 - 0.18X_3 + 0.08X_5 - 0.11X_3^2 \qquad (5)$$

The mathematical model (Eq. 5) enables the maximal point calculation (K

= 2.33) for the coded values of −0.83 and +1.41 corresponding to 8.83% of PEG and 6.02% of NaCl, respectively.

Each of the actual K values (y_o) is compared with the values predicted from the model, y_p, in Table 7. The comparison of the residuals with the error variance S_e^2 (0.23) indicates that none of the individual residuals exceeds twice the square root of the residual variance. All of the above considerations indicate an excellent adequacy of the regression model *(20)*. After the optimum processing conditions were identified by the model derived by RSM, the xylanase partition was performed under the following conditions: PEG = 8.83%; pH 7.0 and NaCl = 6.02%. The experimental xylanase partition coefficient (K) was 2.21, whereas its value predicted by the model was 2.33. This experimental finding is in close agreement with the model prediction. Of the xylanase recovered, 80% remained in the top phase.

CONCLUSIONS

The partition behavior of *P. janthinellum* xylanase has been studied in aqueous two-phase systems. The response surface methodology (RSM) was useful in optimizing the xylanase partition coefficient and graphical analysis aided in locating optimum conditions. The %PEG, %NaCl, and pH were the most important factors affecting the response variable (K). The optimum partition conditions were selected at pH 7.0; PEG = 8.83% and NaCl = 6.02%. Adequacy of the model for predicting optimum response value was tested using these conditions. The experimental xylanase partition coefficient (K) was 2.21, whereas the value predicted by the model was 2.33. These results indicate that the model was adequate for the process. In the range of experimental conditions, MW PEG and phosphate concentration did not affect the xylanase partition coefficient. The results showed the feasibility of xylanase recovery by ATPS and the advantage of employing experimental design for determining optimal extraction conditions.

ACKNOWLEDGMENTS

Silgia A. Costa acknowledges the financial support of CAPES/Brazil, in the form of a Master of Science fellowship, and FAPESP/São Paulo, Brazil. Thanks are also due to Adriane M. F. Milagres for valuable suggestions, and to Maria Eunice M. Coelho for revising this text.

REFERENCES

1. Eriksson, K. E. L., Blanchetter, R. A., Ander, P. (1990), in *Microbial and Enzymatic Degradation of Wood and Wood Components*, Springer-Verlag, New York.
2. Giovanozzi-Sermanni, G., Bertoni, G., Porri, A. (1989), in *Enzyme Systems for Lignocellulose Degradation*, (Coughlan, M. P. ed.) Elsevier New York, pp. 371–382.

3. Mutsaers, J. H. G. M. (1991), in *Xylans and Xylanases International Symposium*, Wageningen, The Netherlands, Novo Nordisk, p. 48.
4. Sharma, H. S. S. (1987), *Appl. Microbiol. Biotechnol.* **26,** 358–362.
5. Milagres, A. M. F. and Durán, N. (1992), *Prog. Biotechnol.* **7,** 539–545.
6. Parthasarathy, V. R. (1990), *Tappi J.* **73,** 243–247.
7. Senior, D. J., Hamilton, J., Bernier, R. L., and Dumanoir, J. R. (1992), *Tappi J.*, **11,** 125–130.
8. Allison, R. W., Clark, T. A., Wrathall, S. H. (1993), *APPITA* **46,** 269–273.
9. Durán, N., Milagres, A. M. F., Sposito, E., and Haun, M. (1995), in *ACS Symposium Series.* (Saddler, J. N. and Penner, N. H., eds.) American Chemical Society, San Diego, CA, pp. 332–338.
10. Palma, M. B., Milagres, A. M. F., Prata, A. M. R., Mancilha, I. M. (1996), *Proc. Biochem.* **31,** 141–145.
11. Zijlstra, G. M., Michielsen, M. J. F., Gooijer, C. D., Pol, L. A., and Tramper, J. (1996), *Biotechnol. Prog.* **12,** 363–370.
12. Schmidt, A. S., Ventom, A. M., and Asenjo, J. A. (1994), *Enzyme Microb. Technol.* **16,** 131–142.
13. Huddleston, J., Veide, A., Köhler, K., Flanagan, J., Enfors, S-O., and Lyddiatt, A. (1991), *TIBTEC* **9,** 381–388.
14. Kula, M-R., Kroner, K. H., and Hustedt, H. (1982), in *Advances in Biochemical Engineering*, Fiechter, A. ed., **24,** 73–118.
15. Box, G. E. P., Hunter, W. G., and Hunter, J. S. (1978), in *Statistics for Experiments*, John Wiley and Sons, New York.
16. Milagres, A. M. F. (1988), MSc. Thesis, Universidade Federal de Viçosa, Viçosa, Brazil.
17. Vogel, H. J. (1956), *Microbial Genet. Bull.* **13,** 42–43.
18. Miller, G. L. (1959), *Anal. Chem.* **31,** 426–428.
19. Albertsson, P. A. (1971), *Partitioning of Cell Particles and Macromolecules*, 2nd ed., Wiley Interscience, New York.
20. Sen, R. (1997), *J. Chem. Tech. Biotechnol.* **68,** 263–270.

Extracellular Proteolytic Processing of *Aspergillus awamori* GAI into GAII is Supported by Physico-Chemical Evidence

HILTON J. NASCIMENTO, VALERIA F. SOARES,
ELBA P. S. BON,* AND JOSÉ G. SILVA, JR

Instituto de Química, Universidade Federal do Rio de Janeiro, CT, Bloco A, Ilha do Fundão, Rio de Janeiro, RJ, Brasil.

ABSTRACT

The proportion of glucoamylases, GAI and GAII, in the culture supernatant of *Aspergillus awamori* fermentations depends on the medium C/N ratio in such a way that the transformation of GAI into GAII is favored by the existence of a surplus of the carbon source in the growth medium. This condition also favors the appearance of the proteolytic activity. The authors report the observation that the shift in the isoenzyme proportion was concomitant to the peak of proteolytic activity. A peptide that may have resulted from the continuous degradation of the GAI C-terminal peptide, Gp-1, was isolated by gel filtration and purified by reverse-phase chromatography. This peptide matched with the region G^{14}-A^{34} of the substrate-binding domain of GAI, thus reinforcing the hypothesis of the extracellular proteolytic processing of GAI.

Index Entries: *Aspergillus awamori*; glucoamylase; GAI and GAII; glucoamylase isoenzymes processing; isoenzymes extracellular processing; isoenzymes proteolytic processing.

INTRODUCTION

Aspergillus awamori 2.B.361 U2/1 produces two glucoamylase isoenzymes, GAI and GAII, whose proportion in the culture supernatant depends on the medium C/N ratio. GAI is prevalent in carbon limited fermentations (C/N 10), whereas GAII predominates under nitrogen limitation (C/N 26) [1]. GAII (54,000 Dalton) differs from GAI (75,000 Dalton) by lacking a C-terminal region of approx 100 amino acids residues [2].

---

*Author to whom all correspondence and reprint requests should be addressed. Email: enzitec@iq.ufrj.br

Because of the specific catalytic properties of the isoenzymes, it would be desirable to elucidate the in vivo mechanism responsible for the generation of the GAII molecule. The limited in vitro degradation of GAI using fungal acid proteases or subtilisin resulted in the production of GAII, by liberating the glycopeptide GpI of GAI (3), suggesting proteolysis as a possible mechanism. A relationship between the conversion of GAI into GAII and the appearance of proteolytic activity during the fermentation was also observed indicating that extracellular proteolysis could be the in vivo mechanism (4,5). This possibility was explored. In consideration of the C/N 26 fermentation where GAII prevails at the end of the cultivation, the isoenzyme profile was also determined at its early stages where no significant proteolytic activity would be expected, and therefore GAI would be the predominant molecular species. For this purpose, samples were collected at selected time intervals during the course of the whole fermentation. These samples were analysed for isoenzyme composition and the presence of peptides derived from the C-terminal region of GAI. For comparison, the same procedure was carried out for C/N 10 fermentations, in which no shift in the pattern of the isoenzymes would be expected. In both cases, the samples were also used for glucoamylase and protease activity determination.

The prevalence of GAI or GAII in the corresponding C/N 10 or C/N 26 fermentations was confirmed in this work by comparing the profiles of the peptides that were obtained after enzymatic cleavage with endoproteinase Lys-C of the reduced and S-alkylated isoenzymes (GAI-RCM and GAII-RCM). The reverse-phase chromatography profiles of the peptides were similar, except for a single peptide that was only present in the digest of GAI-RCM. This mismatched peptide was purified and submitted to amino acid analysis. Its composition was very similar to that of a peptide (residues 554–576) that was expected from the theoretical enzymatic cleavage in the C-terminal region of the GAI isoenzyme.

MATERIAL AND METHODS

Culture and Maintenance, Propagation, and *Aspergillus awamori* Fermentations

All procedures were carried out according to previous work which the same strain, *Aspergillus awamori* 2.B.361.U2/1 was used (1,6,7). During the fermentations, samples were collected and their supernatants were used for glucoamylase and protease activity determination, glucose consumption, GAI and GAII isoenzymes purification and identification, and peptide screening when appropriate.

Crude Glucoamylase and Peptide Isolation

Samples from the culture supernatants were chromatographed in a Biogel P-10 column (120 × 2.5 cm). The elution was performed with

0.5 mM acetic acid using a flow rate of 30 mL/h. The effluent was detected by 280-nm absorption and the relevant fractions lyophilized or dried in SpeedVac system (Savant, Farmingdale, NY).

Glucoamylase Purification by Ion-Exchange Chromatography

The crude glucoamylase activity obtained in the void volume of gel permeation were pooled and chromatographed on an anion-exchange column (Pharmacia, FPLC Mono Q, HR 10/10). The elution was performed at a flow rate of 0.7 mL/min with 5 min of 50 mM Tris-HCl, pH 8.0 followed by a NaCl gradient (0–0.6 M) in the same buffer. The two forms of glucoamylase were detected by a photo dyode array HPLC detector (Waters, Milford, MA) at 215 and 280 nm. The peaks eluting at retention times 21.5 min and 25.2 min were collected, desalted by chromatography on Sephadex G-25 fine column, and dried with a SpeedVac.

Denaturation and S-Carboxymethylation

Both glucoamylase isoenzymes (3 mg) were denatured in 1 mL of 6 M guanidine hydrochloride containing 1 M Tris-HCl buffer, pH 8.0, and 10 mM EDTA for 16 h at 40°C. After reduction with 1.0 mg of DTT for 4 h at 40°C, the two proteins reacted with 1.5 mg of iodoacetamide for 3 h at 40°C. The reaction was stopped by addition of glacial acetic acid to pH 4.0. The mixture was filtered through a Sephadex G-25 fine column (40 × 1.5 cm) eluted with 50 mM amonium hydroxide. The reduced and S-carboxymethylated glucoamylase isoenzymes GAI-RCM and GAII-RCM were detected by absorbance at 280 nm and lyophilized.

Enzymatic Cleavage of the Glucoamylase Isoenzymes with Endoproteinase Lys-C

The reduced and S-carboxymethylated isoenzymes (0.25 mg) were hydrolyzed with endoproteinase Lys-C (Boehringer Mannheim, Mannheim, Germany), as described by the manufacturer, using a 1/50 enzyme/substrate ratio (w/w) in 25 mM Tris-HCl buffer, pH 8.5, for 16 h, at 38°C. The reactions were stopped by addition of 1% trifluoroacetic acid until pH 2.0 and the products separated by HPLC-RP C18.

Peptide Fractionation by HPLC-RP C18

The products of endoproteinase Lys C hydrolysis of the reduced glucoamylase isoenzymes were fractionated by reverse phase high performance liquid chromatography (HPLC-RP) on μBondapak C18 (3.9 × 300 mm, Waters) column. The column was previously equilibrated with 90% of eluant A (0.1% TFA) and 10% of eluant B (70% acetonitrile with 0.08% TFA). The reverse phase chromatography was carried out at room temperature at 0.5 mL/min using a Waters work station equipped with a photo diode array

Fig. 1. Analysis of the endoproteinase Lys C hydrolysates of GAI-RCM (**A**) and GAII-RCM (**B**). The profile corresponding to GAI-RCM hydrolysates presents a mismatched peptide 6a. Detection by absorbance at 220 nm and flow rate 30mL/h.

detector. The gradients used are indicated in the legend of Fig. 1. Samples were injected in 0.35-mL volumes and effluent was monitored at 220 and 280 nm. The relevant samples were dried in a SpeedVac system (Savant).

Purification of the Peptides Separated by Biogel P-10 Chromatography

The peptide fractions collected in the fractionating range of Biogel P-10 chromatography were fractionated by HPLC-RP C4 in a HiPore column (250 × 10 mm, Bio-rad, Richmond, CA). The reverse-phase chromatography was carried out at 1.0 mL/min using a Waters work station. The elution was performed with a acetonitrile gradient in 0.08% trifluoroacetic acid (TFA) for 40 min.

Amino Acid Analysis Of Relevant Peptides

The peptides were hydrolyzed in constant boiling 5.8 N HCI for 22 h at 110°C under vacuum. The hydrolyzate was dried in dessicator and analyzed on a Pharmacia Biotech Biochrom 20 amino acid analyzer. The standard amino acids were from Sigma (2.5 nmoles/µL).

Analytical

Glucoamylase activity and glucose concentration determinations were performed according to previous work (1). Protease activity was measured using azocasein as substrate (8).

RESULTS AND DISCUSSION

Comparing the profiles corresponding to the enzymatic hydrolysis of GAI-RCM and GAII-RCM, which corresponds to the predominant isoenzyme at the end of C/N 10 and C/N 26 fermentations, respectively, it is clear that the peptide 6a from the GAI-RCM hydrolysate is absent in the GAII-RCM hydrolysate (Fig. 1). Table 1 presents a comparison between the amino acid composition of peptide 6a and its theoretical counterpart. Although the composition of the isolated peptide 6a does not precisely match the expected profile, there is a great deal of similarity between the experimental and theoretical data. The results obtained so far support previous results related to the selective production of GAI or GAII depending on the medium composition. The carbon source surplus present in the C/N 26 fermentation is a key element for the GAI transformation into GAII.

The profile of glucoamylase and protease activity during the C/N 10 and C/N 26 fermentation were determined (Fig. 2). The initial levels of glucoamylase and protease were similar in both media conditions up to fourth day. Glucose depletion occured before the fifth day of the cultivation for the C/N 10 media and, therefore, the culture died was of nutrient depletion. Considering the C/N 26 fermentation, a different pattern was observed the carbon source availability beyond this period. Although glucoamylase activity remained stable from the fourth to the seventh day, protease activity kept increasing until a peak was reached within 1 wk of fermentation. Coincidently, glucoamylase activity showed a sudden rise at this point. Considering that GAII shows a higher activity towards maltose in comparison to GAI (1), this increase in enzyme activity could be caused by the generation of GAII from GAI at this point in the course of the fermentation, instead of an increase on enzyme production. Considering this possibility, the isoenzyme proportion in the culture supernatant

Table 1
Amino Acid Composition of the Peptide 6a Isolated by HPLC-Bondapak C18
from Endoproteinase Lys-C Digest of GAI-RCM

Amino acid	Amino acid composition of 6a (T_R = 39 min) peptide		Theoretical composition of peptide 6a no. of mols/mol of peptide
	ρ moles/µL	No. of mols/mol of peptide	
cysteic acid	≤2 (≤0.05)		
S-CMC	≤2 (≤0.05)		
D	60 (1.50)	1–2	1
E	126 (3.16)	3	2
S	125 (3.12)	3	3
G	85 (2.13)	2–3	1
H	4 (0.10)	0	0
R	15 (0.38)	0	0
T	120 (3.00)	3	3
A	50 (1.25)	1	1
P	75 (1.88)	2	2
Y	60 (1.50)	1–2	3
V	76 (1.90)	2	2
M	17 (0.42)	0	0
C	10 (0.25)	0	0
I	22 (0.55)	0–1	0
L	78 (1.95)	2	2
F	40 (1.00)	1	1
K	39 (0.98)	1	1
W[a]	—	—	1
TOTAL		22–25	23

[a] Residue not determined.

The amino acid composition of a theoretical peptide (Y^{554}-K^{576}) present in the C-terminal portion of GAI resulting from the same treatment is also presented.

was examined before and after the seventh day of incubation. The results that were obtained are presented in Figs. 3 and 4. There is a clear inversion of the proportion of GAI and GAII in the analyzed supernatants, suggesting the transformation of GAI into GAII after 7 d of fermentation. This transformation may be related to proteolysis as protease concentration may have reached a threshold concentration on the seventh day that allowed the occurence of glucoamylase processing. Figure 5 presents the isoenzymes profile of the C/N 10 fermentation supernatant. This profile indicates, as expected, the prevalence of GAI.

Although the foregoing evidence supports the hypothesis of extracellular processing for the generation of GAII, the definitive argument would come from the identification of the peptide Gp-1 in the culture supernatant.

Extracellular Processing of GAI into GAII

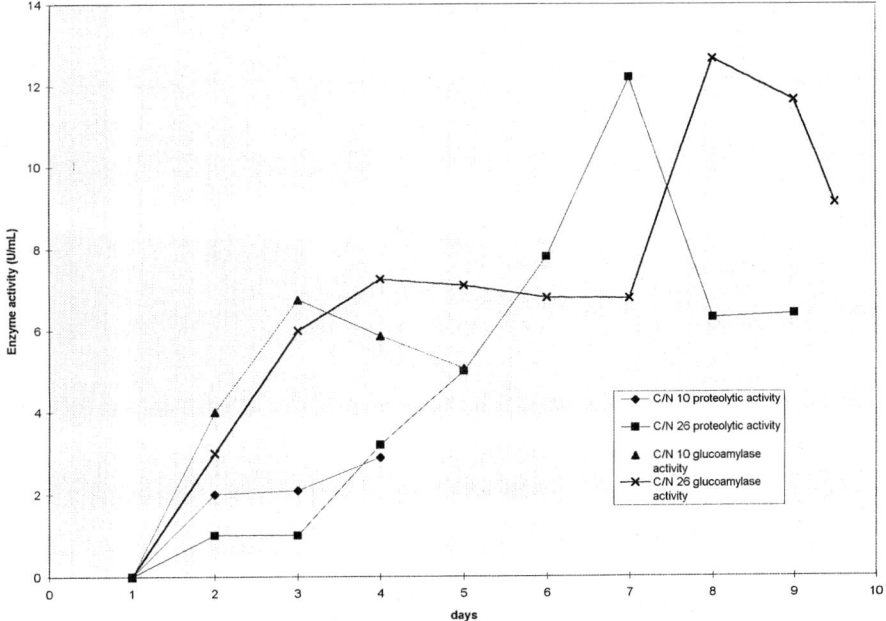

Fig. 2. Profile for glucoamylase and protease activity in C/N 10 and C/N 26 fermentations.

Fig. 3. Ion-exchange chromatography in FPLC-mono Q column of the culture supernatants and GAI from 3 and 4 d of C/N 26 fermentation, showing GAII (peak with retention time 21.5 min) and GAI (peak with retention time of 25.2 min).

Fig. 4. Ion-exchange chromatography in FPLC-mono Q column of the culture supernatants and GAI from 8 and 9 d C/N 26 fermentation, showing GAII (peak with retention time 21.5 min) and GAI (peak with retention time of 25.2 min).

Fig. 5. Ion-exchange chromatography in FPLC-mono Q column of the culture supernatants and GAI from 3 and 4 d of C/N 10 fermentation, showing GAII (peak with retention time 21.5 min) and GAI (peak with retention time of 25.2 min).

Table 2
Retention Time (R_T) of HPLC-RP C4 for Peptide I and Peptide II Fractions Isolated from Biogel P-10 Column

	C/N 26 medium	
Day	Peptide I	Peptide II
4	31 min^a, 38 min, 34 min, and 39 min	24 min, 26 min, 30 min 34 min, and 36 min
7	31 min,^{a} 38 min and 39 min	26 min,^{a} 30–39 min
8	17.5 min,^{a} 38 min, and 39 min^a	20 min^a and 26 min^a
9	18.5 min, 19 min,^{a} 38.5 min, and 36.5 min	18 min and 18.5 min^a, 19 min and 19.5 min

^{a} The major peptide at 220 and 280 nm detection.
Obs.: the peptide II R_T = 26 min related to 7 and 8 d fermentation of C/N 26 fermentation was isolated and had its amino acid composition determined.

Accordingly, Bio-gel P10 chromatography was used in this study to separate the glucoamylase proteins, which eluted in the void volume, and to fractionate peptides with molecular weight within 1500 and 20,000. This methodology could resolve Gp-1, as it shows a molecular weight of 11,000 (100 amino acid residues). Two peptide fractions were resolved from the supernatant of the C/N 26 fermentation mainless of the fermentation time. These fractions, however, showed different V_e/V_o values. The fractions corresponding to the fourth day showed the value of 1.34 (peptide 1) and the value of 1.40 (peptide 2). Peptides 1 and 2 from the seventh eighth, and ninth days presented the same V_{eo} values, i.e., 1.67 and 1.70, respectively. The HPLC-RP-C4 profiles from peptide 1 and peptide 2 showed at fourth and seventh days peptide fractions with high retention times, whereas the eighth and ninth days samplings resulted in fractions with low retention times, suggesting a continuous degradation (Table 2). The main peptide fraction (retention time 26 min) from HPLC-RP C4 profiles of peptide 2 from seventh and eighth day, was collected and subjected to amino acid composition analysis (Table 3). This amino acid composition was similar to a theoretical peptide produced by the cleavage in the residues G^{526} and A^{551} from the C-terminal portion of GAI.

CONCLUSIONS

Structural analysis of GAI and GAII glucoamylase isoenzymes confirmed the effect of the medium C/N ratio on the selective production of GAI or GAII in submerged fermentations of *Aspergillus awamori*. Carbon limited fermentations favors GAI predominance, whereas nitrogen limitation favors GAII formation. The *in vivo* extracellular mechanism responsible for the formation of GAII seems to involve proteolytic processing, because the conversion of GAI into GAII during the fermentation was

Table 3
Amino Acid Composition of a Reverse-Phase Peptide (R_T = 26 min) Isolated from Peptide II Fraction HPLC-RP C4a 7 and 8 d

Amino acid	ρ noles	Σρ noles	Assumed D+E = 6.00	GA_I theoretical peptide (G^{526}-A^{551})
D+E	380+860	1240	6.00 (6)	6
T+S+G	270^a+635^a+960	1865	9.02 (9)	9
A+V+I+L	440+400+250+570	1660	8.03 (8)	8
Y+F	60+150	210	1.02 (1)	1
P	≤50	≤50	≤25 (0)	0
C	≤50	≤50	≤25 (0)	0
M	≤50	≤50	≤25 (0)	0
R	≤50	≤50	≤25 (0)	0
			24	24

^{a} Values corrected considering 8% for T destruction and 15% for S destruction.
An amino acid composition of a theoretical peptide (G^{526}-A^{551}) present in the C-terminal portion of GAI is also showed. The amino acids were grouped according to its similarities. The lysine and histidine residues were lost.

concomitant with high proteolytic activity in the culture medium. The medium C/N ratio also effects protease production that is favored in nitrogen-limited fermentations. The presence of peptide Gp-1 was not identified in the culture supernatant. It was possible, however, to identify a smaller peptide that may have resulted from the continuous degradation of the C-terminal peptide Gp-1. This peptide matched with the region G^{14}-A^{39} of Gp-1, according to the hypothesis of the extracellular proteolytic processing of GAI.

REFERENCES

1. Silva Jr., J. G., Nascimento, H. J., Soares, V. F. and Bon, E. P. S. (1997), *Appl. Biochem. Biotech.* **63/65,** 87–95.
2. Clarke, A. J. and Svensson, B. (1984), *Carlsberg Res. Commun.* **49,** 559–566.
3. Hayashida, S., Nakahara, K., Kuroda, K., Kamachi, T., Otha, K., Iwanaga, S., Miyata, T. and Sakaki, Y. (1988), *Agric. Biol. Chem.* **52(1),** 273–275.
4. Hayashida, S. (1975), *Agric. Biol. Chem.* **39(11),** 2093–2099.
5. Bartoszewitz, K (1986), *Acta Biochim. Polonica* **23,** 17–29.
6. Bon, E. and Weeb, C. (1993), *Appl. Biochem. Biotech.* **39/40,** 349–369.
7. Bon, E. and Weeb, C. (1989), *Enzyme Microb. Technol.* **11,** 495–499.
8. Charney, J. and Tomarelli, R. M. (1947), *J. Biol. Chem.*, **171,** 501–505.

Technical and Economic Evaluation of Different Reactors for Methanotrophic Cultures for Propylene Oxide Production

BHUPENDRA K. SONI*, ROBERT L. KELLEY, AND VIPUL J. SRIVASTAVA

1700 Mount Prospect Road, Des Plaines, IL

ABSTRACT

A two-stage process for the manufacture of propylene oxide is described. The preliminary economics based on use of methanol as a regeneration factor has resulted in a production cost of $12.10/lb of propylene oxide based on propylene oxide production rate of 40 mg/g-cell/h in conventional reactor. Increasing the propylene oxide production from 40 to 500 mg/g-cell/h resulted in a cost reduction from $12.10 to 5.8/lb of propylene oxide. The granular-activated, carbon-fluidized bed reactor (GAC-FBR) absorbs the propylene oxide and when saturated is eluted with ethyl acetate, and the bed is regenerated by steam to drive off the residual solvents. The estimated manufacturing costs are approx 59% lower (from $12.10/lb in conventional reactors to $5.00/lb for GAC-FBRs) for products that are highly inhibitory such as epoxides. In the GAC-FBR reactor, enhancing the propylene oxide production rate from 120 to 1500 mg/g-cell/h has resulted in the cost reduction to $2.00/lb. Enhancing the production capacity from 1 million lb to 10 million lb/yr has further reduced the cost of production to $1.00/lb.

Index Entries: Propylene oxide; granular-activated carbon; fluidized bed reactor.

INTRODUCTION

Most enzymes are very specific in respect to the carbon source. One of the major incentives for research on the biochemistry of methanotrophs is the potential industrial exploitation of the unusual lack of substrate specificity of the methane monooxygenase enzyme. The ability of this enzyme to insert oxygen into a wide variety of chemicals makes the methane-

*Author to whom all correspondence and reprint requests should be addressed.

utilizing bacteria of potential economic importance. A wide range of nongrowth substrates including alkanes, haloalkanes, alkenes, alicyclics, and aromatics have been evaluated (1–3). For example, propylene oxide can be produced from propylene through a co-oxidative process (4,5). Methane is used to grow the organisms and to regenerate them after the production of propylene oxide. Thus, it is possible to have continuous production of the co-oxidized product because the cosubstrate, methane, provides the energy and carbon requirements of the cell.

In the earlier studies conducted at Institute of Gas Technology (IGT, Des Plaines, IL), a continuous process has been developed for enhancement of methane monooxygenase activity so that higher amounts of propylene oxide can be produced (6,7). The production of propylene oxide has been demonstrated in bench-scale studies (8).

The objective of the proposed economic analysis is to develop the cost of propylene oxide for conventional processes, which involve a continuous production of biocatalyst followed by a batch-reactor process for production of propylene oxide. The goal is to compare the conventional process with the GAC-FBR process. The GAC-FBR process has been investigated in various ground-water applications (9–11). GAC-FBR is a single-stage process in which propylene oxide produced by bacteria is adsorbed on the granula-activated carbon, which results in reduced inhibition on enzyme and bacteria producing it. The cost has been estimated for conventional process and GAC-FBR at several different capacities of production under various propylene-oxide production rates. For technical and economic evaluation, the process parameters achieved by IGT were used to design and economically evaluate a schematic process using conventional stirred tank fermentors and air stripping for propylene oxide recovery. Furthermore, the process parameters achieved by the granular-activated carbon-fluidized bed reactor technology (GAC-FBR) for methanotrophic oxidation for waste treatment of chlorinated hydrocarbons (12) were used to design and economically evaluate this GAC-FBR reactor scheme for chemical production by methanotrophs.

RESULTS

Process Description and Economic Evaluation of Conventional/IGT Process for Methanotrophic Co-Oxidation

A schematic of the process is shown in Fig. 1. This process scheme and the yields, rates, and conversion factors were developed, based on the work conducted at IGT.

The stage 1 fermentors produce regenerated cells and oxidations enzyme/cofactors from methane oxidation. In general, it is extremely

Propylene Oxide Production

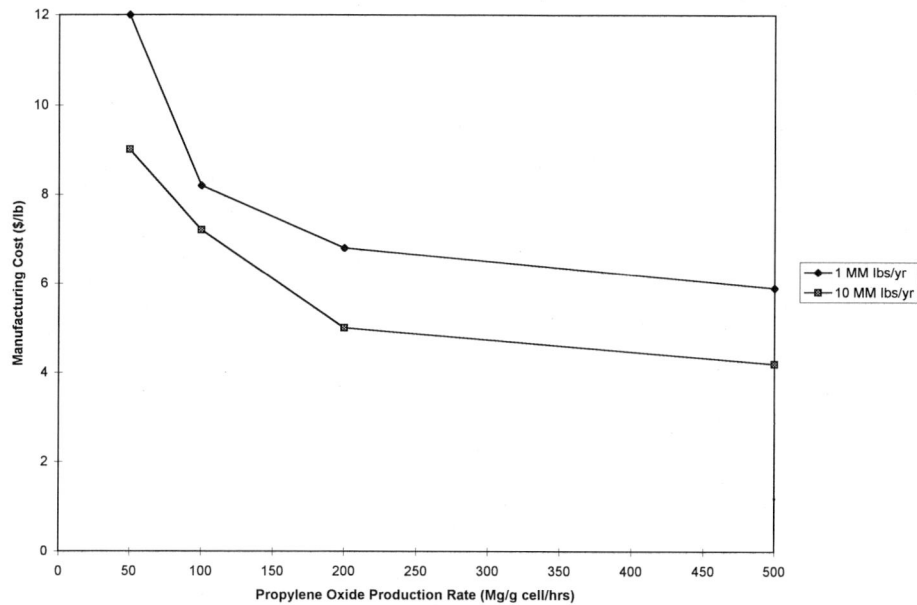

Fig. 1. Schematic of the conventional process of methanotrophic oxidation.

difficult to maintain the sterility of methanogenic cultures, however, under the optimized-nutrient conditions and pH of the reactor, sterile conditions were demonstrated for more than a month of continuous operation (6). In stage 2 fermentors, the cells are used to oxidize propylene to propylene oxide (PO) with methanol as a cofactor regenerant. Air is used in large excess to strip PO from the broth and the stripped PO vapor is adsorbed in granular-activated carbon (GAC) beds. The GAC when saturated with PO is eluted with ethyl acetate (azeotrope composition with water) and the bed is regenerated by steam to drive off the residual solvents. The eluant and regenerant condensate are combined and sent to a series of two distillation columns—the eluant distillation separates and recovers the product PO and the second ethyl acetate column recovers the ethyl acetate azeotrope for recycling and the water, heavies, and some of the unrecoverable ethyl acetate go into the wastewater stream.

The current conventional/IGT process was evaluated. The manufacturing cost to produce one million pounds of propylene oxide was used as a representative example. The estimated manufacturing cost for PO from this process is shown in Table 1 (approx $12.00/lb). This economic analysis was developed based on a material balance from IGT data. Some of the factors important for this analysis are:

Table 1
Manufacturing Base-Cost Process for Methanotrophic Co-oxidation

Item	Quantity, U/lb of product	Price per U, $	Cost per Yr, million	Cost per lb product, $
Raw materials:				
Propylene	0.75	0.14		0.10
Methane	31.88	0.073		2.33
Nutrients	1.28	0.1		0.13
Methanol	0.22	0.06		0.01
Subtotal				2.57
Chemicals, supplies:				
Ethyl acetate	0.21	0.55		0.12
Carbon (GAC)	0.002	0.8		0.00
Miscellaneous			0.10	0.10
Subtotal				0.22
Variable Utilities:				
Steam, 50 psig thousand lb	0.04	3.8		0.15
Water process, thousand gal	0.31	1.5		0.46
Water cooling, thousand gal	4.79	0.1		0.48
Electricity, kWh	65.12	0.05		3.26
Subtotal				4.35
Total variable cost:				7.14
Operating labor	4/shift	@30,000/yr	0.48	0.48
Supervision	0.8/shift	@60,000/yr	0.18	0.18
Maintenance	3% of DFC		0.67	0.67
Plant overhead	75% (labor + supervision)		0.50	0.50
Insurance and taxes	1.5% of DFC		0.33	0.33
Total fixed cost				2.15
Total cash cost (fixed + variable)				9.29
Depreciation (8 yr straight line)			2.78	2.78
Manufacturing cost (cash + depreciation)				12.07

Basis: 1 million lb propylene oxide/yr, 8000 h/yr

1. The large quantity of methane used and the equipment and power needs associated with the process are very important cost factors.
2. This arises primarily because the PO productivity is so low (40 mg/g cell-h as an average productivity achieved by IGT to date). Thus, a large mass of cells have to be regenerated to carry out the reaction.

3. This low productivity is caused by very strong end-product inhibition at low concentration of the PO (or other epoxide products/intermediates).
4. The size of the equipment, particularly, the fermentors is very large. They are to be made of carbon steel and lined where possible and needed (to cut down the cost).
5. These fermentors are conventional and methane and oxygen gas are sparged simultaneously. Mixing of combustible/explosive gases in the fermentors is also a major design and engineering problem.
6. With this large volume and low productivity process it would be very difficult to reduce the costs. It would also be very difficult to scale-up or convince industrial clients to scale-up given the process uncertainties.
7. Process schemes and technical advances that will enable simultaneous separation of the PO end-product, thus reducing the inhibition and also make the equipment volumes lower and less unwidely would have clear advantage.

Process Description and Economic Evaluation of GAC-FBR Process for Methanotrophic Co-Oxidation

A process economic analysis was also conducted based on a GAC/FBR process developed at the Michigan Biotech Institute (MBI) using chlorinated organics (TCE). A schematic for the GAC/FBR process is shown in Fig. 2. Methane and oxygen are dissolved in fresh broth via inverted-cone diffusers, and this broth is pumped up flow through the GAC/FBRs. The methanotrophs form a biofilm on the GAC so that biocatalyst is retained. In the co-oxidation cycle, propylene is oxidized with methanol as a cofactor regenerant. One of the advantages of using GAC-FBR reactor is continuous adsorption of propylene oxide on the bed. This will minimize the inhibition of biocatalyst because of exposure to a reduced amount of propylene oxide. This will lead to a larger conversion cycle of propylene to propylene oxide as compared to the conventional process. The GAC, when saturated with PO, is eluted with ethyl acetate (azeotrope composition with water), and the bed is regenerated by steam to drive off the residual solvents. It is important to mention here that once a bed is in propylene-oxide production mode, the biocatalyst accumulation is done for next bed. The biocatalyst, being dead because of production of propylene oxide, is not recyclable any further. The PO-adsorbed GAC is eluted with ethyl acetate and then regenerated with steam. The ethyl acetate PO separation and ethyl acetate water separation is done by distillation as discussed earlier.

The estimated manufacturing cost for this process scheme is shown in Table 2. This process economic analysis was based on the costs and operation information on GAC/FBRs of MBI. The estimated cost was 41% of the conventional process because of reduced inhibition and the corresponding enhancement in the PO production rates.

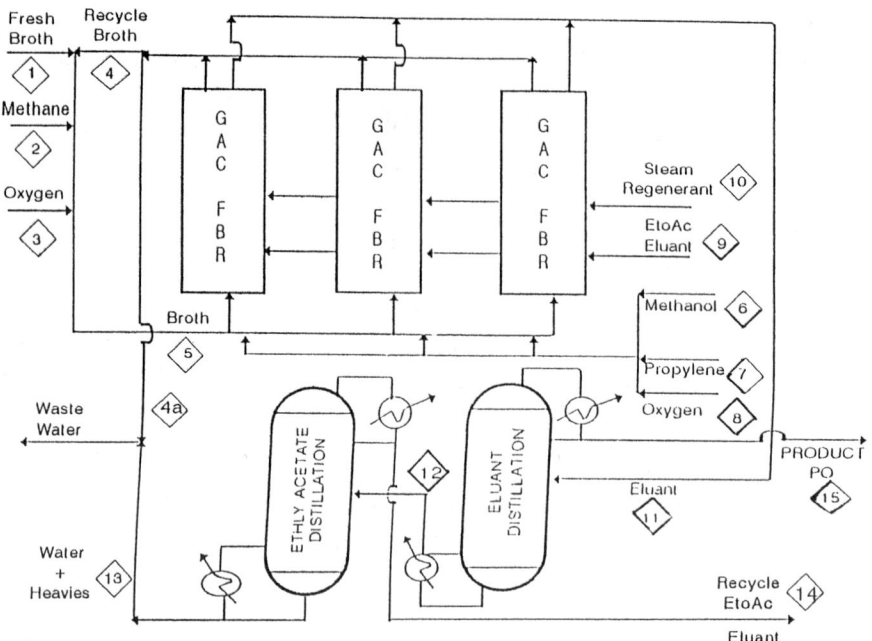

Fig. 2. Schematic of the granular activated carbon/fluid-bed reactor (GAC/FBR) process for methanotroph co-oxidation.

Process Economics—Sensitivity Analysis

Brief sensitivity analyses of the manufacturing cost to changes in production volume 1 to 10 MM lbs/yr and to the organism's rate of production were also conducted. The increase in rate has a very marked effect on the manufacturing cost because the low productivity caused by end-product inhibition increases all costs (Figs. 3 and 4). The increase in production volume reduces the manufacturing cost as expected, because the fixed costs per unit volume are reduced. The GAC-FBR process, because of its ability to adsorb the inhibitory product/intermediate has been assumed to have a threefold increase in corresponding production rate—this assumption is based on a similar decrease in inhibition seen with the methanotrophic oxidation process for waste treatment of toxic halogenated compounds.

This analysis shows that the GAC-FBR process will have a lower cost than the conventional process and at large volumes and high rates (for products that are not very inhibitory), the manufacturing cost can be < $1.50/lb and be as low as $1.00/lb. A wide range of specialty or fine chemicals fall in this range of manufacturing costs. Our results suggest that cost of production of propylene is not competitive with the current cost ($ 0.57/lb). Thus the application of this process needs two important considerations: use higher capacity than demonstrated

Table 2
Manufacturing Cost—GAC/FBR Process for Methanotrophic Co-oxidation

Item	Quantity, U/lb of product	Price per U, $	Cost per Yr, million	Cost per lb product, $
Raw materials:				
Propylene	0.74	0.14		0.10
Methane	7.01	0.073		0.51
Nutrients	0.42	0.1		0.04
Methanol	0.22	0.06		0.01
Subtotal				0.67
Chemicals, supplies:				
Ethyl acetate	0.27	0.55	0.15	
Carbon (GAC)	0.005	0.8		0.00
Oxygen	20.35	0.075		1.53
Miscellaneous			0.05	0.05
Subtotal				1.73
Variable Utilities:				
Steam, 50 psig thousand lb	0.04	4		0.17
Water process, thousand gal	0.05	1.5		0.08
Water Cooling, thousand gal	1.26	0.1		0.13
Electricity, kWh	2.16	0.05		0.11
Subtotal				0.48
Total variable cost:				2.88
Operating labor	3/shift	@30,000/yr	0.36	0.36
Supervision	0.8/shift	@60,000/yr	0.18	0.18
Maintenance	3% of DFC		0.20	0.20
Plant overhead	75% (labor + supervision)		0.41	0.41
Insurance and taxes	1.5% of DFC		0.10	0.10
Total fixed cost				1.25
Total cash cost (fixed + variable)				4.13
Depreciation (8 yr straight line)			0.85	0.85
Manufacturing cost (cash + depreciation)				4.98

in this economic analysis (> 10 million lb/yr.); or determine the ratio of L- and D-forms and purify a single stereospecific form. These products are currently sold at very high prices (> $ 1000.00). Thus, the actual strategy will depend upon the requirement of quantity and type of the product at specific production site.

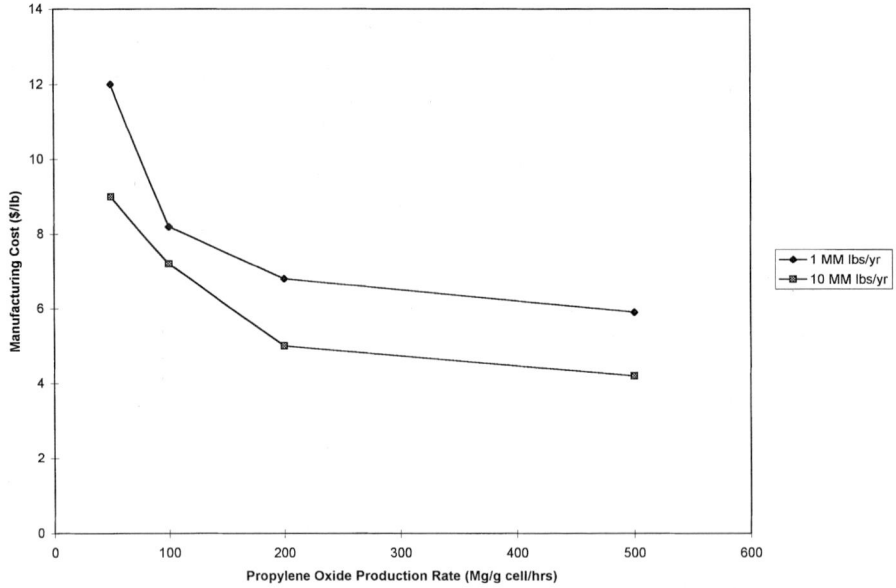

Fig. 3. Effect of propylene-oxide production rates and plant capacity on manufacturing cost in the conventional process.

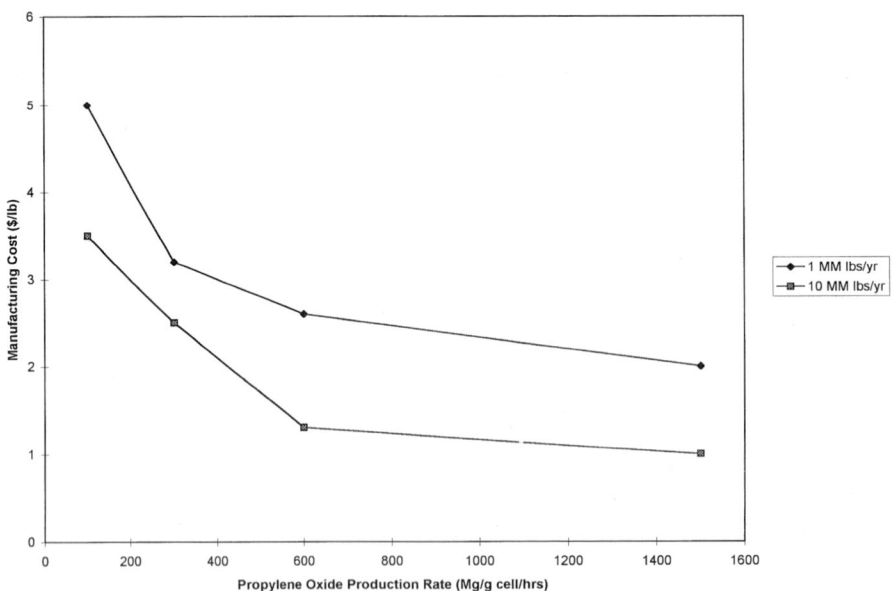

Fig. 4. Effect of propylene oxide production rates and plant capacity on manufacturing cost in the GAC-FBR process.

CONCLUSIONS

The GAC-FBR reactor for propylene oxide production was found to be superior:

1. The estimated manufacturing costs are approx 59% lower (from $12.00/lb in conventional reactors to $5.00/lb) for GAC-FBRs for products that are highly inhibitory such as epoxides, and, considerably lower costs (i.e. < $1.50/lb) are possible for less inhibitory higher productivity products produced in large volumes (> 10 million lbs/yr).
2. The process enables simultaneous conversion and removal of product by adsorption, thus reducing the effect of end-product inhibition on productivity.
3. The capital costs are considerably (3.3-fold) lower in GAC-FBR reactor.
4. The equipment required is not large and unwieldy.
5. Methane and oxygen are dissolved in the media by separated-inverted cone diffusers, hence they do not mix in the gas phase and have explosive/combustible mixtures.
6. The GAC-FBR technology has been scaled-up to very large >2000 GPM reactors and has had a very good track record to-date.

ACKNOWLEDGMENTS

The authors wish to thank Rathin Datta for providing excellence guidance to develop the economic analysis, and John Conrad for the excellent review of the manuscript.

REFERENCES

1. Colby, J., Stirling, D. I., and Dalton, H. (1977), *Biochem. J.* **165,** 395–402.
2. Higgins, I. J., Hammond, R. C., Sariaslani, F. S., Best, D., Davies, M. M, Tryhorn, S. E., and Taylor, F. (1979), *Biochem. Biophy. Res. Commun.* **89,** 671–677.
3. Hou, C. T., Patel, R. N. Laskin, A. I. (1979), *FEMS Micro Letts.* **9,** 267–270.
4. Higgins, I. J., Best, I. J., Hammond, R. C., and Scott, D. (1981), *Microorganism Microbiol. Rev.* **45,** 556–589.
5. Hou, C. T., Patel, R. N., Laskin, A. I., and Barnabe, N. (1982), *J. Appl. Biochem.* **4,** 379–383.
6. Lee, J., Soni, B. K., and Kelley, R. L. (1996a), *Biotechnol. Lett.* **8,** 897–902.
7. Lee, J., Soni, B. K., and Kelley, R. L. (1996b), *Biotechnol. Lett.* **8,** 903–908.
8. Hill, A. H., Kelley, R. L., Srivastava, V. J., Akin, C., and Hayes, T. D. (1991), in *Gas, Oil, Coal and Environmental Biotechnology*, vol. III. pp. 417–435.
9. Voice, T. C., Pak, D., Zhao, J. Shi, and Hickey, R. F. (1992), *Water Res.* **26(10),** 1389–1401.
10. Hickey, R. F., Wagner, D., and Mazewski, G. (1991), *Remediation* **2,** 447–460.
11. Jeris, J. S., Owens, R. W., Hickey, R. F., and Flood, F. (1977), *JWPCF* **49,** 816–831.
12. Hickney, R. F. (personal communication, Michigan Biotechnology Institute, 1994).

Xylanase Recovery by Ethanol and Na_2SO_4 Precipitation

Ely V. Cortez, Adalberto Pessoa Jr,* and Adilson N. Assis

Depto. Biotecnologia/FAENQUIL, Rod. Itajubá-Lorena, Km 74,5. 12.600-000, Lorena/SP, Brazil

ABSTRACT

Xylans are the major components of the hemicellulosic fraction of lignocellulosic biomass and their hydrolysis can be obtained using xylanases from *Penicillium janthinellum*. In this work, sugarcane bagasse hemicellulosic hydrolysate was used as the substrate for producing xylanase. The precipitation of these enzymes was studied using ethanol and Na_2SO_4 as precipitating agents. Ethanol precipitation experiments were performed batchwise in concentrations ranging from 10 to 80%, pH 4.0 to 7.0, at 4°C. The concentrations used in the precipitations with Na_2SO_4 were from 5 to 60% at pH 5.5 and 25°C. Solubility curves as a function of xylanase activity and total protein for both precipitating agents were made. According to the results, Na_2SO_4 is not appropriate for precipitating xylanases in this medium since at salt concentrations higher than 25%, the enzyme was denatured and at this concentration less than 80% of the enzyme and total protein were precipitated. Because of differences in xylanase and total protein solubility, a fractionated precipitation using ethanol can be performed, since with 40% ethanol, 49% of the total protein was precipitated and more than 95% of the enzyme was kept in solution. On the other hand approx 100% of the xylanases were recovered by precipitation after adding 80% ethanol.

Index Entries: Xylanase; ethanol; precipitation; sodium sulfate.

INTRODUCTION

Xylans are the major components of the hemicellulosic fraction of lignocellulosic biomass and their hydrolysis can be obtained using xylanases *(1)*. The enzymatic complex is composed of endoxylanases that cleave internal xylosidic linkages on the xylan backbone and β-xylosidases that release xylosyl residues by endwise attack of xylooligosaccharides *(1,2)*. Efforts have been made in the pulp and paper industries to reduce

*Author to whom all correspondence and reprint requests should be addressed.

the amount of chlorine used for bleaching. Studies have been conducted on the effluent treatment and the use of less toxic bleaching agents are under study *(3)*. The residual lignin removal by enzymatic method is a very active research area *(4,5)*. According to Durán et al. *(6)*, xylanases produced by *Penicillium janthinellum* can be used to reduce the chlorine charge in *Eucalyptus*-pulp bleaching with a simultaneous brightness gain. Xylanases have also been used in bread making *(7)*, clarification of beer and juice *(1)*, and partial xylan hydrolysis in animal feed *(2)*. According to Gattinger et al. *(8)*, xylanases are produced from processed or refined substrates such as sugars, cellulose, and xylan. All of these methods are expensive for industrial-scale production and to lower production costs, cheaper substrates must be employed. Sugarcane bagasse hemicellulosic hydrolysate, composed mainly of xylose oligomers, is a potential substrate. Optimal conditions for cultivation parameters (like agitation and aeration rates) have been investigated *(9)*. However, xylanase recovery from the cultivated medium using a scalable technique has to be studied.

Precipitation and recovery of protein precipitates by centrifugation are widely practiced in the biotechnological industry. Ethanol precipitation is a promising technique since it has been applied to other types of proteins on an industrial scale *(10,11)*. Because of their low dielectric constants (compared to water) organic solvents increase Coulombic attraction between protein molecules. Aggregates are formed until the particle size reaches macroscopic proportions and precipitation occurs. Ethanol is by far the most important of the solvents owing to its good physicochemical properties, like complete miscibility with water, good freezing-point depression, no explosive mixtures, high volatility, chemical inertness, low toxicity, and low cost (especially in Brazil). Protein precipitation by salting out is the oldest type of precipitation and is still used regularly on a laboratory scale. The mechanism is, nevertheless, not completely understood although it is clear that high-salt concentrations remove water associated with protein. The most commonly used salts are ammonium sulfate and sodium sulfate. Ammonium sulfate, although widely used, presents waste disposal problems because of the nitrogen content and corrosive properties. Sodium sulfase is an alternative to salt precipitation since it constitutes a simple means of recycling by reduction of the temperature and separation of the salt crystals. The present study was carried out to examine the xylanase and total protein precipitation using ethanol and sodium sulfate. Ethanol and salt precipitations are used in xylanase precipitations, but the experiments are performed exclusively for laboratory applications *(11–16)*.

MATERIALS AND METHODS

Preparation of Sugarcane Bagasse Acid Hydrolysate

In order to prepare the hydrolysate for cultivation, 800 g of dry-milled bagasse was mixed with 8 L sulfuric acid solution (0.25%) and autoclaved for 45 min at 121°C. The liquid fraction was separated by filtration adjusted to pH 5.5 with NaOH.

Cultivation Medium and Enzyme Production

The isolation of *P. janthinellum* from the decaying wood was described by Milagres *(17)*. This microorganism was identified by the Biosystematic Research Center of Canada (Ottawa, Ontario) and deposited in their collection with the designation of CRC 87M-115. The strain was maintained in silica stocks and, by transfer, on agar slants. The fungus was cultivated at 30°C for 5 d in medium containing 1% glucose, 0.1% yeast extract, 2% (v/v) concentrated complete salts solution based on Vogel's medium (18), and 2% agar-agar. The medium was sterilized at 121°C for 15 min. The spore inocula were obtained by suspending spores in water and filtering through gauze into Erlenmeyer flasks. The final concentration of spores was 10^5/mL. The cultivation medium for enzyme production contained sugarcane bagasse hemicellulosic hydrolysate supplemented with 2% (v/v) concentrated salt solution based on Vogel's medium and 0.1% yeast extract. The medium was then autoclaved for 15 min at 121°C. The cultivation was carried out in Erlenmeyer flasks (125-mL) containing 25 mL of medium. Standard cultivation conditions were: temperature 30°C; initial pH 5.5 (uncontrolled); and 96 h of cultivation time. In general, xylanase produced by *P. janthinellum* in an aqueous solution is stable at a range between pH 4.0 and 8.0; it is rapidly inactivated below pH 4.0 and above pH 8.0 *(17)*. At temperatures above 30°C, the enzyme is also unstable.

Enzyme Activity and Protein Determination

Extracellular xylanase activities were determined by incubating 0.1 mL of diluted culture filtrate with 0.9 mL of a "Birchwood" xylan suspension (10 g/L) in 0.05 M phosphate buffer (pH 5.5) for 5 min at 50°C, according to Bailey et al. *(19)*. The released reducing equivalents were measured by a colorimetric assay *(20)* using xylose solution as a standard reference. Activity units were expressed as micromoles of reducing equivalents released per min at 50°C. The amount of total protein was determined according to the Coomassie blue method described by Bradford *(21)* using bovine serum albumin (BSA) as a protein concentration standard.

Protein and Enzyme Precipitation

The concentrations of Na_2SO_4 used in the precipitations were: 5, 15, 25, 40, and 60% (w/w). The calculated amount of salt in the solid form was added slowly to a 15-mL centrifuge tube containing culture medium. After the addition of salt, all of the tubes contained 5.0 g of the mixture, which was agitated for 15 s in a vortex at pH 5.5 at room temperature (25°C). The precipitate was collected by centrifugation (6000 g, for 20 min) and then dissolved in 0.05 M acetate buffer (pH 5.5) up to the initial mass (5.0 g).

The ethanol concentrations employed in the precipitation experiments were: 10, 20, 30, 40, 50, 60, 70, and 80%. The pH of the precipitation medium was adjusted to the desired value by adding 0.10 M acetate buffer (pH 4.0 and 5.5) or phosphate buffer (pH 7.0). The ethanol was slowly added to the medium under agitation (200 rpm) and the temperature was maintained at 4°C. After the addition of ethanol, the agitation was stopped for 30 min and the mixture was centrifuged (2000 g for 30 min, Centrifuge Jouan Mod. [Saint-Herblain, France] 1812) under refrigetation (4°C). The pellet was resuspended using 0.05 M acetate buffer (pH 5.5) at room temperature (~25°C). The total protein content was determined in triplicate: in the supernatant, in the pellet, and in the initial sample for both precipitations. Enzymatic activities were determined only in the resuspended pellet and in the initial sample since the ethanol and the Na_2SO_4 present in the supernate could interfere with the methodology.

Chemicals

Birchwood 4-O-methyl-D-glucoroxylan (90% xylose) were obtained from Sigma (St. Louis, MO). All of the other chemicals were of analytical grade.

RESULTS AND DISCUSSION

In large-scale enzyme precipitations, organic solvents were more successful than protein salting out (22,23). The efficiencies of two compounds (ethanol and sodium sulfate) to recover *P. janthinellum* xylanase were studied. The cell-free filtrate of acid hemicellulose hydrolyzate culture was used as a crude enzyme mixture. The results shown in Figs. 1 and 2 indicate that recovered xylanase fractions ranged from 18 to 72% in the precipitations conducted with sodium sulfate, and from 0 to 100% in the precipitations with ethanol. In the precipitations performed with sodium sulfate, the maximal xylanase recovery (71.8%) was attained at 25% concentration (Fig. 1). At higher salt concentrations the enzyme redissolved in the supernate and the recovery yield decreased. The total protein precipitation curve showed a different behavior. The highest recovery level (68%) was observed at 40% salt concentration. A comparison between the behaviors of the curves for total protein and xylanase activity reveal that the only benefit derived from the precipitation studies is the concentration factor. A fractionation is not recommended in this case since a satisfactory increase in the purification factor cannot be achieved. As can be seen in Fig. 2, at 40% ethanol concentration, approx 30% of the total protein was precipitated, whereas 95% of the xylanase remained in the solution; however at 80% ethanol concentration, 100% of the xylanase precipitated and approx 30% of the total protein remained in the solution. These results suggest a fractionated precipitation when using ethanol as a precipitating agent. The fractionation can be performed in two steps; the first at 40% and the second

Fig. 1. Xylanase recovery (–■–) and total protein recovery (–●–) as a function of sodium sulfate concentration (%).

Fig. 2. Xylanase recovery (–■–) and total protein recovery (–●–) as a function of ethanol concentration (%).

one at 80% ethanol concentration. The purification factor can increase twofold.

CONCLUSIONS

The precipitation of xylanase from *P. janthinellum* using ethanol and Na_2SO_4 as precipitating agents was studied. Sodium sulfate proved to be inappropriate for enzyme purification, since the xylanase and total protein precipitations occurred similarly in all the concentrations tested. The ethanol precipitation showed promising results. A fractionated precipitation is possible since at 40% ethanol, 30% of the total protein precipitates, whereas 95% of the enzyme keeps soluble. On the other hand, at 80% ethanol concentration, 30% of the total protein keeps soluble and 100% of the enzyme is precipitated. In this way the purification factor can increase twofold for the concentrated enzyme.

ACKNOWLEDGMENTS

Ely V. Cortez acknowledges the financial support of CAPES/Brazil, and FAPESP/São Paulo, in the form of a Master of Science fellowship. Thanks are also due to Maria Eunice M. Coelho for revising this text.

REFERENCES

1. Biely, P. (1985), *Trends Biotechnol.* **3**, 286–290.
2. Wong, K. K. Y., Tan, L. U. L., and Saddler, J. N. (1988), *Microbiol. Rev.* **52**, 305–317.
3. Parthasarathy, V. R. (1990), *Tappi J.* **73**, 243–247.
4. Senior, D. J., Hamilton, J., Bernier, R. L., and Dumanoir, J. R. (1992), *Tappi J.*, **11**, 125–130.
5. Allison, R. W., Clark, T. A., and Wrathall, S. H. (1993), *Appita J.* **46**, 269–273.
6. Durán, N., Milagres, A. M. F., Sposito, E., and Haun, M. in (1993), *ACS Symposium Series*, Saddler, J. N. and Penner, N. H., eds., American Chemical Society, San Diego, CA, pp. 332–338.
7. Mutsaers, J. H. G. M. (1991), in *Xylans and Xylanases International Symposium*. Wageningen, The Netherlands, Novo Nordisk, p. 48.
8. Gattinger, L. D., Duvnjak, Z., and Khan, A. W. (1990), *Appl. Microbiol. Biotechnol.* **33**, 21–25.
9. Palma, M. B., Milagres, A. M. F., Prata, A. M. R., and Mancilha, I. M. (1996), *Process Biochem.* **31**, 141–145.
10. Miranda, E. A. and Berglund, K. A. (1995), *Braz. J. Chem. Eng.* **12**, 1–12.
11. El-Helow, E. R. and El-Gazaerly, M. A. (1996), *J. Basic Microbiol.* **36**, 75–81.
12. Kang, M. K., Maeng, P. J., and Rhee, Y. J. (1996), *Appl. Environ. Microbiol.* **62**, 3480–3482.
13. Chauthaiwale, J. and Rao, M. (1994), *Appl. Environ. Microbiol.* **60**, 4495–4499.
14. Kubata, B. K., Suzuki, T., Horitsu, H., Kawai, K., Takamizawa, K. (1994), *Appl. Environ. Microbiol.* **60**, 531–535.
15. He, L., Bickerstaff, G. F., Paterson, A., and Buswell, J. A. (1993), *Enzyme-Microb. Technol.* **15**, 13–18.
16. Grabski, A. C. and Jeffries, T. W. (1991), *Appl. Environ. Microbiol.* **57**, 987–992.
17. Milagres, A. M. F. (1988), M. Sc. Thesis, UFV, Brazil.
18. Vogel, H. J. (1956), *Microb. Genet. Bull.* **13**, 42–43.
19. Bailey, M. J., Biely, P., and Poutanen, K. (1992), *J. Biotechnol.* **23**, 257–271.
20. Miller, G. L. (1959), *Anal. Biochem.* **72**, 248–254.
21. Bradford, M. A. (1976), *Anal. Biochem.* **71**, 248–254.
22. Volesky, B. and Luong, J. H. T. (1983), *CRC Cri. Rev. Biotechnol.* **2**, 119–146.
23. Dunnill, P. (1983), *Process Biochem.* **18**, 9–13.

Production of Citronellyl Acetate in a Fed-Batch System Using Immobilized Lipase

Scientific Note

HEIZIR F. DE CASTRO,* DIOVANA A. S. NAPOLEÃO, AND PEDRO C. OLIVEIRA

Department of Chemical Engineering, Faculdade de Engenharia Química de Lorena, PO Box 116, 12600-000-Lorena, S. Paulo, Brazil

ABSTRACT

Several reports exist in the literature citing the decrease in conversion rates of organic-phase catalytic synthesis reactions when acetic acid is present as a reaction component. This inhibition is thought to result from damage to either the hydration layer–protein interaction or the overall enzyme structure. In this work, the inhibitory effect of acetic acid on lipase enzyme activity was ameliorated by conducting syntheses under acetic acid-limiting conditions in a fed-batch system, resulting in higher product yields. Periodic additions of acetic acid at levels of 40 mM or less gave maximum yields of 65% conversion for the reaction of citronellol and acetic acid to form citronellyl acetate. The enzyme used was a fungal lipase from *Mucor miehei*, and was immobilized on macroporous synthetic resin (a Novo lipozyme Novo Nordisk, Denmark). These results represent a fourfold improvement over batch runs reported in the literature for direct esterification of terpene alcohol with acetic acid using lipozyme as a catalytic agent.

Index Entries: Citronellyl acetate; lipase; esterification; fed-batch system.

INTRODUCTION

The use of enzymes in nonaqueous media has become a favored research topic and publications are numerous (1–3). At the research level, difficulties associated with selection of enzymes that are more compatible with nonaqueous media and establishing the somewhat unusual optimal conditions (e.g., enzyme hydration, solvent, and substrate polarity) have

*Author to whom all correspondence and reprint requests should be addressed.

been solved, at least for a few processes (4–6). Among these, bioprocessing using lipases is a well-established, useful method for the preparation of esters either by esterification or interesterification reactions (7–9).

Esters of carboxylic acids are important components of natural aromas, contributing to the flavor in most fruits and many other foods. The lipase-catalyzed synthesis of more than 50 flavoring esters has been described to date (10), and, in principle, the reaction can be carried out in a mixture of alcohol and carboxylic acid with or without solvents, resulting in very high productivities and yields (9–12). An exception is found when acetic acid is used as acyl donor because of its inhibitory effects on lipase esterification activity. Although several commercial lipase preparations have been already screened for their ability to promote synthesis of selected low-molecular-weight esters (water soluble, <C4), the majority of them display very low yields (13–16). Previous results (17,18) show that it is more difficult to prepare acetates in similar conditions than their homologs (butyrates, pentanoates, myristates), for which yields of nearly 100% are attainable (11,12). According to several researchers, the presence of acetic acid in the reaction medium can damage either the hydration layer-protein interaction or the overall enzyme structure, causing reaction inhibition (13–15). Since acetate esters are considered important flavors, several attempts have been made to minimize the inhibitory effect of acetic acid on lipase esterification activity, including the use of alcoholysis reactions (19,20).

A different approach is proposed in this work. By starting the synthesis under acetic acid-limiting conditions followed by periodic additions of acetic acid to the organic medium, higher yields are possible. The system chosen for this study was the esterification of citronellol with acetic acid using fungal lipase from *Mucor miehei* immobilized on macroporous synthetic resin (Novo Nordesk). Lipozyme has been described and characterized throughly (4,6,21,22).

MATERIAL AND METHODS

Materials

The enzyme, Lipozyme IM[20], had an activity of 24 BIU/g (1 BIU corresponds to µmol of palmitic acid incorporated into triolein per min, at standard conditions) and was kindly provided by Novo Nordisk. It was used as supplied (10% w/w moisture content). Reactants (R/S- citronellol and acetic acid) were purchased from Sigma (St. Louis, MO). All substrates were dehydrated before use, with 0.32-cm molecular sieves (aluminum sodium silicate, type 13 X, BHD Chemicals).

Citronellyl Acetate Synthesis

The esterification reactions were performed at 30°C in 250-mL round-bottomed flasks with magnetic stirrer (100 rpm). The initial substrates consisted of citronellol at fixed concentration (240 mM) and variable amounts of acetic acid (80–240 mM), using heptane as the solvent. Substrates were inoculated with 25% (w/w reactants) of lipozyme (17). To control and monitor the water level in the reaction media, syntheses were carried out in the presence of molecular sieves, as previously described (23). Acetic acid (40–120 mM) was periodically added to the reaction medium up to an equimolar ratio of 1:1 between the substrate materials.

Analysis

The reactions were monitored by measuring reactant concentrations by gas chromatography using a 6 ft 5% DEGS on Chromosorb WHP, 80/10 mesh column (Hewlett Packard), and heptanol as an internal standard. Water concentrations in the liquid and solid phases were measured by Karl Fischer method using the Karl Fischer Tritrator (Mettler DL 18). The results were evaluated by calculating the citronellol conversion rates, as follows.

$$\text{Conversion rate (\%)} = \left[(C_0 - C)/C_0\right] \times 100 \qquad (1)$$

where: C_0 = initial concentration of the reactant and C = concentration of reactant at a given time.

Estimation of Partition Coefficients

The partition coefficients (lipozyme/external organic solvent) of citronellol and acetic acid were estimated according to the following equation (11,18):

$$\text{Partition coefficient} = \left[(C_0 - C)/C\right] \times \left[(V_0/(V - V_0)\right] \qquad (2)$$

In order to estimate the lipozyme volume ($V - V_0$), a calibration curve volume of lipozyme vs mass of the lipozyme was established (volume of matrix [cm^3] = 0.41 × mass of lipozyme [g] − 0.025).

RESULTS AND DISCUSSION

Although synthesis parameters can be studied as generic factors, some of them will be very process specific. This is especially true when a more hydrophilic organic is used as substrate. In such systems, biocatalyst deactivation may occur, presumably by either the action of a toxic organic substance or by the potential deleterious effects of liquid–liquid interfaces on the structure of the biocatalyst (3). The actual mechanism of the enzyme inhibition caused by some organic media is not well-understood, although it has been suggested that the reduction of the catalytic activity could be

Table 1
Identification of the Experiment Runs for the Production of Citronellyl Acetate

Run Code	Reactants	Initial Conditions	Acetic Acid Additions		
Run 1	Citronellol	240 mM			
	Acetic Acid	240 mM	0	0	0
Run 2	Citronellol	240 mM			
	Acetic Acid	120 mM	0	0	0
Run 3	Citronellol	240 mM			
	Acetic Acid	120 MM	120 mM	0	0
Run 4	Citronellol	240 mM			
	Acetic Acid	120 mM	60 mM	60 mM	0
Run 5	Citronellol	240 mM			
	Acetic Acid	120 mM	40 mM	0	0
Run 6	Citronellol	240 mM			
	Acetic Acid	80 mM	40 mM	40mM	0
Run 7	Citronellol	240 mM			
	Acetic Acid	80 mM	40mM	40mM	40 mM

associated with conformational changes on the enzyme structure (1,3). This phenomenon appears to be related to the protein–solvent interaction and/or to the removal of water and dehydration of enzyme protein.

In the case of acetic acid, which is more soluble in aqueous phase than in organic media, it is expected that most of the acetic acid would be located in the microaqueous environment of the enzyme. Consequently, the local pH decreases, modifying the enzyme active site and making the reaction nearly impossible.

To investigate the acetic acid tolerance of lipozyme, a set of experiments was performed involving a gradual increase of acetic acid concentration in the reaction medium up to a level in which equimolar amounts of the reactants initially present. The experimental conditions and results are given in Table 1 and Fig. 1, respectively.

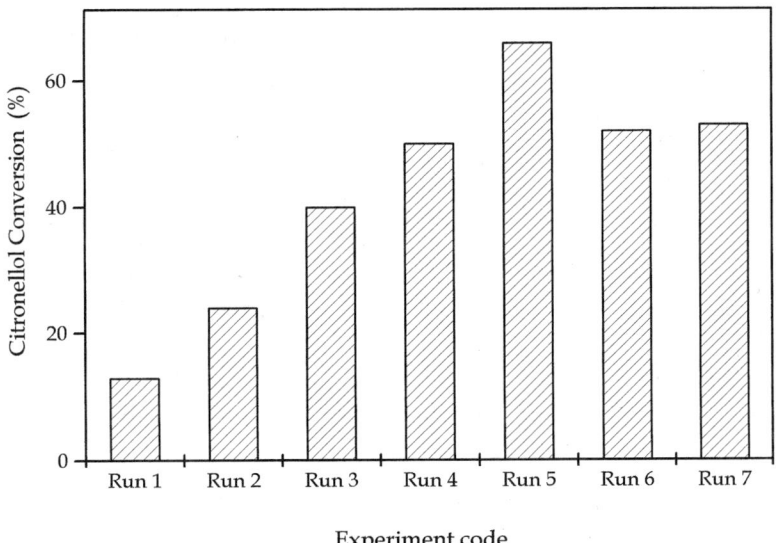

Fig. 1. Final citronellol conversion attained for different proportions of acetic acid added to the reaction medium. Reactions were carried out as described in Table 1. Batch runs 1 and 2, were used as control without any further supplementation of acetic acid.

In the presence of equimolar amounts of acetic acid and citronellol (batch run 1), little ester synthesis occurred. Average citronellol conversions were lower than 15%. Under such conditions, the acetic acid changes the polarity of the reaction medium, which in this turn modifies the partitioning of water between the solid phase (enzyme preparation) and the liquid phase (substrate), resulting in its accumulation on the enzyme solid phase *(17)*. Similar behavior was observed for a different model of study using this enzyme preparation *(24)*. This is also supported by more recent investigations of enzyme activity dependence on substrate concentration in the aqueous layer around the catalyst in which enzymatic reaction occurs *(25)*. According to these authors *(25)*, a correlation between enzyme activity and substrate partition coefficient could be determined. Previous results showed that there is a negative relationship between enzyme activity and substrate-partition coefficient (P_s); that is, the higher the substrate-partition coefficient, the lower the amount of product formed *(18)*. For this particular case, based on the coefficient partitions of each reactant, a substrate partition coefficient value of 11 was estimated, a value favoring the migration of acetic acid to the enzyme preparation (*see* Table 2).

Probably because of this, the esterification performance was improved when acetic acid was used as limiting reactant (batch run 2); however, no further citronellol conversion was attained. Since the esterification is an equimolar reaction, acetic acid should be added up to a molar ratio of 1:1 in order to increase the conversion of citronellol, as performed in runs 3–7.

Table 2
Partition Coefficients Matrix/Heptane of Reactants Estimated According to
Eq. 2 at Initial Bulk Concentration of 0.25 M, 30°C and 150 rpm.

Reactant	log P	Partition coefficient
Citronellol	3.9	0.60
Acetic acid	-0.23	6.80
Citronellol + Acetic Acid	1.3	11[a]

[a] Substrate partition coefficient was calculated on the basis of the relation between the partition coefficients of acetic acid and citronellol (see 18,25).

Fig. 2. Progress of the esterification reactions for different acetic acid additions, runs 3–7.

The response for citronellol consumption when different amounts of acetic acid were added at regular intervals (24 h) is shown in Fig. 2. For these runs, there was a significant increase in the citronellol conversion, ranging from 40 to 65%. Better performance was achieved when small levels of acetic acid (1 × 40 mM of acetic acid) were added in the reaction medium (run 5).

The concentration profiles of citronellol and acetic acid for run 5 (Fig. 3), indicate that citronellol levels gradually decrease whereas acetic acid sharply decreases in the first 24 h and then reaches a constant value. This

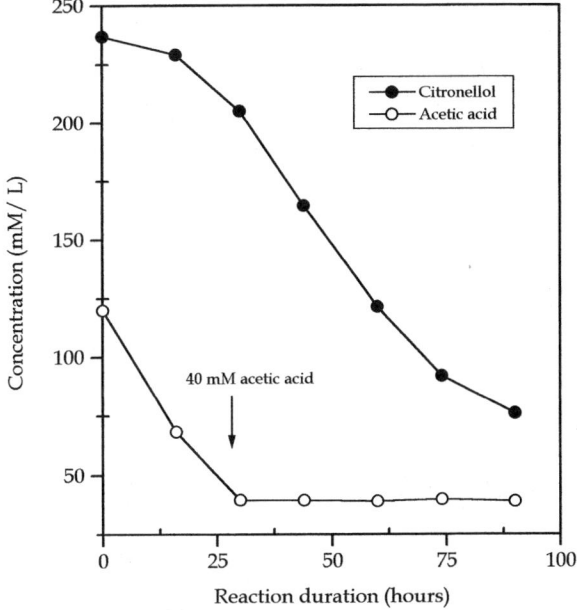

Fig. 3. Concentration profiles of citronellol and acetic acid for run 5 (40 mM of acetic acid was added after 24 h incubation).

suggests that during esterification, the consumption of citronellol is greater than the consumption of acetic acid. However, care should be taken on the interpretation of these data as the actual acetic-acid concentration should be greater than the measured one, because the largest amount of acetic acid can be adsorbed by the reaction solid phase, particularly for this enzyme preparation, which has high water-binding capacity (12).

As already mentioned, acetic acid is a potential inhibitor of the reaction. However, the results suggest that there is a concentration range in which satisfactory enzyme activity could be maintained. The approach investigated in this work provides a means to decrease the inhibitory effect of acetic acid on the enzyme activity and enhance the esterication yields to 65%, which is approx 4 times higher than the one obtained in traditional batch runs.

CONCLUSIONS

The presence of acetic acid in the reaction medium can damage either the hydration layer–protein interaction or the overall enzyme structure, causing reaction inhibition. This negative effect can be related to the high polarity of the acyl group, which promotes its migration to the solid-reaction phase (enzyme preparation). This study investigated the feasibility of carrying out the esterification of citronellyl acetate under fed-batch process. For the conditions used here, the reaction was found to work satisfactorily under acetic acid-limiting conditions, supplemented with periodic additions at levels of 40 mM or less. Such conditions gave maximum esterification yields of 65%. This represents a fourfold improvement over batch runs reported in the literature for direct esterification of terpene alcohol with acetic acid using lipozyme as a catalytic agent. However, taking into consideration the cost of the starting materials, it is still necessary to increase the conversion of citronellol in order to make the application of such a process feasible on an industrial scale.

ACKNOWLEDGMENTS

The authors gratefully acknowledge the Brazilian Research Council (CNPq) for its financial support.

REFERENCES

1. Welsh, F. H., Murray, W. D., and Williams, R. E. (1989), *Crit. Rev. Biotechnol.* **9,** 105–169.
2. Balcão, V. M., Paiva, A. L., and Malcata, F. X. (1996), *Enzyme Microb. Technol.* **18,** 392–416.
3. Scott, C. D., Scott, T. C., Blanch, H. W., Klibanov, A. M., and Russell, A. J. (1996), in *Bioprocessing in Nonaqueous Media: Critical Needs and Opportunities,* Oak Ridge National Laboratory, Report ORNL/TM-12849.
4. de Castro, H. F. and Anderson, W. A. (1995), *Química Nova* **18,** 544–554.
5. Novo Nordisk Bioindustrial Group. (1994), *Biotimes* **3,** 6.
6. Miller, C., Austin, H., Posorske, L., and Gonzalez, T. (1988), *J. Am. Oil Chem. Soc.* **65,** 927.
7. Monot, F., Benoit, Y., Ballerini, D., and Vandescasteele, J. P. (1990), *Appl. Biochem. Biotech.* **24/25,** 375.
8. Langrand, G., Triantaphylides, C., and Baratti, J. (1988), *Biotechnol. Lett.* **10,** 549–554.
9. Langrand, G., Rondot, N., Triantaphylides, C., and Baratti, J. (1990), *Biotechnol. Lett.* **12,** 581–586.
10. Vulfson, E. N. (1993), *Trends Food Sci. Technol.* **4,** 209–215.
11. Dias, S., Vilas-Boas, L., Cabral, J. M. S., and Fonseca, M. M. R. (1991), *Biocatalysis* **5,** 23.
12. de Castro, H. F., Pereira, E. B., and Anderson, W. A. (1996), *J. Braz. Chem. Soc.* **7,** 219–224.
13. Razafindralambo, H., Blecker, C., Lognay, G., Marlier, M., Wathelet, J. P., and Severin, M. (1994), *Biotechnol. Lett.* **16,** 247–253.
14. Fonteyn, F., Blecker, C., Lognay, G., Marlier, M., and Severin, M. (1994), *Biotechnol. Lett.* **16,** 693–696.
15. Manjón, A., Iborra, I. L., and Arocas, A. (1991), *Biotechnol. Lett.* **13,** 339.
16. Claon, P. A. and Akoh, C. C. (1993), *Biotechnol. Lett.* **15,** 1211–1216.
17. Napoleão, D. A. S., Silva, F. J., and Castro, H. F. (1995), in *Anais do 1°Congresso Brasileiro de Engenharia,* (COBEC-IC), São Carlos, Brazil, pp. 243–246.
18. de Castro, H. F., Oliveira, P. C., and Pereira, E. B. (1997), *Biotechnol. Lett.* **19,** 229–232.
19. Claon, P. A. and Akoh, C. C. (1994), *Biotechnol. Lett.* **16,** 235–240.
20. Claon, P. A. and Akoh, C. C. (1994), *J. Agric. Food. Chem.* **42,** 2349–2352.
21. Novo Nordisk Bioindustrials Inc. (1992), *Product Information,* B347c-GB, December.

22. Jensen, B. F. and Eigtved, P. (1990), *Food Biotechnol.* **4,** 699–725.
23. de Castro, H. F. and Jacques, S. S. (1995), *Arq. Biol. Tecnol.* **38,** 339.
24. de Castro, H. F., Anderson, W. A., Moo-Young, M., and Legge, R. L. (1992), in *Biocatalysis in Non-Conventional Media*, J. Tramper, M. H. Vermue, H. H. Beeftink, and U. von Stockar, eds., Elsevier, New York, NY pp. 475–482.
25. Yang, Z and Robb, D, (1994), *Biotechnol. Bioeng.* **43,** 336–370.

Production and Purification of Tartrate Dehydrogenase

Role of Aqueous Two-Phase Extraction

R. Harve and R. K. Bajpai*

University of Missouri-Columbia, Chemical Engineering Department, University of Missouri-Columbia, W2030 EBE, Columbia, MO

ABSTRACT

Tartrate dehydrogenase (TDH) is a stereospecific intracellular enzyme produced by *Pseudomonas putida*. Several methods for separation of nucleic acids from the proteins in cell homogenate were compared in this study. These methods included precipitation (using streptomycin sulfate, manganous sulfate, and protamine sulfate) and aqueous two-phase extraction. Under optimal conditions of separation, a single-step aqueous two-phase extraction followed by back-extraction of the enzyme from enzyme-rich PEG-phase resulted in 77% recovery of enzyme. This compared favorably with 50% enzyme recovery using protamine sulfate treatment. Furthermore, the remaining enzyme activity was accounted in the nucleic acid-rich dextran phase and the spent-PEG phase, suggesting that a multistep extraction process would increase enzyme recovery even more. Under the conditions of aqueous two-phase extraction, the selectivity of proteins over nucleic acids was 30, indicating a high degree of separation of proteins and nucleic acids in this process. The experimental data and their implications are presented.

Index Entries: Nucleic acids; proteins; bioseparation; precipitation; *Pseudomonas putida*.

INTRODUCTION

Production of intracellular enzymes involves a series of operations that start with centrifugation to concentrate the cells, followed by cell disruption to release the intracellular components, removal of nucleic acids,

---

*Author to whom all correspondence and reprint requests should be addressed.

concentration and purification of the enzyme, and finally, polishing. Each of these operations results in some loss of the product; the net result is that the final yield of the product is very low (1). In an analysis of a purification scheme of intracellular enzyme tartrate dehydrogenase produced by *Pseudomonas putida*, the total loss in enzymatic activity ranged from 72 to 84%; 52–65% of the enzymatic activity was lost in the initial stages in which the enzyme was separated from nucleic acids and other undesirable enzymes (2). These steps typically involve sequential precipitation of undesirable components that are somewhat nonspecific for their target molecules. This is in keeping with the standard practice of using low cost/volume, but low-resolution, methods in operations involving large volumes and using higher unit-cost but very specific techniques for final stages of purification (3). Yet, it will be desirable to use methods that minimize the enzyme losses while achieving the target of removing the nucleic acids and the undesirable enzymes. Because of the large losses of enzymatic activity in these initial phases, it was decided to investigate methods of precipitation of nucleic acids from the homogenates of *Pseudomonas putida* and their impact on losses of tartrate dehydrogenase. Three different precipitation agents and aqueous two-phase extraction involving PEG and dextran were evaluated and compared.

LITERATURE SURVEY

The methods for removal of nucleic acids from cell homogenates involve precipitation, extraction, and enzymatic degradation. Precipitation is induced by addition of polycationic-complexing agents that interact with the phosphate residues of nucleic acids. Cetyltrimethyl ammonium bromide (4,5), streptomycin sulfate (6), protamine sulfate (7), manganous sulfate (8), and polyethyleneimine (9,10) are some of the chemicals used as precipitating agents. The complexing between nucleic acids and the precipitating agents is nonspecific and results in simultaneous precipitation of many proteins as well. In general, conditions promoting increased precipitation of nucleic acids cause an increase in loss of enzyme too (8).

Oxenburgh and Snoswell (6) examined the conditions for precipitation of nucleic acids in extracts of *Lactobacillus plantarum* using streptomycin sulfate. The optimum pH range was established to be 6.0–8.0 and the best results were obtained with a ratio of streptomycin to proteins of 10.0 (w/w). At a protein concentration of 10 g/L, 10% (w/v) streptomycin sulfate, pH 7.0, solution conductivity of 0.38 ms/cm, protein losses were determined to be 24%. This compared very favorably with other methods. Use of protamine sulfate caused considerably higher losses of proteins. The streptomycin-sulfate precipitation was highly reproducible. Because of the interference caused by salts, Oxenburgh and Snoswell (6) found it necessary to to dialyze the ammonium-sulfate extracts before addition of streptomycin sulfate. Toxicity of streptomycin sulfate for the operators and the possible development of streptomycin-resistant microbes, were pointed out as potential disadvantages of this method. In another study involving recovery of catalase and oxaloacetate decarboxylase from the homogenates of *Micrococcus lysodiekticus*, protamine-sulfate treatment again resulted in considerable loss in enzyme activity, whereas the removal of nucleic acid was still poor. In comparison, with polyethyleneimine treatment, 90% of the nucleic acids were removed and recovery of the enzymes was 70% *(9)*. Polyethyleneimine was effective in removing large quantities of both DNA and RNA *(9)* and in selective purification of restriction endonuclease EcoRI *(10)*. Its drawbacks include potential carcinogenicity of the unreacted monomer. Higgins et al. *(8)* studied the use of heat treatment and manganous sulfate for precipitation of nucleic acids in production of β-galactosidase from *E. coli*. RNA precipitation was dependent on the concentration of manganous sulfate, but the losses in β-galactosidase activity were as high as 70%.

Precipitation of nucleic acids by the complexing agents is affected by solution pH, salt concentration, and the ratio of nucleic acids to complexing agent *(6,11,12)*. Guerritore and Bellelli *(5)* found that sodium salts (chloride, sulfate, and citrate) interfered with precipitation of nucleic acids by cetyltrimethyl ammonium bromide when present at concentrations above 0.2 M. On the other hand, glycine, glucose, and urea had no effect. In an extract of *E. coli* EM 20031, the effectiveness in precipitating nucleic acids decreased in the following order: polylysine > polyethyleneimine > cetyltrimethyl ammonium bromide > streptomycin sulfate > protamine sulfate > $MnCl_2$ > spermine > spermidine.

Aqueous two-phase extraction has been used for the separation and large-scale purification of enzymes from *Klebsiella pneumoniae (13)* and *E. coli (14)*. These methods depend upon incompatibilities of aqueous solutions of polymers such as polyethylene glycol (PEG) with solutions of salts (such as ammonium or potassium phosphate) or dextran. This results in formation of two separate phases in appropriate ranges of polymer concentrations. Proteins and nucleic acids partition differently in these two phases because of differences in surface charges and hydrophobic/hydrophilic character. As a result, proteins generally prefer the aqueous phase rich in PEG, whereas nucleic acids partition preferentially into dextran-rich phase *(15)*. Other parameters that influence partitioning of a given molecule in the two phases, include its molecular weight, the concentrations and molecular weights of the phase-forming polymers, temperature, pH, ionic strength of the mixture, and presence of polyvalent salts in the mixture. The operating conditions in terms of pH and salt concentration may be manipulated to fractionate proteins as well *(16,17)*. By coupling a ligand to PEG, partitioning of a specific protein in the PEG-rich phase can be enhanced *(18)*. An advantage of this method is that the polymers stabilize the tertiary structure and biological activities *(19)*. Despite these advantages, aqueous-phase partitioning has not become very popular because of the high cost of phase-forming polymers. Use of crude dextran and other less-expensive polymer systems has also been explored with success *(20)*.

Hydrolysis of nucleic acids by nucleases, followed by ammonium-sulfate precipitation of proteins and dialysis of the dissolved precipitate, has been found to be an effective method for removal of nucleic acids *(21–23)*. The hydrolysis can be conducted by ribonucleases and by deoxyribonucleases. The cost of nucleases is a major hindrance in use of this method in removal of nucleic acids from mixtures during protein purification.

MATERIALS AND METHODS

Tartrate dehydrogenase (TDH)-producing strain *Pseudomonas putida* ATCC 17642 were maintained and cultured in a basal salt medium prescribed by Kohn et al. *(24)*. The cells were maintained on agar plates containing the basal medium, stored at 4°C and transferred once a month. The cells were cultured according to the methods described by Tipton and Peisach *(25)*. The cells were centrifuged and homogenized by sonication. Prior to disruption, the cell paste was suspended in an equal weight of 20 mM phosphate buffer (pH 7.2) containing 1 mM dithiotreitol. Protease inhibitors, phenyl methanesulfonyl fluoride in acetone and N$^\alpha$-β-tosyllysine methanesulfonyl fluoride in water, were added to the suspension to a final concentration of 0.5 mM just before cell disruption. The suspension was sonicated thrice, each time for a duration of 2 min with a 2-min cooling period between each sonication. The disrupted-cell suspension was centrifuged at 10°C, 25000g for 30 min, the aqueous phase was recovered, stored at 4°C, and used for studies involving nucleic-acid removal.

Details of the precipitation studies are described along with the results. All the experiments were conducted at 4°C.

Protein concentration in the samples was measured using Bradford Coomassie blue method *(26)* with bovine serum albumin as a standard. The activity of TDH was measured by the method described by Tipton and Peisach *(25)*. The activity is defined as the rate of oxidation of (+) tartrate, one unit being the amount of enzyme that catalyzes the formation of 1 μmol of oxaloglycolate per minute under the standardized conditions of assay *(16)*. The oxidation rate was determined by monitoring the formation of NADH.H$^+$ at 340 nm in the reaction mixture using a spectrophotometer. For determination of the relative contents of proteins and nucleic acids in a mixture, Christian and Waarburg's method *(27)* was used.

All the chemicals used in this study were obtained from Sigma (St. Louis, MO). All salts were of analytical grade. Polyethylene glycol 8000 and dextran 70 were used to form the aqueous two-phase system.

RESULTS AND DISCUSSION

Separation of nucleic acids from proteins in the cell homogenate was studied using three precipitating agents (streptomycin sulfate, manganous sulfate, and protamine sulfate) and in an aqueous two-phase system. The homogenate used in all the experiments came from the same fermentation run. The concentrations of proteins and nucleic acids, and the activity of TDH was monitored in each phase.

Precipitation by Streptomycin Sulfate

According to Linn and Lehman (28), 0.48 g of streptomycin sulfate should be added per gram of RNA in the crude homogenate. Based on the RNA content of the cells (7% of cell dry weight; ref. 29), a value of 4.09 mmol of streptomycin sulfate per L of cell homogenate was calculated. The actual amount was varied by ±20% and by ±50% from this value to study the effect of this variable. Streptomycin sulfate was added to the crude homogenate as a powder and stirred for 10 min. pH of the solution was adjusted to 7.0–7.1 by addition of 1 N KOH. The precipitate was removed by centrifugation in a cold room and the supernate was analyzed for TDH activity and the relative contents of proteins and nucleic acids. The results are presented in Fig. 1A.

Based on these results, addition of 4 mmol streptomycin sulfate per L of crude homogenate is optimal for precipitation of nucleic acids. At this value, 74% of the tartrate dehydrogenase was still in solution. Increasing the ratio of streptomycin sulfate caused more precipitation of enzyme without increasing the precipitation of nucleic acids. More enzyme remained in solution at lower ratios, but so did more nucleic acids. As shown in Fig. 1B, the ratio of loss in enzyme activity to nucleic acid precipitated had a minimum value of 0.4 under the optimal condition. No attempt was made to recover the precipitated enzyme, as it was presumed to have lost its activity.

Precipitation by Protamine Sulfate

Protamine sulfate has been used by Tipton and Peisach (25) in their recovery process to precipitate the nucleic acid from cell homogenate. These authors recommended using 5 mg of protamine sulfate per gram of wet paste used for homogenization. In experiments conducted by Smith and Serafozo (2) TDH activity was reduced 23–30% in this process. However, nucleic-acid removal was not measured. Hence, experiments were conducted in which protamine sulfate was added to the homogenate in increments of 3 mg/g wet paste. The mixture was stirred for 10 min, after which the precipitate was removed by centrifugation and the supernate was analyzed for activity of TDH and concentrations of nucleic acids and proteins. These results are presented in Fig. 2A.

Fig. 1. Precipitation of nucleic acids with streptomycin sulfate. (**A**) Effect on nucleic acid precipitation and residual TDH activity in the supernate. (**B**) Effect on ratio of TDH lost to nucleic acid precipitation.

Fig. 2. Precipitation of nucleic acids with protamine sulfate. (**A**) Effect on nucleic acid precipitation and residual TDH activity in the supernate. (**B**) Effect on ratio of TDH lost to nucleic acid precipitation.

These results essentially duplicated the observation of 23 to 30% loss of TDH activity at 5 mg/g loading of protamine sulfate. At this value, only approx 20% of the nucleic acids would have been precipitated. Obviously, it was not the best choice of protamine-sulfate loading. The ratio of enzyme loss to nucleic acid precipitation has been plotted in Fig. 2B. This ratio decreased as the amount of protamine sulfate in the homogenate was increased and it stabilized at approx 1.0 at a 9 mg/g wet-paste loading of salt. Under these conditions, approx 50% of the nucleic acid precipitated with a concomitant loss of enzyme activity.

Precipitation by Manganous Sulfate

Melling and Atkinson (23) have referred to the high cost and other dangers of using streptomycin and protamine sulfate as complexing agents. Hence, manganous sulfate was used as a precipitating agent, as suggested by Higgins et al. (8) $MnSO_4$, H_2O was added to the crude homogenate at different concentrations ranging from 0.05 to 0.5 mol/L. The mixture was stirred for 10 min followed by centrifugation and analyses of proteins, nucleic acids, and TDH activity in the supernate. The results have been plotted in Fig. 3A. The results show that the loss in TDH activity paralleled that of nucleic acids in the whole range of concentrations of manganous sulfate. As found by Higgins et al. (8) manganous sulfate had little selectivity for nucleic acids over the desired enzyme. Even at the concentration (100 mmol/L) recommended by Higgins et al. (8) only 25% of the nucleic acids were precipitated. About the same percentage of TDH activity was lost because of the treatment. The ratio of loss in TDH activity to nucleic acids precipitated remained almost constant at approx 1.0 for manganous sulfate addition between 0.1 to 0.5 mol/L homogenate (Fig. 3B).

Based on these results, it can be concluded that streptomycin sulfate is the best precipitating agent, not only because it precipitated most nucleic acids but also because the associated losses of enzyme were the least.

Aqueous Two-Phase Extraction

Bajpai et al. (16) have investigated the partitioning of proteins in PEG 8000-dextran 70 aqueous two-phase system and established the effect of nucleic acids on the partition coefficients of proteins (defined as the ratio of protein concentration in PEG-rich phase to that in dextran-rich phase). Their protocol was used in this work to optimize the partitioning of TDH and nucleic acids in the two-phase system. The two-phase system was created by mixing 480 g dextran solution (21.78% w/w) with 133 g PEG solution (45% w/w) and 387 g of phosphate buffer to produce a mixture containing 10.0% dextran and 5.1% PEG. The volume ratio of the two aqueous phases thus produced (PEG-rich phase:dextran-rich phase) was 3:7. pH of the system could be changed by changing the pH of phosphate buffer. In partitioning experiments, phosphate buffer was replaced by the cell homogenate.

The results of preliminary experiments are shown in Table 1. Here, pH of the system was varied between 5.5 and 7.2 and the partition coefficients of total protein, TDH, and of nucleic acids in the two phases have

Table 1
Partitioning of Proteins, TDH, and Nucleic Acids from Crude Homogenate in Aqueous Two-Phase Extraction System

pH	NaCl Concentration (M)	Partition Coefficients			% TDH recovered in PEG phase
		Proteins	TDH	Nucleic acids	
7.2	0	5.28±0.15	2.03±0.38	0.91±0.03	47
6.5	0	9.30±0.07	4.26±0.21	0.72±0.04	65
5.5	0	11.03±0.56	4.61±0.03	0.49±0.08	66
6.5	5	9.51±0.81	9.20±0.04	0.29±0.02	80

Fig. 3. Precipitation of nucleic acids with manganous sulfate. **(A)** Effect on nucleic acid precipitation and residual TDH activity in the supernate. **(B)** Effect on ratio of TDH lost to nucleic acid precipitation.

been reported. Clearly, the partitioning of proteins and of TDH in the PEG phase improved as the pH was reduced. The partitioning of nucleic acids in the PEG phase decreased at the same time, pointing to improved separation of proteins and nucleic acids. Reducing pH from 6.5 to 5.5 resulted in increased losses of total enzyme activity. Moreover, from the partition coefficients, it was clear that the total proteins partitioned more favorably in the PEG phase than did TDH. Earlier work with partitioning of different proteins had suggested that this could be improved by increasing the molarity of NaCl in the solution. An experiment at pH 6.5 and 5 M NaCl concentration, showed significant improvement in this category. Under this condition, partition coefficients of total protein and of TDH were somewhat identical and the partition coefficient of nucleic acids reduced even

Table 2
Aqueous Two-Phase Extraction of Cell Homogenate (pH 6.5, 5 M NaCl) and Back Extraction of PEG Phase with Phosphate Buffer (pH 7.2); Protamine Sulfate Control

Total activity added to the biphasic system	32 units
Activity in PEG phase	27 units
Activity in dextran phase	5 units
Activity in buffer back-extracted from PEG phase	25 units
% yield in the back-extracted buffer	77 %
Total activity used for protamine sulfate control	32 units
Activity in supernate after precipitation	16 units
% yield	50%

(Loading 9 mg/g wet paste); wt of wet paste: 48 g, dry weight: 10.7 g.

further. The selectivity (ratio of partition coefficient of proteins to that of nucleic acids) was 30, suggesting a high extent of removal of nucleic acids from the crude homogenate.

In another experiment, TDH partitioned in the PEG phase (at pH 6.5, 5 M NaCl) was back-extracted by phosphate buffer at pH 7.2. These data are shown in Table 2. Under these conditions, 77% of the activity of TDH was found in the phosphate buffer back-extract. Moreover, the TDH present in dextran phase was also active. Thus the total activity that could be eventually extracted in a multistage or countercurrent aqueous two-phase extraction system could be even higher. By comparison, only 50% of the TDH activity of the crude homogenate was present in the supernate of protamine-sulfate precipitation, used as control (loading 9 mg/g wet paste).

CONCLUSIONS

Among the complexing agents evaluated for precipitation of nucleic acids from cell homogenate of *Pseudomonas putida*, streptomycin sulfate proved to be most effective. Under optimal conditions, it precipitated 65% of nucleic acids, incurring only 30% losses in the activity of desired enzyme, tartrate dehydrogenase. Protamine sulfate and manganous sulfate were not as effective. In an extraction using aqueous two-phase PEG 8000-dextran 70 system, almost all of the enzyme activity was accounted between the two phases. In a single-pass extraction back-extraction, 77% of the enzyme was recovered in the buffer system, making it potentially a very effective means of removing nucleic acids from the homogenate.

ACKNOWLEDGMENTS

Peter Tipton of Department of Biochemistry, University of Missouri, Columbia provided the *Pseudomonas* strain and assistance to Rohit Harve in carrying out this work in his laboratory. M. Smith and P. Serofozo of

Tipton's laboratory performed the initial purifications; data from these were used for evaluation of the scheme of purification. The authors gratefully acknowledge this help.

REFERENCES

1. Dunhill, P. and Lilly, M. D. (1972), *Enzyme Eng.*, **10:1**.
2. Rohit, H. (1994), *Protein purification with tartrate dehydrogenase enzyme as a model*. MS Thesis, University of Missouri-Columbia.
3. Bonnerjea, J., Oh, S., Hoare, M., and Dunhill, P. (1986), *Biotechnology* **4**, 954.
4. Jones, A. S. (1953), *Biochim. Biophys. Acta.* **10**, 607.
5. Guerritore, D. and Bellelli, L. (1969), *Nature* **184**, 1638.
6. Oxenburgh, M. S. and Snoswell, A. N. (1965), *Nature* **203**, 1416.
7. Heppel, L. A. (1955), *Methods Enzymol.* **1**, 137.
8. Higgins, J. J., Lewis, D. J., Daly, W. H., Mosqueira, F. G., Dunhill, P., and Lilly, M. D., (1978), *Biotech. Bioengin.* **20**, 159.
9. Atkinson, A. and Jack, G. (1973), *Biochim. Biophys. Acta* **308**, 41.
10. Bingham, A. H. A., Sharman, A. F., and Atkinson, A. (1977), *FEBS Lett.* **2**, 250.
11. Moskowitz, M. (1963), *Nature* **200**, 335.
12. Dinovick, R. Bayan, A. P., Canales, P., and Pansy, J. (1948), *J. Bacteriol.* **56**, 125.
13. Hustedt, H., Kroner, K. H., and Kula, M. R. (1984), *Proc. Eur. Congr. Biotechnol.* **1**, 597.
14. Takahashi, T. and Adachi, Y. (1982), *J. Biochem. (Tokyo)* **91**, 1719.
15. Albertsson, P.-A. (1985), in *Partitioning in Aqueous Two-Phase Systems: Theory, Methods, Uses, and Applications in Biotechnology*, H. Walter, D. E. Brooks, and D. Fisher eds., Academic, New York, pp. 1–10.
16. Bajpai, R. K., Harve, R., and Tipton, P. (1995), *Appl. Biochem. Biotechnol.* **54**, 193.
17. Johansson, G. and Joelsson, J. (1989), in *Separations Using Aqueous Two Phase Systems*, D. Fisher and I. A. Sutherland eds., Plenum, New York, p. 33.
18. Kopperschlager, G. and Johansson, G. (1982), *Anal. Biochem.* **124**, 117.
19. Mattiasson, B. (1983), *Trends Biotechnol.* **1**, 16.
20. Szlag, D. C. and Guiliano, K. A. (1988), *Biotechnol. Techniques* **2(4)**, 277.
21. Davidson, P. F. (1965), *Proc. Nat. Acad. Sci. USA* **45**, 1560.
22. Burgess, R. R. (1969), *J. Biol. Chem.* **244**, 6160.
23. Melling, J. and Atkinson, A. (1972), *J. Appl. Chem. Biotechnol.* **22**, 739.
24. Kohn, L. D., Packman, P. M., Allen, R. H., and Jakoby, W. B. (1968), *J. Biol. Chem.* **243**, 2469.
25. Tipton, P. A. and Peisach, J. (1990), *Biochemistry* **29**, 1749.
26. Bradford, M. M. (1976). *Analyt. Biochem.* **72**, 248.
27. Thorne, C. J. R. (1978), *Techniques in Protein and Enzyme Biochemistry*, Part 1, B104, Elsevier-North, Holland, p. 451.
28. Linn, A. and Lehman, J. K. (1964), *J. Biol. Chem.* **240**, 3.
29. Wang, D. I. C., Cooney, C. L., Demain, A. L., Dunhill, P., Humphrey, A. E., and Lilly, M. D. (1979), *Fermentation and Enzyme Technology*, Wiley, New York.

Demonstration-Scale Evaluation of a Novel High-Solids Anaerobic Digestion Process for Converting Organic Wastes to Fuel Gas and Compost

CHRISTOPHER J. RIVARD,*,[1] BRIAN W. DUFF,[1] JAMES H. DICKOW,[1] CARLTON C. WILES,[2] NICHOLAS J. NAGLE,[2] JAMES L. GADDY,[3] AND EDGAR C. CLAUSEN[3]

[1]Pinnacle Biotechnologies International, Inc., Golden, CO, [2]National Renewable Energy Laboratory, Golden, CO, and [3]Bioengineering Resources, Inc., Fayetteville, AR

ABSTRACT

Early evaluations of the bioconversion potential for combined wastes such as tuna sludge and sorted municipal solid waste (MSW) were conducted at laboratory scale and compared conventional low-solids, stirred-tank anaerobic systems with the novel, high-solids anaerobic digester (HSAD) design. Enhanced feedstock conversion rates and yields were determined for the HSAD system. In addition, the HSAD system demonstrated superior resiliency to process failure. Utilizing relatively dry feedstocks, the HSAD system is approximately one-tenth the size of conventional low-solids systems. In addition, the HSAD system is capable of organic loading rates (OLRs) on the order of 20–25 g volatile solids per liter digester volume per d (gVS/L/d), roughly 4–5 times those of conventional systems.

Current efforts involve developing a demonstration-scale (pilot-scale) HSAD system. A two-ton/d plant has been constructed in Stanton, CA and is currently in the commissioning/startup phase. The purposes of the project are to verify laboratory- and intermediate-scale process performance; test the performance of large-scale prototype mechanical systems; demonstrate the long-term reliability of the process; and generate the process and economic data required for the design, financing, and construction of full-scale commercial systems. This study presents conformational fermentation data obtained at intermediate-scale and a snapshot of the pilot-scale project.

*Author to whom all correspondence and reprint requests should be addressed.

Index Entries: High solids; anaerobic digestion; biogas; cogeneration; pilot-scale.

INTRODUCTION

The anaerobic digestion of lignocellulosic materials such as municipal solid waste (MSW) and biomass, is limited by the hydrolysis rate (1,2). The primary biodegradable polymer in MSW and biomass—cellulose—may be shielded by lignin, a relatively inert, polyphenylpropane, three-dimensional polymer (3), or by hemicellulose (4). This complex structure dictates that natural biodegradation occurs on a scale of months or years, rather than hours or days. Such slow rates of polymer degradation require long retention times, large reactor volumes (for conventional low-solids systems); thus, they result in high capital costs for large-scale application.

Because the value of the methane produced is relatively low, the anaerobic process must be rather simple in design, require little energy to operate, and have high gas production rates. The conversion process must also result in near-complete digestion to maximize energy production and residue value.

In preliminary economic evaluations of anaerobic digestion processes for producing fuel gas from solid wastes, reactor capital costs have been identified as important factors. If the reactor volume could be reduced significantly and power use maintained or decreased, the economics of the anaerobic digestion process would benefit greatly. Increasing the solids concentration within the reactor would be particularly beneficial because a decreased reactor volume is possible while the same solids-loading rate and retention time are maintained. However, high-solids slurries are very viscous and resemble solid materials more closely than typical fluids. Therefore, conventional mixers such as those employed in continuous stirred-tank reactor (CSTR) systems do not ensure homogeneity within the reactor, and problems develop in providing adequate dispersion of substrate, intermediates, and microorganisms while minimizing power requirements.

Early research on anaerobic digestion conducted at the National Renewable Energy Laboratory (NREL) under funding from the United States Department of Energy focused on enhancing the fundamental understanding of the HSAD process. A novel horizontal-shaft bioreactor was designed for laboratory studies of the process. These laboratory-scale digesters were used to study a wide variety of process parameters, including the most effective agitator designs and best procedures for adapting the anaerobic microbial consortia to high-solids levels (5), the nutrient requirements for optimum anaerobic conversion rates (6), the maximum solids levels and minimum mixing requirements (7), the maximum process organic-loading rate (8), and the minimum retention time for effective high-solids conversion (9). Laboratory-scale digesters were used to study the levels of extracellular hydrolytic enzymes in the anaerobic digestion process and their

High-Solids Organic Waste Processing

Fig. 1. Scale-up efforts detailing laboratory-, intermediate-, and pilot-scale system dimensions and operational parameters.

effects on conversion rates *(10–14)* and the effects of total solids on microbial populations *(15)*. Various organic feed stocks including sorted MSW feedstocks *(16)*, the blending of sorted MSW and food-processing wastes *(17,18)*, various agricultural residues *(19)*, dewatered sewage sludges *(20,21)*, and sewage-derived fat, oil, and grease *(22)*, were evaluated with the laboratory-scale high-solids digester. Finally, the laboratory-scale digesters were used to study the apparent horsepower requirements for mixing at various solids levels *(23)*, as well as the application of high-solids anaerobic residues as a soil amendment *(24)*.

After establishing a firm understanding of the important process parameters that affect the high-solids digestion process, research focus shifted to issues of bioreactor scale-up and the benefits of high-solids operation on commercial-scale system capital and operating costs. Independent review of the HSAD process confirmed the economic advantages of the process application *(25)*, but only laboratory-scale fermentation data were available. These data also formed the basis for a computer-modeling approach to simulating the commercial-scale costs (and benefits) of applying HSAD technology to industrial-waste disposal *(26)*. A step-wise approach to scale-up of the novel horizontal-shaft agitator bioreactor (United States Statutory Invention Record #H1149, March, 1993) is depicted in Fig. 1 and follows 40- to 50-fold scaling increments. Research efforts now focus on confirming high-solids anaerobic digestion process performance determined previously using laboratory-scale bioreactors at intermediate- and pilot-scales to enhance confidence in economic model simulations for commercial-scale technology application. The current work summarizes intermediate-scale HSAD fermentation data and outlines pilot-scale efforts.

METHODS

Intermediate-Scale High-Solids Digestion System

The unique bioreactor design allows for a plug-flow movement of feed solids during the digestion process with zonal mixing to release entrained gas pockets and provide inoculation of feedstock entering the digester. Slow-speed, tine-blade agitation enhances microbial film formation in the digester. A 1000-L (1 m^3) intermediate-scale high-solids digestion system was fabricated by Interpro (Golden, CO). The digester was approx 1.0 m (3.25 ft) in diameter, 1.75 m (5.75 ft) long, and constructed from 304 stainless steel. The vessel was fabricated with five ports for feed introduction, effluent sludge removal, product biogas removal, a 2.5 psi graphite rupture disk, and an inspection window. Feedstocks were blended with a small commercial dough mixer and continuously fed to the intermediate-scale digester with an adjustable-speed screw feeder. The digester was mounted on a 2268-kg (5000-lb) platform scale with digital readout to monitor sludge level on a weight basis. Process-effluent sludge was removed from the digester with a pneumatic-operated pinch valve at the opposite end of the digester from the feed port. Digester biogas head pressure was maintained at approx 0.5 psig with a water-filled gas bubbler. Biogas was measured with both a standard wet-gas meter and a mass-flow meter with totalizer. Digester temperature was maintained at 55°C with an electric heating blanket. Digester temperature was regulated with an internal thermocouple probe and a temperature controller. The horizontal-tine agitator assembly was designed to be similar to laboratory-scale high-solids systems and employed a hydraulic motor with a 16:1 gear reducer to maintain a constant 1 rpm shaft speed. The data acquisition and control system (DACS) used a GE Fanuc model 90/30 programmable logic controller (PLC) and Wonderware software (Intouch, Irvine, CA) on a PC-based server as the man machine interface (MMI). Daily data reports were logged in Excel spreadsheet files for retrieval and data analysis.

Feedstock

The blended feedstock for the intermediate-scale digestion studies was similar to that used in laboratory-scale studies. Analysis of the tuna sludge and MSW used during all experimental studies is shown in Table 1.

Table 1
Feedstock Characteristics

Parameter	Feedstock	
	MSW (RDF)	Tuna sludge
Total solids (%)	96.8 ± 0.2	24.9 ± 1.6
Volatile solids (as % of TS)	88.8 ± 0.5	94.2 ± 0.9
Ash (as % of TS)	11.2 ± 0.5	5.8 ± 0.9
Chemical oxygen demand (g COD/g wet wt)	1.21 ± 0.11	0.35 ± 0.03
pH	—	5.3
Volatile fatty acids (C2–C5, mM)	—	117.9 ± 14.2
Free ammonia (g/L)	—	0.78 ± 0.08

Analytical Analysis

Analysis of feedstock and process-residue samples for total solids, volatile solids, ash, pH, free ammonia, chemical oxygen demand (COD), volatile fatty acids, and biogas composition were as previously described *(21,27)*.

RESULTS

Intermediate-Scale HSAD Development

Design, fabrication, and start-up of the intermediate-scale, high-solids digestion system is documented elsewhere *(28,29)*. HSAD performance data obtained during a 2-y effort that employed the intermediate-scale digestion system are described in Table 2. Data obtained for process organic-loading rates (OLRs) of 18 and 20 gVS/L/d were taken at three retention times and may not fully represent steady-state data. However, the average anaerobic bioconversion (as determined by feedstock COD reduction) is in the low 80% range. This level of conversion at OLRs of 4–20 gVS/L/d confirms previous data determined for laboratory-scale HSAD studies that used the blended tuna sludge and MSW feedstock *(30)*. High conversion rates may also be attributed to longer solids retention times (i.e., 16 d at 20 gVS/L/d loading), which are a consequence of a lower water content in the feedstock.

Pilot-Scale HSAD Development

Confirmation of laboratory-scale high-solids data at intermediate-scale further enhanced confidence in the process economics to pursue pilot-scale system development. In addition, equipment selection for material-handling issues that relate to continuous-feed addition, process monitoring, data collection, and effluent removal increased confidence in the ability to effectively design larger-scale systems. Following com-

Table 2
Intermediate-Scale High-Solids Digester Performance at Various OLRs

Organic loading rate (gVS/L/d)	4	8	10	12	14	16	18	20
Biogas production (L/L/d)	3.19 ± 0.31	5.62 ± 0.52	8.10 ± 0.52	9.88 ± 0.90	10.51 ± 1.10	11.50 ± 1.09	12.61 ± 0.99	12.94 ± 1.46
Methane content (%)	57.7 ± 3.4	63.4 ± 3.1	57.5 ± 1.2	56.0 ± 1.4	58.8 ± 1.7	60.1 ± 2.0	59.7 ± 2.3	59.1 ± 2.8
Methane production (L/L/d)	1.84 ± 0.18	3.56 ± 0.33	4.66 ± 0.36	5.53 ± 0.50	6.18 ± 0.69	6.91 ± 0.81	7.53 ± 0.88	7.65 ± 1.10
COD loading (g COD/L/d)	5.9	11.7	14.6	17.5	20.4	23.3	26.2	29.2
% Bioconversion[a]	85.5 ± 8.5	83.9 ± 7.7	87.7 ± 6.7	86.8 ± 7.9	83.1 ± 7.2	81.2 ± 8.1	78.6 ± 6.9	71.4 ± 9.1
Sludge total solids (%)	26.3 ± 1.3	26.7 ± 1.8	26.2 ± 2.1	26.5 ± 1.9	26.8 ± 3.2	27.2 ± 2.9	27.8 ± 3.1	28.2 ± 4.1
Sludge pH	7.4 ± 1.1	7.4 ± 1.8	7.3 ± 1.6	7.3 ± 1.1	7.3 ± 1.8	7.3 ± 1.4	7.2 ± 2.1	7.2 ± 1.9
Sludge VFAs (C2–C5, mM)	23.1	32.1	29.1	39.1	44.4	59.7	54.7	59.2
Sludge ammonia (g/L)	1.3	1.3	1.3	1.2	1.3	1.3	1.3	1.2

[a] The percent anaerobic bioconversion was determined from the COD loading to the process using the relationship of 1 g COD is equivalent to 0.35 L of methane. All methane production values were first corrected for STP prior to calculating the percent bioconversion.

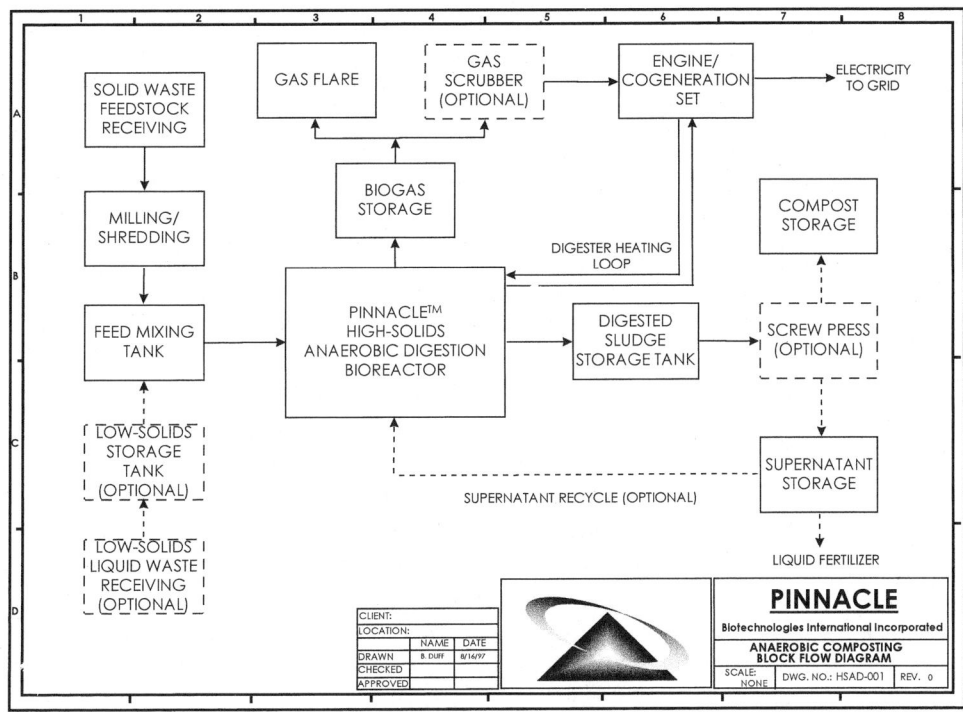

Fig. 2. Block-flow diagram of the pilot-scale HSAD process installed at Stanton, CA and used to confirm smaller-scale fermentation system performance for MSW and tuna sludge feedstocks.

petitive procurement practices, a contract was awarded (December, 1994) to Bioengineering Resources (Fayetteville, AR) to design, fabricate, install, and operate a 2 ton/d anaerobic digestion pilot system for treating a combined feedstock of MSW and tuna sludge. Collaborators on the project included Black and Veatch (Kansas City, MO) for engineering work and Envirex (Waukesha, WI) for design and fabrication of the digester system. Stanton, CA was selected for the pilot plant site to locate the system close to the MSW sorting facility of the project's partner, CR&R. This site is also close to the source of tuna processing sludge, provided by Tri Union International (Terminal Island, CA). Local site permitting assistance was provided by The Planning Center (Newport Beach, CA), and SCEC (Orange, CA).

Early design engineering work consisted of a collaboration by Black & Veatch, Bioengineering Resources, and developers at NREL. The block-flow design for the pilot-scale system is shown in Fig. 2. The design, fabrication, installation, and permitting stages are complete. Commissioning of equipment and start-up (including inoculation of the pilot-scale digester) are underway.

DISCUSSION

Fermentation performance data for applying high-solids anaerobic digestion of MSW, biomass, and food processing wastes were determined to be very successful at laboratory- and intermediate-scale. Economic projections that used the data developed to date also demonstrate that HSAD provides a cost-effective alternative to normal routes for disposal of organic wastes such as landfilling, aerobic composting, and incineration. Early assessments of high-solids anaerobic digestion performance that used the novel, horizontally mixed design developed at NREL indicate substantial product-yield improvements over competing designs (31). Comparison of the HSAD system with conventional, low-solids designs and even with alternative high-solids designs has demonstrated a 10 to 65% greater yield in methane production. This may be attributed to the design of the system, which encourages both biofilm production and plug-flow operation. However, pilot-scale fermentation performance data are necessary to further refine the projected economics for commercial-scale applications (these kinetic data form the basis for computer-model-simulating programs). Operating the pilot-scale system also allows the unique set of equipment chosen for integration to be verified; thereby addressing important solids-handling issues that pertain to applying this technology to a wide variety of solid organic feedstocks. The successful commercialization of the HSAD process would provide many community and industrial leaders with an alternative, environmentally responsible, method for recycling organic wastes.

ACKNOWLEDGMENTS

This work was funded by the Municipal Solid Waste Management Program of the United States Department of Energy.

REFERENCES

1. Boone, D. R., (1982), *Appl. Environ. Microbiol.* **43,** 57–64.
2. Noike, T., Endo, G., Chang, J-E., Yaguchi, J-I., and Matsumoto, J-I., (1985), *Biotechnol. Bioeng.* **27,** 1482–1489.
3. Sarkanen, K. V. and Ludwig, C. H. (1971), *Lignins: Occurrence, Formation, Structure and Reactions,* Wiley-Interscience, New York.
4. Grohmann, K., Torget, R., and Himmel, M. E. (1985), *Biotech. Bioeng. Symp.* **15,** 59–80.
5. Rivard, C. J., Himmel, M. E., Vinzant, T. B., Adney, W. S., Wyman, C. E., and Grohmann, K. (1989), *Appl. Biochem. Biotech.* **20/21,** 461–478.
6. Rivard, C. J., Adney, W. S., Vinzant, T. B., and Grohmann, K. (1989), *J. Environ. Health.* **52,** 96–100.
7. Rivard, C. J., Himmel, M. E., Vinzant, T. B., Adney, W. S., Wyman, C. E., and Grohmann, K. (1990), *Biotech. Lett.* **12,** 235–240.
8. Rivard, C. J. (1993), *Appl. Biochem. Biotech.* **39/40,** 71–82.

9. Vinzant, T. B., Adney, W. S., Grohmann, K., and Rivard, C. J. (1990), *Appl. Biochem. Biotech.* **24/25,** 765–771.
10. Adney, W. S., Rivard, C. J., Grohmann, K., and Himmel, M. E. (1989), *Biotech. Appl. Biochem.* **11,** 387–400.
11. Adney, W. S., Rivard, C. J., Grohmann, K., and Himmel, M. E. (1989), *Biotech. Lett.* **11,** 207–210.
12. Rivard, C. J., Adney, W. S., and Himmel, M. E. (1991), in *Enzymes in Biomass Conversion,* Leatham, G. and Himmel, M. E., eds., American Chemical Society Books, Washington, DC, **460,** pp. 22–34.
13. Adney, W. S., Rivard, C. J., Shiang, M., and Himmel, M. E. (1991), *Appl. Biochem. Biotech.* **30,** 165–183.
14. Rivard, C. J., Nieves, R. A., Nagle, N. J., and Himmel, M. E. (1994), *Appl. Biochem. Biotech.* **45/46,** 453–462.
15. Rivard, C. J., Nagle, N. J., Adney, W. S., and Himmel, M. E. (1993), *Appl. Biochem. Biotech.* **39/40,** 107–117.
16. Rivard, C. J., Vinzant, T. B., Adney, W. S., Grohmann, K., and Himmel, M. E. (1990), *Biomass* **23,** 201–214.
17. Rivard, C. J., (1992), in *Proceedings of the 1992 Food Industry Environmental Conference,* GTRI, Atlanta, GA, 119–127.
18. Rivard, C. J., and Nagle, N. J. (1992), in *Second United States Conference on Municipal Solid Waste Management: Moving Ahead,* U.S. Environmental Protection Agency, Arlington, VA, pp. 1–10.
19. Rivard, C. J. (1992), in *Energy from Biomass and Wastes XVI,* Klass, D. L., ed., Institute of Gas Technology, Chicago, IL, pp. 1025–1041.
20. Nagle, N. J., Rivard, C. J., Adney, W. S., and Himmel, M. E. (1992), *Appl. Biochem. Biotech.* **34/35,** 737–751.
21. Rivard, C. J., and Nagle, N. J. (1996), *Appl. Biochem. Biotech.* **57/58,** 983–991.
22. Rivard, C. J., and Nagle, N. J. (1993), in *Proceedings of the 1993 Food Industry Environmental Conference,* GTRI, Atlanta, GA, pp. 71–80.
23. Rivard, C. J., Kay, B. D., Kerbaugh, D. H., Nagle, N. J., and Himmel, M. E. (1995), *Appl. Biochem. Biotech.* **51/52,** 155–162.
24. Rivard, C. J., Rodriguez, J. B., Nagle, N. J., Self, J. R., Kay, B. D., Soltanpour, P. N., and Nieves, R. A. (1995), *Appl. Biochem. Biotech.* **51/52,** 125–135.
25. Engineering Science, Inc. (1993), *Preliminary Economic Analysis and Comparison for Three Anaerobic Digestion Processes for Treatment of Municipal Solid Waste and Tuna Sludge for Full Scale Technology Application in American Samoa,* NREL Subcontract Report #AE-2-12258-1, Golden, CO.
26. Ruth, M. and Landucci, R. (1994), *Anaerobic Digestion Analysis Model: User's Manual,* NREL/TP-421-6331, Golden, CO.
27. APWA-AWWA-WPCF (1980), *Standard Methods for the Examination of Water and Wastewater Analysis.* 15th ed., APHA, Washington, D.C.
28. Rivard, C. J. (1995), *Intermediate-Scale High-Solids Anaerobic Digestion System: Operational Development,* DOE/R4/10591-1, Golden, CO.
29. Interpro, Inc. (1994), *Intermediate-Scale High Solids Anaerobic Reactor Control System Documentation,* NREL Subcontract Report #941030, Golden, CO.
30. Rivard, C. (1993), *Anaerobic Digestion as a Waste Disposal Option for American Samoa,* NREL/TP-422-5043, Golden, CO.
31. Rivard, C. J. (1996), in *Deploying Anaerobic Digesters: Current Status and Future Possibilities,* Lusk, P., Wheeler, P., and Rivard, C., eds. NREL/TP-427-20558, pp. 28–35.

Recycling of Process Streams in Ethanol Production from Softwoods Based on Enzymatic Hydrolysis

KERSTIN STENBERG,[1] CHARLOTTE TENGBORG,[1] MATS GALBE,[1] GUIDO ZACCHI,*,[1] EVA PALMQVIST,[2] AND BÄRBEL HAHN-HÄGERDAL[2]

[1]Department of Chemical Engineering I and [2]Department of Applied Microbiology, Lund University, P. O. Box 124, SE-221 00 Lund, Sweden

ABSTRACT

In ethanol production from lignocellulose by enzymatic hydrolysis and fermentation, it is desirable to minimize addition of fresh-water and waste-water streams, which leads to an accumulation of substances in the process. This study shows that the amount of fresh water used and the amount of waste water thereby produced in the production of fuel ethanol from softwood, can be reduced to a large extent by recycling of either the stillage stream or part of the liquid stream from the fermenter. A reduction in fresh-water demand of more than 50%, from 3 kg/kg dry raw material to 1.5 kg/kg dry raw material was obtained without any negative effects on either hydrolysis or fermentation. A further decrease in the amount of fresh water, to one-fourth of what was used without recycling of process streams, resulted in a considerable decrease in the ethanol productivity and a slight decrease in the ethanol yield.

Index Entries: Ethanol production; recycling; softwood; inhibition; steam pretreatment.

INTRODUCTION

Ethanol can be produced from lignocellulosic materials by enzymatic hydrolysis and fermentation (1–3). Before efficient hydrolysis can occur, the material must be pretreated to make the lignocellulose more susceptible to enzymatic attack (4,5). Steam pretreatment at high temperatures is a method that is often utilized. The efficiency of the pretreatment can be enhanced by the addition of a catalyst, such as sulfur dioxide or sulfuric

*Author to whom all correspondence and reprint requests should be addressed.

Fig. 1. Schematic flowsheet of ethanol production based on enzymatic hydrolysis.

acid *(6–8)*, which has been shown to be of special importance when softwoods are employed *(9–11)*. During pretreatment, different sugar- and lignin-degradation byproducts are formed that can be inhibitory to hydrolysis *(12–14)* and fermentation *(15,16)*.

In a process for production of ethanol from lignocellulosic materials, it is highly desirable to minimize the addition of fresh-water and wastewater streams. Recycling of process streams decreases the use of fresh water, and minimizes effluent volume. However, this leads to an accumulation of nonfermentable substances and inhibitors in the process *(17)*. Investigations to evaluate the effect of recycling have been performed in previous studies on willow and softwood in a bench-scale unit *(18,19)*. The most inhibiting substances were found to be nonvolatile, which is in agreement with the results from another study *(16)*.

The aim of the present study was to investigate the effects of recirculation of different process streams on hydrolysis and fermentation of a softwood material. Figure 1 shows a schematic flowsheet of the process, including the recirculation alternatives investigated. The fresh-water stream (S1) in the hydrolysis step can be replaced by part of either the stillage stream (S2) or the dilute-ethanol stream from the fermenter (S3). The latter alternative increases the ethanol concentration in the feed to distillation, which results in lower energy costs for distillation.

MATERIALS AND METHODS

The experimental set-up is shown schematically in Fig. 2. The study was performed in a bench-scale unit comprising the following units: a steam-pretreatment reactor, a hydrolysis reactor, a fermenter, a filter press, and an evaporator *(18)*. Five different experiments including hydrolysis, fermentation, and evaporation, (called the base case, R1, R2, R2E, and R3E) were run. In the base-case run, fresh water was used in the hydrolysis for

Recycling of Process Streams

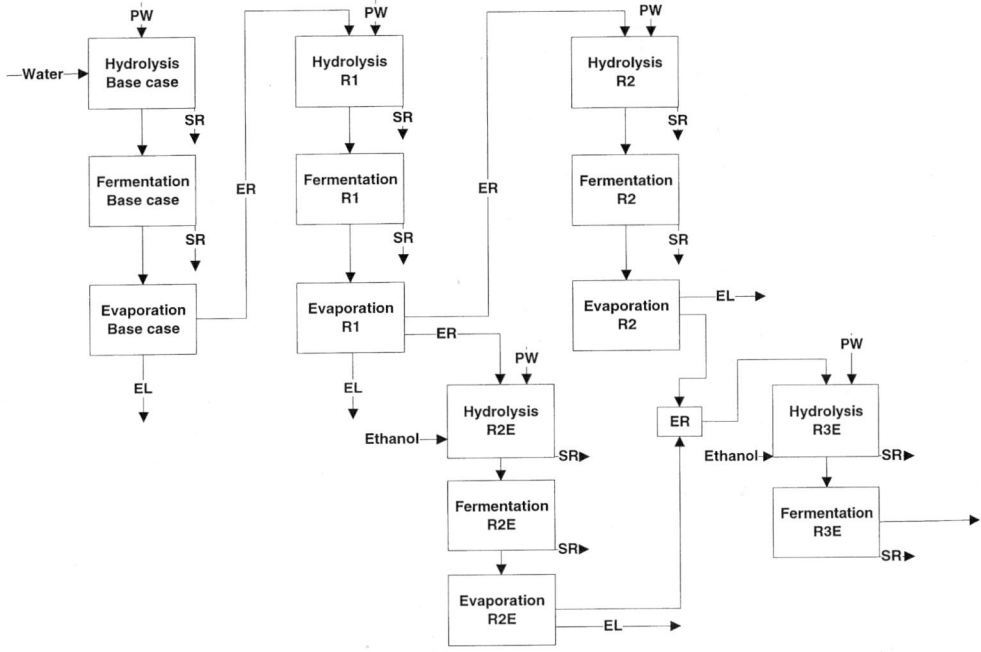

Fig. 2. Experimental procedure. Pretreated wood (PW); Evaporated liquid (EL); Evaporated Residue (ER); Solid Residue (SR).

dilution of the pretreated material. In runs R1 and R2, the fresh water in the hydrolysis stage was replaced by recycled liquid, produced by evaporation of the filtered fermentation broth from a preceding run, thus simulating recycling of the stillage stream (Fig. 1, stream S2). The same simulated recycling was performed in runs R2E and R3E, except that ethanol was added to simulate splitting of the process stream from the fermenter (Fig. 1, stream S3) to the distillation unit into two streams, one of which is recycled to the hydrolysis unit.

Raw Material

Chips of freshly cut spruce, free from bark were generously provided by a sawmill, Höörsågen AB (Höör, Sweden). The wood chips were re-chipped and sieved and the fraction between 2 and 20 mm was used. The fractionated material had a dry matter content of 41.7%. The composition was determined according to the Hägglund method *(20)* (Table 1).

Steam Pretreatment

Prior to pretreatment, the wood chips were impregnated with sulfur dioxide (3.6% SO_2 wt/wt dry matter). The material was placed in a plastic bag, and SO_2 was supplied from a gas cylinder. The amount of SO_2 added

Table 1
Composition of Spruce

Composition	% Dry matter
Extractives	1.0
Galactan	1.8
Glucan	43.9
Mannan	12.0
Arabinan	1.1
Xylan	4.9
Lignin	28.1

to the bag was estimated by weighing the cylinder. The absorbed amount of SO_2 was determined by weighing the bag before and after SO_2 addition. After 20 min at room temperature, the treated material was steam pretreated at 215°C for 5 min. After steam pretreatment, a sample was collected, washed, and the dry matter content and the yield of fibrous material were determined. The liquid fraction was analyzed for glucose, mannose, furfural, 5-hydroxy-2-methylfurfural (HMF), and acetic acid. Steam pretreatment was performed in three separate batches because of the limited capacity of the steam-pretreatment equipment, which has a reactor volume of 10 L. The first batch was used in the base case run, the second in run R1, and the third in the following three runs.

Hydrolysis

Enzymatic hydrolysis was performed at 40°C. The base case run, run R1, and R2 were performed in a stirred tank with a working volume of 20 L. Runs R2E and R3E were carried out in a 10-L fermenter (Bioengineering, Wald, Switzerland) to minimize the evaporation of ethanol. Liquid, either as tap water (base case run) or as nonvolatile evaporation residue from a previous run (runs R1, R2, R2E, and R3E) (see Fig. 2) was added to adjust the dry matter content to 7.5% (wt dry matter/wt total). The added liquid constituted approx one third of the total amount of liquid present in the hydrolysis. In hydrolysis runs R2E and R3E, ethanol was added to the liquid to concentrations of 2.3% (wt/wt liquid) and 4.4%, respectively. This corresponds to the ethanol concentration obtained when fermentation broth is recycled, resulting in concentration factors of 2.2 and 3.1, (Table 2).

The pH was preadjusted to 4.8 with solid calcium hydroxide and 10% (wt/wt) sodium hydroxide was then used to maintain the pH at 4.8 during hydrolysis. Novo Celluclast 2 L, 0.15 g/g fibrous material, supplemented with 0.03 g/g fibrous material of β-glucosidase in the form of Novozym 188 was added to perform the hydrolysis. Both enzyme preparations were

Table 2
Concentration Factors
for Nonvolatile Substances

Run	CF
Base case	1
R1	1.6
R2	2.2
R2E	2.2
R3E	3.1

kind gifts from Novo Industri A/S (Bagsvaerd, Denmark). The activity of Celluclast was 75 FPU/g (21). The β-glucosidase activity in Celluclast was 12 IU/g (22), and in Novozym it was 392 IU/g. Hydrolysis was allowed to proceed for 96 h and samples were withdrawn at regular intervals and analyzed for glucose, mannose, furfural, 5-hydroxy-2-methylfurfural (HMF), glycerol, and acetic acid. The solid residues after hydrolysis and after fermentation were separated from the liquid with a filter press unit, PF 0.1H2 (Larox OY, Helsinki, Finland). A pressure of 15 bar was applied to the slurry.

Fermentation

Fermentation of the filtered hydrolysates was performed in a Bioengineering NL22 fermenter with a working volume of 16 L, using compressed baker's yeast, *Saccharomyces cerevisiae*, from Jästbolaget AB (Rotebro, Sweden). The pH was not adjusted after hydrolysis but maintained at 4.8 during fermentation with 10% (wt/wt) sodium hydroxide. The hydrolysates were supplemented with nutrients to a final concentration of 0.5 g/L $(NH_4)_2HPO_4$ and 0.025 g/L $MgSO_4 \cdot 7H_2O$.

The hydrolysates were inoculated with yeast to 10 g dry weight/L (after inoculation) and incubated at 30°C. The broth was stirred at 300 rpm. Samples were withdrawn at regular intervals, and the fermentation was allowed to proceed until a glucose stick (Boehringer, Mannheim, Germany) was negative, or for a maximum of 25 h. The samples were analyzed for glucose, mannose, furfural, 5-hydroxy-2-methylfurfural (HMF), and acetic acid.

Evaporation

The filtered liquid from the fermentation stage was concentrated in an evaporation unit (18). The evaporation residue was collected and used in the subsequent hydrolysis to replace water. The concentration factor (CF) was defined as the ratio between the amounts of nonvolatile solubles in the actual hydrolysis and in the hydrolysis in the base case run (Table 2). The CF was calculated by weighing both the fermentation liquid prior to evaporation and the amount of evaporated liquid, and then compensating for the dilution of the recycled liquid in the next hydrolysis stage.

Analysis

The liquid fractions after pretreatment, hydrolysis, and fermentation were analysed on an HPLC (Shimadzu, Kyoto, Japan) equipped with a refractive index detector (Waters Millipore, Milford, CT). Glucose, ethanol, furfural, HMF, acetic acid, and glycerol were determined using an Aminex HPX-87H column (Bio-Rad, Hercules, CA) at 45°C, using 5 mM H_2SO_4 as eluant, at a flow rate of 0.6 mL/min. Mannose was determined using a Polymer Labs (Shropshire, UK) PL Hi-Plex Pb column at 80°C, using ultrapure water as eluant, at a flow rate of 0.4 mL/min. The dry-matter content of the washed steam-pretreated material was determined by drying the material at 105°C overnight. The dry weight of the yeast was determined by drying the samples in a microwave oven for 15 min.

RESULTS

All yields are expressed as g/100 g dry raw material, unless otherwise stated. The pretreated material was produced in three individual pretreatment runs, resulting in dry matter contents of 11.4, 10.9, and 12.2%, respectively. The yields of fibrous material after these three pretreatment runs were 59.9, 61.3, and 64.1 g/100 g, respectively. The yields of glucose and mannose in the liquids from pretreated material were approximately the same in all three pretreatment runs: 7 and 8 g/100 g, respectively.

The hydrolysis rates during the first 20 h were approximately the same for the first four runs, 0.63 g/l h, whereas run R3E exhibited a lower hydrolysis rate, 0.35 g/l h, (Fig. 3). The hydrolysis yield of glucose was between 13 and 16 g/100 g and resulted in an overall yield, i.e., for both pretreatment and hydrolysis, of 20–24 g/100 g. Since the pretreated material employed in the base case run, run R1, and runs R2, R2E, and R3E originated from three separate batches, it is difficult to draw any conclusions as to whether the increased recirculation or variations in pretreatment caused the decrease (Table 3). The amount of mannose obtained in the hydrolysis was approx 1 g/100 g.

The formation of ethanol in the fermentation for the five runs is shown in Fig. 4A. The productivity was calculated from the fermentation curves as the average ethanol production rate during five hours, r_{5h}, (Table 4). The productivities were approximately the same for the base case run and run R1, 3 g/L/h. In runs R2, R2E, and R3E they were considerably lower. For run R3E, r_{5h} was difficult to calculate because of the high initial ethanol concentration in the fermentation broth. However, the glucose consumption rates confirmed the trend of declining productivity (Fig. 4B).

Table 3
The Yield of Glucose and Mannose (in g/100 g)

Run	Glucose (in hydrolysis)	Glucose (overall)	Mannose (overall)
Base case	15.2	21.9	8.1
R1	16.1	23.6	8.4
R2	13.5	20.5	8.9
R2E	14.3	21.3	9.3
R3E	13.0	20.0	8.4

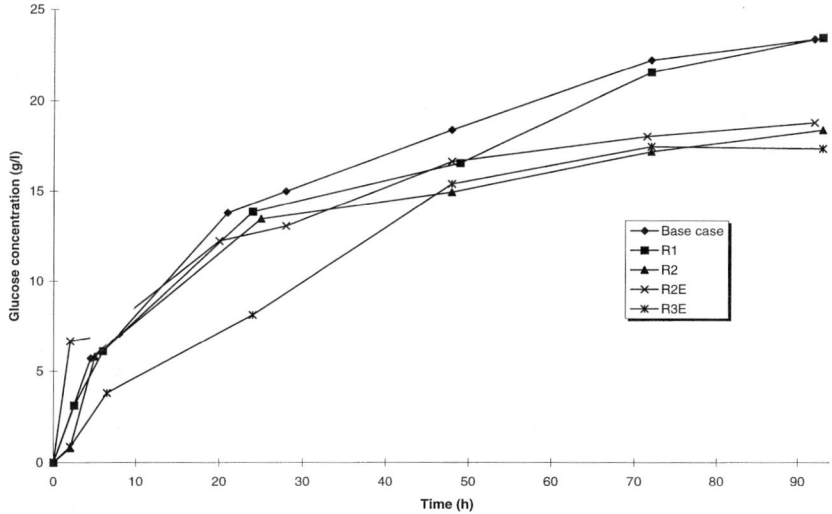

Fig. 3. Hydrolysis rates in the various runs.

The fermentation yield in the base case run was 85% of the theoretical yield, based on fermentable sugars in the hydrolysate (Table 4). The fermentation yields were somewhat lower in runs R1, R2, and R2E, and significantly reduced in run R3E. The overall ethanol yield was approx 13 g/100 g in the base case run and in run R1. In runs R2 and R2E, the overall yield was somewhat lower, approx 10–12 g/100 g, whereas run R3E resulted in only 7.6 g/100 g (Table 4).

The concentrations of HMF and furfural in the hydrolysates were all less than or equal to 2 g/L (Table 5). The acetic acid concentration increased because of the recirculation from 2.2 g/L in the base case fermentation to 7.1 g/L in fermentation R3E. A similar trend was observed for the glycerol concentration.

Table 4
Ethanol Yield in Fermentation, Overall Yield, and Initial Productivity

Run	Productivity, r_{5h} (g/L/h)	Yield of ethanol in fermentation (% of theoretical)	Overall yield of ethanol (g/100 g)
Base case	3.0	85.3	13.1
R1	2.9	79.9	13.0
R2	0.9	66.1	9.9
R2E	1.4	74.8	11.7
R3E	0.6	52.7	7.6

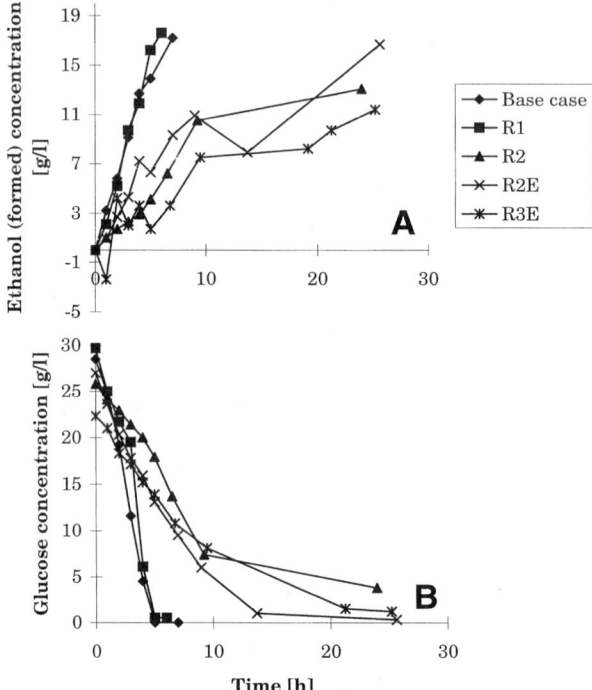

Fig. 4. **(A)** Ethanol formation rates; **(B)** Glucose consumption rates in the various runs.

DISCUSSION

In the present study, the influence of increasing concentration of soluble compounds on hydrolysis and fermentation, when fresh water is replaced by process liquids, has been investigated. However, in the fermentation it was only possible to evaluate the effect of inhibitors on ethanol production, and not on cell growth, because of the high initial cell concentration.

Table 5
The Concentrations of Byproducts in the Hydrolysate

Run	HMF (g/l)	Furfural (g/l)	Acetic acid (g/l)	Glycerol (g/l)
Base case	1.2	0.8	2.2	0.2
R1	1.7	1.0	4.3	2.0
R2	1.4	0.6	4.6	2.6
R2E	1.4	0.7	4.3	2.4
R3E	2.0	0.8	7.1	4.1

Table 6
Amount of Water Used and Required Dry-Matter Content After Pretreatment

Run	Amount of fresh water used (kg/kg dry raw material) In pretreatment (as steam)	In hydrolysis	Required dry matter content after pretreatment (%)
Base case	1.4	1.6	12
R1	1.4	–	12
R2	0.8	–	16.5
R2E	0.8	–	16.5
R3E	0.3	–	23

The base case requires the addition of fresh water corresponding to 3 kg/kg dry raw material, which is added as 1.4 kg/kg steam in the pretreatment stage (flash vapor excluded) and 1.6 kg/kg water in the hydrolysis stage. This is based on a raw material having a moisture content of 50%, a fiber yield after pretreatment of 60%, and an initial dry-matter content in the hydrolysis stage of 7.5%. The amount of fresh water can be reduced by replacing the fresh water in hydrolysis with recycled process liquid. This was simulated experimentally in runs R1, R2, R2E, and R3E. In runs R1 and R2, recycling of the stillage stream from the distillation unit, and in runs R2E and R3E recycling of the outlet stream from the fermenter were simulated (Fig. 1). By replacing all the fresh water used in hydrolysis (run R1) the remaining water requirement will be 1.4 kg/kg for steam pretreatment (Table 6). This corresponds to a concentration of nonvolatile substances 1.6 times higher than in the base case run (see Table 2).

To reduce the amount of fresh water even further, a drier material after pretreatment is required (see Table 6). This can be accomplished in an alternative type of pretreatment equipment, using indirect heating, or by using a drier raw material. Runs R2, R2E, and R3E (which correspond to a higher dry matter content after pretreatment) were simulated by concentrating the recycled process liquid further, to estimate the effect of the increased inhibitor concentrations. The procedure used in the present experimental investigation assumes that the same amount of inhibitors is formed independent of the dry-matter content in the pretreatment step.

The hydrolysis yield was not influenced by recycling, whereas the rate decreased in run R3E. In fermentation, neither the yield nor the productivity was influenced when the fresh water was replaced by process liquid in run R1 (Tables 3 and 4). However, when the amount of fresh water was further decreased the yield, as well as the productivity, decreased.

The concentrations of HMF and furfural were approximately the same in all the runs, 2 g/L and 1 g/L, respectively (Table 5). Furfural was not recycled since it is rather volatile and was removed in the preceding evaporation stage. HMF is nonvolatile and should therefore increase, but analyses of the HMF concentrations before and after fermentation showed that the yeast consumed HMF. This has also been shown previously (23). At these low concentrations, furfural or HMF do not cause any significant inhibition (15,24). Acetic acid is distributed between the volatile and nonvolatile fractions and increases with increasing degree of recycling. In runs R1, R2, and R2E, the acetic acid concentrations were about the same. Various references on the toxicity of acetic acid can be found in the literature. However, the exact inhibiting concentration of acetic acid is difficult to determine (25,26).

The presence of ethanol in run R2E did not affect the yield or the productivity in the fermentation step compared with run R2 in which the same liquid was used except for the addition of ethanol. This is in accordance with another study showing that inhibition starts at 25 g/L and is total at 95 g/L (26). Neither was the presence of ethanol in run R3E the major cause of the decreased yield and productivity, since the ethanol concentration in this case was also low, approx 50 g/L. However, it is not possible to draw any conclusions regarding the synergistic action of ethanol, acetic acid, and other byproducts present during fermentation. It is more likely that the decrease in productivity and yield in fermentation runs R2, R3, and R3E originated from the increased concentration of nonvolatile lignin-degradation products formed at the high temperatures employed during steam pretreatment (27).

Mannose was not completely consumed in any of the fermentation runs, so the initial concentration in the fermentation broth increased from 12 to 16 g/L when process liquids were recycled. The concentration of mannose remaining after fermentation was in the range 2–5 g/L. The affinity of *S. cerevisiae* for mannose is one tenth that for glucose, and mannose will not be utilized until the glucose concentration has reached a sufficiently low level (28). Thus, the ethanol yields might be improved by increasing the fermentation time.

The low overall yield of ethanol (13 g/100 g DM) was mainly caused by inefficient hydrolysis. The yields of fermentable sugars after hydrolysis were lower than expected, also in the absence of recycled liquid. This is most likely caused by inhibitors already present in the pretreated material. In laboratory scale experiments, the hydrolysis-rate and yield increased by approx 50% when the fibrous material was washed and the hydrolysis was performed in buffer solution (data not shown). The inhibitory effect of the liquid was also reduced when the enzyme load was doubled, and the glucose yield increased by approx 25%. A similar result has been obtained in a study in which eucalyptus was used as raw material (29). This suggests that the inhibition may be caused by product inhibition by the sugar-rich liquid from the steam-pretreated material. This is also supported in a study on simultaneous saccharification and fermentation (SSF) of softwoods in which higher yields and productivities have been achieved (data not shown). The relatively low yield may also be caused by difficulty in stirring the material properly. At 7.5% dry-matter content, the mixture is rather viscous, thus hampering mass transfer in the tank. In a previous study, where mixed softwoods were hydrolyzed at 5% dry-matter content, a higher yield was obtained (19). In general, softwoods are more recalcitrant to hydrolysis than hardwoods (4,6).

CONCLUSIONS

This study shows that the amount of fresh-water used, and thus the amount of waste-water produced, in the production of fuel ethanol from softwood, can be reduced to a large extent by recycling of either the stillage stream or part of the liquid stream from the fermenter. A reduction in fresh-water requirement by more than 50% from 3 kg/kg dry raw material was obtained without any negative effects on either hydrolysis or fermentation. Recycling of the liquid stream from fermentation increases the ethanol concentration in the fermentation broth. No negative effects on fermentation were observed with an ethanol concentration in hydrolysis of 2.3 wt% (run R2E). The ethanol concentration in the feed to the distillation unit increased from 1.8 to 3.4 wt% compared with the case of no recycling. This will reduce the energy demand in distillation by 42% (30). A further decrease in the amount of fresh water, to one fourth of what was used without recycling of process streams, resulted in a considerable decrease in the ethanol productivity and a slight decrease in the ethanol yield.

ACKNOWLEDGMENTS

The Swedish National Board for Industrial and Technical Development (NUTEK) is gratefully acknowledged for its financial support.

REFERENCES

1. Vallander, L and Eriksson, K. E. (1990), *Adv. Biochem. Eng/Biotech.* **42**, 63–95.
2. Katzen, R. and Fowler, D. E. (1994), *Appl. Biochem. Biotech.* **45/46**, 697–707.

3. Lynd, L. R., Cushman, J. H., Nichols, R. J, and Wyman, C. E. (1991), *Science* **251**, 1318–1323.
4. Grethlein, H. E. and Converse, A. (1991), *Biores. Technol.* **36**, 77–82.
5. Schell, D. J., Torget, R., Power, A., Walter, P. J., Grohmann, K., and Hinman, N. D. (1991), *Appl. Biochem. Biotechnol.* **28/29**, 87–97.
6. Ramos, L. P., Breuil, C., and Saddler, J. N. (1992), *Appl. Biochem. Biotechnol.* **34/35**, 37–48.
7. Mackie, K. L., Brownell, H. H., West, K. L., and Saddler, J. N. (1985), *J. Wood Chem. Technol.* **5(3)**, 405–425.
8. Eklund, R., Galbe, M., and Zacchi, G (1995), *Bioresour. Eng.* **52**, 225–229.
9. Mamers, H. and Menz, D. N. J. (1984), *Appita* **37(8)**, 644–649.
10. Clark, T. A. and Mackie, K. L. (1987), *J. Wood. Chem. Technol.* **7(3)**, 373–403.
11. Schwald, W., Smaridge, T., Chan, M., Breuil, C., and Saddler, J. N. (1989), Enzyme Syst. Lignocellul. Degrad., [Proc. Workshop Prod., Charact. Appl. Cellul.-, Hemicellul.-LigninDegrading Enzyme Syst.], 231–242. Coughlan, M. P., ed., Elsevier, London, UK.
12. Sinitsyn, A. P., Clesceri, L. S., and Bungay, H. R. (1982), *Appl. Biochem. Biotechnol.* **7(6)**, 455–458.
13. Mes-Hartree, M. and Saddler, J. N. (1983), *Biotechnol. Lett.* **5(8)**, 531–536.
14. Dekker, R. F. H. (1988), *Appl. Microbiol. Biotechnol.* **29**, 593–598.
15. Clark, T. A. and Mackie, K. L. (1984), *J. Chem. Tech. Biotechnol.* **34B**, 101–110.
16. Palmqvist, E, Hahn-Hägerdal, B., Galbe, M., and Zacchi, G. (1995), *Enzym. Microbiol. Technol.* **19**, 470–476.
17. Galbe, M. and Zacchi, G. (1993), Biotechnology in Agriculture 9, Saddler, J. N., CAB International, Wallingford, UK, pp. 291–321.
18. Palmqvist, E., Hahn-Hägerdal, B., Galbe, M., Larsson, M., Stenberg, K., Szengyel, Z., Tengborg, C., and Zacchi, G. (1996) *Biores. Technol.* **58**, 171–179.
19. Larsson, M., Galbe, M, and Zacchi G. (1997), *Biores. Technol.*, **60**, 143–151.
20. Hägglund, E. (1951), in *Chemistry of Wood*, Academic, New York.
21. Mandels, M., Andreotti, R., and Roche, C. (1976) *Biotechnol. Bioeng. Symp.* **6**, 21–33.
22. Berghem, L. E. R. (1974), *Eur. J. Biochem.* **46**, 295–305.
23. Sanchez, B. and Bautista, J. (1988), *Enzyme Microb. Technol.* **10**, 315–318.
24. Boyer, L. J., Vega, J. L., Klasson, K. T., Clausen, E. C., and Gaddy, J. L. (1992), *Biomass Bioenergy* **3(1)**, 41–48.
25. Maiorella, L. J., Blanch, H. W., and Wilke, C. R. (1983), *Biotechnol Bioeng.* **25**, 103–121.
26. Linden, T., Peetre, J., and Hahn-Hägerdal, B. (1992), *Appl. Environ. Microbiol.* **58(5)**, 1661–1669.
27. Buchert, J., Puls, J., and Poutanen, K. (1989), *Appl. Biochem. Biotechnol.* **20/21**, 309–318.
28. Nevado, J., Navarro, A., and Heredio, G. F. (1994), *Yeast* **10**, 59–65.
29. Ramos, L. P., Breuil, C. and Saddler, J. N. (1993), *Enzyme Microb. Technol.* **15**, 19–25.
30. Busche, R. M. (1983), *Biotech. Bioeng. Symp.* **13**, 597–615.

Biological-Chemical Treatment of Soils Contaminated with Exploration and Production Wastes

BHUPENDRA. K. SONI,* J. ROBERT PATEREK, SALIL PRADHAN, AND VIPUL. J. SRIVASTAVA

Institute of Gas Technology, 1700 S. Mount Prospect Road, Des Plaines, Illinois 60018

ABSTRACT

Oil-gas exploration and production (E&P) soils contaminated with total petroleum hydrocarbons (TPHs) have been tested for degradation by two different treatments: biological and chemical. Biological treatment includes the use of native microorganisms for transformation of the various hydrocarbons found in E&P soils. Degradation of TPH of 80 and 86%, was achieved for two different soils, respectively in control experiments. The effect of growth stimulants such as glucose, acetic acid, and valeric acid was examined on TPH degradation. Incorporation of inducer (valerate) enhanced the degradation up to 89 and 93%, for the two soils, respectively. A large portion (> 41%) of contaminant in one soil was comprised of compounds in the carbon range of C_{10}–C_{16} and < 7% constituted carbon range of C_{24}–C_{28}. The degradation of C_{10}–C_{16} compounds was higher (> 98%) as compared to C_{24}–C_{28} compounds (< 75%). Likewise, the degradation rate was also higher (58 mg/kg/d) for lower compounds as compared to higher carbon range compounds (6.7 mg/kg/d). Experiments conducted on chemical treatment included the effect of chelators on stabilization of H_2O_2, comparative studies between buffer and water (used for soil preparation), and the effect of pH on TPH degradation. The rate of oxygen evolution from H_2O_2 was significantly reduced with use of either chelated iron or phosphate buffer using naphthelene as a model compound. Chemical treatment demonstrated a higher degradation of TPH from contaminated soils at pH 4.0 as compared to a pH of 7.0. More degradation was obtained with slurry prepared in phosphate buffer as compared to deionized water.

Index Entries: Biological treatment; chemical treatment; total petroleum hydrocarbons (TPHs).

---

* Author to whom all correspondence and reprint requests should be addressed.

INTRODUCTION

Oil and Gas Exploration and Production Soils

The drilling and operation of gas-exploratory wells and the operation of natural-gas production wells generate a number of waste materials. They are usually stored and/or processed at the drilling/operations site. The waste materials can include: oil-based drilling muds and cuttings; water-based drilling muds and cuttings; site-related soil or sediments contaminated by the drilling and gas recovery activities; brines, including drilling and produced water; production pit and storage tank sludge; produced oily sands and solids; residual gas condensates; vapors and odors; and natural-gas-plant-processing wastes.

The most common contaminants associated with natural-gas wells are hydrocarbons associated with the natural-gas condensates such as: benzene, ethyl benzene, toluene, and the xylenes (para-, ortho-, meta- moieties), which are commonly designated as BTEX; total petroleum hydrocarbons (TPHs); and polynuclear aromatic hydrocarbons (PAHs) from the formation's natural gas deposits.

Institute of Gas Technology (IGT) has developed innovative technology for the treatment of several hydrocarbon wastes such as PAHs *(1)* and PCBs (polychlorinated biphenyls) *(2)*.

Biological-Chemical Treatment Process (BCT)

Bioremediation is the basis of many treatment systems under development at IGT. Microorganisms exhibit abilities under laboratory conditions to degrade nearly all contaminants examined. The rates and extent of these activities might be low as compared to the requirements of the treatment technology. The enhancements under study are designed to accelerate the rate and/or extent of degradation.

IGT has developed and is investigating the application of an integrated biological-chemical treatment (BCT) process for the remediation of soils contaminated with polynuclear aromatic hydrocarbons *(1,3–11)*. A schematic diagram of the process is presented in Fig. 1. The chemical treatment can be performed as a pretreatment before the biological degradation or can be integrated as a step between biological treatments. The process uses a mild chemical treatment with Fenton's reagent that produces hydroxyl radicals that react with the organic contaminants. The contaminants are modified to forms that are more readily degraded by native or supplemented microorganisms. Results with approx 25 MGP (manufactured gas plant) soils show that the CBT process is capable of enhancing the rate, as well as the extent, of PAH degradation. This integrated process generates environmentally benign products, that is, carbon dioxide (CO_2), inorganic salts, biomass, and water.

Fig. 1. TPH-REM process: different alternatives.

The application of a biological-chemical treatment to E&P-contaminated soils has been under study for 2 yr. Preliminary results indicated that this biological-chemical treatment exhibited success levels warranting further research and development (9). The research presented here examined the chemical treatment on two E&P soils under pH-regulated conditions. The laboratory research falls into two major areas of IGT's chemical treatment: oxygen evolution and pH optimization.

MATERIALS AND METHODS

Soil Sample Preparation

Soil samples were collected from E&P sites by various organizations and shipped to IGT for study. These soils with PIT ID no. of 6689 and 1260 were designated as soil 2 and 3 for further reference. The samples were transported to IGT's laboratories and stored at 4°C until used. Soil was sieved (0.25-in screen) prior to use. The soil was supplemented with the nutrient medium containing the following components for biological treatment per liter: NH_4Cl, 1.0 g; KH_2PO_4 0.5 g; and K_2HPO_4 0.5 g.

Determination of Oxygen Evolution

The oxygen evolution, with chelated iron and control (no chelating compound), was conducted in glass bottles (50 mL). The glass bottles were sealed with rubber septum and aluminum crimp. The glass syringes (50 mL) were used to collect the gas produced in the head space. Sodium citrate (10 mM) was used for the chelation of ferrous sulfate (10 mM). Naphthelene (7219 PPM) was used as the sole carbon source for these studies. The test samples were prepared both in phosphate buffer and deionized water (pH 5.0). At least two different pHs, 5.0 and 7.0, were tested using phosphate buffer (0.05 M). The oxygen-evolution studies were started by the addition of H_2O_2 to achieve a final concentration of 2% (w/v).

Effect of pH and Buffer on Degradation of TPH Compounds

The effect of pH was tested by varying the pH in the range of 4.0 to 7.0 using phosphate buffer (0.05 M). A 20% soil slurry was prepared in glass bottles (160 mL) by addition of 10 g of soil in 50 mL of buffer. Sodium citrate and ferrous sulfate were each added at the final concentration of 10 mM. The treatment was started by addition of H_2O_2 (2% w/v). In some microcosoms, soil slurry was prepared in water to compare the performance with the buffer used for slurry preparation.

Soil-Treatability Studies

The soil treatability studies were conducted in glass bottles (160 mL). The nutrient medium, as described above, was used for the biological treatment. Several growth inducers such as acetic acid, glucose, and valeric acid were tested for enhancement in TPH degradation. For chemical treatments, air-dried soils were weighed (4–10 g) and 20–50 mL of phosphate buffer (pH 4.0-7.0) was added to prepare a 20% soil slurry. Ferrous sulfate (10 mM) and sodium citrate (10 mM) were added in all studies. In one set of tests, deionized water was added at the initial pH of 5.0. H_2O_2 was added to achieve a final concentration of 2% (w/v). The treatment was allowed to react for 1–3 wk and liquid samples were evaporated to dryness and extracted with methylene chloride for analysis of TPHs.

Gas Chromatography

All extracts were analyzed by capillary column gas chromatography (Varian 3400 gas chromatograph with flame ionization detector, autosampler/controller). The sieved soil was dried overnight and soil extracts were prepared by shaking the soil vigorously with methylene chloride for 3–5 min. The Wisconsin DNR PUBL-SW-141 (modified DRO method) had three choices of extractant solvents: methylene chloride; hexane, and carbon disulfide. The extraction procedure was repeated three more times and all the aliquots were combined. The column used for separation was a 30 m EC-Wax (Carbowax) with an internal diameter of 0.75 mm and a 0.25 μm film thickness (Alltech, Deerfield, IL). The column temperature was kept at 32°C (1-min holding) followed by a programming of 18°C/min up to 300°C (2-min holding). The injector and detector temperatures were maintained at 270 and 290°C, respectively. The flow rate of make-up gas (nitrogen) was 32 mL/min and air flow was 300 mL/min. The split ratio was maintained at 10:1. The total peaks detected were integrated to generate a total petroleum hydrocarbon (TPH) value. The standardization was done using the external standard method with various hydrocarbon mixtures. The chromatographic parameter, extraction protocols, and standardization methods were taken from the U.S. Environmental Protection Agency's test methods for Evaluating Solid Waste, SW-846. Final concentrations were reported on a soil dry-weight basis.

RESULTS AND DISCUSSIONS

Biological Treatment

Stimulation of Bacterial Consortia by Use of Volatile Fatty Acids and Glucose as Carbon Source

E&P soils 2 and 3 were evaluated for stimulation of indigenous microflora using glucose, acetic acid, and valeric acid (1 g/l) as additional

Table 1. Composition of Contaminants in Soil 3

Compounds	%
C_{10}-C_{16}	41.1
C_{16}-C_{20}	32.0
C_{20}-C_{24}	20.6
C_{24}-C_{28}	6.3

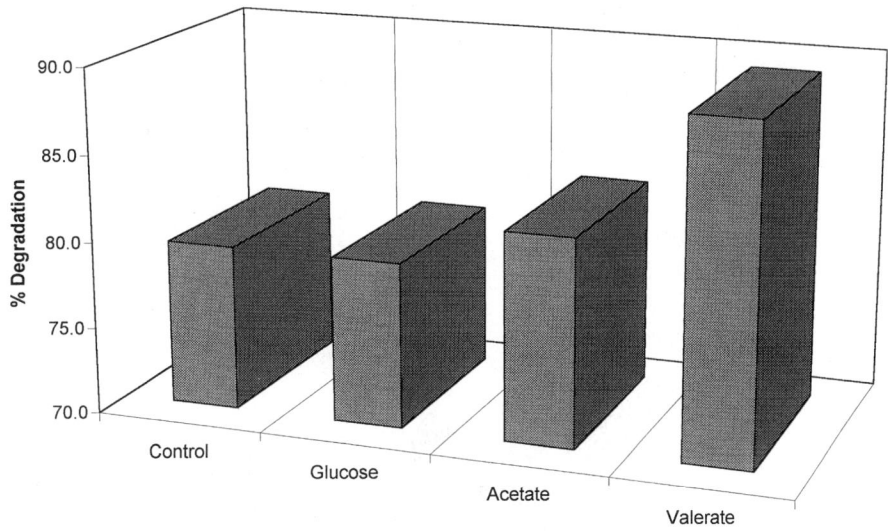

Fig. 2. Effect of various inducers on TPH degradation of soil 2.

growth factors. The studies were continued for 3 wk and samples were analyzed for degradation of TPHs. The results summarized (Figs. 2 and 3) showed that degradation of TPH was 80 and 86% for soils 2 and 3, respectively. Incorporation of the inducer, valerate, did enhance the degradation further to 89 and 93%, for soils 2 and 3, respectively. Statistically this is in the same range as noted without any inducer. Soil 3 was further analyzed for composition of individual TPHs degradation. The compositions (Table 1) of contaminants in soil 3 indicate that over 41% of contaminants fall in the carbon range of C_{10}–C_{16} and less than 7% in the carbon range of C_{24}–C_{28}. Despite the presence of higher proportion of lower-carbon-range compounds, the degradation of lower-carbon-range compounds (C_{10}–C_{16}) was higher (> 98%) as compared to higher (C_{24}–C_{28}) (Fig. 4) carbon-range compounds (> 74%). Likewise, the degradation rate (Fig. 5) was also higher in the lower-carbon-range compounds (58 mg/kg/d) as compared to higher-carbon-range components (6.7 mg/kg/d). These findings indicate that lower-carbon-range compounds are degraded to a

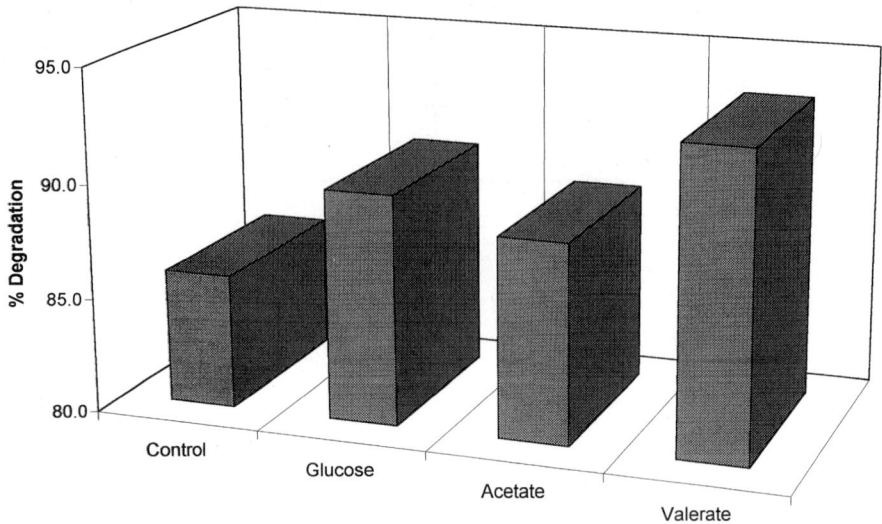

Fig. 3. Effect of various inducers on TPH degradation of soil 3.

Fig. 4. Degradation of different groups of TPH compounds by biological treatment.

larger extent (> 98%) and at faster rates (58 mg/kg/d) compared to higher-carbon compounds. These experiments have suggested to us that none of the inducers were useful for significantly enhancing the degradation of TPHs. One of the reasons for this could be attributed to higher degradation of TPHs without any inducer (> 85%). These results suggest that TPH degradation does not need any externally added inducers.

Fig. 5. Degradation rate of different TPH compounds by biological treatment.

Chemical Treatment

Determination of Oxygen Evolution With and Without Chelating/Stabilizing Agents at Different pHs

Hydrogen peroxide decomposes readily to water and oxygen at room temperature. The decomposition of H_2O_2 can be minimized by the use of stabilizing agents. The evolution of oxygen from H_2O_2 was studied under different conditions: water at pH 7.0 and 5.0; sodium citrate as chelator (10 mM); phosphate buffer (0.05 M) at pH 5.0 and 7.0. The objective was to determine the stabilization/chelation obtained by addition of sodium citrate and phosphate buffers. Naphthalene was used as model substrate for these studies at the initial concentration of 7219 mg/L. The results (Fig. 6) suggested a significant amount of oxygen production at pH 7.0 using deionized water alone with H_2O_2 and $FeSO_4$ (10 mM) addition. Incorporation of sodium citrate (10 mM) resulted in a marked reduction in oxygen production. Experiments conducted at a lower pH of 5.0 did not show oxygen production with or without addition of chelator (results not shown). In other studies with phosphate buffer (pH 5.0 and 7.0), no oxygen was produced at the lower pH of 5.0. The overall degradation of naphthalene was between 6 and 24%. It may be concluded from these experiments that loss of hydrogen peroxide for unproductive oxygen can be minimized by use of different stabilizers/chelators such as citrate or phosphate salts. It is important to mention here that soluble form of iron is important for Fenton's reaction. This soluble form is essentially achieved either at lower pH (< 4.0) or alternatively by using chelators. Our results have suggested that addition of either sodium citrate or phosphate buffer has stabilized the decomposition of H_2O_2 even at higher pH as evidenced by lower oxygen

Fig. 6. Evolution of oxygen with and without sodium citrate in naphthelene-added reactor. (—◆— sodium citrate, —□— control).

evolution. However, at lower pH (5.0), the oxygen evolution was lower suggesting a stabilization without addition of external compounds. Fenton's reaction involves several series of reactions. The production of oxygen from HO_2 radical is dependent on pH. It was suggested that the rate constant for oxygen production from HO_2 is pH dependent with a value of 2×10^4 at lower pH (< 3.0) to a value of 1×10^6 L/mole-sec at higher pH (near neutral) (12).

Optimization of pH for Degradation of TPHs in Contaminated Soils

The objective of this study was to determine the pH for maximum degradation of TPHs. The optimization of pH was done by using phosphate buffers at several different pH (4.0–7.0). The results (Fig. 7) showed that at the lower pH of 4.0, TPH degradation of greater than 64% was achieved. The degradation was significantly reduced at the higher pH of 7.0. Replacing the buffer with deionized water resulted in lower degradation of TPHs in soil-slurry system. The degradation of TPH in buffer was twice as compared to deionized water used for preparation of slurry (Fig. 8).

CONCLUSIONS

The research conducted to-date has successfully demonstrated the removal of TPHs in soils by biological and integrated approaches. The degradation of up to 93% was achieved by biotreatment when valerate was used as a growth inducer. The major proportion ($> 41\%$) of the contaminants were in the carbon range of C_{10}–C_{16}. The degradation of over

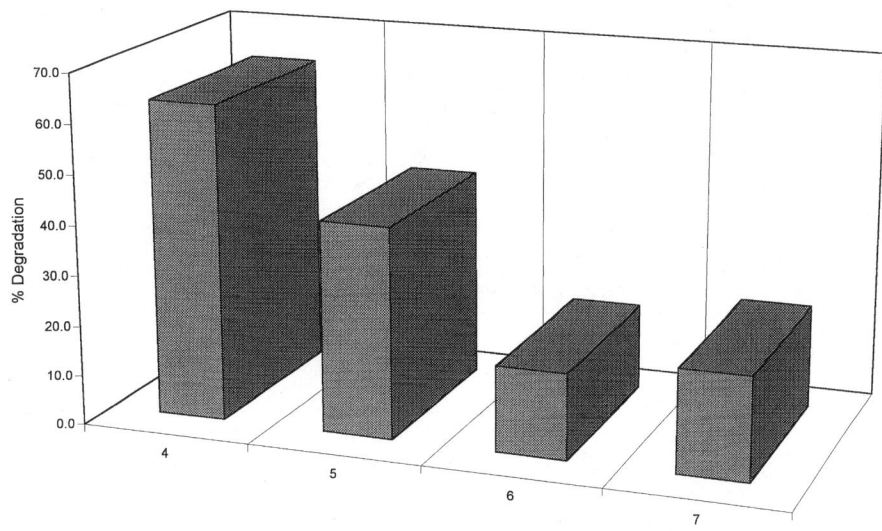

Fig. 7. Effect of pH on degradation of TPHs from soil slurry.

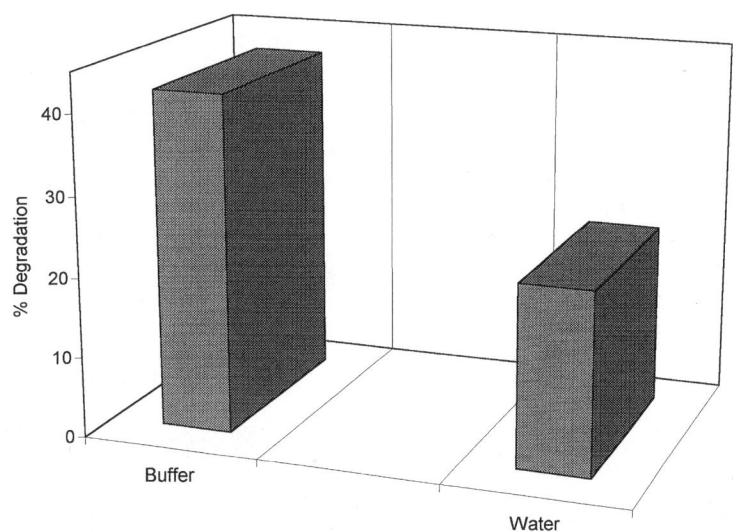

Fig. 8. TPH degradation: a comparison of buffer and water used for preparation of soil slurry.

98% was achieved for the lower-carbon compounds (C_{10}–C_{16}) as compared to 75% for higher-carbon-range compounds (C_{24}–C_{28}). Likewise, the degradation rate was also higher for degradation of lower-carbon-range compounds (58 mg/kg/d) as compared to high carbon compounds (6.7 mg/kg/day). Lower pH range of 4.0–5.0 is more effective for chemical oxidation of TPH components. Use of phosphate buffer has shown better degradation as compared to deionized water.

ACKNOWLEDGMENTS

The research program presented in this paper is possible through funding from the Gas Research Institute; IGT's Sustaining Membership Program; Exxon Corporation; Remediation Technologies, Inc.; and several gas industry members. The authors recognize the significant technical contributions of the staff of IGT, including John Conrad and Vesna Trisic, as well as the staff of the Gas Research Institute; Exxon Corporation; Remediation Technologies, Inc.; and supporting gas companies.

REFERENCES

1. Gauger, W. K., Kilbane, J. J., Kelley, R. L., and Srivastava, V. J. (1989), in *IGT's Second International Symposium on Gas, Oil, Coal, and Environmental Biotechnology*, New Orleans pp. 171–203.
2. Soni, B. K., Srivastava, V. J., Kayser, K., and Kelley, R. L. (1995), in *Eighth International IGT Symposium on Gas, Oil, and Environmental Biotechnology*, Colorado Springs, CO. Institute of Gas Technology, Des Plaines, IL, pp 105–120.
3. Kelley, R. L., Gauger, W. K., and Srivastava, V. J. (1990), in *Third International IGT Symposium on Gas, Oil, and Environmental Biotechnology*, Colorado Springs, CO. Institute of Gas Technology Des Plaines, IL pp. 105–120.
4. Kelley, R. L., Srivastava, V. J., Conrad, J. R., Paterek, J. R., and Liu, B. Y. (1993), in *Sixth International IGT Symposium on Gas, Oil, and Environmental Biotechnology*, Colorado Springs, CO. Institute of Gas Technology, Des Plaines, IL pp. 453–469.
5. Kelley, R. L., Soni, B. K., Conrad, J. R., Misra, B., and Srivastava, V. J. (1996), in *Ninth International IGT Symposium on Gas, Oil, and Environmental Biotechnology*, Colorado Springs, CO. Institute of Gas Technology, Des Plaines, IL.
6. Liu, B. Y., Srivastava, V. J., Paterek, J. R., Pradhan, S. P., Pope, J. R., Hayes, T. D., Linz, D. G., and Jerger, D. E. (1993), in *Sixth International IGT Symposium on Gas, Oil, and Environmental Biotechnology*, Colorado Springs, CO. Institute of Gas Technology, Des Plaines, IL pp. 119–133.
7. Liu, B. Y., Pradhan, S., Srivastava, V. J., Pope, J. R., Hayes, T. D., Linz, D. G., Proulx, C., Jerger, D. E., and Woodhull, P. M. (1994), in *Seventh International IGT Symposium on Gas, Oil, and Environmental Biotechnology*, Colorado Springs, CO. Institute of Gas Technology, Des Plaines, IL pp. 221, 222.
8. Paterek, J. R. (1993), in *Hazardous and Environmental Sensitive Waste Management in the Gas Industry Symposium*, Albuquerque, NM. Institute of Gas Technology, Des Plaines, IL.
9. Paterek, J. R., Aronstein, B., Rice, L. E., Jackowski, K. A., Srivastava, V. J., and Hayes, T. D. (1994), in *Seventh International Biotechnology Symposium on Gas, Oil, and Environmental Biotechnology*, Colorado Springs, CO. Institute of Gas Technology, Des Plaines, IL, pp. 461–469.
10. Paterek, J. R., Aronstein, B. N., Rice, L. E., Srivastava, V. J., Chini, J., and Bohrnerud, A. (1994), in *Seventh International IGT Symposium on Gas, Oil, and Environmental Biotechnology*, Colorado Springs, CO. Institute of Gas Technology, Des Plaines, IL pp. 63–75.
11. Srivastava, V. J., Kelley, R. L., Paterek, J. R., Hayes, T. D., Nelson, G. L., and Golchin, J. (1994), *Appl. Biochem. Biotechnol.* **45/46**, 741–755.
12. Watts, R. J., Udell, M. D., and Leung S. W. (1992), in *Proceedings of the First International Symposium, Chemical Oxidation: Technology for the Nineties*, Vanderbilt University, Nashville, TN, February 20–22, 1991, Eckenfelder W. W., eds. et al. Technomic Publishing, Lancaster, PA, pp. 37–50.

Recovery and Refining of Au by Gold-Cyanide Ion Biosorption Using Animal Fibrous Proteins

SHIN-ICHI ISHIKAWA* AND KYOZO SUYAMA

Laboratory of Applied Biochemistry, Faculty of Agriculture, Tohoku University, Sendai, 981, Japan

ABSTRACT

Animal fibrous proteins (AFPs) such as egg-shell membrane (ESM), chicken feather (CF), wool, silk, or elastin are an intricate network of stable and water-insoluble fibers with high surface area and are abundant bioresources. Every AFP tested was found to accumulate gold-cyanide ion from aqueous solutions in high yield, depending on pH and some other parameters. Gold-cyanide ion is adsorbed by AFP at low pH range, with maximum binding observed at approx pH 2.0. Under the certain conditions, gold-cyanide ion was accumulated up to 8.6, 7.1, 9.8, 2.4, and 3.9% of dry weight on ESM, CF, wool, silk, and elastin, respectively. In the case of ESM, it was found that ESM removed gold-cyanide ion almost quantitatively and almost all the gold uptake by ESM was easily desorbed with 0.1 M NaOH. ESM can be used repeatedly for the process of gold adsorption-desorption. The gold-biosorptive capacity of ESM that was chemically modified with glutaraldehyde was higher than that of control. In column procedure, ESM packed on column removed gold-cyanide ion from the dilute aqueous solution to extremely low concentrations (nondetectable concentration of below 1 ppb).

INTRODUCTION

Gold is a precious metal and used in not only jewelry but extensively in high technology areas, particularly in computer applications that demand the highest reliability *(1)*. Gold recovery from secondary sources such as electronic scrap *(2)* and waste electroplating solutions is therefore an important technology as well as recovery from primary resources such as leach solutions.

*Author to whom all correspondence and reprint requests should be addressed. E-mail: suyama@bios.tohoku.ac.jp

The recovery of gold from dilute solutions generally involves either zinc-dust precipitation, carbon adsorption, or solvent extraction. Recently recovery process alternatives based on ion-exchange resins have received specific attention (3). However, in most cases, their high costs limit the usage of ion-exchange resins. Recently, some investigations have focused on the gold recovery and mining process by means of biosorbents, such as algae (4–7) and microorganisms (8–12). However, few reports appear on the biosorption of gold-cyanide ion.

Cyanidation of gold ores is commonly used for the mobilization of the metal. Although effective, the process poses a number of problems in the recovery of gold.

Recently, we showed that various heavy metals are accumulated in high yield (13) by the hen egg-shell membrane (ESM), and the ESM have removed gold from dilute tetrachloroaurate (III) solutions. Also, we found that chicken feather (CF) was promising to use in the removal/recovery of precious metals as well as water pollution control (14).

This study focuses on the accumulation of gold-cyanide ion by animal fibrous proteins (AFPs), such as egg-shell membrane (ESM), chicken feather (CF), wool, silk, or elastin, and outlines some parameters in the potential use of AFP for removing gold from aqueous solution. The refining of the gold-cyanide ion from the electroplating solution and mining operations are also discussed.

MATERIALS AND METHODS

Sample Preparation

ESM was mechanically stripped from the shells after immersion of the hen (White leghorn) egg shells, which were collected from a local confectionery, in 0.5 M HCl overnight and then further in 0.5 M NaOH for 1 h followed by rinsing with distilled deionized water 10 times. Broiler-chicken feather sample was obtained from the waste of a local chicken industry of Japan. The feather was washed twice with water containing detergent by electric washing machine, then washed with deionized-distilled water, and then dried in air at room temperature. Both Corriedale virgin wool and silk samples were obtained from the university stock farm (Tohoku University). The wool was washed twice with water containing detergent by dipping overnight, air-dried at room temperature, defatted with diethyl ether for 48 h, and washed with methanol and then deionized-distilled water. Silk and cotton were washed with methanol, then fully washed with deionized-distilled water. Elastin was acquired as follows. Bovine *ligamentum nuchae* was cleaned of adhering fat, cut into small cubes, and homogenized in 1 M NaCl. The precipitates were dilapidated with chloroform-methanol (2:1, v/v), and then washed with ethanol and deionized-distilled water.

All samples were desiccated over phosphorus pentaoxide under reduced pressure at room temperature for over 10 h.

Chemical Modifications of ESM

Chemical modification reactions of ESM were performed with formaldehyde and glutaraldehyde as follows:

1. Formaldehyde crosslinking followed a modified procedure of Bullock (15). Approximately 250 mg of ESM and 45 mL of aqueous 3.7% (w/v) formaldehyde containing 0.1% (w/v) hydrochloric acid was heated at 100°C for 12 h. The chemically modified ESM sample was washed with distilled water, 0.5% (w/v) sodium carbonate, and finally deionized-distilled water.
2. Glutaraldehyde crosslinking used a modified procedure of Griffith (16). Approximately 250 mg of ESM was reacted with 45 mL of aqueous 1.25% (w/v) glutaraldehyde containing 0.1% (w/v) hydrochloric acid at 100°C for 12 h. The result was washed first with distilled water, then 0.5% (w/v) sodium carbonate, and finally deionized-distilled water.

Both modified samples were dried over phosphorus pentaoxide under reduced pressure over 10 h in a desiccator.

Chemicals and Metal Solutions

Stock solutions of gold-cyanide ion (18 mM) were prepared by dissolving gold (I) cyanide in aqueous 5% (w/v) sodium-cyanide solution. Dilution was done by addition of deionized-distilled water daily as required. All pH adjustments were carried out with 1 M HCl and 1 M NaOH. For safety precautions, all experiments with cyanide-containing solutions were performed in a hood to avoid exposure to HCN.

Adsorption Experiments

Two procedures were utilized in experiments reported here. In the batch procedure, approximately the same weight of AFPs (approx 25–35 mg) was precisely weighed, then placed directly into the aqueous solution containing gold-cyanide ion in 40-mL test tubes. To prevent metal contamination during experiments, the test tubes were soaked with 10% (w/v) HCl overnight. Then, they were washed and rinsed with deionized-distilled water. This cleaning procedure was used throughout the experiments. The volume of solution was always 10 mL unless otherwise specified. The pH values of reaction mixtures were adjusted with 1 M HCl or 1 M NaOH at the start of the experiment and maintained when possible. Solutions were shaken at elimination in a thermostated water bath with shaker at 25°C. After an appropriate time, aliquots of the solution of reaction mixture were analyzed for the remaining metal ions.

In the column procedure, experimental scale of collection of gold was carried out by the crushed ESM powder-packed mini column (diameter 1 cm, length 10 or 20 cm). The slurry of powdered ESM was poured into a column, then pH-adjusted solution containing metal ions was passed through the column, and then effluent was analyzed for the metal ions. The crushed ESM powder was prepared by crushing by Ostar blending crusher. The column-sorption experiments were performed at room temperature.

Desorption Experiments

ESM was isolated from a gold-ESM reaction mixture of batch method with tweezers and resoaked in an equal volume of solution containing 0.1 M NaOH for 1 h at 25°C. The amount of gold liberated from the ESM was determined from the free-gold concentrations measured in the solution.

Gold Analysis

Gold determinations were performed by atomic-absorption spectrophotometer (AES SAS-727, Seiko) or an inductively coupled plasma mass spectrometer (ELAN 6000 ICP-MS, Perkin Elmer, Norwalk, CT).

Sorption capacity is calculated by $q = (C_0 - C_e)V/m$; where q is the sorption capacity (mg Au/g sorbent), C_0 the initial gold concentration, C_e the residual gold concentration in solution (mg Au/L), V the volume of solution (l), and m the sorbent mass (g).

RESULTS AND DISCUSSION

Figure 1 shows the kinetics of biosorption of gold-cyanide ion on ESM, CF, wool, silk, and elastin at pH 2.0 and 25°C by the batch method. The AFP-gold system reached the equilibrium plateau corresponding to 100% of the gold-uptake capacity of AFP at a contact time in excess of 3 h. Within 1 h of contact, the biosorption system reached approx 80–100% of the total gold uptake.

The effect of pH and temperature on the uptake of gold-cyanide ion by ESM, CF, and wool was studied in the pH range from 1.0 to 12.0 for 3 h of soaking time at 4–40°C. As shown in Figs. 2, 3, and 4, the adsorption of gold-cyanide ion on AFPs was highly dependent on pH, with maximum adsorption occurring near pH 2.0, below and above which the uptake declined greatly. Thus, the adsorption of gold-cyanide ion by AFPs is markedly affected by the pH of the solution. Also, a slight increase in the gold-cyanide ion biosorptive uptake at pH 2.0 was observed when the temperature decreased from 40 to 4°C. However, differences were not considerable for the temperature-range tested.

To screen AFPs for maximal uptake of gold-cyanide ion, ESM, CF, wool, silk, and elastin were examined. The sorption experiments were

Fig. 1. Kinetics of the biosorption of gold-cyanide ion by ESM (○), CF (●), wool (△), silk (▲) and elastin (□). The sorbents were soaked in 10 mL of a solution (pH 2.0) containing 3.0 mM of gold at 25°C.

Fig. 2. Effect of pH and temperature on the biosorption of gold-cyanide ion by ESM. The ESM was soaked in 10 mL of a solution containing 3.8 mM of gold; (○) at 4°C (refrigerator); (●) at 25°C; (△) at 40°C.

performed in the maximum conditions (3 h of contact time at pH 2.0) at 25°C. As shown in Table 1, the ability to uptake gold-cyanide ion differs with different AFPs. ESM, CF, and wool took up more than 70 mg Au/g AFP dry weight from the solution, which suggested that AFPs have excellent ability for accumulation of gold-cyanide ion. Of the AFP tested, extremely high abilities for gold-cyanide uptake were found in ESM and

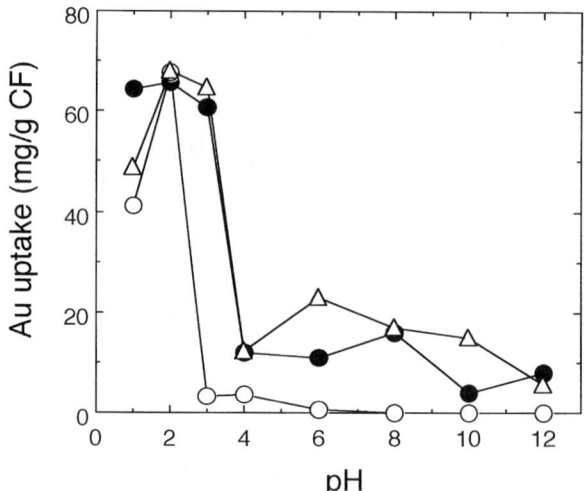

Fig. 3. Effect of pH and temperature on the biosorption of gold-cyanide ion by CF. The CF was soaked in 10 mL of a solution containing 3.8 mM of gold: (○) at 4°C (refrigerator); (●) at 25°C; (△) at 40°C.

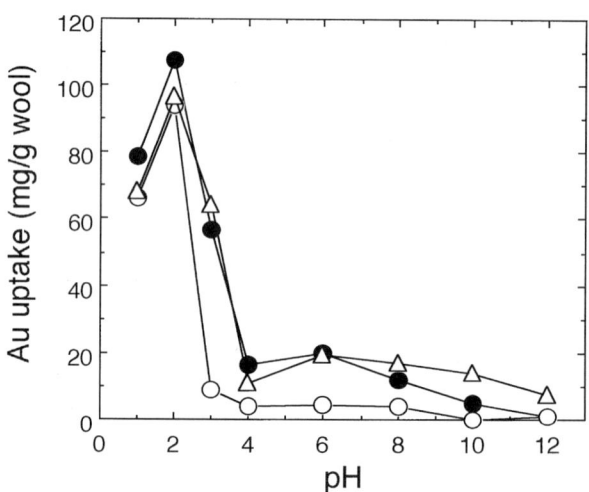

Fig. 4. Effect of pH and temperature on the biosorption of gold-cyanide ion by wool. The wool was soaked in 10 mL of a solution containing 3.8 mM of gold; (○) at 4°C (refrigerator); (●) at 25°C; (△) at 40°C.

wool. No uptake phenomenon was found on cotton. As described previously, ESM have a high ability for accumulating gold-cyanide ion. In order to elucidate of recovering of the gold uptake, we examined five cycles of gold adsorption-desorption by ESM. In batch systems, the gold-uptake ESM (Au: 3.0 mM, pH: 2.0, contact time: 3 h, temp.: 25°C) was desorbed by 0.1 M NaOH solution for 1 h at 25°C. As shown in Fig. 5, ESM recovered

Table 1
Biosorption of Gold-Cyanide Ion by Animal Fibrous Proteins (AFPs) and Cotton. The Sorbents (30 mg dry weight basis) Were Soaked in 10 mL of a Solution Containing 3.4 mM of Gold for 3 h at 25°C

sorbents	Au uptake (mg/g sorbent)
Egg shell membrane	85.8
Chicken feather	70.8
Wool	98.1
Silk	23.6
Elastin	38.9
Cotton	3.3

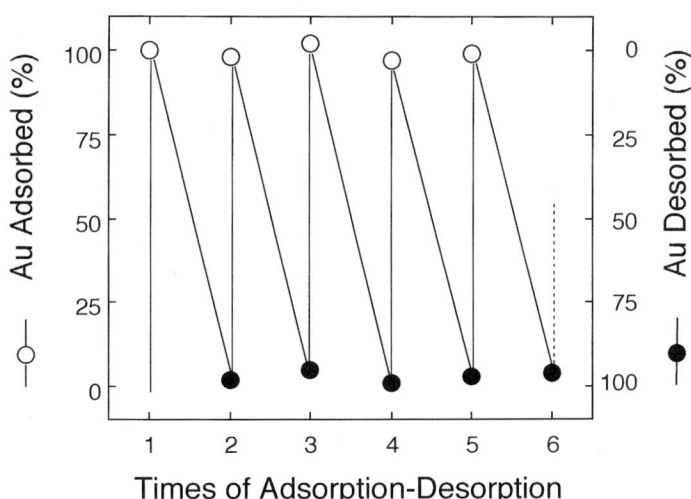

Fig. 5. Repeated test of gold adsorption (○) -desorption (●) by ESM. The gold adsorbed ESM (Au: 3.0 mM, pH: 2.0, contact time: 3 h, temp.: 25°C) was desorbed with 10 mL of 0.1 M NaOH solution for 1 h at 25°C, expressed as a percentage of the first uptake value (73 mg Au/g ESM).

gold almost quantitatively and almost all the gold uptake was desorbed with 0.1 M NaOH through the five cycles tested. The ESM had no damage during five adsorption-desorption cycles. These results show that ESM is very stable and can be used repeatedly.

The adsorption isotherms for gold cyanide ion, plotted in Fig. 6, show the evolution of metal concentration in AFP vs residual metallic concentration in solution. Each isotherm has a similar form, and shows two regions for AFPs: At low concentrations the isotherms are concave, at higher con-

Fig. 6. Gold biosorption isotherms for ESM (○), CF (●), wool (△), silk (▲) and elastin (□). The sorbents were soaked in 10 mL of a solution (pH 2.0) containing from 0.3 to 7.0 mM of gold for 3 h at 25°C.

centrations they are convex toward the concentration axis. These isotherms are not fully fitted by two representative adsorption isotherms, Langmuir and Freundlich models. Such adsorption isotherms are called BET (Brunauer-Emmett-Teller) adsorption isotherm, and it is used in case of the formation of multimolecular adsorbed layers (17). The characteristics of the gold-cyanide ion adsorption profile suggest ionic interactions, possibly involving electrostatic interactions between the negatively charged gold-cyanide ion and positively charged AFP ligands at pH 2.0. Moreover, it is suggested that gold-adsorption isotherms for AFPs are convex to the concentration axis at high concentrations, because an interaction among gold-cyanide ions begins to influence the adsorption phenomenon as gold concentration in solution increases.

ESM protein was chemically modified in two different ways in order to increase its gold sorption capacity. The uptake of gold-cyanide ion by native and modified ESM is compared in Table 2. Although the apparent differences on the uptake of gold were hardly discernible between native and formaldehyde-crosslinked ESM, the sorption capacity increased 40% when crosslinked with glutaraldehyde.

A column was used for the removal of metal from dilute aqueous solutions. Figure 7 shows the gold removal by the column packed with ESM (1.5 g) from very low concentrations of aqueous gold-cyanide ion solution (100 ppb). As shown in Fig. 7, high removal efficiency was maintained for initial 100 mL of effluent to unidentified concentration of gold (below 1 ppb). The results indicate that this biosorption process can deal

Table 2
Effect of the Chemical Modifications of ESM on the
Biosorption of Gold Cyanide Ion

Sorbent type	Au uptake (mg/ g ESM)
ESM	73.4
ESM crosslinked with formaldehyde	70.3
ESM crosslinked with glutaraldehyde	101.0

The modified ESM was soaked in 10 mL of a solution (pH 2.0) containing 3.0 mM of gold at 25°C. The chemical modification details were given in the text.

Fig. 7. Gold removal by the column packed with ESM (1.5 g) from a solution (pH 2.0) containing 100 ppb of gold at room temperature. Flow rate was about 50 ml/h.

with low concentration of metallic effluents such as waste-electroplating solutions. Actually, when the metals cyanized solution (25 ppm of gold with 18 ppm of zinc and 12 ppm of copper as co-ions) from gold electroplating of computer contacts was charged on the same ESM column, gold was efficiently removed from the solution to an undetectable concentration of gold (below 10 ppb; data not shown). Cyanidation was carried out by dissolving in aqueous 5% (w/v) sodium-cyanide solution. These experiments suggested that the application of such biosorption columns have a potential for the recovery and refining of gold from secondary sources such as electronic scrap and waste electroplating solutions.

It was found from the results that AFP have a high affinity for the binding gold-cyanide ions. A knowledge of the chemistry of AFP interactions with gold might provide insight into mechanisms for environmental transport and deposition of gold. It appears that the AFP system might be useful in the recovery of gold from industrial waste waters or from direct mining operations. We have shown that ESM can be used for the purpose of removal/recovery of gold. Thus, it may be possible to use the AFP-adsorption system to recover gold complexes from mining dumps or present-day mining operations.

REFERENCES

1. Wan, R. Y. and Miller, J. D. (1986), *J. Metals*, **38**, pp. 35–40.
2. Dunning, B. W. (1986), In *Proceedings of the Bureau of Mines*, National Western Mining Conference, Denver, Colorado, Washington, DC, February **12**, pp. 112–115.
3. Akser, M., Wan, R. Y., and Miller, J. D. (1986), *Solvent Extr. Ion Exch.* **4**, 531–546.
4. Kuyucak, N. and Volesky, B. (1989), *Biorecovery* **1**, 189–204.
5. Darnall, D. W., Green, B., Henzl, M., Hosea, J. M., McPherson, R., Sneddon, J., and Alexander, M. D. (1986), *Environ. Sci. Technol.* **20**, 206–208.
6. Green, B., Hosea, M., McPherson, R., Henzl, M., Alexander, M. D., and Darnall, D. W. (1986), *Environ. Sci. Technol.* **20**, 627–632.
7. Gardea-Torresdey J. L., Becker-Hapak, M. K., Hosea, J. M., and Darnall, D. W., (1990), *Environ. Sci. Technol.* **24**, 1372–1378.
8. Hosea, M., Green, B., McPherson, R., Henzl, M., Alexander, M. D., and Darnall, D. W. (1986), *Inorg. Chim. Acta, Bioinorg. Chem.* **123**, 161–165.
9. Treen-Sears, M. E., Volesky, B., and Neufeld, E. J. (1984), *Biotechnol. Bioeng.* **26**, 1323–1329.
10. Beveridge, T. J. and Murray, R. G. E. (1976), *J. Bacteriol.* **127**, 1502–1518.
11. Dissanayake, C. B. and Kritsotakis, K. (1984), *Chem. Geol.* **42**, 61–76.
12. Beveridge, T. J. and Murray, R. G. E. (1978), *Can. J. Microbiol.* **24**, 89–104.
13. Suyama, K., Fukazawa, Y., and Umetsu, Y. (1994), *Appl. Biochem. Biotechnol.* **45/46**, 871–879.
14. Suyama, K., Fukazawa, Y., and Suzumura, H. (1996), *Appl. Biochem. Biotechnol.* **57/58**, 67–74.
15. Bullock, A. L., (1965), in *Methods in Carbohydrate Chemistry*, Whistler, R. H. ed., Academic, NY, pp. 409–411.
16. Griffith, I. P., (1972), *Biochem. J.* **126**, 553–560.
17. Brunauer, S., Emmett, P. H., and Teller, E. (1938), *J. Amxer. Chem. Soc.* **60**, 309–319.

Effect of Temperature and Pressure on Growth and Methane Utilization by Several Methanotrophic Cultures

B. K. SONI,* JOHN CONRAD, ROBERT L. KELLEY, AND VIPUL J. SRIVASTAVA

Institute of Gas Technology, 1700 S. Mount Prospect Road, Des Plaines, IL 60018

ABSTRACT

Several methanotrophic microorganisms, i.e., *Methylococcus capsulatus* (Bath), *Methylomonas albus* (BG-8), *Methylosinus trichosporium* OB3b, and *Methylocystis parvus* (OBBP), were evaluated for growth and methane utilization. The effect of temperature was examined in the range of 25 to 45°C for growth and methane utilization. The temperature variations (25–35°C) had minimal effect on growth of *M. albus* and *M. parvus*. Methane consumption varied at different temperatures with a maximum of 0.67 mol%/h and 0.53 mol%/h. at 30 and 35°C, respectively, for *M. albus* and *M. parvus*. The growth and methane consumption was slower for *M. trichosporium* OB3b as a maximum methane consumption of 0.07 mol%/h was obtained at 25°C and growth was inhibited at 35°C. *M. capsulatus* grew the best at 37°C and growth was affected at higher temperature of 45°C. Of the different cultures examined, *M. albus* and *M. capsulatus* grew the best and were further evaluated for the effect of pressure in the range of 10–50 psi. The results obtained using *M. albus* demonstrated an enhancement in methane consumption rate by fourfold and final cell concentration by 40% at a pressure of 20 psi by injecting a methane/oxygen mixture, however further increase in the pressure up to 50 psi inhibited the growth. The inhibition was not seen with nitrogen incorporated mixture of oxygen and methane, which suggest that the high partial pressure of methane and/or oxygen are inhibitory for the growth of *M. albus*. *M. capsulatus* was more sensitive to pressure as evidenced by inhibition at the relatively low pressure of 10 psi.

Index Entries: *Methylococcus capsulatus*; *Methylomonas albus*; temperature; pressure.

*Author to whom all correspondence and reprint requests should be addressed.

INTRODUCTION

Methanotrophic bacteria have in common the ability to utilize methane as a sole source of energy and as a major carbon source (1). The methanotrophic bacteria that have been isolated and characterized to-date have all been Gram-negative, obligatory aerobic, and have intracytoplamic membranes (2–4).

Methanotrophs are classified in several groups based on the metabolic pathway and other characteristics. Types I and X are phylogenetically related to the L-subgroup of the proteobateria (5). Examples of these groups include *Methylomonas albus* and *Methylococcus capsulatus*, respectively. These bacteria utilize the RuMP pathway for formaldehyde assimilation, though other pathways do exist for their metabolism. Type II are also related to L-subgroup of the proteobacteria but uses the serine pathway for formaldehyde fixation. Examples of this group includes *Methylosinus trichosporium* OB 3b and *Methylocystis parvus*. Type X bacteria are able to grow at a higher temperature (> 40°C), however, Type II do not grow at temperatures above 40°C. Examples of Type II include *Methylosinus trichosporium* OB3. This is similar to Type I bacteria, except that it contains enzymes of the Calvin cycle and is capable of carbon dioxide fixation.

Based on this classification, the growth of different classes of methane-utilizing bacteria are dependent on temperature. Therefore, the goal of the present work was to determine the growth and methane utilization rate of different groups of methane-utilizing bacteria at different temperatures. The objective was also to determine the effect of pressure on growth of different organisms and to determine the levels for activation and inhibition of growth.

MATERIALS AND METHODS

Microorganisms

Cultures of *Methylococcus capsulatus* (Bath), *Methylomonas albus* (BG-8), *Methylocystis parvus* (OBBP), and *Methylosinus trichosporium* OB3b (ATCC 35070), was preserved in glycerol vials (15%) at −80°C to minimize the genetic drift caused by repetitive transfers. Working cultures were maintained on nitrate mineral-salts medium (NMS) agar plates under a 1:3 methane/air atmosphere. Seed cultures were started in vials containing 50 mL of NMS medium, charged with methane/air, sealed with a rubber stopper and aluminum crimp, and incubated at 30°C. The composition of medium used in current investigations was the Higgin's nitrate mineral medium (6).

Temperature and Pressure Studies

The temperature studies were conducted at 1 atm in serum bottles charged with different gases (CH_4, O_2, and N_2) at several different temperatures as shown in the results section. Pressure-effect experiments were conducted with *M. albus* (at 30°C) and *M. capsulatus* (at 37°C) (Bath) at various (eight) pressures between 10 and 50 psi. Increased pressure was maintained both by using the gas substrate, a mixture of methane and air, and by using nitrogen gas in the presence of methane and air. This was done in an effort to determine the effects of total pressure, as well as partial pressure, of methane and air on the growth of methanotrophs. In all experiments, care was taken to maintain an approximately constant pressure by the addition of inert nitrogen.

ANALYSIS

Dry-Cell-Weight Estimation

The cell concentration was determined by a predetermined correlation between dry cell weight and optical density at 620 nm (Spectronic 20, Milton Roy, Rochester, NY). The fermentor broth was filtered through a 0.2-micron filter, washed twice with saline solution (0.85%), and dried overnight in the oven.

Gas Analysis

The concentration of methane, oxygen and nitrogen was determined using a Fisher gas partition model 1200 (Pittsburgh, PA) chromatograph equipped with a thermal-conductivity detector. Helium was used as the carrier gas at the flow rate of 30 mL/min. The column temperature was maintained at 50°C.

RESULTS

Selection of Microorganisms and Optimization of Temperature for Enhanced Utilization of Gaseous Substrates

In view of conducting research for co-oxidative products, several microorganisms were selected, and the effect of physical parameters on the mass transfer of methane was evaluated. The effect of temperature on growth and methane utilization was examined, on four selected mesophilic cultures. These cultures were *M. albus*, *M. parvus*, *M. capsulatus*, and *M. trichosporium*. At least three temperatures (25, 30, and 35°C) were evaluated. The results (Figs. 1–4) revealed that growth of *M. albus* and *M. Parvus* were not significantly affected by temperature variations. The maximum biocatalyst concentration of *M. albus* was in the range of 280 to 320 mg/L (dry weight) with a maximum at 30°C. This finding is in agreement with the literature values that showed a similar optimal of 30°C, though conducted at different composition of gas mixture. The methane content was reduced from 24 to 10% with a overall consumption rate of 0.22 mol%/h for this organism. During the initial period (16 h), the methane consumption of 0.67 mol/h was observed. The biocatalyst concentration for *M. parvus* was

Fig. 1. The effect of temperature on the growth of *Methylomonas albus* (BG-8) (2:1 CH_4/O_2). (—▲—, 25; —●—, 30; —■—, 35).

Fig. 2. The effect of temperature on growth of *Methylocystis parbus* (OBBP) (2:1 CH_4/O_2). (—▲—, 25; —●—, 37; —■—, 45).

Fig. 3. The effect of temperature on growth of *Methylosinus trichosporium* (OB3b) (2:1 CH_4/O_2). (—▲—, 25; —●—, 30; —■—, 35).

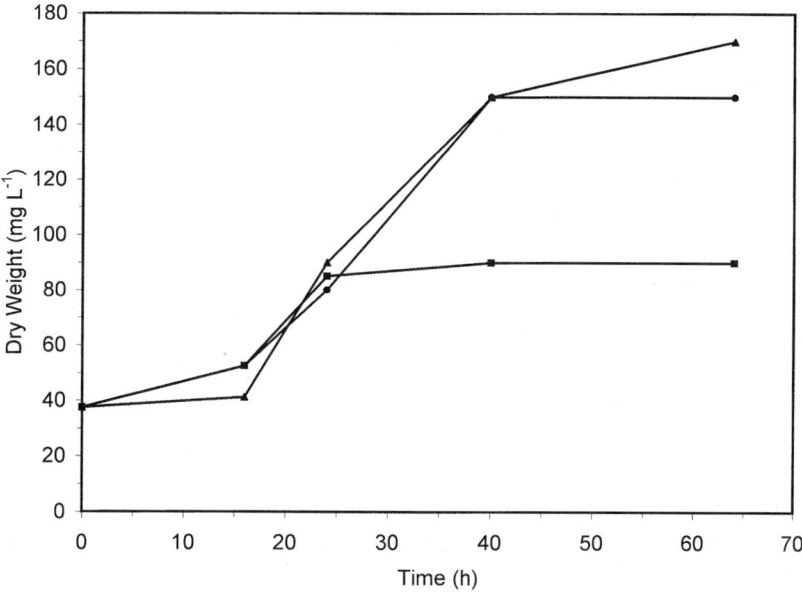

Fig. 4. The effect of temperature on growth of *Methylococcus capsulatus* (Foster and Davis) (2:1 CH_4/O_2). (—▲—, 25; —●—, 30; —■—, 35).

Table 1
Baseline Kinetic Data for Methanotrophs Grown at Optimal Temperature

Cultures	Temperature °C	Growth Rate mg/h	Methane Consumption Rate mg/mol%/h	Cell Yield mg/mol%CH_4
M. albus	30	4.4	0.31	20
M. parvus	30	3.9	0.35	20.7
M. trichosporium	25	1.8	0.05	n.d
M. capsulatus	37	6.2	0.62	24.6

in the range of 260 to 300 mg/L with a maximum of 300 mg/L at 30°C. The methane consumption rate was 0.22 mol%/h at 30°C. During the initial period (16 h), the maximum methane consumption rate of 0.53 mol%/h was observed. The growth of *M. trichosporium* was the same at both 25 and 30°C, but was inhibited at 35°C. The biocatalyst concentration for *M. trichosporium* was in the range of 67 to 127 mg/L with a maximum of 127 mg/L at 25°C. *M. capsulatus* (Foster and Davis) showed a biocatalyst concentration in the range of 235–260 mg/L with a maximum of 260 mg/L at 25°C. The methane consumption rate of 0.275 mol%/h was maximum at 45°C. The effect of temperature is summarized in Table 1. The biocatalyst concentration of 6.2 mg/L was highest for *M. capsulatus*. These results indicate that the maximum growth and methane consumption were achieved with *M. albus* and *M. capsulatus*; therefore, *M. albus* and *M. capsulatus* were selected subsequently for pressure studies.

Effect of Pressure

The purpose of the pressure experiments was to enhance the mass transfer of methane for enhancing the growth. The pressure experiments were conducted with two of the better growing cultures of *M. albus* and *M. capsulatus*. The objective was to determine the growth under control conditions. The reactor were pressurized by increasing the partial pressure of methane and oxygen (2:1). Next, to prove that there was inhibition caused by increased partial pressure of oxygen, the CH_4/O_2 was left constant but pressure was increased with an inert gas like nitrogen. The results (Fig. 5) demonstrated that up to 20 psi, the growth was enhanced as compared to the control. At the high pressure of 50 psi (CH_4/O_2 mixture), a marked inhibition was observed.

The results summarized in Table 2 revealed a significant enhancement in growth rate (40%) and methane consumption rate (up to 20 psi) beyond which a significant inhibition was observed. The reactor first flushed with methane/air and then brought to the desired pressure point with nitrogen did not show the inhibition. These studies indicated that a higher pressure (>25 psi) of methane/oxygen was inhibitory for the growth of *M. albus*. The

Table 2
Effect of Pressure on Growth and Methane
Consumption Rates using *M. albus*

Test Pressure psig	Growth Rate mg/h	Methane Consumption Rate mol% CH_4/h
10	10	0.21
15	10	0.25
20	11	0.35
25	9	0.22
30	5	0.04
35	4	0.04
40	2	0.09
45	2	0.11
50	1	0.08

[a] The vessel were pressurized with CH_4/O_2 mixture in the head space.

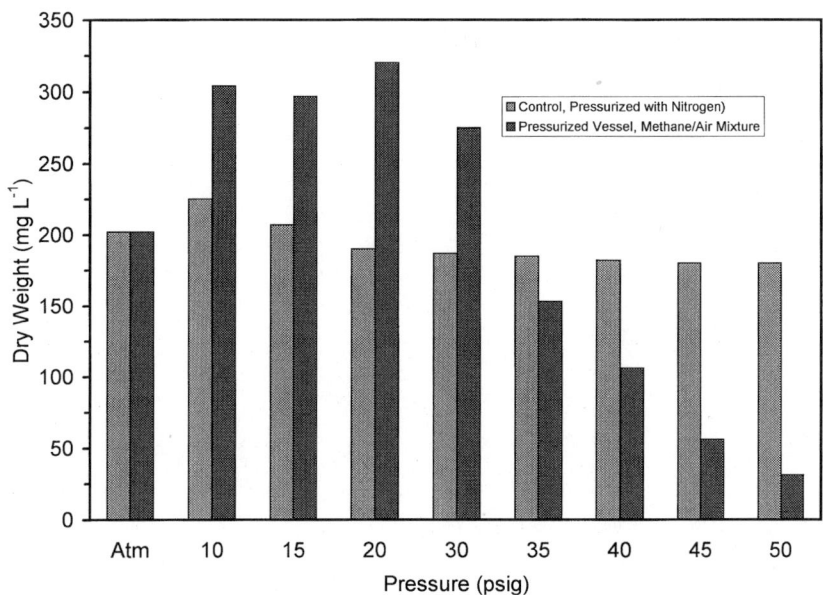

Fig. 5. Comparison of *Methylomonas albus* growth at various pressures (pressurized with CH_4/O_2 2:1).

effect of pressure on growth rate and methane consumption rates suggested maximum growth of *M. albus* was achieved at 20 psi. The maximum methane consumption of 0.35 mol%/CH_4/h was achieved at this pressure.

Experiments were also conducted to determine the effect of pressure on growth of *M. capsulatus*. The results shown indicated (Figs. 6 and 7) no

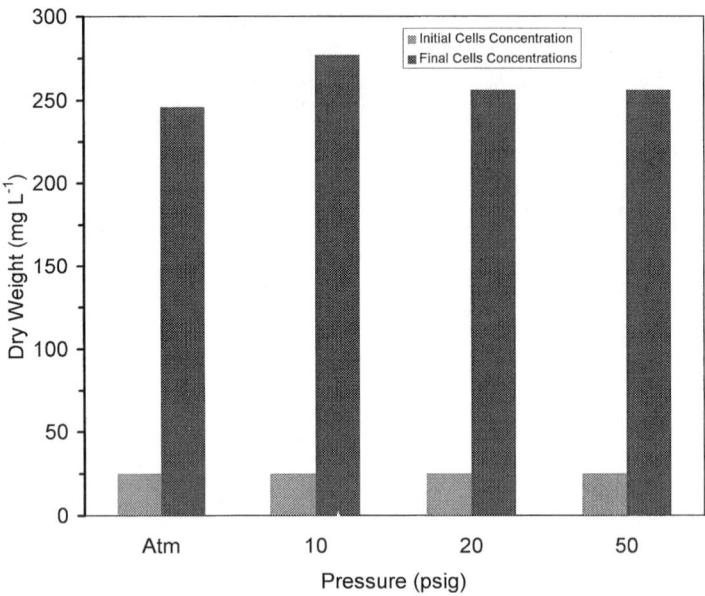

Fig. 6. Comparison of *Methylococcus capsulatus* (Bath) at various pressures (pressurized with CH_4/O_2 2:1).

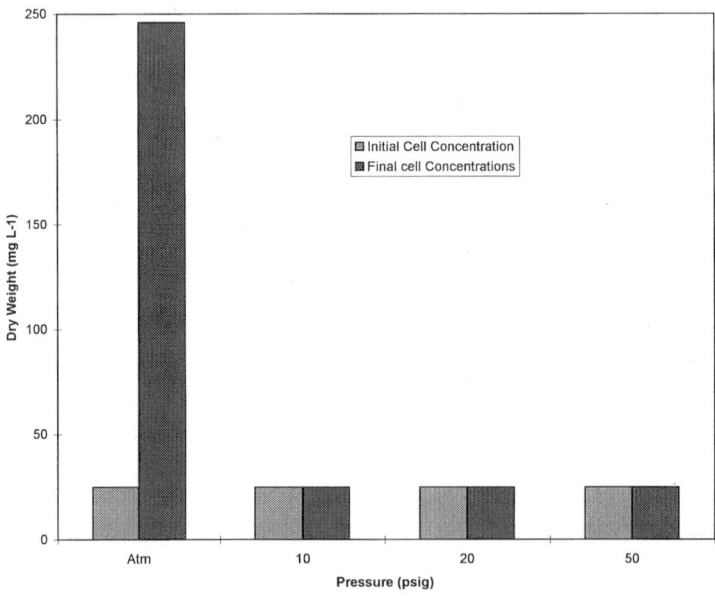

Fig. 7. Comparison of *Methylococcus capsulatus* (Bath) at various pressures (pressurized with nitrogen).

inhibition in reactors pressurized with nitrogen. The growth of *M. capsulatus* was more sensitive to pressure, as a significant inhibition was observed at the relatively low pressure of 10 psi (Fig. 6). However, pressurizing with nitrogen resulted in a better growth, confirming again the inhibition of methane/air (Fig. 7).

The results summarized in Table 2 revealed a significant enhancement in growth rate for *M. albus* (40%) and methane consumption rate (up to 20 psi) beyond which a significant inhibition was observed. The reactor first flushed with methane/air and then brought to the desired pressure point with nitrogen did not show the inhibition of microorganisms. These studies indicated that a higher pressure (>25 psi) of methane/oxygen was inhibitory for the growth of *M. albus*. The effect of pressure on growth rate and methane consumption rates suggested that maximum growth of *M. albus* was achieved at 20 psi. The maximum methane consumption of 0.35 mol%/ CH_4/h was achieved at this pressure.

From the results it is quite evident that different groups of bacterial have different temperatures optimal for growth and methane utilization. The methane utilization was maximum in the initial period of incubation for all the organisms examined. *M. capsulatus* showed higher biocatalyst growth rate (6.2 mg/L/h) as well as methane yield. The higher growth could be attributed possibly to the higher optimal temperature of growth. On the contrary, the growth rate for *M. trichosporium* OB3B was low. This can once again be explained by the fact that optimal temperature of growth is 30°C (7). These finding suggests that biocatalyst growth rates for methane-utilizing bacteria are dependent on temperatures. The sensitivity of *M. albus* and *M. capsulatus* was different as the later showed inhibition at 10 psi reactor pressure. These facts suggests that different methane-utilizing organisms are sensitive to increased partial pressure of oxygen or methane.

Pressure studies conducted with *M. albus* and *M. trichosporium* OB3B demonstrated that enhanced pressure up to 20 psig increases the growth rate. Pressure higher than this with methane/oxygen has resulted in reduction of biocatalyst growth. This reduction in growth could be attributed to the interaction of growth substrate (methane or oxygen) with biocatalyst. The inhibition was not seen when the reactor was pressurized with inert gas (nitrogen).

CONCLUSIONS

Several conclusions can be drawn from these studies as described below:

1. The methane-consumption rate for all the organisms examined was higher in the initial period of growth (< 16 h).
2. *M. capsulatus* showed the highest growth rate and *M. trichosporium* OB3B showed the lowest rates. These rate differences could be ex-

plained on the basis of higher (37°C) and lower optimal growth temperatures (30°C) for these microorganisms.
3. Enhancing the pressure with growth substrate has enhanced the biocatalyst concentration up to 20 psi, beyond which a inhibition of growth of *M. albus* was observed. *M. capsulatus* was more sensitive to growth substrates as seen by growth inhibition at the lower pressure of 10 psi.
4. Neither of the organism tested showed inhibition when inert gas was used for pressure, suggesting that these organisms are sensitive to presence of higher levels of growth substrate (methane or oxygen).

ACKNOWLEDGMENTS

We wish to acknowledge the Gas Research Institute (GRI) and Sustaining Membership Program (SMP) of Institute of Gas Technology for the partial financial support to conduct this research.

REFERENCES

1. Anthony, C. (1982), *The Biochemistry of Methylotroph*, Academic, London.
2. Hanson, R. S., Natrusov, A. I., and Tsuji, K. (1991), in *The Procaryotes*, Balows, A., Truper, H. G., Dworkin, M., Harder, W., and Schleifer, K. H., eds., Springer-Verlag, New York. pp. 661–684.
3. Whittenbury, R. and Dalton, H. (1981), in *The Procaryotes*, Starr, M. P., Stolp, H. Truper, H. G., Balows, H., and Schlegel, H. G., eds., Springer-Verlag KG, Berlin, pp. 894–902.
4. Whittenbury, R. and Krieg, N. R. (1984), in *Bergey's Manual of Determinative Bacteriology*, vol. 1, Krieg, N. R. and Holt, J. G., ed., Williams, and Wilkins, Baltimore, pp. 256–262.
5. Tsuji, K., Tsien, H. C., Hanson, R. S., Depalma, S. R., Scholtz, R., and LaRoche, L. (1990), *J. Gen. Microbiol.* **136,** 1–10.
6. Cornish A, Nicholls, K. M., Scott, D., Hunter, B. K., Higgins, I. S., and Sanders, J. K. M. (1984), *J. Gen. Microbiol.* **130,** 2565–2575.
7. ATCC Catalogue of Bacteria and Bacteria Phages, 17 ed. Gherna, R., Pienta P., and Cote R., eds. p. 152.

The Production of Hydrocarbons from Photoautotrophic Growth of *Dunaliella salina* 1650

DON-HEE PARK,* HWA-WON RUY,[1] KI-YOUNG LEE,[1]
CHOON-HYOUNG KANG,[2] TAE-HO KIM,[2]
AND HYEON-YONG LEE[3]

[1]*Department of Biochemical Engineering,* [2]*Department of Chemical Engineering, Chonnam National University, Kwangju 500-757, Korea, and* [3]*Division of Food and Biotechnology Kangwon National University, Chuncheon 200-701, Korea*

ABSTRACT

Microalga, *Dunaliella salina* 1650 was selected to produce hydrocarbons that may possibly substitute for fossil fuels in the near future. It can produce 0.22 (mg/L) of hydrocarbons over 20 d batch cultivation, maintaining 1.32 (g-dry wt./L) of cell density. Its productivity was similar to that from *Botryococcus braunii*, which was known to economically produce liquid fuels. Optimal growth conditions for the alga were also determined as pH 7.2, 28°C, and 0.00034 (Kcal/cm^2/h) of light intensity. It was shown that the hydrocarbon production from the alga was closely related to cell growth, except for the later periods of batch cultivation. Better hydrocarbon production was observed during light periods in light/dark cycle cultivation. Under chemostat conditions, maximum steady cell concentration was maintained as 1.1 (g-dry wt./L) at 0.12 (1/d) of dilution rate. The system reached to the steady state after 30 d of the cultivation. The maximum specific hydrocarbon production rate, 0.024 (mg/cell/d) was also obtained under this condition. It proves that the hydrocarbon production from *D. salina* 1650 can compete with that from *B. braunii*.

Index Entries: Algal hydrocarbon production; photoautotrophic growth; *Dunaliella salina*.

INTRODUCTION

In algal biotechnology, one of the current research interests in developing new energy resources is to produce usable liquid fuels that can

*Author to whom all correspondence and reprint requests should be addressed. E-mail: dhpark@orion.chonnam.ac.kr

substitute fossil fuels from photosynthetic algae through biological conversion of solar energy (1–3). The algae can produce saturated and unsaturated hydrocarbons by assimilating carbon dioxide. The algal hydrocarbons can also be used directly or indirectly for substituting gasoline by several cracking processes (4,5). It has been reported that the green alga, *Botryococcus braunii* is most promising photosynthetic organism since it can produce economic quantities of hydrocarbons by utilizing artificial or natural light (6–8). It has been intensively investigated to develop a process of cultivating mass amounts of biomass in indoor and outdoor photobioreactors by using various sources of light energy. The development of mass-cultivation technologies is most economic methodology in producing biomass and the products of interest from photosynthetic algae (9–11). However, the outdoor cultivation of photosynthetic algae has been less studied to produce hydrocarbons because an open-pond cultivation requires delicate elaboration of strong light intensity of solar energy, effective pH and temperature controls for long periods of the cultivation, and continuous supply of culture media with limited contamination of other species (12). Therefore, in this work, we investigated the kinetics of hydrocarbon production from green alga, *Dunaliella salina* 1650, using an outdoor cultivation process (13).

MATERIALS AND METHODS

The green alga, *Dunaliella salina* 1650 was originally obtained from the Algal Culture Center (UTEX, USA), and adapted by growing cells in hydrocarbon-producing medium (13) (pH 7.58 and 6.8% (w/v) NaCl) at 25°C with 2.3×10^{-4} (Kcal/cm^2/h) of light intensity. An open-culture system was designed for outdoor batch and continuous cultivations as shown in Fig. 1 (150 × 150 × 25-cm, total working volume was 500 L). To simulate outdoor environment, 12:12 hour light and dark cycle was used by eight 20 W white cool fluorescent lamps illuminated the pond in a dark room. The pH and temperature were not controlled for this experiment. For continuous cultivations, a peristaltic pump was used for feeding fresh medium and the effluent was collected out of the drain at the top of the pond. Two top-driven paddle mixers were also used for the agitation of the media in the pond as shown in Fig. 1. The outdoor cultivation was carried out when the change of seasonal temperature was minimum.

The light intensity was measured by a quantum sensor (Li-Cor LB-125, Yellowsprings, USA) every day at same time (noon). Cell density was measured by filtering 10 mL of the sample through 0.45-m pore-size filter paper and drying it at 105°C for 24 h. To determine the concentration of hydrocarbons within the algae, filtered medium and algae disrupted by a sonicator (Far M-150, Madison, WI) were centrifuged at 1300g for 20 min, then they were extracted by adding two volumes of hexane into the supernatant for 1 h at room temperature (14). The extracts were dried in

1. Substrate reservoir
2. Peristaltic pump
3. Cool-white flourescent lamp
4. Agitator (paddle wheel)
5. Thermometer
6. pH controller
7. Temperature controller
8. Radiometer
9. Effluent reservoir

Fig. 1. A schematic diagram of an open-pond culture system.

a rotary-vacuum evaporator and measured for total crude hydrocarbons. The crude hydrocarbons were purified by 110°C activated silica gel (60 GF-2.5, Merk) chromatography (15), then identified by a thin-layer chromatography (TLC) with diluted commercial gasoline. The energy content of algal hydrocarbons was also estimated by differential scanning calorimeter (DSC) (DuPont 2100, Boston, MA) at 10°C/min of the heating rate up to 400°C.

RESULTS AND DISCUSSION

Figure 2 shows the effect of light intensity on the growth of *D. salina* 1650. An optimal light intensity was observed as 3.4×10^{-4} (Kcal/cm^2/h) when 0.155 (1/d) of specific growth rate was maintained at 1.32 (g-dry wt/L) of maximum cell density. At higher light intensity, cell growth rate was hampered, compared to that at lower light intensity, with 0.063 (1/d) vs 0.104 (1/d) of specific growth rate, respectively. The cell density gradually increased during 17 d of cultivation at relatively low-light intensity. Figure 3 shows the kinetics of cell growth and hydrocarbon production at 3.4×10^{-4} (Kcal/cm^2/h) of constant light intensity in batch cultivation. The hydrocarbon production seems to be closely related to cell growth except for the later periods of cultivation. The concentration of crude hydrocarbons within the cells remained constant even though overall cell density dropped. However, the hydrocarbon yield per g-dry cell was gradually increased according to the cultivation time, which implies that the produc-

Fig. 2. The effect of light intensity on the growth of *D. salina* 1650 as a function of light intensity for batch cutivation.

Fig. 3. The cell growth and hydrocarbon production for batch cultivation of *D. salina* 1650 at 3.4×10^{-4} (Kcal/cm²/h) of the incidence light intensity.

tion of hydrocarbons in the cell can be continued to the last period of cultivation. The maximum specific hydrocarbon production rate was estimated as 0.029 (mg crude hydrocarbon/g-cell/d) at 0.22 (mg/g-cell) of maximum specific hydrocarbon production. 0.22 (mg/g-cell) of hydrocarbon production yield from *D. salina* 1650 in this process is close to the reported value of 0.345 (mg/g-cell) for *Botryococcus braunii* (16).

Fig. 4. The kinetics of cell growth and hydrocarbon yield for 12:12 light/dark diurnal cultivation at the light intensity of 3.4×10^{-4} (Kcal/cm^2/h).

Figure 4 is the result of cultivating the cells by a 12:12 hour light/dark-cycle process. The cell growth seems to be affected by light/dark-cycle cultivation since better cell growth was observed in light periods. However, hydrocarbon yield increased during the light phases and reached maximum value during the third light phase, then it fell. This implies that the metabolism of producing hydrocarbon within the cells lags behind the overall cell growth. This might be the reason that the hydrocarbon yield slowly responds to light/dark cycle.

Figure 5 shows the kinetics of cell growth and hydrocarbon production as well as the changes of pH at 0.12 (1/d) of dilution rate under chemostat condition. The system reached to the steady state after 40 d cultivation, based on relatively constant cell concentration and pH change, whose data well fits to chemostat theory. The hydrocarbon production could also be correlated to cell growth during the steady state. pH of the medium was dropped in later periods of the cultivation because of the decrease of the cell growth. Figure 6 illustrates the relationship between cell growth and specific hydrocarbon-production rate according to dilution rate by 12:12 light/dark-cycle process. The cell growth remained stable at relatively wide ranges of dilution rates. The wash-out dilution rate was observed 0.3 (1/d). The maximum specific hydrocarbon production rate was calculated as 0.024 (mg/g-cell/d) at 0.12 (1/d) of dilution rate. Therefore, for maximum hydrocarbon production an optimal operating condition can be determined as 0.12 (1/d) of dilution rate, having 1.11 (g-dry wt./L) of cell density. It also proves that long-term outdoor cultivation is possible for green alga, *D. salina* 1650 at relatively wide ranges of dilution rate (0.05–0.25 1/d of dilution rate).

Fig. 5. The changes of cell density, hydrocarbon concentration, and pH at 0.12 (1/d) of dilution rate in 12:12 light/dark cycle continuous cultivation.

Fig. 6. The relationship between cell density and specific-hydrocarbon production rate as a function of dilution rate.

Table 1 is to compare the elemental analysis of high-hydrocarbon-producing microalgae since the carbon content in a cell can represent the hydrocarbon productivity during the cultivation. *D. salina* 1650 has relatively high carbon fraction among other algae, even the well-known high-hydrocarbon-producing algae, *B. braunii*. Figure 7 is the result of DSC analysis of hexane extracts for evaluating algal hydrocarbons from *D. salina* as an alternative energy source. Energy contents of the extracts was com-

Table 1
The Results of Elemental Analysis of Hexane-Extracted Hydrocarbons from Several Microalgae

Species	Composition (Wt%)				References
	C	H	N	S	
D. Salina	77.34	10.85	5.36	0.33	
B. Braunii	83.38	11.96	0.17	<0.1	(7)
Spirulina sp.	66.86	10.37	10.05	0.43	(10)

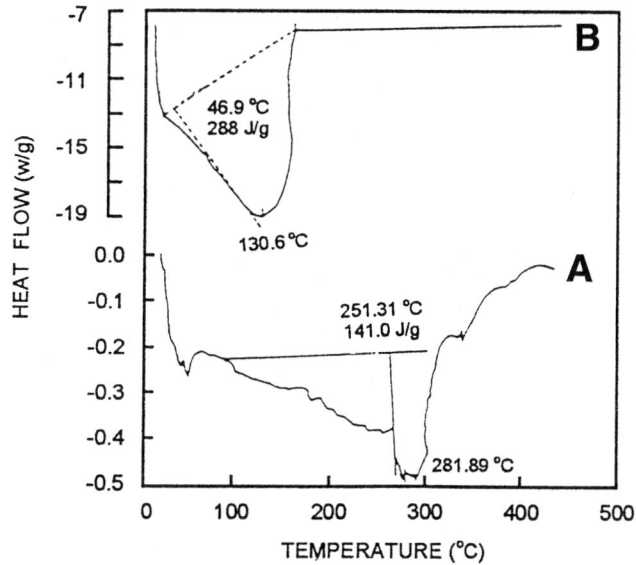

Fig. 7. The thermogram of DSC analysis for algal hydrocarbons from *D. salina* 1650 (**A**) and petroleum (**B**).

pared to that of currently used petroleum. Algal hydrocarbons have good energy level as 141 (J/g) and high heating temperature at 251.3°C. It tells that the hydrocarbons from *D. salina* can be used as a substitute energy directly or through cracking processes.

ACKNOWLEDGMENT

This paper was supported by Special Fund (on the studies of biofuel production) for University Research Institute, Korea Research Foundation. The authors deeply appreciate their financial support.

REFERENCES

1. Bailliez, C., Largeau, C., Berkaloff, C., and Casadevall, E. (1986), *Appl. Microbiol. Biotechnol.* **23**, 361–366.

2. Bailliez, C., Largeau, C., Casadevall, E., Yang. L., and Berkaloff, C. (1988), *Appl. Microbiol. Biotechnol.* **29**, 141–147.
3. Barclay, W. and McIntosch, R, eds., (1986), *Algal Biomass Technologies: An Interdisciplinary Perspective*, Berlin-Stuttgart, Berlin, pp. 126.
4. Benemann, J. R. (1987), *Microalgae Biotechnology: Products, Processes and Opportunities*, vol. 2, OMEC, Washington, DC.
5. Brown, A. C. and Knights, B. A. (1969), *J. Phytochem.* **8**, 543–549.
6. Cane, R. and Albion, P. R. (1973), *Geochim. Cosmochim. Acta.* **37**, 1543–1550.
7. Casadevell, E., Largeau, D. T., Guidin, C., Chaumount, D., and Desanti, O. (1985), *Biotechnol. Bioeng.* **27**, 286–295.
8. Freeman, C. P. and West, D. (1966), *J. Lipid Res.* **7**, 324–329.
9. Frenz, J., Largeau, C., Casadevall, E., Kollerup, F., and Dougulis, A. (1989), *Biotechnol. Bioeng.* **34**, 755–762.
10. Lee, H. Y., Erickson, L. E., and Yang, S. S. (1987), *Biotechnol. Bioeng.* **29**, 832–841.
11. Lee, Y. K. and Low, C. S. (1991), *Biotechnol. Bioeng.* **38**, 995–1000.
12. Meier, R. L. Solar Energy Research. F, Daniels, and J. A. Duffies, eds. (1986), *Microalgal Culture, Crit. Rev. Biotechnol.* **4**, 369–438.
13. Pak, J. H., Lee, S. Y., Shin, W. C., and Lee, H. Y. (1991), *J. Microbiol. Biotechnol.* **1**, 111–115.
14. Schelef, G. and Soeder, C. J., eds., (1980), *Algal Biomass and Production and Use*, Elsevier/North-Holland Biomedical Press, Amsterdam, p. 852.
15. Wolf, F. R., Nonomura, A. M., and Basshan, J. A. (1985), *J. Phycol.* **21**, 388–394.

Fouling and Protein Adsorption

Effect of Low-Temperature Plasma Treatment of Membrane Surfaces

J. JOHANSSON, H. K. YASUDA, AND R. K. BAJPAI*

Chemical Engineering Department, University of Missouri - Columbia, W2030 EBE, Columbia, MO 65211

ABSTRACT

Adsorption of proteins and the effect of the chemical nature of membrane surfaces on protein adsorption were investigated using ^{14}C-tagged albumin and several microporous membranes (polyvinylidene fluoride, PVDF; nylon; polypropylene, PP; and polycarbonate, PC). The membrane surfaces were modified by exposing them to low-temperature plasma of several different monomers (n-butane, oxygen, nitrogen alone or as mixtures) in a radiofrequency plasma reactor. Transients in the permeability of albumin solutions through the membranes and changes in flux of distilled water through the membranes before and after adsorption of albumin were used to investigate the role of protein adsorption on membrane fouling. The results show that the extent of adsorption of albumin on hydrophobic membranes was considerably more than that on hydrophilic membranes. The hydrophilic membranes were susceptible to electrostatic interactions and less prone to fouling. A pore-blocking model was successfully used to correlate the loss of water flux through pores of defined geometry.

Index Entries: Albumin; polymer membranes; permeability; radiofrequency plasma; radiochemical.

INTRODUCTION

Microporous membranes are commonly used in bioprocessing (1). The applications range from filtration of fermentation broth to filter-sterilization of liquid media. Hollow-fiber microporous membranes have potential for use as extraction devices also (2). Wang et al. (3) have demonstrated that such extractors may be used for a single-step purification and concen-

---

*Author to whom all correspondence and reprint requests should be addressed.

tration of lactic acid from broth. However, the process was marked by significant reductions in the extractive flux of lactic acid over time. It was demonstrated that irreversible loss in permeation rate through the membrane was the primary reason for this reduction (4). Similar reductions in flux have been reported for many other applications also involving porous membranes (electrodialysis, reverse osmosis, ultrafiltration, and microfiltration) (5) and their negative economic impact on membrane industry has been estimated to be of the order of $500 million annually (6).

The flux declines in membrane operations are attributed to several distinct phenomena that include membrane compaction and degradation, concentration polarization, and membrane fouling (7). Concentration polarization has been widely studied and is relatively well understood (8). It can be minimized by an appropriate design of the membrane module and by appropriate selection of operating conditions (9). Membrane fouling, on the other hand, is poorly understood. Adsorption of solutes and their aggregates on the exterior and the interior membrane surfaces has been suggested to be the main reason for fouling (10–12). Hence, a number of authors have investigated surface properties of membranes and phenomenon of adsorption of solutes on the surfaces. One of the problems in relating the chemistry of membrane surface with solute adsorption using readily available membranes is that the membranes differ not only in chemistry but also in their pore structures. In these studies, it is desirable to use membranes with different chemistries but the same pore configurations. Low-temperature plasma treatment of surfaces permits creation of surfaces with different properties (13). Preparation of tailor-made membranes using low-temperature plasma treatment of surfaces has been suggested as a potential remedy for fouling (14–16). The objective of this work was to modify the surface characteristics of two polymeric membranes by low-temperature plasma treatment and then to investigate the relationship between surface chemistry, solute adsorption, and fouling characteristics of the surfaces.

Biological fluids contain several solutes such as proteins, carbohydrates, lipids, and their polymers, all of which have been suggested to be potential foulants (6). Of these, proteins have received considerable attention because of their common occurrence and complex interactions with themselves and with membrane surfaces (17–20). Hence, a protein (bovine serum albumin, BSA) was selected as a model foulant in this study.

MATERIALS AND METHODS

Globulin-free, lyophilized, and crystallized unlabeled bovine serum albumin was obtained from Sigma (St. Louis, MO). Its properties are well documented (21). ^{14}C-tagged BSA was obtained from American Radiolabeled Chemical (St. Louis, MO). The protein was dissolved in phosphate buffer (pH 7.0) prepared in high-purity water (distilled, ion-exchanged, and deionized) and equilibrated for at least 3 h. The final solution was prefiltered through a 0.2-μm mixed cellulose-esters hollow-fiber capsule (Fisher, St. Louis, MO) and a 0.1-μm Calyx nylon filter capsule (MSI, Westboro, MA). Sodium azide (0.02%) was added to the albumin solution to prevent bacterial growth. The concentration of albumin in solution was measured by absorbance at 290 nm and by Commassie blue method. Radiolabeled BSA was tracked by counting the rate of disintegration in a liquid-scintillation counter (LS 7000, Beckman, Fullerton, CA) in presence of scintillation cocktail (ScintiSafe, Fisher).

Track-etched polycarbonate (PC) membranes of nominal pore diameters between 0.05 and 0.4 μm nominal pore sizes (diameter 47 mm) were obtained from Poretics (Livermore, CA). These membranes were reported to have straight nontortuous pores and low porosity; maximum deviation of pore axes from perpendicularity to the membrane surface is less than 30% and only up to 10% of the pores exist as doublets or triplets (22). Another hydrophobic membrane, polypropylene (PP), of nominal pore size 0.1 μm was obtained from Gelman Science (Ann Arbor, MI) and that of 0.45 μm nominal pore size from MSI (Westboro, MA). The 0.1 μm PP membrane was a stretched membrane with a very high membrane surface area (22 m^2/g). The 0.45-μm PP membrane was a prefilter membrane made of intertwined fibers. Its surface area (2.3 m^2/g) was low and similar to those of PC membranes (1.4–1.6 m^2/g). The other membranes used were polyvinylidene fluoride (PVDF) from Millipore (Marlborough, MA), and naturally charged nylon membranes from Whatman (Fairfield, NJ). Both of these membranes were obtained in nominal pore sizes of 0.1, 0.2, and 0.45 μm. The PVDF membrane, as supplied, was hydrophilic because of grafting with polyhydroxypropylacrylate. Most of the available surface in the membranes was in pores (the ratio of total surface area to frontal surface area 595:1 for 0.1 μm PP membrane and 190:1 for 0.45 μm PP membrane); the corresponding values in PC membranes ranged from 10:1 to 12:1.

Fouling was quantified by measurements of permeation rates of clean water and of albumin solutions through modified and unmodified membranes in an Amicon 8050 stirred cell (Amicon, Beverly, MA). The experimental set-up is shown in Fig. 1. The transmembrane pressure in this study was varied between 0 and 12 psi. No aggregation of BSA has been reported during passage through microporous membranes at these transmembrane pressures (23). PC membranes (0.1 μm) were prewetted in order to reduce its wetting pressure in the operating range of pressures.

Protein adsorption on the surfaces was measured under static conditions using ^{14}C-bovine serum albumin. Equilibration between protein solution and the membrane(s) was conducted for 16 h followed by measurement of radioactivity on the membrane surface as well as in solution. The detection limit of the counting system and the effect of scintillation cocktail was established before the experiments. Surface area of the membranes was determined using a BET Sorptometer (Porous Material,

Fig. 1. Dead-end filtration set up: **(A)** nitrogen gas cylinder, **(B)** inline gas filter, **(C)** valve, **(D)** testgauge 0–15 psi, **(E)** reservoir, **(F)** pressure relief valve, **(G)** inline filters, **(H)** pressure tranducers, **(I)** stirred cell, **(J)** magnetic stirrer.

Fig. 2. Radiofrequency low-temperature plasma reactor.

Ithaca, NY) and argon gas. The data were analyzed using the Brunauer-Emmett-Teller analysis (24).

The low-temperature plasma treatment of membranes was conducted in a radiofrequency reactor, shown in Fig. 2. It consisted of a center tube (length 69 cm, od 3.9 cm, id 3.5 cm). The power was delivered in the form of 13.46 MHz radio frequency through two ring-shaped copper electrodes (width 1.3 cm, diameter 8 cm) located outside the reactor. The distance between the electrodes was set at 6 cm and the electrode assembly was placed symmetrically at the center of the tube. The unit was connected to a vacuum pump and pressure in the tube could be varied between 20 and 40 mtorr by manipulating a precision needle valve that also controlled the flow rate of the monomers. The monomers used were oxygen and nitrogen without and with *n*-butane. Use of the different monomers (plasma) imparts specific groups and, therefore, different chemical nature to the membrane surface as discussed in the RESULTS AND DISCUSSION subheading without changing the pore structure of the surfaces. This permits elucidation of the effect of chemical nature of the surface on protein adsorption and fouling without undue effect of pore geometry. The membrane sample to be treated was placed at the center of the tube and in the middle of the electrodes. Proper placement of the membrane and the conditions for plasma treatment were established using silicon wafer as a substrate.

Surfaces were characterized by measuring the advancing contact angle of a distilled-water droplet on the membrane surface. The instrument used was a Gonimeter (Model G-I, Kernco, TX) connected to a video camera (Hitachi, Congers, NY). Scanning electron micrographs of the membranes were taken with an Amray scanning electron microscope model 1600 (Amray, Bedford, MA) at an acceleration voltage of 10 kV.

RESULTS AND DISCUSSION:

Protein Adsorption on Membranes

Equilibrium adsorption of BSA on the four membrane surfaces, measured under static conditions, is reported in Fig. 3. The adsorption on hydrophilic surfaces (PVDF and nylon) was measured in the membrane, as whole of the surface, including the pores, would be wetted under static conditions. For the hydrophobic surfaces, the pores would not be wetted under static conditions. Hence, adsorption was measured using flat sheets. In Fig. 3, the adsorbed concentrations have been normalized by the total (internal and external) surface areas that were calculated from BET adsorption isotherms of argon. The adsorption behavior on PVDF and nylon membranes was independent of pore sizes, suggesting that adsorption took place in the pores as well. This is to be expected as the size of the protein molecule ($11.6 \times 2.7 \times 2.7$ nm^3) was considerably smaller than the nominal pore size. For the hydrophilic membranes (PVDF and nylon), a simple calculation of monolayer adsorption of side-on BSA molecules shows that the adsorption was monolayer.

The adsorption behavior on PVDF, nylon, and PC surfaces was linear. Only on polypropylene surface did the adsorption show a saturation behavior. These results are in agreement with those of Persson and Nilsson (25) and of Gök et al. (26). Yui et al. (27) also reported similar anomalies in adsorption of albumin on PP and polystyrene films; adsorption on PP showed a saturation behavior and that on polystyrene was linear in the whole concentration range. The adsorption on the hydrophilic membranes (PVDF and nylon) was considerably lower than that on hydrophobic membranes (PP and PC), as shown in Fig. 4. The higher amount of adsorption

Fig. 3. Adsorption of BSA on untreated membranes.

on PP and PC is attributed to the hydrophobic nature of these surfaces. This too is in agreement with the observations of Persson et al. (28), Bowen and Gan (10) and Pitt (29). Higher adsorption of albumin on nylon than on PVDF suggests that charge interactions played a role in the case of hydrophilic surfaces. Nylon surface has a natural positive charge, whereas albumin (pI value of 4.8) was negatively charged at the operating pH (7.0). On the other hand, the hydrophilic PVDF membrane has been reported to have a zeta potential of -19.5 mV at pH of 7.0 (10). These results were confirmed by the effect of ionic strength of solution on the adsorption phenomenon (results not presented).

Fouling Behavior of Untreated Membranes

Membrane fouling was studied using two methods: by the transient changes in the flux of albumin solutions (50 mg/L and 1 g/L) through the

Fig. 4. Adsorption of BSA on untreated surfaces of 0.45-μm nylon and PVDF, and on flat PP and PC sheets (T = 24°C, pH = 7.0, 0.01 M phosphate buffer).

membranes in a dead-end stirred cell, and by the changes in water-flux through the membranes after static adsorption of BSA on the surfaces. It must be noted that the membranes were microporous and were not expected to show any rejection of protein at the pressure drops used in these experiments (1–10 psi). Water flux through the clean membranes were found to be invariant with time, suggesting that water arriving at the membrane surface had no particulates that could foul the surfaces.

1. The permeation flux of albumin solutions through the different unmodified membranes is shown in Fig. 5. The membrane with the least protein adsorption, PVDF (Fig. 4), showed no sign of flux-reduction for up to 10 h operation. As expected, the fluxes through larger pore-size membrane were higher than those through membrane with smaller nominal pore size. Several authors have reported that adsorption equilibrium is achieved relatively rapidly and is even faster under flow conditions. Hence, it may be concluded that protein solutions did not foul the PVDF membranes.

Fig. 5. Behavior of flux of albumin solution through untreated membranes.

It has been suggested by Persson et al. *(28)* that for these membranes, 10% of the pores control 90% of the solution flux. Hence, even if the smaller pores got fouled, the effect on flux may not have been observed.

Permeation flux through the nylon membranes (also hydrophilic) showed relatively more flux drop than for PVDF membrane for all the pore sizes. Although adsorption of BSA on nylon is also higher, it still was in the range of monolayer adsorption. Since Persson et al. *(28)* have suggested that monolayer sorption should result in only 3 to 4% flux reduction on a 0.2 μm nominal pore size membrane, the flux reductions observed here could signify induced aggregation of albumin *(30)*.

For the hydrophobic membranes (PP and PC), the permeation flux increased with time before reducing again. For these hydrophobic membranes, initial adsorption of albumin perhaps increased the wettability of the surface, thus causing an increase in the flux. Further adsorption resulted in fouling and reduction in flux. The increase was more dominant at lower concentration of BSA than at higher concentrations. The observations of the effect of albumin concentration on fouling were inconsistent with the published reports of Mueller and Davis (23) who found that fouling was more severe with low concentrations of protein than at higher concentrations. The 0.1-μm PP membranes did not show any initial increase in permeation flux since these were prewetted by 50% (V/V) isopropanol in water. This was necessary to get any flux at all through the smaller pore-size membrane under the operating pressures.

2. The losses in permeation flux of water through membranes, on which BSA was preadsorbed, were studied with PC membranes because of their well-defined pore geometry. Since PC is a hydrophobic membrane, adsorption of protein increases its wettability and thus the ease of passage of water through its pores. Hence, PC membranes coated with polyvinylpyrrolidone (PVP) were used in this part of study; membranes treated in such a manner were hydrophilic and permitted water flux through it at low transmembrane pressure drops, without any complications caused by changes in hydrophobicity of the membrane. The water flux through these membranes before and after BSA adsorption were measured and the ratio of the two water fluxes, defined as nominal flux, was plotted as a function of pore-size (Fig. 6). The changes in relative flux were analyzed by a pore-blocking model (21) that assumes that pore size is reduced by adsorption of monolayer of the protein molecules. In such a case, the steady-state expression of relative flux is given by

$$\left(\frac{J}{J_0}\right) = \left(1 - \left[\frac{\Delta r}{r_0}\right]\right)^4 \tag{1}$$

Here r_0 is the pore size before protein adsorption and Δr is the reduction in pore-size caused by to protein adsorption. Norde (21) has estimated the value of Δr for BSA 4.6 nm. Equation (1) is plotted in Fig. 6 as a solid line. An excellent agreement is seen between the predictions and the experimental observations. The dotted line in Fig. 6 represents a case in which protein adsorption would have no effect on water flux. It clearly establishes a relationship between adsorption and fouling.

Effect of Plasma Treatment on Membrane Fouling

The membranes were treated with oxygen, n-butane/oxygen, nitrogen, and n-butane/nitrogen plasma. The conditions for plasma treatments (power, duration, and flow rate) were optimized earlier. The effect of plasma-treatment on membrane fouling was studied only with hydropho-

Fig. 6. Ratio of water flux through hydrophilic PC membranes after and before protein adsorption (nominal flux) as a function of pore sizes.

bic membranes (PP and PC) of large nominal pore sizes (0.4 or 0.45 μm). Fig. 7–9 show the SEM pictures of untreated and plasma-treated PP and PC membranes. Clearly, the membrane structures did not change appreciably. In all the cases, it was noticed that plasma penetrated the membrane pores and caused either the etching or the polymer deposition on the surface (including surface in the pores). Under the conditions of treatment, plasma treatment resulted in making the membrane surface more hydrophilic (i.e., advancing contact angle increased). The water fluxes through the plasma-treated and untreated membranes are listed in Table 1. Pore sizes of the PC membranes, estimated from SEM pictures are also listed in Table 1.

The corresponding water-permeation rates for PP and PC membranes are shown in Fig. 10. The untreated membranes showed a classic hydrophobic-membrane water-breakthrough response. Since the PP membrane has a wide pore-size distribution, water permeation rates through it show a highly nonlinear behavior as smaller and smaller pores are penetrated. It eventually becomes linear when all the pores get wet. PC membrane being of narrow pore-size distribution showed a linear increase in water flux after water breakthrough. The oxygen- and nitrogen-plasma-treated membranes showed water breakthrough even at zero transmembrane pressure drop, suggesting a hydrophilic nature of membranes thus treated. Butane-oxygen plasma treatment also resulted in a hydrophilic membrane and enhanced water flux, even though it was not as high as that for oxygen-treated membrane. The butane-nitrogen plasma resulted in even more hydrophobic membrane (advancing water contact angle increased) with considerably smaller pores. The reduction in pore-sizes was confirmed

Fig. 7. SEM pictures of untreated and oxygen plasma treated 0.45-μm PP (4 W, F_{oxyg} = 0.7 sccm and 20 min).

by measurements of thickness of plasma-film deposited on silicon wafers under identical conditions. Butane-nitrogen plasma-treatment of silicon wafer produced a film thickness of 0.08 μm that corresponds well with a 0.15 μm reduction in pore size. Similarly, butane-oxygen treatment of silicon wafer produced a film thickness of 0.015 μm that would result in a pore-size reduction of 0.03 μm. This also agrees well with the observed reduction of 0.04 to 0.06 μm in SEM pictures.

The changes in permeability of albumin solution through treated and untreated membranes were also measured. These are shown in Fig. 11.

Fig. 8. SEM pictures on: **(A)** untreated 0.40-μm PC **(B)** oxygen plasma treated 0.40-μm PC (4 W, F_{oxyg} = 0.8 sccm and 20 min), **(C)** *n*-butane/oxygen plasma-treated 0.40-μm PC (20 W, F_{but} = 1.1 sccm, 40% oxyg and 20 min.).

Oxygen-plasma treatment of 0.45 μm PP membrane resulted not only in an increase in the solution flux, the flux did not undergo any reduction for 6 h of operation. On the other hand, the untreated membrane exhibited an initial increase in the flux, followed by a gradual decline that continued past the 6-h operation. The PC membranes also exhibited increases in flux after treatment with oxygen or butane-oxygen or nitrogen plasma. In all the cases, no transient increase was observed as was the case with untreated membrane. But the PC membranes demonstrated a continued reduction

Fouling and Protein Adsorption

Fig. 8. *(continued)*

Fig. 9. SEM pictures of: **(A)** untreated 0.4-μm PC, **(B)** Nitrogen plasma treated 0.4-μm PC (7 W, 0.85 sccm nitrogen and 5 min), **(C)** *n*-butane/nitrogen plasma-treated 0.4-μm PC (20 W, 1.4 sccm *n*-butane, 48% nitrogen and 5 min).

of flux of albumin solution after 5 h of operation. Even after all the loss of flux, it was significantly higher than flux through the untreated membrane. The nitrogen-plasma treated membrane also showed a higher initial flux, but it dropped very rapidly. The butane-nitrogen-treated membrane had strongly shrunk pores as a result of treatment as well as higher hydrophobicity and flux through this membrane was never more than a fraction of fluxes in other cases.

Fig. 9. *(continued)*

The stability of these treatments was tested by measurement of advancing contact angles of the treated membranes over an extended time period. The results showed that the treatments were stable.

CONCLUSIONS

Based on the results presented in this work, it may be concluded that hydrophilic membranes are least prone to fouling with bovine serum albumin present in fermentation broth. Under the conditions of operation, hydrophilic membranes were susceptible to electrostatic interactions. On the other hand, a strong absorption was observed on the hydrophobic membranes. A simple pore-blocking model could be used to correlate the flux loss through pores of defined geometry.

Table 1

Membrane	Nominal pore size (n)	Treatment	Water flux mL/(min/psi/cm^2)	Pore-size from SEM-pictures	Advancing water contact angle
PP	0.45	None	3.2		
		4W, F_{oxy} = 0.7 sccm, 20 min.	4.5		
PC	0.40	None	0.3	0.30–0.32	67°
		4W, F_{oxy} = 0.8 sccm, 20 min.	1.9	0.30	43.5°
		20W, F_{but} = 1.1 sccm, 40 vol % oxygen, 20 min.	1.4	0.26	40°
PC	0.40	7W, F_{nit} = 0.85 sccm, 5 min.	0.95	0.34	32°
		20W, F_{but} = 1.4 sccm, 48 vol % oxygen, 5 min.	0.005	0.17	78°

Fig. 10. Water permeation rates through untreated and treated PP (0.45-μm) and PC (0.40-μm) membranes as a function of transmembrane pressure drip.

Fig. 11. Flux of albumin solution as a function of time for untreated and treated (**A**) 0.45-μm PP membrane and (**B**) 0.40-μm PC membrane.

Fig. 12. Decay of advancing contact angle for oxygen plasma-treated 0.45 PP and *n*-butane/oxygen plasma treated 0.40 PC. For the first point in the decay of the oxygen plasma-treated membrane the advancing contact angle of a flat surface with the same treatment was used. Since the membrane was totally wet and no measurement of contact angle was possible.

Results of treatment of the hydrophobic membranes with oxygen, nitrogen, butane-oxygen, and butane-nitrogen plasma suggested that oxygen plasma and butane-oxygen plasma are the method of choice in reducing flux loss. Nitrogen treatment introduced amino groups on the surface. Under the pH of operation, these are positively charged and undergo charge interactions with the negatively charged protein and result in increased fouling. Butane-nitrogen plasma treatment failed to increase the flux and in making the surface more hydrophilic.

REFERENCES

1. Ho, W. S. W. and Sirkar, K. K. (1992), *Membrane Handbook*, Van Nostrand Reinhold, New York.
2. Prasad, R. and Sirkar, K. K. (1988), *AIChE J.* **34**, 177.
3. Wang, C. J., Bajpai, R. K., and Iannotti, E. L. (1991), *Appl. Biochem. Biotechnol.* **28/29**, 589–603.
4. Scheler, C., Popovic, M., Iannotti, E. L., Mueller, R., and Bajpai, R. K. (1996), *J. Appl. Biochem. Biotechnol.* **57/58**, 29–38.
5. Baker, R. W. (1991), *Membrane Separation Systems: Recent Developments and Future Directions*, Noyes Data Corp., Park Ridge, NJ.
6. Smolders, C. A. and van den Boomgaard, T. (1989), *J. Membr. Sci.* **40**, 121–122.
7. van den Berg, G. B. and Smolders, C. A. (1988), *Filtration Separation* 115–122.
8. Aimar, P and Howell, J. A. (1991), *J. Membr. Sci.* **59**, 81–99.
9. Matthiasson, E. and Civik, B. (1980), *Desalination* **35**, 59–103.
10. Bowen, W. R. and Gan, Q. (1991), *Biotechnol. Bioengin.* **38**, 688–696.
11. Gekas, V. (1988), *Desalination* **68**, 77–92.
12. Tracey, E. M. and Davis, R. H. (1994), *J. Colloid Interface Sci.* **167**, 104–116.
13. Yasuda, H. (1985), *Plasma Polymerization*, Academic, New York.
14. Andrade, J. (1985), *Surface Interfacial Aspects of Biomedical Polymers*, Plenum, New York.
15. Michaels, A. (1989), *Chem. Technol.* 162.
16. McGuire, J. and Krishdasima, V. (1991), *Food Technol.* 92–96.
17. Brink, L. E. S. and Romijn, D. J. (1990), *Desalination* **78**, 209–233.
18. Fane, A. G., Fell, C. J. D., and Waters, A. G. (1983), *J. Membr. Sci.* **16**, 211–224.
19. Hlavacek, M. and Bouchet, F. (1993), *J. Membr. Sci.* **82**, 285–295.
20. Marshall, A. D., Munro, P. A., and Tragardh, G. (1993), *Desalination* **91**, 65–108.
21. Norde, W. (1986), *Adv. Colloid Interface Sci.* **25**, 267–340.
22. Chandavarkar, 1990.
23. Mueller, J. and Davis, R. H. (1996), *J. Membr. Sci.* **116(1)**, 47–60.
24. Brunauer, S., Emmet, P. H., and Teller, J. (1938), *Am. Chem. Soc.* **60**, 309–319.
25. Persson, K. M. and Nilsson, J. L. (1991), *Desalination* **80**, 123–138.
26. Gök, E., Kiremitci, M., and Ates, I. S. (1994), *Reactive Polymers* **24**, 41–48.
27. Yui, N., Suzuki, Y., Mori, H., and Terano, M. (1995), *Polymer J.* **27(6)**, 614–622.
28. Persson, K. M., Capannelli, G., Bottino, A., and Trägård, G. (1993), *J. Membr. Sci.* **76**, 61–71.
29. Pitt, A. (1987), *J. Parenteral Sci.* **41(3)**, 110–113.
30. Kelly, S. T., Opong, W. S., and Zydney, A. L. (1993), *J. Membr. Sci.* **80**, 175–187.

Biocatalytic Removal of Nickel and Vanadium from Petroporphyrins and Asphaltenes

L. Mogolloń, R. Rodríguez, W. Larrota, C. Ortiz, and R. Torres*

Colombian Petroleum Institute, Laboratory of Biotechnology, Km 7, Via Piedecuesta, Bucaramanga, Colombia

ABSTRACT

Asphaltenes from a crude oil rich in heavy metals (Castilla crude oil) were fractionated and partially characterized. Biocatalytic modifications of these fractionated asphaltenes by three different hemoproteins: chloroperoxidase (CPO), cytochrome C peroxidase (Cit-C), and lignin peroxidase (LPO) were evaluated in both aqueous buffer and organic solvents. The reactions were carried out in aqueous buffers, ternary systems of toluene:isopropanol:water, and aqueous-miscible organic solvent solutions with petroporphyrins as substrate. The petroporphyrins were more soluble in the ternary systems and aqueous miscible-organic solvent systems than in the aqueous buffer systems. However, only the CPO-mediated reactions were effective in eliminating the Soret peak in both aqueous and organic solvent systems. The effects of CPO-mediated reactions on the release of the metals complexed with the porphyrins and asphaltenes were also determined. Chloroperoxidase was able to alter components in the heavy fractions of petroleum and remove 53 and 27% of total heavy metals (Ni and V, respectively) from petroporphyrin-rich fractions and asphaltenes.

Index Entries: Asphaltenes; chloroperoxidase; biocatalytic modification; organic solvents; demetallation.

INTRODUCTION

Petroleum is a complex mixture containing a wide type of organic compounds. Among these are petroporhyrins, macromolecules of high molecular weights, associated with the pentane-insoluble fraction of petroleum known as asphaltenes *(1)*. Asphaltenes are very high-molecular-

---

*Author to whom all correspondence and reprint requests should be addressed.

weight compounds containing aromatic and aliphatic constituents, heteroatoms (S,O, and N) (1,2) and heavy metals, like Ni and V (3).

Their chemical and average molecular structures may vary considerably with their sizes (4,5). In addition, high-molecular fractions of asphaltenes are more prone to forming methylene chloride-insoluble polymers on heating than are the low-molecular-weight fractions (4). These differences in both chemical structures and chemical reactivities suggest that some fractions of asphaltenes are more susceptible to enzymatic or microbial attack than others (6).

Microorganisms have been shown to associate with bitumens. Although there have been numerous reports demonstrating that microbial cells can degrade hydrocarbons such as polycyclic aromatic hydrocarbons (PAHs) (7) and sulfur heterocycles (8), there is a little chemical or physical evidence that microorganism can modify asphaltenes (9).

The data included in this paper are the result of the separation and partial characterization of asphaltenes from Colombian crude oil rich in heavy metals (Castilla crude oil) and its biocatalytic modification by three different hemoproteins in organic solvents. The investigations showed that only chloroperoxidase (CPO E.C. 1.11.1.10) in a buffer phosphate and organic solvent systems could eliminate the Soret peak in the petroporphyrin fraction of the asphaltenes under the experimental conditions studied. The effect of the CPO-mediated reactions on the fate of the metals complexed with the porphyrins and asphaltenes was also determined.

MATERIAL AND METHODS

Chemicals

Cytochrome C from horse heart was obtained from Sigma (St. Louis, MO). Lignin peroxidase, partially purified was obtained from Tienzyme (State College, PA). Chloroperoxidase from *Caldariomyces fumago* (Rz = 1.4) was a gift from Michael Pickard (Department of Microbiology, University of Alberta). The high-performance liquid chromatography (HPLC)-grade solvents isopropanol, acetonitrile, methylene chloride, chloroform, toluene, and tetrahydrofuran were obtained from Merck (Darmstadt, Germany). Tetrahydrofuran was distilled in the presence of ferrous sulphate to eliminate peroxides. Buffer salts and hydrogen peroxide were purchased from Merck. Asphaltenes was obtained by n-hexane precipitation from Castilla crude oil.

Fractionation of Asphaltenes

Castilla crude oil is an oil with a 25% asphaltene content. Asphaltenes were fractioned by adsorption chromatography on silica gel using chloroform as eluent.

Analytical GPC

The fractions from the silica gel column were analyzed by HPLC using three 45 × 1-cm polystyrene/divinylbenzene columns (100, 500, and 10,000) in series with methylene chloride as eluant. The instrument used was a Perkin Elmer HPLC 410 UV LC-95 System with a UV detector at 254 nm.

Ultraviolet-Visible Spectrophotometry and Calculation of Porphyrin Concentrations

A Varian Cary 1E spectrophotometer was used to obtain UV-visible scans (200–600 nm) of asphaltene fractions dissolved in methylene chloride. The presence of petroporphyrins was detected by the prominent absorbance peak at approx 410 nm (Soret peak) and a weaker absorbance a 573 nm (α peak). Semiquantitative analysis of porphyrins was done according to Semple et al (5). The concentration, in (μg/g) was estimated using an approximated molecular weight for porphyrins of 600 g/mol (Vanadium octaethylporphyrin).

Biocatalytic Modifications with Hemoproteins in Aqueous Buffer Solutions

The enzymatic reactions with hemoproteins in aqueous systems were carried out using petroporphyrins from fractionated asphaltenes as substrate.

Specific activities of lignin peroxidase (LPO) from *Phanerochaete chrysosporium* were estimated in a 2-mL reaction mixture containing 25 nM of enzyme and 50 μg of substrate in 40 mM succinate buffer pH 4.0 with KCl (9 mM). Reactions were started by adding 0.1 mM hydrogen peroxide.

Mixture reactions (2 mL) with horse heart cytochrome c (Cit-C) were carried out with 0.8 $\mu$$M$ of cytochrome C with 50 μg substrate in a 60 mM phosphate buffer pH 6.1 with KCl (9 mM). The reactions were started by adding 1 mM hydrogen peroxide.

For CPO, the reaction mixture (2 mL) contained: 50 μg petroporphyrin, 9 mM KCl in a 3mM KH$_2$PO$_4$ (pH 3.0) buffer. For enzyme reactions, the H$_2$O$_2$ was added first and mixed in the buffer and then CPO (10 μg/mL) was added to start the reaction. Six additions (6X) of CPO 5 μg/mL with 0.25 mM H$_2$O$_2$ were both effective in reducing the Soret peak of petroporphyrin (6).

Methylene chloride (2 mL) was added to the small vial in order to stop the reaction and to extract the porphyrin for all hemoprotein-mediated reactions. This porphyrin layer was transfered into a cuvet to determine changes in the UV-visible spectrum.

In aqueous buffer systems, only CPO was effective in eliminating the Soret peak. For this reason the reactions in organic solvents were mainly carried out with CPO so as to evaluate the biocatalyitic modifications and demetallation of petroporphyrin and asphaltenes.

CPO Reactions in Ternary Solvent Systems

The experiments used a solvent system of toluene:isopropanol:water. Five different ternary systems were prepared using clear microemulsion (normal ternary system). All hemoproteins were evaluated in these systems. However, only CPO could eliminate the Soret peak from petroporphyrins

For all hemoprotein reactions, petroporphyrins were dissolved in toluene and added to the solvent mixture. The reactions were carried out with two different concentrations of petroporphyrins (12.5 and 25 μg/mL).

Aqueous portion for enzymatic reactions was: For LPO, 40 mM succinate buffer pH 4.0, for CPO, 3 mM KH_2PO_4 buffer, pH 3.0, and for Cit-C, 3 mM phosphate buffer pH 6.1. KCl at concentration of 20 mM was added to all buffers. Isopropanol was added at different concentrations (Table 4) and the mixtures were then shaken well to ensure the formation of a clear microemulsion.

The complete reaction mixture contained 0.5 mM H_2O_2 and 1.25 μg CPO/mL. For petroporphyrins, the disappearance of the Soret peak at 407 nm was followed. The maximum rate of the reaction was determined from the linear portion of the curve. The appearance and disappearance of the visible peak at 435 nm was also monitored.

CPO Reactions in Binary Solvent Systems (THF-Aqueous Buffer)

Several water-miscible organic solvents for dissolution of asphaltenes were evaluated (data not shown). The experiments of biocatalytic modifications were carried out with CPO, Cit-C, and LPO by using different tetrahydrofuran concentrations (range 5–30% v/v of THF). As in aqueous buffer and microemulsions experiments, only CPO showed biocatalytic modification on petroporphyrins indicated by changes in the UV-visible spectra (elimination of Soret peak).

For CPO reaction, the buffer solution was 3 mM KH_2PO_4 pH 3.0 and KCl 20 (mM). Petroporphyrins were dissolved in THF and then were added to the reaction mixture. Two different concentrations of petroporphyrins (25 and 50 μg/mL) were evaluated in THF. The complete reaction mixture contained 0.5 mM H_2O_2 and 1.25 μg CPO/mL. The biocatalytic reaction was stopped by adding THF (60% v/v). The reaction mixtures were then transfered to cuvets to quantify the disappearance of the visible peak at approx 410 nm (Soret peak).

Scale-Up of CPO-Mediated Reactions in Ternary Systems and Demetallation

Reactions in ternary systems with CPO were used for scale-up experiments. Petroporphyrins or asphaltenes dissolved in methylene chloride were distributed into 500-mL Erlenmeyer flasks and the solvent was allowed to evaporate. The solid residue in the Erlenmeyer flask was then dissolved in toluene and appropriate volumes of isopropanol and buffer were added. Mixture 5 showed very good enzyme activity, and this was chosen for scale-up experiments. The H_2O_2 was added to give a final concentration of 0.5 mM, and the reaction was started by the addition of CPO to give a concentration of 1.25 µg of enzyme per mL. The progress of the reaction with petroporphyrins was followed by placing a portion of the reaction mixture into a cuvet and following the disappearance of the Soret peak or the appearance and subsequent reduction of the product peaks (A = 435 nm). Further additions of CPO and H_2O_2 were made at 10-min intervals (10X) in order to complete the reaction. An extraction blank, consisting of a ternary system without porphyrin (or asphaltene), and controls with CPO only or with H_2O_2 only were included with each experiment. At the end of the reaction, two volumes of water were added, thus yielding two phases and allowing for the extraction of the porphyrin material.

Analytical Methods

Petroporphyrins and asphaltenes were acid-extracted from the reaction mixtures in a separatory funnel. The organic phase was filtered through sodium sulfate to remove water. The sample was evaporated under a nitrogen atmosphere and analyzed by atomic absorption. The aqueous phase of the reaction was collected in an acid-washed beaker, concentrated more than 20-fold on a steam bath, and then dissolved in 2% HNO_3. The presence of released metals (Ni and V) was analyzed by atomic absorption spectroscopy in a Perkin Elmer 5100 spectrophotometer.

The enzyme concentrations were estimated by protein measurement with the Bio-Rad (Richmond, CA) procedure using bovine serum albumin (BSA) as standard and by spectrophotometry using an absorption coefficient of 168 mM^{-1}/cm at 409 nm for LPO *(10)*, 75.3 mM^{-1}/cm at 403 nm for chloroperoxidase *(11)* and 29.5 mM^{-1}/cm at 550 nm for cytochrome C peroxidase *(12)* reduced with sodium dithionite.

RESULTS AND DISCUSSION

Fractionation of Asphaltenes

Tables 1 and 2 show petroporphyrin contents and heavy metal (nickel and vanadium) composition in asphaltene fractions obtained by adsorption chromatography in silica gel. Petroporphyrins were measured by quantification of Soret peak at approx 410 nm. Almost 65% of the petroporphyrins contained in asphaltenes were concentrated in fractions 5 and 6, with molecular weights close to 600 g/moL (probably vanadium-porphyrins) *(5)*. These petroporphyrin fractions were used to evaluate biocatalytic

Table 1
Characterization of Asphaltene Fractions Obtained from Castilla Crude Oil

Fraction	Porphyrin content[a] μg/g	Porphyrin content[a] μmoles/g[b]	Distribution in asphaltenes (%)	MW[c](g/mol)
1	290	0.48	5.0	940
2	270	0.45	6.6	765
3	450	0.75	7.0	600
4	605	1.01	7.0	540
5	4900	8.17	12.8	535
6	20650	34.42	51.3	520
No Fractionated[c]	295	0.49	10.3	N.D.

[a] Initial content is 1.53 μmoles/g asphaltene and 920 μg/g asphaltene.
[b] A MW of 600 is used for the porphyrins.
[c] No fractionated asphaltenes.
[d] MW obtained from maximum peak. Average MW of asphaltenes = 750 g/mol.
N.D.: Not determined.

Table 2
Nickel and Vanadium Contents in Asphaltene Fractions Using Adsorption Chromatography

Fraction	Metal content (μg/g)[a] Ni	Metal content (μg/g)[a] V	V in porphyrins[b] μg/g	V in porphyrins[b] %[c]
1	543	2071	25	1.2
2	508	1907	23	1.2
3	513	1937	38	2.0
4	507	1845	51	2.8
5	458	1755	416	23.7
6	294	4155	1753	42.2
No Fractioned	N.D.	N.D.	(25)	(26.9)

[a] Metal contents in asphaltenes is 1686 and 43 μg/g for V and Ni, respectively.
[b] Determined from μmol/g porphyrin assuming 1 mol V complexed per porphyrin.
[c] % of V found that is associated with petroporphyrins. In asphaltenes is 4.4%. The value between parentheses was determined by difference.
N.D.: Not determined.

reactions with hemoproteins. This large fraction of vanadium porphyrins was confirmed with atomic absorption analysis. In these asphaltenes, higher concentrations of vanadium were complexed with porphyrins (13,14). Table 2 displays Ni and V contents in fractionated porphyrins. The vanadium content varies with porphyrin concentrations, and this is

Table 3
Biocatalytic Modification of Petroporphyrin in
Aqueous Buffer with Three Different Hemoproteins

Enzyme	Relative Activity (ΔAbs_{410}/min/µg HP)[a]	
	Petroporphyrin (12.5 µg/mL)	Petroporphyrin (25 µg/mL)
LPO	N.D.	N.D.
Cit-C	N.D.	N.D.
CPO	1.71×10^{-3}	3.14×10^{-3}

[a] HP = Hemoprotein.
N.D.: Not detected.

increased in rich-porphyrin fractions. However, nickel contents are decreased as asphaltenes are eluted from the column. These results indicate that vanadium is more associated to light fractions (vanadium porphyrins) *(13)*, whereas nickel appears at higher concentrations in heavy nonporphyrinic fractions. These results are in agreement with characterizations of porphyrinic and nonporphyrin metal complexes in asphaltenes from other oils *(3,13,15)*.

Reactions in Aqueous Buffer

Treatment of petroporphyrin with LPO, Cit-C, and CPO were evaluated in the presence of H_2O_2. Table 3 shows the results obtained in biocatalytic modification of petroporphyrins in aqueous buffer. Only CPO resulted in changes in the petroporphyrin UV-visible spectrum. The Soret peak at 410 nm, characteristic of vanadium-petroporphyrin molecules *(5,16)* was reduced in size (Fig. 1). There was no effect when this fraction was treated with either CPO or H_2O_2 alone. Moreover, chloride ions were required for biocatalytic modification of petroporphyrin in aqueous solutions *(6)*.

No biocatalytic modifications of asphaltenes was observed with Cit-C and LPO. It was not possible eliminate the Soret peak in all reaction mixtures. Although Cit-C and LPO can oxidize polyaromatic hydrocarbons (PAHs) *(17,18)* and sulfur asphaltenes as dibenzotiophene *(19)*, these hemoproteins were not effective in the elimination of the Soret peak.

In order to get most complete reduction of the Soret peak, it was better to use multiple additions of small amounts of enzyme and peroxide rather than a single addition. Successive additions of CPO and H_2O_2 produced a complete reduction of the Soret peak. The results obtained in the aqueous system in differents replicates showed that the reduction of the Soret peak reached a maximum of 80% under the experimental conditions. This incomplete peak removal is mainly caused by mass-transfer limita-

Fig. 1. UV-visible spectra of asphaltenes rich in petroporphyrins with and without CPO.

tions. The results were inconsistent due to the insolubility of petroporhyrin and because mass-transfer limitations were difficult to avoid.

CPO Reactions in Ternary Solvent Systems

Treatment of petroporphyrin in microemulsions with LPO, Cit-C, and CPO were evaluated in the presence of H_2O_2. However only CPO was able to eliminate the Soret peak. In order to eliminate mass-transfer limitations in the system and improve the action of CPO on petroporphyrins, the enzyme reactions were carried out in ternary systems of isopropanol:toluene:water (6,20). Five different mixtures were used to evaluate biocatalytic modifications of petroporphyrin because they were effective in eliminating the Soret peak for nickel octaethylporphine (NiOEP) in presence of CPO and H_2O_2 (6) (Table 4). In this system, CPO showed a higher activity than in aqueous buffer. The rate of disappearance of the Soret peak of petroporphyrins in this ternary system was proportional to the amount of CPO used and the relative composition of compounds in the microemulsion.

The highest enzyme activity was observed in mixture 5. Reaction mixtures with a low water content (approx 15%) showed good results in removal of Soret peak. Some reactions stopped with the formation of a product with a maximum absorption peak of 435 nm. There was no loss of Soret peak when the reaction mixtures lacked CPO or H_2O_2 and chloride was absolutely required for activity of CPO on the petroporphyrins in the microemulsions to take place. These results suggest that free radical

Table 4
Biocatalytic Modification of Petroporphyrins in Normal Ternary Systems

Mixture	Toluene	Isopropanol	Aqueous	Relative Activity (ΔAbs_{410}/min/μg CPO)	
				Petroporphyrin (12.5 μg/mL)	Petroporphyrin (25 μg/mL)
1	20.7 (0.11)	64.9 (0.46)	14.4 (0.43)	5.4×10^{-3}	10.1×10^{-3}
2	30 (0.14)	55 (0.39)	15 (0.46)	12.6×10^{-3}	23.8×10^{-3}
3	5 (0.02)	75 (0.46)	20 (0.52)	9.2×10^{-3}	20.0×10^{-3}
4	20 (0.09)	60 (0.38)	20 (0.53)	17.6×10^{-3}	34.8×10^{-3}
5	15 (0.06)	60 (0.34)	25 (0.60)	21.7×10^{-3}	43.3×10^{-3}

Percent composition by volume (mole fraction).

formation of ClO^- is necessary for the destruction of porphyrin in asphaltenes (6). Free-radical production by some peroxidases has been demonstrated in monophasic organic solvents (21,22).

Moreover, CPO is more selective to oxidize chloride than other hemoproteins (23). Hemoproteins such as horseradish peroxidase, lactoperoxidase, and lignin peroxidase have been reported to catalyze the oxidation of iodide and bromide but only CPO has been shown to be effective in the oxidation of chloride (23,24).

Microemulsions based on toluene or hexane have been used for enzyme reactions on water-insoluble substrates (6,21). In this work, the ternary system of toluene:isopropanol:buffer enhanced the CPO-catalyzed modifications of water insoluble petroporphyrins.

CPO Reactions in Binary Solvent Systems (THF-Aqueous Buffer)

In these experiments, solvent systems containing THF were used because it was the only water-miscible organic solvent able to dissolve appreciable amounts of petroporphyrins and asphaltenes. The biocatalytic modification of petroporphyrins (a water-insoluble substrate) in an organic solvent system requires a good correlation between enzyme activity, stability, and the solubility of the substrate (22).

Figure 2 shows the results obtained with CPO at different concentrations of THF. In this system CPO carried out similar modifications of the petroporphyrins as in microemulsions. However, the enzyme activity was dramatically decreased in the presence of high concentrations of THF. No reaction was observed in a mixture containing 25% of THF. This enzyme inactivation by high concentrations of THF and other aqueous-miscible organic solvents has been observed in other hemoproteins (17,25,26). The enzymatic activity in these water-miscible systems with high organic solvent concentrations is suppressed because the organic solvent replaces water in the protein surface layer (25,27).

Fig. 2. Effect of THF concentration on the activity of CPO using petroporphyrin as substrate.

Demetallation of Petroporphyrins and Asphaltenes

The reactions were carried out with mixture 5, because it was the microemulsion that resulted in the best demetallation of Ni and V from petroporphyrins and asphaltenes. The enzyme reactions were worked with 12 mg of asphaltenes (or porphyirins) and an enzyme concentration of 1.25 µg mL (10X).

Tables 5 and 6 include the results obtained in the demetallation experiments using the ternary system (mixture n°5). Table 5 shows that the CPO reactions increased the amount of Ni and V in aqueous phase with a total heavy metal removal from petroporphyrins of 53%. Metal removal rates of heavy metals from petroporphyrins were 4.1 (mg Ni/g CPO/h) and 35.76 (mg V/g CPO/h).

Table 6 shows the results obtained with demetallation of asphaltenes. Metal removal rates were estimated from these results. Heavy metals removal from asphaltenes were lower than petroporphyrins. The following metal removal rates were reached: 2.4 (mg Ni/g CPO/h) and 15.5 (mg V/g CPO/h). These results confirm metal release from the petroporphyrins by biocatalytic activity with a total removal of 27% of both Ni and V.

Oxidative demetallation by chemical means has been carried out by using oxidating agents such as hypochlorite, chlorine, and sulfuryl chloride. The objective of such studies was the destruction of porphyrins with minor effects on petroleum (28,29).

Treatment of asphaltenes with CPO is an alternative technology to remove heavy metals from crude oils and minimize problems related with poisoning catalysts. Ni and V from organometallic compounds are concen-

Table 5
Metal Analysis of Asphaltene Fraction Rich in Heavy Metals in Mixture 5.
Experiences of Demetallation Using CPO (10×)[a]

	Amount of nickel				Amount of vanadium			
	Organic phase		Aqueous phase		Organic phase		Aqueous phase	
Sample	g	%R[b]	g	%F[c]	g	%R	g	%F
Control without CPO	3.0	0	0.6	20.0	28.4	0	0.2	0.7
Treated with CPO (10×)	1.3	56.7	1.7	53.1	13.5	52.5	4.5	15.8
Control without Asphaltenes	N.D.		0.3		N.D.		0.2	

[a] A sample of 12 mg of petroporphyrins was used, and this expected to contain approx 32.4 V and 3.67 µg Ni.
[b] %R is defined as percentage of metal removal from asphaltene. Based on Ni or V in organic phase control.
[c] %F is defined as percentage of metal found. Based on Ni or V in organic phase control.
N.D.: Not determined.

Table 6
Metal Analysis of Asphaltenes in Mixture 5. Experiences of Demetallation Using CPO (10×)[a]

	Amount of niquel				Amount of vanadium			
	Organic phase		Aqueous phase		Organic phase		Aqueous phase	
Sample	g	%R[b]	g	%F[c]	g	%R	g	%F
Control without CPO	4.3	0	0.1	2.3	18.7	0	<0.1	<0.5
Treatment with CPO (10×)	3.4	20.9	0.3	8.8	12.9	31.0	1.4	7.5
Control without Asphaltenes	N.D.		0.3		N.D.			<0.1

[a] A sample of 12 mg of asphaltenes was used, and this expected to contain approx 20.3 V and 5.2 µg Ni.
[b] %R is defined as percentage of metal removal from asphaltene. Based on Ni or V in organic phase control.
[c] %F is defined as percentage of metal found. Based on Ni or V in organic phase control.
N.D.: Not determined.

trated in the asphaltenes and affect hydrotreatment and cracking catalysts (30). In this perspective, the application of biocatalysis in organic solvents can open new technological processes in petroleum upgrading.

ACKNOWLEDGMENTS

This work was supported by the Colombian Petroleum Institute, Ecopetrol, Colombia. Acknowledgment is made to Michael Pickard from the University of Alberta, Canada for donation of chloroperoxidase vials. The advice of Rafael Vazquez-Duhalt from Biotechnology Institute, UNAM, Mexico and the review of the English version by Luis E. Ortiz from Environmental Department, ECOPETROL, Colombia is gratefully appreciated.

REFERENCES

1. Strauz, O. P., Mojelsky, T. W., and Lown, E. M. (1992), *Fuel* **71**, 1355–1363.
2. Waldo, G. S., Carlson, R. M. K., Moldowan, J. M., Peters, K. E., and Penner-Hahn, J. E. (1991), *Geochim. Cosmochim. Acta*, **55**, 801–814.
3. Fish, R. H., Komlenic, J. J., and Wines, B. K. (1984), *Anal. Chem.* **56**, 2452–2460.
4. Ignasiak, T. M., Kotlyar, L., Samman, N., Montgomery, D. S., and Strausz, O. P. (1983), *Fuel* **62**, 363–369.
5. Semple, K. M., Cyr, N., Fedorak, P. M., and Westlake, D. W. S. (1990), *Can. J. Chem.* **68**, 1092–1099.
6. Fedorak, P., Semple, Vasquez-Duhalt, R., and Westlake, D. (1993), *Enzyme Microb. Technol.* **15**, 429–437.
7. Cerniglia, C. E. (1992), *Biodegradation* **3**, 351–368.
8. Finnerty, W. R., Schokley, K., and Attaway, H. (1983), in *Microbial Enhanced Oil Recovery*, Zajuic, J. E. et al., eds., American Chemical Society, Washington, DC, pp. 2–39.
9. Rontani, J. F., Bosser-Joulak, P. M., Rambeloarisos, E., Bertrand, J. C., and Giusti, G. (1985), *Chemosphere*, **14**, 1413–1422.
10. Farrel, R. L., Murtagh, K. E., Tien, M., Mozuch, M. D., and Kirk (1989), *Enzyme Microb. Technol.* **11**, 322–328.
11. Morris, D. P. and Hager, L. P. (1966), *J. Biol. Chem.* **241**, 1763–1768.
12. Blauer, G., Sreetama, N., and Woody, R. W. (1993), *Biochemistry* **32**, 6674–6679.
13. Yen, T. F. (1975), in *The Role of Trace Metals in Petroleum*, Yen, T. F., ed., Ann Arbor Science Publishers, Ann Arbor, MI, pp. 1–30.
14. Valkovic, V. (1978), *Trace Metals in Petroleum*, Petroleum Publishing, Tulsa, OK.
15. Filby, R. H. and Van Berkel, G. J. (1987), in *Metals Complexes in Fossil Fuels*, Filby, R. H. and Branthaver, J. F., eds., American Chemical Society, Washington, DC, pp. 2–39.
16. Baker, E. W. (1969) in *Organic Geochemistry*, Eglinton, G. and Murphy, M. T. J., eds., Springer-Verlag, Berlin, pp. 464–497.
17. Vazquez-Duhalt, R., Semple, K. M., Westlake, D. W. S., and Fedorak, P. M. (1993), *Enzyme Microbiol. Technol.* **15**, 494–499.
18. Vazquez-Duhalt, R. Westlake, D. W. S., and Fedorak, P. M. (1994), *Appl. Environ. Microbiol.*, **60**, 459–466.
19. Klyachko, N. L. and Klibanov, A. M. (1992), *Appl. Biochem. Biotechnol.* **37**, 53–68.
20. Vulfson, E. N., Ahmed, G., Gill, Y., Kozlov, I. A., Goodenough, P. W., and Law, B. A. (1991), *Biotechnol. Lett.* **13**, 91–96.
21. Khmelnitsky, Y. L., Levanshov, A. V., Klyachko, N. L., and Martinek, K. (1988), *Enzyme Microbiol. Technol.* **10**, 710–724.
22. Dordick, J. S. (1989), *Enzyme Microbiol. Technol.* **11**, 194–211.
23. Doerge, D. (1986), *Arch. Biochem. Biophys.* **244**, 678–685.
24. Farhangrazi, Z. S., Sinclair, R., Yamazaki, Y., and Powers, L. S. (1992), *Biochemistry* **31**, 10,763–10,768.
25. Khmelnitsky, Y. L., Belova, A. B., Levashov, A. V., and Mozhaev, V. V. (1991), *FEBS Lett.* **284**, 267–269.
26. Mozhaev, V. V., Khmelnitsky, Y. L., Sergeeva, M. V., Belova, A. B., Klyachko, N. L., Levashov, A. V., and Martinek, K. (1989). *Eur. J. Biochem.* **184**, 597–602.

27. Gorman, L. A. and Dordick, J. S. (1992), *Biotechnol. Bioeng.* **39,** 392–397.
28. Sugihara, J. M., Branthaver, J. F., and Willcox, K. W. (1973), *Prepr. Am. Chem. Soc. Div. Petrol. Chem.* **18,** 645–647.
29. Gould, K. A. (1980), *Fuel* **59,** 733–736.
30. Pelet, R., Behar, F., and Monin, J. C. (1986), *Org. Geochem.* **10,** 481–498.

Expanded-Bed Adsorption Utilizing Ion-Exchange Resin to Purify Extracellular β-Galactosidase

José Antonio Marques Pereira,[1] Paulo De Tarso Vieira E Rosa,[2] Glaucia Maria Pastore,[3] and Cesar Costapinto Santana*,[2]

[1]*Departamento de Technologia de Alimentos, Universidade Federal de Viçosa,* [2]*Departamento de Processos Biotecnológicos, Faculdade de Engenharia Química,* [3]*Departamento de Ciências de Alimentos, Faculdade de Engenharia de Alimentos, Universidade Estadual de Campinas (UNICAMP), C.P. 6066, CEP 13081-970, Campinas, SP, Brazil*

ABSTRACT

The application of expanded-bed ion-exchange resins allows the elimination of intermediary particulate separation steps like filtration or centrifugation prior to adsorption steps in enzyme-purification processes from crude fermentation broths. This work is concerned with the experimental evaluation data of a process related to the adsorption of an extracellular β-galactosidase from the fungi *Scopulariopsis*. The protein recovery in the ion-exchange resin Accell Plus QMA™ was accomplished using a continuous-monitoring method. The direct adsorption step was followed by a elution step with concentrated NaCl solutions aiming to improve the enzyme-specific activity. Experimental data for fixed and expanded bed were compared.

Index Entries: Extracellular β-galactosidase; adsorption; expanded bed; enzyme recovery.

INTRODUCTION

The preservation of molecular integrity and the pursuit of high-product yield during recovery and purification operations is of paramount concern to the bioprocess engineer. This is best accomplished by the elimination of unnecessary processing operations. The initial steps of the recovery and purification train typically involve time-consuming and inefficient

*Author to whom all correspondence and reprint requests should be addressed.

solid-liquid separation stages to generate the particulate-free feedstream required for subsequent chromatographic operations. Conventional chromatography is operated in downflow fixed-bed adsorption columns. Because of the small inter-particle voidage between the solid-phase components, a particulate-free feedstream is mandatory for their operation. Such problems can be avoided if the direction of the inlet feed is reversed and the linear flow velocity of the liquid is increased. Inter-particle voidage increases and whole-broth processing becomes a possibility. This operation is named expanded-bed adsorption and has been utilized by several authors for enzyme recovery (1–4).

A strain of *Scopulariopsis* that is a potent producer of extracellular β-galactosidase was isolated by Pastore and Park in 1980 and submitted to some preliminary purification and characterization (5). The low yield (4%) obtained in those procedures motivated the application of the new technique of expanded-bed chromatography with relatively high-density resins. This paper describes the use of a purpose-designed expanded-bed adsorption system for the direct capture of the extracellular enzyme from homogenized fermentation broths and the comparison of this approach with traditional packed-bed routes.

MATERIALS AND METHODS

Production of β-Galactosidase from *Scopulariopsis*

The strain of the fungi *Scopulariopsis* selected previously (5) was utilized to produce β-galactosidase in wheat-bran medium (Koji process). After incubation of wheat (that is inoculated with spores) at 30°C for 5 d, 100 mL of deionized water was added to 20 g bran. This was soaked for 1 h at room temperature. The proteic extract containing the enzyme was precipitated by ethanol (70% of final concentration) at 4°C. The precipitate was centrifugated and the solids were subjected to freeze-drying.

Determination of the Protein Concentration

Total protein concentration was monitored trough UV-absorbance by using a medium-pressure chromatographic system (Biologic System, Biorad Richmond, CA). The protein concentration was determined by the Bradford method (6) utilizing bovine serum albumin as standard.

Determination of the Enzyme Activity

As proposed by Pastore and Park (5), β-galactosidase activity was determined by incubating a mixture of 10 μ of the enzyme and 0.15 mL of 0.25% w/w solution of the substrate o-nitrophenyl β-D-galactopyranoside (ONPG) for 15 min at 60°C. The buffer utilized for the control of incubation experiments was 1.69 mL of 100 mM acetate, pH = 5.0. The product of the reaction is o-nitrophenol (ONP), and the reaction is interrupted by the addition of 0.15 mL of 10% sodium carbonate solution. One unit of β-galactosidase activity was defined as the amount of enzyme that liberates 1 μmole of ONP per minute under the conditions described above.

EXPERIMENTAL SET-UP FOR ADSORPTION STUDIES

In Fig. 1 the experimental set-up utilized for the adsorption studies, for fixed and fluidized beds is depicted. A glass column from Pharmacia (Uppsala, Sweden) with a diameter of 1 cm and a total height of 20 cm was utilized both for the packed-bed and fluidized-bed experiments. This column has an appropriate distributor and a mobile piston that enables the variation of the bed height. Systems were equilibrated with the Tris-HCl buffer (pH = 7.5 and ionic strength of 0.03 M), loaded and washed at a flow rate of 1 mL/min. The typical amount of resin for adsorption and elution tests was 1.0 g, which is equivalent to an initial bed height of 2.6 cm. Elution was carried out using a linear gradient with the same buffer to which NaCl was added to form a 0.75 M NaCl solution. Fractions of 2 mL were collected both during the adsorption and elution procedures.

Characterization of the Resin Beds

The adsorbent Accell Plus QMA is an anion-exchange resin commercialized by Waters, a division of Millipore (Bedford, MAP). This resin has a core of silica covered by an acrylamide copolymer containing quaternary amine functional groups. The range of particle size is from 37 to 55 μ and the density when equilibrated in the buffer Tris-HCl, pH = 7.5 and ionic strength of 0.03 M is 1.18 g/mL, as determined by picnometry. These physical properties are convenient for experiments involving expanded beds at low flow rates.

The determination of bed structural characteristics such as bed porosity and particle porosity is essential for scale-up calculations. Aiming at the experimental evaluation of those parameters, a pulse method using the moments theory was developed by Arnold et al. (7) and utilized in the present work. The principles of the method involve the concentration measurement of a tracer pulse as a function of time, being the concentration C(t) related to the first moment (μ_1) by the equations:

$$\mu_1 = \frac{\left(\int_0^\infty C(t) \times t \times dt\right)}{\left(\int_0^\infty C(t)dt\right)} \quad (1)$$

$$\mu_1 = \left(\frac{L}{u}\right)\left\{\varepsilon + (1 - \varepsilon)\beta\left(1 + \left[\frac{\rho_p}{\beta}\right]K\right)\right\} + \left(\frac{t_0}{2}\right) \quad (2)$$

Fig. 1. Scheme of the experimental setup for column adsorption studies. **(A)** Enzyme solution; **(B)** piston pump; **(C)** expanded- or fixed-bed column; **(D)** UV detector; **(E)** data-acquisition system; **(F)** fraction collector.

In equation (2), L is the bed length, u is the fluid superficial velocity, ε is the bed porosity, β is the particle porosity, ρ_p the particles density, t_0 is the injection time, and K is the adsorption constant for a linear isotherm. Therefore the equations above give the relationship among the first moment and bed and hydrodynamics variables.

For the case of a tracer pulse in which the adsorption is avoided (large molecule or high ionic strength of the solution) the above equations can be simplified to:

$$\mu_1 = \left(\frac{L}{u}\right)\{\varepsilon + (1 - \varepsilon)\beta\} + \left(\frac{t_0}{2}\right) \qquad (3)$$

The use of a high-molecular-weight polymer (blue dextran) as a tracer avoids the pore penetration by the solute and implies that $\beta = 0$. Combining this fact with experiments carriedout by injection of a low-molecular-weight tracer (acetone), allows the simultaneous determination of β and ε from the above equations, for a given injection time t_0.

RESULTS AND DISCUSSION

Bed Expansion

The measurement of the bed height as a function of the fluid superficial velocity lead to the bed expansion characteristics as shown in Fig. 2. Straight lines as those depicted are classical in fluidization of solids by using liquid (8). Our results shows a slight modification for the original

Fig. 2. Effect of superficial velocity on the bed expantion ○ - Original resin; ■ - eluted resin; ◇ - loaded resin.

and used resin particles, caused probably by the variation in particle density and to injection of solutions of slightly different viscosities during the adsorption and elution cycle.

Characterization of the Resin Bed

Concentration of both blue dextran and acetone as a function of time in pulse experiments is depicted in Figs. 3 and 4. From equation *(1)*, the first moment μ_1 can be calculated, and bed porosity, ε, and the particles porosity β, can be obtained from equation *(3)*. A summary of the calculations is depicted in Fig. 5. The experiments conducted with blue dextrane and acetone for the bed constituted by Accell Plus QMA resin particles resulted in the values of 0.41 and 0.59 for ε and β, respectively, which are very characteristic for fixed beds and particles of ion-exchange resins. From the fixed-bed porosity value, the expanded bed porosity could be calculated by using the expansion curve (Fig. 2) and equation *(4)* based on the fluidization theory *(9)*:

$$H_1(1 - \varepsilon_1) = H_0(1 - \varepsilon_0) \qquad (4)$$

In equation *(4)* subscript 0 means initial conditions for porosity and bed height and subscript 1 indicates the values for the same variables in expanded bed.

Fig. 3. Response peaks for the injection of 200 µL of a 1.0 mg/mL blue dextran solution. ☐ - First injection; × - second injection.

Fig. 4. Response peaks for the injection of 200 µL of a 1% (v/v) acetone solution. ☐ - First injection; △ - second injection; × - third injection.

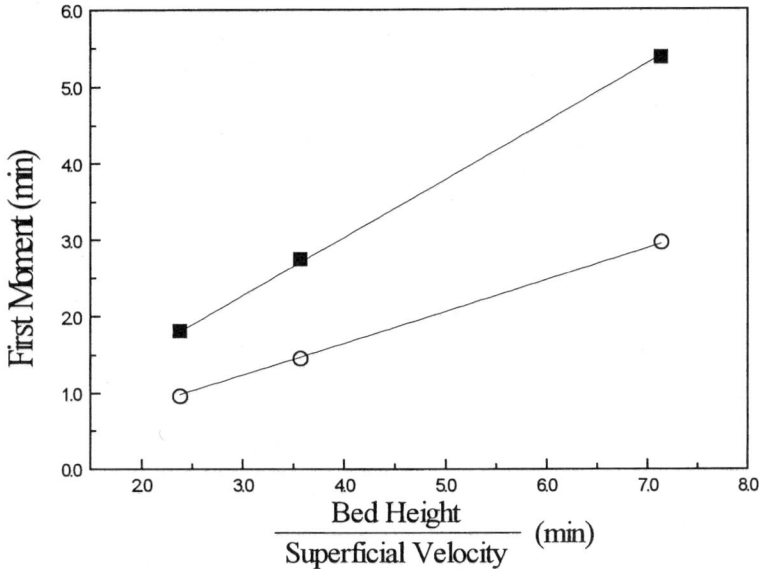

Fig. 5. Determination of bed and particle porosities using statistical moments analysis. ■ - Acetone; ○ - blue dextran.

Recovery of β-Galactosidase: Comparison Between Packed and Expanded Bed Adsorption

The loading flow rate of 1 mL/min, correspondent to a superficial velocity of 1.27 cm/min, resulted in a degree of expansion of approx 3.0. The degree of expansion dropped slightly during loading, but did not require a change in loading flow rate.

The specific activity of the enzyme preparation determined by the assay method described previously was 25.2 U/mg for the packed-bed experiments and 17.2 U/mg for the expanded-bed assays.

Typical chromatograms for the fixed and expanded bed are illustrated in Figs. 6 and 7. In the packed-bed route 92.4% of the total β-galactosidase loaded was recovered compared to 83% for the expanded bed. Both routes led to well-defined sharp β-galactosidase peaks, and Table 1 presents the results from successive loadings onto both the expanded and packed bed in terms of purification factor and yield. Purification factors for the fixed and expanded bed reached values of 3.2 and 2.2, respectively.

CONCLUSIONS

From the experimental procedures carried out, is clear that the direct recovery of extracellular β-galactosidase utilizing an ion-exchange resin utilizing an expanded-bed column is a viable operation. The enzyme yield

Fig. 6. Chromatogram for β-galactosidase recovery using ion-exchange resin Accell Plus in packed bed. ○ - Enzyme activity; ■ - protein concentration.

Fig. 7. Chromatogram for β-galactosidase recovery using ion-exchange resin Accell Plus in expanded bed. ○ - Enzyme activity; ■ - protein concentration.

obtained from the expanded-bed system is lower than the packed bed, but both have comparable performances. The purification factors are 3.2 and 2.2, respectively. These results can be enhanced by optimization of the operational conditions such as flow rate, pH, and ionic strength. It should be remembered that packed beds foul more rapidly than expanded beds,

Table 1
Yield and Purification Factors for Packed- and Expanded-Bed Operations

		Volume (mL)	SA[1] (U/mg)	Purification Factor	Total Activity	yield %
Packed Bed	Initial	144	25.2	1.0	396	100.0
	Final	8	79.6	3.2	366	92.4
Expanded Bed	Initial	144	17.2	1.0	400	100.0
	Final	14	37.3	2.2	333	83.0

[1] SA - Specific Activity of β-Galactosidase

requiring normal solid-liquid operations before the adsorption step. The additional operation required by packed-bed operations results in advantage for the expanded bed in terms of total processing time economics as well as in terms of global efficiencies.

Utilization of the moments theory enabled the characterization of important bed parameters such us the bed porosity and the particle porosity for the resin Accell Plus QMA.

REFERENCES

1. Chase, H. A. (1994), *Trends Biotechnol.* **12**, 296–303.
2. Frej, A. K. B., Hjorth, R., and Hammarstrom, A. (1994), *Biotechnol. Bioengineering* **44**, 922–929.
3. Hansson, M., Stahl, S., Hjorth, R., Uhlén, M., and Moks, T. (1994), *Bio/Technology* **12**, 285–288.
4. Thommes, J., Halfar, M., Lenz, S., and Kula, M. R. (1995), *Biotechnol. Bioengineering* **45**, 205–211.
5. Pastore, G. M. and Park, Y. K. (1980), *J. Ferment. Technol.* **58**, 79–81.
6. Bradford, M. (1976), *Anal. Biochem.* **72**, 248–254.
7. Arnold, F. H., Blanch, H. W., and Wilke, C. R. (1985), *Chem. Eng. J.* **30**, B25–B36.
8. Richardson, J. F. and Zaki, W. N. (1954), *Trans. Inst. Chem. Eng.* **32**, 35–53.
9. Kunii, D. and Levenspiel, O. (1977), *Fluidization Engineering*, 1st ed., R. E. Krieger Publishing, New York.

β-Cyclodextrin Production by Simultaneous Fermentation and Cyclization

Heron O. S. Lima, Flavio F. De Moraes, and Gisella M. Zanin*

State University of Maringá, Chemical Engineering Department, Av. Colombo, 5790 - BL. E46 - 09, 87020-900, Maringá, PR, Brazil

ABSTRACT

Production of β-cyclodextrin (CD) with high-dextrose equivalent (DE) starch hydrolysates by simultaneous fermentation and cyclization (SFC) gives higher yields than using only the enzyme CGTase, because fermentation eliminates glucose and maltose that inhibit CD production, while at the same time, produces ethanol that increases yield. A 10% (w/v) solution of cassava starch, liquefied with α-amylase, was incubated with CGTase using: only the enzyme, added ethanol (from 1 to 5%), and added yeast, *S. cerevisiae* (12% w/v), plus nutrients, the latter being the SFC process. Reaction conditions were: 38°C, pH 6.0, DE from 2 to 25, and 3.3 mL of CGTase/L. The yield of β-CD has decreased with an increase in DE, and maximum reaction yields were found for DE equal to 3.54, reaching 5.6, 14.7, and 11.5 mM β-CD, respectively. For an increase of DE, of approx 6 times (from 3.54 to 23.79), β-CD yield decreased 6 times for the first, and second reaction media with 3% (v/v) ethanol, and only approx 3 times for SFC (from 11.5 to 3.73 mM), showing that this process is less sensitive to variations in the DE.

Index Entries: Cyclodextrin; CGTase; fermentation; cassava starch.

INTRODUCTION

Cyclodextrins (CDs) are cyclic oligosaccharides that are normally formed by 6 to 8 glucopyranose units, linked by α-1,4 bonds. The most common cyclodextrins are: α-CD (cyclohexamilose), β-CD (cycloheptamilose), and γ-CD (cyclooctamilose). These cyclic maltodextrins are produced through the action of the enzyme cyclodextrin glycosyl transferase (CGTase) upon liquefied starch (1). The ring structure formed is highly hydrophilic on the outside because of the great numbers of hydroxyls, but

---

*Author to whom all correspondence and reprint requests should be addressed (E-mail: gisellazanin@cybertelecom.com.br).

relatively hydrophobic inside because of the glycosylic bonds and hydrogen atoms that face the interior of the cavity (2). In aqueous solutions, this structure allows the inclusion of nonpolar molecules of suitable size, inside the cyclodextrin cavity. For long and complex molecules it may be its most hydrophobic parts that can be included inside the CDs (3–5). This property of the CDs makes them valuable microencapsulation products at the molecular level, conferring chemical and physical stabilization to the complexed substances. As a consequence of this property, CDs have a broad actual and potential field of application in various industries such as pharmaceutical, food, cosmetics, and agroindustries (6,7).

The production of β-CD from corn and potato starches with the addition of solvents (ethanol among others) was reviewed recently (8–13). There are at least four mechanisms through which a solvent can enhance CD production (12). The solvent could:

1. Affect starch structure and accessibility by opening the starch molecule.
2. Reduce the concentration of CD products by complexation and precipitation, and therefore shift the reaction towards the formation of products.
3. Change the enzyme conformation and improve affinity.
4. Reduce water activity, and therefore decrease hydrolytic reactions that produce small size oligosaccharides, which inhibit CD production.

Ethanol is emerging as a solvent with greater potential of being applied to enhance CD production, because (11–14):

1. Ethanol has no toxicity restriction if CDs produced in this way would be applied in the food or pharmaceutical industries.
2. CDs form inclusion complexes with ethanol and become less susceptible to reverse decomposition reactions.
3. Ethanol helps control microbial contamination and can be easily removed by distillation and be recycled.
4. Fermentation industries have long experience with handling ethanol, and the risks of flammability are safely controlled.

In these processes, reaction yield is affected by the presence of maltooligosaccharides that not only participate in reactions of intermolecular transglycosylation, but also function as inhibitors of the CGTase enzyme. Noteworthy among these saccharides are glucose and maltose. Mori et al. (13) observed that the presence of ethanol and glucose enhances the reverse decomposition of CDs. Therefore, it seemed plausible that, elimination of glucose and addition of ethanol higher CD yields. This is the concept of the simultaneous fermentation and cyclization process (SFC) proposed in this article. The basic idea is to eliminate glucose and maltose by fermentation, because they inhibit CD production, and at the same time produce ethanol that increases CD yield.

Given the great potential of application of the CDs, and the abundance of cassava starch available at low price in Brazil, it was decided to study in this work the production of β-CD through the process of simultaneous fermentation and cyclization (SFC), offering therefore, another alternative end use for a national resource.

OBJECTIVES

The following objectives were set for this work:
1. Production of liquefied cassava starch solutions of a desired dextrose equivalent (DE) value.
2. Determination of the activity profile for Wacker (Munich, Germany) CGTase as a function of pH using liquefied cassava starch as substrate.
3. To study the influence of enzyme dosage on β-CD yield.
4. To determine the influence of the yeast nutrients upon the CGTase.
5. To select compromise conditions for the production of cyclodextrins using simultaneously CGTase, and fermentation by yeast.
6. To study the influence of ethanol on β-CD production.
7. To test the production of CDs using three comparative conditions:
   a. Only the enzyme CGTase.
   b. Addition of ethanol (CGTase + ethanol). In this condition ethanol is added to the composition of the substrate solution.
   c. SFC process (CGTase + yeast + nutrients). This condition constitutes the simultaneous fermentation and cyclization process (SFC), in which ethanol is produced by yeast fermentation at the same time that CDs are produced by the enzyme CGTase.
8. To evaluate the SFC process in comparison with the other conditions.

MATERIALS AND METHODS

Substrate

The substrate is cassava starch (Copagra) at a nominal concentration of 10% (w/v), and actual 9.33% (w/v) dry-solids content, liquefied with Novo Termamyl α-amylase (4.3 μL of enzyme/g of starch) for different lengths of reaction time at 95°C, pH 6.0, giving maltodextrin solutions of different dextrose equivalent (DE), from 2 to 25. After reaching the DE value desired, the α-amylase was inactivated with 0.27% (v/v) of 1 N HCl and boiling for 10 min. Then the pH of the solution was corrected to 6.0 with 1 N NaOH.

CGTase Enzyme

The CGTase enzyme used was a gift from Wacker. It was obtained from *E. coli* engineered with CGTase from an alkalophilic *Bacillus* sp.1 (15). Specific enzyme activity at 38°C, and pH 6.0 was 70.5 μmol β-CD/min/mg of protein, determined as given below. Enzyme concentration in the stock solution was 0.997 mg of protein/mL.

Yeast and Nutrients

Commercial dried baker's yeast from Fleischmann containing *Saccharomyces cerevisiae* cells was used. Following the recommendation of Lima et al. *(16)* to obtain a well-developed fermentation, the commercial products ammonium sulfate and triple superphosphate were used, as a source of nitrogen and phosphorus, each in the proportion of 0.1% (w/v) of substrate solution. Yeast and nutrients were used together with CGTase in the simultaneous fermentation and cyclization process (SFC).

CGTase Enzyme Activity as a Function of pH

This test was conducted in assay tubes containing a final concentration of 0.5% (w/v) liquefied cassava starch, 0.1 mL CGTase/L, Tris-HCl buffer 0.01 M, and $CaCl_2$ 5 mM, at 38°C, and pH in the range of 5.0 to 8.0.

Influence of Enzyme Dosage on β-CD Yield

For this test, liquefied cassava starch with a DE of 23.5 was used at 38°C, pH 6.0, and the CGTase enzyme was added in one of the following concentrations: 0.16, 3.3, 5, and 6.6 mL of enzyme/L. Samples were taken at regular intervals with the objective of following the production of β-CD during a period of 24 h.

Influence of Yeast Nutrients on β-CD Yields

In this case, the production of β-CD as a function of time was followed in the presence of 0.1% (w/v) of ammonium sulfate and 0.1% (w/v) triple superphosphate that provide the essential nutrients for yeast fermentation. However, during this test, yeast cells were not added, and therefore, there was no fermentation, the objective of the experiment being to test the influence of these nutrients on β-CD yield. Liquefied cassava starch with a DE of 26.06, 10% (w/v) was used at 38°C and pH 6.0, with 1 mL CGTase/L.

Influence of Ethanol on β-CD Yield

This test was conducted with liquefied cassava starch, 10% (w/v), with two DE values, namely 12.88 and 23.37 at 38°C, pH 6.0, and two ethanol concentrations: 3 and 5% (v/v) respectively. CGTase concentration was set to 1 mL/L and the production of β-CD was followed for a period of 24 h.

Production of β-CD by the SFC Process and Other Comparative Conditions

The substrate solution was incubated with either 0.8 or 3.3 mL of CGTase/L in three different test conditions:

1. Only the enzyme CGTase.
2. Addition of ethanol: CGTase + ethanol (1 to 5% v/v).
3. SFC process: CGTase + yeast (12% w/v) + nutrients (0.1% w/v of both ammonium sulfate and triple superphosphate).

Erlenmeyer flasks of 250 mL containing 150 mL of the reaction mixture were incubated and gently shaken in a Dubnoff thermoregulated water bath. The DE of the liquefied cassava starch solution was varied from approx 2 to 25, and reaction conditions were 38°C, pH 6.0. Samples of 0.5 mL were taken at regular intervals, boiled for 10 min, stocked at 4°C, and later assayed for β-CD produced. Each test was run for a period of 24 h.

Assay Methods

Protein concentration was determined by the Coomassie blue method according to Bradford *(17)*. The DE was determined through the reducing-sugar method of Somogyi *(18)*, using glucose as standard. The quantity of β-CD produced in the reaction medium was assayed by the dye-extinction method of Vikmon *(19)*, modified by Mäkelä et al. *(20)* and Hamon and Moraes *(21)*. The concentration of ethanol produced by fermentation was determined by gas chromatography after ethanol separation by distillation.

RESULTS AND DISCUSSION

Production of Liquefied Cassava Starch Solutions of Desired DE

Figure 1 presents the data on dextrose equivalent (DE) as a function of hydrolysis time. The production of liquefied cassava starch, within the specified conditions, gives DE values that increases linearly with hydrolysis time up to 15 min, and for the whole period tested it can be correlated by the following equation:

$$DE = 1.101\ t/(1 + 0.001875\ t^{1.649}) \qquad (1)$$

valid for the interval $0 \leq t \leq 50$ min, where t is the reaction time, and $r^2 = 0.9986$.

These results were used for producing the other cassava starch solutions of desired DE values as used in this work.

Activity as a Function of pH for Wacker CGTase

Figure 2 shows the results for the activity of Wacker CGTase as a function of pH. It can be seen that optimum activity is observed in the pH range of 6.0 to 7.0; at 38°C. For pH 6.0 the activity value is 70.5 μmol β-CD/min/mg of protein. These results are in accord with data obtained by Hamon and Moraes *(21)* with the same enzyme, and Merck soluble starch

Fig. 1. Dextrose equivalent (DE) as a function of hydrolysis time, 10% (w/v) liquefied cassava starch, at 95°C, pH 6.0, 4.3 µmol of Novo Termamyl/g of starch.

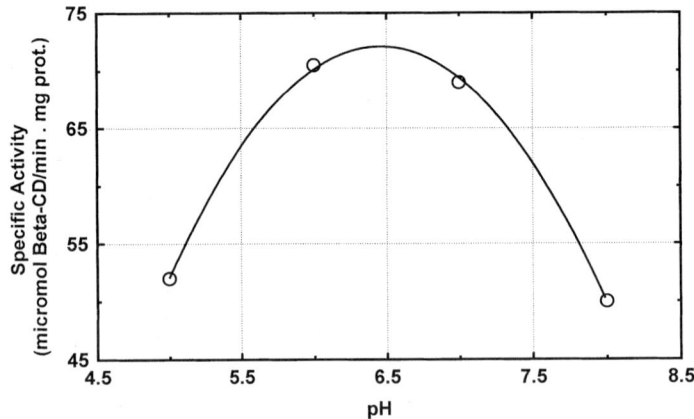

Fig. 2. Activity profile of Wacker CGTase as a function of pH. Reaction conditions: 0.5% (w/v) liquefied cassava starch, 0.1 mL CGTase/L, Tris-HCl buffer 0.01 M and $CaCl_2$ 5mM, at 38°C.

as the substrate. Mäkelä et al. (22) studying the CGTase from alkalophilic bacillus ATCC 21783 also observed maximum activity at pH values of approx 7.0. Collected optimum pH and temperature values for various CGTases are shown in Table 1 for comparison.

Influence of Enzyme Dosage on β-CD Yield

Figure 3 presents the results for β-CD production with different enzyme dosages. It can be seen that β-CD concentration increases with time for a period up to 6 h of reaction, and then the curve flattens, an effect resulting from the reversibility of the reactions that occurs in cyclodextrin

Table 1
Optimum Temperatures and pHs for Various CGTases Produced by Different *Bacillus*

Microorganisms	Optimum pH	Optimum Temperature (°C)	Main CD Produced	Reference
Bacillus macerans (IFO-3490)	6.1-6.2	55	α	23
Bacillus megaterium	5.2-6.2	55	β	23
Bacillus sp (ATCC-21783)	4.5-4.7	50	β	24
Bacillus coagulans	6.5-8.5	65	β	25
Bacillus alkalophilic 290-3	6.0-8.0	60	γ	26
B. thermoanaerobacter sp	5	60	α,β	10
Bacillus sp BE –101	6.0-6.5	45	β	11
Bacillus alkalophilic	6.0-7.0	50	β	21
Bacillus lentus	6.5-8.5	45-55	β	27
Bacillus alkalophilic sp	5.5	50-55	α	28
Brevibacterium sp	10	55	γ	29

Fig. 3. Production of β-CD as a function of Wacker CGTase enzyme dosage: (+) 0.8 mL/L; (∗) 1.6 mL/L; (Δ) 3.3 mL/L; (○) 5.0 mL/L; (□) 6.6 mL/L.

production. The relatively low production of β-CD results from the high DE, as it will become clear with further data. For the same enzyme and using 10% (w/v) Fluka (Buchs, Switzerland) maltodextrin 10, at 50°C, pH 8.0, Hamon and Moraes *(21)* have obtained about 10 mM β-CD in 24 h.

Influence of Yeast Nutrients on β-CD Yield

The influence of the yeast nutrients, ammonium sulfate, and triple superphosphate, upon CGTase capability of producing β-CD, was shown to be negligible as can be seen in Fig. 4, because in both cases, with or without nutrients, the final concentration reached by β-CD is the same.

Compromise Conditions for the Simultaneous Fermentation and Cyclization Process (SFC)

The results of Fig. 2 have permitted to choose pH 6.0 for the following tests, because in this pH, Wacker CGTase enzyme activity is close to its optimum, and in the SFC tests this pH is not too far from the optimum for fermentation with *Saccharomyces (16)*.

Recommended temperature for alcoholic fermentation is within 25 to 36°C, with lower temperatures retarding fermentation and higher temperatures favoring ethanol evaporation *(16)*. However, since Wacker CGTase can work well in temperatures up to 50°C, a compromise temperature was chosen, namely 38°C, for running the subsequent tests that compare the production of β-CD for conditions with or without ethanol and the SFC process.

Enzyme dosage was set for most cases at 3.3 mL of stock Wacker CGTase/L of liquefied cassava starch, because this was considered a good compromise between the yields of β-CD and enzyme expenditure, as seen on Fig. 3.

Influence of Ethanol on β-CD Production

Figure 5 shows that the addition of 5% ethanol increases the production of β-CD from liquefied cassava starch at the DE values tested, namely an increase of 80% for DE = 12.88 and 100% for DE = 23.37. Additional results in this article show that addition of ethanol increases β-CD yield for all DE values, in accordance with results from Lee and Kim *(11)*, and Mattsson et al. *(12)*.

Production of β-CD by the SFC Process and other Comparative Conditions

Figures 6–13 show the evolution of β-CD produced in the three comparative conditions: only the enzyme CGTase, addition of ethanol: CGTase + ethanol, and SFC (simultaneous fermentation and cyclization) process, that is CGTase + yeast + nutrients. It can be seen that the fastest produc-

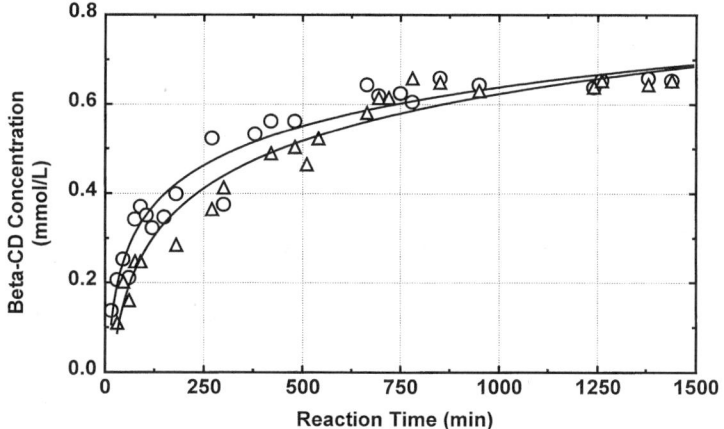

Fig. 4. Production of β-CD as a function of time in the presence of the yeast nutrients; (○) without nutrients; (Δ) with 0.1% (w/v) of triple superphosphate and 0.1% (w/v) ammonium sulfate; 10% (w/v) cassava starch, at 38°C, pH 6.0, 1 mL CGTase/L, and initial DE = 26.06.

Fig. 5. Production of β-CD as a function of time in the presence and absence of ethanol: (○) DE = 12.88 + 5% (v/v) ethanol; (Δ) DE = 12.88 without ethanol; (◇) DE = 23.37 + 5% (v/v) ethanol; (□) DE = 23.37 without ethanol; 10% (w/v) liquefied cassava starch, at 38°C, pH 6.0, with 1 mL CGTase/L.

tion of β-CD occurs when there is addition of ethanol, and the slowest in the case of the SFC process. The SFC process conditions give generally a higher final β-CD yield than using only CGTase, and sometimes even higher than with addition of ethanol, depending on the quantity of ethanol added. This proves the basic idea of the SFC process.

Data on CD production starting with the same DE shows that fermentation that produces endogenous ethanol in the reaction medium, is more

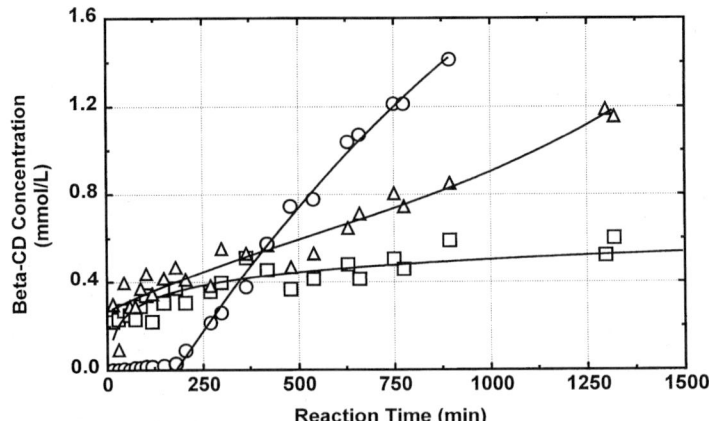

Fig. 6. Production of β-CD with liquefied cassava starch with DE = 23.89; 10% (w/v), at 38°C, pH 6.0 and 0.8 mL CGTase/L: (□) only the enzyme CGTase; (Δ) CGTase + 3% (v/v) ethanol; (○) SFC.

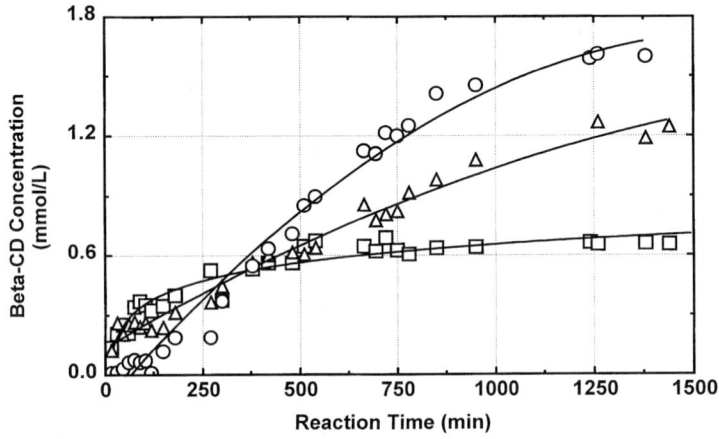

Fig. 7. Production of β-CD with liquefied cassava starch with DE = 26.06; 10% (w/v), at 38°C, pH 6.0 and 0.8 mL CGTase/L: (□) only the enzyme CGTase; (Δ) CGTase + 3% (v/v) ethanol; (○) SFC.

effective for increasing CD yield than the addition of exogenous ethanol to the substrate. Therefore, the elimination of glucose and maltose, which occurs with fermentation, is more important to increase CD yield than the presence of ethanol.

Table 2 allows a rapid comparison of final values for β-CD concentration in the three comparative cases, and the plot of this data clearly shows in Fig. 14 the strong inhibitory effect of high DE on the production of β-CD with liquefied cassava starch solution. These results are in line with those obtained with other starchy materials (30,31).

Cyclodextrin Production

Fig. 8. Production of β-CD with liquefied cassava starch with DE = 2.88; 10% (w/v), at 38°C, pH 6.0 and 0.33% (v/v) CGTase: (□) only the enzyme CGTase; (Δ) CGTase + 1% (v/v) ethanol; (○) SFC.

Fig. 9. Production of β-CD with liquefied cassava starch with DE = 3.54; 10% (w/v), at 38°C, pH 6.0 and 0.33% (v/v) CGTase: (□) only the enzyme CGTase; (Δ) CGTase + 3% (v/v) ethanol; (○) SFC.

The yield of β-CD has decreased with an increase in DE values from 2 to 25 in any of the reaction media, and it appears that there is an optimum for DE values between 2 and 10. Maximum β-CD yields for each reaction media were found for DE equal to 3.54, reaching 5.6 and 14.4 mM β-CD for the first and second medium, and 11.5 mM β-CD for the SFC process conditions. For an increase of DE of approx 6 times (from 3.54 to 23.79), β-CD yield decreased 6 times for reaction media one, and two (with 3% v/v ethanol), whereas for the SFC process, this reduction was only approx

Fig. 10. Production of β-CD with liquefied cassava starch with DE = 7.43; 10% (w/v), at 38°C, pH 6.0 and 0.33% (v/v) CGTase: (□) only the enzyme CGTase; (Δ) CGTase + 3% (v/v) ethanol; (○) SFC.

Fig. 11. Production of β-CD with liquefied cassava starch with DE = 10.58; 10% (w/v), at 38°C, pH 6.0 and 0.33% (v/v) CGTase: (□) only the enzyme CGTase; (Δ) CGTase + 5% (v/v) ethanol; (○) SFC.

3 times (from 11.5 to 3.73 mM), showing that the SFC process is less sensitive to variations in the DE of the liquefied starch solution.

It can also be observed in Fig. 14 that the addition of exogenous ethanol or the SFC process are able to counteract to a certain extent, the inhibitory effects of high DE on β-CD production, but not to the point of producing, for very high DE, the same amount of β-CD as is obtained for example, with very low DE. Therefore, the inhibitory effects of small oligosaccharides during the formation of cyclodextrins are stronger than the enhancement produced by the presence of ethanol.

Cyclodextrin Production

Fig. 12. Production of β-CD with liquefied cassava starch with DE = 18.41; 10% (w/v), at 38°C, pH 6.0 and 0.33% (v/v) CGTase/L: (□) only the enzyme CGTase; (Δ) CGTase + 5% (v/v) ethanol; (○) SFC.

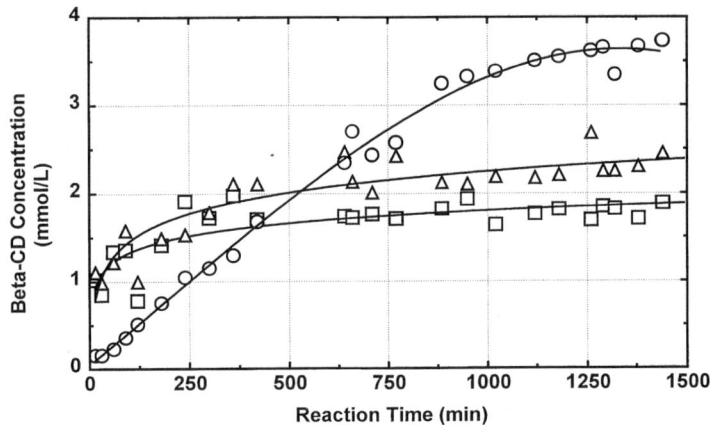

Fig. 13. Production of β-CD with liquefied cassava starch with DE = 23.79; 10% (w/v), at 38°C, pH 6.0 and 0.33% (v/v) CGTase/L: (□) only the enzyme CGTase; (Δ) CGTase + 5% (v/v) ethanol; (○) SFC.

This work has shown that given a cassava starch solution of high DE, it is possible with the SFC process to produce β-CD at higher yields than using only the enzyme CGTase. For example, for DE = 22.2, the SFC process gives 97% higher yields than using only CGTase, and 52% higher than using CGTase plus 5% v/v ethanol. However, at high DE values the CD yield is low enough to possibly preclude the use of the SFC process for industrial application, unless the high DE solutions are byproduct streams that might gain added value by the presence of a small percentage of CDs.

Table 2
Concentration of CD Produced at 38°C, and pH 6.0, with Various Liquefied
Cassava Starch Solutions of Different DE Values, in the Presence
and Absence of Ethanol, and for the Simultaneous Fermentation
and Cyclization (SFC) Process

Volume of CGTase added (mL/L)	1		0.8		3.3					
Dextrose Equivalent (DE)	12.88	23.37	23.89	26.06	2.88	3.54	7.43	10.58	18.41	23.79
Concentration of β-CD(mM) after 24 h; no ethanol	1.83*	0.82*	0.59*	0.66	8.72	11.5	6.50	6.80	2.38	1.89
Ethanol added (% v/v)	5	5	3	5	1	3	3	5	5	5
Concentration of β-CD (mM) after 24 h; with ethanol	2.48*	1.30*	0.85*	1.25	11.1	14.7	8.74	10.8	4.41	2.45
Ethanol produced (SFC, % v/v)	-	-	1.54	2.18	0.84	1.03	1.18	0.86	0.52	1.68
Concentration of β-CD (mM) after 24 h; SFC.	-	-	1.41	1.61	9.27	11.5	9.51	8.6	5.44	3.73

(*) 15 h of reaction.

Fig. 14. Maximum yields of β-CD as a function of initial DE of the liquefied cassava starch 10% (w/v), at 38°C, pH 6.0 and 0.33% (v/v) CGTase: (○) only the enzyme CGTase; (□) CGTase + 3% (v/v) ethanol; (Δ) SFC.

The results of this work have show additionally that: cassava starch is a suitable raw material for the production of β-CD, and additional work is needed in this area to try improve CD yield.

ACKNOWLEDGMENTS

The authors are thankful for the financial support received from CAPES/PADCT, CNPq, FINEP, and the State University of Maringá. The companies that supplied materials (Copagra, Novo, and Wacker) are also acknowledged.

REFERENCES

1. Bender, H. (1986), in *Advances in Biotechnological Process*, vol 6, Liss, A. R., ed., pp. 31–71.
2. Bender, H. and Komiyama, M. (1978), *Cyclodextrin Chemistry*, Springer-Verlag, New York, pp. 2–23.
3. Szejtli, J., (1988), *Cyclodextrin Technology*, Kluwer Academic Publishers, Dordrecht, Netherlands, pp. 79–185.
4. Tabushi, I. (1982), *Chem. Res.* **15**, 66–72.
5. Duchêne, D. and Vaution, C. (1986), *Les Entretiens du Carla -Tome VII*, Conferénce donnée le 24 juin.
6. Horikoshi, K. (1979), *Proc. Biochem.* **14**, 26–30.
7. Lee, J. H., Choi, K. H., Choi, J. Y., Lee, Y. S., Know, I. B., and Yu, J. H., (1992), *Enzyme Microb. Technol.* **14**, 1017–1020.
8. Armbruster, F. C. and Kooi, E. R. (1969), Production of cyclodextrin, U.S. patent No. 3, 425, 910.
9. Armbruster, F. C. (1988), in *Proceedings of the Fourth International Symposium on Cyclodextrins*, Huber, O. and Szejtli, J. eds. Kluwer Academic Publishers, Dordrecht, Netherlands, p. 33–39.
10. Starnes, R. L., Flint, V. M., and Katkocin, D. M. (1990), in *Minutes of the Fifth International Symposium on Cyclodextrins*, Duchêne, D. ed., Editions de Santé, Paris, pp. 55–61.
11. Lee, Y. D. and Kim, H. S., (1991), *Enzyme Microb. Technol.* **13**, 499–503.
12. Mattsson, P., Korpela, T., Paavilainen, S., and Mäkelä, M., (1991), *Appl. Biochem. Biotechnol.* **30**, 17–28.
13. Mori, S., Goto, M., Mase, T., Matsuura, A., Oya, T., and Kitahata, K. (1995), *Biosci. Biotech. Biochem.* **59**, 1012–1015.
14. Morita, T., Yoshida, N., and Karube, I. (1996), *Appl. Biochem. Biotechnol.* **56**, 311–324.
15. Schmid, G., Englbrecht, A., and Schmid, D. (1988), in *Proceedings of the Fourth International Symposium on Cyclodextrins*, Huber O. and Szejtli, J. eds., Kluwer Academic Press, Dordrecht, Netherlands, pp. 71–76.
16. Lima, U., Borzani, W. and Aquarone, E. (1975), *Biotecnologia - Tecnologia das Fermentações*, vol 1, Edgard Blucher, São Paulo, pp. 48–69.
17. Bradford, M. M. (1976), *Anal. Biochem.* **72**, 248–254.
18. Somogyi, M. (1945), *J. Biol. Chem.* **160**, 61–68.
19. Vikmon, M. (1982), in *Proceedings of First International Symposium on Cyclodextrins*, Szejtli, J, ed., D. Riedel Publishing, Dordrecht, Netherlands, pp. 69–74.
20. Mäkelä, M., Korpela, T., and Laaksso, S. (1987), *J. Biochem. Biophys. Meth.* **14**, 85–92.
21. Hamon, V. and Moraes, F. F. de (1990), Etude preliminaire a l'immobilisation de l'enzyme CGTase Wacker, report, Laboratoire de Technologie Enzymatique, Université de Technologie de Compiène.

22. Mäkelä, M., Mattsson, P., Schinina, M. E., and Korpela, T. (1988), *Biotechnol. Appl. Biochem.* **10,** 414–427.
23. Sabioni, J. G. (1991), Ph. D. Thesis, Universidade Estadual de Campinas, pp. 5–27.
24. Nakamura, N. and Horikoshi, K. (1976), *Agr. Biol. Chem.* **40,** 753–757.
25. Kaneko, T., Yoshida, M., Yamamoto, M., Nakamura, N., and Horikoshi, K. (1990), *Starch*, **42,** 277–281.
26. Englbrecht, A., Harrer, G., Lebert, M., and Schmid, G. (1990), in *Minutes of the Fifth International Symposium on Cyclodextrins*, Duchêne, D., ed., Editions de Santé, Paris, pp. 25–31.
27. Sabioni, J. G. and Park, Y. K. (1992), *Starch/Stärke* **44,** 225–229.
28. Kometani, T., Terada, Y., Nishimura, T., Takii, H., and Shigetaka, O. (1994), *Biosci. Biotech. Biochem.* **58,** 517–520.
29. Mori, S., Hirose, S., Oya, T., and Kitahata, S. (1994), *Biosci. Biotech. Biochem.* **58,** 1968–1972.
30. Horikoshi, K., Nakamura, N., Matsuzawa, N., and Yamamoto, M. (1982), in *Proceedings of the First International Symposium on Cyclodextrins*, Szejtli, J., ed., D. Reidel Publishing, Dordrecht, Netherlands, pp. 25–39.
31. Teijin Limited (1975), Procédé de préparation de cyclodextrine. French Patent 2,284,675, 21/July/1975, 19 p.

Session 4

Industrial Needs, Commercialization, and Process Economics

Patrick Foody[1] and J. Rus Miller[2]

[1]Iogen Corp., Ottowa, Ontario, Canada, and [2]Arkenol, Mission-Viejo, CA

Presentations at the "Commercialization" and "Process Economics" section of the 19th Symposium were arranged to cover the spectrum of biomass conversion, from "conceptual" processes through to the most advanced piloted technology. Input to the discussion was also sought from the State of California, regarding specific opportunities in that State, it was considered appropriate that some words of caution on the rigors of the marketplace be included.

The session attracted six papers. It opened with a coauthored presentation by two representatives Of the California Energy Commission. This was followed by an "upbeat" review of the "possible" by Dr. Lee Lynd of Dartmouth. Next came a solid presentation by the Co-Chairman, Rus Miller, on the potential for "Strong Acid" hydrolysis. The "Strong Acid" process was characterized as a well researched and understood technology which had recently become economically feasible because of technical advances by Arkenol in novel chromatographic separations for acid recovery. Next in line were two presentations on enzymatic conversion processes. The first was by SWAN, an AMOCO, Stone & Webster joint venture. This was followed by a National Renewable Energy Laboratory (NREL) paper. The estimated costs of ethanol across the spectrum of processes were not dissimilar when adjusted to an equalized base. The final presentation was by Ray Katzen and dealt mainly with the marketplace and the need for clear business plans and a framework for financing.

The analyses presented were generally encouraging. All the "pro forma cost" estimates for ethanol from lignocellulosics were, for the first time, below those of grain ethanol and this represented an unusual consensus. Concern was expressed about the impact of the "deregulation" of electrical utilities on total plant revenue. Deregulation appears to have reduced the potential value of electrical power by 2 to 3 cents/kWH and, hence, the power "by-product" revenue from plants burning their lignin by 10-15 cents per gallon. This was highlighted as potentially as serious an economic problem as the cost of enzyme. Some discussion from the 'floor' raised the prospect that this might be reversed in the future. By special "green power" allowances given the growing US acceptance of "climate change" as a real problem. This development which could offset the losses from deregulation was seen as possible.

Use of Net Present Value Analysis to Evaluate a Publicly Funded Biomass-to-Ethanol Research, Development, and Demonstration Program and Valuate Expected Private Sector Participation

Norman D. Hinman* and Mark A. Yancey

Center for Renewable Fuels and Biotechnology National Renewable Energy Laboratory Golden, CO 80401-3393

ABSTRACT

One of the functions of government is to invest tax dollars in programs, projects, and properties that will result in greater public benefit than would have resulted from leaving the tax dollars in the private sector or using them to pay off the public debt. This paper describes the use of Net Present Value (NPV) as an approach to analyze and select investment opportunities for government money in public research, development, and demonstration (RD&D) programs and to evaluate potential private sector participation in the programs. This approach is then applied to a specific biomass-to-ethanol opportunity in California.

INTRODUCTION

One of the functions of government is to invest tax dollars in programs, projects, and properties that will result in greater social benefit than would have resulted from leaving those tax dollars in the private sector or using them to pay off the public debt. One traditional area for investment by government is research, development, and demonstration (RD&D) of new technology. According to Battelle, US R&D expenditures reached $164.5 billion in 1994, and federal support represented $69.8 billion (42.4% of the total) *(1)*. If invested wisely, these tax dollars could potentially lead to greater social benefit than would be obtained by leaving them in the private sector or using the money to pay off the federal debt. However, if not invested wisely, this could result in less than optimal benefit or, even worse, in less benefit than could be obtained from the other two

---

*Author to whom all correspondence and reprint requests should be addressed.

options. The purpose of this paper is to describe an approach to analyzing and selecting investment opportunities for government money in public RD&D programs and valuating expected private sector participation in the programs and to apply this approach to a specific biomass-to-ethanol opportunity in California.

BASICS OF INVESTMENT ANALYSIS

For all investment situations there are five basic variables: costs; revenues or benefits; time; discount rate; and risk. In the analysis of investment opportunities, the opportunities under consideration may have differences with respect to costs and revenue or benefits, project lives, and uncertainties. If the effects of these factors are not quantified systematically, correctly assessing opportunities is very difficult.

Many methods are available to decision makers to systematically evaluate investment opportunities. These methods, described in detail in a variety of books and articles (2), include present, annual, and future value; rate of return; and break-even analysis. The application of each method depends on whether the analysis is for a single opportunity, two mutually exclusive opportunities, or several nonmutually exclusive opportunities.

Net present value (NPV) is the tool of choice for a evaluating all these situations because it is much less time consuming and is straightforward, allows direct comparison between projects between widely differing lives, objectives, and scopes, and allows a rational approach to valuating private sector participation in public programs.

THE NPV APPROACH

To apply NPV to a single opportunity situation, the NPV is calculated by determining the present value of the revenue/benefit stream calculated at the minimum rate of return (hurdle rate) and then subtracting the present value of investment dollars and other costs, also calculated at the rate of return.

$$\text{NPV} = \text{present value revenues @ } i^* - \text{present value costs @ } i^*$$

where i^* is the minimum rate of return. If the NPV is zero, there is enough revenue to cover the costs at a rate of return that is equal to the minimum rate of return required by the investor. Projects with NPV less than zero are dropped from consideration. If the NPV is greater than zero, the NPV represents how many present value dollars will be returned to the investor above and beyond those that will be returned at the minimum rate of return.

To apply NPV to mutually exclusive investments, one calculated the NPV for each potential project. The project with the largest NPV is selected. If neither project has a positive NPV, neither is selected.

A nonmutually exclusive investment situation is one for which more than one investment option can be selected, depending on available capital of budget restrictions. The objective is to select projects that maximize the cumulative profitability of benefit from the available investment dollars. Here the NPV for each project is calculated. Projects with an NPV less than zero are dropped from further consideration because their rate of return is less than the minimum required return. Once the NPV for each project is calculated, the decision maker looks at all possible combinations of projects to determine which combination (whose total investment does not exceed the amount of money available) has the largest cumulative NPV. This is the best possible investment portfolio.

SPECIAL CONSIDERATIONS FOR GOVERNMENT INVESTMENTS

Converting Intangible Benefits and Costs into Dollar Values

A basic tenant of this paper is that to make rational investments of pubic dollars one must have some approximate, quantitative idea of the value of critical costs and benefits. Moreover, as a practical matter, the measure of value must be the same for costs and benefits so that direct comparisons between costs and benefits can be made. The most universal measure of value is the dollar. In the private sector this is the measure of cost and benefit. In the public sector, particularly with respect to RD&D programs, it is the established measure of cost. However, on the benefit side, there is no established measure of value. The authors contend that the dollar should be the measure of benefit so that direct comparisons can be made with costs and so that the established and the well recognized investments analysis methodology previously can be employed in the public sector.

In many cases, converting benefits and costs to dollars is fairly straightforward. For example, the US Department of Energy (DOE) is interested in reducing imported petroleum. The dollar value of the yearly benefit can easily be calculated from the present and projected price of petroleum (3). As another example, the net annual increase or decrease in jobs that results from introducing new technology can be estimated. In addition, placing a dollar value on these jobs is fairly straightforward (4). Other possible costs and benefits are environmental and social, which are more difficult to quantify. However, various agencies such as the US Environmental Protection Agency have studied these issues carefully and have given dollar estimates of health costs associated with various types and levels of pollution, as an example.

Minimal Rate of Return for Public Projects

Establishing a minimal rate of return for public projects requires some special considerations, which have been reviewed extensively by Terry

Heaps (5) for Canadian public projects. He concluded that the correct social discount rate for Canada was 3–7%. In another study performed by Wilson Hill Associates (3) a discount rate of 7% was used for projects evaluated for the Office of Transportation Programs in the DOE.

SELECTING PUBLIC RD&D PROGRAMS AND VALUATING EXPECTED PARTICIPATION BY THE IMPLEMENTING INDUSTRY

A government R&D program is usually initiated without the private sector, but the private sector is expected to "come on board" at some point to carry the ball forward into the commercial arena. For these situations, the government and the private sector invest in RD&D to obtain what each desires—maximum overall public benefit in the case of government, and profit in the case of the private companies.

Analysis of the value of these programs demands answers to three questions: What portion of the RD&D cost can the private sector incur and still obtain its minimum return from commercializing the technology?; when this private sector cost allowance is subtracted from the total estimated cost to carry out RD&D so as to obtain an estimate of the RD&D cost that must be borne by government, is the estimated government RD&D cost justified given the expected public benefit from implementing the technology?; and if the answer to questions 2 is positive, does the program represent one of government's best opportunities for its limited investment dollars?

The NPV approach to investments provides the answer to all three questions. For example, to answer the first question one calculates the industry NPV. To do this, one estimates over time the capital and operating costs the industry at large will incur to implement a new technology and, using the average minimum interest rate for the industry, calculates the present value of these costs to industry at the initial time of commercialization. One also estimates over time the present value at the time of commercialization of the expected increased revenues or savings the industry should experience from implementing the technology. Subtracting the present value costs from the present value revenues gives the industry NPV at the time when commercialization is expected to begin. If the NPV is negative, the industry cannot afford to contribute to the RD&D effort and cannot afford the capital and/or operating costs of commercialization. As a result, it will not "come on board" and the government should drop the program from consideration. If the industry NPV is zero, industry cannot afford to contribute to the RD&D costs, but can afford the capital and operating costs to commercialize the technology. In this situation, the government will have to incur all the RD&D costs in order for industry to adopt the technology. If the industry NPV is positive, the government can expect the industry to participate in the RD&D costs at a level equivalent to the NPV.

To answer the second question, one calculates the government NPV. To do this, the expected public benefits are estimated over time and dollar values are assigned. Then the present value of these benefits is calculated at the time of commercialization using the social discount factor. Next, the entire RD&D costs over time are estimated and present-valued to the time of commercialization using the social discount factor. Finally, the expected RD&D contribution from industry, calculated above as industry NPV, is subtracted from the entire RD&D costs to obtain the government's expected RD&D costs present valued to the time of commercialization. These present-valued government RD&D costs are then subtracted from the benefits, also present-valued to the time of commercialization, to obtain the government NPV for the program. If the government NPV is less than zero, the program should not be considered for investment of tax dollars. If the government NPV is zero or greater, it should be thrown into the pot of possible government investments.

To answer the third question, government should list all investment options with a NPV greater than zero and select that combination of projects that will maximize the governments cumulative NPV for the amount of funds available.

VALUATING EXPECTED PARTICIPATION BY INDIVIDUAL COMPANIES

If, from the above analysis, the industry NPV is positive, individual companies that are members of the industry can be expected to cost share in the RD&D phase of a program or purchase licensing arrangements. However, the level of cost sharing or license fees will depend on each company's circumstances. The expected level of cost sharing or the licensing fee for a given company can be calculated using company NPV derived from projected revenues and costs a company will experience in implementing the technology in commercial use. If the company NPV is negative, the particular company cannot afford to commercialize the technology even if the technology is provided free. Such a company is not a viable partner to the government program. If the company NPV is zero, the company may be a partner only in the sense that it will implement the government-developed technology if it is free to the company. If the company NPV is positive, the company can afford to cost share the RD&D effort or purchase a licensing arrangement at a level equal to the company NPV.

BIOMASS-TO-ETHANOL OPPORTUNITY IN CALIFORNIA

Background

California is faced with several issues related to how its forest resources are used and managed. In particular, forest fire suppression has caused large quantities of small-diameter and dead and diseased trees and underbrush to accumulate in the forest, which have in turn resulted in severe fuel loads that have led to ever more intense and destructive wildfires. Unabated, this pattern will continue to worsen, and will create greater risk and loss to natural resources, property, and firefighters. In addition, the costs to government for fire suppression and disaster relief will grow at a time when government is trying to curtail costs and reduce budgets. To deal with this issue, the USDA Forest Service, the California Department of Forestry, and local organizations such as the Quincy Library Group have put forth a plan to strategically thin the forests to reduce total wildfire costs and losses. In addition, such a plan would significantly increase the total amount and value of certain assets such as water and timber. However, a key question is what will be done with the forest thinnings once they are removed from the forests. One potential use of the biomass is to convert it into fuel ethanol and cogenerate electricity.

Public Benefits of Thinning Commercial Timberland in California

There are approx 16 million acres of commercial timberland in California. Of this, about 13 million acres are at a slope of 30° or less, which is a requirement for thinning the forest at a reasonable cost (6). Thinning 5% of the 13 million acres each year results in thinning 650,000 acres of forest. The public benefits associated with thinning these acres are: reduced costs of wildfire protection: reduced loss of assets from wildfires: and increased water and timber asset values.

Reduced Costs of Wildfire Protection

Table 1 summarizes the estimated 1993–1994 state, federal, and local government's costs of California's wild land fire protection system. The chart further identifies wild land fire protection phases—initial attack, major fires, and disaster relief—for each level of government.

There are 40 million acres of forest in California. Since 650,000 acres is approx 2% of the acres of forest, we assume that thinning this number of acres will reduce the cost of fire protection by the same percent or $23/acre thinned, and that this effect will last as long as 5 yr after which the effect is lost as the foliage returns.

Reduced Loss of Assets from Wildfires

A primary purpose of wild land fire protection in California is to protect a wide range of assets found in California wild lands. These assets include: timber; range; water and watershed; buildings; human life and safety; air quality; wildlife, plants, and ecosystem health; scenic areas; recreation; and cultural and historic resources. Table 2 shows the total value

Table 1
Wild Land Fire Protection Budgets by Level of Government by Fire Phase (7)
($ thousand)

	Initial Attack	Major Fire	Disaster Relief	Total
Total Federal	$132.00	$85.00	$92.00	$309.00
Total State	208.00	45.00	13.00	266.00
Total Local	170.00	1.00	1.00	172.00
Total	$510.00	$131.00	$106.00	$747.00

Table 2
Total California Assets at Risk from Wildfire (7)

Asset Type	Total Value
Timber	$105 billion
Range	$138 million/yr.

Water/Watershed
•Increased runoff
  -residential/commercial/industrial use (6 million acre ft.)
  -agriculture (24 million acre ft.)
  -hydropower
	
	$3 billion/yr
	$1.5 billion/yr
	$1.6 billion/yr

Structures	$107 billion
Human life/safety	(1)
Air quality	(1)
Wildlife, habitat, plants, ecosystem health	(1)
Scenic resources and recreation	(1)
Recreation	$1.5 billion/yr
Unique sites	(1)

of each of these assets at risk and Table 3 shows the estimated loss on a yearly basis or per-acre burned basis from wildfire. The estimates for loss for each asset on a per-acre burned basis cover a range of values, which depend on the location of the asset. For example, the timber value lost is estimated to range from $2538/acre in the northern interior to $8823/acre on the central coast.

Table 4 shows the estimate for total aggregate asset loss in California on a yearly basis due to wildfire.

As noted earlier, the total forest land in California is 40 million acres. This study assumes that thinning 650,000 acres will reduce asset loss by 2%/yr or $14/acre thinned and that this effect will last as long as 5 yr, after which the effect is lost as foliage returns.

Table 3
Asset Losses in California from Wildfires

Asset	Asset Losses from Fire
Timber	($2,538-$8,823/acre burned)
Range	($8/acre burned)
Water/Watershed -increased residential/commercial/industrial/agricultural water -increased hydropower -reduced reservoir capacity from sediment -sediment removal -water shed rehabilitation	$3-$12/acre burned (1) $17.50/acre burned (1) ($9-$90/acre burned) ($100-$1,000/acre burned) (1) ($230-$400/acre burned) (1)
Structures	$163 million/yr.
Human life/safety	(2)
Air quality	($1-$15,000/acre burned)
Wildlife, habitat, plants, ecosystem, health	(2)
Scenic resources	(2)
Recreation	$5-$107/acre burned
Unique sites	(2)

Table 4
Total Asset Losses Per Year in California Due
to Wildfire According to Who Pays (7)

Federal	$235 million
State	$54 million
Local	$164 million
Total	$453 million

Increased Water and Timber Asset Values

Thinning forests will probably result in increased water yields. Based on a 1991 report by Ken Turner of the California Department of Water Resources (8), when 75%+ of the vegetation is removed, increased water yields can occur of 0.1 acre feet per acre in areas of 15 in of precipitation annually, up to 0.8 acre feet on acres of 40 in of precipitation annually for the first few years after the vegetation is removed. For the purposes of this report, the average increase is assumed to be 0.25 acre feet per acre thinned.

The value of each new acre foot of water depends on its use. However, on average the value of this water for agricultural, residential, commercial, and industrial use is estimated to be $60 per acre foot and the value for hydropower is estimated to be $70 per acre foot (7). At best, 50% of the increased water yield would be stored for consumption, so the value of the increased water from thinning is estimated to be $16.25/acre thinned. Thus, thinning 650,000 acres will result in $10.5 million in new asset value. Furthermore, this effect is assumed to last as long as 5 yr, after which the effect is low as foliage returns.

Timber growth rates after thinning can return a very high rate of return. Indicators are that annual tree growth can increase by as much as 5–200% or more per acre (9). Although increased timber asset value would clearly be a benefit of thinning, no attempt was made in this study to estimate this increased value.

Expenses and Revenues Associated with Ethanol Facilities That Would Utilize Thinnings from Commercial Timberland in California

Thinning 650,000 acres of forest will yield 3.25 million dry tons of biomass. For this study, we have assumed that this biomass can be converted into 240 million gallons of ethanol. Twelve ethanol plants with an average annual ethanol production capacity of 20 million gallons could produce this amount of ethanol. For this study, certain assumptions were made about the operation and economics of each 20 million gallon per year facility. The key assumptions are: capital costs of $2/annual gallon; O&M costs of $6.1 million/year; and biomass costs at $40/dry ton. These assumptions and others are shown in the block flow diagram in Fig. 1. It is important to note that these assumptions do not presume any particular technology and are not based on detailed engineering and economic analyses.

RD&D Costs

The total RD&D cost is an estimate of the funds needed to extend current knowledge of biomass-to-ethanol technology to the specific opportunity in California. The costs are assumed to be $45 million. The capital contribution to a demonstration plant is an estimate of funds needed to allow the operation of this 20 million gallon per year facility at a 20% rate of return. This contribution is estimated at $20 million. Thinning demonstration costs are funds needed to fully establish thinning methodologies. Total thinning demonstration costs are estimated at $20 million.

Benefits, Revenue, and Cost Streams Over Time

Table 5 shows ethanol and electricity revenue, capital investment, operating costs, increased water values, savings from reduced costs of

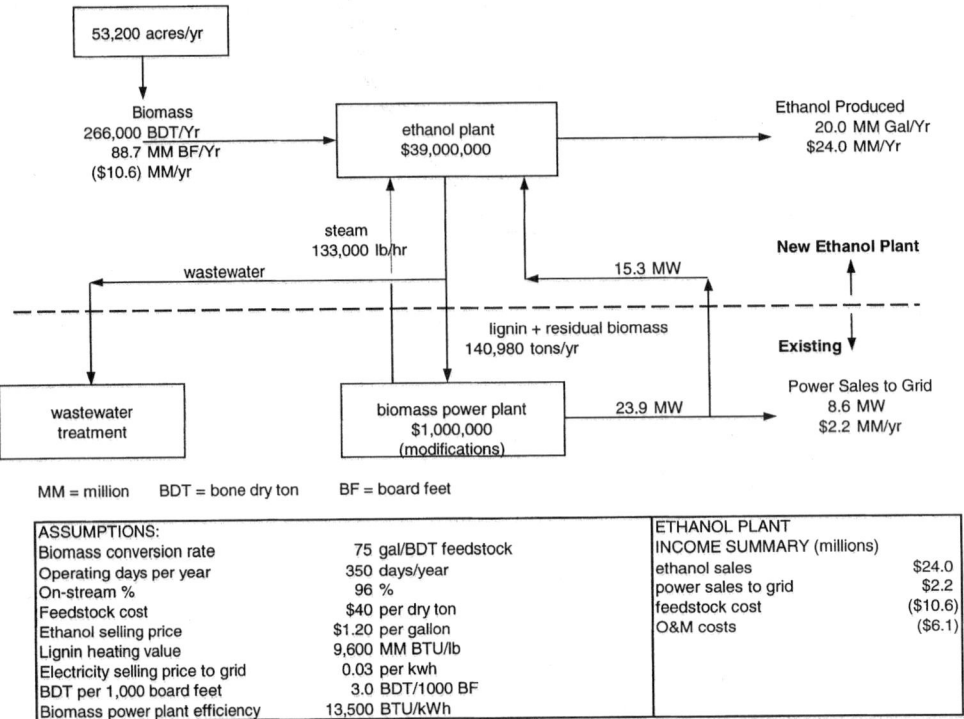

Fig. 1. Biomass to ethanol conversion. Colocation with an Existing Biomass Power Plant and Wastewater Plant in CA.

wildfire protection, savings from reduced loss of assets from wildfires, and RD&D costs associated with developing and operating 12 biomass-to-ethanol plants in California. The first two commercial plants are completed in yr 0. After that, two commercial plants are brought on stream each year for the next 5 yr, starting in yr 1. As noted earlier, the water yield increase and savings from reduced costs of wildfire and reduced loss of assets from thinning are assumed to last as long as 5 yr, after which the benefit begins to decline. For example, for a given plant, the increased water yield is 12,960 per acre foot for the first year from the thinning operations of that year. For the second year, the increased water yield is that from the first year plus that resulting from the thinning operations of that year. For the second year, and so on until yr 5 is reached. After yr 5, it is assumed that the first thinned acres begin to fill in with biomass and water yields from these areas begin to decrease, thereby negating the effects of new thinning operations that continue to provide increased water yields. Thus, for each plant, water yield increases and savings from reduced costs of wildfire and reduced loss of assets increase each year for

Table 5
Public Benefits, Industrial Revenues, Industrial Costs, and RD&D Costs Associated with a California Biomass-to-Ethanol Industry

Years	-4	-3	-2	-1	0	1	2	3	4	5	6	7	8	9	10	11	12	13	14	15
Public Benefits																				
Reduced Wildfire Cost (Million $)						2.5	7.5	15	25	38	50	60	67.5	72.5	75	75	75	75	75	75
Reduced Loss from Wildfire (Million $)						1.5	4.5	9	15	22	30	36	40.5	43.5	45	45	45	45	45	45
Value of Increased Water (Million $)						1.8	5.3	10.5	17.5	26	35	42	47.3	50.8	52.5	52.5	52.5	52.5	52.5	52.5
Total Public Benefit (Million $)						5.8	17.3	34.5	57.5	86	115	138	155.3	166.8	172.5	172.5	172.5	172.5	172.5	172.5
Industrial Revenue																				
Ethanol Revenue (Million $)						48	96	144	192	240	288	288	288	288	288	288	288	288	288	288
Electricity Revenue (Million $)						4.4	8.8	13.2	17.6	22	26.4	26.4	26.4	26.4	26.4	26.4	26.4	26.4	26.4	26.4
Total Ethanol & Electricity Revenue (Million $)						52.4	104.8	157.2	209.6	262	314.4	314.4	314.4	314.4	314.4	314.4	314.4	314.4	314.4	314.4
Industrial Costs																				
Capital Investment for 12 Commercial Plants (Million $)					80	80	80	80	80	80										
Operating Costs for Demonstration and Commerciall Plants (Million $)						33.4	66.8	100.2	133.6	167	200.4	200.4	200.4	200.4	200.4	200.4	200.4	200.4	200.4	200.4
Total Commercial Investment & Operating Costs for Demonstration and Commercial Plants (Million $)					80	113.4	146.8	180.2	213.6	247	200.4	200.4	200.4	200.4	200.4	200.4	200.4	200.4	200.4	200.4
RD&D Costs																				
Technology R&D	10	20	10	5																
Capital Contribution to a Demonstration Plant to Allow the Operator to Obtain a 20% Rate of Return or an NPV = 0			10	10																
Thinning Demonstration	4	4	4	4	4															

Table 6
Industry Net Present Value

	Present Value at Time Zero @ Discount Rate of 20% (Million of $)
Revenue	
• Total ethanol & electricity	943
Costs	
• Total commercial investment for commercial plants and operating costs for 11 commercial plants and demonstration plant	920
Net Present Value = $23 million	

5 yr and then level off. In addition, for this study the public benefits associated with the demonstration plant are ignored.

NPV Analysis

The industry NPV was established by calculating the present value at the end of yr 0 of the total ethanol and electricity revenues at a discount rate of 20% ($943 million) and subtracting the present value at the end of yr 0 of the capital investment for 12 commercial plants and the operating costs for all 12 plants ($920 million). The NPV was +$23 million as shown in Table 6. This means that industry could contribute $23 million to RD&D or construction, or both, of the first demonstration plant and still obtain a 20% return from its commercial operations.

The government NPV was established by calculating the present value of the public benefits produced at a 7% discount rate ($945 million) and then subtracting the present value of the RD&D costs at a 7% discount rate ($95 million) less the expected contribution from industry ($23 million). The government NPV is $873 million as shown in Table 7. In other words, an investment of $72 million ($95 − $23 million) on the part of government and RD&D will potentially provide a 7% return plus $873 million in present value.

Discussion

This NPV analysis suggests that the public benefit realized from thinning the 13 million acres of California's commercial timberland overwhelmingly compensates for the RD&D costs associated with the development of thinning and ethanol production operations. Because not all public benefits were estimated in this analyses, this conclusion might be considered conservation. When other benefits are considered (enhanced

Table 7
Government Net Present Value

	Present Value at Time Zero @ Discount Rate of 7% (Millions of $)
Revenue	
• Public Benefit	945
Costs	
• R&D capital contribution to a demonstration plant to allow it to operate at 20% return, and thinning demonstration costs	95
Net Present Value with Industry Participation = 945-(95-23) = $873 million	
Net Present Value Without Industry Participation = 945-95 = $850 million	

wildlife, habitat and ecosystem balance; increased timber value; jobs; reduced dependence on foreign energy; improved air quality; etc.), the rationale for public investment will become even more compelling.

ACKNOWLEDGMENT

This work was funded by the Biochemical Conversion Element of the Office of Fuels Development of the US Department of Energy.

REFERENCES

1. *Manufacturing Engineering*, (1994), Vol. 112, No. 2.
2. Stermole, F. J. (1984), *Economic Evaluation and Investment Decision Methods*, Fifth Edition, Golden, CO.
3. Santone, L. C. (1981), *Methods for Evaluating and Ranking Transportation Energy Conservation Programs Final Report*, Washington, DC, April 30, 1981.
4. Tyson, K. S., Putsche, V., and Bergeron, P. (1996), *Modeling the Penetration of the Biomass-Ethanol Industry and its Future Benefits*, Golden, CO, March 15, 1996.
5. Heaps, T. and Pratt, B. (1989), FRDA Report 071, *The Social Discount Rate for Silvicultural Investments*, Victoria, B.C., March, 1989.
6. Personal Communication from Loyd Forest of TSS Consultants, Sacramento, CA.
7. California Fire Plan, A Framework for Minimizing Costs and Losses from Wild Land Fires, R. J. Kerstiens, Chairman, California State Board of Forestry, March 1996.
8. Turner, K. M. (1991), Water Salvage from Mediterranean-Type Ecosystems. In: Water Supply and Water Reuse: 1991 and Beyond. Proceedings of the American Water Resources Association Annual Symposium, pp. 83–90.
9. Personal communication by L. Forrest of TSS Consultants, Sacramento, California.

Coordinating California's Efforts to Promote Waste to Alcohol Production*

WILLIAM J. BLACKBURN** AND JONATHAN M. TEAGUE

California Energy Commission, Sacramento, CA 95814

ABSTRACT

Alcohol fuels produced from biomass can improve air quality, enhance energy security, create employment opportunities, and reduce waste disposal problems. Opportunities in California exist to produce alcohols from waste streams from various sectors of the economy. Government agencies have promoted waste-to-alcohol activities, but efforts have been inconsistent and intermittent. Often these efforts have been hindered by contradictory but mandate-driven policies.

A prudent approach to coordinate statewide efforts includes the development of an integrated statewide policy to examine barriers that impede private sector business efforts to produce alcohols from biomass. A multi-agency task force to promote research, development, commercialization, and marketing efforts for biomass-produced alcohols is desirable.

Index Entries: Biomass; ethanol; methanol; partnerships; government.

INTRODUCTION

For a variety of environmental and economic reasons, much attention has been devoted to the development and commercialization of nonpetroleum energy sources. Indeed, few areas within the scope of environmental and energy fields have garnered as much interest as the search for clean burning, renewably-based fuels. Since the 1973–1974 oil embargo, the search for renewable energy sources has often been intense, albeit somewhat sporadic in nature.

Modern industrial economies, such as California's, have become heavily dependent on petroleum-based fuels. Gasoline and diesel make up more than 97% of the state's transportation fuel mix. In California, vehicles travel approx 200 million miles each day, resulting in the consumption of approx 18 billion gallons of gasoline and diesel fuel per year [1,2].

---

*The views and opinions contained in this document do not necessarily reflect those of the California Energy Commission, its staff, management, or the State of California.
**Author to whom all correspondence and reprint requests should be addressed.

The near-exclusive reliance upon petroleum-based fuels increases vulnerability to economic dislocations from oil price instabilities, contributes significantly to air quality degradation, and strongly affects the national balance of trade. These issues, combined with the intrinsic long-term limitations of nonrenewable fuels, have provided the basis for government-led activities devoted to developing viable alternatives to petroleum-based fuels.

Additionally, the recent trend in the manufacturing sector to completely eliminate waste products in order to develop a sustainable business environment may also create opportunities to generate alcohol from waste (3). Many private sector concerns are faced with rising disposal costs and, simultaneously, shortages of critical inputs, such as suitably high quality process water. In some cases, these businesses are seeking the means to convert waste streams into useable and potentially valuable products (e.g., alcohol) as part of the water re-use and recycle process.

Alcohol fuels have the potential to replace petroleum-based fuels. For example, if the United States converted 14% of its farmland to produce crops for ethanol, the results would be sufficient to supply the nation's entire gasoline use (4). Tree farming cultivation of perennial grasses is a plausible means of producing the enormous quantities of biomass needed to replace fossil fuels. Short rotation intensive culture (SRIC) farming techniques can be applied to a variety of tree species. Various nonwoody crops, such as switchgrass, are also efficient producers of lignocellulosic biomass. However, completely replacing petroleum-based fuels in the near term is not practical and may not be desirable. Only the most cost effective and environmentally benign efforts should be pursued.

The conversion of waste biomass to ethanol holds a good chance of becoming both environmentally and economically attractive. Given the potential availability of biomass, one might reasonably ask: Why isn't more ethanol now produced from waste? The answer is that although biomass can be relatively inexpensive to purchase, it is difficult (and therefore costly) to convert into ethanol. Historically, many proposed processes have suffered from inherently low yields and high costs (5). Moreover, biomass streams may be erratic or seasonal in nature, and collection and transportation costs are often significant, due to the low energy density of the materials.

Unlike energy crops, biomass from waste is often available for no or negative cost (e.g., waste generators would be willing to pay to dispose of waste). And although there is currently a significant amount of USDA-designated cropland not in production in the United States, the conversion of arable farmland from production of foodstuffs to energy feedstocks may pose long-term concerns. For these reasons, the focus of this paper is limited to discussions on waste biomass-to-alcohol production.

This paper will examine some of the current waste-to-alcohol activities presently underway in California, with a particular focus on ethanol.

The more promising technologies for converting biomass to alcohol will be reviewed and suggestions will be made to better coordinate state agencies that are involved with or interested in waste-to-alcohol activities.

PROBLEMS WITH CURRENT TRANSPORTATION AND WASTE DISPOSAL SYSTEMS

Consequences of Long-Term Petroleum Dependence

The sharp volatility in oil prices has had a devastating effect on the US national economy. The Oak Ridge National Laboratory estimates this cost to be up to $4 trillion during the period 1973–1990 (6). Similarly, California's economy has suffered substantially during times of high petroleum-based fuel prices.

Because California and other developed economies are heavily dependent on petroleum for transportation, they are vulnerable to oil price and supply constraints. Much like the approach taken by investors to diversify their assets in order to reduce risk, energy strategists advocate a similar approach in regard to transportation fuels. Developing alternatives to petroleum-based fuels could significantly reduce the economic risks of petroleum price swings. Therefore, the development of nonpetroleum fuels represents a prudent approach to energy and transportation planning.

Trends in Sources of Petroleum Supply

California consumes most of its petroleum from domestic sources. Historically, about 50% of the petroleum used in the state came from in-state production, 45% from Alaska, and 5% from foreign sources. However, the percentage supplied from foreign sources will increase as both Alaska and California production decline (7).

California energy markets, however, are thoroughly integrated with national and international markets. By contrast, the United States as a whole currently imports more than 50% of its oil from foreign sources. This fraction is expected to increase in the future. Dependence on imported oil is a major contributor to the unfavorable trade balance that the United States maintains with much of the world. Imported petroleum accounts for approx 45% of the US trade deficit (8).

Fully two thirds of the world's oil reserves are found in the Middle East. Given the area's history of political instability, and regional antiwestern sentiment, access to this resource likely will remain worrisome for decades to come. The extent of US military expense in maintaining commercial access to the oil reserves of this region are a matter of dispute but are enormous.

Environmental Impact of Petroleum Use

The most serious air quality problem in California (and the United States) is the production of ground-level ozone. This reactive form of oxygen (O_3) is produced when hydrocarbons and oxides of nitrogen (considered to be "ozone precursors") chemically react when exposed to the ultraviolet radiation in sunlight. Often referred to as photochemical smog, exposure to ground-level ozone results in a variety of health-related problems and agricultural crop damage. The majority of these ozone precursors are produced from the combustion of gasoline and diesel in motor vehicles.

Ozone and other forms of air pollution, such as carbon monoxide, particulate matter, and toxic air contaminants, cause widespread health problems in California. Some individuals, notably children and the elderly, are particularly susceptible and may be severely affected. Especially at risk are the four million Californians who suffer from heart and lung diseases (9).

In addition to vehicle tailpipe emissions, significant quantities of hydrocarbons escape from vehicles through evaporation from the vehicles' fuel system and during refueling. Although substantial improvements in reducing tailpipe and evaporative emissions have been made in recent years, both deliberate tampering with emission control systems and malfunctioning in vehicles remains problematic.

Emissions (predominantly hydrocarbons) also are produced during the refining and transporting of petroleum. Without stringent controls and continual monitoring, petroleum-based fuel production, transportation, and consumption can generate massive amounts of regulated pollutants.

Greenhouse gas emissions, primarily carbon dioxide from hydrocarbon fuels combustion, is also a cause for concern. While CO_2 emissions are not currently regulated, these are nonetheless of great concern. Strong and accumulating scientific evidence suggests that the increase in atmospheric concentrations of greenhouse gases will induce global climate change. In the United States, approx 32% of the CO_2 released into the atmosphere is from transportation sources (10). Thus, it may be prudent to develop fuels derived from renewable biomass. The production and subsequent use of ethanol and methanol from biomass tends to be "CO_2 neutral" as the production of biomass from plant sources involves CO_2 absorption on the same timescale as its release from combustion and does not contribute to the atmospheric carbon burden.

Problem of Lignocellulosic Waste Disposal in California

Agriculture has been and remains a foundation of the state's economy, contributing some $22 billion dollars annually in gross revenues (11). This sector also produces enormous volumes of residual biomass, which poses significant disposal problems. Forestry activities (thinning, timber production, etc.) also generate large amounts of cellulosic material, as does urban waste in the form of paper, wood waste, and landscaping debris. Altogether California produces abundant cellulose-containing residues, approx 46 million bone dry tons per year (12).

The state's biomass power industry provides some avenues for productive use of woodwaste and some agricultural residues as boiler fuel, but the future of this industry is uncertain at present. Historically, however, much agricultural residue, such as rice straw and orchard prunings, has been disposed of by open-field burning, in part to control pests and plant diseases. The environmental impacts of this activity have been sufficiently severe that restrictions are being imposed on open-field burning as a means of disposal. The existing alternatives, such as incorporating crop residues into the soil, can impose substantial costs on farmers, and may exacerbate some crop diseases. Nevertheless, at least in regard to the burning of rice straw, state law (e.g., Assembly Bill 1378 of 1991) calls for a gradual phase-down of open-field burning in the Sacramento Valley; ultimately, the law envisions that no more than 25% of the straw will be disposed of by burning, but only for purposes of crop disease management.

An economic use for these wastes is preferable and is envisioned under the laws that restrict open burning. Each waste type poses specific disposal problems. For example, more than 1.5 million tons per year of rice straw is generated from rice production in California's Central Valley. This material is high in silica and resistant to decomposition, requiring extensive tillage to incorporate into the soil. Proposals to convert rice straw into ethanol suggest attractive alternatives that could divert a large fraction of the existing supply. Other off-farm uses, such as conversion to construction material, landscaping and erosion control are possible, but do not at present appear likely to absorb more than a small share of the available residue. Accordingly, conversion to alcohols or other chemical feedstocks could offer benefits to this sector of the economy.

Forest waste also represents a significant component of California's biomass resource. In many of the state's forests, excessive fuel loading increases the risk of catastrophic forest fires that destroy life, property, and natural habitats, cause significant air quality impacts (as particulates or PM_{10}) and contribute to severe soil erosion.

Similarly, municipal solid waste is increasingly a problem in the state. Constraints on landfill capacity led to the passage of legislation in 1989—AB 939, the California Integrated Waste Management Act. This Act mandates stringent goals for diverting solid waste from landfills using a hierarchy of reuse, recycling, composting, or transformation to energy products to avoid disposal. Municipalities operating solid waste disposal facilities were required to divert 25% of the waste stream to these uses beginning in 1995 and must attain 50% diversion rate by the year 2000. A subsequent law (AB 688) set forth additional conditions for calculating the credits for diverting waste materials and limited the degree to which biomass transformation (either combustion or fuel production) could be counted to satisfy the diversion requirements.

Apart from yard and construction wastes, cellulosic wastes (primarily paper products) constitute a major fraction (estimated at approx 60%) of the municipal solid waste stream. This portion represents a high value, relatively pure stream of cellulose. Although much of this material can be directly recycled for use as paper products, part of this stream is either too degraded or too contaminated for paper recycling and could be used for fuel production. Like other forms of biomass, conversion of municipal wastes to ethanol and other valued chemicals could be an effective way to reduce this burden.

VALUE AND UTILITY OF ALCOHOL FUELS IN CALIFORNIA TRANSPORTATION

Alcohol in California's Alternative Transportation Fuels Programs

The repeated experience of strong fuel price volatility and concerns over security of access to overseas oil supplies has spurred development of nonpetroleum transportation fuels, often referred to as alternative fuels. Alcohol fuels, including ethanol and methanol, are popular alternative transportation fuels.

The efforts of the California Energy Commission and others have led to the use of methanol by fleet operators and individual citizens as an alternative to gasoline in the state. Developed first in the mid 1980's, the flexible fuel vehicle (FFV) allows the operator to use either gasoline, M85 (85% methanol blended with 15% unleaded gasoline), or any combination of the two fuels. By the end of 1996, approx 14,800 FFVs were operating in California (13).

In contrast to methanol, ethanol as a motor fuel is almost nonexistent in California. And unlike many midwestern states, where corn growers and grain processors support conventional ethanol production, California does not have a comparable economic interest group supporting alcohol fuels. California does possess, however, a nascent methanol fuel distribution system, and the continued development of this alcohol fuel distribution infrastructure could potentially benefit both fuels. Additionally, methanol and ethanol FFVs utilize nearly identical engine technologies.

In California during the 1980's and early 1990's ethanol was used as an octane booster in gasoline (14), typically in blends of up to 10%. At its peak in 1991, over five million gallons of ethanol per month were blended with gasoline in the state, making up approx 5% of California's gasoline supply. The average from 1981 to 1991 was 36 million gallons per year consumed in motor fuels (15).

The growing concern over the role of gasoline-based emissions on urban air quality problems led to the requirement that oxygenates be added to the gasoline pool to reduce carbon monoxide and ozone precursors. These oxygenates typically were in the form of ethanol or alcohol-derived ethers. These ethers, principally methyl tertiary-butyl ether (MTBE), now dominate the oxygenated fuel market, in part because current California air quality regulations for reformulated gasoline limit oxygen content in gasoline to 2% by weight (ethers contain less oxygen by weight than their associated alcohols), and limit fuel volatility (California regulations limit Reid vapor pressure to a maximum of 7.0 pounds per square inch). Refiners would be required to use gasoline blends with very low volatility to utilize ethanol. Because of higher costs associated with this formulation, no refiner has sold gasoline/ethanol blends (e.g., 10% ethanol) in the state since reformulated gasolines became required in early 1996.

Past Experience with Fuel Alcohol Production in California

A number of ethanol production demonstration projects were conducted by state agencies in the early 1980's, under the impetus of state Senate Bill 620. These efforts were intended to exploit agricultural waste products from cultivation and food processing, using then available fermentation and distillation technology. Projects under this program were based on project economics and volatility in the larger fuel markets. The technology of the day, which depended upon reliable and constant sources of high quality sugar and starch feedstocks in order to be economically viable, proved unsuccessful (16). The successful ethanol industry model that has developed in the Midwest, based on rainfall-watered starch crops (primarily corn) and substantial federal tax incentives, was judged to be ill-adapted to irrigated agriculture in California.

These project failures strongly influenced subsequent thinking about in-state fuel alcohol production. However, technological improvements in feedstock processing, biotechnology, membrane-based separation processes, and process engineering have altered the technical and economic landscape to the point where a strong potential now appears to exist for biomass-derived alcohol production. Furthermore, reliance upon feedstocks that are plentiful and that have a low or negative cost may change the equation for industrial-scale alcohol production in the state. Currently, there are two ethanol production plants operating in California, one utilizing beverage industry waste as a feedstock and the other using cheese whey. Total production is on the order of six million gallons per year.

Benefits of Alcohol Fuels

United States and California efforts in the area of biomass-derived alcohol from the early 1980's on were driven primarily by the need to reduce dependence upon imported oil. It has since been recognized that other benefits may accrue from developing this source of primary energy. In particular, both ethanol and methanol fuels hold important potential benefits for California, in the following four principal categories.

Energy

- Can be produced from renewable resources;
- Aids biomass industry by creating demand for biomass feedstocks;
- Could be developed in conjunction with electric power generation, improving overall efficiency of plant operation;

- Allows use of less refined blend stock to achieve requisite octane values in gasoline, thereby reducing energy consumption during refining;
- Domestically produced, thereby reducing oil imports;
- Can be used with advanced transportation technologies such as hybrid electric- or fuel cell-powered vehicles, as ultra low-emission motor fuel (energy and air quality benefit); and
- Potential for combustion efficiency improvements in conventional internal combustion engine vehicles.

Air Quality and Emission Reductions

- Reduces the generation of CO_2, a principal greenhouse gas, since the use of biomass displaces consumption of carbon-based fossil fuels;
- Important oxygenate source for reformulated gasolines (ethanol and its ether ETBE);
- Reductions of toxic air contaminants (e.g., 1–3 butadiene and benzene) commonly found in gasoline; and
- Reduces emissions that would otherwise result from disposal of biomass in landfills (e.g., methane, another greenhouse gas) or by open-field burning (particulates and carbon monoxide).

Waste Disposal

- Potential for reduction in waste volumes through economic use of waste stream components;
- Recovery of economically valuable by products from waste streams, and reduction in process costs by enabling reuse of inputs such as process water; and
- Extend the operating life of existing landfills.

Economic Benefit

- Provides benefits of competition to conventional fuel markets;
- Creates employment opportunities through new economic development (including the economic multiplier effects of new primary economic activities);
- Liquid fuels are convenient to handle, store and use, due to higher energy density; and
- Can produce marketable byproducts in addition to alcohol.

Ethanol Price History

Presently, the cost of ethanol derived from biomass has dropped dramatically from approx $0.95/L ($3.60/gal.) in 1980 to less than $0.32/L ($1.22/gal.) at the pilot plant level. Research is continuing to further improve biomass pretreatment, cellulase enzyme production, cellulose conversion, five-carbon sugar utilization, lignin use, and product recovery, with the goal of reducing the cost to $0.18/L ($0.67/gallon by 2010, National Renewable Energy Laboratory) *(17)*. At the level of $0.18/L, ethanol can compete with gasoline as a neat fuel at an oil price of $25/barrel, or $0.30/L at an oil price of $33/barrel *(18)*.

As the production efficiencies improve and its cost of production correspondingly fall, biomass-derived ethanol will approach competitiveness with fossil-based fuels *(19)*. Although some ethanol operations are now, or will soon be cost competitive, on the whole the fuel's competitiveness as a neat motor fuel is marginal. Because of this, government support will continue to be necessary for the near-term. Government programs to stimulate continued production efficiencies also will be needed to advance ethanol's commercial prospects.

At present, biomass-based alcohol receives a blenders' tax credit of 54 cents per gallon; at the gasoline pump for 10% "gasohol" blends this results in a tax break of 5.4 cents per gallon. Producers can instead opt for a tax credit taken at the time of production. Under the Energy Policy Act of 1992, a tax credit for production of energy from "closed-loop" biomass cultivation was instituted, but this provides little assistance to business seeking to exploit existing waste streams for alcohol feedstocks.

BRIEF CHARACTERIZATION OF OVERALL PROCESS TECHNOLOGIES

Biomass

Many types of biomass, including wood, agricultural residues, herbaceous crops, and municipal solid waste, can be used as feedstocks for ethanol production. The various forms of biomass outwardly appear to be very different, but their chemical makeup is quite similar. About 35% to 50% of the material is cellulose, a polymer of glucose sugar that forms a crystalline structure. Another 15% to 30% of lignocellulosic biomass is hemicellulose, generally a heterogeneous polymer of various sugars often dominated by the five-carbon sugar xylose. The remaining 20% to 30% is composed primarily of lignin, with lesser amounts of extractives, ash, and other components *(20)*.

Both cellulose and hemicellulose are carbohydrates and can be hydrolyzed by enzymes to simple sugars. These sugars are then converted through fermentation to ethanol. Lignin (a heterocyclic phenolic polymer) can not be converted to ethanol, but may be used as a high energy content boiler fuel in the ethanol production process and for electric power generation. Some potential exists to use the lignin as feedstock for chemical synthesis to produce a variety of products.

Technology Approaches

Methanol synthesis from syngas is a well demonstrated, commercially available industrial scale technology. Incremental improvements in efficiency of conversion continue to be made, and liquid phase synthesis processes may hold some promise. Similarly, distillation and refining technologies for ethanol are well-proven. In both cases, the issue for utilizing biomass feedstocks is the production of useful precursors for the upstream synthesis or fermentation processes. In the case of methanol, it is the production of clean, hydrogen-balanced synthesis gas (a mixture of carbon monoxide, carbon dioxide, steam, and hydrogen). For ethanol production, it is the hydrolysis of lignocellulosic substrates into simple sugars that can be efficiently fermented to produce alcohol. Hence, the choice of process for feedstock processing and conversion is at the heart of the various technology options.

Thermochemical processes entail the production of methanol and other products from syngas. These processes include gasification, pyrolysis, and combustion. Thermochemical processes have grown out of considerable past work in the conversion of coal and lignite to synthetic fuels and other products. Specific issues related to the chemical composition of many types of biomass affect the utility and efficiency of these processes when applied to biomass feedstocks. High alkali and silica content of some feedstocks, such as rice straw, can result in severe problems with slagging and fouling of combustors and gasifiers (21,22). Technical measures to overcome these problems have been under development for some time. However, it appears that biologically-based conversion processes may hold significant promise for utilizing other relatively intractable biomass feedstocks.

Ultimately, some combination of these two approaches, such as an initial fermentation-based process with offstreams being fed to a thermochemical conversion process, may prove most effective to capture the full value of the feedstocks.

Biological processes are typically fermentation based, leading to the production of ethanol and other coproducts, depending upon the technology and microorganisms used. Process economics may be strongly influenced by the nature and value of coproducts obtained. These could include sodium silicate; animal feeds; boiler and/or lignin-derived engine fuels. Although the fermentation stage may vary by process design, these processes may all be distinguished primarily by the feedstock pretreatments that are employed to obtain fermentable sugars from the raw biomass. This pretreatment may be seen as consisting of three basic steps: mechanical processing and sizing of the feedstock; disruption of the microstructure of the biomass to expose its components to hydrolysis, and subsequent hydrolysis of these components to simple sugars for fermentation by the alcohol producing organisms.

The various forms of pretreatment may be configured in different ways, with the products of a particular disruption process particularly suited to a subsequent hydrolysis process. Hydrolysis processes include:

- Strong acid hydrolysis;
- Weak acid hydrolysis;

- Enzymatic hydrolysis; and
- Lime-based hydrolysis (laboratory scale only).

CUSTOMER BASE AND ECOSYSTEM MARKET ANALYSIS

Potential beneficiaries of a biomass-based fuels industry in California are diverse and will depend upon the specific feedstocks and processes employed. In general, however, present opportunities for biomass utilization exist that could help the following sectors of the state's economy, loosely characterized in terms of either disposal, or production and consumption.

Disposal

Avoided costs of disposal could be a significant benefit to certain economic sectors, and at some point it might be feasible to achieve revenues from the sale of former waste products. Economic interests that would benefit from a viable biomass-to-alcohol industry in California would potentially include the following.

Various Agricultural Producers

These include crop producers (prunings, orchard clearing for replanting, crop processing wastes, such as almond shells); rice growers (straw disposal for disease control); poultry, livestock and dairy producers (manure and bedding disposal); and cotton growers (gin trash).

Forest Products

Disposal of forest slash, mill residues (such as sawdust, bark and slab wood) can be problematic for the wood products industry. Although new uses for wood fiber can potentially consume much of this material (such as particle board, oriented strand board, pulp, etc.), substantial amounts of biomass remain unutilized and now represent economic costs to the industry.

Natural Resource Managers

Public and private forest land owners confronting the need to reduce fuel loading in fire-prone forests and manage vegetation for watershed and soil conservation purposes. Such essential activities could generate substantial quantities of biomass, which could make them at least partially self-funding.

Municipalities

Cities, counties, and sanitation districts that face the need to divert wastes from landfills, either because of economic pressures or regulatory requirements, such those created by State Assembly Bill 939.

The distribution of total potential of biomass resources in California is as follows (23):

Livestock manure	26%
Chaparral	16%
Field and seed crops	14%
Lumber mill waste	12%
Forest slash	11%
Urban yard waste	7%
Fruit and nut crops	4%
Food processing waste	4%
Urban wood waste	3%
Other	3%
Total	*100%*

Production and Consumption

A number of economic sectors could benefit directly from the production and consumption of fuel alcohols from biomass.

Agricultural Producers

Farmers would benefit from the conversion of problematic wastes into economic resources. Opportunities for vertically integrating alcohol fuels production with existing production and for value-added processing could prove attractive.

Biomass Power Generators

The large biomass power industry in California was developed partially in response to the availability of cheap biomass feedstocks for conversion to electricity. At present, the market price of electricity is too low to support operation of many of these plants, especially as existing power sales contracts with more favorable price terms expire. Operating costs are typically high, owing to problems with boiler fouling and slagging from combusting raw biomass. Combining an ethanol plant with biomass power unit could improve the economic viability of both processes by added value production. Initial processing of biomass fuels for conversion to ethanol would result in a valuable commodity product. The remaining biomass residue would then be usable as a clean-burning fuel stream for power generation, reducing plant operating and maintenance (O&M) costs.

Alternative Fuel Consumers

The Energy Policy Act of 1992 (EPACT) mandates alternative fuel use by fleet operators, in particular state and local government fleets, fuel providers, and possibly private fleets. EPACT and other federal laws also encourage alternative fuel deployment by means of a number of economic incentives. Replacement of diesel fuel with cleaner burning alcohol fuel in heavy duty applications both on and off road would be helped by a domestic alcohol fuel production.

Chemical Manufacturers

Ethanol, methanol, and other possible products from biomass feedstocks are widely used as chemical intermediates and solvents.

Electronics Industry

Manufacturers require silicon for producing semiconductor chips and other electronic components. Certain feedstocks, such as rice straw, may have the potential to provide high purity sodium silicate for this purpose.

The general public would also benefit, in terms of economic growth, and the reduction or elimination of environmental externalities associated with the generation of these biomass wastes. Note that each of these sectors above is not likely to absorb more than a portion of the available biomass feedstocks.

Competing or Displaced Interests

It should be recognized that some economic interests would view the emergence of a fuel alcohol industry as a competitive threat.

Petroleum Producers, Refiners, Distributors, and Retailers

Although petroleum refiners and blenders would benefit from the availability of sources of oxygenates within the state, they may view a largescale alternative fuel production industry as a competitive threat. Retailers could, however, choose to offer the alternative alcohol fuels as part of their product slates. Independent or small refiners and marketers might prefer ethanol as an oxygenate or at least welcome a choice of blending agents.

Wood Products Industries

Biomass feedstock uses could compete for the lower grade fractions of timber and/or forestry wastes, thereby resulting in higher costs for these inputs.

Midwestern Ethanol Producers

As competing suppliers for California ethanol demand, these producers may view the development of biomass ethanol as problematic, even though in the near term, they will probably remain price competitive. In the longer term, the commercialization of biomass-to-ethanol technologies would be likely to benefit firms in agricultural regions outside of California by securing the broad market and public acceptance for this fuel. Moreover, the market developments necessary to establish a commercially viable biomass ethanol industry are likely to benefit all producers. It is worth noting that technologies for converting cellulosic biomass to ethanol may find their first broad application in converting corn fiber (a residue from conventional ethanol production) to alcohol.

Natural Gas-Based Methanol Producers

As with ethanol, to the extent that markets expand for alcohol fuels, this would benefit the conventional natural gas-based methanol industry. The development of biomass-derived methanol could conceivably pose a problem to the methanol industry to the extent that growth in market demand could not absorb the output from biomass-based methanol plants. Given the current low cost of natural gas-derived methanol, however, it seems unlikely that biomass-derived methanol would be a competitive threat in the near-term.

Waste Disposal Companies

Firms operating landfills on a for-profit basis could lose tipping fees as wastes are diverted to biomass conversion facilities.

Note that there may exist possibilities for strategic alliances with some of these competing interests. For example, a firm that operates landfills might also be interested in diverting biomass to a utilization facility (perhaps sited at the landfill), while still deriving revenue on collection and transport of the biomass from the waste generators. Existing commodity alcohol producers also could opt to expand into biomass-based alcohol production, capitalizing upon their established expertise and experience with alcohol process technologies. Because biomass-based alcohol production may be developed in smaller increments of production capacity, the market impacts of new plant additions could be reduced. This is in contrast, for example to the sustained and periodic collapses to which methanol prices have been subject due to the commissioning of new large natural gas-based methanol plants.

GOVERNMENT'S ROLE WITH CURRENT AND FUTURE BIOFUEL ACTIVITIES

Numerous government agencies have developed programs or have participated in waste-to-alcohol related activities.

Federal Agencies

The various federal agencies that have been or may be involved in these efforts include: the US Department of Energy (DOE) and its national labs, especially the National Renewable Energy Laboratory (NREL), the US Environmental Protection Agency (EPA), the US Department of Agriculture (USDA), and others. Unlike California state government, the federal government through NREL has a multitude of nationwide programs to conduct research and development, pilot production biomass-to-alcohol activities and other efforts to aid commercialization activities. NREL has accomplished a great deal of research and technology development in this area. To the authors' knowledge, NREL has not formed regional coalitions to promote these efforts. It may in fact be more appropriate for state and regional governments to act as catalysts in bringing together disparate groups of similar interest.

State Agencies

On the state level a number of agencies have been involved with or have shown interest in waste-to-alcohol, including: the California Integrated Waste Management Board (CIWMB), the California Air Resources Board (CARB) and its parent agency the California Environmental Protection Agency (CalEPA), the California Department of Food and Agriculture (CDFA), the California Department of Forestry and Fire Protection (DFFP), the California Trade and Commerce Agency (TCA), the California Energy Commission (CEC), the California Department of Water Resources (DWR), and other agencies.

Local Government Agencies

Myriad regional, local, and tribal governmental bodies exist in California. Of particular importance in this context are waste disposal agencies and air quality/air pollution districts throughout the state, as well as sewage and reclamation districts, agricultural conservancy boards, and economic development agencies. The Sacramento Metropolitan Air Quality Management District (SMAQMD) has shown strong interest in developing local green waste or rice straw-to-ethanol projects, given its jurisdiction within a major rice straw burning area. In particular, the SMAQMD would like to see locally produced, inexpensive ethanol for use in heavy-duty vehicles in the region and has been studying the potential for waste-to-alcohol in the area (24). Although other air districts may be less enthusiastic about waste-to-alcohol, they do stand to benefit from the work and may similarly act as a support resource around the state. Other local agencies such as waste disposal agencies, agricultural offices, etc., may also have interest in waste-to-alcohol efforts.

Disparate Nature of Organizations Promoting Biofuels

Government-backed programs that have been implemented over the last two decades aimed at encouraging the development of waste-to-alcohol activities have rarely resulted in long-term success stories. Furthermore, regulations developed and enforced by governmental bodies in the pursuit of other objectives often slow or halt hopeful entrepreneurs pursuing biomass conversion efforts.

Quite often government's lack of ability to develop effective programs in this area stems from the approach taken by agencies. Individual governmental agencies tend to approach problems separately and independently. State agencies generally focus on a singular problem or field in response to specific legislative directives. When issues arise that involve multiple agencies, the results are often less than satisfactory. One such example is the complex issue of open-field burning rice straw in the Sacramento Valley region. Burning the remains of rice straw after harvest potentially affects three state agencies: the California Air Resources Board, the California Integrated Waste Management Board, and the Department of Food and Agriculture, as well as local air pollution control districts and county agricultural commissions.

Moreover, if the remaining rice straw was collected and converted to an energy product such as alcohol fuel rather than being burned, yet another state agency—the California Energy Commission—could become involved. Because each agency has its own set of directives—air quality, waste management, agriculture, and energy—and because the state has not created a specific statewide policy, insufficient cooperation exists between these agencies. Also, the independent nature of these state-directed efforts can cause a variety of new problems or result in duplicative efforts. Additionally, these various organizations often maintain conflicting interests. For example, the Energy Commission may promote ethanol as a viable transportation fuel, only to have the California Air Resource Board resist its use because of concerns over potential increases in certain regulated pollutants. Rather than partnering together as sister state agencies, the result is an unintended struggle to move forward mandated agendas.

Therefore, an integrated statewide policy is needed to coordinate efforts by various state agencies and other affected parties. The single most important point made by representatives of the private business community to the authors is the need for an integrated, crossagency policy. Working cooperatively between agencies to remove barriers to commercializing waste-to-alcohol can significantly change the economic outlook for rural and urban development with the potential for creating a substantial number of jobs. It is unlikely that the situation will improve without a unified statewide policy.

BRINGING KEY PARTIES TOGETHER

Government's Role in Facilitating Development

The role of government in promoting waste-to-alcohol activities can assume a variety of forms, ranging from direct involvement in attempting to establish a new technologies in the marketplace, to research, to demonstration, to providing information-only support. The approach considered here suggests that government can function best as a catalyst in promoting introduction of new technologies, but that their successful adoption depends upon market acceptance by both producers and consumers. Private sector investment is more likely to account for the risks and rewards of a new technology. Public involvement is best leveraged in the form of public/private partnerships.

Another role of government in promoting waste-to-alcohol efforts can include what agencies accomplish by removing hindrances to economic development. In the course of exercising essential functions to serve legitimate public purposes, regulatory and development agencies often work at cross purposes. The creation of barriers to market entry by new technologies can be an unintended but important consequence of measures undertaken in pursuit of public goals (e.g., environmental or public health protection). It is desirable that public agencies function with an eye toward coordination and flexibility, so that burdens to private economic activity are minimized without compromising the necessary protection of public interests.

Direct market creation for alcohol fuels has functioned with mixed results. For example, the technical success of the fuel methanol demonstration program in California has not been followed by significant economic penetration of methanol into the transportation fuel market. This can be attributed to a wide variety of factors, most of which focus on the difficulty of establishing a nonfungible transportation fuel in competition with existing, universally accepted petroleum fuels. The present price trends for petroleum and its enormous economies of scale in an integrated worldwide market pose competitive challenges for any alternative motor fuel seeking to gain market share based on pricing alone.

A quick overview of quarterly usage data from the California Energy Commission's fuel methanol program is instructive in this regard. Methanol usage in light duty vehicles peaked in August 1993 at 115,641 gallons of methanol per month, which was blended with gasoline to make M85. Despite the addition of a few thousand more FFVs added to the total population each year, fuel use has declined precipitously. In December 1996, usage had fallen to 22,360 gallons. Ultimately the commercialization of a neat ethanol blend (e.g., E85) and vehicles that operate on such a fuel face similar challenges.

Although the Energy Commission is presently investigating the reasons behind the volume decline, it is unclear what government-driven efforts could be employed to reverse this trend. Contacts with fleet customers have revealed a number of complaints including: inconvenience of fuel access system, too few stations, high fuel price, requirement of unique and expensive motor oil formulations, and others. Because ethanol and methanol are similar in many respects, it appears likely that ethanol may face many of the same issues. It should be noted that ethanol is a less aggressive solvent than methanol, requiring less extreme measures to prevent degradation of fuel-wetted parts. However, at least one automobile manufacturer has indicated experience with some fuel corrosion problems particularly associated with ethanol blends.

Ethanol does hold a long-term potential that is lacking for methanol, in that it can be used directly as a gasoline oxygenate. The potential ability to penetrate the motor fuel market on this basis affords a unique near-term opportunity for the development of biomass-based ethanol.

Coordinating Efforts of State Agencies

A unique example of cooperation among public and private efforts to construct a waste-to-alcohol facility is the Gridley Rice Straw-to-Ethanol Project, initiated in February 1996. Located approx 60 miles north of Sacramento, Gridley is a small agricultural community with a sizable rice industry in and around the area. Originally, project partners included the City of Gridley, the National Renewable Energy Laboratory (NREL), the California Institute of Food and Agricultural Research (CIFAR, at the University of California, Davis), the Rice Research Board of California, Stone and Webster Engineering (as project manager), SWAN Biomass Company, the California Energy Commission and other private and municipal organizations. An energy company, BC International, has recently become a project partner, whereas SWAN Biomass Company is no longer involved.

The impetus for the Gridley project was the requirement under California law to phase down open-field burning of rice straw over a period of several years to improve air quality. Converting the straw to ethanol is seen as one of the more promising ways of disposal (25).

Funding is coming from the US Department of Energy through NREL and from other project participants. Phases I and II are expected to cost $2.2 million to complete. If the project is shown to be feasible, several rice grower's cooperatives may be interested along with private investors in supplying the required capital resources. The project operator will be determined during the feasibility study, and the owners will ultimately be private investors. Although premature to predict the outcome of this project, the effective partnership that has developed can serve as a model for future alcohol production efforts.

Development of Multidisciplinary/Multiagency Consortium

One way of effectively coordinating activities among government and private entities is to create an organization to pull their often diverse interests together. A consortium of interested parties could be established to leverage resources, share information, and jointly address related problems, including the following.

- Legislation;
- Regulatory issues;
- Technology issues;
- Economic barriers;

- Capital investments;
- Partnering projects;
- Biomass resource availability (feedstock issues);
- Market penetration;
- Environmental restraints and permitting; and
- Other issues.

Furthermore, a consortium must work in an environment of cooperation. The following is an example of goals for interagency cooperation efforts.

- Establish a state policy to encourage sustainable economic development;
- Policy carried out through collaboration; coordinate with state, local, tribal governments and public;
- Phase in implementation of cross agency management process—i.e., Interagency Biomass Activities Task Force;
- Establish specific overarching goals and general guidelines for cross agency ecosystem planning and management process; and
- Direct agencies to interpret their existing authorities as broadly as possible to implement the ecosystem management policy and process.

In sum, work should be directed toward understanding, inclusiveness, and breaking away from traditional piecemeal approaches.

One of the advantages of forming a consortium would be its ability to proactively advance the pertinent issues. Ideally, this consortium should include: state air quality management/air pollution control districts; state agencies (i.e., CEC, CARB, CDFA, CIWMB, etc.); local waste disposal offices; federal agencies (NREL, DOE, EPA, etc.); ethanol producers; members representing agricultural, forestry, food processing industries; nongovernmental organizations and other interested parties.

Because the California State Resources Agency and CalEPA are the umbrella organizations that oversee numerous state environmental/resource agencies, it seems most appropriate to seek participation of these two agencies in a multiagency ethanol task force. The Resources Agency and CalEPA could shepherd their constituent agencies and interested local agencies in an efforts to bring a biomass-based alcohol industry to maturity in California.

CONCLUSION

In surveying California ethanol production facilities, it became evident that few exist today, and none of substantial size (e.g., greater than 10 million galls./yr). Furthermore, ethanol will likely remain in a role of fuel oxygenate rather than as a neat fuel, in terms of its potential in near term transportation fuel markets. Ethanol market economics as a fuel oxygenate or as a fuel in flexible fuel vehicles will be tied to those of petroleum fuels, due to the dominance of the latter in the transportation sector. The degree to which air quality and other regulations (e.g., groundwater protection) allow or drive the use of ethanol as a fuel oxygenate will determine the viability of a biomass-derived ethanol production industry in California.

Ethanol's current price, even with the current federal subsidy, is comparatively high, rendering it problematic to market as a neat fuel. However, advances in production efficiencies are likely to continue to drive down the price of biomass-derived alcohols to better compete with conventional petroleum motor fuels.

The diffuse nature of biomass resources hinders their competitiveness with petroleum fuels. Accounting for externalities associated with the choice of petroleum fuels could shift the competitive balance. Such externalities include: resource depletion, air quality impacts, land conservation, carbon emissions, energy security, rural economic development, and balance of trade. If these benefits could be "monetized" and credited to biomass-based alcohols, these fuels could become a significant contributor to the US energy supply.

Numerous federal, state, and local agencies have participated in waste-to-alcohol activities, but many have abandoned these efforts. Development of a coordinated state energy policy, natural resource policy, environmental policy, and economic development policy from the national level down to the local level must be a top priority if a sustainable transportation system, including one incorporating fuels produced from waste biomass, is to emerge. A directive from the executive branch to develop a statewide policy on producing renewable fuels from biomass is a key element.

ACKNOWLEDGMENTS

The authors would like to thank the generous assistance from the following people and organizations: Dr. Sharon Shoemaker, Carsten Vala, and Suanne Klahorst—California Institute of Food and Agricultural Research (CIFAR), University of California, Davis; Dr. Dennis Pendleton—Director, Public Service Research Program, UCD; Dr. Bryan Jenkins—Dept. of Biological & Agricultural Engineering, UCD; Dr. Charles Shoemaker—Food Science and Technology Department, UCD; Dr. Vashek Cervinka and Steve Shaffer—California Department of Food and Agriculture; Alan Jacobson, Loyd Forrest—TSS Consultants; Bill Smith—Stone and Webster Engineering; Neil Koehler—Parallel Products Co.; Raphael Katzen—Raphael Katzen Associates International, Inc.; Cindy Hasenjager—California Renewable Fuels Council; and Dr. Valentino Tiangco, Mary Johannis, George Simons and Susan Patterson—California Energy Commission.

REFERENCES

1. California Energy Commission (1995), *Fuels Report,* December, P300-95-017. The federal Department of Transportation, Federal Highway Administration lists 15.5 billion gallons of gasoline for 1996.
2. California Air Resources Board (1994), *Annual Report.*
3. For example, in April 1995, the first World Congress on Zero Emissions was held in Tokyo. *Chemical & Engineering News;* July 8, 1996; pp. 8–16) This conference, and the subsequent one held in Chattanooga, Tennessee the same year, illustrate that the business community will have to rethink virtually all of our industrial processes if we are to eliminate waste. It may be this effort, to find valuable products from its waste, that will encourage the private sector to convert economically and technically attractive wastes into useable and potentially valuable products such, as alcohol.
4. Foody, B., 1989, *Ethanol from Biomass, The Factors Affecting Its Commercial Feasibility;* Iogen Corporation, 400 Hunt Club Road, Ottawa, Ontario, Canada, K1G3N3
5. Foody, *Ethanol from Biomass, loc. cit.*
6. Greene, D. L., Jones, D. W., and Leiby, P. N. (June 1995), *The Outlook for US Oil Dependence,* Oak Ridge National Laboratory, ORNL-6873.
7. California Energy Commission (1994), *Fuels Report,* P300-95-017.
8. *Office of Industries/US International Trade Commission, Industry and Trade Summary: Crude Petroleum* (November 1992), USITC Publication 2578 (CH-4), P. 8.: as cited in *The Environmental Externality Costs of Petroleum,* June 1994; ENERGETICS.
9. California Air Resources Board (1994), *Motor Vehicle Fuels, Compliance Assistance Program;* pp. 100–5&6.
10. Wyman, C. E. (1995), *Ethanol Production from Lignocellulosic Biomass,* NREL.
11. California Department of Food and Agriculture (April 1997), personal communication.
12. California Energy Commission (1994), Draft *Biomass Resource Assessment Report for California.*
13. California Energy Commission (1996), *The ABCs of AFVs,* P180-96-001(figure includes 1996 model year FFVs).
14. Rask, K. N., Rask, N. *The Economic Characteristics of the US Fuel Ethanol Market,* p. 1177.
15. Shaffer, S. California Dept. of Food and Agriculture, communication May 1997.
16. California Energy Commission (1982), *Senate Bill 620: Ethanol Production and Demonstration Program - Staff Report,* P500-82-002. December 1981; and *California's Ethanol Production Demonstration Program: 1982 Report for Senate Bill 620,* P500-82-057.
17. Wyman (1995), *loc. cit.*
18. Foody and Foody (1991).
19. See recent study by Lynd, Wyman, et. al., updating production cost estimates and equilibrium economics of future ethanol plants: Lynd, Lee R. (Thayer School of Engineering, Dartmouth College, Hanover, NH 03755; Independence Biofuel, Inc., P.O. Box 163, Meriden, NH 03770), Richard T. Elander (National Renewable Energy Laboratory, Golden, CO) and Charles E. Wyman (NREL), Likely Features and Costs of Mature Biomass Ethanol Technology, in *Applied Biochemistry and Biotechnology,* Vol. 57/58, 1996.
20. Wyman (1995), *loc. cit.*
21. Jenkins, B. Baxter, L. L., Miles, Jr. T. R., and Miles, T. R., Combustion Properties of Biomass. *Biomass and Bioenergy.*
22. Katofsky, R., *The Production of Fluid Fuels from Biomass.* PU/CEES Report No. 279, Princeton University, June 1993.
23. California Energy Commission (1991), *Biomass Assessment Report,* P500-94-007, pp. 2–19.
24. Tim Taylor, Sacramento Air Quality Management District, oral communication to Bill Blackburn, CEC, May 1997.
25. California Institute of Food and Agricultural Research (1996), *CIFAR News,* University of California, Davis.

Session 5

Specialty Chemicals with Emphasis on Environmentally Benign Products and Processes

ROBERT DORSCH[1] AND NHUAN NGHIEM[2]

[1]*DuPont Co., Wilmington, DE and* [2]*Oak Ridge National Laboratory, Oak Ridge, TN*

Most industrial chemicals in use today are manufactured by petrochemical processes. These processes are very energy-intensive since they normally are operated at high temperatures and pressures. In addition, the raw materials are petroleum-based. Their uses therefore exert a strong dependence on oil supply. The need for an alternative has resulted in the development of many biological processes for the production of industrial chemicals. These processes generally are very environmentally friendly. The starting materials are sugars which are biodegradable. Any of these which are not converted in the bioreactors can easily be removed in the wastewater treatment plants. The catalysts in most cases are microorganisms. At the end of the production cycle, they can be killed in waste deactivation systems. Biological processes therefore do not release harmful materials to the environment.

The papers in this session discuss the production of various products by biological routes. The first paper concerns with the production of 1,3-propanediol from sugars by genetically engineered microorganisms. The sugars are first converted to glycerol, which subsequently is converted to the final product of interest. The second paper discusses a mathematical model which can be used to optimize an integrated process for the production of lactic acid from wheat starch. Results from the simulation and optimization studies are presented. The third paper presents a fermentation process using a recombinant Escherichia coli strain for dehydroshikimate production. Acetate production was minimized by using controlled glucose feed based on dissolved oxygen concentration measurements. The fourth paper discusses a fermentation process for the production of succinic acid from fumaric acid. In this process, a recombinant E. coli strain with amplified fumarate reductase genes is used. The next paper presents the results of studies on the production of cellulases in transgenic tobacco whole plants and cell cultures. The last paper discusses the production of ethanol from starch derived from babassu coconut, which is the fruit of a Brazilian native palm. The technical and economical aspects of the process are discussed with results obtained in a 500-liter/day industrial plant.

Production of L-Malic Acid via Biocatalysis Employing Wild-Type and Respiratory-Deficient Yeasts

Xiaohai Wang, C. S. Gong, and George T. Tsao*

Laboratory of Renewable Resources Engineering, 1295 Potter Engineering Center, Purdue University, West Lafayette, IN 47907

ABSTRACT

The yeast *Saccharomyces cerevisiae* has been used to efficiently produce L-malic acid from fumaric acid. Fumarase is responsible for the reversible conversion of fumaric and L-malic acids in the TCA cycle. To investigate the function of mitochondrial and cytoplasmic fumarase isoenzymes in L-malic acid bioconversion, a wild-type strain and a cytoplasmic respiratory-deficient mutant devoid of functional mitochondria were employed. The mutant strain, which only contained the cytoplasmic fumarase, was still functional in fumaric acid to L-malic acid bioconversion. However, its specific conversion rate was much lower (0.20 g/g·h) than that of the wild-type strain (0.55 g/g·h).

Index Entries: *Saccharomyces cerevisiae*; fumarase; respiratory deficient; L-malic acid; fumaric acid.

INTRODUCTION

L-Malic acid is an intermediate of cell metabolism that is involved in two respiratory metabolic cycles: the tricarboxylic acid cycle and the glyoxylic acid cycle. It is also the predominant acid component in apple and many other fruits. Having a greater acid taste and better taste retention than citric acid, malic acid is commonly used as a food and beverage acidulent *(1)*, and is also used in pharmaceuticals, cosmetics, metal cleaning, coatings, polymers, and resins *(1)*. Extraction from apple juice (0.4–0.7% L-malic acid) was the traditional method for L-malic acid preparation *(2)*. However, the relatively low concentrations of L-malic acid present in natural sources makes its isolation expensive and impractical. Currently, malic acid can be produced by chemical synthesis (D, L-racemate mixture) via

*Author to whom all correspondence and reprint requests should be addressed.

hydration of maleic acid or fumaric acid at elevated temperature and pressure, or by biosynthesis (L-isomer) from fumaric acid *(1–3)*.

Immobilized *Brevibacterium flavum* and *Brevibacterium ammoniagenes* have been employed for the bioconversion of L-malic acid from fumaric acid *(4,5)*. However, succinic acid was found as an undesirable byproduct. Recently, the yeast *Saccharomyces cerevisiae* has been used to efficiently convert fumaric acid to L-malic acid without producing succinic acid *(6)*.

Fumarase catalyzes the interconversion of fumaric acid and L-malic acid in the tricarboxylic acid cycle. Like most tricarboxylic acid cycle enzymes, fumarase is located in the matrix compartment of mitochondria. In the baker's yeast *S. cerevisiae*, two species of fumarase isoenzymes, with mol wt of 48,000 and 53,000, have been isolated from mitochondria and cytoplasm, respectively *(7)*. The enzymatic function of the cytoplasmic fumarase has not been established.

The respiratory deficient mutants, or petite mutants, of *S. cerevisiae* are mutants which are devoid of functional mitochondria. They can grow on standard agar medium containing glucose as energy source, producing ATP by glycolysis. When glucose is depleted from the medium, colonies stop growing, and their sizes are thus very much smaller than those of normal colonies at stationary phase. Respiratory-deficient cells are unable to grow on nonfermentable medium containing, for example, ethanol as the sole carbon and energy source *(8)*.

As a means to study the in vivo bioconversion function of mitochondrial and cytoplasmic fumarase, we have employed a wild-type *S. cerevisiae* strain that contains both isoenzymes and a corresponding cytoplasmic respiratory-deficient petite strain, which only contains the cytoplasmic fumarase in this study. Their growth patterns and capabilities in bioconversion have been examined.

MATERIALS AND METHODS

Strains

S. cerevisiae strain 43 (ATCC 42510) and its corresponding cytoplasmic respiratory-deficient mutant strain 311 (ATCC 42511) *(9)* were employed in this study.

Triphenyltetrazolium Staining Assay

The test medium consists of 2,3,5-triphenyltetrazolium chloride (1 g/L) and Bacto-agar (15 g/L, Difco, Detroit, MI) in 0.067 M phosphate buffer at pH 7.0. 2,3,5-Triphenyltetrazolium chloride is not sterilized, since it is reduced chemically by autoclaving with the agar. About 50 yeast colonies per plate were grown for 3 d on a conventional agar medium containing yeast extract (10 g/L, Difco), peptone (20 g/L, Difco), Bacto-agar (20 g/L, Difco), and glucose (10 g/L). The assay was performed by pouring 20 mL test medium at 50°C over the colony-bearing plates. Red or white colonies were scored at 3 h after overlay (10,11).

Media and Cultivation

YPD medium (12), used for cell growth, contains yeast extract (10 g/L, Difco), peptone (20 g/L, Difco), and glucose (20 g/L). SD medium contains yeast nitrogen base without amino acids (6.7 g/L, Difco), and glucose (3 g/L). SE medium contains yeast nitrogen base without amino acids (6.7 g/L, Difco), and ethanol (3 g/L). The SD and SE media were buffered at pH 5.0 with 0.05 M citrate buffer. YPF medium, used for bioconversion, contains yeast extract (10 g/L, Difco), peptone (20 g/L, Difco), Triton X-100 (1 g/L), Tween-80 (1 g/L), and fumaric acid (95 g/L) neutralized with sodium hydroxide to pH 7.0.

Batch cultures for cell growth and aerobic bioconversion were conducted in 250-mL Erlenmeyer flasks with silicone sponge closures (Sigma, St. Louis, MO) containing 20 or 30 mL medium. All flasks were incubated in an incubator shaker (model G24, New Brunswick Scientific, Edison, NJ) at 30°C and 250 rpm. Batch cultures for anaerobic bioconversion were conducted in 50-mL Erlenmeyer flasks sealed with rubber plugs containing 25 mL medium, and were incubated in the incubator shaker at 30°C and 80 rpm.

Analytical Methods

Biomass concentrations were determined from turbidimetric measurements at 600 nm and a correlation between biomass and OD 600 nm. Fumaric acid, L-malic acid, succinic acid, glucose, and ethanol were determined and quantified by HPLC, with an Intelligent Pump (Hitachi Instrument, L-6200A), an Intelligent Auto Sampler (Hitachi Instrument, AS-4000), a Bio-Rad (Hercules, CA) Aminex HPX-87H ion-exclusion column (300 × 7.8 mm), a refractive index detector (Hitachi Instrument, L-3350 RI), and a Chromato-Integrator (Hitachi Instrument, D-2500). The column temperature was maintained at 60°C, and the column was eluted with 5 mM sulfuric acid at a flow rate of 0.8 mL/min.

RESULTS AND DISCUSSION

Characterization of Respiratory—Deficient Mutant by Tetrazolium Overlay Assay

Actively respiring yeast cells can rapidly reduce 2,3,5-triphenyltetrazolium chloride to 2,3,5-triphenylformazan, which is an insoluble red pigment. The vital staining by overlaying yeast colonies with 2,3,5-triphenyltetrazolium agar can be used to discriminate respiration-competent wild-type cells and respiration-deficient mutants. The respiration-competent cells turn red, and the respiration-deficient mutants stay white (10,11).

To confirm the respiration capability or deficiency of the strains employed in the study, about 100 colonies each from strains 43 and 311 were tested by the triphenyltetrazolium staining assay as described above. All strain 43 colonies turned red, showing respiration competency; all strain 311 colonies remained white during the test period, indicating their respiration deficiency.

Growth Behavior

The growth of the wild-type strain 43 and the respiratory-deficient mutant strain 311 was studied. Two distinct media were employed: a minimal defined glucose-containing medium SD, and a minimal defined ethanol-containing medium SE. The growth was monitored by measuring OD of the cultures. Glucose and ethanol concentrations were also measured. The strains employed were first grown overnight in the SD medium inoculated from frozen stocks at $-70°C$. All batch cultures were started with 0.012 g/L biomass inoculated from the overnight SD cultures. The results are shown in Figure 1 and Figure 2 for growth in the SD and SE medium, respectively.

In the minimal glucose-containing SD medium (Fig. 1), strain 43 grew much faster than strain 311, as expected. The specific growth rates of strains 43 and 311 were 0.345/h and 0.196/h, respectively. Final biomass concentration of strain 43 reached 0.82 g/L, and was 4.8 × that of strain 311 (0.17 g/L). Glucose concentrations decreased to zero for strain 43 and to 0.13 g/L for strain 311 at 15 h. Ethanol concentrations accumulated to 0.67 g/L and 1.21 g/L at 15 h, and then dropped to zero at 25 h and to 0.97 g/L at 32.5 h for strains 43 and 311, respectively.

When the minimal medium SE, which contained ethanol as the sole carbon and energy source, was employed (Fig. 2), the growth profile of strain 43 showed a distinctive 25-h-long exponential phase that ended with the depletion of ethanol. The specific growth rate was 0.16/h and the maximal biomass concentration achieved 0.74 g/L. The strain 311, however, did not grow at all. This is consistent with the respiratory-deficient phenotype of petite mutants, and further confirms that strain 311 is deficient in respiration. The ethanol profile of strain 311, however, did not stay at the initial level. It decreased to 1.1 g/L at the end of the culture. Ethanol might be converted to other metabolites inside the cell, but it could not be used as the carbon and the energy source to support the growth of strain 311 in the SE medium.

Aerobic Conversion of Fumaric Acid to L-Malic Acid

Strains 43 and 311 were grown in the YPD medium overnight. Cells were precipitated by centrifugation. An equal amount of cell mass (3.62 g/L) of each strain was inoculated into the culture flasks containing YPF medium for the bioconversion of fumaric acid to L-malic acid. The results of aerobic bioconversion by the wild-type strain 43 and the petite strain 311

L-Malic Acid Bioconversion

Fig. 1. Growth curves, glucose and ethanol profiles of the wild-type (wt) strain 43 and the respiratory-deficient (rd) mutant strain 311 in the minimal glucose-containing SD medium: Biomass (wt) (-■-), Biomass (rd) (-□-), Glucose (wt) (-♦-), Glucose (rd) (-◇-), Ethanol (wt) (-★-), and Ethanol (rd) (-×-).

Fig. 2. Growth curves and ethanol profiles of the wild-type (wt) strain 43 and the respiratory-deficient (rd) mutant strain 311 in the minimal ethanol-containing SE medium: Biomass (wt) (-■-), Biomass (rd) (-□-), Ethanol (wt) (-★-), and Ethanol (rd) (-×-).

are shown in Fig. 3. The production of L-malic acid by the wild-type strain 43 was much faster than that by the respiratory-deficient strain 311. Final L-malic acid concentration reached 92.5 g/L in the strain 43 culture, compared to 37.8 g/L in the strain 311 culture at 60 h. Fumaric acid consumption was closely correlated to the production of L-malic acid for both strains.

Table 1
Specific and Volumetric L-Malic Acid Production Rates

	Strain 43		Strain 311	
	SP^a (g/g·h)	VP^b (g/L·h)	SP^a (g/g·h)	VP^b (g/L·h)
Aerobic^c	0.55	2.24	0.20	0.72
Anaerobic^d	0.29	1.13	0.11	0.33

^{a} SP, Average specific production rate of L-malic acid (g L-malic acid/g cells·h).
^{b} VP, Average volumetric production rate of L-malic acid (g L-malic acid/L·h).
^{c} Average rates under aerobic condition, incubation time of 34 h.
^{d} Average rates under anaerobic condition, incubation time of 25 h.

Fig. 3. Kinetic profiles of aerobic L-malic acid production, fumaric acid consumption, and biomass variation employing the wild-type (wt) strain 43 and the respiratory deficient (rd) mutant strain 311: Malic (wt) (-■-), Malic (rd) (-◆-), Fumaric (wt) (-□-), Fumaric (rd) (-◇-), Biomass (wt) (-★-), and Biomass (rd) (-∗-).

Slight variations in biomass were also observed. The specific and volumetric production rates of L-malic acid are listed in Table 1. A substantial difference in fumarase activity was implied in the rate data. Hiraga et al. (13) also reported a higher enzymatic activity for the mitochondrial fumarase. The exact reason for the observed difference is not clear at this point. It is possible that cytoplasmic fumarase represents a premature version of the enzyme, therefore less active, and needs to be further processed in functional mitochondria to yield the mature fumarase. The specific L-malic acid production rate achieved by the wild-type strain 43 is comparable with our previous results employing the laboratory *S. cerevisiae* strain SHY2 (6). It is also better than, or similar to, most of the specific production rates achieved by *B. ammoniagenes* treated with various detergents, except

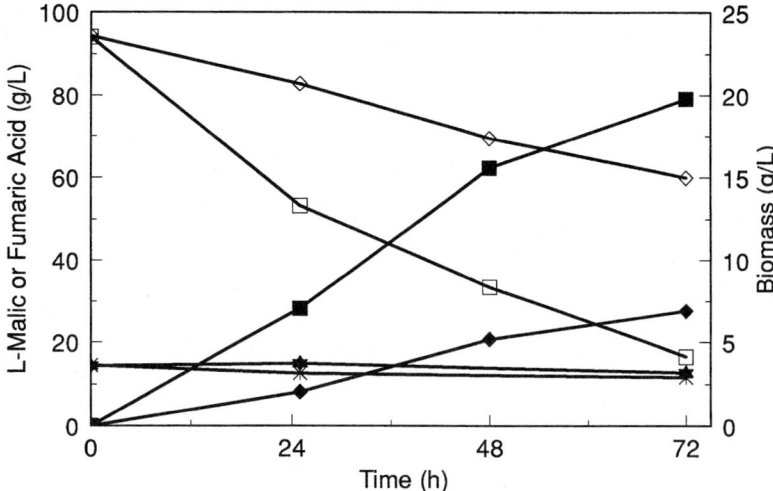

Fig. 4. Kinetic profiles of anaerobic L-malic acid production, fumaric acid consumption, and biomass variation employing the wild-type (wt) strain 43 and the respiratory deficient (rd) mutant strain 311: Malic (wt) (–■–), Malic (rd) (–◆–), Fumaric (wt) (–□–), Fumaric (rd) (–◇–), Biomass (wt) (–★–), and Biomass (rd) (–*–).

bile extract (4). No byproduct, such as succinic acid, was detected in our experiments.

Anaerobic Conversion of Fumaric Acid to L-Malic Acid

Under anaerobic conditions, the respiratory activity of the wild-type strain 43 is also repressed. To investigate any possible effect of respiration on fumarase activity and to place those two strains on the same energetic stage for fair comparison, the authors further conducted the experiment under anaerobic conditions. Strains 43 and 311 were grown and harvested as described above. The same initial biomass concentration (3.62 g/L) was also employed for each strain. The results of anaerobic bioconversion by the wild-type strain and the petite strain are shown in Figure 4. Final L-malic acid concentration achieved at 72 h by strain 43 was much higher, 79.1 g/L, than that achieved by strain 311, 27.6 g/L. Fumaric acid consumption was also closely correlated to the production of L-malic acid for both strains. A slight decrease in final biomass was observed. The specific and volumetric production rates are shown in Table 1. Compared to aerobic bioconversion, anaerobic bioconversion of fumaric acid to L-malic acid was less efficient in both cases employing wild-type and petite mutant strains. Consistent with the data from aerobic bioconversion, the production of L-malic acid by the wild-type strain 43 was also much faster than that by the respiratory-deficient strain 311 under anaerobic conditions.

CONCLUSION

The results indicate that both cytoplasmic and mitochondrial fumarase are functional in the in vivo conversion of fumarate to L-malate. The respiratory-deficient mutant strain containing only the cytoplasmic fumarase, however, appeared to be less active in the bioconversion than the wild-type strain under both aerobic and anaerobic conditions.

ACKNOWLEDGMENTS

This work was supported in part by the National Science Foundation (Grant BES-9412582), and the US Department of Agriculture (Grant 96-35500-3192).

REFERENCES

1. Blair, G. T. and DeFraties, J. J. (1995), in *Kirk-Othmer Encyclopedia of Chemical Technology*, vol. 13, Kroschwitz, J. I. and Howe-Grant, M., eds., John Wiley, New York, pp. 1063–1081.
2. Peleg, Y., Stieglitz, B., and Goldberg, I. (1988), *Appl. Microbiol. Biotechnol.* **28,** 69–75.
3. Battat, E., Peleg, Y., Bercovitz, A., Rokem, J. S., and Goldberg, I. (1991), *Biotechnol. Bioeng.* **37,** 1108–1116.
4. Yamamoto, K., Tosa, T., Yamashita, K., and Chibata, I. (1976), *Eur. J. Appl. Microbiol.* **3,** 169–183.
5. Takata, I., Yamamoto, K., Tosa, T., and Chibata, I. (1980), *Enzyme Microb. Technol.* **2,** 30–36.
6. Wang, X., Gong, C. S., and Tsao, G. T. (1996), *Biotechnol. Lett.* **18,** 1441–1446.
7. Boonyarat, D. and Doonan, S. (1988), *Int. J. Biochem.* **20,** 1125–1132.
8. Wilkie, D. (1983), in *Yeast Genetics: Fundamental and Applied Aspects*, Spencer, J. F. T., Spencer, D. M., and Smith, A. R. W., eds., Springer-Verlag, New York, pp. 255–267.
9. Heerde, E. and Radler, F. (1978), *Arch. Microbiol.* **117,** 269–276.
10. Ogur, M., St. John, R., and Nagai, S. (1957), *Science* **125,** 928–929.
11. Boker-Schmitt, E., Francisci, S., and Schweyen, R. J. (1982), *J. Bacteriol.* **151,** 303–310.
12. Sherman, F., Fink, G. R., and Hicks, J. (1986), *Methods in Yeast Genetics*, Cold Spring Harbor Laboratory, Cold Spring Harbor, NY.
13. Hiraga, K., Inoue, I., Manaka, H., and Tuboi, S. (1984), *Biochem. Int.* **9,** 455–461.

Copyright © 1998 by Humana Press Inc.
All rights of any nature whatsoever reserved.
0273-2289/98/70-72—0853$10.50

Fate of Branched-Chain Fatty Acids in Anaerobic Environment of River Sediment

H. CHUA,*,[1] W. LO[2] AND P. H. F. YU[2]

[1]*Department of Civil and Structural Engineering and* [2]*Department of Applied Biology and Chemical Technology, Hong Kong Polytechnic University, Kowloon, Hong Kong*

ABSTRACT

The fate of six different branched-chain fatty acids (BCFAs) in an anaerobic environment of a river sediment was studied in vitro by culturing enrichment consortia. The anaerobic consortium of BCFA-degrading genus degraded BCFAs with tertiary carbons through β-oxidation, followed by methanogenesis by methane-producing anaerobic bacteria. The consortium could not degrade BCFAs with quaternary carbon. Degree of branching at the alpha or beta position along the carbon chain interfered with the beta-oxidation mechanisms of the branched-chain fatty acid.

Index Entries: Anaerobic consortium; β-oxidation; branched-chain fatty acid; river sediment.

INTRODUCTION

Branched-chain fatty acids (BCFAs) were first isolated from the preen gland waxes of birds, degras, and animal fats. BCFAs were also found as the constituents of various bacteria, namely *Sarcina* sp. and *Bacillus* sp. *(1)*. These compounds are of natural lipid origins that have single-methyl substitutions at up to four separate positions along the carbon (C) chain. In recent years, increased industrial applications of synthetic BCFAs and the somewhat uncertain degradation mechanisms have drawn much attention *(2–4)*. A number of BCFAs are produced as intermediate products through degradation of certain industrial wastes or directly discharged in other industrial effluents. For instance, 2-methylbutanoic acid (2-MBuA) and 3-methylbutanoic acid (3-MBuA) are produced through anaerobic degradation of a number of common amino acids, namely leucine, isoleucine, and valine *(5)*. On the other hand, xenobiotic BCFAs, such as 2,2,-

---

*Author to whom all correspondence and reprint requests should be addressed.

dimethylpropanoic acid (2,2-DMPrA) and 2-ethylhexanoic acid (2-EHeA) are discharged in pharmaceutical wastewaters *(3,6)*.

There has been much controversy about biological degradability and mechanisms of BCFAs degradation. A number of synthetic BCFAs were believed to be persistent in microbial ecosystems *(3)*. In particular, 2,2-DMPrA and 2-EHeA in pharmaceutical wastewaters have been reported to be persistent in on-site aerobic effluent treatment facilities comprised of trickling filters, activated sludge basins, secondary clarifiers, and sludge return *(7)*. These compounds are not readily acclimatized by the microorganisms in biological processes of on-site pretreatment facilities and municipal sewage treatment works, and are discharged largely undegraded to the receiving water bodies. McInerney et al. *(8,9)* reported that an anaerobic bacterial consortium, which degraded straight-chain fatty acids up to C_8, could not degrade the branched-chain 2-MBuA. On the contrary, recent studies by Chua et al. *(10,11)*, Yap et al. *(12)*, and Jimeno et al. *(4)* showed that BCFAs were degradable in anaerobic filters under specific conditions. Although β-oxidation is widely accepted as the mechanism in biodegradation of straight-chain fatty acids *(13)*, the mechanisms in biodegradation of BCFAs are uncertain. Richardson et al. *(14)* isolated 2-MBuA-degrading cultures, composed of an obligate syntroph and methanogens, but the degradation mechanism was not described. β-oxidation was first assumed to be the mechanism of anaerobic degradation of 2-EHeA, and an anaerobic process treating a synthetic wastewater bearing 2-EHeA was mathematically modeled and verified *(15)*.

In this paper, the fate of six selected BCFAs in the anaerobic environment of a river sediment are investigated. In vitro investigation of the anaerobic microbial population, degradation mechanisms and the effect of branching on biological degradability are reported.

MATERIALS AND METHODS

Branched-Chain Fatty Acids

Six different BCFAs were separately used as the sole C source in enrichment cultures 1–6 (Table 1). 2-Ethylhexanoic acid (2-EHeA) represented BCFAs with an even number of C in the main chain and a branching at the alpha position. 2-Ethylpentanoic acid (2-EPeA) represented BCFAs with an odd number of C in the main chain and a branching at the alpha position. 3-Ethylhexanoic acid (3-EHeA) and 3-ethylpentanoic acid represented BCFAs with a branching at the beta position. All four of these BCFAs had a tertiary C. 2,2-Diethylhexanoic acid (2,2-DEHeA) represented BCFAs with two branchings at the alpha position; 3,3-diethylhexanoic acid (3,3-DEHeA) represented BCFAs with two branchings at the beta position, all of which had a quaternary C.

Table 1
Branched-Chain Fatty Acids in Enrichment Consortia

Culture number	Carbon source	Structural formula
1	2-Ethylhexanoic acid (2-EHeA)	$CH_3-CH_2-CH_2-CH_2-CH(C_2H_5)-COOH$
2	2-Ethylpentanoic acid (2-EPeA)	$CH_3-CH_2-CH_2-CH(C_2H_5)-COOH$
3	3-Ethylhexanoic acid (3-EHeA)	$CH_3-CH_2-CH_2-CH(C_2H_5)-CH_2-COOH$
4	3-Ethylpentanoic acid (3-EPeA)	$CH_3-CH_2-CH(C_2H_5)-CH_2-COOH$
5	2,2-Diethylhexanoic acid (2,2-DEHeA)	$CH_3-CH_2-CH_2-CH_2-C(C_2H_5)_2-COOH$
6	3,3-Diethylhexanoic acid (3,3-DEHeA)	$CH_3-CH_2-CH_2-C(C_2H_5)_2-CH_2-COOH$

Enrichment Consortia

A series of enrichment consortia were used to study the bacterial populations and the degradation of BCFAs in an anaerobic environment. The enrichment consortia medium was prepared in six 100-mL serum bottles, with the following formulation, in g/L: NH_4Cl, 0.0159; KH_2PO_4, 0.0037; $MgSO_4 \cdot 7H_2O$, 0.0200; $FeCl_3$, 0.0284; $MnCl_2 \cdot 2H_2O$, 0.003; $Al_2(SO_4)_3 \cdot 18H_2O$, 0.0022; $CaCl_2$, 0.0400; $CoCl_2 \cdot 6H_2O$, 0.0080; $NaSiO_3 \cdot 5H_2O$, 0.0040; H_3BO_3, 0.0040; $ZnSO_4 \cdot 7H_2O$, 0.0020; $CuSO_4 \cdot 5H_2O$, 0.0020; $(NH_4)_2MoO_4$, 0.0020; thiamine hydrogen chloride, 0.0080. Each of the six serum bottles was added with a different BCFA (analytical reagents, Fluka Chemie AG), to an initial concentration of 16 mmol/L. The BCFA acted as the sole C source in each bottle.

The bottles were then inoculated with 3 g of soft clay taken from a deep river sediment. This resulted in initial cell densities between 10^4 and 10^5 cells/mL in the culture medium. The inoculated consortia were maintained at 35°C. The entire procedure for preparing the enrichment consortia was carried out in the oxygen-free environment of an anaerobic chamber (Forma Scientific Model 1029). Redox potentials were not measured.

Bacterial Observation and Enumeration

The enrichment consortia were periodically sampled for observations of the bacterial populations by scanning electron microscopic techniques. One-mL sample from the enrichment consortia was filtered through a Nuclepore (13 mm × 0.4 μ) cellulose nitrate membrane, which was pre wetted with Triton-X 100 surfactant to ensure uniform distribution of bacteria on the membrane. The techniques were similar to that described by

Drier et al. (16). The membrane with the fixed bacterial sample was coated with a 25-nm layer of gold-palladium mixture (Joel Fine-Coat Ion Sputter Type JFC-1100) and observed with a scanning electron microscope (Joel JSM-T220A) at 10 kV accelerating voltage and × 5 000–20,000 magnification. The bacterial cell density was calculated as $X (A_1/A_2)/v$, where X was the number of cells seen on the micrograph, A_1 and A_2 were the areas of filter membrane and field of micrograph, respectively, and v was the volume of sample.

Fluorescence microscopic techniques used for bacterial identification were similar to that described by Birk (17). A Leitz Ortholux 2 microscope, with 4-Lambda Ploem Opak for incident light fluorescence excitation, 250s mirror house, and 250 lamp house were used.

Analytical Methods

The enrichment consortia were also periodically sampled for analysis of BCFA concentrations and intermediate volatile fatty acid (VFA) concentrations by a gas chromatograph (Shimadzu Model GC-14A) with a Chromosorb WAW 100/120 mesh (FFAP 15% and H_3PO_4 1%) column. A sample size of 3 μL was analyzed, and nitrogen (high-purity grade) at a flow rate of 20 mL/min was used as the carrier gas. Biogas quality was examined using a gas chromatograph (Varian Model 3300) with a 2 m Porapak Q 80/100 mesh column. A sample size of 0.5 mL was analyzed, and helium (high-purity grade) at a flow rate of 30 mL/min was used as the carrier gas.

RESULTS AND DISCUSSION

The anaerobic bacteria in enrichment consortia numbers 1–4 were similar and were composed of three morphologically distinctive species (Fig. 1): curved rods with rounded ends (0.3–2.0 μ in diameter and 1.5–5.0 μ in length); rods and filaments with distinctive truncated ends (0.3–0.8 μ in diameter and 3–15 μ in length); and cocci (0.5–1.2 μ in diameter), which autofluoresced when excited at 420 nm. The obligate syntrophic behavior between the first and third species in interspecies hydrogen transfer and the fluorescent property of the cocci, because of the presence of intracellular coenzyme F_{420} for electron transfer in the H_2 reduction of carbon dioxide, were previously described in detail by Chua et al. (2). The three species were respectively identified as the BCFA-degrading and VFA-producings *Syntrophomonas* sp., decarboxylating *Methanothrix* sp., and H_2-utilizing *Methanococcus* sp.

Figure 2 shows the concentration profiles of fatty acids in the 2-EHeA enrichment consortia. Presence of butanoic and ethanoic acids in the 2-EHeA enrichment suggested that 2-EHeA was β-oxidized to butanoic acid (butyric acid) by a cleavage between the α- and β-C along the main chain of the 2-EHeA molecule (Fig. 3). Cleavage at the ethyl side chain

Fig. 1. Scanning electron micrograph of enrichment (20KV, ×5000).

Fig. 2. Fatty acid concentration profiles in 2-EHeA enrichment: 2-EHeA (—), Butanoic acid (—■—), and Ethanoic acid (—*—).

would have otherwise produced hexanoic acid as an intermediate product. The overall 2-EHeA degradation mechanism is shown in Fig. 4. Butanoic acid produced was further β-oxidized to ethanoic acid. Stoichiometrically, degradation of each mole of 2-EHeA should have generated 2 mol of butanoic acid and 4 mol of ethanoic acid (acetic acid) as the intermediate VFAs. However, the generation of VFAs and their subsequent degradation into final products, namely methane and carbon dioxide, were in a dynamic equilibrium. These resulted in the VFA concentrations in the culture medium being lower than the theoretical values calculated from chemical

Fig. 3. Beta-oxidation of 2-EHeA.

Fig. 4. Mechanism of 2-EHeA degradation.

Fig. 5. Fatty acid concentration profiles in 2-EPeA enrichment: 2-EHeA (—), Butanoic acid (—■—), Propanoic acid (—□—), and Ethanoic acid (—✳—).

Fig. 6. Mechanism of 2-EPeA degradation.

reaction stoichiometry (Fig. 2). Because about 16 mmol/L of 2-EHeA was degraded in the enrichment culture, a maximum of only 4 mmol/L of butanoic acid and 33 mmol/L of ethanoic acid were detected in the culture medium. Methane and carbon dioxide were detected at between 3:1 and 2:1 mol ratios in the head space of the enrichment bottle.

Figure 5 shows the concentration profiles of fatty acids in the 2-EPeA enrichment consortia. The concentration profiles of 2-EPeA, butanoic, and ethanoic acids were similar to that in the 2-EHeA consortia. In addition, an accumulation of propanoic acid in the culture medium was also observed. This suggested that 2-EPeA was β-oxidized to butanoic and propanoic acids by cleavages between the α- and and β-C along the main chain of the 2-EPeA molecule (Fig. 6). Consequently, the butanoic and ethanoic

Fig. 7. Fate of branched-chain fatty acids in anaerobic environment.

acid concentrations in the culture medium were lower than that observed in 2-EHeA consortia. Butanoic acid produced was further β-oxidized to ethanoic acid; propanoic acid remained largely undergraded by the anaerobic consortium. *Syntrophomonas* sp. in the consortia could not effectively utilize propanoic acid as a C source. Stoichiometrically, degradation of 2-EPeA and generation of propanoic acid were in a one-to-one ratio, which agreed with the profiles observed in Fig. 5. As 16 mmol/L of 2-EPeA was degraded in the enrichment consortia, a maximum of about 15 mmol/L of propanoic acid was detected in the culture medium.

The degradation of 3-EHeA and 3-EPeA showed similar patterns. In the 3-EHeA culture medium, hexanoic, butanoic, and ethanoic acids were detected as the intermediate VFAs. In the 3-EPeA culture medium, pentanoic, propanoic, and ethanoic acids were detected as the intermediate VFAs. Produced propanoic acid accumulated in the culture medium without being further degraded. These observations agreed with the degradation pathway proposed in Figs. 4 and 6. The fate of the 4 BCFAs with a tertiary C in an anaerobic environment and the roles of various bacterial groups are as summarized in Fig. 7. BCFAs were β-oxidized by the *Syntrophomonas* sp., via intermediate VFAs, to ethanoic acid with concomitant H_2 production. Ethanoic acid was decarboxylated by the *Methanothrix* sp. to CH_4 and CO_2; H_2 was utilized by the *Methanococcus* sp. to reduce CO_2 to CH_4.

The maximum cell densities in enrichment consortia 1 to 4, using 2-EHeA, 2-EPeA, 3-EHeA, and 3-EPeA, respectively, as the sole C, ranged between 4.9 and 5.9×10^5 cells/mL (Table 2). These values were more than an order lower than that in the enrichment consortia on similar BCFAs and inoculated with anaerobic biofilms taken from a BCFA-degrading biofilter (2). The maximum degradation rates between 5.0 and 8.5×10^{-3} mmol/h in enrichment consortia 1 to 4 were also about an order lower

Table 2
Cell Densities and Degradation Rates

Enrichment cultures	Maximum cell density[a] ($\times 10^5$ cells/mL)	Maximum degradation rate ($\times 10^{-3}$ mmol/h)	Maximum specific degradation rate[b] ($\times 10^{-10}$ mmol/h-cell)
2-EHeA	5.0	6.5	1.335
2-EPeA	5.9	8.5	1.410
3-EHeA	4.9	5.0	1.030
3-EPeA	5.7	6.0	1.095
2,2-DEHeA	1.2	0.4	0.378
3,3-DEHeA	1.2	0.3	0.233

[a] Highest cell count observed in each enrichment consortia.
[b] Calculated based on the maximum degradation rate divided by the maximum total number of cells.

Fig. 8. Fatty acid concentration profiles in 2,2-DEHeA and 3,3-DEHeA enrichments. 2,2-DEHeA (—✻—) and 3,3-DEHeA (—□—).

than that reported by Chua et al. (2). These results agree with the findings of Jimeno et al. (4) and Richardson et al. (14) that persistent BCFAs with a tertiary C could be degraded, although at slow degradation rates, by anaerobic consortia isolated by enrichment techniques.

On the other hand, 2,2-DEHeA and 3,3-DEHeA in enrichment consortia 5 and 6 were not readily degradable by, and could hardly support, cell growth in the anaerobic consortia (Table 2). The initial acid concentrations of 16 mmol/L only decreased by less than 1.5 mmol/L during the 360-h culture (Fig. 8), which were equivalent to degradation rates that were an order lower than that in enrichment consortia 1–4.

CONCLUSIONS

In assessing the fate or biodegradability of BCFAs in the anaerobic environment of a river sediment, 2-EHeA, 2-EPeA, 3-EHeA, and 3-EPeA formed a class of persistent BCFAs. These BCFAs have alkyl substituent at the α- or β-C from the carboxylic end of the C chain, resulting in a tertiary C, thus differentiating the compounds from the natural lipid-origin anteiso fatty acids described by Smith (1), which are substituted at the antepenultimate position (third C from the alkyl end). The substituents at the α or β-positions are believed to interfere with the dehydrogenation and cleaving mechanism in β-oxidation, thus slowing down the degradation rates.

2,2-DEHeA and 3,3-DEHeA, on the other hand, form another class of multiple-branching, recalcitrant BCFAs. These recalcitrant BCFAs are different from the those isolated from preen gland waxes, which have single-methyl substitutions at up to four separate positions in the C chain (1). The recalcitrant BCFAs are substituted with two alkyl groups at the α- or β-positions, resulting in a quaternary C. The recalcitrance of these BCFAs was attributed to the presence of quarternary C, which rendered the dehydrogenation and cleavage by β-oxidation impossible.

REFERENCES

1. Smith, C. R. (1970), *Topics in Lipid Chemistry*, Gunstone, F. D., ed., Logos, London, pp. 277–368.
2. Chua, H., Yap, M. G. S., and Ng, W. J. (1996), *Water Research* **30**, 3007–3016.
3. Yap, M. G. S., Relf, R. D., and Tan, S. B. (1990), *Proc. Seminar on NUS-Industry Acheivements in R and D Collaboration*. National University of Singapore, pp. 67–71.
4. Jimeno, A., Bermudez, J. J., Canovas-Diaz, M., Manjon, A., and Iborra, J. L. (1990), *Biol. Wastes* **34**, 241–250.
5. Masey, L. K., Sokatch, J., and Conrad, R. S. (1976), *Bacteriol. Rev.* **40**, 42–54.
6. Chen, Y. F. (1993), *Masters Thesis*, National University of Singapore.
7. Ng, W. J., Yap, M. G. S., and Sivadas, M. (1989), *Biol. Wastes*. **29**, 299–311.
8. McInerney, M. J., Bryant, M. P., and Pfennig, N. (1979), *Arch. Microbiol.* **122**, 129–135.
9. McInerney, M. J., Bryant, M. P., Hespell, R. B., and Costerton, J. W. (1981), *Appl. Env. Microbiol.* **41**, 1029–1039.
10. Chua, H., Yap, M. G. S., and Ng, W. J. (1992), *Appl. Biochem. Biotechnol.* **34/35**, 789–800.
11. Chua, H. and Chen, Y. F. (1995), *Marine Pollut. Bull.* **31**, 313–316.
12. Yap, M. G. S., Ng, W. J., and Chua, H. (1992), *Bioresource Technol.* **41**, 45–51.
13. Novak, J. T. and Carlson, D. A (1970), *J. WPCF* **42**, 1932–1943.
14. Richardson, A. J., Hobson, P. N., and Campbell, G. P. (1987), *Lett. Appl. Microbiol.* **5**, 119–121.
15. Chua, H., Yap, M. G. S., and Ng, W. J. (1995), *Appl. Biochem. Biotechnol.* **51**, 705–716.
16. Drier, T. M. and Thurston, E. L. (1978), *Scanning Electron Microscopy* **11**, 843–848.
17. Birk, G. (1984), *Instrumentation and Techniques for Fluorescence Microscopy*, Wild Leitz, Sydney, Australia.

Simultaneous Enzymatic Synthesis of Gluconic Acid and Sorbitol

Continuous Process Development Using Glucose-Fructose Oxidoreductase from *Zymomonas mobilis*

MARISOL SILVA-MARTINEZ, DIETMAR HALTRICH, SENAD NOVALIC, KLAUS D. KULBE, AND BERND NIDETZKY*

Division of Biochemical Engineering, Institute of Food Technology, Universität für Bodenkultur Wien (BOKU), Muthgasse 18, A-1190 Vienna, Austria

ABSTRACT

The production of sorbitol and gluconic acid by isolated glucose-fructose oxidoreductase (GFOR) from *Zymomonas mobilis* has been studied in a convective, 100–mL loop reactor with tangential ultrafiltration. Using a dilution rate of 0.04/h and 5 kU/L GFOR, substrate conversion (3 M sugar) in a single stage was >85%, and productivities of 126 g sorbitol/(L·d) were obtained. At a constant recycle rate (3/min) and a membrane area of 50 cm^2, the dilution rates (and thus productivities) were however limited by a more than 30-fold reduction of the permeate flow in the presence of high sugar and protein concentrations (5 g/L). Protein was added, together with 10 mM dithiothreitol, to improve the stability of GFOR during substrate turnover and crossflow filtration, thus leading to a stable operation of the enzyme reactor for at least 5 d.

Index Entries: Glucose-fructose oxireductase; sorbitol; gluconic acid; *zymomonas mobilis*.

INTRODUCTION

Glucose-fructose oxidoreductase (GFOR) from *Zymomonas mobilis* (1) converts fructose to sorbitol and, simultaneously, glucose to glucono-δ-

---

*Author to whom all correspondence and reprint request should be addressed. E-mail: nide@mail.boku.ac.at

Fig. 1. Simultaneous synthesis of glucono-lactone and sorbitol by glucose-fructose oxidoreductase from *Zymomonas mobilis*. The enzyme contains tightly bound NADP(H). Glucono-lactone hydrolyzes spontaneously to gluconic acid.

lactone (Fig. 1). The enzyme has a high specific activity (250–300 U/mg) but a very low affinity for fructose (K_m approx 400–450 mM) (2–4). GFOR catalyzes two half reactions, the reduction of fructose and the oxidation of glucose, and NADP(H) serves as the nondissociable coenzyme (1–4). The glucono-δ-lactone product is degraded rapidly and irreversibly to gluconic acid, so that the reaction equilibrium lies completely on the product side (2–4).

Sorbitol and gluconic acid have various food and nonfood applications. Hence, the conversion of mixtures of glucose and fructose by GFOR is thought to be of a significant technical relevance. Cells of *Zymomonas* can be made permeable to substrates and products, whereas proteins like GFOR are retained in the cellular matrix (5,6). Such systems have been used for the continuous production of sorbitol and gluconic acid and are characterized by good operational stabilities at temperatures of up to 39°C (5–11) that lead to rapid inactivation of the isolated enzyme (12–14). The degrees of substrate conversion reported in these studies are in a range of 50–95% (5–11). The high K_m for fructose is clearly responsible for the difficulty to achieve complete substrate conversion in a single reactor, and a two-stage cascade was shown to improve not only the amount of substrate converted but also the productivities (5).

Use of isolated, soluble GFOR in high concentrations would, in principle, offer the opportunity to increase the productivities of such a biocatalytic reactor and could make a significant contribution to the production of gluconic acid and sorbitol. Unfortunately, GFOR is not very stable in the cell-free form, and even at 25–30°C stabilization of the enzyme is necessary during substrate conversion (12,13). Given that the low operational stability of soluble GFOR can be significantly improved (12,13), a continuous enzyme reactor with GFOR retained by ultrafiltration membranes could be an interesting production system. Results obtained in a single-stage laboratory enzyme reactor with dead-end ultrafiltration (flat-membrane configuration) indicated that the half-life of soluble GFOR can be extended beyond 500 h reaction time, and, even at a reaction temperature of 25°C, good productivities can be obtained (13). The aim of the work presented in this communication was to study the performance of GFOR in a reactor

Enzymatic Synthesis of Sorbitol and Gluconic Acid

with a technically more realistic recycle or loop configuration using an external tangential flow membrane module to separate the biocatalyst from products and substrates. Hence, transport data for process development and scale up can be obtained together with information concerning the stability and kinetic properties of GFOR.

MATERIALS AND METHODS

Enzymes

For the production of GFOR, *Zymomonas mobilis* DSM 473 was used *(12,14)*, and the crude cell extract with a specific GFOR activity of approx 2 U/mg was employed in all experiments without further purification. Gluconolactonase was partially purified from cell extracts of *Rhodotorula rubra* DSM 70403 by ammonium sulfate precipitation (50% saturation) followed by Sephadex-G 25 gel filtration *(14)*.

Assays/Analytical

GFOR activity was measured by a reported assay *(2,12)* using excess (10 U/mL) of gluconolactonase from *R. rubra*. One unit of activity refers to 1 μmol gluconic acid formed per min by the action of GFOR together with gluconolactonase. Sorbitol, gluconic acid, fructose, and glucose were quantitated by high-performance liquid chromatography (HPLC) *(13)*.

Continuous Conversions

Reactions were carried out at 30°C in a total volume of 100 mL, including the volume of the ultrafiltration loop (Fig. 2). The pH was constant at 6.2, and 2 M Tris or Na_2CO_3 were used for automatic pH control. Conductivity was recorded. An ultrafiltration membrane cassette with a 30-kDa cut-off and a membrane area of 50 cm^2 (mini-ultrasette; Filtron, Nortborough, MA) was employed. The recycle flow was constant at 300 mL/min (Watson Marlow, Falmouth, UK, 505 S pump), equivalent to a recirculation rate of 3/min, and the corresponding flux of permeate was measured. Substrate (1.5 M glucose, 1.5 M fructose) was fed continuously (Pharmacia, Uppsala, Sweden, P-500 pump), and the flow from the substrate tank together with the titrated alkaline component was adjusted to a variation of the permeate flux with reaction time. The substrate solution contained 10 mM dithiothreitol and 0.05% sodium azide (by weight) as a biocide. The activities of GFOR in the reactor were 5 kU/L, and the total protein concentration was adjusted to 5 g/L with bovine serum albumine (BSA). Samples (0.5 mL) from the permeate were taken in regular intervals and used for sugar analysis *(13)*. GFOR activities were measured in gelfiltered samples (NAP5 columns, Pharmacia) taken from the retentate. Substrate conversion (%) was calculated from the concentrations of sorbitol and fructose as: $100 \times [sorbitol]/\{[sorbitol] + [fructose]\}$.

Fig. 2. Schematic representation of the recycle reactor used in this study, equipped with an external ultrafiltration loop. **(A)** Batch reactor. **(B)** Ultrafiltration loop.

RESULTS AND DISCUSSION

In recent work, the authors have been studying the process properties of isolated GFOR in a continuous 50-mL enzyme reactor that had a flat membrane configuration and thus operated with dead-end filtration *(12,13)*. Whereas this system is well-suited to investigate characteristic properties of the enzymatic conversion such as kinetics and stability of the biocatalyst, transport data for scale-up cannot be obtained. A small 100-mL recycle reactor (Fig. 2) was therefore used here to gather information more directly relevant to process development. To avoid inactivation of GFOR during the course of substrate turnover *(12–14)*, the following conditions were chosen.

1. Use of Tris (2 M) *(13)* or sodium carbonate (2–3 M) for the neutralization of gluconic acid.
2. Addition of dithiothreitol (10 mM) to prevent thiol oxidations, and BSA, to give a total protein concentration of 5 g/L, that is thought to protect against aggregation of GFOR during substrate conversion *(15)*.

Tangential Ultrafiltration

Using a constant recycle flow of 300 mL/min at 30°C, the corresponding permeate flow was measured as a function of the composition of the reaction medium. Compared to water, the reduction in permeate flow was approx 12-fold when a solution of 3 M sugar was employed. In the presence of 5 g/L protein, another threefold decrease in permeate flow was observed (Fig. 3). Hence, limitations to permeate flow have to be considered as factors potentially limiting the range of applicable substrate flow rates at high concentrations of sugar and protein.

Continuous Conversions

The results of a typical conversion reaction using isolated GFOR in the recycle reactor are shown in Fig. 4. Initially, a permeate flow of

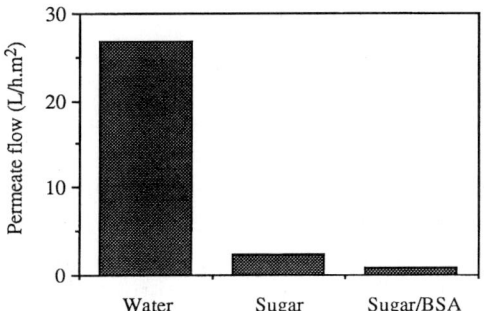

Fig. 3. Permeate flow as a function of the solute composition. The recycle flow was 300 mL/min and the corresponding permeate flow (at 30°C) was measured. The sugar concentration was 3 M (1.5 M glucose, 1.5 fructose), the total protein concentration was 5 g/L.

Fig. 4. Continuous conversion of glucose and fructose in the recycle membrane reactor. The substrate solution contained 3 M sugar, 10 mM dithiothreitol, and 20 mM sodium phosphate, pH 6.2. The reactor was operated at 30°C and contained 5 kU/L GFOR and a total protein concentration of 5 g/L. The pH was constant at 6.2, titration was carried out with 3 M sodium carbonate. The recycle flow was 300 mL/min. The permeate flow is indicated by arrows. The substrate flow is approximately half of the permeate flow because of the addition of alkali.

10 mL/h was obtained, so that the substrate feed was adjusted (5–6 mL/h) to give an average residence time of approx 10 h, taking into consideration the dosage of alkali. The permeate flow was not constant during the reaction and decreased to a lower limit of 4 mL/h after approx 20 h reaction time. Accordingly, the substrate feed was reduced because the recycle flow was set to a constant value in these experiments (300 mL/min). At an average residence time of 25 h (dilution rate of 0.04), the degree of substrate conversion was approx 85%, and the reactor performance was stable for

at least 5 d (not fully shown in Fig. 4). The resulting productivity is 126 g sorbitol/(L·d), a value that is certainly limited by the permeate flow that in turn is expected to increase significantly when a larger membrane area is used and eventually higher recycle flow rates are employed. It is important to notice that the alkaline component used for the neutralization of gluconic acid was sodium carbonate. The advantages of sodium carbonate, compared to Tris (13), are lower costs and a presumably easier separation of the number of components present in the final product mixture. In the enzyme reactor with a flat membrane configuration, carbonate was difficult to use because the formation of carbon dioxide seemed to hamper the dead-end ultrafiltration. In case of the recycle reactor (Fig. 2), the results obtained with Tris or sodium carbonate as alkaline component were identical with regard to enzyme stability, degrees of substrate conversion, and productivity. Current activities are focused on the efficient separation of a typical product mixture, obtained from a conversion experiment like shown in Fig. 4.

REFERENCES

1. Kingston, R. L., Scopes, R. K., and Baker, E. N. (1996), *Structure* **4**, 1413–1428.
2. Zachariou, M. and Scopes, R. K. (1986), *J. Bacteriol.* **167**, 863–869.
3. Hardman, M. J. and Scopes, R. K. (1988), *Eur. J. Biochem.* **173**, 203–209.
4. Hardman, M. J., Tsao, M., and Scopes, R. K. (1992), *Eur. J. Biochem.* **205**, 715–720.
5. Rehr, B., Wilhelm, C., and Sahm, H. (1991), *Appl. Microbiol. Biotechnol.* **35**, 144–148.
6. Ichikawa, Y., Kitamoto, Y., Kato, N., and Mori, N. (1988), EP 0 322 723 A2.
7. Paterson, S. L., Fane, A. G., Fell, C. J. D., Chun, U. H., and Rogers, P. L. (1988), *Biocatalysis* **1**, 217–229.
8. Roh, H. -S. and Kim, H. -S. (1992), *Enzyme Microb. Technol.* **13**, 920–924.
9. Kim, D. -M. and Kim, H. -S. (1992), *Biotechnol. Bioeng.* **39**, 336–342.
10. Chun, U. H., and Rogers, P. L. (1988), *Appl. Microbiol. Biotechnol.* **29**, 19–24.
11. Scopes, R. K., Rogers, P. L., and Leigh, D. A (1988), US Patent 4755467.
12. Gollhofer, D., Nidetzky, B., Fürlinger, M., and Kulbe, K. D. (1995), *Enzyme Microb. Technol.* **17**, 235–240.
13. Nidetzky, B., Fürlinger, M., Gollhofer, D., Scopes, R., Haltrich, D., and Kulbe, K. D. (1997), *Biotechnol. Bioeng.* **53**, 624–629.
14. Nidetzky, B., Fürlinger, M., Gollhofer, D., Haug, I., Haltrich, D., and Kulbe, K. D. (1997), *Appl. Biochem. Biotechnol.*, **63–65**, 173–188.
15. Fürlinger, M., Haltrich, D., Kulbe, K. D., and Nidetzky, B. (1998), *Eur. J. Biochem.*, in press.

Biotechnological Production of Xylitol from Agroindustrial Residues

Evaluation of Bioprocesses

Denise C. G. A. Rodrigues, Silvio S. Silva,* Arnaldo Márcio R. Prata, and Maria das Gracas A. Felipe

Departamento de Biotecnologia, Faculdade de Engenharia Quimica de Lorena, P. Box 116, Lorena, SP, 12600.000, Brazil

ABSTRACT

Batch, fed-batch, and semicontinuous fermentation processes were used for the production of xylitol from sugarcane bagasse hemicellulosic hydrolysate. The best results were achieved by the semicontinuous fermentation process: a xylitol yield of 0.79 g/g with an efficiency of 86% and a volumetric productivity of 0.66 g/L/h.

Index Entries: *Candida guilliermondii*, sugarcane bagasse hemicellulosic hydrolysate, xylitol, batch, fed-batch, semicontinuous.

INTRODUCTION

Xylitol, a five-carbon sugar alcohol, has attracted much attention as a food sweetener because of its anticariogenic and cariostatic properties *(1,2)*. It can be used for the treatment of diabetes and disorders in lipid metabolism *(3,4)*. Although it is a constituent of many fruits and vegetables, its concentration levels are low, making its extraction very uneconomical *(5)*. Xylitol is currently obtained by catalytic hydrogenation of xylose with Raney-Nickel-catalysts *(6)*. This chemical process is very costly, since it demands a very pure xylose solution. An alternative method is the microbial conversion of the xylose present in agroindustrial residues.

A good and available source of D-xylose is sugarcane bagasse *(7,8)*. This residue is abundant in Brazil and represents a low-cost raw material for fermentation processes. It can be hydrolyzed with dilute acid to obtain a mixture of fermentable sugars, xylose being the major component *(9)*. In this hydrolysis, some byproducts are generated, such as acetic acid,

---

*Author to whom all correspondence and reprint requests should be addressed.

furfural, phenolic compounds, and lignin-degradation products, which are potential inhibitors of microbial metabolism *(10,11)*. For the hydrolysate to become a suitable substrate for fermentation, these substances have to be removed by treatment with activated charcoal or cation-exchange resins *(12,13)*.

The yeast *Candida guilliermondii* FTI 20037, selected in our laboratory by Barbosa et al. *(14)*, is able to convert xylose to xylitol with high efficiency (81% of the theoretical value). The maximum theoretical yield of xylitol from D-xylose is 0.917 g/g *(14)*.

For most studies on batchwise production of xylitol by fermentation, Erlenmeyer flasks or bioreactors and synthetic medium were employed. Few reports describe the utilization of agroindustrial residues for xylitol production by biotechnological processes. This work evaluates some of these processes for obtaining xylitol from sugarcane bagasse hemicellulosic hydrolysate with a view to large-scale production.

MATERIALS AND METHODS

Hemicellulosic Hydrolysate

The hemicellulosic hydrolysate was obtained by acid hydrolysis of sugarcane bagasse, in a 250-L steel reactor under the following conditions: 121°C, 10 min reaction time and 100 mg sulfuric acid/g sugarcane bagasse (dry weight). After hydrolysis, the liquid was concentrated by heating at 70°C under vacuum, to obtain xylose at a concentration of 50–60 g/L. The hydrolysate was treated with CaO and aluminium sulfate *(15)* to minimize the inhibition of microbial metabolism.

Microorganism and Inoculum Preparation

A culture of *Candida guilliermondii* FTI 20037 was used. The cells were previously grown in a medium composed of hydrolysate supplemented with 20 g/L of rice bran extract, 0.1 g/L $CaCl_2/2H_2O$ and 5 g/L $(NH_4)_2SO_4$, in 125 mL Erlenmeyer flasks (50 mL of medium) placed on a rotatory shaker set at 200 revolutions/min at 30°C for 48 h. The initial cell concentration in all fermentations was 1.0 g/L (dry weight).

Fermentation Conditions

The fermentation medium used for obtaining the initial culture was the same described for the inoculum cultivation. The experiments were carried out in a 5-L fermenter (BIOFLO III, New Brunswick Scientific, New Brunswick, NJ) at 30°C, 300 revolutions/min, aeration of 0.4 vvm (volume of air per volume of medium per min), initial pH of 5.5.

The processes employed were batch, fed-batch, and semicontinuous. For the fed-batch fermentation a medium containing 79 g/L of xylose (supplemented with the aforementioned nutrients) was continuously fed at a rate of 28 mL/h, using a peristaltic pump (Watson Marlow 505 S). The feeding was initiated after 65 h of batch cultivation. In the semicontinuous process, 67% of the fermented medium was removed after 63 h of batch fermentation. In all cases, samples were collected at different times to analyze cell concentration, xylose, xylitol, and pH.

Analytical Methods

Xylose, glucose, and xylitol were analyzed with a Shimadzu (Kyoto, Japan) high-performance liquid chromatograph (HPLC) using a refractive index (RI) detector and a BioRad (Hercules, CA) Aminex HPX-87H (300 × 7.8 mm) column at 45°C and 0.01 N H_2SO_4 as the eluant at a flow rate of 0.6 mL/min.

Furfural and hydroxymethylfurfural were measured by high-performance liquid chromatography using an ultraviolet (UV-VIS) detector and a Hewelett Packard RP18 (200 × 4.6 mm) column under the following conditions: acetonitrile (1:8) with 1% acetic acid as the eluant, 0.8 mL/min flow rate, column temperature 25°C, wavelength 276 nm, sample volume 20 µL.

Cell concentration was estimated by measuring absorbance at 600 nm. The relationship between absorbance and dry weight (g/L) was given by a standard curve (1 OD unit = 1.55 g dry weight cells/L).

RESULTS AND DISCUSSION

The basic composition of the sugarcane bagasse hydrolysate before and after concentration is shown in Table 1. Under the conditions used, a mixture of monosaccharides containing 78% xylose, 16% glucose, and 6% arabinose was obtained. The level of acetic acid (5.5 g/L) resulting from the decomposition of acetylated sugars was lower than the level normally found in wood hydrolysates (12 g/L) *(16,17)*. This difference was because of the lignocellulosic source and the hydrolysis conditions employed *(18)*. As reported by Felipe et al. *(19)*, acetic acid at concentrations up to 6 g/L is toxic for *Candida guilliermondii*. As the authors explained, the toxic effect of acetic acid is increased by a low pH value in the medium because of the acid's entry into the cell in its nondissociated form. Acetic acid inside the cell, especially under these conditions, may induce cytoplasm acidification *(20)*.

The progress of the fermentation runs is shown in Figs. 1–3. *Candida guilliermondii* was able to grow and accumulate xylitol in all processes (Table 2).

By means of the semicontinuous process, significant increases in the xylitol concentration and productivity were obtained both in relation to the batch process (84 and 128%, respectively) and in relation to the fed-batch process (191 and 275%, respectively). Also regarding the semicontinuous process, the maximum xylitol production was 34 g/L with a yield

Table 1
Basic Composition of the Sugarcane Bagasse
Hemicellulosic Hydrolysate

Components	Hydrolysate (g/L)	
	Original	Concentrated
Glucose	5.5	8.0
Xylose	26.4	62.1
Arabinose	2.1	5.1
Acetic Acid	5.5	8.0
Furfural	<0.5	<0.1
Hydroxymethylfurfural	<0.1	<0.1

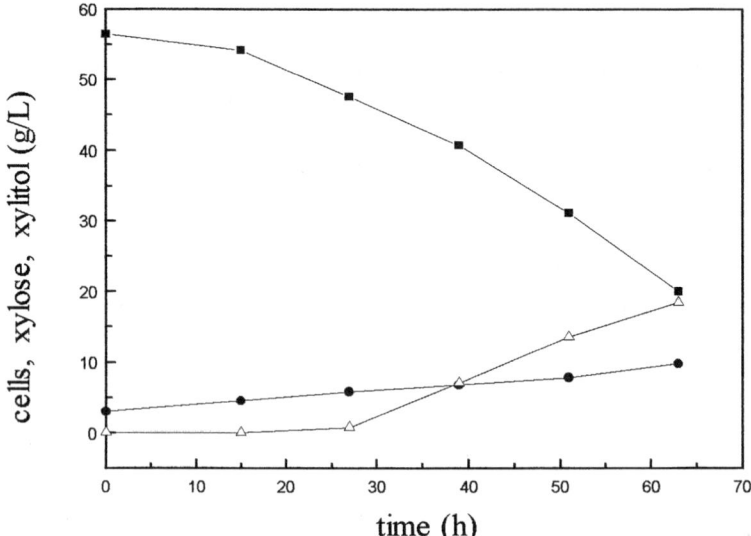

Fig. 1. Progress of the fermentation by batch process: (●) cell mass, (■) xylose, (Δ) xylitol.

of 0.79 g/g, a productivity of 0.66 g/L/h and an efficiency of 86%. Silva et al. (21), using the same yeast in a batch cultivation in synthetic medium, had achieved a xylitol yield of 0.60 g/g and a productivity of 0.55 g/L/h.

Continuous and fed-batch culture techniques often provide better yields and productivities in the production of microbial metabolites than batch culture techniques (22). In this study, however, the outcome of the fed-batch process was not any better, probably because of the fermentation conditions used, like the feeding rate.

For an effective xylitol production, the first critical step is the rapid production of cell mass in the culture medium. This could be achieved by maintaining the medium at a high level of aeration throughout the

Production of Xylitol

Fig. 2. Progress of the fermentation by semicontinuous process: (●) cell mass, (■) xylose, (Δ) xylitol (semicontinuous process started 63 h after batch).

Fig. 3. Progress of the fermentation by fed-batch process: (●) cell mass, (■) xylose, (Δ) xylitol (feeding started 65 h and finished 92 h after batch).

fermentation, but in this case, cell mass would be produced instead of xylitol (23). The semicontinuous fermentation furthers cell adaptation to the fermented medium, thereby giving higher productivity rates than the batch process.

Table 2
Fermentations of the Sugarcane Bagasse Hemicellulosic Hydrolysate to Xylitol Using Different Fermentation Processes

Process	Maximum cell concentration (g/L)	Xylose consumption (%)	Acetic acid consumption (%)	Xylitol (g/L)	$Y_{P/S}$ (g/g)	Q_P (g/L/h)	Q_X (g/g/h)
Batch	9.8	65	58	18.4	0.50	0.29	0.04
Semi-continuous	5.5	98	65	34.0	0.79	0.65	0.12
Fed-batch	1.6	84	66	23.7	0.46	0.16	0.10

$Y_{P/S}$ product/substrate yield coefficient (g/g); Q_P: volumetric productivity (g/L/h); Q_X: specific rate of xylose consumption (g/g/h).

Felipe et al. *(19)*, using a synthetic medium containing xylose, studied the effect of acetic acid on xylitol production by *Candida guilliermondii*. For a medium containing 6 g/L of acetic acid, the xylitol yield was 0.66 g/g and the productivity 0.38 g/L/h. In this work, the concentration of acetic acid in the medium was 7.2–7.7 g/L. The results in Table 2 indicate a reduction in the inhibitory effect of the acetic acid during the semicontinuous process. Both the acetic acid and the xylose were consumed by the yeasts simultaneously, as previously observed by van Zyl *et al.* (17).

The results lead to the conclusion that the hemicellulosic hydrolysate of sugarcane bagasse is a potential biomass for xylitol production by semicontinuous fermentation. Studies for the optimization and scale-up of this process shall be undertaken.

ACKNOWLEDGMENTS

The authors acknowledge the financial support of 'Fundação de Amparo à Pesquisa do Estado de São Paulo' (FAPESP), 'Conselho Nacional de Pesquisa e Desenvolvimento' (CNPq) and 'Coordenação de Aperfeiçoamento de Pessoal de Nível Superior' (CAPES) in Brazil. The authors are also grateful to Maria Eunice Machado Coelho for the revision of this paper.

REFERENCES

1. Pepper, T. and Olinger, P. M. (1988), *Food Technol.* **42,** 98–106.
2. Aguirre-Zero, O., Zero, D. T., and Proskin, H. M. (1993), *Caries Research,* **27,** 55–59.
3. Manz, U., Vanninen, E. and Voirol, F. (1973), in *Food R. A. Symp. Sugar and Replacements* 10 Oct., London.
4. Touster, O. (1974), in *Sugar in Nutrition, Siaple, H. L. and McNutt, K. W.,* eds., Academic Neurosurg, pp. 229–258.
5. Emodi, A. (1978), *Food Technol.* **30,** 28–32.
6. Melaja, A. J. and Hämäläinen, L. (1977), US Patent 4.008.285.

7. Pfeifer, M. J., Silva, S. S., Felipe, M. G. A., Roberto, I. C., and Mancilha, I. M. (1996), *Appl. Biochem. Biotechnol.*, **57/58,** 423–430.
8. Roberto, I. C., Felipe, M. G. A., Mancilha, I. M., Vitolo, M., Sato, S. and Silva, S. S. (1995), *Bioresource Technol.*, **51,** 255–257.
9. Ladisch, M. R. (1979), *Process Biochem.* 21–25.
10. Azhar, A. F., Bery, M. K., Colcord, A. R., Roberts, R. S. and Corbitt, G. V. (1981), *Biotechnol. Bioeng. Symp.* 11, 293–300.
11. Webb, S. R. and Lee, H. (1991), *Appl. Biochem. and Biotechnol.*, **30,** 325–337.
12. Dominguez, J. M., Gong, C. S., and Tsao, G. T. (1996), *Appl. Biochem. Biotechnol.* **57/58,** 49–56.
13. Gong, C. S., Chen, C. S. and Chen, L. F. (1993), *Appl. Biochem. Biotechnol.*, **39–40,** 83–88.
14. Barbosa, M. F. S., Medeiros, M. B., Mancilha, I. M., Schneider, H., and Lee, H. (1988), *J. Ind. Microbiol.*, **3,** 241–251.
15. Ramos, R. M. (1996), tese de mestrado, Universidade Federal de Viçosa, Viçosa, Brazil.
16. Tran, A. V. and Chambers, R. P. (1985), *Biotechnol. Lett.* **7(11),** 841–846.
17. van Zyl, C., Prior, B. A., and du Preez, J. C. (1988), *Appl. Biochem. Biotechnol.* **17,** 357–369.
18. Lee, Y. Y. and Mc Caskey, T. A. (1983), *Tappi J.*, **66(5),** 102–107.
19. Felipe, M. G. A., Vieira, D. C., Vitolo, M., Silva, S. S., Roberto, I. C., Mancilha, I. M. (1995), *J. Basic Microbiol.* **35,** 171–177.
20. Ferrari, M. D., Nerotti, E., Albornoz, C. and Saucedo, E. (1992), *Biotech. Bioeng.*, **40,** 753–759.
21. Silva, S. S., Roberto, I. C., Felipe, M. G. A., Mancilha, I. M. (1996), *Process Biochem.* **31 (6),** 549–553.
22. Kumar, P. K. R., Singh, A., and Schügerl, K. (1991), *Process Biochem.* **26,** 209–216.
23. Horitsu, H., Yahashi, Y., Takamizawa, K., Kawai, K., Suzuki, T., and Watanabe, N. (1992), *Biotechnol. Bioeng.* **40,** 1085–1091.

Ethanol from Babassu Coconut Starch

Technical and Economical Aspects

EDMOND A. BARUQUE FILHO,*[1,2]
MARIA DA GRAÇA A. BARUQUE,[1] DENISE M. G. FREIRE,[2]
AND GERALDO L. SANT'ANNA, JR.[2]

[1]TOBASA-Tocantins Babaçu S.A., Rua Evaristo da Veiga, 35/cj. 1710 CEP 20031-040 Rio de Janeiro, Brazil; and [2]COPPE/Universidade Federal do Rio de Janeiro

ABSTRACT

This study describes a pioneering industrial-scale experience by Tobasa in ethanol production from the amylaceous flour obtained by mechanical processing of the babassu mesocarp. Technical aspects related to enzymatic and fermentation processes, as well as overall economical aspects, are discussed. When produced in a small-size industrial plant (5000 L/d), babassu ethanol has a final cost of about \$218/m^3. The impact of raw materials, production, and processing (enzymes, steam, energy, and so on) on the final product cost is also presented. Babassu coconut ethanol can be produced at low cost, compared with traditional starchy raw materials or sugar cane. The net profitability of ethanol production is about 40% for babassu coconut and just 10% for sugar cane. If the estimated renewable babassu resources were entirely industrially used, 1 billion L/yr of ethanol could be produced, which would roughly correspond to 8% of the current Brazilian ethanol production.

Index Entries: Babassu coconut; amylaceous flour; ethanol; alcohol production.

INTRODUCTION

Babassu coconut is the fruit of a Brazilian native palm (*Orbignya phalerata* Mart.), which is found in the north of the country over a very large area (about 15 million ha) *(1)*. It is a source of fuels and chemicals, mostly lauric oil, starch for ethanol production, and charcoal. The babassu palm exploitation is still carried out on an extractivist basis, but it has a relevant

*Author to whom all correspondence and reprint requests should be addressed.

social and economical role, assuring the subsistence of about 300,000 families (2). The interest in babassu coconut as an energy source started in the 1970s, but practically all the projects in this field were discontinued. The only exception is the project conducted by Tobasa, currently in operation, and its success can be credited to the utilization of an integrated industrialization approach.

Data concerning babassu coconut productivity are controversial, but a conservative value of 2.5 metric ton/ha/yr may be taken as a representative average of the regions in which these fruits are currently exploited. Considering that only 33% of these native palms are productive, the potential productivity can be estimated as 12.4×10^6 metric ton of coconut/yr.

Babassu coconut is 80–140 mm long and consists of three layers: a fibrous external (epicarp), a fibrous-amylaceous intermediate (mesocarp), and a woody internal (endocarp) where the kernels are enclosed. The average weight contents of babassu coconut are: 12% epicarp, 23% mesocarp, 58% endocarp, and 7% kernels. Traditionally, the exploitation of babassu coconuts is oriented for oil production from kernels, wasting about 93% of the fruit biomass (3). If an integral industrial utilization is performed, a significant production of fuels and chemicals will be obtained, as indicated in Table 1, which illustrates the potential of that biomass. It can be observed that the potential for ethanol production is very high, reaching 1 billion L/yr, corresponding to 8% of the current ethanol production in Brazil. Nowadays, the internal annual consumption of lauric oils is about half of the estimated value indicated in Table 1. Thus, considering exportation of oil and the existing market for the other products, it is possible to establish an economical exploitation of babassu coconut in a large scale.

Integral coconut utilization, which is a new concept of babassu fruit processing, is based on the complete separation of its basic components (epicarp, mesocarp, endocarp, and kernels). After this industrial operation, several interesting products can be obtained by diverse processing routes, as illustrated in Fig. 1. The project implanted by Tobasa at its industrial site located at Tocantinopolis, Tocantins State, Brazil, has an integrated infrastructure from the coconut harvest, transportation, and storage until the industrial processing. Mechanical processing of the coconut fruit involves dehusking (which separates epicarp and mesocarp) and cutting of the fruits, leading to the continuous separation of the kernels and endocarp pieces. In the current project stage, only some of the products shown in Fig. 1 are industrially produced; these are lauric oil and animal feedstock (both obtained from the kernel pressing operation), primary fuel for steam generation (fibrous epicarp), charcoal and gas from the endocarp carbonization process, and amylaceous flour and ethanol from the mesocarp. The production of more sophisticated products (shown in Fig. 1) will require the development of technology and high investment costs.

This work describes an industrial process for ethanol production developed by Tobasa. This development was motivated by the significant starch content found in the coconut mesocarp (about 68%, when a manual

Ethanol from Babassu Coconut Starch

Table 1
Estimated Production Potential and Values of Babassu Coconut and Its Derived Products

Product	Product/ha/yr	Total product/yr	Unit value (US$)	Total value (US$/yr)
Coconut (t)	2.5	1.24×10^7	30	3.7×10^8
Alcohol (L)	200	1×10^9	0.42	4.2×10^8
Charcoal (t)	0.36	1.8×10^6	200	3.6×10^8
Oil (t)	0.10	5×10^5	1000	5×10^8
Gas^a (m^3)	435	2.2×10^9	0.02	4.4×10^7
Epicarp (t)	0.30	1.5×10^6	30	4.5×10^7

^{a} Gas from the carbonization process.

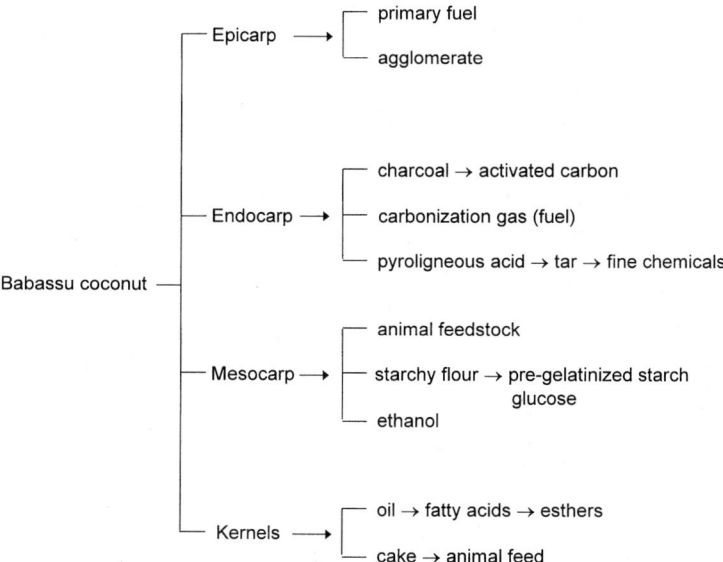

Fig. 1. Potential products obtained from the industrial processing of babassu coconut.

dehusking processing is used), and the relatively low mesocarp cost. Technical results are presented, and an economical evaluation is performed, based on an industrial-scale experience.

INDUSTRIAL PROCESS CHARACTERISTICS

Babassu Starch Characteristics

As mentioned, fruits are dehusked in an industrial machine and the following fractions are obtained: dehusked fruits, fibers (epicarp), and amylaceous flour (mesocarp). When the mechanical process is used, the per-

Table 2
Average Composition of Babassu Mesocarp
(Industrial Amylaceous Flour)

Components	Weight distribution (%)
Moisture	14.0
Starch	50.0
Protein	2.3
Fibers	10.0
Lipids	2.8
Soluble carbohydrates	1.3
Pentosans	3.4
Ash	1.3
Other components	14.8

centage of starch in the flour is about 50% (w/w) and its fiber content is around 10% (w/w). A more detailed composition of industrial babassu amylaceous flour is presented in Table 2. This flour has a brownish color because of tannins. The gelatinization temperature of the starch granules is in the range of 63–73°C, and the Brabender viscosity curves are very similar to those of corn starch. The babassu starch physicochemical properties are close to those of other common cereal starches and are very different from the properties of starches from roots and tubers (cassava, potato, and so on), as remarked by Rosenthal and Espindola (4). Because of its significant amylose content, babassu starch presents a high autoretrogradation trend, thus cooking and liquefaction steps are crucial for the saccharification process, requiring a strict control of cooking and cooling process temperatures (5).

Process Flowsheet

Figure 2 summarizes Tobasa's process for ethanol production from babassu amylaceous flour (6). First, the milled flour is mixed with room temperature water in a 3000-L capacity stirred tank, in a batch process. The resulting slurry has a solid content of 20% w/v. This slurry is supplemented with calcium hydroxide, in order to assure a necessary amount of calcium for enzyme activity, and to adjust the solution to pH 6.0. Some amount of bactericide is also added to prevent contamination. This slurry is pumped to a buffer tank, where 33% of the amount of commercial α-amylase (Termamyl 120 L, Novo Nordisk, Denmark) required for the liquefaction process, is added. This tank assures the continuous operation of the gelatinization step, which is carried out in a specially designed jet-cooker for babassu starch processing, using saturated steam. The gelatinized starch is continuously fed to a flash tank, which promotes an abrupt

Ethanol from Babassu Coconut Starch

Fig. 2. Tobasa's flowsheet for alcohol production from babassu amylaceous flour.

pressure loss, reducing the liquid temperature to 85–90°C. The enzyme α-amylase (67% of the total amount required) is added to the contents of this tank in an intermittent way. To complete the liquefaction process, an additional 6000-L tank, in series with the flash tank, is necessary. This tank has an internal refrigerating coil and an external heating jacket to keep the correct liquefaction temperature. Partial starch saccharification is conducted in a 6000-L stirred tank, where an intermittent addition of the commercial enzyme glucoamylase (AMG 200 L, Novo Nordisk) is performed. This tank maintains the liquid temperature in the range of 55–60°C. The operation pH value (4.5–4.8) is controlled by the addition of commercial hydrochloric acid. The desired degree of dextrinization is achieved in a downstream 10,000-L stirred tank. The content of this tank is continuously pumped through heat exchangers to reduce the liquid temperature to 30°C, which is used in the fermentation step. This latter process step is performed in conventional open batch vessels, using six 100,000-L capacity fermenters, which are mechanically agitated and coil-refrigerated. Finally, the product is fermented by *Saccharomyces cerevisae*, and is continuously fed to a bubble-cap-tray-distillation column.

BABASSU ETHANOL QUALITY

The characteristics of babassu ethanol are similar to those of other cereal alcohols, presenting a density of 0.78 g/mL, total acidity in the range of 3–8 mg/L, and a very pleasant smell. Table 3 compares babassu alcohol produced by Tobasa with other cereal and sugar-cane alcohols found in the Brazilian market, in terms of minor components. These chromatographic results indicate that babassu alcohol has nondetectable levels of propanol and isobutyl alcohol, in contrast with sugar-cane alcohols, which present very high levels of these components. However, it presents higher amounts of ethyl acetate and acetaldehyde, compared with commercial cereal alcohols, because the distillation step at Tobasa industrial plant is not yet completely optimized. Improvements in babassu alcohol quality are expected in the near future.

ECONOMIC ASPECTS OF BABASSU ETHANOL PRODUCTION

Figure 3 shows the contribution of itemized costs on the final product cost. Raw material is the major contribution for alcohol production cost, followed by enzymes, manpower, electricity, chemicals, steam, and mechanical maintenance. Steam generation costs are quite insignificant because of epicarp utilization as solid fuel for boilers. This is a favorable aspect for the net energy balance of the industrial plant, as remarked by Menezes (7).

The results obtained in Tobasa's industrial plant enable us to compare production costs and profitabilities for ethanol production from babassu starch, conventional amylaceous raw materials, and sugar cane. Table 4 summarizes technical and economical data concerning raw material market prices and its starch content, ethanol yield (based on an starch–ethanol production of 0.60 L of ethanol/kg of starch for all the amylaceous materials), conversion costs, processing costs (considered 30% of the ethanol price for amylaceous raw materials and 24% for sugar cane), and final production costs and profitabilities.

Table 3
Minor Components of Commercial and Babassu Coconut Alcohols

Alcohol type	Propanol (mg/L)	Isobutyl alcohol (mg/L)	Methanol (mg/L)	Acetaldehyde (mg/L)	Ethyl acetate (mg/L)
Babassu coconut (Tobasa)	nd	nd	84	51	226
Cereal (Trade mark A)[a]	nd	nd	51	23	10
Cereal (Trade mark B)[b]	20	nd	25	14	80
Sugar-cane (Trade mark B)	640	270	5	52	913
Sugar-cane (Trade mark C)[c]	110	32	8	110	1350

[a,b,c] Correspond to samples of commercial Brazilian alcohols (for ethical reasons, true trade marks were preserved). Component determination was performed by gas chromatography using the following conditions: FID detector, column temperature (75°C), detector temperature (150°C), stainless steel column-PAC 3334, and injection volume (5 µL).

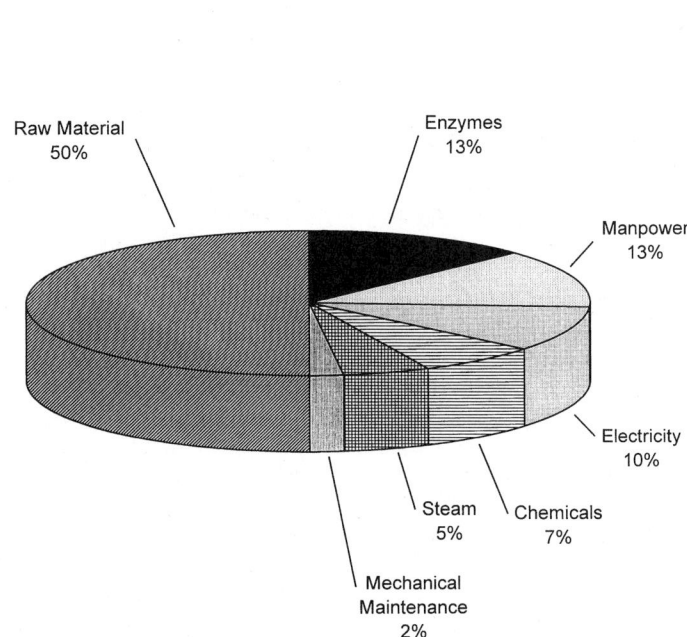

Fig. 3. Contribution of production costs on the final babassu coconut alcohol cost.

Table 4
Ethanol Production from Amylaceous Raw Materials, Babassu Coconut, and Sugar Cane: an Economical Comparative Evaluation

Parameter	Rice	Cassava	Corn	Sorghum	Babassu mesocarp	Sugar cane
FOB price (US$/metric t)[a]	85	30	120	100	32	19
Starch content (w/w%)	35	15	65	63	50	–
Ethanol yield (L/metric t)	208	90	400	370	290	80
Conversion cost (US$/m^3)[c]	408	330	300	270	109	238
Conversion cost/ethanol price (%)[b]	112	91	83	74	30	66
Processing cost (US$/m^3)[d]	109	109	109	109	109	87
Processing cost/ethanol price (%)[b]	30	30	30	30	30	24
Final production cost (US$/m^3)[e]	515	439	410	377	218	326
Final prod. cost/ethanol price (%)[b]	142	121	113	104	60	90
Profitability (US$/m^3)[f]	−152	−76	−47	−15	145	36
Profitability/ethanol price (%)[b]	−42	−21	−13	−4	40	10

[a] Raw materials prices based on current market prices of April, 1996.
[b] Ethanol established government price (free taxes): US$ 362.68/m^3 (April, 1996).
[c] The conversion cost is the result of the division of FOB price per ethanol yield.
[d] The processing cost was estimated as 30% (starchy materials) and 24% (sugar cane) of the ethanol price, based on Tobasa's experience.
[e] The final production cost is the sum of conversion and processing costs.
[f] The profitability value is the difference between the ethanol price and the final production cost.

Data from Table 4 show that ethanol production from conventional amylaceous raw materials does not present economical viability, as indicated by the negative profitability values obtained for rice, cassava, corn, and sorghum. The high profitability of ethanol production from babassu coconut is strongly linked to the pronounced starch content of its mesocarp, and the relatively low price allocated for this coconut fraction. Obviously, if the coconut fruit was purchased only for ethanol production, the profitability would become negative, as for the other starchy raw materials. Thus, only the integral utilization of the fruit allows a profitable production of alcohol. It is important to note that babassu coconut is also a source of oleaginous, proteinaceous, and carbonaceous materials, as well as fibrous material (epicarp), which, used as a primary fuel, has a major contribution for the process energy balance.

Sugar cane is considered the most competitive source for ethanol production in Brazil, mostly because of its low raw material and processing costs, compared with conventional starchy raw materials. However, as shown in Table 4, when these costs are compared with those of babassu mesocarp, a new picture is established, as conversion costs of sugar cane and babassu coconut are 66 and 30% of the alcohol price, respectively. This advantage overcomes the higher processing costs of babassu coconut. Thus, the final ethanol production cost is $326/m^3 for sugar cane and $218/m^3 for babassu coconut ethanol, resulting in profitability values of $36/m^3 and $145/m^3 respectively.

CONCLUSIONS

The production process of ethanol from babassu coconut mesocarp was developed and implanted on a small industrial scale (5,000 L ethanol/d). This process, consisting of physicochemical, enzymatic, and fermentation steps, reaches an ethanol yield of 0.60 L ethanol/kg starch, which is similar to those obtained in conventional plants processing other amylaceous materials. For babassu mesocarp, this yield corresponds to 290 L alcohol/metric ton amylaceous flour.

Alcohol production from amylaceous raw materials seldom is economically viable, because of starchy flour costs, which have increased their market prices. The production of ethanol from babassu coconut starch will be economically viable if, and only if, an integral fruit-processing approach is adopted. Babassu ethanol can be produced at a final cost of $218/m^3 which is a low value in comparison with ethanol produced from other starchy raw materials, and even sugar cane. Furthermore, it can be produced with a profitability that is significantly higher than that currently presented by sugar-cane alcohol.

The results presented in this work indicate that a rational and intensive utilization of babassu coconut can be economically performed on an industrial scale. Furthermore, babassu palms are an important native and renewable forest resource, which assures fuel and chemical production without developing new agricultural frontiers, and avoids the substitution of traditional food crops by sugar cane or other energetic crops.

ACKNOWLEDGMENTS

This work was partially supported by Conselho Nacional de Desenvolvimento Cientifico e Tecnologico (CNPq), Brazil.

REFERENCES

1. Amaral Filho, J. (1990), *A Economia Política do Babaçu*, Sioge, São Luiz.
2. Anderson, A. B., May, P. H., and Balick, M. J. (1991), *The Subsidy from Nature: Palm Forests, Peasantry and Development on an Amazon Frontier*. Columbia University Press, New York.
3. Pinheiro, C. U. B. and Frazão, J. M. F. (1995), *Econ. Botany* **49,** 31–39.
4. Rosenthal, F. and Espindola, A. C. (1975). *Revista Brasileira de Tecnologia* **6,** 307–315.
5. Rosenthal, F., Espindola, L., Nakamura, T., Nakamura, L., Ghiotti, A., Lima, M., and Silva, S. (1978). *Informativo do INT* **XI,** 19, 3–16.
6. Baruque Filho, E. A. (1996), Tocantins Babaçu S. A. (Tobasa), Technical Report, Tocantins, Brazil.
7. Menezes, T. J. B. (1982), *Proc. Biochem.* **17,** 32–35.

Biotechnological Production of Acrylic Acid from Biomass

H. Danner*, M. Ürmös, M. Gartner, and R. Braun

Institute for Agrobiotechnology Tulln (IFA-Tulln) Department for Environmental Biotechnology, Konrad Lorenz Str. 20, A-3430 Tulln, Austria

ABSTRACT

The quantitative conversion of lactic acid to acrylic acid would open a new market for renewable resources within the chemical industry. This paper focuses on the theoretical ways of producing acrylics out of renewable resources. It summarizes possible fermentation routes from carbohydrates to acrylic acid and reviews former research activities in this area. It also illustrates novel approaches involving recombinant microorganisms.

Index Entries: *Clostridium propionicum*; acrylic acid; lactic acid; biomass.

INTRODUCTION

Acrylic acid (CH_2=CH—COOH) is a commodity chemical of considerable value (1.65 ECU/kg) *(1)*. Acrylic acid and its amide and ester derivatives are principle materials in the manufacture of polymeric products. Numerous applications in surface coatings, textiles, adhesives, paper treatment, polishes, leather, fibers, detergents, and super-absorbent materials such as diapers are known *(2)*. The worldwide production of compounds associated with acrylics is estimated to be approx 2.8 million tons/yr. Currently 100% of acrylic acid is produced out of fossil oil, most of it via direct oxidation of propene *(3)*.

This project focuses on the biotechnological production of acrylic acid from organic feed stocks. A few microorganisms have been described to produce acrylic acid as a biochemical intermediate substance *(4–6)*, but observations of free acrylic acid in biological systems are rare. Anerobic formation of acrylic acid is found in the direct reduction pathway of lactic acid (CH_3—CH_2OH—COOH). This conversion is a dehydration reaction. The enzyme responsible for this conversion, lactyl-CoA dehydratase, has

---

*Author to whom all correspondence and reprint requests should be addressed. E-mail: danner@ifa1.boku.ac.at

been partially purified from *Clostridium propionicum* (7). When this microorganism uses lactic acid as the energy source, the main metabolic products are propionic acid (2/3) and acetic acid (1/3).

According to Sanseverino (8) and Akedo (9), the propionic pathway may be blocked with 3-butynoic acid. Nevertheless, acrylate concentrations never exceeded 1% of the initial substrate concentrations. These low yields are due to the enrichment of reduction equivalents like ferredoxin, rubredoxin, and flavodoxin, which inhibit further growth of cells. These reduction equivalents should be regenerated if the cells were provided with an electron acceptor. Within this study, 3-butynoic acid with and without methylenblue as an electron acceptor was investigated.

Possible Strategies for the Production of Acrylic Acid

Acrylic acid normally is not a metabolic end-product. In principle, the following strategies may be followed to obtain high acrylic acid concentrations:

A. Biotechnological production of lactic acid out of biomass, product concentration, purification, and finally chemical conversion of lactic acid to acrylic acid (*see* pathways 1 and 5 in Fig. 1). Unfortunately, this conversion gives quite low yields because of decarbonylation, decarboxylation, and condensation reactions, which mainly leads to acetaldehyde or 2,3-pentanedione (10).

B. Conversion of complex substrates to lactic acid via conventional homofermentative lactic acid fermentation and further conversion of the lactic acid fermentation broth with *Clostridium propionicum* to acrylic acid (pathways 1 and 4 in Fig. 1). This microorganism has been demonstrated to convert 3 mol lactic acid into 1 mol acetate and 2 mol propionate via the acrylyl-CoA pathway. Normally, acrylate is only produced after blocking the direct-reduction pathway. Two PhD studies dealing with the influence of 3-butynoic acid as blocker are available already (8,9). In both works, acrylic acid concentrations never exceeded 1% of the initial substrate concentration.

C. Insertion of lactyl-CoA dehydratase gene into lactic acid bacteria or into *Clostridium butyricum* (pathways 1 and 4 in Fig. 1).

D. Direct conversion of complex substrates into propionic acid with cocultures of *Lactobacilli* and for example, *Propionibacterium shermanii* with further conversion of propionate to acrylate by *Clostridium propionicum* in the presence of an electron acceptor. This approach using methylene blue is described by O'Brien et al. (1). They observed conversion rates of propionate into acrylate up to 18.5% by resting cells (pathways 2 and 6 in Fig. 1).

The proposed strategy 2, above from lactate (or L-alanine) to acrylic acid with *Clostridium propionicum* is shown in Fig. 2.

Production of Acrylic Acid from Biomass

Fig. 1. Possible fermentation routes from carbohydrates to acrylic acid. *Clostridium propionicum* is normally producing propionate from lactate via the direct reduction pathway *(3)*. Modifying this pathway enables the production of acrylic acid from lactic acid *(5)*. Further details of this modification, *see* Fig. 2.

Within this project, the investigations will focus on biotechnological routes to obtain acrylic acid.

MATERIAL AND METHODS

Clostridium propionicum DSM 1682 was obtained from the German Strain Collection. Inoculum preparation and batch experiments were done on standard medium: 1000 mL bidistilled water contain L-alanine (3.0 g), peptone (3.0 g), yeast extract (4.0 g), cysteine-hydrochloride (0.3 g), $MgSO_4 \cdot 7\,H_2O$ (0.1 g), $FeSO_4 \cdot 7\,H_2O$ (0.018 g), $1\,M$ K-PO_4-Puffer (pH 7.1) (5.0 mL), $CaSO_4$ (saturated solution, 2.5 mL), Resazurin (1 mg). Experiments with lactic acid as carbon source contained 3 mL 90% lactic acid solution instead of L-alanine. 3-butynoic acid was synthesized according to Heilbron et al. *(12)* using petrolether instead of recommended ether. Experiments were done strictly anerobic under nitrogen atmosphere in 500-mL flasks.

Analysis of lactose, glucose, galactose, ethanol, acrylate, butyrate, acetate, propionate, and lactic acid were done with HPLC (HP 1100C) using Bio-Rad (Hercules, CA) HPX-87H column and RI (HP1047 A) detector. The mobile phase was $0.01\,N\,H_2SO_4$ (flow 0.45 mL/min, temperature 55°C). Samples were diluted 1:5 with $0.01\,N\,H_2SO_4$ and centrifugated (Beckmann, Fullerton, CA, GS-15, 10 min). 5 µL of the supernatant were injected into the HPLC.

Fig. 2. Direct reduction pathway of *Clostridium propionicum* (pathway 3 of Fig. 1). Blocking the dehydrogenase and inserting a hydrogenase for regeneration of reduction equivalents (e.g., ferredoxin) should lead the microorganism to produce mainly acrylic acid (pathway 4 in Fig. 1). L-alanine undergoes oxidative deamination to pyruvate. A part of the pyruvate is oxidized to acetate and another part of pyruvate is reduced to propionate, balancing the reduction equivalents that are derived from L-alanine deamination and pyruvate oxidation *(9)*.

Protein was detected using Bio-Rad Micro Assay in microtiter plates after base hydrolysis of cell protein.

RESULTS AND DISCUSSIONS

Growth of *Clostridium propionicum* on L-alanine

As indicated in Fig. 3, growth of *C. propionicum* took place within less than 20 h after inoculation. Highest productivities were observed after

Fig. 3. Growth of *Clostridium propionicum* in batch with L-alanine as the main carbon source.

maximum growth. The ratio of propionic acid to acetic acid is between 1.5–2:1. This confirms the suggested metabolism of Fig. 2, according to which the ratio should be 2:1. Almost no acrylic acid could be detected.

Growth of *Clostridium Propionicum* on Lactic Acid

The conversion of renewable resources to acrylic acid is only feasible when cheap and easily available substrates can be used. The production of lactic acid out of various sugars and crop hydrolysates have been demonstrated frequently and may be realized at costs, which are compareable with production costs out of fossil oil. From the economical point of view, biotechnological production of acrylic acid has to be based on the conversion of lactic acid and not on alanine.

Figure 4 demonstrates that *C. propionicum* may be grown successfully on lactic acid. Preculturing of the inocula on lactic acid is necessary to obtain sufficient conversion rates and propionic and acetic acid yields. The ratio of propionate to acetate is again approx 1.5:1. Almost no formation of acrylate was observed.

3-Butynoic Acid as Blocker of Propionate Pathway

In order to achieve an accumulation of acrylic acid, the desired metabolic intermediate, a classical approach is to inhibit a certain reaction or enzyme activity with use of substrate analogs that might be metabolic inhibitors. 3-Butynoic acid (HC≡C—CH$_2$–COOH) has a structure similar

Fig. 4. Growth of *Clostridium propionicum* in batch with lactic acid as the main carbon source.

to acrylic acid. The hypothesis proposed is that this analog may inhibit the activity of propionyl-CoA dehydrogenase so that acrylyl-CoA is not further metabolized and acrylate would accumulate (9).

Figure 5 and Table 1 show the influence of 3-butynoic acid on the growth and product formation of *C. propionicum*. Different amounts of 3-butynoic acid were added after 20 h fermentation time. The formation of acetate is not influenced at all by 3-butynoic acid. The increase in protein and in propionate is significantly lower when butynoic acid is added. Increasing the concentration of 3-butynoic acid decreases the propionate formation. This indicates that 3-butynoic acid is a suitable blocker of propionyl-CoA dehydrogenase. However, no additional acrylate was observed. This is caused by the formation and enrichment of reduction equivalents as NADH and 6-OH-FAD-ETF. Only if a regeneration mechanism for these reduction equivalents is provided, acrylic acid will be accumulated.

Methylene Blue as Electron Acceptor

In alanine fermentation, methylene blue can be used as an alternative electron acceptor. This concept is supported by an observation that *C. propionicum* cells can oxidize propionate to acrylate anaerobically in the presence of methylene blue as an electron acceptor replacing oxygen (7). Although methylene blue is known as a bacteriostatic agent, no inhibition at methylene blue concentrations of 0.1 and 0.3% (wt/vol) for enterococci and lactic acid streptococci was observed (13). Whenever we added meth-

Production of Acrylic Acid from Biomass

Fig. 5. Growth of *C. propionicum* on alanine. 3-Butynoic acid was added after 20 h fermentation, graph on the top with 2.5 m*M* 3 butynoic acid, graph below with 10 m*M* butynoic acid. Both graphs represent the average values of two replicate experiments.

Table 1
Influence of 3-Butynoic Acid on Growth of C. propionicum

	Concentration of 3-Butynoic acid				
	0 mM	2.5 mM	5 mM	7.5 mM	10 mM
Protein (mg/L)	46.3	26.5	27.4	20.3	22.6
Acetate (mM/L)	0.26	0.21	0.19	0.21	0.07
Propionate (mM/L)	0.44	0.37	0.25	0.24	0.12
Acrylate (mM/L)	0.05	0.01	0.02	0.02	0.02

Protein, acetate, propionate, and acrylate indicates the amount of protein or acid formed after addition of 3-butynoic acid after 20 h fermentation time

ylene blue to a final concentration of 0.05 to 0.2% (w/v) prior to inoculation or after 20 h fermentation, no further growth or conversion of substrate could be observed.

CONCLUSION

The major problems of acrylate formation from lactic acid and from renewable resources are the regeneration of reduction equivalents (ferredoxin, NADH) and the sufficient inhibition of propionyl-CoA dehydrogenase. It was demonstrated, that 3-butynoic acid is a suitable inhibitor of propionyl-CoA dehydrogenase, although required concentrations of this blocker are quite high (>10 mM).

It also was demonstrated, that the regeneration of reduction equivalents with methylene blue as the electron acceptor is not an ideal one. Further experiments especially the insertion of a hydrogenase gene into C. propionicum are necessary and in progress.

REFERENCES

1. McCoy, M. (1996), *Chemical Marketing Reporter* September, 1996.
2. Brockinton, L., Savage, P., and Hunter D. (1986), *Chemical Week* October 15, 1986.
3. Falbe, J. and Rebitz, M. (1995), *Roempp Chemielexikon.* 9th ed. Thieme, Stuttgart.
4. Hodgson, B. and McGarry, J. D. (1968), *Biochem. J.* **107,** 7–18.
5. Ladd, J. N. and Walker, D. J. (1959), *Biochem. J.* **71,** 364–373.
6. Leaver, F. W., Wood, H. G., and Stjernholm, R. (1955), *J. Bacteriol.* **70,** 521–530.
7. Schweiger, G. and Buckel, W. (1985), *FEBS.* **185,** 253–256.
8. Sanseverino, J. (1989), Dissertation. Lehigh University.
9. Akedo, M. (1983), Biological formation of acrylic acid by *Clostridium propionicum*. Dissertation. Massachusetts Institute of Technology.
10. Gunter, G. C., Langford, R. H., Jackson, J. E., and Miller, D. J. (1995), *Ind. Eng. Chem. Res.* **34,** 974–980.
11. O'Brien, D. J., Panzer, C. C., and Eisele W. P. (1990), *Biotechnol. Prog.* **6,** 237–242.
12. Heilbron, I., Jones, E. R. H. and Sondheimer, F. (1949), *J. Chem. Soc.* 640–607.
13. Mundt, J. O. (1986), In Seath, P. H. A., Mair, N. S., Sharpe, M. E., and Holt, J. G., ed., *Bergey's Manual of Systematic Bateriology,* vol. 2, Williams & Wilkins, Baltimore, MD pp. 1063–1066.

Bacillus stearothermophilus for Thermophilic Production of L-Lactic Acid

H. Danner,* M. Neureiter, L. Madzingaidzo, M. Gartner, and R. Braun

Institute for Agrobiotechnology Tulln (IFA-Tulln), Department for Environmental Biotechnology, Konrad Lorenz Str. 20, A-3430 Tulln, Austria

ABSTRACT

A process for the continuous production of high purity L-lactic acid in a membrane bioreactor at 65°C has been developed. Two different *Bacillus stearothermophilus* strains have been tested in batch experiments. Lactic acid yields are between 60 and more than 95% of theoretical yields. The amounts of ethanol, acetate, and formate formed varied between 0 and 0.4, 0 and 0.1, and 0 and 0.5, respectively (mol/mol glucose). All byproducts are valuable and may be separated easily by rectification of the fermentation broth. Complete cell retention enables high volumetric productivity (5 g/Lh), and a minimum of growth supplements. The high temperature of 65°C allows the autoselective fermentation without problems with contamination.

Index Entries: *Bacillus stearothermophilus*; L-lactic acid; thermophilic; continuous fermentation.

INTRODUCTION

Lactic acid may be easily fermented from organic substances such as molasses, whey, or potato sap *(1)*. Although it has been called a "commodity chemical sleeping giant" *(2)*, it still has a relatively small world market of 54,500–59,000 t/yr *(1)*. At present, the global market is estimated to be growing at about 3–5% annually *(3)*, but, to gain access to large markets, conversion of lactic acid to other chemicals or polymers is required. At the end of 1996, market prices for both food and technical grade 88% lactic acid were about 1.8 US $/kg *(4)*.

According to Kharas et al. *(5)* physical and biological properties of lactic acid polymers are related to the enantiomeric purity of lactic acid stereocopolymers. The homopolymers have very regular structures and develop a crystalline phase. When copolymerized with D- or L-lactides or

---

*Author to whom all correspondence and reprint requests should be addressed. E-mail: danner@ifa1.boku.ac.at

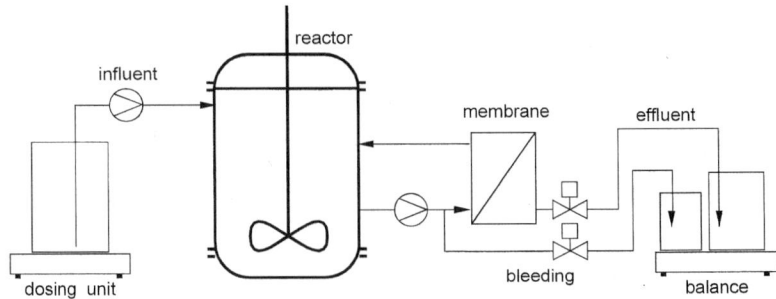

Fig. 1. Schemata of the membrane bioreactor system with cross-flow filtration unit.

lactic acid, their regular structures are interrupted, producing amorphous materials. Therefore, it is desirable to produce high-purity D- or L-lactic acid monomers.

In principle, various methods for the production of lactic acid enantiomers are described in the literature, including chemical synthesis, biotechnological fermentation processes, and enzymatic conversion of pyruvate or mixtures of D- and L-lactic acid. From an economic point of view, direct fermentation of sugars to lactic acid seems the most promising production method. Various microorganisms, including bacteria like lactobacilli, streptococci, or bacilli, and molds like *Rhizopus*, are known to produce either D- or L-lactic acid *(1)*. But fermentation processes suffer from the main disadvantage of possible contaminations. Although *Lactobacillus* fermentations can be performed at quite autoselective conditions at pH 5.0–6.0 and high temperatures of 45°C *(6)*, contamination, of other lactobacilli, which may produce both enantiomers, are possible.

To overcome the problem of contamination, *Bacillus stearothermophilus* was applied in this study. *B. stearothermophilus* is a gram-positive, endosporeforming microorganism, which is capable of growth up to 65°C. The organism may be grown anaerobically where L-lactic acid is the main product. It also grows aerobically when organic acids are further converted to CO_2. Unfortunately, *B. stearothermophilus* requires complex media constituents like yeast extract and peptone for cell growth *(7)*, which raise the production costs. The application of a membrane bioreactor (Fig. 1) should help to overcome this problem.

Membrane bioreactor systems are widely applied in the fermentation industry. Several applications of membranes in combination with bioreactors have been described recently. In all cases an increase of cell density within the bioreactor is the main purpose *(8–11)*. As a result of the high cell concentration, the volumetric productivity may be raised from less than 10 g/Lh lactic acid to more than 85 g/Lh *(12)*.

Because microbes can be recycled completely in membrane bioreactors, cell growth is not a significant parameter, the addition of cost-intensive supplements, such as yeast extract or peptone for growth, can be minimized.

MATERIAL AND METHODS

B. stearothermophilus strain IFA6 was obtained from the German strain collection; strain IFA9 is an isolate which was made available from the high school of chemistry Rosensteingasse in Vienna. Conservation of the strains was performed according to Jones et al. *(13)*, using glass beads at −70°C.

Inocula preparation, strain collection, and batch cultivation experiments were done on the standard synthetic media, consisting of the following substances: 3 g/L meat extract 5 g/L meat peptone, 4 g/L yeast extract, and 30 g/L glucose.

For continuous fermentation, two solutions were applied: solution (a) was pure sucrose from a local sugar company (50 g/L), which was prepared in 1000 L scale. Solution (a) was sterile-filtrated (Durapore CVGL71TP3 from Millipore (Bedford, MA) with pore size of 0.22 μm), and could be stored over weeks. Solution (b) was the supplement solution consisting of (40 g/L casein peptone and 32 g/L yeast extract). This solution was prepared in 10 L scale and sterilized at 120°C. All chemicals were obtained from Sigma. Membranes for cell retention were obtained by Millipore (Ceraflow MSDN 40U50, pore size 50,000 da, 3 modules, membrane surface 0.378 m^2). All the equipment was steam-sterilized at 120°C prior to use.

Analysis of lactose, glucose, galactose, ethanol, formate, and lactic acid were done with HPLC (HP 1050C) using Bio-Rad (Hercules, CA) HPX-87H column and RI (HP1047 A) detectors. The mobile phase was 0.01 N H_2SO_4 (flow 0.6 mL/min, temperature 35°C). Samples were diluted 1:5 with 0.01 N H_2SO_4 and centrifuged (Beckmann GS-15, 10 min). Five μL of the supernatant were injected into the HPLC. Determination of L- and D-lactic acid was done enzymatically according to Gawehn *(14)*.

Cell density was determined by absorbance using a Perkin-Elmer (Norwalk, CT) UV/VIS spectrometer Lambda 2S at 600 nm wavelength. On-line determinations were done using the same spectrometer, with a Perkin-Elmer flow-through cell (model 175, 10 mm light path). Cell dry wt was determined after centrifugation of 5 mL cell suspension, washing the pellet with bidistilled water, and drying at 100°C till weight was constant.

RESULTS AND DISCUSSION

Batch Experiments

Supplement Requirement

Figure 2 shows the results of a batch experiment, which was done in 20 mL flasks with varying substrate constituents. This experiment indicates

Fig. 2. Influence of media constituents on the growth of *B. stearothermophilus*.

that the amount of yeast extract in the media is the most important factor in the growth of *B. stearothermophilus* IFA9, the influence of nutrient broth, which consists of meat extract and peptone, is significant, but less effective. It also indicates that sucrose from sugarbeet is a suitable substrate, although the optical density of the fermentation broth is lower. This is probably a result of preculturing the inocula in glucose.

Growth Kinetics

Experiments for the estimation of growth parameters, such as maximum growth rate, maximum production rate, and cell or product yields, were done in 5-L reactors (Biostat ED), which were obtained from B Braun Biotech International (Germany). Absorbance (600 nm) and base consumption were detected on-line; sugar, ethanol, and organic acids were determined in intervals of 1–12 h by HPLC.

Figures 3 and 4 show typical fermentation curves of *B. stearothermophilus* IFA6 and IFA9. Prior to inoculation, the reactor was sparged with nitrogen to eliminate oxygen.

B. stearothermophilus IFA6 produces considerable amounts of byproducts such as formic acid, acetic acid, and ethanol (Table 1). The sum of produced ethanol and acetate more or less equals the amount of formate formed; the formation of ethanol is about 1.2–3.5 times more than the amount of detectable acetate. The authors assume that these variations are dependent on dissolved oxygen presented during fermentation. The higher the oxygen level, the more acetate will be formed, instead of ethanol. The formation of CO_2 under anaerobic conditions is negligible. This also confirms the proposed pathway for *B. stearothermophilus* (15).

L-Lactic Acid from B. stearothermophilus

Fig. 3. Growth of *B. stearothermophilus* IFA6 on standard synthetic media. pH was maintained at 7.2 by adding 4 N NaOH (alkali). ACR means alkali consumption rate. The calculation of μ is based on the measurements of the optical density (not shown in figure).

Fig. 4. Growth of *B. stearothermophilus* IFA9 on standard synthetic media. pH was maintained at 7.2 by adding 4 N NaOH (alkali). ACR means alkali consumption rate. The calculation of μ is based on the measurements of the absorbance at 600 nm (not shown in figure).

Table 1
Observed Yields of Growth and Metabolism of *B. stearothermophilus* Strain IFA6 and IFA9 Under Comparable Conditions

	IFA6	IFA9
Maximum growth rate μ_{max} (h^{-1})	1.36–1.54	0.863
Cell yield Y_x (g DW/g glucose)	0.023–0.025	0.030
Lactic acid produced Y_{la} (mol/mol glucose)	1.24–1.68	1.974
$Y_{la\ observed}/Y_{theoretical}$ (%)	60.5–84.0	98.7
Optical purity (% L-lactic acid/lactic acid)	99.22–99.85	99.4
Ethanol produced Y_{EtOH} (mol/mol glucose)	0.12–0.39	–
Acetate produced $Y_{acetate}$ (mol/mol glucose)	0.086–0.127	–
Formate produced $Y_{formate}$ (mol/mol glucose)	0.27–0.51	–
Maximum productivity (g lactic acid/lh)	1.2–6.24	1.20
Remaining sugar (g/L)	–	3.67

Given values are the minimum and maximum observed values of various batch experiments. DW, cell dry wt.

B. stearothermophilus IFA9 converts glucose mainly to lactic acid (98.7% of theoretical maximum yield of 2 mol lactic acid/mol glucose). Almost no formation of byproducts could be observed. Unfortunately, not all sugar is converted. The remaining sugar concentration varies between 2.5 and 4 g/L.

Strain Selection for Continuous Fermentation

As already mentioned above, in membrane bioreactor systems, growth of cells is not the key parameter, because cells are kept in the system. According to Boyaval et al. *(16)* and Ferras et al. *(8)*, total cell recycling does not seem suitable: They suggest to install a so-called bleeding of the reactor. This prevents the accumulation of dead cells and a significant decrease in specific productivity. On the other hand continuous growth of biomass is required. Hence, the demands for a suitable strain for continuous conversion of sugar to lactic acid in a membrane reactor are cell growth at high lactic acid concentrations, complete conversion of sugar, and high productivities.

To evaluate the most suitable strain, the lactic acid concentrations (alkali consumed) vs the actual productivity (alkali consumption rate) is shown in Fig. 5. From this, it follows, that IFA6 has higher productivities at higher product concentrations.

Continuous Fermentation in a Membrane Bioreactor

The results of the continuous fermentation (in a membrane bioreactor [Fig. 6]) are shown in Fig. 7. The substrate contains pure sucrose from sugar beets at a concentration of 40 g/L. After a 20 h batch phase, continuous feed

Fig. 5. Alkali (4 N NaOH) consumed vs alkali consumption rate (mL 4 N NaOH/h and 5 L reactor volume). Strain IFA6 has higher productivities at higher product concentrations.

Fig. 6. Flow scheme of the continuous fermentation in membrane bioreactor. The amounts of alkali consumption, growth supplements, substrate, and bleeding were determined by weighing machines. The reactor volume (15 L) was kept constant with a level control sensor, which was connected with the permeate valve.

of 1500 mL sucrose solution/h was started. The feeding rate of sucrose was kept constant over the experiment. Supplement addition (solution [b], consisting of 40 g/L casein peptone and 32 g/L yeast extract) was varied between 10 g/h and 175 g/h. From 20 to 90 h, supplement addition was kept at 10 g/h; from 90 to 138 h, the rate was doubled (20 g/h); and from

Fig. 7. Course of product formation and cell growth during continuous fermentation in a membrane bioreactor.

138 to 231 h, supplement addition was increased to 40 g/h. Complete conversion of sugar was achieved after raising the supplement addition to 80 g/h and stopping the sucrose feeding for 12 hs (big spike at 300 hs). Complete conversion in continuous fermentation was observed only when supplement addition was set to 175 g/h. The membrane cleaning procedure is responsible for the big dip at 550 h, when losses of biomass are inevitable.

Volumetric productivity was 5 g/Lh. This value is comparable with technical fermentation productivity in batch processes.

Byproduct formation was more or less constant over the whole fermentation process. It is not proportional to the formed lactic acid. The observed concentrations of about 40 g/L lactic acid prove that most of the sucrose is converted to lactic acid. This indicates that no contamination of thermophilic, nonlactic acid producers occurred, and that therefore fermentation under nonsterile conditions is possible.

No problems caused by sporulation were observed.

For the technical realization of the process, the involvement of the product recovery process is required. The application of electrodialysis for separation of lactic acid enables the partial reuse of growth supplements, and therefore will reduce the amounts of cost-intensive substances such as yeast extract and peptone. Bipolar electrodialysis may also lower the costs for alkali, which play a major role in the economics of the process. Further experiments are currently in progress at the institute.

CONCLUSION

The experiments described above have shown that complete conversion of sucrose to L-lactic acid is possible. The chief advantage of this process is the opportunity for continuous production of optically pure lactic acid under nonsterile conditions. The experiments in batch mode and in membrane bioreactor have proved, that growth supplements are necessary for complete conversion of sugar.

REFERENCES

1. Litchfield, J. H. (1996), *Adv. Appl. Microbiol.* **42,** 45–96.
2. Lipinsky, E. S. and Sinclair, R. G. (1986). *Chem. Eng. Prog.* **82,** 26.
3. Lerner, M. (1996), *Chem Marketing Reporter* **May 13,** 7–18.
4. McCoy, M. (1996), *Chem Marketing Reporter.* September 16, 1992.
5. Kharas, G. B., Sanchez-Riera, F., and Severson, D. K. (1994), in *Plastic from microbes*, Mobley D. P., ed., Hanser/Gardner, pp. 93–132.
6. Tyagi, R. D., Kluepfel, D., and Couillard, D. (1991), in *Bioconversion of Waste Materials to Industrial Products,* Martin A.M. (ed.), Elsevier, Essex, UK.
7. Sneath, P. H. A., Mair, N. S., Sharpe, M. E., and Holt, J. G. (1986), *Bergey's Manual of Systematic Bacteriology,* Williams & Wilkins, Baltimore.
7. Ferras, E., Minier, M., and Goma, G. (1986), *Biotechnol. Bioeng.* **28,** 523–533.
9. Taniguchi, M., Kotani, N., and Kobayashi, T. (1987), *Appl. Microbiol. Biotechnol.* **25,** 438–441.
10. Blanc, P. and Goma, G. (1987), *Bioprocess Eng.* **2,** 137–139.
11. Borgardts, P., Krischke, W., Chmiel, H., and Trosch, W. (1994), Proceedings ECB 6, in *Progress in Biotechnology* **9,** 905–908. Elsevier.
12. Mehaia, M. and Cheryan, M. (1985), *Enzyme Microb. Technol.* **8,** 289–292.
13. Jones, D., Pell, R., Sneath. R. H. A. (1991), in *Maintenance of Microorganisms and Cultured Cells. A manual of Laboratory Methods. 2nd ed.* Kirsop B. E. and Doyle A. Academic New York, 45–50.
14. Gawehn, K. (1984), in *Methods of Enzymatic Analysis* 3rd ed., vol., Bergmeyer, H. U., ed.), Verlag Chemie, Weinheim, Deerfield Beach Florida, Basel, pp. 588–592.
15. Hartley, B. S., Baghaei-Yazdi, N., Javed, M., Jackson, R. A., San Martin, R., and Leak, D. J. (1993), Straw 93' Conference, Royal Agricultural College, Cirencester, Gloucestershire, UK.
16. Boyaval, P., Corre, C., and Terre, S. (1987), *Biotechnol. Lett* **9/3,** 207–212.

In Situ Mutagenesis and Chemotactic Selection of Microorganisms in a Diffusion Gradient Chamber

MARK R. MIKOLA, MARK T. WIDMAN, R. MARK WORDEN*

Department of Chemical Engineering, Michigan State University, East Lansing, MI 48824

ABSTRACT

A new method has been developed to rapidly generate and select microbial strains having increased resistance to an inhibitory compound. The method combines *in situ* mutagenesis with use of a continuous gradient of the inhibitor to sort cells according to their resistance levels. Microbial chemotaxis is induced to accelerate the selection process. The method was used to develop a strain of *E. coli* having a feedback-resistant DAHP synthase enzyme. An unsteady-state mathematical model of the process has been developed. The model, that can reproduce key trends observed experimentally, was used to explore the effects of chemotaxis on the efficiency of the selection process.

Index Entries: Chemotaxis; selection; *E. coli*; mutagenesis; DGC.

INTRODUCTION

In many applications, it is desirable to increase a microbe's tolerance of an inhibitory substance. In biocatalysis, cells may need to function either in high concentrations of a toxic substrate or product, or under extreme conditions (e.g., low pH). Metabolic engineering involves alteration of the microbe's natural patterns of metabolic regulation so as to increase carbon flux through a desired biosynthetic pathway. Mechanisms for metabolic regulation include inhibition of DNA transcription and inhibition of enzymes. Pathway substrates, intermediates, or products are common inhibitors in such cases. As an example, the first committed step in the common aromatic biosynthetic pathway of *Escherichia coli* is the condensation between erythrose 4-phosphate (E4P) and phosphoenolpyruvate (PEP) to form 3-deoxy-D-*arabino*-heptulosonate 7-phosphate (DAHP). Feedback in-

---

*Author to whom all correspondence and reprint requests should be addressed.

hibition has been demonstrated to be the dominant regulatory mechanism in controlling carbon flow into this pathway *(1)*. The enzyme catalyzing the reaction, DAHP synthase, is strongly inhibited by the products of the pathway. The isozyme of DAHP synthase designated AroF is strongly inhibited by L-tyrosine *(2)*. Also, transcriptional repression of the gene encoding the AroF enzyme by tyrosine can lead to a 20-fold reduction in DAHP synthase activity *(3)*.

Enzyme-level inhibition may be ameliorated by changing the amino-acid sequence of the enzyme, so as to improve the enzyme's kinetic properties. Both random and site-specific mutagenesis of the structural gene for the enzyme have been used for this purpose. Transcription-level regulation may be altered by changing promoters, increasing the copy number of the structural gene, or modifying the regulatory gene *(4,5)*.

Regardless of which approach is taken, an improved strain must be selected after the genetic transformation. Selection can be a challenging task; in many instances, there is no rapid method to screen a large number of cells for improved properties in the presence of the inhibitor. The traditional, brute-force approach is to screen isolates for growth in medium containing high concentrations of the inhibitor *(6)*. One difficulty with this method is that the optimal inhibitor concentration for the screening is typically not known in advance. Too high a concentration may prevent even the best mutant from growing, whereas too low a concentration allows too many strains to grow. Consequently, experiments are generally repeated at different inhibitor concentrations. This redundancy makes the brute-force method inefficient and time intensive.

We have developed a new method that combines *in situ* mutatagenesis with selection based on each strain's relative tolerance to the inhibitor. The method uses a diffusion gradient chamber (DGC) to generate a continuous concentration gradient of the inhibitor across a slab of semisolid agarose. Motile cells can rapidly swim through the agarose into regions of having high inhibitor concentrations. Because only cells tolerant of the inhibitor can grow in these regions, this method sorts cells according to their relative tolerance to the inhibitor.

The new method also uses chemotaxis to enhance the selection efficiency. Chemotaxis is the ability of a microbe to sense a concentration gradient of a chemoattractant and migrate in the direction of the gradient. Many common microbes, including *E. coli*, are chemotactic *(7)*. By overlaying a chemoattractant gradient on the inhibitor gradient, cells can be rapidly drawn by chemotaxis into increasingly higher inhibitor concentrations, thus accelerating the sorting process.

This paper describes the new method and its successful application to isolate *E. coli* strains having a tyrosine-resistant AroF enzyme. A mathematical model of the DGC is presented and used to investigate how the interplay of diffusion and cell transport results in cell selection, and how the selection efficiency is enhanced by chemotaxis.

In Situ Mutagenesis

Fig. 1. Schematic diagram of the diffusion gradient chamber

METHODS

E. coli Strain

The *E. coli* strain was provided by John Frost. It was developed from host AB3248, which lacks all DAHP-synthase activity *(8)*, by the chromosomal addition of a single *aroF* gene. The AroF is 50% inhibited by 20 μM tyrosine *(9)* and is strongly inhibited by the nonmetabolized tyrosine analog *m*-fluorotyrosine (m-FT) *(10)*.

Diffusion Gradient Chamber

The DGC, shown schematically in Fig. 1, has been previously described *(11)*. It was fabricated by Koh Development (Ann Arbor, MI). The polycarbonate body of the DGC contains a slab-shaped layer of dilute (0.15%) agarose. The agarose prevents convective liquid movement within the DGC yet still allows cell swimming. The agarose is surrounded on four sides by liquid reservoirs. A 0.05-μm pore-size polycarbonate membrane allows exchange of small molecules between the agarose and the reservoirs. When solute exchange is not wanted, a silastic silicone sheet (Dow Corning, Midland, MI) is used in place of the membrane. The stainless-steel inlet and outlet ports of the reservoir allow fresh liquid feed to be pumped through the reservoirs to maintain a constant solute concentration. When different solute concentrations are maintained in reservoirs on opposite sides of the DGC, diffusion creates a solute concentration gradient across the agarose.

The experimental protocol was patterned after Emerson et al. *(11)*. A minimal salt medium (M63) supplemented with 5 mM glycerol and 40 mg/L of the amino acids histidine, isoleucine, proline, arginine, valine, and serine was used in the liquid reservoirs and to prepare the agarose. The *E. coli* strain was auxotrophic for these amino acids, as well as the aromatic amino acids. The glycerol served as the carbon source for growth. Two Erlenmeyer flasks, each containing 800 mL of this supplemented M63 medium, were used to provide fresh medium to two liquid reservoirs on opposite sides of the DGC. One of the flasks (designated the source flask) was also supplemented with 125 μM m-FT as an inhibitor and/or 5 mM glucose as a chemoattractant *(12)*. The other was designated the sink flask. The remaining two reservoirs were sealed off using silicone sheets. This arrangement created a one-dimensional glucose gradient that spanned 0 to 125 μM m-FT and/or 0 to 5 mM glucose from the sink reservoir to the source reservoir. The DGC, liquid-feed flasks, and agarose solution were all steam sterilized prior to use. Sterile transfer operations, such as pouring the molten agarose into the DGC, were carried out in a laminar flow hood.

To generate the gradient, liquid medium from the two Erlenmeyer flasks was continuously pumped through the source and sink reservoirs of the DGC. A flow rate of 2.5 mL/h was maintained using a dual-channel peristaltic pump (LKB Bromma Microperpex, Sweden). The effluent from the reservoir was carried by tubing to an effluent chamber. The elevation of the effluent chamber relative to the DGC had to be adjusted carefully, because it regulated the back pressure in the chamber reservoirs. Excessive back pressure causes flooding of the agarose, and insufficient back pressure causes agarose shrinkage because of liquid siphoning. Flow of liquid through the reservoirs was started 6 h before inoculation to initiate the gradients.

A starter culture of *E. coli* was grown in 5 mL of sterile LB medium from a single colony and then incubated at 37°C overnight with shaking. One mL of the starter culture was added to 100 mL of supplemented M63 medium containing 10 mM glycerol. This culture was grown (37°C and 250 rpm shaking) to stationary phase, and four 1-mL aliquots were concentrated by a factor of 16 by microcentrifugation. The center point of the DGC was inoculated with 15 μL of the 16X concentrated culture using a micropipet to disperse the cells evenly throughout the depth of the agarose as the pipet was withdrawn from the gel.

After inoculation, the DGC was stationed on a transilluminator box (TB) located in a 30°C warm room. Inside the TB, two 30-cm fluorescent lights (single 8 W, cool white bulbs) provided diffuse illumination from about a 45° angle beneath the DGC. The bottom of the TB was covered with black felt to provide a dark background. A portion of the light was diffracted by the cells toward a CCD camera (Pulnix TM-7CN, Sunnyvale, CA) mounted directly above the DGC. Consequently, regions of the agarose containing cells appeared bright against a dark background. The TB light was turned on only when an image was being recorded. The images were recorded on a dedicated personal computer using Photofinish software (Zsoft, Marietta, GA).

In Situ Mutagenesis

The removable lid of the DGC allowed mutagenesis by ultraviolet (UV) radiation to be performed *in situ*, while the cells were growing. The method for UV mutagenesis was adapted from Miller *(13)*. Conditions giving the desired survival rate (approx 0.1%) were determined by manipulating the UV-radiation exposure time. *E. coli* cells grown on M63 agar plates were exposed to the UV radiation (Sylvania germicidal 8W lamp) from a distance of 60 cm for a duration of 0 (control), 5, 10, 15, 30, 60, 80, or 100 s immediately after plating. The optimal exposure time was found to be 10 s, which resulted in a 0.6% survival rate. An identical exposure time was used to mutagenize *E. coli* cells growing in the DGC. To confirm that UV radiation would penetrate the agarose gel, the absorbance spectrum of the agarose was determined using a UV spectrophotometer. The gel did not absorb light significantly in the wavelength range of 200 to 800 nm.

Mathematical Model

The mathematical model of the DGC system developed by Widman et al. *(14)* was modified and used to predict the performance of the DGC method for strain selection. The original model consisted of two-dimensional, unsteady-state, material balance equations for a single species of cells, a nutrient (H), and two chemoattractants. One of the chemoattractants (S) is the applied gradient (e.g., glucose), and the other (Q) is oxygen. For this study, a second cell balance was added to account for both the original (A) and the mutant (B) strains. Also, a balance was added to account for diffusion and cellular uptake of the inhibitor (P) by both strains:

$$\frac{\partial P}{\partial t} = D_P \nabla^2 P - \left(\frac{v_{aP}P}{C_{aP} + P}\right)u_a - \left(\frac{v_{bP}P}{C_{bP} + P}\right)u_b \qquad (1)$$

where D_P is the diffusion coefficient of the inhibitor; v_{aP} and v_{bP} are the specific consumption coefficients for populations A and B consuming P;

and C_{aP} and C_{bP} are the saturation constants for consumption of P by populations A and B, respectively.

The chemotactic sensitivity coefficients for both populations responding to the chemoattractants S and Q, the random motility coefficients, and the maximum specific growth rates were modified to incorporate the inhibition effects. The modified chemotactic sensitivity is given by

$$\chi'_{0ij} = \frac{\chi_{0ij}}{\left[1 + \left(\frac{P}{K_{Iij\chi}}\right)\right]} \quad (2)$$

where χ'_{0ij} is the inhibited chemotactic sensitivity for the the ith population ($i = a$ or b) responding to the jth chemoattractant ($j = S$ or Q), χ_{0ij} is the uninhibited chemotactic sensitivity, and $K_{Iij\chi}$ is the inhibition constant for the chemotactic sensitivity of the ith population to the jth chemoattractant. The modified random motility coefficient is given by

$$\mu'_i = \frac{\mu_i}{\left[1 + \left(\frac{P}{K_{Ii\mu}}\right)\right]} \quad (3)$$

where μ'_i is the inhibited random motility coefficient for the ith population, μ'_i is the uninhibited random motility coefficient, and $K_{Iij\mu}$ is the inhibition constant for the random motility of the ith population. The inhibited maximum specific growth rate is given by

$$v'_{iH} = \frac{v_{iH}}{\left[1 + \left(\frac{P}{K_{Iiv}}\right)\right]} \quad (4)$$

where v'_{iH} is the inhibited maximum specific growth rate for the ith population growing on H, v_{iH} is the maximum specific growth rate, and K_{Iiv} is the inhibition constant for the specific growth rate of the ith population. In terms of the variables defined above, the two cell balance equations are given by

$$\frac{\partial u_i}{\partial t} = \mu'_i \nabla^2 u_i - \sum_{j=S}^{Q} \chi'_{0ij} \nabla \cdot \left[\left(\frac{K_{Dij}}{(K_{Dij} + j)^2}\right) u_i \nabla j\right] + \left[\frac{(v'_{iH} H)}{(C_{iH} + H)}\right] u_i \quad (5)$$

where u_i is the ith cell population (either a or b). The nutrient balance was also modified to include the inhibited maximum specific growth rates:

$$\frac{\partial H}{\partial t} = D_H \nabla^2 H \left\{ \left[\frac{v'_{aH} H}{(C_{aH} + H)} \right] \left(\frac{u_a}{Y_{aH}} \right) - \left[\frac{v'_{bH} H}{(C_{bH} + H)} \right] \left(\frac{u_b}{Y_{bH}} \right) \right\} \quad (6)$$

where D_H is the diffusion coefficient for H, C_{iH} is the half-saturation constant, and Y_{iH} is the yield coefficient. Values for the modeling constants were taken from the literature (14) insofar as possible. Values for the inhibition constants included in *equations (2–5)* were chosen that were physically realistic and gave reasonable agreement with the experimental trends.

RESULTS

Chemotaxis Toward Glucose

In this experiment, the source flask contained glucose but not m-FT. Flow of liquid was initiated through the source and sink reservoirs 6 h before the DGC was inoculated. After 72 h, there was a clear chemotactic bias in the cells' growth pattern; at that time, cell growth extended to within 6 mm of the source reservoir, but was no closer than 18 mm from the sink reservoir.

The mathematical model was able to reproduce this chemotactic bias. Fig. 2 shows a time sequence of contour plots depicting cell concentration as a function of position. The chemoattractant (glucose) gradient is indicated by gray shading, with lighter gray indicating a higher glucose concentration. The source reservoir is on the side corresponding to the top of each figure. Bias of the cell's migration in this direction is comparable to that observed experimentally.

Generation and Selection of Feedback-Resistant *E. coli* Mutant

To select for mutants resistant to tyrosine inhibition, a similar experiment was conducted using simultaneous, parallel gradients of glucose and m-FT. The glucose gradient was intended to draw the cells into increasing concentrations of m-FT. Because the growth medium contained no tyrosine, and the native AroF enzyme was strongly inhibited by m-FT, no growth was expected near the m-FT source. After inoculation, the DGC was incubated for 3 d, and the image shown in Fig. 3A was captured. (In Fig. 3, the arrows indicate the source of the glucose and m-FT. The growth pattern is indicative of severe inhibition by m-FT, with no growth occurring in the vicinity of the m-FT source. After Fig. 3A was recorded, UV mutagenesis was performed. As shown in Fig. 3B, putative feedback-resistant

Fig. 2. Simulation of the wild-type population responding to a chemoattractant gradient. No inhibitor is present. The contour lines indicate the cell population, and the shaded gray area represents the chemoattractant diffusion gradient. The whiter areas correspond to higher chemoattractant concentrations. Each graph represents a single time-point, with the earliest time at the top of the figure.

mutants migrated into regions of higher m-FT concentration 7 d after mutagenesis. Nine days after mutagenesis (Fig. 3C), the cell pattern reached the source reservoir, indicating a high level of resistance to m-FT inhibition.

Cell samples taken from regions showing growth in high m-FT concentrations were streaked onto agar plates. One of the colonies tested expressed an AroF enzyme whose activity was unaffected by tyrosine over the range of 0 to 330 µM. Details of the methods for characterizing the mutants have been given elsewhere (15).

The ability of the mathematical model to reproduce the experimental trends shown in Fig. 3 was evaluated. Figure 4 shows the predicted growth-pattern evolution in which gradients of both a chemoattractant and an inhibitor are applied, and only the population sensitive to the inhibitor is present. The effect of chemotaxis, which was significant in Fig. 2,

Fig. 3. **(A)** Cell-growth pattern in the DGC prior to mutagenesis (3 d of growth); **(B)** cell growth pattern in the DGC 7 d after mutagenesis; **(C)** cell growth pattern in the DGC 9 d after mutagenesis.

is overwhelmed by the effect of the inhibitor. As the concentration gradient of the inhibitor becomes established, cell growth occurs predominantly in the sink end of the DGC. This trend was observed experimentally in Fig. 3A.

Figure 5 is a simulation of a mutation event giving rise to a mutant (population B) that is insensitive to the inhibitor. As in Fig. 4, population A, which is sensitive to the inhibitor, grows preferentially near the sink. Population B takes longer to appear in significant concentration, because it was initiated well after population A, and at a lower concentration. The beneficial effect of chemotaxis in separating the two populations is evident, as population B moves preferentially toward the source reservoir, whereas population A remains predominantly near the sink reservoir. These trends are similar to those observed experimentally in Figs. 3B and 3C.

Figure 6 shows a simulation of the case where the mutation occurs in the middle of the growth pattern of population A, and the mutant (population B) was not chemotactic to S. In this case, the two populations were still able to be separated, but the separation is based predominantly on the influence of the inhibitor gradient and spreading caused by chemotaxis to Q. To estimate the benefit obtained by chemotaxis toward S, this simulation was then repeated with chemotaxis to S reinstated. The results, shown in Fig. 7, indicate that chemotaxis toward S does further enhance the rate of separation. Presumably, the more potent the chemoattractant (i.e., the higher the χ_{Oij} value) the greater would be the enhancement.

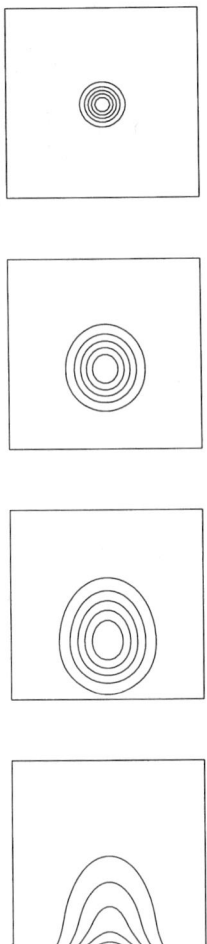

Fig. 4. Simulation of the wild-type population responding to a chemoattractant gradient and an inhibitor gradient. The contour lines indicate the cell population.

DISCUSSION

This new method for *in situ* mutagenesis and strain selection offers several advantages over traditional approaches. First, it allows different strains to be physically sorted according to their maximum tolerances of the inhibitor. The mutant having the highest tolerance would be able to grow closest to the inhibitor source, and could thus be easily selected. In the traditional methods, a fixed inhibitor concentration is used. However, the optimal inhibitor concentration for the screening is not known in advance, so some experiments would be expected to give too many isolates and others none at all.

Second, the method uses chemotaxis to draw cells toward the source of the inhibitor, thus accelerating the sorting process. Although, cell move-

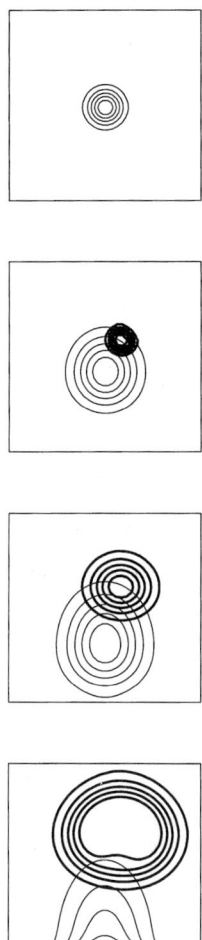

Fig. 5. Simulation of the experiment shown in Fig. 3. Simultaneous gradients of glucose and m-FT are diffusing in from the top. The wild-type strain, represented by the thin contour lines, is initially present. The mutant strain, represented by the thick contour lines, appears later near the upper right hand edge of the wild-type growth pattern.

ment by random motility could eventually result in cell sorting in the presence of the inhibitor gradient, cellular fluxes caused by chemotaxis have been shown to be orders of magnitude higher than those caused by random motility (14). The mathematical model confirmed the beneficial effect of chemotaxis in selection.

Third, because the mutagenesis is performed *in situ* in a nondestructive fashion, multiple rounds of mutagenesis, or even continuous mutagenesis, may be used. As the cells become more tolerant of the inhibitor, they migrate into regions having higher concentrations. In this way, the selection pressure is automatically increased to continue challenging mutants

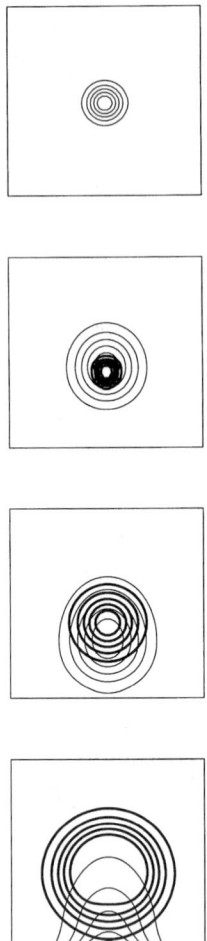

Fig. 6. Same as Fig. 5, except the mutant strain appears in the carbon-source depleted region in the center of the wild-type population, and there is no chemotaxis toward glucose.

as the evolution process continues. As a result, increasingly more tolerant strains could be developed rapidly over time with minimal additional effort.

This new method was developed because a more traditional method consisting of chemical mutagenesis followed by selection on agar plates containing m-FT failed to produce an *E. coli* mutant with a feedback-resistant AroF enzyme. The new method succeeded on the first attempt. To investigate the reproducibility of the method, the process was repeated, and was successful again. The method was found to be reasonably simple to implement, and should thus find numerous applications where enhanced tolerance to an environmental challenge is desired. Examples include metabolic engineering, in which reactants, metabolic intermediates,

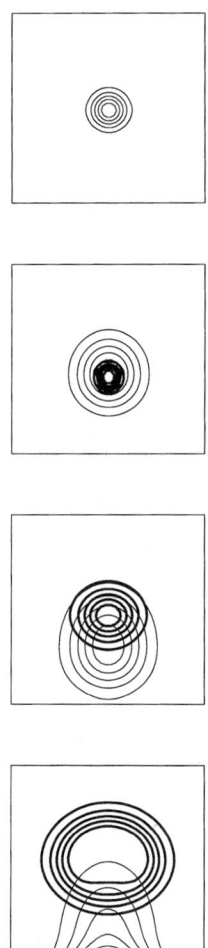

Fig. 7. Same as Fig. 6, except chemotaxis toward glucose has been reinstated.

or products may be inhibitory; environmental engineering, in which tolerance to toxic chemicals is needed; and biocatalysis using extremophiles, in which cell survival under unusual conditions (e.g., high or low pH) is desirable.

ACKNOWLEDGMENTS

We thank John Frost, Karen Draths, and Kai Li for providing the *E. coli* strain and for discussions on mutagenesis methods. Koh Development, provided the diffusion gradient chambers and accessories. Financial support for the research was provided by the Environmental Protection Agency and the MSU Crop and Food Bioprocessing Center/Research Excellence Fund.

REFERENCES

1. Herrmann, K. M. (1983), in *Amino Acids: Biosynthesis and Genetic Regulation*, Herrmann, K. M. and Somerville, R. L., eds., Addison-Wesley, Reading, MA, p. 310.
2. Herrmann, K. M. (1983), in *Amino Acids: Biosynthesis and Genetic Regulation*, Herrmann, K. M. and Somerville, R. L., eds., Addison-Wesley, Reading, MA, p. 305.
3. Cui, J. and Sommerville, R. L. (1993), *J. Bacteriol.* **175**, 303.
4. Lewin, B. (1994), *Genes V* 414.
5. Herrmann, K. M. (1983), in *Amino Acids: Biosynthesis and Genetic Regulation*, Herrmann, K. M. and Somerville, R. L., eds., Addison-Wesley, Reading, MA, p. 309.
6. Elander, R. P. and S. J. Chang (1991), in *Recombinant DNA Technology and Applications*, Prokop, A., Bajpai, R. K., and Ho, C., ed., McGraw-Hill, New York, p. 155.
7. Macnab, R. M. (1987), in *Escherichia coli and Salmonella typhimurium: Cellular and Molecular Biology*, vol. 1, Neidhardt, F. C., Ingraham, J. L., Low, K. B., Magasanik, B., Schaechter, M., and Umbarger, H. E., eds., American Society for Microbiology, Washington, DC, pp. 732–759.
8. Pittard, J. and Wallace B. J. (1966), *J. Bacteriol.* **4**, 1494.
9. Herrmann, K. M. (1983), in *Amino Acids: Biosynthesis and Genetic Regulation*, Herrmann, K. M. and Somerville, R. L., eds., Addison-Wesley, Reading, MA, p. 307.
10. Weaver, L. M. and Herrmann, K. M. (1990), *J. Bacteriol.* **172**, 6581–6584.
11. Emerson, D., Worden, R. M., and Breznak, J. A., (1994), *Appl. Environ. Microbiol.* **60**, 1269–1278.
12. Mesibov, R., Ordal, G. W., and Adler, J. (1973), *J. Gen. Physiol.* **62**, 203.
13. Miller, J. H. (1972), *Experiments in Molecular Genetics*, Cold Spring Harbor Laboratory Press, Plainview, NY.
14. Widman, M. T., Emerson, D., Chiu, C. C., and Worden, R. M. (1997), *Biotechnol. Bioeng.* **55**, 192–205.
15. Mikola, M. R. (1996), M. S. Thesis, Michigan State University.

Bioconversion of Fumaric Acid to Succinic Acid by Recombinant *E. coli*

XIAOHAI WANG, C. S. GONG, AND GEORGE T. TSAO*

Laboratory of Renewable Resources Engineering, 1295 Potter Engineering Center, Purdue University, West Lafayette, IN 47907

ABSTRACT

Succinic acid was produced efficiently from fumaric acid by a recombinant *E. coli* strain DH5α/pGC1002 containing multicopy fumarate reductase genes. The effects of initial fumaric acid and glucose concentration on the production of succinic acid were investigated. Succinic acid reached 41 to over 60 g/L in 48.5 h starting with 50 to 64 g/L fumaric acid. Significant substrate inhibition was observed at initial fumaric acid concentration of 90 g/L. L-Malic acid became the major fermentation product under these conditions. Provision of glucose (5–30 g/L) to the fermentation medium stimulated the initial succinic acid production rate over two folds.

Index Entries: Recombinant *E. coli*; fumarate reductase; succinic acid; fumaric acid.

INTRODUCTION

Succinic acid is an intermediate of cellular metabolism. It has wide applications in agriculture, food, medicine, cosmetics, and polymer synthesis *(1)*. It is currently produced by chemical processes. Production of succinic acid through fermentation represents an alternative synthesis route via the utilization of renewable feedstocks. A strict anaerobe *Anaerobiospirillum succiniciproducens* has been employed to ferment glucose to a mixture of succinic acid and acetic acid *(2,3)*. Metabolic engineering methods have been used to create recombinant *E. coli* for enhanced succinic acid production from glucose by overexpression of the enzyme phosphoenolpyruvate carboxylase *(4)*. In addition to glucose, fumaric acid has been used as a substrate for the synthesis of succinic acid by recombinant *E. coli* strains with amplified fumarate reductase *(5)*. Although *A. succiniciproducens* is a good succinic acid producer, the operation of fermentation pro-

*Author to whom all correspondence and reprint requests should be addressed.

cesses employing this strict anaerobe is not as easy as that employing the facultative organism *E. coli*. Since fumaric acid has been produced very efficiently from glucose at a weight yield of 85% *(6)*, we further explore bioconversion of fumaric acid to succinic acid as an alternative production method.

Fumarate reductase of *E. coli* is a complex flavoprotein enzyme bound to the cytoplasmic membrane *(7)*. It is composed of four nonidentical polypeptides A, B, C, and D. The 66-kDa A subunit contains a covalently bound flavin adenine dinucleotide prosthetic group. The 27-kDa B subunit contains catalytic iron-sulfur centers. The 15-kDa C and the 13-kDa D subunits bind the catalytic AB subunits to the inner surface of the cytoplasmic membrane. Fumarate reductase catalyzes the reduction of fumarate to succinate and is a key enzyme under anaerobic conditions when fumarate is the terminal electron acceptor *(8)*. It is structurally similar to, but functionally distinct from, the tricarboxylic-acid-cycle enzyme succinate dehydrogenase. The two enzymes catalyze the interconversion of fumarate and succinate with different substrate affinities and reaction rates *(9)*. The expression of fumarate reductase gene is induced, whereas that of the succinate dehydrogenase gene is repressed under anaerobic conditions *(8)*. The four genes *(frdABCD)* that code for the corresponding four fumarate reductase subunits have been cloned on a plasmid pGC1002 *(10)*. Total reductase activity was measured to be 31.2 U (μmol succinate oxidized/min at 38°C per 50 mg of cell protein) in a pGC1002-containing *E. coli* strain, whereas that of the host strain alone was only 8.1 U *(10)*.

Recombinant *E. coli* strains with amplified fumarate reductase activity have been shown to produce succinate from fumarate at significantly higher rates and yields than wild-type strain (molar yield ratio of 125 vs 17.6 in 4 d) *(5)*. Previous research mainly focused on the effect of cell concentration on the bioconversion of fumarate to succinate and the analysis of enzyme subunits produced. A detailed investigation of the effects of various levels of initial glucose concentration and initial fumaric acid concentration on succinic acid production have not been reported. In this study, a recombinant *E. coli* strain containing a multicopy plasmid (pGC1002) carrying the fumarate reductase genes *(frdABCD)* has been employed to convert fumaric acid to succinic acid. The effects of initial substrate and glucose concentrations on the bioconversion process have been examined.

MATERIALS AND METHODS

Strain and Plasmid

The laboratory *E. coli* strain DH5α was employed as the host strain in the study. Plasmid pGC1002 *(10)* is a pBR322-based multicopy vector containing the ampicillin-resistance gene and the fumarate reductase genes *frdABCD* coding for all four subunits of the enzyme complex.

E. coli Transformation

The plasmid pGC1002 was introduced into the host strain DH5α. The resulting recombinant *E. coli* strain was denoted as DH5α/pGC1002. Transformation of *E. coli* was performed as described by Sambrook et al. *(11)*.

Media and Cultivation

LB/amp medium for cell growth contains tryptone (10 g/L, Difco, Detroit, MI), yeast extract (5 g/L, Difco), NaCl (10 g/L), and ampicillin (50 mg/L). Fermentation medium contains $K_2HPO_4 \cdot 3H_2O$ (1.3 g/L), $MgSO_4 \cdot 7H_2O$ (0.5 g/L), $Fe(NH_4)_2(SO_4)_2 \cdot 6H_2O$ (0.03 g/L), casamino acid (0.5 g/L, Difco), peptone (0.6 g/L), yeast extract (0.3 g/L), and various concentrations of glucose and fumaric acid neutralized with sodium hydroxide to pH 7.1–7.2.

Batch cultures for cell growth were conducted in 2-L Erlenmeyer flasks containing 500 mL LB/amp medium and were incubated in an incubator shaker at 37°C and 250 rpm. Batch cultures for bioconversion were conducted under microaerobic conditions in 50-mL Erlenmeyer flasks sealed with rubber plugs containing 25 mL fermentation medium and were incubated in an incubator shaker at 37°C and 80 rpm.

Analytical Methods

Fumaric acid, succinic acid, L-malic acid, acetic acid, glucose, and other organic acids were determined and quantified by HPLC with an Intelligent Pump (Hitachi, L-6200A), an Intelligent Auto Sampler (Hitachi, AS-4000, Tokyo, Japan), a Bio-Rad (Hercules, CA) Aminex HPX-87H ion-exclusion column (300 × 7.8 mm), a refractive index detector (Hitachi, L-3350 RI), and a Chromato-Integrator (Hitachi, D-2500). The column temperature was maintained at 60°C and the column was eluted with 5 mM sulfuric acid at a flow rate of 0.8 mL/min.

RESULTS AND DISCUSSION

The recombinant *E. coli* strain DH5α/pGC1002 was grown overnight to stationary phase in LB/amp medium. Cells were precipitated by centrifugation. An equal amount of cell mass (202.5 mg dry weight) was inoculated into each fermentation flask. Samples were taken at different time points during the fermentation and were analyzed by HPLC.

Figure 1 shows the profiles of the fermentation with initial fumaric acid and glucose concentrations of 50.8 and 23.7 g/L, respectively. The production of succinic acid was rapid; the volumetric production rate was 4.38 g/L/h for the first 6 h. The final succinic acid concentration reached 47.5 g/L at 32 h. The weight yield of succinic acid based on the amount of fumaric acid consumed was 0.93. The consumption of fumaric acid was

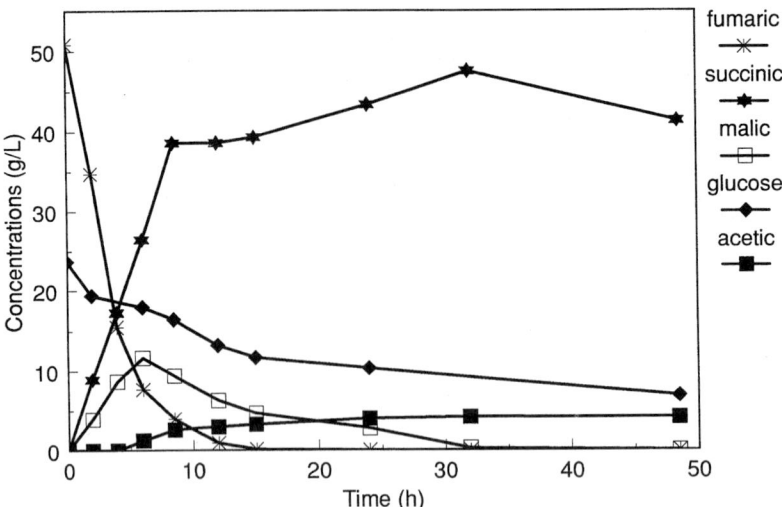

Fig. 1. General profiles of succinic acid fermentation from fumaric acid.

also fast with only 7.7 and 3.9 g/L of fumaric acid left at 6 and 8.5 h, respectively. Glucose was gradually consumed to 10.3 g/L at 24 h and 6.9 g/L at 48.5 h. L-Malic acid concentration first increased from zero to 11.7 g/L at 6 h and then gradually decreased to zero at 32 h. It is very likely that part of fumaric acid was initially converted to L-malic acid by a hydration reaction catalyzed by the enzyme fumarase and then L-malic acid was converted back to fumaric acid when the concentration of fumaric acid dropped to a very low level after 6 h. Acetic acid was also produced as a low-concentration byproduct in the bioconversion process. It was produced slowly after 4 h of incubation and its concentration finally accumulated to a level of about 4.1 g/L in the fermentation broth. The fermentation results are summarized in Table 1.

Effect of Initial Fumaric Acid Concentration

The effect of the initial fumaric acid concentration on the production of succinic acid by the recombinant *E. coli* strain DH5α/pGC1002 was investigated. The concentrations of initial glucose level were fixed at around 21–23 g/L. Four initial fumaric acid levels were examined: 50, 57, 64, and 90 g/L.

As shown in Fig. 2, the production of succinic acid was very rapid with initial fumaric acid concentrations of 50 to 64 g/L. At initial fumaric acid concentration of 90 g/L, however, the production of succinic acid was much slower. The initial volumetric production rates were 4.38, 4.32, 3.01, and 0.24 g/L/h in 6 h for initial fumaric acid concentrations of 50, 57, 64, and 90 g/L, respectively. Final succinic acid concentrations reached 41 to over 60 g/L for those fermentations started with 50 to 64 g/L fumaric acid.

Table 1
Summary of Fermentation Results Obtained at 32 h

Succinic acid concentration (g/L)	47.5
Succinic acid yield (g/g fumaric acid consumed)	93%
Specific productivity of succinic acid (g/g biomass·h)	0.183
Volumetric productivity of succinic acid (g/L·h)	1.48
Succinic acid : acetic acid (g/g)	11.4

Fig. 2. Succinic acid production with various initial fumaric acid concentrations. The initial concentrations of glucose were approx 20 g/L.

When 90 g/L fumaric acid was used, substantial substrate inhibition effect was observed and final succinic acid concentration could only reach 26.8 g/L at 48.5 h.

As evident from Fig. 3A, the consumption rate of fumaric acid at initial concentration of 90 g/L was much slower than those at lower initial fumaric acid concentrations. Glucose profiles also indicated significant difference between the fermentation started with 90 g/L fumaric acid and those started with lower fumaric acid levels (Fig. 3B). L-Malic acid profiles were generally similar to that in Fig. 1 with one exception. For the fermentation with initial fumaric acid concentration of 90 g/L, L-malic acid continued to accumulate and became the major product at a final concentration of 49.7 g/L (Fig. 4A). Acetic acid finally accumulated to 1.4 g/L for the fermentation with 90 g/L initial fumaric acid, and to 4.1–4.4 g/L for the other three cases (Fig. 4B).

Fig. 3. Profiles of fumaric acid consumption (**A**) and glucose utilization (**B**) in the bioconversion of fumaric acid to succinic acid with various initial fumaric acid concentrations. The initial concentrations of glucose were approx 20 g/L.

Effect of Initial Glucose Concentration

The effect of initial glucose concentration on the bioconversion of fumaric acid to succinic acid was studied with initial glucose levels at 0, 5, 10, 20, and 30 g/L. The initial fumaric acid levels were fixed at around 50 g/L.

The production of succinic acid was similar for those fermentations started with at least 5 g/L glucose (Fig. 5). Their initial volumetric production rates ranged from 4.1 to 4.6 g/L/h in the first 6 h. Maximal concentrations of succinic acid achieved were from 41 to 46 g/L at 32 h. In contrast, the initial succinic acid production rate was 2.0 g/L/h and final succinic acid concentration only reached 27 g/L for the fermentation started with 0 g/L glucose (Fig. 5).

The profiles for fumaric acid and glucose consumption are shown in Fig. 6 and those for L-malic acid and acetic acid accumulation are shown in Fig. 7. L-Malic acid accumulated to 12.6 g/L for the fermentation started without glucose. For fermentations with at least 5 g/L initial glucose levels, L-malic acid eventually approached zero at the end of the incubation. Acetic

Fig. 4. Profiles of L-malic acid concentration (**A**) and acetic acid accumulation (**B**) in the bioconversion of fumaric acid to succinic acid with various initial fumaric acid concentrations. The initial concentrations of glucose were approx 20 g/L.

Fig. 5. Succinic acid production with various initial glucose concentrations. The initial concentrations of fumaric acid were approx 50 g/L.

Fig. 6. Profiles of fumaric acid consumption (**A**) and glucose utilization (**B**) in the bioconversion of fumaric acid to succinic acid with various initial glucose concentrations. The initial concentrations of fumaric acid were approx 50 g/L.

acid was not detected for the fermentation started with 0 g/L glucose. For the other fermentations, final acetic acid levels ranged from 2.3 to 4.1 g/L.

The experimental data demonstrated that providing glucose, even at a concentration of 5 g/L, could substantially increase both the production rate and the final concentration of succinic acid in the fumaric acid to succinic acid bioconversion process. Further experiment performed at 2 g/L initial glucose level (with the same initial fumaric acid concentration) only allowed initial succinic acid productivity to reach 3.7 g/L/h, lower than that with 5 g/L initial glucose concentration. A control experiment employing a fermentation medium containing 10 g/L glucose and 0 g/L fumaric acid was also performed. No succinic acid was detected in the fermentation broth. Therefore, glucose alone was not converted to succinic acid. Glucose was possibly used as the hydrogen donor for the conversion of fumaric acid to succinic acid, it provided additional amount of the required cofactor $FADH_2$ to extend the reaction for succinic acid production. Glucose was also converted to acetic acid and possibly to some other cellular metabolites. Biomass was not changed at the end of the fermentation.

Fig. 7. Profiles of L-malic acid concentration (**A**) and acetic acid accumulation (**B**) in the bioconversion of fumaric acid to succinic acid with various initial glucose concentrations. The initial concentrations of fumaric acid were approx 50 g/L.

CONCLUSIONS

Fumaric acid was efficiently converted to succinic acid employing the recombinant *E. coli* strain DH5α/pGC1002. Initial succinic acid production rate could reach 4.3 g/L/h and final succinic acid concentration achieved 41 to over 60 g/L in 2 d under batch operations. For initial fumaric acid concentration over 64 g/L, substrate inhibition effect could be a problem. At 90 g/L initial fumaric acid level, L-malic acid, instead of succinic acid, became the major fermentation product. Fed-batch operations can be employed to optimize succinic acid bioconversion process. Glucose was observed to promote succinic acid bioconversion from fumaric acid. Over twofold increase in initial production rate was obtained by adding glucose (5–30 g/L) to the fermentation medium. Only slight difference in succinic acid synthesis was seen for initial glucose levels from 5 to 30 g/L. Acetic acid appeared to be a low-concentration byproduct in the bioconversion.

ACKNOWLEDGMENTS

This work was supported in part by the National Science Foundation (Grant BES-9412582), and the US Department of Agriculture (Grant 96-

35500-3192). The authors thank Robert P. Gunsalus (Department of Microbiology and the Molecular Biology Institute, University of California, Los Angeles) for providing the plasmid pGC1002.

REFERENCES

1. Winstrom, L. O. (1983), in *Kirk-Othmer Encyclopedia of Chemical Technology*, vol. 21, Grayson, M. and Eckroth, D., eds., Wiley, New York, pp. 848–864.
2. Datta, R. (1992), US patent no. 5,143,833.
3. Zeikus, J. G., Elankovan, P., and Grethlein, A. (1995), *Chem. Processing* **58**, 71–73.
4. Millard, C. S., Chao, Y.-P., Liao, J. C., and Donnelly, M. I. (1996), *Appl. Environ. Microbiol.* **62**, 1808–1810.
5. Goldberg, I., Lonberg-Holm, K., Bagley, E. A., and Stieglitz, B. (1983), *Appl. Environ. Microbiol.* **45**, 1838–1847.
6. Cao, N., Du, J., Gong, C. S., and Tsao, G. T. (1996), *Appl. Environ. Microbiol.* **62**, 2926–2931.
7. Blaut, M., Whittaker, K., Valdovinos, A., Ackrell, B. A. C., Gunsalus, R. P., and Cecchini, G. (1989), *J. Biol. Chem.* **264**, 13,599–13,604.
8. Van Hellemond, J. J. and Tielens, A. G. M. (1994), *Biochem. J.* **304**, 321–331.
9. Hirsch, C. A., Rasminsky, M., Davis, B. D., and Lin, E. C. C. (1963), *J. Biol. Chem.* **238**, 3770–3774.
10. Cecchini, G., Ackrell, B. A. C., Kearney, E. B., and Gunsalus, R. P. (1984), in *Flavins and Flavoproteins*, Bray, R. C., Engel, P. C., and Mayhew, S. G., eds., Walter de Gruyter, New York, pp. 555–558.
11. Sambrook, J., Fritsch, E. F., and Maniatis, T. (1989), *Molecular Cloning: A Laboratory Manual*, 2nd ed., Cold Spring Harbor Laboratory Press, Cold Spring Harbor, NY.

Accumulation of Biodegradable Copolyesters of 3-Hydroxy-Butyrate and 3-Hydroxyvalerate in *Alcaligenes eutrophus*

H. CHUA,*,[1] P. H. F. YU,[2] AND W. LO[2]

Departments of [1]Civil and Structural Engineering, and [2]Applied Biology and Chemical Technology, The Hong Kong Polytechnic University, Hung Hom, Hong Kong

ABSTRACT

Biodegradable copolyesters of 3-hydroxybutyrate-co-3-hydroxyvalerate (3HB-3HV) were produced by *Alcaligenes eutrophus* in a two-staged process, namely growth stage and nitrogen-deficient polyester-accumulation stage. When C_5 was used as the sole carbon source, the copolyester contained 43 mol % of 3HV. A range of copolyesters with 0–43 mol % of 3HV could be produced by using a medium containing different concentration ratios of butyric acid C_4 and C_5. T_m of PHB homopolymer was 177.6°C and that of copolyester with highest 3HV mol fraction of 43% was 99.0°C. C_5 concentration in the medium could be an effective means to control the polymeric composition and mechanical properties of the copolyesters accumulated in *A. eutrophus*.

Index Entries: *Alcaligenes eutrophus* H16; biodegradable plastics; physical properties; poly (3-hydroxybutyrate-co-3-hydroxyvalerate).

INTRODUCTION

Over the past two decades, plastic usage and plastic-waste generation have been on a drastically increasing trend, and are forecast to increase at a rate of 15% a year over the next decade *(1–4)*. Commonly used plastics, including high-density polyethylene (HDPE), low-density polyethylene (LDPE), polyethylene terephthalate (PET), polypropylene (PP), polystyrene (PS), and polyvinyl chloride (PVC), are synthetic polymeric hyrocarbons derived from petroleum, and are not easily decomposed by microorganisms *(5–8)*. Plastic wastes are therefore considered to be among

---

* Author to whom all correspondence and reprint requests should be addressed.

the most environmentally harmful wastes and have led to two environmental concerns.

Microbial polyesters and copolyesters, namely poly-hydroxyalkanoates (PHAs), have been recognized as a potential environment-friendly substitute for traditional plastics *(9–12)*. A number of bacteria, including *Alcaligenes*, *Pseudomonas*, recombinant *Escherichia coli*, and a few filamentous genera, accumulate these polyesters or copolyesters as an intracellular carbon reserve when unfavorable environmental conditions are encountered *(13–16)*. The extracted and processed polymeric substances have properties that are comparable to commonly used PE and PP, namely thermoplastic processability and 100% resistance to water. In addition, PHAs are completely biodegradable in natural environments *(17)*.

In recent years, much effort has been spent in optimizing the PHA production process and reducing costs *(18–20)*. Chua et al. *(21,22)* demonstrated a novel technique to produce PHA copolyesters as a byproduct from conventional activated sludge wastewater treatment process. However, an important aspect that has to be addressed before widespread applications of PHAs in packaging and disposable products are possible is the improvement of the physical and mechanical properties. *Alcaligenes eutrophus* produces PHB homopolyester and 3HB-3HV copolyester from various carbon sources, including glucose, fructose, and organic acids *(23)*. When even-numbered carbon compounds, such as glucose, acetic, butyric, and caproic acids, were used as the carbon source, only PHB was produced. When odd-numbered carbon compounds, such as propionic and valeric acids and propanol, were used as the carbon source, different 3HB-3HV copolyesters could be produced. PHB homopolymer was a relatively stiff and brittle material because of high crystallinity, and the melting point was 177°C. On the other hand, the physical properties of 3HB-3HV copolyesters, including melting point, mechanical strength, and biodegradability, were widely varied and dependent on the mole fractions of monomeric units in the copolyester *(24–28)*. Unfortunately, information on the control of HV fraction in the copolyester production that improve mechanical properties and decrease melting point are seriously lacking.

In this paper, butyric and valeric acids were used as carbon sources to produce PHA in a two-stage cultivation of *A. eutrophus*. The relationship between butyric acid (C_4) to valeric acid (C_5) ratio and the 3HV fraction in and the physical property of the copolyester was investigated.

MATERIALS AND METHODS

Bacterial Cell Line and Growth Medium

A. eutrophus H16 (American Type Culture Collection ATCC 17966) was used to generate copolyesters. The first stage of the two-stage cultivation was a cell-growth stage using a nutrient-rich medium, consisting of

10 g/L yeast extract, 10 g/L polypeptone, 5 g/L meat extract, and 5 g/L ammonium sulphate, in a 500-mL shaker flask at 30°C and shaken at 160 rpm for 24 h.

The second stage was a nutrient-deficient copolyester accumulation stage. Cells from the first stage were harvested by centrifugation and washed with buffer solution to remove any residual nitrogenous matters. About 7.5 g of the harvested and washed cells was transferred into a jar fermentor filled with a nitrogen-free medium. The medium contained C_4 and/or C_5 as the carbon sources, supplementary trace minerals and a growth factor with formulations previously described by Chua et al. (22). In separate batch cultures, the C_4 to C_5 weight ratios in the medium were respectively adjusted to 100:0, 80:20, 60:40, 40:60, 20:80, and 0:100. The initial total concentration of fatty acids was adjusted to 1 g/L to avoid possible growth inhibition as reported by Kim et al. (29). The fermentor was operated in a fed-batch mode by adding fatty acids into the fermentor once every 16 h to maintain the total acid concentration at a level equal to or lower than 1 g/L.

Fermentation Conditions

The automatic jar fermentation (Bioengineering Model ALF, Ruti/Switzerland) was of a 3-L working volume and operated at 300 rpm and 30°C for 48 h. The pH of the culture medium was automatically maintained at 7.0 by the addition of a sterilized 2 M NaOH solution or 2 M H_2SO_4 solutions.

Sampling and Analytical Techniques

The culture broth was periodically sampled and analyzed for residual carbon concentration and dry cell mass. Residual carbon concentration was measured as total organic carbon (TOC) using an Astro 2001 System 2 automatic TOC analyzer. The TOC analytical technique and dry cell-mass measurement procedure were in accordance with the standard methods (30).

The copolyesters accumulated in the cells were extracted by using a Soxhlet extractor with 200 mL of chloroform at 70°C for 24 h. The extracted copolyesters were purified by precipitation with 200 mL of methanol and weighed with an analytical balance. Thermal property of the extracted copolyesters was analyzed by a digital melting-point apparatus (electrothermal digital melting point apparatus Model IA9100). The extraction and analytical procedures were according to that described by Ho (10).

The accumulated copolyesters were also extracted by a different procedure for composition analysis. Culture broth (10 mL) was sampled and centrifuged at 3000 rpm for 15 min. The settled cells were washed with buffer solution and centrifuged again. The settled cells were then resuspended in a mixture containing 2 mL of methanol with 3% of concentrated

H_2SO_4 and 2 mL of chloroform with 2 mg/mL of benzoic acid as the internal standard. The sample was placed in a closed test tube and heated at 100°C for 3 h to convert the 3HB and 3HV constituents into their methylesters. The composition of these extracts was then analyzed by gas chromatographic techniques using a Varian model 3700 gas chromatograph equipped with a Carbowax 20 M column (1 meter) and a Shimazu C-R5A Chromatopac flame-ionization detector. The analytical procedure was according to that described by Ho *(10)*.

RESULTS AND DISCUSSION

Cell Growth and Polymer Production at Various Fatty Acid Ratios

The initially inoculated cell mass in fermentor was approx 7.5 g and the final cell mass ranged from 7.8 to 9.0 g in different batches of cultures with varied fatty acid ratios. The overall cell mass in the nutrient-deficient stage remained almost unchanged throughout the 48-h operation. An increase in C_5 concentration in the medium from 0 to 100 weight % resulted in a decline in both the specific copolyester yield, $Y_{p/x}$, from 0.41 to 0.04 g-copolyester/g-cell, and the copolyester production yield, $Y_{p/s}$, from 0.41 to 0.06 g-copolyester/g-TOC consumed (Table 1). These results indicated inhibitory effect of C_5 on the production of copolyesters. The highest $Y_{p/s}$ of 0.41 g/g achieved was within the theoretical maximum value of 0.65 g/g for *A. eutrophus* *(16)*. However, under the subinhibitory concentration of total fatty acid (less than 1.0 g/L) in the medium, the observed decline of $Y_{p/s}$ with increasing C_5 mol ratio was in contrast to that reported by Ishihara et al. *(31)* that $Y_{p/s}$ remained largely unchanged regardless of C_5 ratio in the medium.

Effect of Fatty Acid Ratio on 3HV Fraction in Copolyesters

Table 1 shows the effect of C_5 concentration in the medium on the 3HV-mole fraction of PHA. When C_4 was used as the sole carbon source, only PHB homopolyester (0 % 3HV) was produced without any 3HB-3HV copolyesters. On the other hand, a highest 3HV-mole fraction of 43 mol % in the copolyester accumulated was obtained when C_5 was used as the sole carbon source. The mole fraction of 3HV in the accumulated copolyesters increased with the increasing C_5 concentration in the medium (Table 1). The relationship between 3HV-mole fraction in the 3HB-3HV copolyesters and the C_5 concentration in the medium was linear. If the 3HV-mole fraction was represented by Y_1 and the C_5 concentration was represented by X_1, then the correlation could be expressed in *equation (1)* as follows.

$$Y_1 = 0.464 X_1 + 0.400, r^2 = 0.956 \qquad (1)$$

These results were in agreement with that observed in other works *(31,32)*. The changes of 3HB and 3HV mole fractions in the copolyesters

Table 1
Copolyester Accumulation Under Different Fatty Acid Ratios

Carbon Conc. (g/L)		C_4 to C_5 Ratio	$Y_{p/x}^1$ (g/g)	$Y_{p/s}^2$ (g/g)	HV Fraction (mol %)	T_m (°C)
C_4	C_5					
3.0	0.0	100:0	0.41	0.41	0	177.6
2.4	0.6	80:20	0.36	0.18	11	149.0
1.8	1.2	60:40	0.32	0.15	27	136.0
1.2	1.8	40:60	0.26	0.12	33	129.5
0.6	2.4	20:80	0.06	0.10	41	116.0
0.0	3.0	0:100	0.04	0.06	43	99.0

accumulated were attributed to the beta-oxidation metabolic pathway for fatty acids. When C_5 was used as the carbon source, acetyl-CoA and propionyl-CoA were produced from 3-keto-valeryl-CoA and both 3HB and 3HV monomeric units were produced. On the other hand, when C_4 was used as the sole carbon, propionyl-CoA was not produced, and only 3HB monomeric units were produced. These results also indicated that C_5 concentration in the medium could be an effective means to control the 3HV-mole fraction in the copolyesters accumulated in *A. eutrophus*.

Melting Temperature of Various Copolyesters

The melting temperature, T_m, of the copolyesters accumulated by *A. eutrophus* with different fatty acid ratios in the medium ranged from 99.0 to 177.6°C (Table 1). The highest value of T_m (177.6°C) was that of PHB, which was produced when C_4 was used as the sole carbon source in the medium. This product was of relatively high crystallinity, resulting in high stiffness and brittleness (Fig. 1). On the other hand, the lowest value of T_m was that of copolyester with the highest 3HV-mole fraction of 43 %, which was produced when C_5 was used as the sole carbon source in the medium. Therefore, the reduced crystallinity with the increasing of 3HV monomeric units in the copolyesters improved such mechanical properties as tensile, shear, compressive, and flexural strengths (Fig. 2). Results in Table 1 show that the T_m decreased with the increasing of 3HV monomeric units in the copolyesters. The relationship between T_m and 3HV-mol % was linear. If T_m was represented by Y_2 and the 3HV-mol % was represented by X_2, then the correlation could be expressed in equation (2) as follows.

$$Y_2 = 1.547 X_2 + 174.470, \quad r^2 = 0.938 \tag{2}$$

These results also indicated that C_5 concentration in the medium could be an effective means to control the mechanical properties of the copolyesters accumulated in *A. eutrophus*.

Fig. 1. Sample of PHB showing high crystallinity resulting in high stiffness and brittleness.

Fig. 2. Sample of 3HB-3HV copolyesters 43% of 3HV mole fraction showing reduced crystallinity resulting in improved mechanical properties.

CONCLUSION

When C_5 was used as the sole carbon source, the copolyester contained 43 mol % of 3HV. A range of copolyesters with 0 to 43 mol % of 3HV could be produced by using a medium containing different concentration ratios of butyric acid C_4 and C_5. T_m of PHB homopolymer was 177.6°C and that of copolyester with highest 3HV mol fraction of 43% was 99.0°C. C_5 concentration in the medium could be an effective means to control the polymeric composition and mechanical properties of the copolyesters accumulated in *A. eutrophus*.

REFERENCES

1. Chua, H., Yu, P. H. F., Xing, S., and Ho, L. Y. (1995), *J. Plast. Technol.* **21,** 65–73.
2. Hong Kong Environmental Protection Department (1994), Environment Hong Kong 1994, Hong Kong Government Press, pp. 51–66.
3. Hong Kong and Kowloon Plastic Product Merchants United Association (1992), Hong Kong *Plast. Ind. Bull.*, 33, December Issue.
4. Hong Kong Government Industry Department (1993), Hong Kong's Manufact. Industry 1993, Hong Kong Government Press.
5. Billmeyer, F. W. (1971), *Polymer Science*, Wiley, New York, pp. 379–490.

6. Huang, T., Zhao, J. Q., and Shen, J. R. (1991), *Plast. Ind.*, **4,** 23–27.
7. Young, R. J. (1981), *Introduction to Polymers*, Chapman and Hall, New York pp. 9–85.
8. Emballage Digest (1988), *Emballage Dig.* **30,** 80.
9. Chua, H., Yu, P. H. F. and Hu, W. F. (1996), *Proceedings of the 12th International Conference on Solid Waste Technology Management* Nov. 17–20, 1996, University of Pennsylvania.
10. Ho, L. Y., (1997), Master Thesis, The Hong Kong Polytechnic University.
11. Industrie-Anzeiger (1987), *Industrie-Anzeiger*, **109,** 26.
12. Pelissero, A. (1987), *Imballaggio*, **38,** 54.
13. Pfeffer, J. T. (1992), *Solid Waste Manage, Eng.* 72–84.
14. Shen J. R., Zhao, J. Q., Huang, T., and Chen, S. M. (1994), *Better Living Through Innovative Biochemical Engineering*, (Teo, W. K. et al, ed., Singapore University Press, Singapore 843–845.
15. Linko, S., Vaheri, H., and Seppala, J. (1993), *Appl. Microbiol. Biotechnol.* **39,** 11–15.
16. Yamane, T. (1993), *Biotechnol. Bioeng.* **41,** 165–170.
17. Kumagai, Y. (1992), *Polym. Degrad. Stabil.* **37,** 253–256.
18. Lee, S. Y., Chang, H. N., and Chang, Y. K. (1994), *Better Living Through Innovative Biochem. Eng.*, Teo, W. K., ed., Singapore University Press, pp. 53–55.
19. Shirai, Y., Yamaguchi, M., Kusubayashi, N., Hibi, K., Uemura, T., and Hashimoto, K. (1994), *Better Living Through Innovative Biochem. Eng.*, Teo, W. K., ed., Singapore University Press, pp. 263–265.
20. Shimizu, H., Sonoo, S., Shioya, S., and Suga, K. (1992), *Biochem. Eng. for 2001*, Furusaki et al. ed., Springer-Verlag, Tokyo, pp. 195–197.
21. Chua, H., Yu, P. H. F., and Ho, L. Y. (1997), *Appl. Biochem. Biotechnol.* **63,** 627–635.
22. Chua, H., Hu, W. F., and Ho, L. Y. (1997), *J. IES*, **37(2),** 9–13.
23. Doi, Y. (1990). *Microb. Polyesters.* VCH. New York.
24. Bluhm, T. L., Hamer, G. K., Marchessault, R. H., Fyfe, C. A., and Veregin, R. P. (1986), *Macromolecules* **19,** 2871–2876.
25. Bloemnergen, S., Holden, D. A., Harmer, G. K., Bluhm, T. L., and Marchessault, R. H. (1986), *Macromolecules* **19,** 2865–2871.
26. Cox, M. K. (1994), *Biodegrad. Plast. Polym.*, Doi et al., ed., Elsevier, Tokyo, pp. 120–135.
27. Doi, Y., Kanesawa, Y., Kunioka, M., and Saito, T. (1990), *Macromolecules* **23,** 26–31.
28. Mergaert, J., Webb, A., Anderson, C., Wouters, A. and Swings, J. (1993), *Appl. Environ. Microbiol.*, **59,** 3233–3238.
29. Kim, G. J., Yun, K. Y., Bae, K. S., and Rhee, Y. H. (1992), *Biotechnol. Lett.* **14(1),** 27–32.
30. APHA (1995), *Standard Methods for Examination of Waste and Wastewater*, 19th ed., APHA, AWWA, WPCF, Washington, DC.
31. Ishihara, Y., Shimizu, H., and Shioya, S. (1996), *J. Ferment. Bioeng.* **81,** 422–428.
32. Doi, Y., Tamaki, A., Kunioka, M., and Soga, K. (1988), *Appl. Microbiol. Biotecnol.* **28,** 330–334.

Session 6

Biotechnology in the Pulp and Paper Industry

J. N. Sandler[1] and E. Chornet[2]

[1]*Chair of Forest Products Biotechnology, University of British Colombia, Vancouver, Canada and* [2]*Sherbrooke University, Toronto, Canada*

One of the areas of biotechnology that has been rapidly expanding is its application in the pulp and paper sector. Many of the microorganisms, enzymes and approaches that have been used to try to completely breakdown lignocellulosic substrates to fermentable sugars have found a related application in fiber modification. For example, combinations of cellulase enzymes used at low concentrations can produce desirable pulp characteristics such as enhanced dewatering or better inter-fiber bonding while leading to very little yield loss. Other enzyme applications in the pulp and paper sector include pitch removal, deinking, reduced fiber coarseness, and biobleaching. Many of these areas have moved rapidly from initial laboratory observation to full scale commercial application. For example, Finnish researchers in 1986 first showed that xylanase treatment of kraft pulps could significantly decrease the bleaching chemicals required to reach a target brightness. A little more than ten years later, many pulp mills routinely operate with an enzyme step in their bleaching sequence to both reduce the level of pollutants in the waste water stream and reduce chemical costs. In the session describing advances in biotechnology in the pulp and paper sector a range of topics were covered. It was shown how classical mutation, selection and the addition of molecular biology have allowed enzyme companies to produce "tailored" xylanases which can operate at elevated temperatures and pH. More fundamental studies described how xylanases could be divided into two major "families" based on their protein sequence and secondary structure. Further subgrouping based on the pI of the enzyme indicated how enzymes with alkaline pI's have a greater affinity for xylan substituted with uronic acid side chains which normally impede xylanase activity. Thus a combination of very fundamental and commercial work has allowed rapid implementation of enzymes while improving the properties of the enzyme at the same time. Other enzymes with potential roles in the bleaching of wood pulp were also described. These include laccases plus redox mediations, cellobiose dehydrogenase (CDH) and manganese peroxidase (MnP).

The mechanism of cellulase action continues to receive considerable attention with the availability of individual, cloned enzymes allowing synergistic studies to be revisited while direct commercial application of these "cellulase cocktails" growing in areas such as "biostoning" of denim jeans. "biopolishing" of dyed cotton fabric and other textile, detergent and food/feed applications.

It is likely that many of the biotechnology applications that might be used in the bioconversion area will first find a related use in the pulp and paper area. This is partly because of the need to find more cost effective, less environmentally polluting ways of producing pulp and paper. However, the main driver is that enzymes will help produce a cellulose based product that usually sells for more than $600 a tonne. Thus the large scale application of enzymes in the processing of biomass will likely be pioneered in the pulp and paper sector with much of the low value fiber then available for bioconversion into other products such as ethanol.

Efficient Production of Mannan-Degrading Enzymes by the Basidiomycete *Sclerotium rolfsii*

ALOIS SACHSLEHNER,[1] DIETMAR HALTRICH,*,[1] GEORG GÜBITZ,[2] BERND NIDETZKY[1] AND KLAUS D. KULBE[1]

[1] *Abteilung Biochemische Technologie, Institut für Lebensmitteltechnologie, Universität für Bodenkultur BOKU (University of Agricultural Sciences Vienna), Muthgasse 18, A-1190 Wien, Austria;* [2] *Institut für Mikrobiologie, Technische Universität Graz, Petersgasse 12, A-8010 Graz, Austria*

ABSTRACT

Sclerotium rolfsii CBS 191.62 was cultivated on a number of carbon (C) sources, including mono- and disaccharides, as well as on polysaccharides, to study the formation of different mannan-degrading enzyme activities. Highest levels of mannanase activity were obtained when α-cellulose-based media were used for growth, but formation of mannanase could not be enhanced by employing galactomannan as the only carbon source. Although both xylanase and cellulase formation was almost completely repressed when *S. rolfsii* was grown on more readily metabolizable carbohydrates, including glucose or mannose, considerable amounts of mannanase activity were secreted under these growth conditions. Enhanced mannanase production only commenced when glucose was depleted in the medium. The maximal mannanase activity of 240 IU/mL obtained in a laboratory fermentation is remarkable. Mannanase activity formed under these derepressed conditions could be mainly attributed to one major, acidic mannanase isoenzyme with a pI value of 2.75.

Index Entries: *Sclerotium rolfsii*; mannanase; xylanase; endoglucanase; regulation.

INTRODUCTION

Hemicellulose forms, together with cellulose and lignin, the main polymeric constituents of lignocellulose, which is a major reservoir of fixed carbon (C) in nature. Hemicellulose comprises several heterogeneous groups

*Author to whom all correspondence and reprint requests should be addressed.

of polysaccharides, which are combined in this group on essentially practical and historical reasons, such as solubility in alkali and application of chemical extraction procedures. These are usually named according to the main sugar residues in the backbone chain of the polymer, e.g., xylans, mannans, and galactans. Depending on the source from which they have been obtained, and the physicochemical extraction procedure, hemicelluloses vary significantly in their structures and mol wt (1).

The major hemicelluloses in softwoods are galactoglucomannans; their content varies between 15–20% of the total dry wt. In hardwood, they are found in quantities up to 5%. Glucomannans, and especially galactomannans, also occur in annual plants, mainly in seeds and tubers, in which they serve as storage carbohydrates. The mannose and glucose units in the backbone of softwood mannans are partially substituted at the O-2 and O-3 position by acetyl groups. Additionally, D-galactosyl units are attached to the main chain by α-(1,6)-bonds to a varying extent, but these side groups are not found in hardwood mannans (1–3).

Owing to the complex structure of mannans, several different enzymes are necessary for their complete enzymatic degradation. The backbone is hydrolyzed by the action of endo-(1,4)-β-D-mannanase, yielding mannobiose and mannotriose, and various mixed oligosaccharides, which are further cleaved by β-D-mannosidase and β-D-glucosidase. The side group substituents are removed by α-D-galactosidase and various esterases, including acetylesterase (4).

β-Mannanases have been reported to be isolated from a wide spectrum of organisms, including bacteria, fungi, germinating seeds of terrestrial plants, marine algae, and animals (5), as well as yeasts and yeast-like microorganisms (6). Similarly, the occurrence of β-mannosidase, α-galactosidase, and various esterases has been described for a large range of plant and animal tissues, as well as for many microorganisms (7).

Mannan-degrading enzymes can find numerous applications in the food, feed, and pulp and paper industries. Their employment is probably most useful when the selective removal of mannans is required. Limited endohydrolysis or removal of side groups may be applied for the modification of mannans used as thickeners in the food industry. Mannanases can also be used in the processing of instant coffee, in which they reduce the viscosity of the coffee extracts by hydrolyzing galactomannans, which decreases the costs for subsequent evaporation and drying (8). In the past few years, the employment of mannan-degrading enzymes in the pulp and paper industry gained much interest. In combination with xylanases, mannanases are used to partially hydrolyze mannan and xylan in kraft pulps. This leads to an increase in brightness and to a significant decrease in the amount of chemicals required for bleaching (9).

Sclerotium rolfsii (or *Athelia rolfsii*, which is used for the teleomorph) is an aggressive plant pathogen of many crops in the tropics and subtropics. The fungus colonizes organic matter in the soil, from where it may parasitize certain plants. During its attack to plant material, it forms several different enzymes that rapidly destroy cell walls, thus enabling it to enter the host. *S. rolfsii* is known as a good producer of cellulolytic and hemicellulolytic enzymes, including mannanases *(10,11)*. It was the aim of our work to investigate the formation of mannan-degrading enzymes in more detail.

MATERIALS AND METHODS

Chemicals

α-Cellulose, *p*-nitrophenyl glycosides, α-naphthyl acetate, locust bean gum (LBG; a galactomannan from *Ceratonia siliqua*, with a mannose-to-galactose ratio of 4:1), and guar gum (a galactomannan from *Cyamopsis tetragonobola*, with a mannose-to-galactose ratio of 2:1) were from Sigma (St. Louis, MO); lactose, L-sorbose, D-mannose, D-melibiose, D-raffinose, and carboxymethylcellulose were from Fluka (Buchs, Switzerland). Xylo-oligosaccharides containing more than 95% xylobiose (xylooligo-95) were a kind gift from Suntory (Tokyo, Japan). The hydrolysate of LBG was prepared as follows. Fifty g of LBG were dissolved in 1 L 0.05 M sodium citrate buffer, pH 4.0, and incubated with a crude culture filtrate of *S. rolfsii* (20 U mannanase activity/g LBG) for 24 h on an orbital shaker (170 rpm, 50°C). The hydrolysate, consisting mostly of mannobiose and mannotriose, was then lyophilized. Azo-carob galactomannan (covalently dyed with Remazol brilliant blue) was purchased from Megazyme (Sydney, Australia). Xylan from birchwood was from Roth (Karlsruhe, Germany), and peptone from meat was from Merck (Darmstadt, Germany). All other chemicals were analytical grade.

Organism and Culture Conditions

Sclerotium (Athelia) rolfsii CBS 191.62 (Centraalbureau voor Schimmelcultures, Baarn, The Netherlands) was used throughout this study. Stock cultures were maintained on glucose-maltose Sabouraud agar, and routinely subcultured every 4 wk. Inoculated plates were incubated at 30°C for 4–6 d, and then stored at 4°C.

The strain was cultivated in unbaffled 300-mL Erlenmeyer flasks at 30°C for 13 d on a medium containing (in g/L) peptone from meat, 80; NH_4NO_3, 2.5; $MgSO_4 \cdot 7H_2O$, 1.5; KH_2PO_4, 1.2; KCl, 0.6, and trace element solution, 0.3 mL/L *(12)*. The trace element solution is comprised of (in g/L): $ZnSO_4 \cdot 7H_2O$, 1.0; $MnCl_2 \cdot 4H_2O$, 0.3; H_3BO_3, 3.0; $CoCl_2 \cdot 6H_2O$, 2.0; $CuSO_4 \cdot 5H_2O$, 0.1; $NiCl_2 \cdot 6H_2O$, 0.2, and H_2SO_4 conc., 4.0 mL/L. Carbon sources were added as indicated at a concentration of 42.6 g/L, unless otherwise stated. All media were prepared with tap water. The pH was adjusted to 5.0 using phosphoric acid prior to sterilization. Flasks were inoculated with a 1-cm² piece from an actively growing, 4–6-d-old culture of *S. rolfsii* on Sabouraud agar. All media not containing insoluble components were homogenized with a laboratory homogenizer (Polytron, Kinematica, Kriens, Switzerland) at 9,500 rpm for 15 s after inoculation. This was necessary to break up the piece of agar with the mycelium to obtain homogeneous growth. The inoculated flasks were incubated at 30°C, with continuous shaking at 150 rpm (stroke 25 mm), for 13 d. Each culture was then centrifuged, and the clear supernatant was used for the estimation of enzyme activities. Results given are the mean of at least duplicate experiments.

Fermentation studies were carried out in a 20-L laboratory fermenter (MBR Bio Reactor, Wetzikon, Switzerland) with a working volume of 15 L, and equipped with four disk turbine impellers, each with six flat blades.

Enzyme Activity Assays

All activity assays were carried out in 0.05 M sodium citrate buffer, pH 4.5, unless otherwise stated. Mannanase (EC 3.2.1.78) activity was assayed using a 0.5% solution of LBG galactomannan in 0.05 M sodium citrate buffer, pH 4.0, as a substrate. The release of reducing sugars in 5 min at 50°C was measured as mannose equivalents, using the dinitrosalicylic acid (DNS) method *(13)*. Xylanase (EC 3.2.1.8) and endoglucanase (carboxymethylcellulase, EC 3.2.1.4) activities were assayed similar to mannanase activity, using a 1% solution of xylan (4-O-methyl glucuronoxylan from birchwood) or of carboxymethylcellulose (sodium salt, ultra-low viscosity) respectively, as the substrates. Reducing sugars were assayed as xylose or glucose, using the DNS method. Filter paper cellulase activity was measured according to IUPAC recommendations, employing filter paper (Whatman No. 1, Maidstone, UK) as a substrate *(14)*. One unit (IU) of enzyme activity is defined as the amount of enzyme releasing 1 µmol of xylose, mannose, or glucose equivalents per min under the given conditions. One IU corresponds to 16.67 nkat.

α-Galactosidase (EC 3.2.1.22), β-glucosidase (EC 3.2.1.21), and β-mannosidase (EC 3.2.1.25) were quantified in a similar manner, using the respective p-nitrophenyl-glycosides (8 mM final concentration) as substrates. Buffer (0.5 mL) was incubated with 0.25 mL of the appropriately diluted enzyme solution and 0.25 mL of substrate solution at 50°C for 10 min. The reaction was stopped by adding 2.0 mL of 1 M Na$_2$CO$_3$, and the absorbance was measured at 405 nm. Activities are expressed on the basis of the liberation of p-nitrophenol.

Acetyl esterase (EC 3.1.1.6) activity was determined using 1 mM α-naphthylacetate as the substrate *(15)*. One unit of enzyme activity is expressed as the amount of enzyme liberating 1 µmol α-naphthol per min.

Protein Assays

Protein concentrations were determined according to the dye-binding method of Bradford *(16)*, using bovine serum albumin (fraction V, United States Biochemical Corp., Cleveland, OH) as standard.

Analytical Isoelectrical Focusing (IEF) and Activity Stains

Isoelectric focusing was carried out on the Pharmacia Phast System using precast, dry gels (PhastGel dry IEF, Pharmacia, Uppsala, Sweden), rehydrated with carrier ampholytes (7.5 parts Pharmalyte, pH 2.5–5, and 2.5 parts Ampholine, pH 3.5–5.0; Pharmacia), as described by the manufacturer. Mannanase activity in IEF gels was detected by active staining (zymogram technique), using covalently dyed mannan overlays, as described by Biely *(17)*. The pI values of the mannanase isoenzymes were determined by comparison with marker proteins (Pharmacia, low pI kit, pH range 2.8–6.5), which were run simultaneously, and were visualized by silver staining, as recommended by the manufacturer.

RESULTS

Growth Experiments

Sclerotium rolfsii CBS 191.62, which was identified as an outstanding producing strain of mannanase activity when grown on cellulose-based media *(11)*, was cultivated on a number of different substrates, to investigate, in detail, the formation of mannan-degrading enzymes. These substrates included various polysaccharides, which structurally resemble the main carbohydrate constituents of lignocellulose, and disaccharides, which are liberated from these polysaccharides by the action of the respective endoglycanases. Furthermore, several well-known inducers of cellulolytic or hemicellulolytic enzymes, including lactose and sorbose *(18,19)*, as well as a number of more easily metabolized sugars were employed. A blank containing no carbohydrate supplemented to the medium was used. After 13 d of growth, the mycelia were separated by centrifugation, and several mannan-degrading enzymes, as well as endoxylanase and endoglucanase, were assayed in the culture supernatants (Tables 1 and 2). Growth of *S. rolfsii* on α-cellulose resulted in the highest activities of all three endoglycanases investigated. This stimulating effect of α-cellulose is especially pronounced for both xylanase and endoglucanase, since activities resulting from growth on the other substrates used in this experiment were lower by at least one order of magnitude. Titers of mannanase could not be enhanced when α-cellulose was substituted by several mannans as the inducing substrate. However, these latter C sources were employed in lower concentrations, because they drastically increased the viscosity of the culture medium. The superiority of cellulose as inducer of mannanase is also reflected when comparing results obtained for cellobiose and the manno-oligosaccharides (LBG hydrolysate). The former substrate clearly stimulated the formation of higher mannanase activities. Although growth of the fungus on more readily metabolized carbohydrates, e.g., glucose or mannose, resulted in only low, presumably constitutive levels of both

Table 1
Formation of Extracellular Protein and Endoglycanase Activities by *S. rolfsii* CBS 191.62 when Grown in Shaken Flasks on Different Substrates (Conditions: 30°C, 150 rpm, 13 d)

Inducing substrate[a] (g/L)	Extracellular protein (mg/mL)	Activities (IU/mL)			Ratios		
		Mannanase	Xylanase	Endoglucanase	Mannanase to endoglucanase	Mannanase to xylanase	Xylanase to endoglucanase
Polysaccharides							
Cellulose	3.83	675	228	1090	0.62	2.96	0.21
Guar gum (21.3)	1.04	169	12.2	41.6	4.06	13.9	0.29
LBG[b] (21.3)	1.01	146	7.8	26.6	5.49	18.7	0.29
Xylan birchwood	0.41	10.2	4.6	2.9	3.52	2.22	1.59
Oligosaccharides							
Cellobiose	1.12	286	11.0	127	2.25	26	0.09
Lactose	0.90	135	1.4	10.6	12.7	96.4	0.13
Xylooligo-95	0.52	98.2	4.2	8.0	12.3	23.4	0.53
Hydrolysate of LBG[b]	0.72	167	1.6	22.1	7.56	104	0.07
Melibiose	0.64	94.4	1.4	8.4	11.2	67.4	0.17
Raffinose	0.76	122	1.0	10.9	11.2	122	0.09
Monosaccharides							
D-Glucose	0.62	81.5	0.59	6.8	12.0	138	0.09
D-Mannose	0.60	113	0.71	6.4	17.7	159	0.11
L-Sorbose	0.30	98.8	7.1	15.8	6.25	13.9	0.45
D-Galactose	0.25	31.9	0.94	3.5	9.11	33.9	0.27
D-Fructose	0.70	25.1	0.24	2.0	12.6	104	0.12
Blank	0.26	6.3	0.16	0.33	19.1	39.4	0.48

[a] Concentration of substrates was 42.6 g/L, unless otherwise indicated.
[b] Locust bean gum.

Table 2
Effect of Different Substrates on Production of Auxiliary Mannan-Degrading Enzymes by S. rolfsii when Grown in Shaken Flasks

Inducing substrate[a] (g/L)	Activities (IU/mL)			
	β-Mannosidase	β-Glucosidase	α-Galactosidase	Acetylesterase
Polysaccharides				
Cellulose	0.50	1.31	7.44	35.8
Guar gum (21.3)	0.61	1.17	5.63	16.5
LBG[b] (21.3)	0.57	1.41	5.44	15.9
Xylan birchwood	0.40	1.57	2.52	5.40
Oligosaccharides				
Cellobiose	0.49	0.72	3.94	13.7
Lactose	0.48	0.57	2.65	7.60
Xylooligo-95	0.43	0.70	2.76	6.10
Hydrolysate of LBG[b]	0.57	0.93	3.99	13.1
Melibiose	0.47	0.70	2.81	9.24
Raffinose	0.56	0.57	3.48	11.7
Monosaccharides				
D-Glucose	0.28	0.57	2.36	8.60
D-Mannose	0.22	0.45	2.40	8.20
L-Sorbose	0.25	1.16	1.66	5.40
D-Galactose	0.20	0.60	1.42	2.80
Blank	0.20	0.63	1.71	1.90

[a] Concentration of inducing substrate was 42.6 g/L, unless otherwise indicated.
[b] Locust bean gum.

xylanase and endoglucanase, considerable activities of mannanase were formed in the cultivations on these monosaccharides.

Even though *S. rolfsii* did not produce increased levels of mannanase when cultured on mannan-based media, the galactomannans specifically provoked the synthesis of mannanase activity, with relatively lower concomitant formation of endoglucanase and xylanase, when compared to cultures using α-cellulose or cellobiose. This is obvious from the ratios of mannanase to endoglucanase or xylanase, which were calculated from the experimental data, and are given in Table 1. The value for the mannanase-to-endoglucanase ratio of 0.62, which was relatively constant for cellulose-based cultures *(11)*, increased significantly for the cultivations on the galactomannans. An even further increase of this ratio, indicating that relatively more mannanase than endoglucanase activity is secreted by the fungus, was observed for several oligosaccharides, including manno-oligosaccharides, as well as for most of the more easily metabolizable monosaccharides. The highest value of this ratio, 17.7, was obtained for the mannose-based medium. A similar conclusion can be drawn when considering the mannanase-to-xylanase ratio. Cultivations on various mannans and manno-oligosaccharides, as well as on a number of monosaccharides, resulted in a relatively increased formation of mannanase, compared to the simultaneously produced xylanase activity (Table 1).

As is evident from Table 2, the effect of the inducing substrate employed in the cultivations is less pronounced with respect to the formation of the auxiliary mannan-degrading enzymes, which are necessary for the complete hydrolysis of substituted mannans, than for the endoglycanases. However, the experimental data suggest that the synthesis of these hydrolases is inducible as well. Highest activities of β-mannosidase were obtained when the fungus was grown on mannan or on the manno-oligosaccharides; the formation of elevated activities of β-glucosidase was provoked by growth on cellulose, cellobiose, or sorbose. The latter is a known inducer of cellulases in fungi *(18,20)*. Similarly, the oligosaccharides melibiose and raffinose, which contain an α-1,6-linked galactosyl moiety, as well as galactomannan or the LBG hydrolysate, when employed as substrates, stimulated the synthesis of α-galactosidase. Interestingly, α-cellulose was clearly the best inducer for the formation of this enzyme which is part of the mannan-degrading enzyme system of the fungus.

Fermentation Studies

To study the effect of the more readily metabolized sugars on the formation of the mannan-degrading enzyme system by *S. rolfsii*, and to evaluate the use of these sugars as cheap substrates for the production of mannanases, *S. rolfsii* was cultivated in a 20-L laboratory fermenter, using 42.6 g/L glucose as the main C and energy source. The inoculum was an

Fig. 1. Time-course of a laboratory cultivation of *S. rolfsii* CBS 191.62 in a 20-L stirred-tank reactor (working volume 15 L) using a medium based on glucose (42.6 g/L). The temperature was controlled at 30°C, and the pH, initially adjusted to 5.0, was allowed to float. Aeration was automatically varied from 0.1 to 1.0 vol of air/ fluid vol/min to maintain a pO_2 of 30% of air saturation. Symbols: (□), glucose; (○), culture pH; (×), extracellular protein; (▲), cell dry wt.

11-d-old shaken culture grown on the glucose-based fermentation medium. The time-course of this cultivation is shown in Fig. 1. Glucose was rapidly consumed during the initial phase of the cultivation, and was depleted after 44 h. This initial phase of glucose consumption was accompanied by a characteristic decrease in the culture pH from 4.7 to 2.5. The maximum amount of biomass formed also coincides with the depletion of glucose. Production of extracellular enzymes, however, was not significant in this first phase of the cultivation (Figs. 2 and 3). It only started after glucose was spent in the medium. The maximum value of mannanase activity of 240 IU/mL was reached after approx 10 d, and remained constant thereafter (Fig. 2). This corresponds to a volumetric productivity of 1,000 IU/L/h. Simultaneously, very low activities of endoglucanase and xylanase were secreted by the organism. The maximum activities of 5.6 and 3.0 IU/mL, respectively, were obtained after approx 140 h, and again remained constant for at least 170 h. Furthermore, filter paper cellulase activity could not be detected in appreciable amounts during the course of this fermentation. Compared to the shaken-flask cultivations, production of acetylesterase and several glycosidases was significantly increased in the fermentation process (Fig. 3). The maximum activities of β-mannosidase (0.65 IU/mL), β-glucosidase (1.95 IU/mL), and α-galactosidase (8.30 IU/mL) produced in this experiment were even higher than the values obtained by growth on galactomannan or α-cellulose in shaken flasks.

Fig. 2. Production of mannanase, xylanase, and endoglucanase by *S. rolfsii* in a laboratory fermenter. Symbols: (▲), mannanase; (□), xylanase; (●), endoglucanase.

Fig. 3. Formation of β-mannosidase, β-glucosidase, α-galactosidase, and acetylesterase by *S. rolfsii*. Symbols: (▲), β-mannosidase; (○), β-glucosidase; (□), α-galactosidase; (▲), acetylesterase.

Analysis of Multiple Mannanases

S. rolfsii has recently been reported to form at least five multiple isoforms of mannanase when grown on a cellulose-based medium supplemented with konjac glucomannan *(21,22)*. To investigate the mannanase isoenzymes secreted by the organism when cultivated on glucose, the culture filtrates obtained at various cultivation times were separated by ana-

Fig. 4. Activity staining of mannanase in culture filtrates obtained at different times during a laboratory cultivation using glucose as the substrates. Lanes 1–7, samples taken at 28, 49, 71, 92, 143, 221, and 309 h, respectively. Equal activities of 0.05 IU mannanase were applied for each of the samples to the separation gel.

lytical IEF and subsequently monitored by active staining with Remazol brilliant blue (RBB) mannan overlays (Fig. 4). Mannanase activity could be attributed mostly to one protein showing a pI value of 2.75. In addition, a second minor mannanase of pI 4.85 could be detected by the zymogram analysis. Judging from the activity staining, in which equal mannanase activity in each sample was applied to the separation gel, the relative amount of these two isoformic mannanases remained relatively constant for the entire cultivation period.

DISCUSSION

In several recent studies (11,12,23) the basidiomycete *S. rolfsii* has been identified as an outstanding producer of mannan-degrading enzymes. Especially, the levels of endomannanase activity secreted by this plant pathogenic fungus are exceptional. In this investigation, the production of mannan-hydrolyzing enzymes was further studied. *S. rolfsii* was cultivated on a number of different substrates, including mono-, oligo-, and polysaccharides, and the resulting mannanolytic enzyme activities, together with endoglucanase and xylanase, were determined. Pure cellulose clearly is the best substrate for the production of mannanase, as well as for the other two endoglycanases by *S. rolfsii*. Levels of mannanase activities could not be enhanced when replacing cellulose by various mannans as the inducing substrate. A similar observation has been made for several other fungi and bacteria *(10,23–26)*; *Trichoderma harzianum* produces almost the same mannanase activities during cultivation on crystalline cellulose or LBG *(27)*. When several of the low-mol-wt compounds that are described as inducers of cellulase or xylanase activity in different fungi, e.g., cellobiose or sorbose *(18,20,28)*, were used as growth substrates, elevated endoglycanase levels were obtained as well. These were, however, significantly lower than those resulting from growth on cellulose. Glucose, an easily metabolizable sugar, was used as a control in these growth experiments, and resulted in only very low activities of xylanase and endoglucanase formed, but levels of mannanase activity were surprisingly high. A similar effect was found when several other easily metabolizable, typically repressive carbohydrates, including fructose or mannose, were used as growth substrates.

From these results, it is evident that the synthesis of mannanase, xylanase, and endoglucanase is inducible in *S. rolfsii*. This is in agreement with reports on several other fungi that have been closely investigated in this respect *(18,19,28,29)*. Contrary to the reports on most other fungi, however, this induction is closely related in *S. rolfsii*, and seems to depend on the presence of cellulose. This assumption is further confirmed when comparing the enzyme activities resulting from growth on cellobiose, xylobiose (xylooligo-95), and manno-oligosaccharides (LBG hydrolysate). Highest titers of all three endoglycanase activities were unequivocally obtained in the cellobiose-based medium. Cellobiose, which is one of the end products of the enzymatic degradation of cellulose, could be a common in vivo inducer of these three endoglycanases. In accordance with the generally accepted model on induction of polysaccharide-degrading enzymes *(18,20,28)*, cellobiose is released from the substrate cellulose by the action of constitutive amounts of cellulases secreted by the organism, is taken up by the fungal cell, and finally triggers the elevated synthesis of the three enzyme activities.

Although the induction is closely related, there seems to be no common regulatory mechanism for mannanase, xylanase, and endoglucanase activity in *S. rolfsii*, since certain significant differences pertaining to the regulation of these three enzyme activities exist. The most obvious one certainly concerns the formation of mannanase, which is synthesized by the organism, even when easily metabolizable sugars are used as substrates. As was shown in a laboratory fermentation using glucose as the substrate, elevated levels of mannanase were formed under derepressed conditions, i.e., when glucose was depleted in the medium; simultaneously, only low, presumably constitutive levels of endoglucanase and xylanase were secreted by the organism. For the enhanced formation of these two latter endoglycanases, the presence of an appropriate inducer seems to be a prerequisite, indicating that the synthesis of both endoglucanase and xylanase is more tightly controlled in *S. rolfsii* than that of mannanase.

This apparent difference in the regulation, i.e., the increased formation of mannanase under derepressed, noninducing conditions, also explains the increase in the ratios of mannanase to endoglucanase or xylanase (Table 1), which indicate that relatively higher activities of mannanase are secreted in the presence of certain carbohydrates as growth substrates. Cellulose is relatively resistant to enzymatic degradation, and therefore cellobiose will only be slowly released, which will ensure the prolonged availability of the inducer, and presumably does not lead to repressive conditions. On the other hand, enzymatic hydrolysis of the disaccharides, or even of the soluble polysaccharides mannan or xylan, will be considerably faster, which results in a more rapid degradation of the low-mol-wt inducers caused by the action of glycosidases. Under these conditions, mannanase formation will be favored.

In contrast to this, the xylanase-to-endoglucanase ratio was found to be relatively constant when different carbohydrates were used as growth substrates for *S. rolfsii* (Table 1). This indicates that the regulation of their synthesis is more closely linked. However, even here certain differences exist. The values of this ratio, calculated for the enzyme activities secreted in response to xylan or xylobiose, are significantly higher, suggesting that xylanase is preferentially formed in their presence.

When *S. rolfsii* was cultivated on glucose, only two isoforms of mannanase could be detected by the activity staining. The major mannanase produced under these growth conditions was found to have a pI value of 2.75. This acidic pI is notable, since, for most of the fungal mannanases characterized to date, this value has been reported to be in the range of 3.2–5.8 *(7)*. In addition, formation of only two isoformic mannanases under derepressed conditions is in contrast to results obtained for a wild-type isolate of *S. rolfsii*, when grown on a cellulose-based medium supplemented with konjac mannan. Under these culture conditions, five multiple mannanase isoforms were formed by this isolate *(21,22)*. Furthermore, it was found that resting mycelia of *S. rolfsii* CBS 191.62 secreted at least seven mannanase isoforms in response to bacterial cellulose, which is formed by the bacterium *Acetobacter xylinum*, and contains no traces of hemicellulose (unpublished results). This indicates that the regulation of the synthesis of mannanases in *S. rolfsii* is even more complicated. Whereas some of the multiple mannanases are formed under derepressed conditions, certain isoforms are apparently separately regulated and only efficiently secreted under inducing conditions. This complex regulation seems to correspond to the different inducibility of endoglucanases, as well as the two major xylanases of *Trichoderma reesei* QM9414 *(30,31)*.

The findings reported in this study are of technological significance. By selecting appropriate culture conditions, mannanase preparations practically free of cellulase activity can be easily produced. In this respect, the maximum mannanase activity of 240 IU/mL, obtained in a laboratory cultivation of *S. rolfsii* employing a glucose-based medium, is remarkable. This value is significantly higher when compared to data reported in the literature on the production of mannanases by other microorganisms, using cellulose or mannan as inducing substrates *(23,25,32)*. Such enzyme preparations have gained increased interest because of new areas of application within the pulp and paper industry. This absence of cellulase activity, which would negatively affect pulp properties, is essential for these applications *(7,33)*. On the other hand, enzyme preparations containing appreciable amounts of cellulase and xylanase can be obtained when cellulose or suitable lignocellulosic material is used as the substrate *(10,11)*. Since *S. rolfsii* also forms considerable amounts of auxiliary hemicellulolytic enzymes, such enzyme preparations can be used when the complete hydrolysis of lignocellulose is the aim.

ACKNOWLEDGMENT

This work was supported by grants from the Fonds zur Förderung der wissenschaftlichen Forschung (Austrian Research Foundation) P10753-MOB, which the authors gratefully acknowledge.

REFERENCES

1. Eriksson, K.-E. L., Blanchette, R. A., and Ander, P. (1990), in *Microbial and Enzymatic Degradation of Wood and Wood Components*, Springer, Berlin, pp. 181–184.
2. Ward, O. P. and Moo-Young, M. (1989), *CRC Crit. Rev. Biotechnol.* **8**, 237–274.
3. Stephen, A. M. (1983), in *The Polysaccharides*, vol. 3, Aspinall, G. O., ed., Academic, New York, pp. 97–193.
4. Puls, J. and Schuseil, J. (1993), in *Hemicellulose and Hemicellulases*, Coughlan, M. P. and Hazlewood, G. P., ed., Portland, London, pp. 1–27.
5. Dekker, R. F. H. and Richards, G. N. (1976), in *Advances in Carbohydrate Chemistry and Biochemistry*, vol. 32, Tipson, R. S. and Horton, D., ed., Academic, New York, pp. 277–352.
6. Kremnicky, L., Slávikóvá, E., Mislovicová, D., and Biely, P. (1996), *Folia Microbiol.* **41**, 43–47.
7. Viikari, L., Tenkanen, M., Buchert, J., Rättö, M., Bailey, M., Siika-aho, M., and Linko, M. (1993), in *Bioconversion of Forest and Agricultural Plant Residues*, Saddler, J. N., ed., C.A.B. International, Wallingford, pp. 131–182.
8. Wong, K. K. Y. and Saddler, J. N. (1993), in *Hemicellulose and Hemicellulases*, Coughlan, M. P. and Hazlewood, G. P., ed., Portland, London, pp. 127–143.
9. Viikari, L., Kantelinen, A., Sundquist, J. and Linko, M. (1994), *FEMS Microbiol. Rev.* **13**, 335–350.
10. Haltrich, D., Laussamayer, B., Steiner, W., Nidetzky, B., and Kulbe, K. D. (1994), *Bioresource Technol.* **50**, 43–50.
11. Sachslehner, A., Haltrich, D., Nidetzky, B., and Kulbe, K. D. (1997), *Appl. Biochem. Biotechnol.* **63–65**, 189–201.
12. Haltrich, D., Laussamayer, B., and Steiner, W. (1994), *Appl. Microbiol. Biotechnol.* **42**, 522–530.
13. Miller, G. L. (1959), *Anal. Chem.* **31**, 426–428.
14. Ghose, T. K. (1987), *Pure Appl. Chem.* **59**, 257–268.
15. Poutanen, K. and Puls, J. (1988), *Appl. Microbiol. Biotechnol.* **28**, 425–432.
16. Bradford, M. M. (1976), *Anal. Biochem.* **72**, 248–254.
17. Biely, P., Markovic, O., and Mislovicová, D. (1985), *Anal. Biochem.* **144**, 147–151.

18. Kubicek, C. P., Messner, R., Gruber, F., Mach, R. L., and Kubicek-Pranz, E. M. (1993), *Enzyme Microb. Technol.* **15,** 90–99.
19. Haltrich, D., Nidetzky, B., Kulbe, K. D., Steiner, W., and Zupancic, S. (1996), *Bioresource Technol.* **58,** 137–161.
20. Bisaria, V. S. and Mishra, S. (1989), *CRC Crit. Rev. Biotechnol.* **9,** 61–103.
21. Gübitz, G. M., Hayn, M., Sommerauer, M., and Steiner, W. (1996), *Bioresource Technol.* **58,** 127–135.
22. Gübitz, G. M., Hayn, M., Urbanz, G., and Steiner, W. (1996), *J. Biotechnol.* **45,** 165–172.
23. Gübitz, G. M. and Steiner, W. (1995), *ACS Symp. Ser.* **618,** 319–331.
24. Reese, E. T. and Shibata, Y. (1965), *Can. J. Microbiol.* **11,** 167–183.
25. Rättö, M. and Poutanen, K. (1988), *Biotechnol. Lett.* **10,** 661–664.
26. Arisan-Atac, I., Hodits, R., Kristufek, D., and Kubicek, C. P. (1993), *Appl. Microbiol. Biotechnol.* **39,** 58–62.
27. Torrie, J. P., Senior, D. J., and Saddler, J. N. (1990), *Appl. Microbiol. Biotechnol.* **34,** 303–307.
28. Biely, P. (1993), in *Hemicellulose and Hemicellulases,* Coughlan, M. P. and Hazlewood, G. P., ed., Portland, London, pp. 29–51.
29. Coughlan, M. P. and Hazlewood, G. P. (1993), *Biotechnol. Appl. Biochem.* **17,** 259–289.
30. Messner, R., Gruber, F., and Kubicek, C. P. (1988), *J. Bacteriol.* **170,** 3689–3693.
31. Zeilinger, S., Mach, R. L., Schindler, M., Herzog, P., and Kubicek, C. P. (1996), *J. Biol. Chem.* **271,** 25,624–25,629.
32. Farrell, R. L., Biely, P., and McKay, D. L. (1996), in *Biotechnology in the Pulp and Paper Industry,* Srebotnik, E. and Messner, K., ed., Facultas-Universitätsverlag, Vienna, pp. 485–489.
33. Biely, P. (1991), *ACS Symp. Ser.* **460,** 408–416.

Production and Characterization of *Phanerochaete chrysosporium* Lignin Peroxidases for Pulp Bleaching

M. E. A. de Carvalho,*[1,2] M. C. Monteiro,[1] E. P. S. Bon,[2] and G. L. Sant'Anna, Jr.[3]

[1]*Departamento de Biotecnologia, Faculdade de Engenharia Química de Lorena, P.O. Box 116, CEP 12600-000, Lorena, SP, Brazil;* [2]*Instituto de Química and* [3]*COPPE, Universidade Federal do Rio de Janeiro, Rio de Janeiro, RJ, Brazil*

ABSTRACT

The production of lignin peroxidase from *Phanerochaete chrysosporium* was studied using immobilized mycelia in nylon-web cubes in semicontinuous fermentation using glucose pulses or ammonium tartrate pulses. Consistent enzyme production was achieved when glucose pulses were used, leading to an average activity of 253 U/L. The crude enzyme was added to eucalyptus kraft pulp before conventional and ECF bleaching sequences. Optimization of the enzymatic pretreatment led to the following operational conditions: enzyme load of 2 U/g of pulp, hydrogen peroxide addition rate of 10 ppm/h, and reaction time of 60 min. Pulp final characteristics were dependent on the chemical treatment sequence that followed enzymatic pretreatment. The chief advantage of enzymatic pretreatment was pulp viscosity preservation, which was observed in most of the experiments carried out with seven different chemical treatment sequences.

Index Entries: *Phanerochaete chrysosporium*; lignin peroxidase; enzymatic prebleaching; eucalyptus kraft pulp; selectivity.

INTRODUCTION

World production of market pulp was estimated at 30 million metric tons in 1995, consisting of 55% softwood pulp and 45% hardwood pulp. The main exporters of bleached sulfate pulp are Canada, United States, Sweden, Brazil, Finland, and Chile *(1)*.

---

*Author to whom all correspondence and reprint requests should be addressed.

Conventional industrial bleaching with chlorine produces effluents with chlorinated organic compounds that adversely effect the environment. As a consequence, pulp and paper industries are under growing pressure from authorities, consumers, and environmental groups to reduce their pollution load (2).

The environmental concerns also contribute to the development of new technologies of delignification, bleaching, and effluent treatment. In the case of bleaching, new technologies include use of ClO_2 instead of chlorine; bleaching without chlorine; use of hydrogen peroxide, oxygen, ozone, peracids; and, finally, bleaching with the fungal enzymes, lignin peroxidases (LiP), manganese-peroxidases (MnP), laccases and xylanases, known as biobleaching (3–6).

Several studies have also dealt with the direct cultivation of the fungus *Phanerochaete chrysosporium* on chips or pulps. In some cases, it was observed reduction of the energy required for refining and improvement of the mechanical properties of the pulp, Kappa number reduction, and brightness increase. Biopulping, however, presents the major drawback of extended time of treatment (7).

The major components of the ligninolytic system of the white rot fungus *P. chrysosporium* are the two extracellular heme peroxidases, LiP and MnP (8,9). The production of these enzymes for industrial purposes, however, has been hindered by their low stability under usual fermentation conditions. Considering that enzyme inactivation can be caused by the simultaneous presence of proteases in the culture medium, some reported strategies for enzyme production involve the use of glucose pulses, which would selectively repress protease production (10–14).

Among the possible applications of LiP and MnP in the pulp and paper industry, the utilization in biobleaching is the most promising because the enzymes may be very efficient, and can be used under industrial conditions. LiP prebleaching may enhance bleachability of the pulp, preserving its strength properties (15). Xylanase has also been studied for prebleaching, and its use is reported to facilitate the subsequent chemical bleaching of kraft pulp, although the mode of the enzyme action is not fully understood. According to Kantelinen et al. (6), hemicellulases remove xylan from the fiber surface and renders it more permeable. This facilitates the further stages of pulp treatment with bleaching chemicals. LiP and xylanase pretreatment can be applied to any traditional or modern bleaching sequence without significant investments in existing plants. The objectives of the enzymatic treatment are to decrease the consumption of chemicals, to reduce the pollution load, and to increase final pulp brightness.

The main objective of this study was to investigate the adequacy of LiP pretreatment for bleaching eucalyptus kraft pulp. Enzyme production and stability under fermentation conditions were also studied. Semicontinuous fermentations were carried out, using glucose pulses in the absence of nitrogen source, and ammonium tartrate pulses in absence of glucose.

MATERIALS AND METHODS

Strain Maintenance and Activation

The fungus *P. chrysosporium* ATCC 24725 was used for enzyme production. The strain was maintained on 2% malt-agar slants, under refrigeration. Slopes of the same agar were used for production of spores, which were used as inoculum in all fermentations. A dense sporulation was observed within 1 wk of incubation at room temperature. To obtain the spore suspension to be used as inoculum, 5.0 mL of sterile dH_2O were carefully added to a slope. The resultant spore suspension was filtered through sterile glass wool to remove hyphal fragments. The concentration of spores was determined using a standard curve that correlated absorbance at 650 nm with spores concentration in spores/mL.

Fermentations

Experiments were performed using immobilized mycelium. For the immobilization step, shake-flask fermentations, using a chemically defined, carbon-limited medium, were carried out in the presence of nylon-web cubes *(16)*. Flasks containing 75 mL of the culture medium were inoculated with 1.65×10^7 spores, incubated at 37°C, and shaken at 150 rpm. The spores were found to germinate preferentially inside the particles *(17)*. After the growth phase, which was coincident to glucose depletion, veratryl alcohol (2.5 m*M*) was added to the culture, and both immobilized mycelium and culture medium were transferred to a packed-bed reactor. The bioreactor consisted of a 120-mL jacketed glass column (L/D ratio, 5:1) containing a fixed bed of nylon-web cubes. Reactor operation was performed at 37°C in a semicontinuous mode. At defined time intervals, one-third of the liquid volume was replaced by fresh medium containing 1 g/L glucose or 0.22 g/L ammonium tartrate. Veratryl alcohol (2.5 m*M*) and Tween-80 (1.3 g/L) were also added to the fresh medium. Glucose or ammonium tartrate were used, in order to perform pulses during the enzyme production stage. Oxygen gas was directly fed to the reactor bottom at a constant specific rate of 0.22 mL O_2/mL/min.

Enzyme Assays

All enzyme concentration measurements were carried out in the supernatant.

LiP activity was measured by oxidation of veratryl alcohol to veratraldehyde, in presence of hydrogen peroxide, according to Tien and Kirk *(18)*. Oxidation of veratryl alcohol was monitored by absorbance at 310 nm ($\epsilon_{310} = 9300 \text{ m}^{-1}\text{cm}^{-1}$). One unit (U) of enzyme activity was defined as the amount of enzyme that oxidizes one µmol of veratryl alcohol to veratraldehyde in 1 min.

MnP activity was determined as described by Kuwahara et al. *(8)*. Reaction was monitored at 610 nm for 60 s ($\epsilon_{610} = 4460$ m^{-1}cm^{-1}). One U of enzyme activity was defined as the amount of enzyme that oxidized 1 μmol of phenol red per min in presence of Mn(II) and hydrogen peroxide.

Endoglucanase activity was measured as described by Mandels et al. *(19)*, using carboxymethyl cellulose (CMC) as substrate. One U of enzyme activity produced 1 μmol of reducing sugars per min.

Protease activity was measured according to Charney and Tomarelli *(20)*, using azocasein as substrate. Samples (0.5 mL) were incubated for 40 min at 37°C, with 0.5 mL of 0.5% (w/v) azocasein solution in 50 mM acetate buffer. The reaction was stopped by the addition of 0.5 mL of 10% (w/v) trichloroacetic acid solution. The residual substrate was removed by centrifugation at 7000×g for 5 min. To 1 mL of supernatant, 1 mL of 5 N KOH was added, and the mixture was spectrophotometrically assayed at 428 nm. The determinations were carried out against both substrate and enzyme blanks. One U of enzyme activity catalyzed the release of azo dye, causing a change in absorbance at 428 nm of 0.001/min.

Analytical

Glucose was measured as reducing sugars by the method of Nelson *(21)*. Hydrogen peroxide was measured, using a specific Peroxide test (Merckoquant 10011).

Pulp Characteristics

The industrial pulp used in the experiments was an unbleached hardwood kraft pulp, which was obtained using a mixture of several eucalyptus varieties. The initial characteristics of the pulp were: Kappa number 15; viscosity 31.7 cp., and brightness 32% ISO.

Enzymatic Pretreatment

The following conditions were investigated: enzyme loadings from 2 to 10 U/g pulp basis, reaction time from 60 to 180 min, and the rate of hydrogen peroxide addition from 10 and 20 ppm/h. Pulp at 10% consistency was acidified to pH 4.5 with lactic acid. Enzyme and hydrogen peroxide at the desired load, and 4 mM veratryl alcohol, were added to the pulp, and the mixture was incubated at 30°C for the required time. As a control, pulp was treated under identical conditions without enzyme. After the enzymatic treatment carried out under the studied conditions, the pulp was submitted to a short bleaching sequence (CE), and analyzed for Kappa number and viscosity. According to the results, optimized conditions for enzymatic prebleaching were defined. These conditions were used before the conventional and ECF bleaching sequences. Both enzymatic and chemical treatments were performed in polyethylene bags, in a water bath, at the desired temperatures and time intervals.

Table 1
Conditions of Conventional and ECF Bleaching Chemical Sequences

Conventional bleaching sequences

Sequence 1

A: C3E sequence
Conditions: Chlorination: 3% chlorine, 30°C, 30 min, 3% consistency
Extraction: 2% NaOH, 60°C, 60 min, 10% consistency

B: C2E sequence
Conditions: Same as sequence 1A, except with 2% chlorine

Sequence 2

CEHH sequence
Conditions: Chlorination: 3% chlorine, 30°C, 30 min, 3% consistency
Extraction: 2% NaOH, 60°C, 60 min, 10% consistency
First and second hypochlorination: 1.7% HClO, 50°C, 120 min, 10% consistency

Sequence 3

A: CEHD sequence
Conditions: Chlorination: 2,5% chlorine, 50°C, 30 min, 4% consistency
Extraction: 2.2% NaOH, 65°C, 65 min, 12% consistency
Hypochlorination: 1.0% HClO, 50°C, 120 min, 12% consistency
Dioxidation: 1.0% ClO_2, 75°C, 180 min, 12% consistency

B: CEpHD sequence
Conditions: Same as sequence 3A, except with 0.5% H_2O_2 in extraction stage

ECF Bleaching sequences

Sequence 4

DED sequence
Conditions: First dioxidation: 1.0% ClO_2, 70°C, 180 min, 6% consistency
Extraction: 2% NaOH, 60°C, 60 min, 10% consistency
Second dioxidation: 0.4% ClO_2, 70°C, 120 min, 10% consistency

Sequence 5

DEDED sequence
Conditions: First dioxidation: 1.0% ClO_2, 70°C, 180 min, 6% consistency
First and second extraction: 2% NaOH, 60°C, 60 min, 10% consistency
Second and third dioxidation: 0.4% ClO_2, 70°C, 120 min, 10% consistency

Bleaching

The conventional and ECF bleaching chemical sequences were performed as indicated in Table 1. All concentrations were expressed in terms of air-dried brownstock pulp. At the end of each stage, the pulp was filtered and washed with dH_2O.

Evaluation of Pulp Treatments

All determinations of brightness, Kappa number, and viscosities were made according to standard Technical Association of the Pulp and Paper

Industry methods. Selectivity was defined as the ratio between the degree of delignification and loss of pulp viscosity. Kappa numbers and pulp viscosities values were determined after the extraction stage and at the end of the bleaching sequences, respectively.

RESULTS AND DISCUSSION

Fermentations

Fermentation experiments were carried out to achieve consistent lignin peroxidase production. In the fermenter, feed scheme pulses of ammonium tartrate, which can be used by the microorganism as carbon (C) and nitrogen (N) source, were compared to glucose pulses. The advantages of using the salt, instead of glucose, include handling simplicity and lower chances of contamination in such a long operation.

According to the data presented in Figs. 1 and 2, which show lignin peroxidase activity and pH variation against time, using glucose or salt pulses every 12 h, overall better results were observed for the glucose scheme. These results confirm previous findings for 24-h frequency glucose pulses *(10)*, although more regular enzyme production was observed for the 12-h time interval. Enzyme production was observed for both the glucose and ammonium tartrate-operated reactors, which were run for 660 h. In the salt feed reactor, however, lignin peroxidase activity dropped at 560 h. This was concomitant to a pH increase that suggested a decrease in the viability of the immobilized mycelia.

Considering the profile of protease production (Fig. 2), the glucose feed mode was more effective in keeping a lower protease accumulation. This would explain the higher LiP activity when glucose was used.

According to the foregoing, the crude enzyme preparation used in the present work was obtained using the glucose-operated reactor. This preparation showed LiP activity within the range of 70–350 U/L. The ratio between LiP and MnP activities in the crude preparation showed an average value of 0.10. No endoglucanase activity was observed.

Bleaching

The factors affecting the efficiency of pulp treatment by LiP are pH, temperature, presence, or absence of veratryl alcohol, hydrogen peroxide concentration, reaction time, and enzyme concentration.

The presence of veratryl alcohol is an important factor for enzymatic treatment performance, because it stabilizes lignin peroxidase under excess of hydrogen peroxide, and acts as mediator in the electron transfer process between enzyme and substrate *(22)*. In this work, a concentration of 4 mM was applied, because, according to the literature, this reagent is used within the range of 2.5–4 mM *(23,24)*.

Eucalyptus kraft pulp was treated with LiP according to the conditions indicated in Table 1. In this first set of experiments, the reaction time

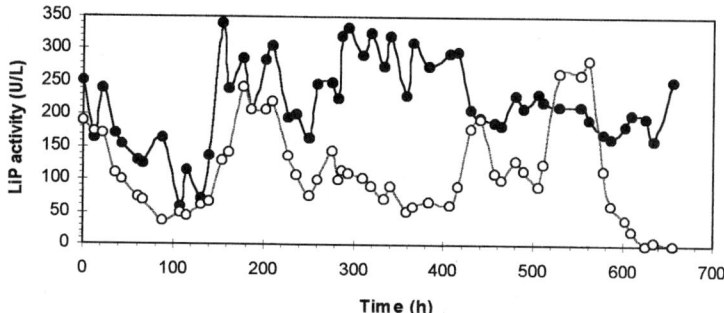

Fig. 1. Lignin Peroxidase production in a packed-bed biorreactor submitted to glucose and ammonium tartrate pulses, symbol: ● glucose pulses, and ○ ammonium tartrate pulses.

Fig. 2. Protease production and pH variation in a packed-bed biorreactor submitted to glucose and ammonium tartrate pulses, symbol: ● glucose pulses, ■ glucose pulses, ○ ammonium tartrate pulses, and □ amonium tartrate pulses.

was fixed at 180 min, which was compatible to both enzymatic reaction and industrial bleaching processes. After enzymatic prebleaching, the pulps were submitted to bleaching sequences 1A and 1B, which are listed in Table 1.

Delignification and percentage of viscosity reduction are shown in Fig. 3. The percentage of delignification remained comparable for pulps treated with enzyme loads of 2 and 5 U/g. Moreover, treatment with 10 U/g did not significantly improve delignification. Beyond 5 U/g, no appreciable viscosity reduction was observed.

Pulps treated with 2 and 5 U/g, and with 10 and 20 ppm/h of H_2O_2, showed delignification levels around 70%, when submitted to a CE bleaching sequence with 3% of chlorine. When the same sequence, with a lower level of chlorine (2%), was used, only 50% of delignification was achieved, and high viscosities were observed. In both cases, the increase of the rate of H_2O_2 addition did not improve pulp delignification. However, viscosity reduction was affected by the increase of H_2O_2 supply in most cases, inde-

Fig. 3. Effect of enzyme and peroxide addition rate on the degree of delignification (**A**) and viscosity reduction (**B**).

pendent of chlorine concentration, as illustrated by the results shown in Table 2. Increase of enzyme load from 2 to 5 U/g did not improve pulp delignification, as expressed by Kappa number values of 5.1 and 5.0, respectively. Final pulp viscosity was affected by both H_2O_2 supply rate and enzyme load, as illustrated in Fig. 4. As shown in Table 2, the percentages of viscosity reduction for all pulps treated with LIP were lower than those of control tests in which no enzyme or no peroxide was used.

These results (Fig. 3 and Table 2) could be explained by modifications of lignin molecules caused by the enzymatic action. The modified lignin would be more resistant to alkaline extraction. The viscosity, however, was favored by the enzymatic treatment. This could be because of the protection of the cellulose fibers by the modified lignin.

Table 2
Effect of Enzyme Load and H_2O_2 Addition Rate on Enzymatic Treatment and on Pulp Properties

LiP (U/g pulp)	H_2O_2 (ppm/h)	Chlorine %	Delignif. %	Viscosity reduction %	Selectivity
2	10	2	48	9	5.3
2	10	3	67	12	5.6
2	20	2	42	11.6	3.6
2	20	3	70	16	4.3
5	10	2	48.4	19	2.5
5	10	3	68	21	2.7
5	20	2	48	19	2.5
5	20	3	70	25	2.8
Standard					
	–	2	56	32	1.8
CE	–	3	76	39	2.0

Pulps treated with LiP were bleached with CE sequence with 3% Cl.

Fig. 4. Viscosities of pulps treated with LiP prebleaching.

Table 3 compares the selectivity values for pulps treated with 2 and 5 U/g to standard pulps treated with CE sequence. The results of the experiments, which were conducted for different periods of time, indicated that optimal enzyme dosage was 2 U/g. Moreover, a short reaction time led to high selectivity values. The delignification range was within 65–72%, and therefore comparable to the control experiment. LiP treatment, however, caused a marked improvement in viscosity, and it also seems to improve bleacheability in the subsequent bleaching stages (results not shown).

Experiments with pulps submitted to enzymatic prebleaching followed by ECF (elemental chlorine-free) bleaching sequences were also conducted. When a DED sequence was used (Table 4), higher selectivity was obtained for the shorter reaction time (60 min) and the lower enzyme

Table 3
Effect of Reaction Time on Kappa Number, Viscosity, and Selectivity of Pulps
Treated with Lignin Peroxidase and C_3E Bleaching Sequence

LiP (U/g pulp)	Reaction time (min)	Kappa number	Delignification %	Viscosity cp	Viscosity reduction %	Selectivity
2	60	4.7	68.7	29.0	8.5	8.0
2	120	4.9	67.4	29.0	8.5	7.9
2	180	5.2	65.5	28.4	10.4	6.3
5	60	4.5	69.9	28.5	10.1	6.9
5	120	4.3	71.8	28.3	10.7	6.7
5	180	4.3	71.8	27.6	12.9	5.6
C_3E^a	180	4.4	70.6	25.5	19.6	3.6
C_2E^a	180	6.1	59.4	26.8	15.4	3.8

[a] Standard: treatment without enzyme and with H_2O_2 (10 ppm/h) and VA (4 mM).

Table 4
Effect of Reaction Time and Enzyme Load on Properties of Pulp Submitted to
Lignin Peroxidase Prebleaching and ECF Bleaching

LiP (U/g pulp)	Reaction time (min)	Delignification (%)	Viscosity reduction (%)	Selectivity
2	60	55.0	12.6	4.4
2	180	55.3	25.8	2.1
5	60	48.0	20.2	2.4
5	180	52.6	18.3	2.9
DED^a	60	53.5	26.5	2.1
DED^a	180	53.3	28.4	1.9

[a] Standard with VA (4 mM) and peroxide (10 ppm/h), without LiP.

concentration (2 U/g) tested. In these experiments, an improvement of selectivity was observed for higher reaction times (180 min) when the enzyme load was 5 U/g.

Selectivity was also affected by the bleaching sequence used. When pulps were submitted to prebleaching (2 U/g) and subsequent bleaching with CE sequence with 3 and 2% of chlorine, those pulps showed selectivities of 5.6 and 5.3, respectively (Table 5). However, when the pulps were submitted to a DED sequence, the selectivity decreased to 4.4. Pulp characteristics resulting from several treatment sequences are also shown in Table 5.

Concerning selectivity, the best results (5.4, 4.4, and 4.0) were obtained when sequences L_2CEHD, L_2*DED, and $L_2*DEDED$ were used, respectively. The low selectivity value observed when the sequence L_2CE_pHD

Table 5
Selectivity and Brightness of Pulp Treated with Lignin Peroxidase and Several Bleaching Sequences

Sequences	Kappa number	Viscosity cp	Selectivity	Brightness %ISO
CEHD	3.8	18.0	1.7	81.3
L_2C_3E	5.1	27.9	5.6	nd
L_2C_2E	7.8	28.8	5.3	nd
L_2CEHH	5.2	11.2	1.1	nd
L_2CEHD	5.6	28.0	5.4	75.3
L_2CE_pHD	4.7	17.7	1.6	81.2
L_2^*DED	6.9	27.7	4.4	nd
L_2^*DEDED	6.7	27.0	4.0	nd

All conditions of bleaching sequences are listed in Table 1.
L_2: 2U LiP/g pulp/180 min.
L_2^*: 2U LiP/g pulp/60 min.

was employed may be attributed to the utilization of hydrogen peroxide, which reduced pulp viscosity. However, the addition of H_2O_2 in the extraction stage of the sequence L_2CE_pHD increased pulp brightness from 75.3% to 81.2% ISO (Table 5).

The L_2CEHH sequence showed very low selectivity values. In this case, severe viscosity reduction occurred in the final bleaching stages. Sodium hypochloride promotes depolymerization of the cellulose chain, reducing pulp viscosity.

When ECF sequences were used (Chlorine dioxide sequences), namely, L_2^*DED, and L_2^*DEDED, selectivity results were similar to those obtained with a conventional chlorine sequence (L_2CEHD). This is an important result, because it allows the use of dioxide as a substitute of elemental chlorine. Table 5 shows that viscosities were preserved in both cases, and, as a consequence, the strength properties of pulp were also maintained. Kappa numbers were higher for pulps treated with dioxide (L_2^*DED and L_2^*DEDED) than for pulps treated with a conventional bleaching sequence using chlorine (L_2CEHD). The use of LiP prebleaching, associated with severe bleaching chemical sequences (L_2^*DED and L_2^*DEDED), may produce pulps with low Kappa numbers, high viscosities, and brightness compatible to or higher than, that of conventional pulps.

CONCLUSION

Consistent lignin peroxidase production was observed using a packed-bed reactor containing *P. chrysosporium* immobilized in sponge cubes. The reactor was operated in a semicontinuous mode, with glucose pulses every 12 h over 660 h. Under these conditions average values of LiP activity of 253 U/L were observed. The crude enzyme preparation was shown to be effective for the biobleaching of eucalyptus kraft pulp.

Pretreatment of the eucalyptus kraft pulp with crude lignin peroxidase from *P. chrysosporium* under optimized conditions, i.e., enzyme load of 2 U/g of pulp, hydrogen peroxide addition rate of 10 ppm/h, during 60 min at 30°C, yielded a great improvement of the pulp selectivity. The delignification degree was not higher in comparison to the conventional treatment, but LiP treatment was beneficial because the pulp viscosity was preserved, even under the drastic conditions of the chemical treatments.

Considering the use of LiP on conventional and ECF sequences, the improvement on the selectivity depended on the conditions used in both the enzymatic pretreatment and the chemical bleaching sequences.

REFERENCES

1. Macedo, A. R. P., Valenca, A. C. V., and Lima, A. S. (1996) *O Papel* **11**, 45–58.
2. Mehta, V. and Gupta, J. K. (1992) *Tappi J.* **75**, 151–152.
3. McDonough, T. J. (1995) *Tappi J.* **78**, 55–62.
4. Liebergoth, N. (1996) *Pulp and Paper Canada*, **97**, 73–75.
5. Reid, I. D. and Paice, M. G. (1994) *FEMS Microbiol. Rev.* **13**, 369–376.
6. Kantelinen, A., Hortling, B., Ranua, M., and Viikari, L. (1993) *Holzforschung* **47**, 29–35.
7. Eriksson, K.-E. (1985) *Tappi J.* **68**, 46–55.
8. Kuwahara, M., Glenn, J. K., Morgan, M. A., and Gold, M. H. (1984) *FEBS Lett.* **169**, 247–250.
9. Paszczynski, A., Huynh, V. B., and Crawford, R. L. (1985) *FEMS Microbiol. Lett.* **29**, 37–41.
10. Linko, S. (1988) *Enzyme Microbiol. Technol.* **10**, 410–417.
11. Linko, S. (1988) *J. Biotechnol.* **6**, 229–243.
12. Dosoretz, C. G., Chen, H. C., and Grethlein, H. E. (1990) *Appl. Environ. Microbiol.* **56**, 395–400.
13. Dosoretz, C. G., Dass, S. B., Reddy, C. A., and Grethlein, H. E. (1990) *Appl. Environ. Microbiol.* **56**, 3429–3434.
14. Feijoo, G., Dosoretz, C. and Lema, J. M. (1995) *J. Biotechnol.* **42**, 247–253.
15. Eriksson, K.-E. (1990) *Wood Sci. Technol.* **24**, 79–101.
16. Linko, S. and Zhong, L.-C. (1987) *Biotechnol. Techn.* **1**, 249–254.
17. Bon, E. P. S. and Webb, C. (1989) *Enzyme Microbiol. Technol.* **11**, 495–499.
18. Tien, M. and Kirk, T. K. (1984) *Proc. Natl. Acad. Sci.* **81**, 2280–2284.
19. Mandels, M., Andreotti, R., and Roche, C. (1976) *Biotech. Bioeng. Symp.* **6**, 17–34.
20. Charney, J. and Tomarelli, R. M. (1947) *J. Biol. Chem.* **171**, 501–505.
21. Nelson, N. (1944) *J. Biol. Chem.* **153**, 375–380.
22. Goodwin, D. C., Aust, S. D., and Grover, T. A. (1995) *Biochemistry* **34**, 5060–5065.
23. Faison, B. D., Kirk, T. K., and Farrel, R. L. (1986) *Appl. Environ. Microbiol.* **52**, 251–254.
24. Arbeloa, M., Leseleuc, J., Goma, G. and Pommier, J. C. (1992) *Tappi J.* **7**, 215–221.

In Vitro Degradation of Insoluble Lignin in Aqueous Media by Lignin Peroxidase and Manganese Peroxidase

DAVID N. THOMPSON,*[1,2] BONNIE R. HAMES,[3] C. A. REDDY,[2,4] AND HANS E. GRETHLEIN[1,2,5]

[1]Department of Chemical Engineering and [2]NSF Center for Microbial Ecology, Michigan State University, East Lansing, MI 48824; [3]Biomass Analysis Group, National Renewable Energy Laboratory, Golden, CO 80401; [4]Department of Microbiology, Michigan State University, East Lansing, MI 48824; and [5]Michigan Biotechnology Institute, Lansing, MI 48909

ABSTRACT

The abilities of lignin peroxidase (LIP) and manganese peroxidase (MNP) from *Phanerochaete chrysosporium* to degrade an insoluble hardwood lignin in vitro in aqueous media were tested. Neither LIP nor MNP appreciably changed the mass or lignin content, although both produced small amounts of unique solubilized lignin fragments. Treatment with both LIP and MNP, however, decreased the mass by 11%, decreased the lignin content by 5.1% (4.2% as total weight), and solubilized unique lignin-derived molecules. These results suggest that LIP and MNP synergistically degrade high molecular weight insoluble lignin, but singly, neither enzyme is sufficient to effect lignin degradation.

Index Entries: Lignin peroxidase (LIP); manganese peroxidase (MNP); *Phanerochaete chrysosporium*.

INTRODUCTION

Billions of tons of lignocellulosic biomass are produced each year worldwide *(1)*. Lignin biodegradation is a rate-limiting step in the mineralization of lignocellulosic biomass in the biosphere. Lignin removal in the pulp and paper industry is costly, energy-intensive, and toxic effluents are produced *(2)*. Thus, microbial delignifying enzymes have important potential applications in the pulp and paper industry, and also in the bio-

* Author to whom all correspondence and reprint requests should be addressed. Present address: Biotechnologies Dept. Idaho National Engineering and Environmental Laboratory P.O. Box 1625 Idaho Falls, ID 83415-2203.

remediation of aromatic xenobiotics (3). Reactions catalyzed by the extracellular peroxidases of *Phanerochaete chrysosporium* have been well characterized on soluble lignin model compounds (4). In contrast, attempts to degrade the insoluble, high molecular weight polymer in completely aqueous media have been largely unsuccessful, casting considerable doubt on the putative roles of these peroxidases in the in vivo degradation of natural lignins (4).

The goal of this study was to determine whether it is possible to degrade a water-insoluble natural lignin polymer in aqueous media using the extracellular ligninolytic enzymes of *P. chrysosporium*. This was done using the two classes of peroxidases singly and in combination to treat an isolated water-insoluble lignin, which was shown by thioacidolysis and syringyl:guaiacyl ratio to be chemically and structurally similar to the initial poplar substrate. Fourier transform infrared spectroscopy (FTIR), gas chromatography/mass spectrometry (GC/MS), and mass balances were used to indicate lignin degradation. The results indicate that LIP and MNP act synergistically to degrade insoluble lignin in aqueous media, but singly, little or no degradation of the solid is observed.

MATERIALS AND METHODS

Lignin Preparation

The lignin substrate used in this study was isolated from hybrid poplar *(Poplus eugeneii)*, as previously described (5,6), using a sequential treatment of dilute-acid hydrolysis, cellulolytic hydrolysis, and exhaustive extractions at 37°C with chloroform, ethyl acetate, and methanol. Carbohydrate and lignin compositions were measured by quantitative saccharification (7). In this technique, the sample was subjected first to hydrolysis with 72 wt% H_2SO_4 for 1 h at 30°C, and then with 4 wt% H_2SO_4 for 1 h at 121°C in an autoclave. The liquid fraction was quantitatively collected, diluted to a known volume, and analyzed for carbohydrates by HPLC, using a Bio-Rad (Hercules, CA) HPX-87P column with deashing system. Carbohydrate values were corrected for losses incurred during the high-temperature step through the use of a carbohydrate recovery experiment at 4 wt% H_2SO_4 and 121°C. The remaining insoluble portion was defined as Klason lignin with extractives and ash.

Enzymes

LIP and MNP enzymes were produced at 37°C from the extracellular fluid of agitated nitrogen-limited, acetate-buffered cultures of *P. chrysosporium*, strain BKM-F-1767 (ATCC 24725), as described by Tien and Kirk (8). Agitated cultures were inoculated with mycelia from stationary cultures containing 11.5 ppm of Mn(II) (8), as previously described (9). Basic media for these cultures were prepared as previously described (5,9), except that Mn(II) was added as $MnSO_4$ to production cultures at 100 ppm, to produce MNP for harvest on d 4, and at 0 ppm to produce LIP for

Fig. 1. Schematic of the dialysis reactor.

harvest on d 5. After concentration by ultrafiltration and dialysis (5,6), aliquots of the dialyzed concentrates for each were stored at −20°C.

Activity Assays

LIP activity was determined spectrophotometrically at 310 nm (8), defining 1 unit (U) of LIP activity as the oxidation of 1 μmol of veratryl alcohol to veratraldehyde per min, using a molar absorptivity of 9300 M/cm. MNP activity was measured spectrophotometrically at 610 nm (2), with a reaction time of 4 min, defining 1 U of MNP activity as the oxidation of 1 μmol of phenol red per min, using a molar absorptivity of 4460 M/cm. Note that 1 U/L of LIP activity represents a much higher amount of heme protein than 1 U/L of MNP.

Lignin Treatment and In Vitro Reactor System

Lignin treatment with peroxidases was done in a small-scale dialysis reactor (Fig. 1) for 12 h at 37°C, 2% (w/v) solids, with approx 50 U LIP/g lignin and/or 2800 U MNP/g lignin in 20 mM tartrate buffer, pH 3.5, at 350 strokes/min, as previously described (5,6). A 6000–8000 mol wt cutoff dialysis membrane, with a measured wet membrane thickness of 61 μm, was used to separate the chambers. Only silanized glassware, including the reactor halves and all sample handling equipment, were contacted with reaction mixtures, samples, or solvent extracts, to prevent compounds released from the lignin from binding to the glass. This reactor system was used to control the levels of H_2O_2 and veratryl alcohol (VA) in contact with the peroxidases over the course of the reaction. Control was achieved through use of a predictive control scheme (5), which allowed adjustments of H_2O_2 and/or VA in the reagent reservoir (1 mL liquid), and LIP and/or

Table 1
Working Concentration Ranges in the Reaction Chamber
Over 12 h of Treatment

Enzyme(s) used	H_2O_2 Concentration range (μM) (Average in parentheses)	VA^a Concentration range (μM) (Average in parentheses)
LIP	10 < [H_2O_2] (15) < 25	60 < [VA] (360) < 550
MNP	15 < [H_2O_2] (40) < 80	$-^b$
LIP + MNP	5.0 < [H_2O_2] (8.0) < 15	20 < [VA] (340) < 550

Brackets indicate concentration.
[a] VA, veratryl alcohol.
[b] VA was not present for this enzyme case.

MNP in the reaction chamber (1 mL slurry). Adjustments were necessary because of the small volume of the system. Concentration ranges maintained in contact with the enzymes over the course of the various runs are listed in Table 1. Initial conditions for each run, including controls, are presented in Table 2. Tartrate was used both as buffer and as the Mn(III) chelator *(10)*. After 12 h of reaction, the slurry in the reaction chamber was quantitatively collected (95% recovery of solids) by centrifugation. The supernatant was stored at $-80°C$ for future GC/MS analyses. The resulting lignin pellet was sequentially washed free of protein by centrifugation with 1 M NaCl, followed by distilled H_2O. The pellet was then suspended in 1 mL of distilled H_2O and stored at $-80°C$.

Mass Balance and FTIR Analyses

Fourier transform infrared spectra were measured for all lignin samples using a Nicolet 5SXC FTIR spectrophotometer equipped with a deuterated triglycine sulfate detector. Transmission FTIR spectra of the pellets, averaged from 50 scans, were measured from 4000 to 650 cm^{-1} with 2 cm^{-1} resolution. A KBr background spectrum was subtracted from each spectrum. Measurements were performed on wafers containing approximately 0.5 wt% of solvent extracted (*see* extraction procedure below), quantitatively lyophilized lignin, pressed under vacuum, as previously described *(5,11)*. The mass balance was calculated from the lyophilized weight prior to pressing the wafer. Spectra were analyzed by partial least-squares (PLS) regression *(11)* for methoxyls/C_9, phenolic hydroxyls/C_9, carbohydrate, and lignin contents for each sample, using methods that were procedurally similar to that for the methoxyl content *(11)*. The regression methods were calibrated separately, as follows: The methoxyl method was calibrated with milled wood lignins *(11)*; the carbohydrate and lignin methods were calibrated with a combination of whole-wood samples and wood pulps in a manner analogous to that for methoxyl content; finally, the phenolic hydroxyl method was calibrated with organic solvent-isolated kraft lignins.

Table 2
Initial Conditions and Experimental Design for Reactor Runs and Controls

Enzyme(s)[a] used	Condition	Lignin (mg)	[LIP] (U/L)	[MNP] (U/L)	Reactant reservoir [H$_2$O$_2$] (mM)	Reactant reservoir [VA] (mM)	Reaction chamber [H$_2$O$_2$] (μM)[c]	Reaction chamber [VA] (μM)[c]	Reaction chamber [Mn(II)] (μM)[d]
LIP	Base case	22.8	1000		38.8	149	100	600	
LIP	No lignin control		1000		38.8	149	100	600	
LIP	No reagent control	23.1	1000						
LIP	No enzyme control	22.9			0.1	0.6	100	600	
LIP	Autoclaved enzyme control	23.1	nd[b]		0.1	0.6	100	600	
LIP	No enzyme, no reagent control	22.9							
MNP	Base case	22.8		56000	11.0		100		100
MNP	No lignin control			56000	11.0		100		100
MNP	No reagent control	23.0		56000					
MNP	No enzyme control	22.8		nd	0.1		100		100
MNP	Autoclaved enzyme control	22.9		nd	0.1		100		100
MNP	No enzyme, no reagent control	23.1							
LIP+MNP	Base case	22.8	1000	56000	44.4	120	100	600	100
LIP+MNP	No lignin control		1000	56000	44.4	120	100	600	100
LIP+MNP	No reagent control	22.8	1000	56000					
LIP+MNP	No enzyme control	23.1			0.1	0.6	100	600	100
LIP+MNP	Autoclaved enzyme control	23.0	nd	nd	0.1	0.6	100	600	100
LIP+MNP	No enzyme, no reagent control	23.0							

Brackets indicate concentration, while blanks indicate the absence of the species.
[a] The LIP concentrate consisted mainly of isoenzymes H2, H6, and H8, and also contained a very small amount of MNP, essentially all H4. The MNP concentrate consisted mainly of H3. Note that 1 U/L of LIP activity represents a much higher relative activity than 1 U/L of MNP activity; i.e., the LIP+MNP mixture contained 56% LIP and 44% MNP measured as heme protein.
[b] nd, None detected.
[c] Note that these values are higher than the maximum levels listed in Table 1 in which enzyme(s) are present. This is because conditions were set initially to quickly drive the H$_2$O$_2$ and VA levels down into the desired ranges (listed in Table 1) and maintain them there.
[d] Added as tartrate-chelated Mn(III) by first mixing it with the active MNP and stoichiometric H$_2$O$_2$.

Extraction of Liquid and Solid Samples for GC/MS

Recovered liquid fractions were thawed and extracted sequentially with three equal volumes of chloroform, recovering the chloroform layer each time. The process was repeated with ethyl acetate. Recovered solid fractions were thawed and pelleted by centrifugation, discarding the supernatant. The pellet was extracted twice at 37°C with 1 mL of chloroform, and then twice with 1 mL of ethyl acetate. The extracted solids were then washed free of solvents by centrifugation with distilled H_2O and stored at -80°C under 1 mL of distilled H_2O. To roughly quantify unknown peaks, internal standard, 2-chloro-5-(trifluoromethyl)-benzoic acid was added to each vial just prior to GC/MS analysis at 1020 ng/vial, using a clean glass microliter syringe. The extracts were then evaporated nearly to dryness under a light stream of nitrogen and derivatized for GC/MS (see below). Once samples were extracted, the extracts were stored at -20°C for no longer than 24 h before analysis, to guard against excessive volatilization.

GC/MS Analyses

Chloroform and ethyl acetate extracts of the liquid and solid samples were analyzed as trimethylsilane (TMS) ethers by GC/MS, as previously described (5). Derivatization was done with N,O-bis-(trimethylsilyl)-trifluoroacetamide (BSTFA). Extracts were analyzed at 70 eV, using a JEOL AX 505 double-focusing magnetic sector mass spectrometer equipped with an HP5890 gas chromatograph. Separation of extracts was achieved with either a 30 m DB5-MS capillary GC column (J and W, Folsom, CA), or with a 30 m DB-1 capillary GC column (SPB-1, Supelco, Bellefonte, PA) (5,6). Run conditions and the column used were taken into account when comparing retention times for comparison of controls and base case ion chromatograms.

Treatment of GC/MS Data

Unique peaks were identified in the base case chromatograms by comparison with the chromatograms of the controls. Peaks that were not unique to the base cases were excluded, unless the relative concentration was at least $3 \times$ higher in the base case. The selected peaks were integrated and their concentrations estimated from the internal standard. Commercially available databases were searched for matches to the mass spectra of the unknowns. Structures were suggested for each remaining unknown, using the mass spectrum, the derivatization pattern, and the following assumptions: It was assumed that the compounds were lignin-derived, and thus contained only carbon, hydrogen, and oxygen. Next, it was assumed that the products retained ring structures in aromatic or quinone forms, which limited the number of rings plus double bonds (12).

RESULTS

Substrate Composition

Substrate compositions and physical characterization data were obtained in order to demonstrate the similarity of this lignin substrate to natural, insoluble lignin as it occurs in wood. The compositions of the poplar residue at each step of the lignin isolation procedure are presented in Table 3. The composition of the untreated poplar was typical for this type of hardwood (13). Cellulose/hemicellulose removal yielded a substrate containing 74.5% lignin and 12.2% carbohydrates, after exhaustive solvent extraction. The final lignin residue was insoluble in dioxane, dimethylformamide, and several aqueous solutions of the two (5). Acetylation (14) did not produce a soluble lignin, and thioacidolysis (15,16) produced yields of syringyl and guaiacyl monomer products of 78 and 87%, relative to those from the native poplar, respectively, indicating no significant condensation of β-O-4 intermonomer linkages. The syringyl:guaiacyl ratios for the isolated lignin and the untreated poplar were not significantly different (16), at 1.02 and 1.15, respectively. These data indicate that the isolated lignin was similar to the untreated poplar lignin in terms of monomer contents, ratio of monomer types, and frequency of intact interunit linkages.

LIP and MNP Production

The LIP concentrate consisted mostly of isoenzymes H2, H6, and H8, and also contained a very small amount of MNP (*see* Table 2), essentially all H4. This probably occurred because of a small amount of Mn(II) carryover in the mycelial inoculum to the production cultures. The effect of this MNP in the LIP-alone lignin treatment runs was negated, however, by excluding Mn(II) from the reaction mixture, so that only LIP was active. The MNP produced consisted mainly of H3 and contained no detectable LIP activity.

Treatment of Lignin

Mass balances on the solid lignin before and after treatment with LIPs and/or MNPs were calculated as a simple measure of whether mass was lost because of enzyme activity. Results of the mass balances are presented in Fig. 2. The statistical significance of the base case measurements were tested using the null hypotheses that they were part of their control sets, since the controls have the same systematic errors associated with their measurement. The significance levels, using the Student's t-distribution with three degrees of freedom as the reference distribution, were 0.376, 0.130, and 0.018 for the LIP, MNP, and LIP + MNP base cases, respectively. Note that each control set was used separately, with its corresponding base cases (with which it was run). Thus, the null hypothesis was rejected

Table 3
Compositions of Untreated Poplar and Residue After Each Step of Lignin Isolation Procedure

Component	Untreated poplar (%)	Acid hydrolyzed (%)	Cellulase hydrolyzed (%)	Solvent extracted (%)
Glucan	42.9	62.9	8.1	10.1
Xylan	13.8	0.9	0.9	1.1
Galactan	0.1	nd[a]	0.3	1.0
Arabinan	0.6	nd	nd	nd
Mannan	1.0	nd	nd	nd
Lignin[b]	25.4	29.3	78.1	74.5
SUM[c]	83.8	93.1	87.4	86.7

Sequential isolation steps proceed from left to right in the table.
[a] nd, none detected.
[b] Includes Klason lignin, extractives, and ash.
[c] Not equal to 100% because of recovery errors and unknown uronic acid content.

for LIP+MNP, meaning that it was not part of its control set. The null hypothesis was not rejected for LIP or MNP alone. Comparatively, the significance level for a control data point showing a 7% loss, which was the maximum deviation seen among the control sets, was 0.208 in its control set, and 0.057 over all the control sets combined. Thus, the LIP+MNP base case data point, at 11% release of mass, was not part of its control set to 98.2% certainty. This indicates that neither LIP nor MNP solubilized the solid singly, but together they were effective in solubilizing the solid. The solid released may have included lignin and/or carbohydrates.

FTIR/PLS Analyses

FTIR/PLS measurements of lignin, carbohydrate, methoxyl, and phenolic hydroxyl contents were performed to provide an independent means of demonstrating lignin loss from the solid. The PLS software used to analyze the data indicated that the carbohydrate spectra for all samples varied significantly from the spectra used for the whole-wood and wood-pulp calibration set (data not shown). This is probably because of the comparatively low carbohydrate content in the residual lignin used. Therefore, the carbohydrate predictions cannot be used with confidence, and are not included.

The FTIR spectra (Fig. 3) for all samples were typical for hardwood lignins (11,17,18). The spectra were normalized so that the absorbance at 1504 cm^{-1} was proportional to the known lignin concentration in the pellets, as suggested by Faix (19), since the aromatic skeletal vibrations at this band correlate well with lignin concentration. Results of the lignin content measurements are presented in Fig. 4. The statistical significance of the

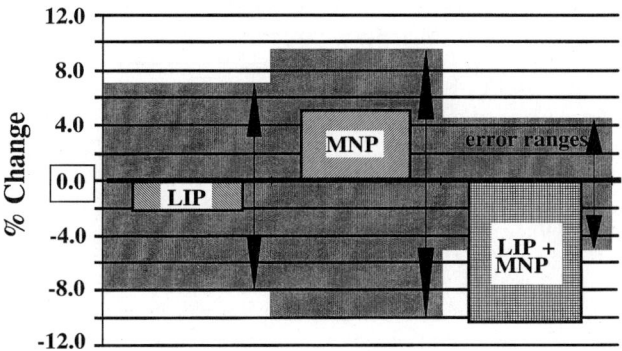

Fig. 2. Results of the mass balances for the lignin treatment runs. Controls are represented by the shaded error ranges, which are the averages ± 2 σ for the control mass balance data sets done with each base case. The LIP + MNP result is statistically different from the control set to 98.2% certainty.

Fig. 3. Raw, unnormalized FTIR spectra of base cases (solid curves) and their ER⁻ controls (dashed curves). **(A)** LIP alone; **(B)** MNP alone; and **(C)** LIP + MNP. Note that 0.5 absorbance units have been added to each base case spectrum to proportionally separate it from its controls.

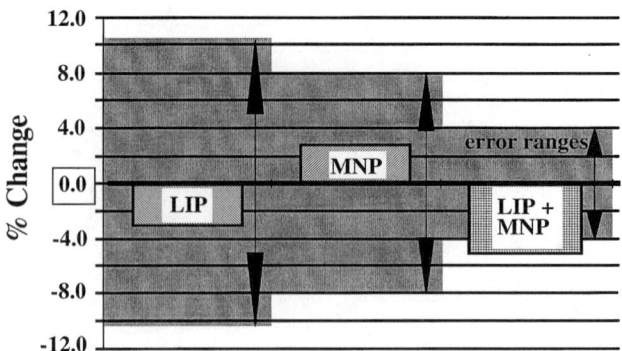

Fig. 4. Results of the FTIR/PLS determination of lignin content in the residue from the lignin treatment runs. Controls are represented by the shaded error ranges, which are the averages ± 2 σ for the control mass balance data sets done with each base case. The LIP and MNP results are clearly part of their control sets; the LIP + MNP result is statistically different from the control set to 98.8 or 92.3% certainty, using the normal distribution or the t-distribution, respectively.

base case measurements were tested as above, using the same null hypothesis, again using each control set separately with its base case. The t-distribution significance levels for LIP, MNP, and LIP + MNP base case mass balances were 0.324, 0.319, and 0.077, respectively, with seven degrees of freedom for each (duplicate measurements). Using the normal distribution, these were 0.301, 0.247, and 0.012, respectively. The null hypothesis was accepted for both LIP and MNP, using either distribution. It was rejected for LIP + MNP using the normal distribution, and borderline rejected using the t-distribution. It thus seems clear that the loss of mass shown above was at least in part a result of lignin release by LIP + MNP.

There was no change in the methoxyl content for any base case or control, with all measurements not statistically different from 1.34 ± 0.03 methoxyls/C_9 (average ± 2σ). This indicates that any lignin removal that may have occurred was not preferential for syringyl or guaiacyl units. The PLS analyses for phenolic hydroxyl contents predicted values well below the range of the Kraft lignin calibration set, for which the lower boundary was 0.70 OH groups per aromatic unit. All samples from this study were indicated at 0.00 ± 0.07 OH groups per C_9, which indicates that the lignin samples were highly polymeric, and also suggests that any lignin removal that may have occurred was at the free-phenolic ends of the lignin polymer.

GC/MS Analyses

The GC/MS analyses of solvent extracts of the aqueous and solid phases after enzymatic treatment were performed to present an additional independent verification of enzymatic lignin solubilization. Syringaldehyde, vanillic acid, and syringic acid were identified in all samples. 2,6-dimethoxyhydroquinone was also present in all samples, although much more was present in the base case extracts, while 2-methoxyhydroquinone was present in some, but not all, controls. Retention times, concentrations, and functional information derived from the mass spectra for unique and higher relative concentration peaks (*see* above for peak selection criteria), identified in the base case extracts, are presented in Table 4. Also included in Table 4 are the possible molecular formulas based on the available data and the procedure outline above. These are not meant to be confirmed identifications of the compounds, but are merely suggestions that fit the available information, since identification was not possible. Please refer to Thompson (5) and Thompson et al. (6) for the mass spectra and some possible structures.

DISCUSSION

The ideal substrate for lignin depolymerization studies would be an unmodified, carbohydrate-free lignin as it occurs in wood (4). The methods chosen for the isolation of the lignin substrate used here represent an attempt to approach this ideal. Although the lignin used in this study has been modified to some extent during the isolation process and still contains at least 12% carbohydrates, thioacidolysis and other tests indicated that it was not significantly different from the original lignin (16). Thus, the lignin used in this study was closer to the ideal substrate than many substrates used previously.

The mass balances for the base cases vs the controls provide some evidence that degradation and perhaps modification occurred for LIP + MNP, but not for either enzyme singly. The loss of mass was statistically different from its controls to 98.2% certainty. An interesting observation, noted during treatment runs containing MNP, was that the base case lignin color darkened over the course of the run from light to dark brown, and remained dark after treatment (5). When LIP + MNP was used, the base case lignin first darkened, and then lightened back to its original color. With LIP singly, no color change was observed. Although this color change could be a result of precipitation of MnO_2, controls containing MNP and reagents, but devoid of lignin, did not form the characteristic MnO_2 precipitate, suggesting that the color changes may have been caused by modification of the lignin by MNP, and then release of the modified fragments by LIP.

The FTIR/PLS analyses indicate that neither LIP nor MNP singly altered the lignin content of the substrate, but together the two enzymes released 5.1% of the lignin (4.2% as total weight). This is an important verification that at least part of the 11% weight loss seen in the mass balance was caused by release of lignin, and not to carbohydrate solubilization. The carbohydrate contents were not well modeled by the calibration

Table 4
Summary of GC Data, Including Concentration and Functional Information, and Possible Molecular Formulas for Peaks of Interest Identified by GC/MS

RT (min:s)[a]	Derivatized mol wt (g/mol)	Base case	Unique? (y/n)	Extr.[f]	Est. conc. (ng/mL)	Highest control conc.[e] (ng/mL)	Possible no. of TMS groups	No. of rings + double bonds	Possible formula
6:22	280	LIP+MNP	Yes	L-EA	0.0235	—	1	7	$C_{10}H_8O_5$
							1	6	$C_{11}H_{12}O_4$
6:32	318	MNP	Yes	S-EA	0.0189	—	1	?	Unknown[c]
8:08	182	LIP	Yes	L-EA	0.1656	—	0	5	$C_9H_{10}O_4$
							0	4	$C_{10}H_{14}O_3$
8:59	226	LIP	Yes	L-EA	0.8820	—	1	4	$C_8H_{10}O_3$[d]
9:17	308	LIP+MNP	No	S-C	0.0532	0.0065	1	6	$C_{13}H_{16}O_4$
10:12	298?	MNP	Yes	L-EA	0.0358	—	2?	?	Unknown[c]
11:10	282	LIP	Yes	L-EA	0.0925	—	2	5	$C_7H_6O_3$[b]
11:15	256	LIP	Yes	L-EA	0.9226	—	0	4	$C_9H_{12}O_4$
12:07	356	LIP+MNP	Yes	S-EA	0.0252	—	2	5	$C_{10}H_{12}O_5$
							2	4	$C_{11}H_{16}O_4$
							3	4	$C_7H_8O_3$
13:36	372	MNP	Yes	L-C	0.0249	—	2	4	$C_{11}H_{16}O_5$
							2	9	$C_{14}H_{12}O_5$
		LIP+MNP	No	S-C	0.0450	0.0112	3	4	$C_7H_8O_4$

Retention time[a]	MW	Source	Identified[b]	Extract[f]	Value 1	Value 2			Formula
15:25	342	MNP	Yes	L-C	0.0445	—	2	5	$C_9H_{10}O_5$
15:54	400	MNP	No	L-EA	0.1499	0.0179	2	4	$C_{10}H_{14}O_4$
			No	S-EA	0.0902	0.0179	1	12	$C_{17}H_{12}O_7$
		LIP+MNP	No	L-EA	0.1933	0.0405	1	11	$C_{18}H_{16}O_6$
							1	10	$C_{19}H_{20}O_5$
							2	6	$C_{11}H_{12}O_7$
							2	5	$C_{12}H_{16}O_6$
							2	4	$C_{13}H_{20}O_5$
							3	5	$C_8H_8O_5$
							3	4	$C_9H_{12}O_4$
16:00	370?	LIP	Yes	L-C	0.0256	—	2?	?	Unknown[c]
21:37	526	MNP	Yes	L-C	0.0498	—	2	14	$C_{20}H_{14}O_8$
							3	7	$C_{15}H_{18}O_7$
							3	6	$C_{16}H_{22}O_6$
22:26	552	MNP	Yes	S-C	0.0988	—	1	16	$C_{25}H_{20}O_{10}$
							1	15	$C_{26}H_{24}O_9$
							2	10	$C_{19}H_{20}O_{10}$
							2	9	$C_{20}H_{24}O_9$
							3	9	$C_{16}H_{16}O_8$
							3	8	$C_{17}H_{20}O_7$

Peaks are listed by increasing retention time, and by the extracts from which they were identified.

[a] Retention time with a DB-1 GC column and 3.75 min solvent hold for LIP and LIP+MNP runs, and retention time with a DB5-MS GC column and 2.00 min solvent hold for MNP runs.

[b] Identified as p-hydroxybenzoic acid by comparison with spectrum in NIST database.

[c] The noise level in the mass spectrum for this peak is too high to identify the molecular ion with a high degree of certainty.

[d] Identified as 3,4-dimethoxyphenol by comparison with spectrum of authentic standard.

[e] Highest concentration in any control extract for that enzyme case.

[f] L or S, Extract of liquid or solid phases, respectively; C, chloroform; EA, ethyl acetate.

method and so are not included here. The lack of differences in the methoxyl contents of all samples indicates that degradation took place in the LIP + MNP mixtures without preference for syringyl or guaiacyl units, and that no significant demethoxylation of the lignin occurred. Finally, the very low phenolic hydroxyl contents for all samples indicate that the lignin samples are highly polymeric, which suggests that intermonomer bond cleavage did not occur to any significant extent in the interior of the polymer, but must have occurred near the end units where the free-phenolic hydroxyl content is highest *(14)*.

The presence of unique products in the base case extracts for all three enzyme cases suggests that there was a release of lignin fragments catalyzed by both enzymes over the course of the 12 h reaction, alone and in combination. Reasonable structures that could be lignin-derived are possible for all unique products found in the extracts of all three enzyme cases *(5)*. It should be noted that no ring-opened structures were considered during the structural analysis procedure. It is possible that the unique products observed are enzyme oxidation products of soluble lignin extractives, such as syringaldehyde, not completely removed during the exhaustive solvent extractions of the dilute-acids/cellulase-treated poplar. The inclusion of a control to eliminate this type of product is, in retrospect, desirable. Although these data do not provide direct proof of release of lignin fragments by identification and quantification of the products, and elucidation of their formation pathways, the data are still useful, because they provide verification that products appear in the liquid phase that have mass spectra and functionalities characteristic of lignin-derived products. As such, they play an important supportive role to the FTIR and mass balance data.

This is the first in vitro evidence with an insoluble natural lignin in aqueous media, that these enzymes take part in lignin degradation. Each piece of evidence is insufficient to verify ligninolytic activity, but, when considered as a whole, it is clear that lignin degradation occurred with LIP and MNP present, but not to any significant extent with LIP or MNP singly. These results support those of Perez and Jeffries *(20)*, that MNP performs the initial stages of depolymerization and that LIP performs the bulk of the depolymerization, and of Tuor et al. *(21)*, that LIP and MNP act synergistically to degrade lignin.

CONCLUSIONS

The results of this study indicate that LIP and MNP from *P. chrysosporium* synergistically effect the degradation of lignin in aqueous media, with MNP perhaps catalyzing the initial steps to make the lignin a more suitable substrate for LIP. Because degradation was observed with control of H_2O_2 and VA levels, but not in previous studies without control, it is plausible that the previous lack of observable aqueous depolymerization of lignin in vitro may have been caused by inactivation of the enzymes by H_2O_2, or to competitive inhibition by H_2O_2. It has also been demonstrated (22) that many peroxidases not generally capable of degrading lignin in aqueous media are able to depolymerize lignin in aqueous organic solvents. It is likely that the previous lack of observable aqueous lignin degradation was caused by a combination of these factors.

ACKNOWLEDGMENTS

The authors wish to thank Jennifer Johnson and Douglas Gage (MSU) for GC/MS analyses, the NREL Biomass Analysis Group for assistance with PLS, and John Obst (USDA Forest Products Laboratory) for thioacidolysis and acetylation analyses. Kenneth Hammel (USDA Forest Products Laboratory) and Michael Gold (Oregon Graduate Center) also provided valuable discussions. This work was supported by the NSF Center for Microbial Ecology at Michigan State University, grant BIR-912-0006. Additional funds were provided by the INEEL under DOE Idaho Operations Office Contract DE-AC07-94ID13223.

REFERENCES

1. Tsao, G. T., Ladisch, M. R., and Bungay, H. R. (1987), in *Advanced Biochemical Engineering*, Bungay, H. R. and Belfort, G., eds. Wiley-Interscience, New York, pp. 79–101.
2. Michel, F. C., Jr., Dass, S. B., Grulke, E. A., and Reddy, C. A. (1991), *Appl. Environ. Microbiol.* **57**, 2368–2375.
3. Boominathan, K. and Reddy, C. A. (1992), in *Handbook of Applied Mycology*, vol 4, *Fungal Biotechnology*, Arora, D. K., Elander, R. P., and Mukerji, K. G., eds., Marcel Dekker, New York, pp. 763–822.
4. Tien, M. (1987), *Crit. Rev. Microbiol.* **15**, 141–168.
5. Thompson, D. N. (1994), PhD thesis, Michigan State University, East Lansing, MI.
6. Thompson, D. N., Hames, B. R., Reddy, C. A., and Grethlein, H. E. (1998), *Biotechnol. Bioeng.*, **57(6)**, 704–717.
7. Saeman, J. F., Bubl, J. L., and Harris, E. E. (1945), *Ind. Eng. Chem.* **17**, 35–37.
8. Tien, M. and Kirk, T. K. (1988), in *Methods in Enzymology*, vol. 161, Colowick, S. P. and Kaplan, N. O., eds., Academic, New York, pp. 238–249.
9. Dosoretz, C. G. and Grethlein, H. E. (1991), *Appl. Biochem. Biotechnol.* **28/29**, 253–265.
10. Wariishi, H., Valli, K., and Gold, M. H. (1992), *J. Biol. Chem.* **267**, 23,688–23,695.
11. Hames, B., Black, S. K., Agblevor, F., Evans, R., Johnson, D. K., and Chum, H. L. (1991), Paper presented at the 45th APPITA General Conference and Exhibition, 6th International Symposium on Wood and Pulping Chemistry, Melbourne, Australia, April 25–May 10.
12. McLafferty, F. W. (1973), *Interpretation of Mass Spectra*, 2nd ed., W. A. Benjamin Reading, MA.
13. Grous, W. R., Converse, A. O., and Grethlein, H. E. (1986), *Enzyme Microb. Technol.* **8**, 274–280.
14. Sarkanen, K. V. and Ludwig, C. H. (1971), in *Lignins: Occurrence, Formation, Structure, and Reactions*. Wiley-Interscience, New York.
15. Obst, J. R. (1982), *Holzforschung* **36**, 143–152.
16. Rolando, C., Monties, B., and Lapierre, C. (1992), in *Methods in Lignin Chemistry*, Lin, S. Y. and Dence, C. W. eds., Springer-Verlag, New York pp. 334–349.

17. Schultz, T. P., Templeton, M. C., and McGinnis, G. D. (1985), *Anal. Chem.* **57,** 2867–2869.
18. Schultz, T. P. and Glasser, W. G. (1986), *Holzforschung* **40,** 37–44.
19. Faix, O. (1992), in *Methods in Lignin Chemistry.*, Lin, S. Y. and Dence, C. W. eds., Springer-Verlag, New York, p. 102.
20. Perez, J. and Jeffries, T. W. (1992), *Appl. Environ. Microbiol.* **58,** 2402–2409.
21. Tuor, U., Wariishi, H., Schoemaker, H. E., and Gold, M. H. (1992), *Biochemistry* **31,** 4986–4995.
22. Dordick, J. S., Marletta, M. A., and Klibanov, A. M. (1986), *Proc. Natl. Acad. Sci. USA* **83,** 6255–6257.

Pulp Bleaching Using Laccase from *Trametes versicolor* Under High Temperature and Alkaline Conditions

M. C. Monteiro and M. E. A. de Carvalho*

Departamento de Biotecnologia, Faculdade de Engenharia Química de Lorena, P.O. Box 116, CEP 12600-000, Lorena, SP, Brazil

ABSTRACT

Kraft pulp was delignified using laccase produced by the white rot fungus *Trametes versicolor* immobilized in solid support under specific conditions. The stability tests showed that this enzyme was stable for 6 h at 55°C and pH 8.0, allowing its use under pH and temperature conditions very close to those used in industrial bleaching. In this work, unbleached hardwood Kraft pulp was submitted to prebleaching using 2 U laccase/g pulp basis. Reaction time, temperature, and pH of the enzymatic treatment were investigated. Good results regarding Kappa number reduction, selectivities, and high viscosities were obtained when prebleaching was performed for 1 h at temperature of 55°C and pH 8.0 followed by alkaline extraction and ECF bleaching sequences.

Index Entries: *Trametes versicolor*; laccase; eucalyptus Kraft pulp; biobleaching; enzyme stability.

INTRODUCTION

Conventional Kraft pulp bleaching, employing chlorine and its derivatives to bleach and remove lignin, produces effluents containing a variety of undesirable colored and chlorinated compounds that are often toxic, mutagenic and carcinogenic. Hence, there is a great interest in eliminating or at least reducing the use of these compounds [1,2].

The use of ligninolytic white-rot fungi and their enzymes in biopulping, biobleaching, and effluent treatment has been studied as potential alternative technologies [3–17]. Direct biobleaching of pulps by using fungi

*Author to whom all correspondence and reprint requests should be addressed.

is economically troublesome, mostly because of the long treatment time required, but the use of enzymes is promising.

The white-rot fungus *Trametes versicolor* has been studied because of its capability to degrade the three major wood components *(18)*. The fungus can delignify and brighten hardwood Kraft pulp effectively. It also produces the three degrading lignin enzymes, laccase, manganese peroxidase, and lignin peroxidase *(19–22)*. The mechanism of action of these enzymes and their role in lignin breakdown is still under investigation.

Laccase is known to oxidize phenols and phenolic substructures of lignin, and to demethylate phenolic and nonphenolic lignin substructures (23–27). The use of 2,2'-azinobis-(3-ethylbenzthiazoline-6-sulfonate) (ABTS), together with laccase, can increase the demethylation of pulps during the bleaching *(28,29)*. Its use has been studied by several authors, and, according to Bourbonnais and Paice *(30)*, methanol release and delignification by laccase are reduced in the absence of ABTS.

The objective of this work was to demonstrate that laccase produced under specific conditions can be used in biobleaching of pulp, under pH and temperature values compatible to those used in industrial bleaching.

MATERIAL AND METHODS

Strain Maintenance and Activation

The fungus *T. versicolor* (ATCC 20869) was used for enzyme production. The strain was maintained on 2% malt agar slants. Slopes of the same agar were used for production of spores that were used as inoculum in all fermentations. The concentration of spores was determined using a standard curve that correlated absorbance at 650 nm with spores concentration in terms of number of spores per mL.

Laccase Production

Experiments were performed using immobilized mycelium. For the immobilization step, shake-flask fermentations using a medium containing glucose (2, 5, or 10 g/L), peptone (2 or 10 g/L), trace metals, and 20 mM ammonium tartrate buffer at pH 5.0 were carried out in the presence of nylon-web cubes. The 500-mL flasks containing 200 mL of the culture medium were inoculated with 1.0×10^7 spores, incubated at 30°C, and shaken at 200 rpm. Laccase production was also performed in a semicontinuous mode, and, in this case, at defined time intervals, the extracellular medium was replaced by fresh medium.

The carbon/nitrogen (C/N) ratio was calculated, assuming the N content of the peptone used (12%), and the C content of the peptone (60%) and glucose (40%).

Enzyme Assays

Enzyme concentration was determined by measuring the level of activity of the cultivation medium. Laccase activity was determined as described by Szklarz (31), with syringaldazine as substrate. The reaction was monitored by change in absorbance at 525 nm ($\epsilon = 65{,}000$ M^{-1} cm^{-1}) for 1 min. One unit (U) of enzyme activity is the amount of enzyme that oxidizes 1 µmol of the substrate/min. Endoglucanase activity was determined according to Mandels et al. (32), using carboxymethyl cellulose (CMC) as substrate. One U of enzyme activity produces 1 µmol reducing sugars/min.

Enzyme stability

To study the effect of pH and temperature on the stability of the enzyme, a crude enzyme preparation with high laccase activity was separated into samples in which pH was adjusted to the desired values. Then, each one of these samples were divided in 3-mL aliquots and stored at different temperatures. The pHs and temperatures ranged from 5 to 8 and from $-10°C$ to $55°C$, respectively. After incubation for the desired time period, the activity was measured according to Szklarz (31).

Analytical

Glucose was determined by the method of Nelson (33).

Pulp Characteristics

The pulp used in the experiments was an unbleached hardwood Kraft pulp (a mixture of various eucalyptus varieties) obtained from Brazilian mills. The initial Kappa number and viscosity values were 15.2 and 28.9 cp., respectively.

Enzymatic Prebleaching

The conditions investigated are listed in Table 1. Both prebleaching and bleaching sequences were performed in polyethylene bags in a water bath at the required time periods and temperatures. Enzymatic treatment assays were followed by a conventional alkaline extraction stage at 60°C for 1 h. After prebleaching, and after each bleaching stage, the pulp was filtered and washed with dH$_2$O.

Bleaching

The chemical bleaching sequences were used as indicated in Table 2. All concentrations are expressed in terms of air-dried brownstock pulp basis. The DED and DEDED bleaching controls were not submitted to prebleaching conditions; EDED control was submitted to prebleaching conditions, however, without enzyme.

Evaluation of Pulp Treatments

Kappa numbers and viscosities of pulp samples were made according to the standard The Clinical Association of the Pulp and Paper Industry

Table 1
Prebleaching Conditions

Conditions	Temperature 30°C	55°C
Consistency (%)	10	10
Time (h)	1 and 3	1
pH	5 and 8	8
ABTS (mM)	0 and 1	0
Laccase (U/g pulp)	2	2

Table 2
ECF Bleaching Conditions

DED sequence Conditions:	1° Dioxidation: 1.0% ClO_2, 70°C, 180 min, 6% consistency Extraction: 2% NaOH, 60°C, 60 min, 10% consistency 2° Dioxidation: 0.4% ClO_2, 70°C, 120 min, 10% consistency
DEDED sequence Conditions:	1° Dioxidation: 1.0% ClO_2, 70°C, 180 min, 6% consistency 1° and 2° Extraction: 2% NaOH, 60°C, 60 min, 10% consistency 2° and 3° Dioxidation: 0.4% ClO_2, 70°C, 120 min, 10% consistency
EDED sequence Conditions:	1° Dioxidation: 1.0% ClO_2, 70°C, 180 min, 6% consistency 1° and 2° Extraction: 2% NaOH, 60°C, 60 min, 10% consistency 2° Dioxidation: 0.4% ClO_2, 70°C, 120 min, 10% consistency

(TAPPI) methods. Selectivity was defined as the ratio of a desirable bleaching effect (delignification) vs an undesirable bleaching effect (loss of pulp viscosity). Kappa numbers and pulp viscosity values were determined after the first alkaline extraction stage and at the end of the bleaching sequences, respectively. All values reported are means of six replicate experiments.

RESULTS AND DISCUSSION

Laccase Production in Batch Cultures

In order to reach a high laccase level, several glucose and peptone concentrations were tested during the growth phase. When culture media containing 10 g/L peptone and 2 or 5 g/L glucose were used, laccase activity was first detected in 5 d after inoculation, and activity levels of about 800 U/L and 1500 U/L, respectively, were reached in 15 d. When glucose

Fig. 1. Laccase production by immobilized *T. versicolor* using 10 g/L peptone and different glucose concentrations.

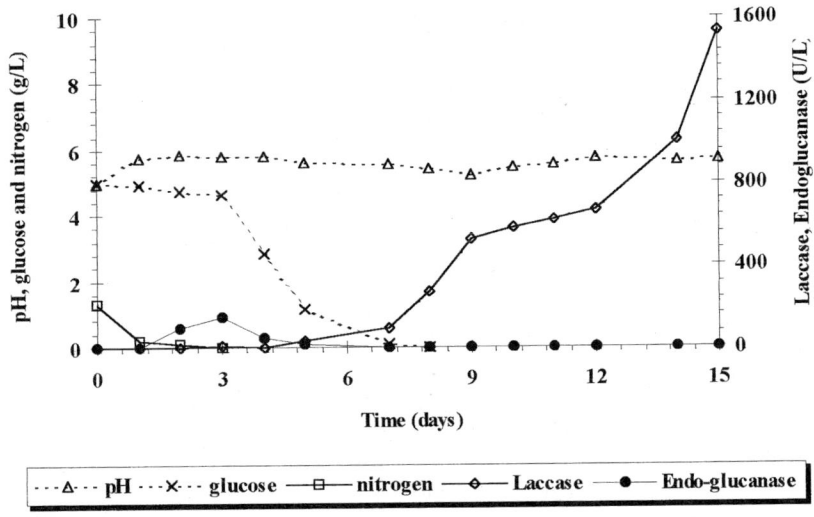

Fig. 2. Laccase production by immobilized *T. versicolor* using 10 g/L peptone and 5 g/L glucose.

concentration was increased to 10 g/L, laccase activity was detected only on d 12, and the activity level was about 500 U/L on d 17 (Fig. 1).

According to the data presented in Fig. 2, which shows enzyme activities, glucose, and N consumption and pH variation against time, using 5 g/L glucose and 10 g/L peptone, the peptone was preferentially consumed. This can be observed once glucose was not consumed until d 3

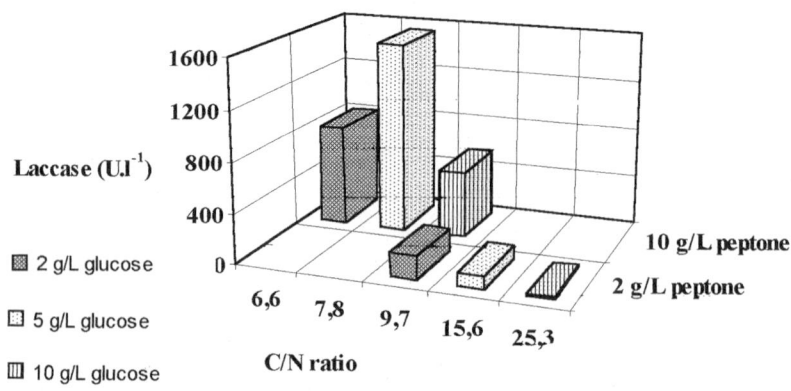

Fig. 3. Effect of carbon/nitrogen ratio on laccase production.

whereas, N from the peptone was used up in 2 d. On the other hand, the pH profile during the fermentation suggested that the buffer was not used by the microorganism as C and N source.

The peptone had an important role in laccase production. The microorganism have mechanisms of metabolism control that are used according to environmental conditions: Medium composition is one of this conditions. This study tested only one source of N (peptone) and two sources of C (peptone and glucose). Figure 3 shows the effect of the initial C/N ratio on laccase production. When 10 g/L peptone was used, high activities were obtained, and the best result was achieved with a C/N ratio value of 7.8. When 2 g/L was used, the increase in the C/N ratio had a negative effect on the amount of enzyme production. Laccase accumulation was affected by C/N ratio, type of substrate (see C/N ratio value of 9.7), and, probably, by the low concentration of N when 2 g/L peptone was used. Thus, other N sources are being tested.

Semicontinuous Laccase Production

The results from repeated batch cultures with immobilized mycelium are shown in Fig. 4. In these experiments, glucose and peptone concentrations were reduced to 2.5 g/L and 5 g/L, respectively, in order to prevent excessive mycelium growth. Under these conditions, it was possible to produce laccase with high activity for at least five successive bath-shake cultures. In this figure are also presented comparative data of enzyme production using immobilized mycelium in shake culture, and free mycelium in stationary culture. The use of bath-shake fermentation made possible an increase in the enzyme production from about 300 U/L to 1500 U/L on a time culture of 15 d. Moreover, the reaction time for successive harvests with approximately the same level of enzyme activity was reduced from 15 to 8 d.

Fig. 4. Laccase production using immobilized mycelium in successive batch-shake cultures and free mycelium in no-agitated cultures.

Enzyme Stability

The enzyme was evaluated regarding storage at different pH and temperature values. As shown in Fig. 5, laccase has good stability at the tested conditions. The best stability was observed in samples maintained under alkaline conditions (pH 8.0) and temperature of $-10°C$. In this case, after 120 d, laccase lost only 20% of initial activity. A good stability was also obtained at pH 5.7, with a loss of about 30% of the initial activity after 4 mo storage. However, at pH 5.0 a drastic activity decrease (about 40%) was observed in the first days, followed by a very slow decrease in the activity that achieved 50% in 120 d. The enzyme also showed good stability when maintained at temperature of 5°C at all tested pH values.

At temperature of 55°C and pH 8.0, the enzyme was stable for 6 h, with 10% reduction of the initial activity. It was observed that, after 6 h, a significant activity reduction occurred, and, after the 24 h, the enzyme had only 30% of initial activity. However, the maintenance of activity for 6 h was sufficient for laccase action in alkaline conditions and high temperature during biobleaching tests of pulps. This result is interesting, because it points out the potential for the use of the enzyme under industrial conditions.

Bleaching

Several biobleaching experiments were made to evaluate the action of laccase on pulp at different conditions of reaction time, pH, temperature, and the use of ABTS, (*see* Fig. 6). The best results were obtained when

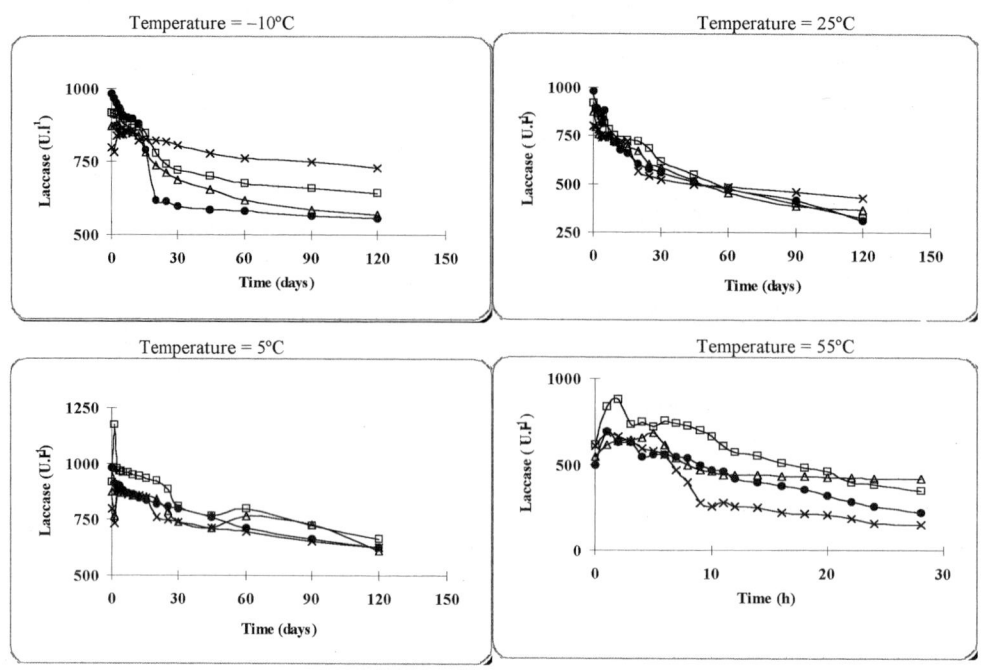

Fig. 5. Laccase stability at different temperatures and pH values.
(●) pH 5.0; (■) pH 5.7; (△) pH 7.0; (×) pH 8.0.

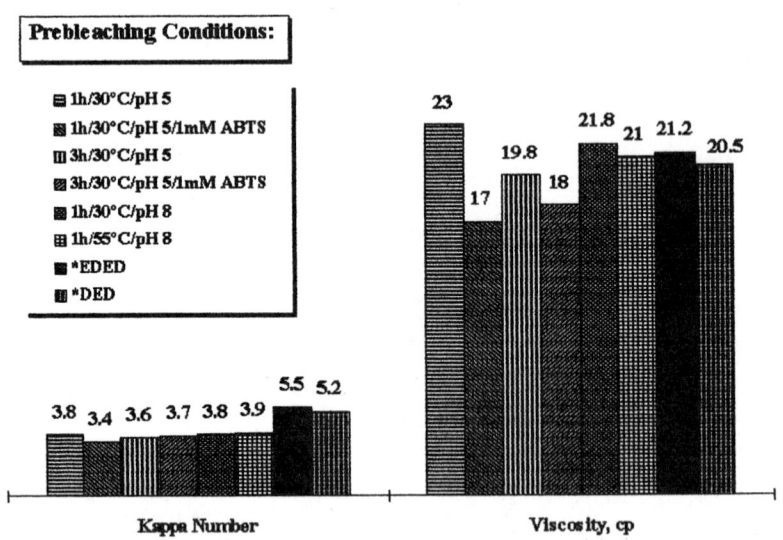

Fig. 6. Biobleaching of hardwood kraft pulp by laccase:
*Bleaching controls were not submitted to the enzymatic delignification. All pulps were prebleached using 2 U laccase/g pulp and submitted to alkaline extraction followed by DED chemical bleaching sequence. Kappa numbers and viscosities were determined after first alkaline extraction-stage (chemical bleaching) and after bleaching, respectively.

Fig. 7. Selectivity of pulps prebleaching by laccase at acid and alkaline conditions. *Controls were not submitted to the enzymatic delignification. All pulps were prebleached using 2 U laccase/g pulp and submitted to alkaline extraction followed by DED and DEDED chemical bleaching sequences.

pulp was treated for 1 h with laccase at 30°C/pH 5.0, and 30°C/pH 8.0, followed by a subsequent bleaching using a DED sequence. In both cases, the Kappa number was 3.8, corresponding to a 75% delignification and a viscosity reduction of 20 and 24.6%, respectively. Results obtained when pulp was treated for 1 h with laccase at 55°C/pH 8.0 confirmed that laccase can act in alkaline conditions and at high temperature. Kappa number and viscosity were 3.9 and 21 cp, corresponding to delignification and viscosity reduction of around 74 and 27.3%, respectively.

ABTS is frequently used as a mediator to increase the delignification and demethylation during the biobleaching of pulps by laccase (28,29). However, for laccase biobleaching at short reaction times, the use of ABTS yielded a Kappa number reduction greater than that obtained by laccase itself, but viscosities were negatively affected. At reaction times of 1 and 3 h, the use of ABTS increased pulp delignification, resulting in Kappa number values of 3.4 and 3.7, respectively. However, viscosities were affected, and their values decreased to 17 and 18 cp, respectively (Fig. 6).

Figure 7 shows selectivities of pulps treated with laccase, followed by DED and DEDED chemical-bleaching sequences. The best selectivities were obtained when pulps were treated with laccase at 30°C/pH 5.0, followed by both chemical bleaching sequences, DED and DEDED, respectively. The selectivity decreased to 3.1 and 2.7 when the pretreatment was performed at 30°C/pH 8.0, followed by DED and DEDED bleaching sequences. It was also observed that pulps treated with laccase at 55°C/pH 8.0 showed selectivities of 2.7 and 2.4 when submitted to DED and DEDED sequences, respectively. All these pretreatments resulted in selectivity values higher than selectivity values obtained for the controls.

One unexpected result was obtained when the pulp was treated for 1 h with laccase at 30°C/pH 5.0 in presence of 1 mM ABTS. In this case, despite the increase in delignification, the viscosity was not preserved, resulting in selectivity value smaller than those observed for the controls.

CONCLUSION

High laccase activity can be obtained with shake-flask fermentation, in a semicontinuous mode, using a chemically defined medium containing 5 g/L of glucose and 10 g/L of peptone. The laccase produced in these conditions can be stored for a long time at low temperatures. Stability tests showed also that laccase was stable at 55°C and pH 8.0, for sufficient time to allow its use in biobleaching at pH and temperature conditions similar to those used in the pulp industry.

The selectivity depended on the conditions used in both the biobleaching and the chemical bleaching sequences. The best selectivity was showed by pulp pretreated using laccase at 30°C/pH 5.0; however, the use of high temperature and alkaline pH (55°C/pH 8.0) also result in selectivities higher than that obtained for the controls.

ACKNOWLEDGMENTS

The authors thank CAPES for the financial support.

REFERENCES

1. Tsai, T. Y., Renard, J. J., and Phillips, R. B. (1994), *Tappi J.* **77**, 149–157.
2. Yin, C., Renard, J. J., and Phillips, R. B. (1994), *Tappi J.* **77**, 158–162.
3. Archibald, F. S. (1992), *Holzforschung* **46**, 305–310.
4. Paice, M. G., Jurasek, L., Ho, C., Bourbonnais, R., and Archibald, F. (1989), *Tappi J.* **72**, 217–221.
5. Leatham, G. F., Myers, G. C., and Wegner, T. H. (1990), *Tappi J.* **73**, 197–200.
6. Mehta, V., Gupta, J. K., Jauhari, M. B. (1992), *Tappi J.* **75**, 151–152.
7. Brown, J., Cheek, M. C., Jameel, H., and Joyce, T. W. (1994), *Tappi J.* **77**, 105–109.
8. Suurnäkki, A., Kantelinen, A., Buchert, J., and Viikari, L. (1994), *Tappi J.* **77**, 111–116.
9. Kantelinen, A., Hortling, B., Ranua, M., and Viikari, L. (1993), *Holzforschung* **47**, 29–35.
10. Hamilton, J., Senior, D. J., Rodriguez, A., Santiago, D., Szwec, J., Ragauskas, A. J. (1996), *Tappi J.* **79**, 231–234.
11. Pham, P. L., Alric, I., Delmas, M. (1996), *Appita* **48**, 213–217.
12. Reid, I. D., and Paice, M. G. (1994), *FEMS Microbiol. Rev.* **13**, 369–376.
13. Davis, S., and Burns, R. G. (1992), *Appl. Microbiol. Biotechnol.* **37**, 474–479.
14. Bergbauer, M., Eggert, C., and Kraepelin, G. (1991), *Appl. Microbiol. Biotechnol.* **35**, 105–109.
15. Davis, S., and Burns, R. G. (1990), *Appl. Microbiol. Biotechnol.* **32**, 721–726.
16. Mehna, A., Bajpai, P., and Bajpai, P. K. (1995), *Enzyme Microb. Technol.* **17**, 18–22.

17. Michel, F. C., Dass, S. B., Grulke, E. A., and Reddy, C. A. (1991), *Appl. Environ. Microbiol.* **57**, 2368–2375.
18. Morohoshi, N. (1991) in *Enzymes in Biomass Conversion*, Leathan, G. F. and Himmel, M. E., eds., American Chemical Society, Washington, DC, pp. 207–223.
19. Jönsson, L., Johansson, T., Sjöström, K., and Nyman, P. O. (1987), *Acta Chem. Scand.* **B41**, 766–769.
20. Johansson, T., and Nyman, P. O. (1987), *Acta Chem. Scand.* **B41**, 762–765.
21. Johansson, T., and Nyman, P. O. (1993), *Arch. Biochem. Biophys.* **300**, 49–56.
22. Paice, M. G., Reid, I. D., Bourbonnais, R., Archibald, F. S., and Jurasek, L. (1993), *Appl. Environ. Microbiol.* **59**, 260–265.
23. Szklarz, G., and Leonowicz, A. (1986), *Phytochemistry* **25**, 2537–2539.
24. Dodson, P. J., Evans, C. S., Harvey, P. J., and Palmer, J. M. (1987), *FEMS Microbiol. Lett.* **42**, 17–22.
25. Bourbonnais, R. and Paice, M. G. (1990), *FEBS Lett.* **267**, 99–102.
26. Muheim, A., Fiechter, A., Harvey, P. J., Schoemaker, H. E. (1992), *Holzforschung* **46**, 121–126.
27. Kawai, S. and Ohashi, H. (1993), *Holzforschung* **47**, 97–102.
28. Kirkpatrick, N., Reid, I. D., Ziomeck, E., and Paice, M. G. (1990), *Appl. Microbiol. Biotechnol.* **33**, 105–108.
29. Bourbonnais, R. and Paice, M. G. (1996), *Tappi J.* **79**, 199–204.
30. Bourbonnais, R. and Paice, M. G. (1992), *Appl. Microbiol. Biotechnol.* **36**, 823–827.
31. Szklarz G., Antibus, R. K., Sinsabaugh, R. L., Linkins, A. E. (1989), *Mycologia* **81**, 234–240.
32. Mandels, M., Andreotti, R., and Roche, C. (1976), *Biotechnol. Bioeng. Symp.* **6**, 21–33.
33. Nelson, N. (1944) *J. Biol. Chem.* **153**, 375–380.

Microbial Oxidation of Mixtures of Methylmercaptan and Hydrogen Sulfide

ANBU SUBRAMANIYAN,[1] RAVINDRA KOLHATKAR,[1] K. L. SUBLETTE*,[1] AND ROBERT BEITLE[2]

[1]Center for Environmental Research and Technology, University of Tulsa, 600 S. College Avenue, Tulsa, OK 74104-3189; [2]Department of Chemical Engineering, University of Arkansas, Fayetteville, AR 72701

ABSTRACT

Refinery spent-sulfidic caustic, containing only inorganic sulfides, has previously been shown to be amenable to biotreatment with *Thiobacillus denitrificans* strain F with complete oxidation of sulfides to sulfate. However, many spent caustics contain mercaptans that cannot be metabolized by this strict autotroph. An aerobic enrichment culture was developed from mixed *Thiobacilli* and activated sludge that was capable of simultaneous oxidation of inorganic sulfide and mercaptans using hydrogen sulfide (H_2S) and methylmercaptan (MeSH) gas feeds used to simulate the inorganic and organic sulfur of a spent-sulfidic caustic. The enrichment culture was also capable of biotreatment of an actual mercaptan-containing, spent-sulfidic caustic but at lower rates than predicted by operation on MeSH and H_2S fed to the culture in the gas phase, indicating that the caustic contained other inhibitory components.

Index Entries: Mercaptan; spent-sulfidic caustic; hydrogen sulfide; biotreatment.

INTRODUCTION

Sodium hydroxide (NaOH) solutions are used in petroleum refining to remove hydrogen sulfide (H_2S) from various hydrocarbon streams. Once H_2S reacts with the majority of NaOH, the solution becomes known as a spent-sulfidic caustic. Spent caustics typically have a pH > 12.0 and sulfide concentrations exceeding 2–3 wt%. Depending on the source, spent caustic may also contain phenols, mercaptans, amines, and other organic compounds that are soluble or emulsified in the caustic (1).

*Author to whom all correspondence and reprint requests should be addressed.

Although biological treatment can be an inexpensive disposal option, many refineries do not have the waste-water treatment capacity to treat the entire amount of spent caustic generated. Additionally, concerns regarding odors and toxicity frequently prohibit on-site treatment. Currently, most spent-sulfidic caustics generated by refineries are either sent off-site to commercial operations for recovery or reuse (pulp and paper mills, for example) or for disposal by deep-well injection.

Future regulatory changes could result in more stringent controls and increased cost for off-site management of spent caustic. In such an event, low-cost on-site treatment options would be desired. Even without regulatory changes, current off-site transportation and disposal costs warrant further investigation of on-site management alternatives. Wet-air oxidation (WAO) for on-site management is commercially available (2), but can result in significant capital investment and high operating costs. WAO can be particularly expensive for spent-caustic streams from small- to medium-size refineries owing to an insufficient economy of scale.

We have previously reported an evaluation of the feasibility of biologically treating mercaptan-free refinery spent-sulfidic caustic using a bioreactor containing a microbial culture augmented with a sulfide-tolerant strain (strain F) of the chemoautotroph *Thiobacillus denitrificans*. It was envisioned that this process could be implemented either by augmenting an existing refinery-activated sludge unit so that it could handle higher concentrations of sulfides without toxicity or odor problems, or by using a relatively small bioreactor that would be specialized for treating spent-sulfidic caustic streams.

Mercaptan-free, spent-sulfidic caustic from two refineries (Table 1) was successfully biotreated at the bench scale (1.5 L) and pilot-scale (3.7 m^3) resulting in neutralization and removal of active sulfides (3,4). Sulfides were completely oxidized to sulfate by *T. denitrificans*. Microbial oxidation of sulfides produced acid that at least partially neutralized the caustic. Mixed heterotrophs in the treatment culture acclimated to methyldiethanolamine (MDEA) present in these samples, resulting also in complete degradation of the amine. A preliminary economic analysis showed that the caustics could be treated for roughly 4–9¢/gal (1–2.3¢/L) plus the cost of any additional acid required to maintain a near-neutral pH over and above that produced by the microbial oxidation of sulfide (5).

As noted above, many refinery spent-sulfidic caustics also contain mercaptans. Although mixotrophic strains of certain *Thiobacilli* have been reported (6), *T. denitrificans* strain F is strictly autotrophic and incapable of using organic sulfur compounds as carbon and energy sources. Therefore, a microbial culture capable of oxidation of both inorganic sulfide and organic sulfur compounds, like mercaptans, will either be mixotrophic, or a coculture of a heterotrophic organism capable of mercaptan oxidation and an autotrophic, sulfide-oxidizer like *T. denitrificans*. We report here the development of an aerobic enrichment culture capable of oxidizing

Table 1
Characteristics of Spent-Sulfidic Caustic Successfully
Biotreated at Bench and Pilot-Scale (3,4)

Sample	Sulfide, M	COD, mg/L	MDEA, wt%	OH, M
D1	1.06	82100	2.37	2.60
D2	1.05	113800	3.17	1.04
D3	1.06	107000	3.81	1.03
PC1	0.60	73300	2.08	2.46
PC2	0.58	40200	---	2.91

mercaptans and sulfides fed simultaneously. The culture was enriched for organisms capable of metabolizing mercaptans and sulfides using MeSH and H_2S as gas feeds. In this way, mercaptan and sulfide oxidation could be studied in the absence of other complicating factors related to the composition of sulfidic caustic. The enrichment culture was then used to biotreat an actual refinery spent-sulfidic caustic sample containing mercaptans.

MATERIALS AND METHODS

Organisms and Stock Cultures

Several species of *Thiobacilli* including *T. thioparus* (ATCC 23647), *T. versutus* (ATCC 25364), *T. thiooxidans* (ATCC 8085), and *T. neopolitanus* (ATCC 23638) were obtained from the American Type Culture Collection (Rockville, MD). *T. denitrificans* strain F was isolated as previously described (7).

Stock cultures of these organisms were grown in partially-filled 10-mL culture tubes at 30°C in the thiosulfate mineral salts medium (8). In this medium, thiosulfate is the energy source; bicarbonate is the source of carbon; and ammonium ion is the source of reduced nitrogen. The medium also contained a phosphate buffer and sources of Mg^{2+}, Ca^{2+}, Fe^{3+}, Mn^{2+}, and trace elements.

Development of Aerobic Enrichment Culture

Each of the above referenced organisms was cultured separately and aerobically in thiosulfate mineral-salts medium in a B. Braun Biostat M (Allentown, PA) (culture volume, 1.45 L) fermenter at 30°C. The pH for each of the cultures was maintained at the optimum for each organism as follows: *T. denitrificans*: 7.0; *T. thioparus*: 7.0; *T. neopolitanus*: 7.0; *T. versutus*: 7.5; *T. thiooxidans*: 5.0. The pH was controlled by addition of 10 N NaOH as needed. The cultures were maintained in a fed-batch mode with a gas feed of 300 mL/min of air + CO_2. The medium used was thiosulfate limiting with respect to the growth of all five organisms. When thiosulfate was depleted, the agitation and aeration were terminated and the medium was centrifuged to recover the biomass which was stored at 4°C.

Activated sludge was obtained from a refinery aerobic waste-water treatment system (aerobic) and from an anaerobic industrial digester. The activated sludge from both the sources were washed free of organics with 0.10 M phosphate buffer (pH 7.0) prior to use.

Biomass from each of the *Thiobacillus* cultures were suspended together in a mineral-salts medium (Table 2) in a B. Braun Biostat M fermenter. About 200 mL of settled activated sludge from the refinery anerobic waste-water treatment system and approx 300 mL of activated sludge from the aerobic industrial digester were also added to the reactor. The volume was then made up to 1.45 L with mineral-salts medium and the mixed culture was maintained at 30°C and pH 7.0 with an aeration of 300 mL/min of air + 5% CO_2 to ensure that the medium would not become carbon limiting. The system was left under aeration overnight to ensure that there was no energy source remaining. Once these conditions were established, the culture began receiving a gas feed of 0.5% methylmercaptan (MeSH) in nitrogen (U.S. Specialty Gas, Tulsa, OK) at an initial rate of 10 mL/min. The outlet gas was passed to an 500-mL Erlenmeyer flask containing 300 mL of 0.3 wt% zinc acetate to trap any fugitive emissions of H_2S from the bioreactor. The MeSH gas was fed to the reactor during the normal working hours for approx 6–9 h/d and the outlet gas routinely monitored for MeSH. Samples of the supernatant (25 mL) were taken twice each day when feed was initiated and terminated and the culture was monitored for sulfate and ammonium-ion concentrations and soluble COD (sCOD). The culture medium was left under aeration overnight when not receiving MeSH feed.

It was anticipated that in the early phases of the enrichment there would be a significant amount of cell death and lysis among bacteria that were not capable of utilizing the mercaptan as a sole carbon source. This lysis would provide substrates for some of the heterotrophs in the culture causing a "bloom" that could deplete the medium of other nutrients. Therefore, the medium was changed every 15 d, to replenish nutrients. To change the medium, agitation, aeration, and mercaptan gas feed were terminated and the biomass allowed to settle under gravity. The supernatant liquid was then discarded and replaced with fresh mineral-salts medium. This process also selected for retention of flocculated biomass.

In the early stages of the development of the culture the ammonium-ion concentration was seen to decline quite rapidly. Initially, when the ammonium ion was depleted, the medium was changed as described above. Subsequently, when the culture became more tolerant of MeSH, NH_4Cl was added directly to the reactor when depleted to give a concentration of approx 0.2 mg/mL and the medium was replenished only when

Table 2
Mineral Salts Medium

Component	Per Liter
Na_2HPO_4	1.2 g
KH_2PO_4	1.8 g
$MgSO_4 \cdot 7H_2O$	0.4 g
NH_4Cl	0.5 g
$CaCl_2$	0.03 g
$MnSO_4$	0.02 g
$NaHCO_3$	1.0 g
Trace mineral solution (8)	15.0 mL

the sulfate concentration exceeded 10,000 mg/L. The enrichment culture was operated with a MeSH feed at increasing feed rates for 4 mo until a target feed rate of 1.2 mmoles MeSH/h (100 mL/min of 0.5% MeSH) was achieved with minimal MeSH breakthrough (<5 ppmv).

The caustic chosen for treatment in this study (Table 3) contained both sulfides and mercaptans. Therefore, once mercaptan oxidation by the enrichment culture had been demonstrated, H_2S (g) was blended with the MeSH feed gas to the culture to see if the enrichment culture was capable of oxidizing both compounds simultaneously. The enrichment culture proved to be intolerant of even low levels of H_2S (10 mL/min of 1.0% H_2S, balance N_2) indicating that the enrichment culture was incapable of oxidizing both organic and inorganic sulfur. At this point, a known sulfide-oxidizing bacterium, *Thiobacillus denitrificans* strain F, was added to the enrichment culture a second time. The flocculated organism was grown separately and harvested by centrifugation as described by Ongcharit, et al., *(9)*. About 2 g of wet-packed cells of flocculated *T. denitrificans* strain F were added to the mercaptan-oxidizing culture reactor and the reactor was kept under aeration for 24 h before initiating a combined MeSH and H_2S feed.

Ultimately a combined feed of 1.2 mmoles/h MeSH and 0.64 mmoles/h H_2S could be achieved with no breakthrough of H_2S and little breakthrough (< 5 ppmv) of MeSH in the reactor outlet gas. These MeSH and H_2S feed rates corresponded to a caustic (Table 3) feed rate of 40 mL/d in terms of mercaptan and sulfide components only. Operation of the enrichment culture initially on gas feed of MeSH and H_2S provided a benchmark

Table 3
Results of Analysis of Mercaptan-Containing Caustic Used
as Feed to the Aerobic Enrichment Culture

Parameter	Value
pH	13.1
COD	104,800 mg/L
Sulfide	0.485 M
Sulfate	281 mg/L
Ammonium ion	0.4 mg/L
Nitrate	306 mg/L
Mercaptans	0.758 M as MeSH
OH^- alkalinity	3.75 N (142.8 g/L as NaOH)
Total alkalinity	212,800 mg/L as $CaCO_3$

for caustic treatment. Treatment of caustic at lesser mercaptan and sulfide feed rates would indicate inhibition of the enrichment culture by other caustic components.

Biotreatability of Spent-Sulfidic Caustic Containing Mercaptans

The objective of this part of the study was to evaluate the ability of the acclimated enrichment culture described above to treat a selected spent-sulfidic caustic containing mercaptans. Samples of a refinery spent-sulfidic caustic were shipped from a major refinery and were sealed until used. Samples were analyzed for sulfides, mercaptans, OH^- alkalinity, sulfates, COD, nitrate, ammonium ion, and carbonates (Table 3).

At start-up the bioreactor (B. Braun Biostat M) was filled to 1.45 L with a suspension of the enrichment culture in mineral-salts medium at an initial mixed liquid suspended solids (MLSS) concentration of approx 3000 mg/L. After the suspension had equilibrated at 30°C and had been aerated for approx 1 h, caustic feed was initiated. The caustic feed was conveyed to the bioreactor using a Harvard Apparatus (Cambridge, MA) syringe pump with a 50-mL syringe with a Teflon seal. The caustic feed was introduced into the reactor at a point approx 2.5 cm from the bottom adjacent to the agitator impeller. The agitation rate was maintained at 200 rpm. The temperature was controlled at 30°C and pH was maintained at 7.0. To control the pH, 5 N HNO_3 was used. This particular acid was chosen to permit the monitoring of acid addition in the bioreactor by following the nitrate concentration. The culture was aerated with 300 mL/min air

Fig. 1. Schematic diagram of equipment used in the biotreatment of spent-sulfidic caustic.

+ 5% CO_2. The outlet gas from the bioreactor was passed to a 500-mL Erlenmeyer flask containing 350 mL of 0.3% zinc acetate to trap any fugitive emissions of H_2S from the bioreactor. A tee connection was located between the bioreactor and the zinc acetate trap for gas sampling. A schematic diagram of the equipment setup is shown in Fig. 1.

Caustic feed was initiated at a rate of 3.8 mL/d and the H_2S and MeSH gases in the outlet monitored periodically. The reactor was successfully operated without upset (MeSH or H_2S breakthrough) at this rate. Over the next 2 mo, the caustic feed rate was increased in steps of 2–3 mL/d to 20.4 mL/d. The culture received caustic feed during normal working hours and only aeration without feed overnight.

Analytical

Thiosulfate, sulfide, sulfate, ammonium ion, nitrate, MLSS, sCOD, hydroxide alkalinity, carbonate alkalinity, and total alkalinity in culture medium samples or caustic were determined as previously described (3,4). Soluble COD was determined on the supernatant of culture medium samples after centrifugation at 5000g for 10 min. Total mercaptan in the caustic was determined by titration with standard 0.10 M lead perchlorate using an Orion Model 94–16 (Cambridge, MA) sulfide/silver electrode to detect the endpoint.

The hydrogen sulfide and methylmercaptan concentrations in the outlet gas of the bioreactor were analyzed using Gastec Analyzer tubes (Gastec, Ayase-City, Japan) with a minimum accuracy of ± 25%. Various ranges of the analyzer tubes in the concentration of 2.5 to 5000 ppmv were used with 100-mL samples. No effect of H_2S as an interferent in MeSH analysis was observed up to H_2S concentration of 500 ppmv.

RESULTS AND DISCUSSION

Development of Aerobic Enrichment Culture

The enrichment culture was operated with 0.5% MeSH (g) feed for a total of 139 h, ultimately achieving a feed rate of 104 mL/min without significant breakthrough of MeSH in the outlet gas. During this time sulfate accumulated in the culture medium as MeSH was removed from the feed gas. A sulfur balance (Table 4) indicated complete oxidation of mercaptan sulfur to sulfate. Ammonium ion was also observed to be utilized as the culture grew on MeSH as a carbon and energy source. During 139 h, of operation the soluble COD was observed to increase to 240 mg/L. This increase in COD has been attributed to the accumulation of cell lysis products from organisms incapable of using MeSH.

As noted above, once mercaptan oxidation had been demonstrated, hydrogen sulfide was blended with the MeSH feed gas to the culture to demonstrate simultaneous removal and oxidation of both compounds. Ultimately, a combined feed of 100 mL/min of 0.5% MeSH and 25 mL/min of 1.0% H_2S was achieved without significant breakthrough of either sulfur compound in the outlet gas. This combined mercaptan and sulfide feed rate was equivalent to 40 mL/d of caustic feed (Table 3). This feed condition was maintained for over 25 h of operating time. During this time, sulfate was observed to accumulate in the culture medium and ammonium ion was consumed as MeSH and H_2S were removed from the feed gas. A sulfur balance showed that at least 87% of MeSH and H_2S sulfur was recovered as sulfate. No other forms of sulfur were detected in the culture medium or outlet gas. It should be noted that when operating with a combined MeSH and H_2S feed, it was important to initiate the H_2S feed before or at the same time that MeSH feed was initiated. If the culture was first exposed to MeSH, H_2S could not be tolerated. However, in this case cessation of MeSH and H_2S feed and overnight aeration of the culture resulted in a renewed capability to treat both simultaneously when initiated in the right sequence.

Biotreatability of Spent Sulfidic-Caustic Containing Mercaptans

Caustic feed was initiated to the reactor at a rate of 3.8 mL/d. During this start-up period, neither H_2S nor MeSH were detected in the outlet gas of the reactor. The caustic feed rate was gradually increased periodically in increments of 2–3 mL/d. Ultimately, it was possible to increase the feed rate to 20.4 mL/d without significant emission of MeSH or H_2S (< 3 ppmv).

As caustic was fed to the culture there was a corresponding increase in the sulfate concentration (Fig. 2). A mass balance on sulfur indicated at least 90% conversion of MeSH sulfur and sulfide from the caustic to sulfate. Ammonium-ion concentrations in the reactor medium fell in response to caustic feed indicating microbial utilization of NH_4^+ as the nitrogen source

Table 4
Sulfur Balance in the Aerobic MeSH Enrichment Culture with a
Gas Feed of 0.5% MeSH in Nitrogen

MeSH Feed Rate (mL/min)	MeSH Feed Rate (mmol/h)	MeSH In (mmol)	MeSH Out (mmol)	Sulfate Produced (mmol)	SO_4^{-2}/MeSH
58–104	0.7–1.0	144.3	3.8	133.1	0.95
74	0.89	68.2	5.3	58.2	0.93

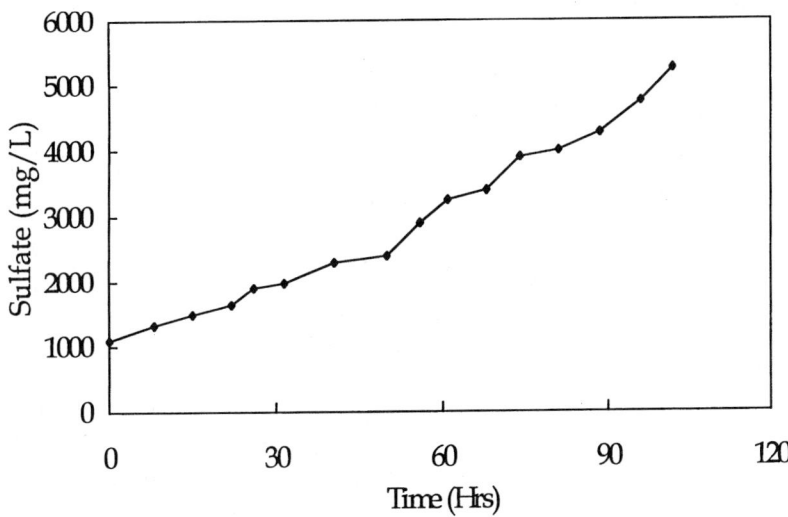

Fig. 2. Sulfate accumulation in enrichment culture during caustic feed.

as the organisms used mercaptans and sulfides in the caustic as energy sources. During caustic feeding there was a small increase in the soluble COD (to 180 mg/L) in the culture suggesting the possible accumulation of some material from the caustic in the culture medium. However, this COD accumulation was very small compared to simple dilution of the caustic in the culture medium (Fig. 3). The MLSS concentration was seen to increase as the cultures grew at the expense of mercaptan and sulfide oxidation in the culture from 3000 mg/L when caustic feed was initiated to 4600 mg/L after a total of 180 h of operation on a caustic feed.

Attempts to increase the flow rate beyond 20.4 mL/d resulted in large emissions of MeSH. The preliminary conclusion from these observations was that the caustic (Table 3) contained some component(s) that was (were) inhibitory to the MeSH-oxidizing organisms in the culture since the culture could tolerate combined feeds of MeSH and H_2S in the gas phase that greatly exceeded the molar feed rates of MeSH and H_2S represented by

Fig. 3. Accumulation of sCOD in enrichment culture during caustic feed.

these caustic feed rates. In equivalent molar terms the culture should have been able to treat 40 mL/d of this caustic.

Based on the sulfide and mercaptan analysis of the spent caustic given in Table 3, and assuming an average MLSS of 4.0 g/L, the highest specific activities for oxidation of sulfide and mercaptan in spent caustic observed in these experiments were 0.102 mmoles sulfide/h/g MLSS and 0.170 moles mercaptan/h/g MLSS, respectively. In terms of sulfide oxidation this is approximately one-tenth the specific activity for sulfide oxidation observed previously in the treatment of mercaptan-free caustic by *T. denitrificans* strain F *(10)*.

CONCLUSIONS

An aerobic enrichment culture has been developed that is capable of biotreatment of refinery spent caustic containing both inorganic sulfide and mercaptans. Treatment rates of refinery caustic were lower than predicted based on tolerance of the culture for combined gas feeds of MeSH and H_2S used to simulate spent-sulfidic caustic feed. These results suggest that the selected caustic contains other components inhibitory to the process culture.

REFERENCES

1. Bechok, M. R. (1967), *Aqueous Wastes from Petroleum and Petrochemical Plants*, Wiley, New York.

2. Schaefer, P. T. (1981), *Hydrocarbon Processing* **60,** 100–104.
3. Rajganesh, B., Sublette, K. L., and Camp, C. (1995), *Biotech. Prog.* **11,** 228–230.
4. Rajganesh, B., Sublette, K. L., and Camp, C. (1995), *Appl. Biochem. Biotech.* **51/52,** 661–670.
5. Sublette, K. L. (1997), *Appl. Biochem. Biotech.*, **63–65,** 695–706.
6. Kangawa, T. and Mikami, E. (1989), *Appl. Environ. Microbid.* **55,** 555–558.
7. Sublette, K. L. and Woolsey, M. E. (1989), *Biotech. Bioeng.* **34,** 565–567.
8. Sublette, K. L. (1987), *Biotech. Bioeng.* **29,** 690–694.
9. Ongcharit, C., Dauben, P., and Sublette, K. L. (1989), *Biotech. Bioeng.* **33,** 1077–1080.
10. Kolhatkar, A. and Sublette, K. L. (1996), *Appl. Biochem. Biotech.* **57/58,** 945–958.

Author Index

A
Adney, W. S., 395
Akano, T., 301
Al-Azmeh, H., 207
Al-Bahra, M., 207
Alves, L. A., 89
Alves, T. L. M., 463
Araujo, M. L. G. C., 493, 579
Assis, A. N., 661

B
Bajpai, R. K., 677, 747
Baker, J. O., 395
Ballesteros, I., 369
Ballesteros, M., 369
Baron, M., 67
Baruque, M. D. G. A., 877
Bates, D., 137
Bautista-Ramirez, M. E., 277
Beitle, R., 995
Belkacemi, K., 441
Berson, R. E., 615
Bienkowski, P. R., 429
Blackburn, W. J., 821
Bollók, M., 225
Bon, E. P. S., 641, 955
Bothast, R. J., 115
Boynton, B., 17, 77
Braun, R., 887, 895
Brewer, M., 99
Brumbauer, A., 225

C
Cabañas, A., 369
Cao, N., 323
Cappelaro, E., 67
Carrasco, J., 369

Chen, G. Q., 603
Chen, M. J., 187
Chen, R., 37
Chetsumon, A., 249
Chowdhury, W. Q., 301
Chua, H., 603, 853, 929
Clark, D. P., 187
Clausen, E. C., 527, 687
Conrad, J., 729
Cortez, E. V., 661
Costa, A. C., 463
Costa, S. A., 629
Cruz, A. J. G., 579
Cruz, F., 215

D
Dale, B. E., 51
Danner, H., 887, 895
Davis, M. W., 257
Davison, B. H., 429
de Carvalho, M. E. A., 955, 983
de Castro, H. F., 667
de Halleux, D., 441
de Moraes, F. F., 267, 383, 789
Desmons, S., 513
Dickow, J. H., 687
Diels, L., 311
Dien, B. S., 115
Donnelly, M. I., 187
Du, J., 323
Duff, B. W., 569, 687

E
Ehrman, C. I., 395
Eiamwat, J., 559
Evans, K., 285
Evrard, P., 513

F

Farmer, J. D., 25
Favela, E., 215
Felipe, M. D. G. A., 89, 127, 331, 869
Filho, E. A. B., 877
Finkelstein, M., 285
Foerster, L. A., 67
Fontana, J. D., 67
Freire, D. M. G., 877
Furutani, T., 301

G

Gaddy, J. L., 527, 687
Galbe, M., 3, 697
Gao, P., 405
Garcia-Kirchner, O., 277
Gartner, M., 887, 895
Gaspar, A., 535
Giordano, R. C., 493, 579
Goldwasser, I., 51
Gong, C. S., 323, 845, 919
Grethlein, H. E., 967
Gübitz, G., 939
Guimarães, M. F., 267
Guyot, J. P., 215

H

Hahn-Hägerdal, B., 3, 697
Haltrich, D., 237, 863, 939
Hames, B. R., 967
Hanley, T. R., 615
Harve, R., 677
Hendrickson, R., 99
Henriques, A. W. S., 463
Himmel, M. E., 395
Hinman, N. D., 807
Hokka, C. O., 493, 579
Huang, A. L., 603

I

Idehara, K., 301
Ikuta, Y., 301
Ishikawa, S.-I., 719

J

Jacques, Ph., 311
Jeffries, T. W., 257
Johansson, J., 747

K

Kang, C.-H., 739
Keller, F. A., 25, 77, 137
Kelley, R. L., 651, 729
Kim, T.-H., 739
Kitano, Y., 341
Klasson, K. T., 527
Ko, S., 547
Kolhatkar, R., 995
Krhouz, H., 513
Kulbe, K. D., 237, 863, 939

L

Ladisch, M. R., 99
Lanças, F. M., 67
Larrota, W., 593, 765
Larsson, S., 3
Lawford, H. G., 161, 173, 353
Lee, H.-Y., 739
Lee, K.-Y., 739
Lee, Y. Y., 37, 479
Leitner, C., 237
Lima, E. L., 463
Lima, H. O. S., 789
Lo, W., 603, 853, 929
Loha, V., 547, 559
Lundbäck, K. M. O., 527

M

Madzingaidzo, L., 895
Maeda, I., 249, 301
Mancilha, I. M., 331
Mane, T. V., 615
Matioli, G., 267
Maul, A. A., 67
Mayerhoff, Z. D. V. L., 149
McMillan, J. D., 353
Mergeay, M., 311
Mikola, M. R., 905
Millard, C. S., 187
Miura, Y., 249, 301, 341
Miyasaka, H., 301
Mizoguchi, T., 249, 301, 341
Mogollón, L., 593, 765
Mohagheghi, A., 285, 353
Monroy, O., 215
Monteiro, M. C., 955, 983
Moon, S.-H., 417
Murakami, Y., 341

N

Nagle, N. J., 569, 687
Napoleão, D. A. S., 667
Nascimento, H. J., 641
Navarro, A. A., 369
Neureiter, M., 895
Nghiem, N. P., 429
Nguyen, Q., 17, 25, 77, 137
Nidetzky, B., 237, 863, 939
Noseda, M., 67
Novalic, S., 863

O

Oliva, J. M., 369
Oliveira, P. C., 667
Ortiz, C., 765

P

Palmqvist, E., 3, 697
Park, D.-H., 739
Passos, M., 67
Pastore, G. M., 779
Paterek, J. R., 709
Pereira, J. A. M., 779
Peres, W. A., 67
Pessoa Jr., A., 505, 629, 661
Pomíllo, A. B., 67
Pradhan, S. 709
Prata, A. M. R., 89, 869
Prillinger, H., 237
Prokop, A., 547, 559

R

Ramirez, F., 215
Rathke, J. W., 187
Réczey, K., 225
Reddy, C. A., 967
Rivard, C. J., 569, 687
Roberto, I. C., 149, 629
Rodrigues, D. C. G. A., 869
Rodríguez, R., 593, 765
Rosa, P. D. T. V. E., 779
Rosa, S. M. A., 127
Rousseau, J. D., 161, 173, 353
Ruiz, R., 137
Ruy, H.-W., 739

S

Sachslehner, A., 939
Saha, B. C., 115
Sant'Anna, Jr., G. L., 877, 955
Santana, C. C., 779
Sarikaya, A., 99
Savoie, P., 441
Schell, D. J., 17, 25, 77

Segura-Granados, M., 277
Shioji, N., 301
Silva Jr., J. G., 641
Silva, J. B. A. E., 89
Silva, S. S., 89, 127, 149, 331, 869
Silva-Martinez, M., 863
Soares, V. F., 641
Soni, B. K., 651, 709, 729
Springael, D., 311
Srivastava, V. J., 651, 709, 729
Stenberg, K., 3, 697
Strodiot, L., 535
Sublette, K. L., 995
Subramaniyan, A., 995
Sun, M. Y., 429
Suyama, K., 719
Svihla, C. K., 615
Szengyel, Zs., 225

T
Tanner, R. D., 547, 559
Teague, J. M., 821
Tengborg, C., 3, 697
Thomas, S. R., 395
Thompson, D. N., 967
Thonart, P., 311, 513, 535
Torres, R., 593, 765
Tsai, S. P., 417
Tsao, G. T., 323, 845, 919
Tucker, M., 17, 25, 77
Turcotte, G., 441

U
Umeda, F., 249, 341

Ürmös, M., 887

V
Vilegas, J., 67
Vitale, A., 67
Vitolo, M., 127, 505

W
Wang, L., 51
Wang, X., 845, 919
Wang, Z., 405
Webb, O. F., 429
Webber, A. C., 67
Weekers, F., 311
Weil, J., 99
Widman, M. T., 905
Wiles, C. C., 687
Worden, R. M., 905
Wu, Z., 37, 479

Y
Yagi, K., 249, 301, 341
Yancey, M. A., 807
Yang, V. W., 257
Yasuda, H. K., 747
Yu, P. H. F., 603, 853, 929
Yurttas, L., 51

Z
Zacchi, G., 3, 225, 697
Zaghloul, T. I., 199, 207
Zanin, G. M., 267, 383, 789
Zhang, M., 285
Zhao, X., 405

Subject Index

A

Accelerated storage test, 513
Acetamide, 215
Acetic acid, 161
Acetonitrile, 67
Acidothermus cellulolyticus, 395
Acrylic acid, 887
Adaptation, 137, 353
Adaptive control, 463
Adsorption, 779
AFEX, 441
Agricultural residues, 441
Albumin, 747
Alcaligenes eutrophus, 603
    H16, 929
Alcaligenes latus, 603
Alcohol production, 877
Algal fermentative products, 301
Algal hydrocarbon production, 739
Alkali, 17
Alkaline protease gene, 199
Alkaline proteases, 277
Alkalophylic bacillus, 267
Ammonia fiber explosion, 51
Amylaceous flour, 877
Amyloglucosidase, 383
Anaerobic consortium, 853
Anaerobic digestion, 687
Annonacea, 67
Antibiotic, 249
Aqueous pretreatment, 99
Aqueous two-phase system, 629
Arabinose, 285
Aspergillus, 225
Aspergillus awamori, 641
Aspergillus niger, 395
Asphaltenes, 765

B

Babassu coconut, 877
Bacillus, 207
Bacillus sphaericus, 215
Bacillus stearothermophilus, 895
Bacillus subtilis, 199
Batch, 869
Batch foam fractionation, 547
Beneficial reuse, 569
Biobleaching, 983
Biocatalytic modification, 765
Bioconversion, 25, 77
Biodegradable plastics, 929
Biodegradation, 311
Biofuel, 441
Biogas, 687
Biological treatment, 709
Biomass, 25, 77, 821, 887
Biomass pretreatment, 51
Bioseparation, 547, 677
Biosolids, 569
Biotreatment, 995
Branched-chain fatty acid, 853
Bubble fermentor, 323

C

Candida guilliermondii, 89, 127, 869
Candida kefyr, 505
Candida mogii, 149
Cassava starch, 383, 789
Caterpillars, 67
Cellulases, 395
Cellulose hydrolysis, 17
Cellulose, 99
Cephalosporin C, 579
Cephalosporin C production, 493

Cephalosporium acremonium, 579
CGTase, 267, 789
Characterization, 405
Chemical treatment, 709
Chemotaxis, 905
Chloroperoxidase, 765
Chromatography, 257
Citronellyl acetate, 667
Clostridium propionicum, 887
Cofermentation, 161, 285
Cogeneration, 687
Column reactor, 479
Computer simulation, 51
Continuous cofermentation, 353
Continuous culture with biomass accumulation, 215
Continuous fermentation, 173, 895
Continuous reactor, 615
Corn cobs/stover mixture, 37
Corn Fiber, 115
Cost estimation and economic analysis, 37
Cross-flow filtration, 505
Culture medium development, 237
Cyanobacteria, 249
Cyclodextrin, 267, 789
Cyclodextringlycosyltransferase, see CGTase

D

Degradation, 257
Demetallation, 765
DGC, 905
Diauxic phenomenon, 579
Dilute-acid pretreatment, 25, 37
Downstream, 67, 505
Drought resistance, 311
Dunaliella salina, 739

E

E. coli, 187, 905
Effectiveness factor, 493
Endoglucanase, 939
Enzymatic hydrolysis, 277, 441
Enzymatic prebleaching, 955
Enzymatic saccharification, 115
Enzyme recovery, 779
Enzyme stability, 983
Escherichia Coli, see *E. coli*
Esterification, 667
Ethanol, 17, 25, 77, 137, 161, 285, 301, 353, 429, 479, 661, 821, 877
  production, 3, 463, 697
  yield, 173
Ethanolic fermentation, 441
Eucalyptus Kraft pulp, 955, 983
Expanded bed, 779
Expanded bed adsorption, 505
Experimental design, 629
Extracellular β-galactosidase, 779

F

Factorial design, 149
Fed-batch system, 667, 869
Fermentation, 115, 137, 187, 479, 513, 789
Fluidized bed reactor, 429, 651
Forages, 441
Freeze-drying, 513
Fuel ethanol, 115
Fumarase, 845
Fumarate reductase, 919
Fumaric acid, 845, 919

G

GAI and GAII, 641
Gas–liquid mixing, 615

Glucoamylase isoenzymes
  processing
    extracellular, 641
    proteolytic, 641
Glucoamylase, 429, 641
Gluconic acid, 863
Glucose fed-batch, 161
Glucose-fructose oxireducatase, 863
β-Glucosidase production, 225
Government, 821
Granular-activated carbon, 651

H
H_2SO_4, 3
Half-life, 383
Hemicellulose hydrolysate of willow, 225
Hemicellulose hydrolysate, 149
Hemicellulosic hydrolysates, 331
High solids, 687
Hybrid neural modeling, 463
Hydrogen production, 301
Hydrogen sulfide, 995
Hydrolysate, 89
Hydrolysis, 99, 215, 479
Hydrothermal, 99

I
Identification, 207
Immobilized cells, 493
Immobilized enzyme, 383
Inhibition, 697
Inulinase, 505
Isoenzymes proteolytic processing, 641
Isolation, 207

K
K_La, 535
Keratinases, 277

Keratinolytic activity, 199, 207

L
L-Lactic acid, 323, 895
L. brevis, 513
Laccase, 983
Lactic acid, 173, 417, 887
Lactic acid dehydrogenase, 173
Lignin extraction, 17
Lignin peroxidase (LIP), 955, 967
Lignocellulose biomass, 369
Lignocellulosic biomass, 17
Lipase, 667

M
M-Zyme, 277
Maintenance of properties, 311
L-Malic acid, 845
Malt waste, 603
Manganese peroxidase (MNP), 967
Mannanase, 939
Mercaptan, 995
Methanol, 821
Methylococcus capsulatus, 729
Methylomonas albus, 729
Microbispora bispora, 395
Mixed culture fermentation, 285
Modeling, 37
Mutagenesis, 905
Mycelial pellet, 323

N
Neural network, 579
Neurospora crassa, 405
Nitrogenase derepression, 301
Nonisothermal, 479
Nucleic acids, 677
Nucleotide sequence, 341

O

Oligosaccharides, 257
Optimal control, 463
Organic solvents, 765
β-Oxidation, 853

P

Partnerships, 821
Penicillium, 225
Penicillium canescens, 535
Penicillium janthinellum, 629
Percolation, 37
Permeability, 747
pH Control, 99
pH Monitoring, 99
Phanerochaete chrysosporium, 955, 967
Phenomenolgical modeling, 579
Photoautotrophic growth, 739
Photobioreactor, 249
Physical properties, 929
Pilot-scale, 687
Polar acetogenins, 67
Poly (3-hydroxybutyrate-co-3-hydroxyvalerate), 929
Poly(3-hydroxybutyrate) (PHB) accumulation, 301
Polyhydroxyalkanoate (PHA), 341, 603
Polyhydroxybutyrate (PHB), 341, 603
Polyhydroxyvalerate (PHV), 603
Polymer membranes, 747
Polyurethane foam, 249
Precipitation, 661, 677
Pressure, 729
Pretreatment, 77, 89, 115
Process design, 51
Propylene oxide, 651
Protein separation, 547
Proteins, 677

Pseudomonas acidophila, 341
Pseudomonas putida, 677
Pulsed amperometric detection, 257
Purification, 405
Pyranose 2-oxidase, 237

R

Radiochemical, 747
Radiofrequency plasma, 747
Recombinant *E. coli*, 919
Recombinant *Zymomonas*, 161, 285
Recycling, 697
Regulation, 939
Respiration rate, 493
Respiratory deficient, 845
Reversed micelles, 505
Rhizopus oryzae, 323
Rice straw, 149
River sediment, 853
Roller bottle reactor, 615

S

Saccharomyces cerevisiae, 845
Sclerotium rolfsii, 939
Screening, 237, 267
Scytonema, 267
Selection, 311, 903
Selectivity, 955
Semicontinuous, 869
Sewage sludge, 569
Shrinking bed, 37
Simultaneous saccharification and fermentation (SSF), 369, 417, 429, 479
SO_2, 3
Sodium sulfate, 661
Softwood, 3, 77, 137, 697
Sorbitol, 863
Soya waste, 603

Specific power, 535
Spent-sulfidic caustic, 995
SSCF, 161
Stability, 383
Starch, 417, 429
Steam pretreatment, 3, 697
Streptomyces sp, 277
Succinic acid, 187, 919
Sugarcane bagasse, 89
Sugarcane bagasse hemicellulosic hydrolysate, 869
Surfactant, 369
Sweet potato proteins, 547
Synthetic biomass prehydrolyzate, 161
System, 667

T
Temperature, 729
Thermal mechanical treatment, 569
Thermomonospora fusca, 395
Thermophillic, 895
Total petroleum hydrocarbons (TPHs), 709

Trametes multicolor, 237
Trametes versicolor, 983
Trichoderma, 225
Trichoderma reesei, 395

U
UASB reactor, 215

X
Xylanase, 257, 535, 629, 661, 939
Xylitol, 89, 149, 331, 869
Xylitol dehydrogenase, 127
Xylose, 161, 285, 353, 405
  fermentation, 331
  reductase, 127, 405
Xylose-fermenting yeasts, 331

Y
Yeast, 137

Z
Zeolite, 369
Zymomonas mobilis, 173, 353, 429, 863